Contents

3 Solving Equations and Problem Solving 161

4 Fractions and Mixed Numbers 212

5 Decimals 329

6 Percent 413

7 Graphs and Triangle Applications 489

8 Geometry and Measurement 553

9 Equations, Inequalities, and Problem Solving 648

10 Exponents and Polynomials 725

11 Factoring Polynomials 801

12 Rational Expressions 870

13 Graphing Equations and Inequalities 946

14 Systems of Equations 1055

15 Roots and Radicals 1105

16 Quadratic Equations 1160

Tools to Help Students Succeed

Your textbook includes a number of features designed to help you succeed in this math course—as well as the next math course you take. These features include:

Feature	Benefit	Page
Well-crafted Exercise Sets: We learn math by doing math	The exercise sets in your text offer an ample number of exercises carefully ordered so you can master basic mathematical skills and concepts while developing all-important problem solving skills. Exercise sets include Mixed Practice exercises to help you master multiple key concepts, as well as Vocabulary and Readiness Check, Writing, Applications, Concept Check, Concept Extension, and Review exercises.	179–181
Study Skills Builders: Maximize your chances for success	Study Skills Builders reinforce the material in ***Section 1.1—Tips for Success in Mathematics.*** Study Skills Builders are a great resource for study ideas and self-assessment to maximize your opportunity for success in this course. Take your new study skills with you to help you succeed in your next math course.	140
The Bigger Picture: Succeed in this math course and the next one you take	The Bigger Picture focuses on the key concept of this course—operations on sets of numbers, simplifying expressions, and solving equations and inequalities—and asks you to keep an ongoing outline so you can recognize and perform operations on different types of numbers, simplify expressions, and solve equations and inequalities. A strong foundation in these topics will help you succeed in this prealgebra course, as well as the next math course you take.	191, 712
Examples: Step-by-step instruction for you	Examples in the text provide you with clear, concise step-by-step instructions to help you learn. Annotations in the examples provide additional instruction.	176
Helpful Hints: Help where you'll need it most	Helpful Hints provide tips and advice at exact locations where students need it most. Strategically placed where you might have the most difficulty, Helpful Hints will help you work through common trouble spots.	176
Practice Problems: Immediate reinforcement	Practice Problems offer immediate reinforcement after every example. Try each Practice Problem after studying the corresponding example to make sure you have a good working knowledge of the concept.	176
Integrated Review: Mid-chapter progress check	To ensure that you understand the key concepts covered in the first sections of the chapter, work the exercises in the Integrated Review before you continue with the rest of the chapter.	182
Vocabulary Check: Key terms and vocabulary **Vocabulary and Readiness Check**	Make sure you understand key terms and vocabulary in each chapter with the Vocabulary Check at the end of the chapter. Use the Vocabulary and Readiness Checks to build your skills in the sections.	200, 188
Chapter Highlights: Study smart	Chapter Highlights outline the key concepts of the chapter along with examples to help you focus your studying efforts as you prepare for your test.	201
Chapter Test: Take a practice test	In preparation for your classroom test, take this practice test to make sure you understand the key topics in the chapter. Be sure to use the **Chapter Test Prep Video CD** included with this text to see the author present a fully worked-out solution to each exercise in the Chapter Test.	208

Martin-Gay's CD VIDEO RESOURCES Help Students Succeed

Martin-Gay's Chapter Test Prep Video CD (available with this text)

- Provides students with help during their most "teachable moment"—while they are studying for a test.
- Text author Elayn Martin-Gay presents step-by-step solutions to the exact exercises found in each Chapter Test in the book.
- Easy video navigation allows students to instantly access the worked-out solutions to the exercises they want to review.
- Captioned in English and Spanish.

Martin-Gay's CD Lecture Series (with Tips for Success in Mathematics)

- Text author Elayn Martin-Gay presents the key concepts from every section of the text in 10–15 minute mini-lectures.
- Students can easily review a section or a specific topic before a homework assignment, quiz, or test.
- Includes fully worked-out solutions to exercises marked with a CD Video icon ( ) in each section.
- Includes ***Section 1.1, Tips for Success in Mathematics.***
- Captioned in English and Spanish.
- Ask your bookstore for information about Martin-Gay's *Prealgebra & Introductory Algebra,* Second Edition, CD Lecture Series, or visit www.prenhall.com.

Additional Resources to Help You Succeed

Student Study Pack
A single, easy-to-use package—available bundled with your textbook or by itself—for purchase through your bookstore. This package contains the following resources to help you succeed:
Student Solutions Manual • Contains worked-out solutions to odd-numbered exercises from each section exercise set, Practice Problems, Mental Math exercises, and all exercises found in the Chapter Review and Chapter Tests. **Prentice Hall Math Tutor Center** • Staffed by qualified math instructors who provide students with tutoring on examples and odd-numbered exercises from the textbook. Tutoring is available via toll-free telephone, toll-free fax, email, or the Internet. **Martin-Gay's CD Lecture Series** • Text author Elayn Martin-Gay presents the key concepts from every section of the text with 10–15 minute mini-lectures. Students can easily review a section or a specific topic before a homework assignment, quiz, or test. • Includes fully worked-out solutions to exercises marked with a CD Video icon ( ) in each section. Also includes ***Section 1.1, Tips for Success in Mathematics.***

Online Homework and Tutorial Resources
MyMathLab® MyMathLab MyMathLab is a series of text specific, easily customizable, online courses for Prentice Hall textbooks in mathematics and statistics. MyMathLab is powered by Course Compass™—Pearson Education's online teaching and learning environment—and by MathXL®—our online homework, tutorial, and assessment system. MyMathLab gives instructors the tools they need to deliver all or a portion of their course online, whether students are in a lab setting or working from home. MyMathLab provides a rich and flexible set of course materials, featuring free-response exercises that are algorithmically generated for unlimited practice and mastery. Students can also use online tools, such as video lectures, animations, and a multimedia textbook, to independently improve their understanding and performance. MyMathLab is available to qualified adopters. For more information, visit our Web site at www.mymathlab.com or contact your Prentice Hall sales representative. (MyMathLab must be set up and assigned by your instructor.)
MathXL® www.mathxl.com MathXL® MathXL is a powerful online homework, tutorial, and assessment system that accompanies the text. With MathXL, instructors can create, edit, and assign online homework and tests using algorithmically generated exercises correlated to your textbook. All student work is tracked in MathXL's online gradebook. Students can take chapter tests in MathXL and receive personalized study plans based on their test results. The study plan diagnoses weaknesses and links students directly to tutorial exercises for the objectives they need to study and retest. Students can also access supplemental animations and video clips directly from selected exercises. MathXL is available to qualified adopters. For more information, visit our Web site at www.mathxl.com, or contact your Prentice Hall sales representative for a product demonstration. (MathXL must be set up and assigned by your instructor.)

Preface

Prealgebra & Introductory Algebra was first written in response to the needs of those teaching combined courses. My goals were to help students make the transition from arithmetic to algebra and to provide a solid foundation in algebra. To achieve these goals, I introduce algebraic concepts early and repeat them often, as traditional arithmetic topics are discussed, thus laying the groundwork for algebra. Specific care was taken to ensure that all core topics of an introductory algebra course are covered and that students have the most up to date relevant text preparation for future courses that require an understanding of algebraic fundamentals.

The many factors that contributed to the success of the previous edition have been retained. In preparing the Second Edition, I considered comments and suggestions of colleagues, students, and many users of the prior edition throughout the country.

What's New in the Second Edition?

Enhanced Exercise Sets

- **NEW!** Three forms of mixed sections of exercises have been added to the Second Edition.
  - **Mixed Practice** exercises combining objectives within a section
  - **Mixed Practice** exercises combining previous sections
  - **Mixed Review** exercises included at the end of the Chapter Review

  These exercises require students to determine the problem type and strategy needed in order to solve it. In doing so, students need to think about key concepts to proceed with a correct method of solving—just as they would need to do on a test.
- **NEW! Vocabulary and Readiness Check** exercises appear at the beginning of most exercise sets. These exercises quickly check a student's understanding of new vocabulary words so that forthcoming instructions in the problem sets will be clear. The **readiness** exercises center on a student's understanding of a concept that is necessary in order to continue with the exercise set. For example, such a readiness exercise might be for students to write out both $\frac{2^3}{7}$ and $\left(\frac{2}{7}\right)^3$.
- **NEW! Concept Check exercises** have been added to the section exercise sets. These exercises are related to the Concept Check(s) found within the section. They help students measure their understanding of key concepts by focusing on common trouble areas. These exercises may ask students to identify a common error, and/or provide an explanation.
- **NEW! Concept Extensions** (formerly Combining Concepts) have been revised. These exercises extend the concepts and require students to combine several skills or concepts to solve the exercises in this section.

Increased Emphasis on Study Skills and Student Success

- **NEW! Study Skills Builders** (formerly Study Skill Reminders) Found at the end of many exercise sets, Study Skills Builders allow instructors to assign exercises that will help students improve their study skills and take responsibility for their part of the learning process. Study Skills Builders reinforce the material found in Section 1.1, "Tips for Success in Mathematics," and serve as an excellent tool for self-assessment.
- **NEW! The Bigger Picture** is a recurring feature that focuses on the key concepts of the course—operations on sets of numbers, simplifying expressions,

and solving equations and inequalities. Students develop an outline to recognize and perform operations on different sets of numbers, simplify expressions, and solve different types of equations and inequalities. By working the exercises and developing this outline throughout the text, students can begin to transition from thinking "section by section" to thinking about how the mathematics in this course is part of the "bigger picture" of mathematics in general. A completed outline is provided in Appendix B so students have a model for their work.

- **Chapter Test Prep Video CD** New captioning options available in both English and Spanish provide students with help during their most "teachable moment"—while they are studying for a test. Included with every copy of the student edition of the text, this video CD provides fully worked-out solutions by the author to every exercise from each Chapter Test in the text. The easy video navigation allows students to instantly access the solutions to the exercises they want to review. The problems are solved by the author in the same manner as in the text.
- **NEW! Chapter Test files in TestGen** provide algorithms specific to each exercise from each Chapter Test in the text. Allows for easy replication of Chapter Tests with consistent, algorithmically generated problem types for additional assignments or assessment purposes.

Content Changes in the Second Edition

- **Enhanced Treatment of Equation Solving** to help students better prepare for their work in Chapter 3, Solving Equations and Problem Solving, and to expose them to prealgebra topics earlier in the course, the idea of equation solving is now introduced in Section 1.8, and revisited in Section 2.6, with dedicated coverage following in Chapter 3.

  This progression helps students gain greater exposure to and experience with equation solving—an essential tool for the prealgebra course as well as future courses. They are better prepared to work with equation solving in dedicated Chapter 3 and throughout the text. The earlier exposure helps reduce overlap with basic math topics and improve student motivation.

  In Section 1.8 (Introduction to Variables, Algebraic Expressions, and Equations) students determine whether a number is indeed a solution of a given equation. In Section 2.6, students are introduced to the multiplication and addition properties of equality and solve equations requiring one application of a property. The properties are reviewed in Section 3.2, and students solve equations requiring more than one property.
- **The global concepts in prealgebra that students struggle with are revisited in the text in the Review exercises, Mixed Practice exercises and Concept Extensions.** These concepts include:
  - Understanding the difference between an equation and an expression and between solving equations and simplifying expressions
  - Translating word phrases and statements to expressions and equations
  - Remembering operations on fractions
- **Adding and Subtracting Whole Numbers are now covered in Section 1.3** versus separate sections, to improve the pace of the chapter.
- To help students remember and retain concepts and operations on fractions, often a side-by-side illustration is used to help those who are visual learners. For example in Section 4.5, Adding and Subtracting Unlike Fractions, page 265, this side-by-side approach is used. On one side, figures illustrating unlike fractions are added. On the other side, the same fractions are added algebraically.
- All exercise sets have been reviewed and updated to ensure that even- and odd-numbered exercises are paired.

Key Pedagogical Features

The following key features have been retained and/or updated for the Second Edition of the text:

Problem Solving Process This is formally introduced in Chapter 3 with a four-step process that is integrated throughout the text. The four steps are **Understand, Translate, Solve,** and **Interpret.** The repeated use of these steps in a variety of examples shows their wide applicability. Reinforcing the steps can increase students' comfort level and confidence in tackling problems.

Exercise Sets Revised and Updated The exercise sets have been carefully examined and extensively revised. Special focus was placed on making sure that even- and odd-numbered exercises are paired.

Examples Detailed step-by-step examples were added, deleted, replaced, or updated as needed. Many of these reflect real life. Additional instructional support is provided in the annotated examples.

Practice Problems Throughout the text, each worked-out example has a parallel Practice Problem. These invite students to be actively involved in the learning process. Students should try each Practice Problem after finishing the corresponding example. Learning by doing will help students grasp ideas before moving on to other concepts. Answers to the Practice Problems are provided at the bottom of each page.

Helpful Hints Helpful Hints contain practical advice on applying mathematical concepts. Strategically placed where students are most likely to need immediate reinforcement, Helpful Hints help students avoid common trouble areas and mistakes.

Concept Checks This feature allows students to gauge their grasp of an idea as it is being presented in the text. Concept Checks stress conceptual understanding at the point-of-use and help suppress misconceived notions before they start. Answers appear at the bottom of the page. Exercises related to Concept Checks are now included in the exercise sets.

Integrated Reviews A unique, mid-chapter exercise set that helps students assimilate new skills and concepts that they have learned separately over several sections. These reviews provide yet another opportunity for students to work with "mixed" exercises as they master the topics.

Vocabulary Check Provides an opportunity for students to become more familiar with the use of mathematical terms as they strengthen their verbal skills. These appear at the end of each chapter before the Chapter Highlights.

Chapter Highlights Found at the end of every chapter, these contain key definitions and concepts with examples to help students understand and retain what they have learned and help them organize their notes and study for tests.

Chapter Review The end of every chapter contains a comprehensive review of topics introduced in the chapter. The Chapter Review offers exercises keyed to every section in the chapter, as well as Mixed Review **(NEW!)** exercises that are not keyed to sections.

Chapter Test and Chapter Test Prep Video CD The Chapter Test is structured to include those problems that involve common student errors. The **Chapter Test Prep Video CD** gives students instant author access to a step-by-step video solution of each exercise in the Chapter Test. New captioning options are available in English and Spanish.

Cumulative Review Follows every chapter in the text (except Chapter 1). Each odd-numbered exercise contained in the Cumulative Review is an earlier worked example in the text that is referenced in the back of the book along with the answer.

Writing Exercises These exercises occur in almost every exercise set and require students to provide a written response to explain concepts or justify their thinking.

Applications Real-world and real-data applications have been thoroughly updated and many new applications are included. These exercises occur in almost every exercise set and show the relevance of mathematics and help students gradually, and continuously develop their problem solving skills.

Review Exercises (formerly Review and Preview exercises) These exercises occur in each exercise set (except in Chapter 1) and are keyed to earlier sections. They review concepts learned earlier in the text that will be needed in the next section or chapter.

Exercise Set Resource Icons Located at the opening of each exercise set, these icons remind students of the resources available for extra practice and support:

See Student Resource descriptions pages xxi–xxii for details on the individual resources available.

Exercise Icons These icons facilitate the assignment of specialized exercises and let students know what resources can support them.

CD Video icon: exercise worked on Martin-Gay's CD Lecture Series.
△ Triangle icon: identifies exercises involving geometric concepts.
Pencil icon: indicates a written response is needed.
Calculator icon: optional exercises intended to be solved using a scientific or graphing calculator.

Group Activities Found at the end of each chapter, these activities are for individual or group completion, and are usually hands-on or data-based activities that extend the concepts found in the chapter allowing students to make decisions and interpretations and to think and write about algebra.

Optional: Calculator Exploration Boxes and Calculator Exercises The optional Calculator Explorations provide key strokes and exercises at appropriate points to provide an opportunity for students to become familiar with these tools. Section exercises that are best completed by using a calculator are identified by for ease of assignment.

A Word about Textbook Design and Student Success

The design of developmental mathematics textbooks has become increasingly important. As students and instructors have told Prentice Hall in focus groups and market research surveys, these textbooks cannot look "cluttered" or "busy." A "busy" design can distract a student from what is most important in the text. It can also heighten math anxiety.

As a result of the conversations and meetings we have had with students and instructors, we concluded the design of this text should be understated and focused on the most important pedagogical elements. Students and instructors helped us to identify the primary elements that are central to student success. These primary elements include:

- Exercise Sets
- Examples and Practice Problems
- Helpful Hints
- Rules, Property, and Definition boxes

As you will notice in this text, these primary features are the most prominent elements in the design. We have made every attempt to make sure these elements are the features the eye is drawn to. The remaining features, the secondary elements in the design, blend into the "fabric" or "grain" of the overall design. These secondary elements complement the primary elements without becoming distractions.

Prentice Hall's thanks goes to all of the students and instructors (as noted by the author in Acknowledgments) who helped us develop the design of this text. At every step in the design process, their feedback proved valuable in helping us to make the right decisions. Thanks to your input, we're confident the design of this text will be both practical and engaging as it serves its educational and learning purposes.

Sincerely,

Paul Murphy

Executive Editor
Developmental Mathematics
Prentice Hall

Instructor and Student Resources

The following resources are available to help instructors and students use this text more effectively.

Instructor Resources

Annotated Instructor's Edition (0-13-231928-4)

- Answers to all exercises printed on the same text page
- Teaching Tips throughout the text placed at key points
- Includes Vocabulary Check Tips at the beginning of relevant sections
- General tips and suggestions for classroom or group activities

Instructor Solutions Manual (0-13-231977-2)

- Solutions to the even-numbered exercises
- Solutions to every Practice Problem
- Solutions to every exercise in the Integrated Reviews, Chapter Reviews, Chapter Tests, and Cumulative Reviews

Instructor's Resource Manual with Tests (0-13-231979-9)

- **NEW!** Includes Mini-Lectures for every section from the text
- Group Activities
- Free Response Test Forms, Multiple Choice Test Forms, Cumulative Tests, and Additional Exercises
- Answers to all items

TestGen (0-13-231935-7)

- Enables instructors to build, edit, print, and administer tests
- Features a computerized bank of questions developed to cover all text objectives
- Available on dual-platform Windows/Macintosh CD-ROM

Instructor Adjunct Resource Kit (0-13-231974-8)

The Martin-Gay Instructor/Adjunct Resource Kit (IARK) contains tools and resources to help adjuncts and instructors succeed in the classroom. The IARK includes:

- Instructor-to-Instructor CD Videos that offer tips, suggestions, and strategies for engaging students and presenting key topics
- PDF files of the Instructor Solutions Manual and the Instructor's Resource Manual
- TestGen

MyMathLab Instructor Version (0-13-147898-2)
MyMathLab® www.mymathlab.com

MyMathLab is a series of text specific, easily customizable, online courses for Prentice Hall textbooks in mathematics and statistics. MyMathLab is powered by CourseCompass™—Pearson Education's online teaching and learning environment—and by MathXL®—our online homework, tutorial, and assessment system. MyMathLab gives instructors the tools they need to deliver all or a portion of their course online, whether students are in a lab setting or working from home. MyMathLab provides a rich and flexible set of course materials, featuring free-response exercises that are algorithmically generated for unlimited practice and mastery. Students can also use online tools, such as video lectures, animations, and a multimedia textbook, to independently improve their understanding and performance. Instructors can use MyMathLab's homework and test managers to select and assign online exercises correlated directly to the text, and they can import TestGen tests into MyMathLab for added flexibility. MyMathLab's online gradebook—designed specifically for mathematics and statistics—automatically tracks students' homework and test results and gives the instructor control over how to calculate final grades. Instructors can also add offline (paper-and-pencil) grades to the gradebook. MyMathLab is available to qualified adopters. For more information, visit our website at www.mymathlab.com or contact your Prentice Hall sales representative.

MathXL Instructor Version (0-13-147895-8)
MathXL® www.mathxl.com

MathXL is a powerful online homework, tutorial, and assessment system that accompanies the text. With MathXL, instructors can create, edit, and assign online homework and tests using algorithmically generated exercises correlated to your textbook. All student work is tracked in MathXL's online gradebook. Students can take chapter tests in MathXL and receive personalized study plans based on their test results. The study plan diagnoses weaknesses and links students directly to tutorial exercises for the objectives they need to study and retest. Students can also access supplemental animations and video clips directly from selected exercises. MathXL is available to qualified adopters. For more information, visit our Web site at www.mathxl.com, or contact your Prentice Hall sales representative for a product demonstration.

Interact Math® Tutorial Web site www.interactmath.com

Get practice and tutorial help online! This interactive tutorial Web site provides algorithmically generated practice exercises that correlate directly to the exercises in your textbook. You can retry an exercise as many times as you like with new values each time for unlimited practice and mastery. Every exercise is accompanied by an interactive guided solution that gives you helpful feedback if you enter an incorrect answer, and you can also view a worked-out sample problem that steps you through an exercise similar to the one you're working on.

Student Resources

Student Solutions Manual (0-13-231975-6)

- Solutions to the odd-numbered section exercises
- Solutions to the Practice Problems
- Solutions to every exercise found in the Chapter Reviews and Chapter Tests

Martin-Gay's CD Lecture Series (0-13-231932-2)

- Perfect for review of a section or a specific topic, these mini-lectures by Elayn Martin-Gay cover the key concepts from each section of the text in approximately 10–15 minutes
- Includes fully worked-out solutions to exercises in each section marked with a
- Includes coverage of Section 1.1, "Tips for Success Mathematics"
- New! Captioning options available in both English and Spanish

New—Worksheets for Classroom and Lab Practice (0-13-235399-7)

- Convenient, ready to use format, with ample work space
- Helps students stay organized and show their work step-by-step. Students work sample examples in class with their instructor. Then solve examples independently on the same page. Practice sets provide immediate reinforcement and extra practice. The worksheets provide an excellent way to check to see if students understand the skills and concepts of each section.

Prentice Hall Math Tutor Center (0-13-064604-0)

- Staffed by qualified math instructors who provide students with tutoring on examples and odd-numbered exercises from the textbook
- Tutoring is available via toll-free telephone, toll-free fax, e-mail, or the Internet
- Whiteboard technology allows tutors and students to see problems worked while they "talk" in real time over the Internet during tutoring sessions

Prealgebra & Introductory Algebra, Second Edition Student Study Pack (0-13-134903-1)

The Student Study Pack includes:

- Martin-Gay's CD Lecture Series
- Student Solutions Manual
- Prentice Hall Math Tutor Center access code

Chapter Test Prep Video CD—Standalone (0-13-231931-4)

- Includes fully worked-out solutions to every problem from each Chapter Test in the text.

MathXL Tutorials on CD—Standalone (0-13-231971-3)

- Provides algorithmically generated practice exercises that correlate to exercises at the end of sections.
- Every exercise is accompanied by an example and a guided solution, selected exercises include a video clip.
- The software recognizes student errors and provides feedback. It can also generate printed summaries of students progress.

Interact Math® Tutorial Web Site www.interactmath.com

Get practice and tutorial help online! This interactive tutorial Web site provides algorithmically generated practice exercises that correlate directly to the exercises in your textbook. You can retry an exercise as many times as you like with new values each time for unlimited practice and mastery. Every exercise is accompanied by an interactive guided solution that gives you helpful feedback if you enter an incorrect answer, and you can also view a worked-out sample problem that steps you through an exercise similar to the one you're working on.

Acknowledgments

There are many people who helped me develop this text, and I will attempt to thank some of them here. Carrie Green and Cindy Trimble were *invaluable* for contributing to the overall accuracy of the text. Elaine Page and Suellen Robinson were *invaluable* for their many suggestions and contributions during the development and writing of this Second Edition. Ingrid Benson provided guidance throughout the production process.

A special thanks to my editor, Paul Murphy, for all of his assistance, support, and contributions to this project. A very special thank you goes to my project manager, Mary Beckwith, for being there 24/7/365, as my students say. Last, my thanks to the staff at Prentice Hall for all their support: Linda Behrens, Maura Zaldivar, Patty Burns, Tom Benfatti, Paul Belfanti, Maureen Eide, Suzanne Behnke, Kate Valentine, Patrice Jones, Chris Hoag, Paul Corey, and Tim Bozik.

I would like to thank the following reviewers for their input and suggestions:

Carole Bergen, *Mercy College*
Laurel Berry, *Bryant & Stratton*
Donald Clayton, *Madisonville Community College*
Anita Collins, *Mesa Community College*

Robert Diaz, *Fullerton College*
Kathryn Gunderson, *Three Rivers Community College*
Elizabeth Hamman, *Cypress College*
Lloyd Harris, *Gulf Coast Community College*
Paul Jones, *University of Cincinnati*
Jeff Koleno, *Lorain County Community College*
Patricia Lamdin, *Chesapeake College*
Judy Langer, *Westchester Community College*
Sandy Lofstock, *St. Petersburg College*
Stan Mattoon, *Merced College*
Dr. Kris Mudunuri, *Long Beach City College*
Greg Nguyen, *Fullerton College*
Jean Olsen, *Pikes Peake Community College*
Darlene Ornelas, *Fullerton College*
Lourdes Pajo, *Pikes Peake Community College*
Elaine Paris, *Mercy College*
Marilyn Platt, *Gaston College*
Warren Powell, *Tyler Junior College*
Jeanette Shea, *Central Texas College*
Jeffrey Simmons, *Ivy Tech State College–Fort Wayne*
Edward Wagner, *Central Texas College*
Jenny Wilson, *Tyler Junior College*

I would also like to thank the following dedicated group of instructors who participated in our focus groups, Martin-Gay Summits, and our design review for the series. Their feedback and insights have helped to strengthen this edition of the text. These instructors include:

Cedric Atkins, *Mott Community College*
Laurel Berry, *Bryant & Stratton*
John Beyers, *University of Maryland*
Bob Brown, *Community College of Baltimore County–Essex*
Lisa Brown, *Community College of Baltimore County–Essex*
Gail Burkett, *Palm Beach Community College*
Cheryl Cantwell, *Seminole Community College*
Jackie Cohen, *Augusta State College*
Julie Dewan, *Mohawk Valley Community College*
Janice Ervin, *Central Piedmont Community College*
Pauline Hall, *Iowa State College*
Sonya Johnson, *Central Piedmont Community College*
Irene Jones, *Fullerton College*
Paul Jones, *University of Cincinnati*
Nancy Lange, *Inver Hills Community College*
Sandy Lofstock, *St. Petersburg College*
Jean McArthur, *Joliet Junior College*
Marica Molle, *Metropolitan Community College*
Greg Nguyen, *Fullerton College*
Linda Padilla, *Joliet Junior College*
Ena Salter, *Manatee Community College*
Carole Shapero, *Oakton Community College*
Anne Smallen, *Mohawk Valley Community College*
Jennifer Strehler, *Oakton Community College*

Tanomo Taguchi, *Fullerton College*
Leigh Ann Wheeler, *Greenville Technical Community College*
Valerie Wright, *Central Piedmont Community College*

A special thank you to those students who participated in our design review: Katherine Browne, Mike Bulfin, Nancy Canipe, Ashley Carpenter, Jeff Chojnachi, Roxanne Davis, Mike Dieter, Amy Dombrowski, Kay Herring, Todd Jaycox, Kaleena Levan, Matt Montgomery, Tony Plese, Abigail Polkinghorn, Harley Price, Eli Robinson, Avery Rosen, Robyn Schott, Cynthia Thomas, and Sherry Ward.

Additional Acknowledgments

As usual, I would like to thank my husband, Clayton, for his constant encouragement. A special thanks to my children, Eric and Bryan, for providing most of the cooking and humor in our household. I would also like to thank my extended family for their help and wonderful sense of humor. Their contributions are too numerous to list. They are Rod and Karen Pasch; Peter, Michael, Christopher, Matthew, and Jessica Callac; Stuart and Earline Martin; Josh, Mandy, Bailey, Ethan, Avery and Mia Barnes; Mark, Sabrina, and Madison Martin; Leo and Barbara Miller; and Jewett Gay.

Elayn Martin-Gay

About the Author

Elayn Martin-Gay has taught mathematics at the University of New Orleans for more than 25 years. Her numerous teaching awards include the local University Alumni Association's Award for Excellence in Teaching, and Outstanding Developmental Educator at University of New Orleans, presented by the Louisiana Association of Developmental Educators.

Prior to writing textbooks, Elayn Martin-Gay developed an acclaimed series of lecture videos to support developmental mathematics students in their quest for success. These highly successful videos originally served as the foundation material for her texts. Today, the videos are specific to each book in the Martin-Gay series. The author has also created Chapter Test Prep Videos to help students during their most "teachable moment"—as they prepare for a test, along with Instructor-to-Instructor videos that provide teaching tips, hints, and suggestions for each developmental mathematics course, including basic mathematics, prealgebra, beginning algebra, and intermediate algebra.

Elayn is the author of 10 published textbooks as well as multimedia interactive mathematics, all specializing in developmental mathematics courses. She has participated as an author across the broadest range of educational materials: textbooks, videos, tutorial software, and courseware. All of these components are designed to work together. This offers an opportunity of various combinations for an integrated teaching and learning package offering great consistency for the student.

Applications Index

1

The Whole Numbers

Whole numbers are the basic building blocks of mathematics. The whole numbers answer the question "How many?"

This chapter covers basic operations on whole numbers. Knowledge of these operations provides a good foundation on which to build further mathematical skills.

A river basin is the geographic area drained by a river and its tributaries. The Mississippi River Basin, drawn to the right, is the third largest in the world. It is divided into the six sub-basins shown. In Section 1.3, Exercises 75 through 78, page 27, we will see how whole numbers can be used to measure the area of the Mississippi River Basin.

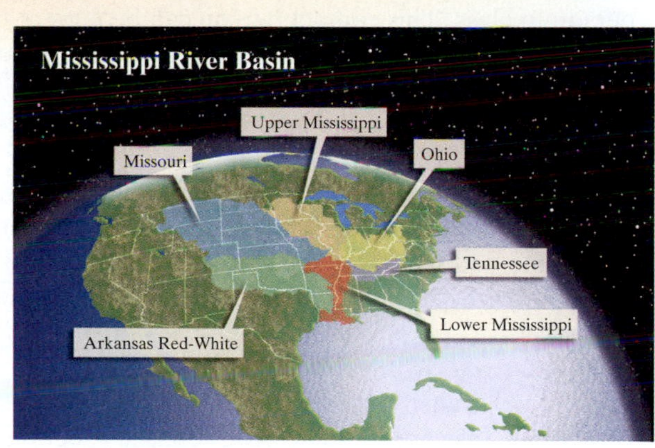

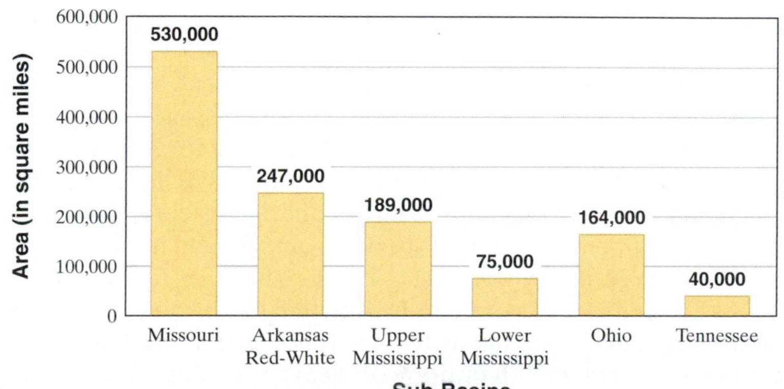

SECTIONS

CHECK YOUR PROGRESS

Objectives

- **A** Get Ready for This Course.
- **B** Understand Some General Tips for Success.
- **C** Understand How to Use This Text.
- **D** Get Help As Soon As You Need It.
- **E** Learn How to Prepare for and Take an Exam.
- **F** Develop Good Time Management.

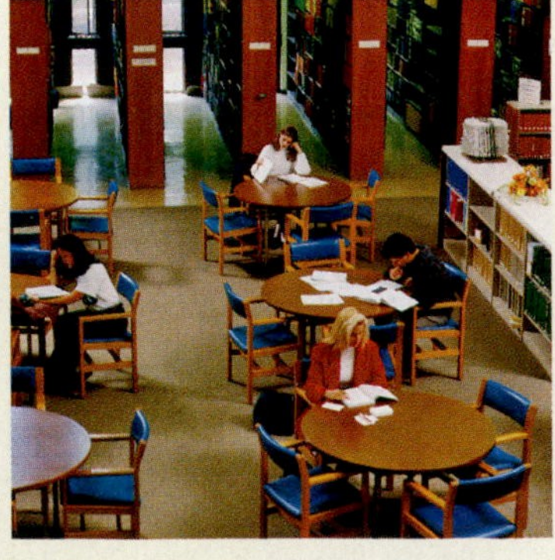

1.1 TIPS FOR SUCCESS IN MATHEMATICS

Before reading this section, remember that your instructor is your best source of information. Please see your instructor for any additional help or information.

Objective A Getting Ready for This Course

Now that you have decided to take this course, remember that a *positive attitude* will make all the difference in the world. Your belief that you can succeed is just as important as your commitment to this course. Make sure you are ready for this course by having the time and positive attitude that it takes to succeed.

Next, make sure that you have scheduled your math course at a time that will give you the best chance for success. For example, if you are also working, you may want to check with your employer to make sure that your work hours will not conflict with your course schedule.

On the day of your first class period, double-check your schedule and allow yourself extra time to arrive on time in case of traffic problems or difficulty locating your classroom. Make sure that you bring at least your textbook, paper, and a writing instrument. Are you required to have a lab manual, graph paper, calculator, or some other supplies besides this text? If so, also bring this material with you.

Objective B General Tips for Success

Below are some general tips that will increase your chance for success in a mathematics class. Many of these tips will also help you in other courses you may be taking.

Exchange names and phone numbers or e-mail addresses with at least one other person in class. This contact person can be a great help if you miss an assignment or want to discuss math concepts or exercises that you find difficult.

Choose to attend all class periods. If possible, sit near the front of the classroom. This way, you will see and hear the presentation better. It may also be easier for you to participate in classroom activities.

Do your homework. You've probably heard the phrase "practice makes perfect" in relation to music and sports. It also applies to mathematics. You will find that the more time you spend solving mathematics exercises, the easier the process becomes. Be sure to schedule enough time to complete your assignments before the next class period.

Check your work. Review the steps you made while working a problem. Learn to check your answers in the original problems. You may also compare your answers with the "Answers to Selected Exercises" section in the back of the book. If you have made a mistake, try to figure out what went wrong. Then correct your mistake. If you can't find what went wrong, don't erase your work or throw it away. Bring your work to your instructor, a tutor in a math lab, or a classmate. It is easier for someone to find where you had trouble if he or she looks at your original work.

Learn from your mistakes. Everyone, even your instructor, makes mistakes. Use your errors to learn and to become a better math student. The key is finding and understanding your errors. Was your mistake a careless one, or did you make it because you can't read your own math writing? If so, try to work more slowly or write more neatly and make a conscious effort to carefully check your work. Did you make a mistake because you don't understand a concept? If so, take the time to review the concept or ask questions to better understand it.

Know how to get help if you need it. It's all right to ask for help. In fact, it's a good idea to ask for help whenever there is something that you don't understand. Make sure you know when your instructor has office hours and how to find his or her office. Find out whether math tutoring services are available on your campus. Check on the hours, location, and requirements of the tutoring service. Know whether software is available and how to access this resource.

Organize your class materials, including homework assignments, graded quizzes and tests, and notes from your class or lab. All of these items will make valuable references throughout your course and when studying for upcoming tests and the final exam. Make sure that you can locate these materials when you need them.

Read your textbook before class. Reading a mathematics textbook is unlike reading a novel or a newspaper. Your pace will be much slower. It is helpful to have paper and a pencil with you when you read. Try to work out examples on your own as you encounter them in your text. You should also write down any questions that you want to ask in class. When you read a mathematics textbook, sometimes some of the information in a section will be unclear. But after you hear a lecture or watch a lecture video on that section, you will understand it much more easily than if you had not read your text beforehand.

Don't be afraid to ask questions. You are not the only person in class with questions. Other students are normally grateful that someone has spoken up.

Hand in assignments on time. This way you can be sure that you will not lose points for being late. Show every step of a problem and be neat and organized. Also be sure that you understand which problems are assigned for homework. If allowed, you can always double-check the assignment with another student in your class.

Objective C **Using This Text**

There are many helpful resources that are available to you in this text. It is important that you become familiar with and use these resources. They should increase your chances for success in this course.

- *Practice Problems*. Each example in every section has a parallel Practice Problem. As you read a section, try each Practice Problem after you've finished the corresponding example. This "learn-by-doing" approach will help you grasp ideas before you move on to other concepts.
- *Chapter Test Prep Video CD*. This book contains a CD. This CD contains all of the Chapter Test exercises worked out by the author. This supplement is very helpful before a classroom chapter test.
- *Lecture Video CDs*. Exercises marked with a ◉ are fully worked out by the author on video CDs. Check with your instructor for the availability of these video CDs.
- *Symbols at the beginning of an exercise set*. If you need help with a particular section, the symbols listed at the beginning of each exercise set will remind you of the numerous supplements available.
- *Objectives*. The main section of exercises in each exercise set is referenced by an objective, such as A or B, and also an example(s). There is also often a section of exercises entitled "Mixed Practice," which is referenced by two or more objectives or sections. These are mixed exercises written to prepare you for your next exam. Use all of this referencing if you have trouble completing an assignment from the exercise set.
- *Icons (Symbols)*. Make sure that you understand the meaning of the icons that are beside many exercises. ◉ tells you that the corresponding exercise may be viewed on the video segment that corresponds to that section. ✎ tells you that this exercise is a writing exercise in which you should answer in complete sentences. △ icon tells you that the exercise involves geometry.
- *Integrated Reviews*. Found in the middle of each chapter, these reviews offer you a chance to practice—in one place—the many concepts that you have learned separately over several sections.
- *End of Chapter Opportunities*. There are many opportunities at the end of each chapter to help you understand the concepts of the chapter.

  Vocabulary Checks contain key vocabulary terms introduced in the chapter.

  Chapter Highlights contain chapter summaries and examples.

  Chapter Reviews contain review problems. The first part is organized section by section and the second part contains a set of mixed exercises.

Chapter Tests are sample tests to help you prepare for an exam. The Chapter Test Prep Video CD, found in this text, contains all the Chapter Test exercises worked by the author.
Cumulative Reviews are reviews consisting of material from the beginning of the book to the end of that particular chapter.

- *The Bigger Picture*. This feature contains the directions for building an outline to be used throughout the course. The purpose of this outline is to help you make the transition from thinking "section by section" to thinking about how the mathematics in this course is part of a bigger picture.
- *Study Skills Builder*. This feature is found at the end of many exercise sets. In order to increase your chance of success in this course, please read and answer the questions in the Study Skills Builder. For your convenience, the table below contains selected Study Skills Builder titles and their location.

Study Skills Builder Title	Page of First Occurence
Learning New Terms	Page 29
What To Do the Day of an Exam?	Page 74
Have You Decided to Complete This Course Successfully?	Page 83
Organizing a Notebook	Page 121
How Are Your Homework Assignments Going?	Page 140
Are You Familiar with Your Textbook Supplements?	Page 172
How Well Do You Know Your Textbook?	Page 199
Are You Organized?	Page 226
Are You Satisfied with Your Performance on a Particular Quiz or Exam?	Page 276
Tips for Studying for an Exam	Page 313
Are You Getting All the Mathematics Help That You Need?	Page 342
How Are You Doing?	Page 442
Are You Preparing for Your Final Exam?	Page 550

See the Preface at the beginning of this text for a more thorough explanation of the features of this text.

Objective D Getting Help

If you have trouble completing assignments or understanding the mathematics, get help as soon as you need it! This tip is presented as an objective on its own because it is so important. In mathematics, usually the material presented in one section builds on your understanding of the previous section. This means that if you don't understand the concepts covered during a class period, there is a good chance that you will not understand the concepts covered during the next class period. If this happens to you, get help as soon as you can.

Where can you get help? Many suggestions have been made in this section on where to get help, and now it is up to you to do it. Try your instructor, a tutoring center, or a math lab, or you may want to form a study group with fellow classmates. If you do decide to see your instructor or go to a tutoring center, make sure that you have a neat notebook and are ready with your questions.

Objective E Preparing for and Taking an Exam

Make sure that you allow yourself plenty of time to prepare for a test. If you think that you are a little "math anxious," it may be that you are not preparing for a test in a way that will ensure success. The way that you prepare for a test in mathematics is important. To prepare for a test:

1. Review your previous homework assignments.
2. Review any notes from class and section-level quizzes you have taken. (If this is a final exam, also review chapter tests you have taken.)
3. Review concepts and definitions by reading the Chapter Highlights at the end of each chapter.
4. Practice working out exercises by completing the Chapter Review found at the end of each chapter. (If this is a final exam, go through a Cumulative Review. There is one found at the end of each chapter except Chapter 1. Choose the review found at the end of the latest chapter that you have covered in your course.) *Don't stop here!*
5. It is important that you place yourself in conditions similar to test conditions to find out how you will perform. In other words, as soon as you feel that you know the material, get a few blank sheets of paper and take a sample test. There is a Chapter Test available at the end of each chapter, or you can work selected problems from the Chapter Review. Your instructor may also provide you with a review sheet. During this sample test, do not use your notes or your textbook. Then check your sample test. If you are not satisfied with the results, study the areas that you are weak in and try again.
6. On the day of the test, allow yourself plenty of time to arrive at where you will be taking your exam.

When taking your test:

1. Read the directions on the test carefully.
2. Read each problem carefully as you take the test. Make sure that you answer the question asked.
3. Watch your time and pace yourself so that you can attempt each problem on your test.
4. If you have time, check your work and answers.
5. Do not turn your test in early. If you have extra time, spend it double-checking your work.

Objective F Managing Your Time

As a college student, you know the demands that classes, homework, work, and family place on your time. Some days you probably wonder how you'll ever get everything done. One key to managing your time is developing a schedule. Here are some hints for making a schedule:

1. Make a list of all of your weekly commitments for the term. Include classes, work, regular meetings, extracurricular activities, etc. You may also find it helpful to list such things as laundry, regular workouts, grocery shopping, etc.
2. Next, estimate the time needed for each item on the list. Also make a note of how often you will need to do each item. Don't forget to include time estimates for the reading, studying, and homework you do outside of your classes. You may want to ask your instructor for help estimating the time needed.
3. In the exercise set that follows, you are asked to block out a typical week on the schedule grid given. Start with items with fixed time slots like classes and work.
4. Next, include the items on your list with flexible time slots. Think carefully about how best to schedule items such as study time.
5. Don't fill up every time slot on the schedule. Remember that you need to allow time for eating, sleeping, and relaxing! You should also allow a little extra time in case some items take longer than planned.
6. If you find that your weekly schedule is too full for you to handle, you may need to make some changes in your workload, classload, or in other areas of your life. You may want to talk to your advisor, manager or supervisor at work, or someone in your college's academic counseling center for help with such decisions.

1.1 EXERCISE SET

FOR EXTRA HELP

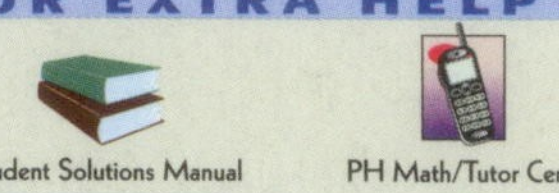

1. What is your instructor's name?
2. What are your instructor's office location and office hours?
3. What is the best way to contact your instructor?
4. Do you have the name and contact information of at least one other student in class?
5. Will your instructor allow you to use a calculator in this class?
6. Is tutorial software available to you? If so, what type and where?
7. Is there a tutoring service available on campus? If so, what are its hours? What services are available?
8. Have you attempted this course before? If so, write down ways that you might improve your chances of success during this second attempt.
9. List some steps that you can take if you begin having trouble understanding the material or completing an assignment.
10. How many hours of studying does your instructor advise for each hour of instruction?
11. What does the ✎ icon in this text mean?
12. What does the ◎ icon in this text mean?
13. What does the △ icon in this text mean?
14. Search the minor columns in your text. What are Practice Problems?
15. When might be the best time to work a Practice Problem?
16. Where are the answers to Practice Problems?
17. What answers are contained in this text and where are they?
18. What solutions are contained in this text and where are they?
19. What and where are Integrated Reviews?
20. What video CD is contained in this book, where is it, and what material is on it?
21. Chapter Highlights are found at the end of each chapter. Find the Chapter 1 Highlights and explain how you might use it and how it might be helpful.
22. Chapter Reviews are found at the end of each chapter. Find the Chapter 1 Review and explain how you might use it and how it might be useful.
23. Chapter Tests are found at the end of each chapter. Find the Chapter 1 Test and explain how you might use it and how it might be helpful when preparing for an exam on Chapter 1. Include how the Chapter Test Prep Video in this book may help.
24. Read or reread objective **F** and fill out the schedule grid below.

	Monday	Tuesday	Wednesday	Thursday	Friday	Saturday	Sunday
7:00 a.m.							
8:00 a.m.							
9:00 a.m.							
10:00 a.m.							
11:00 a.m.							
12:00 noon							
1:00 p.m.							
2:00 p.m.							
3:00 p.m.							
4:00 p.m.							
5:00 p.m.							
6:00 p.m.							
7:00 p.m.							
8:00 p.m.							
9:00 p.m.							

1.2 PLACE VALUE AND NAMES FOR NUMBERS

The **digits** 0, 1, 2, 3, 4, 5, 6, 7, 8, and 9 can be used to write numbers. For example, the **whole numbers** are

0, 1, 2, 3, 4, 5, 6, 7, 8, 9, 10, 11, . . .

and the **natural numbers** are 1, 2, 3, 4, 5, 6, 7, 8, 9, 10, 11, ...

The three dots (. . .) after the 11 mean that this list continues indefinitely. That is, there is no largest whole number. The smallest whole number is 0.

Objective A Finding the Place Value of a Digit in a Whole Number

The position of each digit in a number determines its **place value.** For example, the distance (in miles) between the planet Mercury and the planet Earth can be represented by the whole number 48,337,000. Below is a place-value chart for this whole number.

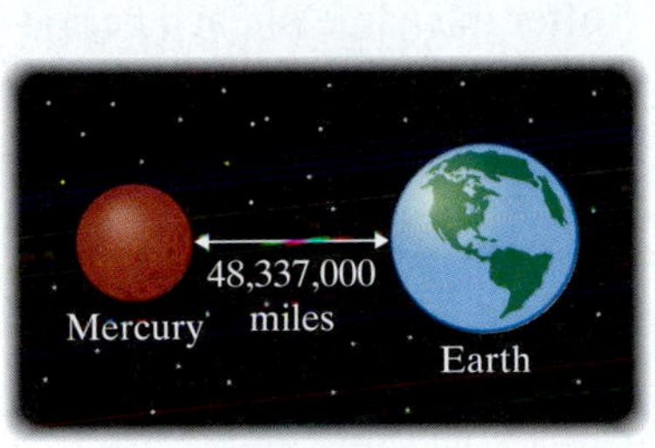

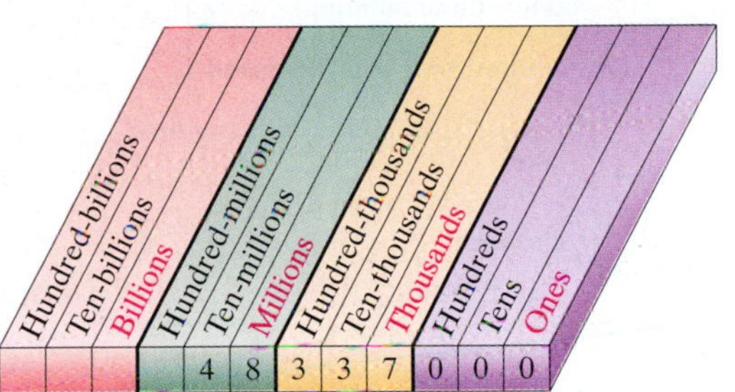

The two 3s in 48,337,000 represent different amounts because of their different placements. The place value of the 3 on the left is hundred-thousands. The place value of the 3 on the right is ten-thousands.

EXAMPLES Find the place value of the digit 3 in each whole number.

1. 396,418 ↑ hundred-thousands

2. 93,192 ↑ thousands

3. 534,275,866 ↑ ten-millions

Work Practice Problems 1–3

Objective B Writing a Whole Number in Words and in Standard Form

A whole number such as 1,083,664,500 is written in **standard form.** Notice that commas separate the digits into groups of three, starting from the right. Each group of three digits is called a **period.** The names of the first four periods are shown in red.

Objectives

A Find the Place Value of a Digit in a Whole Number.

B Write a Whole Number in Words and in Standard Form.

C Write a Whole Number in Expanded Form.

D Read Tables.

PRACTICE PROBLEMS 1–3

Find the place value of the digit 8 in each whole number.

1. 38,760,005

2. 67,890

3. 481,922

Answers

1. millions **2.** hundreds **3.** ten-thousands

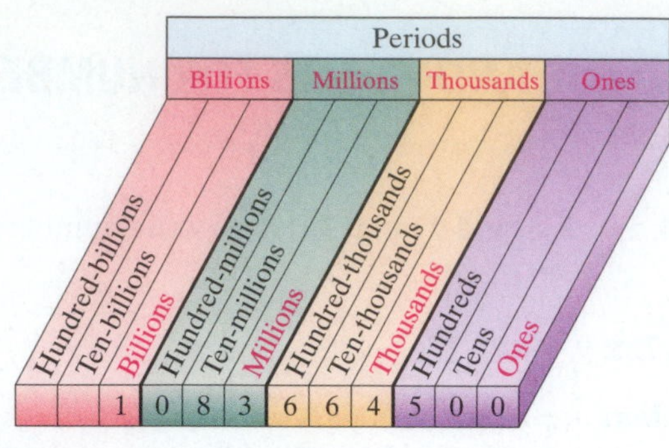

Writing a Whole Number in Words

To write a whole number in words, write the number in each period followed by the name of the period. (The ones period is usually not written.) This same procedure can be used to read a whole number.

For example, we write 1,083,664,500 as

one **billion**,
eighty-three **million**,
six hundred sixty-four **thousand**,
five **hundred**

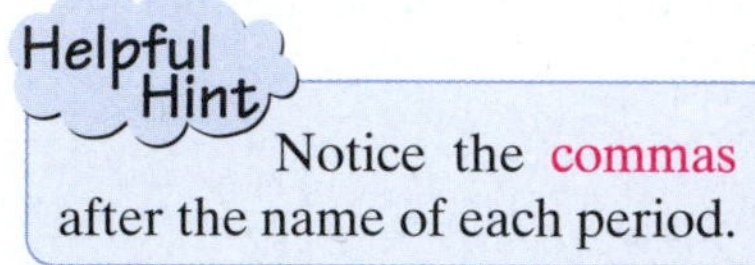

Notice the commas after the name of each period.

Helpful Hint The name of the ones period is not used when reading and writing whole numbers. For example,

9,265

is read as

"nine **thousand**, two **hundred** sixty-five."

PRACTICE PROBLEMS 4–6

Write each number in words.

4. 54

5. 678

6. 93,205

EXAMPLES Write each number in words.

4. 72 seventy-two

5. 546 five hundred forty-six

6. 27,034 twenty-seven thousand, thirty-four

Work Practice Problems 4–6

Helpful Hint The word "and" is *not* used when reading and writing whole numbers. It is used when reading and writing mixed numbers and some decimal values, as shown later in this text.

PRACTICE PROBLEM 7

Write 679,430,105 in words.

EXAMPLE 7 Write 308,063,557 in words.

Solution: 308,063,557 is written as

three hundred eight **million**, sixty-three **thousand**, five **hundred** fifty-seven

Work Practice Problem 7

✔**Concept Check** True or false? When writing a check for \$2600, the word name we write for the dollar amount of the check is "two thousand sixty." Explain your answer.

Answers

4. fifty-four **5.** six hundred seventy-eight **6.** ninety-three thousand, two hundred five **7.** six hundred seventy-nine million, four hundred thirty thousand, one hundred five

✔ Concept Check Answer

false

Writing a Whole Number in Standard Form

To write a whole number in standard form, write the number in each period, followed by a comma.

EXAMPLES Write each number in standard form.

8. forty-one 41 **9.** seven hundred eight 708

10. six thousand, four hundred ninety-three

6,493 or 6493

11. three million, seven hundred forty-six thousand, five hundred twenty-two

3,746,522

Work Practice Problems 8–11

PRACTICE PROBLEMS 8–11

Write each number in standard form.

8. thirty-seven

9. two hundred twelve

10. eight thousand, two hundred seventy-four

11. five million, fifty-seven thousand, twenty-six

Helpful Hint A comma may or may not be inserted in a four-digit number. For example, both

6,493 and 6493

are acceptable ways of writing six thousand, four hundred ninety-three.

Objective C Writing a Whole Number in Expanded Form

The place value of a digit can be used to write a number in expanded form. The **expanded form** of a number shows each digit of the number with its place value. For example, 5672 is written in expanded form as

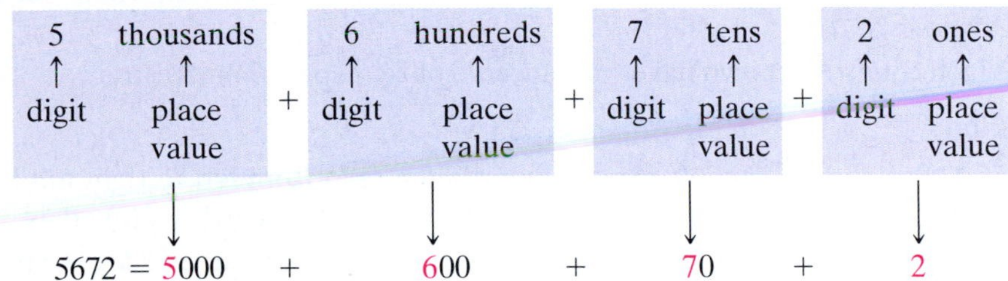

EXAMPLE 12 Write 5,207,034 in expanded form.

Solution: 5,000,000 + 200,000 + 7,000 + 30 + 4

Work Practice Problem 12

PRACTICE PROBLEM 12

Write 4,026,301 in expanded form.

We can visualize whole numbers by points on a line. The line below is called a **number line.** This number line has equally spaced marks for each whole number. The arrow to the right simply means that the whole numbers continue indefinitely. In other words, there is no largest whole number.

Number Line

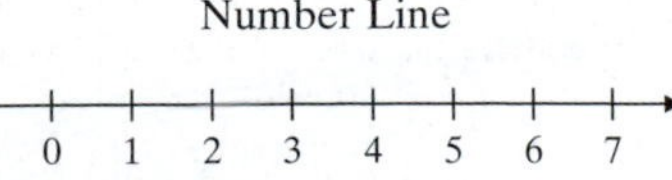

We will study number lines further in Section 1.4.

Answers

8. 37 **9.** 212 **10.** 8,274 **11.** 5,057
12. 4,000,000 + 20,000 + 6000 + 3

Objective D Reading Tables

Now that we know about place value and names for whole numbers, we introduce one way that whole number data may be presented. **Tables** are often used to organize and display facts that contain numbers. The following table shows the countries that have won the most medals during the Olympic Winter Games for the years 1924 through 2006. (Although the medals are truly won by athletes from the various countries, for simplicity we will state that countries have won the medals.)

Most Medals—Olympic Winter (1924–2006) Games

Country	Gold	Silver	Bronze	Total	Country	Gold	Silver	Bronze	Total
Germany*	118	116	92	326	Finland	41	57	52	150
Russia	121	89	86	296	Sweden	46	32	44	122
Norway	96	100	83	279	Canada	37	38	43	118
United States	78	80	58	216	Switzerland	37	37	43	117
Austria	50	64	71	185					

(*Source:* International Olympic Committee)
*Includes West Germany 1952, 1968–1988; East Germany 1968–1988

For example, by reading from left to right along the row marked "United States," we find that the United States won 78 gold, 80 silver, and 58 bronze medals during the Olympic Winter Games from the years 1924 to 2006.

PRACTICE PROBLEM 13

Use the Winter Games table to answer each question.

a. How many gold medals did Norway win during the Olympic Winter Games?

b. Which countries shown have won more than 100 gold medals?

EXAMPLE 13 Use the Winter Games table to answer each question.

a. How many silver medals did Sweden win during the Winter Games of the Olympics?

b. Which countries shown have won fewer bronze medals than Austria?

Solution:

a. Find "Sweden" in the left-hand column. Then read from left to right until the "silver" column is reached. We find that Sweden has won 32 silver medals.

b. Austria has won 71 bronze medals. United States, Finland, Sweden, Canada, and Switzerland have each won fewer than 71 bronze medals.

Work Practice Problem 13

Answers

13. a. 96 **b.** Germany, Russia

Vocabulary and Readiness Check

Use the choices below to fill in each blank.

standard form	period	whole
expanded form	place value	words

1. The numbers 0, 1, 2, 3, 4, 5, 6, 7, 8, 9, 10, 11, 12, … are called ___________ numbers.
2. The number 1,286 is written in ___________.
3. The number "twenty-one" is written in ___________.
4. The number 900 + 60 + 5 is written in ___________.
5. In a whole number, each group of three digits is called a(n) ___________.
6. The ___________ of the digit 4 in the whole number 264 is ones.

1.2 EXERCISE SET

FOR EXTRA HELP

Student Solutions Manual

PH Math/Tutor Center

CD/Video for Review

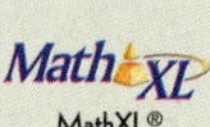

MathXL®

MyMathLab

Objective A *Determine the place value of the digit 5 in each whole number. See Examples 1 through 3.*

1. 657 **2.** 905 **3.** 5423 **4.** 6527

5. 43,526,000 **6.** 79,050,000 **7.** 5,408,092 **8.** 51,682,700

Objective B *Write each whole number in words. See Examples 4 through 7.*

9. 354 **10.** 316 **11.** 8279 **12.** 5445

13. 26,990 **14.** 42,009 **15.** 2,388,000 **16.** 3,204,000

17. 24,350,185 **18.** 47,033,107

Write each number in the sentence in words. See Examples 4 through 7.

19. As of this writing, the population of Bermuda is 65,773. (*Source*: *The World Factbook*)

20. Each Home Depot store in the United States and Canada stocks at least 40,000 different kinds of building materials, home improvement supplies, and lawn and garden products. (*Source:* The Home Depot, Inc.)

21. The world's tallest building, the Taipei 101 building in Taiwan is 1679 feet tall. (*Source:* Council on Tall Buildings and Urban Habitat)

22. In a recent year, there were 6912 patients in the United States waiting for an organ transplant. (*Source:* United Network for Organ Sharing)

23. Each day, FedEx Ground delivers 2,800,000 packages throughout the United States, Canada, and Puerto Rico. (*Source:* FedEx Corporation)

24. In a recent year, the approximate attendance figure for Disney World's Magic Kingdom was 15,200,000 people. (*Source: Amusement Business*)

25. The highest point in Oregon is Mount Hood, at an elevation of 11,239 feet. (*Source:* U.S. Geological Survey)

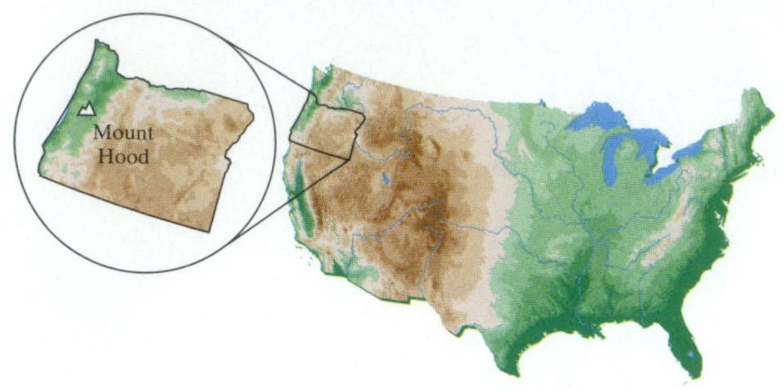

26. The highest point in Colorado is Mount Elbert, at an elevation of 14,433 feet. (*Source:* U.S. Geological Survey)

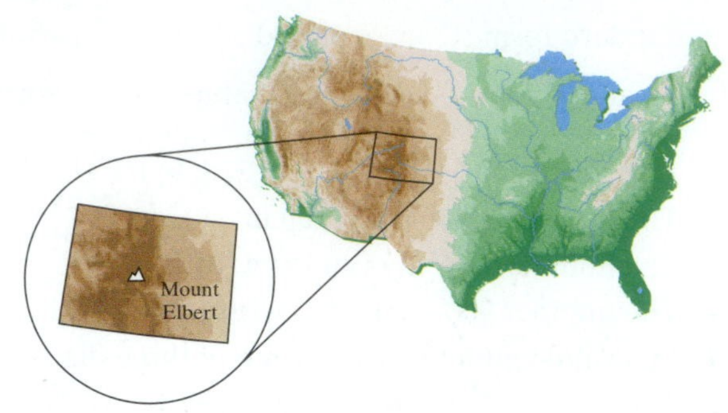

27. The Goodyear blimp *Eagle* holds 202,700 cubic feet of helium. (*Source:* The Goodyear Tire & Rubber Company)

28. In a recent year, zinc mines in the United States mined 799,000 metric tons of zinc. (*Source:* U.S. Dept. of Interior)

Write each whole number in standard form. See Examples 8 through 11.

29. Six thousand, five hundred eighty-seven

30. Four thousand, four hundred sixty-eight

31. Fifty-nine thousand, eight hundred

32. Seventy-three thousand, two

33. Thirteen million, six hundred one thousand, eleven

34. Sixteen million, four hundred five thousand, sixteen

35. Seven million, seventeen

36. Two million, twelve

37. Two hundred sixty thousand, nine hundred ninety-seven

38. Six hundred forty thousand, eight hundred eighty-one

Write the whole number in each sentence in standard form. See Examples 8 through 11.

39. The Hubble Space Telescope orbits above Earth at an altitude of three hundred fifty-three miles. (*Source:* Hubblesite.org)

40. The average distance between the surfaces of the Earth and the Moon is about two hundred thirty-four thousand miles.

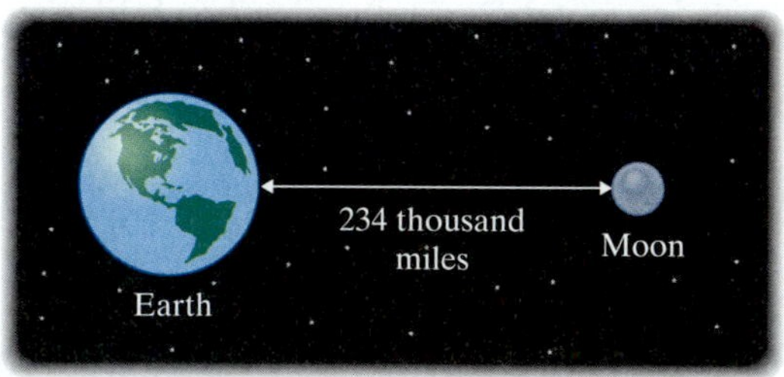

41. The price for a 2006 Porsche Carrera GT is four hundred eighty-four thousand, two hundred thirty-five dollars. (*Source:* Porsche Cars North America)

42. The world's tallest self-supporting structure is the CN Tower in Toronto, Canada. It is one thousand, eight hundred fifteen feet tall. (*Source*: *The World Almanac*, 2006)

43. The Buena Vista film *Pirates of the Caribbean: Dead Man's Chest* set the world record for opening day income when it took in fifty-four million, five hundred thousand dollars on July 7, 2006. (*Source: Chicago Tribune,* 2006)

44. The Twentieth Century Fox film *StarWars: Episode III—Revenge of the Sith* holds the record for second highest opening day income; it took in fifty million, thirteen thousand, eight hundred fifty-nine dollars on May 19, 2005. (*Source: Guinness Book,* 2006)

45. In 2006, Barry Bonds surpassed Babe Ruth's home run record in Major League baseball of seven hundred fourteen home runs.

46. As of this writing, Hank Aaron holds the career record for home runs in Major League baseball since 1974, with a total of seven hundred fifty-five home runs. (*Source:* Major League Baseball)

Objective **C** *Write each whole number in expanded form. See Example 12.*

47. 209 **48.** 789 **49.** 3470 **50.** 6040

51. 80,774 **52.** 20,215 **53.** 66,049 **54.** 99,032

55. 39,680,000 **56.** 47,703,029

Objectives **B** **C** **D** **Mixed Practice** *The table shows the six tallest mountains in New England and their elevations. Use this table to answer Exercises 57 through 62. See Example 13.*

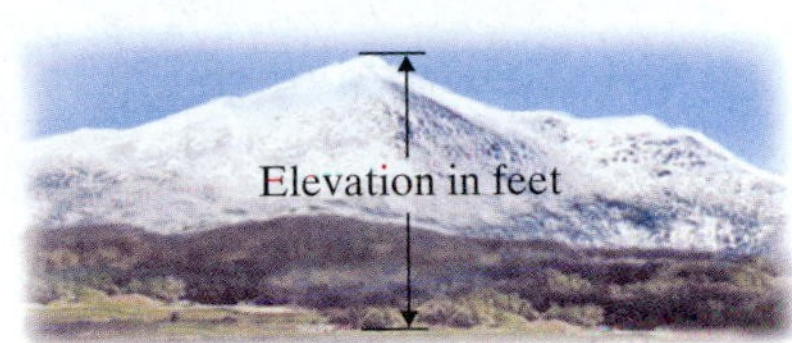

Mountain (State)	Elevation (in feet)
Boott Spur (NH)	5492
Mt. Adams (NH)	5774
Mt. Clay (NH)	5532
Mt. Jefferson (NH)	5712
Mt. Sam Adams (NH)	5584
Mt. Washington (NH)	6288

Source: U.S. Geological Survey

57. Write the elevation of Mt. Clay in standard form and then in words.

58. Write the elevation of Mt. Washington in standard form and then in words.

59. Write the height of Boott Spur in expanded form.

60. Write the height of Mt. Jefferson in expanded form.

61. Which mountain is the tallest in New England?

62. Which mountain is the second tallest in New England?

The table shows the top ten popular breeds of dogs in 2004 according to the American Kennel Club. Use this table to answer Exercises 63 through 68. See Example 13.

Top Ten American Kennel Club Registrations in 2004

Breed	Number of Registered Dogs	Average Dog Maximum Height (in inches)	Average Dog Maximum Weight (in pounds)
Beagle	44,557	15	30
Boxer	37,744	25	70
Chihuahua	24,853	9	6
Dachshund	40,774	9	25
German shepherd dog	46,054	26	95
Golden retriever	52,560	24	80
Labrador retriever	146,714	25	75
Poodle (standard, miniature, and toy)	32,673	standard: 26	standard: 70
Shih Tzu	28,960	11	16
Yorkshire terrier	43,527	9	7

(*Source:* American Kennel Club)

63. Which breed has fewer dogs registered, Chihuahua or Golden retriever?

64. Which breed has more dogs registered, Beagle or Yorkshire terrier?

65. Which breed has the most American Kennel Club registrations? Write the number of registrations for this breed in words.

66. Which of the listed breeds has the fewest registrations? Write the number of registered dogs for this breed in words.

67. What is the maximum weight of an average-size Dachshund?

68. What is the maximum height of an average-size Yorkshire terrier?

Concept Extensions

69. Write the largest four-digit number that can be made from the digits 1, 9, 8, and 6 if each digit must be used once.

____ ____ ____ ____

70. Write the largest five-digit number that can be made using the digits 5, 3, and 7 if each digit must be used at least once.

____ ____ ____ ____ ____

Check to see whether each number written in standard form matches the number written in words. If not, correct the number in words. See the Concept Check in this section.

71.

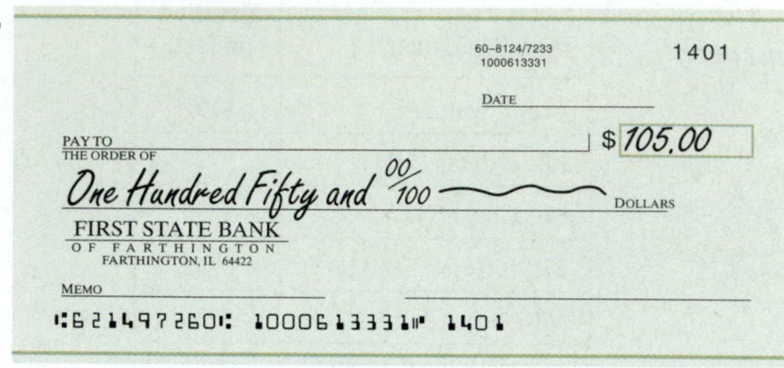

72.

73. If a number is given in words, describe the process used to write this number in standard form.

74. If a number is written in standard form, describe the process used to write this number in expanded form.

75. The world's fastest super computer is Blue Gene/L, a joint project of IBM and the Department of Energy. It can perform 135.5 trillion calculations in a second. Look up "trillion" in a dictionary and use the definition to write this number in standard form (*Source:* IBM).

1.3 ADDING AND SUBTRACTING WHOLE NUMBERS, AND PERIMETER

Objectives

A Add Whole Numbers.

B Subtract Whole Numbers.

C Find the Perimeter of a Polygon.

D Solve Problems by Adding or Subtracting Whole Numbers.

Objective A Adding Whole Numbers

The iPod is a hard drive–based portable audio player. As of 2006, it is the most popular digital music player in the United States.

Suppose that an electronics store received a shipment of two boxes of iPods one day and an additional four boxes of iPods the next day. The **total** shipment in the two days can be found by adding 2 and 4.

2 boxes of iPods + 4 boxes of iPods = 6 boxes of iPods

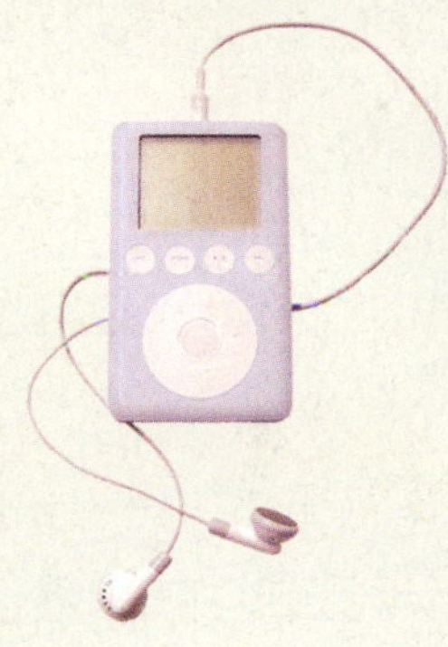

The **sum** (or total) is 6 boxes of iPods. Each of the numbers 2 and 4 is called an **addend,** and the process of finding the sum is called **addition.**

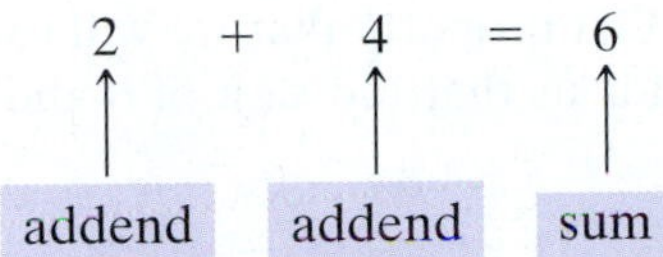

To add whole numbers, we add the digits in the ones place, then the tens place, then the hundreds place, and so on. For example, let's add 2236 + 160.

```
 2236    Line up numbers vertically so that the place values correspond. Then
+ 160    add digits in corresponding place values, starting with the ones place.
 2396
```

2396: sum of ones; sum of tens; sum of hundreds; sum of thousands

EXAMPLE 1 Add: 46 + 713

Solution:

```
   46
+ 713
  759
```

■ Work Practice Problem 1

PRACTICE PROBLEM 1

Add: 4135 + 252

Adding by Carrying

When the sum of digits in corresponding place values is more than 9, **carrying** is necessary. For example, to add 365 + 89, add the ones-place digits first.

Carrying

```
   1
  365
+  89    5 ones + 9 ones = 14 ones or 1 ten + 4 ones
    4    Write the 4 ones in the ones place and carry the 1 ten to the tens place.
```

Next, add the tens-place digits.

```
  11
  365
+  89    1 ten + 6 tens + 8 tens = 15 tens or 1 hundred + 5 tens
   54    Write the 5 tens in the tens place and carry the 1 hundred to the hundreds place.
```

Next, add the hundreds-place digits.

```
  11
  365
+  89    1 hundred + 3 hundreds = 4 hundreds
  454    Write the 4 hundreds in the hundreds place.
```

Answer

1. 4387

PRACTICE PROBLEM 2

Add: 47,364 + 135,898

EXAMPLE 2 Add: 46,278 + 124,931

Solution:

$$\begin{array}{r} {\scriptstyle 1\,1\;\,1} \\ 46{,}278 \\ +\ 124{,}931 \\ \hline 171{,}209 \end{array}$$

Work Practice Problem 2

✔Concept Check What is wrong with the following computation?

$$\begin{array}{r} 394 \\ +\ 283 \\ \hline 577 \end{array}$$

Before we continue adding whole numbers, let's review some properties of addition that you may have already discovered. The first property that we will review is the **addition property of 0.** This property reminds us that the sum of 0 and any number is that same number.

Addition Property of 0

The sum of 0 and any number is that number. For example,

$7 + 0 = 7$
$0 + 7 = 7$

Next, notice that we can add any two whole numbers in any order and the sum is the same. For example,

$4 + 5 = 9$ and $5 + 4 = 9$

We call this special property of addition the **commutative property of addition.**

Commutative Property of Addition

Changing the **order** of two addends does not change their sum. For example,

$2 + 3 = 5$ and $3 + 2 = 5$

Another property that can help us when adding numbers is the **associative property of addition.** This property states that when adding numbers, the grouping of the numbers can be changed without changing the sum. We use parentheses to group numbers. They indicate which numbers to add first. For example, let's use two different groupings to find the sum of 2 + 1 + 5.

$$(2 + 1) + 5 = 3 + 5 = 8$$

Also,

$$2 + (1 + 5) = 2 + 6 = 8$$

Both groupings give a sum of 8.

Answer

2. 183,262

✔ Concept Check Answer

forgot to carry 1 hundred to the hundreds place

Associative Property of Addition

Changing the **grouping** of addends does not change their sum. For example,

$$3 + (5 + 7) = 3 + 12 = 15 \quad \text{and} \quad (3 + 5) + 7 = 8 + 7 = 15$$

The commutative and associative properties tell us that we can add whole numbers using any order and grouping that we want.

When adding several numbers, it is often helpful to look for two or three numbers whose sum is 10, 20, and so on. Why? Adding multiples of 10 such as 10 and 20 is easier.

EXAMPLE 3 Add: $13 + 2 + 7 + 8 + 9$

Solution: $13 + 2 + 7 + 8 + 9 = 39$

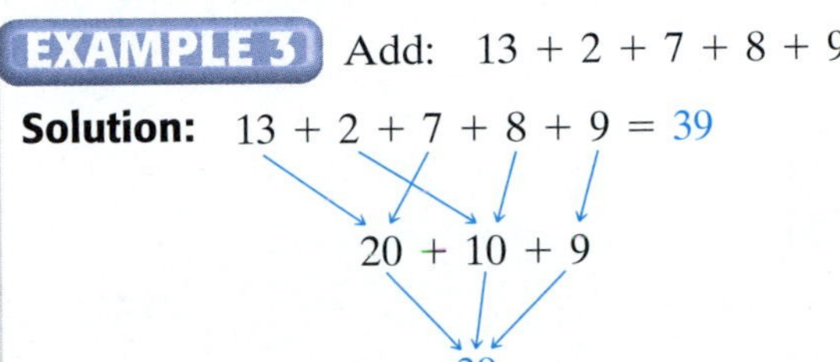

Work Practice Problem 3

PRACTICE PROBLEM 3

Add: $12 + 4 + 8 + 6 + 5$

Feel free to use the process of Example 3 anytime when adding.

EXAMPLE 4 Add: $1647 + 246 + 32 + 85$

Solution:

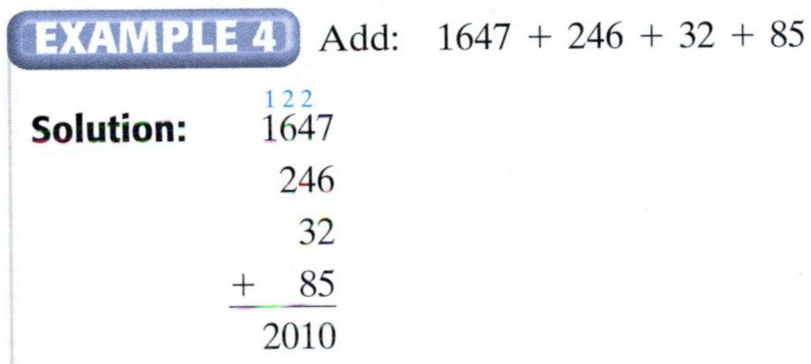

Work Practice Problem 4

PRACTICE PROBLEM 4

Add: $6432 + 789 + 54 + 28$

Objective B Subtracting Whole Numbers

If you have \$5 and someone gives you \$3, you have a total of \$8, since $5 + 3 = 8$. Similarly, if you have \$8 and then someone borrows \$3, you have \$5 left. **Subtraction** is finding the **difference** of two numbers.

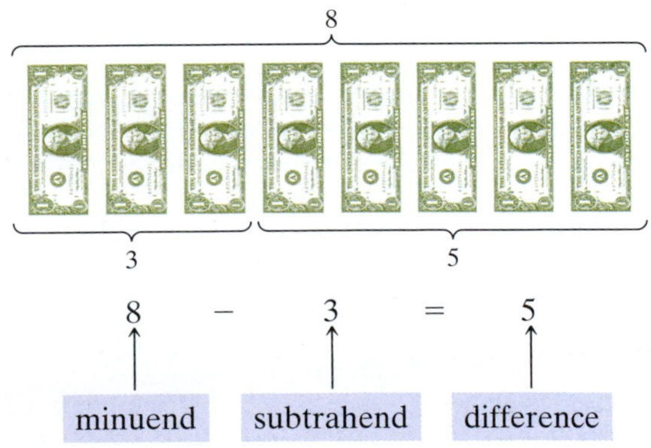

$$8 - 3 = 5$$

minuend subtrahend difference

Answers
3. 35 **4.** 7303

In this example, 8 is the **minuend,** and 3 is the **subtrahend.** The **difference** between these two numbers, 8 and 3, is 5.

Notice that addition and subtraction are very closely related. In fact, subtraction is defined in terms of addition.

$8 - 3 = 5$ because $5 + 3 = 8$

This means that subtraction can be *checked* by addition, and we say that addition and subtraction are reverse operations.

PRACTICE PROBLEM 5

Subtract. Check each answer by adding.

a. $14 - 6$

b. $20 - 8$

c. $93 - 93$

d. $42 - 0$

EXAMPLE 5 Subtract. Check each answer by adding.

a. $12 - 9$ **b.** $22 - 7$ **c.** $35 - 35$ **d.** $70 - 0$

Solution:

a. $12 - 9 = 3$ because $3 + 9 = 12$
b. $22 - 7 = 15$ because $15 + 7 = 22$
c. $35 - 35 = 0$ because $0 + 35 = 35$
d. $70 - 0 = 70$ because $70 + 0 = 70$

Work Practice Problem 5

Look again at Examples 5(c) and 5(d).

5(c) $35 - 35 = 0$ — 35 and 35: same number; 0: difference is 0

5(d) $70 - 0 = 70$ — $70 - 0$: a number minus 0; 70: difference is the same number

These two examples illustrate the subtraction properties of 0.

Subtraction Properties of 0

The difference of any number and that same number is 0. For example,

$11 - 11 = 0$

The difference of any number and 0 is that same number. For example,

$45 - 0 = 45$

To subtract whole numbers we subtract the digits in the ones place, then the tens place, then the hundreds place, and so on. When subtraction involves numbers of two or more digits, it is more convenient to subtract vertically. For example, to subtract $893 - 52$,

```
 893  ← minuend
− 52  ← subtrahend
 841  ← difference
```

(8 − 0, 9 − 5, 3 − 2)

Line up the numbers vertically so that the minuend is on top and the place values correspond. Subtract in corresponding place values, starting with the ones place.

To check, add.

```
  difference     or     841
+ subtrahend          +  52
  minuend               893  ← Since this is the original minuend, the problem checks.
```

Answers

5. a. 8 **b.** 12 **c.** 0 **d.** 42

EXAMPLE 6 Subtract: 7826 − 505. Check by adding.

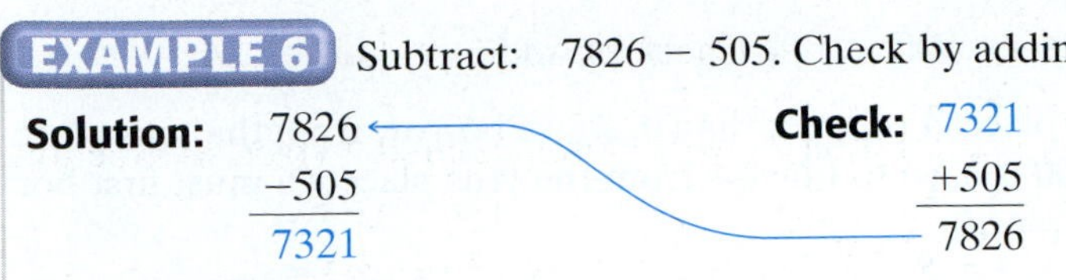

Solution:

$$\begin{array}{r} 7826 \\ -505 \\ \hline 7321 \end{array}$$

Check:

$$\begin{array}{r} 7321 \\ +505 \\ \hline 7826 \end{array}$$

■ **Work Practice Problem 6**

PRACTICE PROBLEM 6

Subtract. Check by adding.

a. 9163 − 142

b. 978 − 851

Subtracting by Borrowing

When subtracting vertically, if a digit in the second number (subtrahend) is larger than the corresponding digit in the first number (minuend), **borrowing** is necessary. For example, consider

$$\begin{array}{r} 81 \\ -63 \end{array}$$

Since the 3 in the ones place of 63 is larger than the 1 in the ones place of 81, borrowing is necessary. We borrow 1 ten from the tens place and add it to the ones place.

Borrowing

8 tens − 1 ten = 7 tens → 7 11 ← 1 ten + 1 one = 11 ones

$$\begin{array}{r} \overset{7}{\cancel{8}}\overset{11}{\cancel{1}} \\ -6\,3 \end{array}$$

Now we subtract the ones-place digits and then the tens-place digits.

$$\begin{array}{r} \overset{7}{\cancel{8}}\overset{11}{\cancel{1}} \\ -63 \\ \hline 18 \end{array}$$

8 ← 11 − 3 = 8

1 ← 7 − 6 = 1

Check:

$$\begin{array}{r} 18 \\ +63 \\ \hline 81 \end{array}$$

81 The original minuend.

EXAMPLE 7 Subtract: 543 − 29. Check by adding.

Solution:

$$\begin{array}{r} 54\overset{3\ 13}{\cancel{3}} \\ -29 \\ \hline 514 \end{array}$$

Check:

$$\begin{array}{r} 514 \\ +29 \\ \hline 543 \end{array}$$

■ **Work Practice Problem 7**

PRACTICE PROBLEM 7

Subtract. Check by adding.

a. $\begin{array}{r} 697 \\ -\ 49 \\ \hline \end{array}$

b. $\begin{array}{r} 326 \\ -245 \\ \hline \end{array}$

c. $\begin{array}{r} 1234 \\ -\ 822 \\ \hline \end{array}$

Sometimes we may have to borrow from more than one place. For example, to subtract 7631 − 152, we first borrow from the tens place.

$$\begin{array}{r} 76\overset{2}{\cancel{3}}\overset{11}{\cancel{1}} \\ -\ 152 \\ \hline 9 \end{array}$$

← 11 − 2 = 9

In the tens place, 5 is greater than 2, so we borrow again. This time we borrow from the hundreds place.

6 hundreds − **1 hundred** = 5 hundreds

1 hundred + 2 tens or 10 tens + 2 tens = 12 tens

$$\begin{array}{r} 7\overset{5}{\cancel{6}}\overset{12}{\cancel{2}}\overset{11}{\cancel{1}} \\ -152 \\ \hline 7479 \end{array}$$

Check:

$$\begin{array}{r} 7479 \\ +152 \\ \hline 7631 \end{array}$$

7631 The original minuend.

Answers

6. a. 9021 **b.** 127

7. a. 648 **b.** 81 **c.** 412

PRACTICE PROBLEM 8

Subtract. Check by adding.

a. $\begin{array}{r} 400 \\ -\ 164 \\ \hline \end{array}$

b. $\begin{array}{r} 1000 \\ -\ 762 \\ \hline \end{array}$

EXAMPLE 8 Subtract: 900 − 174. Check by adding.

Solution: In the ones place, 4 is larger than 0, so we borrow from the tens place. But the tens place of 900 is 0, so to borrow from the tens place we must first borrow from the hundreds place.

$$\begin{array}{r} \scriptstyle 8\ 10\ \ \\ \not{9}\,\not{0}\,0 \\ -\,1\,7\,4 \\ \hline \end{array}$$

Now borrow from the tens place.

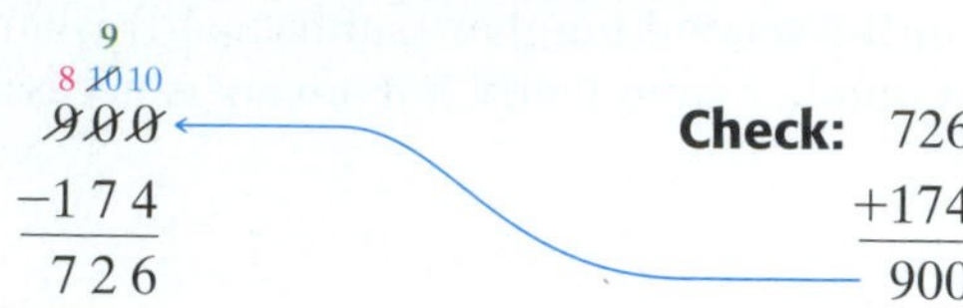

Work Practice Problem 8

Objective C Finding the Perimeter of a Polygon

In geometry, addition is used to find the perimeter of a polygon. A **polygon** can be described as a flat figure formed by line segments connected at their ends. (For more review, see Appendix A.1.) Geometric figures such as triangles, squares, and rectangles are called polygons.

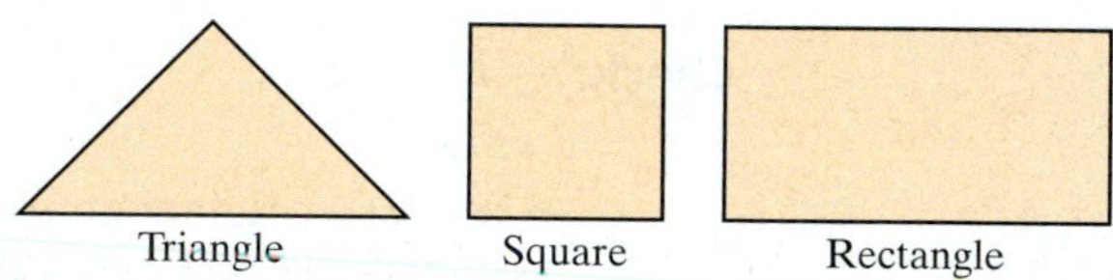

The **perimeter** of a polygon is the *distance around* the polygon. This means that the perimeter of a polygon is the sum of the lengths of its sides.

PRACTICE PROBLEM 9

Find the perimeter of the polygon shown. (A centimeter is a unit of length in the metric system.)

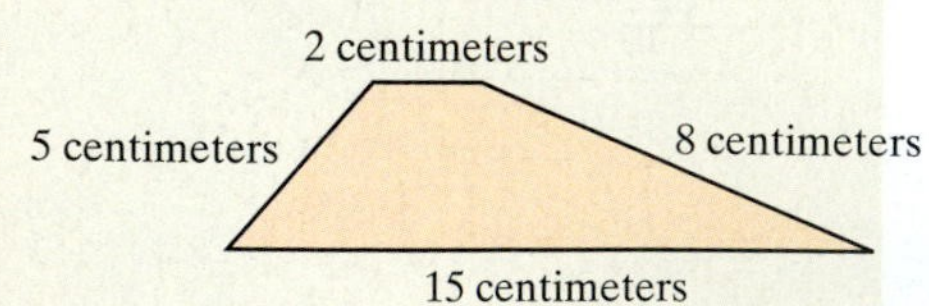

EXAMPLE 9 Find the perimeter of the polygon shown.

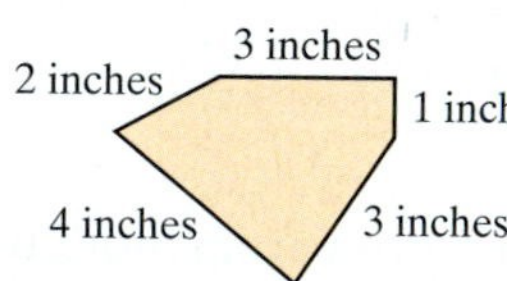

Solution: To find the perimeter (distance around), we add the lengths of the sides.

2 in. + 3 in. + 1 in. + 3 in. + 4 in. = 13 in.

The perimeter is 13 inches.

Work Practice Problem 9

To make the addition appear simpler, we will often not include units with the addends. If you do this, make sure units are included in the final answer.

Answers

8. a. 236 **b.** 238
9. 30 cm

EXAMPLE 10 Calculating the Perimeter of a Building

The largest commercial building in the world under one roof is the flower auction building of the cooperative VBA in Aalsmeer, Netherlands. The floor plan is a rectangle that measures 776 meters by 639 meters. Find the perimeter of this building. (A meter is a unit of length in the metric system.) (*Source: The Handy Science Answer Book,* Visible Ink Press)

Solution: Recall that opposite sides of a rectangle have the same length. To find the perimeter of this building, we add the lengths of the sides. The sum of the lengths of its sides is

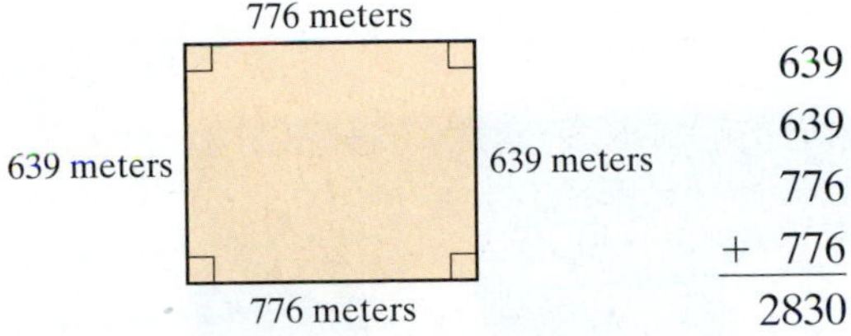

$$\begin{array}{r} 639 \\ 639 \\ 776 \\ +\ 776 \\ \hline 2830 \end{array}$$

The perimeter of the building is 2830 meters.

Work Practice Problem 10

PRACTICE PROBLEM 10

A park is in the shape of a triangle. Each of the park's three sides is 647 feet. Find the perimeter of the park.

Objective D Solving Problems by Adding or Subtracting

Often, real-life problems occur that can be solved by adding or subtracting. The first step in solving any word problem is to *understand* the problem by reading it carefully.

Descriptions of problems solved through addition or subtraction *may* include any of these key words or phrases:

Addition		
Key Words or Phrases	**Examples**	**Symbols**
added to	5 added to 7	7 + 5
plus	0 plus 78	0 + 78
increased by	12 increased by 6	12 + 6
more than	11 more than 25	25 + 11
total	the total of 8 and 1	8 + 1
sum	the sum of 4 and 133	4 + 133

Subtraction		
Key Words or Phrases	**Examples**	**Symbols**
subtract	subtract 5 from 8	8 − 5
difference	the difference of 10 and 2	10 − 2
less	17 less 3	17 − 3
less than	2 less than 20	20 − 2
take away	14 take away 9	14 − 9
decreased by	7 decreased by 5	7 − 5
subtracted from	9 subtracted from 12	12 − 9

Helpful Hint

Be careful when solving applications that suggest subtraction. Although order *does not* matter when adding, order *does* matter when subtracting. For example, 20 − 15 and 15 − 20 do not simplify to the same number.

Answer

10. 1941 ft

✔**Concept Check** In each of the following problems, identify which number is the minuend and which number is the subtrahend.

a. What is the result when 6 is subtracted from 40?

b. What is the difference of 15 and 8?

c. Find a number that is 15 fewer than 23.

To solve a word problem that involves addition or subtraction, we first use the facts given to write an addition or subtraction statement. Then we write the corresponding solution of the real-life problem. It is sometimes helpful to write the statement in words (brief phrases) and then translate to numbers.

PRACTICE PROBLEM 11

The radius of Uranus is 15,759 miles. The radius of Neptune is 458 miles less than the radius of Uranus. What is the radius of Neptune? (*Source:* National Space Science Data Center)

EXAMPLE 11 Finding the Radius of a Planet

The radius of Jupiter is 43,441 miles. The radius of Saturn is 7257 miles less than the radius of Jupiter. Find the radius of Saturn. (*Source:* National Space Science Data Center)

Solution:

In Words		Translate to Numbers
radius of Jupiter	⟶	43,441
− 7257	⟶	− 7257
radius of Saturn	⟶	36,184

The radius of Saturn is 36,184 miles.

Work Practice Problem 11

Helpful Hint Since subtraction and addition are reverse operations, don't forget that a subtraction problem can be checked by adding.

Graphs can be used to visualize data. The graph shown next is called a **bar graph.** For this bar graph, the height of each bar is labeled above the bar. To check this height, follow the top of each bar to the vertical line to the left. For example, the first bar is labeled 370. Follow the top of that bar to the left until the vertical line is reached, not quite halfway between 350 and 400, or 370.

Answer

11. 15,301 miles

✔ Concept Check Answers

a. minuend: 40; subtrahend: 6
b. minuend: 15; subtrahend: 8
c. minuend: 23; subtrahend: 15

EXAMPLE 12 **Reading a Bar Graph**

Airline executives are studying selected aircraft models and their seating capacity for possible equipment expansion and replacement. In the following graph, each bar represents an aircraft model and the height of each bar represents the corresponding seating capacity.

Selected Aircraft Seating Capacity

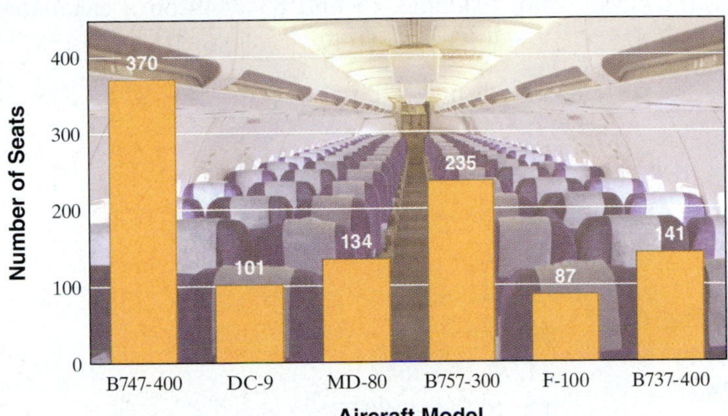

Source: Air Transport Association of America, Inc.

a. Which model shown has the greatest seating capacity?

b. Find the total number of seats for the DC-9, the B757-300, and the MD-80.

Solution:

a. The model with the greatest capacity corresponds to the highest bar, which is the B747-400.

b. The key word here is "total." To find the total number of seats in the DC-9, the B757-300, and the MD-80, we add.

In Words		Translate to Numbers
DC-9	→	101
B757-300	→	235
MD-80	→	+ 134
	Total	470

The total number of seats in the DC-9, the B757-300, and the MD-80 is 470.

Work Practice Problem 12

PRACTICE PROBLEM 12

Use the graph in Example 12 to answer the following:

a. Which aircraft model shown contains the fewest seats?

b. Find the total number of seats for the B747-400, the F-100, and the B737-400.

Answers

12. a. F-100 **b.** 598

CALCULATOR EXPLORATIONS Adding and Subtracting Numbers

Adding Numbers

To add numbers on a calculator, find the keys marked [+] and [=] or [ENTER].

For example, to add 5 and 7 on a calculator, press the keys [5] [+] [7] [=] or [ENTER].

The display will read [12].
Thus, 5 + 7 = 12.

To add 687, 981, and 49 on a calculator, press the keys

[687] [+] [981] [+] [49] [=] or [ENTER].

The display will read [1717].

Thus, 687 + 981 + 49 = 1717. (Although entering 687, for example, requires pressing more than one key, here numbers are grouped together for easier reading.)

Use a calculator to add.

1. 89 + 45

2. 76 + 97

3. 285 + 55

4. 8773 + 652

5.
$$\begin{array}{r} 985 \\ 1210 \\ 562 \\ +\ \ 77 \\ \hline \end{array}$$

6.
$$\begin{array}{r} 465 \\ 9888 \\ 620 \\ +1550 \\ \hline \end{array}$$

Subtracting Numbers

To subtract numbers on a calculator, find the keys marked [−] and [=] or [ENTER].

For example, to find 83 − 49 on a calculator, press the keys [83] [−] [49] [=] or [ENTER].

The display will read [34]. Thus, 83 − 49 = 34.

Use a calculator to subtract.

7. 865 − 95

8. 76 − 27

9. 147 − 38

10. 366 − 87

11. 9625 − 647

12. 10,711 − 8925

Vocabulary and Readiness Check

Use the choices below to fill in each blank.

0	order	addend	associative
sum	number	grouping	commutative
perimeter	minuend	subtrahend	difference

1. The sum of 0 and any number is the same ______________.

2. In 35 + 20 = 55, the number 55 is called the ______________ and 35 and 20 are each called a(n) ______________.

3. The difference of any number and that same number is ______________.

4. The difference of any number and 0 is the same ______________.

5. In 37 − 19 = 18, the number 37 is the ______________, the 19 is the ______________, and the 18 is ______________.

6. The distance around a polygon is called its ______________.

7. Since 7 + 10 = 10 + 7, we say that changing the ______________ in addition does not change the sum. This property is called the ______________ property of addition.

8. Since (3 + 1) + 20 = 3 + (1 + 20), we say that changing the ______________ in addition does not change the sum. This property is called the ______________ property of addition.

1.3 EXERCISE SET

FOR EXTRA HELP

Student Solutions Manual

PH Math/Tutor Center

CD/Video for Review

Math XL
MathXL®

MyMathLab

Objective A *Add. See Examples 1 through 4.*

1. $14 + 22$

2. $27 + 31$

3. $\begin{array}{r} 62 \\ +230 \\ \hline \end{array}$

4. $\begin{array}{r} 37 \\ +542 \\ \hline \end{array}$

5. $\begin{array}{r} 12 \\ 13 \\ +24 \\ \hline \end{array}$

6. $\begin{array}{r} 23 \\ 45 \\ +30 \\ \hline \end{array}$

7. $\begin{array}{r} 5267 \\ +\ \ 132 \\ \hline \end{array}$

8. $\begin{array}{r} 236 \\ +6243 \\ \hline \end{array}$

9. $22{,}781 + 186{,}297$

10. $17{,}427 + 821{,}059$

11. $\begin{array}{r} 8 \\ 9 \\ 2 \\ 5 \\ +1 \\ \hline \end{array}$

12. $\begin{array}{r} 3 \\ 5 \\ 8 \\ 5 \\ +7 \\ \hline \end{array}$

13. $\begin{array}{r} 81 \\ 17 \\ 23 \\ 79 \\ +12 \\ \hline \end{array}$

14. $\begin{array}{r} 64 \\ 28 \\ 56 \\ 25 \\ +32 \\ \hline \end{array}$

15. $24 + 9006 + 489 + 2407$

16. $16 + 1056 + 748 + 7770$

17. $\begin{array}{r} 6820 \\ 4271 \\ +5626 \\ \hline \end{array}$

18. $\begin{array}{r} 6789 \\ 4321 \\ +5555 \\ \hline \end{array}$

19. $\begin{array}{r} 49 \\ 628 \\ 5\,762 \\ +29{,}462 \\ \hline \end{array}$

20. $\begin{array}{r} 26 \\ 582 \\ 4\,763 \\ +62{,}511 \\ \hline \end{array}$

21. $\begin{array}{r} 121{,}742 \\ 57{,}279 \\ 26{,}586 \\ +426{,}782 \\ \hline \end{array}$

22. $\begin{array}{r} 504{,}218 \\ 321{,}920 \\ 38{,}507 \\ +594{,}687 \\ \hline \end{array}$

Objective B *Subtract. Check by adding. See Examples 5 through 8.*

23. $\begin{array}{r} 62 \\ -37 \\ \hline \end{array}$

24. $\begin{array}{r} 55 \\ -29 \\ \hline \end{array}$

25. $\begin{array}{r} 749 \\ -149 \\ \hline \end{array}$

26. $\begin{array}{r} 957 \\ -257 \\ \hline \end{array}$

27. $\begin{array}{r} 922 \\ -634 \\ \hline \end{array}$

28. $\begin{array}{r} 674 \\ -299 \\ \hline \end{array}$

29. $\begin{array}{r} 600 \\ -432 \\ \hline \end{array}$

30. $\begin{array}{r} 300 \\ -149 \\ \hline \end{array}$

31. $\begin{array}{r} 6283 \\ -\ \ 560 \\ \hline \end{array}$

32. $\begin{array}{r} 5349 \\ -\ \ 720 \\ \hline \end{array}$

33. $\begin{array}{r} 533 \\ -\ \ 29 \\ \hline \end{array}$

34. $\begin{array}{r} 724 \\ -\ \ 16 \\ \hline \end{array}$

35. $1983 - 1904$

36. $1983 - 1914$

37. $50{,}000 - 17{,}289$

38. $40{,}000 - 23{,}582$

39. $7020 - 1979$

40. $6050 - 1878$

41. $51{,}111 - 19{,}898$

42. $62{,}222 - 39{,}898$

Objectives A B Mixed Practice *Add or subtract as indicated. See Examples 1 through 8.*

43. $\begin{array}{r} 986 \\ +\ \ 48 \\ \hline \end{array}$

44. $\begin{array}{r} 986 \\ -\ \ 48 \\ \hline \end{array}$

45. $76 - 67$

46. $80 + 93 + 17 + 9 + 2$

47. $\begin{array}{r} 9000 \\ -\ 482 \\ \hline \end{array}$

48. $\begin{array}{r} 10{,}000 \\ -\ 1786 \\ \hline \end{array}$

49. $\begin{array}{r} 10{,}962 \\ 4851 \\ +\ 7063 \\ \hline \end{array}$

50. $\begin{array}{r} 12{,}468 \\ 3211 \\ +\ 1988 \\ \hline \end{array}$

Objective C *Find the perimeter of each figure. See Examples 9 and 10.*

51.

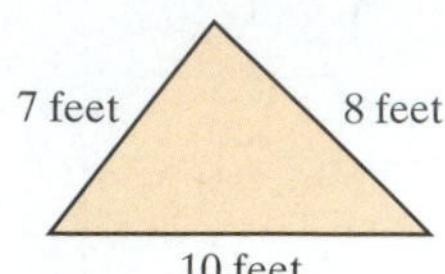

52.

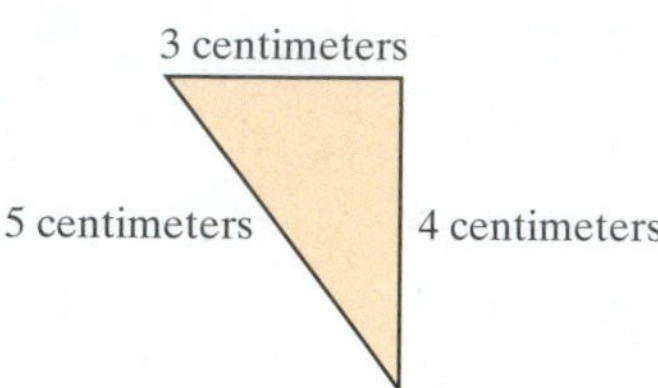

△ **53.**

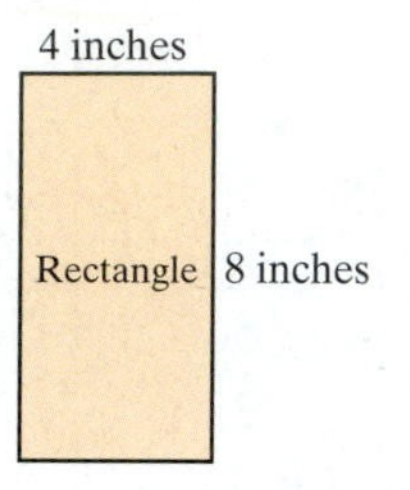

△ **54.**

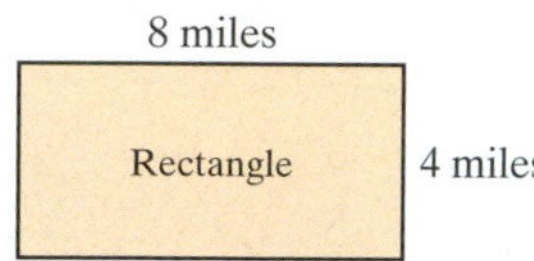

△ **55.**

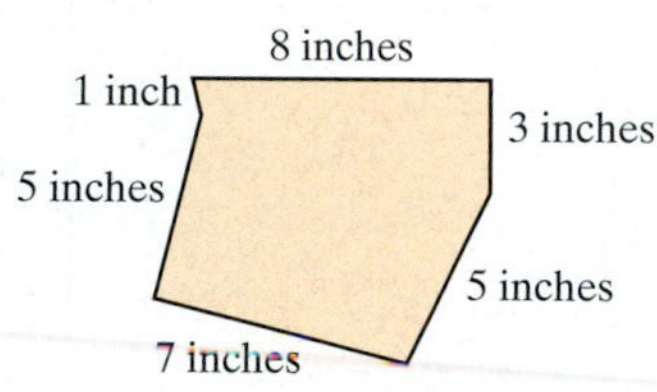

△ **56.**

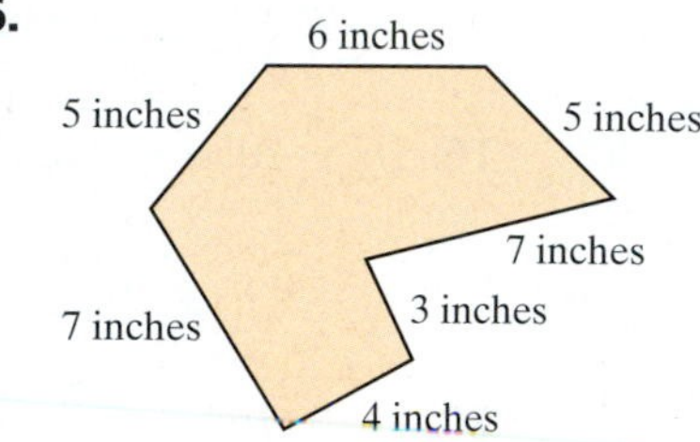

57.

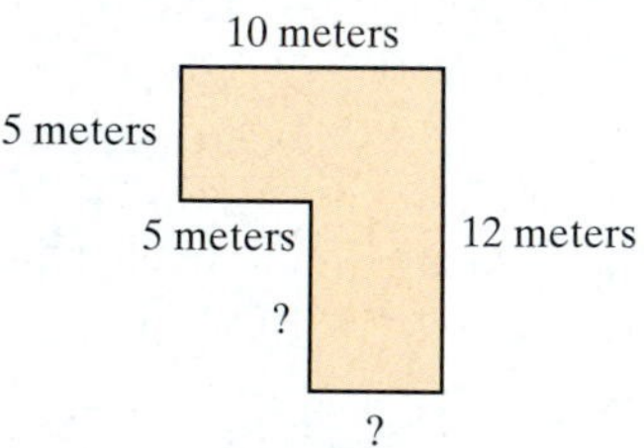

58.

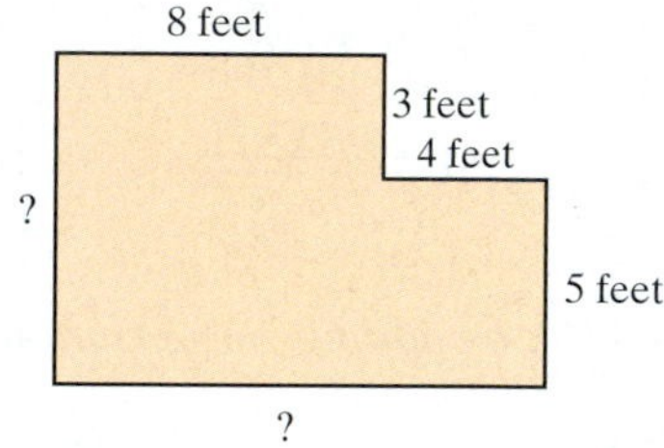

Objectives C D Mixed Practice–Translating *Solve. See Examples 9 through 12.*

59. Find the sum of 297 and 1796.

60. Find the sum of 802 and 6487.

61. Find the total of 76, 39, 8, 17, and 126.

62. Find the total of 89, 45, 2, 19, and 341.

63. Find the difference of 41 and 21.

64. Find the difference of 16 and 5.

65. What is 452 increased by 92?

66. What is 712 increased by 38?

67. Find 108 less 36.

68. Find 25 less 12.

69. Find 12 subtracted from 100.

70. Find 86 subtracted from 90.

71. The population of California is projected to grow from 36,039 thousand in 2005 to 38,067 thousand in 2010. What is California's projected population increase over this time period?

72. The population of Illinois is projected to grow from 12,699 thousand in 2005 to 12,917 thousand in 2010. What is Illinois's projected population increase over this time period?

73. A new DVD player with remote control costs $295. A college student has $914 in her savings account. How much will she have left in her savings account after she buys the DVD player?

74. A stereo that regularly sells for $547 is discounted by $99 in a sale. What is the sale price?

A river basin is the geographic area drained by a river and its tributaries. The Mississippi River Basin is the third largest in the world and is divided into six sub-basins, whose areas are shown in the following bar graph. Use this graph for Exercises 75 through 78.

75. Find the total U.S. land area drained by the Upper Mississippi and Lower Mississippi sub-basins.

76. Find the total U.S. land area drained by the Ohio and Tennessee sub-basins.

77. How many more square miles of land are drained by the Missouri sub-basin than the Arkansas Red-White sub-basin?

78. How many more square miles of land are drained by the Upper Mississippi sub-basin than the Lower Mississippi sub-basin?

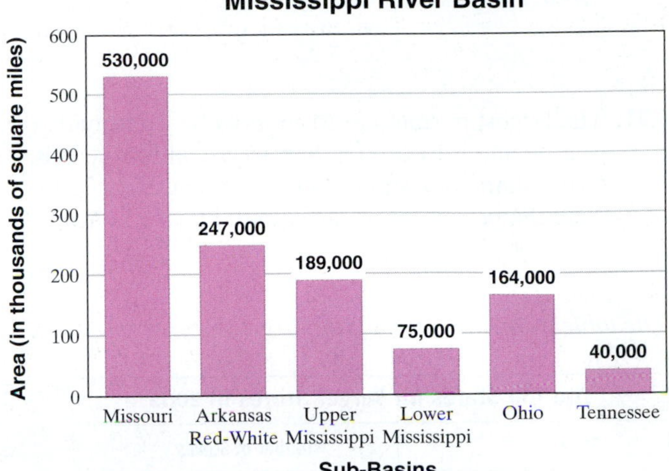

△ **79.** A homeowner is installing a fence in his backyard. How many feet of fencing are needed to enclose the yard below?

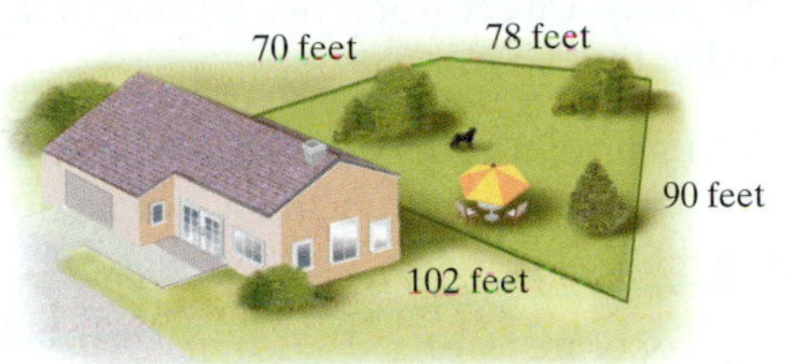

△ **80.** A homeowner is considering installing gutters around her home. Find the perimeter of her rectangular home.

81. Professor Graham is reading a 503-page book. If she has just finished reading page 239, how many more pages must she read to finish the book?

82. When a couple began a trip, the odometer read 55,492. When the trip was over, the odometer read 59,320. How many miles did they drive on their trip?

83. The two top-selling automobiles in the United States are the Honda Accord, with sales of 386,119 cars and the Toyota Camry with sales of 426,990 cars in 2005. What is the total number of Accords and Camrys sold in the United States in 2005? (*Source:* Toyota Corp. and Honda Corp.)

84. In 2005, the country of New Zealand had 35,758,801 more sheep than people. If the human population of New Zealand in 2005 was 4,141,199, what was the sheep population? (*Source: Statistics: New Zealand*)

The decibel (dB) is a unit of measurement for sound. Every increase of 10 dB is a tenfold increase in sound intensity. The following bar graph below shows the decibel levels for some common sounds. Use this graph for Exercises 85 through 88.

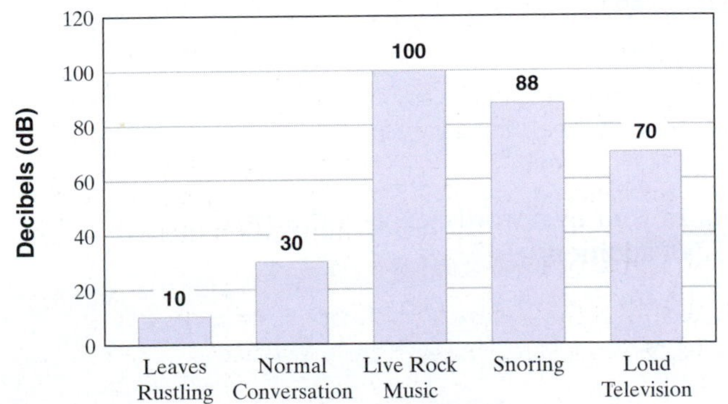

85. What is the dB rating for live rock music?

86. Which is the quietest of all the sounds shown in the graph?

87. How much louder is the sound of snoring than normal conversation?

88. What is the difference is sound intensity between live rock music and loud television?

89. In 2006, there were 5696 Blockbuster video rental stores located in the United States and 3346 located outside the United States. How many Blockbuster video rental stores were located worldwide? (*Source: Blockbuster Inc.*)

90. Automobile classes are defined by the amount of interior room. A subcompact car is defined as a car with a maximum interior space of 99 cubic feet. A midsize car is defined as a car with a maximum interior space of 119 cubic feet. What is the difference in volume between a midsize and a subcompact car?

91. The largest permanent Monopoly board is made of granite and is located in San Jose, California. It is in the shape of a square with side lengths of 31 ft. Find the perimeter of the square playing board.

92. The smallest commercially available jigsaw puzzle is a 1000-piece puzzle manufactured in Spain. It is in the shape of a rectangle with length of 18 inches and width of 12 inches. Find the perimeter of this rectangular-shaped puzzle.

The table shows the number of Target stores in ten states. Use this table to answer Exercises 93 through 98.

The Top States for Target Stores in 2005

State	Number of Stores
Arizona	41
California	205
Florida	95
Georgia	44
Illinois	75
New York	49
Michigan	53
Minnesota	64
Ohio	49
Texas	121

(*Source:* Target Corporation)

93. Which state has the most Target stores?

94. Which of the states listed in the table has the fewest number of Target stores?

95. What is the total number of Target stores located in the three states with the most Target stores?

96. How many Target stores are located in the ten states listed in the table?

97. Which pair of neighboring states have more Target stores combined, Florida and Georgia or Michigan and Ohio?

98. Target operates stores in 47 states. There are 601 Target stores located in the states not listed in the table. How many Target stores are in the United States?

99. The State of Delaware has 2029 miles of urban highways and 3865 miles of rural highways. Find the total highway mileage in Delaware. (*Source:* U.S. Federal Highway Administration)

100. The state of Rhode Island has 5193 miles of urban highways and 1222 miles of rural highways. Find the total highway mileage in Rhode Island. (*Source:* U.S. Federal Highway Administration)

Concept Extensions

For each exercise, identify which number is the minuend and which number is the subtrahend. See the Concept Check in this section.

101. $\begin{array}{r} 48 \\ -\ 1 \\ \hline \end{array}$

102. $\begin{array}{r} 2863 \\ -1904 \\ \hline \end{array}$

103. Subtract 7 from 70.

104. Find 86 decreased by 25.

105. In your own words, explain the commutative property of addition.

106. In your own words, explain the associative property of addition.

Check each addition below. If it is incorrect, find the correct answer. See the Concept Check in this section.

107. $\begin{array}{r} 566 \\ 932 \\ +\ 871 \\ \hline 2369 \end{array}$

108. $\begin{array}{r} 773 \\ 659 \\ +\ 481 \\ \hline 1913 \end{array}$

109. $\begin{array}{r} 14 \\ 173 \\ 86 \\ +\ 257 \\ \hline 520 \end{array}$

110. $\begin{array}{r} 19 \\ 214 \\ 49 \\ +\ 651 \\ \hline 923 \end{array}$

Identify each answer as correct or incorrect. Use addition to check. If the answer is incorrect, then write the correct answer.

111. $\begin{array}{r} 741 \\ -\ 56 \\ \hline 675 \end{array}$

112. $\begin{array}{r} 478 \\ -\ 89 \\ \hline 389 \end{array}$

113. $\begin{array}{r} 1029 \\ -\ 888 \\ \hline 141 \end{array}$

114. $\begin{array}{r} 7615 \\ -\ 547 \\ \hline 7168 \end{array}$

Fill in the missing digits in each problem.

115. $\begin{array}{r} 526\_ \\ -2\_85 \\ \hline 28\_4 \end{array}$

116. $\begin{array}{r} 10,\_4\_ \\ -8\ 5\_4 \\ \hline \_710 \end{array}$

117. Is there a commutative property of subtraction? In other words, does order matter when subtracting? Why or why not?

118. Explain why the phrase "Subtract 7 from 10" translates to "10 − 7."

119. The local college library is having a Million Pages of Reading promotion. The freshmen have read a total of 289,462 pages; the sophomores have read a total of 369,477 pages; the juniors have read a total of 218,287 pages; and the seniors have read a total of 121,685 pages. Have they reached a goal of one million pages? If not, how many more pages need to be read?

STUDY SKILLS BUILDER

Learning New Terms

Many of the terms used in this text may be new to you. It will be helpful to make a list of new mathematical terms and symbols as you encounter them and to review them frequently. Placing these new terms (including page references) on 3 × 5 index cards might help you later when you're preparing for a quiz.

Answer the following.

1. Name one way you might place a word and its definition on a 3 × 5 card.
2. How do new terms stand out in this text so that they can be found?

Objectives

A Round Whole Numbers.

B Use Rounding to Estimate Sums and Differences.

C Solve Problems by Estimating.

1.4 ROUNDING AND ESTIMATING

Objective A Rounding Whole Numbers

Rounding a whole number means approximating it. A rounded whole number is often easier to use, understand, and remember than the precise whole number. For example, instead of trying to remember the South Carolina state population as 4,198,068, it is much easier to remember it rounded to the nearest million: 4,000,000, or 4 million people.

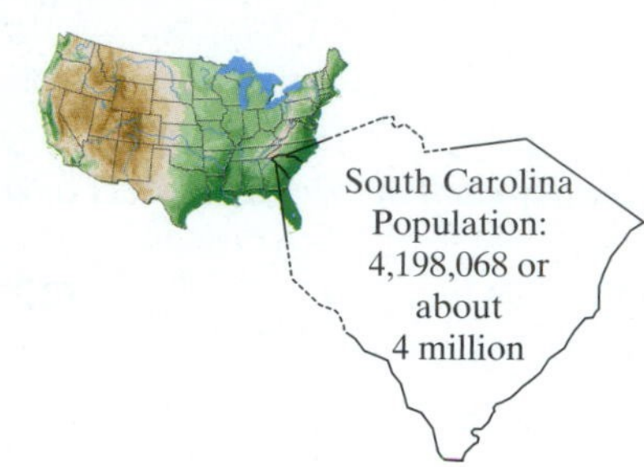

Recall from Section 1.2 that the line below is called a number line. To **graph** a whole number on this number line, we darken the point representing the location of the whole number. For example, the number 4 is graphed below.

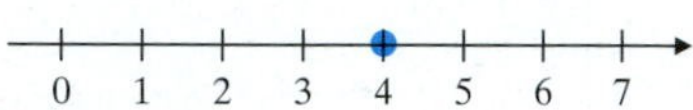

On the number line, the whole number 36 is closer to 40 than 30, so 36 rounded to the nearest ten is 40.

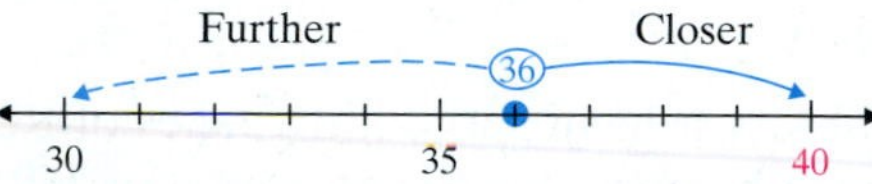

The whole number 52 is closer to 50 than 60, so 52 rounded to the nearest ten is 50.

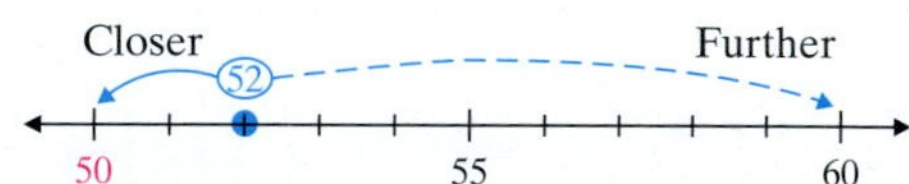

In trying to round 25 to the nearest ten, we see that 25 is halfway between 20 and 30. It is not closer to either number. In such a case, we round to the larger ten, that is, to 30.

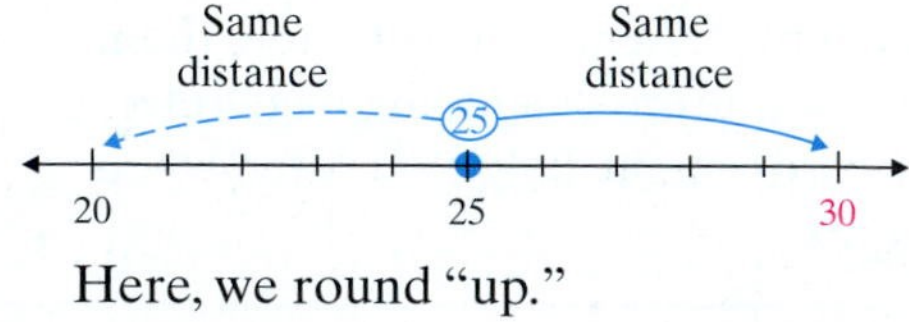

Here, we round "up."

To round a whole number without using a number line, follow these steps:

Rounding Whole Numbers to a Given Place Value

Step 1: Locate the digit to the right of the given place value.

Step 2: If this digit is 5 or greater, add 1 to the digit in the given place value and replace each digit to its right by 0.

Step 3: If this digit is less than 5, replace it and each digit to its right by 0.

EXAMPLE 1 Round 568 to the nearest ten.

Solution: 5 6(8) — The digit to the right of the tens place is the ones place, which is circled.

↑ tens place

5 6(8) — Since the circled digit is 5 or greater, add 1 to the 6 in the tens place and replace the digit to the right by 0.

↑ Add 1. ↖ Replace with 0.

We find that 568 rounded to the nearest ten is 570.

Work Practice Problem 1

PRACTICE PROBLEM 1

Round to the nearest ten.

a. 57

b. 641

c. 325

EXAMPLE 2 Round 278,362 to the nearest thousand.

Solution:

Thousands place ↓; 3 is less than 5.

278,(3)62

Do not add 1. ↑ Replace with zeros. ↑

The number 278,362 rounded to the nearest thousand is 278,000.

Work Practice Problem 2

PRACTICE PROBLEM 2

Round to the nearest thousand.

a. 72,304

b. 9222

c. 671,800

EXAMPLE 3 Round 248,982 to the nearest hundred.

Solution:

Hundreds place ↓; 8 is greater than or equal to 5.

248,9(8)2

↑ Add 1. — $9 + 1 = 10$, so replace the digit 9 by 0 and carry 1 to the place value to the left.

2 4 ~~8~~, ~~9~~ 8 2 (8 + 1 above the 8; 0 above the 9)

Add 1. ↑ Replace with zeros. ↑

The number 248,982 rounded to the nearest hundred is 249,000.

Work Practice Problem 3

PRACTICE PROBLEM 3

Round to the nearest hundred.

a. 3474

b. 76,243

c. 978,865

✔Concept Check Round each of the following numbers to the nearest *hundred*. Explain your reasoning.

a. 59 **b.** 29

Answers

1. a. 60 **b.** 640 **c.** 330
2. a. 72,000 **b.** 9000 **c.** 672,000
3. a. 3500 **b.** 76,200 **c.** 978,900

✔ Concept Check Answers

a. 100 **b.** 0

Objective B Estimating Sums and Differences

By rounding addends, minuends, and subtrahends, we can estimate sums and differences. An estimated sum or difference is appropriate when the exact number is not necessary. Also, an estimated sum or difference can help us determine if we made a mistake in calculating an exact amount. To estimate the sum below, round each number to the nearest hundred and then add.

768	rounds to	800
1952	rounds to	2000
225	rounds to	200
+ 149	rounds to	+ 100
		3100

The estimated sum is 3100, which is close to the **exact** sum of 3094.

PRACTICE PROBLEM 4

Round each number to the nearest ten to find an estimated sum.

$$\begin{array}{r} 49 \\ 25 \\ 32 \\ 51 \\ +98 \\ \hline \end{array}$$

EXAMPLE 4

Round each number to the nearest hundred to find an estimated sum.

$$\begin{array}{r} 294 \\ 625 \\ 1071 \\ +\ 349 \\ \hline \end{array}$$

Solution:

Exact:		Estimate:
294	rounds to	300
625	rounds to	600
1071	rounds to	1100
+ 349	rounds to	+ 300
		2300

The estimated sum is 2300. (The exact sum is 2339.)

Work Practice Problem 4

PRACTICE PROBLEM 5

Round each number to the nearest thousand to find an estimated difference.

$$\begin{array}{r} 3785 \\ -2479 \\ \hline \end{array}$$

EXAMPLE 5

Round each number to the nearest hundred to find an estimated difference.

$$\begin{array}{r} 4725 \\ -2879 \\ \hline \end{array}$$

Solution:

Exact:		Estimate:
4725	rounds to	4700
− 2879	rounds to	− 2900
		1800

The estimated difference is 1800. (The exact difference is 1846.)

Work Practice Problem 5

Objective C Solving Problems by Estimating

Making estimates is often the quickest way to solve real-life problems when solutions do not need to be exact.

Answers
4. 260 **5.** 2000

EXAMPLE 6 Estimating Distances

Jose Guillermo is trying to estimate quickly the distance from Temple, Texas, to Brenham, Texas. Round each distance given on the map to the nearest ten to estimate the total distance.

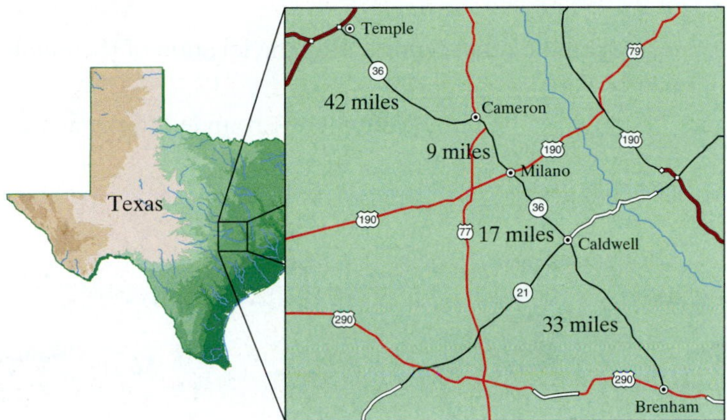

Solution:

Exact Distance:		Estimate:
42	rounds to	40
9	rounds to	10
17	rounds to	20
+33	rounds to	+ 30
		100

It is approximately 100 miles from Temple to Brenham. (The exact distance is 101 miles.)

Work Practice Problem 6

PRACTICE PROBLEM 6

Tasha Kilbey is trying to estimate how far it is from Grove, Kansas, to Hays, Kansas. Round each given distance on the map to the nearest ten to estimate the total distance.

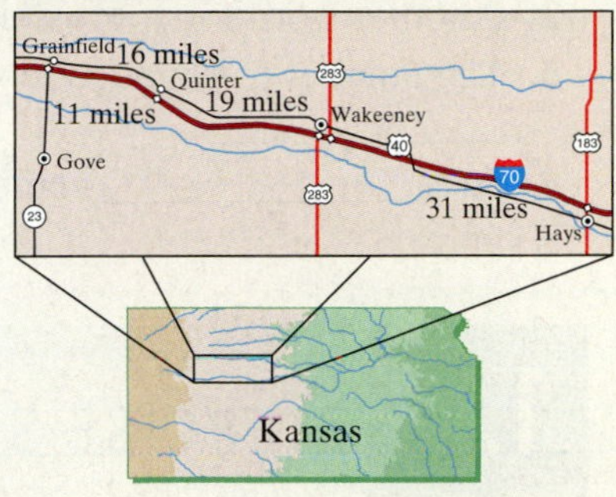

EXAMPLE 7 Estimating Data

In three recent years the numbers of reported cases of mumps in the United States were 906, 1537, and 1692. Round each number to the nearest hundred to estimate the total number of cases reported over this period. (*Source:* Centers for Disease Control and Prevention)

Solution:

Exact Number of Cases:		Estimate:
906	rounds to	900
1537	rounds to	1500
+ 1692	rounds to	+ 1700
		4100

The approximate number of cases reported over this period was 4100. (The exact number of cases was 4135.)

Work Practice Problem 7

PRACTICE PROBLEM 7

In a recent year, there were 120,710 reported cases of chicken pox, 22,878 reported cases of tuberculosis, and 45,974 reported cases of salmonellosis in the United States. Round each number to the nearest ten-thousand to estimate the total number of cases reported for these diseases. (*Source:* Centers for Disease Control and Prevention)

Answers

6. 80 mi **7.** 190,000

Vocabulary and Readiness Check

Use the choices below to fill in each blank.

60	rounding	exact
70	estimate	graph

1. To ______________ a number on a number line, darken the point representing the location of the number.
2. Another word for approximating a whole number is ______________.
3. The number 65 rounded to the nearest ten is ______________, but the number 61 rounded to the nearest ten is ______________.
4. A(n) ______________ number of products is 1265, but a(n) ______________ is 1000.

1.4 EXERCISE SET

FOR EXTRA HELP

Student Solutions Manual

PH Math/Tutor Center

CD/Video for Review

MathXL®

MyMathLab
MyMathLab

Objective A *Round each whole number to the given place. See Examples 1 through 3.*

1. 423 to the nearest ten

2. 273 to the nearest ten

3. 635 to the nearest ten

4. 846 to the nearest ten

5. 2791 to the nearest hundred

6. 8494 to the nearest hundred

7. 495 to the nearest ten

8. 898 to the nearest ten

9. 21,094 to the nearest thousand

10. 82,198 to the nearest thousand

11. 33,762 to the nearest thousand

12. 42,682 to the nearest ten-thousand

13. 328,495 to the nearest hundred

14. 179,406 to the nearest hundred

15. 36,499 to the nearest thousand

16. 96,501 to the nearest thousand

17. 39,994 to the nearest ten

18. 99,995 to the nearest ten

19. 29,834,235 to the nearest ten-million

20. 39,523,698 to the nearest million

Complete the table by estimating the given number to the given place value.

		Tens	Hundreds	Thousands
21.	5281			
22.	7619			
23.	9444			
24.	7777			
25.	14,876			
26.	85,049			

Round each number to the indicated place.

27. In Fall 2005, the University of Texas at El Paso enrolled a student body of 19,264. Round this number to the nearest thousand. (*Source:* University of Texas at El Paso)

28. The number of passengers handled in 2006 by the Hartsfield Atlanta International Airport was 85,907,423. Round this number to the nearest hundred-thousand. (*Source:* Airports Council International)

29. Kareem Abdul-Jabbar holds the NBA record for points scored, a total of 38,387 over his NBA career. Round this number to the nearest thousand. (*Source:* National Basketball Association)

30. It takes 60,149 days for Neptune to make a complete orbit around the Sun. Round this number to the nearest hundred. (*Source:* National Space Science Data Center)

31. The most valuable brand in the world in 2006 was Coca-Cola, with an estimated brand value of $67,520,000,000. Round this to the nearest billion. (*Source:* BBC news)

32. According to the 2000 U.S. Census, the population of the United States was 281,421,906. Round this population figure to the nearest million. (*Source:* U.S. Census Bureau)

33. The average salary for a Boston Red Sox baseball player during the 2006 season was $3,023,894. Round this average salary to the nearest hundred-thousand. (*Source: USA Today*)

34. In 2005, the Procter & Gamble Company had $56,720,000,000 in sales. Round this sales figure to the nearest billion. (*Source:* Procter & Gamble)

35. The United States currently has 207,897,000 cellular mobile phone users, while India has 69,193,300 users. Round each of the user numbers to the nearest million. (*Source*: Cellular Telecommunications Industry Association)

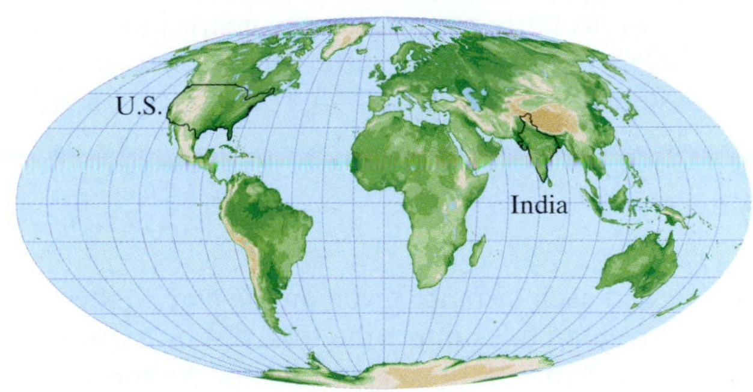

36. In 2005, U.S. farms produced 114,878,000 bushels of oats. Round the oat production figure to the nearest ten-million. (*Source:* U.S. Department of Agriculture)

Objective B *Estimate the sum or difference by rounding each number to the nearest ten. See Examples 4 and 5.*

37. 39 + 45 + 22 + 17

38. 52 + 33 + 15 + 29

39. 449 − 373

40. 555 − 235

Estimate the sum or difference by rounding each number to the nearest hundred. See Examples 4 and 5.

41. 1913 + 1886 + 1925

42. 4050 + 3133 + 1220

43. 1774 − 1492

44. 1989 − 1870

45. 3995 + 2549 + 4944

46. 799 + 1655 + 271

Three of the given calculator answers below are incorrect. Find them.

47. 463 + 219 680

48. 522 + 785 1307

49. 229 + 443 + 606 1278

50. 542 + 789 + 198 2139

51. 7806 + 5150 12,956

52. 5233 + 4988 9011

Helpful Hint

Estimation is useful to check for incorrect answers when using a calculator. For example, pressing a key too hard may result in a double digit, while pressing a key too softly may result in the digit not appearing in the display.

Objective C *Solve each problem by estimating. See Examples 6 and 7.*

53. An appliance store advertises three refrigerators on sale at \$899, \$1499, and \$999. Round each cost to the nearest hundred to estimate the total cost.

54. Suppose you scored 89, 97, 100, 79, 75, and 82 on your biology tests. Round each score to the nearest ten to estimate your total score.

55. The distance from Kansas City to Boston is 1429 miles and from Kansas City to Chicago is 530 miles. Round each distance to the nearest hundred to estimate how much farther Boston is from Kansas City than Chicago is.

56. The Gonzales family took a trip and traveled 588, 689, 277, 143, 59, and 802 miles on six consecutive days. Round each distance to the nearest hundred to estimate the distance they traveled.

57. The peak of Mt. McKinley, in Alaska, is 20,320 feet above sea level. The top of Mt. Rainier, in Washington, is 14,410 feet above sea level. Round each height to the nearest thousand to estimate the difference in elevation of these two peaks. (*Source:* U.S. Geological Survey)

58. A student is pricing new car stereo systems. One system sells for \$1895 and another system sells for \$1524. Round each price to the nearest hundred dollars to estimate the difference in price of these systems.

59. In 2005, the population of Jacksonville, Florida, was 782,623, and the population of Miami was 382,894. Round each population to the nearest ten-thousand to estimate how much larger Jacksonville was than Miami. (*Source:* Internet research)

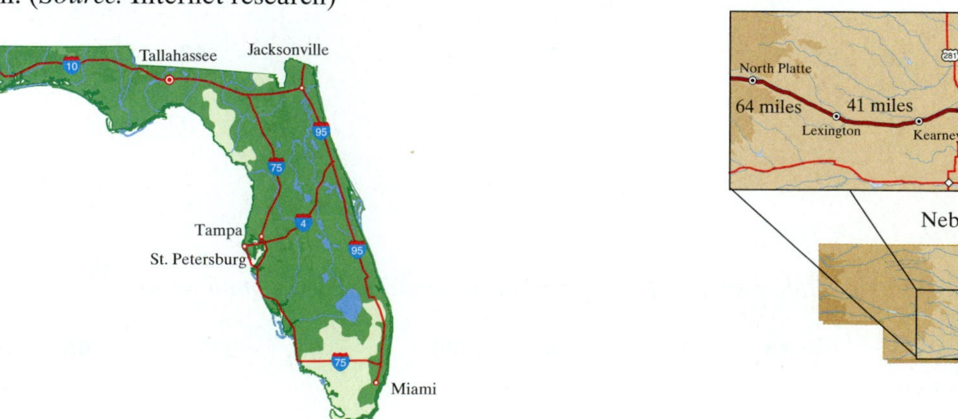

60. Round each distance given on the map to the nearest ten to estimate the total distance from North Platte, Nebraska, to Lincoln, Nebraska.

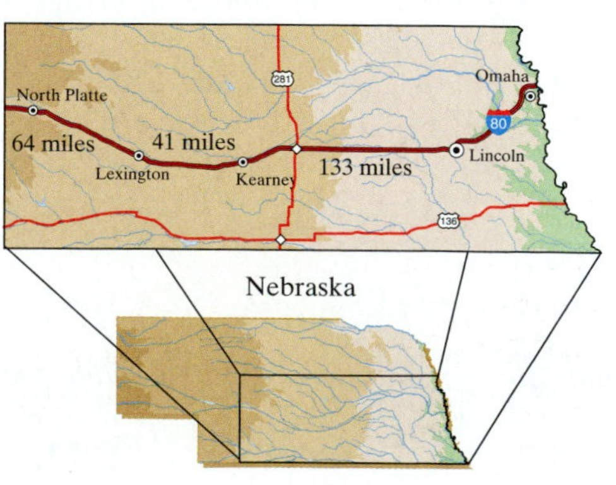

61. Head Start is a national program that provides developmental and social services for America's low-income preschool children ages three to five. Enrollment figures in Head Start programs showed a decrease from 909,608 children in 2003 to 906,993 children in 2005. Round each number of children to the nearest thousand to estimate this decrease. (*Source:* Head Start Bureau)

62. Enrollment figures at a local community college showed an increase from 49,713 credit hours in 2005 to 51,746 credit hours in 2006. Round each number to the nearest thousand to estimate the increase.

Mixed Practice (*Sections 1.3 and 1.4*) *The following table shows a few of the top leading television advertisers in the United States for 2005 and the amount of money spent that year on local and national television advertising. Complete this table. The first line is completed for you.*

	Advertiser	Amount Spent on Television Advertising in 2005 (in millions of dollars)	Amount Written in Standard Form	Standard Form Rounded to Nearest Ten-Million	Standard Form Rounded to Nearest Hundred-Million
	Daimler Chrysler AG	528	$528,000,000	$530,000,000	$500,000,000
63.	Honda	339			
64.	Yum Brands	214			
65.	Federated Department Stores	179			
66.	General Mills	155			
	(*Source:* Television Bureau of Advertising, Inc.)				

Concept Extensions

67. Find one number that when rounded to the nearest hundred is 5700.

68. Find one number that when rounded to the nearest ten is 5700.

69. A number rounded to the nearest hundred is 8600.

a. Determine the smallest possible number.
b. Determine the largest possible number.

70. On August 23, 1989, it was estimated that 1,500,000 people joined hands in a human chain stretching 370 miles to protest the fiftieth anniversary of the pact that allowed what was then the Soviet Union to annex the Baltic nations in 1939. If the estimate of the number of people is to the nearest hundred-thousand, determine the largest possible number of people in the chain.

71. In your own words, explain how to round a number to the nearest thousand.

72. Estimate the perimeter of the triangle by first rounding the length of each side to the nearest hundred.

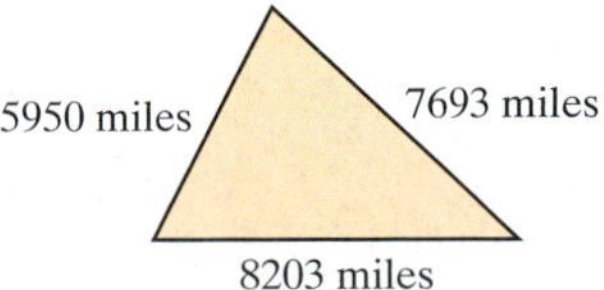

73. Estimate the perimeter of the rectangle by first rounding the length of each side to the nearest ten.

54 meters

Rectangle

17 meters

Objectives

A Use the Properties of Multiplication.

B Multiply Whole Numbers.

C Find the Area of a Rectangle.

D Solve Problems by Multiplying Whole Numbers.

1.5 MULTIPLYING WHOLE NUMBERS AND AREA

Multiplication Shown as Repeated Addition Suppose that we wish to count the number of laptops provided in a computer class. The laptops are arranged in 5 rows, and each row has 6 laptops.

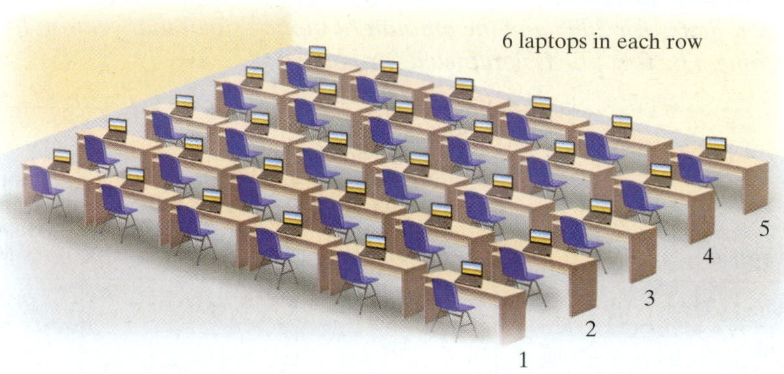

Adding 5 sixes gives the total number of laptops. We can write this as $6 + 6 + 6 + 6 + 6 = 30$ laptops. When each addend is the same, we refer to this as **repeated addition.**

Multiplication is repeated addition but with different notation.

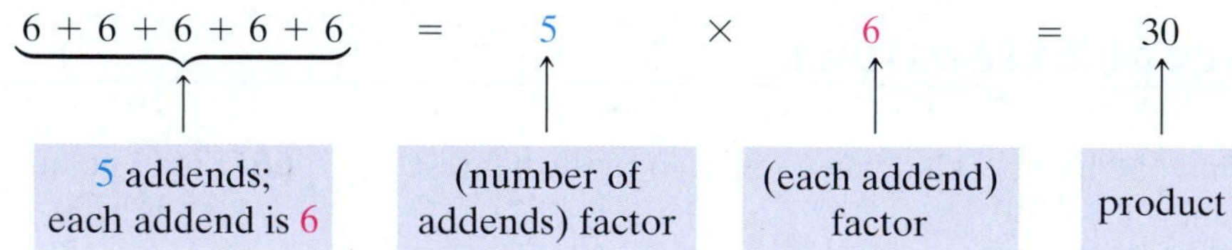

The $\times$ is called a **multiplication sign.** The numbers 5 and 6 are called **factors.** The number 30 is called the **product.** The notation 5×6 is read as "five times six." The symbols $\cdot$ and ( ) can also be used to indicate multiplication.

$$5 \times 6 = 30, \quad 5 \cdot 6 = 30, \quad (5)(6) = 30, \quad \text{and} \quad 5(6) = 30$$

✔ Concept Check

a. Rewrite $5 + 5 + 5 + 5 + 5 + 5 + 5$ using multiplication.

b. Rewrite 3×16 as repeated addition. Is there more than one way to do this? If so, show all ways.

Objective A Using the Properties of Multiplication

As with addition, we memorize products of one-digit whole numbers and then use certain properties of multiplication to multiply larger numbers. (If necessary, review the multiplication of one-digit numbers.)

Notice that when any number is multiplied by 0, the result is always 0. This is called the **multiplication property of 0.**

✔ Concept Check Answers

a. $7 \times 5 = 35$

b. $16 + 16 + 16 = 48$; yes, $3 + 3 + 3 + 3 + 3 + 3 + 3 + 3 + 3 + 3 + 3 + 3 + 3 + 3 + 3 + 3 = 48$

Multiplication Property of 0

The product of 0 and any number is 0. For example,

$5 \cdot 0 = 0$ and $0 \cdot 8 = 0$

Also notice that when any number is multiplied by 1, the result is always the original number. We call this result the **multiplication property of 1.**

Multiplication Property of 1

The product of 1 and any number is that same number. For example,

$1 \cdot 9 = 9$ and $6 \cdot 1 = 6$

EXAMPLE 1 Multiply.

a. 4×1 **b.** $0(3)$ **c.** $1 \cdot 64$ **d.** $(48)(0)$

Solution:

a. $4 \times 1 = 4$ **b.** $0(3) = 0$

c. $1 \cdot 64 = 64$ **d.** $(48)(0) = 0$

Work Practice Problem 1

PRACTICE PROBLEM 1

Multiply.

a. 6×0

b. $(1)8$

c. $(50)(0)$

d. $75 \cdot 1$

Like addition, multiplication is commutative and associative. Notice that when multiplying two numbers, the order of these numbers can be changed without changing the product. For example,

$3 \cdot 5 = 15$ and $5 \cdot 3 = 15$

This property is called the **commutative property of multiplication.**

Commutative Property of Multiplication

Changing the **order** of two factors does not change their product. For example,

$9 \cdot 2 = 18$ and $2 \cdot 9 = 18$

Another property that can help us when multiplying is the **associative property of multiplication.** This property states that when multiplying numbers, the grouping of the numbers can be changed without changing the product. For example,

$$(2 \cdot 3) \cdot 4 = 6 \cdot 4 = 24$$

Also,

$$2 \cdot (3 \cdot 4) = 2 \cdot 12 = 24$$

Both groupings give a product of 24.

Answers

1. a. 0 **b.** 8 **c.** 0 **d.** 75

Associative Property of Multiplication

Changing the **grouping** of factors does not change their product. From the previous work, we know that for example,

$$(2 \cdot 3) \cdot 4 = 2 \cdot (3 \cdot 4)$$

With these properties, along with the **distributive property,** we can find the product of any whole numbers. The distributive property says that multiplication **distributes** over addition. For example, notice that $3(2 + 5)$ simplifies to the same number as $3 \cdot 2 + 3 \cdot 5$.

$$3(2 + 5) = 3(7) = 21$$

$$3 \cdot 2 + 3 \cdot 5 = 6 + 15 = 21$$

Since $3(2 + 5)$ and $3 \cdot 2 + 3 \cdot 5$ both simplify to 21, then

$$3(2 + 5) = 3 \cdot 2 + 3 \cdot 5$$

Notice in $3(2 + 5) = 3 \cdot 2 + 3 \cdot 5$ that each number inside the parentheses is multiplied by 3.

Distributive Property

Multiplication distributes over addition. For example,

$$2(3 + 4) = 2 \cdot 3 + 2 \cdot 4$$

PRACTICE PROBLEM 2

Rewrite each using the distributive property.

a. $6(4 + 5)$

b. $30(2 + 3)$

c. $7(2 + 8)$

EXAMPLE 2 Rewrite each using the distributive property.

a. $5(6 + 5)$ **b.** $20(4 + 7)$ **c.** $2(7 + 9)$

Solution: Using the distributive property, we have

a. $5(6 + 5) = 5 \cdot 6 + 5 \cdot 5$

b. $20(4 + 7) = 20 \cdot 4 + 20 \cdot 7$

c. $2(7 + 9) = 2 \cdot 7 + 2 \cdot 9$

Work Practice Problem 2

Objective B Multiplying Whole Numbers

Let's use the distributive property to multiply 7(48). To do so, we begin by writing the expanded form of 48 (see Section 1.2) and then applying the distributive property.

$7(48) = 7(40 + 8)$	Write 48 in expanded form.
$= 7 \cdot 40 + 7 \cdot 8$	Apply the distributive property.
$= 280 + 56$	Multiply.
$= 336$	Add.

Answers

2. a. $6(4 + 5) = 6 \cdot 4 + 6 \cdot 5$ **b.** $30(2 + 3) = 30 \cdot 2 + 30 \cdot 3$ **c.** $7(2 + 8) = 7 \cdot 2 + 7 \cdot 8$

This is how we multiply whole numbers. When multiplying whole numbers, we will use the following notation.

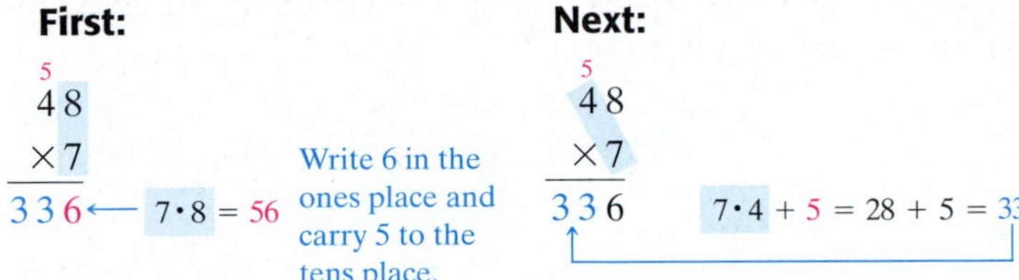

First:
5
48
×7
336 ← 7·8 = 56
Write 6 in the ones place and carry 5 to the tens place.
Next:
5
48
×7
336
7·4 + 5 = 28 + 5 = 33

The product of 48 and 7 is 336.

EXAMPLE 3 Multiply:

a. $\begin{array}{r} 25 \\ \times\ 8 \\ \hline \end{array}$ b. $\begin{array}{r} 246 \\ \times\ 5 \\ \hline \end{array}$

Solution:

a. $\begin{array}{r} {}^{4} \\ 25 \\ \times\ 8 \\ \hline 200 \end{array}$ b. $\begin{array}{r} {}^{23} \\ 246 \\ \times\ 5 \\ \hline 1230 \end{array}$

Work Practice Problem 3

To multiply larger whole numbers, use the following similar notation. Multiply 89×52.

Step 1

$\begin{array}{r} {}^{1} \\ 89 \\ \times\ 52 \\ \hline 178 \end{array}$ ← Multiply 89×2.

Step 2

$\begin{array}{r} {}^{4} \\ 89 \\ \times\ 52 \\ \hline 178 \\ 4450 \\ \hline \end{array}$ ← Multiply 89×50.

Step 3

$\begin{array}{r} 89 \\ \times\ 52 \\ \hline 178 \\ 4450 \\ \hline 4628 \end{array}$ Add.

The numbers 178 and 4450 are called **partial products.** The sum of the partial products, 4628, is the product of 89 and 52.

EXAMPLE 4 Multiply: 236×86

Solution: $\begin{array}{rl} 236 & \\ \times\ \ 86 & \\ \hline 1\,416 & \leftarrow 6(236) \\ 18\,880 & \leftarrow 80(236) \\ \hline 20{,}296 & \text{Add.} \end{array}$

Work Practice Problem 4

EXAMPLE 5 Multiply: 631×125

Solution: $\begin{array}{rl} 631 & \\ \times\ \ 125 & \\ \hline 3\,155 & \leftarrow 5(631) \\ 12\,620 & \leftarrow 20(631) \\ 63\,100 & \leftarrow 100(631) \\ \hline 78{,}875 & \text{Add.} \end{array}$

Work Practice Problem 5

PRACTICE PROBLEM 3

Multiply.

a. $\begin{array}{r} 29 \\ \times\ 6 \\ \hline \end{array}$ b. $\begin{array}{r} 648 \\ \times\ 5 \\ \hline \end{array}$

PRACTICE PROBLEM 4

Multiply.

$\begin{array}{r} 306 \\ \times\ 81 \\ \hline \end{array}$

PRACTICE PROBLEM 5

Multiply.

$\begin{array}{r} 726 \\ \times 142 \\ \hline \end{array}$

Answers

3. a. 174 **b.** 3240
4. 24,786
5. 103,092

✔Concept Check Find and explain the error in the following multiplication problem.

$$\begin{array}{r} 102 \\ \times\ 33 \\ \hline 306 \\ 306 \\ \hline 612 \end{array}$$

Objective C Finding the Area of a Rectangle

A special application of multiplication is finding the **area** of a region. Area measures the amount of surface of a region. For example, we measure a plot of land or the living space of a home by its area. The figures below show two examples of units of area measure. (A centimeter is a unit of length in the metric system.)

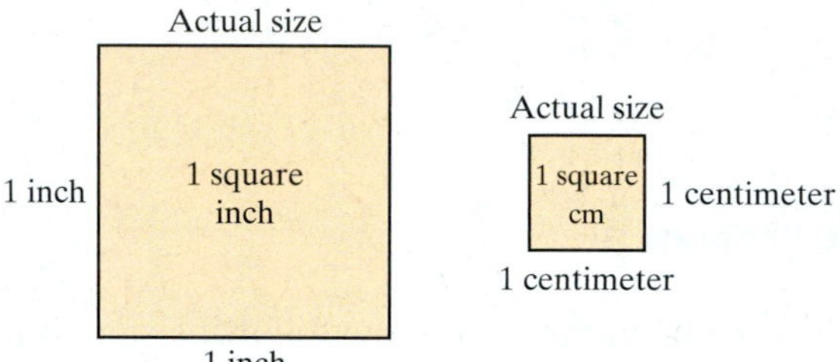

For example, to measure the area of a geometric figure such as the rectangle below, count the number of square units that cover the region.

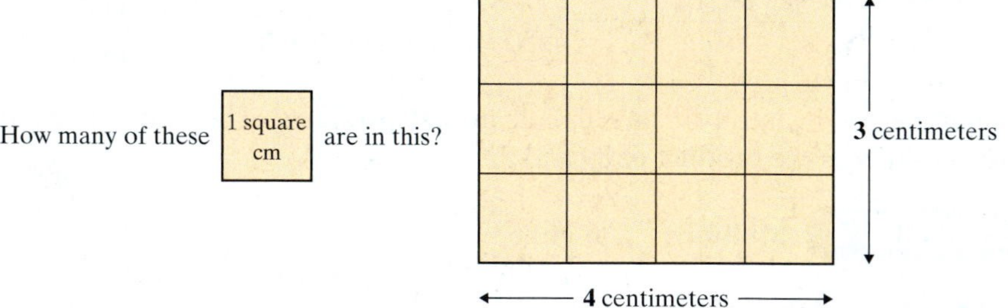

This rectangular region contains 12 square units, each 1 square centimeter. Thus, the area is 12 square centimeters. This total number of squares can be found by counting or by multiplying **4 · 3** (length · width).

$$\begin{aligned} \text{Area of a rectangle} &= \text{length} \cdot \text{width} \\ &= (4 \text{ centimeters})(3 \text{ centimeters}) \\ &= 12 \text{ square centimeters} \end{aligned}$$

In this section, we find the areas of rectangles only. In later sections, we will find the areas of other geometric regions.

✔ Concept Check Answer

$$\begin{array}{r} 102 \\ \times\ 33 \\ \hline 306 \\ 3060 \\ \hline 3366 \end{array}$$

EXAMPLE 6 Finding the Area of a State

The state of Colorado is in the shape of a rectangle whose length is 380 miles and whose width is 280 miles. Find its area.

Solution:

The area of a rectangle is the product of its length and its width.

Area = length · width
= (380 miles)(280 miles)
= 106,400 square miles

The area of Colorado is 106,400 square miles.

Work Practice Problem 6

PRACTICE PROBLEM 6

The state of Wyoming is in the shape of a rectangle whose length is 360 miles and whose width is 280 miles. Find its area.

Objective D Solving Problems by Multiplying

There are several words or phrases that indicate the operation of multiplication. Some of these are as follows:

Multiplication		
Key Words or Phrases	**Examples**	**Symbols**
multiply	multiply 5 by 7	$5 \cdot 7$
product	the product of 3 and 2	$3 \cdot 2$
times	10 times 13	$10 \cdot 13$

Many key words or phrases describing real-life problems that suggest addition might be better solved by multiplication instead. For example, to find the **total** cost of 8 shirts, each selling for $27, we can either add

$$27 + 27 + 27 + 27 + 27 + 27 + 27 + 27$$

or we can multiply 8(27).

EXAMPLE 7 Finding DVD Space

A digital video disc (DVD) can hold about 4800 megabytes (MB) of information. How many megabytes can 12 DVDs hold?

Solution:

Twelve DVDs will hold 12 × 4800 megabytes.

In Words		**Translate to Numbers**
megabytes per disk	→	4800
× DVDs	→	× 12
		9600
		48000
total megabytes		57,600

Twelve DVDs will hold 57,600 megabytes.

Work Practice Problem 7

PRACTICE PROBLEM 7

A particular computer printer can print 16 pages per minute in color. How many pages can it print in 45 minutes?

Answers

6. 100,800 sq mi
7. 720 pages

PRACTICE PROBLEM 8

Ken Shimura purchased DVDs and CDs through a club. Each DVD was marked at \$11 and each CD cost \$9. Ken bought eight DVDs and five CDs. Find the total cost of the order.

EXAMPLE 8 Budgeting Money

Ann Sheridan and two friends agree to take their children to the San Antonio Zoo. The ticket price for each child is \$7 and for each adult, \$9. If 8 children and 3 adults plan to go, how much money is needed for admission? (*Source:* The San Antonio Zoo)

Solution: If the price of one child's ticket is \$7, the price for 8 children is $8 \cdot 7 = \$56$. The price of one adult ticket is \$9, so the price for 3 adults is $9 \cdot 3 = \$27$. The total cost is:

In Words		Translate to Numbers
price of 8 children	→	56
+ price of 3 adults	→	+ 27
total cost		83

The total cost is \$83.

Work Practice Problem 8

PRACTICE PROBLEM 9

If an average page in a book contains 163 words, estimate, rounding each number to the nearest hundred, the total number of words contained on 391 pages.

EXAMPLE 9 Estimating Word Count

The average page of a book contains 259 words. Estimate, rounding each number to the nearest hundred, the total number of words contained on 212 pages.

Solution: The exact number of words is 259×212. Estimate this product by rounding each factor to the nearest hundred.

$$\begin{array}{rcr} 259 & \text{rounds to} & 300 \\ \times 212 & \text{rounds to} & \times 200, \end{array} \quad 300 \times 200 = 60{,}000 \quad (3 \cdot 2 = 6)$$

There are approximately 60,000 words contained on 212 pages.

Work Practice Problem 9

Answers

8. \$133

9. 80,000 words

CALCULATOR EXPLORATIONS Multiplying Numbers

To multiply numbers on a calculator, find the keys marked [×] and [=] or [ENTER]. For example, to find $31 \cdot 66$ on a calculator, press the keys [31] [×] [66] [=] or [ENTER]. The display will read [2046]. Thus, $31 \cdot 66 = 2046$.

Use a calculator to multiply.

1. 72×48

2. 81×92

3. $163 \cdot 94$

4. $285 \cdot 144$

5. 983(277)

6. 1562(843)

Vocabulary and Readiness Check

Use the choices below to fill in each blank.

area	grouping	commutative	1	product	length
factor	order	associative	0	distributive	number

1. The product of 0 and any number is ______________.
2. The product of 1 and any number is the ______________.
3. In $8 \cdot 12 = 96$, the 96 is called the ______________ and 8 and 12 are each called a(n) ______________.
4. Since $9 \cdot 10 = 10 \cdot 9$, we say that changing the ______________ in multiplication does not change the product. This property is called the ______________ property of multiplication.
5. Since $(3 \cdot 4) \cdot 6 = 3 \cdot (4 \cdot 6)$, we say that changing the ______________ in multiplication does not change the product. This property is called the ______________ property of multiplication.
6. ______________ measures the amount of surface of a region.
7. Area of a rectangle = ______________ $\cdot$ width.
8. We know $9(10 + 8) = 9 \cdot 10 + 9 \cdot 8$ by the ______________ property.

1.5 EXERCISE SET

FOR EXTRA HELP

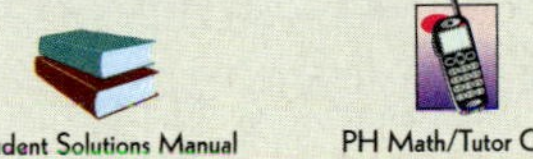
Student Solutions Manual
PH Math/Tutor Center

CD/Video for Review

MathXL®

MyMathLab

Objective A *Multiply. See Example 1.*

1. $1 \cdot 24$ **2.** $55 \cdot 1$ **3.** $0 \cdot 19$ **4.** $27 \cdot 0$

5. $8 \cdot 0 \cdot 9$ **6.** $7 \cdot 6 \cdot 0$ **7.** $87 \cdot 1$ **8.** $1 \cdot 41$

Use the distributive property to rewrite each expression. See Example 2.

9. $6(3 + 8)$ **10.** $5(8 + 2)$ **11.** $4(3 + 9)$

12. $6(1 + 4)$ **13.** $20(14 + 6)$ **14.** $12(12 + 3)$

Objective B *Multiply. See Example 3.*

15. $\begin{array}{r} 64 \\ \times\ 8 \\ \hline \end{array}$ **16.** $\begin{array}{r} 79 \\ \times\ 3 \\ \hline \end{array}$ **17.** $\begin{array}{r} 613 \\ \times\ \ 6 \\ \hline \end{array}$ **18.** $\begin{array}{r} 638 \\ \times\ \ 5 \\ \hline \end{array}$

19. 277×6 **20.** 882×2 **21.** 1074×6 **22.** 9021×3

Objectives **A B Mixed Practice** *Multiply. See Examples 1 through 5.*

23. $\begin{array}{r} 89 \\ \times 13 \\ \hline \end{array}$

24. $\begin{array}{r} 91 \\ \times 72 \\ \hline \end{array}$

25. $\begin{array}{r} 421 \\ \times\ 58 \\ \hline \end{array}$

26. $\begin{array}{r} 526 \\ \times\ 23 \\ \hline \end{array}$

27. $\begin{array}{r} 306 \\ \times\ 81 \\ \hline \end{array}$

28. $\begin{array}{r} 708 \\ \times\ 21 \\ \hline \end{array}$

29. (780)(20)

30. (720)(80)

31. (495)(13)(0)

32. (593)(47)(0)

33. (640)(1)(10)

34. (240)(1)(20)

35. 1234×39

36. 1357×79

37. 609×234

38. 807×127

39. $\begin{array}{r} 8649 \\ \times\ 274 \\ \hline \end{array}$

40. $\begin{array}{r} 1234 \\ \times\ 567 \\ \hline \end{array}$

41. $\begin{array}{r} 589 \\ \times 110 \\ \hline \end{array}$

42. $\begin{array}{r} 426 \\ \times 110 \\ \hline \end{array}$

43. $\begin{array}{r} 1941 \\ \times 2035 \\ \hline \end{array}$

44. $\begin{array}{r} 1876 \\ \times 1407 \\ \hline \end{array}$

Objective **C** *Find the area of each rectangle. See Example 6.*

45.

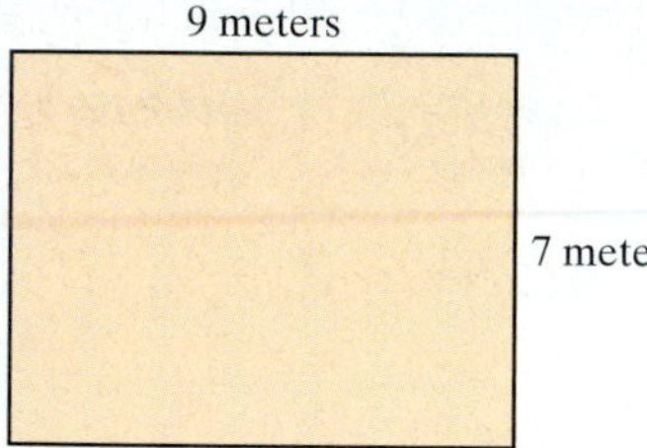

46.

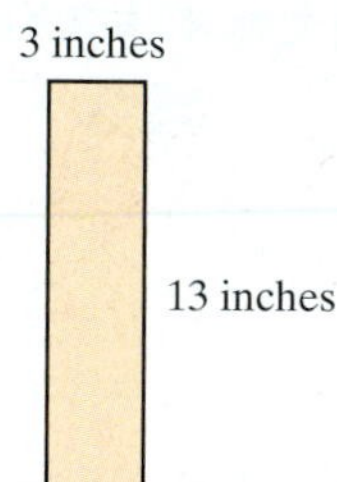

47.

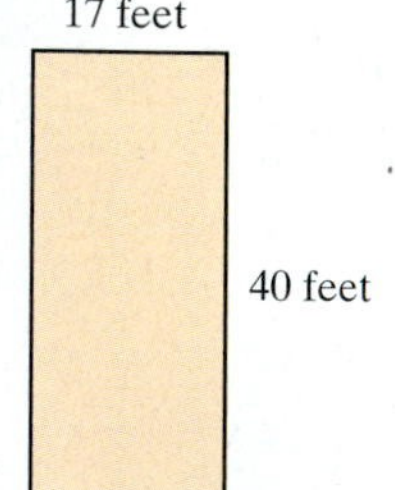

48.

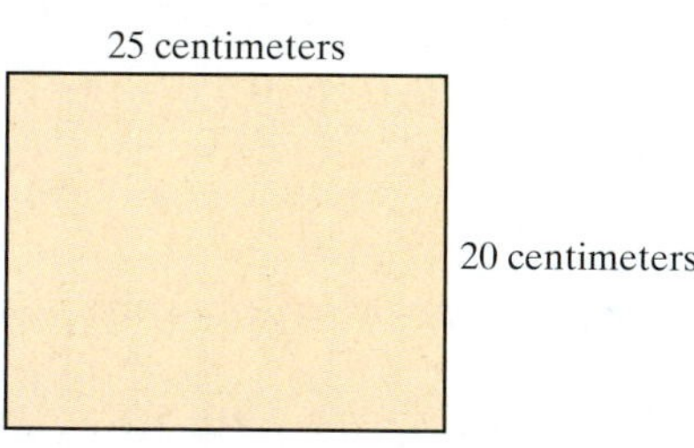

Objectives **C D Mixed Practice** *Estimate the products by rounding each factor to the nearest hundred. See Example 9.*

49. 576×354

50. 982×650

51. 604×451

52. 111×999

Without actually calculating, mentally round, multiply, and choose the best estimate.

53. $38 \times 42 =$
- **a.** 16
- **b.** 160
- **c.** 1600
- **d.** 16,000

54. $2872 \times 12 =$
- **a.** 2872
- **b.** 28,720
- **c.** 287,200
- **d.** 2,872,000

55. $612 \times 29 =$
- **a.** 180
- **b.** 1800
- **c.** 18,000
- **d.** 180,000

56. $706 \times 409 =$
- **a.** 280
- **b.** 2800
- **c.** 28,000
- **d.** 280,000

Translating *Solve. See Examples 6 through 9.*

57. Multiply 80 by 11.

58. Multiply 70 by 12.

59. Find the product of 6 and 700.

60. Find the product of 9 and 900.

61. Find 2 times 2240.

62. Find 3 times 3310.

63. One tablespoon of olive oil contains 125 calories. How many calories are in 3 tablespoons of olive oil? (*Source: Home and Garden Bulletin No. 72,* U.S. Department of Agriculture).

64. One ounce of hulled sunflower seeds contains 14 grams of fat. How many grams of fat are in 8 ounces of hulled sunflower seeds? (*Source: Home and Garden Bulletin No. 72,* U.S. Department of Agriculture).

65. The textbook for a course in biology costs $94. There are 35 students in the class. Find the total cost of the biology books for the class.

66. The seats in a large lecture hall are arranged in 14 rows with 34 seats in each row. Find how many seats are in this room.

67. Cabot Creamery is packing a palet of 20-lb boxes of cheddar cheese to send to a local restaurant. There are five layers of boxes on the pallet, and each layer is four boxes wide by five boxes deep.

a. How many boxes are in one layer?
b. How many boxes are on the pallet?
c. What is the weight of the cheese on the pallet?

68. An apartment building has *three floors*. Each floor has five rows of apartments with four apartments in each row.

a. How many apartments are on 1 floor?
b. How many apartments are in the building?

△ **69.** A plot of land measures 80 feet by 110 feet. Find its area.

△ **70.** A house measures 45 feet by 60 feet. Find the floor area of the house.

△ **71.** The largest hotel lobby can be found at the Hyatt Regency in San Francisco, CA. It is in the shape of a rectangle that measures 350 feet by 160 feet. Find its area.

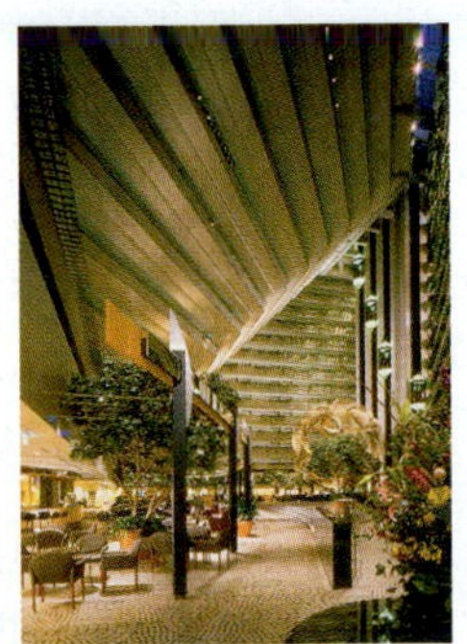

△ **72.** Recall from an earlier section that the largest commercial building in the world under one roof is the flower auction building of the cooperative VBA in Aalsmeer, Netherlands. The floor plan is a rectangle that measures 776 meters by 639 meters. Find the area of this building. (A meter is a unit of length in the metric system.) (*Source: The Handy Science Answer Book,* Visible Ink Press)

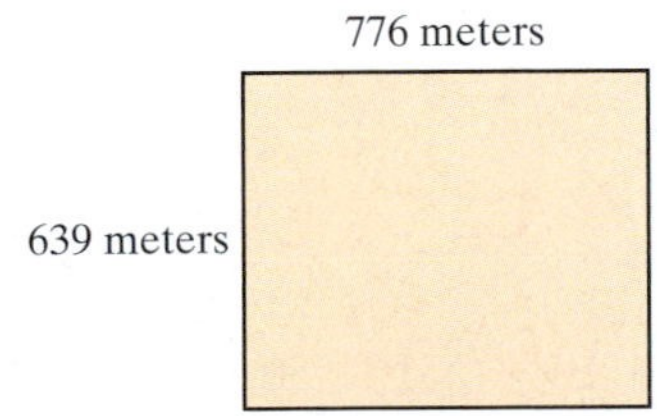

73. A pixel is a rectangular dot on a graphing calculator screen. If a graphing calculator screen contains 62 pixels in a row and 94 pixels in a column, find the total number of pixels on a screen.

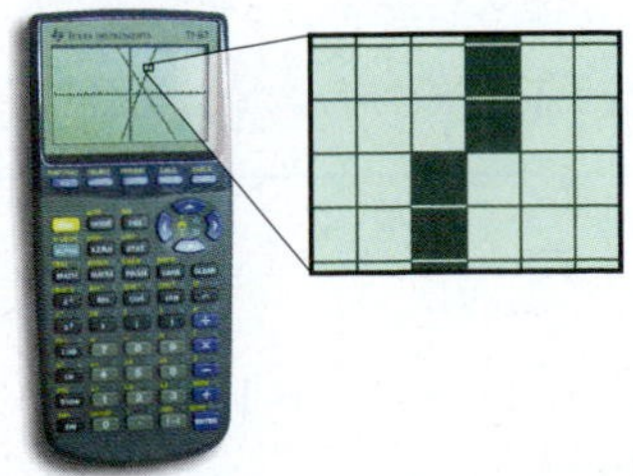

74. A compact disc (CD) can hold 700 megabytes (MB) of information. How many MBs can 17 discs hold?

75. A line of print on a computer contains 60 characters (letters, spaces, punctuation marks). Find how many characters there are in 35 lines.

76. An average cow eats 3 pounds of grain per day. Find how much grain a cow eats in a year. (Assume 365 days in 1 year.)

77. One ounce of Planters® Dry Roasted Peanuts has 160 calories. How many calories are in 8 ounces? (*Source:* RJR Nabisco, Inc.)

78. One ounce of Planters® Dry Roasted Peanuts has 13 grams of fat. How many grams of fat are in 16 ounces? (*Source:* RJR Nabisco, Inc.)

79. The Thespian club at a local community college is ordering T-shirts. T-shirts size S, M, or L cost $10 each and T-shirts size XL or XXL cost $12 each. Use the table below to find the total cost. (The first row is filled in for you.)

T-Shirt Size	Number of Shirts Ordered	Cost per Shirt	Cost per Size Ordered
S	4	$10	$40
M	6		
L	20		
XL	3		
XXL	3		

Total Cost ____

80. The student activities group at North Shore Community College is planning a trip to see the local minor league baseball team. Tickets cost $5 for students, $7 for nonstudents, and $2 for children under 12. Use the following table to find the total cost.

Person	Number of Persons	Cost per Person	Cost per Category
Student	24	$5	$120
Nonstudent	4		
Children under 12	5		

Total Cost ____

81. Hershey's main chocolate factory in Hershey, Pennsylvania, uses 700,000 quarts of milk each day. How many quarts of milk would be used during the month of March, assuming that chocolate is made at the factory every day of the month? (*Source:* Hershey Foods Corp.)

82. Among older Americans (ages 65 and over) there are about 3 times as many widows as widowers. There were about 2,641,000 widowers in 2005. How many widows were there in 2005? (*Source:* Administration on Aging, U.S. Census Bureau)

Mixed Practice (*Sections 1.3, 1.5*) *Perform each indicated operation.*

83. $\begin{array}{r} 128 \\ +\ 7 \\ \hline \end{array}$

84. $\begin{array}{r} 126 \\ -\ 8 \\ \hline \end{array}$

85. $\begin{array}{r} 134 \\ \times\ 16 \\ \hline \end{array}$

86. 47 + 26 + 10 + 231 + 50

87. Find the sum of 19 and 4.

88. Find the product of 19 and 4.

89. Find the difference of 19 and 4.

90. Find the total of 19 and 4.

Concept Extensions

Solve. See the first Concept Check in this section.

91. Rewrite 6 + 6 + 6 + 6 + 6 using multiplication.

92. Rewrite 11 + 11 + 11 + 11 + 11 + 11 using multiplication.

93. **a.** Rewrite $3 \cdot 5$ as repeated addition.
b. Explain why there is more than one way to do this.

94. **a.** Rewrite $4 \cdot 5$ as repeated addition.
b. Explain why there is more than one way to do this.

Find and explain the error in each multiplication problem. See the second Concept Check in this section.

95.
$$\begin{array}{r} 203 \\ \times\ 14 \\ \hline 812 \\ 203 \\ \hline 1015 \end{array}$$

96.
$$\begin{array}{r} 31 \\ \times\ 50 \\ \hline 155 \end{array}$$

Fill in the missing digits in each problem.

97.
$$\begin{array}{r} 4\_ \\ \times\ \_3 \\ \hline 126 \\ 3780 \\ \hline 3906 \end{array}$$

98.
$$\begin{array}{r} \_7 \\ \times\ 6\_ \\ \hline 171 \\ 3420 \\ \hline 3591 \end{array}$$

99. Explain how to multiply two 2-digit numbers using partial products.

100. During the NBA's 2005–2006 regular season, Dirk Nowitzki of the Dallas Mavericks scored 110 three-point field goals, 641 two-point field goals, and 539 free throws (worth one point each). How many points did Nowitzki score during the 2005–2006 regular season? (*Source:* NBA)

101. A window washer in New York City is bidding for a contract to wash the windows of a 23-story building. To write a bid, the number of windows in the building is needed. If there are 7 windows in each row of windows on 2 sides of the building and 4 windows per row on the other 2 sides of the building, find the total number of windows.

Objectives

A **Divide Whole Numbers.**

B **Perform Long Division.**

C **Solve Problems That Require Dividing by Whole Numbers.**

D **Find the Average of a List of Numbers.**

1.6 DIVIDING WHOLE NUMBERS

Suppose three people pooled their money and bought a raffle ticket at a local fund-raiser. Their ticket was the winner and they won a \$75 cash prize. They then divided the prize into three equal parts so that each person received \$25.

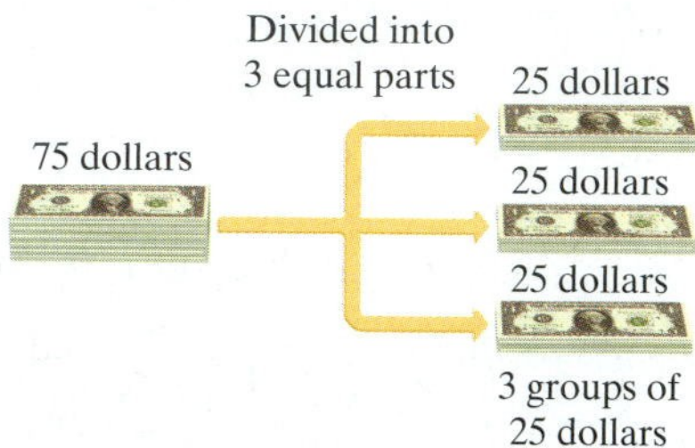

Objective A Dividing Whole Numbers

The process of separating a quantity into equal parts is called **division.** The division above can be symbolized by several notations.

quotient → 25
3)75 ← dividend
↑ divisor

dividend ↓
$\frac{75}{3} = 25$ ← quotient
↑ divisor

$75 \div 3 = 25$ (75: dividend, 3: divisor, 25: quotient)

$75/3 = 25$ (75: dividend, 3: divisor, 25: quotient)

(In the notation $\frac{75}{3}$, the bar separating 75 and 3 is called a **fraction bar.**) Just as subtraction is the reverse of addition, division is the reverse of multiplication. This means that division can be checked by multiplication.

$3\overline{)75}$ with quotient 25 because $25 \cdot 3 = 75$

Quotient · Divisor = Dividend

Since multiplication and division are related in this way, you can use your knowledge of multiplication facts to review quotients of one-digit divisors if necessary.

EXAMPLE 1 Find each quotient. Check by multiplying.

a. $42 \div 7$ **b.** $\frac{64}{8}$ **c.** $3\overline{)21}$

Solution:

a. $42 \div 7 = 6$ because $6 \cdot 7 = 42$

b. $\frac{64}{8} = 8$ because $8 \cdot 8 = 64$

c. $3\overline{)21}$ with quotient 7, because $7 \cdot 3 = 21$

Work Practice Problem 1

EXAMPLE 2 Find each quotient. Check by multiplying.

a. $1\overline{)7}$ **b.** $12 \div 1$ **c.** $\frac{6}{6}$ **d.** $9 \div 9$ **e.** $\frac{20}{1}$ **f.** $18\overline{)18}$

Solution:

a. $1\overline{)7}$ with quotient 7, because $7 \cdot 1 = 7$

b. $12 \div 1 = 12$ because $12 \cdot 1 = 12$

c. $\frac{6}{6} = 1$ because $1 \cdot 6 = 6$

d. $9 \div 9 = 1$ because $1 \cdot 9 = 9$

e. $\frac{20}{1} = 20$ because $20 \cdot 1 = 20$

f. $18\overline{)18}$ with quotient 1, because $1 \cdot 18 = 18$

Work Practice Problem 2

Example 2 illustrates the important properties of division described next:

Division Properties of 1

The quotient of any number (except 0) and that same number is 1. For example,

$8 \div 8 = 1$ $\frac{5}{5} = 1$ $4\overline{)4}$ with quotient 1

The quotient of any number and 1 is that same number. For example,

$9 \div 1 = 9$ $\frac{6}{1} = 6$ $1\overline{)3}$ with quotient 3 $\frac{0}{1} = 0$

EXAMPLE 3 Find each quotient. Check by multiplying.

a. $9\overline{)0}$ **b.** $0 \div 12$ **c.** $\frac{0}{5}$ **d.** $\frac{3}{0}$

Solution:

a. $9\overline{)0}$ with quotient 0, because $0 \cdot 9 = 0$ **b.** $0 \div 12 = 0$ because $0 \cdot 12 = 0$

c. $\frac{0}{5} = 0$ because $0 \cdot 5 = 0$

Continued on next page

PRACTICE PROBLEM 1

Find each quotient. Check by multiplying.

a. $9\overline{)72}$

b. $40 \div 5$

c. $\frac{24}{6}$

PRACTICE PROBLEM 2

Find each quotient. Check by multiplying.

a. $\frac{7}{7}$ **b.** $5 \div 1$

c. $1\overline{)11}$ **d.** $4 \div 1$

e. $\frac{10}{1}$ **f.** $21 \div 21$

PRACTICE PROBLEM 3

Find each quotient. Check by multiplying.

a. $\frac{0}{7}$ **b.** $8\overline{)0}$

c. $7 \div 0$ **d.** $0 \div 14$

Answers

1. a. 8 **b.** 8 **c.** 4 **2. a.** 1 **b.** 5 **c.** 11 **d.** 4 **e.** 10 **f.** 1 **3. a.** 0 **b.** 0 **c.** undefined **d.** 0

d. If $\frac{3}{0}$ = a *number,* then the *number* times $0 = 3$. Recall from Section 1.5 that any number multiplied by 0 is 0 and not 3. We say, then, that $\frac{3}{0}$ is **undefined.**

Work Practice Problem 3

Example 3 illustrates important division properties of 0.

Division Properties of 0

The quotient of 0 and any number (except 0) is 0. For example,

$$0 \div 9 = 0 \quad \frac{0}{5} = 0 \quad 14\overline{)0}\,\text{ with quotient } 0$$

The quotient of any number and 0 is not a number. We say that

$$\frac{3}{0}, \quad 0\overline{)3}, \quad \text{and} \quad 3 \div 0$$

are **undefined.**

Objective B Performing Long Division

When dividends are larger, the quotient can be found by a process called **long division.** For example, let's divide 2541 by 3.

divisor → $3\overline{)2541}$ ← dividend

We can't divide 3 into 2, so we try dividing 3 into the first two digits.

$$\begin{array}{r} 8 \\ 3\overline{)2541} \end{array}$$

25 ÷ 3 = 8 with 1 left, so our best estimate is 8. We place 8 over the 5 in 25.

Next, multiply 8 and 3 and subtract this product from 25. Make sure that this difference is less than the divisor.

$$\begin{array}{rl} 8 & \\ 3\overline{)2541} & \\ -24 & 8(3) = 24 \\ \hline 1 & 25 - 24 = 1\text{, and 1 is less than the divisor 3.} \end{array}$$

Bring down the next digit and go through the process again.

$$\begin{array}{rl} 84 & 14 \div 3 = 4 \text{ with 2 left} \\ 3\overline{)2541} & \\ -24\downarrow & \\ \hline 14 & \\ -12 & 4(3) = 12 \\ \hline 2 & 14 - 12 = 2 \end{array}$$

Once more, bring down the next digit and go through the process.

$$\begin{array}{rl} 847 & 21 \div 3 = 7 \\ 3\overline{)2541} & \\ -24 & \\ \hline 14 & \\ -12\downarrow & \\ \hline 21 & \\ -21 & 7(3) = 21 \\ \hline 0 & 21 - 21 = 0 \end{array}$$

The quotient is 847. To check, see that $847 \times 3 = 2541$.

EXAMPLE 4 Divide: $3705 \div 5$. Check by multiplying.

Solution:

```
     7        37 ÷ 5 = 7 with 2 left. Place this estimate, 7, over the 7 in 37.
  5)3705
   -35↓       7(5) = 35
     20       37 − 35 = 2, and 2 is less than the divisor 5.
     ↑        Bring down the 0.
    74        20 ÷ 5 = 4
  5)3705
   -35
     20
    -20↓      4(5) = 20
     05       20 − 20 = 0, and 0 is less than the divisor 5.
      ↑       Bring down the 5.

    741       5 ÷ 5 = 1
  5)3705
   -35
     20
    -20↓
      5
     -5       1(5) = 5
      0       5 − 5 = 0
```

Check:

$$\begin{array}{r} 741 \\ \times \quad 5 \\ \hline 3705 \end{array}$$

Work Practice Problem 4

Helpful Hint Since division and multiplication are reverse operations, don't forget that a division problem can be checked by multiplying.

EXAMPLE 5 Divide and check: $1872 \div 9$

Solution:

```
    208
  9)1872
   -18↓       2(9) = 18
     07       18 − 18 = 0; bring down the 7.
     -0↓      0(9) = 0
      72      7 − 0 = 7; bring down the 2.
     -72      8(9) = 72
       0      72 − 72 = 0
```

Check: $208 \cdot 9 = 1872$

Work Practice Problem 5

PRACTICE PROBLEM 4

Divide. Check by multiplying.

a. $4908 \div 6$
b. $2212 \div 4$
c. $753 \div 3$

PRACTICE PROBLEM 5

Divide and check by multiplying.

a. $7\overline{)2128}$
b. $9\overline{)45{,}900}$

Answers

4. a. 818 **b.** 553 **c.** 251
5. a. 304 **b.** 5100

Naturally, quotients don't always "come out even." Making 4 rows out of 26 chairs, for example, isn't possible if each row is supposed to have exactly the same number of chairs. Each of 4 rows can have 6 chairs, but 2 chairs are still left over.

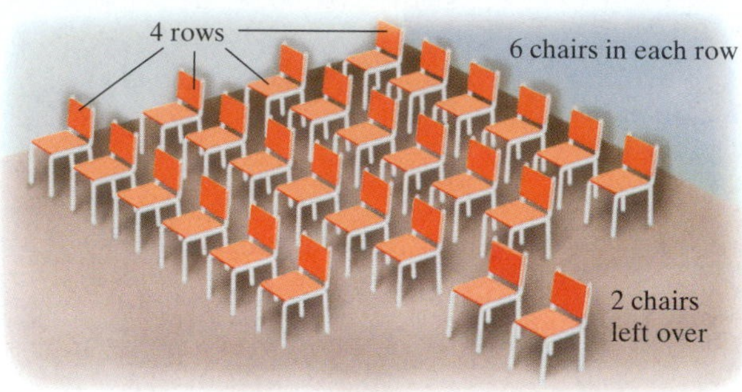

We signify "leftovers" or **remainders** in this way:

$$4\overline{)26}\quad 6 \text{ R } 2$$

The **whole number part of the quotient** is 6; the **remainder part of the quotient** is 2. Checking by multiplying,

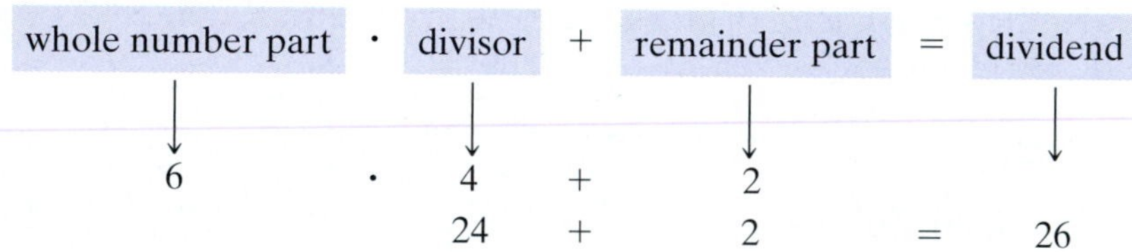

PRACTICE PROBLEM 6

Divide and check.

a. $4\overline{)939}$

b. $5\overline{)3287}$

EXAMPLE 6 Divide and check: $2557 \div 7$

Solution:

```
    365 R 2
  7)2557
   -21
     45
    -42
     37
    -35
      2
```

3(7) = 21

25 − 21 = 4; bring down the 5.

6(7) = 42

45 − 42 = 3; bring down the 7.

5(7) = 35

37 − 35 = 2; the remainder is 2.

Check: 365 · 7 + 2 = 2557

whole number part · divisor + remainder part = dividend

Work Practice Problem 6

Answers

6. a. 234 R 3 **b.** 657 R 2

EXAMPLE 7 Divide and check: 56,717 ÷ 8

Solution:

```
   7089  R 5
8)56717
 -56↓||      7(8) = 56
   07        Subtract and bring down the 7.
   -0↓       0(8) = 0
    71       Subtract and bring down the 1.
   -64↓      8(8) = 64
     77      Subtract and bring down the 7.
    -72      9(8) = 72
      5      Subtract. The remainder is 5.
```

Check:

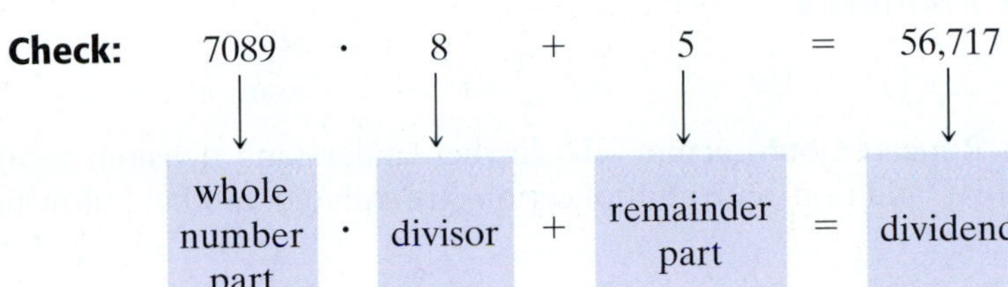

Work Practice Problem 7

PRACTICE PROBLEM 7

Divide and check.

a. 9)81,605

b. 4)23,310

When the divisor has more than one digit, the same pattern applies. For example, let's find 1358 ÷ 23.

```
     5        135 ÷ 23 = 5 with 20 left over. Our estimate is 5.
23)1358
  -115↓       5(23) = 115
   208        135 − 115 = 20. Bring down the 8.
```

Now we continue estimating.

```
    59  R 1   208 ÷ 23 = 9 with 1 left over.
23)1358
  -115
   208
  -207        9(23) = 207
     1        208 − 207 = 1. The remainder is 1.
```

To check, see that 59 · 23 + 1 = 1358.

EXAMPLE 8 Divide: 6819 ÷ 17

Solution:

```
    401  R 2
17)6819
  -68↓|       4(17) = 68
   01         Subtract and bring down the 1.
   -0↓        0(17) = 0
    19        Subtract and bring down the 9.
   -17        1(17) = 17
     2        Subtract. The reminder is 2.
```

To check, see that 401 · 17 + 2 = 6819.

Work Practice Problem 8

PRACTICE PROBLEM 8

Divide: 8920 ÷ 17

Answers

7. a. 9067 R 2 **b.** 5827 R 2
8. 524 R 12

PRACTICE PROBLEM 9

Divide: 33,282 ÷ 678

EXAMPLE 9 Divide: 51,600 ÷ 403

Solution:

$403\overline{)51600}$ = 128 R 16	
−403	1(403) = 403
1130	Subtract and bring down the 0.
−806	2(403) = 806
3240	Subtract and bring down the 0.
−3224	8(403) = 3224
16	Subtract. The remainder is 16.

To check, see that $128 \cdot 403 + 16 = 51{,}600$.

Work Practice Problem 9

Division Shown as Repeated Subtraction To further understand division, recall from Section 1.5 that addition and multiplication are related in the following manner:

$$\underbrace{3 + 3 + 3 + 3}_{\text{4 addends; each addend is 3}} = 4 \times 3 = 12$$

In other words, multiplication is repeated addition. Likewise, division is repeated subtraction.

For example, let's find

$$35 \div 8$$

by repeated subtraction. Keep track of the number of times 8 is subtracted from 35. We are through when we can subtract no more because the difference is less than 8.

35 ÷ 8: Repeated Subtraction

35 −8	1 time
27 −8	2 times
19 −8	3 times
11 −8	4 times
3 ⟵	Remainder (We cannot subtract 8 again.)

35 dollars
8 dollars — 1 time
8 dollars — 2 times
8 dollars — 3 times
8 dollars — 4 times
3 dollars left over

Thus, 35 ÷ 8 = 4 R 3.

To check, perform the same multiplication as usual, but finish by adding in the remainder.

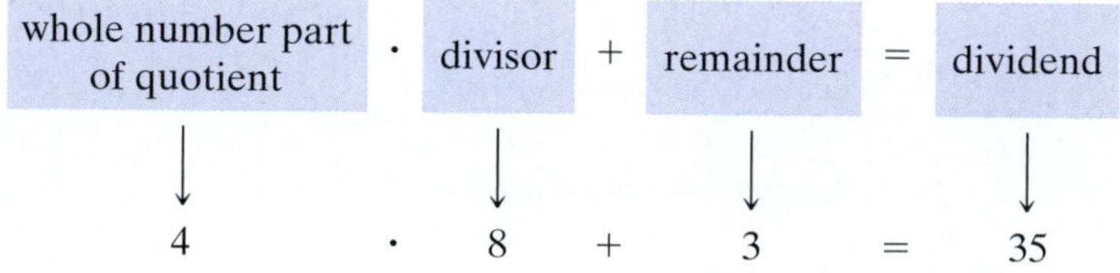

Answer

9. 49 R 60

Objective C Solving Problems by Dividing

Below are some key words and phrases that may indicate the operation of division:

Division		
Key Words or Phrases	**Examples**	**Symbols**
divide	divide 10 by 5	$10 \div 5$ or $\frac{10}{5}$
quotient	the quotient of 64 and 4	$64 \div 4$ or $\frac{64}{4}$
divided by	9 divided by 3	$9 \div 3$ or $\frac{9}{3}$
divided or shared equally among	\$100 divided equally among five people	$100 \div 5$ or $\frac{100}{5}$
per	100 miles per 2 hours	$\frac{100 \text{ miles}}{2 \text{ hours}}$

✔ **Concept Check** Which of the following is the correct way to represent "the quotient of 60 and 12"? Or are both correct? Explain your answer.

a. $12 \div 60$

b. $60 \div 12$

EXAMPLE 10 Finding Shared Earnings

Three college students share a paper route to earn money for expenses. The total in their fund after expenses was \$2895. How much is each person's equal share?

Solution:

In words:	Each person's share	=	total money	÷	number of persons
	↓		↓		↓
Translate:	Each person's share	=	2895	÷	3

Then

$$\begin{array}{r} 965 \\ 3\overline{)2895} \\ -27 \\ \hline 19 \\ -18 \\ \hline 15 \\ -15 \\ \hline 0 \end{array}$$

Each person's share is \$965.

Work Practice Problem 10

PRACTICE PROBLEM 10

Three students bought 171 blank CDs to share equally. How many CDs did each person get?

Answer

10. 57 CDs

✔ Concept Check Answers

a. incorrect **b.** correct

PRACTICE PROBLEM 11

Printers can be packed 12 to a box. If 532 printers are to be packed but only full boxes are shipped, how many full boxes will be shipped? How many printers are left over and not shipped?

EXAMPLE 11 Dividing Number of Downloads

As part of a promotion, Becky Foster receives 238 cards, each good for one free song download from MSN music. If Becky wants to share them evenly with 19 friends, how many download cards will each friend receive? How many will be left over?

Solution:

In words:	Number of cards for each person	=	number of cards	÷	number of friends
	↓		↓		↓
Translate:	Number of cards for each person	=	238	÷	19

$$\begin{array}{r} 12 \quad \text{R } 10 \\ 19\overline{)238} \\ \underline{-19} \\ 48 \\ \underline{-38} \\ 10 \end{array}$$

Each friend will receive 12 download cards. The cards cannot be divided equally among her friends since there is a nonzero remainder. There will be 10 download cards left over.

Work Practice Problem 11

Objective D Finding Averages

A special application of division (and addition) is finding the average of a list of numbers. The **average** of a list of numbers is the sum of the numbers divided by the *number* of numbers.

$$\text{average} = \frac{\text{sum of numbers}}{\textit{number} \text{ of numbers}}$$

PRACTICE PROBLEM 12

To compute a safe time to wait for reactions to occur after allergy shots are administered, a lab technician is given a list of elapsed times between administered shots and reactions. Find the average of the times 4 minutes, 7 minutes, 35 minutes, 16 minutes, 9 minutes, 3 minutes, and 52 minutes.

EXAMPLE 12 Averaging Scores

A mathematics instructor is checking a simple program she wrote for averaging the scores of her students. To do so, she averages a student's scores of 75, 96, 81, and 88 by hand. Find this average score.

Solution: To find the average score, we find the sum of his scores and divide by 4, the number of scores.

$$\begin{array}{r} 75 \\ 96 \\ 81 \\ \underline{+88} \\ 340 \end{array} \text{ sum} \qquad \text{average} = \frac{340}{4} = 85 \qquad \begin{array}{r} 85 \\ 4\overline{)340} \\ \underline{-32} \\ 20 \\ \underline{-20} \\ 0 \end{array}$$

The average score is 85.

Work Practice Problem 12

Answers

11. 44 full boxes; 4 printers left over
12. 18 min

CALCULATOR EXPLORATIONS Dividing Numbers

To divide numbers on a calculator, find the keys marked [÷] and [=] or [ENTER]. For example, to find 435 ÷ 5 on a calculator, press the keys [435] [÷] [5] [=] or [ENTER]. The display will read [87]. Thus, 435 ÷ 5 = 87.

Use a calculator to divide.

1. 848 ÷ 16

2. 564 ÷ 12

3. $95\overline{)5890}$

4. $27\overline{)1053}$

5. $\frac{32{,}886}{126}$

6. $\frac{143{,}088}{264}$

7. 0 ÷ 315

8. 315 ÷ 0

Vocabulary and Readiness Check

Use the choices below to fill in each blank.

1	number	divisor	dividend
0	undefined	average	quotient

1. In 90 ÷ 2 = 45, the answer 45 is called the ______________, 90 is called the ______________, and 2 is called the ______________.

2. The quotient of any number and 1 is the same ______________.

3. The quotient of any number (except 0) and the same number is ______________.

4. The quotient of 0 and any number (except 0) is ______________.

5. The quotient of any number and 0 is ______________.

6. The ______________ of a list of numbers is the sum of the numbers divided by the ______________ of numbers.

1.6 EXERCISE SET

FOR EXTRA HELP

Student Solutions Manual

PH Math/Tutor Center

CD/Video for Review

MathXL®

MyMathLab

Objective A *Find each quotient. See Examples 1 through 3.*

1. 54 ÷ 9

2. 72 ÷ 9

3. 36 ÷ 3

4. 24 ÷ 3

5. 0 ÷ 8

6. 0 ÷ 4

7. 31 ÷ 1

8. 38 ÷ 1

9. $\frac{18}{18}$

10. $\frac{49}{49}$

11. $\frac{24}{3}$

12. $\frac{45}{9}$

13. 26 ÷ 0

14. $\frac{12}{0}$

15. 26 ÷ 26

16. 6 ÷ 6

17. 0 ÷ 14

18. 7 ÷ 0

19. 18 ÷ 2

20. 18 ÷ 3

Objectives **A** **B** **Mixed Practice** *Divide and then check by multiplying. See Examples 1 through 5.*

21. $3\overline{)87}$ **22.** $5\overline{)85}$ **23.** $3\overline{)222}$ **24.** $8\overline{)640}$ **25.** $3\overline{)1014}$ **26.** $4\overline{)2104}$

27. $\frac{30}{0}$ **28.** $\frac{0}{30}$ **29.** $63 \div 7$ **30.** $56 \div 8$ **31.** $150 \div 6$ **32.** $121 \div 11$

Divide and then check by multiplying. See Examples 6 and 7.

33. $7\overline{)479}$ **34.** $7\overline{)426}$ **35.** $6\overline{)1421}$ **36.** $3\overline{)1240}$

37. $305 \div 8$ **38.** $167 \div 3$ **39.** $2286 \div 7$ **40.** $3333 \div 4$

Divide and then check by multiplying. See Examples 8 and 9.

41. $55\overline{)715}$ **42.** $23\overline{)736}$ **43.** $23\overline{)1127}$ **44.** $42\overline{)2016}$ **45.** $97\overline{)9417}$

46. $1938 \div 44$ **47.** $3146 \div 15$ **48.** $7354 \div 12$ **49.** $6578 \div 13$ **50.** $5670 \div 14$

51. $9299 \div 46$ **52.** $2505 \div 64$ **53.** $\frac{12{,}744}{236}$ **54.** $\frac{5781}{123}$ **55.** $\frac{10{,}297}{103}$

56. $\frac{23{,}092}{240}$ **57.** $20{,}619 \div 102$ **58.** $40{,}853 \div 203$ **59.** $244{,}989 \div 423$ **60.** $164{,}592 \div 543$

Divide. See Examples 1 through 9.

61. $7\overline{)119}$ **62.** $8\overline{)104}$ **63.** $7\overline{)3580}$ **64.** $5\overline{)3017}$

65. $40\overline{)85{,}312}$ **66.** $50\overline{)85{,}747}$ **67.** $142\overline{)863{,}360}$ **68.** $214\overline{)650{,}560}$

Objective **C** **Translating** *Solve. See Examples 10 through 11.*

69. Find the quotient of 117 and 5.

70. Find the quotient of 94 and 7.

71. Find 200 divided by 35.

72. Find 116 divided by 32.

73. Find the quotient of 62 and 3.

74. Find the quotient of 78 and 5.

75. Martin Thieme teaches American Sign Language classes for $65 per student for a 7-week session. He collects $2145 from the group of students. Find how many students are in the group.

76. Kathy Gomez teaches Spanish lessons for $85 per student for a 5-week session. From one group of students, she collects $4930. Find how many students are in the group.

77. The gravity of Jupiter is 318 times as strong as the gravity of Earth, so objects on Jupiter weigh 318 times as much as they weigh on Earth. If a person would weigh 52,470 pounds on Jupiter, find how much the person weighs on Earth.

78. Twenty-one people pooled their money and bought lottery tickets. One ticket won a prize of $5,292,000. Find how many dollars each person received.

79. An 18-hole golf course is 5580 yards long. If the distance to each hole is the same, find the distance between holes.

80. A truck hauls wheat to a storage granary. It carries a total of 5768 bushels of wheat in 14 trips. How much does the truck haul each trip if each trip it hauls the same amount?

81. There is a bridge over highway I-35 every three miles. The first bridge is at the beginning of a 265-mile stretch of highway. Find how many bridges there are over 265 miles of I-35.

82. The white stripes dividing the lanes on a highway are 25 feet long, and the spaces between them are 25 feet long. Let's call a "lane divider" a stripe followed by a space. Find how many whole "lane dividers" there are in 1 mile of highway. (A mile is 5280 feet.)

83. Ari Trainor is in the requisitions department of Central Electric Lighting Company. Light poles along a highway are placed 492 feet apart. The first light pole is at the beginning of a 1-mile strip. Find how many poles he should order for the 1-mile strip of highway. (A mile is 5280 feet.)

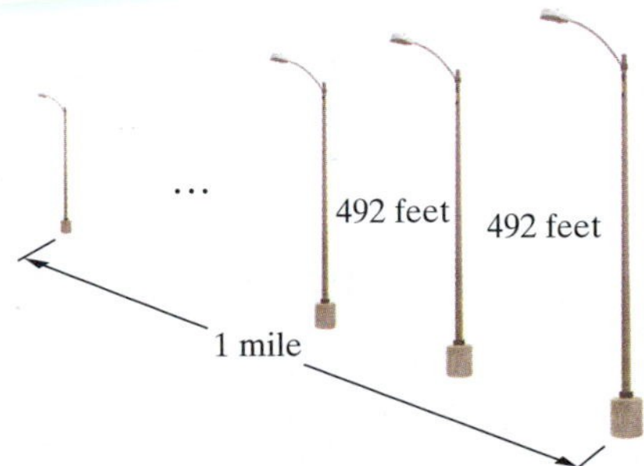

84. Professor Lopez has a piece of rope 185 feet long that she wants to cut into pieces for an experiment in her physics class. Each piece of rope is to be 8 feet long. Determine whether she has enough rope for her 22-student class. Determine the amount extra or the amount short.

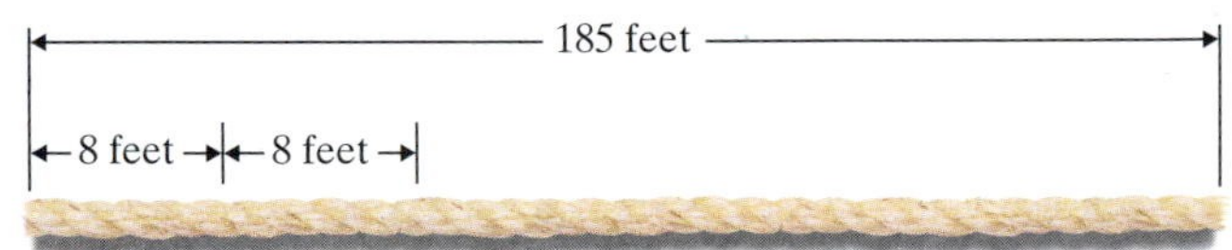

85. Broad Peak in Pakistan is the twelfth-tallest mountain in the world. Its elevation is 26,400 feet. A mile is 5280 feet. How many miles tall is Broad Peak? (*Source:* National Geographic Society)

86. Shaun Alexander of the Seattle Seahawks led the NFL in touchdowns during the 2005 regular football season, scoring a total of 168 points from touchdowns. If a touchdown is worth 6 points, how many touchdowns did Alexander make during the 2005 season? (*Source:* National Football League)

87. Find how many yards are in 1 mile. (A mile is 5280 feet; a yard is 3 feet.)

88. Find how many whole feet are in 1 rod. (A mile is 5280 feet; 1 mile is 320 rods.)

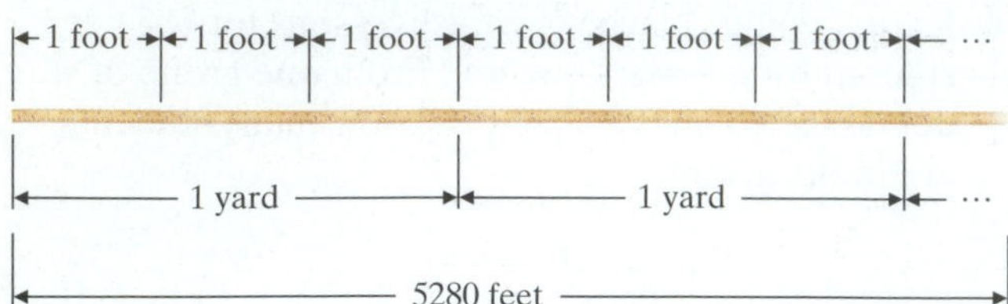

Objective **D** *Find the average of each list of numbers. See Example 12.*

89. 10, 24, 35, 22, 17, 12

90. 37, 26, 15, 29, 51, 22

91. 205, 972, 210, 161

92. 121, 200, 185, 176, 163

93. 86, 79, 81, 69, 80

94. 92, 96, 90, 85, 92, 79

The normal monthly temperature in degrees Fahrenheit for Salt Lake City, Utah, is given in the graph. Use this graph to answer Exercises 95 and 96. (Source: National Climatic Data Center)

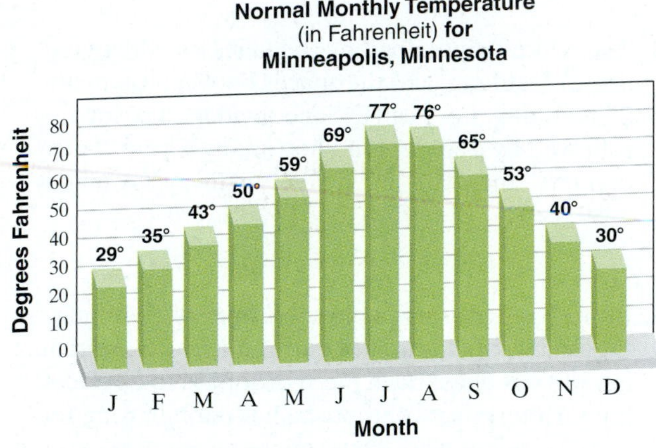

95. Find the average temperature for June, July, and August.

96. Find the average temperature for October, November, and December.

Mixed Practice (Sections 1.3, 1.5, 1.6) *Perform each indicated operation. Watch the operation symbol.*

97. $82 + 463 + 29 + 8704$

98. $23 + 407 + 92 + 7011$

99. $\begin{array}{r} 546 \\ \times\ 28 \\ \hline \end{array}$

100. $\begin{array}{r} 712 \\ \times\ 54 \\ \hline \end{array}$

101. $\begin{array}{r} 722 \\ -\ 43 \\ \hline \end{array}$

102. $\begin{array}{r} 712 \\ -\ 54 \\ \hline \end{array}$

103. $\dfrac{45}{0}$

104. $\dfrac{0}{23}$

105. $228 \div 24$

106. $304 \div 31$

Concept Extensions

Match each word phrase to the correct translation. (Not all letter choices will be used.) See the Concept Check in this section.

107. The quotient of 40 and 8

108. The quotient of 200 and 20

109. 200 divided by 20

110. 40 divided by 8

a. $20 \div 200$

b. $200 \div 20$

c. $40 \div 8$

d. $8 \div 40$

The following table shows the top five leading U.S. television advertisers in 2005 and the amount of money spent that year on television advertising. Use this table to answer Exercises 111 and 112.

Advertiser	Amount Spent on Television Advertising in 2005
General Motors, Dealers and corporate	\$797,000,000
Ford Motors, Dealers and corporate	\$667,000,000
Toyota, Dealers and corporate	\$544,000,000
DaimlerChrysler AG	\$528,000,000
Honda	\$339,000,000

111. Find the average amount of money spent on television ads for the year by the top two advertisers.

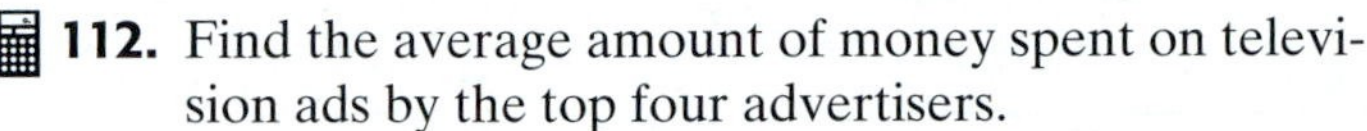

112. Find the average amount of money spent on television ads by the top four advertisers.

In Example 12 in this section, we found that the average of 75, 96, 88, and 81 is 85. Use this information to answer Exercises 113 and 114.

113. If the number 75 is removed from the list of numbers, does the average increase or decrease? Explain why.

114. If the number 96 is removed from the list of numbers, does the average increase or decrease? Explain why.

115. Without computing it, tell whether the average of 126, 135, 198, 113 is 86. Explain why it is or why it is not.

116. If the area of a rectangle is 60 square feet and its width is 5 feet, what is its length?

117. Write down any two numbers whose quotient is 25.

118. Find $26 \div 5$ using the process of repeated subtraction.

THE BIGGER PICTURE Operations on Sets of Numbers

This is a special feature that we begin in this section. Among other concepts introduced later in the text, it is very important for you to be able to perform operations on different sets of numbers. To help you remember these operations, we begin an outline below and continually expand this outline throughout this text. Although suggestions are given, this outline should be in your own words. Once you complete the new portion of your outline, try the exercises below. Remember: Study your outline often as you proceed through this text.

I. Some Operations on Sets of Numbers

A. Whole Numbers

1. Add or Subtract:

$$\begin{array}{r} 13 \\ +29 \\ \hline 42 \end{array} \qquad \begin{array}{r} 400 \\ -\ 38 \\ \hline 362 \end{array}$$

2. Multiply or Divide:

$$\begin{array}{r} 238 \\ \times 47 \\ \hline 1666 \\ 9520 \\ \hline 11186 \end{array} \qquad \begin{array}{r} 127\text{ R }2 \\ 7\overline{)891} \\ -7 \\ \hline 19 \\ -14 \\ \hline 51 \\ -49 \\ \hline 2 \end{array}$$

Perform indicated operations.

1. $82 + 39$

2. $82 - 39$

3. 82×39

4. $2592 \div 29$

5. $0 \cdot 15$

6. $0 \div 11$

7. $26 \cdot 1$

8. $36 \div 0$

9. $82 \div 1$

10. $2000 - 156$

Answers

1. ______
2. ______
3. ______
4. ______
5. ______
6. ______
7. ______
8. ______
9. ______
10. ______
11. ______
12. ______
13. ______
14. ______
15. ______
16. ______
17. ______
18. ______
19. ______
20. ______
21. ______
22. ______
23. ______
24. ______
25. ______
26. ______
27. ______
28. ______
29. ______
30. ______

INTEGRATED REVIEW

Sections 1.1–1.6

Operations on Whole Numbers

Perform each indicated operation.

1. $\begin{array}{r} 42 \\ 63 \\ +89 \\ \hline \end{array}$

2. $\begin{array}{r} 7006 \\ -\ 451 \\ \hline \end{array}$

3. $\begin{array}{r} 87 \\ \times 52 \\ \hline \end{array}$

4. $8\overline{)4496}$

5. $1 \cdot 67$

6. $\dfrac{36}{0}$

7. $16 \div 16$

8. $5 \div 1$

9. $0 \cdot 21$

10. $7 \cdot 0 \cdot 8$

11. $0 \div 7$

12. $12 \div 4$

13. $9 \cdot 7$

14. $45 \div 5$

15. $\begin{array}{r} 207 \\ -\ 69 \\ \hline \end{array}$

16. $\begin{array}{r} 207 \\ +\ 69 \\ \hline \end{array}$

17. $3718 - 2549$

18. $1861 + 7965$

19. $7\overline{)1278}$

20. $\begin{array}{r} 1259 \\ \times\ \ 63 \\ \hline \end{array}$

21. $7\overline{)7695}$

22. $9\overline{)1000}$

23. $32\overline{)21,240}$

24. $65\overline{)70,000}$

25. $4000 - 2963$

26. $10,000 - 101$

27. $\begin{array}{r} 303 \\ \times 101 \\ \hline \end{array}$

28. $(475)(100)$

29. Find the total of 62 and 9.

30. Find the product of 62 and 9.

31. Find the quotient of 62 and 9.

32. Find the difference of 62 and 9.

33. Subtract 17 from 200.

34. Find the difference of 432 and 201.

Complete the table by rounding the given number to the given place value.

		Tens	Hundreds	Thousands
35.	9735			
36.	1429			
37.	20,801			
38.	432,198			

Find the perimeter and area of each figure.

△ **39.**

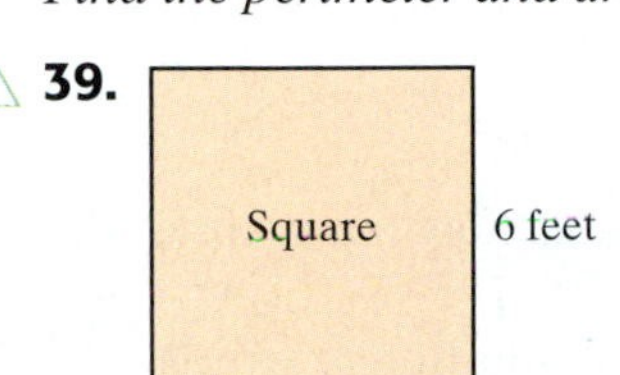

△ **40.**

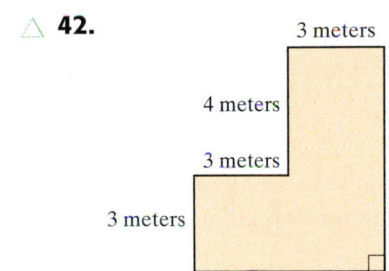

14 inches

Rectangle

7 inches

Find the perimeter of each figure.

△ **41.**

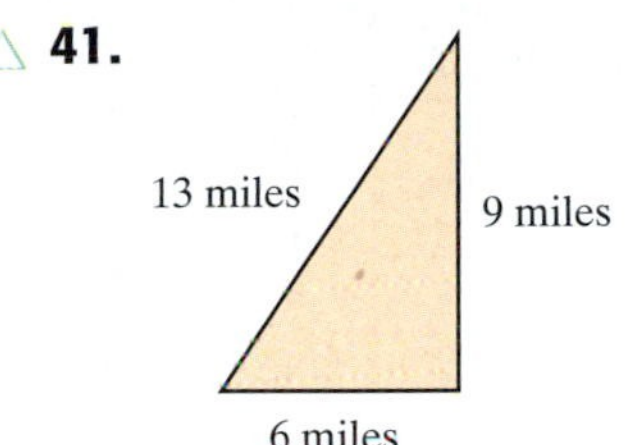

△ **42.**

3 meters

4 meters

3 meters

3 meters

Find the average of each list of numbers.

43. 19, 15, 25, 37, 24

44. 108, 131, 98, 159

45. The Mackinac Bridge is a suspension bridge that connects the lower and upper peninsulas of Michigan across the Straits of Mackinac. Its total length is 26,372 feet. The Lake Pontchartrain Bridge is a twin concrete trestle bridge in Slidell, Louisiana. Its total length is 28,547 feet. Which bridge is longer and by how much? (*Sources:* Mackinac Bridge Authority and Federal Highway Administration, Bridge Division)

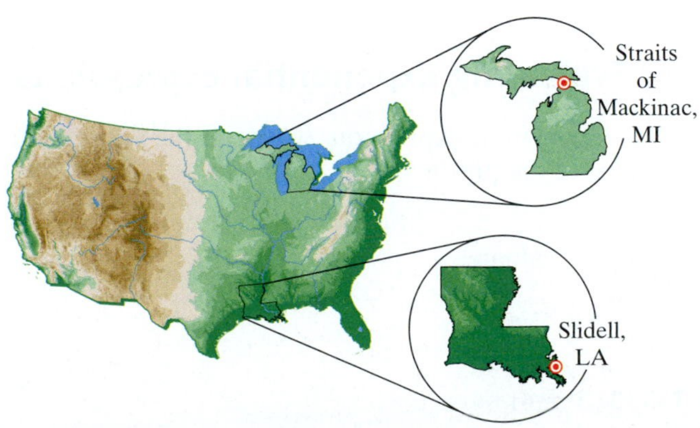

46. In North America, the average toy expenditure per child is $347 per year. On average, how much is spent on toys for a child by the time he or she reaches age 18? (*Source:* The NPD Group Worldwide)

31. ______

32. ______

33. ______

34. ______

35. see table

36. see table

37. see table

38. see table

39. ______

40. ______

41. ______

42. ______

43. ______

44. ______

45. ______

46. ______

1.7 EXPONENTS AND ORDER OF OPERATIONS

Objectives

A Write Repeated Factors Using Exponential Notation.

B Evaluate Expressions Containing Exponents.

C Use the Order of Operations.

D Find the Area of a Square.

Objective A Using Exponential Notation

In the product $3 \cdot 3 \cdot 3 \cdot 3 \cdot 3$, notice that 3 is a factor several times. When this happens, we can use a shorthand notation, called an **exponent,** to write the repeated multiplication.

$\underbrace{3 \cdot 3 \cdot 3 \cdot 3 \cdot 3}_{\text{3 is a factor 5 times}}$ can be written as

exponent

3^5 Read as "three to the fifth power."

base

This is called **exponential notation.** The **exponent,** 5, indicates how many times the **base,** 3, is a factor.

The table below shows examples of reading exponential notation in words.

Expression	In Words
5^2	"five to the second power" or "five squared"
5^3	"five to the third power" or "five cubed"
5^4	"five to the fourth power"

Usually, an exponent of 1 is not written, so when no exponent appears, we assume that the exponent is 1. For example, $2 = 2^1$ and $7 = 7^1$.

PRACTICE PROBLEMS 1–4

Write using exponential notation.

1. $8 \cdot 8 \cdot 8 \cdot 8$
2. $3 \cdot 3 \cdot 3$
3. $10 \cdot 10 \cdot 10 \cdot 10 \cdot 10$
4. $5 \cdot 5 \cdot 4 \cdot 4 \cdot 4 \cdot 4 \cdot 4 \cdot 4$

EXAMPLES Write using exponential notation.

1. $7 \cdot 7 \cdot 7 = 7^3$
2. $3 \cdot 3 = 3^2$
3. $6 \cdot 6 \cdot 6 \cdot 6 \cdot 6 = 6^5$
4. $3 \cdot 3 \cdot 3 \cdot 3 \cdot 9 \cdot 9 \cdot 9 = 3^4 \cdot 9^3$

Work Practice Problems 1–4

Objective B Evaluating Exponential Expressions

To **evaluate** an exponential expression, we write the expression as a product and then find the value of the product.

PRACTICE PROBLEMS 5–8

Evaluate.

5. 4^2
6. 8^3
7. 11^1
8. $2 \cdot 3^2$

EXAMPLES Evaluate.

5. $9^2 = 9 \cdot 9 = 81$
6. $6^1 = 6$
7. $3^4 = 3 \cdot 3 \cdot 3 \cdot 3 = 81$
8. $5 \cdot 6^2 = 5 \cdot 6 \cdot 6 = 180$

Work Practice Problems 5–8

Answers

1. 8^4 **2.** 3^3 **3.** 10^5 **4.** $5^2 \cdot 4^6$
5. 16 **6.** 512 **7.** 11 **8.** 18

Example 8 illustrates an important property: An exponent applies only to its base. The exponent 2, in $5 \cdot 6^2$, applies only to its base, 6.

Helpful Hint An exponent applies only to its base. For example, $4 \cdot 2^3$ means $4 \cdot 2 \cdot 2 \cdot 2$.

Helpful Hint Don't forget that 2^4, for example, is *not* $2 \cdot 4$. The expression 2^4 means repeated multiplication of the same factor.

$$2^4 = 2 \cdot 2 \cdot 2 \cdot 2 = 16, \quad \text{whereas } 2 \cdot 4 = 8$$

✔Concept Check Which of the following statements is correct?

a. 3^5 is the same as $5 \cdot 5 \cdot 5$.
b. "Ten cubed" is the same as 10^2.
c. "Six to the fourth power" is the same as 6^4.
d. 12^2 is the same as $12 \cdot 2$.

Objective C Using the Order of Operations

Suppose that you are in charge of taking inventory at a local cell phone store. An employee has given you the number of a certain cell phone in stock as the expression

$$6 + 2 \cdot 30$$

To calculate the value of this expression, do you add first or multiply first? If you add first, the answer is 240. If you multiply first, the answer is 66.

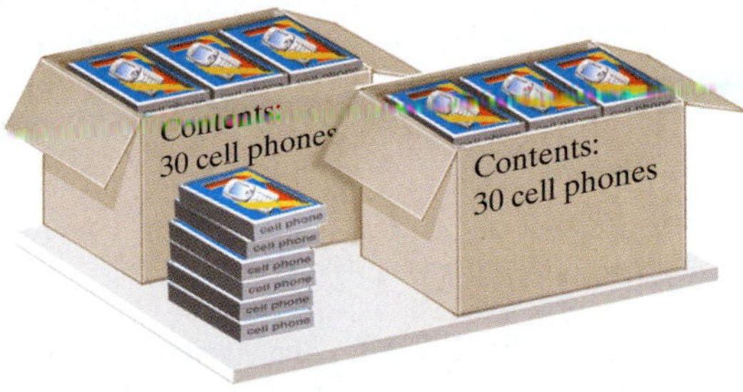

Mathematical symbols wouldn't be very useful if two values were possible for one expression. Thus, mathematicians have agreed that, given a choice, we multiply first.

$$\begin{aligned} 6 + 2 \cdot 30 &= 6 + 60 && \text{Multiply.} \\ &= 66 && \text{Add.} \end{aligned}$$

This agreement is one of several **order of operations** agreements.

✔ Concept Check Answer
c

Order of Operations

1. Perform all operations within parentheses ( ), brackets [ ], or other grouping symbols such as fraction bars, starting with the innermost set.
2. Evaluate any expressions with exponents.
3. Multiply or divide in order from left to right.
4. Add or subtract in order from left to right.

Below we practice using order of operations to simplify expressions.

PRACTICE PROBLEM 9

Simplify: $9 \cdot 3 - 8 \div 4$

EXAMPLE 9 Simplify: $2 \cdot 4 - 3 \div 3$

Solution: There are no parentheses and no exponents, so we start by multiplying and dividing, from left to right.

$$\begin{aligned} 2 \cdot 4 - 3 \div 3 &= 8 - 3 \div 3 && \text{Multiply.} \\ &= 8 - 1 && \text{Divide.} \\ &= 7 && \text{Subtract.} \end{aligned}$$

Work Practice Problem 9

PRACTICE PROBLEM 10

Simplify: $48 \div 3 \cdot 2^2$

EXAMPLE 10 Simplify: $4^2 \div 2 \cdot 4$

Solution: We start by evaluating 4^2.

$$4^2 \div 2 \cdot 4 = 16 \div 2 \cdot 4 \quad \text{Write } 4^2 \text{ as } 16.$$

Next we multiply or divide *in order* from left to right. Since division appears before multiplication from left to right, we divide first, then multiply.

$$\begin{aligned} 16 \div 2 \cdot 4 &= 8 \cdot 4 && \text{Divide.} \\ &= 32 && \text{Multiply.} \end{aligned}$$

Work Practice Problem 10

PRACTICE PROBLEM 11

Simplify: $(10 - 7)^4 + 2 \cdot 3^2$

EXAMPLE 11 Simplify: $(8 - 6)^2 + 2^3 \cdot 3$

Solution:

$$\begin{aligned} (8 - 6)^2 + 2^3 \cdot 3 &= 2^2 + 2^3 \cdot 3 && \text{Simplify inside parentheses.} \\ &= 4 + 8 \cdot 3 && \text{Write } 2^2 \text{ as 4 and } 2^3 \text{ as 8.} \\ &= 4 + 24 && \text{Multiply.} \\ &= 28 && \text{Add.} \end{aligned}$$

Work Practice Problem 11

Answers

9. 25 **10.** 64 **11.** 99

EXAMPLE 12 Simplify: $4^3 + [3^2 - (10 \div 2)] - 7 \cdot 3$

Solution: Here we begin with the innermost set of parentheses.

$$
\begin{aligned}
4^3 + [3^2 - (10 \div 2)] - 7 \cdot 3 &= 4^3 + [3^2 - 5] - 7 \cdot 3 && \text{Simplify inside parentheses.}\\
&= 4^3 + [9 - 5] - 7 \cdot 3 && \text{Write } 3^2 \text{ as 9.}\\
&= 4^3 + 4 - 7 \cdot 3 && \text{Simplify inside brackets.}\\
&= 64 + 4 - 7 \cdot 3 && \text{Write } 4^3 \text{ as 64.}\\
&= 64 + 4 - 21 && \text{Multiply.}\\
&= 47 && \text{Add and subtract from left to right.}
\end{aligned}
$$

Work Practice Problem 12

PRACTICE PROBLEM 12

Simplify: $36 \div [20 - (4 \cdot 2)] + 4^3 - 6$

EXAMPLE 13 Simplify: $\dfrac{7 - 2 \cdot 3 + 3^2}{5(2 - 1)}$

Solution: Here, the fraction bar is like a grouping symbol. We simplify above and below the fraction bar separately.

$$
\begin{aligned}
\frac{7 - 2 \cdot 3 + 3^2}{5(2 - 1)} &= \frac{7 - 2 \cdot 3 + 9}{5(1)} && \text{Evaluate } 3^2 \text{ and } (2 - 1).\\
&= \frac{7 - 6 + 9}{5} && \text{Multiply } 2 \cdot 3 \text{ in the numerator and multiply 5 and 1 in the denominator.}\\
&= \frac{10}{5} && \text{Add and subtract from left to right.}\\
&= 2 && \text{Divide.}
\end{aligned}
$$

Work Practice Problem 13

PRACTICE PROBLEM 13

Simplify: $\dfrac{25 + 8 \cdot 2 - 3^3}{2(3 - 2)}$

EXAMPLE 14 Simplify: $64 \div 8 \cdot 2 + 4$

Solution:

$$
\begin{aligned}
64 \div 8 \cdot 2 + 4 &= 8 \cdot 2 + 4 && \text{Divide.}\\
&= 16 + 4 && \text{Multiply.}\\
&= 20 && \text{Add.}
\end{aligned}
$$

Work Practice Problem 14

PRACTICE PROBLEM 14

Simplify: $36 \div 6 \cdot 3 + 5$

Objective D Finding the Area of a Square

Since a square is a special rectangle, we can find its area by finding the product of its length and its width.

Area of a rectangle = length · width

By recalling that each side of a square has the same measurement, we can use the following procedure to find its area:

$$
\begin{aligned}
\text{Area of a square} &= \text{length} \cdot \text{width}\\
&= \text{side} \cdot \text{side}\\
&= (\text{side})^2
\end{aligned}
$$

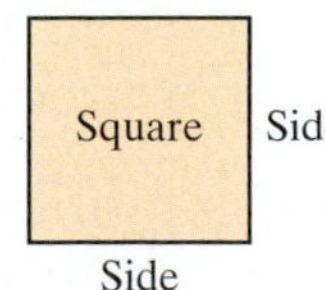

Answers
12. 61 13. 7 14. 23

PRACTICE PROBLEM 15

Find the area of a square whose side measures 12 centimeters.

EXAMPLE 15 Find the area of a square whose side measures 4 inches.

Solution: Area of a square = $(\text{side})^2$
$= (4 \text{ inches})^2$
$= 16$ square inches

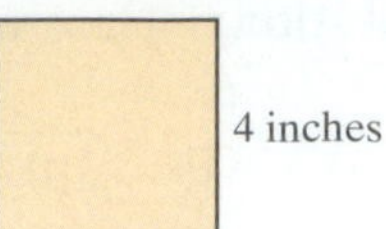

The area of the square is 16 square inches.

Work Practice Problem 15

Answer

15. 144 sq cm

CALCULATOR EXPLORATIONS Exponents

To evaluate an exponent such as 4^7 on a calculator, find the keys marked [y^x] or [∧] and [=] or [ENTER]. To evaluate 4^7, press the keys [4] [y^x] (or [∧]) [7] [=] or [ENTER]. The display will read [16384]. Thus, $4^7 = 16{,}384$.

Use a calculator to evaluate.

1. 3^6 **2.** 5^6 **3.** 4^5

4. 7^6 **5.** 2^{11} **6.** 6^8

Order of Operations

To see whether your calculator has the order of operations built in, evaluate $5 + 2 \cdot 3$ by pressing the keys [5] [+] [2] [×] [3] [=] or [ENTER]. If the display reads [11], your calculator does have the order of operations built in. This means that most of the time you can key in a problem exactly as it is written and the calculator will perform operations in the proper order. When evaluating an expression containing parentheses, key in the parentheses. (If an expression contains brackets, key in parentheses.) For example, to evaluate $2[25 - (8 + 4)] - 11$, press the keys [2] [×] [( ] [25] [−] [( ] [8] [+] [4] [ )] [ )] [−] [11] [=] or [ENTER].

The display will read [15].

Use a calculator to evaluate.

7. $7^4 + 5^3$

8. $12^4 - 8^4$

9. $63 \cdot 75 - 43 \cdot 10$

10. $8 \cdot 22 + 7 \cdot 16$

11. $4(15 \div 3 + 2) - 10 \cdot 2$

12. $155 - 2(17 + 3) + 185$

Vocabulary and Readiness Check

Use the choices below to fill in each blank.

addition	multiplication	exponent
subtraction	division	base

1. In $2^5 = 32$, the 2 is called the ____________ and the 5 is called the ____________.
2. To simplify $8 + 2 \cdot 6$, which operation should be performed first? ____________
3. To simplify $(8 + 2) \cdot 6$, which operation should be performed first? ____________
4. To simplify $9(3 - 2) \div 3 + 6$, which operation should be performed first? ____________
5. To simplify $8 \div 2 \cdot 6$, which operation should be performed first ?__________

1.7 EXERCISE SET

FOR EXTRA HELP

Student Solutions Manual

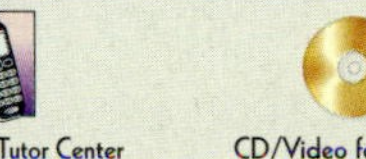
PH Math/Tutor Center

CD/Video for Review

MathXL®

MyMathLab

Objective A *Write using exponential notation. See Examples 1 through 4.*

1. $4 \cdot 4 \cdot 4$ **2.** $5 \cdot 5 \cdot 5 \cdot 5$ **3.** $7 \cdot 7 \cdot 7 \cdot 7 \cdot 7 \cdot 7$ **4.** $6 \cdot 6 \cdot 6 \cdot 6 \cdot 6 \cdot 6 \cdot 6$

5. $12 \cdot 12 \cdot 12$ **6.** $10 \cdot 10 \cdot 10$ **7.** $6 \cdot 6 \cdot 5 \cdot 5 \cdot 5$ **8.** $4 \cdot 4 \cdot 3 \cdot 3 \cdot 3$

9. $9 \cdot 8 \cdot 8$ **10.** $7 \cdot 4 \cdot 4 \cdot 4$ **11.** $3 \cdot 2 \cdot 2 \cdot 2 \cdot 2$ **12.** $4 \cdot 6 \cdot 6 \cdot 6 \cdot 6$

13. $3 \cdot 2 \cdot 2 \cdot 2 \cdot 2 \cdot 5 \cdot 5 \cdot 5 \cdot 5 \cdot 5$ **14.** $6 \cdot 6 \cdot 2 \cdot 9 \cdot 9 \cdot 9 \cdot 9$

Objective B *Evaluate. See Examples 5 through 8.*

15. 8^2 **16.** 6^2 **17.** 5^3 **18.** 6^3 **19.** 2^5 **20.** 3^5

21. 1^{10} **22.** 1^{12} **23.** 7^1 **24.** 8^1 **25.** 2^7 **26.** 5^4

27. 2^8 **28.** 3^3 **29.** 4^4 **30.** 4^3 **31.** 9^3 **32.** 8^3

33. 12^2 **34.** 11^2 **35.** 10^2 **36.** 10^3 **37.** 20^1 **38.** 14^1

39. 3^6 **40.** 4^5 **41.** $3 \cdot 2^6$ **42.** $5 \cdot 3^2$ **43.** $2 \cdot 3^4$ **44.** $2 \cdot 7^2$

Objective C *Simplify. See Examples 9 through 14.*

45. $15 + 3 \cdot 2$ **46.** $24 + 6 \cdot 3$ **47.** $28 \div 7 \cdot 1 + 3$ **48.** $100 \div 10 \cdot 5 + 4$

49. $32 \div 4 - 3$

50. $42 \div 7 - 6$

51. $13 + \frac{24}{8}$

52. $32 + \frac{8}{2}$

53. $6 \cdot 5 + 8 \cdot 2$

54. $3 \cdot 4 + 9 \cdot 1$

55. $\frac{5 + 12 \div 4}{1^7}$

56. $\frac{6 + 9 \div 3}{3^2}$

57. $(7 + 5^2) \div 4 \cdot 2^3$

58. $6^2 \cdot (10 - 8)$

59. $5^2 \cdot (10 - 8) + 2^3 + 5^2$

60. $5^3 \div (10 + 15) + 9^2 + 3^3$

61. $\frac{18 + 6}{2^4 - 2^2}$

62. $\frac{40 + 8}{5^2 - 3^2}$

63. $(3 + 5) \cdot (9 - 3)$

64. $(9 - 7) \cdot (12 + 18)$

65. $\frac{7(9 - 6) + 3}{3^2 - 3}$

66. $\frac{5(12 - 7) - 4}{5^2 - 18}$

67. $8 \div 0 + 37$

68. $18 - 7 \div 0$

69. $2^4 \cdot 4 - (25 \div 5)$

70. $2^3 \cdot 3 - (100 \div 10)$

71. $3^4 - [35 - (12 - 6)]$

72. $[40 - (8 - 2)] - 2^5$

73. $(7 \cdot 5) + [9 \div (3 \div 3)]$

74. $(18 \div 6) + [(3 + 5) \cdot 2]$

75. $8 \cdot [2^2 + (6 - 1) \cdot 2] - 50 \cdot 2$

76. $35 \div [3^2 + (9 - 7) - 2^2] + 10 \cdot 3$

77. $\frac{9^2 + 2^2 - 1^2}{8 \div 2 \cdot 3 \cdot 1 \div 3}$

78. $\frac{5^2 - 2^3 + 1^4}{10 \div 5 \cdot 4 \cdot 1 \div 4}$

79. $\frac{2 + 4^2}{5(20 - 16) - 3^2 - 5}$

80. $\frac{3 + 9^2}{3(10 - 6) - 2^2 - 1}$

81. $9 \div 3 + 5^2 \cdot 2 - 10$

82. $10 \div 2 + 3^3 \cdot 2 - 20$

83. $[13 \div (20 - 7) + 2^5] - (2 + 3)^2$

84. $[15 \div (11 - 6) + 2^2] + (5 - 1)^2$

85. $7^2 - \{18 - [40 \div (5 \cdot 1) + 2] + 5^2\}$

86. $29 - \{5 + 3[8 \cdot (10 - 8)] - 50\}$

Objective D *Find the area of each square. See Example 15.*

△ **87.**

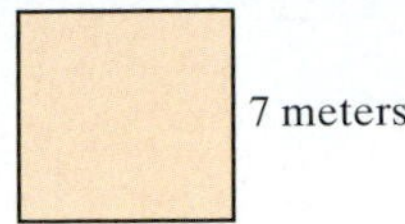

△ **88.**

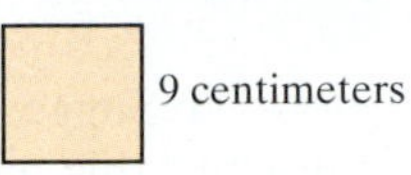

△ **89.**

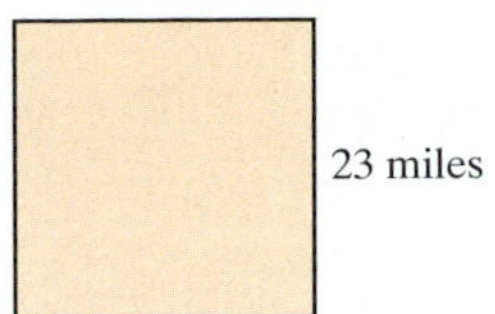

△ **90.**

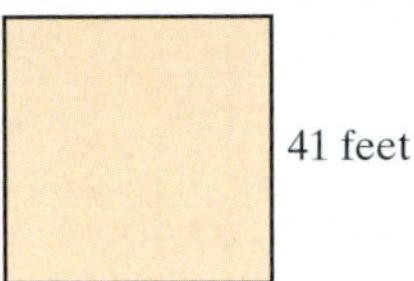

Concept Extensions

Answer the following true or false. See the Concept Check in this section.

91. "Six to the fifth power" is the same as 6^5.

92. "Seven squared" is the same as 7^2.

93. 2^5 is the same as $5 \cdot 5$.

94. 4^9 is the same as $4 \cdot 9$.

Insert grouping symbols (parentheses) so that each given expression evaluates to the given number.

95. $2 + 3 \cdot 6 - 2$; evaluate to 28

96. $2 + 3 \cdot 6 - 2$; evaluate to 20

97. $24 \div 3 \cdot 2 + 2 \cdot 5$; evaluate to 14

98. $24 \div 3 \cdot 2 + 2 \cdot 5$; evaluate to 15

△ **99.** A building contractor is bidding on a contract to install gutters on seven homes in a retirement community, all in the shape shown. To estimate the cost of materials, she needs to know the total perimeter of all seven homes. Find the total perimeter.

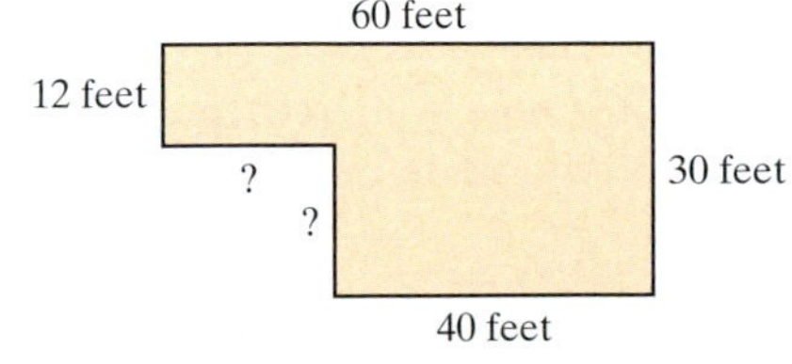

Simplify.

100. $25^3 \cdot (45 - 7 \cdot 5) \cdot 5$

101. $(7 + 2^4)^5 - (3^5 - 2^4)^2$

102. Explain why $2 \cdot 3^2$ is not the same as $(2 \cdot 3)^2$.

103. Write an expression that simplifies to 5. Use multiplication, division, addition, subtraction, and at least one set of parentheses. Explain the process you would use to simplify the expression.

THE BIGGER PICTURE Operations on Sets of Numbers

Continue your outline started in Section 1.6. Suggestions are once again written to help you complete this part of your outline.

I. Some Operations on Sets of Numbers

A. Whole Numbers

1. **Add or Subtract** (Section 1.3)
2. **Multiply or Divide** (Sections 1.5, 1.6)
3. **Exponent:** $3^4 = \overbrace{3 \cdot 3 \cdot 3 \cdot 3}^{\text{4 factors of 3}} = 81$
4. **Order of Operations:**

$$
\begin{aligned}
&24 \div 3 \cdot 2 - (2 + 8) \\
&= 24 \div 3 \cdot 2 - (10) && \text{Parentheses.} \\
&= 8 \cdot 2 - 10 && \text{Multiply or divide from left to right.} \\
&= 16 - 10 && \text{Multiply or divide from left to right.} \\
&= 6 && \text{Add or subtract from left to right.}
\end{aligned}
$$

Perform the indicated operations.

1. 6^3
2. $2^3 \cdot 6^1$
3. $8 \cdot 5^2$
4. $1 + 2(3 + 4)$
5. $2 + 5(10 - 3)$
6. $200 \div 2 \cdot 2$
7. $978 - 179$
8. $\begin{array}{r} 72 \\ \times 30 \\ \hline \end{array}$
9. $614 \div 58$
10. $3[(7 - 3)^2 - (25 - 22)^2] + 6$

STUDY SKILLS BUILDER

What to Do the Day of an Exam?

Your first exam may be soon. On the day of an exam, don't forget to try the following:

- Allow yourself plenty of time to arrive.
- Read the directions on the test carefully.
- Read each problem carefully as you take your test. Make sure that you answer the question asked.
- Watch your time and pace yourself so that you may attempt each problem on your test.
- Check your work and answers.
- ***Do not turn your test in early.*** If you have extra time, spend it double-checking your work.

Good luck!

Answer the following questions based on your most recent mathematics exam, whenever that was.

1. How soon before class did you arrive?
2. Did you read the directions on the test carefully?
3. Did you make sure you answered the question asked for each problem on the exam?
4. Were you able to attempt each problem on your exam?
5. If your answer to question 4 is no, list reasons why.
6. Did you have extra time on your exam?
7. If your answer to question 6 is yes, describe how you spent that extra time.

1.8 INTRODUCTION TO VARIABLES, ALGEBRAIC EXPRESSIONS, AND EQUATIONS

Objectives

A Evaluate Algebraic Expressions Given Replacement Values.

B Identify Solutions of Equations.

C Translate Phrases into Variable Expressions.

Objective A Evaluating Algebraic Expressions

Perhaps the most important quality of mathematics is that it is a science of patterns. Communicating about patterns is often made easier by using a letter to represent all the numbers fitting a pattern. We call such a letter a **variable.** For example, in Section 1.2 we presented the addition property of 0, which states that the sum of 0 and any number is that number. We might write

$$0 + 1 = 1$$
$$0 + 2 = 2$$
$$0 + 3 = 3$$
$$0 + 4 = 4$$
$$0 + 5 = 5$$
$$0 + 6 = 6$$
$$\vdots$$

continuing indefinitely. This is a pattern, and all whole numbers fit the pattern. We can communicate this pattern for all whole numbers by letting a letter, such as a, represent all whole numbers. We can then write

$$0 + a = a$$

Using variable notation is a primary goal of learning **algebra.** We now take some important first steps in beginning to use variable notation.

A combination of operations on letters (variables) and numbers is called an **algebraic expression** or simply an **expression.**

Algebraic Expressions

$3 + x \qquad 5 \cdot y \qquad 2 \cdot z - 1 + x$

If two variables or a number and a variable are next to each other, with no operation sign between them, the operation is multiplication. For example,

$2x$ means $2 \cdot x$

and

xy or $x(y)$ means $x \cdot y$

Also, the meaning of an exponent remains the same when the base is a variable. For example,

$$x^2 = \underbrace{x \cdot x}_{\text{2 factors of } x} \quad \text{and} \quad y^5 = \underbrace{y \cdot y \cdot y \cdot y \cdot y}_{\text{5 factors of } y}$$

Algebraic expressions such as $3x$ have different values depending on replacement values for x. For example, if x is 2, then $3x$ becomes

$$3x = 3 \cdot 2$$
$$= 6$$

If x is 7, then $3x$ becomes

$$3x = 3 \cdot 7$$
$$= 21$$

Replacing a variable in an expression by a number and then finding the value of the expression is called **evaluating the expression** for the variable. When finding the value of an expression, remember to follow the order of operations given in Section 1.7.

PRACTICE PROBLEM 1

Evaluate $x - 2$ if x is 7.

EXAMPLE 1 Evaluate $x + 6$ if x is 8.

Solution: Replace x with 8 in the expression $x + 6$.

$x + 6 = 8 + 6$ Replace x with 8.

$= 14$ Add.

Work Practice Problem 1

When we write a statement such as "x is 5," we can use an equals symbol (=) to represent "is" so that

x is 5 can be written as $x = 5$.

PRACTICE PROBLEM 2

Evaluate $y(x - 3)$ for $x = 8$ and $y = 4$.

EXAMPLE 2 Evaluate $2(x - y)$ for $x = 6$ and $y = 3$.

Solution: $2(x - y) = 2(6 - 3)$ Replace x with 6 and y with 3.

$= 2(3)$ Subtract.

$= 6$ Multiply.

Work Practice Problem 2

PRACTICE PROBLEM 3

Evaluate $\frac{y + 6}{x}$ for $x = 6$ and $y = 18$.

EXAMPLE 3 Evaluate $\frac{x - 5y}{y}$ for $x = 35$ and $y = 5$.

Solution: $\frac{x - 5y}{y} = \frac{35 - 5(5)}{5}$ Replace x with 35 and y with 5.

$= \frac{35 - 25}{5}$ Multiply.

$= \frac{10}{5}$ Subtract.

$= 2$ Divide.

Work Practice Problem 3

PRACTICE PROBLEM 4

Evaluate $25 - z^3 + x$ for $z = 2$ and $x = 1$.

EXAMPLE 4 Evaluate $x^2 + z - 3$ for $x = 5$ and $z = 4$.

Solution: $x^2 + z - 3 = 5^2 + 4 - 3$ Replace x with 5 and z with 4.

$= 25 + 4 - 3$ Evaluate 5^2.

$= 26$ Add and subtract from left to right.

Work Practice Problem 4

Answers

1. 5 **2.** 20 **3.** 4 **4.** 18

Helpful Hint If you are having difficulty replacing variables with numbers, first replace each variable with a set of parentheses, then insert the replacement number within the parentheses.

Example:

$$\begin{aligned} x^2 + z - 3 &= (\ )^2 + (\ ) - 3 \\ &= (5)^2 + (4) - 3 \\ &= 25 + 4 - 3 \\ &= 26 \end{aligned}$$

✔Concept Check What's wrong with the solution to the following problem? Evaluate $3x + 2y$ for $x = 2$ and $y = 3$.

Solution:
$$\begin{aligned} 3x + 2y &= 3(3) + 2(2) \\ &= 9 + 4 \\ &= 13 \end{aligned}$$

EXAMPLE 5 The expression $\frac{5(F - 32)}{9}$ can be used to write degrees Fahrenheit F as degrees Celsius C. Find the value of this expression for $F = 86$.

Solution:
$$\begin{aligned} \frac{5(F - 32)}{9} &= \frac{5(86 - 32)}{9} \\ &= \frac{5(54)}{9} \\ &= \frac{270}{9} \\ &= 30 \end{aligned}$$

Thus 86°F is the same temperature as 30°C.

Work Practice Problem 5

PRACTICE PROBLEM 5

Evaluate $\frac{5(F - 32)}{9}$ for $F = 41$.

Objective B Identifying Solutions of Equations

In Objective A, we learned that a combination of operations on variables and numbers is called an algebraic expression or simply an expression. Frequently in this book, we have written statements like $7 + 4 = 11$ or area = length · width. Each of these statements is called an **equation.** An equation is of the form

expression = expression

Helpful Hint An equation contains "=", while an expression does not.

An equation can be labeled as

$$\underbrace{x + 7}_{\text{left side}} \overset{\text{equal sign}}{=} \underbrace{10}_{\text{right side}}$$

Answer

5. 5

✔ Concept Check Answer

$$\begin{aligned} 3x + 2y &= 3(2) + 2(3) \\ &= 6 + 6 \\ &= 12 \end{aligned}$$

When an equation contains a variable, deciding which values of the variable make an equation a true statement is called **solving** an equation for the variable. A **solution** of an equation is a value for the variable that makes an equation a true statement. For example, 2 is a solution of the equation $x + 5 = 7$, since replacing x with 2 results in the *true* statement $2 + 5 = 7$. Similarly, 3 is not a solution of $x + 5 = 7$, since replacing x with 3 results in the *false* statement $3 + 5 = 7$.

PRACTICE PROBLEM 6

Determine whether 8 is a solution of the equation $3(y - 6) = 6$.

EXAMPLE 6 Determine whether 6 is a solution of the equation $4(x - 3) = 12$.

Solution: We replace x with 6 in the equation.

$$4(x - 3) = 12$$
$$\downarrow$$
$$4(6 - 3) \stackrel{?}{=} 12 \quad \text{Replace } x \text{ with 6.}$$
$$4(3) \stackrel{?}{=} 12$$
$$12 = 12 \quad \text{True}$$

Since $12 = 12$ is a true statement, 6 *is* a solution of the equation.

Work Practice Problem 6

A collection of numbers enclosed by braces is called a set. For example,

$\{0, 1, 2, 3, \ldots\}$

is the set of whole numbers that we are studying about in this chapter. The three dots after the number 3 in the set mean that this list of numbers continues in the same manner indefinitely.

The next example contains set notation.

PRACTICE PROBLEM 7

Determine which numbers in the set $\{10, 6, 8\}$ are solutions of the equation $5n + 4 = 34$.

EXAMPLE 7 Determine which numbers in the set $\{26, 40, 20\}$ are solutions of the equation $2n - 30 = 10$.

Solution: Replace n with each number from the set to see if a true statement results.

Let *n* be 26.

$$2n - 30 = 10$$
$$2 \cdot 26 - 30 \stackrel{?}{=} 10$$
$$52 - 30 \stackrel{?}{=} 10$$
$$22 = 10 \quad \text{False}$$

Let *n* be 40.

$$2n - 30 = 10$$
$$2 \cdot 40 - 30 \stackrel{?}{=} 10$$
$$80 - 30 \stackrel{?}{=} 10$$
$$50 = 10 \quad \text{False}$$

Let *n* be 20.

$$2n - 30 = 10$$
$$2 \cdot 20 - 30 \stackrel{?}{=} 10$$
$$40 - 30 \stackrel{?}{=} 10$$
$$10 = 10 \quad \text{True} \checkmark$$

Thus, 20 is a solution while 26 and 40 are not solutions.

Work Practice Problem 7

Objective C Translating Phrases into Variable Expressions

To aid us in solving problems later, we practice translating verbal phrases into algebraic expressions. Certain key words and phrases suggesting addition, subtraction, multiplication, or division are reviewed next.

Answers

6. yes

7. 6 is a solution.

Addition (+)	Subtraction (−)	Multiplication (·)	Division (÷)
sum	difference	product	quotient
plus	minus	times	divide
added to	subtract	multiply	shared equally among
more than	less than	multiply by	per
increased by	decreased by	of	divided by
total	less	double/triple	divided into

EXAMPLE 8 Write as an algebraic expression. Use x to represent "a number."

a. 7 increased by a number
b. 15 decreased by a number
c. The product of 2 and a number
d. The quotient of a number and 5
e. 2 subtracted from a number

Solution:

a. In words: 7 | increased by | a number
Translate: $7 \quad + \quad x$

b. In words: 15 | decreased by | a number
Translate: $15 \quad - \quad x$

c. In words: The product of 2 and a number
Translate: $2 \cdot x$ or $2x$

d. In words: The quotient of a number and 5
Translate: $x \div 5$ or $\frac{x}{5}$

e. In words: 2 | subtracted from | a number
Translate: $x - 2$

Work Practice Problem 8

Helpful Hint Remember that order is important when subtracting. Study the order of numbers and variables below.

Phrase	Translation
a number *decreased by* 5	$x - 5$
a number *subtracted from* 5	$5 - x$

PRACTICE PROBLEM 8

Write as an algebraic expression. Use x to represent "a number."

a. Twice a number.
b. 8 increased by a number.
c. 10 minus a number.
d. 10 subtracted from a number.
e. The quotient of 6 and a number.

Answers

8. a. $2x$ **b.** $8 + x$ **c.** $10 - x$ **d.** $x - 10$ **e.** $6 \div x$ or $\frac{6}{x}$

Vocabulary and Readiness Check

Use the choices below to fill in each blank. You may use each choice more than once.

evaluating an expression   variable(s)   expression   equation   solution

1. A combination of operations on letters (variables) and numbers is a(n) ______________.
2. A letter that represents a number is a(n) ______________.
3. $3x - 2y$ is called a(n) ______________ and the letters x and y are ______________.
4. Replacing a variable in an expression by a number and then finding the value of the expression is called ______________.
5. A statement of the form "expression = expression" is called a(n) ______________.
6. A value for the variable that makes the equation a true statement is called a(n) ______________.

1.8 EXERCISE SET

FOR EXTRA HELP

Student Solutions Manual

PH Math/Tutor Center

CD/Video for Review

Math XL
MathXL®

MyMathLab

Objective A *Complete the table. The first row has been done for you. See Examples 1 through 5.*

	a	b	a + b	a − b	a · b	a ÷ b
	45	9	54	36	405	5
1.	21	7				
2.	24	6				
3.	152	0				
4.	298	0				
5.	56	1				
6.	82	1				

Evaluate each following expression for $x = 2$, $y = 5$, and $z = 3$. See Examples 1 through 5.

7. $3 + 2z$

8. $7 + 3z$

9. $3xz - 5x$

10. $4yz + 2x$

11. $z - x + y$

12. $x + 5y - z$

13. $4x - z$

14. $2y + 5z$

15. $y^3 - 4x$

16. $y^3 - z$

17. $2xy^2 - 6$

18. $3yz^2 + 1$

19. $8 - (y - x)$

20. $3 + (2y - 4)$

21. $x^5 + (y - z)$

22. $x^4 - (y - z)$

23. $\frac{6xy}{z}$

24. $\frac{8yz}{15}$

25. $\frac{2y - 2}{x}$

26. $\frac{6 + 3x}{z}$

27. $\frac{x + 2y}{z}$

28. $\frac{2z + 6}{3}$

29. $\frac{5x}{y} - \frac{10}{y}$

30. $\frac{70}{2y} - \frac{15}{z}$

31. $2y^2 - 4y + 3$

32. $3x^2 + 2x - 5$

33. $(4y - 5z)^3$

34. $(4y + 3z)^2$

35. $(xy + 1)^2$

36. $(xz - 5)^4$

37. $2y(4z - x)$

38. $3x(y + z)$

39. $xy(5 + z - x)$

40. $xz(2y + x - z)$

41. $\frac{7x + 2y}{3x}$

42. $\frac{6z + 2y}{4}$

43. The expression $16t^2$ gives the distance in feet that an object falls after t seconds. Complete the table by evaluating $16t^2$ for each given value of t.

t	1	2	3	4
$16t^2$				

44. The expression $\frac{5(F - 32)}{9}$ gives the equivalent degrees Celsius for F degrees Fahrenheit. Complete the table by evaluating this expression for each given value of F.

F	50	59	68	77
$\frac{5(F-32)}{9}$				

Objective **B** *Decide whether the given number is a solution of the given equation. See Example 6.*

45. Is 10 a solution of $n - 8 = 2$?

46. Is 9 a solution of $n - 2 = 7$?

47. Is 3 a solution of $24 = 80n$?

48. Is 50 a solution of $250 = 5n$?

49. Is 7 a solution of $3n - 5 = 10$?

50. Is 8 a solution of $11n + 3 = 91$?

51. Is 20 a solution of $2(n - 17) = 6$?

52. Is 0 a solution of $5(n + 9) = 40$?

53. Is 0 a solution of $5x + 3 = 4x + 13$?

54. Is 2 a solution of $3x - 6 = 5x - 10$?

55. Is 8 a solution of $7f = 64 - f$?

56. Is 5 a solution of $8x - 30 = 2x$?

Determine which numbers in each set are solutions to the corresponding equations. See Example 7.

57. $n - 2 = 10$; {10, 12, 14}

58. $n + 3 = 16$; {9, 11, 13}

59. $5n = 30$; {6, 25, 30}

60. $3n = 45$; {15, 30, 45}

61. $6n + 2 = 26$; {0, 2, 4}

62. $4n - 14 = 6$; {0, 5, 10}

63. $3(n - 4) = 10$; {5, 7, 10}

64. $6(n + 2) = 23$; {1, 3, 5}

65. $7x - 9 = 5x + 13$; {3, 7, 11}

66. $9x - 15 = 5x + 1$; {2, 4, 11}

Objective **C** **Translating** *Write each phrase as a variable expression. Use x to represent "a number." See Example 8.*

67. Eight more than a number

68. The sum of three and a number

69. The total of a number and eight

70. The difference of a number and five hundred

71. Twenty decreased by a number

72. A number less thirty

73. The product of 512 and a number

74. A number times twenty

75. The quotient of six and a number

76. A number divided by 11

77. The sum of seventeen and a number added to the product of five and the number

78. The difference of twice a number, and four

79. The product of five and a number

80. The quotient of twenty and a number, decreased by three

81. A number subtracted from 11

82. Twelve subtracted from a number

83. A number less 5

84. The product of a number and 7

85. 6 divided by a number

86. The sum of a number and 7

87. Fifty decreased by eight times a number

88. Twenty decreased by twice a number

Concept Extensions

For Exercises 89 through 92, use a calculator to evaluate each expression for $x = 23$ and $y = 72$.

89. $x^4 - y^2$

90. $2(x + y)^2$

91. $x^2 + 5y - 112$

92. $16y - 20x + x^3$

93. If x is a whole number, which expression is the largest: $2x$, $5x$, or $\frac{x}{3}$? Explain your answer.

94. If x is a whole number, which expression is the smallest: $2x$, $5x$, or $\frac{x}{3}$? Explain your answer.

95. In Exercise 43, what do you notice about the value of $16t^2$ as t gets larger?

96. In Exercise 44, what do you notice about the value of $\frac{5(F - 32)}{9}$ as F gets larger?

STUDY SKILLS BUILDER

Have You Decided to Complete This Course Successfully?

Ask yourself if one of your current goals is to complete this course successfully.

If it is not a goal of yours, ask yourself why? One common reason is fear of failure. Amazingly enough, fear of failure alone can be strong enough to keep many of us from doing our best in any endeavor.

Another common reason is that you simply haven't taken the time to make successfully completing this course one of your goals. How do you do this? Start by writing this goal in your mathematics notebook. Then list steps you will take to ensure success. A great first step is to read or reread Section 1.1 and make a commitment to try the suggestions in that section.

Good luck, and don't forget that a positive attitude will make a big difference.

Let's see how you are doing.

1. Have you decided to make "successfully completing this course" a goal of yours? If no, please list reasons why this has not happened. Study your list and talk to your instructor about this.
2. If your answer to question 1 is yes, take a moment and list in your notebook further specific goals that will help you achieve this major goal of successfully completing this course. (For example, "My goal this semester is not to miss any of my mathematics classes.")
3. Rate your commitment to this course with a number between 1 and 5. Use the diagram below to help.

High Commitment		Average Commitment	Not Committed at All	
5	4	3	2	1

5. If you have rated your personal commitment level (from the exercise above) as a 1, 2, or 3, list the reasons why this is so. Then determine whether it is possible to increase your commitment level to a 4 or 5.

CHAPTER 1 Group Activity

Investigating Endangered and Threatened Species

An **endangered** species is one that is thought to be in danger of becoming extinct throughout all or a major part of its habitat. A **threatened** species is one that may become endangered. The Division of Endangered Species at the U.S. Fish and Wildlife Service keeps close tabs on the state of threatened and endangered wildlife in the United States and around the world. The table on the next page was compiled from 2006 data in the Division of Endangered Species' box score. The "Total Species" column gives the total number of endangered and threatened species for each group.

1. Round each number of *endangered animal species* to the nearest ten to estimate the Animal Total.
2. Round each number of *endangered plant species* to the nearest ten to estimate the Plant Total.
3. Add the exact numbers of endangered animal species to find the exact Animal Total and record it in the table in the Endangered Species column. Add the exact numbers of endangered plant species to find the Plant Total and record it in the table in the Endangered Species column. Then find the total number of endangered species (animals and plants combined) and record this number in the table as the Grand Total in the Endangered Species column.
4. Find the Animal Total, Plant Total, and Grand Total for the Total Species column. Record these values in the table.
5. Use the data in the table to complete the Threatened Species column.
6. Write a paragraph discussing the conclusions that can be drawn from the table.

Continued on next page

	Endangered and Threatened Species Worldwide			
	Group	Endangered Species	Threatened Species	Total Species
Animals	Mammals	323		351
	Birds	251		270
	Reptiles	79		115
	Amphibians	20		30
	Fishes	84		128
	Clams	64		72
	Snails	25		37
	Insects	51		61
	Arachnids	12		12
	Crustaceans	19		22
	Animal Total			
Plants	Flowering Plants	571		714
	Conifers	2		5
	Ferns and Allies	24		26
	Lichens	2		2
	Plant Total			
	Grand Total			

(*Source:* U.S. Fish and Wildlife Service, Division of Endangered Species)

Chapter 1 Vocabulary Check

Fill in each blank with one of the words or phrases listed below.

difference	factor	perimeter	dividend	minuend
place value	whole numbers	equation	divisor	variable
sum	set	addend	exponent	expression
solution	quotient	subtrahend	product	digits
area				

1. The ______________ are 0, 1, 2, 3, . . .
2. The ______________ of a polygon is its distance around or the sum of the lengths of its sides.
3. The position of each digit in a number determines its ______________.
4. An ______________ is a shorthand notation for repeated multiplication of the same factor.
5. To find the ______________ of a rectangle, multiply length times width.
6. The ______________ used to write numbers are 0, 1, 2, 3, 4, 5, 6, 7, 8, and 9.
7. A letter used to represent a number is called a ______________.
8. An ______________ can be written in the form "expression = expression."
9. A combination of operations on variables and numbers is called an ______________.
10. A ______________ of an equation is a value of the variable that makes the equation a true statement.
11. A collection of numbers (or objects) enclosed by braces is called a ______________.

Use the facts below for Exercises 12 through 21.

$2 \cdot 3 = 6$ $\quad 4 + 17 = 21$ $\quad 20 - 9 = 11$ $\quad \begin{array}{r} 7 \\ 5\overline{)35} \end{array}$

12. The 21 above is called the ________________.

13. The 5 above is called the ________________.

14. The 35 above is called the ________________.

15. The 7 above is called the ________________.

16. The 3 above is called a ________________.

17. The 6 above is called the ________________.

18. The 20 above is called the ________________.

19. The 9 above is called the ________________.

20. The 11 above is called the ________________.

21. The 4 above is called an ________________.

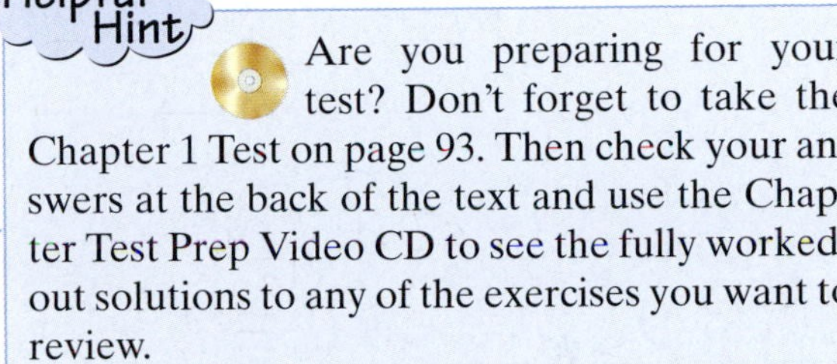

Helpful Hint
Are you preparing for your test? Don't forget to take the Chapter 1 Test on page 93. Then check your answers at the back of the text and use the Chapter Test Prep Video CD to see the fully worked-out solutions to any of the exercises you want to review.

1 Chapter Highlights

Definitions and Concepts	**Examples**
Section 1.2 Place Value and Names for Numbers	
The **whole numbers** are 0, 1, 2, 3, 4, 5, ...	0, 14, 968, 5,268,619
The position of each digit in a number determines its **place value.** A place-value chart is shown next with the names of the periods given. 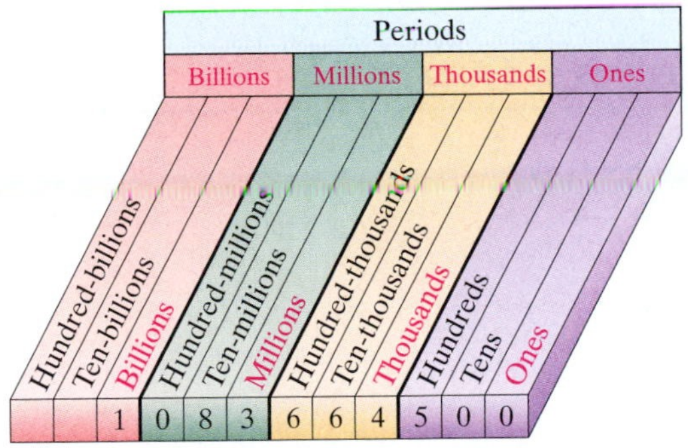	
To write a whole number in words, write the number in each period followed by the name of the period. (The name of the ones period is not included.)	9,078,651,002 is written as nine billion, seventy-eight million, six hundred fifty-one thousand, two.
To write a whole number in standard form, write the number in each period, followed by a comma.	Four million, seven hundred six thousand, twenty-eight is written as 4,706,028.
Section 1.3 Adding and Subtracting Whole Numbers, and Perimeter	
To add whole numbers, add the digits in the ones place, then the tens place, then the hundreds place, and so on, carrying when necessary.	Find the sum: $\begin{array}{rl} {}^{2\,1\,1} & \\ 2689 & \leftarrow \text{addend} \\ 1735 & \leftarrow \text{addend} \\ +\ 662 & \leftarrow \text{addend} \\ \hline 5086 & \leftarrow \text{sum} \end{array}$

DEFINITIONS AND CONCEPTS	**EXAMPLES**

Section 1.3 Adding and Subtracting Whole Numbers, and Perimeter (*continued*)

To subtract whole numbers, subtract the digits in the ones place, then the tens place, then the hundreds place, and so on, borrowing when necessary.

Subtract:

$$\begin{array}{rl} \scriptstyle 8\ 15 & \\ 7\not{9}\not{5}4 & \leftarrow \text{minuend} \\ -5673 & \leftarrow \text{subtrahend} \\ \hline 2281 & \leftarrow \text{difference} \end{array}$$

The **perimeter** of a polygon is its distance around or the sum of the lengths of its sides.

Find the perimeter of the polygon shown.

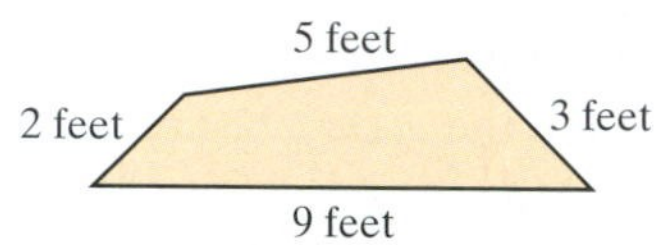

The perimeter is 5 feet + 3 feet + 9 feet + 2 feet = 19 feet.

Section 1.4 Rounding and Estimating

ROUNDING WHOLE NUMBERS TO A GIVEN PLACE VALUE

Step 1. Locate the digit to the right of the given place value.

Step 2. If this digit is 5 or greater, add 1 to the digit in the given place value and replace each digit to its right with 0.

Step 3. If this digit is less than 5, replace it and each digit to its right with 0.

Round 15,721 to the nearest thousand.

15,(7)21 — Add 1 (to the 5); Replace with zeros (the digits 721).

Since the circled digit is 5 or greater, add 1 to the given place value and replace digits to its right with zeros.

15,721 rounded to the nearest thousand is 16,000.

Section 1.5 Multiplying Whole Numbers and Area

To multiply 73 and 58, for example, multiply 73 and 8, then 73 and 50. The sum of these partial products is the product of 73 and 58. Use the notation to the right.

$$\begin{array}{rl} 73 & \leftarrow \text{factor} \\ \times\ 58 & \leftarrow \text{factor} \\ \hline 584 & \leftarrow 73 \times 8 \\ 3650 & \leftarrow 73 \times 50 \\ \hline 4234 & \leftarrow \text{product} \end{array}$$

To find the **area** of a rectangle, multiply length times width.

Find the area of the rectangle shown.

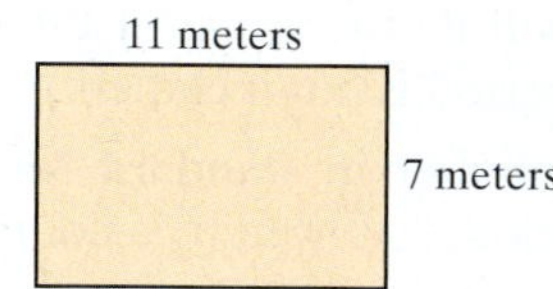

$$\begin{aligned} \text{area of rectangle} &= \text{length} \cdot \text{width} \\ &= (11 \text{ meters})(7 \text{ meters}) \\ &= 77 \text{ square meters} \end{aligned}$$

Definitions and Concepts	**Examples**

Section 1.6 Dividing Whole Numbers

Division Properties of 0

The quotient of 0 and any number (except 0) is 0.

The quotient of any number and 0 is not a number. We say that this quotient is undefined.

$\frac{0}{5} = 0$

$\frac{7}{0}$ is undefined

To divide larger whole numbers, use the process called **long division** as shown to the right.

$$\begin{array}{r} 507 \text{ R } 2 \\ 14\overline{)7100} \\ -70 \\ \hline 10 \\ -0 \\ \hline 100 \\ -98 \\ \hline 2 \end{array}$$

divisor → 14; 507 R 2 ← quotient; 7100 ← dividend

$5(14) = 70$

Subtract and bring down the 0.

$0(14) = 0$

Subtract and bring down the 0.

$7(14) = 98$

Subtract. The remainder is 2.

To check, see that $507 \cdot 14 + 2 = 7100$.

The **average** of a list of numbers is

$$\text{average} = \frac{\text{sum of numbers}}{\text{number of numbers}}$$

Find the average of 23, 35, and 38.

$$\text{average} = \frac{23 + 35 + 38}{3} = \frac{96}{3} = 32$$

Section 1.7 Exponents and Order of Operations

An **exponent** is a shorthand notation for repeated multiplication of the same factor.

$$3^4 = \underbrace{3 \cdot 3 \cdot 3 \cdot 3}_{\text{4 factors of 3}} = 81$$

exponent: 4; base: 3

Order of Operations

1. Perform all operations within parentheses ( ), brackets [ ], or other grouping symbols such as fraction bars, starting with the innermost set.
2. Evaluate any expressions with exponents.
3. Multiply or divide in order from left to right.
4. Add or subtract in order from left to right.

Simplify: $\frac{5 + 3^2}{2(7 - 6)}$

Simplify above and below the fraction bar separately.

$\frac{5 + 3^2}{2(7 - 6)} = \frac{5 + 9}{2(1)}$ Evaluate 3^2 above the fraction bar. Subtract: $7 - 6$ below the fraction bar.

$= \frac{14}{2}$ Add. Multiply.

$= 7$ Divide.

The **area of a square** is $(\text{side})^2$.

Find the area of a square with side length 9 inches.

$$\begin{aligned} \text{Area of the square} &= (\text{side})^2 \\ &= (9 \text{ inches})^2 \\ &= 81 \text{ square inches} \end{aligned}$$

Definitions and Concepts	**Examples**
Section 1.8 Introduction to Variables, Algebraic Expressions, and Equations	
A letter used to represent a number is called a **variable.**	Variables: $x, \quad y, \quad z, \quad a, \quad b$
A combination of operations on variables and numbers is called an **algebraic expression.**	Algebraic expressions: $3 + x, \quad 7y, \quad x^3 + y - 10$
Replacing a variable in an expression by a number, and then finding the value of the expression is called **evaluating the expression** for the variable.	Evaluate $2x + y$ for $x = 22$ and $y = 4$. $2x + y = 2 \cdot 22 + 4$ Replace x with 22 and y with 4. $= 44 + 4$ Multiply. $= 48$ Add.
A statement written in the form "expression = expression" is an **equation.**	Equations: $n - 8 = 12$ $2(20 - 7n) = 32$ Area = length · width
A **solution** of an equation is a value for the variable that makes the equation a true statement.	Determine whether 2 is a solution of the equation $4(x - 1) = 7$. $4(2 - 1) \stackrel{?}{=} 7$ Replace x with 2. $4(1) \stackrel{?}{=} 7$ Subtract. $4 \stackrel{?}{=} 7$ False No, 2 is not a solution.

1 CHAPTER REVIEW

(1.2) *Determine the place value of the digit 4 in each whole number.*

1. 7640

2. 46,200,120

Write each whole number in words.

3. 7640

4. 46,200,120

Write each whole number in expanded form.

5. 3158

6. 403,225,000

Write each whole number in standard form.

7. Eighty-one thousand, nine hundred

8. Six billion, three hundred four million

The following table shows the populations of the ten largest cities in the United States. Use this table to answer Exercises 9 through 12 and other exercises throughout this review.

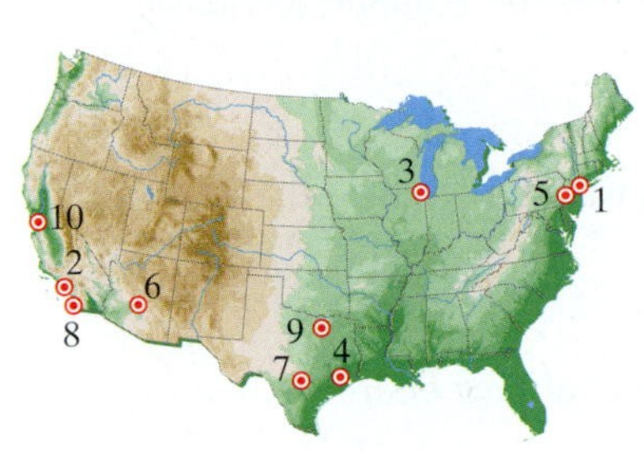

Rank	City	2005	2000	1990
1	New York, NY	8,143,197	8,008,278	7,322,564
2	Los Angeles, CA	3,844,829	3,694,820	3,485,398
3	Chicago, IL	2,842,518	2,896,016	2,783,726
4	Houston, TX	2,016,582	1,953,631	1,630,553
5	Philadelphia, PA	1,463,281	1,517,550	1,585,577
6	Phoenix, AZ	1,461,575	1,321,045	983,403
7	San Antonio, TX	1,256,509	1,144,646	935,933
8	San Diego, CA	1,255,540	1,223,400	1,110,549
9	Dallas, TX	1,213,825	1,188,580	1,006,877
10	San Jose, CA	912,332	894,943	782,248

(*Source:* U.S. Census Bureau)

9. Find the population of Houston, Texas, in 2000.

10. Find the population of Los Angeles, California, in 2005.

11. Which city in the table had the smallest population in 2005?

12. Which city had the largest population in 2005?

(1.3) *Add or subtract as indicated.*

13. 18 + 49

14. 28 + 39

15. 462 − 397

16. 583 − 279

17. 428 + 21

18. 819 + 21

19. 4000 − 86

20. 8000 − 92

21. 91 + 3623 + 497

22. 82 + 1647 + 238

Translating *Solve.*

23. Find the sum of 74, 342, and 918.

24. Find the sum of 49, 529, and 308.

25. Subtract 7965 from 25,862.

26. Subtract 4349 from 39,007.

27. The distance from Washington, DC, to New York City is 205 miles. The distance from New York City to New Delhi, India, is 7318 miles. Find the total distance from Washington, DC, to New Delhi if traveling by air through New York City.

28. Susan Summerline earned salaries of $62,589, $65,340, and $69,770 during the years 2004, 2005, and 2006, respectively. Find her total earnings during those three years.

Find the perimeter of each figure.

△ **29.**

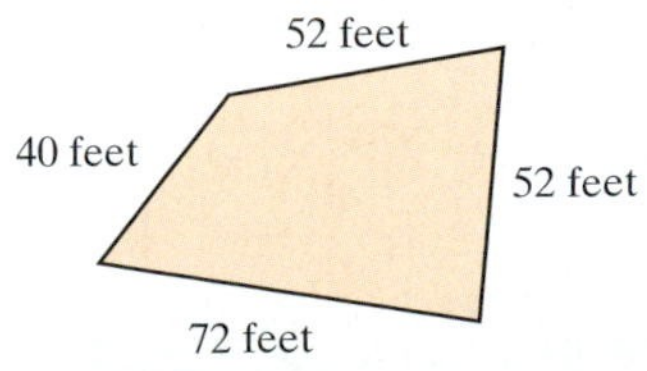

△ **30.**

11 kilometers

20 kilometers

35 kilometers

Use the city population table for Exercises 31 and 32.

31. Find the increase in population for San Antonio, Texas, from 2000 to 2005.

32. Find the decrease in population for Philadelphia, Pennsylvania, from 2000 to 2005.

The following bar graph shows the monthly savings account balance for a freshman attending a local community college. Use this graph to answer Exercises 33 through 36.

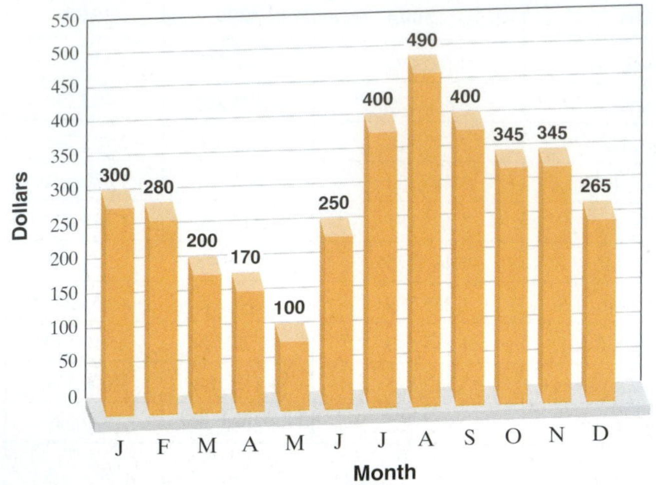

33. During what month was the balance the least?

34. During what month was the balance the greatest?

35. By how much did the balance decrease from February to April?

36. By how much did the balance increase from June to August?

(1.4) *Round to the given place.*

37. 43 to the nearest ten

38. 45 to the nearest ten

39. 876 to the nearest ten

40. 493 to the nearest hundred

41. 3829 to the nearest hundred

42. 57,534 to the nearest thousand

43. 39,583,819 to the nearest million

44. 768,542 to the nearest hundred-thousand

Estimate the sum or difference by rounding each number to the nearest hundred.

45. $3785 + 648 + 2866$

46. $5925 - 1787$

47. A group of students took a week-long driving trip and traveled 630, 192, 271, 56, 703, 454, and 329 miles on seven consecutive days. Round each distance to the nearest hundred to estimate the distance they traveled.

48. According to the city population table, the 2005 population of Phoenix was 1,461,575, and for Dallas it was 1,213,825. Round each number to the nearest hundred-thousand and estimate how much larger Phoenix is than Dallas.

(1.5) *Multiply.*

49. $\begin{array}{r} 276 \\ \times \ 8 \\ \hline \end{array}$

50. $\begin{array}{r} 349 \\ \times \ 4 \\ \hline \end{array}$

51. $\begin{array}{r} 57 \\ \times 40 \\ \hline \end{array}$

52. $\begin{array}{r} 69 \\ \times 42 \\ \hline \end{array}$

53. 20(7)(4)

54. 25(9)(4)

55. $26 \cdot 34 \cdot 0$

56. $62 \cdot 88 \cdot 0$

57. $\begin{array}{r} 586 \\ \times \ 29 \\ \hline \end{array}$

58. $\begin{array}{r} 242 \\ \times \ 37 \\ \hline \end{array}$

59. $\begin{array}{r} 642 \\ \times 177 \\ \hline \end{array}$

60. $\begin{array}{r} 347 \\ \times 129 \\ \hline \end{array}$

61. $\begin{array}{r} 1026 \\ \times \ 401 \\ \hline \end{array}$

62. $\begin{array}{r} 2107 \\ \times \ 302 \\ \hline \end{array}$

Translating *Solve.*

63. Find the product of 6 and 250.

64. Find the product of 6 and 820.

65. A golf pro orders shirts for the company sponsoring a local charity golfing event. Shirts size large cost \$32 while shirts size extra-large cost \$38. If 15 large shirts and 11 extra-large shirts are ordered, find the cost.

66. The cost to attend Black Hills State University full-time is \$4,820 per semester. Determine the cost for 20 students to attend full-time. (*Source: World Almanac*, 2006)

Find the area of each rectangle.

 **67.**

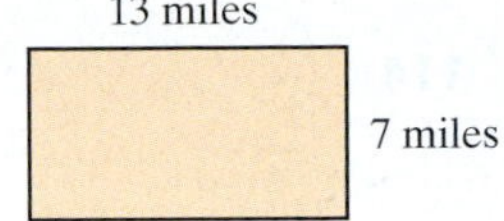

△ **68.**

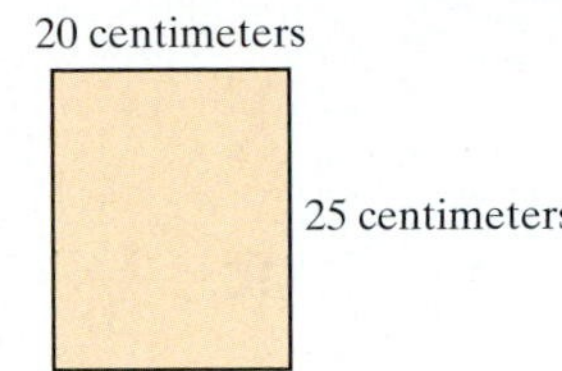

(1.6) *Divide and then check.*

69. $\frac{49}{7}$ **70.** $\frac{36}{9}$ **71.** $27 \div 5$ **72.** $18 \div 4$ **73.** $918 \div 0$

74. $0 \div 668$ **75.** $5\overline{)167}$ **76.** $8\overline{)159}$ **77.** $26\overline{)626}$ **78.** $19\overline{)680}$

79. $47\overline{)23{,}792}$ **80.** $53\overline{)48{,}111}$ **81.** $207\overline{)578{,}291}$ **82.** $306\overline{)615{,}732}$

Translating *Solve.*

83. Find the quotient of 92 and 5.

84. Find the quotient of 86 and 4.

85. A box can hold 24 cans of corn. How many boxes can be filled with 648 cans of corn?

86. One mile is 1760 yards. Find how many miles there are in 22,880 yards.

87. Find the average of the numbers 76, 49, 32, and 47.

88. Find the average of the numbers 23, 85, 62, and 66.

(1.7) *Simplify.*

89. 8^2 **90.** 5^3 **91.** $5 \cdot 9^2$ **92.** $4 \cdot 10^2$

93. $18 \div 2 + 7$ **94.** $12 - 8 \div 4$ **95.** $\frac{5(6^2 - 3)}{3^2 + 2}$ **96.** $\frac{7(16 - 8)}{2^3}$

97. $48 \div 8 \cdot 2$

98. $27 \div 9 \cdot 3$

99. $2 + 3[1^5 + (20 - 17) \cdot 3] + 5 \cdot 2$

100. $21 - [2^4 - (7 - 5) - 10] + 8 \cdot 2$

101. $19 - 2(3^2 - 2^2)$

102. $16 - 2(4^2 - 3^2)$

103. $4 \cdot 5 - 2 \cdot 7$

104. $8 \cdot 7 - 3 \cdot 9$

105. $(6 - 4)^3 \cdot [10^2 \div (3 + 17)]$

106. $(7 - 5)^3 \cdot [9^2 \div (2 + 7)]$

107. $\frac{5 \cdot 7 - 3 \cdot 5}{2(11 - 3^2)}$

108. $\frac{4 \cdot 8 - 1 \cdot 11}{3(9 - 2^3)}$

Find the area of each square.

△ **109.** A square with side length of 7 meters.

△ **110.**

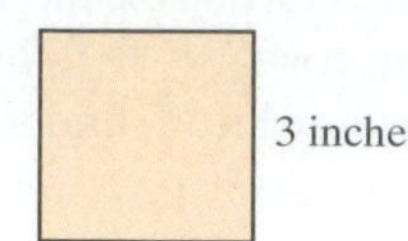

(1.8) *Evaluate each expression for $x = 5$, $y = 0$, and $z = 2$.*

111. $\frac{2x}{z}$

112. $4x - 3$

113. $\frac{x + 7}{y}$

114. $\frac{y}{5x}$

115. $x^3 - 2z$

116. $\frac{7 + x}{3z}$

117. $(y + z)^2$

118. $\frac{100}{x} + \frac{y}{3}$

Translating *Translate each phrase into a variable expression.*

119. Five subtracted from a number

120. Seven more than a number

121. Ten divided by a number

122. The product of 5 and a number

Decide whether the given number is a solution of the given equation.

123. Is 5 a solution of $n + 12 = 20 - 3$?

124. Is 23 a solution of $n - 8 = 10 + 6$?

125. Is 14 a solution of $30 = 3(n - 3)$?

126. Is 20 a solution of $5(n - 7) = 65$?

Determine which numbers in each set are solutions to the corresponding equations.

127. $7n = 77$; {6, 11, 20}

128. $n - 25 = 150$; {125, 145, 175}

129. $5(n + 4) = 90$; {14, 16, 26}

130. $3n - 8 = 28$; {3, 7, 15}

Mixed Review

Perform the indicated operations.

131. $485 - 68$

132. $729 - 47$

133. 732×3

134. 629×4

135. $374 + 29 + 698$

136. $593 + 52 + 766$

137. $13\overline{)5962}$

138. $18\overline{)4267}$

139. 1968×36

140. 5324×18

141. $2000 - 356$

142. $9000 - 519$

Round to the given place.

143. 842 to the nearest ten

144. 258,371 to the nearest hundred-thousand

Simplify.

145. $24 \div 4 \cdot 2$

146. $\frac{(15 + 3) \cdot (8 - 5)}{2^3 + 1}$

Solve.

147. Is 9 a solution of $5n - 6 = 40$?

148. Is 3 a solution of $2n - 6 = 5n - 15$?

149. A manufacturer of drinking glasses ships his delicate stock in special boxes that can hold 32 glasses. If 1714 glasses are manufactured, how many full boxes are filled? Are there any glasses left over?

150. A teacher orders 2 small white boards for $27 each and 8 boxes of dry erase pens for $4 each. What is her total bill before taxes?

1 CHAPTER TEST

Use the Chapter Test Prep Video CD to see the fully worked-out solutions to any of the exercises you want to review.

Simplify.

1. Write 82,426 in words.

2. Write "four hundred two thousand, five hundred fifty" in standard form.

3. $59 + 82$

4. $600 - 487$

5. $\begin{array}{r} 496 \\ \times\ 30 \\ \hline \end{array}$

6. $52{,}896 \div 69$

7. $2^3 \cdot 5^2$

8. $98 \div 1$

9. $0 \div 49$

10. $62 \div 0$

11. $(2^4 - 5) \cdot 3$

12. $16 + 9 \div 3 \cdot 4 - 7$

13. $6^1 \cdot 2^3$

14. $2[(6 - 4)^2 + (22 - 19)^2] + 10$

15. $5698 \cdot 1000$

16. Find the average of 62, 79, 84, 90, and 95.

17. Round 52,369 to the nearest thousand.

Estimate each sum or difference by rounding each number to the nearest hundred.

18. $6289 + 5403 + 1957$

19. $4267 - 2738$

Solve.

20. Subtract 15 from 107.

21. Find the sum of 15 and 107.

22. Find the product of 15 and 107.

23. Find the quotient of 107 and 15.

24. Twenty-nine cans of Sherwin-Williams paint cost \$493. How much was each can?

25. Jo McElory is looking at two new refrigerators for her apartment. One costs \$599 and the other costs \$725. How much more expensive is the higher-priced one?

Answers

1. ______
2. ______
3. ______
4. ______
5. ______
6. ______
7. ______
8. ______
9. ______
10. ______
11. ______
12. ______
13. ______
14. ______
15. ______
16. ______
17. ______
18. ______
19. ______
20. ______
21. ______
22. ______
23. ______
24. ______
25. ______

26. ______

27. ______

28. ______

29. ______

30. ______

31. ______

32. a. ______

b. ______

33. ______

34. ______

26. One tablespoon of white granulated sugar contains 45 calories. How many calories are in 8 tablespoons of white granulated sugar? (*Source: Home and Garden Bulletin No. 72,* U.S. Department of Agriculture)

27. A small business owner recently ordered 16 digital cameras that cost \$430 each and 5 printers that cost \$205 each. Find the total cost for these items.

Find the perimeter and the area of each figure.

△ **28.**

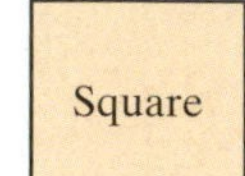

△ **29.**

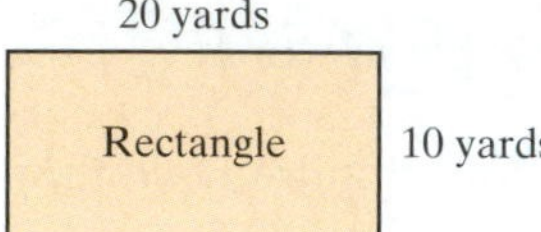

30. Evaluate $5(x^3 - 2)$ for $x = 2$.

31. Evaluate $\frac{3x - 5}{2y}$ for $x = 7$ and $y = 8$.

32. Translate the following phrases into mathematical expressions. Use x to represent "a number."

a. The quotient of a number and 17

b. Twice a number, decreased by 20

33. Is 6 a solution of the equation $5n - 11 = 19$?

34. Determine which (if any) number in the set is a solution to the given equation.

$n + 20 = 4n - 10$; $\{0, 10, 20\}$

2

Integers and Introduction to Solving Equations

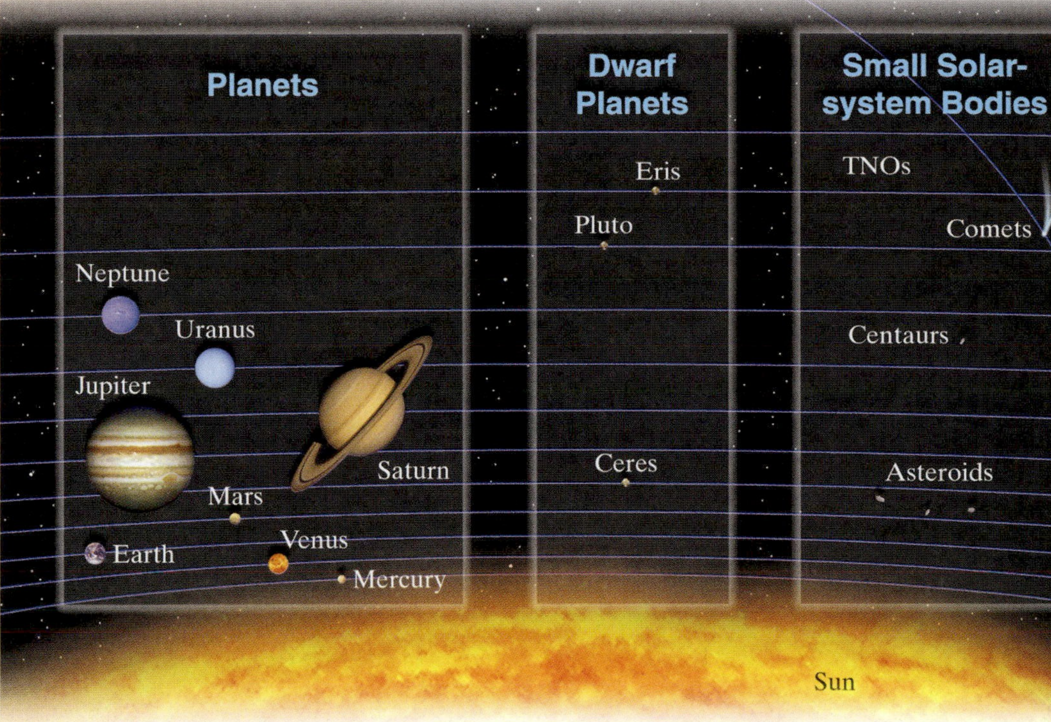

Thus far, we have studied whole numbers, but these numbers are not sufficient for representing many situations in real life. For example, to express 5 degrees below zero or $100 in debt, numbers less than 0 are needed. This chapter is devoted to integers, which include numbers less than 0, and operations on these numbers.

SECTIONS

CHECK YOUR PROGRESS

Members of the International Astronomical Union (IAU) met in August 2006 and approved a resolution that there be three distinct categories of bodies in our Solar System: "Planet," "Dwarf Planet" (completely distinct from a planet), and "Small Solar-System Bodies." See the following chart.

In Section 2.1, Example 8 and Exercises 91 through 94, and Section 2.3, Exercises 81 and 82, we will see how integers can be used to compare the average surface temperatures of several planets in our Solar System.

Planets	Mercury, Venus, Earth, Mars, Jupiter, Saturn, Uranus, and Neptune
Dwarf Planets	Pluto, Ceres (was an asteroid), and 2003 UB313 (newly named Eris)*
Small Solar-System Bodies	Asteroids, comets, and other small bodies

*More "dwarf planets" are expected to be announced by the IAU in the near future.

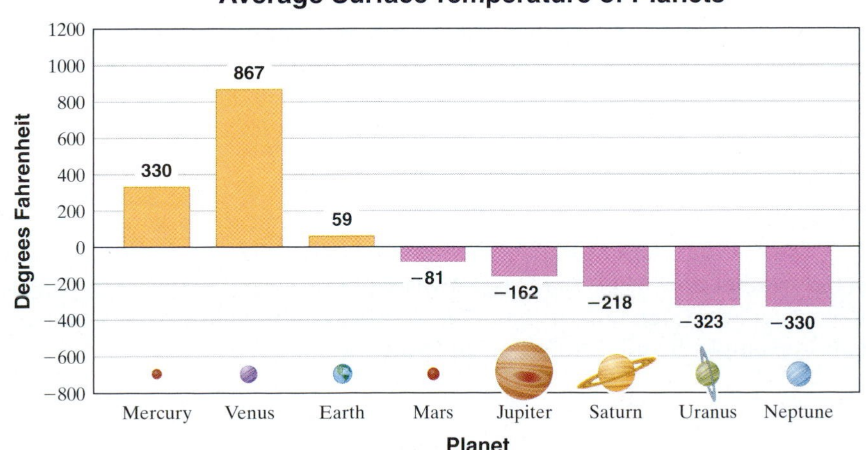

Source: The World Almanac, 2006

* For some planets, the temperature given is the temperature where the atmosphere pressure equals 1 Earth atmosphere.

Objectives

A Represent Real-Life Situations with Integers.

B Graph Integers on a Number Line.

C Compare Integers.

D Find the Absolute Value of a Number.

E Find the Opposite of a Number.

F Read Bar Graphs Containing Integers.

2.1 INTRODUCTION TO INTEGERS

Objective A Representing Real-Life Situations

Thus far in this text, all numbers have been 0 or greater than 0. Numbers greater than 0 are called **positive numbers.** However, sometimes situations exist that cannot be represented by a number greater than 0. For example,

To represent these situations, we need numbers less than 0.

Extending the number line to the left of 0 allows us to picture **negative numbers,** which are numbers that are less than 0.

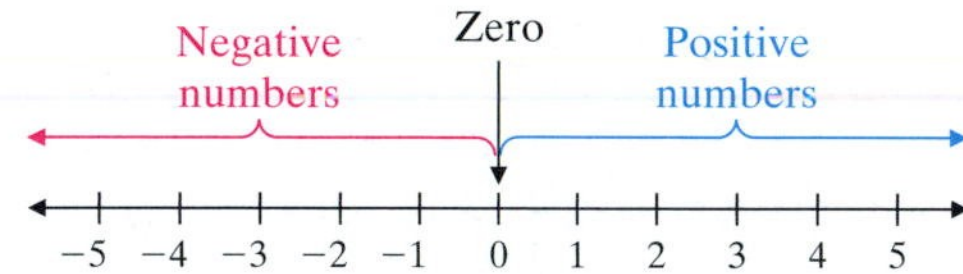

When a single + sign or no sign is in front of a number, the number is a positive number. When a single − sign is in front of a number, the number is a negative number. Together, we call positive numbers, negative numbers, and zero the **signed numbers.**

−5 indicates "negative five."

5 and +5 both indicate "positive five."

The number 0 is neither positive nor negative.

Helpful Hint Notice that 0 is neither positive nor negative.

Some signed numbers are integers. The **integers** consist of the numbers labeled on the number line above. The integers are

$\ldots, -3, -2, -1, 0, 1, 2, 3, \ldots$

Now we have numbers to represent the situations previously mentioned.

5 degrees below 0 $-5°$

20 feet below sea level -20 feet

Helpful Hint A − sign, such as the one in −1, tells us that the number is to the left of 0 on the number line. −1 is read "negative one."

A + sign or no sign tells us that a number lies to the right of 0 on the number line. For example, 3 and +3 both mean "positive three."

EXAMPLE 1 Representing Depth with an Integer

The world's deepest cave is Krubera (or Voronja), in the country of Georgia, located by the Black Sea in Asia. It has been explored to a depth of 6824 feet below the surface of the Earth. Represent this position using an integer. (*Source: Guinness World Records* 2006)

Solution: If 0 represents the surface of the Earth, then 6824 feet below the surface can be represented by −6824.

Work Practice Problem 1

Objective B Graphing Integers

EXAMPLE 2 Graph 0, −3, 2, and −2 on the number line.

Solution:

Work Practice Problem 2

Objective C Comparing Integers

We can compare integers by using a number line. For any two numbers graphed on a number line, the number to the **right** is the **greater number** and the number to the **left** is the **smaller number.** Also, the symbols < and > are called **inequality symbols.**

The inequality symbol > means **"is greater than"** and
the inequality symbol < means **"is less than."**

For example, both −5 and −7 are graphed on the number line below.

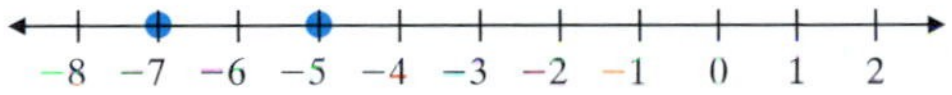

On the graph, −7 is **to the left of** −5, so −7 **is less than** −5, written as

$$-7 < -5$$

We can also write

$$-5 > -7$$

since −5 is **to the right** of −7, so −5 **is greater than** −7.

✔Concept Check Is there a largest positive number? Is there a smallest negative number? Explain.

EXAMPLE 3 Insert < or > between each pair of numbers to make a true statement.

a. −7 7 **b.** 0 −4 **c.** −9 −11

Solution:

a. −7 is to the left of 7 on a number line, so −7 < 7.
b. 0 is to the right of −4 on a number line, so 0 > −4.
c. −9 is to the right of −11 on a number line, so −9 > −11.

Work Practice Problem 3

PRACTICE PROBLEM 1

a. A deep-sea diver is 836 feet below the surface of the ocean. Represent this position using an integer.

b. A company reports a $1 million loss for the year. Represent this amount using an integer.

PRACTICE PROBLEM 2

Graph −4, −1, 2, and −2 on the number line.

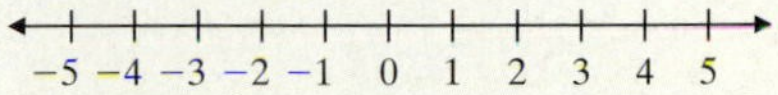

PRACTICE PROBLEM 3

Insert < or > between each pair of numbers to make a true statement.

a. 0 −5 **b.** −3 3
c. −7 −12

Answers

1. a. −836 **b.** −1 million

2.

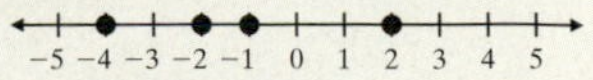

3. a. > **b.** < **c.** >

✔ Concept Check Answer

no

Helpful Hint

If you think of $<$ and $>$ as arrowheads, notice that in a true statement the arrow always points to the smaller number.

$5 > -4$ (−4: smaller number) $\quad -3 < -1$ (−3: smaller number)

Objective D Finding the Absolute Value of a Number

The **absolute value** of a number is the number's distance from 0 on the number line. The symbol for absolute value is $|\ |$. For example, $|3|$ is read as "the absolute value of 3."

$|3| = 3$ because 3 is 3 units from 0.

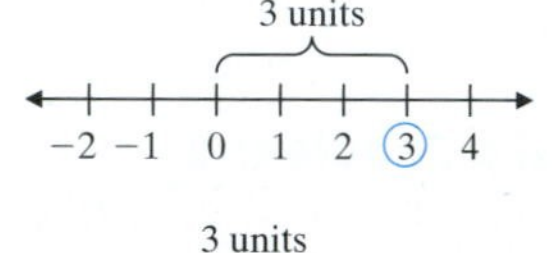

$|-3| = 3$ because -3 is 3 units from 0.

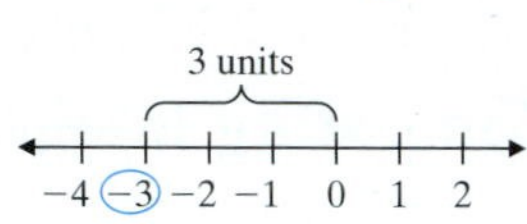

PRACTICE PROBLEM 4

Simplify.

a. $|-6|$
b. $|4|$
c. $|-12|$

EXAMPLE 4 Simplify.

a. $|-9|$ **b.** $|3|$ **c.** $|0|$

Solution:

a. $|-9| = 9$ because -9 is 9 units from 0.
b. $|3| = 3$ because 3 is 3 units from 0.
c. $|0| = 0$ because 0 is 0 units from 0.

Work Practice Problem 4

Helpful Hint

Since the absolute value of a number is that number's *distance* from 0, the absolute value of a number is always 0 or positive. It is never negative.

$|0| = 0$ (zero) $\quad |-6| = 6$ (a positive number)

Objective E Finding Opposites

Two numbers that are the same distance from 0 on the number line but are on opposite sides of 0 are called **opposites.**

4 and −4 are opposites.

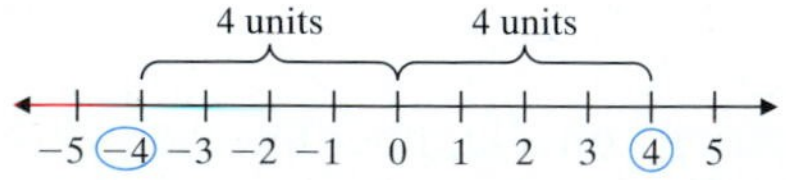

When two numbers are opposites, we say that each is the opposite of the other. Thus **4 is the opposite of** -4 and -4 **is the opposite of 4.**

Answers
4. a. 6 **b.** 4 **c.** 12

The phrase "the opposite of" is written in symbols as "−." For example,

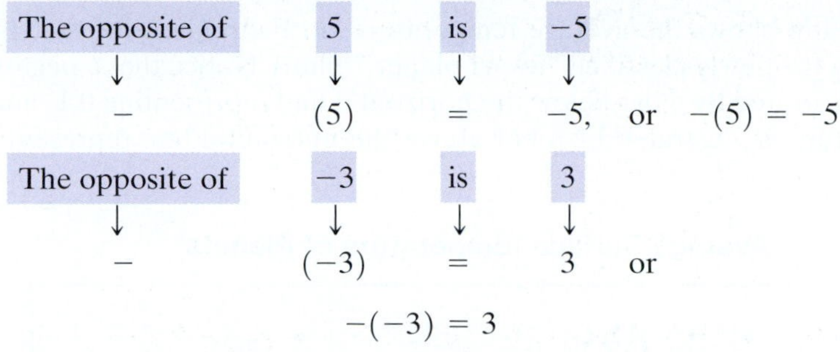

$-(-3) = 3$

In general, we have the following:

Opposites

If a is a number, then $-(-a) = a$.

Notice that because "the opposite of" is written as "−", to find the opposite of a number we place a "−" sign in front of the number.

EXAMPLE 5 Find the opposite of each number.

a. 13 **b.** −2 **c.** 0

Solution:

a. The opposite of 13 is −13.
b. The opposite of −2 is $-(-2)$ or 2.
c. The opposite of 0 is 0.

Helpful Hint Remember that 0 is neither positive nor negative.

Work Practice Problem 5

PRACTICE PROBLEM 5

Find the opposite of each number.

a. 14
b. −9

✓Concept Check True or false? The number 0 is the only number that is its own opposite.

EXAMPLE 6 Simplify.

a. $-(-4)$ **b.** $-|-5|$ **c.** $-|6|$

Solution:

a. $-(-4) = 4$ The opposite of negative 4 is 4.

b. $-|-5| = -5$ The opposite of the absolute value of −5 is the opposite of 5, or −5.

c. $-|6| = -6$ The opposite of the absolute value of 6 is the opposite of 6, or −6.

Work Practice Problem 6

PRACTICE PROBLEM 6

Simplify.

a. $-|-7|$
b. $-|4|$
c. $-(-12)$

EXAMPLE 7 Evaluate $-|-x|$ if $x = -2$.

Solution: Carefully replace x with −2; then simplify.

$-|-x| = -|-(-2)|$ Replace x with −2.

Then $-|-(-)2| = -|2| = -2$.

Work Practice Problem 7

PRACTICE PROBLEM 7

Evaluate $-|x|$ if $x = -6$.

Answers

5. a. −14 **b.** 9
6. a. −7 **b.** −4 **c.** 12 **7.** −6

✓ Concept Check Answer

True

Objective F Reading Bar Graphs Containing Integers

The bar graph below shows the average temperature (in Fahrenheit) of the eight planets, excluding the newly classified "dwarf planet," Pluto. Notice that a negative temperature is illustrated by a bar below the horizontal line representing 0°F, and a positive temperature is illustrated by a bar above the horizontal line representing 0°F.

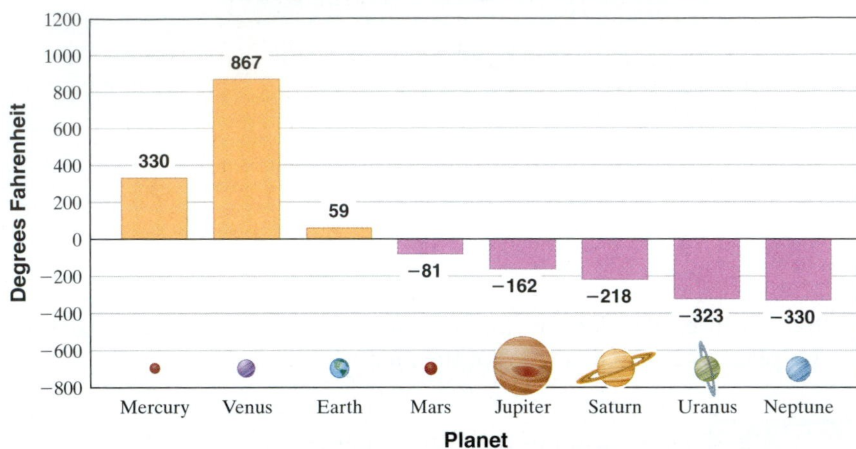

Source: The World Almanac, 2006
* For some planets, the temperature given is the temperature where the atmosphere pressure equals 1 Earth atmosphere.

PRACTICE PROBLEM 8

Which planet has the highest average temperature?

EXAMPLE 8 Which planet has the lowest average temperature?

Solution The planet with the lowest average temperature is the one that corresponds to the bar that extends the furthest in the negative direction (downward.) Neptune has the lowest average temperature of −330°F.

Work Practice Problem 8

Answer

8. Venus; 867°F

Vocabulary and Readiness Check

Use the choices below to fill in each blank. Not all choices will be used.

opposites	absolute value	right	is less than
inequality symbols	negative	positive	left
signed	integers	is greater than	

1. The numbers . . . $-3, -2, -1, 0, 1, 2, 3, \ldots$ are called ______________.

2. Positive numbers, negative numbers, and zero, together, are called ______________ numbers.

3. The symbols “$<$” and “$>$” are called ______________.

4. Numbers greater than 0 are called ______________ numbers while numbers less than 0 are called ______________ numbers.

5. The sign “$<$” means ______________ and “$>$” means ______________.

6. On a number line, the greater number is to the ______________ of the lesser number.

7. A number's distance from 0 on the number line is the number's ______________.

8. The numbers -5 and 5 are called ______________.

2.1 EXERCISE SET

FOR EXTRA HELP: Student Solutions Manual | PH Math/Tutor Center | CD/Video for Review | MathXL® | MyMathLab

Objective A *Represent each quantity by an integer. See Example 1.*

1. A worker in a silver mine in Nevada works 1235 feet underground.

2. A scuba diver is swimming 25 feet below the surface of the water in the Gulf of Mexico.

3. The peak of Mount Elbert in Colorado is 14,433 feet above sea level. (*Source:* U.S. Geological Survey)

4. The lowest elevation in the United States is found at Death Valley, California, at an elevation of 282 feet below sea level. (*Source:* U.S. Geological Survey)

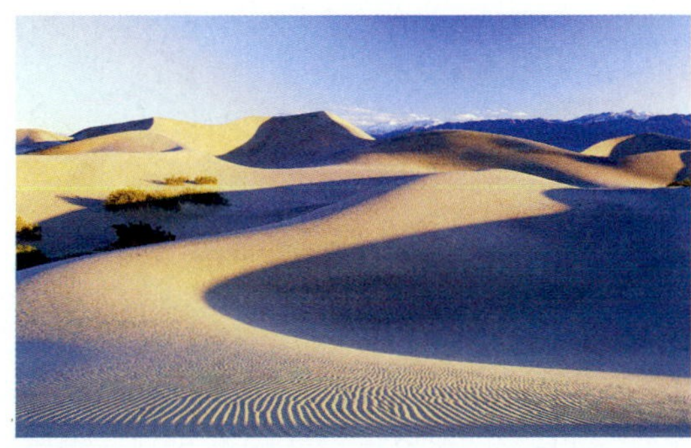

5. The record high temperature in Nevada is 118 degrees above zero Fahrenheit. (*Source:* National Climatic Data Center)

6. The Minnesota Viking football team gained 28 yards on a play.

7. The average depth of the Pacific Ocean is 13,000 feet below the surface of the ocean. (*Source:* U.S. Navy)

8. The Dow Jones stock market average fell 213 points in one day.

9. General Motors Corporation posted a net income loss of $10,458 million for the fiscal year 2005. (*Source:* General Motors)

10. For the fiscal year 2005, General Motors reported a loss of about $19 per share of stock. (*Source:* General Motors)

11. Two divers are exploring the wreck of the *Andrea Doria,* south of Nantucket Island, Massachusetts. Guillermo is 160 feet below the surface of the ocean and Luigi is 147 feet below the surface. Represent each quantity by an integer and determine who is deeper.

12. The temperature on one January day in Chicago was 10° below 0° Celsius. Represent this quantity by an integer and tell whether this temperature is cooler or warmer than 5° below 0° Celsius.

13. In a recent year, the number of music cassette singles shipped to retailers reflected an 81 percent loss from the previous year. Write an integer to represent the percent loss in cassette singles shipped. (*Source:* Recording Industry Association of America)

14. In a recent year, the number of music CDs shipped to retailers reflected a 49 percent loss from the previous year. Write an integer to represent the percent loss in CDs shipped. (*Source:* Recording Industry Association of America)

Objective **B** *Graph each integer in the list on the same number line. See Example 2.*

15. 0, 3, 4, 6

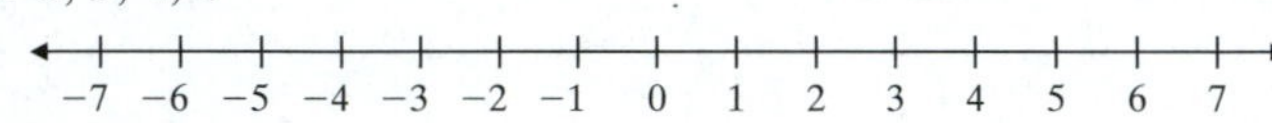

16. 7, 5, 2, 0

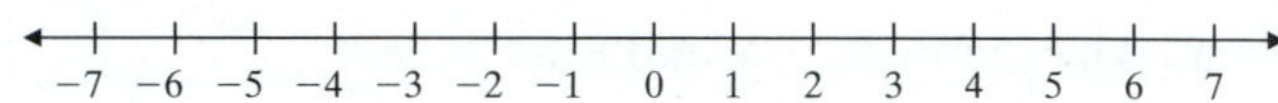

17. 1, −1, 2, −2, −4,

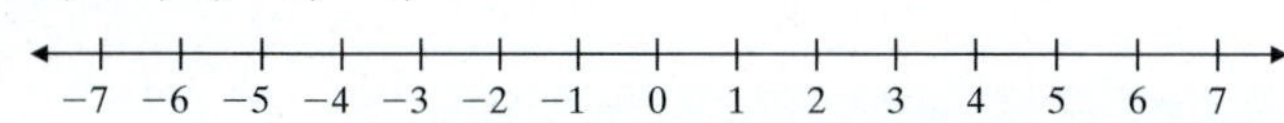

18. 3, −3, 5, −5, 6

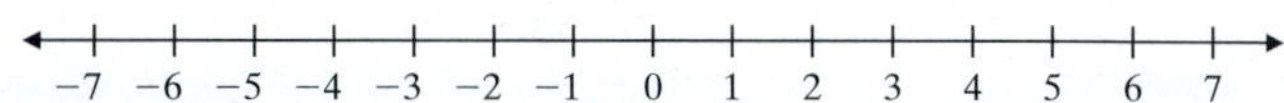

19. 0, 1, 9, 14

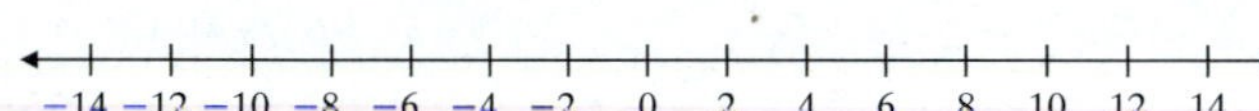

20. 0, 3, 10, 11

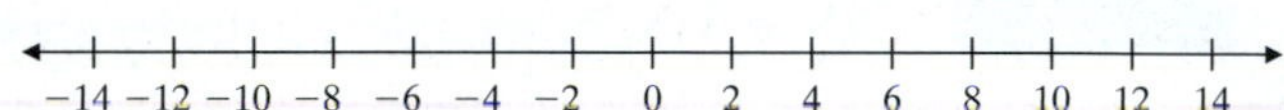

21. 0, −2, −7, −5

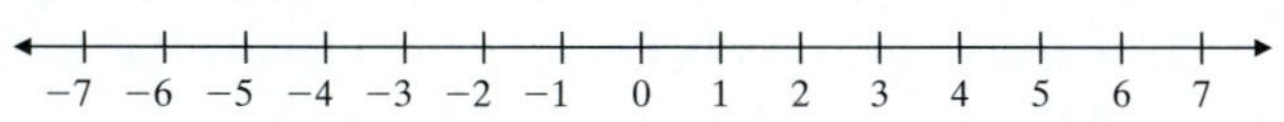

22. 0, −7, 3, −6

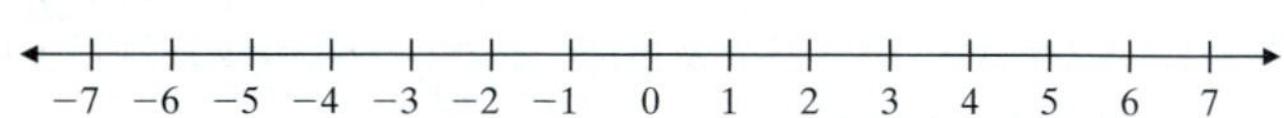

Objective **C** *Insert < or > between each pair of integers to make a true statement. See Example 3.*

23. 0   −7   **24.** −8   0   **25.** −7   −5   **26.** −12   −10

27. −30   −35   **28.** −27   −29   **29.** −26   26   **30.** 13   −13

Objective **D** *Simplify. See Example 4.*

31. $|5|$ **32.** $|7|$ **33.** $|-8|$ **34.** $|-19|$

35. $|0|$ **36.** $|100|$ **37.** $|-55|$ **38.** $|-10|$

Objective **E** *Find the opposite of each integer. See Example 5.*

39. 5   **40.** 8   **41.** −4   **42.** −6

43. 23   **44.** 123   **45.** −85   **46.** −13

Objectives **C D E** **Mixed Practice** *Simplify. See Example 6.*

47. $|-7|$ **48.** $|-11|$ **49.** $-|20|$ **50.** $-|43|$

51. $-|-3|$ **52.** $-|-18|$ **53.** $-(-43)$ **54.** $-(-27)$

55. $|-15|$ **56.** $-(-14)$ **57.** $-(-33)$ **58.** $-|-29|$

Evaluate. See Example 7.

59. $|-x|$ if $x = -6$ **60.** $-|x|$ if $x = -8$ **61.** $-|-x|$ if $x = 2$ **62.** $-|-x|$ if $x = 10$

63. $|x|$ if $x = -32$ **64.** $|x|$ if $x = 32$ **65.** $-|x|$ if $x = 7$ **66.** $|-x|$ if $x = 1$

Insert $<, >,$ *or* $=$ *between each pair of numbers to make a true statement. See Examples 3 through 6.*

67. $-12 \quad -6$ **68.** $-4 \quad -17$ **69.** $|-8| \quad |-11|$ **70.** $|-8| \quad |-4|$

71. $|-47| \quad -(-47)$ **72.** $-|17| \quad -(-17)$ **73.** $-|-12| \quad -(-12)$ **74.** $|-24| \quad -(-24)$

75. $0 \quad -9$ **76.** $-45 \quad 0$ **77.** $|0| \quad |-9|$ **78.** $|-45| \quad |0|$

79. $-|-2| \quad -|-10|$ **80.** $-|-8| \quad -|-4|$ **81.** $-(-12) \quad -(-18)$ **82.** $-22 \quad -(-38)$

Objectives **D** **E** **Mixed Practice** *Fill in the chart. See Examples 4 through 7.*

	Number	Absolute Value of Number	Opposite of Number
83.	31		
85.			−28

	Number	Absolute Value of Number	Opposite of Number
84.	−13		
86.			90

Objective **F** *The bar graph shows the elevations of selected lakes. Use this graph For Exercises 87 through 90 (Source:* U.S. Geological Survey). *See Example 8.*

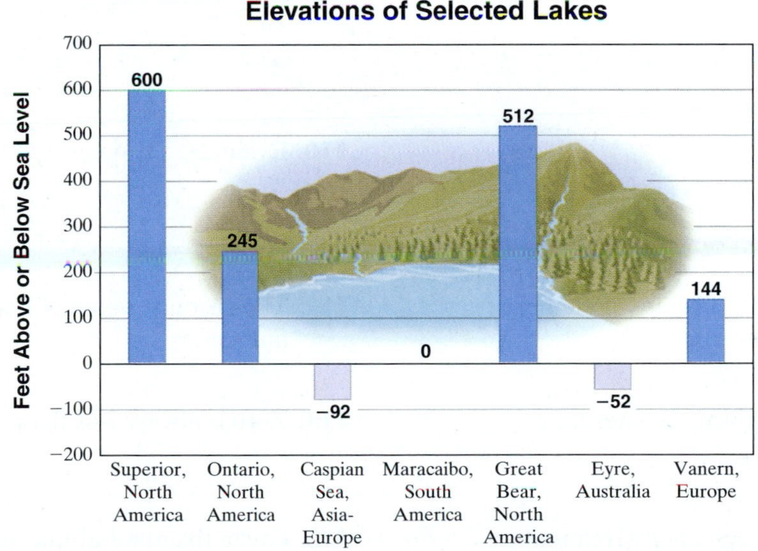

87. Which lake shown has the lowest elevation?

88. Which lake has an elevation at sea level?

89. Which lake shown has the highest elevation?

90. Which lake shown has the second lowest elevation?

Use the bar graph from Example 8 to answer Exercises 91 through 94.

91. Which planet has a negative average temperature closest to 0°F?

92. Which planet has an average temperature closest to 0°F?

93. Which planet has an average temperature closest to −300°F?

94. Which planet has an average temperature closest to −200°F?

Review

Add. See Section 1.3.

95. $0 + 13$

96. $9 + 0$

97. $15 + 20$

98. $20 + 15$

99. $47 + 236 + 77$

100. $362 + 37 + 90$

Concept Extensions

Write the given numbers in order from least to greatest.

101. $2^2, -|3|, -(-5), -|-8|$

102. $|10|, 2^3, -|-5|, -(-4)$

103. $|-1|, -|-6|, -(-6), -|1|$

104. $1^4, -(-3), -|7|, |-20|$

105. $-(-2), 5^2, -10, -|-9|, |-12|$

106. $3^3, -|-11|, -(-10), -4, -|2|$

Choose all numbers for x from each given list that make each statement true.

107. $|x| > 8$

a. -9 **b.** -5 **c.** 8 **d.** -12

108. $|x| > 4$

a. 0 **b.** -4 **c.** 5 **d.** -100

109. Evaluate: $-(-|-8|)$

110. Evaluate: $(-|-(-7)|)$

Answer true or false for Exercises 111 through 115.

111. If $a > b$, then a must be a positive number.

112. The absolute value of a number is *always* a positive number.

113. A positive number is always greater than a negative number.

114. Zero is always less than a positive number.

115. The number $-a$ is always a negative number. (*Hint:* Read "$-$" as "the opposite of.")

116. Given the number line 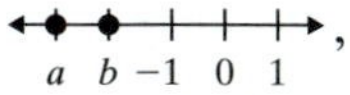, is it true that $b < a$?

117. Write in your own words how to find the absolute value of a signed number.

118. Explain how to determine which of two signed numbers is larger.

For Exercises 119 and 120, see the first Concept Check in this section.

119. Is there a largest negative number? If so, what is it?

120. Is there a smallest positive number? If so, what is it?

STUDY SKILLS BUILDER

Have You Decided to Complete This Course Successfully?*

Ask yourself if one of your current goals is to complete this course successfully.

If it is not a goal of yours, ask yourself why? One common reason is fear of failure. Amazingly enough, fear of failure alone can be strong enough to keep many of us from doing our best in any endeavor.

Another common reason is that you simply haven't taken the time to make successfully completing this course one of your goals. How do you do this? Start by writing this goal in your mathematics notebook. Then list steps you will take to ensure success. A great first step is to read or reread Section 1.1 and make a commitment to try the suggestions in that section.

Good luck, and don't forget that a positive attitude will make a big difference.

Let's see how you are doing.

1. Have you decided to make "successfully completing this course" a goal of yours? If no, please list reasons why this has not happened. Study your list and talk to your instructor about this.
2. If your answer to question 1 is yes, take a moment and list in your notebook further specific goals that will help you achieve this major goal of successfully completing this course. (For example, "My goal this semester is not to miss any of my mathematics classes.")
3. Rate your commitment to this course with a number between 1 and 5. Use the diagram below to help.

High Commitment		Average Commitment		Not committed at all
5	4	3	2	1

4. If you have rated your personal commitment level (from the exercise above) as a 1, 2, or 3, list the reasons why this is so. Then determine whether it is possible to increase your commitment level to a 4 or 5.

*Because of its importance, this is a repeat of the Study Skills Builder on page 83.

2.2 ADDING INTEGERS

Objectives

A Add Integers.

B Evaluate an Algebraic Expression by Adding.

C Solve Problems by Adding Integers.

Objective A Adding Integers

Adding integers can be visualized using a number line. A positive number can be represented on the number line by an arrow of appropriate length pointing to the right, and a negative number by an arrow of appropriate length pointing to the left.

Both arrows represent 2 or +2.
They both point to the right and they are both 2 units long.

Both arrows represent −3.
They both point to the left and they are both 3 units long.

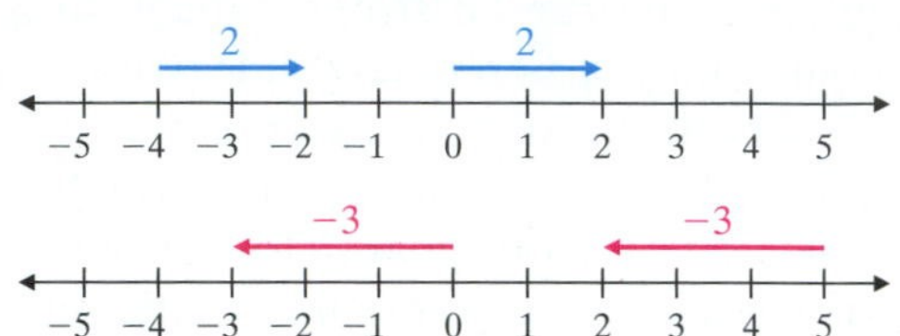

EXAMPLE 1 Add using a number line: $5 + (-2)$

Solution: To add integers on a number line, such as $5 + (-2)$, we start at 0 on the number line and draw an arrow representing 5. From the tip of this arrow, we draw another arrow representing −2. The tip of the second arrow ends at their sum, 3.

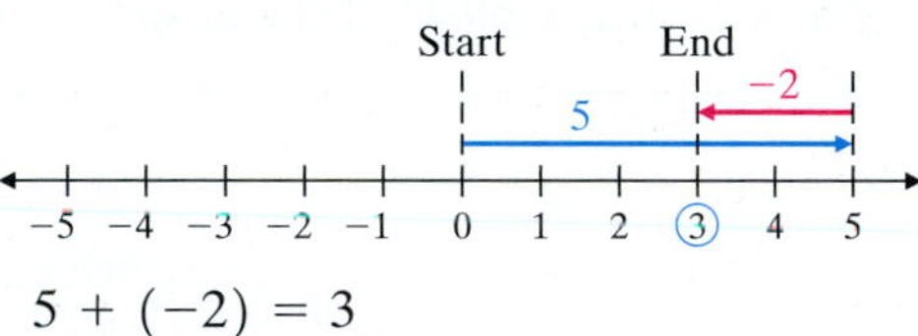

$$5 + (-2) = 3$$

Work Practice Problem 1

PRACTICE PROBLEM 1

Add using a number line:
$5 + (-1)$

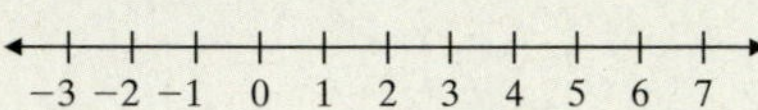

EXAMPLE 2 Add using a number line: $-1 + (-4)$

Start at 0 and draw an arrow representing −1. From the tip of this arrow, we draw another arrow representing −4. The tip of the second arrow ends at their sum, −5.

Solution:

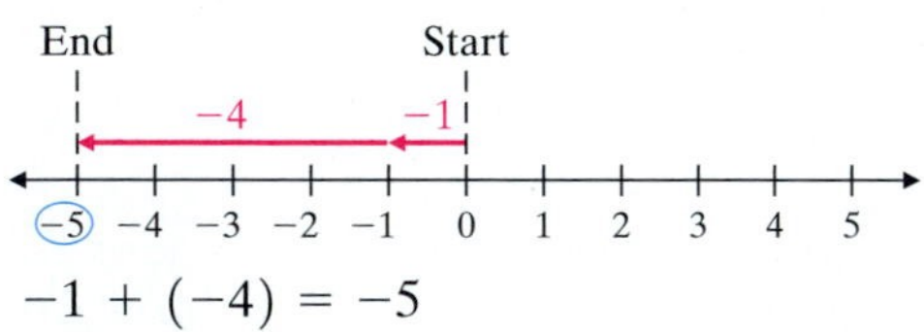

$$-1 + (-4) = -5$$

Work Practice Problem 2

PRACTICE PROBLEM 2

Add using a number line:
$-6 + (-2)$

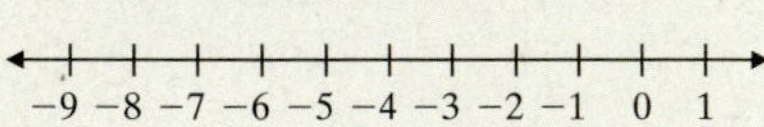

EXAMPLE 3 Add using a number line: $-7 + 3$

Solution:

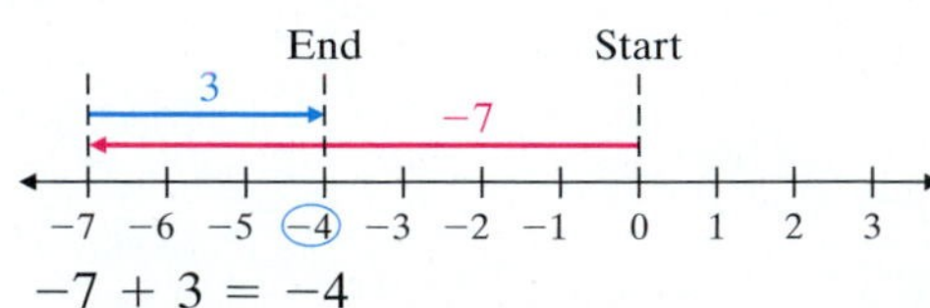

$$-7 + 3 = -4$$

Work Practice Problem 3

PRACTICE PROBLEM 3

Add using a number line:
$-8 + 3$

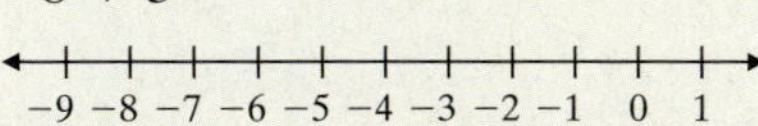

Answers

1. 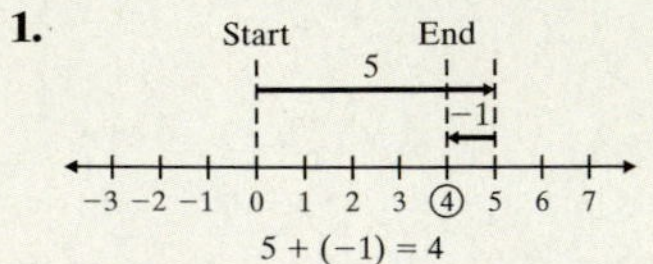

2. 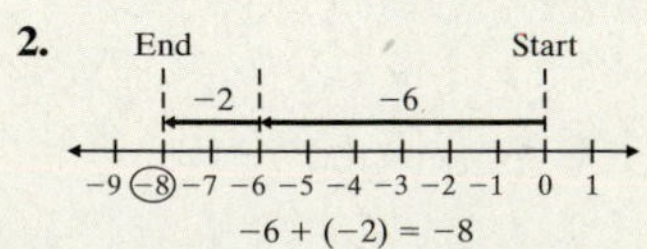

3. 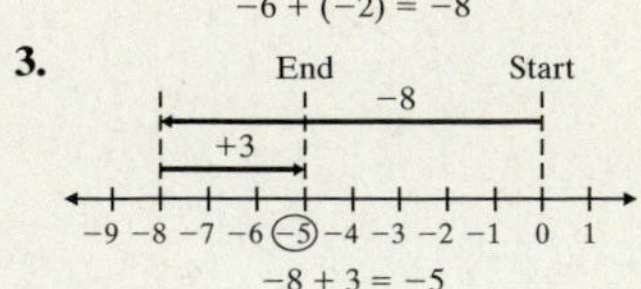

Using a number line each time we add two numbers can be time consuming. Instead, we can notice patterns in the previous examples and write rules for adding signed numbers.

Rules for adding signed numbers depend on whether we are adding numbers with the same sign or different signs. When adding two numbers with the same sign, as in Example 2, notice that the sign of the sum is the same as the sign of the addends.

Adding Two Numbers with the Same Sign

Step 1: Add their absolute values.

Step 2: Use their common sign as the sign of the sum.

EXAMPLE 4 Add: $-2 + (-21)$

Solution:

Step 1: $|-2| = 2, |-21| = 21$, and $2 + 21 = 23$.
Step 2: Their common sign is negative, so the sum is negative:

$$-2 + (-21) = -23$$

Work Practice Problem 4

PRACTICE PROBLEM 4

Add: $(-3) + (-19)$

EXAMPLES Add.

5. $-15 + (-10) = -25$
6. $2 + 6 = 8$

Work Practice Problems 5–6

PRACTICE PROBLEMS 5–6

Add.

5. $-12 + (-30)$
6. $9 + 4$

When adding two numbers with different signs, as in Examples 1 and 3, the sign of the result may be positive, negative, or the result may be 0.

Adding Two Numbers with Different Signs

Step 1: Find the larger absolute value minus the smaller absolute value.

Step 2: Use the sign of the number with the larger absolute value as the sign of the sum.

EXAMPLE 7 Add: $-2 + 5$

Solution:

Step 1: $|-2| = 2, |5| = 5$, and $5 - 2 = 3$.
Step 2: 5 has the larger absolute value and its sign is an understood +:

$$-2 + 5 = +3 \text{ or } 3$$

Work Practice Problem 7

PRACTICE PROBLEM 7

Add: $-1 + 6$

EXAMPLE 8 Add: $3 + (-7)$

Solution:

Step 1: $|3| = 3, |-7| = 7$, and $7 - 3 = 4$.
Step 2: -7 has the larger absolute value and its sign is $-$:

$$3 + (-7) = -4$$

Work Practice Problem 8

PRACTICE PROBLEM 8

Add: $2 + (-8)$

Answers
4. -22 **5.** -42 **6.** 13 **7.** 5 **8.** -6

PRACTICE PROBLEMS 9–11

Add.

9. $-54 + 20$

10. $7 + (-2)$

11. $-3 + 0$

EXAMPLES Add.

9. $-18 + 10 = -8$

10. $12 + (-8) = 4$

11. $0 + (-5) = -5$ The sum of 0 and any number is the number.

Work Practice Problems 9–11

Recall that numbers such as 7 and -7 are called opposites. In general, the sum of a number and its opposite is always 0.

$7 + (-7) = 0$ (opposites) $\quad -26 + 26 = 0$ (opposites) $\quad 1008 + (-1008) = 0$ (opposites)

If a is a number, then

$-a$ is its opposite. Also,

$$\left.\begin{aligned} a + (-a) &= 0 \\ -a + a &= 0 \end{aligned}\right\} \text{The sum of a number and its opposite is 0.}$$

PRACTICE PROBLEMS 12–13

Add.

12. $18 + (-18)$

13. $-64 + 64$

EXAMPLES Add.

12. $-21 + 21 = 0$

13. $36 + (-36) = 0$

Work Practice Problems 12–13

✔Concept Check What is wrong with the following calculation?

$5 + (-22) = 17$

In the following examples, we add three or more integers. Remember that by the associative and commutative properties for addition, we may add numbers in any order that we wish. In Examples 14 and 15, let's add the numbers from left to right.

PRACTICE PROBLEM 14

Add: $6 + (-2) + (-15)$

EXAMPLE 14 Add: $(-3) + 4 + (-11)$

Solution: $(-3) + 4 + (-11) = 1 + (-11)$
$= -10$

Work Practice Problem 14

PRACTICE PROBLEM 15

Add: $5 + (-3) + 12 + (-14)$

Helpful Hint Don't forget that addition is commutative and associative. In other words, numbers may be added in any order.

EXAMPLE 15 Add: $1 + (-10) + (-8) + 9$

Solution: $1 + (-10) + (-8) + 9 = -9 + (-8) + 9$
$= -17 + 9$
$= -8$

The sum will be the same if we add the numbers in any order. To see this, let's add the positive numbers together and then the negative numbers together first.

$1 + 9 = 10$ Add the positive numbers.

$(-10) + (-8) = -18$ Add the negative numbers.

$10 + (-18) = -8$ Add these results.

The sum is -8.

Work Practice Problem 15

Answers

9. -34 **10.** 5 **11.** -3 **12.** 0 **13.** 0 **14.** -11 **15.** 0

✔ Concept Check Answer

$5 + (-22) = -17$

Objective B Evaluating Algebraic Expressions

We can continue our work with algebraic expressions by evaluating expressions given integer replacement values.

EXAMPLE 16 Evaluate $2x + y$ for $x = 3$ and $y = -5$.

Solution: Replace x with 3 and y with -5 in $2x + y$.

$$\begin{aligned} 2x + y &= 2 \cdot 3 + (-5) \\ &= 6 + (-5) \\ &= 1 \end{aligned}$$

Work Practice Problem 16

EXAMPLE 17 Evaluate $x + y$ for $x = -2$ and $y = -10$.

Solution: $x + y = (-2) + (-10)$ Replace x with -2 and y with -10.

$= -12$

Work Practice Problem 17

Objective C Solving Problems by Adding Integers

Next, we practice solving problems that require adding integers.

EXAMPLE 18 **Calculating Temperature**

In Barrow, Alaska, the monthly average temperature for February is −16° Fahrenheit. In March, this average temperature rises 2 degrees, and in April, it rises another 13 degrees. What is the average temperature in April? (*Source:* National Climatic Data Center)

Solution:

In words:	April temperature	=	February temperature	+	rise of 2°	+	rise of 13°
	↓		↓		↓		↓
Translate:	April temperature	=	−16	+	(+2)	+	(+13)

$= -14 + (+13)$

$= -1$

The average temperature in Barrow, Alaska, during the month of April is −1°F.

Work Practice Problem 18

PRACTICE PROBLEM 16

Evaluate $x + 3y$ for $x = -6$ and $y = 2$.

PRACTICE PROBLEM 17

Evaluate $x + y$ for $x = -13$ and $y = -9$.

PRACTICE PROBLEM 18

If the temperature was −7° Fahrenheit at 6 a.m., and it rose 4 degrees by 7 a.m. and then rose another 7 degrees in the hour from 7 a.m. to 8 a.m., what was the temperature at 8 a.m.?

Answers

16. 0 **17.** −22 **18.** 4°F

CALCULATOR EXPLORATIONS Entering Negative Numbers

To enter a negative number on a calculator, find the key marked [+/−]. (Some calculators have a key marked [CHS] and some calculators have a special key [(−)] for entering a negative sign.) To enter the number −2, for example, press the keys [2] [+/−]. The display will read [−2].

To find −32 + (−131), press the keys

[32] [+/−] [+] [131] [+/−] [=] or

[(−)] [32] [+] [(−)] [131] [ENTER]

The display will read [−163]. Thus −32 + (−131) = −163.

Use a calculator to perform each indicated operation.

1. −256 + 97
2. 811 + (−1058)
3. 6(15) + (−46)
4. −129 + 10(48)
5. −108,650 + (−786,205)
6. −196,662 + (−129,856)

Vocabulary and Readiness Check

Use the choices below to fill in each blank. Not all choices will be used.

$-a$ $\quad a \quad$ 0 $\quad$ commutative $\quad$ associative

1. If n is a number, then $-n + n =$ ______________.
2. Since $x + n = n + x$, we say that addition is ______________.
3. If a is a number, then $-(-a) =$ ______________.
4. Since $n + (x + a) = (n + x) + a$, we say that addition is ______________.

2.2 EXERCISE SET

FOR EXTRA HELP

Student Solutions Manual

PH Math/Tutor Center

CD/Video for Review

MathXL®

MyMathLab

Objective A *Add using a number line. See Examples 1 through 3.*

1. $-1 + (-6)$

−7 −6 −5 −4 −3 −2 −1 0 1 2 3 4 5 6 7

2. $9 + (-4)$

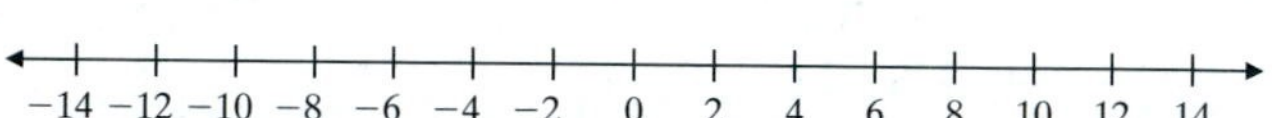

3. $-4 + 7$

−7 −6 −5 −4 −3 −2 −1 0 1 2 3 4 5 6 7

4. $10 + (-3)$

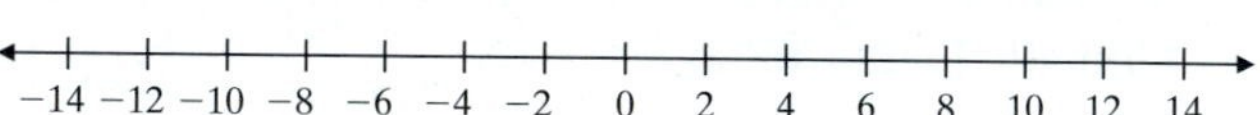

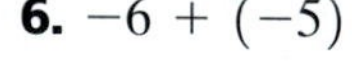

5. $-13 + 7$

−14 −12 −10 −8 −6 −4 −2 0 2 4 6 8 10 12 14

6. $-6 + (-5)$

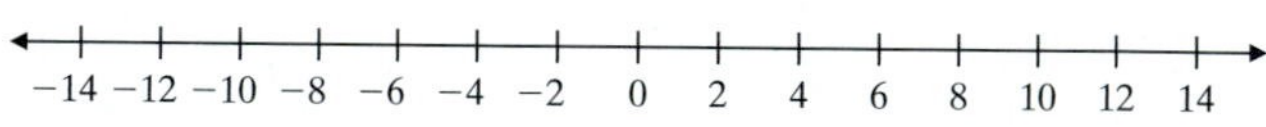

Add. See Examples 4 through 13.

7. $46 + 21$

8. $15 + 42$

9. $-8 + (-2)$

10. $-5 + (-4)$

11. $-43 + 43$

12. $-62 + 62$

13. $6 + (-2)$

14. $8 + (-3)$

15. $-6 + 0$

16. $-8 + 0$

17. $3 + (-5)$

18. $5 + (-9)$

19. $-2 + (-7)$

20. $-6 + (-1)$

21. $-12 + (-12)$

22. $-23 + (-23)$

23. $-640 + (-200)$

24. $-400 + (-256)$

25. $12 + (-5)$

26. $24 + (-10)$

27. $-6 + 3$

28. $-8 + 4$

29. $-56 + 26$

30. $-89 + 37$

31. $-45 + 85$

32. $-32 + 62$

33. $124 + (-144)$

34. $325 + (-375)$

35. $-82 + (-43)$

36. $-56 + (-33)$

Add. See Examples 14 and 15.

37. $-4 + 2 + (-5)$

38. $-1 + 5 + (-8)$

39. $-52 + (-77) + (-117)$

40. $-103 + (-32) + (-27)$

41. $12 + (-4) + (-4) + 12$

42. $18 + (-9) + 5 + (-2)$

43. $(-10) + 14 + 25 + (-16)$

44. $34 + (-12) + (-11) + 213$

Objective **A** **Mixed Practice** *Add. See Examples 1 through 15.*

45. $-6 + (-15) + (-7)$

46. $-12 + (-3) + (-5)$

47. $-26 + 15$

48. $-35 + (-12)$

49. $5 + (-2) + 17$

50. $3 + (-23) + 6$

51. $-13 + (-21)$

52. $-100 + 70$

53. $3 + 14 + (-18)$

54. $(-45) + 22 + 20$

55. $-92 + 92$

56. $-87 + 0$

57. $-13 + 8 + (-10) + (-27)$

58. $-16 + 6 + (-14) + (-20)$

Objective **B** *Evaluate $x + y$ for the given replacement values. See Examples 16 and 17.*

59. $x = -20$ and $y = -50$

60. $x = -1$ and $y = -29$

Evaluate $3x + y$ for the given replacement values. See Examples 16 and 17.

61. $x = 2$ and $y = -3$

62. $x = 7$ and $y = -11$

63. $x = 3$ and $y = -30$

64. $x = 13$ and $y = -17$

Objective **C** **Translating** *Translate each phrase; then simplify. See Example 18.*

65. Find the sum of -6 and 25.

66. Find the sum of -30 and 15.

67. Find the sum of -31, -9, and 30.

68. Find the sum of -49, -2, and 40.

Solve. See Example 18.

69. Suppose a deep-sea diver dives from the surface to 215 feet below the surface. He then dives down 16 more feet. Use positive and negative numbers to represent this situation. Then find the diver's present depth.

70. Suppose a diver dives from the surface to 248 meters below the surface and then swims up 8 meters, down 16 meters, down another 28 meters, and then up 32 meters. Use positive and negative numbers to represent this situation. Then find the diver's depth after these movements.

In golf, it is possible to have positive and negative scores. The following table shows the results of the eighteen-hole playoff between Pat Hurst and Annika Sorenstam at the 2006 U.S. Open. Use the table to answer Exercises 71 and 72.

Player/Hole	1	2	3	4	5	6	7	8	9	10	11	12	13	14	15	16	17	18
Sorenstam	−1	0	−1	0	0	+1	0	0	0	0	0	−1	+1	0	0	0	0	0
Hurst	+1	0	0	0	0	+2	0	0	+1	0	0	0	0	0	0	0	0	−1

(*Source:* Ladies' Professional Golf Association)

71. Find the total score for each of the athletes in the playoff.

72. In golf, the lower score is the winner. Use the result of Exercise 71 to determine who won the 2006 U.S. Open.

The following bar graph shows the yearly net income for Apple, Inc. Net income is one indication of a company's health. It measures revenue (money taken in) minus cost (money spent). Use this graph to answer Exercises 73 through 76. (Source: Apple, Inc.)

73. What was the net income (in dollars) for Apple, Inc. in 2002?

74. What was the net income (in dollars) for Apple, Inc in 2001?

75. Find the total net income for the years 2001 and 2002.

76. Find the total net income for all the years shown.

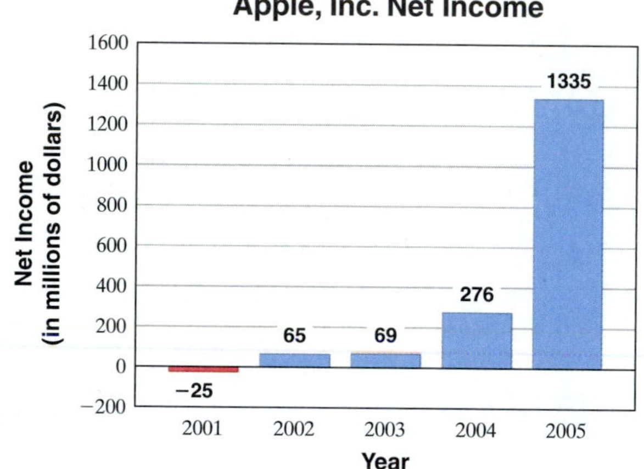

77. The temperature at 4 p.m. on February 2 was $-10°$ Celsius. By 11 p.m. the temperature had risen 12 degrees. Find the temperature at 11 p.m.

78. In some card games, it is possible to have both positive and negative scores. After four rounds of play, Michelle had scores of 14, −5, −8, and 7. What was her total score for the game?

A small business company reports the following net incomes. Use this table to answer Exercises 79 and 80.

Year	Net Income (in Dollars)
2003	−\$10,412
2004	−\$1,786
2005	\$15,395
2006	\$31,418

79. Find the sum of the net incomes for 2004 and 2005.

80. Find the sum of the net incomes for all four years shown.

81. The all-time record low temperature for Wisconsin is −55°F. West Virginia's all-time record low temperature is 8°F higher than Wisconsin record low. What is West Virginia's record low temperature? (*Source:* National Climatic Data Center)

82. The all-time record low temperature for Oklahoma is −27°F. In Georgia, the lowest temperature ever recorded is 10°F higher than Oklahoma's all-time low temperature. What is the all-time record low temperature for Georgia? (*Source:* National Climatic Data Center)

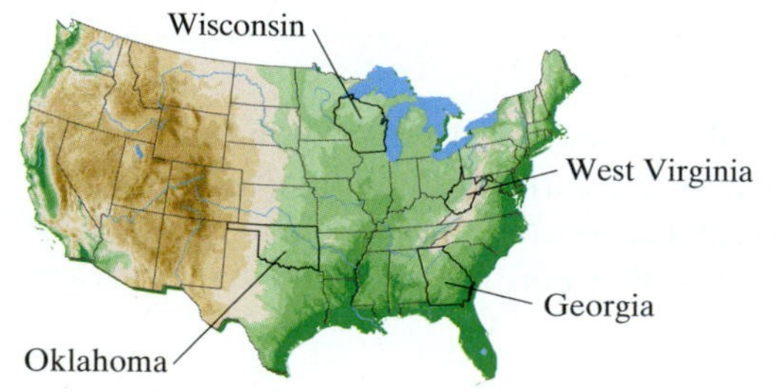

83. The deepest spot in the Pacific Ocean is the Mariana Trench, which has an elevation of 10,924 meters below sea level. The bottom of the Pacific's Aleutian Trench has an elevation 3245 meters higher than that of the Mariana Trench. Use a negative number to represent the depth of the Aleutian Trench. (*Source:* Defense Mapping Agency)

84. The deepest spot in the Atlantic Ocean is the Puerto Rico Trench, which has an elevation of 8605 meters below sea level. The bottom of the Atlantic's Cayman Trench has an elevation 1070 meters above the level of the Puerto Rico Trench. Use a negative number to represent the depth of the Cayman Trench. (*Source:* Defense Mapping Agency)

Review

Subtract. See Section 1.3.

85. $44 - 0$

86. $91 - 0$

87. $200 - 59$

88. $400 - 18$

Concept Extensions

89. Name 2 numbers whose sum is -17.

90. Name 2 numbers whose sum is -30.

Each calculation below is incorrect. Find the error and correct. See the Concept Check in this section.

91. $7 + (-10) \stackrel{?}{=} 17$

92. $-10 + (-12) \stackrel{?}{=} -120$

93. $-4 + 14 \stackrel{?}{=} -18$

94. $-15 + (-17) \stackrel{?}{=} 32$

For Exercises 95 through 98, determine whether each statement is true or false.

95. The sum of two negative numbers is always a negative number.

96. The sum of two positive numbers is always a positive number.

97. The sum of a positive number and a negative number is always a negative number.

98. The sum of zero and a negative number is always a negative number.

99. In your own words, explain how to add two negative numbers.

100. In your own words, explain how to add a positive number and a negative number.

2.3 SUBTRACTING INTEGERS

Objectives

- **A** Subtract Integers.
- **B** Add and Subtract Integers.
- **C** Evaluate an Algebraic Expression by Subtracting.
- **D** Solve Problems by Subtracting Integers.

In Section 2.1, we discussed the opposite of an integer.

The opposite of 3 is -3.
The opposite of -6 is 6.

In this section, we use opposites to subtract integers.

Objective A Subtracting Integers

To subtract integers, we will write the subtraction problem as an addition problem. To see how to do this, study the examples below.

$$10 - 4 = 6$$
$$10 + (-4) = 6$$

Since both expressions simplify to 6, this means that

$$10 - 4 = 10 + (-4) = 6$$

Also,

$$3 - 2 = 3 + (-2) = 1$$
$$15 - 1 = 15 + (-1) = 14$$

Thus, to subtract two numbers, we add the first number to the opposite of the second number. (The opposite of a number is also known as its **additive inverse.**)

Subtracting Two Numbers

If a and b are numbers, then $a - b = a + (-b)$.

PRACTICE PROBLEMS 1–4

Subtract.

1. $13 - 4$
2. $-8 - 2$
3. $11 - (-15)$
4. $-9 - (-1)$

EXAMPLES Subtract.

	subtraction	=	first number	+	opposite of the second number		
1.	$8 - 5$	=	8	+	(-5)	=	3
2.	$-4 - 10$	=	-4	+	(-10)	=	-14
3.	$6 - (-5)$	=	6	+	5	=	11
4.	$-11 - (-7)$	=	-11	+	7	=	-4

Work Practice Problems 1–4

PRACTICE PROBLEMS 5–7

Subtract.

5. $6 - 9$
6. $-14 - 5$
7. $-3 - (-4)$

EXAMPLES Subtract.

5. $-10 - 5 = -10 + (-5) = -15$

6. $8 - 15 = 8 + (-15) = -7$

7. $-4 - (-5) = -4 + 5 = 1$

Work Practice Problems 5–7

Answers

1. 9 2. -10 3. 26 4. -8 5. -3 6. -19 7. 1

Helpful Hint

To visualize subtraction, try the following:

The difference between 5°F and −2°F can be found by subtracting. That is,

$$5 - (-2) = 5 + 2 = 7$$

Can you visually see from the thermometer on the right that there are actually 7 degrees between 5°F and −2°F?

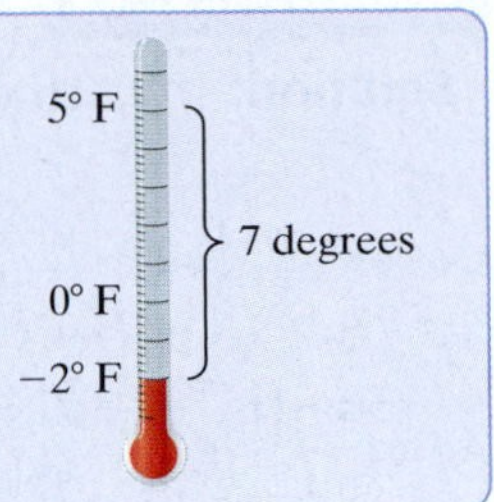

✔Concept Check What is wrong with the following calculation?

$-9 - (-5) = -14$

EXAMPLE 8 Subtract 7 from −3.

Solution: To subtract 7 *from* −3, we find

$$-3 - 7 = -3 + (-7) = -10$$

Work Practice Problem 8

PRACTICE PROBLEM 8

Subtract 6 from −15.

Objective B Adding and Subtracting Integers

If a problem involves adding or subtracting more than two integers, we rewrite differences as sums and add. Recall that by associative and commutative properties, we may add numbers in any order. In Examples 9 and 10, we will add from left to right.

EXAMPLE 9 Simplify: $7 - 8 - (-5) - 1$

Solution:

$$\begin{aligned} 7 - 8 - (-5) - 1 &= \underbrace{7 + (-8)} + 5 + (-1) \\ &= \underbrace{-1 + 5} + (-1) \\ &= \underbrace{4 + (-1)} \\ &= 3 \end{aligned}$$

Work Practice Problem 9

PRACTICE PROBLEM 9

Simplify: $-6 - 5 - 2 - (-3)$

EXAMPLE 10 Simplify: $7 + (-12) - 3 - (-8)$

Solution:

$$\begin{aligned} 7 + (-12) - 3 - (-8) &= \underbrace{7 + (-12)} + (-3) + 8 \\ &= \underbrace{-5 + (-3)} + 8 \\ &= -8 + 8 \\ &= 0 \end{aligned}$$

Work Practice Problem 10

PRACTICE PROBLEM 10

Simplify:

$8 + (-2) - 9 - (-7)$

Objective C Evaluating Expressions

Now let's practice evaluating expressions when the replacement values are integers.

EXAMPLE 11 Evaluate $x - y$ for $x = -3$ and $y = 9$.

Solution: Replace x with −3 and y with 9 in $x - y$.

$$\begin{aligned} & x \quad - \quad y \\ & \downarrow \quad \downarrow \quad \downarrow \\ &= (-3) - 9 \\ &= (-3) + (-9) \\ &= -12 \end{aligned}$$

Work Practice Problem 11

PRACTICE PROBLEM 11

Evaluate $x - y$ for $x = -5$ and $y = 13$.

Answers

8. −21 **9.** −10 **10.** 4 **11.** −18

✔Concept Check Answer

$-9 - (-5) = -4$

PRACTICE PROBLEM 12

Evaluate $3y - z$ for $y = 9$ and $z = -4$.

EXAMPLE 12 Evaluate $2a - b$ for $a = 8$ and $b = -6$.

Solution: Watch your signs carefully!

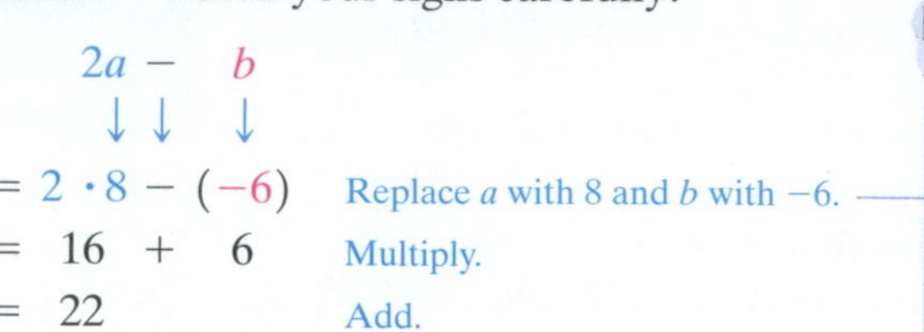

$$2a - b$$
$$= 2 \cdot 8 - (-6) \quad \text{Replace } a \text{ with 8 and } b \text{ with } -6.$$
$$= 16 + 6 \quad \text{Multiply.}$$
$$= 22 \quad \text{Add.}$$

Helpful Hint

Watch carefully when replacing variables in the expression $a - b$. Make sure that all symbols are inserted and accounted for.

Work Practice Problem 12

Objective D Solving Problems by Subtracting Integers

Solving problems often requires subtraction of integers.

PRACTICE PROBLEM 13

The highest point in Asia is the top of Mount Everest, at a height of 29,028 feet above sea level. The lowest point is the Dead Sea, which is 1312 feet below sea level. How much higher is Mount Everest than the Dead Sea? (*Source:* National Geographic Society)

EXAMPLE 13 Finding a Change in Elevation

The highest point in the United States is the top of Mount McKinley, at a height of 20,320 feet above sea level. The lowest point is Death Valley, California, which is 282 feet below sea level. How much higher is Mount McKinley than Death Valley? (*Source:* U.S. Geological Survey)

Solution:

1. UNDERSTAND. Read and reread the problem. To find "how much higher," we subtract. Don't forget that since Death Valley is 282 feet *below* sea level, we represent its height by −282. Draw a diagram to help visualize the problem.

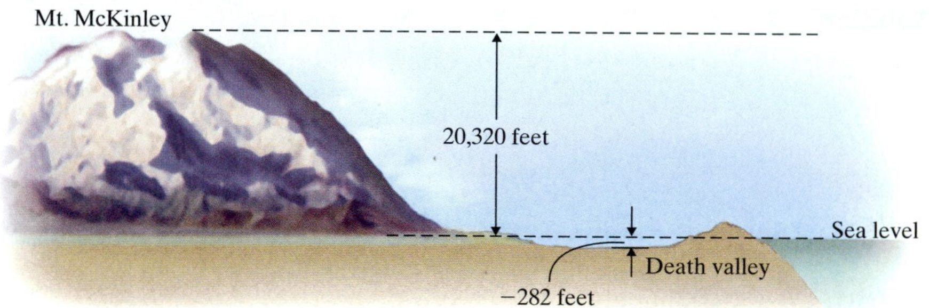

2. TRANSLATE.

In words:	how much higher is Mt. McKinley	=	height of Mt. McKinley	minus	height of Death Valley
	↓	↓	↓	↓	↓
Translate:	how much higher is Mt. McKinley	=	20,320	−	(−282)

3. SOLVE:

$20{,}320 - (-282) = 20{,}320 + 282 = 20{,}602$

4. INTERPRET. Check and state your conclusion: Mount McKinley is 20,602 feet higher than Death Valley.

Work Practice Problem 13

Answers
12. 31 **13.** 30,340 ft

Vocabulary and Readiness Check

Multiple choice: Select the correct lettered response following each exercise.

1. It is true that $a - b =$ ________.

a. $b - a$ **b.** $a + (-b)$ **c.** $a + b$

2. The opposite of n is ______.

a. $-n$ **b.** $-(-n)$ **c.** n

3. To evaluate $x - y$ for $x = -10$ and $y = -14$, we replace x with -10 and y with -14 and evaluate ____________.

a. $10 - 14$ **b.** $-10 - 14$ **c.** $-14 - 10$ **d.** $-10 - (-14)$

4. The expression $-5 - 10$ equals __________.

a. $5 - 10$ **b.** $5 + 10$ **c.** $-5 + (-10)$ **d.** $10 - 5$

2.3 EXERCISE SET

FOR EXTRA HELP

Student Solutions Manual

PH Math/Tutor Center

CD/Video for Review

MathXL®

MyMathLab

Objective A *Subtract. See Examples 1 through 7.*

1. $-8 - (-8)$ **2.** $-6 - (-6)$ **3.** $19 - 16$ **4.** $15 - 12$

5. $3 - 8$ **6.** $2 - 5$ **7.** $11 - (-11)$ **8.** $12 - (-12)$

9. $-4 - (-7)$ **10.** $-25 - (-25)$ **11.** $-16 - 4$ **12.** $-2 - 42$

13. $3 - 15$ **14.** $8 - 9$ **15.** $42 - 55$ **16.** $17 - 63$

17. $478 - (-30)$ **18.** $844 - (-20)$ **19.** $-4 - 10$ **20.** $-5 - 8$

21. $-7 - (-3)$ **22.** $-12 - (-5)$ **23.** $17 - 29$ **24.** $16 - 45$

Translating *Translate each phrase; then simplify. See Example 8.*

25. Subtract 17 from -25.

26. Subtract 10 from -22.

27. Find the difference of -22 and -3.

28. Find the difference of -8 and -13.

29. Subtract -12 from 2.

30. Subtract -50 from -50.

Mixed Practice (Sections 2.2, 2.3) *Add or subtract as indicated.*

31. $-37 + (-19)$ **32.** $-35 + (-11)$ **33.** $8 - 13$ **34.** $4 - 21$

35. $-56 - 89$ **36.** $-105 - 68$ **37.** $30 - 67$ **38.** $86 - 98$

Objective **B** *Simplify. See Examples 9 and 10.*

39. $8 - 3 - 2$

40. $8 - 4 - 1$

41. $13 - 5 - 7$

42. $30 - 18 - 12$

43. $-5 - 8 - (-12)$

44. $-10 - 6 - (-9)$

45. $-11 + (-6) - 14$

46. $-15 + (-8) - 4$

47. $18 - (-32) + (-6)$

48. $23 - (-17) + (-9)$

49. $-(-5) - 21 + (-16)$

50. $-(-9) - 14 + (-23)$

51. $-10 - (-12) + (-7) - 4$

52. $-6 - (-8) + (-12) - 7$

53. $-3 + 4 - (-23) - 10$

54. $5 + (-18) - (-21) - 2$

Objective **C** *Evaluate $x - y$ for the given replacement values. See Examples 11 and 12.*

55. $x = -4$ and $y = 7$

56. $x = -7$ and $y = 1$

57. $x = 8$ and $y = -23$

58. $x = 9$ and $y = -2$

Evaluate $2x - y$ for the given replacement values. See Examples 11 and 12.

59. $x = 4$ and $y = -4$

60. $x = 8$ and $y = -10$

61. $x = 1$ and $y = -18$

62. $x = 14$ and $y = -12$

Objective **D** *Solve. See Example 13.*

The bar graph below shows the monthly average temperature in Fairbanks, Alaska. Notice that a negative temperature is illustrated by a bar below the horizontal line representing 0°F. Use this graph to answer Exercises 63 through 66.

63. Find the difference in temperature between the months of March and February.

64. Find the difference in temperature between the months of November and December.

65. Find the difference in temperature between the two months with the lowest temperatures.

66. Find the difference in temperature between the month with the warmest temperature and the month with the coldest temperature.

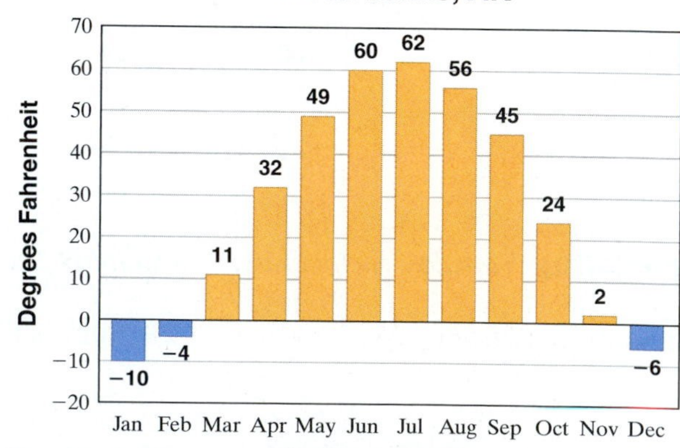

Source: National Climatic Data Center

67. The coldest temperature ever recorded on Earth was −129°F in Antarctica. The warmest temperature ever recorded was 136°F in the Sahara Desert. How many degrees warmer is 136°F than −129°F? (*Source: Questions Kids Ask*, Grolier Limited, 1991, and *The World Almanac*, 2005)

68. The coldest temperature ever recorded in the United States was −80°F in Alaska. The warmest temperature ever recorded was 134°F in California. How many degrees warmer is 134°F than −80°F? (*Source: The World Almanac*, 2005)

Solve.

69. Marta Saarens received a statement of her charge account at Old Navy. She spent \$93 on purchases last month. She returned an \$18 top because she didn't like the color. She also returned a \$26 night shirt because it was damaged. What does she actually owe on her account?

70. Tiger Woods finished the Cialis Western Open golf tournament in 2006 in second place, with a score of -11, or 11 under par. In 82nd place was Nick Watney, with a score of $+8$, or eight over par. What was the difference in scores between Watney and Woods?

71. The temperature on a February morning is $-4°$ Celsius at 6 a.m. If the temperature drops 3 degrees by 7 a.m., rises 4 degrees between 7 a.m. and 8 a.m., and then drops 7 degrees between 8 a.m. and 9 a.m., find the temperature at 9 a.m.

72. Mauna Kea in Hawaii has an elevation of 13,796 feet above sea level. The Mid-America Trench in the Pacific Ocean has an elevation of 21,857 feet below sea level. Find the difference in elevation between those two points. (*Source:* National Geographic Society and Defense Mapping Agency)

Some places on Earth lie below sea level, which is the average level of the surface of the oceans. Use this diagram to answer Exercises 73 through 76. (Source: Fantastic Book of Comparisons, Russell Ash*)*

73. Find the difference in elevation between Death Valley and Quattâra Depression.

74. Find the difference in elevation between Danakil and Turfan Depressions.

75. Find the difference in elevation between the two lowest elevations shown.

76. Find the difference in elevation between the highest elevation shown and the lowest elevation shown.

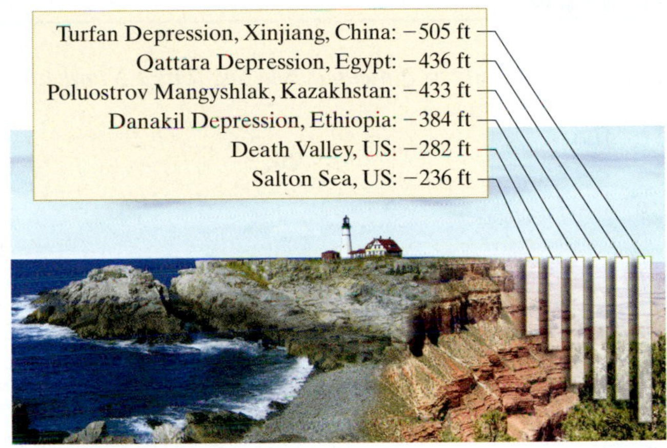

The bar graph shows heights of selected lakes. For Exercises 77 through 80, find the difference in elevation for the lakes listed. (Source: U.S. Geological Survey*)*

77. Lake Superior and Lake Eyre

78. Great Bear Lake and Caspian Sea

79. Lake Maracaibo and Lake Vanern

80. Lake Eyre and Caspian Sea

Elevations of Selected Lakes

Feet Above or Below Sea Level (axis: −200 to 700)

Superior, North America	Ontario, North America	Caspian Sea, Asia-Europe	Maracaibo, South America	Great Bear, North America	Eyre, Australia	Vanern, Europe
600	245	−92	0	512	−52	144

Solve.

81. The surface temperature of the hottest planet, Venus, is 867°F, while the surface temperature of the coldest planet, Neptune, is -330°F. Find the difference in temperatures.

82. The surface temperature of Mercury is 330°F, while the surface temperature of Jupiter is -162°F. Find the difference in temperatures.

83. The difference between a country's exports and imports is called the country's *trade balance*. In 2005, the United States had \$895 billion in exports and \$1677 billion in imports. What was the U.S. trade balance in 2005? (*Source:* U.S. Dept. of Commerce)

84. In 2005, the United States exported 369 million barrels of petroleum products and imported 1267 million barrels of petroleum products. What was the U.S. trade balance for petroleum products in 2005? (*Source:* U.S. Energy Information Administration)

Mixed Practice–Translating (*Sections 2.2, 2.3*) *Translate each phrase to an algebraic expression. Use "x" to represent "a number."*

85. The sum of -5 and a number.

86. The difference of -3 and a number.

87. Subtract a number from -20.

88. Add a number and -36.

Review

Multiply or divide as indicated. See Sections 1.5 and 1.6.

89. $\frac{100}{20}$

90. $\frac{96}{3}$

91. 23×46

92. 51×89

Concept Extensions

93. Name two numbers whose difference is -3.

94. Name two numbers whose difference is -10.

Each calculation below is ***incorrect.*** *Find the error and correct. See the Concept Check in this section.*

95. $9 - (-7) \stackrel{?}{=} 2$

96. $-4 - 8 \stackrel{?}{=} 4$

97. $10 - 30 \stackrel{?}{=} 20$

98. $-3 - (-10) \stackrel{?}{=} -13$

Simplify. (Hint: Find the absolute values first.)

99. $|-3| - |-7|$

100. $|-12| - |-5|$

101. $|-5| - |5|$

102. $|-8| - |8|$

103. $|-15| - |-29|$

104. $|-23| - |-42|$

For Exercises 103 and 104, determine whether each statement is true or false.

105. $|-8 - 3| = 8 - 3$

106. $|-2 - (-6)| = |-2| - |-6|$

107. In your own words, explain how to subtract one signed number from another.

108. A student explains to you that the first step to simplify $8 + 12 \cdot 5 - 100$ is to add 8 and 12. Is the student correct? Explain why or why not.

STUDY SKILLS BUILDER

Organizing a Notebook

It's never too late to get organized. If you need ideas about organizing a notebook for your mathematics course, try some of these:

- Use a spiral or ring binder notebook with pockets and use it for mathematics only.
- Start each page by writing the book's section number you are working on at the top.
- When your instructor is lecturing, take notes. ***Always*** include any examples your instructor works for you.
- Place your worked-out homework exercises in your notebook immediately after the lecture notes from that section. This way, a section's worth of material is together.
- Homework exercises: Attempt all assigned homework. For odd-numbered exercises, you are not through until you check your answers against the back of the book. Correct any exercises with incorrect answers. You may want to place a "?" by any homework exercises or notes that you need to ask questions about. Also, consider placing a "!" by any notes or exercises you feel are important.
- Place graded quizzes in the pockets of your notebook. If you are using a binder, you can place your quizzes in a special section of your binder.

Let's check your notebook organization by answering the following questions.

1. Do you have a spiral or ring binder notebook for your mathematics course only?
2. Have you ever had to flip through several sheets of notes and work in your mathematics notebook to determine what section's work you are in?
3. Are you now writing the textbook's section number at the top of each notebook page?
4. Have you ever lost or had trouble finding a graded quiz or test?
5. Are you now placing all your graded work in a dedicated place in your notebook?
6. Are you attempting all of your homework and placing all of your work in your notebook?
7. Are you checking and correcting your homework in your notebook? If not, why not?
8. Are you writing in your notebook the examples your instructor works for you in class?

Objectives

A Multiply Integers.

B Divide Integers.

C Evaluate an Algebraic Expression by Multiplying or Dividing.

D Solve Problems by Multiplying or Dividing Integers.

2.4 MULTIPLYING AND DIVIDING INTEGERS

Multiplying and dividing integers is similar to multiplying and dividing whole numbers. One difference is that we need to determine whether the result is a positive number or a negative number.

Objective A Multiplying Integers

Consider the following pattern of products.

First factor decreases by 1 each time.

$$3 \cdot 2 = 6$$
$$2 \cdot 2 = 4$$
$$1 \cdot 2 = 2$$
$$0 \cdot 2 = 0$$

Product decreases by 2 each time.

This pattern can be continued, as follows.

$$-1 \cdot 2 = -2$$
$$-2 \cdot 2 = -4$$
$$-3 \cdot 2 = -6$$

This suggests that the product of a negative number and a positive number is a negative number.

What is the sign of the product of two negative numbers? To find out, we form another pattern of products. Again, we decrease the first factor by 1 each time, but this time the second factor is negative.

$$2 \cdot (-3) = -6$$
$$1 \cdot (-3) = -3$$
$$0 \cdot (-3) = 0$$

Product increases by 3 each time.

This pattern continues as:

$$-1 \cdot (-3) = 3$$
$$-2 \cdot (-3) = 6$$
$$-3 \cdot (-3) = 9$$

This suggests that the product of two negative numbers is a positive number. Thus we can determine the sign of a product when we know the signs of the factors.

Multiplying Numbers

The product of two numbers having the same sign is a positive number.

The product of two numbers having different signs is a negative number.

Product of Like Signs

$$(+)(+) = +$$
$$(-)(-) = +$$

Product of Different Signs

$$(-)(+) = -$$
$$(+)(-) = -$$

EXAMPLES Multiply.

1. $-7 \cdot 3 = -21$
2. $-2(-5) = 10$
3. $0 \cdot (-4) = 0$
4. $10(-8) = -80$

Work Practice Problems 1–4

Recall that by the associative and commutative properties for multiplication, we may multiply numbers in any order that we wish. In Example 5, we multiply from left to right.

EXAMPLES Multiply.

5. $7(-6)(-2) = -42(-2)$
$= 84$
6. $(-2)(-3)(-4) = 6(-4)$
$= -24$
7. $(-1)(-2)(-3)(-4) = -1(-24)$ We have -24 from Example 6.
$= 24$

Work Practice Problems 5–7

✔Concept Check What is the sign of the product of five negative numbers? Explain.

Recall from our study of exponents that $2^3 = 2 \cdot 2 \cdot 2 = 8$. We can now work with bases that are negative numbers. For example,

$$(-2)^3 = (-2)(-2)(-2) = -8$$

EXAMPLE 8 Evaluate: $(-5)^2$

Solution: Remember that $(-5)^2$ means 2 factors of -5.

$$(-5)^2 = (-5)(-5) = 25$$

Work Practice Problem 8

Helpful Hint

Have you noticed a pattern when multiplying signed numbers?

If we let $(-)$ represent a negative number and $(+)$ represent a positive number, then

$(-)(-) = (+)$

$(-)(-)(-) = (-)$

$(-)(-)(-)(-) = (+)$

$(-)(-)(-)(-)(-) = (-)$

The product of an even number of negative numbers is a positive result.

The product of an odd number of negative numbers is a negative result.

Notice in Example 8 the parentheses around -5 in $(-5)^2$. With these parentheses, -5 is the base that is squared. Without parentheses, such as -5^2, only the 5 is squared. In other words, $-5^2 = -(5 \cdot 5) = -25$.

PRACTICE PROBLEMS 1–4

Multiply.

1. $-3 \cdot 7$
2. $-5(-2)$
3. $0 \cdot (-20)$
4. $10(-5)$

PRACTICE PROBLEMS 5–7

Multiply.

5. $8(-6)(-2)$
6. $(-9)(-2)(-1)$
7. $(-3)(-4)(-5)(-1)$

PRACTICE PROBLEM 8

Evaluate $(-2)^4$.

Answers

1. -21 2. 10 3. 0 4. -50 5. 96 6. -18 7. 60 8. 16

✔ Concept Check Answer

Negative

PRACTICE PROBLEM 9

Evaluate: -8^2

EXAMPLE 9 Evaluate: -7^2

Solution: Remember that without parentheses, only the 7 is squared.

$-7^2 = -(7 \cdot 7) = -49$

Work Practice Problem 9

Helpful Hint

Make sure you understand the difference between Examples 8 and 9.

parentheses, so -5 is squared

$(-5)^2 = (-5)(-5) = 25$

no parentheses, so only the 7 is squared

$-7^2 = -(7 \cdot 7) = -49$

Objective B Dividing Integers

Division of integers is related to multiplication of integers. The sign rules for division can be discovered by writing a related multiplication problem. For example,

$\frac{6}{2} = 3$ because $3 \cdot 2 = 6$

$\frac{-6}{2} = -3$ because $-3 \cdot 2 = -6$

$\frac{6}{-2} = -3$ because $-3 \cdot (-2) = 6$

$\frac{-6}{-2} = 3$ because $3 \cdot (-2) = -6$

Helpful Hint

Just as for whole numbers, division can be checked by multiplication.

Dividing Numbers

The quotient of two numbers having the same sign is a positive number.

The quotient of two numbers having different signs is a negative number.

Quotient of Like Signs

$$\frac{(+)}{(+)} = + \qquad \frac{(-)}{(-)} = +$$

Quotient of Different Signs

$$\frac{(+)}{(-)} = - \qquad \frac{(-)}{(+)} = -$$

PRACTICE PROBLEMS 10–12

Divide.

10. $\frac{42}{-7}$

11. $-16 \div (-2)$

12. $\frac{-80}{10}$

EXAMPLES Divide.

10. $\frac{-12}{6} = -2$

11. $-20 \div (-4) = 5$

12. $\frac{48}{-3} = -16$

Work Practice Problems 10–12

✔ **Concept Check** What is wrong with the following calculation?

$\frac{-36}{-9} = -4$

Answers

9. -64 10. -6 11. 8 12. -8

✔ Concept Check Answer

$\frac{-36}{-9} = 4$

EXAMPLES Divide, if possible.

13. $\frac{0}{-5} = 0$ because $0 \cdot -5 = 0$

14. $\frac{-7}{0}$ is undefined because there is no number that gives a product of -7 when multiplied by 0.

■ **Work Practice Problems 13–14**

PRACTICE PROBLEMS 13–14

Divide, if possible.

13. $\frac{-6}{0}$ **14.** $\frac{0}{-7}$

Objective C Evaluating Expressions

Next, we practice evaluating expressions given integer replacement values.

EXAMPLE 15 Evaluate xy for $x = -2$ and $y = 7$.

Solution: Recall that xy means $x \cdot y$.

Replace x with -2 and y with 7.

$$xy = -2 \cdot 7$$
$$= -14$$

■ **Work Practice Problem 15**

PRACTICE PROBLEM 15

Evaluate xy for $x = 5$ and $y = -8$.

EXAMPLE 16 Evaluate $\frac{x}{y}$ for $x = -24$ and $y = 6$.

Solution: $\frac{x}{y} = \frac{-24}{6}$ Replace x with -24 and y with 6.

$$= -4$$

■ **Work Practice Problem 16**

PRACTICE PROBLEM 16

Evaluate $\frac{x}{y}$ for $x = -12$ and $y = -3$.

Objective D Solving Problems by Multiplying and Dividing Integers

Many real-life problems involve multiplication and division of signed numbers.

EXAMPLE 17 **Calculating Total Golf Score**

A professional golfer finished seven strokes under par (-7) for each of three days of a tournament. What was her total score for the tournament?

Solution:

1. UNDERSTAND. Read and reread the problem. Although the key word is "total," since this is repeated addition of the same number we multiply.
2. TRANSLATE.

In words:	golfer's total score	=	number of days	·	score each day
	↓	↓	↓	↓	↓
Translate:	golfer's total	=	3	·	(−7)

3. SOLVE: $3 \cdot (-7) = -21$
4. INTERPRET. Check and state your conclusion: The golfer's total score is -21, or 21 strokes under par.

■ **Work Practice Problem 17**

PRACTICE PROBLEM 17

A card player had a score of -13 for each of four games. Find her total score.

Answers

13. undefined **14.** 0 **15.** −40 **16.** 4 **17.** −52

Vocabulary and Readiness Check

Use the choices below to fill in each blank. Each choice may be used more than once.

negative	0
positive	undefined

1. The product of a negative number and a positive number is a ______________ number.
2. The product of two negative numbers is a ______________ number.
3. The quotient of two negative numbers is a ______________ number.
4. The quotient of a negative number and a positive number is a ______________ number.
5. The product of a negative number and zero is ______________.
6. The quotient of 0 and a negative number is ______________.
7. The quotient of a negative number and 0 is ______________.

2.4 EXERCISE SET

FOR EXTRA HELP

Objective A *Multiply. See Examples 1 through 4.*

1. $-6(-2)$ **2.** $5(-3)$ **3.** $-4(9)$ **4.** $-7(-2)$

5. $9(-9)$ **6.** $-9(9)$ **7.** $0(-11)$ **8.** $-6(0)$

Multiply. See Examples 5 through 7.

9. $6(-2)(-4)$ **10.** $-2(3)(-7)$ **11.** $-1(-3)(-4)$ **12.** $-8(-3)(-3)$

13. $-4(4)(-5)$ **14.** $2(-5)(-4)$ **15.** $10(-5)(0)(-7)$ **16.** $3(0)(-4)(-8)$

17. $-5(3)(-1)(-1)$ **18.** $-2(-1)(3)(-2)$ -12

Evaluate. See Examples 8 and 9.

19. -3^2 **20.** -2^4 **21.** $(-3)^3$ **22.** $(-1)^4$

23. -6^2 **24.** -4^3 **25.** $(-4)^3$ **26.** $(-3)^2$

Objective B *Find each quotient. See Examples 10 through 14.*

27. $-24 \div 3$ **28.** $90 \div (-9)$ **29.** $\frac{-30}{6}$ **30.** $\frac{56}{-8}$

31. $\frac{-77}{-11}$ **32.** $\frac{-32}{4}$ **33.** $\frac{0}{-21}$ **34.** $\frac{-13}{0}$

35. $\frac{-10}{0}$ **36.** $\frac{0}{-15}$ **37.** $\frac{56}{-4}$ **38.** $\frac{-24}{-12}$

Objectives **A** **B** **Mixed Practice** *Multiply or divide as indicated.*

39. $-14(0)$ **40.** $0(-100)$ **41.** $-5(3)$ **42.** $-6 \cdot 2$

43. $-9 \cdot 7$ **44.** $-12(13)$ **45.** $-7(-6)$ **46.** $-9(-5)$

47. $-3(-4)(-2)$ **48.** $-7(-5)(-3)$ **49.** $(-7)^2$ **50.** $(-5)^2$

51. $-\frac{25}{5}$ **52.** $-\frac{30}{5}$ **53.** $-\frac{72}{8}$ **54.** $-\frac{49}{7}$

55. $-18 \div 3$ **56.** $-15 \div 3$ **57.** $4(-10)(-3)$ **58.** $6(-5)(-2)$

59. $-30(6)(-2)(-3)$ **60.** $-20 \cdot 5 \cdot (-5) \cdot (-3)$ **61.** $3 \cdot (-8) \cdot 0$ **62.** $-(4)(0)$

63. $\frac{120}{-20}$ **64.** $\frac{63}{-9}$ **65.** $280 \div (-40)$ **66.** $480 \div (-8)$

67. $\frac{-12}{-4}$ **68.** $\frac{-36}{-3}$ **69.** -1^4 **70.** -2^3

71. $(-2)^5$ **72.** $(-11)^2$ **73.** $-2(3)(5)(-6)$ **74.** $-1(2)(7)(-3)$

75. $(-1)^{32}$ **76.** $(-1)^{33}$ **77.** $-2(-3)(-5)$ **78.** $-2(-2)(-3)(-2)$

79. $-48 \cdot 23$ **80.** $-56 \cdot 43$ **81.** $35 \cdot (-82)$ **82.** $70 \cdot (-23)$

Objective **C** *Evaluate ab for the given replacement values. See Example 15.*

83. $a = -8$ and $b = 7$ **84.** $a = 5$ and $b = -1$ **85.** $a = 9$ and $b = -2$

86. $a = -9$ and $b = -6$ **87.** $a = -7$ and $b = -5$ **88.** $a = -8$ and $b = 8$

Evaluate $\frac{x}{y}$ for the given replacement values. See Example 16.

89. $x = 5$ and $y = -5$ **90.** $x = 9$ and $y = -3$ **91.** $x = -15$ and $y = 0$

92. $x = 0$ and $y = -5$ **93.** $x = -36$ and $y = -6$ **94.** $x = -10$ and $y = -10$

Evaluate xy and also $\frac{x}{y}$ for the given replacement values.

95. $x = -8$ and $y = -2$ **96.** $x = 20$ and $y = -5$ **97.** $x = 0$ and $y = -8$ **98.** $x = -3$ and $y = 0$

Objective **D** **Translating** *Translate each phrase; then simplify. See Example 17.*

99. Find the quotient of −54 and 9.

100. Find the quotient of −63 and −3.

101. Find the product of −42 and −6.

102. Find the product of −49 and 5.

Translating *Translate each phrase to an expression. Use x to represent a number. See Example 17.*

103. The product of −71 and a number

104. The quotient of −8 and a number

105. Subtract a number from −16.

106. The sum of a number and −12

107. −29 increased by a number

108. The difference of a number and −10

109. Divide a number by −33.

110. Multiply a number by −17.

Solve. See Example 17.

111. A football team lost four yards on each of three consecutive plays. Represent the total loss as a product of signed numbers and find the total loss.

112. Joe Norstrom lost $400 on each of seven consecutive days in the stock market. Represent his total loss as a product of signed numbers and find his total loss.

113. A deep-sea diver must move up or down in the water in short steps in order to keep from getting a physical condition called the "bends." Suppose a diver moves down from the surface in five steps of 20 feet each. Represent his total movement as a product of signed numbers and find the product.

114. A weather forecaster predicts that the temperature will drop five degrees each hour for the next six hours. Represent this drop as a product of signed numbers and find the total drop in temperature.

The graph shows melting points in degrees Celsius of selected elements. Use this graph to answer Exercises 115 through 118.

115. The melting point of nitrogen is 3 times the melting point of radon. Find the melting point of nitrogen.

116. The melting point of rubidium is −1 times the melting point of mercury. Find the melting point of rubidium.

117. The melting point of argon is −3 times the melting point of potassium. Find the melting point of argon.

118. The melting point of strontium is −11 times the melting point of radon. Find the melting point of strontium.

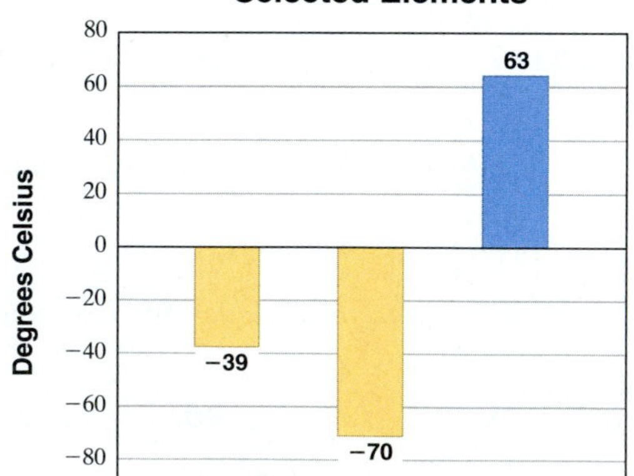

119. During the first quarter of 2006, Ford Motor Company's North American automotive operations posted a pretax income of −$457 million. If this continues, what will Ford's North American automotive operations' net income be after four quarters? (*Source:* Ford Motor Company)

120. At the end of 2005, United Airlines posted a full year net income of −$21,176 million. If the income rate was consistent over the entire year, how much would you expect United's net income to be for each quarter? (*Source:* United Airlines)

121. In 2001, approximately 626 million prerecorded VHS movie cassettes were shipped to retailers in the United States. In 2005, this number dropped to approximately 50 million cassettes. (*Source:* Motion Picture Association: Worldwide Market Research)

a. Find the change in the number of VHS cassettes shipped to retailers from 2001 to 2005.

b. Find the average change per year in the number of VHS cassettes shipped to retailers over this period.

122. In 1987, there were only 27 California Condors in the entire world. Thanks to conservation efforts, in 2006 there were 293 California Condors. (*Source:* Arizona Game and Fish)

a. Find the change in the number of California Condors from 1987 to 2006.

b. Find the average change per year in the California Condor population over the period in part a.

Review

Perform each indicated operation. See Section 1.7.

123. $90 + 12^2 - 5^3$

124. $3 \cdot (7 - 4) + 2 \cdot 5^2$

125. $12 \div 4 - 2 + 7$

126. $12 \div (4 - 2) + 7$

Concept Extensions

Mixed Practice (*Sections 2.2, 2.3, 2.4*) *Perform indicated operations.*

127. $-57 \div 3$

128. $-9(-11)$

129. $-8 - 20$

130. $-4 + (-3) + 21$

131. $-4 - 15 - (-11)$

132. $-16 - (-2)$

Solve. For Exercises 133 and 134, see the first Concept Check in this section.

133. What is the sign of the product of seven negative numbers?

134. What is the sign of the product of ten negative numbers?

Without actually finding the product, write the list of numbers in Exercises 135 and 136 in order from least to greatest. For help, see a helpful hint box in this section.

135. $(-2)^{12}, (-2)^{17}, (-5)^{12}, (-5)^{17}$

136. $(-1)^{50}, (-1)^{55}, 0^{15}, (-7)^{20}, (-7)^{23}$

137. In your own words, explain how to divide two integers.

138. In your own words, explain how to multiply two integers.

THE BIGGER PICTURE Operations on Sets of Numbers

Continue your outline from Sections 1.6 and 1.7. Suggestions are once again written to help you complete this part of your outline. Notice that this part of the outline has to do with operations on integers.

I. Operations on Sets of Numbers

A. Whole Numbers

1. **Add or Subtract** (Section 1.3)
2. **Multiply or Divide** (Sections 1.5, 1.6)
3. **Exponent** (Section 1.7)
4. **Order of Operations** (Section 1.7)

B. Integers

1. **Add:**

   $-5 + (-2) = -7$ Adding like signs. Add absolute value. Attach the common sign.

   $-5 + 2 = -3$ Adding unlike signs. Subtract absolute values. Attach the sign of the number with the larger absolute value.

2. **Subtract:** Add the first number to the opposite of the second number.

   $7 - 10 = 7 + (-10) = -3$

3. **Multiply or Divide:** Multiply or divide as usual. If the signs of the two numbers are the same, the answer is positive. If the signs of the two numbers are different, the answer is negative.

   $-5 \cdot 5 = -25, \quad \frac{-32}{-8} = 4$

Perform the indicated operations.

1. $-9 + 14$
2. $-5(-11)$
3. $5 - 11$
4. $58 - |-70|$
5. $18 + (-30)$
6. $(-9)^2$
7. -9^2
8. $-10 + (-24)$
9. $1 - (-9)$
10. $-15 - 15$
11. $-3(2)(-5)$
12. $\dfrac{|-88|}{-|-8|}$
13. $2 + 4(7 - 9)^3$
14. $-100 - (-20)$
15. $1 + 2(7)$
16. $30 \div 2 \cdot 3$

INTEGRATED REVIEW Sections 2.1–2.4

Integers

Represent each quantity by an integer.

1. The record low temperature in New Mexico is 50 degrees Fahrenheit below zero. The highest temperature in that state is 122 degrees above zero. Represent each quantity by an integer.

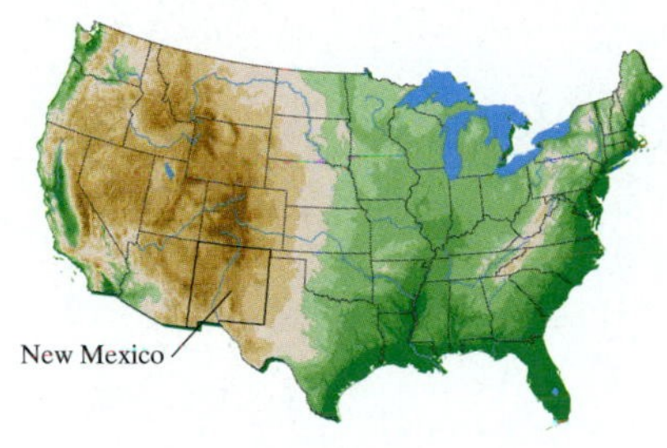

2. Graph the signed numbers on the given number line. $-4, 0, -1, 3$

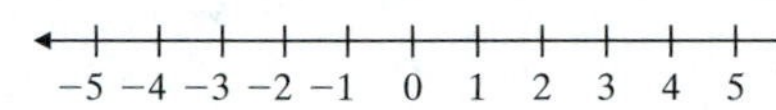

Insert < or > between each pair of numbers to make a true statement.

3. $0 \quad -10$ **4.** $-15 \quad -5$ **5.** $-4 \quad 4$ **6.** $-2 \quad -7$

Simplify.

7. $|-3|$ **8.** $-|-4|$ **9.** $|-9|$ **10.** $-(-5)$

Find the opposite of each number.

11. 11 **12.** -3 **13.** 64 **14.** 0

Perform the indicated operation.

15. $-3 + 15$ **16.** $-9 + (-11)$ **17.** $-8(-6)(-1)$ **18.** $-18 \div 2$

19. $65 + (-55)$ **20.** $1000 - 1002$ **21.** $53 - (-53)$ **22.** $-2 - 1$

Answers

1. ______
2. see number line
3. ______
4. ______
5. ______
6. ______
7. ______
8. ______
9. ______
10. ______
11. ______
12. ______
13. ______
14. ______
15. ______
16. ______
17. ______
18. ______
19. ______
20. ______
21. ______
22. ______

23. ______
24. ______
25. ______
26. ______
27. ______
28. ______
29. ______
30. ______
31. ______
32. ______
33. ______
34. ______
35. ______
36. ______
37. ______
38. ______
39. ______
40. ______
41. ______
42. ______
43. ______
44. ______
45. ______
46. ______

23. $\frac{0}{-47}$

24. $\frac{-36}{-9}$

25. $-17 - (-59)$

26. $-8 + (-6) + 20$

27. $\frac{-95}{-5}$

28. $-9(100)$

29. $-12 - 6 - (-6)$

30. $-4 + (-8) - 16 - (-9)$

31. $\frac{-105}{0}$

32. $7(-16)(0)(-3)$

Translating *Translate each phrase; then simplify.*

33. Subtract -8 from -12.

34. Find the sum of -17 and -27.

35. Find the product of -5 and -25.

36. Find the quotient of -100 and -5.

Translating *Translate each phrase to an expression. Use x to represent a number.*

37. Divide a number by -17

38. The sum of -3 and a number

39. A number decreased by -18

40. The product of -7 and a number

Evaluate the expressions below for $x = -3$ and $y = 12$.

41. $x + y$

42. $x - y$

43. $2y - x$

44. $3y + x$

45. $5x$

46. $\frac{y}{x}$

2.5 ORDER OF OPERATIONS

Objectives

A Simplify Expressions by Using the Order of Operations.

B Evaluate an Algebraic Expression.

C Find the Average of a List of Numbers.

Objective A Simplifying Expressions

We first discussed the order of operations in Chapter 1. In this section, you are given an opportunity to practice using the order of operations when expressions contain signed numbers. The rules for the order of operations from Section 1.7 are repeated here.

Order of Operations

1. Perform all operations within parentheses ( ), brackets [ ], or other grouping symbols such as fraction bars, starting with the innermost set.
2. Evaluate any expressions with exponents.
3. Multiply or divide in order from left to right.
4. Add or subtract in order from left to right.

Before simplifying other expressions, make sure you are confident simplifying Examples 1 through 3.

EXAMPLES Find the value of each expression.

1. $(-3)^2 = (-3)(-3) = 9$ The base of the exponent is -3.
2. $-3^2 = -(3)(3) = -9$ The base of the exponent is 3.
3. $2 \cdot 5^2 = 2 \cdot (5 \cdot 5) = 2 \cdot 25 = 50$ The base of the exponent is 5.

Work Practice Problems 1–3

PRACTICE PROBLEMS 1–3

Find the value of each expression.

1. $(-2)^4$
2. -2^4
3. $3 \cdot 6^2$

Helpful Hint

When simplifying expressions with exponents, remember that parentheses make an important difference.

$(-3)^2$ and -3^2 **do not** mean the same thing.

$(-3)^2$ means $(-3)(-3) = 9$.

-3^2 means the opposite of $3 \cdot 3$, or -9.

Only with parentheses around it is the -3 squared.

EXAMPLE 4 Simplify: $\dfrac{-6(2)}{-3}$

Solution: First we multiply -6 and 2. Then we divide.

$$\frac{-6(2)}{-3} = \frac{-12}{-3}$$
$$= 4$$

Work Practice Problem 4

PRACTICE PROBLEM 4

Simplify: $\dfrac{-25}{5(-1)}$

Answers

1. 16 **2.** -16 **3.** 108 **4.** 5

PRACTICE PROBLEM 5

Simplify: $\dfrac{-18 + 6}{-3 - 1}$

EXAMPLE 5 Simplify: $\dfrac{12 - 16}{-1 + 3}$

Solution: We simplify above and below the fraction bar separately. Then we divide.

$$\frac{12 - 16}{-1 + 3} = \frac{-4}{2}$$
$$= -2$$

Work Practice Problem 5

PRACTICE PROBLEM 6

Simplify: $30 + 50 + (-4)^3$

EXAMPLE 6 Simplify: $60 + 30 + (-2)^3$

Solution: $60 + 30 + (-2)^3 = 60 + 30 + (-8)$ Write $(-2)^3$ as -8.

$= 90 + (-8)$ Add from left to right.

$= 82$

Work Practice Problem 6

PRACTICE PROBLEM 7

Simplify: $-2^3 + (-4)^2 + 1^5$

EXAMPLE 7 Simplify: $-4^2 + (-3)^2 - 1^3$

Solution:

$-4^2 + (-3)^2 - 1^3 = -16 + 9 - 1$ Simplify expressions with exponents.

$= -7 - 1$ Add or subtract from left to right.

$= -8$

Work Practice Problem 7

PRACTICE PROBLEM 8

Simplify:
$2(2 - 9) + (-12) - 3$

EXAMPLE 8 Simplify: $3(4 - 7) + (-2) - 5$

Solution:

$3(4 - 7) + (-2) - 5 = 3(-3) + (-2) - 5$ Simplify inside parentheses.

$= -9 + (-2) - 5$ Multiply.

$= -11 - 5$ Add or subtract from left to right.

$= -16$

Work Practice Problem 8

PRACTICE PROBLEM 9

Simplify:
$(-5) \cdot |-8| + (-3) + 2^3$

EXAMPLE 9 Simplify: $(-3) \cdot |-5| - (-2) + 4^2$

Solution:

$(-3) \cdot |-5| - (-2) + 4^2 = (-3) \cdot 5 - (-2) + 4^2$ Write $|-5|$ as 5.

$= (-3) \cdot 5 - (-2) + 16$ Write 4^2 as 16.

$= -15 - (-2) + 16$ Multiply.

$= -13 + 16$ Add or subtract from left to right.

$= 3$

Work Practice Problem 9

Answers

5. 3 **6.** 16 **7.** 9 **8.** −29 **9.** −35

EXAMPLE 10 Simplify: $-2[-3 + 2(-1 + 6)] - 5$

Solution: Here we begin with the innermost set of parentheses.

$$\begin{aligned} -2[-3 + 2(-1 + 6)] - 5 &= -2[-3 + 2(5)] - 5 && \text{Write } -1 + 6 \text{ as } 5. \\ &= -2[-3 + 10] - 5 && \text{Multiply.} \\ &= -2(7) - 5 && \text{Add.} \\ &= -14 - 5 && \text{Multiply.} \\ &= -19 && \text{Subtract.} \end{aligned}$$

Work Practice Problem 10

PRACTICE PROBLEM 10

Simplify:
$-4[-6 + 5(-3 + 5)] - 7$

✔**Concept Check** True or false? Explain your answer. The result of

$-4(3 - 7) - 8(9 - 6)$

is positive because there are four negative signs.

Objective B Evaluating Expressions

Now we practice evaluating expressions.

EXAMPLE 11 Evaluate x^2 and $-x^2$ for $x = -11$.

Solution: $x^2 = (-11)^2 = (-11)(-11) = 121$

$-x^2 = -(-11)^2 = -(-11)(-11) = -121$

Work Practice Problem 11

PRACTICE PROBLEM 11

Evaluate x^2 and $-x^2$ for $x = -15$.

EXAMPLE 12 Evaluate $6z^2$ for $z = 2$ and $z = -2$.

Solution: $6z^2 = 6(2)^2 = 6(4) = 24$

$6z^2 = 6(-2)^2 = 6(4) = 24$

Work Practice Problem 12

PRACTICE PROBLEM 12

Evaluate $5y^2$ for $y = 4$ and $y = -4$.

EXAMPLE 13 Evaluate $x + 2y - z$ for $x = 3$ and $y = -5$ and $z = -4$.

Solution: Replace x with 3, y with -5, z with -4 and simplify.

Helpful Hint Remember to rewrite the subtraction sign.

$$\begin{aligned} x + 2y - z &= 3 + 2(-5) - (-4) && \text{Let } x = 3, y = -5, \text{ and } z = -4. \\ &= 3 + (-10) + 4 && \text{Replace } 2(-5) \text{ with its product, } -10. \\ &= -3 && \text{Add.} \end{aligned}$$

Work Practice Problem 13

PRACTICE PROBLEM 13

Evaluate $x^2 + y$ for $x = -6$ and $y = -3$.

Answers

10. -23 **11.** 225; -225 **12.** 80; 80 **13.** 33

✔ **Concept Check Answer**

false; $-4(3 - 7) - 8(9 - 6) = -8$

PRACTICE PROBLEM 14

Evaluate $4 - x^2$ for $x = -8$.

EXAMPLE 14 Evaluate $7 - x^2$ for $x = -4$.

Solution: Replace x with -4 and simplify carefully!

$$\begin{aligned} 7 - x^2 &= 7 - (-4)^2 \\ &= 7 - 16 \quad && (-4)^2 = (-4)(-4) = 16 \\ &= -9 && \text{Subtract.} \end{aligned}$$

Work Practice Problem 14

Objective C Finding Averages

Recall from Chapter 1 that the average of a list of numbers is

$$\text{average} = \frac{\text{sum of numbers}}{\textit{number} \text{ of numbers}}$$

PRACTICE PROBLEM 15

Find the average of the temperatures for the months October through April.

EXAMPLE 15 The graph shows the monthly normal temperatures for Barrow, Alaska. Use this graph to find the average of the temperatures for months January through May.

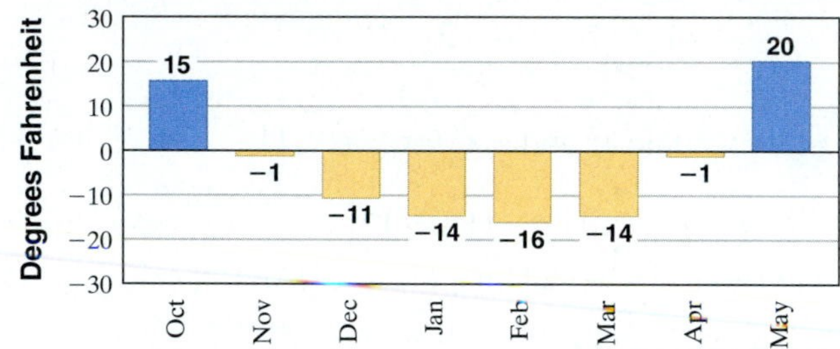

Solution: By reading the graph, we have

$$\begin{aligned} \text{average} &= \frac{-14 + (-16) + (-14) + (-1) + 20}{5} \quad && \text{There are 5 months from January through May.} \\ &= \frac{-25}{5} \\ &= -5 \end{aligned}$$

The average for the temperatures is -5°F.

Work Practice Problem 15

Answers

14. -60 **15.** -6°F

CALCULATOR EXPLORATIONS Simplifying an Expression Containing a Fraction Bar

Recall that even though most calculators follow the order of operations, parentheses must sometimes be inserted. For example, to simplify $\frac{-8 + 6}{-2}$ on a calculator, enter parentheses about the expression above the fraction bar so that it is simplified separately.

To simplify $\frac{-8 + 6}{-2}$, press the keys

[( ] [8] [+/−] [+] [6] [ )] [÷] [2] [+/−] [=] or

[( ] [(−)] [8] [+] [6] [ )] [÷] [(−)] [2] [ENTER]

The display will read [1].

Thus, $\frac{-8 + 6}{-2} = 1$.

Use a calculator to simplify.

1. $\frac{-120 - 360}{-10}$

2. $\frac{4750}{-2 + (-17)}$

3. $\frac{-316 + (-458)}{28 + (-25)}$

4. $\frac{-234 + 86}{-18 + 16}$

Vocabulary and Readiness Check

Use the choices below to fill in each blank. Not all choices will be used.

average	subtraction	division	$-7 - 3(1)$
addition	multiplication	$-7 - 3(-1)$	

1. To simplify $-2 \div 2 \cdot (3)$, which operation should be performed first? ____________
2. To simplify $-9 - 3 \cdot 4$, which operation should be performed first? ____________
3. The ____________ of a list of numbers is $\frac{\text{sum of numbers}}{\textit{number} \text{ of numbers}}$.
4. To simplify $5[-9 + (-3)] \div 4$, which operation should be performed first? ____________
5. To simplify $-2 + 3(10 - 12) \cdot (-8)$, which operation should be performed first? ____________
6. To evaluate $x - 3y$ for $x = -7$ and $y = -1$, replace x with -7 and y with -1 and evaluate ____________.

2.5 EXERCISE SET

FOR EXTRA HELP

Student Solutions Manual

PH Math/Tutor Center

CD/Video for Review

MathXL®

MyMathLab

Objective A *Simplify. See Examples 1 through 10.*

1. $(-5)^3$
2. -2^4
3. -4^3
4. $(-2)^4$
5. $8 \cdot 2^2$
6. $5 \cdot 2^3$
7. $8 - 12 - 4$
8. $10 - 23 - 12$
9. $7 + 3(-6)$
10. $-8 + 4(3)$
11. $5(-9) + 2$
12. $7(-6) + 3$
13. $(-10) + 4 \div 2$
14. $(-12) + 6 \div 3$
15. $6 + 7 \cdot 3 - 10$
16. $5 + 9 \cdot 4 - 20$
17. $\frac{16 - 13}{-3}$
18. $\frac{20 - 15}{-1}$
19. $\frac{24}{10 + (-4)}$
20. $\frac{88}{-8 - 3}$
21. $5(-3) - (-12)$
22. $7(-4) - (-6)$
23. $[8 + (-4)]^2$
24. $[9 + (-2)]^3$
25. $8 \cdot 6 - 3 \cdot 5 + (-20)$
26. $7 \cdot 6 - 6 \cdot 5 + (-10)$
27. $4 - (-3)^4$
28. $7 - (-5)^2$
29. $|7 + 3| \cdot 2^3$
30. $|-3 + 7| \cdot 7^2$
31. $7 \cdot 6^2 + 4$
32. $10 \cdot 5^3 + 7$
33. $7^2 - (4 - 2^3)$
34. $8^2 - (5 - 2)^4$
35. $|3 - 15| \div 3$
36. $|12 - 19| \div 7$
37. $-(-2)^6$
38. $-(-2)^3$
39. $(5 - 9)^2 \div (4 - 2)^2$
40. $(2 - 7)^2 \div (4 - 3)^4$

41. $|8 - 24| \cdot (-2) \div (-2)$

42. $|3 - 15| \cdot (-4) \div (-16)$

43. $(-12 - 20) \div 16 - 25$

44. $(-20 - 5) \div 5 - 15$

45. $5(5 - 2) + (-5)^2 - 6$

46. $3 \cdot (8 - 3) + (-4) - 10$

47. $(2 - 7) \cdot (6 - 19)$

48. $(4 - 12) \cdot (8 - 17)$

49. $(-36 \div 6) - (4 \div 4)$

50. $(-4 \div 4) - (8 \div 8)$

51. $(10 - 4^2)^2$

52. $(11 - 3^2)^3$

53. $2(8 - 10)^2 - 5(1 - 6)^2$

54. $-3(4 - 8)^2 + 5(14 - 16)^3$

55. $3(-10) \div [5(-3) - 7(-2)]$

56. $12 - [7 - (3 - 6)] + (2 - 3)^3$

57. $\dfrac{(-7)(-3) - (4)(3)}{3[7 \div (3 - 10)]}$

58. $\dfrac{10(-1) - (-2)(-3)}{2[-8 \div (-2 - 2)]}$

59. $-3[5 + 2(-4 + 9)] + 15$

60. $-2[6 + 4(2 - 8)] - 25$

Objective **B** *Evaluate each expression for $x = -2$, $y = 4$, and $z = -1$. See Examples 11 through 14.*

61. $x + y + z$

62. $x - y - z$

63. $2x - 3y - 4z$

64. $5x - y + 4z$

65. $x^2 - y$

66. $x^2 + z$

67. $\dfrac{5y}{z}$

68. $\dfrac{4x}{y}$

Evaluate each expression for $x = -3$ and $z = -4$. See Examples 11 through 14.

69. x^2

70. z^2

71. $-z^2$

72. $-x^2$

73. $2z^3$

74. $3x^2$

75. $10 - x^2$

76. $3 - z^2$

77. $2x^3 - z$

78. $3z^2 - x$

Objective **C** *Find the average of each list of numbers. See Example 15.*

79. −10, 8, −4, 2, 7, −5, −12

80. −18, −8, −1, −1, 0, 4

81. −17, −26, −20, −13

82. −40, −20, −10, −15, −5

Scores in golf can be 0 (also called par), a positive integer (also called above par), or a negative integer (also called below par). The bar graph shows scores of selected golfers from a tournament. Use this graph for Exercises 83 through 88.

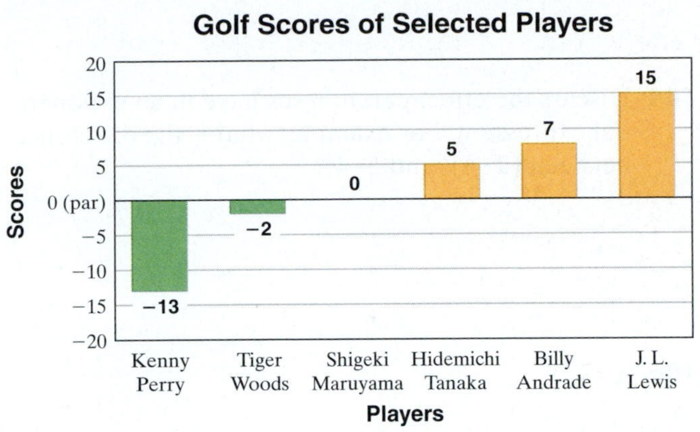

83. Find the difference between the lowest score shown and the highest score shown.

84. Find the difference between the two lowest scores.

85. Find the average of the scores shown. (*Hint*: Here, the average is the sum of the scores divided by the number of players.)

86. Find the average of the scores for Perry, Woods, Maruyama, and Lewis.

87. Can the average for these scores be greater than the highest score, 15? Explain why or why not.

88. Can the average of all the scores shown be less than the lowest score, −13? Explain why or why not.

Review

Perform each indicated operation. See Sections 1.3, 1.5, and 1.6.

89. $45 \cdot 90$

90. $90 \div 45$

91. $90 - 45$

92. $45 + 90$

Find the perimeter of each figure. See Section 1.3.

△ **93.** Square

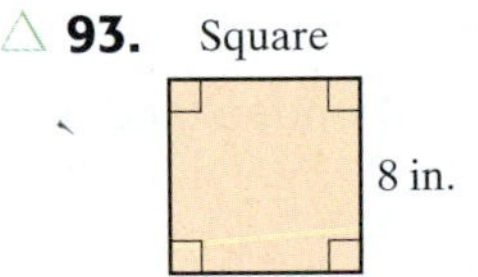

△ **94.** Parallelogram

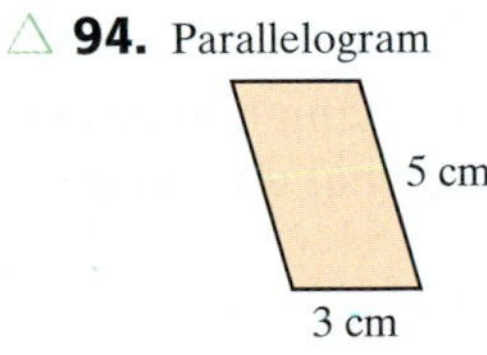

△ **95.** Rectangle

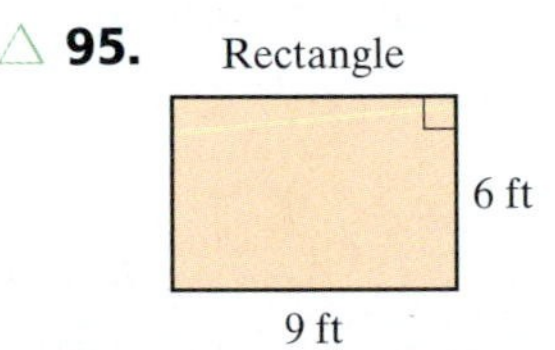

△ **96.** Triangle

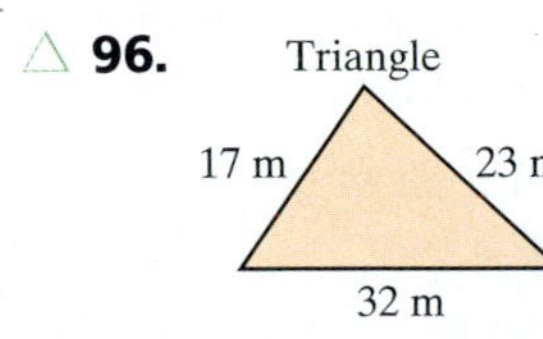

Concept Extensions

Insert parentheses where needed so that each expression evaluates to the given number.

97. $2 \cdot 7 - 5 \cdot 3$; evaluates to 12

98. $7 \cdot 3 - 4 \cdot 2$; evaluates to 34

99. $-6 \cdot 10 - 4$; evaluates to -36

100. $2 \cdot 8 \div 4 - 20$; evaluates to -36

101. Are parentheses necessary in the expression $3 + (4 \cdot 5)$? Explain your answer.

102. Are parentheses necessary in the expression $(3 + 4) \cdot 5$? Explain your answer.

103. Discuss the effect parentheses have in an exponential expression. For example, what is the difference between $(-6)^2$ and -6^2?

104. Discuss the effect parentheses have in an exponential expression. For example, what is the difference between $(2 \cdot 4)^2$ and $2 \cdot 4^2$?

Evaluate.

105. $(-12)^4$

106. $(-17)^6$

107. $x^3 - y^2$ for $x = 21$ and $y = -19$

108. $3x^2 + 2x - y$ for $x = -18$ and $y = 2868$

109. $(xy + z)^x$ for $x = 2$, $y = -5$, and $z = 7$

110. $5(ab + 3)^b$ for $a = -2$, $b = 3$

STUDY SKILLS BUILDER

How Are Your Homework Assignments Going?

It is very important in mathematics to keep up with homework. Why? Many concepts build on each other. Often your understanding of a day's concepts depends on an understanding of the previous day's material.

Remember that completing your homework assignment involves a lot more than attempting a few of the problems assigned.

To complete a homework assignment, remember these four things:

- Attempt all of it.
- Check it.
- Correct it.
- If needed, ask questions about it.

Take a moment and review your completed homework assignments. Answer the questions below based on this review.

1. Approximately how much of your homework have you attempted?
2. Approximately how much of your homework have you checked (if possible)?
3. If you are able to check your homework, have you corrected it when errors have been found?
4. When working homework, if you do not understand a concept, what do you do?

2.6 SOLVING EQUATIONS: THE ADDITION AND MULTIPLICATION PROPERTIES

Objectives

A Identify Solutions of Equations.

B Use the Addition Property of Equality to Solve Equations.

C Use the Multiplication Property of Equality to Solve Equation.

In this section, we introduce properties of equations and we use these properties to begin solving equations. Now that we know how to perform operations on integers, this is an excellent way to practice these operations.

First, let's recall the difference between an equation and an expression. From Section 1.8, a combination of operations on variables and numbers is an expression and an equation is of the form "expression = expression."

Equations	Expressions
$3x - 1 = -17$	$3x - 1$
area = length · width	$5(20-3) + 10$
$8 + 16 = 16 + 8$	y^3
$-9a + 11b = 14b + 3$	$-x^2 + y - 2$

Helpful Hint

Simply stated, an equation contains "=" while an expression does not. Also, we *simplify* expressions and *solve* equations.

Objective A Identifying Solutions of Equations

Let's practice identifying solutions of equations. Recall from Section 1.8 that a solution of an equation is a number that when substituted for a variable makes the equation a true statement.

For example,

-8 is a solution of

$\frac{x}{2} = -4$, because

$\frac{-8}{2} = -4$, or

$-4 = -4$ is true.

Also,

-8 is not a solution of

$x + 6 = 2$, because

$-8 + 6 = 2$ is false.

Let's practice determining whether a number is a solution of an equation. In this section, we will be performing operations on integers.

EXAMPLE 1 Determine whether -1 is a solution of the equation $3y + 1 = 3$.

Solution:

$$3y + 1 = 3$$
$$3(-1) + 1 \stackrel{?}{=} 3$$
$$-3 + 1 \stackrel{?}{=} 3$$
$$-2 = 3 \quad \text{False.}$$

Since $-2 = 3$ is false, -1 is *not* a solution of the equation.

Work Practice Problem 1

PRACTICE PROBLEM 1

Determine whether -2 is a solution of the equation $-4x - 3 = 5$.

Now we know how to check whether a number is a solution. But, given an equation, how do we find its **solution?** In other words, how do we find a number that makes the equation true? How do we solve an equation?

Answer

1. yes

Objective B Using the Addition Property to Solve Equations

To solve an equation, we use properties of equality to write simpler equations, all equivalent to the original equation, until the final equation has the form

$x =$ **number** or **number** $= x$

Equivalent equations have the same solution, so the word "number" above represents the solution of the original equation. The first property of equality to help us write simpler, equivalent equations is the **addition property of equality.**

Addition Property of Equality

Let a, b, and c represent numbers. Then

$a = b$	Also, $a = b$
and $a + c = b + c$	and $a - c = b - c$
are equivalent equations.	are equivalent equations.

In other words, the same number may be added to or subtracted from both sides of an equation without changing the solution of the equation. (Recall in Section 2.3 that we defined subtraction as addition of the first number and the opposite of the second number. Because of this, the addition property of equality also allows us to subtract the same number from both sides.)

A good way to visualize a true equation is to picture a balanced scale. Since it is balanced, each side of the scale weighs the same amount. Similarly, in a true equation the expressions on each side have the same value. Picturing our balanced scale, if we add the same weight to each side, the scale remains balanced.

PRACTICE PROBLEM 2

Solve the equation for y: $y - 6 = -2$

EXAMPLE 2 Solve: $x - 2 = -1$ for x.

Solution: To solve the equation for x, we need to rewrite the equation in the form $x =$ number. In other words, our goal is to get x alone on one side of the equation. To do so, we add 2 to both sides of the equation.

$x - 2 = -1$

$x - 2 + 2 = -1 + 2$ Add 2 to both sides of the equation.

$x + 0 = 1$ Replace $-2 + 2$ with 0.

$x = 1$ Simplify by replacing $x + 0$ with x.

Check: To check, we replace x with 1 in the *original* equation.

$x - 2 = -1$ Original equation

$1 - 2 \stackrel{?}{=} -1$ Replace x with 1.

$-1 = -1$ True

Since $-1 = -1$ is a true statement, 1 is the solution of the equation.

Work Practice Problem 2

Answer

2. 4

Helpful Hint Note that it is always a good idea to check the solution in the *original* equation to see that it makes the equation a true statement.

Let's visualize how we used the addition property of equality to solve an equation. Picture the equation, $x - 2 = 1$, as a balanced scale. The left side of the equation has the same value as the right side.

If the same weight is added to each side of a scale, the scale remains balanced. Likewise, if the same number is added to each side of an equation, the left side continues to have the same value as the right side.

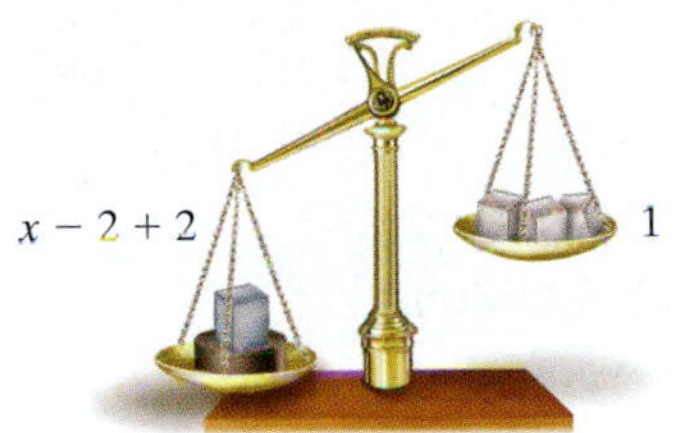

EXAMPLE 3 Solve: $-8 = n + 1$

Solution: To get n alone on one side of the equation, we subtract 1 from both sides of the equation.

$$-8 = n + 1$$
$$-8 - 1 = n + 1 - 1 \quad \text{Subtract 1 from both sides.}$$
$$-9 = n + 0 \quad \text{Replace } 1 - 1 \text{ with } 0.$$
$$-9 = n \quad \text{Simplify.}$$

Check:

$$-8 = n + 1$$
$$-8 \stackrel{?}{=} -9 + 1 \quad \text{Replace } n \text{ with } -9.$$
$$-8 = -8 \quad \text{True}$$

The solution is -9.

Work Practice Problem 3

Helpful Hint Remember that we can get the variable alone on either side of the equation. For example, the equations $-9 = n$ and $n = -9$ both have the solution of -9.

✓Concept Check What number should be added to or subtracted from both sides of the equation in order to solve the equation $-3 = y + 2$?

PRACTICE PROBLEM 3

Solve: $-2 = z + 8$

Answer

3. -10

✓ Concept Check Answer

Subtract 2 from both sides.

PRACTICE PROBLEM 4

Solve: $10x = -2 + 9x$

EXAMPLE 4 Solve: $7x = 6x + 4$

Solution: Subtract $6x$ from both sides so that variable terms will be on the same side of the equation.

$$7x = 6x + 4$$

$$7x - 6x = 6x + 4 - 6x \quad \text{Subtract } 6x \text{ from both sides.}$$

$$1x = 4 \quad \text{or} \quad x = 4 \quad \text{Simplify.}$$

Check to see that 4 is the solution.

Work Practice Problem 4

Objective C Using the Multiplication Property to Solve Equations

Although the addition property of equality is a powerful tool for helping us solve equations, it cannot help us solve all types of equations. For example, it cannot help us solve an equation such as $2x = 6$. To solve this equation, we use a second property of equality called the **multiplication property of equality.**

Multiplication Property of Equality

Let a, b, and c represent numbers and let $c \neq 0$. Then

$a = b$	Also, $a = b$
and $a \cdot c = b \cdot c$	and $\dfrac{a}{c} = \dfrac{b}{c}$
are equivalent equations.	are equivalent equations.

In other words, both sides of an equation may be multiplied or divided by the same nonzero number without changing the solution of the equation. (We will see in Chapter 4 how the multiplication property also allows us to divide both sides of an equation by the same nonzero number.)

To solve an equation like $2x = 6$ for x, notice that 2 is *multiplied* by x. To get x alone, we use the multiplication property of equality to *divide* both sides of the equation by 2, and simplify as follows:

$$2x = 6$$

$$\frac{2 \cdot x}{2} = \frac{6}{2} \quad \text{Divide both sides by 2.}$$

Then it can be shown that an expression such as $\frac{2 \cdot x}{2}$ is equivalent to $\frac{2}{2} \cdot x$, so

$$\frac{2 \cdot x}{2} = \frac{6}{2} \quad \text{can be written as} \quad \frac{2}{2} \cdot x = \frac{6}{2}$$

$$1 \cdot x = 3 \quad \text{or} \quad x = 3$$

Picturing again our balanced scale, if we multiply or divide the weight on each side by the same nonzero number, the scale (or equation) remains balanced.

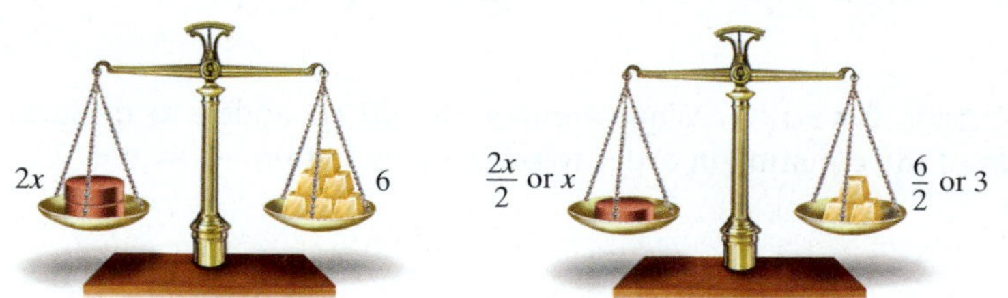

Answer

4. −2

EXAMPLE 5 Solve: $-5x = 15$

Solution: To get x alone, divide both sides by -5.

$$-5x = 15 \quad \text{Original equation}$$
$$\frac{-5x}{-5} = \frac{15}{-5} \quad \text{Divide both sides by } -5.$$
$$\frac{-5}{-5} \cdot x = \frac{15}{-5}$$
$$1x = -3 \quad \text{or} \quad x = -3 \quad \text{Simplify.}$$

Check: To check, replace x with -3 in the original equation.

$$-5x = 15 \quad \text{Original equation}$$
$$-5(-3) \stackrel{?}{=} 15 \quad \text{Let } x = -3.$$
$$15 = 15 \quad \text{True.}$$

The solution is -3.

Work Practice Problem 5

PRACTICE PROBLEM 5

Solve: $3y = -18$

EXAMPLE 6 Solve: $-27 = 3y$

Solution: To get y alone, divide both sides of the equation by 3.

$$-27 = 3y$$
$$\frac{-27}{3} = \frac{3y}{3} \quad \text{Divide both sides by 3.}$$
$$\frac{-27}{3} = \frac{3}{3} \cdot y$$
$$-9 = 1y \quad \text{or} \quad y = -9$$

Check to see that -9 is the solution.

Work Practice Problem 6

PRACTICE PROBLEM 6

Solve: $-32 = 8x$

EXAMPLE 7 Solve: $-12x = -36$

Solution: Divide both sides of the equation by the coefficient of x, which is -12.

$$-12x = -36$$
$$\frac{-12x}{-12} = \frac{-36}{-12}$$
$$\frac{-12}{-12} \cdot x = \frac{-36}{-12}$$
$$x = 3$$

Check: To check, replace x with 3 in the original equation.

$$-12x = -36$$
$$-12(3) \stackrel{?}{=} -36 \quad \text{Let } x = 3.$$
$$-36 = -36 \quad \text{True}$$

Since $-36 = -36$ is a true statement, the solution is 3.

Work Practice Problem 7

PRACTICE PROBLEM 7

Solve: $-3y = -27$

Answers

5. -6 **6.** -4 **7.** 9

✔**Concept Check** Which operation is appropriate for solving each of the following equations, addition or division?

a. $12 = x - 3$
b. $12 = 3x$

The multiplication property also allows us to solve equations like

$$\frac{x}{5} = 2$$

Here, x is *divided* by 5. To get x alone, we use the multiplication property to *multiply* both sides by 5.

$5 \cdot \frac{x}{5} = 5 \cdot 2$ Multiply both sides by 5.

Then it can be shown that

$5 \cdot \frac{x}{5} = 5 \cdot 2$ can be written as $\frac{5}{5} \cdot x = 5 \cdot 2$

$1 \cdot x = 10$ or $x = 10$

PRACTICE PROBLEM 8

Solve: $\frac{x}{-4} = 7$

EXAMPLE 8 Solve: $\frac{x}{3} = -2$

Solution: To get x alone, multiply both sides by 3.

$\frac{x}{3} = -2$

$3 \cdot \frac{x}{3} = 3 \cdot (-2)$ Multiply both sides by 3.

$\frac{3}{3} \cdot x = 3 \cdot (-2)$

$1x = -6$ or $x = -6$ Simplify.

Check: Replace x with -6 in the original equation.

$\frac{x}{3} = -2$ Original equation

$\frac{-6}{3} \stackrel{?}{=} -2$ Let $x = -6$.

$-2 = -2$ True

The solution is -6.

Work Practice Problem 8

Answer

8. -28

✔ Concept Check answers

a. Addition **b.** Division

Vocabulary and Readiness Check

Use the choices below to fill in each blank. Some choices may be used more than once.

equation	multiplication	addition
expression	solution	equivalent

1. A combination of operations on variables and numbers is called an ______________.

2. A statement of the form "expression = expression" is called an ______________.

3. An ______________ contains an equals sign (=) while an ______________ does not.

4. An ______________ may be simplified and evaluated while an ______________ may be solved.

5. A ______________ of an equation is a number that when substituted for a variable makes the equation a true statement.

6. ______________ equations have the same solution.

7. By the ______________ property of equality, the same number may be added to or subtracted from both sides of an equation without changing the solution of the equation.

8. By the ______________ property of equality, the same nonzero number may be multiplied or divided by both sides of an equation without changing the solution of the equation.

2.6 EXERCISE SET

FOR EXTRA HELP

Objective **A** *Decide whether the given number is a solution of the given equation. See Example 1.*

1. Is 6 a solution of $x - 8 = -2$?

2. Is 9 a solution of $y - 16 = -7$?

3. Is -5 a solution of $x + 12 = 17$?

4. Is -7 a solution of $a + 23 = -16$?

5. Is -8 a solution of $-9f = 64 - f$?

6. Is -6 a solution of $-3k = 12 - k$?

7. Is 3 a solution of $5(c - 5) = -10$?

8. Is 1 a solution of $2(b - 3) = 10$?

Objective **B** *Solve. Check each solution. See Examples 2 through 4.*

9. $a + 5 = 23$

10. $f + 4 = -6$

11. $d - 9 = -21$

12. $s - 7 = -15$

13. $7 = y - 2$

14. $1 = y + 7$

15. $11x = 10x - 17$

16. $14z = 13z - 15$

Objective C *Solve. Check each solution. See Examples 5 through 8.*

17. $5x = 20$

18. $6y = 48$

19. $-3z = 12$

20. $-2x = 26$

21. $\frac{n}{7} = -2$

22. $\frac{n}{11} = -5$

23. $2z = -34$

24. $7y = -21$

25. $-4y = 0$

26. $-9x = 0$

27. $-10x = -10$

28. $-31x = -31$

Objectives B C Mixed Practice *Solve. See Examples 2 through 8.*

29. $5x = -35$

30. $3y = -27$

31. $n - 5 = -55$

32. $n - 4 = -48$

33. $-15 = y + 10$

34. $-36 = y + 12$

35. $\frac{x}{-6} = -6$

36. $\frac{x}{-9} = -9$

37. $11n = 10n + 21$

38. $17z = 16z + 8$

39. $-12y = -144$

40. $-11x = -121$

41. $\frac{n}{4} = -20$

42. $\frac{n}{5} = -20$

43. $-64 = 32y$

44. $-81 = 27x$

Mixed Review–Translating *Translate each phrase to an algebraic expression. Use x to represent "a number." See Section 1.8.*

45. A number decreased by -2

46. A number increased by -5

47. The product of -6 and a number

48. The quotient of a number and -20.

49. The sum of -15 and a number

50. -32 multiplied by a number

51. -8 divided by a number

52. Subtract a number from -18.

Concept Extensions

Solve.

53. $n - 42{,}860 = -1{,}286$

54. $\frac{y}{-18} = 1098$

55. $-38x = 15{,}542$

56. $n + 961 = 120$

57. Explain the differences between an equation and an expression.

58. Explain the differences between the addition property of equality and the multiplication property of equality.

59. Write an equation that can be solved using the addition property of equality.

60. Write an equation that can be solved using the multiplication property of equality.

CHAPTER 2 Group Activity

Magic Squares

Sections 2.1–2.3

A magic square is a set of numbers arranged in a square table so that the sum of the numbers in each column, row, and diagonal is the same. For instance, in the magic square below, the sum of each column, row, and diagonal is 15. Notice that no number is used more than once in the magic square.

2	9	4
7	5	3
6	1	8

The properties of magic squares have been known for a very long time and once were thought to be good luck charms. The ancient Egyptians and Greeks understood their patterns. A magic square even made it into a famous work of art. The engraving titled *Melencolia I,* created by German artist Albrecht Dürer in 1514, features the following four-by-four magic square on the building behind the central figure.

16	3	2	13
5	10	11	8
9	6	7	12
4	15	14	1

Source: Melancolia (engraving) by Albrecht Duerer (1471–1528). Guildhall Library, Corporation of London, UK/Bridgeman Art Library.

Exercises

1. Verify that what is shown in the Dürer engraving is, in fact, a magic square. What is the common sum of the columns, rows, and diagonals?
2. Negative numbers can also be used in magic squares. Complete the following magic square:

	−1	
0		−4

3. Use the numbers −16, −12, −8, −4, 0, 4, 8, 12, and 16 to form a magic square:

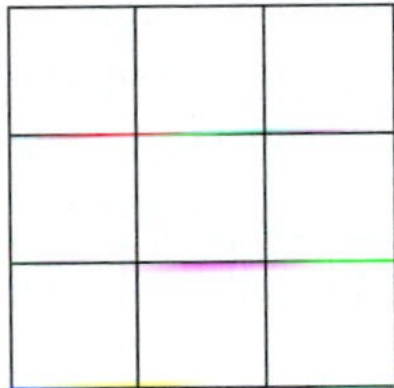

Chapter 2 Vocabulary Check

Fill in each blank with one of the words or phrases listed below.

inequality symbols	addition	solution	is less than	integers
expression	average	negative	absolute value	equation
positive	opposites	is greater than	multiplication	

1. Two numbers that are the same distance from 0 on the number line but are on opposite sides of 0 are called ______________.
2. The ______________ of a number is that number's distance from 0 on a number line.
3. The ______________ are $\ldots, -3, -2, -1, 0, 1, 2, 3, \ldots$.
4. The ______________ numbers are numbers less than zero.
5. The ______________ numbers are numbers greater than zero.
6. The symbols " $<$ " and " $>$ " are called ________________.
7. A ______________ of an equation is a number that when substituted for a variable makes the equation a true statement.
8. The ______________ of a list of numbers is $\dfrac{\text{sum of numbers}}{\textit{number} \text{ of numbers}}$.
9. A combination of operations on variables and numbers is called an ______________.
10. A statement of the form "expression $=$ expression" is called an ______________.
11. The sign "$<$" means ______________ and "$>$" means ______________.
12. By the ______________ property of equality, the same number may be added to or subtracted from both sides of an equation without changing the solution of the equation.
13. By the ______________ property of equality, the same nonzero number may be multiplied or divided by both sides of an equation without changing the solution of the equation.

Helpful Hint

Are you preparing for your test? Don't forget to take the Chapter 2 Test on page 157. Then check your answers at the back of the text and use the Chapter Test Prep Video CD to see the fully worked-out solutions to any of the exercises you want to review.

2 Chapter Highlights

Definitions and Concepts	**Examples**
Section 2.1 Introduction to Integers	
Together, positive numbers, negative numbers, and 0 are called **signed numbers.**	$-432, -10, 0, 15$
The **integers** are $\ldots, -3, -2, -1, 0, 1, 2, 3, \ldots$.	
The **absolute value** of a number is that number's distance from 0 on a number line. The symbol for absolute value is $\lvert\ \rvert$.	$\lvert -2 \rvert = 2$ 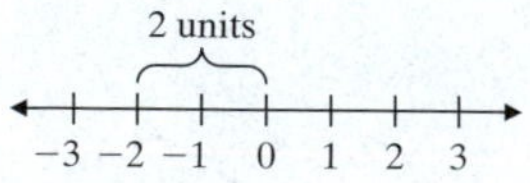 $\lvert 2 \rvert = 2$ 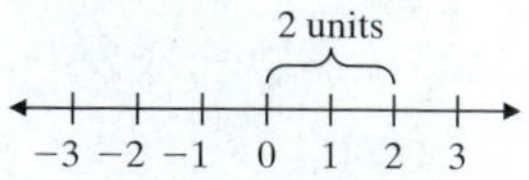

Definitions and Concepts	**Examples**
Section 2.1 Introduction to Integers (*continued*)	
Two numbers that are the same distance from 0 on the number line but are on opposite sides of 0 are called **opposites.**	5 and −5 are opposites. 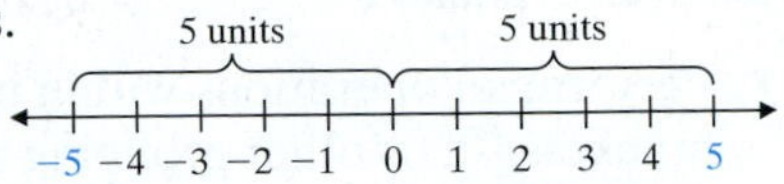
If a is a number, then $-(-a) = a$.	$-(-11) = 11$. Do not confuse with $-\|-3\| = -3$
Section 2.2 Adding Integers	
Adding Two Numbers with the Same Sign **Step 1.** Add their absolute values. **Step 2.** Use their common sign as the sign of the sum.	Add: $-3 + (-2) = -5$ $-7 + (-15) = -22$
Adding Two Numbers with Different Signs **Step 1.** Find the larger absolute value minus the smaller absolute value. **Step 2.** Use the sign of the number with the larger absolute value as the sign of the sum.	$-6 + 4 = -2$ $17 + (-12) = 5$ $-32 + (-2) + 14 = -34 + 14$ $= -20$
Section 2.3 Subtracting Integers	
Subtracting Two Numbers If a and b are numbers, then $a - b = a + (-b)$.	Subtract: $-35 - 4 = -35 + (-4) = -39$ $3 - 8 = 3 + (-8) = -5$ $-10 - (-12) = -10 + 12 = 2$ $7 - 20 - 18 - (-3) = 7 + (-20) + (-18) + (+3)$ $= -13 + (-18) + 3$ $= -31 + 3$ $= -28$
Section 2.4 Multiplying and Dividing Integers	
Multiplying Numbers The product of two numbers having the same sign is a positive number. The product of two numbers having unlike signs is a negative number.	Multiply: $(-7)(-6) = 42$ $9(-4) = -36$ Evaluate: $(-3)^2 = (-3)(-3) = 9$
Dividing Numbers The quotient of two numbers having the same sign is a positive number. The quotient of two numbers having unlike signs is a negative number.	Divide: $-100 \div (-10) = 10$ $\frac{14}{-2} = -7$, $\frac{0}{-3} = 0$, $\frac{22}{0}$ is undefined.

DEFINITIONS AND CONCEPTS	EXAMPLES
Section 2.5 Order of Operations	
ORDER OF OPERATIONS 1. Perform all operations within parentheses ( ), brackets [ ], or other grouping symbols such as fraction bars, starting with the innermost set. 2. Evaluate any expressions with exponents. 3. Multiply or divide in order from left to right. 4. Add or subtract in order from left to right.	Simplify: $3 + 2\cdot(-5) = 3 + (-10)$ $= -7$ $\frac{-2(5-7)}{-7+\lvert -3\rvert} = \frac{-2(-2)}{-7+3}$ $= \frac{4}{-4}$ $= -1$
Section 2.6 Solving Equations: The Addition and Multiplication Properties	
ADDITION PROPERTY OF EQUALITY Let a, b, and c represent numbers. If $a = b$, then $a + c = b + c$ and $a - c = b - c$ In other words, the same number may be added to or subtracted from both sides of an equation without changing the solution of the equation.	Solve: $x + 8 = 1$ $x + 8 - 8 = 1 - 8$ Subtract 8 from both sides. $x = -7$ Simplify. The solution is -7.
MULTIPLICATION PROPERTY OF EQUALITY Let a, b, and c represent numbers and let $c \neq 0$. If $a = b$, then $a\cdot c = b\cdot c$ and $\frac{a}{c} = \frac{b}{c}$ In other words, both sides of an equation may be multiplied or divided by the same nonzero number without changing the solution of the equation.	Solve: $-6y = 30$ $\frac{-6y}{-6} = \frac{30}{-6}$ Divide both sides by -6. $y = -5$ Simplify. The solution is -5.

STUDY SKILLS BUILDER

Are You Prepared for a Test on Chapter 2?

Below I have listed some *common trouble areas* for students in Chapter 2. After studying for your test—but before taking your test—read these.

- Don't forget the difference between $-(-5)$ and $-|-5|$.

$-(-5) = 5$ The opposite of -5 is 5.

$-|-5| = -5$ The opposite of the absolute value of -5 is the opposite of 5, which is -5.

- Remember how to simplify $(-7)^2$ and -7^2.

$(-7)^2 = (-7)(-7) = 49$

$-7^2 = -(7)(7) = -49$

- Don't forget order of operations.

$1 + 3(4 - 6) = 1 + 3(-2)$ Simplify inside parentheses.

$= 1 + (-6)$ Multiply.

$= -5$ Add.

Remember: This is simply a checklist of common trouble spots. For a review of Chapter 2, see the Highlights and Chapter Review at the end of this chapter.

2 CHAPTER REVIEW

(2.1) *Represent each quantity by an integer.*

1. A gold miner is working 1572 feet down in a mine.

2. Mount Hood, in Oregon, has an elevation of 11,239 feet.

Source: Greg Vaughn/ Pacificstock.com

Graph each integer in the list on the same number line.

3. $-3, -5, 0, 7$

−7 −6 −5 −4 −3 −2 −1 0 1 2 3 4 5 6 7

4. $-6, -1, 0, 5$

−7 −6 −5 −4 −3 −2 −1 0 1 2 3 4 5 6 7

Simplify.

5. $|-11|$

6. $|0|$

7. $-|8|$

8. $-(-9)$

9. $-|-16|$

10. $-(-2)$

Insert < or > between each pair of integers to make a true statement.

11. $-18 \quad -20$

12. $-5 \quad 5$

13. $|-123| \quad -|-198|$

14. $8 - |-12| \quad -|-16|$

Find the opposite of each integer.

15. -18

16. 42

Answer true or false for each statement.

17. If $a < b$, then a must be a negative number.

18. The absolute value of an integer is always 0 or a positive number.

19. A negative number is always less than a positive number.

20. If a is a negative number, then $-a$ is a positive number.

Evaluate.

21. $|y|$ if $y = -2$

22. $|-x|$ if $x = -3$

23. $-|-z|$ if $z = -5$

24. $-|-n|$ if $n = -10$

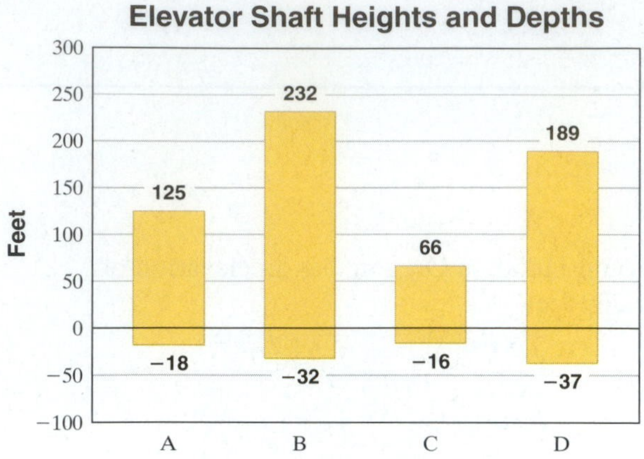

Elevator shafts in some buildings extend not only above ground, but in many cases below ground to accommodate basements, underground parking, etc. The bar graph shows four such elevators and their shaft distance above and below ground. Use the bar graph to answer Exercises 25 and 26.

25. Which elevator shaft extends the farthest below ground?

26. Which elevator shaft extends the highest above ground?

(2.2) *Add.*

27. $5 + (-3)$

28. $18 + (-4)$

29. $-12 + 16$

30. $-23 + 40$

31. $-8 + (-15)$

32. $-5 + (-17)$

33. $-24 + 3$

34. $-89 + 19$

35. $15 + (-15)$

36. $-24 + 24$

37. $-43 + (-108)$

38. $-100 + (-506)$

39. The temperature at 5 a.m. on a day in January was $-15°$ Celsius. By 6 a.m. the temperature had fallen 5 degrees. Use a signed number to represent the temperature at 6 a.m.

40. A diver starts out at 127 feet below the surface and then swims downward another 23 feet. Use a signed number to represent the diver's current depth.

41. During the 2006 PGA Masters tournament, the winner, Phil Mickelson, has scores of -2, 0, -2, and -3 over four rounds of golf. What was his total score for the tournament? (*Source*: Professional Golfer's Association)

42. During the 2006 Women's World Cup of Golf in Sun City, South Africa, the winners, the Swedish team, had a score of -7. The fourth-place finisher, the U.S. team, had a score that was 8 more than the winning score. What was the U.S. team's score in Women's World Cup of Golf? (*Source:* Ladies' Professional Golf Association)

(2.3) *Subtract.*

43. $12 - 4$

44. $-12 - 4$

45. $-7 - 17$

46. $7 - 17$

47. $7 - (-13)$

48. $-6 - (-14)$

49. $16 - 16$

50. $-16 - 16$

51. $-12 - (-12)$

52. $-5 - (-12)$

53. $-(-5) - 12 + (-3)$

54. $-8 + (-12) - 10 - (-3)$

Solve.

55. If the elevation of Lake Superior is 600 feet above sea level and the elevation of the Caspian Sea is 92 feet below sea level, find the difference of the elevations.

56. Josh Weidner has $142 in his checking account. He writes a check for $125, makes a deposit for $43, and then writes another check for $85. Represent the balance in his account by an integer.

57. Some roller coasters travel above and below ground. One such roller coaster is Tremors, located in Silverwood Theme Park, Athol, Idaho. If this coaster rises to a height of 85 feet above ground, then drops 99 feet, how many feet below ground are you at the end of the drop? (*Source:* ultimaterollercoaster.com)

58. Go to the bar graph for Review Exercises 25 and 26 and find the total length of the elevator shaft for Elevator C.

Answer true or false for each statement.

59. $|-5| - |-6| = 5 - 6$

60. $|-5 - (-6)| = 5 + 6$

(2.4) *Multiply.*

61. $-3(-7)$

62. $-6(3)$

63. $-4(16)$

64. $-5(-12)$

65. $(-5)^2$

66. $(-1)^5$

67. $12(-3)(0)$

68. $-1(6)(2)(-2)$

Divide.

69. $-15 \div 3$

70. $\frac{-24}{-8}$

71. $\frac{0}{-3}$

72. $\frac{-46}{0}$

73. $\frac{100}{-5}$

74. $\frac{-72}{8}$

75. $\frac{-38}{-1}$

76. $\frac{45}{-9}$

77. A football team lost 5 yards on each of two consecutive plays. Represent the total loss by a product of integers, and find the product.

78. A race horse bettor lost \$50 on each of four consecutive races. Represent the total loss by a product of integers, and find the product.

79. A person has a debt of \$1024 and is ordered to pay it back in four equal payments. Represent the amount of each payment by a quotient of integers, and find the quotient.

80. Overnight, the temperature dropped 45 degrees Fahrenheit. If this took place over a time period of nine hours, represent the average temperature drop each hour by a quotient of integers. Then find the quotient.

(2.5) *Simplify.*

81. $(-7)^2$

82. -7^2

83. $5 - 8 + 3$

84. $-3 + 12 + (-7) - 10$

85. $-10 + 3 \cdot (-2)$

86. $5 - 10 \cdot (-3)$

87. $16 \div (-2) \cdot 4$

88. $-20 \div 5 \cdot 2$

89. $16 + (-3) \cdot 12 \div 4$

90. $(-12) + 10 \div (-5)$

91. $4^3 - (8 - 3)^2$

92. $(-3)^3 - 90$

93. $\frac{(-4)(-3) - (-2)(-1)}{-10 + 5}$

94. $\frac{4(12 - 18)}{-10 \div (-2 - 3)}$

Find the average of each list of numbers.

95. $-18, 25, -30, 7, 0, -2$

96. $-45, -40, -30, -25$

Evaluate each expression for $x = -2$ and $y = 1$.

97. $2x - y$

98. $y^2 + x^2$

99. $\dfrac{3x}{6}$

100. $\dfrac{5y - x}{-y}$

(2.6) *For Exercises 101 and 102, answer "yes" or "no."*

101. Is -5 a solution of $2n - 6 = 16$?

102. Is -2 a solution of $2(c - 8) = -20$?

Solve.

103. $n - 7 = -20$

104. $-5 = n + 15$

105. $10x = -30$

106. $-8x = 72$

107. $9y = 8y - 13$

108. $6x - 31 = 7x$

109. $\dfrac{n}{-4} = -11$

110. $\dfrac{x}{-2} = 13$

111. $n + 12 = -7$

112. $n - 40 = -2$

113. $-36 = -6x$

114. $-40 = 8y$

Mixed Review

Perform the indicated operations.

115. $-6 + (-9)$

116. $-16 - 3$

117. $-4(-12)$

118. $\dfrac{84}{-4}$

119. $-76 - (-97)$

120. $-9 + 4$

121. Joe owed his mother \$32. He gave her \$23. Write his financial situation as a signed number.

122. The temperature at noon on a Monday in December was $-11°C$. By noon on Tuesday, it had warmed by 17°C. What was the temperature at noon on Tuesday?

123. The top of the mountain has an altitude of 12,923 feet. The bottom of the valley is 195 feet below sea level. Find the difference between these two elevations.

124. Wednesday's lowest temperature was $-18°C$. The cold weather continued and by Friday it had dropped another 9°C. What was the temperature on Friday?

Simplify.

125. $(3 - 7)^2 \div (6 - 4)^3$

126. $3(4 + 2) + (-6) - 3^2$

127. $2 - 4 \cdot 3 + 5$

128. $4 - 6 \cdot 5 + 1$

129. $\dfrac{-|-14| - 6}{7 + 2(-3)}$

130. $5(7 - 6)^3 - 4(2 - 3)^2 + 2^4$

Solve.

131. $n - 9 = -30$

132. $n + 18 = 1$

133. $-4x = -48$

134. $9x = -81$

135. $\dfrac{n}{-2} = 100$

136. $\dfrac{y}{-1} = -3$

2 CHAPTER TEST

Remember to use the Chapter Test Prep Video CD to see the fully worked-out solutions to any of the exercises you want to review.

Answers

Simplify each expression.

1. $-5 + 8$

2. $18 - 24$

3. $5 \cdot (-20)$

4. $(-16) \div (-4)$

5. $(-18) + (-12)$

6. $-7 - (-19)$

7. $(-5) \cdot (-13)$

8. $\dfrac{-25}{-5}$

9. $|-25| + (-13)$

10. $14 - |-20|$

11. $|5| \cdot |-10|$

12. $\dfrac{|-10|}{-|-5|}$

13. $(-8) + 9 \div (-3)$

14. $-7 + (-32) - 12 + 5$

15. $(-5)^3 - 24 \div (-3)$

16. $(5 - 9)^2 \cdot (8 - 2)^3$

17. $-(-7)^2 \div 7 \cdot (-4)$

18. $3 - (8 - 2)^3$

1. __________
2. __________
3. __________
4. __________
5. __________
6. __________
7. __________
8. __________
9. __________
10. __________
11. __________
12. __________
13. __________
14. __________
15. __________
16. __________
17. __________
18. __________

19. ______
20. ______
21. ______
22. ______
23. ______
24. ______
25. ______
26. ______
27. ______
28. ______
29. ______
30. ______
31. a. ______
b. ______
32. ______
33. ______
34. ______
35. ______

19. $\frac{4}{2} - \frac{8^2}{16}$

20. $\frac{-3(-2) + 12}{-1(-4 - 5)}$

21. $\frac{|25 - 30|^2}{2(-6) + 7}$

22. $5(-8) - [6 - (2 - 4)] + (12 - 16)^2$

Evaluate each expression for $x = 0$, $y = -3$, and $z = 2$.

23. $7x + 3y - 4z$

24. $10 - y^2$

25. $\frac{3z}{2y}$

26. Mary Dunstan, a diver, starts at sea level and then makes 4 successive descents of 22 feet. After the descents, what is her elevation?

27. Aaron Hawn has $129 in his checking account. He writes a check for $79, withdraws $40 from an ATM, and then deposits $35. Represent the new balance in his account by an integer.

28. Mt. Washington in New Hampshire has an elevation of 6288 feet above sea level. The Romanche Gap in the Atlantic Ocean has an elevation of 25,354 feet below sea level. Represent the difference in elevation between these two points by an integer. (*Source:* National Geographic Society and Defense Mapping Agency)

29. Lake Baykal in Siberian Russia is the deepest lake in the world with a maximum depth of 5315 feet. The elevation of the lake's surface is 1495 feet above sea level. What is the elevation (with respect to sea level) of the deepest point in the lake? (*Source:* U.S. Geological Survey)

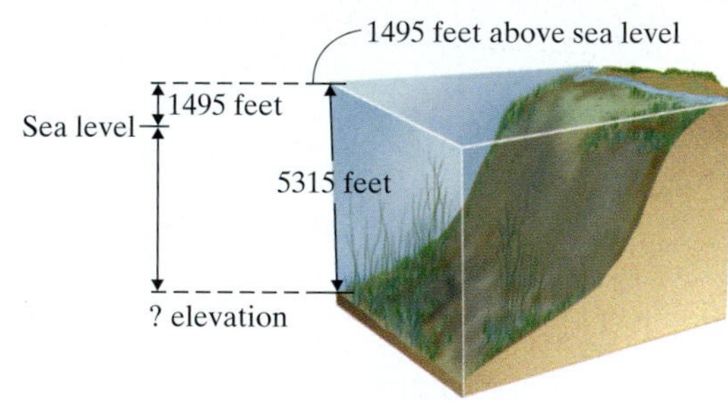

30. Find the average of -12, -13, 0, 9.

31. Translate the following phrases into mathematical expressions. Use x to represent "a number."

a. The product of a number and 17

b. Twice a number subtracted from 20

Solve.

32. $-9n = -45$

33. $\frac{n}{-7} = 4$

34. $x - 16 = -36$

35. $9x = 8x - 4$

CUMULATIVE REVIEW

Chapters 1–2

Find the place value of the digit 3 in each whole number.

1. 396,418

2. 4308

3. 93,192

4. 693,298

5. 534,275,866

6. 267,301,818

7. Insert $<$ or $>$ to make a true statement.

a. $-7 \quad 7$
b. $0 \quad -4$
c. $-9 \quad -11$

8. Insert $<$ or $>$ to make a true statement

a. 12 4
b. 13 31
c. 82 79

9. Add: $13 + 2 + 7 + 8 + 9$

10. Add: $11 + 3 + 9 + 16$

11. Subtract: $7826 - 505$ Check by adding.

12. Subtract: $3285 - 272$ Check by adding.

13. The radius of Jupiter is 43,441 miles. The radius of Saturn is 7257 miles less than the radius of Jupiter. Find the radius of Saturn. (*Source:* National Space Science Data Center)

14. C. J. Dufour wants to buy a digital camera. She has $762 in her savings account. If the camera costs $237, how much money will she have in her account after buying the camera?

15. Round 568 to the nearest ten.

16. Round 568 to the nearest hundred.

17. Round each number to the nearest hundred to find an estimated difference.

$$\begin{array}{r} 4725 \\ -2879 \\ \hline \end{array}$$

18. Round each number to the nearest thousand to find an estimated difference.

$$\begin{array}{r} 8394 \\ -2913 \\ \hline \end{array}$$

19. Rewrite each using the distributive property.

a. $5(6 + 5)$
b. $20(4 + 7)$
c. $2(7 + 9)$

20. Rewrite each using the distributive property.

a. $5(2 + 12)$
b. $9(3 + 6)$
c. $4(8 + 1)$

21. Multiply: 631×125

22. Multiply: 299×104

23. Find each quotient. Check by multiplying.

a. $42 \div 7$
b. $\frac{64}{8}$
c. $3\overline{)21}$

24. Find each quotient. Check by multiplying.

a. $\frac{35}{5}$
b. $64 \div 8$
c. $4\overline{)48}$

Answers

1. ____
2. ____
3. ____
4. ____
5. ____
6. ____
7. a. ____ b. ____ c. ____
8. a. ____ b. ____ c. ____
9. ____
10. ____
11. ____
12. ____
13. ____
14. ____
15. ____
16. ____
17. ____
18. ____
19. a. ____ b. ____ c. ____
20. a. ____ b. ____ c. ____
21. ____
22. ____
23. a. ____ b. ____ c. ____
24. a. ____ b. ____ c. ____

25. ______
26. ______
27. ______
28. ______
29. ______
30. ______
31. ______
32. ______
33. ______
34. ______
35. ______
36. ______
37. ______
38. ______
39. a. ______
b. ______
c. ______
40. a. ______
b. ______
41. ______
42. ______
43. ______
44. ______
45. ______
46. ______
47. ______
48. ______
49. ______
50. ______

25. Divide: $3705 \div 5$
Check by multiplying.

26. Divide: $3648 \div 8$
Check by multiplying.

27. As part of a promotion, Becky Foster receives 238 cards, each good for one free song download from MSN Music. If Becky wants to share them evenly with 19 friends, how many download cards will each friend receive? How many will be left over?

28. Mrs. Mallory's first grade class is going to the zoo. She pays a total of \$324 for 36 admission tickets. How much did each ticket cost?

Evaluate.

29. 9^2

30. 5^3

31. 6^1

32. 4^1

33. $5 \cdot 6^2$

34. $2^3 \cdot 7$

35. Simplify: $\dfrac{7 - 2 \cdot 3 + 3^2}{5(2 - 1)}$

36. Simplify: $\dfrac{6^2 + 4 \cdot 4 + 2^3}{37 - 5^2}$

37. Evaluate $x + 6$ if x is 8.

38. Evaluate $5 + x$ if x is 9.

39. Simplify:
a. $|-9|$
b. $|3|$
c. $|0|$

40. Simplify:
a. $|4|$
b. $|-7|$

41. Add: $-2 + 5$

42. Add: $8 + (-3)$

43. Evaluate $2a - b$ for $a = 8$ and $b = -6$.

44. Evaluate $x - y$ for $x = -2$ and $y = -7$.

45. Multiply: $-7 \cdot 3$

46. Multiply: $5(-2)$

47. Multiply: $0 \cdot (-4)$

48. Multiply: $-6 \cdot 9$

49. Simplify: $3(4 - 7) + (-2) - 5$

50. Simplify: $4 - 8(7 - 3) - (-1)$

3 Solving Equations and Problem Solving

In this chapter, we continue making the transition from arithmetic to algebra. Recall that in algebra, letters (called variables) represent unknown quantities. Using variables is a very powerful method for solving problems that cannot be solved with arithmetic alone. This chapter introduces operations on algebraic expressions, and we continue solving variable equations.

SECTIONS

CHECK YOUR PROGRESS

Roads and highways can lead travelers on an exciting journey through every period of U.S. history. The *Santa Fe Trail Scenic and Historic Byway* in Colorado retraces the famous road taken by wagon trains of the Old West. The *Seward Highway* in Alaska is named for the U.S. Secretary of State who arranged the purchase of Alaska from Russia in 1867. The *Selma to Montgomery March Byway* in Alabama traces the historic route taken by Dr. Martin Luther King, Jr. in 1965. The horizontal bar graph below shows the states with the greatest number of roadway miles. Section 3.4, Example 3, Exercises 33, 34, and the Mixed Review Exercises 93, 94, page 207, all have to do with roadway miles or miles between cities.

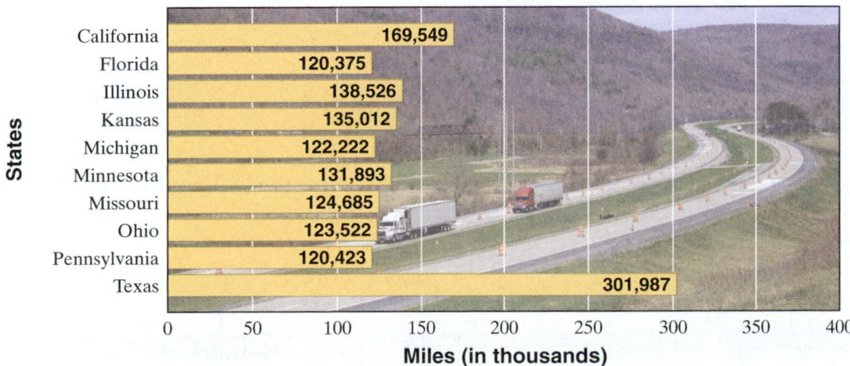

Source: U.S. Federal Highway Administration

3.1 SIMPLIFYING ALGEBRAIC EXPRESSIONS

Objectives

A Use Properties of Numbers to Combine Like Terms.

B Use Properties of Numbers to Multiply Expressions.

C Simplify Expressions by Multiplying and then Combining Like Terms.

D Find the Perimeter and Area of Figures.

Recall from Section 1.8 that a combination of numbers, letters (variables), and operation symbols is called an **algebraic expression** or simply an **expression.** For example,

Algebraic Expressions

$4 \cdot x, \quad n + 7, \quad \text{and} \quad 3y - 5 - x$

are expressions.

Recall that if two variables or a number and a variable are next to each other, with no operation sign between them, the indicated operation is multiplication. For example,

$3y \quad \text{means} \quad 3 \cdot y$

and

$xy \text{ or } x(y) \quad \text{means} \quad x \cdot y$

Also, the meaning of an exponent remains the same when the base is a variable. For example,

$y^2 = \underbrace{y \cdot y}_{\text{2 factors of } y} \quad \text{and} \quad x^4 = \underbrace{x \cdot x \cdot x \cdot x}_{\text{4 factors of } x}$

Just as we can add, subtract, multiply, and divide numbers, we can add, subtract, multiply, and divide algebraic expressions. In previous sections we evaluated algebraic expressions like $x + 3$, $4x$, and $x + 2y$ for particular values of the variables. In this section, we explore working with variable expressions without evaluating them. We begin with a definition of a term.

Objective A Combining Like Terms

The addends of an algebraic expression are called the **terms** of the expression.

$x + 3$ — 2 terms

$3y^2 + (-6y) + 4$ — 3 terms

A term that is only a number has a special name. It is called a **constant term,** or simply a **constant.** A term that contains a variable is called a **variable term.**

x	+	3	and	$3y^2 + (-6y)$	+	4
variable term		constant term		variable terms		constant term

The number factor of a variable term is called the **numerical coefficient.** A numerical coefficient of 1 is usually not written.

Helpful Hint Recall that $1 \cdot$ any number = that number. This means that

$1 \cdot x = x \quad \text{or that } 1x = x$

Thus x can always be replaced by $1x$ or $1 \cdot x$.

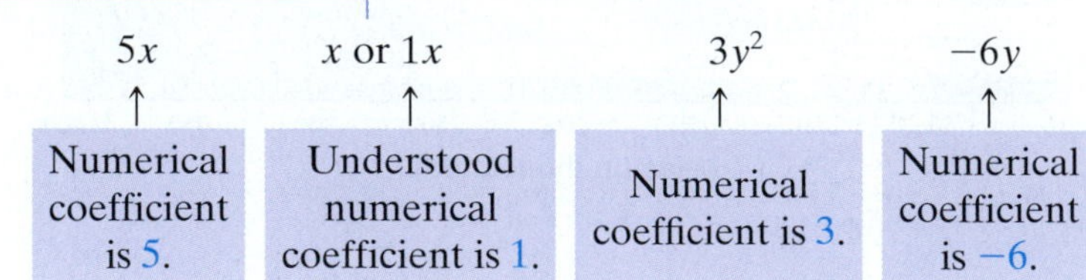

Terms with the same variable factors, except that they may have different numerical coefficients, are called **like terms.**

Like Terms	Unlike Terms
$3x, -4x$	$5x, x^2$
$-6y, 2y, y$	$7x, 7y$

✔**Concept Check** True or false? The terms $-7xz^2$ and $3z^2x$ are like terms. Explain.

A sum or difference of like terms can be simplified using the **distributive property.** Recall from Section 1.5 that the distributive property says that multiplication distributes over addition (and subtraction). Using variables, we can write the distributive property as follows:

$$(a + b)c = ac + bc.$$

If we write the right side of the equation first, then the left side, we have the following:

Distributive Property

If a, b, and c are numbers, then

$$ac + bc = (a + b)c$$

Also,

$$ac - bc = (a - b)c$$

The distributive property guarantees that, no matter what number x is, $7x + 2x$ (for example) has the same value as $(7 + 2)x$, or $9x$. We then have that

$$7x + 2x = (7 + 2)x = 9x$$

This is an example of **combining like terms.** An algebraic expression is **simplified** when all like terms have been combined.

EXAMPLE 1 Simplify each expression by combining like terms.

a. $4x + 6x$ **b.** $y - 5y$ **c.** $3x^2 + 5x^2 - 2$

Solution: Add or subtract like terms.

a. $4x + 6x = (4 + 6)x$
$= 10x$

Understood 1

b. $y - 5y = 1y - 5y$
$= (1 - 5)y$
$= -4y$

c. $3x^2 + 5x^2 - 2 = (3 + 5)x^2 - 2$
$= 8x^2 - 2$

Work Practice Problem 1

PRACTICE PROBLEM 1

Simplify each expression by combining like terms.

a. $8m - 14m$

b. $6a + a$

c. $-y^2 + 3y^2 + 7$

Answers

1. a. $-6m$ **b.** $7a$ **c.** $2y^2 + 7$

✔ Concept Check Answer

True

Helpful Hint

In this section, we are simplifying expressions. Try not to confuse the two processes below.

Expression: $5y - 8y$	Equation: $8n = -40$
Simplify the expression:	Solve the equation:
$5y - 8y = (5 - 8)y$	$8n = -40$
$= -3y$	$\frac{8n}{8} = \frac{-40}{8}$ Divide both sides by 8.
	$n = -5$ The solution is -5.

The commutative and associative properties of addition and multiplication can also help us simplify expressions. We presented these properties in Sections 1.3 and 1.5 and state them again using variables.

Properties of Addition and Multiplication

If a, b, and c are numbers, then

$a + b = b + a$ Commutative property of addition

$a \cdot b = b \cdot a$ Commutative property of multiplication

That is, the **order** of adding or multiplying two numbers can be changed without changing their sum or product.

$(a + b) + c = a + (b + c)$ Associative property of addition

$(a \cdot b) \cdot c = a \cdot (b \cdot c)$ Associative property of multiplication

That is, the **grouping** of numbers in addition or multiplication can be changed without changing their sum or product.

Helpful Hint

- Examples of these properties are

$2 + 3 = 3 + 2$ Commutative property of addition

$7 \cdot 9 = 9 \cdot 7$ Commutative property of multiplication

$(1 + 8) + 10 = 1 + (8 + 10)$ Associative property of addition

$(4 \cdot 2) \cdot 3 = 4 \cdot (2 \cdot 3)$ Associative property of addition

- These properties are not true for subtraction or division.

PRACTICE PROBLEM 2

Simplify: $6z + 5 + z - 4$

EXAMPLE 2 Simplify: $2y - 6 + 4y + 8$

Solution: We begin by writing subtraction as the opposite of addition.

$$\begin{aligned} 2y - 6 + 4y + 8 &= 2y + (-6) + 4y + 8 && \text{Apply the commutative property of addition.} \\ &= 2y + 4y + (-6) + 8 \\ &= (2 + 4)y + (-6) + 8 && \text{Apply the distributive property.} \\ &= 6y + 2 && \text{Simplify.} \end{aligned}$$

Work Practice Problem 2

Answer

2. $7z + 1$

EXAMPLES Simplify each expression by combining like terms.

3. $6x + 2x - 5 = 8x - 5$
4. $4x + 2 - 5x + 3 = 4x + 2 + (-5x) + 3$
$= 4x + (-5x) + 2 + 3$
$= -1x + 5 \quad \text{or} \quad -x + 5$
5. $2x - 5 + 3y + 4x - 10y + 11$
$= 2x + (-5) + 3y + 4x + (-10y) + 11$
$= 2x + 4x + 3y + (-10y) + (-5) + 11$
$= 6x - 7y + 6$

Work Practice Problems 3–5

PRACTICE PROBLEMS 3–5

Simplify each expression by combining like terms.

3. $6y + 12y - 6$
4. $7y - 5 + y + 8$
5. $-7y + 2 - 2y - 9x + 12 - x$

As we practice combining like terms, keep in mind that some of the steps may be performed mentally.

Objective B Multiplying Expressions

We can also use properties of numbers to multiply expressions such as $3(2x)$. By the associative property of multiplication, we can write the product $3(2x)$ as $(3 \cdot 2)x$, which simplifies to $6x$.

EXAMPLES Multiply.

6. $5(3y) = (5 \cdot 3)y$ Apply the associative property of multiplication.
$= 15y$ Multiply.
7. $-2(4x) = (-2 \cdot 4)x$ Apply the associative property of multiplication.
$= -8x$ Multiply.

Work Practice Problems 6–7

PRACTICE PROBLEMS 6–7

Multiply.

6. $6(4a)$
7. $-8(9x)$

We can use the distributive property to combine like terms, which we have done, and also to multiply expressions such as $2(3 + x)$. By the distributive property, we have

$2(3 + x) = 2 \cdot 3 + 2 \cdot x$ Apply the distributive property.
$= 6 + 2x$ Multiply.

EXAMPLE 8 Use the distributive property to multiply: $6(x + 4)$

Solution: By the distributive property,

$5(x + 4) = 5 \cdot x + 5 \cdot 4$ Apply the distributive property.
$= 5x + 20$ Multiply.

Work Practice Problem 8

PRACTICE PROBLEM 8

Use the distributive property to multiply: $8(y + 2)$

✔**Concept Check** What's wrong with the following?

$8(a - b) = 8a - b$

Answers

3. $18y - 6$ 4. $8y + 3$
5. $-9y - 10x + 14$
6. $24a$ 7. $-72x$ 8. $8y + 16$

✔ Concept Check Answer

Did not distribute the 8; $8(a - b) = 8a - 8b$

PRACTICE PROBLEM 9

Multiply: $3(7a - 5)$

EXAMPLE 9 Multiply: $-3(5a + 2)$

Solution: By the distributive property,

$$-3(5a + 2) = -3(5a) + (-3)(2)$$ Apply the distributive property.

$$= (-3 \cdot 5)a + (-6)$$ Use the associative property and multiply.

$$= -15a - 6$$ Multiply.

Work Practice Problem 9

PRACTICE PROBLEM 10

Multiply: $6(5 - y)$

EXAMPLE 10 Multiply: $8(x - 4)$

Solution:

$$8(x - 4) = 8 \cdot x - 8 \cdot 4$$

$$= 8x - 32$$

Work Practice Problem 10

Objective C Simplifying Expressions

Next we will **simplify** expressions by first using the distributive property to multiply and then **combining** any like terms.

PRACTICE PROBLEM 11

Simplify: $5(2y - 3) - 8$

EXAMPLE 11 Simplify: $2(3 + 7x) - 15$

Solution: First we use the distributive property to remove parentheses.

$$2(3 + 7x) - 15 = 2(3) + 2(7x) - 15$$ Apply the distributive property.

$$= 6 + 14x - 15$$ Multiply.

$$= 14x + (-9) \quad \text{or} \quad 14x - 9$$ Combine like terms.

Helpful Hint 2 is *not* distributed to the -15 since it is not within the parentheses.

Work Practice Problem 11

PRACTICE PROBLEM 12

Simplify:
$-7(x - 1) + 5(2x + 3)$

EXAMPLE 12 Simplify: $-2(x - 5) + 4(2x + 2)$

Solution: First we use the distributive property to remove parentheses.

$$-2(x - 5) + 4(2x + 2) = -2(x) - (-2)(5) + 4(2x) + 4(2)$$ Apply the distributive property.

$$= -2x + 10 + 8x + 8$$ Multiply.

$$= 6x + 18$$ Combine like terms.

Work Practice Problem 12

PRACTICE PROBLEM 13

Simplify:
$-(y + 1) + 3y - 12$

EXAMPLE 13 Simplify: $-(x + 4) + 5x + 16$

Solution: The expression $-(x + 4)$ means $-1(x + 4)$.

$$-(x + 4) + 5x + 16 = -1(x + 4) + 5x + 16$$ Apply the distributive property.

$$= -1 \cdot x + (-1)(4) + 5x + 16$$ Multiply.

$$= -x + (-4) + 5x + 16$$ Combine like terms.

$$= 4x + 12$$

Work Practice Problem 13

Answers

9. $21a - 15$ **10.** $30 - 6y$
11. $10y - 23$ **12.** $3x + 22$
13. $2y - 13$

Objective D Finding Perimeter and Area

EXAMPLE 14 Find the perimeter of the triangle.

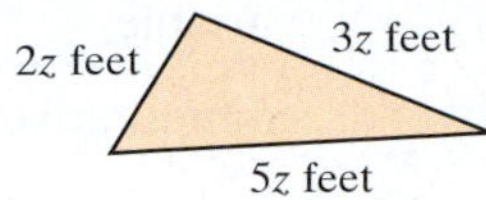

Solution: Recall that the perimeter of a figure is the distance around the figure. To find the perimeter, then, we find the sum of the lengths of the sides. We use the letter P to represent perimeter.

$$P = 2z + 3z + 5z$$
$$= 10z$$

Helpful Hint Don't forget to insert proper units.

The perimeter is $10z$ feet.

Work Practice Problem 14

PRACTICE PROBLEM 14

Find the perimeter of the square.

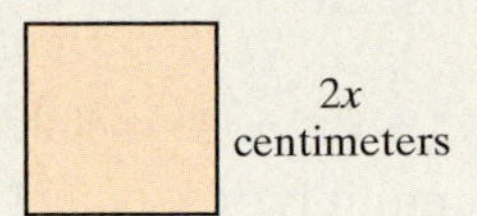

EXAMPLE 15 **Finding the Area of a Deck**

Find the area of the rectangular deck.

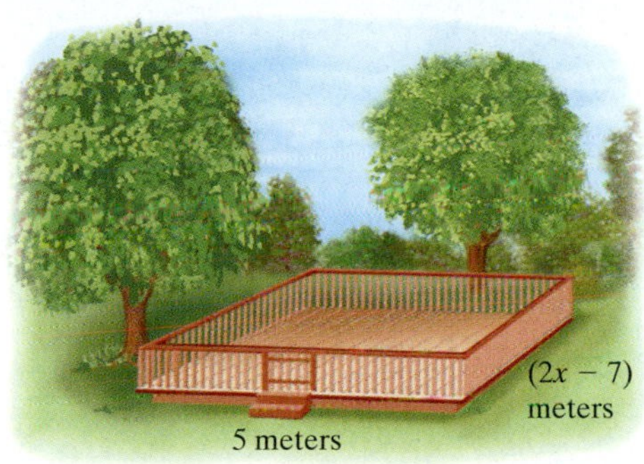

Solution: Recall how to find the area of a rectangle. Area = Length · Width, or if A represents area, l represents length, and w represents width, we have $A = l \cdot w$.

$$A = l \cdot w$$
$$= 5(2x - 7) \quad \text{Let length = 5 and width = } (2x - 7).$$
$$= 10x - 35 \quad \text{Multiply.}$$

The area is $(10x - 35)$ *square* meters.

Work Practice Problem 15

PRACTICE PROBLEM 15

Find the area of the rectangular garden.

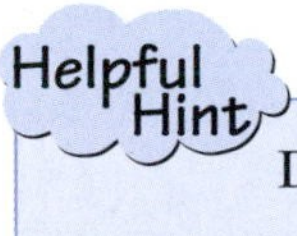

Don't forget . . .

Area:	Perimeter:
• surface enclosed	• distance around
• measured in square units	• measured in units

Answers

14. $8x$ cm
15. $(36y + 27)$ sq yd

Vocabulary and Readiness Check

Use the choices below to fill in each blank. Some choices may be used more than once.

numerical coefficient	combine like terms	like	term	variable	associative
constant	expression	unlike	distributive	commutative	

1. $14y^2 + 2x - 23$ is called a(n) ______________ while $14y^2$, $2x$, and -23 are each called a(n) ______________.
2. To multiply $3(-7x + 1)$, we use the ______________ property.
3. To simplify an expression like $y + 7y$, we ____________________.
4. By the ______________ properties, the *order* of adding or multiplying two numbers can be changed without changing their sum or product.
5. The term $5x$ is called a(n) ______________ term while the term 7 is called a(n) ______________ term.
6. The term z has an understood ____________________ of 1.
7. By the ______________ properties, the grouping of adding or multiplying numbers can be changed without changing their sum or product.
8. The terms $-x$ and $5x$ are ______________ terms and the terms $5x$ and $5y$ are ______________ terms.
9. For the term $-3x^2y$, -3 is called the __________________.

3.1 EXERCISE SET

FOR EXTRA HELP

Objective A *Simplify each expression by combining like terms. See Examples 1 through 5.*

1. $3x + 5x$ **2.** $8y + 3y$ **3.** $2n - 3n$ **4.** $7z - 10z$

5. $4c + c - 7c$ **6.** $5b - 8b - b$ **7.** $4x - 6x + x - 5x$ **8.** $8y + y - 2y - 8y$

9. $3a + 2a + 7a - 5$ **10.** $5b - 4b + b - 15$

Objective B *Multiply. See Examples 6 and 7.*

11. $6(7x)$ **12.** $4(4x)$ **13.** $-3(11y)$ **14.** $-3(21z)$

15. $12(6a)$ **16.** $13(5b)$

Multiply. See Examples 8 through 10.

17. $2(y + 3)$ **18.** $3(x + 1)$ **19.** $3(a - 6)$ **20.** $4(y - 6)$

21. $-4(3x + 7)$ **22.** $-8(8y + 10)$

Objective C *Simplify each expression. First use the distributive property to multiply and remove parentheses. See Examples 11 through 13.*

23. $2(x + 4) - 7$ **24.** $5(6 - y) - 2$ **25.** $8 + 5(3c - 1)$ **26.** $10 + 4(6d - 2)$

27. $-4(6n - 5) + 3n$ **28.** $-3(5 - 2b) - 4b$ **29.** $3 + 6(w + 2) + w$ **30.** $8z + 5(6 + z) + 20$

31. $2(3x + 1) + 5(x - 2)$ **32.** $3(5x - 2) + 2(3x + 1)$ **33.** $-(2y - 6) + 10$ **34.** $-(5x - 1) - 10$

Objectives A B C Mixed Practice *Simplify each expression. See Examples 1 t[illegible]*

35. $18y - 20y$ **36.** $x + 12x$ **37.** $z - 8z$

38. $-12x + 8x$ **39.** $9d - 3c - d$ **40.** $8r + s - 7s$

41. $2y - 6 + 4y - 8$ **42.** $a + 4 - 7a - 5$ **43.** $5q + p - 6q - p$

44. $m - 8n + m + 8n$ **45.** $2(x + 1) + 20$ **46.** $5(x - 1) + 18$

47. $5(x - 7) - 8x$ **48.** $3(x + 2) - 11x$ **49.** $-5(z + 3) + 2z$

50. $-8(1 + v) + 6v$ **51.** $8 - x + 4x - 2 - 9x$ **52.** $5y - 4 + 9y - y + 15$

53. $-7(x + 5) + 5(2x + 1)$ **54.** $-2(x + 4) + 8(3x - 1)$ **55.** $3r - 5r + 8 + r$

56. $6x - 4 + 2x - x + 3$ **57.** $-3(n - 1) - 4n$ **58.** $5(c + 2) + 7c$

59. $4(z - 3) + 5z - 2$ **60.** $8(m + 3) - 20 + m$ **61.** $6(2x - 1) - 12x$

62. $5(2a + 3) - 10a$ **63.** $-(4x - 5) + 5$ **64.** $-(7y - 2) + 6$

65. $-(4xy - 10) + 2(3xy + 5)$ **66.** $-(12ab - 10) + 5(3ab - 2)$ **67.** $3a + 4(a + 3)$

68. $b + 2(b - 5)$ **69.** $5y - 2(y - 1) + 3$ **70.** $3x - 4(x + 2) + 1$

Objective **D** *Find the perimeter of each figure. See Example 14.*

△ **71.**

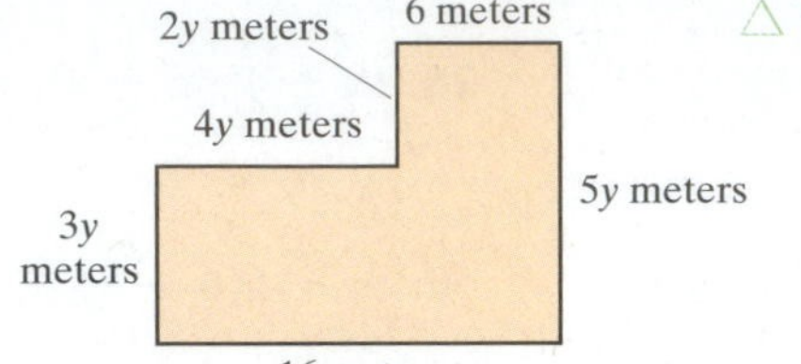

△ **72.**

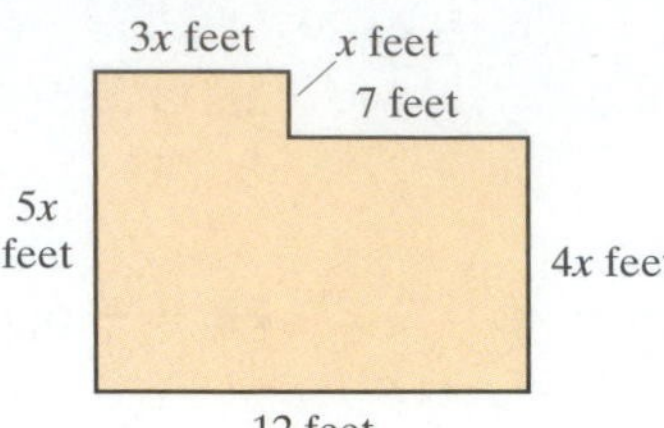

△ **73.**

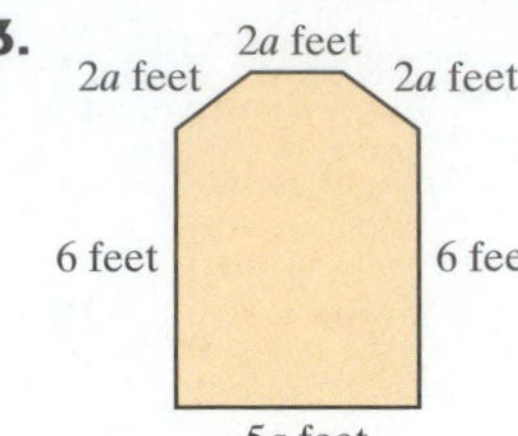

△ **74.**

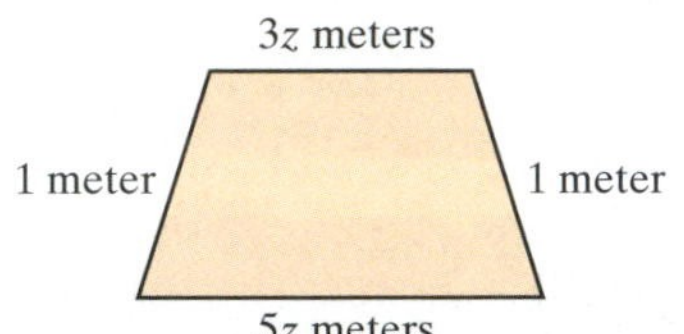

△ **75.**

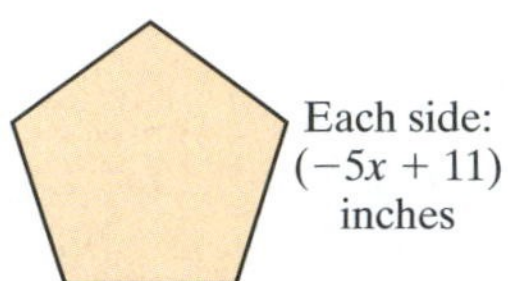

76.

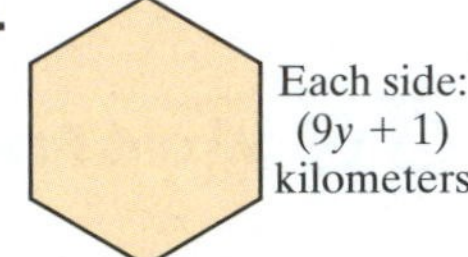

Find the area of each rectangle. See Example 15.

77.

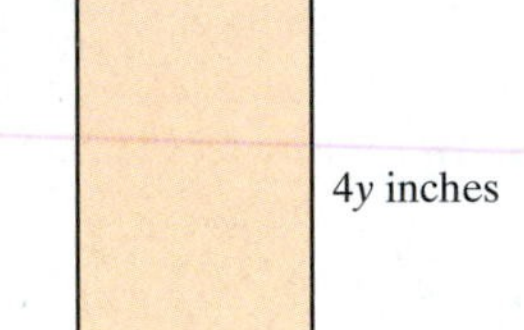

78.

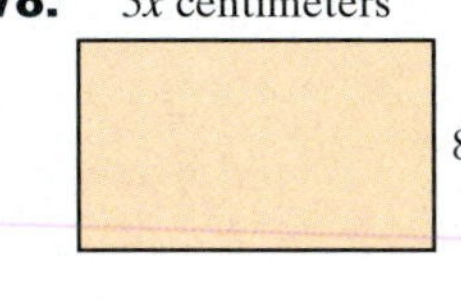

79.

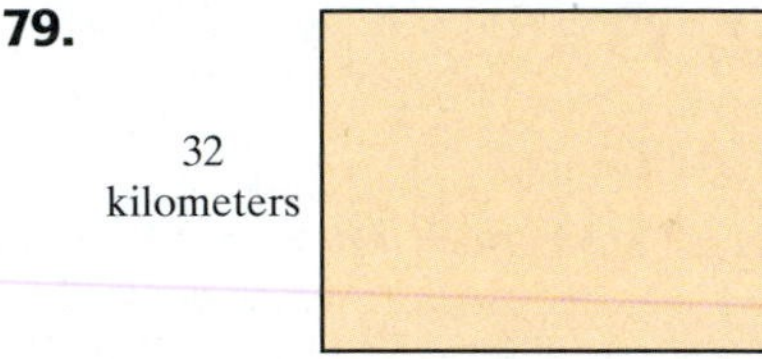

80.

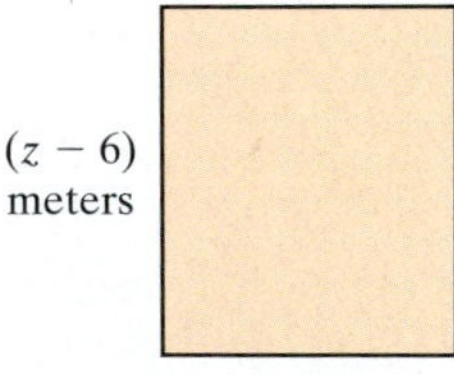

81.

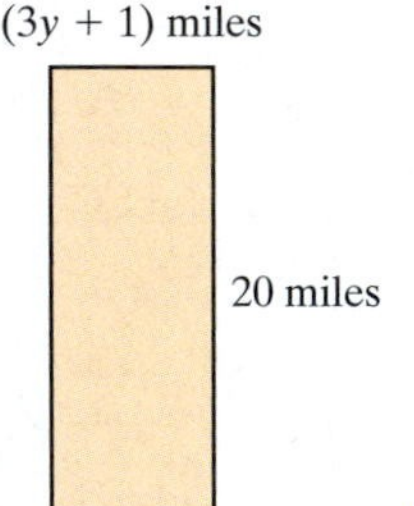

82.

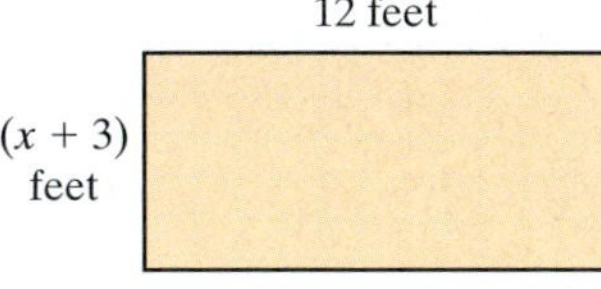

Objectives **A B C D** **Mixed Practice** *Solve. See Examples 1 through 15.*

83. Find the area of a rectangular movie screen that is 50 feet long and 40 feet high.

84. Find the area of a 60-meter by 25-meter rectangular swimming pool.

85. A decorator wishes to put a wallpaper border around a rectangular room that measures 14 feet by 18 feet. Find the room's perimeter.

86. How much fencing will a rancher need for a rectangular cattle lot that measures 80 feet by 120 feet?

87. Find the perimeter of a triangular garden that measures 5 feet by x feet by $(2x + 1)$ feet.

88. Find the perimeter of a triangular picture frame that measures x inches by x inches by $(x - 14)$ inches.

Review

Perform each indicated operation. See Sections 2.2 and 2.3.

89. $-13 + 10$

90. $-7 - (-4)$

91. $-4 - (-12)$

92. $-15 + 23$

93. $-4 + 4$

94. $8 + (-8)$

Concept Extensions

If the expression on the left side of the equal sign is equivalent to the right, write "correct." If not, write "incorrect" and then write an expression that is equivalent to the left side. See the Concept Check in Objective B.

95. $5(3x - 2) \stackrel{?}{=} 15x - 2$

96. $2(xy) \stackrel{?}{=} 2x \cdot 2y$

97. $7x - (x + 2) \stackrel{?}{=} 7x - x - 2$

98. $4(y - 3) + 11 \stackrel{?}{=} 4y - 6 + 11$

Review commutative, associative, and distributive properties. Then identify which property allows us to write the equivalent expression on the right side of the equal sign.

99. $6(2x - 3) + 5 = 12x - 18 + 5$

100. $9 + 7x + (-2) = 7x + 9 + (-2)$

101. $-7 + (4 + y) = (-7 + 4) + y$

102. $(x + y) + 11 = 11 + (x + y)$

Write the expression that represents the area of each composite figure. Then simplify to find the total area.

103.

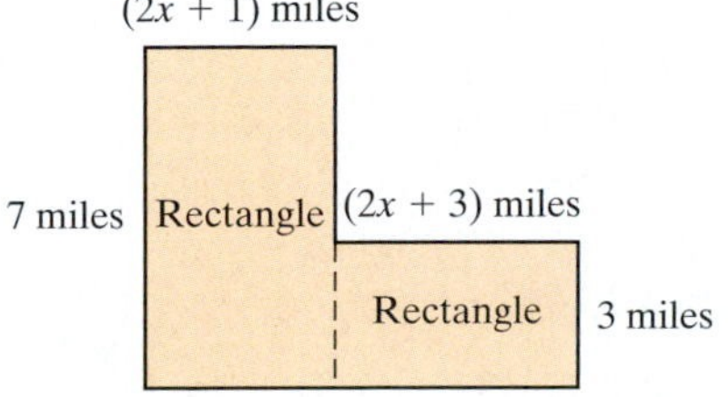

104.

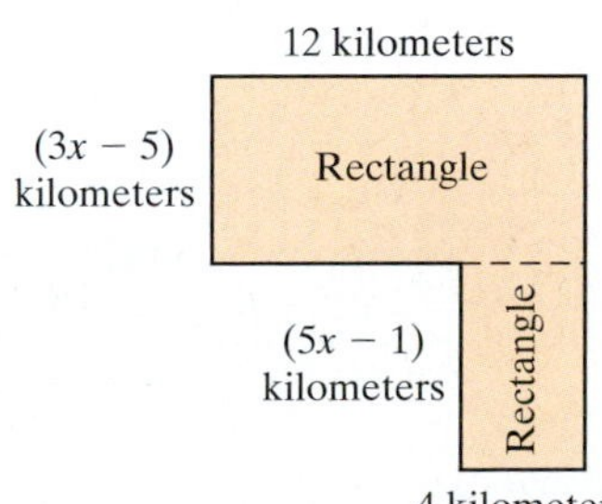

Simplify.

105. $9684q - 686 - 4860q + 12{,}960$

106. $76(268x + 592) - 2960$

107. If x is a whole number, which expression is the greatest: $2x$ or $5x$? Explain your answer.

108. If x is a whole number, which expression is the greatest: $-2x$ or $-5x$? Explain your answer.

109. Explain what makes two terms "like terms."

110. Explain how to combine like terms.

STUDY SKILLS BUILDER

Are You Familiar with Your Textbook Supplements?

There are many student supplements available for additional study. Below, I have listed some of these. See the Preface of this text or your instructor for further information.

Chapter Test Prep Video CD. This material is found in your textbook and is fully explained there. The CD contains video clip solutions to the Chapter Test exercises in this text and are excellent help when studying for chapter tests.

Lecture Video CDs. These video segments are keyed to each sections of the text. The material is presented by me, Elayn Martin-Gay, and I have placed a video icon by the exercises in the text that I have worked on the video.

The Student Solutions Manual. This contains worked out solutions to odd-numbered exercises as well as every exercise in the Integrated Reviews, Chapter Reviews, Chapter Tests, and Cumulative Reviews.

Prentice Hall Tutor Center. Mathematics questions may be phoned, faxed, or emailed to this center.

MyMathLab, MathXL, and Interact Math. These are computer and Internet tutorials. This supplement may already be available to you somewhere on campus, for example at your local learning resource lab. Take a moment and find the name and location of any such lab on campus.

As usual, your instructor is your best source of information.

Let's see how you are doing with textbook supplements:

1. Name one way the Chapter Test Prep Video can help you prepare for a chapter test.
2. List any textbook supplements that you have found useful.
3. Have you located and visited a learning resource lab located on your campus?
4. List the textbook supplements that are currently housed in your campus's learning resource lab.

3.2 SOLVING EQUATIONS: REVIEW OF THE ADDITION AND MULTIPLICATION PROPERTIES

Objectives

A Use the Addition Property or the Multiplication Property to Solve Equations.

B Use Both Properties to Solve Equations.

C Translate Word Phrases to Mathematical Expressions.

Objective A Using the Addition Property or the Multiplication Property

In this section, we continue solving equations using the properties first introduced in Section 2.6.

First, let's recall the difference between an **expression** and an **equation.** Remember—an equation contains an equals sign and an expression does not.

Equations	Expressions
$7x = 6x + 4$	$7x - 6x + 4$
$3(3y - 5) = 10y$	$y - 1 + 11y - 21$

equals signs

no equals signs

Thus far in this text, we have

Solved some equations (Sections 1.8 and 2.6)

and

Simplified some expressions (Section 3.1)

As we will see in this section, the ability to simplify expressions will help us as we solve more equations.

The addition and multiplication properties are reviewed below.

Addition Property of Equality

Let a, b, and c represent numbers. Then

$a = b$ and $a + c = b + c$ are equivalent equations.

Also, $a = b$ and $a - c = b - c$ are equivalent equations.

Multiplication Property of Equality

Let a, b, and c represent numbers and let $c \neq 0$. Then

$a = b$ and $a \cdot c = b \cdot c$ are equivalent equations.

Also, $a = b$ and $\frac{a}{c} = \frac{b}{c}$ are equivalent equations.

In other words, the same number may be added to or subtracted from *both* sides of an equation without changing the solution of the equation. Also, *both* sides of an equation may be multiplied or divided by the same nonzero number without changing the solution of the equation.

Many equations in this section will contain expressions that can be simplified. If one or both sides of an equation can be simplified, do that first.

PRACTICE PROBLEM 1

Solve: $x + 6 = 1 - 3$

EXAMPLE 1 Solve: $y - 5 = -2 - 6$

Solution: First we simplify the right side of the equation.

$$y - 5 = -2 - 6$$
$$y - 5 = -8 \quad \text{Combine like terms.}$$

Next we get y alone by using the addition property of equality. We add 5 to both sides of the equation.

$$y - 5 + 5 = -8 + 5 \quad \text{Add 5 to both sides.}$$
$$y = -3 \quad \text{Simplify.}$$

Check: To see that -3 is the solution, replace y with -3 in the original equation.

$$y - 5 = -2 - 6$$
$$-3 - 5 \stackrel{?}{=} -2 - 6 \quad \text{Replace } y \text{ with } -3.$$
$$-8 = -8 \quad \text{True}$$

Since $-8 = -8$ is true, the solution is -3.

Work Practice Problem 1

PRACTICE PROBLEM 2

Solve: $10 = 2m - 4m$

EXAMPLE 2 Solve: $3y - 7y = 12$

Solution: First, simplify the left side of the equation by combining like terms.

$$3y - 7y = 12$$
$$-4y = 12 \quad \text{Combine like terms.}$$

Next, we get y alone by using the multiplication property of equality and dividing both sides by -4.

$$\frac{-4y}{-4} = \frac{12}{-4} \quad \text{Divide both sides by } -4.$$
$$y = -3 \quad \text{Simplify.}$$

Check: Replace y with -3 in the original equation.

$$3y - 7y = 12$$
$$3(-3) - 7(-3) \stackrel{?}{=} 12$$
$$-9 + 21 \stackrel{?}{=} 12$$
$$12 = 12 \quad \text{True}$$

The solution is -3.

Work Practice Problem 2

PRACTICE PROBLEM 3

Solve: $-8 + 6 = \frac{a}{3}$

EXAMPLE 3 Solve: $\frac{z}{-4} = 11 - 5$

Solution: Simplify the right side of the equation first.

$$\frac{z}{-4} = 11 - 5$$
$$\frac{z}{-4} = 6$$

Answers

1. -8 **2.** -5 **3.** -6

Next, to get z alone, multiply both sides by -4.

$$-4 \cdot \frac{z}{-4} = -4 \cdot 6 \quad \text{Multiply both sides by } -4.$$

$$\frac{-4}{-4} \cdot z = -4 \cdot 6$$

$$1z = -24 \quad \text{or} \quad z = -24$$

Check to see that -24 is the solution.

Work Practice Problem 3

EXAMPLE 4 Solve: $5x + 2 - 4x = 7 - 19$

Solution: First we simplify each side of the equation separately.

$$5x + 2 - 4x = 7 - 19$$

$$\underbrace{5x - 4x} + 2 = \underbrace{7 - 19}$$

$$1x + 2 = -12 \quad \text{Combine like terms.}$$

To get x alone on the left side, we subtract 2 from both sides.

$$1x + 2 - 2 = -12 - 2 \quad \text{Subtract 2 from both sides.}$$

$$1x = -14 \quad \text{or} \quad x = -14 \quad \text{Simplify.}$$

Check to see that -14 is the solution.

Work Practice Problem 4

PRACTICE PROBLEM 4

Solve:
$-6y - 1 + 7y = 17 + 2$

EXAMPLE 5 Solve: $8x - 9x = 12 - 17$

Solution: First combine like terms on each side of the equation.

$$8x - 9x = 12 - 17$$

$$-x = -5$$

Recall that $-x$ means $-1x$ and divide both sides by -1.

$$\frac{-1x}{-1} = \frac{-5}{-1} \quad \text{Divide both sides by } -1.$$

$$x = 5 \quad \text{Simplify.}$$

Check to see that the solution is 5.

Work Practice Problem 5

PRACTICE PROBLEM 5

Solve: $-4 - 10 = 4y - 5y$

EXAMPLE 6 Solve: $3(3x - 5) = 10x$

Solution: First we multiply on the left side to remove the parentheses.

$$3(3x - 5) = 10x$$

$$3 \cdot 3x - 3 \cdot 5 = 10x \quad \text{Use the distributive property.}$$

$$9x - 15 = 10x$$

Now we subtract $9x$ from both sides.

$$9x - 15 - 9x = 10x - 9x \quad \text{Subtract } 9x \text{ from both sides.}$$

$$-15 = 1x \quad \text{or} \quad x = -15 \quad \text{Simplify.}$$

Work Practice Problem 6

PRACTICE PROBLEM 6

Solve: $13x = 4(3x - 1)$

Answers

4. 20 **5.** 14 **6.** -4

Objective B Using Both Properties to Solve Equations

We now solve equations in one variable using more than one property of equality. To solve an equation such as $2x - 6 = 18$, we first get the variable term $2x$ alone on one side of the equation.

PRACTICE PROBLEM 7

Solve: $5y + 2 = 17$

EXAMPLE 7 Solve: $2x - 6 = 18$

Solution: We start by adding 6 to both sides to get the variable term $2x$ alone.

$$2x - 6 = 18$$

$$2x - 6 + 6 = 18 + 6 \quad \text{Add 6 to both sides.}$$

$$2x = 24 \quad \text{Simplify.}$$

To finish solving, we divide both sides by 2.

$$\frac{2x}{2} = \frac{24}{2} \quad \text{Divide both sides by 2.}$$

$$x = 12 \quad \text{Simplify.}$$

Helpful Hint Don't forget to check the proposed solution in the *original* equation.

Check:

$$2x - 6 = 18$$

$$2(12) - 6 \stackrel{?}{=} 18 \quad \text{Replace } x \text{ with 12 and simplify.}$$

$$24 - 6 \stackrel{?}{=} 18$$

$$18 = 18 \quad \text{True}$$

The solution is 12.

Work Practice Problem 7

Don't forget, if one or both sides of an equation can be simplified, do that first.

PRACTICE PROBLEM 8

Solve:

$-4(x + 2) - 60 = 2 - 10$

EXAMPLE 8 Solve: $2 - 6 = -5(x + 4) - 39$

Solution: First, simplify each side of the equation.

$$2 - 6 = -5(x + 4) - 39$$

$$2 - 6 = -5x - 20 - 39 \quad \text{Use the distributive property.}$$

$$-4 = -5x - 59 \quad \text{Combine like terms on each side.}$$

$$-4 + 59 = -5x - 59 + 59 \quad \text{Add 59 to both sides to get the variable term alone.}$$

$$55 = -5x \quad \text{Simplify.}$$

$$\frac{55}{-5} = \frac{-5x}{-5} \quad \text{Divide both sides by } -5.$$

$$-11 = x \quad \text{or} \quad x = -11 \quad \text{Simplify.}$$

Check to see that -11 is the solution.

Work Practice Problem 8

Answers

7. 3 **8.** -15

Objective C Translating Word Phrases into Expressions

Section 3.4 in this chapter contains a formal introduction to problem solving. To prepare for this section, let's once again review writing phrases as algebraic expressions using the following key words and phrases as a guide:

Addition	Subtraction	Multiplication	Division
sum	difference	product	quotient
plus	minus	times	divided
added to	subtracted from	multiply	shared equally among
more than	less than	twice	per
increased by	decreased by	of	divided by
total	less	twice/double/triple	divided into

EXAMPLE 9 Write each phrase as an algebraic expression. Use x to represent "a number."

a. a number increased by -5
b. the product of -7 and a number
c. a number less 20
d. the quotient of -18 and a number
e. a number subtracted from -2

Solution:

a. In words: a number | increased by | -5

Translate: x $+$ (-5) or $x - 5$

b. In words:

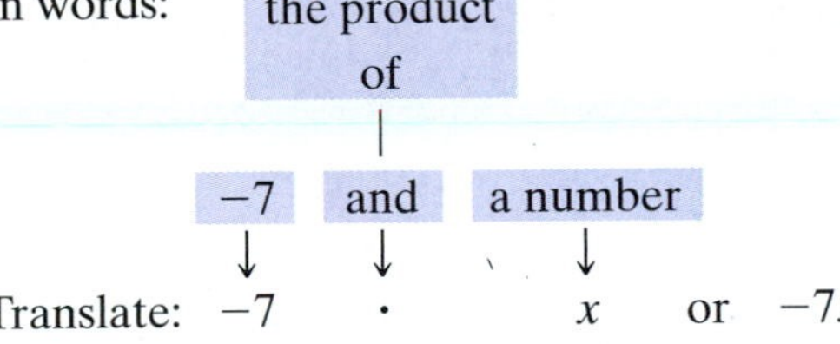

Translate: $-7 \cdot x$ or $-7x$

c. In words: a number | less | 20

Translate: $x - 20$

d. In words: the quotient of — -18 | and | a number

Translate: $-18 \div x$ or $\frac{-18}{x}$ or $-\frac{18}{x}$

e. In words:

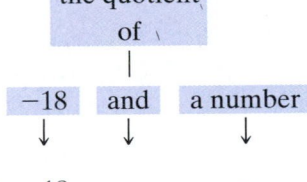

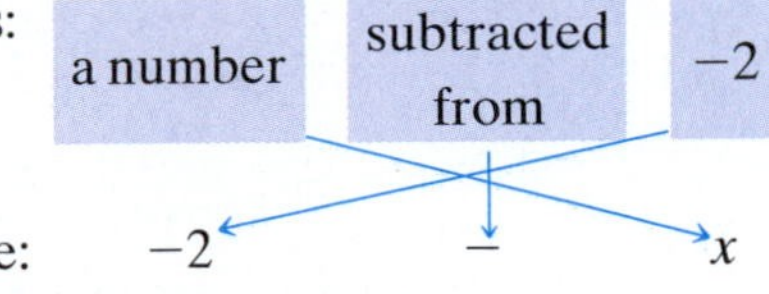

Translate: $-2 - x$

Work Practice Problem 9

PRACTICE PROBLEM 9

Write each phrase as an algebraic expression. Use x to represent "a number."

a. the sum of -3 and a number
b. -5 decreased by a number
c. three times a number
d. a number subtracted from 83
e. the quotient of a number and -4

Answers

9. a. $-3 + x$ **b.** $-5 - x$ **c.** $3x$ **d.** $83 - x$ **e.** $\frac{x}{-4}$ or $-\frac{x}{4}$

Helpful Hint

As we reviewed in Chapter 1, don't forget that order is important when subtracting. Notice the translation order of numbers and variables below.

Phrase	Translation
a number less 9	$x - 9$
a number subtracted from 9	$9 - x$

PRACTICE PROBLEM 10

Translate each phrase into an algebraic expression. Let x be the unknown number.

a. The product of 5 and a number, decreased by 25

b. Twice the sum of a number and 3

c. The quotient of 39 and twice a number

EXAMPLE 10 Write each phrase as an algebraic expression. Let x be the unknown number.

a. Twice a number, increased by -9

b. Three times the difference of a number and 11

c. The quotient of 5 times a number and 17

Solution:

a. In words: Twice a number | increased by | -9

Translate: $2x$ $+$ (-9) or $2x - 9$

b. In words: Three times | the difference of [a number | and | 11]

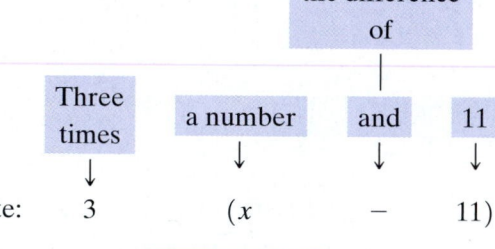

Translate: 3 $(x$ $-$ $11)$

c. In words: The quotient of

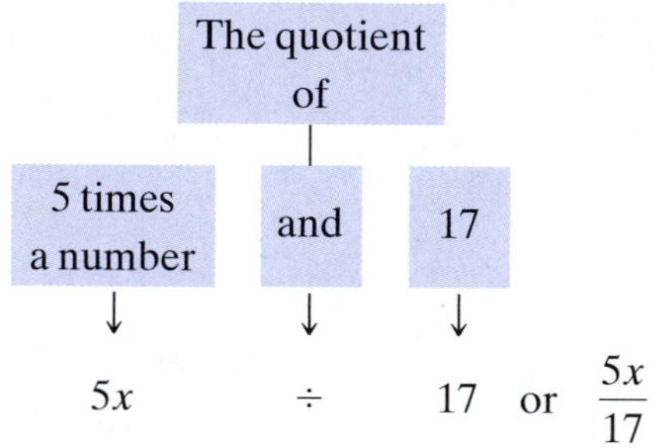

$5x$ $\div$ 17 or $\frac{5x}{17}$

Work Practice Problem 10

Answers

10. a. $5x - 25$ **b.** $2(x + 3)$ **c.** $39 \div 2x$ or $\frac{39}{2x}$

Vocabulary and Readiness Check

Use the choices below to fill in each blank.

equation	multiplication	equivalent	expression
solving	addition	simplifying	

1. The equations $-3x = 51$ and $\frac{-3x}{-3} = \frac{51}{-3}$ are called ______ equations.

2. The difference between an equation and an expression is that a(n) ______ contains an equal sign, while an ______ does not.

3. The process of writing $-3x + 10x$ as $7x$ is called ______ the expression.

4. For the equation $-5x - 1 = -21$, the process of finding that 4 is the solution is called ______ the equation.

5. By the ______ property of equality, $x = -2$ and $x + 7 = -2 + 7$ are equivalent equations.

6. By the ______ property of equality, $y = 8$ and $3 \cdot y = 3 \cdot 8$ are equivalent equations.

3.2 EXERCISE SET

FOR EXTRA HELP

Student Solutions Manual

PH Math/Tutor Center

CD/Video for Review

MathXL®

MyMathLab

Objective A *Solve each equation. First combine any like terms on each side of the equation. See Examples 1 through 5.*

1. $x - 3 = -1 + 4$

2. $x + 7 = 2 + 3$

3. $-7 + 10 = m - 5$

4. $1 - 8 = n + 2$

5. $2w - 12w = 40$

6. $10y - y = 45$

7. $24 = t + 3t$

8. $100 = 15y + 5y$

9. $2z = 12 - 14$

10. $-3x = 11 - 2$

11. $4 - 10 = \frac{z}{-3}$

12. $20 - 22 = \frac{z}{-4}$

13. $-3x - 3x = 50 - 2$

14. $5y - 9y = -14 + (-14)$

15. $\frac{x}{5} = -26 + 16$

16. $\frac{y}{3} = 32 - 52$

17. $7x + 7 - 6x = 10$

18. $-3 + 5x - 4x = 13$

19. $-8 - 9 = 3x + 5 - 2x$

20. $-7 + 10 = 4x - 6 - 3x$

Solve. First multiply to remove parentheses. See Example 6.

21. $2(5x - 3) = 11x$

22. $6(3x + 1) = 19x$

23. $3y = 2(y + 12)$

24. $17x = 4(4x - 6)$

25. $21y = 5(4y - 6)$

26. $28z = 9(3z - 2)$

27. $-1(4 - 6z) = 7z$

28. $-2(-1 - 3y) = 7y$

Objective B *Solve each equation. See Examples 7 and 8.*

29. $2x - 8 = 0$

30. $3y - 12 = 0$

31. $7y + 3 = 24$

32. $5m + 1 = 46$

33. $-7 = 2x - 1$

34. $-11 = 3t - 2$

35. $6(6y - 4) = 12y$

36. $4(3y - 5) = 14y$

37. $11(x - 6) = -4 - 7$

38. $5(x - 6) = -2 - 8$

39. $-3(x - 1) - 10 = 12 + 8$

40. $-2(x - 5) - 42 = -8 - 4$

41. $y - 20 = 6y$

42. $x - 63 = 10x$

43. $22 - 42 = 4(x - 1) - 4$

44. $35 - (-3) = 3(x - 2) + 17$

Objectives **A** **B** **Mixed Practice** *Solve each equation. See Examples 1 through 8.*

45. $-2 - 3 = -4 + x$

46. $7 - (-10) = x - 5$

47. $y + 1 = -3 + 4$

48. $y - 8 = -5 - 1$

49. $3w - 12w = -27$

50. $y - 6y = 20$

51. $-4x = 20 - (-4)$

52. $6x = 5 - 35$

53. $18 - 11 = \dfrac{x}{-5}$

54. $9 - 14 = \dfrac{x}{-12}$

55. $9x - 12 = 78$

56. $8x - 8 = 32$

57. $10 = 7t - 12t$

58. $-30 = t + 9t$

59. $5 - 5 = 3x + 2x$

60. $-42 + 20 = -2x + 13x$

61. $50y = 7(7y + 4)$

62. $65y = 8(8y - 9)$

63. $8x = 2(6x + 10)$

64. $10x = 6(2x - 3)$

65. $7x + 14 - 6x = -4 - 10$

66. $-10x + 11x + 5 = 9 - 5$

67. $\dfrac{x}{-4} = -1 - (-8)$

68. $\dfrac{y}{-6} = 6 - (-1)$

69. $23x + 8 - 25x = 7 - 9$

70. $8x - 4 - 6x = 12 - 22$

71. $-3(x + 9) - 41 = 4 - 60$

72. $-4(x + 7) - 30 = 3 - 37$

Objective **C** **Translating** *Write each phrase as a variable expression. Use x to represent "a number." See Examples 9 and 10.*

73. The sum of −7 and a number

74. Negative eight plus a number

75. Eleven subtracted from a number

76. A number subtracted from twelve

77. The product of −13 and a number

78. Twice a number

79. A number divided by −12

80. The quotient of negative six and a number

81. The product of −11 and a number, increased by 5

82. The quotient of −20 and a number, decreased by three

83. Negative ten decreased by 7 times a number

84. Twice a number decreased by thirty

85. Seven added to the product of 4 and a number

86. The product of 7 and a number, added to 100

87. Twice a number, decreased by 17

88. A number decreased by 14 and increased by 5 times the number

89. The product of -6 and the sum of a number and 15

90. Twice the sum of a number and -5

91. The quotient of 45 and the product of a number and -5

92. The quotient of ten times a number, and -4

93. The quotient of seventeen and a number, increased by -15

94. The difference of -9 times a number, and 1

Review

The trumpeter swan is the largest waterfowl in the United States. Although it was thought to be nearly extinct at the beginning of the twentieth century, recent conservation efforts have been succeeding. Use the bar graph to answer Exercises 95 through 98. See Section 1.3.

95. During what year shown is the number of trumpeter swans the greatest?

96. During what year shown is the number of trumpeter swans the least?

97. Use the length of the bar to estimate the number of trumpeter swans in 2005.

98. Describe any trends shown in this graph.

Trumpeter Swan Population in North America

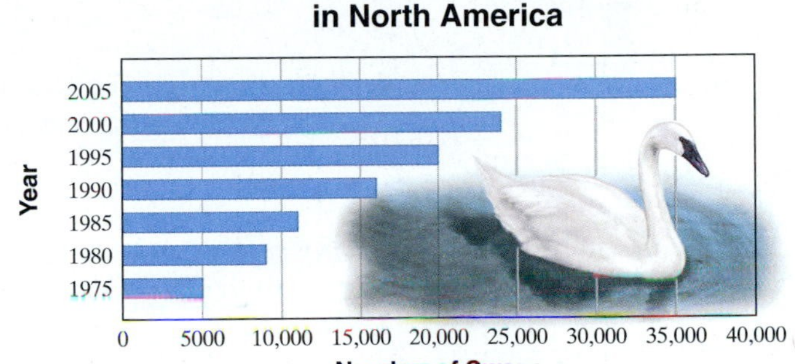

Source: U.S. Fish and Wildlife Service

Concept Extensions

99. In your own words, explain the addition property of equality.

100. Write an equation that can be solved using the addition property of equality.

101. Are the equations below equivalent? Why or why not?

$x + 7 = 4 + (-9)$

$x + 7 = 5$

102. Write 2 equivalent equations.

103. Why does the multiplication property of equality not allow us to divide both sides of an equation by zero?

104. Is the equation $-x = 6$ solved for the variable? Explain why or why not.

Solve.

105. $\frac{y}{72} = -86 - (-1029)$

106. $\frac{x}{-13} = 4^6 - 5^7$

107. $\frac{x}{-2} = 5^2 - |-10| - (-9)$

108. $\frac{y}{10} = (-8)^2 - |20| + (-2)^2$

109. $|-13| + 3^2 = 100y - |-20| - 99y$

110. $4(x - 11) + |90| - |-86| + 2^5 = 5x$

Answers

7. ____________

8. ____________

9. ____________

10. ____________

11. ____________

12. ____________

13. ____________

14. ____________

15. ____________

16. ____________

17. ____________

18. ____________

19. ____________

20. ____________

INTEGRATED REVIEW

Sections 3.1–3.2

Expressions and Equations

For the table below, identify each as an expression or an equation.

		Expression or Equation
1.	$7x - 5y + 14$	
2.	$7x = 35 + 14$	
3.	$3(x - 2) = 5(x + 1) - 17$	
4.	$-9(2x + 1) - 4(x - 2) + 14$	

Fill in each blank with "simplify" or "solve."

5. To ________ an expression, we combine any like terms.

6. To ________ an equation, we use the properties of equality to find any value of the variable that makes the equation a true statement.

Simplify each expression by combining like terms.

7. $7x + x$

8. $6y - 10y$

9. $2a + 5a - 9a - 2$

10. $6a - 12 - a - 14$

Multiply and simplify if possible.

11. $-2(4x + 7)$

12. $-3(2x - 10)$

13. $5(y + 2) - 20$

14. $12x + 3(x - 6) - 13$

15. Find the area of the rectangle.

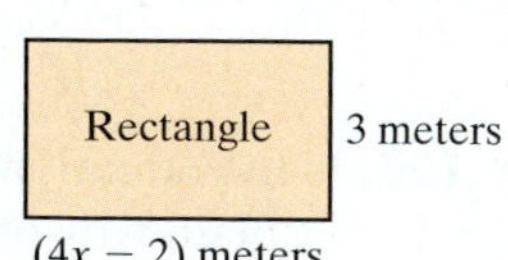

16. Find the perimeter of the triangle.

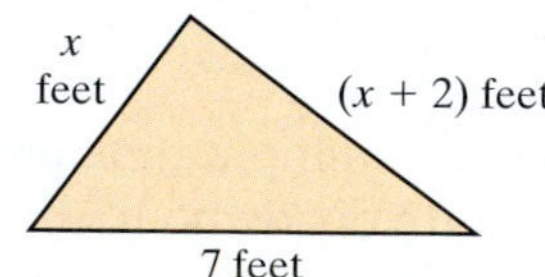

Solve and check.

17. $12 = 11x - 14x$

18. $8y + 7y = -45$

19. $x - 12 = -45 + 23$

20. $6 - (-5) = x + 5$

Solve and check.

21. $\frac{x}{3} = -14 + 9$

22. $\frac{z}{4} = -23 - 7$

23. $-6 + 2 = 4x + 1 - 3x$

24. $5 - 8 = 5x + 10 - 4x$

25. $6(3x - 4) = 19x$

26. $25x = 6(4x - 9)$

27. $-36x - 10 + 37x = -12 - (-14)$

28. $-8 + (-14) = -80y + 20 + 81y$

29. $3x - 16 = -10$

30. $4x - 21 = -13$

31. $-8z - 2z = 26 - (-4)$

32. $-12 + (-13) = 5x - 10x$

33. $-4(x + 8) - 11 = 3 - 26$

34. $-6(x - 2) + 10 = -4 - 10$

Translating *Write each phrase as an algebraic expression. Use x to represent "a number."*

35. The difference of a number and 10

36. The sum of −20 and a number

37. The product of 10 and a number

38. The quotient of 10 and a number

39. Five added to the product of −2 and a number

40. The product of −4 and the difference of a number and 1

21. ____
22. ____
23. ____
24. ____
25. ____
26. ____
27. ____
28. ____
29. ____
30. ____
31. ____
32. ____
33. ____
34. ____
35. ____
36. ____
37. ____
38. ____
39. ____
40. ____

Objectives

A Solve Linear Equations Using the Addition and Multiplication Properties.

B Solve Linear Equations Containing Parentheses.

C Write Numerical Sentences as Equations.

3.3 SOLVING LINEAR EQUATIONS IN ONE VARIABLE

In this chapter, the equations we are solving are called **linear equations in one variable** or **first-degree equations in one variable.** For example, an equation such as $5x - 2 = 6x$ is a linear equation in one variable. It is called linear or first degree because the exponent on each x is 1 and there is no variable below a fraction bar. It is an equation in one variable because it contains one variable, x.

Let's continue solving linear equations in one variable.

Objective A Solving Equations Using the Addition and Multiplication Properties

If an equation contains variable terms on both sides, we use the addition property of equality to get all the variable terms on one side and all the constants or numbers on the other side.

PRACTICE PROBLEM 1

Solve: $7x + 12 = 3x - 4$

EXAMPLE 1 Solve: $3a - 6 = a + 4$

Solution: Although it makes no difference which side you choose, let's move variable terms to the left side and constants to the right side.

$$3a - 6 = a + 4$$

$$3a - 6 + 6 = a + 4 + 6 \quad \text{Add 6 to both sides.}$$

$$3a = a + 10 \quad \text{Simplify.}$$

$$3a - a = a + 10 - a \quad \text{Subtract } a \text{ from both sides.}$$

$$2a = 10 \quad \text{Simplify.}$$

$$\frac{2a}{2} = \frac{10}{2} \quad \text{Divide both sides by 2.}$$

$$a = 5 \quad \text{Simplify.}$$

Check:

$$3a - 6 = a + 4 \quad \text{original equation}$$

$$3 \cdot 5 - 6 \stackrel{?}{=} 5 + 4 \quad \text{Replace } a \text{ with 5.}$$

$$15 - 6 \stackrel{?}{=} 9 \quad \text{Simplify.}$$

$$9 = 9 \quad \text{True}$$

The solution is 5.

Work Practice Problem 1

Helpful Hint

Make sure you understand which property to use to solve an equation.

Addition

$$x + 2 = 10$$

To undo addition of 2, we subtract 2 from both sides.

$$x + 2 - 2 = 10 - 2 \quad \text{Use addition property of equality.}$$

$$x = 8$$

Check: $x + 2 = 10$

$8 + 2 = 10$

$10 = 10$

Understood multiplication

$$2x = 10$$

To undo multiplication of 2, we divide both sides by 2.

$$\frac{2x}{2} = \frac{10}{2} \quad \text{Use multiplication property of equality.}$$

$$x = 5$$

Check: $2x = 10$

$2 \cdot 5 = 10$

$10 = 10$

Answer

1. -4

EXAMPLE 2 Solve: $17 - 7x + 3 = -3x + 21 - 3x$

Solution: First, simplify both sides of the equation.

$$17 - 7x + 3 = -3x + 21 - 3x$$
$$20 - 7x = -6x + 21 \quad \text{Simplify.}$$

Next, move variable terms on one side of the equation and constants, or numbers, to the other side. To begin, let's add $6x$ to both sides.

$$20 - 7x + 6x = -6x + 21 + 6x \quad \text{Add } 6x \text{ to each side.}$$
$$20 - x = 21 \quad \text{Simplify.}$$
$$20 - x - 20 = 21 - 20 \quad \text{Subtract 20 from both sides.}$$
$$-1x = 1 \quad \text{Simplify. Recall that } -x \text{ means } -1x.$$
$$\frac{-1x}{-1} = \frac{1}{-1} \quad \text{Divide both sides by } -1.$$
$$x = -1 \quad \text{Simplify.}$$

Check:
$$17 - 7x + 3 = -3x + 21 - 3x$$
$$17 - 7(-1) + 3 \stackrel{?}{=} -3(-1) + 21 - 3(-1)$$
$$17 + 7 + 3 \stackrel{?}{=} 3 + 21 + 3$$
$$27 = 27 \quad \text{True}$$

The solution is -1.

Work Practice Problem 2

PRACTICE PROBLEM 2

Solve:
$40 - 5y + 5 = -2y - 10 - 4y$

Objective B Solving Equations Containing Parentheses

Recall from the previous section that if an equation contains parentheses, we will first use the distributive property to remove them.

EXAMPLE 3 Solve: $7(x - 2) = 9x - 6$

Solution: First we apply the distributive property.

$$7(x - 2) = 9x - 6$$
$$7x - 14 = 9x - 6 \quad \text{Apply the distributive property.}$$

Next we move variable terms to one side of the equation and constants to the other side.

$$7x - 14 - 9x = 9x - 6 - 9x \quad \text{Subtract } 9x \text{ from both sides.}$$
$$-2x - 14 = -6 \quad \text{Simplify.}$$
$$-2x - 14 + 14 = -6 + 14 \quad \text{Add 14 to both sides.}$$
$$-2x = 8 \quad \text{Simplify.}$$
$$\frac{-2x}{-2} = \frac{8}{-2} \quad \text{Divide both sides by } -2.$$
$$x = -4 \quad \text{Simplify.}$$

Check to see that -4 is the solution.

Work Practice Problem 3

PRACTICE PROBLEM 3

Solve: $6(a - 5) = 4a + 4$

✔**Concept Check** In Example 3, the solution is -4. To check this solution, what equation should you use?

Answers

2. -55 **3.** 17

✔ **Concept Check Answer**

$7(x - 2) = 9x - 6$

You may want to use the following steps to solve equations.

Steps for Solving an Equation

Step 1: If parentheses are present, use the distributive property.

Step 2: Combine any like terms on each side of the equation.

Step 3: Use the addition property of equality to rewrite the equation so that variable terms are on one side of the equation and constant terms are on the other side.

Step 4: Use the multiplication property of equality to divide both sides by the numerical coefficient of the variable to solve.

Step 5: Check the solution in the *original equation*.

PRACTICE PROBLEM 4

Solve: $4(x + 3) + 1 = 13$

EXAMPLE 4 Solve: $3(2x - 6) + 6 = 0$

Solution: $3(2x - 6) + 6 = 0$

Step 1: $6x - 18 + 6 = 0$ Apply the distributive property.

Step 2: $6x - 12 = 0$ Combine like terms on the left side of the equation.

Step 3: $6x - 12 + 12 = 0 + 12$ Add 12 to both sides.

$6x = 12$ Simplify.

Step 4: $\frac{6x}{6} = \frac{12}{6}$ Divide both sides by 6.

$x = 2$ Simplify.

Check:

Step 5: $3(2x - 6) + 6 = 0$

$3(2 \cdot 2 - 6) + 6 \stackrel{?}{=} 0$

$3(4 - 6) + 6 \stackrel{?}{=} 0$

$3(-2) + 6 \stackrel{?}{=} 0$

$-6 + 6 \stackrel{?}{=} 0$

$0 = 0$ True

The solution is 2.

Work Practice Problem 4

Objective C Writing Numerical Sentences as Equations

Next, we practice translating sentences into equations. Below are key words and phrases that translate to an equals sign: (*Note:* For a review of key words and phrases that translate to addition, subtraction, multiplication, and division, see Sections 1.8 and 3.2.)

Key Words or Phrases	Examples	Symbols
equals	3 equals 2 plus 1	$3 = 2 + 1$
gives	the quotient of 10 and −5 gives −2	$\frac{10}{-5} = -2$
is/was	17 minus 12 is 5	$17 - 12 = 5$
yields	11 plus 2 yields 13	$11 + 2 = 13$
amounts to	twice −15 amounts to −30	$2(-15) = -30$
is equal to	−24 is equal to 2 times −12	$-24 = 2(-12)$

Answer

4. 0

EXAMPLE 5 Translate each sentence into an equation.

a. The product of 7 and 6 is 42.
b. Twice the sum of 3 and 5 is equal to 16.
c. The quotient of −45 and 5 yields −9.

Solution:

a. In words:

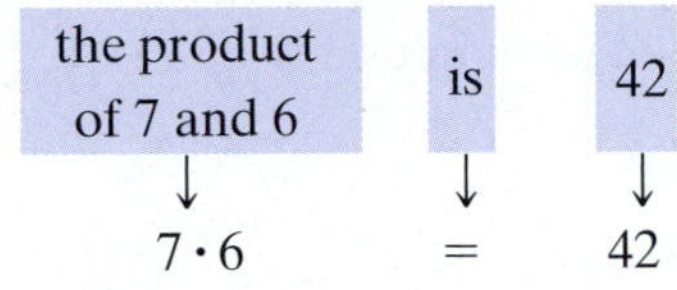

Translate: $7 \cdot 6 = 42$

b. In words:

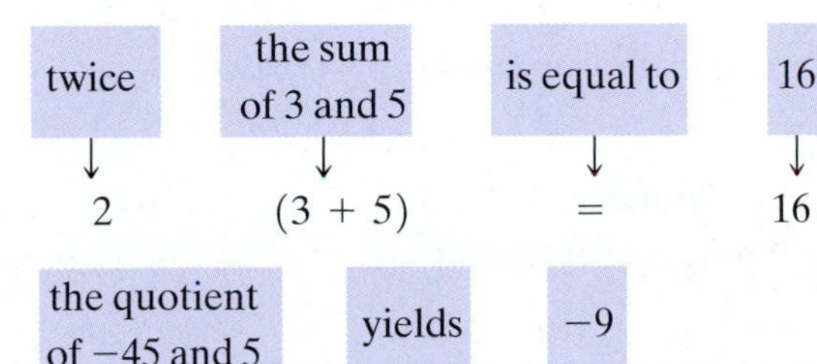

Translate: $2 \quad (3 + 5) \quad = \quad 16$

c. In words:

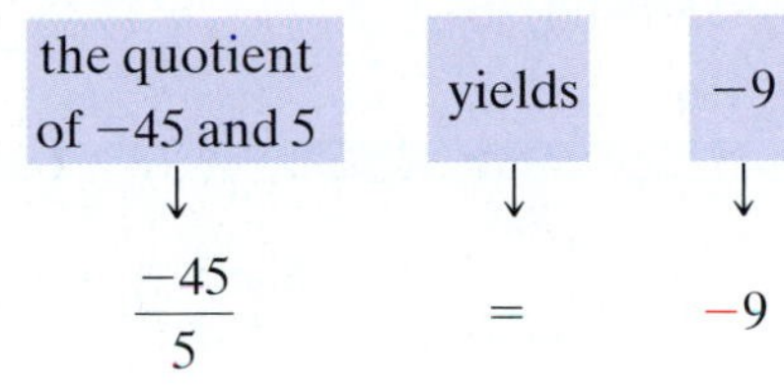

Translate: $\frac{-45}{5} = -9$

Work Practice Problem 5

PRACTICE PROBLEM 5

Translate each sentence into an equation.

a. The difference of 110 and 80 is 30.
b. The product of 3 and the sum of −9 and 11 amounts to 6.
c. The quotient of 24 and −6 yields −4.

Answers

5. a. $110 - 80 = 30$ **b.** $3(-9 + 11) = 6$ **c.** $\frac{24}{-6} = -4$

CALCULATOR EXPLORATIONS Checking Equations

A calculator can be used to check possible solutions of equations. To do this, replace the variable by the possible solution and evaluate each side of the equation separately. For example, to see whether 7 is a solution of the equation $52x = 15x + 259$, replace x with 7 and use your calculator to evaluate each side separately.

Equation: $52x = 15x + 259$

$52 \cdot 7 \stackrel{?}{=} 15 \cdot 7 + 259$ Replace x with 7.

Evaluate left side: [52] [×] [7] [=] or [ENTER]. Display: [364].

Evaluate right side: [15] [×] [7] [+] [259] [=] or [ENTER]. Display: [364].

Since the left side equals the right side, 7 is a solution of the equation $52x = 15x + 259$.

Use a calculator to determine whether the numbers given are solutions of each equation.

1. $76(x - 25) = -988$; 12
2. $-47x + 862 = -783$; 35
3. $x + 562 = 3x + 900$; −170
4. $55(x + 10) = 75x + 910$; −18
5. $29x - 1034 = 61x - 362$; −21
6. $-38x + 205 = 25x + 120$; 25

Vocabulary and Readiness Check

Use the choices below to fill in each blank. Some choices may be used more than once.

addition	multiplication	combine like terms
$5(2x + 6) - 1 = 39$	$3x - 9 + x - 16$	distributive

1. An example of an expression is ________________ while an example of an equation is ________________.

2. To solve $\frac{x}{-7} = -10$, we use the ________________ property of equality.

3. To solve $x - 7 = -10$, we use the ________________ property of equality.

Use the order of the Steps for Solving an Equation in this section to answer Exercises 4 through 6.

4. To solve $9x - 6x = 10 + 6$, first ________________.

5. To solve $5(x - 1) = 25$, first use the ________________ property.

6. To solve $4x + 3 = 19$, first use the ________________ property of equality.

3.3 EXERCISE SET

FOR EXTRA HELP

Student Solutions Manual | PH Math/Tutor Center |  CD/Video for Review | MathXL® | MyMathLab

Objective A *Solve each equation. See Examples 1 and 2.*

1. $3x - 7 = 4x + 5$

2. $7x - 1 = 8x + 4$

3. $10x + 15 = 6x + 3$

4. $5x - 3 = 2x - 18$

5. $19 - 3x = 14 + 2x$

6. $4 - 7m = -3m + 4$

7. $-14x - 20 = -12x + 70$

8. $57y + 140 = 54y - 100$

9. $x + 20 + 2x = -10 - 2x - 15$

10. $2x + 10 + 3x = -12 - x - 20$

11. $40 + 4y - 16 = 13y - 12 - 3y$

12. $19x - 2 - 7x = 31 + 6x - 15$

Objective B *Solve each equation. See Examples 3 and 4.*

13. $35 - 17 = 3(x - 2)$

14. $22 - 42 = 4(x - 1)$

15. $3(x - 1) - 12 = 0$

16. $2(x + 5) + 8 = 0$

17. $2(y - 3) = y - 6$

18. $3(z + 2) = 5z + 6$

19. $-2(y + 4) = 2$

20. $-1(y + 3) = 10$

21. $2t - 1 = 3(t + 7)$

22. $-4 + 3c = 4(c + 2)$

23. $3(5c + 1) - 12 = 13c + 3$

24. $4(3t + 4) - 20 = 3 + 5t$

Mixed Practice (*Sections 1.8, 3.2, 3.3*) *Solve each equation.*

25. $-4x = 44$

26. $-3x = 51$

27. $x + 9 = 2$

28. $y - 6 = -11$

29. $8 - b = 13$

30. $7 - z = 15$

31. $-20 - (-50) = \frac{x}{9}$

32. $-2 - 10 = \frac{z}{10}$

33. $3r + 4 = 19$

34. $7y + 3 = 38$

35. $-7c + 1 = -20$

36. $-2b + 5 = -7$

37. $8y - 13y = -20 - 25$

38. $4x - 11x = -14 - 14$

39. $6(7x - 1) = 43x$

40. $5(3y - 2) = 16y$

41. $-4 + 12 = 16x - 3 - 15x$

42. $-9 + 20 = 19x - 4 - 18x$

43. $-10(x + 3) + 28 = -16 - 16$

44. $-9(x + 2) + 25 = -19 - 19$

45. $4x + 3 = 2x + 11$

46. $6y - 8 = 3y + 7$

47. $-2y - 10 = 5y + 18$

48. $7n + 5 = 12n - 10$

49. $-8n + 1 = -6n - 5$

50. $10w + 8 = w - 10$

51. $9 - 3x = 14 + 2x$

52. $4 - 7m = -3m$

53. $9a + 29 + 7 = 0$

54. $10 + 4v + 6 = 0$

55. $7(y - 2) = 4y - 29$

56. $2(z - 2) = 5z + 17$

57. $12 + 5t = 6(t + 2)$

58. $4 + 3c = 2(c + 2)$

59. $3(5c - 1) - 2 = 13c + 3$

60. $4(2t + 5) - 21 = 7t - 6$

61. $10 + 5(z - 2) = 4z + 1$

62. $14 + 4(w - 5) = 6 - 2w$

63. $7(6 + w) = 6(2 + w)$

64. $6(5 + c) = 5(c - 4)$

Objective **C** **Translating** *Write each sentence as an equation. See Example 5.*

65. The sum of -42 and 16 is -26.

66. The difference of -30 and 10 equals -40.

67. The product of -5 and -29 gives 145.

68. The quotient of -16 and 2 yields -8.

69. Three times the difference of -14 and 2 amounts to -48.

70. Negative 2 times the sum of 3 and 12 is -30.

71. The quotient of 100 and twice 50 is equal to 1.

72. Seventeen subtracted from -12 equals -29.

Review

The following bar graph shows the number of U.S. federal individual income tax returns that are filed electronically during the years shown (some years are projected). Electronically filed returns include Telefile and online returns. Use this graph to answer Exercises 73 through 76. See Section 2.1.

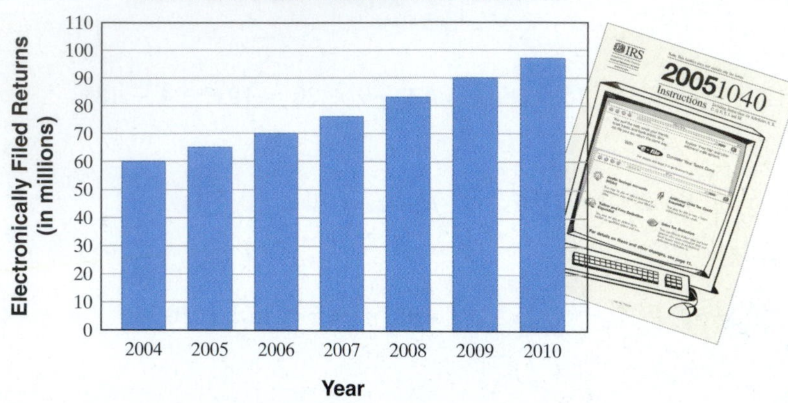

Source: IRS Compliance Research Division

73. Approximate the number of electronically filed returns projected for 2010.

74. Approximate the number of electronically filed returns projected for 2008.

75. By how much is the number of electronically filed returns expected to increase from 2006 to 2009?

76. Describe any trends shown in this graph.

Evaluate each expression for $x = 3$, $y = -1$, and $z = 0$. See Section 2.5.

77. $x^3 - 2xy$

78. $y^3 + 3xyz$

79. $y^5 - 4x^2$

80. $(-y)^3 + 3xyz$

Concept Extensions

Using the steps for solving an equation, choose the next operation for solving the given equation.

81. $2x - 5 = -7$

a. Add 7 to both sides.
b. Add 5 to both sides.
c. Divide both sides by 2.

82. $3x + 2x = -x - 4$

a. Add 4 to both sides.
b. Subtract $2x$ from both sides.
c. Add $3x$ and $2x$.

83. $-3x = -12$

a. Divide both sides by -3.
b. Add 12 to both sides.
c. Add $3x$ to both sides.

84. $9 - 5x = 15$

a. Divide both sides by -5.
b. Subtract 15 from both sides.
c. Subtract 9 from both sides.

85. A classmate shows you his steps for solving the given equation. His solution does not check, but he is unable to find the error. Check this solution, find the error, and correct it.

$$\begin{aligned} 2(3x - 5) &= 5x - 7 \\ 6x - 5 &= 5x - 7 \\ 6x - 5 + 5 &= 5x - 7 + 5 \\ 6x &= 5x - 2 \\ 6x - 5x &= 5x - 2 - 5x \\ x &= -2 \end{aligned}$$

Solve.

86. $(-8)^2 + 3x = 5x + 4^3$

87. $3^2 \cdot x = (-9)^3$

88. $2^3(x + 4) = 3^2(x + 4)$

89. $x + 45^2 = 54^2$

90. A classmate tries to solve $3x = 39$ by subtracting 3 from both sides of the equation. Will this step solve the equation for x? Why or why not?

91. A classmate tries to solve $2 + x = 20$ by dividing both sides by 2. Will this step solve the equation for x? Why or why not?

THE BIGGER PICTURE Operations on Sets of Numbers and Solving Equations

Continue your outline from Sections 1.6, 1.7, and 2.4. Suggestions are once again written to help you complete this part of your outline. Notice that this part of the outline has to do with solving equations.

I. Operations on Sets of Numbers

- **A. Whole Numbers**
  1. **Add or Subtract** (Section 1.3)
  2. **Multiply or Divide** (Sections 1.5, 1.6)
  3. **Exponent** (Section 1.7)
  4. **Order of Operations** (Section 1.7)
- **B. Integers**
  1. **Add** (Section 2.2)
  2. **Subtract** (Section 2.3)
  3. **Multiply or Divide** (Section 2.4)

II. Solving Equations

- **A. Equations in General:** Simplify both sides of the equation by removing parentheses and combining any like terms. Then use the addition property to write variable terms on one side, constants (or numbers) on the other side. Then use the multiplication property to solve for the variable by dividing both sides of the equation by the coefficient of the variable.

Solve: $2(x - 5) = 80$

$2x - 10 = 80$ Use the distributive property.

$2x - 10 + 10 = 80 + 10$ Add 10 to both sides.

$2x = 90$ Simplify.

$\frac{2x}{2} = \frac{90}{2}$ Divide both sides by 2.

$x = 45$ Simplify.

Solve.

1. $-8x = 40$
2. $x - 14 = -3$
3. $2x = -14$
4. $5y + 7 - 4y - 10 = 100$
5. $-3n + n = -100 - 50$
6. $8x + 9 = -79$
7. $4x - 5 = 2x - 17$
8. $30 + 5(2n - 4) = 100$

Objectives

A Write Sentences as Equations.

B Use Problem-Solving Steps to Solve Problems.

3.4 LINEAR EQUATIONS IN ONE VARIABLE AND PROBLEM SOLVING

Objective A Writing Sentences as Equations

Now that we have practiced solving equations for a variable, we can extend considerably our problem-solving skills. We begin by writing sentences as equations using the following key words and phrases as a guide:

Addition	Subtraction	Multiplication	Division	Equals Sign
sum	difference	product	quotient	equals
plus	minus	times	divide	gives
added to	subtracted from	multiply	shared equally among	is/was
more than	less than	twice	per	yields
increased by	decreased by	of	divided by	amounts to
total	less	double	divided into	is equal to

Notice that these sentences contain unknown numbers, which we will represent with x.

PRACTICE PROBLEM 1

Write each sentence as an equation. Use x to represent "a number."

a. Four times a number is 20.

b. The sum of a number and -5 yields 32.

c. Fifteen subtracted from a number amounts to -23.

d. Five times the difference of a number and 7 is equal to -8.

e. The quotient of triple a number and 5 gives 1.

EXAMPLE 1 Write each sentence as an equation. Use x to represent "a number."

a. Twenty increased by a number is 5.

b. Twice a number equals -10.

c. A number minus 11 amounts to 168.

d. Three times the sum of a number and 5 is -30.

e. The quotient of twice a number and 8 is equal to 2.

Solution:

a. In words: twenty | increased by | a number | is | 5

Translate: $20 \quad + \quad x \quad = \quad 5$

b. In words: twice a number | equals | -10

Translate: $2x \quad = \quad -10$

c. In words: a number | minus | 11 | amounts to | 168

Translate: $x \quad - \quad 11 \quad = \quad 168$

d. In words:

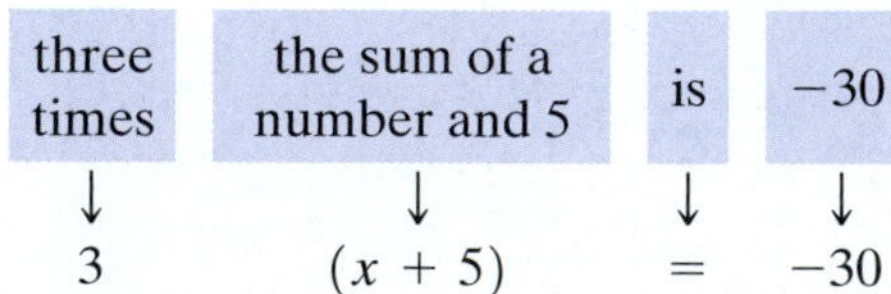

Translate: $3 \quad (x + 5) \quad = \quad -30$

Answers

1. a. $4x = 20$ **b.** $x + (-5) = 32$ **c.** $x - 15 = -23$ **d.** $5(x - 7) = -8$ **e.** $\frac{3x}{5} = 1$

e. In words:

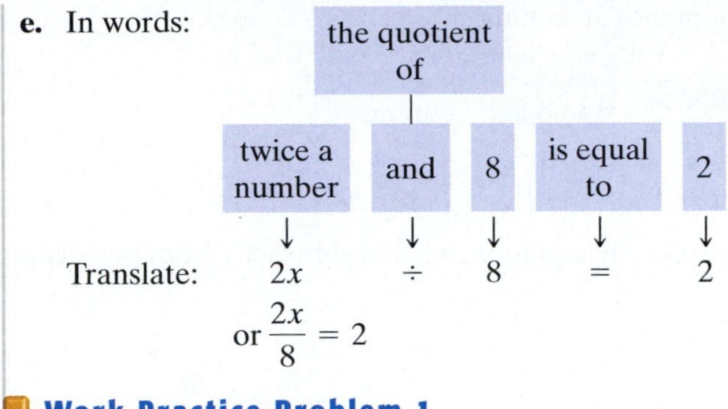

Translate: $2x \quad \div \quad 8 \quad = \quad 2$

or $\frac{2x}{8} = 2$

Work Practice Problem 1

Objective B Using Problem-Solving Steps to Solve Problems

Our main purpose for studying arithmetic and algebra is to solve problems. In previous sections, we have prepared for problem solving by writing phrases as algebraic expressions and sentences as equations. We now draw upon this experience as we solve problems. The following problem-solving steps will be used throughout this text.

Problem-Solving Steps

1. UNDERSTAND the problem. During this step, become comfortable with the problem. Some ways of doing this are as follows:
   - Read and reread the problem.
   - Construct a drawing.
   - Propose a solution and check. Pay careful attention to how you check your proposed solution. This will help when writing an equation to model the problem.
   - Choose a variable to represent the unknown. Use this variable to represent any other unknowns.
2. TRANSLATE the problem into an equation.
3. SOLVE the equation.
4. INTERPRET the results: *Check* the proposed solution in the stated problem and *state* your conclusion.

The first problem that we solve consists of finding an unknown number.

EXAMPLE 2 Finding an Unknown Number

Twice a number plus 3 is the same as the number minus 6. Find the unknown number.

Solution:

1. UNDERSTAND the problem. To do so, we read and reread the problem.
   Let's propose a solution to help us understand. Suppose the unknown number is 5. Twice this number plus 3 is $2 \cdot 5 + 3$ or 13. Is this the same as the number minus 6, or $5 - 6$, or -1? Since 13 is not the same as -1, we know that 5 is not the solution. However, remember that the purpose of proposing a solution is not to guess correctly, but to better understand the problem.
   Now let's choose a variable to represent the unknown. Let's let

   $x =$ unknown number

Continued on next page

PRACTICE PROBLEM 2

Translate "the sum of a number and 2 equals 6 added to three times the number" into an equation and solve.

Answer

2. -2

2. TRANSLATE the problem into an equation.

In words:	twice a number	plus 3	is the same as	the number minus 6
	↓	↓	↓	↓
Translate:	$2x$	$+ 3$	$=$	$x - 6$

3. SOLVE the equation. To solve the equation, we first subtract x from both sides.

$$2x + 3 = x - 6$$
$$2x + 3 - x = x - 6 - x$$
$$x + 3 = -6 \quad \text{Simplify.}$$
$$x + 3 - 3 = -6 - 3 \quad \text{Subtract 3 from both sides.}$$
$$x = -9 \quad \text{Simplify.}$$

4. INTERPRET the results. First, *Check* the proposed solution in the stated problem. Twice "−9" is −18 and −18 + 3 is −15. This is equal to the number minus 6, or "−9" − 6, or −15. Then *state* your conclusion: The unknown number is −9.

Work Practice Problem 2

✔ **Concept Check** Suppose you have solved an equation involving perimeter to find the length of a rectangular table. Explain why you would want to recheck your math if you obtain the result of −5.

PRACTICE PROBLEM 3

The distance by road from Cincinnati, Ohio, to Denver, Colorado, is 71 miles *less* than the distance from Denver to San Francisco, California. If the total of these two distances is 2399 miles, find the distance from Denver to San Francisco.

EXAMPLE 3 Determining Distances

The distance by road from Chicago, Illinois, to Los Angeles, California, is 1091 miles *more* than the distance from Chicago to Boston, Massachusetts. If the total of these two distances is 3017 miles, find the distance from Chicago to Boston. (*Source: World Almanac* 2006)

1. UNDERSTAND the problem. We read and reread the problem.

   Let's propose and check a solution to help us better understand the problem. Suppose the distance from Chicago to Boston is 600 miles. Since the distance from Chicago to Los Angeles is 1091 *more* miles, then this distance is 600 + 1091 = 1691 miles. With these numbers, the total of the distances is 600 + 1691 = 2291 miles. This is less than the given total of 3017 miles, so we are incorrect. But not only do we have a better understanding of this exercise, we also know that the distance from Boston to Chicago is greater than 600 miles since this proposed solution led to a total too small. Now let's choose a variable to represent an unknown. Then use this variable to represent any other unknown quantities. Let

   x = distance from Chicago to Boston

   Then

   $x + 1091$ = distance from Chicago to Los Angeles

   since that distance is 1091 more miles.

2. TRANSLATE the problem into an equation.

In words:	Chicago to Boston distance	+	Chicago to Los Angeles distance	=	total miles
	↓		↓		↓
Translate:	x	$+$	$x + 1091$	$=$	3017

Answer

3. 1235 miles

✔ **Concept Check Answer**

length cannot be negative

3. SOLVE the equation:

$$x + x + 1091 = 3017$$
$$2x + 1091 = 3017 \quad \text{Combine like terms.}$$
$$2x + 1091 - 1091 = 3017 - 1091 \quad \text{Subtract 1091 from both sides.}$$
$$2x = 1926 \quad \text{Simplify.}$$
$$\frac{2x}{2} = \frac{1926}{2} \quad \text{Divide both sides by 2.}$$
$$x = 963 \quad \text{Simplify.}$$

4. INTERPRET the results. First *Check* the proposed solution in the stated problem. Since x represents the distance from Chicago to Boston, this is 963 miles. The distance from Chicago to Los Angeles is $x + 1091 = 963 + 1091 = 2054$ miles. To check, notice that the total number of miles is $963 + 2054 = 3017$ miles, the given total of miles. Also, 2054 is 1091 more miles than 963, so the solution checks. Then, *state* your conclusion: The distance from Chicago to Boston is 963 miles.

Work Practice Problem 3

EXAMPLE 4 Calculating Separate Costs

A sales person at an electronics store sold a computer system and software for \$2100, receiving four times as much money for the computer system as for the software. Find the price of each.

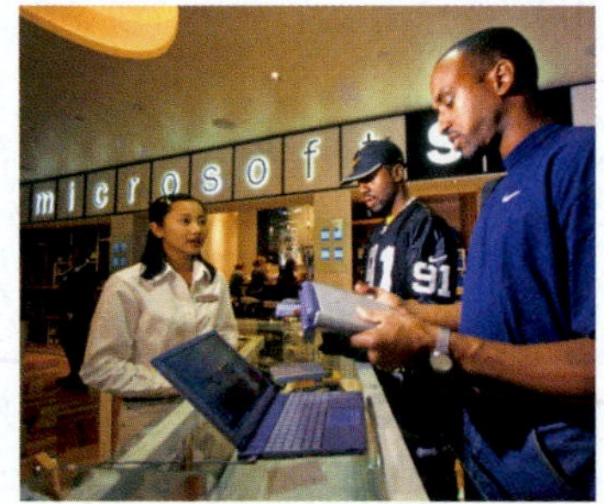

PRACTICE PROBLEM 4

A woman's \$57,000 estate is to be divided so that her husband receives twice as much as her son. How much will each receive?

Solution:

1. UNDERSTAND the problem. We read and reread the problem. Then we choose a variable to represent an unknown. We use this variable to represent any other unknown quantities. We let

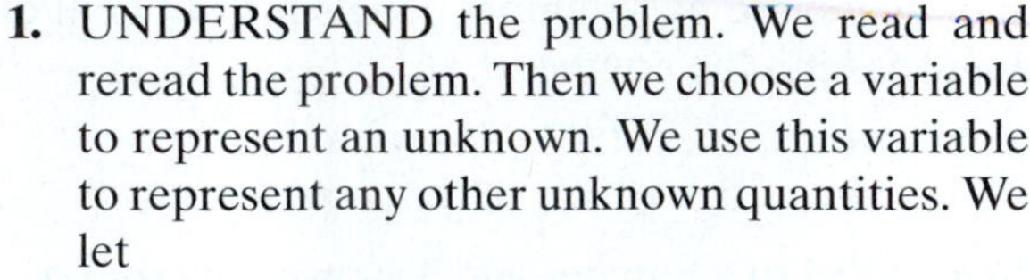

$x =$ the software price

$4x =$ the computer system price

2. TRANSLATE the problem into an equation.

In words:	software price	and	computer price	is	2100
	↓	↓	↓	↓	↓
Translate:	x	$+$	$4x$	$=$	2100

3. SOLVE the equation:

$$x + 4x = 2100$$
$$5x = 2100 \quad \text{Combine like terms.}$$
$$\frac{5x}{5} = \frac{2100}{5} \quad \text{Divide both sides by 5.}$$
$$x = 420 \quad \text{Simplify.}$$

4. INTERPRET the results. *Check* the proposed solution in the stated problem. The software sold for \$420. The computer system sold for $4x = 4(\$420) = \$1680$. Since $\$420 + \$1680 = \$2100$, the total price, and \$1680 is four times \$420, the solution checks. *State* your conclusion: The software sold for \$420, and the computer system sold for \$1680.

Work Practice Problem 4

Answer

4. husband: \$38,000; son: \$19,000

3.4 EXERCISE SET

FOR EXTRA HELP

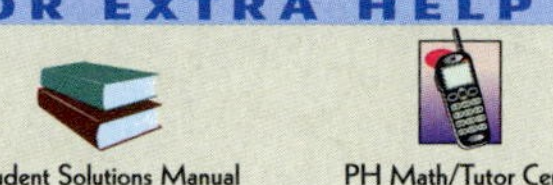
Student Solutions Manual

PH Math/Tutor Center

CD/Video for Review

MathXL®

MyMathLab

Objective A Translating *Write each sentence as an equation. Use x to represent "a number." See Example 1.*

1. A number added to −5 is −7.

2. Five subtracted from a number equals 10.

3. Three times a number yields 27.

4. The quotient of 8 and a number is −2.

5. A number subtracted from −20 amounts to 104.

6. Two added to twice a number gives −14.

7. Twice a number gives 108.

8. Five times a number is equal to −75.

9. The product of 5 and the sum of −3 and a number is −20.

10. Twice the sum of −17 and a number is −14.

Objective B Translating *Translate each to an equation. Then solve the equation. See Example 2.*

11. Three times a number, added to 9 is 33. Find the number.

12. Twice a number, subtracted from 60 is 20. Find the number.

13. The sum of 3, 4, and a number amounts to 16. Find the number.

14. The sum of 7, 9, and a number is 40. Find the number.

15. The difference of a number and 3 is equal to the quotient of 10 and 5. Find the number.

16. Eight decreased by a number equals the quotient of 15 and 5. Find the number.

17. Thirty less a number is equal to the product of 3 and the sum of the number and 6. Find the number.

18. The product of a number and 3 is twice the sum of that number and 5. Find the number.

19. 40 subtracted from five times a number is 8 more than the number. Find the number.

20. Five times the sum of a number and 2 is 11 less than the number times 8. Find the number.

21. Three times the difference of some number and 5 amounts to the quotient of 108 and 12. Find the number.

22. Seven times the difference of some number and 1 gives the quotient of 70 and 10. Find the number.

23. The product of 4 and a number is the same as 30 less twice that same number. Find the number.

24. Twice a number equals 25 less triple that same number. Find the number.

Solve. See Examples 3 and 4.

25. Based on the 2000 Census, Florida has 28 fewer electoral votes for president than California. If the total number of electoral votes for these two states is 82, find the number for each state. (*Source: The World Almanac* 2006)

26. In the 2004 presidential election, George W. Bush received 34 more electoral votes than John Kerry. If a total of 538 electoral votes were cast for the two candidates, find how many votes each candidate received. (*Source:* Voter News Service)

27. A falcon, when diving, can travel five times as fast as a pheasant's top speed. If the total speeds for these two birds is 222 miles per hour, find the fastest speed of the falcon and the fastest speed of the pheasant. (*Source: Fantastic Book of Comparisons*)

28. Norway has had three times as many rulers as Liechtenstein. If the total rulers for both countries is 56, find the number of rulers for Norway and the number for Liechtenstein.

29. The country with the most universities* is India, followed by the United States. If India has 2649 more universities than the United States and their combined total is 14,165, find the number of universities in India and the number in the United States [*Includes all further education establishments.] (*Source: The Top 10 of Everything*)

30. The average life expectancy for an elephant is 24 years longer than the life expectancy for a chimpanzee. If the total of these life expectancies is 130 years, find the life expectancy of each.

31. An Xbox 360 game system and several games are sold for $560. The cost of the Xbox 360 is 3 times as much as the cost of the games. Find the cost of the Xbox 360 and the cost of the games.

32. In a recent year, the two top-selling PC games were *World of Warcraft* and *The Sims 2: University Expansion Pack*. The average price of *World of Warcraft* is $14 more than the average price of *Sims*. If the total of these two prices is $80, find the price of each game. (*Source:* The NPD Group, Inc.)

33. By air, the distance from New York City to London is 2001 miles *less* than the distance from Los Angeles to Tokyo. If the total of these two distances is 8939 miles, find the distance from Los Angeles to Tokyo.

34. By air, the distance from Melbourne, Australia, to Cairo, Egypt, is 2338 miles *more* than the distance from Madrid, Spain, to Bangkok, Thailand. If the total of these distances is 15,012 miles, find the distance from Madrid to Bangkok.

35. The two NCAA stadiums with the largest capacities are Michigan Stadium (Univ. of Michigan) and Neyland Stadium (Univ. of Tennessee). Michigan Stadium has a capacity of 4647 more than Neyland. If the combined capacity for the two stadiums is 210,355, find the capacity for each stadium. (*Source*: National Collegiate Athletic Association)

36. A National Hot Rod Association (NHRA) top fuel dragster has a top speed of 95 mph faster than an Indy Racing League car. If the combined top speeds for these two cars is 565 mph, find the top speed of each car. (*Source: USA Today*)

37. In 2020, China is projected to be the country with the greatest number of visiting tourists. This number is twice the number of tourists projected for Spain. If the total number of tourists for these two countries is projected to be 210 million, find the number projected for each. (*Source: The State of the World Atlas* by Dan Smith)

38. California contains the largest state population of native Americans. This population is three times the native American population of Washington state. If the total of these two populations is 412 thousand, find the native American population in each of these two states. (*Source:* U.S. Census Bureau)

39. In Germany, about twice as many cars are manufactured per day than in Spain. If the total number of these cars manufactured per day is 19,827, find the number manufactured in Spain and the number manufactured in Germany.

40. A Toyota Camry is traveling twice as fast as a Dodge truck. If their combined speed is 105 miles per hour, find the speed of the car and find the speed of the truck.

41. Anthony Tedesco sold his used mountain bike and accessories for $270. If he received five times as much money for the bike as he did for the accessories, find how much money he received for the bike.

42. A tractor and a plow attachment are worth $1200. The tractor is worth seven times as much money as the plow. Find the value of the tractor and the value of the plow.

43. During the 2006 Women's NCAA Division I basketball championship game, the Maryland Terrapins scored 3 more points than the Duke Blue Devils. Together, both teams scored a total of 153 points. How many points did the 2006 Champion Maryland Terrapins score during this game? (*Source:* National Collegiate Athletic Association)

44. During the 2006 Men's NCAA Division I basketball championship game, the UCLA Bruins scored 16 fewer points than the Florida Gators. Together, both teams scored a total of 130 points. How many points did the 2006 Champion Florida Gators score during this game? (*Source:* National Collegiate Athletic Association)

Review

Round each number to the given place value. See Section 1.4.

45. 586 to the nearest ten

46. 82 to the nearest ten

47. 1026 to the nearest hundred

48. 52,333 to the nearest thousand

49. 2986 to the nearest thousand

50. 101,552 to the nearest hundred

Concept Extensions

51. Solve Example 3 again, but this time let x be the distance from Chicago to Los Angeles. Did you get the same results? Explain why or why not.

In real estate, a house's selling price P is found by adding the real estate agent's commission C to the amount A that the seller of the house receives: $P = A + C$.

52. A house sold for $230,000. The owner's real estate agent received a commission of $13,800. How much did the seller receive? (*Hint:* Substitute the known values into the equation, then solve the equation for the remaining unknown.)

53. A homeowner plans to use a real estate agent to sell his house. He hopes to sell the house for $165,000 and keep $156,750 of that. If everything goes as he has planned, how much will his real estate agent receive as a commission?

In retailing, the retail price P of an item can be computed using the equation $P = C + M$, *where C is the wholesale cost of the item and M is the amount of markup.*

54. The retail price of a computer system is $999 after a markup of $450. What is the wholesale cost of the computer system? (*Hint:* Substitute the known values into the equation, then solve the equation for the remaining unknown.)

55. Slidell Feed and Seed sells a bag of cat food for $12. If the store paid $7 for the cat food, what is the markup on the cat food?

STUDY SKILLS BUILDER

How Well Do You Know Your Textbook?

The questions below will determine whether you are familiar with your textbook. For help, see Section 1.1 in this text.

1. What does the icon mean?
2. What does the icon mean?
3. What does the △ icon mean?
4. Where can you find a review for each chapter? What answers to this review can be found in the back of your text?
5. Each chapter contains an overview of the chapter along with examples. What is this feature called?
6. Each chapter contains a review of vocabulary. What is this feature called?
7. There is a CD in your text. What content is contained on this CD?
8. What is the location of the section that is entirely devoted to study skills?
9. There are Practice Problems that are contained in the margin of the text. What are they and how can they be used?

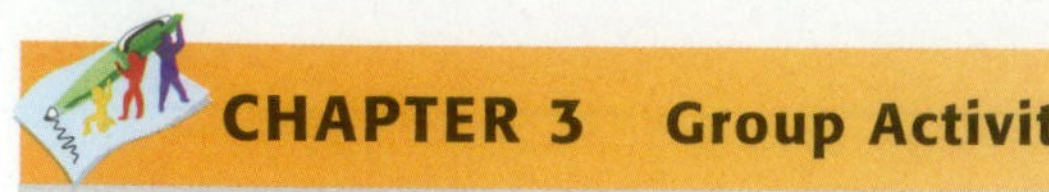

CHAPTER 3 Group Activity

Modeling Equation Solving with Addition and Subtraction

Sections 3.1–3.4

We can use positive counters ● and negative counters ● to help us model the equation-solving process. We also need to use an object that represents a variable. We use small slips of paper with the variable name written on them.

Recall that taking a ● and ● together creates a neutral or zero pair. After a neutral pair has been formed, it can be removed from or added to an equation model without changing the overall value. We also need to remember that we can add or remove the same number of positive or negative counters from both sides of an equation without changing the overall value.

We can represent the equation $x + 5 = 2$ as follows:

To get the variable by itself, we must remove 5 black counters from both sides of the model. Because there are only 2 counters on the right side, we must add 5 negative counters to both sides of the model. Then we can remove neutral pairs: 5 from the left side and 2 from the right side.

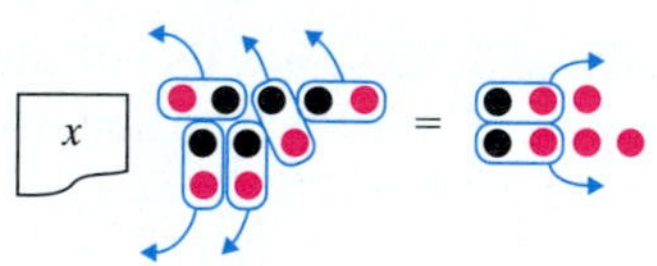

We are left with the following model, which represents the solution, $x = -3$.

Similarly, we can represent the equation $x - 4 = -6$ as follows:

To get the variable by itself, we must remove 4 red counters from both sides of the model

We are left with the following model, which represents the solution, $x = -2$.

Use the counter model to solve each equation.

1. $x - 3 = -7$

2. $x - 1 = -9$

3. $x + 2 = 8$

4. $x + 4 = 5$

5. $x + 8 = 3$

6. $x - 5 = -1$

7. $x - 2 = 1$

8. $x - 5 = 10$

9. $x + 3 = -7$

10. $x + 8 = -2$

Chapter 3 Vocabulary Check

Fill in each blank with one of the words or phrases listed below.

variable	addition	constant	algebraic expression	equation
terms	simplified	multiplication	evaluating the expression	solution
like	combined	numerical coefficient	distributive	

1. An algebraic expression is ______________ when all like terms have been ______________.
2. Terms that are exactly the same, except that they may have different numerical coefficients, are called ______________ terms.
3. A letter used to represent a number is called a ______________.
4. A combination of operations on variables and numbers is called an ______________.
5. The addends of an algebraic expression are called the ______________ of the expression.
6. The number factor of a variable term is called the ______________.
7. Replacing a variable in an expression by a number and then finding the value of the expression is called ______________ for the variable.
8. A term that is a number only is called a ______________.
9. An ______________ is of the form expression = expression.
10. A ______________ of an equation is a value for the variable that makes an equation a true statement.
11. To multiply $-3(2x + 1)$, we use the ______________ property.
12. By the ______________ property of equality, we may multiply or divide both sides of an equation by any nonzero number without changing the solution of the equation.
13. By the ______________ property of equality, the same number may be added to or subtracted from both sides of an equation without changing the solution of the equation.

Are you preparing for your test? Don't forget to take the Chapter 3 Test on page 208. Then check your answers at the back of the text and use the Chapter Test Prep Video CD to see the fully worked-out solutions to any of the exercises you want to review.

3 Chapter Highlights

Definitions and Concepts	**Examples**
Section 3.1 Simplifying Algebraic Expressions	
The addends of an algebraic expression are called the **terms** of the expression.	$5x^2 + (-4x) + (-2)$ — 3 terms
The number factor of a variable term is called the **numerical coefficient.**	**Term** / **Numerical Coefficient**: $7x$ — 7; $-6y$ — -6; x or $1x$ — 1
Terms that are exactly the same, except that they may have different numerical coefficients, are called **like terms.**	$5x + 11x = (5 + 11)x = 16x$ (like terms)
An algebraic expression is **simplified** when all like terms have been **combined.**	$y - 6y = (1 - 6)y = -5y$ (like terms)
Use the **distributive property** to multiply an algebraic expression by a term.	Simplify: $-4(x + 2) + 3(5x - 7)$ $= -4(x) + (-4)(2) + 3(5x) + 3(-7)$ $= -4x + (-8) + 15x + (-21)$ $= 11x + (-29)$ or $11x - 29$

Definitions and Concepts	Examples

Section 3.2 Solving Equations: Review of the Addition and Multiplication Properties

Addition Property of Equality

Let a, b, and c represent numbers. Then

$a = b$ and $a + c = b + c$ are equivalent equations.	Also, $a = b$ and $a - c = b - c$ are equivalent equations.

In other words, the same number may be added to or subtracted from both sides of an equation without changing the solution of the equation.

Solve for x:

$$x + 8 = 2 + (-1)$$
$$x + 8 = 1$$
$$x + 8 - 8 = 1 - 8 \quad \text{Subtract 8 from both sides.}$$
$$x = -7 \quad \text{Simplify.}$$

The solution is -7.

Multiplication Property of Equality

Let a, b, and c represent numbers and let $c \neq 0$. Then

$a = b$ and $a \cdot c = b \cdot c$ are equivalent equations.	Also, $a = b$ and $\frac{a}{c} = \frac{b}{c}$ are equivalent equations.

In other words, both sides of an equation may be multiplied or divided by the same nonzero number without changing the solution of the equation.

Solve: $y - 7y = 30$

$$-6y = 30 \quad \text{Combine like terms.}$$
$$\frac{-6y}{-6} = \frac{30}{-6} \quad \text{Divide both sides by } -6.$$
$$y = -5 \quad \text{Simplify.}$$

The solution is -5.

Section 3.3 Solving Linear Equations in One Variable

Steps for Solving an Equation

Step 1. If parentheses are present, use the distributive property.

Step 2. Combine any like terms on each side of the equation.

Step 3. Use the addition property of equality to rewrite the equation so that variable terms are on one side of the equation and constant terms are on the other side.

Step 4. Use the multiplication property of equality to divide both sides by the numerical coefficient of the variable to solve.

Step 5. Check the solution in the *original equation*.

Solve for x: $5(3x - 1) + 15 = -5$

Step 1. $15x - 5 + 15 = -5$ Apply the distributive property.

Step 2. $15x + 10 = -5$ Combine like terms.

Step 3. $15x + 10 - 10 = -5 - 10$ Subtract 10 from both sides.

$15x = -15$

Step 4. $\frac{15x}{15} = \frac{-15}{15}$ Divide both sides by 15.

$x = -1$

Step 5. Check to see that -1 is the solution.

Section 3.4 Linear Equations in One Variable and Problem Solving

Problem-Solving Steps

1. UNDERSTAND the problem. Some ways of doing this are

   Read and reread the problem.

   Construct a drawing.

   Choose a variable to represent an unknown in the problem.

The incubation period for a golden eagle is three times the incubation period for a hummingbird. If the total of their incubation periods is 60 days, find the incubation period for each bird. (*Source: Wildlife Fact File*, International Masters Publishers)

1. UNDERSTAND the problem. Then choose a variable to represent an unknown. Let

   x = incubation period of a hummingbird

   $3x$ = incubation period of a golden eagle

DEFINITIONS AND CONCEPTS	**EXAMPLES**
Section 3.4 Linear Equations in One Variable and Problem Solving (*continued*)	
2. TRANSLATE the problem into an equation.	**2.** TRANSLATE. incubation of hummingbird + incubation of golden eagle is 60 ↓ ↓ ↓ ↓ $x + 3x = 60$
3. SOLVE the equation.	**3.** SOLVE: $x + 3x = 60$ $4x = 60$ $\frac{4x}{4} = \frac{60}{4}$ $x = 15$
4. INTERPRET the results. *Check* the proposed solution in the stated problem and *state* your conclusion.	**4.** INTERPRET the solution in the stated problem. The incubation period for a hummingbird is 15 days. The incubation period for a golden eagle is $3x = 3 \cdot 15 = 45$ days. Since 15 days + 45 days = 60 days and 45 is 3(15), the solution checks. *State* your conclusion: The incubation period for a hummingbird is 15 days. The incubation period for a golden eagle is 45 days.

STUDY SKILLS BUILDER

Are You Prepared for a Test on Chapter 3?

Below I have listed some *common trouble areas* for students in Chapter 3. After studying for your test, but before taking your test, read these.

- Be careful when evaluating expressions. For example, evaluate $3x - y$ when $x = -2$ and $y = -3$.

$$3x - y = 3(-2) - (-3) \quad \text{Let } x = -2 \text{ and } y = -3.$$
$$= -6 - (-3) \quad \text{Multiply.}$$
$$= -6 + 3$$
$$= -3 \quad \text{Add.}$$

- Remember the distributive property.

$$5(4x - 3) + 2 = 5 \cdot 4x - 5 \cdot 3 + 2 \quad \text{Use the distributive property.}$$
$$= 20x - 15 + 2 \quad \text{Simplify.}$$
$$= 20x - 13 \quad \text{Combine like terms.}$$

- Don't forget the steps for solving a linear equation.

$$2(3x - 2) + 16 = 6$$
$$6x - 4 + 16 = 6 \quad \text{Apply the distributive property.}$$
$$6x + 12 = 6 \quad \text{Combine like terms.}$$
$$6x + 12 - 12 = 6 - 12 \quad \text{Subtract 12 from both sides.}$$
$$6x = -6 \quad \text{Simplify.}$$
$$\frac{6x}{6} = \frac{-6}{6} \quad \text{Divide both sides by 6.}$$
$$1 \cdot x = -1 \quad \text{Simplify.}$$
$$x = -1$$

Remember: This is simply a checklist of common trouble areas. For a review of Chapter 3 see the Highlights and Chapter Review at the end of this chapter.

3 CHAPTER REVIEW

(3.1) *Simplify each expression by combining like terms.*

1. $3y + 7y - 15$

2. $2y - 10 - 8y$

3. $8a + a - 7 - 15a$

4. $y + 3 - 9y - 1$

Multiply.

5. $2(x + 5)$

6. $-3(y + 8)$

Simplify.

7. $7x + 3(x - 4) + x$

8. $-(3m + 2) - m - 10$

9. $3(5a - 2) - 20a + 10$

10. $6y + 3 + 2(3y - 6)$

11. $6y - 7 + 11y - y + 2$

12. $10 - x + 5x - 12 - 3x$

Find the perimeter of each figure.

△ **13.**

$2x$ yards

3 yards

Rectangle

△ **14.**

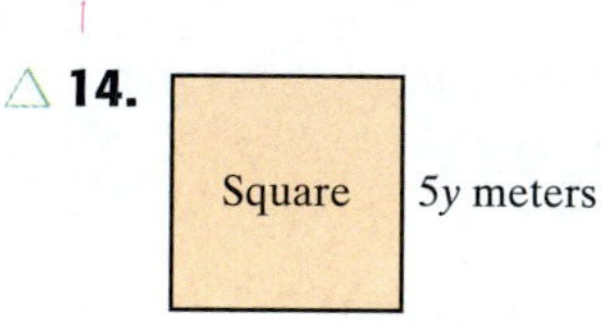

Find the area of each figure.

△ **15.**

$(2x - 1)$ yards

3 yards

Rectangle

△ **16.**

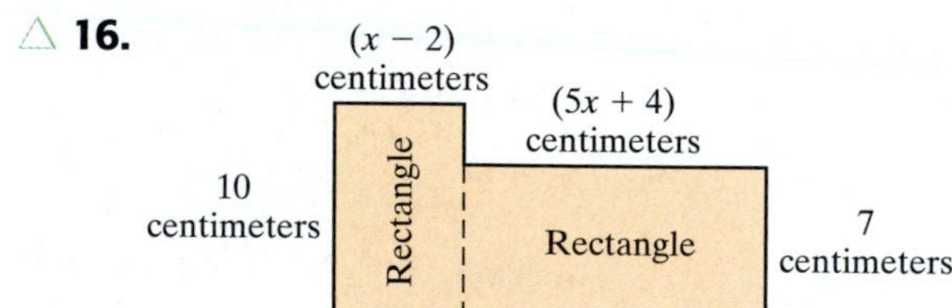

(3.2) *Solve each equation.*

17. $z - 5 = -7$

18. $3x + 10 = 4x$

19. $3y = -21$

20. $-3a = -15$

21. $\frac{x}{-6} = 2$

22. $\frac{y}{-15} = -3$

23. $n + 18 = 10 - (-2)$

24. $c - 5 = -13 + 7$

25. $7x + 5 - 6x = -20$

26. $17x = 2(8x - 4)$

27. $5x + 7 = -3$

28. $-14 = 9y + 4$

29. $\frac{z}{4} = -8 - (-6)$

30. $-1 + (-8) = \frac{x}{5}$

31. $6y - 7y = 100 - 105$

32. $19x - 16x = 45 - 60$

33. $9(2x - 7) = 19x$

34. $-5(3x + 3) = -14x$

35. $3x - 4 = 11$

36. $6y + 1 = 73$

37. $2(x + 4) - 10 = -2(7)$

38. $-3(x - 6) + 13 = 20 - 1$

Translating *Translate each phrase into an algebraic expression. Let x represent "a number."*

39. The product of -5 and a number

40. Three subtracted from a number

41. The sum of -5 and a number

42. The quotient of -2 and a number

43. Eleven added to twice a number

44. The product of -5 and a number, decreased by 50

45. The quotient of 70 and the sum of a number and 6

46. Twice the difference of a number and 13

(3.3) *Solve each equation.*

47. $2x + 5 = 7x - 100$

48. $-6x - 4 = x + 66$

49. $2x + 7 = 6x - 1$

50. $5x - 18 = -4x$

51. $5(n - 3) = 7 + 3n$

52. $7(2 + x) = 4x - 1$

53. $6x + 3 - (-x) = -20 + 5x - 7$

54. $x - 25 + 2x = -5 + 2x - 10$

55. $3(x - 4) = 5x - 8$

56. $4(x - 3) = -2x - 48$

57. $6(2n - 1) + 18 = 0$

58. $7(3y - 2) - 7 = 0$

59. $95x - 14 = 20x - 10 + 10x - 4$

60. $32z + 11 - 28z = 50 + 2z - (-1)$

Translating *Write each sentence as an equation.*

61. The difference of 20 and −8 is 28.

62. Five times the sum of 2 and −6 yields −20.

63. The quotient of −75 and the sum of 5 and 20 is equal to −3.

64. Nineteen subtracted from −2 amounts to −21.

(3.4) Translating *Write each sentence as an equation using x as the variable.*

65. Twice a number minus 8 is 40.

66. The product of some number and 6 is equal to the sum of the number and 2.

67. Twelve subtracted from the quotient of a number and 2 is 10.

68. The difference of a number and 3 is the quotient of the number and 4.

Solve.

69. Five times a number subtracted from 40 is the same as three times the number. Find the number.

70. The product of a number and 3 is twice the difference of that number and 8. Find the number.

71. In an election the incumbent received 14,000 votes of the 18,500 votes cast. Of the remaining votes, the Democratic candidate received 272 more than the Independent candidate. Find how many votes the Democratic candidate received.

72. Rajiv Puri has twice as many movies on DVDs as he has video tapes. Find the number of DVDs if he has a total of 126 movie recordings.

Mixed Review

Simplify.

73. $9x - 20x$

74. $-5(7x)$

75. $12x + 5(2x - 3) - 4$

76. $-7(x + 6) - 2(x - 5)$

Solve.

77. $c - 5 = -13 + 7$

78. $7x + 5 - 6x = -20$

79. $-7x + 3x = -50 - 2$

80. $-x + 8x = -38 - 4$

81. $9x + 12 - 8x = -6 + (-4)$

82. $-17x + 14 + 20x - 2x = 5 - (-3)$

83. $5(2x - 3) = 11x$

84. $\frac{y}{-3} = -1 - 5$

85. $12y - 10 = -70$

86. $-6(x - 3) = x + 4$

87. $4n - 8 = 2n + 14$

88. $9(3x - 4) + 63 = 0$

89. $-5z + 3z - 7 = 8z - 1 - 6$

90. $4x - 3 + 6x = 5x - 3 - 30$

91. Three times a number added to twelve is 27. Find the number.

92. Twice the sum of a number and four is ten. Find the number.

93. Out of the 50 states, Hawaii has the least number of roadway miles followed by Delaware. If Delaware has 1585 more roadway miles than Hawaii and the total number of roadway miles for both states is 10,203, find the number of roadway miles for each state.

94. North and South Dakota both have over 80,000 roadway miles. North Dakota has 3094 more miles and the total number of roadway miles for both states is 170,470. Find the number of roadway miles for North Dakota and for South Dakota.

Answers

1. ____

2. ____

3. ____

4. ____

5. ____

6. ____

7. ____

8. ____

9. ____

10. ____

11. ____

12. ____

13. ____

14. ____

15. ____

16. ____

3 CHAPTER TEST

Remember to check your answers and use the Chapter Test Prep Video to view solutions.

1. Simplify $7x - 5 - 12x + 10$ by combining like terms.

2. Multiply: $-2(3y + 7)$

3. Simplify: $-(3z + 2) - 5z - 18$

△ **4.** Write an expression that represents the perimeter of the equilateral triangle. Simplify the expression. (A triangle with three sides of equal length.)

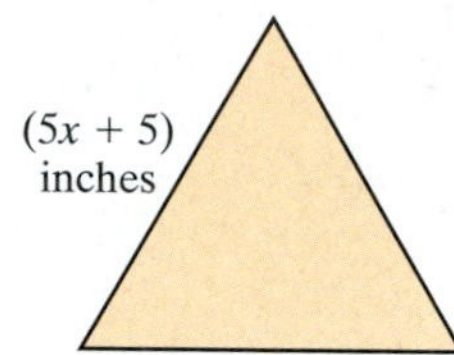

5. Write an expression that represents the area of the rectangle. Simplify the expression.

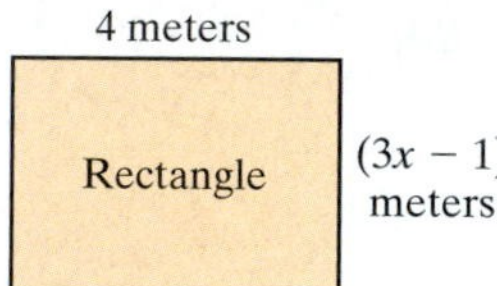

Solve each equation.

6. $12 = y - 3y$

7. $\frac{x}{2} = -5 - (-2)$

8. $5x + 12 - 4x - 14 = 22$

9. $-4x + 7 = 15$

10. $2(x - 6) = 0$

11. $-4(x - 11) - 34 = 10 - 12$

12. $5x - 2 = x - 10$

13. $4(5x + 3) = 2(7x + 6)$

14. $6 + 2(3n - 1) = 28$

Translate the following phrases into mathematical expressions. If needed, use x to represent "a number."

15. The sum of -23 and a number.

16. Three times a number, subtracted from -2.

Translate each sentence into an equation. If needed, use x to represent "a number."

17. The sum of twice 5 and -15 is -5.

18. Six added to three times a number equals -30.

Solve.

19. The difference of three times a number and five times the same number is 4. Find the number.

20. In a championship basketball game, Paula Zimmerman made twice as many free throws as Maria Kaminsky. If the total number of free throws made by both women was 12, find how many free throws Paula made.

21. In a 10-kilometer race, there are 112 more men entered than women. Find the number of female runners if the total number of runners in the race is 600.

17. ____________

18. ____________

19. ____________

20. ____________

21. ____________

CUMULATIVE REVIEW

Chapters 1–3

Answers

1. ______
2. ______
3. ______
4. ______
5. ______
6. ______
7. ______
8. ______
9. ______
10. ______
11. ______
12. ______
13. ______
14. ______
15. ______
16. ______
17. ______
18. a. ______
b. ______
19. ______
20. ______
21. ______
22. ______

1. Write 308,063,557 in words.

2. Write 276,004 in words.

△ **3.** Find the perimeter of the polygon shown.

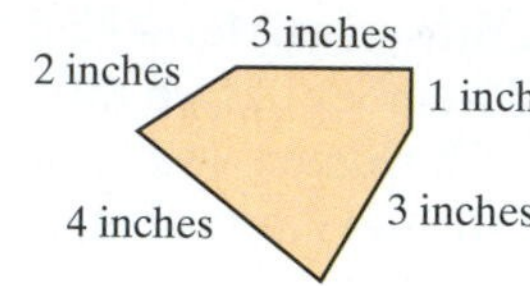

4. Find the perimeter of the rectangle shown.

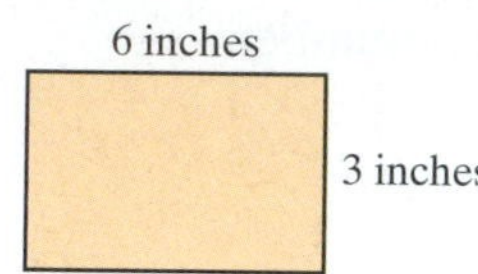

5. Subtract: 900 − 174. Check by adding.

6. Subtract: 17,801 − 8216. Check by adding.

7. Round 248,982 to the nearest hundred.

8. Round 844,497 to the nearest thousand.

9. Multiply: 25×8

10. Multiply: 395×74

11. Divide and check: $1872 \div 9$

12. Divide and check: $3956 \div 46$

13. Simplify: $2 \cdot 4 - 3 \div 3$

14. Simplify: $8 \cdot 4 + 9 \div 3$

15. Evaluate $x^2 + z - 3$ for $x = 5$ and $z = 4$.

16. Evaluate $2a^2 + 5 - c$ for $a = 2$ and $c = 3$.

17. Determine which numbers in the set {26, 40, 20} are solutions of the equation $2n - 30 = 10$.

18. Insert $<$ or $>$ to make a true statement.

a. $-14 \quad 0$
b. $-(-7) \quad -8$

19. Add using a number line: $5 + (-2)$

20. Add using a number line: $-3 + (-4)$

Add.

21. $-15 + (-10)$

22. $3 + (-7)$

23. $-2 + 5$

24. $21 + 15 + (-19)$

Subtract.

25. $-4 - 10$

26. $-2 - 3$

27. $6 - (-5)$

28. $19 - (-10)$

29. $-11 - (-7)$

30. $-16 - (-13)$

Divide.

31. $\frac{-12}{6}$

32. $\frac{-30}{-5}$

33. $-20 \div (-4)$

34. $26 \div (-2)$

35. $\frac{48}{-3}$

36. $\frac{-120}{12}$

Find the value of each expression.

37. $(-3)^2$

38. -2^5

39. -3^2

40. $(-5)^2$

41. Simplify: $2y - 6 + 4y + 8$

42. Simplify: $6x + 2 - 3x + 7$

43. Determine whether -1 is a solution of the equation $3y + 1 = 3$.

44. Determine whether 2 is a solution of $5x - 3 = 7$.

45. Solve: $-12x = -36$

46. Solve: $-3y = 15$

47. Solve: $2x - 6 = 18$

48. Solve: $3a + 5 = -1$

49. A salesperson at an electronics store sold a computer system and software for $2100, receiving four times as much money for the computer system as for the software. Find the price of each.

50. Rose Daunis is thinking of a number. Two times the number plus four is the same amount as three times the number minus seven. Find Rose's number.

23. ______
24. ______
25. ______
26. ______
27. ______
28. ______
29. ______
30. ______
31. ______
32. ______
33. ______
34. ______
35. ______
36. ______
37. ______
38. ______
39. ______
40. ______
41. ______
42. ______
43. ______
44. ______
45. ______
46. ______
47. ______
48. ______
49. ______
50. ______

4 Fractions and Mixed Numbers

Fractions are numbers and, like whole numbers and integers, they can be added, subtracted, multiplied, and divided. Fractions are very useful and appear frequently in everyday language, in common phrases such as "half an hour," "quarter of a pound," and "third of a cup." This chapter reviews the concept of fractions and mixed numbers and demonstrates how to add, subtract, multiply, and divide these numbers.

SECTIONS

CHECK YOUR PROGRESS

The following graph is called a circle graph or pie chart. Each sector (shaped like a piece of pie) shows the fraction of entering college freshmen in the United States who expect to major in each discipline shown. The whole circle represents the entire class of college freshmen. In Section 4.2, Exercises 92–95, Section 4.3, Exercise 85, and Section 4.5, Exercises 95–98, we study this circle graph further and use the information to project the number of college freshmen in a certain major.

College Freshmen Majors

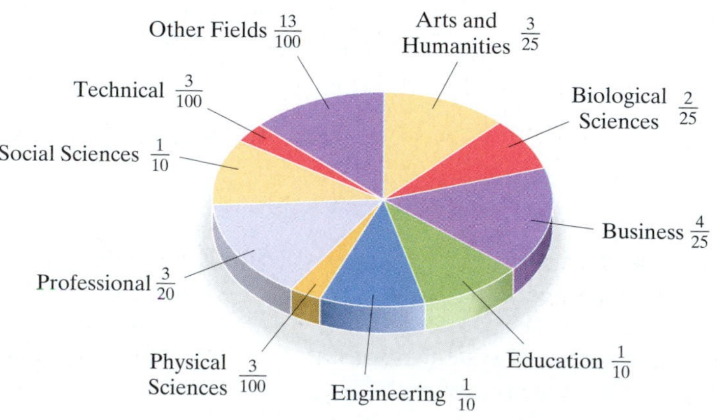

Source: The Higher Education Research Institute

4.1 INTRODUCTION TO FRACTIONS AND MIXED NUMBERS

Objective A Identifying Numerators and Denominators

Whole numbers are used to count whole things or units, such as cars, horses, dollars, and people. To refer to a part of a whole, fractions can be used. Here are some examples of **fractions.** Study these examples for a moment.

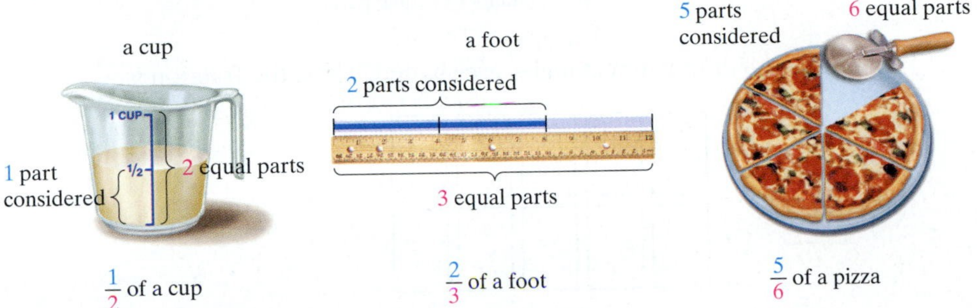

$\frac{1}{2}$ of a cup $\quad$ $\frac{2}{3}$ of a foot $\quad$ $\frac{5}{6}$ of a pizza

In a fraction, the top number is called the **numerator** and the bottom number is called the **denominator.** The bar between the numbers is called the **fraction bar.**

Names	Fraction	Meaning
numerator ⟶	5	⟵ number of parts being considered
denominator ⟶	6	⟵ number of equal parts in the whole

EXAMPLES Identify the numerator and the denominator of each fraction.

1. $\frac{3}{7}$ ← numerator (3), ← denominator (7)

2. $\frac{13}{5x}$ ← numerator (13), ← denominator ($5x$)

Work Practice Problems 1–2

Helpful Hint

$\frac{3}{7}$ ← Remember that the bar in a fraction means division. Since division by 0 is undefined, a fraction with a denominator of 0 is undefined. For example, $\frac{3}{0}$ is undefined.

Objective B Writing Fractions to Represent Parts of Figures or Real-Life Data

One way to become familiar with the concept of fractions is to visualize fractions with shaded figures. We can then write a fraction to represent the shaded area of the figure (or diagram).

Objectives

- **A** Identify the Numerator and the Denominator of a Fraction.
- **B** Write a Fraction to Represent Parts of Figures or Real-Life Data.
- **C** Graph Fractions on a Number Line.
- **D** Review Division Properties for 0 and 1.
- **E** Write Mixed Numbers as Improper Fractions.
- **F** Write Improper Fractions as Mixed Numbers or Whole Numbers.

PRACTICE PROBLEMS 1–2

Identify the numerator and the denominator of each fraction.

1. $\frac{11}{2}$

2. $\frac{10y}{17}$

Answers

1. numerator = 11, denominator = 2
2. numerator = $10y$, denominator = 17

PRACTICE PROBLEMS 3–4

Write a fraction to represent the shaded part of each figure.

3.

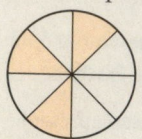

4.

EXAMPLES Write a fraction to represent the shaded part of each figure.

3. In this figure, 2 of the 5 equal parts are shaded. Thus, the fraction is $\frac{2}{5}$.

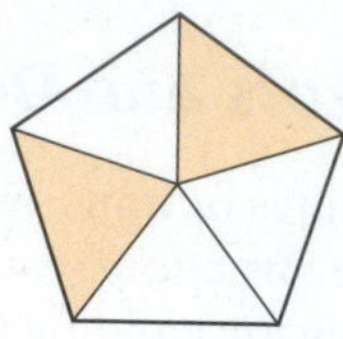

$\frac{2}{5}$ ← number of parts shaded
← number of equal parts

4. In this figure, 3 of the 10 rectangles are shaded. Thus, the fraction is $\frac{3}{10}$.

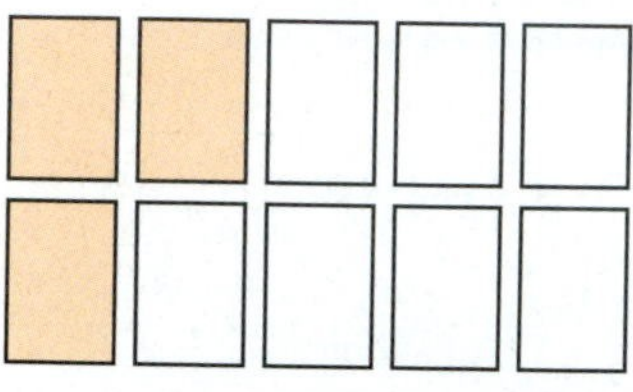

$\frac{3}{10}$ ← number of parts shaded
← number of equal parts

Work Practice Problems 3–4

PRACTICE PROBLEMS 5–6

Write a fraction to represent the part of the whole shown.

5. Just consider this part of the syringe

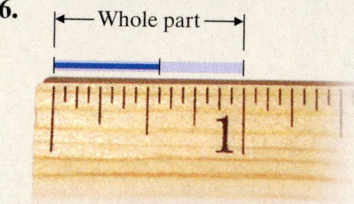

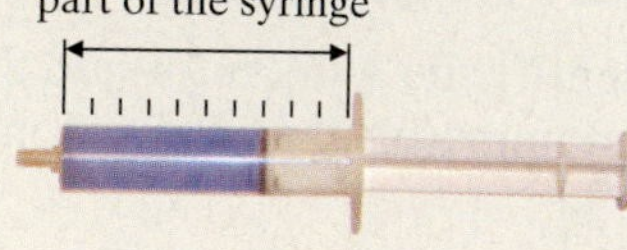

6. Whole part

EXAMPLES Write a fraction to represent the shaded part of the diagram.

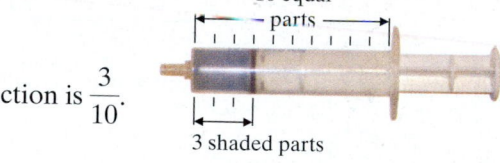

5. The fraction is $\frac{3}{10}$.

10 equal parts

3 shaded parts

6. The fraction is $\frac{1}{3}$.

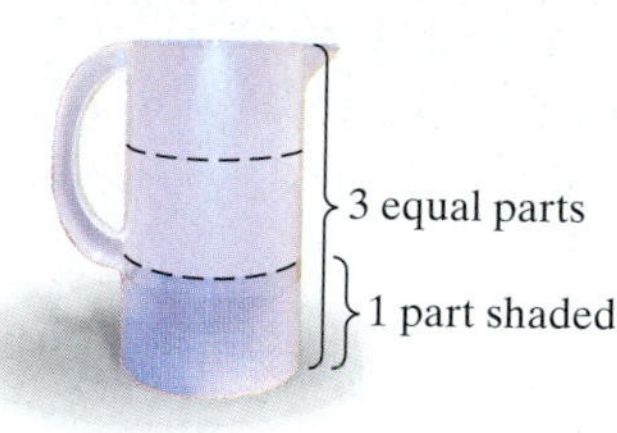

Work Practice Problems 5–6

PRACTICE PROBLEM 7

Draw and shade a part of a figure to represent the fraction.

7. $\frac{2}{3}$ of a figure

EXAMPLES Draw a figure and then shade a part of it to represent each fraction.

7. $\frac{5}{6}$ of a figure

We will use a geometric figure such as a rectangle. Since the denominator is 6, we divide it into 6 equal parts. Then we shade 5 of the equal parts.

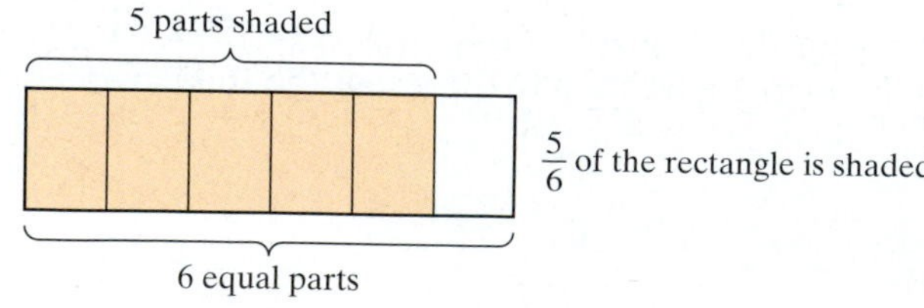

Answers

3. $\frac{3}{8}$ 4. $\frac{1}{6}$ 5. $\frac{7}{10}$ 6. $\frac{9}{16}$

7. answers may vary; for example,

8. $\frac{3}{8}$ of a figure

If you'd like, our figure can consist of 8 triangles of the same size. We will shade 3 of the triangles.

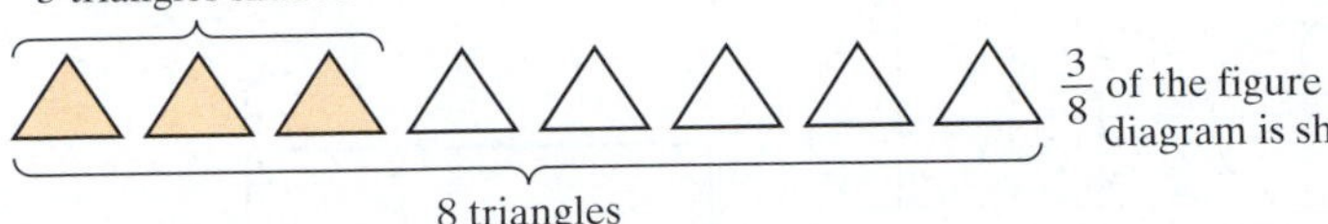

■ **Work Practice Problems 7–8**

✔Concept Check If ▭▭▭▭▭▭ represents $\frac{6}{7}$ of a whole diagram, sketch the whole diagram.

EXAMPLE 9 Writing Fractions from Real-Life Data

Of the eight planets in our solar system, (Pluto is now a dwarf planet), three are closer to the Sun than Mars. What fraction of the planets are closer to the Sun than Mars?

Solution: The fraction of planets closer to the Sun than Mars is:

$\frac{3}{8}$ ← number of planets closer / ← number of planets in our solar system

Thus, $\frac{3}{8}$ of the planets in our solar system are closer to the Sun than Mars.

■ **Work Practice Problem 9**

A **proper fraction** is a fraction whose numerator is less than its denominator. Proper fractions are less than 1. For example, the shaded portion of the triangle is represented by $\frac{2}{3}$.

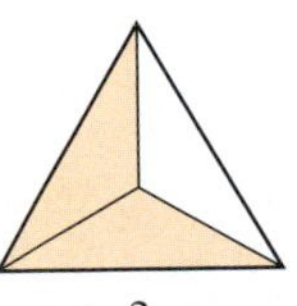

$\frac{2}{3}$

An **improper fraction** is a fraction whose numerator is greater than or equal to its denominator. Improper fractions are greater than or equal to 1.

PRACTICE PROBLEM 8

Draw and shade a part of a figure to represent the fraction.

8. $\frac{7}{11}$ of a figure

PRACTICE PROBLEM 9

Of the eight planets in our solar system, five are farther from the Sun than Earth is. What fraction of the planets are farther from the Sun than Earth is?

Answers

8. answers may vary; for example, ▭▭▭▭▭▭▭▭▭▭▭ (7 of 11 shaded)

9. $\frac{5}{8}$

✔ Concept Check Answer

▭▭▭▭▭▭▭ (6 of 7 shaded)

The shaded part of the group of circles below is $\frac{9}{4}$. The shaded part of the rectangle is $\frac{6}{6}$. Recall that $\frac{6}{6}$ simplifies to 1 and notice that the entire rectangle (1 whole figure) is shaded below.

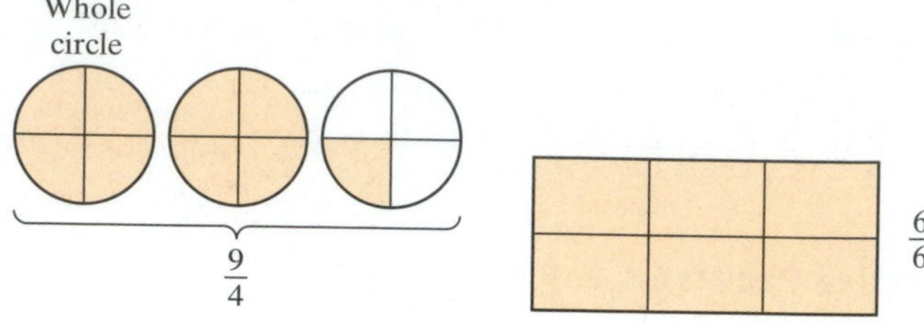

A **mixed number** contains a whole number and a fraction. Mixed numbers are greater than 1. Earlier, we wrote the shaded part of the group of circles below as the improper fraction $\frac{9}{4}$. Now let's write the shaded part as a mixed number. The shaded part of the group of circles' area is $2\frac{1}{4}$. Read this as "two and one-fourth."

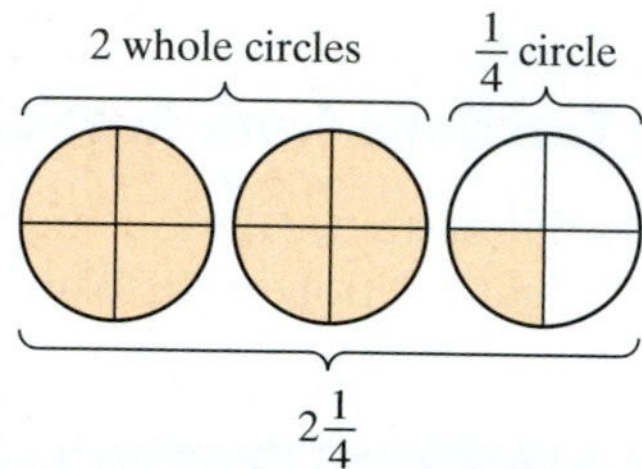

> Note: The mixed number $2\frac{1}{4}$, diagramed above, represents $2 + \frac{1}{4}$.
>
> The mixed number $-3\frac{1}{5}$ represents $-\left(3 + \frac{1}{5}\right)$ or $-3 - \frac{1}{5}$. We review this later in this chapter.

PRACTICE PROBLEMS 10–11

Represent the shaded part of each figure group as both an improper fraction and a mixed number.

10.

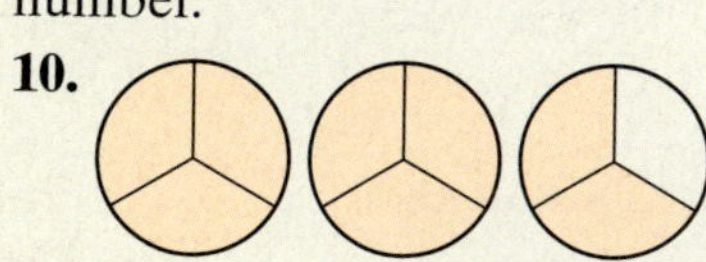

11.

EXAMPLES Represent the shaded part of each figure group as both an improper fraction and a mixed number.

10.

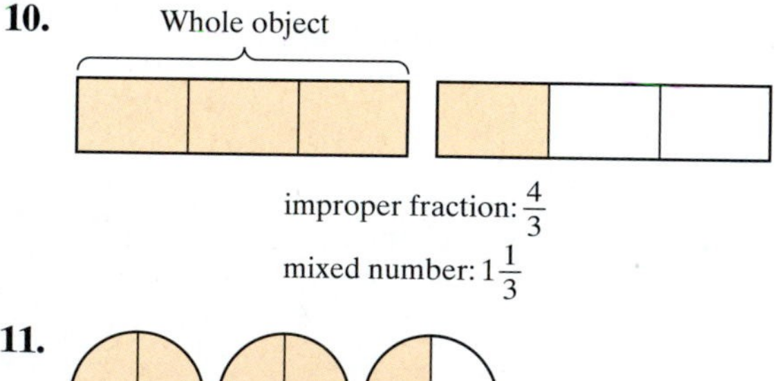

improper fraction: $\frac{4}{3}$

mixed number: $1\frac{1}{3}$

11.

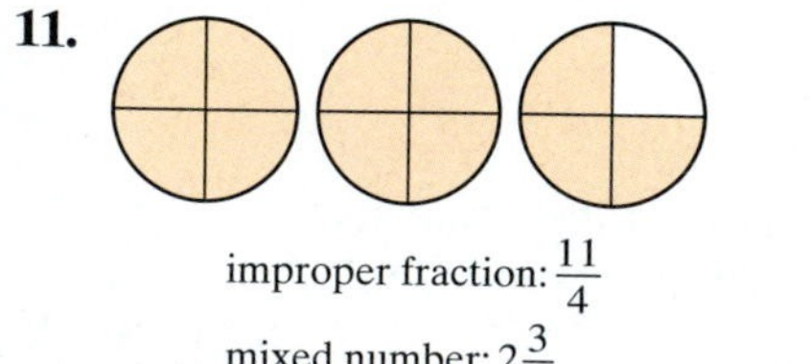

improper fraction: $\frac{11}{4}$

mixed number: $2\frac{3}{4}$

Work Practice Problems 10–11

✔Concept Check If you were to round $2\frac{3}{4}$, shown in Example 11 above, to the nearest a whole number, would you choose 2 or 3? Why?

Answers

10. $\frac{8}{3}, 2\frac{2}{3}$ **11.** $\frac{5}{4}, 1\frac{1}{4}$

✔ Concept Check Answer

3, answers may vary

Objective C Graphing Fractions on a Number Line

Another way to visualize fractions is to graph them on a number line. To do this, think of 1 unit on the number line as a whole. To graph $\frac{2}{5}$, for example, divide the distance from 0 to 1 into 5 equal parts. Then start at 0 and count 2 parts to the right.

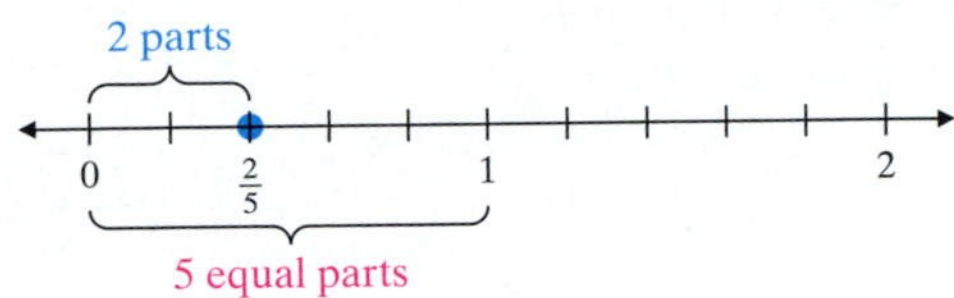

Notice that the graph of $\frac{2}{5}$ lies between 0 and 1. This means

$$0 < \frac{2}{5}\left(\text{or } \frac{2}{5} > 0\right) \text{ and also } \frac{2}{5} < 1$$

EXAMPLE 12 Graph each proper fraction on a number line.

a. $\frac{3}{4}$ **b.** $\frac{1}{2}$ **c.** $\frac{3}{6}$

Solution:

a. To graph $\frac{3}{4}$, divide the distance from 0 to 1 into 4 parts. Then start at 0 and count over 3 parts.

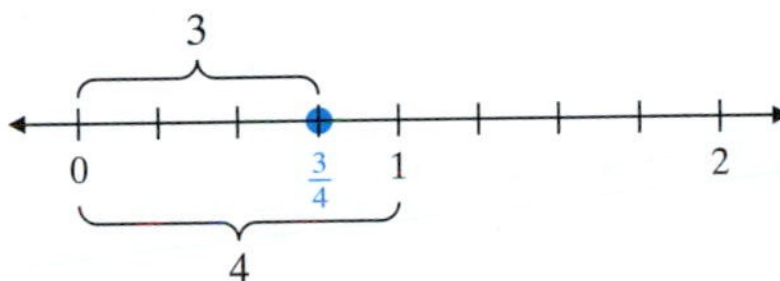

b.

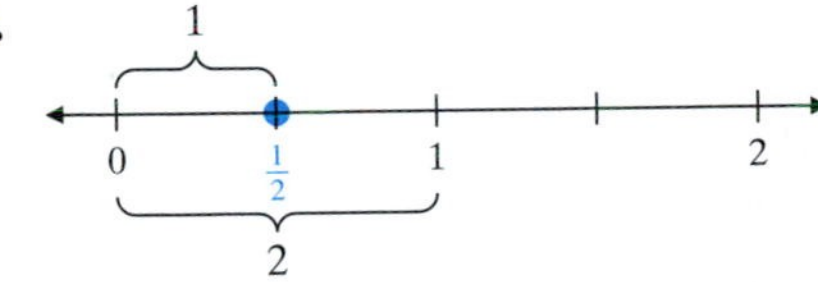

c.

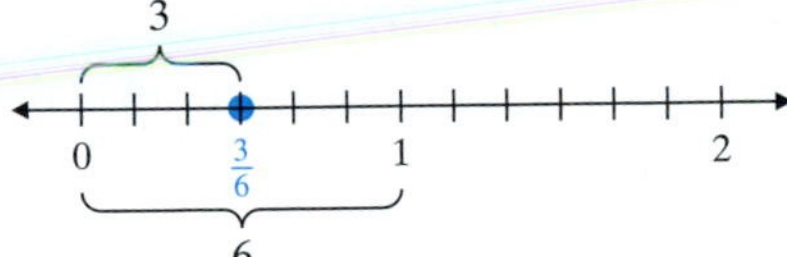

Work Practice Problem 12

The fractions in Example 12 are all proper fractions. Notice that the value of each is less than 1. This is always true for proper fractions since the numerator of a proper fraction is less than the denominator.

On the next page, we graph improper fractions. Notice that improper fractions are greater than or equal to 1. This is always true since the numerator of an improper fraction is greater than or equal to the denominator.

PRACTICE PROBLEM 12

Graph each proper fraction on a number line.

a. $\frac{5}{7}$ **b.** $\frac{2}{3}$ **c.** $\frac{4}{6}$

Answers

12. a. **b.** **c.**

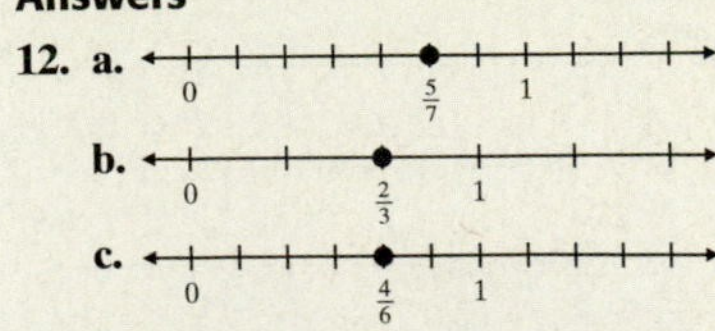

PRACTICE PROBLEM 13

Graph each improper fraction on a number line.

a. $\frac{8}{3}$ **b.** $\frac{5}{4}$ **c.** $\frac{7}{7}$

EXAMPLE 13 Graph each improper fraction on a number line.

a. $\frac{7}{6}$ **b.** $\frac{9}{5}$ **c.** $\frac{6}{6}$ **d.** $\frac{3}{1}$

Solution:

a.

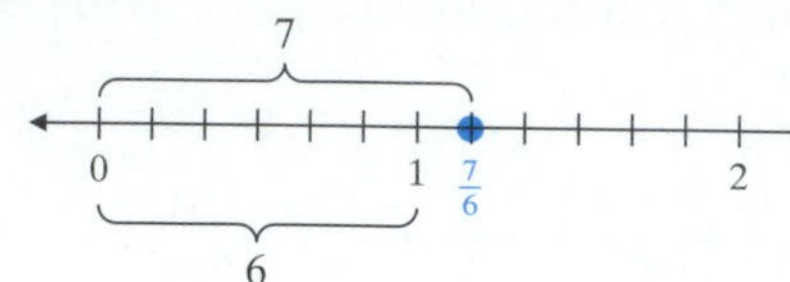

b.

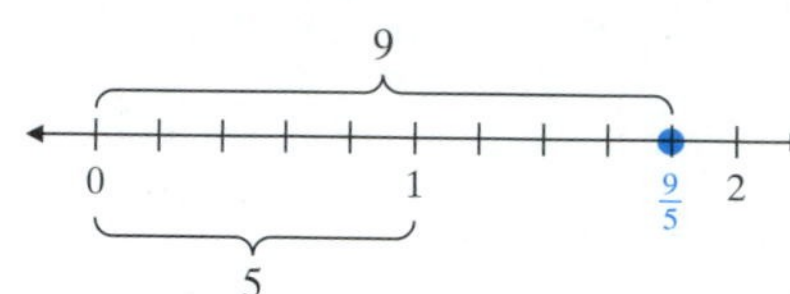

c.

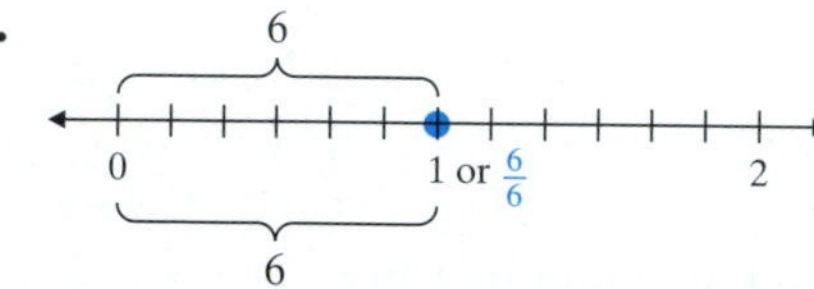

d. Each 1-unit distance has 1 equal part. Count over 3 parts.

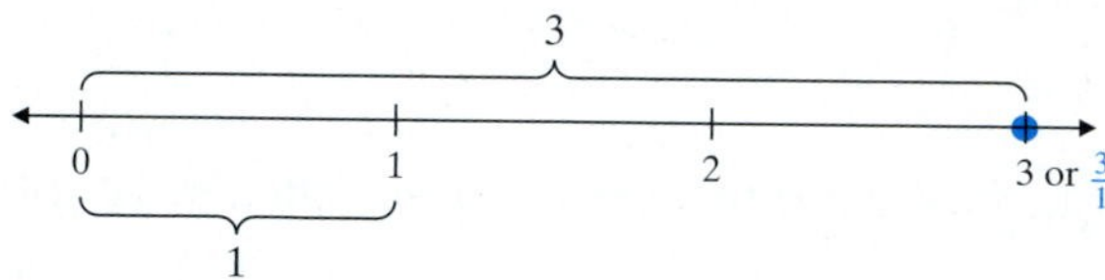

Work Practice Problem 13

Note: We will graph mixed numbers at the end of this chapter.

Objective D Reviewing Division Properties for 0 and 1

Before we continue further, don't forget from Section 1.6 that the fraction bar indicates division. Let's review some division properties for 1 and 0.

$\frac{9}{9} = 1$ because $1 \cdot 9 = 9$ $\frac{-11}{1} = -11$ because $-11 \cdot 1 = -11$

$\frac{0}{6} = 0$ because $0 \cdot 6 = 0$ $\frac{6}{0}$ *is undefined* because there is no number that when multiplied by 0 gives 6.

In general, we can say the following.

Let n be any integer except 0.

$\frac{n}{n} = 1$ $\frac{0}{n} = 0$

$\frac{n}{1} = n$ $\frac{n}{0}$ is undefined.

Answers

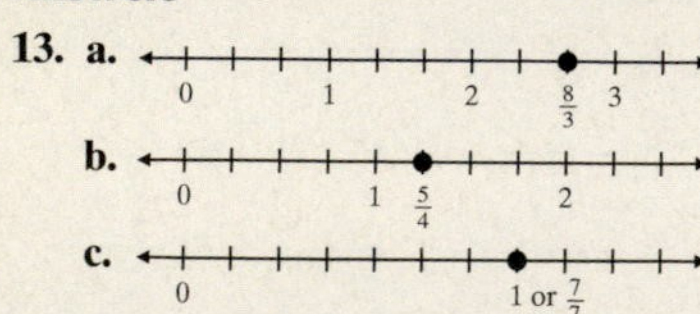

EXAMPLES Simplify.

14. $\frac{5}{5} = 1$ **15.** $\frac{-2}{-2} = 1$ **16.** $\frac{0}{-5} = 0$

17. $\frac{-5}{1} = -5$ **18.** $\frac{41}{1} = 41$ **19.** $\frac{19}{0}$ is undefined

Work Practice Problems 14–19

Notice from Example 18 that we can have negative fractions. In fact,

$$\frac{-5}{1} = -5, \quad \frac{5}{-1} = -5, \quad \text{and } -\frac{5}{1} = -5$$

Because all of the fractions equal -5, we have

$$\frac{-5}{1} = \frac{5}{-1} = -\frac{5}{1}$$

This means that the negative sign in a fraction can be written in the numerator, the denominator, or in front of the fraction. Remember this as we work with negative fractions.

Objective E Writing Mixed Numbers as Improper Fractions

Earlier in this section, mixed numbers and improper fractions were both used to represent the shaded part of the figure groups. For example,

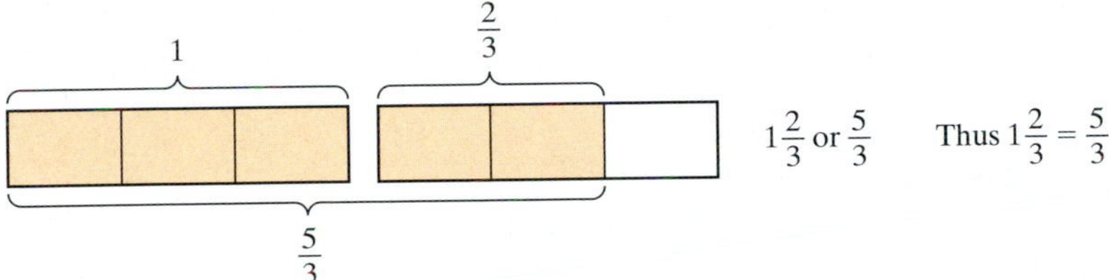

The following steps may be used to write a mixed number as an improper fraction:

Writing a Mixed Number as an Improper Fraction

To write a mixed number as an improper fraction:

Step 1: Multiply the denominator of the fraction by the whole number.

Step 2: Add the numerator of the fraction to the product from Step 1.

Step 3: Write the sum from Step 2 as the numerator of the improper fraction over the original denominator.

For example,

$$1\frac{2}{3} = \frac{\overbrace{3 \cdot 1}^{\text{Step 1}} + \overbrace{2}^{\text{Step 2}}}{\underbrace{3}_{\text{Step 3}}} = \frac{3+2}{3} = \frac{5}{3} \quad \text{or} \quad 1\frac{2}{3} = \frac{5}{3}, \text{ as stated above.}$$

EXAMPLE 20 Write each as an improper fraction.

a. $4\frac{2}{9} = \frac{9 \cdot 4 + 2}{9} = \frac{36 + 2}{9} = \frac{38}{9}$ **b.** $1\frac{8}{11} = \frac{11 \cdot 1 + 8}{11} = \frac{11 + 8}{11} = \frac{19}{11}$

Work Practice Problem 20

PRACTICE PROBLEMS 14–19

Simplify.

14. $\frac{9}{9}$ **15.** $\frac{-6}{-6}$

16. $\frac{0}{-1}$ **17.** $\frac{4}{1}$

18. $\frac{-13}{0}$ **19.** $\frac{-13}{1}$

Helpful Hint

Remember, for example, that

$$-\frac{2}{3} = \frac{-2}{3} = \frac{2}{-3}$$

PRACTICE PROBLEM 20

Write each as an improper fraction.

a. $5\frac{2}{7}$ **b.** $6\frac{2}{3}$ **c.** $10\frac{9}{10}$ **d.** $4\frac{1}{5}$

Answers

14. 1 **15.** 1 **16.** 0 **17.** 4 **18.** undefined **19.** -13 **20. a.** $\frac{37}{7}$ **b.** $\frac{20}{3}$ **c.** $\frac{109}{10}$ **d.** $\frac{21}{5}$

Objective F Writing Improper Fractions as Mixed Numbers or Whole Numbers

Just as there are times when an improper fraction is preferred, sometimes a mixed or a whole number better suits a situation. To write improper fractions as mixed or whole numbers, we use division. Recall once again from Section 1.6 that the fraction bar means division. This means that the fraction

$\frac{5}{3}$ (5 numerator, 3 denominator) means $3\overline{)5}$ (3 denominator, 5 numerator)

Writing an Improper Fraction as a Mixed Number or a Whole Number

To write an improper fraction as a mixed number or a whole number:

Step 1: Divide the denominator into the numerator.

Step 2: The whole number part of the mixed number is the quotient. The fraction part of the mixed number is the remainder over the original denominator.

$$\text{quotient}\frac{\text{remainder}}{\text{original denominator}}$$

For example,

Step 1: $\frac{5}{3}$: $3\overline{)5}$ gives quotient 1; $5 - 3 = 2$ remainder

Step 2: $\frac{5}{3} = 1\frac{2}{3}$ ← 2 remainder, 3 original denominator, 1 quotient

PRACTICE PROBLEM 21

Write each as a mixed number or a whole number.

a. $\frac{9}{5}$ b. $\frac{23}{9}$ c. $\frac{48}{4}$

d. $\frac{62}{13}$ e. $\frac{51}{7}$ f. $\frac{21}{20}$

EXAMPLE 21 Write each as a mixed number or a whole number.

a. $\frac{30}{7}$ b. $\frac{16}{15}$ c. $\frac{84}{6}$

Solution:

a. $\frac{30}{7}$: $7\overline{)30}$ gives quotient 4; $30 - 28 = 2$ remainder. $\frac{30}{7} = 4\frac{2}{7}$

b. $\frac{16}{15}$: $15\overline{)16}$ gives quotient 1; $16 - 15 = 1$ remainder. $\frac{16}{15} = 1\frac{1}{15}$

c. $\frac{84}{6}$: $6\overline{)84}$ gives quotient 14; $84 - 6 = 24$, $24 - 24 = 0$. $\frac{84}{6} = 14$ Since the remainder is 0, the result is the whole number 14.

Helpful Hint

When the remainder is 0, the improper fraction is a whole number. For example, $\frac{92}{4} = 23$.

$4\overline{)92}$ gives quotient 23; $9 - 8 = 1$, bring down 2: $12 - 12 = 0$

Work Practice Problem 21

Answers

21. a. $1\frac{4}{5}$ b. $2\frac{5}{9}$ c. 12 d. $4\frac{10}{13}$ e. $7\frac{2}{7}$ f. $1\frac{1}{20}$

Vocabulary and Readiness Check

Use the choices below to fill in each blank.

improper	fraction	proper
is undefined	mixed number	$= 0$
≥ 1	denominator	$= 1$
< 1	numerator	

1. The number $\frac{17}{31}$ is called a __________. The number 31 is called its __________ and 17 is called its __________.

2. If we simplify each fraction, $\frac{-9}{-9}$ __________, $\frac{0}{-4}$ __________, and we say $\frac{-4}{0}$ __________.

3. The fraction $\frac{8}{3}$ is called a(n) __________ fraction, the fraction $\frac{3}{8}$ is called a(n) __________ fraction, and $10\frac{3}{8}$ is called a(n) __________.

4. The value of an improper fraction is always __________, and the value of a proper fraction is always __________.

4.1 EXERCISE SET

FOR EXTRA HELP

Student Solutions Manual

PH Math/Tutor Center

CD/Video for Review

MathXL®

MyMathLab

Objective A *Identify the numerator and the denominator of each fraction and identify each fraction as proper or improper. See Examples 1 and 2.*

1. $\frac{1}{2}$

2. $\frac{1}{4}$

3. $\frac{10}{3}$

4. $\frac{53}{21}$

5. $\frac{15}{15}$

6. $\frac{26}{26}$

Objective B *Write a proper or improper fraction to represent the shaded part of each. If an improper fraction is appropriate, write the shaded figure as (a) an improper fraction and (b) a mixed number. See Examples 3 through 6 and 10 and 11.*

7.

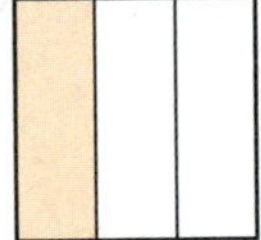

8.

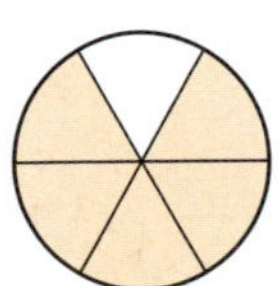

9.

10.

11.

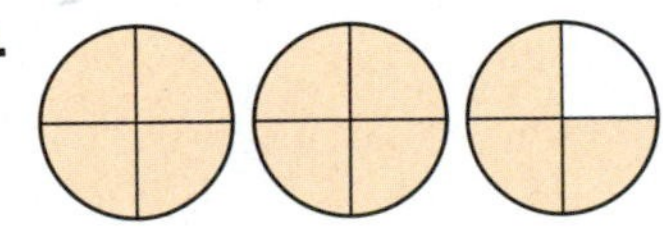

12.

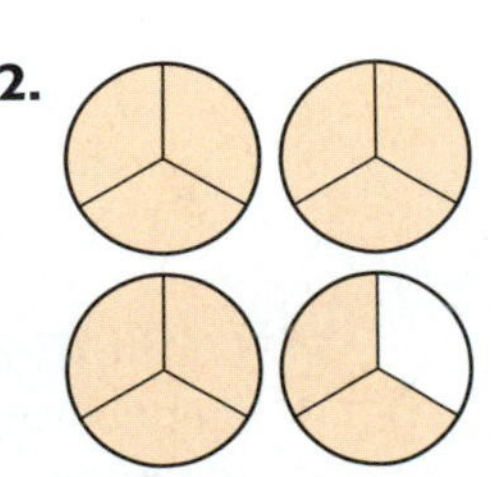

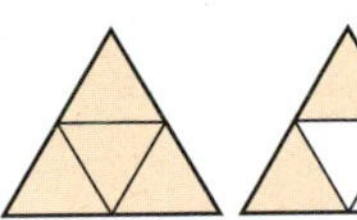

13.

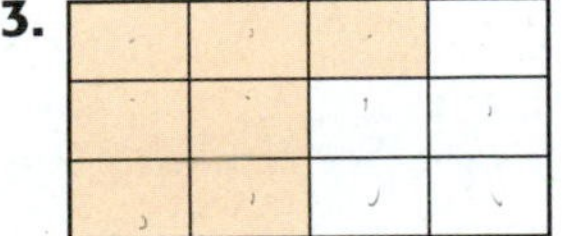

14. 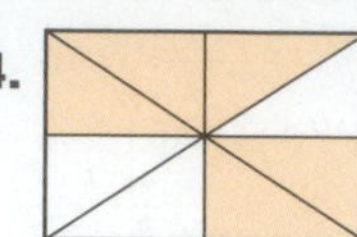

15.

16. 

17. 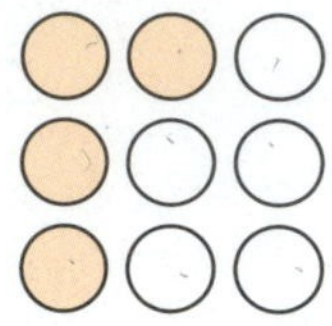

18.

19. 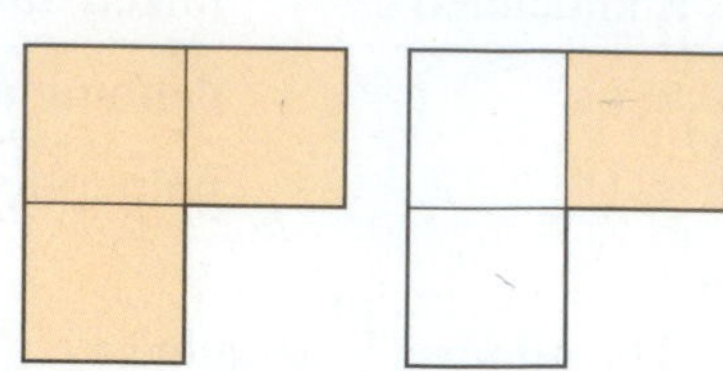

20. 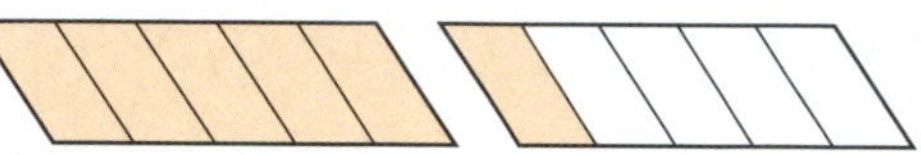

21. 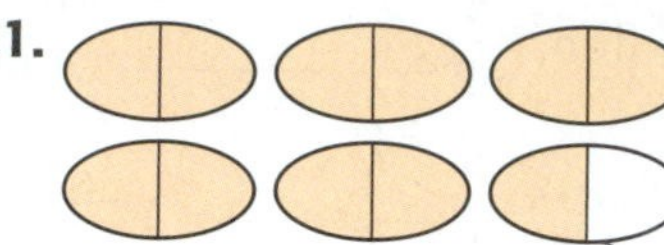

22. 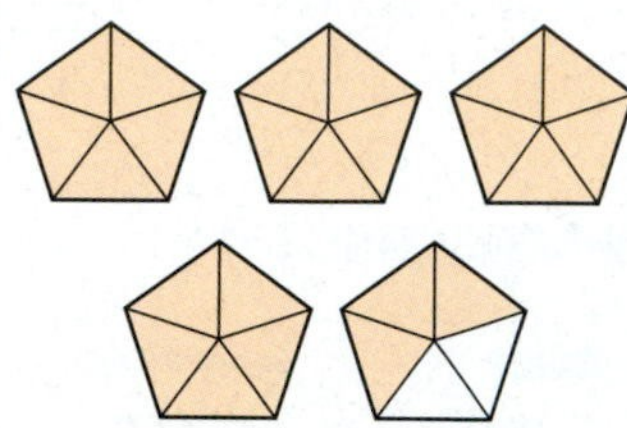

23. 

24. 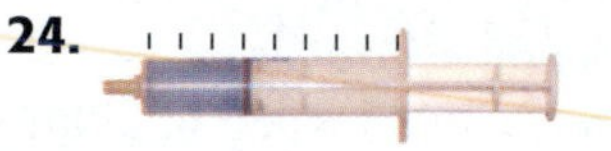

25. 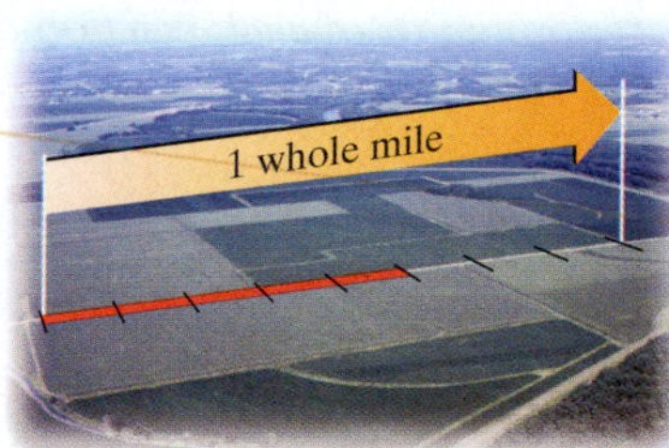

26. 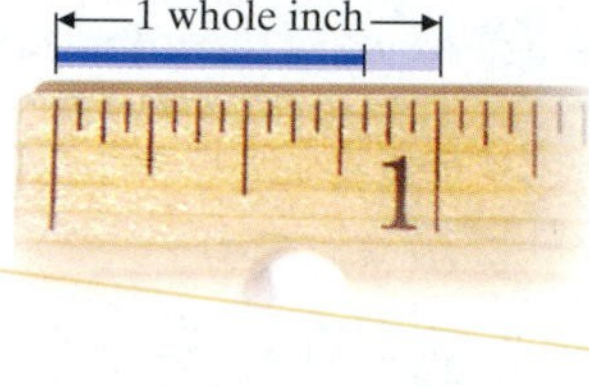

Objective **B** *Draw and shade a part of a diagram to represent each fraction. See Examples 7 and 8.*

27. $\frac{1}{5}$ of a diagram

28. $\frac{1}{16}$ of a diagram

29. $\frac{6}{7}$ of a diagram

30. $\frac{7}{9}$ of a diagram

31. $\frac{4}{4}$ of a diagram

32. $\frac{6}{6}$ of a diagram

Write each fraction. See Example 9.

33. Of the 131 students at a small private school, 42 are freshmen. What fraction of the students are freshmen?

34. Of the 63 employees at a new biomedical engineering firm, 22 are men. What fraction of the employees are men?

35. Use Exercise 33 to answer a and b.

a. How many students are *not* freshmen?

b. What fraction of the students are *not* freshmen?

36. Use Exercise 34 to answer a and b.

a. How many of the employees are women?

b. What fraction of the employees are women?

37. As of 2006, the United States has had 43 different presidents. A total of eight U.S. presidents were born in the state of Virginia, more than any other state. What fraction of U.S. presidents were born in Virginia? (*Source: 2006 World Almanac and Book of Facts*)

38. Of the eight planets in our solar system, four have days that are longer than the 24-hour Earth day. What fraction of the planets have longer days than Earth has? (*Source:* National Space Science Data Center)

39. The Atlantic hurricane season of 2005 rewrote the record books. There were 28 tropical storms, 15 of which turned into hurricanes. What fraction of the 2005 Atlantic tropical storms escalated to hurricanes?

40. There are 12 inches in a foot. What fractional part of a foot does 5 inches represent?

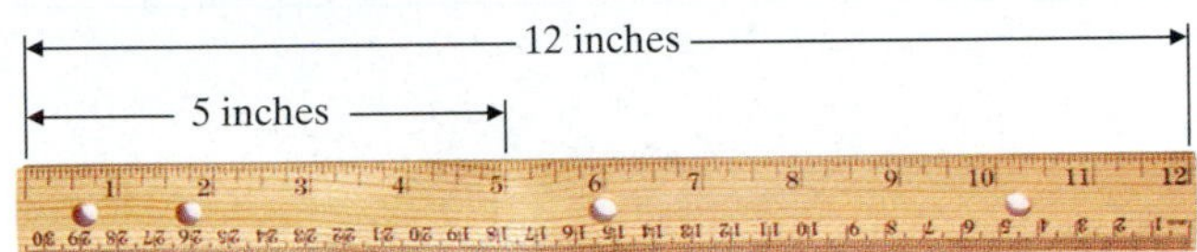

41. There are 31 days in the month of March. What fraction of the month does 11 days represent?

Mon.	Tue.	Wed.	Thu.	Fri.	Sat.	Sun.
					1	2
3	4	5	6	7	8	9
10	11	12	13	14	15	16
17	18	19	20	21	22	23
24	25	26	27	28	29	30
31						

42. There are 60 minutes in an hour. What fraction of an hour does 37 minutes represent?

43. In a prealgebra class containing 31 students, there are 18 freshmen, 10 sophomores, and 3 juniors. What fraction of the class is sophomores?

44. In a sports team with 20 children, there are 9 boys and 11 girls. What fraction of the team is boys?

45. Thirty-three out of the fifty total states in the United States contain federal Indian reservations.

a. What fraction of the states contain federal Indian reservations?

b. How many states do not contain federal Indian reservations?

c. What fraction of the states do not contain federal Indian reservations? (*Source:* Tiller Research, Inc., Albuquerque, NM)

46. Consumer fireworks are legal in 40 out of the 50 total states in the United States.

a. In what fraction of the states are consumer fireworks legal?

b. In how many states are consumer fireworks illegal?

c. In what fraction of the states are consumer fireworks illegal? (*Source:* United States Fireworks Safety Council)

47. A bag contains 50 red or blue marbles. If 21 marbles are blue,

a. What fraction of the marbles are blue?
b. How many marbles are red?
c. What fraction of the marbles are red?

48. An art dealer is taking inventory. His shop contains a total of 37 pieces, which are all sculptures, watercolor paintings, or oil paintings. If there are 15 watercolor paintings and 17 oil paintings, answer each question.

a. What fraction of the inventory is watercolor paintings?
b. What fraction of the inventory is oil paintings?
c. How many sculptures are there?
d. What fraction of the inventory is sculptures?

Objective C *Graph each fraction on a number line. See Examples 12 and 13.*

49. $\frac{1}{4}$

0

50. $\frac{1}{3}$

0

51. $\frac{4}{7}$

0

52. $\frac{5}{6}$

0

53. $\frac{8}{5}$

0

54. $\frac{9}{8}$

0

55. $\frac{7}{3}$

0

56. $\frac{13}{7}$

0

Objective D *Simplify by dividing. See Examples 14 through 19.*

57. $\frac{12}{12}$ **58.** $\frac{-3}{-3}$ **59.** $\frac{-5}{1}$ **60.** $\frac{-20}{1}$

61. $\frac{0}{-2}$ **62.** $\frac{0}{-8}$ **63.** $\frac{-8}{-8}$ **64.** $\frac{-14}{-14}$

65. $\frac{-9}{0}$ **66.** $\frac{-7}{0}$ **67.** $\frac{3}{1}$ **68.** $\frac{5}{5}$

Objective E *Write each mixed number as an improper fraction. See Example 20.*

69. $2\frac{1}{3}$ **70.** $1\frac{13}{17}$ **71.** $3\frac{3}{5}$ **72.** $2\frac{5}{9}$

73. $6\frac{5}{8}$ **74.** $7\frac{3}{8}$ **75.** $11\frac{6}{7}$ **76.** $12\frac{2}{5}$

77. $9\frac{7}{20}$ **78.** $10\frac{14}{27}$ **79.** $166\frac{2}{3}$ **80.** $114\frac{2}{7}$

Objective **F** *Write each improper fraction as a mixed number or a whole number. See Example 21.*

81. $\frac{17}{5}$ **82.** $\frac{13}{7}$ **83.** $\frac{37}{8}$ **84.** $\frac{64}{9}$

85. $\frac{47}{15}$ **86.** $\frac{65}{12}$ **87.** $\frac{225}{15}$ **88.** $\frac{196}{14}$

89. $\frac{182}{175}$ **90.** $\frac{149}{143}$ **91.** $\frac{737}{112}$ **92.** $\frac{901}{123}$

Review

Simplify. See Section 1.7.

93. 3^2 **4.** 4^3 **95.** 5^3 **96.** 3^4

Concept Extensions

Write each fraction in two other equiv- -nt ways by inserting the negative sign in different places.

97. $-\frac{11}{2} = \quad = \quad$

98. $-\frac{21}{4} = \quad = \quad$

99. $\frac{-13}{15} = \quad = \quad$

100. $\frac{45}{-57} = \quad = \quad$

101. In your own words, explain why $\frac{0}{10} = 0$ and $\frac{10}{0}$ are -fined.

Solve. See the Concept Checks in this section.

102. If ◯◯◯◯ represents $\frac{4}{9}$ of a whole diagram, sketch the whole diagram.

1('f △△ represents $\frac{1}{3}$ of a whole diagram, -etch the whole diagram.

104. Round the mixed number $7\frac{1}{8}$ to the nearest whole number.

105. R- -d the mixed number $5\frac{11}{12}$ to the nearest whole nu- -er.

106. The Gap Corporation owns stores with four different brand names, as shown on the bar graph. What fraction of the stores owned by The Gap Corporation are named "Old Navy"?

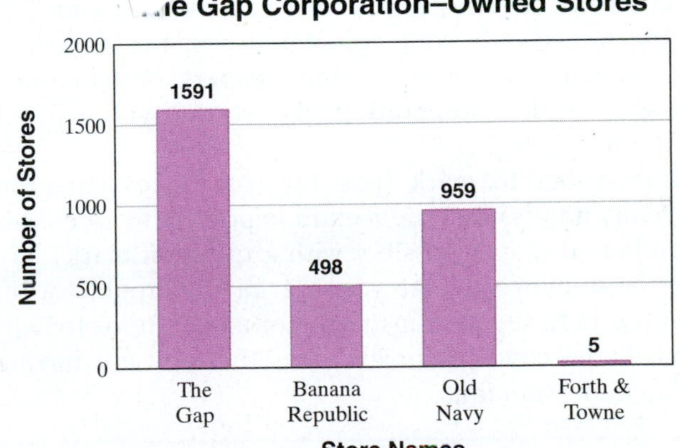

107. The Public Broadcasting Service (PBS) provides programming to the noncommercial public TV stations of the United States. The bar graph shows a breakdown of the public television licensees by type. Each licensee operates one or more PBS member TV stations. What fraction of the public television licensees are universities or colleges? (*Source:* The Public Broadcast Service)

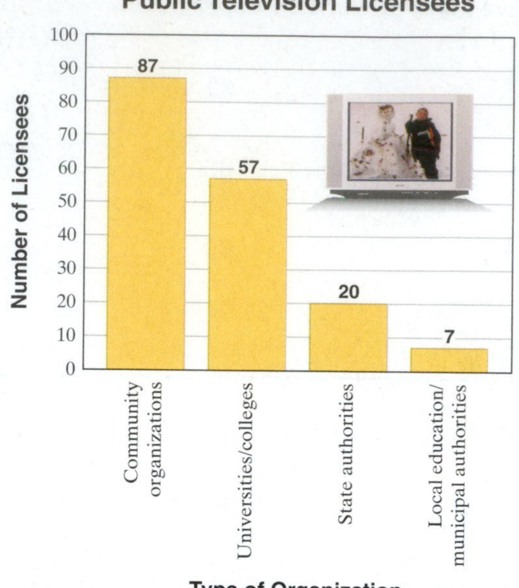

108. Habitat for Humanity is a nonprofit organization that helps provide affordable housing to families in need. Habitat for Humanity does its work of building and renovating houses through 1651 local affiliates in the United States and 634 international affiliates. What fraction of the total Habitat for Humanity affiliates are located in the United States? (*Hint:* First find the total number of affiliates.) (*Source:* Habitat for Humanity International)

109. The United States Marine Corps (USMC) has five principal training centers in California, three in North Carolina, two in South Carolina, one in Arizona, one in Hawaii, and one in Virginia. What fraction of the total USMC principal training centers are located in California? (*Source:* U.S. Department of Defense)

STUDY SKILLS BUILDER

Are You Organized?

Have you ever had trouble finding a completed assignment? When it's time to study for a test, are your notes neat and organized? Have you ever had trouble reading your own mathematics handwriting? (Be honest—I have.)

When any of these things happen, it's time to get organized. Here are a few suggestions:

Write your notes and complete your homework assignment in a notebook with pockets (spiral or ring binder.) Take class notes in this notebook, and then follow the notes with your completed homework assignment. When you receive graded papers or handouts, place them in the notebook pocket so that you will not lose them.

Remember to mark (possibly with an exclamation point) any note(s) that seem extra important to you. Also remember to mark (possibly with a question mark) any notes or homework that you are having trouble with. Don't forget to see your instructor or a math tutor to help you with the concepts or exercises that you are having trouble understanding.

Also, if you are having trouble reading your own handwriting, *slow down* and write your mathematics work clearly!

Exercises

1. Have you been completing your assignments on time?
2. Have you been correcting any exercises you may be having difficulty with?
3. If you are having trouble with a mathematical concept or correcting any homework exercises, have you visited your instructor, a tutor, or your campus math lab?
4. Are you taking lecture notes in your mathematics course? (By the way, these notes should include worked-out examples solved by your instructor.)
5. Is your mathematics course material (handouts, graded papers, lecture notes) organized?
6. If your answer to Exercise 5 is no, take a moment and review your course material. List at least two ways that you might better organize it. Then read the Study Skills Builder on organizing a notebook in Section 2.3.

4.2 FACTORS AND SIMPLEST FORM

Objectives

A Write a Number as a Product of Prime Numbers.

B Write a Fraction in Simplest Form.

C Determine Whether Two Fractions Are Equivalent.

D Solve Problems by Writing Fractions in Simplest Form.

Objective A Writing a Number as a Product of Prime Numbers

Recall from Section 1.5 that since $12 = 2 \cdot 2 \cdot 3$, the numbers 2 and 3 are called *factors* of 12. A **factor** is any number that divides a number evenly (with a remainder of 0).

To perform operations on fractions, it is necessary to be able to factor a number. Remember that factoring a number means writing a number as a product. We first practice writing a number as a product of prime numbers.

A **prime number** is a natural number greater than 1 whose only factors are 1 and itself. The first few prime numbers are 2, 3, 5, 7, 11, 13, 17, 19, 23, 29, ...

A **composite number** is a natural number greater than 1 that is not prime.

The natural number 1 is neither prime nor composite.

When a composite number is written as a product of prime numbers, this product is called the **prime factorization** of the number. For example, the prime factorization of 12 is $2 \cdot 2 \cdot 3$ because

$$12 = \underbrace{2 \cdot 2 \cdot 3}$$

This product is 12 and each number is a prime number.

Because multiplication is commutative, the order of the factors is not important. We can write the factorization $2 \cdot 2 \cdot 3$ as $2 \cdot 3 \cdot 2$ or $3 \cdot 2 \cdot 2$. Any of these is called the prime factorization of 12.

Every whole number greater than 1 has exactly one prime factorization.

One method for finding the prime factorization of a number is by using a factor tree, as shown in the next example.

EXAMPLE 1 Write the prime factorization of 45.

Solution: We can begin by writing 45 as the product of two numbers, say 5 and 9.

```
  45
 /  \
5 · 9
```

The number 5 is prime but 9 is not, so we write 9 as $3 \cdot 3$.

```
      45
     /  \
    5 ·  9
   /    / \
  5 ·  3 · 3
```

A factor tree

Each factor is now a prime number, so the prime factorization of 45 is $3 \cdot 3 \cdot 5$ or $3^2 \cdot 5$.

Work Practice Problem 1

PRACTICE PROBLEM 1

Use a factor tree to find the prime factorization of each number.

a. 30 **b.** 56 **c.** 72

Answers

1. a. $2 \cdot 3 \cdot 5$ **b.** $2^3 \cdot 7$ **c.** $2^3 \cdot 3^2$

✔ **Concept Check** True or false? Two different numbers can have exactly the same prime factorization. Explain your answer.

PRACTICE PROBLEM 2

Write the prime factorization of 60.

EXAMPLE 2 Write the prime factorization of 80.

Solution: Write 80 as a product of two numbers. Continue this process until all factors are prime.

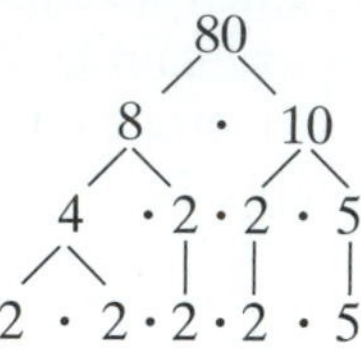

All factors are now prime, so the prime factorization of 80 is

$2 \cdot 2 \cdot 2 \cdot 2 \cdot 5$ or $2^4 \cdot 5$.

Work Practice Problem 2

Helpful Hint

It makes no difference which factors you start with. The prime factorization of a number will be the same.

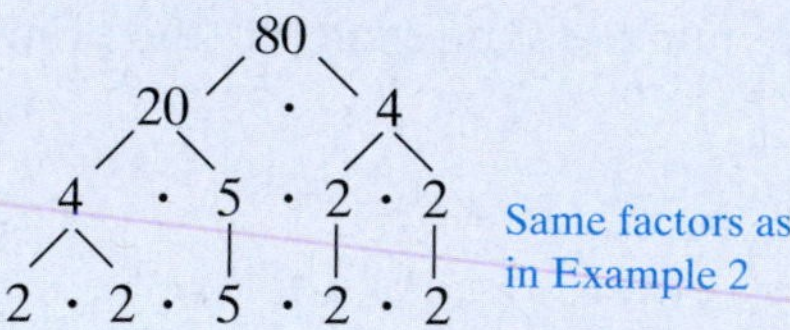

Same factors as in Example 2

There are a few quick **divisibility tests** to determine whether a number is divisible by the primes 2, 3, or 5. (A number is divisible by 2, for example, if 2 divides it evenly so that the remainder is 0.)

Divisibility Tests

A whole number is divisible by:

- **2** if the last digit is 0, 2, 4, 6, or 8.

  132 is divisible by 2 since the last digit is a 2.

- **3** if the sum of the digits is divisible by 3.

  144 is divisible by 3 since $1 + 4 + 4 = 9$ is divisible by 3.

- **5** if the last digit is 0 or 5.

  1115 is divisible by 5 since the last digit is a 5.

Answer

2. $2^2 \cdot 3 \cdot 5$

✔ Concept Check Answer

false; answers may vary

Helpful Hint

Here are a few other divisibility tests you may want to use. A whole number is divisible by:

- 4 if its last two digits are divisible by 4.
  1712 is divisible by 4.
- 6 if it's divisible by 2 and 3.
  9858 is divisible by 6.
- 9 if the sum of its digits is divisible by 9.
  5238 is divisible by 9 since $5 + 2 + 3 + 8 = 18$ is divisible by 9.

When finding the prime factorization of larger numbers, you may want to use the procedure shown in Example 3.

EXAMPLE 3 Write the prime factorization of 252.

Solution: For this method, we divide prime numbers into the given number. Since the ones digit of 252 is 2, we know that 252 is divisible by 2.

$$2\overline{)252}\text{ quotient }126$$

126 is divisible by 2 also.

$$\begin{array}{r} 63 \\ 2\overline{)126} \\ 2\overline{)252} \end{array}$$

63 is not divisible by 2 but is divisible by 3. Divide 63 by 3 and continue in this same manner until the quotient is a prime number.

$$\begin{array}{r} 7 \\ 3\overline{)\ 21} \\ 3\overline{)\ 63} \\ 2\overline{)126} \\ 2\overline{)252} \end{array}$$

Helpful Hint

The order of choosing prime numbers does not matter. For consistency, we use the order 2, 3, 5, 7, ...

The prime factorization of 252 is $2 \cdot 2 \cdot 3 \cdot 3 \cdot 7$ or $2^2 \cdot 3^2 \cdot 7$.

Work Practice Problem 3

In this text, we will write the factorization of a number from the smallest factor to the largest factor.

✔Concept Check True or false? The prime factorization of 117 is $9 \cdot 13$. Explain your reasoning.

PRACTICE PROBLEM 3

Write the prime factorization of 297.

Answer

3. $3^3 \cdot 11$

✔ Concept Check Answer

No; $3 \cdot 3 \cdot 13$

Objective B Writing Fractions in Simplest Form

Fractions that represent the same portion of a whole or the same point on a number line are called **equivalent fractions.** Study the table below to see two ways to visualize equivalent fractions.

Equivalent Fractions	
Figures	**Number Line**
When we shade $\frac{1}{3}$ and $\frac{2}{6}$ on the same-sized figures, 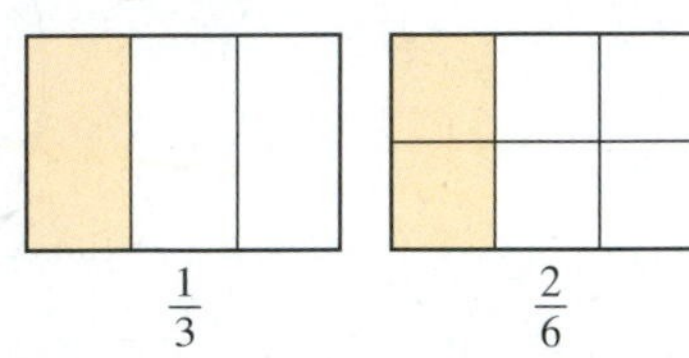 notice that both $\frac{1}{3}$ and $\frac{2}{6}$ represent the same portion of a whole. These fractions are called **equivalent fractions** and we write $\frac{1}{3} = \frac{2}{6}$.	When we graph $\frac{1}{3}$ and $\frac{2}{6}$ on a number line, 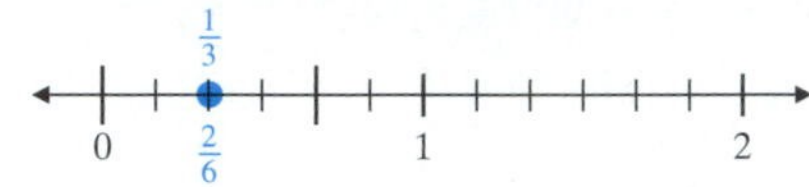 notice that both $\frac{1}{3}$ and $\frac{2}{6}$ correspond to the same point. These fractions are called **equivalent fractions,** and we write $\frac{1}{3} = \frac{2}{6}$.
Thus, $\frac{1}{3} = \frac{2}{6}$ and $\frac{1}{3}$ and $\frac{2}{6}$ are equivalent.	

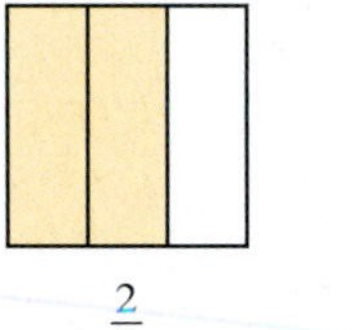

$\frac{2}{3}$

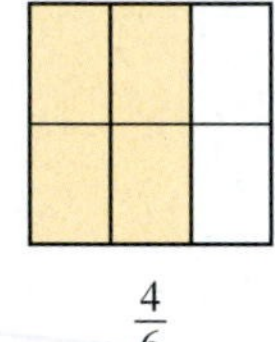

$\frac{4}{6}$

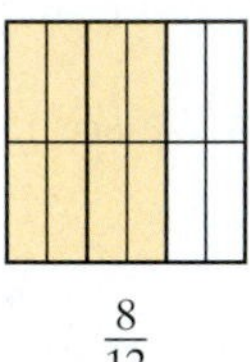

$\frac{8}{12}$

For example, $\frac{2}{3}$, $\frac{4}{6}$, and $\frac{8}{12}$ all represent the same shaded portion of the rectangle's area, so they are equivalent fractions. To show that these fractions are equivalent, we place an equals sign between them. In other words,

$$\frac{2}{3} = \frac{4}{6} = \frac{8}{12}$$

There are many equivalent forms of a fraction. A special equivalent form of a fraction is called **simplest form.**

Simplest Form of a Fraction

A fraction is written in **simplest form** or **lowest terms** when the numerator and the denominator have no common factors other than 1.

For example, the fraction $\frac{2}{3}$ *is* in simplest form because 2 and 3 have no common factor other than 1. The fraction $\frac{4}{6}$ *is not* in simplest form because 4 and 6 both have a factor of 2. That is, 2 is a common factor of 4 and 6. The process of writing a fraction in simplest form is called **simplifying** the fraction.

To simplify $\frac{4}{6}$ and write it as $\frac{2}{3}$, let's first study a few properties. Recall from Section 4.1 that any nonzero whole number n divided by itself is 1.

Any nonzero number n divided by itself is 1.

$$\frac{5}{5} = 1, \ \frac{17}{17} = 1, \ \frac{24}{24} = 1, \text{ or, in general, } \frac{n}{n} = 1$$

Also, in general, if $\frac{a}{b}$ and $\frac{c}{d}$ are fractions (with b and d not 0), the following is true.

$$\frac{a \cdot c}{b \cdot d} = \frac{a}{b} \cdot \frac{c}{d}*$$

These properties allow us to do the following:

$$\frac{4}{6} = \frac{2 \cdot 2}{2 \cdot 3} = \frac{2}{2} \cdot \frac{2}{3} = 1 \cdot \frac{2}{3} = \frac{2}{3}$$

When 1 is multiplied by a number, the result is the same number.

($\frac{2}{2}$: This is 1)

EXAMPLE 4 Write in simplest form: $\frac{12}{20}$

Solution: Notice that 12 and 20 have a common factor of 4.

$$\frac{12}{20} = \frac{4 \cdot 3}{4 \cdot 5} = \frac{4}{4} \cdot \frac{3}{5} = 1 \cdot \frac{3}{5} = \frac{3}{5}$$

Since 3 and 5 have no common factors (other than 1), $\frac{3}{5}$ is in simplest form.

Work Practice Problem 4

PRACTICE PROBLEM 4

Write in simplest form: $\frac{30}{45}$

If you have trouble finding common factors, write the prime factorization of the numerator and the denominator.

EXAMPLE 5 Write in simplest form: $\frac{42x}{66}$

Solution: Let's write the prime factorizations of 42 and 66. Remember that $42x$ means $42 \cdot x$.

$$\frac{42x}{66} = \frac{2 \cdot 3 \cdot 7 \cdot x}{2 \cdot 3 \cdot 11} = \frac{2}{2} \cdot \frac{3}{3} \cdot \frac{7x}{11} = 1 \cdot 1 \cdot \frac{7x}{11} = \frac{7x}{11}$$

Work Practice Problem 5

PRACTICE PROBLEM 5

Write in simplest form: $\frac{39x}{51}$

In the example above, you may have saved time by noticing that 42 and 66 have a common factor of 6.

$$\frac{42x}{66} = \frac{6 \cdot 7x}{6 \cdot 11} = \frac{6}{6} \cdot \frac{7x}{11} = 1 \cdot \frac{7x}{11} = \frac{7x}{11}$$

Helpful Hint

Writing the prime factorizations of the numerator and the denominator is helpful in finding any common factors.

The method for simplifying negative fractions is the same as for positive fractions.

EXAMPLE 6 Write in simplest form: $-\frac{10}{27}$

Solution:

$$-\frac{10}{27} = -\frac{2 \cdot 5}{3 \cdot 3 \cdot 3}$$

Prime factorizations of 10 and 27.

Since 10 and 27 have no common factors, $-\frac{10}{27}$ is already in simplest form.

Work Practice Problem 6

PRACTICE PROBLEM 6

Write in simplest form: $-\frac{9}{50}$

Answers

4. $\frac{2}{3}$ **5.** $\frac{13x}{17}$ **6.** $-\frac{9}{50}$

*Note: We will study this concept further in the next section.

PRACTICE PROBLEM 7

Write in simplest form: $\frac{49}{112}$

EXAMPLE 7 Write in simplest form: $\frac{30}{108}$

Solution:

$$\frac{30}{108} = \frac{2 \cdot 3 \cdot 5}{2 \cdot 2 \cdot 3 \cdot 3 \cdot 3} = \frac{2}{2} \cdot \frac{3}{3} \cdot \frac{5}{2 \cdot 3 \cdot 3} = 1 \cdot 1 \cdot \frac{5}{18} = \frac{5}{18}$$

Work Practice Problem 7

We can use a shortcut procedure with common factors when simplifying.

$$\frac{4}{6} = \frac{\overset{1}{\cancel{2}} \cdot 2}{\underset{1}{\cancel{2}} \cdot 3} = \frac{1 \cdot 2}{1 \cdot 3} = \frac{2}{3}$$

Divide out the common factor of 2 in the numerator and denominator.

This procedure is possible because dividing out a common factor in the numerator and denominator is the same as removing a factor of 1 in the product.

Writing a Fraction in Simplest Form

To write a fraction in simplest form, write the prime factorization of the numerator and the denominator and then divide both by all common factors.

PRACTICE PROBLEM 8

Write in simplest form: $-\frac{64}{20}$

EXAMPLE 8 Write in simplest form: $-\frac{72}{26}$

Solution:

$$-\frac{72}{26} = -\frac{\overset{1}{\cancel{2}} \cdot 2 \cdot 2 \cdot 3 \cdot 3}{\underset{1}{\cancel{2}} \cdot 13} = -\frac{1 \cdot 2 \cdot 2 \cdot 3 \cdot 3}{1 \cdot 13} = -\frac{36}{13}$$

Work Practice Problem 8

✔ **Concept Check** Which is the correct way to simplify the fraction $\frac{15}{25}$? Or are both correct? Explain.

a. $\frac{15}{25} = \frac{3 \cdot \overset{1}{\cancel{5}}}{5 \cdot \underset{1}{\cancel{5}}} = \frac{3}{5}$

b. $\frac{1\overset{1}{\cancel{5}}}{2\underset{1}{\cancel{5}}} = \frac{11}{21}$

In this chapter, we will simplify and perform operations on fractions containing variables. When the denominator of a fraction contains a variable, such as $\frac{6x^2}{60x^3}$, we will assume that the variable does not represent 0. Recall that the denominator of a fraction cannot be 0.

PRACTICE PROBLEM 9

Write in simplest form: $\frac{7a^3}{56a^2}$

EXAMPLE 9 Write in simplest form: $\frac{6x^2}{60x^3}$

Solution: Notice that 6 and 60 have a common factor of 6. Let's also use the definition of an exponent to factor x^2 and x^3.

$$\frac{6x^2}{60x^3} = \frac{\overset{1}{\cancel{6}} \cdot \overset{1}{\cancel{x}} \cdot \overset{1}{\cancel{x}}}{\underset{1}{\cancel{6}} \cdot 10 \cdot \underset{1}{\cancel{x}} \cdot \underset{1}{\cancel{x}} \cdot x} = \frac{1 \cdot 1 \cdot 1}{1 \cdot 10 \cdot 1 \cdot 1 \cdot x} = \frac{1}{10x}$$

Work Practice Problem 9

Answers

7. $\frac{7}{16}$ **8.** $-\frac{16}{5}$ **9.** $\frac{a}{8}$

✔ Concept Check Answers

a. correct **b.** incorrect

Helpful Hint

Be careful when all factors of the numerator or denominator are divided out. In Example 9, the numerator was $1 \cdot 1 \cdot 1 = 1$, so the final result was $\frac{1}{10x}$.

Objective C Determining Whether Two Fractions Are Equivalent

Recall from Objective B that two fractions are equivalent if they represent the same part of a whole. One way to determine whether two fractions are equivalent is to see whether they simplify to the same fraction.

EXAMPLE 10 Determine whether $\frac{16}{40}$ and $\frac{10}{25}$ are equivalent.

Solution: Simplify each fraction.

$$\frac{16}{40} = \frac{\overset{1}{\cancel{8}} \cdot 2}{\underset{1}{\cancel{8}} \cdot 5} = \frac{1 \cdot 2}{1 \cdot 5} = \frac{2}{5}$$

$$\frac{10}{25} = \frac{2 \cdot \overset{1}{\cancel{5}}}{5 \cdot \underset{1}{\cancel{5}}} = \frac{2 \cdot 1}{5 \cdot 1} = \frac{2}{5}$$

Since these fractions are the same, $\frac{16}{40} = \frac{10}{25}$.

Work Practice Problem 10

PRACTICE PROBLEM 10

Determine whether $\frac{7}{9}$ and $\frac{21}{27}$ are equivalent.

There is a shortcut method you may use to check or test whether two fractions are equivalent. In the example above, we learned that the fractions are equivalent, or

$$\frac{16}{40} = \frac{10}{25}$$

In this example above, we call $25 \cdot 16$ and $40 \cdot 10$ **cross products** because they are the products one obtains by multiplying diagonally across the equal sign, as shown below.

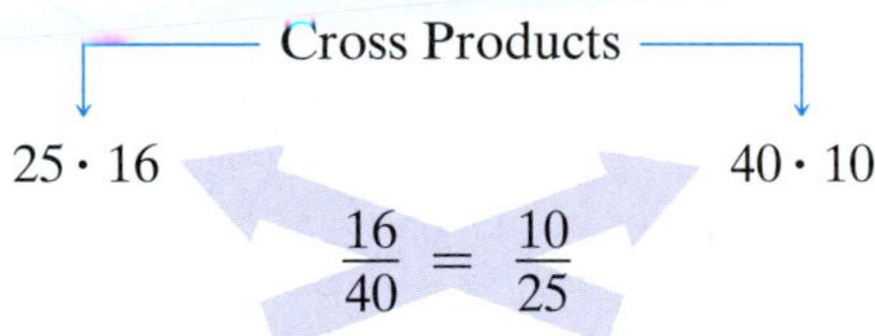

Notice that these cross products are equal

$$25 \cdot 16 = 400, \quad 40 \cdot 10 = 400$$

In general, this is true for equivalent fractions.

Equality of Fractions

$8 \cdot 6$ $\qquad$ $24 \cdot 2$

$$\frac{6}{24} \overset{?}{=} \frac{2}{8}$$

Since the cross products ($8 \cdot 6 = 48$ and $24 \cdot 2 = 48$) are equal, the fractions are equal.

Note: If the cross products are not equal, the fractions are not equal.

Answer

10. equivalent

PRACTICE PROBLEM 11

Determine whether $\frac{4}{13}$ and $\frac{5}{18}$ are equivalent.

EXAMPLE 11 Determine whether $\frac{8}{11}$ and $\frac{19}{26}$ are equivalent.

Solution: Let's check cross products.

$$26 \cdot 8 = 208 \qquad \frac{8}{11} \stackrel{?}{=} \frac{19}{26} \qquad 11 \cdot 19 = 209$$

Since $208 \neq 209$, then $\frac{8}{11} \neq \frac{19}{26}$.

Helpful Hint "Not equal to" symbol

Work Practice Problem 11

Objective D Solving Problems by Writing Fractions in Simplest Form

Many real-life problems can be solved by writing fractions. To make the answers clearer, these fractions should be written in simplest form.

PRACTICE PROBLEM 12

There are six national parks (including historic) in Washington state. See Example 12 and determine what fraction of the United States' national parks are located in Washington. Write the fraction in simplest form.

EXAMPLE 12 **Calculating Fraction of Parks in Wyoming State**

There are currently 58 national parks in the United States. Two of these parks are located in the state of Wyoming. What fraction of the United States' national parks can be found in Wyoming? Write the fraction in simplest form. (*Source:* National Park Service)

Solution: First we determine the fraction of parks found in Wyoming state.

$\frac{2}{58}$ ← national parks in Wyoming / ← total national parks

Next we simplify the fraction.

$$\frac{2}{58} = \frac{\overset{1}{\cancel{2}}}{\underset{1}{\cancel{2}} \cdot 29} = \frac{1}{1 \cdot 29} = \frac{1}{29}$$

Thus, $\frac{1}{29}$ of the United States' national parks are in Wyoming state.

Work Practice Problem 12

Answers
11. not equivalent **12.** $\frac{3}{29}$

CALCULATOR EXPLORATIONS Simplifying Fractions

Scientific Calculator

Many calculators have a fraction key, such as [a b/c], that allows you to simplify a fraction on the calculator. For example, to simplify $\frac{324}{612}$, enter

[3] [2] [4] [a b/c] [6] [1] [2] [=]

The display will read

[9|17]

which represents $\frac{9}{17}$, the original fraction simplified.

Graphing Calculator

Graphing calculators also allow you to simplify fractions. The fraction option on a graphing calculator may be found under the [MATH] menu.

To simplify $\frac{324}{612}$, enter

[3] [2] [4] [÷] [6] [1] [2] [MATH] [ENTER] [ENTER]

The display will read

[324/612 ▶ Frac 9/17]

Helpful Hint

The Calculator Explorations boxes in this chapter provide only an introduction to fraction keys on calculators. Any time you use a calculator, there are both advantages and limitations to its use. Never rely solely on your calculator. It is very important that you understand how to perform all operations on fractions by hand in order to progress through later topics. For further information, talk to your instructor.

Use your calculator to simplify each fraction.

1. $\frac{128}{224}$ **2.** $\frac{231}{396}$ **3.** $\frac{340}{459}$

4. $\frac{999}{1350}$ **5.** $\frac{432}{810}$ **6.** $\frac{225}{315}$

7. $\frac{54}{243}$ **8.** $\frac{455}{689}$

Vocabulary and Readiness Check

Use the choices below to fill in each blank.

cross products	equivalent	composite
simplest form	prime factorization	prime

1. The number 40 equals $2 \cdot 2 \cdot 2 \cdot 5$. Since each factor is prime, we call $2 \cdot 2 \cdot 2 \cdot 5$ the ________________ of 40.
2. A natural number, other than 1, that is not prime is called a ________________ number.
3. A natural number that has exactly two different factors, 1 and itself, is called a ________________ number.
4. In $\frac{11}{48}$, since 11 and 48 have no common factors other than 1, $\frac{11}{48}$ is in ________________.
5. Fractions that represent the same portion of a whole are called ________________ fractions.
6. In the statement $\frac{5}{12} = \frac{15}{36}$, $5 \cdot 36$ and $12 \cdot 15$ are called ________________.

4.2 EXERCISE SET

FOR EXTRA HELP

Student Solutions Manual | PH Math/Tutor Center | CD/Video for Review | MathXL® | MyMathLab

Objective A *Write the prime factorization of each number. See Examples 1 through 3.*

1. 20 **2.** 12 **3.** 48 **4.** 75

5. 81 **6.** 64 **7.** 162 **8.** 128

9. 110 **10.** 130 **11.** 85 **12.** 93

13. 240 **14.** 836 **15.** 828 **16.** 504

Objective B *Write each fraction in simplest form. See Examples 4 through 9.*

17. $\frac{3}{12}$ **18.** $\frac{5}{30}$ **19.** $\frac{4x}{42}$ **20.** $\frac{9y}{48}$ **21.** $\frac{14}{16}$

22. $\frac{22}{34}$ **23.** $\frac{20}{30}$ **24.** $\frac{70}{80}$ **25.** $\frac{35a}{50a}$ **26.** $\frac{25z}{55z}$

27. $-\frac{63}{81}$ **28.** $-\frac{21}{49}$ **29.** $\frac{30x^2}{36x}$ **30.** $\frac{45b}{80b^2}$ **31.** $\frac{27}{64}$

32. $\frac{32}{63}$ **33.** $\frac{25xy}{40y}$ **34.** $\frac{36y}{42yz}$ **35.** $-\frac{40}{64}$ **36.** $-\frac{28}{60}$

37. $\frac{36x^3y^2}{24xy}$ **38.** $\frac{60a^2b}{36ab^3}$ **39.** $\frac{90}{120}$ **40.** $\frac{60}{150}$ **41.** $\frac{40xy}{64xyz}$

42. $\frac{28abc}{60ac}$ **43.** $\frac{66}{308}$ **44.** $\frac{65}{234}$ **45.** $-\frac{55}{85y}$ **46.** $-\frac{78}{90x}$

47. $\frac{189z}{216z}$ **48.** $\frac{144y}{162y}$ **49.** $\frac{224a^3b^4c^2}{16ab^4c^2}$ **50.** $\frac{270x^4y^3z^3}{15x^3y^3z^3}$

Objective C *Determine whether each pair of fractions is equivalent. See Examples 10 and 11.*

51. $\frac{2}{6}$ and $\frac{4}{12}$ **52.** $\frac{3}{6}$ and $\frac{5}{10}$ **53.** $\frac{7}{11}$ and $\frac{5}{8}$

54. $\frac{2}{5}$ and $\frac{4}{11}$ **55.** $\frac{10}{15}$ and $\frac{6}{9}$ **56.** $\frac{4}{10}$ and $\frac{6}{15}$

57. $\frac{3}{9}$ and $\frac{6}{18}$ **58.** $\frac{2}{8}$ and $\frac{7}{28}$ **59.** $\frac{10}{13}$ and $\frac{13}{15}$

60. $\frac{16}{20}$ and $\frac{9}{16}$ **61.** $\frac{8}{18}$ and $\frac{12}{24}$ **62.** $\frac{6}{21}$ and $\frac{14}{35}$

Objective D *Solve. Write each fraction in simplest form. See Example 12.*

63. A work shift for an employee at Starbucks consists of 8 hours. What fraction of the employee's work shift is represented by 2 hours?

64. Two thousand baseball caps were sold one year at the U.S. Open Golf Tournament. What fractional part of this total does 200 caps represent?

65. There are 5280 feet in a mile. What fraction of a mile is represented by 2640 feet?

66. There are 100 centimeters in 1 meter. What fraction of a meter is 20 centimeters?

67. Fifteen out of the total fifty states in the United States have Ritz-Carlton hotels. (*Source:* Ritz-Carlton Hotel Company, LLC)

a. What fraction of states can claim at least one Ritz-Carlton hotel?
b. How many states do not have a Ritz-Carlton hotel?
c. Write the fraction of states without a Ritz-Carlton hotel.

68. There are 78 national monuments in the United States. Ten of these monuments are located in New Mexico. (*Source:* National Park Service)

a. What fraction of the national monuments in the United States can be found in New Mexico?
b. How many of the national monuments in the United States are found outside New Mexico?
c. Write the fraction of national monuments found in states other than New Mexico.

69. The outer wall of the Pentagon is 24 inches wide. Ten inches is concrete, 8 inches is brick, and 6 inches is limestone. What fraction of the wall is concrete?

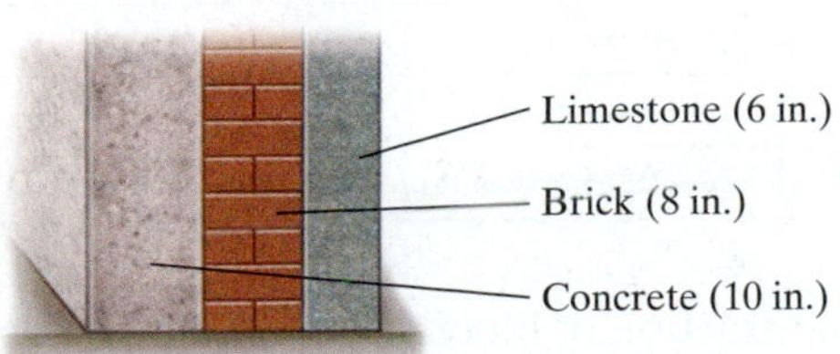

70. There are 35 students in a biology class. If 10 students made an A on the first test, what fraction of the students made an A?

71. As Internet usage grows in the United States, more and more state governments are placing services online. Twenty-two out of the total fifty states have Web sites that allow residents to file their state income tax online directly to the state.

a. How many states do not have this type of Web site?
b. What fraction of states do not have this type of Web site? (*Source:* Federation of Tax Administrators)

72. Chris Callac just bought a brand new 2006 Toyota Camry for $22,000. His old car was traded in for $10,000.

a. How much of his purchase price was not covered by his trade-in?
b. What fraction of the purchase price was not covered by the trade-in?

73. As of March 2006, a total of 449 individuals from around the world had flown in space. Of these, 291 were United States astronauts. What fraction of individuals who have flown into space were Americans? (*Source:* NASA)

74. Worldwide, Hallmark employs 18,000 full-time employees. About 4500 employees work at the Hallmark headquarters in Kansas City, Missouri. What fraction of Hallmark employees work in Kansas City? (*Source:* Hallmark Cards, Inc.)

Review

Evaluate each expression using the given replacement numbers. See Section 2.5.

75. $\frac{x^3}{9}$ when $x = -3$

76. $\frac{y^3}{5}$ when $y = -5$

77. $2y$ when $y = -7$

78. $-5a$ when $a = -4$

Concept Extensions

79. In your own words, define equivalent fractions.

80. Given a fraction, say $\frac{15}{40}$, how many fractions are there that are equivalent to it, but in simplest form or lowest terms? Explain your answer.

Write each fraction in simplest form.

81. $\frac{3975}{6625}$

82. $\frac{9506}{12{,}222}$

There are generally considered to be eight basic blood types. The table shows the number of people with the various blood types in a typical group of 100 blood donors. Use the table to answer Exercises 83 through 86. Write each answer in simplest form.

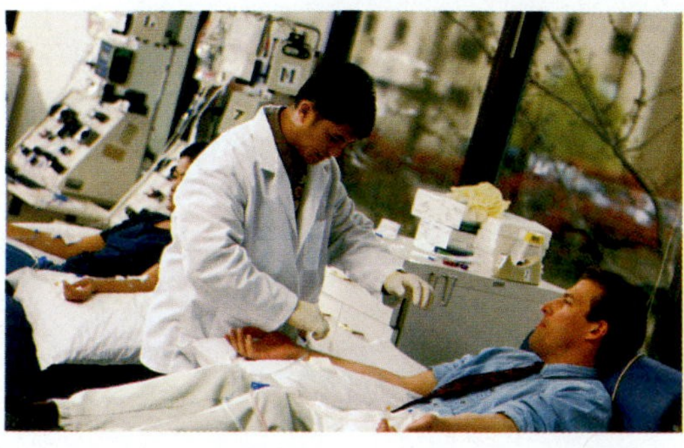

Distribution of Blood Types in Blood Donors

Blood Type	Number of People
O Rh-positive	37
O Rh-negative	7
A Rh-positive	36
A Rh-negative	6
B Rh-positive	9
B Rh-negative	1
AB Rh-positive	3
AB Rh-negative	1
(*Source:* American Red Cross Biomedical Services)	

83. What fraction of blood donors have blood type A Rh-positive?

84. What fraction of blood donors have an O blood type?

85. What fraction of blood donors have an AB blood type?

86. What fraction of blood donors have a B blood type?

Find the prime factorization of each number.

87. 34,020

88. 131,625

89. In your own words, define a prime number.

90. The number 2 is a prime number. All other even natural numbers are composite numbers. Explain why.

91. Two students have different prime factorizations for the same number. Is this possible? Explain.

The following graph is called a circle graph or pie chart. Each sector (shaped like a piece of pie) shows the fraction of entering college freshmen who expect to major in each discipline shown. The whole circle represents the entire class of college freshmen. Use this graph to answer Exercises 92 through 95.

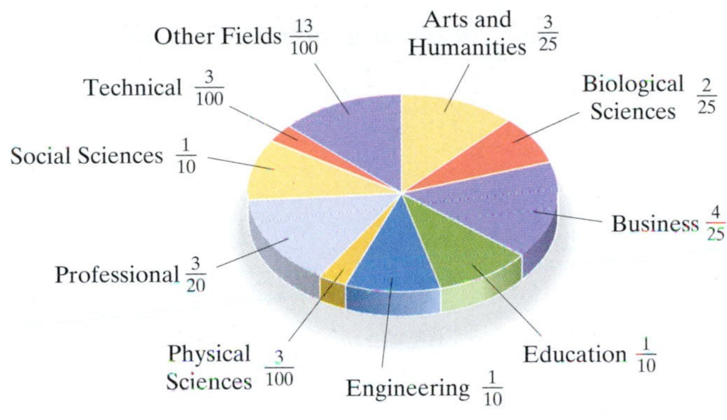

Source: The Higher Education Research Institute

92. What fraction of entering college freshmen plan to major in education?

93. What fraction of entering college freshmen plan to major in biological sciences?

94. Why is the Social Sciences sector the same size as the Engineering sector?

95. Why is the Physical Sciences sector smaller than the Business sector?

Use the following numbers for Exercises 96 through 99.

8691 786 1235 2235 85 105 22 222 900 1470

96. List the numbers divisible by both 2 and 3.

97. List the numbers that are divisible by both 3 and 5.

98. The answers to Exercise 96 are also divisible by what number? Tell why.

99. The answers to Exercise 97 are also divisible by what number? Tell why.

4.3 MULTIPLYING AND DIVIDING FRACTIONS

Objectives

A Multiply Fractions.

B Evaluate Exponential Expressions with Fractional Bases.

C Divide Fractions.

D Multiply and Divide Given Fractional Replacement Values.

E Solve Applications That Require Multiplication of Fractions.

Objective A Multiplying Fractions

Let's use a diagram to discover how fractions are multiplied. For example, to multiply $\frac{1}{2}$ and $\frac{3}{4}$, we find $\frac{1}{2}$ of $\frac{3}{4}$. To do this, we begin with a diagram showing $\frac{3}{4}$ of a rectangle's area shaded.

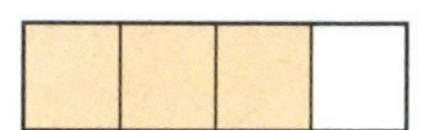

$\frac{3}{4}$ of the rectangle's area is shaded.

To find $\frac{1}{2}$ of $\frac{3}{4}$, we heavily shade $\frac{1}{2}$ of the part that is already shaded.

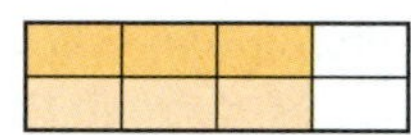

By counting smaller rectangles, we see that $\frac{3}{8}$ of the larger rectangle is now heavily shaded, so that

$\frac{1}{2}$ of $\frac{3}{4}$ is $\frac{3}{8}$, or $\frac{1}{2} \cdot \frac{3}{4} = \frac{3}{8}$ Notice that $\frac{1}{2} \cdot \frac{3}{4} = \frac{1 \cdot 3}{2 \cdot 4} = \frac{3}{8}$.

Multiplying Fractions

To multiply two fractions, multiply the numerators and multiply the denominators.

If a, b, c, and d represent numbers, and b and d are not 0, we have

$$\frac{a}{b} \cdot \frac{c}{d} = \frac{a \cdot c}{b \cdot d}$$

EXAMPLES Multiply.

1. $\frac{2}{3} \cdot \frac{5}{11} = \frac{2 \cdot 5}{3 \cdot 11} = \frac{10}{33}$ Multiply numerators. Multiply denominators.

This fraction is in simplest form since 10 and 33 have no common factors other than 1.

2. $\frac{1}{4} \cdot \frac{1}{2} = \frac{1 \cdot 1}{4 \cdot 2} = \frac{1}{8}$ This fraction is in simplest form.

Work Practice Problems 1–2

PRACTICE PROBLEMS 1–2

Multiply.

1. $\frac{3}{7} \cdot \frac{5}{11}$ **2.** $\frac{1}{3} \cdot \frac{1}{9}$

Answers

1. $\frac{15}{77}$ **2.** $\frac{1}{27}$

EXAMPLE 3 Multiply and simplify: $\frac{6}{7} \cdot \frac{14}{27}$

Solution:

$$\frac{6}{7} \cdot \frac{14}{27} = \frac{6 \cdot 14}{7 \cdot 27}$$

We can simplify by finding the prime factorizations and using our shortcut procedure of dividing out common factors in the numerator and denominator.

$$\frac{6 \cdot 14}{7 \cdot 27} = \frac{2 \cdot \overset{1}{\cancel{3}} \cdot 2 \cdot \overset{1}{\cancel{7}}}{\underset{1}{\cancel{7}} \cdot \underset{1}{\cancel{3}} \cdot 3 \cdot 3} = \frac{2 \cdot 2}{3 \cdot 3} = \frac{4}{9}$$

Work Practice Problem 3

Helpful Hint Remember that the shortcut procedure above is the same as removing factors of 1 in the product.

$$\frac{6 \cdot 14}{7 \cdot 27} = \frac{2 \cdot 3 \cdot 2 \cdot 7}{7 \cdot 3 \cdot 3 \cdot 3} = \frac{7}{7} \cdot \frac{3}{3} \cdot \frac{2 \cdot 2}{3 \cdot 3} = 1 \cdot 1 \cdot \frac{4}{9} = \frac{4}{9}$$

EXAMPLE 4 Multiply and simplify: $\frac{23}{32} \cdot \frac{4}{7}$

Solution: Notice that 4 and 32 have a common factor of 4.

$$\frac{23}{32} \cdot \frac{4}{7} = \frac{23 \cdot 4}{32 \cdot 7} = \frac{23 \cdot \overset{1}{\cancel{4}}}{\underset{1}{\cancel{4}} \cdot 8 \cdot 7} = \frac{23}{8 \cdot 7} = \frac{23}{56}$$

Work Practice Problem 4

After multiplying two fractions, always check to see whether the product can be simplified.

EXAMPLE 5 Multiply: $-\frac{1}{4} \cdot \frac{1}{2}$

Solution: Recall that the product of a negative number and a positive number is a negative number.

$$-\frac{1}{4} \cdot \frac{1}{2} = -\frac{1 \cdot 1}{4 \cdot 2} = -\frac{1}{8}$$

Work Practice Problem 5

EXAMPLES Multiply.

6. $\left(-\frac{6}{13}\right)\left(-\frac{26}{30}\right) = \frac{6 \cdot 26}{13 \cdot 30} = \frac{\overset{1}{\cancel{6}} \cdot \overset{1}{\cancel{13}} \cdot 2}{\underset{1}{\cancel{13}} \cdot \underset{1}{\cancel{6}} \cdot 5} = \frac{2}{5}$ The product of two negative numbers is a positive number.

7. $\frac{1}{3} \cdot \frac{2}{5} \cdot \frac{9}{16} = \frac{1 \cdot 2 \cdot 9}{3 \cdot 5 \cdot 16} = \frac{1 \cdot \overset{1}{\cancel{2}} \cdot \overset{1}{\cancel{3}} \cdot 3}{\underset{1}{\cancel{3}} \cdot 5 \cdot \underset{1}{\cancel{2}} \cdot 8} = \frac{3}{40}$

Work Practice Problems 6–7

PRACTICE PROBLEM 3

Multiply and simplify: $\frac{6}{77} \cdot \frac{7}{8}$

PRACTICE PROBLEM 4

Multiply and simplify: $\frac{4}{27} \cdot \frac{3}{8}$

Helpful Hint Don't forget that we may identify common factors that are not prime numbers.

PRACTICE PROBLEM 5

Multiply.

5. $\frac{1}{2} \cdot \left(-\frac{11}{28}\right)$

PRACTICE PROBLEMS 6–7

Multiply.

6. $\left(-\frac{4}{11}\right)\left(-\frac{33}{16}\right)$

7. $\frac{1}{6} \cdot \frac{3}{10} \cdot \frac{25}{16}$

Answers

3. $\frac{3}{44}$ 4. $\frac{1}{18}$ 5. $-\frac{11}{56}$ 6. $\frac{3}{4}$ 7. $\frac{5}{64}$

We multiply fractions in the same way if variables are involved.

PRACTICE PROBLEM 8

Multiply: $\frac{2}{3} \cdot \frac{3y}{2}$

EXAMPLE 8 Multiply: $\frac{3x}{4} \cdot \frac{8}{5x}$

Solution: Notice that 8 and 4 have a common factor of 4.

$$\frac{3x}{4} \cdot \frac{8}{5x} = \frac{3 \cdot \overset{1}{\cancel{x}} \cdot \overset{1}{\cancel{4}} \cdot 2}{\underset{1}{\cancel{4}} \cdot 5 \cdot \underset{1}{\cancel{x}}} = \frac{3 \cdot 1 \cdot 1 \cdot 2}{1 \cdot 5 \cdot 1} = \frac{6}{5}$$

Work Practice Problem 8

Helpful Hint

Recall that when the denominator of a fraction contains a variable, such as $\frac{8}{5x}$, we assume that the variable does not represent 0.

PRACTICE PROBLEM 9

Multiply: $\frac{a^3}{b^2} \cdot \frac{b}{a^2}$

EXAMPLE 9 Multiply: $\frac{x^2}{y} \cdot \frac{y^3}{x}$

Solution:

$$\frac{x^2}{y} \cdot \frac{y^3}{x} = \frac{x^2 \cdot y^3}{y \cdot x} = \frac{\overset{1}{\cancel{x}} \cdot x \cdot \overset{1}{\cancel{y}} \cdot y \cdot y}{\underset{1}{\cancel{y}} \cdot \underset{1}{\cancel{x}}} = \frac{x \cdot y \cdot y}{1} = xy^2$$

Work Practice Problem 9

Objective B Evaluating Expressions with Fractional Bases

The base of an exponential expression can also be a fraction.

$$\left(\frac{1}{3}\right)^4 = \underbrace{\frac{1}{3} \cdot \frac{1}{3} \cdot \frac{1}{3} \cdot \frac{1}{3}}_{\frac{1}{3}\text{ is a factor 4 times.}} = \frac{1 \cdot 1 \cdot 1 \cdot 1}{3 \cdot 3 \cdot 3 \cdot 3} = \frac{1}{81}$$

PRACTICE PROBLEM 10

Evaluate.

a. $\left(\frac{3}{4}\right)^3$ **b.** $\left(-\frac{4}{5}\right)^2$

EXAMPLE 10 Evaluate.

a. $\left(\frac{2}{5}\right)^4 = \frac{2}{5} \cdot \frac{2}{5} \cdot \frac{2}{5} \cdot \frac{2}{5} = \frac{2 \cdot 2 \cdot 2 \cdot 2}{5 \cdot 5 \cdot 5 \cdot 5} = \frac{16}{625}$

b. $\left(-\frac{1}{4}\right)^2 = \left(-\frac{1}{4}\right) \cdot \left(-\frac{1}{4}\right) = \frac{1 \cdot 1}{4 \cdot 4} = \frac{1}{16}$ The product of two negative numbers is a positive number.

Work Practice Problem 10

Objective C Dividing Fractions

Before we can divide fractions, we need to know how to find the **reciprocal** of a fraction.

Reciprocal of a Fraction

Two numbers are **reciprocals** of each other if their product is 1. The reciprocal of the fraction $\frac{a}{b}$ is $\frac{b}{a}$ because $\frac{a}{b} \cdot \frac{b}{a} = \frac{a \cdot b}{b \cdot a} = 1$.

Answers

8. y **9.** $\frac{a}{b}$ **10. a.** $\frac{27}{64}$ **b.** $\frac{16}{25}$

Helpful Hint
Every number has a reciprocal except 0. The number 0 has no reciprocal because there is no number such that $0 \cdot a = 1$.

For example,

The reciprocal of $\frac{2}{5}$ is $\frac{5}{2}$ because $\frac{2}{5} \cdot \frac{5}{2} = \frac{10}{10} = 1$.

The reciprocal of 5 is $\frac{1}{5}$ because $5 \cdot \frac{1}{5} = \frac{5}{1} \cdot \frac{1}{5} = \frac{5}{5} = 1$.

The reciprocal of $-\frac{7}{11}$ is $-\frac{11}{7}$ because $-\frac{7}{11} \cdot -\frac{11}{7} = \frac{77}{77} = 1$.

Division of fractions has the same meaning as division of whole numbers. For example,

$10 \div 5$ means: How many 5s are there in 10?

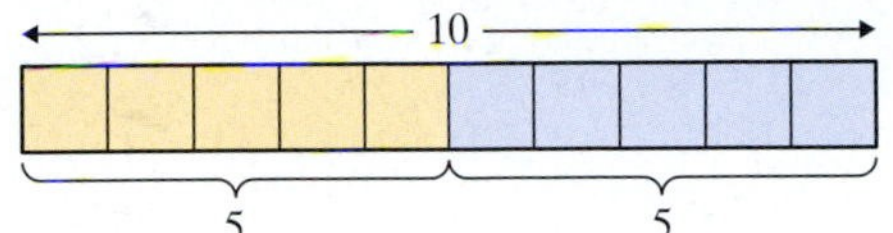

There are two 5s in 10, so $10 \div 5 = 2$.

$\frac{3}{4} \div \frac{1}{8}$ means: How many $\frac{1}{8}$s are there in $\frac{3}{4}$?

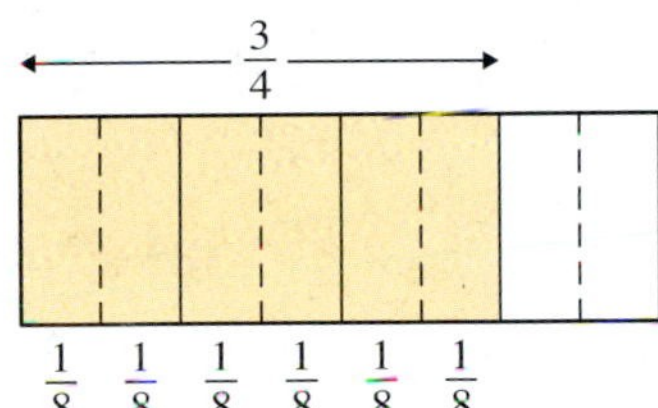

There are six $\frac{1}{8}$s in $\frac{3}{4}$, so $\frac{3}{4} \div \frac{1}{8} = 6$.

We use reciprocals to divide fractions.

Dividing Fractions

To divide two fractions, multiply the first fraction by the reciprocal of the second fraction.

If a, b, c, and d represent numbers, and b, c, and d are not 0, then

$$\frac{a}{b} \div \frac{c}{d} = \frac{a}{b} \cdot \frac{d}{c} = \frac{a \cdot d}{b \cdot c}$$

(reciprocal)

For example,

(multiply by reciprocal)

$$\frac{3}{4} \div \frac{1}{8} = \frac{3}{4} \cdot \frac{8}{1} = \frac{3 \cdot 8}{4 \cdot 1} = \frac{3 \cdot 2 \cdot \cancel{4}^{1}}{\cancel{4}_{1} \cdot 1} = \frac{6}{1} \text{ or } 6$$

After dividing fractions, *always* check to see whether the result can be simplified.

PRACTICE PROBLEMS 11–12

Divide and simplify.

11. $\frac{8}{7} \div \frac{2}{9}$

12. $\frac{4}{9} \div \frac{1}{2}$

EXAMPLES Divide and simplify.

11. $\frac{5}{16} \div \frac{3}{4} = \frac{5}{16} \cdot \frac{4}{3} = \frac{5 \cdot 4}{16 \cdot 3} = \frac{5 \cdot \overset{1}{\cancel{4}}}{\underset{1}{\cancel{4}} \cdot 4 \cdot 3} = \frac{5}{12}$

12. $\frac{2}{5} \div \frac{1}{2} = \frac{2}{5} \cdot \frac{2}{1} = \frac{2 \cdot 2}{5 \cdot 1} = \frac{4}{5}$

Work Practice Problems 11–12

Helpful Hint

When dividing by a fraction, do not look for common factors to divide out until you rewrite the division as multiplication.

Do not try to divide out these two 2s.

$$\frac{1}{\mathbf{2}} \div \frac{\mathbf{2}}{3} = \frac{1}{2} \cdot \frac{3}{2} = \frac{3}{4}$$

PRACTICE PROBLEM 13

Divide: $-\frac{10}{4} \div \frac{2}{9}$

EXAMPLE 13 Divide: $-\frac{5}{16} \div -\frac{3}{4}$

Solution: Recall that the quotient (or product) of two negative numbers is a positive number.

$$-\frac{5}{16} \div -\frac{3}{4} = -\frac{5}{16} \cdot -\frac{4}{3} = \frac{5 \cdot \overset{1}{\cancel{4}}}{\underset{1}{\cancel{4}} \cdot 4 \cdot 3} = \frac{5}{12}$$

Work Practice Problem 13

PRACTICE PROBLEM 14

Divide: $\frac{3y}{4} \div 5y^3$

EXAMPLE 14 Divide: $\frac{2x}{3} \div 3x^2$

Solution:

$$\frac{2x}{3} \div 3x^2 = \frac{2x}{3} \div \frac{3x^2}{1} = \frac{2x}{3} \cdot \frac{1}{3x^2} = \frac{2 \cdot \overset{1}{\cancel{x}} \cdot 1}{3 \cdot 3 \cdot \underset{1}{\cancel{x}} \cdot x} = \frac{2}{9x}$$

Work Practice Problem 14

PRACTICE PROBLEM 15

Simplify: $\left(-\frac{2}{3} \cdot \frac{9}{14}\right) \div \frac{7}{15}$

EXAMPLE 15 Simplify: $\left(\frac{4}{7} \cdot \frac{3}{8}\right) \div -\frac{3}{4}$

Solution: Remember to perform the operations inside the ( ) first.

$$\left(\frac{4}{7} \cdot \frac{3}{8}\right) \div -\frac{3}{4} = \left(\frac{\overset{1}{\cancel{4}} \cdot 3}{7 \cdot 2 \cdot \underset{1}{\cancel{4}}}\right) \div -\frac{3}{4} = \frac{3}{14} \div -\frac{3}{4}$$

Now divide.

$$\frac{3}{14} \div -\frac{3}{4} = \frac{3}{14} \cdot -\frac{4}{3} = -\frac{\overset{1}{\cancel{3}} \cdot \overset{1}{\cancel{2}} \cdot 2}{\underset{1}{\cancel{2}} \cdot 7 \cdot \underset{1}{\cancel{3}}} = -\frac{2}{7}$$

Work Practice Problem 15

Answers

11. $\frac{36}{7}$ **12.** $\frac{8}{9}$ **13.** $-\frac{45}{4}$ **14.** $\frac{3}{20y^2}$ **15.** $-\frac{45}{49}$

✓**Concept Check** Which is the correct way to divide $\frac{3}{5}$ by $\frac{5}{12}$? Explain.

a. $\frac{3}{5} \div \frac{5}{12} = \frac{5}{3} \cdot \frac{5}{12}$ **b.** $\frac{3}{5} \div \frac{5}{12} = \frac{3}{5} \cdot \frac{12}{5}$

Objective D Multiplying and Dividing with Fractional Replacement Values

Recall the difference between an expression and an equation. For example, xy and $x \div y$ are expressions. They contain no equals signs. In Example 16, we practice *simplifying* expressions given fractional replacement values.

EXAMPLE 16 If $x = \frac{7}{8}$ and $y = -\frac{1}{3}$, evaluate **(a)** xy and **(b)** $x \div y$.

Solution: Replace x with $\frac{7}{8}$ and y with $-\frac{1}{3}$.

a. $xy = \frac{7}{8} \cdot -\frac{1}{3} = -\frac{7 \cdot 1}{8 \cdot 3} = -\frac{7}{24}$

b. $x \div y = \frac{7}{8} \div -\frac{1}{3} = \frac{7}{8} \cdot -\frac{3}{1} = -\frac{7 \cdot 3}{8 \cdot 1} = -\frac{21}{8}$

Work Practice Problem 16

PRACTICE PROBLEM 16

If $x = -\frac{3}{4}$ and $y = \frac{9}{2}$, evaluate **(a)** xy, and **(b)** $x \div y$.

Helpful Hint "of" usually translates to multiplication.

EXAMPLE 17 Is $-\frac{2}{3}$ a solution of the equation $-\frac{1}{2}x = \frac{1}{3}$?

Solution: To check whether a number is a solution of an equation, recall that we replace the variable with the given number and see if a true statement results.

$-\frac{1}{2} \cdot x = \frac{1}{3}$ Recall that $-\frac{1}{2}x$ means $-\frac{1}{2} \cdot x$.

$-\frac{1}{2} \cdot -\frac{2}{3} \stackrel{?}{=} \frac{2}{3}$ Replace x with $-\frac{2}{3}$.

$\frac{1 \cdot \cancel{2}^{1}}{\cancel{2}_{1} \cdot 3} \stackrel{?}{=} \frac{1}{3}$ The product of two negative numbers is a positive number.

$\frac{1}{3} = \frac{1}{3}$ True

Since we have a true statement, $-\frac{2}{3}$ is a solution.

Work Practice Problem 17

PRACTICE PROBLEMS 17

Is $-\frac{9}{8}$ a solution of the equation $2x = -\frac{9}{4}$?

Objective E Solving Problems by Multiplying Fractions

To solve real-life problems that involve multiplying fractions, we use our four problem-solving steps from Chapter 3. In Example 18, a new key word that implies multiplication is used. That key word is **"of."**

Answers

16. a. $-\frac{27}{8}$ **b.** $-\frac{1}{6}$ **17.** yes

✓ Concept Check Answers

a. incorrect **b.** correct

PRACTICE PROBLEM 18

Hershey Park is an amusement park in Hershey, Pennsylvania. Of its 60 rides, $\frac{1}{6}$ of them are roller coasters. How many roller coasters are in HersheyPark?

EXAMPLE 18 Finding the Number of Roller Coasters in an Amusement Park

Cedar Point is an amusement park located in Sandusky, Ohio. Its collection of 68 rides is the largest in the world. Of the rides, $\frac{4}{17}$ are roller coasters. How many roller coasters are in Cedar Point's collection of rides? (*Source:* Wikipedia)

Solution:

1. UNDERSTAND the problem. To do so, read and reread the problem. We are told that $\frac{4}{17}$ of Cedar Point's rides are roller coasters. The word "of" here means multiplication.
2. TRANSLATE.

In words:	Number of roller coasters	is	$\frac{4}{17}$	of	total rides at Cedar Point
	↓	↓	↓	↓	↓
Translate:	Number of roller coasters	=	$\frac{4}{17}$	·	68

3. SOLVE: Before we solve, let's estimate a reasonable answer. The fraction $\frac{4}{17}$ is less than $\frac{1}{2}$ (draw a diagram, if needed), and $\frac{1}{2}$ of 68 rides is 34 rides, so the number of roller coasters should be less than 34.

$$\frac{4}{17}\cdot 68 = \frac{4}{17}\cdot\frac{68}{1} = \frac{4\cdot 68}{17\cdot 1} = \frac{4\cdot \overset{1}{\cancel{17}}\cdot 4}{\underset{1}{\cancel{17}}\cdot 1} = \frac{16}{1} \quad \text{or} \quad 16$$

4. INTERPRET. *Check* your work. From our estimate, our answer is reasonable. *State* your conclusion: The number of roller coasters at Cedar Point is 16.

Work Practice Problem 18

Helpful Hint

To help visualize a fractional part of a whole number, look at the diagram below.

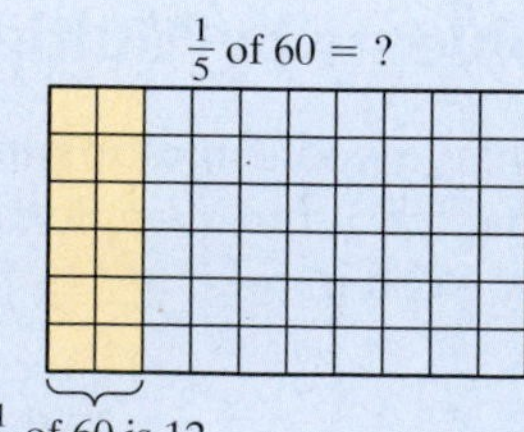

Answer

18. 10 roller coasters

Vocabulary and Readiness Check

Use the choices below to fill in each blank. Not all choices will be used.

multiplication	$\dfrac{a \cdot d}{b \cdot c}$	$\dfrac{a \cdot c}{b \cdot d}$	$\dfrac{2 \cdot 2 \cdot 2}{7}$	$\dfrac{2}{7} \cdot \dfrac{2}{7} \cdot \dfrac{2}{7}$
division	0	reciprocals		

1. To multiply two fractions, we write $\frac{a}{b} \cdot \frac{c}{d} =$ ________.

2. Two numbers are ________ of each other if their product is 1.

3. The expression $\frac{2^3}{7} =$ ________ while $\left(\frac{2}{7}\right)^3 =$ ________.

4. Every number has a reciprocal except ________.

5. To divide two fractions, we write $\frac{a}{b} \div \frac{c}{d} =$ ________.

6. The word "of" indicates ________.

4.3 EXERCISE SET

FOR EXTRA HELP

Student Solutions Manual

PH Math/Tutor Center

CD/Video for Review

MathXL® · MyMathLab

Objective A *Multiply. Write the product in simplest form. See Examples 1 through 9.*

1. $\frac{6}{11} \cdot \frac{3}{7}$

2. $\frac{5}{9} \cdot \frac{7}{4}$

3. $-\frac{2}{7} \cdot \frac{5}{8}$

4. $\frac{4}{15} \cdot -\frac{1}{20}$

5. $-\frac{1}{2} \cdot -\frac{2}{15}$

6. $-\frac{3}{11} \cdot -\frac{11}{12}$

7. $\frac{18x}{20} \cdot \frac{36}{99}$

8. $\frac{5}{32} \cdot \frac{64y}{100}$

9. $3a^2 \cdot \frac{1}{4}$

10. $-\frac{2}{3} \cdot 6y^3$

11. $\frac{x^3}{y^3} \cdot \frac{y^2}{x}$

12. $\frac{a}{b^3} \cdot \frac{b}{a^3}$

13. $0 \cdot \frac{8}{9}$

14. $\frac{11}{12} \cdot 0$

15. $-\frac{17y}{20} \cdot \frac{4}{5y}$

16. $-\frac{13x}{20} \cdot \frac{5}{6x}$

17. $\frac{11}{20} \cdot \frac{1}{7} \cdot \frac{5}{22}$

18. $\frac{27}{32} \cdot \frac{10}{13} \cdot \frac{16}{30}$

Objective B *Evaluate. See Example 10.*

19. $\left(\frac{1}{5}\right)^3$

20. $\left(\frac{8}{9}\right)^2$

21. $\left(-\frac{2}{3}\right)^2$

22. $\left(-\frac{1}{2}\right)^4$

23. $\left(-\frac{2}{3}\right)^3 \cdot \frac{1}{2}$

24. $\left(-\frac{3}{4}\right)^3 \cdot \frac{1}{3}$

Objective C *Divide. Write all quotients in simplest form. See Examples 11 through 14.*

25. $\frac{2}{3} \div \frac{5}{6}$

26. $\frac{5}{8} \div \frac{3}{4}$

27. $-\frac{6}{15} \div \frac{12}{5}$

28. $-\frac{4}{15} \div -\frac{8}{3}$

29. $-\frac{8}{9} \div \frac{x}{2}$

30. $\frac{10}{11} \div -\frac{4}{5x}$

31. $\frac{11y}{20} \div \frac{3}{11}$

32. $\frac{9z}{20} \div \frac{2}{9}$

33. $-\frac{2}{3} \div 4$

34. $-\frac{5}{6} \div 10$

35. $\frac{1}{5x} \div \frac{5}{x^2}$

36. $\frac{3}{y^2} \div \frac{9}{y^3}$

Objectives **A B C Mixed Practice** *Perform each indicated operation. See Examples 1 through 15.*

37. $\frac{2}{3} \cdot \frac{5}{9}$

38. $\frac{8}{15} \cdot \frac{5}{32}$

39. $\frac{3x}{7} \div \frac{5}{6x}$

40. $\frac{2}{5y} \div \frac{5y}{11}$

41. $\frac{16}{27y} \div \frac{8}{15y}$

42. $\frac{12y}{21} \div \frac{4y}{7}$

43. $-\frac{5}{28} \cdot \frac{35}{25}$

44. $\frac{24}{45} \cdot -\frac{5}{8}$

45. $\left(-\frac{3}{4}\right)^2$

46. $\left(-\frac{1}{2}\right)^5$

47. $\frac{x^2}{y} \cdot \frac{y^3}{x}$

48. $\frac{b}{a^2} \cdot \frac{a^3}{b^3}$

49. $7 \div \frac{2}{11}$

50. $-100 \div \frac{1}{2}$

51. $-3x \div \frac{x^2}{12}$

52. $-7x^2 \div \frac{14x}{3}$

53. $\left(\frac{2}{7} \div \frac{7}{2}\right) \cdot \frac{3}{4}$

54. $\frac{1}{2} \cdot \left(\frac{5}{6} \div \frac{1}{12}\right)$

55. $-\frac{19}{63y} \cdot 9y^2$

56. $16a^2 \cdot -\frac{31}{24a}$

57. $-\frac{2}{3} \cdot -\frac{6}{11}$

58. $-\frac{1}{5} \cdot -\frac{6}{7}$

59. $\frac{4}{8} \div \frac{3}{16}$

60. $\frac{9}{2} \div \frac{16}{15}$

61. $\frac{21x^2}{10y} \div \frac{14x}{25y}$

62. $\frac{17y^2}{24x} \div \frac{13y}{18x}$

63. $\left(1 \div \frac{3}{4}\right) \cdot \frac{2}{3}$

64. $\left(33 \div \frac{2}{11}\right) \cdot \frac{5}{9}$

65. $\frac{a^3}{2} \div 30a^3$

66. $15c^3 \div \frac{3c^2}{5}$

67. $\frac{ab^2}{c} \cdot \frac{c}{ab}$

68. $\frac{ac}{b} \cdot \frac{b^3}{a^2c}$

69. $\left(\frac{1}{2} \cdot \frac{2}{3}\right) \div \frac{5}{6}$

70. $\left(\frac{3}{4} \cdot \frac{8}{9}\right) \div \frac{2}{5}$

71. $-\frac{4}{7} \div \left(\frac{4}{5} \cdot \frac{3}{7}\right)$

72. $\frac{5}{8} \div \left(\frac{4}{7} \cdot -\frac{5}{16}\right)$

Objective **D** *Given the following replacement values, evaluate (**a**) xy and (**b**) x ÷ y. See Example 16.*

73. $x = \frac{2}{5}$ and $y = \frac{5}{6}$

a. **b.**

74. $x = \frac{8}{9}$ and $y = \frac{1}{4}$

a. **b.**

75. $x = -\frac{4}{5}$ and $y = \frac{9}{11}$

a. **b.**

76. $x = \frac{7}{6}$ and $y = -\frac{1}{2}$

a. **b.**

Determine whether the given replacement values are solutions of the given equations. See Example 17.

77. Is $-\frac{5}{18}$ a solution to $3x = -\frac{5}{6}$?

78. Is $\frac{9}{11}$ a solution to $\frac{2}{3}y = \frac{6}{11}$?

79. Is $\frac{2}{5}$ a solution to $-\frac{1}{2}z = \frac{1}{10}$?

80. Is $\frac{3}{5}$ a solution to $5x = \frac{1}{3}$?

Objective **E Translating** *Solve. Write each answer in simplest form. For Exercises 81 through 84, recall that "of" translates to multiplication. See Example 18.*

81. Find $\frac{1}{4}$ of 200.

82. Find $\frac{1}{5}$ of 200.

83. Find $\frac{5}{6}$ of 24.

84. Find $\frac{5}{8}$ of 24.

Solve. See Example 18.

85. In the United States, $\frac{4}{25}$ of college freshmen major in business. A community college in Pennsylvania has a freshman enrollment of approximately 800 students. How many of these freshmen might we project are majoring in business?

86. In a recent year, movie theater owners received a total of \$7660 million in movie admission tickets. About $\frac{7}{10}$ of this amount was for R-rated movies. Find the amount of money received from R-rated movies. (*Source:* Motion Picture Association of America)

87. The Oregon National Historic Trail is 2,170 miles long. It begins in Independence, Missouri, and ends in Oregon City, Oregon. Manfred Coulon has hiked $\frac{2}{5}$ of the trail before. How many miles has he hiked? (*Source:* National Park Service)

88. Each turn of a screw sinks it $\frac{3}{16}$ of an inch deeper into a piece of wood. Find how deep the screw is after 8 turns.

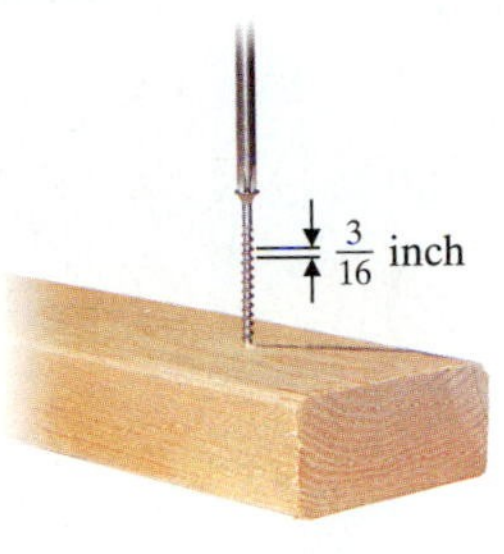

89. The radius of a circle is one-half of its diameter as shown. If the diameter of a circle is $\frac{3}{8}$ of an inch, what is its radius?

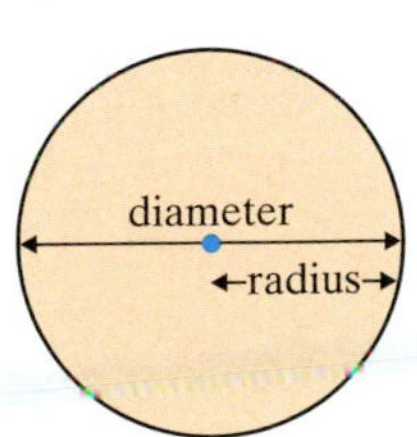

90. A patient was told that no more than $\frac{1}{5}$ of his calories should come from fat. If his diet consists of 3000 calories a day, how many of these calories can come from fat?

91. A special on a cruise to the Bahamas is advertised to be $\frac{2}{3}$ of the regular price. If the regular price is \$2757, what is the sale price?

92. The Gonzales family recently sold their house for \$102,000, but $\frac{3}{50}$ of this amount goes to the real estate companies that helped them sell their house. How much money do the Gonzales pay to the real estate companies?

93. There have been about 180 different contestants on the reality television show *Survivor* over 12 seasons. Some of these contestants have appeared in two different series and/or seasons. If the number of returning contestants was $\frac{1}{9}$ of the total number of participants in the first 12 seasons, how many contestants participated more than once? (*Source:* Survivor.com)

94. In 2005, about $\frac{23}{50}$ of the people who saw movies in theaters were in the 16–24 age bracket. If the local cinema boasted that its total attendance for 2005 was 854,000 people, how many might the cinema owners believe were between the ages of 16 and 24? (*Source:* Motion Picture Association of America)

Find the area of each rectangle. Recall that area = length · width.

△ **95.**

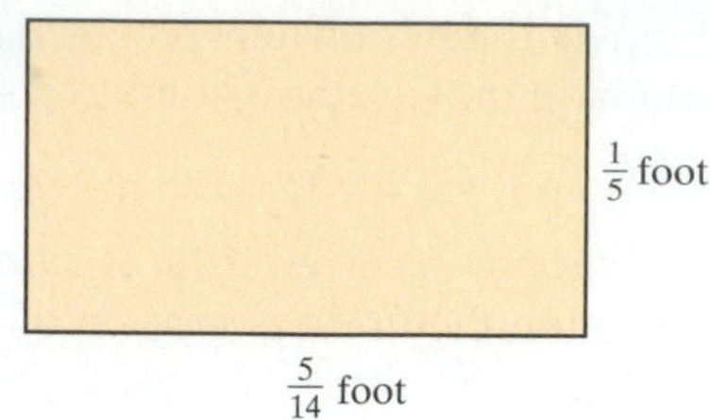

△ **96.**

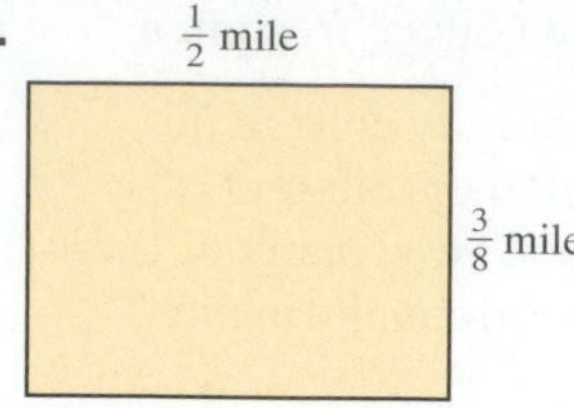

Recall from Section 4.2 that the following graph is called a ***circle graph*** *or* ***pie chart.*** *Each sector (shaped like a piece of pie) shows the fractional part of a car's total mileage that falls into a particular category. The whole circle represents a car's total mileage.*

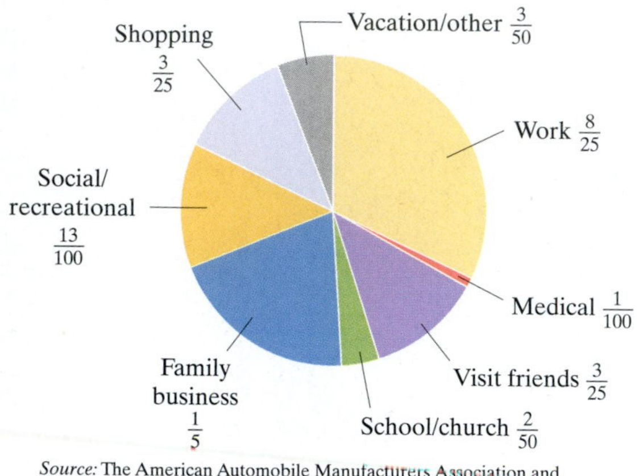

Source: The American Automobile Manufacturers Association and The National Automobile Dealers Association

In one year, a family drove 12,000 miles in the family car. Use the circle graph to determine how many of these miles might be expected to fall in the categories shown in Exercises 97 through 100.

97. Work

98. Shopping

99. Family business

100. Medical

Review

Perform each indicated operation. See Section 1.3.

101. $\begin{array}{r} 27 \\ 76 \\ +\ 98 \\ \hline \end{array}$

102. $\begin{array}{r} 811 \\ 42 \\ +\ 69 \\ \hline \end{array}$

103. $\begin{array}{r} 968 \\ -\ 772 \\ \hline \end{array}$

104. $\begin{array}{r} 882 \\ -\ 773 \\ \hline \end{array}$

Concept Extensions

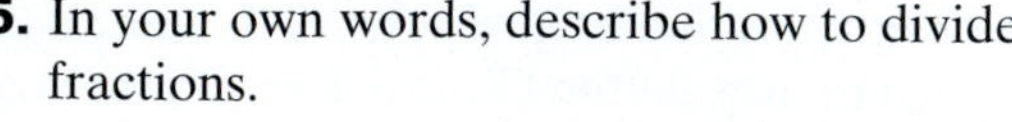

105. In your own words, describe how to divide fractions.

106. In your own words, explain how to multiply fractions.

Simplify.

107. $\frac{42}{25} \cdot \frac{125}{36} \div \frac{7}{6}$

108. $\left(\frac{8}{13} \cdot \frac{39}{16} \cdot \frac{8}{9}\right)^2 \div \frac{1}{2}$

109. Approximately $\frac{3}{25}$ of the U.S. population lives in the state of California. If the U.S. population is approximately 290,810,000, find the approximate population of California. (*Source: World Almanac* 2006)

110. In 2006, there were approximately 10,600 commercial radio stations broadcasting in the United States. Of these stations, $\frac{32}{265}$ were news/talk stations. How many radio stations were news/talk stations in 2006? (*Source:* Corporation for Public Broadcasting)

111. If $\frac{3}{4}$ of 36 students on a first bus are girls and $\frac{2}{3}$ of the 30 students on a second bus are *boys,* how many students on the two buses are girls?

112. The FedEx Express air fleet includes 252 Cessnas. These Cessnas make up $\frac{42}{113}$ of the FedEx fleet. How many aircraft make up the entire FedEx Express air fleet? (*Source:* FedEx Corporation)

THE BIGGER PICTURE Operations on Sets of Numbers

Continue your outline from Sections 1.6, 1.7, 2.4, and 3.3. Suggestions are once again written to help you complete this part of your outline, Section I.C. Fractions.

I. Operations on Sets of Numbers

- **A. Whole Numbers**
  1. **Add or Subtract** (Section 1.3)
  2. **Multiply or Divide** (Sections 1.5, 1.6)
  3. **Exponent** (Section 1.7)
  4. **Order of Operations** (Section 1.7)
- **B. Integers**
  1. **Add** (Section 2.2)
  2. **Subtract** (Section 2.3)
  3. **Multiply or Divide** (Section 2.4)
- **C. Fractions**
  1. **Simplify:** Factor the numerator and denominator. Then divide out factors of 1 by dividing out common factors in the numerator and denominator.

     Simplify: $\frac{20}{28} = \frac{\overset{1}{\cancel{4}} \cdot 5}{\underset{1}{\cancel{4}} \cdot 7} = \frac{5}{7}$

  2. **Multiply:** Numerator times numerator over denominator times denominator. $\frac{5}{9} \cdot \frac{2}{7} = \frac{10}{63}$
  3. **Divide:** First fraction times the reciprocal of the second fraction.

     $\frac{2}{11} \div \frac{3}{4} = \frac{2}{11} \cdot \frac{4}{3} = \frac{8}{33}$

II. Solving Equations

- **A. Equations in General** (Section 3.3)

Perform the indicated operations.

1. $\frac{2}{3} \cdot \frac{8}{9}$

2. $-\frac{2}{3} \div \frac{8}{9}$

3. $\frac{12}{1} \cdot \frac{1}{9}$

4. $\frac{9x}{2} \div \frac{15x}{8}$

5. $80 \div 20 \cdot 2$

6. $3^2 \cdot 2^3$

7. $\frac{11}{20} \cdot \frac{5}{8} \cdot \frac{4}{33}$

8. $20 \div \frac{1}{2}$

9. $3 + 4(18 - 16)^3$

10. $100 - 76$

Solve.

11. $-12x + 7 = 7$

12. $9n - 14 = 6n - 32$

Objectives

A Add or Subtract Like Fractions.

B Add or Subtract Given Fractional Replacement Values.

C Solve Problems by Adding or Subtracting Like Fractions.

D Find the Least Common Denominator of a List of Fractions.

E Write Equivalent Fractions.

4.4 ADDING AND SUBTRACTING LIKE FRACTIONS, LEAST COMMON DENOMINATOR, AND EQUIVALENT FRACTIONS

Fractions with the same denominator are called **like fractions.** Fractions that have different denominators are called **unlike fractions.**

Like Fractions	Unlike Fractions
$\frac{2}{5}$ and $\frac{3}{5}$ ← same denominator	$\frac{2}{5}$ and $\frac{3}{4}$ ← different denominators
$\frac{5}{21}, \frac{16}{21}$, and $\frac{7}{21}$ ← same denominator	$\frac{5}{7}$ and $\frac{5}{9}$ ← different denominators

Objective A Adding or Subtracting Like Fractions

To see how we add like fractions (fractions with the same denominator), study one or both illustrations below.

Add: $\frac{1}{5} + \frac{3}{5}$

Figures	Number Line
$\frac{1}{5} + \frac{3}{5}$ $\frac{1}{5} + \frac{3}{5} = \frac{4}{5}$	To add $\frac{1}{5} + \frac{3}{5}$, start at 0 and draw an arrow $\frac{1}{5}$ of a unit long pointing to the right. From the tip of this arrow, draw an arrow $\frac{3}{5}$ of a unit long also pointing to the right. The tip of the second arrow ends at their sum, $\frac{4}{5}$. Start End $\frac{1}{5}$ $\frac{3}{5}$ 0 $\frac{4}{5}$ 1 $\frac{1}{5} + \frac{3}{5} = \frac{4}{5}$

Thus, $\frac{1}{5} + \frac{3}{5} = \frac{4}{5}$.

Notice that the numerator of the sum is the sum of the numerators. Also, the denominator of the sum is the **common denominator.** This is how we add fractions. Similar illustrations can be shown for subtracting fractions.

Adding or Subtracting Like Fractions (Fractions with the Same Denominator)

If a, b, and c, are numbers and b is not 0, then

$$\frac{a}{b} + \frac{c}{b} = \frac{a + c}{b} \quad \text{and also} \quad \frac{a}{b} - \frac{c}{b} = \frac{a - c}{b}$$

In other words, to add or subtract fractions with the same denominator, add or subtract their numerators and write the sum or difference over the **common** denominator.

For example,

$$\frac{1}{4} + \frac{2}{4} = \frac{1+2}{4} = \frac{3}{4}$$ Add the numerators. Keep the denominator.

$$\frac{4}{5} - \frac{2}{5} = \frac{4-2}{5} = \frac{2}{5}$$ Subtract the numerators. Keep the denominator.

Helpful Hint

As usual, don't forget to write all answers in simplest form.

EXAMPLES Add and simplify.

1. $\frac{2}{7} + \frac{3}{7} = \frac{2+3}{7} = \frac{5}{7}$ ⟵ Add the numerators. ⟵ Keep the common denominator.

2. $\frac{3}{16} + \frac{7}{16} = \frac{3+7}{16} = \frac{10}{16} = \frac{\overset{1}{\cancel{2}} \cdot 5}{\underset{1}{\cancel{2}} \cdot 8} = \frac{5}{8}$

3. $\frac{7}{8} + \frac{6}{8} + \frac{3}{8} = \frac{7+6+3}{8} = \frac{16}{8}$ or 2

Work Practice Problems 1–3

PRACTICE PROBLEMS 1–3

Add and simplify.

1. $\frac{6}{13} + \frac{2}{13}$

2. $\frac{5}{8} + \frac{1}{8}$

3. $\frac{20}{11} + \frac{6}{11} + \frac{7}{11}$

✔Concept Check Find and correct the error in the following:

$$\frac{1}{5} + \frac{1}{5} = \frac{2}{10}$$

EXAMPLES Subtract and simplify.

4. $\frac{8}{9} - \frac{1}{9} = \frac{8-1}{9} = \frac{7}{9}$ ⟵ Subtract the numerators. ⟵ Keep the common denominator.

5. $\frac{7}{8} - \frac{5}{8} = \frac{7-5}{8} = \frac{2}{8} = \frac{\overset{1}{\cancel{2}}}{\underset{1}{\cancel{2}} \cdot 4} = \frac{1}{4}$

Work Practice Problems 4–5

PRACTICE PROBLEMS 4–5

Subtract and simplify.

4. $\frac{11}{12} - \frac{6}{12}$ 5. $\frac{7}{15} - \frac{2}{15}$

From our earlier work, we know that

$$\frac{-12}{6} = \frac{12}{-6} = -\frac{12}{6} \text{ since these all simplify to } -2.$$

In general, the following is true:

$$\frac{-a}{b} = \frac{a}{-b} = -\frac{a}{b} \text{ as long as } b \text{ is not } 0.$$

EXAMPLE 6 Add: $-\frac{11}{8} + \frac{6}{8}$

Solution: $-\frac{11}{8} + \frac{6}{8} = \frac{-11+6}{8}$

$= \frac{-5}{8}$ or $-\frac{5}{8}$

Work Practice Problem 6

PRACTICE PROBLEM 6

Add: $-\frac{8}{17} + \frac{4}{17}$

Answers

1. $\frac{8}{13}$ 2. $\frac{3}{4}$ 3. 3 4. $\frac{5}{12}$ 5. $\frac{1}{3}$ 6. $-\frac{4}{17}$

✔ Concept Check Answer

We don't add denominators together; correct solution: $\frac{1}{5} + \frac{1}{5} = \frac{2}{5}$.

PRACTICE PROBLEM 7

Subtract: $\frac{2}{5} - \frac{7y}{5}$

EXAMPLE 7 Subtract: $\frac{3x}{4} - \frac{7}{4}$

Solution: $\frac{3x}{4} - \frac{7}{4} = \frac{3x - 7}{4}$

Recall from Section 3.1 that the terms in the numerator are unlike terms and cannot be combined.

Work Practice Problem 7

PRACTICE PROBLEM 8

Subtract: $\frac{4}{11} - \frac{6}{11} - \frac{3}{11}$

EXAMPLE 8 Subtract: $\frac{3}{7} - \frac{6}{7} - \frac{3}{7}$

Solution: $\frac{3}{7} - \frac{6}{7} - \frac{3}{7} = \frac{3 - 6 - 3}{7} = \frac{-6}{7}$ or $-\frac{6}{7}$

Work Practice Problem 8

Helpful Hint

Recall that $\frac{-6}{7} = -\frac{6}{7}$ $\left(\text{Also, } \frac{6}{-7} = -\frac{6}{7}, \text{ if needed.}\right)$

Objective B Adding and Subtracting Given Fractional Replacement Values

PRACTICE PROBLEM 9

Evaluate $x + y$ if $x = -\frac{10}{12}$ and $y = \frac{5}{12}$.

EXAMPLE 9 Evaluate $y - x$ if $x = -\frac{3}{10}$ and $y = -\frac{8}{10}$.

Solution: Be very careful when replacing x and y with replacement values.

$$y - x = -\frac{8}{10} - \left(-\frac{3}{10}\right) \quad \text{Replace } x \text{ with } -\frac{3}{10} \text{ and } y \text{ with } -\frac{8}{10}.$$

$$= \frac{-8 - (-3)}{10}$$

$$= \frac{-5}{10} = \frac{-1 \cdot 5}{2 \cdot 5} = \frac{-1}{2} \text{ or } -\frac{1}{2}$$

Work Practice Problem 9

✔Concept Check Fill in each blank with the best choice given.

expression equation simplified solved

An ____________ contains an equals sign and may be ____________ for the variable. An ____________ does not contain an equals sign but may be ____________.

Objective C Solving Problems by Adding or Subtracting Like Fractions

Many real-life problems involve finding the perimeters of square or rectangular-shaped figures such as pastures, swimming pools, and so on. We can use our knowledge of adding fractions to find perimeters.

Answers

7. $\frac{2 - 7y}{5}$ 8. $-\frac{5}{11}$ 9. $-\frac{5}{12}$

✔ Concept Check Answer

equation solved expression simplified

EXAMPLE 10 Find the perimeter of the rectangle.

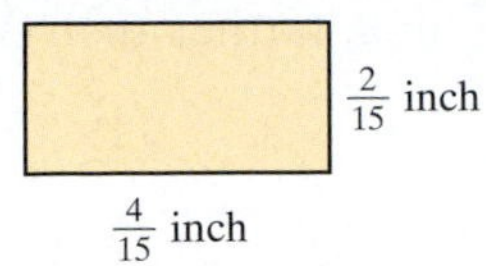

Solution: Recall that perimeter means distance around and that opposite sides of a rectangle are the same length.

$\frac{4}{15}$ inch (top), $\frac{2}{15}$ inch (left), $\frac{2}{15}$ inch (right), $\frac{4}{15}$ inch (bottom)

$$\text{Perimeter} = \frac{2}{15} + \frac{4}{15} + \frac{2}{15} + \frac{4}{15} = \frac{2+4+2+4}{15}$$

$$= \frac{12}{15} = \frac{\overset{1}{\cancel{3}} \cdot 4}{\underset{1}{\cancel{3}} \cdot 5} = \frac{4}{5}$$

The perimeter of the rectangle is $\frac{4}{5}$ inch.

Work Practice Problem 10

We can combine our skills in adding and subtracting fractions with our four problem-solving steps from Section 3.4 to solve many kinds of real-life problems.

EXAMPLE 11 **Calculating Distance**

The distance from home to the World Gym is $\frac{7}{8}$ of a mile and from home to the post office is $\frac{3}{8}$ of a mile. How much farther is it from home to the World Gym than from home to the post office?

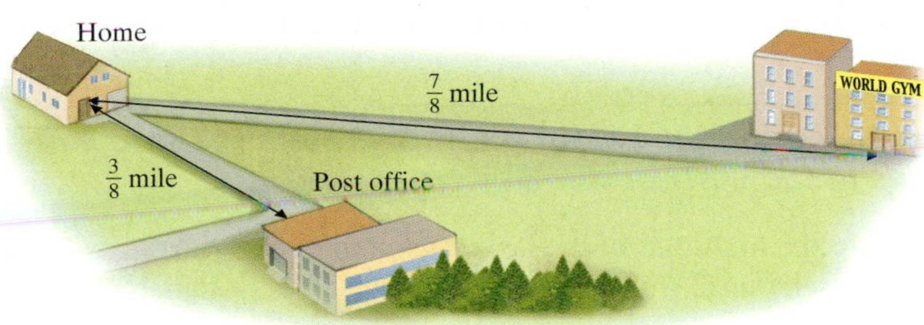

Solution:

1. UNDERSTAND. Read and reread the problem. The phrase "How much farther" tells us to subtract distances.
2. TRANSLATE.

In words:	distance farther	is	home to World Gym distance	minus	home to post office distance
	↓	↓	↓	↓	↓
Translate:	distance farther	=	$\frac{7}{8}$	−	$\frac{3}{8}$

3. SOLVE: $\frac{7}{8} - \frac{3}{8} = \frac{7-3}{8} = \frac{4}{8} = \frac{\overset{1}{\cancel{4}}}{2 \cdot \underset{1}{\cancel{4}}} = \frac{1}{2}$

Continued on next page

PRACTICE PROBLEM 10

Find the perimeter of the square.

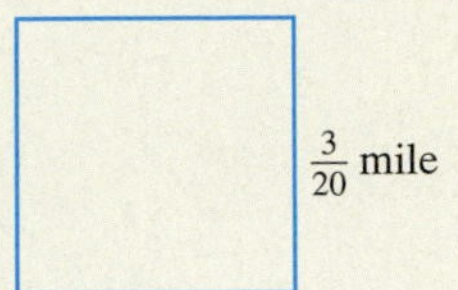

PRACTICE PRO LEM 11

A jogger ran $\frac{13}{4}$ n es on Monday and $\frac{11}{4}$ mi on Wednesday. How m h farther did he run on Mond than on Wednesday?

Answers

10. $\frac{3}{5}$ mi **11.** $\frac{1}{2}$ mi

4. INTERPRET. *Check* your work. *State* your conclusion: The distance from home to the World Gym is $\frac{1}{2}$ mile farther than from home to the post office.

Work Practice Problem 11

Objective D Finding the Least Common Denominator

In the next section, we will add and subtract fractions that have different, or unlike, denominators. To do so, we first write them as equivalent fractions with a common denominator.

Although any common denominator can be used to add or subtract unlike fractions, we will use the **least common denominator (LCD).** The LCD of a list of fractions is the same as the **least common multiple (LCM)** of the denominators. Why do we use this number as the common denominator? Since the LCD is the *smallest* of all common denominators, operations are usually less tedious with this number.

The **least common denominator (LCD)** of a list of fractions is the smallest positive number divisible by all the denominators in the list. (The least common denominator is also the **least common multiple (LCM) of the denominators.**)

For example, the LCD of $\frac{1}{4}$ and $\frac{3}{10}$ is 20 because 20 is the smallest positive number divisible by both 4 and 10.

Finding the LCD: Method 1

One way to find the LCD is to see whether the larger denominator is divisible by the smaller denominator. If so, the larger number is the LCD. If not, then check consecutive multiples of the larger denominator until the LCD is found.

Method 1: Finding the LCD of a List of Fractions Using Multiples of the Largest Number

Step 1: Write the multiples of the largest denominator (starting with the number itself) until a multiple common to all denominators in the list is found.

Step 2: The multiple found in Step 1 is the LCD.

PRACTICE PROBLEM 12

Find the LCD of $\frac{7}{8}$ and $\frac{11}{16}$.

EXAMPLE 12 Find the LCD of $\frac{3}{7}$ and $\frac{5}{14}$.

Solution: We write the multiples of 14 until we find one that is also a multiple of 7.

$14 \cdot 1 = 14$ A multiple of 7

The LCD is 14.

Work Practice Problem 12

Answer
12. 16

EXAMPLE 13 Find the LCD of $\frac{11}{12}$ and $\frac{7}{20}$.

Solution: We write the multiples of 20 until we find one that is also a multiple of 12.

$20 \cdot 1 = 20$ Not a multiple of 12

$20 \cdot 2 = 40$ Not a multiple of 12

$20 \cdot 3 = 60$ A multiple of 12

The LCD is 60.

Work Practice Problem 13

PRACTICE PROBLEM 13

Find the LCD of $\frac{23}{25}$ and $\frac{1}{30}$.

Method 1 for finding multiples works fine for smaller numbers, but may get tedious for larger numbers. For this reason, let's study a second method that uses prime factorization.

Finding the LCD: Method 2

For example, to find the LCD of $\frac{11}{12}$ and $\frac{7}{20}$, such as in Example 13, let's look at the prime factorization of each denominator.

$12 = 2 \cdot 2 \cdot 3$

$20 = 2 \cdot 2 \cdot 5$

Recall that the LCD must be a multiple of both 12 and 20. Thus, to build the LCD, we will circle the greatest number of factors for each different prime number. The LCD is the product of the circled factors.

Prime Number Factors

12 =	(2·2)	(3)	
20 =	2·2		(5)

Circle either pair of 2s, but not both.

$\text{LCD} = 2 \cdot 2 \cdot 3 \cdot 5 = 60$

The number 60 is the smallest number that both 12 and 20 divide into evenly. This method 2 is summarized below:

Method 2: Finding the LCD of a List of Denominators Using Prime Factorization

Step 1: Write the prime factorization of each denominator.

Step 2: For each different prime factor in Step 1, circle the *greatest* number of times that factor occurs in any one factorization.

Step 3: The LCD is the product of the circled factors.

EXAMPLE 14 Find the LCD of $-\frac{23}{72}$ and $\frac{17}{60}$.

Solution: First we write the prime factorization of each denominator.

$72 = 2 \cdot 2 \cdot 2 \cdot 3 \cdot 3$

$60 = 2 \cdot 2 \cdot 3 \cdot 5$

For the prime factors shown, we circle the greatest number of factors found in either factorization.

$72 = (2 \cdot 2 \cdot 2) \cdot (3 \cdot 3)$

$60 = 2 \cdot 2 \cdot 3 \cdot (5)$

Continued on next page

PRACTICE PROBLEM 14

Find the LCD of $\frac{-3}{40}$ and $\frac{11}{108}$.

Answers

13. 150 **14.** 1080

Helpful Hint If you prefer working with exponents, circle the factor with the greatest exponent.

Example 14:

$72 = (2^3) \cdot (3^2)$

$60 = 2^2 \cdot 3 \cdot (5)$

$\text{LCD} = 2^3 \cdot 3^2 \cdot 5 = 360$

The LCD is the product of the circled factors.

$\text{LCD} = 2 \cdot 2 \cdot 2 \cdot 3 \cdot 3 \cdot 5 = 360$

The LCD is 360.

Work Practice Problem 14

Helpful Hint If the number of factors of a prime number are equal, circle either one, but not both. For example,

$12 = (2 \cdot 2) \cdot (3)$

$15 = 3 \cdot (5)$

Circle either 3 but not both.

The LCD is $2 \cdot 2 \cdot 3 \cdot 5 = 60$.

PRACTICE PROBLEM 15

Find the LCD of $\frac{7}{20}, \frac{1}{24}$, and $\frac{13}{45}$.

EXAMPLE 15 Find the LCD of $\frac{1}{15}, \frac{5}{18}$, and $\frac{53}{54}$.

Solution: $15 = 3 \cdot (5)$

$18 = (2) \cdot 3 \cdot 3$

$54 = 2 \cdot (3 \cdot 3 \cdot 3)$

The LCD is $2 \cdot 3 \cdot 3 \cdot 3 \cdot 5$ or 270.

Work Practice Problem 15

PRACTICE PROBLEM 16

Find the LCD of $\frac{7}{y}$ and $\frac{6}{11}$.

EXAMPLE 16 Find the LCD of $\frac{3}{5}, \frac{2}{x}$, and $\frac{7}{x^3}$.

Solution: $5 = (5)$

$x = x$

$x^3 = (x \cdot x \cdot x)$

$\text{LCD} = 5 \cdot x \cdot x \cdot x = 5x^3$

Work Practice Problem 16

✔ **Concept Check** True or false? The LCD of the fractions $\frac{1}{6}$ and $\frac{1}{8}$ is 48.

Objective E Writing Equivalent Fractions

To add or subtract unlike fractions in the next section, we first write equivalent fractions with the LCD as the denominator.

To write $\frac{1}{3}$ as an equivalent fraction with a denominator of 6, we multiply by 1 in the form of $\frac{2}{2}$. Why? Because $3 \cdot 2 = 6$, so the new denominator will become 6, as shown below.

$$\frac{1}{3} = \frac{1}{3} \cdot 1 = \frac{1}{3} \cdot \frac{2}{2} = \frac{1 \cdot 2}{3 \cdot 2} = \frac{2}{6}$$

$\frac{2}{2} = 1$

So $\frac{1}{3} = \frac{2}{6}$.

Answers

15. 360 **16.** $11y$

✔ **Concept Check Answer**

false; it is 24.

To write an equivalent fraction,

$$\frac{a}{b} = \frac{a}{b} \cdot \frac{c}{c} = \frac{a \cdot c}{b \cdot c}$$

where a, b, and c are nonzero numbers.

✔**Concept Check** Which of the following is *not* equivalent to $\frac{3}{4}$?

a. $\frac{6}{8}$ **b.** $\frac{18}{24}$ **c.** $\frac{9}{14}$ **d.** $\frac{30}{40}$

EXAMPLE 17 Write $\frac{3}{4}$ as an equivalent fraction with a denominator of 20.

$$\frac{3}{4} = \frac{}{20}$$

Solution: In the denominators, since $4 \cdot 5 = 20$, we will multiply by 1 in the form of $\frac{5}{5}$.

$$\frac{3}{4} = \frac{3}{4} \cdot \frac{5}{5} = \frac{3 \cdot 5}{4 \cdot 5} = \frac{15}{20}$$

Thus, $\frac{3}{4} = \frac{15}{20}$.

Work Practice Problem 17

Helpful Hint

To check Example 17, write $\frac{15}{20}$ in simplest for[m].

$$\frac{15}{20} = \frac{3 \cdot \overset{1}{\cancel{5}}}{4 \cdot \underset{1}{\cancel{5}}} = \frac{3}{4}, \text{ the original fraction.}$$

If the original fraction is in lowest terms, we can check our work by wri[ting the] equivalent fraction in simplest form. This form should be the original fra[ction].

✔**Concept Check** True or false? When the fraction $\frac{2}{9}$ is rewritten as an equivalent fraction with 27 as the denominator, the result is $\frac{2}{27}$.

EXAMPLE 18 Write an equivalent fraction with the given denominator.

$$\frac{2}{5} = \frac{}{15}$$

Solution: Since $5 \cdot 3 = 15$, we multiply by 1 in the form of $\frac{3}{3}$.

$$\frac{2}{5} = \frac{2}{5} \cdot \frac{3}{3} = \frac{2 \cdot 3}{5 \cdot 3} = \frac{6}{15}$$

Then $\frac{2}{5}$ is equivalent to $\frac{6}{15}$. They both represent the same part of a whole.

Work Practice Problem 18

PRACTICE PROBLEM 17

Write $\frac{7}{8}$ as an equivalent fraction with a denominator of 56.

$$\frac{7}{8} = \frac{}{56}$$

PRA[CTICE PR]OBLEM 18

Write an equivalent fraction with the given denominator.

$$\frac{1}{4} = \frac{}{20}$$

Answers

17. $\frac{49}{56}$ **18.** $\frac{5}{20}$

✔ Concept Check Answers

c

false; the correct result would be $\frac{6}{27}$

PRACTICE PROBLEM 19

Write an equivalent fraction with the given denominator.

$\frac{3x}{7} = \frac{}{42}$

EXAMPLE 19 Write an equivalent fraction with the given denominator.

$\frac{9x}{11} = \frac{}{44}$

Solution: Since $11 \cdot 4 = 44$, we multiply by 1 in the form of $\frac{4}{4}$.

$$\frac{9x}{11} = \frac{9x}{11} \cdot \frac{4}{4} = \frac{9x \cdot 4}{11 \cdot 4} = \frac{36x}{44}$$

Then $\frac{9x}{11}$ is equivalent to $\frac{36x}{44}$.

Work Practice Problem 19

PRACTICE PROBLEM 20

Write an equivalent fraction with the given denominator.

$4 = \frac{}{6}$

EXAMPLE 20 Write an equivalent fraction with the given denominator.

$3 = \frac{}{7}$

Solution: Recall that $3 = \frac{3}{1}$. Since $1 \cdot 7 = 7$, multiply by 1 in the form $\frac{7}{7}$.

$$\frac{3}{1} = \frac{3}{1} \cdot \frac{7}{7} = \frac{3 \cdot 7}{1 \cdot 7} = \frac{21}{7}$$

Work Practice Problem 20

Don't forget that when the denominator of a fraction contains a variable, such as $\frac{8}{3x}$, we will assume that the variable does not represent 0. Recall that the denominator of a fraction cannot be 0.

PRACTICE PROBLEM 21

Write an equivalent fraction with the given denominator.

$\frac{9}{4x} = \frac{}{36x}$

EXAMPLE 21 Write an equivalent fraction with the given denominator.

$\frac{8}{3x} = \frac{}{24x}$

Solution: Since $3x \cdot 8 = 24x$, multiply by 1 in the form $\frac{8}{8}$.

$$\frac{8}{3x} = \frac{8}{3x} \cdot \frac{8}{8} = \frac{8 \cdot 8}{3x \cdot 8} = \frac{64}{24x}$$

Work Practice Problem 21

✔ Concept Check What is the first step in writing $\frac{3}{10}$ as an equivalent fraction whose denominator is 100?

Answers

19. $\frac{18x}{42}$ **20.** $\frac{24}{6}$ **21.** $\frac{81}{36x}$

✔ Concept Check Answer

answers may vary

Vocabulary and Readiness Check

Use the choices below to fill in each blank. Not all choices will be used.

least common denominator (LCD)	like	$-\frac{a}{b}$	$\frac{a-c}{b}$	$\frac{a+c}{b}$	$-\frac{a}{-b}$
perimeter	unlike				
equivalent					

1. The fractions $\frac{9}{11}$ and $\frac{13}{11}$ are called _____________ fractions while $\frac{3}{4}$ and $\frac{1}{3}$ are called _____________ fractions.

2. $\frac{a}{b}+\frac{c}{b}=$ _________ and $\frac{a}{b}-\frac{c}{b}=$ _________.

3. As long as b is not 0, $\frac{-a}{b}=\frac{a}{-b}=$ _____.

4. The distance around a figure is called its _____________.

5. The smallest positive number divisible by all the denominators of a list of fractions is called the ______________________________.

6. Fractions that represent the same portion of a whole are called _____________ fractions.

4.4 EXERCISE SET

FOR EXTRA HELP

Student Solutions Manual | PH Math/Tutor Center  | CD/Video for Review  | MathXL® | MyMathLab

Objective A *Add and simplify. See Examples 1 through 3, and 6.*

1. $\frac{5}{11}+\frac{2}{11}$ **2.** $\frac{9}{17}+\frac{2}{17}$ **3.** $\frac{2}{9}+\frac{4}{9}$ **4.** $\frac{3}{10}+\frac{2}{10}$

5. $-\frac{6}{20}+\frac{1}{20}$ **6.** $-\frac{3}{8}+\frac{1}{8}$ **7.** $-\frac{3}{14}+\left(-\frac{4}{14}\right)$ **8.** $-\frac{5}{24}+\left(-\frac{7}{24}\right)$

9. $\frac{2}{9x}+\frac{4}{9x}$ **10.** $\frac{3}{10y}+\frac{2}{10y}$ **11.** $-\frac{7x}{18}+\frac{3x}{18}+\frac{2x}{18}$ **12.** $-\frac{7z}{15}+\frac{3z}{15}+\frac{1z}{15}$

Subtract and simplify. See Examples 4, 5, 7, and 8.

13. $\frac{10}{11}-\frac{4}{11}$ **14.** $\frac{9}{13}-\frac{5}{13}$ **15.** $\frac{7}{8}-\frac{1}{8}$ **16.** $\frac{5}{6}-\frac{1}{6}$

17. $\frac{1}{y}-\frac{4}{y}$ **18.** $\frac{4}{z}-\frac{7}{z}$ **19.** $-\frac{27}{33}-\left(-\frac{8}{33}\right)$ **20.** $-\frac{37}{45}-\left(-\frac{18}{45}\right)$

21. $\frac{20}{21}-\frac{10}{21}-\frac{17}{21}$ **22.** $\frac{27}{28}-\frac{5}{28}-\frac{28}{28}$ **23.** $\frac{7a}{4}-\frac{3}{4}$ **24.** $\frac{18b}{5}-\frac{3}{5}$

Mixed Practice *Perform the indicated operation. See Examples 1 through 8.*

25. $-\frac{9}{100}+\frac{99}{100}$ **26.** $-\frac{15}{200}+\frac{85}{200}$ **27.** $-\frac{13x}{28}-\frac{13x}{28}$ **28.** $-\frac{15}{26y}-\frac{15}{26y}$

29. $\frac{9x}{15}+\frac{1}{15}$ **30.** $\frac{2x}{15}+\frac{7}{15}$ **31.** $\frac{7x}{16}-\frac{15x}{16}$ **32.** $\frac{3}{16z}-\frac{15}{16z}$

33. $\frac{9}{12}-\frac{7}{12}-\frac{10}{12}$ **34.** $\frac{1}{8}-\frac{15}{8}+\frac{2}{8}$ **35.** $\frac{x}{4}+\frac{3x}{4}-\frac{2x}{4}+\frac{x}{4}$ **36.** $\frac{9y}{8}+\frac{2y}{8}+\frac{5y}{8}-\frac{4y}{8}$

Objective B *Evaluate each expression for the given replacement values. See Example 9.*

37. $x + y;\ x = \frac{3}{4}, y = \frac{2}{4}$

38. $x - y;\ x = \frac{7}{8}, y = \frac{9}{8}$

39. $x - y;\ x = -\frac{1}{5}, y = -\frac{3}{5}$

40. $x + y;\ x = -\frac{1}{6}, y = \frac{5}{6}$

Objective C *Find the perimeter of each figure. (Hint: Recall that perimeter means distance around.) See Example 10.*

△ **41.**

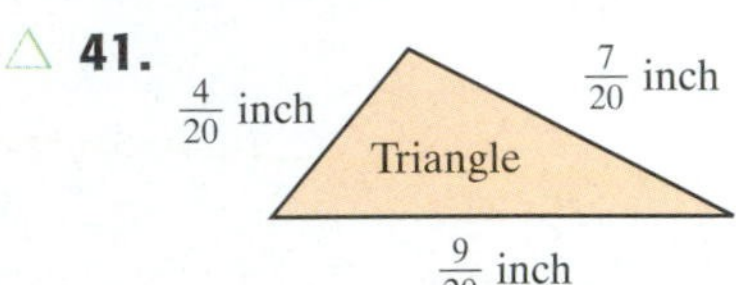

△ **42.**

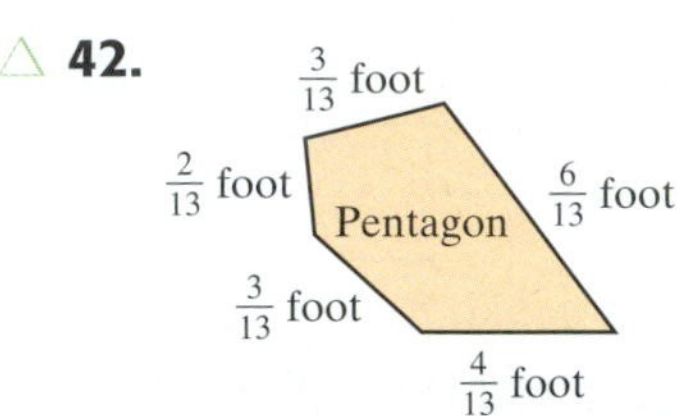

△ **43.** $\frac{5}{12}$ meter | Rectangle | $\frac{7}{12}$ meter

△ **44.** Square | $\frac{1}{6}$ centimeter

Solve. Write each answer in simplest form. See Example 11.

The map of the world below shows the fraction of the world's surface land area taken up by each continent. In other words, the continent of Africa makes up $\frac{20}{100}$ of the land in the world. Use this map for Exercises 45 through 48.

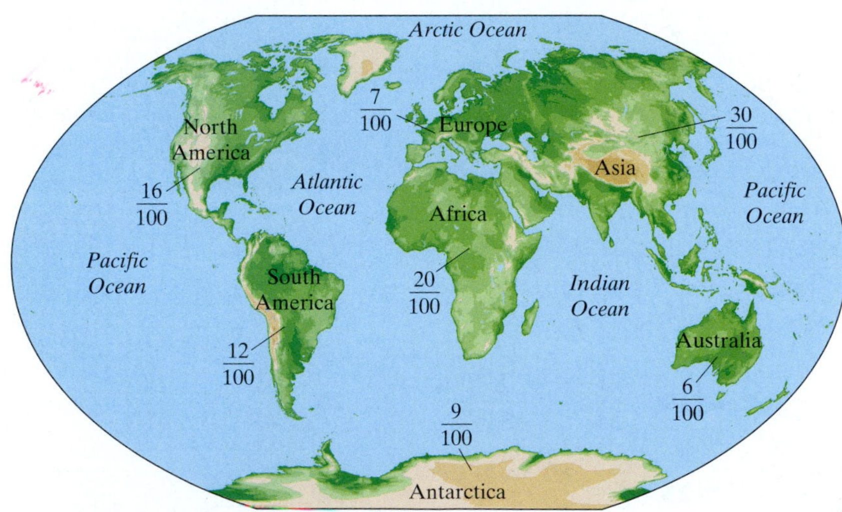

45. Find the fractional part of the world's land area within the continents of North America and South America.

46. Find the fractional part of the world's land area within the continents of Asia and Africa.

47. How much greater is the fractional part of the continent of Antarctica than the fractional part of the continent of Europe?

48. How much greater is the fractional part of the continent of Asia than the continent of Australia?

Solve.

49. A railroad inspector must inspect $\frac{19}{20}$ of a mile of railroad track. If she has already inspected $\frac{5}{20}$ of a mile, how much more does she need to inspect?

50. Scott Davis has run $\frac{11}{8}$ miles already and plans to complete $\frac{16}{8}$ miles. To do this, how much farther must he run?

51. As of 2006, the fraction of states in the United States with maximum interstate highway speed limits up to and including 70 mph was $\frac{37}{50}$. The fraction of states with 70 mph speed limits was $\frac{17}{50}$. What fraction of states had speed limits that were less than 70 mph? (*Source:* Insurance Institute for Highway Safety)

52. When people take aspirin, $\frac{31}{50}$ of the time it is used to treat some type of pain. Approximately $\frac{7}{50}$ of all aspirin use is for treating headaches. What fraction of aspirin use is for treating pain other than headaches? (*Source:* Bayer Market Research)

Objective D *Find the LCD of each list of fractions. See Examples 12 through 16.*

53. $\frac{2}{9}, \frac{6}{15}$

54. $\frac{7}{12}, \frac{3}{20}$

55. $-\frac{1}{36}, \frac{1}{24}$

56. $-\frac{1}{15}, \frac{1}{90}$

57. $\frac{2}{25}, \frac{3}{15}, \frac{5}{6}$

58. $\frac{3}{4}, \frac{1}{14}, \frac{13}{20}$

59. $-\frac{7}{24}, -\frac{5}{x}$

60. $-\frac{11}{y}, -\frac{13}{70}$

61. $\frac{23}{18}, \frac{1}{21}$

62. $\frac{45}{24}, \frac{2}{45}$

63. $\frac{4}{3}, \frac{8}{21}, \frac{3}{56}$

64. $\frac{12}{11}, \frac{20}{33}, \frac{12}{121}$

Objective E *Write each fraction as an equivalent fraction with the given denominator. See Examples 17 through 21.*

65. $\frac{2}{3} = \frac{}{21}$

66. $\frac{5}{6} = \frac{}{24}$

67. $\frac{4}{7} = \frac{}{35}$

68. $\frac{3}{5} = \frac{}{100}$

69. $\frac{1}{2} = \frac{}{50}$

70. $\frac{1}{5} = \frac{}{50}$

71. $\frac{14x}{17} = \frac{}{68}$

72. $\frac{19z}{21} = \frac{}{126}$

73. $\frac{2y}{3} = \frac{}{12}$

74. $\frac{3x}{2} = \frac{}{12}$

75. $\frac{5}{9} = \frac{}{36a}$

76. $\frac{7}{6} = \frac{}{36a}$

The table shows the fraction of the population in selected countries that used cell phones in a recent year. Use this table to answer Exercises 77 through 80.

77. Complete the table by writing each fraction as an equivalent fraction with a denominator of 100.

78. Which of these countries has the largest fraction of cell phone users?

79. Which of these countries shown has the smallest fraction of cell phone users?

80. In which of these countries do over $\frac{3}{4}$ of the population use cell phones? (*Hint:* Write $\frac{3}{4}$ as an equivalent fraction with a denominator of 100.)

Country	Fraction of Population Using Cell Phones	Equivalent Fraction with a Denominator of 100
Denmark	$\frac{89}{100}$	
Finland	$\frac{9}{10}$	
Israel	$\frac{24}{25}$	
Spain	$\frac{23}{25}$	
Japan	$\frac{17}{25}$	
Norway	$\frac{91}{100}$	
Singapore	$\frac{4}{5}$	
South Korea	$\frac{69}{100}$	
India	$\frac{2}{25}$	
United States	$\frac{7}{10}$	

(*Source: International Telecommunication and World Almanac, 2006*)

Review

Simplify. See Section 1.8.

81. 3^2

82. 4^3

83. 5^3

84. 3^4

85. 7^2

86. 5^4

87. $2^3 \cdot 3$

88. $4^2 \cdot 5$

Concept Extensions

Find and correct the error. See the first Concept Check in this section.

89. $\frac{2}{7} + \frac{9}{7} = \frac{11}{14}$

90. $\frac{3}{4} - \frac{1}{4} = \frac{2}{8} = \frac{1}{4}$

Solve.

91. In your own words, explain how to add like fractions.

92. In your own words, explain how to subtract like fractions.

93. Use the map of the world for Exercises 45 through 48 and find the sum of all the continents' fractions. Explain your answer.

94. Mike Cannon jogged $\frac{3}{8}$ of a mile from home and then rested. Then he continued jogging further from home for another $\frac{3}{8}$ of a mile until he discovered his watch had fallen off. He walked back along the same path for $\frac{4}{8}$ of a mile until he found his watch. Find how far he was from his home.

Write each fraction as an equivalent fraction with the indicated denominator.

95. $\frac{37x}{165} = \frac{}{3630}$

96. $\frac{108}{215y} = \frac{}{4085y}$

97. In your own words, explain how to find the LCD of two fractions.

98. In your own words, explain how to write a fraction as an equivalent fraction with a given denominator.

Solve. See the Concept Checks in this section.

99. Which of the following are equivalent to $\frac{2}{3}$?

a. $\frac{10}{15}$ **b.** $\frac{40}{60}$

c. $\frac{16}{20}$ **d.** $\frac{200}{300}$

100. True or False? When the fraction $\frac{7}{12}$ is rewritten with a denominator of 48, the result is $\frac{11}{48}$. If false, give the correct fraction.

4.5 ADDING AND SUBTRACTING UNLIKE FRACTIONS

Objectives

A Add or Subtract Unlike Fractions.

B Write Fractions in Order.

C Evaluate Expressions Given Fractional Replacement Values.

D Solve Problems by Adding or Subtracting Unlike Fractions.

Objective A Adding and Subtracting Unlike Fractions

In this section we add and subtract fractions with unlike denominators. To add or subtract these unlike fractions, we first write the fractions as equivalent fractions with a common denominator and then add or subtract the like fractions. Recall from the previous section that the common denominator we use is called the **least common denominator (LCD).**

To begin, let's add the unlike fractions $\frac{3}{4} + \frac{1}{6}$.

The LCD of these fractions is 12. So we write each fraction as an equivalent fraction with a denominator of 12, then add as usual. This addition process is shown next and also illustrated by figures.

Add: $\frac{3}{4} + \frac{1}{6}$	**The LCD is 12.**
Figures	**Algebra**
$\frac{3}{4} + \frac{1}{6}$ $\frac{9}{12} + \frac{2}{12}$ $\frac{9}{12} + \frac{2}{12} = \frac{11}{12}$	$\frac{3}{4} = \frac{3}{4} \cdot \frac{3}{3} = \frac{9}{12}$ and $\frac{1}{6} = \frac{1}{6} \cdot \frac{2}{2} = \frac{2}{12}$ Remember $\frac{3}{3} = 1$ and $\frac{2}{2} = 1$. Now we can add just as we did in Section 4.4. $\frac{3}{4} + \frac{1}{6} = \frac{9}{12} + \frac{2}{12} = \frac{11}{12}$
Thus, the sum is $\frac{11}{12}$.	

Adding or Subtracting Unlike Fractions

Step 1: Find the least common denominator (LCD) of the fractions.

Step 2: Write each fraction as an equivalent fraction whose denominator is the LCD.

Step 3: Add or subtract the like fractions.

Step 4: Write the sum or difference in simplest form.

PRACTICE PROBLEM 1

Add: $\frac{2}{7} + \frac{8}{21}$

PRACTICE PROBLEM 2

Add: $\frac{5y}{6} + \frac{2y}{9}$

PRACTICE PROBLEM 3

Add: $-\frac{1}{5} + \frac{9}{20}$

Answers

1. $\frac{2}{3}$ **2.** $\frac{19y}{18}$ **3.** $\frac{1}{4}$

EXAMPLE 1 Add: $\frac{2}{5} + \frac{4}{15}$

Solution:

Step 1: The LCD of the fractions is 15. In later examples, we shall simply say, for example, that the LCD of 5 and 15 is 15.

Step 2: $\frac{2}{5} = \frac{2}{5} \cdot \frac{3}{3} = \frac{6}{15}$, $\frac{4}{15} = \frac{4}{15}$ ← This fraction already has a denominator of 15.

Multiply by 1 in the form $\frac{3}{3}$

Step 3: $\frac{2}{5} + \frac{4}{15} = \frac{6}{15} + \frac{4}{15} = \frac{10}{15}$

Step 4: Write in simplest form.

$$\frac{10}{15} = \frac{2 \cdot \overset{1}{\cancel{5}}}{3 \cdot \underset{1}{\cancel{5}}} = \frac{2}{3}$$

Work Practice Problem 1

When the fractions contain variables, we add and subtract the same way.

EXAMPLE 2 Add: $\frac{2x}{15} + \frac{3x}{10}$

Solution:

Step 1: The LCD of the denominators 15 and 10 is 30.

Step 2: $\frac{2x}{15} = \frac{2x}{15} \cdot \frac{2}{2} = \frac{4x}{30}$ $\quad \frac{3x}{10} = \frac{3x}{10} \cdot \frac{3}{3} = \frac{9x}{30}$

Step 3: $\frac{2x}{15} + \frac{3x}{10} = \frac{4x}{30} + \frac{9x}{30} = \frac{13x}{30}$

Step 4: $\frac{13x}{30}$ is in simplest form.

Work Practice Problem 2

EXAMPLE 3 Add: $-\frac{1}{6} + \frac{1}{2}$

Solution: The LCD of the denominators 6 and 2 is 6.

$$-\frac{1}{6} + \frac{1}{2} = \frac{-1}{6} + \frac{1 \cdot 3}{2 \cdot 3}$$
$$= \frac{-1}{6} + \frac{3}{6}$$
$$= \frac{2}{6}$$

Helpful Hint

Recall that

$-\frac{1}{6} = \frac{-1}{6} = \frac{1}{-6}$

Next, simplify $\frac{2}{6}$.

$$\frac{2}{6} = \frac{\overset{1}{\cancel{2}}}{\underset{1}{\cancel{2}} \cdot 3} = \frac{1}{3}$$

Work Practice Problem 3

✔ **Concept Check** Find and correct the error in the following:

$$\frac{2}{9} + \frac{4}{11} = \frac{6}{20} = \frac{3}{10}$$

EXAMPLE 4 Subtract: $\frac{2}{3} - \frac{10}{11}$

Solution:

Step 1: The LCD of 3 and 11 is 33.

Step 2: $\frac{2}{3} = \frac{2}{3} \cdot \frac{11}{11} = \frac{22}{33}$ $\quad \frac{10}{11} = \frac{10}{11} \cdot \frac{3}{3} = \frac{30}{33}$

Step 3: $\frac{2}{3} - \frac{10}{11} = \frac{22}{33} - \frac{30}{33} = \frac{-8}{33}$ or $-\frac{8}{33}$

Step 4: $-\frac{8}{33}$ is in simplest form.

Work Practice Problem 4

EXAMPLE 5 Find: $-\frac{3}{4} - \frac{1}{14} + \frac{6}{7}$

Solution: The LCD of 4, 14, and 7 is 28.

$$-\frac{3}{4} - \frac{1}{14} + \frac{6}{7} = -\frac{3 \cdot 7}{4 \cdot 7} - \frac{1 \cdot 2}{14 \cdot 2} + \frac{6 \cdot 4}{7 \cdot 4}$$

$$= -\frac{21}{28} - \frac{2}{28} + \frac{24}{28}$$

$$= \frac{1}{28}$$

Work Practice Problem 5

✔ **Concept Check** Find and correct the error in the following:

$$\frac{7}{12} - \frac{3}{4} = \frac{4}{8} = \frac{1}{2}$$

EXAMPLE 6 Subtract: $2 - \frac{x}{3}$

Solution: Recall that $2 = \frac{2}{1}$. The LCD of the denominators 1 and 3 is 3.

$$\frac{2}{1} - \frac{x}{3} = \frac{2 \cdot 3}{1 \cdot 3} - \frac{x}{3}$$

$$= \frac{6}{3} - \frac{x}{3}$$

$$= \frac{6 - x}{3}$$

Helpful Hint

The expression $\frac{6 - x}{3}$ from Example 6 *does not simplify* to $2 - x$. The number 3 must be a factor of both terms in the numerator (not just 6) in order to simplify.

The numerator $6 - x$ cannot be simplified further since 6 and $-x$ are unlike terms.

Work Practice Problem 6

PRACTICE PROBLEM 4

Subtract: $\frac{5}{7} - \frac{9}{10}$

PRACTICE PROBLEM 5

Find: $\frac{5}{8} - \frac{1}{3} - \frac{1}{12}$

PRACTICE PROBLEM 6

Subtract: $5 - \frac{y}{4}$

Answers

4. $-\frac{13}{70}$ **5.** $\frac{5}{24}$ **6.** $\frac{20 - y}{4}$

✔ **Concept Check Answer**

When adding unlike fractions, we don't add the denominators. Correct solution:

$\frac{2}{9} + \frac{4}{11} = \frac{22}{99} + \frac{36}{99} = \frac{58}{99}$

$\frac{7}{12} - \frac{3}{4} = \frac{7}{12} - \frac{9}{12} = -\frac{2}{12} = -\frac{1}{6}$

Objective B Writing Fractions in Order

One important application of the least common denominator is to use the LCD to help order or compare fractions.

PRACTICE PROBLEM 7

Insert < or > to form a true sentence.

$\frac{5}{8} \quad \frac{11}{20}$

EXAMPLE 7 Insert < or > to form a true sentence.

$\frac{3}{4} \quad \frac{9}{11}$

Solution: The LCD for these fractions is 44. Let's write each fraction as an equivalent fraction with a denominator of 44.

$$\frac{3}{4} = \frac{3 \cdot 11}{4 \cdot 11} = \frac{33}{44} \qquad \frac{9}{11} = \frac{9 \cdot 4}{11 \cdot 4} = \frac{36}{44}$$

Since $33 < 36$, then

$$\frac{33}{44} < \frac{36}{44} \text{ or}$$
$$\frac{3}{4} < \frac{9}{11}$$

Work Practice Problem 7

PRACTICE PROBLEM 8

Insert < or > to form a true sentence.

$-\frac{17}{20} \quad -\frac{4}{5}$

EXAMPLE 8 Insert < or > to form a true sentence.

$-\frac{2}{7} \quad -\frac{1}{3}$

Solution: The LCD is 21.

$$-\frac{2}{7} = -\frac{2 \cdot 3}{7 \cdot 3} = -\frac{6}{21} \text{ or } \frac{-6}{21} \qquad -\frac{1}{3} = -\frac{1 \cdot 7}{3 \cdot 7} = -\frac{7}{21} \text{ or } \frac{-7}{21}$$

Since $-6 > -7$, then

$$-\frac{6}{21} > -\frac{7}{21} \text{ or}$$
$$-\frac{2}{7} > -\frac{1}{3}$$

Work Practice Problem 8

Objective C Evaluating Expressions Given Fractional Replacement Values

PRACTICE PROBLEM 9

Evaluate $x - y$ if $x = \frac{5}{11}$ and $y = \frac{4}{9}$.

EXAMPLE 9 Evaluate $x - y$ if $x = \frac{7}{18}$ and $y = \frac{2}{9}$.

Solution: Replace x with $\frac{7}{18}$ and y with $\frac{2}{9}$ in the expression $x - y$.

$$x - y = \frac{7}{18} - \frac{2}{9}$$

The LCD of the denominators 18 and 9 is 18. Then

$$\frac{7}{18} - \frac{2}{9} = \frac{7}{18} - \frac{2 \cdot 2}{9 \cdot 2}$$
$$= \frac{7}{18} - \frac{4}{18}$$
$$= \frac{3}{18} = \frac{1}{6} \quad \text{Simplified}$$

Work Practice Problem 9

Answers

7. > **8.** < **9.** $\frac{1}{99}$

Objective D Solving Problems by Adding or Subtracting Unlike Fractions

Very often, real-world problems involve adding or subtracting unlike fractions.

EXAMPLE 10 Finding Total Weight

A freight truck has $\frac{1}{4}$ ton of computers, $\frac{1}{3}$ ton of televisions, and $\frac{3}{8}$ ton of small appliances. Find the total weight of its load.

Solution:

1. UNDERSTAND. Read and reread the problem. The phrase "total weight" tells us to add.
2. TRANSLATE.

In words:	total weight	is	weight of computers	plus	weight of televisions	plus	weight of appliances
	↓	↓	↓	↓	↓	↓	↓
Translate:	total weight	=	$\frac{1}{4}$	+	$\frac{1}{3}$	+	$\frac{3}{8}$

3. SOLVE: The LCD is 24.

$$\begin{aligned}\frac{1}{4}+\frac{1}{3}+\frac{3}{8} &= \frac{1}{4}\cdot\frac{6}{6}+\frac{1}{3}\cdot\frac{8}{8}+\frac{3}{8}\cdot\frac{3}{3}\\ &= \frac{6}{24}+\frac{8}{24}+\frac{9}{24}\\ &= \frac{23}{24}\end{aligned}$$

4. INTERPRET. *Check* the solution. *State* your conclusion: The total weight of the truck's load is $\frac{23}{24}$ ton.

Work Practice Problem 10

PRACTICE PROBLEM 10

To repair her sidewalk, a homeowner must pour small amounts of cement in three different locations. She needs $\frac{3}{5}$ of a cubic yard, $\frac{3}{10}$ of a cubic yard, and $\frac{1}{15}$ of a cubic yard for these locations. Find the total amount of cement the homeowner needs.

Answer

10. $\frac{29}{30}$ cu yd

PRACTICE PROBLEM 11

Find the difference in length of two boards if one board is $\frac{3}{4}$ of a foot long and the other is $\frac{2}{3}$ of a foot long.

EXAMPLE 11 **Calculating Flight Time**

A flight from Tucson to Phoenix, Arizona, requires $\frac{5}{12}$ of an hour. If the plane has been flying $\frac{1}{4}$ of an hour, find how much time remains before landing.

Solution:

1. UNDERSTAND. Read and reread the problem. The phrase "how much time remains" tells us to subtract.
2. TRANSLATE.

In words:	time remaining	is	flight time from Tucson to Phoenix	minus	flight time already passed
	↓	↓	↓	↓	↓
Translate:	time remaining	=	$\frac{5}{12}$	−	$\frac{1}{4}$

3. SOLVE: The LCD is 12.

$$\frac{5}{12} - \frac{1}{4} = \frac{5}{12} - \frac{1}{4} \cdot \frac{3}{3}$$
$$= \frac{5}{12} - \frac{3}{12}$$
$$= \frac{2}{12} = \frac{\overset{1}{\cancel{2}}}{\underset{1}{\cancel{2}} \cdot 6} = \frac{1}{6}$$

4. INTERPRET. *Check* the solution. *State* your conclusion: The flight time remaining is $\frac{1}{6}$ of an hour.

Work Practice Problem 11

Answer

11. $\frac{1}{12}$ ft

CALCULATOR EXPLORATIONS Performing Operations on Fractions

Scientific Calculator

Many calculators have a fraction key, such as [a b/c], that allows you to enter fractions, perform operations on fractions, and will give the result as a fraction. If your calculator has a fraction key, use it to calculate

$$\frac{3}{5} + \frac{4}{7}$$

Enter the keystrokes

[3] [a b/c] [5] [+] [4] [a b/c] [7] [=]

The display should read [1_6 | 35]

which represents the mixed number $1\frac{6}{35}$. Let's write the result as a fraction. To convert from mixed number notation to fractional notation, press

[2nd] [d/c]

The display now reads [41 | 35]

which represents $\frac{41}{35}$, the sum in fractional notation.

Graphing Calculator

Graphing calculators also allow you to perform operations on fractions and will give exact fractional results. The fraction option on a graphing calculator may be found under the [MATH] menu. To perform the addition above, try the keystrokes.

[3] [÷] [5] [+] [4] [÷] [7] [MATH] [ENTER] [ENTER]

The display should read

[3/5 + 4/7 ▶ Frac 41/35]

Use a calculator to add the following fractions. Give each sum as a fraction.

1. $\frac{1}{16} + \frac{2}{5}$ **2.** $\frac{3}{20} + \frac{2}{25}$ **3.** $\frac{4}{9} + \frac{7}{8}$

4. $\frac{9}{11} + \frac{5}{12}$ **5.** $\frac{10}{17} + \frac{12}{19}$ **6.** $\frac{14}{31} + \frac{15}{21}$

Vocabulary and Readiness Check

Use the choices below to fill in each blank. Any numerical answers are not listed.

expression	least common denominator	$>$
equation	equivalent	$<$

1. To add or subtract unlike fractions, we first write the fractions as ________ fractions with a common denominator. The common denominator we use is called the ____________________.

2. The LCD for $\frac{1}{6}$ and $\frac{5}{8}$ is ________.

3. $\frac{1}{6} + \frac{5}{8} = \frac{1}{6} \cdot \frac{4}{4} + \frac{5}{8} \cdot \frac{3}{3} = \underline{\quad} + \underline{\quad} = \underline{\quad}$.

4. $\frac{1}{6} - \frac{5}{8} = \frac{1}{6} \cdot \frac{4}{4} - \frac{5}{8} \cdot \frac{3}{3} = \underline{\quad\quad} - \underline{\quad\quad} = \underline{\quad\quad}$.

5. $x - y$ is an ________ while $3x = \frac{1}{5}$ is an ________.

6. Since $-10 < -1$, we know that $-\frac{10}{13}$ ________ $-\frac{1}{13}$.

4.5 EXERCISE SET

FOR EXTRA HELP

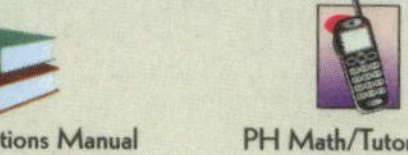

Student Solutions Manual | PH Math/Tutor Center | CD/Video for Review | MathXL® | MyMathLab

Objective A *Add or subtract as indicated. See Examples 1 through 6.*

1. $\frac{2}{3} + \frac{1}{6}$

2. $\frac{5}{6} + \frac{1}{12}$

3. $\frac{1}{2} - \frac{1}{3}$

4. $\frac{2}{3} - \frac{1}{4}$

5. $-\frac{2}{11} + \frac{2}{33}$

6. $-\frac{5}{9} + \frac{1}{3}$

7. $\frac{3}{14} - \frac{3}{7}$

8. $\frac{2}{15} - \frac{2}{5}$

9. $\frac{11x}{35} + \frac{2x}{7}$

10. $\frac{2y}{5} + \frac{3y}{25}$

11. $2 - \frac{y}{12}$

12. $5 - \frac{y}{20}$

13. $\frac{5}{12} - \frac{1}{9}$

14. $\frac{7}{12} - \frac{5}{18}$

15. $-7 + \frac{5}{7}$

16. $-10 + \frac{7}{10}$

17. $\frac{5a}{11} + \frac{4a}{9}$

18. $\frac{7x}{18} + \frac{2x}{9}$

19. $\frac{2y}{3} - \frac{1}{6}$

20. $\frac{5z}{6} - \frac{1}{12}$

21. $\frac{1}{2} + \frac{3}{x}$

22. $\frac{2}{5} + \frac{3}{x}$

23. $-\frac{2}{11} - \frac{2}{33}$

24. $-\frac{5}{9} - \frac{1}{3}$

25. $\frac{9}{14} - \frac{3}{7}$

26. $\frac{4}{5} - \frac{2}{15}$

27. $\frac{11y}{35} - \frac{2}{7}$

28. $\frac{2b}{5} - \frac{3}{25}$

29. $\frac{1}{9} - \frac{5}{12}$

30. $\frac{5}{18} - \frac{7}{12}$

31. $\frac{7}{15} - \frac{5}{12}$

32. $\frac{5}{8} - \frac{3}{20}$

33. $\frac{5}{7} - \frac{1}{8}$

34. $\frac{10}{13} - \frac{7}{10}$

35. $\frac{7}{8} + \frac{3}{16}$

36. $\frac{7}{18} + \frac{2}{9}$

37. $\frac{3}{9} - \frac{5}{9}$

38. $\frac{1}{13} - \frac{4}{13}$

39. $-\frac{2}{5} + \frac{1}{3} - \frac{3}{10}$

40. $-\frac{1}{3} - \frac{1}{4} + \frac{2}{5}$

41. $\frac{5}{11} + \frac{y}{3}$

42. $\frac{5z}{13} + \frac{3}{26}$

43. $-\frac{5}{6} - \frac{3}{7}$

44. $-\frac{1}{2} - \frac{3}{29}$

45. $\frac{x}{2} + \frac{x}{4} + \frac{2x}{16}$

46. $\frac{z}{4} + \frac{z}{8} + \frac{2z}{16}$

47. $\frac{7}{9} - \frac{1}{6}$

48. $\frac{9}{16} - \frac{3}{8}$

49. $\frac{2a}{3} + \frac{6a}{13}$

50. $\frac{3y}{4} + \frac{y}{7}$

51. $\frac{7}{30} - \frac{5}{12}$ **52.** $\frac{7}{30} - \frac{3}{20}$ **53.** $\frac{5}{9} + \frac{1}{y}$ **54.** $\frac{1}{12} - \frac{5}{x}$ **55.** $\frac{6}{5} - \frac{3}{4} + \frac{1}{2}$

56. $\frac{6}{5} + \frac{3}{4} - \frac{1}{2}$ **57.** $\frac{4}{5} + \frac{4}{9}$ **58.** $\frac{11}{12} - \frac{7}{24}$ **59.** $\frac{5}{9x} + \frac{1}{8}$ **60.** $\frac{3}{8} + \frac{5}{12x}$

61. $-\frac{9}{12} + \frac{17}{24} - \frac{1}{6}$ **62.** $-\frac{5}{14} + \frac{3}{7} - \frac{1}{2}$ **63.** $\frac{3x}{8} + \frac{2x}{7} - \frac{5}{14}$ **64.** $\frac{9x}{10} - \frac{1}{2} + \frac{x}{5}$

Objective B *Insert $<$ or $>$ to form a true sentence. See Examples 7 and 8.*

65. $\frac{2}{7} \quad \frac{3}{10}$ **66.** $\frac{5}{9} \quad \frac{6}{11}$ **67.** $\frac{5}{6} \quad -\frac{13}{15}$

68. $-\frac{7}{8} \quad -\frac{5}{6}$ **69.** $-\frac{3}{4} \quad -\frac{11}{14}$ **70.** $-\frac{2}{9} \quad -\frac{3}{13}$

Objective C *Evaluate each expression if $x = \frac{1}{3}$ and $y = \frac{3}{4}$. See Example 9.*

71. $x + y$ **72.** $x - y$ **73.** xy

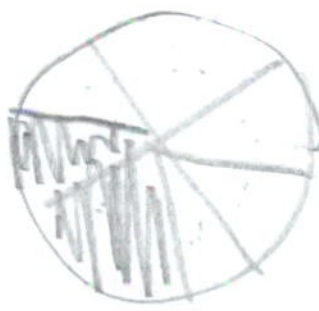

74. $x \div y$ **75.** $2y + x$ **76.** $2x + y$

Objective D *Find the perimeter of each geometric figure. (Hint: Recall that perimeter means distance around.)*

 △ **77.**

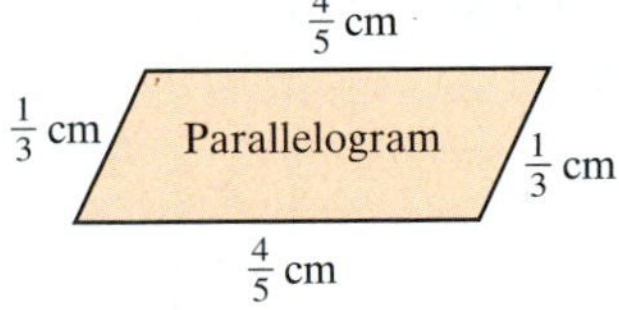

△ **78.**

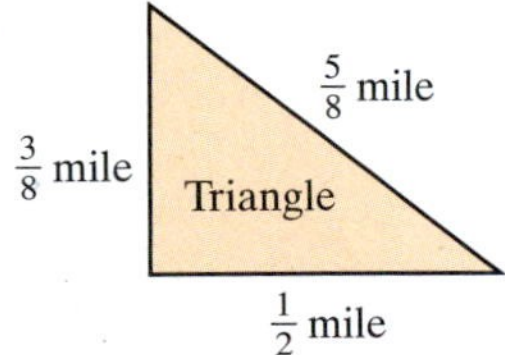

 **79.**

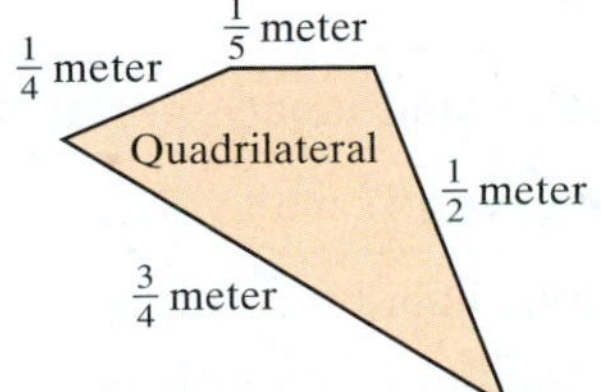

△ **80.**

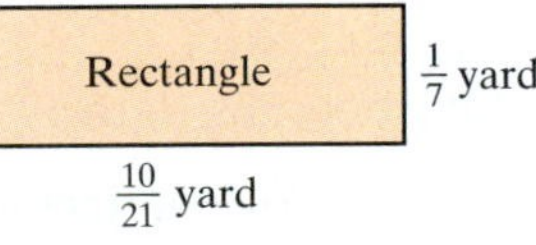

Translating *Translate each phrase to an algebraic expression. Use "x" to represent "a number." See Examples 10 and 11.*

81. The sum of a number and $\frac{1}{2}$

82. A number increased by $-\frac{2}{5}$

83. A number subtracted from $-\frac{3}{8}$

84. The difference of a number and $\frac{7}{20}$

Solve. See Examples 10 and 11.

85. Find the inner diameter of the washer.

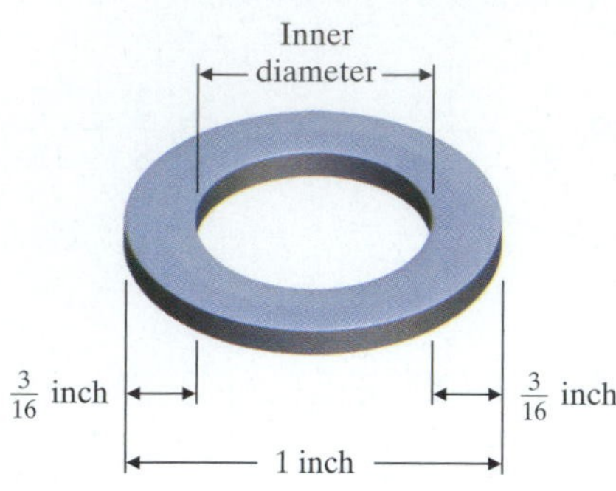

86. Find the inner diameter of the tubing.

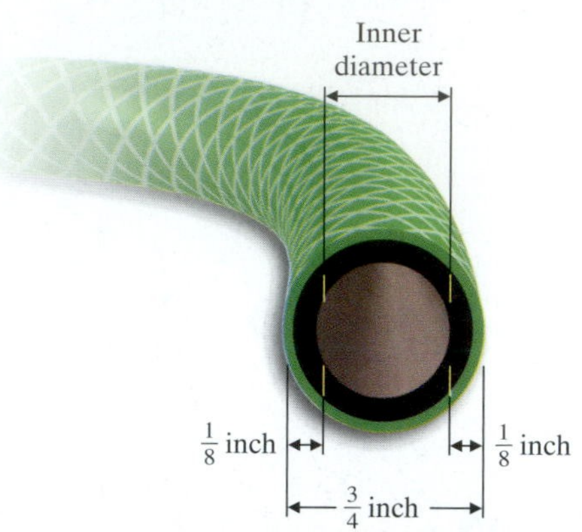

87. The slowest mammal is the three-toed sloth from South America. The sloth has an average ground speed of $\frac{1}{10}$ mph. In the trees, it can accelerate to $\frac{17}{100}$ mph. How much faster can a sloth travel in the trees? (*Source: The Guiness Book of World Records*)

88. Killer bees have been known to chase people for up to $\frac{1}{4}$ of a mile, while domestic European honeybees will normally chase a person for no more than 100 feet, or $\frac{5}{264}$ of a mile. How much farther will a killer bee chase a person than a domestic honeybee? (*Source:* Coachella Valley Mosquito & Vector Control District)

89. Given the following diagram, find its total length.

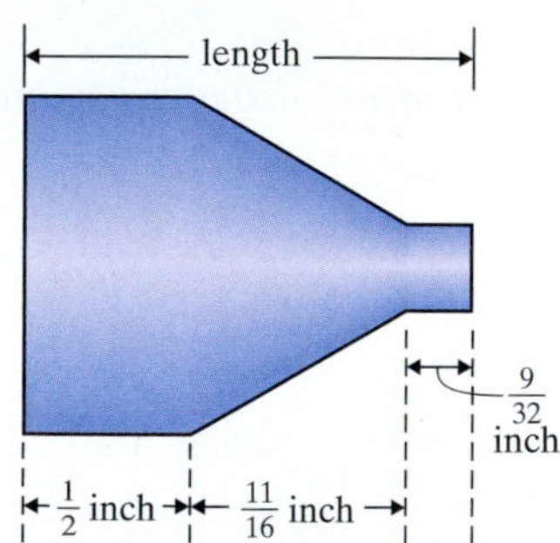

90. Given the following diagram, find its total width.

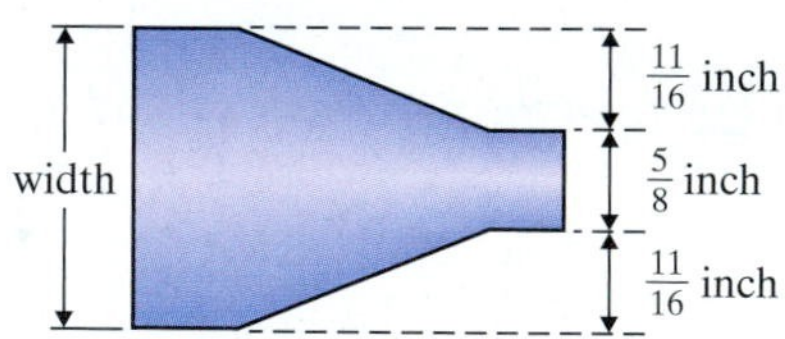

91. About $\frac{13}{20}$ of American students ages 10 to 17 name math, science, or art as their favorite subject in school. Art is the favorite subject for about $\frac{4}{25}$ of the American students ages 10 to 17. For what fraction of students this age is math or science their favorite subject? (*Source:* Peter D. Hart Research Associates for the National Science Foundation)

92. Together, the United States' and Japan's postal services handle $\frac{49}{100}$ of the world's mail volume. Japan's postal service alone handles $\frac{3}{50}$ of the world's mail. What fraction of the world's mail is handled by the postal service of the United States? (*Source:* United States Postal Service)

The table below shows the fraction of the Earth's water area taken up by each ocean. Use this table for Exercises 93 and 94.

Fraction of Earth's Water Area per Ocean	
Ocean	**Fraction**
Arctic	$\frac{1}{25}$
Atlantic	$\frac{13}{50}$
Pacific	$\frac{1}{2}$
Indian	$\frac{1}{5}$

93. What fraction of the world's water surface area is accounted for by the Pacific and Atlantic Oceans?

94. What fraction of the world's water surface area is accounted for by the Arctic and Indian Oceans?

Each sector (shaped like a piece of pie) in the circle graph below shows the fraction of entering college freshmen who expect to major in each discipline shown. The whole circle represents the entire class of college freshmen. Use this graph to answer Exercises 95 through 98.

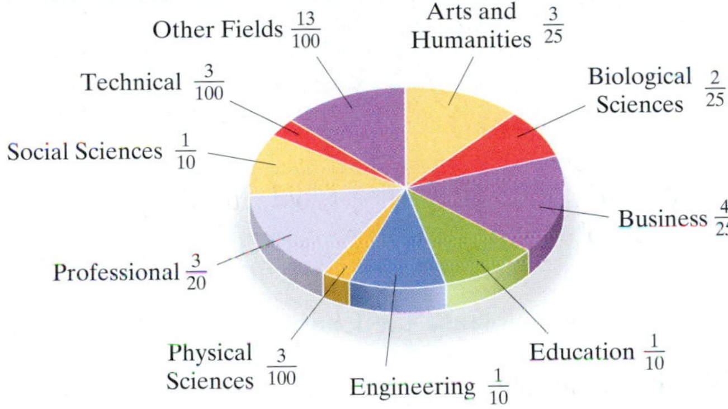

Source: The Higher Education Research Institute

95. What fraction of beginning freshmen plan to major in arts and humanities or education?

96. What fraction of beginning freshmen plan to major in a technical or professional field?

97. What fraction of beginning freshmen do *not* plan to major in business?

98. What fraction of beginning freshmen do *not* plan to major in engineering?

Review

Use order of operations to simplify.

99. $-50 \div 5 \cdot 2$

100. $8 - 6 \cdot 4 - 7$

101. $(8 - 6) \cdot (4 - 7)$

102. $50 \div (5 \cdot 2)$

Concept Extensions

For Exercises 103 and 104 below, do the following:

a. *Draw three rectangles of the same size and represent each fraction in the sum or difference, one fraction per rectangle, by shading.*
b. *Using these rectangles as estimates, determine whether there is an error in the sum.*
c. *If there is an error, correctly calculate the sum.*

See the Concept Check in this section.

103. $\frac{3}{5} + \frac{4}{5} \stackrel{?}{=} \frac{7}{10}$

104. $\frac{5}{8} - \frac{3}{4} \stackrel{?}{=} \frac{2}{4}$

Subtract from left to right.

105. $\frac{2}{3} - \frac{1}{4} - \frac{2}{540}$

106. $\frac{9}{10} - \frac{7}{200} - \frac{1}{3}$

Perform each indicated operation.

107. $\frac{30}{55} + \frac{1000}{1760}$

108. $\frac{19}{26} - \frac{968}{1352}$

109. In your own words, describe how to add or subtract two fractions with different denominators.

110. Find the sum of the fractions in the circle graph on page 275. Did the sum surprise you? Why or why not?

111. In a recent year, about $\frac{24}{53}$ of the total number of pieces of mail delivered by the United States Postal Service was first class mail. That same year, about $\frac{51}{106}$ of the total number of pieces of mail delivered by the United States Postal Service was standard mail. Which of these two categories account for a greater portion of the mail handled by volume? (*Source*: United States Postal Service)

STUDY SKILLS BUILDER

Are You Satisfied with Your Performance on a Particular Quiz or Exam?

If not, don't forget to analyze your quiz or exam and look for common errors. Were most of your errors a result of:

- *Carelessness?* Did you turn in your quiz or exam before the allotted time expired? If so, resolve next time to use the entire time allotted. Any extra time can be spent checking your work.
- *Running out of time?* If so, make a point to better manage your time on your next quiz or exam. Try completing any questions that you are unsure of last and delay checking your work until all questions have been answered.
- *Not understanding a concept?* If so, review that concept and correct your work. Try to understand how this happened so that you make sure it doesn't happen before the next quiz or exam.
- *Test conditions?* When studying for a quiz or exam, make sure you place yourself in conditions similar to test conditions. For example, before your next quiz or exam, use a few sheets of blank paper and take a sample test without the aid of your notes or text. (See your instructor or use the Chapter Test at the end of each chapter.)

Exercises

1. Have you corrected all your previous quizzes and exams?
2. List any errors you have found common to two or more of your graded papers.
3. Is one of your common errors not understanding a concept? If so, are you making sure you understand all the concepts for the next quiz or exam?
4. Is one of your common errors making careless mistakes? If so, are you now taking all the time allotted to check over your work so that you can minimize the number of careless mistakes?
5. Are you satisfied with your grades thus far on quizzes and tests?
6. If your answer to Exercise 5 is no, are there any more suggestions you can make to your instructor or yourself to help? If so, list them here and share these with your instructor.

INTEGRATED REVIEW Sections 4.1–4.5

Summary on Fractions and Operations on Fractions

Use a fraction to represent the shaded area of each figure. If the fraction is improper, also write the fraction as a mixed number.

1. 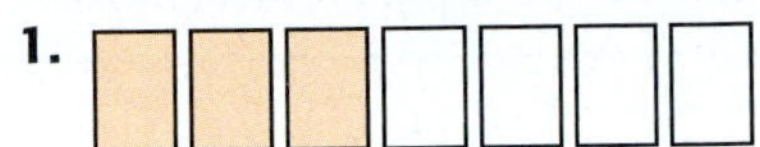

2. 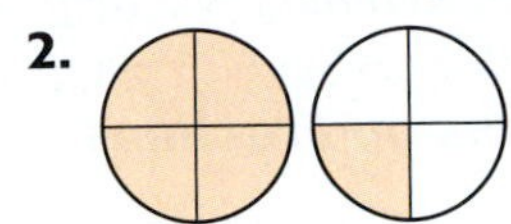

Solve.

3. In a survey, 73 people out of 85 get fewer than 8 hours of sleep each night. What fraction of people in the survey get fewer than 8 hours of sleep?

4. Sketch a diagram to represent $\frac{9}{13}$.

Simplify.

5. $\frac{11}{-11}$ 6. $\frac{17}{1}$ 7. $\frac{0}{-3}$ 8. $\frac{7}{0}$

Write the prime factorization of each composite number. Write any repeated factors using exponents.

9. 65 10. 70 11. 315 12. 441

Write each fraction in simplest form.

13. $\frac{2}{14}$ 14. $\frac{24}{20}$ 15. $-\frac{56}{60}$ 16. $-\frac{72}{80}$

17. $\frac{54x}{135}$ 18. $\frac{90}{240y}$ 19. $\frac{165z^3}{210z}$ 20. $\frac{245ab}{385a^2b^3}$

Determine whether each pair of fractions is equivalent.

21. $\frac{7}{8}$ and $\frac{9}{10}$ 22. $\frac{10}{12}$ and $\frac{15}{18}$

23. Of the 50 states, 2 states are not adjacent to any other states.
   - **a.** What fraction of the states are not adjacent to other states?
   - **b.** How many states are adjacent to other states?
   - **c.** What fraction of the states are adjacent to other states?

24. In a recent year, 460 new films were released. Of these, 92 were rated R. (*Source:* Motion Picture Association of America)
   - **a.** What fraction were rated R?
   - **b.** How many films were rated other than R?
   - **c.** What fraction of films were rated other than R?

Answers

1. ____
2. ____
3. ____
4. ____
5. ____
6. ____
7. ____
8. ____
9. ____
10. ____
11. ____
12. ____
13. ____
14. ____
15. ____
16. ____
17. ____
18. ____
19. ____
20. ____
21. ____
22. ____
23. a. b. c.
24. a. b. c.

25. ______ 26. ______
27. ______ 28. ______
29. ______ 30. ______
31. ______ 32. ______
33. ______ 34. ______
35. ______ 36. ______
37. ______
38. ______
39. ______
40. ______
41. ______
42. ______
43. ______
44. ______
45. ______
46. ______
47. ______
48. ______
49. ______
50. ______
51. ______
52. ______
53. ______

Find the LCM of each list of numbers.

25. 5, 6 **26.** 2, 14 **27.** 6, 18, 30

Write each fraction as an equivalent fraction with the indicated denominator.

28. $\frac{7}{9} = \frac{}{36}$ **29.** $\frac{11}{15} = \frac{}{75}$ **30.** $\frac{5}{6} = \frac{}{48}$

The following summary will help you with the following review of operations on fractions.

Operations on Fractions

Let a, b, c, and d be integers.

Addition: $\frac{a}{b} + \frac{c}{b} = \frac{a+c}{b}$ $(b \neq 0)$ ↑ ↑ **common denominator**

Subtraction: $\frac{a}{b} - \frac{c}{b} = \frac{a-c}{b}$ $(b \neq 0)$ ↑ ↑ **common denominator**

Multiplication: $\frac{a}{b} \cdot \frac{c}{d} = \frac{a \cdot c}{b \cdot d}$ $(b \neq 0, d \neq 0)$

Division: $\frac{a}{b} \div \frac{c}{d} = \frac{a}{b} \cdot \frac{d}{c} = \frac{a \cdot d}{b \cdot c}$ $(b \neq 0, d \neq 0, c \neq 0)$

Perform each indicated operation.

31. $\frac{1}{5} + \frac{3}{5}$ **32.** $\frac{1}{5} - \frac{3}{5}$ **33.** $\frac{1}{5} \cdot \frac{3}{5}$ **34.** $\frac{1}{5} \div \frac{3}{5}$

35. $\frac{2}{3} \div \frac{5}{6}$ **36.** $\frac{2a}{3} \cdot \frac{5}{6a}$ **37.** $\frac{2}{3y} - \frac{5}{6y}$ **38.** $\frac{2x}{3} + \frac{5x}{6}$

39. $-\frac{1}{7} \cdot -\frac{7}{18}$ **40.** $-\frac{4}{9} \cdot -\frac{3}{7}$ **41.** $-\frac{7z}{8} \div 6z^2$ **42.** $-\frac{9}{10} \div 5$

43. $\frac{7}{8} + \frac{1}{20}$ **44.** $\frac{5}{12} - \frac{1}{9}$ **45.** $\frac{2}{9} + \frac{1}{18} + \frac{1}{3}$ **46.** $\frac{3y}{10} + \frac{y}{5} + \frac{6}{25}$

Translating *Translate each to an expression. Use x to represent a number.*

47. $\frac{2}{3}$ of a number

48. The quotient of a number and $-\frac{1}{5}$

49. A number subtracted from $-\frac{8}{9}$

50. $\frac{6}{11}$ increased by a number

Solve.

51. Find $\frac{2}{3}$ of 1530.

52. A contractor is using 18 acres of his land to sell $\frac{3}{4}$-acre lots. How many lots can he sell?

53. Suppose that the cross-section of a piece of pipe looks like the diagram shown. What is the inner diameter?

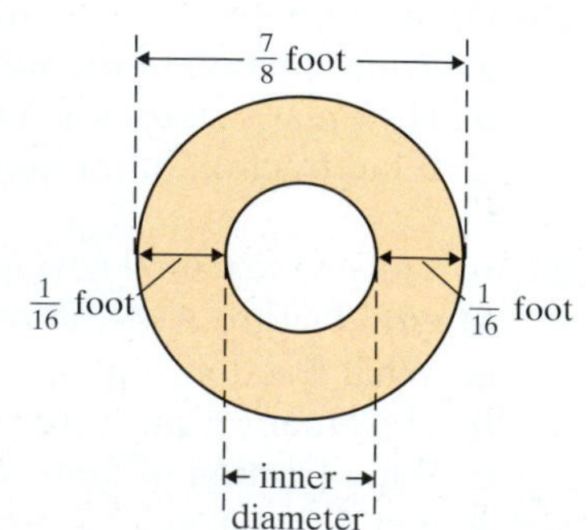

4.6 COMPLEX FRACTIONS AND REVIEW OF ORDER OF OPERATIONS

Objectives

A Simplify Complex Fractions.

B Review the Order of Operations.

C Evaluate Expressions Given Replacement Values.

Objective A Simplifying Complex Fractions

Thus far, we have studied operations on fractions. We now practice simplifying fractions whose numerators or denominators themselves contain fractions. These fractions are called **complex fractions.**

Complex Fraction

A fraction whose numerator or denominator or both numerator and denominator contain fractions is called a **complex fraction.**

Examples of complex fractions are

$$\frac{\frac{x}{4}}{\frac{3}{2}} \qquad \frac{\frac{1}{2}+\frac{3}{8}}{\frac{3}{4}-\frac{1}{6}} \qquad \frac{\frac{y}{5}-2}{\frac{3}{10}} \qquad \frac{-4z}{\frac{3}{5}}$$

Method 1 for Simplifying Complex Fractions

Two methods are presented to simplify complex fractions. The first method makes use of the fact that a fraction bar means division.

EXAMPLE 1 Simplify: $\frac{\frac{x}{4}}{\frac{3}{2}}$

Solution: Since a fraction bar means division, the complex fraction $\frac{\frac{x}{4}}{\frac{3}{2}}$ can be written as $\frac{x}{4} \div \frac{3}{2}$. Then divide as usual to simplify.

$$\frac{x}{4} \div \frac{3}{2} = \frac{x}{4} \cdot \frac{2}{3} \quad \text{Multiply by the reciprocal.}$$

$$= \frac{x \cdot \overset{1}{\cancel{2}}}{\underset{1}{\cancel{2}} \cdot 2 \cdot 3}$$

$$= \frac{x}{6}$$

Work Practice Problem 1

PRACTICE PROBLEM 1

Simplify: $\frac{\frac{7y}{10}}{\frac{1}{5}}$

EXAMPLE 2 Simplify: $\frac{\frac{1}{2}+\frac{3}{8}}{\frac{3}{4}-\frac{1}{6}}$

Solution: Recall the order of operations. Since the fraction bar is considered a grouping symbol, we simplify the numerator and the denominator of the complex fraction separately. Then we divide.

$$\frac{\frac{1}{2}+\frac{3}{8}}{\frac{3}{4}-\frac{1}{6}} = \frac{\frac{1 \cdot 4}{2 \cdot 4}+\frac{3}{8}}{\frac{3 \cdot 3}{4 \cdot 3}-\frac{1 \cdot 2}{6 \cdot 2}} = \frac{\frac{4}{8}+\frac{3}{8}}{\frac{9}{12}-\frac{2}{12}} = \frac{\frac{7}{8}}{\frac{7}{12}}$$

Continued on next page

PRACTICE PROBLEM 2

Simplify: $\frac{\frac{1}{2}+\frac{1}{6}}{\frac{3}{4}-\frac{2}{3}}$

Answers

1. $\frac{7y}{2}$ 2. $\frac{8}{1}$ or 8

Thus,

$$\frac{\frac{1}{2}+\frac{3}{8}}{\frac{3}{4}-\frac{1}{6}} = \frac{\frac{7}{8}}{\frac{7}{12}}$$

$$= \frac{7}{8} \div \frac{7}{12} \quad \text{Rewrite the quotient using the} \div \text{sign.}$$

$$= \frac{7}{8} \cdot \frac{12}{7} \quad \text{Multiply by the reciprocal.}$$

$$= \frac{\overset{1}{\cancel{7}} \cdot 3 \cdot \overset{1}{\cancel{4}}}{2 \cdot \underset{1}{\cancel{4}} \cdot \underset{1}{\cancel{7}}} \quad \text{Multiply.}$$

$$= \frac{3}{2} \quad \text{Simplify.}$$

Work Practice Problem 2

Method 2 for Simplifying Complex Fractions The second method for simplifying complex fractions is to multiply the numerator and the denominator of the complex fraction by the LCD of all the fractions in its numerator and its denominator. This has the effect of leaving sums and differences of integers in the numerator and the denominator as we shall see in the example below.

Let's use this second method to simplify the complex fraction in Example 2 again.

PRACTICE PROBLEM 3

Use Method 2 to simplify:

$$\frac{\frac{1}{2}+\frac{1}{6}}{\frac{3}{4}-\frac{2}{3}}$$

EXAMPLE 3 Simplify: $\frac{\frac{1}{2}+\frac{3}{8}}{\frac{3}{4}-\frac{1}{6}}$

Solution: The complex fraction contains fractions with denominators 2, 8, 4, and 6. The LCD is 24. By the fundamental property of fractions, we can multiply the numerator and the denominator of the complex fraction by 24. Notice below that by the distributive property, this means that we multiply each term in the numerator and denominator by 24.

$$\frac{\frac{1}{2}+\frac{3}{8}}{\frac{3}{4}-\frac{1}{6}} = \frac{24\left(\frac{1}{2}+\frac{3}{8}\right)}{24\left(\frac{3}{4}-\frac{1}{6}\right)}$$

$$= \frac{\left(\overset{12}{\cancel{24}} \cdot \frac{1}{\underset{1}{\cancel{2}}}\right) + \left(\overset{3}{\cancel{24}} \cdot \frac{3}{\underset{1}{\cancel{8}}}\right)}{\left(\overset{6}{\cancel{24}} \cdot \frac{3}{\underset{1}{\cancel{4}}}\right) - \left(\overset{4}{\cancel{24}} \cdot \frac{1}{\underset{1}{\cancel{6}}}\right)} \quad \text{Apply the distributive property. Then divide out common factors to aid in multiplying.}$$

$$= \frac{12+9}{18-4} \quad \text{Multiply.}$$

$$= \frac{21}{14}$$

$$= \frac{\overset{1}{\cancel{7}} \cdot 3}{\underset{1}{\cancel{7}} \cdot 2} = \frac{3}{2} \quad \text{Simplify.}$$

Work Practice Problem 3

The simplified result is the same, of course, no matter which method is used.

Answer

3. $\frac{8}{1}$ or 8

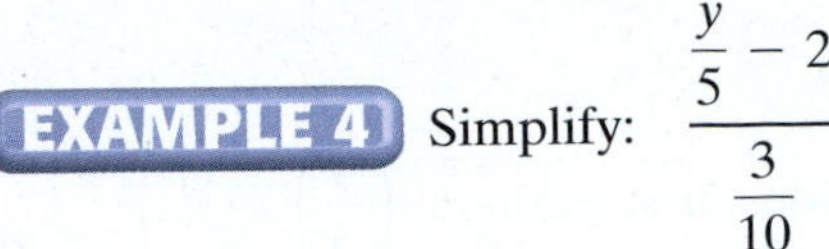

EXAMPLE 4 Simplify: $\dfrac{\frac{y}{5} - 2}{\frac{3}{10}}$

Solution: Use the second method and multiply the numerator and the denominator of the complex fraction by the LCD of all fractions. Recall that $2 = \frac{2}{1}$. The LCD of the denominators 5, 1, and 10 is 10.

$$\frac{\frac{y}{5} - \frac{2}{1}}{\frac{3}{10}} = \frac{10\left(\frac{y}{5} - \frac{2}{1}\right)}{10\left(\frac{3}{10}\right)}$$ Multiply the numerator and denominator by 10.

$$= \frac{\left(\overset{2}{\cancel{10}} \cdot \frac{y}{\underset{1}{\cancel{5}}}\right) - \left(10 \cdot \frac{2}{1}\right)}{\overset{1}{\cancel{10}} \cdot \frac{3}{\underset{1}{\cancel{10}}}}$$ Apply the distributive property. Then divide out common factors to aid in multiplying.

$$= \frac{2y - 20}{3}$$ Multiply.

Work Practice Problem 4

PRACTICE PROBLEM 4

Simplify: $\dfrac{\frac{3}{4}}{\frac{x}{5} - 1}$

Helpful Hint

Don't forget to multiply the numerator and the denominator of the complex fraction by the same number—the LCD.

Objective B Reviewing the Order of Operations

At this time, it is probably a good idea to review the order of operations on expressions containing fractions. Before we do so, let's review how we perform operations on fractions.

Review of Operations on Fractions

Operation	Procedure	Example
Multiply	Multiply the numerators and multiply the denominators.	$\frac{5}{9} \cdot \frac{1}{2} = \frac{5 \cdot 1}{9 \cdot 2} = \frac{5}{18}$
Divide	Multiply the first fraction by the reciprocal of the second fraction.	$\frac{2}{3} \div \frac{11}{13} = \frac{2}{3} \cdot \frac{13}{11} = \frac{2 \cdot 13}{3 \cdot 11} = \frac{26}{33}$
Add or Subtract	**1.** Write each fraction as an equivalent fraction whose denominator is the LCD. **2.** Add or subtract numerators and write the result over the common denominator.	$\frac{3}{4} + \frac{1}{8} = \frac{3}{4} \cdot \frac{2}{2} + \frac{1}{8} = \frac{6}{8} + \frac{1}{8} = \frac{7}{8}$

Now let's review order of operations.

Order of Operations

1. Perform all operations within parentheses ( ), brackets [ ], or other grouping symbols such as fraction bars, starting with the innermost set.
2. Evaluate any expressions with exponents.
3. Multiply or divide in order from left to right.
4. Add or subtract in order from left to right.

Answer

4. $\frac{15}{4x - 20}$

PRACTICE PROBLEM 5

Simplify: $\left(\frac{2}{3}\right)^3 - 2$

PRACTICE PROBLEM 6

Simplify: $\left(-\frac{1}{2} + \frac{1}{5}\right)\left(\frac{7}{8} + \frac{1}{8}\right)$

PRACTICE PROBLEM 7

Evaluate $-\frac{3}{5} - xy$ if $x = \frac{3}{10}$ and $y = \frac{2}{3}$.

Answers

5. $-\frac{46}{27}$ **6.** $-\frac{3}{10}$ **7.** $-\frac{4}{5}$

✔ Concept Check Answer

Add inside parentheses.

EXAMPLE 5 Simplify: $\left(\frac{4}{5}\right)^2 - 1$

Solution: According to the order of operations, first evaluate $\left(\frac{4}{5}\right)^2$.

$$\left(\frac{4}{5}\right)^2 - 1 = \frac{16}{25} - 1 \quad \text{Write } \left(\frac{4}{5}\right)^2 \text{ as } \frac{16}{25}.$$

Next, combine the fractions. The LCD of 25 and 1 is 25.

$$\frac{16}{25} - 1 = \frac{16}{25} - \frac{25}{25} \quad \text{Write 1 as } \frac{25}{25}.$$

$$= \frac{-9}{25} \text{ or } -\frac{9}{25} \quad \text{Subtract.}$$

Work Practice Problem 5

EXAMPLE 6 Simplify: $\left(\frac{1}{4} + \frac{2}{3}\right)\left(\frac{11}{12} + \frac{1}{4}\right)$

Solution: First perform operations inside parentheses. Then multiply.

$$\left(\frac{1}{4} + \frac{2}{3}\right)\left(\frac{11}{12} + \frac{1}{4}\right) = \left(\frac{1 \cdot 3}{4 \cdot 3} + \frac{2 \cdot 4}{3 \cdot 4}\right)\left(\frac{11}{12} + \frac{1 \cdot 3}{4 \cdot 3}\right) \quad \text{Each LCD is 12.}$$

$$= \left(\frac{3}{12} + \frac{8}{12}\right)\left(\frac{11}{12} + \frac{3}{12}\right)$$

$$= \left(\frac{11}{12}\right)\left(\frac{14}{12}\right) \quad \text{Add.}$$

$$= \frac{11 \cdot \overset{1}{\cancel{2}} \cdot 7}{\underset{1}{\cancel{2}} \cdot 6 \cdot 12} \quad \text{Multiply.}$$

$$= \frac{77}{72} \quad \text{Simplify.}$$

Work Practice Problem 6

✔Concept Check What should be done first to simplify the expression $\frac{1}{5} \cdot \frac{5}{2} - \left(\frac{2}{3} + \frac{4}{5}\right)^2$?

Objective C Evaluating Algebraic Expressions

EXAMPLE 7 Evaluate $2x + y^2$ if $x = -\frac{1}{2}$ and $y = \frac{1}{3}$.

Solution: Replace x and y with the given values and simplify.

$$2x + y^2 = 2\left(-\frac{1}{2}\right) + \left(\frac{1}{3}\right)^2 \quad \text{Replace } x \text{ with } -\frac{1}{2} \text{ and } y \text{ with } \frac{1}{3}.$$

$$= 2\left(-\frac{1}{2}\right) + \frac{1}{9} \quad \text{Write } \left(\frac{1}{3}\right)^2 \text{ as } \frac{1}{9}.$$

$$= -1 + \frac{1}{9} \quad \text{Multiply.}$$

$$= -\frac{9}{9} + \frac{1}{9} \quad \text{The LCD is 9.}$$

$$= -\frac{8}{9} \quad \text{Add.}$$

Work Practice Problem 7

Helpful Hint

If you find it difficult replacing a variable with a number, try the following. First, replace the variable with a set of parentheses, then place the replacement number between the parentheses.

If $x = \frac{4}{5}$, find $2x + x^2$.

$$2x + x^2 = 2(\ ) + (\ )^2$$
$$= 2\left(\frac{4}{5}\right) + \left(\frac{4}{5}\right)^2 \ldots$$

then continue simplifying.

Vocabulary and Readiness Check

Use the choices below to fill in each blank.

addition	multiplication	evaluate the exponential expression
subtraction	division	complex

1. A fraction whose numerator or denominator or both numerator and denominator contain fractions is called a __________ fraction.

2. To simplify $-\frac{1}{2} + \frac{2}{3} \cdot \frac{7}{8}$, which operation do we perform first? __________

3. To simplify $-\frac{1}{2} \div \frac{2}{3} \cdot \frac{7}{8}$, which operation do we perform first? __________

4. To simplify $\frac{7}{8} \cdot \left(\frac{1}{2} - \frac{2}{3}\right)$, which operation do we perform first? __________

5. To simplify $\frac{1}{3} \div \frac{1}{4} \cdot \left(\frac{9}{11} + \frac{3}{8}\right)^3$, which operation do we perform first? __________

6. To simplify $9 - \left(-\frac{3}{4}\right)^2$, which operation do we perform first? __________

4.6 EXERCISE SET

FOR EXTRA HELP

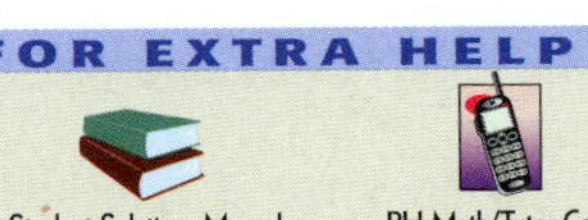

Objective A *Simplify each complex fraction. See Examples 1 through 4.*

1. $\dfrac{\frac{1}{8}}{\frac{3}{4}}$ **2.** $\dfrac{\frac{5}{12}}{\frac{15}{12}}$ **3.** $\dfrac{\frac{2}{3}}{\frac{2}{7}}$ **4.** $\dfrac{\frac{9}{25}}{\frac{6}{25}}$ **5.** $\dfrac{\frac{2x}{27}}{\frac{4}{9}}$

6. $\dfrac{\frac{3y}{11}}{\frac{1}{2}}$ **7.** $\dfrac{\frac{3}{4} + \frac{2}{5}}{\frac{1}{2} + \frac{3}{5}}$ **8.** $\dfrac{\frac{7}{6} + \frac{2}{3}}{\frac{3}{2} - \frac{8}{9}}$ **9.** $\dfrac{\frac{3x}{4}}{5 - \frac{1}{8}}$ **10.** $\dfrac{\frac{3}{10} + 2}{\frac{2}{5y}}$

Objective B *Use the order of operations to simplify each expression. See Examples 5 and 6.*

11. $\frac{1}{5}+\frac{1}{3}\cdot\frac{1}{4}$

12. $\frac{1}{2}+\frac{1}{6}\cdot\frac{1}{3}$

13. $\frac{5}{6}\div\frac{1}{3}\cdot\frac{1}{4}$

14. $\frac{7}{8}\div\frac{1}{4}\cdot\frac{1}{7}$

15. $2^2-\left(\frac{1}{3}\right)^2$

16. $3^2-\left(\frac{1}{2}\right)^2$

17. $\left(\frac{2}{9}+\frac{4}{9}\right)\left(\frac{1}{3}-\frac{9}{10}\right)$

18. $\left(\frac{1}{5}-\frac{1}{10}\right)\left(\frac{1}{5}+\frac{1}{10}\right)$

19. $\left(\frac{7}{8}-\frac{1}{2}\right)\div\frac{3}{11}$

20. $\left(-\frac{2}{3}-\frac{7}{3}\right)\div\frac{4}{9}$

21. $2\cdot\left(\frac{1}{4}+\frac{1}{5}\right)+2$

22. $\frac{2}{5}\cdot\left(5-\frac{1}{2}\right)-1$

23. $\left(\frac{3}{4}\right)^2\div\left(\frac{3}{4}-\frac{1}{12}\right)$

24. $\left(\frac{8}{9}\right)^2\div\left(2-\frac{2}{3}\right)$

25. $\left(\frac{2}{5}-\frac{3}{10}\right)^2$

26. $\left(\frac{3}{2}-\frac{4}{3}\right)^3$

27. $\left(\frac{3}{4}+\frac{1}{8}\right)^2-\left(\frac{1}{2}+\frac{1}{8}\right)$

28. $\left(\frac{1}{6}+\frac{1}{3}\right)^3+\left(\frac{2}{5}\cdot\frac{3}{4}\right)^2$

Objective C *Evaluate each expression if* $x=-\frac{1}{3}$, $y=\frac{2}{5}$, *and* $z=\frac{5}{6}$. *See Example 7.*

29. $5y-z$

30. $2z-x$

31. $\frac{x}{z}$

32. $\frac{y+x}{z}$

33. x^2-yz

34. $(1+x)(1+z)$

Objectives A B Mixed Practice *Simplify the following. See Examples 1 through 6.*

35. $\dfrac{\frac{5a}{24}}{\frac{1}{12}}$

36. $\dfrac{\frac{7}{10}}{\frac{14z}{25}}$

37. $\left(\frac{3}{2}\right)^3+\left(\frac{1}{2}\right)^3$

38. $\left(\frac{5}{21}\div\frac{1}{2}\right)+\left(\frac{1}{7}\cdot\frac{1}{3}\right)$

39. $\left(-\frac{1}{2}\right)^2+\frac{1}{5}$

40. $\left(-\frac{3}{4}\right)^2+\frac{3}{8}$

41. $\dfrac{2+\frac{1}{6}}{1-\frac{4}{3}}$

42. $\dfrac{3-\frac{1}{2}}{4+\frac{1}{5}}$

43. $\left(1-\frac{2}{5}\right)^2$

44. $\left(-\frac{1}{2}\right)^2-\left(\frac{3}{4}\right)^2$

45. $\left(\frac{3}{4}-1\right)\left(\frac{1}{8}+\frac{1}{2}\right)$

46. $\left(\frac{1}{10}+\frac{3}{20}\right)\left(\frac{1}{5}-1\right)$

47. $\left(-\frac{2}{9}-\frac{7}{9}\right)^4$

48. $\left(\frac{5}{9}-\frac{2}{3}\right)^2$

49. $\dfrac{\frac{1}{3}-\frac{5}{6}}{\frac{3}{4}+\frac{1}{2}}$

50. $\dfrac{\frac{7}{10}+\frac{1}{2}}{\frac{4}{5}+\frac{3}{4}}$

51. $\left(\frac{3}{4}\div\frac{6}{5}\right)-\left(\frac{3}{4}\cdot\frac{6}{5}\right)$

52. $\left(\frac{1}{2}\cdot\frac{2}{7}\right)-\left(\frac{1}{2}\div\frac{2}{7}\right)$

53. $\dfrac{\frac{x}{3}+2}{5+\frac{1}{3}}$

54. $\dfrac{1-\frac{x}{4}}{2+\frac{3}{8}}$

Review

Perform each indicated operation. If the result is an improper fraction, also write the improper fraction as a mixed number. See Sections 4.1 and 4.5.

55. $3+\frac{1}{2}$

56. $2+\frac{2}{3}$

57. $9-\frac{5}{6}$

58. $4-\frac{1}{5}$

Concept Extensions

59. Calculate $\frac{2^3}{3}$ and $\left(\frac{2}{3}\right)^3$. Do both of these expressions simplify to the same number? Explain why or why not.

60. Calculate $\left(\frac{1}{2}\right)^2\cdot\left(\frac{3}{4}\right)^2$ and $\left(\frac{1}{2}\cdot\frac{3}{4}\right)^2$. Do both of these expressions simplify to the same number? Explain why or why not.

Recall that to find the average of two numbers, find their sum and divide by 2. For example, the average of $\frac{1}{2}$ and $\frac{3}{4}$ is $\dfrac{\frac{1}{2}+\frac{3}{4}}{2}$. Find the average of each pair of numbers.

61. $\frac{1}{2}, \frac{3}{4}$

62. $\frac{3}{5}, \frac{9}{10}$

63. $\frac{1}{4}, \frac{2}{14}$

64. $\frac{5}{6}, \frac{7}{9}$

65. Two positive numbers, a and b, are graphed below. Where should the graph of their average lie?

Answer true or false for each statement.

66. It is possible for the average of two numbers to be greater than both numbers.

67. It is possible for the average of two numbers to be less than both numbers.

68. The sum of two negative fractions is always a negative number.

69. The sum of a negative fraction and a positive fraction is always a positive number.

70. It is possible for the sum of two fractions to be a whole number.

71. It is possible for the difference of two fractions to be a whole number.

72. What operation should be performed first to simplify

$$\frac{1}{5}\cdot\frac{5}{2}-\left(\frac{2}{3}+\frac{4}{5}\right)^2$$

Explain your answer.

73. A student is to evaluate $x - y$ when $x = \frac{1}{5}$ and $y = -\frac{1}{7}$. This student is asking you if he should evaluate $\frac{1}{5} - \frac{1}{7}$. What do you tell this student and why?

Each expression contains one addition, one subtraction, one multiplication, and one division. Write the operations in the order that they should be performed. Do not actually simplify. See the Concept Check in this section.

74. $[9 + 3(4 - 2)] \div \frac{10}{21}$

75. $[30 - 4(3 + 2)] \div \frac{5}{2}$

76. $\frac{1}{3} \div \left(\frac{2}{3}\right)\left(\frac{4}{5}\right) - \frac{1}{4} + \frac{1}{2}$

77. $\left(\frac{5}{6} - \frac{1}{3}\right)\cdot\frac{1}{3} + \frac{1}{2} \div \frac{9}{8}$

Evaluate each expression if $x = \frac{3}{4}$ *and* $y = -\frac{4}{7}$.

78. $\frac{2 + x}{y}$

79. $4x + y$

80. $x^2 + 7y$

81. $\frac{\frac{9}{14}}{x + y}$

4.7 OPERATIONS ON MIXED NUMBERS

Objective A Graphing Fractions and Mixed Numbers

Let's review graphing fractions and practice graphing mixed numbers on a number line. This will help us visualize rounding and estimating operations with mixed numbers.

Recall that $5\frac{2}{3}$ means $5 + \frac{2}{3}$ and

$-4\frac{1}{6}$ means $-4 - \frac{1}{6}$ or $-4 + \left(-\frac{1}{6}\right)$

EXAMPLE 1 Graph the numbers on a number line:

$\frac{1}{2}, -\frac{3}{4}, 2\frac{2}{3}, -3, -3\frac{1}{8}$

Solution: Remember that $2\frac{2}{3}$ means $2 + \frac{2}{3}$.

Also, $-3\frac{1}{8}$ means $-3 - \frac{1}{8}$, so $-3\frac{1}{8}$ lies to the left of -3.

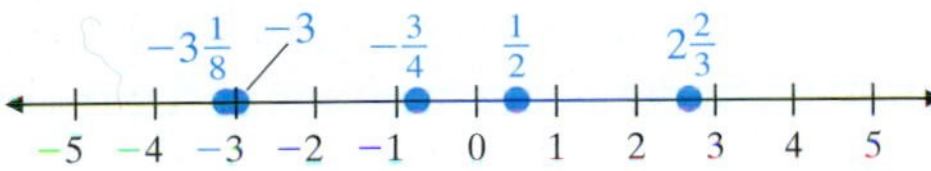

Work Practice Problem 1

✔Concept Check Which of the following are equivalent to 9?

a. $7\frac{6}{3}$ **b.** $8\frac{4}{4}$ **c.** $8\frac{9}{9}$ **d.** $\frac{18}{2}$ **e.** all of these

Objective B Multiplying or Dividing with Mixed Numbers or Whole Numbers

When multiplying or dividing a fraction and a mixed or a whole number, remember that mixed and whole numbers can be written as improper fractions.

Multiplying or Dividing Fractions and Mixed Numbers or Whole Numbers

To multiply or divide with mixed numbers or whole numbers, first write any mixed or whole numbers as improper fractions and then multiply or divide as usual.

(*Note:* If an exercise contains a mixed number, we will write the answer as a mixed number, if possible.)

Objectives

A Graph Positive and Negative Fractions and Mixed Numbers.

B Multiply or Divide Mixed or Whole Numbers.

C Add or Subtract Mixed Numbers.

D Solve Problems Containing Mixed Numbers.

E Perform Operations on Negative Mixed Numbers.

PRACTICE PROBLEM 1

Graph the numbers on a number line.

$-5, -4\frac{1}{2}, 2\frac{3}{4}, \frac{1}{8}, -\frac{1}{2}$

−5 −4 −3 −2 −1 0 1 2 3 4 5

Answer

1. 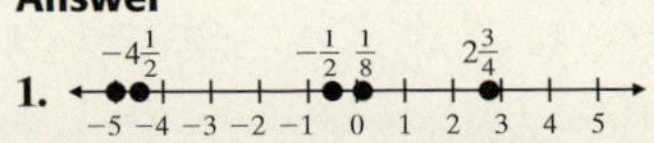

✔ Concept Check Answer

e

PRACTICE PROBLEM 2

Multiply and simplify: $1\frac{2}{3}\cdot\frac{11}{15}$

EXAMPLE 2 Multiply: $3\frac{1}{3}\cdot\frac{7}{8}$

Solution: Recall from Section 4.1 that the mixed number $3\frac{1}{3}$ can be written as the fraction $\frac{10}{3}$. Then

$$3\frac{1}{3}\cdot\frac{7}{8}=\frac{10}{3}\cdot\frac{7}{8}=\frac{\overset{1}{\cancel{2}}\cdot5\cdot7}{3\cdot\underset{1}{\cancel{2}}\cdot4}=\frac{35}{12}\quad\text{or}\quad 2\frac{11}{12}$$

Work Practice Problem 2

Don't forget that a whole number can be written as a fraction by writing the whole number over 1. For example,

$$20=\frac{20}{1}\qquad\text{and}\qquad 7=\frac{7}{1}$$

PRACTICE PROBLEM 3

Multiply: $\frac{5}{6}\cdot18$

EXAMPLE 3 Multiply: $\frac{3}{4}\cdot20$

$$\frac{3}{4}\cdot20=\frac{3}{4}\cdot\frac{20}{1}=\frac{3\cdot20}{4\cdot1}=\frac{3\cdot\overset{1}{\cancel{4}}\cdot5}{\underset{1}{\cancel{4}}\cdot1}=\frac{15}{1}\quad\text{or}\quad15$$

Work Practice Problem 3

When both numbers to be multiplied are mixed or whole numbers, it is a good idea to estimate the product to see if your answer is reasonable. To do this, we first practice rounding mixed numbers to the nearest whole. If the fraction part of the mixed number is $\frac{1}{2}$ or greater, we round the whole number part up. If the fraction part of the mixed number is less than $\frac{1}{2}$, then we do not round the whole number part up. Study the table below for examples.

Mixed Number	Rounding
$5\frac{1}{4}$ — $\frac{1}{4}$ is less than $\frac{1}{2}$	Thus, $5\frac{1}{4}$ rounds to 5.
$3\frac{9}{16}$ — ←9 is greater than 8; →Half of 16 is 8.	Thus, $3\frac{9}{16}$ rounds to 4.
$1\frac{3}{7}$ — ←3 is less than $3\frac{1}{2}$. →Half of 7 is $3\frac{1}{2}$.	Thus, $1\frac{3}{7}$ rounds to 1.

PRACTICE PROBLEM 4

Multiply. Check by estimating.

$3\frac{1}{5}\cdot2\frac{3}{4}$

EXAMPLE 4 Multiply. Check by estimating.

$$1\frac{2}{3}\cdot2\frac{1}{4}=\frac{5}{3}\cdot\frac{9}{4}=\frac{5\cdot9}{3\cdot4}=\frac{5\cdot\overset{1}{\cancel{3}}\cdot3}{\underset{1}{\cancel{3}}\cdot4}=\frac{15}{4}\text{ or }3\frac{3}{4}\quad\text{Exact}$$

Let's check by estimating.

$$1\frac{2}{3}\text{ rounds to 2, }2\frac{1}{4}\text{ rounds to 2, and }2\cdot2=4\quad\text{Estimate}$$

The estimate is close to the exact value, so our answer is reasonable.

Work Practice Problem 4

Answers

2. $\frac{11}{9}$ or $1\frac{2}{9}$ **3.** 15 **4.** $\frac{44}{5}$ or $8\frac{4}{5}$

EXAMPLE 5 Multiply: $7 \cdot 2\frac{11}{14}$. Check by estimating.

$$7 \cdot 2\frac{11}{14} = \frac{7}{1} \cdot \frac{39}{14} = \frac{7 \cdot 39}{1 \cdot 14} = \frac{\overset{1}{\cancel{7}} \cdot 39}{1 \cdot 2 \cdot \underset{1}{\cancel{7}}} = \frac{39}{2} \text{ or } 19\frac{1}{2} \quad \text{Exact}$$

To estimate,

$2\frac{11}{14}$ rounds to 3 and $7 \cdot 3 = 21$. Estimate

The estimate is close to the exact value, so our answer is reasonable.

Work Practice Problem 5

PRACTICE PROBLEM 5

Multiply. Check by estimating.

$3 \cdot 6\frac{7}{15}$

✓Concept Check Find the error.

$$2\frac{1}{4} \cdot \frac{1}{2} = 2\frac{1 \cdot 1}{4 \cdot 2} = 2\frac{1}{8}$$

EXAMPLES Divide.

6. $\frac{3}{4} \div 5 = \frac{3}{4} \div \frac{5}{1} = \frac{3}{4} \cdot \frac{1}{5} = \frac{3 \cdot 1}{4 \cdot 5} = \frac{3}{20}$

7. $\frac{11}{18} \div 2\frac{5}{6} = \frac{11}{18} \div \frac{17}{6} = \frac{11}{18} \cdot \frac{6}{17} = \frac{11 \cdot 6}{18 \cdot 17} = \frac{11 \cdot \overset{1}{\cancel{6}}}{\underset{1}{\cancel{6}} \cdot 3 \cdot 17} = \frac{11}{51}$

8. $5\frac{2}{3} \div 2\frac{5}{9} = \frac{17}{3} \div \frac{23}{9} = \frac{17}{3} \cdot \frac{9}{23} = \frac{17 \cdot 9}{3 \cdot 23} = \frac{17 \cdot \overset{1}{\cancel{3}} \cdot 3}{\underset{1}{\cancel{3}} \cdot 23} = \frac{51}{23} \text{ or } 2\frac{5}{23}$

Work Practice Problems 6–8

PRACTICE PROBLEMS 6–8

Divide.

6. $\frac{4}{9} \div 7$

7. $\frac{8}{15} \div 3\frac{4}{5}$

8. $3\frac{2}{7} \div 2\frac{3}{14}$

Objective C Adding or Subtracting Mixed Numbers

We can add or subtract mixed numbers, too, by first writing each mixed number as an improper fraction. But it is often easier to add or subtract the whole-number parts and add or subtract the proper-fraction parts vertically.

Adding or Subtracting Mixed Numbers

To add or subtract mixed numbers, add or subtract the fraction parts and then add or subtract the whole number parts.

EXAMPLE 9 Add: $2\frac{1}{3} + 5\frac{3}{8}$. Check by estimating.

Solution: The LCD of 3 and 8 is 24.

$$2\frac{1 \cdot 8}{3 \cdot 8} = 2\frac{8}{24}$$

$$+5\frac{3 \cdot 3}{8 \cdot 3} = 5\frac{9}{24}$$

$$7\frac{17}{24}$$

← Add the fractions

↑ Add the whole numbers

To check by estimating, we round as usual. The fraction $2\frac{1}{3}$ rounds to 2, $5\frac{3}{8}$ rounds to 5, and $2 + 5 = 7$, our estimate.

Our exact answer is close to 7, so our answer is reasonable.

Work Practice Problem 9

PRACTICE PROBLEM 9

Add: $2\frac{1}{6} + 4\frac{2}{5}$.

Check by estimating.

Answers

5. $19\frac{2}{5}$ 6. $\frac{4}{63}$ 7. $\frac{8}{57}$ 8. $\frac{46}{31}$ or $1\frac{15}{31}$ 9. $6\frac{17}{30}$

✓ Concept Check Answer

forgot to change mixed number to fraction

Helpful Hint When adding or subtracting mixed numbers and whole numbers, it is a good idea to estimate to see if your answer is reasonable.

For the rest of this section, we leave most of the checking by estimating to you.

PRACTICE PROBLEM 10

Add: $3\frac{5}{14} + 2\frac{6}{7}$

EXAMPLE 10 Add: $3\frac{4}{5} + 1\frac{4}{15}$

Solution: The LCD of 5 and 15 is 15.

$$3\frac{4}{5} = 3\frac{12}{15}$$
$$+1\frac{4}{15} = 1\frac{4}{15} \quad \text{Add the fractions; then add the whole numbers.}$$
$$4\frac{16}{15} \quad \text{Notice that the fraction part is improper.}$$

Since $\frac{16}{15}$ is $1\frac{1}{15}$ we can write the sum as

$$4\frac{16}{15} = 4 + 1\frac{1}{15} = 5\frac{1}{15}$$

Work Practice Problem 10

✔Concept Check Explain how you could estimate the following sum: $5\frac{1}{9} + 14\frac{10}{11}$.

PRACTICE PROBLEM 11

Add: $12 + 3\frac{6}{7} + 2\frac{1}{5}$

EXAMPLE 11 Add: $2\frac{4}{5} + 5 + 1\frac{1}{2}$

Solution: The LCD of 5 and 2 is 10.

$$2\frac{4}{5} = 2\frac{8}{10}$$
$$5 = 5$$
$$+1\frac{1}{2} = 1\frac{5}{10}$$
$$8\frac{13}{10} = 8 + 1\frac{3}{10} = 9\frac{3}{10}$$

Work Practice Problem 11

PRACTICE PROBLEM 12

Subtract: $32\frac{7}{9} - 16\frac{5}{18}$

EXAMPLE 12 Subtract: $8\frac{3}{7} - 5\frac{2}{21}$. Check by estimating.

Solution: The LCD of 7 and 21 is 21.

$$8\frac{3}{7} = 8\frac{9}{21} \quad \leftarrow \text{The LCD of 7 and 21 is 21.}$$
$$-5\frac{2}{21} = -5\frac{2}{21}$$
$$3\frac{7}{21} \quad \leftarrow \text{Subtract the fractions.}$$

↑ Subtract the whole numbers.

Answers

10. $6\frac{3}{14}$ **11.** $18\frac{2}{35}$ **12.** $16\frac{1}{2}$

✔ Concept Check Answer

Round each mixed number to the nearest whole number and add. $5\frac{1}{9}$ rounds to 5 and $14\frac{10}{11}$ rounds to 15, and the estimated sum is $5 + 15 = 20$.

Then $3\frac{7}{21}$ simplifies to $3\frac{1}{3}$. The difference is $3\frac{1}{3}$.

To check, $8\frac{3}{7}$ rounds to 8, $5\frac{2}{21}$ rounds to 5, and $8 - 5 = 3$, our estimate.

Our exact answer is close to 3, so our answer is reasonable.

Work Practice Problem 12

When subtracting mixed numbers, borrowing may be needed, as shown in the next example.

EXAMPLE 13 Subtract: $7\frac{3}{14} - 3\frac{6}{7}$

Solution: The LCD of 7 and 14 is 14.

$$\begin{array}{rcr} 7\frac{3}{14} & = & 7\frac{3}{14} \\ -3\frac{6}{7} & = & -3\frac{12}{14} \end{array}$$

Notice that we cannot subtract $\frac{12}{14}$ from $\frac{3}{14}$, so we borrow from the whole number 7.

borrow 1 from 7

$$7\frac{3}{14} = 6 + 1\frac{3}{14} = 6 + \frac{17}{14} \text{ or } 6\frac{17}{14}$$

Now subtract.

$$\begin{array}{rcrcr} 7\frac{3}{14} & = & 7\frac{3}{14} & = & 6\frac{17}{14} \\ -3\frac{6}{7} & = & -3\frac{12}{14} & = & -3\frac{12}{14} \\ \hline & & & & 3\frac{5}{14} \end{array}$$

$\leftarrow$ Subtract the fractions.

$\uparrow$ Subtract the whole numbers.

Work Practice Problem 13

PRACTICE PROBLEM 13

Subtract: $9\frac{7}{15} - 4\frac{3}{5}$

✔ **Concept Check** In the subtraction problem $5\frac{1}{4} - 3\frac{3}{4}$, $5\frac{1}{4}$ must be rewritten because $\frac{3}{4}$ cannot be subtracted from $\frac{1}{4}$. Why is it incorrect to rewrite $5\frac{1}{4}$ as $5\frac{5}{4}$?

EXAMPLE 14 Subtract: $14 - 8\frac{3}{7}$

Solution:

$$\begin{array}{rcr} 14 & = & 13\frac{7}{7} \\ -8\frac{3}{7} & = & -8\frac{3}{7} \\ \hline & & 5\frac{4}{7} \end{array}$$

Borrow 1 from 14 and write it as $\frac{7}{7}$.

$\leftarrow$ Subtract the fractions.

$\uparrow$ Subtract the whole numbers.

Work Practice Problem 14

PRACTICE PROBLEM 14

Subtract: $25 - 10\frac{2}{9}$

Answers

13. $4\frac{13}{15}$ **14.** $14\frac{7}{9}$

✔ Concept Check Answer

Rewrite $5\frac{1}{4}$ as $4\frac{5}{4}$ by borrowing from the 5.

Objective D Solving Problems Containing Mixed Numbers

Now that we know how to perform operations on mixed numbers, we can solve real-life problems.

PRACTICE PROBLEM 15

The measurement around the trunk of a tree just below shoulder height is called its girth. The largest known American beech tree in the United States has a girth of $23\frac{1}{4}$ feet. The largest known sugar maple tree in the United States has a girth of $19\frac{5}{12}$ feet. How much larger is the girth of the largest known American beech tree than the girth of the largest known sugar maple tree? (*Source: American Forests*)

PRACTICE PROBLEM 16

A designer of women's clothing designs a woman's dress that requires $3\frac{1}{7}$ yards of material. How many dresses can be made from a 44-yard bolt of material?

Answers

15. $3\frac{5}{6}$ ft **16.** 14 dresses

EXAMPLE 15 Finding Legal Lobster Size

Lobster fishermen must measure the upper body shells of the lobsters they catch. Lobsters that are too small are thrown back into the ocean. Each state has its own size standard for lobsters to help control the breeding stock. In 1988, Massachusetts increased its legal lobster size from $3\frac{3}{16}$ inches to $3\frac{7}{32}$ inches. How much of an increase was this? (*Source:* Peabody Essex Museum, Salem, Massachusetts)

Solution:

1. UNDERSTAND. Read and reread the problem carefully. The word "increase" found in the problem might make you think that we add to solve the problem. But the phrase "how much of an increase" tells us to subtract to find the increase.
2. TRANSLATE.

In words:	increase	is	new lobster size	minus	old lobster size
	↓	↓	↓	↓	↓
Translate:	increase	=	$3\frac{7}{32}$	−	$3\frac{3}{16}$

3. SOLVE: Before we solve, let's estimate. The fraction $3\frac{7}{32}$ rounds to 3, $3\frac{3}{16}$ rounds to 3, and $3 - 3 = 0$. The increase is not 0, but will be very small.

$$\begin{array}{rcr} 3\frac{7}{32} & = & 3\frac{7}{32} \\ -3\frac{3}{16} & = & 3\frac{6}{32} \\ \hline & & \frac{1}{32} \end{array}$$

4. INTERPRET. *Check* your work. Our estimate tells us that the exact increase of $\frac{1}{32}$ inch is reasonable. *State* your conclusion: The increase in lobster size is $\frac{1}{32}$ of an inch.

Work Practice Problem 15

EXAMPLE 16 Calculating Manufacturing Materials Needed

In a manufacturing process, a metal-cutting machine cuts strips $1\frac{3}{5}$ inches long from a piece of metal stock. How many such strips can be cut from a 48-inch piece of stock?

Solution:

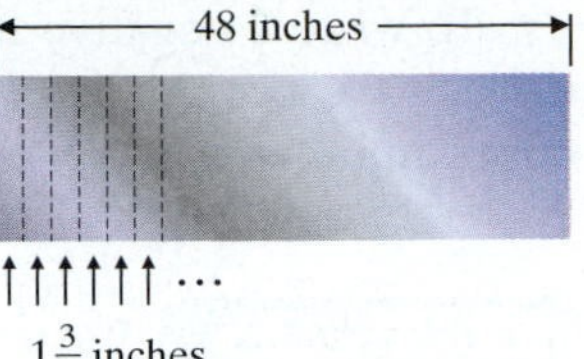

1. UNDERSTAND the problem. To do so, read and reread the problem. Then draw a diagram:

   We want to know how many $1\frac{3}{5}$s there are in 48.

2. TRANSLATE.

In words:	Number of strips	is	48	divided by	$1\frac{3}{5}$
	↓	↓	↓	↓	↓
Translate:	Number of strips	=	48	÷	$1\frac{3}{5}$

3. SOLVE: Let's estimate a reasonable answer. The mixed number $1\frac{3}{5}$ rounds to 2 and $48 \div 2 = 24$.

$$48 \div 1\frac{3}{5} = 48 \div \frac{8}{5} = \frac{48}{1} \cdot \frac{5}{8} = \frac{48 \cdot 5}{1 \cdot 8} = \frac{\overset{1}{\cancel{8}} \cdot 6 \cdot 5}{1 \cdot \underset{1}{\cancel{8}}} = \frac{30}{1} \text{ or } 30$$

4. INTERPRET. *Check* your work. Since the exact answer of 30 is close to our estimate of 24, our answer is reasonable. *State* your conclusion: Thirty strips can be cut from the 48-inch piece of stock.

Work Practice Problem 16

Objective E Operating on Negative Mixed Numbers

To perform operations on negative mixed numbers, let's first practice writing these numbers as negative fractions and negative fractions as negative mixed numbers.

To understand negative mixed numbers, we simply need to know that, for example,

$$-3\frac{2}{5} \text{ means } -\left(3\frac{2}{5}\right)$$

Thus, to write a negative mixed number as a fraction, we do the following.

$$-3\frac{2}{5} = -\left(3\frac{2}{5}\right) = -\left(\frac{5 \cdot 3 + 2}{5}\right) = -\left(\frac{17}{5}\right) \quad \text{or} \quad -\frac{17}{5}$$

EXAMPLES Write each as a fraction.

17. $-1\frac{7}{8} = -\frac{8 \cdot 1 + 7}{8} = -\frac{15}{8}$ Write $1\frac{7}{8}$ as an improper fraction and keep the negative sign.

18. $-23\frac{1}{2} = -\frac{2 \cdot 23 + 1}{2} = -\frac{47}{2}$ Write $23\frac{1}{2}$ as an improper fraction and keep the negative sign.

Work Practice Problems 17–18

PRACTICE PROBLEMS 17–18

Write each as a fraction.

17. $-9\frac{3}{7}$ **18.** $-5\frac{10}{11}$

Answers

17. $-\frac{66}{7}$ **18.** $-\frac{65}{11}$

To write a negative fraction as a negative mixed number, we use a similar procedure. We simply disregard the negative sign, convert the improper fraction to a mixed number, then reinsert the negative sign.

PRACTICE PROBLEMS 19–20

Write each as a mixed number.

19. $-\frac{37}{8}$ **20.** $-\frac{46}{5}$

EXAMPLES Write each as a mixed number.

19. $-\frac{22}{5} = -4\frac{2}{5}$

$$5\overline{)22} \text{ gives } 4,\quad -20,\quad \text{remainder } 2 \qquad \frac{22}{5} = 4\frac{2}{5}$$

20. $-\frac{9}{4} = -2\frac{1}{4}$

$$4\overline{)9} \text{ gives } 2,\quad -8,\quad \text{remainder } 1 \qquad \frac{9}{4} = 2\frac{1}{4}$$

Work Practice Problems 19–20

We multiply or divide with negative mixed numbers the same way that we multiply or divide with positive mixed numbers. We first write each mixed number as a fraction.

PRACTICE PROBLEMS 21–22

21. $2\frac{3}{4}\cdot\left(-3\frac{3}{5}\right)$

22. $-4\frac{2}{7} \div 1\frac{1}{4}$

EXAMPLES Perform the indicated operations.

21. $-4\frac{2}{5}\cdot 1\frac{3}{11} = -\frac{22}{5}\cdot\frac{14}{11} = -\frac{22\cdot 14}{5\cdot 11} = -\frac{2\cdot \overset{1}{\cancel{11}}\cdot 14}{5\cdot \underset{1}{\cancel{11}}} = -\frac{28}{5}$ or $-5\frac{3}{5}$

22. $-2\frac{1}{3} \div \left(-2\frac{1}{2}\right) = -\frac{7}{3} \div \left(-\frac{5}{2}\right) = -\frac{7}{3}\cdot\left(-\frac{2}{5}\right) = \frac{7\cdot 2}{3\cdot 5} = \frac{14}{15}$

Work Practice Problems 21–22

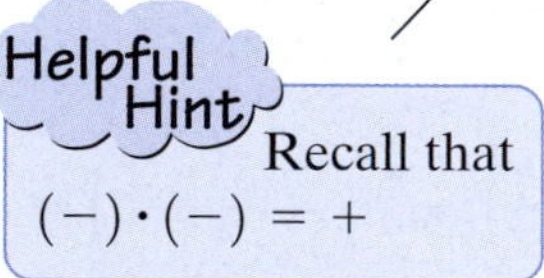
Helpful Hint
Recall that
(−)·(−) = +

To add or subtract with negative mixed numbers, we must be very careful! Problems arise because recall that

$$-3\frac{2}{5} \text{ means } -\left(3\frac{2}{5}\right)$$

This means that

$$-3\frac{2}{5} = -\left(3\frac{2}{5}\right) = -\left(3 + \frac{2}{5}\right) = -3 - \frac{2}{5}$$ This can sometimes be easily overlooked.

To avoid problems, we will add or subtract negative mixed numbers by rewriting as addition and recalling how to add signed numbers.

Answers

19. $-4\frac{5}{8}$ **20.** $-9\frac{1}{5}$

21. $-9\frac{9}{10}$ **22.** $-3\frac{3}{7}$

EXAMPLE 23 Add: $6\frac{3}{5} + \left(-9\frac{7}{10}\right)$

Solution: Here we are adding two numbers with different signs. Recall that we then subtract the absolute values and keep the sign of the larger absolute value.

Since $-9\frac{7}{10}$ has the larger absolute value, the answer is negative.

First, subtract absolute values:

$$\begin{aligned} 9\frac{7}{10} &= 9\frac{7}{10} \\ -6\frac{3\cdot 2}{5\cdot 2} &= -6\frac{6}{10} \\ \hline & \quad 3\frac{1}{10} \end{aligned}$$

Thus,

$$6\frac{3}{5} + \left(-9\frac{7}{10}\right) = -3\frac{1}{10}$$ The result is negative since $-9\frac{7}{10}$ has the larger absolute value.

Work Practice Problem 23

PRACTICE PROBLEM 23

Add: $6\frac{2}{3} + \left(-12\frac{3}{4}\right)$

EXAMPLE 24 Subtract: $-11\frac{5}{6} - 20\frac{4}{9}$

Solution: Let's write as an equivalent addition: $-11\frac{5}{6} + \left(-20\frac{4}{9}\right)$. Here, we are adding two numbers with like signs. Recall that we add their absolute values and keep the common negative sign.

First, add absolute values:

$$\begin{aligned} 11\frac{5\cdot 3}{6\cdot 3} &= 11\frac{15}{18} \\ +20\frac{4\cdot 2}{9\cdot 2} &= +20\frac{8}{18} \\ \hline & \quad 31\frac{23}{18} \text{ or } 32\frac{5}{18} \end{aligned}$$

Since $\frac{23}{18} = 1\frac{5}{18}$

Thus,

$$-11\frac{5}{6} - 20\frac{4}{9} = -32\frac{5}{18}$$

Keep the common sign.

Work Practice Problem 24

PRACTICE PROBLEM 24

Subtract: $-9\frac{2}{7} - 30\frac{11}{14}$

Answers

23. $-6\frac{1}{12}$ **24.** $-40\frac{1}{14}$

CALCULATOR EXPLORATIONS Converting Between Mixed-Number and Fraction Notation

If your calculator has a fraction key, such as [a b/c], you can use it to convert between mixed-number notation and fraction notation.

To write $13\frac{7}{16}$ as an improper fraction, press

[1] [3] [a b/c] [7] [a b/c] [1] [6] [2nd] [d/c]

The display will read

[215 | 16]

which represents $\frac{215}{16}$. Thus $13\frac{7}{16} = \frac{215}{16}$.

To convert $\frac{190}{13}$ to a mixed number, press

[1] [9] [0] [a b/c] [1] [3] [=]

The display will read

[14_8/13]

which represents $14\frac{8}{13}$. Thus $\frac{190}{13} = 14\frac{8}{13}$.

Write each mixed number as a fraction and each fraction as a mixed number.

1. $25\frac{5}{11}$ **2.** $67\frac{14}{15}$ **3.** $107\frac{31}{35}$

4. $186\frac{17}{21}$ **5.** $\frac{365}{14}$ **6.** $\frac{290}{13}$

7. $\frac{2769}{30}$ **8.** $\frac{3941}{17}$

Vocabulary and Readiness Check

Use the choices below to fill in each blank.

round	fraction	whole number
improper	mixed number	

1. The number $5\frac{3}{4}$ is called a(n) ______________.

2. For $5\frac{3}{4}$, the 5 is called the ______________ part and $\frac{3}{4}$ is called the ______________ part.

3. To estimate operations on mixed numbers, we ______________ mixed numbers to the nearest whole number.

4. The mixed number $2\frac{5}{8}$ written as an ______________ fraction is $\frac{21}{8}$.

4.7 EXERCISE SET

FOR EXTRA HELP

Student Solutions Manual

PH Math/Tutor Center

CD/Video for Review

MathXL®

MyMathLab

Objective A *Graph each list of numbers on the given number line. See Example 1.*

1. $-2, -2\frac{2}{3}, 0, \frac{7}{8}, -\frac{1}{3}$

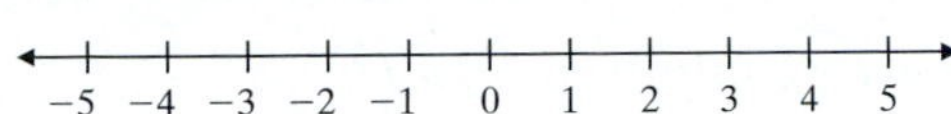

2. $-1, -1\frac{1}{4}, -\frac{1}{4}, 3\frac{1}{4}, 3$

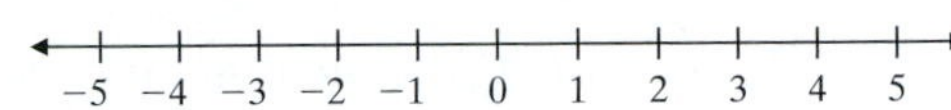

3. $4, \frac{1}{3}, -3, -3\frac{4}{5}, 1\frac{1}{3}$

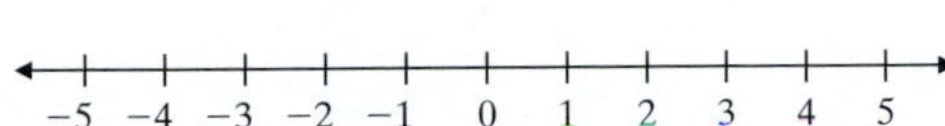

4. $3, \frac{3}{8}, -4, -4\frac{1}{3}, -\frac{9}{10}$

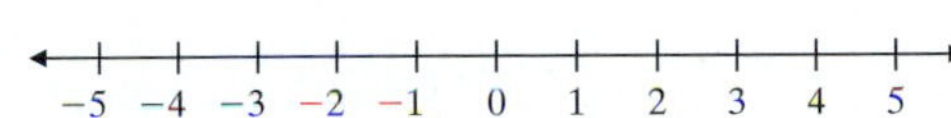

Objective B *Choose the best estimate for each product or quotient. See Examples 4 through 8.*

5. $2\frac{11}{12} \cdot 1\frac{1}{4}$

a. 2 **b.** 3 **c.** 1 **d.** 12

6. $5\frac{1}{6} \cdot 3\frac{5}{7}$

a. 9 **b.** 15 **c.** 8 **d.** 20

7. $12\frac{2}{11} \div 3\frac{9}{10}$

a. 3 **b.** 4 **c.** 36 **d.** 9

8. $20\frac{3}{14} \div 4\frac{8}{11}$

a. 5 **b.** 80 **c.** 4 **d.** 16

Multiply or divide. For Exercises 13 through 16, find an exact answer and an estimated answer. See Examples 2 through 8.

9. $2\frac{2}{3} \cdot \frac{1}{7}$

10. $\frac{5}{9} \cdot 4\frac{1}{5}$

11. $7 \div 1\frac{3}{5}$

12. $9 \div 1\frac{2}{3}$

13. $2\frac{1}{5} \cdot 3\frac{1}{2}$

Exact:

Estimate:

14. $2\frac{1}{4} \cdot 7\frac{1}{8}$

Exact:

Estimate:

15. $3\frac{4}{5} \cdot 6\frac{2}{7}$

Exact:

Estimate:

16. $5\frac{5}{6} \cdot 7\frac{3}{5}$

Exact:

Estimate:

17. $5 \cdot 2\frac{1}{2}$

18. $6 \cdot 3\frac{1}{3}$

19. $3\frac{2}{3} \cdot 1\frac{1}{2}$

20. $2\frac{4}{5} \cdot 2\frac{5}{8}$

21. $2\frac{2}{3} \div \frac{1}{7}$

22. $\frac{5}{9} \div 4\frac{1}{5}$

Objective C *Choose the best estimate for each sum or difference. See Examples 9 through 13.*

23. $3\frac{7}{8} + 2\frac{1}{5}$

a. 6 **b.** 5 **c.** 1 **d.** 2

24. $3\frac{7}{8} - 2\frac{1}{5}$

a. 6 **b.** 5 **c.** 1 **d.** 2

25. $8\frac{1}{3} + 1\frac{1}{2}$

a. 4 **b.** 10 **c.** 6 **d.** 16

26. $8\frac{1}{3} - 1\frac{1}{2}$

a. 4 **b.** 10 **c.** 6 **d.** 16

Add. For Exercises 27 through 30, find an exact sum and an estimated sum. See Examples 9 through 11.

27. $4\frac{7}{12} + 2\frac{1}{12}$

Exact:

Estimate:

28. $7\frac{4}{11} + 3\frac{2}{11}$

Exact:

Estimate:

29. $10\frac{3}{14} + 3\frac{4}{7}$

Exact:

Estimate:

30. $12\frac{5}{12} + 4\frac{1}{6}$

Exact:

Estimate:

31. $9\frac{1}{5} + 8\frac{2}{25}$

32. $6\frac{2}{13} + 8\frac{7}{26}$

33. $12\frac{3}{14}$, 10, $+25\frac{5}{12}$

34. $8\frac{2}{9}$, 32, $+9\frac{10}{21}$

35. $15\frac{4}{7} + 9\frac{11}{14}$

36. $23\frac{3}{5} + 8\frac{8}{15}$

37. $3\frac{5}{8}$, $2\frac{1}{6}$, $+7\frac{3}{4}$

38. $4\frac{1}{3}$, $9\frac{2}{5}$, $+3\frac{1}{6}$

Subtract. For Exercises 39 through 42, find an exact difference and an estimated difference. See Examples 12 through 14.

39. $4\frac{7}{10} - 2\frac{1}{10}$

Exact:

Estimate:

40. $7\frac{4}{9} - 3\frac{2}{9}$

Exact:

Estimate:

41. $10\frac{13}{14} - 3\frac{4}{7}$

Exact:

Estimate:

42. $12\frac{5}{12} - 4\frac{1}{6}$

Exact:

Estimate:

43. $9\frac{1}{5} - 8\frac{6}{25}$

44. $5\frac{2}{13} - 4\frac{7}{26}$

45. $6 - 2\frac{4}{9}$

46. $8 - 1\frac{7}{10}$

47. $63\frac{1}{6} - 47\frac{5}{12}$

48. $86\frac{2}{15} - 27\frac{3}{10}$

Objectives B C Mixed Practice *Perform each indicated operation. See Examples 2 through 14.*

49. $2\frac{3}{4} + 1\frac{1}{4}$

50. $5\frac{5}{8} + 2\frac{3}{8}$

51. $15\frac{4}{7} - 9\frac{11}{14}$

52. $23\frac{3}{5} - 8\frac{8}{15}$

53. $3\frac{1}{9} \cdot 2$

54. $4\frac{1}{2} \cdot 3$

55. $1\frac{2}{3} \div 2\frac{1}{5}$

56. $5\frac{1}{5} \div 3\frac{1}{4}$

57. $22\frac{4}{9} + 13\frac{5}{18}$

58. $15\frac{3}{25} - 5\frac{2}{5}$

59. $5\frac{2}{3} - 3\frac{1}{6}$

60. $5\frac{3}{8} - 2\frac{13}{16}$

61. $15\frac{1}{5}$, $20\frac{3}{10}$, $+37\frac{2}{15}$

62. $7\frac{3}{7}$, 15, $+20\frac{1}{2}$

63. $6\frac{4}{7} - 5\frac{11}{14}$

64. $47\frac{5}{12} - 23\frac{19}{24}$

65. $4\frac{2}{7} \cdot 1\frac{3}{10}$

66. $6\frac{2}{3} \cdot 2\frac{3}{4}$

67. $6\frac{2}{11}$, 3, $+4\frac{10}{33}$

68. $3\frac{7}{16}$, $6\frac{1}{2}$, $+9\frac{3}{8}$

Objective D Translating *Translate each phrase to an algebraic expression. Use x to represent a number. See Examples 15 and 16.*

69. $-5\frac{2}{7}$ decreased by a number

70. The sum of $8\frac{3}{4}$ and a number

71. Multiply $1\frac{9}{10}$ by a number.

72. Divide a number by $-6\frac{1}{11}$.

Solve. Write each answer in simplest form. See Examples 15 and 16.

73. The Gauge Act of 1846 set the standard gauge for U.S. railroads at $56\frac{1}{2}$ inches. (See figure.) If the standard gauge in Spain is $65\frac{9}{10}$ inches, how much wider is Spain's standard gauge than the U.S. standard gauge? (*Source:* San Diego Railroad Museum)

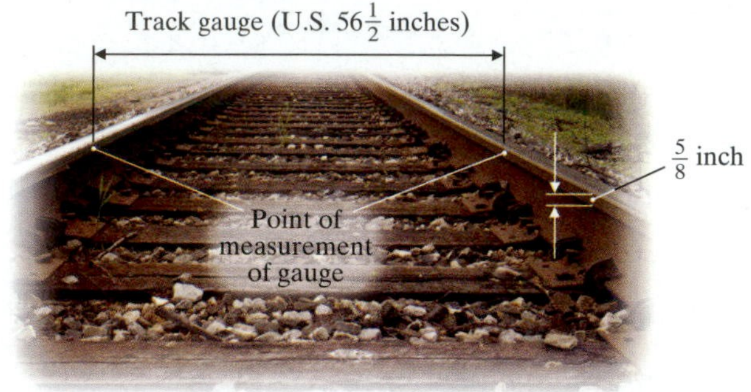

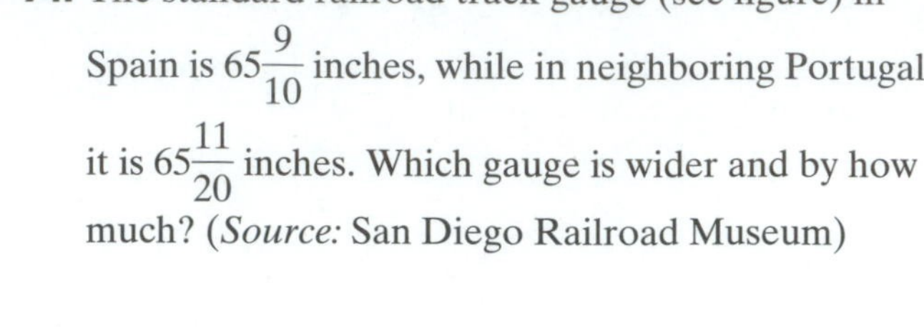

74. The standard railroad track gauge (see figure) in Spain is $65\frac{9}{10}$ inches, while in neighboring Portugal it is $65\frac{11}{20}$ inches. Which gauge is wider and by how much? (*Source:* San Diego Railroad Museum)

75. If Tucson's average rainfall is $11\frac{1}{4}$ inches and Yuma's is $3\frac{3}{5}$ inches, how much more rain, on the average, does Tucson get than Yuma?

76. A pair of crutches needs adjustment. One crutch is 43 inches and the other is $41\frac{5}{8}$ inches. Find how much the short crutch should be lengthened to make both crutches the same length.

For Exercises 77 and 78, find the area of each figure.

△**77.**

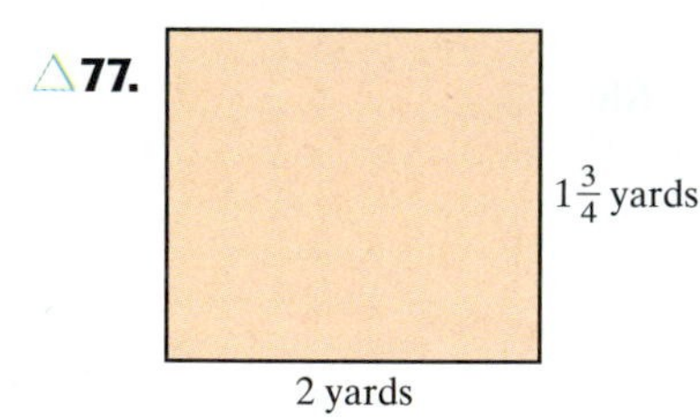

△**78.**

5 inches

$3\frac{1}{2}$ inches

△**79.** A model for a proposed computer chip measures $\frac{3}{4}$ inch by $1\frac{1}{4}$ inches. Find its area.

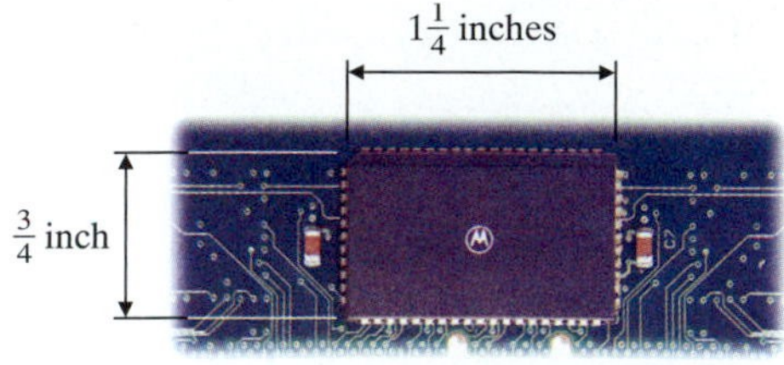

△**80.** The Saltalamachios are planning to build a deck that measures $4\frac{1}{2}$ yards by $6\frac{1}{3}$ yards. Find the area of their proposed deck.

For Exercises 81 and 82, find the perimeter of each figure.

△**81.**

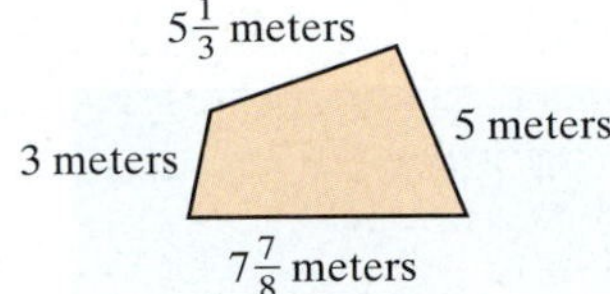

△**82.**

$3\frac{1}{4}$ yards
$3\frac{1}{4}$ yards
$3\frac{1}{4}$ yards
$3\frac{1}{4}$ yards
$3\frac{1}{4}$ yards

83. A homeowner has $15\frac{2}{3}$ feet of plastic pipe. She cuts off a $2\frac{1}{2}$-foot length and then a $3\frac{1}{4}$-foot length. If she now needs a 10-foot piece of pipe, will the remaining piece do? If not, by how much will the piece be short?

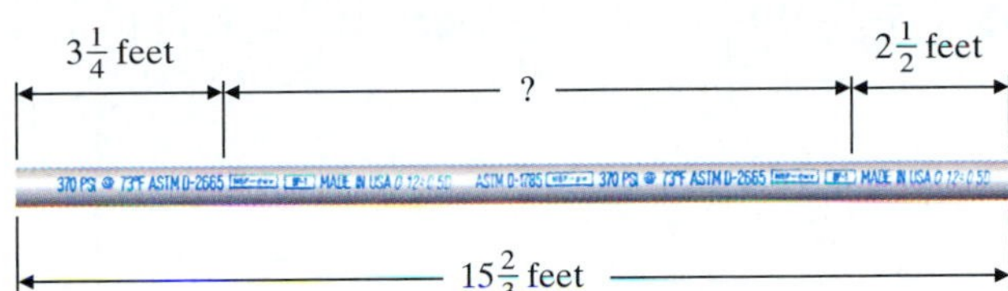

84. A trim carpenter cuts a board $3\frac{3}{8}$ feet long from one 6 feet long. How long is the remaining piece?

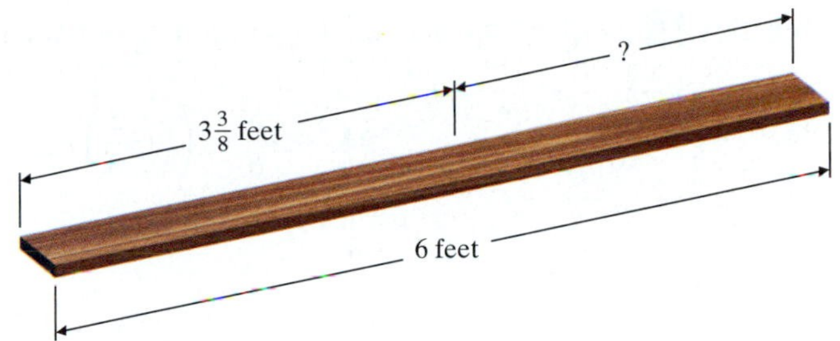

85. A heart attack patient in rehabilitation walked on a treadmill $12\frac{3}{4}$ miles over 4 days. How many miles is this per day?

86. A local restaurant is selling hamburgers from a booth on Memorial Day. A total of $27\frac{3}{4}$ pounds of hamburger have been ordered. How many quarter-pound hamburgers can this make?

△**87.** The area of the rectangle below is 12 square meters. If its width is $2\frac{4}{7}$ meters, find its length.

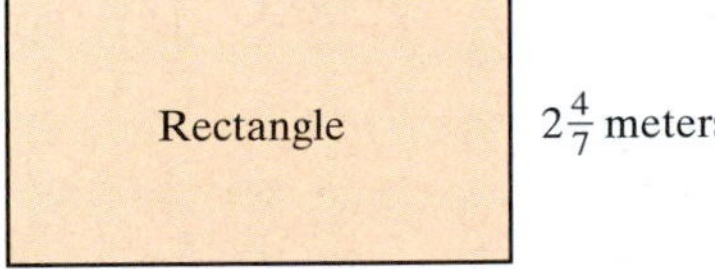

△**88.** The perimeter of the square below is $23\frac{1}{2}$ feet. Find the length of each side.

Square

The following table lists some upcoming total eclipses of the Sun that will be visible in North America. The duration of each eclipse is listed in the table. Use the table to answer Exercises 89 through 92.

Total Solar Eclipses Visible from North America

Date of Eclipse	Duration (in Minutes)
August 1, 2008	$2\frac{9}{20}$
August 21, 2017	$2\frac{2}{3}$
April 8, 2024	$4\frac{7}{15}$

(*Source:* NASA/Goddard Space Flight Center)

89. What is the total duration for the three eclipses?

90. What is the total duration for the two eclipses occuring in even-numbered years?

91. How much longer will the April 8, 2024, eclipse be than the August 21, 2017, eclipse?

92. How much longer will the August 21, 2017, eclipse be than the August 1, 2008, eclipse?

Objective **E** *Perform the indicated operations. See Examples 17 through 24.*

93. $-4\frac{2}{5} \cdot 2\frac{3}{10}$

94. $-3\frac{5}{6} \div \left(-3\frac{2}{3}\right)$

95. $-5\frac{1}{8} - 19\frac{3}{4}$

96. $17\frac{5}{9} + \left(-14\frac{2}{3}\right)$

97. $-31\frac{2}{15} + 17\frac{3}{20}$

98. $-1\frac{5}{7} \cdot \left(-2\frac{1}{2}\right)$

99. $1\frac{3}{4} \div \left(-3\frac{1}{2}\right)$

100. $-31\frac{7}{8} - \left(-26\frac{5}{12}\right)$

101. $11\frac{7}{8} - 13\frac{5}{6}$

102. $-20\frac{2}{5} + \left(-30\frac{3}{10}\right)$

103. $-7\frac{3}{10} \div (-100)$

104. $-4\frac{1}{4} \div 2\frac{3}{8}$

Review

Multiply. See Section 4.3.

105. $\frac{1}{3}(3x)$

106. $\frac{1}{5}(5y)$

107. $\frac{2}{3}\left(\frac{3}{2}a\right)$

108. $-\frac{9}{10}\left(-\frac{10}{9}m\right)$

Concept Extensions

Solve. See the first Concept Check in this section.

109. Which of the following are equivalent to 10?

a. $9\frac{5}{5}$ **b.** $9\frac{100}{100}$ **c.** $6\frac{44}{11}$ **d.** $8\frac{13}{13}$

110. Which of the following are equivalent to $7\frac{3}{4}$?

a. $6\frac{7}{4}$ **b.** $5\frac{11}{4}$ **c.** $7\frac{12}{16}$ **d.** all of them

Solve. See the second Concept Check in this section.

111. A student asked you to check her work below. Is it correct? If not, where is the error?

$$20\frac{2}{3} \div 10\frac{1}{2} \stackrel{?}{=} 2\frac{1}{3}$$

112. In your own words, describe how to divide mixed numbers.

113. In your own words, explain how to multiply
a. fractions
b. mixed numbers

114. In your own words, explain how to round a mixed number to the nearest whole number.

Solve. See the second Concept Check in this section.

115. A student asked you to check his work below. Is it correct? If not, where is the error?

$$3\frac{2}{3} \cdot 1\frac{1}{7} \stackrel{?}{=} 3\frac{2}{21}$$

Solve.

116. Explain in your own words why $9\frac{13}{9}$ is equal to $10\frac{4}{9}$.

117. In your own words, explain
a. when to borrow when subtracting mixed numbers, and
b. how to borrow when subtracting mixed numbers.

THE BIGGER PICTURE Operations on Sets of Numbers

Continue your outline from Sections 1.6, 1.7, 2.4, 3.3, and 4.3. Suggestions are once again written to help you complete this part of your outline.

I. Operations on Sets of Numbers

A. Whole Numbers

1. **Add or Subtract** (Section 1.3)
2. **Multiply or Divide** (Sections 1.5, 1.6)
3. **Exponent** (Section 1.7)
4. **Order of Operations** (Section 1.7)

B. Integers

1. **Add** (Section 2.2)
2. **Subtract** (Section 2.3)
3. **Multiply or Divide** (Section 2.4)

C. Fractions

1. **Simplify** (Section 4.2)
2. **Multiply** (Section 4.3)
3. **Divide** (Section 4.3)
4. **Add or Subtract:** Must have same denominators. If not, find the LCD, and write each fraction as an equivalent fraction with the LCD as denominator.

$$\frac{2}{5} + \frac{1}{15} = \frac{2}{5}\cdot\frac{3}{3} + \frac{1}{15} = \frac{6}{15} + \frac{1}{15} = \frac{7}{15}$$

II. Solving Equations

A. Equations in General (Section 3.3)

Perform indicated operations.

1. $\frac{3}{17} + \frac{2}{17}$

2. $\frac{9}{10} - \frac{1}{10}$

3. $\frac{2}{3} + \frac{3}{10}$

4. $\frac{23}{24} - \frac{11}{12}$

5. $\frac{7}{8} + \frac{19}{20}$

6. $\frac{3^3}{4^3}$

7. $\begin{array}{r} 16 \\ -\ 3\frac{4}{7} \\ \hline \end{array}$

8. $\begin{array}{r} 2\frac{5}{8} \\ 1\frac{1}{6} \\ +5\frac{3}{4} \\ \hline \end{array}$

9. $\frac{6}{11} \cdot \frac{8}{9}$

10. $2\frac{4}{15} \div 1\frac{4}{5}$

Objectives

A Solve Equations Containing Fractions.

B Solve Equations by Multiplying by the LCD.

C Review Adding and Subtracting Fractions.

4.8 SOLVING EQUATIONS CONTAINING FRACTIONS

Objective A Solving Equations Containing Fractions

In Chapter 3, we solved linear equations in one variable. In this section, we practice this skill by solving linear equations containing fractions. To help us solve these equations, let's review the properties of equality.

Addition Property of Equality

Let a, b, and c represent numbers. Then

$a = b$	Also, $a = b$
and $a + c = b + c$	and $a - c = b - c$
are equivalent equations.	are equivalent equations.

Multiplication Property of Equality

Let a, b, and c represent numbers and let $c \neq 0$. Then

$a = b$	Also, $a = b$
and $a \cdot c = b \cdot c$	and $\frac{a}{c} = \frac{b}{c}$
are equivalent equations.	are equivalent equations.

In other words, the same number may be added to or subtracted from both sides of an equation without changing the solution of the equation. Also, the same nonzero number may be multiplied to or divided by both sides of an equation without changing the solution.

Also, don't forget that to solve an equation in x, our goal is to use properties of equality to write simpler equations, all equivalent to the original equation, until the final equation has the form

$x =$ **number** or **number** $= x$

PRACTICE PROBLEM 1

Solve: $y - \frac{2}{3} = \frac{5}{12}$

EXAMPLE 1 Solve: $x - \frac{3}{4} = \frac{1}{20}$

Solution: To get x by itself, add $\frac{3}{4}$ to both sides.

$$x - \frac{3}{4} = \frac{1}{20}$$

$$x - \frac{3}{4} + \frac{3}{4} = \frac{1}{20} + \frac{3}{4} \quad \text{Add } \tfrac{3}{4} \text{ to both sides.}$$

$$x = \frac{1}{20} + \frac{3 \cdot 5}{4 \cdot 5} \quad \text{The LCD of 20 and 4 is 20.}$$

$$x = \frac{1}{20} + \frac{15}{20}$$

$$x = \frac{16}{20}$$

$$x = \frac{\overset{1}{\cancel{4}} \cdot 4}{\underset{1}{\cancel{4}} \cdot 5} = \frac{4}{5} \quad \text{Write } \tfrac{16}{20} \text{ in simplest form}$$

Continued on next page

Answer

1. $\frac{13}{12}$

Check: To check, replace x with $\frac{4}{5}$ in the original equation.

$$x - \frac{3}{4} = \frac{1}{20}$$

$$\frac{4}{5} - \frac{3}{4} \stackrel{?}{=} \frac{1}{20} \quad \text{Replace } x \text{ with } \frac{4}{5}.$$

$$\frac{4 \cdot 4}{5 \cdot 4} - \frac{3 \cdot 5}{4 \cdot 5} \stackrel{?}{=} \frac{1}{20} \quad \text{The LCD of 5 and 4 is 20.}$$

$$\frac{16}{20} - \frac{15}{20} \stackrel{?}{=} \frac{1}{20}$$

$$\frac{1}{20} = \frac{1}{20} \quad \text{True}$$

Thus $\frac{4}{5}$ is the solution of $x - \frac{3}{4} = \frac{1}{20}$.

Work Practice Problem 1

EXAMPLE 2 Solve: $\frac{1}{3}x = 7$

Solution: Recall that isolating x means that we want the coefficient of x to be 1. To do so, we use the multiplication property of equality and multiply both sides of the equation by the reciprocal of $\frac{1}{3}$, or 3. Since $\frac{1}{3} \cdot 3 = 1$, we will have isolated x.

$$\frac{1}{3}x = 7$$

$$3 \cdot \frac{1}{3}x = 3 \cdot 7 \quad \text{Multiply both sides by 3.}$$

$$1 \cdot x = 21 \text{ or } x = 21 \quad \text{Simplify.}$$

To check, replace x with 21 in the original equation.

$$\frac{1}{3}x = 7 \quad \text{Original equation}$$

$$\frac{1}{3} \cdot 21 \stackrel{?}{=} 7 \quad \text{Replace } x \text{ with 21.}$$

$$7 = 7 \quad \text{True}$$

Since $7 = 7$ is a true statement, 21 is the solution of $\frac{1}{3}x = 7$.

Work Practice Problem 2

PRACTICE PROBLEM 2

Solve: $\frac{1}{5}y = 2$

EXAMPLE 3 Solve: $\frac{3}{5}a = 9$

Solution: Multiply both sides by $\frac{5}{3}$, the reciprocal of $\frac{3}{5}$, so that the coefficient of a is 1.

$$\frac{3}{5}a = 9$$

$$\frac{5}{3} \cdot \frac{3}{5}a = \frac{5}{3} \cdot 9 \quad \text{Multiply both sides by } \frac{5}{3}.$$

$$1a = \frac{5 \cdot 9}{3} \quad \text{Multiply.}$$

$$a = 15 \quad \text{Simplify.}$$

Continued on next page

PRACTICE PROBLEM 3

Solve: $\frac{5}{7}b = 25$

Answers

2. 10 **3.** 35

To check, replace a with 15 in the original equation.

$$\frac{3}{5}a = 9$$

$$\frac{3}{5} \cdot 15 \stackrel{?}{=} 9 \quad \text{Replace } a \text{ with 15.}$$

$$\frac{3 \cdot 15}{5} \stackrel{?}{=} 9 \quad \text{Multiply.}$$

$$9 = 9 \quad \text{True}$$

Since $9 = 9$ is true, 15 is the solution of $\frac{3}{5}a = 9$.

Work Practice Problem 3

PRACTICE PROBLEM 4

Solve: $-\frac{7}{10}x = \frac{2}{5}$

EXAMPLE 4 Solve: $\frac{3}{4}x = -\frac{1}{8}$

Solution: Multiply both sides of the equation by $\frac{4}{3}$, the reciprocal of $\frac{3}{4}$.

$$\frac{3}{4}x = -\frac{1}{8}$$

$$\frac{4}{3} \cdot \frac{3}{4}x = \frac{4}{3} \cdot -\frac{1}{8} \quad \text{Multiply both sides by } \frac{4}{3}.$$

$$1x = -\frac{4 \cdot 1}{3 \cdot 8} \quad \text{Multiply.}$$

$$x = -\frac{1}{6} \quad \text{Simplify.}$$

To check, replace x with $-\frac{1}{6}$ in the original equation.

$$\frac{3}{4}x = -\frac{1}{8} \quad \text{Original equation}$$

$$\frac{3}{4} \cdot -\frac{1}{6} \stackrel{?}{=} -\frac{1}{8} \quad \text{Replace } x \text{ with } -\frac{1}{6}.$$

$$-\frac{1}{8} = -\frac{1}{8} \quad \text{True}$$

Since we arrived at a true statement, $-\frac{1}{6}$ is the solution of $\frac{3}{4}x = -\frac{1}{8}$.

Work Practice Problem 4

PRACTICE PROBLEM 5

Solve: $5x = -\frac{3}{4}$

EXAMPLE 5 Solve: $3y = -\frac{2}{11}$

Solution: We can either divide both sides by 3 or multiply both sides by the reciprocal of 3, which is $\frac{1}{3}$.

$$3y = -\frac{2}{11}$$

$$\frac{1}{3} \cdot 3y = \frac{1}{3} \cdot -\frac{2}{11} \quad \text{Multiply both sides by } \frac{1}{3}.$$

$$1y = -\frac{1 \cdot 2}{3 \cdot 11} \quad \text{Multiply.}$$

$$y = -\frac{2}{33} \quad \text{Simplify.}$$

Check to see that the solution is $-\frac{2}{33}$.

Work Practice Problem 5

Answers

4. $-\frac{4}{7}$ **5.** $-\frac{3}{20}$

Objective B Solving Equations by Multiplying by the LCD

Solving equations with fractions can be tedious. If an equation contains fractions, it is often helpful to first multiply both sides of the equation by the LCD of the fractions. This has the effect of eliminating the fractions in the equation, as shown in the next example.

Let's solve the equation in Example 4 again. This time, we will multiply both sides by the LCD.

EXAMPLE 6 Solve: $\frac{3}{4}x = -\frac{1}{8}$

Solution: First, multiply both sides of the equation by the LCD of the fractions $\frac{3}{4}$ and $-\frac{1}{8}$. The LCD of the denominators is 8.

$$\frac{3}{4}x = -\frac{1}{8}$$

$$8 \cdot \frac{3}{4}x = 8 \cdot -\frac{1}{8} \quad \text{Multiply both sides by 8.}$$

$$\frac{8 \cdot 3}{1 \cdot 4}x = -\frac{\overset{1}{\cancel{8}} \cdot 1}{1 \cdot \underset{1}{\cancel{8}}} \quad \text{Multiply the fractions.}$$

$$\frac{2 \cdot \overset{1}{\cancel{4}} \cdot 3}{1 \cdot \underset{1}{\cancel{4}}}x = -\frac{1 \cdot 1}{1 \cdot 1} \quad \text{Simplify.}$$

$$6x = -1$$

$$\frac{6x}{6} = \frac{-1}{6} \quad \text{Divide both sides by 6.}$$

$$x = \frac{-1}{6} \text{ or } -\frac{1}{6} \quad \text{Simplify.}$$

As seen in Example 4, the solution is $-\frac{1}{6}$.

Work Practice Problem 6

PRACTICE PROBLEM 6

Solve: $\frac{11}{15}x = -\frac{3}{5}$

EXAMPLE 7 Solve: $\frac{x}{6} + 1 = \frac{4}{3}$

Solution: First multiply both sides of the equation by the LCD of the fractions. The LCD of the denominators 6 and 3 is 6.

$$\frac{x}{6} + 1 = \frac{4}{3}$$

$$6\left(\frac{x}{6} + 1\right) = 6\left(\frac{4}{3}\right) \quad \text{Multiply both sides by 6.}$$

$$\overset{1}{\cancel{6}}\left(\frac{x}{\underset{1}{\cancel{6}}}\right) + 6(1) = \overset{2}{\cancel{6}}\left(\frac{4}{\underset{1}{\cancel{3}}}\right) \quad \text{Apply the distributive property.}$$

$$x + 6 = 8 \quad \text{Simplify.}$$

$$x + 6 + (-6) = 8 + (-6) \quad \text{Add } -6 \text{ to both sides.}$$

$$x = 2 \quad \text{Simplify.}$$

Continued on next page

PRACTICE PROBLEM 7

Solve: $\frac{y}{8} + \frac{3}{4} = 2$

Answers

6. $-\frac{9}{11}$ 7. 10

To check, replace x with 2 in the original equation.

$$\frac{x}{6} + 1 = \frac{4}{3} \quad \text{Original equation}$$

$$\frac{2}{6} + 1 \stackrel{?}{=} \frac{4}{3} \quad \text{Replace } x \text{ with 2.}$$

$$\frac{1}{3} + \frac{3}{3} \stackrel{?}{=} \frac{4}{3} \quad \text{Simplify } \frac{2}{6}\text{. The LCD of 3 and 1 is 3.}$$

$$\frac{4}{3} = \frac{4}{3} \quad \text{True}$$

Since we arrived at a true statement, 2 is the solution of $\frac{x}{6} + 1 = \frac{4}{3}$.

Work Practice Problem 7

Let's review the steps for solving equations in x. An extra step is now included to handle equations containing fractions.

Solving an Equation in x

Step 1: If fractions are present, multiply both sides of the equation by the LCD of the fractions.

Step 2: If parentheses are present, use the distributive property.

Step 3: Combine any like terms on each side of the equation.

Step 4: Use the addition property of equality to rewrite the equation so that variable terms are on one side of the equation and constant terms are on the other side.

Step 5: Divide both sides of the equation by the numerical coefficient of x to solve.

Step 6: Check the answer in the **original equation.**

PRACTICE PROBLEM 8

Solve: $\frac{x}{5} - x = \frac{1}{5}$

EXAMPLE 8 Solve: $\frac{z}{5} - \frac{z}{3} = 6$

Solution: $\frac{z}{5} - \frac{z}{3} = 6$

$$15\left(\frac{z}{5} - \frac{z}{3}\right) = 15(6) \quad \text{Multiply both sides by the LCD, 15.}$$

$$\overset{3}{\cancel{15}}\left(\frac{z}{\underset{1}{\cancel{5}}}\right) - \overset{5}{\cancel{15}}\left(\frac{z}{\underset{1}{\cancel{3}}}\right) = 15(6) \quad \text{Apply the distributive property.}$$

$$3z - 5z = 90 \quad \text{Simplify.}$$

$$-2z = 90 \quad \text{Combine like terms.}$$

$$\frac{-2z}{-2} = \frac{90}{-2} \quad \text{Divide both sides by } -2\text{, the coefficient of } z.$$

$$z = -45 \quad \text{Simplify.}$$

Helpful Hint Don't forget to multiply *both* sides of the equation by the LCD.

To check, replace z with -45 in the **original equation** to see that a true statement results.

Work Practice Problem 8

Answer

8. $-\frac{1}{4}$

EXAMPLE 9 Solve: $\frac{x}{2} = \frac{x}{3} + \frac{1}{2}$

Solution: First multiply both sides by the LCD, 6.

$$\frac{x}{2} = \frac{x}{3} + \frac{1}{2}$$

$$6\left(\frac{x}{2}\right) = 6\left(\frac{x}{3} + \frac{1}{2}\right) \quad \text{Multiply both sides by the LCD, 6.}$$

$$\overset{3}{\cancel{6}}\left(\frac{x}{\underset{1}{\cancel{2}}}\right) = \overset{2}{\cancel{6}}\left(\frac{x}{\underset{1}{\cancel{3}}}\right) + \overset{3}{\cancel{6}}\left(\frac{1}{\underset{1}{\cancel{2}}}\right) \quad \text{Apply the distributive property.}$$

$$3x = 2x + 3 \quad \text{Simplify.}$$

$$3x + (-2x) = 2x + 3 + (-2x) \quad \text{Add } (-2x) \text{ to both sides.}$$

$$x = 3 \quad \text{Simplify.}$$

To check, replace x with 3 in the original equation to see that a true statement results.

Work Practice Problem 9

PRACTICE PROBLEM 9

Solve: $\frac{y}{2} = \frac{y}{5} + \frac{3}{2}$

Objective C Review of Adding and Subtracting Fractions

Make sure you understand the difference between **solving an equation** containing fractions and **adding or subtracting two fractions.** To solve an equation containing fractions, we use the multiplication property of equality and multiply both sides by the LCD of the fractions, thus eliminating the fractions. This method does not apply to adding or subtracting fractions. The multiplication property of equality applies only to equations. To add or subtract unlike fractions, we write each fraction as an equivalent fraction using the LCD of the fractions as the denominator. See the next example for a review.

EXAMPLE 10 Add: $\frac{x}{3} + \frac{2}{5}$

Solution: This expression is not an equation. Here, we are adding two unlike fractions. To add unlike fractions, we need to find the LCD. The LCD of the denominators 3 and 5 is 15. Write each fraction as an equivalent fraction with a denominator of 15.

$$\frac{x}{3} + \frac{2}{5} = \frac{x}{3} \cdot \frac{5}{5} + \frac{2}{5} \cdot \frac{3}{3} = \frac{x \cdot 5}{3 \cdot 5} + \frac{2 \cdot 3}{5 \cdot 3}$$

$$= \frac{5x}{15} + \frac{6}{15}$$

$$= \frac{5x + 6}{15}$$

Work Practice Problem 10

PRACTICE PROBLEM 10

Subtract: $\frac{9}{10} - \frac{y}{3}$

✔ **Concept Check** Which of the following are equations and which are expressions?

a. $\frac{1}{2} + 3x = 5$ **b.** $\frac{2}{3}x - \frac{x}{5}$

c. $\frac{x}{12} + \frac{5x}{24}$ **d.** $\frac{x}{5} = \frac{1}{10}$

Answers

9. 5 **10.** $\frac{27 - 10y}{30}$

✔ **Concept Check Answers**

equations: a, d; expressions: b, c

4.8 EXERCISE SET

FOR EXTRA HELP

Student Solutions Manual

PH Math/Tutor Center

CD/Video for Review

MathXL®

MyMathLab

Objective A *Solve each equation. Check your proposed solution. See Example 1.*

1. $x + \frac{1}{3} = -\frac{1}{3}$

2. $x + \frac{1}{9} = -\frac{7}{9}$

3. $y - \frac{3}{13} = -\frac{2}{13}$

4. $z - \frac{5}{14} = \frac{4}{14}$

5. $3x - \frac{1}{5} - 2x = \frac{1}{5} + \frac{2}{5}$

6. $5x + \frac{1}{11} - 4x = \frac{2}{11} - \frac{5}{11}$

7. $x - \frac{1}{12} = \frac{5}{6}$

8. $y - \frac{8}{9} = \frac{1}{3}$

9. $\frac{2}{5} + y = -\frac{3}{10}$

10. $\frac{1}{2} + a = -\frac{3}{8}$

11. $7z + \frac{1}{16} - 6z = \frac{3}{4}$

12. $9x - \frac{2}{7} - 8x = \frac{11}{14}$

13. $-\frac{2}{9} = x - \frac{5}{6}$

14. $-\frac{1}{4} = y - \frac{7}{10}$

Solve each equation. See Examples 2 through 5.

15. $7x = 2$

16. $-5x = 4$

17. $\frac{1}{4}x = 3$

18. $\frac{1}{3}x = 6$

19. $\frac{2}{9}y = -6$

20. $\frac{4}{7}x = -8$

21. $-\frac{4}{9}z = -\frac{3}{2}$

22. $-\frac{11}{10}x = -\frac{2}{7}$

23. $7a = \frac{1}{3}$

24. $2z = -\frac{5}{12}$

25. $-3x = -\frac{6}{11}$

26. $-4z = -\frac{12}{25}$

Objective B *Solve each equation. See Examples 6 through 9.*

27. $\frac{5}{9}x = -\frac{3}{18}$

28. $\frac{3}{5}y = -\frac{7}{20}$

29. $\frac{x}{3} + 2 = \frac{7}{3}$

30. $\frac{x}{5} - 1 = \frac{7}{5}$

31. $\frac{x}{5} - \frac{5}{10} = 1$

32. $\frac{x}{3} - x = -6$

33. $\frac{1}{2} - \frac{3}{5} = \frac{x}{10}$

34. $\frac{2}{3} - \frac{1}{4} = \frac{x}{12}$

35. $\frac{x}{3} = \frac{x}{5} - 2$

36. $\frac{a}{2} = \frac{a}{7} + \frac{5}{2}$

Objective **C** *Add or subtract as indicated. See Example 10.*

37. $\frac{x}{7} - \frac{4}{3}$

38. $-\frac{5}{9} + \frac{y}{8}$

39. $\frac{y}{2} + 5$

40. $2 + \frac{7x}{3}$

41. $\frac{3x}{10} + \frac{x}{6}$

42. $\frac{9x}{8} - \frac{5x}{6}$

Objectives **A B C** **Mixed Practice** *Solve. If no equation is given, perform the indicated operation.*

43. $\frac{3}{8}x = \frac{1}{2}$

44. $\frac{2}{5}y = \frac{3}{10}$

45. $\frac{2}{3} - \frac{x}{5} = \frac{4}{15}$

46. $\frac{4}{5} + \frac{x}{4} = \frac{21}{20}$

47. $\frac{9}{14}z = \frac{27}{20}$

48. $\frac{5}{16}a = \frac{5}{6}$

49. $-3m - 5m = \frac{4}{7}$

50. $30n - 34n = \frac{3}{20}$

51. $\frac{x}{4} + 1 = \frac{1}{4}$

52. $\frac{y}{7} - 2 = \frac{1}{7}$

53. $\frac{5}{9} - \frac{2}{3}$

54. $\frac{8}{11} - \frac{1}{2}$

55. $\frac{1}{5}y = 10$

56. $\frac{1}{4}x = -2$

57. $\frac{5}{7}y = -\frac{15}{49}$

58. $-\frac{3}{4}x = \frac{9}{2}$

59. $\frac{x}{2} - x = -2$

60. $\frac{y}{3} = -4 + y$

61. $-\frac{5}{8}y = \frac{3}{16} - \frac{9}{16}$

62. $-\frac{7}{9}x = -\frac{5}{18} - \frac{4}{18}$

63. $17x - 25x = \frac{1}{3}$

64. $27x - 30x = \frac{4}{9}$

65. $\frac{7}{6}x = \frac{1}{4} - \frac{2}{3}$

66. $\frac{5}{4}y = \frac{1}{2} - \frac{7}{10}$

67. $\frac{b}{4} = \frac{b}{12} + \frac{2}{3}$

68. $\frac{a}{6} = \frac{a}{3} + \frac{1}{2}$

69. $\frac{x}{3} + 2 = \frac{x}{2} + 8$

70. $\frac{y}{5} - 2 = \frac{y}{3} - 4$

Review

Round each number to the given place value. See Section 1.4.

71. 57,236 to the nearest hundred

72. 576 to the nearest hundred

73. 327 to the nearest ten

74. 2333 to the nearest ten

Concept Extensions

75. Explain why the method for eliminating fractions in an **equation** does not apply to simplifying **expressions** containing fractions.

Solve.

76. $\frac{14}{11} + \frac{3x}{8} = \frac{x}{2}$

77. $\frac{19}{53} = \frac{353x}{1431} + \frac{23}{27}$

78. The area of the rectangle is $\frac{5}{12}$ square inches. Find its length, x.

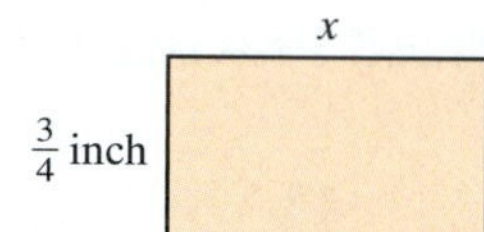

CHAPTER 4 Group Activity

Lobster Classification

Sections 4.1, 4.7, 4.8

This activity may be completed by working in groups or individually.

Lobsters are normally classified by weight. Use the weight classification table to answer the questions in this activity.

Classification of Lobsters	
Class	**Weight (in Pounds)**
Chicken	1 to $1\frac{1}{8}$
Quarter	$1\frac{1}{4}$
Half	$1\frac{1}{2}$ to $1\frac{3}{4}$
Select	$1\frac{3}{4}$ to $2\frac{1}{2}$
Large select	$2\frac{1}{2}$ to $3\frac{1}{2}$
Jumbo	Over $3\frac{1}{2}$
(*Source:* The Maine Lobster Promotion Council)	

1. A lobster fisher has kept four lobsters from a lobster trap. Classify each lobster if they have the following weights:

a. $1\frac{7}{8}$ pounds

b. $1\frac{9}{16}$ pounds

c. $2\frac{3}{4}$ pounds

d. $2\frac{3}{8}$ pounds

2. A recipe requires 5 pounds of lobster. Using the minimum weight for each class, decide whether a chicken, half, and select lobster will be enough for the recipe, and explain your reasoning. If not, suggest a better choice of lobsters to meet the recipe requirements.

3. A lobster market customer has selected two chickens, a select, and a large select. What is the most that these four lobsters could weigh? What is the least that these four lobsters could weigh?

4. A lobster market customer wishes to buy three quarters. If lobsters sell for \$7 per pound, how much will the customer owe for her purchase?

5. Why do you think there is no classification for lobsters weighing under 1 pound?

STUDY SKILLS BUILDER

Tips for Studying for an Exam

To prepare for an exam, try the following study techniques:

- Start the study process days before your exam.
- Make sure that you are up-to-date on your assignments.
- If there is a topic that you are unsure of, use one of the many resources that are available to you. For example,

  See your instructor.

  Visit a learning resource center on campus.

  Read the textbook material and examples on the topic.

  View a video on the topic.

- Reread your notes and carefully review the Chapter Highlights at the end of any chapter.
- Work the review exercises at the end of the chapter. Check your answers and correct any mistakes. If you have trouble, use a resource listed above.
- Find a quiet place to take the Chapter Test found at the end of the chapter. Do not use any resources when taking this sample test. This way, you will have a clear indication of how prepared you are for your exam. Check your answers and make sure that you correct any missed exercises.
- Get lots of rest the night before the exam. It's hard to show how well you know the material if your brain is foggy from lack of sleep.

Good luck, and keep a positive attitude.

Let's see how you did on your last exam.

1. How many days before your last exam did you start studying for that exam?
2. Were you up-to-date on your assignments at that time or did you need to catch up on assignments?
3. List the most helpful text supplement (if you used one).
4. List the most helpful campus supplement (if you used one).
5. List your process for preparing for a mathematics test.
6. Was this process helpful? In other words, were you satisfied with your performance on your exam?
7. If not, what changes can you make in your process that will make it more helpful to you?

Chapter 4 Vocabulary Check

Fill in each blank with one of the words or phrases listed below.

mixed number	complex fraction	like	numerator	prime factorization
composite number	equivalent	cross products	least common denominator	denominator
prime number	improper fraction	simplest form	undefined	0
reciprocals	proper fraction			

1. Two numbers are ______________ of each other if their product is 1.
2. A ______________ is a natural number greater than 1 that is not prime.
3. Fractions that represent the same portion of a whole are called ______________ fractions.
4. An ______________ is a fraction whose numerator is greater than or equal to its denominator.
5. A ______________ is a natural number greater than 1 whose only factors are 1 and itself.
6. A fraction is in ______________ when the numerator and the denominator have no factors in common other than 1.
7. A ______________ is one whose numerator is less than its denominator.
8. A ______________ contains a whole number part and a fraction part.
9. In the fraction $\frac{7}{9}$, the 7 is called the ______________ and the 9 is called the ______________.
10. The ______________ of a number is the factorization in which all the factors are prime numbers.
11. The fraction $\frac{3}{0}$ is ______________.
12. The fraction $\frac{0}{5}$ = ______________.
13. Fractions that have the same denominator are called ______________ fractions.
14. The LCM of the denominators in a list of fractions is called the ______________.
15. A fraction whose numerator or denominator or both numerator and denominator contain fractions is called a ______________.
16. In $\frac{a}{b} = \frac{c}{d}$, $a \cdot d$ and $b \cdot c$ are called ______________.

Helpful Hint

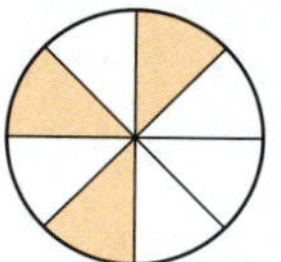

Are you preparing for your test? Don't forget to take the Chapter 4 Test on page 325. Then check your answers at the back of the text and use the Chapter Test Prep Video CD to see the fully worked-out solutions to any of the exercises you want to review.

4 Chapter Highlights

Definitions and Concepts	**Examples**
Section 4.1 Introduction to Fractions and Mixed Numbers	
A **fraction** is of the form $\frac{\text{numerator}}{\text{denominator}}$ ← number of parts being considered; ← number of equal parts in the whole	Write a fraction to represent the shaded part of the figure. $\frac{3}{8}$ ← number of parts shaded; ← number of equal parts
A fraction is called a **proper fraction** if its numerator is less than its denominator.	Proper Fractions: $\frac{1}{3}, \frac{2}{5}, \frac{7}{8}, \frac{100}{101}$
A fraction is called an **improper fraction** if its numerator is greater than or equal to its denominator.	Improper Fractions: $\frac{5}{4}, \frac{2}{2}, \frac{9}{7}, \frac{101}{100}$
A **mixed number** contains a whole number and a fraction.	Mixed Numbers: $1\frac{1}{2}, 5\frac{7}{8}, 25\frac{9}{10}$

Definitions and Concepts	**Examples**
Section 4.1 Introduction to Fractions and Mixed Numbers (*continued*)	
To Write a Mixed Number as an Improper Fraction **1.** Multiply the denominator of the fraction by the whole number. **2.** Add the numerator of the fraction to the product from Step 1. **3.** Write this sum from Step 2 as the numerator of the improper fraction over the original denominator.	$5\frac{2}{7} = \frac{5 \cdot 7 + 2}{7} = \frac{35 + 2}{7} = \frac{37}{7}$
To Write an Improper Fraction as a Mixed Number or a Whole Number **1.** Divide the denominator into the numerator. **2.** The whole number part of the mixed number is the quotient. The fraction is the remainder over the original denominator. $\text{quotient}\frac{\text{remainder}}{\text{original denominator}}$	$\frac{17}{3} = 5\frac{2}{3}$ $3\overline{)17}$ quotient 5; -15; remainder 2
Section 4.2 Factors and Simplest Form	
A **prime number** is a natural number that has exactly two different factors, 1 and itself.	2, 3, 5, 7, 11, 13, 17, . . .
A **composite number** is any natural number other than 1 that is not prime.	4, 6, 8, 9, 10, 12, 14, 15, 16, . . .
The **prime factorization** of a number is the factorization in which all the factors are prime numbers.	Write the prime factorization of 60. $60 = 6 \cdot 10$ $= 2 \cdot 3 \cdot 2 \cdot 5$ or $2^2 \cdot 3 \cdot 5$
Fractions that represent the same portion of a whole are called **equivalent fractions.**	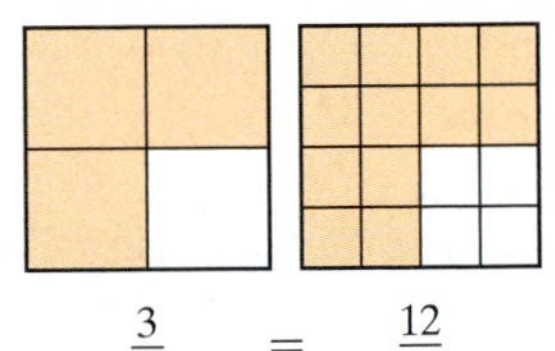 $\frac{3}{4} = \frac{12}{16}$
A fraction is in **simplest form** or **lowest terms** when the numerator and the denominator have no common factors other than 1.	The fraction $\frac{2}{3}$ is in simplest form.
To write a fraction in simplest form, write the prime factorizations of the numerator and the denominator and then divide both by all common factors.	Write in simplest form: $\frac{30}{36}$ $\frac{30}{36} = \frac{2 \cdot 3 \cdot 5}{2 \cdot 2 \cdot 3 \cdot 3} = \frac{2}{2} \cdot \frac{3}{3} \cdot \frac{5}{2 \cdot 3} = 1 \cdot 1 \cdot \frac{5}{6} = \frac{5}{6}$ or $\frac{30}{36} = \frac{\overset{1}{\cancel{2}} \cdot \overset{1}{\cancel{3}} \cdot 5}{\underset{1}{\cancel{2}} \cdot 2 \cdot \underset{1}{\cancel{3}} \cdot 3} = \frac{5}{6}$

DEFINITIONS AND CONCEPTS	**EXAMPLES**
Section 4.2 Factors and Simplest Form (*continued*)	
Two fractions are equivalent if **Method 1.** They simplify to the same fraction. **Method 2.** Their cross products are equal. $24 \cdot 7 = 168$ $\quad \frac{7}{8} = \frac{21}{24} \quad$ $8 \cdot 21 = 168$ Since $168 = 168, \frac{7}{8} = \frac{21}{24}$	Determine whether $\frac{7}{8}$ and $\frac{21}{24}$ are equivalent. $\frac{7}{8}$ is in simplest form $\frac{21}{24} = \frac{\overset{1}{\cancel{3}} \cdot 7}{\underset{1}{\cancel{3}} \cdot 8} = \frac{1 \cdot 7}{1 \cdot 8} = \frac{7}{8}$ Since both simplify to $\frac{7}{8}$, then $\frac{7}{8} = \frac{21}{24}$.
Section 4.3 Multiplying and Dividing Fractions	
To multiply two fractions, multiply the numerators and multiply the denominators.	Multiply. $\frac{2x}{3} \cdot \frac{5}{7} = \frac{2x \cdot 5}{3 \cdot 7} = \frac{10x}{21}$ $\frac{3}{4} \cdot \frac{1}{6} = \frac{3 \cdot 1}{4 \cdot 6} = \frac{\overset{1}{\cancel{3}} \cdot 1}{4 \cdot \underset{1}{\cancel{3}} \cdot 2} = \frac{1}{8}$
To find the **reciprocal** of a fraction, interchange its numerator and denominator.	The reciprocal of $\frac{3}{5}$ is $\frac{5}{3}$.
To divide two fractions, multiply the first fraction by the reciprocal of the second fraction.	Divide. $-\frac{3}{10} \div \frac{7}{9} = -\frac{3}{10} \cdot \frac{9}{7} = -\frac{3 \cdot 9}{10 \cdot 7} = -\frac{27}{70}$
Section 4.4 Adding and Subtracting Like Fractions, Least Common Denominator, and Equivalent Fractions	
Fractions that have the same denominator are called **like fractions.**	$-\frac{1}{3}$ and $\frac{2}{3}$; $\frac{5x}{7}$ and $\frac{6}{7}$
To add or subtract like fractions, combine the numerators and place the sum or difference over the common denominator.	$\frac{2}{7} + \frac{3}{7} = \frac{5}{7}$ ← Add the numerators. ← Keep the common denominator. $\frac{7}{8} - \frac{4}{8} = \frac{3}{8}$ ← Subtract the numerators. ← Keep the common denominator.
The least common denominator (LCD) of a list of fractions is the smallest positive number divisible by all the denominators in the list.	The LCD of $\frac{1}{2}$ and $\frac{5}{6}$ is 6 because 6 is the smallest positive number that is divisible by both 2 and 6.

Definitions and Concepts	Examples
Section 4.4 Adding and Subtracting Like Fractions, Least Common Denominator, and Equivalent Fractions	
Method 1 for Finding the LCD of a List of Fractions Using Multiples **Step 1.** Write the multiples of the largest denominator (starting with the number itself) until a multiple common to all denominators in the list is found. **Step 2.** The multiple found in Step 1 is the LCD.	Find the LCD of $\frac{1}{4}$ and $\frac{5}{6}$ using Method 1. $6 \cdot 1 = 6$ Not a multiple of 4 $6 \cdot 2 = 12$ A multiple of 4 The LCD is 12.
Method 2 for Finding the LCD of a List of Fractions Using Prime Factorization **Step 1.** Write the prime factorization of each denominator. **Step 2.** For each different prime factor in Step 1, circle the greatest number of times that factor occurs in any one factorization. **Step 3.** The LCD is the product of the circled factors.	Find the LCD of $\frac{5}{6}$ and $\frac{11}{20}$ using Method 2. $6 = 2 \cdot \textcircled{3}$ $20 = \boxed{2 \cdot 2} \cdot \textcircled{5}$ The LCD is $2 \cdot 2 \cdot 3 \cdot 5 = 60$
Equivalent fractions represent the same portion of a whole.	Write an equivalent fraction with the indicated denominator. $\frac{2}{8} = \frac{}{16}$ $\frac{2 \cdot 2}{8 \cdot 2} = \frac{4}{16}$
Section 4.5 Adding and Subtracting Unlike Fractions	
To Add or Subtract Fractions with Unlike Denominators **Step 1.** Find the LCD.	Add: $\frac{3}{20} + \frac{2}{5}$ **Step 1.** The LCD of 20 and 5 is 20.
Step 2. Write each fraction as an equivalent fraction whose denominator is the LCD.	**Step 2.** $\frac{3}{20} = \frac{3}{20}$; $\frac{2}{5} = \frac{2}{5} \cdot \frac{4}{4} = \frac{8}{20}$
Step 3. Add or subtract the like fractions.	**Step 3.** $\frac{3}{20} + \frac{2}{5} = \frac{3}{20} + \frac{8}{20} = \frac{11}{20}$
Step 4. Write the sum or difference in simplest form.	**Step 4.** $\frac{11}{20}$ is in simplest form.
Section 4.6 Complex Fractions and Review of Order of Operations	
A fraction whose numerator or denominator or both contain fractions is called a **complex fraction.**	Complex Fractions: $\frac{\frac{11}{4}}{\frac{7}{10}}, \frac{\frac{y}{6} - 11}{\frac{4}{3}}$

Definitions and Concepts	Examples
Section 4.6 Complex Fractions and Review of Order of Operations (*continued*)	
One method for simplifying complex fractions is to multiply the numerator and the denominator of the complex fraction by the LCD of all fractions in its numerator and its denominator.	$\dfrac{\frac{y}{6}-11}{\frac{4}{3}} = \dfrac{6\left(\frac{y}{6}-11\right)}{6\left(\frac{4}{3}\right)} = \dfrac{6\left(\frac{y}{6}\right)-6(11)}{6\left(\frac{4}{3}\right)}$ $= \dfrac{y-66}{8}$
Section 4.7 Operations on Mixed Numbers	
To multiply with mixed numbers or whole numbers, first write any mixed or whole numbers as improper fractions and then multiply as usual.	$2\frac{1}{3}\cdot\frac{1}{9} = \frac{7}{3}\cdot\frac{1}{9} = \frac{7\cdot 1}{3\cdot 9} = \frac{7}{27}$
To divide with mixed numbers or whole numbers, first write any mixed or whole numbers as fractions and then divide as usual.	$2\frac{5}{8} \div 3\frac{7}{16} = \frac{21}{8} \div \frac{55}{16} = \frac{21}{8}\cdot\frac{16}{55} = \frac{21\cdot 16}{8\cdot 55}$ $= \frac{21\cdot 2\cdot \overset{1}{\cancel{8}}}{\underset{1}{\cancel{8}}\cdot 55} = \frac{42}{55}$
To add or subtract with mixed numbers, add or subtract the fractions and then add or subtract the whole numbers.	Add: $2\frac{1}{2} + 5\frac{7}{8}$ $2\frac{1}{2} = 2\frac{4}{8}$ $+5\frac{7}{8} = 5\frac{7}{8}$ $7\frac{11}{8} = 7 + 1\frac{3}{8} = 8\frac{3}{8}$
Section 4.8 Solving Equations Containing Fractions	
To Solve an Equation in *x*	
Step 1. If fractions are present, multiply both sides of the equation by the LCD of the fractions.	Solve: $\frac{x}{15} + 2 = \frac{7}{3}$
Step 2. If parentheses are present, use the distributive property.	$15\left(\frac{x}{15} + 2\right) = 15\left(\frac{7}{3}\right)$ Multiply by the LCD 15.
Step 3. Combine any like terms on each side of the equation.	$15\left(\frac{x}{15}\right) + 15\cdot 2 = 15\left(\frac{7}{3}\right)$
Step 4. Use the addition property of equality to rewrite the equation so that variable terms are on one side of the equation and constant terms are on the other side.	$x + 30 = 35$ $x + 30 + (-30) = 35 + (-30)$ $x = 5$
Step 5. Divide both sides by the numerical coefficient of x to solve.	
Step 6. Check the answer in the *original equation*.	Check to see that 5 is the solution.

STUDY SKILLS BUILDER

Are you Prepared for a Test on Chapter 4?

Below I have listed some *common trouble areas* for students in Chapter 4. After studying for your test—but before taking your test—read these.

Make sure you remember how to perform different operations on fractions!!! Try to add, subtract, multiply, then divide $\frac{3}{5}$ and $\frac{7}{15}$. Check your results below.

$$\frac{3}{5}+\frac{7}{15}=\frac{3}{5}\cdot\frac{3}{3}+\frac{7}{15}=\frac{9}{15}+\frac{7}{15}=\frac{16}{15} \text{ or } 1\frac{1}{15}$$

To add or subtract, the fractions must have a common denominator.

$$\frac{3}{5}-\frac{7}{15}=\frac{3}{5}\cdot\frac{3}{3}-\frac{7}{15}=\frac{9}{15}-\frac{7}{15}=\frac{2}{15}$$

$$\frac{3}{5}\cdot\frac{7}{15}=\frac{3\cdot 7}{5\cdot 15}=\frac{\overset{1}{\cancel{3}}\cdot 7}{5\cdot\underset{1}{\cancel{3}}\cdot 5}=\frac{7}{25}$$

To multiply, multiply numerators and multiply denominators.

$$\frac{3}{5}\div\frac{7}{15}=\frac{3}{5}\cdot\frac{15}{7}=\frac{3\cdot 15}{5\cdot 7}=\frac{3\cdot 3\cdot\overset{1}{\cancel{5}}}{\underset{1}{\cancel{5}}\cdot 7}=\frac{9}{7} \text{ or } 1\frac{2}{7}$$

To divide, multiply by the reciprocal.

4 CHAPTER REVIEW

(4.1) *Write a fraction to represent the shaded area. If the fraction is improper, write the shaded area as a mix number, also.*

1. 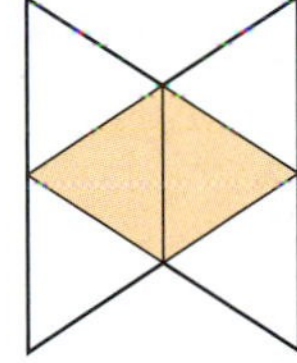

2. 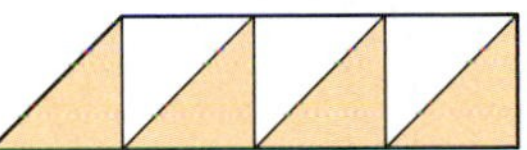

3. 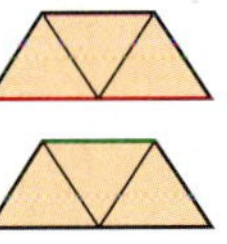
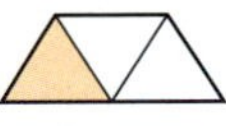

4. 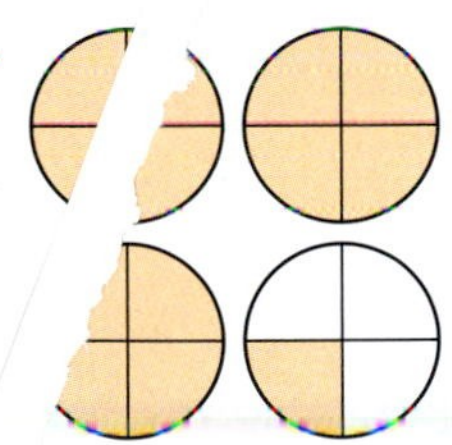

5. A basketball player made 11 free throws out of 12 during a game. What fraction of free throws did the player make?

6. A new car lot contai 23 blue cars out of a total of 131 cars.

a. How many cars the lot are not blue?
b. What fraction cars on the lot are not blue?

Simplify by dividing.

7. $\frac{3}{-3}$

8. $\frac{-20}{-20}$

9. $\frac{0}{-1}$

10. $\frac{4}{0}$

Graph each fraction on a number line.

11. $\frac{7}{9}$

12. $\frac{4}{7}$

13. $\frac{5}{4}$

14. $\frac{7}{5}$

Write each improper fraction as a mixed number or a whole number.

15. $\frac{15}{4}$

16. $\frac{39}{13}$

Write each mixed number as an improper fraction.

17. $2\frac{1}{5}$

18. $3\frac{8}{9}$

(4.2) *Write each fraction in simplest form.*

19. $\frac{12}{28}$

20. $\frac{15}{27}$

21. $-\frac{25x}{75x^2}$

22. $-\frac{36y^3}{72y}$

23. $\frac{29ab}{32abc}$

24. $\frac{18xyz}{23xy}$

25. $\frac{45x^2y}{27xy^3}$

26. $\frac{42ab^2c}{30abc^3}$

27. There are 12 inches in a foot. What fractional part of a foot does 8 inches represent?

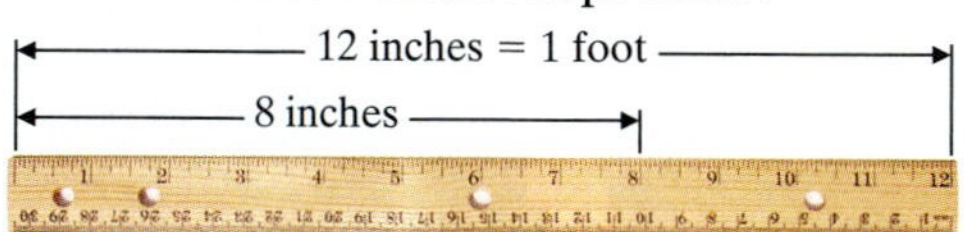

28. Six out of 15 cars are white. What fraction of cars are *not* white?

Determine whether each two fractions are equivalent.

29. $\frac{10}{34}$ and $\frac{4}{14}$

30. $\frac{30}{50}$ and $\frac{9}{15}$

(4.3) *Multiply.*

31. $\frac{3}{5} \cdot \frac{1}{2}$

32. $-\frac{6}{7} \cdot \frac{5}{12}$

33. $-\frac{24x}{5} \cdot -\frac{15}{8x^3}$

34. $\frac{27y^3}{21} \cdot \frac{7}{18y^2}$

35. $\left(-\frac{1}{3}\right)^3$

36. $\left(-\frac{5}{12}\right)^2$

Divide.

37. $-\frac{3}{4} \div \frac{3}{8}$

38. $\frac{21a}{4} \div \frac{7a}{5}$

39. $-\frac{9}{2} \div -\frac{1}{3}$

40. $-\frac{5}{3} \div 2y$

41. Evaluate $x \div y$ if $x = \frac{9}{7}$ and $y = \frac{3}{4}$.

42. Evaluate ab if $a = -7$ and $b = \frac{9}{10}$.

Find the area of each figure.

△ **43.**

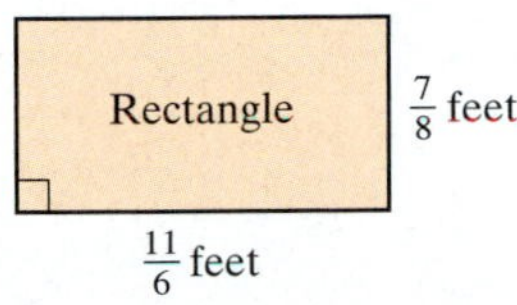

△ **44.**

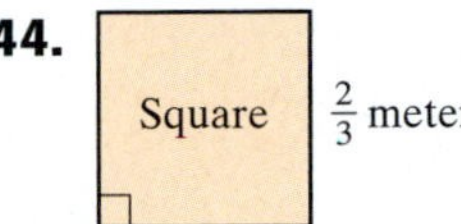

(4.4) *Add or subtract as indicated.*

45. $\frac{7}{11} + \frac{3}{11}$

46. $\frac{4}{9} + \frac{2}{9}$

47. $\frac{1}{12} - \frac{5}{12}$

48. $\frac{11x}{15} + \frac{x}{15}$

49. $\frac{4y}{21} - \frac{3}{21}$

50. $\frac{4}{15} - \frac{3}{15} - \frac{2}{15}$

Find the LCD of each list of fractions.

51. $\frac{2}{3}, \frac{5}{x}$

52. $\frac{3}{4}, \frac{3}{8}, \frac{7}{12}$

Write each fraction as an equivalent fraction with the given denominator.

53. $\frac{2}{3} = \frac{?}{30}$

54. $\frac{5}{8} = \frac{?}{56}$

55. $\frac{7a}{6} = \frac{?}{42}$

56. $\frac{9b}{4} = \frac{?}{20}$

57. $\frac{4}{5x} = \frac{?}{50x}$

58. $\frac{5}{9y} = \frac{?}{18y}$

Solve.

59. One evening Mark Alorenzo did $\frac{3}{8}$ of his homework before supper, another $\frac{2}{8}$ of it while his children did their homework, and $\frac{1}{8}$ after his children went to bed. What part of his homework did he do that evening?

△ **60.** The Simpsons will be fencing in their land, which is in the shape of a rectangle. In order to do this, they need to find its perimeter. Find the perimeter of their land.

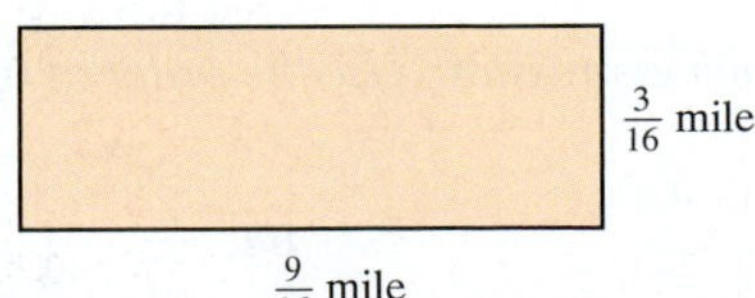

(4.5) *Add or subtract as indicated.*

61. $\frac{7}{18} + \frac{2}{9}$

62. $\frac{4}{13} - \frac{1}{26}$

63. $-\frac{1}{3} + \frac{1}{4}$

64. $-\frac{2}{3} + \frac{1}{4}$

65. $\frac{5x}{11} + \frac{2}{55}$

66. $\frac{4}{15} + \frac{b}{5}$

67. $\frac{5y}{12} - \frac{2y}{9}$

68. $\frac{7x}{18} + \frac{2x}{9}$

69. $\frac{4}{9} + \frac{5}{y}$

70. $-\frac{9}{14} - \frac{3}{7}$

71. $\frac{4}{25} + \frac{23}{75} + \frac{7}{50}$

72. $\frac{2}{3} - \frac{2}{9} - \frac{1}{6}$

Find the perimeter of each figure.

△ **73.**

△ **74.**

75. In a group of 100 blood donors, typically $\frac{9}{25}$ have type A Rh-positive blood and $\frac{3}{50}$ have type A Rh-negative blood. What fraction have type A blood?

76. Find the difference in length of two scarves if one scarf is $\frac{5}{12}$ of a yard long and the other is $\frac{2}{3}$ of a yard long.

(4.6) *Simplify each complex fraction.*

77. $\dfrac{\frac{2x}{5}}{\frac{7}{10}}$

78. $\dfrac{\frac{3y}{7}}{\frac{11}{7}}$

79. $\dfrac{\frac{2}{5} - \frac{1}{2}}{\frac{3}{4} - \frac{7}{10}}$

80. $\dfrac{\frac{5}{6} - \frac{1}{4}}{\frac{-1}{12y}}$

Evaluate each expression if $x = \frac{1}{2}$, $y = -\frac{2}{3}$, *and* $z = \frac{4}{5}$.

81. $\frac{x}{y + z}$

82. $\frac{x + y}{z}$

Evaluate each expression. Use the order of operations to simplify.

83. $\frac{5}{13} \div \frac{1}{2} \cdot \frac{4}{5}$

84. $\frac{2}{27} - \left(\frac{1}{3}\right)^2$

85. $\frac{9}{10} \cdot \frac{1}{3} - \frac{2}{5} \cdot \frac{1}{11}$

86. $-\frac{2}{7} \cdot \left(\frac{1}{5} + \frac{3}{10}\right)$

(4.7) *Perform operations as indicated. Simplify your answers. Estimate where noted.*

87. $\begin{array}{r} 7\frac{3}{8} \\ 9\frac{5}{6} \\ +3\frac{1}{12} \\ \hline \end{array}$

88. $\begin{array}{r} 8\frac{1}{5} \\ -5\frac{3}{11} \\ \hline \end{array}$

Exact:

Estimate:

89. $1\frac{5}{8} \cdot 3\frac{1}{5}$

Exact:

Estimate:

90. $6\frac{3}{4} \div 1\frac{2}{7}$

91. A truck traveled 341 miles on $15\frac{1}{2}$ gallons of gas. How many miles might we expect the truck to travel on 1 gallon of gas?

92. There are $7\frac{1}{3}$ grams of fat in each ounce of hamburger. How many grams of fat are in a 5-ounce hamburger patty?

Find the unknown measurements.

 **93.**

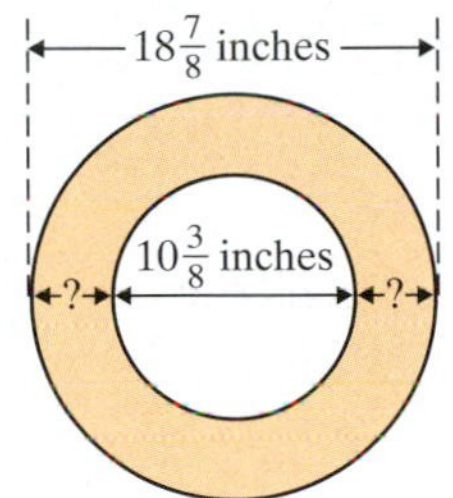

△ **94.**

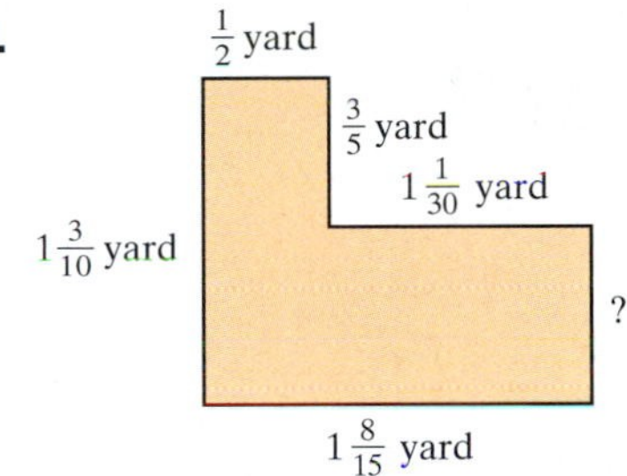

Perform the indicated operations.

95. $-12\frac{1}{7} + \left(-15\frac{3}{14}\right)$

96. $-3\frac{1}{5} \div \left(-2\frac{7}{10}\right)$

97. $-2\frac{1}{4} \cdot 1\frac{3}{4}$

98. $23\frac{7}{8} - 24\frac{7}{10}$

(4.8) *Solve each equation.*

99. $a - \frac{2}{3} = \frac{1}{6}$

100. $9x + \frac{1}{5} - 8x = -\frac{7}{10}$

101. $-\frac{3}{5}x = 6$

102. $\frac{2}{9}y = -\frac{4}{3}$

103. $\frac{x}{7} - 3 = -\frac{6}{7}$

104. $\frac{y}{5} + 2 = \frac{11}{5}$

105. $\frac{1}{6} + \frac{x}{4} = \frac{17}{12}$

106. $\frac{x}{5} - \frac{5}{4} = \frac{x}{2} - \frac{1}{20}$

Mixed Review

Perform indicated operations. Write each answer in simplest form. Estimate where noted.

107. $\frac{6}{15} \cdot \frac{5}{8}$

108. $\frac{5x^2}{y} \div \frac{10x^3}{y^3}$

109. $\frac{3}{10} - \frac{1}{10}$

110. $\frac{7}{8x} \cdot -\frac{2}{3}$

111. $\frac{2x}{3} + \frac{x}{4}$

112. $-\frac{5}{11} + \frac{2}{55}$

113. $-1\frac{3}{5} \div \frac{1}{4}$

114. $2\frac{7}{8} + 9\frac{1}{2}$
Exact:
Estimate:

115. $12\frac{1}{7} - 9\frac{3}{5}$
Exact:
Estimate:

116. Simplify: $\dfrac{2 + \frac{3}{4}}{1 - \frac{1}{8}}$

Solve.

117. Evaluate: $-\frac{3}{8} \cdot \left(\frac{2}{3} - \frac{4}{9}\right)$

118. $11x - \frac{2}{7} - 10x = -\frac{13}{14}$

119. $-\frac{3}{5}x = \frac{4}{15}$

120. $\frac{x}{12} + \frac{5}{6} = \frac{3}{4}$

Solve.

121. A ribbon $5\frac{1}{2}$ yards long is cut from a reel of ribbon with 50 yards on it. Find the length of the piece remaining on the reel.

△ **122.** A slab of natural granite is purchased and a rectangle with length $5\frac{1}{2}$ feet and width $7\frac{4}{11}$ feet is cut from it. Find the area of the rectangle.

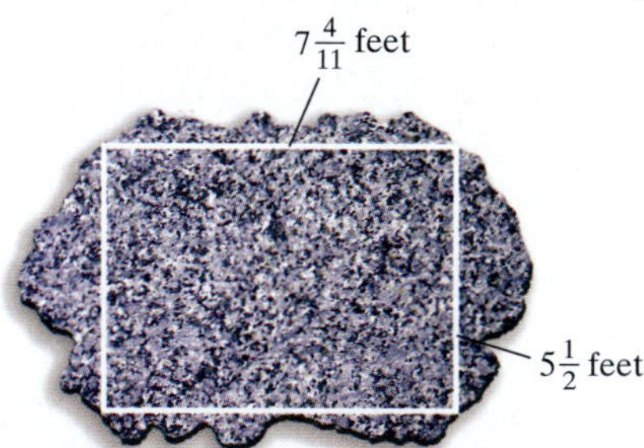

4 CHAPTER TEST

Remember to use the Chapter Test Prep Video CD to see the fully worked-out solutions to any of the exercises you want to review.

Write a fraction to represent the shaded area.

1.

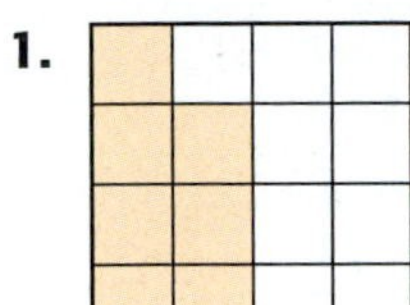

Write the mixed number as an improper fraction.

2. $7\frac{2}{3}$

Write the improper fraction as a mixed number.

3. $\frac{75}{4}$

Write each fraction in simplest form.

4. $\frac{24}{210}$

5. $-\frac{42x}{70}$

Determine whether these fractions are equivalent.

6. $\frac{5}{7}$ and $\frac{8}{11}$

7. $\frac{6}{27}$ and $\frac{14}{63}$

Find the prime factorization of each number.

8. 84

9. 495

Perform each indicated operation and write the answers in simplest form.

10. $\frac{4}{4} \div \frac{3}{4}$

11. $-\frac{4}{3} \cdot \frac{4}{4}$

12. $\frac{7x}{9} + \frac{x}{9}$

13. $\frac{1}{7} - \frac{3}{x}$

14. $\frac{xy^3}{z} \cdot \frac{z}{xy}$

15. $-\frac{2}{3} \cdot -\frac{8}{15}$

16. $\frac{9a}{10} + \frac{2}{5}$

17. $-\frac{8}{15y} - \frac{2}{15y}$

18. $\frac{3a}{8} \cdot \frac{16}{6a^3}$

19. $\frac{11}{12} - \frac{3}{8} + \frac{5}{24}$

20.
$$\begin{array}{r} 3\frac{7}{8} \\ 7\frac{2}{5} \\ +2\frac{3}{4} \\ \hline \end{array}$$

21.
$$\begin{array}{r} 19 \\ -2\frac{3}{11} \\ \hline \end{array}$$

Answers

1. ______
2. ______
3. ______
4. ______
5. ______
6. ______
7. ______
8. ______
9. ______
10. ______
11. ______
12. ______
13. ______
14. ______
15. ______
16. ______
17. ______
18. ______
19. ______
20. ______
21. ______

22. ______
23. ______
24. ______
25. ______
26. ______
27. ______
28. ______
29. ______
30. ______
31. ______
32. ______
33. ______
34. ______
35. ______
36. ______
37. ______
38. ______
39. ______
40. ______

22. $-\frac{16}{3} \div -\frac{3}{12}$

23. $3\frac{1}{3} \cdot 6\frac{3}{4}$

24. $-\frac{2}{7} \cdot \left(6 - \frac{1}{6}\right)$

25. $\frac{1}{2} \div \frac{2}{3} \cdot \frac{3}{4}$

26. $\left(-\frac{3}{4}\right)^2 \div \left(\frac{2}{3} + \frac{5}{6}\right)$

27. Find the average of $\frac{5}{6}$, $\frac{4}{3}$, and $\frac{7}{12}$.

Simplify each complex fraction.

28. $\dfrac{\frac{5x}{7}}{\frac{20x^2}{21}}$

29. $\dfrac{5 + \frac{3}{7}}{2 - \frac{1}{2}}$

Solve.

30. $-\frac{3}{8}x = \frac{3}{4}$

31. $\frac{x}{5} + x = -\frac{24}{5}$

32. $\frac{2}{3} + \frac{x}{4} = \frac{5}{12} + \frac{x}{2}$

Evaluate each expression for the given replacement values.

33. $-5x$; $x = -\frac{1}{2}$

34. $x \div y$; $x = \frac{1}{2}$, $y = 3\frac{7}{8}$

Solve.

35. A carpenter cuts a piece $2\frac{3}{4}$ feet long from a cedar plank that is $6\frac{1}{2}$ feet long. How long is the remaining piece?

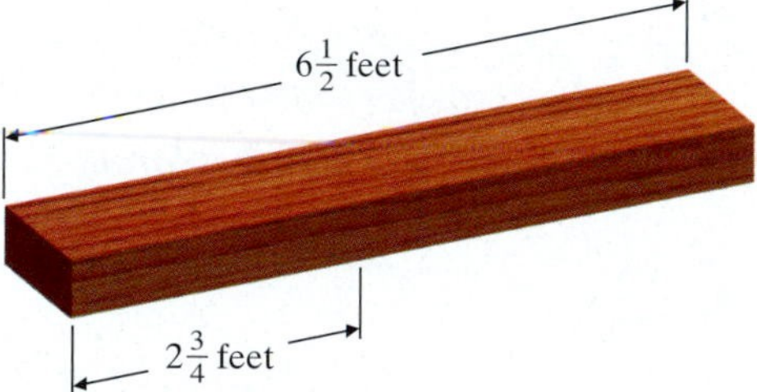

The circle graph below shows us how the average consumer spends money. For example, $\frac{7}{50}$ of your spending goes for food. Use this information for Exercises 36 through 38.

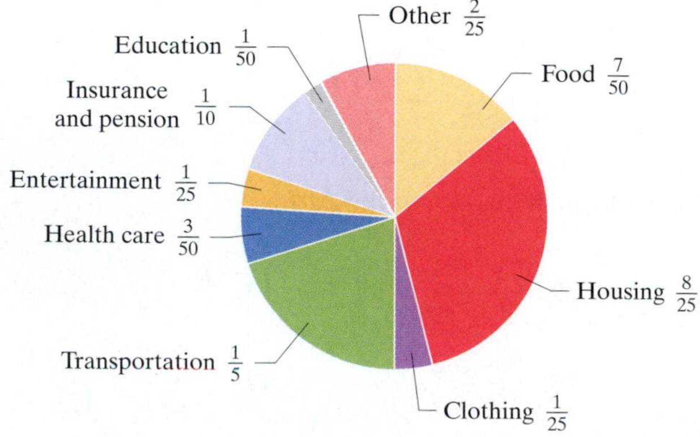

Source: U.S. Bureau of Labor Statistics; based on survey

36. What fraction of spending goes for housing and food combined?

37. What fraction of spending goes for education, transportation, and clothing?

38. Suppose your family spent \$47,000 on the items in the graph. How much might we expect was spent on health care?

Find the perimeter and area of the figure.

△ **39.**

Rectangle	$\frac{2}{3}$ foot
1 foot	

40. During a 258-mile trip, a car used $10\frac{3}{4}$ gallons of gas. How many miles would we expect the car to travel on 1 gallon of gas?

CUMULATIVE REVIEW

Chapters 1–4

Write each number in words.

1. 546

2. 115

3. 27,034

4. 6573

5. Add: $46 + 713$

6. Add: $587 + 44$

7. Subtract: $543 - 29$. Check by adding.

8. Subtract: $995 - 62$. Check by adding.

9. Round 278,362 to the nearest thousand.

10. Round 1436 to the nearest ten.

11. A digital video disc (DVD) can hold about 4800 megabytes (MB) of information. How many megabytes can 12 DVDs hold?

12. On a trip across country, Daniel Daunis travels 435 miles per day. How many total miles does he travel in 3 days?

13. Divide and check: $56{,}717 \div 8$

14. Divide and check: $4558 \div 12$

Write using exponential notation.

15. $7 \cdot 7 \cdot 7$

16. $7 \cdot 7$

17. $3 \cdot 3 \cdot 3 \cdot 3 \cdot 9 \cdot 9 \cdot 9$

18. $9 \cdot 9 \cdot 9 \cdot 9 \cdot 5 \cdot 5$

19. Evaluate $2(x - y)$ if $x = 6$ and $y = 3$.

20. Evaluate $8a + 3(b - 5)$ if $a = 5$ and $b = 9$.

21. The world's deepest cave is Krubera (or Voronja) in the country of Georgia, located by the Black Sea in Asia. It has been explored to a depth of 6824 feet below the surface of the Earth. Represent this position using an integer. (*Source: Guinness World Records* 2006)

22. The temperature on a cold day in Minneapolis, MN, is 21°F below zero. Represent this temperature using an integer.

Answers

1. ______
2. ______
3. ______
4. ______
5. ______
6. ______
7. ______
8. ______
9. ______
10. ______
11. ______
12. ______
13. ______
14. ______
15. ______
16. ______
17. ______
18. ______
19. ______
20. ______
21. ______
22. ______

23. ____
24. ____
25. ____
26. ____
27. ____
28. ____
29. ____
30. ____
31. ____
32. ____
33. ____
34. ____
35. ____
36. ____
37. ____
38. ____
39. a. ____ b. ____
40. ____
41. ____
42. ____
43. ____
44. ____
45. ____
46. ____

23. Add using a number line: $-7 + 3$

24. Add using a number line: $-3 + 8$

25. Simplify: $7 - 8 - (-5) - 1$

26. Simplify: $6 + (-8) - (-9) + 3$

27. Evaluate: $(-5)^2$

28. Evaluate: -2^4

29. Simplify: $3(4 - 7) + (-2) - 5$

30. Simplify: $(20 - 5^2)^2$

31. Simplify: $2y - 6 + 4y + 8$

32. Simplify: $5x - 1 + x + 10$

Solve.

33. $5x + 2 - 4x = 7 - 19$

34. $9y + 1 - 8y = 3 - 20$

35. $17 - 7x + 3 = -3x + 21 - 3x$

36. $9x - 2 = 7x - 24$

37. Write a fraction to represent the shaded part of the figure.

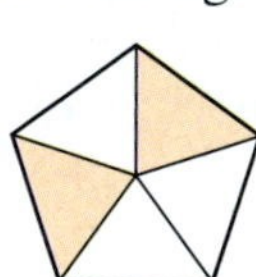

38. Write the prime factorization of 156.

39. Write each as an improper fraction.

a. $4\frac{2}{9}$ **b.** $1\frac{8}{11}$

40. Write $7\frac{4}{5}$ as an improper fraction.

41. Write in simplest form: $\frac{42x}{66}$

42. Write in simplest form: $\frac{70}{105y}$

43. Multiply: $3\frac{1}{3} \cdot \frac{7}{8}$

44. Multiply: $\frac{2}{3} \cdot 4$

45. Divide and simplify: $\frac{5}{16} \div \frac{3}{4}$

46. Divide: $1\frac{1}{10} \div 5\frac{3}{5}$

5

Decimals

Decimal numbers represent parts of a whole, just like fractions. For example, one penny is 0.01 or $\frac{1}{100}$ of a dollar. In this chapter, we learn to perform arithmetic operations on decimals and to analyze the relationship between fractions and decimals. We also learn how decimals are used in the real world.

SECTIONS

CHECK YOUR PROGRESS

A twinkling star that we see from Earth is actually steady starlight that has been bent by traveling through Earth's atmosphere. Telescopes here on Earth, which also require visual focusing through our atmosphere, reveal these same distortions. That's why the Hubble Space Telescope (HST) was conceived and built, to detect light from objects in space before it is absorbed or distorted. The HST was launched in 1990 from the space shuttle *Discovery,* and it orbits at a height of 380 miles above Earth's surface. (*Source:* NASA)

In Section 5.3, Exercise 41, and Section 5.5, Exercises 37 and 38, we examine some facts about the Hubble Space Telescope.

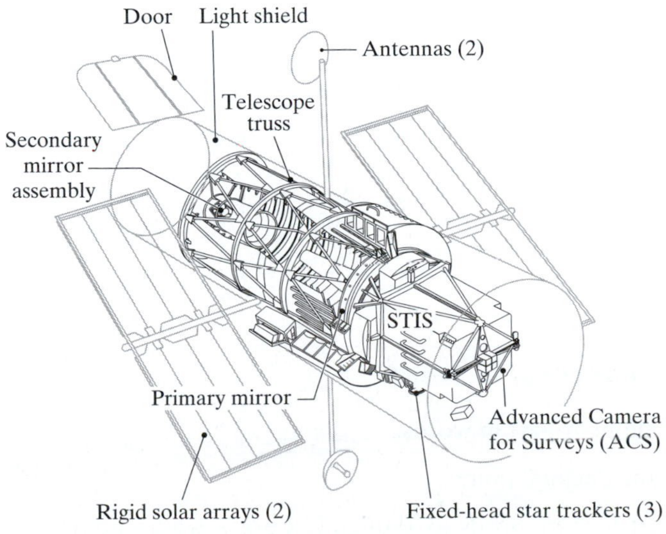

Length

13.3 meters or 43.5 feet (It is nearly the size of a large school bus.)

Diameter:

4.2 meters or 14 feet

Orbit:

28.5 degrees from the equator

Primary Mirror Diameter:

2.4 meters or 94.5 inches

Secondary Mirror Diameter:

0.3 meters or 12 inches

Cost at Launch:

$1.5 billion

Source: National Aeronautics and Space Administration

Objectives

A Know the Meaning of Place Value for a Decimal Number and Write Decimals in Words.

B Write Decimals in Standard Form.

C Write Decimals as Fractions.

D Compare Decimals.

E Round Decimals to a Given Place Value.

5.1 INTRODUCTION TO DECIMALS

Objective A Decimal Notation and Writing Decimals in Words

Like fractional notation, decimal notation is used to denote a part of a whole. Numbers written in decimal notation are called **decimal numbers,** or simply **decimals.** The decimal 17.758 has three parts.

17 . 758

Whole number part — Decimal point — Decimal part

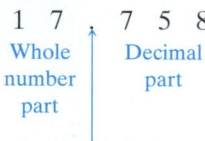

In Section 1.2, we introduced place value for whole numbers. Place names and place values for the whole number part of a decimal number are exactly the same. Place names and place values for the decimal part are shown below.

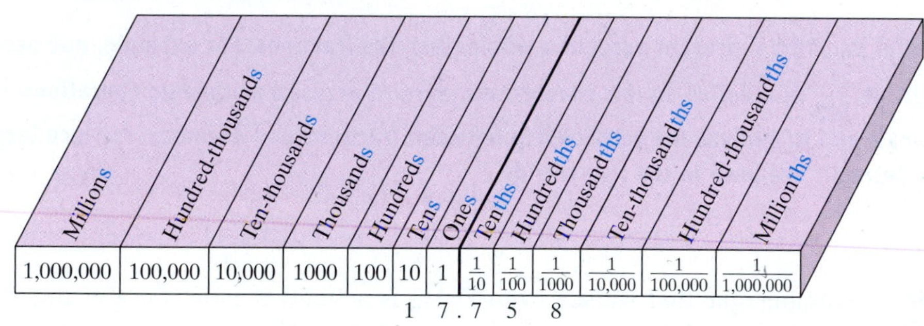

Helpful Hint Notice that place values to the left of the decimal point end in "s." Place values to the right of the decimal point end in "ths."

Notice that the value of each place is $\frac{1}{10}$ of the value of the place to its left. For example,

$$\underset{\text{ones}}{1} \cdot \frac{1}{10} = \underset{\text{tenths}}{\frac{1}{10}} \quad \text{and} \quad \underset{\text{tenths}}{\frac{1}{10}} \cdot \frac{1}{10} = \underset{\text{hundredths}}{\frac{1}{100}}$$

The decimal number 17.758 means

1 ten	+	7 ones	+	7 tenths	+	5 hundredths	+	8 thousandths
or $1 \cdot 10$	+	$7 \cdot 1$	+	$7 \cdot \frac{1}{10}$	+	$5 \cdot \frac{1}{100}$	+	$8 \cdot \frac{1}{1000}$
or 10	+	7	+	$\frac{7}{10}$	+	$\frac{5}{100}$	+	$\frac{8}{1000}$

Writing (or Reading) a Decimal in Words

Step 1: Write the whole number part in words.

Step 2: Write "and" for the decimal point.

Step 3: Write the decimal part in words as though it were a whole number, followed by the place value of the last digit.

EXAMPLE 1 Write each decimal in words.

a. 0.7 **b.** −50.82 **c.** 21.093

Solution:

a. seven tenths
b. negative fifty and eighty-two hundredths
c. twenty-one and ninety-three thousandths

Work Practice Problem 1

PRACTICE PROBLEM 1

Write each decimal in words.

a. 0.06
b. −200.073
c. 0.0829

EXAMPLE 2

Write the decimal in the following sentence in words: The Golden Jubilee Diamond is a 545.67-carat cut diamond. (*Source: The Guinness Book of Records*)

Solution: five hundred forty-five and sixty-seven hundredths

Work Practice Problem 2

PRACTICE PROBLEM 2

Write the decimal 87.31 in words.

EXAMPLE 3

Write the decimal in the following sentence in words: The oldest known fragments of the Earth's crust are Zircon crystals; they were discovered in Australia and are thought to be 4.276 billion years old. (*Source: The Guinness Book of Records*)

Solution: four and two hundred seventy-six thousandths

Work Practice Problem 3

PRACTICE PROBLEM 3

Write the decimal 52.1085 in words.

Suppose that you are paying for a purchase of $368.42 at Circuit City by writing a check. Checks are usually written using the following format.

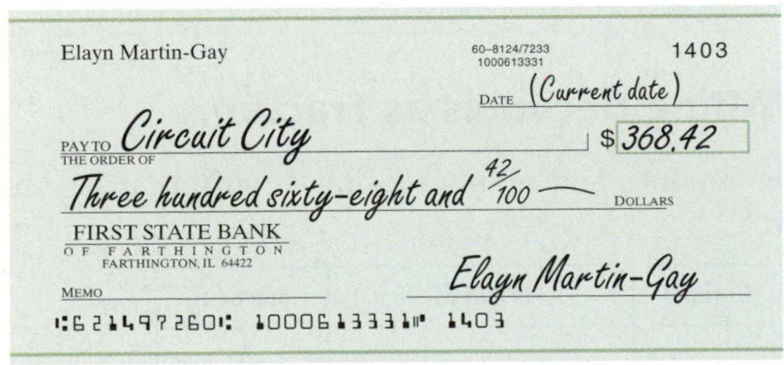

Answers

1. a. six hundredths **b.** negative two hundred and seventy-three thousandths **c.** eight hundred twenty-nine ten-thousandths
2. eighty-seven and thirty-one hundredths **3.** fifty-two and one thousand eighty-five ten-thousandths

PRACTICE PROBLEM 4

Fill in the check to CLECO (Central Louisiana Electric Company) to pay for your monthly electric bill of $207.40.

Your Preprinted Name
Your Preprinted Address
60-8124/7233 1000613331 1406
DATE
PAY TO THE ORDER OF $
DOLLARS
FIRST STATE BANK OF FARTHINGTON FARTHINGTON, IL 64422
MEMO
⑆621497260⑆ 1000613331⑈ 1406

EXAMPLE 4 Fill in the check to Camelot Music to pay for your purchase of $92.98.

Solution:

Your Preprinted Name
Your Preprinted Address
60-8124/7233 1000613331 1404
DATE *(Current date)*
PAY TO THE ORDER OF *Camelot Music* $ *92.98*
Ninety-two and 98/100 DOLLARS
FIRST STATE BANK OF FARTHINGTON FARTHINGTON, IL 64422
MEMO *(Your signature)*
⑆621497260⑆ 1000613331⑈ 1404

Work Practice Problem 4

Objective B Writing Decimals in Standard Form

A decimal written in words can be written in standard form by reversing the procedure in Objective A.

PRACTICE PROBLEMS 5–6

Write each decimal in standard form.

5. Five hundred and ninety-six hundredths
6. Thirty-nine and forty-two thousandths

EXAMPLES Write each decimal in standard form.

5. Forty-eight and twenty-six hundredths is

6. Six and ninety-five thousandths is

Work Practice Problems 5–6

Helpful Hint When converting a decimal from words to decimal notation, make sure the last digit is in the correct place by inserting 0s if necessary. For example,

Two and thirty-eight thousandths is 2.038
thousandths place

Objective C Writing Decimals as Fractions

Once you master reading and writing decimals, writing a decimal as a fraction follows naturally.

Decimal	In Words	Fraction
0.7	seven tenths	$\frac{7}{10}$
0.51	fifty-one hundredths	$\frac{51}{100}$
0.009	nine thousandths	$\frac{9}{1000}$
0.05	five hundredths	$\frac{5}{100} = \frac{1}{20}$

Answers

4.
Your Preprinted Name
Your Preprinted Address
60-8124/7233 1000613331 1405
DATE *(Current date)*
PAY TO THE ORDER OF *CLECO* $ *207.40*
Two hundred seven and 40/100 DOLLARS
FIRST STATE BANK OF FARTHINGTON FARTHINGTON, IL 64422
MEMO *(Your signature)*
⑆621497260⑆ 1000613331⑈ 1405

5. 500.96 **6.** 39.042

Notice that the number of decimal places in a decimal number is the same as the number of zeros in the denominator of the equivalent fraction. We can use this fact to write decimals as fractions.

$$0.31 = \frac{31}{100} \qquad 0.007 = \frac{7}{1000}$$

(0.31: 2 decimal places; 100: 2 zeros. 0.007: 3 decimal places; 1000: 3 zeros.)

EXAMPLE 7 Write 0.47 as a fraction.

Solution: $0.47 = \frac{47}{100}$

(2 decimal places; 2 zeros)

Work Practice Problem 7

EXAMPLE 8 Write 5.9 as a mixed number.

Solution: $5.9 = 5\frac{9}{10}$

(1 decimal place; 1 zero)

Work Practice Problem 8

EXAMPLES Write each decimal as a fraction or a mixed number. Write your answer in simplest form.

9. $0.125 = \frac{125}{1000} = \frac{1}{8}$

10. $43.5 = 43\frac{5}{10} = 43\frac{\overset{1}{\cancel{5}}}{2 \cdot \underset{1}{\cancel{5}}} = 43\frac{1}{2 \cdot 1} = 43\frac{1}{2}$

11. $-105.083 = -105\frac{83}{1000}$

Work Practice Problems 9–11

Later in the chapter, we write fractions as decimals. If you study Examples 7–11, you already know how to write fractions with denominators of 10, 100, 1000, and so on, as decimals.

Objective D Comparing Decimals

One way to compare positive decimals is by comparing digits in corresponding places. To see why this works, let's compare 0.5 or $\frac{5}{10}$ and 0.8 or $\frac{8}{10}$. We know

$$\frac{5}{10} < \frac{8}{10} \text{ since } 5 < 8\text{, so}$$

$$0.5 < 0.8 \text{ since } 5 < 8$$

This leads to the following.

PRACTICE PROBLEM 7

Write 0.051 as a fraction.

PRACTICE PROBLEM 8

Write 29.97 as a mixed number.

PRACTICE PROBLEMS 9–11

Write each decimal as a fraction or mixed number. Write your answer in simplest form.

9. 0.12
10. 64.8
11. −209.986

Answers

7. $\frac{51}{1000}$ 8. $29\frac{97}{100}$ 9. $\frac{3}{25}$ 10. $64\frac{4}{5}$ 11. $-209\frac{493}{500}$

Comparing Two Positive Decimals

Compare digits in the same places from left to right. When two digits are not equal, the number with the larger digit is the larger decimal. If necessary, insert 0s after the last digit to the right of the decimal point to continue comparing.

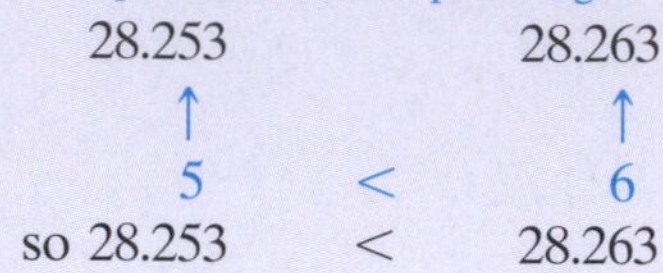

Helpful Hint

For any decimal, writing 0s after the last digit to the right of the decimal point does not change the value of the number.

$7.6 = 7.60 = 7.600$, and so on

When a whole number is written as a decimal, the decimal point is placed to the right of the ones digit.

$25 = 25.0 = 25.00$, and so on

PRACTICE PROBLEM 12

Insert $<$, $>$, or $=$ to form a true statement.

26.208 26.28

EXAMPLE 12 Insert $<$, $>$, or $=$ to form a true statement.

0.378 0.368

Solution: 0.3 78 0.3 68 The tenths places are the same.

0.37 8 0.36 8 The hundredths places are different.

Since $7 > 6$, then $0.378 > 0.368$.

Work Practice Problem 12

PRACTICE PROBLEM 13

Insert $<$, $>$, or $=$ to form a true statement.

0.12 0.026

EXAMPLE 13 Insert $<$, $>$, or $=$ to form a true statement.

0.052 0.236

Solution: $0.052 < 0.236$ 0 is smaller than 2 in the tenths place.

Work Practice Problem 13

We can also use a number line to compare decimals. This is especially helpful when comparing negative decimals. Remember, the number whose graph is to the left is smaller, and the number whose graph is to the right is larger.

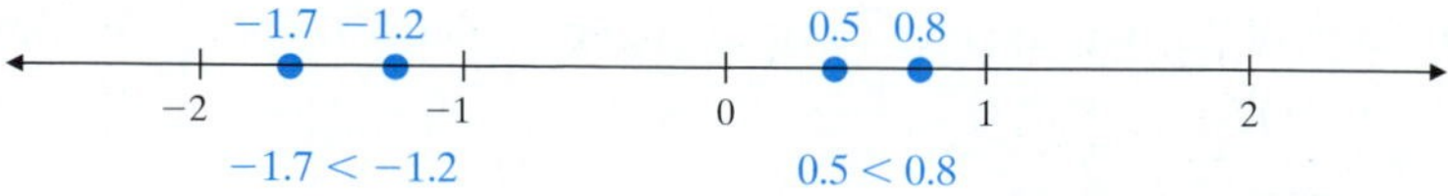

Answers

12. $<$ 13. $>$

Helpful Hint

If you have trouble comparing two negative decimals, try the following: Compare their absolute values. Then to correctly compare the negative decimals, reverse the direction of the inequality symbol.

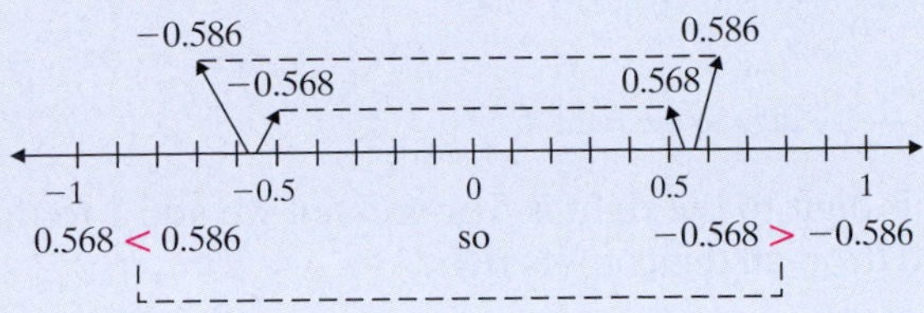

0.568 < 0.586 so −0.568 > −0.586

EXAMPLE 14 Insert <, >, or = to form a true statement.

−0.0101 −0.00109

Solution: Since 0.0101 > 0.00109, then −0.0101 < −0.00109

Work Practice Problem 14

PRACTICE PROBLEM 14

Insert <, >, or = to form a true statement.

−0.039 −0.0309

Objective E Rounding Decimals

We **round the decimal part** of a decimal number in nearly the same way as we round whole numbers. The only difference is that we drop digits to the right of the rounding place, instead of replacing these digits with 0s. For example,

36.954 rounded to the nearest hundredth is 36.95.

Rounding Decimals to a Place Value to the Right of the Decimal Point

Step 1: Locate the digit to the right of the given place value.

Step 2: If this digit is 5 or greater, add 1 to the digit in the given place value and drop all digits to its right. If this digit is less than 5, drop all digits to the right of the given place.

EXAMPLE 15 Round 736.2359 to the nearest tenth.

Solution:

Step 1: We locate the digit to the right of the tenths place.

736.2 3 59 (tenths place: 2; digit to the right: 3)

Step 2: Since this digit to the right is less than 5, we drop it and all digits to its right.

Thus, 736.2359 rounded to the nearest tenth is 736.2.

Work Practice Problem 15

PRACTICE PROBLEM 15

Round 482.7817 to the nearest thousandth.

The same steps for rounding can be used when the decimal is negative.

Answers

14. −0.039 < −0.0309 **15.** 482.782

PRACTICE PROBLEM 16

Round −0.032 to the nearest hundredth.

EXAMPLE 16 Round −0.027 to the nearest hundredth.

Solution:

Step 1: Locate the digit to the right of the hundredths place.

hundredths place

−0.02 7

digit to the right

Step 2: Since this digit to the right is 5 or greater, we add 1 to the hundredths digit and drop all digits to its right.

Thus, −0.027 is −0.03 rounded to the nearest hundredth.

Work Practice Problem 16

The following number line illustrates the rounding of negative decimals.

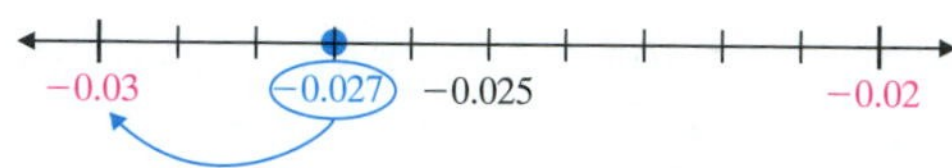

In Section 5.3, we will introduce a formula for the distance around a circle. This distance around a circle is given a special name called **circumference.**

The symbol π is the Greek letter pi, pronounced "pie." We use π to denote the following constant:

$$\pi = \frac{\text{circumference of a circle}}{\text{diameter of a circle}}$$

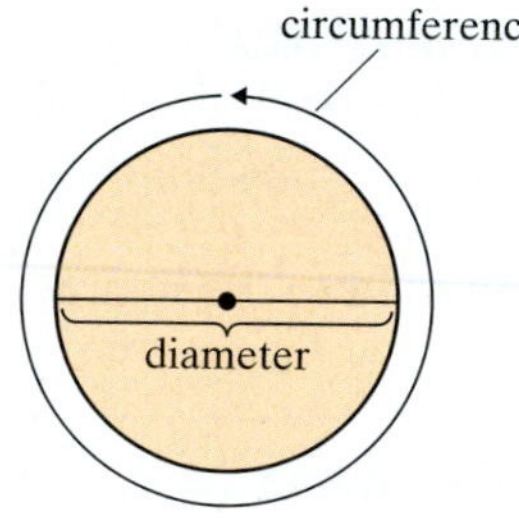

The value π is an **irrational number.** This means if we try to write it as a decimal, it neither ends nor repeats in a pattern.

PRACTICE PROBLEM 17

$\pi \approx 3.14159265$. Round π to the nearest ten thousandth.

EXAMPLE 17 $\pi \approx 3.14159265$. Round π to the nearest hundredth.

hundredths place — 1 is less than 5.

3.14159265

Delete these digits.

Thus, 3.14159265 rounded to the nearest hundredth is 3.14. In other words, $\pi \approx 3.14$.

Work Practice Problem 17

Answers

16. −0.03 **17.** $\pi \approx 3.1416$

Rounding often occurs with money amounts. Since there are 100 cents in a dollar, each cent is $\frac{1}{100}$ of a dollar. This means that if we want to round to the nearest cent, we round to the nearest hundredth of a dollar.

✔ **Concept Check** 1756.0894 rounded to the nearest *ten* is

a. 1756.1 **b.** 1760.0894 **c.** 1760 **d.** 1750

EXAMPLE 18 Determining State Taxable Income

A high school teacher's taxable income is $41,567.72. The tax tables in the teacher's state use amounts rounded to the nearest dollar. Round the teacher's income to the nearest whole dollar.

Solution: Rounding to the nearest whole dollar means rounding to the ones place.

ones place — 7 is greater than 5.

$41,567.72

Add 1. Delete these digits.

Thus, the teacher's income rounded to the nearest dollar is $41,568.

Work Practice Problem 18

PRACTICE PROBLEM 18

Water bills in Mexia are always rounded to the nearest dollar. Round a water bill of $24.62 to the nearest dollar.

Answer

18. $25

✔ Concept Check Answer

c

Vocabulary and Readiness Check

Use the choices below to fill in each blank.

words	decimals	tenths	after
tens	circumference	and	standard form

1. The number "twenty and eight hundredths" is written in __________ and "20.08" is written in __________.
2. Another name for the distance around a circle is its __________.
3. Like fractions, __________ are used to denote part of a whole.
4. When writing a decimal number in words, the decimal point is written as __________.
5. The place value __________ is to the right of the decimal point while __________ is to the left of the decimal point.
6. The decimal point in a whole number is __________ the last digit.

5.1 EXERCISE SET

FOR EXTRA HELP

Student Solutions Manual

PH Math/Tutor Center

CD/Video for Review

MathXL®

MyMathLab

Objective A *Write each decimal number in words. See Examples 1 through 3.*

1. 5.62
2. 9.57
3. 16.23
4. 47.65
5. −0.205
6. −0.495
7. 167.009
8. 233.056
9. 3000.04
10. 5000.02
11. 105.6
12. 410.3
13. The Akashi Kaikyo Bridge, between Kobe and Awaji-Shima, Japan, is approximately 2.43 miles long.
14. The English Channel Tunnel is 31.04 miles long. (*Source: Railway Directory & Year Book*)

Fill in each check for the described purchase. See Example 4.

15. Your monthly car loan of $321.42 to R. W. Financial.

Your Preprinted Name
Your Preprinted Address
60-8124/7233
1000613331
1407
DATE
PAY TO THE ORDER OF $
DOLLARS
FIRST STATE BANK
OF FARTHINGTON
FARTHINGTON, IL 64422
MEMO
⑆621497260⑆ 1000613331⑈ 1407

16. Your part of the monthly apartment rent, which is $213.70. You pay this to Amanda Dupre.

Your Preprinted Name
Your Preprinted Address
60-8124/7233
1000613331
1408
DATE
PAY TO THE ORDER OF $
DOLLARS
FIRST STATE BANK
OF FARTHINGTON
FARTHINGTON, IL 64422
MEMO
⑆621497260⑆ 1000613331⑈ 1408

17. Your cell phone bill of $59.68 to Bell South.

Your Preprinted Name
Your Preprinted Address
60-8124/7233
1000613331
1409
DATE
PAY TO THE ORDER OF $
DOLLARS
FIRST STATE BANK
OF FARTHINGTON
FARTHINGTON, IL 64422
MEMO
⑆621497260⑆ 1000613331⑈ 1409

18. Your grocery bill of $87.49 at Albertsons.

Your Preprinted Name
Your Preprinted Address
60-8124/7233
1000613331
1410
DATE
PAY TO THE ORDER OF $
DOLLARS
FIRST STATE BANK
OF FARTHINGTON
FARTHINGTON, IL 64422
MEMO
⑆621497260⑆ 1000613331⑈ 1410

Objective **B** *Write each decimal number in standard form. See Examples 5 and 6.*

19. Two and eight tenths

20. Five and one tenth

21. Nine and eight hundredths

22. Twelve and six hundredths

23. Negative seven hundred five and six hundred twenty-five thousandths

24. Negative eight hundred four and three hundred ninety-nine thousandths

25. Forty-six ten-thousandths

26. Eighty-three ten-thousandths

Objective **C** *Write each decimal as a fraction or a mixed number. Write your answer in simplest form. See Examples 7 through 11.*

27. 0.7

28. 0.9

29. 0.27

30. 0.39

31. 0.4

32. 0.8

33. 5.4

34. 6.8

35. −0.058

36. −0.024

37. 7.008

38. 9.005

39. 15.802

40. 11.406

41. 0.3005

42. 0.2006

Objectives **A** **B** **C** **Mixed Practice** *Fill in the chart. The first row is completed for you.*

	Decimal Number in Standard Form	In Words	Fraction
	0.37	thirty-seven hundredths	$\frac{37}{100}$
43.		eight tenths	
44.		five tenths	
45.	0.077		
46.	0.019		

Objective D *Insert* $<$, $>$, *or* $=$ *between each pair of numbers to form a true statement. See Examples 12 through 14.*

47. 0.15 0.16

48. 0.12 0.15

49. −0.57 −0.54

50. −0.59 −0.52

51. 0.098 0.1

52. 0.0756 0.2

53. 0.54900 0.549

54. 0.98400 0.984

55. 167.908 167.980

56. 519.3405 519.3054

57. −1.062 −1.07

58. −18.1 −18.01

59. −7.052 7.0052

60. 0.01 −0.1

61. −0.023 −0.024

62. −0.562 −0.652

Objective E *Round each decimal to the given place value. See Examples 15 through 18.*

63. 0.57, nearest tenth

64. 0.64, nearest tenth

65. 98,207.23, nearest ten

66. 68,934.543, nearest ten

67. −0.234, nearest hundredth

68. −0.892, nearest hundredth

69. 0.5942, nearest thousandth

70. 63.4523, nearest thousandth

Recall that the number π*, written as a decimal, neither ends nor repeats in a pattern. Given that* $\pi \approx 3.14159265$*, round* π *to the given place values below. (We study* π *further in Section 5.3.) See Example 17.*

71. tenth

72. ones

73. thousandth

74. hundred thousandth

Round each monetary amount to the nearest cent or dollar as indicated. See Example 18.

75. $26.95, to the nearest dollar

76. $14,769.52, to the nearest dollar

77. $0.1992, to the nearest cent

78. $0.7633, to the nearest cent

Round each number to the given place value. See Example 18.

79. At this writing, the smallest personal computer, developed by OQO in the United States, measures 10.414 centimeters across. Round this number to the nearest tenth. (*Source: Guinness World Records* 2006)

80. A large tropical cockroach of the family Dictyoptera is the fastest-moving insect. This insect was clocked at a speed of 3.36 miles per hour. Round this number to the nearest tenth. (*Source:* University of California, Berkeley)

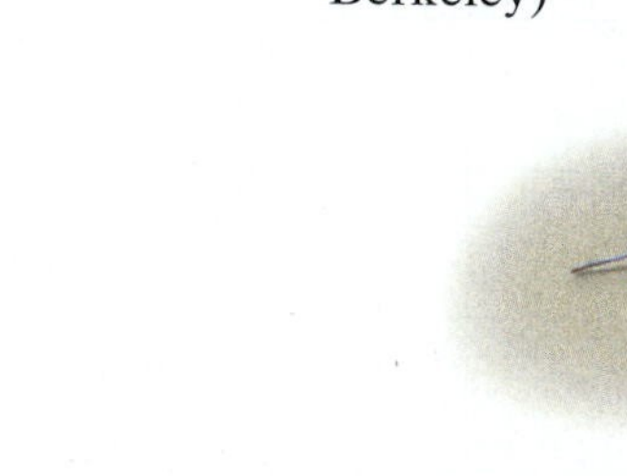

81. During the 2006 Boston Marathon, Edith Hunkeler of Switzerland was the winner in the women's wheelchair division, with a finishing time of 1.7283 hours. Round this time to the nearest hundredth of an hour. (*Source:* Boston Athletic Association)

82. The population density of the state of Louisiana is 102.5794 people per square mile. Round this population density to the nearest tenth. (*Source:* U.S. Census Bureau)

83. A used biology textbook is priced at \$47.89. Round this price to the nearest dollar.

84. A used office desk is advertised at \$19.95 by Drawley's Office Furniture. Round this price to the nearest dollar.

85. The length of a day on Mars, a full rotation about its axis, is 24.6229 hours. Round this figure to the nearest thousandth. (*Source:* National Space Science Data Center)

86. Venus makes a complete orbit around the Sun every 224.695 days. Round this figure to the nearest whole day. (*Source:* National Space Science Data Center)

Review

Perform each indicated operation. See Section 1.3.

87. $3452 + 2314$

88. $8945 + 4536$

89. $82 - 47$

90. $4002 - 3897$

Concept Extensions

Solve. See the Concept Check in this section.

91. 2849.1738 rounded to the nearest hundred is

a. 2849.17 **b.** 2800 **c.** 2850 **d.** 2849.174

92. 146.059 rounded to the nearest ten is

a. 146.0 **b.** 146.1 **c.** 140 **d.** 150

93. 2849.1738 rounded to the nearest hundredth is

a. 2849.17 **b.** 2800 **c.** 2850 **d.** 2849.18

94. 146.059 rounded to the nearest tenth is

a. 146.0 **b.** 146.1 **c.** 140 **d.** 150

95. In your own words, describe how to write a decimal as a fraction or a mixed number.

96. Explain how to identify the value of the 9 in the decimal 486.3297.

97. Write $7\frac{12}{100}$ as a decimal.

98. Write $17\frac{268}{1000}$ as a decimal.

99. Write 0.00026849576 as a fraction.

100. Write 0.00026849576 in words.

101. Write a 5-digit number that rounds to 1.7.

102. Write a 4-digit number that rounds to 26.3.

103. Write a decimal number that is greater than 48.1, but less than 48.2.

104. Which number(s) rounds to 0.26?
0.26559 0.26499 0.25786 0.25186

105. Which number(s) rounds to 0.06?
0.0612 0.066 0.0586 0.0506

Write these numbers from smallest to largest.

106. 0.9
0.1038
0.10299
0.1037

107. 0.01
0.0839
0.09
0.1

108. The all-time top six movies (those that earned the most money in the United States) along with the approximate amount of money they have earned are listed in the table. Estimate the total amount of money that these movies have earned by first rounding each earning to the nearest hundred million. (*Source:* Movie Web)

Top All-Time American Movies

Movie	Gross Domestic Earnings
Titanic (1997)	\$600.8 million
Star Wars: A New Hope (1977)	\$461.0 million
Shrek 2 (2004)	\$437.2 million
E.T. (1982)	\$433.0 million
Star Wars: The Phantom Menace (1999)	\$431.1 million
Pirates of the Caribbean: Dead Man's Chest (2006)	\$421.4 million

109. In 2005, American manufacturers shipped approximately 33.8 million music videos to retailers. The value of these shipments was approximately \$602.2 million. Estimate the value of an individual music video by rounding 602.2 and 33.8 to the nearest ten, then dividing. (*Source:* Recording Industry Association of America)

STUDY SKILLS BUILDER

Are You Getting All the Mathematics Help That You Need?

Remember that, in addition to your instructor, there are many places to get help with your mathematics course. For example,

- This text has an accompanying video lesson for every section and worked out solutions to every Chapter Test exercise on video.
- The back of the book contains answers to odd-numbered exercises and selected solutions.
- A *Student Solutions Manual* is available that contains worked-out solutions to odd-numbered exercises as well as solutions to every exercise in the Integrated Reviews, Chapter Reviews, Chapter Tests, and Cumulative Reviews.
- Don't forget to check with your instructor for other local resources available to you, such as a tutor center.

Exercises

1. List items you find helpful in the text and all student supplements to this text.
2. List all the campus help that is available to you for this course.
3. List any help (besides the textbook) from Exercises 1 and 2 above that you are using.
4. List any help (besides the textbook) that you feel you should try.
5. Write a goal for yourself that includes trying anything you listed in Exercise 4 during the next week.

5.2 ADDING AND SUBTRACTING DECIMALS

Objectives

A Add or Subtract Decimals.

B Estimate when Adding or Subtracting Decimals.

C Evaluate Expressions with Decimal Replacement Values.

D Simplify Expressions Containing Decimals.

E Solve Problems That Involve Adding or Subtracting Decimals.

Objective A Adding or Subtracting Decimals

Adding or subtracting decimals is similar to adding or subtracting whole numbers. We add or subtract digits in corresponding place values from right to left, carrying or borrowing if necessary. To make sure that digits in corresponding place values are added or subtracted, we line up the decimal points vertically.

Adding or Subtracting Decimals

Step 1: Write the decimals so that the decimal points line up vertically.

Step 2: Add or subtract as with whole numbers.

Step 3: Place the decimal point in the sum or difference so that it lines up vertically with the decimal points in the problem.

In this section, we will insert zeros in decimals numbers so that place value digits line up neatly. For instance, see Example 1.

EXAMPLE 1 Add: 23.85 + 1.604

Solution: First we line up the decimal points vertically.

$$\begin{array}{r} 23.850 \\ +\ 1.604 \\ \hline \end{array}$$

Insert one 0 so that digits line up neatly.

↑ line up decimal points

Then we add the digits from right to left as for whole numbers.

$$\begin{array}{r} \overset{1}{2}3.850 \\ +\ 1.604 \\ \hline 25.454 \end{array}$$

Place the decimal point in the sum so that all decimal points line up.

Work Practice Problem 1

PRACTICE PROBLEM 1

Add.

a. 19.52 + 5.371

b. 40.08 + 17.612

c. 0.125 + 422.8

Helpful Hint

Recall that 0s may be placed after the last digit to the right of the decimal point without changing the value of the decimal. This may be used to help line up place values when adding decimals.

3.2	becomes	3.200	Insert two 0s.
15.567		15.567	
+ 0.11		+ 0.110	Insert one 0.
		18.877	Add.

Answers

1. a. 24.891 **b.** 57.692 **c.** 422.925

PRACTICE PROBLEM 2

Add.

a. 34.567 + 129.43 + 2.8903

b. 11.21 + 46.013 + 362.526

EXAMPLE 2 Add: 763.7651 + 22.001 + 43.89

Solution: First we line up the decimal points.

$$\begin{array}{r} \overset{1\ 1\ 1}{763.7651} \\ 22.0010 \\ +\ 43.8900 \\ \hline 829.6561 \end{array}$$

22.0010 — Insert one 0.
43.8900 — Insert two 0s.
829.6561 — Add.

Work Practice Problem 2

Helpful Hint Don't forget that the decimal point in a whole number is positioned after the last digit.

PRACTICE PROBLEM 3

Add: 19 + 26.072

EXAMPLE 3 Add: 45 + 2.06

Solution:

$$\begin{array}{r} 45.00 \\ +\ 2.06 \\ \hline 47.06 \end{array}$$

45.00 — Insert a decimal point and two 0s.
+ 2.06 — Line up decimal points.
47.06 — Add.

Work Practice Problem 3

✔Concept Check What is wrong with the following calculation of the sum of 7.03, 2.008, 19.16, and 3.1415?

$$\begin{array}{r} 7.03 \\ 2.008 \\ 19.16 \\ +3.1415 \\ \hline 3.6042 \end{array}$$

PRACTICE PROBLEM 4

Add: 7.12 + (−9.92)

EXAMPLE 4 Add: 3.62 + (−4.78)

Solution: Recall from Chapter 2 that to add two numbers with different signs we find the difference of the larger absolute value and the smaller absolute value. The sign of the answer is the same as the sign of the number with the larger absolute value.

$$\begin{array}{r} 4.78 \\ -3.62 \\ \hline 1.16 \end{array}$$

Subtract the absolute values.

Thus, 3.62 + (−4.78) = −1.16

The sign of the number with the larger absolute value; −4.78 has the larger absolute value.

Work Practice Problem 4

Subtracting decimals is similar to subtracting whole numbers. We line up digits and subtract from right to left, borrowing when needed.

Answers

2. a. 166.8873 **b.** 419.749
3. 45.072 **4.** −2.8

✔ Concept Check Answer

The decimal places are not lined up properly.

EXAMPLE 5 Subtract: 3.5 − 0.068. Check your answer.

Solution:

$$\begin{array}{r} 3.500 \\ -\,0.068 \\ \hline 3.432 \end{array}$$

Insert two 0s. Line up decimal points. Subtract.

Check: Recall that we can check a subtraction problem by adding.

$$\begin{array}{rl} 3.432 & \text{Difference} \\ +\,0.068 & \text{Subtrahend} \\ \hline 3.500 & \text{Minuend} \end{array}$$

Work Practice Problem 5

PRACTICE PROBLEM 5

Subtract. Check your answers.

a. 6.7 − 3.92
b. 9.72 − 4.068

EXAMPLE 6 Subtract: 85 − 17.31. Check your answer.

Solution:

$$\begin{array}{r} 85.00 \\ -\,17.31 \\ \hline 67.69 \end{array}$$

Check:

$$\begin{array}{rl} 67.69 & \text{Difference} \\ +\,17.31 & \text{Subtrahend} \\ \hline 85.00 & \text{Minuend} \end{array}$$

Work Practice Problem 6

PRACTICE PROBLEM 6

Subtract. Check your answers.

a. 73 − 29.31
b. 210 − 68.22

EXAMPLE 7 Subtract 3 from 6.98.

Solution:

$$\begin{array}{r} 6.98 \\ -3.00 \\ \hline 3.98 \end{array}$$

Insert two 0s.

Check:

$$\begin{array}{rl} 3.98 & \text{Difference} \\ +\,3.00 & \text{Subtrahend} \\ \hline 6.98 & \text{Minuend} \end{array}$$

Work Practice Problem 7

PRACTICE PROBLEM 7

Subtract: 19 from 25.91

EXAMPLE 8 Subtract: −5.8 − 1.7

Solution: Recall from Chapter 2 that to subtract 1.7 we add the opposite of 1.7, or −1.7. Thus

$-5.8 - 1.7 = -5.8 + (-1.7)$ To subtract, add the opposite of 1.7 which is −1.7.

Add the absolute values.

$= -7.5.$

Use the common negative sign.

Work Practice Problem 8

PRACTICE PROBLEM 8

Subtract: −5.4 − 9.6

EXAMPLE 9 Subtract: −2.56 − (−4.01)

Solution: $-2.56 - (-4.01) = -2.56 + 4.01$ To subtract, add the opposite of −4.01, which is 4.01.

Add the absolute values.

$= 1.45$

The answer is positive since 4.01 has the larger absolute value.

Work Practice Problem 9

PRACTICE PROBLEM 9

Subtract: −1.05 − (−7.23)

Answers

5. a. 2.78 **b.** 5.652
6. a. 43.69 **b.** 141.78
7. 6.91 **8.** −15 **9.** 6.18

Objective B Estimating when Adding or Subtracting Decimals

To help avoid errors, we can also estimate to see if our answer is reasonable when adding or subtracting decimals. Although only one estimate is needed per operation, we show two for variety.

PRACTICE PROBLEM 10

Add or subtract as indicated. Then estimate to see if the answer is reasonable by rounding the given numbers and adding or subtracting the rounded numbers.

a. 58.1 + 326.97

b. 16.08 − 0.925

EXAMPLE 10 Add or subtract as indicated. Then estimate to see if the answer is reasonable by rounding the given numbers and adding or subtracting the rounded numbers.

a. 27.6 + 519.25

Exact		Estimate 1		Estimate 2
27.60	rounds to	30		30
+ 519.25	rounds to	+ 500	or	+ 520
546.85		530		550

Since the exact answer is close to either estimate, it is reasonable. (In the first estimate, each number is rounded to the place value of the leftmost digit. In the second estimate, each number is rounded to the nearest ten.)

b. 11.01 − 0.862

Exact		Estimate 1		Estimate 2
11.010	rounds to	10		11
− 0.862	rounds to	− 1	or	− 1
10.148		9		10

In the first estimate, we rounded the first number to the nearest ten and the second number to the nearest one. In the second estimate, we rounded both numbers to the nearest one. Both estimates show us that our answer is reasonable.

Work Practice Problem 10

Helpful Hint Remember that estimates are used for our convenience to quickly check the reasonableness of an answer.

✔ Concept Check Why shouldn't the sum 21.98 + 42.36 be estimated as 30 + 50 = 80?

Objective C Using Decimals as Replacement Values

Let's review evaluating expressions with given replacement values. This time the replacement values are decimals.

PRACTICE PROBLEM 11

Evaluate $y - z$ for $y = 11.6$ and $z = 10.8$.

EXAMPLE 11 Evaluate $x - y$ for $x = 2.8$ and $y = 0.92$.

Solution: Replace x with 2.8 and y with 0.92 and simplify.

$$x - y = 2.8 - 0.92$$
$$= 1.88$$

$$\begin{array}{r} 2.80 \\ -0.92 \\ \hline 1.88 \end{array}$$

Work Practice Problem 11

Answers

10. a. Exact: 385.07 **b.** Exact: 15.155
11. 0.8

✔ Concept Check Answer

Each number is rounded incorrectly. The estimate is too high.

EXAMPLE 12 Is 2.3 a solution of the equation $6.3 = x + 4$?

Solution: Replace x with 2.3 in the equation $6.3 = x + 4$ to see if the result is a true statement.

$6.3 = x + 4$
$6.3 \stackrel{?}{=} 2.3 + 4$ Replace x with 2.3.
$6.3 = 6.3$ True

Since $6.3 = 6.3$ is a true statement, 2.3 is a solution of $6.3 = x + 4$.

Work Practice Problem 12

PRACTICE PROBLEM 12

Is 12.1 a solution of the equation $y - 4.3 = 7.8$?

Objective D Simplify Expressions Containing Decimals

EXAMPLE 13 Simplify by combining like terms:

$11.1x - 6.3 + 8.9x - 4.6$

Solution:
$$11.1x - 6.3 + 8.9x - 4.6 = 11.1x + 8.9x + (-6.3) + (-4.6)$$
$$= 20x + (-10.9)$$
$$= 20x - 10.9$$

Work Practice Problem 13

PRACTICE PROBLEM 13

Simplify by combining like terms:
$-4.3y + 7.8 - 20.1y + 14.6$

Objective E Solving Problems by Adding or Subtracting Decimals

Decimals are very common in real-life problems.

EXAMPLE 14 **Calculating the Cost of Owning an Automobile**

Find the total monthly cost of owning and operating a certain automobile given the expenses shown.

Monthly car payment:	\$256.63
Monthly insurance cost:	\$47.52
Average gasoline bill per month:	\$95.33

Solution:

1. UNDERSTAND. Read and reread the problem. The phrase "total monthly cost" tells us to add.
2. TRANSLATE.

In words:	total monthly cost	is	car payment	plus	insurance cost	plus	gasoline bill
	↓	↓	↓	↓	↓	↓	↓
Translate:	total monthly cost	=	\$256.63	+	\$47.52	+	\$95.33

3. SOLVE: Let's also estimate by rounding each number to the nearest ten.

256.63	rounds to	260	
47.52	rounds to	50	
+ 95.33	rounds to	100	
\$399.48	Exact	410	Estimate

Continued on next page

PRACTICE PROBLEM 14

Find the total monthly cost of owning and operating a certain automobile given the expenses shown.

Monthly car payment:	\$563.52
Monthly insurance cost:	\$52.68
Average gasoline bill per month:	\$127.50

Answers
12. yes **13.** $-24.4y + 22.4$ **14.** \$743.70

4. INTERPRET. *Check* your work. Since our estimate is close to our exact answer, our answer is reasonable. *State* your conclusion: The total monthly cost is $399.48.

Work Practice Problem 14

The next bar graph has horizontal bars. To visualize the value represented by a bar, see how far it extends to the right. The value of each bar is labeled, and we will study bar graphs further in a later chapter.

PRACTICE PROBLEM 15

Use the bar graph in Example 15. How much greater is the average height in the Netherlands than the average height in Czechoslovakia?

EXAMPLE 15 Comparing Average Heights

The bar graph shows the current average heights for adults in various countries. How much greater is the average height in Denmark than the average height in the United States?

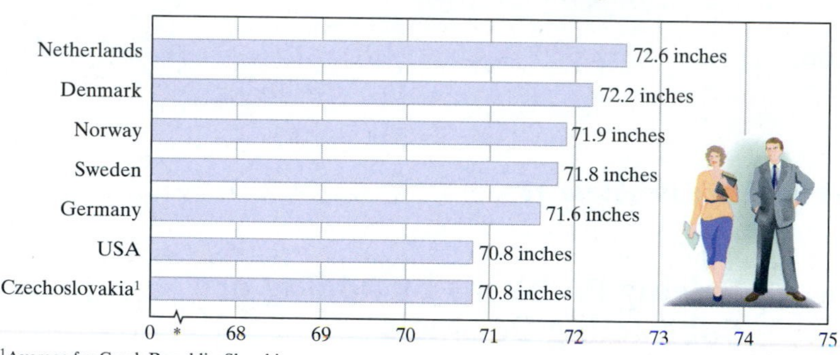

[1]Average for Czech Republic, Slovakia
Source: USA Today

* The -√- means that some numbers are purposefully missing on the axis.

Solution:

1. UNDERSTAND. Read and reread the problem. Since we want to know "how much greater," we subtract.
2. TRANSLATE.

In words:	How much greater	is	Denmark's average height	minus	U.S. average height
	↓	↓	↓	↓	↓
Translate:	How much greater	=	72.2	−	70.8

3. SOLVE: We estimate by rounding each number to the nearest whole.

72.2 (borrow: 1 12)	rounds to	72	
− 70.8	rounds to	− 71	
1.4 Exact		1	Estimate

4. INTERPRET. *Check* your work. Since our estimate is close to our exact answer, 1.4 inches is reasonable. *State* your conclusion: The average height in Denmark is 1.4 inches greater than the average U.S. height.

Work Practice Problem 15

Answer
15. 1.8 in.

CALCULATOR EXPLORATIONS Decimals

Entering Decimal Numbers

To enter a decimal number, find the key marked [·]. To enter the number 2.56, for example, press the keys [2] [·] [5] [6].
The display will read [2.56].

Operations on Decimal Numbers

Operations on decimal numbers are performed in the same way as operations on whole or signed numbers. For example, to find 8.625 − 4.29, press the keys [8.625] [−] [4.29] [=] or [ENTER].
The display will read [4.335]. (Although entering 8.625, for example, requires pressing more than one key, we group numbers together here for easier reading.)

Use a calculator to perform each indicated operation.

1. 315.782 + 12.96

2. 29.68 + 85.902

3. 6.249 − 1.0076

4. 5.238 − 0.682

5.
```
   12.555
  224.987
    5.2
 +622.65
```

6.
```
  47.006
   0.17
 313.259
 139.088
```

Vocabulary and Readiness Check

Use the choices below to fill in each blank. Not all choices will be used.

minuend	vertically	like	true
difference	subtrahend	last	false

1. The decimal point in a whole number is positioned after the ______________ digit.

2. In 89.2 − 14.9 = 74.3, the number 74.3 is called the ______________, 89.2 is the ______________, and 14.9 is the ______________.

3. To simplify an expression, we combine any ______________ terms.

4. True or false: If we replace x with 11.2 and y with −8.6 in the expression $x - y$, we have 11.2 − 8.6. ______________

5. To add or subtract decimals, we line up the decimal points ______________.

5.2 EXERCISE SET

FOR EXTRA HELP

Student Solutions Manual

PH Math/Tutor Center

CD/Video for Review

Math XL
MathXL®

MyMathLab

Objectives A B Mixed Practice *Add or subtract. See Examples 1 through 4, and 10. For those exercises marked, also estimate to see if the answer is reasonable.*

1. $5.6 + 2.1$

2. $3.6 + 4.1$

3. $8.2 + 2.15$

4. $5.17 + 3.7$

5. $24.6 + 2.39 + 0.0678$

6. $32.4 + 1.58 + 0.0934$

7. $-2.6 + (-5.97)$

8. $-18.2 + (-10.8)$

9. $18.56 + (-8.23)$

10. $4.38 + (-6.05)$

11. $\begin{array}{r} 234.89 \\ +\ 230.67 \\ \hline \end{array}$

Exact: Estimate:

12. $\begin{array}{r} 734.89 \\ +\ 640.56 \\ \hline \end{array}$

Exact: Estimate:

13. $\begin{array}{r} 100.009 \\ 6.08 \\ +\ 9.034 \\ \hline \end{array}$

Exact: Estimate:

14. $\begin{array}{r} 200.89 \\ 7.49 \\ +\ 62.83 \\ \hline \end{array}$

Exact: Estimate:

15. Find the sum of 39, 3.006, and 8.403

16. Find the sum of 65, 5.0903, and 6.9003

Subtract and check. See Examples 5 through 10. For those exercises marked, also estimate to see if the answer is reasonable.

17. $12.6 - 8.2$

18. $8.9 - 3.1$

19. $18 - 2.7$

20. $28 - 3.3$

21. $\begin{array}{r} 654.9 \\ -\ 56.67 \\ \hline \end{array}$

22. $\begin{array}{r} 863.23 \\ -\ 39.453 \\ \hline \end{array}$

23. $5.9 - 4.07$
Exact:
Estimate:

24. $6.4 - 3.04$
Exact:
Estimate:

25. $\begin{array}{r} 1000 \\ -\ 123.4 \\ \hline \end{array}$

Exact:

Estimate:

26. $\begin{array}{r} 2000 \\ -\ 327.47 \\ \hline \end{array}$

Exact:

Estimate:

27. $200 - 5.6$

28. $800 - 8.9$

29. $-1.12 - 5.2$

30. $-8.63 - 5.6$

31. $5.21 - 11.36$

32. $8.53 - 17.84$

33. $-2.6 - (-5.7)$

34. $-9.4 - (-10.4)$

35. $3 - 0.0012$

36. $7 - 0.097$

37. Subtract 6.7 from 23.

38. Subtract 9.2 from 45.

Objective A *Perform the indicated operation. See Examples 1 through 10.*

39. $0.9 + 2.2$

40. $0.7 + 3.4$

41. $-6.06 + 0.44$

42. $-5.05 + 0.88$

43. $500.21 - 136.85$

44. $600.47 - 254.68$

45. $50.2 - 600$

46. $40.3 - 700$

47. Subtract 61.9 from 923.5

48. Subtract 45.8 from 845.9

49. Add 100.009 and 6.08 and 9.034

50. Add 200.89 and 7.49 and 62.83

51. $-0.003 + 0.091$

52. $-0.004 + 0.085$

53. $-102.4 - 78.04$

54. $-36.2 - 10.02$

55. $-2.9 - (-1.8)$

56. $-6.5 - (-3.3)$

Objective **C** *Evaluate each expression for $x = 3.6$, $y = 5$, and $z = 0.21$. See Example 11.*

57. $x + z$

58. $y + x$

59. $x - z$

60. $y - z$

61. $y - x + z$

62. $x + y + z$

Determine whether the given values are solutions to the given equations. See Example 12.

63. Is 7 a solution to $x + 2.7 = 9.3$?

64. Is 3.7 a solution to $x + 5.9 = 8.6$?

65. Is -11.4 a solution to $27.4 + y = 16$?

66. Is -22.9 a solution to $45.9 + z = 23$?

67. Is 1 a solution to $2.3 + x = 5.3 - x$?

68. Is 0.9 a solution to $1.9 - x = x + 0.1$?

Objective **D** *Simplify by combining like terms. See Example 13.*

69. $30.7x + 17.6 - 23.8x - 10.7$

70. $14.2z + 11.9 - 9.6z - 15.2$

71. $-8.61 + 4.23y - 2.36 - 0.76y$

72. $-8.96x - 2.31 - 4.08x + 9.68$

Objective **E** *Solve. See Examples 14 and 15.*

73. Microsoft stock opened the day at \$21.90 per share, and the closing price the same day was \$22.07. By how much did the price of each share change?

74. A pair of eyeglasses costs a total of \$347.89. The frames of the glasses are \$97.23. How much do the lenses of the eyeglasses cost?

75. Find the perimeter.

Square 7.14 meters

76. Find the perimeter.

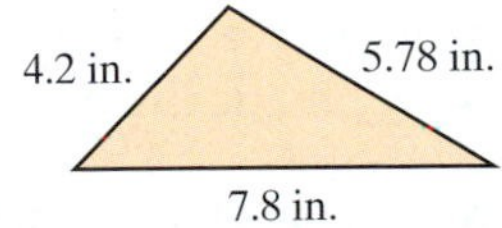

The iPod mini is a miniature version of Apple Computer's popular iPod portable audio player. This mini was introduced in January 2004 with a storage capacity of 4 gigabytes. (This is about 1000 3-minute or 3-megabyte songs.)

77. The top face of the iPod mini shown measures 3.6 inches by 2.0 inches. Find the perimeter of the rectangular face.

78. The face of the larger Apple iPod measures 4.1 inches by 2.4 inches. Find the perimeter of this rectangular face.

Solve.

79. Ann-Margaret Tober bought a book for $32.48. If she paid with two $20 bills, what was her change?

80. Phillip Guillot bought a car part for $18.26. If he paid with two $10 bills, what was his change?

81. The average wind speed at the weather station on Mt. Washington in New Hampshire is 35.2 miles per hour. The highest speed ever recorded at the station is 321.0 miles per hour. How much faster is the highest speed than the average wind speed? (*Source:* National Climatic Data Center)

82. The average annual rainfall in Omaha, Nebraska, is 30.22 inches. The average annual rainfall in New Orleans, Louisiana, is 61.88 inches. On average, how much more rain does New Orleans receive annually than Omaha? (*Source:* National Climatic Data Center)

This bar graph shows the predicted decrease in home video rental revenue for chains such as Blockbuster and Hollywood Video. Use this graph for Exercise 83.

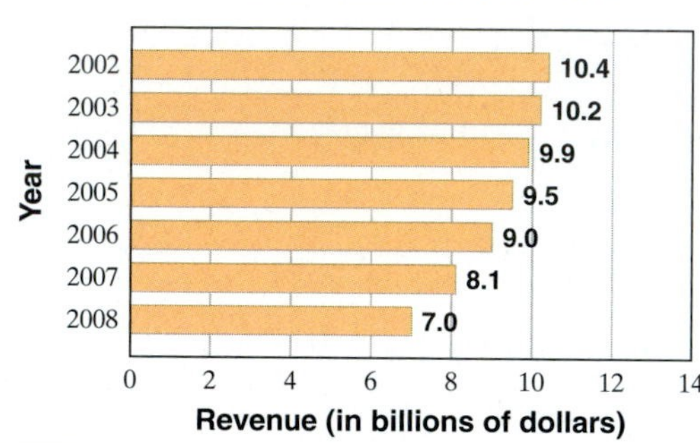

Source: Forrester Research; *Note:* Many of these years are projections.

83. Find the decrease in video rental revenue from the year 2002 to 2008.

84. It is predicted that home video *sales* (not shown on the bar graph) will increase from $15.3 billion in 2002 to $24.6 billion in 2008. Find the amount of increase.

85. The top three U.S. movies that made the most money through movie ticket sales are *Titanic* (1997) $600.78 million, *Star Wars IV: A New Hope* (1997) $460.99 million, and *Shrek 2* (2004) $436.72 million. What was the total amount of ticket sales for these three movies? (*Source:* MovieWeb)

86. In 1995, the average credit-card late fee was $12.53. By 2005, the average late fee had increased to $29.45. By how much did the average credit-card late fee increase from 1995 to 2005? (*Source: Money magazine*)

87. The snowiest city in the United States is Blue Canyon, California, which receives an average of 111.6 more inches of snow than the second snowiest city. The second snowiest city in the United States is Marquette, Michigan. Marquette receives an average of 129.2 inches of snow annually. How much snow does Blue Canyon receive on average each year? (*Source:* National Climatic Data Center)

88. The driest city in the world is Aswan, Egypt, which receives an average of only 0.02 inch of rain per year. Yuma, Arizona, is the driest city in the United States. Yuma receives an average of 2.63 more inches of rain each year than Aswan. What is the average annual rainfall in Yuma? (*Source:* National Climatic Data Center)

△ **89.** A landscape architect is planning a border for a flower garden shaped like a triangle. The sides of the garden measure 12.4 feet, 29.34 feet, and 25.7 feet. Find the amount of border material needed.

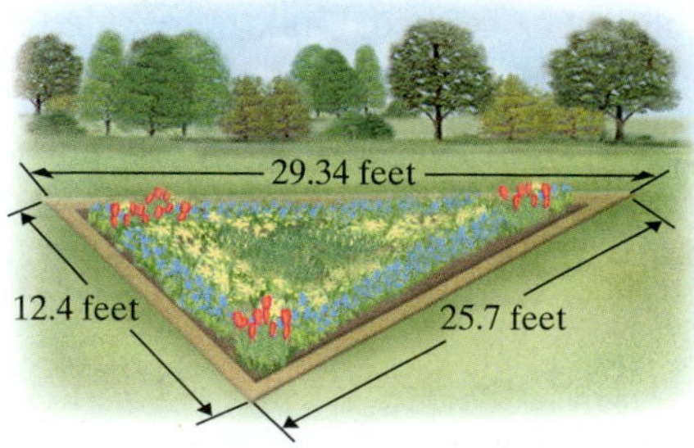

△ **90.** A contractor purchased enough railing to completely enclose the newly built deck shown below. Find the amount of railing purchased.

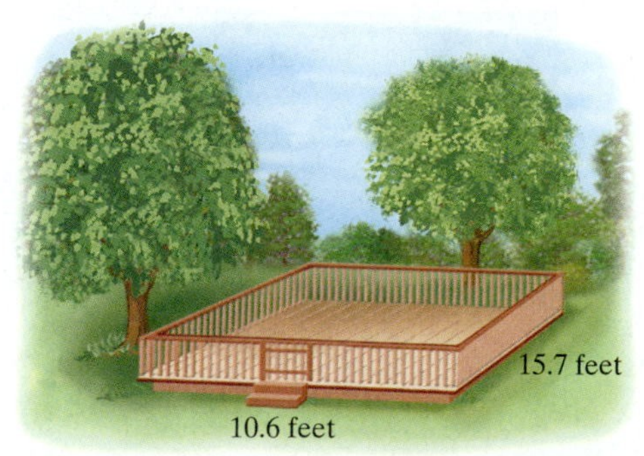

The table shows the average speeds for the Daytona 500 Winners for the years shown. Use this table to answer Exercises 91 and 92.

Daytona 500 Winners		
Year	**Winner**	**Average Speed**
1966	Richard Petty	160.927
1976	David Pearson	152.181
1986	Geoff Bodine	148.124
1996	Dale Jarrett	154.308
2006	Jimmie Johnson	142.734
(*Source:* Daytona International Speedway)		

91. How much slower was the average Daytona 500 winning speed in 2006 than in 1966?

92. How much faster was the average Daytona 500 winning speed in 1996 than in 1986?

The bar graph shows the top five chocolate-consuming nations in the world. Use this table to answer Exercises 93 through 97.

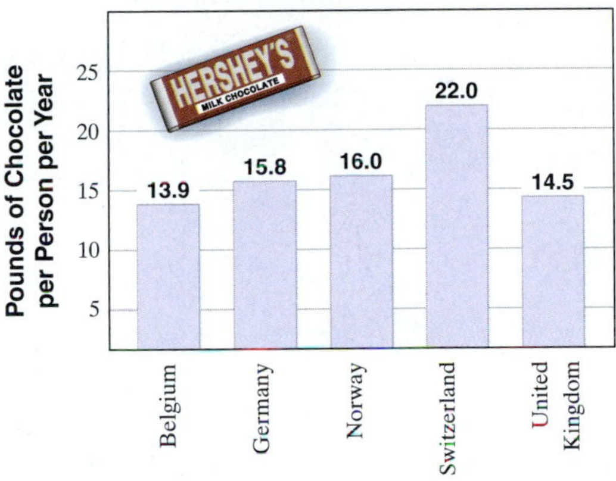

Source: Hershey Foods Corporation

93. Which country in the table has the greatest chocolate consumption per person?

94. Which country in the table has the least chocolate consumption per person?

95. How much more is the greatest chocolate consumption than the least chocolate consumption shown in the table?

96. How much more chocolate does the average German consume than the average citizen of the United Kingdom?

97. Make a new chart listing the countries and their corresponding chocolate consumptions in order from greatest to least.

Review

Multiply. See Sections 1.5 and 4.3.

98. $23 \cdot 2$

99. $46 \cdot 3$

100. $\left(\frac{2}{3}\right)^2$

101. $\left(\frac{1}{5}\right)^3$

Concept Extensions

Solve. See the first Concept Check in this section.

102. A friend asks you to check his calculation to the right. Is it correct? If not, explain your friends' error and correct the calculation.

$$\begin{array}{r} \overset{1}{9.2} \\ \overset{1}{8.63} \\ +4.005 \\ \hline 4.960 \end{array}$$

Find the unknown length in each figure.

△ **103.**

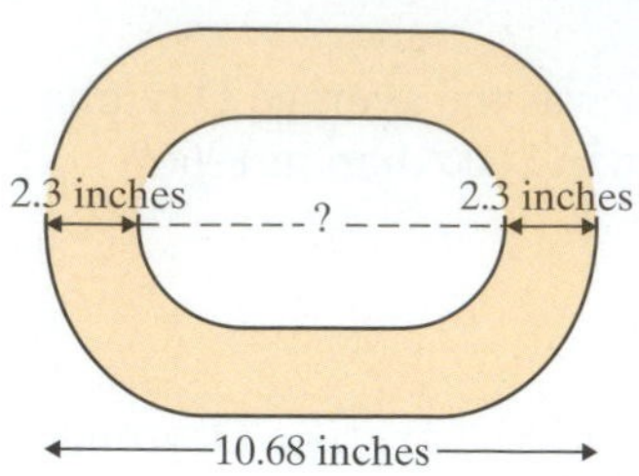

△ **104.**

5.26 meters | 7.82 meters | ? meters

17.67 meters

Let's review the values of these common U.S. coins in order to answer the following exercises.

Penny Nickel Dime Quarter

$0.01 $0.05 $0.10 $0.25

For Exercises 105 and 106, write the value of each group of coins. To do so, it is usually easiest to start with the coin(s) of greatest value and end with the coin(s) of least value.

105.

106.

107. Name the different ways that coins can have a value of \$0.17 given that you may use no more than 10 coins.

108. Name the different ways that coin(s) can have a value of \$0.25 given that there are no pennies.

109. Why shouldn't the sum

$$82.95 + 51.26$$

be estimated as $90 + 60 = 150$? See the second Concept Check in this section.

110. Explain how adding or subtracting decimals is similar to adding or subtracting whole numbers.

111. Laser beams can be used to measure the distance to the moon. One measurement showed the distance to the moon to be 256,435.235 miles. A later measurement showed that the distance is 256,436.012 miles. Find how much farther away the moon is in the second measurement as compared to the first.

Combine like terms and simplify.

112. $-8.689 + 4.286x - 14.295 - 12.966x + 30.861x$

113. $14.271 - 8.968x + 1.333 - 201.815x + 101.239x$

114. Can the sum of two negative decimals ever be a positive decimal? Why or why not?

5.3 MULTIPLYING DECIMALS AND CIRCUMFERENCE OF A CIRCLE

Objective A Multiplying Decimals

Multiplying decimals is similar to multiplying whole numbers. The only difference is that we place a decimal point in the product. To discover where a decimal point is placed in the product, let's multiply 0.6×0.03. We first write each decimal as an equivalent fraction and then multiply.

$$\underset{\text{1 decimal place}}{0.6} \times \underset{\text{2 decimal places}}{0.03} = \frac{6}{10} \times \frac{3}{100} = \frac{18}{1000} = \underset{\text{3 decimal places}}{0.018}$$

Notice that $1 + 2 = 3$, the number of decimal places in the product. Now let's multiply 0.03×0.002.

$$\underset{\text{2 decimal places}}{0.03} \times \underset{\text{3 decimal places}}{0.002} = \frac{3}{100} \times \frac{2}{1000} = \frac{6}{100{,}000} = \underset{\text{5 decimal places}}{0.00006}$$

Again, we see that $2 + 3 = 5$, the number of decimal places in the product.

Instead of writing decimals as fractions each time we want to multiply, we notice a pattern from these examples and state a rule that we can use:

Multiplying Decimals

Step 1: Multiply the decimals as though they are whole numbers.

Step 2: The decimal point in the product is placed so that the number of decimal places in the product is equal to the *sum* of the number of decimal places in the factors.

EXAMPLE 1 Multiply: 23.6×0.78

Solution:

$$\begin{array}{r} 23.6 \\ \times 0.78 \\ \hline 1888 \\ 16520 \\ \hline 18.408 \end{array}$$

23.6 — 1 decimal place
×0.78 — 2 decimal places

Since $1 + 2 = 3$, insert the decimal point in the product so that there are 3 decimal places.

Work Practice Problem 1

EXAMPLE 2 Multiply: 0.0531×16

Solution:

$$\begin{array}{r} 0.0531 \\ \times \quad 16 \\ \hline 3186 \\ 5310 \\ \hline 0.8496 \end{array}$$

0.0531 — 4 decimal places
× 16 — 0 decimal places

4 decimal places $(4 + 0 = 4)$

Work Practice Problem 2

Objectives

A Multiply Decimals.

B Estimate when Multiplying Decimals.

C Multiply Decimals by Powers of 10.

D Evaluate Expressions with Decimal Replacement Values

E Find the Circumference of Circle.

F Solve Problems by Multiplying Decimals.

PRACTICE PROBLEM 1

Multiply: 34.8×0.62

PRACTICE PROBLEM 2

Multiply: 0.0641×27

Answers

1. 21.576 **2.** 1.7307

✔Concept Check True or false? The number of decimal places in the product of 0.261 and 0.78 is 6. Explain.

PRACTICE PROBLEM 3

Multiply: $(7.3)(-0.9)$

EXAMPLE 3 Multiply: $(-2.6)(0.8)$

Solution: Recall that the product of a negative number and a positive number is a negative number.

$(-2.6)(0.8) = -2.08$

Work Practice Problem 3

Objective B Estimating when Multiplying Decimals

Just as for addition and subtraction, we can estimate when multiplying decimals to check the reasonableness of our answer.

PRACTICE PROBLEM 4

Multiply: 30.26×2.89. Then estimate to see whether the answer is reasonable.

EXAMPLE 4 Multiply: 28.06×1.95. Then estimate to see whether the answer is reasonable by rounding each factor, then multiplying the rounded numbers.

Solution:

Exact:	Estimate 1	Estimate 2
28.06	28 Rounded to ones or	30 Rounded to tens
$\times$ 1.95	$\times$ 2	$\times$ 2
14030	56	60
252540		
280600		
54.7170		

The answer 54.7170 is reasonable.

Work Practice Problem 4

As shown in Example 4, estimated results will vary depending on what estimates are used. Notice that estimating results is a good way to see whether the decimal point has been correctly placed.

Objective C Multiplying Decimals by Powers of 10

There are some patterns that occur when we multiply a number by a power of 10 such as 10, 100, 1000, 10,000, and so on.

$23.6951 \times 10 = 236.951$ (1 zero) Move the decimal point *1 place* to the *right.*

$23.6951 \times 100 = 2369.51$ (2 zeros) Move the decimal point *2 places* to the *right.*

$23.6951 \times 100{,}000 = 2{,}369{,}510.$ (5 zeros) Move the decimal point *5 places* to the *right* (insert a 0).

Answers

3. −6.57 4. 87.4514

✔ Concept Check Answer

false: 3 decimal places and 2 decimal places means 5 decimal places in the product

Notice that we move the decimal point the same number of places as there are zeros in the power of 10.

Multiplying Decimals by Powers of 10 such as 10, 100, 1000, 10,000 . . .

Move the decimal point to the *right* the same number of places as there are *zeros* in the power of 10.

EXAMPLES Multiply.

5. $7.68 \times 10 = 76.8$ — 7.68
6. $23.702 \times 100 = 2370.2$ — 23.702
7. $(-76.3)(1000) = -76{,}300$ — 76.300

Work Practice Problems 5–7

PRACTICE PROBLEMS 5–7

Multiply.

5. 46.8×10
6. 203.004×100
7. $(-2.33)(1000)$

There are also powers of 10 that are less than 1. The decimals 0.1, 0.01, 0.001, 0.0001, and so on are examples of powers of 10 less than 1. Notice the pattern when we multiply by these powers of 10:

$569.2 \times 0.1 = 56.92$ — Move the decimal point *1 place* to the *left.*
(1 decimal place)

$569.2 \times 0.01 = 5.692$ — Move the decimal point 2 *places* to the *left.*
(2 decimal places)

$569.2 \times 0.0001 = 0.05692$ — Move the decimal point 4 *places* to the *left* (insert one 0).
(4 decimal places)

Multiplying Decimals by Powers of 10 such as 0.1, 0.01, 0.001, 0.0001 . . .

Move the decimal point to the *left* the same number of places as there are *decimal places* in the power of 10.

EXAMPLES Multiply.

8. $42.1 \times 0.1 = 4.21$ — 42.1
9. $76{,}805 \times 0.01 = 768.05$ — 76,805.
10. $(-9.2)(-0.001) = 0.0092$ — 0009.2

Work Practice Problems 8–10

PRACTICE PROBLEMS 8–10

Multiply.

8. 6.94×0.1
9. 3.9×0.01
10. $(-7682)(-0.001)$

Many times we see large numbers written, for example, in the form 297.9 million rather than in the longer standard notation. The next example shows us how to interpret these numbers.

Answers

5. 468 6. 20,300.4 7. −2330
8. 0.694 9. 0.039 10. 7.682

PRACTICE PROBLEM 11

In 2004, there were 76.2 million families in the United States. Write this number in standard notation. (*Source:* U.S. Census Bureau)

EXAMPLE 11

In 2010, the population of the United States is projected to be 308.9 million. Write this number in standard notation. (*Source:* U.S. Census Bureau)

Solution: $308.9 \text{ million} = 308.9 \times 1 \text{ million}$

$= 308.9 \times 1{,}000{,}000 = 308{,}900{,}000$

Work Practice Problem 11

Objective D Using Decimals as Replacement Values

Now let's practice working with variables.

PRACTICE PROBLEM 12

Evaluate $7y$ for $y = -0.028$.

EXAMPLE 12 Evaluate xy for $x = 2.3$ and $y = 0.44$.

Solution: Recall that xy means $x \cdot y$.

$xy = (2.3)(0.44)$

$$\begin{array}{r} 2.3 \\ \times\ 0.44 \\ \hline 92 \\ 920 \\ \hline 1.012 \end{array}$$

$= 1.012 \longleftarrow$

Work Practice Problem 12

PRACTICE PROBLEM 13

Is -5.5 a solution of the equation $-6x = 33$?

EXAMPLE 13 Is -9 a solution of the equation $3.7y = -3.33$?

Solution: Replace y with -9 in the equation $3.7y = -3.33$ to see if a true equation results.

$3.7y = -3.33$

$3.7(-9) \stackrel{?}{=} -3.33$ Replace y with -9.

$-33.3 = -3.33$ False

Since $-33.3 = -3.33$ is a false statement, -9 is **not** a solution of $3.7y = -3.33$.

Work Practice Problem 13

Objective E Finding the Circumference of a Circle

Recall from Section 1.3 that the distance around a polygon is called its perimeter. The distance around a circle is given a special name called the **circumference,** and this distance depends on the radius or the diameter of the circle.

Circumference of a Circle

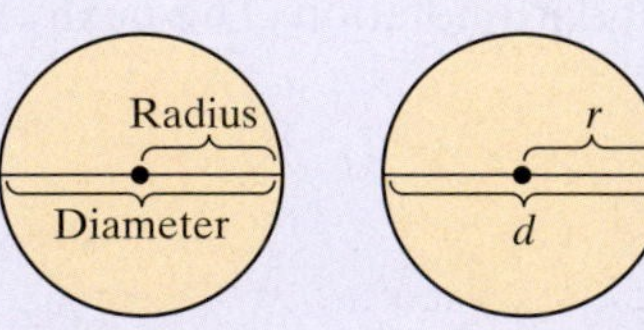

Circumference $= 2 \cdot \pi \cdot$ **radius** or Circumference $= \pi \cdot$ **diameter**

$C = 2\pi r$ or $C = \pi d$

Answers

11. 76,200,000 **12.** -0.196 **13.** yes

In Section 5.1, we learned about the symbol π as the Greek letter pi, pronounced "pie." It is a constant between 3 and 4. A decimal approximation for π is 3.14. Also, a fraction approximation for π is $\frac{22}{7}$.

EXAMPLE 14 Circumference of a Circle

Find the circumference of a circle whose radius is 5 inches. Then use the approximation 3.14 for π to approximate the circumference.

Solution: Let $r = 5$ in the formula $C = 2\pi r$.

$$C = 2\pi r$$
$$= 2\pi \cdot 5$$
$$= 10\pi$$

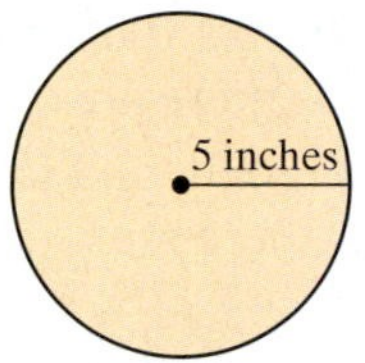

Next, replace π with the approximation 3.14.

$$C = 10\pi$$
(is approximately) ⟶ $\approx 10(3.14)$
$$= 31.4$$

The **exact** circumference or distance around the circle is 10π inches, which is **approximately** 31.4 inches.

Work Practice Problem 14

PRACTICE PROBLEM 14

Find the circumference of a circle whose radius is 11 meters. Then use the approximation 3.14 for π to approximate this circumference.

Objective F Solving Problems by Multiplying Decimals

The solutions to many real-life problems are found by multiplying decimals. We continue using our four problem-solving steps to solve such problems.

EXAMPLE 15 Finding the Total Cost of Materials for a Job

A college student is hired to paint a billboard with paint costing \$2.49 per quart. If the job requires 3 quarts of paint, what is the total cost of the paint?

Solution:

1. UNDERSTAND. Read and reread the problem. The phrase "total cost" might make us think addition, but since this problem requires repeated addition, let's multiply.
2. TRANSLATE.

In words:	Total cost	is	cost per quart of paint	times	number of quarts
	↓	↓	↓	↓	↓
Translate:	Total cost	=	2.49	×	3

3. SOLVE. We can estimate to check our calculations. The number 2.49 rounds to 2 and $2 \times 3 = 6$.

$$\begin{array}{r} \overset{1\,2}{2.49} \\ \times \quad 3 \\ \hline 7.47 \end{array}$$

4. INTERPRET. *Check* your work. Since 7.47 is close to our estimate of 6, our answer is reasonable. *State* your conclusion: The total cost of the paint is \$7.47.

Work Practice Problem 15

PRACTICE PROBLEM 15

A biology major is fertilizing her personal garden. She uses 5.6 ounces of fertilizer per square yard. The garden measures 60.5 square yards. How much fertilizer does she need?

Answers

14. 22π m $\approx$ 69.08 m
15. 338.8 oz

Vocabulary and Readiness Check

Use the choices below to fill in each blank.

circumference	left	sum	zeros
decimal places	right	product	factor

1. When multiplying decimals, the number of decimal places in the product is equal to the ____________ of the number of decimal places in the factors.
2. In $8.6 \times 5 = 43$, the number 43 is called the ____________, while 8.6 and 5 are each called a ____________.
3. When multiplying a decimal number by powers of 10, such as 10, 100, 1000, and so on, we move the decimal point in the number to the ____________ the same number of places as there are ____________ in the power of 10.
4. When multiplying a decimal number by powers of 10, such as 0.1, 0.01, and so on, we move the decimal point in the number to the ____________ the same number of places as there are ____________ in the power of 10.
5. The distance around a circle is called its ____________.

5.3 EXERCISE SET

FOR EXTRA HELP

Student Solutions Manual | PH Math/Tutor Center | CD/Video for Review | MathXL® | MyMathLab

Objectives A B Mixed Practice *Multiply. See Examples 1 through 4. For those exercises marked, also estimate to see if the answer is reasonable.*

1. 0.17×8

2. 0.23×9

3. 1.2×0.5

4. 6.8×0.3

5. $(-2.3)(7.65)$

6. $(4.7)(-9.02)$

7. $(-5.73)(-9.6)$

8. $(-7.84)(-3.5)$

9. 6.8×4.2
Exact:
Estimate:

10. 8.3×2.7
Exact:
Estimate:

11. 0.347×0.3

12. 0.864×0.4

13. 1.0047×8.2
Exact: Estimate:

14. 2.0005×5.5
Exact: Estimate:

15. 490.2×0.023

16. 300.9×0.032

Objective C *Multiply. See Examples 5 through 10.*

17. 6.5×10

18. 7.2×100

19. 8.3×0.1

20. 23.4×0.1

21. $(-7.093)(1000)$

22. $(-1.123)(1000)$

23. 7.093×100

24. 0.5×100

25. $(-9.83)(-0.01)$

26. $(-4.72)(-0.01)$

27. 25.23×0.001

28. 36.41×0.001

Objectives A B C Mixed Practice *Multiply. See Examples 1 through 10.*

29. 0.123×0.4

30. 0.216×0.3

31. $(147.9)(100)$

32. $(345.2)(100)$

33. 8.6×0.15

34. 0.42×5.7

35. $(937.62)(-0.01)$

36. $(-0.001)(562.01)$

37. 562.3×0.001

38. 993.5×0.001

39. $\begin{array}{r} 6.32 \\ \times\ 5.7 \\ \hline \end{array}$

40. $\begin{array}{r} 9.21 \\ \times\ 3.8 \\ \hline \end{array}$

Write each number in standard notation. See Example 11.

41. The cost of the Hubble Space Telescope at launch was \$1.5 billion. (*Source:* NASA)

42. About 40.4 million American households own at least one dog. (*Source:* American Pet Products Manufacturers Association)

43. The Blue Streak is the oldest roller coaster at Cedar Point, an amusement park in Sandusky, Ohio. Since 1964, it has given more than 49.8 million rides. (*Source:* Cedar Fair, L.P.)

44. There are 166.2 thousand full service restaurants in the United States. (*Source:* National Restaurant Association)

Objective D *Evaluate each expression for $x = 3$, $y = -0.2$, and $z = 5.7$. See Example 12.*

45. xy

46. yz

47. $xz - y$

48. $-5y + z$

Determine whether the given value is a solution of each given equation. See Example 13.

49. Is 14.2 a solution of $0.6x = 4.92$?

50. Is 1414 a solution of $100z = 14.14$?

51. Is -4 a solution of $3.5y = -14$?

52. Is -3.6 a solution of $0.7x = -2.52$?

Objective E *Find the circumference of each circle. Then use the approximation 3.14 for π and approximate each circumference. See Example 14.*

△ **53.**

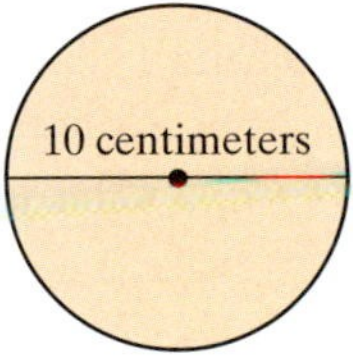

△ **54.**

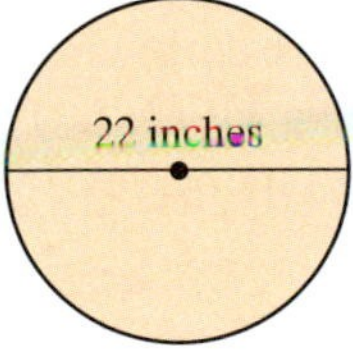

△ **55.**

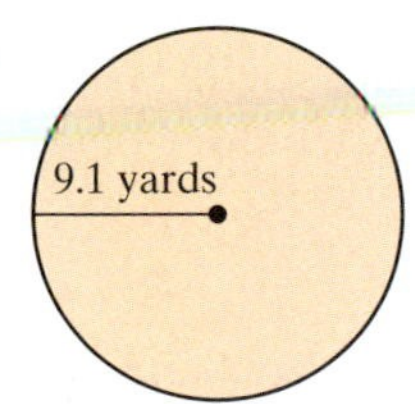

△ **56.**

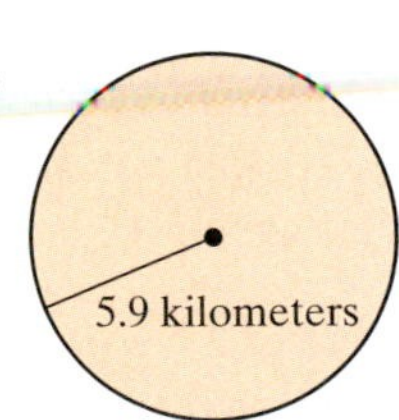

Objectives E F Mixed Practice *Solve. See Examples 14 and 15. For circumference applications find the exact circumference and then use 3.14 for π to approximate the circumference.*

57. A 1-ounce serving of cream cheese contains 6.2 grams of saturated fat. How much saturated fat is in 4 ounces of cream cheese? (*Source: Home and Garden Bulletin No. 72;* U.S. Department of Agriculture)

58. A 3.5-ounce serving of lobster meat contains 0.1 gram of saturated fat. How much saturated fat do 3 servings of lobster meat contain? (*Source:* The National Institute of Health)

59. Recall that the top face of the Apple iPod mini (see Section 5.2) measures 3.6 inches by 2.0 inches. Find the area of the face of the iPod mini.

60. Recall from Section 5.2 that the face of the regular Apple iPod measures 4.1 inches by 2.4 inches. Find the area of this iPod.

△ **61.** In 1893, the first ride called a Ferris wheel was constructed by Washington Gale Ferris. Its diameter was 250 feet. Find its circumference. Give an exact answer and an approximation using 3.14 for π. (*Source: The Handy Science Answer Book,* Visible Ink Press, 1994)

△ **62.** The radius of Earth is approximately 3950 miles. Find the distance around Earth at the equator. Give an exact answer and an approximation using 3.14 for π. (*Hint:* Find the circumference of a circle with radius 3950 miles.)

△ **63.** The London Eye, built for the Millennium celebration in London, resembles a gigantic ferris wheel with a diameter of 135 meters. If Adam Hawn rides the Eye for one revolution, find how far he travels. Give an exact answer and an approximation using 3.14 for π. (*Source:* Londoneye.com)

△ **64.** The world's longest suspension bridge is the Akashi Kaikyo Bridge in Japan. This bridge has two circular caissons, which are underwater foundations. If the diameter of a caisson is 80 meters, find its circumference. Give an exact answer and an approximation using 3.14 for π. (*Source: Scientific American;* How Things Work Today)

65. A meter is a unit of length in the metric system that is approximately equal to 39.37 inches. Sophia Wagner is 1.65 meters tall. Find her approximate height in inches.

66. The doorway to a room is 2.15 meters tall. Approximate this height in inches. (*Hint:* See Exercise 65.)

67. An electrician for Central Power and Light worked 40 hours last week. Calculate his pay before taxes for last week if his hourly wage is $17.88.

68. An assembly line worker worked 20 hours last week. Her hourly rate is $19.52 per hour. Calculate her pay before taxes.

△ **69. a.** Approximate the circumference of each circle.

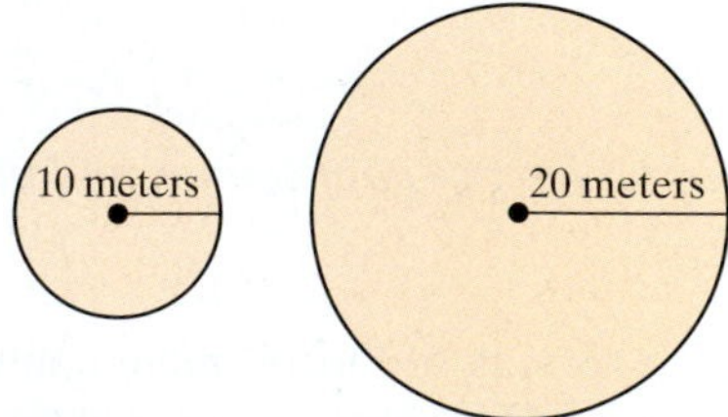

b. If the radius of a circle is doubled, is its corresponding circumference doubled?

△ **70. a.** Approximate the circumference of each circle.

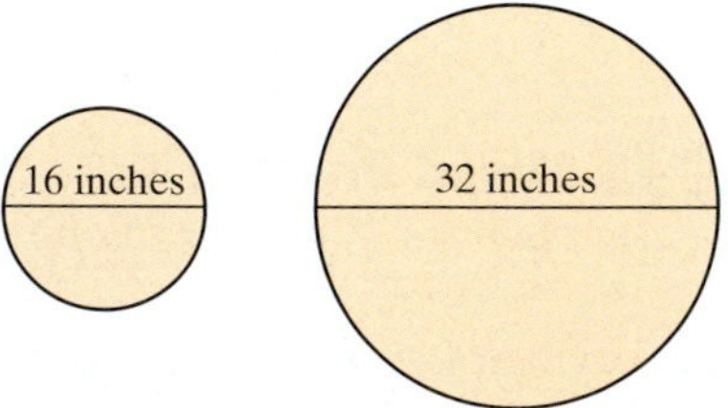

b. If the diameter of a circle is doubled, is its corresponding circumference doubled?

71. In 2005, American farmers received an average of $3.40 per bushel of wheat. How much did a farmer receive for selling 100 bushels of wheat? (*Source:* National Agricultural Statistics Service)

72. In 2005, American farmers received an average of $5.50 per bushel of soybeans. How much did a farmer receive for selling 10,000 bushels of soybeans? (*Source:* National Agricultural Statistics Service)

This table shows currency exchange rates for various countries on June 19, 2006. To find the amount of foreign currency equivalent to an amount of U.S. dollars, multiply the U.S. dollar amount by the exchange rate listed in the table. Use this table to answer Exercises 73 through 76.

Foreign Currency Exchange Rates

Country	Exchange Rate
Australian Dollar	1.3559
Canadian Dollar	1.1232
Sri Lanka Rupee	103.550
Japanese Yen	115.26
European Union Euro	0.7951
Chinese Yuan	8.0035

73. How many Japanese yen are equivalent to 675 U.S. dollars?

74. Suppose you wish to exchange 500 U.S. dollars into Chinese yuan. How much money, in Chinese yuan, would you receive?

75. The Robinson family is taking a vacation to Europe. How much European Union euros can they "buy" with 900 U.S. dollars?

76. An Australian tourist spent \$85 for souvenir tee shirts in San Francisco. What did he pay for the tee shirts in Australian dollars?

Review

Divide. See Sections 1.6 and 4.3.

77. $2920 \div 365$

78. $2916 \div 6$

79. $-\frac{24}{7} \div \frac{8}{21}$

80. $\frac{162}{25} \div -\frac{9}{75}$

Concept Extensions

Mixed Practice (*Sections 5.2, 5.3*) *Perform the indicated operations.*

81. $3.6 + 0.04$

82. -3.6×0.04

83. $3.6 - 0.04$

84. $100 - 48.6$

85. -0.221×0.5

86. $7.2 + 0.14 + 98.6$

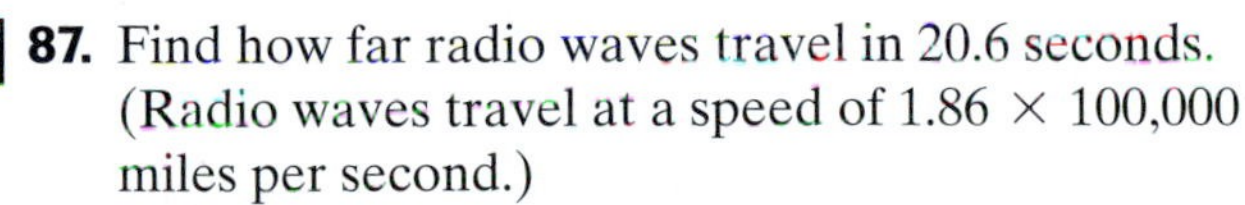

87. Find how far radio waves travel in 20.6 seconds. (Radio waves travel at a speed of $1.86 \times 100,000$ miles per second.)

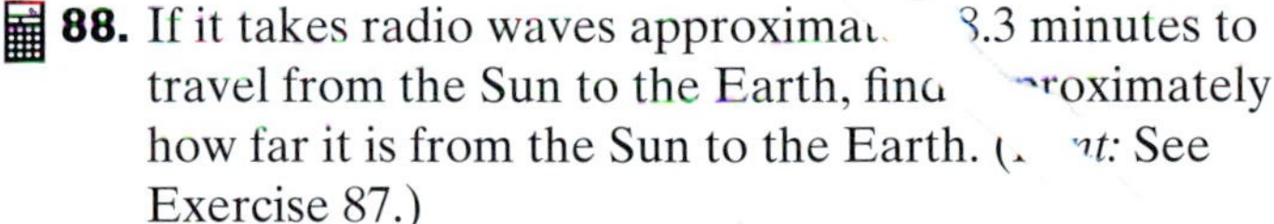

88. If it takes radio waves approximately 8.3 minutes to travel from the Sun to the Earth, find approximately how far it is from the Sun to the Earth. (*Hint:* See Exercise 87.)

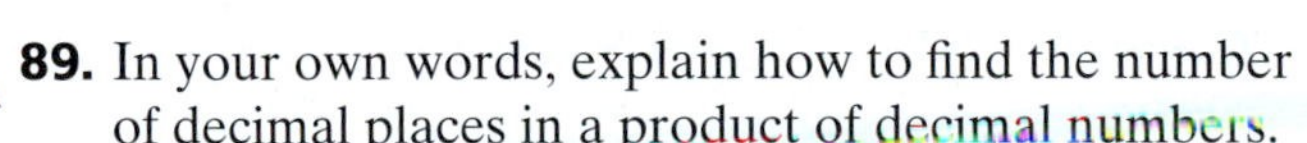

89. In your own words, explain how to find the number of decimal places in a product of decimal numbers.

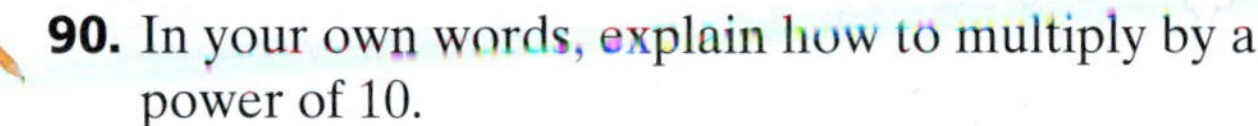

90. In your own words, explain how to multiply by a power of 10.

91. Write down two decimal numbers whose product will contain 5 decimal places. Without multiplying, explain how you know your answer is correct.

STUDY SKILLS BUILDER

How Are Your Homework Assignments Going?

Remember that it is important to keep up with homework. Why? Many concepts in mathematics build on each other. Often, your understanding of a day's lecture depends on an understanding of the previous day's material.

To complete a homework assignment, remember these four things:

- Attempt all of it.
- Check it.
- Correct it.
- If needed, ask questions about it.

Take a moment and review your completed homework assignments. Answer the exercises below based on this review.

1. Approximate the fraction of your homework you have attempted.
2. Approximate the fraction of your homework you have checked (if possible).
3. If you are able to check your homework, have you corrected it when errors have been found?
4. When working homework, if you do not understand a concept, what do you personally do?

5.4 DIVIDING DECIMALS

Objectives

A Divide Decimals.

B Estimate when Dividing Decimals.

C Divide Decimals by Powers of 10.

D Evaluate Expressions with Decimal Replacement Values.

E Solve Problems by Dividing Decimals.

Objective A Dividing Decimals

Dividing decimal numbers is similar to dividing whole numbers. The only difference is that we place a decimal point in the quotient. If the divisor is a whole number, we place the decimal point in the quotient directly above the decimal point in the dividend, and then divide as with whole numbers. Recall that division can be checked by multiplication.

PRACTICE PROBLEM 1

Divide: 370.4 ÷ 8. Check your answer.

EXAMPLE 1 Divide: 270.2 ÷ 7. Check your answer.

Solution: We divide as usual. The decimal point in the quotient is directly above the decimal point in the dividend.

Write the decimal point.

$$\begin{array}{r} 38.6 \leftarrow \text{quotient} \\ \text{divisor} \rightarrow 7\overline{)270.2} \leftarrow \text{dividend} \\ -21 \\ \hline 60 \\ -56 \\ \hline 4\,2 \\ -4\,2 \\ \hline 0 \end{array}$$

Check:

$$\begin{array}{r} \overset{6\,4}{38.6} \\ \times\quad 7 \\ \hline 270.2 \end{array}$$

The quotient is 38.6.

Work Practice Problem 1

PRACTICE PROBLEM 2

Divide: $48\overline{)34.08}$. Check your answer.

EXAMPLE 2 Divide: $32\overline{)8.32}$

Solution: We divide as usual. The decimal point in the quotient is directly above the decimal point in the dividend.

$$\begin{array}{r} 0.26 \leftarrow \text{quotient} \\ \text{divisor} \rightarrow 32\overline{)8.32} \leftarrow \text{dividend} \\ -6\,4 \\ \hline 1\,92 \\ -1\,92 \\ \hline 0 \end{array}$$

Check:

$$\begin{array}{rl} 0.26 & \text{quotient} \\ \times\ 32 & \text{divisor} \\ \hline 52 & \\ 7\,80 & \\ \hline 8.32 & \text{dividend} \end{array}$$

Work Practice Problem 2

Sometimes to continue dividing we need to insert zeros after the last digit in the dividend.

Answers
1. 46.3 **2.** 0.71

EXAMPLE 3 Divide: $-5.98 \div 115$

Solution: Recall that a negative number divided by a positive number gives a negative quotient.

$$\begin{array}{r} 0.052 \\ 115\overline{)5.980} \\ \underline{-575} \\ 230 \\ \underline{-230} \\ 0 \end{array}$$

← Insert one 0.

Thus $-5.98 \div 115 = -0.052$.

Work Practice Problem 3

PRACTICE PROBLEM 3

Divide and check.

a. $-15.89 \div 14$

b. $-2.808 \div (-104)$

If the divisor is not a whole number, before we divide we need to move the decimal point to the right until the divisor is a whole number.

$$1.5\overline{)64.85}$$

divisor ↑ ↑ dividend

To understand how this works, let's rewrite

$1.5\overline{)64.85}$ as $\dfrac{64.85}{1.5}$

and then multiply by 1 in the form of $\dfrac{10}{10}$. We use the form $\dfrac{10}{10}$ so that the denominator (divisor) becomes a whole number.

$$\frac{64.85}{1.5} = \frac{64.85}{1.5} \cdot 1 = \frac{64.85}{1.5} \cdot \frac{10}{10} = \frac{64.85 \cdot 10}{1.5 \cdot 10} = \frac{648.5}{15},$$

which can be written as $15.\overline{)648.5}$. Notice that

$1.5\overline{)64.85}$ is equivalent to $15.\overline{)648.5}$

The decimal points in the dividend and the divisor were both moved one place to the right, and the divisor is now a whole number. This procedure is summarized next:

Dividing by a Decimal

Step 1: Move the decimal point in the divisor to the right until the divisor is a whole number.

Step 2: Move the decimal point in the dividend to the right the *same number of places* as the decimal point was moved in Step 1.

Step 3: Divide. Place the decimal point in the quotient directly over the moved decimal point in the dividend.

Answers

3. a. −1.135 **b.** 0.027

PRACTICE PROBLEM 4

Divide: 166.88 ÷ 5.6

EXAMPLE 4 Divide: 10.764 ÷ 2.3

Solution: We move the decimal points in the divisor and the dividend one place to the right so that the divisor is a whole number.

$2.3\overline{)10.764}$ becomes

$$\begin{array}{r} 4.68 \\ 23.\overline{)107.64} \\ -92 \\ \hline 15\,6 \\ -13\,8 \\ \hline 1\,84 \\ -1\,84 \\ \hline 0 \end{array}$$

■ **Work Practice Problem 4**

PRACTICE PROBLEM 5

Divide: 1.976 ÷ 0.16

EXAMPLE 5 Divide: 5.264 ÷ 0.32

Solution:

$0.32\overline{)5.264}$ becomes

$$\begin{array}{r} 16.45 \\ 32\overline{)526.40} \\ -32 \\ \hline 206 \\ -192 \\ \hline 14\,4 \\ -12\,8 \\ \hline 1\,60 \\ -1\,60 \\ \hline 0 \end{array}$$

Insert one 0.

■ **Work Practice Problem 5**

✔**Concept Check** Is it always true that the number of decimal places in a quotient equals the sum of the decimal places in the dividend and divisor?

PRACTICE PROBLEM 6

Divide 23.4 ÷ 0.57. Round the quotient to the nearest hundredth.

EXAMPLE 6 Divide: 17.5 ÷ 0.48. Round the quotient to the nearest hundredth.

Solution: First we move the decimal points in the divisor and the dividend two places. Then we divide and round the quotient to the nearest hundredth.

hundredths place

$$\begin{array}{r} 36.458 \approx 36.46 \\ 48.\overline{)1750.000} \\ -144 \\ \hline 310 \\ -288 \\ \hline 22\,0 \\ -19\,2 \\ \hline 2\,80 \\ -2\,40 \\ \hline 400 \\ -384 \\ \hline 16 \end{array}$$

"is approximately"

When rounding to the nearest hundredth, carry the division process out to one more decimal place, the thousandths place.

■ **Work Practice Problem 6**

Answers

4. 29.8 **5.** 12.35 **6.** 41.05

✔ Concept Check Answer

no

✔**Concept Check** If a quotient is to be rounded to the nearest thousandth, to what place should the division be carried out? (Assume that the division carries out to your answer.)

Objective B Estimating when Dividing Decimals

Just as for addition, subtraction, and multiplication of decimals, we can estimate when dividing decimals to check the reasonableness of our answer.

EXAMPLE 7 Divide: 272.356 ÷ 28.4. Then estimate to see whether the proposed result is reasonable.

Solution:

Exact:	**Estimate 1**		**Estimate 2**
$\begin{array}{r} 9.59 \\ 284.\overline{)2723.56} \\ -\,2556 \\ \hline 1675 \\ -\,1420 \\ \hline 25 \\ -\,2 \end{array}$	$\begin{array}{r} 9 \\ 30\overline{)270} \end{array}$	or	$\begin{array}{r} 10 \\ 30\overline{)300} \end{array}$

The estir ıs 9 or 10, so 9.59 is reasonable.

Work ctice Problem 7

PRACTICE PROBLEM 7

Divide: 713.7 ÷ 91.5. Then estimate to see whether the proposed answer is reasonable.

Objec e C Dividing Decimals by Powers of 10

As ı multiplication, there are patterns that occur when we divide decimals by p rs of 10 such as 10, 100, 1000, and so on.

$\frac{569.2}{10} = 56.92$ Move the decimal point *1 place* to the *left*. (1 zero)

$\frac{569.2}{10{,}000} = 0.05692$ Move the decimal point *4 places* to the *left*. (4 zeros)

This pattern suggests the following rule:

Dividing Decimals by Powers of 10 such as 10, 100, or 1000

Move the decimal point of the dividend to the *left* the same number of places as there are *zeros* in the power of 10.

EXAMPLES Divide.

8. $\frac{786.1}{1000} = 0.7861$ Move the decimal point *3 places* to the *left*. (3 zeros)

9. $-\frac{0.12}{10} = -0.012$ Move the decimal point *1 place* to the *left*. (1 zero)

Work Practice Problems 8–9

PRACTICE PROBLEMS 8–9

Divide.

8. $\frac{362.1}{1000}$ 9. $-\frac{0.49}{10}$

Answers

7. 7.8 8. 0.3621 9. −0.049

✔ Concept Check Answer

ten-thousandths place

Objective D Using Decimals as Replacement Values

PRACTICE PROBLEM 10

Evaluate $x \div y$ for $x = 0.035$ and $y = 0.02$.

EXAMPLE 10 Evaluate $x \div y$ for $x = 2.5$ and $y = 0.05$.

Solution: Replace x with 2.5 and y with 0.05.

$$x \div y = 2.5 \div 0.05 \qquad 0.05\overline{)2.5} \text{ becomes } 5\overline{)250} = 50$$
$$= 50$$

Work Practice Problem 10

PRACTICE PROBLEM 11

Is 39 a solution of the equation $\frac{x}{100} = 3.9$?

EXAMPLE 11 Is 720 a solution of the equation $\frac{y}{100} = 7.2$?

Solution: Replace y with 720 to see if a true statement results.

$\frac{y}{100} = 7.2$ Original equation

$\frac{720}{100} \stackrel{?}{=} 7.2$ Replace y with 720.

$7.2 = 7.2$ True

Since $7.2 = 7.2$ is a true statement, 720 is a solution of the equation.

Work Practice Problem 11

Objective E Solving Problems by Dividing Decimals

Many real-life problems involve dividing decimals.

PRACTICE PROBLEM 12

A bag of fertilizer covers 1250 square feet of lawn. Tim Parker's lawn measures 14,800 square feet. How many bags of fertilizer does he need? If he can buy only whole bags of fertilizer, how many whole bags does he need?

EXAMPLE 12 **Calculating Materials Needed for a Job**

A gallon of paint covers a 250-square-foot area. If Betty Adkins wishes to paint a wall that measures 1450 square feet, how many gallons of paint does she need? If she can buy only gallon containers of paint, how many gallon containers does she need?

Solution:

1. UNDERSTAND. Read and reread the problem. We need to know how many 250s are in 1450, so we divide.
2. TRANSLATE.

In words:	number of gallons	is	square feet	divided by	square feet per gallon
	↓	↓	↓	↓	↓
Translate:	number of gallons	=	1450	÷	250

3. SOLVE. Let's see if our answer is reasonable by estimating. The dividend 1450 rounds to 1500 and the divisor 250 rounds to 300. Then $1500 \div 300 = 5$.

$$\begin{array}{r} 5.8 \\ 250\overline{)1450.0} \\ -1250 \\ \hline 200\ 0 \\ -200\ 0 \\ \hline 0 \end{array}$$

4. INTERPRET. *Check* your work. Since our estimate is close to our answer of 5, our answer is reasonable. *State* your conclusion: Betty needs 5.8 gallons of paint. If she can buy only gallon containers of paint, she needs 6 gallon containers of paint to complete the job.

Work Practice Problem 12

Answers

10. 1.75 11. no
12. 11.84 bags; 12 bags

CALCULATOR EXPLORATIONS Estimation

Calculator errors can easily be made by pressing an incorrect key or by not pressing a correct key hard enough. Estimation is a valuable tool that can be used to check calculator results.

EXAMPLE Use estimation to determine whether the calculator result is reasonable or not. (For example, a result that is not reasonable can occur if proper keys are not pressed.)

Simplify: $82.064 \div 23$
Calculator display: 35.68

Solution: Round each number to the nearest 10. Since $80 \div 20 = 4$, the calculator display 35.68 is not reasonable.

Use estimation to determine whether each result is reasonable or not.

1. 102.62×41.8 Result: 428.9516

2. $174.835 \div 47.9$ Result: 3.65

3. $1025.68 - 125.42$ Result: 900.26

4. $562.781 + 2.96$ Result: 858.781

Vocabulary and Readiness Check

Use the choices below to fill in each blank. Some choices may be used more than once, and some not used at all.

dividend	divisor	quotient	true
zeros	left	right	false

1. In $6.5 \div 5 = 1.3$, the number 1.3 is called the ______________, 5 is the ______________, and 6.5 is the ______________.

2. To check a division exercise, we can perform the following multiplication: quotient · __________ = __________.

3. To divide a decimal number by a power of 10, such as 10, 100, 1000, and so on, we move the decimal point in the number to the ______________ the same number of places as there are ______________ in the power of 10.

4. True or false: If we replace x with -12.6 and y with 0.3 in the expression $y \div x$, we have $0.3 \div (-12.6)$. ______________

5.4 EXERCISE SET

FOR EXTRA HELP

Student Solutions Manual

PH Math/Tutor Center

CD/Video for Review

MathXL®

MyMathLab

Objectives A B Mixed Practice *Divide. See Examples 1 through 5 and 7. For those exercises marked, also estimate to see if the answer is reasonable.*

1. $6\overline{)27.6}$

2. $4\overline{)23.6}$

3. $5\overline{)0.47}$

4. $6\overline{)0.51}$

5. $0.06\overline{)18}$

6. $0.04\overline{)20}$

7. $0.82\overline{)4.756}$

8. $0.92\overline{)3.312}$

9. $5.5\overline{)36.3}$
Exact:
Estimate:

10. $2.2\overline{)21.78}$
Exact:
Estimate:

11. $7.434 \div 18$

12. $8.304 \div 16$

13. $36 \div (-0.06)$

14. $36 \div (-0.04)$

15. Divide -4.2 by -0.6.

16. Divide -3.6 by -0.9.

17. $0.27\overline{)1.296}$

18. $0.34\overline{)2.176}$

19. $0.02\overline{)42}$

20. $0.03\overline{)24}$

21. $4.756 \div 0.82$

22. $3.312 \div 0.92$

23. $-36.3 \div -6.6$

24. $-21.78 \div -9.9$

25. $7.2\overline{)70.56}$
Exact:
Estimate:

26. $6.3\overline{)54.18}$
Exact:
Estimate:

27. $5.4\overline{)51.84}$

28. $7.7\overline{)33.88}$

29. $\dfrac{1.215}{0.027}$

30. $\dfrac{3.213}{0.051}$

31. $0.25\overline{)13.648}$

32. $0.75\overline{)49.866}$

33. $3.78\overline{)0.02079}$

34. $2.96\overline{)0.01332}$

Divide. Round the quotients as indicated. See Example 6.

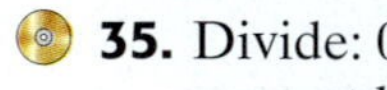

35. Divide: $0.549 \div 0.023$. Round the quotient to the nearest hundredth.

36. Divide: $0.0453 \div 0.98$. Round the quotient to the nearest thousandth.

37. Divide: $68.39 \div 0.6$. Round the quotient to the nearest tenth.

38. Divide: $98.83 \div 3.5$. Round the quotient to the nearest tenth.

Objective C *Divide. See Examples 8 and 9.*

39. $\dfrac{83.397}{100}$

40. $\dfrac{64.423}{100}$

41. $\dfrac{26.87}{10}$

42. $\dfrac{13.49}{10}$

43. $12.9 \div (-1000)$

44. $13.49 \div (-10{,}000)$

Objectives A C Mixed Practice *Divide. See Examples 1 through 5, 8, and 9.*

45. $7\overline{)88.2}$

46. $9\overline{)130.5}$

47. $\frac{13.1}{10}$

48. $\frac{17.7}{10}$

49. $\frac{456.25}{10,000}$

50. $\frac{986.11}{10,000}$

51. $1.239 \div 3$

52. $0.54 \div 12$

53. Divide 4.2 by -0.6.

54. Divide 3.6 by -0.9.

55. $-1.296 \div 0.27$

56. $-2.176 \div 0.34$

57. Divide 42 by 0.02.

58. Divide 24 by 0.03.

59. Divide -18 by -0.6.

60. Divide 20 by 0.4.

61. Divide 87 by -0.0015.

62. Divide 35 by -0.0007.

63. $-1.104 \div 1.6$

64. $-2.156 \div 0.98$

65. $-2.4 \div (-100)$

66. $-86.79 \div (-1000)$

67. $\frac{4.615}{0.071}$

68. $\frac{23.8}{0.035}$

Objective D *Evaluate each expression for $x = 5.65$, $y = -0.8$, and $z = 4.52$. See Example 10.*

69. $z \div y$

70. $z \div x$

71. $x \div y$

72. $y \div 2$

Determine whether the given values are solutions of the given equations. See Example 11.

73. $\frac{x}{4} = 3.04$; $x = 12.16$

74. $\frac{y}{8} = 0.89$; $y = 7.12$

75. $\frac{z}{100} = 0.8$; $z = 8$

76. $\frac{x}{10} = 0.23$; $x = 23$

Objective E *Solve. See Example 12.*

77. Josef Jones is painting the walls of a room. The walls have a total area of 546 square feet. A quart of paint covers 52 square feet. If he must buy paint in whole quarts, how many quarts does he need?

78. A page of a book contains about 1.5 kilobytes of information. If a computer disk can hold 740 kilobytes of information, how many pages of a book can be stored on one computer disk? Round to the nearest tenth of a page.

79. There are approximately 39.37 inches in 1 meter. How many meters, to the nearest tenth of a meter, are there in 200 inches?

← 1 meter →

← ≈39.37 inches →

80. There are 2.54 centimeters in 1 inch. How many inches are there in 50 centimeters? Round to the nearest tenth.

← 1 inch →

← 2.54 cm →

81. In the United States, an average child will wear down 730 crayons by his or her tenth birthday. Find the number of boxes of 64 crayons this is equivalent to. Round to the nearest tenth. (*Source:* Binney & Smith Inc.)

82. In 2005, American farmers received an average of $41.96 per hundred pounds of turkey. What was the average price per pound for turkeys? (*Source:* National Agricultural Statistics Service)

A child is to receive a dose of 0.5 teaspoon of cough medicine every 4 hours. If the bottle contains 4 fluid ounces, answer Exercises 83 through 86.

83. A fluid ounce equals 6 teaspoons. How many teaspoons are in 4 fluid ounces?

84. The bottle of medicine contains how many doses for the child?

85. If the child takes a dose every four hours, how many days will the medicine last?

86. If the child takes a dose every six hours, how many days will the medicine last?

87. Americans aged 18–22 drive, on average, 12,900 miles per year. About how many miles each week is that? Round to the nearest tenth. (*Note:* There are 52 weeks in a year.) (*Source:* U.S. Department of Energy)

88. Javier Xeron was interested in the gas mileage on his "new" used car. He filled the tank, drove 423.8 miles, and filled the tank again. When he refilled the tank, it took 19.35 gallons of gas. Calculate the miles per gallon for Javier's car. Round to the nearest tenth

89. The leading money winner in men's professional golf in 2005 was Tiger Woods. He earned approximately \$10,628,000. Suppose he had earned this working 40 hours each week for a year. Determine his hourly wage to the nearest cent. (*Note:* There are 52 weeks in a year.) (*Source:* Professional Golf Association)

90. The book *Harry Potter and the Half-Blood Prince* was released to the public on July 16, 2005. Booksellers in the United State sold approximately 6900 thousand copies in the first 24 hours after release. If the same number of books were sold each hour, calculate the number of books sold each hour in the United States for that first day.

Review

Perform the indicated operation. See Sections 4.3 and 4.5.

91. $\frac{3}{5} \cdot \frac{7}{10}$

92. $\frac{3}{5} \div \frac{7}{10}$

93. $\frac{3}{5} - \frac{7}{10}$

94. $-\frac{3}{4} - \frac{1}{14}$

Concept Extensions

Mixed Practice (*Sections 5.2, 5.3, 5.4*) *Perform the indicated operation.*

95. $1.278 \div 0.3$

96. 1.278×0.3

97. $1.278 + 0.3$

98. $1.278 - 0.3$

99. $(-8.6)(3.1)$

100. $7.2 + 0.05 + 49.1$

101. $\begin{array}{r} 1000 \\ -\ \ 95.71 \\ \hline \end{array}$

102. $\frac{87.2}{-10,000}$

Choose the best estimate.

103. 8.62×41.7
- **a.** 36
- **b.** 32
- **c.** 360
- **d.** 3.6

104. $1.437 + 20.69$
- **a.** 34
- **b.** 22
- **c.** 3.4
- **d.** 2.2

105. $78.6 \div 97$
- **a.** 7.86
- **b.** 0.786
- **c.** 786
- **d.** 7860

106. $302.729 - 28.697$
- **a.** 270
- **b.** 20
- **c.** 27
- **d.** 300

Recall from Section 1.7 that the average of a list of numbers is their total divided by how many numbers there are in the list. Use this procedure to find the average of the test scores listed in Exercises 107 and 108. If necessary, round to the nearest tenth.

107. 86, 78, 91, 87

108. 56, 75, 80

△ **109.** The area of a rectangle is 38.7 square feet. If its width is 4.5 feet, find its length.

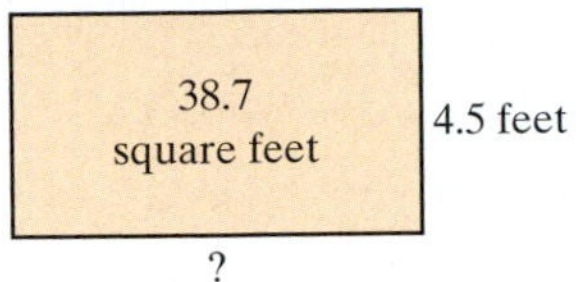

△ **110.** The perimeter of a square is 180.8 centimeters. Find the length of a side.

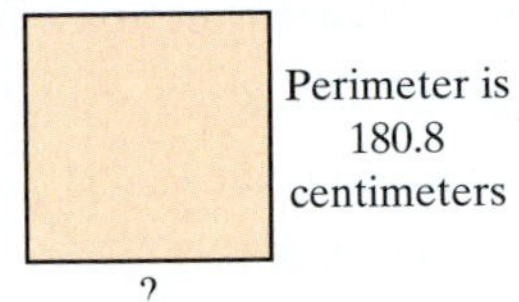

111. When dividing decimals, describe the process you use to place the decimal point in the quotient.

112. In your own words, describe how to quickly divide a number by a power of 10 such as 10, 100, 1000, etc.

To convert wind speeds in miles per hour to knots, divide by 1.15. Use this information and the Saffir-Simpson Hurricane Intensity chart below to answer Exercises 113 and 114. Round to the nearest tenth.

Saffir-Simpson Hurricane Intensity Scale

Category	Wind Speed	Barometric Pressure [inches of mercury (Hg)]	Storm Surge	Damage Potential
1 (Weak)	75–95 mph	≥28.94 in.	4–5 ft	Minimal damage to vegetation
2 (Moderate)	96–110 mph	28.50–28.93 in.	6–8 ft	Moderate damage to houses
3 (Strong)	111–130 mph	27.91–28.49 in.	9–12 ft	Extensive damage to small buildings
4 (Very Strong)	131–155 mph	27.17–27.90 in.	13–18 ft	Extreme structural damage
5 (Devastating)	>155 mph	<27.17 in.	>18 ft	Catastrophic building failures possible

113. The chart gives wind speeds in miles per hour. What is the range of wind speeds for a Category 1 hurricane in knots?

114. What is the range of wind speeds for a Category 4 hurricane in knots?

△ **115.** Don Larson is building a horse corral that's shaped like a rectangle with dimensions of 24.28 meters by 15.675 meters. He plans to make a four-wire fence; that is, he will string four wires around the corral. How much wire will he need?

116. Takeesha Bethel signed up for a new credit card that guarantees her no interest charges on transferred balances for a year. She transferred over a $2523.86 balance from her old credit card. Her minimum payment is $185.35 per month. If she only pays the minimum, will she pay off her balance before interest charges start again?

THE BIGGER PICTURE Operations on Sets of Numbers

Continue your outline from Sections 1.6, 1.7, 2.4, 3.3, 4.3, and 4.7. Suggestions are once again written to help you complete this part of your outline.

I. Operations on Sets of Numbers

A. Whole Numbers

1. **Add or Subtract** (Section 1.3)
2. **Multiply or Divide** (Sections 1.5, 1.6)
3. **Exponent** (Section 1.7)
4. **Order of Operations** (Section 1.7)

B. Integers

1. **Add** (Section 2.2)
2. **Subtract** (Section 2.3)
3. **Multiply or Divide** (Section 2.4)

C. Fractions

1. **Simplify** (Section 4.2)
2. **Multiply** (Section 4.3)
3. **Divide** (Section 4.3)
4. **Add or Subtract** (Sections 4.4, 4.5)

D. Decimals

1. **Add or Subtract:** Line up decimal points.

$$\begin{array}{r} 1.27 \\ +0.6 \\ \hline 1.87 \end{array}$$

2. **Multiply:**

$$\begin{array}{rl} 2.56 & \text{2 decimal places} \\ \times\ 3.2 & \text{1 decimal place} \\ \hline 512 & 2 + 1 = 3 \\ 7680 & \\ \hline 8.192 & \text{3 decimal places} \end{array}$$

3. **Divide:**

$$8\overline{)5.6}\ \ \text{quotient } 0.7 \qquad 0.6\overline{)0.786}\ \ \text{quotient } 1.31$$

II. Solving Equations

A. Equations in General (Section 3.3)

Perform the indicated operations.

1. $3.6 + 8.092 + 10.48$
2. $7 - 3.049$
3. 91.332×100
4. $-\frac{68}{10}$
5. $\begin{array}{r} 5.2 \\ \times\ 0.27 \\ \hline \end{array}$
6. $9\overline{)77.94}$
7. $0.35\overline{)0.01785}$
8. $2.3 - (0.4)^2$
9. $\frac{8}{15} - \frac{2}{5}$
10. $-\frac{8}{15} \cdot \frac{2}{5}$

INTEGRATED REVIEW

Sections 5.1–5.4

Operations on Decimals

Perform the indicated operation.

1. $1.6 + 0.97$

2. $3.2 + 0.85$

3. $9.8 - 0.9$

4. $10.2 - 6.7$

5. $\begin{array}{r} 0.8 \\ \times 0.2 \\ \hline \end{array}$

6. $\begin{array}{r} 0.6 \\ \times 0.4 \\ \hline \end{array}$

7. $8\overline{)2.16}$

8. $6\overline{)3.12}$

9. $(9.6)(-0.5)$

10. $(-8.7)(-0.7)$

11. $\begin{array}{r} 123.6 \\ -\ 48.04 \\ \hline \end{array}$

12. $\begin{array}{r} 325.2 \\ -\ 36.08 \\ \hline \end{array}$

13. $-25 + 0.026$

14. $0.125 + (-44)$

15. $29.24 \div (-3.4)$

16. $-10.26 \div (-1.9)$

17. -2.8×100

18. 1.6×1000

19. $\begin{array}{r} 96.21 \\ 7.028 \\ +121.7 \\ \hline \end{array}$

20. $\begin{array}{r} 0.268 \\ 1.93 \\ +142.881 \\ \hline \end{array}$

21. $-25.76 \div -46$

22. $-27.09 \div 43$

23. $\begin{array}{r} 12.004 \\ \times \quad 2.3 \\ \hline \end{array}$

24. $\begin{array}{r} 28.006 \\ \times \quad 5.2 \\ \hline \end{array}$

Answers

1. ______
2. ______
3. ______
4. ______
5. ______
6. ______
7. ______
8. ______
9. ______
10. ______
11. ______
12. ______
13. ______
14. ______
15. ______
16. ______
17. ______
18. ______
19. ______
20. ______
21. ______
22. ______
23. ______
24. ______

25. ______
26. ______
27. ______
28. ______
29. ______
30. ______
31. ______
32. ______
33. ______
34. ______
35. ______
36. ______
37. ______
38. ______
39. ______

25. Subtract 4.6 from 10.

26. Subtract 18 from 0.26.

27. $-268.19 - 146.25$

28. $-860.18 - 434.85$

29. $\dfrac{2.958}{-0.087}$

30. $\dfrac{-1.708}{0.061}$

31. $160 - 43.19$

32. $120 - 101.21$

33. 15.62×10

34. $15.62 \div 10$

35. $15.62 + 10$

36. $15.62 - 10$

37. Estimate the distance in miles between Garden City, Kansas, and Wichita, Kansas, by rounding each given distance to the nearest ten.

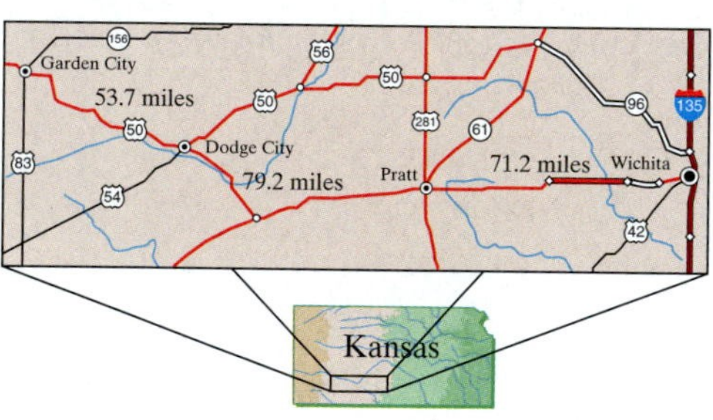

38. It costs \$3.30 to send a 2-pound package locally via parcel post at a U.S. Post Office. To send the same package as Priority Mail, it costs \$4.20. How much more does it cost to send a package as Priority Mail? (*Source:* United States Postal Service)

39. In 2005, Americans spent a total of \$23.324 billion on DVD home entertainment and \$1.556 on VHS home entertainment. Find the total spent on DVD or VHS home entertainment. Write the total in billions of dollars and also in standard notation.

5.5 FRACTIONS, DECIMALS, AND ORDER OF OPERATIONS

Objectives

A Write Fractions as Decimals.

B Compare Fractions and Decimals.

C Simplify Expressions Containing Decimals and Fractions Using Order of Operations

D Solve Area Problems Containing Fractions and Decimals.

E Evaluate Expressions Given Decimal Replacement Values.

Objective A Writing Fractions as Decimals

To write a fraction as a decimal, we interpret the fraction bar to mean division and find the quotient.

Writing Fractions as Decimals

To write a fraction as a decimal, divide the numerator by the denominator.

EXAMPLE 1 Write $\frac{1}{4}$ as a decimal.

Solution: $\frac{1}{4} = 1 \div 4$

$$\begin{array}{r} 0.25 \\ 4\overline{)1.00} \\ -\ 8 \\ \hline 20 \\ -20 \\ \hline 0 \end{array}$$

Thus, $\frac{1}{4}$ written as a decimal is 0.25.

Work Practice Problem 1

PRACTICE PROBLEM 1

a. Write $\frac{2}{5}$ as a decimal.

b. Write $\frac{9}{40}$ as a decimal.

EXAMPLE 2 Write $-\frac{5}{8}$ as a decimal.

Solution: $-\frac{5}{8} = -(5 \div 8) = -0.625$

$$\begin{array}{r} 0.625 \\ 8\overline{)5.000} \\ -4\ 8 \\ \hline 20 \\ -16 \\ \hline 40 \\ -40 \\ \hline 0 \end{array}$$

Work Practice Problem 2

PRACTICE PROBLEM 2

Write $-\frac{3}{8}$ as a decimal.

EXAMPLE 3 Write $\frac{2}{3}$ as a decimal.

Solution:

$$\begin{array}{r} 0.666\ldots \\ 3\overline{)2.000} \\ -1\ 8 \\ \hline 20 \\ -18 \\ \hline 20 \\ -18 \\ \hline 2 \end{array}$$

This pattern will continue because $\frac{2}{3} = 0.6666\ldots$

Remainder is 2, then 0 is brought down.

Remainder is 2, then 0 is brought down.

Remainder is 2.

Continued on next page

PRACTICE PROBLEM 3

a. Write $\frac{5}{6}$ as a decimal.

b. Write $\frac{2}{9}$ as a decimal.

Answers

1. a. 0.4 **b.** 0.225 **2.** −0.375
3. a. $0.8\overline{3}$ **b.** $0.\overline{2}$

Notice that the digit 2 keeps occurring as the remainder. This will continue so that the digit 6 will keep repeating in the quotient. We place a bar over the digit 6 to indicate that it repeats.

$$\frac{2}{3} = 0.666\ldots = 0.\overline{6}$$

We can also write a decimal approximation for $\frac{2}{3}$. For example, $\frac{2}{3}$ rounded to the nearest hundredth is 0.67. This can be written as $\frac{2}{3} \approx 0.67$.

Work Practice Problem 3

PRACTICE PROBLEM 4

Write $\frac{28}{13}$ as a decimal. Round to the nearest thousandth.

EXAMPLE 4 Write $\frac{22}{7}$ as a decimal. (Recall that the fraction $\frac{22}{7}$ is an approximation for π.) Round to the nearest hundredth.

Solution:

$$\begin{array}{r} 3.142 \\ 7\overline{)22.000} \\ -21 \\ \hline 1\,0 \\ -7 \\ \hline 30 \\ -28 \\ \hline 20 \\ -14 \\ \hline 6 \end{array} \approx 3.14$$

Carry the division out to the thousandths place.

The fraction $\frac{22}{7}$ in decimal form is approximately 3.14.

Work Practice Problem 4

PRACTICE PROBLEM 5

Write $3\frac{5}{16}$ as a decimal.

EXAMPLE 5 Write $2\frac{3}{16}$ as a decimal.

Solution:

Option 1. Write the fractional part only as a decimal.

$$\frac{3}{16} \longrightarrow \begin{array}{r} 0.1875 \\ 16\overline{)3.0000} \\ -1\,6 \\ \hline 1\,40 \\ -1\,28 \\ \hline 120 \\ -112 \\ \hline 80 \\ -80 \\ \hline 0 \end{array}$$

Thus $2\frac{3}{16} = 2.1875$

Option 2. Write $2\frac{3}{16}$ as an improper fraction, and divide.

$$2\frac{3}{16} = \frac{35}{16} \longrightarrow \begin{array}{r} 2.1875 \\ 16\overline{)35.0000} \\ -32 \\ \hline 3\,0 \\ -1\,6 \\ \hline 1\,40 \\ -1\,28 \\ \hline 120 \\ -112 \\ \hline 80 \\ -80 \\ \hline 0 \end{array}$$

Thus $2\frac{3}{16} = 2.1875$

Work Practice Problem 5

Answers

4. 2.154 **5.** 3.3125

Some fractions may be written as decimals using our knowledge of decimals. From Section 5.1, we know that if the denominator of a fraction is 10, 100, 1000, or so on, we can immediately write the fraction as a decimal. For example,

$$\frac{4}{10} = 0.4, \quad \frac{12}{100} = 0.12, \text{ and so on}$$

EXAMPLE 6 Write $\frac{4}{5}$ as a decimal.

Solution: Let's write $\frac{4}{5}$ as an equivalent fraction with a denominator of 10.

$$\frac{4}{5} = \frac{4}{5} \cdot \frac{2}{2} = \frac{8}{10} = 0.8$$

Work Practice Problem 6

PRACTICE PROBLEM 6

Write $\frac{3}{5}$ as a decimal.

EXAMPLE 7 Write $\frac{1}{25}$ as a decimal.

Solution: $\frac{1}{25} = \frac{1}{25} \cdot \frac{4}{4} = \frac{4}{100} = 0.04$

Work Practice Problem 7

PRACTICE PROBLEM 7

Write $\frac{3}{50}$ as a decimal.

✔ **Concept Check** Suppose you are writing the fraction $\frac{9}{16}$ as a decimal. How do you know you have made a mistake if your answer is 1.735?

Objective B Comparing Decimals and Fractions

Now we can compare decimals and fractions by writing fractions as equivalent decimals.

EXAMPLE 8 Insert <, >, or = to form a true statement.

$\frac{1}{8} \quad 0.12$

Solution: First we write $\frac{1}{8}$ as an equivalent decimal. Then we compare decimal places.

$$\begin{array}{r} 0.125 \\ 8\overline{)1.000} \\ -8 \\ \hline 20 \\ -16 \\ \hline 40 \\ -40 \\ \hline 0 \end{array}$$

Original numbers	$\frac{1}{8}$	0.12
Decimals	0.125	0.120
Compare	0.125 > 0.12	

Thus, $\frac{1}{8} > 0.12$

Work Practice Problem 8

PRACTICE PROBLEM 8

Insert <, >, or = to form a true statement.

$\frac{1}{5} \quad 0.25$

Answers

6. 0.6 7. 0.06 8. <

✔ **Concept Check Answer**

$\frac{9}{16}$ is less than 1 while 1.735 is greater than 1.

PRACTICE PROBLEM 9

Insert <, >, or = to form a true statement.

a. $\frac{1}{2}$ 0.54 **b.** $0.\overline{4}$ $\frac{4}{9}$

c. $\frac{5}{7}$ 0.72

EXAMPLE 9 Insert <, >, or = to form a true statement.

$0.\overline{7}$ $\frac{7}{9}$

Solution: We write $\frac{7}{9}$ as a decimal and then compare.

$$\begin{array}{r} 0.77\ldots = 0.\overline{7} \\ 9\overline{)7.00} \\ -6\,3 \\ \hline 70 \\ -63 \\ \hline 7 \end{array}$$

Original numbers	$0.\overline{7}$	$\frac{7}{9}$
Decimals	$0.\overline{7}$	$0.\overline{7}$
Compare	$0.\overline{7} = 0.\overline{7}$	

Thus, $0.\overline{7} = \frac{7}{9}$

Work Practice Problem 9

PRACTICE PROBLEM 10

Write the numbers in order from smallest to largest.

a. $\frac{1}{3}$, 0.302, $\frac{3}{8}$ **b.** 1.26, $1\frac{1}{4}$, $1\frac{2}{5}$

c. 0.4, 0.41, $\frac{5}{7}$

EXAMPLE 10 Write the numbers in order from smallest to largest.

$\frac{9}{20}$, $\frac{4}{9}$, 0.456

Solution:

Original numbers	$\frac{9}{20}$	$\frac{4}{9}$	0.456
Decimals	0.450	0.444...	0.456
Compare in order	2nd	1st	3rd

Written in order, we have

1st 2nd 3rd

$\frac{4}{9}$, $\frac{9}{20}$, 0.456

Work Practice Problem 10

Objective C Simplifying Expressions with Decimals and Fractions

In the remaining examples, we will review the order of operations by simplifying expressions that contain decimals.

Order of Operations

1. Perform all operations within parentheses ( ), brackets [ ], or other grouping symbols such as fraction bars.
2. Evaluate any expressions with exponents.
3. Multiply or divide in order from left to right.
4. Add or subtract in order from left to right.

Answers

9. a. < **b.** = **c.** <

10. a. 0.302, $\frac{1}{3}$, $\frac{3}{8}$ **b.** $1\frac{1}{4}$, 1.26, $1\frac{2}{5}$

c. 0.4, 0.41, $\frac{5}{7}$

EXAMPLE 11 Simplify: $723.6 \div 1000 \times 10$

Solution: Multiply or divide in order from left to right.

$$\begin{aligned} 723.6 \div 1000 \times 10 &= 0.7236 \times 10 && \text{Divide.} \\ &= 7.236 && \text{Multiply.} \end{aligned}$$

Work Practice Problem 11

PRACTICE PROBLEM 11

Simplify: $897.8 \div 100 \times 10$

EXAMPLE 12 Simplify: $-0.5(8.6 - 1.2)$

Solution: According to the order of operations, we simplify inside the parentheses first.

$$\begin{aligned} -0.5(8.6 - 1.2) &= -0.5(7.4) && \text{Subtract.} \\ &= -3.7 && \text{Multiply.} \end{aligned}$$

Work Practice Problem 12

PRACTICE PROBLEM 12

Simplify: $-8.69(3.2 - 1.8)$

EXAMPLE 13 Simplify: $(-1.3)^2 + 2.4$

Solution: Recall the meaning of an exponent.

$$\begin{aligned} (-1.3)^2 &= (-1.3)(-1.3) + 2.4 && \text{Use the definition of an exponent.} \\ &= 1.69 + 2.4 && \text{Multiply. The product of two negative numbers is a positive number} \\ &= 4.09 && \text{Add.} \end{aligned}$$

Work Practice Problem 13

PRACTICE PROBLEM 13

Simplify: $(-0.7)^2 + 2.1$

EXAMPLE 14 Simplify: $\dfrac{5.68 + (0.9)^2 \div 100}{0.2}$

Solution: First we simplify the numerator of the fraction. Then we divide.

$$\begin{aligned} \frac{5.68 + (0.9)^2 \div 100}{0.2} &= \frac{5.68 + 0.81 \div 100}{0.2} && \text{Simplify } (0.9)^2. \\ &= \frac{5.68 + 0.0081}{0.2} && \text{Divide.} \\ &= \frac{5.6881}{0.2} && \text{Add.} \\ &= 28.4405 && \text{Divide.} \end{aligned}$$

Work Practice Problem 14

PRACTICE PROBLEM 14

Simplify: $\dfrac{20.06 - (1.2)^2 \div 10}{0.02}$

Objective D Solving Area Problems Containing Fractions and Decimals

Sometimes real-life problems contain both fractions and decimals. In the next example, we review the area of a triangle, and when values are substituted, the result may be an expression containing both fractions and decimals.

Answers

11. 89.78 **12.** -12.166 **13.** 2.59 **14.** 995.8

PRACTICE PROBLEM 15

Find the area of the triangle.

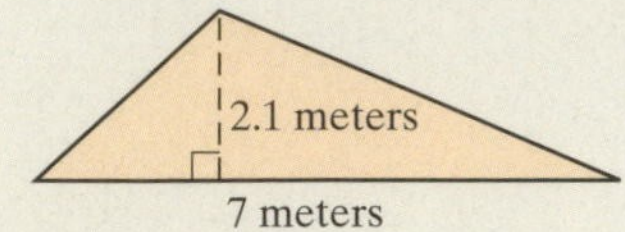

△ **EXAMPLE 15** The area of a triangle is Area $= \frac{1}{2} \cdot \text{base} \cdot \text{height}$. Find the area of the triangle shown.

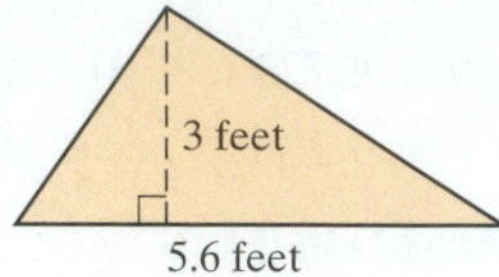

Solution:

$$\begin{aligned} \text{Area} &= \frac{1}{2} \cdot \text{base} \cdot \text{height} \\ &= \frac{1}{2} \cdot 5.6 \cdot 3 \\ &= 0.5 \cdot 5.6 \cdot 3 && \text{Write } \frac{1}{2} \text{ as the decimal 0.5.} \\ &= 8.4 \end{aligned}$$

The area of the triangle is 8.4 square feet.

Work Practice Problem 15

Objective E Using Decimals as Replacement Values

PRACTICE PROBLEM 16

Evaluate $1.7y - 2$ for $y = 2.3$.

EXAMPLE 16 Evaluate $-2x + 5$ for $x = 3.8$.

Solution: Replace x with 3.8 in the expression $-2x + 5$ and simplify.

$$\begin{aligned} -2x + 5 &= -2(3.8) + 5 && \text{Replace } x \text{ with 3.8.} \\ &= -7.6 + 5 && \text{Multiply.} \\ &= -2.6 && \text{Add.} \end{aligned}$$

Work Practice Problem 16

Answers

15. 7.35 sq m **16.** 1.91

Vocabulary and Readiness Check

Answer each exercise "true" or "false."

1. The number $0.\overline{5}$ means 0.555.
2. To write $\frac{9}{19}$ as a decimal, perform the division $19\overline{)9}$.
3. $(-1.2)^2$ means $(-1.2)(-1.2)$ or -1.44.
4. To simplify $8.6(4.8 - 9.6)$, we first subtract.

5.5 EXERCISE SET

FOR EXTRA HELP

Student Solutions Manual

PH Math/Tutor Center

CD/Video for Review
MathXL®
MyMathLab

Objective A *Write each number as a decimal. See Examples 1 through 7.*

1. $\frac{1}{5}$ **2.** $\frac{1}{20}$ **3.** $\frac{17}{25}$ **4.** $\frac{13}{25}$ **5.** $\frac{3}{4}$ **6.** $\frac{3}{8}$

7. $-\frac{2}{25}$ **8.** $-\frac{3}{25}$ **9.** $\frac{9}{4}$ **10.** $\frac{8}{5}$ **11.** $\frac{11}{12}$ **12.** $\frac{5}{12}$

13. $\frac{17}{40}$ **14.** $\frac{19}{25}$ **15.** $\frac{9}{20}$ **16.** $\frac{31}{40}$ **17.** $-\frac{1}{3}$ **18.** $-\frac{7}{9}$

19. $\frac{7}{16}$ **20.** $\frac{9}{16}$ **21.** $\frac{7}{11}$ **22.** $\frac{9}{11}$ **23.** $5\frac{17}{20}$ **24.** $4\frac{7}{8}$

25. $\frac{78}{125}$ **26.** $\frac{159}{375}$

Round each number as indicated. See Example 4.

27. Round your decimal answer to Exercise 17 to the nearest hundredth.

28. Round your decimal answer to Exercise 18 to the nearest hundredth.

29. Round your decimal answer to Exercise 19 to the nearest hundredth.

30. Round your decimal answer to Exercise 20 to the nearest hundredth.

31. Round your decimal answer to Exercise 21 to the nearest tenth.

32. Round your decimal answer to Exercise 22 to the nearest tenth.

Write each fraction as a decimal. If necessary, round to the nearest hundredth. See Examples 1 through 7.

33. Of the U.S. mountains that are over 14,000 feet in elevation, $\frac{56}{91}$ are located in Colorado. (*Source:* U.S. Geological Survey)

34. About $\frac{21}{50}$ of all blood donors have type A blood. (*Source:* American Red Cross Biomedical Services)

35. The United States contains the greatest fraction of people who use the Internet, with about $\frac{67}{94}$ people using it. (*Source:* UCLA Center for Communication Policy)

36. Hungary has the lowest fraction of people using the Internet, with only $\frac{7}{40}$ people using it. (*Source:* UCLA Center for Communication Policy)

37. When first launched, the Hubble Space Telescope's primary mirror was out of shape on the edges by $\frac{1}{50}$ of a human hair. This very small defect made it difficult to focus faint objects being viewed. Because the HST was in low Earth orbit, it was serviced by a shuttle and the defect was corrected.

38. The two mirrors currently in use in the Hubble Space Telescope were ground so that they do not deviate from a perfect curve by more than $\frac{1}{800{,}000}$ of an inch. Do not round this off.

Objective B *Insert <, >, or = to form a true statement. See Examples 8 and 9.*

39. 0.562 0.569

40. 0.983 0.988

41. 0.215 $\frac{43}{200}$

42. $\frac{29}{40}$ 0.725

43. −0.0932 −0.0923

44. −0.00563 −0.00536

45. $0.\overline{6}$ $\frac{5}{6}$

46. $0.\overline{1}$ $\frac{2}{17}$

47. $\frac{51}{91}$ $0.56\overline{4}$

48. $0.58\overline{3}$ $\frac{6}{11}$

49. $\frac{4}{7}$ 0.14

50. $\frac{5}{9}$ 0.557

51. 1.38 $\frac{18}{13}$

52. 0.372 $\frac{22}{59}$

53. 7.123 $\frac{456}{64}$

54. 12.713 $\frac{89}{7}$

Write the numbers in order from smallest to largest. See Example 10.

55. 0.34, 0.35, 0.32

56. 0.47, 0.42, 0.40

57. 0.49, 0.491, 0.498

58. 0.72, 0.727, 0.728

59. 5.23, $\frac{42}{8}$, 5.34

60. 7.56, $\frac{67}{9}$, 7.562

61. $\frac{5}{8}$, 0.612, 0.649

62. $\frac{5}{6}$, 0.821, 0.849

Objective C *Simplify each expression. See Examples 11 through 14.*

63. $(0.3)^2 + 0.5$

64. $(-2.5)(3) - 4.7$

65. $\frac{1 + 0.8}{-0.6}$

66. $(-0.05)^2 + 3.13$

67. $(-2.3)^2(0.3 + 0.7)$

68. $(8.2)(100) - (8.2)(10)$

69. $(5.6 - 2.3)(2.4 + 0.4)$

70. $\frac{0.222 - 2.13}{12}$

71. $\frac{(4.5)^2}{100}$

72. $0.9(5.6 - 6.5)$

73. $\frac{7 + 0.74}{-6}$

74. $(1.5)^2 + 0.5$

Find the value of each expression. Give the result as a decimal. See Examples 11 through 14.

75. $\frac{1}{5} - 2(7.8)$

76. $\frac{3}{4} - (9.6)(5)$

77. $\frac{1}{4}(-9.6 - 5.2)$

78. $\frac{3}{8}(4.7 - 5.9)$

Objective **D** *Find the area of each triangle or rectangle. See Example 15.*

79.

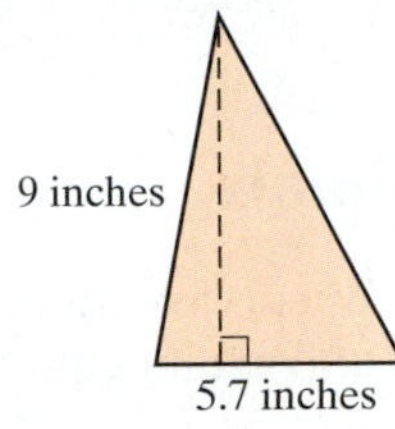

80.

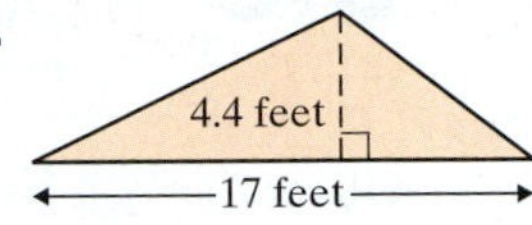

81.

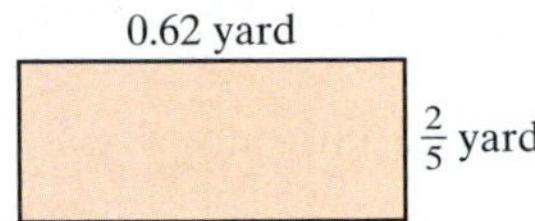

82.

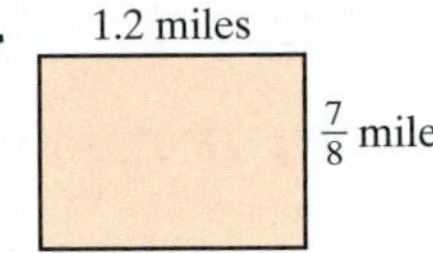

Objective **E** *Evaluate each expression for $x = 6$, $y = 0.3$, and $z = -2.4$. See Example 16.*

83. z^2 **84.** y^2 **85.** $x - y$ **86.** $x - z$ **87.** $4y - z$ **88.** $\frac{x}{y} + 2z$

Review

Simplify. See Sections 1.7 and 4.6.

89. $\frac{9}{10} + \frac{16}{25}$ **90.** $\frac{4}{11} - \frac{19}{22}$ **91.** $\left(\frac{2}{5}\right)\left(\frac{5}{2}\right)^2$ **92.** $\left(\frac{2}{3}\right)^2\left(\frac{3}{2}\right)$

Concept Extensions

Without calculating, describe each number as < 1, $= 1$, or > 1. See the Concept Check in this section.

93. 1.0 **94.** 1.0000 **95.** 1.00001 **96.** $\frac{101}{99}$ **97.** $\frac{99}{100}$ **98.** $\frac{99}{99}$

In 2005, there were 10,661 commercial radio stations in the United States. The most popular formats are listed in the table along with their counts. Use this graph to answer Exercises 99 through 102.

99. Write the fraction of radio stations with a country music format as a decimal. Round to the nearest thousandth.

100. Write the fraction of radio stations with a news/talk format as a decimal. Round to the nearest hundredth.

101. Estimate, by rounding each number in the table to the nearest hundred, the total number of stations with the top six formats in 2005.

102. Use your estimate from Exercise 101 to write the fraction of radio stations accounted for by the top six formats as a decimal. Round to the nearest hundredth.

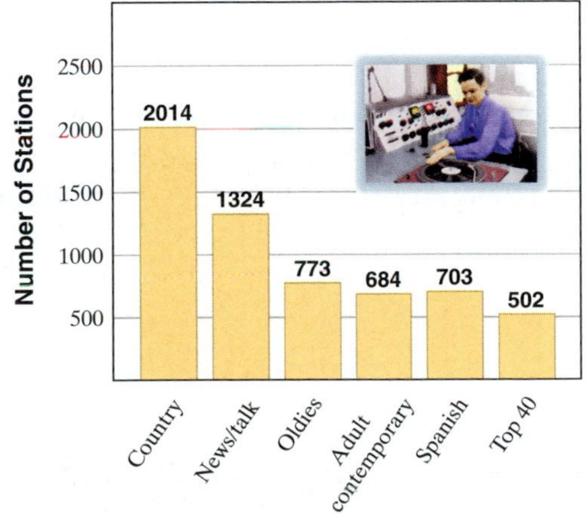

103. Describe two ways to write fractions as decimals.

104. Describe two ways to write mixed numbers as decimals.

Objective

A Solve Equations Containing Decimals.

5.6 EQUATIONS CONTAINING DECIMALS

Objective A Solving Equations Containing Decimals

In this section, we continue our work with decimals and algebra by solving equations containing decimals. First, we review the steps given earlier for solving an equation.

Steps for Solving an Equation in x

Step 1: If fractions are present, multiply both sides of the equation by the LCD of the fractions.

Step 2: If parentheses are present, use the distributive property.

Step 3: Combine any like terms on each side of the equation.

Step 4: Use the addition property of equality to rewrite the equation so that variable terms are on one side of the equation and constant terms are on the other side.

Step 5: Divide both sides by the numerical coefficient of x to solve.

Step 6: Check the answer in the **original equation.**

PRACTICE PROBLEM 1

Solve: $z + 0.9 = 1.3$

EXAMPLE 1 Solve: $x - 1.5 = 8$

Solution: Steps 1 through 3 are not needed for this equation, so we begin with Step 4. To get x alone on one side of the equation, add 1.5 to both sides.

$$x - 1.5 = 8 \quad \text{Original equation}$$
$$x - 1.5 + 1.5 = 8 + 1.5 \quad \text{Add 1.5 to both sides.}$$
$$x = 9.5 \quad \text{Simplify.}$$

Check: To check, replace x with 9.5 in the *original equation*.

$$x - 1.5 = 8 \quad \text{Original equation}$$
$$9.5 - 1.5 \stackrel{?}{=} 8 \quad \text{Replace } x \text{ with 9.5.}$$
$$8 = 8 \quad \text{True}$$

Since $8 = 8$ is a true statement, 9.5 is a solution of the equation.

Work Practice Problem 1

PRACTICE PROBLEM 2

Solve: $0.17x = -0.34$

EXAMPLE 2 Solve: $-2y = 6.7$

Solution: Steps 1 through 4 are not needed for this equation, so we begin with Step 5. To solve for y, divide both sides by the coefficient of y, which is -2.

$$-2y = 6.7 \quad \text{Original equation}$$
$$\frac{-2y}{-2} = \frac{6.7}{-2} \quad \text{Divide both sides by } -2.$$
$$y = -3.35 \quad \text{Simplify.}$$

Check: To check, replace y with -3.35 in the original equation.

$$-2y = 6.7 \quad \text{Original equation}$$
$$-2(-3.35) \stackrel{?}{=} 6.7 \quad \text{Replace } y \text{ with } -3.35.$$
$$6.7 = 6.7 \quad \text{True}$$

Thus -3.35 is a solution of the equation $-2y = 6.7$.

Work Practice Problem 2

Answers
1. 0.4 **2.** −2

EXAMPLE 3 Solve: $1.2x + 5.8 = 8.2$

Solution: We begin with Step 4 and get the variable term alone by subtracting 5.8 from both sides.

$$1.2x + 5.8 = 8.2$$
$$1.2x + 5.8 - 5.8 = 8.2 - 5.8 \quad \text{Subtract 5.8 from both sides.}$$
$$1.2x = 2.4 \quad \text{Simplify.}$$
$$\frac{1.2x}{1.2} = \frac{2.4}{1.2} \quad \text{Divide both sides by 1.2.}$$
$$x = 2 \quad \text{Simplify.}$$

To check, replace x with 2 in the original equation. The solution is 2.

Work Practice Problem 3

PRACTICE PROBLEM 3

Solve: $2.9 = 1.7 + 0.3x$

EXAMPLE 4 Solve: $7x + 3.2 = 4x - 1.6$

Solution: We start with Step 4 to get variable terms on one side and numerical terms on the other.

$$7x + 3.2 = 4x - 1.6$$
$$7x + 3.2 - 3.2 = 4x - 1.6 - 3.2 \quad \text{Subtract 3.2 from both sides.}$$
$$7x = 4x - 4.8 \quad \text{Simplify.}$$
$$7x - 4x = 4x - 4.8 - 4x \quad \text{Subtract } 4x \text{ from both sides.}$$
$$3x = -4.8 \quad \text{Simplify.}$$
$$\frac{\overset{1}{\cancel{3}}x}{\underset{1}{\cancel{3}}} = -\frac{4.8}{3} \quad \text{Divide both sides by 3.}$$
$$x = -1.6 \quad \text{Simplify.}$$

Check to see that -1.6 is the solution.

Work Practice Problem 4

PRACTICE PROBLEM 4

Solve: $8x + 4.2 = 10x + 11.6$

EXAMPLE 5 Solve: $5(x - 0.36) = -x + 2.4$

Solution: First use the distributive property to distribute the factor 5.

$$5(x - 0.36) = -x + 2.4 \quad \text{Original equation}$$
$$5x - 1.8 = -x + 2.4 \quad \text{Apply the distributive property.}$$

Next, get x alone on the left side of the equation by adding 1.8 to both sides of the equation and then adding x to both sides of the equation.

$$5x - 1.8 + 1.8 = -x + 2.4 + 1.8 \quad \text{Add 1.8 to both sides.}$$
$$5x = -x + 4.2 \quad \text{Simplify.}$$
$$5x + x = -x + 4.2 + x \quad \text{Add } x \text{ to both sides.}$$
$$6x = 4.2 \quad \text{Simplify.}$$
$$\frac{6x}{6} = \frac{4.2}{6} \quad \text{Divide both sides by 6.}$$
$$x = 0.7 \quad \text{Simplify.}$$

To verify that 0.7 is the solution, replace x with 0.7 in the original equation.

Work Practice Problem 5

PRACTICE PROBLEM 5

Solve: $6.3 - 5x = 3(x + 2.9)$

Instead of solving equations with decimals, sometimes it may be easier to first rewrite the equation so that it contains integers only. Recall that multiplying a decimal by a power of 10, such as 10, 100, or 1000, has the effect of moving the decimal point to the right. We can use the multiplication property of equality to multiply both sides of the equation through by an appropriate power of 10. The resulting equivalent equation will contain integers only.

Answers

3. 4 **4.** -3.7 **5.** -0.3

PRACTICE PROBLEM 6

Solve: $0.2y + 2.6 = 4$

EXAMPLE 6 Solve: $0.5y + 2.3 = 1.65$

Solution: Multiply the equation through by 100. This will move the decimal point in each term two places to the right.

$$\begin{aligned} 0.5y + 2.3 &= 1.65 && \text{Original equation} \\ 100(0.5y + 2.3) &= 100(1.65) && \text{Multiply both sides by 100.} \\ 100(0.5y) + 100(2.3) &= 100(1.65) && \text{Apply the distributive property.} \\ 50y + 230 &= 165 && \text{Simplify.} \end{aligned}$$

Now the equation contains integers only. Finish solving by subtracting 230 from both sides.

$$\begin{aligned} 50y + 230 &= 165 \\ 50y + 230 - 230 &= 165 - 230 && \text{Subtract 230 from both sides.} \\ 50y &= -65 && \text{Simplify.} \\ \frac{50y}{50} &= \frac{-65}{50} && \text{Divide both sides by 50.} \\ y &= -1.3 && \text{Simplify.} \end{aligned}$$

Check to see that -1.3 is the solution by replacing y with -1.3 in the original equation.

Work Practice Problem 6

Answer

6. 7

✔ Concept Check Answer

Multiply by 1000.

✔Concept Check By what number would you multiply both sides of the following equation to make calculations easier? Explain your choice.

$1.7x + 3.655 = -14.2$

5.6 EXERCISE SET

FOR EXTRA HELP: Student Solutions Manual · PH Math/Tutor Center · CD/Video for Review · MathXL® · MyMathLab

Objective A *Solve each equation. See Examples 1 and 2.*

1. $x + 1.2 = 7.1$ **2.** $y - 0.5 = 9$ **3.** $-5y = 2.15$ **4.** $-0.4x = 50$

5. $6.2 = y - 4$ **6.** $9.7 = x + 11.6$ **7.** $3.1x = -13.95$ **8.** $3y = -25.8$

Solve each equation. See Examples 3 through 5.

9. $-3.5x + 2.8 = -11.2$ **10.** $7.1 - 0.2x = 6.1$ **11.** $6x + 8.65 = 3x + 10$

12. $7x - 9.64 = 5x + 2.32$  **13.** $2(x - 1.3) = 5.8$ **14.** $5(x + 2.3) = 19.5$

Solve each equation by first multiplying both sides through by an appropriate power of 10 so that the equation contains integers only. See Example 6.

15. $0.4x + 0.7 = -0.9$ **16.** $0.7x + 0.1 = 1.5$ **17.** $7x - 10.8 = x$

18. $3y = 7y + 24.4$ **19.** $2.1x + 5 - 1.6x = 10$ **20.** $1.5x + 2 - 1.2x = 12.2$

Solve. See Examples 1 through 6.

21. $y - 3.6 = 4$

22. $x + 5.7 = 8.4$

23. $-0.02x = -1.2$

24. $-9y = -0.162$

25. $6.5 = 10x + 7.2$

26. $2x - 4.2 = 8.6$

27. $2.7x - 25 = 1.2x + 5$

28. $9y - 6.9 = 6y - 11.1$

29. $200x - 0.67 = 100x + 0.81$

30. $2.3 + 500x = 600x - 0.2$

31. $3(x + 2.71) = 2x$

32. $7(x + 8.6) = 6x$

33. $8x - 5 = 10x - 8$

34. $24y - 10 = 20y - 17$

35. $1.2 + 0.3x = 0.9$

36. $1.5 = 0.4x + 0.5$

37. $-0.9x + 2.65 = -0.5x + 5.45$

38. $-50x + 0.81 = -40x - 0.48$

39. $4x + 7.6 = 2(3x - 3.2)$

40. $4(2x - 1.6) = 5x - 6.4$

41. $0.7x + 13.8 = x - 2.16$

42. $y - 5 = 0.3y + 4.1$

Review

Simplify each expression by combining like terms. If parentheses are present, first use the distributive property. See Section 3.1.

43. $2x - 7 + x - 9$

44. $x + 14 - 5x - 17$

Perform the indicated operation. See Sections 4.3 and 4.5.

45. $\frac{6x}{5} \cdot \frac{1}{2x^2}$

46. $\frac{x}{3} + \frac{2x}{7}$

47. $5\frac{1}{3} \div 9\frac{1}{6}$

48. $50 - 14\frac{9}{13}$

Concept Extensions

Mixed Practice (*Sections 5.2 and 5.6*) *This section of exercises contains equations and expressions. If the exercise contains an equation, solve it for the variable. If the exercise contains an expression, simplify it by combining any like terms.*

49. $b + 4.6 = 8.3$

50. $y - 15.9 = -3.8$

51. $2x - 0.6 + 4x - 0.01$

52. $-x - 4.1 - x - 4.02$

53. $5y - 1.2 - 7y + 8$

54. $9a - 5.6 - 3a + 6$

55. $2.8 = z - 6.3$

56. $9.7 = x + 4.3$

57. $4.7x + 8.3 = -5.8$

58. $2.8x + 3.4 = -13.4$

59. $7.76 + 8z - 12z + 8.91$

60. $9.21 + x - 4x + 11.33$

61. $5(x - 3.14) = 4x$

62. $6(x + 1.43) = 5x$

63. $2.6y + 8.3 = 4.6y - 3.4$

64. $8.4z - 2.6 = 5.4z + 10.3$

65. $9.6z - 3.2 - 11.7z - 6.9$

66. $-3.2x + 12.6 - 8.9x - 15.2$

67. Explain in your own words the property of equality that allows us to multiply an equation through by a power of 10.

68. Construct an equation whose solution is 1.4.

Solve.

69. $-5.25x = -40.33575$

70. $7.68y = -114.98496$

71. $1.95y + 6.834 = 7.65y - 19.8591$

72. $6.11x + 4.683 = 7.51x + 18.235$

5.7 DECIMAL APPLICATIONS: MEAN, MEDIAN, AND MODE

Objectives

A Find the Mean of a List of Numbers.

B Find the Median of a List of Numbers.

C Find the Mode of a List of Numbers.

Objective A Finding the Mean

Sometimes we want to summarize data by displaying them in a graph, but sometimes it is also desirable to be able to describe a set of data, or a set of numbers, by a single "middle" number. Three such **measures of central tendency** are the **mean,** the **median,** and the **mode.**

The most common measure of central tendency is the mean (sometimes called the "arithmetic mean" or the "average"). Recall that we first introduced finding the average of a list of numbers in Section 1.6.

The **mean (average)** of a set of number items is the sum of the items divided by the number of items.

$$\text{mean} = \frac{\text{sum of items}}{\text{number of items}}$$

EXAMPLE 1 Finding the Mean Time in an Experiment

Seven students in a psychology class conducted an experiment on mazes. Each student was given a pencil and asked to successfully complete the same maze. The timed results are below:

Student	Ann	Thanh	Carlos	Jesse	Melinda	Ramzi	Dayni
Time (Seconds)	13.2	11.8	10.7	16.2	15.9	13.8	18.5

a. Who completed the maze in the shortest time? Who completed the maze in the longest time?

b. Find the mean time.

c. How many students took longer than the mean time? How many students took shorter than the mean time?

Solution:

a. Carlos completed the maze in 10.7 seconds, the shortest time. Dayni completed the maze in 18.5 seconds, the longest time.

b. To find the mean (or average), we find the sum of the items and divide by 7, the number of items.

$$\text{mean} = \frac{13.2 + 11.8 + 10.7 + 16.2 + 15.9 + 13.8 + 18.5}{7}$$

$$= \frac{100.1}{7} = 14.3$$

c. Three students, Jesse, Melinda, and Dayni, had times longer than the mean time. Four students, Ann, Thanh, Carlos, and Ramzi, had times shorter than the mean time.

Work Practice Problem 1

PRACTICE PROBLEM 1

Find the mean of the following test scores: 87, 75, 96, 91, and 78.

✔ **Concept Check** Estimate the mean of the following set of data:

5, 10, 10, 10, 10, 15

Often in college, the calculation of a **grade point average** (GPA) is a **weighted mean** and is calculated as shown in Example 2.

Answer

1. 85.4

✔ **Concept Check Answer**

10

PRACTICE PROBLEM 2

Find the grade point average if the following grades were earned in one semester.

Grade	Credit Hours
A	2
B	4
C	5
D	2
A	2

EXAMPLE 2 **Calculating Grade Point Average (GPA)**

The following grades were earned by a student during one semester. Find the student's grade point average.

Course	Grade	Credit Hours
College mathematics	A	3
Biology	B	3
English	A	3
PE	C	1
Social studies	D	2

Solution: To calculate the grade point average, we need to know the point values for the different possible grades. The point values of grades commonly used in colleges and universities are given below:

A: 4, B: 3, C: 2, D: 1, F: 0

Now, to find the grade point average, we multiply the number of credit hours for each course by the point value of each grade. The grade point average is the sum of these products divided by the sum of the credit hours.

Course	Grade	Point Value of Grade	Credit Hours	Point Value Credit Hours
College mathematics	A	4	3	12
Biology	B	3	3	9
English	A	4	3	12
PE	C	2	1	2
Social studies	D	1	2	2
		Totals:	12	37

$$\text{grade point average} = \frac{37}{12} \approx 3.08 \text{ rounded to two decimal places}$$

The student earned a grade point average of 3.08.

Work Practice Problem 2

Objective B Finding the Median

You may have noticed that a very low number or a very high number can affect the mean of a list of numbers. Because of this, you may sometimes want to use another measure of central tendency. A second measure of central tendency is called the **median.** The median of a list of numbers is not affected by a low or high number in the list.

The **median** of a set of numbers in numerical order is the middle number. If the number of items is odd, the median is the middle number. If the number of items is even, the median is the mean of the two middle numbers.

PRACTICE PROBLEM 3

Find the median of the list of numbers: 5, 11, 14, 23, 24, 35, 38, 41, 43

EXAMPLE 3 Find the median of the following list of numbers:

25, 54, 56, 57, 60, 71, 98

Solution: Because this list is in numerical order, the median is the middle number, 57.

Work Practice Problem 3

Answers
2. 2.67 **3.** 24

EXAMPLE 4 Find the median of the following list of scores: 67, 91, 75, 86, 55, 91

Solution: First we list the scores in numerical order and then find the middle number.

55, 67, 75, 86, 91, 91

Since there is an even number of scores, there are two middle numbers, 75 and 86. The median is the mean of the two middle numbers.

$$\text{median} = \frac{75 + 86}{2} = 80.5$$

The median is 80.5.

Work Practice Problem 4

Helpful Hint Don't forget to write the numbers in order from smallest to largest before finding the median.

PRACTICE PROBLEM 4

Find the median of the list of scores:
36, 91, 78, 65, 95, 95, 88, 71

Objective C Finding the Mode

The last common measure of central tendency is called the **mode.**

The **mode** of a set of numbers is the number that occurs most often. (It is possible for a set of numbers to have more than one mode or to have no mode.)

EXAMPLE 5 Find the mode of the list of numbers:

11, 14, 14, 16, 31, 56, 65, 77, 77, 78, 79

Solution: There are two numbers that occur the most often. They are 14 and 77. This list of numbers has two modes, 14 and 77.

Work Practice Problem 5

PRACTICE PROBLEM 5

Find the mode of the list of numbers:
14, 10, 10, 13, 15, 15, 15, 17, 18, 18, 20

EXAMPLE 6 Find the median and the mode of the following set of numbers. These numbers were high temperatures for 14 consecutive days in a city in Montana.

76, 80, 85, 86, 89, 87, 82, 77, 76, 79, 82, 89, 89, 92

Solution: First we write the numbers in numerical order.

76, 76, 77, 79, 80, 82, 82, 85, 86, 87, 89, 89, 89, 92

Since there is an even number of items, the median is the mean of the two middle numbers, 82 and 85.

$$\text{median} = \frac{82 + 85}{2} = 83.5$$

The mode is 89, since 89 occurs most often.

Work Practice Problem 6

PRACTICE PROBLEM 6

Find the median and the mode of the list of numbers:
26, 31, 15, 15, 26, 30, 16, 18, 15, 35

✔ **Concept Check** True or false? Every set of numbers *must* have a mean, median, and mode. Explain your answer.

Helpful Hint Don't forget that it is possible for a list of numbers to have no mode. For example, the list

2, 4, 5, 6, 8, 9

has no mode. There is no number or numbers that occur more often than the others.

Answers
4. 83 **5.** 15 **6.** median: 22; mode: 15

✔ Concept Check Answer
false; a set of numbers may have no mode.

Vocabulary and Readiness Check

Use the choices below to fill in each blank. Some choices may be used more than once.

mean	mode	grade point average
median	average	

1. Another word for "mean" is ______________.
2. The number that occurs most often in a set of numbers is called the ______________.
3. The ________________ of a set of number items is $\frac{\text{sum of items}}{\text{number of items}}$.
4. The ______________ of a set of numbers is the middle number. If the number of numbers is even, it is the ________________ of the two middle numbers.
5. An example of weighted mean is a calculation of ________________.

5.7 EXERCISE SET

FOR EXTRA HELP

Student Solutions Manual

PH Math/Tutor Center

CD/Video for Review

MathXL®

MyMathLab

Objectives **A** **B** **C** **Mixed Practice** *For each set of numbers, find the mean, median, and mode. If necessary, round the mean to one decimal place. See Examples 1 and 3 through 6.*

1. 15, 23, 24, 18, 25

2. 45, 36, 28, 46, 52

3. 7.6, 8.2, 8.2, 9.6, 5.7, 9.1

4. 4.9, 7.1, 6.8, 6.8, 5.3, 4.9

5. 0.5, 0.2, 0.2, 0.6, 0.3, 1.3, 0.8, 0.1, 0.5

6. 0.6, 0.6, 0.8, 0.4, 0.5, 0.3, 0.7, 0.8, 0.1

7. 231, 543, 601, 293, 588, 109, 334, 268

8. 451, 356, 478, 776, 892, 500, 467, 780

The ten tallest buildings in the world are listed in the following table. Use this table to answer Exercises 9 through 14. If necessary, round results to one decimal place. See Examples 1 and 3 through 6.

9. Find the mean height of the five tallest buildings.

10. Find the median height of the five tallest buildings.

11. Find the median height of the eight tallest buildings.

12. Find the mean height of the eight tallest buildings.

Building	Height (in Feet)
Taipei 101	1679
Petronas Tower 1, Kuala Lumpur	1483
Petronas Tower 2, Kuala Lumpur	1483
Sears Tower, Chicago	1450
Jin Mao Building, Shanghai	1380
Two International Finance Centre	1362
Citic Plaza, Guangzhou	1283
Shun Hing Square, Shenzhen	1260
Empire State Building, New York	1250
Central Plaza, Hong Kong	1227
(*Source:* Council on Tall Buildings and Urban Habitat)	

13. Given the building heights, explain how you know, without calculating, that the answer to Exercise 10 is more than the answer to Exercise 11.

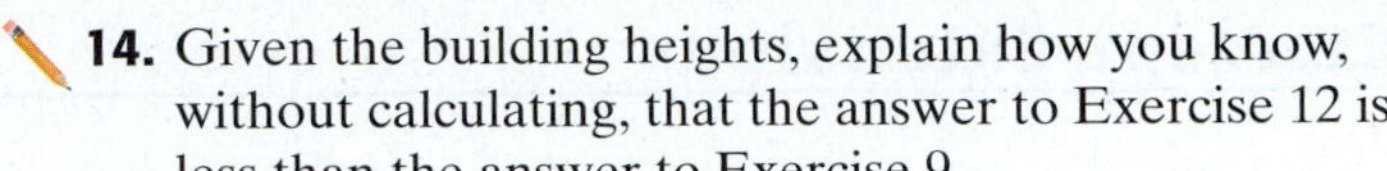

14. Given the building heights, explain how you know, without calculating, that the answer to Exercise 12 is less than the answer to Exercise 9.

For Exercises 15 through 18, the grades are given for a student for a particular semester. Find the grade point average. If necessary, round the grade point average to the nearest hundredth. See Example 2.

15.

Grade	Credit Hours
B	3
C	3
A	4
C	4

16.

Grade	Credit Hours
D	1
F	1
C	4
B	5

17.

Grade	Credit Hours
A	3
A	3
A	4
B	3
C	1

18.

Grade	Credit Hours
B	2
B	2
C	3
A	3
B	3

During an experiment, the following times (in seconds) were recorded:

7.8, 6.9, 7.5, 4.7, 6.9, 7.0

19. Find the mean. Round to the nearest tenth.

20. Find the median.

21. Find the mode.

In a mathematics class, the following test scores were recorded for a student: 93, 85, 89, 79, 88, 91.

22. Find the mean. Round to the nearest hundredth.

23. Find the median.

24. Find the mode.

The following pulse rates were recorded for a group of 15 students:

78, 80, 66, 68, 71, 64, 82, 71, 70, 65, 70, 75, 77, 86, 72.

25. Find the mean.

26. Find the median.

27. Find the mode.

28. How many pulse rates were higher than the mean?

29. How many pulse rates were lower than the mean?

Review

Write each fraction in simplest form. See Section 4.2.

30. $\frac{12}{20}$

31. $\frac{6}{18}$

32. $\frac{4x}{36}$

33. $\frac{18}{30y}$

34. $\frac{35a^3}{100a^2}$

35. $\frac{55y^2}{75y^2}$

Concept Extensions

Find the missing numbers in each set of numbers.

36. 16, 18, _____, _____, _____. The mode is 21. The median is 20.

37. _____, _____, _____, 40, _____. The mode is 35. The median is 37. The mean is 38.

38. Write a list of numbers for which you feel the median would be a better measure of central tendency than the mean.

39. Without making any computations, decide whether the median of the following list of numbers will be a whole number. Explain your reasoning.

36, 77, 29, 58, 43

STUDY SKILLS BUILDER

Tips for Studying for an Exam

To prepare for an exam, try the following study techniques.

- Start the study process days before your exam.
- Make sure that you are up-to-date on your assignments.
- If there is a topic that you are unsure of, use one of the many resources that are available to you. For example,

  See your instructor.

  Visit a learning resource center on campus.

  Read the textbook material and examples on the topic.

  View a video on the topic.
- Reread your notes and carefully review the Chapter Highlights at the end of any chapter.
- Work the review exercises at the end of the chapter. Check your answers and correct any mistakes. If you have trouble, use a resource listed above.
- Find a quiet place to take the Chapter Test found at the end of the chapter. Do not use any resources when taking this sample test. This way, you will have a clear indication of how prepared you are for your exam. Check your answers and make sure that you correct any missed exercises.
- Get lots of rest the night before the exam. It's hard to show how well you know the material if your brain is foggy from lack of sleep.

Good luck and keep a positive attitude.

Let's see how you did on your last exam.

1. How many days before your last exam did you start studying?
2. Were you up-to-date on your assignments at that time or did you need to catch up on assignments?
3. List the most helpful text supplement (if you used one).
4. List the most helpful campus supplement (if you used one).
5. List your process for preparing for a mathematics test.
6. Was this process helpful? In other words, were you satisfied with your performance on your exam?
7. If not, what changes can you make in your process that will make it more helpful to you?

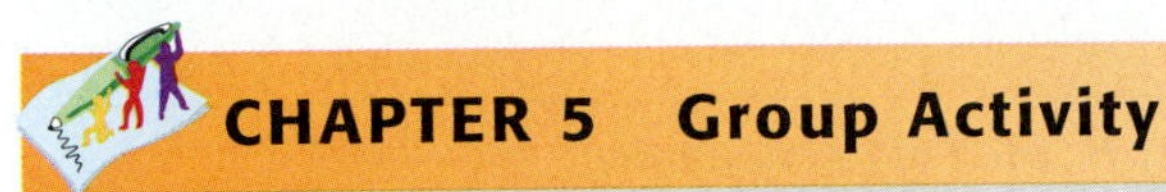

CHAPTER 5 Group Activity

Maintaining a Checking Account
(Sections 5.1, 5.2, 5.3, 5.4)

This activity may be completed by working in groups or individually.

A checking account is a convenient way of handling money and paying bills. To open a checking account, the bank or savings and loan association requires a customer to make a deposit. Then the customer receives a checkbook that contains checks, deposit slips, and a register for recording checks written and deposits made. It is important to record all payments and deposits that affect the account. It is also important to keep the checkbook balance current by subtracting checks written and adding deposits made.

About once a month checking customers receive a statement from the bank listing all activity that the account has had in the last month. The statement lists a beginning balance, all checks and deposits, any service charges made against the account, and an ending balance. Because it may take several days for checks that a customer has written to clear the banking system, the check register may list checks that do not appear on the monthly bank statement. These checks are called **outstanding checks.** Deposits that are recorded in the check register but do not appear on the statement are called **deposits in transit.** Because of these differences, it is important to balance, or reconcile, the checkbook against the monthly statement. The steps for doing so are listed below.

Balancing or Reconciling a Checkbook

Step 1: Place a check mark in the checkbook register next to each check and deposit listed on the monthly bank statement. Any entries in the register without a check mark are outstanding checks or deposits in transit.

Step 2: Find the ending checkbook register balance and add to it any outstanding checks and any interest paid on the account.

Step 3: From the total in Step 2, subtract any deposits in transit and any service charges.

Step 4: Compare the amount found in Step 3 with the ending balance listed on the bank statement. If they are the same, the checkbook balances with the bank statement. Be sure to update the check register with service charges and interest.

Step 5: If the checkbook does not balance, recheck the balancing process. Next, make sure that the running checkbook register balance was calculated correctly. Finally, compare the checkbook register with the statement to make sure that each check was recorded for the correct amount.

For the checkbook register and monthly bank statement given:

a. *update the checkbook register* **b.** *list the outstanding checks and deposits in transit*
c. *balance the checkbook—be sure to update the register with any interest or service fees*

Checkbook Register

#	Date	Description	Payment	✔	Deposit	Balance 425.86
114	4/1	Market Basket	30.27			
115	4/3	May's Texaco	8.50			
	4/4	Cash at ATM	50.00			
116	4/6	UNO Bookstore	121.38			
	4/7	Deposit			100.00	
117	4/9	MasterCard	84.16			
118	4/10	Blockbuster	6.12			
119	4/12	Kroger	18.72			
120	4/14	Parking sticker	18.50			
	4/15	Direct deposit			294.36	
121	4/20	Rent	395.00			
122	4/25	Student fees	20.00			
	4/28	Deposit			75.00	

First National Bank Monthly Statement 4/30

BEGINNING BALANCE:		425.86
Date	Number	Amount
CHECKS AND ATM WITHDRAWALS		
4/3	114	30.27
4/4	ATM	50.00
4/11	117	84.16
4/13	115	8.50
4/15	119	18.72
4/22	121	395.00
DEPOSITS		
4/7		100.00
4/15	Direct deposit	294.36
SERVICE CHARGES		
Low balance fee		7.50
INTEREST		
Credited 4/30		1.15
ENDING BALANCE:		227.22

Chapter 5 Vocabulary Check

Fill in each blank with one of the choices listed below. Some choices may be used more than once.

vertically	decimal	and	right triangle
standard form	mean	median	circumference
sum	denominator	numerator	mode

1. Like fractional notation, ______________ notation is used to denote a part of a whole.
2. To write fractions as decimals, divide the ______________ by the ______________.
3. To add or subtract decimals, write the decimals so that the decimal points line up ______________.
4. When writing decimals in words, write "______________" for the decimal point.
5. When multiplying decimals, the decimal point in the product is placed so that the number of decimal places in the product is equal to the ______________ of the number of decimal places in the factors.
6. The ______________ of a set of numbers is the number that occurs most often.
7. The distance around a circle is called the ______________.
8. The ______________ of a set of numbers in numerical order is the middle number. If there are an even number of numbers, the median is the ______________ of the two middle numbers.
9. The ______________ of a list of numbers of items is $\frac{\text{sum of items}}{\text{number of items}}$.
10. When 2 million is written as 2,000,000, we say it is written in ______________.

Helpful Hint

Are you preparing for your test? Don't forget to take the Chapter 5 Test on page 408. Then check your answers at the back of the text and use the Chapter Test Prep Video CD to see the fully worked-out solutions to any of the exercises you want to review.

5 Chapter Highlights

Definitions and Concepts	Examples
Section 5.1 Introduction to Decimals	

Place-Value Chart

hundreds	tens	ones	.	tenths	hundredths	thousandths	ten-thousandths	hundred-thousandths
		4		2	6	5		
100	10	1	↑ decimal point	$\frac{1}{10}$	$\frac{1}{100}$	$\frac{1}{1000}$	$\frac{1}{10,000}$	$\frac{1}{100,000}$

4.265 means

$$4 \cdot 1 + 2 \cdot \frac{1}{10} + 6 \cdot \frac{1}{100} + 5 \cdot \frac{1}{1000}$$

or

$$4 + \frac{2}{10} + \frac{6}{100} + \frac{5}{1000}$$

DEFINITIONS AND CONCEPTS	EXAMPLES
Section 5.1 Introduction to Decimals (*continued*)	
WRITING (OR READING) A DECIMAL IN WORDS **Step 1.** Write the whole number part in words. **Step 2.** Write "and" for the decimal point. **Step 3.** Write the decimal part in words as though it were a whole number, followed by the place value of the last digit.	Write 3.08 in words. Three and eight hundredths
A decimal written in words can be written in standard form by reversing the above procedure.	Write "negative four and twenty-one thousandths" in standard form. -4.021
TO ROUND DECIMALS TO A PLACE VALUE TO THE RIGHT OF THE DECIMAL POINT **Step 1.** Locate the digit to the right of the given place value. **Step 2.** If this digit is 5 or greater, add 1 to the digit in the given place value and delete all digits to its right. If this digit is less than 5, delete all digits to the right of the given place value.	Round 86.1256 to the nearest hundredth. **Step 1.** 86.12 5 6 (hundredths place: 2; digit to the right: 5) **Step 2.** Since the digit to the right is 5 or greater, we add 1 to the digit in the hundredths place and delete all digits to its right. 86.1256 rounded to the nearest hundredth is 86.13.
Section 5.2 Adding and Subtracting Decimals	
TO ADD OR SUBTRACT DECIMALS **Step 1.** Write the decimals so that the decimal points line up vertically. **Step 2.** Add or subtract as with whole numbers. **Step 3.** Place the decimal point in the sum or difference so that it lines up vertically with the decimal points in the problem.	Add: 4.6 + 0.28 $\begin{array}{r} 4.60 \\ +\ 0.28 \\ \hline 4.88 \end{array}$ Subtract: 2.8 − 1.04 $\begin{array}{r} \overset{7\,10}{2.8\not{0}} \\ -\ 1.04 \\ \hline 1.76 \end{array}$
Section 5.3 Multiplying Decimals and Circumference of a Circle	
TO MULTIPLY DECIMALS **Step 1.** Multiply the decimals as though they are whole numbers. **Step 2.** The decimal point in the product is placed so that the number of decimal places in the product is equal to the *sum* of the number of decimal places in the factors.	Multiply: 1.48 × 5.9 $\begin{array}{r} 1.48 \\ \times\ 5.9 \\ \hline 1332 \\ 7400 \\ \hline 8.732 \end{array}$ 1.48 ← 2 decimal places × 5.9 ← 1 decimal place 8.732 ← 3 decimal places

Definitions and Concepts	Examples

Section 5.3 Multiplying Decimals and Circumference of a Circle (*continued*)

The **circumference** of a circle is the distance around the circle.

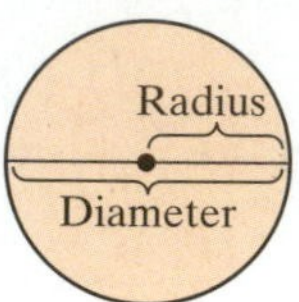

$C = \pi d$ or $C = 2\pi r$

where $\pi \approx 3.14$ or $\pi \approx \frac{22}{7}$.

or

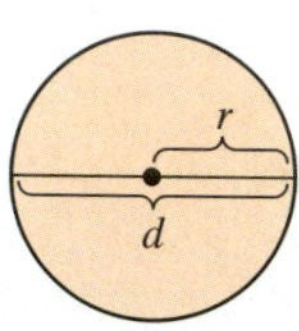

Find the exact circumference of a circle with radius 5 miles and an approximation by using 3.14 for π.

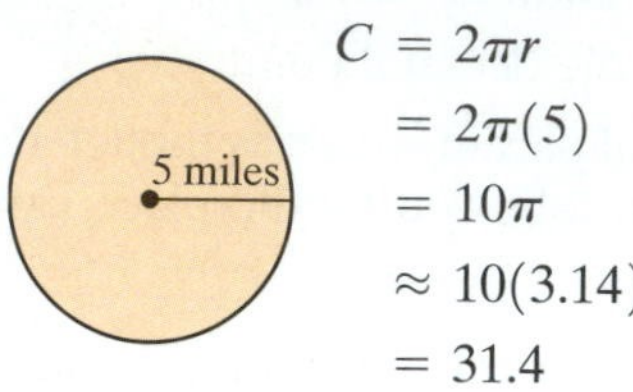

$$\begin{aligned} C &= 2\pi r \\ &= 2\pi(5) \\ &= 10\pi \\ &\approx 10(3.14) \\ &= 31.4 \end{aligned}$$

The circumference is exactly 10π miles and approximately 31.4 miles.

Section 5.4 Dividing Decimals

To Divide Decimals

Step 1. If the divisor is not a whole number, move the decimal point in the divisor to the right until the divisor is a whole number.

Step 2. Move the decimal point in the dividend to the right the *same number of places* as the decimal point was moved in Step 1.

Step 3. Divide. The decimal point in the quotient is directly over the moved decimal point in the dividend.

Divide: $1.118 \div 2.6$

```
         0.43
    2.6)1.118
       - 1 04
         ----
           78
         - 78
         ----
            0
```

Section 5.5 Fractions, Decimals, and Order of Operations

To **write fractions as decimals,** divide the numerator by the denominator.

Write $\frac{3}{8}$ as a decimal.

```
      0.375
    8)3.000
     -24
     ----
       60
      -56
      ---
       40
      -40
      ---
        0
```

Order of Operations

1. Perform all operations within parentheses ( ), brackets [ ], or grouping symbols such as fraction bars.
2. Evaluate any expressions with exponents.
3. Multiply or divide in order from left to right.
4. Add or subtract in order from left to right.

Simplify.

$-1.9(12.8 - 4.1) = -1.9(8.7)$ Subtract.

$= -16.53$ Multiply.

DEFINITIONS AND CONCEPTS	**EXAMPLES**
Section 5.6 Equations Containing Decimals	
STEPS FOR SOLVING AN EQUATION IN *X* **Step 1.** If fractions are present, multiply both sides of the equation by the LCD of the fractions. **Step 2.** If parentheses are present, use the distributive property. **Step 3.** Combine any like terms on each side of the equation. **Step 4.** Use the addition property of equality to rewrite the equation so that variable terms are on one side of the equation and constant terms are on the other side. **Step 5.** Divide both sides by the numerical coefficient of x to solve. **Step 6.** Check your answer in the *original equation.*	Solve: $3(x + 2.6) = 10.92$ $3x + 7.8 = 10.92$ Apply the distributive property. $3x + 7.8 - 7.8 = 10.92 - 7.8$ Subtract 7.8 from both sides. $3x = 3.12$ Simplify. $\frac{3x}{3} = \frac{3.12}{3}$ Divide both sides by 3. $x = 1.04$ Simplify. Check 1.04 in the original equation.
Section 5.7 Decimal Applications: Mean, Median, and Mode	
The **mean** (or **average**) of a set of number items is $\text{mean} = \frac{\text{sum of items}}{\text{number of items}}$ The **median** of a set of numbers in numerical order is the middle number. If the number of items is even, the median is the mean of the two middle numbers. The **mode** of a set of numbers is the number that occurs most often. (A set of numbers may have no mode or more than one mode.)	Find the mean, median, and mode of the following set of numbers: 33, 35, 35, 43, 68, 68 $\text{mean} = \frac{33 + 35 + 35 + 43 + 68 + 68}{6} = 47$ The median is the mean of the two middle numbers, 35 and 43 $\text{median} = \frac{35 + 43}{2} = 39$ There are two modes because there are two numbers that occur twice: 35 and 68

STUDY SKILLS BUILDER

Are You Prepared for a Test on Chapter 5?

Below I have listed some *common trouble areas* for students in Chapter 5. After studying for your test—but before taking your test—read these.

- Don't forget the order of operations. To simplify $-0.7 + 1.3(5 - 0.1)$, should you add, subtract, or multiply first? First, perform the subtraction within parentheses, then multiply, and finally add.

$$\begin{aligned} -0.7 + 1.3(5 - 0.1) &= -0.7 + 1.3(4.9) && \text{Subtract.} \\ &= -0.7 + 6.37 && \text{Multiply.} \\ &= 5.67 && \text{Add.} \end{aligned}$$

- If you are having trouble with ordering or operations on decimals, don't forget that you can insert 0s after the last digit to the right of the decimal point as needed.

Addition:
$$\begin{array}{r} 8.1 \\ 0.6 \\ +\,23.003 \\ \hline \end{array}$$

Addition with zeros inserted:
$$\begin{array}{r} 8.100 \\ 0.600 \\ +\,23.003 \\ \hline 31.703 \end{array}$$

Subtraction:
$$\begin{array}{r} 7 \\ -\,0.28 \\ \hline \end{array}$$

Subtraction with zeros inserted:
$$\begin{array}{r} \overset{6\ \overset{9}{10}\ 10}{\cancel{7}.\cancel{0}\cancel{0}} \\ -\,0.28 \\ \hline 6.72 \end{array}$$

- Place in order from smallest to largest: 0.108, 0.18, 0.0092

  If we insert zeros, we have: 0.1080, 0.1800, 0.0092

  The decimals in order are: 0.0092, 0.1080, 0.1800 or 0.0092, 0.108, 0.18.

- Do you remember that a set of numbers can have no mode, 1 mode, or even more than 1 mode?

2, 5, 8, 9	no mode
2, 2, 8, 9	mode: 2
2, 2, 3, 3, 5, 7, 7	mode: 2, 3, 7

- Do you remember how to find the median of an even-numbered set of numbers?

$$2, 5, 8, 9 \qquad \frac{5 + 8}{2} = 6.5$$ The median is the average of the two "middle" numbers.

5 CHAPTER REVIEW

(5.1) *Determine the place value of the number 4 in each decimal.*

1. 23.45

2. 0.000345

Write each decimal in words.

3. −23.45

4. 0.00345

5. 109.23

6. 200.000032

Write each decimal in standard form.

7. Eight and six hundredths

8. Negative five hundred three and one hundred two thousandths

9. Sixteen thousand twenty-five and fourteen ten-thousandths

10. Fourteen and eleven thousandths

Write each decimal as a fraction or a mixed number.

11. 0.16

12. −12.023

Write each fraction or mixed number as a decimal.

13. $\frac{231}{100{,}000}$

14. $25\frac{1}{4}$

Insert <, >, or = between each pair of numbers to make a true statement.

15. 0.49 0.43

16. 0.973 0.9730

17. −38.0027 −38.00056

18. −0.230505 −0.23505

Round each decimal to the given place value.

19. 0.623, nearest tenth

20. 0.9384, nearest hundredth

21. −42.895, nearest hundredth

22. 16.34925, nearest thousandth

Write each number in standard notation.

23. Saturn is a distance of about 887 million miles from the Sun.

24. The tail of a comet can be over 600 thousand miles long.

(5.2) *Add.*

25. 8.6 + 9.5

26. 3.9 + 1.2

27. −6.4 + (−0.88)

28. −19.02 + 6.98

29. 200.49 + 16.82 + 103.002

30. 0.00236 + 100.45 + 48.29

Subtract.

31. 4.9 − 3.2

32. 5.23 − 2.74

33. −892.1 − 432.4

34. 0.064 − 10.2

35. 100 − 34.98

36. 200 − 0.00198

Solve.

37. Find the total distance between Grove City and Jerome.

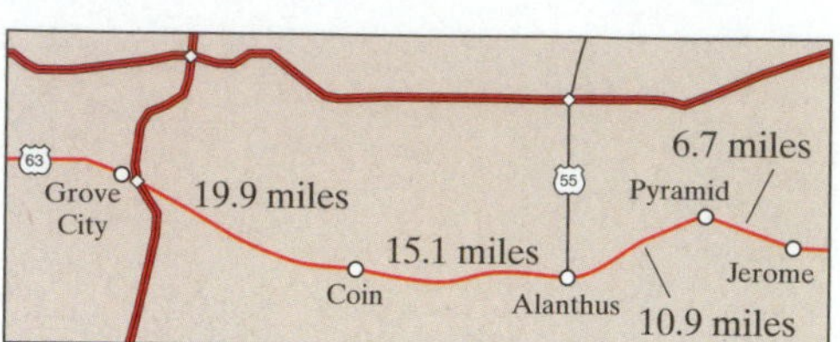

38. Evaluate $x - y$ for $x = 1.2$ and $y = 6.9$.

39. Find the perimeter.

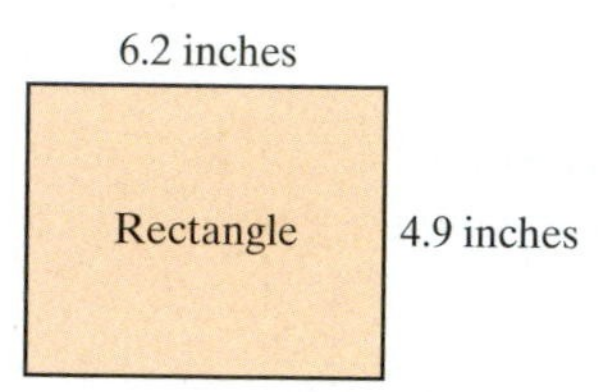

40. Find the perimeter.

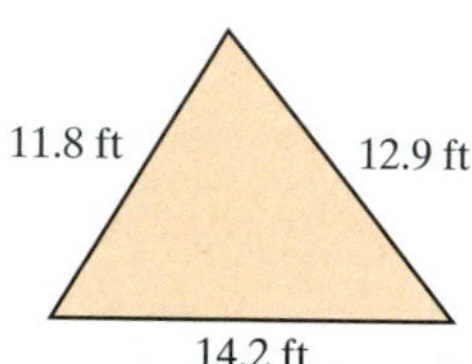

(5.3) *Multiply.*

41. 7.2×10

42. 9.345×1000

43. -34.02×2.3

44. $-839.02 \times (-87.3)$

Find the exact circumference of each circle. Then use the approximation 3.14 for π and approximate the circumference.

45.

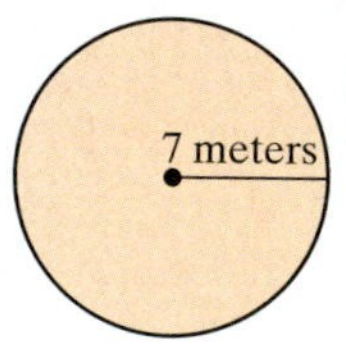

46.

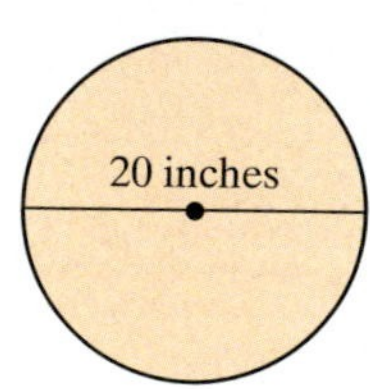

(5.4) *Divide. Round the quotient to the nearest thousandth if necessary.*

47. $3\overline{)0.2631}$

48. $20\overline{)316.5}$

49. $-21 \div (-0.3)$

50. $-0.0063 \div 0.03$

51. $0.34\overline{)2.74}$

52. $19.8\overline{)601.92}$

53. $\frac{23.65}{1000}$

54. $\frac{93}{-10}$

55. There are approximately 3.28 feet in 1 meter. Find how many meters are in 24 feet to the nearest tenth of a meter.

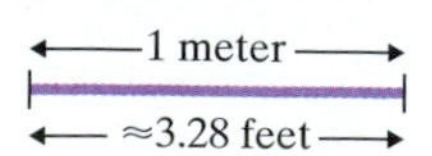

56. George Strait pays $69.71 per month to pay back a loan of $3136.95. In how many months will the loan be paid off?

(5.5) *Write each fraction as a decimal. Round to the nearest thousandth if necessary.*

57. $\frac{4}{5}$

58. $-\frac{12}{13}$

59. $2\frac{1}{3}$

60. $\frac{13}{60}$

Insert <, >, or = to make a true statement.

61. 0.392 0.392 **62.** $\frac{4}{7}$ 0.625 **63.** 0.293 $\frac{5}{17}$ **64.** −0.0231 −0.0221

Write the numbers in order from smallest to largest.

65. 0.837, 0.839, 0.832 **66.** 0.685, 0.626, $\frac{5}{8}$ **67.** $\frac{3}{7}$, 0.42, 0.43 **68.** $\frac{18}{11}$, 1.63, $\frac{19}{12}$

Simplify each expression.

69. $-7.6 \times 1.9 + 2.5$ **70.** $(-2.3)^2 - 1.4$ **71.** $\dfrac{7 + 0.74}{-0.06}$

72. $0.9(6.5 - 5.6)$ **73.** $\dfrac{(1.5)^2 + 0.5}{0.05}$ **74.** $0.0726 \div 10 \times 1000$

Find each area.

75.

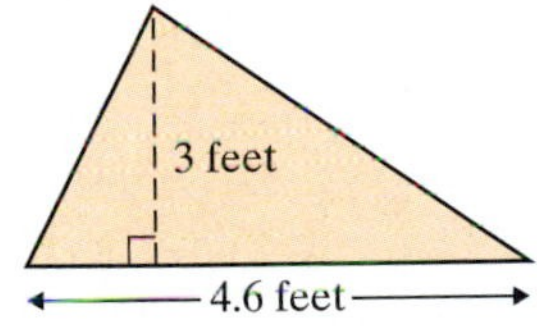

76.

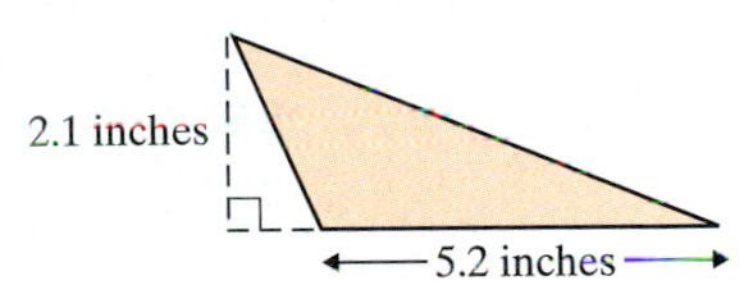

(5.6) *Solve.*

77. $x + 3.9 = 4.2$ **78.** $70 = y - 22.81$ **79.** $2x = 17.2$

80. $-1.1y = 88$ **81.** $3x - 0.78 = 1.2 + 2x$ **82.** $-x + 0.6 - 2x = -4x - 0.9$

83. $-1.3x - 9.4 = -0.4x + 8.6$ **84.** $3(x - 1.1) = 5x - 5.3$

(5.7) *Find the mean, median, and any mode(s) for each list of numbers. If necessary, round to the nearest tenth.*

85. 13, 23, 33, 14, 6 **86.** 45, 86, 21, 60, 86, 64, 45

87. 14,000, 20,000, 12,000, 20,000, 36,000, 45,000 **88.** 560, 620, 123, 400, 410, 300, 400, 780, 430, 450

For Exercises 89 and 90, the grades are given for a student for a particular semester. Find each grade point average. If necessary, round the grade point average to the nearest hundredth.

89.

Grade	Credit Hours
A	3
A	3
C	2
B	3
C	1

90.

Grade	Credit Hours
B	3
B	4
C	2
D	2
B	3

Mixed Review

91. Write 200.0032 in words.

92. Write negative sixteen and nine hundredths in standard form.

93. Write 0.0847 as a fraction or a mixed number.

94. Write the numbers $\frac{6}{7}, \frac{8}{9}$, 0.75 in order from smallest to largest.

Write each fraction as a decimal. Round to the nearest thousandth, if necessary.

95. $-\frac{7}{100}$

96. $\frac{9}{80}$ (Do not round.)

97. $\frac{8935}{175}$

Insert <, >, or = to make a true statement.

98. $-402.000032 \quad -402.00032$

99. $\frac{6}{11} \quad 0.55$

Round each decimal to the given place value.

100. 86.905, nearest hundredth

101. 3.11526, nearest thousandth

Round each money amount to the nearest dollar.

102. $123.46, nearest dollar

103. $3645.52, nearest dollar

Add or subtract as indicated.

104. $3.2 - 4.9$

105. $9.12 - 3.86$

106. $-102.06 + 89.3$

107. $-4.021 + (-10.83) + (-0.056)$

Multiply or divide as indicated. Round to the nearest thousandth, if necessary.

108. $\begin{array}{r} 2.54 \\ \times\ 3.2 \\ \hline \end{array}$

109. $(-3.45)(2.1)$

110. $0.005\overline{)24.5}$

111. $2.3\overline{)54.98}$

Solve.

△ **112.** Tomaso is going to fertilize his lawn, a rectangle that measures 77.3 feet by 115.9 feet. Approximate the area of the lawn by rounding each measurement to the nearest ten feet.

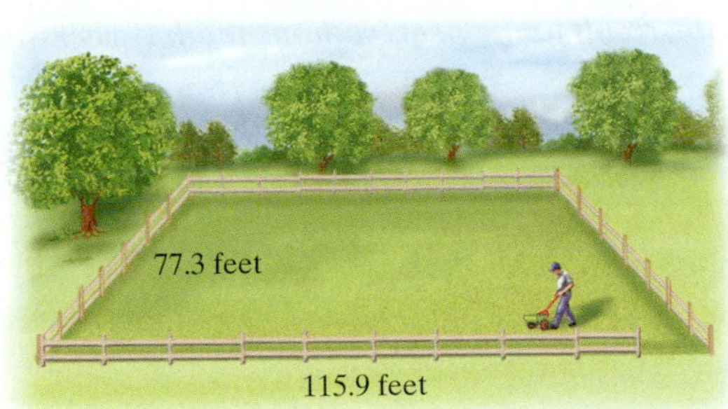

113. Estimate the cost of the items to see whether the groceries can be purchased with a $5 bill.

Simplify each expression.

114. $\dfrac{(3.2)^2}{100}$

115. $(2.6 + 1.4)(4.5 - 3.6)$

Find the mean, median, and any mode(s) for each list of numbers. If needed round answers to two decimal places.

116. 73, 82, 95, 68, 54

117. 952, 327, 566, 814, 327, 729

Answers

1. ______

2. ______

3. ______

4. ______

5. ______

6. ______

7. ______

8. ______

9. ______

10. ______

11. ______

12. ______

13. ______

14. ______

15. ______

16. ______

17. ______

18. ______

5 CHAPTER TEST

Remember to use the Chapter Test Prep Video CD to see the fully worked-out solutions to any of the exercises you want to review.

Write each decimal as indicated.

1. 45.092, in words

2. Three thousand and fifty-nine thousandths, in standard form

Perform each indicated operation. Round the result to the nearest thousandth if necessary.

3. $2.893 + 4.21 + 10.492$

4. $-47.92 - 3.28$

5. $9.83 - 30.25$

6. 10.2×4.01

7. $(-0.00843) \div (-0.23)$

Round each decimal to the indicated place value.

8. 34.8923, nearest tenth

9. 0.8623, nearest thousandth

Insert $<$, $>$, or $=$ between each pair of numbers to form a true statement.

10. 25.0909   25.9090

11. $\frac{4}{9}$ 0.445

Write each decimal as a fraction or a mixed number.

12. 0.345

13. -24.73

Write each fraction as a decimal. If necessary, round to the nearest thousandth.

14. $-\frac{13}{26}$

15. $\frac{16}{17}$

Simplify.

16. $(-0.6)^2 + 1.57$

17. $\frac{0.23 + 1.63}{-0.3}$

18. $2.4x - 3.6 - 1.9x - 9.8$

Solve.

19. $0.2x + 1.3 = 0.7$

20. $2(x + 5.7) = 6x - 3.4$

Find the mean, median, and mode of each list of numbers.

21. 26, 32, 42, 43, 49

22. 8, 10, 16, 16, 14, 12, 12, 13

Find the grade point average. If necessary, round to the nearest hundredth.

23.

Grade	Credit Hours
A	3
B	3
C	3
B	4
A	1

Solve.

24. At its farthest, Pluto is 4,583 million miles from the Sun. Write this number using standard notation.

△ **25.** Find the area.

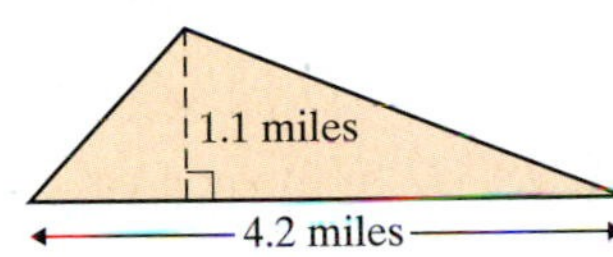

△ **26.** Find the exact circumference of the circle. Then use the approximation 3.14 for π and approximate the circumference.

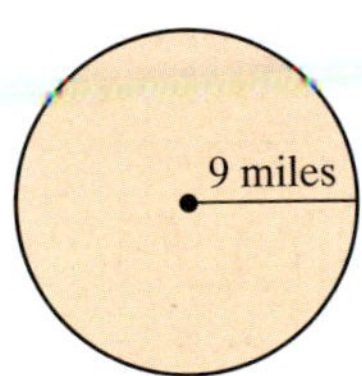

27. Vivian Thomas is going to put insecticide on her lawn to control grubworms. The lawn is a rectangle that measures 123.8 feet by 80 feet. The amount of insecticide required is 0.02 ounces per square foot.

a. Find the area of her lawn.

b. Find how much insecticide Vivian needs to purchase.

28. Find the total distance from Bayette to Center City.

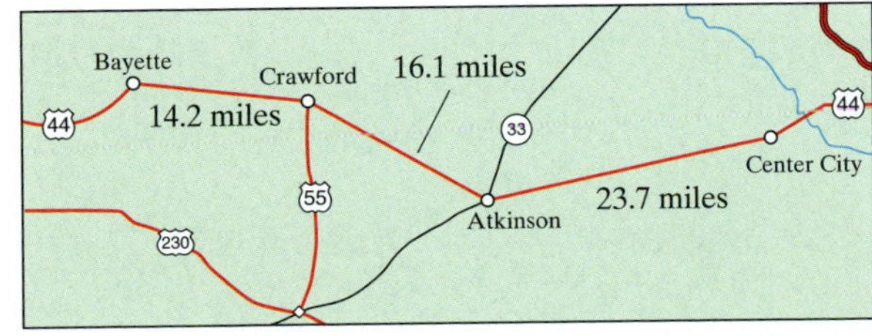

19. ______

20. ______

21. ______

22. ______

23. ______

24. ______

25. ______

26. ______

27. a. ______

b. ______

28. ______

CUMULATIVE REVIEW

Chapters 1–5

Answers

1. ______
2. ______
3. ______
4. ______
5. ______
6. ______
7. ______
8. ______
9. ______
10. ______
11. ______
12. ______
13. a. ______
b. ______
c. ______
d. ______
e. ______
f. ______
14. ______
15. ______
16. ______
17. ______
18. ______

Write each number in words.

1. 72

2. 107

3. 546

4. 5026

5. Add: $46 + 713$

6. Find the perimeter.

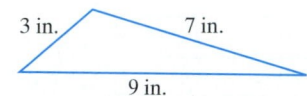

3 in. 7 in. 9 in.

7. Subtract: $543 - 29$. Then check by adding.

8. Divide: $3268 \div 27$

9. Round 278,362 to the nearest thousand.

10. Find all the factors of 30.

11. Multiply: 236×86

12. Multiply: $236 \times 86 \times 0$

13. Find each quotient and then check the answer by multiplying.

a. $1\overline{)7}$

b. $12 \div 1$

c. $\frac{6}{6}$

d. $9 \div 9$

e. $\frac{20}{1}$

f. $18\overline{)18}$

14. Find the average of 25, 17, 19, and 39.

15. Simplify: $2 \cdot 4 - 3 \div 3$

16. Simplify: $77 \div 11 \cdot 7$

Evaluate.

17. 9^2

18. 5^3

19. 3^4

20. 10^3

21. Evaluate $\frac{x - 5y}{y}$ for $x = 35$ and $y = 5$.

22. Evaluate $\frac{2a + 4}{c}$ for $a = 7$ and $c = 3$.

23. Find the opposite of each number.
a. 13 **b.** -2 **c.** 0

24. Find the opposite of each number.
a. -7 **b.** 4 **c.** -1

25. Add: $-2 + (-21)$

26. Add: $-7 + (-15)$

Find the value of each expression.

27. $5 \cdot 6^2$

28. $4 \cdot 2^3$

29. -7^2

30. $(-2)^5$

31. $(-5)^2$

32. -3^2

Represent the shaded part as an improper fraction and a mixed number.

33. 

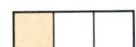

34. 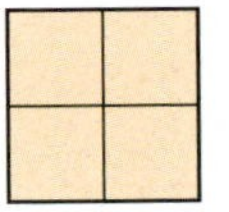 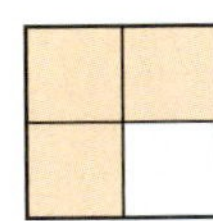

35.

36.

37. Find the prime factorization of 252.

38. Find the difference of 87 and 25.

19. ____
20. ____
21. ____
22. ____
23. a. ____
b. ____
c. ____
24. a. ____
b. ____
c. ____
25. ____
26. ____
27. ____
28. ____
29. ____
30. ____
31. ____
32. ____
33. ____
34. ____
35. ____
36. ____
37. ____
38. ____

39. ______
40. ______
41. ______
42. ______
43. ______
44. ______
45. ______
46. ______
47. ______
48. ______
49. ______
50. ______
51. ______
52. ______

39. Write $-\frac{72}{26}$ in simplest form.

40. Write $9\frac{7}{8}$ as an improper fraction.

41. Determine whether $\frac{16}{40}$ and $\frac{10}{25}$ are equivalent.

42. Insert $<$ or $>$ to form a true statement. $\frac{4}{7} \quad \frac{5}{9}$

Multiply.

43. $\frac{2}{3} \cdot \frac{5}{11}$

44. $2\frac{5}{8} \cdot \frac{4}{7}$

45. $\frac{1}{4} \cdot \frac{1}{2}$

46. $7 \cdot 5\frac{2}{7}$

Solve.

47. $\frac{z}{-4} = 11 - 5$

48. $6x - 12 - 5x = -20$

49. Add: $763.7651 + 22.001 + 43.89$

50. Add: $89.27 + 14.361 + 127.2318$

51. Multiply: 23.6×0.78

52. Multiply: 43.8×0.645

6

Percent

This chapter is devoted to percent, a concept used virtually every day in ordinary and business life. Understanding percent and using it efficiently depends on understanding ratios, because a percent is a ratio whose denominator is 100. We present techniques to write percents as fractions and as decimals. We then solve problems relating to interest rates, sales tax, discounts, and other real-life situations by writing percent equations.

No one is sure who first concocted ice cream. Historians suspect that ice cream was probably developed independently at various times all over the globe. The current popularity of ice cream is due to two major developments in the nineteenth and early twentieth centuries: the invention of the ice cream freezer and mechanical refrigeration. Ice cream became widely available after 1846, when Nancy Johnson invented the hand-cranked freezer. Turning the freezer handle agitates a container of ice cream mix in a bed of salt and ice until the mix is frozen. By 1926, the first commercially successful continuous process freezer was developed, making the mass production of ice cream feasible. In Section 6.5, Example 1, we will examine some of the current tastes in ice cream.

Favorite Ice Cream Flavors of Adults

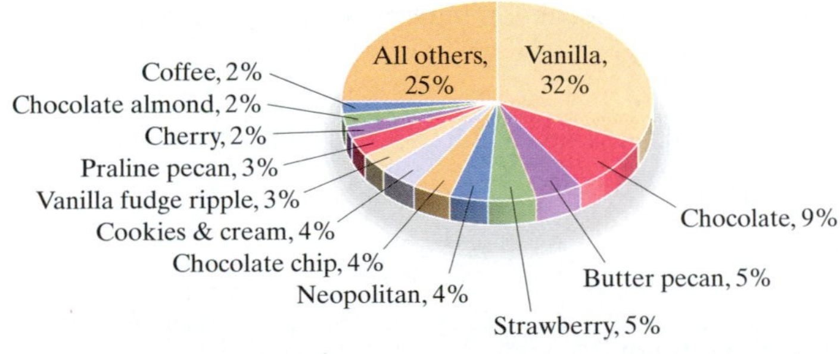

Source: International Ice Cream Association

SECTIONS

CHECK YOUR PROGRESS

Objectives

A Write Ratios as Fractions.

B Solve Proportions.

C Solve Problems Modeled by Proportions.

6.1 RATIO AND PROPORTION

Objective A Writing Ratios as Fractions

A **ratio** is the quotient of two numbers or two quantities. A ratio, in fact, is no different from a fraction, except that a ratio is sometimes written using notation other than fractional notation. For example, the ratio of 1 to 2 can be written as

$$1 \text{ to } 2 \quad \text{or} \quad \frac{1}{2} \quad \text{or} \quad 1:2$$

fractional notation (pointing to $\frac{1}{2}$) colon notation (pointing to $1:2$)

These ratios are all read as, "the ratio of 1 to 2."

Ratio

The ratio of a number a to a number b is their quotient. Ways of writing ratios are

$$a \text{ to } b, \qquad a:b, \quad \text{and} \quad \frac{a}{b}$$

Whenever possible, we will convert quantities in a ratio to the same unit of measurement.

PRACTICE PROBLEM 1

Write a ratio for each phrase. Use fractional notation.

a. The ratio of 3 parts oil to 7 parts gasoline

b. The ratio of 40 minutes to 3 hours

EXAMPLE 1 Write a ratio for each phrase. Use fractional notation.

a. The ratio of 2 parts salt to 5 parts water

b. The ratio of 18 inches to 2 feet

Solution:

a. The ratio of 2 parts salt to 5 parts water is $\frac{2}{5}$.

b. First we convert to the same unit of measurement. For example,

$$2 \text{ feet} = 2 \cdot 12 \text{ inches} = 24 \text{ inches}$$

The ratio of 18 inches to 2 feet is then $\frac{18}{24}$, or $\frac{3}{4}$ in lowest terms.

Work Practice Problem 1

If a ratio compares two decimal numbers, we will write the simplified ratio as a ratio of whole numbers.

PRACTICE PROBLEM 2

Write the ratio of 1.68 to 4.8 as a fraction in simplest form.

EXAMPLE 2 Write the ratio of 2.5 to 3.15 as a fraction in simplest form.

Solution: The ratio is

$$\frac{2.5}{3.15}$$

Now let's clear the ratio of decimals.

$$\frac{2.5}{3.15} = \frac{2.5 \cdot 100}{3.15 \cdot 100} = \frac{250}{315} = \frac{50}{63} \quad \text{Simplest form}$$

Work Practice Problem 2

Answers

1. a. $\frac{3}{7}$ **b.** $\frac{2}{9}$ **2.** $\frac{7}{20}$

EXAMPLE 3 Given the rectangle shown:

a. Find the ratio of its width (shorter side) to its length (longer side).
b. Find the ratio of its length to its perimeter.

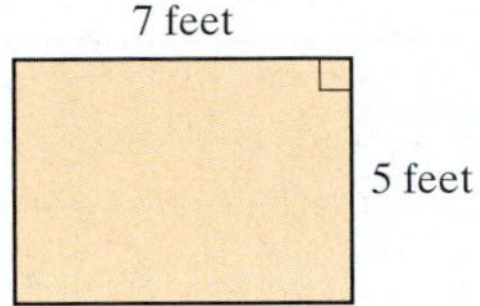

Solution:

a. The ratio of its width to its length is

$$\frac{\text{width}}{\text{length}} = \frac{5 \cancel{\text{feet}}}{7 \cancel{\text{feet}}} = \frac{5}{7}$$

b. Recall that the perimeter of the rectangle is the distance around the rectangle: $7 + 5 + 7 + 5 = 24$ feet. The ratio of its length to its perimeter is

$$\frac{\text{length}}{\text{perimeter}} = \frac{7 \cancel{\text{feet}}}{24 \cancel{\text{feet}}} = \frac{7}{24}$$

Work Practice Problem 3

PRACTICE PROBLEM 3

Given the triangle shown:

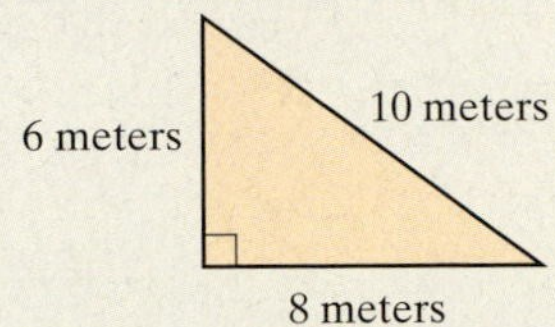

a. Find the ratio of the length of the shortest side to the length of the longest side in simplest form.
b. Find the ratio of the length of the longest side to the perimeter of the triangle in simplest form.

✔Concept Check Explain why the answer $\frac{7}{5}$ would be incorrect for part (a) of Example 3.

Objective B Solving Proportions

Ratios can be used to form proportions. A **proportion** is a mathematical statement that two ratios are equal.

For example, the equation

$$\frac{1}{2} = \frac{4}{8}$$

is a proportion that says that the ratios $\frac{1}{2}$ and $\frac{4}{8}$ are equal.

Notice that a proportion contains four numbers. If any three numbers are known, we can solve and find the fourth number. One way to do so is to use cross products. To understand cross products, which were introduced in Section 4.2, let's start with the proportion

$$\frac{a}{b} = \frac{c}{d}$$

and multiply both sides by the LCD, bd.

$$\frac{a}{b} = \frac{c}{d}$$

$$bd\left(\frac{a}{b}\right) = bd\left(\frac{c}{d}\right) \quad \text{Multiply both sides by the LCD, } bd.$$

$$ad = bc \quad \text{Simplify.}$$

ad — cross product; bc — cross product

Answers

3. a. $\frac{3}{5}$ **b.** $\frac{5}{12}$

✔ Concept Check Answer

$\frac{7}{5}$ is the ratio of the rectangle's length to its width.

Notice why ad and bc are called cross products.

$$\frac{a}{b} = \frac{c}{d}$$

ad bc

Cross Products

If $\frac{a}{b} = \frac{c}{d}$, then $ad = bc$.

PRACTICE PROBLEM 4

Solve for x: $\frac{3}{8} = \frac{63}{x}$

EXAMPLE 4 Solve for x: $\frac{45}{x} = \frac{5}{7}$

Solution: To solve, we set cross products equal.

$$\frac{45}{x} = \frac{5}{7}$$

$45 \cdot 7 = x \cdot 5$ Set cross products equal.

$315 = 5x$ Multiply.

$\frac{315}{5} = \frac{5x}{5}$ Divide both sides by 5.

$63 = x$ Simplify.

Check: To check, substitute 63 for x in the original proportion. The solution is 63.

Work Practice Problem 4

PRACTICE PROBLEM 5

Solve for x: $\frac{2x + 1}{7} = \frac{x - 3}{5}$

EXAMPLE 5 Solve for x: $\frac{x - 5}{3} = \frac{x + 2}{5}$

Solution:

$$\frac{x - 5}{3} = \frac{x + 2}{5}$$

$5(x - 5) = 3(x + 2)$ Set cross products equal.

$5x - 25 = 3x + 6$ Multiply.

$5x = 3x + 31$ Add 25 to both sides.

$2x = 31$ Subtract $3x$ from both sides.

$\frac{2x}{2} = \frac{31}{2}$ Divide both sides by 2.

$x = \frac{31}{2}$

Check: Verify that $\frac{31}{2}$ is the solution.

Work Practice Problem 5

✔Concept Check For which of the following equations can we immediately use cross products to solve for x?

a. $\frac{2 - x}{5} = \frac{1 + x}{3}$ **b.** $\frac{2}{5} - x = \frac{1 + x}{3}$

Answers

4. $x = 168$ **5.** $x = -\frac{26}{3}$

✔ Concept Check Answer

a

Objective C Solving Problems Modeled by Proportions

Writing proportions is a powerful tool for solving problems in almost every field, including business, chemistry, biology, health sciences, and engineering, as well as in daily life. Given a specified ratio (or rate) of two quantities, a proportion can be used to determine an unknown quantity.

EXAMPLE 6 Determining Distances from a Map

On a Chamber of Commerce map of Abita Springs, 5 miles corresponds to 2 inches. How many miles correspond to 7 inches?

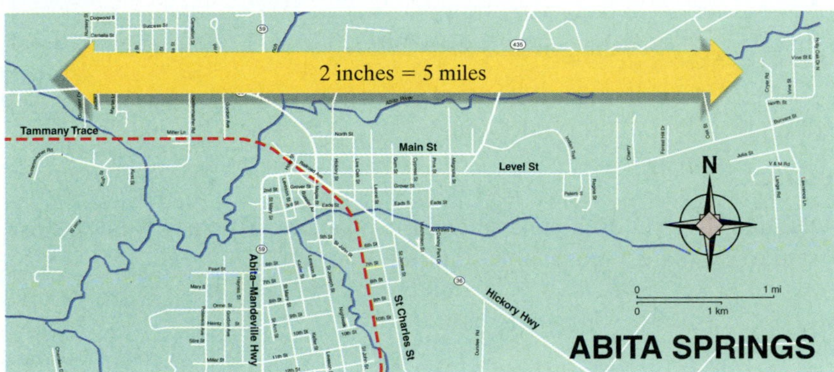

Solution:

1. UNDERSTAND. Read and reread the problem. You may want to draw a diagram.

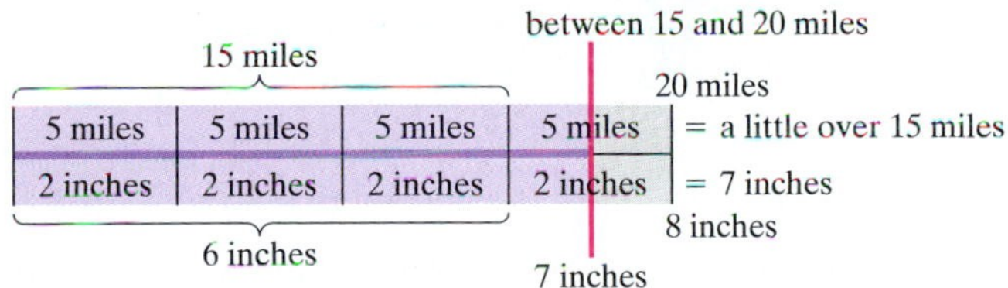

From the diagram we can see that a reasonable solution should be between 15 and 20 miles.

2. TRANSLATE. We will let x be our unknown number. Since 5 miles corresponds to 2 inches as x miles corresponds to 7 inches, we have the proportion

$$\begin{matrix} \text{miles} & \rightarrow \\ \text{inches} & \rightarrow \end{matrix} \ \frac{5}{2} = \frac{x}{7} \ \begin{matrix} \leftarrow & \text{miles} \\ \leftarrow & \text{inches} \end{matrix}$$

3. SOLVE: In earlier sections, we estimated to obtain a reasonable answer. Notice we did this in Step 1 above.

$$\frac{5}{2} = \frac{x}{7}$$

$5 \cdot 7 = 2 \cdot x$	Set the cross products equal to each other.
$35 = 2x$	Multiply.
$\frac{35}{2} = \frac{2x}{2}$	Divide both sides by 2.
$17\frac{1}{2} = x$ or $x = 17.5$	Simplify.

4. INTERPRET. *Check* your work. This result is reasonable since it is between 15 and 20 miles. *State* your conclusion: 7 inches corresponds to 17.5 miles.

Work Practice Problem 6

PRACTICE PROBLEM 6

On an architect's blueprint, 1 inch corresponds to 4 feet. How long is a wall represented by a $4\frac{1}{4}$-inch line on the blueprint?

Answer

6. 17 ft

Helpful Hint

We can also solve Example 6 by writing the proportion

$$\frac{2 \text{ inches}}{5 \text{ miles}} = \frac{7 \text{ inches}}{x \text{ miles}}$$

Although other proportions may be used to solve Example 6, we will solve by writing proportions so that the numerators have the same unit measures and the denominators have the same unit measures.

PRACTICE PROBLEM 7

An auto mechanic recommends that 5 ounces of isopropyl alcohol be mixed with a tankful of gas (16 gallons) to increase the octane of the gasoline for better engine performance. At this rate, how many gallons of gas can be treated with a 8-ounce bottle of alcohol?

EXAMPLE 7 **Finding Medicine Dosage**

The standard dose of an antibiotic is 4 cc (cubic centimeters) for every 25 pounds (lb) of body weight. At this rate, find the standard dose for a 140-lb woman.

Solution:

1. UNDERSTAND. Read and reread the problem. You may want to draw a diagram to estimate a reasonable solution.

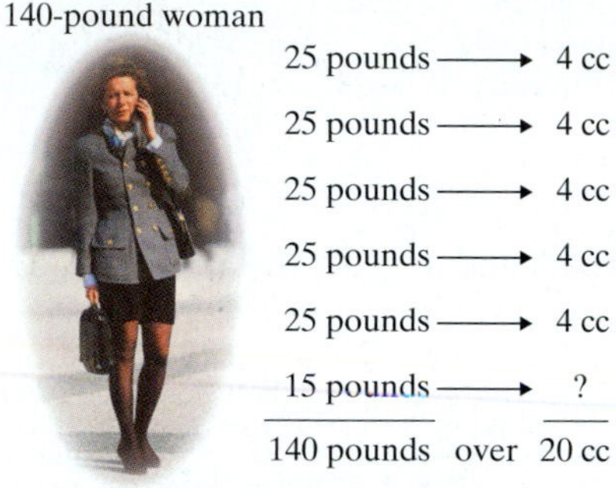

From the diagram, we can see that a reasonable solution is a little over 20 cc.

2. TRANSLATE. We will let x be the unknown number. From the problem, we know that 4 cc is to 25 pounds as x cc is to 140 pounds, or

$$\begin{array}{rcl} \text{cubic centimeters} \rightarrow & \dfrac{4}{25} = \dfrac{x}{140} & \leftarrow \text{cubic centimeters} \\ \text{pounds} \rightarrow & & \leftarrow \text{pounds} \end{array}$$

3. SOLVE:

$$\frac{4}{25} = \frac{x}{140}$$

$4 \cdot 140 = 25 \cdot x$ Set the cross products equal to each other.

$560 = 25x$ Multiply.

$\frac{560}{25} = \frac{25x}{25}$ Divide both sides by 25.

$22\frac{2}{5} = x$ or $x = 22.4$ Simplify.

4. INTERPRET. *Check* your work. This result is reasonable since it is a little over 20 cc. *State* your conclusion: The standard dose for a 140-lb woman is 22.4 cc.

Work Practice Problem 7

Answer

7. $25\frac{3}{5}$ or 25.6 gal

Vocabulary and Readiness Check

Use the words and phrases below to fill in each blank.

ratio   cross products   proportion

1. $\frac{4.2}{8.4} = \frac{1}{2}$ is called a ______________ while $\frac{7}{8}$ is called a ______________.

2. In $\frac{a}{b} = \frac{c}{d}$, $a \cdot d$ and $b \cdot c$ are called ______________.

6.1 EXERCISE SET

FOR EXTRA HELP

Student Solutions Manual

PH Math/Tutor Center

CD/Video for Review

MathXL®

MyMathLab

Objective A *Write each ratio in fractional notation in lowest terms. (For Exercises 9 and 10, note that 1 meter = 100 centimeters.) See Examples 1 through 3.*

1. 2 megabytes to 15 megabytes

2. 18 disks to 41 disks

3. 10 inches to 12 inches

4. 15 miles to 40 miles

5. 5 quarts to 3 gallons

6. 8 inches to 3 feet

7. 4 nickels to 2 dollars

8. 12 quarters to 2 dollars

9. 175 centimeters to 5 meters

10. 90 centimeters to 4 meters

11. 190 minutes to 3 hours

12. 60 hours to 2 days

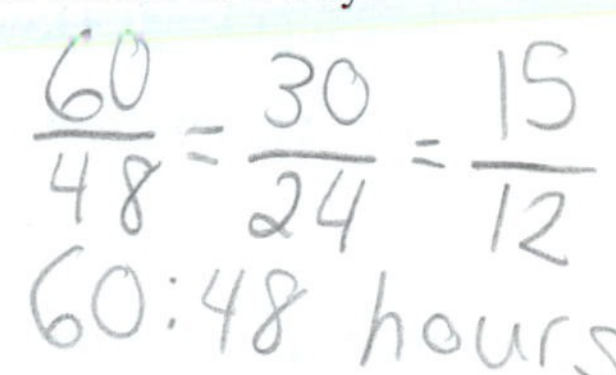

△ **13.** Find the ratio of the length to the width of a regulation size basketball court.

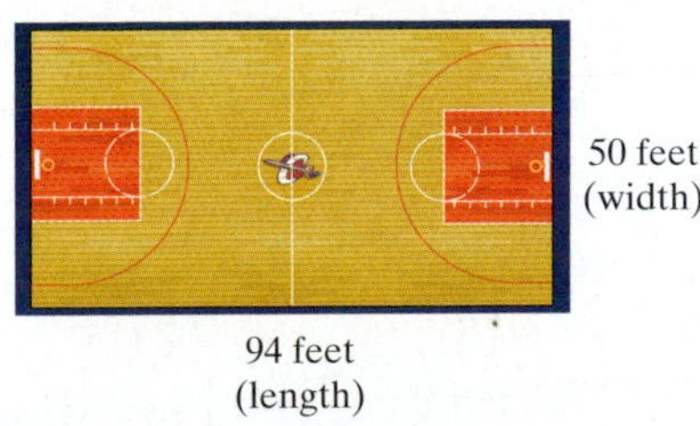

△ **14.** Find the ratio of the base to the height of the triangular mainsail.

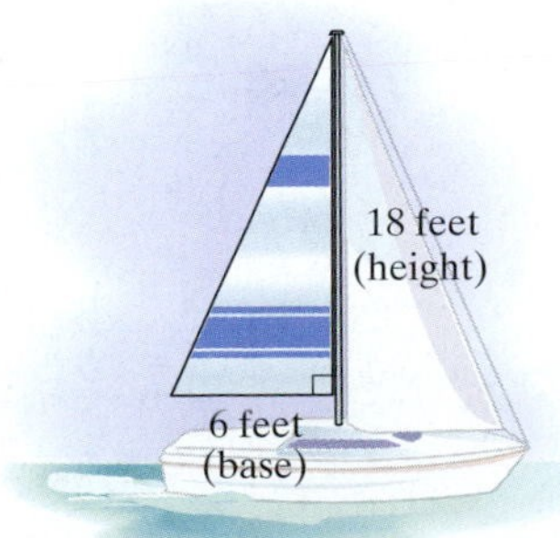

△ **15.** Find the ratio of the longest side to the perimeter of the right-triangular-shaped billboard.

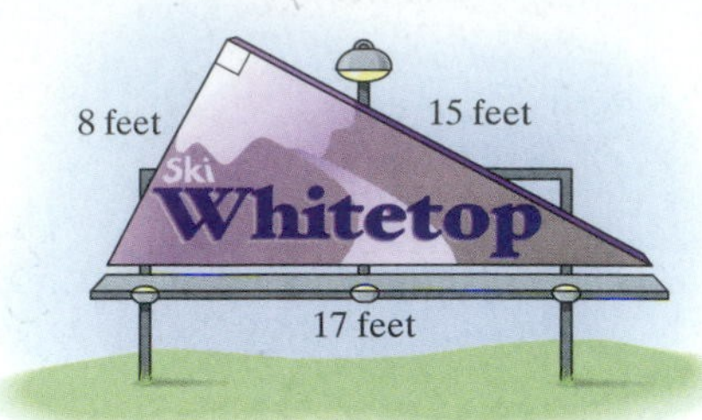

△ **16.** Find the ratio of the width to the perimeter of the rectangular vegetable garden.

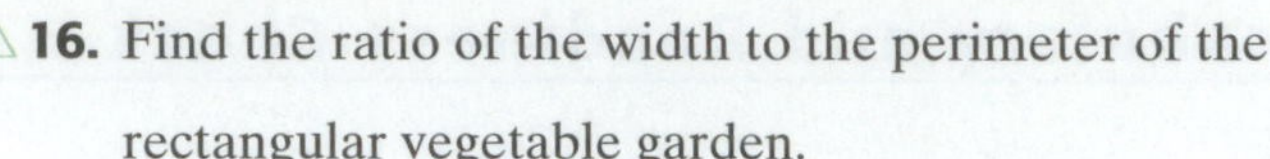

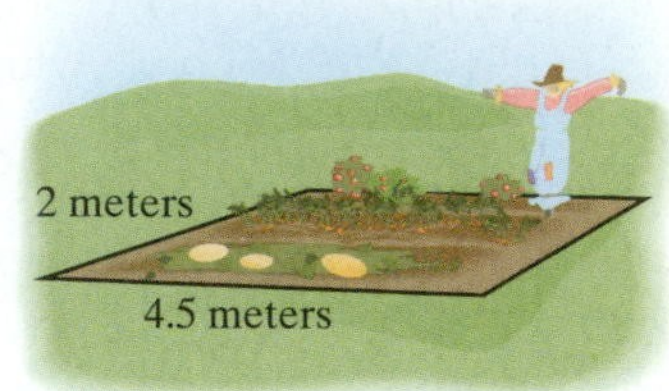

17. A large order of McDonald's french fries has 450 calories. Of this total, 200 calories are from fat. Find the ratio of calories from fat to total calories in a large order of McDonald's french fries. (*Source:* McDonald's Corporation)

18. A McDonald's Quarter Pounder® with Cheese contains 30 grams of fat. A McDonald's Grilled Chicken™ sandwich contains 20 grams of fat. Find the ratio of the amount of fat in a Quarter Pounder with Cheese to the amount of fat in a Grilled Chicken sandwich. (*Source:* McDonald's Corporation)

Blood contains three types of cells: red blood cells, white blood cells, and platelets. For approximately every 600 red blood cells in healthy humans, there are 40 platelets and 1 white blood cell. (*Source:* American Red Cross Biomedical Services) *Use this for Exercises 19 and 20.*

19. Write the ratio of red blood cells to platelet cells.

20. Write the ratio of white blood cells to red blood cells.

21. Suppose someone tells you that the ratio of 11 inches to 2 feet is $\frac{11}{2}$. How would you correct that person and explain the error?

22. Write a ratio that can be written in fractional notation as $\frac{3}{2}$.

Objective **B** *Solve each proportion. See Examples 4 and 5.*

23. $\frac{2}{3} = \frac{x}{6}$

24. $\frac{x}{2} = \frac{16}{6}$

25. $\frac{x}{10} = \frac{5}{9}$

26. $\frac{9}{4x} = \frac{6}{2}$

27. $\frac{4x}{6} = \frac{7}{2}$

28. $\frac{a}{5} = \frac{3}{2}$

29. $\frac{x - 3}{x} = \frac{4}{7}$

30. $\frac{y}{y - 16} = \frac{5}{3}$

31. $\frac{x + 1}{2x + 3} = \frac{2}{3}$

32. $\frac{x + 1}{x + 2} = \frac{5}{3}$

33. $\frac{9}{5} = \frac{12}{3x + 2}$

34. $\frac{6}{11} = \frac{27}{3x - 2}$

35. $\frac{3}{x + 1} = \frac{5}{2x}$

36. $\frac{7}{x - 3} = \frac{8}{2x}$

37. $\frac{15}{3x - 4} = \frac{5}{x}$

38. $\frac{x}{3} = \frac{2x + 5}{6}$

Objective **C** *Solve. See Examples 6 and 7.*

39. The ratio of the weight of an object on Earth to the weight of the same object on Pluto is 100 to 3. If an elephant weighs 4100 pounds on Earth, find the elephant's weight on Pluto.

40. If a 170-pound person weighs approximately 65 pounds on Mars, how much does a 9000-pound satellite weigh on Mars? Round to the nearest pound.

41. There are 110 calories per 28.4 grams of Crispy Rice cereal. Find how many calories are in 42.6 grams of this cereal.

42. On an architect's blueprint, 1 inch corresponds to 4 feet. Find the length of a wall represented by a line that is $3\frac{7}{8}$ inches long on the blueprint.

43. A 16-oz grande Tazo Black Iced Tea at Starbucks has 80 calories. How many calories are there in a 24-oz venti Tazo Black Iced Tea? (*Source:* Starbucks Coffee Company)

44. A 16-oz nonfat Caramel Macchiato at Starbucks has 220 calories. How many calories are there in a 12-oz nonfat Caramel Macchiato? (*Source:* Starbucks Coffee Company)

45. Mosquitos are annoying insects. To eliminate mosquito larvae, a certain granular substance can be applied to standing water in a ratio of 1 tsp per 25 sq ft of standing water.

a. At this rate, find how many teaspoons of granules must be used for 450 square feet.

b. If 3 tsp = 1 tbsp, how many tablespoons of granule must be used?

46. Another type of mosquito control is liquid, where 3 oz of pesticide is mixed with 100 oz of water. This mixture is sprayed on roadsides to control mosquito breeding grounds hidden by tall grass.

a. If one mixture of water with this pesticide can treat 150 feet of roadway, how many ounces of pesticide are needed to treat one mile? (*Hint:* 1 mile = 5280 feet)

b. If 8 liquid ounces equals one cup, write your answer to part **a** in cups. Round to the nearest cup.

△ **47.** The daily supply of oxygen for one person is provided by 625 square feet of lawn. A total of 3750 square feet of lawn would provide the daily supply of oxygen for how many people? (*Source:* Professional Lawn Care Association of America)

48. In the United States, approximately 71 million of the 200 million cars and light trucks in service have driver-side air bags. In a parking lot containing 800 cars and light trucks, how many would be expected to have driver-side air bags? (*Source:* Insurance Institute for Highway Safety)

49. A student would like to estimate the height of the Statue of Liberty in New York City's harbor. The length of the Statue of Liberty's right arm is 42 feet. The student's right arm is 2 feet long and her height is $5\frac{1}{3}$ feet. Use this information to estimate the height of the Statue of Liberty. How close is your estimate to the statue's actual height of 111 feet, 1 inch from heel to top of head? (*Source:* National Park Service)

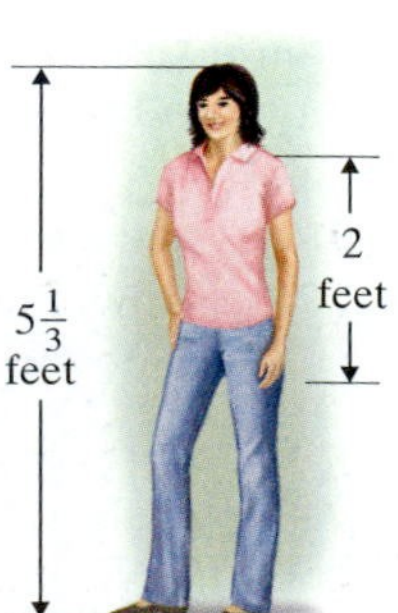

50. The length of the Statue of Liberty's index finger is 8 feet while the height to the top of the head is about 111 feet. Suppose your measurements are proportionaly the same as this statue and your height is 5 feet.

a. Use this information to find the proposed length of your index finger. Give an exact measurement and then a decimal rounded to the nearest hundredth.

b. Measure your index finger and write it as decimal in feet rounded to the nearest hundredth. How close is the length of your index finger to the answer to **a**? Explain why.

51. There are 72 milligrams of cholesterol in a 3.5 ounce serving of lobster. How much cholesterol is in 5 ounces of lobster? Round to the nearest tenth of a milligram. (*Source:* The National Institute of Health)

52. There are 76 milligrams of cholesterol in a 3-ounce serving of skinless chicken. How much cholesterol is in 8 ounces of chicken? (*Source:* USDA)

53. Trump World Tower in New York City is 881 feet tall and contains 72 stories. The Empire State Building contains 102 stories. If the Empire State Building has the same number of feet per floor as the Trump World Tower, approximate its height rounded to the nearest foot. (*Source:* skyscrapers.com)

54. Two out of every 5 men blame their poor eating habits on too much fast food. In a room of 40 men, how many would you expect to blame their not eating well on fast food? (*Source:* Healthy Choice Mixed Grills survey)

55. Medication is prescribed in 7 out of every 10 hospital emergency room visits that involve an injury. If a large urban hospital had 620 emergency room visits involving an injury in the past month, how many of these visits would you expect included a prescription for medication? (*Source:* National Center for Health Statistics)

56. Currently in the American population of people aged 65 years old and older, there are 145 women for every 100 men. In a nursing home with 280 male residents over the age of 65, how many female residents over the age of 65 would be expected? (*Source:* U.S. Bureau of the Census)

57. One out of three American adults has worked in the restaurant industry at some point during his or her life. In an office of 84 workers, how many of these people would you expect to have worked in the restaurant industry at some point? (*Source:* National Restaurant Association)

58. One pound of firmly packed brown sugar yields $2\frac{1}{4}$ cups. How many pounds of brown sugar will be required in a recipe that calls for 6 cups of firmly packed brown sugar? (*Source:* Based on data from *Family Circle* magazine)

When making homemade ice cream in a hand-cranked freezer, the tub containing the ice cream mix is surrounded by a brine (water/salt) solution. To freeze the ice cream mix rapidly so that smooth and creamy ice cream results, the brine solution should combine crushed ice and rock salt in a ratio of 5 to 1. Use this for Exercises 59 and 60. (*Source:* White Mountain Freezers, The Rival Company)

59. A small ice cream freezer requires 12 cups of crushed ice. How much rock salt should be mixed with the ice to create the necessary brine solution?

60. A large ice cream freezer requires $18\frac{3}{4}$ cups of crushed ice. How much rock salt will be needed?

61. The gas/oil ratio for a certain chainsaw is 50 to 1.

a. How much oil (in gallons) should be mixed with 5 gallons of gasoline?

b. If 1 gallon equals 128 fluid ounces, write the answer to part **a** in fluid ounces. Round to the nearest whole ounce.

62. The gas/oil ratio for a certain tractor mower is 20 to 1.

a. How much oil (in gallons) should be mixed with 10 gallons of gas?

b. If 1 gallon equals 4 quarts, write the answer to part **a** in quarts.

63. The adult daily dosage for a certain medicine is 150 mg (milligrams) of medicine for every 20 pounds of body weight.

a. At this rate, find the daily dose for a man who weighs 275 pounds.

b. If the man is to receive 500 mg of this medicine every 8 hours, is he receiving the proper dosage?

64. The adult daily dosage for a certain medicine is 80 mg (milligrams) for every 25 pounds of body weight.

a. At this rate, find the daily dose for a woman who weighs 190 pounds.

b. If she is to receive this medicine every 6 hours, find the amount to be given every 6 hours.

Review

Find the prime factorization of each number. See Section 4.2.

65. 200 **66.** 300 **67.** 32 **68.** 81

Concept Extensions

As we have seen earlier, proportions are often used in medicine dosage calculations. The exercises below have to do with liquid drug preparations, where the weight of the drug is contained in a volume of solution. The description of mg and ml below will help. We will study metric units further in Chapter 8.

mg means milligrams (A paper clip weighs about a gram. A milligram is about the weight of $\frac{1}{1000}$ *of a paper clip.)*

ml means milliliter (A liter is about a quart. A milliliter is about the amount of liquid in $\frac{1}{1000}$ *of a quart.)*

One way to solve the applications below is to set up the proportion $\frac{\text{mg}}{\text{ml}} = \frac{\text{mg}}{\text{ml}}$.

A solution strength of 15 mg of medicine in 1 ml of solution is available.

69. If a patient needs 12 mg of medicine, how many ml do you administer?

70. If a patient needs 33 mg of medicine, how many ml do you administer?

A solution strength of 8 mg of medicine in 1 ml of solution is available.

71. If a patient needs 10 mg of medicine, how many ml do you administer?

72. If a patient needs 6 mg of medicine, how many ml do you administer?

THE BIGGER PICTURE Operations on Sets of Numbers and Solving Equations

Continue your outline from Sections 1.6, 1.7, 2.4, 3.3, 4.3, 4.7, and 5.4. Suggestions are once again written to help you complete this part of your outline. Notice that this part of the outline has to do with solving a certain type of equation, proportions.

I. Operations on Sets of Numbers

- **A. Whole Numbers**
  1. **Add or Subtract** (Section 1.3)
  2. **Multiply or Divide** (Sections 1.5, 1.6)
  3. **Exponent** (Section 1.7)
  4. **Order of Operations** (Section 1.7)
- **B. Integers**
  1. **Add** (Section 2.2)
  2. **Subtract** (Section 2.3)
  3. **Multiply or Divide** (Section 2.4)
- **C. Fractions**
  1. **Simplify** (Section 4.2)
  2. **Multiply** (Section 4.3)
  3. **Divide** (Section 4.3)
  4. **Add or Subtract** (Sections 4.4, 4.5)
- **D. Decimals**
  1. **Add or Subtract** (Section 5.2)
  2. **Multiply** (Section 5.3)
  3. **Divide** (Section 5.4)

II. Solving Equations

- **A. Equations in General** (Section 3.3)
- **B. Proportions:** Set cross products equal to each other. Then solve.

$$\frac{14}{3} = \frac{2}{n}, \text{ or } 14 \cdot n = 3 \cdot 2, \text{ or } 14n = 6, \text{ or } n = \frac{6}{14} = \frac{3}{7}$$

Perform the indicated operations.

1. $\frac{7}{20} - \frac{1}{10}$

2. $\frac{7}{20} \cdot \frac{1}{10}$

3. $\frac{7}{20} \div \frac{1}{10}$

4. $\frac{7}{20} + \frac{1}{10}$

5. $7.6 + 0.02$

6. $7.6(0.02)$

For each proportion, find the unknown number, n.

7. $\frac{4}{n} = \frac{50}{100}$

8. $\frac{60}{10} = \frac{15}{n}$

9. $\frac{n}{0.8} = \frac{0.06}{12}$

10. $\frac{\frac{7}{8}}{\frac{1}{4}} = \frac{n}{\frac{5}{6}}$

6.2 PERCENTS, DECIMALS, AND FRACTIONS

Objective A Understanding Percent

The word **percent** comes from the Latin phrase *per centum,* which means **"per 100."** For example, 53% (53 percent) means 53 per 100. In the square below, 53 of the 100 squares are shaded. Thus, 53% of the figure is shaded.

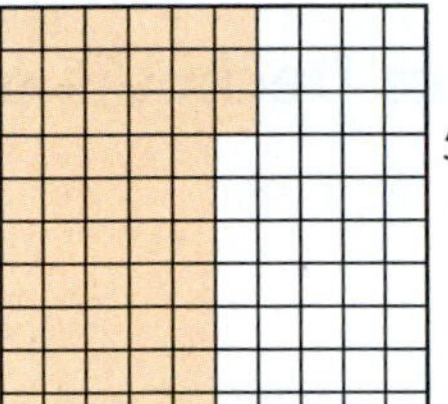

53 of 100 squares are shaded or 53% is shaded.

Since 53% means 53 per 100, 53% is the ratio of 53 to 100, or $\frac{53}{100}$.

$$53\% = \frac{53}{100}$$

Also,

$7\% = \frac{7}{100}$ 7 parts per 100 parts

$73\% = \frac{73}{100}$ 73 parts per 100 parts

$109\% = \frac{109}{100}$ 109 parts per 100 parts

Percent

Percent means **per one hundred.** The "%" symbol is used to denote percent.

Percent is used in a variety of everyday situations. For example,

- 58.7% of the U.S. population uses the Internet.
- The store is having a 25%-off sale.
- 78% of us trust our local fire department.
- The enrollment in community colleges is predicted to increase 1.3% each year.

EXAMPLE 1

In a survey of 100 people, 17 people drive blue cars. What percent of people drive blue cars?

Solution: Since 17 people out of 100 drive blue cars, the fraction is $\frac{17}{100}$. Then

$$\frac{17}{100} = 17\%$$

Work Practice Problem 1

Objectives

A Understand Percent.

B Write Percents as Decimals or Fractions.

C Write Decimals or Fractions as Percents.

D Applications with Percents, Decimals, and Fractions.

PRACTICE PROBLEM 1

Of 100 students in a club, 27 are freshmen. What percent of the students are freshmen?

Answer

1. 27%

PRACTICE PROBLEM 2

31 out of 100 executives are in their forties. What percent of executives are in their forties?

EXAMPLE 2 46 out of every 100 college students live at home. What percent of students live at home? (*Source:* Independent Insurance Agents of America)

Solution:

$$\frac{46}{100} = 46\%$$

Work Practice Problem 2

Objective B Writing Percents as Decimals or Fractions

Since percent means "per hundred," we have that

$$1\% = \frac{1}{100} = 0.01$$

In other words, the percent symbol means "per hundred" or, equivalently, "$\frac{1}{100}$" or "0.01." Thus

Write 87% as a fraction: $87\% = 87 \times \frac{1}{100} = \frac{87}{100}$

or

Write 87% as a decimal: $87\% = 87 \times (0.01) = 0.87$

Results are the same.

Of course, we know that the end results are the same, that is,

$$\frac{87}{100} = 0.87$$

The above gives us two options for converting percents. We can replace the percent symbol, %, by $\frac{1}{100}$ or 0.01 and then multiply.

For consistency, when we

- convert from a percent to a *decimal,* we will drop the % symbol and multiply by 0.01.
- convert from a percent to a *fraction,* we will drop the % symbol and multiply by $\frac{1}{100}$.

Let's practice writing percents as decimals, then writing percents as fractions.

Writing a Percent as a Decimal

Replace the percent symbol with its decimal equivalent, 0.01; then multiply.

$$43\% = 43(0.01) = 0.43$$

Helpful Hint

If it helps, think of writing a percent as a decimal by

Percent → | Remove the % symbol and move decimal point 2 places to the left | → Decimal

Answer

2. 31%

EXAMPLES Write each percent as a decimal.

3. $23\% = 23(0.01) = 0.23$ Replace the percent symbol with 0.01, then multiply.

4. $4.6\% = 4.6(0.01) = 0.046$ Replace the percent symbol with 0.01. Then multiply.

5. $190\% = 190(0.01) = 1.90$ or 1.9

6. $0.74\% = 0.74(0.01) = 0.0074$

7. $100\% = 100(0.01) = 1.00$ or 1

We just learned that $100\% = 1$

Work Practice Problems 3–7

PRACTICE PROBLEMS 3–7

Write each percent as a decimal.

3. 49% **4.** 3.1%
5. 175% **6.** 0.46%
7. 600%

✔Concept Check Why is it incorrect to write the percent 0.033% as 3.3 in decimal form?

Now let's write percents as fractions.

Writing a Percent as a Fraction

Replace the percent symbol with its fraction equivalent, $\frac{1}{100}$; then multiply. Don't forget to simplify the fraction if possible.

$$43\% = 43 \cdot \frac{1}{100} = \frac{43}{100}$$

EXAMPLES Write each percent as a fraction or mixed number in simplest form.

8. $40\% = 40 \cdot \frac{1}{100} = \frac{40}{100} = \frac{2 \cdot \cancel{20}^{1}}{5 \cdot \cancel{20}_{1}} = \frac{2}{5}$

9. $1.9\% = 1.9 \cdot \frac{1}{100} = \frac{1.9}{100}$. We don't want the numerator of the fraction to contain a decimal, so we multiply by 1 in the form of $\frac{10}{10}$.

$$= \frac{1.9}{100} \cdot \frac{10}{10} = \frac{1.9 \cdot 10}{100 \cdot 10} = \frac{19}{1000}$$

10. $125\% = 125 \cdot \frac{1}{100} = \frac{125}{100} = \frac{5 \cdot \cancel{25}^{1}}{4 \cdot \cancel{25}_{1}} = \frac{5}{4}$ or $1\frac{1}{4}$

11. $33\frac{1}{3}\% = 33\frac{1}{3} \cdot \frac{1}{100} = \frac{100}{3} \cdot \frac{1}{100} = \frac{\cancel{100}^{1} \cdot 1}{3 \cdot \cancel{100}_{1}} = \frac{1}{3}$

Write as an improper fraction.

12. $100\% = 100 \cdot \frac{1}{100} = \frac{100}{100} = 1$

Helpful Hint

Just as in Example 7, we confirm that $100\% = 1$.

Work Practice Problems 8–12

PRACTICE PROBLEMS 8–12

Write each percent as a fraction or mixed number in simplest form.

8. 50%

9. 2.3%

10. 150%

11. $66\frac{2}{3}\%$

12. 12%

Answers

3. 0.49 **4.** 0.031 **5.** 1.75 **6.** 0.0046 **7.** 6 **8.** $\frac{1}{2}$ **9.** $\frac{23}{1000}$ **10.** $\frac{3}{2}$ or $1\frac{1}{2}$ **11.** $\frac{2}{3}$ **12.** $\frac{3}{25}$

✔ Concept Check Answer

To write a percent as a decimal, the decimal point should be moved two places to the left, not to the right. So the correct answer is 0.00033.

Objective C Writing Decimals or Fractions as Percents

To write a decimal or fraction as a percent, we use the result of Examples 7 and 12. In these examples, we found that $1 = 100\%$.

Write 0.38 as a percent: $0.38 = 0.38(1) = 0.38(100\%) = 38.\%$

Write $\frac{1}{4}$ as a percent: $\frac{1}{4} = \frac{1}{4}(1) = \frac{1}{4} \cdot 100\% = \frac{100}{4}\% = 25\%$

First, let's practice writing decimals as percents.

Writing a Decimal as a Percent

Multiply by 1 in the form of 100%.

$0.27 = 0.27(100\%) = 27.\%$

Helpful Hint If it helps, think of writing a decimal as a percent by reversing the steps in the Helpful Hint on page 426.

Percent ← | Move the decimal point 2 places to the right and attach a % symbol. | ← Decimal

PRACTICE PROBLEMS 13–16

Write each decimal as a percent.

13. 0.14 **14.** 1.75
15. 0.057 **16.** 0.5

EXAMPLES Write each decimal as a percent.

13. $0.65 = 0.65(100\%) = 65.\%$ or 65% Multiply by 100%.

14. $1.25 = 1.25(100\%) = 125.\%$ or 125%

15. $0.012 = 0.012(100\%) = 001.2\%$ or 1.2%

16. $0.6 = 0.6(100\%) = 060.\%$ or 60%

Helpful Hint A zero was inserted as a placeholder.

Work Practice Problems 13–16

✔Concept Check Why is it incorrect to write the decimal 0.0345 as 34.5% in percent form?

Now let's write fractions as percents.

Writing a Fraction as a Percent

Multiply by 1 in the form of 100%.

$$\frac{1}{8} = \frac{1}{8} \cdot 100\% = \frac{1}{8} \cdot \frac{100}{1}\% = \frac{100}{8}\% = 12\frac{1}{2}\% \quad \text{or} \quad 12.5\%$$

Helpful Hint From Examples 7 and 12, we know that

$100\% = 1$

Recall that when we multiply a number by 1, we are not changing the value of that number. This means that when we multiply a number by 100%, we are not changing its value but rather writing the number as an equivalent percent.

Answers

13. 14% **14.** 175% **15.** 5.7%
16. 50%

✔ Concept Check Answer

To change a decimal to a percent, multiply by 100%, or move the decimal point *only* two places to the right. So the correct answer is 3.45%.

EXAMPLES Write each fraction or mixed number as a percent.

17. $\frac{7}{20} = \frac{7}{20} \cdot 100\% = \frac{7}{20} \cdot \frac{100}{1}\% = \frac{700}{20}\% = 35\%$

18. $\frac{2}{3} = \frac{2}{3} \cdot 100\% = \frac{2}{3} \cdot \frac{100}{1}\% = \frac{200}{3}\% = 66\frac{2}{3}\%$

19. $2\frac{1}{4} = \frac{9}{4} \cdot 100\% = \frac{9}{4} \cdot \frac{100}{1}\% = \frac{900}{4}\% = 225\%$

Helpful Hint

$\frac{200}{3} = 66.\overline{6}$. Thus, another way to write $\frac{200}{3}\%$ is $66.\overline{6}\%$.

Work Practice Problems 17–19

PRACTICE PROBLEMS 17–19

Write each fraction or mixed number as a percent.

17. $\frac{1}{4}$ **18.** $\frac{9}{40}$ **19.** $5\frac{1}{2}$

✔Concept Check Which digit in the percent 76.4582% represents

a. A tenth percent?
b. A thousandth percent?
c. A hundredth percent?
d. A ten percent?

EXAMPLE 20 Write $\frac{1}{12}$ as a percent. Round to the nearest hundredth percent.

Solution:

$$\frac{1}{12} = \frac{1}{12} \cdot 100\% = \frac{1}{12} \cdot \frac{100\%}{1} = \frac{100}{12}\% \approx 8.33\%$$

("approximately")

```
     8.333 ≈ 8.33
12)100.000
   -96
     4 0
    -3 6
      40
     -36
      40
     -36
       4
```

Thus, $\frac{1}{12}$ is approximately 8.33%.

Work Practice Problem 20

PRACTICE PROBLEM 20

Write $\frac{3}{17}$ as a percent. Round to the nearest hundredth percent.

Objective D Applications with Percents, Decimals, and Fractions

Let's summarize what we have learned so far about percents, decimals, and fractions:

Summary of Converting Percents, Decimals, and Fractions

- *To write a percent as a decimal,* replace the % symbol with its decimal equivalent, 0.01; then multiply.
- *To write a percent as a fraction,* replace the % symbol with its fraction equivalent, $\frac{1}{100}$; then multiply.
- *To write a decimal or fraction as a percent,* multiply by 100%.

Answers

17. 25% **18.** $22\frac{1}{2}\%$ **19.** 550% **20.** 17.65%

✔ Concept Check Answers

a. 4 **b.** 8 **c.** 5 **d.** 7

If we let x represent a number, below we summarize using symbols.

Write a percent as a decimal:	Write a percent as a fraction:	Write a number as a percent:
$x\% = x(0.01)$	$x\% = x \cdot \frac{1}{100}$	$x = x \cdot 100\%$

PRACTICE PROBLEM 21

A family decides to spend no more than 27.5% of its monthly income on rent. Write 27.5% as a decimal and as a fraction.

EXAMPLE 21 17.8% of automobile thefts in the continental United States occur in the Midwest. Write this percent as a decimal and as a fraction. (*Source:* The American Automobile Manufacturers Association)

Solution:

As a decimal: $17.8\% = 17.8(0.01) = 0.178.$

As a fraction: $17.8\% = 17.8 \cdot \frac{1}{100} = \frac{17.8}{100} = \frac{17.8}{100} \cdot \frac{10}{10} = \frac{178}{1000} = \frac{\cancel{2}^{1} \cdot 89}{\cancel{2}_{1} \cdot 500} = \frac{89}{500}.$

Thus, 17.8% written as a decimal is 0.178, and written as a fraction is $\frac{89}{500}$.

Work Practice Problem 21

PRACTICE PROBLEM 22

Provincetown's budget for waste disposal increased by $1\frac{3}{4}$ times over the budget from last year. What percent increase is this?

EXAMPLE 22 An advertisement for a stereo system reads "$\frac{1}{4}$ off." What percent off is this?

Solution: Write $\frac{1}{4}$ as a percent.

$$\frac{1}{4} = \frac{1}{4} \cdot 100\% = \frac{1}{4} \cdot \frac{100\%}{1} = \frac{100}{4}\% = 25\%$$

Thus, "$\frac{1}{4}$ off" is the same as "25% off."

Work Practice Problem 22

Note: It is helpful to know a few basic percent conversions. Appendix A.2 contains a handy reference of percent, decimal, and fraction equivalencies.

Also, Appendix A.3 shows how to find common percents of a number.

Answers

21. 0.275, $\frac{11}{40}$ **22.** 175%

Vocabulary and Readiness Check

Use the choices below to fill in each blank. Some choices may be used more than once.

$\frac{1}{100}$ 0.01 100% percent

1. ________ means "per hundred."
2. ___________ = 1.
3. The % symbol is read as ________.
4. To write a decimal or a fraction as a percent, multiply by 1 in the form of ___________.
5. To write a percent as a *decimal,* drop the % symbol and multiply by _________.
6. To write a percent as a *fraction,* drop the % symbol and multiply by _________.

6.2 EXERCISE SET

FOR EXTRA HELP

Student Solutions Manual

PH Math/Tutor Center

CD/Video for Review

MathXL®

MyMathLab

Objective A *Solve. See Examples 1 and 2.*

1. In a survey of 100 college students, 96 use the Internet. What percent use the Internet?

2. A basketball player makes 81 out of 100 attempted free throws. What percent of free throws are made?

One hundred adults were asked to name their favorite sport, and the results are shown in the circle graph.

3. What sport was preferred by most adults? What percent preferred this sport?

4. What sport was preferred by the least number of adults? What percent preferred this sport?

5. What percent of adults preferred football or soccer?

6. What percent of adults preferred basketball or baseball?

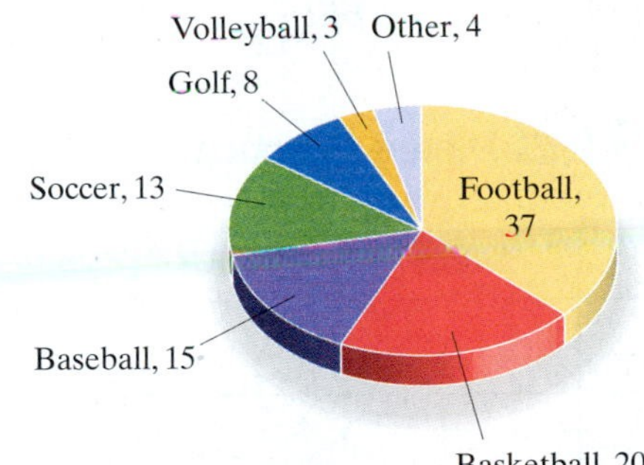

Objective B *Write each percent as a decimal. See Examples 3 through 7.*

7. 41% **8.** 62% **9.** 6% **10.** 3%

11. 100% **12.** 136% **13.** 73.6% **14.** 45.7%

15. 2.8% **16.** 1.4% **17.** 0.6% **18.** 0.9%

19. 300% **20.** 500% **21.** 32.58% **22.** 72.18%

Write each percent as a fraction or mixed number in simplest form. See Examples 8 through 12.

23. 8% **24.** 22% **25.** 4% **26.** 2%

27. 4.5%
28. 7.5%
29. 175%
30. 275%
31. 6.25%

32. 8.75%
33. $10\frac{1}{3}\%$
34. $7\frac{3}{4}\%$
35. $22\frac{3}{8}\%$
36. $21\frac{7}{8}\%$

Objective C *Write each decimal as a percent. See Examples 13 through 16.*

37. 0.006
38. 0.008
39. 0.22
40. 0.44
41. 5.3

42. 2.7
43. 0.056
44. 0.019
45. 0.2228
46. 0.1115

47. 3.00
48. 9.00
49. 0.7
50. 0.8

Write each fraction or mixed number as a percent. See Examples 17 through 19.

51. $\frac{7}{10}$
52. $\frac{9}{10}$
53. $\frac{4}{5}$
54. $\frac{2}{5}$
55. $\frac{34}{50}$

56. $\frac{41}{50}$
57. $\frac{3}{8}$
58. $\frac{5}{16}$
59. $\frac{1}{3}$
60. $\frac{2}{3}$

61. $4\frac{1}{2}$
62. $6\frac{1}{5}$
63. $1\frac{9}{10}$
64. $2\frac{7}{10}$

Write each fraction as a percent. Round to the nearest hundredth percent. See Example 20.

65. $\frac{9}{11}$
66. $\frac{11}{12}$
67. $\frac{4}{15}$
68. $\frac{10}{11}$

Objectives A B C Mixed Practice *Complete each table. See Examples 1 through 20.*

69.

Percent	Decimal	Fraction
60%		
	0.235	
		$\frac{4}{5}$
$33\frac{1}{3}\%$		
		$\frac{7}{8}$
7.5%		

70.

Percent	Decimal	Fraction
	0.525	
		$\frac{3}{4}$
$66\frac{1}{2}\%$		
		$\frac{5}{6}$
100%		
		$\frac{7}{50}$

71.

Percent	Decimal	Fraction
200%		
	2.8	
705%		
		$4\frac{27}{50}$

72.

Percent	Decimal	Fraction
800%		
	3.2	
608%		
		$9\frac{13}{50}$

Objective D *Write each percent as a decimal and a fraction. See Examples 21 and 22.*

73. People take aspirin for a variety of reasons. The most common use of aspirin is to prevent heart disease, accounting for 38% of all aspirin use. (*Source:* Bayer Market Research)

74. Approximately 56% of all U.S. workers use computers on the job. (*Source:* National Center for Education Statistics)

75. In the United States, about 25.2% of the female population age 25 and over graduated from college. (*Source:* U.S. Census Bureau)

76. In the United States, 28.5% of the male population age 25 and over graduated from college. (*Source:* U.S. Census Bureau)

77. Approximately 32.2% of all new cars sold in the United States were manufactured in Japan. (*Source:* American Automobile Manufacturers Association)

78. Approximately 14.8% of new luxury cars are silver, making silver the most popular new vehicle color for that class. (*Source:* Ward's Communications)

In Exercises 79 through 82, write the percent from the circle graph as a decimal and a fraction.

World Population by Continent

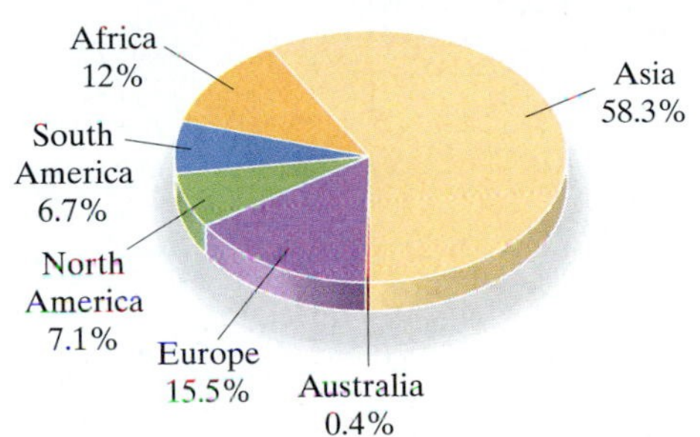

79. Australia: 0.4%

80. Asia: 58.3%

81. Africa: 12%

82. North America: 7.1%

Solve. See Examples 21 and 22.

83. In a recent year, the U.S. Postal Service handled approximately 0.442 of the world's card and letter mail volume. Write this decimal as a percent. (*Source:* U.S. Postal Service)

84. In the state of California, the Hispanic population is 0.347 of the total population. Write this decimal as a percent. (*Source:* U.S. Census Bureau)

85. The mirrors on the Hubble Space Telescope are able to lock onto a target without deviating more than $\frac{7}{1000}$ of an arc-second. Write this fraction as a percent. (*Source:* NASA)

86. In 2005, $\frac{3}{250}$ of all new cars sold in the United States were hybrids. Write this fraction as a percent. (*Source:* American Automobile Manufacturers Association)

87. In 2005, 3.9% of all households with television sets subscribed to TiVo®. Write this percent as a fraction. (*Source:* Plunkett Research)

88. For the 2006 season, the Tuesday broadcasts of *American Idol* had a 12.9% Nielsen rating. Write this percent as a fraction. (*Source: Hollywood Reporter*)

Review

Perform the indicated operation. See Section 4.6.

89. $\frac{3}{4} - \frac{1}{2} \cdot \frac{8}{9}$

90. $\left(\frac{2}{11} + \frac{5}{11}\right)\left(\frac{2}{11} - \frac{5}{11}\right)$

91. $6\frac{2}{3} - 4\frac{5}{6}$

92. $6\frac{2}{3} \div 4\frac{5}{6}$

Concept Extensions

Solve. See the Concept Checks in this section.

93. Given the percent 52.8647%, round as indicated.

a. Round to a tenth of a percent.
b. Round to a hundredth of a percent.

94. Given the percent 0.5269%, round as indicated.

a. Round to a tenth of a percent.
b. Round to a hundredth of a percent.

95. Which of the following are correct?

a. 6.5% = 0.65 **b.** 7.8% = 0.078
c. 120% = 0.12 **d.** 0.35% = 0.0035

96. Which of the following are correct?

a. 0.231 = 23.1% **b.** 5.12 = 0.0512%
c. 3.2 = 320% **d.** 0.0175 = 0.175%

Recall that 1 = 100%. This means that 1 whole is 100%. Use this for Exercises 97 and 98. (Source: Some Body, by Dr. Pete Rowen)

97. The four blood types are A, B, O, and AB. (Each blood type can also be further classified as Rh-positive or Rh-negative depending upon whether your blood contains protein or not.) Given the percent blood types for the United States below, calculate the percent of U.S. population with AB blood type.

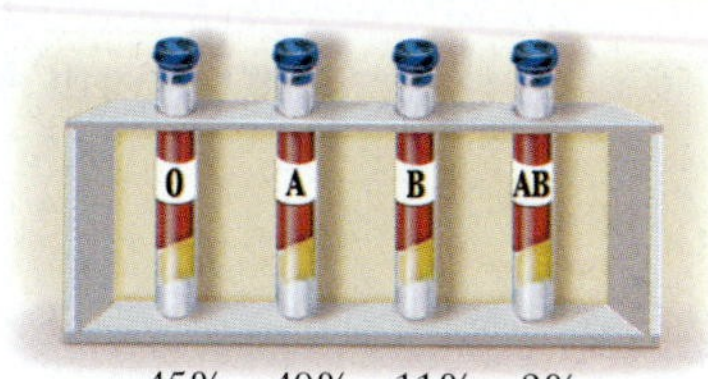

45% 40% 11% ?%

98. The top four components of bone are below. Find the missing percent.

1. Minerals—45%
2. Living tissue—30%
3. Water—20%
4. Other—?

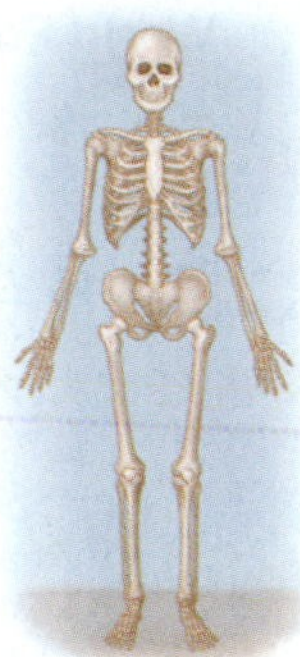

What percent of the figure is shaded?

99.

100. 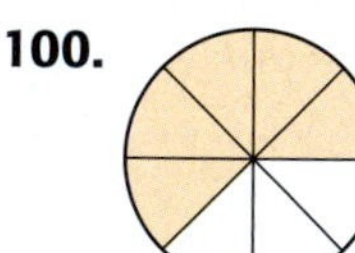

Fill in the blanks.

101. A fraction written as a percent is greater than 100% when the numerator is _______ than the denominator. (greater/less)

102. A decimal written as a percent is less than 100% when the decimal is _______ than 1. (greater/less)

Write each fraction as a decimal and then write each decimal as a percent. Round the decimal to three decimal places (nearest thousandth) and the percent to the nearest tenth of a percent.

 **103.** $\frac{21}{79}$

104. $\frac{56}{102}$

The bar graph shows the predicted fastest-growing occupations. Use this graph for Exercises 105 through 108.

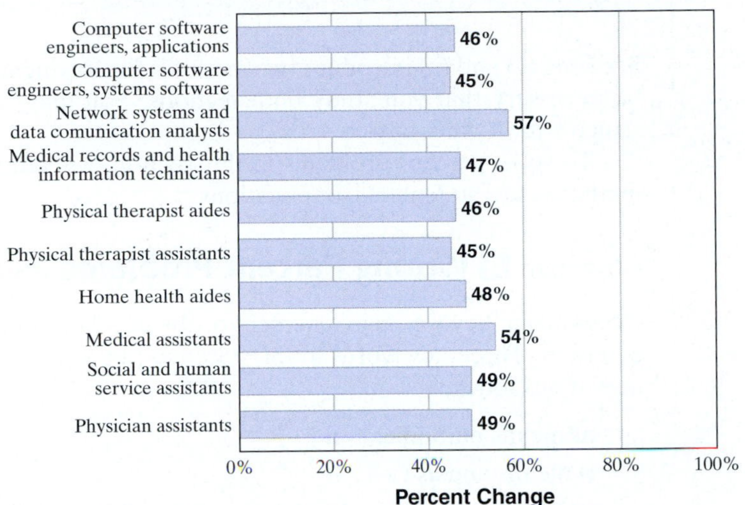

Source: Bureau of Labor Statistics

105. What occupation is predicted to be the fastest growing?

106. What occupation is predicted to be the second fastest growing?

107. Write the percent change for physician assistants as a decimal.

108. Write the percent change for medical assistants as a decimal.

109. In your own words, explain how to write a percent as a decimal.

110. In your own words, explain how to write a decimal as a percent.

STUDY SKILLS BUILDER

Are You Satisfied with Your Performance in This Course thus Far?

To see if there is room for improvement, answer these questions:

1. Am I attending all classes and arriving on time?
2. Am I working and checking my homework assignments on time?
3. Am I getting help (from my instructor or a campus learning resource lab) when I need it?
4. In addition to my instructor, am I using the text supplements that might help me?
5. Am I satisfied with my performance on quizzes and exams?

If you answered no to any of these questions, read or reread Section 1.1 for suggestions in these areas. Also, you might want to contact your instructor for additional feedback.

Objectives

A Write Percent Problems as Equations.

B Solve Percent Problems.

6.3 SOLVING PERCENT PROBLEMS WITH EQUATIONS

Sections 6.3 and 6.4 introduce two methods for solving percent problems. It may not be necessary that you study both sections. You may want to check with your instructor for further advice.

To solve percent problems in this section, we will translate the problems into mathematical statements, or equations.

Objective A Writing Percent Problems as Equations

Recognizing key words in a percent problem is helpful in writing the problem as an equation. Three key words in the statement of a percent problem and their meanings are as follows:

of means **multiplication** (·)
is means **equals** (=)
what (or some equivalent) means **the unknown number.**

In our examples, we will let the letter x stand for the unknown number.

PRACTICE PROBLEM 1

Translate: 8 is what percent of 48?

EXAMPLE 1 Translate to an equation:

5 is what percent of 20?

Solution: 5 is what percent of 20?

$$5 = x \cdot 20$$

Work Practice Problem 1

Helpful Hint

Remember that an equation is simply a mathematical statement that contains an equals sign (=).

$$5 = 20x$$

↑ equals sign

PRACTICE PROBLEM 2

Translate: 2.6 is 40% of what number?

EXAMPLE 2 Translate to an equation:

1.2 is 30% of what number?

Solution: 1.2 is 30% of what number?

$$1.2 = 30\% \cdot x$$

Work Practice Problem 2

PRACTICE PROBLEM 3

Translate: What number is 90% of 0.045?

Answers

1. $8 = x \cdot 48$ **2.** $2.6 = 40\% \cdot x$
3. $x = 90\% \cdot 0.045$

EXAMPLE 3 Translate to an equation:

What number is 25% of 0.008?

Solution: What number is 25% of 0.008?

$$x = 25\% \cdot 0.008$$

Work Practice Problem 3

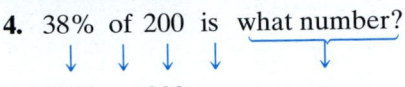

EXAMPLES Translate each question to an equation.

4. 38% of 200 is what number?

$38\% \cdot 200 = x$

5. 40% of what number is 80?

$40\% \cdot x = 80$

6. What percent of 85 is 34?

$x \cdot 85 = 34$

Work Practice Problems 4–6

PRACTICE PROBLEMS 4–6

Translate each question to an equation.

4. 56% of 180 is what number?
5. 12% of what number is 21?
6. What percent of 95 is 76?

✔Concept Check In the equation $2x = 10$, what step is taken to solve the equation?

Objective B Solving Percent Problems

You may have noticed by now that each percent problem has contained three numbers—in our examples, two are known and one is unknown. Each of these numbers is given a special name.

15% of 60 is 9

15% percent	·	60 base	=	9 amount

We call this equation the **percent equation.**

Percent Equation

percent · base = amount

Once a percent problem has been written as a percent equation, we can use the equation to find the unknown number, whether it is the percent, the base, or the amount.

Solving Percent Equations for the Amount

EXAMPLE 7 What number is 35% of 60?

Solution:

$x = 35\% \cdot 60$ Translate to an equation.

$x = 0.35 \cdot 60$ Write 35% as 0.35.

$x = 21$ Multiply:

$$\begin{array}{r} 60 \\ \times 0.35 \\ \hline 300 \\ 1800 \\ \hline 21.00 \end{array}$$

Then 21 is 35% of 40. Is this reasonable? To see, round 35% to 40%. Then 40% or 0.40(60) is 24. Our result is reasonable since 21 is close to 24.

Work Practice Problem 7

PRACTICE PROBLEM 7

What number is 25% of 90?

Answers

4. $56\% \cdot 180 = x$ 5. $12\% \cdot x = 21$
6. $x \cdot 95 = 76$ 7. 22.5

✔ Concept Check Answer

Divide both sides of the equation by 2.

Helpful Hint When solving a percent equation, write the percent as a decimal or fraction.

PRACTICE PROBLEM 8

95% of 400 is what number?

EXAMPLE 8 85% of 300 is what number?

Solution:

$$85\% \cdot 300 = x \quad \text{Translate to an equation.}$$
$$0.85 \cdot 300 = x \quad \text{Write 85\% as 0.85.}$$
$$255 = x \quad \text{Multiply: } 0.85 \cdot 300 = 255.$$

Then 85% of 300 is 255. Is this result reasonable? To see, round 85% to 90%. Then 90% of 300 or 0.90(300) = 270, which is close to 255.

Work Practice Problem 8

Solving Percent Equations for the Base

PRACTICE PROBLEM 9

15% of what number is 2.4?

EXAMPLE 9 12% of what number is 0.6?

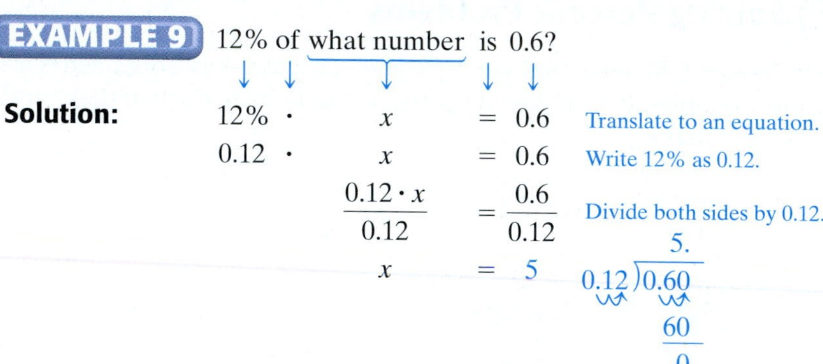

Solution:

$$12\% \cdot x = 0.6 \quad \text{Translate to an equation.}$$
$$0.12 \cdot x = 0.6 \quad \text{Write 12\% as 0.12.}$$
$$\frac{0.12 \cdot x}{0.12} = \frac{0.6}{0.12} \quad \text{Divide both sides by 0.12.}$$
$$x = 5$$

$$0.12\overline{)0.60} = 5. \qquad 60 \qquad 0$$

Then 12% of 5 is 0.6. Is this reasonable? To see, round 12% to 10%. Then 10% of 0.6 or 0.10 (0.6) is 6, which is close to 5.

Work Practice Problem 9

PRACTICE PROBLEM 10

18 is $4\frac{1}{2}\%$ of what number?

EXAMPLE 10 13 is $6\frac{1}{2}\%$ of what number?

Solution:

$$13 = 6\frac{1}{2}\% \cdot x \quad \text{Translate to an equation.}$$
$$13 = 0.065 \cdot x \quad 6\frac{1}{2}\% = 6.5\% = 0.065.$$
$$\frac{13}{0.065} = \frac{0.065 \cdot x}{0.065} \quad \text{Divide both sides by 0.065.}$$
$$200 = x$$

$$0.065\overline{)13.000} = 200. \qquad 130 \qquad 0$$

Then 13 is $6\frac{1}{2}\%$ of 200. Check to see if this result is reasonable.

Work Practice Problem 10

Answers

8. 380 **9.** 16 **10.** 400

Solving Percent Equations for the Percent

EXAMPLE 11 What percent of 12 is 9?

Solution:

$$x \cdot 12 = 9 \quad \text{Translate to an equation.}$$

$$\frac{x \cdot 12}{12} = \frac{9}{12} \quad \text{Divide both sides by 12.}$$

$$\text{or} \quad x = 0.75$$

Next, since we are looking for percent, we can write $\frac{9}{12}$ or 0.75 as a percent.

$x = 75\%$

Then 75% of 12 is 9. To check, see that $75\% \cdot 12 = 9$.

Work Practice Problem 11

PRACTICE PROBLEM 11

What percent of 90 is 27?

Helpful Hint If your unknown in the percent equation is percent, don't forget to convert your answer to a percent.

EXAMPLE 12 78 is what percent of 65?

Solution:

$$78 = x \cdot 65 \quad \text{Translate to an equation.}$$

$$\frac{78}{65} = \frac{x \cdot 65}{65} \quad \text{Divide both sides by 65.}$$

$$1.2 = x$$

$$120\% = x \quad \text{Write 1.2 as a percent.}$$

Then 78 is 120% of 65. Check this result.

Work Practice Problem 12

PRACTICE PROBLEM 12

63 is what percent of 45?

✔ Concept Check Consider these problems

1. 75% of 50 =
   a. 50 b. a number greater than 50 c. a number less than 50
2. 40% of a number is 10. Is the number
   a. 10 b. less than 10 c. greater than 10?
3. 800 is 120% of what number? Is the number
   a. 800 b. less than 800 c. greater than 800?

Helpful Hint Use the following to see if your answers are reasonable.

$$(100\%) \text{ of a number} = \text{the number}$$

$$\left(\begin{array}{c}\text{a percent}\\ \text{greater than}\\ 100\%\end{array}\right) \text{ of a number} = \begin{array}{l}\text{a number greater}\\ \text{than the original number}\end{array}$$

$$\left(\begin{array}{c}\text{a percent}\\ \text{less than } 100\%\end{array}\right) \text{ of a number} = \begin{array}{l}\text{a number less}\\ \text{than the original number}\end{array}$$

Answers

11. 30% **12.** 140%

✔ Concept Check Answers

1. c **2.** c **3.** b

Vocabulary and Readiness Check

Use the choices below to fill in each blank.

percent	amount	of	less
base	the number	is	greater

1. The word ________ translates to "=".
2. The word ________ usually translates to "multiplication."
3. In the statement "10% of 90 is 9," the number 9 is called the ________, 90 is called the ________, and 10 is called the ________.
4. 100% of a number = ________.
5. Any "percent greater than 100%" of "a number" = "a number ________ than the original number."
6. Any "percent less than 100%" of "a number" = "a number ________ than the original number."

6.3 EXERCISE SET

FOR EXTRA HELP

Student Solutions Manual

PH Math/Tutor Center

CD/Video for Review

MathXL®

MyMathLab

Objective A Translating *Translate each to an equation. Do not solve. See Examples 1 through 6.*

1. 18% of 81 is what number?
2. 36% of 72 is what number?
3. 20% of what number is 105?
4. 40% of what number is 6?
5. 0.6 is 40% of what number?
6. 0.7 is 20% of what number?
7. What percent of 80 is 3.8?
8. 9.2 is what percent of 92?
9. What number is 9% of 43?
10. What number is 25% of 55?
11. What percent of 250 is 150?
12. What percent of 375 is 300?

Objective B *Solve. See Examples 7 and 8.*

13. 10% of 35 is what number?
14. 25% of 68 is what number?
15. What number is 14% of 205?
16. What number is 18% of 425?

Solve. See Examples 9 and 10.

17. 1.2 is 12% of what number?
18. 0.22 is 44% of what number?
19. $8\frac{1}{2}\%$ of what number is 51?
20. $4\frac{1}{2}\%$ of what number is 45?

Solve. See Examples 11 and 12.

21. What percent of 80 is 88?
22. What percent of 40 is 60?
23. 17 is what percent of 50?
24. 48 is what percent of 50?

Objectives A B Mixed Practice *Solve. See Examples 1 through 12.*

25. 0.1 is 10% of what number?

26. 0.5 is 5% of what number?

27. 150% of 430 is what number?

28. 300% of 56 is what number?

29. 82.5 is $16\frac{1}{2}\%$ of what number?

30. 7.2 is $6\frac{1}{4}\%$ of what number?

31. 2.58 is what percent of 50?

32. 2.64 is what percent of 25?

33. What number is 42% of 60?

34. What number is 36% of 80?

35. What percent of 184 is 64.4?

36. What percent of 120 is 76.8?

37. 120% of what number is 42?

38. 160% of what number is 40?

39. 2.4% of 26 is what number?

40. 4.8% of 32 is what number?

41. What percent of 600 is 3?

42. What percent of 500 is 2?

43. 6.67 is 4.6% of what number?

44. 9.75 is 7.5% of what number?

45. 1575 is what percent of 2500?

46. 2520 is what percent of 3500?

47. 2 is what percent of 50?

48. 2 is what percent of 40?

Review

Find the value of x in each proportion. See Section 6.1.

49. $\frac{27}{x} = \frac{9}{10}$

50. $\frac{35}{x} = \frac{7}{5}$

51. $\frac{x}{5} = \frac{8}{11}$

52. $\frac{x}{3} = \frac{6}{13}$

Write each sentence as a proportion.

53. 17 is to 12 as x is to 20.

54. 20 is to 25 as x is to 10.

55. 8 is to 9 as 14 is to x.

56. 5 is to 6 as 15 is to x.

Concept Extensions

For each equation, determine the next step taken to find the value of n. See the first Concept Check in this section.

57. $5 \cdot n = 32$

a. $n = 5 \cdot 32$ **b.** $n = \frac{5}{32}$ **c.** $n = \frac{32}{5}$ **d.** none of these

58. $n = 0.7 \cdot 12$

a. $n = 8.4$ **b.** $n = \frac{12}{0.7}$ **c.** $n = \frac{0.7}{12}$ **d.** none of these

59. $0.06 = n \cdot 7$

a. $n = 0.06 \cdot 7$ **b.** $n = \frac{0.06}{7}$ **c.** $n = \frac{7}{0.06}$ **d.** none of these

60. Write a word statement for the equation $20\% \cdot x = 18.6$. Use the phrase "what number" for "x."

61. Write a word statement for the equation $x = 33\frac{1}{3}\% \cdot 24$. Use the phrase "what number" for "x."

For each exercise, determine whether the percent, x, is (a) 100%, (b) greater than 100%, or (c) less than 100%. See the last Concept Check in this section.

62. x% of 20 is 30

63. x% of 98 is 98

64. x% of 120 is 85

For each exercise, determine whether the number, y, is (a) equal to 45, (b) greater than 45, or (c) less than 45.

65. 55% of 45 is y

66. 230% of 45 is y

67. 100% of 45 is y

68. 30% of y is 45

69. 100% of y is 45

70. 180% of y is 45

Solve.

71. In your own words, explain how to solve a percent equation.

72. Write a percent problem that uses the percent 50%.

73. 1.5% of 45,775 is what number?

74. What percent of 75,528 is 27,945.36?

75. 22,113 is 180% of what number?

THE BIGGER PICTURE Operations on Sets of Numbers and Solving Equations

Continue your outline from Sections 1.6, 1.7, 2.4, 3.3, 4.3, 4.7, 5.4, and 6.1. Suggestions are once again written to help you complete this part of your outline. Notice that this part of the outline has to do with solving equations.

I. Operations on Sets of Numbers

- **A. Whole Numbers**
  1. **Add or Subtract** (Section 1.3)
  2. **Multiply or Divide** (Sections 1.5, 1.6)
  3. **Exponent** (Section 1.7)
  4. **Order of Operations** (Section 1.7)
- **B. Integers**
  1. **Add** (Section 2.2)
  2. **Subtract** (Section 2.3)
  3. **Multiply or Divide** (Section 2.4)
- **C. Fractions**
  1. **Simplify** (Section 4.2)
  2. **Multiply** (Section 4.3)
  3. **Divide** (Section 4.3)
  4. **Add or Subtract** (Sections 4.4, 4.5)
- **D. Decimals**
  1. **Add or Subtract** (Section 5.2)
  2. **Multiply** (Section 5.3)
  3. **Divide** (Section 5.4)

II. Solving Equations

- **A. Equations in General** (Section 3.3)
- **B. Proportions** (Section 6.1)
- **C. Percent Problems**
  1. **Solved by Equations:** Remember that "of" means multiplication and "is" means equals.

     12% of some number is 6 translates to

     $12\% \cdot n = 6$ *or* $0.12n = 6$ *or* $n = \dfrac{6}{0.12}$ *or* $n = 50$

Perform the indicated operations.

1. $\frac{2}{9} + \frac{1}{5}$

2. $42 \div 2 \cdot 3$

3. $-0.03(0.7)$

4. $7.9 + 0.1$

Solve.

5. $\frac{3}{8} = \frac{n}{128}$

6. $\frac{7.2}{n} = \frac{36}{8}$

7. 215 is what percent of 86?

8. 95% of 48 is what number?

9. 4.2 is what percent of 15?

10. 93.6 is 52% of what number?

6.4 SOLVING PERCENT PROBLEMS WITH PROPORTIONS

Objectives

A Write Percent Problems as Proportions.

B Solve Percent Problems.

There is more than one method that can be used to solve percent problems. (See the note at the beginning of Section 6.3.) In the last section, we used the percent equation. In this section, we will use proportions.

Objective A Writing Percent Problems as Proportions

To understand the proportion method, recall that 70% means the ratio of 70 to 100, or $\frac{70}{100}$.

$$70\% = \frac{70}{100} = \frac{7}{10}$$

$\frac{7}{10}$ shaded

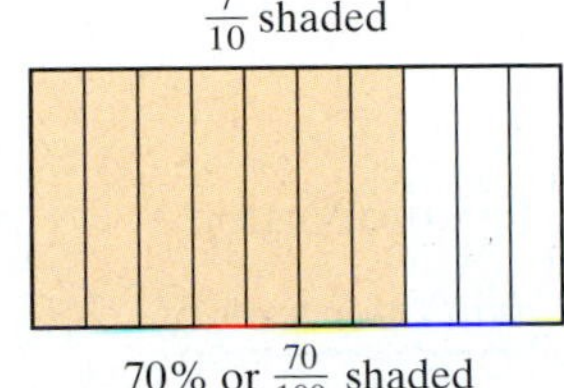

70% or $\frac{70}{100}$ shaded

Since the ratio $\frac{70}{100}$ is equal to the ratio $\frac{7}{10}$, we have the proportion

$$\frac{7}{10} = \frac{70}{100}$$

We call this proportion the **percent proportion.** In general, we can name the parts of this proportion as follows:

Percent Proportion

$$\frac{\text{amount}}{\text{base}} = \frac{\text{percent}}{100} \leftarrow \text{always 100}$$

or

$$\text{amount} \rightarrow \frac{a}{b} = \frac{p}{100} \leftarrow \text{percent}$$
$$\text{base} \rightarrow$$

When we translate percent problems to proportions, the **percent,** p, can be identified by looking for the symbol % or the word *percent*. The **base,** b, usually follows the word *of*. The **amount,** a, is the part compared to the whole.

Helpful Hint This table may be useful when identifying the parts of a proportion.

Part of Proportion	How It's Identified
Percent	% or percent
Base	Appears after *of*
Amount	Part compared to whole

PRACTICE PROBLEM 1

Translate to a proportion.
27% of what number is 54?

EXAMPLE 1 Translate to a proportion.

12% of what number is 47?

Solution:

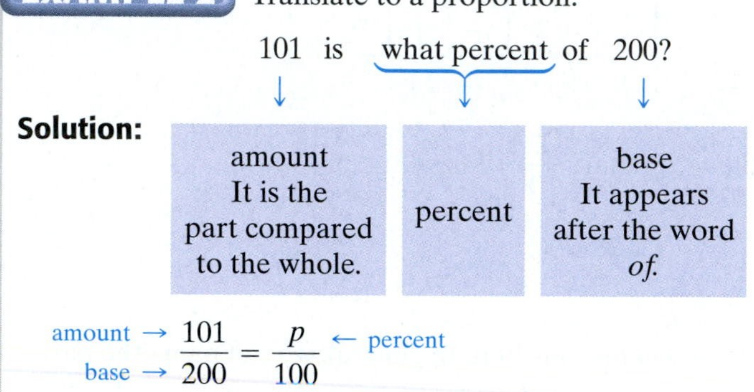

$$\text{amount} \rightarrow \frac{47}{b} = \frac{12}{100} \leftarrow \text{percent}$$
base → b

Work Practice Problem 1

PRACTICE PROBLEM 2

Translate to a proportion.
30 is what percent of 90?

EXAMPLE 2 Translate to a proportion.

101 is what percent of 200?

Solution:

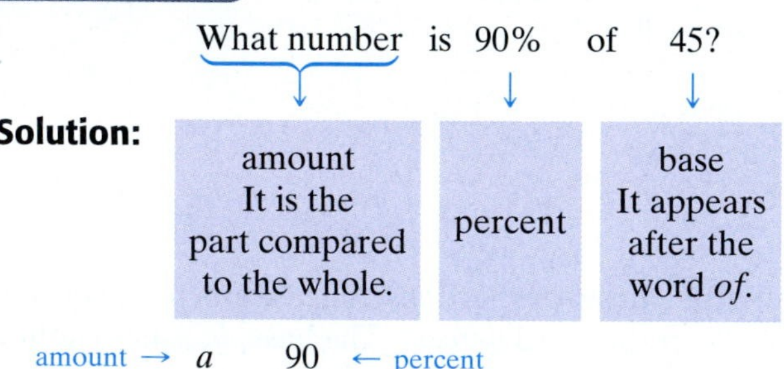

$$\text{amount} \rightarrow \frac{101}{200} = \frac{p}{100} \leftarrow \text{percent}$$
base → 200

Work Practice Problem 2

PRACTICE PROBLEM 3

Translate to a proportion.
What number is 25% of 116?

EXAMPLE 3 Translate to a proportion.

What number is 90% of 45?

Solution:

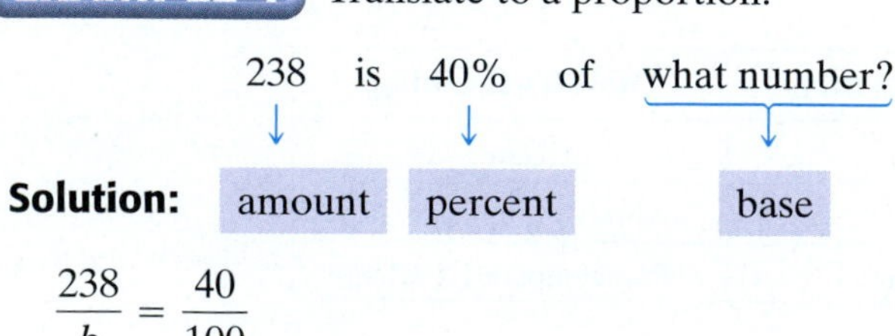

$$\text{amount} \rightarrow \frac{a}{45} = \frac{90}{100} \leftarrow \text{percent}$$
base → 45

Work Practice Problem 3

PRACTICE PROBLEM 4

Translate to a proportion.
680 is 65% of what number?

EXAMPLE 4 Translate to a proportion.

238 is 40% of what number?

Solution: amount percent base

$$\frac{238}{b} = \frac{40}{100}$$

Work Practice Problem 4

Answers

1. $\frac{54}{b} = \frac{27}{100}$ 2. $\frac{30}{90} = \frac{p}{100}$
3. $\frac{a}{116} = \frac{25}{100}$ 4. $\frac{680}{b} = \frac{65}{100}$

EXAMPLE 5 Translate to a proportion.

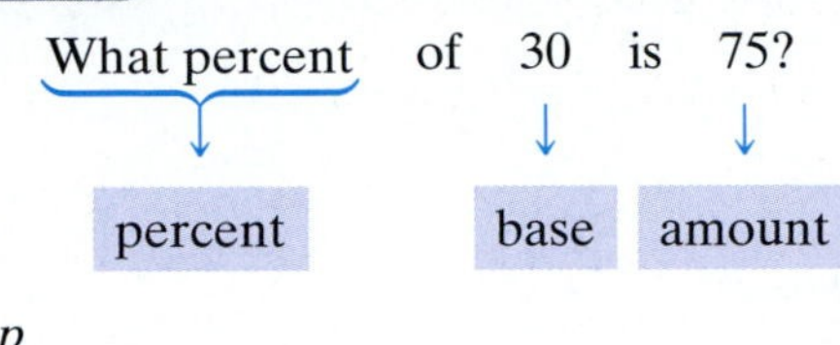

Solution: percent base amount

$$\frac{75}{30} = \frac{p}{100}$$

Work Practice Problem 5

PRACTICE PROBLEM 5

Translate to a proportion.
What percent of 40 is 75?

EXAMPLE 6 Translate to a proportion.

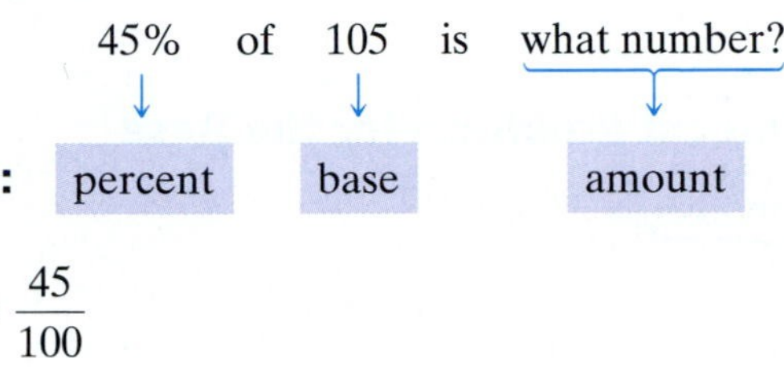

Solution:

$$\frac{a}{105} = \frac{45}{100}$$

Work Practice Problem 6

PRACTICE PROBLEM 6

Translate to a proportion.
46% of 80 is what number?

✔ **Concept Check** Consider the statement "78 is what percent of 350?" Which part of the percent proportion is unknown?

a. the amount
b. the base
c. the percent

Consider another statement: "14 is 10% of some number." Which part of the percent proportion is unknown?

a. the amount
b. the base
c. the percent

Objective B Solving Percent Problems

The proportions that we have written in this section contain three values that can change: the percent, the base, and the amount. If any two of these values are known, we can find the third (the unknown value). To do this, we write a percent proportion and find the unknown value as we did in Section 6.1.

EXAMPLE 7 **Solving Percent Proportions for the Amount**

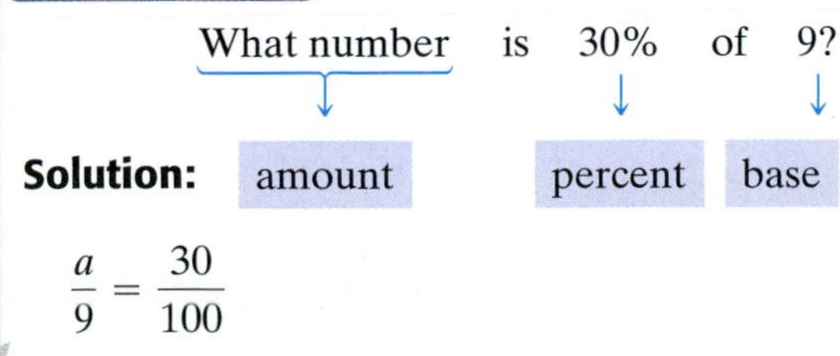

Solution:

$$\frac{a}{9} = \frac{30}{100}$$

Continued on next page

PRACTICE PROBLEM 7

What number is 8% of 120?

Answers

5. $\frac{75}{40} = \frac{p}{100}$ **6.** $\frac{a}{80} = \frac{46}{100}$ **7.** 9.6

✔ **Concept Check Answers**

c, b

Helpful Hint The proportion in Example 7 contains the ratio $\frac{30}{100}$. A ratio in a proportion may be simplified before solving the proportion. The unknown number in both $\frac{a}{9} = \frac{30}{100}$ and $\frac{a}{9} = \frac{3}{10}$ is 2.7.

To solve, we set cross products equal to each other.

$$\frac{a}{9} = \frac{30}{100}$$

$a \cdot 100 = 9 \cdot 30$ Set cross products equal.

$100a = 270$ Multiply.

$\frac{100a}{100} = \frac{270}{100}$ Divide both sides by 100, the coefficient of a.

$a = 2.7$ Simplify.

Thus, 2.7 is 30% of 9.

Work Practice Problem 7

PRACTICE PROBLEM 8

65% of what number is 52?

EXAMPLE 8 Solving Percent Problems for the Base

150% of what number is 30?

Solution: 150% → percent; what number → base; 30 → amount

$\frac{30}{b} = \frac{150}{100}$ Write the proportion.

$\frac{30}{b} = \frac{3}{2}$ Simplify $\frac{150}{100}$ and write as $\frac{3}{2}$.

$30 \cdot 2 = b \cdot 3$ Set cross products equal.

$60 = 3b$ Multiply.

$\frac{60}{3} = \frac{3b}{3}$ Divide both sides by 3.

$20 = b$ Simplify.

Thus, 150% of 20 is 30.

Work Practice Problem 8

✔Concept Check When solving a percent problem by using a proportion, describe how you can check the result.

PRACTICE PROBLEM 9

15.4 is 5% of what number?

EXAMPLE 9

20.8 is 40% of what number?

Solution: 20.8 → amount; 40% → percent; what number → base

$\frac{20.8}{b} = \frac{40}{100}$ or $\frac{20.8}{b} = \frac{2}{5}$ Write the proportion and simplify $\frac{40}{100}$.

$20.8 \cdot 5 = b \cdot 2$ Set cross products equal.

$104 = 2b$ Multiply.

$\frac{104}{2} = \frac{2b}{2}$ Divide both sides by 2.

$52 = b$ Simplify.

So, 20.8 is 40% of 52.

Work Practice Problem 9

Answers

8. 80 **9.** 308

✔ Concept Check Answer

By putting the result into the proportion and checking that the proportion is true

EXAMPLE 10 **Solving Percent Problems for the Percent**

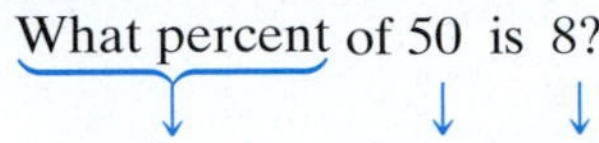

What percent of 50 is 8?

Solution: percent → What percent; base → 50; amount → 8

$$\frac{8}{50} = \frac{p}{100} \quad \text{or} \quad \frac{4}{25} = \frac{p}{100}$$ Write the proportion and simplify $\frac{8}{50}$.

$4 \cdot 100 = 25 \cdot p$ Set cross products equal.

$400 = 25p$ Multiply.

$\frac{400}{25} = \frac{25p}{25}$ Divide both sides by 25.

$16 = p$ Simplify.

So, 16% of 50 is 8.

Work Practice Problem 10

Helpful Hint Recall from our percent proportion that this number already is a percent. Just keep the number as is and attach a % symbol.

PRACTICE PROBLEM 10

What percent of 40 is 8?

EXAMPLE 11

504 is what percent of 360?

Solution: amount → 504; percent → what percent; base → 360

$$\frac{504}{360} = \frac{p}{100}$$

Let's choose not to simplify the ratio $\frac{504}{360}$.

$504 \cdot 100 = 360 \cdot p$ Set cross products equal.

$50{,}400 = 360p$ Multiply.

$\frac{50{,}400}{360} = \frac{360p}{360}$ Divide both sides by 360.

$140 = p$ Simplify.

Notice that by choosing not to simplify $\frac{504}{360}$, we had larger numbers in our equation. Either way, we find that 504 is 140% of 360.

Work Practice Problem 11

PRACTICE PROBLEM 11

414 is what percent of 180?

Helpful Hint Use the following to see whether your answers to the above examples and practice problems are reasonable.

100% of a number = the number

(a percent greater than 100%) of a number = a number larger than the original number

(a percent less than 100%) of a number = a number less than the original number

Answers

10. 20% **11.** 230%

Vocabulary and Readiness Check

Use the choices below to fill in each blank. These choices will be used more than once.

amount base percent

1. When translating the statement "20% of 15 is 3" to a proportion, the number 3 is called the ____________, 15 is the ____________, and 20 is the ____________.
2. In the question "50% of what number is 28?", which part of the percent proportion is unknown? ____________
3. In the question "What number is 25% of 200?", which part of the percent proportion is unknown? ____________
4. In the question "38 is what percent of 380?", which part of the percent proportion is unknown? ____________

6.4 EXERCISE SET

FOR EXTRA HELP

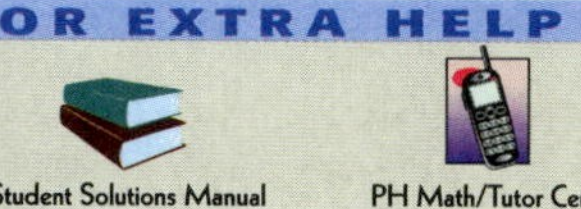

MyMathLab

MyMathLab

Objective A Translating *Translate each to a proportion. Do not solve. See Examples 1 through 6.*

1. 98% of 45 is what number?
2. 92% of 30 is what number?
3. What number is 4% of 150?
4. What number is 7% of 175?
5. 14.3 is 26% of what number?
6. 1.2 is 47% of what number?
7. 35% of what number is 84?
8. 85% of what number is 520?
9. What percent of 400 is 70?
10. What percent of 900 is 216?
11. 8.2 is what percent of 82?
12. 9.6 is what percent of 96?

Objective B *Solve. See Example 7.*

13. 40% of 65 is what number?
14. 25% of 84 is what number?
15. What number is 18% of 105?
16. What number is 60% of 29?

Solve. See Examples 8 and 9.

17. 15% of what number is 90?
18. 55% of what number is 55?
19. 7.8 is 78% of what number?
20. 1.1 is 44% of what number?

Solve. See Examples 10 and 11.

21. 42 is what percent of 35?
22. 147 is what percent of 98?
23. 14 is what percent of 50?
24. 24 is what percent of 50?

Objectives **A** **B** **Mixed Practice** *Solve. See Examples 1 through 11.*

25. 3.7 is 10% of what number?

26. 7.4 is 5% of what number?

27. 2.4% of 70 is what number?

28. 2.5% of 90 is what number?

29. 160 is 16% of what number?

30. 30 is 6% of what number?

31. 394.8 is what percent of 188?

32. 550.4 is what percent of 172?

33. What number is 89% of 62?

34. What number is 53% of 130?

35. What percent of 6 is 2.7?

36. What percent of 5 is 1.6?

37. 140% of what number is 105?

38. 170% of what number is 221?

39. 1.8% of 48 is what number?

40. 7.8% of 24 is what number?

41. What percent of 800 is 4?

42. What percent of 500 is 3?

43. 3.5 is 2.5% of what number?

44. 9.18 is 6.8% of what number?

45. 20% of 48 is what number?

46. 75% of 14 is what number?

47. 2486 is what percent of 2200?

48. 9310 is what percent of 3800?

Review

Add or subtract the fractions. See Sections 4.4, 4.5, and 4.7.

49. $-\frac{11}{16} + \left(-\frac{3}{16}\right)$

50. $\frac{7}{12} - \frac{5}{8}$

51. $3\frac{1}{2} - \frac{11}{30}$

52. $2\frac{2}{3} + 4\frac{1}{2}$

Add or subtract the decimals. See Section 5.2.

53. $\begin{array}{r} 0.41 \\ +0.29 \\ \hline \end{array}$

54. $\begin{array}{r} 10.78 \\ 4.3 \\ +\ 0.21 \\ \hline \end{array}$

55. $\begin{array}{r} 2.38 \\ -0.19 \\ \hline \end{array}$

56. $\begin{array}{r} 16.37 \\ -\ 2.61 \\ \hline \end{array}$

Concept Extensions

57. Write a word statement for the proportion $\frac{x}{28} = \frac{25}{100}$. Use the phrase "what number" for "x."

58. Write a percent statement that translates to $\frac{16}{80} = \frac{20}{100}$

Solve. See the Concept Checks in this section.

Suppose you have finished solving three percent problems using proportions that you set up correctly. Check each answer to see if each makes the proportion a true proportion. If any proportion is not true, solve it to find the correct solution.

59. $\frac{a}{64} = \frac{25}{100}$

Is the amount equal to 17?

60. $\frac{520}{b} = \frac{65}{100}$

Is the base equal to 800?

61. $\frac{36}{12} = \frac{p}{100}$

Is the percent equal to 50 (50%)?

 **62.** In your own words, describe how to identify the percent, the base, and the amount in a percent problem.

 **63.** In your own words, explain how to use a proportion to solve a percent problem.

Solve. Round to the nearest tenth, if necessary.

64. What number is 22.3% of 53,862?

 **65.** What percent of 110,736 is 88,542?

66. 8652 is 119% of what number?

THE BIGGER PICTURE Operations on Sets of Numbers and Solving Equations

Continue your outline from Sections 1.6, 1.7, 2.4, 3.3, 4.3, 4.7, 5.4, 6.1, and 6.3. (If you did not cover Section 6.3, pay no attention to the part of the outline numbered II.C.1.) Suggestions are once again written to help you complete this part of your outline. Notice that this part of the outline has to do with solving proportions.

I. Operations on Sets of Numbers

- **A. Whole Numbers**
  1. **Add or Subtract** (Section 1.3)
  2. **Multiply or Divide** (Sections 1.5, 1.6)
  3. **Exponent** (Section 1.7)
  4. **Order of Operations** (Section 1.7)
- **B. Integers**
  1. **Add** (Section 2.2)
  2. **Subtract** (Section 2.3)
  3. **Multiply or Divide** (Section 2.4)
- **C. Fractions**
  1. **Simplify** (Section 4.2)
  2. **Multiply** (Section 4.3)
  3. **Divide** (Section 4.3)
  4. **Add or Subtract** (Sections 4.4, 4.5)
- **D. Decimals**
  1. **Add or Subtract** (Section 5.2)
  2. **Multiply** (Section 5.3)
  3. **Divide** (Section 5.4)

II. Solving Equations

- **A. Equations in general** (Section 3.3)
- **B. Proportions** (Section 6.1)
- **C. Percent Problems**
  1. **Solved by Equations** (Section 6.3—you may not have covered this section)
  2. **Solved by Proportions:** Remember that percent, p, is identified by % or percent, base, b, usually appears after "of" and amount, a, is the part compared to the whole. 12% of some number is 6 translates to

$$\frac{6}{b} = \frac{12}{100} \text{ or } 6 \cdot 100 = b \cdot 12 \text{ or } \frac{600}{12} = b \text{ or } 50 = b$$

Perform the indicated operations.

1. $\frac{2}{9} + \frac{1}{5}$

2. $42 \div 2 \cdot 3$

3. $-0.03\,(0.7)$

4. $7.9 + 0.1$

Solve.

5. $\frac{3}{8} = \frac{n}{128}$

6. $\frac{7.2}{n} = \frac{36}{8}$

7. 215 is what percent of 86?

8. 95% of 48 is what number?

9. 4.2 is what percent of 15?

10. 93.6 is 52% of what number?

INTEGRATED REVIEW (Sections 6.1–6.4)

Ratio, Proportion, and Percent

Write each ratio as a ratio of whole numbers using fractional notation. Write the fraction in simplest form.

1. 18 to 20 **2.** 36 to 100 **3.** 8.6 to 10 **4.** 1.6 to 4.6

Find the ratio described in each problem.

5. Find the ratio of the width to the length of the sign below.

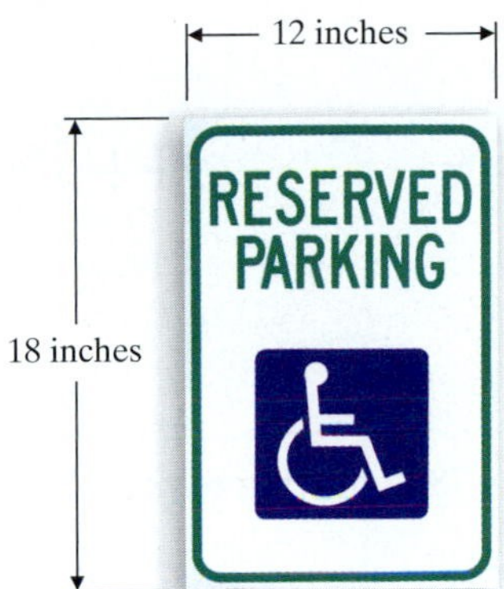

6. At the end of 2002 Lockheed Martin Corporation had \$26 hundred million in assets and \$8 hundred million in debts. Find the ratio of assets to debt. (*Source:* Lockheed Martin Corporation)

Solve.

7. $\frac{3.5}{12.5} = \frac{7}{z}$

8. $\frac{x + 7}{3} = \frac{2x}{5}$

An office uses 5 boxes of envelopes every 3 weeks.

9. Find how long a gross of envelope boxes is likely to last. (A gross of boxes is 144 boxes.) Round to the nearest week.

10. Find how many boxes should be purchased to last a month. Round to the nearest box.

Write each number as a percent.

11. 0.94 **12.** 0.17 **13.** $\frac{3}{8}$ **14.** $\frac{7}{2}$

15. 4.7 **16.** 8 **17.** $\frac{9}{20}$ **18.** $\frac{53}{50}$

19. $6\frac{3}{4}$ **20.** $3\frac{1}{4}$ **21.** 0.02 **22.** 0.06

Answers

1. ______
2. ______
3. ______
4. ______
5. ______
6. ______
7. ______
8. ______
9. ______
10. ______
11. ______
12. ______
13. ______
14. ______
15. ______
16. ______
17. ______
18. ______
19. ______
20. ______
21. ______
22. ______

23. ______ 24. ______
25. ______ 26. ______
27. ______ 28. ______
29. ______ 30. ______
31. ______
32. ______
33. ______
34. ______
35. ______
36. ______
37. ______
38. ______
39. ______ 40. ______
41. ______ 42. ______
43. ______ 44. ______
45. ______ 46. ______
47. ______ 48. ______
49. ______ 50. ______

Write each percent as a decimal.

23. 71% **24.** 31% **25.** 3% **26.** 4%

27. 224% **28.** 700% **29.** 2.9% **30.** 6.6%

Write each percent as a decimal and as a fraction or mixed number in simplest form. (If necessary when writing as a decimal, round to the nearest thousandth.)

31. 7% **32.** 5% **33.** 6.8% **34.** 11.25%

35. 74% **36.** 45% **37.** $16\frac{1}{3}\%$ **38.** $12\frac{2}{3}\%$

Solve each percent problem.

39. 15% of 90 is what number?

40. 78 is 78% of what number?

41. 297.5 is 85% of what number?

42. 78 is what percent of 65?

43. 23.8 is what percent of 85?

44. 38% of 200 is what number?

45. What number is 40% of 85?

46. What percent of 99 is 128.7?

47. What percent of 250 is 115?

48. What number is 45% of 84?

49. 42% of what number is 63?

50. 95% of what number is 58.9?

6.5 APPLICATIONS OF PERCENT

Objectives

A Solve Applications Involving Percent.

B Find Percent Increase and Percent Decrease.

Objective A Solving Applications Involving Percent

Percent is used in a variety of everyday situations. The next examples show just a few ways that percent occurs in real-life settings. (Each of these examples shows two ways of solving these problems. If you studied Section 6.3 only, see *Method 1*. If you studied Section 6.4 only, see *Method 2*.)

EXAMPLE 1

Recall that the circle graph from the beginning of this chapter shows the favorite ice cream flavors of adults in the United States. The state of Delaware has an adult population of 637,000. How many of these adults might we predict will choose butter pecan as their favorite ice cream flavor? (*Source:* U.S. Census Bureau)

Solution: *Method 1.* First, we state the problem in words.

In words: What number is 5% of 637,000?

Translate: $x = 5\% \cdot 637{,}000$

To solve for x, we find $5\% \cdot 637{,}000$.

$x = 0.05 \cdot 637{,}000$ Write 5% as a decimal.

$x = 31{,}850$

We predict that 31,850 adults in Delaware will choose butter pecan as their favorite ice cream flavor.

Method 2. State the problem in words; then translate.

In words: What number (amount) is 5% (percent) of 637,000 (base)?

Translate: amount → $\frac{a}{637{,}000} = \frac{5}{100}$ ← percent; base → 637,000

Next, we solve for a.

$a \cdot 100 = 637{,}000 \cdot 5$ Set cross products equal.

$100a = 3{,}185{,}000$ Multiply.

$\frac{100a}{100} = \frac{3{,}185{,}000}{100}$ Divide both sides by 100.

$a = 31{,}850$ Simplify.

We predict that 31,850 adults in Delaware will choose butter pecan as their favorite ice cream flavor.

Work Practice Problem 1

PRACTICE PROBLEM 1

The state of Rhode Island has an adult population of 837,000. How many of these adults might we predict will choose chocolate chip as their favorite ice cream flavor?

Favorite Ice Cream Flavors of Adults

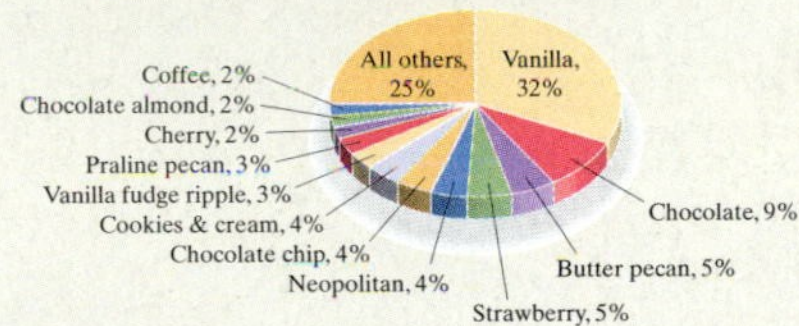

Source: International Ice Cream Association

Answer

1. 33,480

PRACTICE PROBLEM 2

In Florida, about 34,000 new nurses were needed in 2006. If there are currently 130,000 nurses, what percent of new nurses are needed in Florida? Round to the nearest whole percent. (*Source*: *St. Petersburg Times* and *The Registered Nurse Population*)

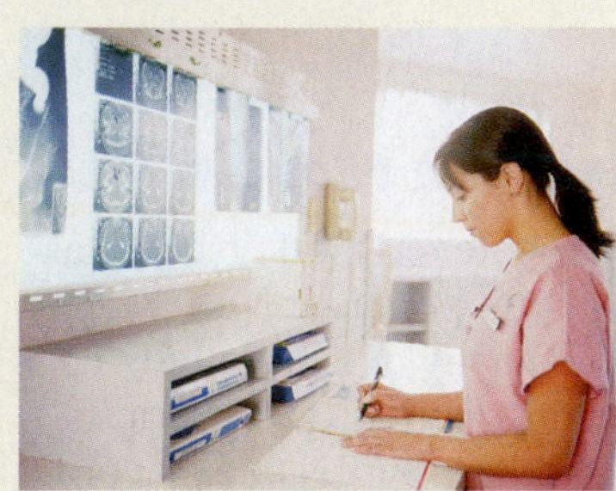

EXAMPLE 2 Finding Percent of Nursing Schools with Increases in Enrollment

There is a worldwide shortage of nurses that is projected to be 20% below requirements by 2020. Until 2001, there had also been a continual decline in enrollment in nursing schools. That has recently changed.

Of the total 2593 nursing schools in the United States, 2178 had an increase in applications or enrollment. What percent of nursing schools had an increase? Round to the nearest whole percent. (*Source:* CNN and *Nurse Week*)

Solution: *Method 1.* First, we state the problem in words.

In words: 2178 is what percent of 2593?

Translate: $2178 = x \cdot 2593$

or $2178 = 2593x$

Next, solve for x.

$$\frac{2178}{2593} = \frac{2593x}{2593} \quad \text{Divide both sides by 2593.}$$

$$0.84 \approx x \quad \text{Divide and round to the nearest hundredth.}$$

$$84\% \approx x \quad \text{Write as a percent.}$$

About 84% of nursing schools had an increase in applications or enrollment.

Method 2.

In words: 2178 is what percent of 2593?

2178 → amount; what percent → percent; 2593 → base

Translate: amount → 2178 / base → 2593 = p / 100 ← percent

$$\frac{2178}{2593} = \frac{p}{100}$$

Next, solve for p.

$$2178 \cdot 100 = 2593 \cdot p \quad \text{Set cross products equal.}$$

$$217{,}800 = 2593p \quad \text{Multiply.}$$

$$\frac{217{,}800}{2593} = \frac{2593p}{2593} \quad \text{Divide both sides by 2593.}$$

$$84 \approx p$$

About 84% of nursing schools had an increase in applications or enrollment.

Work Practice Problem 2

PRACTICE PROBLEM 3

The freshmen class of 864 students is 32% of all students at Euclid University. How many students go to Euclid University?

EXAMPLE 3 Finding the Base Number of Absences

Mr. Percy, the principal at Slidell High School, counted 31 freshmen absent during a particular day. If this is 4% of the total number of freshmen, how many freshmen are there at Slidell High School?

Solution: *Method 1.* First we state the problem in words; then we translate.

In words: 31 is 4% of what number?

Translate: $31 = 4\% \cdot x$

Answers

2. 26% 3. 2700

Next, we solve for x.

$$31 = 0.04 \cdot x \quad \text{Write 4\% as a decimal.}$$

$$\frac{31}{0.04} = \frac{0.04x}{0.04} \quad \text{Divide both sides by 0.04.}$$

$$775 = x \quad \text{Simplify.}$$

There are 775 freshmen at Slidell High School.

Method 2. First we state the problem in words; then we translate.

In words: 31 is 4% of what number?

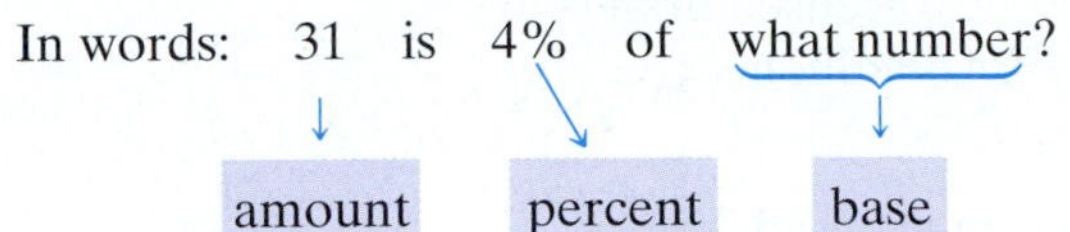

Translate: amount → 31 / base → b = 4 / 100 ← percent

$$\frac{31}{b} = \frac{4}{100}$$

Next, we solve for b.

$$31 \cdot 100 = b \cdot 4 \quad \text{Set cross products equal.}$$

$$3100 = 4b \quad \text{Multiply.}$$

$$\frac{3100}{4} = \frac{4b}{4} \quad \text{Divide both sides by 4.}$$

$$775 = b \quad \text{Simplify.}$$

There are 775 freshmen at Slidell High School.

Work Practice Problem 3

EXAMPLE 4 Finding the Base Increase in Licensed Drivers

From 1994 to 2004, the number of licensed drivers on the road in the United States increased by 14%. In 1994, there were 175 million licensed drivers on the road.

a. Find the increase in licensed drivers from 1994 to 2004.

b. Find the number of licensed drivers on the road in 2004. (*Source:* Federal Highway Administration)

Solution: *Method 1.* First we find the increase in licensed drivers.

In words: What number is 14% of 175?

Translate: $x = 14\% \cdot 175$

Continued on next page

PRACTICE PROBLEM 4

From 1994 to 2004, the number of registered vehicles on the road in the United States increased by 20%. In 1994, the number of vehicles on the road was 202 million.

a. Find the increase in the number of vehicles on the road in 2004.

b. Find the total number of registered vehicles on the road in 2004.

(*Source:* Federal Highway Administration)

Answers

4. a. 40.4 million **b.** 242.4 million

Next, we solve for x.

$x = 0.14 \cdot 175$ Write 14% as a decimal.

$x = 24.5$ Multiply.

a. The increase in licensed drivers is 24.5 million.

b. This means that the number of licensed drivers in 2004 was

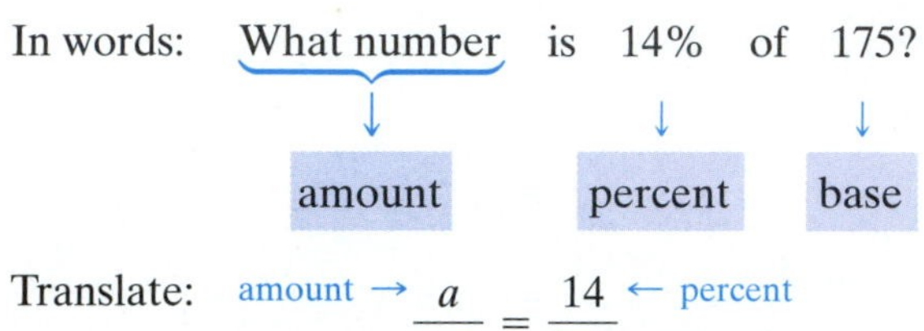

Method 2. First we find the increase in licensed drivers.

In words: What number is 14% of 175?

What number → amount; 14% → percent; 175 → base

Translate: $\frac{a}{175} = \frac{14}{100}$ (amount → a, base → 175, 14 ← percent)

Next, we solve for a.

$a \cdot 100 = 175 \cdot 14$ Set cross products equal.

$100a = 2450$ Multiply.

$\frac{100a}{100} = \frac{2450}{100}$ Divide both sides by 100.

$a = 24.5$ Simplify.

a. The increase in licensed drivers is 24.5 million.

b. This means that the number of licensed drivers in 2004 was

Number of licensed drivers in 2004 = Number of licensed drivers in 1994 + Increase in number of licensed drivers

= 175 million + 24.5 million

= 199.5 million

Work Practice Problem 4

Objective B Finding Percent Increase and Percent Decrease

We often use percents to show how much an amount has increased or decreased.

Suppose that the population of a town is 10,000 people and then it increases by 2000 people. The **percent of increase** is

$$\frac{2000}{10{,}000} = 0.2 = 20\%$$

(amount of increase → 2000; original amount → 10,000)

In general, we have the following.

Percent of Increase

$$\text{percent of increase} = \frac{\text{amount of increase}}{\text{original amount}}$$

Then write the quotient as a percent.

EXAMPLE 5 Finding Percent Increase

The number of applications for a mathematics scholarship at Yale increased from 34 to 45 in one year. What is the percent increase? Round to the nearest whole percent.

Solution: First we find the amount of increase by subtracting the original number of applicants from the new number of applicants.

amount of increase = 45 − 34 = 11

The amount of increase is 11 applicants. To find the percent of increase,

$$\text{percent of increase} = \frac{\text{amount of increase}}{\text{original amount}} = \frac{11}{34} \approx 0.32 = 32\%$$

The number of applications increased by about 32%.

Work Practice Problem 5

PRACTICE PROBLEM 5

The number of people attending the local play, *Peter Pan,* increased from 285 on Friday to 333 on Saturday. Find the percent increase in attendance. Round to the nearest tenth of a percent.

Helpful Hint Make sure that this number is the original number and not the new number.

✔Concept Check A student is calculating the percent increase in enrollment from 180 students one year to 200 students the next year. Explain what is wrong with the following calculations:

$$\text{Amount of increase} = 200 - 180 = 20$$

$$\text{Percent of increase} = \frac{20}{200} = 0.1 = 10\%$$

Suppose that your income was \$300 a week and then it decreased by \$30. The **percent of decrease** is

$$\begin{matrix}\text{amount of decrease} \rightarrow \\ \text{original amount} \rightarrow\end{matrix} \frac{\$30}{\$300} = 0.1 = 10\%$$

Percent of Decrease

$$\text{percent of decrease} = \frac{\text{amount of decrease}}{\text{original amount}}$$

Then write the quotient as a percent.

EXAMPLE 6 Finding Percent Decrease

In response to a decrease in sales, a company with 1500 employees reduces the number of employees to 1230. What is the percent decrease?

Solution: First we find the amount of decrease by subtracting 1230 from 1500.

amount of decrease = 1500 − 1230 = 270

The amount of decrease is 270. To find the percent of decrease,

$$\text{percent of decrease} = \frac{\text{amount of decrease}}{\text{original amount}} = \frac{270}{1500} = 0.18 = 18\%$$

The number of employees decreased by 18%.

Work Practice Problem 6

PRACTICE PROBLEM 6

A town with a population of 20,200 in 1995 decreased to 18,483 in 2005. What was the percent decrease?

✔Concept Check An ice cream stand sold 6000 ice cream cones last summer. This year the same stand sold 5400 cones. Was there a 10% increase, a 10% decrease, or neither? Explain.

Answers
5. 16.8% **6.** 8.5%

✔ Concept Check Answers
To find the percent of increase, you have to divide the amount of increase (20) by the original amount (180);
For the second Concept Check, there is a 10% decrease since this year's sales decreased.

6.5 EXERCISE SET

FOR EXTRA HELP

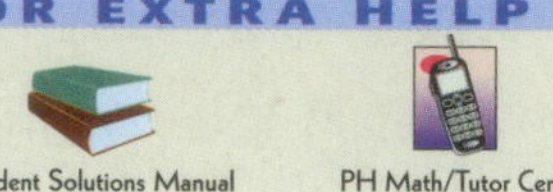

Objective A *Solve. See Examples 1 through 4. If necessary, round percents to the nearest tenth and all other answers to the nearest whole.*

1. The Total Gym® provides weight resistance through adjustments of incline. The minimum weight resistance is 4% of the weight of the person using the Total Gym. Find the minimum weight resistance possible for a 220-pound man. (*Source:* Total Gym)

2. The maximum weight resistance for the Total Gym is 60% of the weight of the person using it. Find the maximum weight resistance possible for a 220-pound man. (See Exercise 1 if needed.)

3. An inspector found 24 defective bolts during an inspection. If this is 1.5% of the total number of bolts inspected, how many bolts were inspected?

4. A day care worker found 28 children absent one day during an epidemic of chicken pox. If this was 35% of the total number of children attending the day care center, how many children attend this day care center?

5. A student's cost for last semester at her Community College was \$2700. She spent \$378 of that on books. What percent of last semester's college costs was spent on books?

6. Pierre Sampeau belongs to his local food cooperative, where he receives a percentage of what he spends each year as a dividend. He spent \$3850 last year at the food cooperative store and received a dividend of \$154. What percent of his total spending at the food cooperative did he receive as a dividend?

7. Approximately 58% of films are rated R. If 940 films were recently rated, how many were rated R? (*Source*: Motion Picture Association of America)

8. Approximately 11% of films are rated PG-13. If 940 films were recently rated, how many were rated PG-13? (*Source*: Motion Picture Association of America)

9. Of the 535 members of the 109th U.S. Congress, 35 have attended a community college. Determine the percent of the members of the 109th Congress who attended a community college. (*Source:* American Association of Community Colleges)

10. Of the 54,300 veterinarians in private practice in the United States, 23,200 of them are female. Find the percent of female veterinarians in private practice in the United States. (*Source:* American Veterinary Medical Association)

11. A furniture company currently produces 6200 chairs per month. If production decreases by 8%, find the decrease and the new number of chairs produced each month.

12. The enrollment at a local college decreased by 5% over last year's enrollment of 7640. Find the decrease in enrollment and the current enrollment.

13. From 2002 to 2012, the number of people employed as physician assistants in the United States is expected to increase by 49%. The number of people employed as physician assistants in 2002 was 63,000. Find the predicted number of physician assistants in 2012. (*Source:* Bureau of Labor Statistics)

14. The state of North Dakota had the smallest percent increase in population, 0.5%, from the 1990 census to the 2000 census. In 1990, the population of North Dakota was 638,800. What was the population of North Dakota in 2000? (*Source:* U.S. Census Bureau)

For each food described, find the percent of total calories from fat. If necessary, round to the nearest tenth of a percent.

15. Ranch dressing serving size of 2 tablespoons

	Calories
Total	40
From fat	20

16. Unsweetened cocoa powder serving size of 1 tablespoon

	Calories
Total	20
From fat	5

17.

Nutrition Facts
Serving Size 1 pouch (20g)
Servings Per Container 6

Amount Per Serving	
Calories	80
Calories from fat	10
	% Daily Value*
Total Fat 1g	**2%**
Sodium 45mg	**2%**
Total Carbohydrate 17g	**6%**
Sugars 9g	
Protein 0g	
Vitamin C	25%

Not a significant source of saturated fat, cholesterol, dietary fiber, vitamin A, calcium and iron.

*Percent Daily Values are based on a 2,000 calorie diet.

Artificial Fruit Snacks

18.

Nutrition Facts
Serving Size $\frac{1}{4}$ cup (33g)
Servings Per Container About 9

Amount Per Serving	
Calories 190 **Calories from Fat** 130	
	% Daily Value
Total Fat 16g	**24%**
Saturated Fat 3g	**16%**
Cholesterol 0mg	**0%**
Sodium 135mg	**6%**
Total Carbohydrate 9g	**3%**
Dietary Fiber 1g	**5%**
Sugars 2g	
Protein 5g	

Vitamin A 0% • Vitamin C 0%
Calcium 0% • Iron 8%

Peanut Mixture

19.

Nutrition Facts
Serving Size 18 crackers (29g)
Servings Per Container About 9

Amount Per Serving	
Calories 120 Calories from Fat 35	
	% Daily Value*
Total Fat 4g	**6%**
Saturated Fat 0.5g	**3%**
Polyunsaturated Fat 0g	
Monounsaturated Fat 1.5g	
Cholesterol 0mg	**0%**
Sodium 220mg	**9%**
Total Carbohydrate 21g	**7%**
Dietary Fiber 2g	**7%**
Sugars 3g	
Protein 2g	

Vitamin A 0% • Vitamin C 0%
Calcium 2% • Iron 4%
Phosphorus 10%

Snack Crackers

20.

Nutrition Facts
Serving Size 28 crackers (31g)
Servings Per Container About 6

Amount Per Serving	
Calories 130 Calories from Fat 35	
	% Daily Value*
Total Fat 4g	**6%**
Saturated Fat 2g	**10%**
Polyunsaturated Fat 1g	
Monounsaturated Fat 1g	
Cholesterol 0mg	**0%**
Sodium 470mg	**20%**
Total Carbohydrate 23g	**8%**
Dietary Fiber 1g	**4%**
Sugars 4g	
Protein 2g	

Vitamin A 0% • Vitamin C 0%
Calcium 0% • Iron 2%

Snack Crackers

Solve. Round money amounts to the nearest cent and all other amounts to the nearest tenth. See Example 4.

21. A family paid \$26,250 as a down payment for a home. If this represents 15% of the price of the home, find the price of the home.

22. A banker learned that \$842.40 is withheld from his monthly check for taxes and insurance. If this represents 18% of his total pay, find the total pay.

23. An owner of a repair service company estimates that for every 40 hours a repairperson is on the job, he can bill for only 78% of the hours. The remaining hours, the repairperson is idle or driving to or from a job. Determine the number of hours per 40-hour week the owner can bill for a repairperson.

24. A manufacturer of electronic components expects 1.04% of its products to be defective. Determine the number of defective components expected in a batch of 28,350 components. Round to the nearest whole component.

25. A car manufacturer announced that next year the price of a certain model of car would increase by 4.5%. This year the price is $19,286. Find the increase in price and the new price.

26. A union contract calls for a 6.5% salary increase for all employees. Determine the increase and the new salary that a worker currently making $58,500 under this contract can expect.

27. The city of Buckeye has a population of approximately 6950 adults. Use the circle graph for Example 1 of this section and predict the number of adults in Buckeye who prefer Vanilla ice cream.

28. The city of Hamlet has a population of approximately 4380 adults. Use the circle graph for Example 1 of this section and predict the number of adults in Hamlet who prefer Strawberry ice cream.

29. The population of Americans aged 65 and older was 35 million in 2000. That population is projected to increase by 80% by 2025. Find the increase and the projected 2025 population. (*Source:* Bureau of the Census)

30. From 2000 to 2010, the number of masters degrees awarded to women is projected to increase by 8.3%. The number of women who received masters degrees in 2000 was 265,000. Find the increase and the predicted number of women to be awarded masters degrees in 2010. (*Source:* U.S. National Center for Education Statistics)

Objective **B** *Find the amount of increase and the percent increase. See Example 5.*

	Original Amount	New Amount	Amount of Increase	Percent Increase
31.	50	80		
32.	8	12		
33.	65	117		
34.	68	170		

Find the amount of decrease and the percent decrease. See Example 6.

	Original Amount	New Amount	Amount of Decrease	Percent Decrease
35.	8	6		
36.	25	20		
37.	160	40		
38.	200	162		

Solve. Round percents to the nearest tenth, if necessary. See Examples 5 and 6.

39. There are 150 calories in a cup of whole milk and only 84 in a cup of skim milk. In switching to skim milk, find the percent decrease in number of calories per cup.

40. In reaction to a slow economy, the number of employees at a soup company decreased from 530 to 477. What was the percent decrease in the number of employees?

41. The number of cable TV systems recently decreased from 10,845 to 10,700. Find the percent decrease.

42. Before taking a typing course, Geoffry Landers could type 32 words per minute. By the end of the course, he was able to type 76 words per minute. Find the percent increase.

43. In 1940, the average size of a privately owned farm in the United States was 174 acres. By 2005, the average size of a privately owned farm in the United States had increased to 444 acres. What was the percent increase? (*Source:* National Agricultural Statistics Service)

44. In 1995, 722.9 million recorded music CDs were shipped to retailers in the United States. By 2005, this number had decreased to 705.4 million CDs. What was the percent decrease? (*Source:* Recording Industry Association of America)

45. In 2004, approximately 455,000 computer programmers were employed in the United States. By 2014, this number is expected to increase to 464,000 computer programmers. What is the percent increase? (*Source:* Bureau of Labor Statistics)

46. In 1994, there were 784 deaths from boating accidents in the United States. By 2004, the number of deaths from boating accidents had decreased to 676. What was the percent decrease? (*Source:* U.S. Coast Guard)

47. In 2006, there were 3,570 thousand elementary and secondary teachers employed in the United States. This number is expected to increase to 3,769 thousand teachers in 2012. What is the percent increase? (*Source:* National Center for Education Statistics)

48. In 2005, approximately 484,000 correctional officers were employed in the United States. By 2008, this number is expected to increase to 518,000 correctional officers. What is the percent increase? (*Source:* Bureau of Labor Statistics)

49. In 1995, there were 7151 indoor cinema sites in the United States. By 2005, this number had decreased to 5713 sites. What was the percent decrease? (*Source:* National Association of Theater Owners)

50. In 1994, approximately 16,000 occupational therapy assistants and aides were employed in the United States. According to one survey, by 2005, this number is expected to increase to 29,000 assistants and aides. What is the percent increase? (*Source:* Bureau of Labor Statistics)

51. The average soft-drink size has increased from 13.1 oz to 19.9 oz over the past two decades. Find the percent increase. (*Source:* University of North Carolina at Chapel Hill, *Journal for American Medicine*)

52. In 1999, discarded electronics, including obsolete computer equipment, accounted for 75,000 tons of solid waste per year in Massachusetts. By 2006, discarded electronic waste increased to 300,000 tons of waste per year in the state. Find the percent increase. (*Source:* Massachusetts Department of Environmental Protection)

53. The number of Americans who subscribed to cellular phone service was 110,000 thousand in 2000. By 2006, this number had increased to about 212,000 thousand. What was the percent increase? (*Source:* CTIA—The Wireless Association®)

54. The population of Tokyo is expected to decrease from 127,400 thousand in 2005 to 99,900 thousand in 2050. Find the percent decrease. (*Source:* International Programs Center, Bureau of the Census, U.S. Dept. of Commerce)

Review

Perform each indicated operation. See Chapters 4 and 5.

55. 0.12×38

56. $29.4 \div 0.7$

57. $9.20 + 1.98$

58. $78 - 19.46$

59. $-\frac{3}{8} - \frac{5}{12}$

60. $\left(-\frac{3}{8}\right)\left(-\frac{5}{12}\right)$

61. $2\frac{4}{5} \div 3\frac{9}{10}$

62. $2\frac{4}{5} - 3\frac{9}{10}$

Concept Extensions

63. If a number is increased by 100%, how does the increased number compare with the original number? Explain your answer.

64. In your own words, explain what is wrong with the following statement. "Last year we had 80 students attend. This year we have a 50% increase or a total of 160 students attend."

Explain what errors were made by each student when solving percent of increase or decrease problems and then correct the errors. "The population of a certain rural town was 150 in 1980, 180 in 1990, and 150 in 2000."

65. Find the percent of increase in population from 1980 to 1990.

Miranda's solution: Percent of increase $= \frac{30}{180} = 0.1\overline{6} \approx 16.7\%$

66. Find the percent of decrease in population from 1990 to 2000.

Jeremy's solution: Percent of decrease $= \frac{30}{150} = 0.20 = 20\%$

67. The percent of increase from 1980 to 1990 is the same as the percent of decrease from 1990 to 2000. True or false. Chris's answer: True because they had the same amount of increase as the amount of decrease.

6.6 PERCENT AND PROBLEM SOLVING: SALES TAX, COMMISSION, AND DISCOUNT

Objectives

A Calculate Sales Tax and Total Price.

B Calculate Commissions.

C Calculate Discount and Sale Price.

Objective A Calculating Sales Tax and Total Price

Percents are frequently used in the retail trade. For example, most states charge a tax on certain items when purchased. This tax is called a **sales tax,** and retail stores collect it for the state. Sales tax is almost always stated as a percent of the purchase price.

A 9% sales tax rate on a purchase of a \$10 calculator gives a sales tax of

sales tax = 9% of \$10 = 0.09 · \$10.00 = \$0.90

The total price to the customer would be

purchase price plus sales tax
↓ ↓ ↓
\$10.00 + \$0.90 = \$10.90

This example suggests the following equations:

Sales Tax and Total Price

sales tax = tax rate · purchase price
total price = purchase price + sales tax

In this section we round dollar amounts to the nearest cent.

EXAMPLE 1 Finding Sales Tax and Purchase Price

Find the sales tax and the total price on the purchase of an \$85.50 atlas in a city where the sales tax rate is 7.5%.

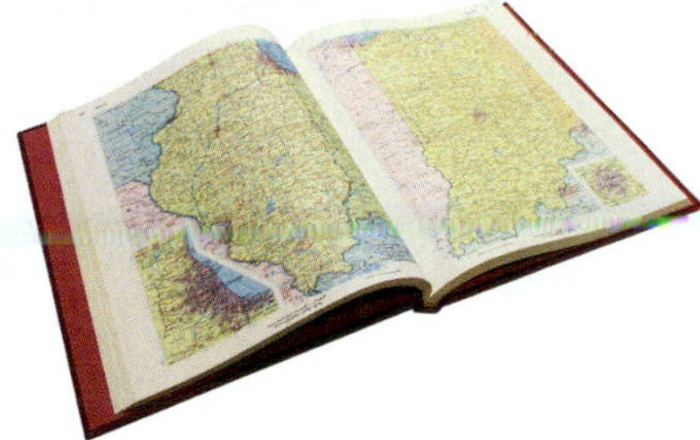

Solution: The purchase price is \$85.50 and the tax rate is 7.5%.

sales tax = tax rate · purchase price
↓ ↓ ↙
sales tax = 7.5% · \$85.50
= 0.075 · \$85.5 Write 7.5% as a decimal.
≈ \$6.41 Rounded to the nearest cent

Thus, the sales tax is \$6.41. Next find the total price.

total price = purchase price + sales tax
↓ ↓ ↙
total price = \$85.50 + \$6.41
= \$91.91

The sales tax on \$85.50 is \$6.41, and the total price is \$91.91.

Work Practice Problem 1

PRACTICE PROBLEM 1

If the sales tax rate is 8.5%, what is the sales tax and the total amount due on a \$59.90 Goodgrip tire? (Round the sales tax to the nearest cent.)

Answer

1. tax: \$5.09; total: \$64.99

✔Concept Check The purchase price of a textbook is \$50 and sales tax is 10%. If you are told by the cashier that the total price is \$75, how can you tell that a mistake has been made?

PRACTICE PROBLEM 2

The sales tax on an \$18,500 automobile is \$1665. Find the sales tax rate.

EXAMPLE 2 Finding a Sales Tax Rate

The sales tax on a \$406 Sony flat screen digital 27-inch television is \$34.51. Find the sales tax rate.

Solution: Let r represent the unknown sales tax rate. Then

sales tax = tax rate · purchase price

$$\$34.51 = r \cdot \$406$$

$$\frac{34.51}{406} = \frac{r \cdot 406}{406} \quad \text{Divide both sides by 300.}$$

$$0.085 = r \quad \text{Simplify.}$$

$$8.5\% = r \quad \text{Write 0.085 as a percent.}$$

The sales tax rate is 8.5%.

Work Practice Problem 2

Objective B Calculating Commissions

A **wage** is payment for performing work. Hourly wage, commissions, and salary are some of the ways wages can be paid. Many people who work in sales are paid a commission. An employee who is paid a **commission** is paid a percent of his or her total sales.

Commission

commission = commission rate · sales

PRACTICE PROBLEM 3

A sales representative for Office Product Copiers sold \$47,632 worth of copy equipment and supplies last month. What is his commission for the month if he is paid a commission of 6.6% of his total sales for the month?

EXAMPLE 3 Finding the Amount of Commission

Sherry Souter, a real estate broker for Wealth Investments, sold a house for \$214,000 last week. If her commission is 1.5% of the selling price of the home, find the amount of her commission.

Solution:

commission = commission rate · sales

$$\text{commission} = 1.5\% \cdot \$214{,}000$$

$$= 0.015 \cdot \$214{,}000 \quad \text{Write 1.5\% as 0.015.}$$

$$= \$3210 \quad \text{Multiply.}$$

Answers

2. 9% **3.** \$3143.71

✔ Concept Check Answer

Since $10\% = \frac{1}{10}$, the sales tax is $\frac{\$50}{10} = \$5$. The total price should have been \$55.

Her commission on the house is $3210.

Work Practice Problem 3

EXAMPLE 4 Finding a Commission Rate

A salesperson earned $1560 for selling $13,000 worth of electronics equipment. Find the commission rate.

Solution: Let r stand for the unknown commission rate. Then

$$\text{commission} = \text{commission rate} \cdot \text{sales}$$

$$\$1560 = r \cdot \$13{,}000$$

$$\frac{1560}{13{,}000} = r \quad \text{Divide 1560 by 13,000, the number multiplied by } r.$$

$$0.12 = r \quad \text{Simplify.}$$

$$12\% = r \quad \text{Write 0.12 as a percent.}$$

The commission rate is 12%.

Work Practice Problem 4

PRACTICE PROBLEM 4

A salesperson earns $645 for selling $4300 worth of appliances. Find the commission rate.

Objective C Calculating Discount and Sale Price

Suppose that an item that normally sells for $40 is on sale for 25% off. This means that the **original price** of $40 is reduced, or **discounted,** by 25% of $40, or $10. The **discount rate** is 25%, the **amount of discount** is $10, and the **sale price** is $40 − $10, or $30. Study the diagram below to visualize these terms.

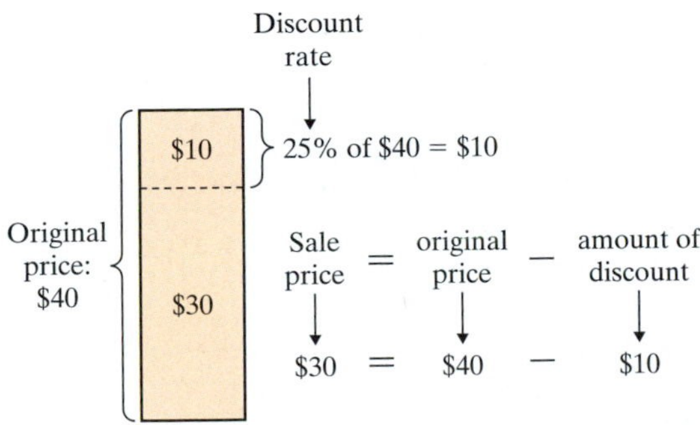

To calculate discounts and sale prices, we can use the following equations:

Answer

4. 15%

Discount and Sale Price

amount of discount = discount rate · original price

sale price = original price − amount of discount

PRACTICE PROBLEM 5

A discontinued washer and dryer combo is advertised on sale for 35% off the regular price of $700. Find the amount of discount and the sale price.

Answer

5. $245; $455

EXAMPLE 5 Finding a Discount and a Sale Price

An electric rice cooker that normally sells for $65 is on sale for 25% off. What is the amount of discount and what is the sale price?

Solution: First we find the amount of discount, or simply the discount.

amount of discount	=	discount rate	·	original price	
↓		↓		↓	
amount of discount	=	25%	·	$65	
	=	0.25 · $65			Write 25% as 0.25.
	=	$16.25			Multiply.

The discount is $16.25. Next, find the sale price.

sale price	=	original price	−	discount	
↓		↓		↓	
sale price	=	$65	−	$16.25	
	=	$48.75			Subtract.

The sale price is $48.75.

Work Practice Problem 5

Vocabulary and Readiness Check

Use the choices below to fill in each blank.

amount of discount	sale price	sales tax
commission	total price	

1. ______________ = tax rate · purchase price.

2. ______________ = purchase price + sales tax.

3. ______________ = commission rate · sales.

4. ________________ = discount rate · original price.

5. ______________ = original price − amount of discount.

6.6 EXERCISE SET

FOR EXTRA HELP

Student Solutions Manual

PH Math/Tutor Center

CD/Video for Review

MathXL®

MyMathLab

Objective **A** *Solve. See Examples 1 and 2.*

1. What is the sales tax on a jacket priced at $150 if the sales tax rate is 5%?

2. If the sales tax rate is 6%, find the sales tax on a microwave oven priced at $188.

3. The purchase price of a camcorder is $799. What is the total price if the sales tax rate is 7.5%?

4. A stereo system has a purchase price of $426. What is the total price if the sales tax rate is 8%?

5. A new large-screen television has a purchase price of $4790. If the sales tax on this purchase is $335.30, find the sales tax rate.

6. The sales tax on the purchase of a $6800 used car is $374. Find the sales tax rate.

7. The sales tax on a table saw is $10.20.

a. What is the purchase price of the table saw (before tax) if the sales tax rate is 8.5%? (*Hint:* Use the sales tax equation and insert the replacement values.)

b. Find the total price of the table saw.

8. The sales tax on a one-half-carat diamond ring is $76.

a. Find the purchase price of the ring (before tax) if the sales tax rate is 9.5%. (See the hint for Exercise 7a.)

b. Find the total price of the ring.

9. A gold and diamond bracelet sells for $1800. Find the sales tax and the total price if the sales tax rate is 6.5%.

10. The purchase price of a personal computer is $1890. If the sales tax rate is 8%, what is the sales tax and the total price?

11. The sales tax on the purchase of a futon is $24.25. If the tax rate is 5%, find the purchase price of the futon.

12. The sales tax on the purchase of a TV-DVD combination is $32.85. If the tax rate is 9%, find the purchase price of the TV-DVD.

13. The sales tax is $98.70 on a stereo sound system purchase of $1645. Find the sales tax rate.

14. The sales tax is $103.50 on a necklace purchase of $1150. Find the sales tax rate.

15. A cell phone costs $210, a battery recharger costs $15, and batteries cost $5. What is the sales tax and total price for purchasing these items if the sales tax rate is 7%?

16. Ms. Warner bought a blouse for $35, a skirt for $55, and a blazer for $95. Find the sales tax and the total price she paid, given a sales tax rate of 6.5%.

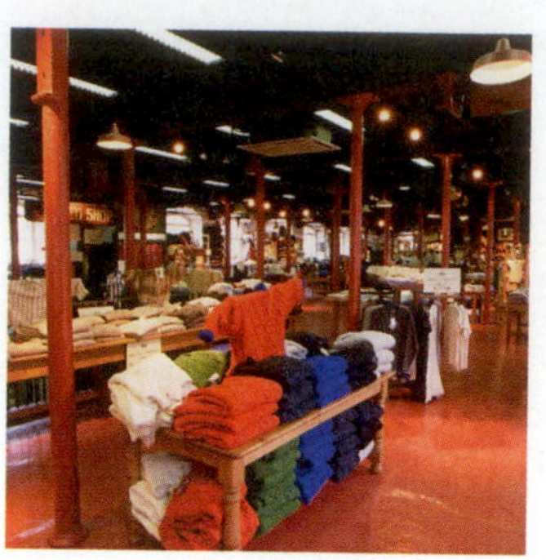

Objective **B** *Solve. See Examples 3 and 4.*

17. A sales representative for a large furniture warehouse is paid a commission rate of 4%. Find her commission if she sold $1,329,401 worth of furniture last year.

18. Rosie Davis-Smith is a beauty consultant for a home cosmetic business. She is paid a commission rate of 12.8%. Find her commission if she sold $1638 in cosmetics last month.

19. A salesperson earned a commission of $1380.40 for selling $9860 worth of paper products. Find the commission rate.

20. A salesperson earned a commission of $3575 for selling $32,500 worth of books to various bookstores. Find the commission rate.

21. How much commission will Jack Pruet make on the sale of a $325,900 house if he receives 1.5% of the selling price?

22. Frankie Lopez sold $9638 of jewelry this week. Find her commission for the week if she receives a commission rate of 5.6%.

23. A real estate agent earned a commission of $5565 for selling a house. If his rate is 3%, find the selling price of the house. (*Hint:* Use the commission equation and insert the replacement values.)

24. A salesperson earned $1750 for selling fertilizer. If her commission rate is 7%, find the selling price of the fertilizer (*Hint:* See Exercise 23.).

Objective **C** *Find the amount of discount and the sale price. See Example 5.*

	Original Price	Discount Rate	Amount of Discount	Sale Price
25.	$89	10%		
26.	$74	20%		
27.	$196.50	50%		
28.	$110.60	40%		
29.	$410	35%		
30.	$370	25%		
31.	$21,700	15%		
32.	$17,800	12%		

33. A $300 fax machine is on sale for 15% off. Find the amount of discount and the sale price.

34. A $4295 designer dress is on sale for 30% off. Find the amount of discount and the sale price.

Objectives **A** **B** **Mixed Practice** *Complete each table.*

	Purchase Price	Tax Rate	Sales Tax	Total Price
35.	\$305	9%		
36.	\$243	8%		
37.	\$56	5.5%		
38.	\$65	8.4%		

	Sale	Commission Rate	Commission
39.	\$235,800	3%	
40.	\$195,450	5%	
41.	\$17,900		\$1432
42.	\$25,600		\$2304

Review

Multiply. See Sections 4.3, 5.3, and 5.5.

43. $2000 \cdot \frac{3}{10} \cdot 2$

44. $500 \cdot \frac{2}{25} \cdot 3$

45. $400 \cdot \frac{3}{100} \cdot 11$

46. $1000 \cdot \frac{1}{20} \cdot 5$

47. $600 \cdot 0.04 \cdot \frac{2}{3}$

48. $6000 \cdot 0.06 \cdot \frac{3}{4}$

Concept Extensions

Solve. See the Concept Check in this section.

49. Your purchase price is \$68 and the sales tax rate is 9.5%. Round each amount and use the rounded amounts to estimate the total price. Choose the best estimate.

a. \$105 **b.** \$58 **c.** \$93 **d.** \$77

50. Your purchase price is \$200 and the tax rate is 10%. Choose the best estimate of the total price.

a. \$190 **b.** \$210 **c.** \$220 **d.** \$300

Tipping *One very useful application of percent is mentally calculating a tip. Recall that to find 10% of a number, simply move the decimal point one place to the left. To find 20% of a number, just double 10% of the number. To find 15% of a number, find 10% and then add to that number half of the 10% amount. Mentally fill in the chart below. To do so, start by rounding the bill amount to the nearest dollar.*

	Tipping Chart			
	Bill Amount	10%	15%	20%
51.	\$40.21			
52.	\$15.89			
53.	\$72.17			
54.	\$9.33			

55. Suppose that the original price of a shirt is \$50. Which is better, a 60% discount or a discount of 30% followed by a discount of 35% of the reduced price? Explain your answer.

56. Which is better, a 30% discount followed by an additional 25% off or a 20% discount followed by an additional 40% off? To see, suppose an item costs \$100 and calculate each discounted price. Explain your answer.

57. A diamond necklace sells for \$24,966. If the tax rate is 7.5%, find the total price.

58. A house recently sold for \$562,560. The commission rate on the sale is 5.5%. If the real estate agent is to receive 60% of the commission, find the amount received by the agent.

STUDY SKILLS BUILDER

Are You Familiar with Your Textbook Supplements?

Below is a review of some of the student supplements available for additional study. Check to see if you are using the ones most helpful to you.

- Chapter Test Prep Videos on CD. This material is found with your textbook and is fully explained there. The CD contains video clip solutions to the Chapter Test exercises in this text and are excellent help when studying for chapter tests.
- Lecture Videos on CD-ROM. These video segments are keyed to each section of the text. The material is presented by me, Elayn Martin-Gay, and I have placed a by the exercises in the text that I have worked on the video.
- The *Student Solutions Manual*. This contains worked out solutions to odd-numbered exercises as well as every exercise in the Integrated Reviews, Chapter Reviews, Chapter Tests, and Cumulative Reviews.
- Prentice Hall Tutor Center. Mathematic questions may be phoned, faxed, or e-mailed to this center.
- MyMathLab, MathXL, and Interact Math. These are computer and Internet tutorials. This supplement may already be available to you somewhere on campus, for example at your local learning resource lab. Take a moment and find the name and location of any such lab on campus.

As usual, your instructor is your best source of information.

Let's see how you are doing with textbook supplements.

1. Name one way the Lecture Videos can be helpful to you.

2. Name one way the Chapter Test Prep Video can help you prepare for a chapter test.

3. List any textbook supplements that you have found useful.

4. Have you located and visited a learning resource lab located on your campus?

5. List the textbook supplements that are currently housed in your campus's learning resource lab.

6.7 PERCENT AND PROBLEM SOLVING: INTEREST

Objectives

A Calculate Simple Interest.

B Calculate Compound Interest.

Objective A Calculating Simple Interest

Interest is money charged for using other people's money. When you borrow money, you pay interest. When you loan or invest money, you earn interest. The money borrowed, loaned, or invested is called the **principal amount,** or simply **principal.** Interest is normally stated in terms of a percent of the principal for a given period of time. The **interest rate** is the percent used in computing the interest. Unless stated otherwise, *the rate is understood to be per year.* When the interest is computed on the original principal, it is called **simple interest.** Simple interest is calculated using the following equation:

Simple Interest

$$\text{Simple Interest} = \text{Principal} \cdot \text{Rate} \cdot \text{Time}$$

$$I = P \cdot R \cdot T$$

where the rate is understood to be per year and time is in years.

EXAMPLE 1 Finding Simple Interest

Find the simple interest after 2 years on \$500 at an interest rate of 12%.

Solution: In this example, $P = \$500$, $R = 12\%$, and $T = 2$ years. Replace the variables with values in the formula $I = PRT$.

$$\begin{aligned} I &= P \cdot R \cdot T \\ I &= \$500 \cdot 12\% \cdot 2 && \text{Let } P = \$500,\ R = 12\%, \text{ and } T = 2. \\ &= \$500 \cdot (0.12) \cdot 2 && \text{Write 12\% as a decimal.} \\ &= \$120 && \text{Multiply.} \end{aligned}$$

The simple interest is \$120.

Work Practice Problem 1

PRACTICE PROBLEM 1

Find the simple interest after 5 years on \$875 at an interest rate of 7%.

If time is not given in years, we need to convert the given time to years.

EXAMPLE 2 Finding Simple Interest

Ivan Borski borrowed \$2400 at 10% simple interest for 8 months to buy a used Toyota Corolla. Find the simple interest he paid.

Solution: Since there are 12 months in a year, we first find what part of a year 8 months is.

$$8 \text{ months} = \frac{8}{12} \text{ year} = \frac{2}{3} \text{ year}$$

Now we find the simple interest.

$$\begin{aligned} I &= P \cdot R \cdot T \\ &= \$2400 \cdot (0.10) \cdot \frac{2}{3} && \text{Let } P = \$2400,\ R = 10\% \text{ or } 0.10, \text{ and } T = \frac{2}{3}. \\ &= \$160 \end{aligned}$$

The interest on Ivan's loan is \$160.

Work Practice Problem 2

PRACTICE PROBLEM 2

A student borrowed \$1500 for 9 months on her credit card at a simple interest rate of 20%. How much interest did she pay?

✔ **Concept Check** Suppose in Example 2 you had obtained an answer of \$16,000. How would you know that you had made a mistake in this problem?

Answers

1. \$306.25 **2.** \$225

✔ Concept Check Answer

\$16,000 is too much interest. Answers may vary.

When money is borrowed, the borrower pays the original amount borrowed, or the principal, as well as the interest. When money is invested, the investor receives the original amount invested, or the principal, as well as the interest. In either case, the **total amount** is the sum of the principal and the interest.

Finding the Total Amount of a Loan or Investment

total amount (paid or received) = principal + interest

PRACTICE PROBLEM 3

If $2100 is borrowed at a simple interest rate of 13% for 6 months, find the total amount paid.

EXAMPLE 3 Finding the Total Amount of an Investment

An accountant invested $2000 at a simple interest rate of 10% for 2 years. What total amount of money will she have from her investment in 2 years?

Solution: First we find her interest.

$$I = P \cdot R \cdot T$$
$$= \$2000 \cdot (0.10) \cdot 2 \quad \text{Let } P = \$2000,\ R = 10\% \text{ or } 0.10, \text{ and } T = 2.$$
$$= \$400$$

The interest is $400.

Next, we add the interest to the principal.

total amount	=	principal	+	interest
↓		↓		↓
total amount	=	$2000	+	$400
	=	$2400		

After 2 years, she will have a total amount of $2400.

Work Practice Problem 3

✔**Concept Check** Which investment would earn more interest: an amount of money invested at 8% interest for 2 years, or the same amount of money invested at 8% for 3 years? Explain.

Objective B Calculating Compound Interest

Recall that simple interest depends on the original principal only. Another type of interest is compound interest. **Compound interest** is computed on not only the principal, but also on the interest already earned in previous compounding periods. Compound interest is used more often than simple interest.

Let's see how compound interest differs from simple interest. Suppose that $2000 is invested at 7% interest **compounded annually** for 3 years. This means that interest is added to the principal at the end of each year and that next year's interest is computed on this new amount. In this section, we round dollar amounts to the nearest cent.

	Amount at Beginning of Year	Principal	·	Rate	·	Time	= Interest	Amount at End of Year
1st year	$2000	$2000	·	0.07	·	1	= $140	$2000 + 140 = $2140
2nd year	$2140	$2140	·	0.07	·	1	= $149.80	$2140 + 149.80 = $2289.80
3rd year	$2289.80	$2289.80	·	0.07	·	1	= $160.29	$2289.80 + 160.29 = $2450.09

The compound interest earned can be found by

total amount	−	original principal	=	compound interest
↓		↓		↓
$2450.09	−	$2000	=	$450.09

Answer

3. $2236.50

✔ **Concept Check Answer**

8% for 3 years. Since the interest rate is the same, the longer you keep the money invested, the more interest you earn.

The simple interest earned would have been

principal	·	rate	·	time	=	interest
↓		↓		↓		↓
\$2000	·	0.07	·	3	=	\$420

Since compound interest earns "interest on interest," compound interest earns more than simple interest.

Computing compound interest using the method above can be tedious. We can use a calculator and the compound interest formula below to compute compound interest more quickly.

Compound Interest Formula

The total amount A in an account is given by

$$A = P\left(1 + \frac{r}{n}\right)^{n \cdot t}$$

where P is the principal, r is the interest rate written as a decimal, t is the length of time in years, and n is the number of times compounded per year.

Let's use this formula to check our earlier compound interest calculations.

EXAMPLE 4 \$2000 is invested at 7% interest compounded annually. Find the total amount after 3 years.

Solution: "Compounded annually" means 1 time a year, so

$n = 1$. Also, $P = \$2000$, $r = 7\% = 0.07$, and $t = 3$ years.

$$A = P\left(1 + \frac{r}{n}\right)^{n \cdot t}$$

$$= 2000\left(1 + \frac{0.07}{1}\right)^{1 \cdot 3}$$

$$= 2000(1.07)^3$$

$$\approx 2450.09 \quad \text{Round to 2 decimal places.}$$

The total amount at the end of 3 years is \$2450.09.

Work Practice Problem 4

PRACTICE PROBLEM 4

\$3000 is invested at 4% interest compounded annually. Find the total amount after 6 years.

Helpful Hint Remember order of operations. **First** evaluate. $(1.07)^3$, then multiply by 2000.

EXAMPLE 5 **Finding Total Amount Received on an Investment**

\$4000 is invested at 5.3% compounded quarterly for 10 years. Find the total amount at the end of 10 years.

Solution: "Compounded quarterly" means 4 times a year, so

$n = 4$. Also, $P = \$4000$, $r = 5.3\% = 0.053$, and $t = 10$ years.

$$A = P\left(1 + \frac{r}{n}\right)^{n \cdot t}$$

$$= 4000\left(1 + \frac{0.053}{4}\right)^{4 \cdot 10}$$

$$= 4000(1.01325)^{40}$$

$$\approx 6772.12$$

The total amount after 10 years is \$6772.12.

Work Practice Problem 5

PRACTICE PROBLEM 5

\$5500 is invested at $6\frac{1}{4}\%$ compounded *daily* for 5 years. Find the total amount at the end of 5 years. (Use 1 year = 365 days.)

Answers
4. \$3795.96 **5.** \$7517.41

CALCULATOR EXPLORATIONS Compound Interest Formula

For a review of using your calculator to evaluate compound interest, see this box.

Let's review the calculator keys pressed to evaluate the expression in Example 5,

$$4000\left(1 + \frac{0.053}{4}\right)^{40}$$

To evaluate, press the keys [4000] [×] [(] [1] [+] [0.053] [÷] [4] [)] [y^x] or [∧] [40] [=] or [ENTER]. The display will read [6772.117549]. Rounded to 2 decimal places, this is 6772.12.

Find the compound interest.

1. \$600, 5 years, 9%, compounded quarterly

2. \$10,000, 15 years, 4%, compounded daily

3. \$1200, 20 years, 11%, compounded annually

4. \$5800, 1 year, 7%, compounded semiannually

5. \$500, 4 years, 6%, compounded quarterly.

6. \$2500, 19 years, 5%, compounded daily.

Vocabulary and Readiness Check

Use the choices below to fill in each blank. Choices may be used more than once.

total amount   simple   compound

1. To calculate ________ interest, use $I = P \cdot R \cdot T$.

2. To calculate ________ interest, use $A = P\left(1 + \frac{r}{n}\right)^{n \cdot t}$.

3. ________ interest is computed on not only the original principal, but on interest already earned in previous compounding periods.

4. When interest is computed on the original principal only, it is called ________ interest.

5. ________ (paid or received) = principal + interest.

6.7 EXERCISE SET

FOR EXTRA HELP

Objective **A** *Find the simple interest. See Examples 1 and 2.*

	Principal	Rate	Time
1.	\$200	8%	2 years
3.	\$160	11.5%	4 years
5.	\$5000	10%	$1\frac{1}{2}$ years
7.	\$375	18%	6 months
9.	\$2500	16%	21 months

	Principal	Rate	Time
2.	\$800	9%	3 years
4.	\$950	12.5%	5 years
6.	\$1500	14%	$2\frac{1}{4}$ years
8.	\$775	15%	8 months
10.	\$1000	10%	18 months

Solve. See Examples 1 through 3.

11. A company borrows \$162,500 for 5 years at a simple interest of 12.5% to buy an airplane. Find the interest paid on the loan.

12. \$265,000 is borrowed to buy a house. If the simple interest rate on the 30-year loan is 8.25%, find the interest paid on the loan.

13. A money market fund advertises a simple interest rate of 9%. Find the total amount received on an investment of $5000 for 15 months.

14. The Real Service Company takes out a 270-day (9-month) short-term, simple interest loan of $4500 to finance the purchase of some new equipment. If the interest rate is 14%, find the total amount that the company pays back.

15. Marsha borrows $8500 and agrees to pay it back in 4 years. If the simple interest rate is 17%, find the total amount she pays back.

16. An 18-year-old is given a high school graduation gift of $2000. If this money is invested at 8% simple interest for 5 years, find the total amount.

Objective B *Find the total amount in each compound interest account. See Examples 4 and 5.*

17. $6150 is compounded semiannually at a rate of 14% for 15 years.

18. $2060 is compounded annually at a rate of 15% for 10 years.

19. $1560 is compounded daily at a rate of 8% for 5 years.

20. $1450 is compounded quarterly at a rate of 10% for 15 years.

21. $10,000 is compounded semiannually at a rate of 9% for 20 years.

22. $3500 is compounded daily at a rate of 8% for 10 years.

23. $2675 is compounded annually at a rate of 9% for 1 year.

24. $6375 is compounded semiannually at a rate of 10% for 1 year.

25. $2000 is compounded annually at a rate of 8% for 5 years.

26. $2000 is compounded semiannually at a rate of 8% for 5 years.

27. $2000 is compounded quarterly at a rate of 8% for 5 years.

28. $2000 is compounded daily at a rate of 8% for 5 years.

Review

Find the perimeter of each figure. See Section 1.3.

29.

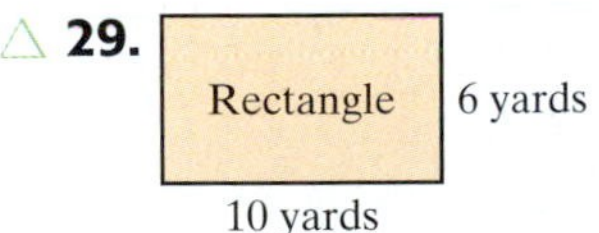

30.

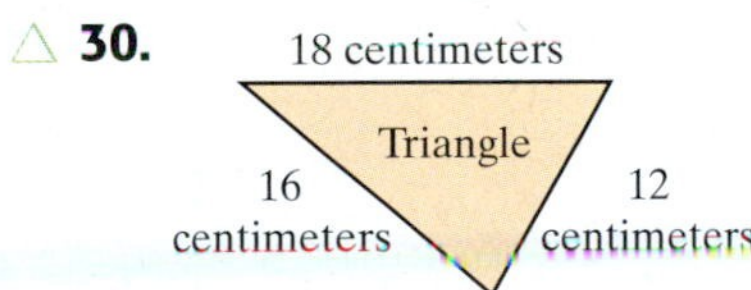

31.

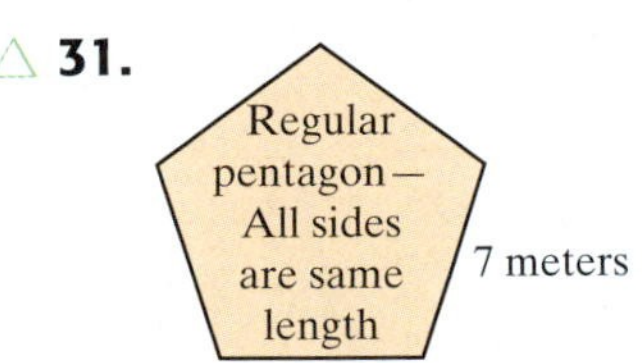

32.

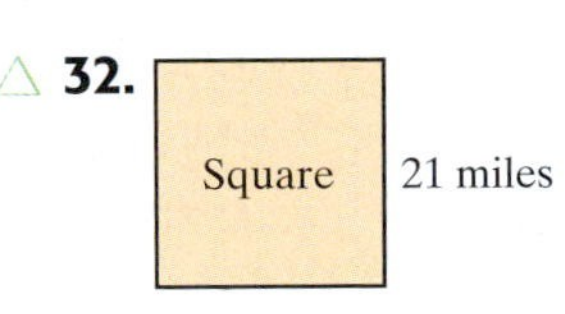

Perform the indicated operations. See Sections 4.3 through 4.6.

33. $\frac{x}{4} + \frac{x}{5}$

34. $-\frac{x}{4} \div \left(-\frac{x}{5}\right)$

35. $\left(\frac{2}{3}\right)\left(-\frac{1}{3}\right) - \left(\frac{9}{10}\right)\left(\frac{2}{5}\right)$

36. $\frac{3}{11} \div \frac{9}{22} \cdot \frac{1}{3}$

Concept Extensions

37. Review Exercises 25 through 28. As the number of compoundings increases per year, how is the total amount affected?

38. Explain how to find the amount of interest in a compounded account.

39. Compare the following accounts: Account 1: $1000 is invested for 10 years at a simple interest rate of 6%. Account 2: $1000 is compounded semiannually at a rate of 6% for 10 years. Discuss how the interest is computed for each account. Determine which account earns more interest. Why?

CHAPTER 6 Group Activity

Investigating Scale Drawings

Section 6.1

Materials:

- ruler
- tape measure
- grid paper (optional)

This activity may be completed by working in groups or individually.

Scale drawings are used by architects, engineers, interior designers, ship builders, and others. In a scale drawing, each unit measurement on the drawing represents a fixed length on the object being drawn. For instance, in an architect's scale drawing, 1 inch on the drawing may represent 10 feet on a building. The scale describes the relationship between the measurements. If the measurements have the same units, the scale can be expressed as a ratio. In this case, the ratio would be 1 : 120, representing 1 inch to 120 inches (or 10 feet).

Use a ruler and the scale drawing of a college building below to answer the following questions.

1. How wide are each of the front doors of the college building?
2. How long is the front of the college building?
3. How tall is the front of the college building?

Now you will draw your own scale floor plan. First choose a room to draw—it can be your math classroom, your living room, your dormitory room, or any room that can be easily measured. Start by using a tape measure to measure the distances around the base of the walls in the room you are drawing.

4. Choose a scale for your floor plan.
5. Convert each measurement in the room you are drawing to the corresponding lengths needed for the scale drawing.
6. Complete your floor plan (you may find it helpful to use grid paper). Mark the locations of doors and windows on your floor plan. Be sure to indicate on the drawing the scale used in your floor plan.

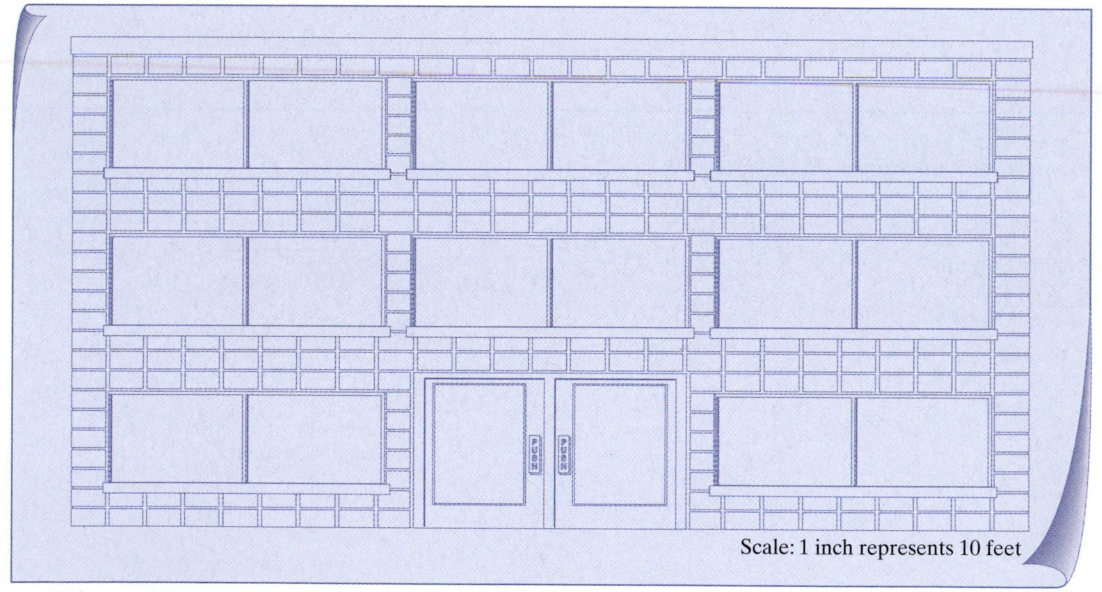

Chapter 6 Vocabulary Check

Fill in each blank with one of the words or phrases listed below. Some words may be used more than once.

percent	sales tax	is	0.01	$\frac{1}{100}$	amount of discount
percent of decrease	total price	ratio	proportion	base	of
amount	100%	compound interest	percent of increase	sale price	commission

1. In a mathematical statement, ______________ usually means "multiplication."
2. In a mathematical statement, ______________ means "equals."
3. ______________ means "per hundred."
4. ______________ is computed not only on the principal, but also on interest already earned in previous compounding periods.

5. In the percent proportion $\underline{\hspace{3cm}} = \frac{\text{percent}}{100}$.

6. To write a decimal or fraction as a percent, multiply by ______________.

7. The decimal equivalent of the % symbol is ______________.

8. The fraction equivalent of the % symbol is ______________.

9. The percent equation is ______________ · percent = ______________.

10. $\underline{\hspace{3cm}} = \frac{\text{amount of decrease}}{\text{original amount}}$.

11. $\underline{\hspace{3cm}} = \frac{\text{amount of increase}}{\text{original amount}}$.

12. ______________ = tax rate · purchase price.

13. ______________ = purchase price + sales tax.

14. ______________ = commission rate · sales.

15. ______________ = discount rate · original price.

16. ______________ = original price − amount of discount.

17. A ______________ is a mathematical statement that two ratios are equal.

18. A ______________ is the quotient of two numbers or two quantities.

Are you preparing for your test? Don't forget to take the Chapter 6 Test on page 485. Then check your answers at the back of the text and use the Chapter Test Prep Video CD to see the fully worked-out solutions to any of the exercises you want to review.

6 Chapter Highlights

Definitions and Concepts	**Examples**
Section 6.1 Ratio and Proportion	
A **ratio** is the quotient of two numbers or two quantities.	Write the ratio of 5 hours to 1 day using fractional notation. $\frac{5\text{ hours}}{1\text{ day}} = \frac{5\text{ hours}}{24\text{ hours}} = \frac{5}{24}$
A **proportion** is a mathematical statement that two ratios are equal. In the proportion $\frac{a}{b} = \frac{c}{d}$, the products ad and bc are called **cross products.**	$\frac{2}{3} = \frac{8}{12} \quad \frac{x}{7} = \frac{15}{35}$ $\frac{2}{3} = \frac{8}{12}$ → $3 \cdot 8$ or 24; → $2 \cdot 12$ or 24
If $\frac{a}{b} = \frac{c}{d}$ then $ad = bc$.	Solve: $\frac{3}{4} = \frac{x}{x-1}$ $\frac{3}{4} = \frac{x}{x-1}$ $3(x-1) = 4x$ Set cross products equal. $3x - 3 = 4x$ $-3 = x$

Definitions and Concepts	Examples

Section 6.2 Percents, Decimals, and Fractions

Percent means "per hundred." The % symbol denotes percent.

$$51\% = \frac{51}{100} \quad 51 \text{ per } 100$$

$$7\% = \frac{7}{100} \quad 7 \text{ per } 100$$

To write a percent as a decimal, replace the % symbol with its decimal equivalent, 0.01, and multiply.

To write a decimal as a percent, multiply by 100%.

$$32\% = 32(0.01) = 0.32$$

$$0.08 = 0.08(100\%) = 08.\% = 8\%$$

To write a percent as a fraction, replace the % symbol with its fraction equivalent, $\frac{1}{100}$, and multiply.

$$25\% = 25 \cdot \frac{1}{100} = \frac{25}{100} = \frac{\cancel{25}^{1}}{4 \cdot \cancel{25}_{1}} = \frac{1}{4}$$

To write a fraction as a percent, multiply by 100%.

$$\frac{1}{6} = \frac{1}{6} \cdot 100\% = \frac{1}{6} \cdot \frac{100}{1}\% = \frac{100}{6}\% = 16\frac{2}{3}\%$$

Section 6.3 Solving Percent Problems with Equations

Three key words in the statement of a percent problem are

of, which means multiplication (·)

is, which means equals (=)

what (or some equivalent word or phrase), which stands for the unknown (x)

Solve:

6 is 12% of what number?

$$6 = 12\% \cdot x$$

$$6 = 0.12 \cdot x \quad \text{Write 12\% as a decimal.}$$

$$\frac{6}{0.12} = \frac{0.12 \cdot x}{0.12} \quad \text{Divide both sides by 0.12.}$$

$$50 = x$$

Thus, 6 is 12% of 50.

Section 6.4 Solving Percent Problems with Proportions

Percent Proportion

$$\frac{\text{amount}}{\text{base}} = \frac{\text{percent}}{100} \leftarrow \text{always 100}$$

or

$$\text{amount} \rightarrow \frac{a}{b} = \frac{p}{100} \leftarrow \text{percent}$$
(base → b)

Solve:

20.4 is what percent of 85?

amount | percent | base

$$\text{amount} \rightarrow \frac{20.4}{85} = \frac{p}{100} \leftarrow \text{percent}$$
(base → 85)

$$20.4 \cdot 100 = 85 \cdot p \quad \text{Set cross products equal.}$$

$$2040 = 85 \cdot p \quad \text{Multiply.}$$

$$\frac{2040}{85} = \frac{85 \cdot p}{85} \quad \text{Divide both sides by 85.}$$

$$24 = p \quad \text{Simplify.}$$

Thus, 20.4 is 24% of 85.

DEFINITIONS AND CONCEPTS	EXAMPLES
Section 6.5 Applications of Percent	
PERCENT INCREASE $\text{percent increase} = \dfrac{\text{amount of increase}}{\text{original amount}}$ **PERCENT DECREASE** $\text{percent decrease} = \dfrac{\text{amount of decrease}}{\text{original amount}}$	A town with a population of 16,480 decreased to 13,870 over a 12-year period. Find the percent decrease. Round to the nearest whole percent. $\text{amount of decrease} = 16{,}480 - 13{,}870 = 2610$ $\text{percent decrease} = \dfrac{\text{amount of decrease}}{\text{original amount}} = \dfrac{2610}{16{,}480} \approx 0.16 = 16\%$ The town's population decreased by about 16%.
Section 6.6 Percent and Problem Solving: Sales Tax, Commission, and Discount	
SALES TAX sales tax = sales tax rate · purchase price total price = purchase price + sales tax	Find the sales tax and the total price of a purchase of \$42.00 if the sales tax rate is 9%. sales tax = sales tax rate · purchase price ↓ ↓ ↓ sales tax = 9% · \$42 = 0.09 · \$42 = \$3.78 The total price is total price = purchase price + sales tax ↓ ↓ ↓ total price = \$42.00 + \$3.78 = \$45.78 The total price is \$45.78.
COMMISSION commission = commission rate · sales	A salesperson earns a commission of 3%. Find the commission from sales of \$12,500 worth of appliances. commission = commission rate · sales ↓ ↓ ↓ commission = 3% · \$12,500 = 0.03 · 12,500 = \$375 The commission is \$375.

Definitions and Concepts	**Examples**

Section 6.6 Percent and Problem Solving: Sales Tax, Commission, and Discount (*continued*)

Discount and Sale Price

amount of discount = discount rate · original price

sale price = original price − amount of discount

A suit is priced at \$320 and is on sale today for 25% off. What is the sale price?

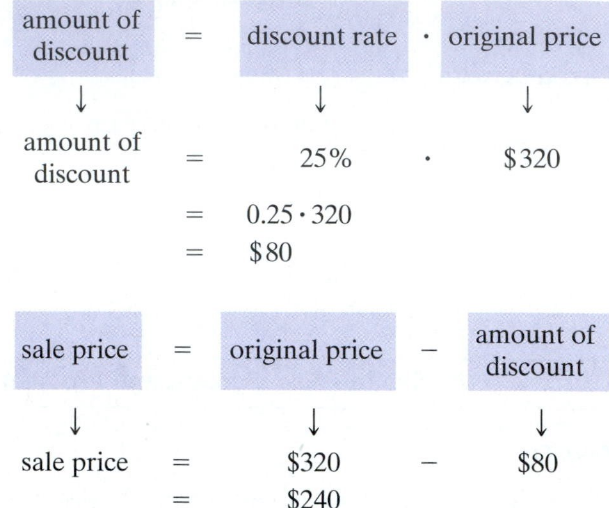
amount of discount = discount rate · original price
amount of discount = 25% · \$320
= 0.25 · 320
= \$80
sale price = original price − amount of discount
sale price = \$320 − \$80
= \$240

The sale price is \$240.

Section 6.7 Percent and Problem Solving: Interest

Simple Interest

interest = principal · rate · time

where the rate is understood to be per year, unless told otherwise.

Find the simple interest after 3 years on \$800 at an interest rate of 5%.

interest = principal · rate · time

↓ ↓ ↓ ↓

interest = \$800 · 5% · 3

= \$800 · 0.05 · 3   Write 5% as 0.05.

= \$120   Multiply.

The interest is \$120.

Compound Interest Formula

The total amount A in an account is

$$A = P\left(1 + \frac{r}{n}\right)^{n \cdot t}$$

where P is the principal, r is the interest rate written as a decimal, t is time in years, and n is the number of times compounded per year.

\$800 is invested at 5% compounded quarterly for 10 years. Find the total amount at the end of 10 years.

"Compounded quarterly" means $n = 4$. Also, $P = \$800$, $r = 5\% = 0.05$, and $t = 10$ years.

$$\begin{aligned} A &= P\left(1 + \frac{r}{n}\right)^{n \cdot t} \\ &= 800\left(1 + \frac{0.05}{4}\right)^{4 \cdot 10} \\ &= 800(1.0125)^{40} \\ &\approx 1314.90 \end{aligned}$$

The total amount is \$1314.90.

STUDY SKILLS BUILDER

Are You Prepared for a Test on Chapter 6?

Below I have listed some *common trouble areas* for students in Chapter 6. After studying for your test—but before taking your test—read these.

- Can you convert from percents to fractions or decimals and from fractions or decimals to percents?

  Percent to decimal: $7.5\% = 7.5(0.01) = 0.075$

  Percent to fraction: $11\% = 11 \cdot \frac{1}{100} = \frac{11}{100}$

  Decimal to percent: $0.36 = 0.36(100\%) = 36\%$

  Fraction to percent: $\frac{6}{7} = \frac{6}{7} \cdot 100\% = \frac{6}{7} \cdot \frac{100}{1}\%$

  $= \frac{600}{7}\% = 85\frac{5}{7}\%$

- Do you remember how to find percent increase or percent decrease? The number of CDs increased from 40 to 48. Find the percent increase.

  $$\frac{\text{percent}}{\text{increase}} = \frac{\text{increase}}{\text{original number}} = \frac{8}{40} = 0.20 = 20\%$$

Remember: This is simply a checklist of common trouble areas. For a review of Chapter 6, see the Highlights and Chapter Review.

6 CHAPTER REVIEW

(6.1) *Write each phrase as a ratio in fractional notation.*

1. 20 cents to 1 dollar

2. four parts red to six parts white

Solve each proportion.

3. $\frac{x}{2} = \frac{12}{4}$

4. $\frac{20}{1} = \frac{x}{25}$

5. $\frac{32}{100} = \frac{100}{x}$

6. $\frac{20}{2} = \frac{c}{5}$

7. $\frac{2}{x-1} = \frac{3}{x+3}$

8. $\frac{4}{y-3} = \frac{3}{y+2}$

9. $\frac{y+2}{y} = \frac{5}{3}$

10. $\frac{x-3}{3x+2} = \frac{2}{5}$

11. A machine can process 300 parts in 20 minutes. Find how many parts can be processed in 45 minutes.

12. As his consulting fee, Mr. Visconti charges $90.00 per day. Find how much he charges for 3 hours of consulting. Assume an 8-hour work day.

(6.2) *Solve.*

13. In a survey of 100 adults, 37 preferred pepperoni on their pizzas. What percent preferred pepperoni?

14. A basketball player made 77 out 100 attempted free throws. What percent of free throws was made?

Write each percent as a decimal.

15. 26%

16. 75%

17. 3.5%

18. 1.5%

19. 275%

20. 400%

21. 47.85%

22. 85.34%

Write each decimal as a percent.

23. 1.6

24. 0.055

25. 0.076

26. 0.085

27. 0.71

28. 0.65

29. 6

30. 9

Write each percent as a fraction or mixed number in simplest form.

31. 7%

32. 15%

33. 25%

34. 8.5%

35. 10.2%

36. $16\frac{2}{3}\%$

37. $33\frac{1}{3}\%$

38. 110%

Write each fraction or mixed number as a percent.

39. $\frac{2}{5}$

40. $\frac{7}{10}$

41. $\frac{7}{12}$

42. $1\frac{2}{3}$

43. $1\frac{1}{4}$

44. $\frac{3}{5}$

45. $\frac{1}{16}$

46. $\frac{5}{8}$

(6.3) Translating *Translate each to an equation and solve.*

47. 1250 is 1.25% of what number?

48. What number is $33\frac{1}{3}\%$ of 24,000?

49. 124.2 is what percent of 540?

50. 22.9 is 20% of what number?

51. What number is 17% of 640?

52. 693 is what percent of 462?

(6.4) Translating *Translate each to a proportion and solve.*

53. 104.5 is 25% of what number?

54. 16.5 is 5.5% of what number?

55. What number is 30% of 532?

56. 63 is what percent of 35?

57. 93.5 is what percent of 85?

58. What number is 33% of 500?

(6.5) *Solve.*

59. In a survey of 2000 people, it was found that 1320 have a microwave oven. Find the percent of people who own microwaves.

60. Of the 12,360 freshmen entering County College, 2000 are enrolled in prealgebra. Find the percent of entering freshmen who are enrolled in prealgebra. Round to the nearest whole percent.

61. The number of violent crimes in a city decreased from 675 to 534. Find the percent decrease. Round to the nearest tenth of a percent.

62. The current charge for dumping waste in a local landfill is \$16 per cubic foot. To cover new environmental costs, the charge will increase to \$33 per cubic foot. Find the percent increase.

63. This year the fund drive for a charity collected \$215,000. Next year, a 4% decrease is expected. Find how much is expected to be collected in next year's drive.

64. A local union negotiated a new contract that increases the hourly pay 15% over last year's pay. The old hourly rate was \$11.50. Find the new hourly rate rounded to the nearest cent.

(6.6) *Solve.*

65. If the sales tax rate is 5.5%, what is the total amount charged for a \$250 coat?

66. Find the sales tax paid on a \$25.50 purchase if the sales tax rate is 4.5%.

67. Russ James is a sales representative for a chemical company and is paid a commission rate of 5% on all sales. Find his commission if he sold \$100,000 worth of chemicals last month.

68. Carol Sell is a sales clerk in a clothing store. She receives a commission of 7.5% on all sales. Find her commission for the week if her sales for the week were \$4005. Round to the nearest cent.

69. A \$3000 mink coat is on sale for 30% off. Find the discount and the sale price.

70. A \$90 calculator is on sale for 10% off. Find the discount and the sale price.

(6.7) *Solve.*

71. Find the simple interest due on \$4000 loaned for 4 months at 12% interest.

72. Find the simple interest due on \$6500 loaned for 3 months at 20%.

73. Find the total amount in an account if \$5500 is compounded annually at 12% for 15 years.

74. Find the total amount in an account if \$6000 is compounded semiannually at 11% for 10 years.

75. Find the total amount in an account if \$100 is compounded quarterly at 12% for 5 years.

76. Find the total amount in an account if \$1000 is compounded quarterly at 18% for 20 years.

Mixed Review

Write each percent as a decimal.

77. 3.8%

78. 124.5%

Write each decimal as a percent.

79. 0.54

80. 95.2

Write each percent as a fraction or mixed number in simplest form.

81. 47%

82. 5.6%

Write each fraction or mixed number as a percent.

83. $\frac{3}{8}$

84. $\frac{6}{5}$

Translating *Translate each into an equation and solve.*

85. 43 is 16% of what number?

86. 27.5 is what percent of 25?

87. What number is 36% of 1968?

88. 67 is what percent of 50?

Translating *Translate each into a proportion and solve.*

89. 75 is what percent of 25?

90. What number is 16% of 240?

91. 28 is 5% of what number?

92. 52 is what percent of 16?

Solve.

93. The total number of cans in a soft drink machine is 300. If 78 soft drinks have been sold, find the percent of soft drink cans that have been sold.

94. A home valued at $96,950 last year has lost 7% of its value this year. Find the loss in value.

95. A dinette set sells for $568.00. If the sales tax rate is 8.75%, find the purchase price of the dinette set.

96. The original price of a video game is $23.00. It is on sale for 15% off. What is the amount of the discount?

97. A candy salesman makes a commission of $1.60 from each case of candy he sells. If a case of candy costs $12.80, what is his rate of commission?

98. Find the total amount due on a 6 month loan of $1400 at a simple interest rate of 13%.

99. Find the total amount due on a loan of $5,500 for 9 years at 12.5% simple interest.

6 CHAPTER TEST

Use the Chapter Test Prep Video CD to see the fully worked-out solutions to any of the exercises you want to review.

Answers

Write each percent as a decimal.

1. 85%

2. 500%

3. 0.6%

Write each decimal as a percent.

4. 0.056

5. 6.1

6. 0.35

Write each percent as a fraction or a mixed number in simplest form.

7. 120%

8. 38.5%

9. 0.2%

Write each fraction or mixed number as a percent.

10. $\frac{11}{20}$

11. $\frac{3}{8}$

12. $1\frac{3}{4}$

13. Sales of bottled water have recently surged. Bottled water accounts for $\frac{1}{5}$ of the total noncarbonated beverage category in convenience stores. Write $\frac{1}{5}$ as a percent. (*Source:* Grocery Manufacturers of America)

14. In small firms in the United States, 64% of full-time employees receive medical insurance benefits. Write 64% as a fraction. (*Source:* U.S. Bureau of Labor Statistics)

Solve.

15. What number is 42% of 80?

16. 0.6% of what number is 7.5?

17. 567 is what percent of 756?

1. __________
2. __________
3. __________
4. __________
5. __________
6. __________
7. __________
8. __________
9. __________
10. __________
11. __________
12. __________
13. __________
14. __________
15. __________
16. __________
17. __________

18. ______
19. ______
20. ______
21. ______
22. ______
23. ______
24. ______
25. ______
26. ______
27. ______
28. ______
29. ______
30. ______
31. ______

Solve. If necessary, round percents to the nearest tenth, dollar amounts to the nearest cent, and all other numbers to the nearest whole.

18. An alloy is 12% copper. How much copper is contained in 320 pounds of this alloy?

19. A farmer in Nebraska estimates that 20% of his potential crop, or \$11,350, has been lost to a hard freeze. Find the total value of his potential crop.

20. If the local sales tax rate is 1.25%, find the total amount charged for a stereo system priced at \$354.

21. A town's population increased from 25,200 to 26,460. Find the percent increase.

22. A \$120 framed picture is on sale for 15% off. Find the discount and the sale price.

23. Randy Nguyen is paid a commission rate of 4% on all sales. Find Randy's commission if his sales were \$9875.

24. A sales tax of \$1.53 is added to an item's price of \$152.99. Find the sales tax rate. Round to the nearest whole percent.

25. Find the simple interest earned on \$2000 saved for $3\frac{1}{2}$ years at an interest rate of 9.25%.

26. \$1365 is compounded annually at 8%. Find the total amount in the account after 5 years.

27. A couple borrowed \$400 from a bank at 13.5% simple interest for 6 months for car repairs. Find the total amount due the bank at the end of the 6-month period.

28. The number of crimes reported in New York City was 125,587 in the first half of 2001 and 118,346 in the first half of 2002. Find the percent decrease in the number of crimes in New York City from 2001 to 2002. (*Source:* Federal Bureau of Investigation. Uniform Crime Reports, January–June 2002)

29. Write the ratio \$75 to \$10 as a fraction in simplest form.

30. Solve: $\frac{5}{y + 1} = \frac{4}{y + 2}$

31. In a sample of 85 fluorescent bulbs, 3 were found to be defective. At this rate, how many defective bulbs should be found in 510 bulbs?

CUMULATIVE REVIEW

Chapters 1–6

1. Multiply: 236×86

2. Multiply: 409×76

3. Subtract: 7 from -3.

4. Subtract: -2 from 8.

5. Solve: $x - 2 = -1$

6. Solve: $x + 4 = 3$

7. Solve: $3(2x - 6) + 6 = 0$

8. Solve: $5(x - 2) = 3x$

9. Write an equivalent fraction with the given denominator.

$3 = \frac{}{7}$

10. Write an equivalent fraction with the given denominator.

$8 = \frac{}{5}$

11. Simplify: $-\frac{10}{27}$

12. Simplify: $\frac{10y}{32}$

13. Divide: $-\frac{5}{16} \div -\frac{3}{4}$

14. Divide: $\frac{-2}{5} \div \frac{7}{10}$

15. Evaluate $y - x$ if $x = -\frac{3}{10}$ and $y = -\frac{8}{10}$.

16. Evaluate $2x + 3y$ if $x = \frac{2}{5}$ and $y = \frac{-1}{5}$.

17. Find: $-\frac{3}{4} - \frac{1}{14} + \frac{6}{7}$

18. Find: $\frac{2}{9} + \frac{7}{15} - \frac{1}{3}$

19. Simplify: $\dfrac{\frac{1}{2} + \frac{3}{8}}{\frac{3}{4} - \frac{1}{6}}$

20. Simplify: $\dfrac{\frac{2}{3} + \frac{1}{6}}{\frac{3}{4} - \frac{3}{5}}$

21. Solve: $\frac{x}{2} = \frac{x}{3} + \frac{1}{2}$

22. Solve: $\frac{x}{2} + \frac{1}{5} = 3 - \frac{x}{5}$

Answers

1. ______
2. ______
3. ______
4. ______
5. ______
6. ______
7. ______
8. ______
9. ______
10. ______
11. ______
12. ______
13. ______
14. ______
15. ______
16. ______
17. ______
18. ______
19. ______
20. ______
21. ______
22. ______

23. a. ______
b. ______
24. a. ______
b. ______
25. ______
26. ______
27. ______
28. ______
29. ______
30. ______
31. ______
32. ______
33. ______
34. ______
35. ______
36. ______
37. ______
38. ______
39. ______
40. ______
41. ______
42. ______
43. ______
44. ______
45. ______
46. ______
47. ______
48. ______
49. ______
50. ______

23. Write each mixed number as an improper fraction.

a. $4\frac{2}{9}$

b. $1\frac{8}{11}$

24. Write each mixed number as an improper fraction.

a. $3\frac{2}{5}$

b. $6\frac{2}{7}$

Write each decimal as a fraction or mixed number. Write your answer in simplest form.

25. 0.125

26. 0.85

27. −105.083

28. 17.015

29. Subtract: $85 - 17.31$

30. Subtract: $38 - 10.06$

Multiply.

31. 7.68×10

32. 12.483×100

33. $(-76.3)(1000)$

34. -853.75×10

35. Evaluate $x \div y$ for $x = 2.5$ and $y = 0.05$.

36. Is 470 a solution of the equation $\frac{x}{100} = 4.75$?

37. Find the median of the scores: 67, 91, 75, 86, 55, 91

38. Find the mean of 36, 40, 86, and 30.

39. Write the ratio of 2.5 to 3.15 as a fraction in simplest form.

40. Write the ratio of 5.8 to 7.6 as a fraction in simplest form.

41. Solve for x: $\frac{45}{x} = \frac{5}{7}$

42. Solve for x: $\frac{x - 1}{3x + 1} = \frac{3}{8}$

43. On a Chamber of Commerce map of Abita Springs, 5 miles corresponds to 2 inches. How many miles correspond to 7 inches?

44. A student doing math homework can complete 7 problems in about 6 minutes. At this rate, how many problems can be completed in 30 minutes?

Write each percent as a fraction or mixed number in simplest form.

45. 1.9%

46. 2.3%

47. $33\frac{1}{3}\%$

48. 108%

49. What number is 35% of 60?

50. What number is 42% of 85?

7 Graphs and Triangle Applications

This chapter reviews presenting data in a usable form on a graph and the basic ideas of probability. Also in this chapter, we introduce important triangle applications, such as the Pythagorean theorem and the concept of congruent and similar triangles.

SECTIONS

CHECK YOUR PROGRESS

Hybrid cars are becoming more widespread in America each year. New models are being offered by an ever-growing number of auto manufacturers. Concern for the environment, rising costs of gasoline, and financial incentives offered by the federal and state governments make hybrid cars a more attractive alternative to regular gasoline engine vehicles. While these vehicles are not the all-electric cars that futurists used to predict, the innovative use of electric and gasoline power makes these hybrids very energy efficient. We explore hybrid car data in Section 7.2, Exercise 27 and the Chapter Review, Exercises 71–74.

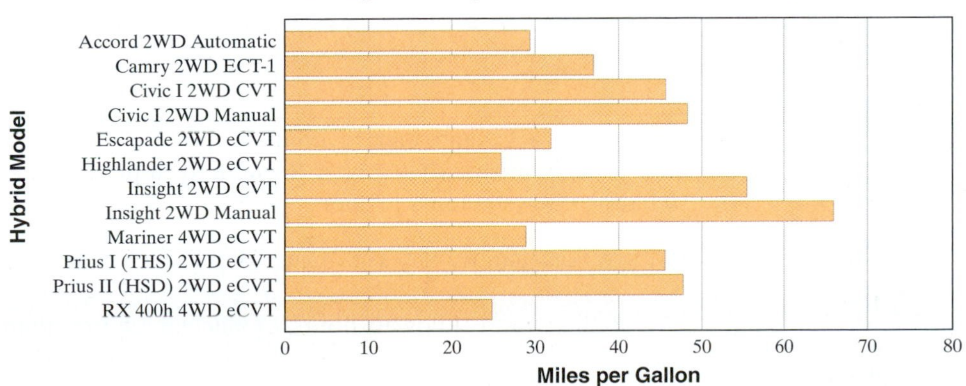

Source: http://www.greenhybrid.com

7.1 READING PICTOGRAPHS, BAR GRAPHS, HISTOGRAMS, AND LINE GRAPHS

Objectives

A Read Pictographs.

B Read and Construct Bar Graphs.

C Read and Construct Histograms.

D Read Line Graphs.

Often data is presented visually in a graph. In this section, we practice reading several kinds of graphs including pictographs, bar graphs, and line graphs.

Objective A Reading Pictographs

A **pictograph** such as the one below is a graph in which pictures or symbols are used. This type of graph contains a key that explains the meaning of the symbol used. An advantage of using a pictograph to display information is that comparisons can easily be made. A disadvantage of using a pictograph is that it is often hard to tell what fractional part of a symbol is shown. For example, in the pictograph below, Sweden shows a part of a symbol, but it's hard to read with any accuracy what fractional part of a symbol is shown.

PRACTICE PROBLEM 1

Use the pictograph shown in Example 1 to answer the following questions:

a. Approximate the amount of nuclear energy that was generated in Sweden.

b. Approximate the total nuclear energy generated in Sweden and Germany.

EXAMPLE 1 Calculating Nuclear Energy Generated

The following pictograph shows the approximate amount of nuclear energy generated by selected countries in the year 2005. Use this pictograph to answer the questions.

Nuclear Energy Generated by Selected Countries (2005)

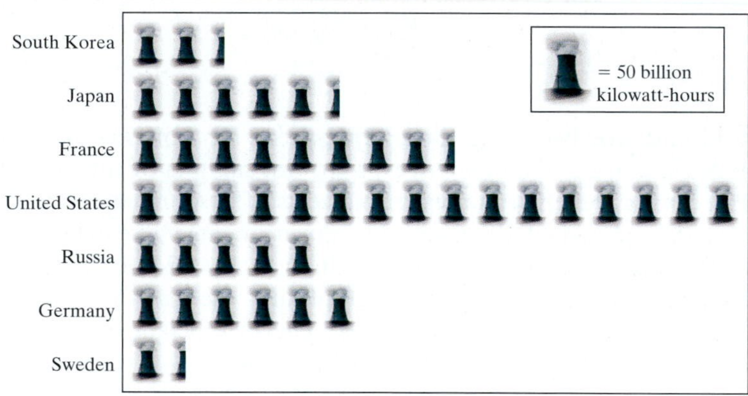

a. Approximate the amount of nuclear energy that was generated in Russia.

b. Approximate how much more nuclear energy was generated in France than in Russia.

Solution:

a. Russia corresponds to 5 symbols, and each symbol represents 50 billion kilowatt-hours of energy. This means that Russia generated approximately $5 \cdot (50 \text{ billion})$ or 250 billion kilowatt-hours of energy.

b. France shows $3\frac{1}{2}$ more symbols than Russia. This means that France generated $3\frac{1}{2} \cdot (50 \text{ billion})$ or 175 billion more kilowatt-hours of nuclear energy than Russia.

Work Practice Problem 1

Answers

1. a. 75 billion kilowatt-hours **b.** 375 billion kilowatt-hours

Objective B Reading and Constructing Bar Graphs

Another way to visually present data is with a **bar graph.** Bar graphs can appear with vertical bars or horizontal bars. Although we have studied bar graphs in previous sections, we now practice reading the height or length of the bars contained in a bar graph. An advantage to using bar graphs is that a scale is usually included for greater accuracy. Care must be taken when reading bar graphs, as well as other types of graphs—they may be misleading, as shown later in this section.

EXAMPLE 2 Finding the Number of Endangered Species

The following bar graph shows the number of endangered species in the United States in 2005. Use this graph to answer the questions.

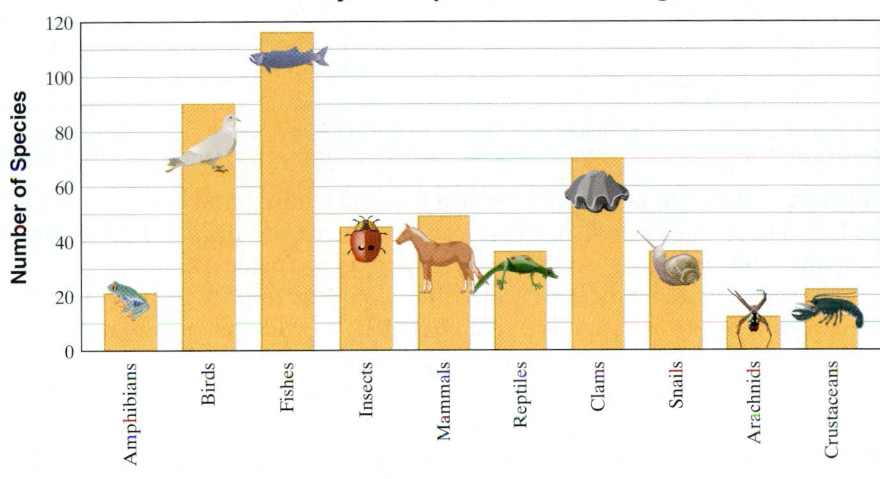

Source: U.S. Fish and Wildlife Service

a. Approximate the number of endangered species that are reptiles.

b. Which category has the most endangered species?

Solution:

a. To approximate the number of endangered species that are reptiles, we go to the top of the bar that represents reptiles. From the top of this bar, we move horizontally to the left until the scale is reached. We read the height of the bar on the scale as approximately 36. There are approximately 36 reptile species that are endangered, as shown.

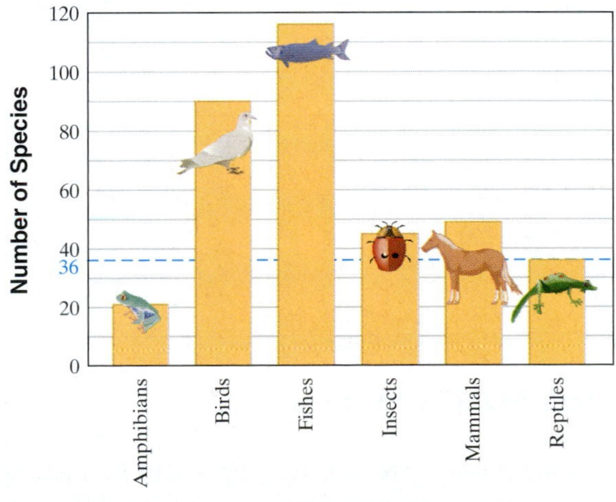

Source: U.S. Fish and Wildlife Service

b. The most endangered species is represented by the tallest (longest) bar. The tallest bar corresponds to fishes.

Work Practice Problem 2

PRACTICE PROBLEM 2

Use the bar graph in Example 2 to answer the following questions:

a. Approximate the number of endangered species that are insects.

b. Which category shows the fewest endangered species?

Answers

2. a. 45 **b.** arachnids

PRACTICE PROBLEM 3

Draw a vertical bar graph using the information in the table about electoral votes for selected states.

Total Electoral Votes by Selected States	
State	**Electoral Votes**
Texas	34
California	55
Florida	27
Nebraska	5
Indiana	11
Georgia	15
(*Source: World Almanac* 2006)	

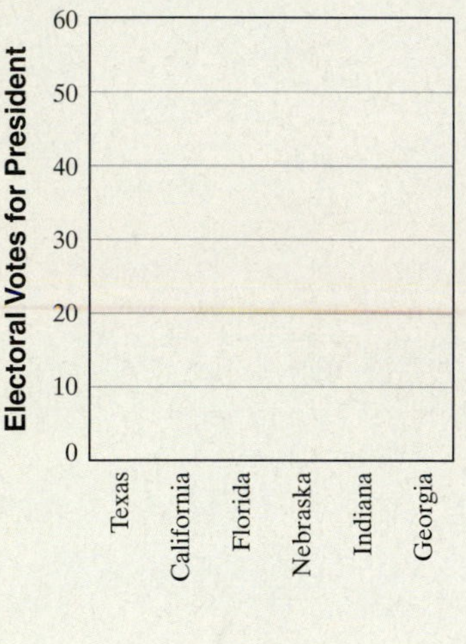

Answer

3.

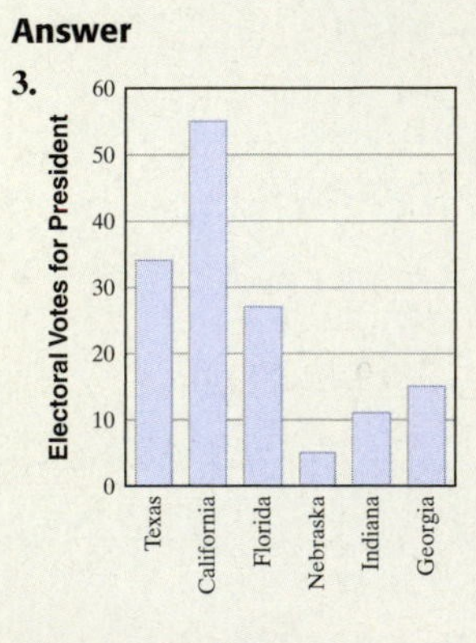

Next, we practice constructing a bar graph.

EXAMPLE 3 Draw a vertical bar graph using the information in the table below that gives the caffeine content of selected foods.

Average Caffeine Content of Selected Foods			
Food	**Milligrams**	**Food**	**Milligrams**
Brewed coffee (percolator, 8 ounces)	124	Instant coffee (8 ounces)	104
Brewed decaffeinated coffee (8 ounces)	3	Brewed tea (U.S. brands, 8 ounces)	64
Coca-Cola Classic (8 ounces)	31	Mr. Pibb (8 ounces)	27
Dark chocolate (semisweet, $1\frac{1}{2}$ ounces)	30	Milk chocolate (8 ounces)	9
(*Sources:* International Food Information Council and the Coca-Cola Company)			

Solution: We draw and label a vertical line and a horizontal line as shown next to the left. These lines are also called axes. We place the different food categories along the horizontal axis. Along the vertical axis, we place a scale.

There are many choices of scales that would be appropriate. Notice that the milligrams range from a low of 3 to a high of 124. From this information, we use a scale that starts at 0 and then shows multiples of 20 so that the scale is not too cluttered. The scale stops at 140, the smallest multiple of 20 that will allow all milligrams to be graphed. It may also be helpful to draw horizontal lines along the scale markings to help draw the vertical bars at the correct height. The finished bar graph is shown below.

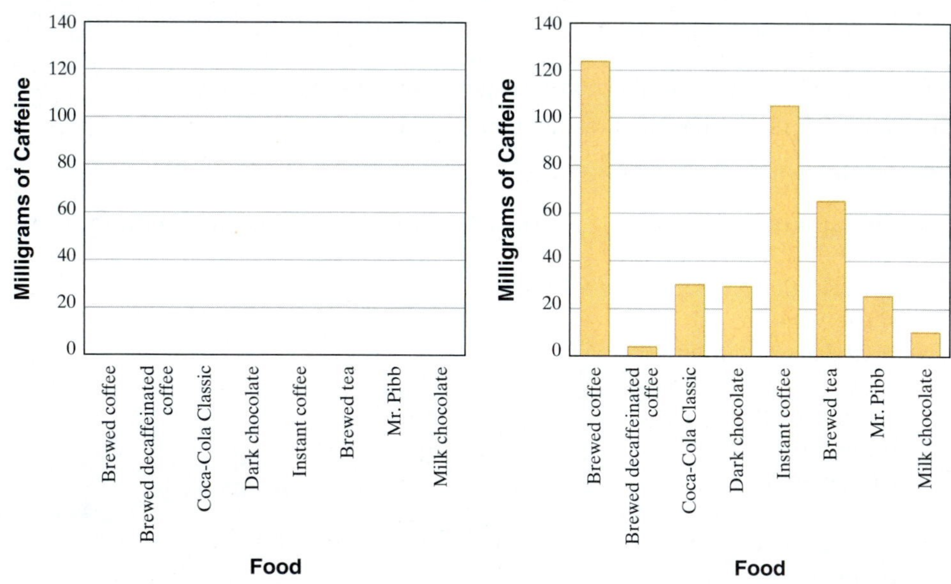

Work Practice Problem 3

As mentioned previously, graphs can be misleading. Both graphs on the next page show the same information, but with different scales. Special care should be taken when forming conclusions from the appearance of a graph.

Notice the ⚡ symbol on each vertical scale on the graphs. This symbol alerts us that numbers are missing on that scale

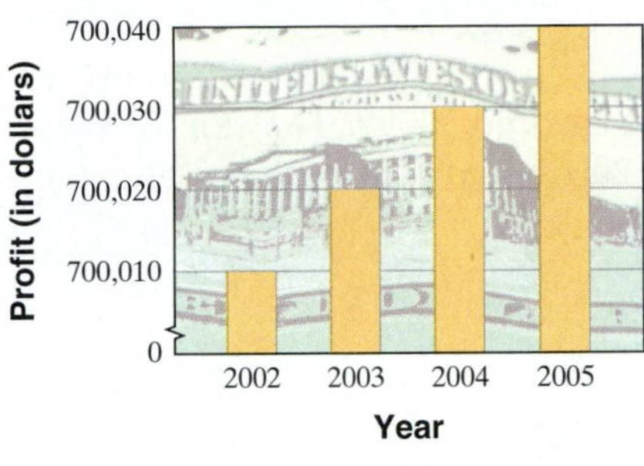

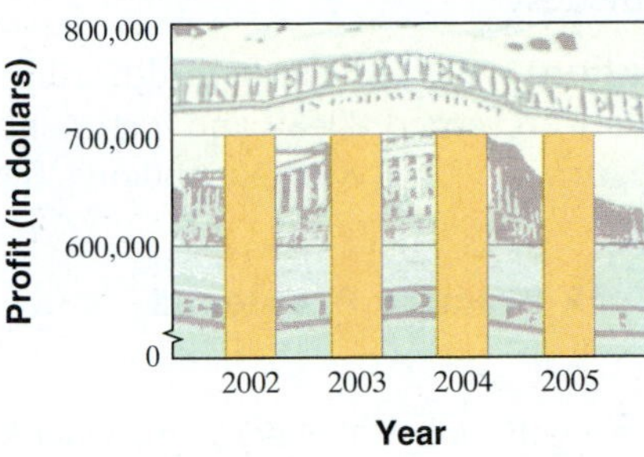

Are profits shown in the graphs above greatly increasing, or are they remaining about the same?

Objective C Reading and Constructing Histograms

Suppose that the test scores of 36 students are summarized in the table below:

Student Scores	Frequency (Number of Students)
40–49	1
50–59	3
60–69	2
70–79	10
80–89	12
90–99	8

The results in the table can be displayed in a histogram. A **histogram** is a special bar graph. The width of each bar represents a range of numbers called a **class interval.** The height of each bar corresponds to how many times a number in the class interval occurs and is called the **class frequency.** The bars in a histogram lie side by side with no space between them.

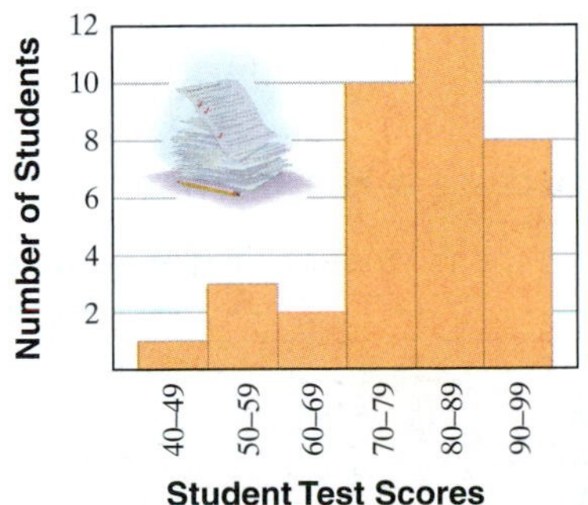

EXAMPLE 4 Reading a Histogram on Student Test Scores

Use the preceding histogram to determine how many students scored 50–59 on the test.

Solution: We find the bar representing 50–59. The height of this bar is 3, which means 3 students scored 50–59 on the test.

Work Practice Problem 4

PRACTICE PROBLEM 4

Use the histogram on the left to determine how many students scored 80–89 on the test.

Answer

4. 12

PRACTICE PROBLEM 5

Use the histogram from Example 4 to determine how many students scored less than 80 on the test.

PRACTICE PROBLEM 6

Complete the frequency distribution table for the data below. Each number represents a credit card owner's unpaid balance for one month.

0	53	89	125
265	161	37	76
62	201	136	42

Class Intervals (Credit Card Balances)	Tally	Class Frequency (Number of Months)
\$0–\$49	___	___
\$50–\$99	___	___
\$100–\$149	___	___
\$150–\$199	___	___
\$200–\$249	___	___
\$250–\$299	___	___

PRACTICE PROBLEM 7

Construct a histogram from the frequency distribution table above.

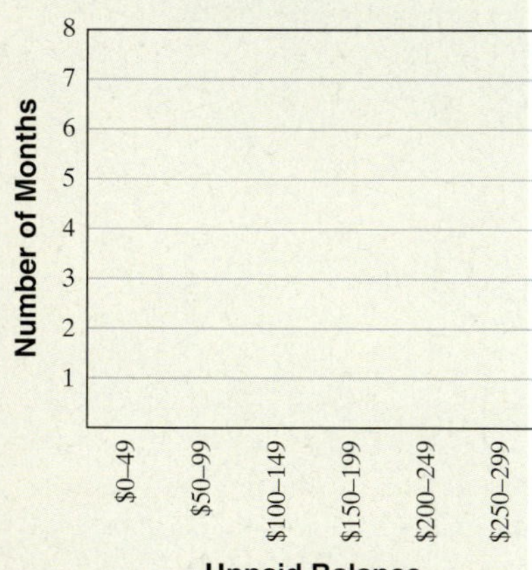

Answers

5. 16

6.

Tally	Class Frequency (Number of Months)
|||	3
||||	4
||	2
|	1
|	1
|	1

7.

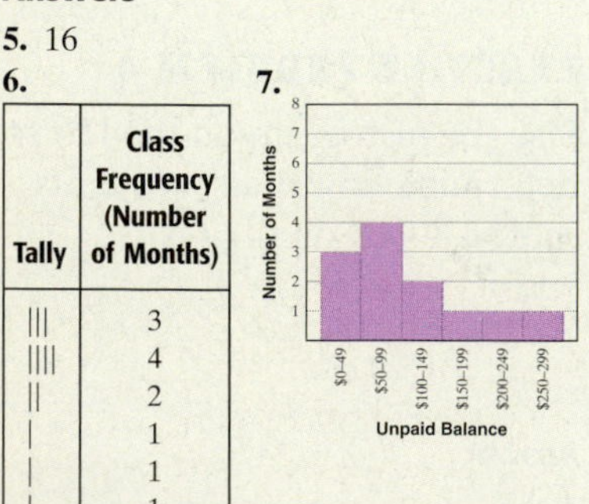

EXAMPLE 5 Reading a Histogram on Student Test Scores

Use the preceding histogram to determine how many students scored 80 or above on the test.

Solution: We see that two different bars fit this description. There are 12 students who scored 80–89 and 8 students who scored 90–99. The sum of these two categories is 12 + 8 or 20 students. Thus, 20 students scored 80 or above on the test.

Work Practice Problem 5

Now we will look at a way to construct histograms.

The daily high temperatures for 1 month in New Orleans, Louisiana, are recorded in the following list:

85°	90°	95°	89°	88°	94°
87°	90°	95°	92°	95°	94°
82°	92°	96°	91°	94°	92°
89°	89°	90°	93°	95°	91°
88°	90°	88°	86°	93°	89°

The data in this list have not been organized and can be hard to interpret. One way to organize the data is to place it in a **frequency distribution table.** We will do this in Example 6.

EXAMPLE 6 Completing a Frequency Distribution on Temperature

Complete the frequency distribution table for the preceding temperature data.

Solution: Go through the data and place a tally mark in the second column of the table next to the class interval. Then count the tally marks and write each total in the third column of the table.

Class Intervals (Temperatures)	Tally	Class Frequency (Number of Days)
82°–84°	|	1
85°–87°	|||	3
88°–90°	~~||||~~ ~~||||~~ |	11
91°–93°	~~||||~~ ||	7
94°–96°	~~||||~~ |||	8

Work Practice Problem 6

EXAMPLE 7 Constructing a Histogram

Construct a histogram from the frequency distribution table in Example 6.

Solution:

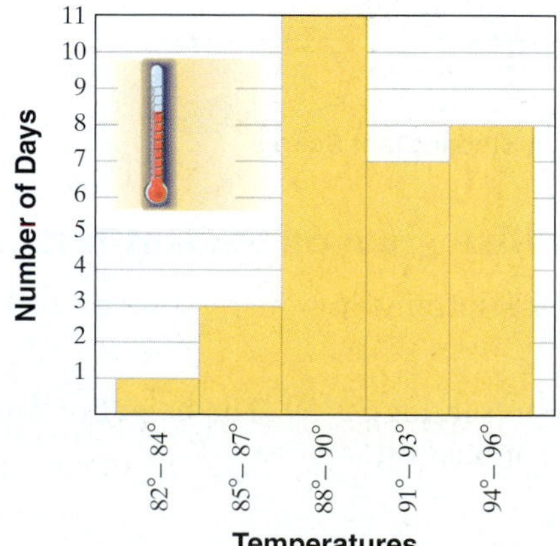

Work Practice Problem 7

✔Concept Check Which of the following sets of data is better suited to representation by a histogram? Explain.

Set 1		Set 2	
Grade on Final	# of Students	Section Number	Avg. Grade on Final
51–60	12	150	78
61–70	18	151	83
71–80	29	152	87
81–90	23	153	73
91–100	25		

Objective D Reading Line Graphs

Another common way to display information with a graph is by using a **line graph.** An advantage of a line graph is that it can be used to visualize relationships between two quantities. A line graph can also be very useful in showing a change over time.

EXAMPLE 8 Reading Temperatures from Line Graph

The following line graph shows the average daily temperature for each month for Omaha, Nebraska. Use this graph to answer the questions below.

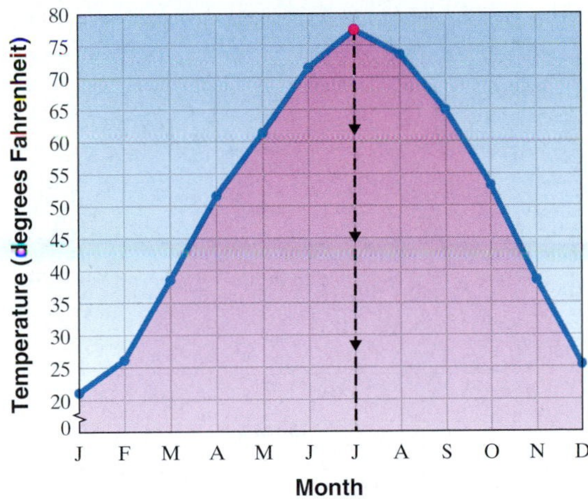

Source: National Climatic Data Center

a. During what month is the average daily temperature the highest?

b. During what month, from July through December, is the average daily temperature 65°F?

c. During what months is the average daily temperature less than 30°F?

Solution:

a. The month with the highest temperature corresponds to the highest point. This is the red point shown on the graph above. We follow this highest point downward to the horizontal month scale and see that this point corresponds to July.

Continued on next page

PRACTICE PROBLEM 8

Use the temperature graph in Example 8 to answer the following questions:

a. During what month is the average daily temperature the lowest?

b. During what month is the average daily temperature 25°F?

c. During what months is the average daily temperature greater than 70°F?

Answers

8. a. January **b.** December **c.** June, July, and August

✔ Concept Check Answer

Set 1; the grades are arranged in ranges of scores.

b. The months July through December correspond to the right side of the graph. We find the 65°F mark on the vertical temperature scale and move to the right until a point on the right side of the graph is reached. From that point, we move downward to the horizontal month scale and read the corresponding month. During the month of September, the average daily temperature is 65°F.

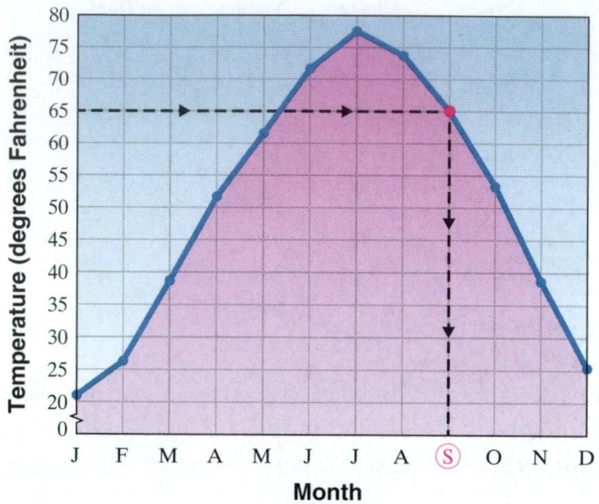

Source: National Climatic Data Center

c. To see what months the temperature is less than 30°F, we find what months correspond to points that fall below the 30°F mark on the vertical scale. These months are January, February, and December.

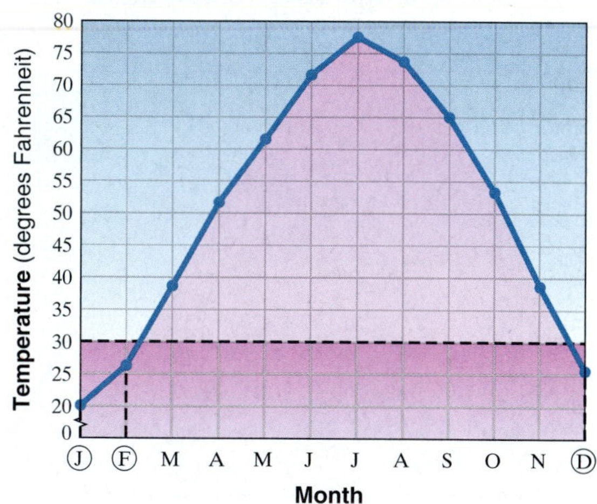

Source: National Climatic Data Center

Work Practice Problem 8

Vocabulary and Readiness Check

Fill in each blank with one of the choices below.

pictograph	bar	class frequency
histogram	line	class interval

1. A ___________ graph presents data using vertical or horizontal bars.
2. A __________ is a graph in which pictures or symbols are used to visually present data.
3. A ___________ graph displays information with a line that connects data points.
4. A __________ is a special bar graph in which the width of each bar represents a ___________ and the height of each bar represents the _____________.

7.1 EXERCISE SET

FOR EXTRA HELP

Objective A *The following pictograph shows the annual apple production in 2005 by the top apple-producing states. Use this graph to answer Exercises 1 through 8. See Example 1.* (*Source:* United States Apple Association)

1. Which state produced the greatest quantity of apples?

2. In which of the states shown were the least amount of apples grown?

3. Approximate the number of bushels of apples grown in New York.

4. Approximate the number of bushels of apples grown in California.

5. Which two states together produced as many apples as New York?

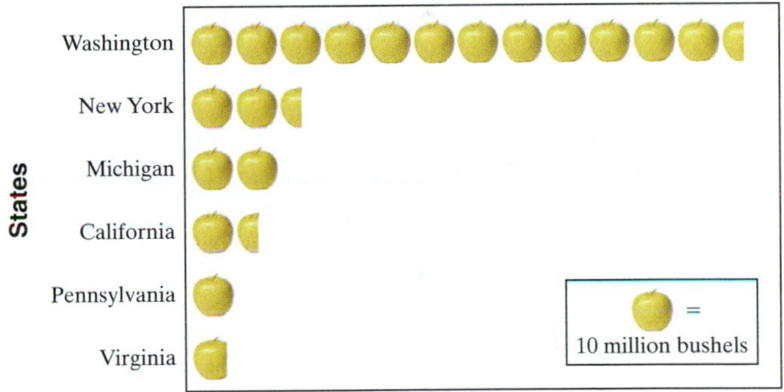

6. Washington produces about 53% of the apples grown in the United States. About how many bushels of apples were grown in the United States in 2005? Round to the nearest whole.

7. In 2001, New York produced about 23.8 million bushels of apples. What was the percent increase in apple production in New York from 2001 to 2005?

8. Which state produced about 20 million bushels of apples?

The following pictograph shows the average number of wildfires in the United States between 2000 and 2005. Use this graph to answer Exercises 9 through 16. See Example 1. (*Source:* National Interagency Fire Center)

9. Approximate the number of wildfires in 2002.

10. Which year, of the years shown, had the most wildfires?

11. Approximately how many wildfires were there in 2000?

12. In what years were the number of wildfires greater than 80,000?

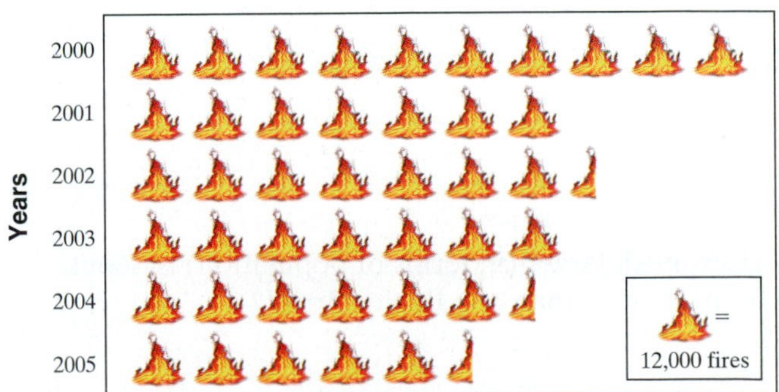

13. What was the amount of decrease in wildfires from 2004 to 2005?

14. What was the amount of decrease in wildfires from 2000 to 2001?

15. What was the average annual number of wildfires from 2003 to 2005?

16. Describe a possible trend in wildfires based on this graph.

Objective B *The National Weather Service has exacting definitions for hurricanes; they are tropical storms with winds in excess of 74 mph. The following bar graph shows the number of hurricanes, by month, that have made landfall on the mainland United States between 1851 and 2005. Use this graph to answer Exercises 17 through 22. See Example 2.* (*Source:* National Weather Service: National Hurricane Center)

17. In which month did the most hurricanes make landfall in the United States?

18. In which month did the fewest hurricanes make landfall in the United States?

19. Approximate the number of hurricanes that made landfall in the United States during the month of August.

20. Approximate the number of hurricanes that made landfall in the United States in September.

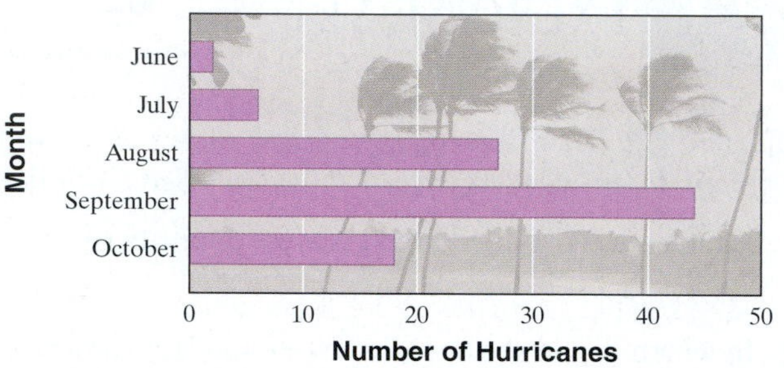

21. In 2005 alone, two hurricanes made landfall during the month of July. What fraction of all hurricanes that made landfall during July is this?

22. In 2005 alone, there were five hurricanes that made landfall on the United States mainland. If there have been 97 hurricanes to make landfall in the United States since 1851, approximately what percent of these arrived in 2005?

The following horizontal bar graph shows the 2005 population of the world's largest cities (including their suburbs). Use this graph to answer Exercises 23 through 28. See Example 2. (*Source:* CityPopulation)

23. Name the city with the largest population, and estimate its population.

24. Name the city whose population is between 16 million and 17 million, and estimate its population.

25. Name the city in the United States with the largest population, and estimate its population.

26. Name the two cities whose population is between 19 and 20 million.

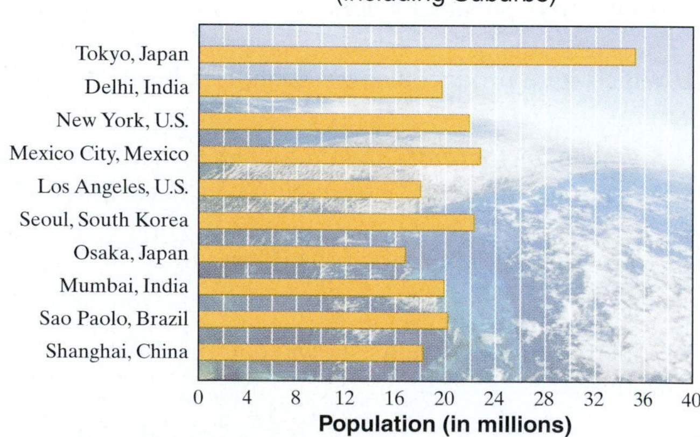

27. How much larger (in terms of population) is Seoul, South Korea, than São Paolo, Brazil?

28. How much larger (in terms of population) is Mexico City, Mexico, than Osaka, Japan?

Use the information given to draw a vertical bar graph. Clearly label the bars. See Example 3.

29.

Fiber Content of Selected Foods	
Food	**Grams of Total Fiber**
Kidney beans $\left(\frac{1}{2}\text{c}\right)$	4.5
Oatmeal, cooked $\left(\frac{3}{4}\text{c}\right)$	3.0
Peanut butter, chunky (2 tbsp)	1.5
Popcorn (1 c)	1.0
Potato, baked with skin (1 med)	4.0
Whole wheat bread (1 slice)	2.5

(*Sources:* American Dietetic Association and National Center for Nutrition and Dietetics)

Fiber Content of Selected Foods

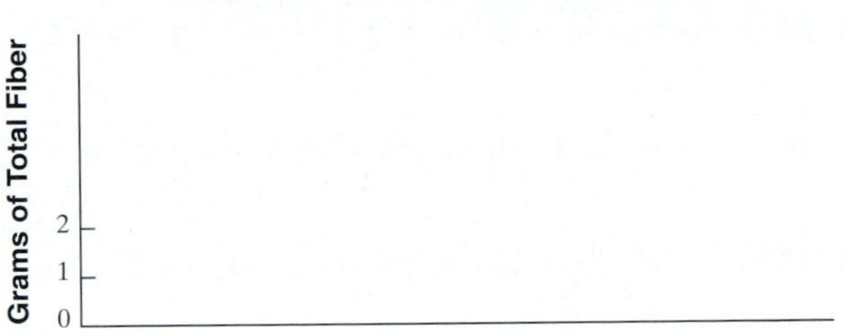

Food

30.

U.S. Annual Food Sales	
Year	**Sales in Billions of Dollars**
1990	557
1995	655
2000	826
2003	947

(*Source:* U.S. Department of Agriculture)

U.S. Annual Food Sales

Year

31.

Best-Selling Albums of All Time (U.S. Sales)	
Album	**Estimated Sales (in millions)**
Shania Twain: *Come on Over* (1997)	20
Pink Floyd: *The Wall* (1979)	23
Michael Jackson: *Thriller* (1982)	27
AC/DC: *Back in Black* (1980)	21
Billy Joel: *Greatest Hits Volumes I & II* (1985)	21
Eagles: *Their Greatest Hits* (1976)	28
Led Zeppelin: *Led Zeppelin IV* (1971)	22

(*Source:* Recording Industry Association of America)

Best-selling Albums of All Time
(U.S. sales)

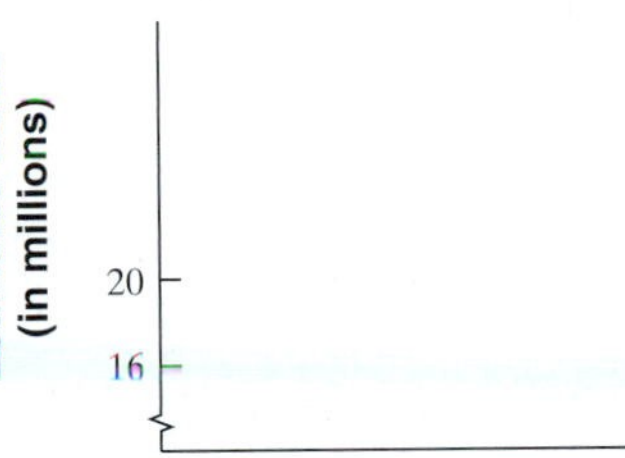

Album

32.

Fuel Economy of Selected 2007 Vehicles in the United States*	
Vehicle (sales rank)	**Highway Fuel Economy* (in miles per gallon)**
Ford F-Series (1)	20
Chevrolet Silverado (2)	19
Dodge Nitro (3)	24
Toyota Camry (4)	34
Honda Accord (5)	34
Ford Explorer (6)	21

*Maximum fuel economy available among all model trims, 2007 models.
(*Sources:* Internet research)

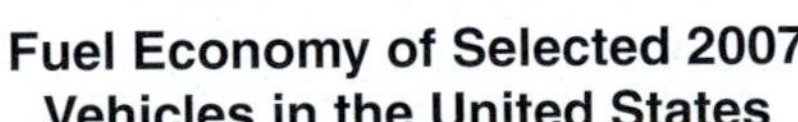

Fuel Economy of Selected 2007 Vehicles in the United States

Vehicle

Objective C *The following histogram shows the number of miles that each adult, from a survey of 100 adults, drives per week. Use this histogram to answer Exercises 33 through 42. See Examples 4 and 5.*

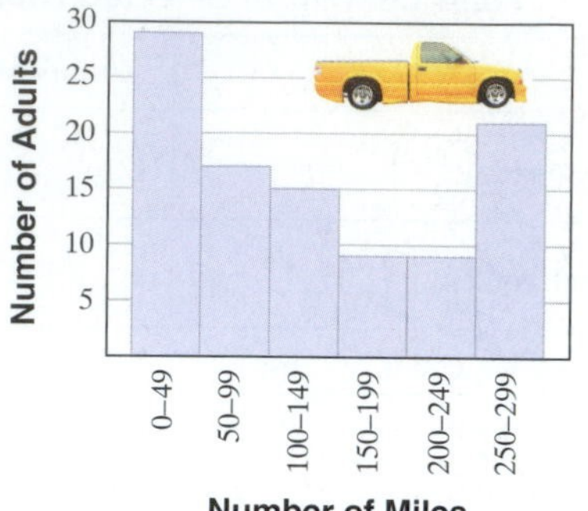

33. How many adults drive 100–149 miles per week?

34. How many adults drive 200–249 miles per week?

35. How many adults drive fewer than 150 miles per week?

36. How many adults drive 200 miles or more per week?

37. How many adults drive 100–199 miles per week?

38. How many adults drive 150–249 miles per week?

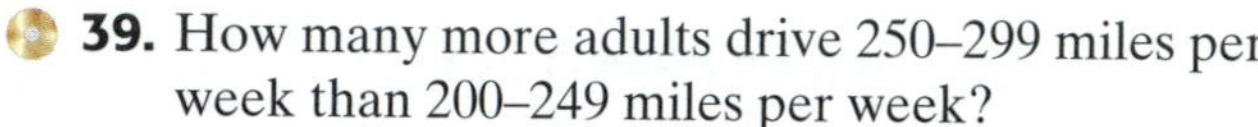

39. How many more adults drive 250–299 miles per week than 200–249 miles per week?

40. How many more adults drive 0–49 miles per week than 50–99 miles per week?

41. What is the ratio of adults who drive 150–199 miles per week to the total number of adults surveyed?

42. What is the ratio of adults who drive 50–99 miles per week to the total number of adults surveyed?

The following histogram shows the projected ages of householders for the year 2010. Use this histogram to answer Exercises 43 through 50. For Exercises 45 through 48, estimate to the nearest whole million. See Examples 4 and 5.

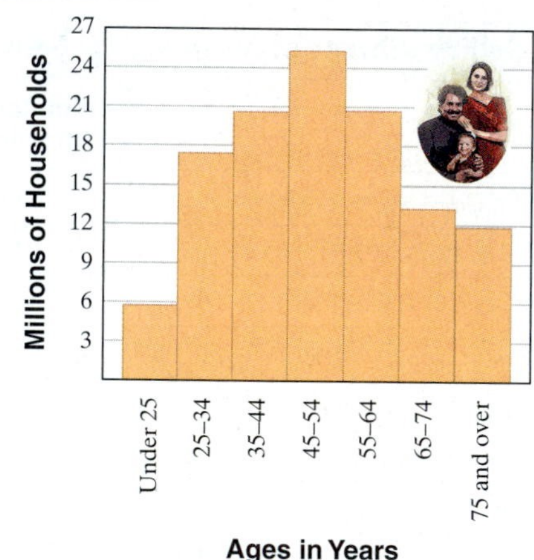

Source: U.S. Bureau of the Census, *Current Population Reports*

43. The most householders will be in what age range?

44. The least number of householders will be in what age range?

45. How many householders will be 55–64 years old?

46. How many householders will be 35–44 years old?

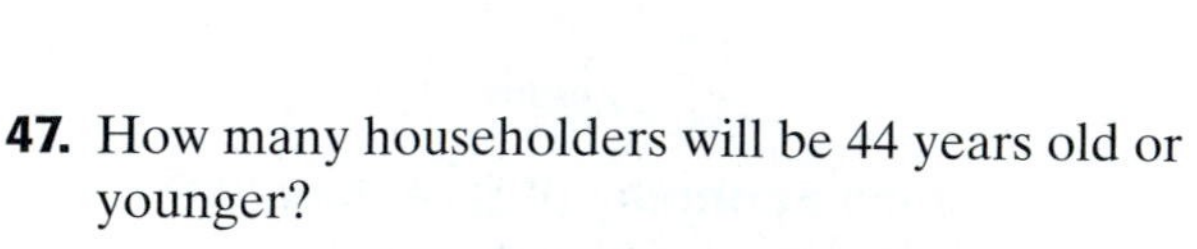

47. How many householders will be 44 years old or younger?

48. How many householders will be 55 years old or older?

49. How many more householders will be 45–54 years old than 55–64 years old?

50. Which age-interval will describe you in the year 2010?

The following list shows the golf scores for an amateur golfer. Use this list to complete the frequency distribution table to the right. See Example 6.

78	84	91	93	97
97	95	85	95	96
101	89	92	89	100

	Class Intervals (Scores)	Tally	Class Frequency (Number of Games)
51.	70–79		
52.	80–89		
53.	90–99		
54.	100–109		

Twenty-five people in a survey were asked to give their current checking account balances. Use the balances shown in the following list to complete the frequency distribution table to the right. See Example 6.

$53	$105	$162	$443	$109
$468	$47	$259	$316	$228
$207	$357	$15	$301	$75
$86	$77	$512	$219	$100
$192	$288	$352	$166	$292

	Class Intervals (Account Balances)	Tally	Class Frequency (Number of People)
55.	$0–$99		
56.	$100–$199		
57.	$200–$299		
58.	$300–$399		
59.	$400–$499		
60.	$500–$599		

61. Use the frequency distribution table from Exercises 51 through 54 to construct a histogram. See Example 7.

Number of Games

2

1

Golf Scores

62. Use the frequency distribution table from Exercises 55 through 60 to construct a histogram. See Example 7.

Account Balances

Objective **D** *The following line graph shows the World Cup goals per game average during the years shown. Use this graph to answer Exercises 63 through 70. See Example 8.*

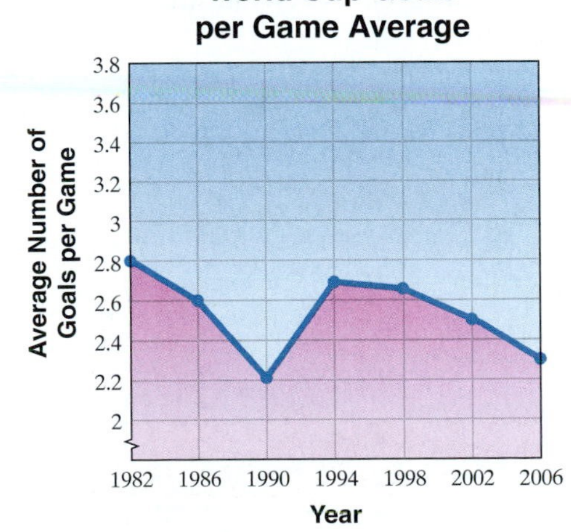

Source: Soccer America Magazine

63. Find the average number of goals per game in 1994.

64. Find the average number of goals per game in 2006.

65. During what year shown was the average number of goals per game the highest?

66. During what year shown was the average number of goals per game the lowest?

67. Between 2002 and 2006, did the average number of goals per game increase or decrease?

68. Between 1990 and 1994, did the average number of goals per game increase or decrease?

69. During what year(s) was the average goals per game less than 2.5?

70. During what year(s) was the average goals per game greater than 2.6?

Review

Find each percent. See Sections 6.3 and 6.4.

71. 30% of 12

72. 45% of 120

73. 10% of 62

74. 95% of 50

Write each fraction as a percent. See Section 6.2.

75. $\frac{1}{4}$

76. $\frac{2}{5}$

77. $\frac{17}{50}$

78. $\frac{9}{10}$

Concept Extensions

The following double-line graph shows temperature highs and lows for a week. Use this graph to answer Exercises 79 through 84.

79. What was the high temperature reading on Thursday?

80. What was the low temperature reading on Thursday?

81. What day was the temperature the lowest? What was this low temperature?

82. What day of the week was the temperature the highest? What was this high temperature?

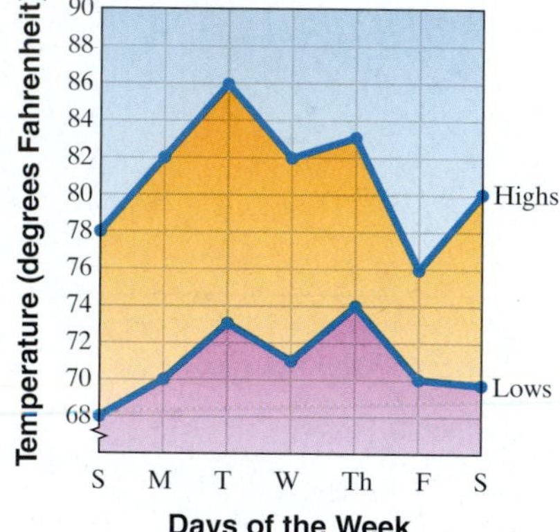

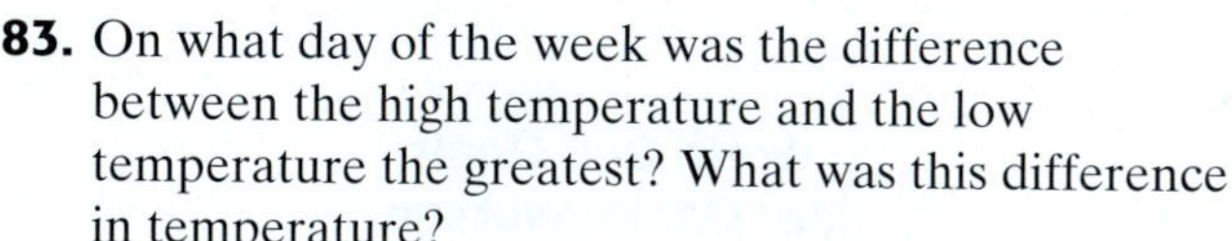

83. On what day of the week was the difference between the high temperature and the low temperature the greatest? What was this difference in temperature?

84. On what day of the week was the difference between the high temperature and the low temperature the least? What was this difference in temperature?

85. True or false? With a bar graph, the width of the bar is just as important as the height of the bar. Explain your answer.

7.2 READING CIRCLE GRAPHS

Objectives

A Read Circle Graphs.

B Draw Circle Graphs.

Objective A Reading Circle Graphs

In Exercise Set 6.2, the following **circle graph** was shown. This particular graph shows the favorite sport for 100 adults.

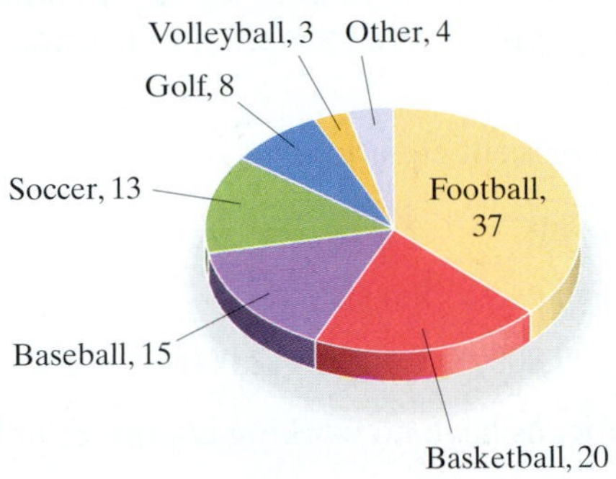

Each sector of the graph (shaped like a piece of pie) shows a category and the relative size of the category. In other words, the most popular sport is football, and it is represented by the largest sector.

EXAMPLE 1 Find the ratio of adults preferring basketball to total adults. Write the ratio as a fraction in simplest form.

Solution: The ratio is

$$\frac{\text{people preferring basketball}}{\text{total adults}} = \frac{20}{100} = \frac{1}{5}$$

Work Practice Problem 1

PRACTICE PROBLEM 1

Find the ratio of adults preferring golf to total adults. Write the ratio as a fraction in simplest form.

A circle graph is often used to show percents in different categories, with the whole circle representing 100%.

EXAMPLE 2 **Using a Circle Graph**

The following circle graph shows the percent of Americans with various numbers of working computers at home. Using the circle graph shown, determine the percent of Americans who have one or more working computers at home.

Number of Working Computers at Home

Three 6%
Four or more 3%
Two 15%
Zero 29%
One 47%

Source: UCLA Center for Communication Policy, 2003

Solution: To find this percent, we add the percents corresponding to one, two, three, and four or more working computers at home. The percent of Americans that have one or more working computers at home is

$$47\% + 15\% + 6\% + 3\% = 71\%$$

Work Practice Problem 2

PRACTICE PROBLEM 2

Using the circle graph shown in Example 2, determine the percent of Americans that have three or more working computers at home.

Answers

1. $\frac{2}{25}$ **2.** 9%

Helpful Hint Since a circle graph represents a whole, the percents should add to 100% or 1. Notice this is true for Example 2.

PRACTICE PROBLEM 3

Using the circle graph from Example 2, find the number of Americans that have four or more working computers at home.

EXAMPLE 3 **Finding Percent of Population**

In 2006, the population of the United States was approximately 299,800,000. Using the circle graph from Example 2, find the number of Americans that have no working computers at home.

Solution: We use the percent equation.

$$\text{amount} = \text{percent} \cdot \text{base}$$
$$\text{amount} = 0.29 \cdot 299{,}800{,}000$$
$$= 0.29(299{,}800{,}000) = 86{,}942{,}000$$

Thus, 86,942,000 Americans have no working computer at home.

Work Practice Problem 3

✔Concept Check Can the following data be represented by a circle graph? Why or why not?

Responses to the Question, "In Which Activities Are You Involved?"	
Intramural sports	60%
On-campus job	42%
Fraternity/sorority	27%
Academic clubs	21%
Music programs	14%

Objective B Drawing Circle Graphs

To draw a circle graph, we use the fact that a whole circle contains 360° (degrees).

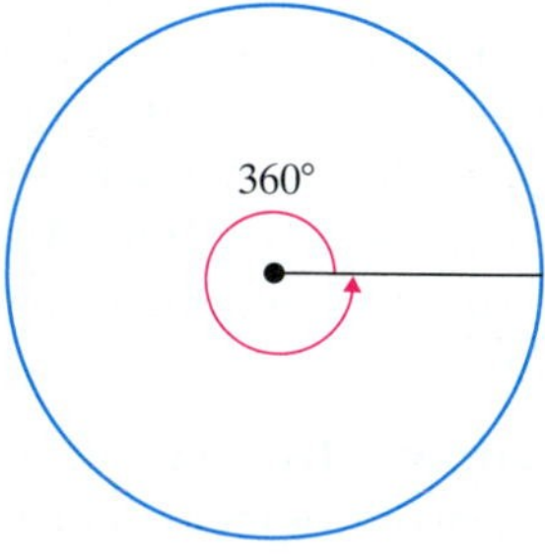

Answer

3. 8,994,000 Americans

✔ Concept Check Answer

no; the percents add up to more than 100%

EXAMPLE 4 Drawing a Circle Graph for U.S. Armed Forces Personnel

The following table shows the percent of U.S. armed forces personnel that are in each branch of service in 2006. (*Source:* U.S. Department of Defense)

Branch of Service	Percent
Army	43
Navy	22
Marine Corps	11
Air Force	22
Coast Guard	2

Draw a circle graph showing this data.

Solution: First we find the number of degrees in each sector representing each branch of service. Remember that the whole circle contains 360°. (We will round degrees to the nearest whole.)

Sector	Degrees in Each Sector
Army	$43\% \times 360° = 0.43 \times 360° = 154.8° \approx 155°$
Navy	$22\% \times 360° = 0.22 \times 360° = 79.2° \approx 79°$
Marine Corps	$11\% \times 360° = 0.11 \times 360° = 39.6° \approx 40°$
Air Force	$22\% \times 360° = 0.22 \times 360° = 79.2° = 79°$
Coast Guard	$2\% \times 360° = 0.02 \times 360° = 7.2° \approx 7°$

Check your calculations by finding the sum of the degrees.

$$155° + 79° + 40° + 79° + 7° = 360°$$

The sum should be 360°. (It may vary only slightly because of rounding.)

Next we draw a circle and mark its center. Then we draw a line from the center of the circle to the circle itself.

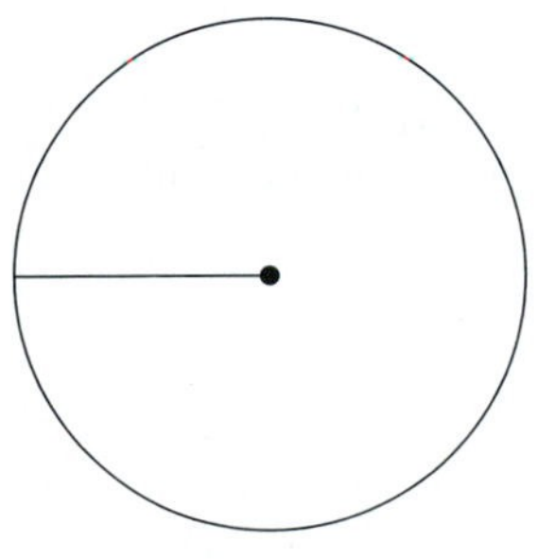

To construct the sectors, we will use a **protractor.** A protractor measures the number of degrees in an angle. We place the hole in the protractor over the center of the circle. Then we adjust the protractor so that 0° on the protractor is aligned with the line that we drew.

Continued on next page

PRACTICE PROBLEM 4

Use the data shown to draw a circle graph.

Freshmen	30%
Sophomores	27%
Juniors	25%
Seniors	18%

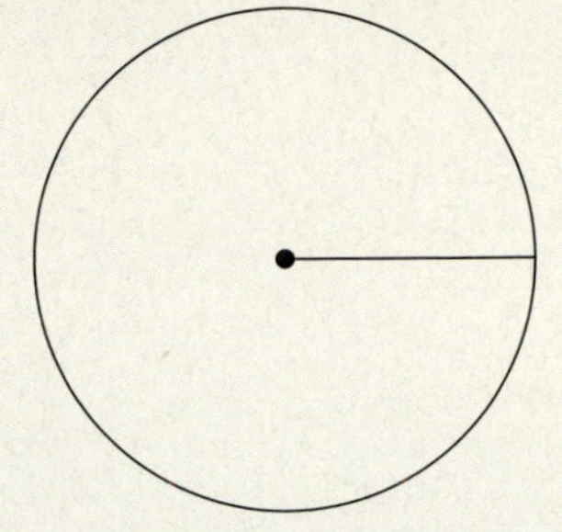

Answer

4.

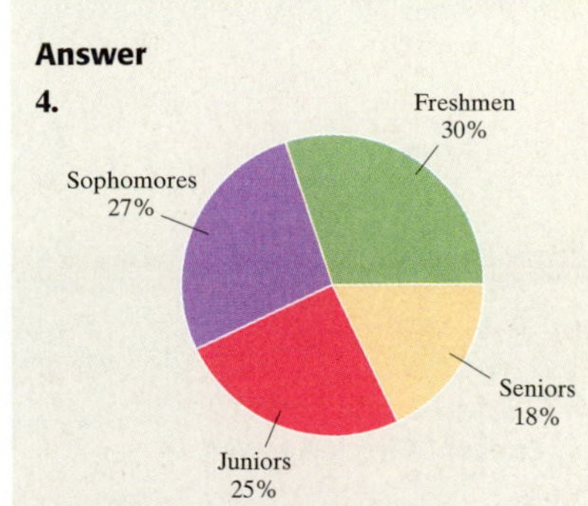

It makes no difference which sector we draw first. To construct the "Army" sector, we find 155° on the protractor and mark our circle. Then we remove the protractor and use this mark to draw a second line from the center to the circle itself.

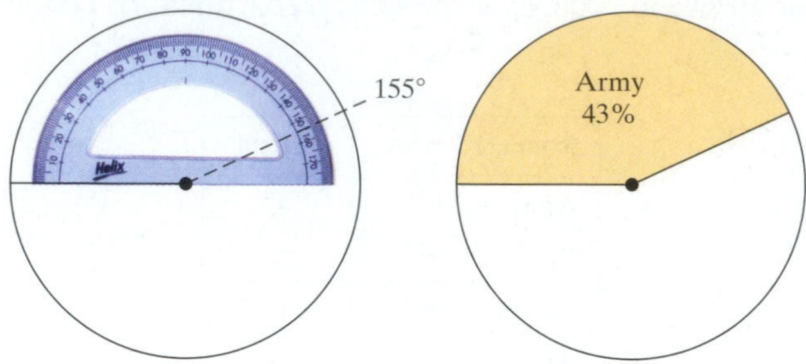

To construct the "Navy" sector, we follow the same procedure as above, except that we line up 0° with the second line we drew and mark the protractor at 79°.

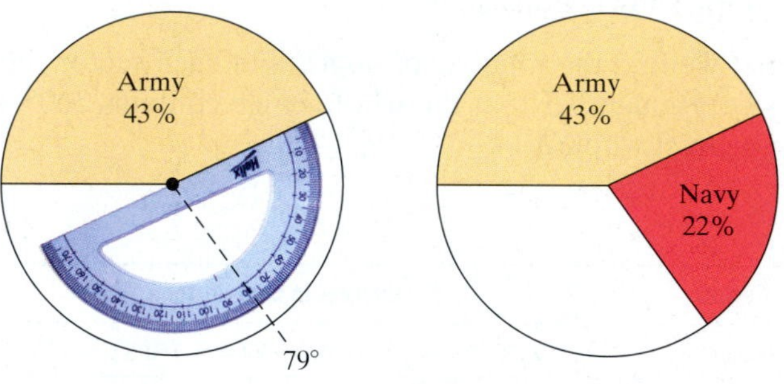

We continue in this manner until the circle graph is complete.

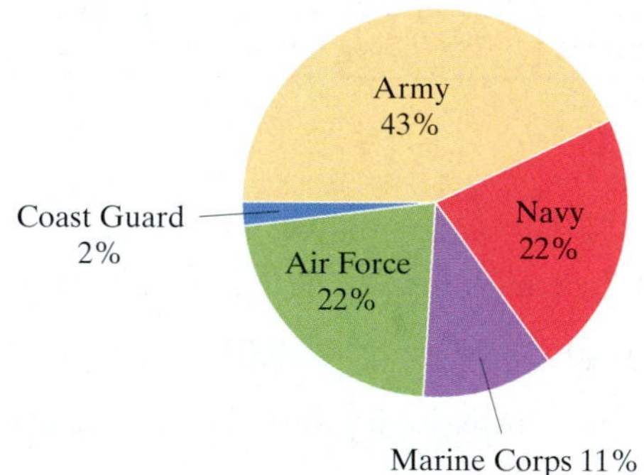

Work Practice Problem 4

✔Concept Check True or false? The larger a sector in a circle graph, the larger the percent of the total it represents. Explain your answer.

✔ Concept Check Answer
true

Vocabulary and Readiness Check

Use the choices below to fill in each blank.

sector circle 100 360

1. In a _______ graph, each section (shaped like a piece of pie) shows a category and the relative size of the category.
2. A circle graph contains pie-shaped sections, each called a ______.
3. The number of degrees in a whole circle is ________.
4. If a circle graph has percent labels, the percents should add up to ________.

7.2 EXERCISE SET

FOR EXTRA HELP

Student Solutions Manual PH Math/Tutor Center

CD/Video for Review

MathXL®

MyMathLab

Objective A *The following circle graph is a result of surveying 700 college students. They were asked where they live while attending college. Use this graph to answer Exercises 1 through 6. Write all ratios as fractions in simplest form. See Example 1.*

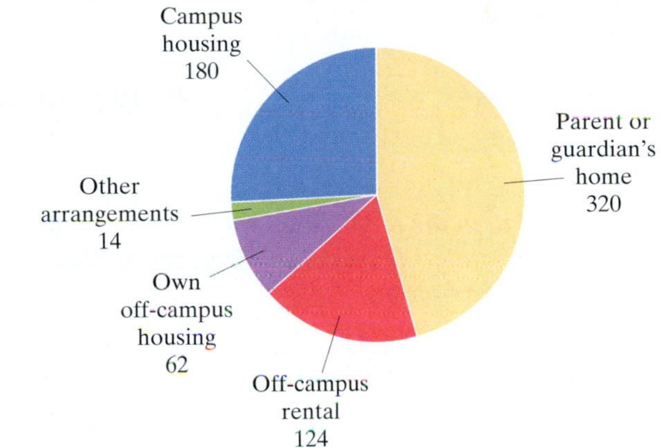

1. Where do most of these college students live?
2. Besides the category "Other Arrangements," where do least of these college students live?
3. Find the ratio of students living in campus housing to total students.
4. Find the ratio of students living in off-campus rentals to total students.
5. Find the ratio of students living in campus housing to students living in a parent or guardian's home.
6. Find the ratio of students living in off-campus rentals to students living in a parent or guardian's home.

The following circle graph shows the percent of the land area of the continents of Earth. Use this graph for Exercises 7 through 14. See Examples 2 and 3.

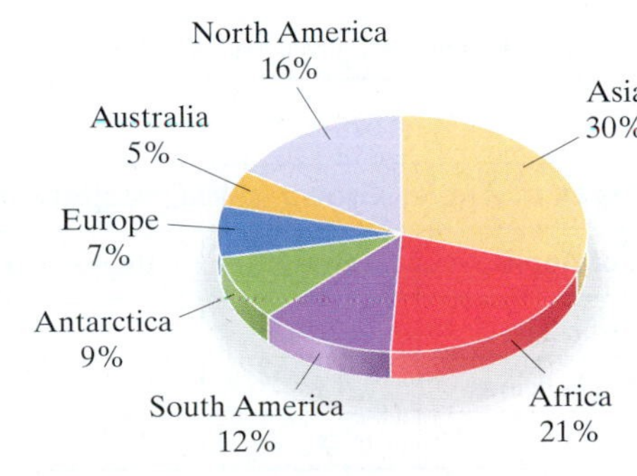

Source: National Geographic Society

7. Which continent is the largest?
8. Which continent is the smallest?
9. What percent of the land on Earth is accounted for by Asia and Europe together?
10. What percent of the land on Earth is accounted for by North and South America?

The total amount of land on Earth is approximately 57,000,000 square miles. Use the graph to find the area of the continents given in Exercises 11 through 14.

11. Asia
12. South America
13. Australia
14. Europe

The following circle graph shows the percent of the types of books available at Midway Memorial Library. Use this graph for Exercises 15 through 24. See Examples 2 and 3.

15. What percent of books are classified as some type of fiction?

16. What percent of books are nonfiction or reference?

17. What is the second-largest category of books?

18. What is the third-largest category of books?

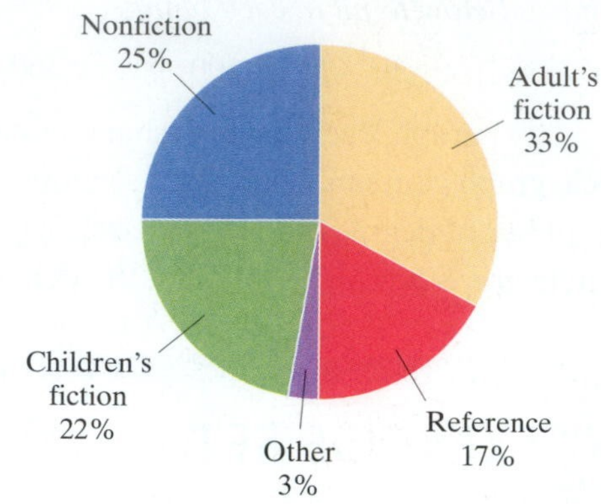

If this library has 125,600 books, find how many books are in each category given in Exercises 19 through 24.

19. Nonfiction

20. Reference

21. Children's fiction

22. Adult's fiction

23. Reference or other

24. Nonfiction or other

Objective B *Fill in the table. Round to the nearest degree. Then draw a circle graph to represent the information given in each table. (Remember: The total of "Degrees in Sector" column should equal 360° or very close to 360° because of rounding.) See Example 4.*

25.

U.S. Electricity Generation by Energy Source: 2006		
Energy Source	**Percent**	**Degrees in Sector**
Coal	50.6%	
Natural gas	15.8%	
Hydroelectric	8.2%	
Nuclear	20.6%	
Petroleum	1.6%	
Other	3.2%	
(*Source:* Energy Information Administration)		

26.

Types of Apples Grown in Washington State		
Type of Apple	**Percent**	**Degrees in Sector**
Red delicious	37%	
Golden delicious	13%	
Fuji	14%	
Gala	15%	
Granny Smith	12%	
Other varieties	6%	
Braeburn	3%	
(*Source:* U.S. Apple Association)		

27.

2006* Hybrid Sales by Make of Car		
Company	**Percent**	**Degrees in Sector**
Toyota	63%	
Honda	17%	
Lexus	10%	
Ford/Mercury	10%	
*Through six months of 2006		

28.

Number of Times the "Are We There Yet?" Question Is Asked to Parents During Road Trips:		
	Percent	**Degrees in Sector**
Never	20%	
Once	11%	
2–5 times	36%	
6–10 times	14%	
More than 10 times	19%	
(*Source:* KRC Research for Goodyear Tire & Rubber Co.)		

Review

Write the prime factorization of each number. See Section 4.2.

29. 20 **30.** 25 **31.** 40

32. 16 **33.** 85 **34.** 105

Concept Extensions

The following circle graph shows the relative sizes of the great oceans.

35. Without calculating, determine which ocean is the largest. How can you answer this question by looking at the circle graph?

36. Without calculating, determine which ocean is the smallest. How can you answer this question by looking at the circle graph?

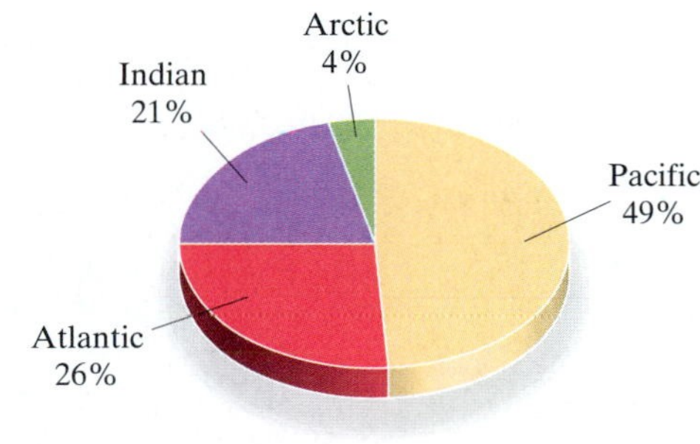

Source: Philip's World Atlas

These oceans together make up 264,489,800 square kilometers of the Earth's surface. Find the square kilometers for each ocean.

 **37.** Pacific Ocean **38.** Atlantic Ocean **39.** Indian Ocean **40.** Arctic Ocean

The following circle graph summarizes the results of a survey of 2800 Internet users who make purchases online. Use this graph for Exercises 41 through 45. Round to the nearest whole.

41. How many of the survey respondents said that they spend \$0–\$15 online each month?

42. How many of the survey repondents said that they spend \$15–\$175 online each month?

43. How many of the survey respondents said that they spend at least \$15 or over \$175 online each month?

Online Spending per Month

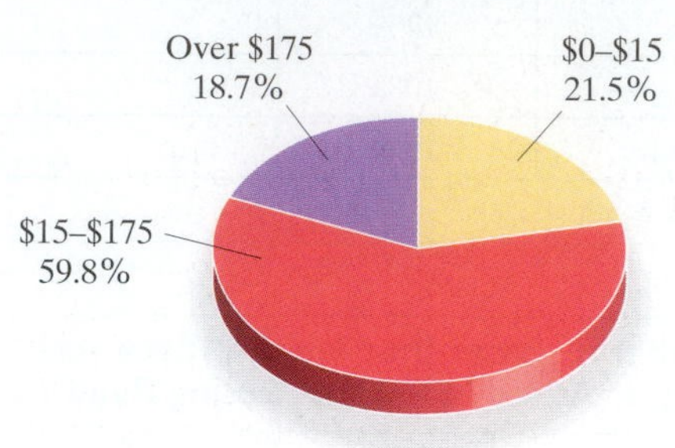

Source: UCLA Center for Communication Policy

44. Find the ratio of *percent* of respondents who spend \$0–\$15 online to *percent* of those who spend \$15–\$175. Write the ratio as a fraction with integers in the numerator and denominator.

45. Find the ratio of *number* of respondents who spend \$0–\$15 online to *number* of respondents who spend \$15–\$175 online. Write the ratio as a fraction with integers in the numerator and denominator.

See the Concept Checks in this section.

46. True or false? The smaller a sector in a circle graph, the smaller the percent of the total it represents. Explain why.

47. Can the data below be represented by a circle graph? Why or why not?

Responses to the Question, "In Which Activities Are You Involved?"	
Intramural sports	60%
On-campus job	42%
Fraternity/sorority	27%
Academic clubs	21%
Music programs	14%

INTEGRATED REVIEW Sections 7.1–7.2

Reading Graphs

The following pictograph shows the six occupations with the largest estimated numerical increase in employment in the United States between 2004 and 2014. Use this graph to answer Exercises 1 through 4.

Jobs with Projected Highest Numerical Increase: 2004–2014

Jobs

Retail Salespeople

Registered Nurses

Post-secondary Teachers

Customer Service Reps

Janitors and Cleaners

Waitstaff

(icon) = 50,000 workers

Source: Bureau of Labor Statistics

1. Approximate the increase in the number of customer service representatives from 2004 to 2014.

2. Approximate the increase in number of post-secondary teachers from 2004 to 2014.

3. Which occupation is expected to show the greatest increase in numbers of employees between the years shown?

4. Which of the listed occupations is expected to show the least increase in numbers of employees between the years shown?

The following bar graph shows the highest U.S. dams. Use this graph to answer Exercises 5 through 8.

Highest U.S. Dams

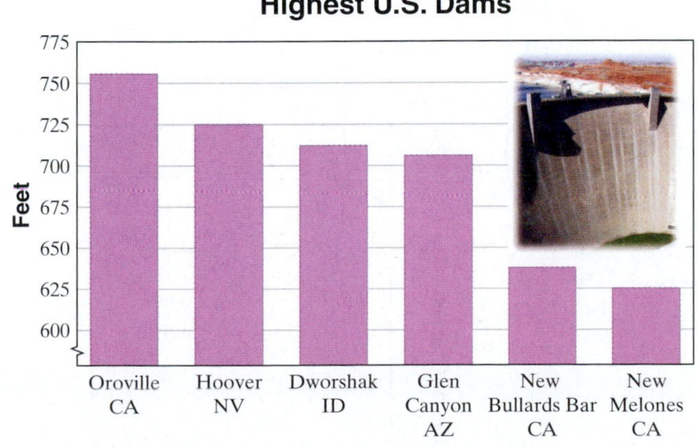

Source: Committee on Register of Dams

5. Name the U.S. dam with the greatest height and estimate its height.

6. Name the U.S. dam whose height is between 625 and 650 feet and estimate its height.

7. Estimate how much higher the Hoover Dam is than the Glen Canyon Dam.

8. How many U.S. dams have heights over 700 feet?

Answers

1. ______

2. ______

3. ______

4. ______

5. ______

6. ______

7. ______

8. ______

9. ______

10. ______

11. ______

12. ______

13. ______

14. ______

15. ______

16. ______

17. see table

18. see table

19. see table

20. see table

21. see table

22. see graph

The following line graph shows the daily high temperatures for one week in Annapolis, Maryland. Use this graph to answer Exercises 9 through 12.

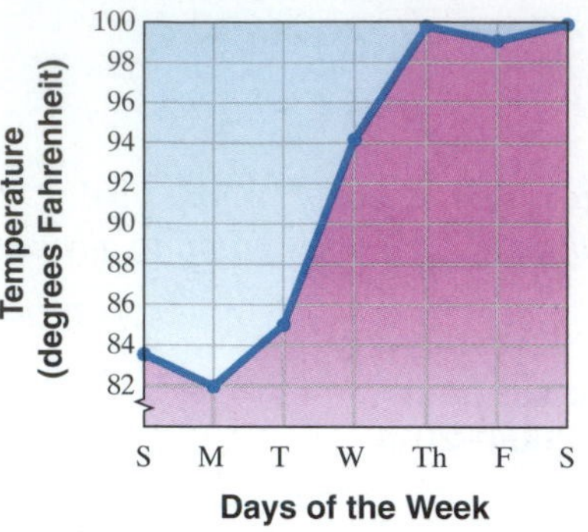

9. Name the day(s) of the week with the highest temperature and give that high temperature.

10. Name the day(s) of the week with the lowest temperature and give that low temperature.

11. On what days of the week was the temperature less than 90° Fahrenheit?

12. On what days of the week was the temperature greater than 90° Fahrenheit?

The following circle graph shows the type of beverage milk consumed in the United States. Use this graph for Exercises 13 through 16. If a store in Kerrville, Texas, sells 200 quart containers of milk per week, estimate how many quart containers are sold in each category below.

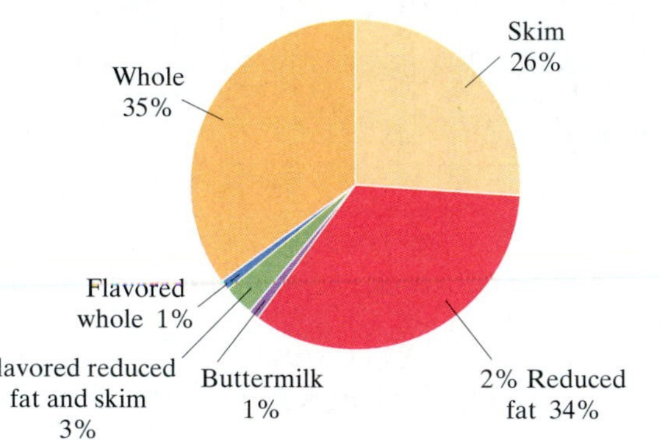

Source: U.S. Department of Agriculture

13. Whole milk

14. Skim milk

15. Buttermilk

16. Flavored reduced fat and skim milk

The following list shows weekly quiz scores for a student in prealgebra. Use this list to complete the frequency distribution table below.

50	80	71	83	86
67	89	93	88	97
	53	90		
75	80	78	93	99

	Class Intervals (Scores)	Tally	Class Frequency (Number of Quizzes)
17.	50–59		
18.	60–69		
19.	70–79		
20.	80–89		
21.	90–99		

22. Use the frequency distribution table from Exercises 17 through 21 to construct a histogram.

Quiz Scores

△ 7.3 SQUARE ROOTS AND THE PYTHAGOREAN THEOREM

Now that we know how to solve proportions, in Section 7.4, we use proportions to help us find unknown sides of similar triangles. In this section, we prepare for work on triangles by studying right triangles and their applications.

First, let's review square roots.

Objective A Finding Square Roots

The **square** of a number is the number times itself. For example,

The square of 5 is 25 because 5^2 or $5 \cdot 5 = 25$.
The square of -5 is also 25 because $(-5)^2$ or $(-5)(-5) = 25$.

The reverse process of squaring is finding a **square root.** For example,

A square root of 25 is 5 because $5 \cdot 5$ or $\ ^2 = 25$.
A square root of 25 is also   because $(-5)(-5)$ or $(-5)^2 = 25$.

Every positive number has two square roots. We see above that the square roots of 25 are 5 and -5.

We use the symbol $\sqrt{\ }$, called a **radical sign,** to indicate the positive square root of a nonnegative number. For example,

$\sqrt{25} = 5$ because $5^2 = 25$ and 5 is positive.
$\sqrt{9} = 3$ because $3^2 = 9$ and 3 is positive.

Square Root of a Number

The square root, $\sqrt{\ }$, of a positive number a is the positive number b whose square is a. In symbols,

$$\sqrt{a} = b, \qquad \text{if } b^2 = a$$

Also, $\sqrt{0} = 0$.

Helpful Hint

Remember that the radical sign $\sqrt{\ }$ is used to indicate the **positive square root** of a nonnegative number.

EXAMPLES Find each square root.

1. $\sqrt{49} = 7$ because $7^2 = 49$.
2. $\sqrt{36} = 6$ because $6^2 = 36$.
3. $\sqrt{1} = 1$ because $1^2 = 1$.
4. $\sqrt{81} = 9$ because $9^2 = 81$.
5. Find $\sqrt{\frac{1}{36}} = \frac{1}{6}$ because $\left(\frac{1}{6}\right)^2$ or $\frac{1}{6} \cdot \frac{1}{6} = \frac{1}{36}$.
6. Find $\sqrt{\frac{4}{25}} = \frac{2}{5}$ because $\left(\frac{2}{5}\right)^2$ or $\frac{2}{5} \cdot \frac{2}{5} = \frac{4}{25}$.

Work Practice Problems 1–6

Objectives

A Find the Square Root of a Number.

B Approximate Square Roots.

C Use the Pythagorean Theorem.

PRACTICE PROBLEMS

Find each square root.

1. $\sqrt{100}$
2. $\sqrt{64}$
3. $\sqrt{121}$
4. $\sqrt{0}$
5. $\sqrt{\frac{1}{4}}$
6. $\sqrt{\frac{9}{16}}$

Answers

1. 10 **2.** 8 **3.** 11 **4.** 0 **5.** $\frac{1}{2}$ **6.** $\frac{3}{4}$

Objective B Approximating Square Roots

Thus far, we have found square roots of perfect squares. Numbers like $\frac{1}{4}$, 36, $\frac{4}{25}$, and 1 are called **perfect squares** because their square root is a whole number or a fraction. A square root such as $\sqrt{5}$ cannot be written as a whole number or a fraction since 5 is not a perfect square.

Although $\sqrt{5}$ cannot be written as a whole number or a fraction, it can be approximated by estimating, by using a table (as in Appendix A.4), or by using a calculator.

PRACTICE PROBLEM 7

Use Appendix A.4 or a calculator to approximate each square root of 11 to the nearest thousandth.

a. $\sqrt{10}$

b. $\sqrt{62}$

EXAMPLE 7 Use Appendix A.4 or a calculator to approximate each square root to the nearest thousandth.

a. $\sqrt{43} \approx 6.557$ ($\approx$ is approximately)

b. $\sqrt{80} \approx 8.944$

Work Practice Problem 7

Helpful Hint

$\sqrt{80}$, above, is *approximately* 8.944. This means that if we multiply 8.944 by 8.944, the product is *close* to 80.

$$8.944 \times 8.944 \approx 79.995$$

Objective C Using the Pythagorean Theorem

One important application of square roots has to do with right triangles. Recall that a **right triangle** is a triangle in which one of the angles is a right angle, or measures 90° (degrees). The **hypotenuse** of a right triangle is the side opposite the right angle. The **legs** of a right triangle are the other two sides. These are shown in the following figure. The right angle in the triangle is indicated by the small square drawn in that angle.

The following theorem is true for all right triangles.

Pythagorean Theorem

If a and b are the lengths of the legs of a right triangle and c is the length of the hypotenuse, then

$$a^2 + b^2 = c^2$$

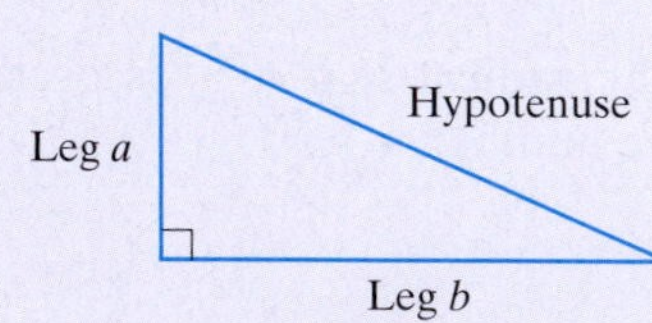

In other words, $(\text{leg})^2 + (\text{other leg})^2 = (\text{hypotenuse})^2$.

PRACTICE PROBLEM 8

Find the length of the hypotenuse of the given right triangle.

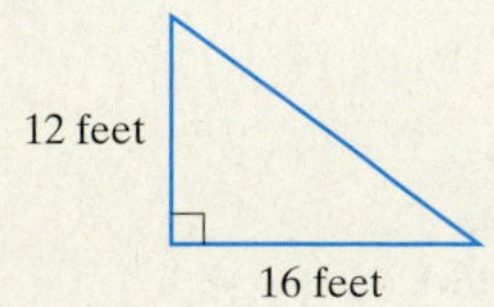

EXAMPLE 8 Find the length of the hypotenuse of the given right triangle.

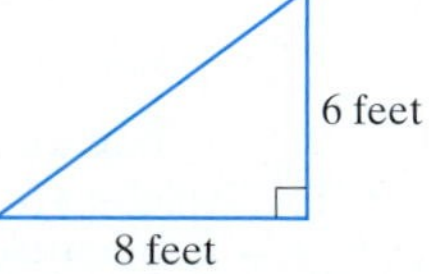

Solution: Let $a = 6$ and $b = 8$. According to the Pythagorean theorem,

$$a^2 + b^2 = c^2$$

$$6^2 + 8^2 = c^2 \quad \text{Let } a = 6 \text{ and } b = 8.$$

$$36 + 64 = c^2 \quad \text{Evaluate } 6^2 \text{ and } 8^2.$$

$$100 = c^2 \quad \text{Add.}$$

Answers

7. a. 3.162 b. 7.874 8. 20 feet

In the equation $c^2 = 100$, the solutions of c are the square roots of 100. Since $10 \cdot 10 = 100$ and $(-10)(-10) = 100$, both 10 and -10 are square roots of 100. Since c represents a length, we are only interested in the positive square root of c^2.

$$c = \sqrt{100}$$
$$= 10$$

The hypotenuse is 10 feet long.

Work Practice Problem 8

EXAMPLE 9 Approximate the length of the hypotenuse of the given right triangle. Round the length to the nearest whole unit.

Solution: Let $a = 17$ and $b = 10$.

$$a^2 + b^2 = c^2$$
$$17^2 + 10^2 = c^2$$
$$289 + 100 = c^2$$
$$389 = c^2$$
$$\sqrt{389} = c \text{ or } c \approx 20 \quad \text{From Appendix A.4 or a calculator}$$

10 meters

17 meters

The hypotenuse is exactly $\sqrt{389}$ meters, which is approximately 20 meters.

Work Practice Problem 9

PRACTICE PROBLEM 9

Approximate the length of the hypotenuse of the given right triangle. Round to the nearest whole unit.

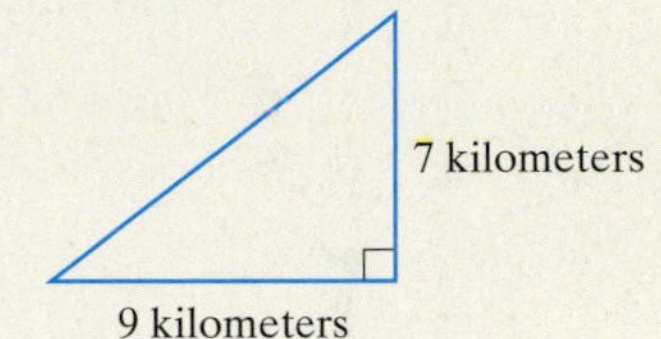

EXAMPLE 10 Find the length of the leg in the given right triangle. Give the exact length and a two-decimal-place approximation.

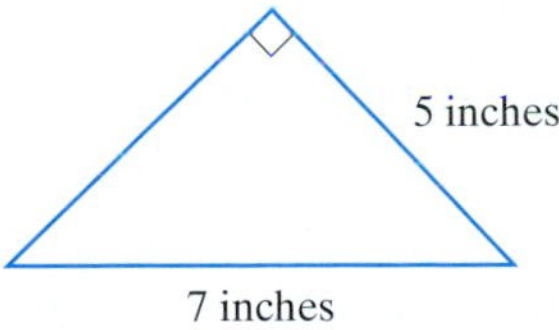

Solution: Notice that the hypotenuse measures 7 inches and that the length of one leg measures 5 inches. Thus, let $c = 7$ and a or b be 5. We will let $a = 5$.

$$a^2 + b^2 = c^2$$
$$5^2 + b^2 = 7^2 \quad \text{Let } a = 5 \text{ and } c = 7.$$
$$25 + b^2 = 49 \quad \text{Evaluate } 5^2 \text{ and } 7^2.$$
$$b^2 = 24 \quad \text{Subtract 25 from both sides.}$$
$$b = \sqrt{24} \approx 4.90$$

The length of the leg is exactly $\sqrt{24}$ inches and approximately 4.90 inches.

Work Practice Problem 10

PRACTICE PROBLEM 10

Find the length of the leg in the given right triangle. Give the exact length and a two-decimal-place approximation.

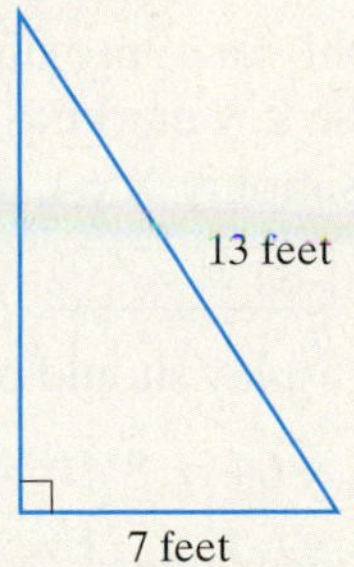

✔ **Concept Check** The following lists are the lengths of the sides of two triangles. Which set forms a right triangle?

a. 8, 15, 17

b. 24, 30, 40

Answers

9. 11 kilometers

10. $\sqrt{120}$ feet $\approx$ 10.95 feet

✔ **Concept Check Answer**

a

PRACTICE PROBLEM 11

A football field is a rectangle measuring 100 yards by 53 yards. Draw a diagram and find the length of the diagonal of a football field to the nearest yard.

Answer

11. ≈113 yards

EXAMPLE 11 Finding the Dimensions of a Park

An inner-city park is in the shape of a square that measures 300 feet on a side. A sidewalk is to be constructed along the diagonal of the park. Find the length of the sidewalk rounded to the nearest whole foot.

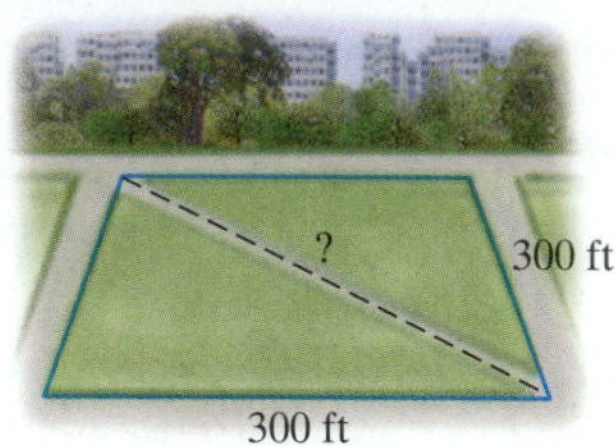

Solution: The diagonal is the hypotenuse of a right triangle, which we label c.

$$\begin{aligned} a^2 + b^2 &= c^2 \\ 300^2 + 300^2 &= c^2 && \text{Let } a = 300 \text{ and } b = 300. \\ 90{,}000 + 90{,}000 &= c^2 && \text{Evaluate } (300)^2. \\ 180{,}000 &= c^2 && \text{Add.} \\ \sqrt{180{,}000} &= c \text{ or } c \approx 424 \end{aligned}$$

The length of the sidewalk is approximately 424 feet.

Work Practice Problem 11

CALCULATOR EXPLORATIONS Finding and Approximating Square Roots

To simplify or approximate square roots using a calculator, locate the key marked $\boxed{\sqrt{\ }}$.

To simplify $\sqrt{64}$, for example, press the keys

$\boxed{64}\ \boxed{\sqrt{\ }}$ or $\boxed{\sqrt{\ }}\ \boxed{64}$

The display should read $\boxed{8}$. Then

$\sqrt{64} = 8$

To *approximate* $\sqrt{10}$, press the keys

$\boxed{10}\ \boxed{\sqrt{\ }}$ or $\boxed{\sqrt{\ }}\ \boxed{10}$

The display should read $\boxed{3.16227766}$. This is an *approximation* for $\sqrt{10}$. A three-decimal-place approximation is

$\sqrt{10} \approx 3.162$

Is this answer reasonable? Since 10 is between the perfect squares 9 and 16, $\sqrt{10}$ is between $\sqrt{9} = 3$ and $\sqrt{16} = 4$. Our answer is reasonable since 3.162 is between 3 and 4.

Simplify.

1. $\sqrt{1024}$

2. $\sqrt{676}$

Approximate each square root. Round each answer to the nearest thousandth.

3. $\sqrt{15}$

4. $\sqrt{19}$

5. $\sqrt{97}$

6. $\sqrt{56}$

Vocabulary and Readiness Check

Use the choices below to fill in each blank. Not all choices will be used, and some choices will be used more than once.

squaring	b^2	radical	-10	leg
hypotenuse	perfect squares	10	c^2	

1. A square root of 100 is ______________ and ______________ because $10 \cdot 10 = 100$ and $(-10)(-10) = 100$.
2. $\sqrt{100} =$ ______________ only because $10 \cdot 10 = 100$ and 10 is positive.
3. The ______________ sign is used to denote the positive square root of a nonnegative number.
4. The reverse process of ______________ a number is finding a square root of a number.
5. The numbers 9, 1, and $\frac{1}{25}$ are called ______________.
6. Label the parts of the right triangle.

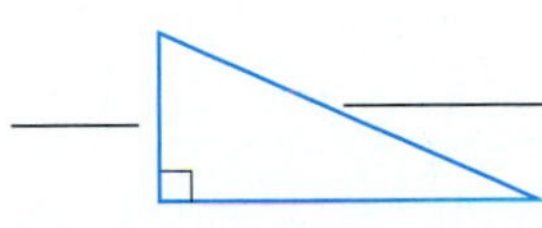

7. In the given triangle,

$a^2 +$ ________ $=$ ________.

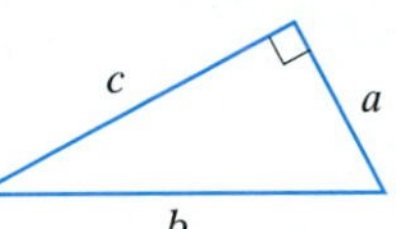

7.3 EXERCISE SET

Objective A *Find each square root. See Examples 1 through 6.*

1. $\sqrt{4}$
2. $\sqrt{9}$
3. $\sqrt{121}$
4. $\sqrt{144}$
5. $\sqrt{\frac{1}{81}}$
6. $\sqrt{\frac{1}{64}}$
7. $\sqrt{\frac{16}{64}}$
8. $\sqrt{\frac{36}{81}}$

Objective B *Use Appendix A.4 or a calculator to approximate each square root. Round the square root to the nearest thousandth. See Example 7.*

9. $\sqrt{3}$
10. $\sqrt{5}$
11. $\sqrt{15}$
12. $\sqrt{17}$
13. $\sqrt{31}$
14. $\sqrt{85}$
15. $\sqrt{26}$
16. $\sqrt{35}$

Objectives A B Mixed Practice *Find each square root. If necessary, round the square root to the nearest thousandth. See Examples 1 through 7.*

17. $\sqrt{256}$
18. $\sqrt{625}$
19. $\sqrt{92}$
20. $\sqrt{18}$
21. $\sqrt{\frac{49}{144}}$
22. $\sqrt{\frac{121}{169}}$
23. $\sqrt{71}$
24. $\sqrt{62}$

Objective **C** *Find the unknown length in each right triangle. If necessary, approximate the length to the nearest thousandth. See Examples 8 through 11.*

25.

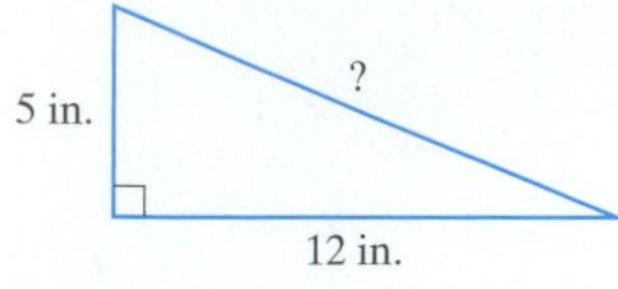

26.

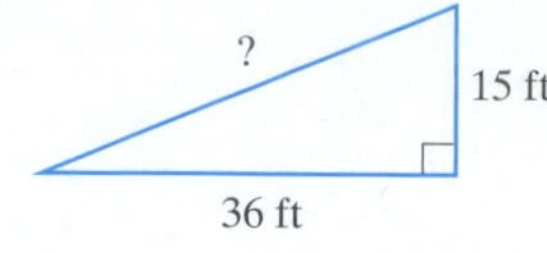

27.

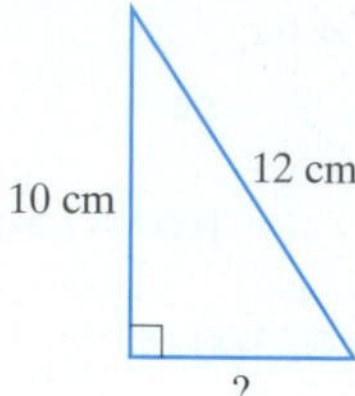

28.

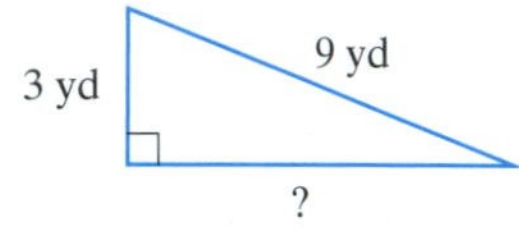

29.

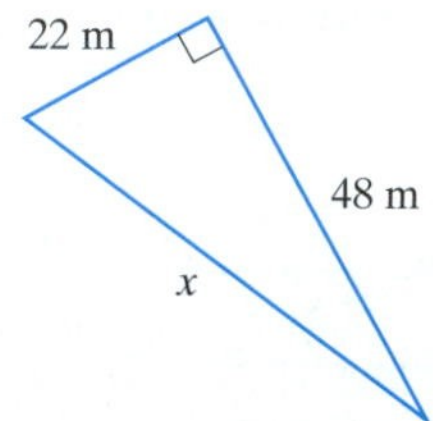

30.

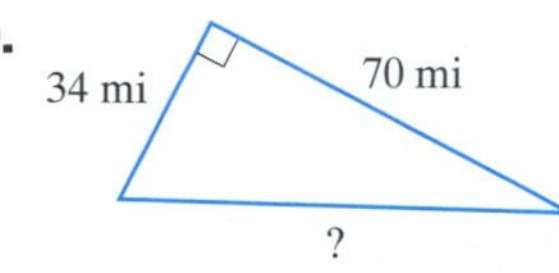

31.

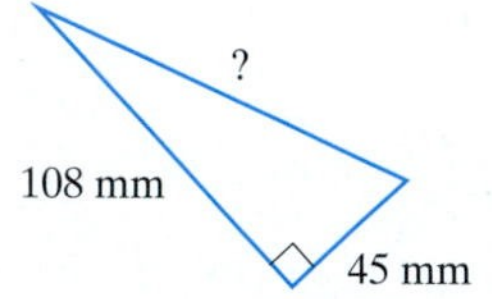

32.

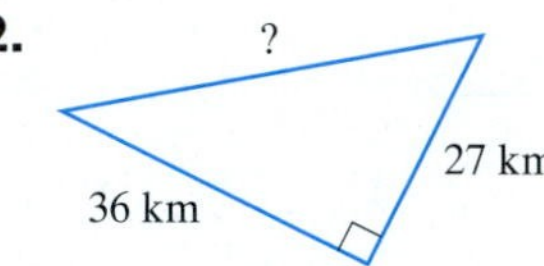

Sketch each right triangle and find the length of the side not given. If necessary, approximate the length to the nearest thousandth. (Each length is in units.) See Examples 8 through 11.

33. leg = 3, leg = 4

34. leg = 9, leg = 12

35. leg = 5, hypotenuse = 13

36. leg = 6, hypotenuse = 10

37. leg = 10, leg = 14

38. leg = 2, leg = 16

39. leg = 35, leg = 28

40. leg = 30, leg = 15

41. leg = 30, leg = 30

42. leg = 21, leg = 21

43. hypotenuse = 2, leg = 1

44. hypotenuse = 9, leg = 8

45. leg = 7.5, leg = 4

46. leg = 12, leg = 22.5

Solve. See Example 12.

47. A standard city block is a square with each side measuring 100 yards. Find the length of the diagonal of a city block to the nearest hundredth yard.

48. A section of land is a square with each side measuring 1 mile. Find the length of the diagonal of the section of land to the nearest thousandth mile.

49. Find the height of the tree. Round the height to one decimal place.

50. Find the height of the antenna. Round the height to one decimal place.

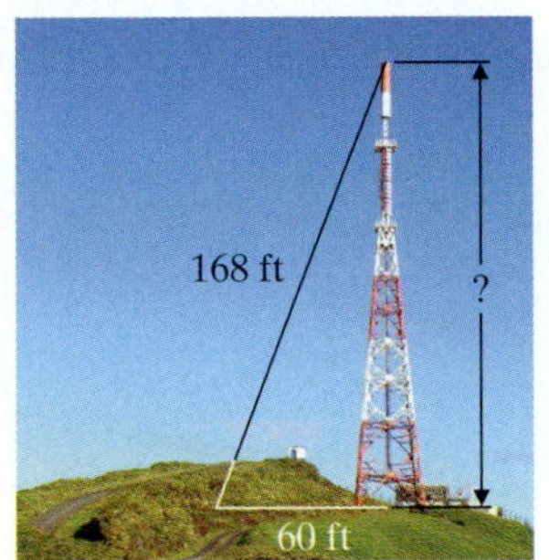

51. The playing field for football is a rectangle that is 300 feet long by 160 feet wide. Find, to the nearest foot, the length of a straight-line run that started at one corner and went diagonally to end at the opposite corner.

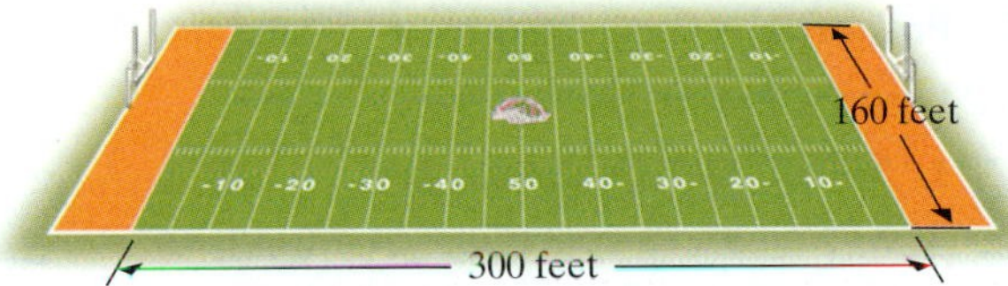

52. A soccer field is in the shape of a rectangle and its dimensions depend on the age of the players. The dimensions of the soccer field below are the minimum dimensions for international play. Find the length of the diagonal of this rectangle. Round answer to the nearest tenth of a yard.

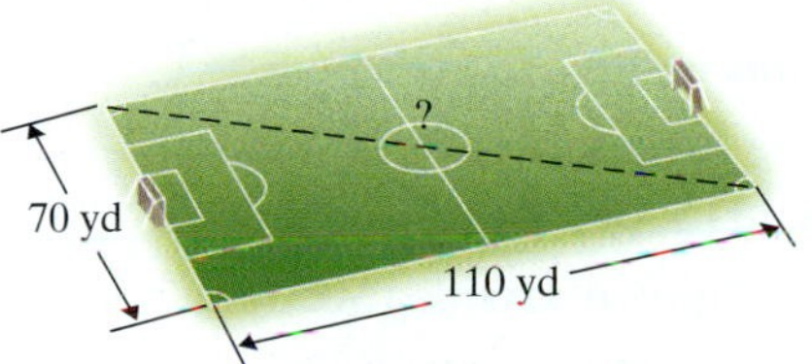

Review

Write each fraction in simplest form. See Section 4.2.

53. $\frac{10}{12}$ **54.** $\frac{10}{15}$ **55.** $\frac{2x}{60}$ **56.** $\frac{35}{75y}$

Perform indicated operations. See Sections 4.3 through 4.5.

57. $\frac{3x}{9} - \frac{5}{9}$ **58.** $\frac{9}{13y} - \frac{12}{13y}$ **59.** $\frac{9}{8} \cdot \frac{x}{8}$ **60.** $\frac{7x}{11} \div \frac{8x}{11}$

Concept Extensions

Determine what two whole numbers each square root is between without using a calculator or table. Then use a calculator or appendix to check.

61. $\sqrt{38}$ **62.** $\sqrt{27}$ **63.** $\sqrt{101}$ **64.** $\sqrt{85}$

65. Without using a calculator, explain how you know that $\sqrt{105}$ is *not* approximately 9.875.

Does the set form the lengths of the sides of a right triangle? See the Concept Check in this section.

66. 25, 60, 65

67. 20, 45, 50

△ **68.** Find the exact length of x. Then give a two-decimal-place approximation.

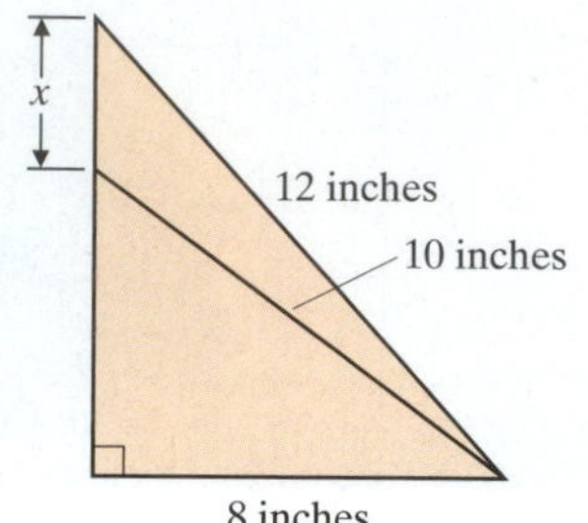

THE BIGGER PICTURE Operations on Sets of Numbers and Solving Equations

Continue your outline from Sections 1.6, 1.7, 2.4, 3.3, 4.3, 4.7, 5.4, 6.1, 6.3, and 6.4. Suggestions are once again written to help you complete this part of your outline.

I. Operations on Sets of Numbers
- **A. Whole Numbers**
  1. **Add or Subtract** (Section 1.3)
  2. **Multiply or Divide** (Sections 1.5, 1.6)
  3. **Exponent** (Section 1.7)
  4. **Order of Operations** (Section 1.7)
  5. **Square Root:** $\sqrt{25} = 5$ *because* $5 \cdot 5 = 25$
- **B. Integers**
  1. **Add** (Section 2.2)
  2. **Subtract** (Section 2.3)
  3. **Multiply or Divide** (Section 2.4)
- **C. Fractions**
  1. **Simplify** (Section 4.2)
  2. **Multiply** (Section 4.3)
  3. **Divide** (Section 4.3)
  4. **Add or Subtract** (Sections 4.4, 4.5)
- **D. Decimals**
  1. **Add or Subtract** (Section 5.2)
  2. **Multiply** (Section 5.3)
  3. **Divide** (Section 5.4)

II. Solving Equations
- **A. Equations in General** (Section 3.3)
- **B. Proportions** (Section 6.1)
- **C. Percent Problems**
  1. **Solved by Equations** (Section 6.3)
  2. **Solved by Proportions** (Section 6.4)

Perform the indicated operations.

1. $\sqrt{81}$

2. $\sqrt{\frac{4}{49}}$

3. $\sqrt{9} \cdot \sqrt{25}$

4. $\sqrt{16} \cdot \sqrt{4}$

5. $6 \cdot \sqrt{9} + 3 \cdot \sqrt{4}$

6. $3 \cdot \sqrt{25} + 2 \cdot \sqrt{81}$

7. $4 \cdot \sqrt{49} - 0 \div \sqrt{100}$

8. $7 \cdot \sqrt{36} - 0 \div \sqrt{64}$

9. $\frac{\sqrt{4}}{3} + \frac{\sqrt{25}}{6}$

10. $\frac{\sqrt{9}}{16} + \frac{\sqrt{49}}{8}$

7.4 CONGRUENT AND SIMILAR TRIANGLES

Objectives

A Decide Whether Two Triangles Are Congruent.

B Find the Ratio of Corresponding Sides in Similar Triangles.

C Find Unknown Lengths of Sides in Similar Triangles.

Objective A Deciding Whether Two Triangles Are Congruent

Two triangles are **congruent** when they have the same shape and the same size. In congruent triangles, the measures of corresponding angles are equal and the lengths of corresponding sides are equal. The following triangles are congruent:

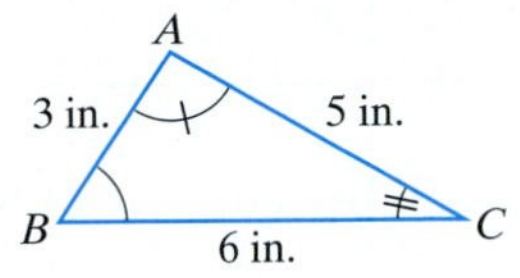

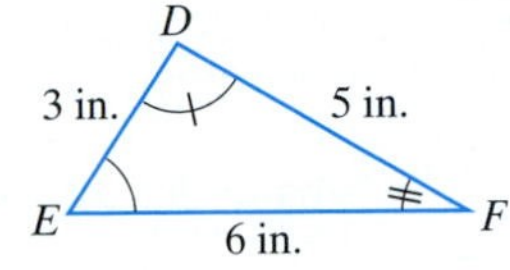

Since these triangles are congruent, the measures of corresponding angles are equal.

Angles with equal measure: $\angle A$ and $\angle D$, $\angle B$ and $\angle E$, $\angle C$ and $\angle F$. Also, the lengths of corresponding sides are equal.

Equal corresponding sides: $\overline{AB}$ and $\overline{DE}$, $\overline{BC}$ and $\overline{EF}$, $\overline{CA}$ and $\overline{FD}$

Any one of the following may be used to determine whether two triangles are congruent:

Congruent Triangles

Angle-Side-Angle (ASA)

If the measures of two angles of a triangle equal the measures of two angles of another triangle, and the lengths of the sides between each pair of angles are equal, the triangles are congruent.

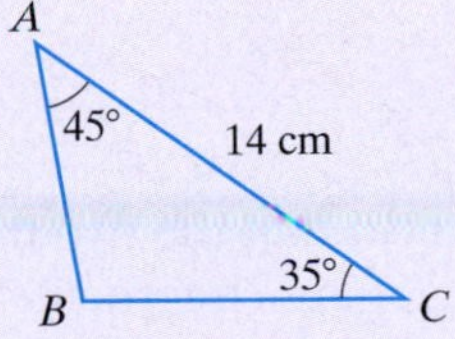

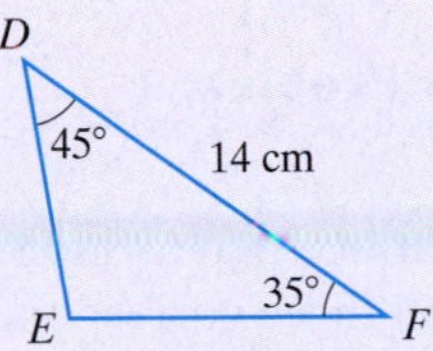

For example, these two triangles are congruent by Angle-Side-Angle.

Side-Side-Side (SSS)

If the lengths of the three sides of a triangle equal the lengths of the corresponding sides of another triangle, the triangles are congruent.

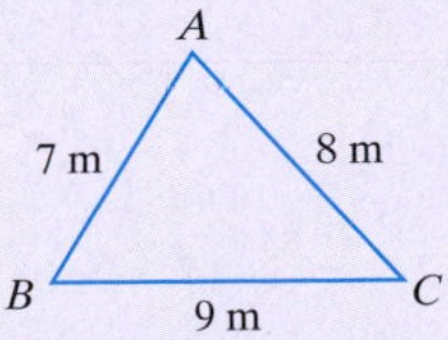

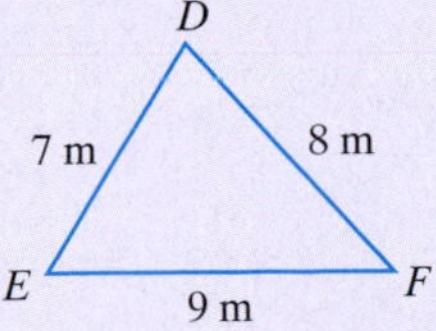

For example, these two triangles are congruent by Side-Side-Side.

(continued)

Congruent Triangles (continued)

Side-Angle-Side (SAS)

If the lengths of two sides of a triangle equal the lengths of corresponding sides of another triangle, and the measures of the angles between each pair of sides are equal, the triangles are congruent.

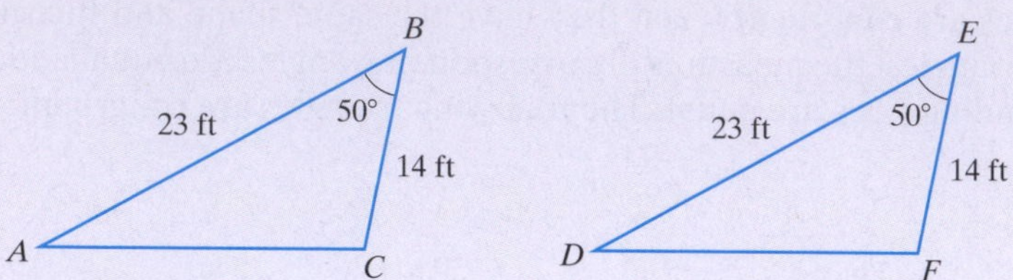

For example, these two triangles are congruent by Side-Angle-Side.

PRACTICE PROBLEM 1

a. Determine whether triangle MNO is congruent to triangle RQS.

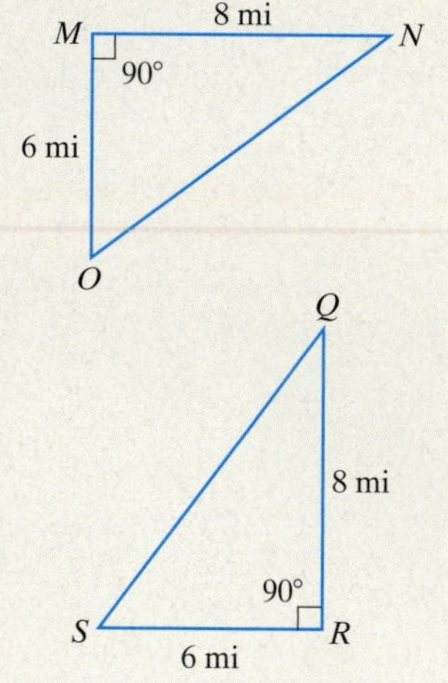

b. Determine whether triangle GHI is congruent to triangle JKL.

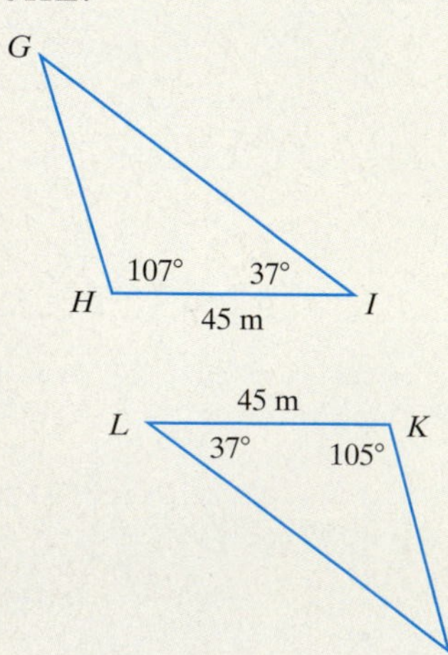

Answers

1. a. congruent **b.** not congruent

EXAMPLE 1 Determine whether triangle ABC is congruent to triangle DEF.

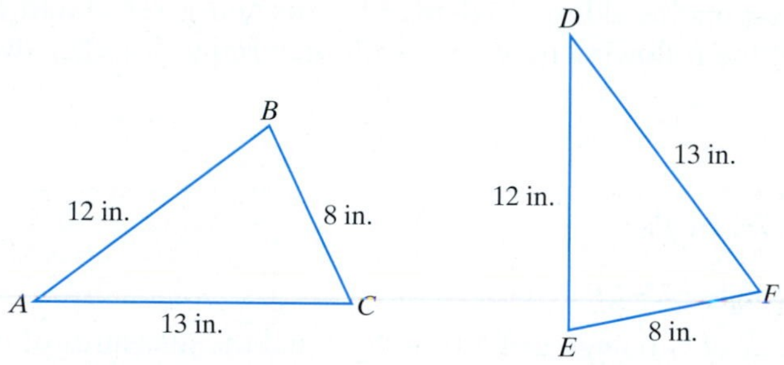

Solution: Since the lengths of all three sides of triangle ABC equal the lengths of all three sides of triangle DEF, the triangles are congruent.

Work Practice Problem 1

In Example 1, notice that as soon as we know that the two triangles are congruent, we know that all three corresponding angles are congruent.

Objective B Finding the Ratios of Corresponding Sides in Similar Triangles

Two triangles are **similar** when they have the same shape but not necessarily the same size. In similar triangles, the measures of corresponding angles are equal and corresponding sides are in proportion. The following triangles are similar:

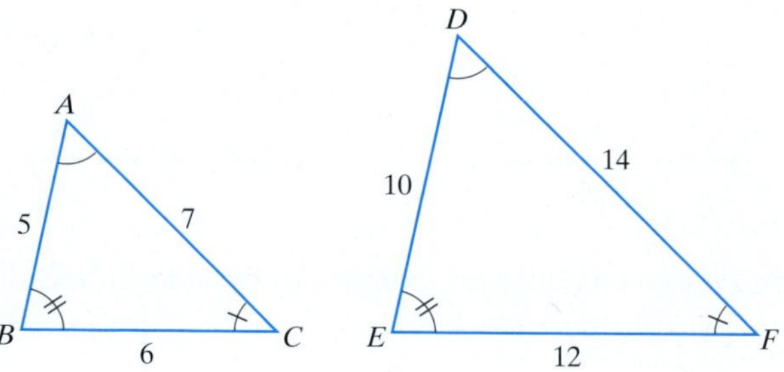

Since these triangles are similar, the measures of corresponding angles are equal.

Angles with equal measure: $\angle A$ and $\angle D$, $\angle B$ and $\angle E$, $\angle C$ and $\angle F$. Also, the lengths of corresponding sides are in proportion.

Sides in proportion: $\frac{AB}{DE} = \frac{BC}{EF} = \frac{CA}{FD}$ or, in this particular case,

$$\frac{AB}{DE} = \frac{5}{10} = \frac{1}{2}, \frac{BC}{EF} = \frac{6}{12} = \frac{1}{2}, \frac{CA}{FD} = \frac{7}{14} = \frac{1}{2}$$

The ratio of corresponding sides is $\frac{1}{2}$.

EXAMPLE 2 Find the ratio of corresponding sides for the similar triangles ABC and DEF.

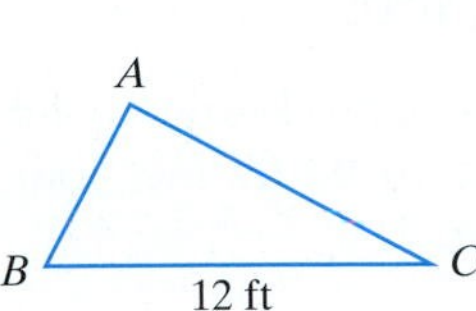

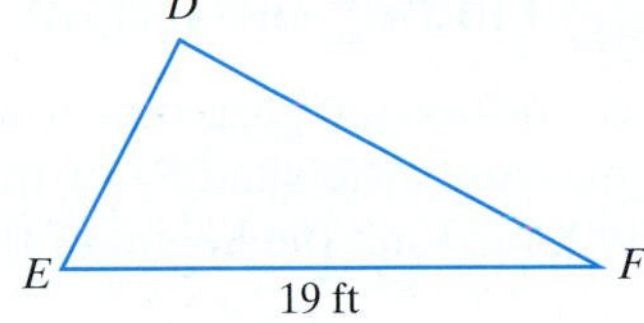

Solution: We are given the lengths of two corresponding sides. Their ratio is

$$\frac{12 \text{ feet}}{19 \text{ feet}} = \frac{12}{19}$$

Work Practice Problem 2

PRACTICE PROBLEM 2

Find the ratio of corresponding sides for the similar triangles QRS and XYZ.

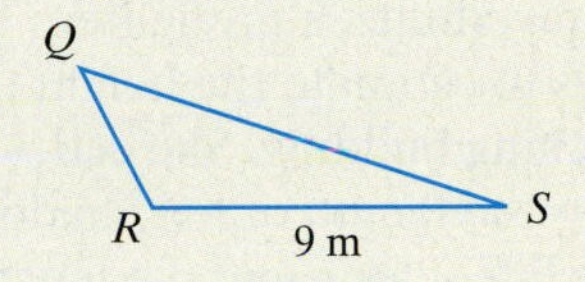

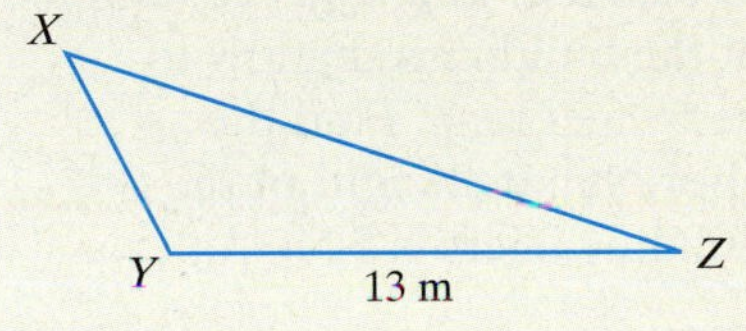

Objective C Finding Unknown Lengths of Sides in Similar Triangles

Because the ratios of lengths of corresponding sides are equal, we can use proportions to find unknown lengths in similar triangles.

EXAMPLE 3 Given that the triangles are similar, find the missing length y.

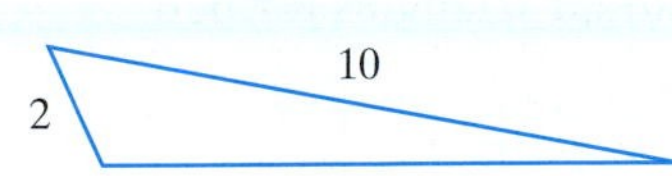

Solution: Since the triangles are similar, corresponding sides are in proportion. Thus, the ratio of 2 to 3 is the same as the ratio of 10 to y, or

$$\frac{2}{3} = \frac{10}{y}$$

To find the unknown length y, we set cross products equal.

$$\frac{2}{3} = \frac{10}{y}$$

$2 \cdot y = 3 \cdot 10$ Set cross products equal.

$2y = 30$ Multiply.

$\frac{2y}{2} = \frac{30}{2}$ Divide both sides by 2.

$y = 15$ Simplify.

The missing length is 15 units.

Work Practice Problem 3

PRACTICE PROBLEM 3

Given that the triangles are similar, find the missing length x.

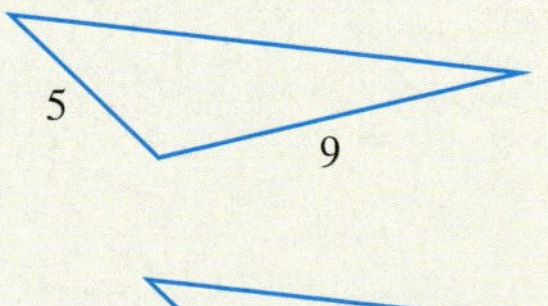

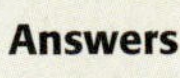

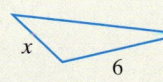

Answers

2. $\frac{9}{13}$ **3.** $x = \frac{10}{3}$ or $3\frac{1}{3}$

✔**Concept Check** The following two triangles are similar. Which vertices of the first triangle appear to correspond to which vertices of the second triangle?

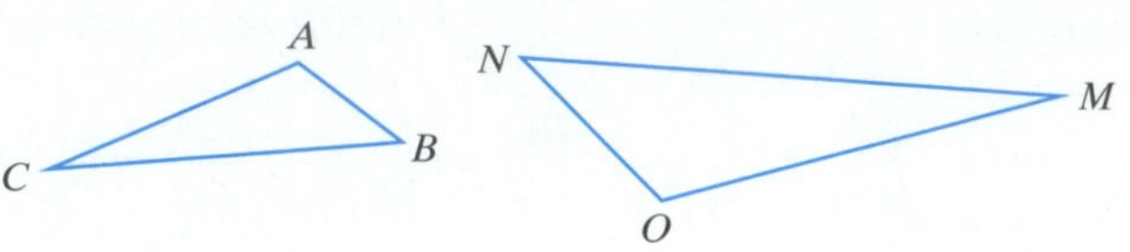

Many applications involve a diagram containing similar triangles. Surveyors, astronomers, and many other professionals continually use similar triangles in their work.

PRACTICE PROBLEM 4

Tammy Shultz, a firefighter, needs to estimate the height of a burning building. She estimates the length of her shadow to be 8 feet long and the length of the building's shadow to be 60 feet long. Find the approximate height of the building if she is 5 feet tall.

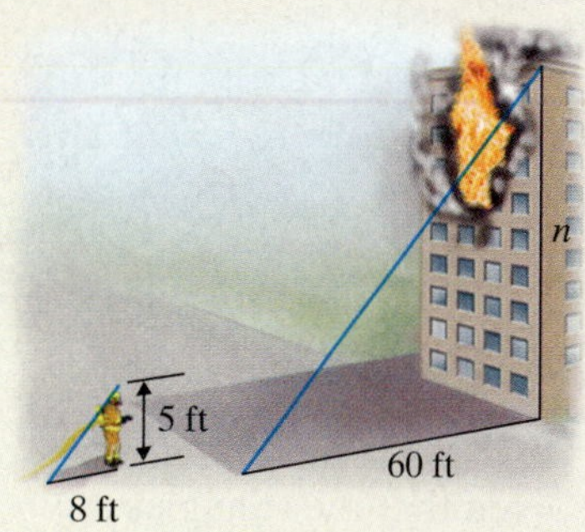

EXAMPLE 4 Finding the Height of a Tree

Mel Wagstaff is a 6-foot-tall park ranger who needs to know the height of a particular tree. He measures the shadow of the tree to be 69 feet long when his own shadow is 9 feet long. Find the height of the tree.

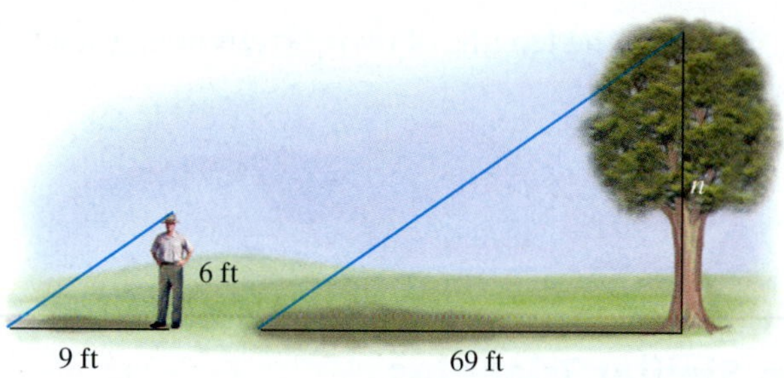

Solution:

1. UNDERSTAND. Read and reread the problem. Notice that the triangle formed by the Sun's rays, Mel, and his shadow is similar to the triangle formed by the Sun's rays, the tree, and its shadow.
2. TRANSLATE. Write a proportion from the similar triangles formed.

$$\frac{\text{Mel's height}}{\text{height of tree}} \rightarrow \frac{6}{x} = \frac{9}{69} \leftarrow \frac{\text{length of Mel's shadow}}{\text{length of tree's shadow}}$$

$$\text{or } \frac{6}{x} = \frac{3}{23} \quad \text{Simplify } \frac{9}{69}. \text{ (ratio in lowest terms)}$$

3. SOLVE for x:

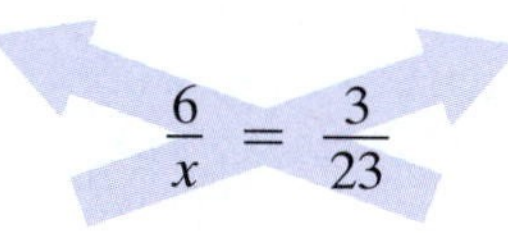

$$6 \cdot 23 = x \cdot 3 \quad \text{Set cross products equal.}$$
$$138 = 3x \quad \text{Multiply.}$$
$$\frac{138}{3} = \frac{3x}{3} \quad \text{Divide both sides by 3.}$$
$$46 = x$$

4. INTERPRET. *Check* to see that replacing x with 46 in the proportion makes the proportion true. *State* your conclusion: The height of the tree is 46 feet.

Work Practice Problem 4

Answer

4. approximately 37.5 ft

✔ Concept Check Answer

A corresponds to *O*; *B* corresponds to *N*; *C* corresponds to *M*

Vocabulary and Readiness Check

Answer each question true or false.

1. Two triangles that have the same shape but not necessarily the same sign are congruent.
2. Two triangles are congruent if they have the same shape and size.
3. Congruent triangles are also similar.
4. Similar triangles are also congruent.
5. For the two similar triangles, the ratio of corresponding sides is $\frac{5}{6}$.

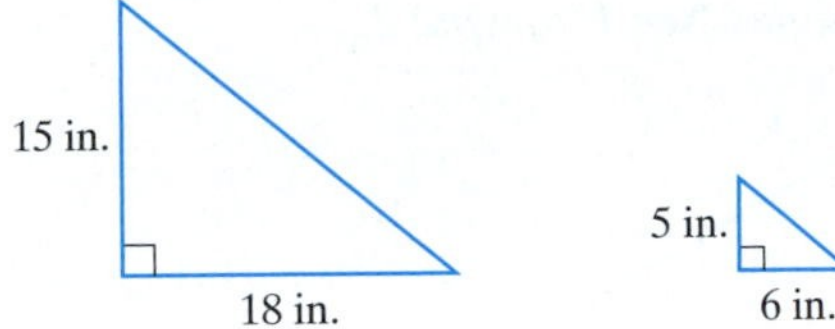

Objective A *Determine whether each pair of triangles is congruent. See Example 1.*

1.

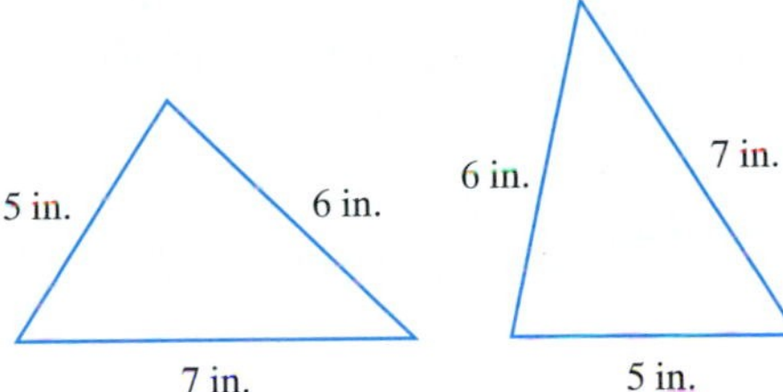

2.

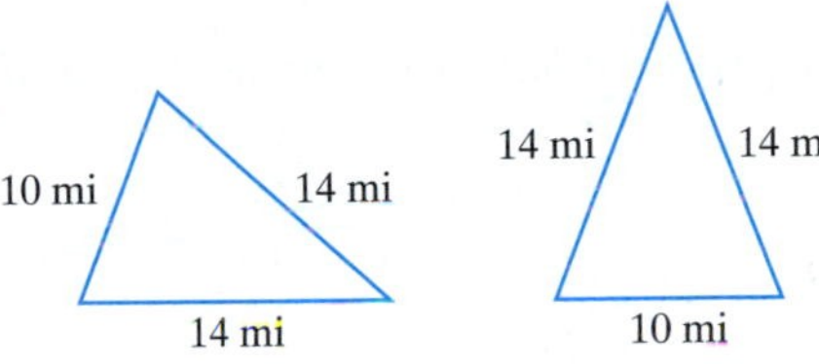

3.

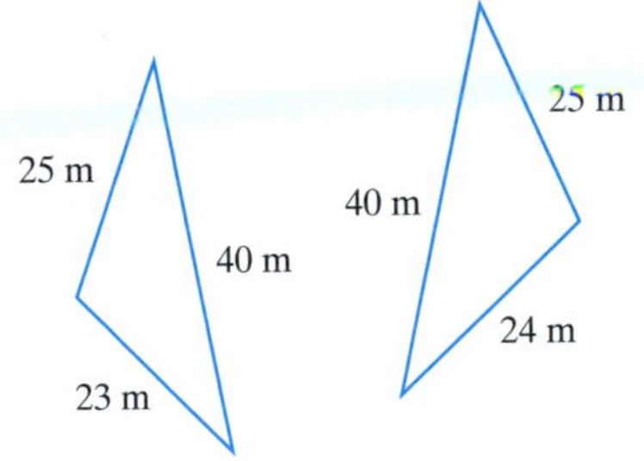

4.

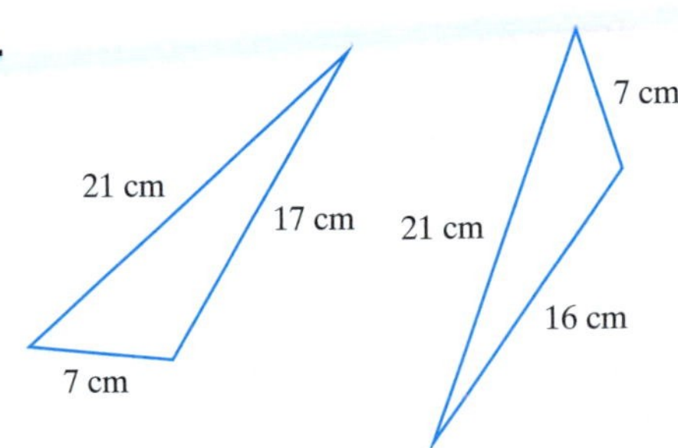

5.

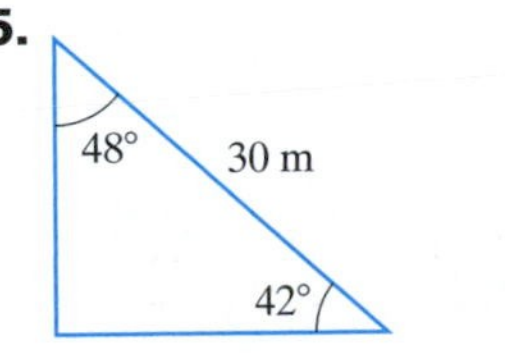

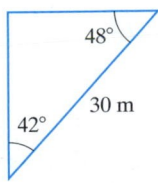

6.

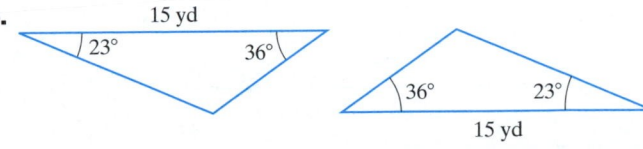

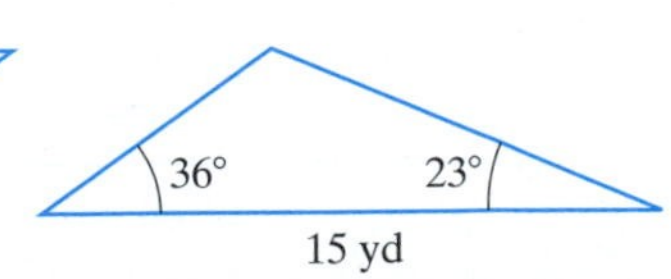

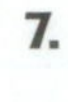

7.

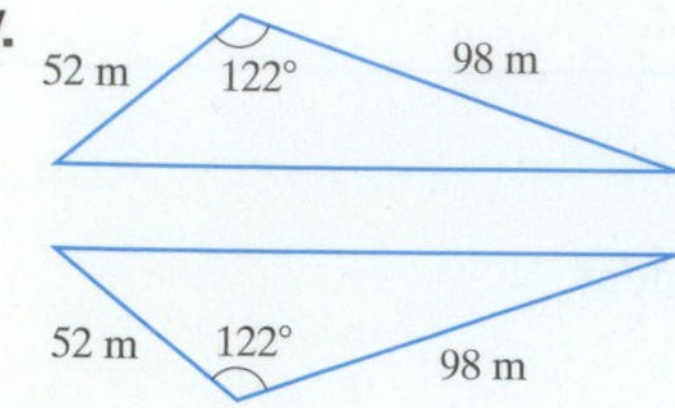

8.

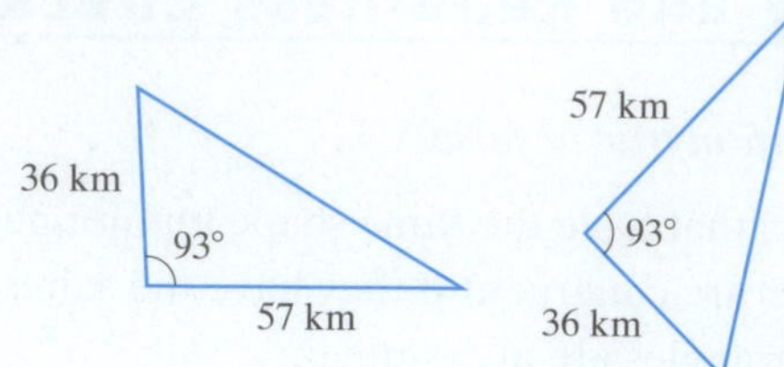

Objective B *Find each ratio of the corresponding sides of the given similar triangles. See Example 2.*

9.

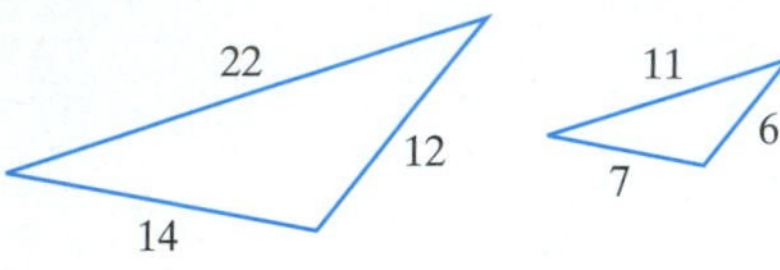

10.

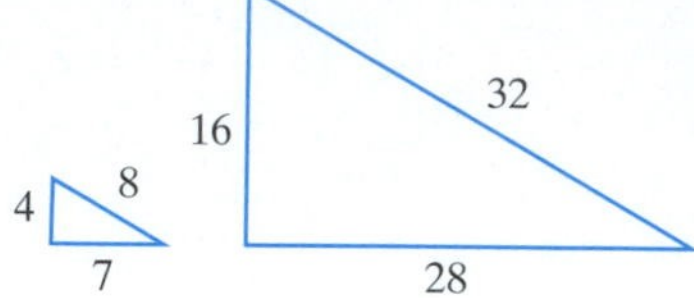

11.

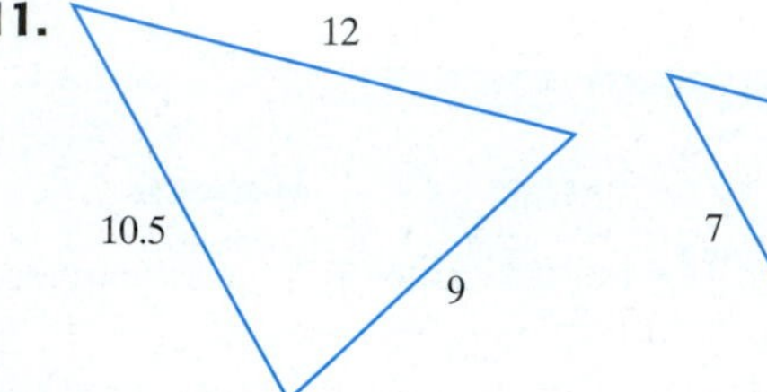

12.

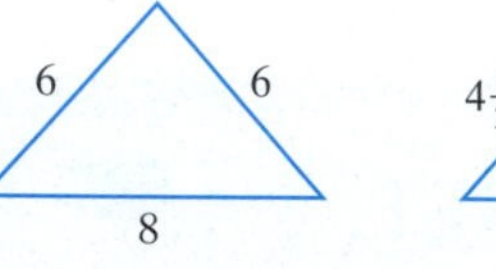

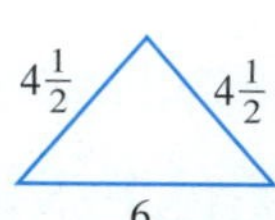

Objective C *Given that the pairs of triangles are similar, find the unknown length of the side labeled with a variable. See Example 3.*

13.

14.

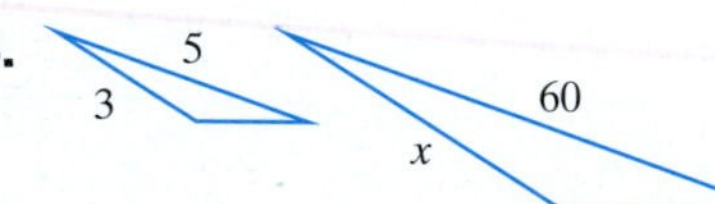

15.

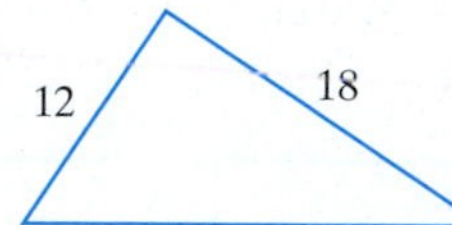

16.

4

7

y

14

17.

y

12

18.

9

15

19.

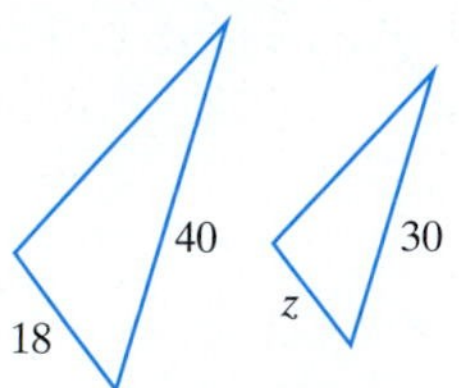

20.

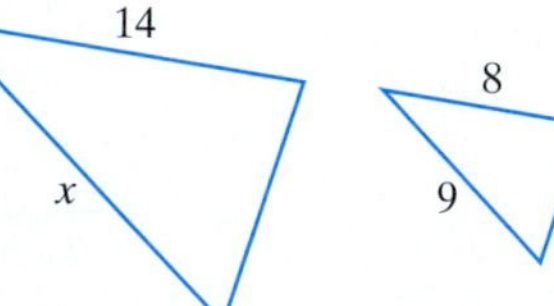

21.

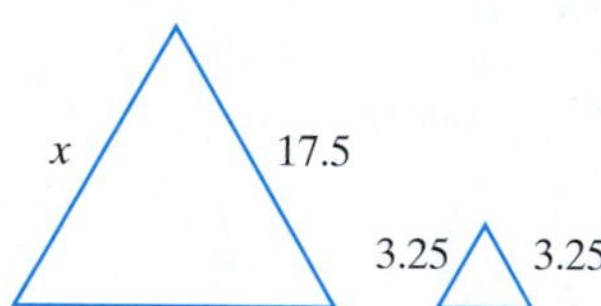

22.

8.3

9.6

23.

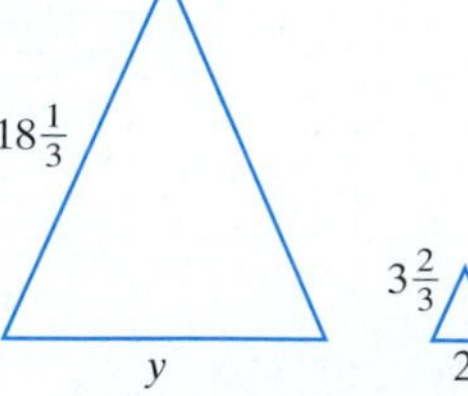

24.

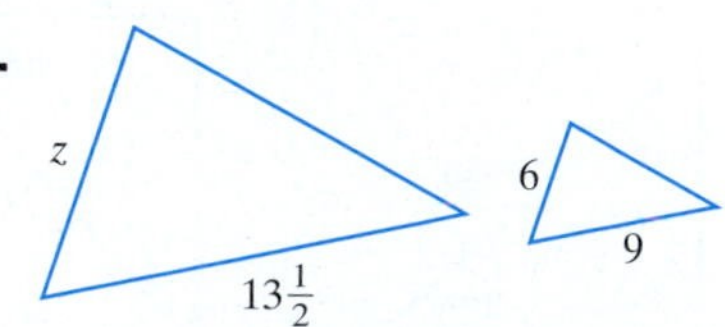

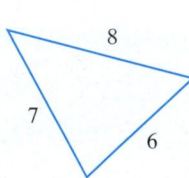

25.

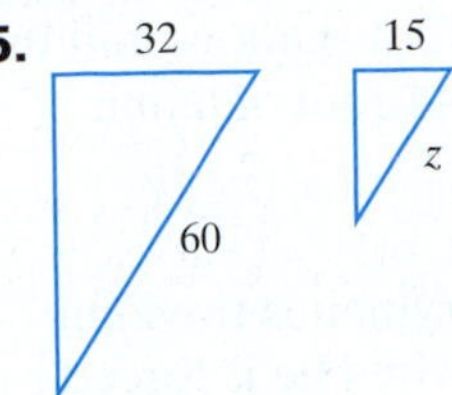

26.

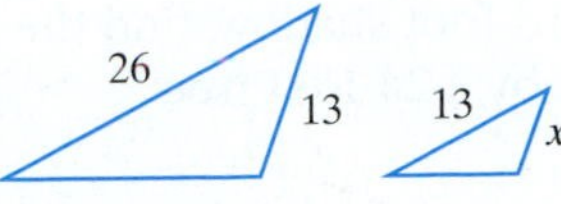

27.

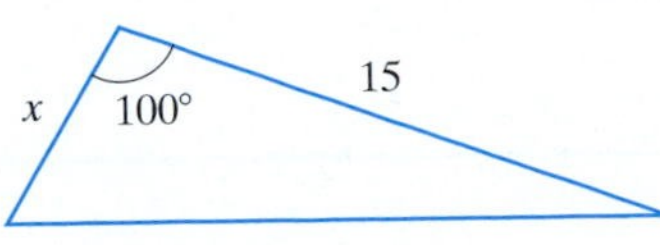

28.

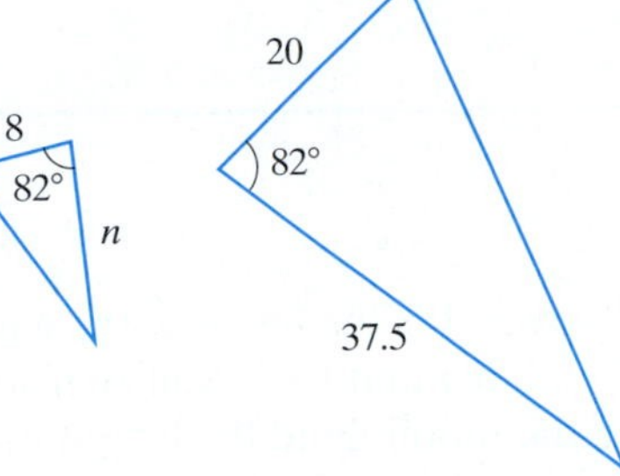

Solve. See Example 4.

29. Given the following diagram, approximate the height of the Bank One Tower in Oklahoma City, OK. (*Source: The World Almanac,* 2006)

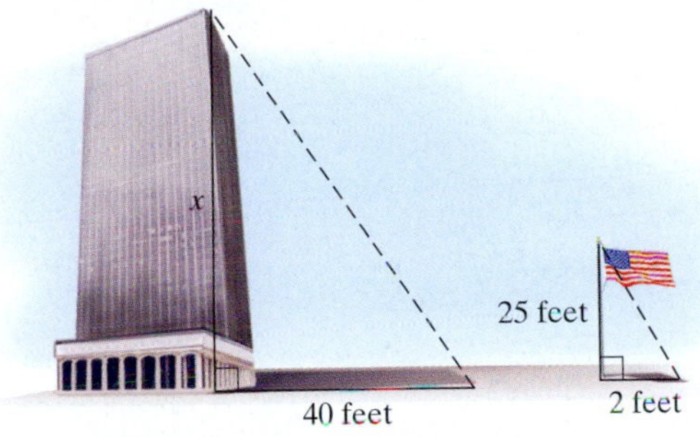

30. The tallest tree standing today is a redwood located in the Humboldt Redwoods State Park near Ukiah, California. Given the following diagram, approximate its height. (*Source: Guinness World Records,* 2006)

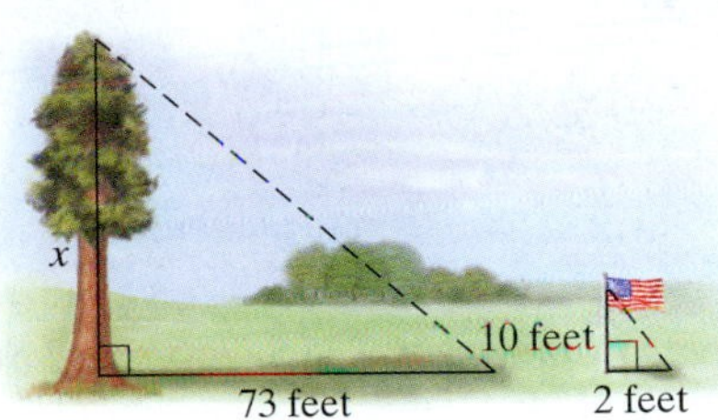

31. Fountain Hills, Arizona, boasts the tallest fountain in the world. The fountain sits in a 28-acre lake and shoots up a column of water every hour. Based on the diagram below, what is the approximate height of the fountain?

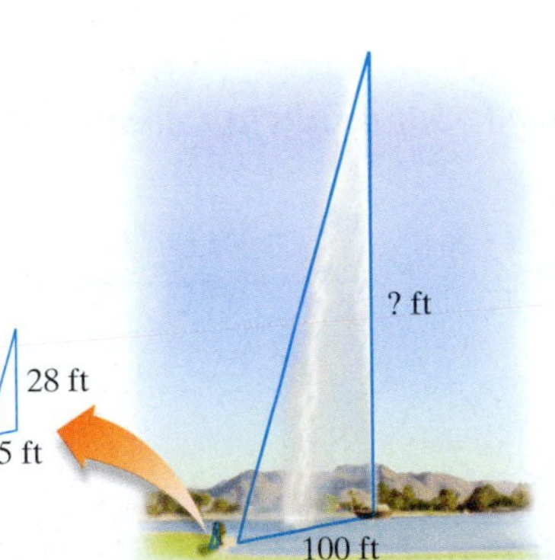

32. Lloyd White, a firefighter, needs to estimate the height of a burning building. He estimates the length of his shadow to be 9 feet long and the length of the building's shadow to be 75 feet long. Find the approximate height of the building if he is 6 feet tall.

33. If a 30-foot tree casts an 18-foot shadow, find the length of the shadow cast by a 24-foot tree.

34. If a 24-foot flagpole casts a 32-foot shadow, find the length of the shadow cast by a 44-foot antenna. Round to the nearest tenth.

35. For the health of his fish, Pete's Sea World uses the standard that a 20-gallon tank should only house 19 neon tetras. Find the number of neon tetras that Pete would place into a 55-gallon tank.

36. A local package express deliveryman is traveling the city expressway at 45 mph when he is forced to slow down due to traffic ahead. His truck slows at the rate of 3 mph every 5 seconds. Find his speed 8 seconds after braking.

Review

Solve. See Section 7.3.

37. Launch Umbilical Tower 1 is the name of the gantry used for the *Apollo* launch that took Neil Armstrong and Buzz Aldrin to the moon. Find the height of the gantry to the nearest whole number.

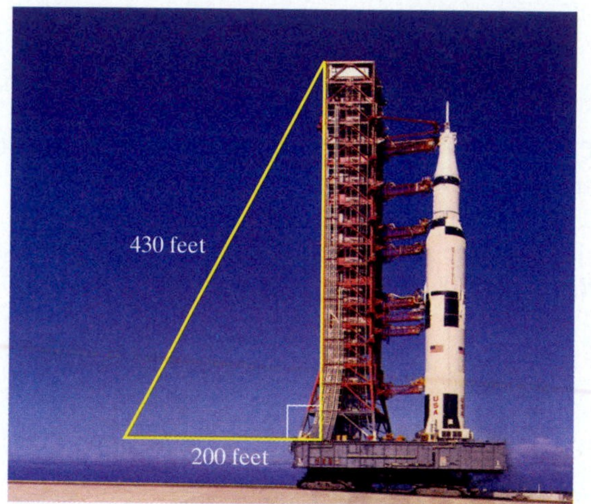

38. A regulation NCAA basketball court is a 94'-by-50' rectangle. Find, to the nearest foot, the length of a straight-line break that starts at one corner and moves diagonally to end at the opposite corner.

Perform the indicated operation. See Sections 5.2 through 5.4.

39. $3.6 + 0.41$

40. $0.41 - 3.6$

41. $(0.41)(-3)$

42. $-0.48 \div 3$

Concept Extensions

43. The print area on a particular page measures 7 inches by 9 inches. A printing shop is to copy the page and reduce the print area so that its length is 5 inches. What will its width be? Will the print now fit on a 3-by-5-inch index card?

Given that the pairs of triangles are similar, find the length of the side labeled n. Round your results to 1 decimal place.

 **44.**

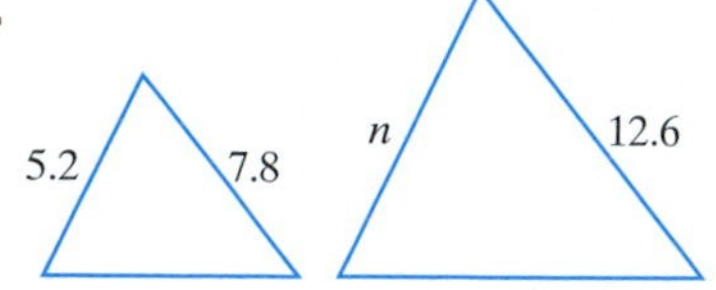

45.

46. In your own words, describe any differences in similar triangles and congruent triangles.

7.5 COUNTING AND INTRODUCTION TO PROBABILITY

Objectives

A Use a Tree Diagram to Count Outcomes.

B Find the Probability of an Event.

Objective A Using a Tree Diagram

In our daily conversations, we often talk about the likelihood or **probability** of a given result occurring. For example:

The *chance* of thundershowers is 70 percent.

What are the *odds* that the New Orleans Saints will go to the Super Bowl?

What is the *probability* that you will finish cleaning your room today?

Each of these chance happenings—thundershowers, the New Orleans Saints playing in the Super Bowl, and cleaning your room today—is called an **experiment.** The possible results of an experiment are called **outcomes.** For example, flipping a coin is an experiment, and the possible outcomes are heads (H) or tails (T).

One way to picture the outcomes of an experiment is to draw a **tree diagram.** Each outcome is shown on a separate branch. For example, the outcomes of flipping a coin are

Heads

Tails

EXAMPLE 1 Draw a tree diagram for tossing a coin twice. Then use the diagram to find the number of possible outcomes.

Solution: There are 4 possible outcomes when tossing a coin twice.

First Coin Toss	Second Coin Toss	Outcomes
H	H	H, H
	T	H, T
T	H	T, H
	T	T, T

Work Practice Problem 1

PRACTICE PROBLEM 1

Draw a tree diagram for tossing a coin three times. Then use the diagram to find the number of possible outcomes.

Answer

1.

H	H	H
		T
	T	H
		T
T	H	H
		T
	T	H
		T

8 outcomes

PRACTICE PROBLEM 2

Draw a tree diagram for an experiment consisting of tossing a coin and then rolling a die. Then use the diagram to find the number of possible outcomes.

EXAMPLE 2 Draw a tree diagram for an experiment consisting of rolling a die and then tossing a coin. Then use the diagram to find the number of possible outcomes.

Die

Solution: Recall that a die has six sides and that each side represents a number, 1 through 6.

Roll a Die	Toss a coin	Outcomes
1	H	1, H
	T	1, T
2	H	2, H
	T	2, T
3	H	3, H
	T	3, T
4	H	4, H
	T	4, T
5	H	5, H
	T	5, T
6	H	6, H
	T	6, T

There are 12 possible outcomes for rolling a die and then tossing a coin.

Work Practice Problem 2

Any number of outcomes considered together are called an **event.** For example, when tossing a coin twice, H, H is an event. The event is tossing heads first and tossing heads second. Another event would be tossing tails first and then heads (T, H), and so on.

Objective B Finding the Probability of an Event

As we mentioned earlier, the **probability of an event is a measure of the chance or likelihood of it occurring.** For example, if a coin is tossed, what is the probability that heads occurs? Since one of two equally likely possible outcomes is heads, the probability is $\frac{1}{2}$.

The Probability of an Event

$$\text{probability of an event} = \frac{\text{number of ways that the event can occur}}{\text{number of possible outcomes}}$$

Note from the definition of probability that the probability of an event is always between 0 and 1, inclusive (i.e., including 0 and 1). A probability of 0 means that an event won't occur, and a probability of 1 means that an event is certain to occur.

Answer

2.

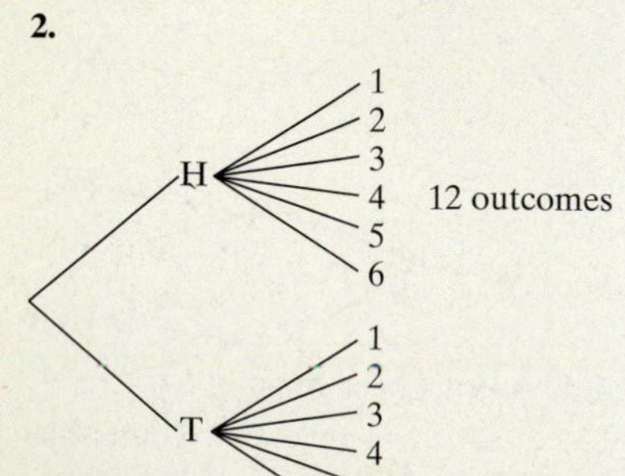

EXAMPLE 3 If a coin is tossed twice, find the probability of tossing heads on the first toss and then heads again on the second toss (H, H).

Solution: 1 way the event can occur

H, T, H, H, T, H, T, T

4 possible outcomes

$$\text{probability} = \frac{1}{4} \quad \begin{matrix}\text{Number of ways the event can occur}\\ \text{Number of possible outcomes}\end{matrix}$$

The probability of tossing heads and then heads is $\frac{1}{4}$.

Work Practice Problem 3

PRACTICE PROBLEM 3

If a coin is tossed three times, find the probability of tossing tails, then heads, then tails (T, H, T).

EXAMPLE 4 If a die is rolled one time, find the probability of rolling a 3 or a 4.

Solution: Recall that there are 6 possible outcomes when rolling a die.

2 ways that the event can occur

possible outcomes: 1, 2, 3, 4, 5, 6

6 possible outcomes

$$\text{probability of a 3 or a 4} = \frac{2}{6} \quad \begin{matrix}\text{Number of ways the event can occur}\\ \text{Number of possible outcomes}\end{matrix}$$

$$= \frac{1}{3} \quad \text{Simplest form}$$

Work Practice Problem 4

PRACTICE PROBLEM 4

If a die is rolled one time, find the probability of rolling a 2 or a 5.

✓Concept Check Suppose you have calculated a probability of $\frac{11}{9}$. How do you know that you have made an error in your calculation?

EXAMPLE 5 Find the probability of choosing a red marble from a box containing 1 red, 1 yellow, and 2 blue marbles.

Solution: 1 way that event can occur

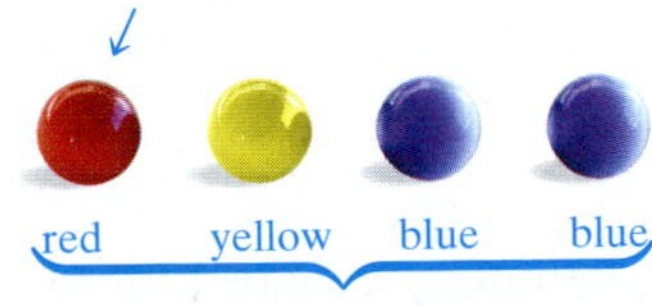

red yellow blue blue

4 possible outcomes

$$\text{probability} = \frac{1}{4}$$

Work Practice Problem 5

PRACTICE PROBLEM 5

Use the diagram and information from Example 5 and find the probability of choosing a blue marble from the box.

Answers

3. $\frac{1}{8}$ **4.** $\frac{1}{3}$ **5.** $\frac{1}{2}$

✓ Concept Check Answer

The number of ways an event can occur can't be larger than the number of possible outcomes.

Vocabulary and Readiness Check

Use the choices below to fill in each blank.

0	probability	tree diagram
1	outcome	

1. A possible result of an experiment is called a(n) ______________.

2. A ______________ shows each outcome of an experiment as a separate branch.

3. The ______________ of an event is a measure of the likelihood of it occurring.

4. ______________ is calculated by the number of ways that the event can occur divided by the number of possible outcomes.

5. A probability of ______________ means that an event won't occur.

6. A probability of ______________ means that an event is certain to occur.

7.5 EXERCISE SET

FOR EXTRA HELP

Student Solutions Manual

PH Math/Tutor Center

CD/Video for Review

MathXL®

MyMathLab

Objective A *Draw a tree diagram for each experiment. Then use the diagram to find the number of possible outcomes. See Examples 1 and 2.*

1. Choosing the letters in the word MATH, then a number (1, 2, or 3)

2. Choosing a number (1 or 2) and then a vowel (a, e, i, o, u)

Spinner A

Spinner B

3. Spinning Spinner A once

4. Spinning Spinner B once

5. Spinning Spinner B twice

6. Spinning Spinner A twice

7. Spinning Spinner A and then Spinner B

8. Spinning Spinner B and then Spinner A

9. Tossing a coin and then spinning Spinner B

10. Spinning Spinner A and then tossing a coin

Objective **B** *If a single die is tossed once, find the probability of each event. See Examples 3 through 5.*

11. A 5

12. A 9

13. A 1 or a 6

14. A 2 or a 3

15. An even number

16. An odd number

17. A number greater than 2

18. A number less than 6

Suppose the spinner shown is spun once. Find the probability of each event. See Examples 3 through 5.

19. The result of the spin is 2.

20. The result of the spin is 3.

21. The result of a spin is 1, 2, or 3.

22. The result of a spin is not 3.

23. The result of the spin is an odd number.

24. The result of the spin is an even number.

If a single choice is made from the bag of marbles shown, find the probability of each event. See Examples 3 through 5.

25. A red marble is chosen.

26. A blue marble is chosen.

27. A yellow marble is chosen.

28. A green marble is chosen.

29. A green or red marble is chosen.

30. A blue or yellow marble is chosen.

A new drug is being tested that is supposed to lower blood pressure. This drug was given to 200 people and the results are shown below.

Lower Blood Pressure	Higher Blood Pressure	Blood Pressure Not Changed
152	38	10

31. If a person is testing this drug, what is the probability that their blood pressure will be higher?

32. If a person is testing this drug, what is the probability that their blood pressure will be lower?

33. If a person is testing this drug, what is the probability that their blood pressure will not change?

34. What is the sum of the answers to Exercises 31, 32, and 33? In your own words, explain why.

Review

Perform each indicated operation. See Sections 4.3 and 4.5.

35. $\frac{1}{2} + \frac{1}{3}$ **36.** $\frac{7}{10} - \frac{2}{5}$ **37.** $\frac{1}{2} \cdot \frac{1}{3}$ **38.** $\frac{7}{10} \div \frac{2}{5}$ **39.** $5 \div \frac{3}{4}$ **40.** $\frac{3}{5} \cdot 10$

Concept Extensions

Recall that a deck of cards contains 52 cards. These cards consist of four suits (hearts, spades, clubs, and diamonds) of each of the following: 2, 3, 4, 5, 6, 7, 8, 9, 10, jack, queen, king, and ace. If a card is chosen from a deck of cards, find the probability of each event.

41. The king of hearts

42. The 10 of spades

43. A king

44. A 10

45. A heart

46. A club

47. A card in black ink

48. A queen or ace

Two dice are tossed. Find the probability of each sum of the dice. (Hint: Draw a tree diagram of the possibilities of two tosses of a die, and then find the sum of the numbers on each branch.)

49. A sum of 6

50. A sum of 10

51. A sum of 13

52. A sum of 2

Solve. See the Concept Check in this section.

53. In your own words, explain why the probability of an event cannot be greater than 1.

54. In your own words, explain when the probability of an event is 0.

CHAPTER 7 Group Activity

Stem-and-Leaf Displays

Sections 7.3, 7.4

Stem-and-leaf displays are another way to organize data. After data are logically organized, it can be much easier to draw conclusions from them.

Suppose we have collected the following set of data. It could represent the test scores for an algebra class or the pulse rates of a group of small children.

90	73	93	99	79	95	69	78	93	80
89	85	97	78	75	79	72	76	97	88
83	98	72	94	92	79	70	98	85	99

In a stem-and-leaf display, the last digit of each number forms the *leaf,* and the remaining digits to the left form the *stem*. For the first number in the list, 90, 9 is the stem and 0 is the leaf. To make the stem-and-leaf display, we write all of the stems in numerical order in a column. Then we write each leaf on the horizontal line next to its stem, aligning leaves in vertical columns. In this case, because the data range from 69 to 99, we use the stems 6, 7, 8, and 9. Each line of the display represents an interval of data; for instance, the line corresponding to the stem **7** represents all data that fall in the interval **70** to **79,** inclusive. After the data have been divided into stems and leaves on the display as shown in the table below, we simply rearrange the leaves on each line to appear in numerical order, as shown in the table in the right column.

Stem	Leaf	Stem	Leaf
6	9	6	9
7	39885926290 ⟶	7	02235688999
8	095835	8	035589
9	039537784289	9	023345778899

Now that the data have been organized into a stem-and-leaf display, it is easy to answer questions about the data such as: What are the least and greatest values in the set of data? Which data interval contains the most items from the data set? How many values fall between 74 and 84? Which data value occurs most frequently in the data set? What patterns or trends do you see in the data?

Make a stem-and-leaf display of the weekend emergency room admission data shown below. Then answer the following questions:

1. What is the difference between the least number and the greatest number of weekend ER admissions?
2. How many weekends had between 125 and 165 ER admissions?
3. What number of weekend ER admissions occurred most frequently?
4. Which interval contains the most weekend ER admissions?

Number of Emergency Room Admissions on Weekends				
198	168	117	185	159
160	177	169	112	175
170	188	137	117	145
198	169	154	163	192
167	179	155	133	121
162	188	124	145	146
128	181	198	149	140
122	162	161	180	177

Chapter 7 Vocabulary Check

Fill in each blank with one of the words or phrases listed below.

outcomes	class interval	experiment	probability	Pythagorean	congruent
pictograph	bar	class frequency	hypotenuse	unit rate	leg
histogram	circle	tree diagram	right	similar	

1. ______________ triangles have the same shape and the same size.

2. ______________ triangles have exactly the same shape but not necessarily the same size.

3–5. Label the sides of the right triangle.

3. ______ 4. ______ 5. ________

6. A triangle with one right angle is called a ______________ triangle.

7. In the right triangle, (sides a, b, c), $a^2 + b^2 = c^2$ is called the ______________ theorem.

8. A ______________ graph presents data using vertical or horizontal bars.

9. The possible results of an experiment are the ______________.

10. A ______________ is a graph in which pictures or symbols are used to visually present data.

11. A ______________ is one way to picture and count outcomes.

12. An ______________ is an activity being considered, such as tossing a coin or rolling a die.

13. In a ______________ graph, each section (shaped like a piece of pie) shows a category and the relative size of the category.

14. The ______________ of an event is $\dfrac{\text{number of ways that the event can occur}}{\text{number of possible outcomes}}$.

15. A ______________ is a special bar graph in which the width of each bar represents a ______________ and the height of each bar represents the ______________.

Helpful Hint

Are you preparing for your test? Don't forget to take the Chapter 7 Test on page 547. Then check your answers at the back of the text and use the Chapter Test Prep Video CD to see the fully worked-out solutions to any of the exercises you want to review.

7 Chapter Highlights

DEFINITIONS AND CONCEPTS	EXAMPLES

Section 7.1 Reading Pictographs, Bar Graphs, Histograms, and Line Graphs

A **pictograph** is a graph in which pictures or symbols are used to visually present data.

A **line graph** displays information with a line that connects data points.

A **bar graph** presents data using vertical or horizontal bars.

The bar graph on the right shows the number of acres of wheat harvested in 1996 for leading states.

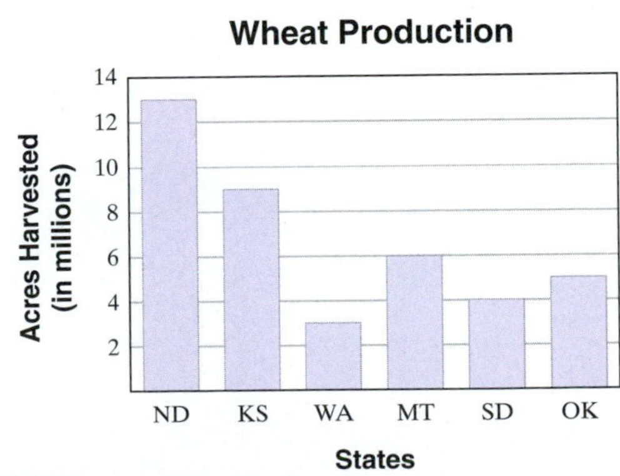

Definitions and Concepts	**Examples**
Section 7.1 Reading Pictographs, Bar Graphs, Histograms, and Line Graphs (*continued*)	
A **histogram** is a special bar graph in which the width of each bar represents a **class interval** and the height of each bar represents the **class frequency.** The histogram on the right shows student quiz scores.	**1.** Approximately how many acres of wheat were harvested in Kansas? 9,000,000 acres **2.** About how many more acres of wheat were harvested in North Dakota than South Dakota? 13 million − 4 million 9 million or 9,000,000 acres 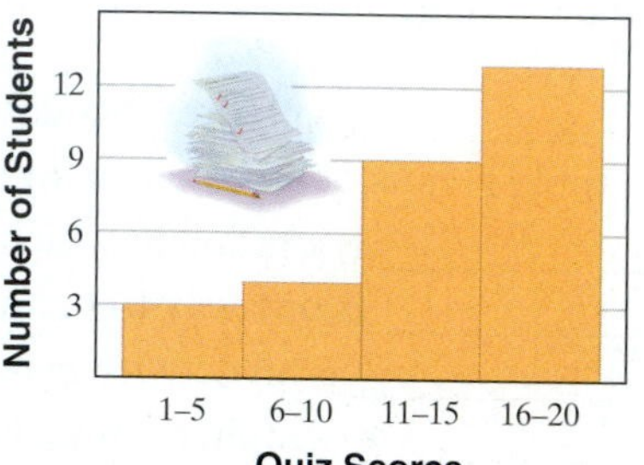 **1.** How many students received a score of 6–10? 4 students **2.** How many students received a score of 11–20? $9 + 13 = 22$ students
Section 7.2 Reading Circle Graphs	
In a **circle graph,** each section (shaped like a piece of pie) shows a category and the relative size of the category. The circle graph on the right classifies tornadoes by wind speed.	**Tornado Wind Speeds** 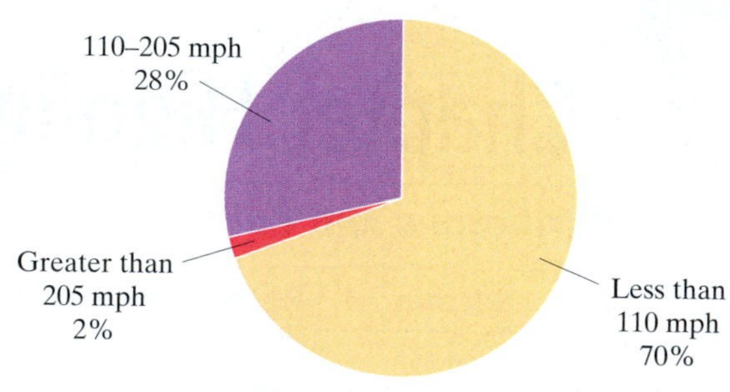 *Source:* National Oceanic and Atmospheric Administration **1.** What percent of tornadoes have wind speeds of 110 mph or greater? $28\% + 2\% = 30\%$ **2.** If there were 1235 tornadoes in the United States in 1995, how many of these might we expect to have had wind speeds less than 110 mph? Find 70% of 1235. $70\%(1235) = 0.70(1235) = 864.5 \approx 865$ Around 865 tornadoes would be expected to have had wind speeds of less than 110 mph.

Definitions and Concepts / Examples

Section 7.3 Square Roots and the Pythagorean Theorem

Square Root of a Number

The square root of a positive number a is the positive number b whose square is a. In symbols,

$$\sqrt{a} = b, \qquad \text{if} \qquad b^2 = a$$

Also, $\sqrt{0} = 0$.

Examples:

$\sqrt{9} = 3, \quad \sqrt{100} = 10, \quad \sqrt{1} = 1$

Pythagorean Theorem

If a and b are the lengths of the legs of a right triangle and c is the length of the hypotenuse, then

$$a^2 + b^2 = c^2$$

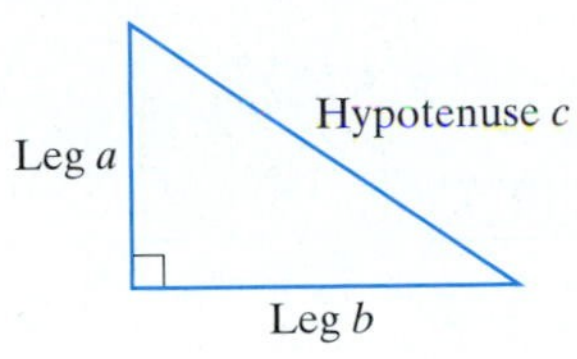

Examples:

Find c. $a = 3$, c, $b = 8$

$$a^2 + b^2 = c^2$$
$$3^2 + 8^2 = c^2 \qquad \text{Let } a = 3 \text{ and } b = 8.$$
$$9 + 64 = c^2 \qquad \text{Multiply.}$$
$$73 = c^2 \qquad \text{Simplify.}$$
$$\sqrt{73} = c \text{ or } c \approx 8.5$$

Section 7.4 Congruent and Similar Triangles

Congruent triangles have the same shape and the same size. Corresponding angles are equal, and corresponding sides are equal.

Similar triangles have exactly the same shape but not necessarily the same size. Corresponding angles are equal, and the ratios of the lengths of corresponding sides are equal.

Examples:

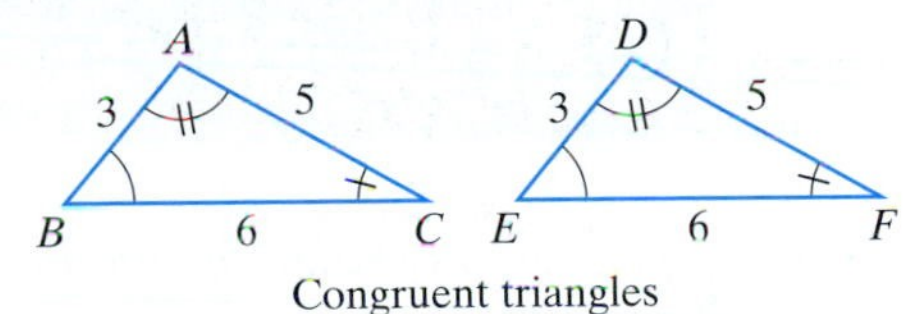

Congruent triangles

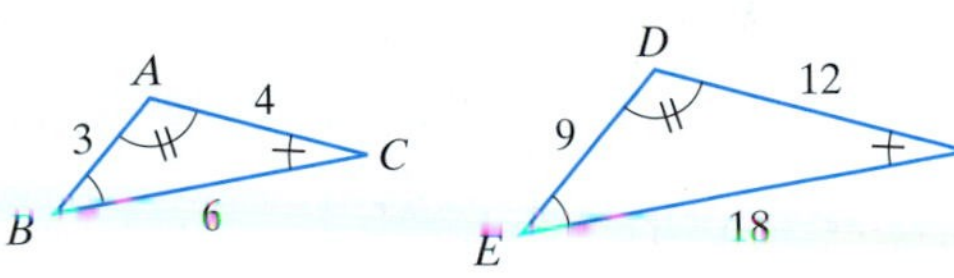

$$\frac{AB}{DE} = \frac{3}{9} = \frac{1}{3}, \frac{BC}{EF} = \frac{6}{18} = \frac{1}{3},$$

$$\frac{CA}{FD} = \frac{4}{12} = \frac{1}{3}$$

DEFINITIONS AND CONCEPTS	EXAMPLES
Section 7.5 Counting and Introduction to Probability	
An **experiment** is an activity being considered, such as tossing a coin or rolling a die. The possible results of an experiment are the **outcomes.** A **tree diagram** is one way to picture and count outcomes.	Draw a tree diagram for tossing a coin and then choosing a number from 1 to 4. 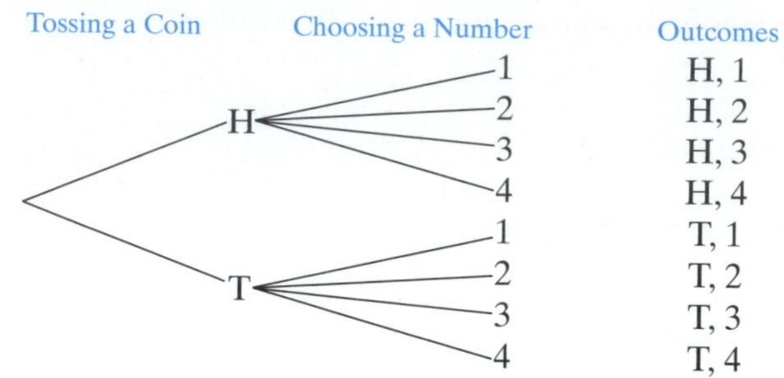
Any number of outcomes considered together is called an **event.** The **probability** of an event is a measure of the chance or likelihood of it occurring. $$\text{probability of an event} = \frac{\text{number of ways that the event can occur}}{\text{number of possible outcomes}}$$	Find the probability of tossing a coin twice and tails occurring each time. 1 way the event can occur (H,H), (H,T), (T,H), (T,T) 4 possible outcomes $$\text{probability} = \frac{1}{4}$$

7 CHAPTER REVIEW

(7.1) *The following pictograph shows the number of new homes constructed from June, 2005, to June, 2006, by region. Use this graph to answer Exercises 1 through 6.*

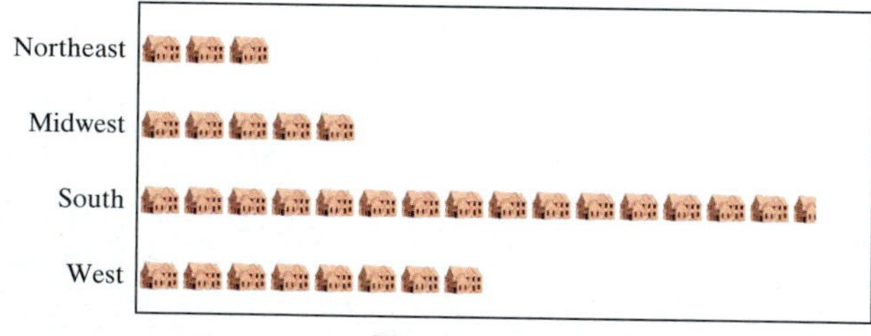

Source: U.S. Census Bureau

1. How many housing starts were there in the Midwest during the given year?

2. How many housing starts were there in the Northeast during the given year?

3. Which region had the most housing starts?

4. Which region had the fewest housing starts?

5. Which region(s) had 6,400,000 or more housing starts?

6. Which region(s) has fewer than 6,400,000 housing starts?

The following bar graph shows the percent of persons age 25 or over who completed four or more years of college. Use this graph to answer Exercises 7 through 10.

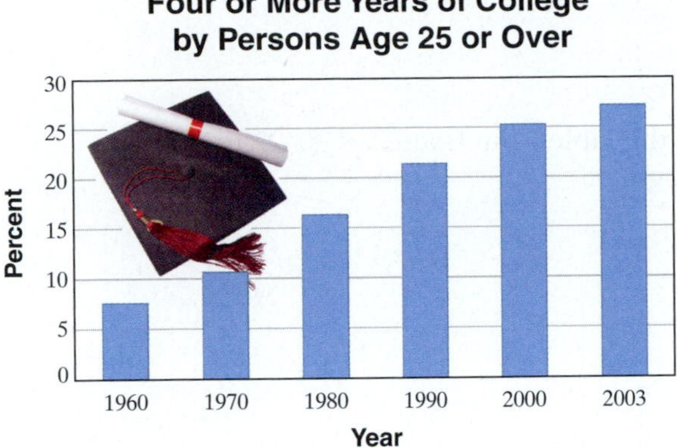

Source: U.S. Census Bureau

7. Approximate the percent of persons who completed four or more years of college in 1960.

8. What year shown had the greatest percent of persons completing four or more years of college?

9. What years shown had 15% or more of persons completing four or more years of college?

10. Describe any patterns you notice in this graph.

The following line graph shows the total number of Olympic Medal Events for Winter Olympics since 1992. Use this graph to answer Exercises 11 through 15.

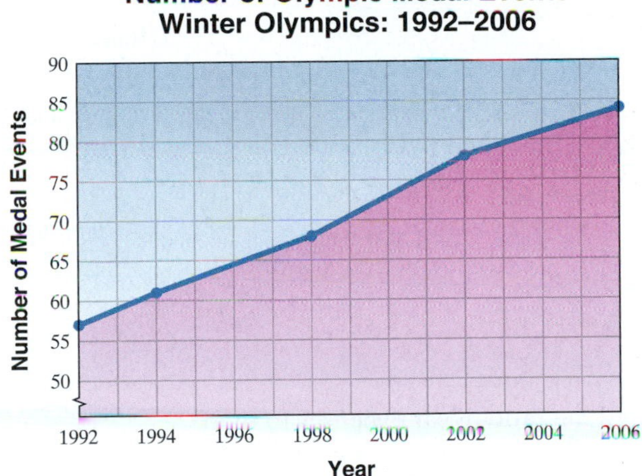

Source: International Olympic Committee

11. Approximate the number of medal events at the Winter Olympics of 1998.

12. Approximate the number of medal events at the Winter Olympics of 2002.

13. Approximate the number of medal events at the Winter Olympics of 2006.

14. How many more medal events were there at the Winter Olympics of 2006 than at the Winter Olympics of 2002?

15. How many more medal events were there at the Winter Olympics of 2006 than the Winter Olympics of 1992?

The following histogram shows the hours worked per week by the employees of Southern Star Furniture. Use this histogram to answer Exercises 16 through 19.

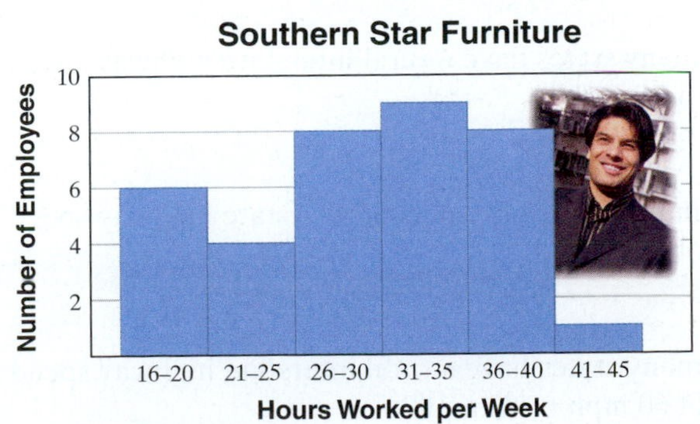

16. How many employees work 21–25 hours per week?

17. How many employees work 41–45 hours per week?

18. How many employees work 36 hours or more per week?

19. How many employees work 30 hours or less per week?

Following is a list of monthly record high temperatures for New Orleans, Louisiana. Use this list to complete the frequency distribution table below.

83	96	101	92
85	100	92	102
89	101	87	84

	Class Intervals (Temperatures)	Tally	Class Frequency (Number of Months)
20.	80°–89°	___	___
21.	90°–99°	___	___
22.	100°–109°	___	___

23. Use the table from Exercises 20–22 to draw a histogram.

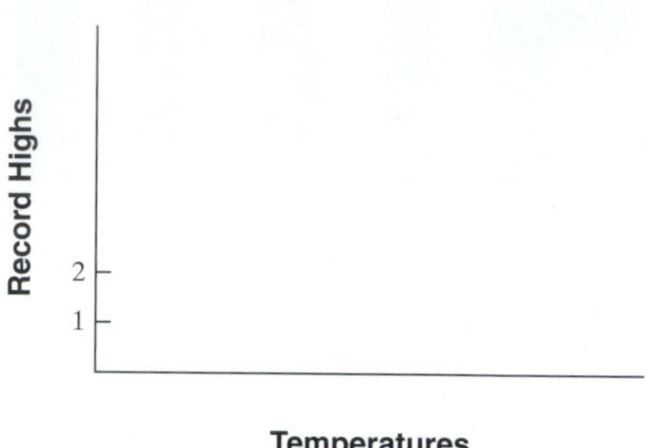

(7.2) *The following circle graph shows a family's $4000 monthly budget. Use this graph to answer Exercises 24 through 30. Write all ratios as fractions in simplest form.*

24. What is the largest budget item?

25. What is the smallest budget item?

26. How much money is budgeted for the mortgage payment and utilities?

27. How much money is budgeted for savings and contributions?

28. Find the ratio of the mortage payment to the total monthly budget.

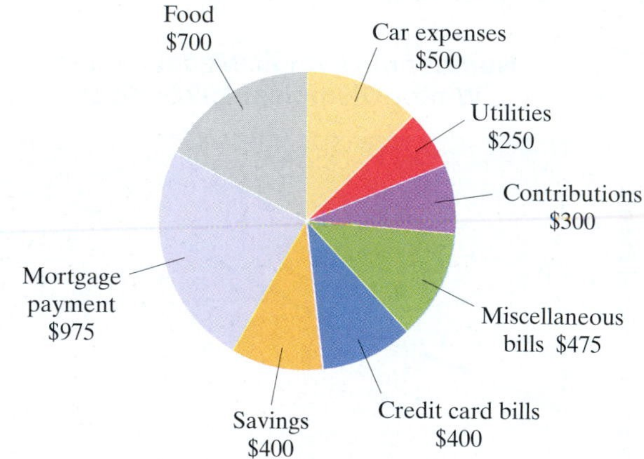

29. Find the ratio of food to the total monthly budget.

30. Find the ratio of car expenses to food.

The following circle graph shows the percent of states with various rural interstate highway speed limits in 2006. Use this graph to determine the number of states with each speed limit in Exercises 31 through 34. All states are included in this data.

Percent of States with Rural Interstate Highway Speed Limit

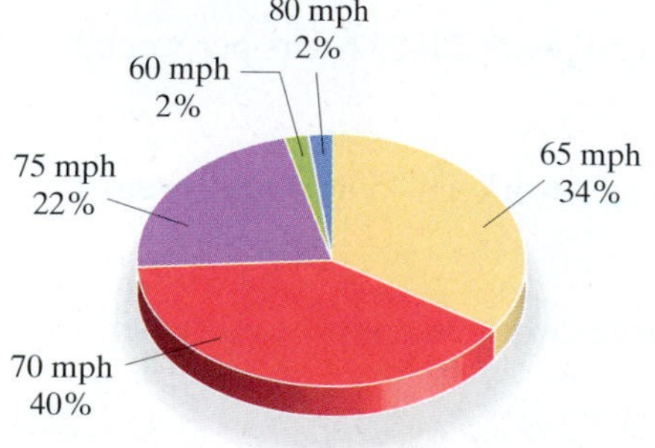

Source: Insurance Institute for Highway Safety

31. How many states have a rural interstate highway speed limit of 70 mph?

32. How many states have a rural interstate highway speed limit of 75 mph?

33. How many states have a rural interstate highway speed limit of 65 mph?

34. How many states have a rural interstate highway speed limit of 60 mph or 80 mph?

(7.3) *Find each square root. If necessary, round the square root to the nearest thousandth.*

35. $\sqrt{64}$ **36.** $\sqrt{144}$ **37.** $\sqrt{12}$ **38.** $\sqrt{15}$ **39.** $\sqrt{0}$

40. $\sqrt{1}$ **41.** $\sqrt{50}$ **42.** $\sqrt{65}$ **43.** $\sqrt{\frac{4}{25}}$ **44.** $\sqrt{\frac{1}{100}}$

Find the unknown length in each given right triangle. If necessary, round to the nearest tenth.

△ **45.** leg = 12, leg = 5

△ **46.** leg = 20, leg = 21

△ **47.** leg = 9, hypotenuse = 14

△ **48.** leg = 66, leg = 56

△ **49.** Find the length to the nearest hundredth of the diagonal of a square that has a side of length 20 centimeters.

△ **50.** Find the height of the building rounded to the nearest tenth.

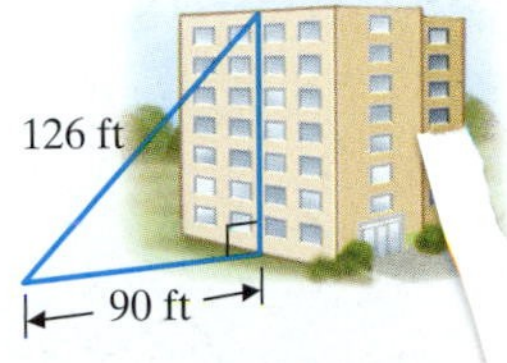

(7.4) *Determine whether each pair of triangles is congruent. If congruent, state the reason why, such as SSS, SAS, or ASA.*

51.

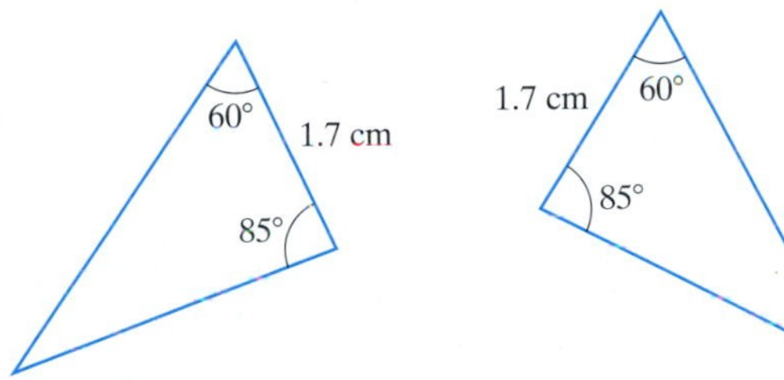

52.

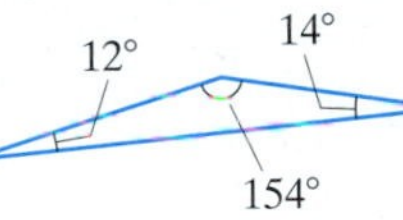

Given that the pairs of triangles are similar, find the unknown length x.

53.

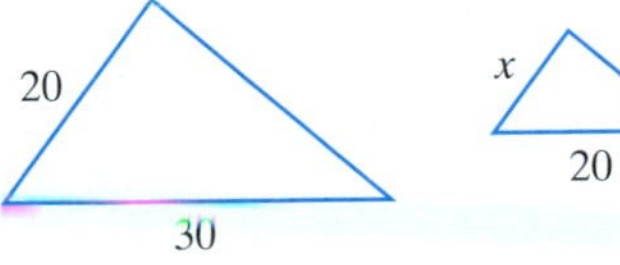

54.

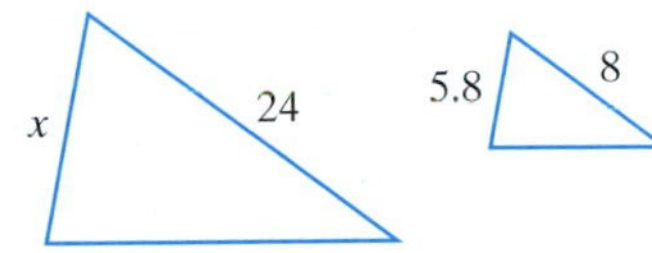

Solve.

55. A housepainter needs to estimate the height of a condominium. He estimates the length of his shadow to be 7 feet long and the length of the building's shadow to be 42 feet long. Find the approximate height of the building if the housepainter is $5\frac{1}{2}$ feet tall.

56. A design company is making a triangular sail for a model sailboat. The model sail is to be the same shape as a life-size sailboat's sail. Use the following diagram to find the unknown lengths x and y.

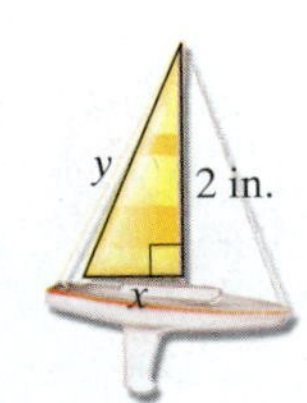

(7.5) *Draw a tree diagram for each experiment. Then use the diagram to determine the number of outcomes.*

Spinner 1

Spinner 2

57. Tossing a coin and then spinning Spinner 1

58. Spinning Spinner 2 and then tossing a coin

59. Spinning Spinner 1 twice

60. Spinning Spinner 1 and then Spinner 2

Find the probability of each event.

Die

61. Rolling a 4 on a die

62. Rolling a 3 on a die

63. Spinning a 4 on Spinner 1

64. Spinning a 3 on Spinner 1

65. Spinning either a 1, 3, or 5 on Spinner 1

66. Spinning either a 2 or a 4 on Spinner 1

Mixed Review

Given a bag containing 2 red marbles, 2 blue marbles, 3 yellow marbles, and 1 green marble, find the following:

67. The probability of choosing a blue marble from the bag

68. The probability of choosing a yellow marble from the bag

69. The probability of choosing a red marble from the bag

70. The probability of choosing a green marble from the bag

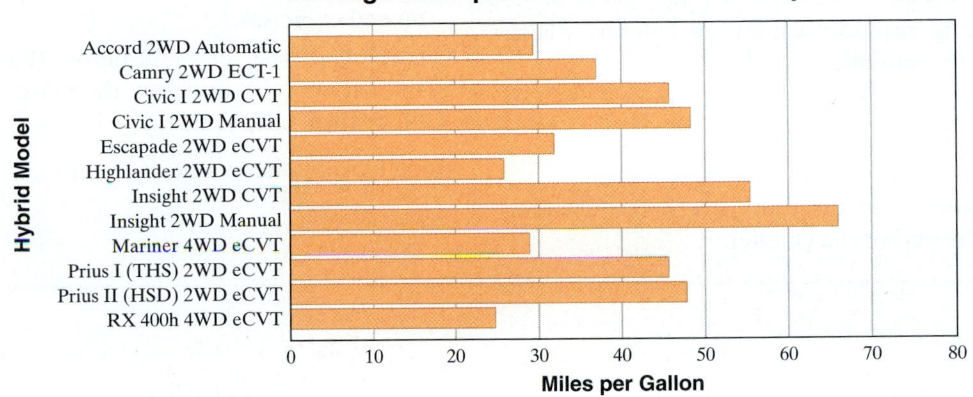

Source: http://www.greenhybrid.com

71. Name the hybrid model with the greatest number of miles per gallon, and estimate the miles per gallon.

72. Name the hybrid model with the least number of miles per gallon, and estimate the miles per gallon.

73. Name the hybrid model whose miles per gallon is between 50 and 60, and estimate the miles per gallon.

74. Name the hybrid model whose miles per gallon is the closest to 40, and estimate the miles per gallon.

Find each square root. If necessary, approximate and round to the nearest thousandth.

75. $\sqrt{36}$

76. $\sqrt{\frac{16}{81}}$

77. $\sqrt{105}$

78. $\sqrt{32}$

Find the unknown length of each given right triangle. If necessary, round to the nearest tenth.

79. leg $= 66$, leg $= 56$

80. leg $= 12$, hypotenuse $= 24$

Given that the pairs of triangles are similar, find the unknown length n.

81.

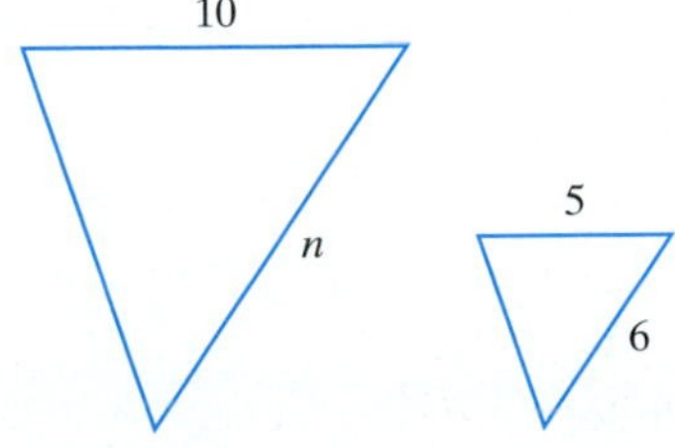

82.

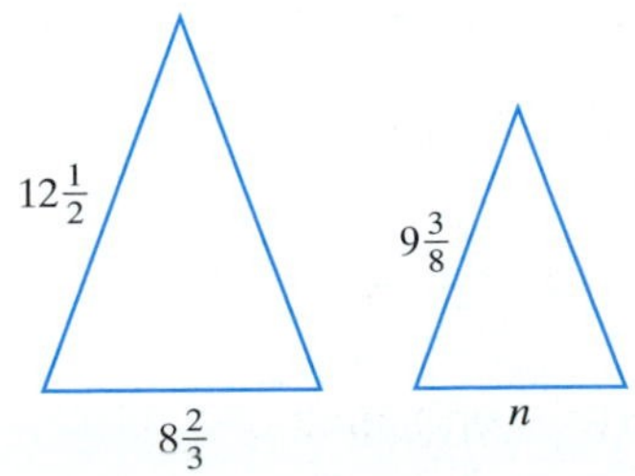

STUDY SKILLS BUILDER

Tips for Studying for an Exam

To prepare for an exam, try the following study techniques.

- Start the study process days before your exam.
- Make sure that you are current and up to date on your assignments.
- If there is a topic that you are unsure of, use one of the many resources that are available to you. For example,

  See your instructor.

  Visit a learning resource center on campus where math tutors are available.

  Read the textbook material and examples on the topic.

  View a videotape on the topic.
- Reread your notes and carefully review the Chapter Highlights at the end of the chapter.
- Work the review exercises at the end of the chapter and check your answers. Make sure that you correct any missed exercises. If you have trouble on a topic, use a resource listed above.
- Find a quiet place to take the Chapter Test found at the end of the chapter. Do not use any resources when taking this sample test. This way you will have a clear indication of how prepared you are for your exam. Check your answers and make sure that you correct any missed exercises.
- Get lots of rest the night before the exam. It's hard to show how well you know the material if your brain is foggy from lack of sleep.

Good luck and keep a positive attitude.

7 CHAPTER TEST

Use the Chapter Test Prep Video CD to see the fully worked-out solutions to any of the exercises you want to review.

Answers

The following pictograph shows the money collected each week from a wrapping paper fundraiser. Use this graph to answer Exercises 1 through 3.

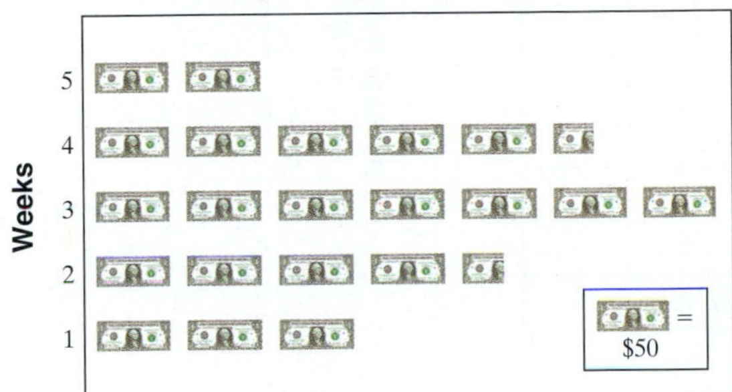

1. How much money was collected during the second week?

2. During which week was the most money collected? How much money was collected during that week?

3. What was the total money collected for the fundraiser?

The bar graph shows the normal monthly precipitation in centimeters for Chicago, Illinois. Use this graph to answer Exercises 4 through 6.

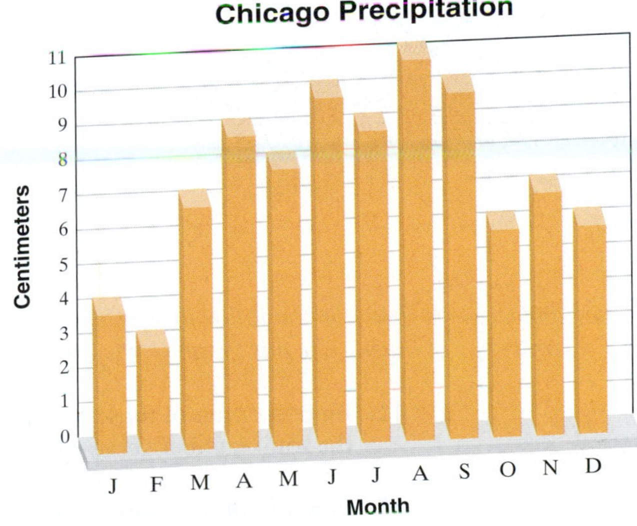

Source: U.S. National Oceanic and Atmospheric Administration, *Climatography of the United States*, No. 81

4. During which month(s) does Chicago normally have more than 9 centimeters of precipitation?

5. During which month does Chicago normally have the least amount of precipitation? How much precipitation occurs during that month?

1. ______

2. ______

3. ______

4. ______

5. ______

6. ______

6. During which month(s) does 7 centimeters of precipitation normally occur?

7. Use the information in the table to draw a bar graph. Clearly label each bar.

7. see graph ______

Countries with the Highest Newspaper Circulations	
Country	**Average Daily Circulation (in millions)**
Japan	72
U.S.	56
China	50
India	31
Germany	24
Russia	24
U.K.	19
(*Source:* World Association of Newspapers)	

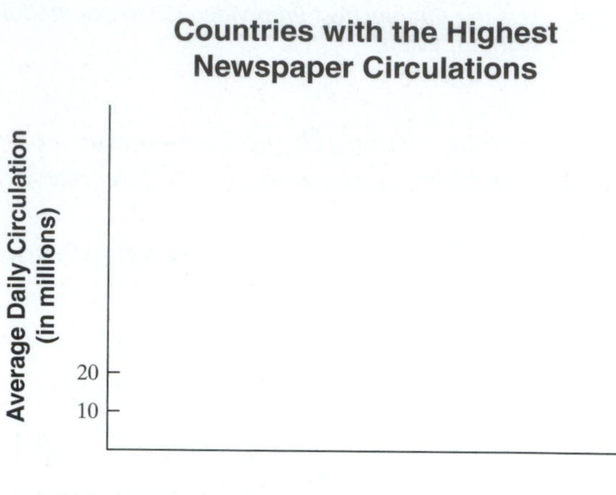

8. ______

The following line graph shows the annual inflation rate in the United States for the years 1990–2003. Use this graph to answer Exercises 8 through 10.

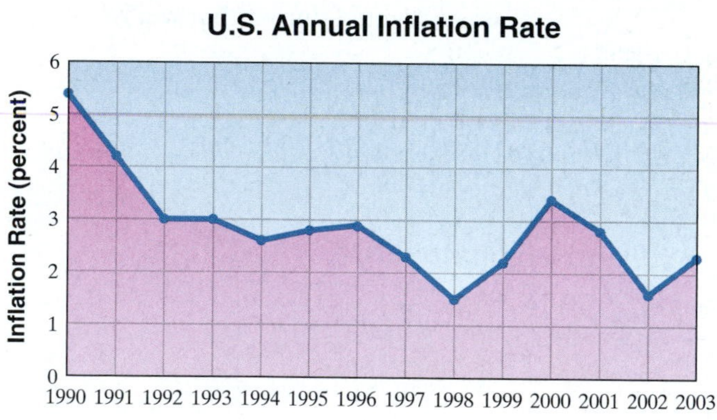

Source: Bureau of Labor Statistics

8. Approximate the annual inflation rate in 2002.

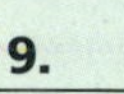

9. ______

9. During which of the years shown was the inflation rate greater than 3%?

10. During which sets of years was the inflation rate increasing?

10. ______

The result of a survey of 200 people is shown in the following circle graph. Each person was asked to tell his or her favorite type of music. Use this graph to answer Exercises 11 and 12.

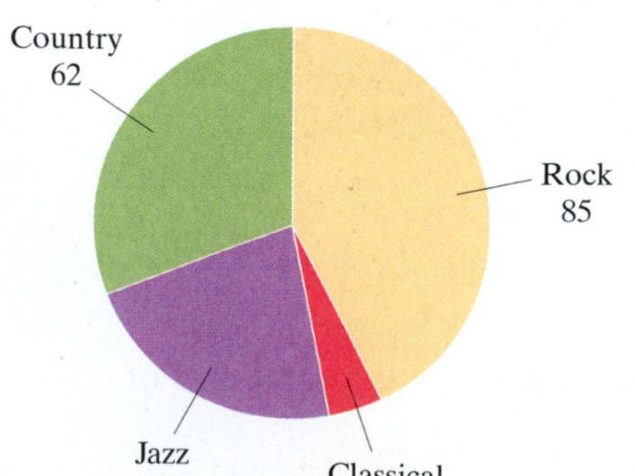

11. Find the ratio of those who prefer rock music to the total number surveyed.

11. ______

12. Find the ratio of those who prefer country music to those who prefer jazz.

12. ______

The following circle graph shows the U.S. labor force employment by industry for 2000. There were approximately 132,000,000 people employed by these industries in the United States in 2000. Use the graph to find how many people were employed by the industries given in Exercises 13 and 14.

13. ______

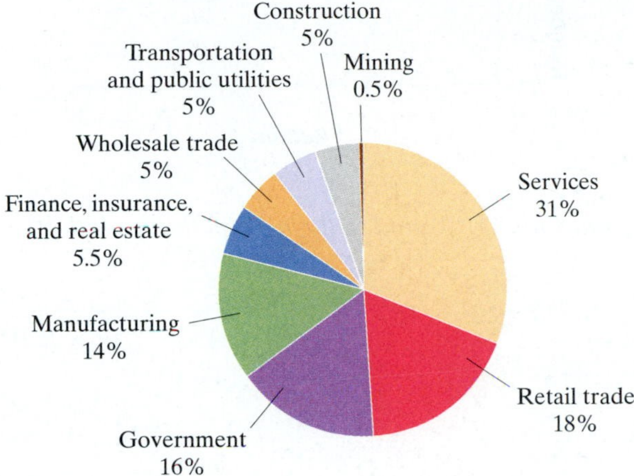

14. ______

13. Services

14. Government

A professor measures the heights of the students in her class. The results are shown in the following histogram. Use this histogram to answer Exercises 15 and 16.

15. ______

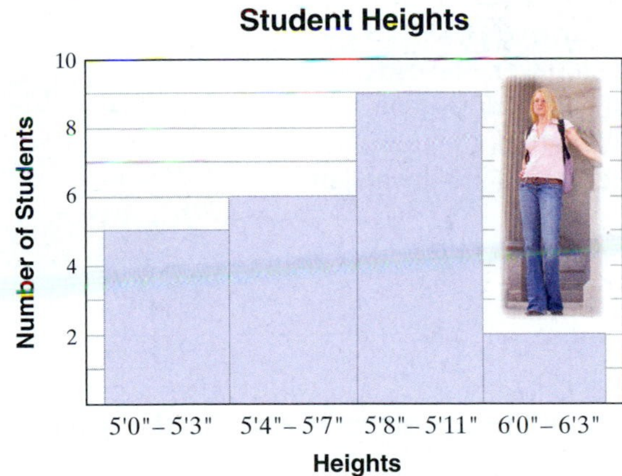

15. How many students are 5′8″–5′11″ tall?

16. How many students are 5′7″ or shorter?

16. ______

17. The history test scores of 25 students are shown below. Use these scores to complete the frequency distribution table.

70	86	81	65	92
43	72	85	69	97
82	51	75	50	68
88	83	85	77	99
77	63	59	84	90

Class Intervals (Scores)	Tally	Class Frequency (Number of Students)
40–49		
50–59		
60–69		
70–79		
80–89		
90–99		

17. see table ______

18. see graph ______

19. ______

20. ______

21. ______

22. ______

23. ______

24. ______

25. ______

26. ______

27. ______

28. ______

18. Use the results of Exercise 17 to draw a histogram.

Number of Students

2
1
0

Scores

Find each square root and simplify. Round to the nearest thousandth if necessary.

19. $\sqrt{49}$

20. $\sqrt{157}$

21. $\sqrt{\frac{64}{100}}$

Solve.

△ **22.** Approximate to the nearest hundredth of a centimeter the length of the missing side of a right triangle with legs of 4 centimeters each.

△ **23.** Given that the following triangles are similar, find the unknown length n.

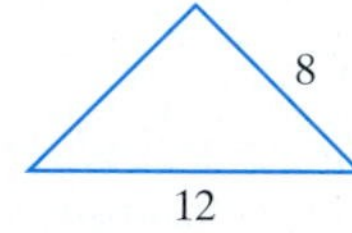

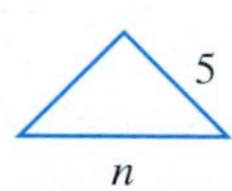

△ **24.** Tamara Watford, a surveyor, needs to estimate the height of a tower. She estimates the length of her shadow to be 4 feet long and the length of the tower's shadow to be 48 feet long. Find the height of the tower if she is $5\frac{3}{4}$ feet tall.

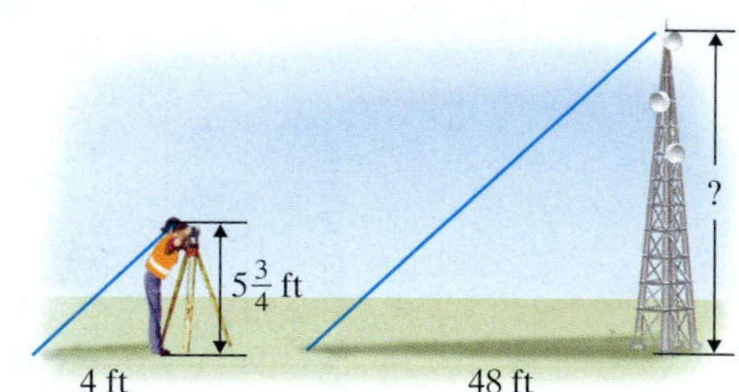

25. Draw a tree diagram for the experiment of spinning the spinner twice.

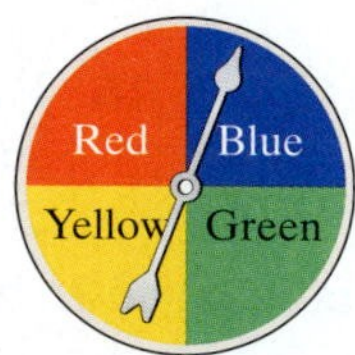

26. Draw a tree diagram for the experiment of tossing a coin twice.

Suppose that the numbers 1 through 10 are each written on the same size sheet of paper and placed in a bag. You then select one sheet of paper from the bag.

27. What is the probability of choosing a 6 from the bag?

28. What is the probability of choosing a 3 or a 4 from the bag?

CUMULATIVE REVIEW

Chapters 1–7

1. Simplify: $4^3 + [3^2 - (10 \div 2)] - 7 \cdot 3$

2. $7^2 - [5^3 + (6 \div 3)] + 4 \cdot 2$

3. Evaluate $x - y$ for $x = -3$ and $y = 9$.

4. Evaluate $x - y$ for $x = 7$ and $y = -2$.

5. Solve: $3y - 7y = 12$

6. Solve: $2x - 6x = 24$

7. Solve: $\frac{x}{6} + 1 = \frac{4}{3}$

8. Solve: $\frac{7}{2} + \frac{a}{4} = 1$

9. Add: $2\frac{1}{3} + 5\frac{3}{8}$

10. Add: $3\frac{2}{5} + 4\frac{3}{4}$

11. Write 5.9 as a mixed number.

12. Write 2.8 as a mixed number.

13. Subtract: $3.5 - 0.068$

14. Subtract: $7.4 - 0.073$.

15. Multiply: 0.0531×16

16. Multiply: 0.147×0.2.

17. Divide: $-5.98 \div 115$

18. Divide: $27.88 \div 205$.

19. Simplify: $(-1.3)^2 + 2.4$

20. Simplify: $(-2.7)^2$

21. Write $\frac{1}{4}$ as a decimal.

22. Write $\frac{3}{8}$ as a decimal.

23. Solve: $5(x - 0.36) = -x + 2.4$

24. Solve: $4(0.35 - x) = x - 7$

25. Approximate $\sqrt{80}$ to the nearest thousandth.

26. Approximate $\sqrt{60}$ to the nearest thousandth.

Answers

1. ______
2. ______
3. ______
4. ______
5. ______
6. ______
7. ______
8. ______
9. ______
10. ______
11. ______
12. ______
13. ______
14. ______
15. ______
16. ______
17. ______
18. ______
19. ______
20. ______
21. ______
22. ______
23. ______
24. ______
25. ______
26. ______

27. ______
28. ______
29. ______
30. ______
31. ______
32. ______
33. ______
34. ______
35. ______
36. ______
37. ______
38. ______
39. ______
40. ______
41. ______
42. ______
43. ______
44. ______
45. ______
46. ______
47. ______
48. ______
49. ______
50. ______

27. Find $\sqrt{\frac{1}{36}}$.

28. Find $\sqrt{\frac{16}{49}}$.

Sketch each right triangle and find the length of the side not given. Round to two decimal places.

29. leg = 5 in., hypotenuse = 7 in.

30. leg = 2, leg = 16.

31. Solve $\frac{x-5}{3} = \frac{x+2}{5}$ for x and then check.

32. Solve $\frac{8}{5} = \frac{x}{10}$ for x and then check.

△**33.** Find the ratio of corresponding sides for the similar triangles ABC and DEF.

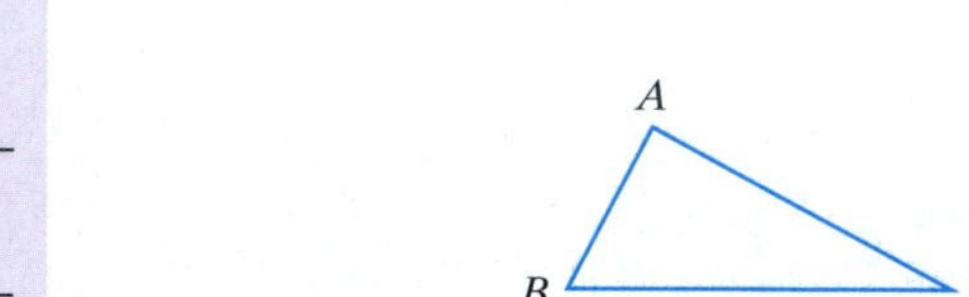

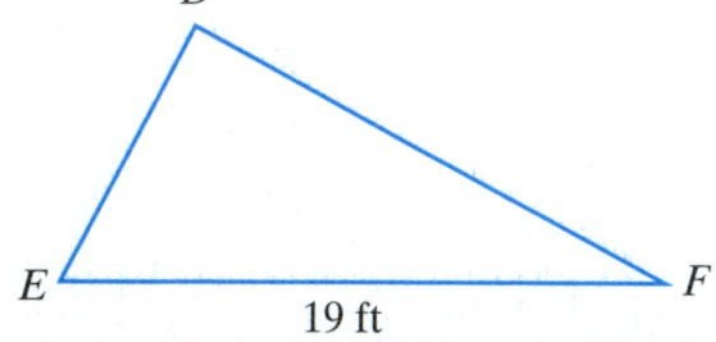

△**34.** Find the ratio of corresponding sides for the similar triangles GHJ and KLM.

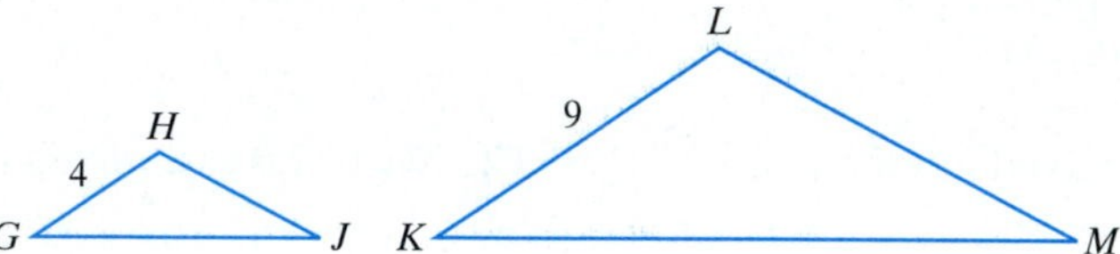

Write each percent as a decimal.

35. 4.6%

36. 32%

37. 0.74%

38. 2.7%

39. What number is 35% of 60?

40. What number is 40% of 36?

41. 20.8 is 40% of what number?

42. 9.5 is 25% of what number?

43. The sales tax on a $406 Sony flat screen digital 27-inch television is $34.51. Find the sales tax rate.

44. The sales tax on a $2.00 yo-yo is $0.13. Find the sales tax rate.

45. Four thousand dollars is invested at 5.3% compounded quarterly for 10 years. Find the total amount at the end of 10 years.

46. Linda Bonnett borrows $1600 for 1 year. If the interest is $128.60, find the monthly payment.

47. Find the median of the list of scores: 67, 91, 75, 86, 55, 91

48. Find the median of the list of numbers. 43, 46, 47, 50, 52, 83

49. If a die is rolled one time, find the probability of rolling a 3 or a 4.

50. If a die is rolled once, find the probability of rolling an even number.

8

Geometry and Measurement

The word *geometry* is formed from the Greek words *geo*, meaning "Earth," and *metron*, meaning "measure." Geometry literally means to measure the Earth. In this chapter we learn about various geometric figures and their properties, such as perimeter, area, and volume. Knowledge of geometry can help us solve practical problems in real-life situations. For instance, knowing certain measures of a circular swimming pool allows us to calculate how much water it can hold.

Did you know there are many more "grand canyons" in the United States in addition to the popular Grand Canyon of the Colorado River? One such canyon is called the Black Canyon of the Gunnison River. The walls of this canyon are deep and narrow, letting in only slanting rays of sunlight, thus the name "Black Canyon." The Gunnison River within the canyon is also unique in that it drops an average of 95 feet per mile (18 meters per kilometer). This is one of the greatest rates of fall for a river in North America. The diagram below shows canyon walls and river elevations for 12 miles (19 kilometers). In Section 8.4, Exercises 85 and 86, we explore the dimensions of some "grand canyons."

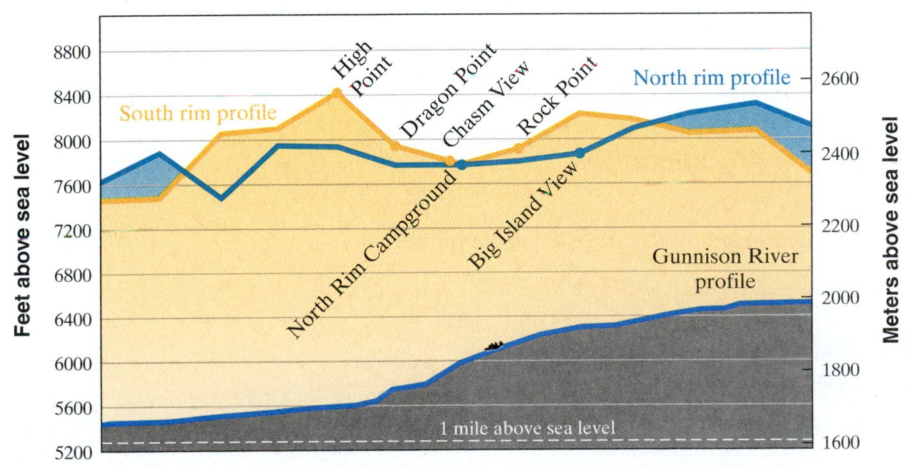

SECTIONS

CHECK YOUR PROGRESS

Objectives

A Identify Lines, Line Segments, Rays, and Angles.

B Classify Angles as Acute, Right, Obtuse, or Straight.

C Identify Complementary and Supplementary Angles.

D Find Measures of Angles.

8.1 LINES AND ANGLES

Objective A Identifying Lines, Line Segments, Rays, and Angles

Let's begin with a review of two important concepts—space and plane.

Space extends in all directions indefinitely. Examples of objects in space are houses, grains of salt, bushes, your *Prealgebra and Introductory Algebra* textbook, and you.

A **plane** is a flat surface that extends indefinitely. Surfaces like a plane are a classroom floor or a blackboard or whiteboard.

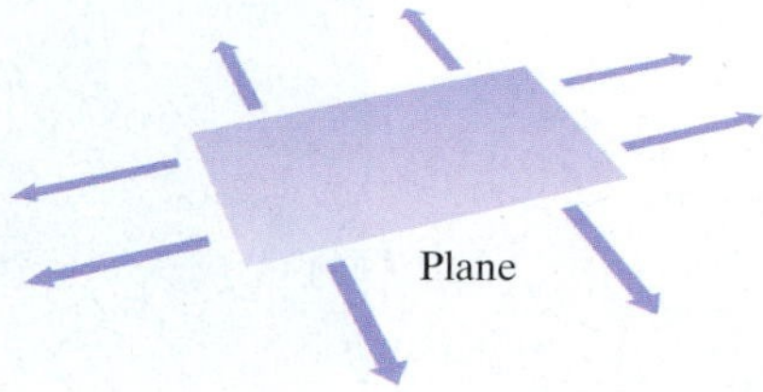

The most basic concept of geometry is the idea of a point in space. A **point** has no length, no width, and no height, but it does have location. We represent a point by a dot, and we usually label points with capital letters.

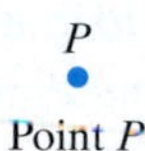

Point P

A **line** is a set of points extending indefinitely in two directions. A line has no width or height, but it does have length. We can name a line by any two of its points or by a single lowercase letter. A **line segment** is a piece of a line with two endpoints.

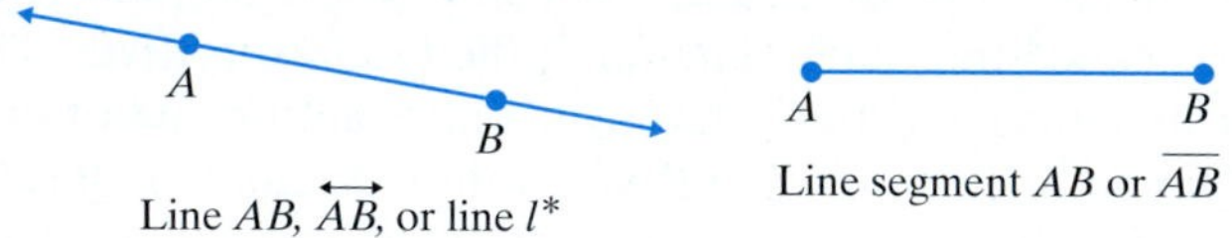

Line AB, $\overleftrightarrow{AB}$, or line l*

Line segment AB or $\overline{AB}$

A **ray** is a part of a line with one endpoint. A ray extends indefinitely in one direction. An **angle** is made up of two rays that share the same endpoint. The common endpoint is called the **vertex.**

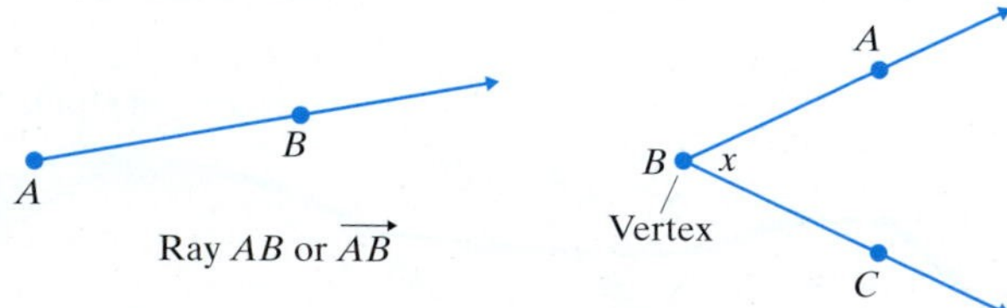

Ray AB or $\overrightarrow{AB}$

The angle in the figure above can be named

$\angle ABC \quad \angle CBA \quad \angle B \quad \text{or} \quad \angle x$

↑ ↑ The vertex is the middle point.

Rays BA and BC are **sides** of the angle.

* Although line l is also line BA or $\overleftrightarrow{BA}$, we will use one order of points only to name a line or line segment.

Helpful Hint

Naming an Angle

When there is no confusion as to what angle is being named, you may use the vertex alone.

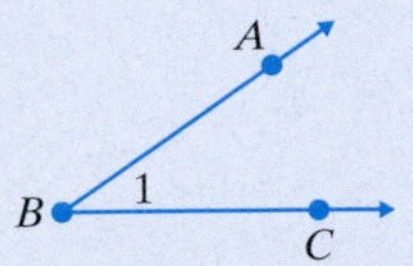

Name of $\angle B$ is all right.
There is no confusion. $\angle B$ means $\angle 1$.

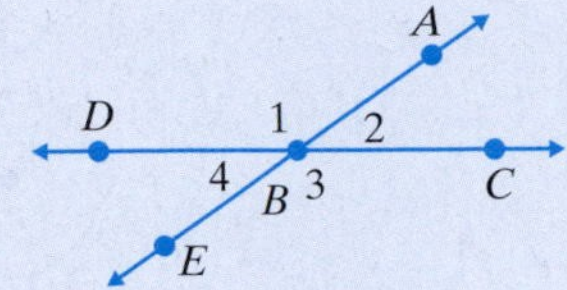

Name of $\angle B$ is *not* all right.
There is confusion. Does $\angle B$ mean $\angle 1$, $\angle 2$, $\angle 3$, or $\angle 4$?

EXAMPLE 1 Identify each figure as a line, a ray, a line segment, or an angle. Then name the figure using the given points.

a.

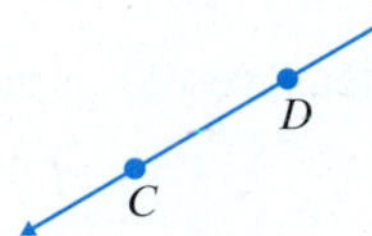

b.

c.

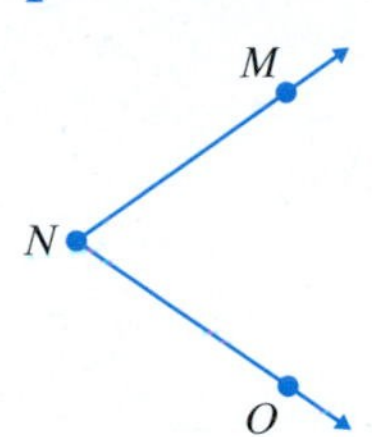

d.

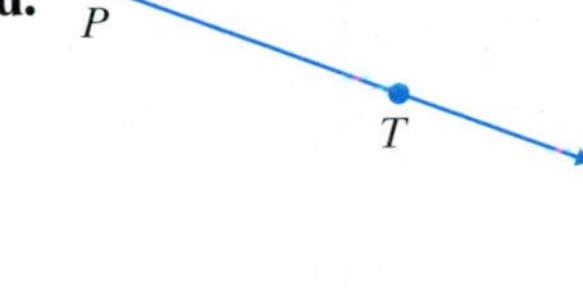

Solution:

Figure (a) extends indefinitely in two directions. It is line CD or $\overleftrightarrow{CD}$.
Figure (b) has two endpoints. It is line segment EF or $\overline{EF}$.
Figure (c) has two rays with a common endpoint. It is $\angle MNO$, $\angle ONM$, or $\angle N$.
Figure (d) is part of a line with one endpoint. It is ray PT or $\overrightarrow{PT}$.

Work Practice Problem 1

EXAMPLE 2 List other ways to name $\angle y$.

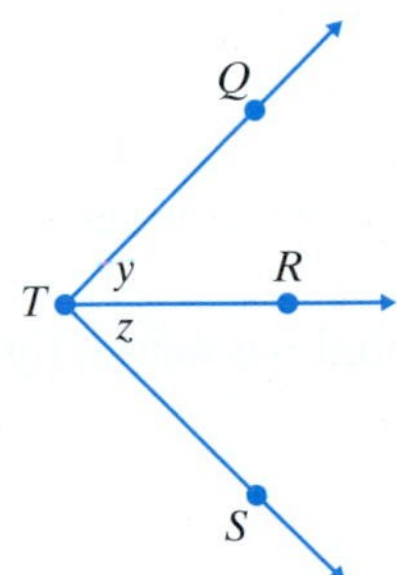

Solution: Two other ways to name $\angle y$ are $\angle QTR$ and $\angle RTQ$. We may *not* use the vertex alone to name this angle because three different angles have T as their vertex.

Work Practice Problem 2

PRACTICE PROBLEM 1

Identify each figure as a line, a ray, a line segment, or an angle. Then name the figure using the given points.

a. **b.**

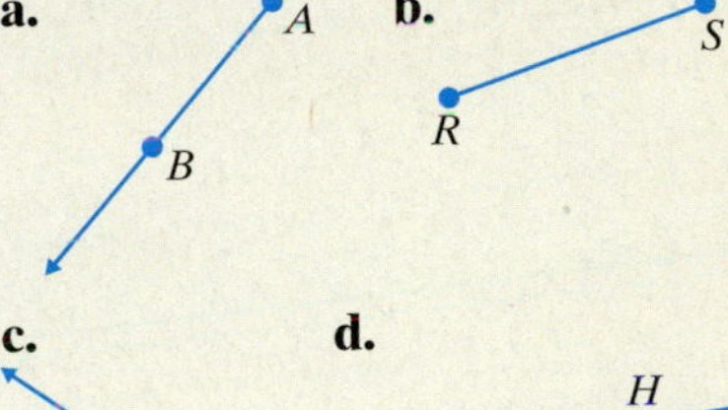

c. **d.**

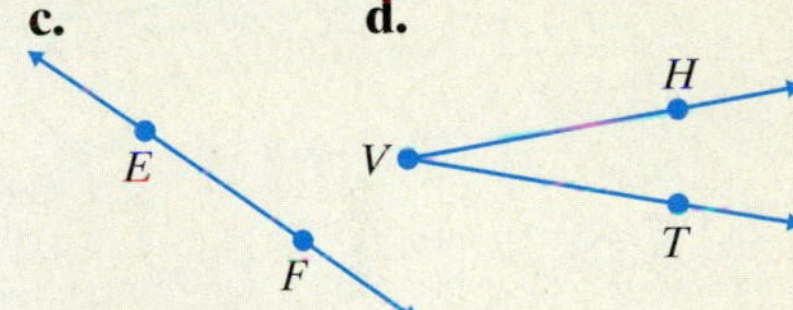

PRACTICE PROBLEM 2

Use the figure in Example 2 to list other ways to name $\angle z$.

Answers

1. a. ray; ray AB or $\overrightarrow{AB}$
b. line segment; line segment RS or $\overline{RS}$
c. line; line EF or $\overleftrightarrow{EF}$
d. angle; $\angle TVH$, or $\angle HVT$ or $\angle V$
2. $\angle RTS$, $\angle STR$

Objective B Classifying Angles as Acute, Right, Obtuse, or Straight

An angle can be measured in **degrees.** The symbol for degrees is a small, raised circle, °. There are 360° in a full revolution, or a full circle.

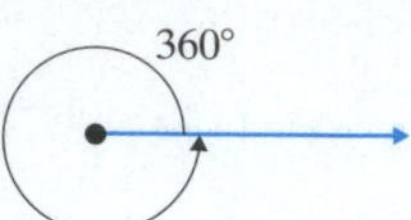

$\frac{1}{2}$ of a revolution measures $\frac{1}{2}(360°) = 180°$. An angle that measures 180° is called a **straight angle.**

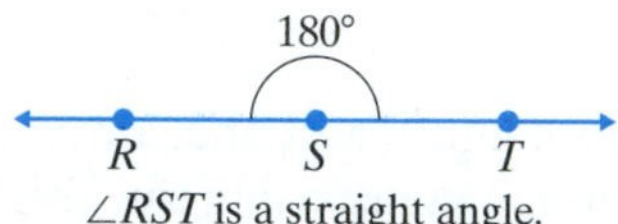

∠*RST* is a straight angle.

$\frac{1}{4}$ of a revolution measures $\frac{1}{4}(360°) = 90°$. An angle that measures 90° is called a **right angle.** The symbol ∟ is used to denote a right angle.

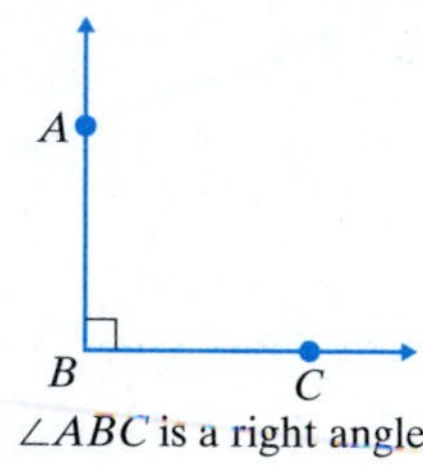

∠*ABC* is a right angle.

An angle whose measure is between 0° and 90° is called an **acute angle.**

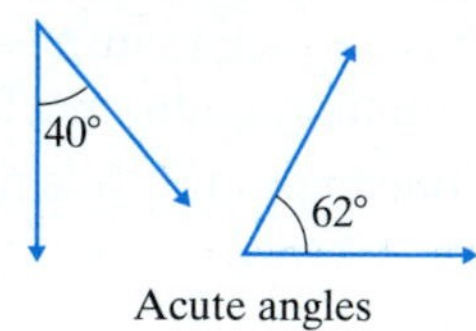

Acute angles

An angle whose measure is between 90° and 180° is called an **obtuse angle.**

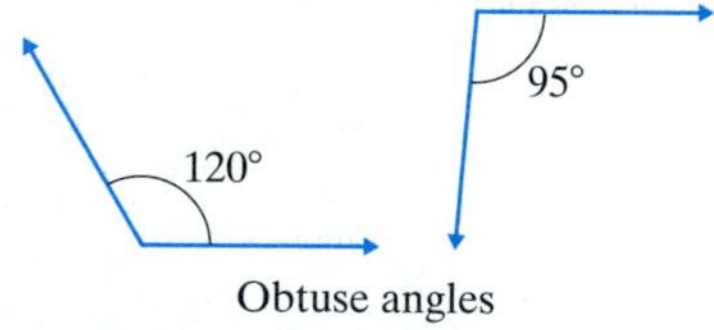

Obtuse angles

EXAMPLE 3 Classify each angle as acute, right, obtuse, or straight.

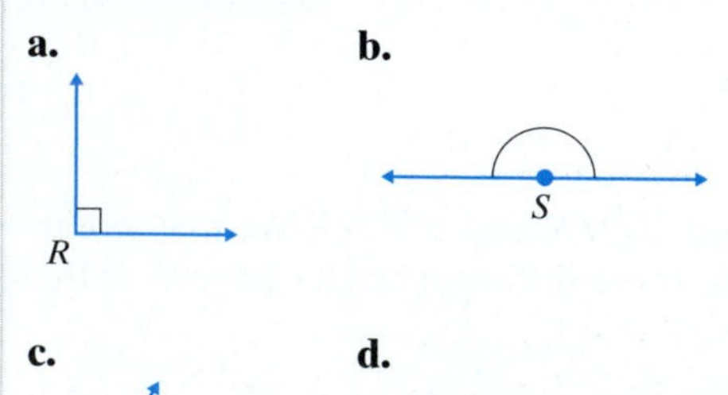

PRACTICE PROBLEM 3

Classify each angle as acute, right, obtuse, or straight.

a. *R* b. *N*

c. *M* d. *Q*

Answers

3. **a.** obtuse **b.** straight **c.** acute **d.** right

Solution:

a. $\angle R$ is a right angle, denoted by ∟. It measures 90°.
b. $\angle S$ is a straight angle. It measures 180°.
c. $\angle T$ is an acute angle. It measures between 0° and 90°.
d. $\angle Q$ is an obtuse angle. It measures between 90° and 180°.

Work Practice Problem 3

Let's look at $\angle B$ below, whose measure is 62°.

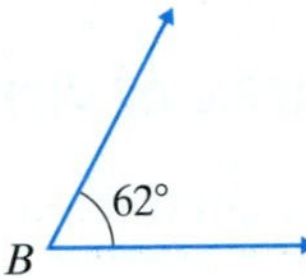

There is a shorthand notation for writing the measure of this angle. To write "The measure of $\angle B$ is 62°," we can write,

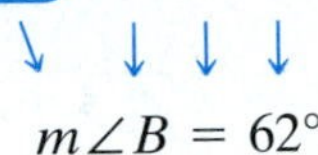

$$m\angle B = 62°.$$

By the way, note that $\angle B$ is an acute angle because $m\angle B$ is between 0° and 90°.

Objective C Identifying Complementary and Supplementary Angles

Two angles that have a sum of 90° are called **complementary angles.** We say that each angle is the **complement** of the other.

$\angle R$ and $\angle S$ are complementary angles because

$$m\angle R + m\angle S = 60° + 30° = 90°$$

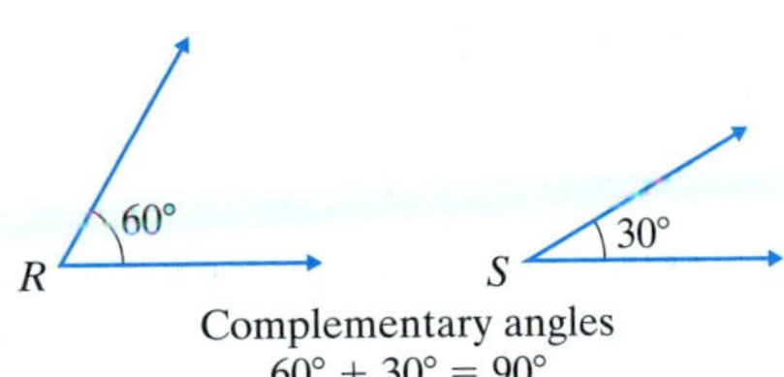

Complementary angles
60° + 30° = 90°

Two angles that have a sum of 180° are called **supplementary angles.** We say that each angle is the **supplement** of the other.

$\angle M$ and $\angle N$ are supplementary angles because

$$m\angle M + m\angle N = 125° + 55° = 180°$$

125° M 55° N

Supplementary angles
125° + 55° = 180°

EXAMPLE 4 Find the complement of a 48° angle.

Solution: Two angles that have a sum of 90° are complementary. This means that the complement of an angle that measures 48° is an angle that measures $90° - 48° = 42°$.

Work Practice Problem 4

PRACTICE PROBLEM 4

Find the complement of a 29° angle.

Answer

4. 61°

PRACTICE PROBLEM 5

Find the supplement of a 67° angle.

EXAMPLE 5 Find the supplement of a 107° angle.

Solution: Two angles that have a sum of 180° are supplementary. This means that the supplement of an angle that measures 107° is an angle that measures 180° − 107° = 73°.

Work Practice Problem 5

✔**Concept Check** True or false? The supplement of a 48° angle is 42°. Explain.

Objective D Finding Measures of Angles

Measures of angles can be added or subtracted to find measures of related angles.

PRACTICE PROBLEM 6

a. Find the measure of $\angle y$.

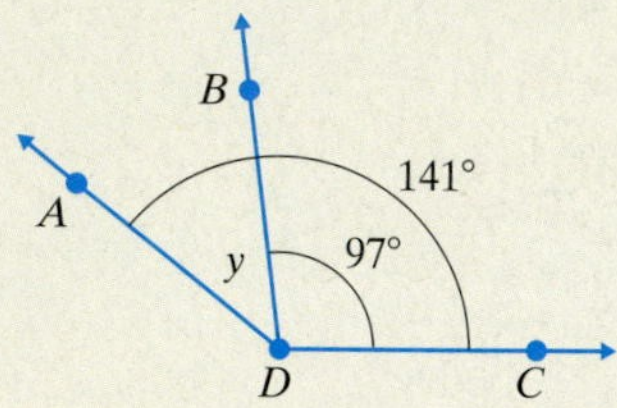

b. Find the measure of $\angle x$.

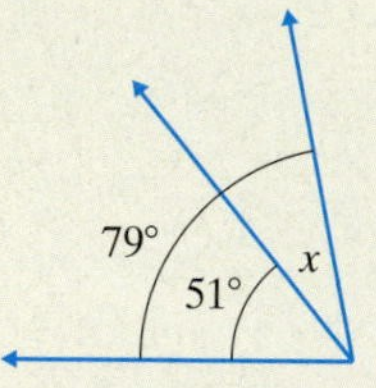

c. Classify $\angle x$ and $\angle y$ as acute, obtuse, or right angles.

EXAMPLE 6 Find the measure of $\angle x$. Then classify $\angle x$ as an acute, obtuse, or right angle.

Solution:

$$m\angle x = m\angle QTS - m\angle RTS$$
$$= 87° - 52°$$
$$= 35°$$

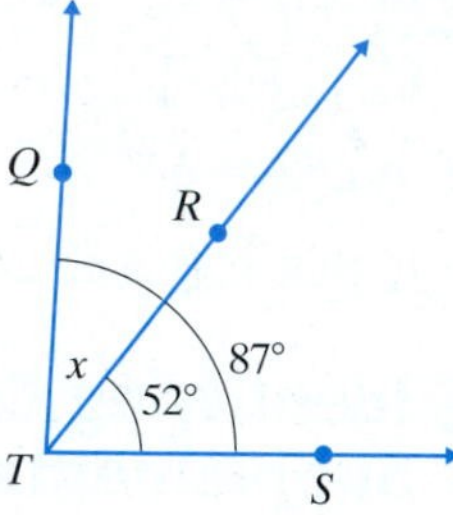

Thus, the measure of $\angle x$ ($m\angle x$) is 35°.

Since $\angle x$ measures between 0° and 90°, it is an acute angle.

Work Practice Problem 6

Two lines in a plane can be either parallel or intersecting. **Parallel lines** never meet. **Intersecting lines** meet at a point. The symbol ∥ is used to indicate "is parallel to." For example, in the figure $p \parallel q$.

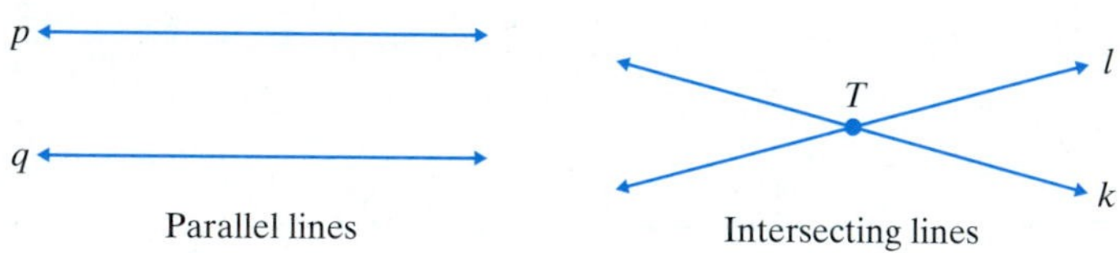

Parallel lines　　Intersecting lines

Some intersecting lines are perpendicular. Two lines are **perpendicular** if they form right angles when they intersect. The symbol ⊥ is used to denote "is perpendicular to." For example, in the figure below, $n \perp m$.

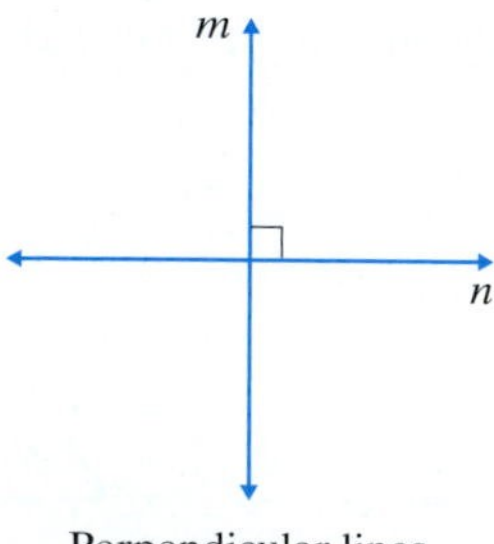

Perpendicular lines

When two lines intersect, four angles are formed. Two angles that are opposite each other are called **vertical angles.** Vertical angles have the same measure.

Two angles that share a common side are called **adjacent angles.** Adjacent angles formed by intersecting lines are supplementary. That is, they have a sum of 180°.

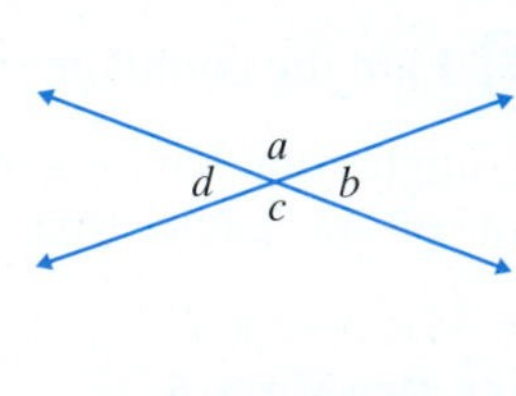

Vertical angles:
$\angle a$ and $\angle c$
$\angle d$ and $\angle b$

Adjacent angles:
$\angle a$ and $\angle b$
$\angle b$ and $\angle c$
$\angle c$ and $\angle d$
$\angle d$ and $\angle a$

Answers

5. 113° 6. a. 44° b. 28° c. acute

✔ **Concept Check Answer**

false; the *complement* of a 48° angle is 42°; the *supplement* of a 48° angle is 132°

Here are a few real-life examples of the lines we just discussed.

Parallel lines

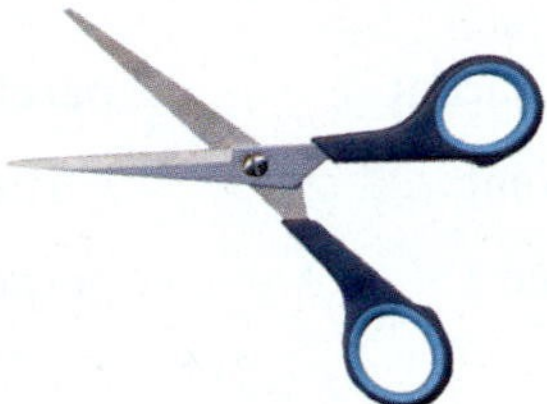
Vertical angles

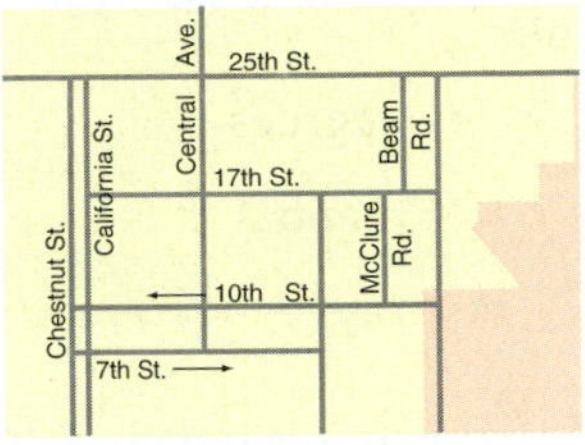

Perpendicular lines

EXAMPLE 7 Find the measure of $\angle x$, $\angle y$, and $\angle z$ if the measure of $\angle t$ is 42°.

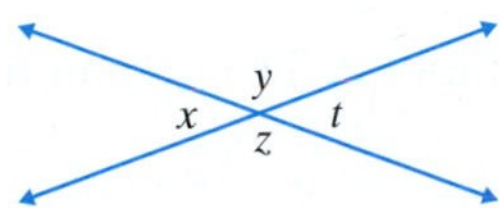

Solution: Since $\angle t$ and $\angle x$ are vertical angles, they have the same measure, so $\angle x$ measures 42°.

Since $\angle t$ and $\angle y$ are adjacent angles, their measures have a sum of 180°. So $\angle y$ measures $180° - 42° = 138°$.

Since $\angle y$ and $\angle z$ are vertical angles, they have the same measure. So $\angle z$ measures 138°.

Work Practice Problem 7

PRACTICE PROBLEM 7

Find the measure of $\angle a$, $\angle b$, and $\angle c$.

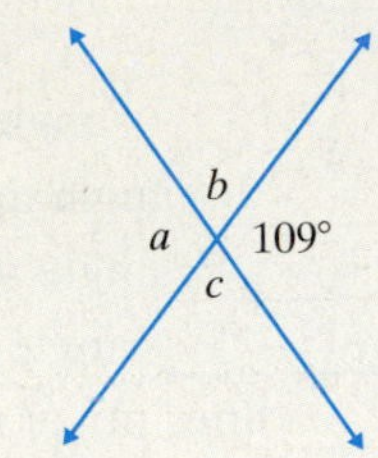

A line that intersects two or more lines at different points is called a **transversal.** Line l is a transversal that intersects lines m and n. The eight angles formed have special names. Some of these names are:

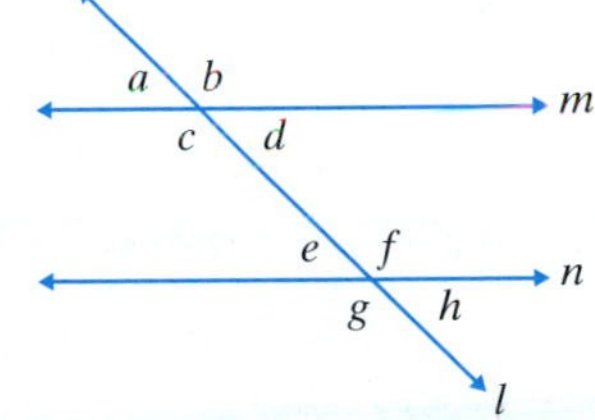

Corresponding angles: $\angle a$ and $\angle e$, $\angle c$ and $\angle g$, $\angle b$ and $\angle f$, $\angle d$ and $\angle h$

Alternate interior angles: $\angle c$ and $\angle f$, $\angle d$ and $\angle e$

When two lines cut by a transversal are *parallel,* the following statement is true:

Parallel Lines Cut by a Transversal

If two parallel lines are cut by a transversal, then the measures of **corresponding angles are equal** and the measures of the **alternate interior angles are equal.**

PRACTICE PROBLEM 8

Given that $m \parallel n$ and that the measure of $\angle w = 45°$, find the measures of all the angles shown.

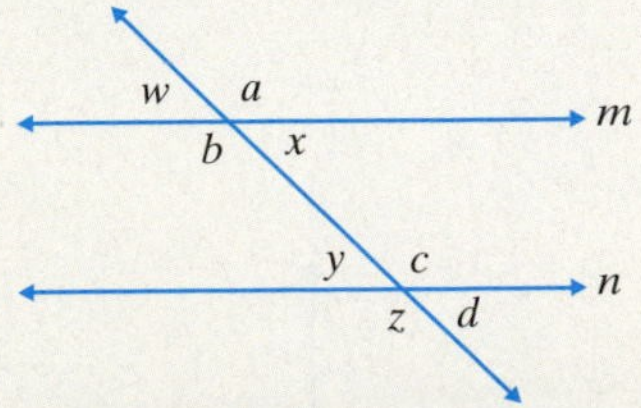

EXAMPLE 8 Given that $m \parallel n$ and that the measure of $\angle w$ is 100°, find the measures of $\angle x$, $\angle y$, and $\angle z$.

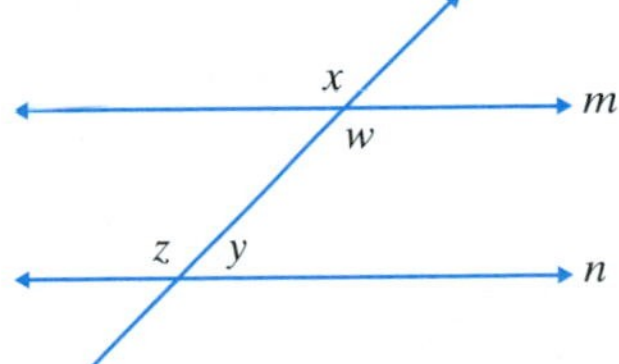

Solution:

$m\angle x = 100°$. $\angle x$ and $\angle w$ are vertical angles.

$m\angle z = 100°$. $\angle x$ and $\angle z$ are corresponding angles.

$m\angle y = 180° - 100° = 80°$. $\angle z$ and $\angle y$ are supplementary angles.

Work Practice Problem 8

Answers

7. $m\angle a = 109°$; $m\angle b = 71°$; $m\angle c = 71°$

8. $m\angle x = 45°$; $m\angle y = 45°$; $m\angle z = 135°$; $m\angle a = 135°$; $m\angle b = 135°$; $m\angle c = 135°$; $m\angle d = 45°$

Vocabulary and Readiness Check

Use the choices below to fill in each blank.

acute	straight	degrees	adjacent	parallel	intersecting
obtuse	space	plane	point	vertical	vertex
right	angle	ray	line	perpendicular	transversal

1. $A(n)$ ___________ is a flat surface that extends indefinitely.
2. $A(n)$ __________ has no length, no width, and no height.
3. __________ extends in all directions indefinitely.
4. $A(n)$___________ is a set of points extending indefinitely in two directions.
5. $A(n)$ ___________ is part of a line with one endpoint.
6. $A(n)$ _________ is made up of two rays that share a common endpoint. The common endpoint is called the __________.
7. $A(n)$ ___________ angle measures 180°.
8. $A(n)$ ________ angle measures 90°.
9. $A(n)$ ________ angle measures between 0° and 90°.
10. $A(n)$ __________ angle measures between 90° and 180°.
11. __________ lines never meet and ___________ lines meet at a point.
12. Two intersecting lines are _____________ if they form right angles when they intersect.
13. An angle can be measured in __________.
14. A line that intersects two or more lines at different points is called a __________.
15. When two lines intersect, four angles are formed, called _________ angles.
16. Two angles that share a common side are called _________ angles.

8.1 EXERCISE SET

FOR EXTRA HELP: Student Solutions Manual · PH Math/Tutor Center · CD/Video for Review · MathXL® · MyMathLab

Objective A *Identify each figure as a line, a ray, a line segment, or an angle. Then name the figure using the given points. See Examples 1 and 2.*

1.
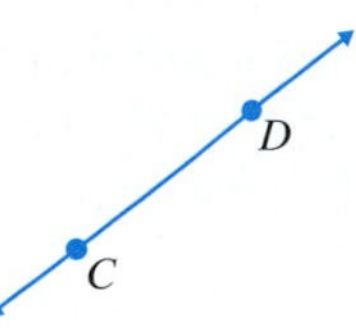

2.
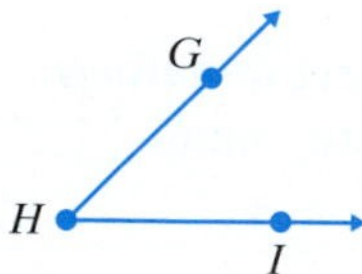

3.

4.
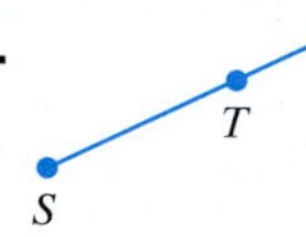

5.
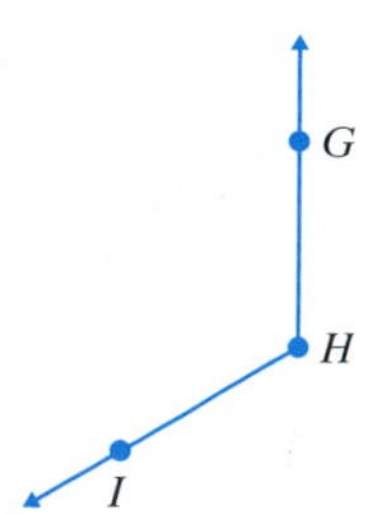

6.
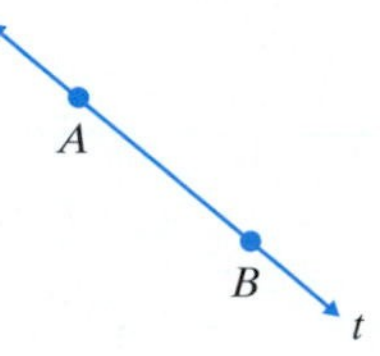

7.
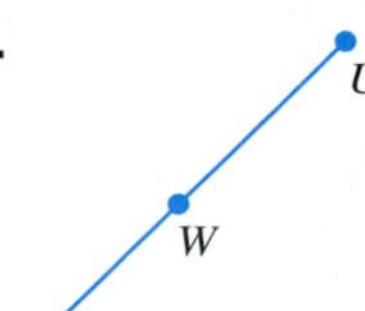

8.
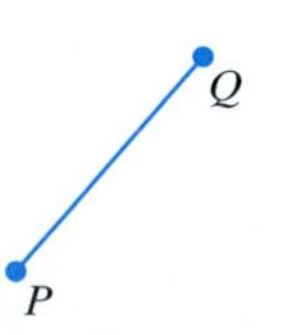

List two other ways to name each angle. See Example 2.

9. $\angle x$

10. $\angle w$

11. $\angle z$

12. $\angle y$

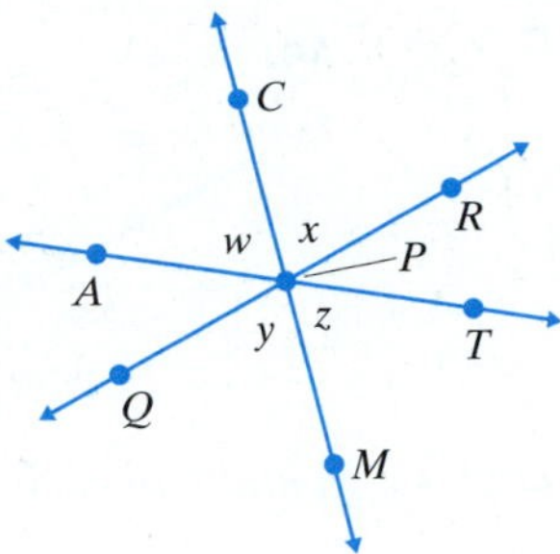

Objective B *Classify each angle as acute, right, obtuse, or straight. See Example 3.*

13. 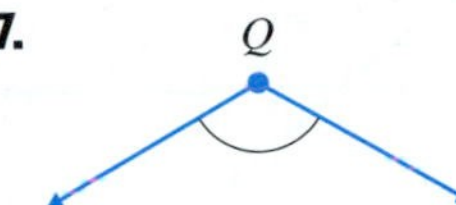

14. 

15. 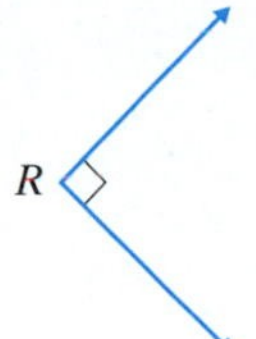

16. 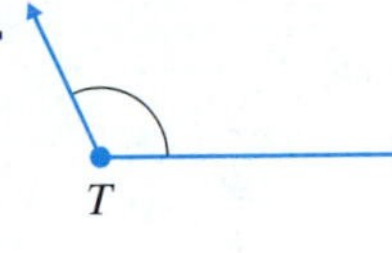

17. 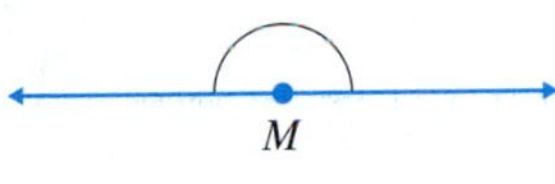

18. 

19. 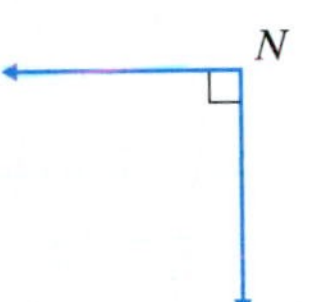

20. 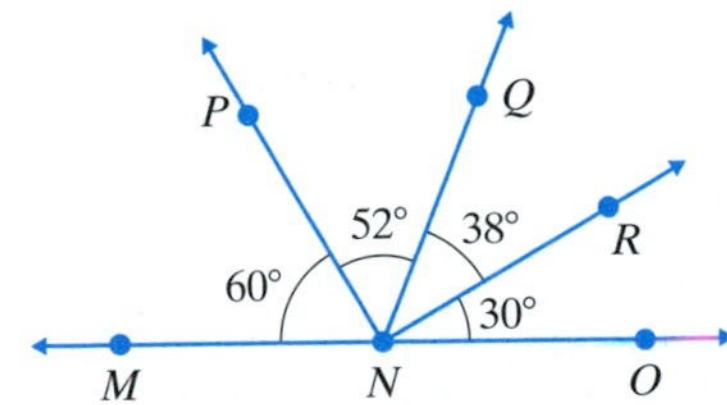

Objective C *Find each complementary or supplementary angle as indicated. See Examples 4 and 5.*

21. Find the complement of a 23° angle.

22. Find the complement of an 77° angle.

23. Find the supplement of a 17° angle.

24. Find the supplement of an 77° angle.

25. Find the complement of a 58° angle.

26. Find the complement of a 22° angle.

27. Find the supplement of a 150° angle.

28. Find the supplement of a 130° angle.

29. Identify the pairs of complementary angles.

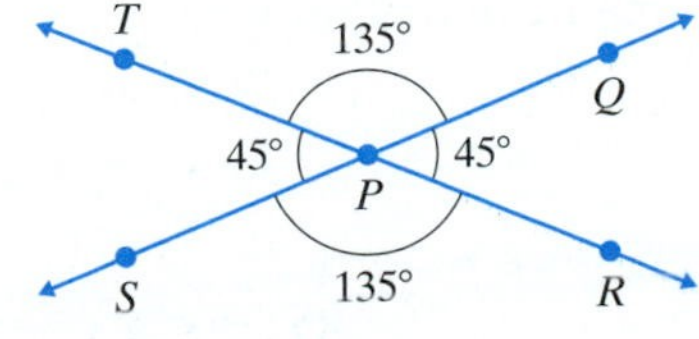

30. Identify the pairs of complementary angles.

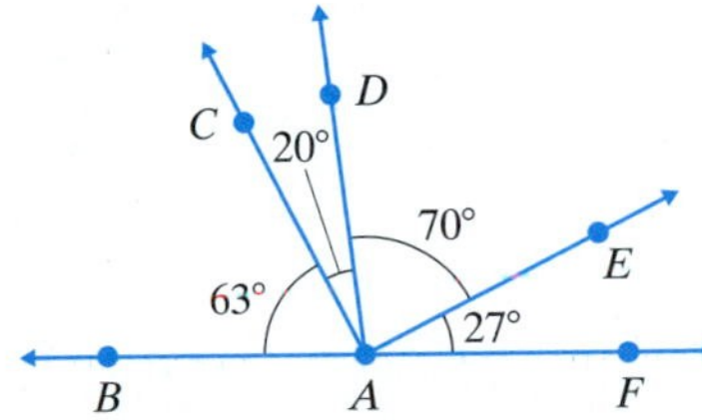

31. Identify the pairs of supplementary angles.

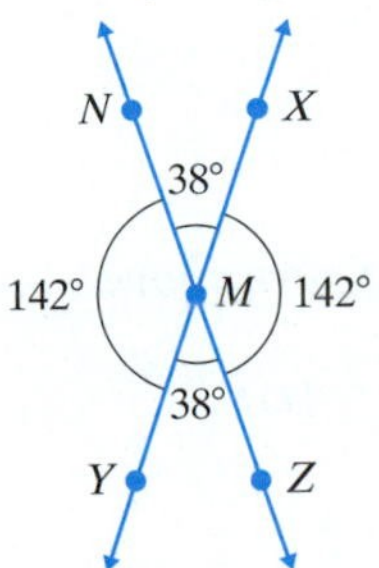

32. Identify the pairs of supplementary angles.

Objective D *Find the measure of* $\angle x$ *in each figure. See Example 6.*

33.

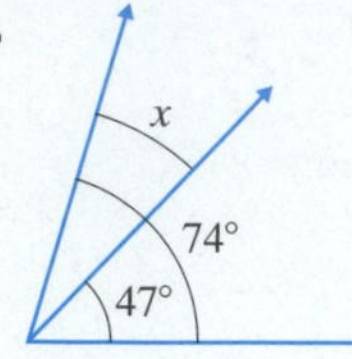

34.

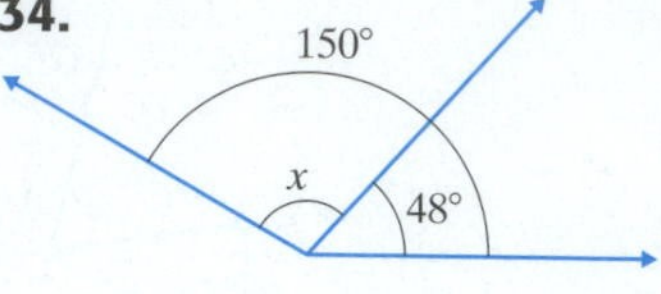

35.

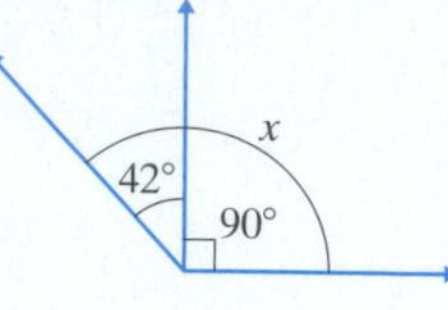

36.

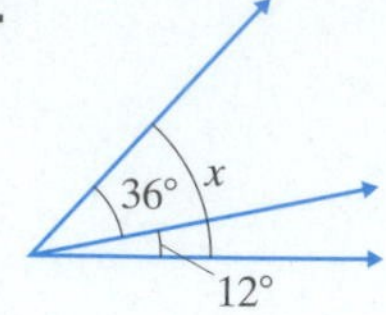

Find the measures of angles x, y, and z in each figure. See Examples 7 and 8.

37.

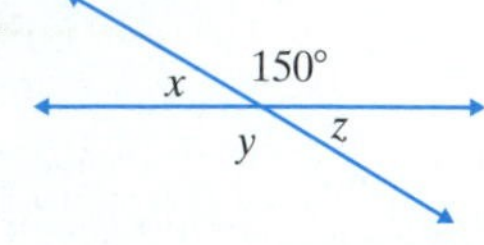

38.

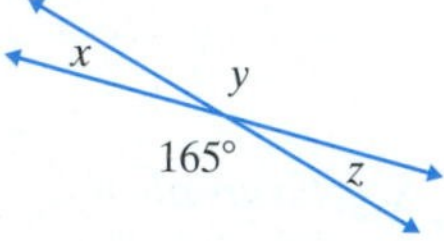

39.

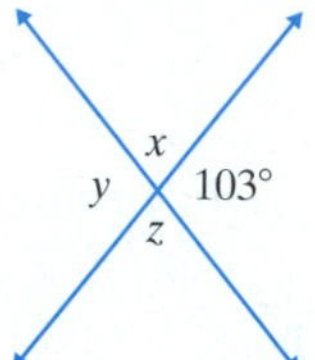

40.

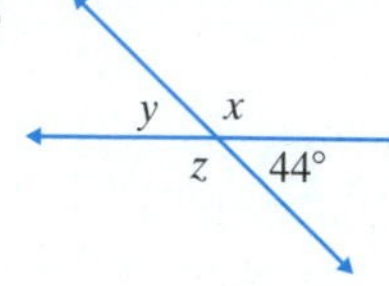

41. $m \parallel n$

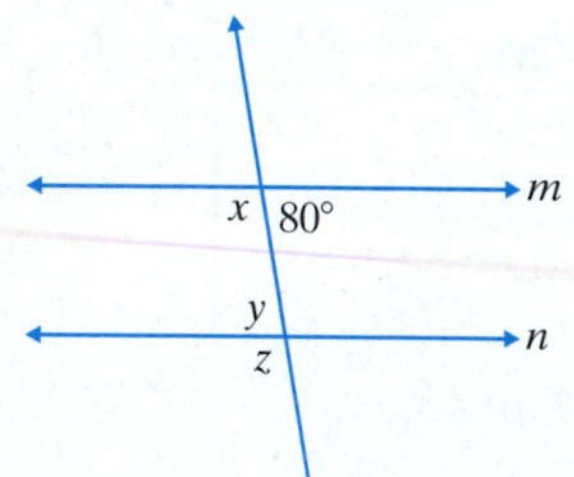

42. $m \parallel n$

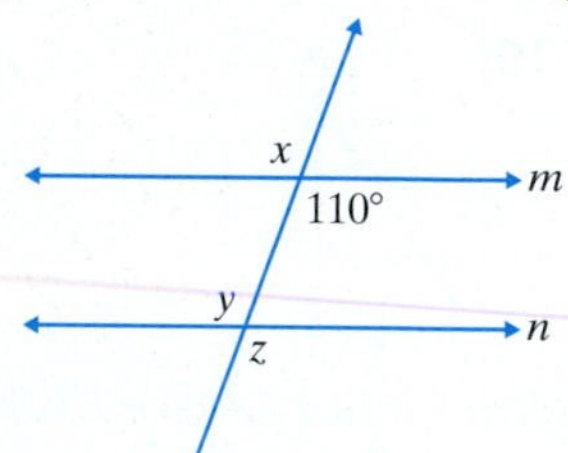

43. $m \parallel n$

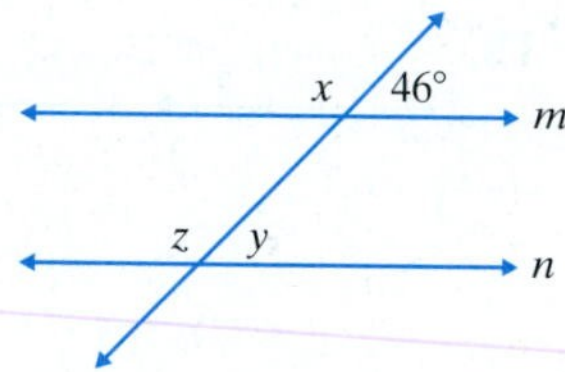

44. $m \parallel n$

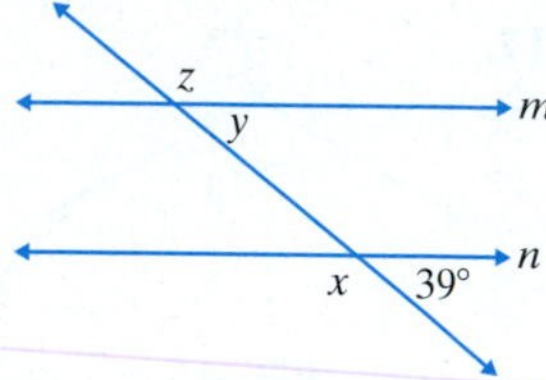

Objectives A D Mixed Practice *Find two other ways of naming each angle. See Example 2.*

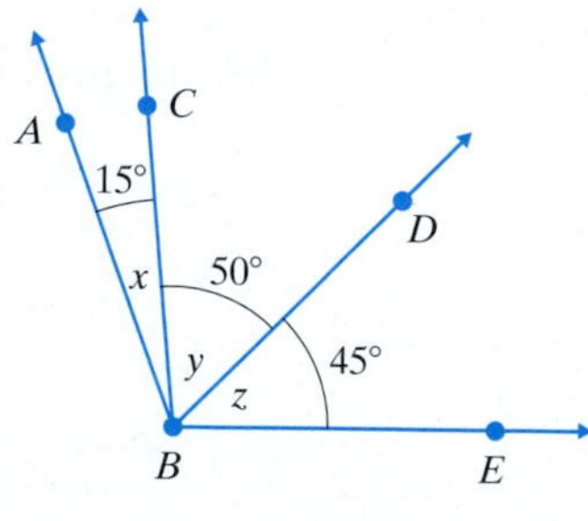

45. $\angle x$

46. $\angle y$

47. $\angle z$

48. $\angle ABE$ (just name one other way)

Find the measure of each angle in the figure above.

49. $\angle ABC$

50. $\angle EBD$

51. $\angle CBD$

52. $\angle CBA$

53. $\angle DBA$

54. $\angle EBC$

55. $\angle CBE$

56. $\angle ABE$

Review

Perform each indicated operation. See Sections 4.3, 4.5, and 4.7.

57. $\frac{7}{8} + \frac{1}{4}$

58. $\frac{7}{8} - \frac{1}{4}$

59. $\frac{7}{8} \cdot \frac{1}{4}$

60. $\frac{7}{8} \div \frac{1}{4}$

61. $3\frac{1}{3} - 2\frac{1}{2}$

62. $3\frac{1}{3} + 2\frac{1}{2}$

63. $3\frac{1}{3} \div 2\frac{1}{2}$

64. $3\frac{1}{3} \cdot 2\frac{1}{2}$

Concept Extensions

65. The angle between the two walls of the Vietnam Veterans Memorial in Washington, D.C., is 125.2°. Find the supplement of this angle. (*Source:* National Park Service)

66. The faces of Khafre's Pyramid at Giza, Egypt, are inclined at an angle of 53.13°. Find the complement of this angle. (*Source:* PBS *NOVA* Online)

Answer true or false for Exercises 67 through 70. See the Concept Check in this section. If false, explain why.

67. The complement of a 100° angle is an 80° angle.

68. It is possible to find the complement of a 120° angle.

69. It is possible to find the supplement of a 120° angle.

70. The supplement of a 5° angle is a 175° angle.

71. If lines m and n are parallel, find the measures of angles a through e.

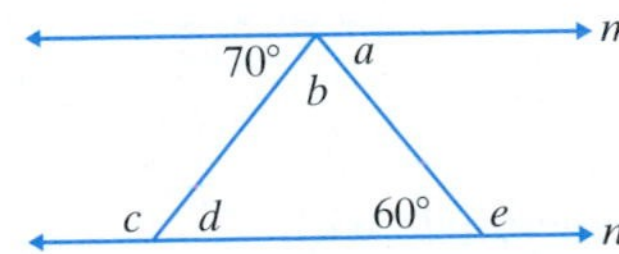

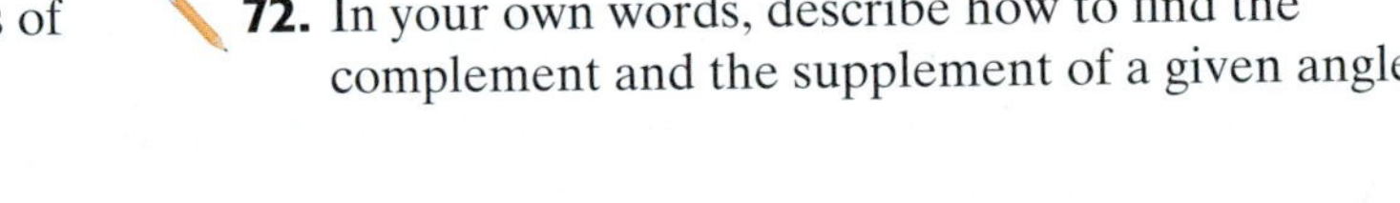

72. In your own words, describe how to find the complement and the supplement of a given angle.

73. Can two supplementary angles both be acute? Explain why or why not.

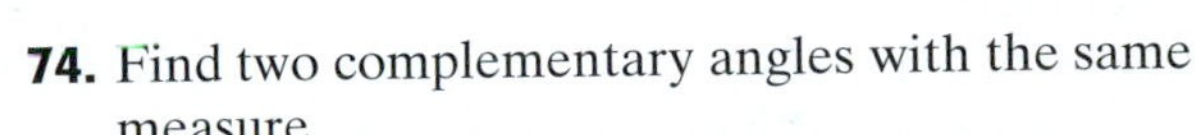

74. Find two complementary angles with the same measure.

75. Is the figure below possible? Why or why not?

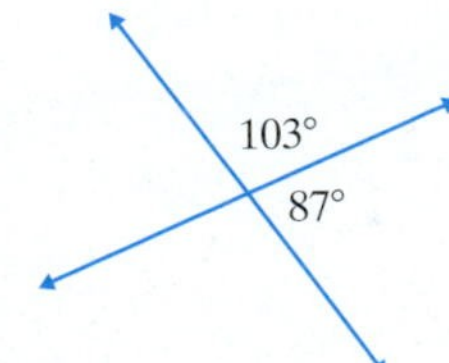

76. Below is a rectangle. List which segments, if extended, would be parallel lines.

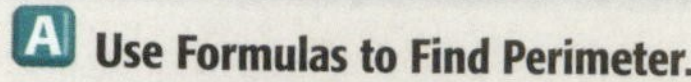

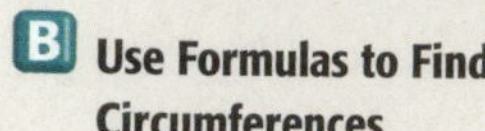

8.2 PERIMETER

Objectives

A Use Formulas to Find Perimeter.

B Use Formulas to Find Circumferences.

Objective A Using Formulas to Find Perimeters

Recall from Section 1.3 that the perimeter of a polygon is the distance around the polygon. This means that the perimeter of a polygon is the sum of the lengths of its sides.

EXAMPLE 1 Find the perimeter of the rectangle below.

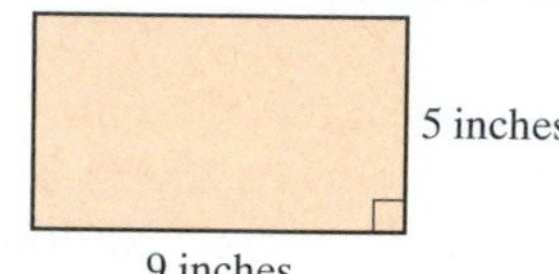

Solution:

perimeter = 9 inches + 9 inches + 5 inches + 5 inches
= 28 inches

Work Practice Problem 1

PRACTICE PROBLEM 1

a. Find the perimeter of the rectangle.

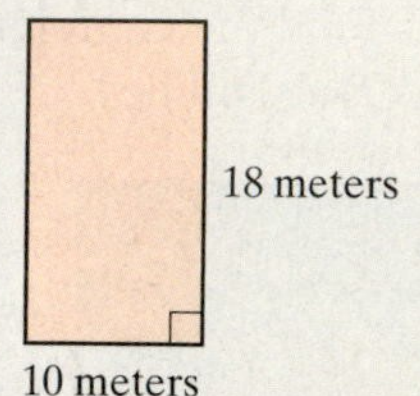

b. Find the perimeter of the rectangular lot shown below:

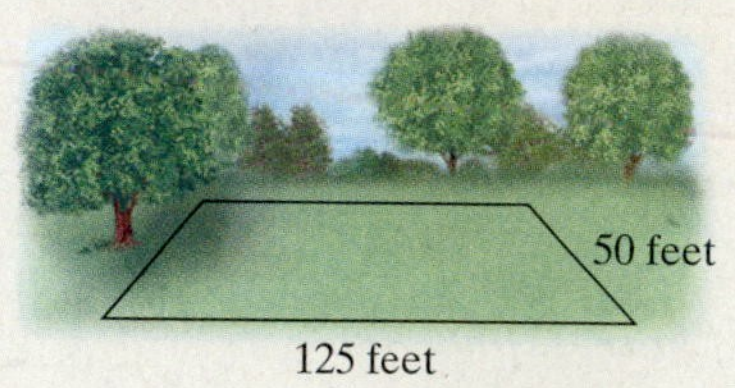

Notice that the perimeter of the rectangle in Example 1 can be written as $2 \cdot (9 \text{ inches}) + 2 \cdot (5 \text{ inches})$.

↑ length ↑ width

In general, we can say that the perimeter of a rectangle is always

$2 \cdot \text{length} + 2 \cdot \text{width}$

As we have just seen, the perimeter of some special figures such as rectangles form patterns. These patterns are given as **formulas.** The formula for the perimeter of a rectangle is shown next:

Perimeter of a Rectangle

Perimeter = 2 · **length** + 2 · **width**

In symbols, this can be written as

$$P = 2l + 2w$$

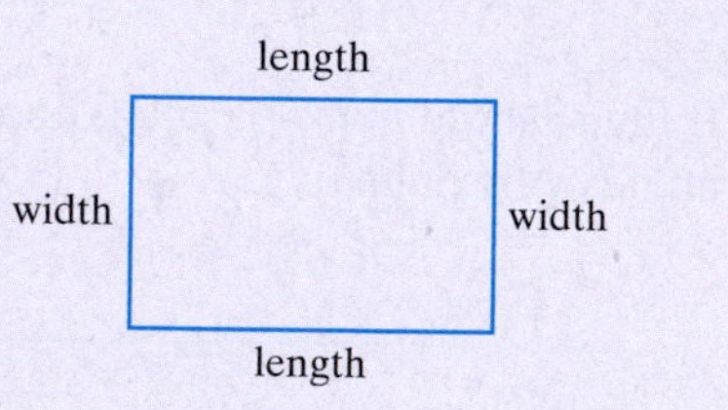

EXAMPLE 2 Find the perimeter of a rectangle with a length of 11 inches and a width of 3 inches.

11 in.

3 in.

Solution: We use the formula for perimeter and replace the letters by their known lengths.

$P = 2l + 2w$

$= 2 \cdot 11 \text{ in.} + 2 \cdot 3 \text{ in.}$ Replace l with 11 in. and w with 3 in.

$= 22 \text{ in.} + 6 \text{ in.}$

$= 28 \text{ in.}$

The perimeter is 28 inches.

Work Practice Problem 2

PRACTICE PROBLEM 2

Find the perimeter of a rectangle with a length of 32 centimeters and a width of 15 centimeters.

Answers

1. a. 56 m **b.** 350 ft **2.** 94 cm

Recall that a square is a special rectangle with all four sides the same length. The formula for the perimeter of a square is shown next:

Perimeter of a Square

Perimeter = **side** + **side** + **side** + **side**
= 4 · **side**

In symbols,

$$P = 4s$$

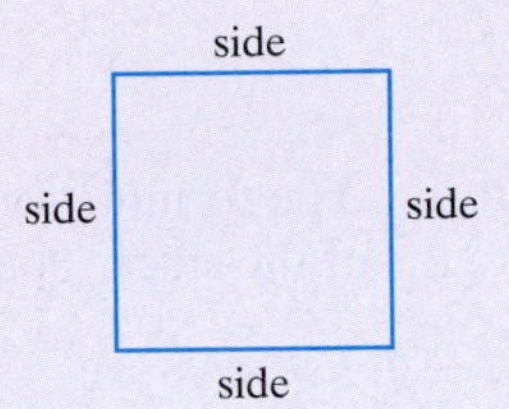

EXAMPLE 3 **Finding the Perimeter of a Field**

How much fencing is needed to enclose a square field 50 yards on a side?

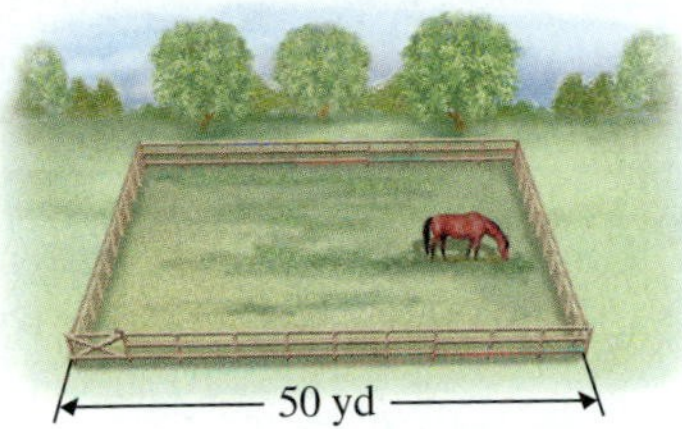

Solution: To find the amount of fencing needed, we find the distance around, or perimeter. The formula for the perimeter of a square is $P = 4 \cdot s$. We use this formula and replace s by 50 yards.

$$P = 4 \cdot s$$
$$= 4 \cdot 50 \text{ yd}$$
$$= 200 \text{ yd}$$

The amount of fencing needed is 200 yards.

Work Practice Problem 3

PRACTICE PROBLEM 3

Find the perimeter of a square tabletop if each side is 4 feet long.

The formula for the perimeter of a triangle with sides of lengths a, b, and c is given next:

Perimeter of a Triangle

Perimeter = side $\boldsymbol{a}$ + side $\boldsymbol{b}$ + side $\boldsymbol{c}$

In symbols,

$$P = a + b + c$$

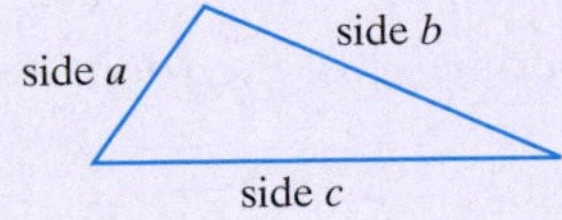

Answer

3. 16 ft

PRACTICE PROBLEM 4

Find the perimeter of a triangle if the sides are 6 centimeters, 10 centimeters, and 8 centimeters in length.

EXAMPLE 4 Find the perimeter of a triangle if the sides are 3 inches, 7 inches, and 6 inches.

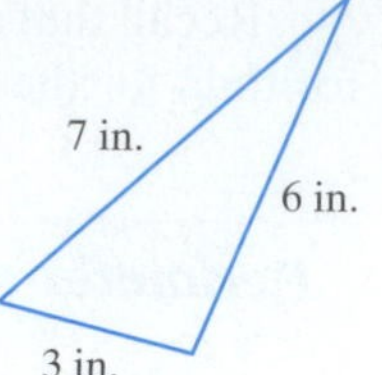

Solution: The formula for the perimeter is $P = a + b + c$, where a, b, and c are the lengths of the sides. Thus,

$$P = a + b + c$$
$$= 3 \text{ in.} + 7 \text{ in.} + 6 \text{ in.}$$
$$= 16 \text{ in.}$$

The perimeter of the triangle is 16 inches.

Work Practice Problem 4

The method for finding the perimeter of any polygon is given next:

Perimeter of a Polygon

The perimeter of a polygon is the sum of the lengths of its sides.

PRACTICE PROBLEM 5

Find the perimeter of the trapezoid shown.

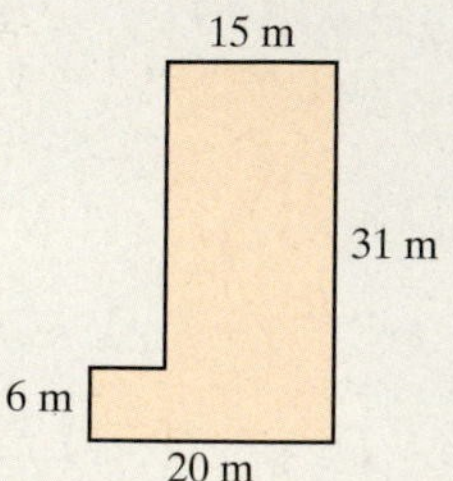

EXAMPLE 5 Find the perimeter of the trapezoid shown below:

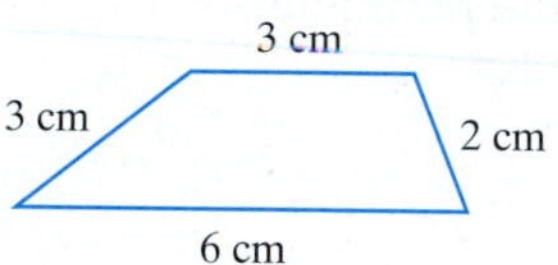

Solution: To find the perimeter, we find the sum of the lengths of its sides.

perimeter $= 3 \text{ cm} + 2 \text{ cm} + 6 \text{ cm} + 3 \text{ cm} = 14 \text{ cm}$

The perimeter is 14 centimeters.

Work Practice Problem 5

PRACTICE PROBLEM 6

Find the perimeter of the room shown.

15 m
31 m
6 m
20 m

EXAMPLE 6 **Finding the Perimeter of a Room**

Find the perimeter of the room shown below:

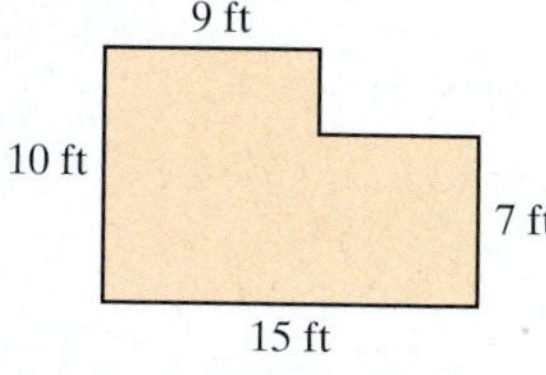

Solution: To find the perimeter of the room, we first need to find the lengths of all sides of the room.

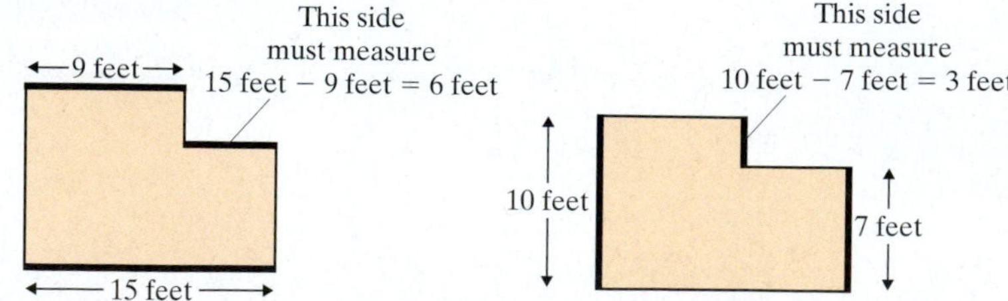

Answers

4. 24 cm **5.** 23 km **6.** 102 m

Now that we know the measures of all sides of the room, we can add the measures to find the perimeter.

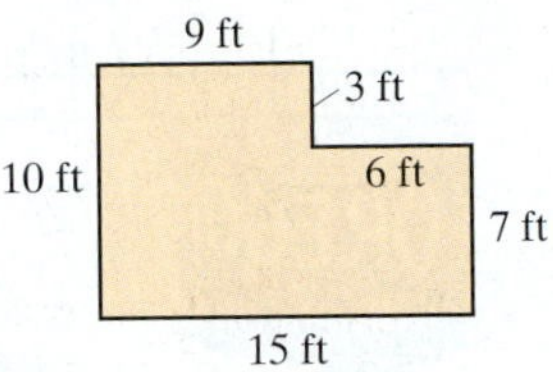

$$\begin{aligned} \text{perimeter} &= 10 \text{ ft} + 9 \text{ ft} + 3 \text{ ft} + 6 \text{ ft} + 7 \text{ ft} + 15 \text{ ft} \\ &= 50 \text{ ft} \end{aligned}$$

The perimeter of the room is 50 feet.

Work Practice Problem 6

EXAMPLE 7 **Calculating the Cost of Wallpaper Border**

A rectangular room measures 10 feet by 12 feet. Find the cost to hang a wallpaper border on the walls close to the ceiling if the cost of the wallpaper border is \$1.09 per foot.

Solution: First we find the perimeter of the room.

$$\begin{aligned} P &= 2 \cdot l + 2 \cdot w \\ &= 2 \cdot 12 \text{ ft} + 2 \cdot 10 \text{ ft} \quad \text{Replace } l \text{ with 12 feet and } w \text{ with 10 feet.} \\ &= 24 \text{ ft} + 20 \text{ ft} \\ &= 44 \text{ ft} \end{aligned}$$

The cost of the wallpaper is

$$\text{cost} = \$1.09 \cdot 44 \text{ ft} = 47.96$$

The cost of the wallpaper is \$47.96.

Work Practice Problem 7

PRACTICE PROBLEM 7

A rectangular lot measures 60 feet by 120 feet. Find the cost to install fencing around the lot if the cost of fencing is \$1.90 per foot.

Objective B Using Formulas to Find Circumferences

Recall from Section 5.3 that the distance around a circle is called the **circumference.** This distance depends on the radius or the diameter of the circle.

The formulas for circumference are shown next:

Circumference of a Circle

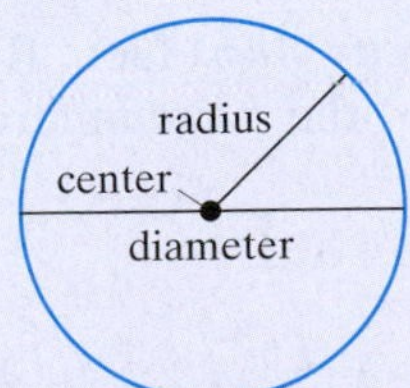

Circumference = $2 \cdot \pi \cdot$ radius or Circumference = $\pi \cdot$ diameter

In symbols,

$$C = 2\pi r \quad \text{or} \quad C = \pi d$$

where $\pi \approx 3.14$ or $\pi \approx \frac{22}{7}$.

Answer

7. \$684

To better understand circumference and π(pi), try the following experiment. Take any can and measure its circumference and its diameter.

The can in the figure above has a circumference of 23.5 centimeters and a diameter of 7.5 centimeters. Now divide the circumference by the diameter.

$$\frac{\text{circumference}}{\text{diameter}} = \frac{23.5 \text{ cm}}{7.5 \text{ cm}} \approx 3.13$$

Try this with other sizes of cylinders and circles—you should always get a number close to 3.1. The exact ratio of circumference to diameter is π. (Recall that $\pi \approx 3.14$ or $\approx \frac{22}{7}$.)

PRACTICE PROBLEM 8

An irrigation device waters a circular region with a diameter of 20 yards. Find the exact circumference of the watered region, then use $\pi \approx 3.14$ to give an approximation.

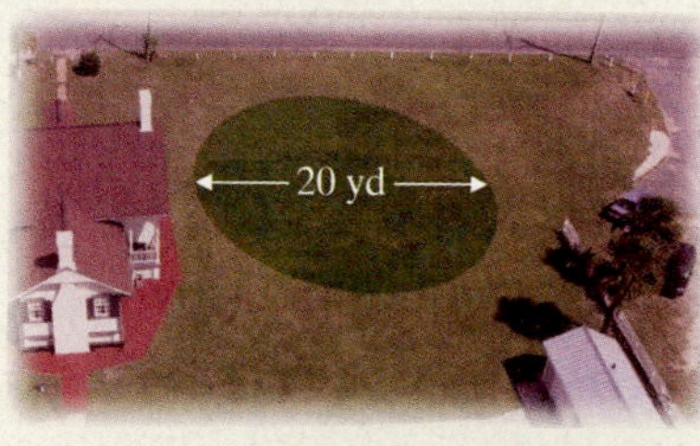

EXAMPLE 8 Finding Circumference of a Circular Spa

A homeowner plans to install a border of new tiling around the circumference of her circular spa. If her spa has a diameter of 14 feet, find its exact circumference. Then use the approximation 3.14 for π to approximate the circumference.

Solution: Because we are given the diameter, we use the formula $C = \pi \cdot d$.

$C = \pi \cdot d$

$= \pi \cdot 14 \text{ ft}$ Replace d with 14 feet.

$= 14\pi \text{ ft}$

The circumference of the spa is *exactly* 14π feet. By replacing π with the *approximation* 3.14, we find that the circumference is *approximately* $14 \text{ feet} \cdot 3.14 = 43.96$ feet.

Work Practice Problem 8

✔Concept Check The distance around which figure is greater: a square with side length 5 inches or a circle with radius 3 inches?

Answer

8. exactly 20π yd ≈ 62.8 yd

✔ Concept Check Answer

a square with side length 5 in.

Vocabulary and Readiness Check

Use the choices below to fill in each blank.

circumference	radius	π	$\frac{22}{7}$
diameter	perimeter	3.14	

1. The ____________ of a polygon is the sum of the lengths of its sides.

2. The distance around a circle is called the ____________.

3. The exact ratio of circumference to diameter is ____________.

4. The diameter of a circle is double its ____________.

5. Both ____________ and ____________ are approximations for π.

6. The radius of a circle is half its ____________.

8.2 EXERCISE SET

FOR EXTRA HELP

Student Solutions Manual | PH Math/Tutor Center | CD/Video for Review | MathXL® | MyMathLab

Objective A *Find the perimeter of each figure. See Examples 1 through 6.*

1.

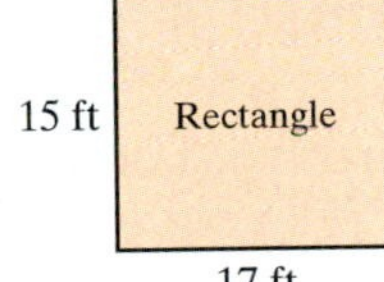

2.

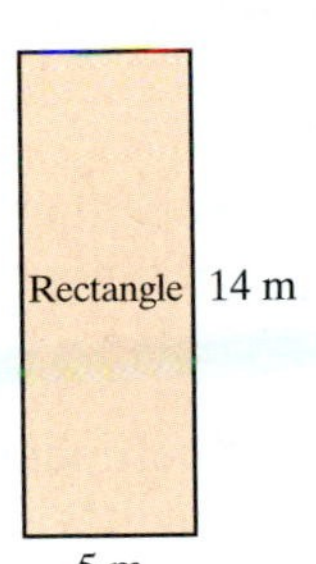

3.

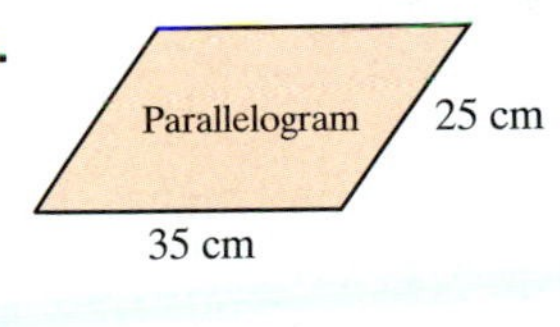

4.

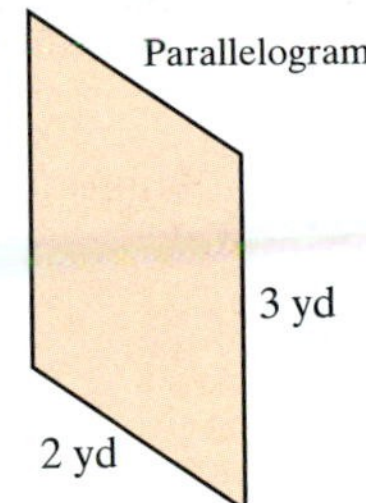

5.

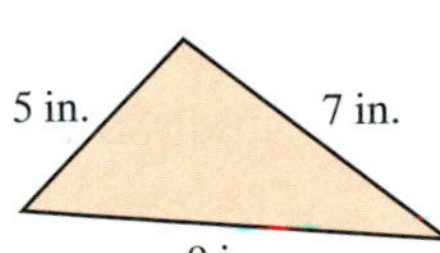

6.

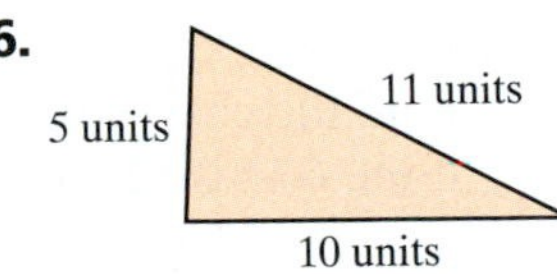

7.

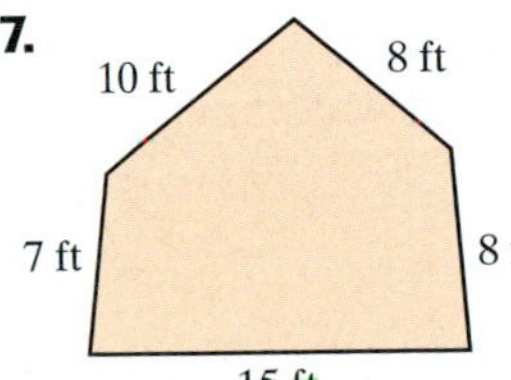

8.

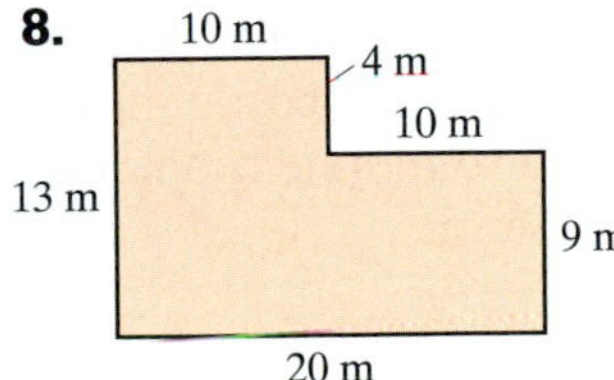

Find the perimeter of each regular polygon. (The sides of a regular polygon have the same length.)

9.

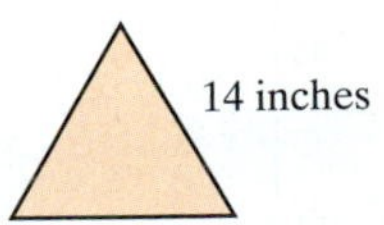

10.

11.

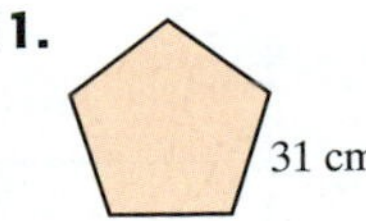

12.

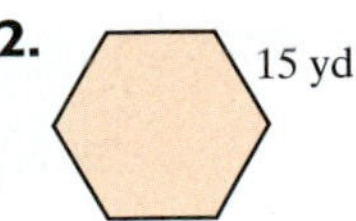

Solve. See Examples 1 through 7.

13. A polygon has sides of length 5 feet, 3 feet, 2 feet, 7 feet, and 4 feet. Find its perimeter.

14. A triangle has sides of length 8 inches, 12 inches, and 10 inches. Find its perimeter.

15. A line marking machine lays down lime powder to mark the base path on a new baseball diamond. If a baseball diamond has 90 feet between each base, how many feet of lime will be deposited on the diamond?

16. Find how much fencing is needed to enclose a rectangular rose garden 85 feet by 15 feet.

17. If a football field is 53 yards wide and 120 yards long, what is the perimeter?

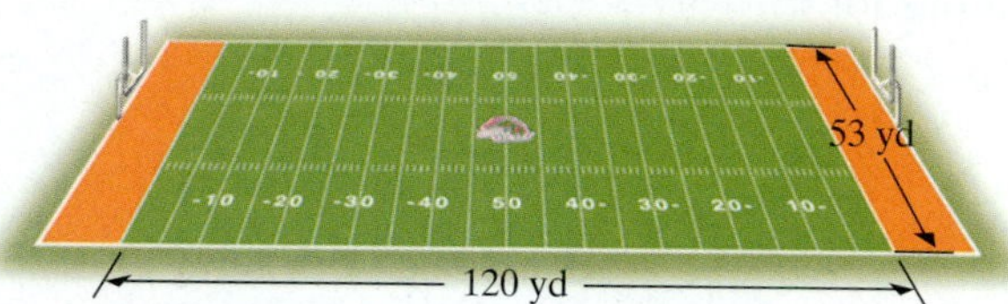

18. A stop sign has eight equal sides of length 12 inches. Find its perimeter.

19. A metal strip is being installed around a workbench that is 8 feet long and 3 feet wide. Find how much stripping is needed.

20. Find how much fencing is needed to enclose a rectangular garden 70 feet by 21 feet.

21. If the stripping in Exercise 19 costs \$2.50 per foot, find the total cost of the stripping.

22. If the fencing in Exercise 20 costs \$2 per foot, find the total cost of the fencing.

23. A regular octagon has a side length of 9 inches. Find its perimeter.

24. A regular pentagon has a side length of 14 meters. Find its perimeter.

25. Find the perimeter of the top of a square compact disc case if the length of one side is 7 inches.

26. Find the perimeter of a square ceramic tile with a side of length 3 inches.

27. A rectangular room measures 10 feet by 11 feet. Find the cost of installing a strip of wallpaper around the room if the wallpaper costs \$0.86 per foot.

28. A rectangular house measures 85 feet by 70 feet. Find the cost of installing gutters around the house if the cost is \$2.36 per foot.

Find the perimeter of each figure. See Example 6.

29.

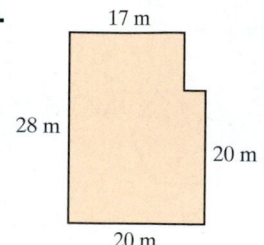

17 m
28 m
20 m
20 m

30.

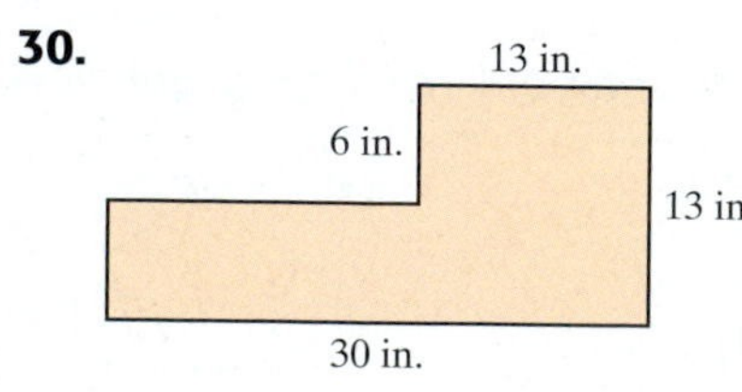

31.

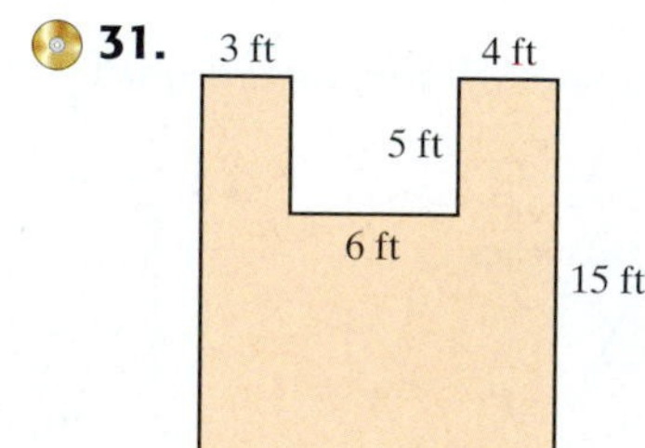

32. 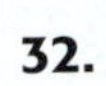

16 cm, 2 cm, 11 cm, 4 cm, 3 cm, 9 cm

33. 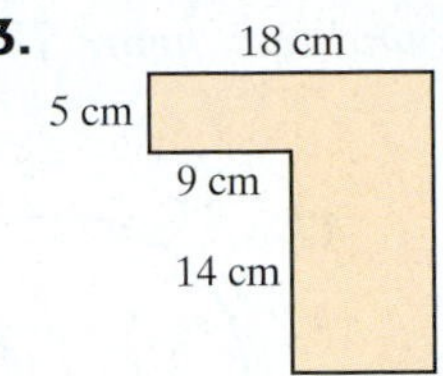

34. 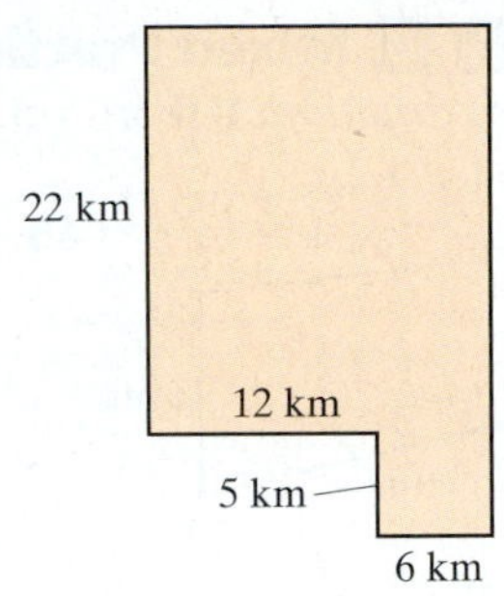

Objective B *Find the circumference of each circle. Give the exact circumference and then an approximation. Use $\pi \approx 3.14$. See Example 8.*

35. 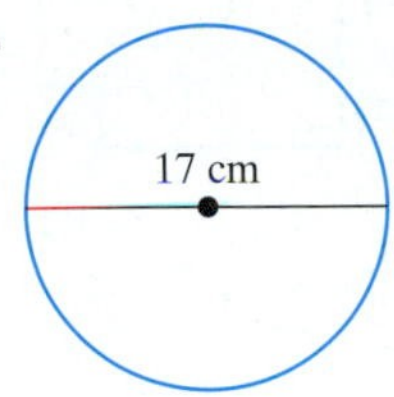

36. 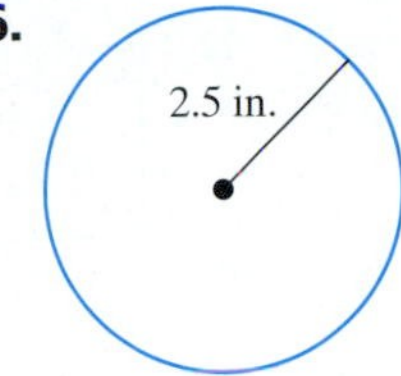

37. 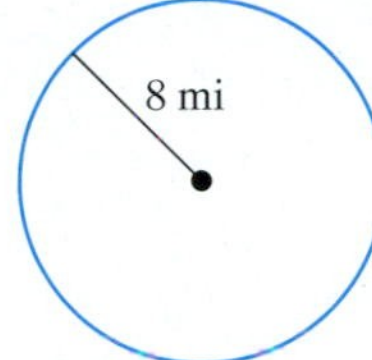

38. 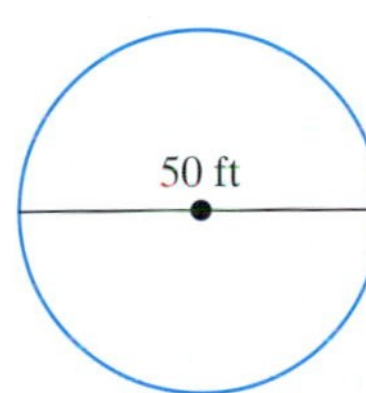

39. 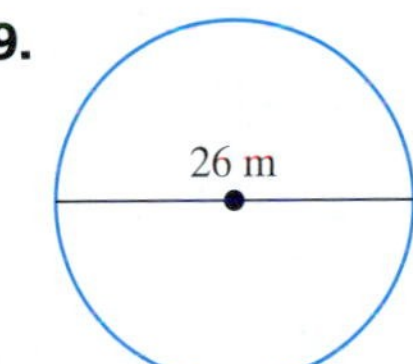

40. 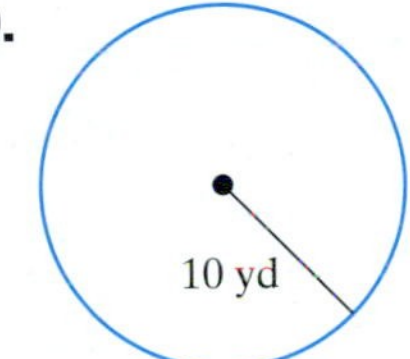

41. Wyley Robinson just bought a trampoline for his children to use. The trampoline has a diameter of 15 feet. If Wyley wishes to buy netting to go around the outside of the trampoline, how many feet of netting does he need?

42. The largest round barn in the world is located at the Marshfield Fairgrounds in Wisconsin. The barn has a diameter of 150 ft. What is the circumference of the barn? (*Source: The Milwaukee Journal Sentinel*)

43. Meteor Crater, near Winslow, Arizona, is 4000 feet in diameter. Approximate the distance around the crater. Use 3.14 for π. (*Source: The Handy Science Answer Book*)

44. The largest pearl, the *Pearl of Lao-tze,* has a diameter of $5\frac{1}{2}$ inches. Approximate the distance around the pearl. Use $\frac{22}{7}$ for π. (*Source: The Guinness World Records*)

Objectives A B Mixed Practice *Find the distance around each figure. For circles, give the exact circumference and then an approximation. Use $\pi \approx 3.14$.*

45.

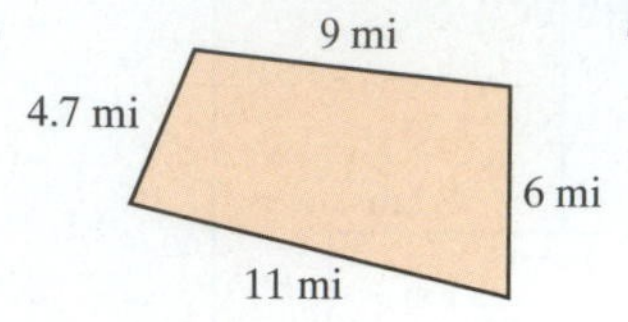

46.

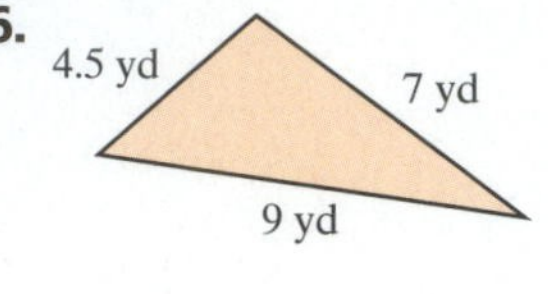

47.

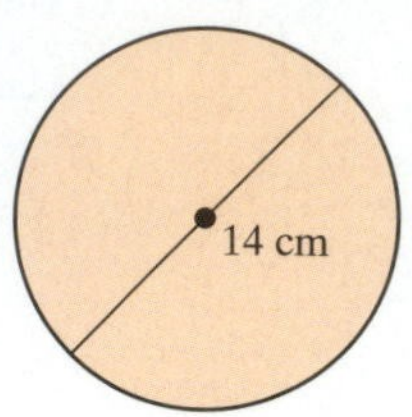

48.

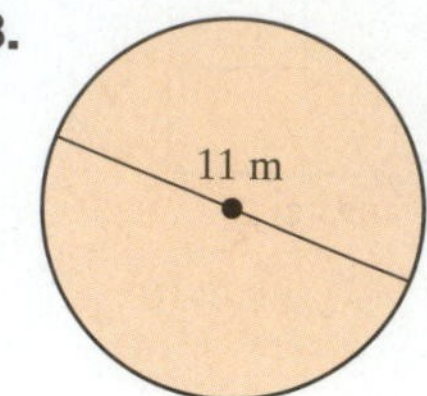

49.

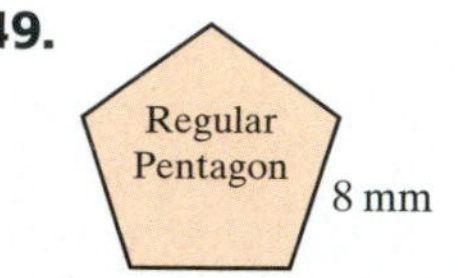

50.

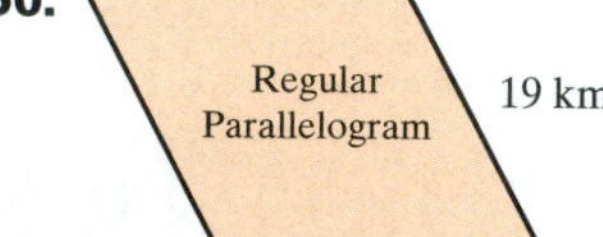

51.

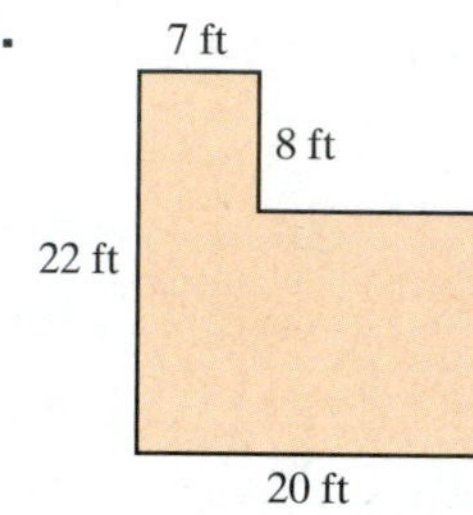

52.

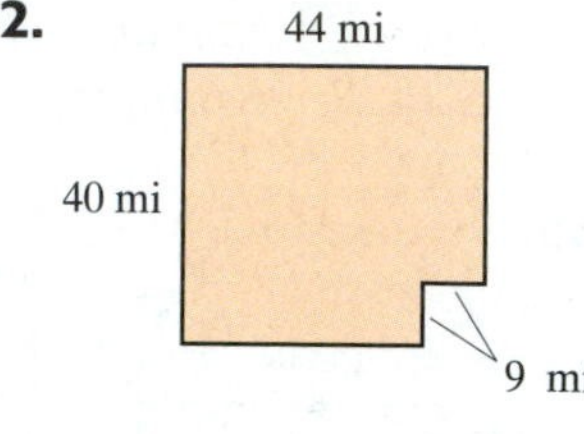

Review

Simplify. See Section 1.7.

53. $5 + 6 \cdot 3$

54. $25 - 3 \cdot 7$

55. $(20 - 16) \div 4$

56. $6 \cdot (8 + 2)$

57. $(18 + 8) - (12 + 4)$

58. $72 \div (2 \cdot 6)$

59. $(72 \div 2) \cdot 6$

60. $4^1 \cdot (2^3 - 8)$

Concept Extensions

There are a number of factors that determine the dimensions of a rectangular soccer field. Use the table below to answer Exercises 61 and 62.

Soccer Field Width and Length		
Age	**Width Min–Max**	**Length Min–Max**
Under 6/7:	15–20 yards	25–30 yards
Under 8:	20–25 yards	30–40 yards
Under 9:	30–35 yards	40–50 yards
Under 10:	40–50 yards	60–70 yards
Under 11:	40–50 yards	70–80 yards
Under 12:	40–55 yards	100–105 yards
Under 13:	50–60 yards	100–110 yards
International:	70–80 yards	110–120 yards

61. a. Find the minimum length and width of a soccer field for 8-year-old children. (Carefully consider the age.)

b. Find the perimeter of this field.

62. a. Find the maximum length and width of a soccer field for 12-year-old children.

b. Find the perimeter of this field.

Solve. See the Concept Check in this section. Choose the figure that has greater distance around.

63. a. A square with side length 3 inches

b. A circle with diameter 4 inches

64. a. A circle with diameter 7 inches

b. A square with side length 7 inches

65. a. Find the circumference of each circle. Approximate the circumference by using 3.14 for π.

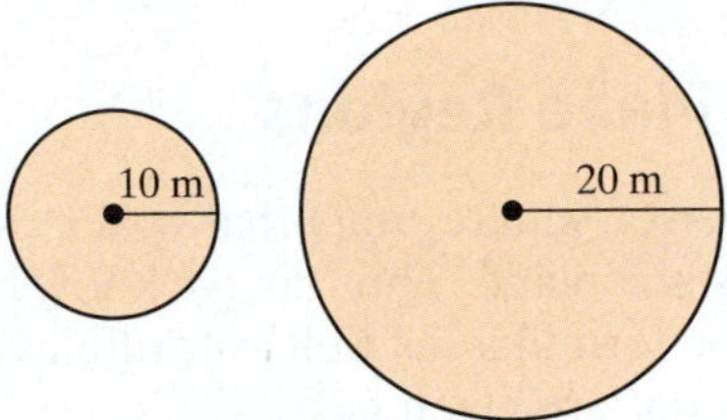

b. If the radius of a circle is doubled, is its corresponding circumference doubled?

66. a. Find the circumference of each circle. Approximate the circumference by using 3.14 for π.

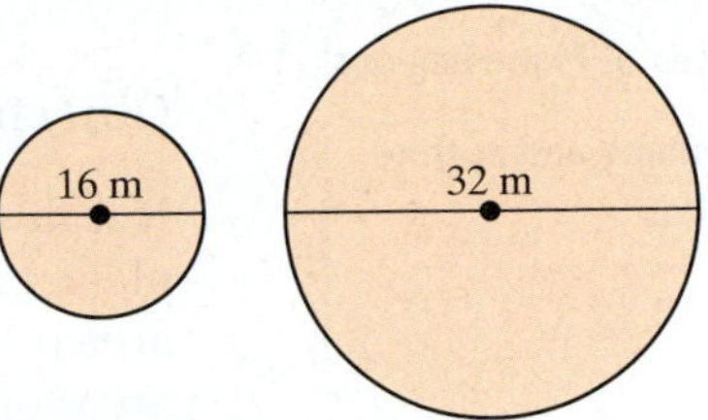

b. If the diameter of a circle is doubled, is its corresponding circumference doubled?

67. Find the perimeter of the skating rink.

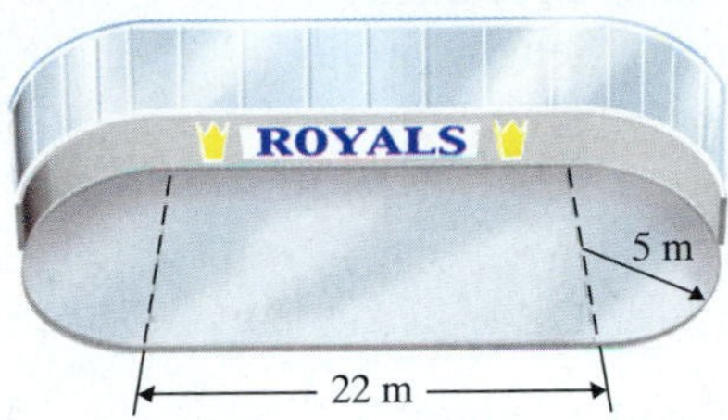

68. In your own words, explain how to find the perimeter of any polygon.

Find the perimeter. Round your results to the nearest tenth.

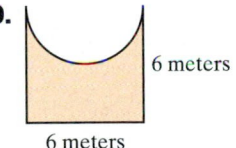

69.

70.

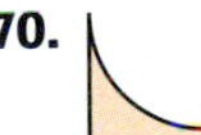

Objectives

A Find the Area of Plane Regions.

B Find the Volume and Surface Area of Solids.

8.3 AREA, VOLUME, AND SURFACE AREA

Objective A Finding Area of Plane Regions

Recall that area measures the number of square units that cover the surface of a plane region; that is, a region that lies in a plane. Thus far, we know how to find the area of a rectangle and a square. These formulas, as well as formulas for finding the areas of other common geometric figures, are given next.

Area Formulas of Common Geometric Figures

Geometric Figure | **Area Formula**

RECTANGLE

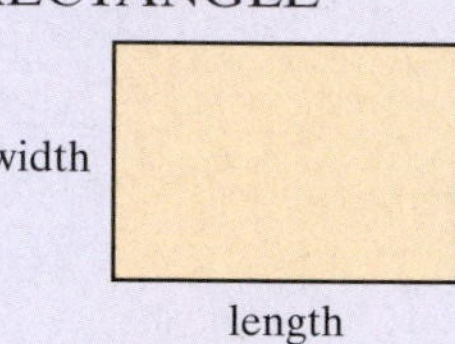

Area of a rectangle:

Area = **l**ength · **w**idth

$A = lw$

SQUARE

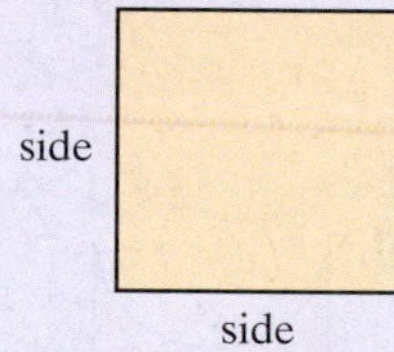

Area of a square:

Area = **s**ide · **s**ide

$A = s \cdot s = s^2$

TRIANGLE

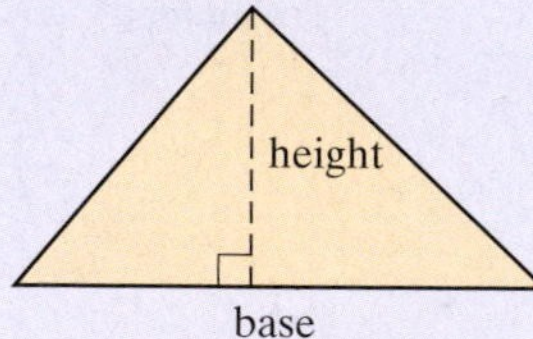

Area of a triangle:

Area $= \frac{1}{2} \cdot$ **b**ase · **h**eight

$A = \frac{1}{2}bh$

PARALLELOGRAM

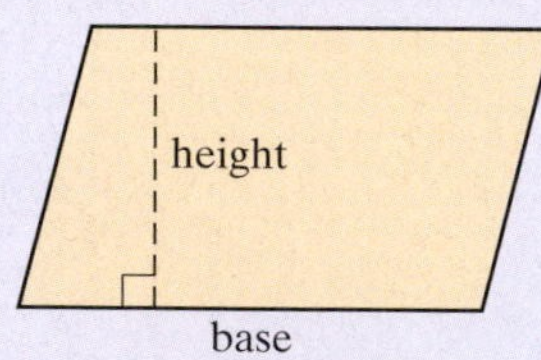

Area of a parallelogram:

Area = **b**ase · **h**eight

$A = bh$

TRAPEZOID

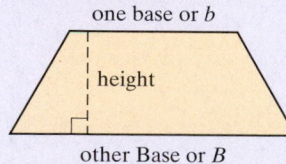

one base or b

height

other Base or B

Area of a trapezoid:

Area $= \frac{1}{2} \cdot$ (one **b**ase + other **b**ase) · **h**eight

$A = \frac{1}{2}(b + B)h$

Use these formulas for the following examples.

Helpful Hint Area is always measured in square units.

EXAMPLE 1 Find the area of the triangle.

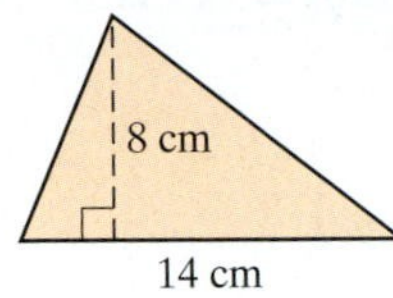

Solution: $A = \frac{1}{2}bh$

$= \frac{1}{2} \cdot 14 \text{ cm} \cdot 8 \text{ cm}$ — Replace b, base, with 14 cm and h, height, with 8 cm.

$= \frac{\overset{1}{\cancel{2}} \cdot 7 \cdot 8}{\underset{1}{\cancel{2}}}$ square centimeters — Write 14 as $2 \cdot 7$.

= 56 square centimeters

The area is 56 square centimeters.

Work Practice Problem 1

EXAMPLE 2 Find the area of the parallelogram.

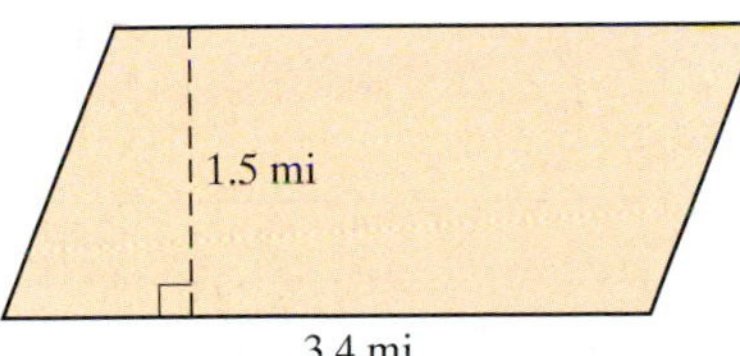

Solution: $A = bh$

$= 3.4 \text{ miles} \cdot 1.5 \text{ miles}$ — Replace b, base, with 3.4 miles and h, height, with 1.5 miles.

= 5.1 square miles

The area is 5.1 square miles.

Work Practice Problem 2

Helpful Hint When finding the area of figures, check to make sure that all measurements are in the same units before calculations are made.

EXAMPLE 3 Find the area of the figure.

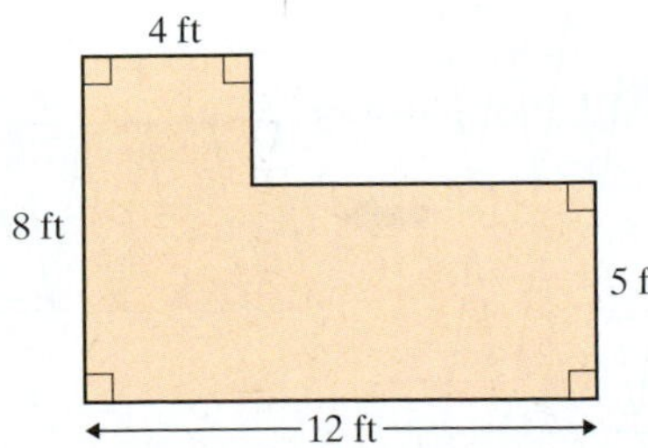

Continued on next page

PRACTICE PROBLEM 1

Find the area of the triangle.

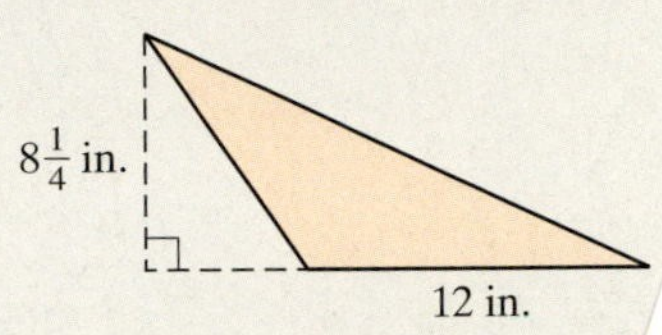

PRACTICE PROBLEM 2

Find the area of the trapezoid.

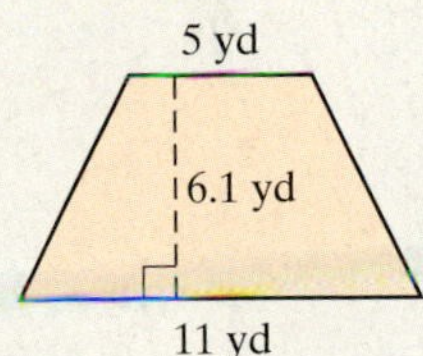

PRACTICE PROBLEM 3

Find the area of the figure.

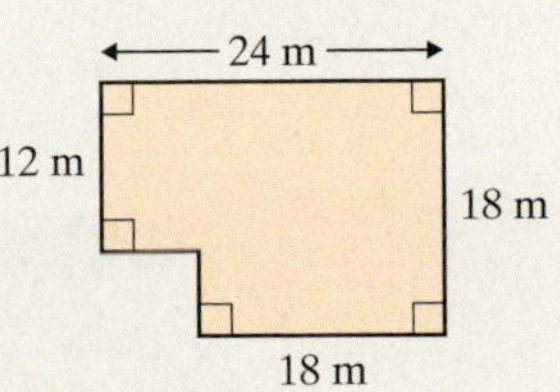

Answers

1. $49\frac{1}{2}$ sq in. **2.** 48.8 sq yd **3.** 396 sq m

Solution: Split the figure into two rectangles. To find the area of the figure, we find the sum of the areas of the two rectangles.

$$\begin{aligned}\text{area of Rectangle 1} &= lw \\ &= 8 \text{ feet} \cdot 4 \text{ feet} \\ &= 32 \text{ square feet}\end{aligned}$$

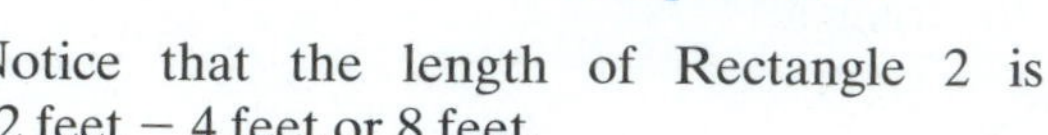

Notice that the length of Rectangle 2 is 12 feet − 4 feet or 8 feet.

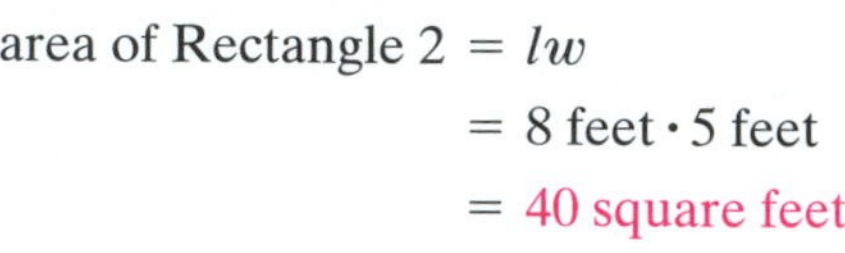

$$\begin{aligned}\text{area of Rectangle 2} &= lw \\ &= 8 \text{ feet} \cdot 5 \text{ feet} \\ &= 40 \text{ square feet}\end{aligned}$$

$$\begin{aligned}\text{area of the figure} &= \text{area of Rectangle 1} + \text{area of Rectangle 2} \\ &= 32 \text{ square feet} + 40 \text{ square feet} \\ &= 72 \text{ square feet}\end{aligned}$$

Work Practice Problem 3

Helpful Hint

The figure in Example 3 could also be split into two rectangles as shown.

To better understand the formula for area of a circle, try the following. Cut a circle into many pieces, as shown.

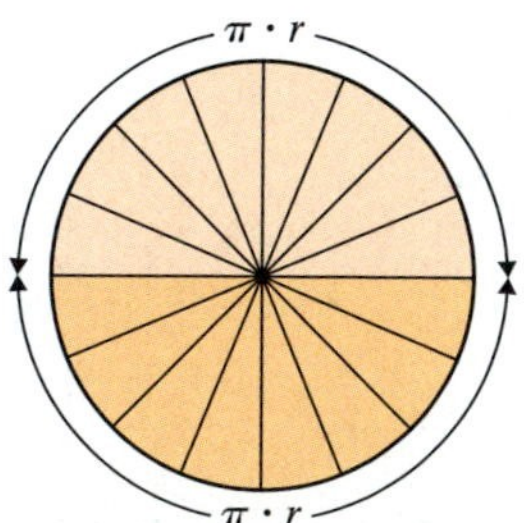

The circumference of a circle is $2\pi r$. This means that the circumference of half a circle is half of $2\pi r$, or πr.

Then unfold the two halves of the circle and place them together, as shown.

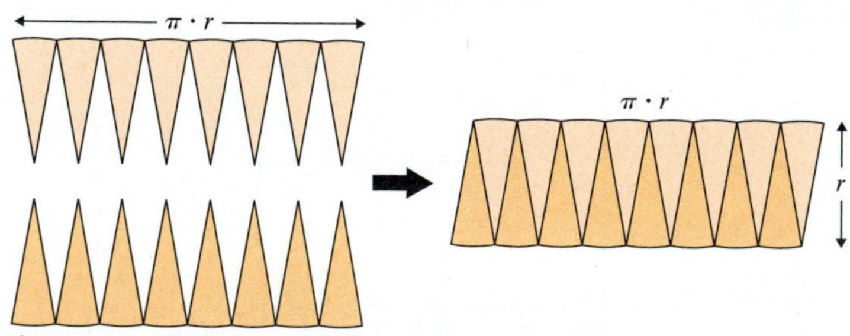

The figure on the right is almost a parallelogram with a base of πr and a height of r. The area is

$$
\begin{aligned}
A &= \text{base} \cdot \text{height} \\
&= (\pi r) \cdot r \\
&= \pi r^2
\end{aligned}
$$

This is the formula for area of a circle.

Area Formula of a Circle

Circle

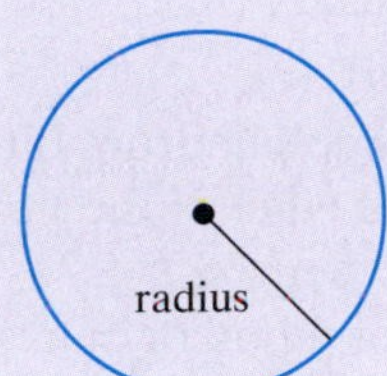

Area of a circle

Area $= \pi \cdot (\text{radius})^2$

$A = \pi r^2$

(A fraction approximation for π is $\frac{22}{7}$.)

(A decimal approximation for π is 3.14.)

EXAMPLE 4 Find the area of a circle with a radius of 3 feet. Find the exact area and an approximation. Use 3.14 as an approximation for π.

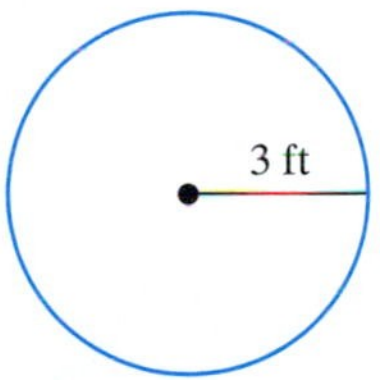

Solution: We let $r = 3$ feet and use the formula

$A = \pi r^2$

$= \pi \cdot (3\text{ feet})^2$ Replace r with 3 feet.

$= 9 \cdot \pi$ square feet Replace $(3\text{ feet})^2$ with 9 sq ft.

To approximate this area, we substitute 3.14 for π.

$9 \cdot \pi$ square feet $\approx 9 \cdot 3.14$ square feet

$= 28.26$ square feet

The *exact* area of the circle is 9π square feet, which is *approximately* 28.26 square feet.

Work Practice Problem 4

✔Concept Check Use estimation to decide which figure would have a larger area: a circle of diameter 10 in. or a square 10 in. long on each side.

PRACTICE PROBLEM 4

Find the area of the given circle. Find the exact area and an approximation. Use 3.14 as an approximation for π.

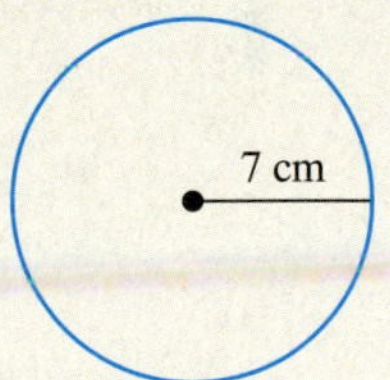

Answer

4. 49π sq cm $\approx$ 153.86 sq cm

✔ Concept Check Answer

A square 10 in. long on each side would have a larger area.

Objective B Finding Volume and Surface Area of Solids

A convex solid is a set of points, *S*, not all in one plane, such that for any two points *A* and *B* in *S*, all points between *A* and *B* are also in *S*. In this section, we will find the volume and surface area of special types of solids called polyhedrons. A solid formed by the intersection of a finite number of planes is called a **polyhedron.** The box below is an example of a polyhedron.

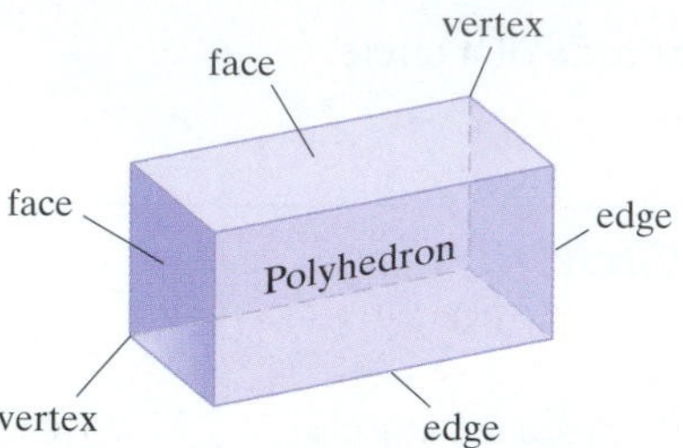

Each of the plane regions of the polyhedron is called a **face** of the polyhedron. If the intersection of two faces is a line segment, this line segment is an **edge** of the polyhedron. The intersections of the edges are the **vertices** of the polyhedron.

Volume is a measure of the space of a region. The volume of a box or can, for example, is the amount of space inside. Volume can be used to describe the amount of juice in a pitcher or the amount of concrete needed to pour a foundation for a house.

The volume of a solid is the number of **cubic units** in the solid. A cubic centimeter and a cubic inch are illustrated.

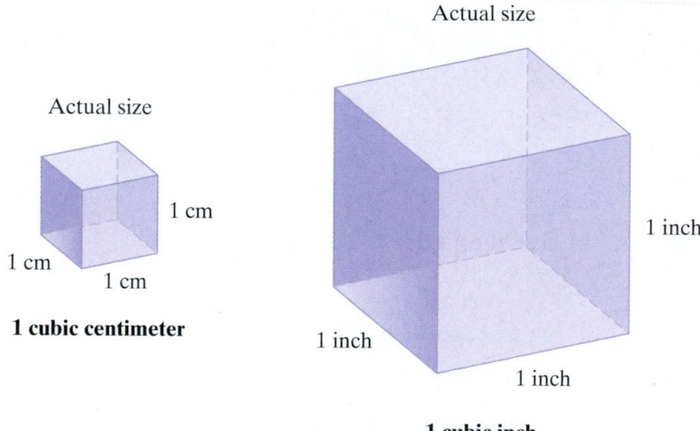

The **surface area** of a polyhedron is the sum of the areas of the faces of the polyhedron. For example, each face of the cube to the left above has an area of 1 square centimeter. Since there are 6 faces of the cube, the sum of the areas of the faces is 6 square centimeters. Surface area can be used to describe the amount of material needed to cover a solid. Surface area is measured in square units.

Formulas for finding the volumes, V, and surface areas, SA, of some common solids are given next.

Volume and Surface Area Formulas of Common Solids

Solid	Formulas
RECTANGULAR SOLID 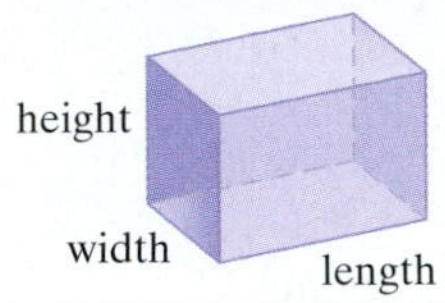	$V = lwh$ $SA = 2lh + 2wh + 2lw$ where h = height, w = width, l = length
CUBE 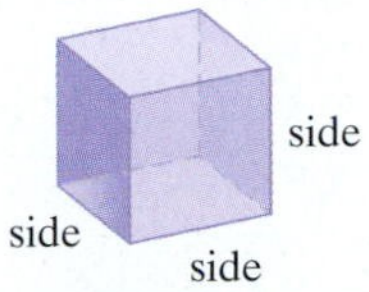	$V = s^3$ $SA = 6s^2$ where s = side
SPHERE 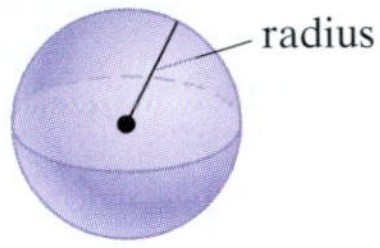	$V = \frac{4}{3}\pi r^3$ $SA = 4\pi r^2$ where r = radius
CIRCULAR CYLINDER 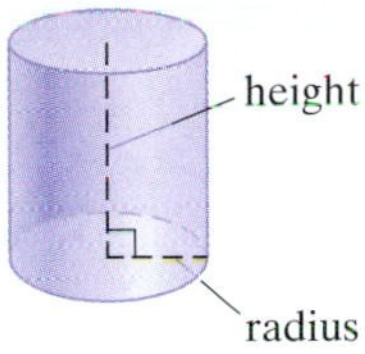	$V = \pi r^2 h$ $SA = 2\pi rh + 2\pi r^2$ where h = height, r = radius
CONE 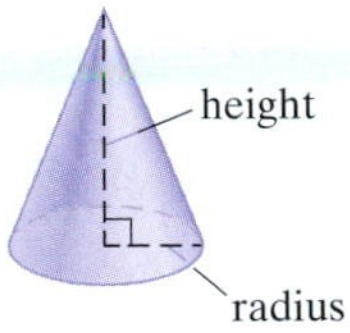	$V = \frac{1}{3}\pi r^2 h$ $SA = \pi r\sqrt{r^2 + h^2} + \pi r^2$ where h = height, r = radius
SQUARE-BASED PYRAMID 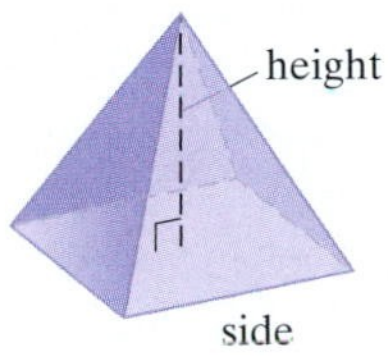	$V = \frac{1}{3}s^2 h$ $SA = B + \frac{1}{2}pl$ where B = area of base; p = perimeter of base, h = height, s = side, l = slant height

PRACTICE PROBLEM 5

Find the volume and surface area of a rectangular box that is 7 feet long, 3 feet wide, and 4 feet deep.

EXAMPLE 5 Find the volume and surface area of a rectangular box that is 12 inches long, 6 inches wide, and 3 inches high.

Solution: Let $h = 3$ in., $l = 12$ in., and $w = 6$ in.

$$V = lwh$$

$$V = 12 \text{ inches} \cdot 6 \text{ inches} \cdot 3 \text{ inches} = 216 \text{ cubic inches}$$

The volume of the rectangular box is 216 cubic inches.

$$\begin{aligned} SA &= 2lh + 2wh + 2lw \\ &= 2(12 \text{ in.})(3 \text{ in.}) + 2(6 \text{ in.})(3 \text{ in.}) + 2(12 \text{ in.})(6 \text{ in.}) \\ &= 72 \text{ sq in.} + 36 \text{ sq in.} + 144 \text{ sq in.} \\ &= 252 \text{ sq in.} \end{aligned}$$

The surface area of rectangular box is 252 square inches.

Work Practice Problem 5

✔ **Concept Check** Juan is calculating the volume of the following rectangular solid. Find the error in his calculation.

$$\begin{aligned} \text{Volume} &= l + w + h \\ &= 14 \text{ cm} + 8 \text{ cm} + 5 \text{ cm} \\ &= 27 \text{ cu cm} \end{aligned}$$

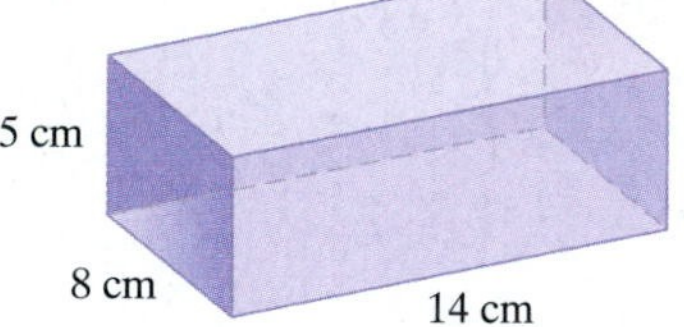

PRACTICE PROBLEM 6

Find the volume and surface area of a ball of radius $\frac{1}{2}$ centimeter. Give the exact volume and surface area. Then use $\frac{22}{7}$ for π.

EXAMPLE 6 Find the volume and surface area of a ball of radius 2 inches. Give the exact volume and surface area. Then use the approximation $\frac{22}{7}$ for π.

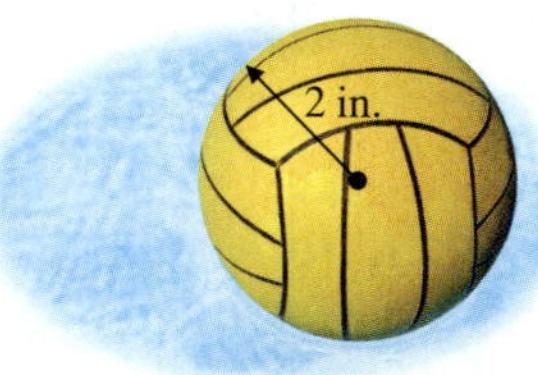

Solution:

$V = \frac{4}{3}\pi r^3$ — Formula for volume of a sphere

$V = \frac{4}{3} \cdot \pi(2 \text{ in.})^3$ — Let $r = 2$ inches.

$= \frac{32}{3}\pi$ cu in. — Exact volume

$\approx \frac{32}{3} \cdot \frac{22}{7}$ cu in. — Approximate π with $\frac{22}{7}$.

$= \frac{704}{21}$ or $33\frac{11}{21}$ cu in. — Approximate volume

Answers

5. $V = 84$ cu ft; $SA = 122$ sq ft

6. $V = \frac{1}{6}\pi$ cu cm $\approx \frac{11}{21}$ cu cm; $SA = \pi$ sq cm $\approx 3\frac{1}{7}$ sq cm

✔ Concept Check Answer

$$\begin{aligned} \text{Volume} &= l \cdot w \cdot h \\ &= 14 \cdot 8 \cdot 5 \\ &= 560 \text{ cu cm} \end{aligned}$$

The volume of the sphere is exactly $\frac{32}{3}\pi$ cubic inches or approximately $33\frac{11}{21}$ cubic inches.

$$
\begin{aligned}
SA &= 4\pi r^2 && \text{Formula for surface area} \\
SA &= 4 \cdot \pi (2\,\text{in.})^2 && \text{Let } r = 2 \text{ inches.} \\
&= 16\pi \text{ sq in.} && \text{Exact surface area} \\
&\approx 16 \cdot \frac{22}{7} \text{ sq in.} && \text{Approximate } \pi \text{ with } \frac{22}{7}. \\
&= \frac{352}{7} \text{ or } 50\frac{2}{7} \text{ sq in.} && \text{Approximate surface area}
\end{aligned}
$$

The surface area of the sphere is exactly 16π square inches or approximately $50\frac{2}{7}$ square inches.

Work Practice Problem 6

EXAMPLE 7 Approximate the volume of a can that has a $3\frac{1}{2}$-inch radius and a height of 6 inches. Use $\frac{22}{7}$ for π. Give an exact volume and an approximate volume.

Solution: Using the formula for a circular cylinder, we have

$$
\begin{aligned}
V &= \pi \cdot r^2 \cdot h && 3\tfrac{1}{2} = \tfrac{7}{2} \\
&= \pi \cdot \left(\frac{7}{2}\text{ in.}\right)^2 \cdot 6 \text{ in.} \\
&= \pi \cdot \frac{49}{4} \text{ sq in.} \cdot 6 \text{ in.} \\
&= \frac{\pi \cdot 49 \cdot \cancel{2} \cdot 3}{\cancel{2} \cdot 2} \text{ cu in.} \\
&= 73\frac{1}{2}\pi \text{ cu in. or } 73.5\pi \text{ cu in.}
\end{aligned}
$$

This is the exact volume. To approximate the volume, use the approximation $\frac{22}{7}$ for π.

$$
\begin{aligned}
V &= 73\frac{1}{2}\pi \text{ or } \frac{147}{2} \cdot \frac{22}{7} \text{ cu in.} && \text{Replace } \pi \text{ with } \frac{22}{7}. \\
&= \frac{21 \cdot \cancel{7} \cdot \cancel{2} \cdot 11}{\cancel{2} \cdot \cancel{7}} \text{ cu in.} \\
&= 231 \text{ cubic in.}
\end{aligned}
$$

The volume is approximately 231 cubic inches.

Work Practice Problem 7

PRACTICE PROBLEM 7

Approximate the volume of a cylinder of radius 5 inches and height 9 inches. Use 3.14 for π. Give an exact answer and an approximate answer.

Answer

7. 225π cu in. ≈ 706.5 cu in.

PRACTICE PROBLEM 8

Find the volume of a square-based pyramid that has a 3-meter side and a height of 5.1 meters.

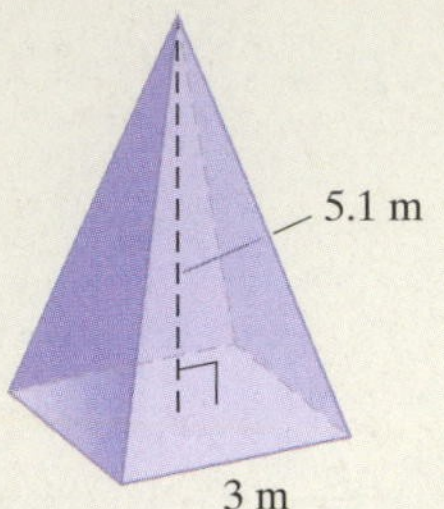

EXAMPLE 8 Approximate the volume of a cone that has a height of 14 centimeters and a radius of 3 centimeters. Use 3.14 for π. Give an exact answer and an approximate answer.

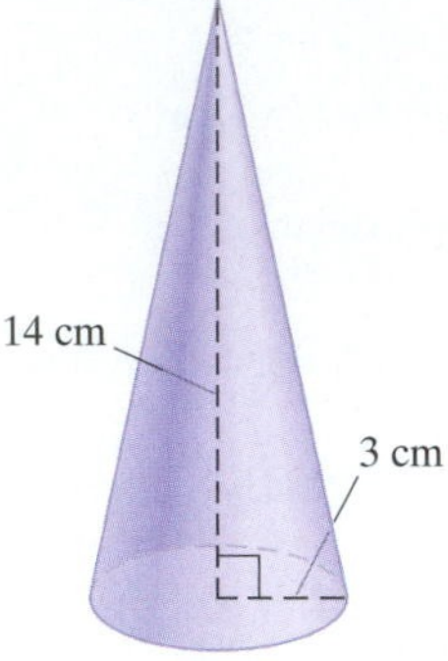

Solution: Using the formula for volume of a cone, we have

$$V = \frac{1}{3} \cdot \pi \cdot r^2 \cdot h$$

$$= \frac{1}{3} \cdot \pi \cdot (3 \text{ cm})^2 \cdot 14 \text{ cm} \quad \text{Replace } r \text{ with 3 cm and } h \text{ with 14 cm.}$$

$$= 42\pi \text{ cu cm}$$

Thus, 42π cubic centimeters is the exact volume. To approximate the volume, use the approximation 3.14 for π.

$$V \approx 42 \cdot 3.14 \text{ cu cm} \quad \text{Replace } \pi \text{ with 3.14.}$$

$$= 131.88 \text{ cu cm}$$

The volume is approximately 131.88 cubic centimeters.

Work Practice Problem 8

Answer

8. 15.3 cu m

Vocabulary and Readiness Check

Use the choices below to fill in each blank. Some choices may be used more than once.

area	surface area	cubic
volume	square	

1. The __________ of a polyhedron is the sum of the areas of its faces.

2. The measure of the space of a solid is its __________.

3. __________ measures the amount of surface enclosed by a region.

4. Volume is measured in __________ units.

5. Area is measured in __________ units.

6. Surface area is measured in __________ units.

8.3 EXERCISE SET

FOR EXTRA HELP

Student Solutions Manual | PH Math/Tutor Center | CD/Video for Review | MathXL® | MyMathLab

Objective A *Find the area of each geometric figure. If the figure is a circle, give an exact area and then use the given* **approximation** *for π to approximate the area. See Examples 1 through 4.*

1.

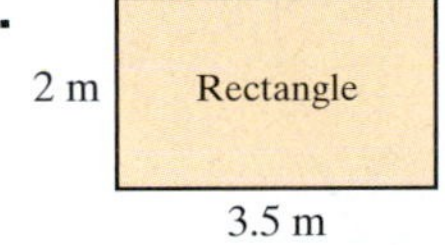

2.

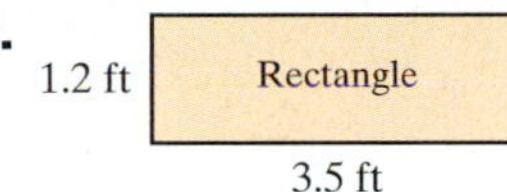

3.

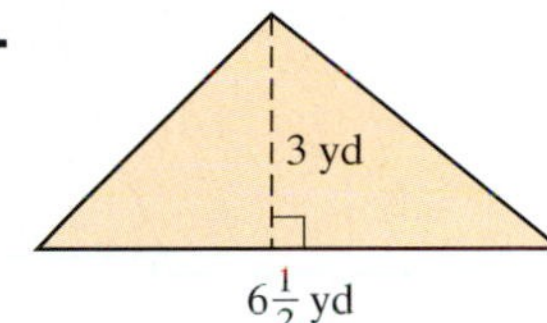

4.

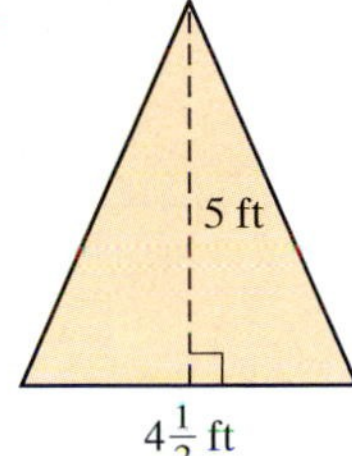

5.

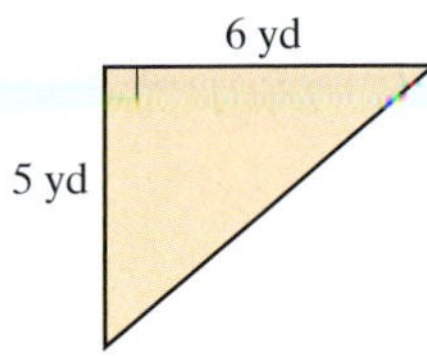

6.

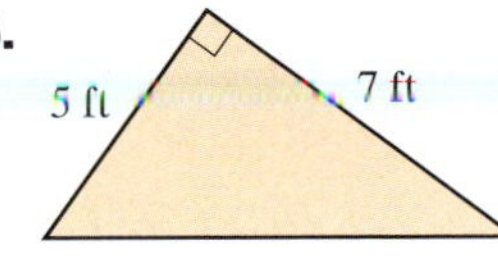

7. Use 3.14 for π.

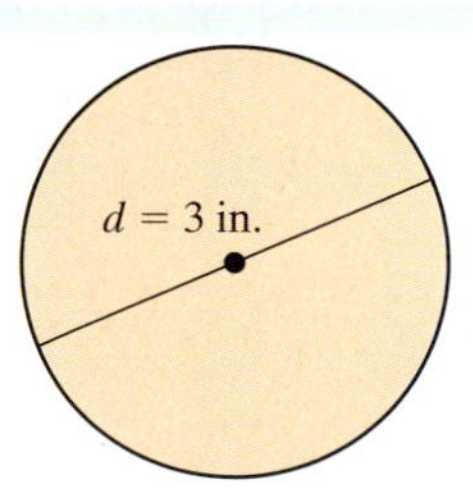

8. Use $\frac{22}{7}$ for π.

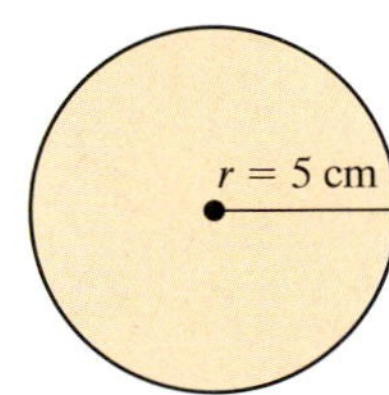

9.

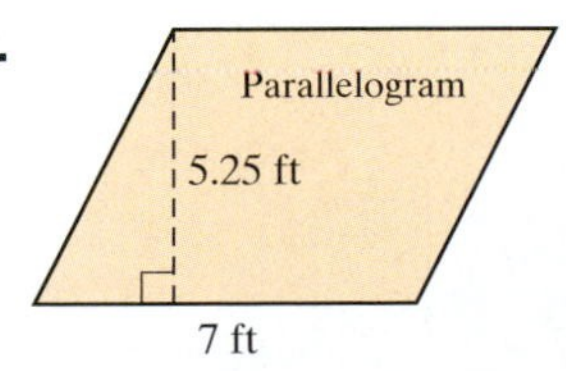

10.

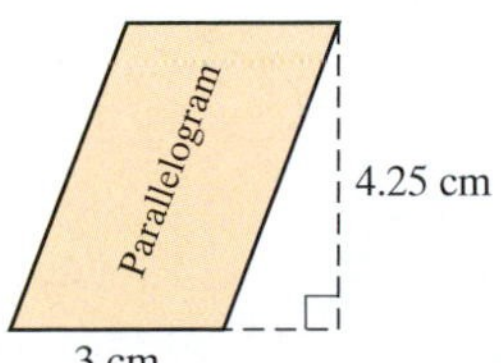

11.

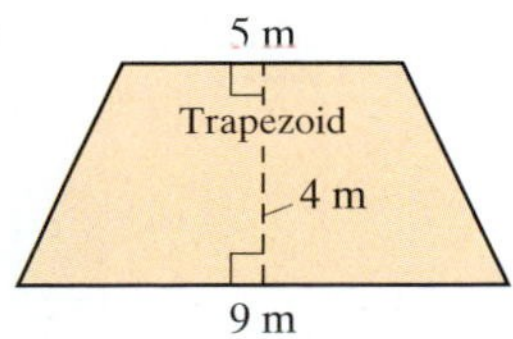

12.

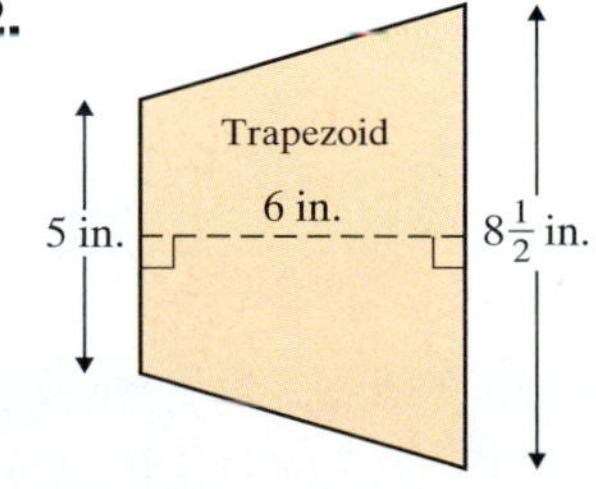

13.

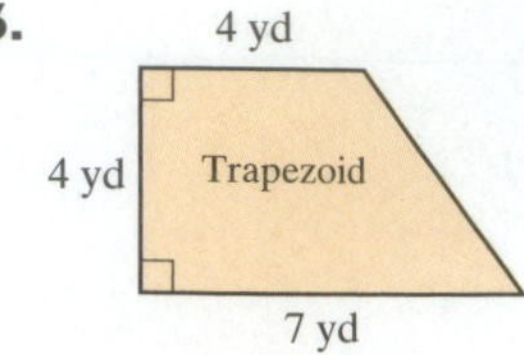

14.

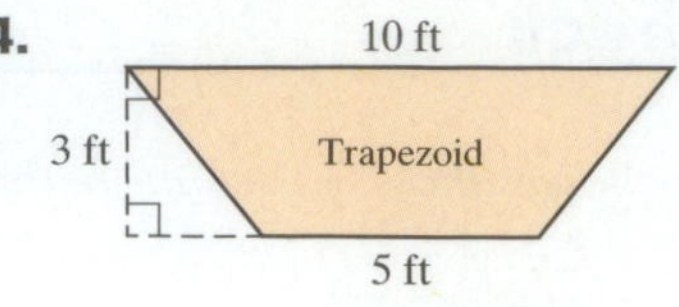

15.

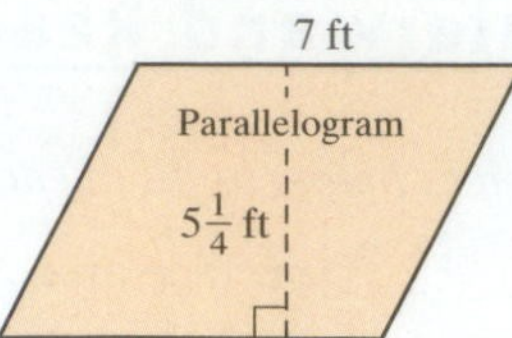

16.

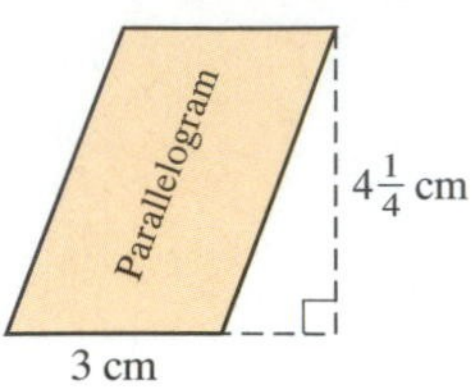

17.

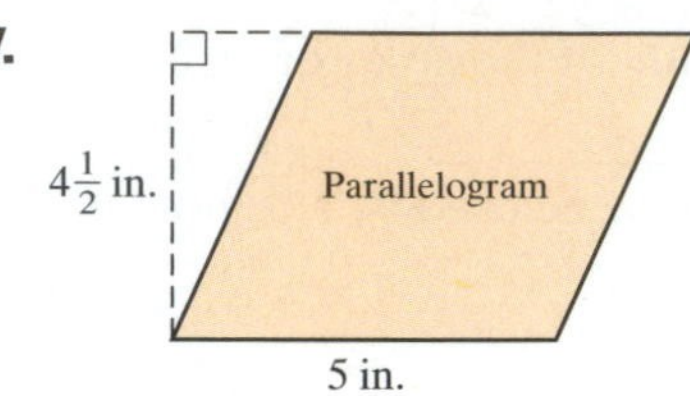

18.

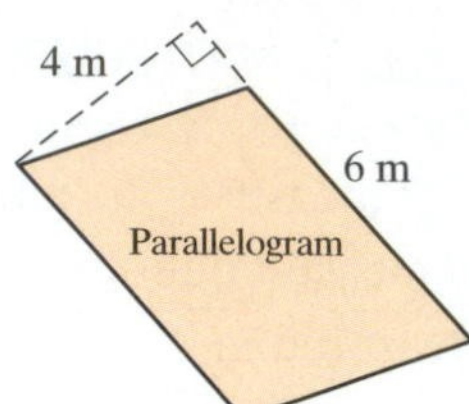

19.

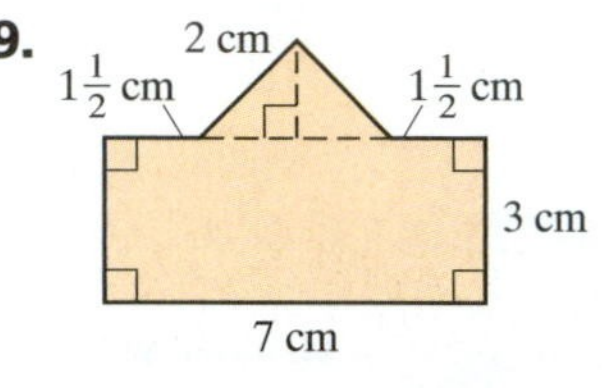

20.

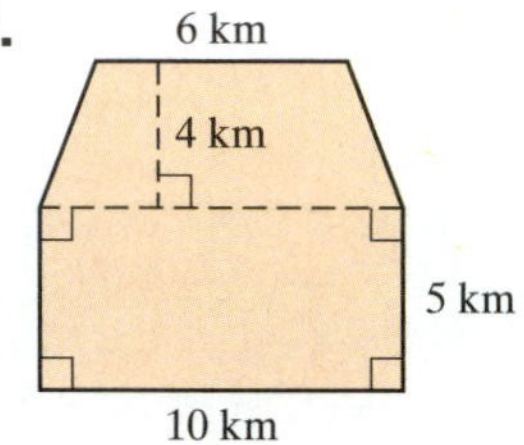

21.

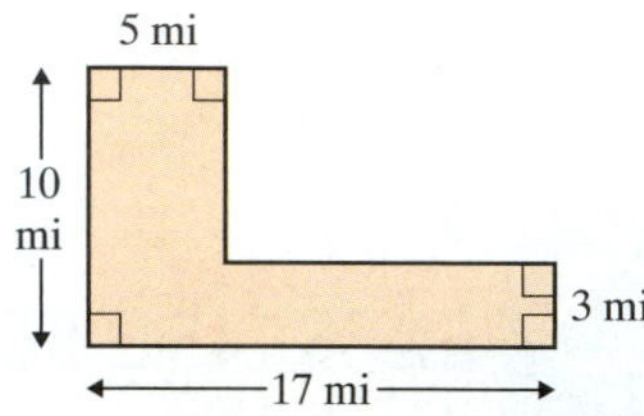

22.

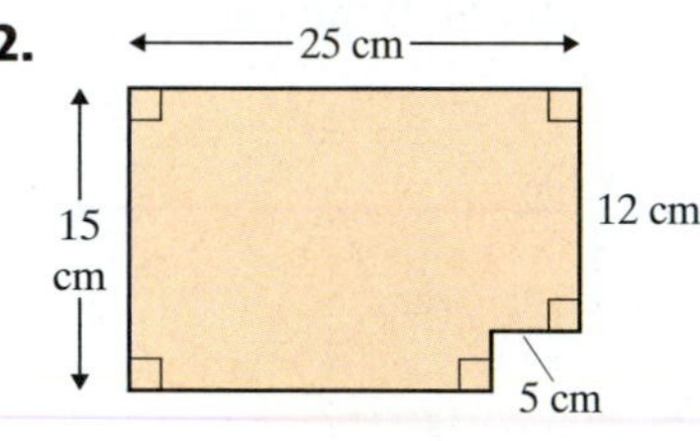

23.

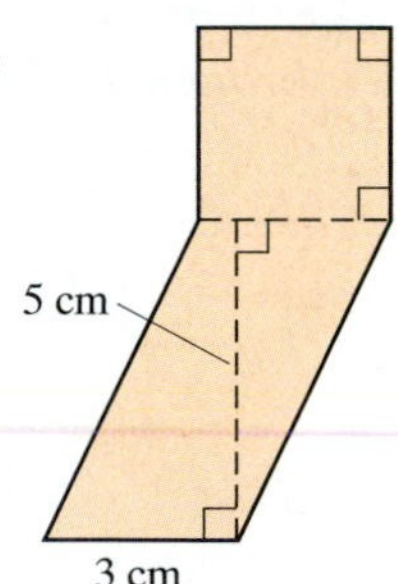

24.

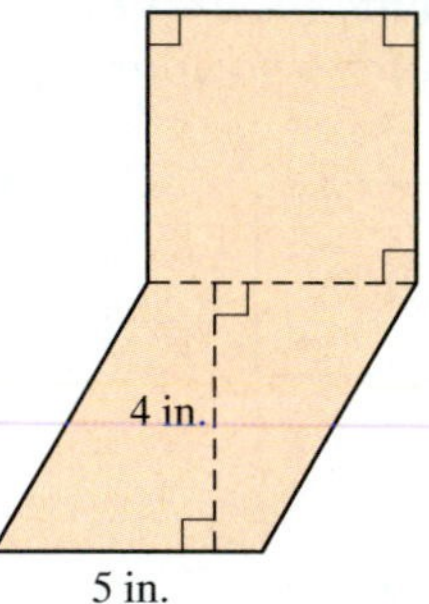

25. Use $\frac{22}{7}$ for π.

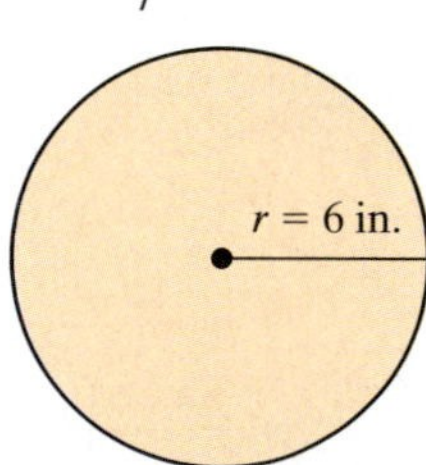

26. Use 3.14 for π.

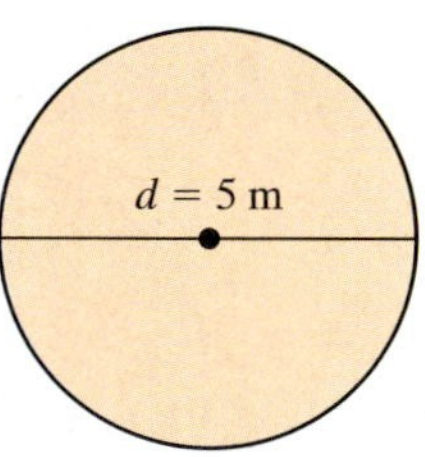

Objective B *Find the volume and surface area of each solid. See Examples 5 through 8. For formulas containing π, give an exact answer and then approximate using $\frac{22}{7}$ for π.*

27.

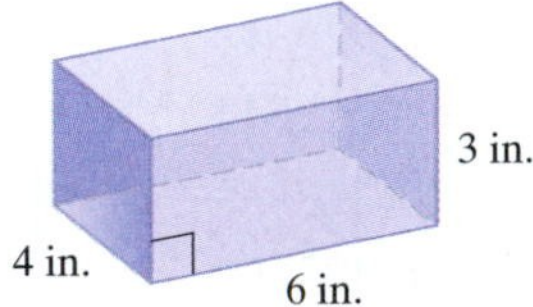

28.

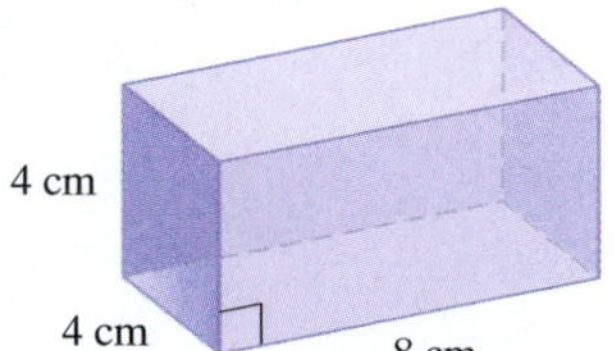

29.

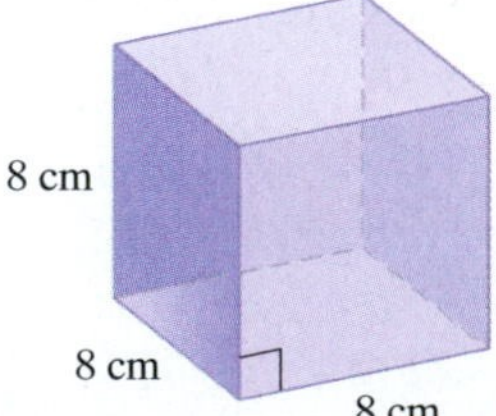

30.

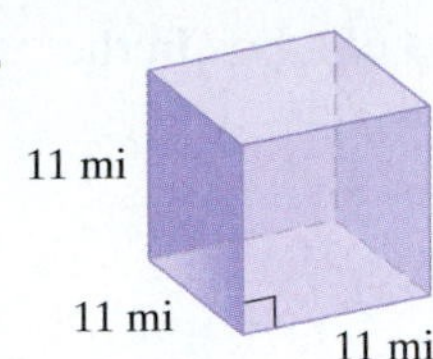

31. For surface area, use $\pi = 3.14$.

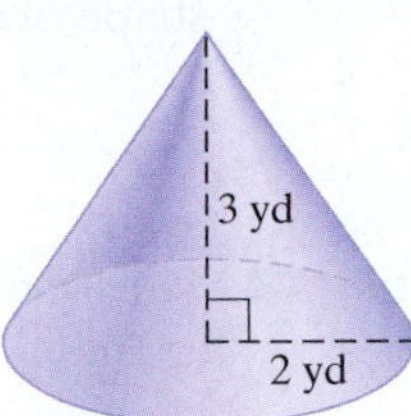

32. For surface area, use $\pi = 3.14$.

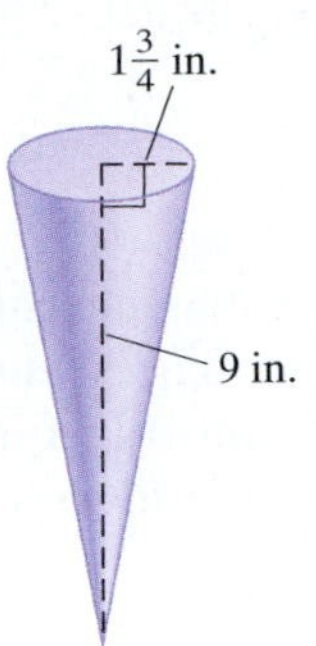

33.

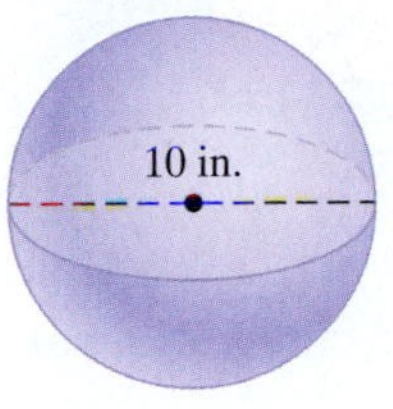

34.

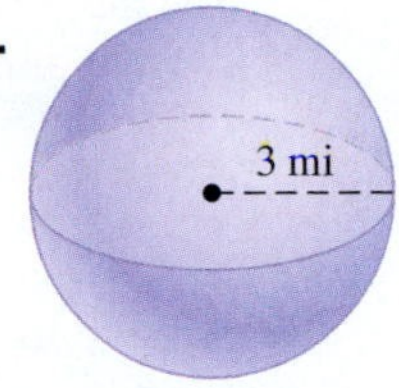

35. Find the volume only.

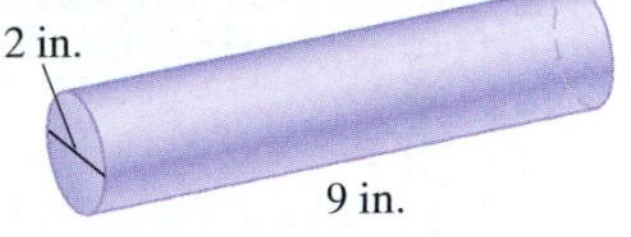

36. Find the volume only.

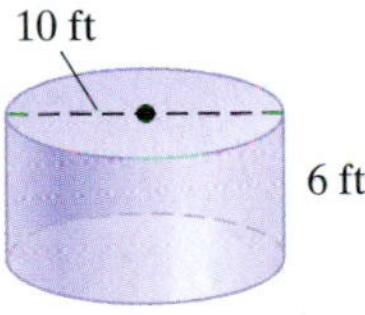

37. Find the volume only.

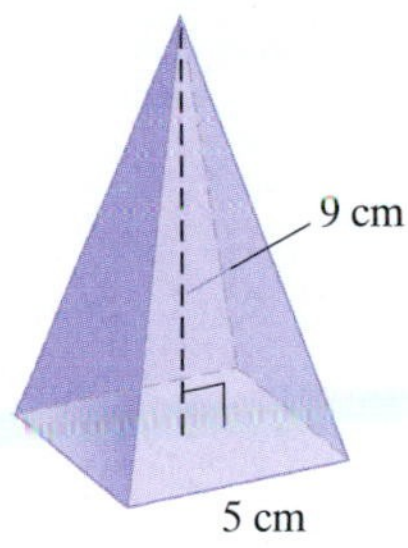

38. Find the volume only.

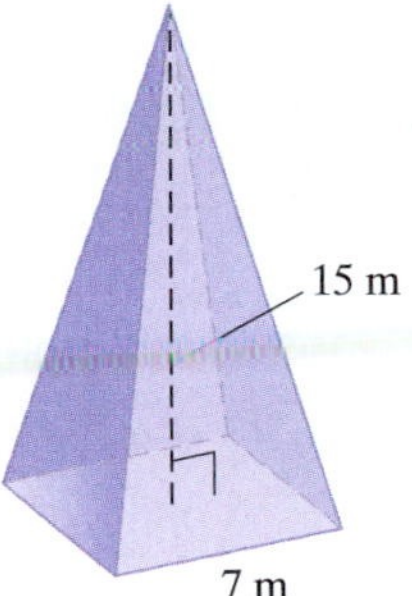

Objectives **A B Mixed Practice** *Solve. See Examples 1 through 8.*

39. Find the volume of a cube with edges of $1\dfrac{1}{3}$ inches.

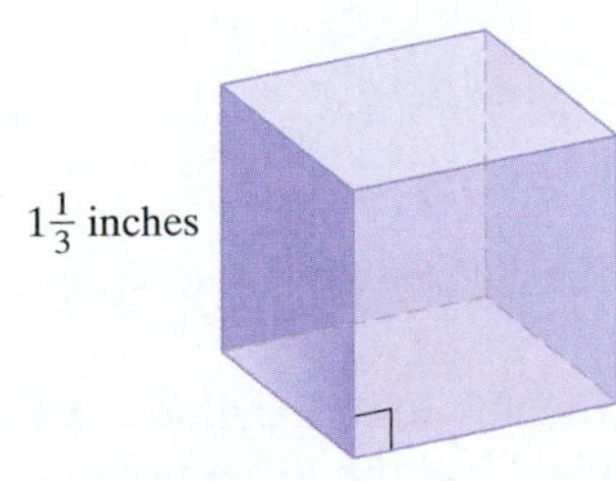

40. A water storage tank is in the shape of a cone with the pointed end down. If the radius is 14 ft and the depth of the tank is 15 ft, approximate the volume of the tank in cubic feet. Use $\dfrac{22}{7}$ for π.

41. Find the volume and surface area of a rectangular box 2 ft by 1.4 ft by 3 ft.

42. Find the volume and surface area of a box in the shape of a cube that is 5 ft on each side.

43. The world's largest flag measures 505 feet by 225 feet. It's the U.S. "Super flag" owned by "Ski" Demski of Long Beach, California. Find its area. (*Source: Guinness World Records,* 2005)

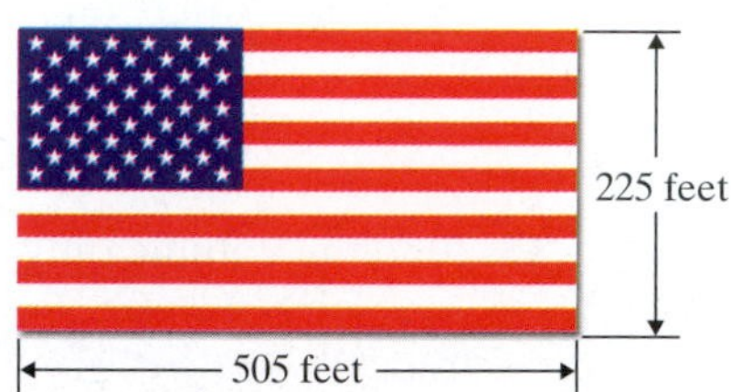

44. The longest illuminated sign is in Ramat Gan, Israel, and measures 197 feet by 66 feet. Find its area. (*Source: The Guinness Book of World Records*)

45. A drapery panel measures 6 ft by 7 ft. Find how many square feet of material are needed for *four* panels.

46. A page in this book measures 27.5 cm by 21.8 cm. Find its area.

47. A paperweight is in the shape of a square-based pyramid 20 centimeters tall. If an edge of the base is 12 centimeters, find the volume of the paperweight.

48. A birdbath is made in the shape of a hemisphere (half-sphere). If its radius is 10 inches, approximate the volume. Use $\frac{22}{7}$ for π.

49. Find how many square feet of land are in the following plot:

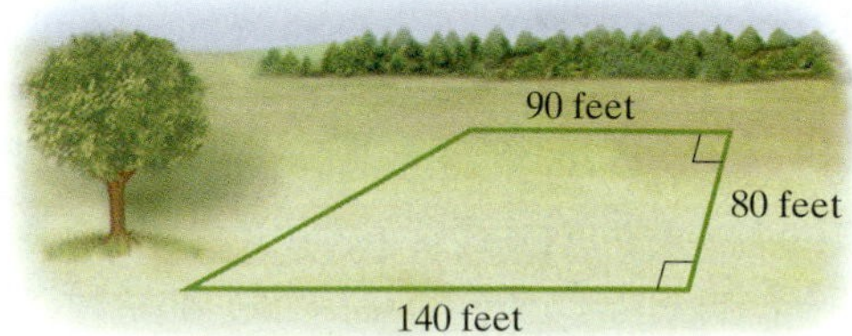

50. For Gerald Gomez to determine how much grass seed he needs to buy, he must know the size of his yard. Use the drawing to determine how many square feet are in his yard.

96 feet
48 feet
48 feet
24 feet
132 feet

51. Find the exact volume and surface area of a sphere with a radius of 7 inches.

52. A tank is in the shape of a cylinder 8 feet tall and 3 feet in radius. Find the exact volume and surface area of the tank.

53. The shaded part of the roof shown is in the shape of a trapezoid and needs to be shingled. The number of packages of shingles to buy depends on the area. Use the dimensions given to find the area of the shaded part of the roof to the nearest whole square foot.

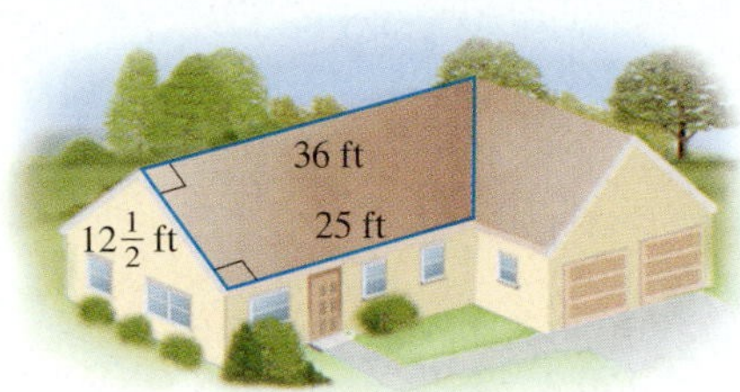

54. The end of the building shaded in the drawing is to be bricked. The number of bricks to buy depends on the area.

a. Find the area.

b. If the side area of each brick (including mortar room) is $\frac{1}{6}$ square ft, find the number of bricks needed to buy.

55. A snow globe has a diameter of 6 inches. Find its exact volume. Then approximate its volume using 3.14 for π.

56. Find the exact volume of a waffle ice cream cone with a 3-in. diameter and a height of 7 inches.

57. Paul Revere's Pizza in the USA will bake and deliver a round pizza with a 4-foot diameter. This pizza is called the "Ultimate Party Pizza" and its current price is $99.99. Find the exact area of the top of the pizza and an approximation. Use 3.14 as an approximation for π.

58. The face of a circular watch has a diameter of 2 centimeters. What is its area? Find the exact area and an approximation. Use 3.14 as an approximation for π.

59. Zorbing is an extreme sport invented by two New Zealanders who joke that they were looking for a way to walk on water. A Zorb is a large sphere inside a second sphere with the space between the spheres pumped full of air. There is a tunnel-like opening so a person can crawl into the inner sphere. You are strapped in and sent down a zorbing hill. A standard zorb is approximately 3 m in diameter. Find the exact volume of a zorb, and approximate the volume using 3.14 for π.

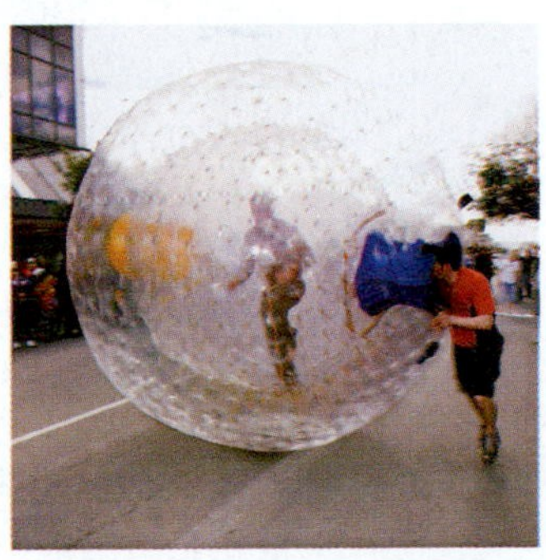

60. Mount Fuji, in Japan, is considered the most beautiful composite volcano in the world. The mountain is in the shape of a cone whose height is about 3.5 kilometers and whose base radius is about 3 kilometers. Approximate the volume of Mt. Fuji in cubic kilometers. Use $\frac{22}{7}$ for π.

61. A $10\frac{1}{2}$-foot by 16-foot concrete wall is to be built using concrete blocks. Find the area of the wall.

62. The floor of Terry's attic is 24 feet by 35 feet. Find how many square feet of insulation are needed to cover the attic floor.

63. Find the volume of a pyramid with a square base 5 inches on a side and a height of $1\frac{3}{10}$ inches.

64. Approximate to the nearest hundredth the volume of a sphere with a radius of 2 centimeters. Use 3.14 for π.

Review

Evaluate. See Section 1.7.

65. 5^2

66. 7^2

67. 3^2

68. 20^2

69. $1^2 + 2^2$

70. $5^2 + 3^2$

71. $4^2 + 2^2$

72. $1^2 + 6^2$

Concept Extensions

Given the following situations, tell whether you are more likely to be concerned with area or perimeter.

73. ordering fencing to fence a yard

74. ordering grass seed to plant in a yard

75. buying carpet to install in a room

76. buying gutters to install on a house

77. ordering paint to paint a wall

78. ordering baseboards to install a room

79. buying a wallpaper border to go on the walls around a room

80. buying fertilizer for your yard

81. A pizza restaurant recently advertised two specials. The first special was a 12-inch pizza for \$10. The second special was two 8-inch pizzas for \$9. Determine the better buy. (*Hint:* First compare the areas of the two specials and then find a price per square inch for both specials.)

82. Find the approximate area of the state of Utah.

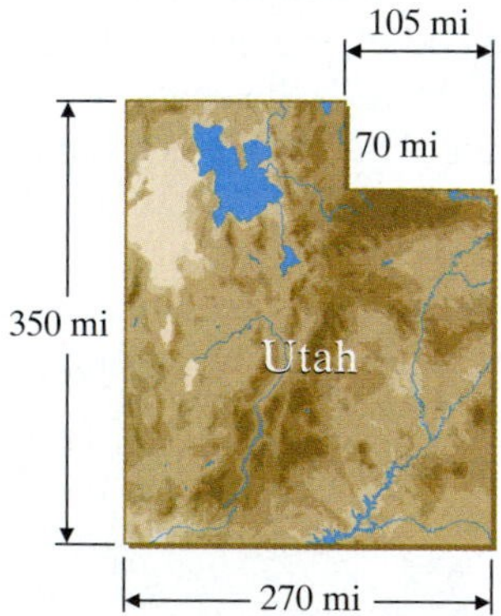

83. The centerpiece of the New England Aquarium in Boston is the Caribbean Coral Reef tank. The centerpiece of this four-story giant tank is a human-made coral reef with over 300 species of coral displayed. This giant reef, with its many caves, is the perfect backdrop for sharks, sea turtles, barracudas, moray eels, and hundreds of smaller tropical fish. The radius of the tank is 16.3 feet and its height is 32 feet. What is the volume of the Caribbean Coral Reef Tank? Use $\pi = 3.14$ and round to the nearest tenth of a cubic foot. (*Source:* New England Aquarium)

84. The Great Pyramid of Khufu at Giza is the largest of the ancient Egyptian pyramids. Its original height was 146.5 meters. The length of each side of its square base was originally 230 meters. Find the volume of the Great Pyramid of Khufu as it was originally built. Round to the nearest whole cubic meter. (*Source:* PBS *NOVA* Online)

85. Can you compute the volume of a rectangle? Why or why not?

86. In your own words, explain why perimeter is measured in units and area is measured in square units. (*Hint:* See Section 1.5 for an introduction on the meaning of area.)

87. Find the area of the shaded region. Use the approximation 3.14 for π.

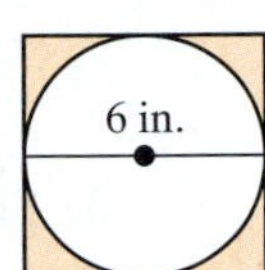

88. The largest pumpkin pie was made and served in Windsor, California. The pie had a diameter of 72 inches. Find the exact area of the top of the pie and an approximation. Use $\pi \approx 3.14$. (*Source: Guinness Book of World Records*)

Find the area of each figure. If needed, use $\pi \approx 3.14$ and round results to the nearest tenth.

89. Find the skating area.

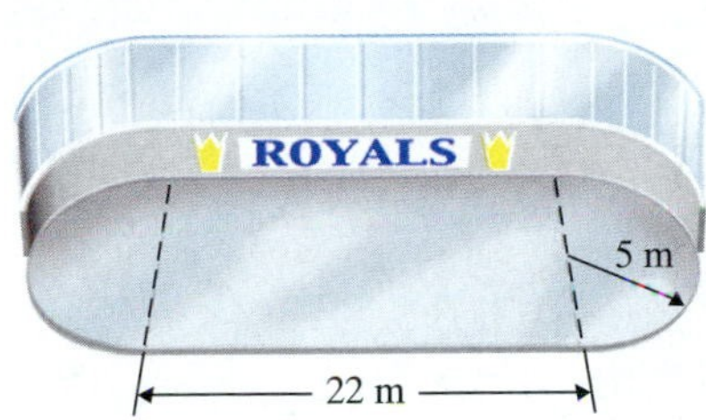

90.

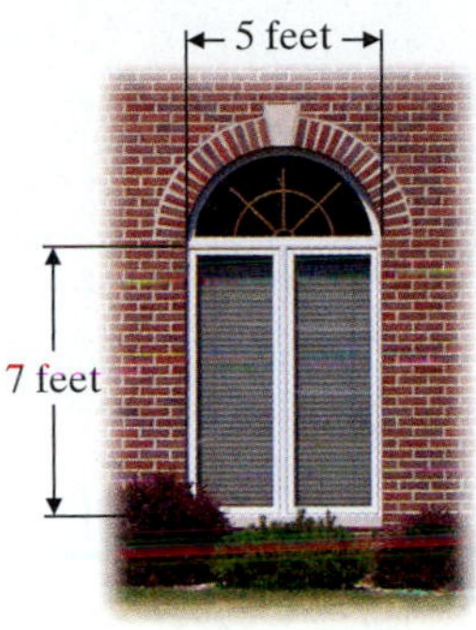

91. Do two rectangles with the same perimeter have the same area? To see, find the perimeter and the area of each rectangle.

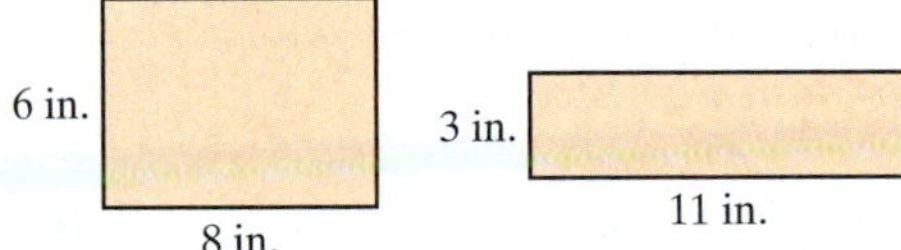

STUDY SKILLS BUILDER

How Well Do You Know Your Textbook?

Let's check to see whether you are familiar with your textbook yet. Remember, for help, see Section 1.1 in this text.

1. What does the icon mean?
2. What does the icon mean?
3. What does the △ icon mean?
4. Where can you find a review for each chapter? What answers to this review can be found in the back of your text?
5. Each chapter contains an overview of the chapter along with examples. What is this feature called?
6. Each chapter contains a review of vocabulary. What is this feature called?
7. There are free CDs in your text. What content is contained on these CDs?
8. What is the location of the section that is entirely devoted to study skills?
9. There are Practice Problems that are contained in the margin of the text. What are they and how can they be used?

INTEGRATED REVIEW Sections 8.1–8.3

Geometry Concepts

Answers

1. ________
2. ________
3. ________
4. ________
5. ________
6. ________
7. ________
8. ________
9. ________
10. ________
11. ________
12. ________
13. ________
14. ________
15. ________
16. ________

△ **1.** Find the supplement and the complement of a 27° angle.

Find the measures of angles x, y, and z in each figure.

2.

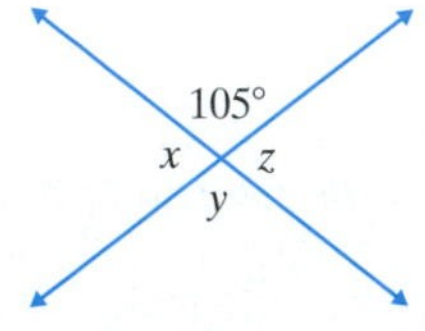

3. $m \| n$

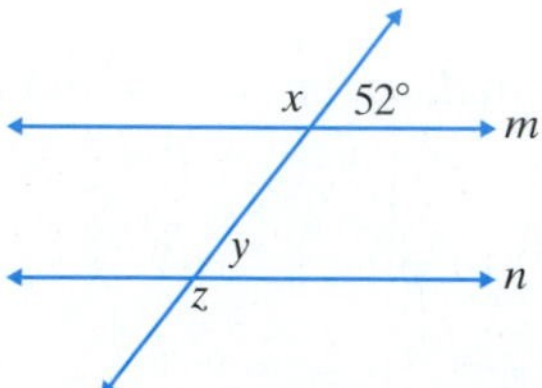

4. Find the measure of $\angle x$.

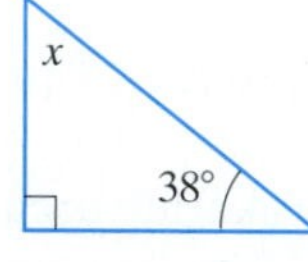

5. Find the diameter.

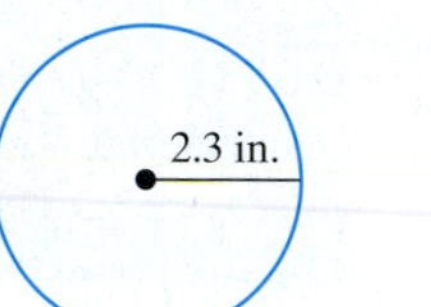

6. Find the radius.

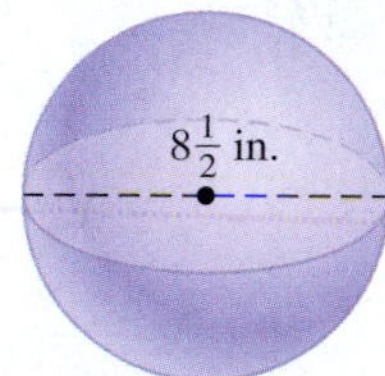

For Exercises 7 through 11, find the perimeter (or circumference) and area of each figure. For the circle give an exact circumference and area. Then use $\pi \approx 3.14$ to approximate each. Don't forget to attach correct units.

7.

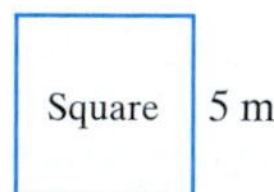

8.

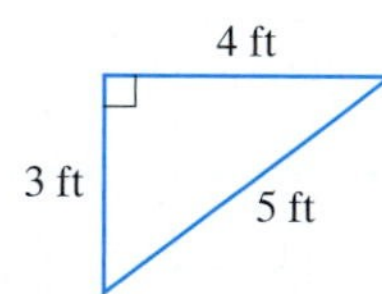

9.

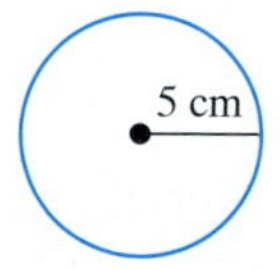

10.

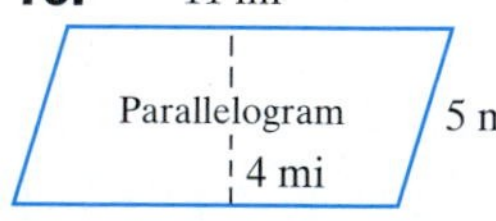

11.

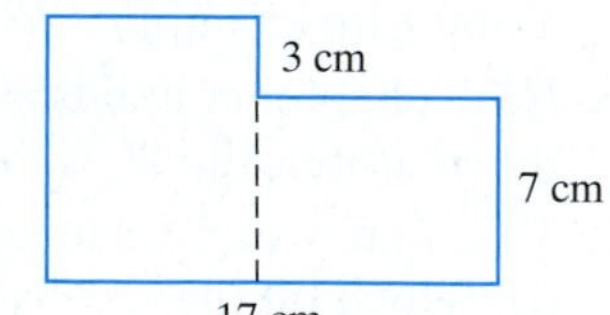

12. The smallest cathedral is in Highlandville, Missouri. The rectangular floor of the cathedral measures 14 feet by 17 feet. Find its perimeter and its area. (*Source: The Guinness Book of World Records*)

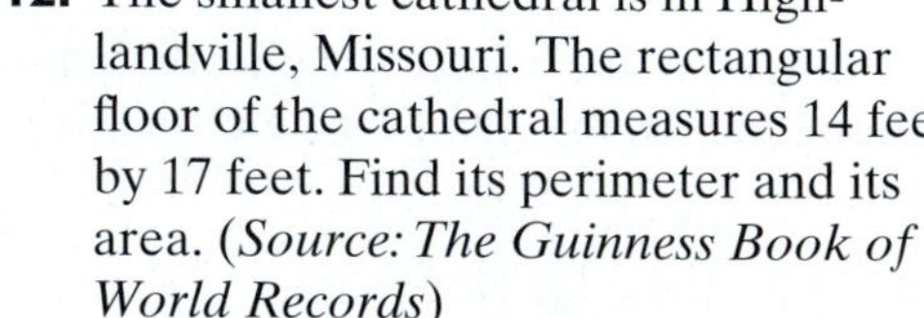

Find the volume of each solid. Don't forget to attach correct units. For Exercises 13 and 14, find the surface area, also.

13. A cube with edges of 4 inches each.

14. A rectangular box 2 feet by 3 feet by 5.1 feet.

15. A pyramid with a square base 10 centimeters on a side and a height of 12 centimeters.

16. A sphere with a diameter of 3 miles. Give the exact volume and then use $\pi \approx \frac{22}{7}$ to approximate.

8.4 LINEAR MEASUREMENT

Objectives

A Define U.S. Units of Length and Convert from One Unit to Another.

B Use Mixed Units of Length.

C Perform Arithmetic Operations on U.S. Units of Length.

D Define Metric Units of Length and Convert from One Unit to Another.

E Perform Arithmetic Operations on Metric Units of Length.

Objective A Defining and Converting U.S. System Units of Length

In the United States, two systems of measurement are commonly used. They are the **United States (U.S.), or English, measurement system** and the **metric system.** The U.S. measurement system is familiar to most Americans. Units such as feet, miles, ounces, and gallons are used. However, the metric system is also commonly used in fields such as medicine, sports, international marketing, and certain physical sciences. We are accustomed to buying 2-liter bottles of soft drinks, watching televised coverage of the 100-meter dash at the Olympic Games, or taking a 200-milligram dose of pain reliever.

The U.S. system of measurement uses the **inch, foot, yard,** and **mile** to measure **length.** The following is a summary of equivalencies between units of length:

U.S. Units of Length

12 inches (in.) = 1 foot (ft)
3 feet = 1 yard (yd)
36 inches = 1 yard
5280 feet = 1 mile (mi)

To convert from one unit of length to another, we will use **unit fractions.** We define a unit fraction to be a fraction that is equivalent to 1. Examples of unit fractions are as follows:

Unit Fractions

$$\frac{12\text{ in.}}{1\text{ ft}} = 1 \text{ or } \frac{1\text{ ft}}{12\text{ in.}} = 1 \text{ (since 12 in. = 1 ft)}$$

$$\frac{3\text{ ft}}{1\text{ yd}} = 1 \text{ or } \frac{1\text{ yd}}{3\text{ ft}} = 1 \text{ (since 3 ft = 1 yd)}$$

$$\frac{5280\text{ ft}}{1\text{ mi}} = 1 \text{ or } \frac{1\text{ mi}}{5280\text{ ft}} = 1 \text{ (since 5280 ft = 1 mi)}$$

Remember that multiplying a number by 1 does not change the value of the number.

EXAMPLE 1 Convert 8 feet to inches.

Solution: We multiply 8 feet by a unit fraction that uses the equality 12 inches = 1 foot. The unit fraction should be in the form $\frac{\text{units to convert to}}{\text{original units}}$ or in this case $\frac{12\text{ inches}}{1\text{ foot}}$. We do this so that like units will divide out to 1, as shown.

$$\begin{aligned} 8\text{ ft} &= \frac{8\text{ ft}}{1} \cdot 1 && \text{Multiply by 1 in the form of } \tfrac{12\text{ in.}}{1\text{ ft}}. \\ &= \frac{8\text{ ft}}{1} \cdot \frac{12\text{ in.}}{1\text{ ft}} \\ &= 8 \cdot 12\text{ in.} \\ &= 96\text{ in.} && \text{Multiply.} \end{aligned}$$

Continued on next page

PRACTICE PROBLEM 1

Convert 6 feet to inches.

Answer

1. 72 in.

Thus, 8 ft = 96 in., as shown in the diagram:

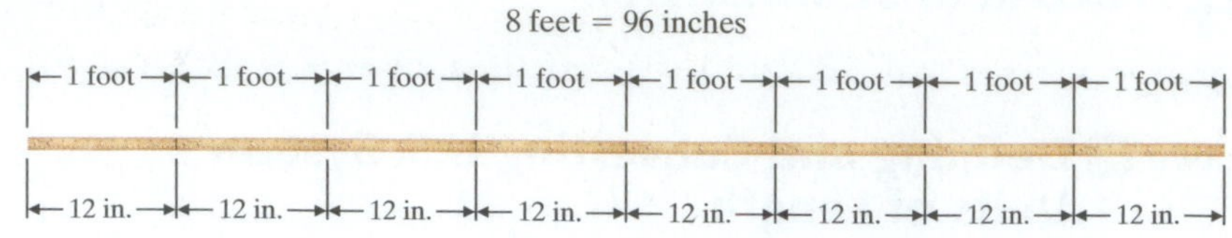

Work Practice Problem 1

PRACTICE PROBLEM 2

Convert 8 yards to feet.

EXAMPLE 2 Convert 7 feet to yards.

Solution: We multiply by a unit fraction that compares 1 yard to 3 feet.

$$7 \text{ ft} = \frac{7 \text{ ft}}{1} \cdot 1$$

$$= \frac{7 \cancel{\text{ft}}}{1} \cdot \frac{1 \text{ yd}}{3 \cancel{\text{ft}}} \quad \begin{array}{l} \leftarrow \text{Units to convert to} \\ \leftarrow \text{Original units} \end{array}$$

$$= \frac{7}{3} \text{ yd}$$

$$= 2\frac{1}{3} \text{ yd} \quad \text{Divide.}$$

Thus, 7 ft $= 2\frac{1}{3}$ yd, as shown in the diagram.

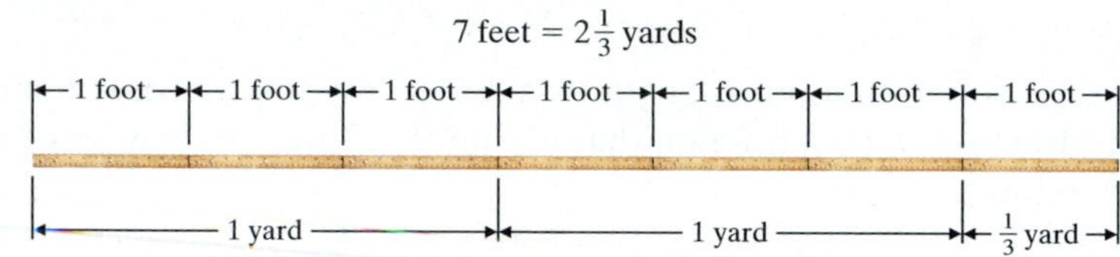

Work Practice Problem 2

Helpful Hint

When converting from one unit to another, select a unit fraction with the properties below:

$$\frac{\text{units you are converting to}}{\text{original units}}$$

By using this unit fraction, the original units will divide out, as wanted.

PRACTICE PROBLEM 3

Suppose the bill in the photo measures 18 inches. Convert 18 inches to feet, using decimals.

EXAMPLE 3 **Finding Length of Pelican's Bill**

The Australian pelican has the longest bill, measuring from 13 to 18.5 inches long. The pelican in the photo has a 15-inch bill. Convert 15 inches to feet, using decimals in your final answer.

Solution:

$$15 \text{ in.} = \frac{15 \cancel{\text{in.}}}{1} \cdot \frac{1 \text{ ft}}{12 \cancel{\text{in.}}} \quad \begin{array}{l} \leftarrow \text{Units to convert to} \\ \leftarrow \text{Original units} \end{array}$$

$$= \frac{15}{12} \text{ ft}$$

$$= \frac{5}{4} \text{ ft} \quad \text{Simplify } \frac{15}{12}.$$

$$= 1.25 \text{ ft} \quad \text{Divide.}$$

Thus, 15 in. = 1.25 ft, as shown in the diagram.

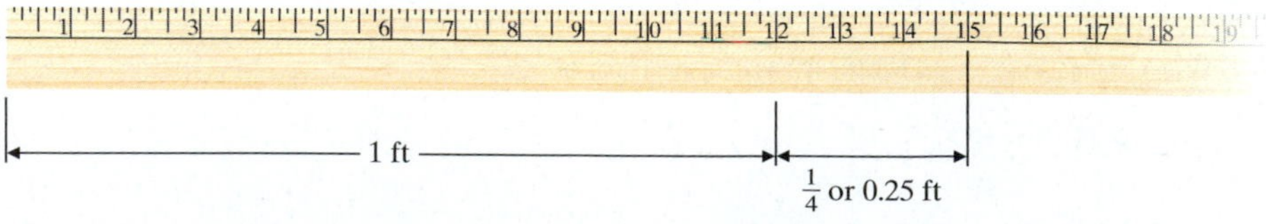

Work Practice Problem 3

Answers

2. 24 ft **3.** 1.5 ft

Objective B Using Mixed U.S. System Units of Length

Sometimes it is more meaningful to express a measurement of length with mixed units such as 1 ft and 5 in. We usually condense this and write 1 ft 5 in.

In Example 2, we found that 7 feet was the same as $2\frac{1}{3}$ yards. The measurement can also be written as a mixture of yards and feet. That is,

7 ft = ____ yd ____ ft

Because 3 ft = 1 yd, we divide 3 into 7 to see how many whole yards are in 7 feet. The quotient is the number of yards, and the remainder is the number of feet.

$$\begin{array}{r} 2 \text{ yd } 1 \text{ ft} \\ 3\overline{)7} \\ -6 \\ \hline 1 \end{array}$$

Thus, 7 ft = 2 yd 1 ft, as seen in the diagram:

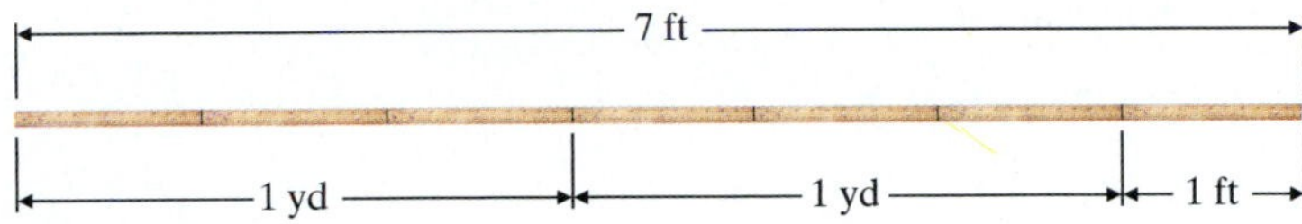

EXAMPLE 4 Convert: 134 in. = _____ ft _____ in.

Solution: Because 12 in. = 1 ft, we divide 12 into 134. The quotient is the number of feet. The remainder is the number of inches. To see why we divide 12 into 134, notice that

$$134 \text{ in.} = \frac{134 \text{ in.}}{1} \cdot \frac{1 \text{ ft}}{12 \text{ in.}} = \frac{134}{12} \text{ ft}$$

$$\begin{array}{r} 11 \text{ ft } 2 \text{ in.} \\ 12\overline{)134} \\ -12 \\ \hline 14 \\ -12 \\ \hline 2 \end{array}$$

Thus, 134 in. = 11 ft 2 in.

Work Practice Problem 4

PRACTICE PROBLEM 4

Convert:
68 in. = _____ ft _____ in.

EXAMPLE 5 Convert 3 feet 7 inches to inches.

Solution: First, we convert 3 feet to inches. Then we add 7 inches.

$$3 \text{ ft} = \frac{3 \text{ ft}}{1} \cdot \frac{12 \text{ in.}}{1 \text{ ft}} = 36 \text{ in.}$$

Then

$$3 \text{ ft } 7 \text{ in.} = 36 \text{ in.} + 7 \text{ in.} = 43 \text{ in.}$$

Work Practice Problem 5

PRACTICE PROBLEM 5

Convert 5 yards 2 feet to feet.

Answers

4. 5 ft 8 in. **5.** 17 ft

Objective C Performing Operations on U.S. System Units of Length

Finding sums or differences of measurements often involves converting units, as shown in the next example. Just remember that, as usual, only like units can be added or subtracted.

PRACTICE PROBLEM 6

Add 4 ft 8 in. to 8 ft 11 in.

EXAMPLE 6 Add 3 ft 2 in. and 5 ft 11 in.

Solution: To add, we line up the similar units.

$$\begin{array}{r} 3\text{ ft } 2\text{ in.} \\ +\ 5\text{ ft } 11\text{ in.} \\ \hline 8\text{ ft } 13\text{ in.} \end{array}$$

Since 13 inches is the same as 1 ft 1 in., we have

$$\begin{aligned} 8\text{ ft } 13\text{ in.} &= 8\text{ ft} + \underbrace{1\text{ ft } 1\text{ in.}} \\ &= 9\text{ ft } 1\text{ in.} \end{aligned}$$

Work Practice Problem 6

Concept Check How could you estimate the following sum?

$$\begin{array}{r} 7\text{ yd } 4\text{ in.} \\ +\ 3\text{ yd } 27\text{ in.} \\ \hline \end{array}$$

PRACTICE PROBLEM 7

Multiply 4 ft 7 in. by 4.

EXAMPLE 7 Multiply 8 ft 9 in. by 3.

Solution: By the distributive property, we multiply 8 ft by 3 and 9 in. by 3.

$$\begin{array}{r} 8\text{ ft } 9\text{ in.} \\ \times 3 \\ \hline 24\text{ ft } 27\text{ in.} \end{array}$$

Since 27 in. is the same as 2 ft 3 in., we simplify the product as

$$\begin{aligned} 24\text{ ft } 27\text{ in.} &= 24\text{ ft} + \underbrace{2\text{ ft } 3\text{ in.}} \\ &= 26\text{ ft } 3\text{ in.} \end{aligned}$$

We divide in a similar manner as above.

Work Practice Problem 7

PRACTICE PROBLEM 8

A carpenter cuts 1 ft 9 in. from a board of length 5 ft 8 in. Find the remaining length of the board.

EXAMPLE 8 **Finding the Length of a Piece of Rope**

A rope of length 6 yd 1 ft has 2 yd 2 ft cut from one end. Find the length of the remaining rope.

Solution: Subtract 2 yd 2 ft from 6 yd 1 ft.

$$\begin{array}{rcr} \text{beginning length} & \rightarrow & 6\text{ yd } 1\text{ ft} \\ -\ \text{amount cut} & \rightarrow & -\ 2\text{ yd } 2\text{ ft} \\ \hline \text{remaining length} & & \end{array}$$

We cannot subtract 2 ft from 1 ft, so we borrow 1 yd from the 6 yd. One yard is converted to 3 ft and combined with the 1 ft already there.

Answers

6. 13 ft 7 in. 7. 18 ft 4 in.
8. 3 ft 11 in.

✔ Concept Check Answer

round each to the nearest yard: 7 yd + 4 yd = 11 yd

Borrow 1 yd = 3 ft

$$\begin{array}{rcr} & & 5\text{ yd} + 1\text{ yd}\ 3\text{ ft} \\ \not{6}\text{ yd } 1\text{ ft} & = & 5\text{ yd } 4\text{ ft} \\ -\,2\text{ yd } 2\text{ ft} & = & -\,2\text{ yd } 2\text{ ft} \\ \hline & & 3\text{ yd } 2\text{ ft} \end{array}$$

The remaining rope is 3 yd 2 ft long.

Work Practice Problem 8

Objective D Defining and Converting Metric System Units of Length

The basic unit of length in the metric system is the **meter.** A meter is slightly longer than a yard. It is approximately 39.37 inches long. Recall that a yard is 36 inches long.

1 yard = 36 inches

1 meter ≈ 39.37 inches

All units of length in the metric system are based on the meter. The following is a summary of the prefixes used in the metric system. Also shown are equivalencies between units of length. Like the decimal system, the metric system uses powers of 10 to define units.

Metric Unit of Length
1 **kilo**meter (km) = 1000 meters (m)
1 **hecto**meter (hm) = 100 m
1 **deka**meter (dam) = 10 m
1 meter (m) = 1 m
1 **deci**meter (dm) = 1/10 m or 0.1 m
1 **centi**meter (cm) = 1/100 m or 0.01 m
1 **milli**meter (mm) = 1/1000 m or 0.001 m

The figure below will help you with decimeters, centimeters, and millimeters.

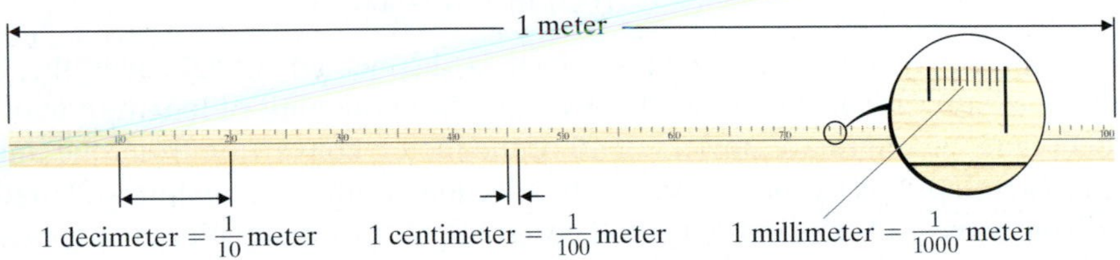

Helpful Hint

Study the figure above for other equivalencies between metric units of length.

10 decimeters = 1 meter

100 centimeters = 1 meter

1000 millimeters = 1 meter

10 millimeters = 1 centimeter

10 centimeters = 1 decimeter

These same prefixes are used in the metric system for mass and capacity. The most commonly used measurements of length in the metric system are the **meter, millimeter, centimeter,** and **kilometer.**

✔**Concept Check** Is this statement reasonable? "The screen of a home television set has a 30-meter diagonal." Why or why not?

Being comfortable with the metric units of length means gaining a "feeling" for metric lengths, just as you have a "feeling" for the length of an inch, a foot, and a mile. To help you accomplish this, study the following examples:

A millimeter is about the thickness of a large paper clip.

A centimeter is about the width of a large paper clip.

A meter is slightly longer than a yard.

A kilometer is about two-thirds of a mile.

The width of this book is approximately 21.5 centimeters.

The distance between New York and Philadelphia is about 160 kilometers.

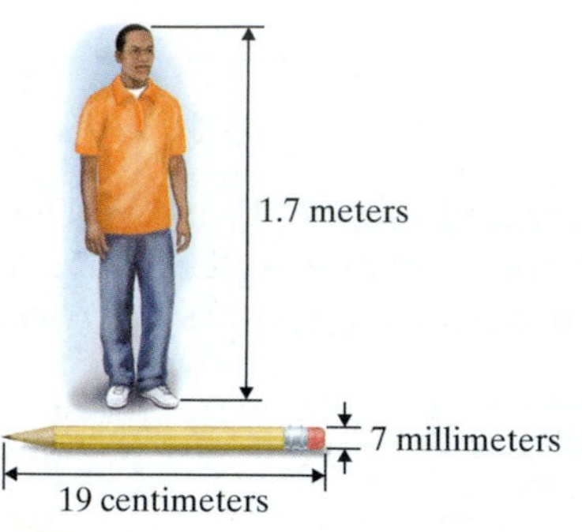

$2\frac{1}{2}$ centimeters is about 1 inch.

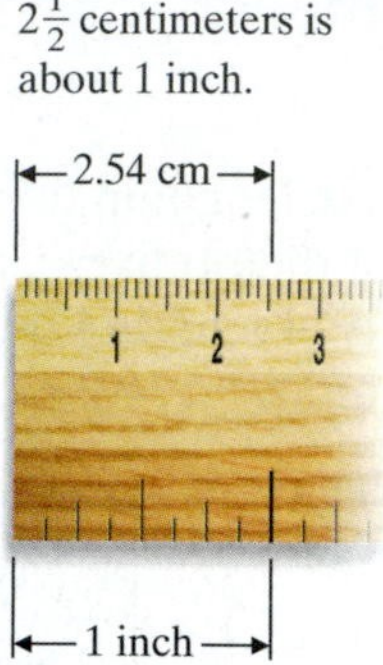

As with the U.S. system of measurement, unit fractions may be used to convert from one unit of length to another. For example, let's convert 1200 meters to kilometers. To do so, we will multiply by 1 in the form of the unit fraction

$$\frac{1\text{ km}}{1000\text{ m}} \quad \begin{matrix}\leftarrow \text{Units to convert to} \\ \leftarrow \text{Original units}\end{matrix}$$

$$1200\text{ m} = \frac{1200\text{ m}}{1} \cdot 1 = \frac{1200\text{ }\cancel{\text{m}}}{1} \cdot \overbrace{\frac{1\text{ km}}{1000\text{ }\cancel{\text{m}}}}^{\text{Unit fraction}} = \frac{1200\text{ km}}{1000} = 1.2\text{ km}$$

The metric system does, however, have a distinct advantage over the U.S. system of measurement: The ease of converting from one unit of length to another. Since all units of length are powers of 10 of the meter, converting from one unit of length to another is as simple as moving the decimal point. Listing units of length in order from largest to smallest helps to keep track of how many places to move the decimal point when converting.

Let's again convert 1200 meters to kilometers. This time, to convert from meters to kilometers, we move along the chart shown 3 units to the left, from meters to kilometers. This means that we move the decimal point 3 places to the left.

km hm dam **m** dm cm mm

3 units to the left

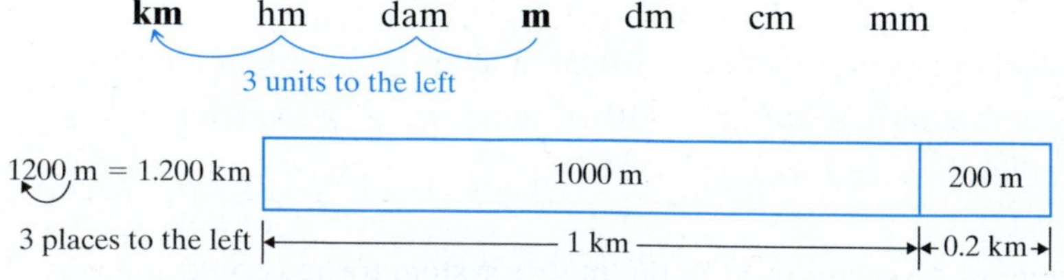

Thus, 1200 m = 1.2 km as shown in the diagram.

✔ **Concept Check Answer**

no; answers may vary

EXAMPLE 9 Convert 2.3 m to centimeters.

Solution: First we will convert by using a unit fraction.

$$2.3 \text{ m} = \frac{2.3 \cancel{\text{m}}}{1} \cdot \overbrace{\frac{100 \text{ cm}}{1 \cancel{\text{m}}}}^{\text{Unit fraction}} = 230 \text{ cm}$$

Now we will convert by listing the units of length in order from left to right and moving from meters to centimeters.

km hm dam m dm cm mm

2 units to the right

2.30 m = 230. cm

2 places to the right

With either method, we get 230 cm.

Work Practice Problem 9

PRACTICE PROBLEM 9

Convert 2.5 m to millimeters.

EXAMPLE 10 Convert 450,000 mm to meters.

Solution: We list the units of length in order from left to right and move from millimeters to meters.

km hm dam m dm cm mm

3 units to the left

Thus, move the decimal point 3 places to the left.

450,000 mm = 450.000 m or 450 m

Work Practice Problem 10

PRACTICE PROBLEM 10

Convert 3500 m to kilometers.

✔Concept Check What is wrong with the following conversion of 150 cm to meters?

150.00 cm = 15,000 m

Objective E Performing Operations on Metric System Units of Length

To add, subtract, multiply, or divide with metric measurements of length, we write all numbers using the same unit of length and then add, subtract, multiply, or divide as with decimals.

EXAMPLE 11 Subtract 430 m from 1.3 km.

Solution: First we convert both measurements to kilometers or both to meters.

430 m = 0.43 km or 1.3 km = 1300 m

$$\begin{array}{r} 1.30 \text{ km} \\ -\ 0.43 \text{ km} \\ \hline 0.87 \text{ km} \end{array} \qquad \begin{array}{r} 1300 \text{ m} \\ -\ 430 \text{ m} \\ \hline 870 \text{ m} \end{array}$$

The difference is 0.87 km or 870 m.

Work Practice Problem 11

PRACTICE PROBLEM 11

Subtract 640 m from 2.1 km.

Answers

9. 2500 mm **10.** 3.5 km **11.** 1.46 km or 1460 m

✔ Concept Check Answer

decimal point should be moved two places to the left: 1.5 m

PRACTICE PROBLEM 12

Multiply 18.3 hm by 5.

EXAMPLE 12 Multiply 5.7 mm by 4.

Solution: Here we simply multiply the two numbers. Note that the unit of measurement remains the same.

$$\begin{array}{r} 5.7 \text{ mm} \\ \times \quad 4 \\ \hline 22.8 \text{ mm} \end{array}$$

Work Practice Problem 12

PRACTICE PROBLEM 13

Doris Blackwell is knitting a scarf that is currently 0.8 meter long. If she knits an additional 45 centimeters, how long will the scarf be?

EXAMPLE 13 **Finding a Person's Height**

Fritz Martinson was 1.2 meters tall on his last birthday. Since then, he has grown 14 centimeters. Find his current height in meters.

Solution:

$$\begin{array}{rcr} \text{original height} & \rightarrow & 1.20 \text{ m} \\ + \text{ height grown} & \rightarrow & + \ 0.14 \text{ m} \\ \hline \text{current height} & & 1.34 \text{ m} \end{array}$$

(Since 14 cm = 0.14 m)

Fritz is now 1.34 meters tall.

Work Practice Problem 13

Answers

12. 91.5 hm **13.** 125 cm or 1.25 m

Vocabulary and Readiness Check

Use the choices below to fill in each blank. Some choices may be used more than once.

inches	yard	unit fraction
feet	meter	

1. The basic unit of length in the metric system is the ______________.
2. The expression $\frac{1 \text{ foot}}{12 \text{ inches}}$ is an example of a ______________.
3. A meter is slightly longer than a ______________.
4. One foot equals 12 ______________.
5. One yard equals 3 ______________.
6. One yard equals 36 ______________.
7. One mile equals 5280 ______________.

8.4 EXERCISE SET

FOR EXTRA HELP

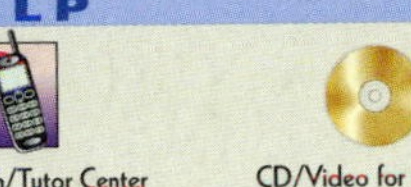

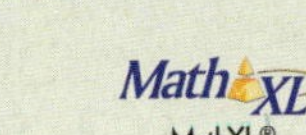

MyMathLab

Objective A *Convert each measurement as indicated. See Examples 1 through 3.*

1. 60 in. to feet
2. 84 in. to feet
3. 12 yd to feet
4. 18 yd to feet
5. 42,240 ft to miles
6. 36,960 ft to miles
7. $8\frac{1}{2}$ ft to inches
8. $12\frac{1}{2}$ ft to inches
9. 10 ft to yards
10. 25 ft to yards
11. 6.4 mi to feet
12. 3.8 mi to feet
13. 162 in. to yd (Write answer as a decimal.)
14. 7216 yd to mi (Write answer as a decimal.)
15. 3 in. to ft (Write answer as a decimal.)
16. 129 in. to ft (Write answer as a decimal.)

Objective B *Convert each measurement as indicated. See Examples 4 and 5.*

17. 40 ft = _____ yd _____ ft
18. 100 ft = _____ yd _____ ft
19. 85 in. = _____ ft _____ in.
20. 59 in. = _____ ft _____ in.
21. 10,000 ft = _____ mi _____ ft
22. 25,000 ft = _____ mi _____ ft
23. 5 ft 2 in. = _____ in.
24. 4 ft 11 in. = _____ in.
25. 8 yd 2 ft = _____ ft
26. 4 yd 1 ft = _____ ft
27. 2 yd 1 ft = _____ in.
28. 1 yd 2 ft = _____ in.

Objective **C** *Perform each indicated operation. Simplify the result if possible. See Examples 6 through 8.*

29. 3 ft 10 in. + 7 ft 4 in.

30. 12 ft 7 in. + 9 ft 11 in.

31. 12 yd 2 ft + 9 yd 2 ft

32. 16 yd 2 ft + 8 yd 2 ft

33. 22 ft 8 in. − 16 ft 3 in.

34. 15 ft 5 in. − 8 ft 2 in.

35. 18 ft 3 in. − 10 ft 9 in.

36. 14 ft 8 in. − 3 ft 11 in.

37. 28 ft 8 in. ÷ 2

38. 34 ft 6 in. ÷ 2

39. 16 yd 2 ft × 5

40. 15 yd 1 ft × 8

Objective **D** *Convert as indicated. See Examples 9 and 10.*

41. 60 m to centimeters

42. 46 m to centimeters

43. 40 mm to centimeters

44. 14 mm to centimeters

45. 500 m to kilometers

46. 400 m to kilometers

47. 1700 mm to meters

48. 6400 mm to meters

49. 1500 cm to meters

50. 6400 cm to meters

51. 0.42 km to centimeters

52. 0.95 km to centimeters

53. 7 km to meters

54. 5 km to meters

55. 8.3 cm to millimeters

56. 4.6 cm to millimeters

57. 20.1 mm to decimeters

58. 140.2 mm to decimeters

59. 0.04 m to millimeters

60. 0.2 m to millimeters

Objective **E** *Perform each indicated operation. See Examples 11 through 13.*

61. 8.6 m + 0.34 m

62. 14.1 cm + 3.96 cm

63. 2.9 m + 40 mm

64. 30 cm + 8.9 m

65. 24.8 mm − 1.19 cm

66. 45.3 m − 2.16 dam

67. 15 km − 2360 m

68. 14 cm − 15 mm

69. 18.3 m × 3

70. 14.1 m × 4

71. 6.2 km ÷ 4

72. 9.6 m ÷ 5

Objectives A C D E Mixed Practice *Solve. Remember to insert units when writing your answers. For Exercises 73 through 82, complete the charts. See Examples 1 through 13.*

		Yards	Feet	Inches
73.	Chrysler Building in New York City		1046	
74.	4-story building			792
75.	Python length		35	
76.	Ostrich height			108

Complete the chart.

		Meters	Millimeters	Kilometers	Centimeters
77.	Length of elephant	5			
78.	Height of grizzly bear	3			
79.	Tennis ball diameter				6.5
80.	Golf ball diameter				4.6
81.	Distance from London to Paris			342	
82.	Distance from Houston to Dallas			396	

83. The National Zoo maintains a small patch of bamboo, which it grows as a food supply for its pandas. Two weeks ago, the bamboo was 6 ft 10 in. tall. Since then, the bamboo has grown 3 ft 8 in. How tall is the bamboo now?

84. While exploring in the Marianas Trench, a submarine probe was lowered to a point 1 mile 1400 feet below the ocean's surface. Later it was lowered an additional 1 mile 4000 feet below this point. How far is the probe below the surface of the Pacific?

85. At its deepest point, the Grand Canyon of the Colorado River in Arizona is about 6000 ft. The Grand Canyon of the Yellowstone River, which is in Yellowstone National Park in Wyoming, is at most 900 feet deep. How much deeper is the Grand Canyon of the Colorado River than the Grand Canyon of the Yellowstone River? (*Source:* National Park Service)

86. The Grand Canyon of the Gunnison River, in Colorado, is often called the Black Canyon of the Gunnison because it is so steep that light rarely penetrates the depth of the canyon. The Black Canyon of the Gunnison is only 1150 ft wide at its narrowest point. At its narrowest, the Grand Canyon of the Yellowstone is $\frac{1}{2}$ mile wide. Find the difference in width between the Grand Canyon of the Yellowstone and the Black Canyon of the Gunnison. (*Note:* Notice that the dimensions are different.) (*Source:* National Park Service)

87. The tallest man in the world is recorded as Robert Pershing Wadlow of Alton, Illinois. Born is 1918, he measured 8 ft 11 in. at his tallest. The shortest man in the world is Gul Mohammed of India, who measures 22.5 in. How many times taller than Gul is Robert? Round to one decimal place. (*Source: Guinness World Records*)

88. A 3.4-m rope is attached to a 5.8-m rope. However, when the ropes are tied, 8 cm of length is lost to form the knot. What is the length of the tied ropes?

89. The ice on a pond is 5.33 cm thick. For safe skating, the owner of the pond insists that it must be 80 mm thick. How much thicker must the ice be before skating is allowed?

90. The sediment on the bottom of the Towamencin Creek is normally 14 cm thick, but the recent flood washed away 22 mm of sediment. How thick is it now?

91. The Amana Corporation stacks up its microwave ovens in a distribution warehouse. Each stack is 1 ft 9 in. wide. How far from the wall would 9 of these stacks extend?

92. The highway commission is installing concrete sound barriers along a highway. Each barrier is 1 yd 2 ft long. Find the total length of 25 barriers placed end to end.

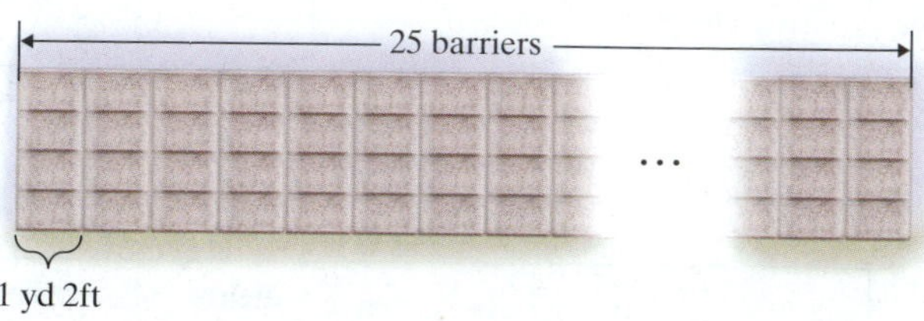

93. A logging firm needs to cut a 67-m-long redwood log into 20 equal pieces before loading it onto a truck for shipment. How long will each piece be?

94. An 18.3-m-tall flagpole is mounted on a 65-cm-high pedestal. How far is the top of the flagpole from the ground?

95. The world's longest Coca-Cola truck is in Sweden and is 79 feet long. How many *yards* long are 4 of these trucks? (*Source: Coca-Cola Today*)

△ **96.** The world's largest Coca-Cola sign is in Arica, Chile. It is in the shape of a rectangle whose length is $133\frac{1}{3}$ yards and whose width is 131 feet. Find the area of the sign in square feet. (*Source: Coca-Cola Today*) (*Hint:* Recall that area of a rectangle is the product: length times width.)

Review

Write each decimal as a fraction and each fraction as a decimal. See Section 5.1.

97. 0.21 **98.** 0.86 **99.** $\frac{13}{100}$ **100.** $\frac{47}{100}$ **101.** $\frac{1}{4}$ **102.** $\frac{3}{20}$

Concept Extensions

Determine whether the measurement in each statement is reasonable.

103. The width of a twin-size bed is 20 meters.

104. A window measures 1 meter by 0.5 meter.

105. A drinking glass is made of glass 2 millimeters thick.

106. A paper clip is 4 kilometers long.

107. The distance across the Colorado River is 50 kilometers.

108. A model's hair is 30 centimeters long.

Estimate each sum or difference. See the first Concept Check in this section.

109. 5 yd 2 in.
+ 7 yd 30 in.

110. 45 ft 1 in.
− 10 ft 11 in.

111. Using a unit other than the foot, write a length that is equivalent to 4 feet. (*Hint:* There are many possibilities.)

112. Using a unit other than the meter, write a length that is equivalent to 7 meters. (*Hint:* There are many possibilities.)

113. To convert from meters to centimeters, the decimal point is moved two places to the right. Explain how this relates to the fact that the prefix *centi* means $\frac{1}{100}$.

114. Explain why conversions in the metric system are easier to make than conversions in the U.S. system of measurement.

115. An advertisement sign outside Fenway Park in Boston measures 18.3 m by 18.3 m. What is the area of this sign?

Objectives

A Define U.S. Units of Weight and Convert from One Unit to Another.

B Perform Arithmetic Operations on Units of Weight.

C Define Metric Units of Mass and Convert from One Unit to Another.

D Perform Arithmetic Operations on Units of Mass.

8.5 WEIGHT AND MASS

Objective A Defining and Converting U.S. System Units of Weight

Whenever we talk about how heavy an object is, we are concerned with the object's **weight.** We discuss weight when we refer to a 12-ounce box of Rice Krispies, a 15-pound tabby cat, or a barge hauling 24 tons of garbage.

12 ounces

15 pounds

24 tons of garbage

The most common units of weight in the U.S. measurement system are the **ounce,** the **pound,** and the **ton.** The following is a summary of equivalencies between units of weight:

U.S. Units of Weight	*Unit Fractions*
16 ounces (oz) = 1 pound (lb)	$\frac{16 \text{ oz}}{1 \text{ lb}} = \frac{1 \text{ lb}}{16 \text{ oz}} = 1$
2000 pounds = 1 ton	$\frac{2000 \text{ lb}}{1 \text{ ton}} = \frac{1 \text{ ton}}{2000 \text{ lb}} = 1$

✔ **Concept Check** If you were describing the weight of a fully-loaded semi-trailer, which type of unit would you use: ounce, pound, or ton? Why?

Unit fractions that equal 1 are used to convert between units of weight in the U.S. system. When converting using unit fractions, recall that the numerator of a unit fraction should contain the units we are converting to and the denominator should contain the original units.

PRACTICE PROBLEM 1

Convert 6500 pounds to tons.

Answer

1. $3\frac{1}{4}$ tons

✔ Concept Check Answer

ton

EXAMPLE 1 Convert 9000 pounds to tons.

Solution: We multiply 9000 lb by a unit fraction that uses the equality

2000 pounds = 1 ton.

Remember, the unit fraction should be $\frac{\text{units to convert to}}{\text{original units}}$ or $\frac{1 \text{ ton}}{2000 \text{ lb}}$.

$$9000 \text{ lb} = \frac{9000 \text{ lb}}{1} \cdot 1 = \frac{9000 \not{\text{lb}}}{1} \cdot \frac{1 \text{ ton}}{2000 \not{\text{lb}}} = \frac{9000 \text{ tons}}{2000} = \frac{9}{2} \text{ tons or } 4\frac{1}{2} \text{ tons}$$

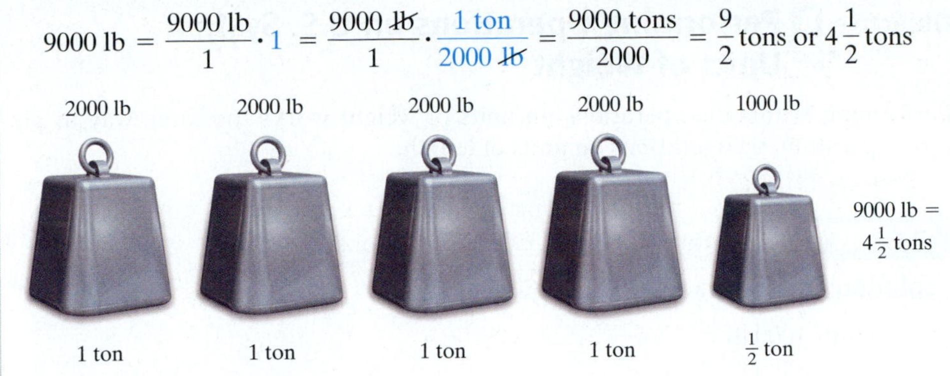

Work Practice Problem 1

EXAMPLE 2 Convert 3 pounds to ounces.

Solution: We multiply by the unit fraction $\frac{16 \text{ oz}}{1 \text{ lb}}$ to convert from pounds to ounces.

$$3 \text{ lb} = \frac{3 \text{ lb}}{1} \cdot 1 = \frac{3 \not{\text{lb}}}{1} \cdot \frac{16 \text{ oz}}{1 \not{\text{lb}}} = 3 \cdot 16 \text{ oz} = 48 \text{ oz}$$

1 pound 1 pound 1 pound

3 lb = 48 oz

16 ounces 16 ounces 16 ounces

Work Practice Problem 2

PRACTICE PROBLEM 2

Convert 72 ounces to pounds.

As with length, it is sometimes useful to simplify a measurement of weight by writing it in terms of mixed units.

EXAMPLE 3 Convert: 33 ounces = _____ lb _____ oz

Solution: Because 16 oz = 1 lb, divide 16 into 33 to see how many pounds are in 33 ounces. The quotient is the number of pounds, and the remainder is the number of ounces. To see why we divide 16 into 33, notice that

$$33 \text{ oz} = 33 \not{\text{oz}} \cdot \frac{1 \text{ lb}}{16 \not{\text{oz}}} = \frac{33}{16} \text{ lb}$$

$$\begin{array}{r} 2 \text{ lb } 1 \text{ oz} \\ 16\overline{)33} \\ -32 \\ \hline 1 \end{array}$$

Thus, 33 ounces is the same as 2 lb 1 oz.

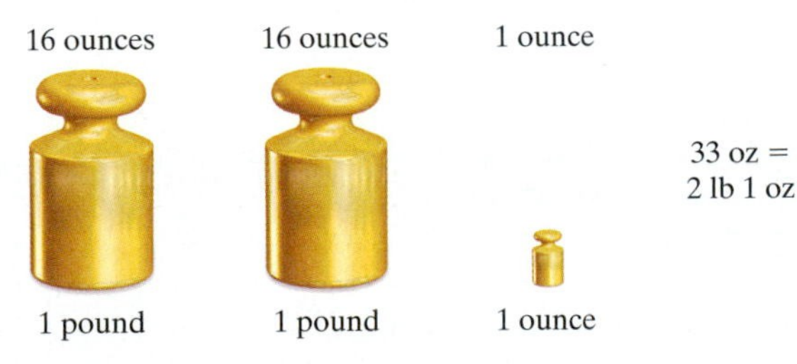

Work Practice Problem 3

PRACTICE PROBLEM 3

Convert:
47 ounces = _____ lb _____ oz

Answers

2. $4\frac{1}{2}$ lb **3.** 2 lb 15 oz

Objective B Performing Operations on U.S. System Units of Weight

Performing arithmetic operations on units of weight works the same way as performing arithmetic operations on units of length.

PRACTICE PROBLEM 4

Subtract 5 tons 1200 lb from 8 tons 100 lb.

EXAMPLE 4 Subtract 3 tons 1350 lb from 8 tons 1000 lb.

Solution: To subtract, we line up similar units.

$$\begin{array}{r} 8 \text{ tons } 1000 \text{ lb} \\ -3 \text{ tons } 1350 \text{ lb} \\ \hline \end{array}$$

Since we cannot subtract 1350 lb from 1000 lb, we borrow 1 ton from the 8 tons. To do so, we write 1 ton as 2000 lb and combine it with the 1000 lb.

7 tons + 1 ton → 2000 lb

$$\begin{array}{r} \not{8} \text{ tons } 1000 \text{ lb} \\ -3 \text{ tons } 1350 \text{ lb} \\ \hline \end{array} \quad = \quad \begin{array}{r} 7 \text{ tons } 3000 \text{ lb} \\ -3 \text{ tons } 1350 \text{ lb} \\ \hline 4 \text{ tons } 1650 \text{ lb} \end{array}$$

To check, see that the sum of 4 tons 1650 lb and 3 tons 1350 lb is 8 tons 1000 lb.

Work Practice Problem 4

PRACTICE PROBLEM 5

Divide 5 lb 8 oz by 4.

EXAMPLE 5 Divide 9 lb 6 oz by 2.

Solution: We divide each of the units by 2.

$$\begin{array}{rl} & 4 \text{ lb} \quad 11 \text{ oz} \\ 2 \overline{)} & 9 \text{ lb} \quad 6 \text{ oz} \\ & \underline{-8} \\ & 1 \text{ lb} = \underline{16 \text{ oz}} \\ & \qquad\quad 22 \text{ oz} \end{array}$$

Divide 2 into 22 oz to get 11 oz.

To check, multiply 4 pounds 11 ounces by 2. The result is 9 pounds 6 ounces.

Work Practice Problem 5

PRACTICE PROBLEM 6

A 5-lb 14-oz batch of cookies is packed into a 6-oz container before it is mailed. Find the total weight.

EXAMPLE 6 **Finding the Weight of a Child**

Bryan weighed 8 lb 8 oz at birth. By the time he was 1 year old, he had gained 11 lb 14 oz. Find his weight at age 1 year.

Solution:

$$\begin{array}{rcr} \text{birth weight} & \rightarrow & 8 \text{ lb } \ 8 \text{ oz} \\ \underline{+ \text{ weight gained}} & \rightarrow & \underline{+ 11 \text{ lb } 14 \text{ oz}} \\ \text{total weight} & \rightarrow & 19 \text{ lb } 22 \text{ oz} \end{array}$$

Since 22 oz equals 1 lb 6 oz,

$$\begin{aligned} 19 \text{ lb } 22 \text{ oz} &= 19 \text{ lb} + 1 \text{ lb } 6 \text{ oz} \\ &= 20 \text{ lb } 6 \text{ oz} \end{aligned}$$

Bryan weighed 20 lb 6 oz on his first birthday.

Work Practice Problem 6

Answers

4. 2 tons 900 lb **5.** 1 lb 6 oz
6. 6 lb 4 oz

Objective C Defining and Converting Metric System Units of Mass

In scientific and technical areas, a careful distinction is made between **weight** and **mass. Weight** is really a measure of the pull of gravity. The farther from Earth an object gets, the less it weighs. However, **mass** is a measure of the amount of substance in the object and does not change. Astronauts orbiting Earth weigh much less than they weigh on Earth, but they have the same mass in orbit as they do on Earth. Here on Earth weight and mass are the same, so either term may be used.

The basic unit of mass in the metric system is the **gram.** It is defined as the mass of water contained in a cube 1 centimeter (cm) on each side.

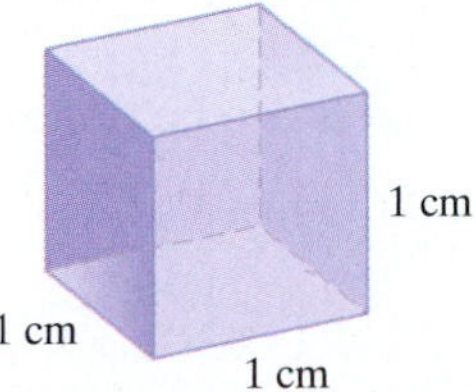

The following examples may help you get a feeling for metric masses:

A tablet contains 200 milligrams of ibuprofen.

A large paper clip weighs approximately 1 gram.

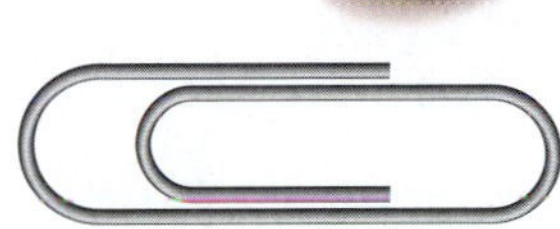

A box of crackers weighs 453 grams.

A kilogram is slightly over 2 pounds. An adult woman may weigh 60 kilograms.

The prefixes for units of mass in the metric system are the same as for units of length, as shown in the following table:

Metric Unit of Mass
1 **kilo**gram (kg) = 1000 grams (g)
1 **hecto**gram (hg) = 100 g
1 **deka**gram (dag) = 10 g
1 gram (g) = 1 g
1 **deci**gram (dg) = 1/10 g or 0.1 g
1 **centi**gram (cg) = 1/100 g or 0.01 g
1 **milli**gram (mg) = 1/1000 g or 0.001 g

✔ **Concept Check** True or false? A decigram is larger than a dekagram. Explain.

The **milligram,** the **gram,** and the **kilogram** are the three most commonly used units of mass in the metric system.

As with lengths, all units of mass are powers of 10 of the gram, so converting from one unit of mass to another involves moving only the decimal point. To convert

✔ **Concept Check Answer**

false

from one unit of mass to another in the metric system, list the units of mass in order from largest to smallest.

Let's convert 4300 milligrams to grams. To convert from milligrams to grams, we move along the list 3 units to the left.

kg hg dag **g** dg cg **mg**

3 units to the left

This means that we move the decimal point 3 places to the left to convert from milligrams to grams.

$4300 \text{ mg} = 4.3 \text{ g}$

Don't forget, the same conversion can be done with unit fractions.

$$4300 \text{ mg} = \frac{4300 \text{ mg}}{1} \cdot 1 = \frac{4300 \cancel{\text{mg}}}{1} \cdot \frac{0.001 \text{ g}}{1 \cancel{\text{mg}}}$$

$$= 4300 \cdot 0.001 \text{ g}$$

$$= 4.3 \text{ g}$$ To multiply by 0.001, move the decimal point 3 places to the left.

To see that this is reasonable, study the diagram:

Thus, 4300 mg = 4.3 g

PRACTICE PROBLEM 7

Convert 3.41 g to milligrams.

EXAMPLE 7 Convert 3.2 kg to grams.

Solution: First we convert by using a unit fraction.

$$3.2 \text{ kg} = 3.2 \text{ kg} \cdot 1 = 3.2 \cancel{\text{kg}} \cdot \overbrace{\frac{1000 \text{ g}}{1 \cancel{\text{kg}}}}^{\text{Unit fraction}} = 3200 \text{ g}$$

Now let's list the units of mass in order from left to right and move from kilograms to grams.

kg hg dag g dg cg mg

3 units to the right

$3.200 \text{ kg} = 3200. \text{ g}$

3 places to the right

1 kg 1 kg 1 kg 0.2 kg

3.2 kg = 3200 g

1000 g 1000 g 1000 g 200 g

Work Practice Problem 7

Answer

7. 3410 mg

EXAMPLE 8 Convert 2.35 cg to grams.

Solution: We list the units of mass in a chart and move from centigrams to grams.

kg hg dag g dg cg mg

2 units to the left

02.35 cg = 0.0235 g

2 places to the left

Work Practice Problem 8

PRACTICE PROBLEM 8

Convert 56.2 cg to grams.

Objective D Performing Operations on Metric System Units of Mass

Arithmetic operations can be performed with metric units of mass just as we performed operations with metric units of length. We convert each number to the same unit of mass and add, subtract, multiply, or divide as with decimals.

EXAMPLE 9 Subtract 5.4 dg from 1.6 g.

Solution: We convert both numbers to decigrams or to grams before subtracting.

5.4 dg = 0.54 g or 1.6 g = 16 dg

$$\begin{array}{r} 1.60\text{ g} \\ -\ 0.54\text{ g} \\ \hline 1.06\text{ g} \end{array} \qquad \begin{array}{r} 16.0\text{ dg} \\ -\ 5.4\text{ dg} \\ \hline 10.6\text{ dg} \end{array}$$

The difference is 1.06 g or 10.6 dg.

Work Practice Problem 9

PRACTICE PROBLEM 9

Subtract 3.1 dg from 2.5 g.

EXAMPLE 10 **Calculating Allowable Weight in an Elevator**

An elevator has a weight limit of 1400 kg. A sign posted in the elevator indicates that the maximum capacity of the elevator is 17 persons. What is the average allowable weight for each passenger, rounded to the nearest kilogram?

Solution: To solve, notice that the total weight of 1400 kilograms ÷ 17 = average weight

$$\begin{array}{r} 82.3\text{ kg} \approx 82\text{ kg} \\ 17\overline{)\ 1400.0\text{ kg}} \\ -136\phantom{.0\text{ kg}} \\ \hline 40\phantom{\text{ kg}} \\ -34\phantom{\text{ kg}} \\ \hline 60 \\ -51 \\ \hline 9 \end{array}$$

Each passenger can weigh an average of 82 kg. (Recall that a kilogram is slightly over 2 pounds, so 82 kilograms is over 164 pounds.)

Work Practice Problem 10

PRACTICE PROBLEM 10

Twenty-four bags of cement weigh a total of 550 kg. Find the average weight of 1 bag, rounded to the nearest kilogram.

Answers

8. 0.562 g **9.** 2.19 g or 21.9 dg
10. 23 kg

Vocabulary and Readiness Check

Use the choices below to fill in each blank.

mass weight gram

1. ______________ is a measure of the amount of substance in an object. This measure does not change.
2. ______________ is the measure of the pull of gravity.
3. The basic unit of mass in the metric system is the ______________.

Fill in these blanks with the correct number. Choices for these blanks are not shown in the list of terms above.

4. One pound equals ______________ ounces.
5. One ton equals ______________ pounds.

8.5 EXERCISE SET

FOR EXTRA HELP

CD/Video for Review

MathXL®

MyMathLab

Objective A *Convert as indicated. See Examples 1 through 3.*

1. 2 pounds to ounces

2. 5 pounds to ounces

3. 5 tons to pounds

4. 7 tons to pounds

5. 18,000 pounds to tons

6. 28,000 pounds to tons

7. 60 ounces to pounds

8. 90 ounces to pounds

9. 3500 pounds to tons

10. 11,000 pounds to tons

11. 12.75 pounds to ounces

12. 9.5 pounds to ounces

13. 4.9 tons to pounds

14. 8.3 tons to pounds

15. $4\frac{3}{4}$ pounds to ounces

16. $9\frac{1}{8}$ pounds to ounces

17. 2950 pounds to the nearest tenth of a ton

18. 51 ounces to the nearest tenth of a pound

19. $\frac{4}{5}$ oz to pounds

20. $\frac{1}{4}$ oz to pounds

21. $5\frac{3}{4}$ lb to ounces

22. $2\frac{1}{4}$ lb to ounces

23. 10 lb 1 oz to ounces

24. 7 lb 6 oz to ounces

25. 89 oz to _____ lb _____ oz

26. 100 oz = _____ lb _____ oz

Objective **B** *Perform each indicated operation. See Examples 4 through 6.*

27. 34 lb 12 oz + 18 lb 14 oz

28. 6 lb 10 oz + 10 lb 8 oz

29. 3 tons 1820 lb + 4 tons 930 lb

30. 1 ton 1140 lb + 5 tons 1200 lb

31. 5 tons 1050 lb − 2 tons 875 lb

32. 4 tons 850 lb − 1 ton 260 lb

33. 12 lb 4 oz − 3 lb 9 oz

34. 45 lb 6 oz − 26 lb 10 oz

35. 5 lb 3 oz × 6

36. 2 lb 5 oz × 5

37. 6 tons 1500 lb ÷ 5

38. 5 tons 400 lb ÷ 4

Objective **C** *Convert as indicated. See Examples 7 and 8.*

39. 500 g to kilograms

40. 820 g to kilograms

41. 4 g to milligrams

42. 9 g to milligrams

43. 25 kg to grams

44. 18 kg to grams

45. 48 mg to grams

46. 112 mg to grams

47. 6.3 g to kilograms

48. 4.9 g to kilograms

49. 15.14 g to milligrams

50. 16.23 g to milligrams

51. 6.25 kg to grams

52. 3.16 kg to grams

53. 35 hg to centigrams

54. 4.26 cg to dekagrams

Objective **D** *Perform each indicated operation. See Examples 9 and 10.*

55. 3.8 mg + 9.7 mg

56. 41.6 g + 9.8 g

57. 205 mg + 5.61 g

58. 2.1 g + 153 mg

59. 9 g − 7150 mg

60. 4 kg − 2410 g

61. 1.61 kg − 250 g

62. 6.13 g − 418 mg

63. 5.2 kg × 2.6

64. 4.8 kg × 9.3

65. 17 kg ÷ 8

66. 8.25 g ÷ 6

Objectives A B C D Mixed Practice *Solve. Remember to insert units when writing your answers. For Exercises 67 through 74, complete the chart. See Examples 1 through 10.*

	Object	Tons	Pounds	Ounces
67.	Statue of Liberty—weight of copper sheeting	100		
68.	Statue of Liberty—weight of steel	125		
69.	A 12-inch cube of osmium (heaviest metal)		1,345	
70.	A 12-inch cube of lithium (lightest metal)		32	

	Object	Grams	Kilograms	Milligrams	Centigrams
71.	Capsule of Amoxicillin (antibiotic)			500	
72.	Tablet of Topamax (epilepsy and migraine uses)			25	
73.	A six-year-old boy		21		
74.	A golf ball	45			

75. A can of 7-Up weighs 336 grams. Find the weight in kilograms of 24 cans.

76. Guy Green normally weighs 73 kg, but he lost 2800 grams after being sick with the flu. Find Guy's new weight.

77. Sudafed is a decongestant that comes in two strengths. Regular strength contains 60 mg of medication. Extra strength contains 0.09 g of medication. How much extra medication is in the extra-strength tablet?

78. A small can of Planters sunflower seeds weighs 177 g. If each can contains 6 servings, find the weight of one serving.

79. Doris Johnson has two open containers of Uncle Ben's rice. If she combines 1 lb 10 oz from one container with 3 lb 14 oz from the other container, how much total rice does she have?

80. Dru Mizel maintains the records of the amount of coal delivered to his department in the steel mill. In January, 3 tons 1500 lb were delivered. In February, 2 tons 1200 lb were delivered. Find the total amount delivered in these two months.

81. Carla Hamtini was amazed when she grew a 28 lb 10 oz zucchini in her garden, but later she learned that the heaviest zucchini ever grown weighed 64 lb 8 oz in Llanharry, Wales, by B. Lavery in 1990. How far below the record weight was Carla's zucchini? (*Source: Guinness World Records*)

82. The heaviest baby born in good health weighed an incredible 22 lb 8 oz. He was born in Italy in September, 1955. How much heavier is this than a 7 lb 12 oz baby? (*Source: Guinness World Records*)

83. The smallest baby born in good health weighed only 8.6 ounces, less than a can of soda. She was born in Chicago in December, 2004. How much lighter was she than an average baby, who weighs about 7 lb 8 ounces?

84. A large bottle of Hire's Root Beer weighs 1900 grams. If a carton contains 6 large bottles of root beer, find the weight in kilograms of 5 cartons.

85. Three milligrams of preservatives are added to a 0.5-kg box of dried fruit. How many milligrams of preservatives are in 3 cartons of dried fruit if each carton contains 16 boxes?

86. One box of Swiss Miss Cocoa Mix weighs 0.385 kg, but 39 grams of this weight is the packaging. Find the actual weight of the cocoa in 8 boxes.

87. A carton of 12 boxes of Quaker Oats Oatmeal weighs 6.432 kg. Each box includes 26 grams of packaging material. What is the actual weight of the oatmeal in the carton?

88. The supermarket prepares hamburger in 85-gram market packages. When Leo Gonzalas gets home, he divides the package in half before refrigerating the meat. How much will each package weigh?

89. The Shop 'n Bag supermarket chain ships hamburger meat by placing 10 packages of hamburger in a box, with each package weighing 3 lb 4 oz. How much will 4 boxes of hamburger weigh?

90. The Quaker Oats Company ships its 1-lb 2-oz boxes of oatmeal in cartons containing 12 boxes of oatmeal. How much will 3 such cartons weigh?

91. A carton of Del Monte Pineapple weighs 55 lb 4 oz, but 2 lb 8 oz of this weight is due to packaging. Subtract the weight of the packaging to find the actual weight of the pineapple in 4 cartons.

92. The Hormel Corporation ships cartons of canned ham weighing 43 lb 2 oz each. Of this weight, 3 lb 4 oz is due to packaging. Find the actual weight of the ham found in 3 cartons.

Review

Write each fraction as a decimal. See Section 5.5.

93. $\frac{4}{25}$

94. $\frac{3}{5}$

95. $\frac{7}{8}$

96. $\frac{3}{16}$

Concept Extensions

Determine whether the measurement in each statement is reasonable.

97. The doctor prescribed a pill containing 2 kg of medication.

98. A full-grown cat weighs approximately 15 g.

99. A bag of flour weighs 4.5 kg.

100. A staple weighs 15 mg.

101. A professor weighs less than 150 g.

102. A car weighs 2000 mg.

103. Use a unit other than centigram and write a mass that is equivalent to 25 centigrams. (*Hint:* There are many possibilities.)

104. Use a unit other than pound and write a weight that is equivalent to 4000 pounds. (*Hint:* There are many possibilities.)

True or False? See the Concept Check in this section.

105. A kilogram is larger than a gram.

106. A decigram is larger than a milligram.

107. Why is the decimal point moved to the right when grams are converted to milligrams?

108. To change 8 pounds to ounces, multiply by 16. Why is this the correct procedure?

STUDY SKILLS BUILDER

How Are Your Homework Assignments Going?

Remember that it is important to keep up with homework. Why? Many concepts in mathematics build on each other. Often, your understanding of a day's lecture depends on an understanding of the previous day's material.

To complete a homework assignment, remember these 4 things:

- Attempt all of it.
- Check it.
- Correct it.
- If needed, ask questions about it.

Take a moment and review your completed homework assignments. Answer the exercises below based on this review.

1. Approximate the fraction of your homework you have attempted.
2. Approximate the fraction of your homework you have checked (if possible).
3. If you are able to check your homework, have you corrected it when errors have been found?
4. When working homework, if you do not understand a concept, what do you personally do?

8.6 CAPACITY

Objectives

A Define U.S. Units of Capacity and Convert from One Unit to Another.

B Perform Arithmetic Operations on U.S. Units of Capacity.

C Define Metric Units of Capacity and Convert from One Unit to Another.

D Perform Arithmetic Operations on Metric Units of Capacity.

Objective A Defining and Converting U.S. System Units of Capacity

Units of **capacity** are generally used to measure liquids. The number of gallons of gasoline needed to fill a gas tank in a car, the number of cups of water needed in a bread recipe, and the number of quarts of milk sold each day at a supermarket are all examples of using units of capacity. The following summary shows equivalencies between units of capacity:

U.S. Units of Capacity

8 fluid ounces (fl oz) = 1 cup (c)
2 cups = 1 pint (pt)
2 pints = 1 quart (qt)
4 quarts = 1 gallon (gal)

Just as with units of length and weight, we can form unit fractions to convert between different units of capacity. For instance,

$$\frac{2\text{ c}}{1\text{ pt}} = \frac{1\text{ pt}}{2\text{ c}} = 1 \quad \text{and} \quad \frac{2\text{ pt}}{1\text{ qt}} = \frac{1\text{ qt}}{2\text{ pt}} = 1$$

EXAMPLE 1 Convert 9 quarts to gallons.

Solution: We multiply by the unit fraction $\frac{1\text{ gal}}{4\text{ qt}}$.

$$\begin{aligned} 9\text{ qt} &= \frac{9\text{ qt}}{1} \cdot 1 \\ &= \frac{9\not{\text{qt}}}{1} \cdot \frac{1\text{ gal}}{4\not{\text{qt}}} \\ &= \frac{9\text{ gal}}{4} \\ &= 2\frac{1}{4}\text{ gal} \end{aligned}$$

Thus, 9 quarts is the same as $2\frac{1}{4}$ gallons, as shown in the diagram:

Work Practice Problem 1

PRACTICE PROBLEM 1

Convert 43 pints to quarts.

Answer

1. $21\frac{1}{2}$ qt

PRACTICE PROBLEM 2

Convert 26 quarts to cups.

EXAMPLE 2 Convert 14 cups to quarts.

Solution: Our equivalency table contains no direct conversion from cups to quarts. However, from this table we know that

$$1\text{ qt} = 2\text{ pt} = \frac{2\text{ pt}}{1}\cdot 1 = \frac{2\text{ pt}}{1}\cdot\frac{2\text{ c}}{1\text{ pt}} = 4\text{ c}$$

so 1 qt = 4 c. Now we have the unit fraction $\frac{1\text{ qt}}{4\text{ c}}$. Thus,

$$14\text{ c} = \frac{14\text{ c}}{1}\cdot 1 = \frac{14\not{c}}{1}\cdot\frac{1\text{ qt}}{4\not{c}} = \frac{14\text{ qt}}{4} = \frac{7}{2}\text{qt} \quad\text{or}\quad 3\frac{1}{2}\text{qt}$$

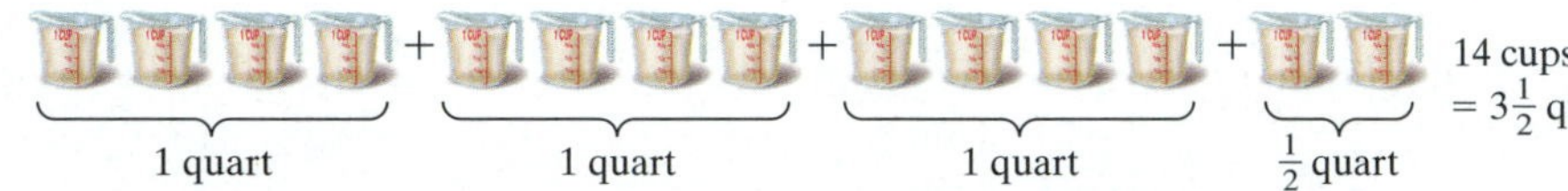

Work Practice Problem 2

✔Concept Check If 50 cups are converted to quarts, will the equivalent number of quarts be less than or greater than 50? Explain.

Objective B Performing Operations on U.S. System Units of Capacity

As is true of units of length and weight, units of capacity can be added, subtracted, multiplied, and divided.

PRACTICE PROBLEM 3

Subtract 2 qt from 1 gal 1 qt.

EXAMPLE 3 Subtract 3 qt from 4 gal 2 qt.

Solution: To subtract, we line up similar units.

$$\begin{array}{r} 4\text{ gal }2\text{ qt} \\ -\quad 3\text{ qt} \\ \hline \end{array}$$

We cannot subtract 3 qt from 2 qt. We need to borrow 1 gallon from the 4 gallons, convert it to 4 quarts, and then combine it with the 2 quarts.

3 gal + (1 gal) 4 qt

$$\begin{array}{r} 4\text{ gal }2\text{ qt} \\ -\quad 3\text{ qt} \\ \hline \end{array} \quad = \quad \begin{array}{r} 3\text{ gal }6\text{ qt} \\ -\quad 3\text{ qt} \\ \hline 3\text{ gal }3\text{ qt} \end{array}$$

To check, see that the sum of 3 gal 3 qt and 3 qt is 4 gal 2 qt.

Work Practice Problem 3

Answers

2. 104 c **3.** 3 qt

✔ Concept Check Answer

less than 50

EXAMPLE 4 **Finding the Amount of Water in an Aquarium**

An aquarium contains 6 gal 3 qt of water. If 2 gal 2 qt of water is added, what is the total amount of water in the aquarium?

Solution:

beginning water	→	6 gal 3 qt
+ water added	→	+ 2 gal 2 qt
total water	→	8 gal 5 qt

Since 5 qt = 1 gal 1 qt, we have

$$\overbrace{8 \text{ gal}}^{8 \text{ gal}} + \overbrace{1 \text{ gal } 1 \text{ qt}}^{5 \text{ qt}}$$
$$= 9 \text{ gal } 1 \text{ qt}$$

The total amount of water is 9 gal 1 qt.

Work Practice Problem 4

PRACTICE PROBLEM 4

A large oil drum contains 15 gal 3 qt of oil. How much will be in the drum if an additional 4 gal 3 qt of oil is poured into it?

Objective C Defining and Converting Metric System Units of Capacity

Thus far, we know that the basic unit of length in the metric system is the meter and that the basic unit of mass in the metric system is the gram. What is the basic unit of capacity? The **liter.** By definition, a **liter** is the capacity or volume of a cube measuring 10 centimeters on each side.

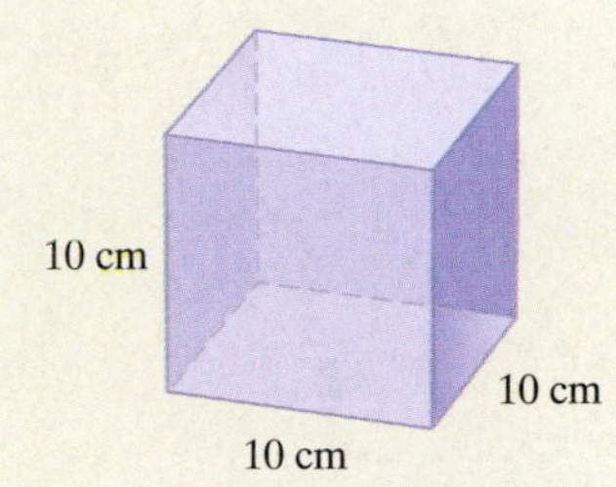

The following examples may help you get a feeling for metric capacities:

One liter of liquid is slightly more than one quart.

Many soft drinks are packaged in 2-liter bottles.

The metric system was designed to be a consistent system. Once again, the prefixes for metric units of capacity are the same as for metric units of length and mass, as summarized in the following table:

Metric Unit of Capacity
1 **kilo**liter (kl) = 1000 liters (L)
1 **hecto**liter (hl) = 100 L
1 **deka**liter (dal) = 10 L
1 liter (L) = 1 L
1 **deci**liter (dl) = 1/10 L or 0.1 L
1 **centi**liter (cl) = 1/100 L or 0.01 L
1 **milli**liter (ml) = 1/1000 L or 0.001 L

The **milliliter** and the **liter** are the two most commonly used metric units of capacity.

Converting from one unit of capacity to another involves multiplying by powers of 10 or moving the decimal point to the left or to the right. Listing units of capacity in order from largest to smallest helps to keep track of how many places to move the decimal point when converting.

Let's convert 2.6 liters to milliliters. To convert from liters to milliliters, we move along the chart 3 units to the right.

kl hl dal **L** dl cl **ml**

3 units to the right

Answer

4. 20 gal 2 qt

This means that we move the decimal point 3 places to the right to convert from liters to milliliters.

2.600 L = 2600. ml

This same conversion can be done with unit fractions.

$$\begin{aligned} 2.6\text{ L} &= \frac{2.6\text{ L}}{1}\cdot 1 \\ &= \frac{2.6\text{ }\cancel{\text{L}}}{1}\cdot\frac{1000\text{ ml}}{1\text{ }\cancel{\text{L}}} \\ &= 2.6\cdot 1000\text{ ml} \\ &= 2600\text{ ml} \end{aligned}$$

To multiply by 1000, move the decimal point 3 places to the right.

To visualize the result, study the diagram below:

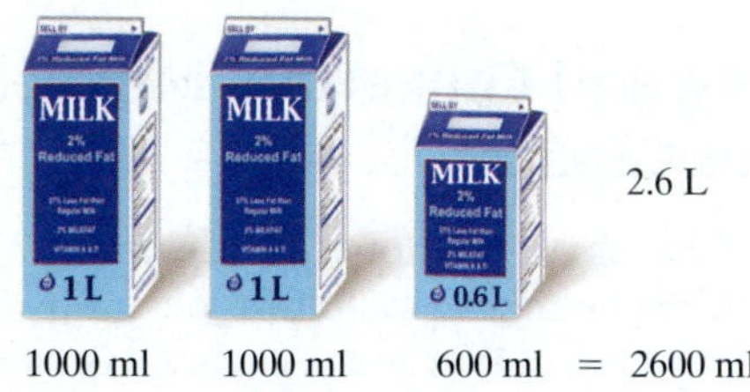

Thus, 2.6 L = 2600 ml.

PRACTICE PROBLEM 5

Convert 2100 ml to liters.

EXAMPLE 5 Convert 3210 ml to liters.

Solution: Let's use the unit fraction method first.

$$3210\text{ ml} = \frac{3210\text{ ml}}{1}\cdot 1 = 3210\text{ }\cancel{\text{ml}}\cdot\overbrace{\frac{1\text{ L}}{1000\text{ }\cancel{\text{ml}}}}^{\text{Unit fraction}} = 3.21\text{ L}$$

Now let's list the unit measures in order from left to right and move from milliliters to liters.

kl hl dal L dl cl ml

3 units to the left

3210 ml = 3.210 L, the same results as before and shown below in the diagram.

3 places to the left

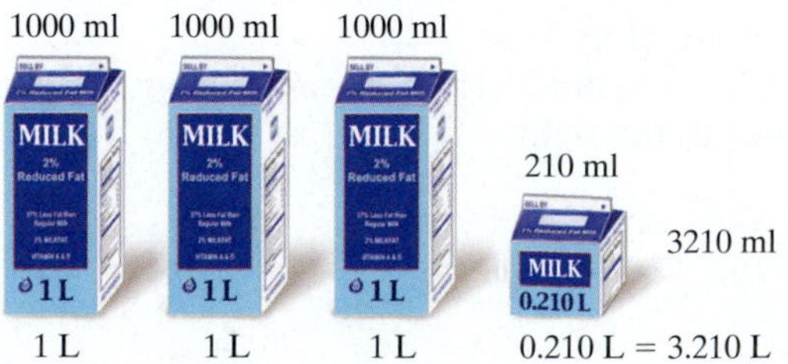

Work Practice Problem 5

Answer

5. 2.1 L

EXAMPLE 6 Convert 0.185 dl to milliliters.

Solution: We list the unit measures in order from left to right and move from deciliters to milliliters.

kl hl dal L dl cl ml

2 units to the right

0.185 dl = 18.5 ml

2 places to the right

Work Practice Problem 6

PRACTICE PROBLEM 6

Convert 2.13 dal to liters.

Objective D Performing Operations on Metric System Units of Capacity

As was true for length and weight, arithmetic operations involving metric units of capacity can also be performed. Make sure that the metric units of capacity are the same before adding, subtracting, multiplying, or dividing.

EXAMPLE 7 Add 2400 ml to 8.9 L.

Solution: We must convert both to liters or both to milliliters before adding the capacities together.

2400 ml = 2.4 L or 8.9 L = 8900 ml

$$\begin{array}{r} 2.4\text{ L} \\ +\ 8.9\text{ L} \\ \hline 11.3\text{ L} \end{array} \qquad \begin{array}{r} 2400\text{ ml} \\ +\ 8900\text{ ml} \\ \hline 11{,}300\text{ ml} \end{array}$$

The total is 11.3 L or 11,300 ml. They both represent the same capacity.

Work Practice Problem 7

PRACTICE PROBLEM 7

Add 1250 ml to 2.9 L.

✔ **Concept Check** How could you estimate the following operation? Subtract 950 ml from 7.5 L.

EXAMPLE 8 **Finding the Amount of Medication a Person Has Received**

A patient hooked up to an IV unit in the hospital is to receive 12.5 ml of medication every hour. How much medication does the patient receive in 3.5 hours?

Solution: We multiply 12.5 ml by 3.5.

medication per hour	→	12.5 ml
× hours	→	× 3.5
total medication		625
		3750
		43.75 ml

The patient receives 43.75 ml of medication.

Work Practice Problem 8

PRACTICE PROBLEM 8

If 28.6 L of water can be pumped every minute, how much water can be pumped in 85 minutes?

Answers

6. 21.3 L **7.** 4150 ml or 4.15 L **8.** 2431 L

✔ Concept Check Answer

950 ml = 0.95 L; round 0.95 to 1; 7.5 − 1 = 6.5 L

Vocabulary and Readiness Check

Use the choices below to fill in each blank. Some choices may be used more than once.

cups	pints	liter
quarts	fluid ounces	capacity

1. Units of __________ are generally used to measure liquids.
2. The basic unit of capacity in the metric system is the ______.
3. One cup equals 8 _____________.
4. One quart equals 2 _______.
5. One pint equals 2 _______.
6. One quart equals 4 _______.
7. One gallon equals 4 ________.

8.6 EXERCISE SET

FOR EXTRA HELP

Student Solutions Manual

PH Math/Tutor Center

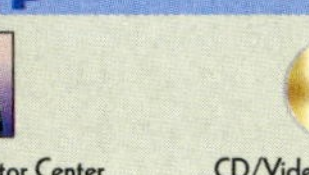
CD/Video for Review

MathXL®

MyMathLab

Objective A *Convert each measurement as indicated. See Examples 1 and 2.*

1. 32 fluid ounces to cups
2. 16 quarts to gallons
3. 8 quarts to pints
4. 9 pints to quarts
5. 14 quarts to gallons
6. 11 cups to pints
7. 80 fluid ounces to pints
8. 18 pints to gallons
9. 2 quarts to cups
10. 3 pints to fluid ounces
11. 120 fluid ounces to quarts
12. 20 cups to gallons
13. 42 cups to quarts
14. 7 quarts to cups
15. $4\frac{1}{2}$ pints to cups
16. $6\frac{1}{2}$ gallons to quarts
17. 5 gal 3 qt to quarts
18. 4 gal 1 qt to quarts
19. $\frac{1}{2}$ cup to pint
20. $\frac{1}{2}$ pint to quarts
21. 58 qt = _____ gal _____ qt
22. 70 qt = _____ gal _____ qt
23. 39 pt = _____ gal _____ qt _____ pt
24. 29 pt = _____ gal _____ qt _____ pt
25. $2\frac{3}{4}$ gallons to pints
26. $3\frac{1}{4}$ quarts to cups

Objective B *Perform each indicated operation. See Examples 3 and 4.*

27. 5 gal 3 qt + 7 gal 3 qt
28. 2 gal 2 qt + 9 gal 3 qt
29. 1 c 5 fl oz + 2 c 7 fl oz

30. 2 c 3 fl oz + 2 c 6 fl oz

31. 3 gal − 1 gal 3 qt

32. 2 pt − 1 pt 1 c

33. 3 gal 1 qt − 1 qt 1 pt

34. 3 qt 1 c − 1 c 4 fl oz

35. 8 gal 2 qt × 2

36. 6 gal 1 pt × 2

37. 9 gal 2 qt ÷ 2

38. 5 gal 6 fl oz ÷ 2

Objective **C** *Convert as indicated. See Examples 5 and 6.*

39. 5 L to milliliters

40. 8 L to milliliters

41. 0.16 L to kiloliters

42. 0.127 L to kiloliters

43. 5600 ml to liters

44. 1500 ml to liters

45. 3.2 L to centiliters

46. 1.7 L to centiliters

47. 410 L to kiloliters

48. 250 L to kiloliters

49. 64 ml to liters

50. 39 ml to liters

51. 0.16 kl to liters

52. 0.48 kl to liters

53. 3.6 L to milliliters

54. 1.9 L to milliliters

Objective **D** *Perform each indicated operation. See Examples 7 and 8.*

55. 3.4 L + 15.9 L

56. 18.5 L + 4.6 L

57. 2700 ml + 1.8 L

58. 4.6 L + 1600 ml

59. 8.6 L − 190 ml

60. 4.8 L − 283 ml

61. 17,500 ml − 0.9 L

62. 6850 ml − 0.3 L

63. 480 ml × 8

64. 290 ml × 6

65. 81.2 L ÷ 0.5

66. 5.4 L ÷ 3.6

Objectives **A B C D** **Mixed Practice** *Solve. Remember to insert units when writing your answers. For Exercises 67 through 70, complete the chart.*

	Capacity	Cups	Gallons	Quarts	Pints
67.	An average-size bath of water		21		
68.	A dairy cow's daily milk yield				38
69.	Your kidneys filter about this amount of blood every minute	4			
70.	The amount of water needed in a punch recipe	2			

71. Mike Schaferkotter drank 410 ml of Mountain Dew from a 2-liter bottle. How much Mountain Dew remains in the bottle?

72. The Werners' Volvo has a 54.5-L gas tank. Only 3.8 liters of gasoline still remain in the tank. How much is needed to fill it?

73. Margie Phitts added 354 ml of Prestone dry gas to the 18.6 L of gasoline in her car's tank. Find the total amount of gasoline in the tank.

74. Chris Peckaitis wishes to share a 2-L bottle of Coca Cola equally with 7 of his friends. How much will each person get?

75. A garden tool engine requires a 30 to 1 gas to oil mixture. This means that $\frac{1}{30}$ of a gallon of oil should be mixed with 1 gallon of gas. Convert $\frac{1}{30}$ gallon to ounces. Round to the nearest tenth.

76. Henning's Supermarket sells homemade soup in 1 qt 1 pt containers. How much soup is contained in three such containers?

77. Can 5 pt 1 c of fruit punch and 2 pt 1 c of ginger ale be poured into a 1-gal container without it overflowing?

78. Three cups of prepared Jell-O are poured into 6 dessert dishes. How many fluid ounces of Jell-O are in each dish?

79. Stanley Fisher paid $14 to fill his car with 44.3 liters of gasoline. Find the price per liter of gasoline to the nearest thousandth of a dollar.

80. A student carelessly misread the scale on a cylinder in the chemistry lab and added 40 cl of water to a mixture instead of 40 ml. Find the excess amount of water.

Review

Write each fraction in simplest form. See Section 4.2.

81. $\frac{20}{25}$ **82.** $\frac{75}{100}$ **83.** $\frac{27}{45}$ **84.** $\frac{56}{60}$ **85.** $\frac{72}{80}$ **86.** $\frac{18}{20}$

Concept Extensions

Determine whether the measurement in each statement is reasonable.

87. Clair took a dose of 2 L of cough medicine to cure her cough.

88. John drank 250 ml of milk for lunch.

89. Jeannie likes to relax in a tub filled with 3000 ml of hot water.

90. Sarah pumped 20 L of gasoline into her car yesterday.

Solve. See the Concept Checks in this section.

91. If 70 pints are converted to gallons, will the equivalent number of gallons be less than or greater than 70? Explain why.

92. If 30 gallons are converted to quarts, will the equivalent number of quarts be less than or greater than 30? Explain why.

93. Explain how to estimate the following operation: Add 986 ml to 6.9 L.

94. Find the number of fluid ounces in 1 gallon.

95. Explain how to borrow in order to subtract 1 gal 2 qt from 3 gal 1 qt.

A cubic centimeter (cc) is the amount of space that a volume of 1 ml occupies. Because of this, we will say that 1 cc = 1 ml.

A common syringe is one with a capacity of 3 cc. Use the diagram and give the measurement indicated by each arrow.

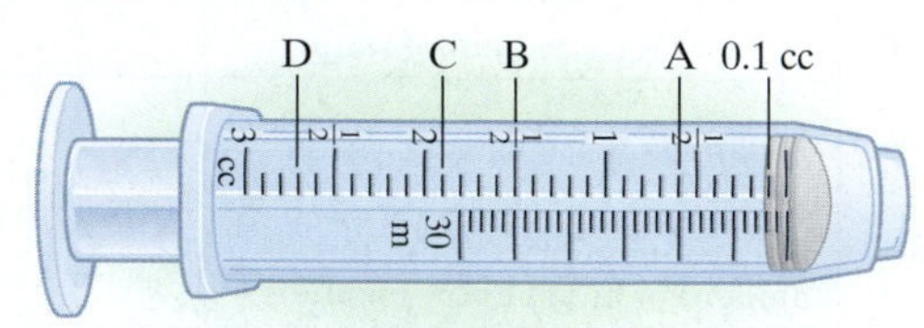

96. A **97.** B **98.** C **99.** D

In order to measure small dosages, such as for insulin, u-100 syringes are used. For these syringes, 1 cc has been divided into 100 equal units (u). Use the diagram and give the measurement indicated by each arrow in units (u) and then cubic centimeters. Use 100 u = 1 cc *and round to the nearest hundredth.*

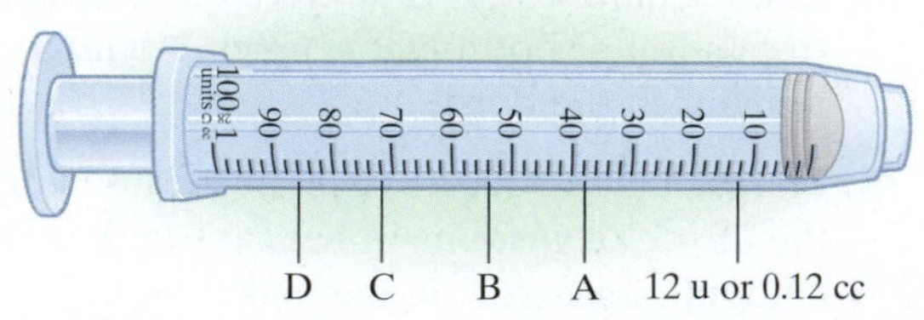

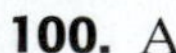

100. A **101.** B

102. C **103.** D

8.7 TEMPERATURE AND CONVERSIONS BETWEEN THE U.S. AND METRIC SYSTEMS

Objective

A Convert Between the U.S. and Metric Systems.

B Convert Temperatures from Degrees Celsius to Degrees Fahrenheit.

C Convert Temperatures from Degrees Fahrenheit to Degrees Celsius.

Objective A Converting Between the U.S. and Metric Systems

The metric system probably had its beginnings in France in the 1600s, but it was the Metric Act of 1866 that made the use of this system legal (but not mandatory) in the United States. Other laws have followed that allow for a slow, but deliberate, transfer to the modernized metric system. In April, 2001, for example, the U.S. Stock Exchanges completed their change to decimal trading instead of fractions. By the end of 2009, all products sold in Europe (with some exceptions) will be required to have only metric units on their labels. (*Source:* U.S. Metric Association and National Institute of Standards and Technology)

You may be surprised at the number of everyday items we use that are already manufactured in metric units. We easily recognize 1L and 2L soda bottles, but what about the following?

Pencil leads (0.5 mm or 0.7 mm)
Camera film (35 mm)
Sporting events (5 km or 10 km races)
Medicines (500 mg capsules)
Labels on retail goods (dual-labeled since 1994)

Since the United States has not completely converted to the metric system, we need to practice converting from one system to the other. Below is a table of mostly approximate conversions.

Length:		**Capacity:**		**Weight** (mass):	
metric	U.S. System	metric	U.S. System	metric	U.S. System
1 m	≈ 1.09 yd	1 L	≈ 1.06 qt	1 kg	≈ 2.20 lb
1 m	≈ 3.28 ft	1 L	≈ 0.26 gal	1 g	≈ 0.04 oz
1 km	≈ 0.62 mi	3.79 L	≈ 1 gal	0.45 kg	≈ 1 lb
2.54 cm	= 1 in.	0.95 L	≈ 1 qt	28.35 g	≈ 1 oz
0.30 m	≈ 1 ft	29.57 ml	≈ 1 fl oz		
1.61 km	≈ 1 mi				

There are many ways to perform these metric to U.S. conversions. We will do so by using unit fractions.

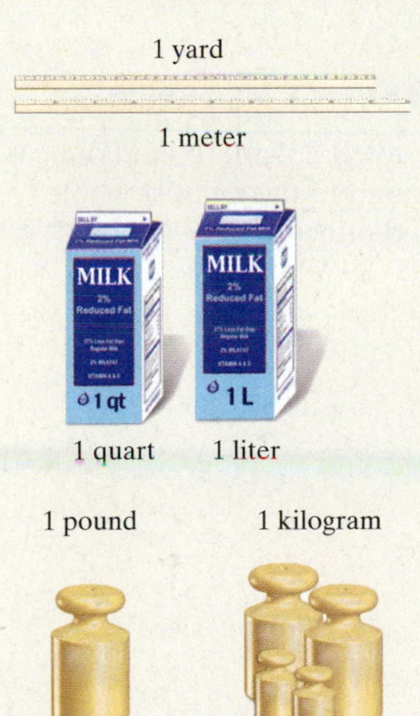

EXAMPLE 1 Compact Discs

Standard-sized compact discs are 12 centimeters in diameter. Convert this length to inches. Round the result to two decimal places. (*Source:* usByte.com)

Solution: From our length conversion table, we know that 2.54 cm = 1 in. This fact gives us two unit fractions: $\frac{2.54 \text{ cm}}{1 \text{ in.}}$ and $\frac{1 \text{ in.}}{2.54 \text{ cm}}$. We use the unit fraction with cm in the denominator so that these units divide out.

1.5 cm

12 cm

$$12 \text{ cm} = \frac{12 \text{ cm}}{1} \cdot 1 = \frac{12 \cancel{\text{cm}}}{1} \cdot \overbrace{\frac{1 \text{ in.}}{2.54 \cancel{\text{cm}}}}^{\text{Unit fraction}} \begin{matrix} \leftarrow \text{Units to convert to} \\ \leftarrow \text{Original units} \end{matrix}$$

$$= \frac{12 \text{ in.}}{2.54}$$

$$\approx 4.72 \text{ in.} \quad \text{Divide.}$$

Continued on next page

PRACTICE PROBLEM 1

The center hole of a standard-sized compact disc is 1.5 centimeters in diameter. Convert this length to inches. Round the result to 2 decimal places.

Answer

1. 0.59 in.

Thus, the diameter of a standard compact disc is exactly 12 cm or approximately 4.72 inches. For a dimension this size, you can use a ruler to check. Another method is to approximate. Our result, 4.72 in., is close to 5 inches. Since 1 in. is about 2.5 cm, then 5 in. is about $5(2.5 \text{ cm}) = 12.5$ cm, which is close to 12 cm.

Work Practice Problem 1

PRACTICE PROBLEM 2

A full-grown human heart weighs about 8 ounces. Convert this weight to grams. If necessary, round your result to the nearest tenth of a gram.

EXAMPLE 2 Liver

The liver is your largest internal organ. It weighs about 3.5 pounds in a grown man. Convert this weight to kilograms. Round to the nearest tenth. (*Source: Some Body!* by Dr. Pete Rowan)

Solution: $3.5 \text{ lb} \approx \frac{3.5 \cancel{\text{lb}}}{1} \cdot \overbrace{\frac{0.45 \text{ kg}}{1 \cancel{\text{lb}}}}^{\text{Unit fraction}} = 3.5(0.45 \text{ kg}) \approx 1.6 \text{ kg}$

Thus 3.5 pounds are approximately 1.6 kilograms. From the table of conversions, we know that $1 \text{ kg} \approx 2.2$ lb. So that $0.5 \text{ kg} \approx 1.1$ lb and after adding, we have $1.5 \text{ kg} \approx 3.3$ lb. Our result is reasonable.

Work Practice Problem 2

PRACTICE PROBLEM 3

Convert 237 ml to fluid ounces. Round to the nearest whole fluid ounce.

EXAMPLE 3 Postage Stamp

Australia converted to the metric system in 1973. In that year, four postage stamps were issued to publicize this conversion. One such stamp is shown to the right. Let's check the mathematics on the stamp by converting 7 fluid ounces to milliliters. Round to the nearest hundred.

Solution: $7 \text{ fl oz} \approx \frac{7 \cancel{\text{fl oz}}}{1} \cdot \overbrace{\frac{29.57 \text{ ml}}{1 \cancel{\text{fl oz}}}}^{\text{Unit fraction}} = 7(29.57 \text{ ml}) = 206.99 \text{ ml}$

Rounded to the nearest hundred, $7 \text{ fl oz} \approx 200$ ml.

Work Practice Problem 3

Now that we have practiced converting between two measurement systems, let's practice converting between two temperature scales.

Temperature When Gabriel Fahrenheit and Anders Celsius independently established units for temperature scales, each based his unit on the heat of water the moment it boils compared to the moment it freezes. One degree Celsius is $\frac{1}{100}$ of the difference in heat. One degree Fahrenheit is $\frac{1}{180}$ of the difference in heat. Celsius arbitrarily labeled the temperature at the freezing point at 0°C, making the boiling point 100°C; Fahrenheit labeled the freezing point 32°F, making the boiling point 212°F. Water boils at 212°F or 100°C.

By comparing the two scales in the figure, we see that a 20°C day is as warm as a 68°F day. Similarly, a sweltering 104°F day in the Mojave desert corresponds to a 40°C day.

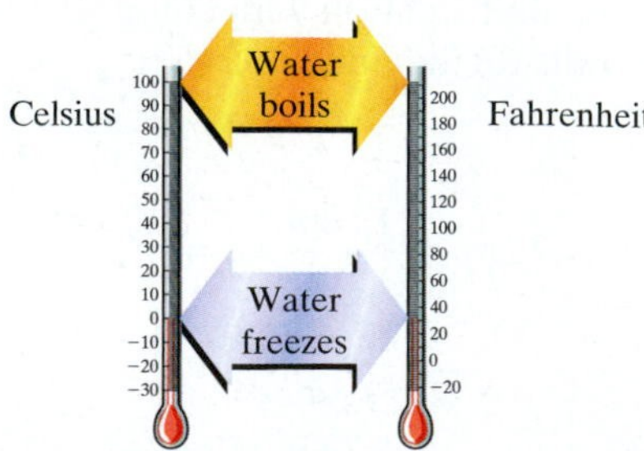

Answers
2. 226.8 g **3.** 8 fl oz

✔ **Concept Check** Which of the following statements is correct? Explain.

a. 6°C is below the freezing point of water.

b. 6°F is below the freezing point of water.

Objective B Converting Degrees Celsius to Degrees Fahrenheit

To convert from Celsius temperatures to Fahrenheit temperatures, see the box below. In this box, we use the symbol F to represent degrees Fahrenheit and the symbol C to represent degrees Celsius.

Converting Celsius to Fahrenheit

$$F = \frac{9}{5}C + 32 \quad \text{or} \quad F = 1.8C + 32$$

(To convert to Fahrenheit temperature, multiply the Celsius temperature by $\frac{9}{5}$ or 1.8, and then add 32.)

EXAMPLE 4 Convert 15°C to degrees Fahrenheit.

Solution:

$$\begin{aligned} F &= \frac{9}{5}C + 32 \\ &= \frac{9}{5} \cdot 15 + 32 && \text{Replace C with 15.} \\ &= 27 + 32 && \text{Simplify.} \\ &= 59 && \text{Add.} \end{aligned}$$

Thus, 15°C is equivalent to 59°F.

Work Practice Problem 4

PRACTICE PROBLEM 4

Convert 60°C to degrees Fahrenheit.

EXAMPLE 5 Convert 29°C to degrees Fahrenheit.

Solution:

$$\begin{aligned} F &= 1.8C + 32 \\ &= 1.8 \cdot 29 + 32 && \text{Replace C with 29.} \\ &= 52.2 + 32 && \text{Multiply 1.8 by 29.} \\ &= 84.2 && \text{Add.} \end{aligned}$$

Therefore, 29°C is the same as 84.2°F.

Work Practice Problem 5

PRACTICE PROBLEM 5

Convert 32°C to degrees Fahrenheit.

Objective C Converting Degrees Fahrenheit to Degrees Celsius

To convert from Fahrenheit temperatures to Celsius temperatures, see the box below. The symbol C represents degrees Celsius and the symbol F represents degrees Fahrenheit.

Converting Fahrenheit to Celsius

$$C = \frac{5}{9}(F - 32)$$

(To convert to Celsius temperature, subtract 32 from the Fahrenheit temperature, and then multiply by $\frac{5}{9}$.)

Answers

4. 140°F **5.** 89.6°F

✔ Concept Check Answer

a. false **b.** true

PRACTICE PROBLEM 6

Convert 68°F to degrees Celsius.

EXAMPLE 6 Convert 59°F to degrees Celsius.

Solution: We evaluate the formula $C = \frac{5}{9}(F - 32)$ when F is 59.

$$C = \frac{5}{9}(F - 32)$$

$$= \frac{5}{9} \cdot (59 - 32) \quad \text{Replace F with 59.}$$

$$= \frac{5}{9} \cdot (27) \quad \text{Subtract inside parentheses.}$$

$$= 15 \quad \text{Multiply.}$$

Therefore, 59°F is the same temperature as 15°C.

Work Practice Problem 6

PRACTICE PROBLEM 7

Convert 113°F to degrees Celsius. If necessary, round to the nearest tenth of a degree.

EXAMPLE 7 Convert 114°F to degrees Celsius. If necessary, round to the nearest tenth of a degree.

Solution:

$$C = \frac{5}{9}(F - 32)$$

$$= \frac{5}{9}(114 - 32) \quad \text{Replace F with 114.}$$

$$= \frac{5}{9} \cdot (82) \quad \text{Subtract inside parentheses.}$$

$$\approx 45.6 \quad \text{Multiply.}$$

Therefore, 114°F is approximately 45.6°C.

Work Practice Problem 7

PRACTICE PROBLEM 8

During a bout with the flu, Albert's temperature reaches 102.8°F. What is his temperature measured in degrees Celsius? Round to the nearest tenth of a degree.

EXAMPLE 8 **Body Temperature**

Normal body temperature is 98.6°F. What is this temperature in degrees Celsius?

Solution: We evaluate the formula $C = \frac{5}{9}(F - 32)$ when F is 98.6.

$$C = \frac{5}{9}(F - 32)$$

$$= \frac{5}{9}(98.6 - 32) \quad \text{Replace F with 98.6.}$$

$$= \frac{5}{9} \cdot (66.6) \quad \text{Subtract inside parentheses.}$$

$$= 37 \quad \text{Multiply.}$$

Therefore, normal body temperature is 37°C.

Work Practice Problem 8

✔ **Concept Check** Clarissa must convert 40°F to degrees Celsius. What is wrong with her work shown below?

$$F = 1.8 \cdot C + 32$$
$$F = 1.8 \cdot 40 + 32$$
$$F = 72 + 32$$
$$F = 104$$

Answers

6. 20°C **7.** 45°C **8.** 39.3°C

✔ Concept Check Answer

She used the conversion for Celsius to Fahrenheit instead of Fahrenheit to Celsius.

8.7 EXERCISE SET

Note: Because approximations are used, your answers may vary slightly from the answers given in the back of the book.

Objective **A** *Convert as indicated. If necessary, round answers to two decimal places. See Examples 1 through 3.*

1. 756 milliliters to fluid ounces

2. 18 liters to quarts

3. 86 inches to centimeters

4. 86 miles to kilometers

5. 1000 grams to ounces

6. 100 kilograms to pounds

7. 93 kilometers to miles

8. 9.8 meters to feet

9. 14.5 liters to gallons

10. 150 milliliters to fluid ounces

11. 30 pounds to kilograms

12. 15 ounces to grams

Fill in the chart. Give exact answers or round to 1 decimal place.

		Meters	Yards	Centimeters	Feet	Inches
13.	**The Height of a Woman**				5	
14.	**Statue of Liberty Length of Nose**	1.37				
15.	**Leaning Tower of Pisa**		60			
16.	**Blue Whale**		36			

Solve. If necessary, round answers to two decimal places. See Examples 1 through 3.

17. The balance beam for female gymnasts is 10 centimeters wide. Convert this width to inches.

18. In men's gymnastics, the rings are 250 centimeters from the floor. Convert this height to inches, then to feet.

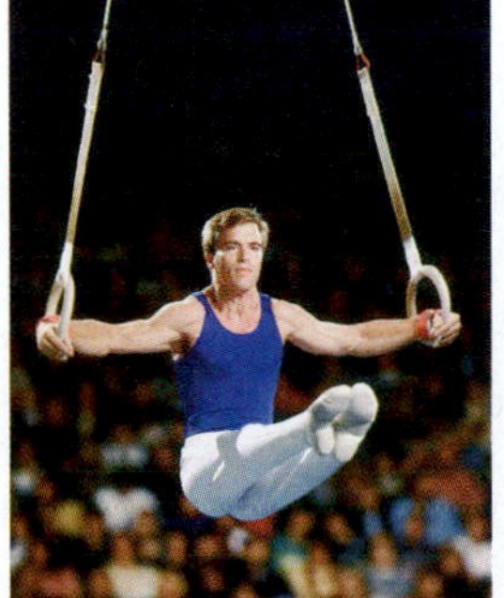

19. In many states, the maximum speed limit for recreational vehicles is 50 miles per hour. Convert this to kilometers per hour.

20. The speed limit is 70 miles per hour. Convert this to kilometers per hour.

21. Ibuprofen comes in 200 milligram tablets. Convert this to ounces. (Round your answer to this exercise to 3 decimal places.)

22. Vitamin C tablets come in 500 milligram caplets. Convert this to ounces.

23. A stone is a unit in the British customary system. Use the conversion: 14 pounds = 1 stone to check the equivalencies in this 1973 Australian stamp. Is 100 kilograms approximately 15 stone 10 pounds?

24. Convert 5 feet 11 inches to centimeters and check the conversion on this 1973 Australian stamp. Is it correct?

25. The Monarch butterfly migrates annually between the northern United States and central Mexico. The trip is about 4500 km long. Convert this to miles.

26. There is a species of African termite that builds nests up to 18 ft high. Convert this to meters.

27. A $3\frac{1}{2}$-inch diskette is not really $3\frac{1}{2}$ inches. To find its actual width, convert this measurement to centimeters, then to millimeters. Round the result to the nearest ten.

28. The average two-year-old is 84 centimeters tall. Convert this to feet and inches.

29. For an average adult, the weight of a right lung is greater than the weight of a left lung. If the right lung weighs 1.5 pounds and the left lung weighs 1.25 pounds, find the difference in grams. (*Source: Some Body!*)

30. The skin of an average adult weighs 9 pounds and is the heaviest organ. Find the weight in grams. (*Source: Some Body!*)

31. A fast sneeze has been clocked at about 167 kilometers per hour. Convert this to miles per hour. Round to the nearest whole.

32. A Boeing 747 has a cruising speed of about 980 kilometers per hour. Convert this to miles per hour. Round to the nearest whole.

33. The General Sherman giant sequoia tree has a diameter of about 8 meters at its base. Convert this to feet. (*Source: Fantastic Book of Comparisions*)

34. The largest crater on the near side of the moon is Billy Crater. It has a diameter of 303 kilometers. Convert this to miles. (*Source: Fantastic Book of Comparisions*)

35. The total length of the track on a CD is about 4.5 kilometers. Convert this to miles. Round to the nearest whole mile.

36. The distance between Mackinaw City, Michigan, and Cheyenne, Wyoming, is 2079 kilometers. Convert this to miles. Round to the nearest whole mile.

37. A doctor orders a dosage of 5 ml of medicine every 4 hours for 1 week. How many fluid ounces of medicine should be purchased? Round up to the next whole fluid ounce.

38. A doctor orders a dosage of 12 ml of medicine every 6 hours for 10 days. How many fluid ounces of medicine should be purchased? Round up to the next whole fluid ounce.

Without actually converting, choose the most reasonable answer.

39. This math book has a height of about _____.

a. 28 mm **b.** 28 cm
c. 28 m **d.** 28 km

40. A mile is _____ a kilometer.

a. shorter than **b.** longer than
c. the same length as

41. A liter has _____ capacity than a quart.

a. less **b.** greater
c. the same

42. A foot is _____ a meter.

a. shorter than **b.** longer than
c. the same length as

43. A kilogram weighs _____ a pound.

a. the same as **b.** less than
c. greater than

44. A football field is 100 yards, which is about _____.

a. 9 m **b.** 90 m
c. 900 m **d.** 9000 m

45. An $8\frac{1}{2}$ ounce glass of water has a capacity of about _____.

a. 250 L **b.** 25 L
c. 2.5 L **d.** 250 ml

46. A 5-gallon gasoline can has a capacity of about _____.

a. 19 L **b.** 1.9 L
c. 19 ml **d.** 1.9 ml

47. The weight of an average man is about _____.

a. 700 kg **b.** 7 kg
c. 0.7 kg **d.** 70 kg

48. The weight of a pill is about _____.

a. 200 kg **b.** 20 kg
c. 2 kg **d.** 200 mg

Objectives **B** **C** *Convert as indicated. When necessary, round to the nearest tenth of a degree. See Examples 4 through 8.*

49. 77°F to degrees Celsius

50. 86°F to degrees Celsius

51. 104°F to degrees Celsius

52. 140°F to degrees Celsius

53. 50°C to degrees Fahrenheit

54. 80°C to degrees Fahrenheit

55. 115°C to degrees Fahrenheit

56. 225°C to degrees Fahrenheit

57. 20°F to degrees Celsius

58. 26°F to degrees Celsius

59. 142.1°F to degrees Celsius

60. 43.4°F to degrees Celsius

61. 92°C to degrees Fahrenheit

62. 75°C to degrees Fahrenheit

63. 12.4°C to degrees Fahrenheit

64. 48.6°C to degrees Fahrenheit

65. The hottest temperature ever recorded in the United States, in Death Valley, was 134°F. Convert this temperature to degrees Celsius. (*Source:* National Climatic Data Center)

66. The hottest temperature ever recorded in the United States in January was 95°F in Los Angeles. Convert this temperature to degrees Celsius. (*Source:* National Climatic Data Center)

67. A weather forecaster in Caracas predicts a high temperature of 27°C. Find this measurement in degrees Fahrenheit.

68. While driving to work, Alan Olda notices a temperature of 18°C flash on the local bank's temperature display. Find the corresponding temperature in degrees Fahrenheit.

69. At Mack Trucks' headquarters, the room temperature is to be set at 70°F, but the thermostat is calibrated in degrees Celsius. Find the temperature to be set.

70. The computer room at Merck, Sharp, and Dohm is normally cooled to 66°F. Find the corresponding temperature in degrees Celsius.

71. In a European cookbook, a recipe requires the ingredients for caramels to be heated to 118°C, but the cook has access only to a Fahrenheit thermometer. Find the temperature in degrees Fahrenheit that should be used to make the caramels.

72. The ingredients for divinity should be heated to 127°C, but the candy thermometer that Myung Kim has is calibrated to degrees Fahrenheit. Find how hot he should heat the ingredients.

73. The temperature of Earth's core is estimated to be 4000°C. Find the corresponding temperature in degrees Fahrenheit.

74. In 2005, the average temperature of the Earth's surface was 58.1°F, the second warmest in recorded history. Convert this temperature to degrees Celsius.

Review

Perform the indicated operations. See Section 1.7.

75. $6 \cdot 4 + 5 \div 1$

76. $10 \div 2 + 9(8)$

77. $3[(1 + 5) \cdot (8 - 6)]$

78. $5[(18 - 8) - 9]$

Concept Extensions

Determine whether the measurement in each statement is reasonable.

79. A 72°F room feels comfortable.

80. Water heated to 110°F will boil.

81. Josiah has a fever if a thermometer shows his temperature to be 40°F.

82. An air temperature of 20°F on a Vermont ski slope can be expected in the winter.

83. When the temperature is 30°C outside, an overcoat is needed.

84. An air-conditioned room at 60°C feels quite chilly.

85. Barbara has a fever when a thermometer records her temperature at 40°C.

86. Water cooled to 32°C will freeze.

Body surface area (BSA) is often used to calculate dosages for some drugs. BSA is calculated in square meters using a person's weight and height.

$$\text{BSA} = \sqrt{\frac{(\text{weight in kg}) \times (\text{height in cm})}{3600}}$$

For Exercises 87 through 92, calculate the BSA for each person. Round to the nearest hundredth. You will need to use the square root key on your calculator.

87. An adult whose height is 182 cm and weight is 90 kg.

88. An adult whose height is 157 cm and weight is 63 kg.

89. A child whose height is 40 in. and weight is 50 kg. (*Hint:* Don't forget to first convert inches to centimeters.)

90. A child whose height is 26 in. and weight is 13 kg.

91. An adult whose height is 60 in. and weight is 150 lb.

92. An adult whose height is 69 in. and weight is 172 lb.

93. On February 17, 1995, in the Tokamak Fusion Test Reactor at Princeton University, the highest temperature produced in a laboratory was achieved. This temperature was 918,000,000°F. Convert this temperature to degrees Celsius. Round your answer to the nearest ten million degrees. (*Source: Guinness World Records*)

94. The hottest-burning substance known is carbon subnitride. Its flame at one atmospheric pressure reaches 9010°F. Convert this temperature to degrees Celsius. (*Source: Guinness World Records*)

95. In your own words, describe how to convert from degrees Celsius to degrees Fahrenheit.

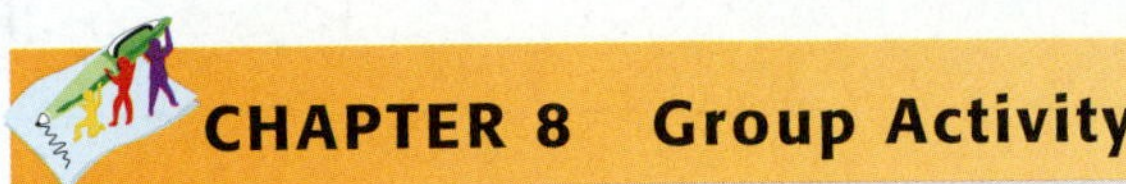

CHAPTER 8 Group Activity

Map Reading

Sections 8.1 and 8.4

Materials:

- ruler
- string
- calculator

This activity may be completed by working in groups or individually.

Investigate the route you would take from Santa Rosa, New Mexico, to San Antonio, New Mexico. Use the map in the figure to answer the following questions. You may find that using string to match the roads on the map is useful when measuring distances.

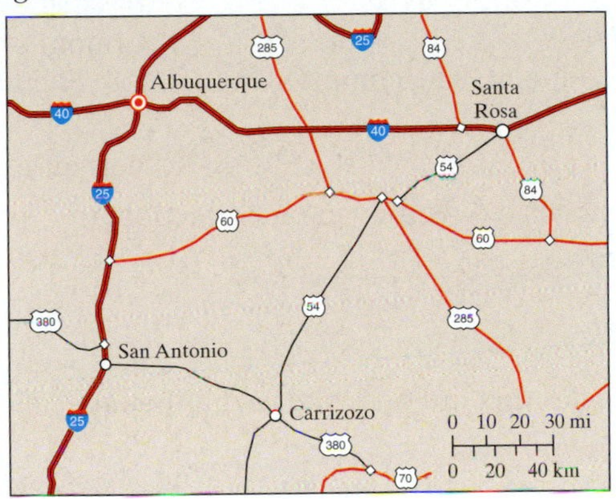

1. How many miles is it from Santa Rosa to San Antonio via Interstate 40 and Interstate 25? Convert this distance to kilometers.

2. How many miles is it from Santa Rosa to San Antonio via U.S. 54 and U.S. 380? Convert this distance to kilometers.

3. Assume that the speed limit on Interstates 40 and 25 is 65 miles per hour. How long would the trip take if you took this route and traveled 65 miles per hour the entire trip?

4. At what average speed would you have to travel on the U.S. routes to make the trip from Santa Rosa to San Antonio in the same amount of time that it would take on the interstate routes? Do you think this speed is reasonable on this route? Explain your reasoning.

5. Discuss in general the factors that might affect your decision among the different routes.

6. Explain which route you would choose in this case and why.

Chapter 8 Vocabulary Check

Fill in each blank with one of the words or phrases listed below.

transversal	line segment	obtuse	straight	adjacent	right	volume	area
acute	perimeter	vertical	supplementary	ray	angle	line	complementary
vertex	mass	unit fractions	gram	weight	meter	liter	surface area

1. ______________ is a measure of the pull of gravity.
2. ______________ is a measure of the amount of substance in an object. This measure does not change.
3. The basic unit of length in the metric system is the ______________.
4. To convert from one unit of length to another, ______________ may be used.
5. A ______________ is the basic unit of mass in the metric system.
6. The ______________ is the basic unit of capacity in the metric system.
7. A ______________ is a piece of a line with two endpoints.
8. Two angles that have a sum of 90° are called ______________ angles.
9. A ______________ is a set of points extending indefinitely in two directions.
10. The ______________ of a polygon is the distance around the polygon.
11. An ______________ is made up of two rays that share the same endpoint. The common endpoint is called the ______________.
12. ______________ measures the amount of surface of a region.
13. A ______________ is a part of a line with one endpoint. A ray extends indefinitely in one direction.
14. A line that intersects two or more lines at different points is called a ______________.
15. An angle that measures 180° is called a ______________ angle.
16. The measure of the space of a solid is called its ______________.
17. When two lines intersect, four angles are formed. Two of these angles that are opposite each other are called ______________ angles.
18. Two of the angles from #17 that share a common side are called ______________ angles.
19. An angle whose measure is between 90° and 180° is called an ______________ angle.
20. An angle that measures 90° is called a ______________ angle.
21. An angle whose measure is between 0° and 90° is called an ______________ angle.
22. Two angles that have a sum of 180° are called ______________ angles.
23. The ______________ of a polyhedron is the sum of the areas of the faces of the polyhedrons.

STUDY SKILLS BUILDER

Are You Prepared for a Test on Chapter 8?

Below I have listed some common trouble areas for topics covered in Chapter 8. After studying for your test—but before taking your test—read these.

- Don't forget the difference between complementary and supplementary angles.

  Complementary angles have a sum of 90°.

  Supplementary angles have a sum of 180°.

  The complement of a 15° angle measures 90° − 15° = 75°.

  The supplement of a 15° angle measures 180° − 15° = 165°.

- Remember:

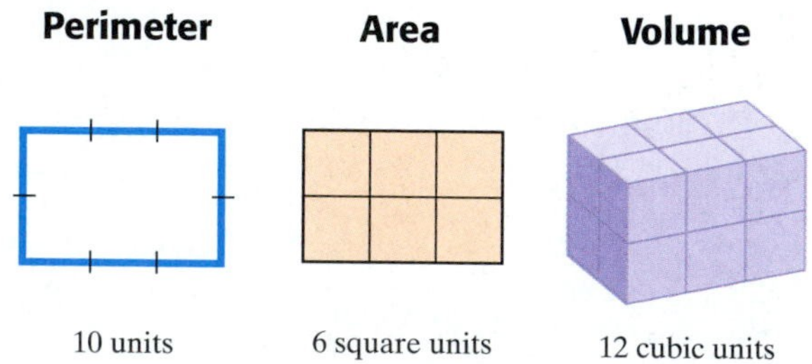

Remember: This is simply a checklist of common trouble areas. For a review of Chapter 8, see the Highlights and Chapter Review at the end of this chapter.

Are you preparing for your test? Don't forget to take the Chapter 8 Test on page 643. Then check your answers at the back of the text and use the Chapter Test Prep Video CD to see the fully worked-out solutions to any of the exercises you want to review.

8 Chapter Highlights

DEFINITIONS AND CONCEPTS	EXAMPLES

Section 8.1 Lines and Angles

A **line** is a set of points extending indefinitely in two directions. A line has no width or height, but it does have length. We name a line by any two of its points.

Line AB or $\overleftrightarrow{AB}$

A **line segment** is a piece of a line with two endpoints.

Line segment AB or $\overline{AB}$

A **ray** is a part of a line with one endpoint. A ray extends indefinitely in one direction.

Ray AB or $\overrightarrow{AB}$

A   B

An **angle** is made up of two rays that share the same endpoint. The common endpoint is called the **vertex.**

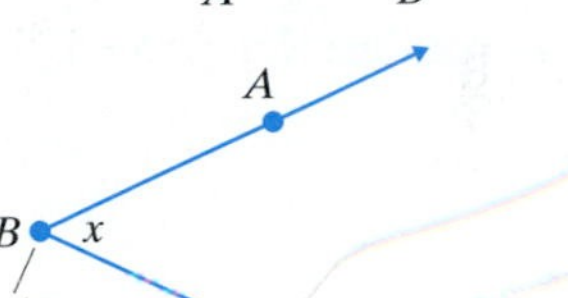

Section 8.2 Perimeter

PERIMETER FORMULAS

Rectangle: $P = 2l + 2w$

Square: $P = 4s$

Triangle: $P = a + b + c$

Circumference of a Circle: $C = 2\pi r$ or $C = \pi d$

where $\pi \approx 3.14$ or $\pi \approx \frac{22}{7}$

Find the perimeter of the rectangle.

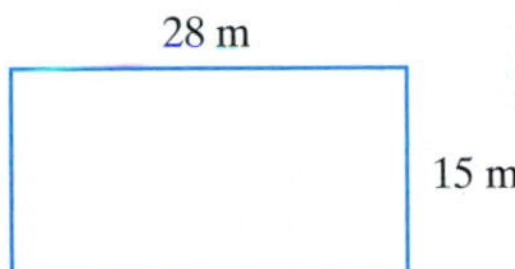

$$\begin{aligned} P &= 2l + 2w \\ &= 2 \cdot 28 \text{ meters} + 2 \cdot 15 \text{ meters} \\ &= 56 \text{ meters} + 30 \text{ meters} \\ &= 86 \text{ meters} \end{aligned}$$

The perimeter is 86 meters.

Section 8.3 Area, Volume, and Surface Area

AREA FORMULAS

Rectangle: $A = lw$

Square: $A = s^2$

Triangle: $A = \frac{1}{2}bh$

Parallelogram: $A = bh$

Trapezoid: $A = \frac{1}{2}(b + B)h$

Circle: $A = \pi r^2$

Find the area of the square.

$$\begin{aligned} A &= s^2 \\ &= (8 \text{ centimeters})^2 \\ &= 64 \text{ square centimeters} \end{aligned}$$

The area of the square is 64 square centimeters.

DEFINITIONS AND CONCEPTS	EXAMPLES

Section 8.3 Area, Volume, and Surface Area (*continued*)

VOLUME FORMULAS

Rectangular Solid: $V = lwh$

Cube: $V = s^3$

Sphere: $V = \frac{4}{3}\pi r^3$

Right Circular Cylinder: $V = \pi r^2 h$

Cone: $V = \frac{1}{3}\pi r^2 h$

Square-Based Pyramid: $V = \frac{1}{3}s^2 h$

Surface Area Formulas: See page 579.

Find the volume of the sphere. Use $\frac{22}{7}$ for π.

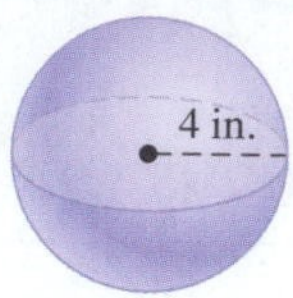

$$V = \frac{4}{3}\pi r^3$$
$$\approx \frac{4}{3}\cdot\frac{22}{7}\cdot(4\text{ inches})^3$$
$$= \frac{4\cdot 22\cdot 64}{3\cdot 7}\text{ cubic inches}$$
$$= \frac{5632}{21} \quad\text{or}\quad 268\frac{4}{21}\text{ cubic inches}$$

Section 8.4 Linear Measurement

To convert from one unit of length to another, multiply by a **unit fraction** in the form

$$\frac{\text{units to convert to}}{\text{original units}}$$

LENGTH: U.S. SYSTEM OF MEASUREMENT

12 inches (in.) = 1 foot (ft)
3 feet = 1 yard (yd)
5280 feet = 1 mile (mi)

The basic unit of length in the metric system is the **meter.** A meter is slightly longer than a yard.

LENGTH: METRIC SYSTEM OF MEASUREMENT

Metric Unit of Length
1 **kilo**meter (km) = 1000 meters (m)
1 **hecto**meter (hm) = 100 m
1 **deka**meter (dam) = 10 m
1 meter (m) = 1 m
1 **deci**meter (dm) = 1/10 m or 0.1 m
1 **centi**meter (cm) = 1/100 m or 0.01 m
1 **milli**meter (mm) = 1/1000 m or 0.001 m

$$\frac{12\text{ inches}}{1\text{ foot}}, \frac{1\text{ foot}}{12\text{ inches}}, \frac{3\text{ feet}}{1\text{ yard}}$$

Convert 6 feet to inches.

$$6\text{ ft} = \frac{6\text{ ft}}{1}\cdot 1$$
$$= \frac{6\not{\text{ft}}}{1}\cdot\frac{12\text{ in.}}{1\not{\text{ft}}} \quad \begin{array}{l}\leftarrow \text{units to convert to}\\ \leftarrow \text{original units}\end{array}$$
$$= 6\cdot 12\text{ in.}$$
$$= 72\text{ in.}$$

Convert 3650 centimeters to meters.

$$3650\text{ cm} = 3650\text{ cm}\cdot 1$$
$$= \frac{3650\not{\text{cm}}}{1}\cdot\frac{0.01\text{ m}}{1\not{\text{cm}}} = 36.5\text{ m}$$

or

km hm dam m dm cm mm
2 units to the left

3650 cm = 36.5 m
2 places to the left

DEFINITIONS AND CONCEPTS	EXAMPLES

Section 8.5 Weight and Mass

Weight is really a measure of the pull of gravity. **Mass** is a measure of the amount of substance in an object and does not change.

Convert 5 pounds to ounces.

$$5 \text{ lb} = 5 \text{ lb} \cdot 1 = \frac{5 \cancel{\text{lb}}}{1} \cdot \frac{16 \text{ oz}}{1 \cancel{\text{lb}}} = 80 \text{ oz}$$

WEIGHT: U.S. SYSTEM OF MEASUREMENT

16 ounces (oz) = 1 pound (1b)
2000 pounds = 1 ton

A **gram** is the basic unit of mass in the metric system. It is the mass of water contained in a cube 1 centimeter on each side. A paper clip weighs about 1 gram.

Convert 260 grams to kilograms.

$$260 \text{ g} = \frac{260 \text{ g}}{1} \cdot 1 = \frac{260 \cancel{\text{g}}}{1} \cdot \frac{1 \text{ kg}}{1000 \cancel{\text{g}}} = 0.26 \text{ kg}$$

or

kg hg dag g dg cg mg
3 units to the left

260 g = 0.260 kg
3 places to the left

MASS: METRIC SYSTEM OF MEASUREMENT

Metric Unit of Mass
1 kilogram (kg) = 1000 grams (g)
1 hectogram (hg) = 100 g
1 dekagram (dag) = 10 g
1 gram (g) = 1 g
1 decigram (dg) = 1/10 g or 0.1 g
1 centigram (cg) = 1/100 g or 0.01 g
1 milligram (mg) = 1/1000 g or 0.001 g

Section 8.6 Capacity

CAPACITY: U.S. SYSTEM OF MEASUREMENT

8 fluid ounces (fl oz) = 1 cup (c)
2 cups = 1 pint (pt)
2 pints = 1 quart (qt)
4 quarts = 1 gallon (gal)

Convert 5 pints to gallons.

1 gal = 4 qt = 8 pt

$$5 \text{ pt} = 5 \text{ pt} \cdot 1 = \frac{5 \cancel{\text{pt}}}{1} \cdot \frac{1 \text{ gal}}{8 \cancel{\text{pt}}} = \frac{5}{8} \text{ gal}$$

The **liter** is the basic unit of capacity in the metric system. It is the capacity or volume of a cube measuring 10 centimeters on each side. A liter of liquid is slightly more than 1 quart.

Convert 1.5 liters to milliliters.

$$1.5 \text{ L} = \frac{1.5 \text{ L}}{1} \cdot 1 = \frac{1.5 \cancel{\text{L}}}{1} \cdot \frac{1000 \text{ ml}}{1 \cancel{\text{L}}} = 1500 \text{ ml}$$

or

kl hl dal L dl cl ml
3 units to the right

1.500 L = 1500 ml
3 places to the right

CAPACITY: METRIC SYSTEM OF MEASUREMENT

Metric Unit of Capacity
1 kiloliter (kl) = 1000 liters (L)
1 hectoliter (hl) = 100 L
1 dekaliter (dal) = 10 L
1 liter (L) = 1 L
1 deciliter (dl) = 1/10 L or 0.1 L
1 centiliter (cl) = 1/100 L or 0.01 L
1 milliliter (ml) = 1/1000 L or 0.001 L

Definitions and Concepts	**Examples**
Section 8.7 Temperature and Conversions Between the U.S. and Metric Systems	
To convert between systems, use approximate unit fractions.	Convert 7 feet to meters. $$7 \text{ ft} \approx \frac{7 \cancel{\text{ft}}}{1} \cdot \frac{0.30 \text{ m}}{1 \cancel{\text{ft}}} = 2.1 \text{ m}$$ Convert 8 liters to quarts. $$8 \text{ L} \approx \frac{8 \cancel{\text{L}}}{1} \cdot \frac{1.06 \text{ qt}}{1 \cancel{\text{L}}} = 8.48 \text{ qt}$$ Convert 363 grams to ounces. $$363 \text{ g} \approx \frac{363 \cancel{\text{g}}}{1} \cdot \frac{0.04 \text{ oz}}{1 \cancel{\text{g}}} = 14.52 \text{ oz}$$
Celsius to Fahrenheit $$F = \frac{9}{5}C + 32 \quad \text{or} \quad F = 1.8C + 32$$	Convert 35°C to degrees Fahrenheit. $$F = \frac{9}{5} \cdot 35 + 32 = 63 + 32 = 95$$ 35°C = 95°F
Fahrenheit to Celsius $$C = \frac{5}{9}(F - 32)$$	Convert 50°F to degrees Celsius. $$C = \frac{5}{9} \cdot (50 - 32) = \frac{5}{9} \cdot (18) = 10$$ 50°F = 10°C

8 CHAPTER REVIEW

(8.1) *Classify each angle as acute, right, obtuse, or straight.*

1.

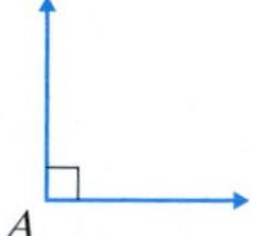

2.

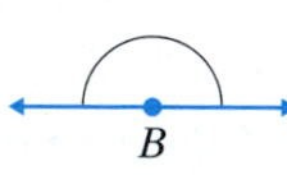

3.

4.

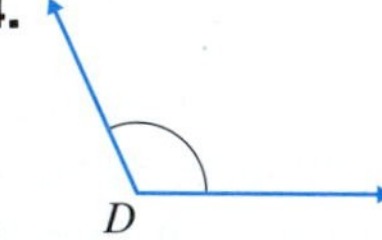

5. Find the complement of a 25° angle.

6. Find the supplement of a 105° angle.

Find the measure of angle x in each figure.

7.

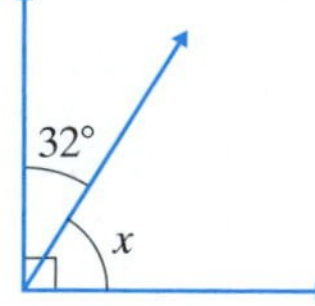

8.

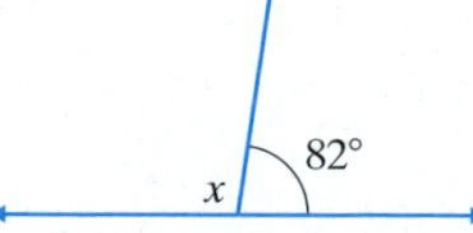

9.

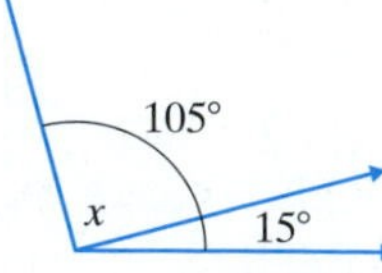

10.

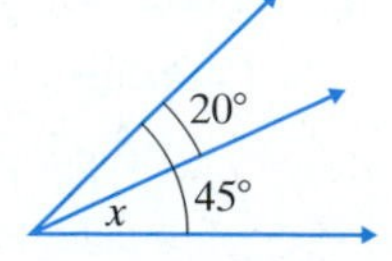

11. Identify the pairs of supplementary angles.

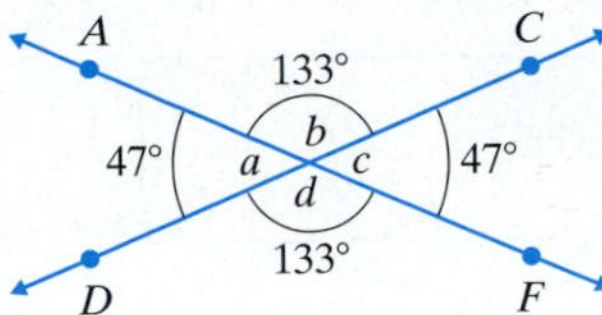

12. Identify the pairs of complementary angles.

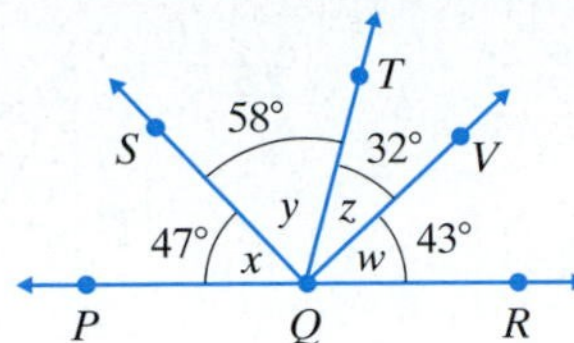

Find the measures of angles x, y, and z in each figure.

13.

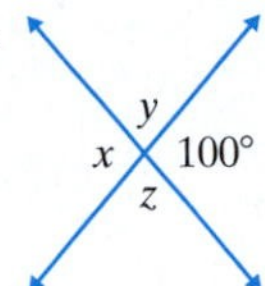

14.

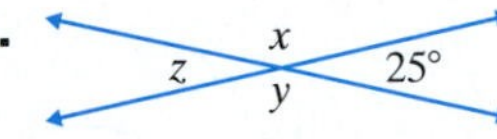

15. Given that $m \parallel n$.

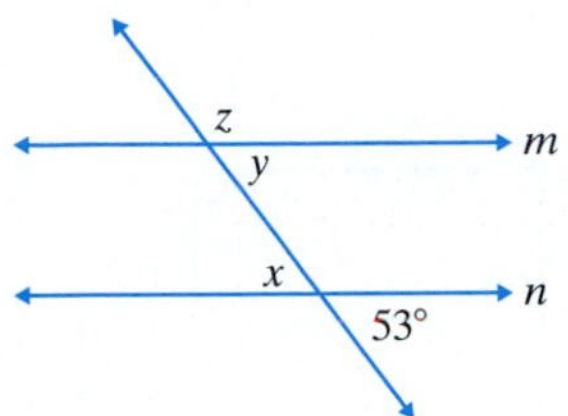

16. Given that $m \parallel n$.

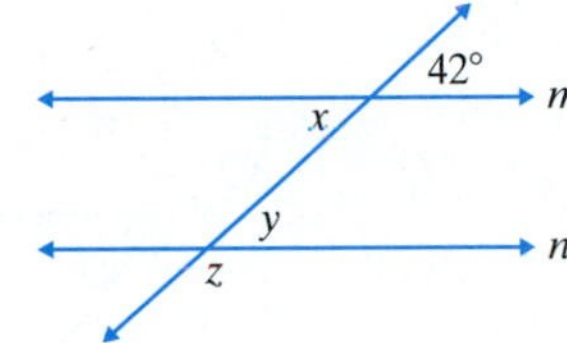

(8.2) *Find the perimeter of each figure.*

17.

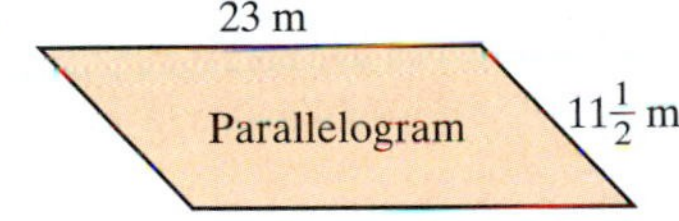

18.

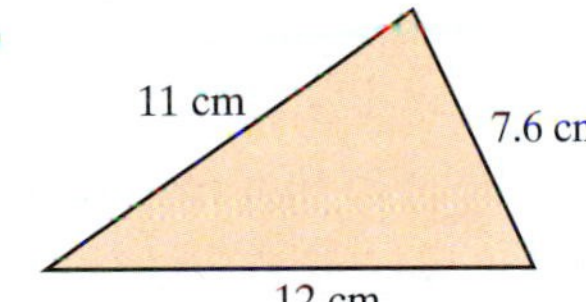

19.

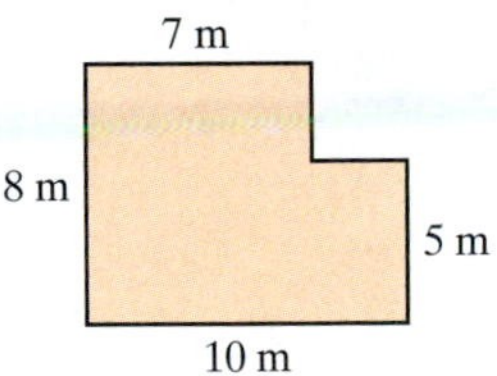

20.

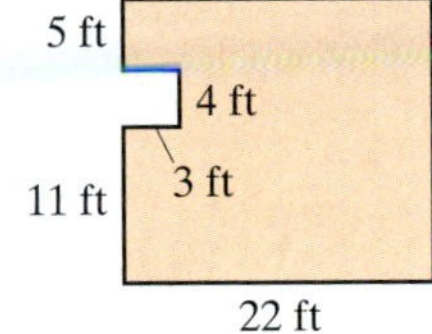

Solve.

21. Find the perimeter of a rectangular sign that measures 6 feet by 10 feet.

22. Find the perimeter of a town square that measures 110 feet on a side.

Find the circumference of each circle. Use $\pi \approx 3.14$.

23.

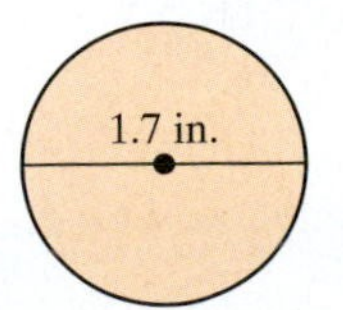

24.

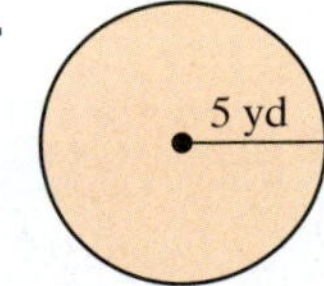

(8.3) *Find the area of each figure. For the circles, find the exact area and then use $\pi \approx 3.14$ to approximate the area.*

25.

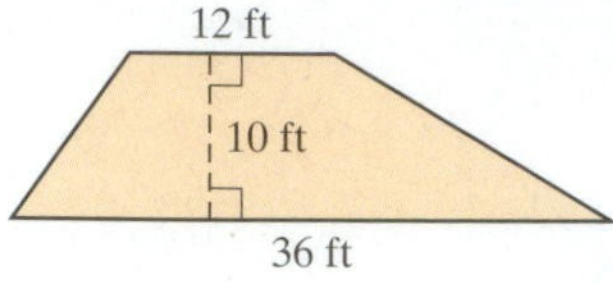

26.

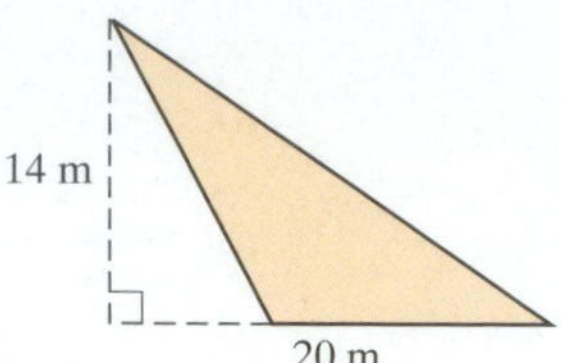

27.

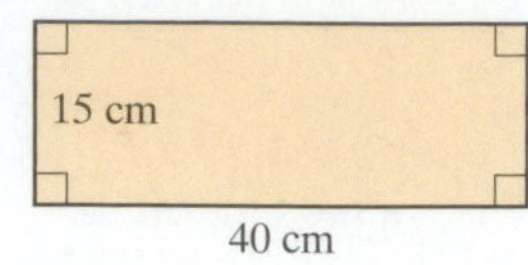

28.

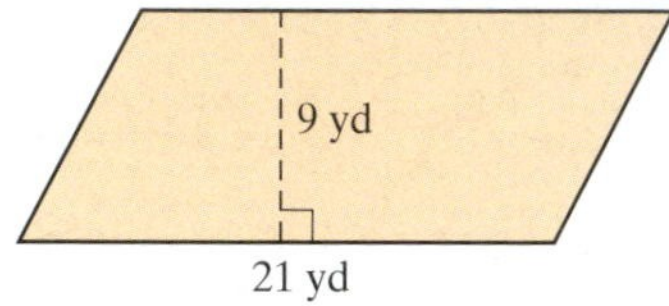

29.

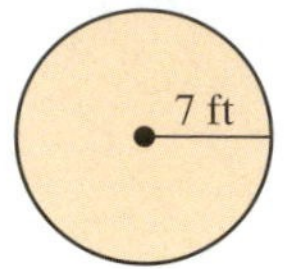

30.

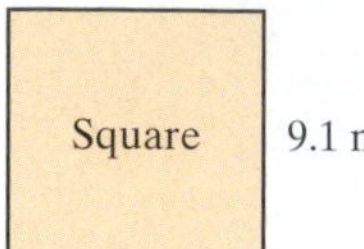

31.

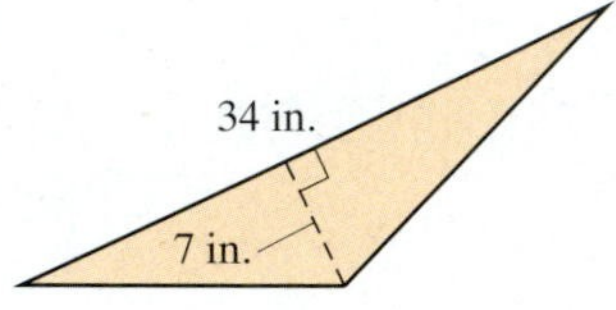

32.

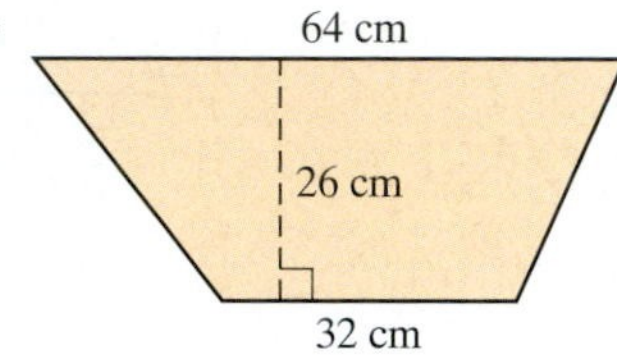

33.

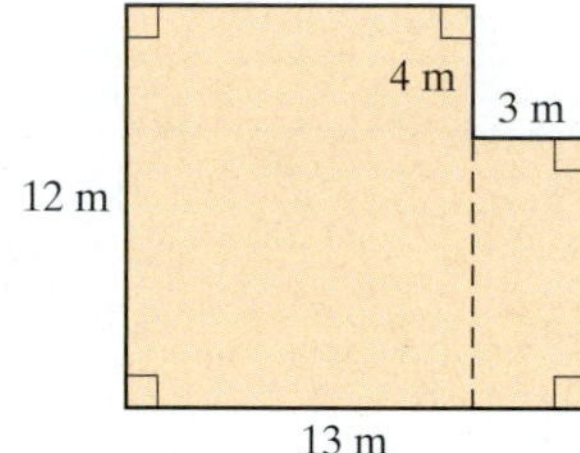

34. The amount of sealer necessary to seal a driveway depends on the area. Find the area of a rectangular driveway 36 feet by 12 feet.

35. Find how much carpet is necessary to cover the floor of the room shown.

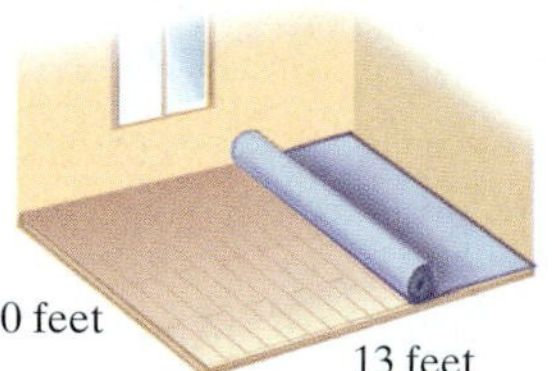

Find the volume and surface area of the solids in Exercises 36 and 37. For Exercises 38 and 39, give an exact volume and an approximation.

36.

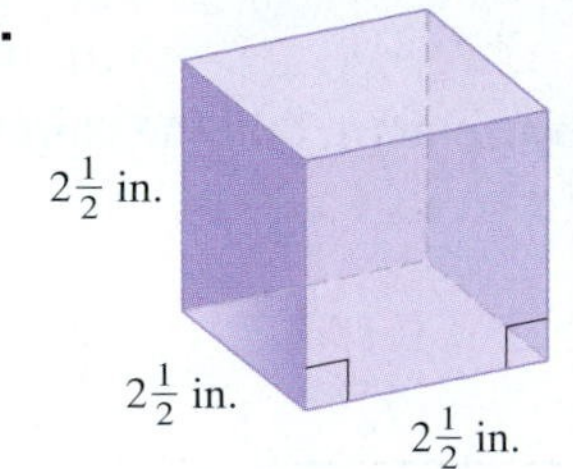

37.

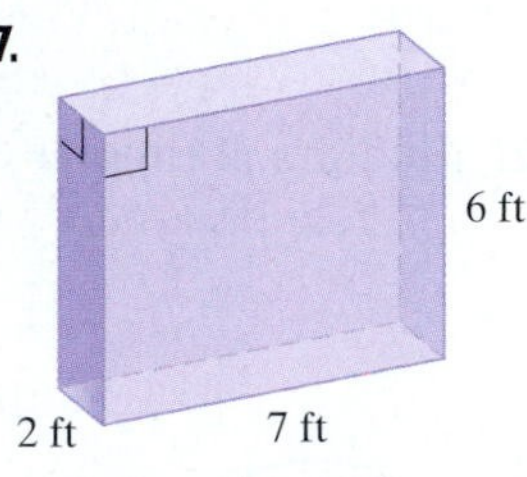

38. Use $\pi \approx 3.14$.

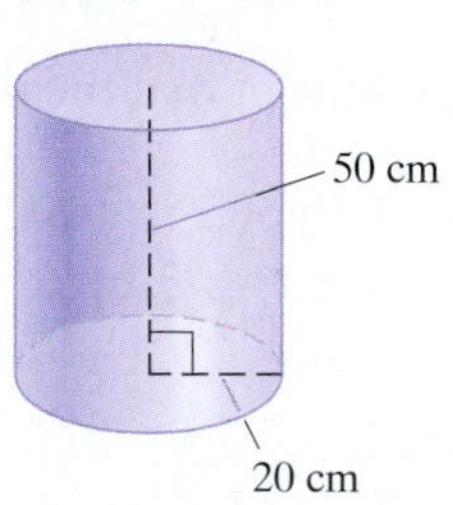

39. Use $\pi \approx \frac{22}{7}$.

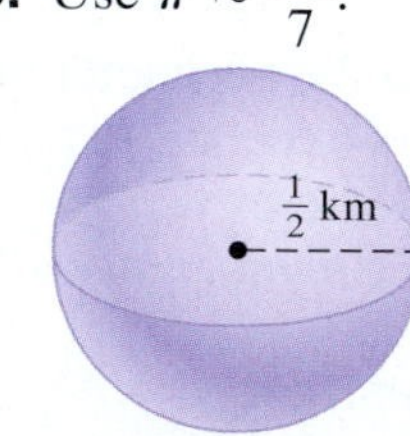

40. Find the volume of a pyramid with a square base 2 feet on a side and a height of 2 feet.

41. Approximate the volume of a tin can 8 inches high and 3.5 inches in radius. Use 3.14 for π.

42. A chest has 3 drawers. If each drawer has inside measurements of $2\frac{1}{2}$ feet by $1\frac{1}{2}$ feet by $\frac{2}{3}$ foot, find the total volume of the 3 drawers.

43. A cylindrical canister for a shop vacuum is 2 feet tall and 1 foot in *diameter*. Find its exact volume.

(8.4) *Convert.*

44. 108 in. to feet

45. 72 ft to yards

46. 1.5 mi to feet

47. $\frac{1}{2}$ yd to inches

48. 52 ft = _____ yd _____ ft

49. 46 in. = _____ ft _____ in.

50. 42 m to centimeters

51. 82 cm to millimeters

52. 12.18 mm to meters

53. 2.31 m to kilometers

Perform each indicated operation.

54. 4 yd 2 ft + 16 yd 2 ft

55. 7 ft 4 in. ÷ 2

56. 8 cm + 15 mm

57. 4 m − 126 cm

Solve.

58. A bolt of cloth contains 333 yd 1 ft of cotton ticking. Find the amount of material that remains after 163 yd 2 ft is removed from the bolt.

59. The student activities club is sponsoring a walk for hunger, and all students who participate will receive a sash with the name of the school to wear on the walk. If each sash requires 5 ft 2 in. of material and there are 50 students participating in the walk, how much material will the student activities club need?

60. The trip from Philadelphia to Washington, D.C., is 217 km each way. Four friends agree to share the driving equally. How far must each drive on this round-trip vacation?

△ **61.** The college has ordered that NO SMOKING signs be placed above the doorway of each classroom. Each sign is 0.8 m long and 30 cm wide. Find the area of each sign. (*Hint:* Recall that the area of a rectangle = width · length.)

0.8 meter

NO SMOKING

30 centimeters

(8.5) *Convert.*

62. 66 oz to pounds

63. 2.3 tons to pounds

64. 52 oz = _____ lb _____ oz.

65. 10,300 lb = _____ tons _____ lb

66. 27 mg to grams

67. 40 kg to grams

68. 2.1 hg to dekagrams

69. 0.03 mg to decigrams

Perform each indicated operation.

70. 6 lb 5 oz − 2 lb 12 oz

71. 8 lb 6 oz × 4

72. 4.3 mg × 5

73. 4.8 kg − 4200 g

Solve.

74. Donshay Berry ordered 1 lb 12 oz of soft-center candies and 2 lb 8 oz of chewy-center candies for his party. Find the total weight of the candy ordered.

75. Four local townships jointly purchase 38 tons 300 lb of cinders to spread on their roads during an ice storm. Determine the weight of the cinders each township receives if they share the purchase equally.

(8.6) *Convert.*

76. 28 pints to quarts

77. 40 fluid ounces to cups

78. 3 qt 1 pt to pints

79. 18 quarts to cups

80. 9 pt = _____ qt _____ pt

81. 15 qt = _____ gal _____ qt

82. 3.8 L to milliliters

83. 14 hl to kiloliters

84. 30.6 L to centiliters

Perform each indicated operation.

85. 1 qt 1 pt + 3 qt 1 pt

86. 3 gal 2 qt × 2

87. 0.946 L − 210 ml

88. 6.1 L + 9400 ml

Solve.

89. Carlos Perez prepares 4 gal 2 qt of iced tea for a block party. During the first 30 minutes of the party, 1 gal 3 qt of the tea is consumed. How much iced tea remains?

90. A recipe for soup stock calls for 1 c 4 fl oz of beef broth. How much should be used if the recipe is cut in half?

91. Each bottle of Kiwi liquid shoe polish holds 85 ml of the polish. Find the number of liters of shoe polish contained in 8 boxes if each box contains 16 bottles.

92. Ivan Miller wants to pour three separate containers of saline solution into a single vat with a capacity of 10 liters. Will 6 liters of solution in the first container combined with 1300 milliliters in the second container and 2.6 liters in the third container fit into the larger vat?

(8.7) *Note: Because approximations are used in this section, your answers may vary slightly from the answers given in the back of the book.*

Convert as indicated. If necessary, round to two decimal places.

93. 7 meters to feet

94. 11.5 yards to meters

95. 17.5 liters to gallons

96. 7.8 liters to quarts

97. 15 ounces to grams

98. 23 pounds to kilograms

99. A compact disc is 1.2 mm thick. Find the height (in inches) of 50 discs.

100. If a person weighs 82 kilograms, how many pounds is this?

Convert. Round to the nearest tenth of a degree, if necessary.

101. 42°C to degrees Fahrenheit

102. 160°C to degrees Fahrenheit

103. 41.3°F to degrees Celsius

104. 80°F to degrees Celsius

Solve. Round to the nearest tenth of a degree, if necessary.

105. A sharp dip in the jet stream caused the temperature in New Orleans to drop to 35°F. Find the corresponding temperature in degrees Celsius.

106. The recipe for meat loaf calls for a 165°C oven. Find the setting used if the oven has a Fahrenheit thermometer.

Mixed Review

Find the following.

107. Find the supplement of a 72° angle.

108. Find the complement of a 1° angle.

Find the measure of angle x in each figure.

109.

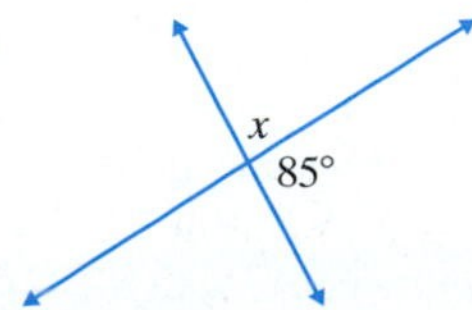

110.

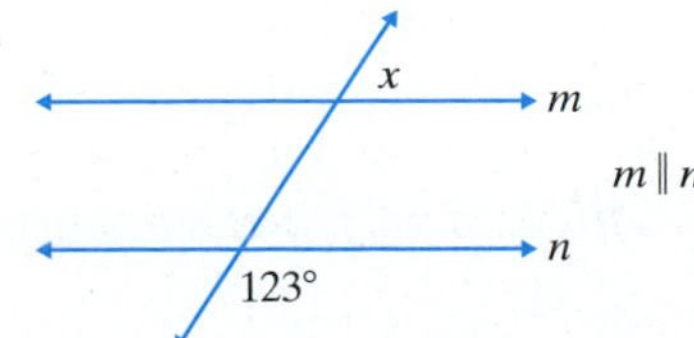

Find the perimeter of each figure.

111.

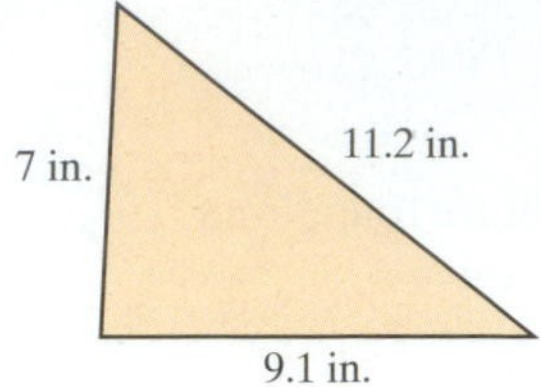

112.

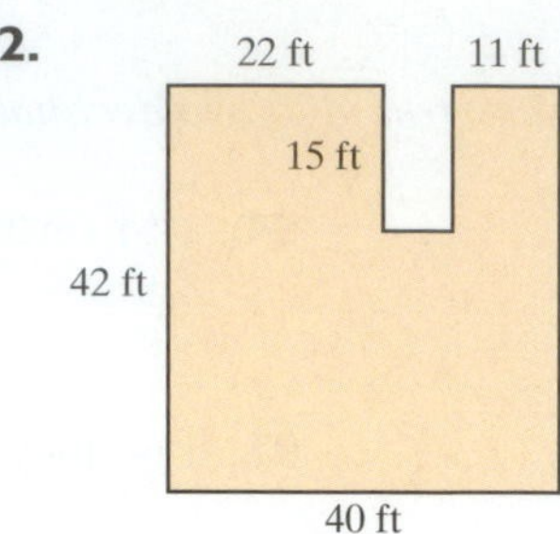

Find the area of each figure. For the circle, find the exact area and then use $\pi = 3.14$ to approximate the area.

113.

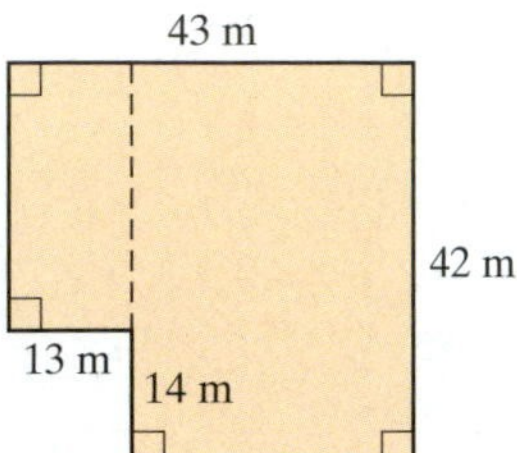

114.

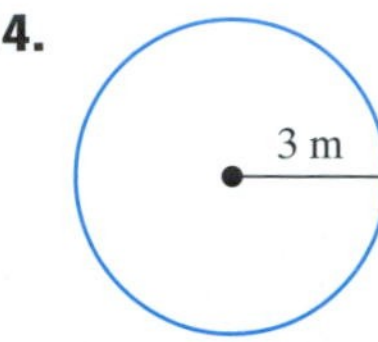

Find the volume of each solid.

115. Give an approximation using $\frac{22}{7}$ for π.

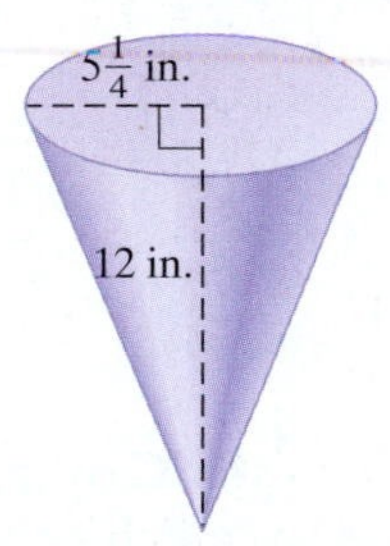

116. Find the surface area, also.

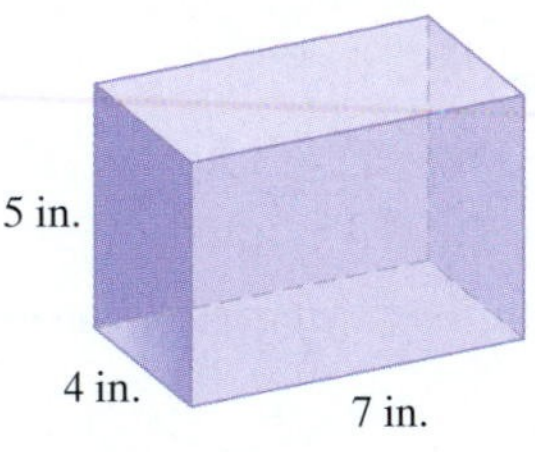

Convert the following.

117. 6.25 ft to inches

118. 8200 lb = _____ tons _____ lb

119. 5 m to centimeters

120. 286 mm to kilometers

121. 1400 mg to grams

122. 6.75 gallons to quarts

123. 86°C to degrees Fahrenheit

124. 51.8°F to degrees Celsius

Perform the indicated operations and simplify.

125. 9.3 km − 183 m

126. 35 L + 700 ml

127. 3 gal 3 qt + 4 gal 2 qt

128. 3.2 kg × 4

8 CHAPTER TEST

Answers

Use the Chapter Test Prep Video CD to see the fully worked-out solutions to any of the exercises you want to review.

△ **1.** Find the complement of a 78° angle.

△ **2.** Find the supplement of a 124° angle.

△ **3.** Find the measure of $\angle x$.

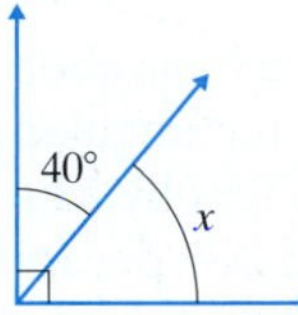

Find the measure of x, y, and z in each figure.

△ **4.**

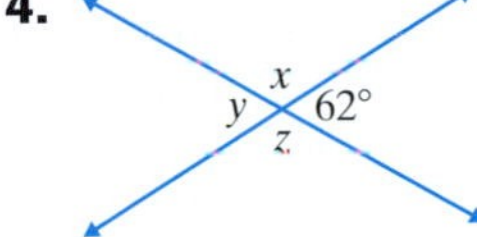

△ **5.**

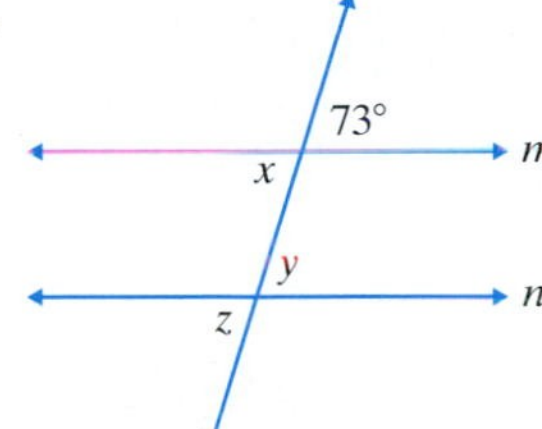

Find the unknown diameter or radius as indicated.

△ **6.**

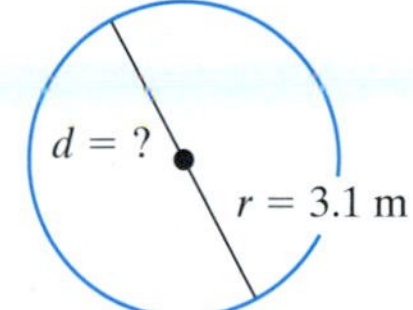

△ **7.**

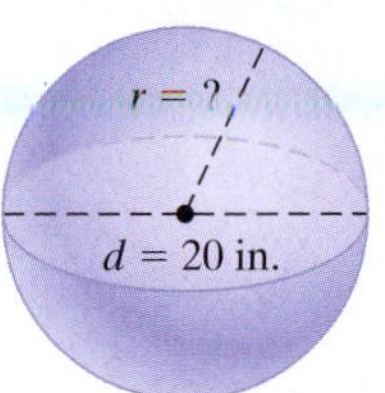

Find the perimeter (or circumference) and area of each figure. For the circle, give the exact value and then use $\pi \approx 3.14$.

△ **8.**

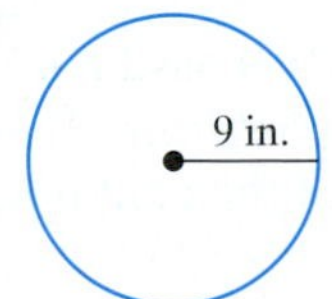

△ **9.**

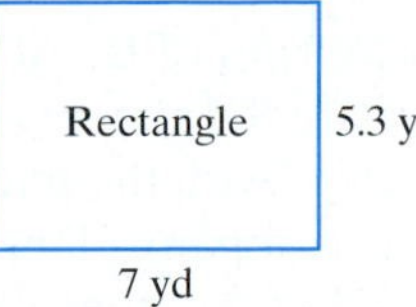

△ **10.**

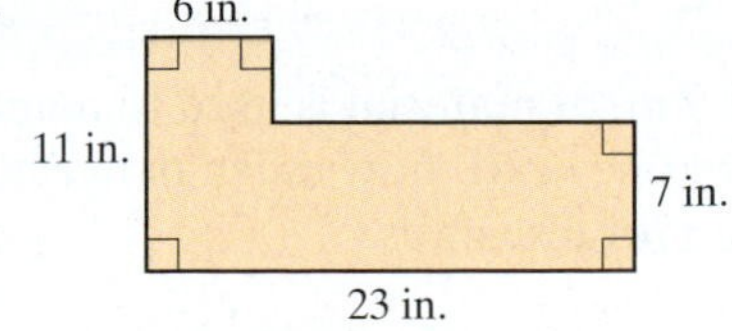

1. ______

2. ______

3. ______

4. ______

5. ______

6. ______

7. ______

8. ______

9. ______

10. ______

Find the volume of each solid. For the cylinder, use $\pi \approx \frac{22}{7}$.

△**11.**

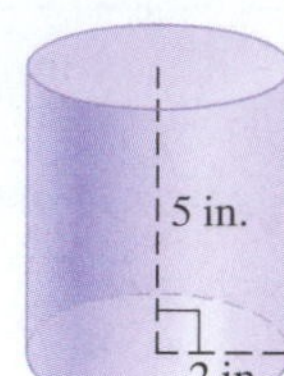

△**12.**

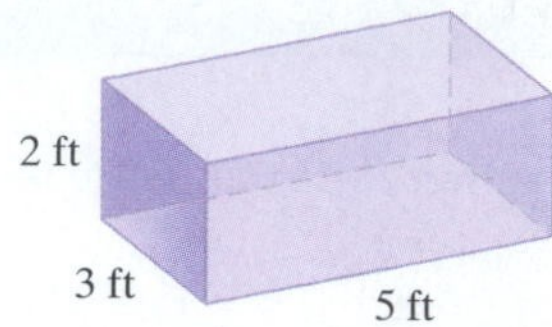

Solve.

△**13.** Find the perimeter of a square frame with a side length of 4 inches.

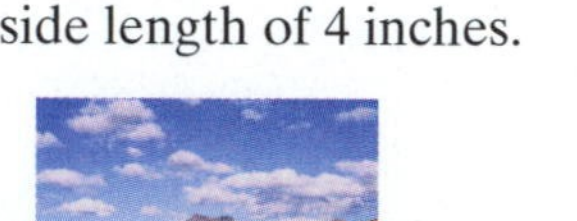

△**14.** How much soil is needed to fill a rectangular hole 3 feet by 3 feet by 2 feet?

15. Find how much baseboard is needed to go around a rectangular room that measures 18 feet by 13 feet. If baseboard costs \$1.87 per foot, also calculate the total cost needed for materials.

Convert.

16. 280 in. = _____ ft _____ in.

17. $2\frac{1}{2}$ gal to quarts

18. 30 oz to pounds

19. 2.8 tons to pounds

20. 38 pt to gallons

21. 40 mg to grams

22. 2.4 kg to grams

23. 3.6 cm to millimeters

24. 4.3 dg to grams

25. 0.83 L to milliliters

Perform each indicated operation.

26. 3 qt 1 pt + 2 qt 1 pt

27. 8 lb 6 oz − 4 lb 9 oz

28. 2 ft 9 in. × 3

29. 5 gal 2 qt ÷ 2

30. 8 cm − 14 mm

31. 1.8 km + 456 m

Convert. Round to the nearest tenth of a degree, if necessary.

32. 84°F to degrees Celsius

33. 12.6°C to degrees Fahrenheit

34. The sugar maples in front of Bette MacMillan's house are 8.4 meters tall. Because they interfere with the phone lines, the telephone company plans to remove the top third of the trees. How tall will the maples be after they are shortened?

35. A total of 15 gal 1 qt of oil has been removed from a 20-gallon drum. How much oil still remains in the container?

36. The engineer in charge of bridge construction said that the span of a certain bridge would be 88 m. But the actual construction required it to be 340 cm longer. Find the span of the bridge, in meters.

37. If 2 ft 9 in. of material is used to manufacture one scarf, how much material is needed for 6 scarves?

11. ______
12. ______
13. ______
14. ______
15. ______
16. ______
17. ______
18. ______
19. ______
20. ______
21. ______
22. ______
23. ______
24. ______
25. ______
26. ______
27. ______
28. ______
29. ______
30. ______
31. ______
32. ______
33. ______
34. ______
35. ______
36. ______
37. ______

CUMULATIVE REVIEW Chapters 1–8

1. Solve: $3a - 6 = a + 4$

2. Solve: $2x + 1 = 3x - 5$.

3. Evaluate:
a. $\left(\frac{2}{5}\right)^4$ **b.** $\left(-\frac{1}{4}\right)^2$

4. Evaluate:
a. $\left(-\frac{1}{3}\right)^3$ **b.** $\left(\frac{3}{7}\right)^2$.

5. Add: $2\frac{4}{5} + 5 + 1\frac{1}{2}$

6. Add: $2\frac{1}{3} + 4\frac{2}{5} + 3$.

7. Simplify by combining like terms:
$11.1x - 6.3 + 8.9x - 4.6$

8. Simplify by combining like terms:
$2.5y + 3.7 - 1.3y - 1.9$

9. Simplify: $\dfrac{5.68 + (0.9)^2 \div 100}{0.2}$

10. Simplify: $\dfrac{0.12 + 0.96}{0.5}$

11. Insert <, >, or = to form a true statement.
$0.\overline{7} \quad \frac{7}{9}$

12. Insert <, >, or = to form a true statement.
$0.43 \quad \frac{2}{5}$

13. Solve: $0.5y + 2.3 = 1.65$

14. Solve: $0.4x - 9.3 = 2.7$

△ **15.** An inner-city park is in the shape of a square that measures 300 feet on a side. Find the length of the diagonal of the park rounded to the nearest whole foot.

△ **16.** A rectangular field is 200 feet by 125 feet. Find the length of the diagonal of the field, rounded to the nearest whole foot.

Answers

1. ______
2. ______
3. a. ______
   b. ______
4. a. ______
   b. ______
5. ______
6. ______
7. ______
8. ______
9. ______
10. ______
11. ______
12. ______
13. ______
14. ______
15. ______
16. ______

17. a. ________

b. ________

18. a. ________

b. ________

19. ________

20. ________

21. ________

22. ________

23. ________

24. ________

25. ________

26. ________

27. ________

28. ________

29. ________

30. ________

31. ________

32. ________

△ **17.** Given the rectangle shown:

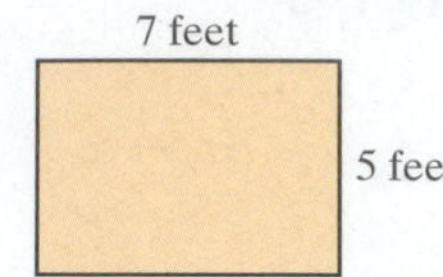

a. Find the ratio of its width to its length.
b. Find the ratio of its length to its perimeter.

△ **18.** A square is 9 inches by 9 inches.

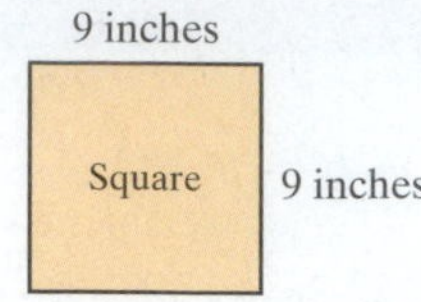

a. Find the ratio of a side to its perimeter.
b. Find the ratio of its perimeter to its area.

19. Determine whether -1 is a solution of the equation $3y + 1 = 3$.

20. Determine whether -5 is a solution of the equation $4(x + 7) = 8$.

21. Solve: $\frac{x}{3} = -2$

22. Solve: $9x = 8x + 1.02$

23. The standard dose of an antibiotic is 4 cc (cubic centimeters) for every 25 pounds (lb) of body weight. At this rate, find the standard dose for a 140-lb woman.

24. A recipe that makes 2 pie crusts calls for 3 cups of flour. How much flour is needed to make 5 pie crusts?

25. In a survey of 100 people, 17 people drive blue cars. What percent of people drive blue cars?

26. Of 100 shoppers surveyed at a mall, 38 paid for their purchases using only cash. What percent of shoppers used only cash to pay for their purchases?

27. 13 is $6\frac{1}{2}\%$ of what number?

28. 54 is $4\frac{1}{2}\%$ of what number?

29. What number is 30% of 9?

30. What number is 42% of 30?

31. The number of applications for a mathematics scholarship at Yale increased from 34 to 45 in one year. What is the percent increase? Round to the nearest whole percent.

32. The price of a gallon of paint rose from \$15 to \$19. Find the percent increase, rounded to the nearest whole percent.

33. Find the sales tax and the total price on a purchase of an $85.50 atlas in a city where the sales tax rate is 7.5%.

34. A sofa has a purchase price of $375. If the sales tax rate is 8%, find the amount of sales tax and the total cost of the sofa.

35. Find the median of the list of numbers: 25, 54, 56, 57, 60, 71, 98

36. Find the median in the list of scores. 60, 95, 89, 72, 83

37. Find the probability of choosing a red marble from a box containing 1 red, 1 yellow, and 2 blue marbles.

38. Find the probability of choosing a nickel at random in a coin purse that contains 2 pennies, 2 nickels, and 3 quarters.

39. Find the complement of a 48° angle.

40. Find the supplement of a 137° angle.

41. Convert 8 feet to inches.

42. Convert 7 yards to feet.

43. Subtract 3 tons 1350 lb from 8 tons 1000 lb.

44. Add 8 lb 15 oz to 9 lb 3 oz.

45. Convert 59°F to degrees Celsius.

46. Convert 86°F to degrees Celsius.

33. ____________

34. ____________

35. ____________

36. ____________

37. ____________

38. ____________

39. ____________

40. ____________

41. ____________

42. ____________

43. ____________

44. ____________

45. ____________

46. ____________

9

Equations, Inequalities, and Problem Solving

In this chapter, we solve equations and inequalities. Once we know how to solve equations and inequalities, we may solve word problems. Of course, problem solving is an integral topic in algebra and its discussion is continued throughout this text.

Since 1948, when NASCAR began, the cars have been transformed from the original "stock" cars, or road models, into the technologically advanced racing machines on the tracks today. In fact, auto manufacturers are creating more advanced street vehicles that can also be used for racing. NASCAR is an increasingly popular sport, with the audience growing daily. In Exercise 35, Section 9.4, you will find the number of points accumulated by the top two finishers for a recent Winston Cup.

SECTIONS

CHECK YOUR PROGRESS

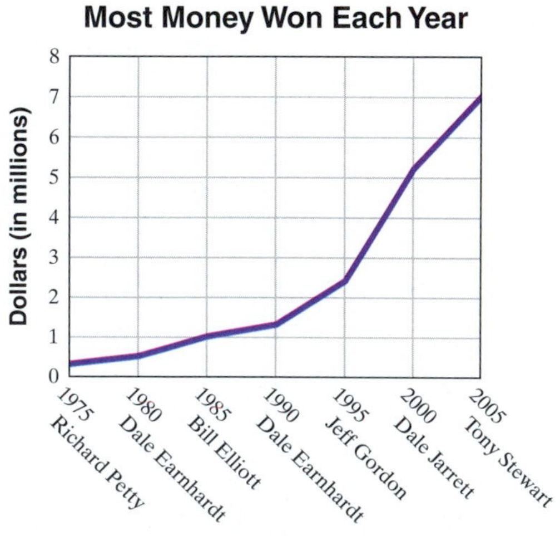

Source: NASCAR

9.1 SYMBOLS AND SETS OF NUMBERS

Objectives

A Define the Meaning of the Symbols $=, \neq, <, >, \leq$, and $\geq$.

B Translate Sentences into Mathematical Statements.

C Identify Integers, Rational Numbers, Irrational Numbers, and Real Numbers.

Throughout the previous chapters, we have studied different sets of numbers. In this section, we review these sets of numbers. We also introduce a few new sets of numbers in order to show the relationships among these common sets of real numbers. We begin with a review of the set of natural numbers and the sets of whole numbers and how we use symbols to compare these numbers. A **set** is a collection of objects, each of which is called a **member** or **element** of the set. A pair of brace symbols { } encloses the list of elements and is translated as "the set of" or "the set containing."

Natural Numbers

$\{1, 2, 3, 4, 5, 6, \ldots\}$

Whole Numbers

$\{0, 1, 2, 3, 4, 5, 6, \ldots\}$

Helpful Hint

The three dots (an ellipsis) at the end of the list of elements of a set means that the list continues in the same manner indefinitely.

Objective A Equality and Inequality Symbols

Picturing natural numbers and whole numbers on a number line helps us to see the order of the numbers. Symbols can be used to describe in writing the order of two quantities. We will use equality symbols and inequality symbols to compare quantities.

Below is a review of these symbols. The letters a and b are used to represent quantities. Letters such as a and b that are used to represent numbers or quantities are called **variables.**

Equality and Inequality Symbols

		Meaning
Equality symbol:	$a = b$	a is equal to b.
Inequality symbols:	$a \neq b$	a is not equal to b.
	$a < b$	a is less than b.
	$a > b$	a is greater than b.
	$a \leq b$	a is less than or equal to b.
	$a \geq b$	a is greater than or equal to b.

These symbols may be used to form **mathematical statements** such as

$2 = 2$ and $2 \neq 6$

Recall that on the number line, we see that a number **to the right of** another number is **larger.** Similarly, a number **to the left of** another number is **smaller.** For example, 3 is to the left of 5 on the number line, which means that 3 is less than 5, or $3 < 5$. Similarly, 2 is to the right of 0 on the number line, which means 2 is greater than 0, or $2 > 0$. Since 0 is to the left of 2, we can also say that 0 is less than 2, or $0 < 2$.

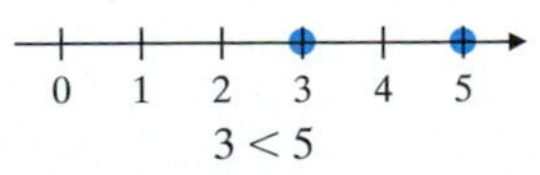

$3 < 5$

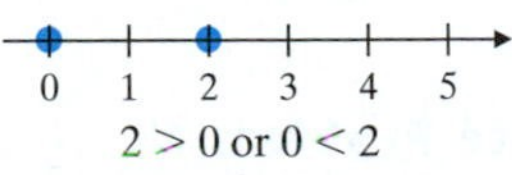

$2 > 0$ or $0 < 2$

Helpful Hint

Recall that $2 > 0$ has exactly the same meaning as $0 < 2$. Switching the order of the numbers and reversing the "direction of the inequality symbol" does not change the meaning of the statement.

$5 > 3$ has the same meaning as $3 < 5$.

Also notice that when the statement is true, the inequality arrow points to the smaller number.

Our discussion above can be generalized in the order property below.

Order Property for Real Numbers

For any two real numbers a and b, a is less than b if a is to the left of b on the number line.

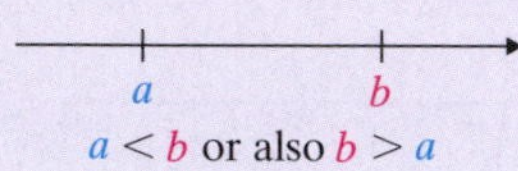

$a < b$ or also $b > a$

PRACTICE PROBLEMS 1–6

Determine whether each statement is true or false.

1. $8 < 6$ **2.** $100 > 10$
3. $21 \leq 21$ **4.** $21 \geq 21$
5. $0 \geq 5$ **6.** $25 \geq 22$

If either $3 < 3$ or $3 = 3$ is true, then $3 \leq 3$ is true.

EXAMPLES Determine whether each statement is true or false.

1. $2 < 3$ True. Since 2 is to the left of 3 on the number line
2. $72 < 27$ False. 72 is to the right of 27 on the number line, so $72 > 27$.
3. $8 \geq 8$ True. Since $8 = 8$ is true
4. $8 \leq 8$ True. Since $8 = 8$ is true
5. $23 \leq 0$ False. Since neither $23 < 0$ nor $23 = 0$ is true
6. $0 \leq 23$ True. Since $0 < 23$ is true

Work Practice Problems 1–6

Objective B Translating Sentences into Mathematical Statements

Now, let's use the symbols discussed above to translate sentences into mathematical statements.

PRACTICE PROBLEM 7

Translate each sentence into a mathematical statement.

a. Fourteen is greater than or equal to fourteen.
b. Zero is less than five.
c. Nine is not equal to ten.

EXAMPLE 7 Translate each sentence into a mathematical statement.

a. Nine is less than or equal to eleven. **b.** Eight is greater than one.
c. Three is not equal to four.

Solution:

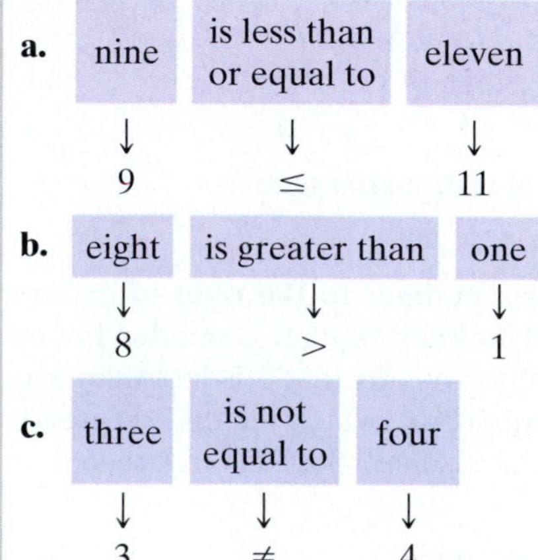

Work Practice Problem 7

Answers

1. false **2.** true **3.** true
4. true **5.** false **6.** true
7. a. $14 \geq 14$ **b.** $0 < 5$ **c.** $9 \neq 10$

Objective C Identifying Common Sets of Numbers

Whole numbers are not sufficient to describe many situations in the real world. For example, quantities smaller than zero must sometimes be represented, such as temperatures less than 0 degrees.

Recall that we can place numbers less than zero on the number line as follows: Numbers less than 0 are to the left of 0 and are labeled −1, −2, −3, and so on. The numbers we have labeled on the number line below are called the set of **integers.**

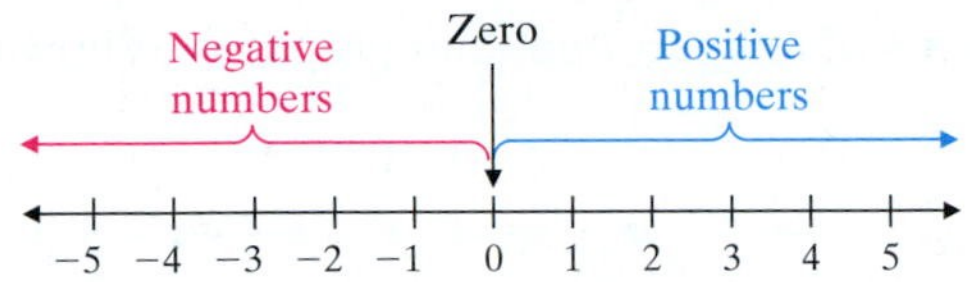

Integers to the left of 0 are called **negative integers;** integers to the right of 0 are called **positive integers.** The integer 0 is neither positive nor negative.

Integers

$\{\ldots, -3, -2, -1, 0, 1, 2, 3, \ldots\}$

Helpful Hint

A − sign, such as the one in −2, tells us that the number is to the left of 0 on the number line.

−2 is read "negative two."

A + sign or no sign tells us that a number lies to the right of 0 on the number line. For example, 3 and +3 both mean positive three.

EXAMPLE 8

Use an integer to express the number in the following. "The lowest temperature ever recorded at South Pole Station, Antarctica, occurred during the month of June. The record-low temperature was 117 degrees below zero." (*Source:* The National Oceanic and Atmospheric Administration)

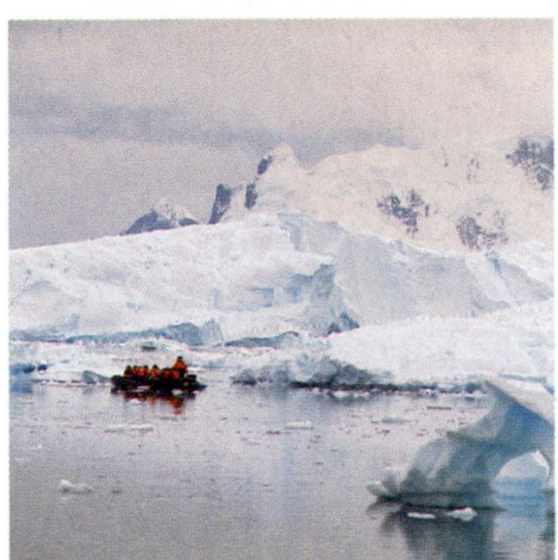

Solution: The integer −117 represents 117 degrees below zero.

Work Practice Problem 8

PRACTICE PROBLEM 8

Use an integer to express the number in the following. Due to hurricane Katrina, the city of New Orleans suffered massive flooding. Under normal conditions, its elevation is 8 feet below sea level. (*Source: The World Almanac*)

Answer

8. −8

A problem with integers in real-life settings arises when quantities are smaller than some integer but greater than the next smallest integer. On the number line, these quantities may be visualized by points between integers. Some of these quantities between integers can be represented as a quotient of integers. For example,

The point on the number line halfway between 0 and 1 can be represented by $\frac{1}{2}$, a quotient of integers.

The point on the number line halfway between 0 and −1 can be represented by $-\frac{1}{2}$. Other quotients of integers and their graphs are shown below.

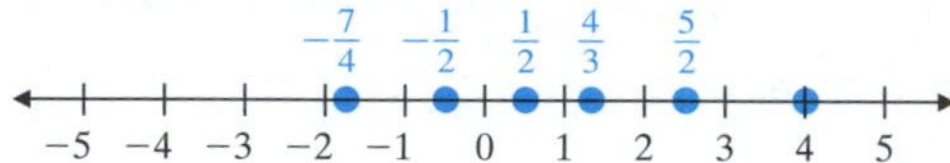

These numbers, each of which can be represented as a quotient of integers, are examples of **rational numbers.** It's not possible to list the set of rational numbers using the notation that we have been using. For this reason, we will use a different notation.

Rational Numbers

$$\left\{\frac{a}{b}\,\middle|\, a \text{ and } b \text{ are integers and } b \neq 0\right\}$$

We read this set as "the set of numbers $\frac{a}{b}$ such that a and b are integers and ***b* is not equal to 0.**"

We commonly refer to rational numbers as fractions.

Notice that every integer is also a rational number since each integer can be written as a quotient of integers. For example, the integer 5 is also a rational number since $5 = \frac{5}{1}$. For the rational number $\frac{5}{1}$, recall that the top number, 5, is called the numerator and the bottom number, 1, is called the denominator.

Let's practice **graphing** numbers on a number line.

PRACTICE PROBLEM 9

Graph the numbers on the number line.

$$-2\frac{1}{2}, \quad -\frac{2}{3}, \quad \frac{1}{5}, \quad \frac{5}{4}, \quad 2.25$$

−5 −4 −3 −2 −1 0 1 2 3 4 5

EXAMPLE 9 Graph the numbers on a number line.

$$-\frac{4}{3}, \quad \frac{1}{4}, \quad \frac{3}{2}, \quad -2\frac{1}{8}, \quad 3.5$$

Solution: To help graph the improper fractions in the list, we first write them as mixed numbers.

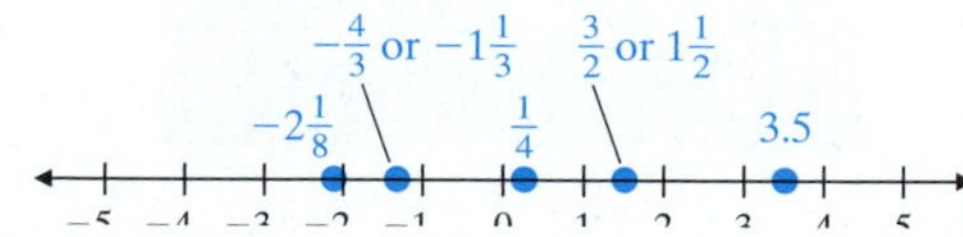

Work Practice Problem 9

Every rational number has a point on the number line that corresponds to it. But not every point on the number line corresponds to a rational number. Those points that do not correspond to rational numbers correspond instead to **irrational numbers.**

Answer

9. $-2\frac{1}{2}$ $-\frac{2}{3}$ $\frac{1}{5}$ $\frac{5}{4}$ 2.25
−5 −4 −3 −2 −1 0 1 2 3 4 5

Irrational Numbers

{Nonrational numbers that correspond to points on the number line}

An irrational number that you have probably seen is π. Also, $\sqrt{2}$, the length of the diagonal of the square shown below, is an irrational number.

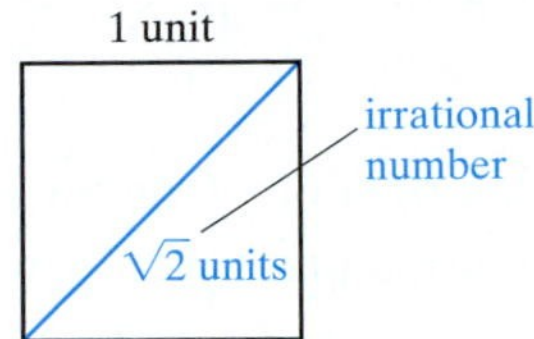

Both rational and irrational numbers can be written as decimal numbers. The decimal equivalent of a rational number will either terminate or repeat in a pattern. For example, upon dividing we find that

$\frac{3}{4} = 0.75$ (Decimal number terminates or ends.)

$\frac{2}{3} = 0.66666\ldots$ (Decimal number repeats in a pattern.)

The decimal representation of an irrational number will neither terminate nor repeat. (For further review of decimals, see Section 6.2.)

The set of numbers, each of which corresponds to a point on the number line, is called the set of **real numbers.** One and only one point on the number line corresponds to each real number.

Real Numbers

{All numbers that correspond to points on the number line}

Several different sets of numbers have been discussed in this section. The following diagram shows the relationships among these sets of real numbers. Notice that, together, the rational numbers and the irrational numbers make up the real numbers.

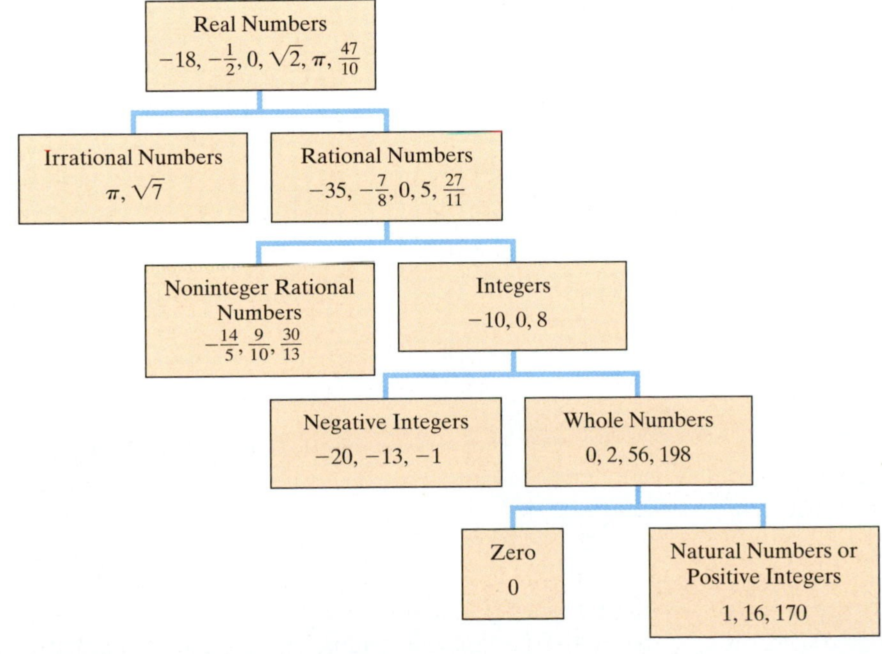

Now that other sets of numbers have been reviewed, let's continue our practice of comparing numbers.

PRACTICE PROBLEM 10

Insert <, >, or = between pairs of numbers to form true statements.

a. $-11 \quad -9$

b. $4.511 \quad 4.151$

c. $\frac{7}{8} \quad \frac{2}{3}$

EXAMPLE 10 Insert <, >, or = between the pairs of numbers to form true statements.

a. $-5 \quad -6$ **b.** $3.195 \quad 3.2$ **c.** $\frac{1}{4} \quad \frac{1}{3}$

Solution:

a. $-5 > -6$ since -5 lies to the right of -6 on the number line.

b. By comparing digits in the same place values, we find that $3.195 < 3.2$. Since $0.1 < 0.2$.

c. By dividing, we find that $\frac{1}{4} = 0.25$ and $\frac{1}{3} = 0.33\ldots$. Since $0.25 < 0.33\ldots$, then $\frac{1}{4} < \frac{1}{3}$.

Work Practice Problem 10

PRACTICE PROBLEM 11

Given the set $\{-100, -\frac{2}{5}, 0, \pi, 6, 913\}$, list the numbers in this set that belong to the set of:

a. Natural numbers

b. Whole numbers

c. Integers

d. Rational numbers

e. Irrational numbers

f. Real numbers

EXAMPLE 11 Given the set $\left\{-2, 0, \frac{1}{4}, 112, -3, 11, \sqrt{2}\right\}$, list the numbers in this set that belong to the set of:

a. Natural numbers **b.** Whole numbers **c.** Integers
d. Rational numbers **e.** Irrational numbers **f.** Real numbers

Solution:

a. The natural numbers are 11 and 112.

b. The whole numbers are 0, 11, and 112.

c. The integers are $-3, -2, 0, 11$, and 112.

d. Recall that integers are rational numbers also. The rational numbers are $-3, -2, 0, \frac{1}{4}, 11$, and 112.

e. The irrational number is $\sqrt{2}$.

f. All numbers in the given set are real numbers.

Work Practice Problem 11

Answers

10. a. < **b.** > **c.** >
11. a. 6, 913 **b.** 0, 6, 913
c. $-100, 0, 6, 913$
d. $-100, -\frac{2}{5}, 0, 6, 913$ **e.** π
f. all numbers in the given set

Vocabulary and Readiness Check

Use the choices below to fill in each blank.

real	natural	whole
rational	inequality	integers

1. The ________ numbers are $\{0, 1, 2, 3, 4, \ldots\}$.
2. The ________ numbers are $\{1, 2, 3, 4, 5, \ldots\}$.
3. The symbols $\neq$, $\leq$, and $>$ are called ____________ symbols.
4. The ________ are $\{\ldots, -3, -2, -1, 0, 1, 2, 3, \ldots\}$.
5. The ________ numbers are {all numbers that correspond to points on the number line}.
6. The __________ numbers are $\left\{ \frac{a}{b} \middle| a \text{ and } b \text{ are integers}, b \neq 0 \right\}$.

9.1 EXERCISE SET

FOR EXTRA HELP

Objectives **A C Mixed Practice** *Insert $<$, $>$, or $=$ in the space between the paired numbers to make each statement true. See Examples 1 through 6, and 10.*

1. 4 10 **2.** 8 5 **3.** 7 3 **4.** 9 15

5. 6.26 6.26 **6.** 1.13 1.13 **7.** 0 7 **8.** 20 0

9. The freezing point of water is 32° Fahrenheit. The boiling point of water is 212° Fahrenheit. Write an inequality statement using $<$ or $>$ comparing the numbers 32 and 212.

10. The freezing point of water is 0° Celsius. The boiling point of water is 100° Celsius. Write an inequality statement using $<$ or $>$ comparing the numbers 0 and 100.

△ **11.** An angle measuring 30° and an angle measuring 45° are shown. Use the inequality symbols $\leq$ or $\geq$ to write a statement comparing the numbers 30 and 45.

30° 45°

△ **12.** The sum of the measures of the angles of a triangle is 180°. The sum of the measures of the angles of a parallelogram is 360°. Use the inequality symbols $\leq$ or $\geq$ to write a statement comparing the numbers 360 and 180.

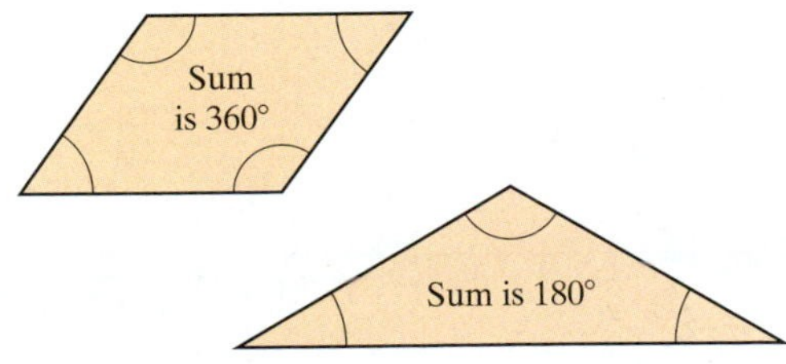

Determine whether each statement is true or false. See Examples 1 through 6 and 10.

13. $11 \leq 11$ **14.** $8 \geq 9$ **15.** $-11 > -10$ **16.** $-16 > -17$

17. $5.092 < 5.902$ **18.** $1.02 > 1.021$ **19.** $\frac{9}{10} \leq \frac{8}{9}$ **20.** $\frac{4}{5} \leq \frac{9}{11}$

Rewrite each inequality so that the inequality symbol points in the opposite direction and the resulting statement has the same meaning as the given one.

21. $25 \geq 20$ **22.** $-13 \leq 13$ **23.** $0 < 6$

24. $5 > 3$ **25.** $-10 > -12$ **26.** $-4 < -2$

Objectives B C Mixed Practice–Translating *Write each sentence as a mathematical statement. See Examples 7 and 10.*

27. Seven is less than eleven.

28. Twenty is greater than two.

29. Five is greater than or equal to four.

30. Negative ten is less than or equal to thirty-seven.

31. Fifteen is not equal to negative two.

32. Negative seven is not equal to seven.

Use integers to represent the values in each statement. See Example 8.

33. The highest elevation in California is Mt. Whitney with an altitude of 14,494 feet. The lowest elevation in California is Death Valley with an altitude of 282 feet below sea level. (*Source:* U.S. Geological Survey)

34. Driskill Mountain, in Louisiana, has an altitude of 535 feet. New Orleans, Louisiana, lies 8 feet below sea level. (*Source:* U.S. Geological Survey)

35. The number of students admitted to the Class of 2008 at UCLA was 43,413 fewer students than the number that applied. (*Source:* UCLA)

36. From 1990 to 2000, the population of Washington, D.C., decreased by 34,841. (*Source:* U.S. Census Bureau)

37. Gretchen Bertani deposited \$475 in her savings account. She later withdrew \$195.

38. David Lopez was deep-sea diving. During his dive, he ascended 17 feet and later descended 15 feet.

Graph each set of numbers on the number line. See Example 9.

39. $-4, 0, 2, -2$

−5 −4 −3 −2 −1 0 1 2 3 4 5

40. $-3, 0, 1, -5$

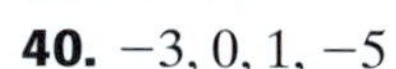

−5 −4 −3 −2 −1 0 1 2 3 4 5

41. $-2, 4, \frac{1}{3}, -\frac{1}{4}$

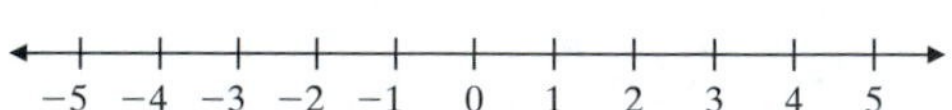

42. $-5, 3, -\frac{1}{3}, \frac{7}{8}$

−5 −4 −3 −2 −1 0 1 2 3 4 5

43. $-4.5, \frac{7}{4}, 3.25, -\frac{3}{2}$

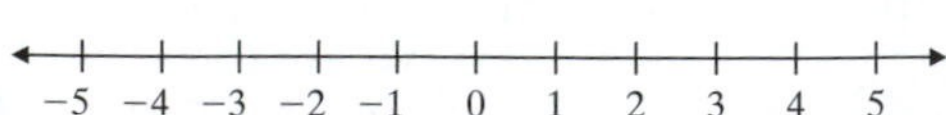

44. $4.5, -\frac{9}{4}, 1.75, -\frac{7}{2}$

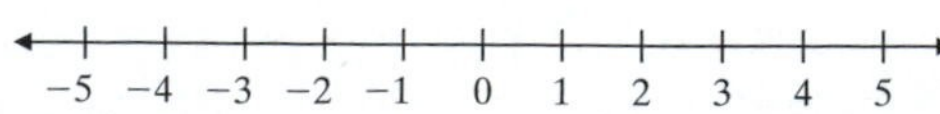

Tell which set or sets each number belongs to: natural numbers, whole numbers, integers, rational numbers, irrational numbers, and real numbers. See Example 11.

45. 0

46. $\frac{1}{4}$

47. -7

48. $-\frac{1}{7}$

49. 265

50. 7941

51. $\frac{2}{3}$

52. $\sqrt{3}$

Determine whether each statement is true or false.

53. Every rational number is also an integer.

54. Every natural number is positive.

55. 0 is a real number.

56. $\frac{1}{2}$ is an integer.

57. Every negative number is also a rational number.

58. Every rational number is also a real number.

59. Every real number is also a rational number.

60. Every whole number is an integer.

Review

Insert <, >, or = in the appropriate space to make each statement true. See Section 2.1.

61. $|-5| \quad -4$

62. $|-12| \quad |0|$

63. $\left|-\frac{5}{8}\right| \quad \left|\frac{5}{8}\right|$

64. $\left|\frac{2}{5}\right| \quad \left|-\frac{2}{5}\right|$

65. $|-2| \quad |-2.7|$

66. $|-5.01| \quad |-5|$

67. $|0| \quad |-8|$

68. $|-12| \quad \frac{-24}{2}$

Concept Extensions

The graph below is called a bar graph. This graph shows apple production in Massachusetts from 1998 through 2003. Each bar represents a different year, and the height of each bar represents the apple production for that year in thousands of bushels. (A bushel is 42 lb.)

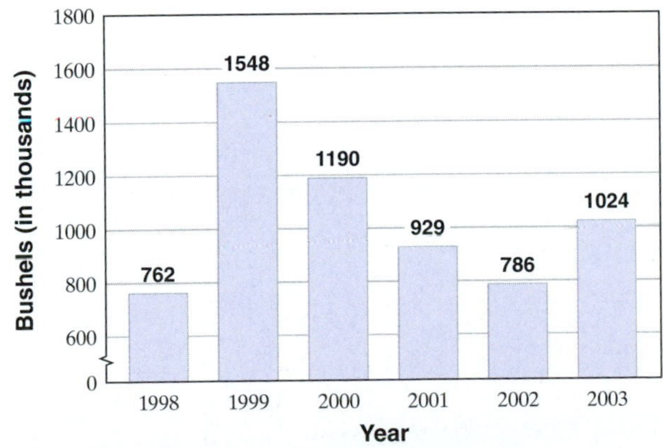

(Note: The ⚡ symbol means that some numbers are missing. Along the vertical data line, notice the numbers between 0 and 600 are missing or not shown.) (Source: New England Agriculture Statistical Service.)

69. Write an inequality comparing the apple production in 1998 with the apple production in 1999.

70. Write an inequality comparing the apple production in 2003 with the apple production in 2002.

71. Determine the change in apple production between 2000 and 2001.

72. According to the bar graph, which year produced the largest crops?

The apparent magnitude of a star is the measure of its brightness as seen by someone on Earth. The smaller the apparent magnitude, the brighter the star. Below, the apparent magnitudes of some stars are listed. Use this table to answer Exercises 73 through 78.

Star	Apparent Magnitude	Star	Apparent Magnitude
Arcturus	−0.04	Spica	0.98
Sirius	−1.46	Rigel	0.12
Vega	0.03	Regulus	1.35
Antares	0.96	Canopus	−0.72
Sun	−26.7	Hadar	0.61

(*Source: Norton's 2000.0: Star Atlas and Reference Handbook,* 18th ed., Longman Group, UK, 1989)

73. The apparent magnitude of the sun is −26.7. The apparent magnitude of the star Arcturus is −0.04. Write an inequality statement comparing the numbers −0.04 and −26.7.

74. The apparent magnitude of Antares is 0.96. The apparent magnitude of Spica is 0.98. Write an inequality statement comparing the numbers 0.96 and 0.98.

75. Which is brighter, the sun or Arcturus?

76. Which is dimmer, Antares or Spica?

77. Which star listed is the brightest?

78. Which star listed is the dimmest?

79. In your own words, explain how to find the absolute value of a number.

80. Give an example of a real-life situation that can be described with integers but not with whole numbers.

9.2 PROPERTIES OF REAL NUMBERS

Objectives

A Use the Commutative and Associative Properties.

B Use the Distributive Property.

C Use the Identity and Inverse Properties.

Objective A Using the Commutative and Associative Properties

In this section we review names to properties of real numbers with which we are already familiar. Throughout this section, the variables a, b, and c represent real numbers.

We know that order does not matter when adding numbers. For example, we know that $7 + 5$ is the same as $5 + 7$. This property is given a special name—the **commutative property of addition.** We also know that order does not matter when multiplying numbers. For example, we know that $-5(6) = 6(-5)$. This property means that multiplication is commutative also and is called the **commutative property of multiplication.**

Commutative Properties

Addition: $a + b = b + a$

Multiplication: $a \cdot b = b \cdot a$

These properties state that the *order* in which any two real numbers are added or multiplied does not change their sum or product. For example, if we let $a = 3$ and $b = 5$, then the commutative properties guarantee that

$3 + 5 = 5 + 3$ and $3 \cdot 5 = 5 \cdot 3$

Helpful Hint

Is subtraction also commutative? Try an example. Is $3 - 2 = 2 - 3$? **No!** The left side of this statement equals 1; the right side equals -1. There is no commutative property of subtraction. Similarly, there is no commutative property of division. For example, $10 \div 2$ does not equal $2 \div 10$.

EXAMPLE 1 Use a commutative property to complete each statement.

a. $x + 5 =$ _____ **b.** $3 \cdot x =$ _____

Solution:

a. $x + 5 = 5 + x$ By the commutative property of addition

b. $3 \cdot x = x \cdot 3$ By the commutative property of multiplication

Work Practice Problem 1

PRACTICE PROBLEM 1

Use a commutative property to complete each statement.

a. $7 \cdot y =$ _____

b. $4 + x =$ _____

✔**Concept Check** Which of the following pairs of actions are commutative?

a. "raking the leaves" and "bagging the leaves"

b. "putting on your left glove" and "putting on your right glove"

c. "putting on your coat" and "putting on your shirt"

d. "reading a novel" and "reading a newspaper"

Answers

1. a. $y \cdot 7$ **b.** $x + 4$

✔ **Concept Check Answer**

b, d

Let's now discuss grouping numbers. When we add three numbers, the way in which they are grouped or associated does not change their sum. For example, we know that $2 + (3 + 4) = 2 + 7 = 9$. This result is the same if we group the numbers differently. In other words, $(2 + 3) + 4 = 5 + 4 = 9$, also. Thus, $2 + (3 + 4) = (2 + 3) + 4$. This property is called the **associative property of addition.**

In the same way, changing the grouping of numbers when multiplying does not change their product. For example, $2 \cdot (3 \cdot 4) = (2 \cdot 3) \cdot 4$ (check it). This is the **associative property of multiplication.**

Associative Properties

Addition: $(a + b) + c = a + (b + c)$

Multiplication: $(a \cdot b) \cdot c = a \cdot (b \cdot c)$

These properties state that the way in which three numbers are *grouped* does not change their sum or their product.

PRACTICE PROBLEM 2

Use an associative property to complete each statement.

a. $5 \cdot (-3 \cdot 6) =$ _____

b. $(-2 + 7) + 3 =$ _____

c. $(q + r) + 17 =$ _____

d. $(ab) \cdot 21 =$ _____

EXAMPLE 2 Use an associative property to complete each statement.

a. $5 + (4 + 6) =$ _____ **b.** $(-1 \cdot 2) \cdot 5 =$ _____

c. $(m + n) + 9 =$ _____ **d.** $(xy) \cdot 12 =$ _____

Solution:

a. $5 + (4 + 6) = (5 + 4) + 6$ By the associative property of addition

b. $(-1 \cdot 2) \cdot 5 = -1 \cdot (2 \cdot 5)$ By the associative property of multiplication

c. $(m + n) + 9 = m + (n + 9)$ By the associative property of addition

d. $(xy) \cdot 12 = x \cdot (y \cdot 12)$ Recall that xy means $x \cdot y$.

Work Practice Problem 2

Helpful Hint

Remember the difference between the commutative properties and the associative properties. The commutative properties have to do with the *order* of numbers and the associative properties have to do with the *grouping* of numbers.

PRACTICE PROBLEMS 3–4

Determine whether each statement is true by an associative property or a commutative property.

3. $5 \cdot (4 \cdot 7) = 5 \cdot (7 \cdot 4)$

4. $-2 + (4 + 9) = (-2 + 4) + 9$

EXAMPLES

Determine whether each statement is true by an associative property or a commutative property.

3. $(7 + 10) + 4 = (10 + 7) + 4$ Since the order of two numbers was changed and their grouping was not, this is true by the commutative property of addition.

4. $2 \cdot (3 \cdot 1) = (2 \cdot 3) \cdot 1$ Since the grouping of the numbers was changed and their order was not, this is true by the associative property of multiplication.

Work Practice Problems 3–4

Let's now illustrate how these properties can help us simplify expressions.

Answers

2. a. $(5 \cdot -3) \cdot 6$ **b.** $-2 + (7 + 3)$ **c.** $q + (r + 17)$ **d.** $a \cdot (b \cdot 21)$ **3.** commutative **4.** associative

EXAMPLES Simplify each expression.

5. $10 + (x + 12) = 10 + (12 + x)$ By the commutative property of addition
$= (10 + 12) + x$ By the associative property of addition
$= 22 + x$ Add.

6. $-3(7x) = (-3 \cdot 7)x$ By the associative property of multiplication
$= -21x$ Multiply.

Work Practice Problems 5–6

PRACTICE PROBLEMS 5–6

Simplify each expression.

5. $(-3 + x) + 17$
6. $4(5x)$

Objective B Using the Distributive Property

The **distributive property of multiplication over addition** is used repeatedly throughout algebra. It is useful because it allows us to write a product as a sum or a sum as a product.

We know that $7(2 + 4) = 7(6) = 42$. Compare that with

$$7(2) + 7(4) = 14 + 28 = 42$$

Since both original expressions equal 42, they must equal each other, or

$$7(2 + 4) = 7(2) + 7(4)$$

This is an example of the distributive property. The product on the left side of the equals sign is equal to the sum on the right side. We can think of the 7 as being distributed to each number inside the parentheses.

Distributive Property of Multiplication Over Addition

$$a(b + c) = ab + ac$$

Since multiplication is commutative, this property can also be written as

$$(b + c)a = ba + ca$$

The distributive property can also be extended to more than two numbers inside the parentheses. For example,

$$3(x + y + z) = 3(x) + 3(y) + 3(z)$$
$$= 3x + 3y + 3z$$

Since we define subtraction in terms of addition, the distributive property is also true for subtraction. For example,

$$2(x - y) = 2(x) - 2(y)$$
$$= 2x - 2y$$

EXAMPLES Use the distributive property to write each expression without parentheses. Then simplify the result.

7. $2(x + y) = 2(x) + 2(y)$
$= 2x + 2y$

8. $-5(-3 + 2z) = -5(-3) + (-5)(2z)$
$= 15 - 10z$

9. $5(x + 3y - z) = 5(x) + 5(3y) - 5(z)$
$= 5x + 15y - 5z$

Continued on next page

PRACTICE PROBLEMS 7–12

Use the distributive property to write each expression without parentheses. Then simplify the result.

7. $5(x + y)$
8. $-3(2 + 7x)$
9. $4(x + 6y - 2z)$
10. $-1(3 - a)$
11. $-(8 + a - b)$
12. $\frac{1}{2}(2x + 4) + 9$

Answers

5. $14 + x$ **6.** $20x$ **7.** $5x + 5y$ **8.** $-6 - 21x$ **9.** $4x + 24y - 8z$ **10.** $-3 + a$ **11.** $-8 - a + b$ **12.** $x + 11$

10. $-1(2 - y) = (-1)(2) - (-1)(y)$
$= -2 + y$

11. $-(3 + x - w) = -1(3 + x - w)$
$= (-1)(3) + (-1)(x) - (-1)(w)$
$= -3 - x + w$

Helpful Hint
Notice in Example 11 that $-(3 + x - w)$ is first rewritten as $-1(3 + x - w)$.

12. $\frac{1}{2}(6x + 14) + 10 = \frac{1}{2}(6x) + \frac{1}{2}(14) + 10$ Apply the distributive property.
$= 3x + 7 + 10$ Multiply.
$= 3x + 17$ Add.

Work Practice Problems 7–12

The distributive property can also be used to write a sum as a product.

PRACTICE PROBLEMS 13–14

Use the distributive property to write each sum as a product.

13. $9 \cdot 3 + 9 \cdot y$

14. $4x + 4y$

EXAMPLES Use the distributive property to write each sum as a product.

13. $8 \cdot 2 + 8 \cdot x = 8(2 + x)$

14. $7s + 7t = 7(s + t)$

Work Practice Problems 13–14

Objective C Using the Identity and Inverse Properties

Next, we look at the **identity properties.**

The number 0 is called the identity for addition because when 0 is added to any real number, the result is the same real number. In other words, the *identity* of the real number is not changed.

The number 1 is called the identity for multiplication because when a real number is multiplied by 1, the result is the same real number. In other words, the *identity* of the real number is not changed.

Identities for Addition and Multiplication

0 is the identity element for addition.

$$a + 0 = a \quad \text{and} \quad 0 + a = a$$

1 is the identity element for multiplication.

$$a \cdot 1 = a \quad \text{and} \quad 1 \cdot a = a$$

Notice that 0 is the *only* number that can be added to any real number with the result that the sum is the same real number. Also, 1 is the *only* number that can be multiplied by any real number with the result that the product is the same real number.

Additive inverses or **opposites** were introduced in Section 2.1. Two numbers are called additive inverses or opposites if their sum is 0. The additive inverse or opposite of 6 is -6 because $6 + (-6) = 0$. The additive inverse or opposite of -5 is 5 because $-5 + 5 = 0$.

Reciprocals or **multiplicative inverses** were introduced in Section 4.3. Two nonzero numbers are called reciprocals or multiplicative inverses if their product is 1. The reciprocal or multiplicative inverse of $\frac{2}{3}$ is $\frac{3}{2}$ because $\frac{2}{3} \cdot \frac{3}{2} = 1$. Likewise, the reciprocal of -5 is $-\frac{1}{5}$ because $-5\left(-\frac{1}{5}\right) = 1$.

Answers
13. $9(3 + y)$ **14.** $4(x + y)$

Additive or Multiplicative Inverses

The numbers a and $-a$ are additive inverses or opposites of each other because their sum is 0; that is,

$$a + (-a) = 0$$

The numbers b and $\frac{1}{b}$ (for $b \neq 0$) are reciprocals or multiplicative inverses of each other because their product is 1; that is,

$$b \cdot \frac{1}{b} = 1$$

✔Concept Check Which of the following is the

a. opposite of $-\frac{3}{10}$, and which is the

b. reciprocal of $-\frac{3}{10}$?

$$1, -\frac{10}{3}, \frac{3}{10}, 0, \frac{10}{3}, -\frac{3}{10}$$

EXAMPLES Name the property illustrated by each true statement.

15. $3(x + y) = 3 \cdot x + 3 \cdot y$ — Distributive property

16. $(x + 7) + 9 = x + (7 + 9)$ — Associative property of addition (grouping changed)

17. $(b + 0) + 3 = b + 3$ — Identity element for addition

18. $2 \cdot (z \cdot 5) = 2 \cdot (5 \cdot z)$ — Commutative property of multiplication (order changed)

19. $-2 \cdot \left(-\frac{1}{2}\right) = 1$ — Multiplicative inverse property

20. $-2 + 2 = 0$ — Additive inverse property

21. $-6 \cdot (y \cdot 2) = (-6 \cdot 2) \cdot y$ — Commutative and associative properties of multiplication (order and grouping changed)

Work Practice Problems 15–21

PRACTICE PROBLEMS 15–21

Name the property illustrated by each true statement.

15. $7(a + b) = 7 \cdot a + 7 \cdot b$

16. $12 + y = y + 12$

17. $-4 \cdot (6 \cdot x) = (-4 \cdot 6) \cdot x$

18. $6 + (z + 2) = 6 + (2 + z)$

19. $3\left(\frac{1}{3}\right) = 1$

20. $(x + 0) + 23 = x + 23$

21. $(7 \cdot y) \cdot 10 = y \cdot (7 \cdot 10)$

Answers

15. distributive property **16.** commutative property of addition **17.** associative property of multiplication **18.** commutative property of addition **19.** multiplicative inverse property **20.** identity element for addition **21.** commutative and associative properties of multiplication

✔ Concept Check Answers

a. $\frac{3}{10}$ **b.** $-\frac{10}{3}$

Vocabulary and Readiness Check

Use the choices below to fill in each blank.

distributive property	associative property of multiplication	commutative property of addition
opposites or additive inverses	associative property of addition	
reciprocals or multiplicative inverses	commutative property of multiplication	

1. $x + 5 = 5 + x$ is a true statement by the ____________________.

2. $x \cdot 5 = 5 \cdot x$ is a true statement by the ____________________.

3. $3(y + 6) = 3 \cdot y + 3 \cdot 6$ is true by the ____________________.

4. $2 \cdot (x \cdot y) = (2 \cdot x) \cdot y$ is a true statement by the ____________________.

5. $x + (7 + y) = (x + 7) + y$ is a true statement by the ____________________.

6. The numbers $-\frac{2}{3}$ and $-\frac{3}{2}$ are called ____________________.

7. The numbers $-\frac{2}{3}$ and $\frac{2}{3}$ are called ____________________.

9.2 EXERCISE SET

FOR EXTRA HELP

Objective A *Use a commutative property to complete each statement. See Examples 1 and 3.*

1. $x + 16 =$ ________ **2.** $8 + y =$ ________ **3.** $-4 \cdot y =$ ________ **4.** $-2 \cdot x =$ ________

5. $xy =$ ________ **6.** $ab =$ ________ **7.** $2x + 13 =$ ________ **8.** $19 + 3y =$ ________

Use an associative property to complete each statement. See Examples 2 and 4.

9. $(xy) \cdot z =$ ________ **10.** $3 \cdot (x \cdot y) =$ ________ **11.** $2 + (a + b) =$ ________

12. $(y + 4) + z =$ ________ **13.** $4 \cdot (ab) =$ ________ **14.** $(-3y) \cdot z =$ ________

15. $(a + b) + c =$ ________ **16.** $6 + (r + s) =$ ________

Use the commutative and associative properties to simplify each expression. See Examples 5 and 6.

17. $8 + (9 + b)$ **18.** $(r + 3) + 11$ **19.** $4(6y)$ **20.** $2(42x)$ **21.** $\frac{1}{5}(5y)$

22. $\frac{1}{8}(8z)$ **23.** $(13 + a) + 13$ **24.** $7 + (x + 4)$ **25.** $-9(8x)$ **26.** $-3(12y)$

27. $\frac{3}{4}\left(\frac{4}{3}s\right)$ **28.** $\frac{2}{7}\left(\frac{7}{2}r\right)$ **29.** $-\frac{1}{2}(5x)$ **30.** $-\frac{1}{3}(7x)$

Objective B *Use the distributive property to write each expression without parentheses. Then simplify the result, if possible. See Examples 7 through 12.*

31. $4(x + y)$

32. $7(a + b)$

33. $9(x - 6)$

34. $11(y - 4)$

35. $2(3x + 5)$

36. $5(7 + 8y)$

37. $7(4x - 3)$

38. $3(8x - 1)$

39. $3(6 + x)$

40. $2(x + 5)$

41. $-2(y - z)$

42. $-3(z - y)$

43. $-\frac{1}{3}(3y + 5)$

44. $-\frac{1}{2}(2r + 11)$

45. $5(x + 4m + 2)$

46. $8(3y + z - 6)$

47. $-4(1 - 2m + n) + 4$

48. $-4(4 + 2p + 5) + 16$

49. $-(5x + 2)$

50. $-(9r + 5)$

51. $-(r - 3 - 7p) + 3$

52. $-(q - 2 + 6r) + 2$

53. $\frac{1}{2}(6x + 7) + \frac{1}{2}$

54. $\frac{1}{4}(4x - 2) - \frac{7}{2}$

55. $-\frac{1}{3}(3x - 9y)$

56. $-\frac{1}{5}(10a - 25b)$

57. $3(2r + 5) - 7$

58. $10(4s + 6) - 40$

59. $-9(4x + 8) + 2$

60. $-11(5x + 3) + 10$

61. $-0.4(4x + 5) - 0.5$

62. $-0.6(2x + 1) - 0.1$

Use the distributive property to write each sum as a product. See Examples 13 and 14.

63. $4 \cdot 1 + 4 \cdot y$

64. $14 \cdot z + 14 \cdot 5$

65. $11x + 11y$

66. $9a + 9b$

67. $(-1) \cdot 5 + (-1) \cdot x$

68. $(-3)a + (-3)y$

69. $30a + 30b$

70. $25x + 25y$

Objectives A C *Name the properties illustrated by each true statement. See Examples 15 through 21.*

71. $3 \cdot 5 = 5 \cdot 3$

72. $4(3 + 8) = 4 \cdot 3 + 4 \cdot 8$

73. $2 + (x + 5) = (2 + x) + 5$

74. $9 \cdot (x \cdot 7) = (9 \cdot x) \cdot 7$

75. $(x + 9) + 3 = (9 + x) + 3$

76. $1 \cdot 9 = 9$

77. $(4 \cdot y) \cdot 9 = 4 \cdot (y \cdot 9)$

78. $-4 \cdot (8 \cdot 3) = (8 \cdot 3) \cdot (-4)$

79. $0 + 6 = 6$

80. $(a + 9) + 6 = a + (9 + 6)$

81. $-4(y + 7) = -4 \cdot y + (-4) \cdot 7$

82. $(11 + r) + 8 = (r + 11) + 8$

83. $6 \cdot \frac{1}{6} = 1$

84. $r + 0 = r$

85. $-6 \cdot 1 = -6$

86. $-\frac{3}{4}\left(-\frac{4}{3}\right) = 1$

Review

Evaluate each expression for the given values. See Section 2.5.

87. If $x = -1$ and $y = 3$, find $y - x^2$

88. If $g = 0$ and $h = -4$, find $gh - h^2$

89. If $a = 2$ and $b = -5$, find $a - b^2$

90. If $x = -3$, find $x^3 - x^2 + 4$

91. If $y = -5$ and $z = 0$, find $yz - y^2$

92. If $x = -2$, find $x^3 - x^2 - x$

Concept Extensions

Fill in the table with the opposite (additive inverse), the reciprocal (multiplicative inverse), or the expression. Assume that the value of each expression is not 0.

	93.	94.	95.	96.	97.	98.
Expression	8	$-\frac{2}{3}$	x	$4y$		
Opposite						$7x$
Reciprocal					$\frac{1}{2x}$	

Decide whether each statement is true or false. See the second Concept Check in this section.

99. The opposite of $-\frac{a}{2}$ is $-\frac{2}{a}$.

100. The reciprocal of $-\frac{a}{2}$ is $\frac{a}{2}$.

Determine which pairs of actions are commutative. See the first Concept Check in this section.

101. "taking a test" and "studying for the test"

102. "putting on your shoes" and "putting on your socks"

103. "putting on your left shoe" and "putting on your right shoe"

104. "reading the sports section" and "reading the comics section"

105. "mowing the lawn" and "trimming the hedges"

106. "baking a cake" and "eating the cake"

107. "feeding the dog" and "feeding the cat"

108. "dialing a number" and "turning on the cell phone"

Name the property illustrated by each step.

109. a. $\triangle + (\square + \bigcirc) = (\square + \bigcirc) + \triangle$

b. $= (\bigcirc + \square) + \triangle$

c. $= \bigcirc + (\square + \triangle)$

110. a. $(x + y) + z = x + (y + z)$

b. $= (y + z) + x$

c. $= (z + y) + x$

111. Explain why 0 is called the identity element for addition.

112. Explain why 1 is called the identity element for multiplication.

113. Write an example that shows that division is not commutative.

114. Write an example that shows that subtraction is not commutative.

STUDY SKILLS BUILDER

Are You Familiar with Your Textbook Supplements?

There are many student supplements available for additional study. Below, I have listed some of these. See the preface of this text or your instructor for further information.

Chapter Test Prep Video CD. This material is found in your textbook and is fully explained there. The CD contains video clips of solutions to the Chapter Test exercises in this text and is excellent help when studying for chapter tests.

Lecture Video CDs. These video segments are keyed to each section of the text. The material is presented by me, Elayn Martin-Gay, and I have placed a video icon by the exercises in the text that I have worked on the video.

The Student Solutions Manual. This contains worked out solutions to odd-numbered exercises as well as every exercise in the Integrated Reviews, Chapter Reviews, Chapter Tests, and Cumulative Reviews.

Prentice Hall Tutor Center. Mathematics questions may be phoned, faxed, or emailed to this center.

MyMathLab, MathXL, and Interact Math. These are computer and Internet tutorials. This supplement may already be available to you somewhere on campus, for example at your local learning resource lab. Take a moment and find the name and location of any such lab on campus.

As usual, your instructor is your best source of information.

Let's see how you are doing with textbook supplements:

1. Name one way the Chapter Test Prep Video can help you prepare for a chapter test.
2. List any textbook supplements that you have found useful.
3. Have you located and visited a learning resource lab located on your campus?
4. List the textbook supplements that are currently housed in your campus' learning resource lab.

Objectives

A Apply the General Strategy for Solving a Linear Equation.

B Solve Equations Containing Fractions or Decimals.

C Recognize Identities and Equations with No Solution.

9.3 FURTHER SOLVING LINEAR EQUATIONS

Objective A Solving Linear Equations

Let's begin with a formal definition of a linear equation in one variable.

A linear equation in one variable can be written in the form

$$Ax + B = C$$

where A, B, and C, are real numbers and $A \neq 0$.

We now combine our knowledge from the previous chapters and review solving linear equations.

To Solve Linear Equations in One Variable

Step 1: If an equation contains fractions, multiply both sides by the LCD to clear the equation of fractions.

Step 2: Use the distributive property to remove parentheses if they occur.

Step 3: Simplify each side of the equation by combining like terms.

Step 4: Get all variable terms on one side and all numbers on the other side by using the addition property of equality.

Step 5: Get the variable alone by using the multiplication property of equality.

Step 6: Check the solution by substituting it into the original equation.

PRACTICE PROBLEM 1

Solve:
$5(3x - 1) + 2 = 12x + 6$

EXAMPLE 1 Solve: $4(2x - 3) + 7 = 3x + 5$

Solution: There are no fractions, so we begin with Step 2.

$$4(2x - 3) + 7 = 3x + 5$$

Step 2: $8x - 12 + 7 = 3x + 5$ Use the distributive property.

Step 3: $8x - 5 = 3x + 5$ Combine like terms.

Step 4: Get all variable terms on one side of the equation and all numbers on the other side. One way to do this is by subtracting $3x$ from both sides and then adding 5 to both sides.

$8x - 5 - 3x = 3x + 5 - 3x$ Subtract $3x$ from both sides.

$5x - 5 = 5$ Simplify.

$5x - 5 + 5 = 5 + 5$ Add 5 to both sides.

$5x = 10$ Simplify.

Step 5: Use the multiplication property of equality to get x alone.

$\frac{5x}{5} = \frac{10}{5}$ Divide both sides by 5.

$x = 2$ Simplify.

Step 6: Check.

$4(2x - 3) + 7 = 3x + 5$ Original equation

$4[2(2) - 3] + 7 \stackrel{?}{=} 3(2) + 5$ Replace x with 2.

$4(4 - 3) + 7 \stackrel{?}{=} 6 + 5$

$4(1) + 7 \stackrel{?}{=} 11$

$4 + 7 \stackrel{?}{=} 11$

$11 = 11$ True

The solution is 2.

Work Practice Problem 1

Answer

1. $x = 3$

EXAMPLE 2 Solve: $8(2 - t) = -5t$

Solution: First, we apply the distributive property.

$$8(2 - t) = -5t$$

Step 2: $16 - 8t = -5t$ Use the distributive property.

Step 4: $16 - 8t + 8t = -5t + 8t$ Add $8t$ to both sides.

$16 = 3t$ Combine like terms.

Step 5: $\frac{16}{3} = \frac{3t}{3}$ Divide both sides by 3.

$\frac{16}{3} = t$ Simplify.

Step 6: Check.

$8(2 - t) = -5t$ Original equation

$8\left(2 - \frac{16}{3}\right) \stackrel{?}{=} -5\left(\frac{16}{3}\right)$ Replace t with $\frac{16}{3}$.

$8\left(\frac{6}{3} - \frac{16}{3}\right) \stackrel{?}{=} -\frac{80}{3}$ The LCD is 3.

$8\left(-\frac{10}{3}\right) \stackrel{?}{=} -\frac{80}{3}$ Subtract fractions.

$-\frac{80}{3} = -\frac{80}{3}$ True

The solution is $\frac{16}{3}$.

Work Practice Problem 2

PRACTICE PROBLEM 2

Solve: $9(5 - x) = -3x$

Helpful Hint
When checking solutions, use the original equation.

Objective B Solving Equations Containing Fractions or Decimals

If an equation contains fractions, we can clear the equation of fractions by multiplying both sides by the LCD of all denominators. By doing this, we avoid working with time-consuming fractions.

EXAMPLE 3 Solve: $\frac{x}{2} - 1 = \frac{2}{3}x - 3$

Solution: We begin by clearing fractions. To do this, we multiply both sides of the equation by the LCD of 2 and 3, which is 6.

$$\frac{x}{2} - 1 = \frac{2}{3}x - 3$$

Step 1: $6\left(\frac{x}{2} - 1\right) = 6\left(\frac{2}{3}x - 3\right)$ Multiply both sides by the LCD, 6.

Step 2: $6\left(\frac{x}{2}\right) - 6(1) = 6\left(\frac{2}{3}x\right) - 6(3)$ Use the distributive property.

$3x - 6 = 4x - 18$ Simplify.

There are no longer grouping symbols and no like terms on either side of the equation, so we continue with Step 4.

Continued on next page

PRACTICE PROBLEM 3

Solve: $\frac{5}{2}x - 1 = \frac{3}{2}x - 4$

Helpful Hint
Don't forget to multiply *each* term by the LCD.

Answers

2. $x = \frac{15}{2}$ **3.** $x = -3$

$$3x - 6 = 4x - 18$$

Step 4: $3x - 6 - 3x = 4x - 18 - 3x$ Subtract $3x$ from both sides.

$-6 = x - 18$ Simplify.

$-6 + 18 = x - 18 + 18$ Add 18 to both sides.

$12 = x$ Simplify.

Step 5: The variable is now alone, so there is no need to apply the multiplication property of equality.

Step 6: Check.

$\frac{x}{2} - 1 = \frac{2}{3}x - 3$ Original equation

$\frac{12}{2} - 1 \stackrel{?}{=} \frac{2}{3} \cdot 12 - 3$ Replace x with 12.

$6 - 1 \stackrel{?}{=} 8 - 3$ Simplify.

$5 = 5$ True

The solution is 12.

Work Practice Problem 3

PRACTICE PROBLEM 4

Solve: $\frac{3(x-2)}{5} = 3x + 6$

EXAMPLE 4 Solve: $\frac{2(a+3)}{3} = 6a + 2$

Solution: We clear the equation of fractions first.

$$\frac{2(a+3)}{3} = 6a + 2$$

Step 1: $3 \cdot \frac{2(a+3)}{3} = 3(6a + 2)$ Clear the fraction by multiplying both sides by the LCD, 3.

$2(a + 3) = 3(6a + 2)$ Simplify.

Step 2: Next, we use the distributive property to remove parentheses.

$2a + 6 = 18a + 6$ Use the distributive property.

Step 4: $2a + 6 - 18a = 18a + 6 - 18a$ Subtract $18a$ from both sides.

$-16a + 6 = 6$ Simplify.

$-16a + 6 - 6 = 6 - 6$ Subtract 6 from both sides.

$-16a = 0$

Step 5: $\frac{-16a}{-16} = \frac{0}{-16}$ Divide both sides by -16.

$a = 0$ Simplify.

Step 6: To check, replace a with 0 in the original equation. The solution is 0.

Work Practice Problem 4

Helpful Hint

Remember: When solving an equation, it makes no difference on which side of the equation variable terms lie. Just make sure that constant terms lie on the other side.

When solving a problem about money, you may need to solve an equation containing decimals. If you choose, you may multiply to clear the equation of decimals.

Answer

4. $x = -3$

EXAMPLE 5 Solve: $0.25x + 0.10(x - 3) = 1.1$

Solution: First we clear this equation of decimals by multiplying both sides of the equation by 100. Recall that multiplying a decimal number by 100 has the effect of moving the decimal point 2 places to the right.

$$0.25x + 0.10(x - 3) = 1.1$$

Step 1: $0.25x + 0.10(x - 3) = 1.10$ Multiply both sides by 100

$25x + 10(x - 3) = 110$

Step 2: $25x + 10x - 30 = 110$ Apply the distributive property.

Step 3: $35x - 30 = 110$ Combine like terms.

Step 4: $35x - 30 + 30 = 110 + 30$ Add 30 to both sides.

$35x = 140$ Combine like terms.

Step 5: $\frac{35x}{35} = \frac{140}{35}$ Divide both sides by 35.

$x = 4$

Step 6: To check, replace x with 4 in the original equation. The solution is 4.

Work Practice Problem 5

PRACTICE PROBLEM 5

Solve:
$0.06x - 0.10(x - 2) = -0.16$

Helpful Hint

If you have trouble with this step, try removing parentheses first.

$0.25x + 0.10(x - 3) = 1.1$

$0.25x + 0.10x - 0.3 = 1.1$

$0.25x + 0.10x - 0.30 = 1.10$

$25x + 10x - 30 = 110$

Then continue.

Objective C Recognizing Identities and Equations with No Solution

So far, each equation that we have solved has had a single solution. However, not every equation in one variable has a single solution. Some equations have no solution, while others have an infinite number of solutions. For example,

$$x + 5 = x + 7$$

has **no solution** since no matter which real number we replace x with, the equation is false.

real number + 5 = same real number + 7 FALSE

On the other hand,

$$x + 6 = x + 6$$

has infinitely many solutions since x can be replaced by any real number and the equation is always true.

real number + 6 = same real number + 6 TRUE

The equation $x + 6 = x + 6$ is called an **identity.** The next few examples illustrate special equations like these.

EXAMPLE 6 Solve: $-2(x - 5) + 10 = -3(x + 2) + x$

Solution:

$-2(x - 5) + 10 = -3(x + 2) + x$

$-2x + 10 + 10 = -3x - 6 + x$ Apply the distributive property on both sides.

$-2x + 20 = -2x - 6$ Combine like terms.

$-2x + 20 + 2x = -2x - 6 + 2x$ Add $2x$ to both sides.

$20 = -6$ Combine like terms.

The final equation contains no variable terms, and the result is the false statement $20 = -6$. This means that there is no value for x that makes $20 = -6$ a true equation. Thus, we conclude that there is **no solution** to this equation.

Work Practice Problem 6

PRACTICE PROBLEM 6

Solve:
$5(2 - x) + 8x = 3(x - 6)$

Answers

5. $x = 9$ **6.** no solution

PRACTICE PROBLEM 7

Solve:

$-6(2x + 1) - 14 = -10(x + 2) - 2x$

EXAMPLE 7 Solve: $3(x - 4) = 3x - 12$

Solution: $3(x - 4) = 3x - 12$

$3x - 12 = 3x - 12$ Apply the distributive property.

The left side of the equation is now identical to the right side. Every real number may be substituted for x and a true statement will result. We arrive at the same conclusion if we continue.

$3x - 12 = 3x - 12$

$3x - 12 - 3x = 3x - 12 - 3x$ Subtract $3x$ from both sides.

$-12 = -12$ Combine like terms.

Again, the final equation contains no variables, but this time the result is the true statement $-12 = -12$. This means that one side of the equation is identical to the other side. Thus, $3(x - 4) = 3x - 12$ is an **identity** and **every real number** is a solution.

Work Practice Problem 7

Answer

7. Every real number is a solution.

✔ Concept Check Answer

a. Every real number is a solution.
b. The solution is 0.
c. There is no solution.

✔Concept Check Suppose you have simplified several equations and obtain the following results. What can you conclude about the solutions to the original equation?

a. $7 = 7$ **b.** $x = 0$ **c.** $7 = -4$

CALCULATOR EXPLORATIONS Checking Equations

We can use a calculator to check possible solutions of equations. To do this, replace the variable by the possible solution and evaluate both sides of the equation separately.

Equation: $3x - 4 = 2(x + 6)$ Solution: $x = 16$

$3x - 4 = 2(x + 6)$ Original equation

$3(16) - 4 \stackrel{?}{=} 2(16 + 6)$ Replace x with 16.

Now evaluate each side with your calculator.

Evaluate left side: [3] [×] [16] [−] [4] [=] or [ENTER]

Display: [44]

Evaluate right side: [2] [(] [16] [+] [6] [)] [=] or [ENTER]

Display: [44]

Since the left side equals the right side, the equation checks.

Use a calculator to check the possible solutions to each equation.

1. $2x = 48 + 6x;\quad x = -12$
2. $-3x - 7 = 3x - 1;\quad x = -1$
3. $5x - 2.6 = 2(x + 0.8);\quad x = 4.4$
4. $-1.6x - 3.9 = -6.9x - 25.6;\quad x = 5$
5. $\frac{564x}{4} = 200x - 11(649);\quad x = 121$
6. $20(x - 39) = 5x - 432;\quad x = 23.2$

9.3 EXERCISE SET

FOR EXTRA HELP

Student Solutions Manual

PH Math/Tutor Center

CD/Video for Review

MathXL®

MyMathLab

Objective A *Solve each equation. See Examples 1 and 2.*

1. $-4y + 10 = -2(3y + 1)$

2. $-3x + 1 = -2(4x + 2)$

3. $15x - 8 = 10 + 9x$

4. $15x - 5 = 7 + 12x$

5. $-2(3x - 4) = 2x$

6. $-(5x - 10) = 5x$

7. $5(2x - 1) - 2(3x) = 1$

8. $3(2 - 5x) + 4(6x) = 12$

9. $-6(x - 3) - 26 = -8$

10. $-4(n - 4) - 23 = -7$

11. $8 - 2(a + 1) = 9 + a$

12. $5 - 6(2 + b) = b - 14$

13. $4x + 3 = -3 + 2x + 14$

14. $6y - 8 = -6 + 3y + 13$

15. $-2y - 10 = 5y + 18$

16. $-7n + 5 = 8n - 10$

Objective B *Solve each equation. See Examples 3 through 5.*

17. $\frac{2}{3}x + \frac{4}{3} = -\frac{2}{3}$

18. $\frac{4}{5}x - \frac{8}{5} = -\frac{16}{5}$

19. $\frac{3}{4}x - \frac{1}{2} = 1$

20. $\frac{2}{9}x - \frac{1}{3} = 1$

21. $0.50x + 0.15(70) = 35.5$

22. $0.40x + 0.06(30) = 9.8$

23. [illegible] / 4 $= 3x - 2$

24. [illegible] / 5 = [illegible]

25. $x + \frac{7}{6} = 2x - \frac{7}{6}$

26. $\frac{5}{2}x - 1 = x + \frac{1}{4}$

27. $0.12(y - 6) + 0.06y = 0.08y - 0.7$

28. $0.60(z - 300) + 0.05z = 0.70z - 205$

Objective C *Solve each equation. See Examples 6 and 7.*

29. $4(3x + 2) = 12x + 8$

30. $14x + 7 = 7(2x + 1)$

31. $\frac{x}{4} + 1 = \frac{x}{4}$

32. $\frac{x}{3} - 2 = \frac{x}{3}$

33. $3x - 7 = 3(x + 1)$

34. $2(x - 5) = 2x + 10$

35. $-2(6x - 5) + 4 = -12x + 14$

36. $-5(4y - 3) + 2 = -20y + 17$

Objectives **A B C Mixed Practice** *Solve. See Examples 1 through 7.*

37. $\frac{6(3-z)}{5} = -z$

38. $\frac{4(5-w)}{3} = -w$

39. $-3(2t-5) + 2t = 5t - 4$

40. $-(4a-7) - 5a = 10 + a$

41. $5y + 2(y-6) = 4(y+1) - 2$

42. $9x + 3(x-4) = 10(x-5) + 7$

43. $\frac{3(x-5)}{2} = \frac{2(x+5)}{3}$

44. $\frac{5(x-1)}{4} = \frac{3(x+1)}{2}$

45. $0.7x - 2.3 = 0.5$

46. $0.9x - 4.1 = 0.4$

47. $5x - 5 = 2(x+1) + 3x - 7$

48. $3(2x-1) + 5 = 6x + 2$

49. $4(2n+1) = 3(6n+3) + 1$

50. $4(4y+2) = 2(1+6y) + 8$

51. $x + \frac{5}{4} = \frac{3}{4}x$

52. $\frac{7}{8}x + \frac{1}{4} = \frac{3}{4}x$

53. $\frac{x}{2} - 1 = \frac{x}{5} + 2$

54. $\frac{x}{5} - 7 = \frac{x}{3} - 5$

55. $2(x+3) - 5 = 5x - 3(1+x)$

56. $4(2+x) + 1 = 7x - 3(x-2)$

57. $0.06 - 0.01(x+1) = -0.02(2-x)$

58. $-0.01(5x+4) = 0.04 - 0.01(x+4)$

59. $\frac{9}{2} + \frac{5}{2}y = 2y - 4$

60. $3 - \frac{1}{2}x = 5x - 8$

Review

Translating *Write each algebraic expression described. See Section 3.1.*

61. A plot of land is in the shape of a triangle. If one side is x meters, a second side is $(2x-3)$ meters, and a third side is $(3x-5)$ meters, express the perimeter of the lot as a simplified expression in x.

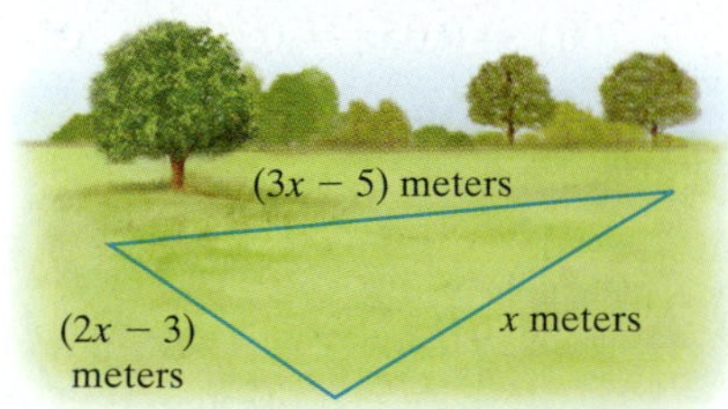

62. A portion of a board has length x feet. The other part has length $(7x-9)$ feet. Express the total length of the board as a simplified expression in x.

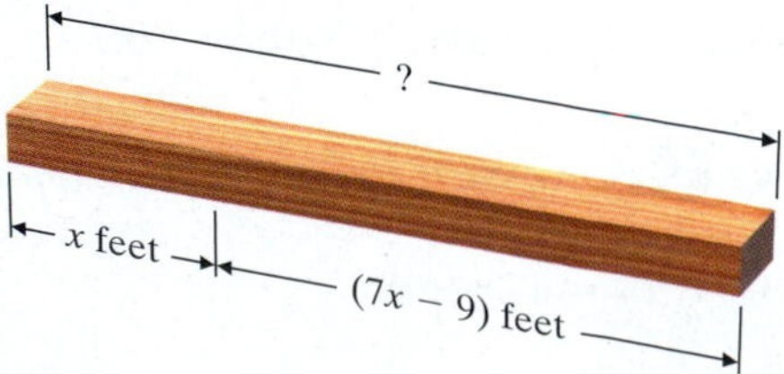

Translating *Write each phrase as an algebraic expression. Use x for the unknown number. See Section 3.2.*

63. A number subtracted from -8

64. Three times a number

65. The sum of -3 and twice a number

66. The difference of 8 and twice a number

67. The product of 9 and the sum of a number and 20

68. The quotient of -12 and the difference of a number and 3

Concept Extensions

See the Concept Check in this section.

69. a. Solve: $x + 3 = x + 3$
b. If you simplify an equation and get $0 = 0$, what can you conclude about the solution(s) of the original equation?
c. On your own, construct an equation for which every real number is a solution.

70. a. Solve: $x + 3 = x + 5$
b. If you simplify an equation and get $3 = 5$, what can you conclude about the solution(s) of the original equation?
c. On your own, construct an equation that has no solution.

Match each equation in the first column with its solution in the second column. Items in the second column may be used more than once.

71. $5x + 1 = 5x + 1$

72. $3x + 1 = 3x + 2$

73. $2x - 6x - 10 = -4x + 3 - 10$

74. $x - 11x - 3 = -10x - 1 - 2$

75. $9x - 20 = 8x - 20$

76. $-x + 15 = x + 15$

a. all real numbers
b. no solution
c. 0

77. Explain the difference between simplifying an expression and solving an equation.

78. On your own, write an expression and then an equation. Label each.

For Exercises 79 and 80, **a.** *Write an equation for perimeter.* **b.** *Solve the equation in part (a).* **c.** *Find the length of each side.*

79. The perimeter of a geometric figure is the sum of the lengths of its sides. If the perimeter of the following pentagon (five-sided figure) is 28 centimeters, find the length of each side.

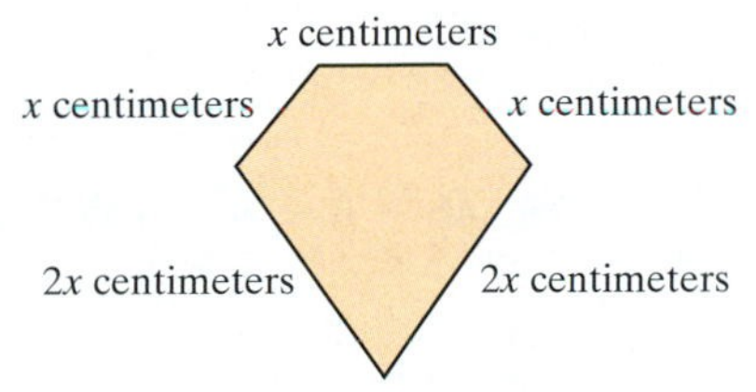

80. The perimeter of the following triangle is 35 meters. Find the length of each side.

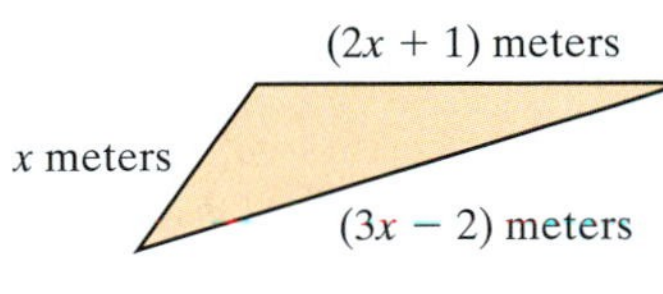

Fill in the blanks with numbers of your choice so that each equation has the given solution. Note: *Each blank may be replaced by a different number.*

81. $x + \_\_\_\_ = 2x - \_\_\_\_$; solution: 9

82. $-5x - \_\_\_\_ = \_\_\_\_$; solution: 2

Solve.

83. $1000(7x - 10) = 50(412 + 100x)$

84. $1000(x + 40) = 100(16 + 7x)$

85. $0.035x + 5.112 = 0.010x + 5.107$

86. $0.127x - 2.685 = 0.027x - 2.38$

INTEGRATED REVIEW Sections 9.1–9.3

Real Numbers and Solving Linear Equations

Answers

1. ______
2. ______
3. ______
4. ______
5. ______
6. ______
7. ______
8. ______
9. ______
10. ______
11. ______
12. ______
13. ______
14. ______
15. ______
16. ______
17. ______
18. ______
19. ______
20. ______

Tell which set or sets each number belongs to: natural numbers, whole numbers, integers, rational numbers, irrational numbers, and real numbers.

1. 0 **2.** 143 **3.** $\frac{3}{8}$ **4.** 1

5. -13 **6.** $\frac{9}{10}$ **7.** $-\frac{1}{9}$ **8.** $\sqrt{5}$

Remove parentheses and simplify each expression.

9. $7(d - 3) + 10$ **10.** $9(z + 7) - 15$

11. $-4(3y - 4) + 12y$ **12.** $-3(2x + 5) - 6x$

Solve. Feel free to use the steps given in Section 9.3.

13. $2x - 7 = 6x - 27$ **14.** $3 + 8y = 3y - 2$

15. $-3a + 6 + 5a = 7a - 8a$ **16.** $4b - 8 - b = 10b - 3b$

17. $-\frac{2}{3}x = \frac{5}{9}$ **18.** $-\frac{3}{8}y = -\frac{1}{16}$

19. $10 = -6n + 16$ **20.** $-5 = -2m + 7$

21. $3(5c - 1) - 2 = 13c + 3$

22. $4(3t + 4) - 20 = 3 + 5t$

23. $\frac{2(z + 3)}{3} = 5 - z$

24. $\frac{3(w + 2)}{4} = 2w + 3$

25. $-2(2x - 5) = -3x + 7 - x + 3$

26. $-4(5x - 2) = -12x + 4 - 8x + 4$

27. $0.02(6t - 3) = 0.04(t - 2) + 0.02$

28. $0.03(m + 7) = 0.02(5 - m) + 0.03$

29. $-3y = \frac{4(y - 1)}{5}$

30. $-4x = \frac{5(1 - x)}{6}$

31. $\frac{5}{3}x - \frac{7}{3} = x$

32. $\frac{7}{5}n + \frac{3}{5} = -n$

21. ______
22. ______
23. ______
24. ______
25. ______
26. ______
27. ______
28. ______
29. ______
30. ______
31. ______
32. ______

Objective

A **Translate a Problem to an Equation, Then Use the Equation to Solve the Problem.**

9.4 FURTHER PROBLEM SOLVING

In this section, we review our problem-solving steps first introduced in Section 3.4 and continue to solve problems that are modeled by linear equations..

General Strategy for Problem Solving

1. UNDERSTAND the problem. During this step, become comfortable with the problem. Some ways of doing this are:

Read and reread the problem.

Choose a variable to represent the unknown.

Construct a drawing.

Propose a solution and check. Pay careful attention to how you check your proposed solution. This will help when writing an equation to model the problem.

2. TRANSLATE the problem into an equation.

3. SOLVE the equation.

4. INTERPRET the results: *Check* the proposed solution in the stated problem and *state* your conclusion.

Objective A Translating and Solving Problems

Much of problem solving involves a direct translation from a sentence to an equation.

PRACTICE PROBLEM 1

Three times the difference of a number and 5 is the same as twice the number decreased by 3. Find the number.

EXAMPLE 1 Finding an Unknown Number

Twice the sum of a number and 4 is the same as four times the number, decreased by 12. Find the number.

Solution:

1. UNDERSTAND. Read and reread the problem. If we let x = the unknown number, then
"the sum of a number and 4" translates to "$x + 4$" and
"four times the number" translates to "$4x$"

2. TRANSLATE.

twice	sum of a number and 4	is the same as	four times the number	decreased by	12
↓	↓	↓	↓	↓	↓
2	$(x + 4)$	$=$	$4x$	$-$	12

3. SOLVE

$$2(x + 4) = 4x - 12$$

$$2x + 8 = 4x - 12 \quad \text{Apply the distributive property.}$$

$$2x + 8 - 4x = 4x - 12 - 4x \quad \text{Subtract } 4x \text{ from both sides.}$$

$$-2x + 8 = -12$$

$$-2x + 8 - 8 = -12 - 8 \quad \text{Subtract 8 from both sides.}$$

$$-2x = -20$$

$$\frac{-2x}{-2} = \frac{-20}{-2} \quad \text{Divide both sides by } -2.$$

$$x = 10$$

Answer

1. The number is 12.

4. INTERPRET.

Check: Check this solution in the problem as it was originally stated. To do so, replace "number" with 10. Twice the sum of "10" and 4 is 28, which is the same as 4 times "10" decreased by 12.

State: The number is 10.

Work Practice Problem 1

The next example has to do with consecutive integers.

EXAMPLE 2

Some states have a single area code for the entire state. Two such states have area codes that are consecutive odd integers. If the sum of these integers is 1208, find the two area codes. (*Source: World Almanac*)

Solution:

1. UNDERSTAND. Read and reread the problem. If we let

x = the first odd integer, then

$x + 2$ = the next odd integer

Helpful Hint

Remember, the 2 here means that odd integers are 2 units apart, for example, the odd integers 13 and $13 + 2 = 15$.

2. TRANSLATE.

first odd integer	the sum of	next odd integer	is	1208
↓	↓	↓		
x	$+$	$(x + 2)$	$=$	1208

3. SOLVE.

$$x + x + 2 = 1208$$
$$2x + 2 = 1208$$
$$2x + 2 - 2 = 1208 - 2$$
$$2x = 1206$$
$$\frac{2x}{2} = \frac{1206}{2}$$
$$x = 603$$

4. INTERPRET.

Check: If $x = 603$, then the next odd integer $x + 2 = 603 + 2 = 605$. Notice their sum, $603 + 605 = 1208$, as needed.

State: The area codes are 603 and 605.

Note: New Hampshire's area code is 603 and South Dakota's area code is 605.

Work Practice Problem 2

During the next example, we expand our discussion of the UNDERSTAND part of the problem-solving process.

PRACTICE PROBLEM 2

The sum of three consecutive even integers is 144. Find the integers.

Answer

2. 46, 48, 50

PRACTICE PROBLEM 3

An 18-foot wire is to be cut so that the longer piece is 5 times longer than the shorter piece. Find the length of each piece.

EXAMPLE 3 Finding the Length of a Board

A 10-foot board is to be cut into two pieces so that the longer piece is 4 times the shorter. Find the length of each piece.

Solution:

1. UNDERSTAND the problem. To do so, read and reread the problem. You may also want to propose a solution. For example, if 3 feet represents the length of the shorter piece, then $4(3) = 12$ feet is the length of the longer piece, since it is 4 times the length of the shorter piece. This guess gives a total board length of 3 feet + 12 feet = 15 feet, which is too long. However, the purpose of proposing a solution is not to guess correctly, but to help better understand the problem and how to model it.

   In general, if we let

   x = length of shorter piece, then
   $4x$ = length of longer piece

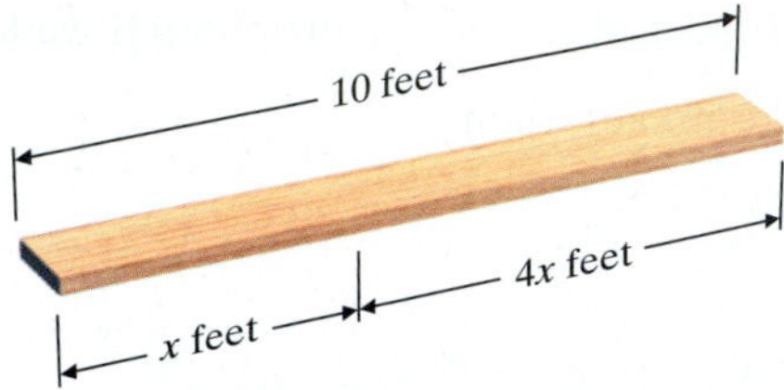

2. TRANSLATE the problem. First, we write the equation in words.

length of shorter piece	added to	length of longer piece	equals	total length of board
↓	↓	↓	↓	↓
x	$+$	$4x$	$=$	10

3. SOLVE.

$$x + 4x = 10$$
$$5x = 10 \quad \text{Combine like terms.}$$
$$\frac{5x}{5} = \frac{10}{5} \quad \text{Divide both sides by 5.}$$
$$x = 2$$

4. INTERPRET.

Check: Check the solution in the stated problem. If the shorter piece of board is 2 feet, the longer piece is $4 \cdot (2 \text{ feet}) = 8$ feet and the sum of the two pieces is 2 feet + 8 feet = 10 feet.

State: The shorter piece of board is 2 feet and the longer piece of board is 8 feet.

Work Practice Problem 3

Make sure that units are included in your answer, if appropriate.

Answer

3. shorter piece: 3 feet; longer piece: 15 feet

EXAMPLE 4 Finding the Number of Republican and Democratic Senators

In a recent year, the U.S. House of Representatives had a total of 431 Democrats and Republicans. There were 15 more Republican representatives than Democratic. Find the number of representatives from each party. (*Source:* Office of the Clerk of the U.S. House of Representatives)

Solution:

1. UNDERSTAND the problem. Read and reread the problem. Let's suppose that there are 200 Democratic representatives. Since there are 15 more Republicans than Democrats, there must be $200 + 15 = 215$ Republicans. The total number of Democrats and Republicans is then $200 + 215 = 415$. This is incorrect since the total should be 431, but we now have a better understanding of the problem.

   In general, if we let

   $x =$ number of Democrats, then

   $x + 15 =$ number of Republicans

2. TRANSLATE the problem. First, we write the equation in words.

number of Democrats	added to	number of Republicans	equals	431
↓	↓	↓	↓	↓
x	$+$	$(x + 15)$	$=$	431

3. SOLVE.

$$x + (x + 15) = 431$$
$$2x + 15 = 431 \quad \text{Combine like terms.}$$
$$2x + 15 - 15 = 431 - 15 \quad \text{Subtract 15 from both sides.}$$
$$2x = 416$$
$$\frac{2x}{2} = \frac{416}{2} \quad \text{Divide both sides by 2.}$$
$$x = 208$$

4. INTERPRET.

Check: If there were 208 Democratic representatives, then there were $208 + 15 = 223$ Republican representatives. The total number of representatives is then $208 + 223 = 431$. The results check.

State: There were 208 Democratic and 223 Republican representatives in Congress.

Work Practice Problem 4

EXAMPLE 5 Calculating Hours on the Job

A computer science major at a local university has a part-time job working on computers for his clients. He charges \$20 to come to your home or office and then \$25 per hour. During one month he visited 10 homes or offices and his total income was \$575. How many hours did he spend working on computers?

Continued on next page

PRACTICE PROBLEM 4

Through the year 2010, the state of California will have 21 more electoral votes for president than the state of Texas. If the total electoral votes for these two states is 89, find the number of electoral votes for each state.

PRACTICE PROBLEM 5

A car rental agency charges \$28 a day and \$0.15 a mile. If you rent a car for a day and your bill (before taxes) is \$52, how many miles did you drive?

Answers

4. Texas: 34 electoral votes; California: 55 electoral votes
5. 160 miles

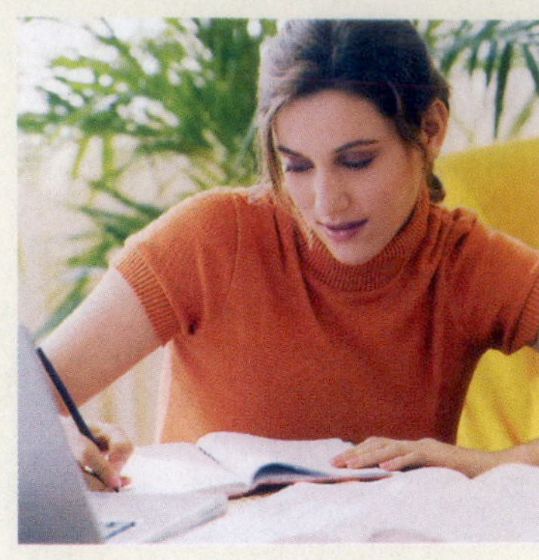

Solution:

1. UNDERSTAND. Read and reread the problem. Let's propose that the student spent 20 hours working on computers. Pay careful attention as to how his income is calculated. For 20 hours and 10 visits, his income is $20(\$25) + 10(\$20) = \$700$, more than \$575. We now have a better understanding of the problem and know that the time working on computers is less than 20 hours.

   Let's let

   $x =$ hours working on computers. Then

   $25x =$ amount of money made while working on computers

2. TRANSLATE.

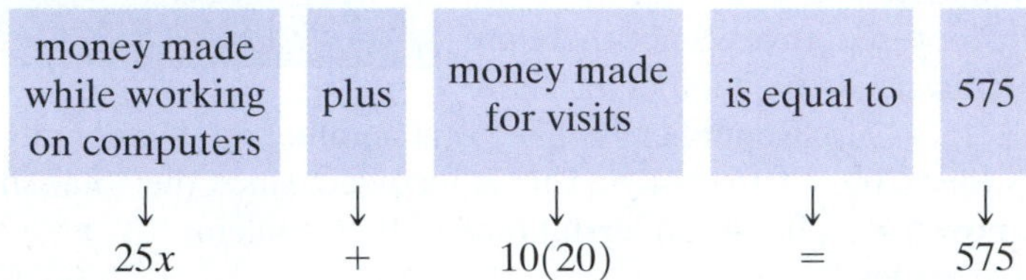

money made while working on computers	plus	money made for visits	is equal to	575
↓	↓	↓	↓	↓
$25x$	$+$	$10(20)$	$=$	575

3. SOLVE.

$$25x + 200 = 575$$

$$25x + 200 - 200 = 575 - 200 \quad \text{Subtract 200 from both sides.}$$

$$25x = 375 \quad \text{Simplify.}$$

$$\frac{25x}{25} = \frac{375}{25} \quad \text{Divide both sides by 25.}$$

$$x = 15 \quad \text{Simplify.}$$

4. INTERPRET.

Check: If the student works 15 hours and makes 10 visits, his income is $15(\$25) + 10(\$20) = \$575$.

State: The student spent 15 hours working on computers.

Work Practice Problem 5

PRACTICE PROBLEM 6

The measure of the second angle of a triangle is twice the measure of the smallest angle. The measure of the third angle of the triangle is three times the measure of the smallest angle. Find the measures of the angles.

EXAMPLE 6 Finding Angle Measures

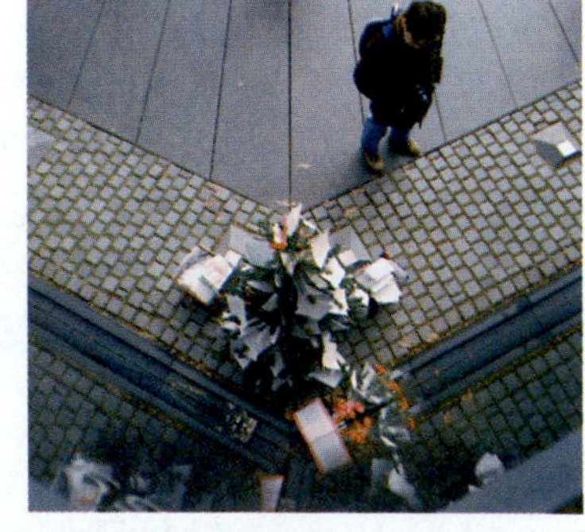

If the two walls of the Vietnam Veterans Memorial in Washington, D.C., were connected, an isosceles triangle would be formed. The measure of the third angle is 97.5° more than the measure of either of the two equal angles. Find the measure of the third angle. (*Source:* National Park Service)

Solution:

1. UNDERSTAND. Read and reread the problem. We then draw a diagram (recall that an isosceles triangle has two angles with the same measure) and let

   $x =$ degree measure of one angle

   $x =$ degree measure of the second equal angle

   $x + 97.5 =$ degree measure of the third angle

Answer

6. smallest: 30°; second: 60°; third: 90°

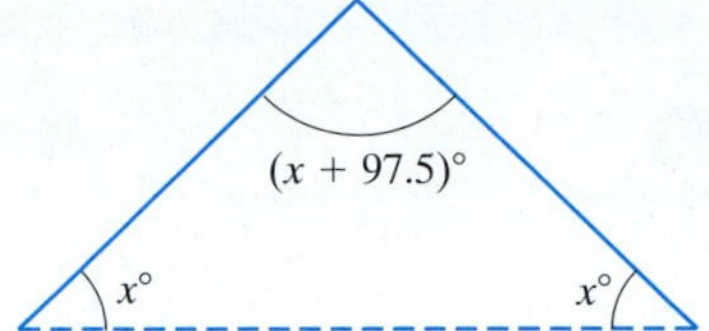

2. TRANSLATE. Recall that the sum of the measures of the angles of a triangle equals 180.

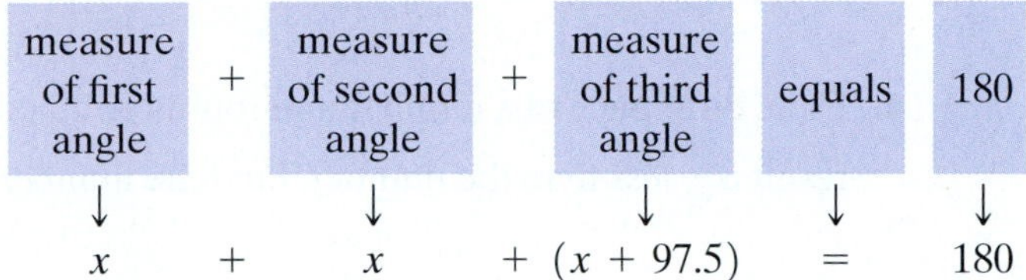

3. SOLVE.

$$x + x + (x + 97.5) = 180$$

$$3x + 97.5 = 180 \quad \text{Combine like terms.}$$

$$3x + 97.5 - 97.5 = 180 - 97.5 \quad \text{Subtract 97.5 from both sides.}$$

$$3x = 82.5$$

$$\frac{3x}{3} = \frac{82.5}{3} \quad \text{Divide both sides by 3.}$$

$$x = 27.5$$

4. INTERPRET.

Check: If $x = 27.5$, then the measure of the third angle is $x + 97.5 = 125$. The sum of the angles is then $27.5 + 27.5 + 125 = 180$, the correct sum.

State: The third angle measures 125°.*

Work Practice Problem 6

*The two walls actually meet at an angle of 125 degrees 12 minutes. The measurement of 97.5° given in the problem is an approximation.

9.4 EXERCISE SET

FOR EXTRA HELP

Student Solutions Manual

PH Math/Tutor Center

CD/Video for Review

MathXL®

MyMathLab

Objective A *Solve. See Example 1.*

1. Twice the difference of a number and 8 is equal to three times the sum of the number and 3. Find the number.

2. Five times the sum of a number and -1 is the same as 6 times the number. Find the number.

3. The product of twice a number and three is the same as the difference of five times the number and $\frac{3}{4}$. Find the number.

4. If the difference of a number and four is doubled, the result is $\frac{1}{4}$ less than the number. Find the number.

Solve. See Example 2.

5. The left and right page numbers of an open book are two consecutive integers whose sum is 469. Find these page numbers.

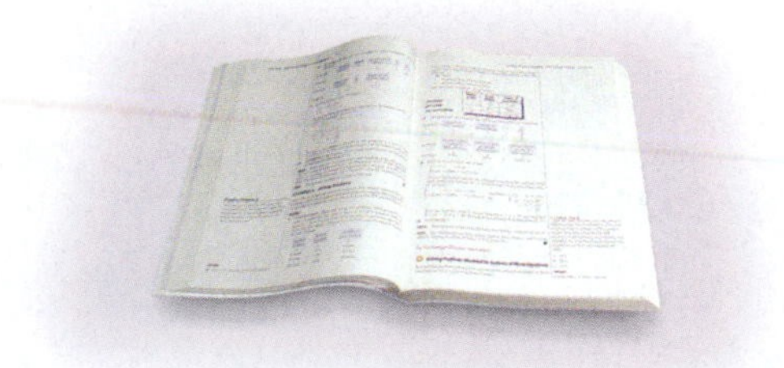

6. The room numbers of two adjacent classrooms are two consecutive even numbers. If their sum is 654, find the classroom numbers.

7. To make an international telephone call, you need the code for the country you are calling. The codes for Belgium, France, and Spain are three consecutive integers whose sum is 99. Find the code for each country. (*Source: The World Almanac and Book of Facts*)

8. The code to unlock a student's combination lock happens to be three consecutive odd integers whose sum is 51. Find the integers.

Solve. See Examples 3 and 4.

9. A 25-inch piece of steel is cut into three pieces so that the second piece is twice as long as the first piece, and the third piece is one inch more than five times the length of the first piece. Find the lengths of the pieces.

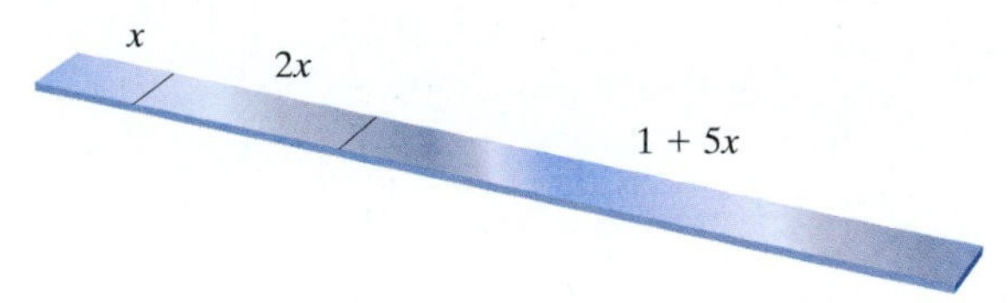

10. A 46-foot piece of rope is cut into three pieces so that the second piece is three times as long as the first piece, and the third piece is two feet more than seven times the length of the first piece. Find the lengths of the pieces.

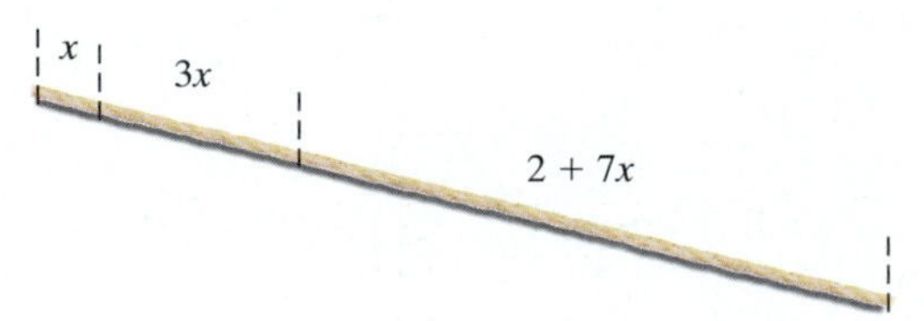

11. A 40-inch board is to be cut into three pieces so that the second piece is twice as long as the first piece and the third piece is 5 times as long as the first piece. If x represents the length of the first piece, find the lengths of all three pieces.

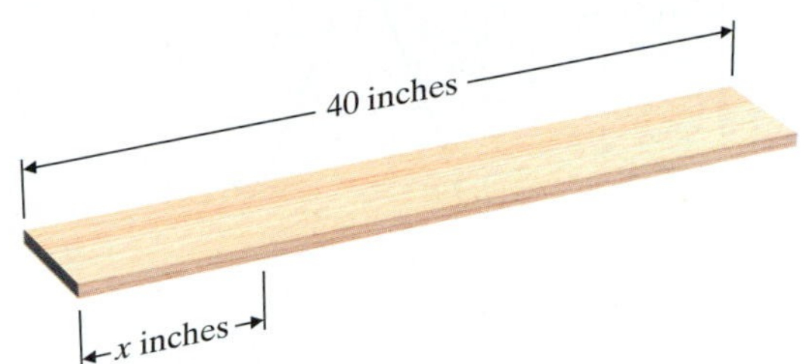

12. A 21-foot beam is to be divided so that the longer piece is 1 foot more than 3 times the shorter piece. If x represents the length of the shorter piece, find the lengths of both pieces.

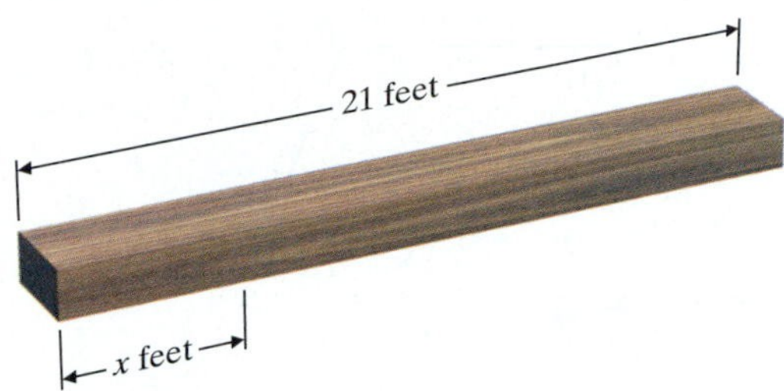

13. The governor of California earns \$50,425 more than the governor of Florida. If the total of their salaries is \$299,575, find the salaries of each. (*Source: The World Almanac,* 2005)

14. In the 2004 Summer Olympics, the United States team won 3 more gold medals than the Chinese team. If the total number of gold medals won by both teams was 67, find the number of gold medals won by each team. (*Source:* Wikipedia)

Solve. See Example 5.

15. A car rental agency advertised renting a Buick Century for \$24.95 per day and \$0.29 per mile. If you rent this car for 2 days, how many whole miles can you drive on a \$100 budget?

16. A plumber gave an estimate for the renovation of a kitchen. Her hourly pay is \$27 per hour and the plumbing parts will cost \$80. If her total estimate is \$404, how many hours does she expect this job to take?

17. In one U.S. city, the taxi cost is \$3 plus \$0.80 per mile. If you are traveling from the airport, there is an additional charge of \$4.50 for tolls. How far can you travel from the airport by taxi for \$27.50?

18. A professional carpet cleaning service charges \$30 plus \$25.50 per hour to come to your home. If your total bill from this company is \$119.25 before taxes, for how many hours were you charged?

Solve. See Example 6.

△ 19. The flag of Equatorial Guinea contains an isosceles triangle. (Recall that an isosceles triangle contains two angles with the same measure.) If the measure of the third angle of the triangle is 30° more than twice the measure of either of the other two angles, find the measure of each angle of the triangle. (*Hint:* Recall that the sum of the measures of the angles of a triangle is 180°.)

△ 20. The flag of Brazil contains a parallelogram. One angle of the parallelogram is 15° less than twice the measure of the angle next to it. Find the measure of each angle of the parallelogram. (*Hint:* Recall that opposite angles of a parallelogram have the same measure and that the sum of the measures of the angles is 360°.)

21. The sum of the measures of the angles of a parallelogram is 360°. In the parallelogram below, angles A and D have the same measure as well as angles C and B. If the measure of angle C is twice the measure of angle A, find the measure of each angle.

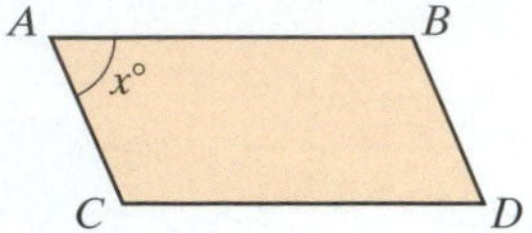

22. Recall that the sum of the measures of the angles of a triangle is 180°. In the triangle below, angle C has the same measure as angle B, and angle A measures 42° less than angle B. Find the measure of each angle.

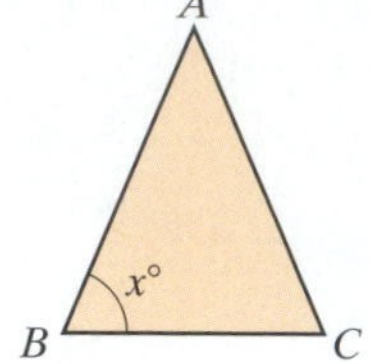

Mixed Practice

23. A 17-foot piece of string is cut into two pieces so that the longer piece is 2 feet longer than twice the shorter piece. Find the lengths of both pieces.

24. A 25-foot wire is to be cut so that the longer piece is one foot longer than 5 times the shorter piece. Find the length of each piece.

25. From 1997 to 2001, the number of prescriptions written for ADHD drugs increased by 5.5 million. If the sum of the number of prescriptions for these two years is 35.7 million, find the number of prescriptions for each year. Check to see that your results agree with the heights of the bars in the graph.

Number of Prescriptions Written for ADHD Drugs

Number of prescriptions written (in millions)

0 10 20 30

1997 1998 1999 2000 2001

Year

Source: IMS Health

26. The Pentagon Building in Washington, D.C., is the headquarters for the U.S. Department of Defense. The Pentagon is also the world's largest office building in terms of ground space with a floor area of over 6.5 million square feet. This is three times the floor area of the Empire State Building. About how much floor space does the Empire State Building have? Round to the nearest tenth.

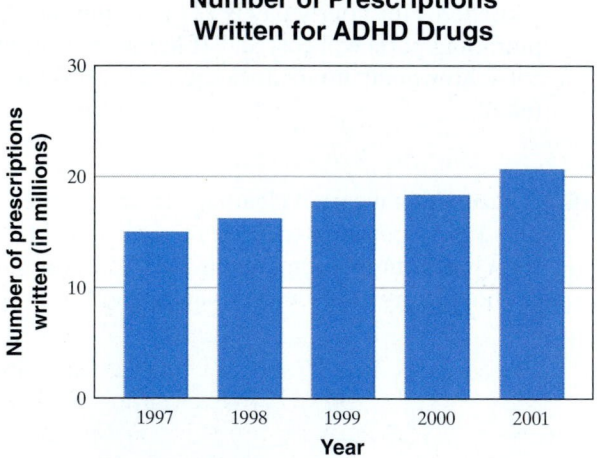

27. Two angles are supplementary if their sum is 180°. One angle measures three times the measure of a smaller angle. If x represents the measure of the smaller angle and these two angles are supplementary, find the measure of each angle.

△ **28.** Two angles are complementary if their sum is 90°. Given the measures of the complementary angles shown, find the measure of each angle.

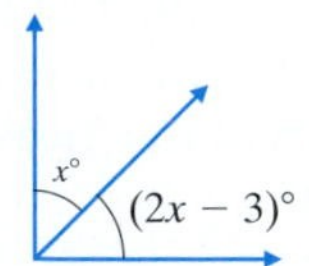

△ **29.** The measures of the angles of a triangle are 3 consecutive even integers. Find the measure of each angle.

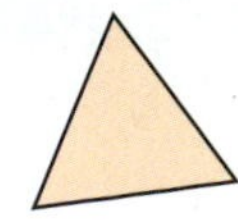

△ **30.** A quadrilateral is a polygon with 4 sides. The sum of the measures of the 4 angles in a quadrilateral is 360°. If the measures of the angles of a quadrilateral are consecutive odd integers, find the measures.

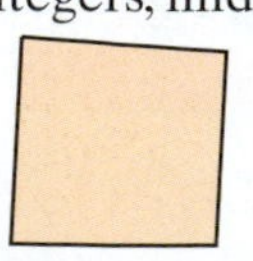

31. The sum of $\frac{1}{5}$ and twice a number is equal to $\frac{4}{5}$ subtracted from three times the number. Find the number.

32. The sum of $\frac{2}{3}$ and four times a number is equal to $\frac{5}{6}$ subtracted from five times the number. Find the number.

33. Hertz Car Rental charges a daily rate of \$39 plus \$0.20 per mile for a certain car. Suppose that you rent that car for a day and your bill (before taxes) is \$95. How many miles did you drive?

34. A woman's \$15,000 estate is to be divided so that her husband receives twice as much as her son. Find the amount of money that her husband receives and the amount of money that her son receives.

35. The winner of the NASCAR Winston Cup in 2003 was Matt Kenseth. Kenseth earned 90 more points than his closest rival, Jimmie Johnson. Together they earned 9954 points. How many points did each driver accumulate during the 2003 Winston Cup Series?

36. During the 2004 Houston Bowl, University of Colorado beat University of Texas–El Paso by 5 points. If their combined scores totaled 61, find the individual team scores. (*Source:* ESPN)

37. The number of counties in California and the number of counties in Montana are consecutive even integers whose sum is 114. If California has more counties than Montana, how many counties does each state have? (*Source: The World Almanac and Book of Facts,* 2006)

38. After a recent election, there were 2 more Republican governors than Democratic governors in the United States. How many Democrats and how many Republicans held governor's offices after this election? (*Source: The World Almanac and Book of Facts*)

39. Over the past few years the satellite Voyager II has passed by the planets Saturn, Uranus, and Neptune, continually updating information about these planets, including the number of moons for each. Uranus is now believed to have 13 more moons than Neptune. Also, Saturn is now believed to have 2 more than twice the number of moons of Neptune. If the total number of moons for these planets is 47, find the number of moons for each planet. (*Source:* National Space Science Data Center)

40. On April 7, 2001, the Mars Odyssey spacecraft was launched, beginning a multi-year mission to observe and map the planet Mars. Mars Odyssey was launched on Boeing's Delta II 7925 launch vehicle using nine strap-on solid rocket motors. Each solid rocket motor has a height that is 8 meters more than 5 times its diameter. If the sum of the height and the diameter for a single solid rocket motor is 14 meters, find each dimension. (*Source:* NASA)

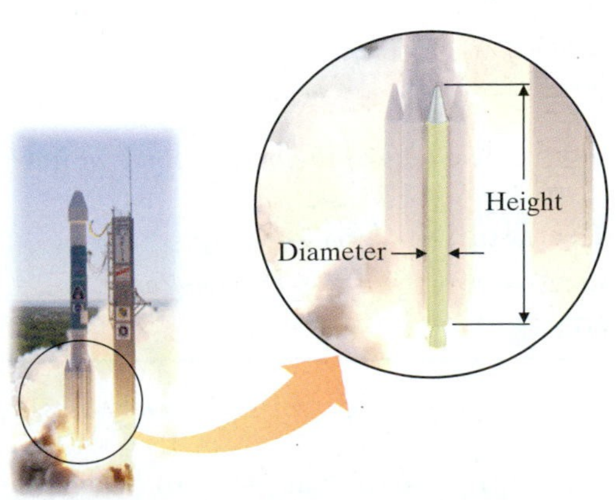

41. If the sum of a number and five is tripled, the result is one less than twice the number. Find the number.

42. Twice the sum of a number and six equals three times the sum of the number and four. Find the number.

43. The area of the Sahara Desert is 7 times the area of the Gobi Desert. If the sum of their areas is 4,000,000 square miles, find the area of each desert.

44. The largest meteorite in the world is the Hoba West located in Namibia. Its weight is 3 times the weight of the Armanty meteorite located in Outer Mongolia. If the sum of their weights is 88 tons, find the weight of each.

45. In the 2004 summer Olympics, France won more gold medals than Italy, who won more gold medals than Korea. If the total number of gold medals won by these three countries is three consecutive integers whose sum is 30, find the number of gold medals won by each. (*Source: The World Almanac*, 2006)

46. To make an international telephone call, you need the code for the country you are calling. The codes for Mali Republic, Côte d'Ivoire, and Niger are three consecutive odd integers whose sum is 675. Find the code for each country.

47. In a recent election in Florida for a seat in the United States House of Representatives, Corrine Brown received 13,288 more votes than Bill Randall. If the total number of votes was 119,436, find the number of votes for each candidate.

48. In a recent election in Texas for a seat in the United States House of Representatives, Max Sandlin received 25,557 more votes than opponent Dennis Boerner. If the total number of votes was 135,821, find the number of votes for each candidate. (*Source:* Voter News Service)

The graph below shows the states with the highest tourism budgets. Use the graph for Exercises 49 through 52.

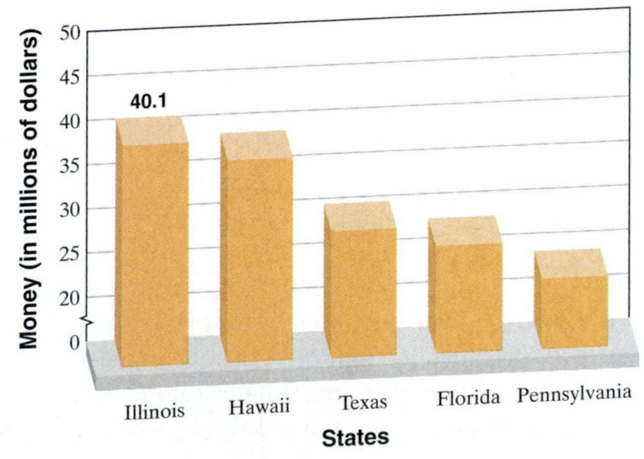

49. Which state spends the most money on tourism?

50. Which states spend between \$25 and \$30 million on tourism?

51. The states of Texas and Florida spend a total of \$56.6 million for tourism. The state of Texas spends \$2.2 million more than the state of Florida. Find the amount that each state spends on tourism.

52. The states of Hawaii and Pennsylvania spend a total of \$60.9 million for tourism. The state of Hawaii spends \$8.1 million less than twice the amount of money that the state of Pennsylvania spends. Find the amount that each state spends on tourism.

Compare the heights of the bars in the graph with your results of the exercises below. Are your answers reasonable?

53. Exercise 51

54. Exercise 52

Review

Evaluate each expression for the given values. See Section 1.8.

55. $2W + 2L$; $W = 7$ and $L = 10$

56. $\frac{1}{2}Bh$; $B = 14$ and $h = 22$

57. πr^2; $r = 15$

58. $r \cdot t$; $r = 15$ and $t = 2$

Concept Extensions

△ **59.** A golden rectangle is a rectangle whose length is approximately 1.6 times its width. The early Greeks thought that a rectangle with these dimensions was the most pleasing to the eye and examples of the golden rectangle are found in many early works of art. For example, the Parthenon in Athens contains many examples of golden rectangles.

Mike Hallahan would like to plant a rectangular garden in the shape of a golden rectangle. If he has 78 feet of fencing available, find the dimensions of the garden.

60. Dr. Dorothy Smith gave the students in her geometry class at the University of New Orleans the following question. Is it possible to construct a triangle such that the second angle of the triangle has a measure that is twice the measure of the first angle and the measure of the third angle is 5 times the measure of the first? If so, find the measure of each angle. (*Hint:* Recall that the sum of the measures of the angles of a triangle is 180°.)

61. The human eye blinks once every 5 seconds on average. How many times does the average eye blink in one hour? In one 16-hour day while awake? In one year?

62. Give an example of how you recently solved a problem using mathematics.

63. In your own words, explain why a solution of a word problem should be checked using the original wording of the problem and not the equation written from the wording.

Recall from Exercise 59 that a golden rectangle is a rectangle whose length is approximately 1.6 times its width.

△ **64.** It is thought that for about 75% of adults, a rectangle in the shape of the golden rectangle is the most pleasing to the eye. Draw three rectangles, one in the shape of the golden rectangle, and poll your class. Do the results agree with the percentage given above?

△ **65.** Examples of golden rectangles can be found today in architecture and manufacturing packaging. Find an example of a golden rectangle in your home. A few suggestions: the front face of a book, the floor of a room, the front of a box of food.

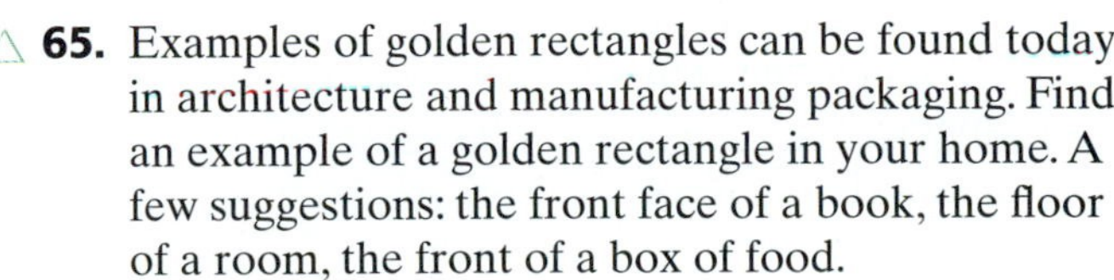

△ **66.** Measure the dimensions of each rectangle and decide which one best approximates the shape of a golden rectangle.

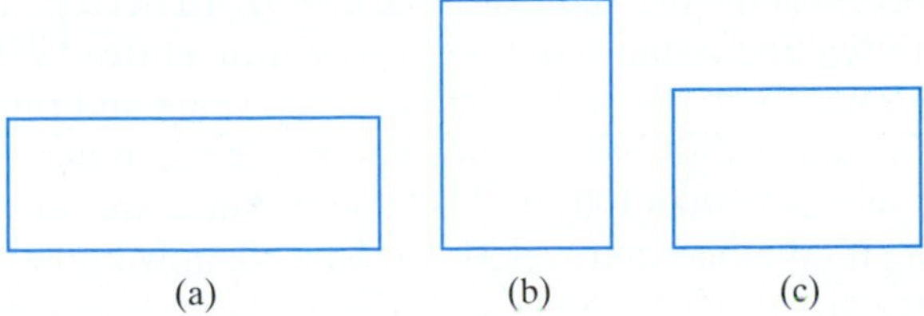

Objectives

A Use Formulas to Solve Problems.

B Solve a Formula or Equation for One of Its Variables.

9.5 FORMULAS AND PROBLEM SOLVING

Objective A Using Formulas to Solve Problems

A **formula** describes a known relationship among quantities. Many formulas are given as equations. For example, the formula

$$d = r \cdot t$$

stands for the relationship

$$\text{distance} = \text{rate} \cdot \text{time}$$

Let's look at one way that we can use this formula.

If we know we traveled a distance of 100 miles at a rate of 40 miles per hour, we can replace the variables d and r in the formula $d = rt$ and find our travel time, t.

$d = rt$ Formula

$100 = 40t$ Replace d with 100 and r with 40.

To solve for t, we divide both sides of the equation by 40.

$\frac{100}{40} = \frac{40t}{40}$ Divide both sides by 40.

$\frac{5}{2} = t$ Simplify.

The travel times was $\frac{5}{2}$ hours, or $2\frac{1}{2}$ hours, or 2.5 hours.

In this section, we solve problems that can be modeled by known formulas. We use the same problem-solving strategy that was introduced in the previous section.

PRACTICE PROBLEM 1

A family is planning their vacation to visit relatives. They will drive from Cincinnati, Ohio, to Rapid City, South Dakota, a distance of 1180 miles. They plan to average a rate of 50 miles per hour. How much time will they spend driving?

EXAMPLE 1 Finding Time Given Rate and Distance

A glacier is a giant mass of rocks and ice that flows downhill like a river. Portage Glacier in Alaska is about 6 miles, or 31,680 *feet,* long and moves 400 *feet* per year. Icebergs are created when the front end of the glacier flows into Portage Lake. How long does it take for ice at the head (beginning) of the glacier to reach the lake?

Solution:

1. UNDERSTAND. Read and reread the problem. The appropriate formula needed to solve this problem is the distance formula, $d = rt$. To become familiar with this formula, let's find the distance that ice traveling at a rate of 400 feet per year travels in 100 years. To do so, we let time t be 100 years and rate r be the given 400 feet per year, and substitute these values into the formula $d = rt$. We then have that distance $d = 400(100) = 40{,}000$ feet. Since we are interested in finding how long it takes ice to travel 31,680 feet, we now know that it is less than 100 years.

Answer

1. 23.6 hours

Since we are using the formula $d = rt$, we let

t = the time in years for ice to reach the lake
r = rate or speed of ice
d = distance from beginning of glacier to lake

2. TRANSLATE. To translate to an equation, we use the formula $d = rt$ and let distance $d = 31{,}680$ feet and rate $r = 400$ feet per year.

$$d = r \cdot t$$
$$31{,}680 = 400 \cdot t \quad \text{Let } d = 31{,}680 \text{ and } r = 400.$$

3. SOLVE. Solve the equation for t. To solve for t, divide both sides by 400.

$$\frac{31{,}680}{400} = \frac{400 \cdot t}{400} \quad \text{Divide both sides by 400.}$$
$$79.2 = t \quad \text{Simplify.}$$

4. INTERPRET.

Check: To check, substitute 79.2 for t and 400 for r in the distance formula and check to see that the distance is 31,680 feet.

State: It takes 79.2 years for the ice at the head of Portage Glacier to reach the lake.

Helpful Hint Don't forget to include units, if appropriate.

Work Practice Problem 1

EXAMPLE 2 Calculating the Length of a Garden

Charles Pecot can afford enough fencing to enclose a rectangular garden with a perimeter of 140 feet. If the width of his garden is to be 30 feet, find the length.

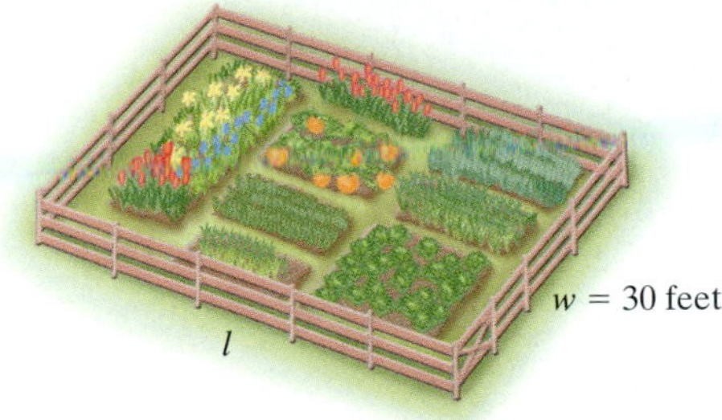

Solution:

1. UNDERSTAND. Read and reread the problem. The formula needed to solve this problem is the formula for the perimeter of a rectangle, $P = 2l + 2w$. Before continuing, let's become familar with this formula.

l = the length of the rectangular garden
w = the width of the rectangular garden
P = perimeter of the garden

2. TRANSLATE. To translate to an equation, we use the formula $P = 2l + 2w$ and let perimeter $P = 140$ feet and width $w = 30$ feet.

$$P = 2l + 2w \quad \text{Let } P = 140 \text{ and } w = 30.$$
$$140 = 2l + 2(30)$$

Continued on next page

PRACTICE PROBLEM 2

A wood deck is being built behind a house. The width of the deck must be 18 feet because of the shape of the house. If there is 450 square feet of decking material, find the length of the deck.

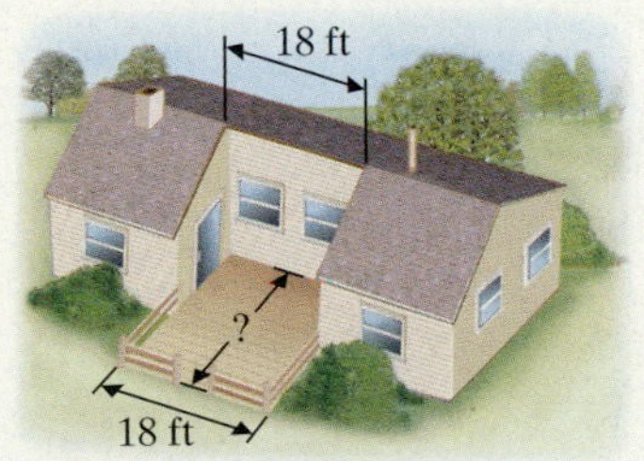

Answer
2. 25 feet

3. SOLVE.

$$140 = 2l + 2(30)$$
$$140 = 2l + 60 \quad \text{Multiply } 2(30).$$
$$140 - 60 = 2l + 60 - 60 \quad \text{Subtract 60 from both sides.}$$
$$80 = 2l \quad \text{Combine like terms.}$$
$$40 = l \quad \text{Divide both sides by 2.}$$

4. INTERPRET.

Check: Substitute 40 for l and 30 for w in the perimeter formula and check to see that the perimeter is 140 feet.

State: The length of the rectangular garden is 40 feet.

Work Practice Problem 2

PRACTICE PROBLEM 3

Convert the temperature 5°C to Fahrenheit.

EXAMPLE 3 Finding an Equivalent Temperature

The average maximum temperature for January in Algiers, Algeria, is 59° Fahrenheit. Find the equivalent temperature in degrees Celsius.

Solution:

1. UNDERSTAND. Read and reread the problem. A formula that can be used to solve this problem is the formula for converting degrees Celsius to degrees Fahrenheit, $F = \frac{9}{5}C + 32$. Before continuing, become familiar with this formula. Using this formula, we let

   C = temperature in degrees Celsius, and

   F = temperature in degrees Fahrenheit.

2. TRANSLATE. To translate to an equation, we use the formula $F = \frac{9}{5}C + 32$ and let degrees Fahrenheit $F = 59$.

   Formula: $F = \frac{9}{5}C + 32$

   Substitute: $59 = \frac{9}{5}C + 32$ Let $F = 59$.

3. SOLVE.

$$59 = \frac{9}{5}C + 32$$
$$59 - 32 = \frac{9}{5}C + 32 - 32 \quad \text{Subtract 32 from both sides.}$$
$$27 = \frac{9}{5}C \quad \text{Combine like terms.}$$
$$\frac{5}{9} \cdot 27 = \frac{5}{9} \cdot \frac{9}{5}C \quad \text{Multiply both sides by } \frac{5}{9}.$$
$$15 = C \quad \text{Simplify.}$$

4. INTERPRET.

Check: To check, replace C with 15 and F with 59 in the formula and see that a true statement results.

State: Thus, 59° Fahrenheit is equivalent to 15° Celsius.

Work Practice Problem 3

Answer

3. 41°F

In the next example, we again use the formula for perimeter of a rectangle as in Example 2. In Example 2, we knew the width of the rectangle. In this example, both the length and width are unknown.

△ EXAMPLE 4 Finding Road Sign Dimensions

The length of a rectangular road sign is 2 feet less than three times its width. Find the dimensions if the perimeter is 28 feet.

Solution:

1. UNDERSTAND. Read and reread the problem. Recall that the formula for the perimeter of a rectangle is $P = 2l + 2w$. Draw a rectangle and guess the solution. If the width of the rectangular sign is 5 feet, its length is 2 feet less than 3 times the width or 3(5 feet) − 2 feet = 13 feet. The perimeter P of the rectangle is then 2(13 feet) + 2(5 feet) = 36 feet, too much. We now know that the width is less than 5 feet.

   Proposed rectangle:

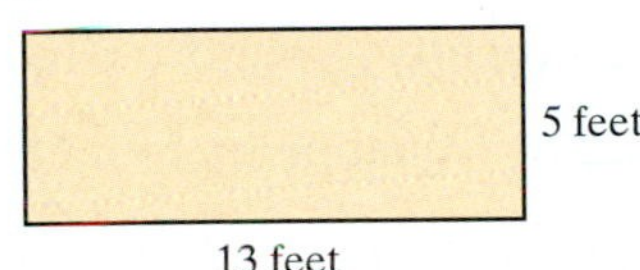

Let

$w =$ the width of the rectangular sign; then

$3w - 2 =$ the length of the sign.

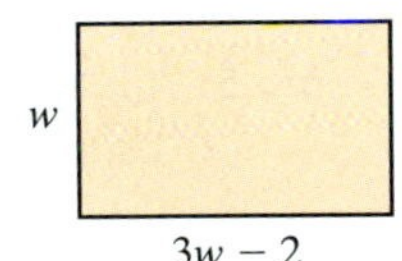

Draw a rectangle and label it with the assigned variables.

2. TRANSLATE.

Formula: $P = 2l + 2w$

Substitute: $28 = 2(3w - 2) + 2w$.

3. SOLVE.

$$28 = 2(3w - 2) + 2w$$
$$28 = 6w - 4 + 2w \quad \text{Apply the distributive property.}$$
$$28 = 8w - 4$$
$$28 + 4 = 8w - 4 + 4 \quad \text{Add 4 to both sides.}$$
$$32 = 8w$$
$$\frac{32}{8} = \frac{8w}{8} \quad \text{Divide both sides by 8.}$$
$$4 = w$$

4. INTERPRET.

Check: If the width of the sign is 4 feet, the length of the sign is 3(4 feet) − 2 feet = 10 feet. This gives a perimeter of P = 2(4 feet) + 2(10 feet) = 28 feet, the correct perimeter.

State: The width of the sign is 4 feet and the length of the sign is 10 feet.

Work Practice Problem 4

PRACTICE PROBLEM 4

The length of a rectangle is one more meter than 4 times its width. Find the dimensions if the perimeter is 52 meters.

Answer

4. length: 21m; width: 5m

Objective B Solving a Formula for a Variable

We say that the formula

$$d = rt$$

is solved for d because d is alone on one side of the equation and the other side contains no d's. Suppose that we have a large number of problems to solve where we are given distance d and rate r and asked to find time t. In this case, it may be easier to first solve the formula $d = rt$ for t. To solve for t, we divide both sides of the equation by r.

$$d = rt$$

$$\frac{d}{r} = \frac{rt}{r} \quad \text{Divide both sides by } r.$$

$$\frac{d}{r} = t \quad \text{Simplify.}$$

To solve a formula or an equation for a specified variable, we use the same steps as for solving a linear equation except that we treat the specified variable as the only variable in the equation. These steps are listed next.

Solving Equations for a Specified Variable

Step 1: Multiply on both sides to clear the equation of fractions if they occur.

Step 2: Use the distributive property to remove parentheses if they occur.

Step 3: Simplify each side of the equation by combining like terms.

Step 4: Get all terms containing the specified variable on one side and all other terms on the other side by using the addition property of equality.

Step 5: Get the specified variable alone by using the multiplication property of equality.

PRACTICE PROBLEM 5

Solve $C = 2\pi r$ for r. (This formula is used to find the circumference C of a circle given its radius r.)

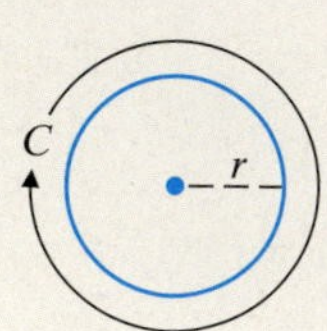

EXAMPLE 5 Solve $V = lwh$ for l.

Solution: This formula is used to find the volume of a box. To solve for l, we divide both sides by wh.

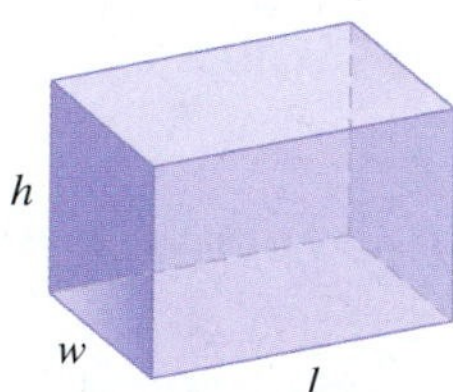

$$V = lwh$$

$$\frac{V}{wh} = \frac{lwh}{wh} \quad \text{Divide both sides by } wh.$$

$$\frac{V}{wh} = l \quad \text{Simplify.}$$

Since we have l alone on one side of the equation, we have solved for l in terms of V, w, and h. Remember that it does not matter on which side of the equation we get the variable alone.

Work Practice Problem 5

Answer

5. $r = \frac{C}{2\pi}$

EXAMPLE 6 Solve $y = mx + b$ for x.

Solution: First we get mx alone by subtracting b from both sides.

$$y = mx + b$$

$$y - b = mx + b - b \quad \text{Subtract } b \text{ from both sides.}$$

$$y - b = mx \quad \text{Combine like terms.}$$

Next we solve for x by dividing both sides by m.

$$\frac{y - b}{m} = \frac{mx}{m}$$

$$\frac{y - b}{m} = x \quad \text{Simplify.}$$

Work Practice Problem 6

PRACTICE PROBLEM 6

Solve $P = 2l + 2w$ for l.

✔Concept Check Solve:

a. ● = ■ − ■ for ■

b. ● = ■ · ▲ − ■ for ■

EXAMPLE 7 Solve $P = 2l + 2w$ for w.

Solution: This formula relates the perimeter of a rectangle to its length and width. Find the term containing the variable w. To get this term, $2w$, alone subtract $2l$ from both sides.

$$P = 2l + 2w$$

$$P - 2l = 2l + 2w - 2l \quad \text{Subtract } 2l \text{ from both sides.}$$

$$P - 2l = 2w \quad \text{Combine like terms.}$$

$$\frac{P - 2l}{2} = \frac{2w}{2} \quad \text{Divide both sides by 2.}$$

$$\frac{P - 2l}{2} = w \quad \text{Simplify.}$$

l w

Work Practice Problem 7

PRACTICE PROBLEM 7

Solve $P = 2a + b - c$ for a.

Helpful Hint

The 2's may *not* be divided out here. Although 2 is a factor of the denominator, 2 is *not* a factor of the numerator since it is not a factor of both terms in the numerator.

The next example has an equation containing a fraction. We will first clear the equation of fractions and then solve for the specified variable.

EXAMPLE 8 Solve $F = \frac{9}{5}C + 32$ for C.

Solution:

$$F = \frac{9}{5}C + 32$$

$$5(F) = 5\left(\frac{9}{5}C + 32\right) \quad \text{Clear the fraction by multiplying both sides by the LCD.}$$

$$5F = 9C + 160 \quad \text{Distribute the 5.}$$

$$5F - 160 = 9C + 160 - 160 \quad \text{To get the term containing the variable } C \text{ alone, subtract 160 from both sides.}$$

$$5F - 160 = 9C \quad \text{Combine like terms.}$$

$$\frac{5F - 160}{9} = \frac{9C}{9} \quad \text{Divide both sides by 9.}$$

$$\frac{5F - 160}{9} = C \quad \text{Simplify.}$$

Work Practice Problem 8

PRACTICE PROBLEM 8

Solve $A = \frac{a + b}{2}$ for b.

Answers

6. $l = \frac{P - 2w}{2}$ 7. $a = \frac{P - b + c}{2}$

8. $b = 2A - a$

✔ Concept Check Answer

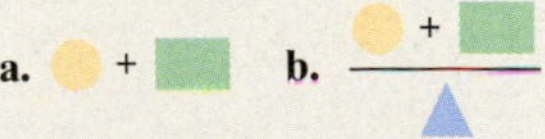

9.5 EXERCISE SET

FOR EXTRA HELP

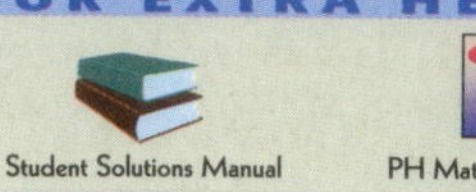

PH Math/Tutor Center

Objective A *Substitute the given values into each given formula and solve for the unknown variable. If necessary, round to one decimal place. See Examples 1 through 4.*

△ **1.** $A = bh$; $A = 45, b = 15$ (Area of a parallelogram)

2. $d = rt$; $d = 195, t = 3$ (Distance formula)

△ **3.** $S = 4lw + 2wh$; $S = 102, l = 7, w = 3$ (Surface area of a special rectangular box)

△ **4.** $V = lwh$; $l = 14, w = 8, h = 3$ (Volume of a rectangular box)

△ **5.** $A = \frac{1}{2}h(B + b)$; $A = 180, B = 11, b = 7$ (Area of a trapezoid)

△ **6.** $A = \frac{1}{2}h(B + b)$; $A = 60, B = 7, b = 3$ (Area of a trapezoid)

△ **7.** $P = a + b + c$; $P = 30, a = 8, b = 10$ (Perimeter of a triangle)

△ **8.** $V = \frac{1}{3}Ah$; $V = 45, h = 5$ (Volume of a pyramid)

△ **9.** $C = 2\pi r$; $C = 15.7$ (Circumference of a circle) (Use the approximation 3.14 for π.)

△ **10.** $A = \pi r^2$; $r = 4$ (Area of a circle) (Use the approximation 3.14 for π.)

Objective B *Solve each formula for the specified variable. See Examples 5 through 8.*

11. $f = 5gh$ for h

△ **12.** $x = 4\pi y$ for y

13. $V = lwh$ for w

14. $T = mnr$ for n

15. $3x + y = 7$ for y

16. $-x + y = 13$ for y

17. $A = P + PRT$ for R

18. $A = P + PRT$ for T

19. $V = \frac{1}{3}Ah$ for A

20. $D = \frac{1}{4}fk$ for k

21. $P = a + b + c$ for a

22. $PR = x + y + z + w$ for z

23. $S = 2\pi rh + 2\pi r^2$ for h

△ **24.** $S = 4lw + 2wh$ for h

Solve. See Examples 1 through 4.

25. For the purpose of purchasing new baseboard and carpet,

a. Find the area and perimeter of the room below (neglecting doors).

b. Identify whether baseboard has to do with area or perimeter and the same with carpet.

26. For the purpose of purchasing lumber for a new fence and seed to plant grass,

a. Find the area and perimeter of the yard below.

b. Identify whether a fence has to do with area or perimeter and the same with grass seed.

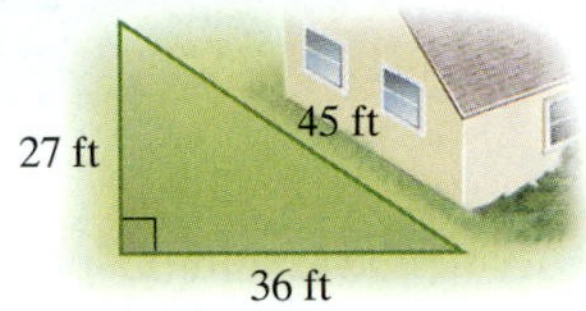

27. A frame shop charges according to both the amount of framing needed to surround the picture and the amount of glass needed to cover the picture.

a. Find the area and perimeter of the picture below.

b. Identify whether the frame has to do with perimeter or area and the same with the glass.

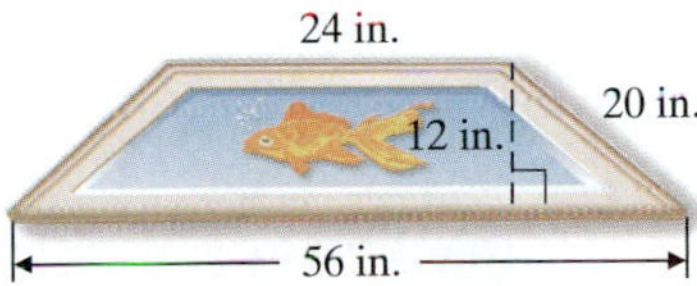

28. A decorator is painting and placing a border completely around the parallelogram-shaped wall.

a. Find the area and perimeter of the wall below.

b. Identify whether the border has to do with perimeter or area and the same with paint.

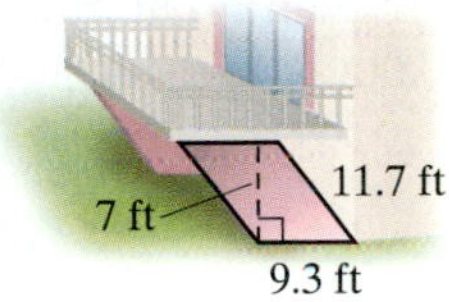

29. The world's largest pink ribbon, the sign of the fight against breast cancer, was erected out of pink post-it notes on a billboard in New York City in October, 2004. If the area of the rectangular billboard covered by the ribbon is approximately 3990 sq ft, and the width of the ribbon was approximately 57 ft, what was the height of this gigantic symbol?

△ **30.** The world's largest sign for Coca-Cola is located in Arica, Chile. The rectangular sign has a length of 400 feet and has an area of 52,400 square feet. Find the width of the sign. (*Source:* Fabulous Facts about Coca-Cola, Atlanta, GA)

31. Convert Nome, Alaska's 14°F high temperature to Celsius.

32. Convert Paris, France's low temperature of −5°C to Fahrenheit.

33. The X-30 is a "space plane" that skims the edge of space at 4000 miles per hour. Neglecting altitude, if the circumference of the Earth is approximately 25,000 miles, how long will it take for the X-30 to travel around the Earth?

34. In the United States, a notable hang glider flight was a 303-mile, $8\frac{1}{2}$ hour flight from New Mexico to Kansas. What was the average rate during this flight?

△ **35.** An architect designs a rectangular flower garden such that the width is exactly two-thirds of the length. If 260 feet of antique picket fencing are to be used to enclose the garden, find the dimensions of the garden.

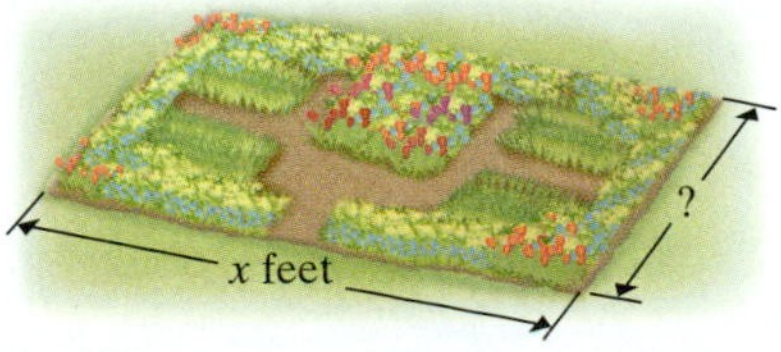

△ **36.** If the length of a rectangular parking lot is 10 meters less than twice its width, and the perimeter is 400 meters, find the length of the parking lot.

△ **37.** A flower bed is in the shape of a triangle with one side twice the length of the shortest side, and the third side is 30 feet more than the length of the shortest side. Find the dimensions if the perimeter is 102 feet.

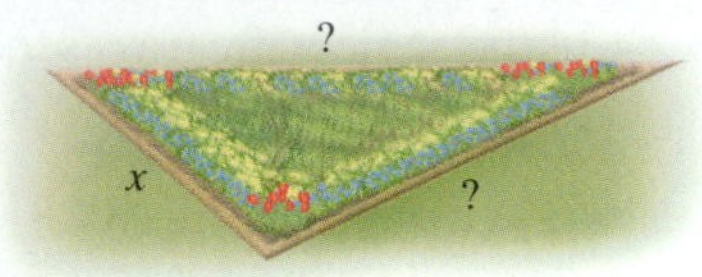

△ **38.** The perimeter of a yield sign in the shape of an isosceles triangle is 22 feet. If the shortest side is 2 feet less than the other two sides, find the length of the shortest side. (*Hint:* An isosceles triangle has two sides the same length.)

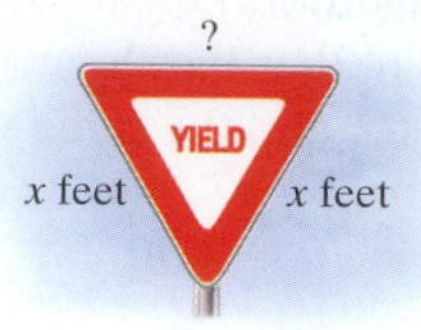

39. The Cat is a high-speed catamaran auto ferry that operates between Bar Harbor, Maine, and Yarmouth, Nova Scotia. The Cat can make the trip in about $2\frac{1}{2}$ hours at a speed of 55 mph. About how far apart are Bar Harbor and Yarmouth? (*Source:* Bay Ferries)

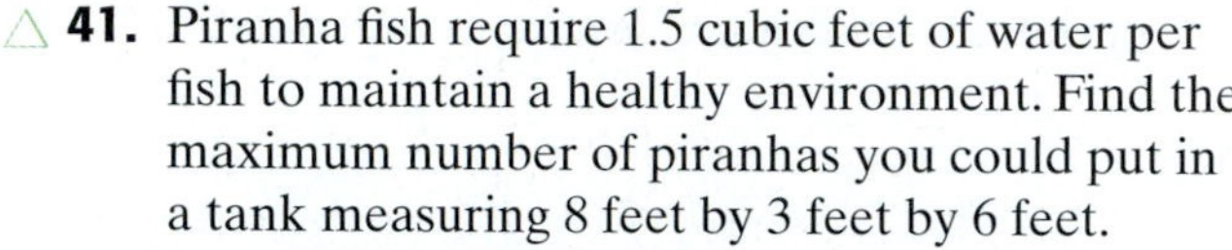

40. A family is planning their vacation to Disney World. They will drive from a small town outside New Orleans, Louisiana, to Orlando, Florida, a distance of 700 miles. They plan to average a rate of 55 mph. How long will this trip take?

△ **41.** Piranha fish require 1.5 cubic feet of water per fish to maintain a healthy environment. Find the maximum number of piranhas you could put in a tank measuring 8 feet by 3 feet by 6 feet.

△ **42.** Find the maximum number of goldfish you can put in a cylindrical tank whose diameter is 8 meters and whose height is 3 meters if each goldfish needs 2 cubic meters of water.

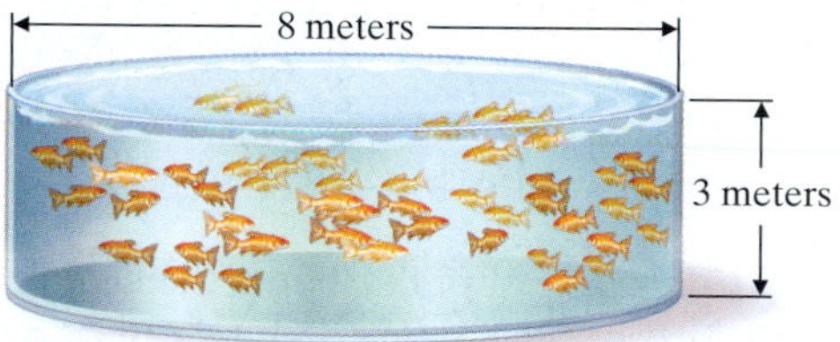

△ **43.** A lawn is in the shape of a trapezoid with a height of 60 feet and bases of 70 feet and 130 feet. How many bags of fertilizer must be purchased to cover the lawn if each bag covers 4000 square feet?

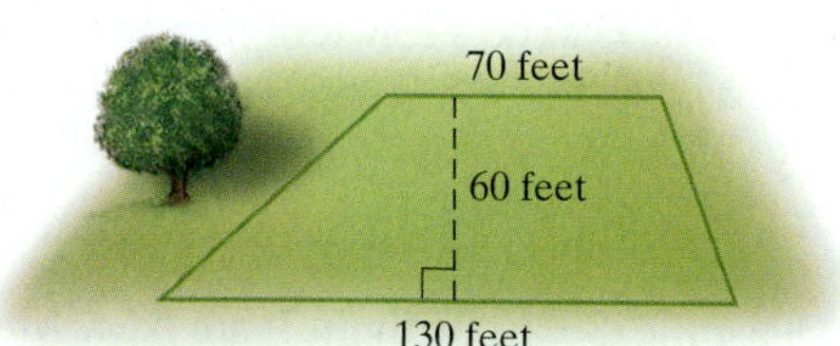

△ **44.** If the area of a right-triangularly shaped sail is 20 square feet and its base is 5 feet, find the height of the sail.

△ **45.** Maria's Pizza sells one 16-inch cheese pizza or two 10-inch cheese pizzas for $9.99. Determine which size gives more pizza.

△ **46.** Find how much rope is needed to wrap around the Earth at the equator, if the radius of the Earth is 4000 miles. (*Hint:* Use 3.14 for π and the formula for circumference.)

47. A Japanese "bullet" train set a new world record for train speed at 552 kilometers per hour during a manned test run on the Yamanashi Maglev Test Line in April 1999. The Yamanashi Maglev Test Line is 42.8 kilometers long. How many *minutes* would a test run on the Yamanashi Line last at this record-setting speed? Round to the nearest hundredth of a minute. (*Source:* Japan Railways Central Co.)

48. In 1983, the Hawaiian volcano Kilauea began erupting in a series of episodes still occurring at the time of this writing. At times, the lava flows advanced at speeds of up to 0.5 kilometer per hour. In 1983 and 1984 lava flows destroyed 16 homes in the Royal Gardens subdivision, about 6 km away from the eruption site. Roughly how long did it take the lava to reach Royal Gardens? (*Source:* U.S. Geological Survey Hawaiian Volcano Observatory)

△ **49.** The perimeter of an equilateral triangle is 7 inches more than the perimeter of a square, and the side of the triangle is 5 inches longer than the side of the square. Find the side of the triangle. (*Hint:* An equilateral triangle has three sides the same length.)

△ **50.** A square animal pen and a pen shaped like an equilateral triangle have equal perimeters. Find the length of the sides of each pen if the sides of the triangular pen are fifteen less than twice a side of the square pen.

51. Find how long it takes Tran Nguyen to drive 135 miles on I-10 if he merges onto I-10 at 10 A.M. and drives nonstop with his cruise control set on 60 mph.

52. Beaumont, Texas, is about 150 miles from Toledo Bend. If Leo Miller leaves Beaumont at 4 A.M. and averages 45 mph, when should he arrive at Toledo Bend?

△ **53.** The longest runway at Los Angeles International Airport has the shape of a rectangle and an area of 1,813,500 square feet. This runway is 150 feet wide. How long is the runway? (*Source:* Los Angeles World Airports)

54. The return stroke of a bolt of lightning can travel at a speed of 87,000 miles per second (almost half the speed of light). At this speed, how many times can an object travel around the world in one second? (See Exercise 46. Round to the nearest tenth.) (*Source: The Handy Science Answer Book*)

55. The highest temperature ever recorded in Europe was 122°F in Seville, Spain, in August of 1881. Convert this record high temperature to Celsius. (*Source:* National Climatic Data Center)

56. The lowest temperature ever recorded in Oceania was −10°C at the Haleakala Summit in Maui, Hawaii, in January 1961. Convert this record low temperature to Fahrenheit. (*Source:* National Climatic Data Center)

△ **57.** The CART FedEx Championship Series is an open-wheeled race car competition based in the United States. A CART car has a maximum length of 199 inches, a maximum width of 78.5 inches, and a maximum height of 33 inches. When the CART series travels to another country for a grand prix, teams must ship their cars. Find the volume of the smallest shipping crate needed to ship a CART car of maximum dimensions. (*Source:* Championship Auto Racing Teams, Inc.)

CART Racing Car

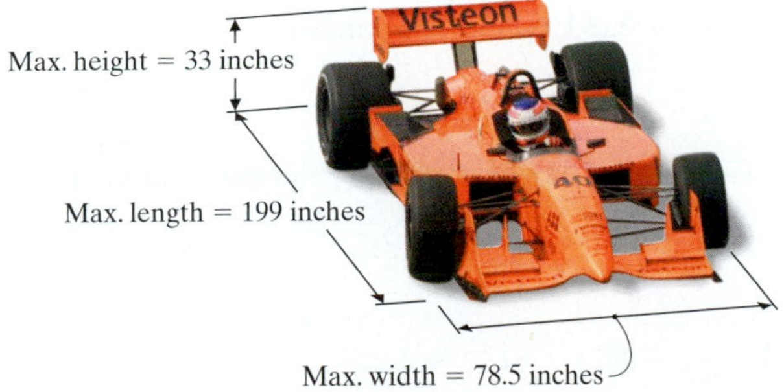

58. On a road course, a CART car's speed can average up to around 105 mph. Based on this speed, how long would it take a CART driver to travel from Los Angeles to New York City, a distance of about 2810 miles by road, without stopping? Round to the nearest tenth of an hour.

△ **59.** The Hoberman Sphere is a toy ball that expands and contracts. When it is completely closed, it has a diameter of 9.5 inches. Find the volume of the Hoberman Sphere when it is completely closed. Use 3.14 for π. Round to the nearest whole cubic inch. (*Hint:* volume of a sphere $= \frac{4}{3}\pi r^3$. *Source:* Hoberman Designs, Inc.)

△ **60.** When the Hoberman Sphere (see Exercise 59) is completely expanded, its diameter is 30 inches. Find the volume of the Hoberman Sphere when it is completely expanded. Use 3.14 for π. Round to the nearest whole cubic inch. (*Source:* Hoberman Designs, Inc.)

61. The average temperature on the planet Mercury is 167°C. Convert this temperature to degrees Fahrenheit. Round to the nearest degree. (*Source:* National Space Science Data Center)

62. The average temperature on the planet Jupiter is −227°F. Convert this temperature to degrees Celsius. Round to the nearest degree. (*Source:* National Space Science Data Center)

Review

Write each percent as a decimal. See Section 6.2.

63. 32%

64. 8%

65. 200%

66. 0.5%

Write each decimal as a percent. See Section 6.2.

67. 0.17

68. 0.03

69. 7.2

70. 5

Concept Extensions

Solve.

71. $N = R + \frac{V}{G}$ for V (Urban forestry: tree plantings per year)

72. $B = \frac{F}{P - V}$ for V (Business: break-even point)

△ **73.** The formula $V = lwh$ is used to find the volume of a box. If the length of a box is doubled, the width is doubled, and the height is doubled, how does this affect the volume? Explain your answer.

△ **74.** The formula $A = bh$ is used to find the area of a parallelogram. If the base of a parallelogram is doubled and its height is doubled, how does this affect the area? Explain your answer.

75. Find the temperature at which the Celsius measurement and Fahrenheit measurement are the same number.

Solve. See the Concept Check in this section.

76. ⬟ · ■ + ▲ = ● for ■

77. ▲ − ● · ▮ = ■ for ●

78. A glacier is a giant mass of rocks and ice that flows downhill like a river. Exit Glacier, near Seward, Alaska, moves at a rate of 20 inches a day. Find the distance in feet the glacier moves in a year. (Assume 365 days a year. Round to two decimal places.)

79. Flying fish do not *actually* fly, but glide. They have been known to travel a distance of 1300 feet at a rate of 20 miles per hour. How many seconds did it take to travel this distance? (*Hint:* First convert miles per hour to feet per second. Recall that 1 mile = 5280 feet. Round to the nearest tenth of a second.)

Substitute the given values into each given formula and solve for the unknown variable. If necessary, round to one decimal place.

80. $I = PRT$; $I = 3750, P = 25{,}000, R = 0.05$ (Simple interest formula)

81. $I = PRT$; $I = 1{,}056{,}000, R = 0.055, T = 6$ (Simple interest formula)

82. $V = \frac{1}{3}\pi r^2 h$; $V = 565.2, r = 6$ (Use a calculator approximation for π.)(Volume of a cone)

83. $V = \frac{4}{3}\pi r^3$; $r = 3$ (Use a calculator approximation for π.) (Volume of a sphere)

STUDY SKILLS BUILDER

Organizing a Notebook

It's never too late to get organized. If you need ideas about organizing a notebook for your mathematics course, try some of these:

- Use a spiral or ring binder notebook with pockets and use it for mathematics only.
- Start each page by writing the book's section number you are working on at the top.
- When your instructor is lecturing, take notes. *Always* include any examples your instructor works for you.
- Place your worked-out homework exercises in your notebook immediately after the lecture notes from that section. This way, a section's worth of material is together.
- Homework exercises: Attempt all assigned homework. For odd-numbered exercises, you are not through until you check your answers against the back of the book. Correct any exercises with incorrect answers. You may want to place a "?" by any homework exercises or notes that you need to ask questions about. Also, consider placing a "!" by any notes or exercises you feel are important.
- Place graded quizzes in the pockets of your notebook. If you are using a binder, you can place your quizzes in a special section of your binder.

Let's check your notebook organization by answering the following questions.

1. Do you have a spiral or ring binder notebook for your mathematics course only?
2. Have you ever had to flip through several sheets of notes and work in your mathematics notebook to determine what section's work you are in?
3. Are you now writing the textbook's section number at the top of each notebook page?
4. Have you ever lost or had trouble finding a graded quiz or test?
5. Are you now placing all your graded work in a dedicated place in your notebook?
6. Are you attempting all of your homework and placing all of your work in your notebook?
7. Are you checking and correcting your homework in your notebook? If not, why not?
8. Are you writing in your notebook the examples your instructor works for you in class?

Objectives

A Graph Inequalities on a Number Line.

B Use the Addition Property of Inequality to Solve Inequalities.

C Use the Multiplication Property of Inequality to Solve Inequali s.

D Use Both Properties to Solve Inequalities.

E Solve Problems Modeled by Inequalities.

9.6 LINEAR INEQUALITIES AND PROBLEM SOLVING

In Section 9.1, we reviewed these inequality symbols and their meanings:

$<$ means "is less than" $\qquad \leq$ means "is less than or equal to"
$>$ means "is greater than" $\qquad \geq$ means "is greater than or equal to"

An **inequality** is a statement that contains one of the symbols above.

Equations	Inequalities
$x = 3$	$x \leq 3$
$5n - 6 = 14$	$5n - 6 > 14$
$12 = 7 - 3y$	$12 \leq 7 - 3y$
$\frac{x}{4} - 6 = 1$	$\frac{x}{4} - 6 > 1$

Objecti **A** Graphing Inequalities on a Number Line

Recall that e single solution to the equation $x = 3$ is 3. The solutions of the inequality $x \leq$ nclude 3 and *all real numbers less than 3* (for example, $-10, \frac{1}{2}, 2,$ and 2.9). Because can't list all numbers less than 3, we show instead a picture of the solutions by gra ng them on a number line.

To graph the solutions of $x \leq 3$, we shade the numbers to the left of 3 since they are less than 3. Then we place a closed circle on the point representing 3. The closed circle indicates that 3 *is* a solution: 3 *is* less than or equal to 3.

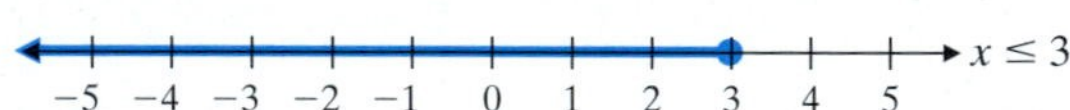

To graph the solutions of $x < 3$, we shade the numbers to the left of 3. Then we place an open circle on the point representing 3. The open circle indicates that 3 *is not* a solution: 3 *is not* less than 3.

PRACTICE PROBLEM 1

Graph: $x \geq -2$

−5 −4 −3 −2 −1 0 1 2 3 4 5

EXAMPLE 1 Graph: $x \geq -1$

Solution: To graph the solutions of $x \geq -1$, we place a closed circle at -1 since the inequality symbol is $\geq$ and -1 is greater than or equal to -1. Then we shade to the right of -1.

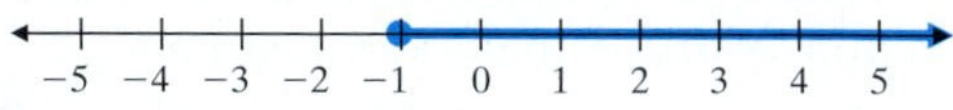

Work Practice Problem 1

PRACTICE PROBLEM 2

Graph: $5 > x$

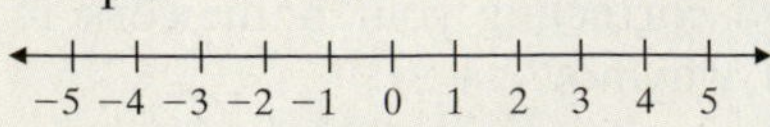

EXAMPLE 2 Graph: $-1 > x$

Solution: Recall from Section 9.1 that $-1 > x$ means the same as $x < -1$. The graph of the solutions of $x < -1$ is shown below.

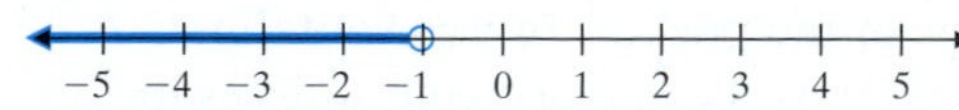

Work Practice Problem 2

Answers

1. 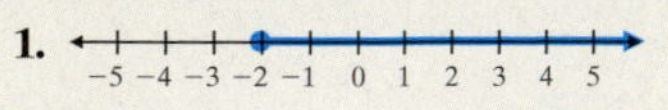

2. −5 −4 −3 −2 −1 0 1 2 3 4 5

EXAMPLE 3 Graph: $-4 < x \le 2$

Solution: We read $-4 < x \le 2$ as "-4 is less than x and x is less than or equal to 2," or as "x is greater than -4 and x is less than or equal to 2." To graph the solutions of this inequality, we place an open circle at -4 (-4 is not part of the graph), a closed circle at 2 (2 is part of the graph), and we shade all numbers between -4 and 2. Why? All numbers between -4 and 2 are greater than -4 *and* also less than 2.

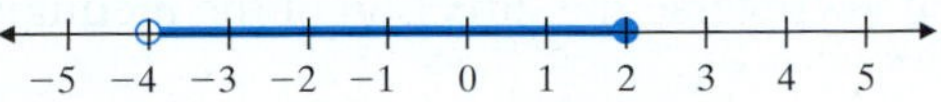

Work Practice Problem 3

PRACTICE PROBLEM 3

Graph: $-3 \le x < 1$

−5 −4 −3 −2 −1 0 1 2 3 4 5

Objective B Using the Addition Property

When solutions of a linear inequality are not immediately obvious, they are found through a process similar to the one used to solve a linear equation. Our goal is to get the variable alone on one side of the inequality. We use properties of inequality similar to properties of equality.

Addition Property of Inequality

If a, b, and c are real numbers, then

$$a < b \quad \text{and} \quad a + c < b + c$$

are equivalent inequalities.

This property also holds true for subtracting values, since subtraction is defined in terms of addition. In other words, adding or subtracting the same quantity from both sides of an inequality does not change the solutions of the inequality.

EXAMPLE 4 Solve $x + 4 \le -6$. Graph the solutions.

Solution: To solve for x, subtract 4 from both sides of the inequality.

$x + 4 \le -6$	Original inequality
$x + 4 - 4 \le -6 - 4$	Subtract 4 from both sides.
$x \le -10$	Simplify.

The graph of the solutions is shown below.

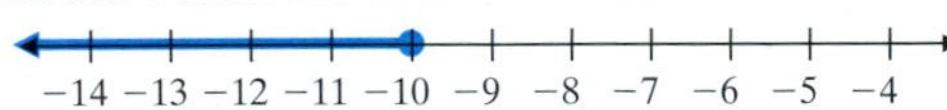

Work Practice Problem 4

PRACTICE PROBLEM 4

Solve $x - 6 \ge -11$. Graph the solutions.

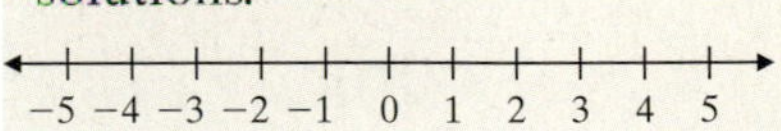

Helpful Hint

Notice that any number less than or equal to -10 is a solution to $x \le -10$. For example, solutions include

$$-10, \quad -200, \quad -11\frac{1}{2}, \quad -\sqrt{130}, \quad \text{and} \quad -50.3$$

Objective C Using the Multiplication Property

An important difference between solving linear equations and solving linear inequalities is shown when we multiply or divide both sides of an inequality by a nonzero real number. For example, start with the true statement $6 < 8$ and multiply both sides by 2. As we see below, the resulting inequality is also true.

$6 < 8$	True
$2(6) < 2(8)$	Multiply both sides by 2.
$12 < 16$	True

Answers

3. 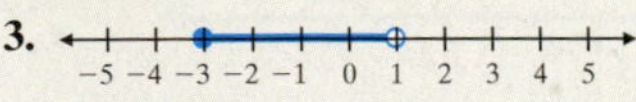

4. $x \ge -5$, 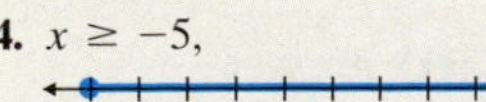

But if we start with the same true statement $6 < 8$ and multiply both sides by -2, the resulting inequality is not a true statement.

$$6 < 8 \quad \text{True}$$
$$-2(6) < -2(8) \quad \text{Multiply both sides by } -2.$$
$$-12 < -16 \quad \text{False}$$

Notice, however, that if we reverse the direction of the inequality symbol, the resulting inequality is true.

$$-12 < -16 \quad \text{False}$$
$$-12 > -16 \quad \text{True}$$

This demonstrates the multiplication property of inequality.

Multiplication Property of Inequality

1. If a, b, and c are real numbers, and c is **positive,** then

$$a < b \quad \text{and} \quad ac < bc$$

are equivalent inequalities.

2. If a, b, and c are real numbers, and c is **negative,** then

$$a < b \quad \text{and} \quad ac > bc$$

are equivalent inequalities.

Because division is defined in terms of multiplication, this property also holds true when dividing both sides of an inequality by a nonzero number: If we multiply or divide both sides of an inequality by a negative number, **the direction of the inequality sign must be reversed for the inequalities to remain equivalent.**

✔ Concept Check Fill in the box with $<$, $>$, $\leq$, or $\geq$.

a. Since $-8 < -4$, then $3(-8) \square 3(-4)$.

b. Since $5 \geq -2$, then $\frac{5}{-7} \square \frac{-2}{-7}$.

c. If $a < b$, then $2a \square 2b$.

d. If $a \geq b$, then $\frac{a}{-3} \square \frac{b}{-3}$.

PRACTICE PROBLEM 5

Solve $-3x \leq 12$. Graph the solutions.

−5 −4 −3 −2 −1 0 1 2 3 4 5

EXAMPLE 5 Solve $-2x \leq -4$. Graph the solutions.

Solution: Remember to reverse the direction of the inequality symbol when dividing by a negative number.

$$-2x \leq -4$$
$$\frac{-2x}{-2} \geq \frac{-4}{-2} \quad \text{Divide both sides by } -2 \text{ and reverse the inequality sign.}$$
$$x \geq 2 \quad \text{Simplify.}$$

The graph of the solutions is shown.

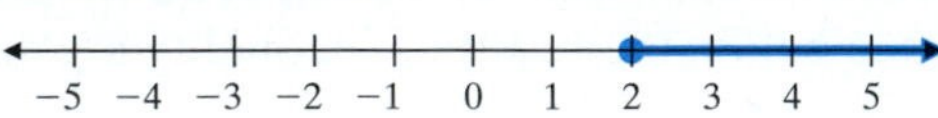

Work Practice Problem 5

Answer

5. $x \geq -4$

−5 −4 −3 −2 −1 0 1 2 3 4 5

✔ Concept Check Answer

a. $<$ b. $\leq$ c. $<$ d. $\leq$

EXAMPLE 6 Solve $2x < -4$. Graph the solutions.

Solution: $2x < -4$

$$\frac{2x}{2} < \frac{-4}{2}$$ Divide both sides by 2. Do not reverse the inequality sign.

$$x < -2$$ Simplify.

The graph of the solutions is shown.

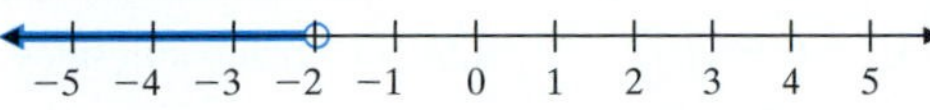

Work Practice Problem 6

PRACTICE PROBLEM 6

Solve $5x > -20$. Graph the solutions.

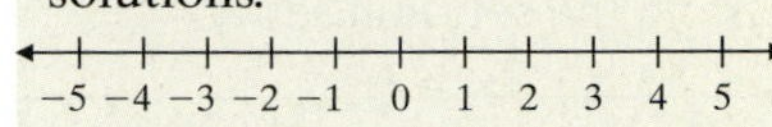

Since we cannot list all solutions to an inequality such as $x < -2$, we will use the set notation $\{x|x < -2\}$. Recall from Section 9.1 that this is read "the set of all x such that x is less than -2." We will use this notation when solving inequalities.

Objective D Using Both Properties of Inequality

The following steps may be helpful when solving inequalities in one variable. Notice that these steps are similar to the ones given in Section 9.3 for solving equations.

To Solve Linear Inequalities in One Variable

Step 1: If an inequality contains fractions, multiply both sides by the LCD to clear the inequality of fractions.

Step 2: Use the distributive property to remove parentheses if they occur.

Step 3: Simplify each side of the inequality by combining like terms.

Step 4: Get all variable terms on one side and all numbers on the other side by using the addition property of inequality.

Step 5: Get the variable alone by using the multiplication property of inequality.

Helpful Hint

Don't forget that if both sides of an inequality are multiplied or divided by a negative number, the direction of the inequality sign must be reversed.

EXAMPLE 7 Solve $-4x + 7 \geq -9$. Graph the solution set.

Solution: $-4x + 7 \geq -9$

$$-4x + 7 - 7 \geq -9 - 7$$ Subtract 7 from both sides.

$$-4x \geq -16$$ Simplify.

$$\frac{-4x}{-4} \leq \frac{-16}{-4}$$ Divide both sides by -4 and reverse the direction of the inequality sign.

$$x \leq 4$$ Simplify.

The graph of the solution set $\{x|x \leq 4\}$ is shown.

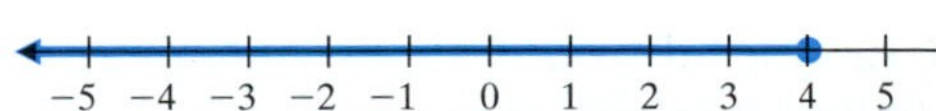

Work Practice Problem 7

PRACTICE PROBLEM 7

Solve $-3x + 11 \leq -13$. Graph the solution set.

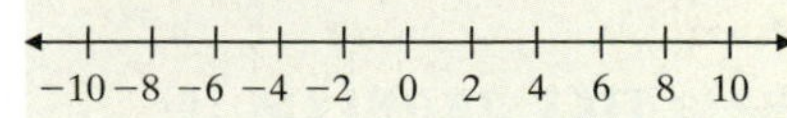

Answers

6. $x > -4$

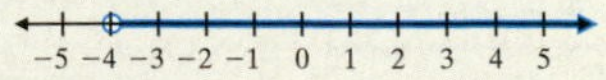

7. $\{x|x \geq 8\}$

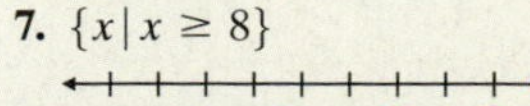

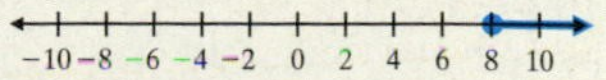

PRACTICE PROBLEM 8

Solve $2x - 3 > 4(x - 1)$. Graph the solution set.

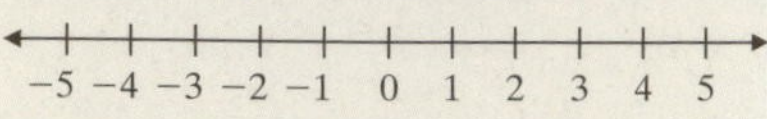

EXAMPLE 8 Solve $-5x + 7 < 2(x - 3)$. Graph the solution set.

Solution: $-5x + 7 < 2(x - 3)$

$-5x + 7 < 2x - 6$ Apply the distributive property.

$-5x + 7 - 2x < 2x - 6 - 2x$ Subtract $2x$ from both sides.

$-7x + 7 < -6$ Combine like terms.

$-7x + 7 - 7 < -6 - 7$ Subtract 7 from both sides.

$-7x < -13$ Combine like terms.

$\frac{-7x}{-7} > \frac{-13}{-7}$ Divide both sides by -7 and reverse the direction of the inequality sign.

$x > \frac{13}{7}$ Simplify.

The graph of the solution set $\left\{x \middle| x > \frac{13}{7}\right\}$ is shown.

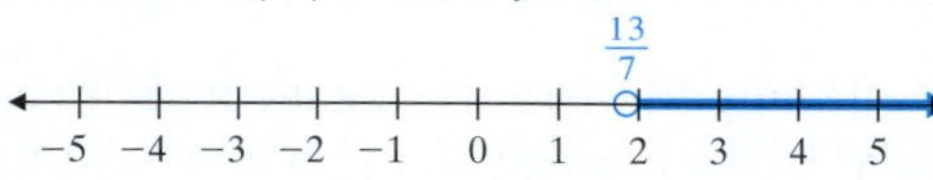

Work Practice Problem 8

PRACTICE PROBLEM 9

Solve:
$3(x + 5) - 1 \geq 5(x - 1) + 7$

EXAMPLE 9 Solve: $2(x - 3) - 5 \leq 3(x + 2) - 18$

Solution: $2(x - 3) - 5 \leq 3(x + 2) - 18$

$2x - 6 - 5 \leq 3x + 6 - 18$ Apply the distributive property.

$2x - 11 \leq 3x - 12$ Combine like terms.

$-x - 11 \leq -12$ Subtract $3x$ from both sides.

$-x \leq -1$ Add 11 to both sides.

$\frac{-x}{-1} \geq \frac{-1}{-1}$ Divide both sides by -1 and reverse the direction of the inequality sign.

$x \geq 1$ Simplify.

The solution set is $\{x | x \geq 1\}$.

Work Practice Problem 9

Objective E Solving Problems Modeled by Inequalities

Problems containing words such as "at least," "at most," "between," "no more than," and "no less than" usually indicate that an inequality should be solved instead of an equation. In solving applications involving linear inequalities, we use the same procedure we used to solve applications involving linear equations.

Some Inequality Translations			
$\geq$	$\leq$	$<$	$>$
at least	at most	is less than	is greater than
no less than	no more than		

PRACTICE PROBLEM 10

Twice a number, subtracted from 35 is greater than 15. Find all numbers that make this true.

EXAMPLE 10 12 subtracted from 3 times a number is less than 21. Find all numbers that make this true.

Solution:

1. UNDERSTAND. Read and reread the problem. This is a direct translation problem, and let's let

$x =$ the unknown number

Answers

8. $\left\{x \middle| x < \frac{1}{2}\right\}$,

9. $\{x | x \leq 6\}$
10. all numbers less than 10

2. TRANSLATE.

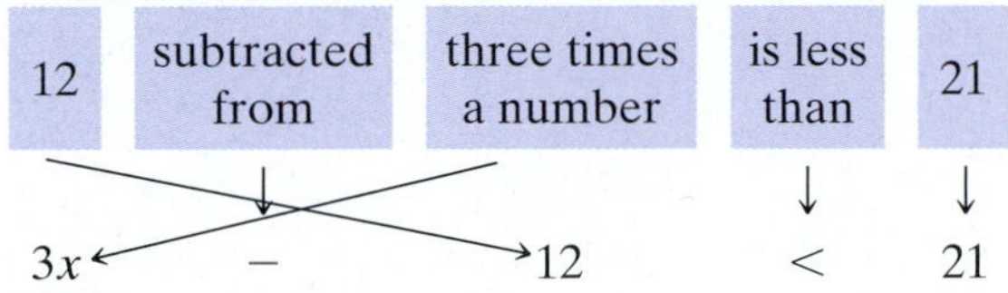

3. SOLVE. $3x - 12 < 21$

$3x < 33$ Add 12 to both sides.

$\frac{3x}{3} < \frac{33}{3}$ Divide both sides by 3 and do not reverse the direction of the inequality sign.

$x < 11$ Simplify.

4. INTERPRET.

Check: Check the translation; then let's choose a number less than 11 to see if it checks. For example, let's check 10. 12 subtracted from 3 times 10 is 12 subtracted from 30, or 18. Since 18 is less than 21, the number 10 checks.

State: All numbers less than 11 make the original statement true.

Work Practice Problem 10

EXAMPLE 11 Budgeting for a Wedding

Marie Chase and Jonathan Edwards are having their wedding reception at the Gallery reception hall. They may spend at most \$1000 for the reception. If the reception hall charges a \$100 cleanup fee plus \$14 per person, find the greatest number of people that they can invite and still stay within their budget.

Solution:

1. UNDERSTAND. Read and reread the problem. Suppose that 50 people attend the reception. The cost is then $\$100 + \$14(50) = \$100 + \$700 = \$800$.

   Let $x =$ the number of people who attend the reception.

2. TRANSLATE.

cleanup fee	+	cost per person	times	number of people	must be less than or equal to	\$1000
↓		↓	↓	↓	↓	↓
100	+	14	·	x	$\leq$	1000

3. SOLVE.

$100 + 14x \leq 1000$

$14x \leq 900$ Subtract 100 from both sides.

$x \leq 64\frac{2}{7}$ Divide both sides by 14.

4. INTERPRET.

Check: Since x represents the number of people, we round down to the nearest whole, or 64. Notice that if 64 people attend, the cost is $\$100 + \$14(64) = \$996$. If 65 people attend, the cost is $\$100 + \$14(65) = \$1010$, which is more than the given \$1000.

State: Marie Chase and Jonathan Edwards can invite at most 64 people to the reception.

Work Practice Problem 11

PRACTICE PROBLEM 11

Alex earns \$600 per month plus 4% of all his sales. Find the minimum sales that will allow Alex to earn at least \$3000 per month.

Answer

11. \$60,000

Vocabulary and Readiness Check

Use the choices below to fill in each blank.

expression   inequality   equation

1. $6x - 7(x + 9)$ ____________

2. $6x = 7(x + 9)$ ____________

3. $6x < 7(x + 9)$ ____________

4. $5y - 2 \geq -38$ ____________

5. $\frac{9}{7} = \frac{x + 2}{14}$ ____________

6. $\frac{9}{7} - \frac{x + 2}{14}$ ____________

Decide which number listed is not a solution to each given inequality.

7. $x \geq -3$; $-3, 0, -5, \pi$ ________

8. $x < 6$; $-6, |-6|, 0, -3.2$ ________

9. $x < 4.01$; $4, -4.01, 4.1, -4.1$ ________

10. $x \geq -3$; $-4, -3, -2, -(-2)$ ________

9.6 EXERCISE SET

FOR EXTRA HELP

Student Solutions Manual

PH Math/Tutor Center

CD/Video for Review

MathXL®

MyMathLab

Objective A *Graph each inequality on a number line. See Examples 1 and 2.*

1. $x \leq -1$

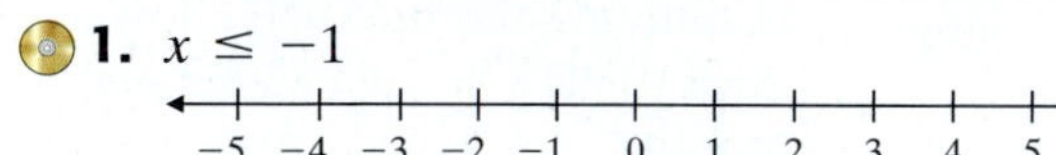

2. $y < 0$

3. $x > \frac{1}{2}$

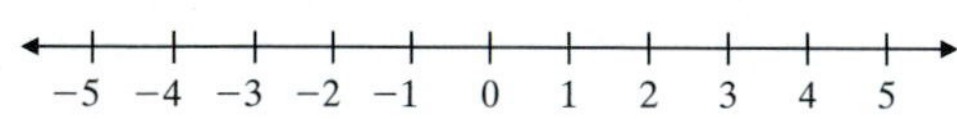

4. $z \geq -\frac{2}{3}$

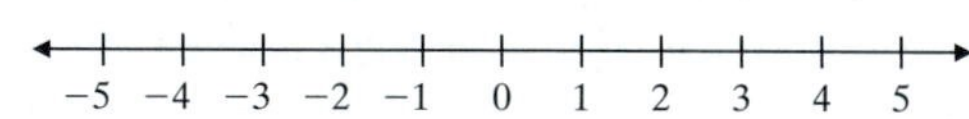

5. $y < 4$

6. $x > 3$

7. $-2 \leq m$

8. $-5 \geq x$

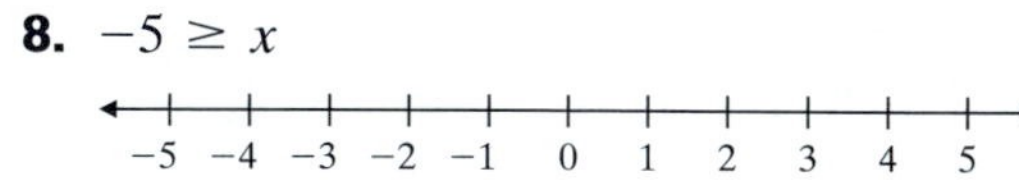

Graph each inequality on a number line. See Example 3.

9. $-1 < x < 3$

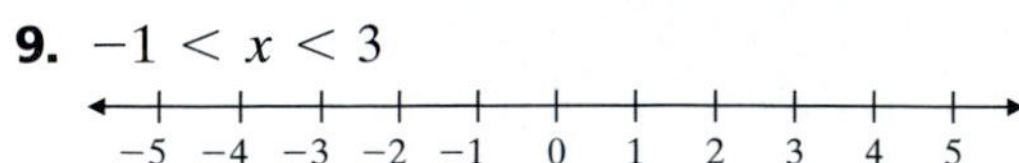

10. $-2 \leq x \leq 3$

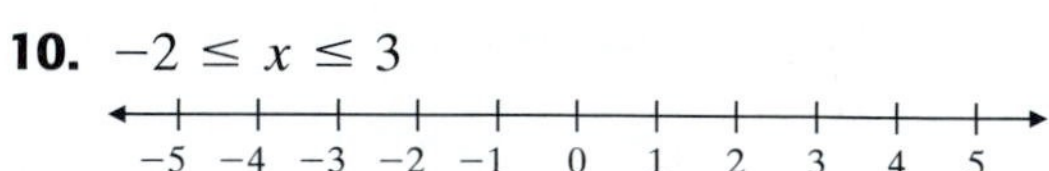

11. $0 \leq y < 2$

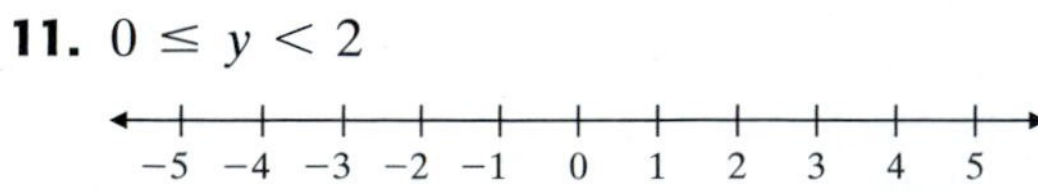

12. $-4 < x \leq 0$

Objective B *Solve each inequality. Graph the solution set. See Example 4.*

13. $x - 2 \geq -7$

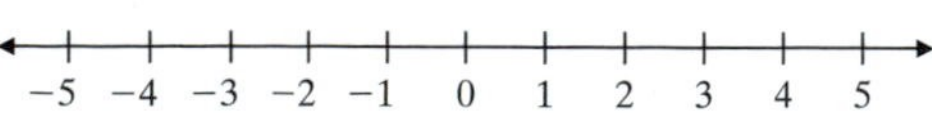

14. $x + 4 \leq 1$

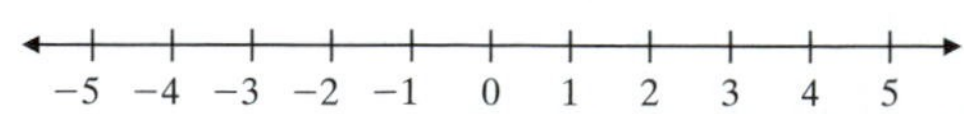

15. $-9 + y < 0$

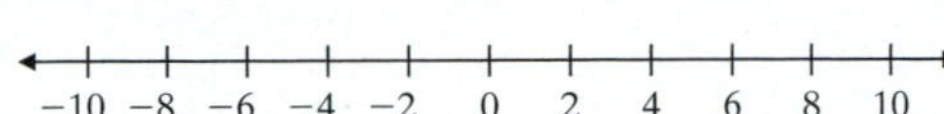

16. $-3 + m > 5$

−10 −8 −6 −4 −2 0 2 4 6 8 10

17. $3x - 5 > 2x - 8$

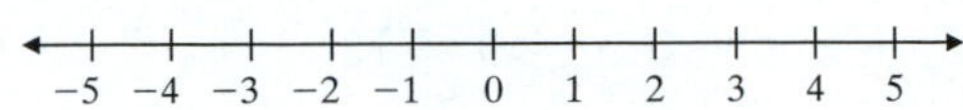

18. $3 - 7x \geq 10 - 8x$

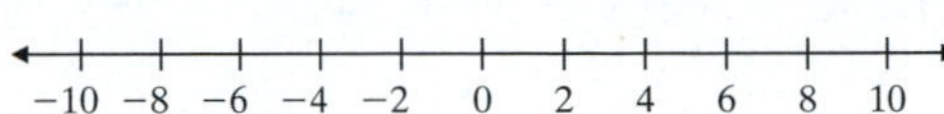

19. $4x - 1 \leq 5x - 2x$

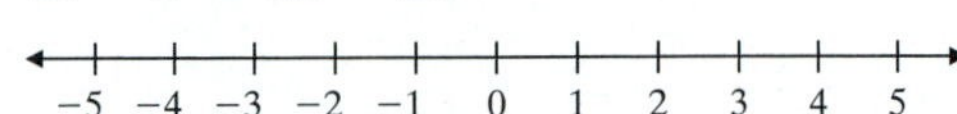

20. $7x + 3 < 9x - 3x$

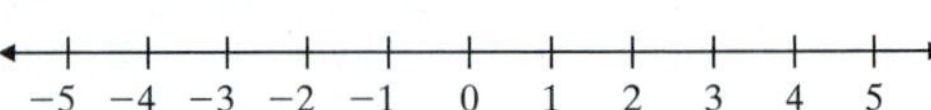

Objective **C** *Solve each inequality. Graph the solution set. See Examples 5 and 6.*

21. $2x < -6$

−5 −4 −3 −2 −1 0 1 2 3 4 5

22. $3x > -9$

−5 −4 −3 −2 −1 0 1 2 3 4 5

23. $-8x \leq 16$

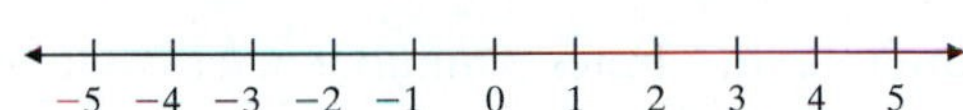

24. $-5x < 20$

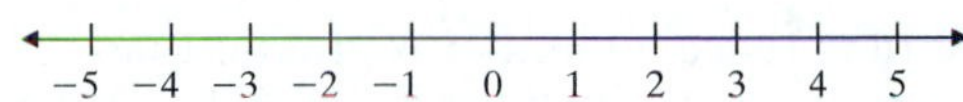

25. $-x > 0$

−5 −4 −3 −2 −1 0 1 2 3 4 5

26. $-y \geq 0$

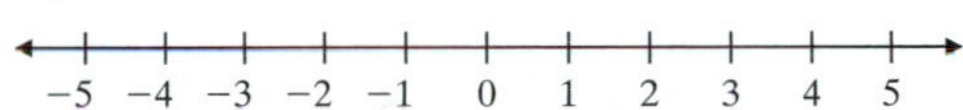

27. $\frac{3}{4}y \geq -2$

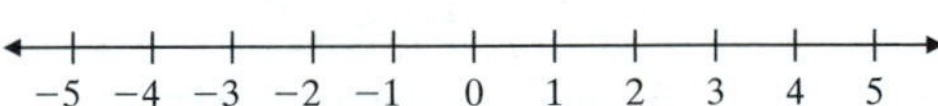

28. $\frac{5}{6}x \leq -8$

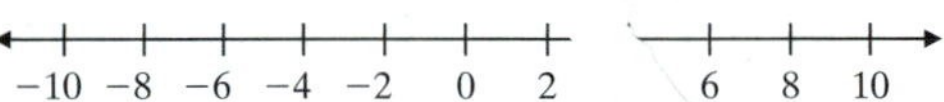

29. $-0.6y < -1.8$

−5 −4 −3 −2 −1 0 1 2 3 4 5

30. $-0.3x > -2.4$

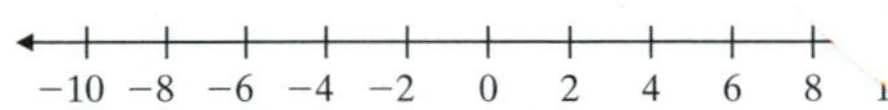

Objectives **B** **C** **D** **Mixed Practice** *Solve each inequality. See Examples 4 through 9.*

31. $-8 < x + 7$

32. $-11 > x + 4$

33. $7(x + 1) - 6x \geq -4$

34. $10(x + 2) - 9x \leq -$

35. $4x > 1$

36. $6x < 5$

37. $-\frac{2}{3}y \leq 8$

38. $-\frac{3}{4}y \geq 9$

39. $4(2z + 1) < 4$

40. $6(2 - z) \geq 12$

41. $3x - 7 < 6x + 2$

42. $2x - 1 \geq 4x - 5$

43. $5x - 7x \leq x + 2$

44. $4 - x < 8x + 2x$

45. $-6x + 2 \geq 2(5 - x)$

46. $-7x + 4 > 3(4 - x)$

47. $3(x - 5) < 2(2x - 1)$

48. $5(x - 2) \leq 3(2x - 1)$

49. $4(3x - 1) \leq 5(2x - 4)$

50. $3(5x - 4) \leq 4(3x - 2)$

51. $3(x + 2) - 6 > -2(x - 3) + 14$

52. $7(x - 2) + x \leq -4(5 - x) - 12$

53. $-5(1 - x) + x \leq -(6 - 2x) + 6$

54. $-2(x - 4) - 3x < -(4x + 1) + 2x$

55. $\frac{1}{4}(x + 4) < \frac{1}{5}(2x + 3)$

56. $\frac{1}{2}(x - 5) < \frac{1}{3}(2x - 1)$

57. $-5x + 4 \leq -4(x - 1)$

58. $-6x + 2 < -3(x + 4)$

Objective E *Solve the following. See Examples 10 and 11.*

59. Six more than twice a number is greater than negative fourteen. Find all numbers that make this statement true.

60. One more than five times a number is less than or equal to ten. Find all such numbers.

△ **61.** The perimeter of a rectangle is to be no greater than 100 centimeters and the width must be 15 centimeters. Find the maximum length of the rectangle.

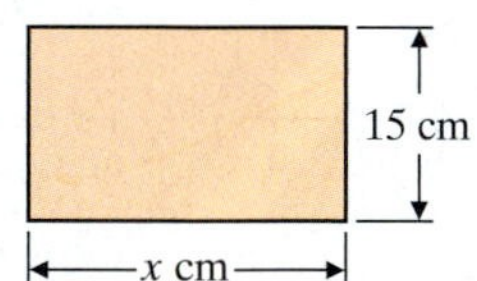

△ **62.** One side of a triangle is four times as long as another side, and the third side is 12 inches long. If the perimeter can be no longer than 87 inches, find the maximum lengths of the other two sides.

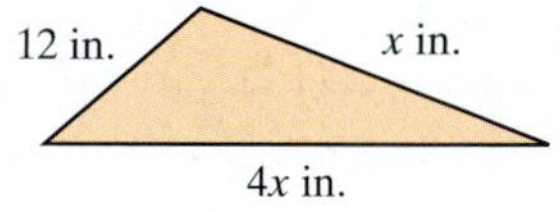

63. Ben Holladay bowled 146 and 201 in his first two games. What must he bowl in his third game to have an average of at least 180? (*Hint:* The average of a list of numbers is their sum divided by the number of numbers in the list.)

64. On an NBA team the two forwards measure 6′8″ and 6′6″ tall and the two guards measure 6′0″ and 5′9″ tall. How tall a center should they hire if they wish to have a starting team average height of at least 6′5″?

65. Dennis and Nancy Wood are celebrating their 30th wedding anniversary by having a reception at Tiffany Oaks reception hall. They have budgeted $3000 for their reception. If the reception hall charges a $50.00 cleanup fee plus $34 per person, find the greatest number of people that they may invite and still stay within their budget.

66. A surprise retirement party is being planned for Pratap Puri. A total of $860 has been collected for the event, which is to be held at a local reception hall. This reception hall charges a cleanup fee of $40 and $15 per person for drinks and light snacks. Find the greatest number of people that may be invited and still stay within the $860 budget.

67. A 150-pound person uses 5.8 calories per minute when walking at a speed of 4 mph. How long must a person walk at this speed to use at least 200 calories? (Round up to the nearest minute.) (*Source:* Home & Garden Bulletin No. 72)

68. A 170-pound person uses 5.3 calories per minute when bicycling at a speed of 5.5 mph. How long must a person ride a bike at this speed in order to use at least 200 calories? (Round up to the nearest minute.) (*Source:* Same as Exercise 67.)

Review

Evaluate each expression. See Sections 1.7 and 4.3.

69. 3^4 **70.** 4^3 **71.** 1^8 **72.** 0^7 **73.** $\left(\frac{7}{8}\right)^2$ **74.** $\left(\frac{2}{3}\right)^3$

The graph shows the number of Krispy Kreme Doughnut locations from 1996 to 2004. The height of the graph for each year shown corresponds to the number of Krispy Kreme locations. Use this graph to answer Exercises 75 through 80. See Section 7.1.

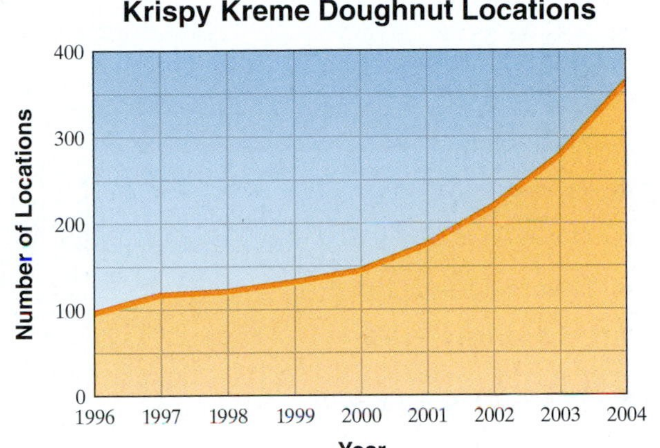

75. How many Krispy Kreme locations were there in 1998?

76. How many Krispy Kreme locations were there in 2003?

77. Between which two years did the greatest increase in the number of Krispy Kreme locations occur?

78. In what year were there approximately 150 Krispy Kreme locations?

79. During which year did the number of Krispy Kreme locations rise above 200?

80. During which year did the number of Krispy Kreme locations rise above 300?

Concept Extensions

Fill in the box with $<$, $>$, $\leq$, or $\geq$. See the Concept Check in this section.

81. Since $3 < 5$, then $3(-4) \square 5(-4)$.

82. If $m \leq n$, then $2m \square 2n$.

83. If $m \leq n$, then $-2m \square -2n$.

84. If $-x < y$, then $x \square -y$.

85. When solving an inequality, when must you reverse the direction of an inequality symbol?

86. If both sides of the inequality $-3x < -30$ are divided by 3, do you reverse the direction of the inequality symbol? Why or why not?

Solve.

87. Eric Daly has scores of 75, 83, and 85 on his history tests. Use an inequality to find the scores he can make on his final exam to receive a B in the class. The final exam counts as **two** tests, and a B is received if the final course average is greater than or equal to 80.

88. Maria Lipco has scores of 85, 95, and 92 on her algebra tests. Use an inequality to find the scores she can make on her final exam to receive an A in the course. The final exam counts as three tests, and an A is received if the final course average is greater than or equal to 90. Round to one decimal place.

THE BIGGER PICTURE Simplifying Expressions and Solving Equations

Starting a New Outline:

In this section, we begin with a new outline in which we now concentrate on not only solving equations, but also simplifying expressions. As usual, once you complete this new portion, try the exercises below.

Remember: Study this outline often as you proceed through the remainder of this text.

I. Simplifying Expressions

Yet to come

II. Solving Equations and Inequalities

A. Linear Equations: power on variable is 1 and there are no variables in the denominator

$7(x - 3) = 4x + 6$	Linear equation. Simplify both sides, then get variable terms on one side, numbers on the other side.
$7x - 21 = 4x + 6$	Use the distributive property.
$7x = 4x + 27$	Add 21 to both sides.
$3x = 27$	Subtract $4x$ from both sides.
$x = 9$	Divide both sides by 3.

B. Linear Inequalities: same as linear equation, except there are inequality symbols, $\leq, <, \geq, >$ Remember, if you multiply or divide by a negative number, then reverse the direction of the inequality symbol.

$-4x - 11 \leq 1$	Linear inequality
$-4x \leq 12$	Add 11 to both sides.
$\dfrac{-4x}{-4} \geq \dfrac{12}{-4}$	Divide both sides by -4 and reverse the direction of the inequality symbol.
$x \geq -3$	Simplify.

Solve each equation or inequality.

1. $-5x = 15$
2. $-5x > 15$
3. $9y - 14 = -12$
4. $9x - 3 = 5x - 4$
5. $4(x - 2) \leq 5x + 7$
6. $5(4x - 1) = 2(10x - 1)$
7. $-5.4 = 0.6x - 9.6$
8. $\frac{1}{3}(x - 4) < \frac{1}{4}(x + 7)$
9. $3y - 5(y - 4) = -2(y - 10)$
10. $\dfrac{7(x - 1)}{3} = \dfrac{2(x + 1)}{5}$

CHAPTER 9 Group Activity

Investigating Averages

Sections 9.1–9.6

Materials:

- small rubber ball or crumpled paper ball
- bucket or waste can

This activity may be completed by working in groups or individually.

1. Try shooting the ball into the bucket or waste can 5 times. Record your results below.

Shots Made	Shots Missed

2. Find your shooting percent for the 5 shots (that is, the percent of the shots you actually made out of the number you tried).
3. Suppose you are going to try an additional 5 shots. How many of the next 5 shots will you have to make to have a 50% shooting percent for all 10 shots? An 80% shooting percent?
4. Did you solve an equation in Question 3? If so, explain what you did. If not, explain how you could use an equation to find the answers.
5. Now suppose you are going to try an additional 22 shots. How many of the next 22 shots will you have to make to have at least a 50% shooting percent for all 27 shots? At least a 70% shooting percent?
6. Choose one of the sports played at your college that is currently in season. How many regular-season games are scheduled? What is the team's current percent of games won?
7. Suppose the team has a goal of finishing the season with a winning percent better than 110% of their current wins. At least how many of the remaining games must they win to achieve their goal?

Chapter 9 Vocabulary Check

Fill in each blank with one of the words or phrases listed below.

no solution	all real numbers	linear equation in one variable
reciprocals	formula	reversed
equivalent equations	opposites	
linear inequality in one variable	the same	

1. A ______________________________ can be written in the form $ax + b = c$.
2. Equations that have the same solution are called ______________________.
3. An equation that describes a known relationship among quantities is called a ___________.
4. A _______________________________ can be written in the form $ax + b < c$, (or $>$, $\le$, $\ge$).
5. The solution(s) to the equation $x + 5 = x + 5$ is/are __________________.
6. The solution(s) to the equation $x + 5 = x + 4$ is/are _____________.
7. If both sides of an inequality are multiplied or divided by the same positive number, the direction of the inequality symbol is ___________.
8. If both sides of an inequality are multiplied by the same negative number, the direction of the inequality symbol is ___________.
9. Two numbers whose sum is 0 are called ____________.
10. Two numbers whose product is 1 are called _____________.

Are you preparing for your test? Don't forget to take the Chapter 9 Test on page 721. Then check your answers at the back of the text and use the Chapter Test Prep Video CD to see the fully worked-out solutions to any of the exercises you want to review.

9 Chapter Highlights

DEFINITIONS AND CONCEPTS	EXAMPLES
Section 9.1 Symbols and Sets of Numbers	
A **set** is a collection of objects, called **elements**, enclosed in braces.	$\{a, c, e\}$
Natural numbers: $\{1, 2, 3, 4, \ldots\}$ **Whole numbers:** $\{0, 1, 2, 3, 4, \ldots\}$ **Integers:** $\{\ldots, -3, -2, -1, 0, 1, 2, 3, \ldots\}$ **Rational numbers:** {real numbers that can be expressed as a quotient of integers} **Irrational numbers:** {real numbers that cannot be expressed as a quotient of integers} **Real numbers:** {all numbers that correspond to a point on the number line}	Given the set $\left\{-3.4, \sqrt{3}, 0, \frac{2}{3}, 5, -4\right\}$ list the numbers that belong to the set of Natural numbers 5 Whole numbers $0,\ 5$ Integers $-4, 0, 5$ Rational numbers $-3.4, 0, \frac{2}{3}, 5, -4$ Irrational numbers $\sqrt{3}$ Real numbers $-3.4, \sqrt{3}, 0, \frac{2}{3}, 5, -4$
A line used to picture numbers is called a **number line.**	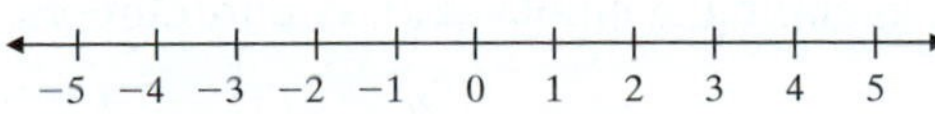
The **absolute value** of a real number a denoted by $\lvert a \rvert$ is the distance between a and 0 on the number line.	$\lvert 5 \rvert = 5 \quad \lvert 0 \rvert = 0 \quad \lvert -2 \rvert = 2$

DEFINITIONS AND CONCEPTS	EXAMPLES
Section 9.1 Symbols and Sets of Numbers (*continued*)	
SYMBOLS: $=$ is equal to	$-7 = -7$
$\neq$ is not equal to	$3 \neq -3$
$>$ is greater than	$4 > 1$
$<$ is less than	$1 < 4$
$\leq$ is less than or equal to	$6 \leq 6$
$\geq$ is greater than or equal to	$18 \geq -\frac{1}{3}$
Section 9.2 Properties of Real Numbers	
COMMUTATIVE PROPERTIES	
Addition: $a + b = b + a$	$3 + (-7) = -7 + 3$
Multiplication: $a \cdot b = b \cdot a$	$-8 \cdot 5 = 5 \cdot (-8)$
ASSOCIATIVE PROPERTIES	
Addition: $(a + b) + c = a + (b + c)$	$(5 + 10) + 20 = 5 + (10 + 20)$
Multiplication: $(a \cdot b) \cdot c = a \cdot (b \cdot c)$	$(-3 \cdot 2) \cdot 11 = -3 \cdot (2 \cdot 11)$
Two numbers whose product is 1 are called **multiplicative inverses** or **reciprocals.** The reciprocal of a nonzero number a is $\frac{1}{a}$ because $a \cdot \frac{1}{a} = 1$.	The reciprocal of 3 is $\frac{1}{3}$. The reciprocal of $-\frac{2}{5}$ is $-\frac{5}{2}$.
DISTRIBUTIVE PROPERTY	
$a(b + c) = a \cdot b + a \cdot c$	$5(6 + 10) = 5 \cdot 6 + 5 \cdot 10$ $-2(3 + x) = -2 \cdot 3 + (-2)(x)$
IDENTITIES	
$a + 0 = a$ $\quad 0 + a = a$	$5 + 0 = 5$ $\quad 0 + (-2) = -2$
$a \cdot 1 = a$ $\quad 1 \cdot a = a$	$-14 \cdot 1 = -14$ $\quad 1 \cdot 27 = 27$
INVERSES	
Additive or opposite: $a + (-a) = 0$	$7 + (-7) = 0$
Multiplicative or reciprocal: $b \cdot \frac{1}{b} = 1, \quad b \neq 0$	$3 \cdot \frac{1}{3} = 1$
Section 9.3 Further Solving Linear Equations	
TO SOLVE LINEAR EQUATIONS	*Solve:* $\frac{5(-2x + 9)}{6} + 3 = \frac{1}{2}$
1. Clear the equation of fractions.	**1.** $6 \cdot \frac{5(-2x + 9)}{6} + 6 \cdot 3 = 6 \cdot \frac{1}{2}$
2. Remove any grouping symbols such as parentheses.	**2.** $5(-2x + 9) + 18 = 3$ $-10x + 45 + 18 = 3$ Apply the distributive property.
3. Simplify each side by combining like terms.	**3.** $-10x + 63 = 3$ Combine like terms.
4. Get all variable terms on one side and all numbers on the other side by using the addition property of equality.	**4.** $-10x + 63 - 63 = 3 - 63$ Subtract 63. $-10x = -60$

Definitions and Concepts	Examples
Section 9.3 Further Solving Linear Equations (*continued*)	
5. Get the variable alone by using the multiplication property of equality. **6.** Check the solution by substituting it into the original equation.	**5.** $\frac{-10x}{-10} = \frac{-60}{-10}$ Divide by -10. $x = 6$
Section 9.4 Further Problem Solving	
Problem-Solving Steps	The height of the Hudson volcano in Chile is twice the height of the Kiska volcano in the Aleutian Islands. If the sum of their heights is 12,870 feet, find the height of each.
1. UNDERSTAND the problem.	**1.** Read and reread the problem. Guess a solution and check your guess. Let x be the height of the Kiska volcano. Then $2x$ is the height of the Hudson volcano. x $2x$
2. TRANSLATE the problem.	**2.** height of Kiska → x; added to → $+$; height of Hudson → $2x$; is → $=$; 12,870 → 12,870
3. SOLVE the equation.	**3.** $x + 2x = 12{,}870$ $3x = 12{,}870$ $x = 4290$
4. INTERPRET the results.	**4.** *Check:* If x is 4290, then $2x$ is 2(4290) or 8580. Their sum is 4290 + 8580 or 12,870, the required amount. *State:* The Kiska volcano is 4290 feet high, and the Hudson volcano is 8580 feet high.
Section 9.5 Formulas and Problem Solving	
An equation that describes a known relationship among quantities is called a **formula.** **To solve a formula for a specified variable,** use the same steps as for solving a linear equation. Treat the specified variable as the only variable of the equation.	$A = lw$ (area of a rectangle) $I = PRT$ (simple interest) *Solve:* $P = 2l + 2w$ for l. $P = 2l + 2w$ $P - 2w = 2l + 2w - 2w$ Subtract $2w$. $P - 2w = 2l$ $\frac{P - 2w}{2} = \frac{2l}{2}$ Divide by 2. $\frac{P - 2w}{2} = l$

DEFINITIONS AND CONCEPTS	EXAMPLES

Section 9.6 Solving Linear Inequalities

Properties of inequalities are similar to properties of equations. However, if you multiply or divide both sides of an inequality by the same *negative* number, you must reverse the direction of the inequality symbol.

$$-2x \le 4$$

$$\frac{-2x}{-2} \ge \frac{4}{-2} \quad \text{Divide by } -2\text{; reverse the inequality symbol.}$$

$$x \ge -2$$

TO SOLVE LINEAR INEQUALITIES

1. Clear the inequality of fractions.
2. Remove grouping symbols.
3. Simplify each side by combining like terms.
4. Write all variable terms on one side and all numbers on the other side using the addition property of inequality.
5. Get the variable alone by using the multiplication property of inequality.

Solve: $3(x + 2) \le -2 + 8$

1. $3(x + 2) \le -2 + 8$ No fractions to clear.
2. $3x + 6 \le -2 + 8$ Apply the distributive property.
3. $3x + 6 \le 6$ Combine like terms.
4. $3x + 6 - 6 \le 6 - 6$ Subtract 6.
   $3x \le 0$
5. $\frac{3x}{3} \le \frac{0}{3}$ Divide by 3.
   $x \le 0$

The solution set is $\{x \mid x \le 0\}$.

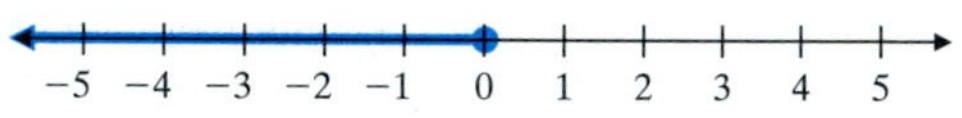

9 CHAPTER REVIEW

(9.1) *Insert <, >, or = in the appropriate space to make each statement true.*

1. 8 10

2. 7 2

3. -4 -5

4. $\frac{12}{2}$ -8

5. $|-7|$ $|-8|$

6. $|-9|$ -9

7. $-|-1|$ -1

8. $|-14|$ $-(-14)$

9. 1.2 1.02

10. $-\frac{3}{2}$ $-\frac{3}{4}$

Translating *Translate each statement into symbols.*

11. Four is greater than or equal to negative three.

12. Six is not equal to five.

13. 0.03 is less than 0.3.

14. New York City has 155 museums and 400 art galleries. Write an inequality statement comparing the numbers 155 and 400. (*Source*: Absolute Trivia.com)

Given the sets of numbers below, list the numbers in each set that also belong to the set of:

a. Natural numbers
b. Whole numbers
c. Integers
d. Rational numbers
e. Irrational numbers
f. Real numbers

15. $\{-6, 0, 1, 1\frac{1}{2}, 3, \pi, 9.62\}$

16. $\{-3, -1.6, 2, 5, \frac{11}{2}, 15.1, \sqrt{5}, 2\pi\}$

The following chart shows the gains and losses in dollars of Density Oil and Gas stock for a particular week. Use this chart to answer Exercises 17 and 18.

Day	Gain or Loss (in dollars)
Monday	+1
Tuesday	−2
Wednesday	+5
Thursday	+1
Friday	−4

17. Which day showed the greatest loss?

18. Which day showed the greatest gain?

(9.2) *Name the property illustrated in each equation.*

19. $-6 + 5 = 5 + (-6)$

20. $6 \cdot 1 = 6$

21. $3(8 - 5) = 3 \cdot 8 + 3 \cdot (-5)$

22. $4 + (-4) = 0$

23. $2 + (3 + 9) = (2 + 3) + 9$

24. $2 \cdot 8 = 8 \cdot 2$

25. $6(8 + 5) = 6 \cdot 8 + 6 \cdot 5$

26. $(3 \cdot 8) \cdot 4 = 3 \cdot (8 \cdot 4)$

27. $4 \cdot \frac{1}{4} = 1$

28. $8 + 0 = 8$

(9.3) *Solve each equation.*

29. $\frac{5}{3}x + 4 = \frac{2}{3}x$

30. $\frac{7}{8}x + 1 = \frac{5}{8}x$

31. $-(5x + 1) = -7x + 3$

32. $-4(2x + 1) = -5x + 5$

33. $-6(2x - 5) = -3(9 + 4x)$

34. $3(8y - 1) = 6(5 + 4y)$

35. $\frac{3(2 - z)}{5} = z$

36. $\frac{4(n + 2)}{5} = -n$

37. $0.5(2n - 3) - 0.1 = 0.4(6 + 2n)$

38. $1.72y - 0.04y = 0.42$

39. $\frac{5(c + 1)}{6} = 2c - 3$

40. $-9 - 5a = 3(6a - 1)$

(9.4) *Solve each of the following.*

41. The height of the Washington Monument is 50.5 inches more than 10 times the length of a side of its square base. If the sum of these two dimensions is 7327 inches, find the height of the Washington Monument. (*Source:* National Park Service)

42. A 12-foot board is to be divided into two pieces so that one piece is twice as long as the other. If x represents the length of the shorter piece, find the length of each piece.

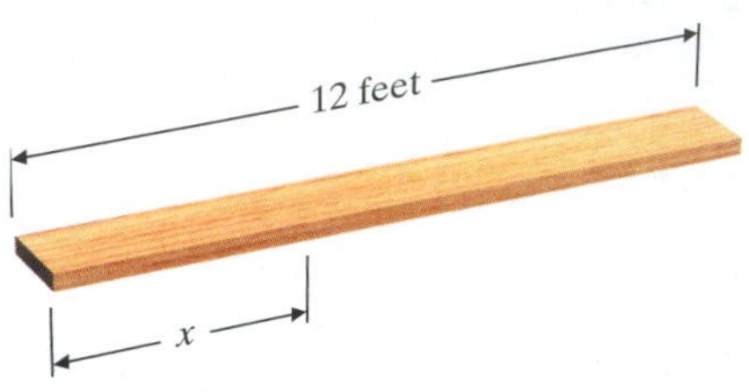

43. In a recent year, Kellogg Company acquired Keebler Foods Company. After the merger, the total number of Kellogg and Keebler manufacturing plants was 53. The number of Kellogg plants was one less than twice the number of Keebler plants. How many of each type of plant were there? (*Source: Kellogg Company 2000 Annual Report*)

44. Find three consecutive integers whose sum is -114.

45. The quotient of a number and 3 is the same as the difference of the number and two. Find the number.

46. Double the sum of a number and 6 is the opposite of the number. Find the number.

(9.5) *Substitute the given values into the given formulas and solve for the unknown variable.*

47. $P = 2l + 2w$; $P = 46, l = 14$

48. $V = lwh$; $V = 192, l = 8, w = 6$

Solve each equation for the indicated variable.

49. $y = mx + b$ for m

50. $r = vst - 5$ for s

51. $2y - 5x = 7$ for x

52. $3x - 6y = -2$ for y

△ **53.** $C = \pi d$ for π

△ **54.** $C = 2\pi r$ for π

△ **55.** A swimming pool holds 900 cubic meters of water. If its length is 20 meters and its height is 3 meters, find its width.

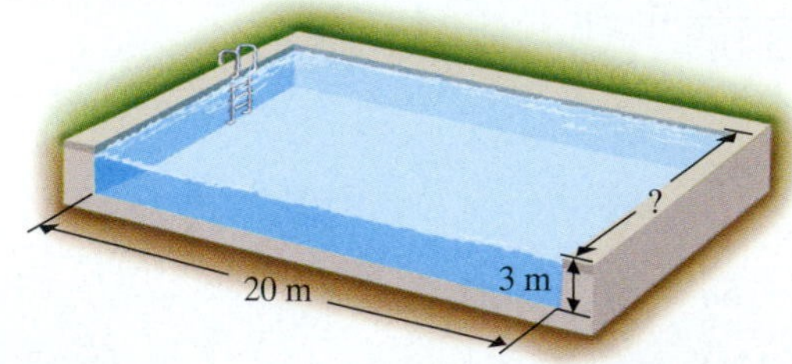

56. The perimeter of a rectangular billboard is 60 feet and has a length 6 feet longer than its width. Find the dimensions of the billboard.

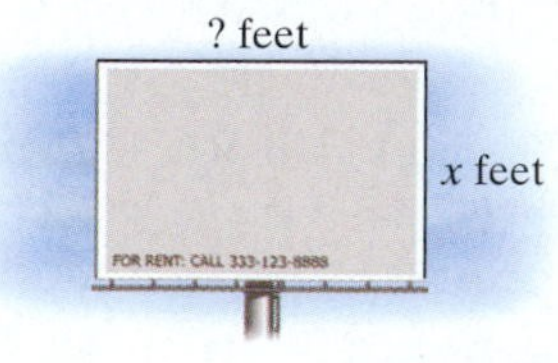

57. A charity 10K race is given annually to benefit a local hospice organization. How long will it take to run/walk a 10K race (10 kilometers or 10,000 meters) if your average pace is 125 **meters** per minute? Give your time in hours and minutes.

58. On April 28, 2001, the highest temperature recorded in the United States was 104°F, which occurred in Death Valley, California. Convert this temperature to degrees Celsius. (*Source:* National Weather Service)

(9.6) *Graph on a number line.*

59. $x \leq -2$

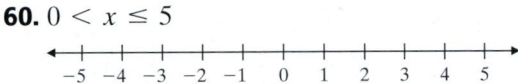

−5 −4 −3 −2 −1 0 1 2 3 4 5

60. $0 < x \leq 5$

−5 −4 −3 −2 −1 0 1 2 3 4 5

Solve each inequality.

61. $x - 5 \leq -4$

62. $x + 7 > 2$

63. $-2x \geq -20$

64. $-3x > 12$

65. $5x - 7 > 8x + 5$

66. $x + 4 \geq 6x - 16$

67. $\frac{2}{3}y > 6$

68. $-0.5y \leq 7.5$

69. $-2(x - 5) > 2(3x - 2)$

70. $4(2x - 5) \leq 5x - 1$

71. Carol Abolafia earns $175 per week plus a 5% commission on all her sales. Find the minimum amount of sales she must make to ensure that she earns at least $300 per week.

72. Joseph Barrow shot rounds of 76, 82, and 79 golfing. What must he shoot on his next round so that his average will be below 80?

Mixed Review

Solve each equation.

73. $6x + 2x - 1 = 5x + 11$

74. $2(3y - 4) = 6 + 7y$

75. $4(3 - a) - (6a + 9) = -12a$

76. $\frac{x}{3} - 2 = 5$

77. $2(y + 5) = 2y + 10$

78. $7x - 3x + 2 = 2(2x - 1)$

Solve.

79. The sum of six and twice a number is equal to seven less than the number. Find the number.

80. A 23-inch piece of string is to be cut into two pieces so that the length of the longer piece is three more than four times the shorter piece. If x represents the length of the shorter piece, find the lengths of both pieces.

Solve for the specified variable.

81. $V = \frac{1}{3}Ah$ for h

Solve each inequality. Graph the solution set.

82. $4x - 7 > 3x + 2$

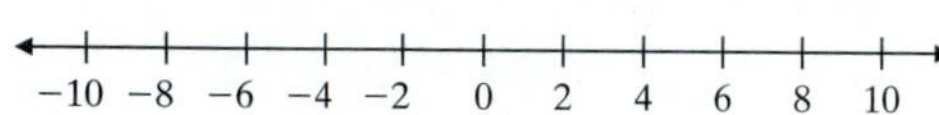

83. $-5x < 20$

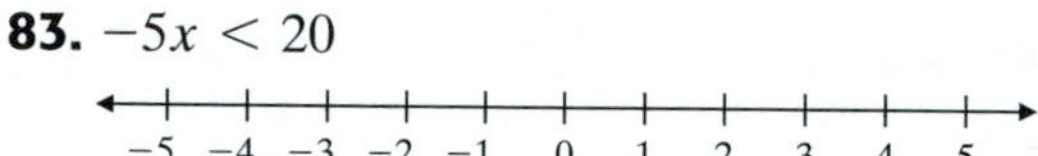

84. $-3(1 + 2x) + x \geq -(3 - x)$

−5 −4 −3 −2 −1 0 1 2 3 4 5

9 CHAPTER TEST

Answers

Use the Chapter Test Prep Video CD to see the fully worked-out solutions to any of the exercises you want to review.

Translate each statement into symbols.

1. The absolute value of negative seven is greater than five.

2. The sum of nine and five is greater than or equal to four.

3. Given

$\{-5, -1, \frac{1}{4}, 0, 1, 7, 11.6, \sqrt{7}, 3\pi\}$, list the numbers in this set that also belong to the set of:

a. Natural numbers
b. Whole numbers
c. Integers
d. Rational numbers
e. Irrational numbers
f. Real numbers

Identify the property illustrated by each expression.

4. $8 + (9 + 3) = (8 + 9) + 3$

5. $6 \cdot 8 = 8 \cdot 6$

6. $-6(2 + 4) = -6 \cdot 2 + (-6) \cdot 4$

7. $\frac{1}{6}(6) = 1$

8. Find the opposite of -9.

9. Find the reciprocal of $-\frac{1}{3}$.

Use the distributive property to write each expression without parentheses. Then simplify if possible.

10. $7 + 2(5y - 3)$

11. $4(x - 2) - 3(2x - 6)$

Solve each equation.

12. $4(n - 5) = -(4 - 2n)$

13. $-2(x - 3) = x + 5 - 3x$

14. $4z + 1 - z = 1 + z$

15. $\frac{2(x + 6)}{3} = x - 5$

16. $\frac{1}{2} - x + \frac{3}{2} = x - 4$

17. $-0.3(x - 4) + x = 0.5(3 - x)$

18. $-4(a + 1) - 3a = -7(2a - 3)$

1. ______
2. ______
3. a. ______
b. ______
c. ______
d. ______
e. ______
f. ______
4. ______
5. ______
6. ______
7. ______
8. ______
9. ______
10. ______
11. ______
12. ______
13. ______
14. ______
15. ______
16. ______
17. ______
18. ______

19. ______

20. ______

21. ______

22. ______

23. ______

24. ______

25. ______

26. ______

27. ______

28. ______

Solve each application.

19. A number increased by two-thirds of the number is 35. Find the number.

△ **20.** A gallon of water seal covers 200 square feet. How many gallons are needed to paint two coats of water seal on a deck that measures 20 feet by 35 feet?

21. Find the value of x if $y = -14$, $m = -2$, and $b = -2$ in the formula $y = mx + b$.

Solve the equation for the indicated variable.

22. $V = \pi r^2 h$ for h

23. $3x - 4y = 10$ for y

Solve the inequality. Graph the solution set.

24. $3x - 5 > 7x + 3$

−5 −4 −3 −2 −1 0 1 2 3 4 5

Solve each inequality.

25. $-0.3x \geq 2.4$

26. $-5(x - 1) + 6 \leq -3(x + 4) + 1$

27. $\dfrac{2(5x + 1)}{3} > 2$

28. New York State has more public libraries than any other state. It has 650 more public libraries than Indiana. If the total number of public libraries for these states is 1504, find the number of public libraries in New York and the number in Indiana. (*Source: The World Almanac and Book of Facts*)

CUMULATIVE REVIEW

Chapters 1–9

Answers

1\. a. ______ b. ______ c. ______
2\. a. ______ b. ______ c. ______ d. ______ e. ______
3\. ______ 4. ______ 5. ______ 6. ______ 7. ______ 8. ______ 9. ______ 10. ______ 11. ______ 12. ______ 13. ______ 14. ______ 15. ______ 16. ______ 17. ______ 18. ______ 19. ______ 20. ______ 21. ______ 22. ______

Simplify.

1. a. $-(-4)$ **b.** $-|-5|$ **c.** $-|6|$

2. Insert $<$, $>$, or $=$ in the appropriate space to make each statement true.

a. $|0| \quad 2$
b. $|-5| \quad 5$
c. $|-3| \quad |-2|$
d. $|5| \quad |6|$
e. $|-7| \quad |6|$

3. Simplify $-2[-3 + 2(-1 + 6)] - 5$

4. Simplify the expression $\dfrac{3 + |4 - 3| + 2^2}{6 - 3}$.

Add.

5. $-18 + 10$ **6.** $(-8) + (-11)$ **7.** $12 + (-8)$

8. $(-2) + 10$ **9.** $(-3) + 4 + (-11)$ **10.** $0.2 + (-0.5)$

Perform the indicated operation.

11. $\dfrac{2}{3} - \dfrac{10}{11}$ **12.** $\dfrac{2}{5} - \dfrac{39}{40}$ **13.** $2 - \dfrac{x}{3}$ **14.** $5 + \dfrac{x}{2}$

Evaluate. Let $x = -\dfrac{1}{2}$ *and* $y = \dfrac{1}{3}$.

15. $2x + y^2$ **16.** $2y + x^2$

Perform the indicated operation.

17. $-4\dfrac{2}{5} \cdot 1\dfrac{3}{11}$ **18.** $-2\dfrac{1}{2} \cdot \left(-2\dfrac{1}{2}\right)$ **19.** $-2\dfrac{1}{3} \div \left(-2\dfrac{1}{2}\right)$

20. $3\dfrac{3}{5} \div \left(-3\dfrac{1}{3}\right)$ **21.** $(-1.3)^2 + 2.4$ **22.** $1.2(7.3 - 9.3)$

23. ______
24. ______
25. ______
26. ______
27. ______
28. ______
29. ______
30. ______
31. ______
32. ______
33. ______
34. ______
35. ______
36. ______
37. ______
38. ______
39. ______
40. ______
41. ______
42. ______
43. ______
44. ______
45. ______
46. ______
47. ______
48. ______
49. ______
50. ______

23. $\sqrt{49}$ **24.** $\sqrt{64}$ **25.** $\sqrt{\frac{4}{25}}$ **26.** $\sqrt{\frac{9}{100}}$

Solve for x.

27. $1.2x + 5.8 = 8.2$

28. $1.3x - 2.6 = -9.1$

Determine whether each statement is true or false.

29. $8 \geq 8$ **30.** $-4 \leq -4$ **31.** $8 \leq 8$ **32.** $-4 \geq -4$

33. $23 \leq 0$ **34.** $-8 \leq 0$ **35.** $0 \leq 23$ **36.** $0 \leq -8$

Use the distributive property to write each expression without parentheses. Then simplify the result.

37. $-5(-3 + 2z)$ **38.** $-4(2x-1)$ **39.** $\frac{1}{2}(6x + 14) + 10$ **40.** $9 + 2(5x + 6)$

Solve.

41. $\frac{2(a + 3)}{3} = 6a + 2$

42. $\frac{x}{2} + \frac{x}{5} = 3$

43. In a recent year, the U.S. House of Representatives had a total of 431 Democrats and Republicans. There were 15 more Republican representatives than Democratic. Find the number of representatives from each party. (*Source:* Office of the Clerk of the U.S. House of Representatives)

44. Three times the sum of a number and 2 is the same as 9 times the number.

45. A glacier is a giant mass of rocks and ice that flows downhill like a river. Portage Glacier in Alaska is about 6 miles, or 31,680 feet, long and moves 400 feet per year. Icebergs are created when the front end of the glacier flows into Portage Lake. How long does it take for ice at the head (beginning) of the glacier to reach the lake?

46. Solve: $V = lwh$ for w.

47. Graph $-1 > x$.

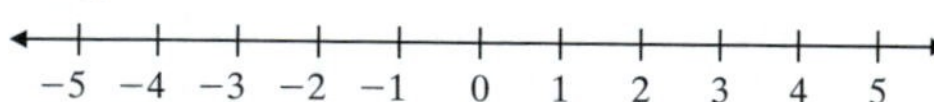

48. Solve: $-3x < -30$

49. Solve: $2(x - 3) - 5 \leq 3(x + 2) - 18$

50. Solve: $10 + x < 6x - 10$

10 Exponents and Polynomials

Recall from Chapter 1 that an exponent is a shorthand notation for repeated factors. This chapter explores additional concepts about exponents and exponential expressions. An especially useful type of exponential expression is a polynomial. Polynomials model many real-world phenomena. Our goal in this chapter is to become proficient with operations on polynomials.

SECTIONS

CHECK YOUR PROGRESS

A popular use of the Internet is the World Wide Web. The World Wide Web was invented in 1989–1990 as an environment originally by which scientists could share information. It has grown into a medium containing text, graphics, audio, animation, and video. Each of the locations, or Web sites below, has an address and can be used to locate other Web sites. In Section 10.2, Exercise 95, you will have the opportunity to estimate the number of visitors to the most popular Web sites.

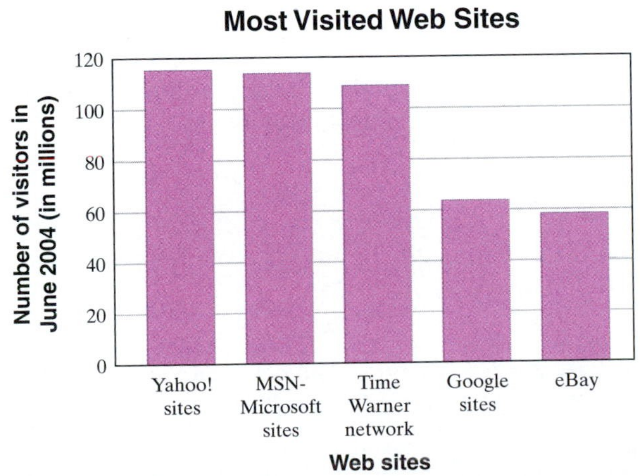

Objectives

A Evaluate Exponential Expressions.

B Use the Product Rule for Exponents.

C Use the Power Rule for Exponents.

D Use the Power Rules for Products and Quotients.

E Use the Quotient Rule for Exponents, and Define a Number Raised to the 0 Power.

F Decide Which Rule(s) to Use to Simplify an Expression.

10.1 EXPONENTS

Objective A Evaluating Exponential Expressions

In this section, we continue our work with integer exponents. Recall from Section 1.7 that repeated multiplication of the same factor can be written using exponents. For example,

$$2 \cdot 2 \cdot 2 \cdot 2 \cdot 2 = 2^5$$

The exponent 5 tells us how many times that 2 is a factor. The expression 2^5 is called an **exponential expression.** It is also called the fifth **power** of 2, or we can say that 2 is **raised** to the fifth power.

$$5^6 = \underbrace{5 \cdot 5 \cdot 5 \cdot 5 \cdot 5 \cdot 5}_{\text{6 factors; each factor is 5}} \quad \text{and} \quad (-3)^4 = \underbrace{(-3) \cdot (-3) \cdot (-3) \cdot (-3)}_{\text{4 factors; each factor is } -3}$$

The base of an exponential expression is the repeated factor. The exponent is the number of times that the base is used as a factor.

$$a^n = \underbrace{a \cdot a \cdot a \cdots a}_{n \text{ factors; each factor is } a}$$

(n: exponent or power; a: base)

PRACTICE PROBLEMS 1–6

Evaluate each expression.

1. 3^4
2. 7^1
3. $(-2)^3$
4. -2^3
5. $\left(\frac{2}{3}\right)^2$
6. $5 \cdot 6^2$

EXAMPLES Evaluate each expression.

1. $2^3 = 2 \cdot 2 \cdot 2 = 8$
2. $3^1 = 3$. To raise 3 to the first power means to use 3 as a factor only once. When no exponent is shown, the exponent is assumed to be 1.
3. $(-4)^2 = (-4)(-4) = 16$
4. $-4^2 = -(4 \cdot 4) = -16$
5. $\left(\frac{1}{2}\right)^4 = \frac{1}{2} \cdot \frac{1}{2} \cdot \frac{1}{2} \cdot \frac{1}{2} = \frac{1}{16}$
6. $4 \cdot 3^2 = 4 \cdot 9 = 36$

Work Practice Problems 1–6

Notice how similar -4^2 is to $(-4)^2$ in the examples above. The difference between the two is the parentheses. In $(-4)^2$, the parentheses tell us that the base, or the repeated factor, is -4. In -4^2, only 4 is the base.

Helpful Hint

Be careful when identifying the base of an exponential expression. Pay close attention to the use of parentheses.

$(-3)^2$	-3^2	$2 \cdot 3^2$
The base is -3.	The base is 3.	The base is 3.
$(-3)^2 = (-3)(-3) = 9$	$-3^2 = -(3 \cdot 3) = -9$	$2 \cdot 3^2 = 2 \cdot 3 \cdot 3 = 18$

An exponent has the same meaning whether the base is a number or a variable. If x is a real number and n is a positive integer, then x^n is the product of n factors, each of which is x.

$$x^n = \underbrace{x \cdot x \cdot x \cdot x \cdot x \cdots x}_{n \text{ factors; each factor is } x}$$

Answers

1. 81 2. 7 3. −8 4. −8 5. $\frac{4}{9}$ 6. 180

EXAMPLE 7 Evaluate each expression for the given value of x.

a. $2x^3$ when x is 5 **b.** $\frac{9}{x^2}$ when x is -3

Solution:

a. When x is 5, $2x^3 = 2 \cdot 5^3$

$$= 2 \cdot (5 \cdot 5 \cdot 5)$$
$$= 2 \cdot 125$$
$$= 250$$

b. When x is -3, $\frac{9}{x^2} = \frac{9}{(-3)^2}$

$$= \frac{9}{(-3)(-3)}$$
$$= \frac{9}{9} = 1$$

Work Practice Problem 7

PRACTICE PROBLEM 7

Evaluate each expression for the given value of x.

a. $3x^2$ when x is 4

b. $\frac{x^4}{-8}$ when x is -2

Objective B Using the Product Rule

Exponential expressions can be multiplied, divided, added, subtracted, and themselves raised to powers. Let's see if we can discover a shortcut method for multiplying exponential expressions with the same base. By our definition of an exponent,

$$5^4 \cdot 5^3 = \underbrace{(5 \cdot 5 \cdot 5 \cdot 5)}_{4 \text{ factors of } 5} \cdot \underbrace{(5 \cdot 5 \cdot 5)}_{3 \text{ factors of } 5}$$
$$= \underbrace{5 \cdot 5 \cdot 5 \cdot 5 \cdot 5 \cdot 5 \cdot 5}_{7 \text{ factors of } 5}$$
$$= 5^7$$

Also,

$$x^2 \cdot x^3 = (x \cdot x) \cdot (x \cdot x \cdot x)$$
$$= x \cdot x \cdot x \cdot x \cdot x$$
$$= x^5$$

In both cases, notice that the result is exactly the same if the exponents are added.

$$5^4 \cdot 5^3 = 5^{4+3} = 5^7 \quad \text{and} \quad x^2 \cdot x^3 = x^{2+3} = x^5$$

This suggests the following rule.

Product Rule for Exponents

If m and n are positive integers and a is a real number, then

$$a^m \cdot a^n = a^{m+n}$$ ← Add exponents.
↑ Keep common base.

For example,

$$3^5 \cdot 3^7 = 3^{5+7} = 3^{12}$$ ← Add exponents.
↑ Keep common base.

Answers

7. a. 48 **b.** -2

Helpful Hint

Don't forget that

$3^5 \cdot 3^7 \neq 9^{12}$ ← Add exponents.
↑ Common base *not* kept.

$$3^5 \cdot 3^7 = \underbrace{3 \cdot 3 \cdot 3 \cdot 3 \cdot 3}_{\text{5 factors of 3}} \cdot \underbrace{3 \cdot 3 \cdot 3 \cdot 3 \cdot 3 \cdot 3 \cdot 3}_{\text{7 factors of 3}}$$

$= 3^{12}$ 12 factors of 3, *not* 9.

In other words, to multiply two exponential expressions with the **same base,** we keep the base and add the exponents. We call this **simplifying** the exponential expression.

PRACTICE PROBLEMS 8–12

Use the product rule to simplify each expression.

8. $7^3 \cdot 7^2$ **9.** $x^4 \cdot x^9$
10. $r^5 \cdot r$ **11.** $s^6 \cdot s^2 \cdot s^3$
12. $(-3)^9 \cdot (-3)$

EXAMPLES Use the product rule to simplify each expression.

8. $4^2 \cdot 4^5 = 4^{2+5} = 4^7$ ← Add exponents.
↑ Keep common base.

9. $x^2 \cdot x^5 = x^{2+5} = x^7$

10. $y^3 \cdot y = y^3 \cdot y^1$
$= y^{3+1}$
$= y^4$

Helpful Hint

Don't forget that if no exponent is written, it is assumed to be 1.

11. $y^3 \cdot y^2 \cdot y^7 = y^{3+2+7} = y^{12}$

12. $(-5)^7 \cdot (-5)^8 = (-5)^{7+8} = (-5)^{15}$

■ **Work Practice Problems 8–12**

✔**Concept Check** Where possible, use the product rule to simplify the expression.

a. $z^2 \cdot z^{14}$ **b.** $x^2 \cdot z^{14}$ **c.** $9^8 \cdot 9^3$ **d.** $9^8 \cdot 2^7$

PRACTICE PROBLEM 13

Use the product rule to simplify $(6x^3)(-2x^9)$.

EXAMPLE 13 Use the product rule to simplify $(2x^2)(-3x^5)$.

Solution: Recall that $2x^2$ means $2 \cdot x^2$ and $-3x^5$ means $-3 \cdot x^5$.

$(2x^2)(-3x^5) = (2 \cdot x^2) \cdot (-3 \cdot x^5)$
$= (2 \cdot -3) \cdot (x^2 \cdot x^5)$ Group factors with common bases (using commutative and associative properties.)
$= -6x^7$ Simplify.

■ **Work Practice Problem 13**

PRACTICE PROBLEMS 14–15

Simplify.

14. $(m^5n^{10})(mn^8)$
15. $(-x^9y)(4x^2y^{11})$

EXAMPLES Simplify.

14. $(x^2y)(x^3y^2) = (x^2 \cdot x^3) \cdot (y^1 \cdot y^2)$ Group like bases and write y as y^1.
$= x^5 \cdot y^3$ or x^5y^3 Multiply.

15. $(-a^7b^4)(3ab^9) = (-1 \cdot 3) \cdot (a^7 \cdot a^1) \cdot (b^4 \cdot b^9)$
$= -3a^8b^{13}$

■ **Work Practice Problems 14–15**

Answers

8. 7^5 **9.** x^{13} **10.** r^6 **11.** s^{11}
12. $(-3)^{10}$ **13.** $-12x^{12}$
14. m^6n^{18} **15.** $-4x^{11}y^{12}$

✔ Concept Check Answers

a. z^{16} **b.** cannot be simplified
c. 9^{11} **d.** cannot be simplified

Helpful Hint

These examples will remind you of the difference between adding and multiplying terms.

Addition

$5x^3 + 3x^3 = (5 + 3)x^3 = 8x^3$ By the distributive property.

$7x + 4x^2 = 7x + 4x^2$ Cannot be combined.

Multiplication

$(5x^3)(3x^3) = 5 \cdot 3 \cdot x^3 \cdot x^3 = 15x^{3+3} = 15x^6$ By the product rule.

$(7x)(4x^2) = 7 \cdot 4 \cdot x \cdot x^2 = 28x^{1+2} = 28x^3$ By the product rule.

Objective C Using the Power Rule

Exponential expressions can themselves be raised to powers. Let's try to discover a rule that simplifies an expression like $(x^2)^3$. By the definition of a^n,

$(x^2)^3 = (x^2)(x^2)(x^2)$ $(x^2)^3$ means 3 factors of (x^2).

which can be simplified by the product rule for exponents.

$$(x^2)^3 = (x^2)(x^2)(x^2) = x^{2+2+2} = x^6$$

Notice that the result is exactly the same if we multiply the exponents.

$$(x^2)^3 = x^{2 \cdot 3} = x^6$$

The following rule states this result.

Power Rule for Exponents

If m and n are positive integers and a is a real number, then

$(a^m)^n = a^{mn}$ ← Multiply exponents. Keep common base.

For example,

$(7^2)^5 = 7^{2 \cdot 5} = 7^{10}$ ← Multiply exponents. Keep common base.

In other words, to raise an exponential expression to a power, we keep the base and multiply the exponents.

EXAMPLES Use the power rule to simplify each expression.

16. $(5^3)^6 = 5^{3 \cdot 6} = 5^{18}$

17. $(y^8)^2 = y^{8 \cdot 2} = y^{16}$

Work Practice Problems 16–17

PRACTICE PROBLEMS 16–17

Use the power rule to simplify each expression.

16. $(9^4)^{10}$ **17.** $(z^6)^3$

Helpful Hint

Take a moment to make sure that you understand when to apply the product rule and when to apply the power rule.

Product Rule → Add Exponents	Power Rule → Multiply Exponents
$x^5 \cdot x^7 = x^{5+7} = x^{12}$	$(x^5)^7 = x^{5 \cdot 7} = x^{35}$
$y^6 \cdot y^2 = y^{6+2} = y^8$	$(y^6)^2 = y^{6 \cdot 2} = y^{12}$

Answers

16. 9^{40} **17.** z^{18}

Objective D Using the Power Rules for Products and Quotients

When the base of an exponential expression is a product, the definition of a^n still applies. For example, simplify $(xy)^3$ as follows.

$$\begin{aligned}(xy)^3 &= (xy)(xy)(xy) && (xy)^3 \text{ means 3 factors of } (xy).\\ &= x\cdot x\cdot x\cdot y\cdot y\cdot y && \text{Group factors with common bases.}\\ &= x^3y^3 && \text{Simplify.}\end{aligned}$$

Notice that to simplify the expression $(xy)^3$, we raise each factor within the parentheses to a power of 3.

$$(xy)^3 = x^3y^3$$

In general, we have the following rule.

Power of a Product Rule

If n is a positive integer and a and b are real numbers, then

$$(ab)^n = a^nb^n$$

For example,

$$(3x)^5 = 3^5x^5$$

In other words, to raise a product to a power, we raise each factor to the power.

PRACTICE PROBLEMS 18–21

Simplify each expression.

18. $(xy)^7$ **19.** $(3y)^4$

20. $(-2p^4q^2r)^3$ **21.** $(-a^4b)^7$

EXAMPLES Simplify each expression.

18. $(st)^4 = s^4\cdot t^4 = s^4t^4$ Use the power of a product rule.

19. $(2a)^3 = 2^3\cdot a^3 = 8a^3$ Use the power of a product rule.

20. $(-5x^2y^3z)^2 = (-5)^2\cdot(x^2)^2\cdot(y^3)^2\cdot(z^1)^2$ Use the power of a product rule.

$= 25x^4y^6z^2$ Use the power rule for exponents.

21. $(-xy^3)^5 = (-1xy^3)^5 = (-1)^5\cdot x^5\cdot(y^3)^5$

$= -1x^5y^{15}$ or $-x^5y^{15}$

Work Practice Problems 18–21

Let's see what happens when we raise a quotient to a power. For example, we simplify $\left(\frac{x}{y}\right)^3$ as follows.

$$\begin{aligned}\left(\frac{x}{y}\right)^3 &= \left(\frac{x}{y}\right)\left(\frac{x}{y}\right)\left(\frac{x}{y}\right) && \left(\frac{x}{y}\right)^3 \text{ means 3 factors of } \left(\frac{x}{y}\right).\\ &= \frac{x\cdot x\cdot x}{y\cdot y\cdot y} && \text{Multiply fractions.}\\ &= \frac{x^3}{y^3} && \text{Simplify.}\end{aligned}$$

Notice that to simplify the expression, $\left(\frac{x}{y}\right)^3$, we raise both the numerator and the denominator to a power of 3.

$$\left(\frac{x}{y}\right)^3 = \frac{x^3}{y^3}$$

In general, we have the following rule.

Answers

18. x^7y^7 **19.** $81y^4$ **20.** $-8p^{12}q^6r^3$ **21.** $-a^{28}b^7$

Power of a Quotient Rule

If n is a positive integer and a and c are real numbers, then

$$\left(\frac{a}{c}\right)^n = \frac{a^n}{c^n}, \quad c \neq 0$$

For example,

$$\left(\frac{y}{7}\right)^3 = \frac{y^3}{7^3}$$

In other words, to raise a quotient to a power, we raise both the numerator and the denominator to the power.

EXAMPLES Simplify each expression.

22. $\left(\frac{m}{n}\right)^7 = \frac{m^7}{n^7}, \quad n \neq 0$ Use the power of a quotient rule.

23. $\left(\frac{2x^4}{3y^5}\right)^4 = \frac{2^4 \cdot (x^4)^4}{3^4 \cdot (y^5)^4}$ Use the power of a quotient rule.

$= \frac{16x^{16}}{81y^{20}}, \quad y \neq 0$ Use the power rule for exponents.

Work Practice Problems 22–23

PRACTICE PROBLEMS 22–23

Simplify each expression.

22. $\left(\frac{r}{s}\right)^6$ **23.** $\left(\frac{5x^6}{9y^3}\right)^2$

Objective E Using the Quotient Rule and Defining the Zero Exponent

Another pattern for simplifying exponential expressions involves quotients.

$$\begin{aligned}\frac{x^5}{x^3} &= \frac{x \cdot x \cdot x \cdot x \cdot x}{x \cdot x \cdot x} \\ &= \frac{x \cdot x \cdot x \cdot x \cdot x}{x \cdot x \cdot x} \\ &= 1 \cdot 1 \cdot 1 \cdot x \cdot x \\ &= x \cdot x \\ &= x^2\end{aligned}$$

Notice that the result is exactly the same if we subtract exponents of the common bases.

$$\frac{x^5}{x^3} = x^{5-3} = x^2$$

The following rule states this result in a general way.

Quotient Rule for Exponents

If m and n are positive integers and a is a real number, then

$$\frac{a^m}{a^n} = a^{m-n}, \quad a \neq 0$$

For example,

$$\frac{x^6}{x^2} = x^{6-2} = x^4, \quad x \neq 0$$

Answers

22. $\frac{r^6}{s^6}, \quad s \neq 0$ **23.** $\frac{25x^{12}}{81y^6}, \quad y \neq 0$

In other words, to divide one exponential expression by another with a common base, we keep the base and subtract the exponents.

PRACTICE PROBLEMS 24–27

Simplify each quotient.

24. $\frac{y^7}{y^3}$ **25.** $\frac{5^9}{5^6}$

26. $\frac{(-2)^{14}}{(-2)^{10}}$ **27.** $\frac{7a^4b^{11}}{ab}$

EXAMPLES Simplify each quotient.

24. $\frac{x^5}{x^2} = x^{5-2} = x^3$ Use the quotient rule.

25. $\frac{4^7}{4^3} = 4^{7-3} = 4^4 = 256$ Use the quotient rule.

26. $\frac{(-3)^5}{(-3)^2} = (-3)^3 = -27$ Use the quotient rule.

27. $\frac{2x^5y^2}{xy} = 2 \cdot \frac{x^5}{x^1} \cdot \frac{y^2}{y^1}$

$= 2 \cdot (x^{5-1}) \cdot (y^{2-1})$ Use the quotient rule.

$= 2x^4y^1$ or $2x^4y$

Work Practice Problems 24–27

Let's now give meaning to an expression such as x^0. To do so, we will simplify $\frac{x^3}{x^3}$ in two ways and compare the results.

$\frac{x^3}{x^3} = x^{3-3} = x^0$ Apply the quotient rule.

$\frac{x^3}{x^3} = \frac{x \cdot x \cdot x}{x \cdot x \cdot x} = 1$ Apply the fundamental principle for fractions.

Since $\frac{x^3}{x^3} = x^0$ and $\frac{x^3}{x^3} = 1$, we define that $x^0 = 1$ as long as x is not 0.

Zero Exponent

$a^0 = 1$, as long as a is not 0.

For example, $5^0 = 1$.

In other words, a base raised to the 0 power is 1, as long as the base is not 0.

PRACTICE PROBLEMS 28–32

Simplify each expression.

28. 8^0 **29.** $(2r^2s)^0$

30. $(-7)^0$ **31.** -7^0

32. $7y^0$

EXAMPLES Simplify each expression.

28. $3^0 = 1$

29. $(5x^3y^2)^0 = 1$

30. $(-4)^0 = 1$

31. $-4^0 = -1 \cdot 4^0 = -1 \cdot 1 = -1$

32. $5x^0 = 5 \cdot x^0 = 5 \cdot 1 = 5$

Work Practice Problems 28–32

Answers

24. y^4 **25.** 125 **26.** 16 **27.** $7a^3b^{10}$ **28.** 1 **29.** 1 **30.** 1 **31.** −1 **32.** 7

✔**Concept Check** Suppose you are simplifying each expression. Tell whether you would *add* the exponents, *subtract* the exponents, *multiply* the exponents, *divide* the exponents, or *none of these.*

a. $(x^{63})^{21}$ **b.** $\frac{y^{15}}{y^3}$ **c.** $z^{16} + z^8$ **d.** $w^{45} \cdot w^9$

Objective F Deciding Which Rule to Use

Let's practice deciding which rule to use to simplify. We will continue this discussion with more examples in the next section.

EXAMPLE 33 Simplify each expression.

a. $x^7 \cdot x^4$ **b.** $\left(\frac{t}{2}\right)^4$ **c.** $(9y^5)^2$

Solution:

a. Here, we have a product, so we use the product rule to simplify.

$x^7 \cdot x^4 = x^{7+4} = x^{11}$

b. This is a quotient raised to a power, so we use the power of a quotient rule.

$\left(\frac{t}{2}\right)^4 = \frac{t^4}{2^4} = \frac{t^4}{16}$

c. This is a product raised to a power, so we use the power of a product rule.

$(9y^5)^2 = 9^2(y^5)^2 = 81y^{10}$

Work Practice Problem 33

PRACTICE PROBLEM 33

Simplify each expression.

a. $\frac{x^7}{x^4}$ **b.** $(3y^4)^4$ **c.** $\left(\frac{x}{4}\right)^3$

Answers

33. a. x^3 **b.** $81y^{16}$ **c.** $\frac{x^3}{64}$

✔ Concept Check Answers

a. multiply **b.** subtract **c.** none of these **d.** add

Vocabulary and Readiness Check

Use the choices below to fill in each blank. Some choices may be used more than once.

0	base	add
1	exponent	multiply

1. Repeated multiplication of the same factor can be written using a(n) ____________.
2. In 5^2, the 2 is called the ____________ and the 5 is called the ________.
3. To simplify $x^2 \cdot x^7$, keep the base and _______ the exponents.
4. To simplify $(x^3)^6$, keep the base and ___________ the exponents.
5. The understood exponent on the term y is _____.
6. If $x^{\square} = 1$, the exponent is _____.

10.1 EXERCISE SET

FOR EXTRA HELP

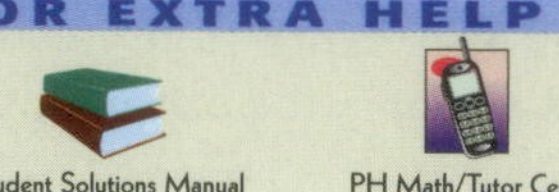

Objective A *Evaluate each expression. See Examples 1 through 6.*

1. 7^2 **2.** -3^2 **3.** $(-5)^1$ **4.** $(-3)^2$ **5.** -2^4 **6.** -4^3

7. $(-2)^4$ **8.** $(-4)^3$ **9.** $\left(\frac{1}{3}\right)^3$ **10.** $\left(-\frac{1}{9}\right)^2$ **11.** $7 \cdot 2^4$ **12.** $9 \cdot 2^2$

Evaluate each expression with the given replacement values. See Example 7.

13. x^2 when $x = -2$

14. x^3 when $x = -2$

15. $5x^3$ when $x = 3$

16. $4x^2$ when $x = 5$

17. $2xy^2$ when $x = 3$ and $y = -5$

18. $-4x^2y^3$ when $x = 2$ and $y = -1$

19. $\frac{5z^4}{7}$ when $z = -2$

20. $\frac{10}{3y^3}$ when $y = -3$

Objective B *Use the product rule to simplify each expression. Write the results using exponents. See Examples 8 through 15.*

21. $x^2 \cdot x^5$ **22.** $y^2 \cdot y$ **23.** $(-3)^3 \cdot (-3)^9$ **24.** $(-5)^7 \cdot (-5)^6$

25. $(5y^4)(3y)$ **26.** $(-2z^3)(-2z^2)$ **27.** $(x^9y)(x^{10}y^5)$ **28.** $(a^2b)(a^{13}b^{17})$

29. $(-8mn^6)(9m^2n^2)$ **30.** $(-7a^3b^3)(7a^{19}b)$ **31.** $(4z^{10})(-6z^7)(z^3)$ **32.** $(12x^5)(-x^6)(x^4)$

△ **33.** The rectangle below has width $4x^2$ feet and length $5x^3$ feet. Find its area as an expression in x.

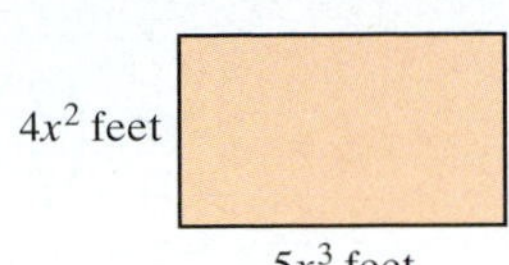

△ **34.** The parallelogram below has base length $9y^7$ meters and height $2y^{10}$ meters. Find its area as an expression in y.

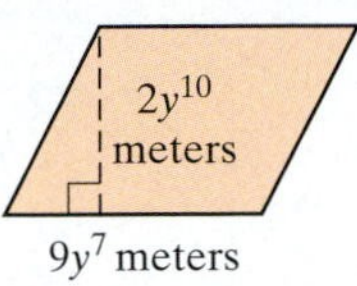

Objectives **C** **D** **Mixed Practice** *Use the power rule and the power of a product or quotient rule to simplify each expression. See Examples 16 through 23.*

35. $(x^9)^4$ **36.** $(y^7)^5$ **37.** $(pq)^8$ **38.** $(ab)^6$

39. $(2a^5)^3$ **40.** $(4x^6)^2$ **41.** $(x^2y^3)^5$ **42.** $(a^4b)^7$

43. $(-7a^2b^5c)^2$ **44.** $(-3x^7yz^2)^3$ **45.** $\left(\frac{r}{s}\right)^9$ **46.** $\left(\frac{q}{t}\right)^{11}$

47. $\left(\frac{mp}{n}\right)^9$ **48.** $\left(\frac{xy}{7}\right)^2$ **49.** $\left(\frac{-2xz}{y^5}\right)^2$ **50.** $\left(\frac{xy^4}{-3z^3}\right)^3$

△ **51.** The square shown has sides of length $8z^5$ decimeters. Find its area.

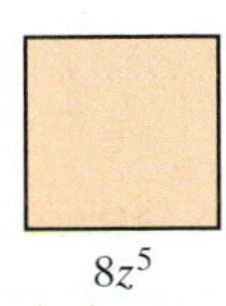

△ **52.** Given the circle below with radius $5y$ centimeters, find its area. Do not approximate π.

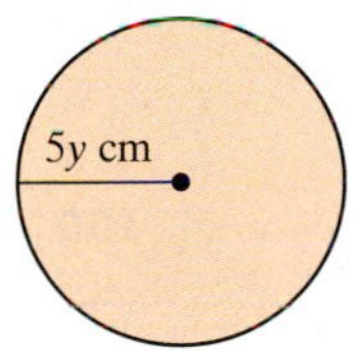

△ **53.** The vault below is in the shape of a cube. If each side is $3y^4$ feet, find its volume.

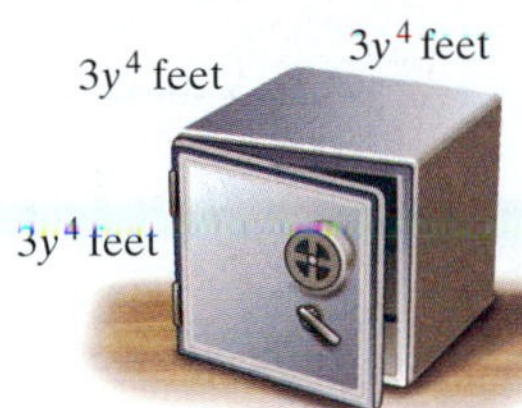

△ **54.** The silo shown is in the shape of a cylinder. If its radius is $4x$ meters and its height is $5x^3$ meters, find its volume. Do not approximate π.

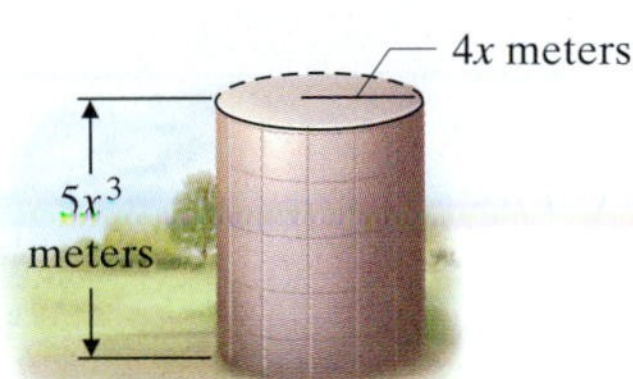

Objective **E** *Use the quotient rule and simplify each expression. See Examples 24 through 28.*

55. $\frac{x^3}{x}$ **56.** $\frac{y^{10}}{y^9}$ **57.** $\frac{(-4)^6}{(-4)^3}$ **58.** $\frac{(-6)^{13}}{(-6)^{11}}$

59. $\frac{p^7q^{20}}{pq^{15}}$ **60.** $\frac{x^8y^6}{xy^5}$ **61.** $\frac{7x^2y^6}{14x^2y^3}$ **62.** $\frac{9a^4b^7}{27ab^2}$

Simplify each expression. See Examples 28 through 32.

63. 7^0 **64.** 23^0 **65.** $(2x)^0$ **66.** $(4y)^0$

67. $-7x^0$ **68.** $-2x^0$ **69.** $5^0 + y^0$ **70.** $-3^0 + 4^0$

Objectives A B C D E F Mixed Practice *Simplify each expression. See Examples 1 through 6, and 8 through 33.*

71. -9^2 **72.** $(-9)^2$ **73.** $\left(\frac{1}{4}\right)^3$ **74.** $\left(\frac{2}{3}\right)^3$

75. b^4b^2 **76.** y^4y **77.** $a^2a^3a^4$ **78.** $x^2x^{15}x^9$

79. $(2x^3)(-8x^4)$ **80.** $(3y^4)(-5y)$ **81.** $(a^7b^{12})(a^4b^8)$ **82.** $(y^2z^2)(y^{15}z^{13})$

83. $(-2mn^6)(-13m^8n)$ **84.** $(-3s^5t)(-7st^{10})$ **85.** $(z^4)^{10}$ **86.** $(t^5)^{11}$

87. $(4ab)^3$ **88.** $(2ab)^4$ **89.** $(-6xyz^3)^2$ **90.** $(-3xy^2a^3)^3$

91. $\frac{z^{12}}{z^4}$ **92.** $\frac{b^4}{b}$ **93.** $\frac{3x^5}{x^4}$ **94.** $\frac{5x^9}{x^3}$

95. $(6b)^0$ **96.** $(5ab)^0$ **97.** $(9xy)^2$ **98.** $(2ab)^5$

99. $2^3 + 2^5$ **100.** $7^2 - 7^0$ **101.** $\left(\frac{3y^5}{6x^4}\right)^3$ **102.** $\left(\frac{2ab}{6yz}\right)^4$

103. $\frac{2x^3y^2z}{xyz}$ **104.** $\frac{x^{12}y^{13}}{x^5y^7}$

Review

Subtract. See Section 2.3.

105. $5 - 7$ **106.** $9 - 12$ **107.** $3 - (-2)$

108. $5 - (-10)$ **109.** $-11 - (-4)$ **110.** $-15 - (-21)$

Concept Extensions

Solve. See the Concept Checks in this section. For Exercises 111 through 114, match the expression with the operation needed to simplify each. A letter may be used more than once and a letter may not be used at all.

111. $(x^{14})^{23}$

112. $x^{14} \cdot x^{23}$

113. $x^{14} + x^{23}$

114. $\frac{x^{35}}{x^{17}}$

a. Add the exponents
b. Subtract the exponents
c. Multiply the exponents
d. Divide the exponents
e. None of these

Fill in the boxes so that each statement is true. (More than one answer is possible for each exercise.)

115. $x^{\square} \cdot x^{\square} = x^{12}$

116. $(x^{\square})^{\square} = x^{20}$

117. $\dfrac{y^{\square}}{y^{\square}} = y^7$

118. $(y^{\square})^{\square} \cdot (y^{\square})^{\square} = y^{30}$

△ **119.** The formula $V = x^3$ can be used to find the volume V of a cube with side length x. Find the volume of a cube with side length 7 meters. (Volume is measured in cubic units.)

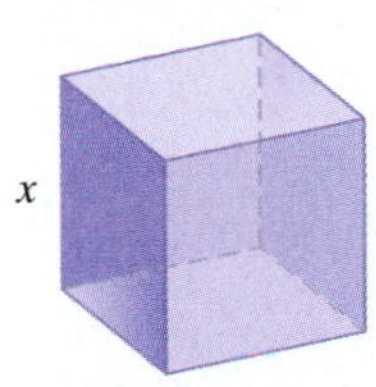

△ **120.** The formula $S = 6x^2$ can be used to find the surface area S of a cube with side length x. Find the surface area of a cube with side length 5 meters. (Surface area is measured in square units.)

△ **121.** To find the amount of water that a swimming pool in the shape of a cube can hold, do we use the formula for volume of the cube or surface area of the cube? (See Exercises 119 and 120.)

△ **122.** To find the amount of material needed to cover an ottoman in the shape of a cube, do we use the formula for volume of the cube or surface area of the cube? (See Exercises 119 and 120.)

✎ **123.** Explain why $(-5)^4 = 625$, while $-5^4 = -625$.

✎ **124.** Explain why $5 \cdot 4^2 = 80$, while $(5 \cdot 4)^2 = 400$.

✎ **125.** In your own words, explain why $5^0 = 1$.

✎ **126.** In your own words, explain when $(-3)^n$ is positive and when it is negative.

Simplify each expression. Assume that variables represent positive integers.

127. $x^{5a}x^{4a}$

128. $b^{9a}b^{4a}$

129. $(a^b)^5$

130. $(2a^{4b})^4$

131. $\dfrac{x^{9a}}{x^{4a}}$

132. $\dfrac{y^{15b}}{y^{6b}}$

STUDY SKILLS BUILDER

How Well Do You Know Your Textbook?

The questions below will determine whether you are familiar with your textbook. For help, see Section 1.1 in this text.

1. What does the ◉ icon mean?
2. What does the ✎ icon mean?
3. What does the △ icon mean?
4. Where can you find a review for each chapter? What answers to this review can be found in the back of your text?
5. Each chapter contains an overview of the chapter along with examples. What is this feature called?
6. Each chapter contains a review of vocabulary. What is this feature called?
7. There is a CD in your text. What content is contained on this CD?
8. What is the location of the section that is entirely devoted to study skills?
9. There are Practice Problems that are contained in the margin of the text. What are they and how can they be used?

Objectives

A Simplify Expressions Containing Negative Exponents.

B Use the Rules and Definitions for Exponents to Simplify Exponential Expressions.

C Write Numbers in Scientific Notation.

D Convert Numbers in Scientific Notation to Standard Form.

10.2 NEGATIVE EXPONENTS AND SCIENTIFIC NOTATION

Objective A Simplifying Expressions Containing Negative Exponents

Our work with exponential expressions so far has been limited to exponents that are positive integers or 0. Here we will also give meaning to an expression like x^{-3}.

Suppose that we wish to simplify the expression $\frac{x^2}{x^5}$. If we use the quotient rule for exponents, we subtract exponents:

$$\frac{x^2}{x^5} = x^{2-5} = x^{-3}, \quad x \neq 0$$

But what does x^{-3} mean? Let's simplify $\frac{x^2}{x^5}$ using the definition of a^n.

$$\frac{x^2}{x^5} = \frac{x \cdot x}{x \cdot x \cdot x \cdot x \cdot x}$$

$$= \frac{x \cdot x}{x \cdot x \cdot x \cdot x \cdot x}$$ Divide numerator and denominator by common factors by applying the fundamental principle for fractions.

$$= \frac{1}{x^3}$$

If the quotient rule is to hold true for negative exponents, then x^{-3} must equal $\frac{1}{x^3}$.

From this example, we state the definition for negative exponents.

Negative Exponents

If a is a real number other than 0 and n is an integer, then

$$a^{-n} = \frac{1}{a^n}$$

For example,

$$x^{-3} = \frac{1}{x^3}$$

In other words, another way to write a^{-n} is to take its reciprocal and change the sign of its exponent.

EXAMPLES Simplify by writing each expression with positive exponents only.

1. $3^{-2} = \frac{1}{3^2} = \frac{1}{9}$ Use the definition of negative exponents.
2. $2x^{-3} = 2^1 \cdot \frac{1}{x^3} = \frac{2^1}{x^3}$ or $\frac{2}{x^3}$ Use the definition of negative exponents.
3. $2^{-1} + 4^{-1} = \frac{1}{2} + \frac{1}{4} = \frac{2}{4} + \frac{1}{4} = \frac{3}{4}$
4. $(-2)^{-4} = \frac{1}{(-2)^4} = \frac{1}{(-2)(-2)(-2)(-2)} = \frac{1}{16}$

Helpful Hint

Don't forget that since there are no parentheses, only x is the base for the exponent -3.

Work Practice Problems 1–4

PRACTICE PROBLEMS 1–4

Simplify by writing each expression with positive exponents only.

1. 5^{-3}
2. $7x^{-4}$
3. $5^{-1} + 3^{-1}$
4. $(-3)^{-4}$

Answers

1. $\frac{1}{125}$ 2. $\frac{7}{x^4}$ 3. $\frac{8}{15}$ 4. $\frac{1}{81}$

Helpful Hint

A negative exponent *does not affect* the sign of its base.
Remember: Another way to write a^{-n} is to take its reciprocal and change the sign of its exponent: $a^{-n} = \frac{1}{a^n}$. For example,

$$x^{-2} = \frac{1}{x^2}, \qquad 2^{-3} = \frac{1}{2^3} \text{ or } \frac{1}{8}$$

$$\frac{1}{y^{-4}} = \frac{1}{\frac{1}{y^4}} = y^4, \qquad \frac{1}{5^{-2}} = 5^2 \text{ or } 25$$

From the preceding Helpful Hint, we know that $x^{-2} = \frac{1}{x^2}$ and $\frac{1}{y^{-4}} = y^4$. We can use this to include another statement in our definition of negative exponents.

Negative Exponents

If a is a real number other than 0 and n is an integer, then

$$a^{-n} = \frac{1}{a^n} \quad \text{and} \quad \frac{1}{a^{-n}} = a^n$$

EXAMPLES Simplify each expression. Write each result using positive exponents only.

5. $\left(\frac{2}{x}\right)^{-3} = \frac{2^{-3}}{x^{-3}} = \frac{2^{-3}}{1} \cdot \frac{1}{x^{-3}} = \frac{1}{2^3} \cdot \frac{x^3}{1} = \frac{x^3}{2^3} = \frac{x^3}{8}$ Use the negative exponents rule.

6. $\frac{y}{y^{-2}} = \frac{y^1}{y^{-2}} = y^{1-(-2)} = y^3$ Use the quotient rule.

7. $\frac{p^{-4}}{q^{-9}} = p^{-4} \cdot \frac{1}{q^{-9}} = \frac{1}{p^4} \cdot q^9 = \frac{q^9}{p^4}$ Use the negative exponents rule.

8. $\frac{x^{-5}}{x^7} = x^{-5-7} = x^{-12} = \frac{1}{x^{12}}$

Work Practice Problems 5–8

PRACTICE PROBLEMS 5–8

Simplify each expression. Write each result using positive exponents only.

5. $\left(\frac{6}{7}\right)^{-2}$ **6.** $\frac{x}{x^{-4}}$

7. $\frac{y^{-9}}{z^{-5}}$ **8.** $\frac{y^{-4}}{y^6}$

Objective B Simplifying Exponential Expressions

All the previously stated rules for exponents apply for negative exponents also. Here is a summary of the rules and definitions for exponents.

Summary of Exponent Rules

If m and n are integers and a, b, and c are real numbers, then

Product rule for exponents:	$a^m \cdot a^n = a^{m+n}$
Power rule for exponents:	$(a^m)^n = a^{m \cdot n}$
Power of a product:	$(ab)^n = a^n b^n$
Power of a quotient:	$\left(\frac{a}{c}\right)^n = \frac{a^n}{c^n}, \quad c \neq 0$
Quotient rule for exponents:	$\frac{a^m}{a^n} = a^{m-n}, \quad a \neq 0$
Zero exponent:	$a^0 = 1, \quad a \neq 0$
Negative exponent:	$a^{-n} = \frac{1}{a^n}, \quad a \neq 0$

Answers

5. $\frac{49}{36}$ **6.** x^5 **7.** $\frac{z^5}{y^9}$ **8.** $\frac{1}{y^{10}}$

PRACTICE PROBLEMS 9–16

Simplify each expression. Write each result using positive exponents only.

9. $\dfrac{(x^5)^3x}{x^4}$ 10. $\left(\dfrac{9x^3}{y}\right)^{-2}$

11. $(a^{-4}b^7)^{-5}$ 12. $\dfrac{(2x)^4}{x^8}$

13. $\dfrac{y^{-10}}{(y^5)^4}$ 14. $(4a^2)^{-3}$

15. $-\dfrac{32x^{-3}y^{-6}}{8x^{-5}y^{-2}}$ 16. $\dfrac{(3x^{-2}y)^{-2}}{(2x^7y)^3}$

EXAMPLES Simplify each expression. Write each result using positive exponents only.

9. $\dfrac{(x^3)^4x}{x^7} = \dfrac{x^{12}\cdot x}{x^7} = \dfrac{x^{12+1}}{x^7} = \dfrac{x^{13}}{x^7} = x^{13-7} = x^6$ Use the power rule.

10. $\left(\dfrac{3a^2}{b}\right)^{-3} = \dfrac{3^{-3}(a^2)^{-3}}{b^{-3}}$ Raise each factor in the numerator and the denominator to the -3 power.

$= \dfrac{3^{-3}a^{-6}}{b^{-3}}$ Use the power rule.

$= \dfrac{b^3}{3^3a^6}$ Use the negative exponent rule.

$= \dfrac{b^3}{27a^6}$ Write 3^3 as 27.

11. $(y^{-3}z^6)^{-6} = (y^{-3})^{-6}(z^6)^{-6}$ Raise each factor to the -6 power.

$= y^{18}z^{-36} = \dfrac{y^{18}}{z^{36}}$

12. $\dfrac{(2x)^5}{x^3} = \dfrac{2^5\cdot x^5}{x^3} = 2^5\cdot x^{5-3} = 32x^2$ Raise each factor in the numerator to the fifth power.

13. $\dfrac{x^{-7}}{(x^4)^3} = \dfrac{x^{-7}}{x^{12}} = x^{-7-12} = x^{-19} = \dfrac{1}{x^{19}}$

14. $(5y^3)^{-2} = 5^{-2}(y^3)^{-2} = 5^{-2}y^{-6} = \dfrac{1}{5^2y^6} = \dfrac{1}{25y^6}$

15. $-\dfrac{22a^7b^{-5}}{11a^{-2}b^3} = -\dfrac{22}{11}\cdot a^{7-(-2)}b^{-5-3} = -2a^9b^{-8} = -\dfrac{2a^9}{b^8}$

16. $\dfrac{(2xy)^{-3}}{(x^2y^3)^2} = \dfrac{2^{-3}x^{-3}y^{-3}}{(x^2)^2(y^3)^2} = \dfrac{2^{-3}x^{-3}y^{-3}}{x^4y^6} = 2^{-3}x^{-3-4}y^{-3-6}$

$= 2^{-3}x^{-7}y^{-9} = \dfrac{1}{2^3x^7y^9}$ or $\dfrac{1}{8x^7y^9}$

Work Practice Problems 9–16

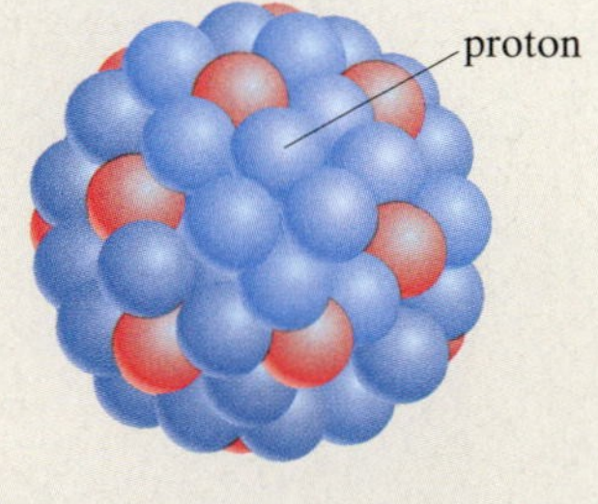

Mass of proton is approximately 0.000 000 000 000 000 000 000 001 65 gram

Objective C Writing Numbers in Scientific Notation

Both very large and very small numbers frequently occur in many fields of science. For example, the distance between the Sun and the dwarf planet Pluto is approximately 5,906,000,000 kilometers, and the mass of a proton is approximately 0.00000000000000000000000165 gram. It can be tedious to write these numbers in this standard decimal notation, so **scientific notation** is used as a convenient shorthand for expressing very large and very small numbers.

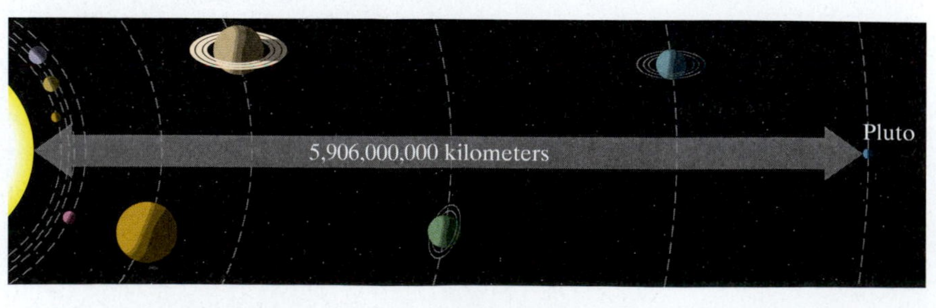

Answers

9. x^{12} 10. $\dfrac{y^2}{81x^6}$ 11. $\dfrac{a^{20}}{b^{35}}$ 12. $\dfrac{16}{x^4}$ 13. $\dfrac{1}{y^{30}}$ 14. $\dfrac{1}{64a^6}$ 15. $-\dfrac{4x^2}{y^4}$ 16. $\dfrac{1}{72x^{17}y^5}$

Scientific Notation

A positive number is written in scientific notation if it is written as the product of a number a, where $1 \le a < 10$, and an integer power r of 10: $a \times 10^r$

The following numbers are written in scientific notation. The $\times$ sign for multiplication is used as part of the notation.

2.03×10^2 7.362×10^7 5.906×10^9 (Distance between the Sun and Pluto)

1×10^{-3} 8.1×10^{-5} 1.65×10^{-24} (Mass of a proton)

The following steps are useful when writing numbers in scientific notation.

To Write a Number in Scientific Notation

Step 1: Move the decimal point in the original number so that the new number has a value between 1 and 10.

Step 2: Count the number of decimal places the decimal point is moved in Step 1. If the original number is 10 or greater, the count is positive. If the original number is less than 1, the count is negative.

Step 3: Multiply the new number in Step 1 by 10 raised to an exponent equal to the count found in Step 2.

EXAMPLE 17 Write each number in scientific notation.

a. 367,000,000
b. 0.000003
c. 20,520,000,000
d. 0.00085

Solution:

a. Step 1: Move the decimal point until the number is between 1 and 10.

Step 2: The decimal point is moved 8 places and the original number is 10 or greater, so the count is positive 8.

Step 3: $367{,}000{,}000 = 3.67 \times 10^8$.

b. Step 1: Move the decimal point until the number is between 1 and 10.

0.000003
6 places

Step 2: The decimal point is moved 6 places and the original number is less than 1, so the count is -6.

Step 3: $0.000003 = 3.0 \times 10^{-6}$

c. $20{,}520{,}000{,}000 = 2.052 \times 10^{10}$

d. $0.00085 = 8.5 \times 10^{-4}$

Work Practice Problem 17

PRACTICE PROBLEM 17

Write each number in scientific notation.

a. 420,000 **b.** 0.00017
c. 9,060,000,000 **d.** 0.000007

Objective D Converting Numbers to Standard Form

A number written in scientific notation can be rewritten in standard form. For example, to write 8.63×10^3 in standard form, recall that $10^3 = 1000$.

$$8.63 \times 10^3 = 8.63(1000) = 8630$$

Notice that the exponent on the 10 is positive 3, and we moved the decimal point 3 places to the right.

Answers

17. a. 4.2×10^5 **b.** 1.7×10^{-4} **c.** 9.06×10^9 **d.** 7×10^{-6}

To write 7.29×10^{-3} in standard form, recall that $10^{-3} = \frac{1}{10^3} = \frac{1}{1000}$.

$$7.29 \times 10^{-3} = 7.29\left(\frac{1}{1000}\right) = \frac{7.29}{1000} = 0.00729$$

The exponent on the 10 is negative 3, and we moved the decimal to the left 3 places.

In general, **to write a scientific notation number in standard form,** move the decimal point the same number of places as the exponent on 10. If the exponent is positive, move the decimal point to the right; if the exponent is negative, move the decimal point to the left.

✔**Concept Check** Which number in each pair is larger?

a. 7.8×10^3 or 2.1×10^5

b. 9.2×10^{-2} or 2.7×10^4

c. 5.6×10^{-4} or 6.3×10^{-5}

PRACTICE PROBLEM 18

Write the numbers in standard notation, without exponents.

a. 3.062×10^{-4}

b. 5.21×10^4

c. 9.6×10^{-5}

d. 6.002×10^6

EXAMPLE 18 Write each number in standard notation, without exponents.

a. 1.02×10^5

b. 7.358×10^{-3}

c. 8.4×10^7

d. 3.007×10^{-5}

Solution:

a. Move the decimal point 5 places to the right.

$1.02 \times 10^5 = 102{,}000.$

b. Move the decimal point 3 places to the left.

$7.358 \times 10^{-3} = 0.007358$

c. $8.4 \times 10^7 = 84{,}000{,}000.$ 7 places to the right

d. $3.007 \times 10^{-5} = 0.00003007$ 5 places to the left

Work Practice Problem 18

Performing operations on numbers written in scientific notation makes use of the rules and definitions for exponents.

PRACTICE PROBLEM 19

Perform each indicated operation. Write each result in standard decimal notation.

a. $(9 \times 10^7)(4 \times 10^{-9})$

b. $\frac{8 \times 10^4}{2 \times 10^{-3}}$

EXAMPLE 19 Perform each indicated operation. Write each result in standard decimal notation.

a. $(8 \times 10^{-6})(7 \times 10^3)$

b. $\frac{12 \times 10^2}{6 \times 10^{-3}}$

Solution:

a.
$$\begin{aligned}(8 \times 10^{-6})(7 \times 10^3) &= 8 \cdot 7 \cdot 10^{-6} \cdot 10^3 \\ &= 56 \times 10^{-3} \\ &= 0.056\end{aligned}$$

b. $\frac{12 \times 10^2}{6 \times 10^{-3}} = \frac{12}{6} \times 10^{2-(-3)} = 2 \times 10^5 = 200{,}000$

Work Practice Problem 19

Answers

18. a. 0.0003062 **b.** 52,100 **c.** 0.000096 **d.** 6,002,000
19. a. 0.36 **b.** 40,000,000

✔ **Concept Check Answer**

a. 2.1×10^5 **b.** 2.7×10^4 **c.** 5.6×10^{-4}

CALCULATOR EXPLORATIONS Scientific Notation

To enter a number written in scientific notation on a scientific calculator, locate the scientific notation key, which may be marked [EE] or [EXP]. To enter 3.1×10^7, press [3.1] [EE] [7]. The display should read [3.1 07].

Enter each number written in scientific notation on your calculator.

1. 5.31×10^3
2. -4.8×10^{14}
3. 6.6×10^{-9}
4. -9.9811×10^{-2}

Multiply each of the following on your calculator. Notice the form of the result.

5. $3{,}000{,}000 \times 5{,}000{,}000$
6. $230{,}000 \times 1{,}000$

Multiply each of the following on your calculator. Write the product in scientific notation.

7. $(3.26 \times 10^6)(2.5 \times 10^{13})$
8. $(8.76 \times 10^{-4})(1.237 \times 10^9)$

Vocabulary and Readiness Check

Fill in each blank with the correct choice.

1. The expression x^{-3} equals ______.

   a. $-x^3$ **b.** $\frac{1}{x^3}$ **c.** $\frac{-1}{x^3}$ **d.** $\frac{1}{x^{-3}}$

2. The expression 5^{-4} equals ______.

   a. -20 **b.** -625 **c.** $\frac{1}{20}$ **d.** $\frac{1}{625}$

3. The number 3.021×10^{-3} is written in ______________.

   a. standard form **b.** expanded form **c.** scientific notation

4. The number 0.0261 is written in ______________.

   a. standard form **b.** expanded form **c.** scientific notation

10.2 EXERCISE SET

FOR EXTRA HELP: Student Solutions Manual | PH Math/Tutor Center | CD/Video for Review | MathXL® | MyMathLab

Objective A *Simplify each expression. Write each result using positive exponents only. See Examples 1 through 8.*

1. 4^{-3} **2.** 6^{-2} **3.** $7x^{-3}$ **4.** $(7x)^{-3}$ **5.** $\left(-\frac{1}{4}\right)^{-3}$ **6.** $\left(-\frac{1}{8}\right)^{-2}$

7. $3^{-1} + 2^{-1}$ **8.** $4^{-1} + 4^{-2}$ **9.** $\frac{1}{p^{-3}}$ **10.** $\frac{1}{q^{-5}}$ **11.** $\frac{p^{-5}}{q^{-4}}$ **12.** $\frac{r^{-5}}{s^{-2}}$

13. $\frac{x^{-2}}{x}$ **14.** $\frac{y}{y^{-3}}$ **15.** $\frac{z^{-4}}{z^{-7}}$ **16.** $\frac{x^{-4}}{x^{-1}}$ **17.** $3^{-2} + 3^{-1}$ **18.** $4^{-2} - 4^{-3}$

19. $(-3)^{-2}$ **20.** $(-2)^{-6}$ **21.** $\frac{-1}{p^{-4}}$ **22.** $\frac{-1}{y^{-6}}$ **23.** $-2^0 - 3^0$ **24.** $5^0 + (-5)^0$

Objective B *Simplify each expression. Write each result using positive exponents only. See Examples 9 through 16.*

25. $\frac{x^2x^5}{x^3}$ **26.** $\frac{y^4y^5}{y^6}$ **27.** $\frac{p^2p}{p^{-1}}$ **28.** $\frac{y^3y}{y^{-2}}$ **29.** $\frac{(m^5)^4m}{m^{10}}$ **30.** $\frac{(x^2)^8x}{x^9}$

31. $\frac{r}{r^{-3}r^{-2}}$
32. $\frac{p}{p^{-3}q^{-5}}$
33. $(x^5y^3)^{-3}$
34. $(z^5x^5)^{-3}$
35. $\frac{(x^2)^3}{x^{10}}$
36. $\frac{(y^4)^2}{y^{12}}$

37. $\frac{(a^5)^2}{(a^3)^4}$
38. $\frac{(x^2)^5}{(x^4)^3}$
39. $\frac{8k^4}{2k}$
40. $\frac{27r^6}{3r^4}$
41. $\frac{-6m^4}{-2m^3}$
42. $\frac{15a^4}{-15a^5}$

43. $\frac{-24a^6b}{6ab^2}$
44. $\frac{-5x^4y^5}{15x^4y^2}$
45. $\frac{6x^2y^3}{-7x^2y^5}$
46. $\frac{-8xa^2b}{-5xa^5b}$
47. $(3a^2b^{-4})^3$
48. $(5x^3y^{-2})^2$

49. $(a^{-5}b^2)^{-6}$
50. $(4^{-1}x^5)^{-2}$
51. $\left(\frac{x^{-2}y^4}{x^3y^7}\right)^2$
52. $\left(\frac{a^5b}{a^7b^{-2}}\right)^{-3}$
53. $\frac{4^2z^{-3}}{4^3z^{-5}}$
54. $\frac{5^{-1}z^7}{5^{-2}z^9}$

55. $\frac{3^{-1}x^4}{3^3x^{-7}}$
56. $\frac{2^{-3}x^{-4}}{2^2x}$
57. $\frac{7ab^{-4}}{7^{-1}a^{-3}b^2}$
58. $\frac{6^{-5}x^{-1}y^2}{6^{-2}x^{-4}y^4}$
59. $\frac{-12m^5n^{-7}}{4m^{-2}n^{-3}}$
60. $\frac{-15r^{-6}s}{5r^{-4}s^{-3}}$

61. $\left(\frac{a^{-5}b}{ab^3}\right)^{-4}$
62. $\left(\frac{r^{-2}s^{-3}}{r^{-4}s^{-3}}\right)^{-3}$
63. $(5^2)(8)(2^0)$
64. $(3^4)(7^0)(2)$
65. $\frac{(xy^3)^5}{(xy)^{-4}}$
66. $\frac{(rs)^{-3}}{(r^2s^3)^2}$

67. $\frac{(-2xy^{-3})^{-3}}{(xy^{-1})^{-1}}$
68. $\frac{(-3x^2y^2)^{-2}}{(xyz)^{-2}}$
69. $\frac{(a^4b^{-7})^{-5}}{(5a^2b^{-1})^{-2}}$
70. $\frac{(a^6b^{-2})^4}{(4a^{-3}b^{-3})^3}$

△ **71.** Find the volume of the cube.

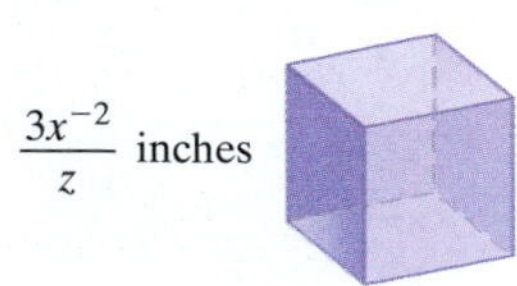

△ **72.** Find the area of the triangle.

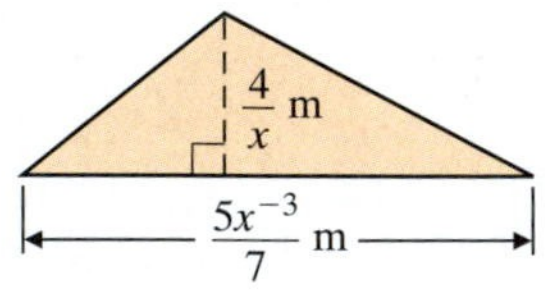

Objective C *Write each number in scientific notation. See Example 17.*

73. 78,000
74. 9,300,000,000
75. 0.00000167
76. 0.00000017

77. 0.00635
78. 0.00194
79. 1,160,000
80. 700,000

81. At this writing, the world's largest optical telescopes are the twin Keck Telescopes located near the summit of Mauna Kea in Hawaii. The elevation of the Keck Telescopes is about 13,600 feet above sea level. Write 13,600 in scientific notation. (*Source:* W.M. Keck Observatory)

82. After more than 30 years, the *Pioneer 10* spacecraft sent its last signal to Earth. Launched on March 2, 1972, it became the first spacecraft to leave our solar system. When it transmitted its last signal, in January 2003, it was approximately 8,000,000,000 miles from Earth. Write 8,000,000,000 in scientific notation. (*Source:* NASA Ames Research Center)

Objective **D** *Write each number in standard notation. See Example 18.*

83. 8.673×10^{-10}

84. 9.056×10^{-4}

85. 3.3×10^{-2}

86. 4.8×10^{-6}

87. 2.032×10^{4}

88. 9.07×10^{10}

89. Each second, the Sun converts 7.0×10^8 tons of hydrogen into helium and energy in the form of gamma rays. Write this number in standard notation. (*Source:* Students for the Exploration and Development of Space)

90. In chemistry, Avogadro's number is the number of atoms in one mole of an element. Avogadro's number is $6.02214199 \times 10^{23}$. Write this number in standard notation. (*Source:* National Institute of Standards and Technology)

Objectives **C** **D** **Mixed Practice** *See Examples 17 and 18. Below are some interesting facts about the Internet. If a number is written in standard form, write it in scientific notation. If a number is written in scientific notation, write it in standard form.*

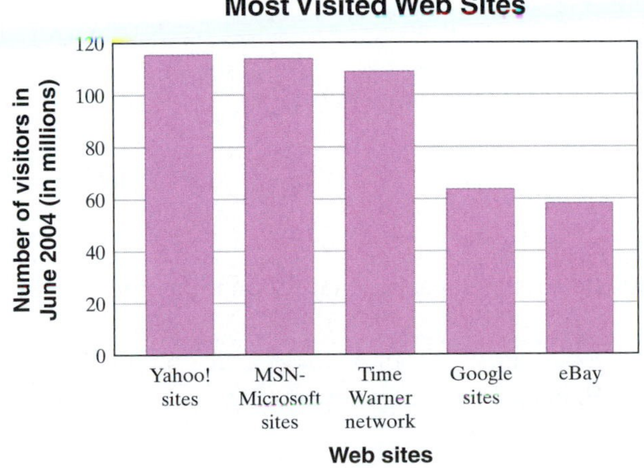

91. The total number of Internet users is 940,000,000.

92. The total number of Internet hosts (sites) is 233,000,000.

93. In a recent year, the revenue generated by the Internet was 1.23×10^{12} dollars.

94. The estimated number of e-mail boxes is 1.2×10^9.

95. The bar graph above shows the most visited Web sites on the computer. Estimate the height of the tallest bar and the shortest bar. Then write each number in scientific notation.

96. Junk e-mail (SPAM) costs consumers and businesses an estimated \$23,000,000,000.

Objective **D** *Evaluate each expression using exponential rules. Write each result in standard notation. See Example 19.*

97. $(1.2 \times 10^{-3})(3 \times 10^{-2})$

98. $(2.5 \times 10^{6})(2 \times 10^{-6})$

99. $(4 \times 10^{-10})(7 \times 10^{-9})$

100. $(5 \times 10^{6})(4 \times 10^{-8})$

101. $\frac{8 \times 10^{-1}}{16 \times 10^{5}}$

102. $\frac{25 \times 10^{-4}}{5 \times 10^{-9}}$

103. $\frac{1.4 \times 10^{-2}}{7 \times 10^{-8}}$

104. $\frac{0.4 \times 10^{5}}{0.2 \times 10^{11}}$

105. Although the actual amount varies by season and time of day, the average volume of water that flows over Niagara Falls (the American and Canadian falls combined) each second is 7.5×10^5 gallons. How much water flows over Niagara Falls in an hour? Write the result in scientific notation. (*Hint:* 1 hour equals 3600 seconds) (*Source:* niagarafallslive.com)

106. A beam of light travels 9.460×10^{12} kilometers per year. How far does light travel in 10,000 years? Write the result in scientific notation.

Review

Simplify each expression by combining any like terms. See Section 3.1.

107. $3x - 5x + 7$

108. $7w + w - 2w$

109. $y - 10 + y$

110. $-6z + 20 - 3z$

111. $7x + 2 - 8x - 6$

112. $10y - 14 - y - 14$

Concept Extensions

Simplify.

113. $(2a^3)^3a^4 + a^5a^8$

114. $(2a^3)^3a^{-3} + a^{11}a^{-5}$

Fill in the boxes so that each statement is true. (More than one answer is possible for these exercises.)

115. $x^{\square} = \frac{1}{x^5}$

116. $7^{\square} = \frac{1}{49}$

117. $z^{\square} \cdot z^{\square} = z^{-10}$

118. $(x^{\square})^{\square} = x^{-15}$

119. Which is larger? See the Concept Check in this section.

a. 9.7×10^{-2} or 1.3×10^{1}

b. 8.6×10^{5} or 4.4×10^{7}

c. 6.1×10^{-2} or 5.6×10^{-4}

120. It was stated earlier that for an integer n,

$$x^{-n} = \frac{1}{x^n}, \quad x \neq 0$$

Explain why x may not equal 0.

121. Determine whether each statement is true or false.

a. $5^{-1} < 5^{-2}$

b. $\left(\frac{1}{5}\right)^{-1} < \left(\frac{1}{5}\right)^{-2}$

c. $a^{-1} < a^{-2}$ for all nonzero numbers.

Simplify each expression. Assume that variables represent positive integers.

122. $a^{-4m} \cdot a^{5m}$

123. $(x^{-3s})^3$

124. $(3y^{2z})^3$

125. $a^{4m+1} \cdot a^4$

10.3 INTRODUCTION TO POLYNOMIALS

Objective A Defining Term and Coefficient

In this section, we introduce a special algebraic expression called a polynomial. Let's first review some definitions presented in Section 3.1.

Recall that a term is a number or the product of a number and variables raised to powers. The terms of an expression are separated by plus signs. The terms of the expression $4x^2 + 3x$ are $4x^2$ and $3x$. The terms of the expression $9x^4 - 7x - 1$, or $9x^4 + (-7x) + (-1)$, are $9x^4$, $-7x$, and -1.

Expression	Terms
$4x^2 + 3x$	$4x^2, 3x$
$9x^4 - 7x - 1$	$9x^4, -7x, -1$
$7y^3$	$7y^3$

The **numerical coefficient** of a term, or simply the **coefficient,** is the numerical factor of each term. If no numerical factor appears in the term, then the coefficient is understood to be 1. If the term is a number only, it is called a **constant term** or simply a **constant.**

Term	Coefficient
x^5	1
$3x^2$	3
$-4x$	-4
$-x^2y$	-1
3 (constant)	3

EXAMPLE 1

Complete the table for the expression $7x^5 - 8x^4 + x^2 - 3x + 5$.

Term	Coefficient
x^2	
	-8
$-3x$	
	7
5	

Solution: The completed table is shown below.

Term	Coefficient
x^2	1
$-8x^4$	-8
$-3x$	-3
$7x^5$	7
5	5

Work Practice Problem 1

Objectives

A Define Term and Coefficient of a Term.

B Define Polynomial, Monomial, Binomial, Trinomial, and Degree.

C Evaluate Polynomials for Given Replacement Values.

D Simplify a Polynomial by Combining Like Terms.

E Simplify a Polynomial in Several Variables.

F Write a Polynomial in Descending Powers of the Variable and with No Missing Powers of the Variable.

PRACTICE PROBLEM 1

Complete the table for the expression $-6x^6 + 4x^5 + 7x^3 - 9x^2 - 1$.

Term	Coefficient
$7x^3$	
	-9
$-6x^6$	
	4
-1	

Answer

1. term: $-9x^2$; $4x^5$, coefficient: 7; -6; -1

Objective B Defining Polynomial, Monomial, Binomial, Trinomial, and Degree

Now we are ready to define what we mean by a polynomial.

Polynomial

A **polynomial in x** is a finite sum of terms of the form ax^n, where a is a real number and n is a whole number.

For example,

$$x^5 - 3x^3 + 2x^2 - 5x + 1$$

is a polynomial in x. Notice that this polynomial is written in **descending powers** of x because the powers of x decrease from left to right. (Recall that the term 1 can be thought of as $1x^0$.)

On the other hand,

$$x^{-5} + 2x - 3$$

is **not** a polynomial because one of its terms contains a variable with an exponent, -5, that is not a whole number.

Types of Polynomials

A **monomial** is a polynomial with exactly one term.

A **binomial** is a polynomial with exactly two terms.

A **trinomial** is a polynomial with exactly three terms.

The following are examples of monomials, binomials, and trinomials. Each of these examples is also a polynomial.

Polynomials			
Monomials	**Binomials**	**Trinomials**	**More Than Three Terms**
ax^2	$x + y$	$x^2 + 4xy + y^2$	$5x^3 - 6x^2 + 3x - 6$
$-3z$	$3p + 2$	$x^5 + 7x^2 - x$	$-y^5 + y^4 - 3y^3 - y^2 + y$
4	$4x^2 - 7$	$-q^4 + q^3 - 2q$	$x^6 + x^4 - x^3 + 1$

Each term of a polynomial has a degree. The **degree of a term in one variable** is the exponent on the variable.

PRACTICE PROBLEM 2

Identify the degree of each term of the trinomial $-15x^3 + 2x^2 - 5$.

EXAMPLE 2 Identify the degree of each term of the trinomial $12x^4 - 7x + 3$.

Solution: The term $12x^4$ has degree 4.

The term $-7x$ has degree 1 since $-7x$ is $-7x^1$.

The term 3 has degree 0 since 3 is $3x^0$.

Work Practice Problem 2

Each polynomial also has a degree.

Degree of a Polynomial

The **degree of a polynomial** is the greatest degree of any term of the polynomial.

Answer

2. 3; 2; 0

EXAMPLE 3 Find the degree of each polynomial and tell whether the polynomial is a monomial, binomial, trinomial, or none of these.

a. $-2t^2 + 3t + 6$ **b.** $15x - 10$ **c.** $7x + 3x^3 + 2x^2 - 1$

Solution:

a. The degree of the trinomial $-2t^2 + 3t + 6$ is 2, the greatest degree of any of its terms.

b. The degree of the binomial $15x - 10$ or $15x^1 - 10$ is 1.

c. The degree of the polynomial $7x + 3x^3 + 2x^2 - 1$ is 3. The polynomial is neither a monomial, binomial, nor trinomial.

Work Practice Problem 3

PRACTICE PROBLEM 3

Find the degree of each polynomial and tell whether the polynomial is a monomial, binomial, trinomial, or none of these.

a. $-6x + 14$

b. $9x - 3x^6 + 5x^4 + 2$

c. $10x^2 - 6x - 6$

Objective C Evaluating Polynomials

Polynomials have different values depending on the replacement values for the variables. When we find the value of a polynomial for a given replacement value, we are evaluating the polynomial for that value.

EXAMPLE 4 Evaluate each polynomial when $x = -2$.

a. $-5x + 6$ **b.** $3x^2 - 2x + 1$

Solution:

a. $-5x + 6 = -5(-2) + 6$ Replace x with -2.

$= 10 + 6$

$= 16$

b. $3x^2 - 2x + 1 = 3(-2)^2 - 2(-2) + 1$ Replace x with -2.

$= 3(4) + 4 + 1$

$= 12 + 4 + 1$

$= 17$

Work Practice Problem 4

PRACTICE PROBLEM 4

Evaluate each polynomial when $x = -1$.

a. $-2x + 10$

b. $6x^2 + 11x - 20$

Many physical phenomena can be modeled by polynomials.

EXAMPLE 5 **Finding Free-Fall Time**

The Swiss Re Building, completed in London in 2003, is a unique building. Londoners often refer to it as the "pickle building." The building is 592.1 feet tall. An object is dropped from the highest point of this building. Neglecting air resistance, the height in feet of the object above ground at time t seconds is given by the polynomial $-16t^2 + 592.1$. Find the height of the object when $t = 1$ second, and when $t = 6$ seconds.

Solution: To find each height, we evaluate the polynomial when $t = 1$ and when $t = 6$.

$-16t^2 + 592.1 = -16(1)^2 + 592.1$ Replace t with 1.

$= -16(1) + 592.1$

$= -16 + 592.1$

$= 576.1$

The height of the object at 1 second is 576.1 feet.

$-16t^2 + 592.1 = -16(6)^2 + 592.1$ Replace t with 6.

$= -16(36) + 592.1$

$= -576 + 592.1 = 16.1$

Continued on next page

PRACTICE PROBLEM 5

Find the height of the object in example 5 when $t = 2$ seconds and $t = 4$ seconds.

Answers

3. a. binomial, 1 **b.** none of these, 6 **c.** trinomial, 2 **4. a.** 12 **b.** −25 **5.** 528.1 feet, 336.1 feet

The height of the object at 6 seconds is 16.1 feet.

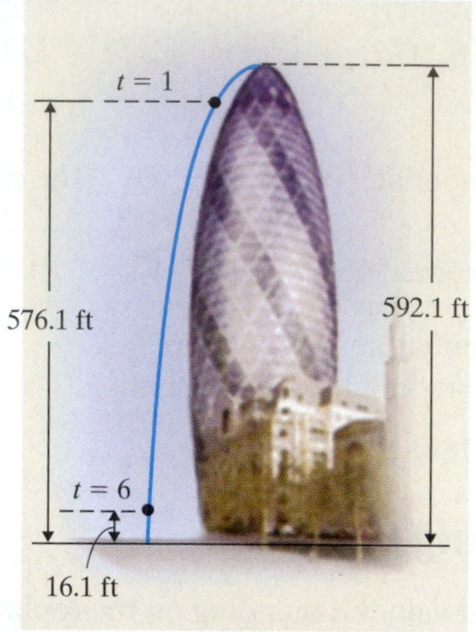

Work Practice Problem 5

Objective D Simplifying Polynomials by Combining Like Terms

We can simplify polynomials with like terms by combining the like terms. Recall from Section 3.1 that like terms are terms that contain exactly the same variables raised to exactly the same powers.

Like Terms	Unlike Terms
$5x^2, -7x^2$	$3x, 3y$
$y, 2y$	$-2x^2, -5x$
$\frac{1}{2}a^2b, -a^2b$	$6st^2, 4s^2t$

Only like terms can be combined. We combine like terms by applying the distributive property.

PRACTICE PROBLEMS 6–10

Simplify each polynomial by combining any like terms.

6. $-6y + 8y$
7. $14y^2 + 3 - 10y^2 - 9$
8. $7x^3 + x^3$
9. $23x^2 - 6x - x - 15$
10. $\frac{2}{7}x^3 - \frac{1}{4}x + 2 - \frac{1}{2}x^3 + \frac{3}{8}x$

EXAMPLES Simplify each polynomial by combining any like terms.

6. $-3x + 7x = (-3 + 7)x = 4x$

7. $11x^2 + 5 + 2x^2 - 7 = 11x^2 + 2x^2 + 5 - 7$
$= 13x^2 - 2$

8. $9x^3 + x^3 = 9x^3 + 1x^3$ Write x^3 as $1x^3$.
$= 10x^3$

9. $5x^2 + 6x - 9x - 3 = 5x^2 - 3x - 3$ Combine like terms $6x$ and $-9x$.

10. $\frac{2}{5}x^4 + \frac{2}{3}x^3 - x^2 + \frac{1}{10}x^4 - \frac{1}{6}x^3$

$$= \left(\frac{2}{5} + \frac{1}{10}\right)x^4 + \left(\frac{2}{3} - \frac{1}{6}\right)x^3 - x^2$$
$$= \left(\frac{4}{10} + \frac{1}{10}\right)x^4 + \left(\frac{4}{6} - \frac{1}{6}\right)x^3 - x^2$$
$$= \frac{5}{10}x^4 + \frac{3}{6}x^3 - x^2$$
$$= \frac{1}{2}x^4 + \frac{1}{2}x^3 - x^2$$

Work Practice Problems 6–10

Answers

6. $2y$ 7. $4y^2 - 6$ 8. $8x^3$
9. $23x^2 - 7x - 15$
10. $-\frac{3}{14}x^3 + \frac{1}{8}x + 2$

EXAMPLE 11 Write a polynomial that describes the total area of the squares and rectangles shown below. Then simplify the polynomial.

Solution: Recall that the area of a rectangle is length times width.

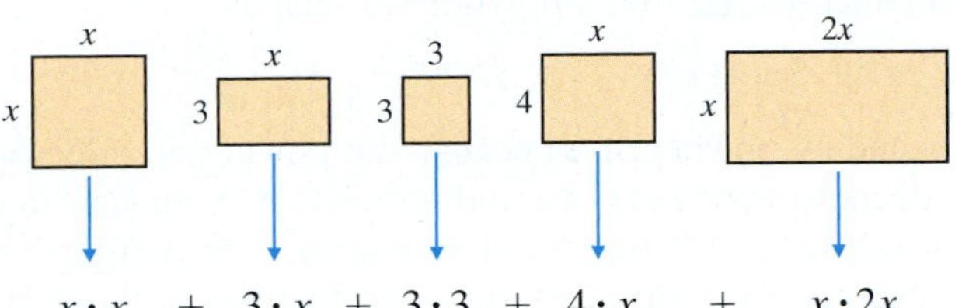

Area: $x \cdot x + 3 \cdot x + 3 \cdot 3 + 4 \cdot x + x \cdot 2x$

$= x^2 + 3x + 9 + 4x + 2x^2$

$= 3x^2 + 7x + 9$ Combine like terms.

Work Practice Problem 11

PRACTICE PROBLEM 11

Write a polynomial that describes the total area of the squares and rectangles shown below. Then simplify the polynomial.

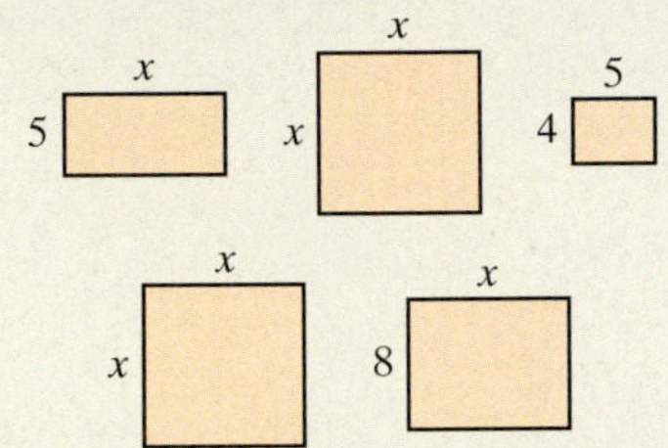

Objective E Simplifying Polynomials Containing Several Variables

A polynomial may contain more than one variable. One example is

$$5x + 3xy^2 - 6x^2y^2 + x^2y - 2y + 1$$

We call this expression a polynomial in several variables.

The **degree of a term** with more than one variable is the sum of the exponents on the variables. The **degree of the polynomial** in several variables is still the greatest degree of the terms of the polynomial.

EXAMPLE 12 Identify the degrees of the terms and the degree of the polynomial $5x + 3xy^2 - 6x^2y^2 + x^2y - 2y + 1$.

Solution: To organize our work, we use a table.

Terms of Polynomial	Degree of Term	Degree of Polynomial
$5x$	1	
$3xy^2$	1 + 2 or 3	
$-6x^2y^2$	2 + 2 or 4	4 (greatest degree)
x^2y	2 + 1 or 3	
$-2y$	1	
1	0	

Work Practice Problem 12

PRACTICE PROBLEM 12

Identify the degrees of the terms and the degree of the polynomial $-2x^3y^2 + 4 - 8xy + 3x^3y + 5xy^2$.

To simplify a polynomial containing several variables, we combine any like terms.

EXAMPLES Simplify each polynomial by combining any like terms.

13. $3xy - 5y^2 + 7yx - 9x^2 = (3 + 7)xy - 5y^2 - 9x^2$

$= 10xy - 5y^2 - 9x^2$

14. $9a^2b - 6a^2 + 5b^2 + a^2b - 11a^2 + 2b^2$

$= 10a^2b - 17a^2 + 7b^2$

Helpful Hint This term can be written as $7yx$ or $7xy$.

Work Practice Problems 13–14

PRACTICE PROBLEMS 13–14

Simplify each polynomial by combining any like terms.

13. $11ab - 6a^2 - ba + 8b^2$

14. $7x^2y^2 + 2y^2 - 4y^2x^2 + x^2 - y^2 + 5x^2$

Answers

11. $2x^2 + 13x + 20$
12. 5, 0, 2, 4, 3; 5
13. $10ab - 6a^2 + 8b^2$
14. $3x^2y^2 + y^2 + 6x^2$

Objective F Inserting "Missing" Terms

To prepare for dividing polynomials in Section 10.7, let's practice writing a polynomial in descending powers of the variable and with no "missing" powers.

Recall from Objective B that a polynomial such as

$$x^5 - 3x^3 + 2x^2 - 5x + 1$$

is written in descending powers of x because the powers of x decrease from left to right. Study the decreasing powers of x and notice that there is a "missing" power of x. This missing power is x^4. Writing a polynomial in decreasing powers of the variable helps you immediately determine important features of the polynomial, such as its degree. It is also sometimes helpful to write a polynomial so that there are no "missing" powers of x. For our polynomial above, if we simply insert a term of $0x^4$, which equals 0, we have an equivalent polynomial with no missing powers of x.

$$x^5 - 3x^3 + 2x^2 - 5x + 1 = x^5 + 0x^4 - 3x^3 + 2x^2 - 5x + 1$$

PRACTICE PROBLEM 15

Write each polynomial in descending powers of the variable with no missing powers.

a. $x^2 + 9$

b. $9m^3 + m^2 - 5$

c. $-3a^3 + a^4$

EXAMPLE 15 Write each polynomial in descending powers of the variable with no missing powers.

a. $x^2 - 4$

b. $3m^3 - m + 1$

c. $2x + x^4$

Solution:

a. $x^2 - 4 = x^2 + 0x^1 - 4$ or $x^2 + 0x - 4$ Insert a missing term of $0x^1$ or $0x$.

b. $3m^3 - m + 1 = 3m^3 + 0m^2 - m + 1$ Insert a missing term of $0m^2$.

c. $2x + x^4 = x^4 + 2x$ Write in descending power of variable.

$= x^4 + 0x^3 + 0x^2 + 2x + 0x^0$ Insert missing terms of $0x^3$, $0x^2$, and $0x^0$ (or 0).

Work Practice Problem 15

Helpful Hint

Since there is no constant as a last term, we insert a $0x^0$. This $0x^0$ (or 0) is the final power of x in our polynomial.

Answers

15. a. $x^2 + 0x + 9$ b. $9m^3 + m^2 + 0m - 5$ c. $a^4 - 3a^3 + 0a^2 + 0a + 0a^0$

Vocabulary and Readiness Check

Use the choices below to fill in each blank. Not all choices will be used.

least	monomial	trinomial	coefficient
greatest	binomial	constant	

1. A(n) ______ is a polynomial with exactly 2 terms.
2. A(n) ______ is a polynomial with exactly one term.
3. A(n) ______ is a polynomial with exactly three terms.
4. The numerical factor of a term is called the ______.
5. A number term is also called a ______.
6. The degree of a polynomial is the ______ degree of any term of the polynomial.

10.3 EXERCISE SET

FOR EXTRA HELP

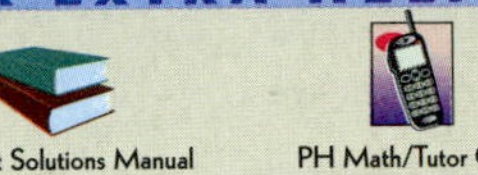
Student Solutions Manual

PH Math/Tutor Center

CD/Video for Review

MathXL®

MyMathLab

Objective A *Complete each table for each polynomial. See Example 1.*

1. $x^2 - 3x + 5$

Term	Coefficient
x^2	
	-3
5	

2. $2x^3 - x + 4$

Term	Coefficient
	2
$-x$	
4	

3. $-5x^4 + 3.2x^2 + x - 5$

Term	Coefficient
$-5x^4$	
$3.2x^2$	
x	
-5	

4. $9.7x^7 - 3x^5 + x^3 - \frac{1}{4}x^2$

Term	Coefficient
$9.7x^7$	
$-3x^5$	
x^3	
$-\frac{1}{4}x^2$	

Objective B *Find the degree of each polynomial and determine whether it is a monomial, binomial, trinomial, or none of these. See Examples 2 and 3.*

5. $x + 2$

6. $-6y + 4$

7. $9m^3 - 5m^2 + 4m - 8$

8. $a + 5a^2 + 3a^3 - 4a^4$

9. $12x^4 - x^6 - 12x^2$

10. $7r^2 + 2r - 3r^5$

11. $3z - 5z^4$

12. $5y^6 + 2$

Objective C *Evaluate each polynomial when* **(a)** $x = 0$ *and* **(b)** $x = -1$. *See Examples 4 and 5.*

13. $5x - 6$

14. $2x - 10$

15. $x^2 - 5x - 2$

16. $x^2 + 3x - 4$

17. $-x^3 + 4x^2 - 15$

18. $-2x^3 + 3x^2 - 6$

A rocket is fired upward from the ground with an initial velocity of 200 feet per second. Neglecting air resistance, the height of the rocket at any time t can be described in feet by the polynomial $-16t^2 + 200t$. Find the height of the rocket at the time given in Exercises 19 through 22. See Example 5.

	Time, t (in seconds)	Height $-16t^2 + 200t$
19.	1	
20.	5	
21.	7.6	
22.	10.3	

23. The polynomial $-24x^2 + 336x - 132$ represents the average number of visitors (in thousands) per day to National Park Service areas, where x represents the month of the year. Use this model to predict the average daily attendance at our national parks for the month of July. (*Hint:* July is the seventh month.) (*Source:* Based on data from the National Park Service)

24. The number of wireless telephone subscribers (in millions) x years after 1994 is given by the polynomial $0.56x^2 + 10x + 15.25$ for 1994 through 2004. Use this model to predict the number of wireless telephone subscribers in 2010 ($x = 16$). (*Source:* Based on data from Cellular Telecommunications & Internet Association)

Objective D *Simplify each expression by combining like terms. See Examples 6 through 10.*

25. $9x - 20x$

26. $14y - 30y$

27. $14x^3 + 9x^3$

28. $18x^3 + 4x^3$

29. $7x^2 + 3 + 9x^2 - 10$

30. $8x^2 + 4 + 11x^2 - 20$

31. $15x^2 - 3x^2 - 13$

32. $12k^3 - 9k^3 + 11$

33. $8s - 5s + 4s$

34. $5y + 7y - 6y$

35. $0.1y^2 - 1.2y^2 + 6.7 - 1.9$

36. $7.6y + 3.2y^2 - 8y - 2.5y^2$

37. $\frac{2}{3}x^4 + 12x^3 + \frac{1}{6}x^4 - 19x^3 - 19$

38. $\frac{2}{5}x^4 - 23x^2 + \frac{1}{15}x^4 + 5x^2 - 5$

39. $\frac{3}{20}x^3 + \frac{1}{10} - \frac{3}{10}x - \frac{1}{5} - \frac{7}{20}x + 6x^2$

40. $\frac{5}{16}x^3 - \frac{1}{8} + \frac{3}{8}x + \frac{1}{4} - \frac{9}{16}x - 14x^2$

Objective E *Identify the degrees of the terms and the degree of the polynomial. See Example 12.*

41. $9ab - 6a + 5b - 3$

42. $y^4 - 6y^3x + 2x^2y^2 - 5y^2 + 3$

43. $x^3y - 6 + 2x^2y^2 + 5y^3$

44. $2a^2b + 10a^4b - 9ab + 6$

Simplify each polynomial by combining any like terms. See Examples 13 and 14.

45. $3ab - 4a + 6ab - 7a$

46. $-9xy + 7y - xy - 6y$

47. $4x^2 - 6xy + 3y^2 - xy$

48. $3a^2 - 9ab + 4b^2 - 7ab$

49. $5x^2y + 6xy^2 - 5yx^2 + 4 - 9y^2x$

50. $17a^2b - 16ab^2 + 3a^3 + 4ba^3 - b^2a$

51. $14y^3 - 9 + 3a^2b^2 - 10 - 19b^2a^2$

52. $18x^4 + 2x^3y^3 - 1 - 2y^3x^3 - 17x^4$

Objective **F** *Write each polynomial in descending powers of the variable and with no missing powers. See Example 15.*

53. $7x^2 + 3$

54. $5x^2 - 2$

55. $x^3 - 64$

56. $x^3 - 8$

57. $5y^3 + 2y - 10$

58. $6m^3 - 3m + 4$

59. $8y + 2y^4$

60. $11z + 4z^4$

61. $6x^5 + x^3 - 3x + 15$

62. $9y^5 - y^2 + 2y - 11$

Objective **D** *Write a polynomial that describes the total area of each set of rectangles and squares shown in Exercises 63 and 64. Then simplify the polynomial. See Example 11.*

△ **63.**

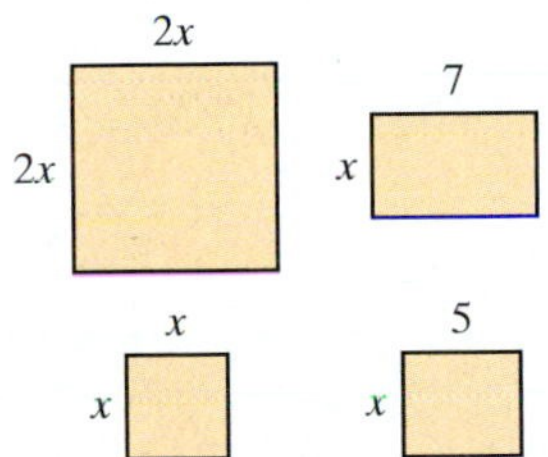

△ **64.**

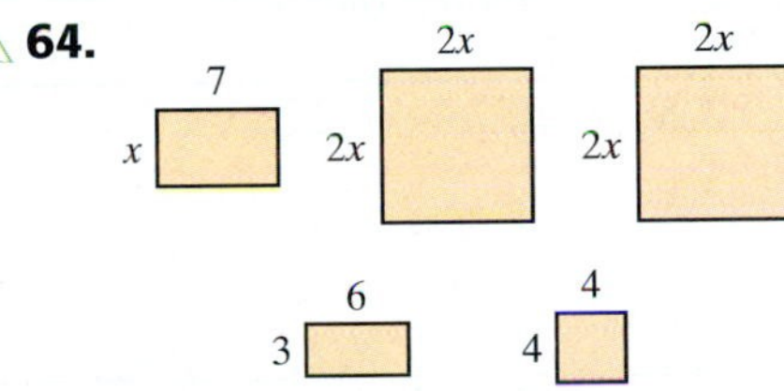

Recall that the perimeter of a figure such as the ones shown in Exercises 65 and 66 is the sum of the lengths of its sides. Write each perimeter as a polynomial. Then simplify the polynomial.

△ **65.**

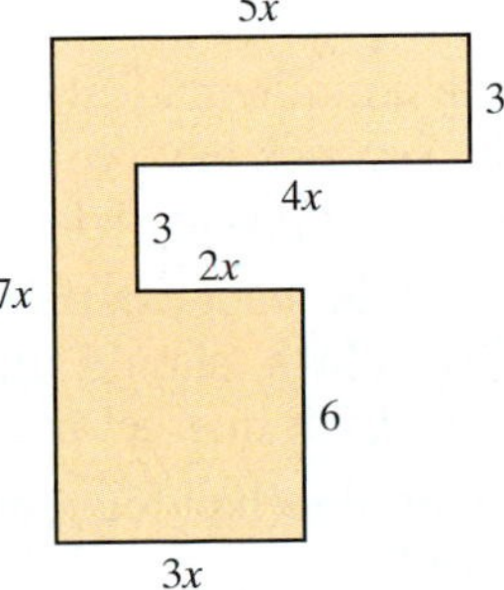

△ **66.**

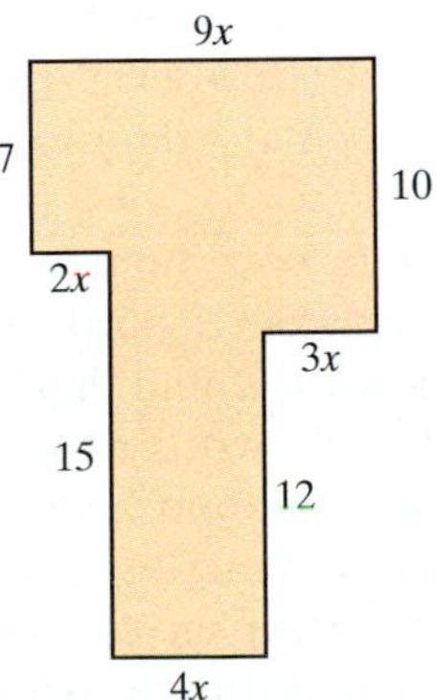

Review

Simplify each expression. See Section 3.1.

67. $4 + 5(2x + 3)$

68. $9 - 6(5x + 1)$

69. $2(x - 5) + 3(5 - x)$

70. $-3(w + 7) + 5(w + 1)$

Concept Extensions

71. Describe how to find the degree of a term.

72. Describe how to find the degree of a polynomial.

73. Explain why xyz is a monomial while $x + y + z$ is a trinomial.

74. Explain why the degree of the term $5y^3$ is 3 and the degree of the polynomial $2y + y + 2y$ is 1.

Simplify, if possible.

75. $x^4 \cdot x^9$

76. $x^4 + x^9$

77. $a \cdot b^3 \cdot a^2 \cdot b^7$

78. $a + b^3 + a^2 + b^7$

79. $(y^5)^4 + (y^2)^{10}$

80. $x^5y^2 + y^2x^5$

Fill in the boxes so that the terms in each expression can be combined. Then simplify. Each exercise has more than one solution.

81. $7x^{\square} + 2x^{\square}$

82. $(3y^2)^{\square} + (4y^3)^{\square}$

83. Explain why the height of the rocket in Exercises 19 through 22 increases and then decreases as time passes.

84. Approximate (to the nearest tenth of a second) how long before the rocket in Exercises 19 through 22 hits the ground.

Simplify each polynomial by combining like terms.

85. $1.85x^2 - 3.76x + 9.25x^2 + 10.76 - 4.21x$

86. $7.75x + 9.16x^2 - 1.27 - 14.58x^2 - 18.34$

STUDY SKILLS BUILDER

Are You Organized?

Have you ever had trouble finding a completed assignment? When it's time to study for a test, are your notes neat and organized? Have you ever had trouble reading your own mathematics handwriting? (Be honest—I have.)

When any of these things happen, it's time to get organized. Here are a few suggestions:

Write your notes and complete your homework assignment in a notebook with pockets (spiral or ring binder.) Take class notes in this notebook, and then follow the notes with your completed homework assignment. When you receive graded papers or handouts, place them in the notebook pocket so that you will not lose them.

Remember to mark (possibly with an exclamation point) any note(s) that seem extra important to you. Also remember to mark (possibly with a question mark) any notes or homework that you are having trouble with. Don't forget to see your instructor or a math tutor to help you with the concepts or exercises that you are having trouble understanding.

Also, if you are having trouble reading your own handwriting, *slow down* and write your mathematics work clearly!

Exercises

1. Have you been completing your assignments on time?
2. Have you been correcting any exercises you may be having difficulty with?
3. If you are having trouble with a mathematical concept or correcting any homework exercises, have you visited your instructor, a tutor, or your campus math lab?
4. Are you taking lecture notes in your mathematics course? (By the way, these notes should include worked-out examples solved by your instructor.)
5. Is your mathematics course material (handouts, graded papers, lecture notes) organized?
6. If your answer to Exercise 5 is no, take a moment and review your course material. List at least two ways that you might better organize it. Then read the Study Skills Builder on organizing a notebook in Chapter 2.

10.4 ADDING AND SUBTRACTING POLYNOMIALS

Objectives

A Add Polynomials.

B Subtract Polynomials.

C Add or Subtract Polynomials in One Variable.

D Add or Subtract Polynomials in Several Variables.

Objective A Adding Polynomials

To add polynomials, we use commutative and associative properties and then combine like terms. To see if you are ready to add polynomials, try the Concept Check.

✔**Concept Check** When combining like terms in the expression $5x - 8x^2 - 8x$, which of the following is the proper result?

a. $-11x^2$ **b.** $-3x - 8x^2$ **c.** $-11x$ **d.** $-11x^4$

To Add Polynomials

To add polynomials, combine all like terms.

EXAMPLES Add.

1. $(4x^3 - 6x^2 + 2x + 7) + (5x^2 - 2x)$

$= 4x^3 - 6x^2 + 2x + 7 + 5x^2 - 2x$ Remove parentheses.

$= 4x^3 + (-6x^2 + 5x^2) + (2x - 2x) + 7$ Combine like terms.

$= 4x^3 - x^2 + 7$ Simplify.

2. $(-2x^2 + 5x - 1) + (-2x^2 + x + 3)$

$= -2x^2 + 5x - 1 - 2x^2 + x + 3$ Remove parentheses.

$= (-2x^2 - 2x^2) + (5x + 1x) + (-1 + 3)$ Combine like terms.

$= -4x^2 + 6x + 2$ Simplify.

Work Practice Problems 1–2

PRACTICE PROBLEMS 1–2

Add.

1. $(3x^5 - 7x^3 + 2x - 1) + (3x^3 - 2x)$

2. $(5x^2 - 2x + 1) + (-6x^2 + x - 1)$

Just as we can add numbers vertically, polynomials can be added vertically if we line up like terms underneath one another.

EXAMPLE 3 Add $(7y^3 - 2y^2 + 7)$ and $(6y^2 + 1)$ using a vertical format.

Solution: Vertically line up like terms and add.

$$\begin{array}{r} 7y^3 - 2y^2 + 7 \\ 6y^2 + 1 \\ \hline 7y^3 + 4y^2 + 8 \end{array}$$

Work Practice Problem 3

PRACTICE PROBLEM 3

Add $(9y^2 - 6y + 5)$ and $(4y + 3)$ using a vertical format.

Objective B Subtracting Polynomials

To subtract one polynomial from another, recall the definition of subtraction. To subtract a number, we add its opposite: $a - b = a + (-b)$. To subtract a polynomial, we also add its opposite. Just as $-b$ is the opposite of b, $-(x^2 + 5)$ is the opposite of $(x^2 + 5)$.

To Subtract Polynomials

To subtract two polynomials, change the signs of the terms of the polynomial being subtracted and then add.

Answers

1. $3x^5 - 4x^3 - 1$ **2.** $-x^2 - x$
3. $9y^2 - 2y + 8$

✔ Concept Check Answer

b

PRACTICE PROBLEM 4

Subtract:
$(9x + 5) - (4x - 3)$

EXAMPLE 4 Subtract: $(5x - 3) - (2x - 11)$

Solution: From the definition of subtraction, we have

$(5x - 3) - (2x - 11) = (5x - 3) + [-(2x - 11)]$ Add the opposite.
$= (5x - 3) + (-2x + 11)$ Apply the distributive property.
$= 5x - 3 - 2x + 11$ Remove parentheses.
$= 3x + 8$ Combine like terms.

Work Practice Problem 4

PRACTICE PROBLEM 5

Subtract:
$(4x^3 - 10x^2 + 1) - (-4x^3 + x^2 - 11)$

EXAMPLE 5 Subtract: $(2x^3 + 8x^2 - 6x) - (2x^3 - x^2 + 1)$

Solution: First, we change the sign of each term of the second polynomial; then we add.

$(2x^3 + 8x^2 - 6x) - (2x^3 - x^2 + 1)$
$= (2x^3 + 8x^2 - 6x) + (-2x^3 + x^2 - 1)$
$= 2x^3 + 8x^2 - 6x - 2x^3 + x^2 - 1$
$= 2x^3 - 2x^3 + 8x^2 + x^2 - 6x - 1$
$= 9x^2 - 6x - 1$ Combine like terms.

Work Practice Problem 5

Just as polynomials can be added vertically, so can they be subtracted vertically.

PRACTICE PROBLEM 6

Subtract $(6y^2 - 3y + 2)$ from $(2y^2 - 2y + 7)$ using a vertical format.

EXAMPLE 6 Subtract $(5y^2 + 2y - 6)$ from $(-3y^2 - 2y + 11)$ using a vertical format.

Solution: Arrange the polynomials in a vertical format, lining up like terms.

$$\begin{array}{r} -3y^2 - 2y + 11 \\ \underline{-(5y^2 + 2y - 6)} \end{array} \qquad \begin{array}{r} -3y^2 - 2y + 11 \\ \underline{-5y^2 - 2y + 6} \\ -8y^2 - 4y + 17 \end{array}$$

Work Practice Problem 6

Helpful Hint

Don't forget to change the sign of each term in the polynomial being subtracted.

Objective C Adding and Subtracting Polynomials in One Variable

Let's practice adding and subtracting polynomials in one variable.

PRACTICE PROBLEM 7

Subtract $(3x + 1)$ from the sum of $(4x - 3)$ and $(12x - 5)$.

EXAMPLE 7 Subtract $(5z - 7)$ from the sum of $(8z + 11)$ and $(9z - 2)$.

Solution: Notice that $(5z - 7)$ is to be subtracted **from** a sum. The translation is

$[(8z + 11) + (9z - 2)] - (5z - 7)$
$= 8z + 11 + 9z - 2 - 5z + 7$ Remove grouping symbols.
$= 8z + 9z - 5z + 11 - 2 + 7$ Group like terms.
$= 12z + 16$ Combine like terms.

Work Practice Problem 7

Answers
4. $5x + 8$ **5.** $8x^3 - 11x^2 + 12$
6. $-4y^2 + y + 5$ **7.** $13x - 9$

Objective D Adding and Subtracting Polynomials in Several Variables

Now that we know how to add or subtract polynomials in one variable, we can also add and subtract polynomials in several variables.

EXAMPLES Add or subtract as indicated.

8. $(3x^2 - 6xy + 5y^2) + (-2x^2 + 8xy - y^2)$
$= 3x^2 - 6xy + 5y^2 - 2x^2 + 8xy - y^2$
$= x^2 + 2xy + 4y^2$ Combine like terms.

9. $(9a^2b^2 + 6ab - 3ab^2) - (5b^2a + 2ab - 3 - 9b^2)$
$= 9a^2b^2 + 6ab - 3ab^2 - 5b^2a - 2ab + 3 + 9b^2$
$= 9a^2b^2 + 4ab - 8ab^2 + 9b^2 + 3$ Combine like terms.

Work Practice Problems 8–9

✔Concept Check If possible, simplify each expression by performing the indicated operation.

a. $2y + y$
b. $2y \cdot y$
c. $-2y - y$
d. $(-2y)(-y)$
e. $2x + y$

PRACTICE PROBLEMS 8–9

Add or subtract as indicated.

8. $(2a^2 - ab + 6b^2) + (-3a^2 + ab - 7b^2)$

9. $(5x^2y^2 + 3 - 9x^2y + y^2) - (-x^2y^2 + 7 - 8xy^2 + 2y^2)$

Answers

8. $-a^2 - b^2$
9. $6x^2y^2 - 4 - 9x^2y + 8xy^2 - y^2$

✔ Concept Check Answers

a. $3y$ **b.** $2y^2$ **c.** $-3y$ **d.** $2y^2$
e. cannot be simplified

10.4 EXERCISE SET

FOR EXTRA HELP

Student Solutions Manual

PH Math/Tutor Center

CD/Video for Review

MathXL®

MyMathLab

Objective A *Add. See Examples 1 and 2.*

1. $(3x + 7) + (9x + 5)$

2. $(-y - 2) + (3y + 5)$

3. $(-7x + 5) + (-3x^2 + 7x + 5)$

4. $(3x - 8) + (4x^2 - 3x + 3)$

5. $(-5x^2 + 3) + (2x^2 + 1)$

6. $(3x^2 + 7) + (3x^2 + 9)$

7. $(-3y^2 - 4y) + (2y^2 + y - 1)$

8. $(7x^2 + 2x - 9) + (-3x^2 + 5)$

9. $(1.2x^3 - 3.4x + 7.9) + (6.7x^3 + 4.4x^2 - 10.9)$

10. $(9.6y^3 + 2.7y^2 - 8.6) + (1.1y^3 - 8.8y + 11.6)$

11. $\left(\frac{3}{4}m^2 - \frac{2}{5}m + \frac{1}{8}\right) + \left(-\frac{1}{4}m^2 - \frac{3}{10}m + \frac{11}{16}\right)$

12. $\left(-\frac{4}{7}n^2 + \frac{5}{6}m - \frac{1}{20}\right) + \left(\frac{3}{7}n^2 - \frac{5}{12}m - \frac{3}{10}\right)$

Add using a vertical format. See Example 3.

13. $\begin{array}{r} 3t^2 + 4 \\ \underline{5t^2 - 8} \end{array}$

14. $\begin{array}{r} 7x^3 + 3 \\ \underline{2x^3 + 1} \end{array}$

15. $\begin{array}{r} 10a^3 - 8a^2 + 4a + 9 \\ \underline{5a^3 + 9a^2 - 7a + 7} \end{array}$

16. $\begin{array}{r} 2x^3 - 3x^2 + x - 4 \\ \underline{5x^3 + 2x^2 - 3x + 2} \end{array}$

Objective B *Subtract. See Examples 4 and 5.*

17. $(2x + 5) - (3x - 9)$

18. $(4 + 5a) - (-a - 5)$

19. $(5x^2 + 4) - (-2y^2 + 4)$

20. $(-7y^2 + 5) - (-8y^2 + 12)$

21. $3x - (5x - 9)$

22. $4 - (-y - 4)$

23. $(2x^2 + 3x - 9) - (-4x + 7)$

24. $(-7x^2 + 4x + 7) - (-8x + 2)$

25. $(5x + 8) - (-2x^2 - 6x + 8)$

26. $(-6y^2 + 3y - 4) - (9y^2 - 3y)$

27. $(0.7x^2 + 0.2x - 0.8) - (0.9x^2 + 1.4)$

28. $(-0.3y^2 + 0.6y - 0.3) - (0.5y^2 + 0.3)$

29. $\left(\frac{1}{4}z^2 - \frac{1}{5}z\right) - \left(-\frac{3}{20}z^2 + \frac{1}{10}z - \frac{7}{20}\right)$

30. $\left(\frac{1}{3}x^2 - \frac{2}{7}x\right) - \left(\frac{4}{21}x^2 + \frac{1}{21}x - \frac{2}{3}\right)$

Subtract using a vertical format. See Example 6.

31. $\begin{array}{r} 4z^2 - 8z + 3 \\ -(6z^2 + 8z - 3) \\ \hline \end{array}$

32. $\begin{array}{r} 7a^2 - 9a + 6 \\ -(11a^2 - 4a + 2) \\ \hline \end{array}$

33. $\begin{array}{r} 5u^5 - 4u^2 + 3u - 7 \\ -(3u^5 + 6u^2 - 8u + 2) \\ \hline \end{array}$

34. $\begin{array}{r} 5x^3 - 4x^2 + 6x - 2 \\ -(3x^3 - 2x^2 - x - 4) \\ \hline \end{array}$

Objectives **A** **B** **C** **Mixed Practice** *Add or subtract as indicated. See Examples 1 through 7.*

35. $(3x + 5) + (2x - 14)$

36. $(2y + 20) + (5y - 30)$

37. $(9x - 1) - (5x + 2)$

38. $(7y + 7) - (y - 6)$

39. $(14y + 12) + (-3y - 5)$

40. $(26y + 17) + (-20y - 10)$

41. $(x^2 + 2x + 1) - (3x^2 - 6x + 2)$

42. $(5y^2 - 3y - 1) - (2y^2 + y + 1)$

43. $(3x^2 + 5x - 8) + (5x^2 + 9x + 12) - (8x^2 - 14)$

44. $(2x^2 + 7x - 9) + (x^2 - x + 10) - (3x^2 - 30)$

45. $(-a^2 + 1) - (a^2 - 3) + (5a^2 - 6a + 7)$

46. $(-m^2 + 3) - (m^2 - 13) + (6m^2 - m + 1)$

Translating *Perform each indicated operation. See Examples 3, 6, and 7.*

47. Subtract $4x$ from $7x - 3$.

48. Subtract y from $y^2 - 4y + 1$.

49. Add $(4x^2 - 6x + 1)$ and $(3x^2 + 2x + 1)$.

50. Add $(-3x^2 - 5x + 2)$ and $(x^2 - 6x + 9)$.

51. Subtract $(5x + 7)$ from $(7x^2 + 3x + 9)$.

52. Subtract $(5y^2 + 8y + 2)$ from $(7y^2 + 9y - 8)$.

53. Subtract $(4y^2 - 6y - 3)$ from the sum of $(8y^2 + 7)$ and $(6y + 9)$.

54. Subtract $(4x^2 - 2x + 2)$ from the sum of $(x^2 + 7x + 1)$ and $(7x + 5)$.

55. Subtract $(3x^2 - 4)$ from the sum of $(x^2 - 9x + 2)$ and $(2x^2 - 6x + 1)$.

56. Subtract $(y^2 - 9)$ from the sum of $(3y^2 + y + 4)$ and $(2y^2 - 6y - 10)$.

Objective **D** *Add or subtract as indicated. See Examples 8 and 9.*

57. $(9a + 6b - 5) + (-11a - 7b + 6)$

58. $(3x - 2 + 6y) + (7x - 2 - y)$

59. $(4x^2 + y^2 + 3) - (x^2 + y^2 - 2)$

60. $(7a^2 - 3b^2 + 10) - (-2a^2 + b^2 - 12)$

61. $(x^2 + 2xy - y^2) + (5x^2 - 4xy + 20y^2)$

62. $(a^2 - ab + 4b^2) + (6a^2 + 8ab - b^2)$

63. $(11r^2s + 16rs - 3 - 2r^2s^2) - (3sr^2 + 5 - 9r^2s^2)$

64. $(3x^2y - 6xy + x^2y^2 - 5) - (11x^2y^2 - 1 + 5yx^2)$

For Exercises 65 through 68, find the perimeter of each figure.

65.

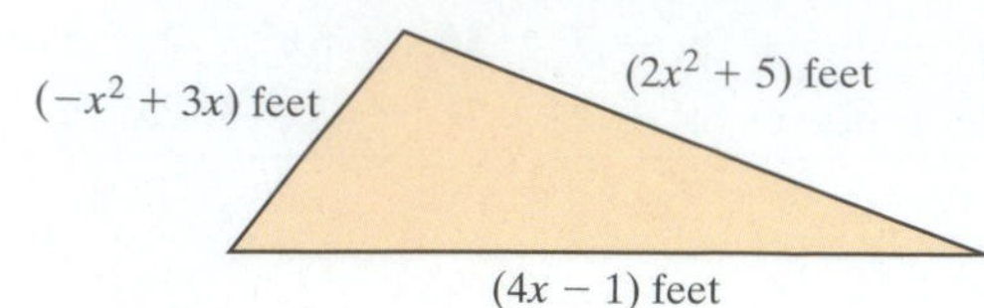

66.

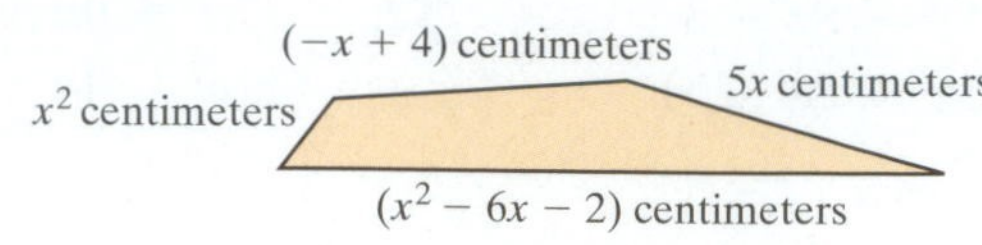

67.

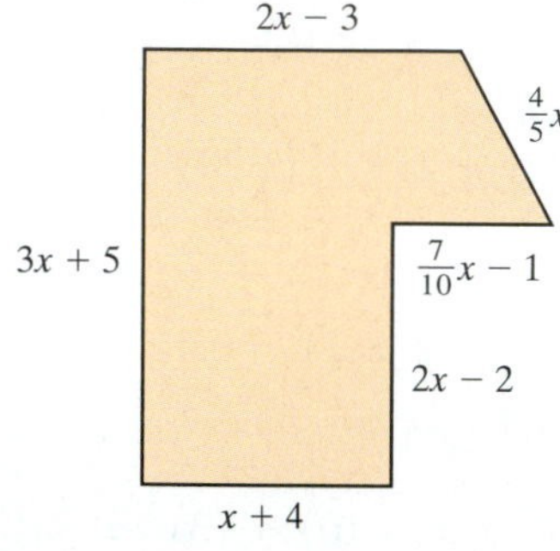

68.

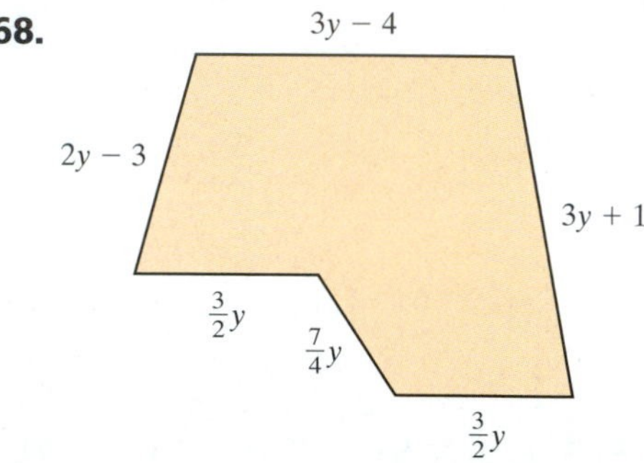

69. A wooden beam is $(4y^2 + 4y + 1)$ meters long. If a piece $(y^2 - 10)$ meters is cut, express the length of the remaining piece of beam as a polynomial in y.

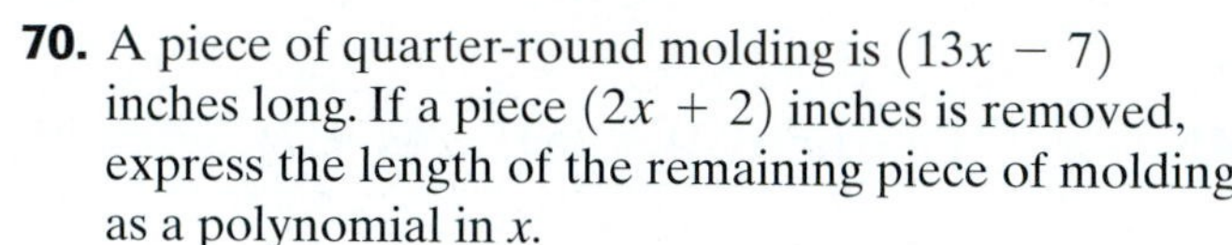

70. A piece of quarter-round molding is $(13x - 7)$ inches long. If a piece $(2x + 2)$ inches is removed, express the length of the remaining piece of molding as a polynomial in x.

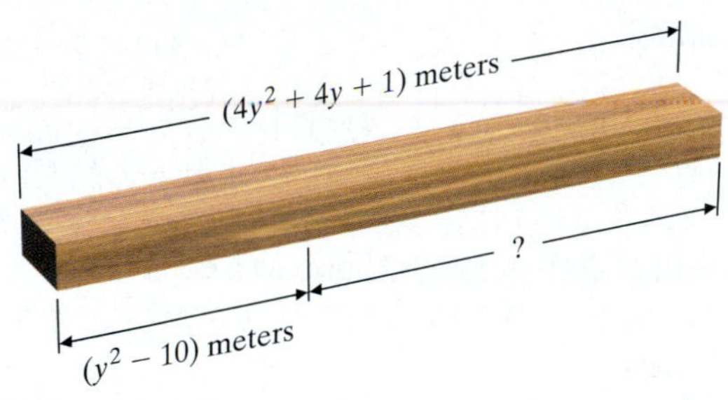

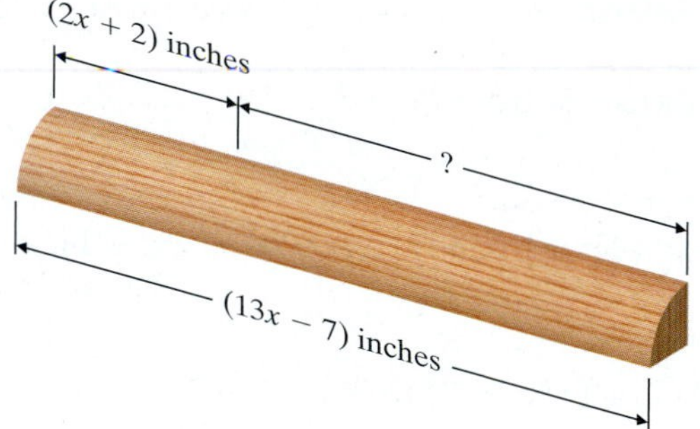

Perform each indicated operation.

71. $[(1.2x^2 - 3x + 9.1) - (7.8x^2 - 3.1 + 8)] + (1.2x - 6)$

72. $[(7.9y^4 - 6.8y^3 + 3.3y) + (6.1y^3 - 5)] - (4.2y^4 + 1.1y - 1)$

Review

Multiply. See Section 10.1.

73. $3x(2x)$ **74.** $-7x(x)$ **75.** $(12x^3)(-x^5)$ **76.** $6r^3(7r^{10})$ **77.** $10x^2(20xy^2)$ **78.** $-z^2y(11zy)$

Concept Extensions

Fill in the squares so that each is a true statement.

79. $3x^{\square} + 4x^2 = 7x^{\square}$

80. $9y^7 + 3y^{\square} = 12y^7$

81. $2x^{\square} + 3x^{\square} - 5x^{\square} + 4x^{\square} = 6x^4 - 2x^3$

82. $3y^{\square} + 7y^{\square} - 2y^{\square} - y^{\square} = 10y^5 - 3y^2$

Match each expression on the left with its simplification on the right. Not all letters on the right must be used and a letter may be used more than once.

83. $10y - 6y^2 - y$

84. $5x + 5x$

85. $(5x - 3) + (5x - 3)$

86. $(15x - 3) - (5x - 3)$

a. $3y$
b. $9y - 6y^2$
c. $10x$
d. $25x^2$
e. $10x - 6$
f. none of these

Simplify each expression by performing the indicated operation. Explain how you arrived at each answer. See the last Concept Check in this section.

87. a. $z + 3z$
b. $z \cdot 3z$
c. $-z - 3z$
d. $(-z)(-3z)$

88. a. $x + x$
b. $x \cdot x$
c. $-x - x$
d. $(-x)(-x)$

89. a. $m \cdot m \cdot m$
b. $m + m + m$
c. $(-m)(-m)(-m)$
d. $-m - m - m$

90. The polynomial $0.0005x^2 + 0.0303x + 1.156$ represents the sale of electricity (in trillion kilowatt-hours) in the U.S. residential sector during 1999–2003. The polynomial $0.0215x^2 - 0.1073x + 2.31$ represents the sale of electricity (in trillion kilowatt-hours) in all other U.S. sectors during 1999–2003. In both polynomials, x represents the number of years after 1999. Find a polynomial for the total sales of electricity (in trillion kilowatt hours) to all sectors in the United States during this period. (*Source:* Based on data from the Energy Information Administration)

91. The polynomial $-0.35x^2 + 0.49x + 71.75$ represents the percent of Americans under age 65 covered by private health insurance during 1999–2003. The polynomial $0.025x^2 + 9.65x + 11.83$ represents the percent of Americans under age 65 covered by public health programs during 1999–2003. In both polynomials, x represents the number of years since 1999. Find a polynomial for the total percent of Americans under age 65 with some form of health coverage during this period. (*Source:* Based on data from the Public Health Service)

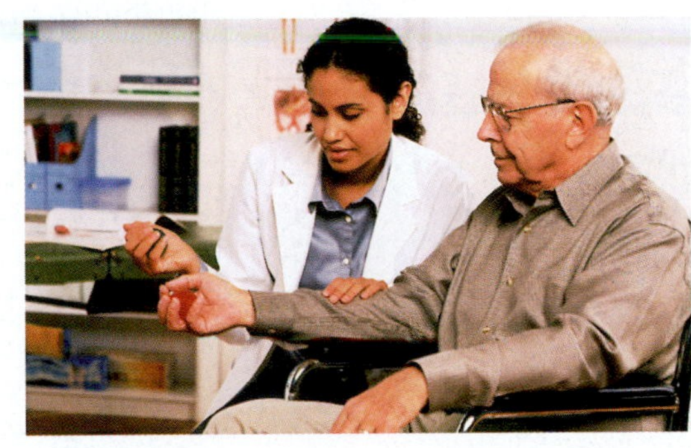

Objectives

A Multiply Monomials.

B Multiply a Monomial by a Polynomial.

C Multiply Two Polynomials.

D Multiply Polynomials Vertically.

10.5 MULTIPLYING POLYNOMIALS

Objective A Multiplying Monomials

Recall from Section 10.1 that to multiply two monomials such as $(-5x^3)$ and $(-2x^4)$, we use the associative and commutative properties and regroup. Remember also that to multiply exponential expressions with a common base, we use the product rule for exponents and add exponents.

$$(-5x^3)(-2x^4) = (-5)(-2)(x^3 \cdot x^4) \quad \text{Use the commutative and associative properties.}$$
$$= 10x^7 \quad \text{Multiply.}$$

PRACTICE PROBLEMS 1–3

Multiply.

1. $10x \cdot 9x$
2. $8x^3(-11x^7)$
3. $(-5x^4)(-x)$

EXAMPLES Multiply.

1. $6x \cdot 4x = (6 \cdot 4)(x \cdot x)$ Use the commutative and associative properties.
   $= 24x^2$ Multiply.
2. $-7x^2 \cdot 2x^5 = (-7 \cdot 2)(x^2 \cdot x^5)$
   $= -14x^7$
3. $(-12x^5)(-x) = (-12x^5)(-1x)$
   $= (-12)(-1)(x^5 \cdot x)$
   $= 12x^6$

Work Practice Problems 1–3

✔Concept Check Simplify.

a. $3x \cdot 2x$ **b.** $3x + 2x$

Objective B Multiplying Monomials by Polynomials

To multiply a monomial such as $7x$ by a trinomial such as $x^3 + 2x + 5$, we use the distributive property.

PRACTICE PROBLEMS 4–6

Multiply.

4. $4x(x^2 + 4x + 3)$
5. $8x(7x^4 + 1)$
6. $-2x^3(3x^2 - x + 2)$

EXAMPLES Multiply.

4. $7x(x^2 + 2x + 5) = 7x(x^2) + 7x(2x) + 7x(5)$ Apply the distributive property.
   $= 7x^3 + 14x^2 + 35x$ Multiply.
5. $5x(2x^3 + 6) = 5x(2x^3) + 5x(6)$ Apply the distributive property.
   $= 10x^4 + 30x$ Multiply.
6. $-3x^2(5x^2 + 6x - 1)$
   $= (-3x^2)(5x^2) + (-3x^2)(6x) + (-3x^2)(-1)$ Apply the distributive property.
   $= -15x^4 - 18x^3 + 3x^2$ Multiply.

Work Practice Problems 4–6

Answers

1. $90x^2$ 2. $-88x^{10}$ 3. $5x^5$ 4. $4x^3 + 16x^2 + 12x$ 5. $56x^5 + 8x$ 6. $-6x^5 + 2x^4 - 4x^3$

✔ Concept Check Answers

a. $6x^2$ b. $5x$

Objective C Multiplying Two Polynomials

We also use the distributive property to multiply two binomials.

EXAMPLE 7 Multiply.

a. $(m + 4)(m + 6)$ **b.** $(3x + 2)(2x - 5)$

Solution:

a. $(m + 4)(m + 6) = m(m + 6) + 4(m + 6)$ Use the distributive property.

$= m \cdot m + m \cdot 6 + 4 \cdot m + 4 \cdot 6$ Use the distributive property.

$= m^2 + 6m + 4m + 24$ Multiply.

$= m^2 + 10m + 24$ Combine like terms.

b. $(3x + 2)(2x - 5) = 3x(2x - 5) + 2(2x - 5)$ Use the distributive property.

$= 3x(2x) + 3x(-5) + 2(2x) + 2(-5)$

$= 6x^2 - 15x + 4x - 10$ Multiply.

$= 6x^2 - 11x - 10$ Combine like terms.

Work Practice Problem 7

PRACTICE PROBLEM 7

Multiply:

a. $(x + 5)(x + 10)$

b. $(4x + 5)(3x - 4)$

This idea can be expanded so that we can multiply any two polynomials.

To Multiply Two Polynomials

Multiply each term of the first polynomial by each term of the second polynomial, and then combine like terms.

EXAMPLES Multiply.

8. $(2x - y)^2$

$= (2x - y)(2x - y)$ Using the meaning of an exponent, we have 2 factors of $(2x - y)$.

$= 2x(2x) + 2x(-y) + (-y)(2x) + (-y)(-y)$

$= 4x^2 - 2xy - 2xy + y^2$ Multiply.

$= 4x^2 - 4xy + y^2$ Combine like terms.

9. $(t + 2)(3t^2 - 4t + 2)$

$= t(3t^2) + t(-4t) + t(2) + 2(3t^2) + 2(-4t) + 2(2)$

$= 3t^3 - 4t^2 + 2t + 6t^2 - 8t + 4$

$= 3t^3 + 2t^2 - 6t + 4$ Combine like terms.

Work Practice Problems 8–9

PRACTICE PROBLEMS 8–9

Multiply.

8. $(3x - 2y)^2$

9. $(x + 3)(2x^2 - 5x + 4)$

✔ **Concept Check** Square where indicated. Simplify if possible.

a. $(4a)^2 + (3b)^2$ **b.** $(4a + 3b)^2$

Objective D Multiplying Polynomials Vertically

Another convenient method for multiplying polynomials is to multiply vertically, similar to the way we multiply real numbers. This method is shown in the next examples.

Answers

7. a. $x^2 + 15x + 50$

b. $12x^2 - x - 20$

8. $9x^2 - 12xy + 4y^2$

9. $2x^3 + x^2 - 11x + 12$

✔ Concept Check Answers

a. $16a^2 + 9b^2$ **b.** $16a^2 + 24ab + 9b^2$

PRACTICE PROBLEM 10

Multiply vertically:
$(3y^2 + 1)(y^2 - 4y + 5)$

EXAMPLE 10 Multiply vertically: $(2y^2 + 5)(y^2 - 3y + 4)$

Solution:

$$
\begin{array}{rrrrrl}
 & & y^2 & -\ 3y & +\ 4 & \\
 & & 2y^2 & & +\ 5 & \\
\hline
 & & 5y^2 & -\ 15y & +\ 20 & \text{Multiply } y^2 - 3y + 4 \text{ by } 5 \\
2y^4 & -\ 6y^3 & +\ 8y^2 & & & \text{Multiply } y^2 - 3y + 4 \text{ by } 2y^2 \\
\hline
2y^4 & -\ 6y^3 & +\ 13y^2 & -\ 15y & +\ 20 & \text{Combine like terms.}
\end{array}
$$

Work Practice Problem 10

PRACTICE PROBLEM 11

Find the product of $(4x^2 - x - 1)$ and $(3x^2 + 6x - 2)$ using a vertical format.

EXAMPLE 11 Find the product of $(2x^2 - 3x + 4)$ and $(x^2 + 5x - 2)$ using a vertical format.

Solution: First, we arrange the polynomials in a vertical format. Then we multiply each term of the second polynomial by each term of the first polynomial.

$$
\begin{array}{rrrrrl}
 & & 2x^2 & -\ 3x & +\ 4 & \\
 & & x^2 & +\ 5x & -\ 2 & \\
\hline
 & & -4x^2 & +\ 6x & -\ 8 & \text{Multiply } 2x^2 - 3x + 4 \text{ by } -2. \\
 & 10x^3 & -\ 15x^2 & +\ 20x & & \text{Multiply } 2x^2 - 3x + 4 \text{ by } 5x. \\
2x^4 & -\ 3x^3 & +\ 4x^2 & & & \text{Multiply } 2x^2 - 3x + 4 \text{ by } x^2. \\
\hline
2x^4 & +\ 7x^3 & -\ 15x^2 & +\ 26x & -\ 8 & \text{Combine like terms.}
\end{array}
$$

Work Practice Problem 11

Answers

10. $3y^4 - 12y^3 + 16y^2 - 4y + 5$
11. $12x^4 + 21x^3 - 17x^2 - 4x + 2$

Vocabulary and Readiness Check

Fill in each blank with the correct choice.

1. The expression $5x(3x + 2)$ equals $5x \cdot 3x + 5x \cdot 2$ by the ____________ property.
a. commutative **b.** associative **c.** distributive

2. The expression $(x + 4)(7x - 1)$ equals $x(7x - 1) + 4(7x - 1)$ by the ____________ property.
a. commutative **b.** associative **c.** distributive

3. The expression $(5y - 1)^2$ equals ____________.
a. $2(5y - 1)$ **b.** $(5y - 1)(5y + 1)$ **c.** $(5y - 1)(5y - 1)$

4. The expression $9x \cdot 3x$ equals ____________.
a. $27x$ **b.** $27x^2$ **c.** $12x$ **d.** $12x^2$

10.5 EXERCISE SET

FOR EXTRA HELP

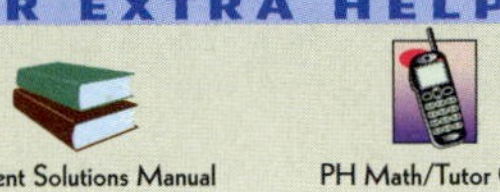

Objective **A** *Multiply. See Examples 1 through 3.*

1. $8x^2 \cdot 3x$

2. $6x \cdot 3x^2$

3. $(-x^3)(-x)$

4. $(-x^6)(-x)$

5. $-4n^3 \cdot 7n^7$

6. $9t^6(-3t^5)$

7. $(-3.1x^3)(4x^9)$

8. $(-5.2x^4)(3x^4)$

9. $\left(-\frac{1}{3}y^2\right)\left(\frac{2}{5}y\right)$

10. $\left(-\frac{3}{4}y^7\right)\left(\frac{1}{7}y^4\right)$

11. $(2x)(-3x^2)(4x^5)$

12. $(x)(5x^4)(-6x^7)$

Objective **B** *Multiply. See Examples 4 through 6.*

13. $3x(2x + 5)$

14. $2x(6x + 3)$

15. $7x(x^2 + 2x - 1)$

16. $5y(y^2 + y - 10)$

17. $-2a(a + 4)$

18. $-3a(2a + 7)$

19. $3x(2x^2 - 3x + 4)$

20. $4x(5x^2 - 6x - 10)$

21. $3a^2(4a^3 + 15)$

22. $9x^3(5x^2 + 12)$

23. $-2a^2(3a^2 - 2a + 3)$

24. $-4b^2(3b^3 - 12b^2 - 6)$

25. $3x^2y(2x^3 - x^2y^2 + 8y^3)$

26. $4xy^2(7x^3 + 3x^2y^2 - 9y^3)$

27. $-y(4x^3 - 7x^2y + xy^2 + 3y^3)$

28. $-x(6y^3 - 5xy^2 + x^2y - 5x^3)$

29. $\frac{1}{2}x^2(8x^2 - 6x + 1)$

30. $\frac{1}{3}y^2(9y^2 - 6y + 1)$

Objective **C** *Multiply. See Examples 7 through 9.*

31. $(x + 4)(x + 3)$

32. $(x + 2)(x + 9)$

33. $(a + 7)(a - 2)$

34. $(y - 10)(y + 11)$

35. $\left(x + \frac{2}{3}\right)\left(x - \frac{1}{3}\right)$

36. $\left(x + \frac{3}{5}\right)\left(x - \frac{2}{5}\right)$

37. $(3x^2 + 1)(4x^2 + 7)$

38. $(5x^2 + 2)(6x^2 + 2)$

39. $(4x - 3)(3x - 5)$

40. $(8x - 3)(2x - 4)$

41. $(1 - 3a)(1 - 4a)$

42. $(3 - 2a)(2 - a)$

43. $(2y - 4)^2$

44. $(6x - 7)^2$

45. $(x - 2)(x^2 - 3x + 7)$

46. $(x + 3)(x^2 + 5x - 8)$

47. $(x + 5)(x^3 - 3x + 4)$

48. $(a + 2)(a^3 - 3a^2 + 7)$

49. $(2a - 3)(5a^2 - 6a + 4)$

50. $(3 + b)(2 - 5b - 3b^2)$

51. $(7xy - y)^2$

52. $(x^2 - 4)^2$

Objective **D** *Multiply vertically. See Examples 10 and 11.*

53. $(2x - 11)(6x + 1)$

54. $(4x - 7)(5x + 1)$

55. $(x + 3)(2x^2 + 4x - 1)$

56. $(4x - 5)(8x^2 + 2x - 4)$

57. $(x^2 + 5x - 7)(2x^2 - 7x - 9)$

58. $(3x^2 - x + 2)(x^2 + 2x + 1)$

Objectives **A B C D** **Mixed Practice** *Multiply. See Examples 1 through 11.*

59. $-1.2y(-7y^6)$

60. $-4.2x(-2x^5)$

61. $-3x(x^2 + 2x - 8)$

62. $-5x(x^2 - 3x + 10)$

63. $(x + 19)(2x + 1)$

64. $(3y + 4)(y + 11)$

65. $\left(x + \frac{1}{7}\right)\left(x - \frac{3}{7}\right)$

66. $\left(m + \frac{2}{9}\right)\left(m - \frac{1}{9}\right)$

67. $(3y + 5)^2$

68. $(7y + 2)^2$

69. $(a + 4)(a^2 - 6a + 6)$

70. $(t + 3)(t^2 - 5t + 5)$

Express as the product of polynomials. Then multiply.

△ **71.** Find the area of the rectangle.

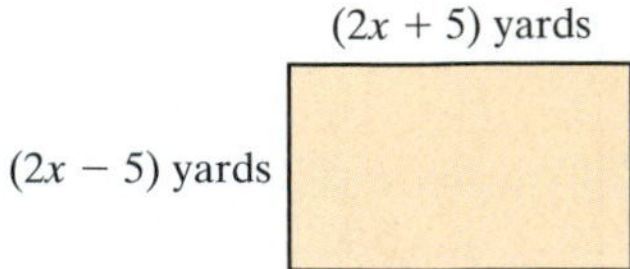

△ **72.** Find the area of the square field.

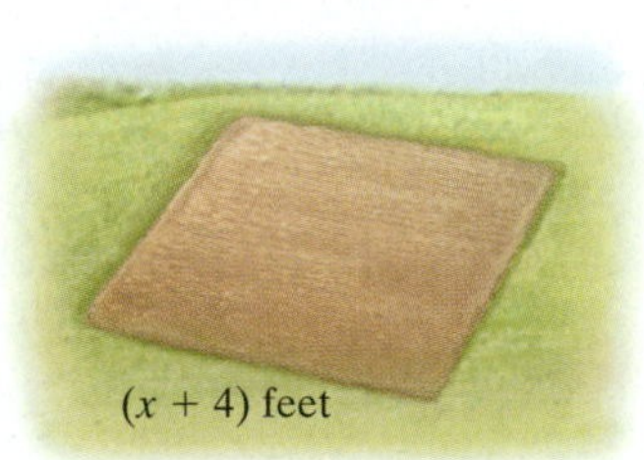

10.6 SPECIAL PRODUCTS

Objective A Using the FOIL Method

In this section, we multiply binomials using special products. First, we introduce a special order for multiplying binomials called the FOIL order or method. This order, or pattern, is a result of the distributive property. We demonstrate by multiplying $(3x + 1)$ by $(2x + 5)$.

The FOIL Method

F stands for the product of the **First** terms. $(3x + 1)(2x + 5)$
$(3x)(2x) = 6x^2$ F

O stands for the product of the **Outer** terms. $(3x + 1)(2x + 5)$
$(3x)(5) = 15x$ O

I stands for the product of the **Inner** terms. $(3x + 1)(2x + 5)$
$(1)(2x) = 2x$ I

L stands for the product of the **Last** terms. $(3x + 1)(2x + 5)$
$(1)(5) = 5$ L

$$\begin{aligned}(3x + 1)(2x + 5) &= \overset{F}{6x^2} + \overset{O}{15x} + \overset{I}{2x} + \overset{L}{5} \\ &= 6x^2 + 17x + 5 \quad \text{Combine like terms.}\end{aligned}$$

Let's practice multiplying binomials using the FOIL method.

EXAMPLE 1 Multiply: $(x - 3)(x + 4)$

Solution:

$$\begin{aligned}(x - 3)(x + 4) &= \overset{F}{(x)(x)} + \overset{O}{(x)(4)} + \overset{I}{(-3)(x)} + \overset{L}{(-3)(4)} \\ &= x^2 + 4x - 3x - 12 \\ &= x^2 + x - 12 \quad \text{Combine like terms.}\end{aligned}$$

Work Practice Problem 1

EXAMPLE 2 Multiply. $(5x - 7)(x - 2)$

Solution:

$$\begin{aligned}(5x - 7)(x - 2) &= \overset{F}{5x(x)} + \overset{O}{5x(-2)} + \overset{I}{(-7)(x)} + \overset{L}{(-7)(-2)} \\ &= 5x^2 - 10x - 7x + 14 \\ &= 5x^2 - 17x + 14 \quad \text{Combine like terms.}\end{aligned}$$

Work Practice Problem 2

Objectives

A Multiply Two Binomials Using the FOIL Method.

B Square a Binomial.

C Multiply the Sum and Difference of Two Terms.

D Use Special Products to Multiply Binomials.

PRACTICE PROBLEM 1

Multiply: $(x + 7)(x - 5)$

PRACTICE PROBLEM 2

Multiply: $(6x - 1)(x - 4)$

Helpful Hint

Remember that the FOIL order for multiplying can only be used for the product of 2 binomials.

Answers

1. $x^2 + 2x - 35$ 2. $6x^2 - 25x + 4$

PRACTICE PROBLEM 3

Multiply: $(2y^2 + 3)(y - 4)$

EXAMPLE 3 Multiply: $(y^2 + 6)(2y - 1)$

Solution: F O I L

$(y^2 + 6)(2y - 1) = 2y^3 - 1y^2 + 12y - 6$

Notice in this example that there are no like terms that can be combined, so the product is $2y^3 - y^2 + 12y - 6$.

Work Practice Problem 3

Objective B Squaring Binomials

An expression such as $(3y + 1)^2$ is called the square of a binomial. Since $(3y + 1)^2 = (3y + 1)(3y + 1)$, we can use the FOIL method to find this product.

PRACTICE PROBLEM 4

Multiply: $(2x + 9)^2$

EXAMPLE 4 Multiply: $(3y + 1)^2$

Solution: $(3y + 1)^2 = (3y + 1)(3y + 1)$

F O I L

$$= (3y)(3y) + (3y)(1) + 1(3y) + 1(1)$$
$$= 9y^2 + 3y + 3y + 1$$
$$= 9y^2 + 6y + 1$$

Work Practice Problem 4

Notice the pattern that appears in Example 4.

$(3y + 1)^2 = 9y^2 + 6y + 1$

- $9y^2$ is the first term of the binomial squared: $(3y)^2 = 9y^2$.
- $6y$ is 2 times the product of both terms of the binomial: $(2)(3y)(1) = 6y$.
- 1 is the second term of the binomial squared: $(1)^2 = 1$.

This pattern leads to the formulas below, which can be used when squaring a binomial. We call these **special products.**

Squaring a Binomial

A binomial squared is equal to the square of the first term plus or minus twice the product of both terms plus the square of the second term.

$$(a + b)^2 = a^2 + 2ab + b^2$$
$$(a - b)^2 = a^2 - 2ab + b^2$$

This product can be visualized geometrically.

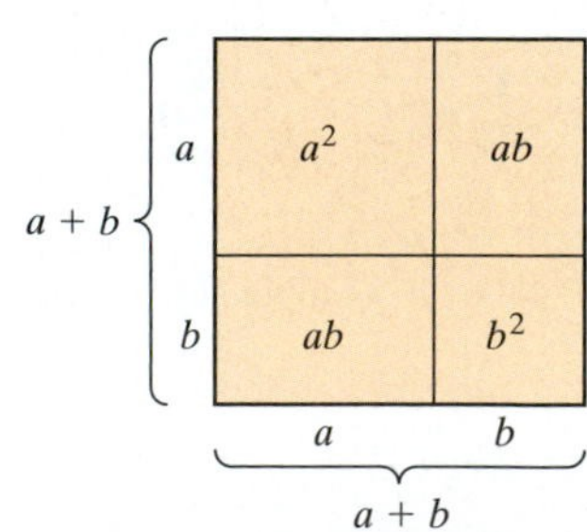

The area of the large square is side · side.

Area $= (a + b)(a + b) = (a + b)^2$

The area of the large square is also the sum of the areas of the smaller rectangles.

Area $= a^2 + ab + ab + b^2 = a^2 + 2ab + b^2$

Thus, $(a + b)^2 = a^2 + 2ab + b^2$.

Answers

3. $2y^3 - 8y^2 + 3y - 12$
4. $4x^2 + 36x + 81$

EXAMPLES Use a special product to square each binomial.

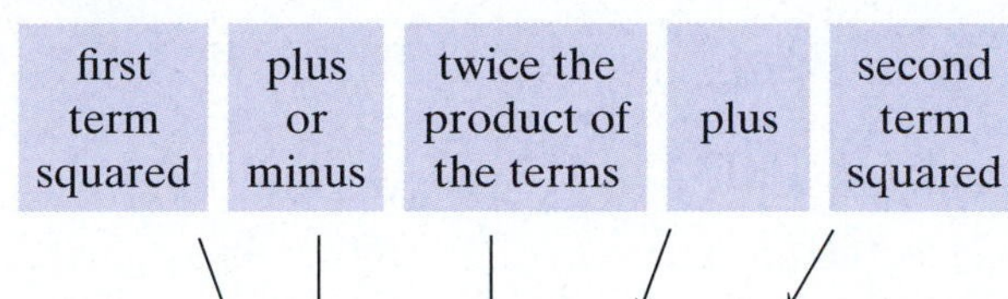

5. $(t + 2)^2 = t^2 + 2(t)(2) + 2^2 = t^2 + 4t + 4$

6. $(p - q)^2 = p^2 - 2(p)(q) + q^2 = p^2 - 2pq + q^2$

7. $(2x + 5)^2 = (2x)^2 + 2(2x)(5) + 5^2 = 4x^2 + 20x + 25$

8. $(x^2 - 7y)^2 = (x^2)^2 - 2(x^2)(7y) + (7y)^2 = x^4 - 14x^2y + 49y^2$

Work Practice Problems 5–8

PRACTICE PROBLEMS 5–8

Use a special product to square each binomial.

5. $(y + 3)^2$

6. $(r - s)^2$

7. $(6x + 5)^2$

8. $(x^2 - 3y)^2$

Notice that

$(a + b)^2 \neq a^2 + b^2$ The middle term $2ab$ is missing.

$(a + b)^2 = (a + b)(a + b) = a^2 + 2ab + b^2$

Likewise,

$(a - b)^2 \neq a^2 - b^2$

$(a - b)^2 = (a - b)(a - b) = a^2 - 2ab + b^2$

Objective C Multiplying the Sum and Difference of Two Terms

Another special product is the product of the sum and difference of the same two terms, such as $(x + y)(x - y)$. Finding this product by the FOIL method, we see a pattern emerge.

$$(x + y)(x - y) = \overset{F}{x^2} - \overset{O}{xy} + \overset{I}{xy} - \overset{L}{y^2}$$
$$= x^2 - y^2$$

Notice that the two middle terms subtract out. This is because the **O**uter product is the opposite of the **I**nner product. Only the **difference of squares** remains.

Multiplying the Sum and Difference of Two Terms

The product of the sum and difference of two terms is the square of the first term minus the square of the second term.

$$(a + b)(a - b) = a^2 - b^2$$

Answers

5. $y^2 + 6y + 9$ **6.** $r^2 - 2rs + s^2$
7. $36x^2 + 60x + 25$
8. $x^4 - 6x^2y + 9y^2$

PRACTICE PROBLEMS 9–13

Use a special product to multiply.

9. $(x + 9)(x - 9)$

10. $(5 + 4y)(5 - 4y)$

11. $\left(x - \frac{1}{3}\right)\left(x + \frac{1}{3}\right)$

12. $(3a - b)(3a + b)$

13. $(2x^2 - 6y)(2x^2 + 6y)$

EXAMPLES Use a special product to multiply.

first term squared — minus — second term squared

9. $(x + 4)(x - 4) = x^2 - 4^2 = x^2 - 16$

10. $(6t + 7)(6t - 7) = (6t)^2 - 7^2 = 36t^2 - 49$

11. $\left(x - \frac{1}{4}\right)\left(x + \frac{1}{4}\right) = x^2 - \left(\frac{1}{4}\right)^2 = x^2 - \frac{1}{16}$

12. $(2p - q)(2p + q) = (2p)^2 - q^2 = 4p^2 - q^2$

13. $(3x^2 - 5y)(3x^2 + 5y) = (3x^2)^2 - (5y)^2 = 9x^4 - 25y^2$

Work Practice Problems 9–13

✔**Concept Check** Match each expression on the left to the equivalent expression or expressions in the list on the right.

$(a + b)^2$

$(a + b)(a - b)$

a. $(a + b)(a + b)$

b. $a^2 - b^2$

c. $a^2 + b^2$

d. $a^2 - 2ab + b^2$

e. $a^2 + 2ab + b^2$

Objective D Using Special Products

Let's now practice using our special products on a variety of multiplication problems. This practice will help us recognize when to apply what special product formula.

PRACTICE PROBLEMS 14–17

Use a special product to multiply, if possible.

14. $(7x - 1)^2$

15. $(5y + 3)(2y - 5)$

16. $(2a - 1)(2a + 1)$

17. $\left(5y - \frac{1}{9}\right)^2$

EXAMPLES Use a special product to multiply, if possible.

14. $(4x - 9)(4x + 9)$ — This is the sum and difference of the same two terms.

$= (4x)^2 - 9^2 = 16x^2 - 81$

15. $(3y + 2)^2$ — This is a binomial squared.

$= (3y)^2 + 2(3y)(2) + 2^2$

$= 9y^2 + 12y + 4$

16. $(6a + 1)(a - 7)$ — No special product applies. Use the FOIL method.

F O I L

$= 6a \cdot a + 6a(-7) + 1 \cdot a + 1(-7)$

$= 6a^2 - 42a + a - 7$

$= 6a^2 - 41a - 7$

17. $\left(4x - \frac{1}{11}\right)^2$ — This is a binomial squared.

$= (4x)^2 - 2(4x)\left(\frac{1}{11}\right) + \left(\frac{1}{11}\right)^2$

$= 16x^2 - \frac{8}{11}x + \frac{1}{121}$

Work Practice Problems 14–17

Helpful Hint

- When multiplying two binomials, you may always use the FOIL order or method.
- When multiplying any two polynomials, you may always use the distributive property to find the product.

Answers

9. $x^2 - 81$ **10.** $25 - 16y^2$ **11.** $x^2 - \frac{1}{9}$ **12.** $9a^2 - b^2$ **13.** $4x^4 - 36y^2$ **14.** $49x^2 - 14x + 1$ **15.** $10y^2 - 19y - 15$ **16.** $4a^2 - 1$ **17.** $25y^2 - \frac{10}{9}y + \frac{1}{81}$

✔ Concept Check Answer

a or e, b

10.6 EXERCISE SET

FOR EXTRA HELP

Student Solutions Manual

PH Math/Tutor Center

CD/Video for Review

MathXL®

MyMathLab

Objective **A** *Multiply using the FOIL method. See Examples 1 through 3.*

1. $(x + 3)(x + 4)$

2. $(x + 5)(x + 1)$

3. $(x - 5)(x + 10)$

4. $(y - 12)(y + 4)$

5. $(5x - 6)(x + 2)$

6. $(3y - 5)(2y - 7)$

7. $(y - 6)(4y - 1)$

8. $(2x - 9)(x - 11)$

9. $(2x + 5)(3x - 1)$

10. $(6x + 2)(x - 2)$

11. $(y^2 + 7)(6y + 4)$

12. $(y^2 + 3)(5y + 6)$

13. $\left(x - \frac{1}{3}\right)\left(x + \frac{2}{3}\right)$

14. $\left(x - \frac{2}{5}\right)\left(x + \frac{1}{5}\right)$

15. $(0.4 - 3a)(0.2 - 5a)$

16. $(0.3 - 2a)(0.6 - 5a)$

17. $(x + 5y)(2x - y)$

18. $(x + 4y)(3x - y)$

Objective **B** *Multiply. See Examples 4 through 8.*

19. $(x + 2)^2$

20. $(x + 7)^2$

21. $(2x - 1)^2$

22. $(7x - 3)^2$

23. $(3a - 5)^2$

24. $(5a + 2)^2$

25. $(x^2 + 0.5)^2$

26. $(x^2 + 0.3)^2$

27. $\left(y - \frac{2}{7}\right)^2$

28. $\left(y - \frac{3}{4}\right)^2$

29. $(2a - 3)^2$

30. $(5b - 4)^2$

31. $(5x + 9)^2$

32. $(6s + 2)^2$

33. $(3x - 7y)^2$

34. $(4s - 2y)^2$

35. $(4m + 5n)^2$

36. $(3n + 5m)^2$

37. $(5x^4 - 3)^2$

38. $(7x^3 - 6)^2$

Objective **C** *Multiply. See Examples 9 through 13.*

39. $(a - 7)(a + 7)$

40. $(b + 3)(b - 3)$

41. $(x + 6)(x - 6)$

42. $(x - 8)(x + 8)$

43. $(3x - 1)(3x + 1)$

44. $(4x - 5)(4x + 5)$

45. $(x^2 + 5)(x^2 - 5)$

46. $(a^2 + 6)(a^2 - 6)$

47. $(2y^2 - 1)(2y^2 + 1)$ **48.** $(3x^2 + 1)(3x^2 - 1)$ **49.** $(4 - 7x)(4 + 7x)$ **50.** $(8 - 7x)(8 + 7x)$

51. $\left(3x - \frac{1}{2}\right)\left(3x + \frac{1}{2}\right)$ **52.** $\left(10x + \frac{2}{7}\right)\left(10x - \frac{2}{7}\right)$ **53.** $(9x + y)(9x - y)$ **54.** $(2x - y)(2x + y)$

55. $(2m + 5n)(2m - 5n)$ **56.** $(5m + 4n)(5m - 4n)$

Objective **D Mixed Practice** *Multiply. See Examples 14 through 17.*

57. $(a + 5)(a + 4)$ **58.** $(a + 5)(a + 7)$ **59.** $(a - 7)^2$ **60.** $(b - 2)^2$

61. $(4a + 1)(3a - 1)$ **62.** $(6a + 7)(6a + 5)$ **63.** $(x + 2)(x - 2)$ **64.** $(x - 10)(x + 10)$

65. $(3a + 1)^2$ **66.** $(4a + 2)^2$ **67.** $(x + y)(4x - y)$ **68.** $(3x + 2)(4x - 2)$

69. $\left(a - \frac{1}{2}y\right)\left(a + \frac{1}{2}y\right)$ **70.** $\left(\frac{a}{2} + 4y\right)\left(\frac{a}{2} - 4y\right)$ **71.** $(3b + 7)(2b - 5)$ **72.** $(3y - 13)(y - 3)$

73. $(x^2 + 10)(x^2 - 10)$ **74.** $(x^2 + 8)(x^2 - 8)$ **75.** $(4x + 5)(4x - 5)$ **76.** $(3x + 5)(3x - 5)$

77. $(5x - 6y)^2$ **78.** $(4x - 9y)^2$ **79.** $(2r - 3s)(2r + 3s)$ **80.** $(6r - 2x)(6r + 2x)$

Express each as a product of polynomials in x. Then multiply and simplify.

△ **81.** Find the area of the square rug if its side is $(2x + 1)$ feet.

△ **82.** Find the area of the rectangular canvas if its length is $(3x - 2)$ inches and its width is $(x - 4)$ inches.

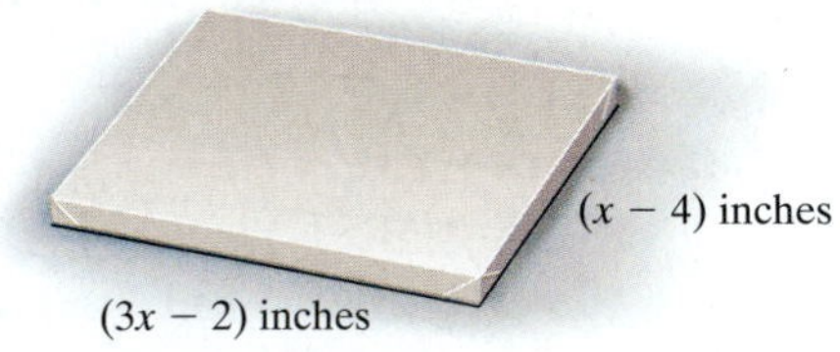

Review

Simplify each expression. See Sections 10.1 and 10.2.

83. $\frac{50b^{10}}{70b^5}$ **84.** $\frac{60y^6}{80y^2}$ **85.** $\frac{8a^{17}b^5}{-4a^7b^{10}}$ **86.** $\frac{-6a^8y}{3a^4y}$ **87.** $\frac{2x^4y^{12}}{3x^4y^4}$ **88.** $\frac{-48ab^6}{32ab^3}$

Concept Extensions

Match each expression on the left to the equivalent expression on the right. See the Concept Check in this section.

89. $(a - b)^2$

90. $(a - b)(a + b)$

91. $(a + b)^2$

92. $(a + b)^2(a - b)^2$

a. $a^2 - b^2$

b. $a^2 + b^2$

c. $a^2 - 2ab + b^2$

d. $a^2 + 2ab + b^2$

e. none of these

Fill in the squares so that a true statement forms.

93. $(x^{\square} + 7)(x^{\square} + 3) = x^4 + 10x^2 + 21$

94. $(5x^{\square} - 2)^2 = 25x^6 - 20x^3 + 4$

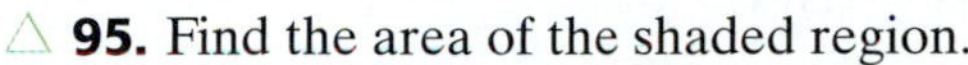

△ **95.** Find the area of the shaded region.

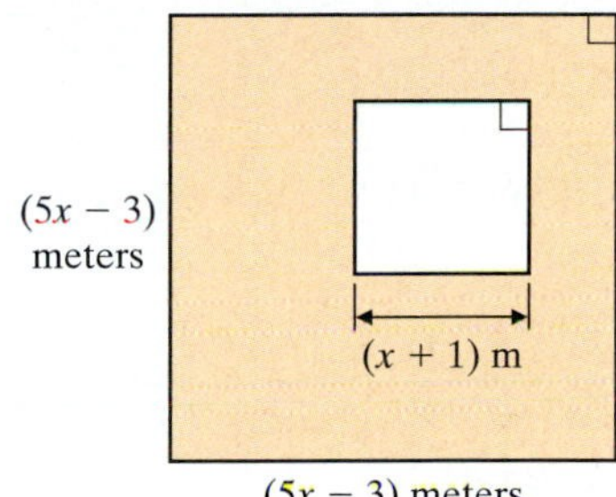

△ **96.** Find the area of the shaded region.

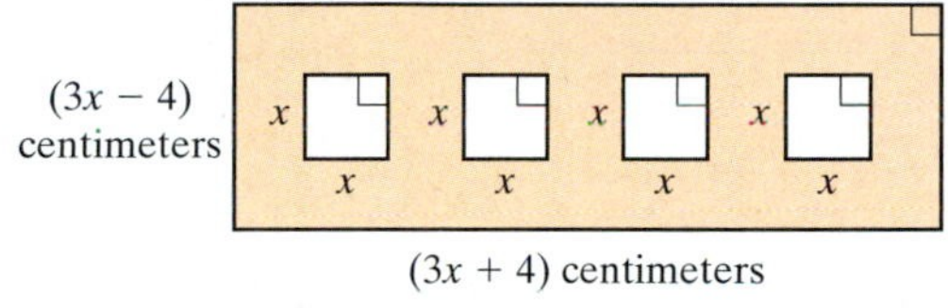

97. In your own words, describe the different methods that can be used to find the product: $(2x - 5)(3x + 1)$.

98. In your own words, describe the different methods that can be used to find the product: $(5x + 1)^2$.

INTEGRATED REVIEW

Sections 10.1–10.6

Exponents and Operations on Polynomials

Answers

1. ______
2. ______
3. ______
4. ______
5. ______
6. ______
7. ______
8. ______
9. ______
10. ______
11. ______
12. ______
13. ______
14. ______
15. ______
16. ______
17. ______
18. ______
19. ______
20. ______

Perform operations and simplify.

1. $(5x^2)(7x^3)$

2. $(4y^2)(-8y^7)$

3. -4^2

4. $(-4)^2$

5. $(x - 5)(2x + 1)$

6. $(3x - 2)(x + 5)$

7. $(x - 5) + (2x + 1)$

8. $(3x - 2) + (x + 5)$

9. $\frac{7x^9y^{12}}{x^3y^{10}}$

10. $\frac{20a^2b^8}{14a^2b^2}$

11. $(12m^7n^6)^2$

12. $(4y^9z^{10})^3$

13. $(4y - 3)(4y + 3)$

14. $(7x - 1)(7x + 1)$

15. $(x^{-7}y^5)^9$

16. 8^{-2}

17. $(3^{-1}x^9)^3$

18. $\frac{(r^7s^{-5})^6}{(2r^{-4}s^{-4})^4}$

19. $(7x^2 - 2x + 3) - (5x^2 + 9)$

20. $(10x^2 + 7x - 9) - (4x^2 - 6x + 2)$

21. $0.7y^2 - 1.2 + 1.8y^2 - 6y + 1$

22. $7.8x^2 - 6.8x - 3.3 + 0.6x^2 - 0.9$

23. Subtract $y^2 + 2$ from $3y^2 - 6y + 1$

24. $(z^2 + 5) - (3z^2 - 1) + \left(8z^2 + 2z - \frac{1}{2}\right)$

25. $(x + 4)^2$

26. $(y - 9)^2$

27. $(x + 4) + (x + 4)$

28. $(y - 9) + (y - 9)$

29. $7x^2 - 6xy + 4(y^2 - xy)$

30. $5a^2 - 3ab + 6(b^2 - a^2)$

31. $(x - 3)(x^2 + 5x - 1)$

32. $(x + 1)(x^2 - 3x - 2)$

33. $(2x - 7)(3x + 10)$

34. $(5x - 1)(4x + 5)$

35. $(2x - 7)(x^2 - 6x + 1)$

36. $(5x - 1)(x^2 + 2x - 3)$

37. $\left(2x + \frac{5}{9}\right)\left(2x - \frac{5}{9}\right)$

38. $\left(12y + \frac{3}{7}\right)\left(12y - \frac{3}{7}\right)$

21. ______

22. ______

23. ______

24. ______

25. ______

26. ______

27. ______

28. ______

29. ______

30. ______

31. ______

32. ______

33. ______

34. ______

35. ______

36. ______

37. ______

38. ______

10.7 DIVIDING POLYNOMIALS

Objectives

A Divide a Polynomial by a Monomial.

B Use Long Division to Divide a Polynomial by a Polynomial Other Than a Monomial.

Objective A Dividing by a Monomial

To divide a polynomial by a monomial, recall addition of fractions. Fractions that have a common denominator are added by adding the numerators:

$$\frac{a}{c} + \frac{b}{c} = \frac{a + b}{c}$$

If we read this equation from right to left and let a, b, and c be monomials, $c \neq 0$, we have the following.

To Divide a Polynomial by a Monomial

Divide each term of the polynomial by the monomial.

$$\frac{a + b}{c} = \frac{a}{c} + \frac{b}{c}, \quad c \neq 0$$

Throughout this section, we assume that denominators are not 0.

PRACTICE PROBLEM 1

Divide: $(25x^3 + 5x^2) \div 5x^2$

EXAMPLE 1 Divide: $(6m^2 + 2m) \div 2m$

Solution: We begin by writing the quotient in fraction form. Then we divide each term of the polynomial $6m^2 + 2m$ by the monomial $2m$ and use the quotient rule for exponents to simplify.

$$\frac{6m^2 + 2m}{2m} = \frac{6m^2}{2m} + \frac{2m}{2m}$$

$$= 3m + 1 \quad \text{Simplify.}$$

Check: To check, we multiply.

$$2m(3m + 1) = 2m(3m) + 2m(1) = 6m^2 + 2m$$

The quotient $3m + 1$ checks.

Work Practice Problem 1

✔ **Concept Check** In which of the following is $\frac{x + 5}{5}$ simplified correctly?

a. $\frac{x}{5} + 1$ **b.** x **c.** $x + 1$

PRACTICE PROBLEM 2

Divide: $\frac{24x^7 + 12x^2 - 4x}{4x^2}$

EXAMPLE 2 Divide: $\frac{9x^5 - 12x^2 + 3x}{3x^2}$

Solution: $\frac{9x^5 - 12x^2 + 3x}{3x^2} = \frac{9x^5}{3x^2} - \frac{12x^2}{3x^2} + \frac{3x}{3x^2}$ Divide each term by $3x^2$.

$$= 3x^3 - 4 + \frac{1}{x} \quad \text{Simplify.}$$

Notice that the quotient is not a polynomial because of the term $\frac{1}{x}$. This expression is called a rational expression—we will study rational expressions in Chapter 12. Although the quotient of two polynomials is not always a polynomial, we may still check by multiplying.

Answers

1. $5x + 1$ **2.** $6x^5 + 3 - \frac{1}{x}$

✔ Concept Check Answer

a

Check: $3x^2\left(3x^3 - 4 + \frac{1}{x}\right) = 3x^2(3x^3) - 3x^2(4) + 3x^2\left(\frac{1}{x}\right)$

$= 9x^5 - 12x^2 + 3x$

Work Practice Problem 2

EXAMPLE 3 Divide: $\dfrac{8x^2y^2 - 16xy + 2x}{4xy}$

Solution: $\dfrac{8x^2y^2 - 16xy + 2x}{4xy} = \dfrac{8x^2y^2}{4xy} - \dfrac{16xy}{4xy} + \dfrac{2x}{4xy}$ Divide each term by $4xy$.

$= 2xy - 4 + \dfrac{1}{2y}$ Simplify.

Check: $4xy\left(2xy - 4 + \frac{1}{2y}\right) = 4xy(2xy) - 4xy(4) + 4xy\left(\frac{1}{2y}\right)$

$= 8x^2y^2 - 16xy + 2x$

Work Practice Problem 3

PRACTICE PROBLEM 3

Divide: $\dfrac{12x^3y^3 - 18xy + 6y}{3xy}$

Objective B Dividing by a Polynomial Other Than a Monomial

To divide a polynomial by a polynomial other than a monomial, we use a process known as long division. Polynomial long division is similar to number long division, so we review long division by dividing 13 into 3660.

```
    281
13)3660
   26↓     2·13 = 26
   106     Subtract and bring down the next digit in the dividend.
   104↓    8·13 = 104
     20    Subtract and bring down the next digit in the dividend.
     13    1·13 = 13
      7    Subtract. There are no more digits to bring down, so the remainder is 7.
```

Helpful Hint Recall that 3660 is called the dividend.

The quotient is 281 R 7, which can be written as $281\,\frac{7}{13}$ ← remainder / ← divisor

Recall that division can be checked by multiplication. To check this division problem, we see that

$13 \cdot 281 + 7 = 3660$, the dividend.

Now we demonstrate long division of polynomials.

EXAMPLE 4 Divide $x^2 + 7x + 12$ by $x + 3$ using long division.

Solution:

To subtract, change the signs of these terms and add.

```
                x              How many times does x divide x²?  x²/x = x.
x + 3)x² + 7x + 12
   →  x² + 3x    ↓             Multiply: x(x + 3).
      ─   ─
           4x + 12             Subtract and bring down the next term.
```

Continued on next page

PRACTICE PROBLEM 4

Divide $x^2 + 12x + 35$ by $x + 5$ using long division.

Answers

3. $4x^2y^2 - 6 + \frac{2}{x}$ 4. $x + 7$

Now we repeat this process.

$$\begin{array}{r} x + 4 \\ x + 3\overline{)x^2 + 7x + 12} \\ \underline{x^2 + 3x} \\ 4x + 12 \\ \underline{\overset{-}{4x} \overset{-}{+} 12} \\ 0 \end{array}$$

How many times does x divide $4x$? $\frac{4x}{x} = 4$.

To subtract, change the signs of these terms and add. → Multiply: $4(x + 3)$.

Subtract. The remainder is 0.

The quotient is $x + 4$.

Check: We check by multiplying.

divisor · quotient + remainder = dividend

or

$$(x + 3) \cdot (x + 4) + 0 = x^2 + 7x + 12$$

The quotient checks.

Work Practice Problem 4

PRACTICE PROBLEM 5

Divide: $8x^2 + 2x - 7$ by $2x - 1$

EXAMPLE 5 Divide $6x^2 + 10x - 5$ by $3x - 1$ using long division.

Solution:

$$\begin{array}{r} 2x + 4 \\ 3x - 1\overline{)6x^2 + 10x - 5} \\ \underline{\overset{-}{6x^2} \overset{+}{-} 2x} \\ 12x - 5 \\ \underline{\overset{-}{12x} \overset{+}{-} 4} \\ -1 \end{array}$$

$\frac{6x^2}{3x} = 2x$, so $2x$ is a term of the quotient.

Multiply: $2x(3x - 1)$.

Subtract and bring down the next term.

$\frac{12x}{3x} = 4$. Multiply: $4(3x - 1)$.

Subtract. The remainder is -1.

Thus $(6x^2 + 10x - 5)$ divided by $(3x - 1)$ is $(2x + 4)$ with a remainder of -1. This can be written as follows.

$$\frac{6x^2 + 10x - 5}{3x - 1} = 2x + 4 + \frac{-1}{3x - 1}$$ ← remainder / ← divisor

$$\text{or } 2x + 4 - \frac{1}{3x - 1}$$

Check: To check, we multiply $(3x - 1)(2x + 4)$. Then we add the remainder, -1, to this product.

$$(3x - 1)(2x + 4) + (-1) = (6x^2 + 12x - 2x - 4) - 1$$
$$= 6x^2 + 10x - 5$$

The quotient checks.

Work Practice Problem 5

Notice that the division process is continued until the degree of the remainder polynomial is less than the degree of the divisor polynomial.

Recall in Section 10.3 that we practiced writing polynomials in descending order of powers and with no missing terms. For example, $2 - 4x^2$ written in this form is $-4x^2 + 0x + 2$. Writing the dividend and divisor in this form is helpful when dividing polynomials.

Answer

5. $4x + 3 + \frac{-4}{2x - 1}$ or $4x + 3 - \frac{4}{2x - 1}$

EXAMPLE 6 Divide: $(2 - 4x^2) \div (x + 1)$

Solution: We use the rewritten form of $2 - 4x^2$ from the previous page.

$$
\begin{array}{r|l}
\begin{array}{r}
-4x + 4 \\
x + 1\overline{)-4x^2 + 0x + 2} \\
\underline{-4x^2 - 4x} \\
4x + 2 \\
\underline{4x + 4} \\
-2
\end{array}
&
\begin{array}{l}
\frac{-4x^2}{x} = -4x, \text{ so } -4x \text{ is a term of the quotient.} \\
\text{Multiply: } -4x(x + 1). \\
\text{Subtract and bring down the next term.} \\
\frac{4x}{x} = 4. \text{ Multiply: } 4(x + 1). \\
\text{Remainder.}
\end{array}
\end{array}
$$

Thus, $\dfrac{-4x^2 + 0x + 2}{x + 1}$ or $\dfrac{2 - 4x^2}{x + 1} = -4x + 4 + \dfrac{-2}{x + 1}$ or $-4x + 4 - \dfrac{2}{x + 1}$.

Check: To check, see that $(x + 1)(-4x + 4) + (-2) = 2 - 4x^2$.

Work Practice Problem 6

PRACTICE PROBLEM 6

Divide: $(15 - 2x^2) \div (x - 3)$

EXAMPLE 7 Divide: $\dfrac{4x^2 + 7 + 8x^3}{2x + 3}$

Solution: Before we begin the division process, we rewrite $4x^2 + 7 + 8x^3$ as $8x^3 + 4x^2 + 0x + 7$. Notice that we have written the polynomial in descending order and have represented the missing x term by $0x$.

$$
\begin{array}{r}
4x^2 - 4x + 6 \\
2x + 3\overline{)8x^3 + 4x^2 + 0x + 7} \\
\underline{8x^3 + 12x^2} \\
-8x^2 + 0x \\
\underline{-8x^2 - 12x} \\
12x + 7 \\
\underline{12x + 18} \\
-11
\end{array}
\quad \text{Remainder.}
$$

Thus, $\dfrac{4x^2 + 7 + 8x^3}{2x + 3} = 4x^2 - 4x + 6 + \dfrac{-11}{2x + 3}$ or $4x^2 - 4x + 6 - \dfrac{11}{2x + 3}$.

Work Practice Problem 7

PRACTICE PROBLEM 7

Divide: $\dfrac{5 - x + 9x^3}{3x + 2}$

EXAMPLE 8 Divide $x^3 - 8$ by $x - 2$.

Solution: Notice that the polynomial $x^3 - 8$ is missing an x^2 term and an x term. We'll represent these terms by inserting $0x^2$ and $0x$.

$$
\begin{array}{r}
x^2 + 2x + 4 \\
x - 2\overline{)x^3 + 0x^2 + 0x - 8} \\
\underline{x^3 - 2x^2} \\
2x^2 + 0x \\
\underline{2x^2 - 4x} \\
4x - 8 \\
\underline{4x - 8} \\
0
\end{array}
$$

Thus, $\dfrac{x^3 - 8}{x - 2} = x^2 + 2x + 4$.

Check: To check, see that $(x^2 + 2x + 4)(x - 2) = x^3 - 8$.

Work Practice Problem 8

PRACTICE PROBLEM 8

Divide: $x^3 - 1$ by $x - 1$

Answers

6. $-2x - 6 + \dfrac{-3}{x - 3}$ or $-2x - 6 - \dfrac{3}{x - 3}$

7. $3x^2 - 2x + 1 + \dfrac{3}{3x + 2}$

8. $x^2 + x + 1$

10.7 EXERCISE SET

FOR EXTRA HELP

Student Solutions Manual | PH Math/Tutor Center | CD/Video for Review | MathXL® | MyMathLab

Objective A *Perform each division. See Examples 1 through 3.*

1. $\dfrac{12x^4 + 3x^2}{x}$

2. $\dfrac{15x^2 - 9x^5}{x}$

3. $\dfrac{20x^3 - 30x^2 + 5x + 5}{5}$

4. $\dfrac{8x^3 - 4x^2 + 6x + 2}{2}$

5. $\dfrac{15p^3 + 18p^2}{3p}$

6. $\dfrac{14m^2 - 27m^3}{7m}$

7. $\dfrac{-9x^4 + 18x^5}{6x^5}$

8. $\dfrac{6x^5 + 3x^4}{3x^4}$

9. $\dfrac{-9x^5 + 3x^4 - 12}{3x^3}$

10. $\dfrac{6a^2 - 4a + 12}{-2a^2}$

11. $\dfrac{4x^4 - 6x^3 + 7}{-4x^4}$

12. $\dfrac{-12a^3 + 36a - 15}{3a}$

Objective B *Find each quotient using long division. See Examples 4 and 5.*

13. $\dfrac{x^2 + 4x + 3}{x + 3}$

14. $\dfrac{x^2 + 7x + 10}{x + 5}$

15. $\dfrac{2x^2 + 13x + 15}{x + 5}$

16. $\dfrac{3x^2 + 8x + 4}{x + 2}$

17. $\dfrac{2x^2 - 7x + 3}{x - 4}$

18. $\dfrac{3x^2 - x - 4}{x - 1}$

19. $\dfrac{9a^3 - 3a^2 - 3a + 4}{3a + 2}$

20. $\dfrac{4x^3 + 12x^2 + x - 14}{2x + 3}$

21. $\dfrac{8x^2 + 10x + 1}{2x + 1}$

22. $\dfrac{3x^2 + 17x + 7}{3x + 2}$

23. $\dfrac{2x^3 + 2x^2 - 17x + 8}{x - 2}$

24. $\dfrac{4x^3 + 11x^2 - 8x - 10}{x + 3}$

Find each quotient using long division. Don't forget to write the polynomials in descending order and fill in any missing terms. See Examples 6 through 8.

25. $\dfrac{x^2 - 36}{x - 6}$

26. $\dfrac{a^2 - 49}{a - 7}$

27. $\dfrac{x^3 - 27}{x - 3}$

28. $\dfrac{x^3 + 64}{x + 4}$

29. $\dfrac{1 - 3x^2}{x + 2}$

30. $\dfrac{7 - 5x^2}{x + 3}$

31. $\dfrac{-4b + 4b^2 - 5}{2b - 1}$

32. $\dfrac{-3y + 2y^2 - 15}{2y + 5}$

Objectives **A** **B** **Mixed Practice** *Divide. If the divisor contains 2 or more terms, use long division. See Examples 1 through 8.*

33. $\dfrac{a^2b^2 - ab^3}{ab}$

34. $\dfrac{m^3n^2 - mn^4}{mn}$

35. $\dfrac{8x^2 + 6x - 27}{2x - 3}$

36. $\dfrac{18w^2 + 18w - 8}{3w + 4}$

37. $\dfrac{2x^2y + 8x^2y^2 - xy^2}{2xy}$

38. $\dfrac{11x^3y^3 - 33xy + x^2y^2}{11xy}$

39. $\dfrac{2b^3 + 9b^2 + 6b - 4}{b + 4}$

40. $\dfrac{2x^3 + 3x^2 - 3x + 4}{x + 2}$

41. $\dfrac{y^3 + 3y^2 + 4}{y - 2}$

42. $\dfrac{3x^3 + 11x + 12}{x + 4}$

43. $\dfrac{5 - 6x^2}{x - 2}$

44. $\dfrac{3 - 7x^2}{x - 3}$

45. $\dfrac{x^5 + x^2}{x^2 + x}$

46. $\dfrac{x^6 - x^4}{x^3 + 1}$

Review

Fill in each blank. See Section 10.1

47. $12 = 4 \cdot$ ____

48. $12 = 2 \cdot$ ____

49. $20 = -5 \cdot$ ____

50. $20 = -4 \cdot$ ____

51. $9x^2 = 3x \cdot$ ____

52. $9x^2 = 9x \cdot$ ____

53. $36x^2 = 4x \cdot$ ____

54. $36x^2 = 2x \cdot$ ____

Concept Extensions

Solve.

△ **55.** The perimeter of a square is $(12x^3 + 4x - 16)$ feet. Find the length of its side.

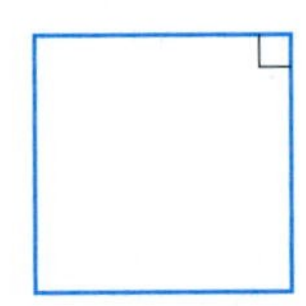

Perimeter is $(12x^3 + 4x - 16)$ feet

△ **56.** The volume of the swimming pool shown is $(36x^5 - 12x^3 + 6x^2)$ cubic feet. If its height is $2x$ feet and its width is $3x$ feet, find its length.

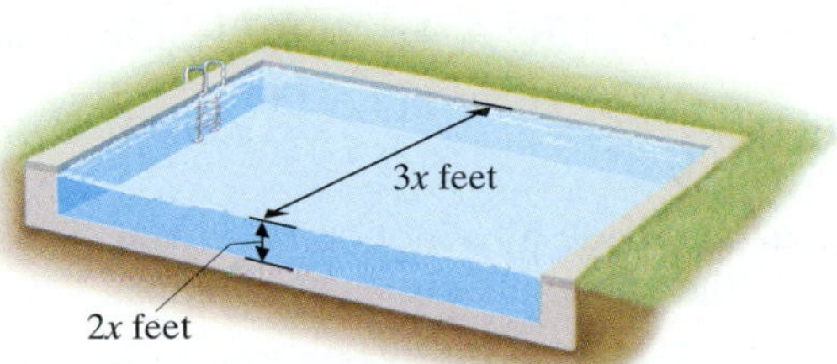

△ **57.** The area of the parallelogram shown is $(10x^2 + 31x + 15)$ square meters. If its base is $(5x + 3)$ meters, find its height.

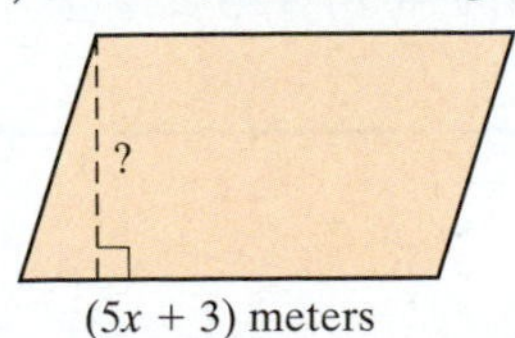

△ **58.** The area of the top of the Ping-Pong table shown is $(49x^2 + 70x - 200)$ square inches. If its length is $(7x + 20)$ inches, find its width.

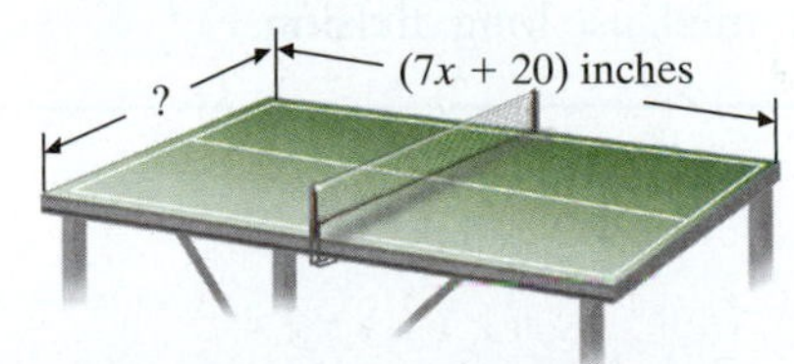

59. Explain how to check a polynomial long division result when the remainder is 0.

60. Explain how to check a polynomial long division result when the remainder is not 0.

61. In which of the following is $\frac{a+7}{7}$ simplified correctly? See the Concept Check in this section.

a. $a + 1$
b. a
c. $\frac{a}{7} + 1$

THE BIGGER PICTURE Simplifying Expressions and Solving Equations

Now we continue our outline from Section 9.6. Although suggestions are given, this outline should be in your own words. Once you complete this new portion, try the exercises below.

I. Simplifying Expressions

A. Exponents — $x^7 \cdot x^5 = x^{12}$; $(x^7)^5 = x^{35}$; $\frac{x^7}{x^5} = x^2$; $x^0 = 1$; $8^{-2} = \frac{1}{8^2} = \frac{1}{64}$

B. Polynomials

1. Add: Combine like terms.

$$(3y^2 + 6y + 7) + (9y^2 - 11y - 15)$$
$$= 3y^2 + 6y + 7 + 9y^2 - 11y - 15$$
$$= 12y^2 - 5y - 8$$

2. Subtract: Change the sign of the terms of the polynomial being subtracted, then add.

$$(3y^2 + 6y + 7) - (9y^2 - 11y - 15)$$
$$= 3y^2 + 6y + 7 - 9y^2 + 11y + 15$$
$$= -6y^2 + 17y + 22$$

3. Multiply: Multiply each term of one polynomial by each term of the other polynomial.

$$(x + 5)(2x^2 - 3x + 4)$$
$$= x(2x^2 - 3x + 4) + 5(2x^2 - 3x + 4)$$
$$= 2x^3 - 3x^2 + 4x + 10x^2 - 15x + 20$$
$$= 2x^3 + 7x^2 - 11x + 20$$

4. Divide:

a. To divide by a monomial, divide each term of the polynomial by the monomial.

$$\frac{8x^2 + 2x - 6}{2x} = \frac{8x^2}{2x} + \frac{2x}{2x} - \frac{6}{2x}$$
$$= 4x + 1 - \frac{3}{x}$$

b. To divide by a polynomial other than a monomial, use long division.

$$\begin{array}{r} x - 6 + \frac{40}{2x+5} \\ 2x + 5\overline{)2x^2 - 7x + 10} \\ \underline{\overset{-}{2x^2} \overset{-}{\cancel{+}} 5x} \\ -12x + 10 \\ \underline{\overset{+}{\cancel{-}}12x \overset{+}{\cancel{-}} 30} \\ 40 \end{array}$$

II. Solving Equations and Inequalities

A. Linear Equations (Section 9.3)

B. Linear Inequalities (Section 9.6)

Simplify the expressions.

1. $-5.7 + (-0.23)$

2. $\frac{1}{2} - \frac{9}{10}$

3. $(-5x^2y^3)(-x^7y)$

4. $2^{-3}a^{-7}a^3$

5. $(7y^3 - 6y + 2) - (y^3 + 2y^2 + 2)$

6. Subtract $(y^2 + 7)$ from $(9y^2 - 3y)$

7. Multiply: $(x - 3)(4x^2 - x + 7)$

8. Multiply: $(6m - 5)^2$

9. Divide: $\frac{20n^2 - 5n + 10}{5n}$

10. Divide: $\frac{6x^2 - 20x + 20}{3x - 1}$

Solve the equations or inequalities.

11. $-6x = 3.6$

12. $-6x < 3.6$

13. $6x + 6 \geq 8x + 2$

14. $7y + 3(y - 1) = 4(y + 1) - 3$

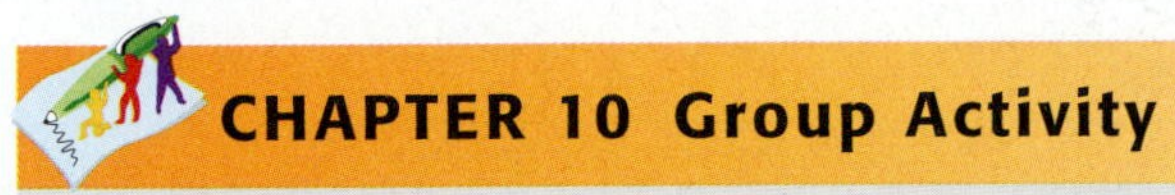

CHAPTER 10 Group Activity

Modeling with Polynomials

Materials

Calculator

This activity may be completed by working in groups or individually.

The polynomial model $-13x^2 + 221x + 8476$ gives the average daily total supply of motor gasoline (in thousand barrels per day) in the United States for the period 2000–2003. The polynomial model $-23x^2 + 192x + 7825$ gives the average daily supply of domestically produced motor gasoline (in thousand barrels per day) in the United States for the same period. In both models, x is the number of years after 2000. The other source of motor gasoline in the United States, contributing to the total supply, is imported motor gasoline. (*Source:* Based on data from the Energy Information Administration)

1. Use the given polynomials to complete the following table showing the average daily supply (both total and domestic) over the period 2000–2003 by evaluating each polynomial at the given values of x. Then subtract each value in the fourth column from the corresponding value in the third column. Record the result in the last column, titled "Difference." What do you think these values represent?

Year	x	Average Daily Total Supply (thousand barrels per day)	Average Daily Domestic Supply (thousand barrels per day)	Difference
2000	0			
2001	1			
2002	2			
2003	3			

2. Use the polynomial models to find a new polynomial model representing the average daily supply of imported motor gasoline. Then evaluate your new polynomial model to complete the accompanying table.

Year	x	Average Daily Imported Supply (thousand barrels per day)
2000	0	
2001	1	
2002	2	
2003	3	

3. Compare the values in the last column of the table in question 1 to the values in the last column of the table in question 2. What do you notice? What can you conclude?

4. Make a bar graph of the data in the table in question 2. Describe what you see.

Average Daily Imported Supply (thousand barrels per day)

0

Year

Chapter 10 Vocabulary Check

Fill in each blank with one of the words or phrases listed below.

term	coefficient	monomial	binomial	trinomial
polynomials	degree of a term	degree of a polynomial	FOIL	

1. A ________ is a number or the product of a number and variables raised to powers.
2. The ________ method may be used when multiplying two binomials.
3. A polynomial with exactly 3 terms is called a ___________.
4. The ____________________ is the greatest degree of any term of the polynomial.
5. A polynomial with exactly 2 terms is called a ___________.
6. The ___________ of a term is its numerical factor.
7. The ________________ is the sum of the exponents on the variables in the term.
8. A polynomial with exactly 1 term is called a ___________.
9. Monomials, binomials, and trinomials are all examples of _____________.

Helpful Hint

Are you preparing for your test? Don't forget to take the Chapter 10 Test on page 796. Then check your answers at the back of the text and use the Chapter Test Prep Video CD to see the fully worked-out solutions to any of the exercises you want to review.

10 Chapter Highlights

Definitions and Concepts	**Examples**
Section 10.1 Exponents	
a^n means the product of n factors, each of which is a.	$3^2 = 3 \cdot 3 = 9$ $(-5)^3 = (-5)(-5)(-5) = -125$ $\left(\frac{1}{2}\right)^4 = \frac{1}{2}\cdot\frac{1}{2}\cdot\frac{1}{2}\cdot\frac{1}{2} = \frac{1}{16}$
Let m and n be integers and no denominators be 0. **Product Rule:** $a^m \cdot a^n = a^{m+n}$ **Power Rule:** $(a^m)^n = a^{mn}$ **Power of a Product Rule:** $(ab)^n = a^nb^n$ **Power of a Quotient Rule:** $\left(\frac{a}{b}\right)^n = \frac{a^n}{b^n}$ **Quotient Rule:** $\frac{a^m}{a^n} = a^{m-n}$ **Zero Exponent:** $a^0 = 1, a \neq 0$	 $x^2 \cdot x^7 = x^{2+7} = x^9$ $(5^3)^8 = 5^{3\cdot 8} = 5^{24}$ $(7y)^4 = 7^4y^4$ $\left(\frac{x}{8}\right)^3 = \frac{x^3}{8^3}$ $\frac{x^9}{x^4} = x^{9-4} = x^5$ $5^0 = 1; x^0 = 1, x \neq 0$

Definitions and Concepts	**Examples**
Section 10.2 Negative Exponents and Scientific Notation	
If $a \neq 0$ and n is an integer, $$a^{-n} = \frac{1}{a^n}$$	$3^{-2} = \frac{1}{3^2} = \frac{1}{9}$; $5x^{-2} = \frac{5}{x^2}$ Simplify: $\left(\frac{x^{-2}y}{x^5}\right)^{-2} = \frac{x^4y^{-2}}{x^{-10}} = x^{4-(-10)}y^{-2} = \frac{x^{14}}{y^2}$
A positive number is written in scientific notation if it is written as the product of a number a, where $1 \leq a < 10$, and an integer power r of 10. $$a \times 10^r$$	$1200 = 1.2 \times 10^3$ $0.000000568 = 5.68 \times 10^{-7}$
Section 10.3 Introduction to Polynomials	
A **term** is a number or the product of a number and variables raised to powers.	$-5x, 7a^2b, \frac{1}{4}y^4, 0.2$
The **numerical coefficient** or **coefficient** of a term is its numerical factor.	**Term** / **Coefficient**: $7x^2$ — 7; y — 1; $-a^2b$ — -1
A **polynomial** is a finite sum of terms of the form ax^n where a is a real number and n is a whole number.	$5x^3 - 6x^2 + 3x - 6$ (Polynomial)
A **monomial** is a polynomial with exactly 1 term.	$\frac{5}{6}y^3$ (Monomial)
A **binomial** is a polynomial with exactly 2 terms.	$-0.2a^2b - 5b^2$ (Binomial)
A **trinomial** is a polynomial with exactly 3 terms.	$3x^2 - 2x + 1$ (Trinomial)
The **degree of a polynomial** is the greatest degree of any term of the polynomial.	**Polynomial** / **Degree**: $5x^2 - 3x + 2$ — 2; $7y + 8y^2z^3 - 12$ — $2 + 3 = 5$
Section 10.4 Adding and Subtracting Polynomials	
To add polynomials, combine like terms.	Add. $(7x^2 - 3x + 2) + (-5x - 6)$ $= 7x^2 - 3x + 2 - 5x - 6$ $= 7x^2 - 8x - 4$
To subtract two polynomials, change the signs of the terms of the second polynomial, and then add.	Subtract. $(17y^2 - 2y + 1) - (-3y^3 + 5y - 6)$ $= (17y^2 - 2y + 1) + (3y^3 - 5y + 6)$ $= 17y^2 - 2y + 1 + 3y^3 - 5y + 6$ $= 3y^3 + 17y^2 - 7y + 7$

Definitions and Concepts	Examples
Section 10.5 Multiplying Polynomials	
To multiply two polynomials, multiply each term of one polynomial by each term of the other polynomial, and then combine like terms.	Multiply. $(2x+1)(5x^2-6x+2)$ $= 2x(5x^2-6x+2) + 1(5x^2-6x+2)$ $= 10x^3 - 12x^2 + 4x + 5x^2 - 6x + 2$ $= 10x^3 - 7x^2 - 2x + 2$
Section 10.6 Special Products	
The **FOIL method** may be used when multiplying two binomials.	Multiply: $(5x-3)(2x+3)$ First, Last, Inner, Outer: $(5x-3)(2x+3)$ F O I L $= (5x)(2x) + (5x)(3) + (-3)(2x) + (-3)(3)$ $= 10x^2 + 15x - 6x - 9$ $= 10x^2 + 9x - 9$
Squaring a Binomial $(a+b)^2 = a^2 + 2ab + b^2$ $(a-b)^2 = a^2 - 2ab + b^2$	Square each binomial. $(x+5)^2 = x^2 + 2(x)(5) + 5^2$ $= x^2 + 10x + 25$ $(3x-2y)^2 = (3x)^2 - 2(3x)(2y) + (2y)^2$ $= 9x^2 - 12xy + 4y^2$
Multiplying the Sum and Difference of Two Terms $(a+b)(a-b) = a^2 - b^2$	Multiply. $(6y+5)(6y-5) = (6y)^2 - 5^2$ $= 36y^2 - 25$
Section 10.7 Dividing Polynomials	
To divide a polynomial by a monomial, $\frac{a+b}{c} = \frac{a}{c} + \frac{b}{c}, c \neq 0$	Divide. $\frac{15x^5 - 10x^3 + 5x^2 - 2x}{5x^2}$ $= \frac{15x^5}{5x^2} - \frac{10x^3}{5x^2} + \frac{5x^2}{5x^2} - \frac{2x}{5x^2}$ $= 3x^3 - 2x + 1 - \frac{2}{5x}$
To divide a polynomial by a polynomial other than a monomial, use long division.	$(10x^2 + 13x - 7) \div (2x+3) = 5x - 1 + \frac{-4}{2x+3}$ or $5x - 1 - \frac{4}{2x+3}$ Long division: $10x^2 + 15x$ subtracted gives $-2x - 7$; $-2x - 3$ subtracted gives -4

STUDY SKILLS BUILDER

Are You Prepared for a Test on Chapter 10?

Below is a list of some *common trouble areas* for students in Chapter 10. After studying for your test—but before taking your test—read these.

- Do you know that a negative exponent does not make the base a negative number? For example,

  $3^{-2} = \frac{1}{3^2} = \frac{1}{9}$

- Make sure you remember that x has an understood coefficient of 1 and an understood exponent of 1. For example,

  $2x + x = 2x + 1x = 3x; \quad x^5 \cdot x = x^5 \cdot x^1 = x^6$

- Do you know the difference between $5x^2$ and $(5x)^2$?

  $5x^2$ is $5 \cdot x^2$; $(5x)^2 = 5^2 \cdot x^2$ or $25 \cdot x^2$

- Can you evaluate $x^2 - x$ when $x = -2$?

  $x^2 - x = (-2)^2 - (-2) = 4 - (-2) = 4 + 2 = 6$

- Can you subtract $5x^2 + 1$ from $3x^2 - 6$?

  $$(3x^2 - 6) - (5x^2 + 1) = 3x^2 - 6 - 5x^2 - 1$$
  $$= -2x^2 - 7$$

- Make sure you are familiar with squaring a binomial.

  $$(3x - 4)^2 = (3x)^2 - 2(3x)(4) + 4^2$$
  $$= 9x^2 - 24x + 16$$

  or

  $$(3x - 4)^2 = (3x - 4)(3x - 4)$$
  $$= 9x^2 - 24x + 16$$

Remember: This is simply a checklist of common trouble areas. For a review of Chapter 10, see the Highlights and Chapter Review.

10 CHAPTER REVIEW

(10.1) *State the base and the exponent for each expression.*

1. 3^2 **2.** $(-5)^4$ **3.** -5^4 **4.** x^6

Evaluate each expression.

5. 8^3 **6.** $(-6)^2$ **7.** -6^2 **8.** $-4^3 - 4^0$ **9.** $(3b)^0$ **10.** $\frac{8b}{8b}$

Simplify each expression.

11. $y^2 \cdot y^7$ **12.** $x^9 \cdot x^5$ **13.** $(2x^5)(-3x^6)$ **14.** $(-5y^3)(4y^4)$ **15.** $(x^4)^2$

16. $(y^3)^5$ **17.** $(3y^6)^4$ **18.** $(2x^3)^3$ **19.** $\frac{x^9}{x^4}$ **20.** $\frac{z^{12}}{z^5}$

21. $\frac{a^5b^4}{ab}$ **22.** $\frac{x^4y^6}{xy}$ **23.** $\frac{12xy^6}{3x^4y^{10}}$ **24.** $\frac{2x^7y^8}{8xy^2}$ **25.** $5a^7(2a^4)^3$

26. $(2x)^2(9x)$ **27.** $(-5a)^0 + 7^0 + 8^0$ **28.** $8x^0 + 9^0$

Simplify the given expression and choose the correct result.

29. $\left(\frac{3x^4}{4y}\right)^3$

a. $\frac{27x^{64}}{64y^3}$ **b.** $\frac{27x^{12}}{64y^3}$ **c.** $\frac{9x^{12}}{12y^3}$ **d.** $\frac{3x^{12}}{4y^3}$

30. $\left(\frac{5a^6}{b^3}\right)^2$

a. $\frac{10a^{12}}{b^6}$ **b.** $\frac{25a^{36}}{b^9}$ **c.** $\frac{25a^{12}}{b^6}$ **d.** $25a^{12}b^6$

(10.2) *Simplify each expression.*

31. 7^{-2}

32. -7^{-2}

33. $2x^{-4}$

34. $(2x)^{-4}$

35. $\left(\frac{1}{5}\right)^{-3}$

36. $\left(\frac{-2}{3}\right)^{-2}$

37. $2^0 + 2^{-4}$

38. $6^{-1} - 7^{-1}$

Simplify each expression. Write each answer using positive exponents only.

39. $\frac{x^5}{x^{-3}}$

40. $\frac{z^4}{z^{-4}}$

41. $\frac{r^{-3}}{r^{-4}}$

42. $\frac{y^{-2}}{y^{-5}}$

43. $\left(\frac{bc^{-2}}{bc^{-3}}\right)^4$

44. $\left(\frac{x^{-3}y^{-4}}{x^{-2}y^{-5}}\right)^{-3}$

45. $\frac{x^{-4}y^{-6}}{x^2y^7}$

46. $\frac{a^5b^{-5}}{a^{-5}b^5}$

Write each number in scientific notation.

47. 0.00027

48. 0.8868

49. 80,800,000

50. 868,000

51. In August 2004, the United States imported approximately 112,400,000 kilograms of coffee. Write this number in scientific notation. (*Source:* International Coffee Organization)

52. The approximate diameter of the Milky Way galaxy is 150,000 light years. Write this number in scientific notation. (*Source:* NASA IMAGE/POETRY Education and Public Outreach Program)

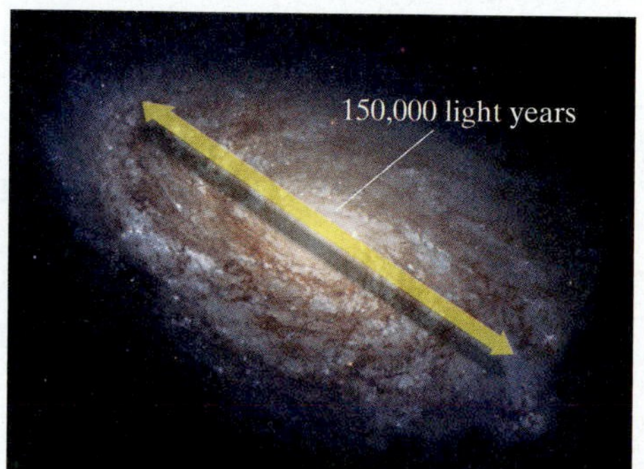

Write each number in standard form.

53. 8.67×10^5

54. 3.86×10^{-3}

55. 8.6×10^{-4}

56. 8.936×10^5

57. The volume of the planet Jupiter is 1.43128×10^{15} cubic kilometers. Write this number in standard notation. (*Source:* National Space Science Data Center)

58. An angstrom is a unit of measure, equal to 1×10^{-10} meter, used for measuring wavelengths or the diameters of atoms. Write this number in standard notation. (*Source:* National Institute of Standards and Technology)

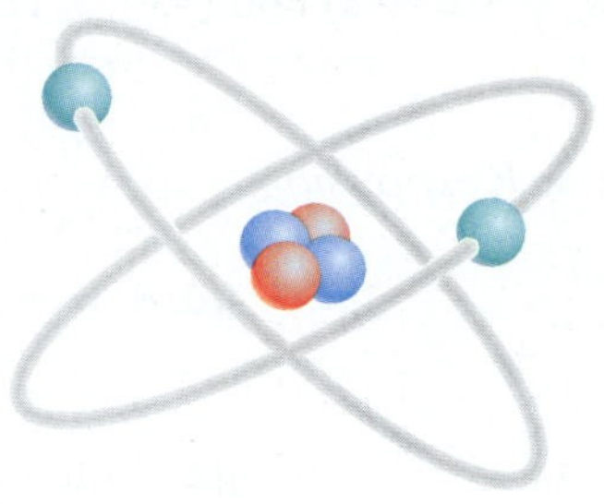

Simplify. Express each result in standard form.

59. $(8 \times 10^4)(2 \times 10^{-7})$

60. $\dfrac{8 \times 10^4}{2 \times 10^{-7}}$

(10.3) *Find the degree of each polynomial.*

61. $y^5 + 7x - 8x^4$

62. $9y^2 + 30y + 25$

63. $-14x^2y - 28x^2y^3 - 42x^2y^2$

64. $6x^2y^2z^2 + 5x^2y^3 - 12xyz$

△ **65.** The surface area of a box with a square base and a height of 5 units is given by the polynomial $2x^2 + 20x$. Fill in the table below by evaluating $2x^2 + 20x$ for the given values of x.

x	1	3	5.1	10
$2x^2 + 20x$				

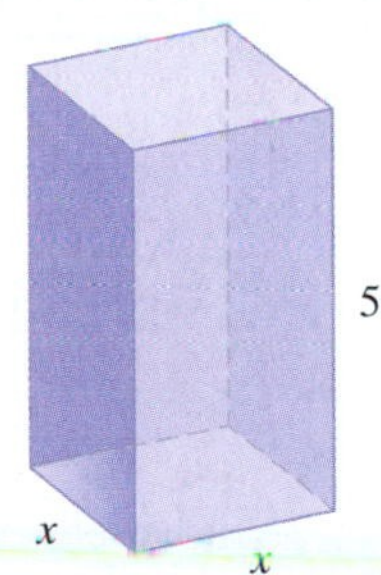

Combine like terms in each expression.

66. $7a^2 - 4a^2 - a^2$

67. $9y + y - 14y$

68. $6a^2 + 4a + 9a^2$

69. $21x^2 + 3x + x^2 + 6$

70. $4a^2b - 3b^2 - 8q^2 - 10a^2b + 7q^2$

71. $2s^{14} + 3s^{13} + 12s^{12} - s^{10}$

(10.4) *Add or subtract as indicated.*

72. $(3x^2 + 2x + 6) + (5x^2 + x)$

73. $(2x^5 + 3x^4 + 4x^3 + 5x^2) + (4x^2 + 7x + 6)$

74. $(-5y^2 + 3) - (2y^2 + 4)$

75. $(2m^7 + 3x^4 + 7m^6) - (8m^7 + 4m^2 + 6x^4)$

76. $(3x^2 - 7xy + 7y^2) - (4x^2 - xy + 9y^2)$

Translating *Perform the indicated operations.*

77. Add $(-9x^2 + 6x + 2)$ and $(4x^2 - x - 1)$.

78. Subtract $(4x^2 + 8x - 7)$ from the sum of $(x^2 + 7x + 9)$ and $(x^2 + 4)$.

(10.5) *Multiply each expression.*

79. $6(x + 5)$

80. $9(x - 7)$

81. $4(2a + 7)$

82. $9(6a - 3)$

83. $-7x(x^2 + 5)$

84. $-8y(4y^2 - 6)$

85. $-2(x^3 - 9x^2 + x)$

86. $-3a(a^2b + ab + b^2)$

87. $(3a^3 - 4a + 1)(-2a)$

88. $(6b^3 - 4b + 2)(7b)$

89. $(2x + 2)(x - 7)$

90. $(2x - 5)(3x + 2)$

91. $(4a - 1)(a + 7)$

92. $(6a - 1)(7a + 3)$

93. $(x + 7)(x^3 + 4x - 5)$

94. $(x + 2)(x^5 + x + 1)$

95. $(x^2 + 2x + 4)(x^2 + 2x - 4)$

96. $(x^3 + 4x + 4)(x^3 + 4x - 4)$

97. $(x + 7)^3$

98. $(2x - 5)^3$

(10.6) *Use special products to multiply each of the following.*

99. $(x + 7)^2$

100. $(x - 5)^2$

101. $(3x - 7)^2$

102. $(4x + 2)^2$

103. $(5x - 9)^2$

104. $(5x + 1)(5x - 1)$

105. $(7x + 4)(7x - 4)$

106. $(a + 2b)(a - 2b)$

107. $(2x - 6)(2x + 6)$

108. $(4a^2 - 2b)(4a^2 + 2b)$

Express each as a product of polynomials in x. Then multiply and simplify.

△ **109.** Find the area of the square if its side is $(3x - 1)$ meters.

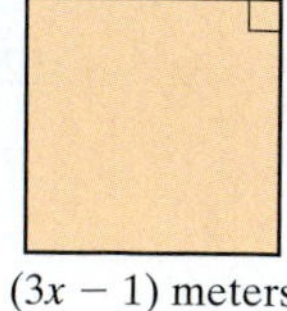

△ **110.** Find the area of the rectangle.

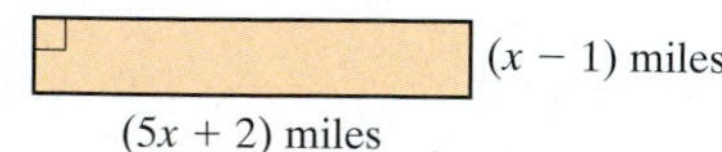

(10.7) *Divide.*

111. $\frac{x^2 + 21x + 49}{7x^2}$

112. $\frac{5a^3b - 15ab^2 + 20ab}{-5ab}$

113. $(a^2 - a + 4) \div (a - 2)$

114. $(4x^2 + 20x + 7) \div (x + 5)$

115. $\dfrac{a^3 + a^2 + 2a + 6}{a - 2}$

116. $\dfrac{9b^3 - 18b^2 + 8b - 1}{3b - 2}$

117. $\dfrac{4x^4 - 4x^3 + x^2 + 4x - 3}{2x - 1}$

118. $\dfrac{-10x^2 - x^3 - 21x + 18}{x - 6}$

△ **119.** The area of the rectangle below is $(15x^3 - 3x^2 + 60)$ square feet. If its length is $3x^2$ feet, find its width.

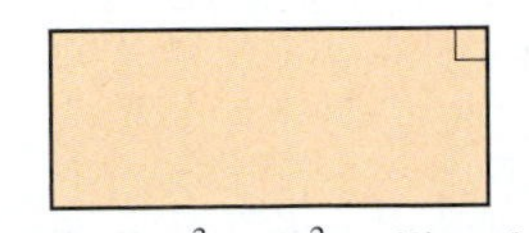

Area is $(15x^3 - 3x^2 + 60)$ sq feet

△ **120.** The perimeter of the equilateral triangle below is $(21a^3b^6 + 3a - 3)$ units. Find the length of a side.

Perimeter is $(21a^3b^6 + 3a - 3)$ units

Mixed Review

Evaluate.

121. $\left(-\dfrac{1}{2}\right)^3$

Simplify each expression. Write each answer using positive exponents only.

122. $(4xy^2)(x^3y^5)$

123. $\dfrac{18x^9}{27x^3}$

124. $\left(\dfrac{3a^4}{b^2}\right)^3$

125. $(2x^{-4}y^3)^{-4}$

126. $\dfrac{a^{-3}b^6}{9^{-1}a^{-5}b^{-2}}$

Perform the indicated operations and simplify.

127. $(6x + 2) + (5x - 7)$

128. $(-y^2 - 4) + (3y^2 - 6)$

129. $(8y^2 - 3y + 1) - (3y^2 + 2)$

130. $(5x^2 + 2x - 6) - (-x - 4)$

131. $4x(7x^2 + 3)$

132. $(2x + 5)(3x - 2)$

133. $(x - 3)(x^2 + 4x - 6)$

134. $(7x - 2)(4x - 9)$

Use special products to multiply.

135. $(5x + 4)^2$

136. $(6x + 3)(6x - 3)$

Divide.

137. $\dfrac{8a^4 - 2a^3 + 4a - 5}{2a^3}$

138. $\dfrac{x^2 + 2x + 10}{x + 5}$

139. $\dfrac{4x^3 + 8x^2 - 11x + 4}{2x - 3}$

Answers

1. ______
2. ______
3. ______
4. ______
5. ______
6. ______
7. ______
8. ______
9. ______
10. ______
11. ______
12. ______
13. ______
14. ______
15. a. see table
b. ______
16. ______
17. ______
18. ______
19. ______

10 CHAPTER TEST

Remember to use the Chapter Test Prep Video CD to see the fully worked-out solutions to any of the exercises you want to review.

Evaluate each expression.

1. 2^5 **2.** $(-3)^4$ **3.** -3^4 **4.** 4^{-3}

Simplify each exponential expression.

5. $(3x^2)(-5x^9)$ **6.** $\frac{y^7}{y^2}$ **7.** $\frac{r^{-8}}{r^{-3}}$

Simplify each expression. Write the result using only positive exponents.

8. $\left(\frac{x^2y^3}{x^3y^{-4}}\right)^2$ **9.** $\frac{6^2x^{-4}y^{-1}}{6^3x^{-3}y^7}$

Express each number in scientific notation.

10. 563,000 **11.** 0.0000863

Write each number in standard form.

12. 1.5×10^{-3} **13.** 6.23×10^4

14. Simplify. Write the answer in standard form.
$(1.2 \times 10^5)(3 \times 10^{-7})$

15. a. Complete the table for the polynomial $4xy^2 + 7xyz + x^3y - 2$.

Term	Numerical Coefficient	Degree of Term
$4xy^2$		
$7xyz$		
x^3y		
-2		

b. What is the degree of the polynomial?

16. Simplify by combining like terms.
$5x^2 + 4x - 7x^2 + 11 + 8x$

Perform each indicated operation.

17. $(8x^3 + 7x^2 + 4x - 7) + (8x^3 - 7x - 6)$

18.
$$\begin{array}{r} 5x^3 + x^2 + 5x - 2 \\ -(8x^3 - 4x^2 + x - 7) \\ \hline \end{array}$$

19. Subtract $(4x + 2)$ from the sum of $(8x^2 + 7x + 5)$ and $(x^3 - 8)$.

Multiply. Exercises 20 through 26.

20. $(3x + 7)(x^2 + 5x + 2)$

21. $3x^2(2x^2 - 3x + 7)$

22. $(x + 7)(3x - 5)$

23. $\left(3x - \frac{1}{5}\right)\left(3x + \frac{1}{5}\right)$

24. $(4x - 2)^2$

25. $(8x + 3)^2$

26. $(x^2 - 9b)(x^2 + 9b)$

27. The height of the Bank of China in Hong Kong is 1001 feet. Neglecting air resistance, the height of an object dropped from this building at time t seconds is given by the polynomial $-16t^2 + 1001$. Find the height of the object at the given times below.

t	0 seconds	1 second	3 seconds	5 seconds
$-16t^2 + 1001$				

△ **28.** Find the area of the top of the table. Express the area as a product, then multiply and simplify.

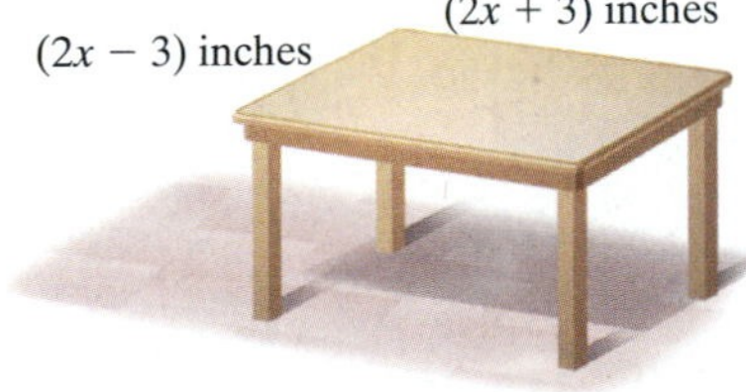

Divide.

29. $\dfrac{4x^2 + 2xy - 7x}{8xy}$

30. $(x^2 + 7x + 10) \div (x + 5)$

31. $\dfrac{27x^3 - 8}{3x + 2}$

20. ______

21. ______

22. ______

23. ______

24. ______

25. ______

26. ______

27. see table ______

28. ______

29. ______

30. ______

31. ______

CUMULATIVE REVIEW

Chapters 1–10

Answers

1. ______
2. ______
3. ______
4. ______
5. a. ______
b. ______
c. ______
d. ______
e. ______
6. a. ______
b. ______
c. ______
d. ______
e. ______
7. ______
8. ______
9. ______
10. ______
11. ______
12. ______
13. ______
14. ______
15. ______

Evaluate.

1. 9^2 **2.** 5^3 **3.** 3^4 **4.** 3^3

Write an algebraic expression. Use x to represent "a number."

5. **a.** 7 increased by a number
b. 15 decreased by a number
c. the product of 2 and a number
d. the quotient of a number and 5
e. 2 subtracted from a number

6. **a.** the sum of a number and 3
b. the product of 3 and a number
c. twice a number
d. 10 decreased by a number
e. 5 times a number increased by 7

Use the distributive property to remove parentheses. Then simplify if possible.

7. $2(3 + 7x) - 15$

8. $5(x + 2)$

9. $-2(x - 5) + 4(2x + 2)$

10. $-2(y + 0.3z - 1)$

Solve.

11. $y - 5 = -2 - 6$

12. $\frac{y}{7} = 20$

13. $7(x - 2) = 9x - 6$

14. $6(2a - 1) - (11a + 6) = 7$

15. $\frac{3}{5}a = 9$

16. $\frac{2}{3}y = 16$

17. $3y = -\frac{2}{11}$

18. $5y = -\frac{1}{5}$

Write each decimal as a fraction or a mixed number.

19. 0.125

20. 0.250

21. 43.5

22. 10.75

23. −105.083

24. −31.07

Evaluate $x - y$ for the given replacement values.

25. $x = 2.8, y = 0.92$

26. $x = -1.2, y = 7.6$

Evaluate xy for the given replacement values.

27. $x = 2.3, y = 0.44$

28. $x = -6.1, y = 0.5$

29. Given the set $\left\{-2, 0, \frac{1}{4}, 112, -3, 11, \sqrt{2}\right\}$, list the numbers in this set that belong to the set of:

a. natural numbers
b. whole numbers
c. integers
d. rational numbers
e. irrational numbers
f. real numbers

30. Tell which set(s) the number $-2\frac{1}{2}$ belongs to. Use the set of numbers from Exercise 29.

Solve.

31. $0.25x + 0.10(x - 3) = 1.1$

32. $0.6x - 10 = 1.4x - 14$

16. ____
17. ____
18. ____
19. ____
20. ____
21. ____
22. ____
23. ____
24. ____
25. ____
26. ____
27. ____
28. ____
29. a. ____
b. ____
c. ____
d. ____
e. ____
f. ____
30. ____
31. ____
32. ____

33. ______
34. ______
35. ______
36. ______
37. ______
38. ______
39. a. ______
b. ______
c. ______
40. a. ______
b. ______
c. ______
41. ______
42. ______
43. ______
44. ______
45. ______
46. ______
47. ______
48. ______
49. ______
50. ______
51. ______
52. ______

33. Twice the sum of a number and 4 is the same as four times the number, decreased by 12. Find the number.

34. Three times the difference of a number and 2 is the same as five times the number, decreased by 10.

△ **35.** Charles Pecot can afford enough fencing to enclose a rectangular garden with a perimeter of 140 feet. If the width of his garden is to be 30 feet, find the length.

36. A house is in the shape of a rectangle and has 2016 square feet of living area. If the length is 63 feet, find its width.

37. Solve: $-4x + 7 \geq -9$. Graph the solutions.

−5 −4 −3 −2 −1 0 1 2 3 4 5

38. Solve: $3x + 4 \geq 2x - 6$

39. Simplify each expression.

a. $x^7 \cdot x^4$ **b.** $\left(\frac{t}{2}\right)^4$ **c.** $(9y^5)^2$

40. Simplify each expression.

a. $y \cdot y^5$ **b.** $\left(\frac{2}{3}\right)^3$ **c.** $(8x^3)^2$

Simplify the following expressions. Write each result using positive exponents only.

41. $\left(\frac{3a^2}{b}\right)^{-3}$ **42.** $\left(\frac{2x^3}{y}\right)^{-2}$ **43.** $(5y^3)^{-2}$ **44.** $(3x^7)^{-3}$

Simplify each polynomial by combining any like terms.

45. $9x^3 + x^3$

46. $9x^3 - x^3$

47. $5x^2 + 6x - 9x - 3$

48. $2x - x^2 + 5x - 4x^2$

Multiply.

49. $7x(x^2 + 2x + 5)$

50. $-2x(x^2 - x + 1)$

Divide.

51. $\dfrac{9x^5 - 12x^2 + 3x}{3x^2}$

52. $\dfrac{4x^7 - 12x^2 + 2x}{2x}$

11

Factoring Polynomials

In Chapter 10, we learned how to multiply polynomials. Now we will deal with an operation that is the reverse process of multiplying–factoring. Factoring is an important algebraic skill because it allows us to write a sum as a product. As we will see in Sections 11.6 and 11.7, factoring can be used to solve equations other than linear equations. In Chapter 12, we will also use factoring to simplify and perform arithmetic operations on rational expressions.

When recently completed, the Taipei 101 building in Taipei, Taiwan, became the world's tallest building. At a height of 1671 feet, it is the world's first super tall building to be built in an active earthquake zone. In Exercise 107, Section 11.5, a polynomial expression for the height of an object dropped from Taipei 101 is factored. (*Source:* Council on Tall Buildings and Urban Habitats)

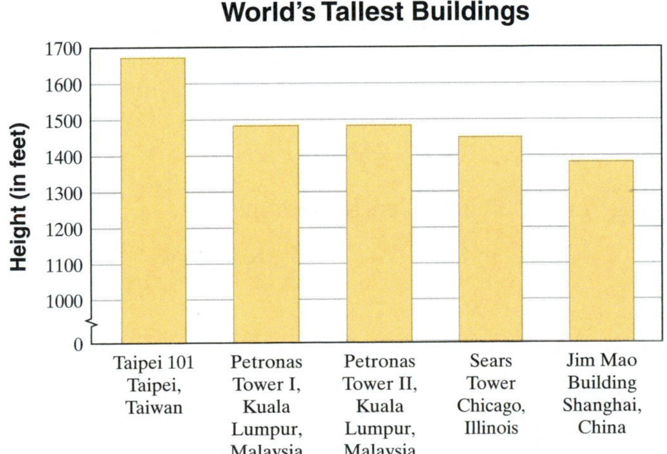

SECTIONS

CHECK YOUR PROGRESS

11.1 THE GREATEST COMMON FACTOR

Objectives

A Find the Greatest Common Factor of a List of Numbers.

B Find the Greatest Common Factor of a List of Terms.

C Factor Out the Greatest Common Factor from the Terms of a Polynomial.

D Factor by Grouping.

In the product $2 \cdot 3 = 6$, the numbers 2 and 3 are called **factors** of 6 and $2 \cdot 3$ is a **factored form** of 6. This is true of polynomials also. Since $(x + 2)(x + 3) = x^2 + 5x + 6$, then $(x + 2)$ and $(x + 3)$ are factors of $x^2 + 5x + 6$, and $(x + 2)(x + 3)$ is a factored form of the polynomial.

> The process of writing a polynomial as a product is called **factoring** the polynomial.

Study the examples below and look for a pattern.

Multiplying: $5(x^2 + 3) = 5x^2 + 15$ $\quad 2x(x - 7) = 2x^2 - 14x$

Factoring: $5x^2 + 15 = 5(x^2 + 3)$ $\quad 2x^2 - 14x = 2x(x - 7)$

Do you see that factoring is the reverse process of multiplying?

$$x^2 + 5x + 6 = (x + 2)(x + 3)$$

factoring →; ← multiplying

✔ **Concept Check** Multiply: $2(x - 4)$
What do you think the result of factoring $2x - 8$ would be? Why?

Objective A Finding the Greatest Common Factor of a List of Numbers

The first step in factoring a polynomial is to see whether the terms of the polynomial have a common factor. If there is one, we can write the polynomial as a product by **factoring out** the common factor. We will usually factor out the *greatest* common factor (GCF).

The GCF of a list of integers is the largest integer that is a factor of all the integers in the list. For example, the GCF of 12 and 20 is 4 because 4 is the largest integer that is a factor of both 12 and 20. With large integers, the GCF may not be easily found by inspection. When this happens, we will write each integer as a product of prime numbers. Recall that a prime number is a whole number other than 1, whose only factors are 1 and itself.

PRACTICE PROBLEM 1

Find the GCF of each list of numbers.

a. 45 and 75 **b.** 32 and 33

c. 14, 24, and 60

EXAMPLE 1 Find the GCF of each list of numbers.

a. 28 and 40 **b.** 55 and 21 **c.** 15, 18, and 66

Solution:

a. Write each number as a product of primes.

$28 = 2 \cdot 2 \cdot 7 = 2^2 \cdot 7$

$40 = 2 \cdot 2 \cdot 2 \cdot 5 = 2^3 \cdot 5$

There are two common factors, each of which is 2, so the GCF is

$\text{GCF} = 2 \cdot 2 = 4$

Answers

1. a. 15 **b.** 1 **c.** 2

✔ **Concept Check Answer**

$2x - 8$; The result would be $2(x - 4)$ because factoring is the reverse process of multiplying.

b. $55 = 5 \cdot 11$

$21 = 3 \cdot 7$

There are no common prime factors; thus, the GCF is 1.

c. $15 = 3 \cdot 5$

$18 = 2 \cdot 3 \cdot 3 = 2 \cdot 3^2$

$66 = 2 \cdot 3 \cdot 11$

The only prime factor common to all three numbers is 3, so the GCF is

GCF = 3

Work Practice Problem 1

Objective B Finding the Greatest Common Factor of a List of Terms

The greatest common factor of a list of variables raised to powers is found in a similar way. For example, the GCF of x^2, x^3, and x^5 is x^2 because each term contains a factor of x^2 and no higher power of x is a factor of each term.

$$x^2 = x \cdot x$$
$$x^3 = x \cdot x \cdot x$$
$$x^5 = x \cdot x \cdot x \cdot x \cdot x$$

There are two common factors, each of which is x, so the GCF $= x \cdot x$ or x^2. From this example, we see that **the GCF of a list of common variables raised to powers is the variable raised to the smallest exponent in the list.**

EXAMPLE 2 Find the GCF of each list of terms.

a. x^3, x^7, and x^5

b. y, y^4, and y^7

Solution:

a. The GCF is x^3, since 3 is the smallest exponent to which x is raised.

b. The GCF is y^1 or y, since 1 is the smallest exponent on y.

Work Practice Problem 2

PRACTI E PROBLEM 2

Find the ;CF of each list of terms.

a. y^4, y ınd y^8

b. x ar x^{10}

The **greatest common factor (GCF) of a list of terms** is the product of the GCF of the numerical coefficients and the GCF of the variable factors.

$$20x^2y^2 = 2 \cdot 2 \cdot 5 \cdot x \cdot x \cdot y \cdot y$$
$$6xy^3 = 2 \cdot 3 \cdot x \cdot y \cdot y \cdot y$$
$$\text{GCF} = 2 \cdot x \cdot y \cdot y = 2xy^2$$

Helpful Hint

Remember that the GCF of a list of terms contains the smallest exponent on each common variable.

The GCF of x^5y^6, x^2y^7 and x^3y^4 is x^2y^4. — Smallest exponent on x. — Smallest exponent on y.

Answers

2. a. y^4 **b.** x

PRACTICE PROBLEM 3

Find the greatest common factor of each list of terms.

a. $6x^2, 9x^4$, and $-12x^5$

b. $-16y, -20y^6$, and $40y^4$

c. a^5b^4, ab^3, and a^3b^2

EXAMPLE 3 Find the greatest common factor of each list of terms.

a. $6x^2, 10x^3$, and $-8x$

b. $-18y^2, -63y^3$, and $27y^4$

c. a^3b^2, a^5b, and a^6b^2

Solution:

a.
$$6x^2 = 2 \cdot 3 \cdot x^2$$
$$10x^3 = 2 \cdot 5 \cdot x^3$$
$$-8x = -1 \cdot 2 \cdot 2 \cdot 2 \cdot x^1$$
→ The GCF of x^2, x^3, and x^1 is x^1 or x.
$$\text{GCF} = 2 \cdot x^1 \quad \text{or} \quad 2x$$

b.
$$-18y^2 = -1 \cdot 2 \cdot 3 \cdot 3 \cdot y^2$$
$$-63y^3 = -1 \cdot 3 \cdot 3 \cdot 7 \cdot y^3$$
$$27y^4 = 3 \cdot 3 \cdot 3 \cdot y^4$$
→ The GCF of y^2, y^3, and y^4 is y^2.
$$\text{GCF} = 3 \cdot 3 \cdot y^2 \quad \text{or} \quad 9y^2$$

c. The GCF of a^3, a^5, and a^6 is a^3.
The GCF of b^2, b, and b^2 is b. Thus,
the GCF of a^3b^2, a^5b, and a^6b^2 is a^3b.

Work Practice Problem 3

Objective C Factoring Out the Greatest Common Factor

To factor a polynomial such as $8x + 14$, we first see whether the terms have a greatest common factor other than 1. In this case, they do: The GCF of $8x$ and 14 is 2.

We factor out 2 from each term by writing each term as the product of 2 and the term's remaining factors.

$$8x + 14 = 2 \cdot 4x + 2 \cdot 7$$

Using the distributive property, we can write

$$8x + 14 = 2 \cdot 4x + 2 \cdot 7$$
$$= 2(4x + 7)$$

Thus, a factored form of $8x + 14$ is $2(4x + 7)$. We can check by multiplying:

$$2(4x + 7) = 2 \cdot 4x + 2 \cdot 7 = 8x + 14.$$

Helpful Hint

A factored form of $8x + 14$ is *not*

$$2 \cdot 4x + 2 \cdot 7$$

Although the *terms* have been factored (written as products), the *polynomial* $8x + 14$ has not been factored. A factored form of $8x + 14$ is the *product* $2(4x + 7)$.

✔**Concept Check** Which of the following is/are factored form(s) of $6t + 18$?

a. 6

b. $6 \cdot t + 6 \cdot 3$

c. $6(t + 3)$

d. $3(t + 6)$

Answers

3. a. $3x^2$ **b.** $4y$ **c.** ab^2

✔ Concept Check Answer

c

EXAMPLE 4 Factor each polynomial by factoring out the greatest common factor (GCF).

a. $5ab + 10a$ **b.** $y^5 - y^{12}$

Solution:

a. The GCF of terms $5ab$ and $10a$ is $5a$. Thus,

$$5ab + 10a = 5a \cdot b + 5a \cdot 2$$
$$= 5a(b + 2) \quad \text{Apply the distributive property.}$$

We can check our work by multiplying $5a$ and $(b + 2)$.
$5a(b + 2) = 5a \cdot b + 5a \cdot 2 = 5ab + 10a$, the original polynomial.

b. The GCF of y^5 and y^{12} is y^5. Thus,

$$y^5 - y^{12} = y^5(1) - y^5(y^7)$$
$$= y^5(1 - y^7)$$

Don't forget the 1.

Work Practice Problem 4

PRACTICE PROBLEM 4

Factor each polynomial by factoring out the greatest common factor (GCF).

a. $10y + 25$

b. $x^4 - x^9$

EXAMPLE 5 Factor: $-9a^5 + 18a^2 - 3a$

Solution:

$$-9a^5 + 18a^2 - 3a = 3a(-3a^4) + 3a(6a) + 3a(-1)$$
$$= 3a(-3a^4 + 6a - 1)$$

Helpful Hint: Don't forget the -1.

Work Practice Problem 5

PRACTICE PROBLEM 5

Factor: $-10x^3 + 8x^2 - 2x$

In Example 5, we could have chosen to factor out $-3a$ instead of $3a$. If we factor out $-3a$, we have

$$-9a^5 + 18a^2 - 3a = (-3a)(3a^4) + (-3a)(-6a) + (-3a)(1)$$
$$= -3a(3a^4 - 6a + 1)$$

Notice the changes in signs when factoring out $-3a$.

EXAMPLES Factor.

6. $6a^4 - 12a = 6a(a^3 - 2)$

7. $\frac{3}{7}x^4 + \frac{1}{7}x^3 - \frac{5}{7}x^2 = \frac{1}{7}x^2(3x^2 + x - 5)$

8. $15p^2q^4 + 20p^3q^5 + 5p^3q^3 = 5p^2q^3(3q + 4pq^2 + p)$

Work Practice Problems 6–8

PRACTICE PROBLEMS 6–8

Factor.

6. $4x^3 + 12x$

7. $\frac{2}{5}a^5 - \frac{4}{5}a^3 + \frac{1}{5}a^2$

8. $6a^3b + 3a^3b^2 + 9a^2b^4$

EXAMPLE 9 Factor: $5(x + 3) + y(x + 3)$

Solution: The binomial $(x + 3)$ is present in both terms and is the greatest common factor. We use the distributive property to factor out $(x + 3)$.

$$5(x + 3) + y(x + 3) = (x + 3)(5 + y)$$

Work Practice Problem 9

PRACTICE PROBLEM 9

Factor: $7(p + 2) + q(p + 2)$

Answers

4. a. $5(2y + 5)$ b. $x^4(1 - x^5)$
5. $2x(-5x^2 + 4x - 1)$
6. $4x(x^2 + 3)$ 7. $\frac{1}{5}a^2(2a^3 - 4a + 1)$
8. $3a^2b(2a + ab + 3b^3)$
9. $(p + 2)(7 + q)$

Objective D Factoring by Grouping

Once the GCF is factored out, we can often continue to factor the polynomial, using a variety of techniques. We discuss here a technique called **factoring by grouping.** This technique can be used to factor some polynomials with four terms.

PRACTICE PROBLEM 10

Factor $ab + 7a + 2b + 14$ by grouping.

Helpful Hint

Notice that this form, $x(y + 2) + 3(y + 2)$, is *not* a factored form of the original polynomial. It is a sum, not a product.

EXAMPLE 10 Factor $xy + 2x + 3y + 6$ by grouping.

Solution: Notice that the first two terms of this polynomial have a common factor of x and the second two terms have a common factor of 3. Because of this, group the first two terms, then the last two terms, and then factor out these common factors.

$$xy + 2x + 3y + 6 = (xy + 2x) + (3y + 6) \quad \text{Group terms.}$$
$$= x(y + 2) + 3(y + 2) \quad \text{Factor out GCF from each grouping.}$$

Next we factor out the common binomial factor, $(y + 2)$.

$$x(y + 2) + 3(y + 2) = (y + 2)(x + 3)$$

Now the result is a factored form because it is a product. We were able to write the polynomial as a product because of the common binomial factor, $(y + 2)$, that appeared. If this does not happen, try rearranging the terms of the original polynomial.

Check: Multiply $(y + 2)$ by $(x + 3)$.

$$(y + 2)(x + 3) = xy + 2x + 3y + 6,$$

the original polynomial.
Thus, the factored form of $xy + 2x + 3y + 6$ is the product $(y + 2)(x + 3)$.

Work Practice Problem 10

You may want to try these steps when factoring by grouping.

To Factor by Grouping

Step 1: Group the terms in two groups so that each group has a common factor.

Step 2: Factor out the GCF from each group.

Step 3: If there is a common binomial factor, factor it out.

Step 4: If not, rearrange the terms and try these steps again.

PRACTICE PROBLEMS 11–13

Factor by grouping.

11. $28x^3 - 7x^2 + 12x - 3$

12. $2xy + 5y^2 - 4x - 10y$

13. $3x^2 + 4xy + 3x + 4y$

EXAMPLES Factor by grouping.

11. $15x^3 - 10x^2 + 6x - 4$

$$= (15x^3 - 10x^2) + (6x - 4) \quad \text{Group the terms.}$$
$$= 5x^2(3x - 2) + 2(3x - 2) \quad \text{Factor each group.}$$
$$= (3x - 2)(5x^2 + 2) \quad \text{Factor out the common factor, } (3x - 2).$$

12. $3x^2 + 4xy - 3x - 4y$

$$= (3x^2 + 4xy) + (-3x - 4y)$$
$$= x(3x + 4y) - 1(3x + 4y) \quad \text{Factor each group. A } -1 \text{ is factored from the second pair of terms so that there is a common factor, } (3x + 4y).$$
$$= (3x + 4y)(x - 1) \quad \text{Factor out the common factor, } (3x + 4y).$$

Answers

10. $(b + 7)(a + 2)$
11. $(4x - 1)(7x^2 + 3)$
12. $(2x + 5y)(y - 2)$
13. $(3x + 4y)(x + 1)$

13. $2a^2 + 5ab + 2a + 5b$

$= (2a^2 + 5ab) + (2a + 5b)$ Factor each group. An understood 1 is written before $(2a + 5b)$ to help remember that $(2a + 5b)$ is $1(2a + 5b)$.

$= a(2a + 5b) + 1(2a + 5b)$

$= (2a + 5b)(a + 1)$ Factor out the common factor, $(2a + 5b)$.

Helpful Hint

Notice the factor of 1 is written when $(2a + 5b)$ is factored out.

Work Practice Problems 11–13

EXAMPLES Factor by grouping.

14. $3x^3 - 2x - 9x^2 + 6$

$= x(3x^2 - 2) - 3(3x^2 - 2)$ Factor each group. A -3 is factored from the second pair of terms so that there is a common factor, $(3x^2 - 2)$.

$= (3x^2 - 2)(x - 3)$ Factor out the common factor, $(3x^2 - 2)$.

15. $3xy + 2 - 3x - 2y$

Notice that the first two terms have no common factor other than 1. However, if we rearrange these terms, a grouping emerges that does lead to a common factor.

$3xy + 2 - 3x - 2y$

$= (3xy - 3x) + (-2y + 2)$

$= 3x(y - 1) - 2(y - 1)$ Factor -2 from the second group.

$= (y - 1)(3x - 2)$ Factor out the common factor, $(y - 1)$.

16. $5x - 10 + x^3 - x^2 = 5(x - 2) + x^2(x - 1)$

There is no common binomial factor that can now be factored out. No matter how we rearrange the terms, no grouping will lead to a common factor. Thus, this polynomial is not factorable by grouping.

Work Practice Problems 14–16

PRACTICE PROBLEMS 14–16

Factor by grouping.

14. $4x^3 + x - 20x^2 - 5$

15. $3xy - 4 + x - 12y$

16. $2x - 2 + x^3 - 3x^2$

Answers

14. $(4x^2 + 1)(x - 5)$
15. $(3y + 1)(x - 4)$
16. cannot be factored by grouping

Helpful Hint

Throughout this chapter, we will be factoring polynomials. Even when the instructions do not so state, it is always a good idea to check your answers by multiplying.

Vocabulary and Readiness Check

Use the choices below to fill in each blank. Some choices may be used more than once and some may not be used at all.

greatest common factor   factors   factoring   true   false   least   greatest

1. Since $5 \cdot 4 = 20$, the numbers 5 and 4 are called ________ of 20.
2. The ____________________ of a list of integers is the largest integer that is a factor of all the integers in the list.
3. The greatest common factor of a list of common variables raised to powers is the variable raised to the ________ exponent in the list.
4. The process of writing a polynomial as a product is called ___________.
5. True or false: A factored form of $7x + 21 + xy + 3y$ is $7(x + 3) + y(x + 3)$. ________
6. True or false: A factored form of $3x^3 + 6x + x^2 + 2$ is $3x(x^2 + 2)$. ________

11.1 EXERCISE SET

FOR EXTRA HELP

Student Solutions Manual

PH Math/Tutor Center

CD/Video for Review

MathXL®

MyMathLab

Objectives **A** **B** **Mixed Practice** *Find the GCF for each list. See Examples 1 through 3.*

1. 32, 36

2. 36, 90

3. 18, 42, 84

4. 30, 75, 135

5. 24, 14, 21

6. 15, 25, 27

7. y^2, y^4, y^7

8. x^3, x^2, x^5

9. z^7, z^9, z^{11}

10. y^8, y^{10}, y^{12}

11. $x^{10}y^2, xy^2, x^3y^3$

12. p^7q, p^8q^2, p^9q^3

13. $14x, 21$

14. $20y, 15$

15. $12y^4, 20y^3$

16. $32x^5, 18x^2$

17. $-10x^2, 15x^3$

18. $-21x^3, 14x$

19. $12x^3, -6x^4, 3x^5$

20. $15y^2, 5y^7, -20y^3$

21. $-18x^2y, 9x^3y^3, 36x^3y$

22. $7x^3y^3, -21x^2y^2, 14xy^4$

23. $20a^6b^2c^8, 50a^7b$

24. $40x^7y^2z, 64x^9y$

Objective **C** *Factor out the GCF from each polynomial. See Examples 4 through 9.*

25. $3a + 6$

26. $18a + 12$

27. $30x - 15$

28. $42x - 7$

29. $x^3 + 5x^2$

30. $y^5 + 6y^4$

31. $6y^4 + 2y^3$

32. $5x^2 + 10x^6$

33. $32xy - 18x^2$

34. $10xy - 15x^2$

35. $4x - 8y + 4$

36. $7x + 21y - 7$

37. $6x^3 - 9x^2 + 12x$

38. $12x^3 + 16x^2 - 8x$

39. $a^7b^6 - a^3b^2 + a^2b^5 - a^2b^2$

40. $x^9y^6 + x^3y^5 - x^4y^3 + x^3y^3$

41. $5x^3y - 15x^2y + 10xy$

42. $14x^3y + 7x^2y - 7xy$

43. $8x^5 + 16x^4 - 20x^3 + 12$

44. $9y^6 - 27y^4 + 18y^2 + 6$

45. $\frac{1}{3}x^4 + \frac{2}{3}x^3 - \frac{4}{3}x^5 + \frac{1}{3}x$

46. $\frac{2}{5}y^7 - \frac{4}{5}y^5 + \frac{3}{5}y^2 - \frac{2}{5}y$

47. $y(x^2 + 2) + 3(x^2 + 2)$

48. $x(y^2 + 1) - 3(y^2 + 1)$

49. $z(y + 4) + 3(y + 4)$

50. $8(x + 2) - y(x + 2)$

51. $r(z^2 - 6) + (z^2 - 6)$

52. $q(b^3 - 5) + (b^3 - 5)$

Factor a −1 from each polynomial. See Example 5.

53. $-x - 7$

54. $-y - 3$

55. $-2 + z$

56. $-5 + y$

57. $3a - b + 2$

58. $2y - z - 11$

Objective **D** *Factor each four-term polynomial by grouping. See Examples 10 through 16.*

59. $x^3 + 2x^2 + 5x + 10$

60. $x^3 + 4x^2 + 3x + 12$

61. $5x + 15 + xy + 3y$

62. $xy + y + 2x + 2$

63. $6x^3 - 4x^2 + 15x - 10$

64. $16x^3 - 28x^2 + 12x - 21$

65. $5m^3 + 6mn + 5m^2 + 6n$

66. $8w^2 + 7wv + 8w + 7v$

67. $2y - 8 + xy - 4x$

68. $6x - 42 + xy - 7y$

69. $2x^3 + x^2 + 8x + 4$

70. $2x^3 - x^2 - 10x + 5$

71. $4x^2 - 8xy - 3x + 6y$

72. $5xy - 15x - 6y + 18$

73. $5q^2 - 4pq - 5q + 4p$

74. $6m^2 - 5mn - 6m + 5n$

Factor out the GCF from each polynomial. Then factor by grouping.

75. $12x^2y - 42x^2 - 4y + 14$

76. $90 + 15y^2 - 18x - 3xy^2$

Review

Multiply. See Section 10.5.

77. $(x + 2)(x + 5)$

78. $(y + 3)(y + 6)$

79. $(b + 1)(b - 4)$

80. $(x - 5)(x + 10)$

Fill in the chart by finding two numbers that have the given product and sum. The first column is filled in for you.

		81.	82.	83.	84.	85.	86.	87.	88.
Two Numbers	4, 7								
Their Product	28	12	20	8	16	−10	−9	−24	−36
Their Sum	11	8	9	−9	−10	3	0	−5	−5

Concept Extensions

See the Concept Checks in this section.

89. Which of the following is/are factored form(s) of $8a - 24$?

a. $8 \cdot a - 24$ **b.** $8(a - 3)$ **c.** $4(2a - 12)$ **d.** $8 \cdot a - 2 \cdot 12$

Which of the following expressions are factored?

90. $(a + 6)(a + 2)$

91. $(x + 5)(x + y)$

92. $5(2y + z) - b(2y + z)$

93. $3x(a + 2b) + 2(a + 2b)$

94. The polynomial $-24x^2 + 336x - 132$ represents the average number of visitors (in thousands) per day to National Park Service areas, where x represents the month of the year. (*Source:* Based on data from National Park Service)

a. Find the average daily number of visitors to National Park Service areas during the month of August. To do so, let $x = 8$ and evaluate $-24x^2 + 336x - 132$.

b. Find the average daily number of visitors in May.

c. Factor the polynomial $-24x^2 + 336x - 132$.

95. The average total daily supply of motor gasoline (in thousands of barrels per day) in the United States for the period 2000–2003 can be approximated by the polynomial $-13x^2 + 221x + 8476$, where x is the number of years after 2000. (*Source:* Based on data from Energy Information Administration)

a. Find the average daily total supply of motor gasoline in 2001. To do so, let $x = 1$ and evaluate $-13x^2 + 221x + 8476$.

b. Find the average daily total supply of motor gasoline in 2003.

c. Factor the polynomial $-13x^2 + 221x + 8476$.

Write an expression for the area of each shaded region. Then write the expression as a factored polynomial.

96.

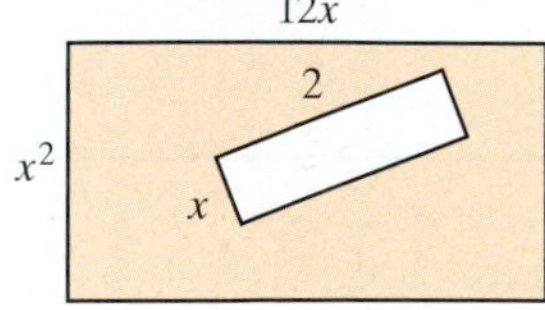

△ **97.**

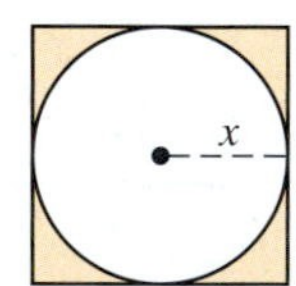

Write an expression for the length of each rectangle. (Hint: Factor the area binomial and recall that Area = width · length.)

△ **98.**

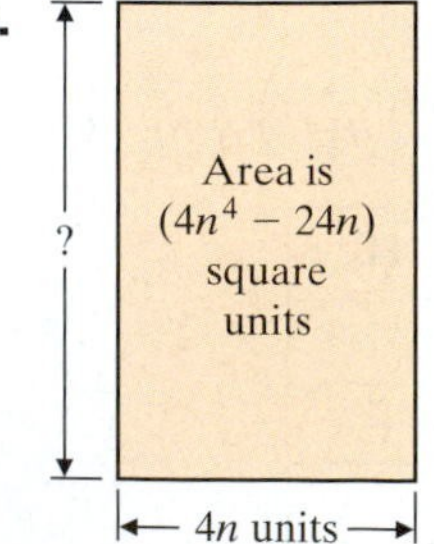

△ **99.**

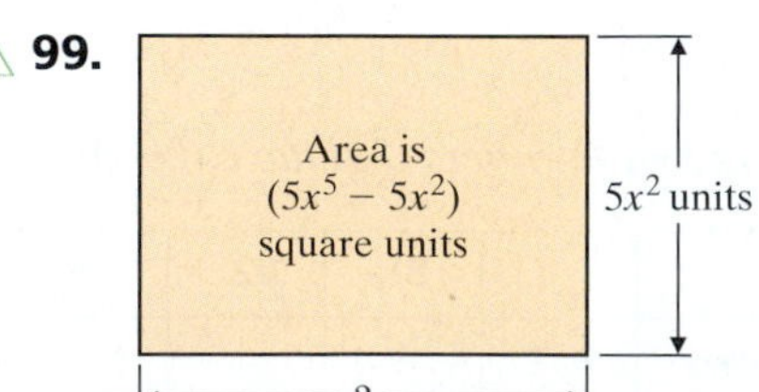

100. Construct a binomial whose greatest common factor is $5a^3$. (*Hint:* Multiply $5a^3$ by a binomial whose terms contain no common factor other than 1. $5a^3(\square + \square)$.)

101. Construct a trinomial whose greatest common factor is $2x^2$. See the hint for Exercise 100.

102. Explain how you can tell whether a polynomial is written in factored form.

103. Construct a four-term polynomial that can be factored by grouping.

STUDY SKILLS BUILDER

Are You Getting All the Mathematics Help That You Need?

Remember that, in addition to your instructor, there are many places to get help with your mathematics course. For example.

- This text has an accompanying video lesson for every section and worked out solutions to every Chapter Test exercise on video.
- The back of the book contains answers to odd-numbered exercises and selected solutions.
- A student *Solutions Manual* is available that contains worked-out solutions to odd-numbered exercises as well as solutions to every exercise in the Integrated Reviews, Chapter Reviews, Chapter Tests, and Cumulative Reviews.
- Don't forget to check with your instructor for other local resources available to you, such as a tutor center.

Exercises

1. List items you find helpful in the text and all student supplements to this text.
2. List all the campus help that is available to you for this course.
3. List any help (besides the textbook) from Exercises 1 and 2 above that you are using.
4. List any help (besides the textbook) that you feel you should try.
5. Write a goal for yourself that includes trying anything you listed in Exercise 4 during the next week.

Objectives

A Factor Trinomials of the Form $x^2 + bx + c$.

B Factor Out the Greatest Common Factor and Then Factor a Trinomial of the Form $x^2 + bx + c$.

11.2 FACTORING TRINOMIALS OF THE FORM $x^2 + bx + c$

Objective A Factoring Trinomials of the Form $x^2 + bx + c$

In this section, we factor trinomials of the form $x^2 + bx + c$, such as

$$x^2 + 7x + 12, \quad x^2 - 12x + 35, \quad x^2 + 4x - 12, \quad \text{and} \quad r^2 - r - 42$$

Notice that for these trinomials, the coefficient of the squared variable is 1.

Recall that factoring means to write as a product and that factoring and multiplying are reverse processes. Using the FOIL method of multiplying binomials, we have the following.

$$\begin{aligned}(x + 3)(x + 1) &= \overset{F}{x^2} + \overset{O}{1x} + \overset{I}{3x} + \overset{L}{3} \\ &= x^2 + 4x + 3\end{aligned}$$

Thus, a factored form of $x^2 + 4x + 3$ is $(x + 3)(x + 1)$.

Notice that the product of the first terms of the binomials is $x \cdot x = x^2$, the first term of the trinomial. Also, the product of the last two terms of the binomials is $3 \cdot 1 = 3$, the third term of the trinomial. The sum of these same terms is $3 + 1 = 4$, the coefficient of the middle, x, term of the trinomial.

The product of these numbers is 3.

$$x^2 + 4x + 3 = (x + 3)(x + 1)$$

The sum of these numbers is 4.

Many trinomials, such as the one above, factor into two binomials. To factor $x^2 + 7x + 10$, let's assume that it factors into two binomials and begin by writing two pairs of parentheses. The first term of the trinomial is x^2, so we use x and x as the first terms of the binomial factors.

$$x^2 + 7x + 10 = (x + \square)(x + \square)$$

To determine the last term of each binomial factor, we look for two integers whose product is 10 and whose sum is 7. The integers are 2 and 5. Thus,

$$x^2 + 7x + 10 = (x + 2)(x + 5)$$

Check: To see if we have factored correctly, we multiply.

$$\begin{aligned}(x + 2)(x + 5) &= x^2 + 5x + 2x + 10 \\ &= x^2 + 7x + 10 \qquad \text{Combine like terms.}\end{aligned}$$

Helpful Hint

Since multiplication is commutative, the factored form of $x^2 + 7x + 10$ can be written as either $(x + 2)(x + 5)$ or $(x + 5)(x + 2)$.

To Factor a Trinomial of the Form $x^2 + bx + c$

The product of these numbers is c.

$$x^2 + bx + c = (x + \square)(x + \square)$$

The sum of these numbers is b.

EXAMPLE 1 Factor: $x^2 + 7x + 12$

Solution: We begin by writing the first terms of the binomial factors.

$(x + \square)(x + \square)$

Next we look for two numbers whose product is 12 and whose sum is 7. Since our numbers must have a positive product and a positive sum, we look at pairs of positive factors of 12 only.

Factors of 12	Sum of Factors
1, 12	13
2, 6	8
3, 4	7

Correct sum, so the numbers are 3 and 4.

Thus, $x^2 + 7x + 12 = (x + 3)(x + 4)$

Check: $(x + 3)(x + 4) = x^2 + 4x + 3x + 12 = x^2 + 7x + 12$.

Work Practice Problem 1

PRACTICE PROBLEM 1

Factor: $x^2 + 12x + 20$

EXAMPLE 2 Factor: $x^2 - 12x + 35$

Solution: Again, we begin by writing the first terms of the binomials.

$(x + \square)(x + \square)$

Now we look for two numbers whose product is 35 and whose sum is -12. Since our numbers must have a positive product and a negative sum, we look at pairs of negative factors of 35 only.

Factors of 35	Sum of Factors
−1, −35	−36
−5, −7	−12

Correct sum, so the numbers are −5 and −7.

$x^2 - 12x + 35 = (x - 5)(x - 7)$

Check: To check, multiply $(x - 5)(x - 7)$.

Work Practice Problem 2

PRACTICE PROBLEM 2

Factor each trinomial.

a. $x^2 - 23x + 22$

b. $x^2 - 27x + 50$

EXAMPLE 3 Factor: $x^2 + 4x - 12$

Solution: $x^2 + 4x - 12 = (x + \square)(x + \square)$

We look for two numbers whose product is -12 and whose sum is 4. Since our numbers must have a negative product, we look at pairs of factors with opposite signs.

Factors of −12	Sum of Factors
−1, 12	11
1, −12	−11
−2, 6	4
2, −6	−4
−3, 4	1
3, −4	−1

Correct sum, so the numbers are −2 and 6.

$x^2 + 4x - 12 = (x - 2)(x + 6)$

Work Practice Problem 3

PRACTICE PROBLEM 3

Factor: $x^2 + 5x - 36$

Answers

1. $(x + 10)(x + 2)$
2. a. $(x - 1)(x - 22)$ b. $(x - 2)(x - 25)$
3. $(x + 9)(x - 4)$

PRACTICE PROBLEM 4

Factor each trinomial.

a. $q^2 - 3q - 40$
b. $y^2 + 2y - 48$

EXAMPLE 4 Factor: $r^2 - r - 42$

Solution: Because the variable in this trinomial is r, the first term of each binomial factor is r.

$r^2 - r - 42 = (r + \square)(r + \square)$

Now we look for two numbers whose product is -42 and whose sum is -1, the numerical coefficient of r. The numbers are 6 and -7. Therefore,

$r^2 - r - 42 = (r + 6)(r - 7)$

Work Practice Problem 4

PRACTICE PROBLEM 5

Factor: $x^2 + 6x + 15$

EXAMPLE 5 Factor: $a^2 + 2a + 10$

Solution: Look for two numbers whose product is 10 and whose sum is 2. Neither 1 and 10 nor 2 and 5 give the required sum, 2. We conclude that $a^2 + 2a + 10$ is not factorable with integers. A polynomial such as $a^2 + 2a + 10$ is called a **prime polynomial.**

Work Practice Problem 5

PRACTICE PROBLEM 6

Factor each trinomial.

a. $x^2 + 9xy + 14y^2$
b. $a^2 - 13ab + 30b^2$

EXAMPLE 6 Factor: $x^2 + 5xy + 6y^2$

Solution: $x^2 + 5xy + 6y^2 = (x + \square)(x + \square)$

Recall that the middle term $5xy$ is the same as $5yx$. Thus, we can see that $5y$ is the "coefficient" of x. We then look for two terms whose product is $6y^2$ and whose sum is $5y$. The terms are $2y$ and $3y$ because $2y \cdot 3y = 6y^2$ and $2y + 3y = 5y$. Therefore,

$x^2 + 5xy + 6y^2 = (x + 2y)(x + 3y)$

Work Practice Problem 6

PRACTICE PROBLEM 7

Factor: $x^4 + 8x^2 + 12$

EXAMPLE 7 Factor: $x^4 + 5x^2 + 6$

Solution: As usual, we begin by writing the first terms of the binomials. Since the greatest power of x in this polynomial is x^4, we write

$(x^2 + \square)(x^2 + \square)$ since $x^2 \cdot x^2 = x^4$

Now we look for two factors of 6 whose sum is 5. The numbers are 2 and 3. Thus,

$x^4 + 5x^2 + 6 = (x^2 + 2)(x^2 + 3)$

Work Practice Problem 7

If the terms of a polynomial are not written in descending powers of the variable, you may want to do so before factoring.

PRACTICE PROBLEM 8

Factor: $48 - 14x + x^2$

EXAMPLE 8 Factor: $40 - 13t + t^2$

Solution: First, we rearrange terms so that the trinomial is written in descending powers of t.

$40 - 13t + t^2 = t^2 - 13t + 40$

Next, try to factor.

$t^2 - 13t + 40 = (t + \square)(t + \square)$

Now we look for two factors of 40 whose sum is -13. The numbers are -8 and -5. Thus,

$t^2 - 13t + 40 = (t - 8)(t - 5)$

Work Practice Problem 8

Answers

4. a. $(q - 8)(q + 5)$
b. $(y + 8)(y - 6)$
5. prime polynomial
6. a. $(x + 2y)(x + 7y)$
b. $(a - 3b)(a - 10b)$
7. $(x^2 + 6)(x^2 + 2)$
8. $(x - 6)(x - 8)$

The following sign patterns may be useful when factoring trinomials.

Helpful Hint

A positive constant in a trinomial tells us to look for two numbers with the same sign. The sign of the coefficient of the middle term tells us whether the signs are both positive or both negative.

both positive ↓ same sign ↓

$$x^2 + 10x + 16 = (x + 2)(x + 8)$$

both negative ↓ same sign ↓

$$x^2 - 10x + 16 = (x - 2)(x - 8)$$

A negative constant in a trinomial tells us to look for two numbers with opposite signs.

opposite signs ↓

$$x^2 + 6x - 16 = (x + 8)(x - 2)$$

opposite signs ↓

$$x^2 - 6x - 16 = (x - 8)(x + 2)$$

Objective B Factoring Out the Greatest Common Factor

Remember that the first step in factoring any polynomial is to factor out the greatest common factor (if there is one other than 1 or -1).

EXAMPLE 9 Factor: $3m^2 - 24m - 60$

Solution: First we factor out the greatest common factor, 3, from each term.

$$3m^2 - 24m - 60 = 3(m^2 - 8m - 20)$$

Now we factor $m^2 - 8m - 20$ by looking for two factors of -20 whose sum is -8. The factors are -10 and 2. Therefore, the complete factored form is

$$3m^2 - 24m - 60 = 3(m + 2)(m - 10)$$

Work Practice Problem 9

Remember to write the common factor 3 as part of the factored form.

EXAMPLE 10 Factor: $2x^4 - 26x^3 + 84x^2$

Solution:

$$2x^4 - 26x^3 + 84x^2 = 2x^2(x^2 - 13x + 42) \quad \text{Factor out common factor, } 2x^2.$$

$$= 2x^2(x - 6)(x - 7) \quad \text{Factor } x^2 - 13x + 42.$$

Work Practice Problem 10

PRACTICE PROBLEM 9

Factor each trinomial.

a. $4x^2 - 24x + 36$

b. $x^3 + 3x^2 - 4x$

PRACTICE PROBLEM 10

Factor: $5x^5 - 25x^4 - 30x^3$

Answers

9. a. $4(x - 3)(x - 3)$
b. $x(x + 4)(x - 1)$
10. $5x^3(x + 1)(x - 6)$

Vocabulary and Readiness Check

Fill in each blank with "true" or "false."

1. To factor $x^2 + 7x + 6$, we look for two numbers whose product is 6 and whose sum is 7. ________
2. We can write the factorization $(y + 2)(y + 4)$ also as $(y + 4)(y + 2)$. ________
3. The factorization $(4x - 12)(x - 5)$ is completely factored. ________
4. The factorization $(x + 2y)(x + y)$ may also be written as $(x + 2y)^2$. ________

11.2 EXERCISE SET

FOR EXTRA HELP

Student Solutions Manual

PH Math/Tutor Center

CD/Video for Review

MathXL®

MyMathLab

Objective A *Factor each trinomial completely. If a polynomial can't be factored, write "prime." See Examples 1 through 8.*

1. $x^2 + 7x + 6$

2. $x^2 + 6x + 8$

3. $y^2 - 10y + 9$

4. $y^2 - 12y + 11$

5. $x^2 - 6x + 9$

6. $x^2 - 10x + 25$

7. $x^2 - 3x - 18$

8. $x^2 - x - 30$

9. $x^2 + 3x - 70$

10. $x^2 + 4x - 32$

11. $x^2 + 5x + 2$

12. $x^2 - 7x + 5$

13. $x^2 + 8xy + 15y^2$

14. $x^2 + 6xy + 8y^2$

15. $a^4 - 2a^2 - 15$

16. $y^4 - 3y^2 - 70$

17. $13 + 14m + m^2$

18. $17 + 18n + n^2$

19. $10t - 24 + t^2$

20. $6q - 27 + q^2$

21. $a^2 - 10ab + 16b^2$

22. $a^2 - 9ab + 18b^2$

Objectives A B Mixed Practice *Factor each trinomial completely. Some of these trinomials contain a greatest common factor (other than 1). Don't forget to factor out the GCF first. See Examples 1 through 10.*

23. $2z^2 + 20z + 32$

24. $3x^2 + 30x + 63$

25. $2x^3 - 18x^2 + 40x$

26. $3x^3 - 12x^2 - 36x$

27. $x^2 - 3xy - 4y^2$

28. $x^2 - 4xy - 77y^2$

29. $x^2 + 15x + 36$

30. $x^2 + 19x + 60$

31. $x^2 - x - 2$

32. $x^2 - 5x - 14$

33. $r^2 - 16r + 48$

34. $r^2 - 10r + 21$

35. $x^2 + xy - 2y^2$

36. $x^2 - xy - 6y^2$

37. $3x^2 + 9x - 30$

38. $4x^2 - 4x - 48$

39. $3x^2 - 60x + 108$

40. $2x^2 - 24x + 70$

41. $x^2 - 18x - 144$

42. $x^2 + x - 42$

43. $r^2 - 3r + 6$

44. $x^2 + 4x - 10$

45. $x^2 - 8x + 15$

46. $x^2 - 9x + 14$

47. $6x^3 + 54x^2 + 120x$

48. $3x^3 + 3x^2 - 126x$

49. $4x^2y + 4xy - 12y$

50. $3x^2y - 9xy + 45y$

51. $x^2 - 4x - 21$

52. $x^2 - 4x - 32$

53. $x^2 + 7xy + 10y^2$

54. $x^2 - 3xy - 4y^2$

55. $64 + 24t + 2t^2$

56. $50 + 20t + 2t^2$

57. $x^3 - 2x^2 - 24x$

58. $x^3 - 3x^2 - 28x$

59. $2t^5 - 14t^4 + 24t^3$

60. $3x^6 + 30x^5 + 72x^4$

61. $5x^3y - 25x^2y^2 - 120xy^3$

62. $7a^3b - 35a^2b^2 + 42ab^3$

63. $162 - 45m + 3m^2$

64. $48 - 20n + 2n^2$

65. $-x^2 + 12x - 11$ (Factor out -1 first.)

66. $-x^2 + 8x - 7$ (Factor out -1 first.)

67. $\frac{1}{2}y^2 - \frac{9}{2}y - 11$ (Factor out $\frac{1}{2}$ first.)

68. $\frac{1}{3}y^2 - \frac{5}{3}y - 8$ (Factor out $\frac{1}{3}$ first.)

69. $x^3y^2 + x^2y - 20x$

70. $a^2b^3 + ab^2 - 30b$

Review

Multiply. See Section 10.5.

71. $(2x + 1)(x + 5)$

72. $(3x + 2)(x + 4)$

73. $(5y - 4)(3y - 1)$

74. $(4z - 7)(7z - 1)$

75. $(a + 3b)(9a - 4b)$

76. $(y - 5x)(6y + 5x)$

Concept Extensions

77. Write a polynomial that factors as $(x - 3)(x + 8)$.

78. To factor $x^2 + 13x + 42$, think of two numbers whose _______ is 42 and whose _______ is 13.

Complete each sentence in your own words.

79. If $x^2 + bx + c$ is factorable and c is negative, then the signs of the last-term factors of the binomials are opposite because . . .

80. If $x^2 + bx + c$ is factorable and c is positive, then the signs of the last-term factors of the binomials are the same because . . .

Remember that perimeter means distance around. Write the perimeter of each rectangle as a simplified polynomial. Then factor the polynomial.

△ **81.**

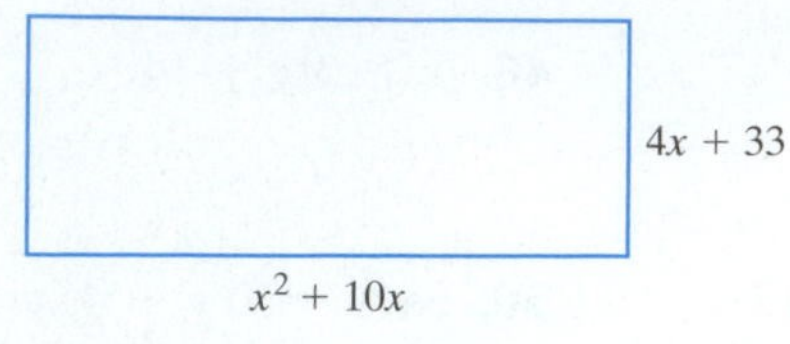

△ **82.**

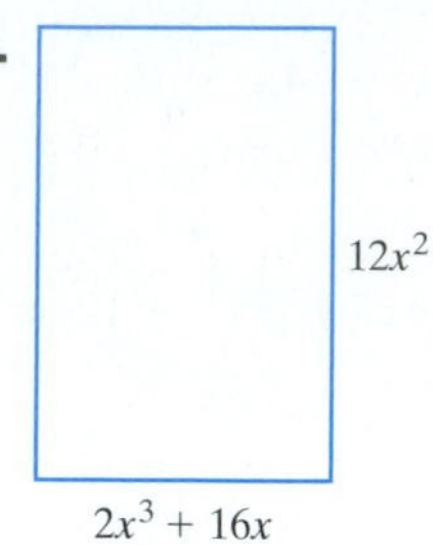

83. An object is thrown upward from the top of an 80-foot building with an initial velocity of 64 feet per second. The height of the object after t seconds is given by $-16t^2 + 64t + 80$. Factor this polynomial.

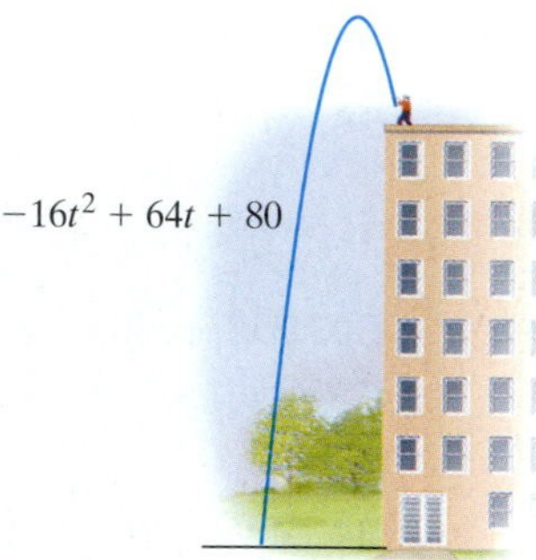

Factor each trinomial completely.

84. $x^2 + x + \frac{1}{4}$

85. $x^2 + \frac{1}{2}x + \frac{1}{16}$

86. $y^2(x + 1) - 2y(x + 1) - 15(x + 1)$

87. $z^2(x + 1) - 3z(x + 1) - 70(x + 1)$

Find a positive value of c so that each trinomial is factorable.

88. $y^2 - 4y + c$

89. $n^2 - 16n + c$

Find a positive value of b so that each trinomial is factorable.

90. $x^2 + bx + 15$

91. $y^2 + by + 20$

Factor each trinomial. (Hint: *Notice that* $x^{2n} + 4x^n + 3$ *factors as* $(x^n + 1)(x^n + 3)$. Remember: $x^n \cdot x^n = x^{n+n}$ *or* x^{2n}.)

92. $x^{2n} + 5x^n + 6$

93. $x^{2n} + 8x^n - 20$

11.3 FACTORING TRINOMIALS OF THE FORM $ax^2 + bx + c$

Objectives

A Factor Trinomials of the Form $ax^2 + bx + c$, where $a \neq 1$.

B Factor Out the GCF before Factoring a Trinomial of the Form $ax^2 + bx + c$.

Objective A Factoring Trinomials of the Form $ax^2 + bx + c$

In this section, we factor trinomials of the form $ax^2 + bx + c$, such as

$$3x^2 + 11x + 6, \quad 8x^2 - 22x + 5, \quad \text{and} \quad 2x^2 + 13x - 7$$

Notice that the coefficient of the squared variable in these trinomials is a number other than 1. We will factor these trinomials using a trial-and-check method based on our work in the last section.

To begin, let's review the relationship between the numerical coefficients of the trinomial and the numerical coefficients of its factored form. For example, since $(2x + 1)(x + 6) = 2x^2 + 13x + 6$,

a factored form of $2x^2 + 13x + 6$ is $(2x + 1)(x + 6)$

Notice that $2x$ and x are factors of $2x^2$, the first term of the trinomial. Also, 6 and 1 are factors of 6, the last term of the trinomial, as shown:

$$2x^2 + 13x + 6 = (2x + 1)(x + 6)$$

$2x \cdot x$

$1 \cdot 6$

Also notice that $13x$, the middle term, is the sum of the following products:

$$2x^2 + 13x + 6 = (2x + 1)(x + 6)$$

$$\begin{array}{r} 1x \\ +12x \\ \hline 13x \end{array}$$ Middle term

Let's use this pattern to factor $5x^2 + 7x + 2$. First, we find factors of $5x^2$. Since all numerical coefficients in this trinomial are positive, we will use factors with positive numerical coefficients only. Thus, the factors of $5x^2$ are $5x$ and x. Let's try these factors as first terms of the binomials. Thus far, we have

$$5x^2 + 7x + 2 = (5x + \square)(x + \square)$$

Next, we need to find positive factors of 2. Positive factors of 2 are 1 and 2. Now we try possible combinations of these factors as second terms of the binomials until we obtain a middle term of $7x$.

$$(5x + 1)(x + 2) = 5x^2 + 11x + 2$$

$$\begin{array}{r} 1x \\ +10x \\ \hline 11x \end{array}$$ → **Incorrect** middle term

Let's try switching factors 2 and 1.

$$(5x + 2)(x + 1) = 5x^2 + 7x + 2$$

$$\begin{array}{r} 2x \\ +5x \\ \hline 7x \end{array}$$ → **Correct** middle term

Thus a factored form of $5x^2 + 7x + 2$ is $(5x + 2)(x + 1)$. To check, we multiply $(5x + 2)$ and $(x + 1)$. The product is $5x^2 + 7x + 2$.

PRACTICE PROBLEM 1

Factor each trinomial.

a. $5x^2 + 27x + 10$

b. $4x^2 + 12x + 5$

EXAMPLE 1 Factor: $3x^2 + 11x + 6$

Solution: Since all numerical coefficients are positive, we use factors with positive numerical coefficients. We first find factors of $3x^2$.

Factors of $3x^2$: $3x^2 = 3x \cdot x$

If factorable, the trinomial will be of the form

$$3x^2 + 11x + 6 = (3x + \square)(x + \square)$$

Next we factor 6.

Factors of 6: $6 = 1 \cdot 6$, $\quad 6 = 2 \cdot 3$

Now we try combinations of factors of 6 until a middle term of $11x$ is obtained. Let's try 1 and 6 first.

$$(3x + 1)(x + 6) = 3x^2 + 19x + 6$$

$1x$
$+18x$
$19x$ ⟶ **Incorrect** middle term

Now let's next try 6 and 1.

$$(3x + 6)(x + 1)$$

Before multiplying, notice that the terms of the factor $3x + 6$ have a common factor of 3. The terms of the original trinomial $3x^2 + 11x + 6$ have no common factor other than 1, so the terms of its factors will also contain no common factor other than 1. This means that $(3x + 6)(x + 1)$ is not a factored form.

Next let's try 2 and 3 as last terms.

$$(3x + 2)(x + 3) = 3x^2 + 11x + 6$$

$2x$
$+9x$
$11x$ ⟶ **Correct** middle term

Thus a factored form of $3x^2 + 11x + 6$ is $(3x + 2)(x + 3)$.

Work Practice Problem 1

Helpful Hint

This is true in general: If the terms of a trinomial have no common factor (other than 1), then the terms of each of its binomial factors will contain no common factor (other than 1).

✔Concept Check Do the terms of $3x^2 + 29x + 18$ have a common factor? Without multiplying, decide which of the following factored forms could not be a factored form of $3x^2 + 29x + 18$.

a. $(3x + 18)(x + 1)$ **b.** $(3x + 2)(x + 9)$
c. $(3x + 6)(x + 3)$ **d.** $(3x + 9)(x + 2)$

PRACTICE PROBLEM 2

Factor each trinomial.

a. $2x^2 - 11x + 12$

b. $6x^2 - 5x + 1$

EXAMPLE 2 Factor: $8x^2 - 22x + 5$

Solution: Factors of $8x^2$: $8x^2 = 8x \cdot x$, $\quad 8x^2 = 4x \cdot 2x$

We'll try $8x$ and x.

$$8x^2 - 22x + 5 = (8x + \square)(x + \square)$$

Since the middle term, $-22x$, has a negative numerical coefficient, we factor 5 into negative factors.

Factors of 5: $5 = -1 \cdot -5$

Answers

1. a. $(5x + 2)(x + 5)$ **b.** $(2x + 5)(2x + 1)$
2. a. $(2x - 3)(x - 4)$ **b.** $(3x - 1)(2x - 1)$

✔ Concept Check Answer

no; a, c, d

Let's try −1 and −5.

$$(8x - 1)(x - 5) = 8x^2 - 41x + 5$$

$$\begin{array}{r} -1x \\ +(-40x) \\ \hline -41x \end{array} \longrightarrow \text{Incorrect middle term}$$

Now let's try −5 and −1.

$$(8x - 5)(x - 1) = 8x^2 - 13x + 5$$

$$\begin{array}{r} -5x \\ +(-8x) \\ \hline -13x \end{array} \longrightarrow \text{Incorrect middle term}$$

Don't give up yet! We can still try other factors of $8x^2$. Let's try $4x$ and $2x$ with −1 and −5.

$$(4x - 1)(2x - 5) = 8x^2 - 22x + 5$$

$$\begin{array}{r} -2x \\ +(-20x) \\ \hline -22x \end{array} \longrightarrow \text{Correct middle term}$$

A factored form of $8x^2 - 22x + 5$ is $(4x - 1)(2x - 5)$.

Work Practice Problem 2

EXAMPLE 3 Factor: $2x^2 + 13x - 7$

Solution: Factors of $2x^2$: $2x^2 = 2x \cdot x$

Factors of −7: $-7 = -1 \cdot 7, \quad -7 = 1 \cdot -7$

We try possible combinations of these factors:

$(2x + 1)(x - 7) = 2x^2 - 13x - 7$ **Incorrect** middle term

$(2x - 1)(x + 7) = 2x^2 + 13x - 7$ **Correct** middle term

A factored form of $2x^2 + 13x - 7$ is $(2x - 1)(x + 7)$.

Work Practice Problem 3

EXAMPLE 4 Factor: $10x^2 - 13xy - 3y^2$

Solution: Factors of $10x^2$: $10x^2 = 10x \cdot x, \quad 10x^2 = 2x \cdot 5x$

Factors of $-3y^2$: $-3y^2 = -3y \cdot y, \quad -3y^2 = 3y \cdot -y$

We try some combinations of these factors:

(Correct ↓ first and last terms)

$(10x - 3y)(x + y) = 10x^2 + 7xy - 3y^2$

$(x + 3y)(10x - y) = 10x^2 + 29xy - 3y^2$

$(5x + 3y)(2x - y) = 10x^2 + xy - 3y^2$

$(2x - 3y)(5x + y) = 10x^2 - 13xy - 3y^2$ **Correct** middle term

A factored form of $10x^2 - 13xy - 3y^2$ is $(2x - 3y)(5x + y)$.

Work Practice Problem 4

EXAMPLE 5 Factor: $3x^4 - 5x^2 - 8$

Solution: Factors of $3x^4$: $3x^4 = 3x^2 \cdot x^2$

Factors of −8: $-8 = -2 \cdot 4, 2 \cdot -4, -1 \cdot 8, 1 \cdot -8$

Continued on next page

PRACTICE PROBLEM 3

Factor each trinomial.

a. $3x^2 + 14x - 5$

b. $35x^2 + 4x - 4$

PRACTICE PROBLEM 4

Factor each trinomial.

a. $14x^2 - 3xy - 2y^2$

b. $12a^2 - 16ab - 3b^2$

PRACTICE PROBLEM 5

Factor: $2x^4 - 5x^2 - 7$

Answers

3. a. $(3x - 1)(x + 5)$ b. $(5x + 2)(7x - 2)$
4. a. $(7x + 2y)(2x - y)$ b. $(6a + b)(2a - 3b)$
5. $(2x^2 - 7)(x^2 + 1)$

Try combinations of these factors:

Correct ↓ (middle term) Correct ↓ (last term)

$(3x^2 - 2)(x^2 + 4) = 3x^4 + 10x^2 - 8$

$(3x^2 + 4)(x^2 - 2) = 3x^4 - 2x^2 - 8$

$(3x^2 + 8)(x^2 - 1) = 3x^4 + 5x^2 - 8$ **Incorrect** sign on middle term, so switch signs in binomial factors.

$(3x^2 - 8)(x^2 + 1) = 3x^4 - 5x^2 - 8$ **Correct** middle term.

Work Practice Problem 5

Helpful Hint

Study the last two lines of Example 5. If a factoring attempt gives you a middle term whose numerical coefficient is the opposite of the desired numerical coefficient, try switching the signs of the last terms in the binomials.

Switched signs

$(3x^2 + 8)(x^2 - 1) = 3x^4 + 5x^2 - 8$ Middle term: +5x

$(3x^2 - 8)(x^2 + 1) = 3x^4 - 5x^2 - 8$ Middle term: −5x

Objective B Factoring Out the Greatest Common Factor

Don't forget that the first step in factoring any polynomial is to look for a common factor to factor out.

PRACTICE PROBLEM 6

Factor each trinomial.

a. $3x^3 + 17x^2 + 10x$

b. $6xy^2 + 33xy - 18x$

EXAMPLE 6 Factor: $24x^4 + 40x^3 + 6x^2$

Solution: Notice that all three terms have a common factor of $2x^2$. Thus we factor out $2x^2$ first.

$$24x^4 + 40x^3 + 6x^2 = 2x^2(12x^2 + 20x + 3)$$

Next we factor $12x^2 + 20x + 3$.

Factors of $12x^2$: $12x^2 = 4x \cdot 3x$, $12x^2 = 12x \cdot x$, $12x^2 = 6x \cdot 2x$

Since all terms in the trinomial have positive numerical coefficients, we factor 3 using positive factors only.

Factors of 3: $3 = 1 \cdot 3$

We try some combinations of the factors.

$2x^2(4x + 3)(3x + 1) = 2x^2(12x^2 + 13x + 3)$

$2x^2(12x + 1)(x + 3) = 2x^2(12x^2 + 37x + 3)$

$2x^2(2x + 3)(6x + 1) = 2x^2(12x^2 + 20x + 3)$ **Correct** middle term

A factored form of $24x^4 + 40x^3 + 6x^2$ is $2x^2(2x + 3)(6x + 1)$.

Helpful Hint

Don't forget to include the common factor in the factored form.

Work Practice Problem 6

When the term containing the squared variable has a negative coefficient, you may want to first factor out a common factor of −1.

PRACTICE PROBLEM 7

Factor: $-5x^2 - 19x + 4$

EXAMPLE 7 Factor: $-6x^2 - 13x + 5$

Solution: We begin by factoring out a common factor of −1.

$-6x^2 - 13x + 5 = -1(6x^2 + 13x - 5)$ Factor out −1.

$= -1(3x - 1)(2x + 5)$ Factor $6x^2 + 13x - 5$.

Work Practice Problem 7

Answers

6. a. $x(3x + 2)(x + 5)$ b. $3x(2y - 1)(y + 6)$ 7. $-1(x + 4)(5x - 1)$

11.3 EXERCISE SET

FOR EXTRA HELP

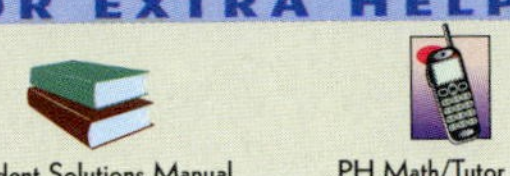
Student Solutions Manual

PH Math/Tutor Center

CD/Video for Review

MathXL®

MyMathLab

Objective A *Complete each factored form. See Examples 1 through 5.*

1. $5x^2 + 22x + 8 = (5x + 2)(\quad)$

2. $2y^2 + 15y + 25 = (2y + 5)(\quad)$

3. $50x^2 + 15x - 2 = (5x + 2)(\quad)$

4. $6y^2 + 11y - 10 = (2y + 5)(\quad)$

5. $20x^2 - 7x - 6 = (5x + 2)(\quad)$

6. $8y^2 - 2y - 55 = (2y + 5)(\quad)$

Factor each trinomial completely. See Examples 1 through 5.

7. $2x^2 + 13x + 15$

8. $3x^2 + 8x + 4$

9. $8y^2 - 17y + 9$

10. $21x^2 - 41x + 10$

11. $2x^2 - 9x - 5$

12. $36r^2 - 5r - 24$

13. $20r^2 + 27r - 8$

14. $3x^2 + 20x - 63$

15. $10x^2 + 17x + 3$

16. $2x^2 + 7x + 5$

17. $x + 3x^2 - 2$

18. $y + 8y^2 - 9$

19. $6x^2 - 13xy + 5y^2$

20. $8x^2 - 14xy + 3y^2$

21. $15m^2 - 16m - 15$

22. $25n^2 - 5n - 6$

23. $-9x + 20 + x^2$

24. $-7x + 12 + x^2$

25. $2x^2 - 7x - 99$

26. $2x^2 + 7x - 72$

27. $-27t + 7t^2 - 4$

28. $-3t + 4t^2 - 7$

29. $3a^2 + 10ab + 3b^2$

30. $2a^2 + 11ab + 5b^2$

31. $49p^2 - 7p - 2$

32. $3r^2 + 10r - 8$

33. $18x^2 - 9x - 14$

34. $42a^2 - 43a + 6$

35. $2m^2 + 17m + 10$

36. $3n^2 + 20n + 5$

37. $24x^2 + 41x + 12$

38. $24x^2 - 49x + 15$

Objectives A B Mixed Practice *Factor each trinomial completely. See Examples 1 through 7.*

39. $12x^3 + 11x^2 + 2x$

40. $8a^3 + 14a^2 + 3a$

41. $21b^2 - 48b - 45$

42. $12x^2 - 14x - 10$

43. $7z + 12z^2 - 12$

44. $16t + 15t^2 - 15$

45. $6x^2y^2 - 2xy^2 - 60y^2$

46. $8x^2y + 34xy - 84y$

47. $4x^2 - 8x - 21$

48. $6x^2 - 11x - 10$

49. $3x^2 - 42x + 63$

50. $5x^2 - 75x + 60$

51. $8x^2 + 6xy - 27y^2$

52. $54a^2 + 39ab - 8b^2$

53. $-x^2 + 2x + 24$

54. $-x^2 + 4x + 21$

55. $4x^3 - 9x^2 - 9x$

56. $6x^3 - 31x^2 + 5x$

57. $24x^2 - 58x + 9$

58. $36x^2 + 55x - 14$

59. $40a^2b + 9ab - 9b$
60. $24y^2x + 7yx - 5x$
61. $30x^3 + 38x^2 + 12x$
62. $6x^3 - 28x^2 + 16x$

63. $6y^3 - 8y^2 - 30y$
64. $12x^3 - 34x^2 + 24x$
65. $10x^4 + 25x^3y - 15x^2y^2$
66. $42x^4 - 99x^3y - 15x^2y^2$

67. $-14x^2 + 39x - 10$
68. $-15x^2 + 26x - 8$
69. $16p^4 - 40p^3 + 25p^2$
70. $9q^4 - 42q^3 + 49q^2$

71. $-2x^2 + 9x + 5$
72. $-3x^2 + 8x + 16$
73. $-4 + 52x - 48x^2$
74. $-5 + 55x - 50x^2$

75. $2t^4 + 3t^2 - 27$
76. $4r^4 - 17r^2 - 15$
77. $5x^2y^2 + 20xy + 1$
78. $3a^2b^2 + 12ab + 1$

79. $6a^5 + 37a^3b^2 + 6ab^4$
80. $5m^5 + 26m^3h^2 + 5mh^4$

Review

Multiply. See Section 10.6.

81. $(x - 4)(x + 4)$
82. $(2x - 9)(2x + 9)$
83. $(x + 2)^2$

84. $(x + 3)^2$
85. $(2x - 1)^2$
86. $(3x - 5)^2$

Concept Extensions

See the Concept Check in this section.

87. Do the terms of $4x^2 + 19x + 12$ have a common factor (other than 1)?

88. Without multiplying, decide which of the following factored forms is not a factored form of $4x^2 + 19x + 12$.

a. $(2x + 4)(2x + 3)$
b. $(4x + 4)(x + 3)$
c. $(4x + 3)(x + 4)$
d. $(2x + 2)(2x + 6)$

Write the perimeter of each figure as a simplified polynomial. Then factor the polynomial.

89.

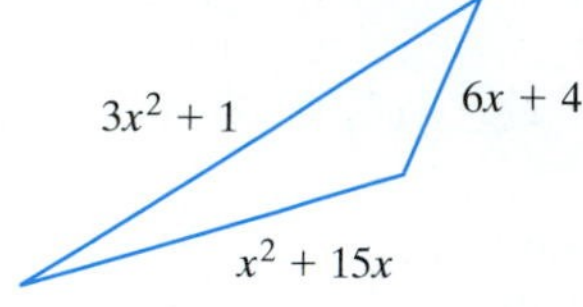

90.

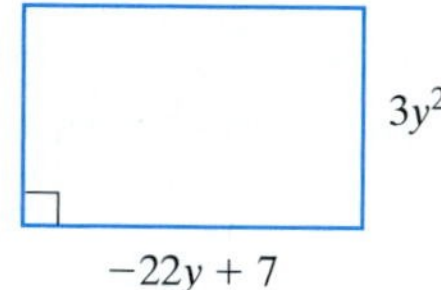

Factor each trinomial completely.

91. $4x^2 + 2x + \frac{1}{4}$

92. $27x^2 + 2x - \frac{1}{9}$

93. $4x^2(y - 1)^2 + 10x(y - 1)^2 + 25(y - 1)^2$

94. $3x^2(a + 3)^3 - 28x(a + 3)^3 + 25(a + 3)^3$

Find a positive value of b so that each trinomial is factorable.

95. $3x^2 + bx - 5$

96. $2z^2 + bz - 7$

Find a positive value of c so that each trinomial is factorable.

97. $5x^2 + 7x + c$

98. $3x^2 - 8x + c$

99. In your own words, describe the steps you will use to factor a trinomial.

STUDY SKILLS BUILDER

Are You Satisfied with Your Performance on a Particular Quiz or Exam?

If not, don't forget to analyze your quiz or exam and look for common errors. Were most of your errors a result of:

- *Carelessness?* Did you turn in your quiz or exam before the allotted time expired? If so, resolve to use the entire time allotted next time. Any extra time can be spent checking your work.
- *Running out of time?* If so, make a point to better manage your time on your next quiz or exam. Try completing any questions that you are unsure of last and delay checking your work until all questions have been answered.
- *Not understanding a concept?* If so, review that concept and correct your work. Try to understand how this happened so that you make sure it doesn't happen before the next quiz or exam.
- *Test conditions?* When studying for a quiz or exam, make sure you place yourself in conditions similar to test conditions. For example, before your next quiz or exam, use a few sheets of blank paper and take a sample test without the aid of your notes or text.

(See your instructor or use the Chapter Test at the end of each chapter.)

Exercises

1. Have you corrected all your previous quizzes and exams?
2. List any errors you have found common to two or more of your graded papers.
3. Is one of your common errors not understanding a concept? If so, are you making sure you understand all the concepts for the next quiz or exam?
4. Is one of your common errors making careless mistakes? If so, are you now taking all the time allotted to check over your work so that you can minimize the number of careless mistakes?
5. Are you satisfied with your grades thus far on quizzes and tests?
6. If your answer to Exercise 5 is no, are there any more suggestions you can make to your instructor or yourself to help? If so, list them here and share these with your instructor.

Objective

A Use the Grouping Method to Factor Trinomials of the Form $ax^2 + bx + c$.

11.4 FACTORING TRINOMIALS OF THE FORM $ax^2 + bx + c$ BY GROUPING

Objective A Using the Grouping Method

There is an alternative method that can be used to factor trinomials of the form $ax^2 + bx + c, a \neq 1$. This method is called the **grouping method** because it uses factoring by grouping as we learned in Section 11.1.

To see how this method works, recall from Section 11.1 that to factor a trinomial such as $x^2 + 11x + 30$, we find two numbers such that

Product is 30

$$x^2 + 11x + 30$$

Sum is 11.

To factor a trinomial such as $2x^2 + 11x + 12$ by grouping, we use an extension of the method in Section 11.1. Here we look for two numbers such that

Product is $2 \cdot 12 = 24$

$$2x^2 + 11x + 12$$

Sum is 11.

This time, we use the two numbers to write

$$2x^2 + 11x + 12 \text{ as}$$
$$= 2x^2 + \square x + \square x + 12$$

Then we factor by grouping. Since we want a positive product, 24, and a positive sum, 11, we consider pairs of positive factors of 24 only.

Factors of 24	Sum of Factors	
1, 24	25	
2, 12	14	
3, 8	11	Correct sum

The factors are 3 and 8. Now we use these factors to write the middle term $11x$ as $3x + 8x$ (or $8x + 3x$). We replace $11x$ with $3x + 8x$ in the original trinomial and then we can factor by grouping.

$$\begin{aligned} 2x^2 + 11x + 12 &= 2x^2 + 3x + 8x + 12 \\ &= (2x^2 + 3x) + (8x + 12) && \text{Group the terms.} \\ &= x(2x + 3) + 4(2x + 3) && \text{Factor each group.} \\ &= (2x + 3)(x + 4) && \text{Factor out } (2x + 3). \end{aligned}$$

In general, we have the following procedure.

To Factor Trinomials by Grouping

Step 1: Factor out a greatest common factor, if there is one other than 1.

Step 2: For the resulting trinomial $ax^2 + bx + c$, find two numbers whose product is $a \cdot c$ and whose sum is b.

Step 3: Write the middle term, bx, using the factors found in Step 2.

Step 4: Factor by grouping.

EXAMPLE 1 Factor $8x^2 - 14x + 5$ by grouping.

Solution:

Step 1: The terms of this trinomial contain no greatest common factor other than 1.

Step 2: This trinomial is of the form $ax^2 + bx + c$ with $a = 8$, $b = -14$, and $c = 5$. Find two numbers whose product is $a \cdot c$ or $8 \cdot 5 = 40$, and whose sum is b or -14.

The numbers are -4 and -10.

Factors of 40	Sum of Factors
−40, −1	−41
−20, −2	−22
−10, −4	−14

Correct sum

Step 3: Write $-14x$ as $-4x - 10x$ so that

$$8x^2 - 14x + 5 = 8x^2 - 4x - 10x + 5$$

Step 4: Factor by grouping.

$$8x^2 - 4x - 10x + 5 = 4x(2x - 1) - 5(2x - 1)$$
$$= (2x - 1)(4x - 5)$$

Work Practice Problem 1

PRACTICE PROBLEM 1

Factor each trinomial by grouping.

a. $3x^2 + 14x + 8$

b. $12x^2 + 19x + 5$

EXAMPLE 2 Factor $6x^2 - 2x - 20$ by grouping.

Solution:

Step 1: First factor out the greatest common factor, 2.

$$6x^2 - 2x - 20 = 2(3x^2 - x - 10)$$

Step 2: Next notice that $a = 3$, $b = -1$, and $c = -10$ in the resulting trinomial. Find two numbers whose product is $a \cdot c$ or $3(-10) = -30$ and whose sum is b, -1. The numbers are -6 and 5.

Step 3: $3x^2 - x - 10 = 3x^2 - 6x + 5x - 10$

Step 4: $3x^2 - 6x + 5x - 10 = 3x(x - 2) + 5(x - 2)$
$$= (x - 2)(3x + 5)$$

The factored form of $6x^2 - 2x - 20 = 2(x - 2)(3x + 5)$.

Don't forget to include the common factor of 2.

Work Practice Problem 2

PRACTICE PROBLEM 2

Factor each trinomial by grouping.

a. $30x^2 - 26x + 4$

b. $6x^2y - 7xy - 5y$

EXAMPLE 3 Factor $18y^4 + 21y^3 - 60y^2$ by grouping.

Solution:

Step 1: First factor out the greatest common factor, $3y^2$.

$$18y^4 + 21y^3 - 60y^2 = 3y^2(6y^2 + 7y - 20)$$

Step 2: Notice that $a = 6$, $b = 7$, and $c = -20$ in the resulting trinomial. Find two numbers whose product is $a \cdot c$ or $6(-20) = -120$ and whose sum is 7. It may help to factor -120 as a product of primes and -1.

$$-120 = 2 \cdot 2 \cdot 2 \cdot 3 \cdot 5 \cdot (-1)$$

Then choose pairings of factors until you have two pairings whose sum is 7.

$$2 \cdot 2 \cdot 2 \cdot 3 \cdot 5 \cdot (-1)$$

(−8 from $2 \cdot 2 \cdot 2 \cdot (-1)$; 15 from $3 \cdot 5$) The numbers are −8 and 15.

Step 3: $6y^2 + 7y - 20 = 6y^2 - 8y + 15y - 20$

Step 4: $6y^2 - 8y + 15y - 20 = 2y(3y - 4) + 5(3y - 4)$
$$= (3y - 4)(2y + 5)$$

The factored form of $18y^4 + 21y^3 - 60y^2$ is $3y^2(3y - 4)(2y + 5)$

Don't forget to include the common factor of $3y^2$.

Work Practice Problem 3

PRACTICE PROBLEM 3

Factor $12y^5 + 10y^4 - 42y^3$ by grouping.

Answers

1. a. $(x + 4)(3x + 2)$
b. $(4x + 5)(3x + 1)$
2. a. $2(5x - 1)(3x - 2)$
b. $y(2x + 1)(3x - 5)$
3. $2y^3(3y + 7)(2y - 3)$

11.4 EXERCISE SET

FOR EXTRA HELP

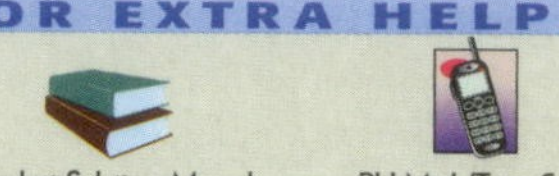
Student Solutions Manual
PH Math/Tutor Center

CD/Video for Review

MathXL®

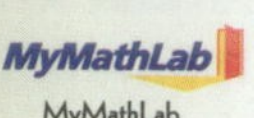
MyMathLab

Objective A *Factor each polynomial by grouping. Notice that Step 3 has already been done in these exercises. See Examples 1 through 3.*

1. $x^2 + 3x + 2x + 6$ **2.** $x^2 + 5x + 3x + 15$ **3.** $y^2 + 8y - 2y - 16$ **4.** $z^2 + 10z - 7z - 70$

5. $8x^2 - 5x - 24x + 15$ **6.** $4x^2 - 9x - 32x + 72$ **7.** $5x^4 - 3x^2 + 25x^2 - 15$ **8.** $2y^4 - 10y^2 + 7y^2 - 35$

Factor each trinomial by grouping. Exercises 9 through 12 are broken into parts to help you get started. See Examples 1 through 3.

9. $6x^2 + 11x + 3$
- **a.** Find two numbers whose product is $6 \cdot 3 = 18$ and whose sum is 11.
- **b.** Write $11x$ using the factors from part (a).
- **c.** Factor by grouping.

10. $8x^2 + 14x + 3$
- **a.** Find two numbers whose product is $8 \cdot 3 = 24$ and whose sum is 14.
- **b.** Write $14x$ using the factors from part (a).
- **c.** Factor by grouping.

11. $15x^2 - 23x + 4$
- **a.** Find two numbers whose product is $15 \cdot 4 = 60$ and whose sum is -23.
- **b.** Write $-23x$ using the factors from part (a).
- **c.** Factor by grouping.

12. $6x^2 - 13x + 5$
- **a.** Find two numbers whose product is $6 \cdot 5 = 30$ and whose sum is -13.
- **b.** Write $-13x$ using the factors from part (a).
- **c.** Factor by grouping.

13. $21y^2 + 17y + 2$ **14.** $15x^2 + 11x + 2$ **15.** $7x^2 - 4x - 11$ **16.** $8x^2 - x - 9$

17. $10x^2 - 9x + 2$ **18.** $30x^2 - 23x + 3$ **19.** $2x^2 - 7x + 5$ **20.** $2x^2 - 7x + 3$

21. $12x + 4x^2 + 9$ **22.** $20x + 25x^2 + 4$ **23.** $4x^2 - 8x - 21$ **24.** $6x^2 - 11x - 10$

25. $10x^2 - 23x + 12$ **26.** $21x^2 - 13x + 2$ **27.** $2x^3 + 13x^2 + 15x$ **28.** $3x^3 + 8x^2 + 4x$

29. $16y^2 - 34y + 18$ **30.** $4y^2 - 2y - 12$ **31.** $-13x + 6 + 6x^2$ **32.** $-25x + 12 + 12x^2$

33. $54a^2 - 9a - 30$ **34.** $30a^2 + 38a - 20$ **35.** $20a^3 + 37a^2 + 8a$ **36.** $10a^3 + 17a^2 + 3a$

37. $12x^3 - 27x^2 - 27x$ **38.** $30x^3 - 155x^2 + 25x$ **39.** $3x^2y + 4xy^2 + y^3$ **40.** $6r^2t + 7rt^2 + t^3$

41. $20z^2 + 7z + 1$

42. $36z^2 + 6z + 1$

43. $24a^2 - 6ab - 30b^2$

44. $30a^2 + 5ab - 25b^2$

45. $15p^4 + 31p^3q + 2p^2q^2$

46. $20s^4 + 61s^3t + 3s^2t^2$

47. $35 + 12x + x^2$

48. $33 + 14x + x^2$

49. $6 - 11x + 5x^2$

50. $5 - 12x + 7x^2$

Review

Multiply. See Section 10.6.

51. $(x - 2)(x + 2)$

52. $(y - 5)(y + 5)$

53. $(y + 4)(y + 4)$

54. $(x + 7)(x + 7)$

55. $(9z + 5)(9z - 5)$

56. $(8y + 9)(8y - 9)$

57. $(4x - 3)^2$

58. $(2z - 1)^2$

Concept Extensions

Write the perimeter of each figure as a simplified polynomial. Then factor the polynomial.

59.

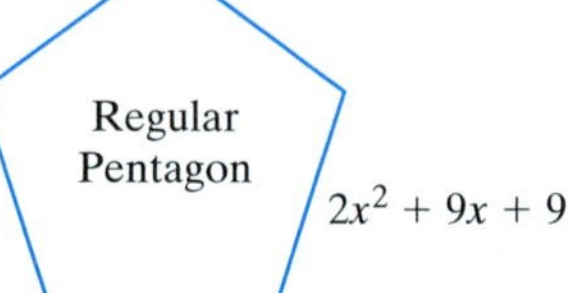

60.

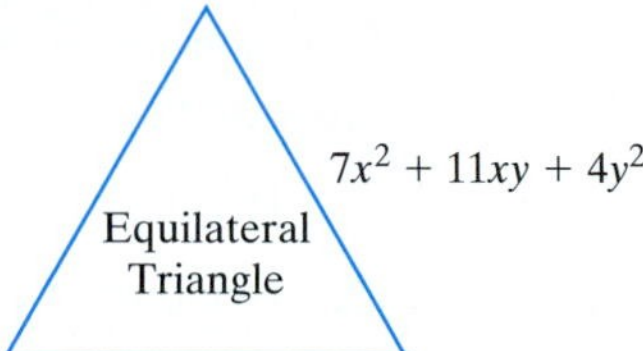

Factor each polynomial by grouping.

61. $x^{2n} + 2x^n + 3x^n + 6$
(*Hint:* Don't forget that $x^{2n} = x^n \cdot x^n$.)

62. $x^{2n} + 6x^n + 10x^n + 60$

63. $3x^{2n} + 16x^n - 35$

64. $12x^{2n} - 40x^n + 25$

65. In your own words, explain how to factor a trinomial by grouping.

11.5 FACTORING PERFECT SQUARE TRINOMIALS AND THE DIFFERENCE OF TWO SQUARES

Objectives

A Recognize Perfect Square Trinomials.

B Factor Perfect Square Trinomials.

C Factor the Difference of Two Squares.

Objective A Recognizing Perfect Square Trinomials

A trinomial that is the square of a binomial is called a **perfect square trinomial.** For example,

$$(x + 3)^2 = (x + 3)(x + 3)$$
$$= x^2 + 6x + 9$$

Thus $x^2 + 6x + 9$ is a perfect square trinomial.

In Chapter 10, we discovered special product formulas for squaring binomials.

$$(a + b)^2 = a^2 + 2ab + b^2 \quad \text{and} \quad (a - b)^2 = a^2 - 2ab + b^2$$

Because multiplication and factoring are reverse processes, we can now use these special products to help us factor perfect square trinomials. If we reverse these equations, we have the following.

Factoring Perfect Square Trinomials

$$a^2 + 2ab + b^2 = (a + b)^2$$
$$a^2 - 2ab + b^2 = (a - b)^2$$

Helpful Hint

Notice that for both given forms of a perfect square trinomial, the last term is positive. This is because the last term is a square.

To use these equations to help us factor, we must first be able to recognize a perfect square trinomial. A trinomial is a perfect square when

1. two terms, a^2 and b^2, are squares and
2. another term is $2 \cdot a \cdot b$ or $-2 \cdot a \cdot b$. That is, this term is twice the product of a and b, or its opposite.

PRACTICE PROBLEM 1

Decide whether each trinomial is a perfect square trinomial.

a. $x^2 + 12x + 36$

b. $x^2 + 20x + 100$

EXAMPLE 1 Decide whether $x^2 + 8x + 16$ is a perfect square trinomial.

Solution:

1. Two terms, x^2 and 16, are squares $(16 = 4^2)$.
2. Twice the product of x and 4 is the other term of the trinomial.

   $2 \cdot x \cdot 4 = 8x$

Thus, $x^2 + 8x + 16$ is a perfect square trinomial.

Work Practice Problem 1

PRACTICE PROBLEM 2

Decide whether each trinomial is a perfect square trinomial.

a. $9x^2 + 20x + 25$

b. $4x^2 + 8x + 11$

EXAMPLE 2 Decide whether $4x^2 + 10x + 9$ is a perfect square trinomial.

Solution:

1. Two terms, $4x^2$ and 9, are squares.

   $4x^2 = (2x)^2 \quad \text{and} \quad 9 = 3^2$

2. Twice the product of $2x$ and 3 is *not* the other term of the trinomial.

   $2 \cdot 2x \cdot 3 = 12x$, *not* $10x$

The trinomial is *not* a perfect square trinomial.

Work Practice Problem 2

Answers

1. a. yes **b.** yes **2. a.** no **b.** no

EXAMPLE 3 Decide whether $9x^2 - 12xy + 4y^2$ is a perfect square trinomial.

Solution:

1. Two terms, $9x^2$ and $4y^2$, are squares.

   $9x^2 = (3x)^2$ and $4y^2 = (2y)^2$

2. Twice the product of $3x$ and $2y$ is the opposite of the other term of the trinomial.

   $2 \cdot 3x \cdot 2y = 12xy$, the opposite of $-12xy$

Thus, $9x^2 - 12xy + 4y^2$ is a perfect square trinomial.

Work Practice Problem 3

PRACTICE PROBLEM 3

Decide whether each trinomial is a perfect square trinomial.

a. $25x^2 - 10x + 1$

b. $9x^2 - 42x + 49$

Objective B Factoring Perfect Square Trinomials

Now that we can recognize perfect square trinomials, we are ready to factor them.

EXAMPLE 4 Factor: $x^2 + 12x + 36$

Solution:

$$x^2 + 12x + 36 = x^2 + 2 \cdot x \cdot 6 + 6^2 \quad 36 = 6^2 \text{ and } 12x = 2 \cdot x \cdot 6$$

$$a^2 + 2 \cdot a \cdot b + b^2$$

$$= (x + 6)^2$$

$$(a + b)^2$$

Work Practice Problem 4

PRACTICE PROBLEM 4

Factor: $x^2 + 16x + 64$

EXAMPLE 5 Factor: $25x^2 + 20xy + 4y^2$

Solution:

$$25x^2 + 20xy + 4y^2 = (5x)^2 + 2 \cdot 5x \cdot 2y + (2y)^2$$
$$= (5x + 2y)^2$$

Work Practice Problem 5

PRACTICE PROBLEM 5

Factor: $9r^2 + 24rs + 16s^2$

EXAMPLE 6 Factor: $4m^4 - 4m^2 + 1$

Solution:

$$4m^4 - 4m^2 + 1 = (2m^2)^2 - 2 \cdot 2m^2 \cdot 1 + 1^2$$

$$a^2 \quad - 2 \cdot a \cdot b + b^2$$

$$= (2m^2 - 1)^2$$

$$(a \quad - b)^2$$

Work Practice Problem 6

PRACTICE PROBLEM 6

Factor: $9n^4 - 6n^2 + 1$

EXAMPLE 7 Factor: $25x^2 + 50x + 9$

Solution: Notice that this trinomial is not a perfect square trinomial.

$25x^2 = (5x)^2, 9 = 3^2$

but

$2 \cdot 5x \cdot 3 = 30x$

and $30x$ is not the middle term $50x$.

Continued on next page

PRACTICE PROBLEM 7

Factor: $9x^2 + 15x + 4$

Answers

3. a. yes **b.** yes **4.** $(x + 8)^2$ **5.** $(3r + 4s)^2$ **6.** $(3n^2 - 1)^2$ **7.** $(3x + 1)(3x + 4)$

Helpful Hint

A perfect square trinomial can also be factored by the methods found in Sections 11.2 through 11.4.

PRACTICE PROBLEM 8

Factor:

a. $8n^2 + 40n + 50$

b. $12x^3 - 84x^2 + 147x$

Although $25x^2 + 50x + 9$ is not a perfect square trinomial, it is factorable. Using techniques we learned in Sections 11.3 or 11.4, we find that

$$25x^2 + 50x + 9 = (5x + 9)(5x + 1)$$

Work Practice Problem 7

EXAMPLE 8 Factor: $162x^3 - 144x^2 + 32x$

Solution: Don't forget to first look for a common factor. There is a greatest common factor of $2x$ in this trinomial.

$$\begin{aligned} 162x^3 - 144x^2 + 32x &= 2x(81x^2 - 72x + 16) \\ &= 2x[(9x)^2 - 2 \cdot 9x \cdot 4 + 4^2] \\ &= 2x(9x - 4)^2 \end{aligned}$$

Work Practice Problem 8

Objective C Factoring the Difference of Two Squares

In Chapter 10, we discovered another special product, the product of the sum and difference of two terms a and b:

$$(a + b)(a - b) = a^2 - b^2$$

Reversing this equation gives us another factoring pattern, which we use to factor the difference of two squares.

Factoring the Difference of Two Squares

$$a^2 - b^2 = (a + b)(a - b)$$

To use this equation to help us factor, we must first be able to recognize the difference of two squares. A binomial is a difference of two squares if

1. both terms are squares and
2. the signs of the terms are different.

Let's practice using this pattern.

PRACTICE PROBLEMS 9–12

Factor each binomial.

9. $x^2 - 9$ **10.** $a^2 - 16$

11. $c^2 - \frac{9}{25}$ **12.** $s^2 + 9$

EXAMPLES Factor each binomial.

9. $z^2 - 4 = z^2 - 2^2 = (z + 2)(z - 2)$

$$\begin{matrix} \uparrow & \uparrow & \uparrow & \uparrow & \uparrow & \uparrow \\ a^2 - & b^2 = & (a + & b)(& a - & b) \end{matrix}$$

10. $y^2 - 25 = y^2 - 5^2 = (y + 5)(y - 5)$

11. $y^2 - \frac{4}{9} = y^2 - \left(\frac{2}{3}\right)^2 = \left(y + \frac{2}{3}\right)\left(y - \frac{2}{3}\right)$

12. $x^2 + 4$

Note that the binomial $x^2 + 4$ is the *sum* of two squares since we can write $x^2 + 4$ as $x^2 + 2^2$. We might try to factor using $(x + 2)(x + 2)$ or $(x - 2)(x - 2)$. But when we multiply to check, we find that neither factoring is correct.

$$(x + 2)(x + 2) = x^2 + 4x + 4$$
$$(x - 2)(x - 2) = x^2 - 4x + 4$$

In both cases, the product is a trinomial, not the required binomial. In fact, $x^2 + 4$ is a prime polynomial.

Work Practice Problems 9–12

Helpful Hint

After the greatest common factor has been removed, the *sum* of two squares cannot be factored further using real numbers.

Answers

8. a. $2(2n + 5)^2$ **b.** $3x(2x - 7)^2$
9. $(x - 3)(x + 3)$
10. $(a - 4)(a + 4)$
11. $\left(c - \frac{3}{5}\right)\left(c + \frac{3}{5}\right)$
12. prime polynomial

EXAMPLES Factor each difference of two squares.

13. $4x^2 - 1 = (2x)^2 - 1^2 = (2x + 1)(2x - 1)$

14. $25a^2 - 9b^2 = (5a)^2 - (3b)^2 = (5a + 3b)(5a - 3b)$

15. $y^4 - 16 = (y^2)^2 - 4^2$

$= (y^2 + 4)(y^2 - 4)$ Factor the difference of two squares.

$= (y^2 + 4)(y + 2)(y - 2)$ Factor the difference of two squares.

Work Practice Problems 13–15

PRACTICE PROBLEMS 13–15

Factor each difference of two squares.

13. $9s^2 - 1$

14. $16x^2 - 49y^2$

15. $p^4 - 81$

Helpful Hint

1. Don't forget to first see whether there's a greatest common factor (other than 1) that can be factored out.

2. Factor completely. In other words, check to see whether any factors can be factored further (as in Example 15).

EXAMPLES Factor each difference of two squares.

16. $4x^3 - 49x = x(4x^2 - 49)$ Factor out the common factor, x.

$= x[(2x)^2 - 7^2]$

$= x(2x + 7)(2x - 7)$ Factor the difference of two squares.

17. $162x^4 - 2 = 2(81x^4 - 1)$ Factor out the common factor, 2.

$= 2(9x^2 + 1)(9x^2 - 1)$ Factor the difference of two squares.

$= 2(9x^2 + 1)(3x + 1)(3x - 1)$ Factor the difference of two squares.

18. $-49x^2 + 16 = -1(49x^2 - 16)$ Factor out -1.

$= -1(7x + 4)(7x - 4)$ Factor the difference of two squares.

Work Practice Problems 16–18

PRACTICE PROBLEMS 16–18

Factor each difference of two squares.

16. $9x^3 - 25x$

17. $48x^4 - 3$

18. $-9x^2 + 100$

EXAMPLE 19 Factor: $36 - x^2$.

Solution: This is the difference of two squares. Factor as is or if you like, first write the binomial with variable term first.

Factor as is: $36 - x^2 = 6^2 - x^2 = (6 + x)(6 - x)$.

Rewrite binomial: $36 - x^2 = -x^2 + 36 = -1(x^2 - 36)$

$= -1(x + 6)(x - 6)$.

Both factorizations are correct and are equal. To see this, factor -1 from $(6 - x)$ in the first factorization.

Work Practice Problem 19

PRACTICE PROBLEM 19

Factor: $121 - m^2$

Helpful Hint

When rearranging terms, keep in mind that the sign of a term is in front of the term.

Answers

13. $(3s - 1)(3s + 1)$
14. $(4x - 7y)(4x + 7y)$
15. $(p^2 + 9)(p + 3)(p - 3)$
16. $x(3x - 5)(3x + 5)$
17. $3(4x^2 + 1)(2x + 1)(2x - 1)$
18. $-1(3x - 10)(3x + 10)$
19. $(11 + m)(11 - m)$ or $-1(m + 11)(m - 11)$

CALCULATOR EXPLORATIONS Graphing

A graphing calculator is a convenient tool for evaluating an expression at a given replacement value. For example, let's evaluate $x^2 - 6x$ when $x = 2$. To do so, store the value 2 in the variable x and then enter and evaluate the algebraic expression.

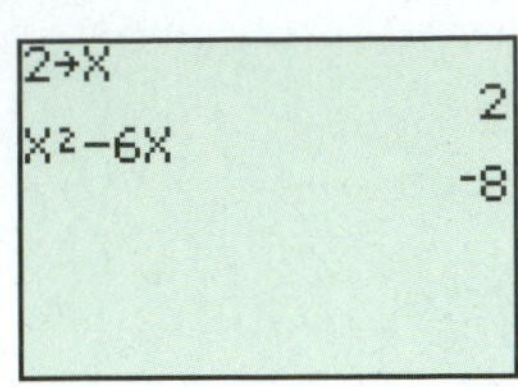

The value of $x^2 - 6x$ when $x = 2$ is -8. You may want to use this method for evaluating expressions as you explore the following.

We can use a graphing calculator to explore factoring patterns numerically. Use your calculator to evaluate $x^2 - 2x + 1$, $x^2 - 2x - 1$, and $(x - 1)^2$ for each value of x given in the table. What do you observe?

	$x^2 - 2x + 1$	$x^2 - 2x - 1$	$(x - 1)^2$
$x = 5$			
$x = -3$			
$x = 2.7$			
$x = -12.1$			
$x = 0$			

Notice in each case that $x^2 - 2x - 1 \neq (x - 1)^2$. Because for each x in the table the value of $x^2 - 2x + 1$ and the value of $(x - 1)^2$ are the same, we might guess that $x^2 - 2x + 1 = (x - 1)^2$. We can verify our guess algebraically with multiplication:

$$(x - 1)(x - 1) = x^2 - x - x + 1 = x^2 - 2x + 1$$

Vocabulary and Readiness Check

Use the choices below to fill in each blank. Some choices may be used more than once and some choices may not be used at all.

perfect square trinomial	true	$(5y)^2$	$(x + 5y)^2$
difference of two squares	false	$(x - 5y)^2$	$5y^2$

1. A(n) ______________ is a trinomial that is the square of a binomial.
2. The term $25y^2$ written as a square is ________.
3. The expression $x^2 + 10xy + 25y^2$ is called a(n) ______________.
4. The expression $x^2 - 49$ is called a(n) ______________.
5. The factorization $(x + 5y)(x + 5y)$ may also be written as ________.
6. True or false: The factorization $(x - 5y)(x + 5y)$ may also be written as $(x - 5y)^2$. ________
7. The trinomial $x^2 - 6x - 9$ is a perfect square trinomial. ________
8. The binomial $y^2 + 9$ factors as $(y + 3)^2$. ________

11.5 EXERCISE SET

FOR EXTRA HELP

Student Solutions Manual

PH Math/Tutor Center

CD/Video for Review

MathXL®

MyMathLab

Objective A *Determine whether each trinomial is a perfect square trinomial. See Examples 1 through 3.*

1. $x^2 + 16x + 64$ **2.** $x^2 + 22x + 121$ **3.** $y^2 + 5y + 25$ **4.** $y^2 + 4y + 16$

5. $m^2 - 2m + 1$ **6.** $p^2 - 4p + 4$ **7.** $a^2 - 16a + 49$ **8.** $n^2 - 20n + 144$

9. $4x^2 + 12xy + 8y^2$ **10.** $25x^2 + 20xy + 2y^2$ **11.** $25a^2 - 40ab + 16b^2$ **12.** $36a^2 - 12ab + b^2$

Objective **B** *Factor each trinomial completely. See Examples 4 through 8.*

13. $x^2 + 22x + 121$ **14.** $x^2 + 18x + 81$ **15.** $x^2 - 16x + 64$ **16.** $x^2 - 12x + 36$

17. $16a^2 - 24a + 9$ **18.** $25x^2 - 20x + 4$ **19.** $x^4 + 4x^2 + 4$ **20.** $m^4 + 10m^2 + 25$

21. $2n^2 - 28n + 98$ **22.** $3y^2 - 6y + 3$ **23.** $16y^2 + 40y + 25$ **24.** $9y^2 + 48y + 64$

25. $x^2y^2 - 10xy + 25$ **26.** $4x^2y^2 - 28xy + 49$ **27.** $m^3 + 18m^2 + 81m$ **28.** $y^3 + 12y^2 + 36y$

29. $1 + 6x^2 + x^4$ **30.** $1 + 16x^2 + x^4$ **31.** $9x^2 - 24xy + 16y^2$ **32.** $25x^2 - 60xy + 36y^2$

Objective **C** *Factor each binomial completely. See Examples 9 through 19.*

33. $x^2 - 4$ **34.** $x^2 - 36$ **35.** $81 - p^2$ **36.** $100 - t^2$

37. $-4r^2 + 1$ **38.** $-9t^2 + 1$ **39.** $9x^2 - 16$ **40.** $36y^2 - 25$

41. $16r^2 + 1$ **42.** $49y^2 + 1$ **43.** $-36 + x^2$ **44.** $-1 + y^2$

45. $m^4 - 1$ **46.** $n^4 - 16$ **47.** $x^2 - 169y^2$ **48.** $x^2 - 225y^2$

49. $18r^2 - 8$ **50.** $32t^2 - 50$ **51.** $9xy^2 - 4x$ **52.** $36x^2y - 25y$

53. $16x^4 - 64x^2$ **54.** $25y^4 - 100y^2$ **55.** $xy^3 - 9xyz^2$ **56.** $x^3y - 4xy^3$

57. $36x^2 - 64y^2$ **58.** $225a^2 - 81b^2$ **59.** $144 - 81x^2$ **60.** $12x^2 - 27$

61. $25y^2 - 9$ **62.** $49a^2 - 16$ **63.** $121m^2 - 100n^2$ **64.** $169a^2 - 49b^2$

65. $x^2y^2 - 1$ **66.** $a^2b^2 - 16$ **67.** $x^2 - \frac{1}{4}$

68. $y^2 - \frac{1}{16}$ **69.** $49 - \frac{9}{25}m^2$ **70.** $100 - \frac{4}{81}n^2$

Objectives **B** **C** **Mixed Practice** *Factor each binomial or trinomial completely. See Examples 4 through 19.*

71. $81a^2 - 25b^2$

72. $49y^2 - 100z^2$

73. $x^2 + 14xy + 49y^2$

74. $x^2 + 10xy + 25y^2$

75. $32n^4 - 112n^2 + 98$

76. $162a^4 - 72a^2 + 8$

77. $x^6 - 81x^2$

78. $n^9 - n^5$

79. $64p^3q - 81pq^3$

80. $100x^3y - 49xy^3$

Review

Solve each equation. See Section 3.3.

81. $x - 6 = 0$

82. $y + 5 = 0$

83. $2m + 4 = 0$

84. $3x - 9 = 0$

85. $5z - 1 = 0$

86. $4a + 2 = 0$

Concept Extensions

Factor each expression completely.

87. $x^2 - \frac{2}{3}x + \frac{1}{9}$

88. $x^2 - \frac{1}{25}$

89. $(x + 2)^2 - y^2$

90. $(y - 6)^2 - z^2$

91. $a^2(b - 4) - 16(b - 4)$

92. $m^2(n + 8) - 9(n + 8)$

93. $(x^2 + 6x + 9) - 4y^2$ (*Hint:* Factor the trinomial in parentheses first.)

94. $(x^2 + 2x + 1) - 36y^2$

95. $x^{2n} - 100$

96. $x^{2n} - 81$

97. Fill in the blank so that $x^2 +$ _____ $x + 16$ is a perfect square trinomial.

98. Fill in the blank so that $9x^2 +$ _____ $x + 25$ is a perfect square trinomial.

99. Describe a perfect square trinomial.

100. Write a perfect square trinomial that factors as $(x + 3y)^2$.

101. What binomial multiplied by $(x - 6)$ gives the difference of two squares?

102. What binomial multiplied by $(5 + y)$ gives the difference of two squares?

The area of the largest square in the figure is $(a + b)^2$. Use this figure to answer Exercises 103 and 104.

103. Write the area of the largest square as the sum of the areas of the smaller squares and rectangles.

104. What factoring formula from this section is visually represented by this square?

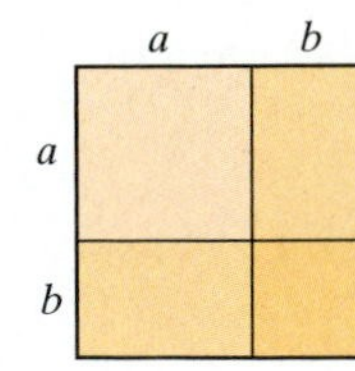

105. An object is dropped from the top of Pittsburgh's USX Towers, which is 841 feet tall. (*Source: World Almanac* research) The height of the object after t seconds is given by the expression $841 - 16t^2$.

a. Find the height of the object after 2 seconds.

b. Find the height of the object after 5 seconds.

c. To the nearest whole second, estimate when the object hits the ground.

d. Factor $841 - 16t^2$.

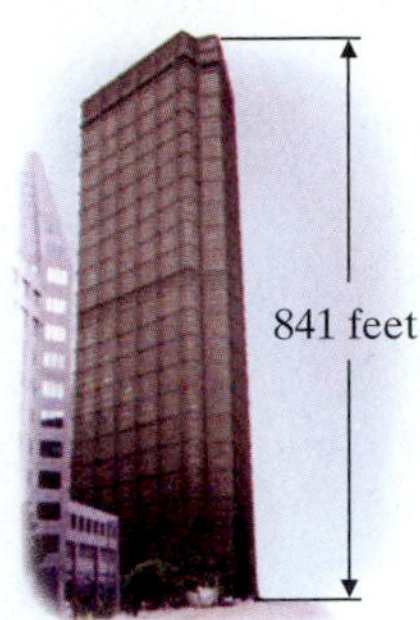

106. A worker on the top of the Aetna Life Building in San Francisco accidentally drops a bolt. The Aetna Life Building is 529 feet tall. (*Source: World Almanac* research) The height of the bolt after t seconds is given by the expression $529 - 16t^2$.

a. Find the height of the bolt after 1 second.

b. Find the height of the bolt after 4 seconds.

c. To the nearest whole second, estimate when the bolt hits the ground.

d. Factor $529 - 16t^2$.

107. At this writing, the world's tallest building is the Taipei 101 in Taipei, Taiwan, at a height of 1671 feet. (*Source:* Council on Tall Buildings and Urban Habitat) Suppose a worker is suspended 71 feet below the top of the pinnacle atop the building, at a height of 1600 feet above the ground. If the worker accidentally drops a bolt, the height of the bolt after t seconds is given by the expression $1600 - 16t^2$.

a. Find the height of the bolt after 3 seconds.

b. Find the height of the bolt after 7 seconds.

c. To the nearest whole second, estimate when the bolt hits the ground.

d. Factor $1600 - 16t^2$.

108. A performer with the Moscow Circus is planning a stunt involving a free fall from the top of the Moscow State University building, which is 784 feet tall. (*Source:* Council on Tall Buildings and Urban Habitat) Neglecting air resistance, the performer's height above gigantic cushions positioned at ground level after t seconds is given by the expression $784 - 16t^2$.

a. Find the performer's height after 2 seconds.

b. Find the performer's height after 5 seconds.

c. To the nearest whole second, estimate when the performer reaches the cushions positioned at ground level.

d. Factor $784 - 16t^2$.

Answers

1. ______
2. ______
3. ______
4. ______
5. ______
6. ______
7. ______
8. ______
9. ______
10. ______
11. ______
12. ______
13. ______
14. ______
15. ______
16. ______
17. ______
18. ______
19. ______
20. ______
21. ______
22. ______
23. ______
24. ______
25. ______
26. ______
27. ______
28. ______
29. ______
30. ______
31. ______
32. ______
33. ______
34. ______
35. ______
36. ______

INTEGRATED REVIEW Sections 11.1–11.5

Choosing a Factoring Strategy

The following steps may be helpful when factoring polynomials.

To Factor a Polynomial

Step 1: Are there any common factors? If so, factor out the GCF.

Step 2: How many terms are in the polynomial?

a. Two terms: Is it the difference of two squares? $a^2 - b^2 = (a - b)(a + b)$

b. Three terms: Try one of the following.

i. Perfect square trinomial: $a^2 + 2ab + b^2 = (a + b)^2$

$a^2 - 2ab + b^2 = (a - b)^2$

ii. If not a perfect square trinomial, factor using the methods presented in Sections 11.2 through 11.4.

c. Four terms: Try factoring by grouping.

Step 3: See if any factors in the factored polynomial can be factored further.

Step 4: Check by multiplying.

Factor each polynomial completely.

1. $x^2 + x - 12$

2. $x^2 - 10x + 16$

3. $x^2 - x - 6$

4. $x^2 + 2x + 1$

5. $x^2 - 6x + 9$

6. $x^2 + x - 2$

7. $x^2 + x - 6$

8. $x^2 + 7x + 12$

9. $x^2 - 7x + 10$

10. $x^2 - x - 30$

11. $2x^2 - 98$

12. $3x^2 - 75$

13. $x^2 + 3x + 5x + 15$

14. $3y - 21 + xy - 7x$

15. $x^2 + 6x - 16$

16. $x^2 - 3x - 28$

17. $4x^3 + 20x^2 - 56x$

18. $6x^3 - 6x^2 - 120x$

19. $12x^2 + 34x + 24$

20. $8a^2 + 6ab - 5b^2$

21. $4a^2 - b^2$

22. $x^2 - 25y^2$

23. $28 - 13x - 6x^2$

24. $20 - 3x - 2x^2$

25. $4 - 2x + x^2$

26. $a + a^2 - 3$

27. $6y^2 + y - 15$

28. $4x^2 - x - 5$

29. $18x^3 - 63x^2 + 9x$

30. $12a^3 - 24a^2 + 4a$

31. $16a^2 - 56a + 49$

32. $25p^2 - 70p + 49$

33. $14 + 5x - x^2$

34. $3 - 2x - x^2$

35. $3x^4y + 6x^3y - 72x^2y$

36. $2x^3y + 8x^2y^2 - 10xy^3$

37. $12x^3y + 243xy$ **38.** $6x^3y^2 + 8xy^2$ **39.** $2xy - 72x^3y$

40. $2x^3 - 18x$ **41.** $x^3 + 6x^2 - 4x - 24$ **42.** $x^3 - 2x^2 - 36x + 72$

43. $6a^3 + 10a^2$ **44.** $4n^2 - 6n$ **45.** $3x^3 - x^2 + 12x - 4$

46. $x^3 - 2x^2 + 3x - 6$ **47.** $6x^2 + 18xy + 12y^2$ **48.** $12x^2 + 46xy - 8y^2$

49. $5(x + y) + x(x + y)$ **50.** $7(x - y) + y(x - y)$ **51.** $14t^2 - 9t + 1$

52. $3t^2 - 5t + 1$ **53.** $-3x^2 - 2x + 5$ **54.** $-7x^2 - 19x + 6$

55. $1 - 8a - 20a^2$ **56.** $1 - 7a - 60a^2$ **57.** $x^4 - 10x^2 + 9$

58. $x^4 - 13x^2 + 36$ **59.** $x^2 - 23x + 120$ **60.** $y^2 + 22y + 96$

61. $x^2 - 14x - 48$ **62.** $16a^2 - 56ab + 49b^2$ **63.** $25p^2 - 70pq + 49q^2$

64. $7x^2 + 24xy + 9y^2$ **65.** $-x^2 - x + 30$ **66.** $-x^2 + 6x - 8$

67. $3rs - s + 12r - 4$ **68.** $x^3 - 2x^2 + x - 2$ **69.** $4x^2 - 8xy - 3x + 6y$

70. $4x^2 - 2xy - 7yz + 14xz$ **71.** $x^2 + 9xy - 36y^2$ **72.** $3x^2 + 10xy - 8y^2$

73. $x^4 - 14x^2 - 32$ **74.** $x^4 - 22x^2 - 75$

75. Explain why it makes good sense to factor out the GCF first, before using other methods of factoring.

76. The sum of two squares usually does not factor. Is the sum of two squares $9x^2 + 81y^2$ factorable?

37. ______
38. ______
39. ______
40. ______
41. ______
42. ______
43. ______
44. ______
45. ______
46. ______
47. ______
48. ______
49. ______
50. ______
51. ______
52. ______
53. ______
54. ______
55. ______
56. ______
57. ______
58. ______
59. ______
60. ______
61. ______
62. ______
63. ______
64. ______
65. ______
66. ______
67. ______
68. ______
69. ______
70. ______
71. ______
72. ______
73. ______
74. ______
75. ______
76. ______

11.6 SOLVING QUADRATIC EQUATIONS BY FACTORING

Objectives

A Solve Quadratic Equations by Factoring.

B Solve Equations with Degree Greater Than 2 by Factoring.

In this section, we introduce a new type of equation—the **quadratic equation.**

Quadratic Equation

A quadratic equation is one that can be written in the form

$$ax^2 + bx + c = 0$$

where a, b, and c are real numbers and $a \neq 0$.

Some examples of quadratic equations are shown below.

$$x^2 - 9x - 22 = 0 \qquad 4x^2 - 28 = -49 \qquad x(2x - 7) = 4$$

The form $ax^2 + bx + c = 0$ is called the **standard form** of a quadratic equation. The quadratic equation $x^2 - 9x - 22 = 0$ is the only equation above that is in standard form.

Quadratic equations model many real-life situations. For example, let's suppose we want to know how long before a person diving from a 144-foot cliff reaches the ocean. The answer to this question is found by solving the quadratic equation $-16t^2 + 144 = 0$. (See Example 1 in Section 11.7.)

Objective **A** Solving Quadratic Equations by Factoring

Some quadratic equations can be solved by making use of factoring and the **zero factor property.**

Zero Factor Property

If a and b are real numbers and if $ab = 0$, then $a = 0$ or $b = 0$.

In other words, if the product of two numbers is 0, then at least one of the numbers must be 0.

PRACTICE PROBLEM 1

Solve: $(x - 7)(x + 2) = 0$

EXAMPLE 1 Solve: $(x - 3)(x + 1) = 0$

Solution: If this equation is to be a true statement, then either the factor $x - 3$ must be 0 or the factor $x + 1$ must be 0. In other words, either

$$x - 3 = 0 \qquad \text{or} \qquad x + 1 = 0$$

If we solve these two linear equations, we have

$$x = 3 \qquad \text{or} \qquad x = -1$$

Answer

1. 7 and −2

Thus, 3 and -1 are both solutions of the equation $(x - 3)(x + 1) = 0$. To check, we replace x with 3 in the original equation. Then we replace x with -1 in the original equation.

Check:

$(x - 3)(x + 1) = 0$
$(3 - 3)(3 + 1) \stackrel{?}{=} 0$ Replace x with 3.
$0(4) = 0$ True

$(x - 3)(x + 1) = 0$
$(-1 - 3)(-1 + 1) \stackrel{?}{=} 0$ Replace x with -1.
$(-4)(0) = 0$ True

The solutions are 3 and -1.

Work Practice Problem 1

Helpful Hint

The zero factor property says that *if a product is 0, then a factor is 0.*

If $a \cdot b = 0$, then $a = 0$ or $b = 0$.
If $x(x + 5) = 0$, then $x = 0$ or $x + 5 = 0$.
If $(x + 7)(2x - 3) = 0$, then $x + 7 = 0$ or $2x - 3 = 0$.

Use this property only when the product is 0. For example, if $a \cdot b = 8$, we do not know the value of a or b. The values may be $a = 2, b = 4$ or $a = 8, b = 1$, or any other two numbers whose product is 8.

EXAMPLE 2 Solve: $(x - 5)(2x + 7) = 0$

Solution: The product is 0. By the zero factor property, this is true only when a factor is 0. To solve, we set each factor equal to 0 and solve the resulting linear equations.

$(x - 5)(2x + 7) = 0$
$x - 5 = 0$ or $2x + 7 = 0$
$x = 5$ $\qquad 2x = -7$
$x = -\frac{7}{2}$

Check: Let $x = 5$.

$(x - 5)(2x + 7) = 0$
$(5 - 5)(2 \cdot 5 + 7) \stackrel{?}{=} 0$ Replace x with 5.
$0 \cdot 17 \stackrel{?}{=} 0$
$0 = 0$ True

Let $x = -\frac{7}{2}$.

$(x - 5)(2x + 7) = 0$
$\left(-\frac{7}{2} - 5\right)\left(2\left(-\frac{7}{2}\right) + 7\right) \stackrel{?}{=} 0$ Replace x with $-\frac{7}{2}$.
$\left(-\frac{17}{2}\right)(-7 + 7) \stackrel{?}{=} 0$
$\left(-\frac{17}{2}\right) \cdot 0 \stackrel{?}{=} 0$
$0 = 0$ True

The solutions are 5 and $-\frac{7}{2}$.

Work Practice Problem 2

PRACTICE PROBLEM 2

Solve: $(x - 10)(3x + 1) = 0$

Answer

2. 10 and $-\frac{1}{3}$

PRACTICE PROBLEM 3

Solve each equation.

a. $y(y + 3) = 0$

b. $x(4x - 3) = 0$

EXAMPLE 3 Solve: $x(5x - 2) = 0$

Solution: $x(5x - 2) = 0$

$$x = 0 \quad \text{or} \quad 5x - 2 = 0 \qquad \text{Use the zero factor property.}$$
$$5x = 2$$
$$x = \frac{2}{5}$$

Check these solutions in the original equation. The solutions are 0 and $\frac{2}{5}$.

Work Practice Problem 3

PRACTICE PROBLEM 4

Solve: $x^2 - 3x - 18 = 0$

EXAMPLE 4 Solve: $x^2 - 9x - 22 = 0$

Solution: One side of the equation is 0. However, to use the zero factor property, one side of the equation must be 0 *and* the other side must be written as a product (must be factored). Thus, we must first factor this polynomial.

$$x^2 - 9x - 22 = 0$$
$$(x - 11)(x + 2) = 0 \qquad \text{Factor.}$$

Now we can apply the zero factor property.

$$x - 11 = 0 \quad \text{or} \quad x + 2 = 0$$
$$x = 11 \qquad\qquad x = -2$$

Check: Let $x = 11$.

$$x^2 - 9x - 22 = 0$$
$$11^2 - 9 \cdot 11 - 22 \stackrel{?}{=} 0$$
$$121 - 99 - 22 \stackrel{?}{=} 0$$
$$22 - 22 \stackrel{?}{=} 0$$
$$0 = 0 \qquad \text{True}$$

Let $x = -2$.

$$x^2 - 9x - 22 = 0$$
$$(-2)^2 - 9(-2) - 22 \stackrel{?}{=} 0$$
$$4 + 18 - 22 \stackrel{?}{=} 0$$
$$22 - 22 \stackrel{?}{=} 0$$
$$0 = 0 \qquad \text{True}$$

The solutions are 11 and -2.

Work Practice Problem 4

PRACTICE PROBLEM 5

Solve: $9x^2 - 24x = -16$

EXAMPLE 5 Solve: $4x^2 - 28x = -49$

Solution: First we rewrite the equation in standard form so that one side is 0. Then we factor the polynomial.

$$4x^2 - 28x = -49$$
$$4x^2 - 28x + 49 = 0 \qquad \text{Write in standard form by adding 49 to both sides.}$$
$$(2x - 7)(2x - 7) = 0 \qquad \text{Factor.}$$

Next we use the zero factor property and set each factor equal to 0. Since the factors are the same, the related equations will give the same solution.

$$2x - 7 = 0 \quad \text{or} \quad 2x - 7 = 0 \qquad \text{Set each factor equal to 0.}$$
$$2x = 7 \qquad\qquad 2x = 7 \qquad \text{Solve.}$$
$$x = \frac{7}{2} \qquad\qquad x = \frac{7}{2}$$

Check: Check this solution in the original equation. The solution is $\frac{7}{2}$.

Work Practice Problem 5

Answers

3. a. 0 and -3 **b.** 0 and $\frac{3}{4}$

4. 6 and -3 **5.** $\frac{4}{3}$

The following steps may be used to solve a quadratic equation by factoring.

To Solve Quadratic Equations by Factoring

Step 1: Write the equation in standard form so that one side of the equation is 0.

Step 2: Factor the quadratic equation completely.

Step 3: Set each factor containing a variable equal to 0.

Step 4: Solve the resulting equations.

Step 5: Check each solution in the original equation.

Since it is not always possible to factor a quadratic polynomial, not all quadratic equations can be solved by factoring. Other methods of solving quadratic equations are presented in Chapter 16.

EXAMPLE 6 Solve: $x(2x - 7) = 4$

Solution: First we write the equation in standard form; then we factor.

$$x(2x - 7) = 4$$

$$2x^2 - 7x = 4 \quad \text{Multiply.}$$

$$2x^2 - 7x - 4 = 0 \quad \text{Write in standard form.}$$

$$(2x + 1)(x - 4) = 0 \quad \text{Factor.}$$

$$2x + 1 = 0 \quad \text{or} \quad x - 4 = 0 \quad \text{Set each factor equal to zero.}$$

$$2x = -1 \qquad x = 4 \quad \text{Solve.}$$

$$x = -\frac{1}{2}$$

Check the solutions in the original equation. The solutions are $-\frac{1}{2}$ and 4.

Work Practice Problem 6

PRACTICE PROBLEM 6

Solve each equation.

a. $x(x - 4) = 5$

b. $x(3x + 7) = 6$

Helpful Hint

To solve the equation $x(2x - 7) = 4$, do **not** set each factor equal to 4. Remember that to apply the zero factor property, one side of the equation must be 0 and the other side of the equation must be in factored form.

✔ Concept Check Explain the error and solve the equation correctly.

$$(x - 3)(x + 1) = 5$$

$$x - 3 = 0 \quad \text{or} \quad x + 1 = 0$$

$$x = 3 \quad \text{or} \quad x = -1$$

Objective B Solving Equations with Degree Greater Than Two by Factoring

Some equations with degree greater than 2 can be solved by factoring and then using the zero factor property.

Answers

6. a. 5 and −1 **b.** $\frac{2}{3}$ and −3

✔ Concept Check Answer

To use the zero factor property, one side of the equation must be 0, not 5. Correctly, $(x - 3)(x + 1) = 5$, $x^2 - 2x - 3 = 5$, $x^2 - 2x - 8 = 0$, $(x - 4)(x + 2) = 0$, $x - 4 = 0$ or $x + 2 = 0$, $x = 4$ or $x = -2$.

PRACTICE PROBLEM 7

Solve: $2x^3 - 18x = 0$

EXAMPLE 7 Solve: $3x^3 - 12x = 0$

Solution: To factor the left side of the equation, we begin by factoring out the greatest common factor, $3x$.

$$3x^3 - 12x = 0$$

$$3x(x^2 - 4) = 0 \quad \text{Factor out the GCF, } 3x.$$

$$3x(x + 2)(x - 2) = 0 \quad \text{Factor } x^2 - 4 \text{, a difference of two squares.}$$

$$3x = 0 \quad \text{or} \quad x + 2 = 0 \quad \text{or} \quad x - 2 = 0 \quad \text{Set each factor equal to 0.}$$

$$x = 0 \qquad x = -2 \qquad x = 2 \quad \text{Solve.}$$

Thus, the equation $3x^3 - 12x = 0$ has three solutions: 0, -2, and 2.

Check: Replace x with each solution in the original equation.

Let $x = 0$.

$$3(0)^3 - 12(0) \stackrel{?}{=} 0$$

$$0 = 0 \quad \text{True}$$

Let $x = -2$.

$$3(-2)^3 - 12(-2) \stackrel{?}{=} 0$$

$$3(-8) + 24 \stackrel{?}{=} 0$$

$$0 = 0 \quad \text{True}$$

Let $x = 2$.

$$3(2)^3 - 12(2) \stackrel{?}{=} 0$$

$$3(8) - 24 \stackrel{?}{=} 0$$

$$0 = 0 \quad \text{True}$$

The solutions are 0, -2, and 2.

Work Practice Problem 7

PRACTICE PROBLEM 8

Solve:
$(x + 3)(3x^2 - 20x - 7) = 0$

EXAMPLE 8 Solve: $(5x - 1)(2x^2 + 15x + 18) = 0$

Solution:

$$(5x - 1)(2x^2 + 15x + 18) = 0$$

$$(5x - 1)(2x + 3)(x + 6) = 0 \quad \text{Factor the trinomial.}$$

$$5x - 1 = 0 \quad \text{or} \quad 2x + 3 = 0 \quad \text{or} \quad x + 6 = 0 \quad \text{Set each factor equal to 0.}$$

$$5x = 1 \qquad 2x = -3 \qquad x = -6 \quad \text{Solve.}$$

$$x = \frac{1}{5} \qquad x = -\frac{3}{2}$$

Check each solution in the original equation. The solutions are $\frac{1}{5}, -\frac{3}{2}$, and -6.

Work Practice Problem 8

Answers

7. 0, 3, and -3 **8.** $-3, -\frac{1}{3}$, and 7

Vocabulary and Readiness Check

Use the choices below to fill in each blank. Not all choices will be used.

$-3, 5$	$a = 0$ or $b = 0$	0	linear
$3, -5$	quadratic	1	

1. An equation that can be written in the form $ax^2 + bx + c = 0$, (with $a \neq 0$), is called a(n) ____________ equation.

2. If the product of two numbers is 0, then at least one of the numbers must be _____.

3. The solutions to $(x - 3)(x + 5) = 0$ are ________.

4. If $a \cdot b = 0$, then __________________.

11.6 EXERCISE SET

FOR EXTRA HELP

Student Solutions Manual

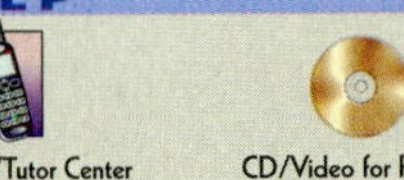
PH Math/Tutor Center

CD/Video for Review

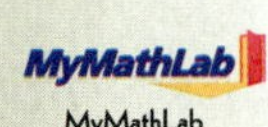
MathXL®

MyMathLab

Objective A *Solve each equation. See Examples 1 through 3.*

1. $(x - 2)(x + 1) = 0$

2. $(x + 3)(x + 2) = 0$

3. $(x - 6)(x - 7) = 0$

4. $(x + 4)(x - 10) = 0$

5. $(x + 9)(x + 17) = 0$

6. $(x - 11)(x - 1) = 0$

7. $x(x + 6) = 0$

8. $x(x - 7) = 0$

9. $3x(x - 8) = 0$

10. $2x(x + 12) = 0$

11. $(2x + 3)(4x - 5) = 0$

12. $(3x - 2)(5x + 1) = 0$

13. $(2x - 7)(7x + 2) = 0$

14. $(9x + 1)(4x - 3) = 0$

15. $\left(x - \frac{1}{2}\right)\left(x + \frac{1}{3}\right) = 0$

16. $\left(x + \frac{2}{9}\right)\left(x - \frac{1}{4}\right) = 0$

17. $(x + 0.2)(x + 1.5) = 0$

18. $(x + 1.7)(x + 2.3) = 0$

Solve. See Examples 4 through 6.

19. $x^2 - 13x + 36 = 0$

20. $x^2 + 2x - 63 = 0$

21. $x^2 + 2x - 8 = 0$

22. $x^2 - 5x + 6 = 0$

23. $x^2 - 7x = 0$

24. $x^2 - 3x = 0$

25. $x^2 + 20x = 0$

26. $x^2 + 15x = 0$

27. $x^2 = 16$

28. $x^2 = 9$

29. $x^2 - 4x = 32$

30. $x^2 - 5x = 24$

31. $(x + 4)(x - 9) = 4x$

32. $(x + 3)(x + 8) = x$

33. $x(3x - 1) = 14$

34. $x(4x - 11) = 3$

35. $3x^2 + 19x - 72 = 0$

36. $36x^2 + x - 21 = 0$

Objectives **A** **B** **Mixed Practice** *Solve each equation. See Examples 1 through 8. (A few exercises are linear equations.)*

37. $4x^3 - x = 0$

38. $4y^3 - 36y = 0$

39. $4(x - 7) = 6$

40. $5(3 - 4x) = 9$

41. $(4x - 3)(16x^2 - 24x + 9) = 0$

42. $(2x + 5)(4x^2 + 20x + 25) = 0$

43. $4y^2 - 1 = 0$

44. $4y^2 - 81 = 0$

45. $(2x + 3)(2x^2 - 5x - 3) = 0$

46. $(2x - 9)(x^2 + 5x - 36) = 0$

47. $x^2 - 15 = -2x$

48. $x^2 - 26 = -11x$

49. $30x^2 - 11x = 30$

50. $12x^2 + 7x - 12 = 0$

51. $5x^2 - 6x - 8 = 0$

52. $9x^2 + 7x = -2$

53. $6y^2 - 22y - 40 = 0$

54. $3x^2 - 6x - 9 = 0$

55. $(y - 2)(y + 3) = 6$

56. $(y - 5)(y - 2) = 28$

57. $x^3 - 12x^2 + 32x = 0$

58. $x^3 - 14x^2 + 49x = 0$

59. $x^2 + 14x + 49 = 0$

60. $x^2 + 22x + 121 = 0$

61. $12y = 8y^2$

62. $9y = 6y^2$

63. $7x^3 - 7x = 0$

64. $3x^3 - 27x = 0$

65. $3x^2 + 8x - 11 = 13 - 6x$

66. $2x^2 + 12x - 1 = 4 + 3x$

67. $3x^2 - 20x = -4x^2 - 7x - 6$

68. $4x^2 - 20x = -5x^2 - 6x - 5$

Review

Perform each indicated operation. Write all results in lowest terms. See Sections 4.3 and 4.5.

69. $\frac{3}{5} + \frac{4}{9}$

70. $\frac{2}{3} + \frac{3}{7}$

71. $\frac{7}{10} - \frac{5}{12}$

72. $\frac{5}{9} - \frac{5}{12}$

73. $\frac{4}{5} \cdot \frac{7}{8}$

74. $\frac{3}{7} \cdot \frac{12}{17}$

Concept Extensions

For Exercises 75 and 76, see the Concept Check in this section.

75. Explain the error and solve correctly:

$x(x - 2) = 8$

$x = 8$ or $x - 2 = 8$

$x = 10$

76. Explain the error and solve correctly:

$(x - 4)(x + 2) = 0$

$x = -4$ or $x = 2$

77. Write a quadratic equation that has two solutions, 6 and -1. Leave the polynomial in the equation in factored form.

78. Write a quadratic equation that has two solutions, 0 and -2. Leave the polynomial in the equation in factored form.

79. Write a quadratic equation in standard form that has two solutions, 5 and 7.

80. Write an equation that has three solutions, 0, 1, and 2.

81. A compass is accidentally thrown upward and out of an air balloon at a height of 300 feet. The height, y, of the compass at time x is given by the equation $y = -16x^2 + 20x + 300$.

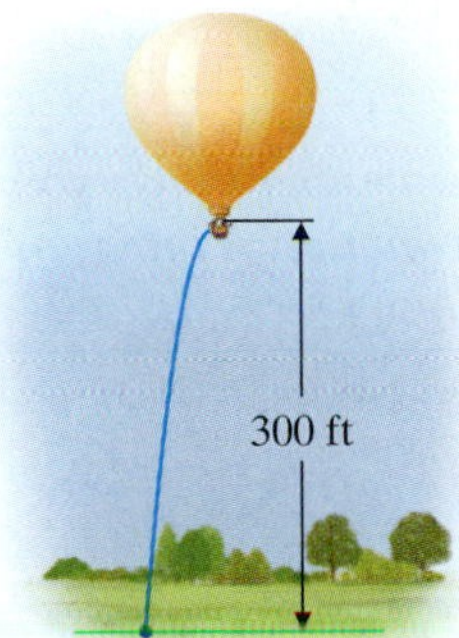

a. Find the height of the compass at the given times by filling in the table below.

Time, x (in seconds)	0	1	2	3	4	5	6
Height, y (in feet)							

b. Use the table to determine when the compass strikes the ground.

c. Use the table to approximate the maximum height of the compass.

82. A rocket is fired upward from the ground with an initial velocity of 100 feet per second. The height, y, of the rocket at any time x is given by the equation $y = -16x^2 + 100x$.

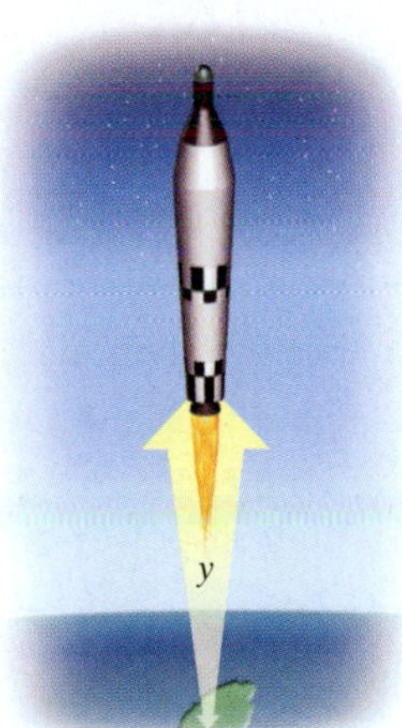

a. Find the height of the rocket at the given times by filling in the table below.

Time, x (in seconds)	0	1	2	3	4	5	6	7
Height, y (in feet)								

b. Use the table to determine between what two whole-numbered seconds the rocket strikes the ground.

c. Use the table to approximate the maximum height of the rocket.

Solve each equation.

83. $(x - 3)(3x + 4) = (x + 2)(x - 6)$

84. $(2x - 3)(x + 6) = (x - 9)(x + 2)$

85. $(2x - 3)(x + 8) = (x - 6)(x + 4)$

86. $(x + 6)(x - 6) = (2x - 9)(x + 4)$

THE BIGGER PICTURE Simplifying Expressions and Solving Equations

Now we continue our outline from Sections 9.6 and 10.7. Although suggestions are given, this outline should be in your own words. Once you complete this new portion, try the exercises below.

I. Simplifying Expressions

A. Exponents (Sections 10.1, 10.2)

B. Polynomials

1. Add (Section 10.4)
2. Subtract (Section 10.4)
3. Multiply (Section 10.5)
4. Divide (Section 10.7)

C. Factoring Polynomials—see the Chapter 11 Integrated Review for steps.

$3x^4 - 78x^2 + 75$

$= 3(x^4 - 26x^2 + 25)$ Factor out GCF—always first step.

$= 3(x^2 - 25)(x^2 - 1)$ Factor trinomial.

$= 3(x + 5)(x - 5)(x + 1)(x - 1)$ Factor further—each difference of squares.

II. Solving Equations and Inequalities

A. Linear Equations (Section 9.3)

B. Linear Inequalities (Section 9.6)

C. Quadratic & Higher Degree Equations (Solving by Factoring)—highest power on variable is at least 2 when equation is written in standard form (set equal to 0).

$x^2 + x = 6$

$x^2 + x - 6 = 0$ Write the equation in standard form (set it equal to 0).

$(x - 2)(x + 3) = 0$ Factor.

$x = 2$ or $x = -3$ Set each factor equal to 0 and solve.

Simplify each expression.

1. $-7 + (-27)$

2. $\dfrac{(x^3)^4}{(x^{-2})^5}$

3. $(x^3 - 6x^2 + 2) - (5x^3 - 6)$

4. $\dfrac{3y^3 - 3y^2 + 9}{3y^2}$

Factor each expression.

5. $10x^3 - 250x$

6. $x^2 - 36x + 35$

7. $6xy + 15x - 6y - 15$

8. $5xy^2 - 2xy - 7x$

Solve each equation. Remember to use your outline to determine whether the equation is linear or quadratic and how to proceed with solving.

9. $(x - 5)(2x + 1) = 0$

10. $5x - 5 = 0$

11. $x(x - 12) = 28$

12. $7(x - 3) + 2(5x + 1) = 14$

11.7 QUADRATIC EQUATIONS AND PROBLEM SOLVING

Objective

A Solve Problems That Can Be Modeled by Quadratic Equations.

Objective A Solving Problems Modeled by Quadratic Equations

Some problems may be modeled by quadratic equations. To solve these problems, we use the same problem-solving steps that were introduced in Section 3.4. When solving these problems, keep in mind that a solution of an equation that models a problem may not be a solution to the problem. For example, a person's age or the length of a rectangle is always a positive number. Thus we discard solutions that do not make sense as solutions of the problem.

EXAMPLE 1 Finding Free-Fall Time

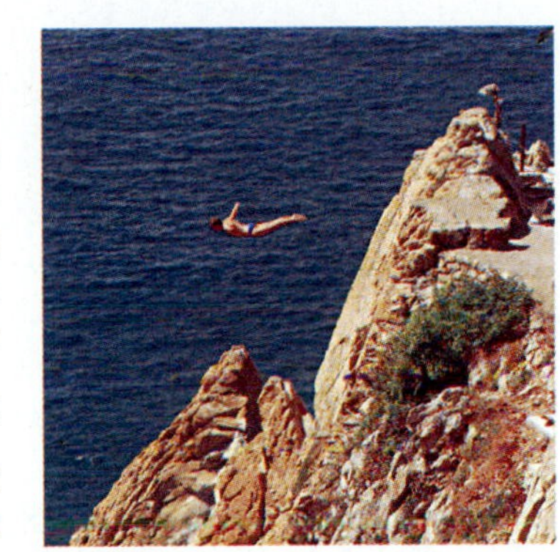

Since the 1940s, one of the top tourist attractions in Acapulco, Mexico, is watching the cliff divers off the La Quebrada. The divers' platform is about 144 feet above the sea. These divers must time their descent just right, since they land in the crashing Pacific, in an inlet that is at most $9\frac{1}{2}$ feet deep. Neglecting air resistance, the height h in feet of a cliff diver above the ocean after t seconds is given by the quadratic equation $h = -16t^2 + 144$.

Find out how long it takes the diver to reach the ocean.

Solution:

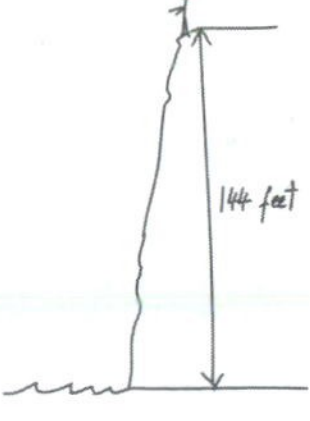

1. UNDERSTAND. Read and reread the problem. Then draw a picture of the problem.

   The equation $h = -16t^2 + 144$ models the height of the falling diver at time t. Familiarize yourself with this equation by find the height of the diver at time $t = 1$ second and $t = 2$ seconds.

   When $t = 1$ second, the height of the diver is $h = -16(1)^2 + 144 = 128$ feet.
   When $t = 2$ seconds, the height of the diver is $h = -16(2)^2 + 144 = 80$ feet.

2. TRANSLATE. To find out how long it takes the diver to reach the ocean, we want to know the value of t for which $h = 0$.

$$0 = -16t^2 + 144$$

$$0 = -16(t^2 - 9) \quad \text{Factor out } -16.$$

$$0 = -16(t - 3)(t + 3) \quad \text{Factor completely.}$$

$$t - 3 = 0 \quad \text{or} \quad t + 3 = 0 \quad \text{Set each factor containing a variable equal to 0.}$$

$$t = 3 \quad \text{or} \quad t = -3 \quad \text{Solve.}$$

3. INTERPRET. Since the time t cannot be negative, the proposed solution is 3 seconds.

Check: Verify that the height of the diver when t is 3 seconds is 0.

When $t = 3$ seconds, $h = -16(3)^2 + 144 = -144 + 144 = 0$.

Work Practice Problem 1

PRACTICE PROBLEM 1

Cliff divers also frequent the falls at Waimea Falls Park in Oahu, Hawaii. Here, a diver can jump from a ledge 64 feet up the waterfall into a rocky pool below. Neglecting air resistance, the height of a diver above the pool after t seconds is $h = -16t^2 + 64$. Find how long it takes the diver to reach the pool.

Answer

1. 2 sec

PRACTICE PROBLEM 2

The square of a number minus twice the number is 63. Find the number.

EXAMPLE 2 Finding a Number

The square of a number plus three times the number is 70. Find the number.

Solution:

1. UNDERSTAND. Read and reread the problem. Suppose that the number is 5. The square of 5 is 5^2 or 25. Three times 5 is 15. Then $25 + 15 = 40$, not 70, so the number must be greater than 5. Remember, the purpose of proposing a number, such as 5, is to better understand the problem. Now that we do, we will let $x =$ the number.
2. TRANSLATE.

the square of a number	plus	three times the number	is	70
↓	↓	↓	↓	↓
x^2	$+$	$3x$	$=$	70

3. SOLVE.

$$x^2 + 3x = 70$$
$$x^2 + 3x - 70 = 0 \quad \text{Subtract 70 from both sides.}$$
$$(x + 10)(x - 7) = 0 \quad \text{Factor.}$$
$$x + 10 = 0 \quad \text{or} \quad x - 7 = 0 \quad \text{Set each factor equal to 0.}$$
$$x = -10 \qquad x = 7 \quad \text{Solve.}$$

4. INTERPRET.

Check: The square of -10 is $(-10)^2$, or 100. Three times -10 is $3(-10)$ or -30. Then $100 + (-30) = 70$, the correct sum, so -10 checks.

The square of 7 is 7^2 or 49. Three times 7 is 3(7), or 21. Then $49 + 21 = 70$, the correct sum, so 7 checks.

State: There are two numbers. They are -10 and 7.

Work Practice Problem 2

PRACTICE PROBLEM 3

The length of a rectangular garden is 5 feet more than its width. The area of the garden is 176 square feet. Find the length and the width of the garden.

Answers

2. 9 and −7
3. length: 16 ft; width: 11 ft

EXAMPLE 3 Finding the Dimensions of a Sail

The height of a triangular sail is 2 meters less than twice the length of the base. If the sail has an area of 30 square meters, find the length of its base and the height.

Solution:

1. UNDERSTAND. Read and reread the problem. Since we are finding the length of the base and the height, we let

   $x =$ the length of the base

   Since the height is 2 meters less than twice the length of the base,

   $2x - 2 =$ the height

   An illustration is shown on the next page.
2. TRANSLATE. We are given that the area of the triangle is 30 square meters, so we use the formula for area of a triangle.

area of triangle	=	$\frac{1}{2}$	·	base	·	height
↓		↓		↓		↓
30	=	$\frac{1}{2}$	·	x	·	$(2x - 2)$

3. SOLVE. Now we solve the quadratic equation.

$$30 = \frac{1}{2}x(2x - 2)$$

$$30 = x^2 - x \quad \text{Multiply.}$$

$$0 = x^2 - x - 30 \quad \text{Write in standard form.}$$

$$0 = (x - 6)(x + 5) \quad \text{Factor.}$$

$$x - 6 = 0 \quad \text{or} \quad x + 5 = 0 \quad \text{Set each factor equal to 0.}$$

$$x = 6 \qquad x = -5$$

4. INTERPRET. Since x represents the length of the base, we discard the solution -5. The base of a triangle cannot be negative. The base is then 6 meters and the height is $2(6) - 2 = 10$ meters.

Check: To check this problem, we recall that

$$\text{area} = \frac{1}{2}\text{base} \cdot \text{height or}$$

$$30 \stackrel{?}{=} \frac{1}{2}(6)(10)$$

$$30 = 30 \quad \text{True}$$

State: The base of the triangular sail is 6 meters and the height is 10 meters.

Work Practice Problem 3

The next examples make use of the **Pythagorean theorem** and consecutive integers. Before we review this theorem, recall that a **right triangle** is a triangle that contains a 90° or right angle. The **hypotenuse** of a right triangle is the side opposite the right angle and is the longest side of the triangle. The **legs** of a right triangle are the other sides of the triangle.

Pythagorean Theorem

In a right triangle, the sum of the squares of the lengths of the two legs is equal to the square of the length of the hypotenuse.

$$(\text{leg})^2 + (\text{leg})^2 = (\text{hypotenuse})^2 \quad \text{or} \quad a^2 + b^2 = c^2$$

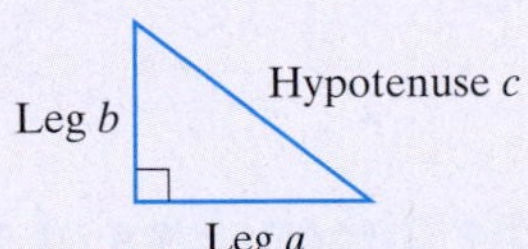

Study the following diagrams for a review of consecutive integers.

Examples

If x is the first integer, then consecutive integers are $x, x + 1, x + 2, \ldots$

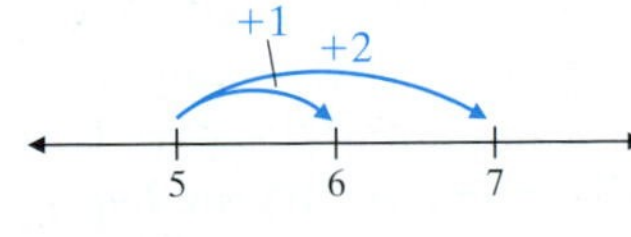

If x is the first even integer, then consecutive even integers are $x, x + 2, x + 4, \ldots$

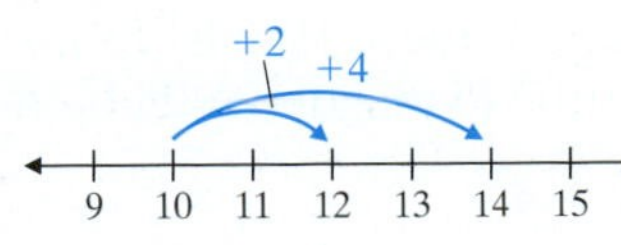

If x is the first odd integer, then consecutive odd integers are $x, x + 2, x + 4, \ldots$

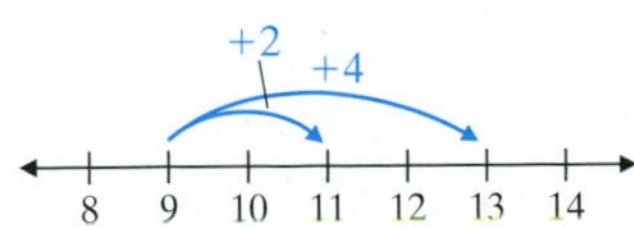

PRACTICE PROBLEM 4

Find two consecutive odd integers whose product is 23 more than their sum.

EXAMPLE 4 Finding Consecutive Even Integers

Find two consecutive even integers whose product is 34 more than their sum.

Solution:

1. UNDERSTAND. Read and reread the problem. Let's just choose two consecutive even integers to help us better understand the problem. Let's choose 10 and 12. Their product is $10(12) = 120$ and their sum is $10 + 12 = 22$. The product is $120 - 22$, or 98 greater than the sum. Thus our guess is incorrect, but we have a better understanding of this example.

   Let's let x and $x + 2$ be the consecutive even integers.

2. TRANSLATE.

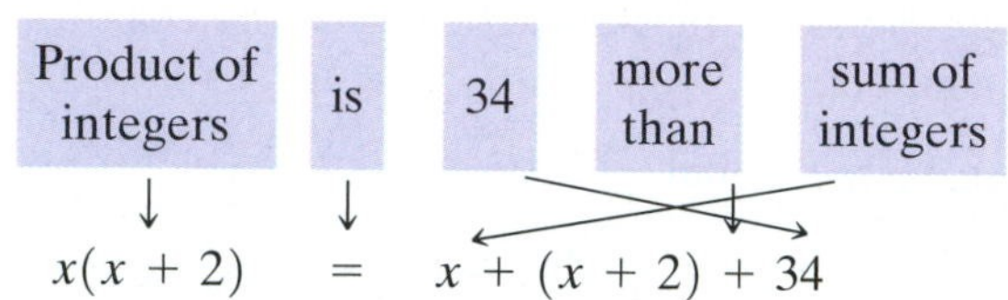

$$x(x + 2) \quad = \quad x + (x + 2) + 34$$

3. SOLVE. Now we solve the equation.

$$x(x + 2) = x + (x + 2) + 34$$
$$x^2 + 2x = x + x + 2 + 34 \quad \text{Multiply.}$$
$$x^2 + 2x = 2x + 36 \quad \text{Combine like terms.}$$
$$x^2 - 36 = 0 \quad \text{Write in standard form.}$$
$$(x + 6)(x - 6) = 0 \quad \text{Factor.}$$
$$x + 6 = 0 \quad \text{or} \quad x - 6 = 0 \quad \text{Set each factor equal to 0.}$$
$$x = -6 \qquad\qquad x = 6 \quad \text{Solve.}$$

4. INTERPRET. If $x = -6$, then $x + 2 = -6 + 2$, or -4.
   If $x = 6$, then $x + 2 = 6 + 2$, or 8.

Check: $-6, -4$

$$-6(-4) \stackrel{?}{=} -6 + (-4) + 34$$
$$24 \stackrel{?}{=} -10 + 34$$
$$24 = 24 \quad \text{True}$$

$6, 8$

$$6(8) \stackrel{?}{=} 6 + 8 + 34$$
$$48 \stackrel{?}{=} 14 + 34$$
$$48 = 48 \quad \text{True}$$

State: The two consecutive even integers are -6 and -4 or 6 and 8.

Work Practice Problem 4

PRACTICE PROBLEM 5

The length of one leg of a right triangle is 7 meters less than the length of the other leg. The length of the hypotenuse is 13 meters. Find the lengths of the legs.

EXAMPLE 5 Finding the Dimensions of a Triangle

Find the lengths of the sides of a right triangle if the lengths can be expressed as three consecutive even integers.

Solution:

1. UNDERSTAND. Read and reread the problem. Let's suppose that the length of one leg of the right triangle is 4 units. Then the other leg is the next even integer, or 6 units, and the hypotenuse of the triangle is the next even integer, or 8 units. Remember that the hypotenuse is the longest side. Let's see if a triangle with sides of these lengths forms a right triangle. To do this, we check to see whether the Pythagorean theorem holds true.

$$4^2 + 6^2 \stackrel{?}{=} 8^2$$
$$16 + 36 \stackrel{?}{=} 64$$
$$52 = 64 \quad \text{False}$$

4 units

8 units

6 units

Answers

4. 5 and 7 or -5 and -3
5. 5 meters, 12 meters

Our proposed numbers do not check, but we now have a better understanding of the problem.

We let $x, x + 2$, and $x + 4$ be three consecutive even integers. Since these integers represent lengths of the sides of a right triangle, we have the following.

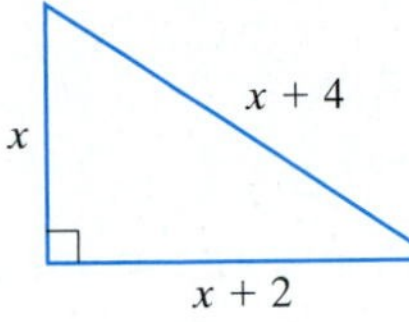

x = one leg

$x + 2$ = other leg

$x + 4$ = hypotenuse (longest side)

2. TRANSLATE. By the Pythagorean theorem, we have that

$$(\text{leg})^2 + (\text{leg})^2 = (\text{hypotenuse})^2$$
$$(x)^2 + (x + 2)^2 = (x + 4)^2$$

3. SOLVE. Now we solve the equation.

$$x^2 + (x + 2)^2 = (x + 4)^2$$
$$x^2 + x^2 + 4x + 4 = x^2 + 8x + 16 \quad \text{Multiply.}$$
$$2x^2 + 4x + 4 = x^2 + 8x + 16 \quad \text{Combine like terms.}$$
$$x^2 - 4x - 12 = 0 \quad \text{Write in standard form.}$$
$$(x - 6)(x + 2) = 0 \quad \text{Factor.}$$
$$x - 6 = 0 \quad \text{or} \quad x + 2 = 0 \quad \text{Set each factor equal to 0.}$$
$$x = 6 \qquad x = -2$$

4. INTERPRET. We discard $x = -2$ since length cannot be negative. If $x = 6$, then $x + 2 = 8$ and $x + 4 = 10$.

Check: Verify that

$$(\text{leg})^2 + (\text{leg})^2 = (\text{hypotenuse})^2$$
$$6^2 + 8^2 \stackrel{?}{=} 10^2$$
$$36 + 64 \stackrel{?}{=} 100$$
$$100 = 100 \quad \text{True}$$

State: The sides of the right triangle have lengths 6 units, 8 units, and 10 units.

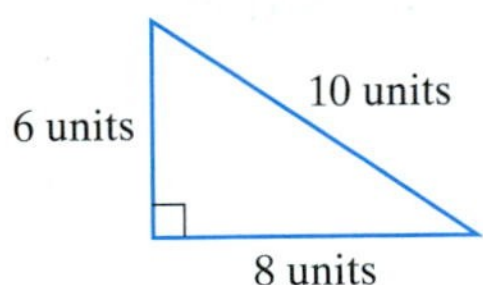

Work Practice Problem 5

11.7 EXERCISE SET

FOR EXTRA HELP

Student Solutions Manual | PH Math/Tutor Center | CD/Video for Review | MathXL® | MyMathLab

Objective A *See Examples 1 through 5 for all exercises.*

Translating *For Exercises 1 through 6, represent each given condition using a single variable, x.*

△ **1.** The length and width of a rectangle whose length is 4 centimeters more than its width

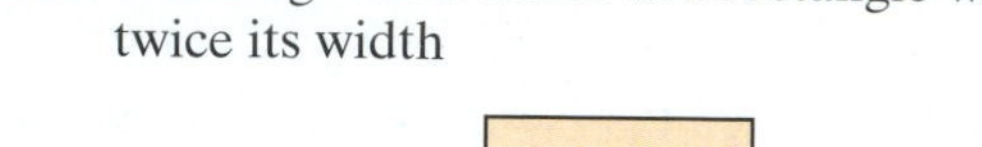

△ **2.** The length and width of a rectangle whose length is twice its width

3. Two consecutive odd integers

4. Two consecutive even integers

△ **5.** The base and height of a triangle whose height is one more than four times its base

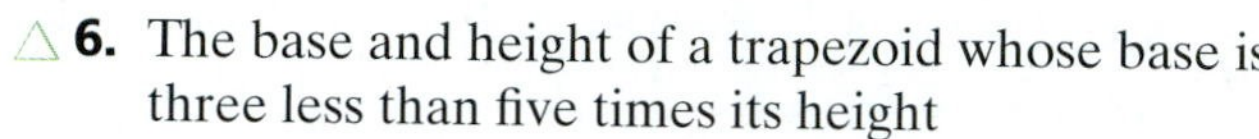

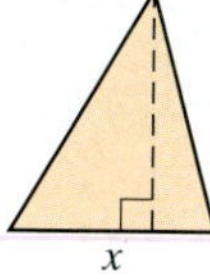

△ **6.** The base and height of a trapezoid whose base is three less than five times its height

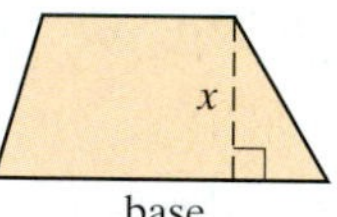

Use the information given to find the dimensions of each figure.

△ **7.**

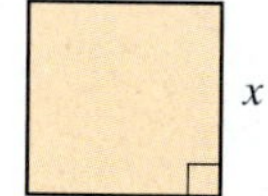

The *area* of the square is 121 square units. Find the length of its sides.

△ **8.**

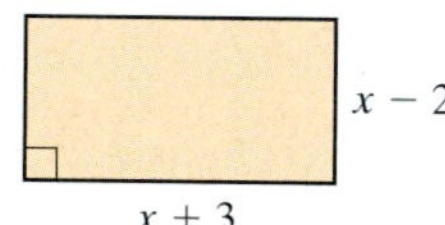

The *area* of the rectangle is 84 square inches. Find its length and width.

△ **9.**

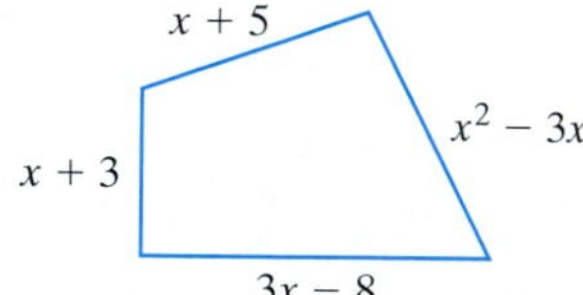

The *perimeter* of the quadrilateral is 120 centimeters. Find the lengths of the sides.

△ **10.**

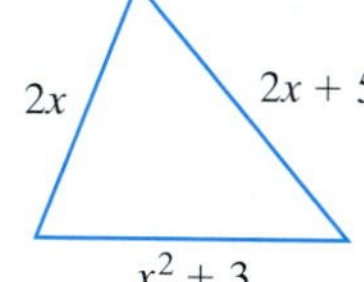

The *perimeter* of the triangle is 85 feet. Find the lengths of its sides.

△ **11.**

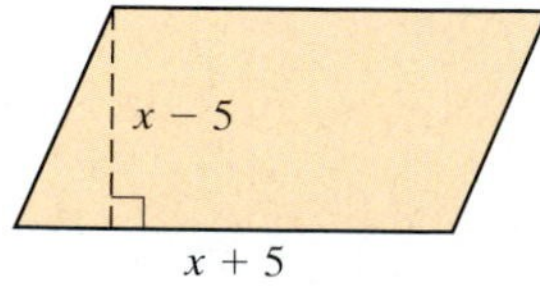

The *area* of the parallelogram is 96 square miles. Find its base and height.

△ **12.**

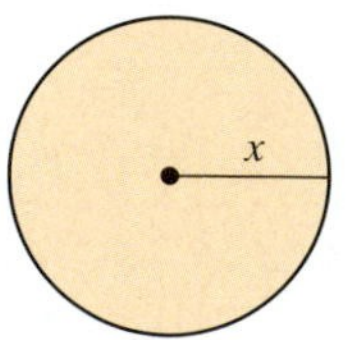

The *area* of the circle is 25π square kilometers. Find its radius.

Solve.

13. An object is thrown upward from the top of an 80-foot building with an initial velocity of 64 feet per second. The height h of the object after t seconds is given by the quadratic equation $h = -16t^2 + 64t + 80$. When will the object hit the ground?

14. A hang glider accidentally drops her compass from the top of a 400-foot cliff. The height h of the compass after t seconds is given by the quadratic equation $h = -16t^2 + 400$. When will the compass hit the ground?

15. The width of a rectangle is 7 centimeters less than twice its length. Its area is 30 square centimeters. Find the dimensions of the rectangle.

16. The length of a rectangle is 9 inches more than its width. Its area is 112 square inches. Find the dimensions of the rectangle.

△ *The equation* $D = \frac{1}{2}n(n - 3)$ *gives the number of diagonals D for a polygon with n sides. For example, a polygon with 6 sides has* $D = \frac{1}{2} \cdot 6(6 - 3)$ *or* $D = 9$ *diagonals. (See if you can count all 9 diagonals. Some are shown in the figure.)*

Use this equation, $D = \frac{1}{2}n(n - 3)$, *for Exercises 17 through 20.*

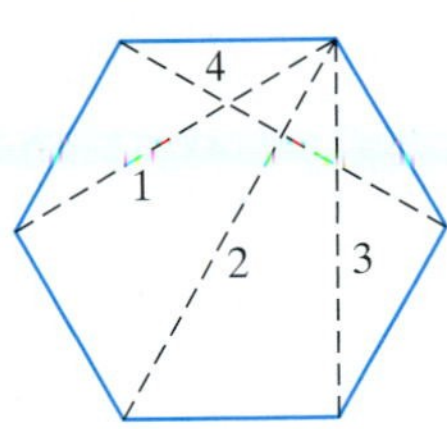

17. Find the number of diagonals for a polygon that has 12 sides.

18. Find the number of diagonals for a polygon that has 15 sides.

19. Find the number of sides n for a polygon that has 35 diagonals.

20. Find the number of sides n for a polygon that has 14 diagonals.

21. The sum of a number and its square is 132. Find the number.

22. The sum of a number and its square is 182. Find the number.

23. The product of two consecutive room numbers is 210. Find the room numbers.

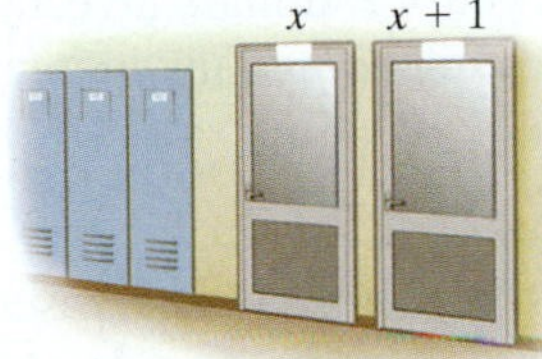

24. The product of two consecutive page numbers is 420. Find the page numbers.

25. A ladder is leaning against a building so that the distance from the ground to the top of the ladder is one foot less than the length of the ladder. Find the length of the ladder if the distance from the bottom of the ladder to the building is 5 feet.

26. Use the given figure to find the length of the guy wire.

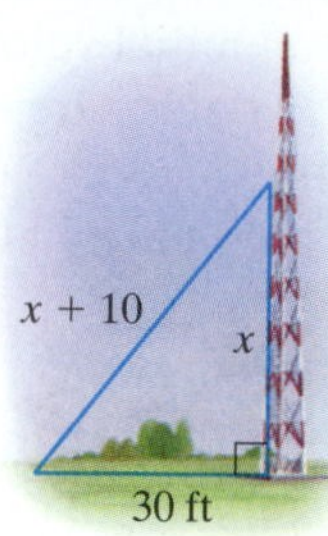

△ **27.** If the sides of a square are increased by 3 inches, the area becomes 64 square inches. Find the length of the sides of the original square.

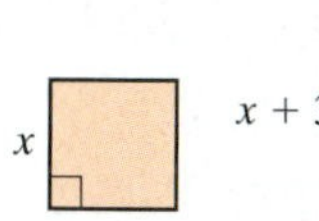

△ **28.** If the sides of a square are increased by 5 meters, the area becomes 100 square meters. Find the length of the sides of the original square.

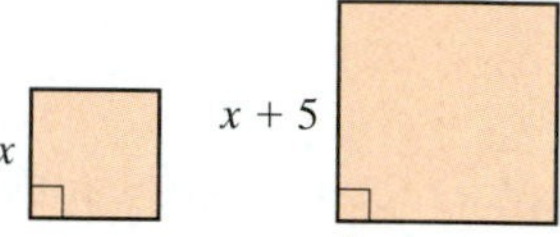

△ **29.** One leg of a right triangle is 4 millimeters longer than the smaller leg and the hypotenuse is 8 millimeters longer than the smaller leg. Find the lengths of the sides of the triangle.

△ **30.** One leg of a right triangle is 9 centimeters longer than the other leg and the hypotenuse is 45 centimeters. Find the lengths of the legs of the triangle.

△ **31.** The length of the base of a triangle is twice its height. If the area of the triangle is 100 square kilometers, find the height.

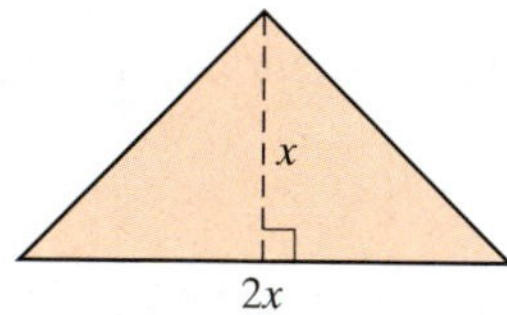

△ **32.** The height of a triangle is 2 millimeters less than the base. If the area is 60 square millimeters, find the base.

△ **33.** Find the length of the shorter leg of a right triangle if the longer leg is 12 feet more than the shorter leg and the hypotenuse is 12 feet less than twice the shorter leg.

△ **34.** Find the length of the shorter leg of a right triangle if the longer leg is 10 miles more than the shorter leg and the hypotenuse is 10 miles less than twice the shorter leg.

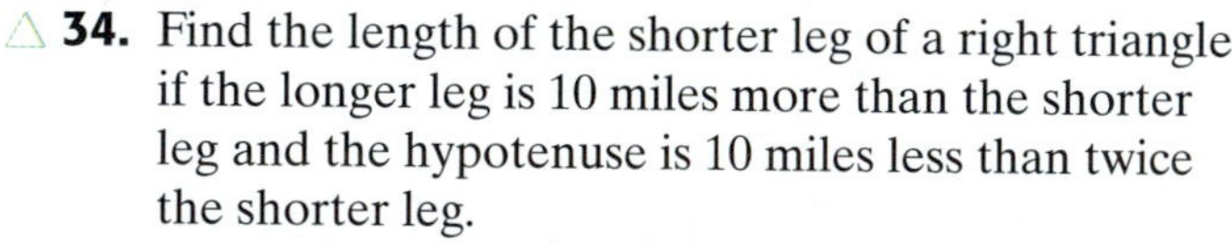

35. An object is dropped from 39 feet below the tip of the pinnacle atop one of the 1483-foot-tall Petronas Twin Towers in Kuala Lumpur, Malaysia. (*Source:* Council on Tall Buildings and Urban Habitat) The height h of the object after t seconds is given by the equation $h = -16t^2 + 1444$. Find how many seconds pass before the object reaches the ground.

36. An object is dropped from the top of 311 South Wacker Drive, a 961-foot-tall office building in Chicago. (*Source:* Council on Tall Buildings and Urban Habitat) The height h of the object after t seconds is given by the equation $h = -16t^2 + 961$. Find how many seconds pass before the object reaches the ground.

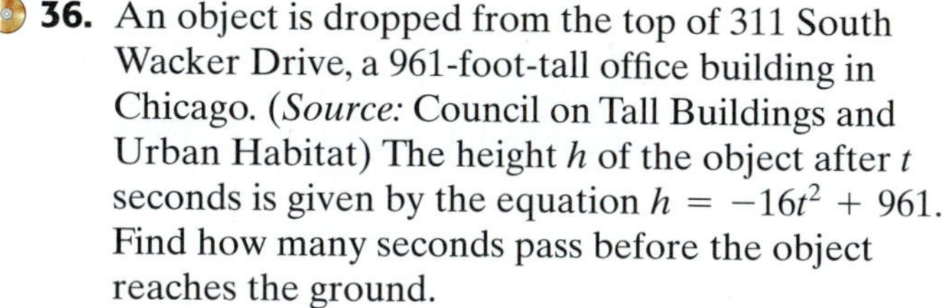

37. At the end of 2 years, P dollars invested at an interest rate r compounded annually increases to an amount, A dollars, given by

$$A = P(1 + r)^2$$

Find the interest rate if \$100 increased to \$144 in 2 years. Write your answer as a percent.

38. At the end of 2 years, P dollars invested at an interest rate r compounded annually increases to an amount, A dollars, given by

$$A = P(1 + r)^2$$

Find the interest rate if \$2000 increased to \$2420 in 2 years. Write your answer as a percent.

△ **39.** Find the dimensions of a rectangle whose width is 7 miles less than its length and whose area is 120 square miles.

△ **40.** Find the dimensions of a rectangle whose width is 2 inches less than half its length and whose area is 160 square inches.

41. If the cost, C, for manufacturing x units of a certain product is given by $C = x^2 - 15x + 50$, find the number of units manufactured at a cost of \$9500.

42. If a switchboard handles n telephones, the number C of telephone connections it can make simultaneously is given by the equation $C = \frac{n(n-1)}{2}$. Find how many telephones are handled by a switchboard making 120 telephone connections simultaneously.

Review

The following double line graph shows a comparison of the amount of land (in thousand acres) occupied by farms in Florida during the years shown with the amount of land occupied by farms in Georgia. Use this graph to answer Exercises 43 through 49. See Section 7.1.

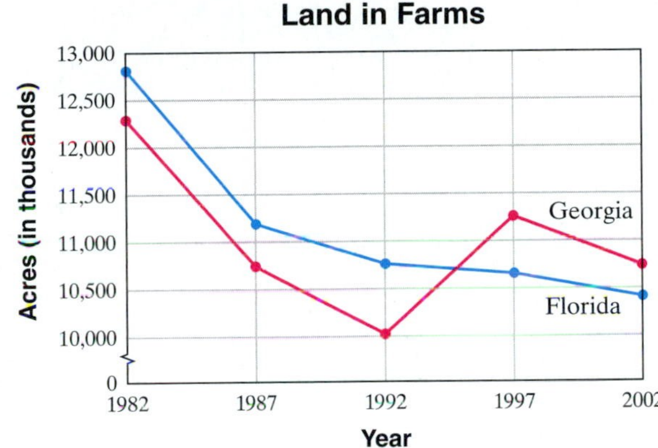

43. Approximate the amount of land occupied by farms in Georgia in 1997.

44. Approximate the amount of land occupied by farms in Florida in 1997.

45. Approximate the amount of land occupied by farms in Georgia in 1987.

46. Approximate the amount of land occupied by farms in Florida in 1987.

47. Approximate the year that the colored lines in this graph intersect.

48. In your own words, explain the meaning of the point of intersection in the graph.

49. Describe the trends shown in this graph and speculate as to why these trends have occurred.

Concept Extensions

△ **50.** Two boats travel at right angles to each other after leaving the same dock at the same time. One hour later the boats are 17 miles apart. If one boat travels 7 miles per hour faster than the other boat, find the rate of each boat.

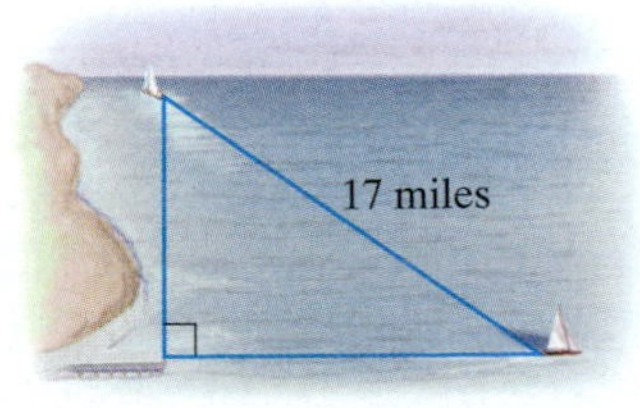

△ **51.** The side of a square equals the width of a rectangle. The length of the rectangle is 6 meters longer than its width. The sum of the areas of the square and the rectangle is 176 square meters. Find the side of the square.

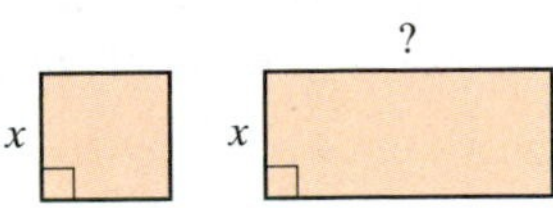

52. The sum of two numbers is 20, and the sum of their squares is 218. Find the numbers.

53. The sum of two numbers is 25, and the sum of their squares is 325. Find the numbers.

△ **54.** According to the International America's Cup Class (IACC) rule, a sailboat competing in the America's Cup match must have a 110-foot-tall mast and a combined mainsail and jib sail area of 3000 square feet. (*Source:* America's Cup Organizing Committee) A design for an IACC-class sailboat calls for the mainsail to be 60% of the combined sail area. If the height of the triangular mainsail is 28 feet more than twice the length of the boom, find the length of the boom and the height of the mainsail.

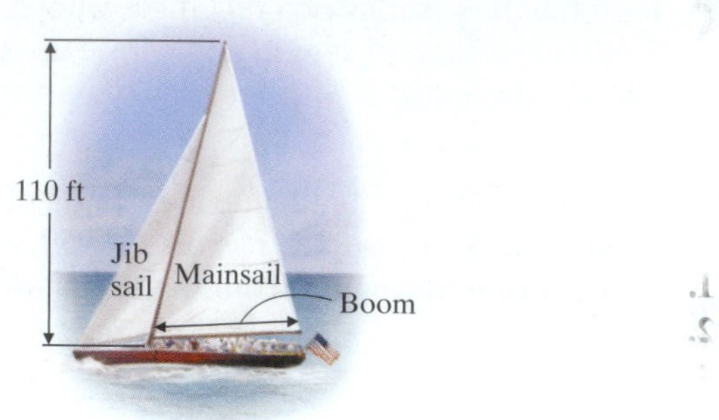

△ **55.** A rectangular pool is surrounded by a walk 4 meters wide. The pool is 6 meters longer than its width. If the total area of the pool and walk is 576 square meters more than the area of the pool, find the dimensions of the pool.

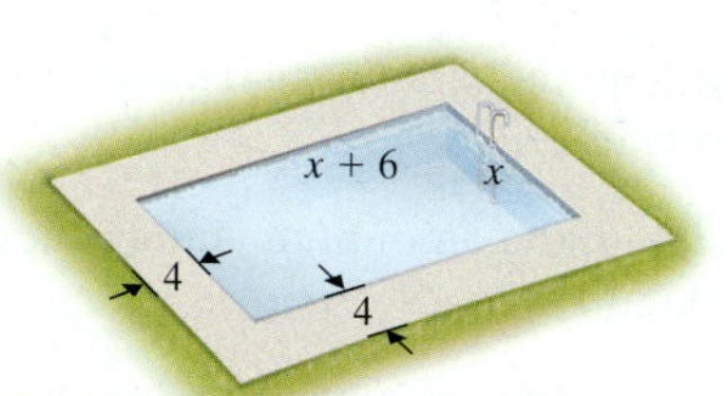

△ **56.** A rectangular garden is surrounded by a walk of uniform width. The area of the garden is 180 square yards. If the dimensions of the garden plus the walk are 16 yards by 24 yards, find the width of the walk.

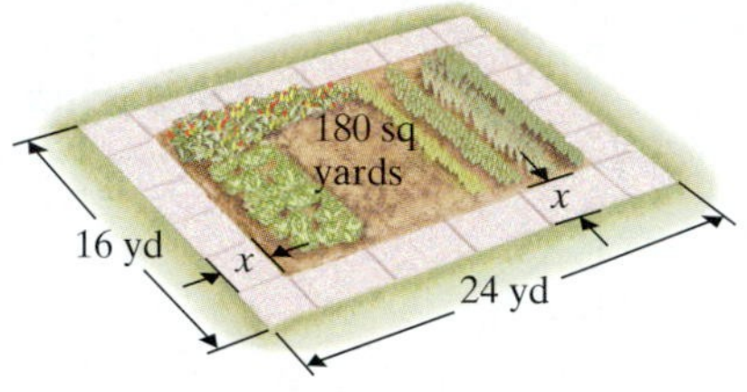

CHAPTER 11 Group Activity

Factoring polynomials can be visualized using areas of rectangles. To see this, let's first find the areas of the following squares and rectangles. (Recall that Area = Length · Width.)

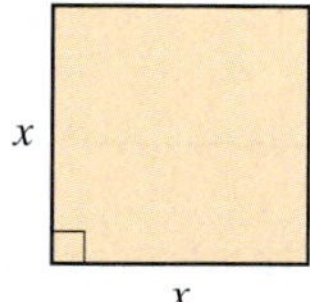

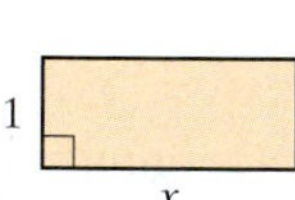

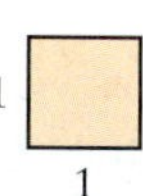

To use these areas to visualize factoring the polynomial $x^2 + 3x + 2$, for example, use the shapes below to form a rectangle. The factored form is found by reading the length and the width of the rectangle as shown below.

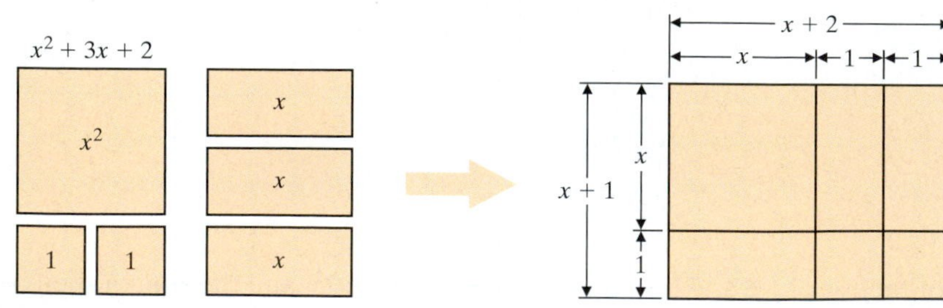

Thus, $x^2 + 3x + 2 = (x + 2)(x + 1)$.

Try using this method to visualize the factored form of each polynomial below.

Work in a group and use tiles to find the factored form of the polynomials below. (Tiles can be hand made from index cards.)

1. $x^2 + 6x + 5$

2. $x^2 + 5x + 6$

3. $x^2 + 5x + 4$

4. $x^2 + 4x + 3$

5. $x^2 + 6x + 9$

6. $x^2 + 4x + 4$

Chapter 11 Vocabulary Check

Fill in each blank with one of the words or phrases listed below.

factoring	quadratic equation
greatest common factor	perfect square trinomial

1. An equation that can be written in the form $ax^2 + bx + c = 0$ (with a not 0) is called a ________________.

2. ____________ is the process of writing an expression as a product.

3. The ________________________ of a list of terms is the product of all common factors.

4. A trinomial that is the square of some binomial is called a ________________________.

Helpful Hint

Are you preparing for your test? Don't forget to take the Chapter 11 Test on page 866. Then check your answers at the back of the text and use the Chapter Test Prep Video CD to see the fully worked-out solutions to any of the exercises you want to review.

11 Chapter Highlights

DEFINITIONS AND CONCEPTS	**EXAMPLES**
Section 11.1 The Greatest Common Factor	
Factoring is the process of writing an expression as a product.	Factor: $6 = 2 \cdot 3$ Factor: $x^2 + 5x + 6 = (x + 2)(x + 3)$
The GCF of a list of variable terms contains the smallest exponent on each common variable.	The GCF of z^5, z^3, and z^{10} is z^3.
The GCF of a list of terms is the product of all common factors.	Find the GCF of $8x^2y$, $10x^3y^2$, and $50x^2y^3$. $8x^2y = 2 \cdot 2 \cdot 2 \cdot x^2 \cdot y$ $10x^3y^2 = 2 \cdot 5 \cdot x^3 \cdot y^2$ $50x^2y^3 = 2 \cdot 5 \cdot 5 \cdot x^2 \cdot y^3$ $\text{GCF} = 2 \cdot x^2 \cdot y$ or $2x^2y$
TO FACTOR BY GROUPING **Step 1.** Group the terms in two groups so that each group has a common factor. **Step 2.** Factor out the GCF from each group. **Step 3.** If there is a common binomial factor, factor it out. **Step 4.** If not, rearrange the terms and try these steps again.	Factor: $10ax + 15a - 6xy - 9y$ **Step 1.** $(10ax + 15a) + (-6xy - 9y)$ **Step 2.** $5a(2x + 3) - 3y(2x + 3)$ **Step 3.** $(2x + 3)(5a - 3y)$
Section 11.2 Factoring Trinomials of the Form $x^2 + bx + c$	
The product of these numbers is c. $x^2 + bx + c = (x + \square)(x + \square)$ The sum of these numbers is b.	Factor: $x^2 + 7x + 12$ $3 + 4 = 7$ $3 \cdot 4 = 12$ $x^2 + 7x + 12 = (x + 3)(x + 4)$

DEFINITIONS AND CONCEPTS	EXAMPLES
Section 11.3 Factoring Trinomials of the Form $ax^2 + bx + c$	
To factor $ax^2 + bx + c$, try various combinations of factors of ax^2 and c until a middle term of bx is obtained when checking.	Factor: $3x^2 + 14x - 5$ Factors of $3x^2$: $3x, x$ Factors of -5: $-1, 5$ and $1, -5$. $(3x - 1)(x + 5)$ $-1x$ $+15x$ $14x$ Correct middle term
Section 11.4 Factoring Trinomials of the Form $ax^2 + bx + c$ by Grouping	
TO FACTOR $ax^2 + bx + c$ BY GROUPING	Factor: $3x^2 + 14x - 5$
Step 1. Find two numbers whose product is $a \cdot c$ and whose sum is b.	**Step 1.** Find two numbers whose product is $3 \cdot (-5)$ or -15 and whose sum is 14. They are 15 and -1.
Step 2. Rewrite bx, using the factors found in Step 1.	**Step 2.** $3x^2 + 14x - 5$ $= 3x^2 + 15x - 1x - 5$
Step 3. Factor by grouping.	**Step 3.** $= 3x(x + 5) - 1(x + 5)$ $= (x + 5)(3x - 1)$
Section 11.5 Factoring Perfect Square Trinomials and the Difference of Two Squares	
A **perfect square trinomial** is a trinomial that is the square of some binomial.	**PERFECT SQUARE TRINOMIAL = SQUARE OF BINOMIAL** $x^2 + 4x + 4 = (x + 2)^2$ $25x^2 - 10x + 1 = (5x - 1)^2$
Factoring Perfect Square Trinomials $a^2 + 2ab + b^2 = (a + b)^2$ $a^2 - 2ab + b^2 = (a - b)^2$	Factor. $x^2 + 6x + 9 = x^2 + 2(x \cdot 3) + 3^2 = (x + 3)^2$ $4x^2 - 12x + 9 = (2x)^2 - 2(2x \cdot 3) + 3^2$ $= (2x - 3)^2$
Difference of Two Squares $a^2 - b^2 = (a + b)(a - b)$	Factor. $x^2 - 9 = x^2 - 3^2 = (x + 3)(x - 3)$
Section 11.6 Solving Quadratic Equations by Factoring	
A **quadratic equation** is an equation that can be written in the form $ax^2 + bx + c = 0$ with a not 0. The form $ax^2 + bx + c = 0$ is called the **standard form** of a quadratic equation.	**Quadratic Equation** — **Standard Form** $x^2 = 16$ — $x^2 - 16 = 0$ $y = -2y^2 + 5$ — $2y^2 + y - 5 = 0$
Zero Factor Property If a and b are real numbers and if $ab = 0$, then $a = 0$ or $b = 0$.	If $(x + 3)(x - 1) = 0$, then $x + 3 = 0$ or $x - 1 = 0$.

DEFINITIONS AND CONCEPTS	**EXAMPLES**
Section 11.6 Solving Quadratic Equations by Factoring *(continued)*	
TO SOLVE QUADRATIC EQUATIONS BY FACTORING	Solve: $3x^2 = 13x - 4$
Step 1. Write the equation in standard form so that one side of the equation is 0.	**Step 1.** $3x^2 - 13x + 4 = 0$
Step 2. Factor completely.	**Step 2.** $(3x - 1)(x - 4) = 0$
Step 3. Set each factor containing a variable equal to 0.	**Step 3.** $3x - 1 = 0$ or $x - 4 = 0$
Step 4. Solve the resulting equations.	**Step 4.** $3x = 1$ $\qquad x = 4$ $x = \frac{1}{3}$
Step 5. Check solutions in the original equation.	**Step 5.** Check both $\frac{1}{3}$ and 4 in the original equation.
Section 11.7 Quadratic Equations and Problem Solving	
PROBLEM-SOLVING STEPS	A garden is in the shape of a rectangle whose length is two feet more than its width. If the area of the garden is 35 square feet, find its dimensions.
1. UNDERSTAND the problem.	**1.** Read and reread the problem. Guess a solution and check your guess. Draw a diagram. Let x be the width of the rectangular garden. Then $x + 2$ is the length. 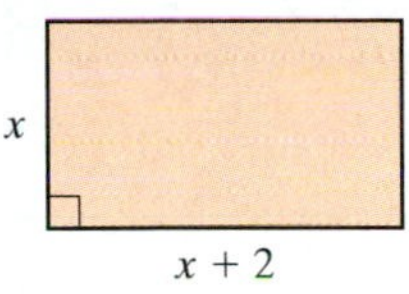
2. TRANSLATE.	**2.** length · width = area $(x + 2) \cdot x = 35$
3. SOLVE.	**3.** $(x + 2)x = 35$ $x^2 + 2x - 35 = 0$ $(x - 5)(x + 7) = 0$ $x - 5 = 0$ or $x + 7 = 0$ $x = 5$ $\qquad x = -7$
4. INTERPRET.	**4.** Discard the solution $x = -7$ since x represents width. *Check:* If x is 5 feet, then $x + 2 = 5 + 2 = 7$ feet. The area of a rectangle whose width is 5 feet and whose length is 7 feet is (5 feet)(7 feet) or 35 square feet. *State:* The garden is 5 feet by 7 feet.

STUDY SKILLS BUILDER

Are You Prepared for a Test on Chapter 11?

Below is a list of some *common trouble areas* for students in Chapter 11. After studying for your test—but before taking your test—read these.

- The difference of two squares such as $x^2 - 25$ factors as $x^2 - 25 = (x + 5)(x - 5)$.
- The sum of two squares, for example, $x^2 + 25$, cannot be factored using real numbers.
- Don't forget that the first step to factor any polynomial is to first factor out any common factors.

$$9x^2 - 36 = 9(x^2 - 4) = 9(x + 2)(x - 2)$$

- Can you completely factor $x^4 - 24x^2 - 25$?

$$x^4 - 24x^2 - 25 = (x^2 - 25)(x^2 + 1) = (x + 5)(x - 5)(x^2 + 1)$$

- Remember that to use the zero factor property to solve a quadratic equation, one side of the equation must be 0 and the other side must be a factored polynomial.

$x(x - 2) = 3$ Cannot use zero factor property.

$x^2 - 2x - 3 = 0$

$(x - 3)(x + 1) = 0$ Now we can use zero factor property.

$x - 3 = 0$ or $x + 1 = 0$

$x = 3$ or $x = -1$

Remember: This is simply a sampling of selected topics given to check your understanding. For a review of Chapter 11 in your text, see the material at the end of this chapter.

11 CHAPTER REVIEW

(11.1) *Complete each factoring.*

1. $6x^2 - 15x = 3x(\quad)$

2. $4x^5 + 2x - 10x^4 = 2x(\quad)$

Factor out the GCF from each polynomial.

3. $5m + 30$

4. $20x^3 + 12x^2 + 24x$

5. $3x(2x + 3) - 5(2x + 3)$

6. $5x(x + 1) - (x + 1)$

Factor each polynomial by grouping.

7. $3x^2 - 3x + 2x - 2$

8. $6x^2 + 10x - 3x - 5$

9. $3a^2 + 9ab + 3b^2 + ab$

(11.2) *Factor each trinomial.*

10. $x^2 + 6x + 8$

11. $x^2 - 11x + 24$

12. $x^2 + x + 2$

13. $x^2 - 5x - 6$

14. $x^2 + 2x - 8$

15. $x^2 + 4xy - 12y^2$

16. $x^2 + 8xy + 15y^2$

17. $72 - 18x - 2x^2$

18. $32 + 12x - 4x^2$

19. $5y^3 - 50y^2 + 120y$

20. To factor $x^2 + 2x - 48$, think of two numbers whose product is _____ and whose sum is _____.

21. What is the first step to factoring $3x^2 + 15x + 30$?

(11.3) or (11.4) *Factor each trinomial.*

22. $2x^2 + 13x + 6$ **23.** $4x^2 + 4x - 3$ **24.** $6x^2 + 5xy - 4y^2$ **25.** $x^2 - x + 2$

26. $2x^2 - 23x - 39$ **27.** $18x^2 - 9xy - 20y^2$ **28.** $10y^3 + 25y^2 - 60y$

Write the perimeter of each figure as a simplified polynomial. Then factor each polynomial.

 **29.**

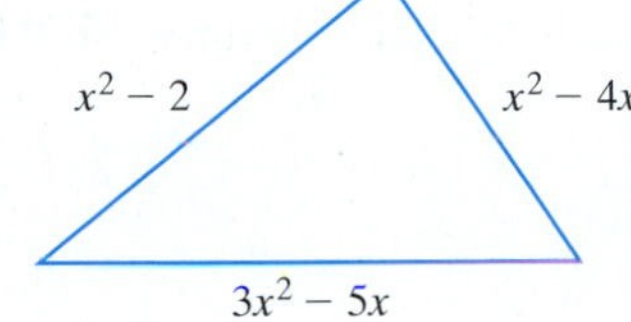

30.

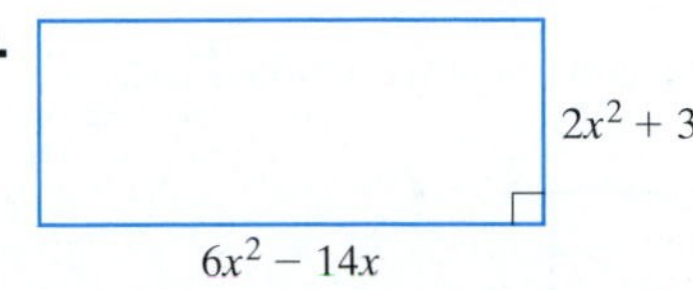

(11.5) *Determine whether each polynomial is a perfect square trinomial.*

31. $x^2 + 6x + 9$ **32.** $x^2 + 8x + 64$ **33.** $9m^2 - 12m + 16$ **34.** $4y^2 - 28y + 49$

Determine whether each binomial is a difference of two squares.

35. $x^2 - 9$ **36.** $x^2 + 16$ **37.** $4x^2 - 25y^2$ **38.** $9a^3 - 1$

Factor each polynomial completely.

39. $x^2 - 81$ **40.** $x^2 + 12x + 36$ **41.** $4x^2 - 9$ **42.** $9t^2 - 25s^2$

43. $16x^2 + y^2$ **44.** $n^2 - 18n + 81$ **45.** $3r^2 + 36r + 108$ **46.** $9y^2 - 42y + 49$

47. $5m^8 - 5m^6$ **48.** $4x^2 - 28xy + 49y^2$ **49.** $3x^2y + 6xy^2 + 3y^3$ **50.** $16x^4 - 1$

(11.6) *Solve each equation.*

51. $(x + 6)(x - 2) = 0$ **52.** $3x(x + 1)(7x - 2) = 0$ **53.** $4(5x + 1)(x + 3) = 0$

54. $x^2 + 8x + 7 = 0$ **55.** $x^2 - 2x - 24 = 0$ **56.** $x^2 + 10x = -25$

57. $x(x - 10) = -16$ **58.** $(3x - 1)(9x^2 + 3x + 1) = 0$ **59.** $56x^2 - 5x - 6 = 0$

60. $m^2 = 6m$

61. $r^2 = 25$

62. Write a quadratic equation that has the two solutions 4 and 5.

(11.7) *Use the given information to choose the correct dimensions.*

△ **63.** The perimeter of a rectangle is 24 inches. The length is twice the width. Find the dimensions of the rectangle.

a. 5 inches by 7 inches
b. 5 inches by 10 inches
c. 4 inches by 8 inches
d. 2 inches by 10 inches

△ **64.** The area of a rectangle is 80 meters. The length is one more than three times the width. Find the dimensions of the rectangle.

a. 8 meters by 10 meters
b. 4 meters by 13 meters
c. 4 meters by 20 meters
d. 5 meters by 16 meters

Use the given information to find the dimensions of each figure.

△ **65.** The *area* of the square is 81 square units. Find the length of a side.

△ **66.** The *perimeter* of the quadrilateral is 47 units. Find the lengths of the sides.

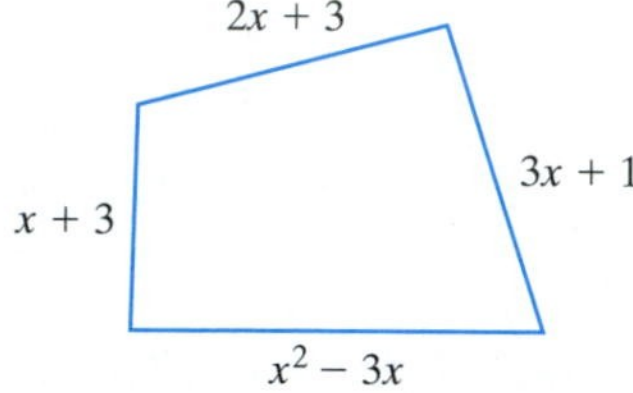

Solve.

△ **67.** A flag for a local organization is in the shape of a rectangle whose length is 15 inches less than twice its width. If the area of the flag is 500 square inches, find its dimensions.

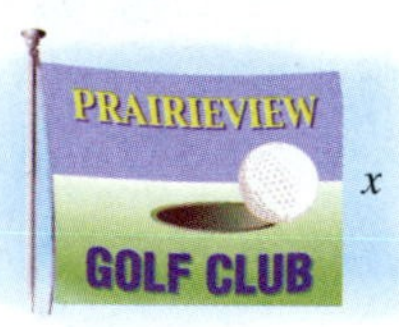

△ **68.** The base of a triangular sail is four times its height. If the area of the triangle is 162 square yards, find the base.

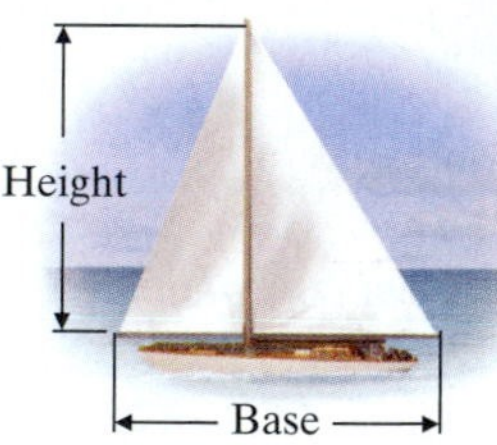

69. Find two consecutive positive integers whose product is 380.

70. A rocket is fired from the ground with an initial velocity of 440 feet per second. Its height h after t seconds is given by the equation $h = -16t^2 + 440t$.

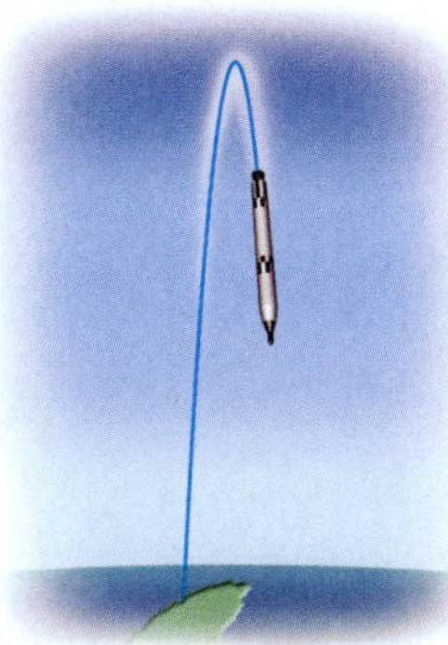

a. Find how many seconds pass before the rocket reaches a height of 2800 feet. Explain why two answers are obtained.

b. Find how many seconds pass before the rocket reaches the ground again.

△ **71.** An architect's squaring instrument is in the shape of a right triangle. Find the length of the longer leg of the right triangle if the hypotenuse is 8 centimeters longer than the longer leg and the shorter leg is 8 centimeters shorter than the longer leg.

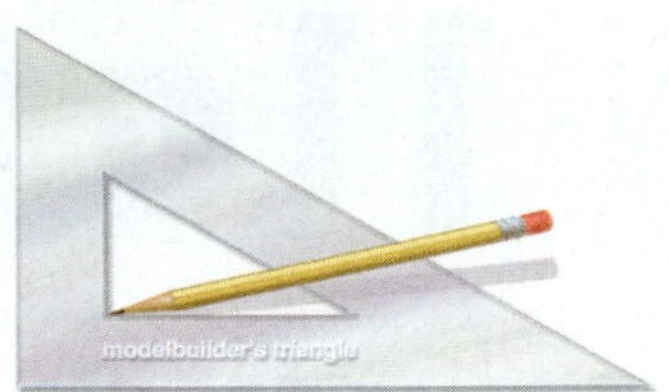

Mixed Review

Factor completely.

72. $6x + 24$

73. $7x - 63$

74. $11x(4x - 3) - 6(4x - 3)$

75. $2x(x - 5) - (x - 5)$

76. $3x^3 - 4x^2 + 6x - 8$

77. $xy + 2x - y - 2$

78. $2x^2 + 2x - 24$

79. $3x^3 - 30x^2 + 27x$

80. $4x^2 - 81$

81. $2x^2 - 18$

82. $16x^2 - 24x + 9$

83. $5x^2 + 20x + 20$

Solve.

84. $2x^2 - x - 28 = 0$

85. $x^2 - 2x = 15$

86. $2x(x + 7)(x + 4) = 0$

87. $x(x - 5) = -6$

88. $x^2 = 16x$

89. The perimeter of the following triangle is 48 inches. Find the lengths of its sides.

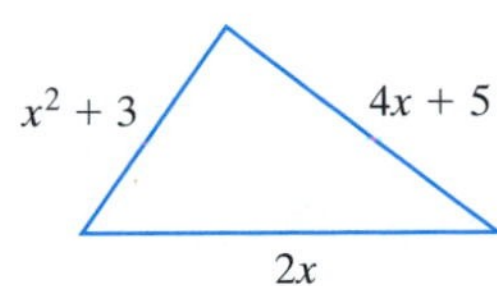

90. The width of a rectangle is 4 inches less than its length. Its area is 12 square inches. Find the dimensions of the rectangle.

11 CHAPTER TEST

Answers

Remember to use the Chapter Test Prep Video CD to see the fully worked-out solutions to any of the exercises you want to review.

Factor each polynomial completely. If a polynomial cannot be factored, write "prime."

1. $9x^2 - 3x$

2. $x^2 + 11x + 28$

3. $49 - m^2$

4. $y^2 + 22y + 121$

5. $x^4 - 16$

6. $4(a + 3) - y(a + 3)$

7. $x^2 + 4$

8. $y^2 - 8y - 48$

9. $3a^2 + 3ab - 7a - 7b$

10. $3x^2 - 5x + 2$

11. $180 - 5x^2$

12. $3x^3 - 21x^2 + 30x$

13. $6t^2 - t - 5$

14. $xy^2 - 7y^2 - 4x + 28$

15. $x - x^5$

16. $x^2 + 14xy + 24y^2$

Solve each equation.

17. $(x - 3)(x + 9) = 0$

18. $x^2 + 5x = 14$

19. $x(x + 6) = 7$

20. $3x(2x - 3)(3x + 4) = 0$

21. $5t^3 - 45t = 0$

22. $t^2 - 2t - 15 = 0$

23. $6x^2 = 15x$

1. ______
2. ______
3. ______
4. ______
5. ______
6. ______
7. ______
8. ______
9. ______
10. ______
11. ______
12. ______
13. ______
14. ______
15. ______
16. ______
17. ______
18. ______
19. ______
20. ______
21. ______
22. ______
23. ______

Solve.

△ **24.** A deck for a home is in the shape of a triangle. The length of the base of the triangle is 9 feet longer than its height. If the area of the triangle is 68 square feet find the length of the base.

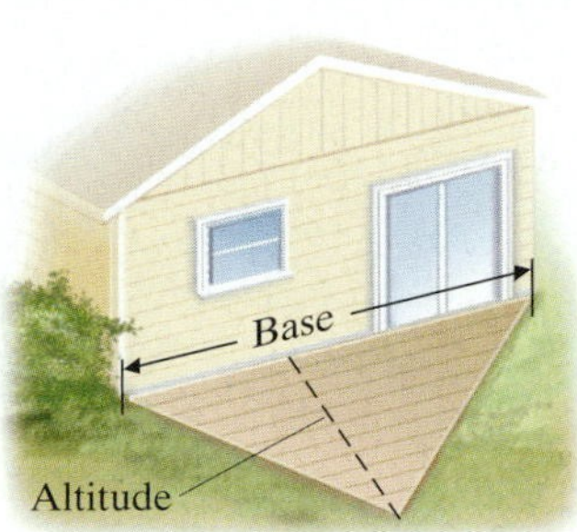

△ **25.** The *area* of the rectangle is 54 square units. Find the dimensions of the rectangle.

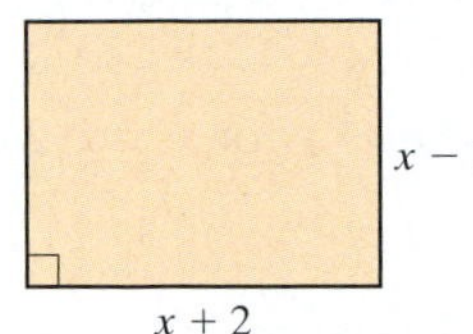

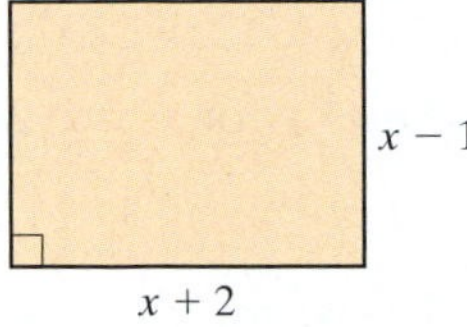

26. An object is dropped from the top of the Woolworth Building on Broadway in New York City. The height h of the object after t seconds is given by the equation

$$h = -16t^2 + 784$$

Find how many seconds pass before the object reaches the ground.

△ **27.** Find the lengths of the sides of a right triangle if the hypotenuse is 10 centimeters longer than the shorter leg and 5 centimeters longer than the longer leg.

28. A window washer is suspended 38 feet below the roof of the 1127-foot-tall John Hancock Center in Chicago. (*Source:* Council on Tall Buildings and Urban Habitat) If the window washer drops an object from this height, the object's height h after t seconds is given by the equation $h = -16t^2 + 1089$. Find how many seconds pass before the object reaches the ground.

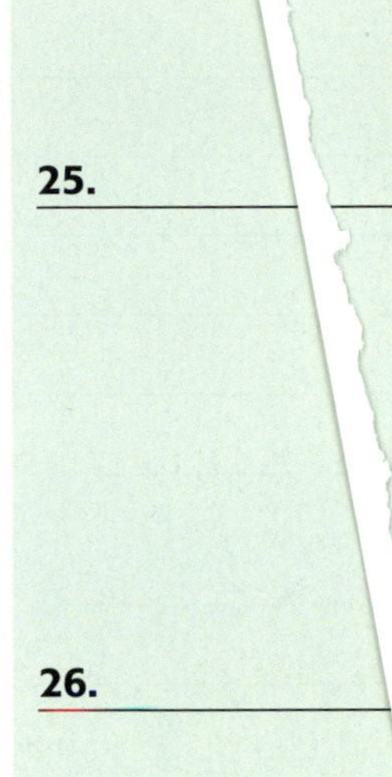

24. ____________

25. ____________

26. ____________

27. ____________

28. ____________

CUMULATIVE REVIEW

Chapters 1–11

Answers

1. ______
2. ______
3. ______
4. ______
5. ______
6. ______
7. ______
8. ______
9. ______
10. ______
11. ______
12. ______
13. ______
14. ______
15. ______
16. ______
17. a. ______
    b. ______
    c. ______
18. a. ______
    b. ______
19. ______
20. ______
21. ______
22. ______

1. Evaluate $x + 2y - z$ for $x = 3$, $y = -5$, and $z = -4$.

2. Evaluate $5x - y$ for $x = -2$ and $y = 4$.

3. Evaluate $7 - x^2$ for $x = -4$.

4. Evaluate $x^3 - y^3$ for $x = -2$ and $y = 4$.

Solve.

5. $-8 = x + 1$

6. $-8x = 72$

Simplify.

7. $2y - 6 + 4y + 8$

8. $-5a - 3 + a + 2$

9. $4x + 2 - 5x + 3$

10. $2.3x + 5x - 6$

Solve.

11. $3(3x - 5) = 10x$

12. $2(7x - 1) = 15x$

13. $3(2x - 6) + 6 = 0$

14. $4(x + 3) - 12 = 0$

15. $\dfrac{x - 5}{3} = \dfrac{x + 2}{5}$

16. $\dfrac{x + 1}{2} = \dfrac{x - 7}{3}$

17. Translate each sentence into a mathematical statement.

a. Nine is less than or equal to eleven.
b. Eight is greater than one.
c. Three is not equal to four.

18. Translate each sentence into a mathematical statement.

a. Five is greater than or equal to one.
b. Two is not equal to negative four.

Solve.

19. $3(x - 4) = 3x - 12$

20. $2(x + 5) = 2x + 9$

21. Solve for l: $V = lwh$

22. Solve for x: $3x - 2y = 5$

Simplify each expression.

23. $(5^3)^6$ **24.** $(7^9)^2$ **25.** $(y^8)^2$ **26.** $(x^{11})^3$

Simplify the following expressions. Write each result using positive exponents only.

27. $\dfrac{(x^3)^4x}{x^7}$ **28.** $\dfrac{(y^3)^9 \cdot y}{y^6}$ **29.** $(y^{-3}z^6)^{-6}$

30. $(xy^{-4})^{-2}$ **31.** $\dfrac{x^{-7}}{(x^4)^3}$ **32.** $\dfrac{(y^2)^5}{y^{-3}}$

Simplify each polynomial by combining any like terms.

33. $-3x + 7x$ **34.** $y + y$ **35.** $11x^2 + 5 + 2x^2 - 7$

36. $8y - y^2 + 4y - y^2$

Multiply.

37. $(2x - y)^2$ **38.** $(3x + 1)^2$

Use a special product to square each binomial.

39. $(t + 2)^2$ **40.** $(x - 4)^2$ **41.** $(x^2 - 7y)^2$ **42.** $(x^2 + 7y)^2$

Divide.

43. $\dfrac{8x^2y^2 - 16xy + 2x}{4xy}$ **44.** $\dfrac{20a^2b^3 - 5ab + 10b}{5ab}$

Factor.

45. $5(x + 3) + y(x + 3)$ **46.** $9(y - 2) + x(y - 2)$ **47.** $x^4 + 5x^2 + 6$

48. $x^4 - 4x^2 - 5$ **49.** $6x^2 - 2x - 20$ **50.** $10x^2 + 25x + 10$

51. Since the 1940s, one of the top tourist attractions in Acapulco, Mexico, is watching the cliff divers off the La Quebrada. The divers' platform is about 144 feet above the sea. These divers must time their descent just right, since they land in the crashing Pacific, in an inlet that is at most $9\frac{1}{2}$ feet deep. Neglecting air resistance, the height h in feet of a cliff diver above the ocean after t seconds is given by the quadratic equation $h = -16t^2 + 144$.

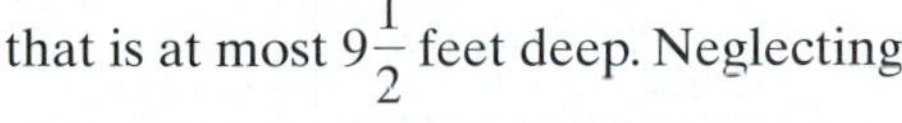

Find out how long it takes the diver to reach the ocean.

52. The square of a number plus twice the number is 120. Find the number.

23. ______
24. ______
25. ______
26. ______
27. ______
28. ______
29. ______
30. ______
31. ______
32. ______
33. ______
34. ______
35. ______
36. ______
37. ______
38. ______
39. ______
40. ______
41. ______
42. ______
43. ______
44. ______
45. ______
46. ______
47. ______
48. ______
49. ______
50. ______
51. ______
52. ______

12 Rational Expressions

In this chapter, we expand our knowledge of algebraic expressions to include algebraic fractions, called *rational expressions*. We explore the operations of addition, subtraction, multiplication, and division using principles similar to the principles for numerical fractions.

SECTIONS

CHECK YOUR PROGRESS

American football is one of this nation's most followed sports. It has its roots in English rugby, a game played with the same shaped ball, but where the ball can only advance through running. American college players were the first to add advancing the ball by throwing or kicking it past the opponents. In 1867, both Rutgers and Princeton established a basic set of rules and played the first intercollegiate football game. In 1876, Walter Camp, the coach at Yale, helped establish that teams were to consist of 11 men, standardized the size of the field, and generally instituted the first cohesive set of rules.

In Exercise 87, Section 12.1, you will have the opportunity to calculate a quarterback's rating.

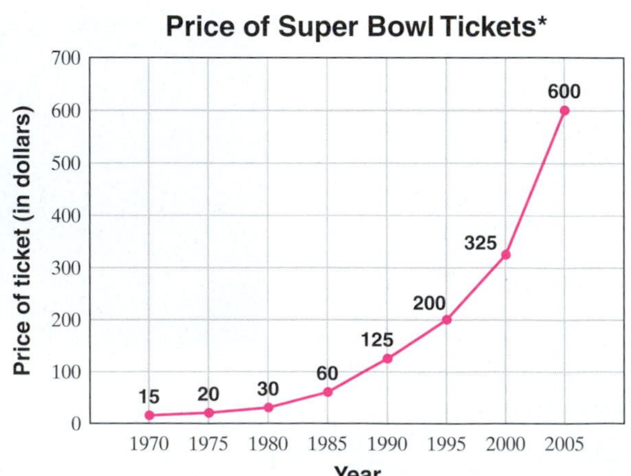

Source: JS Online: Super Bowl Ticket Prices and NFL research
* For years with multiple ticket prices, highest price is shown

12.1 SIMPLIFYING RATIONAL EXPRESSIONS

Objective A Evaluating Rational Expressions

A rational number is a number that can be written as a quotient of integers. A *rational expression* is also a quotient; it is a quotient of polynomials. Examples are

$$\frac{2}{3}, \quad \frac{3y^3}{8}, \quad \frac{-4p}{p^3 + 2p + 1}, \quad \text{and} \quad \frac{5x^2 - 3x + 2}{3x + 7}$$

Rational Expression

A **rational expression** is an expression that can be written in the form

$$\frac{P}{Q}$$

where P and Q are polynomials and $Q \neq 0$.

Rational expressions have different numerical values depending on what values replace the variables.

EXAMPLE 1 Find the numerical value of $\frac{x + 4}{2x - 3}$ for each replacement value.

a. $x = 5$ **b.** $x = -2$

Solution:

a. We replace each x in the expression with 5 and then simplify.

$$\frac{x + 4}{2x - 3} = \frac{5 + 4}{2(5) - 3} = \frac{9}{10 - 3} = \frac{9}{7}$$

b. We replace each x in the expression with -2 and then simplify.

$$\frac{x + 4}{2x - 3} = \frac{-2 + 4}{2(-2) - 3} = \frac{2}{-7} \quad \text{or} \quad -\frac{2}{7}$$

Work Practice Problem 1

In the example above, we wrote $\frac{2}{-7}$ as $-\frac{2}{7}$. For a negative fraction such as $\frac{2}{-7}$, recall from Section 4.1 that

$$\frac{2}{-7} = \frac{-2}{7} = -\frac{2}{7}$$

In general, for any fraction,

$$\frac{-a}{b} = \frac{a}{-b} = -\frac{a}{b}, \quad b \neq 0$$

This is also true for rational expressions. For example,

$$\underbrace{\frac{-(x + 2)}{x}}_{\text{Notice the parentheses.}} = \frac{x + 2}{-x} = -\frac{x + 2}{x}$$

Objectives

A Find the Value of a Rational Expression Given a Replacement Number.

B Identify Values for Which a Rational Expression Is Undefined.

C Simplify, or Write Rational Expressions in Lowest Terms.

D Write Equivalent Forms of Rational Expressions.

PRACTICE PROBLEM 1

Find the value of $\frac{x - 3}{5x + 1}$ for each replacement value.

a. $x = 4$

b. $x = -3$

Answers

1. a. $\frac{1}{21}$ **b.** $\frac{3}{7}$

Objective B Identifying When a Rational Expression Is Undefined

Helpful Hint

Do you recall why division by 0 is not defined? Remember, for example, that $\frac{8}{4} = 2$ because $2 \cdot 4 = 8$. Thus, if $\frac{8}{0} =$ *a number*, then *the number* $\cdot 0 = 8$. There is no number that when multiplied by 0 equals 8; thus $\frac{8}{0}$ is undefined. This is true in general for fractions and rational expressions.

In the definition of rational expression (first "box" in this section), notice that we wrote $Q \neq 0$ for the denominator Q. The denominator of a rational expression must not equal 0 since division by 0 is not defined. (See the Helpful Hint to the left.) This means we must be careful when replacing the variable in a rational expression by a number. For example, suppose we replace x with 5 in the rational expression $\frac{3 + x}{x - 5}$. The expression becomes

$$\frac{3 + x}{x - 5} = \frac{3 + 5}{5 - 5} = \frac{8}{0}$$

But division by 0 is undefined. Therefore, in this expression we can allow x to be any real number *except* 5. **A rational expression is undefined for values that make the denominator 0.** Thus,

> To find values for which a rational expression is undefined, find values for which the denominator is 0.

PRACTICE PROBLEM 2

Are there any values for x for which each rational expression is undefined?

a. $\frac{x}{x + 8}$

b. $\frac{x - 3}{x^2 + 5x + 4}$

c. $\frac{x^2 - 3x + 2}{5}$

EXAMPLE 2 Are there any values for x for which each expression is undefined?

a. $\frac{x}{x - 3}$ **b.** $\frac{x^2 + 2}{x^2 - 3x + 2}$ **c.** $\frac{x^3 - 6x^2 - 10x}{3}$

Solution: To find values for which a rational expression is undefined, we find values that make the denominator 0.

a. The denominator of $\frac{x}{x - 3}$ is 0 when $x - 3 = 0$ or when $x = 3$. Thus, when $x = 3$, the expression $\frac{x}{x - 3}$ is undefined.

b. We set the denominator equal to 0.

$$x^2 - 3x + 2 = 0$$

$$(x - 2)(x - 1) = 0 \quad \text{Factor.}$$

$$x - 2 = 0 \quad \text{or} \quad x - 1 = 0 \quad \text{Set each factor equal to 0.}$$

$$x = 2 \qquad x = 1 \quad \text{Solve.}$$

Thus, when $x = 2$ or $x = 1$, the denominator $x^2 - 3x + 2$ is 0. So the rational expression $\frac{x^2 + 2}{x^2 - 3x + 2}$ is undefined when $x = 2$ or when $x = 1$.

c. The denominator of $\frac{x^3 - 6x^2 - 10x}{3}$ is never 0, so there are no values of x for which this expression is undefined.

Note: Unless otherwise stated, we will now assume that variables in rational expressions are only replaced by values for which the expressions are defined.

Work Practice Problem 2

Objective C Simplifying Rational Expressions

A fraction is said to be written in lowest terms or simplest form when the numerator and denominator have no common factors other than 1 (or -1). For example, the fraction $\frac{7}{10}$ is written in lowest terms since the numerator and denominator have no common factors other than 1 (or -1).

The process of writing a rational expression in lowest terms or simplest form is called **simplifying** a rational expression.

Answers

2. a. $x = -8$ **b.** $x = -4, x = -1$ **c.** no

Simplifying a rational expression is similar to simplifying a fraction. Recall from Section 4.2 that to simplify a fraction, we essentially "remove factors of 1." Our ability to do this comes from these facts:

- Any nonzero number over itself simplifies to 1 $\left(\frac{5}{5} = 1, \frac{-7.26}{-7.26} = 1, \text{ or } \frac{c}{c} = 1 \text{ as long as } c \text{ is not } 0\right)$, and
- The product of any number and 1 is that number $\left(19 \cdot 1 = 19, -8.9 \cdot 1 = -8.9, \frac{a}{b} \cdot 1 = \frac{a}{b}\right)$.

In other words, we have the following:

$$\frac{a \cdot c}{b \cdot c} = \frac{a}{b} \cdot \frac{c}{c} = \frac{a}{b}$$

Since $\frac{a}{b} \cdot 1 = \frac{a}{b}$

Simplify: $\frac{15}{20}$

$$\frac{15}{20} = \frac{3 \cdot 5}{2 \cdot 2 \cdot 5}$$ Factor the numerator and the denominator.

$$= \frac{3 \cdot 5}{2 \cdot 2 \cdot 5}$$ Look for common factors.

$$= \frac{3}{2 \cdot 2} \cdot \frac{5}{5}$$ Common factors in the numerator and denominator form factors of 1.

$$= \frac{3}{2 \cdot 2} \cdot 1$$ Write $\frac{5}{5}$ as 1.

$$= \frac{3}{2 \cdot 2} = \frac{3}{4}$$ Multiply to remove a factor of 1.

Before we use the same technique to simplify a rational expression, remember that as long as the denominator is not 0, $\frac{a^3b}{a^3b} = 1$, $\frac{x + 3}{x + 3} = 1$, and $\frac{7x^2 + 5x - 100}{7x^2 + 5x - 100} = 1$.

Simplify: $\frac{x^2 - 9}{x^2 + x - 6}$

$$\frac{x^2 - 9}{x^2 + x - 6} = \frac{(x - 3)(x + 3)}{(x - 2)(x + 3)}$$ Factor the numerator and the denominator.

$$= \frac{(x - 3)(x + 3)}{(x - 2)(x + 3)}$$ Look for common factors.

$$= \frac{x - 3}{x - 2} \cdot \frac{x + 3}{x + 3}$$

$$= \frac{x - 3}{x - 2} \cdot 1$$ Write $\frac{x + 3}{x + 3}$ as 1.

$$= \frac{x - 3}{x - 2}$$ Multiply to remove a factor of 1.

Just as for numerical fractions, we can use a shortcut notation. Remember that as long as exact factors in both the numerator and denominator are divided out, we are "removing a factor of 1." We will use the following notation to show this:

$$\frac{x^2 - 9}{x^2 + x - 6} = \frac{(x - 3)(x + 3)}{(x - 2)(x + 3)}$$ A factor of 1 is identified by the shading.

$$= \frac{x - 3}{x - 2}$$ Remove a factor of 1.

Thus, the rational expression $\frac{x^2 - 9}{x^2 + x - 6}$ has the same value as the rational expression $\frac{x - 3}{x - 2}$ for all values of x except 2 and -3. (Remember that when x is 2, the denominator of both rational expressions is 0 and when x is -3, the original rational expression has a denominator of 0.)

As we simplify rational expressions, we will assume that the simplified rational expression is equal to the original rational expression for all real numbers except those for which either denominator is 0. The following steps may be used to simplify rational expressions.

To Simplify a Rational Expression

Step 1: Completely factor the numerator and denominator.

Step 2: Divide out factors common to the numerator and denominator. (This is the same as "removing a factor of 1.")

PRACTICE PROBLEM 3

Simplify: $\frac{x^4 + x^3}{5x + 5}$

EXAMPLE 3 Simplify: $\frac{5x - 5}{x^3 - x^2}$

Solution: To begin, we factor the numerator and denominator if possible. Then we look for common factors.

$$\frac{5x - 5}{x^3 - x^2} = \frac{5(x - 1)}{x^2(x - 1)} = \frac{5}{x^2}$$

Work Practice Problem 3

PRACTICE PROBLEM 4

Simplify: $\frac{x^2 + 11x + 18}{x^2 + x - 2}$

EXAMPLE 4 Simplify: $\frac{x^2 + 8x + 7}{x^2 - 4x - 5}$

Solution: We factor the numerator and denominator and then look for common factors.

$$\frac{x^2 + 8x + 7}{x^2 - 4x - 5} = \frac{(x + 7)(x + 1)}{(x - 5)(x + 1)} = \frac{x + 7}{x - 5}$$

Work Practice Problem 4

PRACTICE PROBLEM 5

Simplify: $\frac{x^2 + 10x + 25}{x^2 + 5x}$

EXAMPLE 5 Simplify: $\frac{x^2 + 4x + 4}{x^2 + 2x}$

Solution: We factor the numerator and denominator and then look for common factors.

$$\frac{x^2 + 4x + 4}{x^2 + 2x} = \frac{(x + 2)(x + 2)}{x(x + 2)} = \frac{x + 2}{x}$$

Work Practice Problem 5

Helpful Hint

When simplifying a rational expression, we look for **common *factors*, not common *terms*.**

$$\frac{x \cdot (x + 2)}{x \cdot x} = \frac{x + 2}{x}$$

Common factors. These can be divided out.

$$\frac{x + 2}{x}$$

Common terms. There is no factor of 1 that can be generated.

Answers

3. $\frac{x^3}{5}$ 4. $\frac{x + 9}{x - 1}$ 5. $\frac{x + 5}{x}$

✔Concept Check Recall that we can only remove *factors* of 1. Which of the following are *not* true? Explain why.

a. $\frac{3-1}{3+5}$ simplifies to $-\frac{1}{5}$ **b.** $\frac{2x+10}{2}$ simplifies to $x+5$

c. $\frac{37}{72}$ simplifies to $\frac{3}{2}$ **d.** $\frac{2x+3}{2}$ simplifies to $x+3$

EXAMPLE 6 Simplify: $\frac{x+9}{x^2-81}$

Solution: We factor and then apply the fundamental principle.

$$\frac{x+9}{x^2-81} = \frac{x+9}{(x+9)(x-9)} = \frac{1}{x-9}$$

Work Practice Problem 6

PRACTICE PROBLEM 6

Simplify: $\frac{x+5}{x^2-25}$

EXAMPLE 7 Simplify each rational expression.

a. $\frac{x+y}{y+x}$ **b.** $\frac{x-y}{y-x}$

Solution:

a. The expression $\frac{x+y}{y+x}$ can be simplified by using the commutative property of addition to rewrite the denominator $y+x$ as $x+y$.

$$\frac{x+y}{y+x} = \frac{x+y}{x+y} = 1$$

b. The expression $\frac{x-y}{y-x}$ can be simplified by recognizing that $y-x$ and $x-y$ are opposites. In other words, $y-x=-1(x-y)$. We proceed as follows:

$$\frac{x-y}{y-x} = \frac{1\cdot(x-y)}{(-1)(x-y)} = \frac{1}{-1} = -1$$

Work Practice Problem 7

PRACTICE PROBLEM 7

Simplify each rational expression.

a. $\frac{x+4}{4+x}$

b. $\frac{x-4}{4-x}$

Objective D Writing Equivalent Forms of Rational Expressions

From Example 7a, we have $y+x=x+y$. $y+x$ and $x+y$ are equivalent.

From Example 7b, we have $y-x=-1(x-y)$. $y-x$ and $x-y$ are opposites.

Thus, $\frac{x+y}{y+x} = \frac{x+y}{x+y} = 1$ and $\frac{x-y}{y-x} = \frac{x-y}{-1(x-y)} = \frac{1}{-1} = -1.$

When performing operations on rational expressions, equivalent forms of answers often result. For this reason, it is very important to be able to recognize equivalent answers.

Answers

6. $\frac{1}{x-5}$ **7. a.** 1 **b.** -1

✔ Concept Check Answer

a, c, d

PRACTICE PROBLEM 8

List 4 equivalent forms of $-\frac{3x+7}{x-6}$.

Helpful Hint

Remember, a negative sign in front of a fraction or rational expression may be moved to the numerator or the denominator, but *not* both.

Answer

8. $\frac{-(3x+7)}{x-6}$; $\frac{-3x-7}{x-6}$; $\frac{3x+7}{-(x-6)}$; $\frac{3x+7}{-x+6}$

EXAMPLE 8 List some equivalent forms of $-\frac{5x-1}{x+9}$.

Solution: To do so, recall that $-\frac{a}{b} = \frac{-a}{b} = \frac{a}{-b}$. Thus

$$-\frac{5x-1}{x+9} = \frac{-(5x-1)}{x+9} = \frac{-5x+1}{x+9} \quad \text{or} \quad \frac{1-5x}{x+9}$$

Also,

$$-\frac{5x-1}{x+9} = \frac{5x-1}{-(x+9)} = \frac{5x-1}{-x-9} \quad \text{or} \quad \frac{5x-1}{-9-x}$$

Thus $-\frac{5x-1}{x+9} = \frac{-(5x-1)}{x+9} = \frac{-5x+1}{x+9} = \frac{5x-1}{-(x+9)} = \frac{5x-1}{-x-9}$

Work Practice Problem 8

Keep in mind that many rational expressions may look different, but in fact be equivalent.

Vocabulary and Readiness Check

Use the choices below to fill in each blank. Not all choices will be used.

−1	0	simplifying	$\frac{-a}{-b}$	$\frac{-a}{b}$	$\frac{a}{-b}$
1	2	rational expression			

1. A ________________ is an expression that can be written in the form $\frac{P}{Q}$ where P and Q are polynomials and $Q \neq 0$.
2. The expression $\frac{x+3}{3+x}$ simplifies to ____________.
3. The expression $\frac{x-3}{3-x}$ simplifies to ____________.
4. A rational expression is undefined for values that make the denominator ____________.
5. The expression $\frac{7x}{x-2}$ is undefined for $x =$ ____________.
6. The process of writing a rational expression in lowest terms is called ____________.
7. For a rational expression, $-\frac{a}{b} =$ ______ $=$ ______.

12.1 EXERCISE SET

Student Solutions Manual

PH Math/Tutor Center

CD/Video for Review

MathXL®

MyMathLab

Objective A *Find the value of the following expressions when $x = 2$, $y = -2$, and $z = -5$. See Example 1.*

1. $\dfrac{x + 5}{x + 2}$

2. $\dfrac{x + 8}{x + 1}$

3. $\dfrac{y^3}{y^2 - 1}$

4. $\dfrac{z}{z^2 - 5}$

5. $\dfrac{x^2 + 8x + 2}{x^2 - x - 6}$

6. $\dfrac{x + 5}{x^2 + 4x - 8}$

7. The average cost per DVD, in dollars, for a company to produce x DVDs on exercising is given by the formula: $A = \dfrac{3x + 400}{x}$, where A is the average cost per DVD, and x is the number of DVDs produced.

a. Find the cost for producing 1 DVD.

b. Find the average cost for producing 100 DVDs.

c. Does the cost per DVD decrease or increase when more DVDs are produced? Explain your answer.

8. For a certain model of fax machine, the manufacturing cost C per machine is given by the equation

$$C = \frac{250x + 10{,}000}{x}$$

where x is the number of fax machines manufactured and cost C is in dollars per machine.

a. Find the cost per fax machine when manufacturing 100 fax machines.

b. Find the cost per fax machine when manufacturing 1000 fax machines.

c. Does the cost per machine decrease or increase when more machines are manufactured? Explain why this is so.

Objective B *Find any numbers for which each rational expression is undefined. See Example 2.*

9. $\dfrac{7}{2x}$

10. $\dfrac{3}{5x}$

11. $\dfrac{x + 3}{x + 2}$

12. $\dfrac{5x + 1}{x - 9}$

13. $\dfrac{x - 4}{2x - 5}$

14. $\dfrac{x + 1}{5x - 2}$

15. $\dfrac{9x^3 + 4}{15x^2 + 30x}$

16. $\dfrac{19x^3 + 2}{x^2 - x}$

17. $\dfrac{x^2 - 5x - 2}{4}$

18. $\dfrac{9y^5 + y^3}{9}$

19. $\frac{3x^2 + 9}{x^2 - 5x - 6}$

20. $\frac{11x^2 + 1}{x^2 - 5x - 14}$

21. $\frac{x}{3x^2 + 13x + 14}$

22. $\frac{x}{2x^2 + 15x + 27}$

Objective C *Simplify each expression. See Examples 3 through 7.*

23. $\frac{x + 7}{7 + x}$

24. $\frac{y + 9}{9 + y}$

25. $\frac{x - 7}{7 - x}$

26. $\frac{y - 9}{9 - y}$

27. $\frac{2}{8x + 16}$

28. $\frac{3}{9x + 6}$

29. $\frac{x - 2}{x^2 - 4}$

30. $\frac{x + 5}{x^2 - 25}$

31. $\frac{2x - 10}{3x - 30}$

32. $\frac{3x - 9}{4x - 16}$

33. $\frac{-5a - 5b}{a + b}$

34. $\frac{-4x - 4y}{x + y}$

35. $\frac{7x + 35}{x^2 + 5x}$

36. $\frac{9x + 99}{x^2 + 11x}$

37. $\frac{x + 5}{x^2 - 4x - 45}$

38. $\frac{x - 3}{x^2 - 6x + 9}$

39. $\frac{5x^2 + 11x + 2}{x + 2}$

40. $\frac{12x^2 + 4x - 1}{2x + 1}$

41. $\frac{x^3 + 7x^2}{x^2 + 5x - 14}$

42. $\frac{x^4 - 10x^3}{x^2 - 17x + 70}$

43. $\frac{14x^2 - 21x}{2x - 3}$

44. $\frac{4x^2 + 24x}{x + 6}$

45. $\frac{x^2 + 7x + 10}{x^2 - 3x - 10}$

46. $\frac{2x^2 + 7x - 4}{x^2 + 3x - 4}$

47. $\frac{3x^2 + 7x + 2}{3x^2 + 13x + 4}$

48. $\frac{4x^2 - 4x + 1}{2x^2 + 9x - 5}$

49. $\frac{2x^2 - 8}{4x - 8}$

50. $\frac{5x^2 - 500}{35x + 350}$

51. $\frac{4 - x^2}{x - 2}$

52. $\frac{49 - y^2}{y - 7}$

53. $\frac{x^2 - 1}{x^2 - 2x + 1}$

54. $\frac{x^2 - 16}{x^2 - 8x + 16}$

Simplify each expression. Each exercise contains a four-term polynomial that should be factored by grouping.

55. $\frac{x^2 + xy + 2x + 2y}{x + 2}$

56. $\frac{ab + ac + b^2 + bc}{b + c}$

57. $\frac{5x + 15 - xy - 3y}{2x + 6}$

58. $\frac{xy - 6x + 2y - 12}{y^2 - 6y}$

59. $\frac{2xy + 5x - 2y - 5}{3xy + 4x - 3y - 4}$

60. $\frac{2xy + 2x - 3y - 3}{2xy + 4x - 3y - 6}$

Objective D *Study Example 8. Then list four equivalent forms for each rational expression.*

61. $-\frac{x - 10}{x + 8}$

62. $-\frac{x + 11}{x - 4}$

63. $-\frac{5y - 3}{y - 12}$

64. $-\frac{8y - 1}{y - 15}$

Objectives **C** **D** **Mixed Practice** *Simplify each expression. Then determine whether the given answer is correct. See Examples 3 through 8.*

65. $\frac{9 - x^2}{x - 3}$; Answer: $-3 - x$

66. $\frac{100 - x^2}{x - 10}$; Answer: $-10 - x$

67. $\frac{7 - 34x - 5x^2}{25x^2 - 1}$; Answer: $\frac{x + 7}{-5x - 1}$

68. $\frac{2 - 15x - 8x^2}{64x^2 - 1}$; Answer: $\frac{x + 2}{-8x - 1}$

Review

Perform each indicated operation. See Section 4.3.

69. $\frac{1}{3} \cdot \frac{9}{11}$

70. $\frac{5}{27} \cdot \frac{2}{5}$

71. $\frac{1}{3} \div \frac{1}{4}$

72. $\frac{7}{8} \div \frac{1}{2}$

73. $\frac{13}{20} \div \frac{2}{9}$

74. $\frac{8}{15} \div \frac{5}{8}$

Concept Extensions

Which of the following are incorrect and why? See the Concept Check in this section.

75. $\frac{5a - 15}{5}$ simplifies to $a - 3$

76. $\frac{7m - 9}{7}$ simplifies to $m - 9$

77. $\frac{1 + 2}{1 + 3}$ simplifies to $\frac{2}{3}$

78. $\frac{46}{54}$ simplifies to $\frac{6}{5}$

79. Explain how to write a fraction in lowest terms

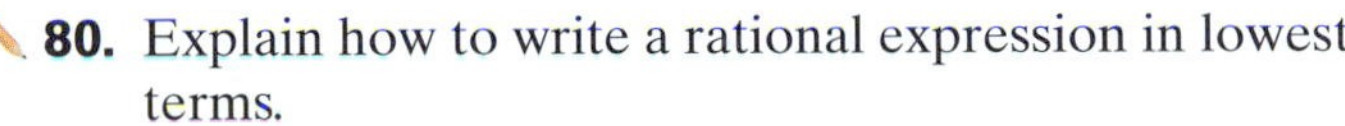

80. Explain how to write a rational expression in lowest terms.

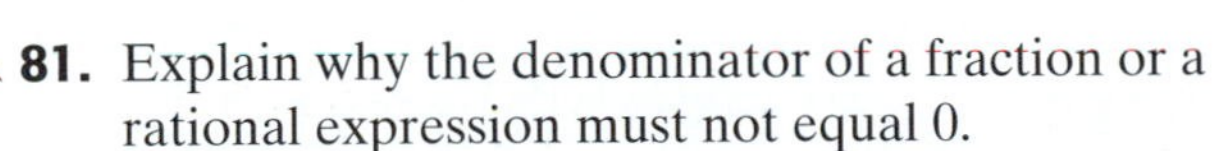

81. Explain why the denominator of a fraction or a rational expression must not equal 0.

82. Does $\frac{(x - 3)(x + 3)}{x - 3}$ have the same value as $x + 3$ for all real numbers? Explain why or why not.

83. The dose of medicine prescribed for a child depends on the child's age A in years and the adult dose D for the medication. Young's Rule is a formula used by pediatricians that gives a child's dose C as

$$C = \frac{DA}{A + 12}$$

Suppose that an 8-year-old child needs medication, and the normal adult dose is 1000 mg. What size dose should the child receive?

84. Calculating body-mass index is a way to gauge whether a person should lose weight. Doctors recommend that body-mass index values fall between 19 and 25. The formula for body-mass index B is

$$B = \frac{705w}{h^2}$$

where w is weight in pounds and h is height in inches. Should a 148-pound person who is 5 feet 6 inches tall lose weight?

85. A baseball player's slugging percentage S can be calculated with the following formula: $S = \frac{h + d + 2t + 3r}{b}$, where h = number of hits, d = number of doubles, t = number of triples, r = number of home runs, and b = number of at-bats. In 2004, Ichiro Suzuki of the Seattle Mariners led the American League in slugging percentage. During the 2004 season, Suzuki had 704 at-bats, 262 hits, 24 doubles, 5 triples, and 8 home runs. (*Source:* Major League Baseball) Calculate Suzuki's 2004 slugging percentage. Round to the nearest tenth of a percent.

86. A company's gross profit margin P can be computed with the formula $P = \frac{R - C}{R}$, where R = the company's revenue and C = cost of goods sold. For fiscal year 2004, consumer electronics retailer Best Buy had revenues of \$24.5 billion and cost of goods sold of \$18.3 billion. (*Source:* Best Buy Company, Inc.) What was Best Buy's gross profit margin in 2004? Express the answer as a percent, rounded to the nearest tenth of a percent.

87. To calculate a quarterback's rating in football, you may use the formula $\left[\frac{20C + 0.5A + Y + 80T - 100I}{A}\right]\left(\frac{25}{6}\right)$, where C = the number of completed passes, A = the number of attempted passes, Y = total yards thrown for passes, T = the number of touchdown passes, and I = the number of interceptions. The New England Patriots were the winners of the Super Bowl in 2005. Their quarterback, Tom Brady, boasted the final season totals of 527 attempts, 317 completions, 3620 yards, 23 touchdown passes, and 12 interceptions. Calculate Brady's quarterback rating for the 2004–2005 season. Round the answer to the nearest tenth. (*Source:* The NFL)

88. Anthropologists and forensic scientists use a measure called the cephalic index to help classify skulls. The cephalic index of a skull with width W and length L from front to back is given by the formula

$$C = \frac{100W}{L}$$

A long skull has an index value less than 75, a medium skull has an index value between 75 and 85, and a broad skull has an index value over 85. Find the cephalic index of a skull that is 5 inches wide and 6.4 inches long. Classify the skull.

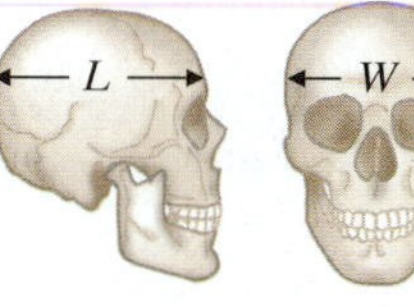

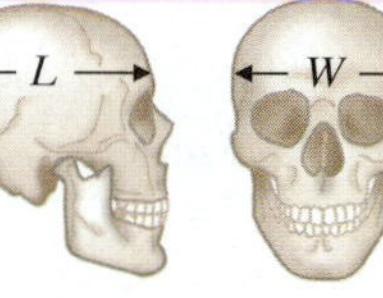

STUDY SKILLS BUILDER

Is Your Notebook Still Organized?

It's never too late to organize your material in a course. Let's see how you are doing.

1. Are all your graded papers in one place in your math notebook or binder?
2. Flip through the pages of your notebook. Are your notes neat and readable?
3. Are your notes complete with no sections missing?
4. Are important notes marked in some way (like an exclamation point) so that you will know to review them before a quiz or task?
5. Are your assignments complete?
6. Do exercises that have given you trouble have a mark (like a question mark) so that you will remember to talk to your instructor or a tutor about them?
7. Describe your attitude toward this course.
8. List ways your attitude can improve and make a commitment to work on at least one of those during the next week.

12.2 MULTIPLYING AND DIVIDING RATIONAL EXPRESSIONS

Objectives

A Multiply Rational Expressions.

B Divide Rational Expressions.

C Multiply and Divide Rational Expressions.

D Convert between Units of Measure.

Objective A Multiplying Rational Expressions

Just as simplifying rational expressions is similar to simplifying number fractions, multiplying and dividing rational expressions is similar to multiplying and dividing number fractions.

Fractions	**Rational Expressions**
Multiply: $\frac{3}{5} \cdot \frac{10}{11}$	Multiply: $\frac{x-3}{x+5} \cdot \frac{2x+10}{x^2-9}$

Multiply numerators and then multiply denominators.

$\frac{3}{5} \cdot \frac{10}{11} = \frac{3 \cdot 10}{5 \cdot 11}$	$\frac{x-3}{x+5} \cdot \frac{2x+10}{x^2-9} = \frac{(x-3) \cdot (2x+10)}{(x+5) \cdot (x^2-9)}$

Simplify by factoring numerators and denominators.

$= \frac{3 \cdot 2 \cdot 5}{5 \cdot 11}$	$= \frac{(x-3) \cdot 2(x+5)}{(x+5)(x+3)(x-3)}$

Apply the fundamental principle.

$= \frac{3 \cdot 2}{11}$ or $\frac{6}{11}$	$= \frac{2}{x+3}$

Multiplying Rational Expressions

If $\frac{P}{Q}$ and $\frac{R}{S}$ are rational expressions, then

$$\frac{P}{Q} \cdot \frac{R}{S} = \frac{PR}{QS}$$

To multiply rational expressions, multiply the numerators and then multiply the denominators.

Note: Recall that for Sections 12.1 through 12.4, we assume variables in rational expressions have only those replacement values for which the expressions are defined.

EXAMPLE 1 Multiply.

a. $\frac{25x}{2} \cdot \frac{1}{y^3}$ **b.** $\frac{-7x^2}{5y} \cdot \frac{3y^5}{14x^2}$

Solution: To multiply rational expressions, we first multiply the numerators and then multiply the denominators of both expressions. Then we write the product in lowest terms.

a. $\frac{25x}{2} \cdot \frac{1}{y^3} = \frac{25x \cdot 1}{2 \cdot y^3} = \frac{25x}{2y^3}$

The expression $\frac{25x}{2y^3}$ is in lowest terms.

b. $\frac{-7x^2}{5y} \cdot \frac{3y^5}{14x^2} = \frac{-7x^2 \cdot 3y^5}{5y \cdot 14x^2}$ Multiply.

Continued on next page

PRACTICE PROBLEM 1

Multiply.

a. $\frac{16y}{3} \cdot \frac{1}{x^2}$

b. $\frac{-5a^3}{3b^3} \cdot \frac{2b^2}{15a}$

Answers

1. a. $\frac{16y}{3x^2}$ **b.** $-\frac{2a^2}{9b}$

The expression $\frac{-7x^2 \cdot 3y^5}{5y \cdot 14x^2}$ is not in lowest terms, so we factor the numerator and the denominator and apply the fundamental principle and "remove factors of 1."

$$= \frac{-1 \cdot 7 \cdot 3 \cdot x^2 \cdot y \cdot y^4}{5 \cdot 2 \cdot 7 \cdot x^2 \cdot y}$$ Common factors in the numerator and denominator form factors of 1.

$$= -\frac{3y^4}{10}$$ Divide out common factors. (This is the same as "removing a factor of 1.")

Work Practice Problem 1

When multiplying rational expressions, it is usually best to factor each numerator and denominator first. This will help us when we apply the fundamental principle to write the product in lowest terms.

PRACTICE PROBLEM 2

Multiply: $\frac{3x + 6}{14} \cdot \frac{7x^2}{x^3 + 2x^2}$

EXAMPLE 2 Multiply: $\frac{x^2 + x}{3x} \cdot \frac{6}{5x + 5}$

Solution:

$$\frac{x^2 + x}{3x} \cdot \frac{6}{5x + 5} = \frac{x(x + 1)}{3x} \cdot \frac{2 \cdot 3}{5(x + 1)}$$ Factor numerators and denominators.

$$= \frac{x(x + 1) \cdot 2 \cdot 3}{3x \cdot 5(x + 1)}$$ Multiply.

$$= \frac{2}{5}$$ Divide out common factors.

Work Practice Problem 2

The following steps may be used to multiply rational expressions.

To Multiply Rational Expressions

Step 1: Completely factor numerators and denominators.

Step 2: Multiply numerators and multiply denominators.

Step 3: Simplify or write the product in lowest terms by dividing out common factors.

✔ **Concept Check** Which of the following is a true statement?

a. $\frac{1}{3} \cdot \frac{1}{2} = \frac{1}{5}$ **b.** $\frac{2}{x} \cdot \frac{5}{x} = \frac{10}{x}$ **c.** $\frac{3}{x} \cdot \frac{1}{2} = \frac{3}{2x}$ **d.** $\frac{x}{7} \cdot \frac{x + 5}{4} = \frac{2x + 5}{28}$

PRACTICE PROBLEM 3

Multiply:

$\frac{4x + 8}{7x^2 - 14x} \cdot \frac{3x^2 - 5x - 2}{9x^2 - 1}$

EXAMPLE 3 Multiply: $\frac{3x + 3}{5x^2 - 5x} \cdot \frac{2x^2 + x - 3}{4x^2 - 9}$

Solution:

$$\frac{3x + 3}{5x^2 - 5x} \cdot \frac{2x^2 + x - 3}{4x^2 - 9} = \frac{3(x + 1)}{5x(x - 1)} \cdot \frac{(2x + 3)(x - 1)}{(2x - 3)(2x + 3)}$$ Factor.

$$= \frac{3(x + 1)(2x + 3)(x - 1)}{5x(x - 1)(2x - 3)(2x + 3)}$$ Multiply.

$$= \frac{3(x + 1)}{5x(2x - 3)}$$ Simplify.

Work Practice Problem 3

Answers

2. $\frac{3}{2}$ **3.** $\frac{4(x + 2)}{7x(3x - 1)}$

✔ **Concept Check Answer**

c

Objective B Dividing Rational Expressions

We can divide by a rational expression in the same way we divide by a number fraction. Recall that to divide by a fraction, we multiply by its reciprocal.

For example, to divide $\frac{3}{2}$ by $\frac{7}{8}$, we multiply $\frac{3}{2}$ by $\frac{8}{7}$.

$$\frac{3}{2} \div \frac{7}{8} = \frac{3}{2} \cdot \frac{8}{7} = \frac{3 \cdot 4 \cdot 2}{2 \cdot 7} = \frac{12}{7}$$

Don't forget how to find reciprocals. The reciprocal of $\frac{a}{b}$ is $\frac{b}{a}$, $a \neq 0$, $b \neq 0$.

Dividing Rational Expressions

If $\frac{P}{Q}$ and $\frac{R}{S}$ are rational expressions and $\frac{R}{S}$ is not 0, then

$$\frac{P}{Q} \div \frac{R}{S} = \frac{P}{Q} \cdot \frac{S}{R} = \frac{PS}{QR}$$

To divide two rational expressions, multiply the first rational expression by the reciprocal of the second rational expression.

EXAMPLE 4 Divide: $\frac{3x^3}{40} \div \frac{4x^3}{y^2}$

Solution:

$$\frac{3x^3}{40} \div \frac{4x^3}{y^2} = \frac{3x^3}{40} \cdot \frac{y^2}{4x^3} \quad \text{Multiply by the reciprocal of } \frac{4x^3}{y^2}.$$

$$= \frac{3x^3 \cdot y^2}{160x^3}$$

$$= \frac{3y^2}{160} \quad \text{Simplify.}$$

Work Practice Problem 4

PRACTICE PROBLEM 4

Divide: $\frac{7x^2}{6} \div \frac{x}{2y}$

EXAMPLE 5 Divide: $\frac{(x+2)^2}{10} \div \frac{2x+4}{5}$

Solution:

$$\frac{(x+2)^2}{10} \div \frac{2x+4}{5} = \frac{(x+2)^2}{10} \cdot \frac{5}{2x+4} \quad \text{Multiply by the reciprocal of } \frac{2x+4}{5}.$$

$$= \frac{(x+2)(x+2) \cdot 5}{5 \cdot 2 \cdot 2 \cdot (x+2)} \quad \text{Factor and multiply.}$$

$$= \frac{x+2}{4} \quad \text{Simplify.}$$

Work Practice Problem 5

PRACTICE PROBLEM 5

Divide: $\frac{(x-4)^2}{6} \div \frac{3x-12}{2}$

Helpful Hint

Remember, **to Divide by a Rational Expression,** multiply by its reciprocal.

Answers

4. $\frac{7xy}{3}$ **5.** $\frac{x-4}{9}$

PRACTICE PROBLEM 6

Divide: $\frac{10x + 4}{x^2 - 4} \div \frac{5x^3 + 2x^2}{x + 2}$

EXAMPLE 6 Divide: $\frac{6x + 2}{x^2 - 1} \div \frac{3x^2 + x}{x - 1}$

Solution:

$$\frac{6x + 2}{x^2 - 1} \div \frac{3x^2 + x}{x - 1} = \frac{6x + 2}{x^2 - 1} \cdot \frac{x - 1}{3x^2 + x}$$ Multiply by the reciprocal.

$$= \frac{2(3x + 1)(x - 1)}{(x + 1)(x - 1) \cdot x(3x + 1)}$$ Factor and multiply.

$$= \frac{2}{x(x + 1)}$$ Simplify.

Work Practice Problem 6

PRACTICE PROBLEM 7

Divide:

$\frac{3x^2 - 10x + 8}{7x - 14} \div \frac{9x - 12}{21}$

EXAMPLE 7 Divide: $\frac{2x^2 - 11x + 5}{5x - 25} \div \frac{4x - 2}{10}$

Solution:

$$\frac{2x^2 - 11x + 5}{5x - 25} \div \frac{4x - 2}{10} = \frac{2x^2 - 11x + 5}{5x - 25} \cdot \frac{10}{4x - 2}$$ Multiply by the reciprocal.

$$= \frac{(2x - 1)(x - 5) \cdot 2 \cdot 5}{5(x - 5) \cdot 2(2x - 1)}$$ Factor and multiply.

$$= \frac{1}{1} \text{ or } 1$$ Simplify.

Work Practice Problem 7

Objective C Multiplying and Dividing Rational Expressions

Let's make sure that we understand the difference between multiplying and dividing rational expressions.

Rational Expressions	
Multiplication	Multiply the numerators and multiply the denominators.
Division	Multiply by the reciprocal of the divisor.

PRACTICE PROBLEM 8

Multiply or divide as indicated.

a. $\frac{x + 3}{x} \cdot \frac{7}{x + 3}$

b. $\frac{x + 3}{x} \div \frac{7}{x + 3}$

c. $\frac{3 - x}{x^2 + 6x + 5} \cdot \frac{2x + 10}{x^2 - 7x + 12}$

EXAMPLE 8 Multiply or divide as indicated.

a. $\frac{x - 4}{5} \cdot \frac{x}{x - 4}$

b. $\frac{x - 4}{5} \div \frac{x}{x - 4}$

c. $\frac{x^2 - 4}{2x + 6} \cdot \frac{x^2 + 4x + 3}{2 - x}$

Solution:

a. $\frac{x - 4}{5} \cdot \frac{x}{x - 4} = \frac{(x - 4) \cdot x}{5 \cdot (x - 4)} = \frac{x}{5}$

b. $\frac{x - 4}{5} \div \frac{x}{x - 4} = \frac{x - 4}{5} \cdot \frac{x - 4}{x} = \frac{(x - 4)^2}{5x}$

c. $\frac{x^2 - 4}{2x + 6} \cdot \frac{x^2 + 4x + 3}{2 - x} = \frac{(x - 2)(x + 2) \cdot (x + 1)(x + 3)}{2(x + 3) \cdot (2 - x)}$ Factor and multiply.

Answers

6. $\frac{2}{x^2(x - 2)}$ **7.** 1

8. a. $\frac{7}{x}$ **b.** $\frac{(x + 3)^2}{7x}$

c. $-\frac{2}{(x + 1)(x - 4)}$

Recall from Section 12.1 that $x - 2$ and $2 - x$ are opposites. This means that $\frac{x-2}{2-x} = -1$. Thus,

$$\frac{(x-2)(x+2)\cdot(x+1)(x+3)}{2(x+3)\cdot(2-x)} = \frac{-1(x+2)(x+1)}{2}$$
$$= -\frac{(x+2)(x+1)}{2}$$

Work Practice Problem 8

Objective D Converting between Units of Measure

How many square inches are in 1 square foot?

How many cubic feet are in a cubic yard?

If you have trouble answering these questions, this section will be helpful to you.

Now that we know how to multiply fractions and rational expressions, we can use this knowledge to help us convert between units of measure. To do so, we will use **unit fractions.** A unit fraction is a fraction that equals 1. For example, since 12 in. = 1 ft, we have the unit fractions

$$\frac{12 \text{ in.}}{1 \text{ ft}} = 1 \quad \text{and} \quad \frac{1 \text{ ft}}{12 \text{ in.}} = 1$$

EXAMPLE 9 18 square feet = _____ square yards

Solution: Let's multiply 18 square feet by a unit fraction that has square feet in denominator and square yards in the numerator. From the diagram, you can see that

1 square yard = 9 square feet

Thus,

$$18 \text{ sq ft} = \frac{18 \text{ sq ft}}{1}\cdot 1 = \frac{\overset{2}{\cancel{18}}\ \cancel{\text{sq ft}}}{1}\cdot\frac{1 \text{ sq yd}}{\underset{1}{\cancel{9}}\ \cancel{\text{sq ft}}}$$
$$= \frac{2\cdot 1}{1\cdot 1}\text{ sq yd} = 2 \text{ sq yd}$$

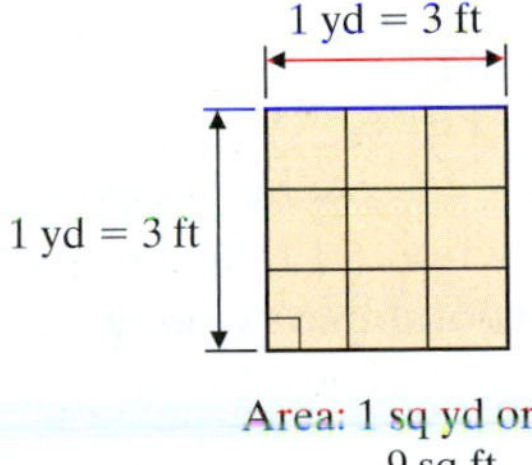

Area: 1 sq yd or 9 sq ft

Thus, 18 sq ft = 2 sq yd.

Draw a diagram of 18 sq ft to help you see that this is reasonable.

Work Practice Problem 9

PRACTICE PROBLEM 9

288 square inches = _____ square feet

EXAMPLE 10 5.2 square yards = _____ square feet

Solution:

$$5.2 \text{ sq yd} = \frac{5.2 \text{ sq yd}}{1}\cdot 1 = \frac{5.2\ \cancel{\text{sq yd}}}{1}\cdot\frac{9 \text{ sq ft}}{1\ \cancel{\text{sq yd}}} \quad \begin{matrix}\leftarrow \text{Units converting to}\\ \leftarrow \text{Units given}\end{matrix}$$
$$= \frac{5.2\cdot 9}{1\cdot 1}\text{ sq ft}$$
$$= 46.8 \text{ sq ft}$$

Thus, 5.2 sq yd = 46.8 sq ft.

Draw a diagram to see that this is reasonable.

Work Practice Problem 10

PRACTICE PROBLEM 10

3.5 square feet = _____ square inches

Answers

9. 2 sq ft **10.** 504 sq in.

PRACTICE PROBLEM 11

The largest casino in the world is the Foxwoods Resort Casino in Ledyard, CT. The gaming area for this casino is approximately 35,000 *square yards*. Find the size of the gaming area in *square feet*. (*Source:* Foxwoods Resort)

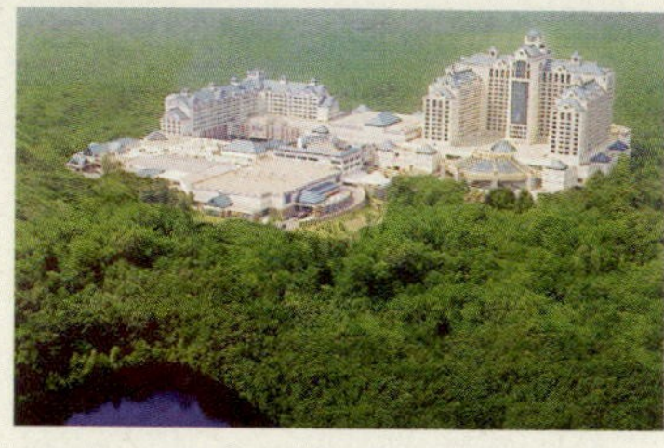

EXAMPLE 11 Converting from Cubic Feet to Cubic Yards

The largest building in the world by volume is The Boeing Company's Everett, Washington, factory complex where Boeing's wide-body jetliners, the 747, 767, and 777, are built. The volume of this factory complex is 472,370,319 cubic feet. Find the volume of this Boeing facility in cubic yards. (*Source:* The Boeing Company)

Solution: There are 27 cubic feet in 1 cubic yard. (See the diagram.)

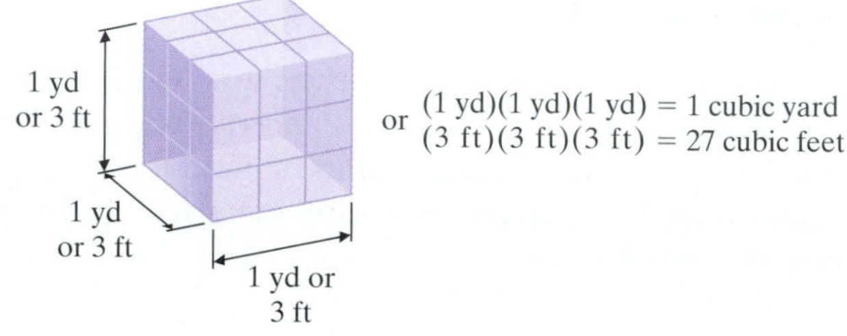

$$472{,}370{,}319 \text{ cu ft} = 472{,}370{,}319 \cancel{\text{cu ft}} \cdot \frac{1 \text{ cu yd}}{27 \cancel{\text{cu ft}}}$$

$$= \frac{472{,}370{,}319}{27} \text{ cu yd}$$

$$= 17{,}495{,}197 \text{ cu yd}$$

Work Practice Problem 11

Helpful Hint

When converting among units of measurement, if possible write the unit fraction so that **the numerator contains the units you are converting to** and **the denominator contains the original units.**

$$48 \text{ in.} = \frac{48 \cancel{\text{in.}}}{1} \cdot \overbrace{\frac{1 \text{ ft}}{12 \cancel{\text{in.}}}}^{\text{Unit fraction}} \begin{array}{l} \leftarrow \text{Units converting to} \\ \leftarrow \text{Original units} \end{array}$$

$$= \frac{48}{12} \text{ ft} = 4 \text{ ft}$$

PRACTICE PROBLEM 12

The cheetah is the fastest land animal, being clocked at about 102.7 feet per second. Convert this to miles per hour. Round to the nearest tenth. (*Source: World Almanac and Book of Facts*)

EXAMPLE 12

At the 2004 Summer Olympics, U.S. athlete Justin Gatlin won the gold medal in the men's 100-meter track event. He ran the distance at an average speed of 33.3 feet per second. Convert this speed to miles per hour. (*Source:* Athens Olympic Committee)

Solution: Recall that 1 mile = 5280 feet and 1 hour = 3600 seconds $(60 \cdot 60)$.

$$33.3 \text{ feet/second} = \frac{33.3 \text{ feet}}{1 \text{ second}} \cdot \overbrace{\frac{3600 \text{ seconds}}{1 \text{ hour}} \cdot \frac{1 \text{ mile}}{5280 \text{ feet}}}^{\text{Unit fractions}}$$

$$= \frac{33.3 \cdot 3600}{5280} \text{ miles/hour}$$

$$\approx 22.7 \text{ miles/hour (rounded to the nearest tenth)}$$

Work Practice Problem 12

Answers

11. 315,000 sq ft
12. 70.0 miles per hour

Vocabulary and Readiness Check

Use the choices below to fill in each blank. Not all choices will be used.

opposites	$\frac{a \cdot d}{b \cdot c}$	$\frac{a \cdot c}{b \cdot d}$	$\frac{x}{42}$	$\frac{x^2}{42}$	$\frac{2x}{42}$	$\frac{6}{7}$	$\frac{7}{6}$
reciprocals							

1. The expressions $\frac{x}{2y}$ and $\frac{2y}{x}$ are called ______________.

2. $\frac{a}{b} \cdot \frac{c}{d} =$ ______________

3. $\frac{a}{b} \div \frac{c}{d} =$ ______________

4. $\frac{x}{7} \cdot \frac{x}{6} =$ ______________

5. $\frac{x}{7} \div \frac{x}{6} =$ ______________

12.2 EXERCISE SET

FOR EXTRA HELP

Student Solutions Manual

PH Math/Tutor Center

CD/Video for Review

MathXL®

MyMathLab

Objective A *Find each product and simplify if possible. See Examples 1 through 3.*

1. $\frac{3x}{y^2} \cdot \frac{7y}{4x}$

2. $\frac{9x^2}{y} \cdot \frac{4y}{3x^3}$

3. $\frac{8x}{2} \cdot \frac{x^5}{4x^2}$

4. $\frac{6x^2}{10x^3} \cdot \frac{5x}{12}$

5. $-\frac{5a^2b}{30a^2b^2} \cdot b^3$

6. $-\frac{9x^3y^2}{18xy^5} \cdot y^3$

7. $\frac{x}{2x - 14} \cdot \frac{x^2 - 7x}{5}$

8. $\frac{4x - 24}{20x} \cdot \frac{5}{x - 6}$

9. $\frac{6x + 6}{5} \cdot \frac{10}{36x + 36}$

10. $\frac{x^2 + x}{8} \cdot \frac{16}{x + 1}$

11. $\frac{(m + n)^2}{m - n} \cdot \frac{m}{m^2 + mn}$

12. $\frac{(m - n)^2}{m + n} \cdot \frac{m}{m^2 - mn}$

13. $\frac{x^2 - 25}{x^2 - 3x - 10} \cdot \frac{x + 2}{x}$

14. $\frac{a^2 - 4a + 4}{a^2 - 4} \cdot \frac{a + 3}{a - 2}$

15. $\frac{x^2 + 6x + 8}{x^2 + x - 20} \cdot \frac{x^2 + 2x - 15}{x^2 + 8x + 16}$

16. $\frac{x^2 + 9x + 20}{x^2 - 15x + 44} \cdot \frac{x^2 - 11x + 28}{x^2 + 12x + 35}$

Objective B *Find each quotient and simplify. See Examples 4 through 7.*

17. $\frac{5x^7}{2x^5} \div \frac{15x}{4x^3}$

18. $\frac{9y^4}{6y} \div \frac{y^2}{3}$

19. $\frac{8x^2}{y^3} \div \frac{4x^2y^3}{6}$

20. $\frac{7a^2b}{3ab^2} \div \frac{21a^2b^2}{14ab}$

21. $\frac{(x - 6)(x + 4)}{4x} \div \frac{2x - 12}{8x^2}$

22. $\frac{(x + 3)^2}{5} \div \frac{5x + 15}{25}$

23. $\frac{3x^2}{x^2 - 1} \div \frac{x^5}{(x + 1)^2}$

24. $\frac{9x^5}{a^2 - b^2} \div \frac{27x^2}{3b - 3a}$

25. $\frac{m^2 - n^2}{m + n} \div \frac{m}{m^2 + nm}$

26. $\frac{(m - n)^2}{m + n} \div \frac{m^2 - mn}{m}$

27. $\dfrac{x+2}{7-x} \div \dfrac{x^2-5x+6}{x^2-9x+14}$

28. $\dfrac{x-3}{2-x} \div \dfrac{x^2+3x-18}{x^2+2x-8}$

29. $\dfrac{x^2+7x+10}{x-1} \div \dfrac{x^2+2x-15}{x-1}$

30. $\dfrac{x+1}{(x+1)(2x+3)} \div \dfrac{20x+100}{2x+3}$

Objective C Mixed Practice *Multiply or divide as indicated. See Example 8.*

31. $\dfrac{5x-10}{12} \div \dfrac{4x-8}{8}$

32. $\dfrac{6x+6}{5} \div \dfrac{9x+9}{10}$

33. $\dfrac{x^2+5x}{8} \cdot \dfrac{9}{3x+15}$

34. $\dfrac{3x^2+12x}{6} \cdot \dfrac{9}{2x+8}$

35. $\dfrac{7}{6p^2+q} \div \dfrac{14}{18p^2+3q}$

36. $\dfrac{3x+6}{20} \div \dfrac{4x+8}{8}$

37. $\dfrac{3x+4y}{x^2+4xy+4y^2} \cdot \dfrac{x+2y}{2}$

38. $\dfrac{x^2-y^2}{3x^2+3xy} \cdot \dfrac{3x^2+6x}{3x^2-2xy-y^2}$

39. $\dfrac{(x+2)^2}{x-2} \div \dfrac{x^2-4}{2x-4}$

40. $\dfrac{x+3}{x^2-9} \div \dfrac{5x+15}{(x-3)^2}$

41. $\dfrac{x^2-4}{24x} \div \dfrac{2-x}{6xy}$

42. $\dfrac{3y}{3-x} \div \dfrac{12xy}{x^2-9}$

43. $\dfrac{a^2+7a+12}{a^2+5a+6} \cdot \dfrac{a^2+8a+15}{a^2+5a+4}$

44. $\dfrac{b^2+2b-3}{b^2+b-2} \cdot \dfrac{b^2-4}{b^2+6b+8}$

45. $\dfrac{5x-20}{3x^2+x} \cdot \dfrac{3x^2+13x+4}{x^2-16}$

46. $\dfrac{9x+18}{4x^2-3x} \cdot \dfrac{4x^2-11x+6}{x^2-4}$

47. $\dfrac{8n^2-18}{2n^2-5n+3} \div \dfrac{6n^2+7n-3}{n^2-9n+8}$

48. $\dfrac{36n^2-64}{3n^2-10n+8} \div \dfrac{3n^2-13n+12}{n^2-5n-14}$

Objective D *Convert as indicated. See Examples 9 through 12.*

49. 10 square feet = ________ square inches.

50. 1008 square inches = _____ square feet.

51. 45 square feet = _____ square yards.

52. 2 square yards = ________ square inches.

53. 3 cubic yards = ______ cubic feet.

54. 2 cubic yards = _________ cubic inches.

55. 50 miles per hour = ______ feet per second (round to the nearest whole).

56. 10 feet per second = ______ miles per hour (round to the nearest tenth).

57. 6.3 square yards = _______ square feet.

58. 3.6 square yards = _______ square feet.

59. The Pentagon, headquarters for the Department of Defense, contains 3,705,793 square feet of office and storage space. Convert this to square yards. Round to the nearest square yard. (*Source:* U.S. Department of Defense)

60. The world's tallest building, Taipei 101 in Taipei, Taiwan, has 427,831 square yards of floor space. Convert this to square feet. (*Source:* Taipei 101)

61. On October 4, 2004, the rocket plane *SpaceShipOne* shot to an altitude of more than 100 km for the second time inside a week to claim the $10 million Ansari X-Prize. At one point in its flight, *SpaceShipOne* was traveling past Mach 1, about 930 miles per hour. Find this speed in feet per second. Round to the nearest whole. (*Source:* Space.com)

62. In 2002, Tim Montgomery of the United States held the current world record for the men's 100-meter track event. In that year, he covered the distance at an average speed of 33.55 feet per second. Convert this speed to miles per hour. Round to the nearest tenth. (*Source:* International Amateur Athletic Association)

Review

Perform each indicated operation. See Section 4.5.

63. $\frac{1}{5} + \frac{4}{5}$

64. $\frac{3}{15} + \frac{6}{15}$

65. $\frac{9}{9} - \frac{19}{9}$

66. $\frac{4}{3} - \frac{8}{3}$

67. $\frac{6}{5} + \left(\frac{1}{5} - \frac{8}{5}\right)$

68. $-\frac{3}{2} + \left(\frac{1}{2} - \frac{3}{2}\right)$

Concept Extensions

Identify each statement as true or false. If false, correct the multiplication. See the Concept Check in this section.

69. $\frac{4}{a} \cdot \frac{1}{b} = \frac{4}{ab}$

70. $\frac{2}{3} \cdot \frac{2}{4} = \frac{2}{7}$

71. $\frac{x}{5} \cdot \frac{x+3}{4} = \frac{2x+3}{20}$

72. $\frac{7}{a} \cdot \frac{3}{a} = \frac{21}{a}$

73. Find the area of the rectangle.

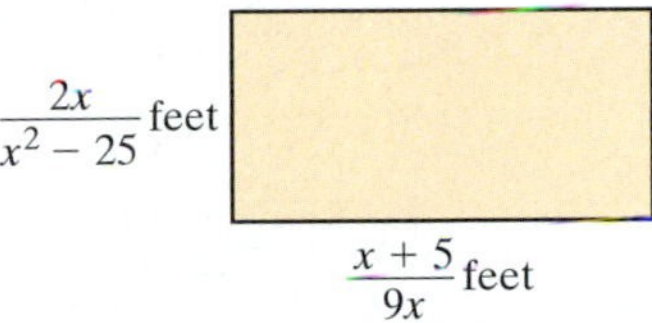

74. Find the area of the square.

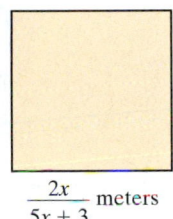

Multiply or divide as indicated.

75. $\left(\frac{x^2 - y^2}{x^2 + y^2} \div \frac{x^2 - y^2}{3x}\right) \cdot \frac{x^2 + y^2}{6}$

76. $\left(\frac{x^2 - 9}{x^2 - 1} \cdot \frac{x^2 + 2x + 1}{2x^2 + 9x + 9}\right) \div \frac{2x + 3}{1 - x}$

77. $\left(\frac{2a + b}{b^2} \cdot \frac{3a^2 - 2ab}{ab + 2b^2}\right) \div \frac{a^2 - 3ab + 2b^2}{5ab - 10b^2}$

78. $\left(\frac{x^2y^2 - xy}{4x - 4y} \div \frac{3y - 3x}{8x - 8y}\right) \cdot \frac{y - x}{8}$

79. In your own words, explain how you multiply rational expressions.

80. Explain how dividing rational expressions is similar to dividing rational numbers.

81. On November 14, 2004, 1 euro was equivalent to 1.2955 U.S. dollars. If you had wanted to exchange \$2000 U.S. for euros on that day for a European vacation, how much would you have received? Round to the nearest hundredth. (*Source:* International Monetary Fund)

82. An environmental technician finds that warm water from an industrial process is being discharged into a nearby pond at a rate of 30 gallons per minute. Plant regulations state that the flow rate should be no more than 0.1 cubic feet per second. Is the flow rate of 30 gallons per minute in violation of the plant regulations? (*Hint:* 1 cubic foot is equivalent to 7.48 gallons.)

12.3 ADDING AND SUBTRACTING RATIONAL EXPRESSIONS WITH THE SAME DENOMINATOR AND LEAST COMMON DENOMINATOR

Objectives

A Add and Subtract Rational Expressions with Common Denominators.

B Find the Least Common Denominator of a List of Rational Expressions.

C Write a Rational Expression as an Equivalent Expression Whose Denominator Is Given.

Objective A Adding and Subtracting Rational Expressions with the Same Denominator

Like multiplication and division, addition and subtraction of rational expressions is similar to addition and subtraction of rational numbers. In this section, we add and subtract rational expressions with a common denominator.

Add: $\frac{6}{5} + \frac{2}{5}$ | Add: $\frac{9}{x+2} + \frac{3}{x+2}$

Add the numerators and place the sum over the common denominator.

$$\frac{6}{5} + \frac{2}{5} = \frac{6+2}{5}$$
$$= \frac{8}{5} \quad \text{Simplify.}$$

$$\frac{9}{x+2} + \frac{3}{x+2} = \frac{9+3}{x+2}$$
$$= \frac{12}{x+2} \quad \text{Simplify.}$$

Adding and Subtracting Rational Expressions with Common Denominators

If $\frac{P}{R}$ and $\frac{Q}{R}$ are rational expressions, then

$$\frac{P}{R} + \frac{Q}{R} = \frac{P+Q}{R} \quad \text{and} \quad \frac{P}{R} - \frac{Q}{R} = \frac{P-Q}{R}$$

To add or subtract rational expressions, add or subtract numerators and place the sum or difference over the common denominator.

PRACTICE PROBLEM 1

Add: $\frac{8x}{3y} + \frac{x}{3y}$

EXAMPLE 1 Add: $\frac{5m}{2n} + \frac{m}{2n}$

Solution:

$$\frac{5m}{2n} + \frac{m}{2n} = \frac{5m+m}{2n} \quad \text{Add the numerators.}$$
$$= \frac{6m}{2n} \quad \text{Simplify the numerator by combining like terms.}$$
$$= \frac{3m}{n} \quad \text{Simplify by applying the fundamental principle.}$$

Work Practice Problem 1

PRACTICE PROBLEM 2

Subtract: $\frac{3x}{3x-7} - \frac{7}{3x-7}$

EXAMPLE 2 Subtract: $\frac{2y}{2y-7} - \frac{7}{2y-7}$

Solution:

$$\frac{2y}{2y-7} - \frac{7}{2y-7} = \frac{2y-7}{2y-7} \quad \text{Subtract the numerators.}$$
$$= \frac{1}{1} \text{ or } 1 \quad \text{Simplify.}$$

Work Practice Problem 2

Answers

1. $\frac{3x}{y}$ 2. 1

EXAMPLE 3 Subtract: $\frac{3x^2 + 2x}{x - 1} - \frac{10x - 5}{x - 1}$

Solution:

$$\frac{3x^2 + 2x}{x - 1} - \frac{10x - 5}{x - 1} = \frac{3x^2 + 2x - (10x - 5)}{x - 1}$$ Subtract the numerators. Notice the parentheses.

$$= \frac{3x^2 + 2x - 10x + 5}{x - 1}$$ Use the distributive property.

$$= \frac{3x^2 - 8x + 5}{x - 1}$$ Combine like terms.

$$= \frac{(x - 1)(3x - 5)}{x - 1}$$ Factor.

$$= 3x - 5$$ Simplify.

Work Practice Problem 3

PRACTICE PROBLEM 3

Subtract: $\frac{2x^2 + 5x}{x + 2} - \frac{4x + 6}{x + 2}$

Helpful Hint

Notice how the numerator $10x - 5$ was subtracted in Example 3.

This $-$ sign applies to the entire numerator $10x - 5$.

So parentheses are inserted here to indicate this.

$$\frac{3x^2 + 2x}{x - 1} - \frac{10x - 5}{x - 1} = \frac{3x^2 + 2x - (10x - 5)}{x - 1}$$

Objective B Finding the Least Common Denominator

Recall from Chapter 4 that to add and subtract fractions with different denominators, we first find a least common denominator (LCD). Then we write all fractions as equivalent fractions with the LCD.

For example, suppose we want to add $\frac{3}{8}$ and $\frac{1}{6}$. To find the LCD of the denominators, factor 8 and 6. Remember, the LCD is the same as the least common multiple LCM. It is the smallest number that is a multiple of 6 and also 8.

$8 = 2 \cdot 2 \cdot 2$

$6 = 2 \cdot 3$

The LCM is a multiple of 6.

$\text{LCM} = 2 \cdot 2 \cdot 2 \cdot 3 = 24$

The LCM is a multiple of 8.

In the next section, we will continue and find the sum: $\frac{3}{8} + \frac{1}{6}$, but for now, let's concentrate on the LCD.

To add or subtract rational expressions with different denominators, we also first find an LCD and then write all rational expressions as equivalent expressions with the LCD. The **least common denominator (LCD) of a list of rational expressions** is a polynomial of least degree whose factors include all the factors of the denominators in the list.

To Find the Least Common Denominator (LCD)

Step 1: Factor each denominator completely.

Step 2: The least common denominator (LCD) is the product of all unique factors found in Step 1, each raised to a power equal to the greatest number of times that the factor appears in any one factored denominator.

Answer

3. $2x - 3$

PRACTICE PROBLEM 4

Find the LCD for each pair.

a. $\frac{2}{9}, \frac{7}{15}$

b. $\frac{5}{6x^3}, \frac{11}{8x^5}$

EXAMPLE 4 Find the LCD for each pair.

a. $\frac{1}{8}, \frac{3}{22}$ **b.** $\frac{7}{5x}, \frac{6}{15x^2}$

Solution:

a. We start by finding the prime factorization of each denominator.

$$8 = 2^3 \quad \text{and}$$
$$22 = 2 \cdot 11$$

Next we write the product of all the unique factors, each raised to a power equal to the greatest number of times that the factor appears.

The greatest number of times that the factor 2 appears is 3.

The greatest number of times that the factor 11 appears is 1.

$$\text{LCD} = 2^3 \cdot 11^1 = 8 \cdot 11 = 88$$

b. We factor each denominator.

$$5x = 5 \cdot x \quad \text{and}$$
$$15x^2 = 3 \cdot 5 \cdot x^2$$

The greatest number of times that the factor 5 appears is 1.

The greatest number of times that the factor 3 appears is 1.

The greatest number of times that the factor x appears is 2.

$$\text{LCD} = 3^1 \cdot 5^1 \cdot x^2 = 15x^2$$

Work Practice Problem 4

PRACTICE PROBLEM 5

Find the LCD of $\frac{3a}{a+5}$ and $\frac{7a}{a-5}$.

EXAMPLE 5 Find the LCD of $\frac{7x}{x+2}$ and $\frac{5x^2}{x-2}$.

Solution: The denominators $x + 2$ and $x - 2$ are completely factored already. The factor $x + 2$ appears once and the factor $x - 2$ appears once.

$$\text{LCD} = (x+2)(x-2)$$

Work Practice Problem 5

PRACTICE PROBLEM 6

Find the LCD of $\frac{7x^2}{(x-4)^2}$ and $\frac{5x}{3x-12}$.

EXAMPLE 6 Find the LCD of $\frac{6m^2}{3m+15}$ and $\frac{2}{(m+5)^2}$.

Solution: We factor each denominator.

$$3m + 15 = 3(m+5)$$
$$(m+5)^2 = (m+5)^2 \quad \text{This denominator is already factored.}$$

The greatest number of times that the factor 3 appears is 1.

The greatest number of times that the factor $m + 5$ appears *in any one denominator* is 2.

$$\text{LCD} = 3(m+5)^2$$

Work Practice Problem 6

✔ **Concept Check** Choose the correct LCD of $\frac{x}{(x+1)^2}$ and $\frac{5}{x+1}$.

a. $x + 1$ **b.** $(x+1)^2$ **c.** $(x+1)^3$ **d.** $5x(x+1)^2$

Answers

4. a. 45 **b.** $24x^5$

5. $(a+5)(a-5)$ **6.** $3(x-4)^2$

✔ Concept Check Answer

b

EXAMPLE 7 Find the LCD of $\frac{t-10}{2t^2+t-6}$ and $\frac{t+5}{t^2+3t+2}$.

Solution:

$$2t^2 + t - 6 = (2t-3)(t+2)$$
$$t^2 + 3t + 2 = (t+1)(t+2)$$
$$\text{LCD} = (2t-3)(t+2)(t+1)$$

Work Practice Problem 7

PRACTICE PROBLEM 7

Find the LCD of $\frac{y+5}{y^2+2y-3}$ and $\frac{y+4}{y^2-3y+2}$.

EXAMPLE 8 Find the LCD of $\frac{2}{x-2}$ and $\frac{10}{2-x}$.

Solution: The denominators $x-2$ and $2-x$ are opposites. That is, $2-x = -1(x-2)$. We can use either $x-2$ or $2-x$ as the LCD.

$\text{LCD} = x - 2$ or $\text{LCD} = 2 - x$

Work Practice Problem 8

PRACTICE PROBLEM 8

Find the LCD of $\frac{6}{x-4}$ and $\frac{9}{4-x}$.

Objective C Writing Equivalent Rational Expressions

Next we practice writing a rational expression as an equivalent rational expression with a given denominator. To do this, we multiply by a form of 1. Recall that multiplying an expression by 1 produces an equivalent expression. In other words,

$$\frac{P}{Q} = \frac{P}{Q}\cdot 1 = \frac{P}{Q}\cdot\frac{R}{R} = \frac{PR}{QR}$$

EXAMPLE 9 Write each rational expression as an equivalent rational expression with the given denominator.

a. $\frac{4b}{9a} = \frac{\quad}{27a^2b}$ **b.** $\frac{7x}{2x+5} = \frac{\quad}{6x+15}$

Solution:

a. We can ask ourselves: "What do we multiply $9a$ by to get $27a^2b$?" The answer is $3ab$, since $9a(3ab) = 27a^2b$. So we multiply by 1 in the form of $\frac{3ab}{3ab}$.

$$\frac{4b}{9a} = \frac{4b}{9a}\cdot 1 = \frac{4b}{9a}\cdot\frac{3ab}{3ab}$$
$$= \frac{4b(3ab)}{9a(3ab)} = \frac{12ab^2}{27a^2b}$$

b. First, factor the denominator on the right.

$$\frac{7x}{2x+5} = \frac{\quad}{3(2x+5)}$$

To obtain the denominator on the right from the denominator on the left, we multiply by 1 in the form of $\frac{3}{3}$.

$$\frac{7x}{2x+5} = \frac{7x}{2x+5}\cdot\frac{3}{3} = \frac{7x\cdot 3}{(2x+5)\cdot 3} = \frac{21x}{3(2x+5)}$$

Work Practice Problem 9

PRACTICE PROBLEM 9

Write the rational expression as an equivalent rational expression with the given denominator.

$$\frac{2x}{5y} = \frac{\quad}{20x^2y^2}$$

Answers

7. $(y+3)(y-2)(y-1)$
8. $x-4$ or $4-x$
9. $\frac{8x^3y}{20x^2y^2}$

PRACTICE PROBLEM 10

Write the rational expression as an equivalent rational expression with the given denominator.

$$\frac{3}{x^2 - 25} = \frac{}{(x + 5)(x - 5)(x - 3)}$$

Answer

10. $\frac{3x - 9}{(x + 5)(x - 5)(x - 3)}$

EXAMPLE 10 Write the rational expression as an equivalent rational expression with the given denominator.

$$\frac{5}{x^2 - 4} = \frac{}{(x - 2)(x + 2)(x - 4)}$$

Solution: First we factor the denominator $x^2 - 4$ as $(x - 2)(x + 2)$. If we multiply the original denominator $(x - 2)(x + 2)$ by $x - 4$, the result is the new denominator $(x - 2)(x + 2)(x - 4)$. Thus, we multiply by 1 in the form of $\frac{x - 4}{x - 4}$.

$$\frac{5}{x^2 - 4} = \frac{5}{(x - 2)(x + 2)} = \frac{5}{(x - 2)(x + 2)} \cdot \frac{x - 4}{x - 4}$$

$$= \frac{5(x - 4)}{(x - 2)(x + 2)(x - 4)}$$

$$= \frac{5x - 20}{(x - 2)(x + 2)(x - 4)}$$

Work Practice Problem 10

Vocabulary and Readiness Check

Use the choices below to fill in each blank. Not all choices will be used.

$\frac{9}{22}$ $\frac{5}{22}$ $\frac{9}{11}$ $\frac{5}{11}$ $\frac{ac}{b}$ $\frac{a - c}{b}$ $\frac{a + c}{b}$ $\frac{5 - 6 + x}{x}$ $\frac{5 - (6 + x)}{x}$

1. $\frac{7}{11} + \frac{2}{11} =$ ________

2. $\frac{7}{11} - \frac{2}{11} =$ ________

3. $\frac{a}{b} + \frac{c}{b} =$ ________

4. $\frac{a}{b} - \frac{c}{b} =$ ________

5. $\frac{5}{x} - \frac{6 + x}{x} =$ ________

12.3 EXERCISE SET

FOR EXTRA HELP

Student Solutions Manual

PH Math/Tutor Center

CD/Video for Review
MathXL®

MyMathLab

Objective A *Add or subtract as indicated. Simplify the result if possible. See Examples 1 through 3.*

1. $\frac{a}{13} + \frac{9}{13}$

2. $\frac{x+1}{7} + \frac{6}{7}$

3. $\frac{4m}{3n} + \frac{5m}{3n}$

4. $\frac{3p}{2q} + \frac{11p}{2q}$

5. $\frac{4m}{m-6} - \frac{24}{m-6}$

6. $\frac{8y}{y-2} - \frac{16}{y-2}$

7. $\frac{9}{3+y} + \frac{y+1}{3+y}$

8. $\frac{9}{y+9} + \frac{y-5}{y+9}$

9. $\frac{5x^2+4x}{x-1} - \frac{6x+3}{x-1}$

10. $\frac{x^2+9x}{x+7} - \frac{4x+14}{x+7}$

11. $\frac{4a}{a^2+2a-15} - \frac{12}{a^2+2a-15}$

12. $\frac{3y}{y^2+3y-10} - \frac{6}{y^2+3y-10}$

13. $\frac{2x+3}{x^2-x-30} - \frac{x-2}{x^2-x-30}$

14. $\frac{3x-1}{x^2+5x-6} - \frac{2x-7}{x^2+5x-6}$

Objective B *Find the LCD for each list of rational expressions. See Examples 4 through 8.*

15. $\frac{19}{2x}, \frac{5}{4x^3}$

16. $\frac{17x}{4y^5}, \frac{2}{8y}$

17. $\frac{9}{8x}, \frac{3}{2x+4}$

18. $\frac{1}{6y}, \frac{3x}{4y+12}$

19. $\frac{2}{x+3}, \frac{5}{x-2}$

20. $\frac{-6}{x-1}, \frac{4}{x+5}$

21. $\frac{x}{x+6}, \frac{10}{3x+18}$

22. $\frac{12}{x+5}, \frac{x}{4x+20}$

23. $\frac{8x^2}{(x-6)^2}, \frac{13x}{5x-30}$

24. $\frac{9x^2}{7x-14}, \frac{6x}{(x-2)^2}$

25. $\frac{1}{3x+3}, \frac{8}{2x^2+4x+2}$

26. $\frac{19x+5}{4x-12}, \frac{3}{2x^2-12x+18}$

27. $\frac{5}{x-8}, \frac{3}{8-x}$

28. $\frac{2x+5}{3x-7}, \frac{5}{7-3x}$

29. $\frac{5x+1}{x^2+3x-4}, \frac{3x}{x^2+2x-3}$

30. $\frac{4}{x^2+4x+3}, \frac{4x-2}{x^2+10x+21}$

31. $\frac{2x}{3x^2+4x+1}, \frac{7}{2x^2-x-1}$

32. $\frac{3x}{4x^2+5x+1}, \frac{5}{3x^2-2x-1}$

33. $\frac{1}{x^2-16}, \frac{x+6}{2x^3-8x^2}$

34. $\frac{5}{x^2-25}, \frac{x+9}{3x^3-15x^2}$

Objective C *Rewrite each rational expression as an equivalent rational expression with the given denominator. See Examples 9 and 10.*

35. $\frac{3}{2x} = \frac{\quad}{4x^2}$

36. $\frac{3}{9y^5} = \frac{\quad}{72y^9}$

37. $\frac{6}{3a} = \frac{\quad}{12ab^2}$

38. $\frac{5}{4y^2x} = \frac{\quad}{32y^3x^2}$

39. $\frac{9}{2x+6} = \frac{\quad}{2y(x+3)}$

40. $\frac{4x+1}{3x+6} = \frac{\quad}{3y(x+2)}$

41. $\frac{9a+2}{5a+10} = \frac{\quad}{5b(a+2)}$

42. $\frac{5+y}{2x^2+10} = \frac{\quad}{4(x^2+5)}$

43. $\frac{x}{x^3+6x^2+8x} = \frac{\quad}{x(x+4)(x+2)(x+1)}$

44. $\frac{5x}{x^3+2x^2-3x} = \frac{\quad}{x(x-1)(x-5)(x+3)}$

45. $\frac{9y-1}{15x^2-30} = \frac{\quad}{30x^2-60}$

46. $\frac{6m-5}{3x^2-9} = \frac{\quad}{12x^2-36}$

Mixed Practice (*Sections 12.2, 12.3*) *Perform the indicated operations.*

47. $\frac{5x}{7} + \frac{9x}{7}$

48. $\frac{5x}{7} \cdot \frac{9x}{7}$

49. $\frac{x+3}{4} \div \frac{2x-1}{4}$

50. $\frac{x+3}{4} - \frac{2x-1}{4}$

51. $\frac{x^2}{x-6} - \frac{5x+6}{x-6}$

52. $\frac{x^2+5x}{x^2-25} \cdot \frac{3x-15}{x^2}$

53. $\frac{-2x}{x^3-8x} + \frac{3x}{x^3-8x}$

54. $\frac{-2x}{x^3-8x} \div \frac{3x}{x^3-8x}$

55. $\frac{12x-6}{x^2+3x} \cdot \frac{4x^2+13x+3}{4x^2-1}$

56. $\frac{x^3+7x^2}{3x^3-x^2} \div \frac{5x^2+36x+7}{9x^2-1}$

Review

Perform each indicated operation. See Section 4.5.

57. $\frac{2}{3} + \frac{5}{7}$ **58.** $\frac{9}{10} - \frac{3}{5}$ **59.** $\frac{2}{6} - \frac{3}{4}$ **60.** $\frac{11}{15} + \frac{5}{9}$ **61.** $\frac{1}{12} + \frac{3}{20}$ **62.** $\frac{7}{30} + \frac{3}{18}$

Concept Extensions

63. Choose the correct LCD of $\frac{11a^3}{4a - 20}$ and $\frac{15a^3}{(a-5)^2}$. See the Concept Check in this section.

a. $4a(a - 5)(a + 5)$ **b.** $a - 5$
c. $(a - 5)^2$ **d.** $4(a - 5)^2$
e. $(4a - 20)(a - 5)^2$

64. An algebra student approaches you with a problem. He's tried to subtract two rational expressions, but his result does not match the book's. Check to see if the student has made an error. If so, correct his work shown below.

$$\frac{2x - 6}{x - 5} - \frac{x + 4}{x - 5}$$
$$= \frac{2x - 6 - x + 4}{x - 5}$$
$$= \frac{x - 2}{x - 5}$$

△ **65.** A square has a side of length $\frac{5}{x - 2}$ meters. Express its perimeter as a rational expression.

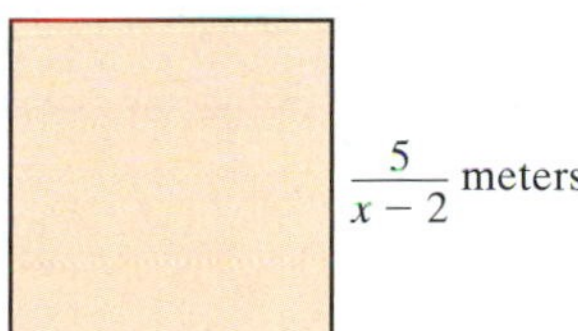

△ **66.** A trapezoid has sides of the indicated lengths. Find its perimeter.

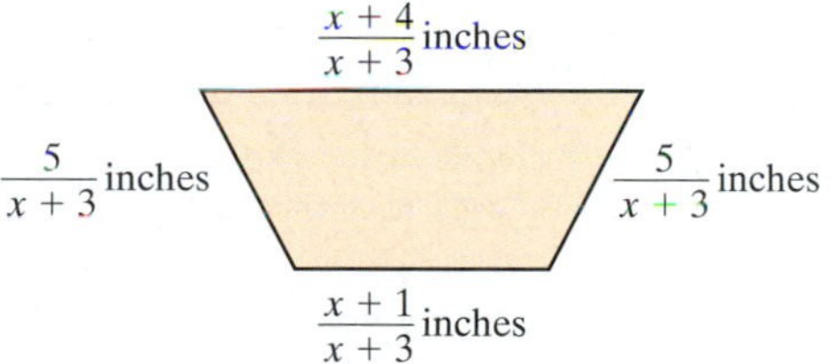

67. Write two rational expressions with the same denominator whose sum is $\frac{5}{3x - 1}$.

68. Write two rational expressions with the same denominator whose difference is $\frac{x - 7}{x^2 + 1}$.

69. You are throwing a barbecue and you want to make sure that you purchase the same number of hot dogs as hot dog buns. Hot dogs come 8 to a package and hot dog buns come 12 to a package. What is the least number of each type of package you should buy?

70. The planet Mercury revolves around the sun in 88 Earth days. It takes Jupiter 4332 Earth days to make one revolution around the Sun. (*Source:* National Space Science Data Center) If the two planets are aligned as shown in the figure, how long will it take for them to align again?

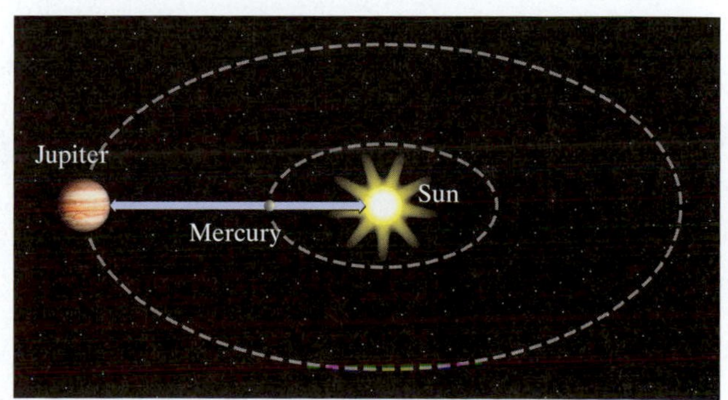

71. Write some instructions to help a friend who is having difficulty finding the LCD of two rational expressions.

72. Explain why the LCD of the rational expressions $\frac{7}{x+1}$ and $\frac{9x}{(x+1)^2}$ is $(x+1)^2$ and not $(x+1)^3$.

73. In your own words, describe how to add or subtract two rational expressions with the same denominators.

74. Explain the similarities between subtracting $\frac{3}{8}$ from $\frac{7}{8}$ and subtracting $\frac{6}{x+3}$ from $\frac{9}{x+3}$.

STUDY SKILLS BUILDER

How Are You Doing?

If you haven't done so yet, take a few moments and think about how you are doing in this course. Are you working toward your goal of successfully completing this course? Is your performance on homework, quizzes, and tests satisfactory? If not, you might want to see your instructor to see if he/she has any suggestions on how you can improve your performance. Reread Section 1.1 for ideas on places to get help with your mathematics course.

Answer the following

1. List any textbook supplements you are using to help you through this course.
2. List any campus resources you are using to help you through this course.
3. Write a short paragraph describing how you are doing in your mathematics course.
4. If improvement is needed, list ways that you can work toward improving your situation as described in Exercise 3.

12.4 ADDING AND SUBTRACTING RATIONAL EXPRESSIONS WITH DIFFERENT DENOMINATORS

Objective

A **Add and Subtract Rational Expressions with Different Denominators.**

Objective A Adding and Subtracting Rational Expressions with Different Denominators

Let's add $\frac{3}{8}$ and $\frac{1}{6}$. In the previous section, we found the LCD to be 24. Now let's write equivalent fractions with denominators of 24 by multiplying by different forms of 1.

$$\frac{3}{8} = \frac{3}{8} \cdot 1 = \frac{3}{8} \cdot \frac{3}{3} = \frac{3 \cdot 3}{8 \cdot 3} = \frac{9}{24}.$$

$$\frac{1}{6} = \frac{1}{6} \cdot 1 = \frac{1}{6} \cdot \frac{4}{4} = \frac{1 \cdot 4}{6 \cdot 4} = \frac{4}{24}.$$

Now that the denominators are the same, we may add.

$$\frac{3}{8} + \frac{1}{6} = \frac{9}{24} + \frac{4}{24} = \frac{9+4}{24} = \frac{13}{24}$$

We add or subtract rational expressions the same way. You may want to use the steps below.

To Add or Subtract Rational Expressions with Different Denominators

Step 1: Find the LCD of the rational expressions.

Step 2: Rewrite each rational expression as an equivalent expression whose denominator is the LCD found in Step 1.

Step 3: Add or subtract numerators and write the sum or difference over the common denominator.

Step 4: Simplify or write the rational expression in lowest terms.

EXAMPLE 1 Perform each indicated operation.

a. $\frac{a}{4} - \frac{2a}{8}$ **b.** $\frac{3}{10x^2} + \frac{7}{25x}$

Solution:

a. First, we must find the LCD. Since $4 = 2^2$ and $8 = 2^3$, the LCD $= 2^3 = 8$. Next we write each fraction as an equivalent fraction with the denominator 8, and then we subtract.

$$\frac{a}{4} = \frac{a}{4} \cdot 1 = \frac{a}{4} \cdot \frac{2}{2} = \frac{a \cdot 2}{4 \cdot 2} = \frac{2a}{8}$$

$$\frac{a}{4} - \frac{2a}{8} = \frac{a(2)}{4(2)} - \frac{2a}{8} = \frac{2a}{8} - \frac{2a}{8} = \frac{2a - 2a}{8} = \frac{0}{8} = 0$$

Notice that we wrote $\frac{a}{4}$ as the equivalent expression $\frac{2a}{8}$. Multiplying by a form of 1 means we multiply the numerator and the denominator by the same number. Since this is so, we will start using the shorthand notation on the next page.

Continued on next page

PRACTICE PROBLEM 1

Perform each indicated operation.

a. $\frac{y}{5} - \frac{3y}{15}$ **b.** $\frac{5}{8x} + \frac{11}{10x^2}$

Answers

1. a. 0 **b.** $\frac{25x + 44}{40x^2}$

$$\frac{a}{4} = \frac{a(2)}{4(2)} = \frac{2a}{8}$$

Multiplying the numerator and denominator by 2 is the same as multiplying by $\frac{2}{2}$ or 1.

b. Since $10x^2 = 2 \cdot 5 \cdot x \cdot x$ and $25x = 5 \cdot 5 \cdot x$, the LCD $= 2 \cdot 5^2 \cdot x^2 = 50x^2$. We write each fraction as an equivalent fraction with a denominator of $50x^2$.

$$\frac{3}{10x^2} + \frac{7}{25x} = \frac{3(5)}{10x^2(5)} + \frac{7(2x)}{25x(2x)}$$

$$= \frac{15}{50x^2} + \frac{14x}{50x^2}$$

$$= \frac{15 + 14x}{50x^2}$$ Add numerators. Write the sum over the common denominator.

Work Practice Problem 1

PRACTICE PROBLEM 2

Subtract: $\frac{10x}{x^2 - 9} - \frac{5}{x + 3}$

EXAMPLE 2 Subtract: $\frac{6x}{x^2 - 4} - \frac{3}{x + 2}$

Solution: Since $x^2 - 4 = (x + 2)(x - 2)$, the LCD $= (x + 2)(x - 2)$. We write equivalent expressions with the LCD as denominators.

$$\frac{6x}{x^2 - 4} - \frac{3}{x + 2} = \frac{6x}{(x + 2)(x - 2)} - \frac{3(x - 2)}{(x + 2)(x - 2)}$$

$$= \frac{6x - 3(x - 2)}{(x + 2)(x - 2)}$$ Subtract numerators. Write the difference over the common denominator.

$$= \frac{6x - 3x + 6}{(x + 2)(x - 2)}$$ Apply the distributive property in the numerator.

$$= \frac{3x + 6}{(x + 2)(x - 2)}$$ Combine like terms in the numerator.

Next we factor the numerator to see if this rational expression can be simplified.

$$\frac{3x + 6}{(x + 2)(x - 2)} = \frac{3(x + 2)}{(x + 2)(x - 2)}$$ Factor.

$$= \frac{3}{x - 2}$$ Apply the fundamental principle to simplify.

Work Practice Problem 2

PRACTICE PROBLEM 3

Add: $\frac{5}{7x} + \frac{2}{x + 1}$

EXAMPLE 3 Add: $\frac{2}{3t} + \frac{5}{t + 1}$

Solution: The LCD is $3t(t + 1)$. We write each rational expression as an equivalent rational expression with a denominator of $3t(t + 1)$.

$$\frac{2}{3t} + \frac{5}{t + 1} = \frac{2(t + 1)}{3t(t + 1)} + \frac{5(3t)}{(t + 1)(3t)}$$

$$= \frac{2(t + 1) + 5(3t)}{3t(t + 1)}$$ Add numerators. Write the sum over the common denominator.

$$= \frac{2t + 2 + 15t}{3t(t + 1)}$$ Apply the distributive property in the numerator.

$$= \frac{17t + 2}{3t(t + 1)}$$ Combine like terms in the numerator.

Work Practice Problem 3

Answers

2. $\frac{5}{x - 3}$ **3.** $\frac{19x + 5}{7x(x + 1)}$

EXAMPLE 4 Subtract: $\frac{7}{x-3} - \frac{9}{3-x}$

Solution: To find a common denominator, we notice that $x - 3$ and $3 - x$ are opposites. That is, $3 - x = -(x - 3)$. We write the denominator $3 - x$ as $-(x - 3)$ and simplify.

$$\frac{7}{x-3} - \frac{9}{3-x} = \frac{7}{x-3} - \frac{9}{-(x-3)}$$

$$= \frac{7}{x-3} - \frac{-9}{x-3} \quad \text{Apply } \frac{a}{-b} = \frac{-a}{b}.$$

$$= \frac{7-(-9)}{x-3} \quad \text{Subtract numerators. Write the difference over the common denominator.}$$

$$= \frac{16}{x-3}$$

Work Practice Problem 4

PRACTICE PROBLEM 4

Subtract: $\frac{10}{x-6} - \frac{15}{6-x}$

EXAMPLE 5 Add: $1 + \frac{m}{m+1}$

Solution: Recall that 1 is the same as $\frac{1}{1}$. The LCD of $\frac{1}{1}$ and $\frac{m}{m+1}$ is $m + 1$.

$$1 + \frac{m}{m+1} = \frac{1}{1} + \frac{m}{m+1} \quad \text{Write 1 as } \frac{1}{1}.$$

$$= \frac{1(m+1)}{1(m+1)} + \frac{m}{m+1} \quad \text{Multiply both the numerator and the denominator of } \frac{1}{1} \text{ by } m+1.$$

$$= \frac{m+1+m}{m+1} \quad \text{Add numerators. Write the sum over the common denominator.}$$

$$= \frac{2m+1}{m+1} \quad \text{Combine like terms in the numerator.}$$

Work Practice Problem 5

PRACTICE PROBLEM 5

Add: $2 + \frac{x}{x+5}$

EXAMPLE 6 Subtract: $\frac{3}{2x^2+x} - \frac{2x}{6x+3}$

Solution: First, we factor the denominators.

$$\frac{3}{2x^2+x} - \frac{2x}{6x+3} = \frac{3}{x(2x+1)} - \frac{2x}{3(2x+1)}$$

The LCD is $3x(2x + 1)$. We write equivalent expressions with denominators of $3x(2x + 1)$.

$$\frac{3}{x(2x+1)} - \frac{2x}{3(2x+1)} = \frac{3(3)}{x(2x+1)(3)} - \frac{2x(x)}{3(2x+1)(x)}$$

$$= \frac{9-2x^2}{3x(2x+1)} \quad \text{Subtract numerators. Write the difference over the common denominator.}$$

Work Practice Problem 6

PRACTICE PROBLEM 6

Subtract: $\frac{4}{3x^2+2x} - \frac{3x}{12x+8}$

Answers

4. $\frac{25}{x-6}$ **5.** $\frac{3x+10}{x+5}$ **6.** $\frac{16-3x^2}{4x(3x+2)}$

PRACTICE PROBLEM 7

Add: $\frac{6x}{x^2 + 4x + 4} + \frac{x}{x^2 - 4}$

EXAMPLE 7 Add: $\frac{2x}{x^2 + 2x + 1} + \frac{x}{x^2 - 1}$

Solution: First we factor the denominators.

$$\frac{2x}{x^2 + 2x + 1} + \frac{x}{x^2 - 1}$$

$$= \frac{2x}{(x + 1)(x + 1)} + \frac{x}{(x + 1)(x - 1)}$$

Rewrite each expression with LCD $(x + 1)(x + 1)(x - 1)$.

$$= \frac{2x(x - 1)}{(x + 1)(x + 1)(x - 1)} + \frac{x(x + 1)}{(x + 1)(x - 1)(x + 1)}$$

$$= \frac{2x(x - 1) + x(x + 1)}{(x + 1)^2(x - 1)}$$ Add numerators. Write the sum over the common denominator.

$$= \frac{2x^2 - 2x + x^2 + x}{(x + 1)^2(x - 1)}$$ Apply the distributive property in the numerator.

$$= \frac{3x^2 - x}{(x + 1)^2(x - 1)} \text{ or } \frac{x(3x - 1)}{(x + 1)^2(x - 1)}$$

The numerator was factored as a last step to see if the rational expression could be simplified further. Since there are no factors common to the numerator and the denominator, we can't simplify further.

Work Practice Problem 7

Answer

7. $\frac{x(7x - 10)}{(x + 2)^2(x - 2)}$

12.4 EXERCISE SET

MathXL®

Objective **A** *Perform each indicated operation. Simplify if possible. See Examples 1 through 7.*

1. $\frac{4}{2x} + \frac{9}{3x}$

2. $\frac{15}{7a} + \frac{8}{6a}$

3. $\frac{15a}{b} + \frac{6b}{5}$

4. $\frac{4c}{d} - \frac{8d}{5}$

5. $\frac{3}{x} + \frac{5}{2x^2}$

6. $\frac{14}{3x^2} + \frac{6}{x}$

7. $\frac{6}{x+1} + \frac{10}{2x+2}$

8. $\frac{8}{x+4} - \frac{3}{3x+12}$

9. $\frac{3}{x+2} - \frac{2x}{x^2-4}$

10. $\frac{5}{x-4} + \frac{4x}{x^2-16}$

11. $\frac{3}{4x} + \frac{8}{x-2}$

12. $\frac{5}{y^2} - \frac{y}{2y+1}$

13. $\frac{6}{x-3} + \frac{8}{3-x}$

14. $\frac{15}{y-4} + \frac{20}{4-y}$

15. $\frac{9}{x-3} + \frac{9}{3-x}$

16. $\frac{5}{a-7} + \frac{5}{7-a}$

17. $\frac{-8}{x^2-1} - \frac{7}{1-x^2}$

18. $\frac{-9}{25x^2-1} + \frac{7}{1-25x^2}$

19. $\frac{5}{x} + 2$

20. $\frac{7}{x^2} - 5x$

21. $\frac{5}{x-2} + 6$

22. $\frac{6y}{y+5} + 1$

23. $\frac{y+2}{y+3} - 2$

24. $\frac{7}{2x-3} - 3$

25. $\frac{-x+2}{x} - \frac{x-6}{4x}$

26. $\frac{-y+1}{y} - \frac{2y-5}{3y}$

27. $\frac{5x}{x+2} - \frac{3x-4}{x+2}$

28. $\frac{7x}{x-3} - \frac{4x+9}{x-3}$

29. $\frac{3x^4}{7} - \frac{4x^2}{21}$

30. $\frac{5x}{6} + \frac{11x^2}{2}$

31. $\frac{1}{x+3} - \frac{1}{(x+3)^2}$

32. $\frac{5x}{(x-2)^2} - \frac{3}{x-2}$

33. $\frac{4}{5b} + \frac{1}{b-1}$

34. $\frac{1}{y+5} + \frac{2}{3y}$

35. $\frac{2}{m} + 1$

36. $\frac{6}{x} - 1$

37. $\frac{2x}{x-7} - \frac{x}{x-2}$

38. $\frac{9x}{x-10} - \frac{x}{x-3}$

39. $\frac{6}{1-2x} - \frac{4}{2x-1}$

40. $\frac{10}{3n-4} - \frac{5}{4-3n}$

41. $\frac{7}{(x+1)(x-1)} + \frac{8}{(x+1)^2}$

42. $\frac{5}{(x+1)(x+5)} - \frac{2}{(x+5)^2}$

43. $\frac{x}{x^2-1} - \frac{2}{x^2-2x+1}$

44. $\frac{x}{x^2-4} - \frac{5}{x^2-4x+4}$

45. $\frac{3a}{2a+6} - \frac{a-1}{a+3}$

46. $\frac{1}{x+y} - \frac{y}{x^2-y^2}$

47. $\frac{y-1}{2y+3} + \frac{3}{(2y+3)^2}$

48. $\frac{x-6}{5x+1} + \frac{6}{(5x+1)^2}$

49. $\frac{5}{2-x} + \frac{x}{2x-4}$

50. $\frac{-1}{a-2} + \frac{4}{4-2a}$

51. $\frac{15}{x^2+6x+9} + \frac{2}{x+3}$

52. $\frac{2}{x^2+4x+4} + \frac{1}{x+2}$

53. $\frac{13}{x^2-5x+6} - \frac{5}{x-3}$

54. $\frac{-7}{y^2-3y+2} - \frac{2}{y-1}$

55. $\frac{70}{m^2-100} + \frac{7}{2(m+10)}$

56. $\frac{27}{y^2-81} + \frac{3}{2(y+9)}$

57. $\frac{x+8}{x^2-5x-6} + \frac{x+1}{x^2-4x-5}$

58. $\frac{x+4}{x^2+12x+20} + \frac{x+1}{x^2+8x-20}$

59. $\frac{5}{4n^2-12n+8} - \frac{3}{3n^2-6n}$

60. $\frac{6}{5y^2-25y+30} - \frac{2}{4y^2-8y}$

Mixed Practice (*Sections 12.2, 12.3, 12.4*) *Perform the indicated operations. Addition, subtraction, multiplication, and division of rational expressions are included here.*

61. $\frac{15x}{x+8} \cdot \frac{2x+16}{3x}$

62. $\frac{9z+5}{15} \cdot \frac{5z}{81z^2-25}$

63. $\frac{8x+7}{3x+5} - \frac{2x-3}{3x+5}$

64. $\frac{2z^2}{4z-1} - \frac{z-2z^2}{4z-1}$

65. $\frac{5a+10}{18} \div \frac{a^2-4}{10a}$

66. $\frac{9}{x^2-1} \div \frac{12}{3x+3}$

67. $\frac{5}{x^2-3x+2} + \frac{1}{x-2}$

68. $\frac{4}{2x^2+5x-3} + \frac{2}{x+3}$

Review

Solve each linear or quadratic equation. See Sections 9.3 and 11.6.

69. $3x + 5 = 7$

70. $5x - 1 = 8$

71. $2x^2 - x - 1 = 0$

72. $4x^2 - 9 = 0$

73. $4(x + 6) + 3 = -3$

74. $2(3x + 1) + 15 = -7$

Concept Extensions

Perform each indicated operation.

75. $\frac{3}{x} - \frac{2x}{x^2 - 1} + \frac{5}{x + 1}$

76. $\frac{5}{x - 2} + \frac{7x}{x^2 - 4} - \frac{11}{x}$

77. $\frac{5}{x^2 - 4} + \frac{2}{x^2 - 4x + 4} - \frac{3}{x^2 - x - 6}$

78. $\frac{8}{x^2 + 6x + 5} - \frac{3x}{x^2 + 4x - 5} + \frac{2}{x^2 - 1}$

79. $\frac{9}{x^2 + 9x + 14} - \frac{3x}{x^2 + 10x + 21} + \frac{x + 4}{x^2 + 5x + 6}$

80. $\frac{x + 10}{x^2 - 3x - 4} - \frac{8}{x^2 + 6x + 5} - \frac{9}{x^2 + x - 20}$

81. A board of length $\frac{3}{x + 4}$ inches was cut into two pieces. If one piece is $\frac{1}{x - 4}$ inches, express the length of the other board as a rational expression.

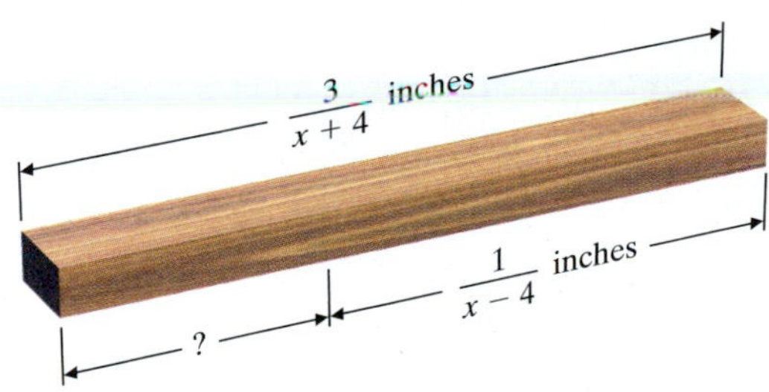

△ **82.** The length of a rectangle is $\frac{3}{y - 5}$ feet, while its width is $\frac{2}{y}$ feet. Find its perimeter and then find its area.

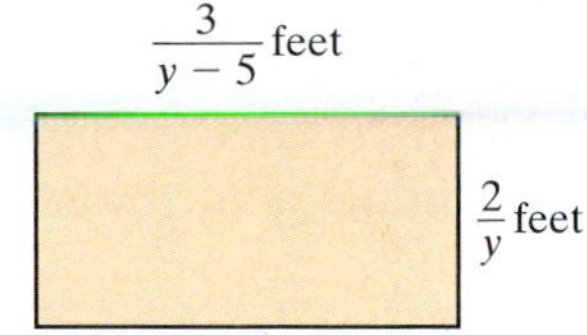

83. In ice hockey, penalty killing percentage is a statistic calculated as $1 - \frac{G}{P}$, where G = opponent's power play goals and P = opponent's power play opportunities. Simplify this expression.

84. The dose of medicine prescribed for a child depends on the child's age A in years and the adult dose D for the medication. Two expressions that give a child's dose are Young's Rule, $\frac{DA}{A + 12}$, and Cowling's Rule, $\frac{D(A + 1)}{24}$. Find an expression for the difference in the doses given by these expressions.

85. Explain when the LCD of the rational expressions in a sum is the product of the denominators.

86. Explain when the LCD is the same as one of the denominators of a rational expression to be added or subtracted.

△ **87.** Two angles are said to be complementary if the sum of their measures is 90°. If one angle measures $\frac{40}{x}$ degrees, find the measure of its complement.

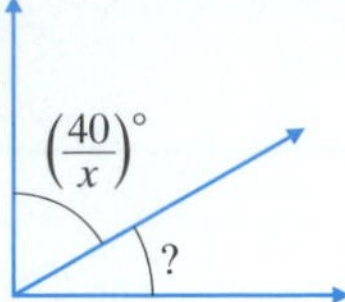

△ **88.** Two angles are said to be supplementary if the sum of their measures is 180°. If one angle measures $\frac{x+2}{x}$ degrees, find the measure of its supplement.

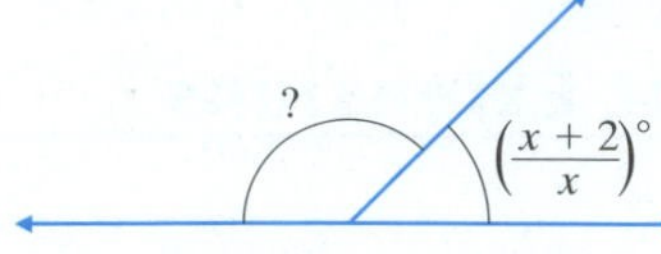

89. In your own words, explain how to add two rational expressions with different denominators.

90. In your own words, explain how to subtract two rational expressions with different denominators.

THE BIGGER PICTURE Simplifying Expressions and Solving Equations

Now we continue our outline from Sections 9.6, 10.7, and 11.6. Although suggestions are given, this outline should be in your own words. Once you complete this new portion, try the exercises below.

I. Simplifying Expressions

A. Exponents (Sections 10.1, 10.2)

B. Polynomials

1. Add (Section 10.4)
2. Subtract (Section 10.4)
3. Multiply (Section 10.5)
4. Divide (Section 10.7)

C. Factoring Polynomials (Chapter 11 Integrated Review)

D. Rational Expressions

1. Simplify: Factor the numerator and denominator. Then divide out factors of 1 by dividing out common factors in the numerator and denominator.

$$\frac{x^2-9}{7x^2-21x} = \frac{(x+3)(x-3)}{7x(x-3)} = \frac{x+3}{7x}$$

2. Multiply: Multiply numerators, then multiply denominators.

$$\frac{5z}{2z^2-9z-18}\cdot\frac{22z+33}{10z} = \frac{5\cdot z}{(2z+3)(z-6)}\cdot\frac{11(2z+3)}{2\cdot 5\cdot z} = \frac{11}{2(z-6)}$$

3. Divide: First fraction times the reciprocal of the second fraction.

$$\frac{14}{x+5}\div\frac{x+1}{2} = \frac{14}{x+5}\cdot\frac{2}{x+1} = \frac{28}{(x+5)(x+1)}$$

4. Add or Subtract: Must have same denominator. If not find the LCD and write each fraction as an equivalent fraction with the LCD as denominator.

$$\frac{9}{10}-\frac{x+1}{x+5} = \frac{9(x+5)}{10(x+5)}-\frac{10(x+1)}{10(x+5)} = \frac{9x+45-10x-10}{10(x+5)} = \frac{-x+35}{10(x+5)}$$

II. Solving Equations and Inequalities

A. Linear Equations (Section 9.3)

B. Linear Inequalities (Section 9.6)

C. Quadratic & Higher Degree Equations (Section 11.6)

Perform indicated operations and simplify.

1. $-8.6+(-9.1)$

2. $(-8.6)(-9.1)$

3. $14-(-14)$

4. $3x^4-7+x^4-x^2-10$

5. $\frac{5x^2-5}{25x+25}$

6. $\frac{7x}{x^2+4x+3}\div\frac{x}{2x+6}$

7. $\frac{2}{9}-\frac{5}{6}$

8. $\frac{x}{9}-\frac{x+3}{5}$

Factor.

9. $9x^3-2x^2-11x$

10. $12xy-21x+4y-7$

Solve.

11. $7x-14=5x+10$

12. $\frac{-x+2}{5}<\frac{3}{10}$

13. $1+4(x+4)=3^2+x$

14. $x(x-2)=24$

12.5 SOLVING EQUATIONS CONTAINING RATIONAL EXPRESSIONS

Objectives

A Solve Equations Containing Rational Expressions.

B Solve Equations Containing Rational Expressions for a Specified Variable.

Objective A Solving Equations Containing Rational Expressions

In Chapters 4 and 9, we solved equations containing fractions. In this section, we continue the work we began in these chapters by solving equations containing rational expressions. For example,

$$\frac{x}{2} + \frac{8}{3} = \frac{1}{6} \quad \text{and} \quad \frac{4x}{x^2 + x - 30} + \frac{2}{x - 5} = \frac{1}{x + 6}$$

are equations containing rational expressions. To solve equations such as these, we use the multiplication property of equality to clear the equation of fractions by multiplying both sides of the equation by the LCD.

EXAMPLE 1 Solve: $\frac{x}{2} + \frac{8}{3} = \frac{1}{6}$

Solution: The LCD of denominators 2, 3, and 6 is 6, so we multiply both sides of the equation by 6.

$$6\left(\frac{x}{2} + \frac{8}{3}\right) = 6\left(\frac{1}{6}\right)$$

$$6\left(\frac{x}{2}\right) + 6\left(\frac{8}{3}\right) = 6\left(\frac{1}{6}\right) \quad \text{Apply the distributive property.}$$

$$3 \cdot x + 16 = 1 \quad \text{Multiply and simplify.}$$

$$3x = -15 \quad \text{Subtract 16 from both sides.}$$

$$x = -5 \quad \text{Divide both sides by 3.}$$

Check: To check, we replace x with -5 in the original equation.

$$\frac{-5}{2} + \frac{8}{3} \stackrel{?}{=} \frac{1}{6} \quad \text{Replace } x \text{ with } -5.$$

$$\frac{1}{6} = \frac{1}{6} \quad \text{True}$$

This number checks, so the solution is -5.

Work Practice Problem 1

PRACTICE PROBLEM 1

Solve: $\frac{x}{4} + \frac{4}{5} = \frac{1}{20}$

Helpful Hint

Make sure that *each* term is multiplied by the LCD.

EXAMPLE 2 Solve: $\frac{t - 4}{2} - \frac{t - 3}{9} = \frac{5}{18}$

Solution: The LCD of denominators 2, 9, and 18 is 18, so we multiply both sides of the equation by 18.

$$18\left(\frac{t - 4}{2} - \frac{t - 3}{9}\right) = 18\left(\frac{5}{18}\right)$$

$$18\left(\frac{t - 4}{2}\right) - 18\left(\frac{t - 3}{9}\right) = 18\left(\frac{5}{18}\right) \quad \text{Apply the distributive property.}$$

$$9(t - 4) - 2(t - 3) = 5 \quad \text{Simplify.}$$

$$9t - 36 - 2t + 6 = 5 \quad \text{Use the distributive property.}$$

$$7t - 30 = 5 \quad \text{Combine like terms.}$$

$$7t = 35$$

$$t = 5 \quad \text{Solve for } t.$$

Continued on next page

PRACTICE PROBLEM 2

Solve: $\frac{x + 2}{3} - \frac{x - 1}{5} = \frac{1}{15}$

Helpful Hint

Multiply *each* term by 18.

Answers

1. $x = -3$ 2. $x = -6$

Check:
$$\frac{t-4}{2} - \frac{t-3}{9} = \frac{5}{18}$$
$$\frac{5-4}{2} - \frac{5-3}{9} \stackrel{?}{=} \frac{5}{18} \quad \text{Replace } t \text{ with 5.}$$
$$\frac{1}{2} - \frac{2}{9} \stackrel{?}{=} \frac{5}{18} \quad \text{Simplify.}$$
$$\frac{5}{18} = \frac{5}{18} \quad \text{True}$$

The solution is 5.

Work Practice Problem 2

Recall from Section 12.1 that a rational expression is defined for all real numbers except those that make the denominator of the expression 0. This means that if an equation contains *rational expressions with variables in the denominator,* we must be certain that the proposed solution does not make the denominator 0. If replacing the variable with the proposed solution makes the denominator 0, the rational expression is undefined and this proposed solution must be rejected.

PRACTICE PROBLEM 3

Solve: $2 + \frac{6}{x} = x + 7$

EXAMPLE 3 Solve: $3 - \frac{6}{x} = x + 8$

Solution: In this equation, 0 cannot be a solution because if x is 0, the rational expression $\frac{6}{x}$ is undefined. The LCD is x, so we multiply both sides of the equation by x.

$$x\left(3 - \frac{6}{x}\right) = x(x+8)$$
$$x(3) - x\left(\frac{6}{x}\right) = x \cdot x + x \cdot 8 \quad \text{Apply the distributive property.}$$
$$3x - 6 = x^2 + 8x \quad \text{Simplify.}$$

Helpful Hint

Multiply *each* term by x.

Now we write the quadratic equation in standard form and solve for x.

$$0 = x^2 + 5x + 6$$
$$0 = (x+3)(x+2) \quad \text{Factor.}$$
$$x + 3 = 0 \quad \text{or} \quad x + 2 = 0 \quad \text{Set each factor equal to 0 and solve.}$$
$$x = -3 \qquad\qquad x = -2$$

Notice that neither -3 nor -2 makes the denominator in the original equation equal to 0.

Check: To check these solutions, we replace x in the original equation by -3, and then by -2.

If $x = -3$:

$$3 - \frac{6}{x} = x + 8$$
$$3 - \frac{6}{-3} \stackrel{?}{=} -3 + 8$$
$$3 - (-2) \stackrel{?}{=} 5$$
$$5 = 5 \quad \text{True}$$

If $x = -2$:

$$3 - \frac{6}{x} = x + 8$$
$$3 - \frac{6}{-2} \stackrel{?}{=} -2 + 8$$
$$3 - (-3) \stackrel{?}{=} 6$$
$$6 = 6 \quad \text{True}$$

Both -3 and -2 are solutions.

Work Practice Problem 3

The following steps may be used to solve an equation containing rational expressions.

Answer

3. $x = -6, x = 1$

To Solve an Equation Containing Rational Expressions

Step 1: Multiply both sides of the equation by the LCD of all rational expressions in the equation.

Step 2: Remove any grouping symbols and solve the resulting equation.

Step 3: Check the solution in the original equation.

EXAMPLE 4 Solve: $\dfrac{4x}{x^2 + x - 30} + \dfrac{2}{x - 5} = \dfrac{1}{x + 6}$

Solution: The denominator $x^2 + x - 30$ factors as $(x + 6)(x - 5)$. The LCD is then $(x + 6)(x - 5)$, so we multiply both sides of the equation by this LCD.

$$(x + 6)(x - 5)\left(\frac{4x}{x^2 + x - 30} + \frac{2}{x - 5}\right) = (x + 6)(x - 5)\left(\frac{1}{x + 6}\right) \quad \text{Multiply by the LCD.}$$

$$(x + 6)(x - 5)\cdot\frac{4x}{x^2 + x - 30} + (x + 6)(x - 5)\cdot\frac{2}{x - 5} \quad \text{Apply the distributive property.}$$

$$= (x + 6)(x - 5)\cdot\frac{1}{x + 6}$$

$$4x + 2(x + 6) = x - 5 \quad \text{Simplify.}$$

$$4x + 2x + 12 = x - 5 \quad \text{Apply the distributive property.}$$

$$6x + 12 = x - 5 \quad \text{Combine like terms.}$$

$$5x = -17$$

$$x = -\frac{17}{5} \quad \text{Divide both sides by 5.}$$

Check: Check by replacing x with $-\dfrac{17}{5}$ in the original equation. The solution is $-\dfrac{17}{5}$.

Work Practice Problem 4

PRACTICE PROBLEM 4

Solve:

$\dfrac{2}{x + 3} + \dfrac{3}{x - 3} = \dfrac{-2}{x^2 - 9}$

EXAMPLE 5 Solve: $\dfrac{2x}{x - 4} = \dfrac{8}{x - 4} + 1$

Solution: Multiply both sides by the LCD, $x - 4$.

$$(x - 4)\left(\frac{2x}{x - 4}\right) = (x - 4)\left(\frac{8}{x - 4} + 1\right) \quad \text{Multiply by the LCD.}$$

$$(x - 4)\cdot\frac{2x}{x - 4} = (x - 4)\cdot\frac{8}{x - 4} + (x - 4)\cdot 1 \quad \text{Use the distributive property.}$$

$$2x = 8 + (x - 4) \quad \text{Simplify.}$$

$$2x = 4 + x$$

$$x = 4$$

Notice that 4 makes the denominator 0 in the original equation. Therefore, 4 is *not* a solution and this equation has *no solution*.

Work Practice Problem 5

PRACTICE PROBLEM 5

Solve: $\dfrac{5x}{x - 1} = \dfrac{5}{x - 1} + 3$

✔ **Concept Check** When can we clear fractions by multiplying through by the LCD?

a. When adding or subtracting rational expressions
b. When solving an equation containing rational expressions
c. Both of these
d. Neither of these

Answers

4. $x = -1$ **5.** No solution

✔ Concept Check Answer

b

Helpful Hint

As we can see from Example 5, it is important to check the proposed solution(s) in the original equation.

PRACTICE PROBLEM 6

Solve:

$$x - \frac{6}{x+3} = \frac{2x}{x+3} + 2$$

EXAMPLE 6 Solve: $x + \frac{14}{x-2} = \frac{7x}{x-2} + 1$

Solution: Notice the denominators in this equation. We can see that 2 can't be a solution. The LCD is $x - 2$, so we multiply both sides of the equation by $x - 2$.

$$(x-2)\left(x + \frac{14}{x-2}\right) = (x-2)\left(\frac{7x}{x-2} + 1\right)$$

$$(x-2)(x) + (x-2)\left(\frac{14}{x-2}\right) = (x-2)\left(\frac{7x}{x-2}\right) + (x-2)(1)$$

$$x^2 - 2x + 14 = 7x + x - 2 \quad \text{Simplify.}$$

$$x^2 - 2x + 14 = 8x - 2 \quad \text{Combine like terms.}$$

$$x^2 - 10x + 16 = 0 \quad \text{Write the quadratic equation in standard form.}$$

$$(x-8)(x-2) = 0 \quad \text{Factor.}$$

$$x - 8 = 0 \quad \text{or} \quad x - 2 = 0 \quad \text{Set each factor equal to 0.}$$

$$x = 8 \qquad x = 2 \quad \text{Solve.}$$

As we have already noted, 2 can't be a solution of the original equation. So we need only replace x with 8 in the original equation. We find that 8 is a solution; the only solution is 8.

Work Practice Problem 6

Objective B Solving Equations for a Specified Variable

The last example in this section is an equation containing several variables, and we are directed to solve for one of the variables. The steps used in the preceding examples can be applied to solve equations for a specified variable as well.

PRACTICE PROBLEM 7

Solve $\frac{1}{a} + \frac{1}{b} = \frac{1}{x}$ for a.

EXAMPLE 7 Solve $\frac{1}{a} + \frac{1}{b} = \frac{1}{x}$ for x.

Solution: (This type of equation often models a work problem, as we shall see in the next section.) The LCD is abx, so we multiply both sides by abx.

$$abx\left(\frac{1}{a} + \frac{1}{b}\right) = abx\left(\frac{1}{x}\right)$$

$$abx\left(\frac{1}{a}\right) + abx\left(\frac{1}{b}\right) = abx \cdot \frac{1}{x}$$

$$bx + ax = ab \quad \text{Simplify.}$$

$$x(b + a) = ab \quad \text{Factor out } x \text{ from each term on the left side.}$$

$$\frac{x(b+a)}{b+a} = \frac{ab}{b+a} \quad \text{Divide both sides by } b + a.$$

$$x = \frac{ab}{b+a} \quad \text{Simplify.}$$

This equation is now solved for x.

Work Practice Problem 7

Answers

6. $x = 4$ 7. $a = \frac{bx}{b - x}$

12.5 EXERCISE SET

FOR EXTRA HELP

Student Solutions Manual

PH Math/Tutor Center

CD/Video for Review
MathXL®

MyMathLab

Objective A *Solve each equation and check each solution. See Examples 1 through 3.*

1. $\frac{x}{5} + 3 = 9$

2. $\frac{x}{5} - 2 = 9$

3. $\frac{x}{2} + \frac{5x}{4} = \frac{x}{12}$

4. $\frac{x}{6} + \frac{4x}{3} = \frac{x}{18}$

5. $2 - \frac{8}{x} = 6$

6. $5 + \frac{4}{x} = 1$

7. $2 + \frac{10}{x} = x + 5$

8. $6 + \frac{5}{y} = y - \frac{2}{y}$

9. $\frac{a}{5} = \frac{a-3}{2}$

10. $\frac{b}{5} = \frac{b+2}{6}$

11. $\frac{x-3}{5} + \frac{x-2}{2} = \frac{1}{2}$

12. $\frac{a+5}{4} + \frac{a+5}{2} = \frac{a}{8}$

Solve each equation and check each proposed solution. See Examples 4 through 6.

13. $\frac{3}{2a-5} = -1$

14. $\frac{6}{4-3x} = -3$

15. $\frac{4y}{y-4} + 5 = \frac{5y}{y-4}$

16. $\frac{2a}{a+2} - 5 = \frac{7a}{a+2}$

17. $2 + \frac{3}{a-3} = \frac{a}{a-3}$

18. $\frac{2y}{y-2} - \frac{4}{y-2} = 4$

19. $\frac{1}{x+3} + \frac{6}{x^2-9} = 1$

20. $\frac{1}{x+2} + \frac{4}{x^2-4} = 1$

21. $\frac{2y}{y+4} + \frac{4}{y+4} = 3$

22. $\frac{5y}{y+1} - \frac{3}{y+1} = 4$

23. $\frac{2x}{x+2} - 2 = \frac{x-8}{x-2}$

24. $\frac{4y}{y-3} - 3 = \frac{3y-1}{y+3}$

Solve each equation. See Examples 1 through 6.

25. $\frac{2}{y} + \frac{1}{2} = \frac{5}{2y}$

26. $\frac{6}{3y} + \frac{3}{y} = 1$

27. $\frac{a}{a-6} = \frac{-2}{a-1}$

28. $\frac{5}{x-6} = \frac{x}{x-2}$

29. $\frac{11}{2x} + \frac{2}{3} = \frac{7}{2x}$

30. $\frac{5}{3} - \frac{3}{2x} = \frac{3}{2}$

31. $\frac{2}{x-2} + 1 = \frac{x}{x+2}$

32. $1 + \frac{3}{x+1} = \frac{x}{x-1}$

33. $\frac{x+1}{3} - \frac{x-1}{6} = \frac{1}{6}$

34. $\frac{3x}{5} - \frac{x-6}{3} = -\frac{2}{5}$

35. $\frac{t}{t-4} = \frac{t+4}{6}$

36. $\frac{15}{x+4} = \frac{x-4}{x}$

37. $\frac{y}{2y+2} + \frac{2y-16}{4y+4} = \frac{2y-3}{y+1}$

38. $\frac{1}{x+2} = \frac{4}{x^2-4} - \frac{1}{x-2}$

39. $\frac{4r-4}{r^2+5r-14} + \frac{2}{r+7} = \frac{1}{r-2}$

40. $\frac{3}{x+3} = \frac{12x+19}{x^2+7x+12} - \frac{5}{x+4}$

41. $\frac{x+1}{x+3} = \frac{x^2-11x}{x^2+x-6} - \frac{x-3}{x-2}$

42. $\frac{2t+3}{t-1} - \frac{2}{t+3} = \frac{5-6t}{t^2+2t-3}$

Objective B *Solve each equation for the indicated variable. See Example 7.*

43. $R = \frac{E}{I}$ for I (Electronics: resistance of a circuit)

44. $T = \frac{V}{Q}$ for Q (Water purification: settling time)

45. $T = \frac{2U}{B+E}$ for B (Merchandising: stock turnover rate)

46. $i = \frac{A}{t+B}$ for t (Hydrology: rainfall intensity)

47. $B = \frac{705w}{h^2}$ for w (Health: body-mass index)

△ **48.** $\frac{A}{W} = L$ for W (Geometry: area of a rectangle)

49. $N = R + \frac{V}{G}$ for G (Urban forestry: tree plantings per year)

50. $C = \frac{D(A+1)}{24}$ for A (Medicine: Cowling's Rule for child's dose)

△ **51.** $\frac{C}{\pi r} = 2$ for r (Geometry: circumference of a circle)

52. $W = \frac{CE^2}{2}$ for C (Electronics: energy stored in a capacitor)

53. $\frac{1}{y} + \frac{1}{3} = \frac{1}{x}$ for x

54. $\frac{1}{5} + \frac{2}{y} = \frac{1}{x}$ for x

Review

Translating *Write each phrase as an expression. See Sections 3.2 and 9.2.*

55. The reciprocal of x

56. The reciprocal of $x + 1$

57. The reciprocal of x, added to the reciprocal of 2

58. The reciprocal of x, subtracted from the reciprocal of 5

Answer each question.

59. If a tank is filled in 3 hours, what part of the tank is filled in 1 hour?

60. If a strip of beach is cleaned in 4 hours, what part of the beach is cleaned in 1 hour?

Concept Extensions

Solve each equation.

61. $\dfrac{4}{a^2 + 4a + 3} + \dfrac{2}{a^2 + a - 6} - \dfrac{3}{a^2 - a - 2} = 0$

62. $\dfrac{-4}{a^2 + 2a - 8} + \dfrac{1}{a^2 + 9a + 20} = \dfrac{-4}{a^2 + 3a - 10}$

Recall that two angles are supplementary if the sum of their measures is 180°. Find the measures of the supplementary angles.

△ **63.**

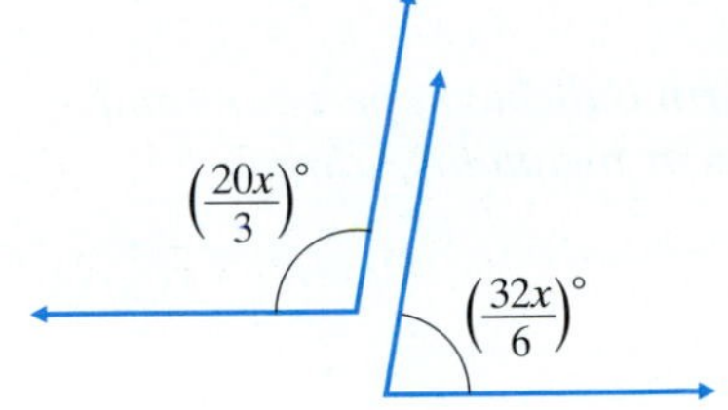

△ **64.**

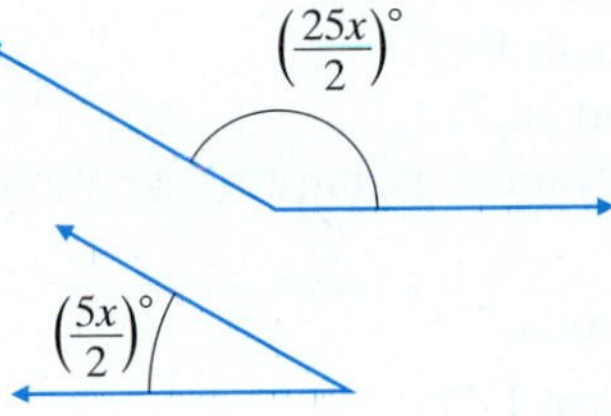

Recall that two angles are complementary if the sum of their measures is 90°. Find the measures of the complementary angles.

△ **65.**

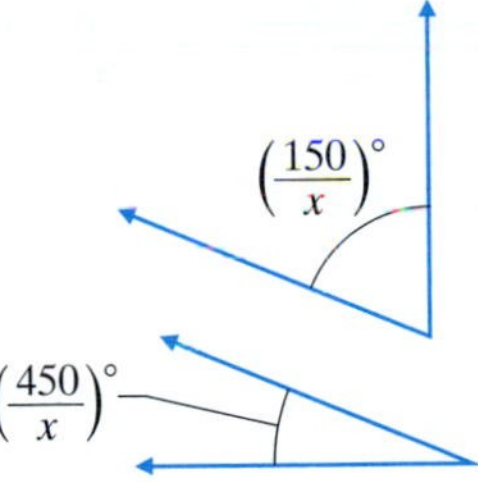

△ **66.**

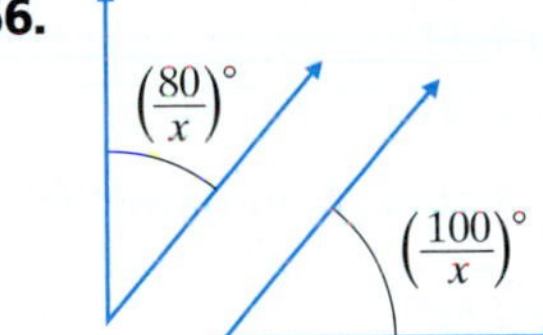

67. When adding the expressions in $\dfrac{3x}{2} + \dfrac{x}{4}$, can you multiply each term by 4? Why or why not?

68. When solving the equation $\dfrac{3x}{2} + \dfrac{x}{4} = 1$, can you multiply both sides of the equation by 4? Why or why not?

THE BIGGER PICTURE Simplifying Expressions and Solving Equations

Now we continue our outline from Sections 9.6, 10.7, 11.6, and 12.4. Although suggestions are given, this outline should be in your own words. Once you complete this new portion, try the exercises below.

I. Simplifying Expressions

- **A.** Exponents (Sections 10.1, 10.2)
- **B.** Polynomials
  1. Add (Section 10.4)
  2. Subtract (Section 10.4)
  3. Multiply (Section 10.5)
  4. Divide (Section 10.7)
- **C.** Factoring Polynomials (Chapter 11 Integrated Review)
- **D.** Rational Expressions
  1. Simplify (Section 12.1)
  2. Multiply (Section 12.2)
  3. Divide (Section 12.2)
  4. Add or Subtract (Section 12.4)

II. Solving Equations and Inequalities

- **A.** Linear Equations (Section 9.3)
- **B.** Linear Inequalities (Section 9.6)
- **C.** Quadratic and Higher Degree Equations (Section 11.6)
- **D.** Equations with Rational Expressions—solving equations with rational expressions

$$\frac{3}{x} - \frac{1}{x-1} = \frac{4}{x-1}$$ Equation with rational expressions.

$$x(x-1)\cdot\frac{3}{x} - x(x-1)\frac{1}{x-1} = x(x-1)\frac{4}{x-1}$$ Multiply through by $x(x-1)$.

$$3(x-1) - x\cdot 1 = x\cdot 4$$ Simplify.

$$3x - 3 - x = 4x$$ Use the distributive property.

$$-3 = 2x$$ Simplify and move variable terms to right side.

$$-\frac{3}{2} = x$$ Divide both sides by 2.

Multiply.

1. $(3x - 2)(4x^2 - x - 5)$
2. $(2x - y)^2$

Factor.

3. $8y^3 - 20y^5$
4. $9m^2 - 11mn + 2n^2$

Simplify or solve.

If an expression, perform indicated operations and simplify. If an equation or inequality, solve it.

5. $\frac{7}{x} = \frac{9}{x-10}$
6. $\frac{7}{x} + \frac{9}{x-10}$
7. $(-3x^5)\left(\frac{1}{2}x^7\right)(8x)$
8. $5x - 1 = |-4| + |-5|$
9. $\frac{8-12}{12 \div 3\cdot 2}$
10. $-2(3y - 4) \leq 5y - 7 - 7y - 1$
11. $\frac{7}{x} + \frac{5}{2x+3} = \frac{-2}{x}$
12. $\frac{(a^{-3}b^2)^{-5}}{ab^4}$

INTEGRATED REVIEW Sections 12.1–12.5

Summary on Rational Expressions

It is important to know the difference between performing operations with rational expressions and solving an equation containing rational expressions. Study the examples below.

Performing Operations with Rational Expressions

Adding: $\frac{1}{x} + \frac{1}{x+5} = \frac{1 \cdot (x+5)}{x(x+5)} + \frac{1 \cdot x}{x(x+5)} = \frac{x+5+x}{x(x+5)} = \frac{2x+5}{x(x+5)}$

Subtracting: $\frac{3}{x} - \frac{5}{x^2y} = \frac{3 \cdot xy}{x \cdot xy} - \frac{5}{x^2y} = \frac{3xy-5}{x^2y}$

Multiplying: $\frac{2}{x} \cdot \frac{5}{x-1} = \frac{2 \cdot 5}{x(x-1)} = \frac{10}{x(x-1)}$

Dividing: $\frac{4}{2x+1} \div \frac{x-3}{x} = \frac{4}{2x+1} \cdot \frac{x}{x-3} = \frac{4x}{(2x+1)(x-3)}$

Solving an Equation Containing Rational Expressions

To solve an equation containing rational expressions, we clear the equation of fractions by multiplying both sides by the LCD.

$$\frac{3}{x} - \frac{5}{x-1} = \frac{1}{x(x-1)}$$ Note that x can't be 0 or 1.

$$x(x-1)\left(\frac{3}{x}\right) - x(x-1)\left(\frac{5}{x-1}\right) = x(x-1) \cdot \frac{1}{x(x-1)}$$ Multiply both sides by the LCD.

$$3(x-1) - 5x = 1$$ Simplify.

$$3x - 3 - 5x = 1$$ Use the distributive property.

$$-2x - 3 = 1$$ Combine like terms.

$$-2x = 4$$ Add 3 to both sides.

$$x = -2$$ Divide both sides by -2.

Don't forget to check to make sure our proposed solution of -2 does not make any denominators 0. If it does, this proposed solution is *not* a solution of the equation. -2 checks and is the solution.

Determine whether each of the following is an equation or an expression. If it is an equation, solve it for its variable. If it is an expression, perform the indicated operation.

1. $\frac{1}{x} + \frac{2}{3}$

2. $\frac{3}{a} + \frac{5}{6}$

3. $\frac{1}{x} + \frac{2}{3} = \frac{3}{x}$

4. $\frac{3}{a} + \frac{5}{6} = 1$

5. $\frac{2}{x+1} - \frac{1}{x}$

6. $\frac{4}{x-3} - \frac{1}{x}$

7. $\frac{2}{x+1} - \frac{1}{x} = 1$

8. $\frac{4}{x-3} - \frac{1}{x} = \frac{6}{x(x-3)}$

9. $\frac{15x}{x+8} \cdot \frac{2x+16}{3x}$

10. $\frac{9z+5}{15} \cdot \frac{5z}{81z^2-25}$

Answers

1. ________

2. ________

3. ________

4. ________

5. ________

6. ________

7. ________

8. ________

9. ________

10. ________

11. ______
12. ______
13. ______
14. ______
15. ______
16. ______
17. ______
18. ______
19. ______
20. ______
21. ______
22. ______
23. ______
24. ______

11. $\dfrac{2x+1}{x-3} + \dfrac{3x+6}{x-3}$

12. $\dfrac{4p-3}{2p+7} + \dfrac{3p+8}{2p+7}$

13. $\dfrac{x+5}{7} = \dfrac{8}{2}$

14. $\dfrac{1}{2} = \dfrac{x+1}{8}$

15. $\dfrac{5a+10}{18} \div \dfrac{a^2-4}{10a}$

16. $\dfrac{9}{x^2-1} \div \dfrac{12}{3x+3}$

17. $\dfrac{x+2}{3x-1} + \dfrac{5}{(3x-1)^2}$

18. $\dfrac{4}{(2x-5)^2} + \dfrac{x+1}{2x-5}$

19. $\dfrac{x-7}{x} - \dfrac{x+2}{5x}$

20. $\dfrac{9}{x^2-4} + \dfrac{2}{x+2} = \dfrac{-1}{x-2}$

21. $\dfrac{3}{x+3} = \dfrac{5}{x^2-9} - \dfrac{2}{x-3}$

22. $\dfrac{10x-9}{x} - \dfrac{x-4}{3x}$

23. Explain the difference between solving an equation such as $\dfrac{x}{2} + \dfrac{3}{4} = \dfrac{x}{4}$ for x and performing an operation such as adding $\dfrac{x}{2} + \dfrac{3}{4}$.

24. When solving an equation such as $\dfrac{y}{4} = \dfrac{y}{2} - \dfrac{1}{4}$, we may multiply all terms by 4. When subtracting two rational expressions such as $\dfrac{y}{2} - \dfrac{1}{4}$, we may not. Explain why.

12.6 RATIONAL EQUATIONS AND PROBLEM SOLVING

Objectives

A Solve Problems about Numbers.

B Solve Problems about Work.

C Solve Problems about Distance.

Objective A Solving Problems about Numbers

In this section, we solve problems that can be modeled by equations containing rational expressions. To solve these problems, we use the same problem-solving steps that were first introduced in Section 3.4. In our first example, our goal is to find an unknown number.

EXAMPLE 1 Finding an Unknown Number

The quotient of a number and 6, minus $\frac{5}{3}$, is the quotient of the number and 2. Find the number.

Solution:

1. UNDERSTAND. Read and reread the problem. Suppose that the unknown number is 2, then we see if the quotient of 2 and 6, or $\frac{2}{6}$, minus $\frac{5}{3}$ is equal to the quotient of 2 and 2, or $\frac{2}{2}$.

   $\frac{2}{6} - \frac{5}{3} = \frac{1}{3} - \frac{5}{3} = -\frac{4}{3}$, not $\frac{2}{2}$

   Don't forget that the purpose of a proposed solution is to better understand the problem.

   Let x = the unknown number.

2. TRANSLATE.

In words:	the quotient of x and 6	minus	$\frac{5}{3}$	is	the quotient of x and 2
	↓	↓	↓	↓	↓
Translate:	$\frac{x}{6}$	$-$	$\frac{5}{3}$	$=$	$\frac{x}{2}$

3. SOLVE. Here, we solve the equation $\frac{x}{6} - \frac{5}{3} = \frac{x}{2}$. We begin by multiplying both sides of the equation by the LCD, 6.

$$6\left(\frac{x}{6} - \frac{5}{3}\right) = 6\left(\frac{x}{2}\right)$$

$$6\left(\frac{x}{6}\right) - 6\left(\frac{5}{3}\right) = 6\left(\frac{x}{2}\right) \quad \text{Apply the distributive property.}$$

$$x - 10 = 3x \quad \text{Simplify.}$$

$$-10 = 2x \quad \text{Subtract } x \text{ from both sides.}$$

$$\frac{-10}{2} = \frac{2x}{2} \quad \text{Divide both sides by 2.}$$

$$-5 = x \quad \text{Simplify.}$$

4. INTERPRET.

Check: To check, we verify that "the quotient of −5 and 6 minus $\frac{5}{3}$ is the quotient of −5 and 2," or $-\frac{5}{6} - \frac{5}{3} = -\frac{5}{2}$.

State: The unknown number is −5.

Work Practice Problem 1

PRACTICE PROBLEM 1

The quotient of a number and 2, minus $\frac{1}{3}$, is the quotient of the number and 6. Find the number.

Answer

1. 1

Objective B Solving Problems about Work

The next example is often called a work problem. Work problems usually involve people or machines doing a certain task.

PRACTICE PROBLEM 2

Andrew and Timothy Larson volunteer at a local recycling plant. Andrew can sort a batch of recyclables in 2 hours alone while his brother Timothy needs 3 hours to complete the same job. If they work together, how long will it take them to sort one batch?

EXAMPLE 2 Finding Work Rates

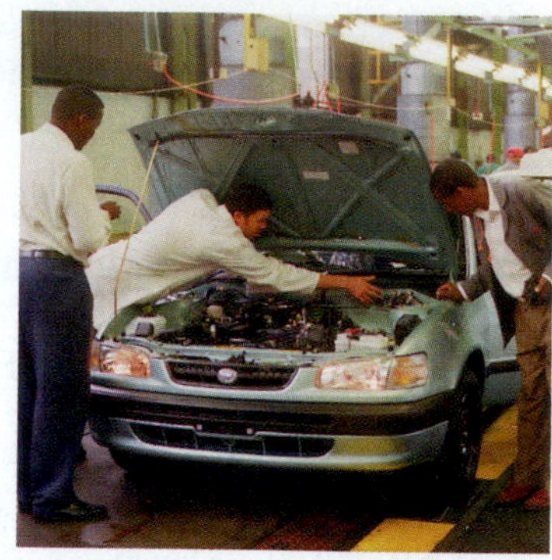

Sam Waterton and Frank Schaffer work in a plant that manufactures automobiles. Sam can complete a quality control tour of the plant in 3 hours while his assistant, Frank, needs 7 hours to complete the same job. The regional manager is coming to inspect the plant facilities, so both Sam and Frank are directed to complete a quality control tour together. How long will this take?

Solution:

1. UNDERSTAND. Read and reread the problem. The key idea here is the relationship between the **time** (hours) it takes to complete the job and the **part of the job** completed in 1 unit of time (hour). For example, if the **time** it takes Sam to complete the job is 3 hours, the **part of the job** he can complete in 1 hour is $\frac{1}{3}$. Similarly, Frank can complete $\frac{1}{7}$ of the job in 1 hour.

   Let $x =$ the **time** in hours it takes Sam and Frank to complete the job together. Then $\frac{1}{x} =$ the **part of the job** they complete in 1 hour.

	Hours to Complete Total Job	Part of Job Completed in 1 Hour
Sam	3	$\frac{1}{3}$
Frank	7	$\frac{1}{7}$
Together	x	$\frac{1}{x}$

2. TRANSLATE.

In words:	part of job Sam completed in 1 hour	added to	part of job Frank completed in 1 hour	is equal to	part of job they completed together in 1 hour
	↓	↓	↓	↓	↓
Translate:	$\frac{1}{3}$	$+$	$\frac{1}{7}$	$=$	$\frac{1}{x}$

3. SOLVE. Here, we solve the equation $\frac{1}{3} + \frac{1}{7} = \frac{1}{x}$. We begin by multiplying both sides of the equation by the LCD, $21x$.

$$21x\left(\frac{1}{3}\right) + 21x\left(\frac{1}{7}\right) = 21x\left(\frac{1}{x}\right)$$

$$7x + 3x = 21 \quad \text{Simplify.}$$

$$10x = 21$$

$$x = \frac{21}{10} \text{ or } 2\frac{1}{10} \text{ hours}$$

Answer

2. $1\frac{1}{5}$ hours

4. INTERPRET.

Check: Our proposed solution is $2\frac{1}{10}$ hours. This proposed solution is reasonable since $2\frac{1}{10}$ hours is more than half of Sam's time and less than half of Frank's time. Check this solution in the originally *stated* problem.

State: Sam and Frank can complete the quality control tour in $2\frac{1}{10}$ hours.

Work Practice Problem 2

✔Concept Check Solve $E = mc^2$

a. for m **b.** for c^2.

Objective C Solving Problems about Distance

Next we look at a problem solved by the distance formula,

$$d = r \cdot t$$

EXAMPLE 3 Finding Speeds of Vehicles

A car travels 180 miles in the same time that a truck travels 120 miles. If the car's speed is 20 miles per hour faster than the truck's, find the car's speed and the truck's speed.

Solution:

1. UNDERSTAND. Read and reread the problem. Suppose that the truck's speed is 45 miles per hour. Then the car's speed is 20 miles per hour more, or 65 miles per hour.

We are given that the car travels 180 miles in the same time that the truck travels 120 miles. To find the time it takes the car to travel 180 miles, remember that since $d = rt$, we know that $\frac{d}{r} = t$.

Car's Time	*Truck's Time*
$t = \frac{d}{r} = \frac{180}{65} = 2\frac{50}{65} = 2\frac{10}{13}$ hours	$t = \frac{d}{r} = \frac{120}{45} = 2\frac{30}{45} = 2\frac{2}{3}$ hours

Since the times are not the same, our proposed solution is not correct. But we have a better understanding of the problem.

Let x = the speed of the truck.

Since the car's speed is 20 miles per hour faster than the truck's, then

$x + 20$ = the speed of the car

Use the formula $d = r \cdot t$ or **d**istance = **r**ate · **t**ime. Prepare a chart to organize the information in the problem.

	Distance	=	Rate	·	Time
Truck	120		x		$\frac{120}{x}$ ← distance / ← rate
Car	180		$x + 20$		$\frac{180}{x+20}$ ← distance / ← rate

Continued on next page

PRACTICE PROBLEM 3

A car travels 600 miles in the same time that a motorcycle travels 450 miles. If the car's speed is 15 miles per hour more than the motorcycle's, find the speed of the car and the speed of the motorcycle.

Helpful Hint

If $d = r \cdot t$,

then $t = \frac{d}{r}$

or *time* $= \frac{\textit{distance}}{\textit{rate}}$.

Answer

3. car: 60 mph; motorcycle: 45 mph

✔ Concept Check Answer

a. $m = \frac{E}{c^2}$ **b.** $c^2 = \frac{E}{m}$

2. TRANSLATE. Since the car and the truck traveled the same amount of time, we have that

In words: car's time = truck's time

Translate: $\frac{180}{x + 20} = \frac{120}{x}$

3. SOLVE. We begin by multiplying both sides of the equation by the LCD, $x(x + 20)$, or cross multiplying.

$$\frac{180}{x + 20} = \frac{120}{x}$$

$$180x = 120(x + 20)$$

$$180x = 120x + 2400 \quad \text{Use the distributive property.}$$

$$60x = 2400 \quad \text{Subtract } 120x \text{ from both sides.}$$

$$x = 40 \quad \text{Divide both sides by 60.}$$

4. INTERPRET. The speed of the truck is 40 miles per hour. The speed of the car must then be $x + 20$ or 60 miles per hour.

Check: Find the time it takes the car to travel 180 miles and the time it takes the truck to travel 120 miles.

Car's Time

$$t = \frac{d}{r} = \frac{180}{60} = 3 \text{ hours}$$

Truck's Time

$$t = \frac{d}{r} = \frac{120}{40} = 3 \text{ hours}$$

Since both travel the same amount of time, the proposed solution is correct.

State: The car's speed is 60 miles per hour and the truck's speed is 40 miles per hour.

Work Practice Problem 3

Vocabulary and Readiness Check

Without solving algebraically, select the best choice for each exercise.

1. One person can complete a job in 7 hours. A second person can complete the same job in 5 hours. How long will it take them to complete the job if they work together?

a. more than 7 hours
b. between 5 and 7 hours
c. less than 5 hours

2. One inlet pipe can fill a pond in 30 hours. A second inlet pipe can fill the same pond in 25 hours. How long before the pond is filled if both inlet pipes are on?

a. less than 25 hours
b. between 25 and 30 hours
c. more than 30 hours

12.6 EXERCISE SET

FOR EXTRA HELP

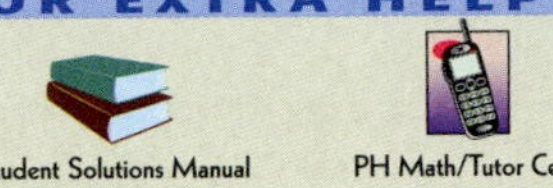
Student Solutions Manual

PH Math/Tutor Center

CD/Video for Review

MathXL®

MyMathLab

Objective A *Solve the following. See Example 1.*

1. Three times the reciprocal of a number equals 9 times the reciprocal of 6. Find the number.

2. Twelve divided by the sum of x and 2 equals the quotient of 4 and the difference of x and 2. Find x.

3. If twice a number added to 3 is divided by the number plus 1, the result is three halves. Find the number.

4. A number added to the product of 6 and the reciprocal of the number equals -5. Find the number.

Objective B *See Example 2.*

5. Smith Engineering found that an experienced surveyor surveys a roadbed in 4 hours. An apprentice surveyor needs 5 hours to survey the same stretch of road. If the two work together, find how long it takes them to complete the job.

6. An experienced bricklayer constructs a small wall in 3 hours. The apprentice completes the job in 6 hours. Find how long it takes if they work together.

7. In 2 minutes, a conveyor belt moves 300 pounds of recyclable aluminum from the delivery truck to a storage area. A smaller belt moves the same quantity of cans the same distance in 6 minutes. If both belts are used, find how long it takes to move the cans to the storage area.

8. Find how long it takes the conveyor belts described in Exercise 7 to move 1200 pounds of cans. (*Hint:* Think of 1200 pounds as four 300-pound jobs.)

Objective C *See Example 3.*

9. A jogger begins her workout by jogging to the park, a distance of 12 miles. She then jogs home at the same speed but along a different route. This return trip is 18 miles and her time is one hour longer. Find her jogging speed. Complete the accompanying chart and use it to find her jogging speed.

	Distance	=	Rate	·	Time
Trip to Park	12				
Return Trip	18				

10. A boat can travel 9 miles upstream in the same amount of time it takes to travel 11 miles downstream. If the current of the river is 3 miles per hour, complete the chart below and use it to find the speed of the boat in still water.

	Distance	=	Rate	·	Time
Upstream	9		$r - 3$		
Downstream	11		$r + 3$		

11. A cyclist rode the first 20-mile portion of his workout at a constant speed. For the 16-mile cooldown portion of his workout, he reduced his speed by 2 miles per hour. Each portion of the workout took the same time. Find the cyclist's speed during the first portion and find his speed during the cooldown portion.

12. A semi-truck travels 300 miles through the flatland in the same amount of time that it travels 180 miles through mountains. The rate of the truck is 20 miles per hour slower in the mountains than in the flatland. Find both the flatland rate and mountain rate.

Objectives A B C Mixed Practice *Solve the following. See Examples 1 through 3. (Note: Some exercises can be modeled by equations without rational expressions.)*

13. One-fourth equals the quotient of a number and 8. Find the number.

14. Four times a number added to 5 is divided by 6. The result is $\frac{7}{2}$. Find the number.

15. Marcus and Tony work for Lombardo's Pipe and Concrete. Mr. Lombardo is preparing an estimate for a customer. He knows that Marcus lays a slab of concrete in 6 hours. Tony lays the same size slab in 4 hours. If both work on the job and the cost of labor is $45.00 per hour, decide what the labor estimate should be.

16. Mr. Dodson can paint his house by himself in 4 days. His son needs an additional day to complete the job if he works by himself. If they work together, find how long it takes to paint the house.

17. A pilot can travel 400 miles with the wind in the same amount of time as 336 miles against the wind. Find the speed of the wind if the pilot's speed in still air is 230 miles per hour.

18. A fisherman on Pearl River rows 9 miles downstream in the same amount of time he rows 3 miles upstream. If the current is 6 miles per hour, find how long it takes him to cover the 12 miles.

19. Two divided by the difference of a number and 3 minus 4 divided by a number plus 3, equals 8 times the reciprocal of the difference of the number squared and 9. What is the number?

20. If 15 times the reciprocal of a number is added to the ratio of 9 times a number minus 7 and the number plus 2, the result is 9. What is the number?

21. A pilot flies 630 miles with a tail wind of 35 miles per hour. Against the wind, he flies only 455 miles in the same amount of time. Find the rate of the plane in still air.

22. A marketing manager travels 1080 miles in a corporate jet and then an additional 240 miles by car. If the car ride takes one hour longer than the jet ride takes, and if the rate of the jet is 6 times the rate of the car, find the time the manager travels by jet and find the time the manager travels by car.

23. A boater travels 16 miles per hour on the water on a still day. During one particular windy day, he finds that he travels 48 miles with the wind behind him in the same amount of time that he travels 16 miles into the wind. Find the rate of the wind.

24. The current on a portion of the Mississippi River is 3 miles per hour. A barge can go 6 miles upstream in the same amount of time it takes to go 10 miles downstream. Find the speed of the boat in still water.

25. Two hikers are 11 miles apart and walking toward each other. They meet in 2 hours. Find the rate of each hiker if one hiker walks 1.1 mph faster than the other.

26. On a 255-mile trip, Gary Alessandrini traveled at an average speed of 70 mph, got a speeding ticket, and then traveled at 60 mph for the remainder of the trip. If the entire trip took 4.5 hours and the speeding ticket stop took 30 minutes, how long did Gary speed before getting stopped?

27. One custodian cleans a suite of offices in 3 hours. When a second worker is asked to join the regular custodian, the job takes only $1\frac{1}{2}$ hours. How long does it take the second worker to do the same job alone?

28. One person proofreads a copy for a small newspaper in 4 hours. If a second proofreader is also employed, the job can be done in $2\frac{1}{2}$ hours. How long does it take for the second proofreader to do the same job alone?

29. A jet plane traveling at 500 mph overtakes a propeller plane traveling at 200 mph that had a 2-hour head start. How far from the starting point are the planes?

30. How long will it take a bus traveling at 60 miles per hour to overtake a car traveling at 40 mph if the car had a 1.5-hour head start?

31. One pipe fills a storage pool in 20 hours. A second pipe fills the same pool in 15 hours. When a third pipe is added and all three are used to fill the pool, it takes only 6 hours. Find how long it takes the third pipe to do the job.

32. One pump fills a tank 2 times as fast as another pump. If the pumps work together, they fill the tank in 18 minutes. How long does it take for each pump to fill the tank?

33. A car travels 280 miles in the same time that a motorcycle travels 240 miles. If the car's speed is 10 miles per hour more than the motorcycle's, find the speed of the car and the speed of the motorcycle.

34. A bus traveled on a level road for 3 hours at an average speed 20 miles per hour faster than it traveled on a winding road. The time spent on the winding road was 4 hours. Find the average speed on the level road if the entire trip was 305 miles.

35. In 6 hours, an experienced cook prepares enough pies to supply a local restaurant's daily order. Another cook prepares the same number of pies in 7 hours. Together with a third cook, they prepare the pies in 2 hours. Find how long it takes the third cook to prepare the pies alone.

36. Mrs. Smith balances the company books in 8 hours. It takes her assistant 12 hours to do the same job. If they work together, find how long it takes them to balance the books.

Review

Simplify. Follow the circled steps in the order shown. See Sections 4.3 and 4.4.

37. $$\frac{\frac{3}{4}+\frac{1}{4}}{\frac{3}{8}+\frac{13}{8}}$$ ← ① Add. (numerator) ← ② Add. (denominator)

38. $$\frac{\frac{9}{5}+\frac{6}{5}}{\frac{17}{6}+\frac{7}{6}}$$ ← ① Add. (numerator) ← ② Add. (denominator)

39. $$\frac{\frac{2}{5}+\frac{1}{5}}{\frac{7}{10}+\frac{7}{10}}$$ ① Add. ← ③ Divide. ② Add.

40. $$\frac{\frac{1}{4}+\frac{5}{4}}{\frac{3}{8}+\frac{7}{8}}$$ ① Add. ← ③ Divide. ② Add.

Concept Extensions

41. One pump fills a tank 3 times as fast as another pump. If the pumps work together, they fill the tank in 21 minutes. How long does it take for each pump to fill the tank?

42. For which of the following equations can we immediately use cross products to solve for x?

a. $\frac{2-x}{5}=\frac{1+x}{3}$

b. $\frac{2}{5}-x=\frac{1+x}{3}$

43. Person A can complete a job in 5 hours, and person B can complete the same job in 3 hours. Without solving algebraically, discuss reasonable and unreasonable answers for how long it would take them to complete the job together.

44. For what value of x is $\frac{x}{x-1}$ in proportion to $\frac{x+1}{x}$? Explain your result.

Solve. See the Concept Check in this section.

Solve $D = RT$

45. for R

46. for T

47. A hyena spots a giraffe 0.5 mile away and begins running toward it. The giraffe starts running away from the hyena just as the hyena begins running toward it. A hyena can run at a speed of 40 mph and a giraffe can run at 32 mph. How long will it take for the hyena to overtake the giraffe? (*Source: The World Almanac and Book of Facts*)

48. The Andretti Green Racing team boasts the proud name of one of the best known Indy car drivers, Mario Andretti. Two of its drivers, Tony Kanaan and Bryan Herta, placed second and fourth, respectively, in the 2004 Indianapolis 500. The track is 2.5 miles long. When traveling at their fastest lap speeds, Herta drove 2.479 miles in the same time that Kanaan completed an entire 2.5-mile lap. Kanaan's fastest lap speed was 1.822 mph faster than Herta's fastest lap speed. Find each driver's fastest lap speed. Round each speed to the nearest tenth. (*Source:* Indy Racing League)

12.7 SIMPLIFYING COMPLEX FRACTIONS

Objectives

A Simplify Complex Fractions Using Method 1.

B Simplify Complex Fractions Using Method 2.

A rational expression whose numerator or denominator or both numerator and denominator contain fractions is called a **complex rational expression** or a **complex fraction.** Some examples are

$$\frac{4}{2-\frac{1}{2}} \qquad \frac{\frac{3}{2}}{\frac{4}{7}-x} \qquad \frac{\frac{1}{x+2}}{x+2-\frac{1}{x}}$$

← Numerator of complex fraction
← Main fraction bar
← Denominator of complex fraction

Our goal in this section is to write complex fractions in simplest form. A complex fraction is in simplest form when it is in the form $\frac{P}{Q}$, where P and Q are polynomials that have no common factors.

Objective A Simplifying Complex Fractions—Method 1

In this section, two methods of simplifying complex fractions are represented. The first method presented uses the fact that the main fraction bar indicates division.

Method 1: To Simplify a Complex Fraction

Step 1: Add or subtract fractions in the numerator or denominator so that the numerator is a single fraction and the denominator is a single fraction.

Step 2: Perform the indicated division by multiplying the numerator of the complex fraction by the reciprocal of the denominator of the complex fraction.

Step 3: Write the rational expression in lowest terms.

EXAMPLE 1 Simplify the complex fraction $\frac{\frac{5}{8}}{\frac{2}{3}}$.

Solution: Since the numerator and denominator of the complex fraction are already single fractions, we proceed to Step 2: perform the indicated division by multiplying the numerator $\frac{5}{8}$ by the reciprocal of the denominator $\frac{2}{3}$.

$$\frac{\frac{5}{8}}{\frac{2}{3}} = \frac{5}{8} \div \frac{2}{3} = \frac{5}{8} \cdot \frac{3}{2} = \frac{15}{16}$$

The reciprocal of $\frac{2}{3}$ is $\frac{3}{2}$.

Work Practice Problem 1

PRACTICE PROBLEM 1

Simplify the complex fraction $\frac{\frac{3}{7}}{\frac{5}{9}}$.

Answer

1. $\frac{27}{35}$

PRACTICE PROBLEM 2

Simplify: $\dfrac{\frac{3}{4} - \frac{2}{3}}{\frac{1}{2} + \frac{3}{8}}$

EXAMPLE 2 Simplify: $\dfrac{\frac{2}{3} + \frac{1}{5}}{\frac{2}{3} - \frac{2}{9}}$

Solution: We simplify the numerator and denominator of the complex fraction separately. First we add $\frac{2}{3}$ and $\frac{1}{5}$ to obtain a single fraction in the numerator. Then we subtract $\frac{2}{9}$ from $\frac{2}{3}$ to obtain a single fraction in the denominator.

$$\frac{\frac{2}{3} + \frac{1}{5}}{\frac{2}{3} - \frac{2}{9}} = \frac{\frac{2(5)}{3(5)} + \frac{1(3)}{5(3)}}{\frac{2(3)}{3(3)} - \frac{2}{9}}$$ The LCD of the numerator's fractions is 15. The LCD of the denominator's fractions is 9.

$$= \frac{\frac{10}{15} + \frac{3}{15}}{\frac{6}{9} - \frac{2}{9}}$$ Simplify.

$$= \frac{\frac{13}{15}}{\frac{4}{9}}$$ Add the numerator's fractions. Subtract the denominator's fractions.

Next we perform the indicated division by multiplying the numerator of the complex fraction by the reciprocal of the denominator of the complex fraction.

$$\frac{\frac{13}{15}}{\frac{4}{9}} = \frac{13}{15} \cdot \frac{9}{4}$$ The reciprocal of $\frac{4}{9}$ is $\frac{9}{4}$.

$$= \frac{13 \cdot 3 \cdot 3}{3 \cdot 5 \cdot 4} = \frac{39}{20}$$

Work Practice Problem 2

PRACTICE PROBLEM 3

Simplify: $\dfrac{\frac{2}{5} - \frac{1}{x}}{\frac{2x}{15} - \frac{1}{3}}$

EXAMPLE 3 Simplify: $\dfrac{\frac{1}{z} - \frac{1}{2}}{\frac{1}{3} - \frac{z}{6}}$

Solution: Subtract to get a single fraction in the numerator and a single fraction in the denominator of the complex fraction.

$$\frac{\frac{1}{z} - \frac{1}{2}}{\frac{1}{3} - \frac{z}{6}} = \frac{\frac{2}{2z} - \frac{z}{2z}}{\frac{2}{6} - \frac{z}{6}}$$ The LCD of the numerator's fractions is $2z$. The LCD of the denominator's fractions is 6.

$$= \frac{\frac{2 - z}{2z}}{\frac{2 - z}{6}}$$

$$= \frac{2 - z}{2z} \cdot \frac{6}{2 - z}$$ Multiply by the reciprocal of $\frac{2 - z}{6}$.

$$= \frac{2 \cdot 3 \cdot (2 - z)}{2 \cdot z \cdot (2 - z)}$$ Factor.

$$= \frac{3}{z}$$ Write in lowest terms.

Work Practice Problem 3

Answers

2. $\frac{2}{21}$ 3. $\frac{3}{x}$

Objective B Simplifying Complex Fractions—Method 2

Next we study a second method for simplifying complex fractions. In this method, we multiply the numerator and the denominator of the complex fraction by the LCD of all fractions in the complex fraction.

Method 2: To Simplify a Complex Fraction

Step 1: Find the LCD of all the fractions in the complex fraction.

Step 2: Multiply both the numerator and the denominator of the complex fraction by the LCD from Step 1.

Step 3: Perform the indicated operations and write the result in lowest terms.

We use method 2 to rework Example 2.

EXAMPLE 4 Simplify: $\dfrac{\dfrac{2}{3}+\dfrac{1}{5}}{\dfrac{2}{3}-\dfrac{2}{9}}$

Solution: The LCD of $\frac{2}{3}, \frac{1}{5}, \frac{2}{3}$, and $\frac{2}{9}$ is 45, so we multiply the numerator and the denominator of the complex fraction by 45. Then we perform the indicated operations, and write in lowest terms.

$$\frac{\dfrac{2}{3}+\dfrac{1}{5}}{\dfrac{2}{3}-\dfrac{2}{9}} = \frac{45\left(\dfrac{2}{3}+\dfrac{1}{5}\right)}{45\left(\dfrac{2}{3}-\dfrac{2}{9}\right)}$$

$$= \frac{45\left(\dfrac{2}{3}\right)+45\left(\dfrac{1}{5}\right)}{45\left(\dfrac{2}{3}\right)-45\left(\dfrac{2}{9}\right)} \quad \text{Apply the distributive property}$$

$$= \frac{30+9}{30-10} = \frac{39}{20} \quad \text{Simplify.}$$

Work Practice Problem 4

Helpful Hint

The same complex fraction was simplified using two different methods in Examples 2 and 4. Notice that the simplified results are the same.

PRACTICE PROBLEM 4

Use method 2 to simplify the complex fraction in Practice Problem 2:

$$\frac{\dfrac{3}{4}-\dfrac{2}{3}}{\dfrac{1}{2}+\dfrac{3}{8}}$$

Answer

4. $\dfrac{2}{21}$

PRACTICE PROBLEM 5

Simplify: $\dfrac{1 + \dfrac{x}{y}}{\dfrac{2x + 1}{y}}$

EXAMPLE 5 Simplify: $\dfrac{\dfrac{x + 1}{y}}{\dfrac{x}{y} + 2}$

Solution: The LCD of $\frac{x + 1}{y}$ and $\frac{x}{y}$ is y, so we multiply the numerator and the denominator of the complex fraction by y.

$$\frac{\dfrac{x + 1}{y}}{\dfrac{x}{y} + 2} = \frac{y\left(\dfrac{x + 1}{y}\right)}{y\left(\dfrac{x}{y} + 2\right)}$$

$$= \frac{y\left(\dfrac{x + 1}{y}\right)}{y\left(\dfrac{x}{y}\right) + y \cdot 2}$$ Apply the distributive property in the denominator.

$$= \frac{x + 1}{x + 2y}$$ Simplify.

Work Practice Problem 5

PRACTICE PROBLEM 6

Simplify: $\dfrac{\dfrac{5}{6y} + \dfrac{y}{x}}{\dfrac{y}{3} - x}$

EXAMPLE 6 Simplify: $\dfrac{\dfrac{x}{y} + \dfrac{3}{2x}}{\dfrac{x}{2} + y}$

Solution: The LCD of $\frac{x}{y}, \frac{3}{2x}, \frac{x}{2}$, and $\frac{y}{1}$ is $2xy$, so we multiply both the numerator and the denominator of the complex fraction by $2xy$.

$$\frac{\dfrac{x}{y} + \dfrac{3}{2x}}{\dfrac{x}{2} + y} = \frac{2xy\left(\dfrac{x}{y} + \dfrac{3}{2x}\right)}{2xy\left(\dfrac{x}{2} + y\right)}$$

$$= \frac{2xy\left(\dfrac{x}{y}\right) + 2xy\left(\dfrac{3}{2x}\right)}{2xy\left(\dfrac{x}{2}\right) + 2xy(y)}$$ Apply the distributive property.

$$= \frac{2x^2 + 3y}{x^2y + 2xy^2}$$

or $\dfrac{2x^2 + 3y}{xy(x + 2y)}$

Work Practice Problem 6

Answers

5. $\dfrac{y + x}{2x + 1}$

6. $\dfrac{5x + 6y^2}{2xy^2 - 6x^2y}$ or $\dfrac{5x + 6y^2}{2xy(y - 3x)}$

12.7 EXERCISE SET

FOR EXTRA HELP

Student Solutions Manual

PH Math/Tutor Center

CD/Video for Review

MathXL® MathXL®

MyMathLab MyMathLab

Objectives A B Mixed Practice *Simplify each complex fraction. See Examples 1 through 6.*

1. $\dfrac{\frac{1}{2}}{\frac{3}{4}}$

2. $\dfrac{\frac{1}{8}}{-\frac{5}{12}}$

3. $\dfrac{-\frac{4x}{9}}{-\frac{2x}{3}}$

4. $\dfrac{-\frac{6y}{11}}{\frac{4y}{9}}$

5. $\dfrac{\frac{1+x}{6}}{\frac{1+x}{3}}$

6. $\dfrac{\frac{6x-3}{5x^2}}{\frac{2x-1}{10x}}$

7. $\dfrac{\frac{1}{2}+\frac{2}{3}}{\frac{5}{9}-\frac{5}{6}}$

8. $\dfrac{\frac{3}{4}-\frac{1}{2}}{\frac{3}{8}+\frac{1}{6}}$

9. $\dfrac{2+\frac{7}{10}}{1+\frac{3}{5}}$

10. $\dfrac{4-\frac{11}{12}}{5+\frac{1}{4}}$

11. $\dfrac{\frac{1}{3}}{\frac{1}{2}-\frac{1}{4}}$

12. $\dfrac{\frac{7}{10}-\frac{3}{5}}{\frac{1}{2}}$

13. $\dfrac{-\frac{2}{9}}{-\frac{14}{3}}$

14. $\dfrac{\frac{3}{8}}{\frac{4}{15}}$

15. $\dfrac{-\frac{5}{12x^2}}{\frac{25}{16x^3}}$

16. $\dfrac{-\frac{7}{8y}}{\frac{21}{4y}}$

17. $\dfrac{\frac{m}{n}-1}{\frac{m}{n}+1}$

18. $\dfrac{\frac{x}{2}+2}{\frac{x}{2}-2}$

19. $\dfrac{\frac{1}{5}-\frac{1}{x}}{\frac{7}{10}+\frac{1}{x^2}}$

20. $\dfrac{\frac{1}{y^2}+\frac{2}{3}}{\frac{1}{y}-\frac{5}{6}}$

21. $\dfrac{1+\frac{1}{y-2}}{y+\frac{1}{y-2}}$

22. $\dfrac{x-\frac{1}{2x+1}}{1-\frac{x}{2x+1}}$

23. $\dfrac{\frac{4y-8}{16}}{\frac{6y-12}{4}}$

24. $\dfrac{\frac{7y+21}{3}}{\frac{3y+9}{8}}$

25. $\dfrac{\frac{x}{y}+1}{\frac{x}{y}-1}$

26. $\dfrac{\frac{3}{5y}+8}{\frac{3}{5y}-8}$

27. $\dfrac{1}{2+\frac{1}{3}}$

28. $\dfrac{3}{1-\frac{4}{3}}$

29. $\dfrac{\frac{ax+ab}{x^2-b^2}}{\frac{x+b}{x-b}}$

30. $\dfrac{\frac{m+2}{m-2}}{\frac{2m+4}{m^2-4}}$

31. $\dfrac{\frac{-3+y}{4}}{\frac{8+y}{28}}$

32. $\dfrac{\frac{-x+2}{18}}{\frac{8}{9}}$

33. $\dfrac{3+\frac{12}{x}}{1-\frac{16}{x^2}}$

34. $\dfrac{2+\frac{6}{x}}{1-\frac{9}{x^2}}$

35. $\dfrac{\frac{8}{x+4}+2}{\frac{12}{x+4}-2}$

36. $\dfrac{\frac{25}{x+5}+5}{\frac{3}{x+5}-5}$

37. $\dfrac{\dfrac{s}{r}+\dfrac{r}{s}}{\dfrac{s}{r}-\dfrac{r}{s}}$

38. $\dfrac{\dfrac{2}{x}+\dfrac{x}{2}}{\dfrac{2}{x}-\dfrac{x}{2}}$

39. $\dfrac{\dfrac{6}{x-5}+\dfrac{x}{x-2}}{\dfrac{3}{x-6}-\dfrac{2}{x-5}}$

40. $\dfrac{\dfrac{4}{x}+\dfrac{x}{x+1}}{\dfrac{1}{2x}+\dfrac{1}{x+6}}$

Review

Use the bar graph below to answer Exercises 41 through 44. See Section 7.1.

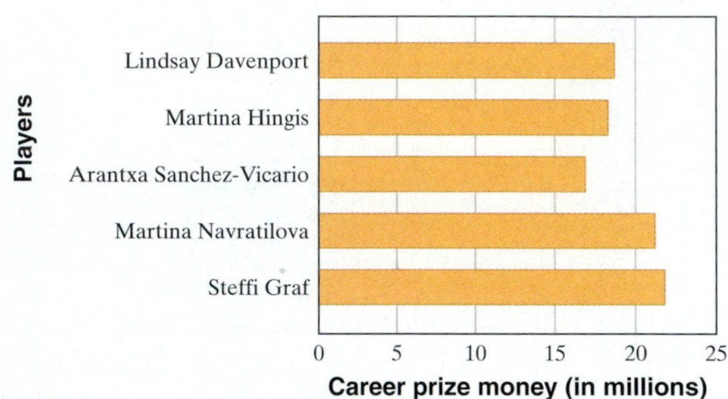

Source: Sanex WTA Tour Media Information System

41. Which women's tennis player has earned the most prize money in her career?

42. Estimate how much more prize money Lindsay Davenport has earned over her career than Arantxa Sanchez-Vicario.

43. Which of the players shown have earned over \$20 million in prize money over their careers?

44. During her career, through July 4, 2004, Martina Navratilova has won 347 singles and doubles tournaments. Assuming her prize money was earned only for tournament titles, how much prize money did she earn per tournament title, on average?

Concept Extensions

45. Explain how to simplify a complex fraction using method 1.

46. Explain how to simplify a complex fraction using method 2.

To find the average of two numbers, we find their sum and divide by 2. For example, the average of 65 and 81 is found by simplifying $\dfrac{65+81}{2}$. *This simplifies to* $\dfrac{146}{2}=73$.

47. Find the average of $\dfrac{1}{3}$ and $\dfrac{3}{4}$.

48. Write the average of $\dfrac{3}{n}$ and $\dfrac{5}{n^2}$ as a simplified rational expression.

49. In electronics, when two resistors R_1 (read R sub 1) and R_2 (read R sub 2) are connected in parallel, the total resistance is given by the complex fraction

$$\dfrac{1}{\dfrac{1}{R_1}+\dfrac{1}{R_2}}.$$

Simplify this expression.

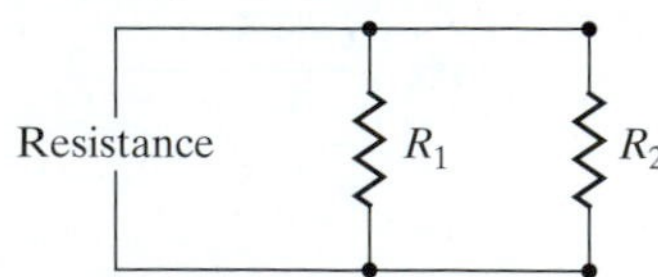

50. Astronomers occasionally need to know the day of the week a particular date fell on. The complex fraction

$$\dfrac{J+\dfrac{3}{2}}{7}$$

where J is the *Julian day number,* is used to make this calculation. Simplify this expression.

Simplify each of the following. First, write each expression with positive exponents. Then simplify the complex fraction. The first step has been completed for Exercise 51.

51. $\dfrac{x^{-1} + 2^{-1}}{x^{-2} - 4^{-1}} = \dfrac{\frac{1}{x} + \frac{1}{2}}{\frac{1}{x^2} - \frac{1}{4}}$

52. $\dfrac{3^{-1} - x^{-1}}{9^{-1} - x^{-2}}$

53. $\dfrac{y^{-2}}{1 - y^{-2}}$

54. $\dfrac{4 + x^{-1}}{3 + x^{-1}}$

55. If the distance formula $d = r \cdot t$ is solved for t, then $t = \dfrac{d}{r}$. Use this formula to find t if distance d is $\dfrac{20x}{3}$ miles and rate r is $\dfrac{5x}{9}$ miles per hour. Write t in simplified form.

△ **56.** If the formula for area of a rectangle, $A = l \cdot w$, is solved for w, then $w = \dfrac{A}{l}$. Use this formula to find w if area A is $\dfrac{4x - 2}{3}$ square meters and length l is $\dfrac{6x - 3}{5}$ meters. Write w in simplified form.

CHAPTER 12 Group Activity

Fast-Growing Careers

According to U.S. Bureau of Labor Statistics projections, the careers listed below will have the largest job growth in the next decade.

Occupation	Employment (number in thousands)		
	2002	**2012**	**Change**
1 Registered nurses	2284	2908	+623
2 Postsecondary teachers	1581	2184	+603
3 Retail salespersons	4076	4672	+596
4 Customer service representatives	1894	2354	+460
5 Combined food preparation and serving workers, including fast food	1990	2444	+454
6 Cashiers, except gaming	3432	3886	+454
7 Janitors and cleaners, except maids and housekeeping cleaners	2267	2681	+414
8 General and operations managers	2049	2425	+376
9 Waiters and waitresses	2097	2464	+367
10 Nursing aides, orderlies, and attendants	1375	1718	+343

What do all of these in-demand occupations have in common? They all require a knowledge of math! For some careers, like nurses, postsecondary teachers, and salespersons, the ways math is used on the job may be obvious. For other occupations, the use of math may not be quite as obvious. However, tasks common to many jobs, such as filling in a time sheet or a medication log, writing up an expense report, planning a budget, figuring a bill, ordering supplies, and even making a work schedule, all require math.

Activity

Suppose that your college placement office is planning to publish an occupational handbook on math in popular occupations. Choose one of the occupations from the given list that interests you. Research the occupation. Then write a brief entry for the occupational handbook that describes how a person in that career would use math in his or her job. Include an example if possible.

Chapter 12 Vocabulary Check

Fill in each blank with one of the words or phrases listed below.

rational expression   complex fraction

1. A ________________ is an expression that can be written in the form $\frac{P}{Q}$, where P and Q are polynomials and Q is not 0.
2. In a ________________, the numerator or denominator or both may contain fractions.

Helpful Hint

Are you preparing for your test? Don't forget to take the Chapter 12 Test on page 941. Then check your answers at the back of the text and use the Chapter Test Prep Video CD to see the fully worked-out solutions to any of the exercises you want to review.

12 Chapter Highlights

DEFINITIONS AND CONCEPTS	EXAMPLES
Section 12.1 Simplifying Rational Expressions	
A **rational expression** is an expression that can be written in the form $\frac{P}{Q}$, where P and Q are polynomials and Q does not equal 0.	$\frac{7y^3}{4}, \frac{x^2+6x+1}{x-3}, \frac{-5}{s^3+8}$
To find values for which a rational expression is undefined, find values for which the denominator is 0.	Find any values for which the expression $\frac{5y}{y^2-4y+3}$ is undefined. $y^2 - 4y + 3 = 0$ Set the denominator equal to 0. $(y-3)(y-1) = 0$ Factor. $y - 3 = 0$ or $y - 1 = 0$ Set each factor equal to 0. $y = 3$ $y = 1$ Solve. The expression is undefined when y is 3 and when y is 1.
TO SIMPLIFY A RATIONAL EXPRESSION **Step 1.** Factor the numerator and denominator. **Step 2.** Divide out factors common to the numerator and denominator. (This is the same as removing a factor of 1.)	Simplify: $\frac{4x+20}{x^2-25}$ $\frac{4x+20}{x^2-25} = \frac{4(x+5)}{(x+5)(x-5)} = \frac{4}{x-5}$

DEFINITIONS AND CONCEPTS	EXAMPLES
Section 12.2 Multiplying and Dividing Rational Expressions	
TO MULTIPLY RATIONAL EXPRESSIONS **Step 1.** Factor numerators and denominators. **Step 2.** Multiply numerators and multiply denominators. **Step 3.** Write the product in lowest terms. $\frac{P}{Q}\cdot\frac{R}{S}=\frac{PR}{QS}$	Multiply: $\frac{4x+4}{2x-3}\cdot\frac{2x^2+x-6}{x^2-1}$ $\frac{4x+4}{2x-3}\cdot\frac{2x^2+x-6}{x^2-1}$ $=\frac{4(x+1)}{2x-3}\cdot\frac{(2x-3)(x+2)}{(x+1)(x-1)}$ $=\frac{4(x+1)(2x-3)(x+2)}{(2x-3)(x+1)(x-1)}$ $=\frac{4(x+2)}{x-1}$
To divide by a rational expression, multiply by the reciprocal. $\frac{P}{Q}\div\frac{R}{S}=\frac{P}{Q}\cdot\frac{S}{R}=\frac{PS}{QR}$	Divide: $\frac{15x+5}{3x^2-14x-5}\div\frac{15}{3x-12}$ $\frac{15x+5}{3x^2-14x-5}\div\frac{15}{3x-12}$ $=\frac{5(3x+1)}{(3x+1)(x-5)}\cdot\frac{3(x-4)}{3\cdot 5}$ $=\frac{x-4}{x-5}$
Section 12.3 Adding and Subtracting Rational Expressions with the Same Denominator and Least Common Denominator	
To add or subtract rational expressions with the same denominator, add or subtract numerators, and place the sum or difference over the common denominator. $\frac{P}{R}+\frac{Q}{R}=\frac{P+Q}{R}$ $\frac{P}{R}-\frac{Q}{R}=\frac{P-Q}{R}$	Perform each indicated operation. $\frac{5}{x+1}+\frac{x}{x+1}=\frac{5+x}{x+1}$ $\frac{2y+7}{y^2-9}-\frac{y+4}{y^2-9}$ $=\frac{(2y+7)-(y+4)}{y^2-9}$ $=\frac{2y+7-y-4}{y^2-9}$ $=\frac{y+3}{(y+3)(y-3)}$ $=\frac{1}{y-3}$
TO FIND THE LEAST COMMON DENOMINATOR (LCD) **Step 1.** Factor the denominators. **Step 2.** The LCD is the product of all unique factors, each raised to a power equal to the greatest number of times that it appears in any one factored denominator.	Find the LCD for $\frac{7x}{x^2+10x+25}$ and $\frac{11}{3x^2+15x}$ $x^2+10x+25=(x+5)(x+5)$ $3x^2+15x=3x(x+5)$ LCD $=3x(x+5)(x+5)$ or $3x(x+5)^2$

DEFINITIONS AND CONCEPTS	EXAMPLES

Section 12.4 Adding and Subtracting Rational Expressions with Different Denominators

TO ADD OR SUBTRACT RATIONAL EXPRESSIONS WITH DIFFERENT DENOMINATORS

Step 1. Find the LCD.

Step 2. Rewrite each rational expression as an equivalent expression whose denominator is the LCD.

Step 3. Add or subtract numerators and place the sum or difference over the common denominator.

Step 4. Write the result in lowest terms.

Perform the indicated operation.

$$\frac{9x+3}{x^2-9} - \frac{5}{x-3}$$

$$= \frac{9x+3}{(x+3)(x-3)} - \frac{5}{x-3}$$

LCD is $(x+3)(x-3)$.

$$= \frac{9x+3}{(x+3)(x-3)} - \frac{5(x+3)}{(x-3)(x+3)}$$

$$= \frac{9x+3-5(x+3)}{(x+3)(x-3)}$$

$$= \frac{9x+3-5x-15}{(x+3)(x-3)}$$

$$= \frac{4x-12}{(x+3)(x-3)}$$

$$= \frac{4(x-3)}{(x+3)(x-3)} = \frac{4}{x+3}$$

Section 12.5 Solving Equations Containing Rational Expressions

TO SOLVE AN EQUATION CONTAINING RATIONAL EXPRESSIONS

Step 1. Multiply both sides of the equation by the LCD of all rational expressions in the equation.

Step 2. Remove any grouping symbols and solve the resulting equation.

Step 3. Check the solution in the original equation.

Solve: $\frac{5x}{x+2} + 3 = \frac{4x-6}{x+2}$ The LCD is $x+2$.

$$(x+2)\left(\frac{5x}{x+2} + 3\right) = (x+2)\left(\frac{4x-6}{x+2}\right)$$

$$(x+2)\left(\frac{5x}{x+2}\right) + (x+2)(3) = (x+2)\left(\frac{4x-6}{x+2}\right)$$

$$5x + 3x + 6 = 4x - 6$$

$$4x = -12$$

$$x = -3$$

The solution checks; the solution is -3.

Section 12.6 Rational Equations and Problem Solving

PROBLEM-SOLVING STEPS

1. UNDERSTAND. Read and reread the problem.

A small plane and a car leave Kansas City, Missouri, and head for Minneapolis, Minnesota, a distance of 450 miles. The speed of the plane is 3 times the speed of the car, and the plane arrives 6 hours ahead of the car. Find the speed of the car.

Let x = the speed of the car.
Then $3x$ = the speed of the plane.

	Distance	= Rate	· Time
Car	450	x	$\frac{450}{x}$ $\left(\frac{\text{distance}}{\text{rate}}\right)$
Plane	450	$3x$	$\frac{450}{3x}$ $\left(\frac{\text{distance}}{\text{rate}}\right)$

DEFINITIONS AND CONCEPTS	**EXAMPLES**
Section 12.6 Rational Equations and Problem Solving (*continued*)	
2. TRANSLATE.	In words: plane's time + 6 hours = car's time Translate: $\frac{450}{3x} + 6 = \frac{450}{x}$
3. SOLVE.	$\frac{450}{3x} + 6 = \frac{450}{x}$ $3x\left(\frac{450}{3x}\right) + 3x(6) = 3x\left(\frac{450}{x}\right)$ $450 + 18x = 1350$ $18x = 900$ $x = 50$
4. INTERPRET.	**Check** this solution in the originally stated problem. **State** the conclusion: The speed of the car is 50 miles per hour.
Section 12.7 Simplifying Complex Fractions	
METHOD 1: TO SIMPLIFY A COMPLEX FRACTION **Step 1.** Add or subtract fractions in the numerator and the denominator of the complex fraction. **Step 2.** Perform the indicated division. **Step 3.** Write the result in lowest terms.	Simplify: $\frac{\frac{1}{x}+2}{\frac{1}{x}-\frac{1}{y}} = \frac{\frac{1}{x}+\frac{2x}{x}}{\frac{y}{xy}-\frac{x}{xy}}$ $= \frac{\frac{1+2x}{x}}{\frac{y-x}{xy}}$ $= \frac{1+2x}{x} \cdot \frac{xy}{y-x}$ $= \frac{y(1+2x)}{y-x}$
METHOD 2: TO SIMPLIFY A COMPLEX FRACTION **Step 1.** Find the LCD of all fractions in the complex fraction. **Step 2.** Multiply the numerator and the denominator of the complex fraction by the LCD. **Step 3.** Perform the indicated operations and write the result in lowest terms.	$\frac{\frac{1}{x}+2}{\frac{1}{x}-\frac{1}{y}} = \frac{xy\left(\frac{1}{x}+2\right)}{xy\left(\frac{1}{x}-\frac{1}{y}\right)}$ $= \frac{xy\left(\frac{1}{x}\right)+xy(2)}{xy\left(\frac{1}{x}\right)-xy\left(\frac{1}{y}\right)}$ $= \frac{y+2xy}{y-x}$ or $\frac{y(1+2x)}{y-x}$

STUDY SKILLS BUILDER

Are You Prepared for a Test on Chapter 12?

Below I have listed *a common trouble* area for students in Chapter 12. After studying for your test, but before taking your test, read this.

Do you know the differences between how to perform operations such as $\frac{4}{x} + \frac{2}{3}$ or $\frac{4}{x} \div \frac{2}{x}$ and how to solve an equation such as $\frac{4}{x} + \frac{2}{3} = 1$?

$$\frac{4}{x} + \frac{2}{3} = \frac{4 \cdot 3}{x \cdot 3} + \frac{2 \cdot x}{3 \cdot x}$$

Addition—write each expression as an equivalent expression with the same LCD denominator.

$$= \frac{12}{3x} + \frac{2x}{3x} = \frac{12 + 2x}{3x} \quad \text{or} \quad \frac{2(6 + x)}{3x}, \text{ the sum.}$$

$$\frac{4}{x} \div \frac{2}{x} = \frac{4}{x} \cdot \frac{x}{2} = \frac{4 \cdot x}{x \cdot 2} = \frac{4}{2} = 2, \text{ the quotient.}$$

Division—multiply the first rational expression by the reciprocal of the second.

$$\frac{4}{x} + \frac{2}{3} = 1 \quad \text{Equation to be solved.}$$

$$3x\left(\frac{4}{x} + \frac{2}{3}\right) = 3x \cdot 1 \quad \text{Multiply both sides of the equation by the LCD, } 3x.$$

$$3x\left(\frac{4}{x}\right) + 3x\left(\frac{2}{3}\right) = 3x \cdot 1 \quad \text{Use the distributive property.}$$

$$12 + 2x = 3x \quad \text{Multiply and simplify.}$$

$$12 = x \quad \text{Subtract } 2x \text{ from both sides.}$$

The solution is 12.

For more examples and exercises, see the Chapter 12 Integrated Review.

12 CHAPTER REVIEW

(12.1) *Find any real number for which each rational expression is undefined.*

1. $\dfrac{x + 5}{x^2 - 4}$

2. $\dfrac{5x + 9}{4x^2 - 4x - 15}$

Find the value of each rational expression when $x = 5$, $y = 7$, and $z = -2$.

3. $\dfrac{2 - z}{z + 5}$

4. $\dfrac{x^2 + xy - y^2}{x + y}$

Simplify each rational expression.

5. $\dfrac{2x + 6}{x^2 + 3x}$

6. $\dfrac{3x - 12}{x^2 - 4x}$

7. $\dfrac{x + 2}{x^2 - 3x - 10}$

8. $\dfrac{x + 4}{x^2 + 5x + 4}$

9. $\dfrac{x^3 - 4x}{x^2 + 3x + 2}$

10. $\dfrac{5x^2 - 125}{x^2 + 2x - 15}$

11. $\dfrac{x^2 - x - 6}{x^2 - 3x - 10}$

12. $\dfrac{x^2 - 2x}{x^2 + 2x - 8}$

Simplify each expression. First, factor the four-term polynomials by grouping.

13. $\dfrac{x^2 + xa + xb + ab}{x^2 - xc + bx - bc}$

14. $\dfrac{x^2 + 5x - 2x - 10}{x^2 - 3x - 2x + 6}$

(12.2) *Perform each indicated operation and simplify.*

15. $\dfrac{15x^3y^2}{z} \cdot \dfrac{z}{5xy^3}$

16. $\dfrac{-y^3}{8} \cdot \dfrac{9x^2}{y^3}$

17. $\dfrac{x^2 - 9}{x^2 - 4} \cdot \dfrac{x - 2}{x + 3}$

18. $\dfrac{2x + 5}{x - 6} \cdot \dfrac{2x}{-x + 6}$

19. $\dfrac{x^2 - 5x - 24}{x^2 - x - 12} \div \dfrac{x^2 - 10x + 16}{x^2 + x - 6}$

20. $\dfrac{4x + 4y}{xy^2} \div \dfrac{3x + 3y}{x^2y}$

21. $\dfrac{x^2 + x - 42}{x - 3} \cdot \dfrac{(x - 3)^2}{x + 7}$

22. $\dfrac{2a + 2b}{3} \cdot \dfrac{a - b}{a^2 - b^2}$

23. $\dfrac{2x^2 - 9x + 9}{8x - 12} \div \dfrac{x^2 - 3x}{2x}$

24. $\dfrac{x^2 - y^2}{x^2 + xy} \div \dfrac{3x^2 - 2xy - y^2}{3x^2 + 6x}$

(12.3) *Perform each indicated operation and simplify.*

25. $\frac{x}{x^2 + 9x + 14} + \frac{7}{x^2 + 9x + 14}$

26. $\frac{x}{x^2 + 2x - 15} + \frac{5}{x^2 + 2x - 15}$

27. $\frac{4x - 5}{3x^2} - \frac{2x + 5}{3x^2}$

28. $\frac{9x + 7}{6x^2} - \frac{3x + 4}{6x^2}$

Find the LCD of each pair of rational expressions.

29. $\frac{x + 4}{2x}, \frac{3}{7x}$

30. $\frac{x - 2}{x^2 - 5x - 24}, \frac{3}{x^2 + 11x + 24}$

Rewrite each rational expression as an equivalent expression whose denominator is the given polynomial.

31. $\frac{5}{7x} = \frac{}{14x^3y}$

32. $\frac{9}{4y} = \frac{}{16y^3x}$

33. $\frac{x + 2}{x^2 + 11x + 18} = \frac{}{(x + 2)(x - 5)(x + 9)}$

34. $\frac{3x - 5}{x^2 + 4x + 4} = \frac{}{(x + 2)^2(x + 3)}$

(12.4) *Perform each indicated operation and simplify.*

35. $\frac{4}{5x^2} - \frac{6}{y}$

36. $\frac{2}{x - 3} - \frac{4}{x - 1}$

37. $\frac{4}{x + 3} - 2$

38. $\frac{3}{x^2 + 2x - 8} + \frac{2}{x^2 - 3x + 2}$

39. $\frac{2x - 5}{6x + 9} - \frac{4}{2x^2 + 3x}$

40. $\frac{x - 1}{x^2 - 2x + 1} - \frac{x + 1}{x - 1}$

(12.5) *Solve each equation.*

41. $\frac{n}{10} = 9 - \frac{n}{5}$

42. $\frac{2}{x + 1} - \frac{1}{x - 2} = -\frac{1}{2}$

43. $\frac{y}{2y + 2} + \frac{2y - 16}{4y + 4} = \frac{y - 3}{y + 1}$

44. $\frac{2}{x - 3} - \frac{4}{x + 3} = \frac{8}{x^2 - 9}$

45. $\frac{x - 3}{x + 1} - \frac{x - 6}{x + 5} = 0$

46. $x + 5 = \frac{6}{x}$

Solve.

47. Five times the reciprocal of a number equals the sum of $\frac{3}{2}$ the reciprocal of the number and $\frac{7}{6}$. What is the number?

48. The reciprocal of a number equals the reciprocal of the difference of 4 and the number. Find the number.

49. A car travels 90 miles in the same time that a car traveling 10 miles per hour slower travels 60 miles. Find the speed of each car.

50. The current in a bayou near Lafayette, Louisiana, is 4 miles per hour. A paddle boat travels 48 miles upstream in the same amount of time it takes to travel 72 miles downstream. Find the speed of the boat in still water.

51. When Mark and Maria manicure Mr. Stergeon's lawn, it takes them 5 hours. If Mark works alone, it takes 7 hours. Find how long it takes Maria alone.

52. It takes pipe A 20 days to fill a fish pond. Pipe B takes 15 days. Find how long it takes both pipes together to fill the pond.

(12.7) *Simplify each complex fraction.*

53. $\dfrac{\frac{5x}{27}}{-\frac{10xy}{21}}$

54. $\dfrac{\frac{3}{5}+\frac{2}{7}}{\frac{1}{5}+\frac{5}{6}}$

55. $\dfrac{3-\frac{1}{y}}{2-\frac{1}{y}}$

56. $\dfrac{\frac{6}{x+2}+4}{\frac{8}{x+2}-4}$

Mixed Review

Simplify each rational expression.

57. $\dfrac{4x+12}{8x^2+24x}$

58. $\dfrac{x^3-6x^2+9x}{x^2+4x-21}$

Perform the indicated operations and simplify.

59. $\dfrac{x^2+9x+20}{x^2-25}\cdot\dfrac{x^2-9x+20}{x^2+8x+16}$

60. $\dfrac{x^2-x-72}{x^2-x-30}\div\dfrac{x^2+6x-27}{x^2-9x+18}$

61. $\dfrac{x}{x^2-36}+\dfrac{6}{x^2-36}$

62. $\dfrac{5x-1}{4x}-\dfrac{3x-2}{4x}$

63. $\dfrac{4}{3x^2+8x-3}+\dfrac{2}{3x^2-7x+2}$

64. $\dfrac{3x}{x^2+9x+14}-\dfrac{6x}{x^2+4x-21}$

Solve.

65. $\dfrac{4}{a-1} + 2 = \dfrac{3}{a-1}$

66. $\dfrac{x}{x+3} + 4 = \dfrac{x}{x+3}$

Solve.

67. The quotient of twice a number and three, minus one-sixth is the quotient of the number and two. Find the number.

68. Mr. Crocker can paint his shed by himself in three days. His son will need an additional day to complete the job if he works alone. If they work together, find how long it takes to paint the house.

Simplify each complex fraction.

69. $\dfrac{\frac{1}{4}}{\frac{1}{3} + \frac{1}{2}}$

70. $\dfrac{4 + \frac{2}{x}}{6 + \frac{3}{x}}$

12 CHAPTER TEST

Answers

Remember to use the Chapter Test Prep Video CD to see the fully worked-out solutions to any of the exercises you want to review.

1. Find any real numbers for which the following expression is undefined.

$$\frac{x+5}{x^2+4x+3}$$

2. For a certain computer desk, the average manufacturing cost C per desk (in dollars) is

$$C = \frac{100x + 3000}{x}$$

where x is the number of desks manufactured.

a. Find the average cost per desk when manufacturing 200 computer desks.

b. Find the average cost per desk when manufacturing 1000 computer desks.

Simplify each rational expression.

3. $\dfrac{3x-6}{5x-10}$

4. $\dfrac{x+6}{x^2+12x+36}$

5. $\dfrac{7-x}{x-7}$

6. $\dfrac{y-x}{x^2-y^2}$

7. $\dfrac{2m^3-2m^2-12m}{m^2-5m+6}$

8. $\dfrac{ay+3a+2y+6}{ay+3a+5y+15}$

Perform each indicated operation and simplify if possible.

9. $\dfrac{x^2-13x+42}{x^2+10x+21} \div \dfrac{x^2-4}{x^2+x-6}$

10. $\dfrac{3}{x-1} \cdot (5x-5)$

11. $\dfrac{y^2-5y+6}{2y+4} \cdot \dfrac{y+2}{2y-6}$

12. $\dfrac{5}{2x+5} - \dfrac{6}{2x+5}$

13. $\dfrac{5a}{a^2-a-6} - \dfrac{2}{a-3}$

14. $\dfrac{6}{x^2-1} + \dfrac{3}{x+1}$

15. $\dfrac{x^2-9}{x^2-3x} \div \dfrac{x^2+4x+1}{2x+10}$

16. $\dfrac{x+2}{x^2+11x+18} + \dfrac{5}{x^2-3x-10}$

17. $\dfrac{4y}{y^2+6y+5} - \dfrac{3}{y^2+5y+4}$

1. ______
2. a. ______
   b. ______
3. ______
4. ______
5. ______
6. ______
7. ______
8. ______
9. ______
10. ______
11. ______
12. ______
13. ______
14. ______
15. ______
16. ______
17. ______

18. ______
19. ______
20. ______
21. ______
22. ______
23. ______
24. ______
25. ______
26. ______
27. ______
28. ______

Solve each equation.

18. $\frac{4}{y} - \frac{5}{3} = \frac{-1}{5}$

19. $\frac{5}{y+1} = \frac{4}{y+2}$

20. $\frac{a}{a-3} = \frac{3}{a-3} - \frac{3}{2}$

21. $\frac{10}{x^2-25} = \frac{3}{x+5} + \frac{1}{x-5}$

22. $x - \frac{14}{x-1} = 4 - \frac{2x}{x-1}$

Simplify each complex fraction.

23. $\dfrac{\frac{5x^2}{yz^2}}{\frac{10x}{z^3}}$

24. $\dfrac{\frac{b}{a} - \frac{a}{b}}{\frac{1}{b} + \frac{1}{a}}$

25. $\dfrac{5 - \frac{1}{y^2}}{\frac{1}{y} + \frac{2}{y^2}}$

26. One number plus five times its reciprocal is equal to six. Find the number.

27. A pleasure boat traveling down the Red River takes the same time to go 14 miles upstream as it takes to go 16 miles downstream. If the current of the river is 2 miles per hour, find the speed of the boat in still water.

28. An inlet pipe can fill a tank in 12 hours. A second pipe can fill the tank in 15 hours. If both pipes are used, find how long it takes to fill the tank.

CUMULATIVE REVIEW

Chapters 1–12

Write each fraction or mixed number as a percent.

1. $\frac{7}{20}$ **2.** $\frac{4}{5}$ **3.** $\frac{2}{3}$ **4.** $\frac{1}{9}$ **5.** $2\frac{1}{4}$ **6.** $3\frac{3}{4}$

Name the property illustrated by each true statement.

7. $2 \cdot (z \cdot 5) = 2 \cdot (5 \cdot z)$

8. $3 + y = y + 3$

9. $(x + 7) + 9 = x + (7 + 9)$

10. $(x \cdot 7) \cdot 9 = x \cdot (7 \cdot 9)$

11. A 10-foot board is to be cut into two pieces so that the longer piece is 4 times longer than the shorter. Find the length of each piece.

12. A 45-foot piece of rope is to be cut into three pieces. Two pieces are to have equal length and the third piece is to be 3 feet shorter than the longest pieces. Find the length of each piece.

13. Solve $y = mx + b$ for x.

14. Solve $y = mx + b$ for b.

15. Solve $x + 4 \leq -6$. Graph the solutions.

−10 −8 −6 −4 −2 0 2 4 6 8 10

16. Solve $x - 1 > -10$.

Simplify each quotient.

17. $\frac{x^5}{x^2}$ **18.** $\frac{x^9}{x}$ **19.** $\frac{4^7}{4^3}$ **20.** $\frac{7^{12}}{7^4}$

21. $\frac{(-3)^5}{(-3)^2}$ **22.** $\frac{(-4)^9}{(-4)^7}$ **23.** $\frac{2x^5y^2}{xy}$ **24.** $\frac{13a^5b}{a^4}$

Answers

1. ______
2. ______
3. ______
4. ______
5. ______
6. ______
7. ______
8. ______
9. ______
10. ______
11. ______
12. ______
13. ______
14. ______
15. ______
16. ______
17. ______
18. ______
19. ______
20. ______
21. ______
22. ______
23. ______
24. ______

25. ______
26. ______
27. ______
28. ______
29. ______
30. ______
31. ______
32. ______
33. ______
34. ______
35. ______
36. ______
37. ______
38. ______
39. ______
40. ______
41. ______
42. ______
43. ______
44. ______

Simplify by writing each expression with positive exponents only.

25. $2x^{-3}$ **26.** $9x^{-2}$ **27.** $(-2)^{-4}$ **28.** $(-3)^{-3}$

Multiply.

29. $5x(2x^3 + 6)$ **30.** $3y(4y^2 - 2)$

31. $-3x^2(5x^2 + 6x - 1)$ **32.** $-5y(7y^2 - 3y + 1)$

Divide.

33. $\dfrac{4x^2 + 7 + 8x^3}{2x + 3}$ **34.** $\dfrac{6x^2 - 7x + 4}{2x+1}$

Factor.

35. $x^2 + 7x + 12$ **36.** $x^2 + 17x + 70$

37. $25x^2 + 20xy + 4y^2$ **38.** $36a^2 - 48ab + 16b^2$

39. Solve: $x^2 - 9x - 22 = 0$ **40.** Solve: $x^2 + 2x - 15 = 0$

41. Multiply: $\dfrac{x^2 + x}{3x} \cdot \dfrac{6}{5x + 5}$ **42.** Divide: $\dfrac{3x - 12}{2} \div \dfrac{5x - 20}{4x}$

43. Subtract: $\dfrac{3x^2 + 2x}{x - 1} - \dfrac{10x - 5}{x - 1}$ **44.** Add: $\dfrac{4x}{x + 2} + \dfrac{8}{x + 2}$

45. Subtract: $\dfrac{6x}{x^2 - 4} - \dfrac{3}{x + 2}$

46. Add: $\dfrac{2}{x - 3} + \dfrac{5x}{x^2 - 9}$

47. Solve: $\dfrac{t - 4}{2} - \dfrac{t - 3}{9} = \dfrac{5}{18}$

48. $\dfrac{y}{2} - \dfrac{y}{4} = \dfrac{1}{6}$

49. Sam Waterton and Frank Schaffer work in a plant that manufactures automobiles. Sam can complete a quality control tour of the plant in 3 hours while his assistant, Frank, needs 7 hours to complete the same job. The regional manager is coming to inspect the plant facilities, so both Sam and Frank are directed to complete a quality control tour together. How long will this take?

50. One machine can complete a task in 18 hours. A second machine can do the same task in 12 hours. How long would it take to complete the task if both machines are able to work on it?

Simplify.

51. $\dfrac{\dfrac{1}{z} - \dfrac{1}{2}}{\dfrac{1}{3} - \dfrac{z}{6}}$

52. $\dfrac{\dfrac{x}{9} - \dfrac{1}{x}}{1 + \dfrac{3}{x}}$

45. __________

46. __________

47. __________

48. __________

49. __________

50. __________

51. __________

52. __________

13 Graphing Equations and Inequalities

In Chapter 9 we learned to solve and graph the solutions of linear equations and inequalities in one variable on number lines. Now we define and present techniques for solving and graphing linear equations and inequalities in two variables on grids. Two-variable equations lead directly to the concept of *function*, perhaps the most important concept in all mathematics. Functions are introduced in Section 13.6.

SECTIONS

CHECK YOUR PROGRESS

Americans enjoy pets more than ever before. Currently 62% of all U.S. households, or about 64.2 million households, have at least one pet. According to an American Pet Products Manufacturing Association survey, companionship, love, company, and affection eclipse all other benefits of pet ownership and are cited as the primary benefits of sharing their lives with their pet.

In Section 13.1, Exercise 17 we will examine the growth of these pet-related expenditures, such as food, veterinary care, supplies, and pet care and grooming.

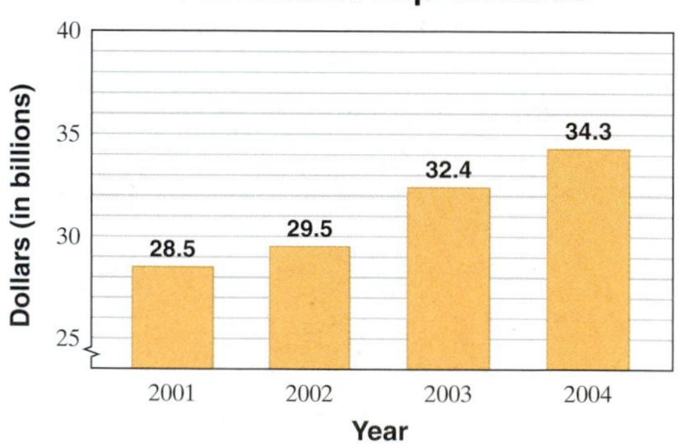

13.1 THE RECTANGULAR COORDINATE SYSTEM

Objectives

A Plot Ordered Pairs of Numbers on the Rectangular Coordinate System.

B Graph Paired Data to Create a Scatter Diagram.

C Find the Missing Coordinate of an Ordered Pair Solution, Given One Coordinate of the Pair.

In Sections 7.1 and 7.2, we learned how to read graphs. The broken line graph below shows the relationship between time spent smoking a cigarette and pulse rate. The horizontal line or axis shows time in minutes and the vertical line or axis shows the pulse rate in heartbeats per minute. Notice that there are two numbers associated with each point of the graph. For example, the graph shows that 15 minutes after "lighting up," the pulse rate is 80 beats per minute. If we agree to write the time first and the pulse rate second, we can say there is a point on the graph corresponding to the **ordered pair** of numbers (15, 80). A few more ordered pairs are shown alongside their corresponding points.

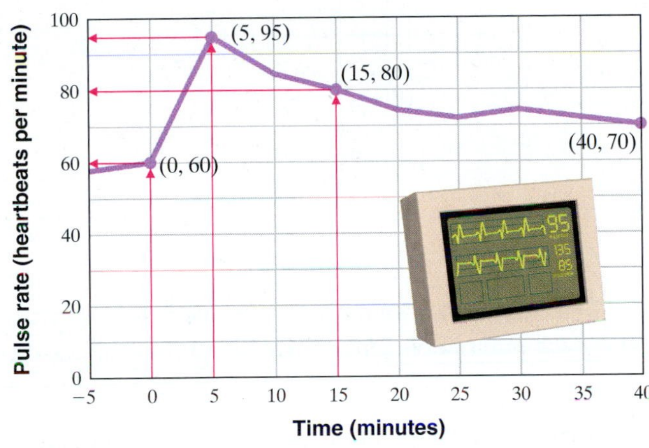

Objective A Plotting Ordered Pairs of Numbers

In general, we use the idea of ordered pairs to describe the location of a point in a plane (such as a piece of paper). We start with a horizontal and a vertical axis. Each axis is a number line, and for the sake of consistency we construct our axes to intersect at the 0 coordinate of both. This point of intersection is called the **origin.** Notice that these two number lines or axes divide the plane into four regions called **quadrants.** The quadrants are usually numbered with Roman numerals as shown. The axes are not considered to be in any quadrant.

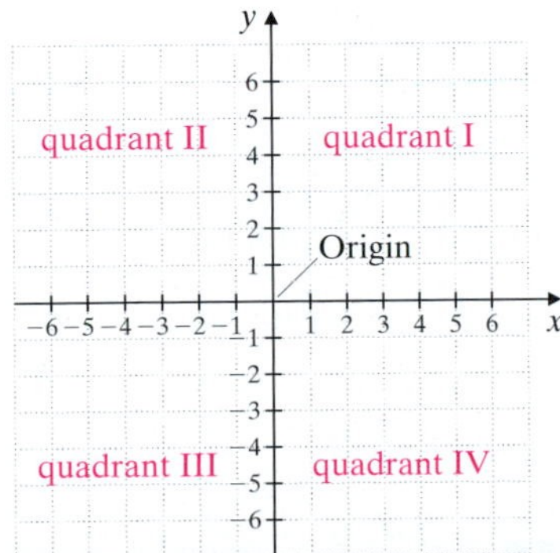

It is helpful to label axes, so we label the horizontal axis the ***x*-axis** and the vertical axis the ***y*-axis.** We call the system described above the **rectangular coordinate system,** or the **coordinate plane.** Just as with other graphs shown, we can then describe the locations of points by ordered pairs of numbers. We list the horizontal ***x*-axis** measurement first and the vertical ***y*-axis** measurement second.

To plot or graph the point corresponding to the ordered pair

(a, b)

we start at the origin. We then move a units left or right (right if a is positive, left if a is negative). From there, we move b units up or down (up if b is positive, down if b is negative). For example, to plot the point corresponding to the ordered pair $(3, 2)$, we start at the origin, move 3 units right, and from there move 2 units up. (See the figure below.) The x-value, 3, is also called the ***x*-coordinate** and the y-value, 2, is also called the ***y*-coordinate.** From now on, we will call the point with coordinates $(3, 2)$ simply the point $(3, 2)$. The point $(-2, 5)$ is also graphed below.

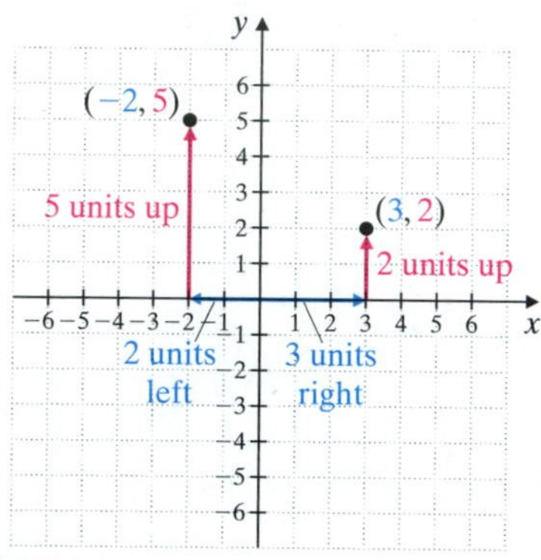

PRACTICE PROBLEM 1

On a single coordinate system, plot each ordered pair. State in which quadrant, or on which axis each point lies.

a. $(4, 2)$ **b.** $(-1, -3)$
c. $(2, -2)$ **d.** $(-5, 1)$
e. $(0, 3)$ **f.** $(3, 0)$
g. $(0, -4)$ **h.** $\left(-2\frac{1}{2}, 0\right)$
i. $\left(1, -3\frac{3}{4}\right)$

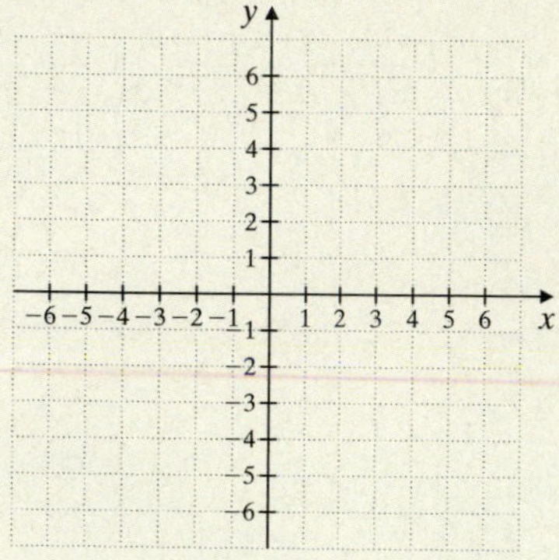

Helpful Hint

Don't forget that **each ordered pair corresponds to exactly one point in the plane and that each point in the plane corresponds to exactly one ordered pair.**

✔Concept Check Is the graph of the point $(-5, 1)$ in the same location as the graph of the point $(1, -5)$? Explain.

EXAMPLE 1 On a single coordinate system, plot each ordered pair. State in which quadrant, or on which axis each point lies.

a. $(5, 3)$ **b.** $(-2, -4)$ **c.** $(1, -2)$ **d.** $(-5, 3)$ **e.** $(0, 0)$
f. $(0, 2)$ **g.** $(-5, 0)$ **h.** $\left(0, -5\frac{1}{2}\right)$ **i.** $\left(4\frac{2}{3}, -3\right)$

Solution:

a. Point $(5, 3)$ lies in quadrant I.
b. Point $(-2, -4)$ lies in quadrant III.
c. Point $(1, -2)$ lies in quadrant IV.
d. Point $(-5, 3)$ lies in quadrant II.
e.–h. Points $(0, 0)$, $(0, 2)$, and $\left(0, -5\frac{1}{2}\right)$ lie on the y-axis. Points $(0, 0)$ and $(-5, 0)$ lie on the x-axis.
i. Point $\left(4\frac{2}{3}, -3\right)$ lies in quadrant IV.

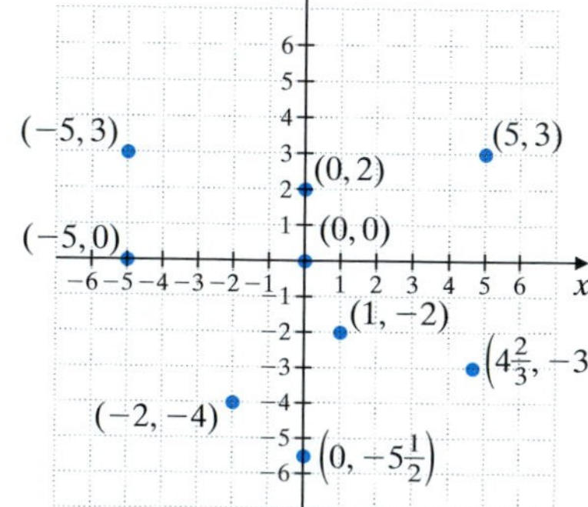

Work Practice Problem 1

Helpful Hint

In Example 1, notice that the point $(0, 0)$ lies on both the x-axis and the y-axis. It is the only point in the entire rectangular coordinate system that has this feature. Why? It is the only point of intersection of the x-axis and the y-axis.

Answers

1.

a. Point $(4, 2)$ lies in quadrant I.
b. Point $(-1, -3)$ lies in quadrant III.
c. Point $(2, -2)$ lies in quadrant IV.
d. Point $(-5, 1)$ lies in quadrant II.
e.–h. Points $(3, 0)$ and $\left(-2\frac{1}{2}, 0\right)$ lie on the x-axis. Points $(0, 3)$ and $(0, -4)$ lie on the y-axis.
i. Point $\left(1, -3\frac{3}{4}\right)$ lies in quadrant IV.

✔ Concept Check Answer

The graph of point $(-5, 1)$ lies in quadrant II and the graph of point $(1, -5)$ lies in quadrant IV. They are *not* in the same location.

✔Concept Check For each description of a point in the rectangular coordinate system, write an ordered pair that represents it.

a. Point A is located three units to the left of the y-axis and five units above the x-axis.

b. Point B is located six units below the origin.

From Example 1, notice that the y-coordinate of any point on the x-axis is 0. For example, the point $(-5, 0)$ lies on the x-axis. Also, the x-coordinate of any point on the y-axis is 0. For example, the point $(0, 2)$ lies on the y-axis.

Objective B Creating Scatter Diagrams

Data that can be represented as ordered pairs are called **paired data.** Many types of data collected from the real world are paired data. For instance, the annual measurements of a child's height can be written as ordered pairs of the form (year, height in inches) and are paired data. The graph of paired data as points in the rectangular coordinate system is called a **scatter diagram.** Scatter diagrams can be used to look for patterns and trends in paired data.

EXAMPLE 2 The table gives the annual net sales for Target Stores for the years shown. (*Source:* TargetCorp.com)

a. Write this paired data as a set of ordered pairs of the form (year, net sales in billions of dollars).

b. Create a scatter diagram of the paired data.

c. What trend in the paired data does the scatter diagram show?

Year	Target Net Sales (in billions of dollars)
1999	34
2000	37
2001	40
2002	44
2003	48

Solution:

a. The ordered pairs are (1999, 34), (2000, 37), (2001, 40), (2002, 44), and (2003, 48).

b. We begin by plotting the ordered pairs. Because the x-coordinate in each ordered pair is a year, we label the x-axis "Year" and mark the horizontal axis with the years given. Then we label the y-axis or vertical axis "Net Sales (in billions of dollars)." In this case, it is convenient to mark the vertical axis in multiples of 5, starting with 0. In Practice Problem 2, since there are no years when the number of tornadoes is less than 900, we use the notation ⸗ to skip to 900, then proceed by multiples of 100.

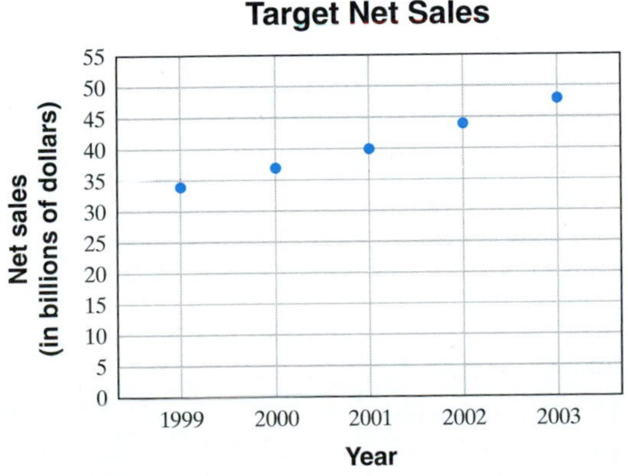

c. The scatter diagram shows that Target net sales steadily increased over the years 1999–2003.

Work Practice Problem 2

PRACTICE PROBLEM 2

The table gives the number of tornadoes that have occurred in the United States for the years shown. (*Source:* Storm Prediction Center, National Weather Service)

Year	Tornadoes
1998	1424
1999	1343
2000	997
2001	1216
2002	941
2003	1376

a. Write this paired data as a set of ordered pairs of the form (year, number of tornadoes).

b. Create a scatter diagram of the paired data.

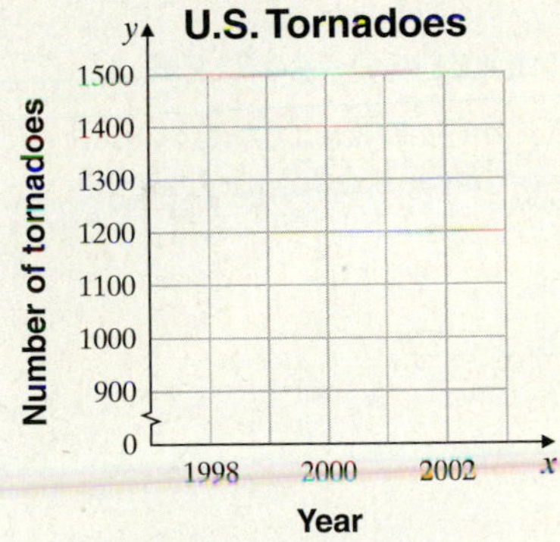

c. What trend in the paired data, if any, does the scatter diagram show?

Answers

2. a. (1998, 1424), (1999, 1343), (2000, 997), (2001, 1216), (2002, 941), (2003, 1376)

b.

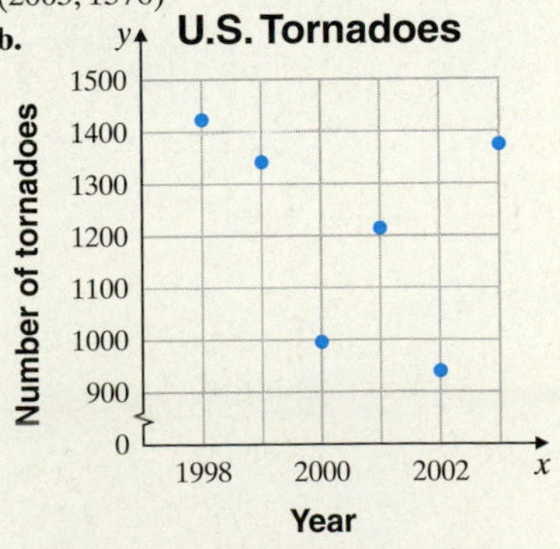

c. The number of tornadoes varies greatly from year to year.

✔ Concept Check Answers

a. $(-3, 5)$ **b.** $(0, -6)$

Objective C Completing Ordered Pair Solutions

Let's see how we can use ordered pairs to record solutions of equations containing two variables. An equation in one variable such as $x + 1 = 5$ has one solution, 4: the number 4 is the value of the variable x that makes the equation true.

An equation in two variables, such as $2x + y = 8$, has solutions consisting of two values, one for x and one for y. For example, $x = 3$ and $y = 2$ is a solution of $2x + y = 8$ because, if x is replaced with 3 and y with 2, we get a true statement.

$$2x + y = 8$$

$$2(3) + 2 \stackrel{?}{=} 8 \quad \text{Replace } x \text{ with 3 and } y \text{ with 2.}$$

$$8 = 8 \quad \text{True}$$

The solution $x = 3$ and $y = 2$ can be written as $(3, 2)$, an ordered pair of numbers.

In general, an ordered pair is a **solution** of an equation in two variables if replacing the variables by the values of the ordered pair results in a *true statement.*

For example, another ordered pair solution of $2x + y = 8$ is $(5, -2)$. Replacing x with 5 and y with -2 results in a true statement.

$$2x + y = 8$$

$$2(5) + (-2) \stackrel{?}{=} 8 \quad \text{Replace } x \text{ with 5 and } y \text{ with } -2.$$

$$10 - 2 \stackrel{?}{=} 8$$

$$8 = 8 \quad \text{True}$$

PRACTICE PROBLEM 3

Complete each ordered pair so that it is a solution to the equation $x + 2y = 8$.

a. $(0, \ )$

b. $(\ , 3)$

c. $(-4, \ )$

EXAMPLE 3 Complete each ordered pair so that it is a solution to the equation $3x + y = 12$.

a. $(0, \ )$ **b.** $(\ , 6)$ **c.** $(-1, \ )$

Solution:

a. In the ordered pair $(0, \ )$, the x-value is 0. We let $x = 0$ in the equation and solve for y.

$$3x + y = 12$$

$$3(0) + y = 12 \quad \text{Replace } x \text{ with 0.}$$

$$0 + y = 12$$

$$y = 12$$

The completed ordered pair is $(0, 12)$.

b. In the ordered pair $(\ , 6)$, the y-value is 6. We let $y = 6$ in the equation and solve for x.

$$3x + y = 12$$

$$3x + 6 = 12 \quad \text{Replace } y \text{ with 6.}$$

$$3x = 6 \quad \text{Subtract 6 from both sides.}$$

$$x = 2 \quad \text{Divide both sides by 3.}$$

The ordered pair is $(2, 6)$.

c. In the ordered pair $(-1, \ )$, the x-value is -1. We let $x = -1$ in the equation and solve for y.

$$3x + y = 12$$

$$3(-1) + y = 12 \quad \text{Replace } x \text{ with } -1.$$

$$-3 + y = 12$$

$$y = 15 \quad \text{Add 3 to both sides.}$$

The ordered pair is $(-1, 15)$.

Work Practice Problem 3

Answers

3. a. $(0, 4)$ **b.** $(2, 3)$ **c.** $(-4, 6)$

Solutions of equations in two variables can also be recorded in a **table of paired values,** as shown in the next example.

EXAMPLE 4 Complete the table for the equation $y = 3x$.

	x	y
a.	−1	
b.		0
c.		−9

Solution:

a. We replace x with −1 in the equation and solve for y.

$y = 3x$

$y = 3(-1)$ Let $x = -1$.

$y = -3$

The ordered pair is $(-1, -3)$.

b. We replace y with 0 in the equation and solve for x.

$y = 3x$

$0 = 3x$ Let $y = 0$.

$0 = x$ Divide both sides by 3.

The ordered pair is $(0, 0)$.

c. We replace y with −9 in the equation and solve for x.

$y = 3x$

$-9 = 3x$ Let $y = -9$.

$-3 = x$ Divide both sides by 3.

The ordered pair is $(-3, -9)$. The completed table is shown to the right.

x	y
−1	−3
0	0
−3	−9

Work Practice Problem 4

PRACTICE PROBLEM 4

Complete the table for the equation $y = -2x$.

	x	y
a.	−3	
b.		0
c.		10

EXAMPLE 5 Complete the table for the equation

$$y = \frac{1}{2}x - 5.$$

	x	y
a.	−2	
b.	0	
c.		0

Solution:

a. Let $x = -2$.

$y = \frac{1}{2}x - 5$

$y = \frac{1}{2}(-2) - 5$

$y = -1 - 5$

$y = -6$

b. Let $x = 0$.

$y = \frac{1}{2}x - 5$

$y = \frac{1}{2}(0) - 5$

$y = 0 - 5$

$y = -5$

c. Let $y = 0$.

$y = \frac{1}{2}x - 5$

$0 = \frac{1}{2}x - 5$ Now, solve for x.

$5 = \frac{1}{2}x$ Add 5.

$10 = x$ Multiply by 2.

Ordered Pairs: $(-2, -6)$ $(0, -5)$ $(10, 0)$

The completed table is

x	−2	0	10
y	−6	−5	0

Work Practice Problem 5

PRACTICE PROBLEM 5

Complete the table for the equation $y = \frac{1}{3}x - 1$.

	x	y
a.	−3	
b.	0	
c.		0

Answers

4.

	x	y
a.	−3	6
b.	0	0
c.	−5	10

5.

	x	y
a.	−3	−2
b.	0	−1
c.	3	0

By now, you have noticed that equations in two variables often have more than one solution. We discuss this more in the next section.

A table showing ordered pair solutions may be written vertically, or horizontally as shown in the next example.

PRACTICE PROBLEM 6

A company purchased a fax machine for $400. The business manager of the company predicts that the fax machine will be used for 7 years and the value in dollars y of the machine in x years is $y = -50x + 400$. Complete the table.

x	1	2	3	4	5	6	7
y							

EXAMPLE 6 A small business purchased a computer for $2000. The business predicts that the computer will be used for 5 years and the value in dollars y of the computer in x years is $y = -300x + 2000$. Complete the table.

x	0	1	2	3	4	5
y						

Solution:

To find the value of y when x is 0, we replace x with 0 in the equation. We use this same procedure to find y when x is 1 and when x is 2.

When $x = 0$,

$y = -300x + 2000$
$y = -300 \cdot 0 + 2000$
$y = 0 + 2000$
$y = 2000$

When $x = 1$,

$y = -300x + 2000$
$y = -300 \cdot 1 + 2000$
$y = -300 + 2000$
$y = 1700$

When $x = 2$,

$y = -300x + 2000$
$y = -300 \cdot 2 + 2000$
$y = -600 + 2000$
$y = 1400$

We have the ordered pairs (0, 2000), (1, 1700), and (2, 1400). This means that in 0 years the value of the computer is $2000, in 1 year the value of the computer is $1700, and in 2 years the value is $1400. To complete the table of values, we continue the procedure for $x = 3$, $x = 4$, and $x = 5$.

When $x = 3$,

$y = -300x + 2000$
$y = -300 \cdot 3 + 2000$
$y = -900 + 2000$
$y = 1100$

When $x = 4$,

$y = -300x + 2000$
$y = -300 \cdot 4 + 2000$
$y = -1200 + 2000$
$y = 800$

When $x = 5$,

$y = -300x + 2000$
$y = -300 \cdot 5 + 2000$
$y = -1500 + 2000$
$y = 500$

The completed table is shown below.

x	0	1	2	3	4	5
y	2000	1700	1400	1100	800	500

Work Practice Problem 6

The ordered pair solutions recorded in the completed table for Example 6 are another set of paired data. They are graphed next. Notice that this scatter diagram gives a visual picture of the decrease in value of the computer.

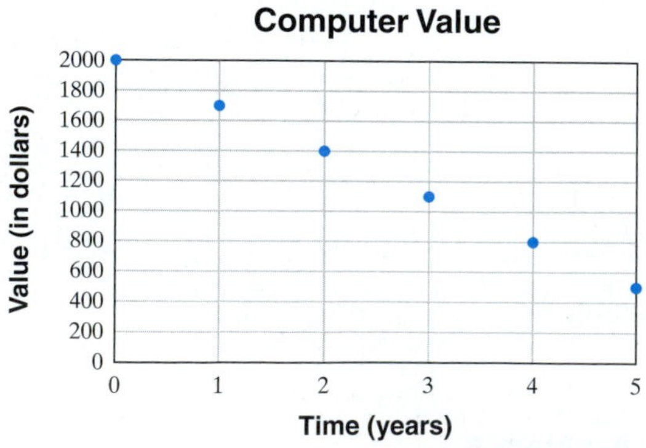

Computer Value

Answer

6.

x	1	2	3	4	5	6	7
y	350	300	250	200	150	100	50

Vocabulary and Readiness Check

Use the choices below to fill in each blank. The exercises below all have to do with the rectangular coordinate system.

origin	x-coordinate	x-axis	one	four
quadrants	y-coordinate	y-axis	solution	

1. The horizontal axis is called the ________.

2. The vertical axis is called the ________.

3. The intersection of the horizontal axis and the vertical axis is a point called the ________.

4. The axes divide the plane into regions, called ___________. There are _______ of these regions.

5. In the ordered pair of numbers $(-2, 5)$, the number -2 is called the ______________ and the number 5 is called the ______________.

6. Each ordered pair of numbers corresponds to _______ point in the plane.

7. An ordered pair is a __________ of an equation in two variables if replacing the variables by the coordinates of the ordered pair results in a true statement.

13.1 EXERCISE SET

FOR EXTRA HELP

Student Solutions Manual

PH Math/Tutor Center

CD/Video for Review

MathXL®

MyMathLab

Objective A *Plot each ordered pair. State in which quadrant or on which axis each point lies. See Example 1.*

1. **a.** $(1, 5)$ **b.** $(-5, -2)$ **c.** $(-3, 0)$ **d.** $(0, -1)$
e. $(2, -4)$ **f.** $\left(-1, 4\frac{1}{2}\right)$ **g.** $(3.7, 2.2)$ **h.** $\left(\frac{1}{2}, -3\right)$

2. **a.** $(2, 4)$ **b.** $(0, 2)$ **c.** $(-2, 1)$ **d.** $(-3, -3)$
e. $\left(3\frac{3}{4}, 0\right)$ **f.** $(5, -4)$ **g.** $(-3.4, 4.8)$ **h.** $\left(\frac{1}{3}, -5\right)$

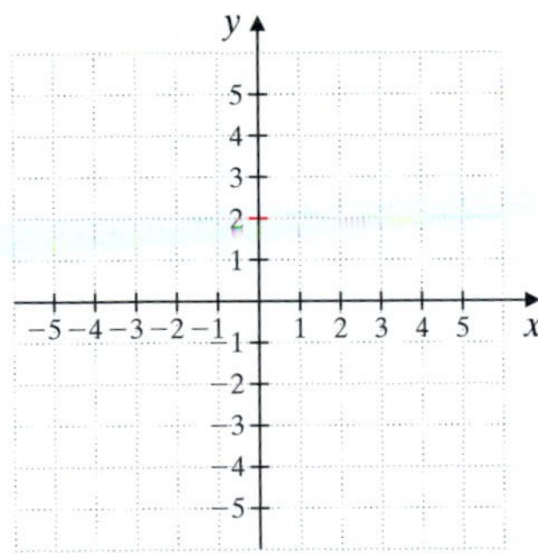

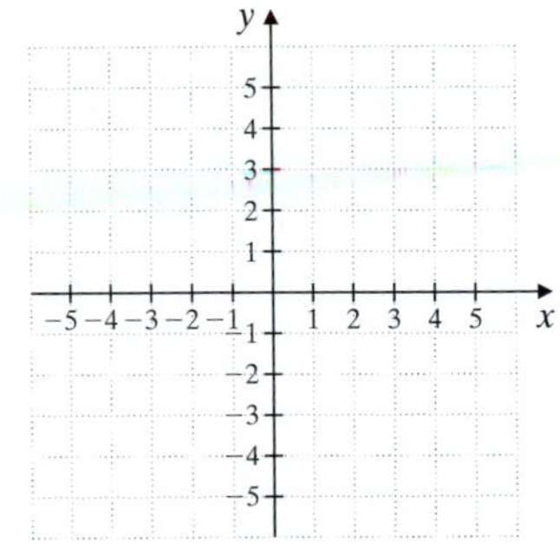

Find the x- and y-coordinates of each labeled point. See Example 1.

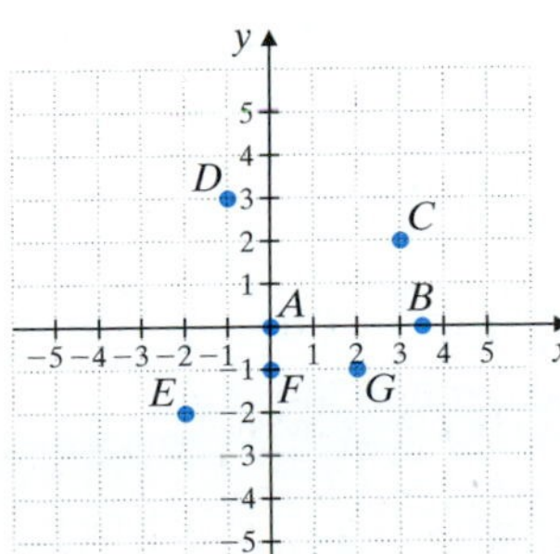

3. A **4.** B **5.** C

6. D **7.** E **8.** F

9. G

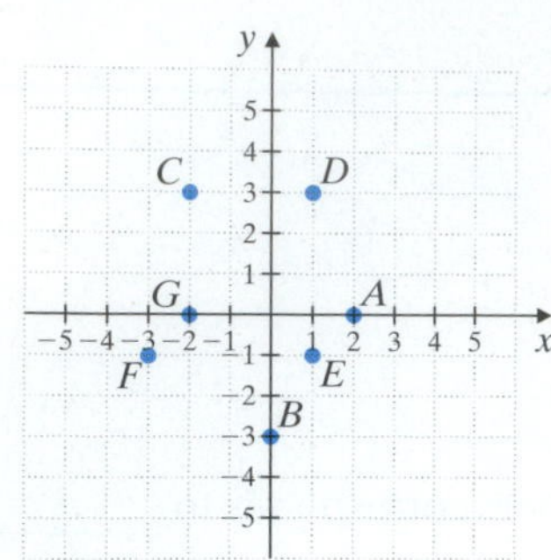

10. A **11.** B **12.** C

13. D **14.** E **15.** F

16. G

Objective B *Solve. See Example 2.*

17. The table shows the amount of money (in billions of dollars) that Americans spent on their pets for the years shown. (*Source:* American Pet Products Manufacturers Association)

Year	Pet-Related Expenditures (in billions of dollars)
2001	28.5
2002	29.5
2003	32.4
2004	34.3

a. Write this paired data as a set of ordered pairs of the form (year, pet-related expenditures).

b. In your own words, write the meaning of the ordered pair (2004, 34.3).

c. Create a scatter diagram of the paired data. Be sure to label the axes appropriately.

Pet-Related Expenditures

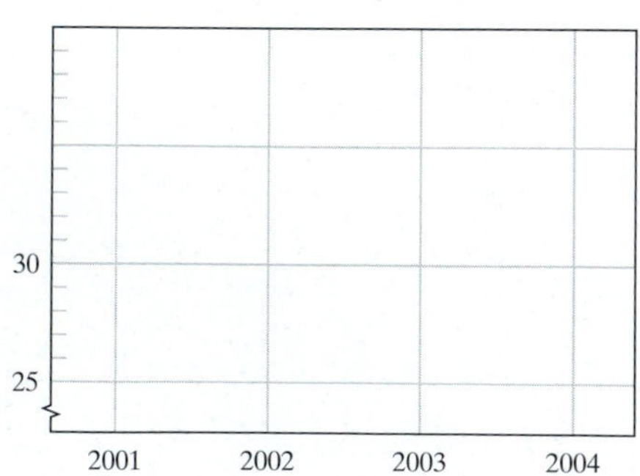

d. What trend in the paired data does the scatter diagram show?

18. The table shows the average farm size (in acres) in the United States during the years shown. (*Source:* National Agricultural Statistics Service)

Year	Average Farm Size (in acres)
1998	435
1999	432
2000	434
2001	438
2002	440
2003	441

a. Write this paired data as a set of ordered pairs of the form (year, average farm size).

b. In your own words, write the meaning of the ordered pair (2003, 441).

c. Create a scatter diagram of the paired data. Be sure to label the axes appropriately.

U.S. Average Farm Size

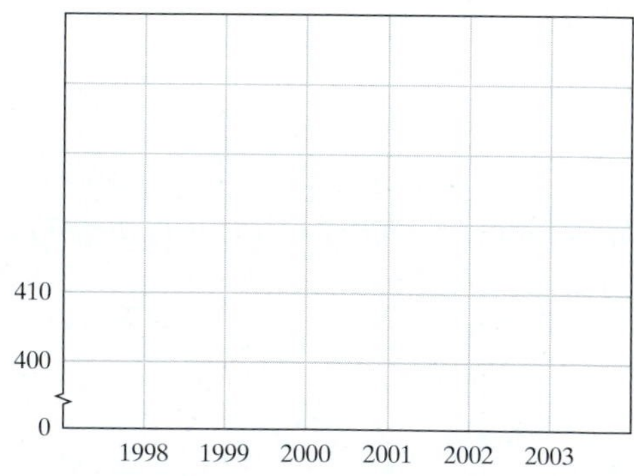

19. Minh, a psychology student, kept a record of how much time she spent studying for each of her 20-point psychology quizzes and her score on each quiz.

Hours Spent Studying	0.50	0.75	1.00	1.25	1.50	1.50	1.75	2.00
Quiz Score	10	12	15	16	18	19	19	20

a. Write each paired data as an ordered pair of the form (hours spent studying, quiz score).

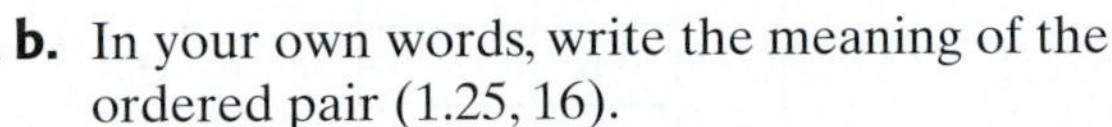

b. In your own words, write the meaning of the ordered pair (1.25, 16).

c. Create a scatter diagram of the paired data. Be sure to label the axes appropriately.

d. What might Minh conclude from the scatter diagram?

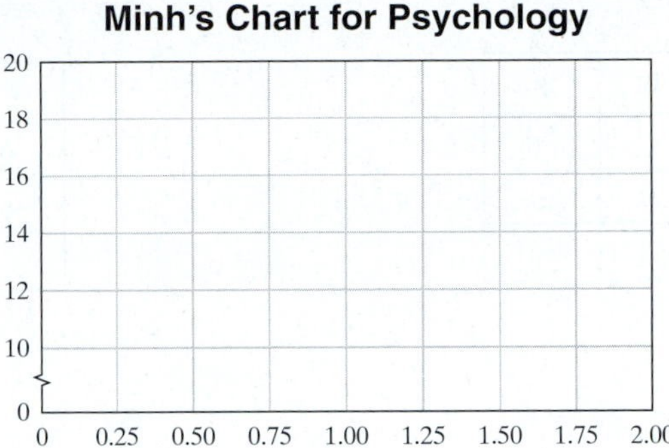

20. A local lumberyard uses quantity pricing. The table shows the price per board for different amounts of lumber purchased.

Price per Board (in dollars)	Number of Boards Purchased
8.00	1
7.50	10
6.50	25
5.00	50
2.00	100

a. Write each paired data as an ordered pair of the form (price per board, number of boards purchased).

b. In your own words, write the meaning of the ordered pair (2.00, 100).

c. Create a scatter diagram of the paired data. Be sure to label the axes appropriately.

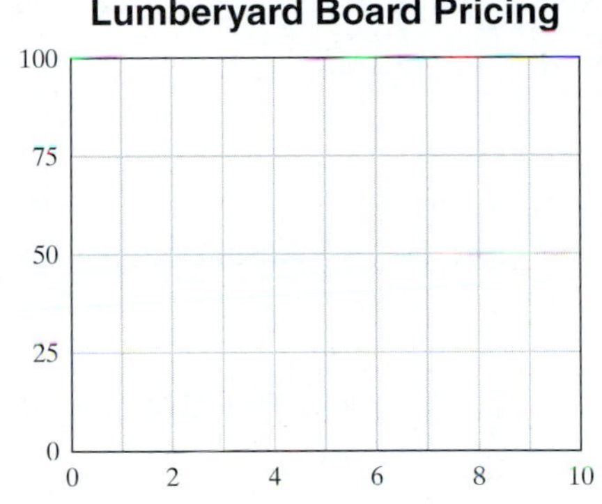

d. What trend in the paired data does the scatter diagram show?

Objective **C** *Complete each ordered pair so that it is a solution of the given linear equation. See Example 3.*

21. $x - 4y = 4$; (, −2), (4,)

22. $x - 5y = -1$; (, −2), (4,)

23. $y = \frac{1}{4}x - 3$; (−8,), (, 1)

24. $y = \frac{1}{5}x - 2$; (−10,), (, 1)

Complete the table of ordered pairs for each linear equation. See Examples 4 and 5.

25. $y = -7x$

x	y
0	
−1	
	2

26. $y = -9x$

x	y
	0
−3	
	2

27. $y = -x + 2$

x	y
0	
	0
−3	

28. $x = -y + 4$

x	y
	0
0	
	−3

29. $y = \frac{1}{2}x$

x	y
0	
−6	
	1

30. $y = \frac{1}{3}x$

x	y
0	
−6	
	1

31. $x + 3y = 6$

x	y
0	
	0
	1

32. $2x + y = 4$

x	y
0	
	0
	2

33. $y = 2x - 12$

x	y
0	
	−2
3	

34. $y = 5x + 10$

x	y
	0
	5
0	

35. $2x + 7y = 5$

x	y
0	
	0
	1

36. $x - 6y = 3$

x	y
0	
1	
	−1

Objectives **A B C Mixed Practice** *Complete the table of ordered pairs for each equation. Then plot the ordered pair solutions. See Examples 1 through 5.*

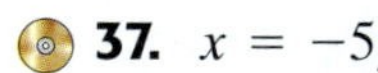

37. $x = -5y$

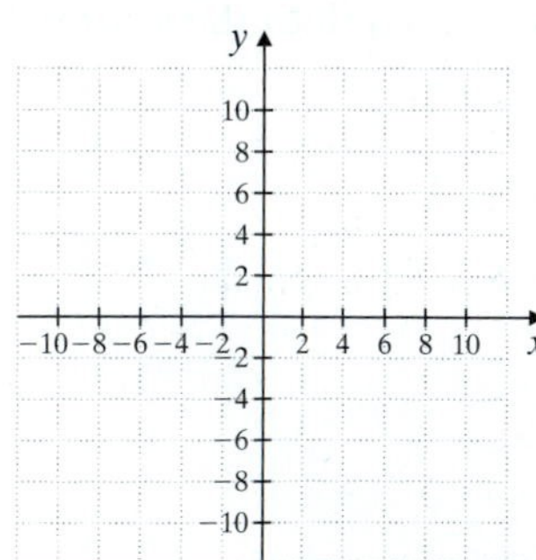

x	y
	0
	1
10	

38. $y = -3x$

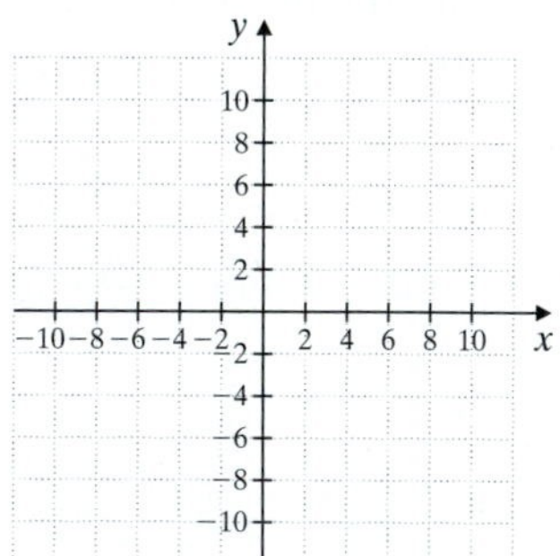

x	y
0	
−2	
	9

39. $y = \frac{1}{3}x + 2$

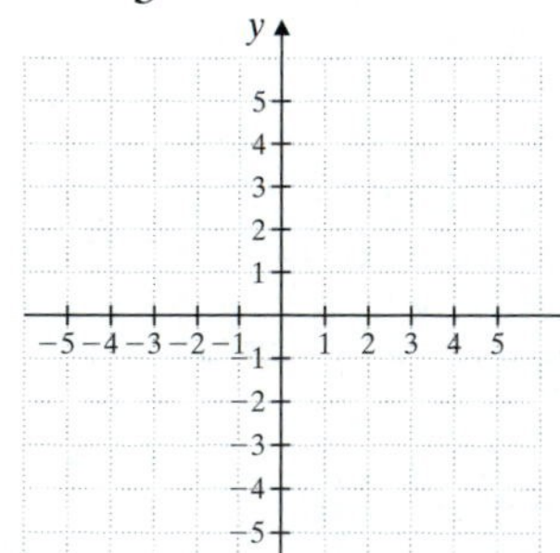

x	y
0	
−3	
	0

40. $y = \frac{1}{2}x + 3$

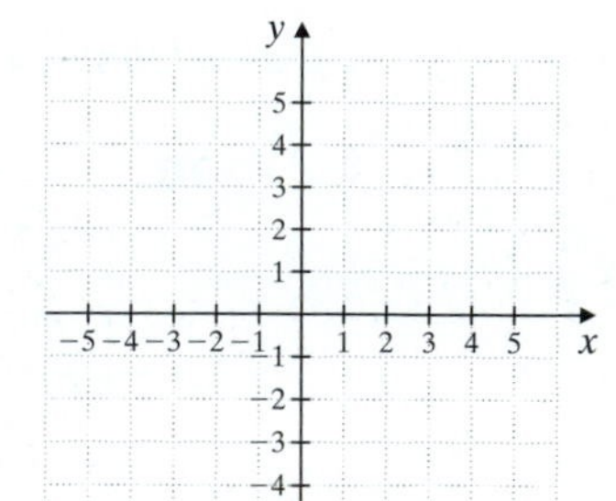

x	y
0	
−4	
	0

Solve. See Example 6.

41. The cost in dollars y of producing x computer desks is given by $y = 80x + 5000$.

a. Complete the table.

x	100	200	300
y			

b. Find the number of computer desks that can be produced for \$8600. (*Hint:* Find x when $y = 8600$.)

42. The hourly wage y of an employee at a certain production company is given by $y = 0.25x + 9$ where x is the number of units produced by the employee in an hour.

a. Complete the table.

x	0	1	5	10
y				

b. Find the number of units that an employee must produce each hour to earn an hourly wage of \$12.25. (*Hint:* Find x when $y = 12.25$.)

43. The percent y of recorded music sales that were in cassette format from 1998 through 2003 is given by $y = -2.4x + 13$. In the equation, x represents the number of years after 1998. (*Source:* Recording Industry Association of America)

a. Complete the table.

x	1	3	5
y			

b. Find the year in which approximately 3% of recorded music sales were cassettes. (*Hint:* Find x when $y = 3$ and round to the nearest whole number.)

44. The amount y of land occupied by farms in the United States (in million acres) from 1993 through 2003 is given by $y = -3.2x + 968$. In the equation, x represents the number of years after 1993. (*Source:* National Agricultural Statistics Service)

a. Complete the table.

x	4	7	10
y			

b. Find the year in which there were approximately 943 million acres of land occupied by farms. (*Hint:* Find x when $y = 943$ and round to the nearest whole number.)

Review

Solve each equation for y. See Section 9.5.

45. $x + y = 5$

46. $x - y = 3$

47. $2x + 4y = 5$

48. $5x + 2y = 7$

49. $10x = -5y$

50. $4y = -8x$

Concept Extensions

Answer each exercise with true or false.

51. Point $(-1, 5)$ lies in quadrant IV.

52. Point $(3, 0)$ lies on the y-axis.

53. For the point $\left(-\frac{1}{2}, 1.5\right)$, the first value, $-\frac{1}{2}$, is the x-coordinate and the second value, 1.5, is the y-coordinate.

54. The ordered pair $\left(2, \frac{2}{3}\right)$ is a solution of $2x - 3y = 6$.

For Exercises 55 through 59, fill in each blank with "0," "positive," or "negative." For Exercises 60 and 61, fill in each blank with "x" or "y."

	Point	Location
55.	(,)	quadrant III
56.	(,)	quadrant I
57.	(,)	quadrant IV
58.	(,)	quadrant II
59.	(,)	origin
60.	(number, 0)	__-axis
61.	(0, number)	__-axis

62. Give an example of an ordered pair whose location is in (or on)
a. quadrant I **b.** quadrant II **c.** quadrant III
d. quadrant IV **e.** x-axis **f.** y-axis

Solve. See the Concept Checks in this section.

63. Is the graph of $(3, 0)$ in the same location as the graph of $(0, 3)$? Explain why or why not.

64. Give the coordinates of a point such that if the coordinates are reversed, their location is the same.

65. In general, what points can have coordinates reversed and still have the same location?

66. In your own words, describe how to plot or graph an ordered pair of numbers.

Write an ordered pair for each point described.

67. Point C is four units to the right of the y-axis and seven units below the x-axis.

68. Point D is three units to the left of the origin.

△ **69.** Find the perimeter of the rectangle whose vertices are the points with coordinates $(-1, 5)$, $(3, 5)$, $(3, -4)$, and $(-1, -4)$.

△ **70.** Find the area of the rectangle whose vertices are the points with coordinates $(5, 2)$, $(5, -6)$, $(0, -6)$, and $(0, 2)$.

The scatter diagram below shows Walt Disney Company's annual revenues. The horizontal axis represents the number of years after 1999.

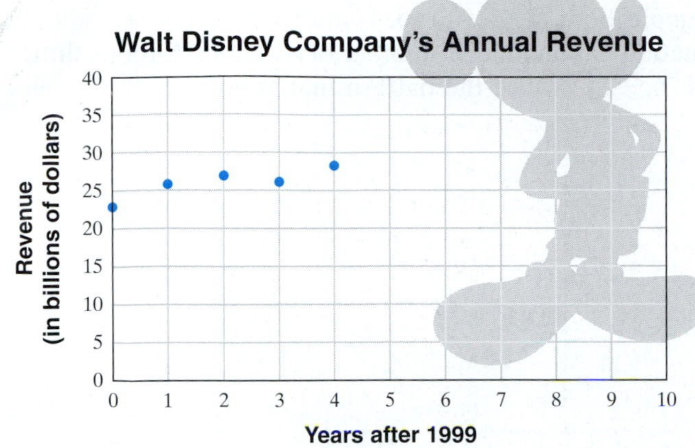

71. Estimate the annual revenues for years 1, 2, 3, and 4.

72. Use a straight edge or ruler and this scatter diagram to predict Disney's revenue in the year 2008.

73. Discuss any similarities in the graphs of the ordered pair solutions for Exercises 37–40.

STUDY SKILLS BUILDER

Are You Satisfied with Your Performance in This Course Thus Far?

To see if there is room for improvement, answer these questions:

1. Am I attending all classes and arriving on time?
2. Am I working and checking my homework assignments on time?
3. Am I getting help (from my instructor or a campus learning resource lab) when I need it?
4. In addition to my instructor, am I using the text supplements that might help me?
5. Am I satisfied with my performance on quizzes and exams?

If you answered no to any of these questions, read or reread Section 1.1 for suggestions in these areas. Also, you might want to contact your instructor for additional feedback.

Objective

A **Graph a Linear Equation by Finding and Plotting Ordered Pair Solutions.**

13.2 GRAPHING LINEAR EQUATIONS

In the previous section, we found that equations in two variables may have more than one solution. For example, both $(2, 2)$ and $(0, 4)$ are solutions of the equation $x + y = 4$. In fact, this equation has an infinite number of solutions. Other solutions include $(-2, 6)$, $(4, 0)$, and $(6, -2)$. Notice the pattern that appears in the graph of these solutions.

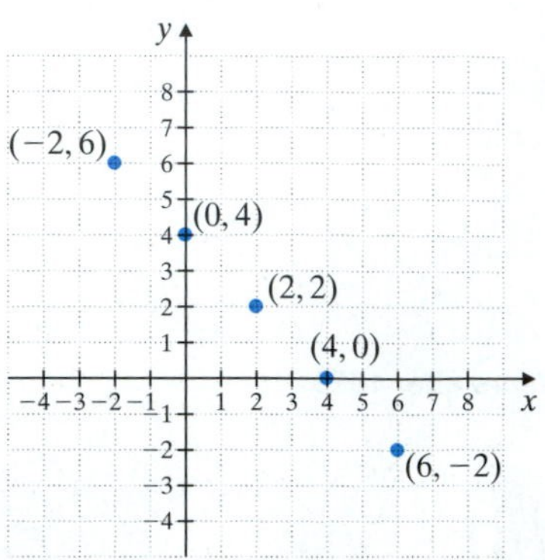

These solutions all appear to lie on the same line, as seen in the second graph. It can be shown that every ordered pair solution of the equation corresponds to a point on this line, and every point on this line corresponds to an ordered pair solution. Thus, we say that this line is the **graph of the equation** $x + y = 4$. Notice that we can only show a part of a line on a graph. The arrowheads on each end of the line below remind us that the line actually extends indefinitely in both directions.

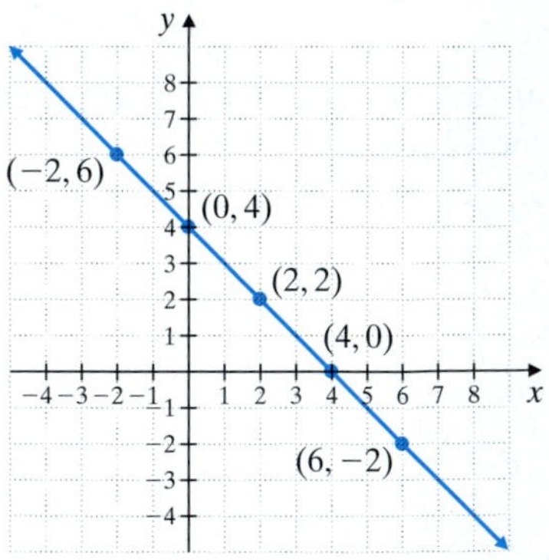

The equation $x + y = 4$ is called a *linear equation in two variables* and *the graph of every linear equation in two variables is a straight line*.

Linear Equation in Two Variables

A **linear equation in two variables** is an equation that can be written in the form

$$Ax + By = C$$

where A, B, and C are real numbers and A and B are not both 0. This form is called **standard form. The graph of a linear equation in two variables is a straight line.**

A linear equation in two variables may be written in many forms. Standard form, $Ax + By = C$, is just one of many of these forms.

Following are examples of linear equations in two variables.

$2x + y = 8$ (Standard Form) $\quad -2x = 7y \quad y = \frac{1}{3}x + 2 \quad y = 7$

Objective A Graphing Linear Equations

From geometry, we know that a straight line is determined by just two points. Thus, to graph a linear equation in two variables we need to find just two of its infinitely many solutions. Once we do so, we plot the solution points and draw the line connecting the points. Usually, we find a third solution as well, as a check.

EXAMPLE 1 Graph the linear equation $2x + y = 5$.

Solution: To graph this equation, we find three ordered pair solutions of $2x + y = 5$. To do this, we choose a value for one variable, x or y, and solve for the other variable. For example, if we let $x = 1$, then $2x + y = 5$ becomes

$$2x + y = 5$$
$$2(1) + y = 5 \quad \text{Replace } x \text{ with 1.}$$
$$2 + y = 5 \quad \text{Multiply.}$$
$$y = 3 \quad \text{Subtract 2 from both sides.}$$

Since $y = 3$ when $x = 1$, the ordered pair $(1, 3)$ is a solution of $2x + y = 5$. Next, we let $x = 0$.

$$2x + y = 5$$
$$2(0) + y = 5 \quad \text{Replace } x \text{ with 0.}$$
$$0 + y = 5$$
$$y = 5$$

The ordered pair $(0, 5)$ is a second solution.

The two solutions found so far allow us to draw the straight line that is the graph of all solutions of $2x + y = 5$. However, we will find a third ordered pair as a check. Let $y = -1$.

$$2x + y = 5$$
$$2x + (-1) = 5 \quad \text{Replace } y \text{ with } -1.$$
$$2x - 1 = 5$$
$$2x = 6 \quad \text{Add 1 to both sides.}$$
$$x = 3 \quad \text{Divide both sides by 2.}$$

The third solution is $(3, -1)$. These three ordered pair solutions are listed in the table and plotted on the coordinate plane. The graph of $2x + y = 5$ is the line through the three points.

x	y
1	3
0	5
3	−1

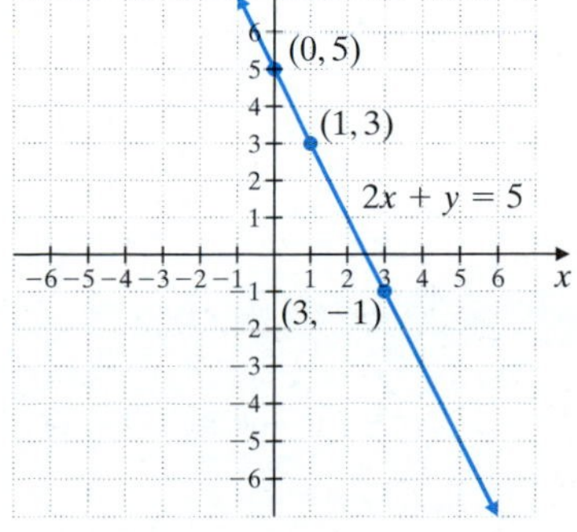

Work Practice Problem 1

PRACTICE PROBLEM 1

Graph the linear equation $x + 3y = 6$.

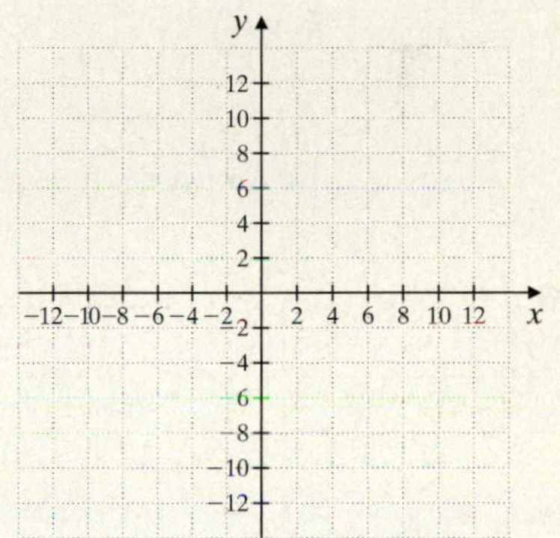

Helpful Hint

All three points should fall on the same straight line. If not, check your ordered pair solutions for a mistake.

Answer

1.

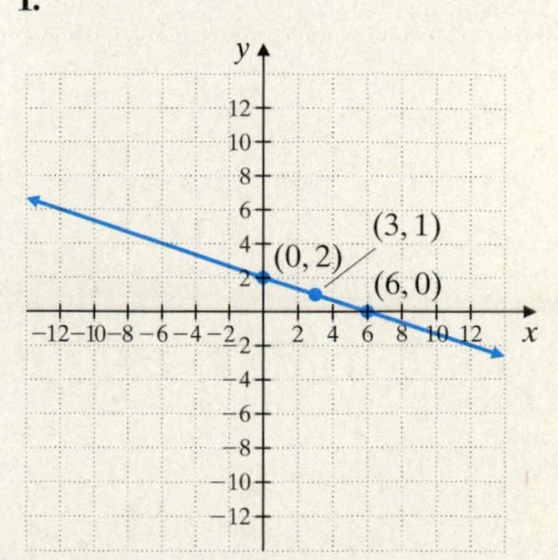

PRACTICE PROBLEM 2

Graph the linear equation $-2x + 4y = 8$.

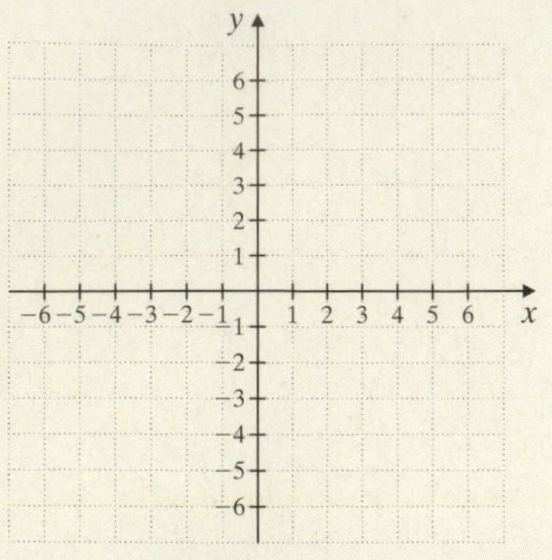

PRACTICE PROBLEM 3

Graph the linear equation $y = 2x$.

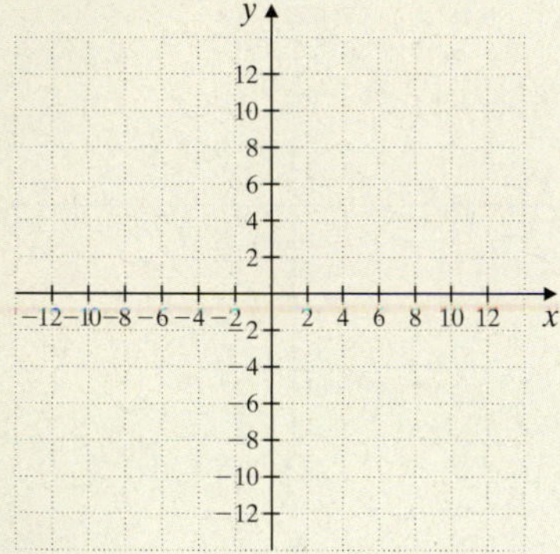

Answers

2.

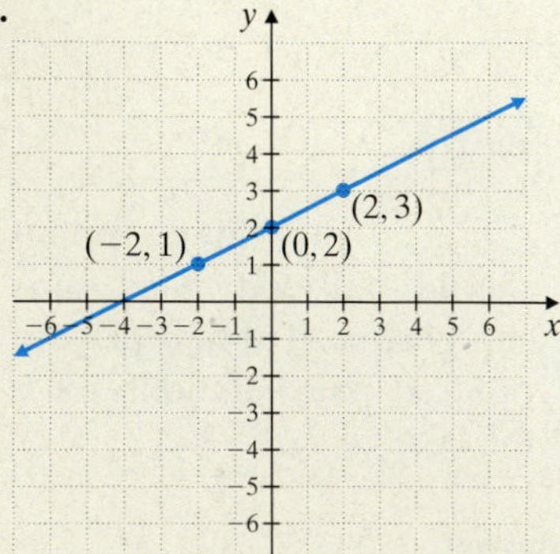

3.

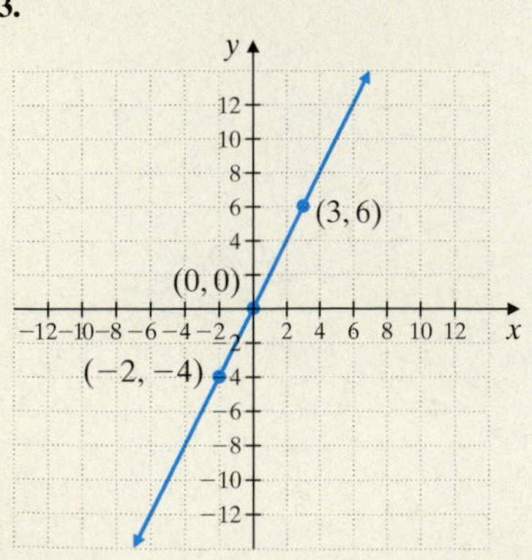

EXAMPLE 2 Graph the linear equation $-5x + 3y = 15$.

Solution: We find three ordered pair solutions of $-5x + 3y = 15$.

Let $x = 0$.

$$-5x + 3y = 15$$
$$-5 \cdot 0 + 3y = 15$$
$$0 + 3y = 15$$
$$3y = 15$$
$$y = 5$$

Let $y = 0$.

$$-5x + 3y = 15$$
$$-5x + 3 \cdot 0 = 15$$
$$-5x + 0 = 15$$
$$-5x = 15$$
$$x = -3$$

Let $x = -2$.

$$-5x + 3y = 15$$
$$-5 \cdot -2 + 3y = 15$$
$$10 + 3y = 15$$
$$3y = 5$$
$$y = \frac{5}{3} \text{ or } 1\frac{2}{3}$$

The ordered pairs are $(0, 5)$, $(-3, 0)$, and $\left(-2, 1\frac{2}{3}\right)$. The graph of $-5x + 3y = 15$ is the line through the three points.

x	y
0	5
−3	0
−2	$1\frac{2}{3}$

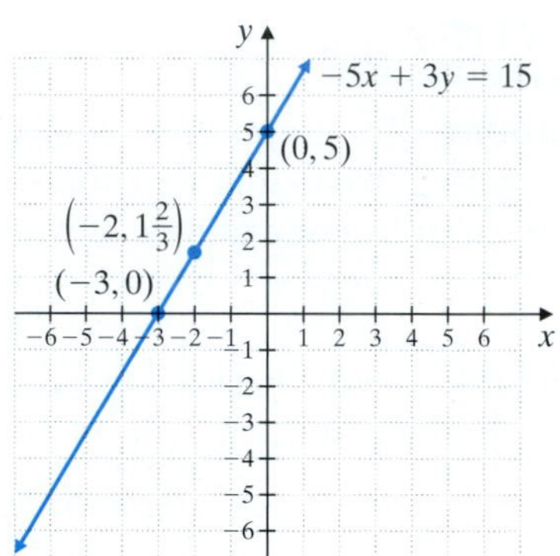

Work Practice Problem 2

EXAMPLE 3 Graph the linear equation $y = 3x$.

Solution: We find three ordered pair solutions. Since this equation is solved for y, we'll choose three x values.

If $x = 2$, $y = 3 \cdot 2 = 6$.
If $x = 0$, $y = 3 \cdot 0 = 0$.
If $x = -1$, $y = 3 \cdot -1 = -3$.

Next, we plot the ordered pair solutions and draw a line through the plotted points. The line is the graph of $y = 3x$.

Think about the following for a moment: A line is made up of an infinite number of points. Every point on the line defined by $y = 3x$ represents an ordered pair solution of the equation and every ordered pair solution is a point on this line.

x	y
2	6
0	0
−1	−3

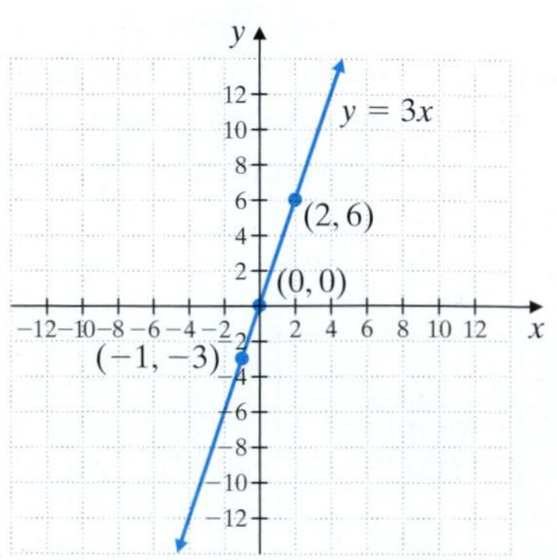

Work Practice Problem 3

Helpful Hint

When graphing a linear equation in two variables, if it is

- solved for y, it may be easier to find ordered-pair solutions by choosing x-values. If it is
- solved for x, it may be easier to find ordered-pair solutions by choosing y-values.

EXAMPLE 4 Graph the linear equation $y = -\frac{1}{3}x + 2$.

Solution: We find three ordered pair solutions, plot the solutions, and draw a line through the plotted solutions. To avoid fractions, we'll choose x values that are multiples of 3 to substitute into the equation.

If $x = 6$, then $y = -\frac{1}{3} \cdot 6 + 2 = -2 + 2 = 0$

If $x = 0$, then $y = -\frac{1}{3} \cdot 0 + 2 = 0 + 2 = 2$

If $x = -3$, then $y = -\frac{1}{3} \cdot -3 + 2 = 1 + 2 = 3$

x	y
6	0
0	2
−3	3

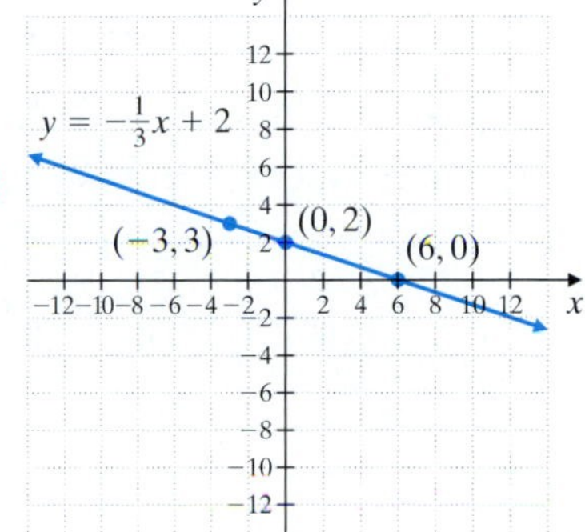

■ **Work Practice Problem 4**

Let's take a moment and compare the graphs in Examples 3 and 4. The graph of $y = 3x$ tilts upward (as we follow the line from left to right) and the graph of $y = -\frac{1}{3}x + 2$ tilts downward (as we follow the line from left to right). We will learn more about the tilt, or slope, of a line in Section 13.4.

EXAMPLE 5 Graph the linear equation $y = -2$.

Solution: The equation $y = -2$ can be written in standard form as $0x + y = -2$. No matter what value we replace x with, y is always -2.

x	y
0	−2
3	−2
−2	−2

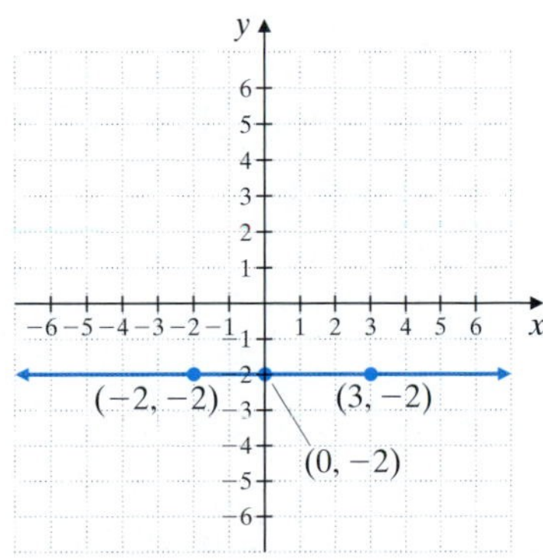

Notice that the graph of $y = -2$ is a horizontal line.

■ **Work Practice Problem 5**

Linear equations are often used to model real data, as seen in the next example.

PRACTICE PROBLEM 4

Graph the linear equation $y = -\frac{1}{2}x + 4$.

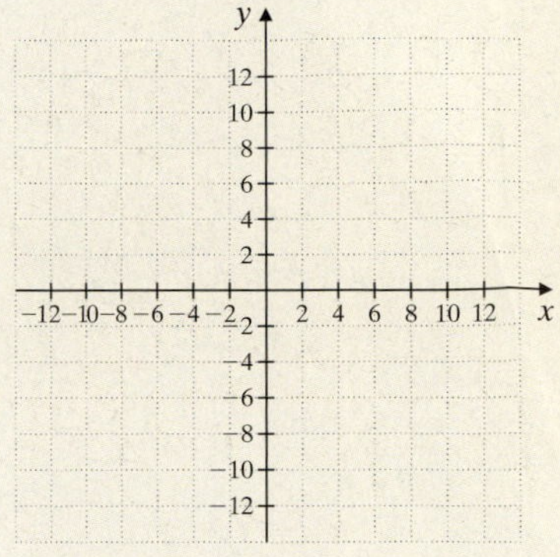

PRACTICE PROBLEM 5

Graph the linear equation $x = 3$.

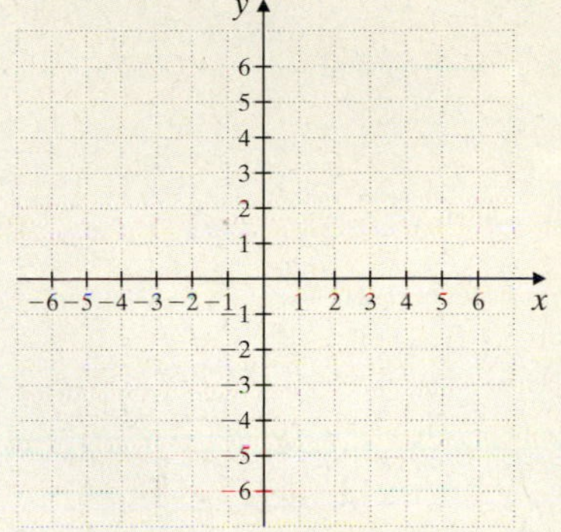

Answers

4.

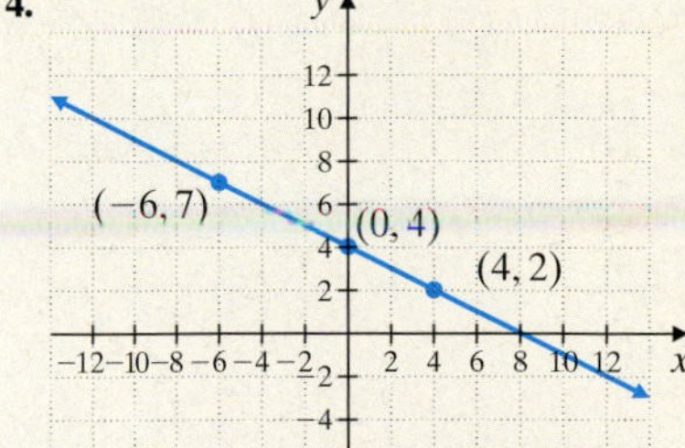

5.

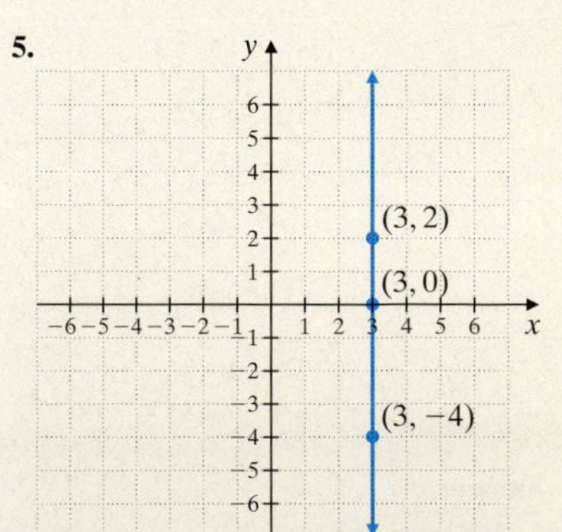

PRACTICE PROBLEM 6

Use the graph in Example 6 to predict the number of medical assistants in 2004.

EXAMPLE 6 Estimating the Number of Medical Assistants

One of the occupations expected to have the most growth in the next few years is medical assistant. The number of people y (in thousands) employed as medical assistants in the United States can be estimated by the linear equation $y = 31.8x + 180$, where x is the number of years after the year 1995. (*Source:* Based on data from the Bureau of Labor Statistics)

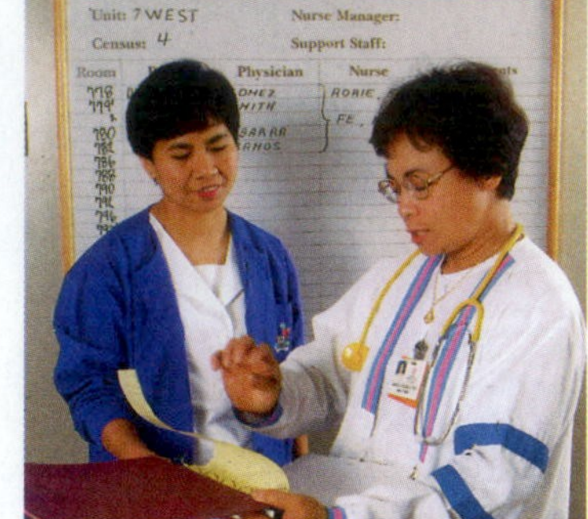

a. Graph the equation.

b. Use the graph to predict the number of medical assistants in the year 2010.

Solution:

a. To graph $y = 31.8x + 180$, choose x-values and substitute in the equation.

If $x = 0$, then $y = 31.8(0) + 180 = 180$.

If $x = 2$, then $y = 31.8(2) + 180 = 243.6$.

If $x = 7$, then $y = 31.8(7) + 180 = 402.6$.

x	y
0	180
2	243.6
7	402.6

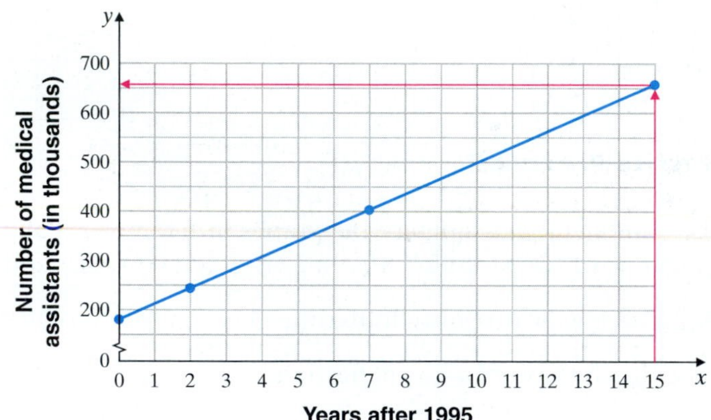

b. To use the graph to *predict* the number of medical assistants in the year 2010, we need to find the y-coordinate that corresponds to $x = 15$. (15 years after 1995 is the year 2010.) To do so, find 15 on the x-axis. Move vertically upward to the graphed line and then horizontally to the left. We approximate the number on the y-axis to be 655. Thus in the year 2010, we predict that there will be 655 thousand medical assistants. (The actual value, using 15 for x, is 657.)

Work Practice Problem 6

Helpful Hint

Make sure you understand that models are mathematical approximations of the data for the known years. (For example, see the model in Example 6.) Any number of unknown factors can affect future years, so be cautious when using models to predict.

Answer

6. 465 thousand

CALCULATOR EXPLORATIONS Graphing

In this section, we begin an optional study of graphing calculators and graphing software packages for computers. These graphers use the same point plotting technique that was introduced in this section. The advantage of this graphing technology is, of course, that graphing calculators and computers can find and plot ordered pair solutions much faster than we can. Note, however, that the features described in these boxes may not be available on all graphing calculators.

The rectangular screen where a portion of the rectangular coordinate system is displayed is called a **window.** We call it a **standard window** for graphing when both the x- and y-axes show coordinates between -10 and 10. This information is often displayed in the window menu on a graphing calculator as follows.

Xmin = -10
Xmax = 10
Xscl = 1 The scale on the x-axis is one unit per tick mark.
Ymin = -10
Ymax = 10
Yscl = 1 The scale on the y-axis is one unit per tick mark.

To use a graphing calculator to graph the equation $y = 2x + 3$, press the [Y=] key and enter the keystrokes [2] [x] [+] [3]. The top row should now read $Y_1 = 2x + 3$. Next press the [GRAPH] key, and the display should look like this:

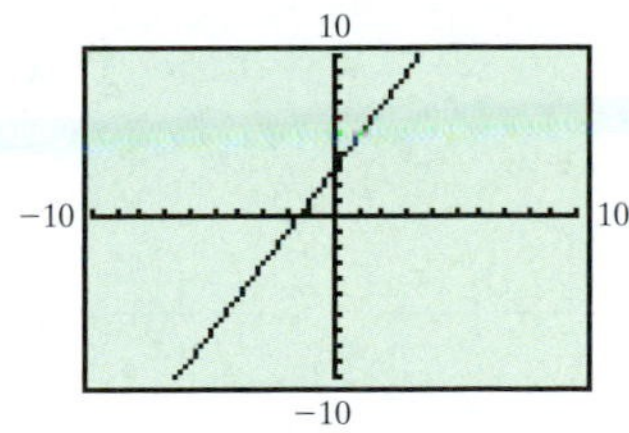

Graph the following linear equations. (Unless otherwise stated, use a standard window when graphing.)

1. $y = -3x + 7$

2. $y = -x + 5$

3. $y = 2.5x - 7.9$

4. $y = -1.3x + 5.2$

5. $y = -\frac{3}{10}x + \frac{32}{5}$

6. $y = \frac{2}{9}x - \frac{22}{3}$

13.2 EXERCISE SET

Student Solutions Manual

PH Math/Tutor Center

CD/Video for Review

MathXL®

MyMathLab

Objective A *For each equation, find three ordered pair solutions by completing the table. Then use the ordered pairs to graph the equation. See Examples 1 through 5.*

1. $x - y = 6$

x	y
	0
4	
	−1

2. $x - y = 4$

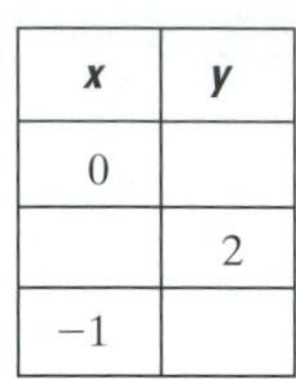

x	y
0	
	2
−1	

3. $y = -4x$

x	y
1	
0	
−1	

4. $y = -5x$

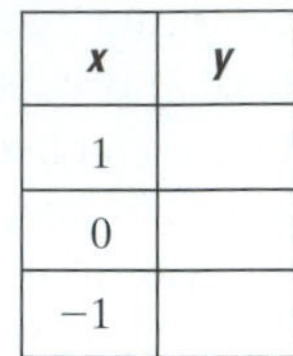

x	y
1	
0	
−1	

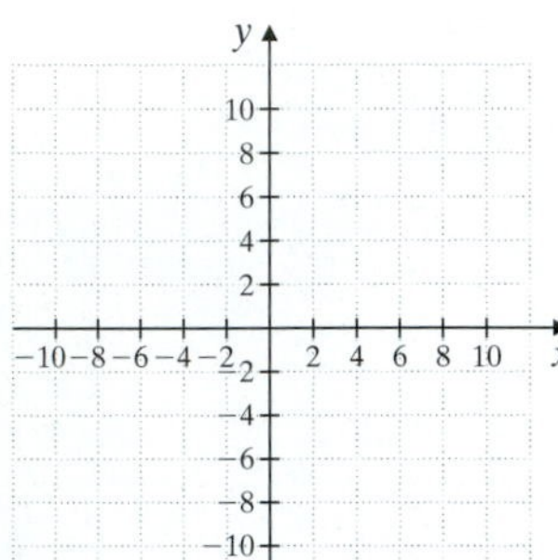

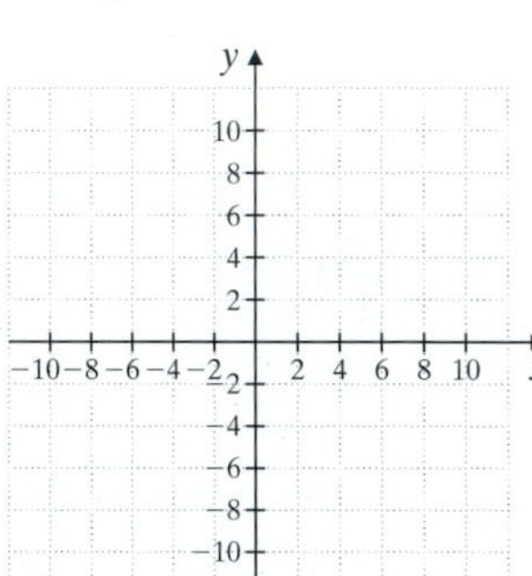

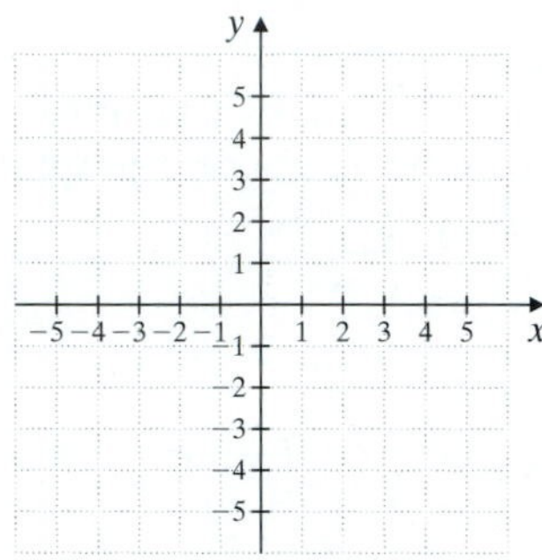

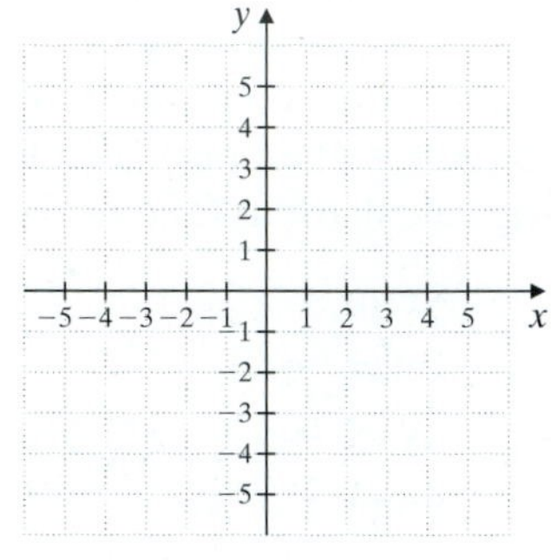

5. $y = \frac{1}{3}x$

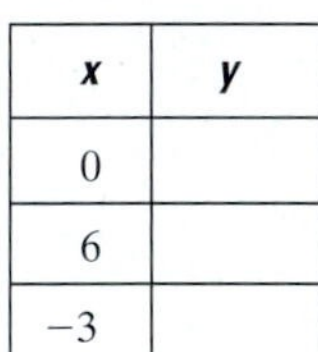

x	y
0	
6	
−3	

6. $y = \frac{1}{2}x$

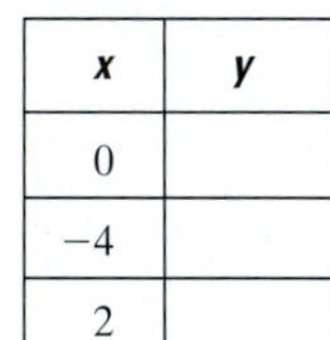

x	y
0	
−4	
2	

7. $y = -4x + 3$

x	y
0	
1	
2	

8. $y = -5x + 2$

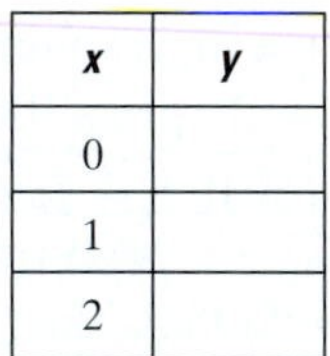

x	y
0	
1	
2	

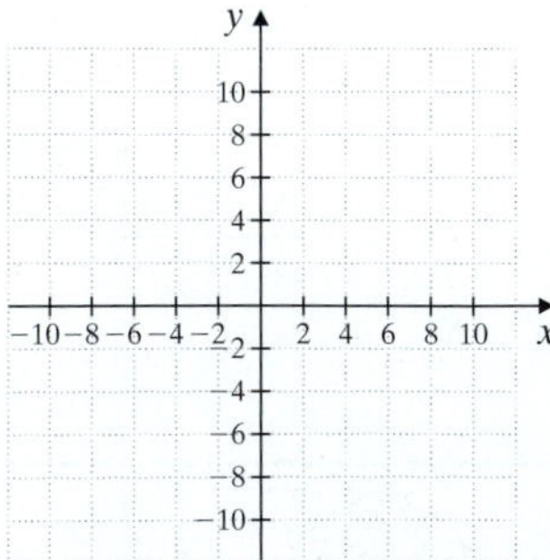

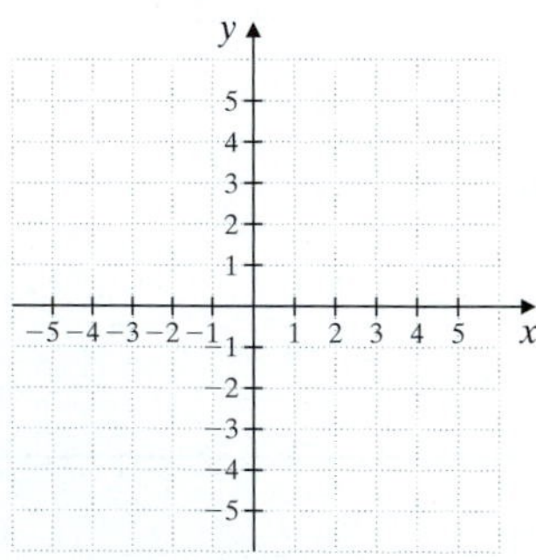

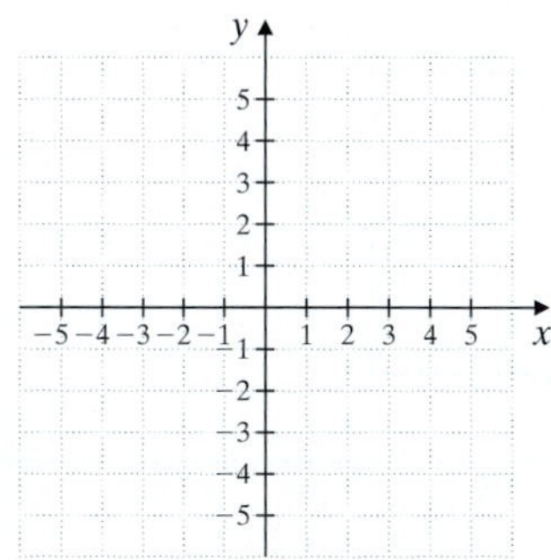

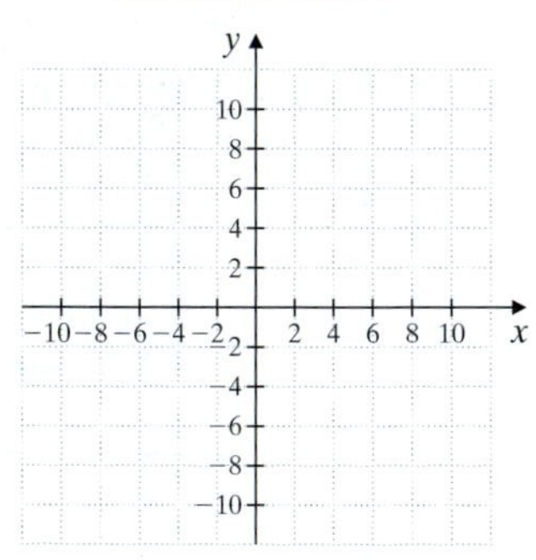

Graph each linear equation. See Examples 1 through 5.

9. $x + y = 1$

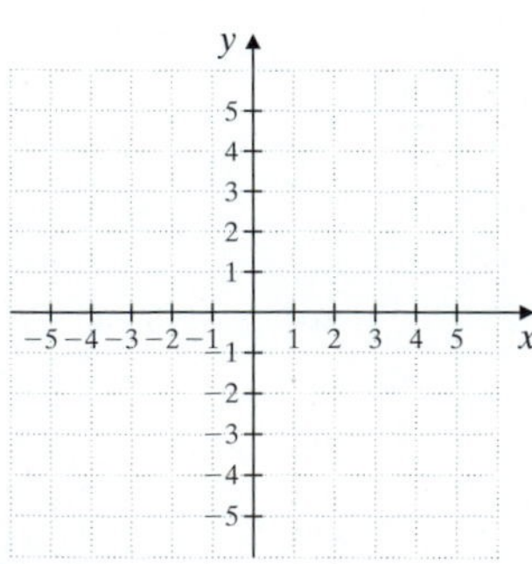

10. $x + y = 7$

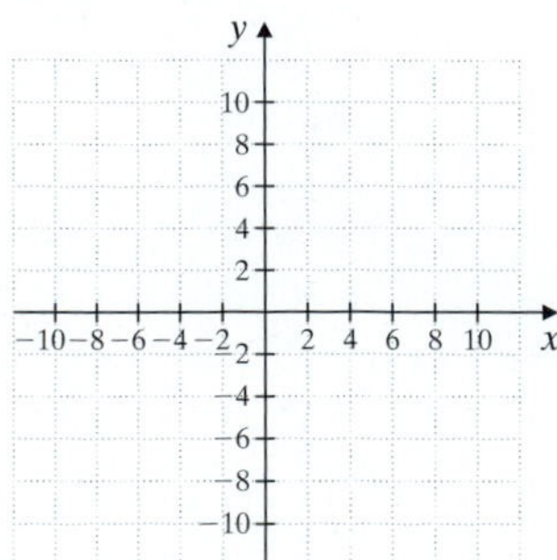

11. $x - y = -2$

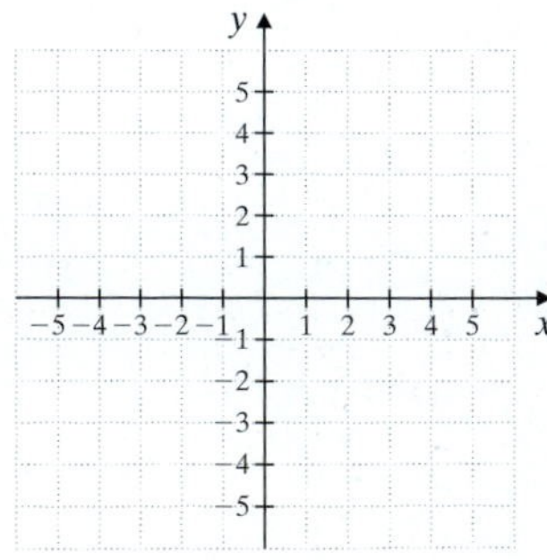

12. $-x + y = 6$

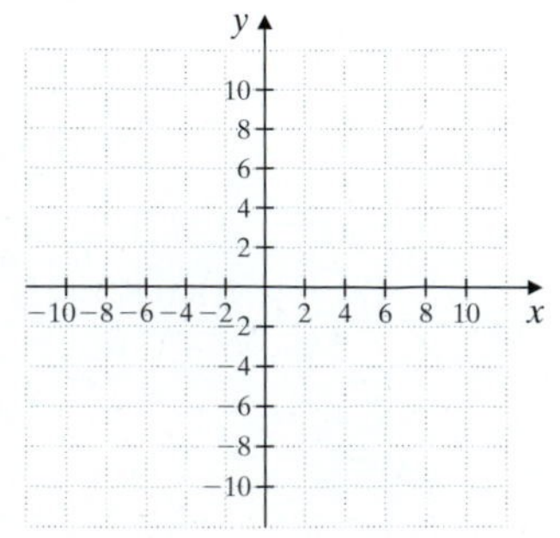

13. $x - 2y = 6$

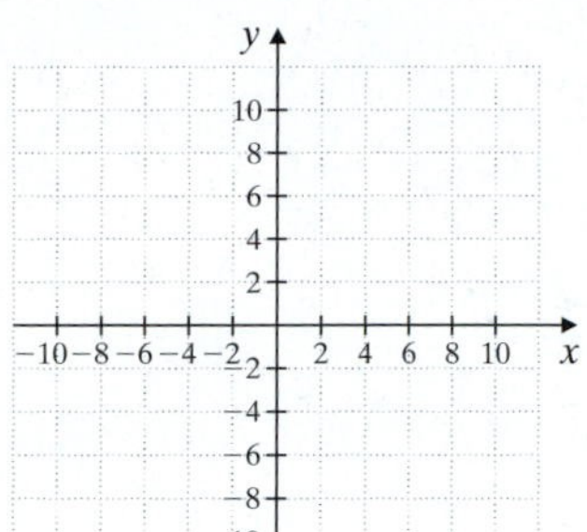

14. $-x + 5y = 5$

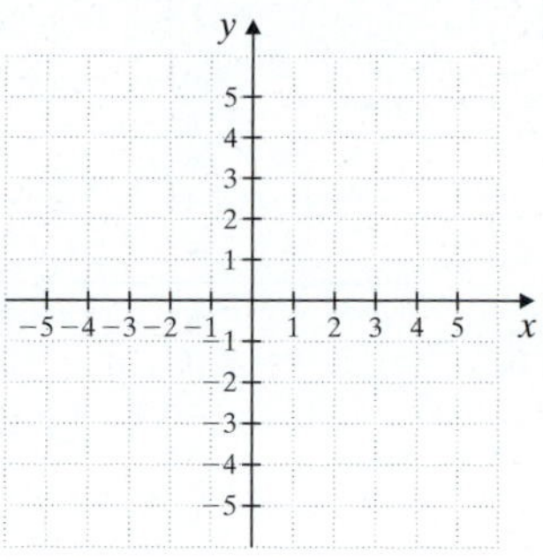

15. $y = 6x + 3$

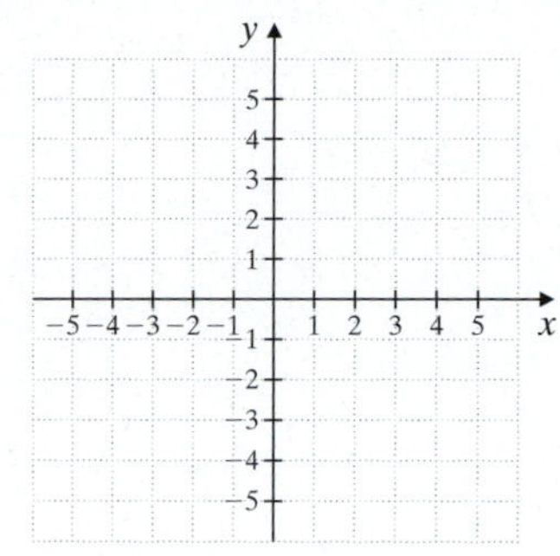

16. $y = -2x + 7$

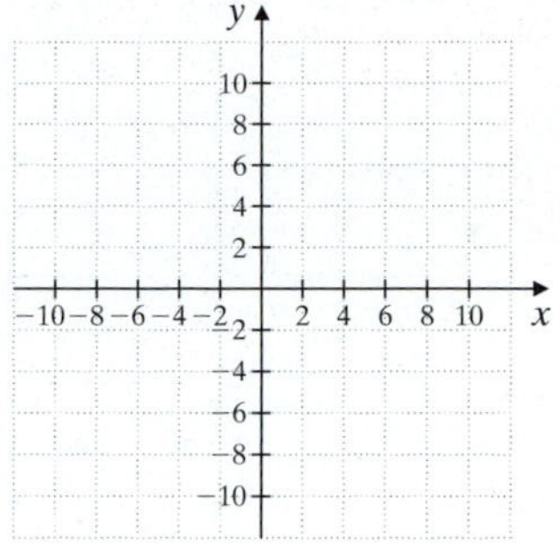

17. $x = -4$

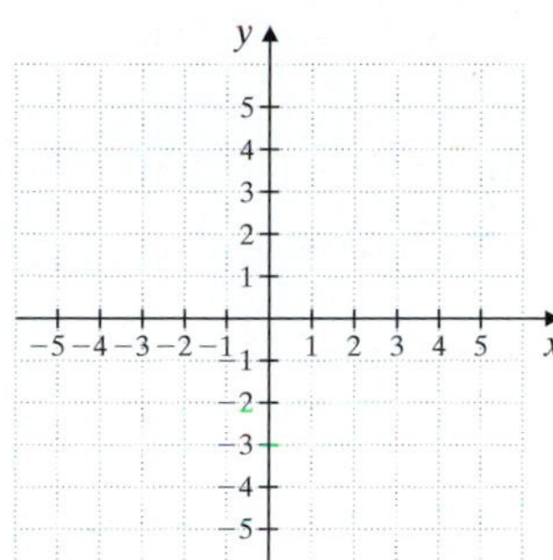

18. $y = 5$

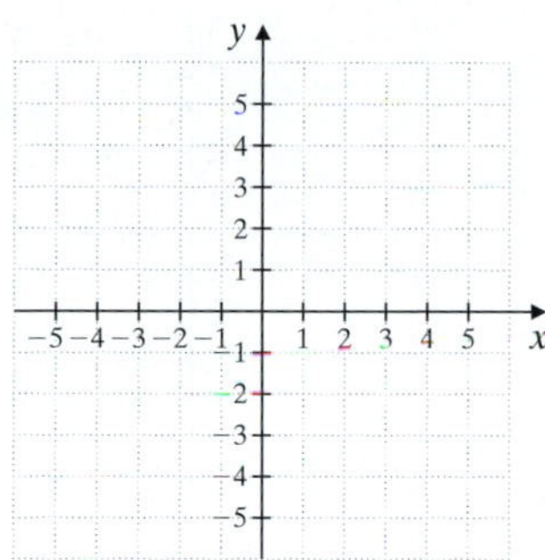

19. $y = 3$

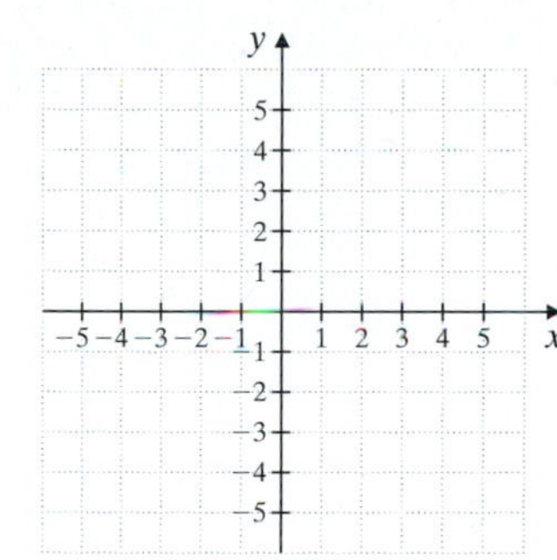

20. $x = -1$

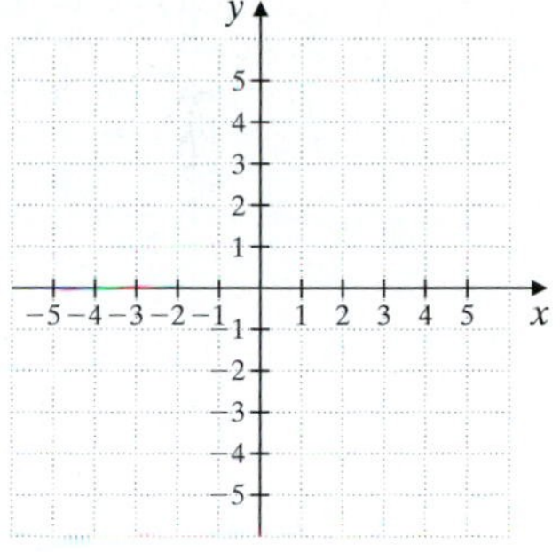

21. $y = x$

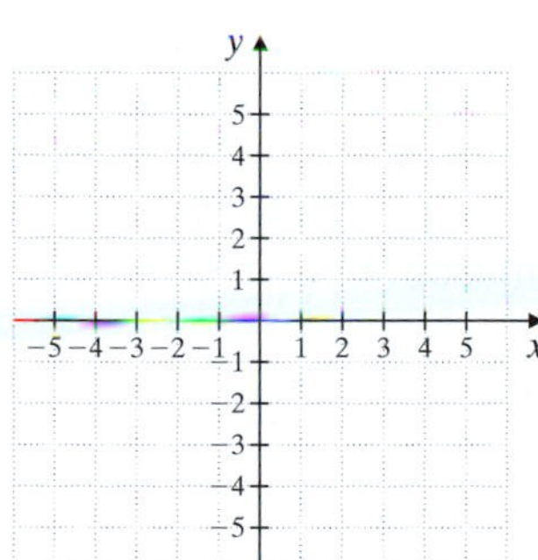

22. $y = -x$

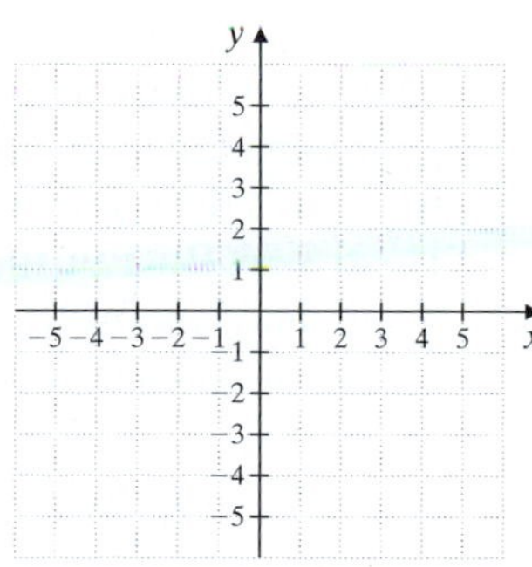

23. $y = 5x$

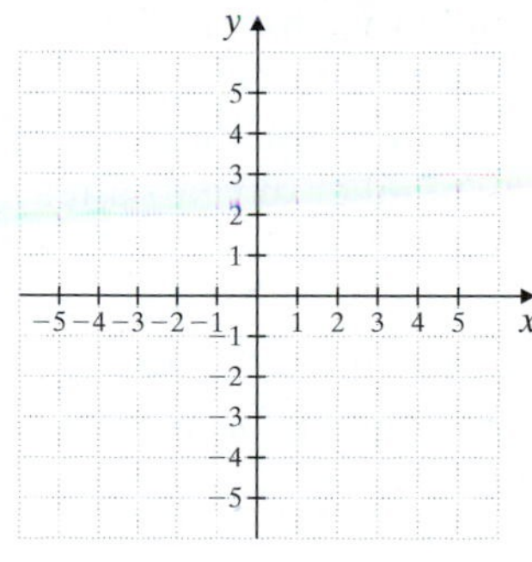

24. $y = 4x$

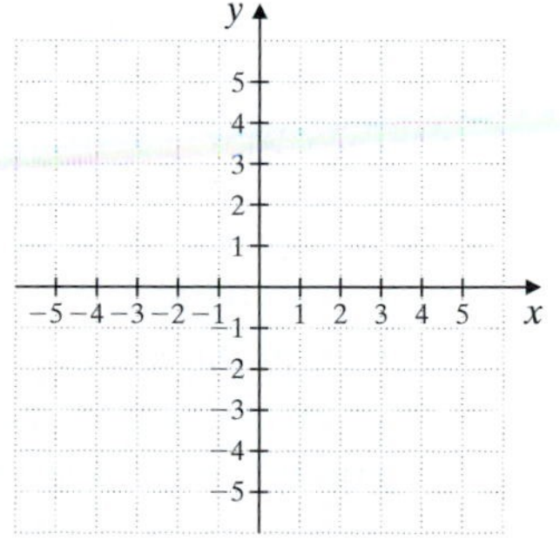

25. $x + 3y = 9$

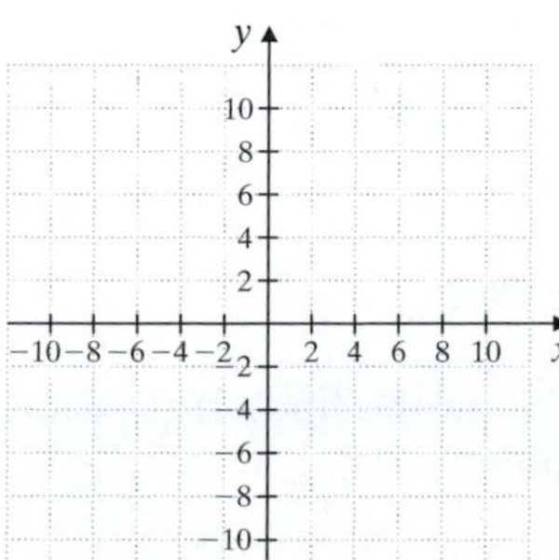

26. $2x + y = 2$

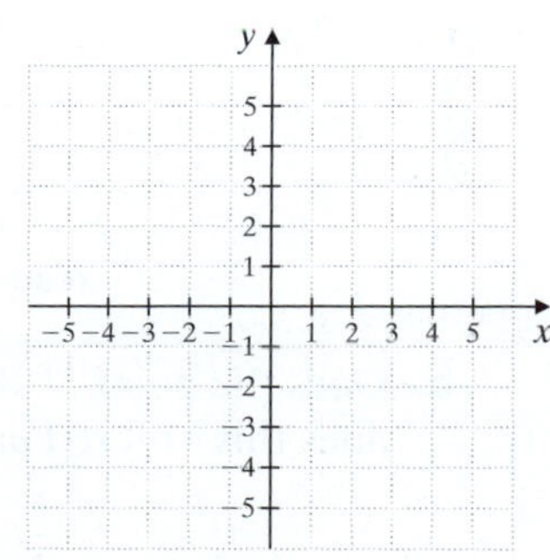

27. $y = \frac{1}{2}x - 1$

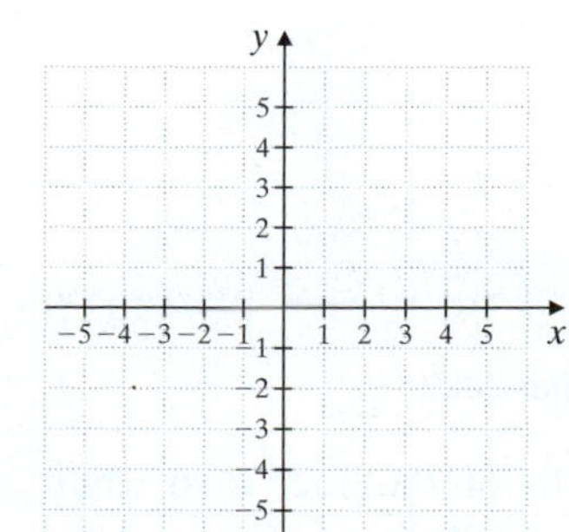

28. $y = \frac{1}{4}x + 3$

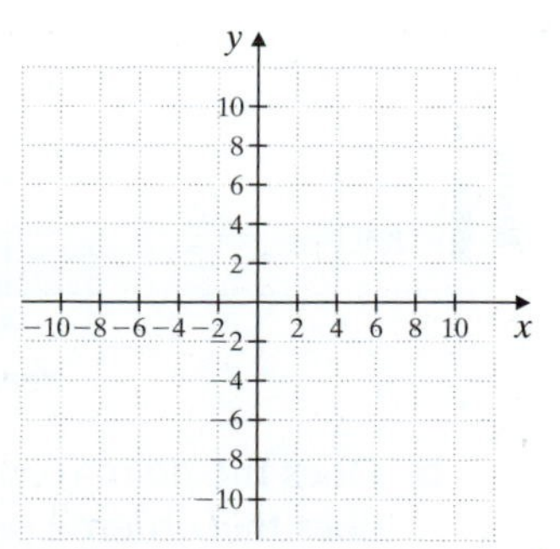

29. $3x - 2y = 12$

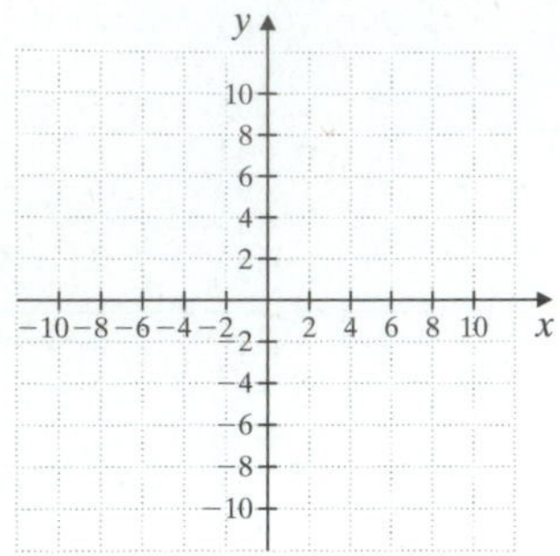

30. $2x - 7y = 14$

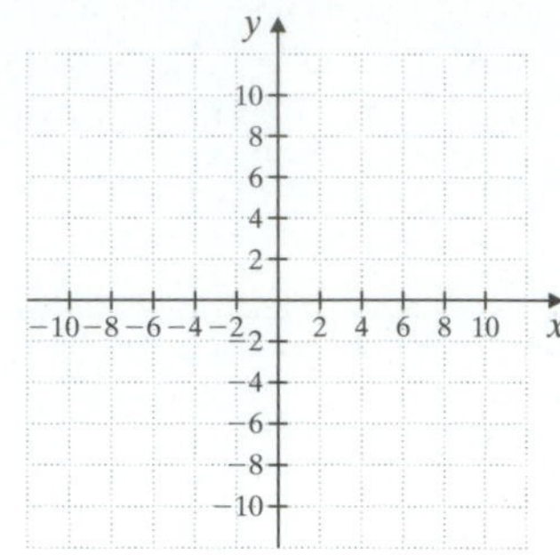

31. $y = -3.5x + 4$

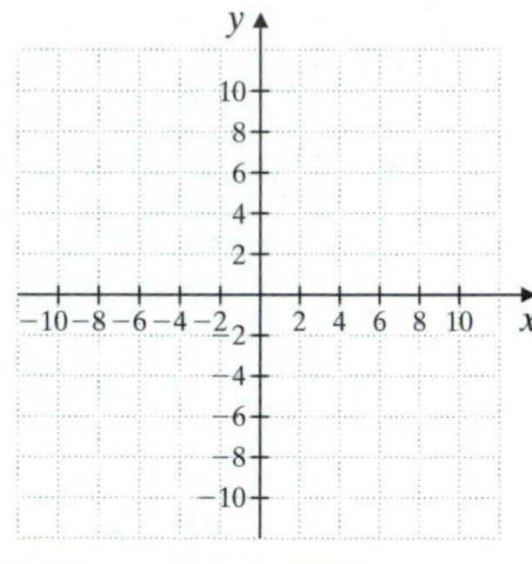

32. $y = -1.5x - 3$

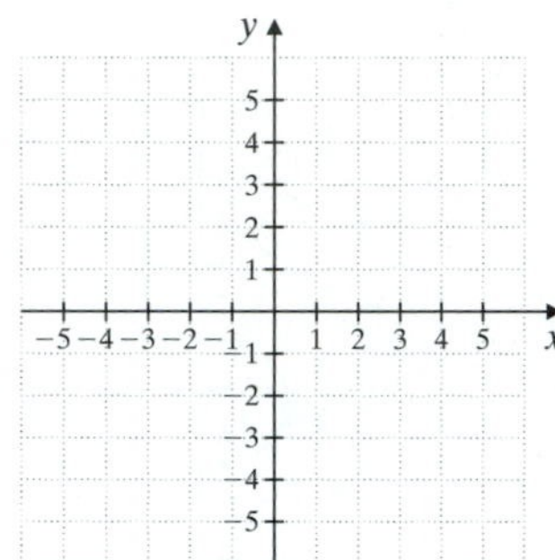

Solve. See Example 6.

33. One of the top five occupations in terms of growth in the next few years is expected to be physician's assistants. The number of people y (in hundreds) employed as physician's assistants in the United States can be estimated by the linear equation $y = 31x + 630$ where x is the number of years after 2002. (*Source:* Based on data from the Bureau of Labor Statistics)

a. Graph the linear equation. The break in the vertical axis means that the numbers between 0 and 600 have been skipped.

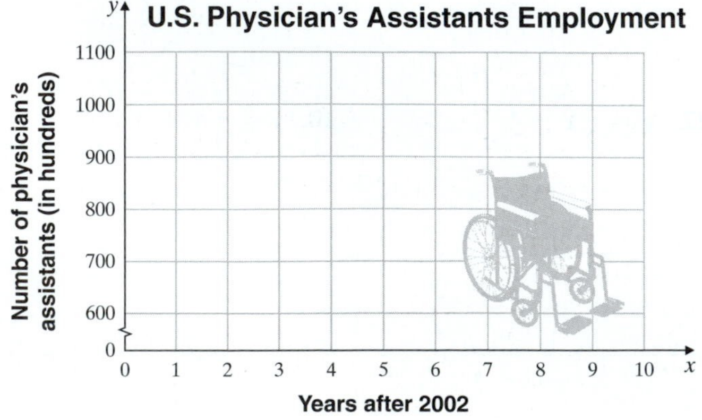

b. Does the point (6, 816) lie on the line? If so, what does this ordered pair mean?

34. Head Start is a comprehensive child development program serving young children in low-income families. The number of children y (in thousands) enrolled in Head Start from 1998 to 2003 can be approximated by the linear equation $y = 21x + 822$, where x is the number of years after 1998. (*Source:* Head Start Bureau, the Administration on Children, Youth and Families)

a. Graph the linear equation.

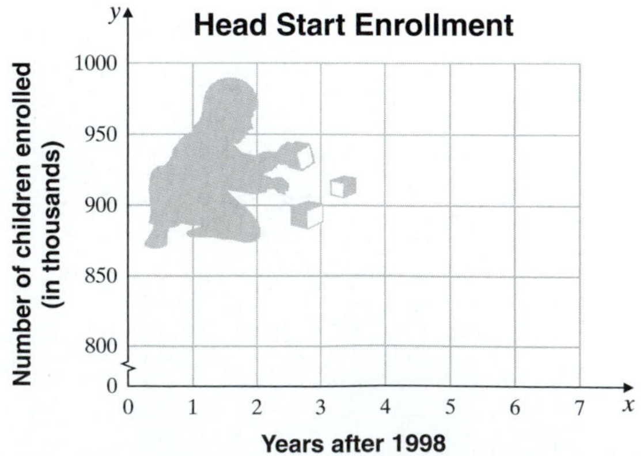

b. Does the point (3, 885) lie on the line? If so, what does this ordered pair mean?

35. The number of U.S. households y in millions that have at least one television set can be estimated by the linear equation $y = 1.5x + 99$ where x is the number of years after 1999. (*Source:* Nielsen Media Research)

a. Graph the linear equation.

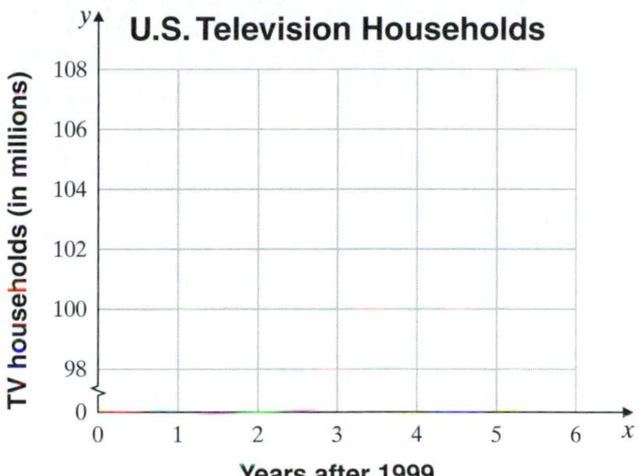

b. Complete the ordered pair (5,).

c. Write a sentence explaining the meaning of the ordered pair found in part b.

36. The restaurant industry is busier than ever. The yearly revenue for restaurants in the United States can be estimated by $y = 11.9x + 284$ where x is the number of years after 2001 and y is the revenue in billions of dollars. (*Source:* National Restaurant Assn.)

a. Graph the linear equation.

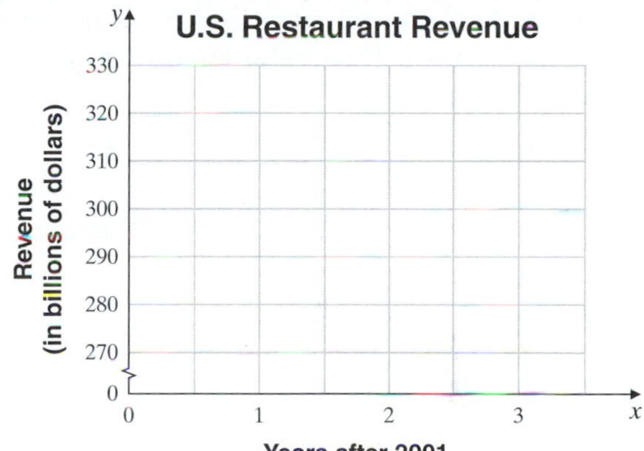

b. Complete the ordered pair (3,).

c. Write a sentence explaining the meaning of the ordered pair found in part b.

Review

△ **37.** The coordinates of three vertices of a rectangle are $(-2, 5)$, $(4, 5)$, and $(-2, -1)$. Find the coordinates of the fourth vertex. See Section 13.1.

△ **38.** The coordinates of two vertices of a square are $(-3, -1)$ and $(2, -1)$. Find the coordinates of two pairs of points possible for the third and fourth vertices. See Section 13.1.

Complete each table. See Section 13.1.

39. $x - y = -3$

x	y
0	
	0

40. $y - x = 5$

x	y
0	
	0

41. $y = 2x$

x	y
0	
	0

42. $x = -3y$

x	y
0	
	0

Concept Extensions

Graph each pair of linear equations on the same set of axes. Discuss how the graphs are similar and how they are different.

43. $y = 5x$
$y = 5x + 4$

44. $y = 2x$
$y = 2x + 5$

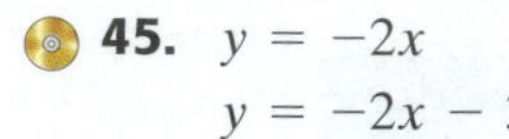
45. $y = -2x$
$y = -2x - 3$

46. $y = x$
$y = x - 7$

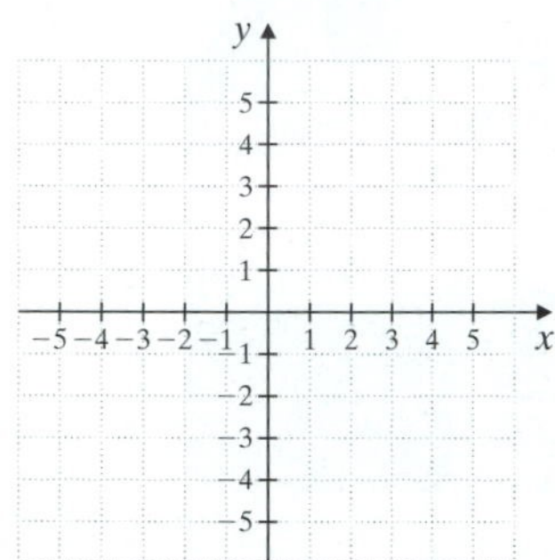

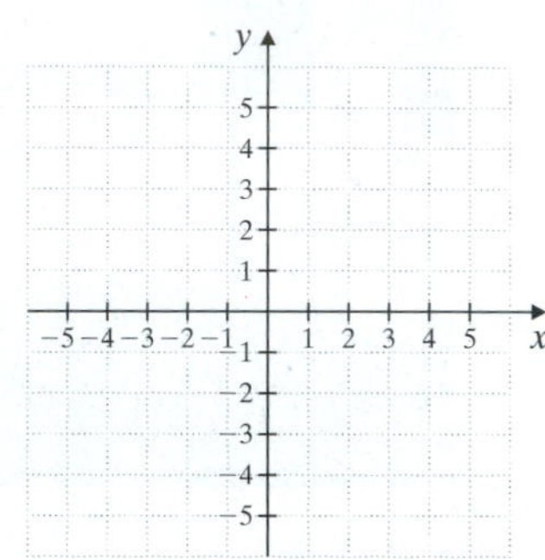

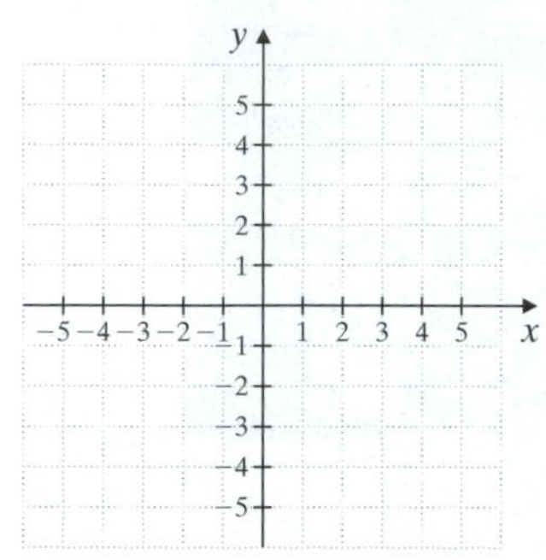

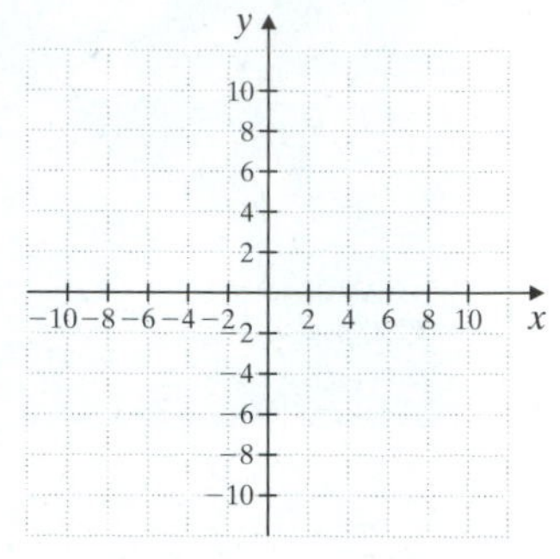

47. Graph the nonlinear equation $y = x^2$ by completing the table shown. Plot the ordered pairs and connect them with a smooth curve.

x	y
0	
1	
−1	
2	
−2	

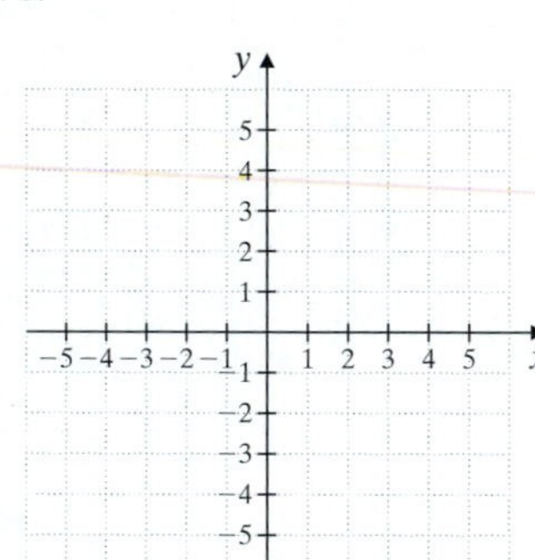

48. Graph the nonlinear equation $y = |x|$ by completing the table shown. Plot the ordered pairs and connect them. This curve is "V" shaped.

x	y
0	
1	
−1	
2	
−2	

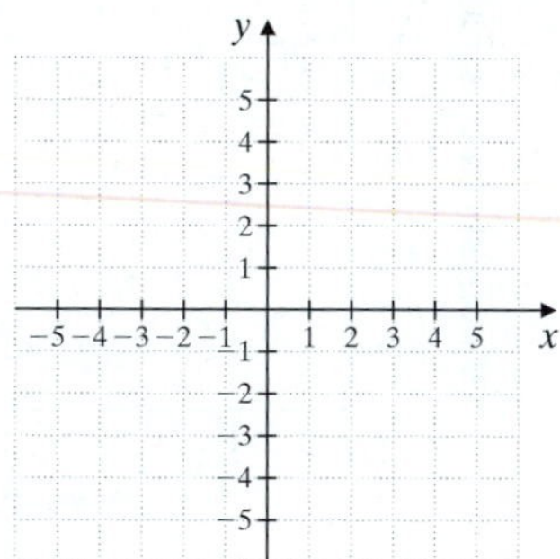

△ **49.** The perimeter of the trapezoid below is 22 centimeters. Write a linear equation in two variables for the perimeter. Find y if x is 3 centimeters.

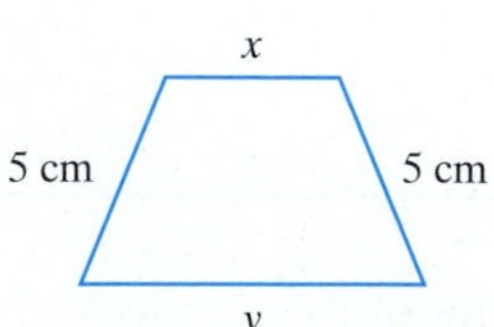

△ **50.** The perimeter of the rectangle below is 50 miles. Write a linear equation in two variables for the perimeter. Use this equation to find x when y is 20.

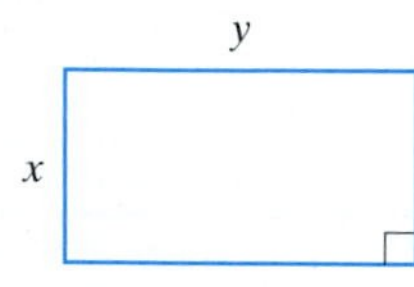

51. If (a, b) is an ordered pair solution of $x + y = 5$, is (b, a) also a solution? Explain why or why not.

13.3 INTERCEPTS

Objective A Identifying Intercepts

The graph of $y = 4x - 8$ is shown below. Notice that this graph crosses the y-axis at the point $(0, -8)$. This point is called the ***y*-intercept.** Likewise the graph crosses the x-axis at $(2, 0)$. This point is called the ***x*-intercept.**

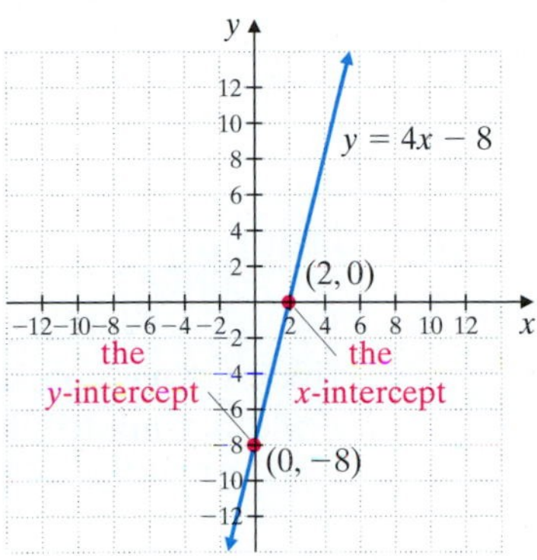

The intercepts are $(2, 0)$ and $(0, -8)$.

Helpful Hint

If a graph crosses the x-axis at $(2, 0)$ and the y-axis at $(0, -8)$, then

$(2, 0)$ ↑ x-intercept $(0, -8)$ ↑ y-intercept

Notice that for the x-intercept, the y-value is 0 and for the y-intercept, the x-value is 0.

Note: Sometimes in mathematics, you may see just the number -8 stated as the y-intercept, and 2 stated as the x-intercept.

EXAMPLES Identify the x- and y-intercepts.

1.

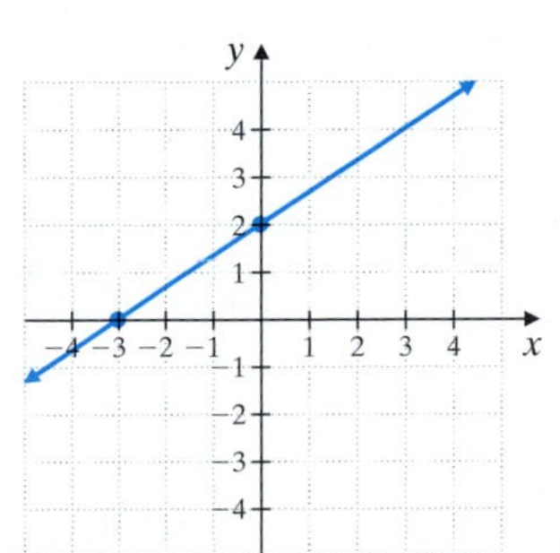

Solution:

x-intercept: $(-3, 0)$

y-intercept: $(0, 2)$

Continued on next page

Objectives

A Identify Intercepts of a Graph.

B Graph a Linear Equation by Finding and Plotting Intercept Points.

C Identify and Graph Vertical and Horizontal Lines.

PRACTICE PROBLEM 1

Identify the x- and y-intercepts.

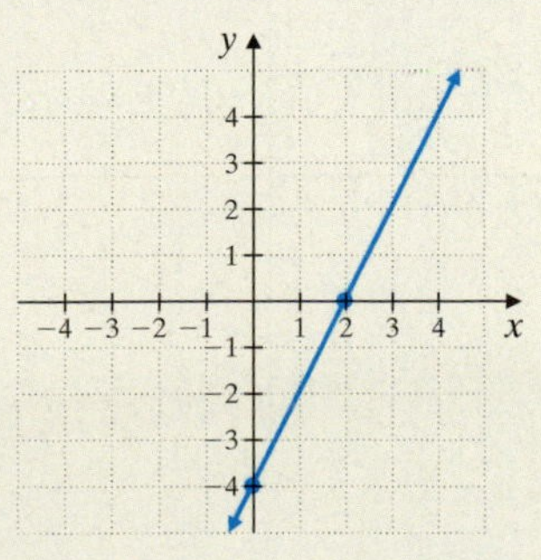

Answer

1. x-intercept: $(2, 0)$; y-intercept: $(0, -4)$

PRACTICE PROBLEMS 2–3

Identify the x- and y-intercepts.

2.

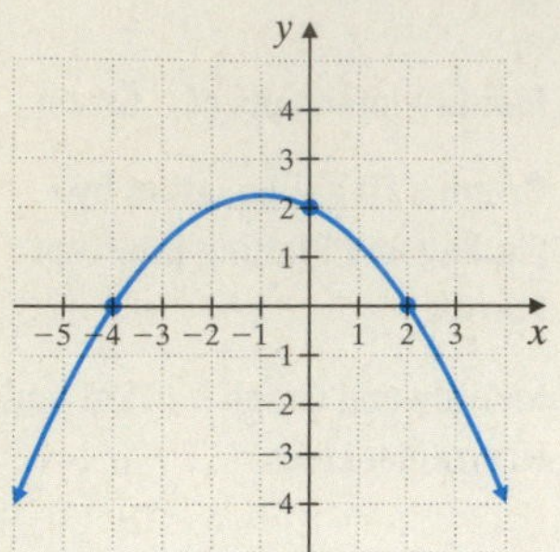

3.

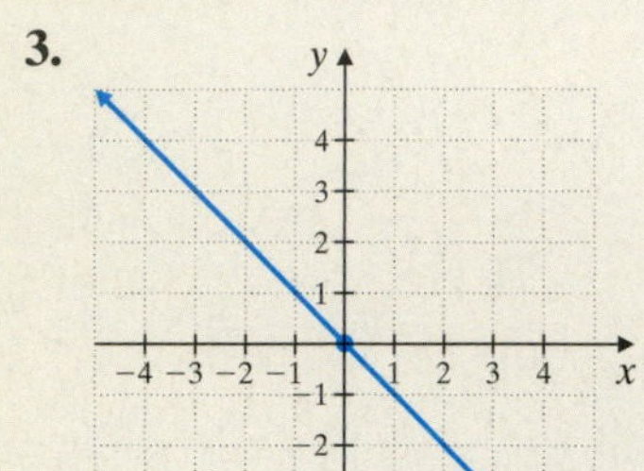

2.

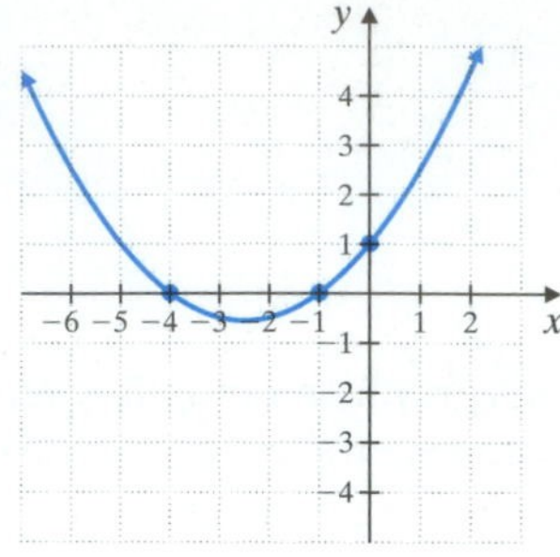

Solution:

x-intercepts: $(-4, 0), (-1, 0)$

y-intercept: $(0, 1)$

3.

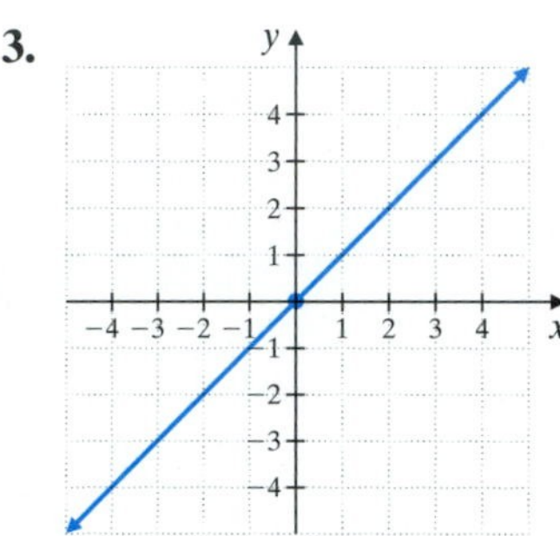

Helpful Hint

Notice that any time $(0, 0)$ is a point of a graph, then it is an x-intercept and a y-intercept. Why? It is the *only* point that lies on both axes.

Solution:

x-intercept: $(0, 0)$

y-intercept: $(0, 0)$

Here, the x- and y-intercept happen to be the same point.

Work Practice Problems 1–3

Objective B Finding and Plotting Intercepts

Given an equation of a line, we can usually find intercepts easily since one coordinate is 0.

To find the x-intercept of a line from its equation, let $y = 0$, since a point on the x-axis has a y-coordinate of 0. To find the y-intercept of a line from its equation, let $x = 0$, since a point on the y-axis has an x-coordinate of 0.

PRACTICE PROBLEM 4

Graph $2x - y = 4$ by finding and plotting its intercepts.

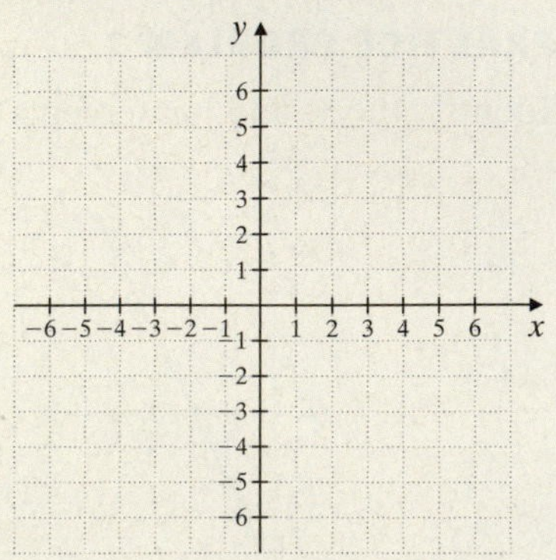

Finding x- and y-Intercepts

To find the x-intercept, let $y = 0$ and solve for x.
To find the y-intercept, let $x = 0$ and solve for y.

EXAMPLE 4 Graph $x - 3y = 6$ by finding and plotting its intercepts.

Solution: We let $y = 0$ to find the x-intercept and $x = 0$ to find the y-intercept.

Let $y = 0$.	Let $x = 0$.
$x - 3y = 6$	$x - 3y = 6$
$x - 3(0) = 6$	$0 - 3y = 6$
$x - 0 = 6$	$-3y = 6$
$x = 6$	$y = -2$

The x-intercept is $(6, 0)$ and the y-intercept is $(0, -2)$. We find a third ordered pair solution to check our work. If we let $y = -1$, then $x = 3$. We plot the points $(6, 0)$,

Answers

2. x-intercepts: $(-4, 0)$ $(2, 0)$; y-intercept: $(0, 2)$
3. x-intercept and y-intercept: $(0, 0)$
4. See page 973.

(0, −2), and (3, −1). The graph of $x - 3y = 6$ is the line drawn through these points as shown.

x	y
6	0
0	−2
3	−1

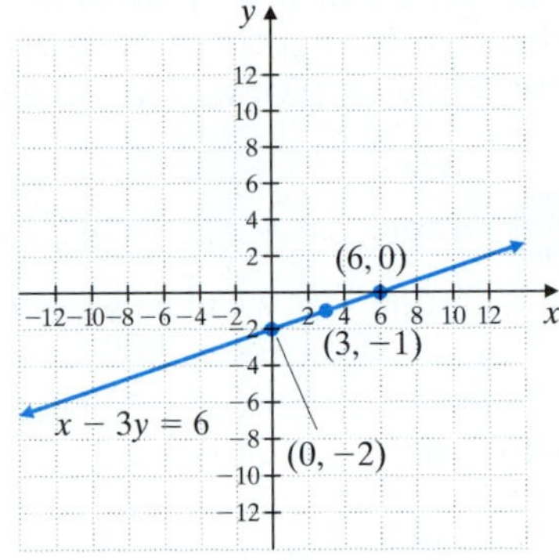

Work Practice Problem 4

EXAMPLE 5 Graph $x = -2y$ by finding and plotting its intercepts.

Solution: We let $y = 0$ to find the x-intercept and $x = 0$ to find the y-intercept.

Let $y = 0$.
$$x = -2y$$
$$x = -2(0)$$
$$x = 0$$

Let $x = 0$.
$$x = -2y$$
$$0 = -2y$$
$$0 = y$$

Both the x-intercept and y-intercept are (0, 0). In other words, when $x = 0$, then $y = 0$, which gives the ordered pair (0, 0). Also, when $y = 0$, then $x = 0$, which gives the same ordered pair (0, 0). This happens when the graph passes through the origin. Since two points are needed to determine a line, we must find at least one more ordered pair that satisfies $x = -2y$. Since the equation is solved for x, we choose y-values so that there is no need to solve to find the corresponding x-value. We let $y = -1$ to find a second ordered pair solution and let $y = 1$ as a check point.

Let $y = -1$.
$$x = -2(-1)$$
$$x = 2 \quad \text{Multiply.}$$

Let $y = 1$.
$$x = -2(1)$$
$$x = -2 \quad \text{Multiply.}$$

The ordered pairs are (0, 0), (2, −1), and (−2, 1). We plot these points to graph $x = -2y$.

x	y
0	0
2	−1
−2	1

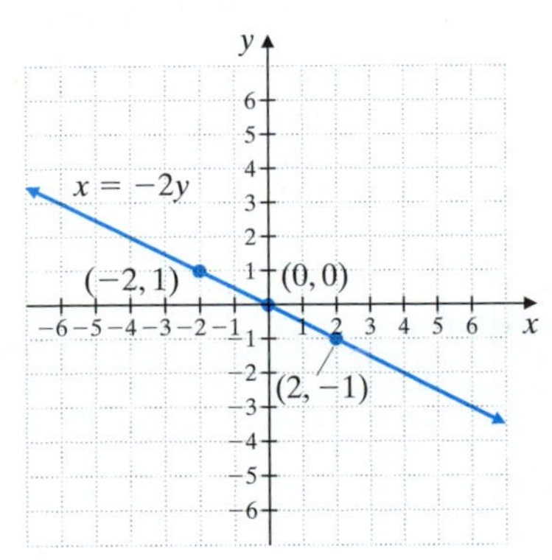

Work Practice Problem 5

PRACTICE PROBLEM 5

Graph $y = 3x$ by finding and plotting its intercepts.

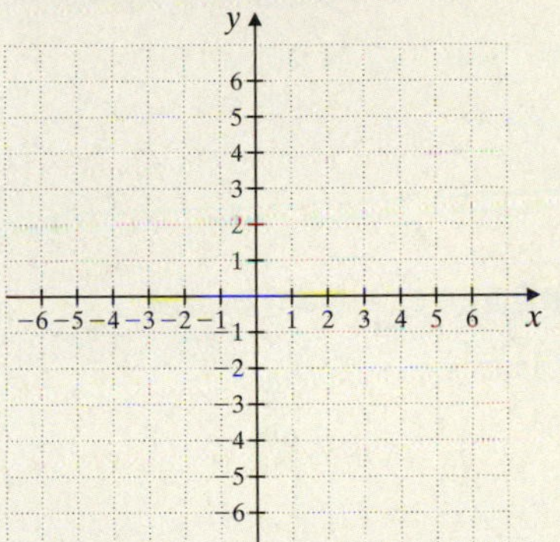

Answers

4.

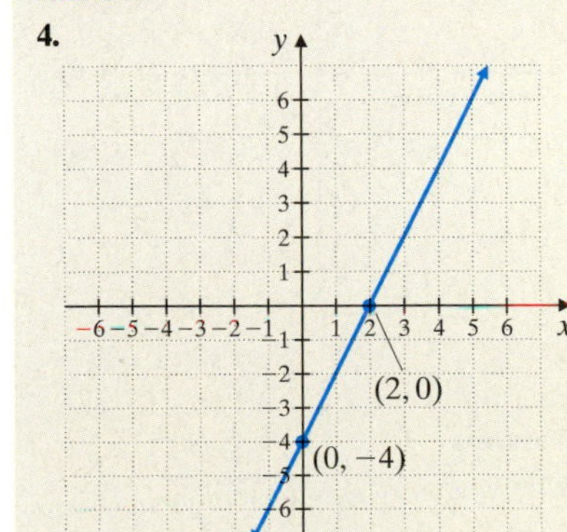

5.

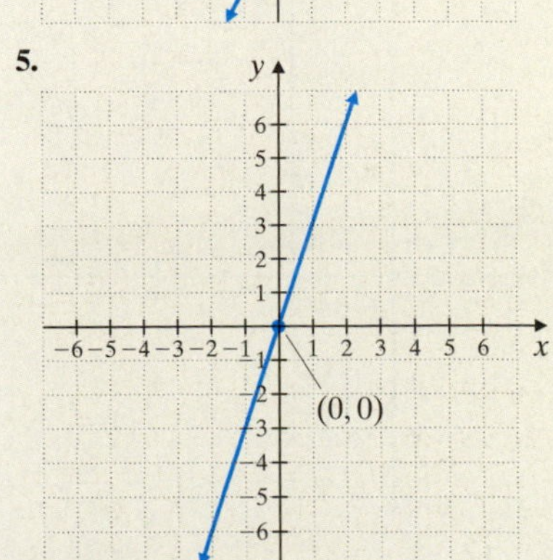

Objective C Graphing Vertical and Horizontal Lines

From Section 13.2, recall that the equation $x = 2$, for example, is a linear equation in two variables because it can be written in the form $x + 0y = 2$. The graph of this equation is a vertical line, as reviewed in the next example.

PRACTICE PROBLEM 6

Graph: $x = -3$

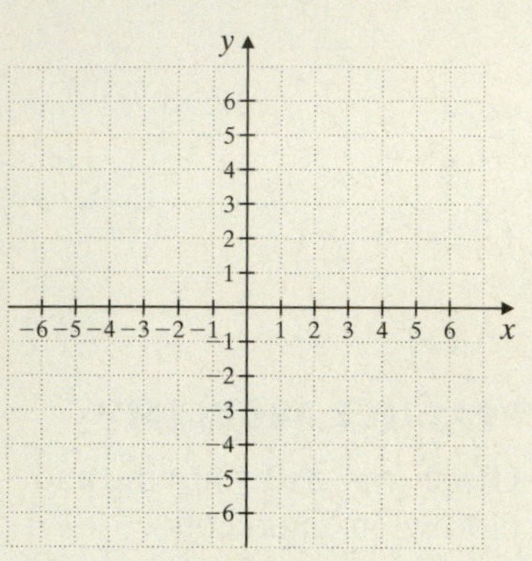

EXAMPLE 6 Graph: $x = 2$

Solution: The equation $x = 2$ can be written as $x + 0y = 2$. For any y-value chosen, notice that x is 2. No other value for x satisfies $x + 0y = 2$. Any ordered pair whose x-coordinate is 2 is a solution of $x + 0y = 2$. We will use the ordered pair solutions (2, 3), (2, 0), and (2, −3) to graph $x = 2$.

x	y
2	3
2	0
2	−3

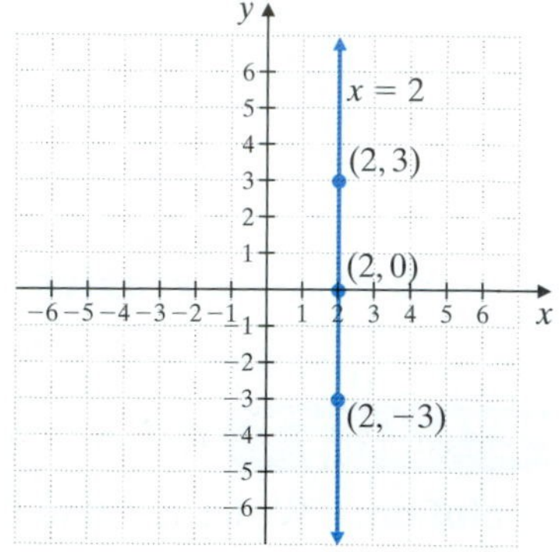

The graph is a vertical line with x-intercept 2. Note that this graph has no y-intercept because x is never 0.

Work Practice Problem 6

PRACTICE PROBLEM 7

Graph: $y = 4$

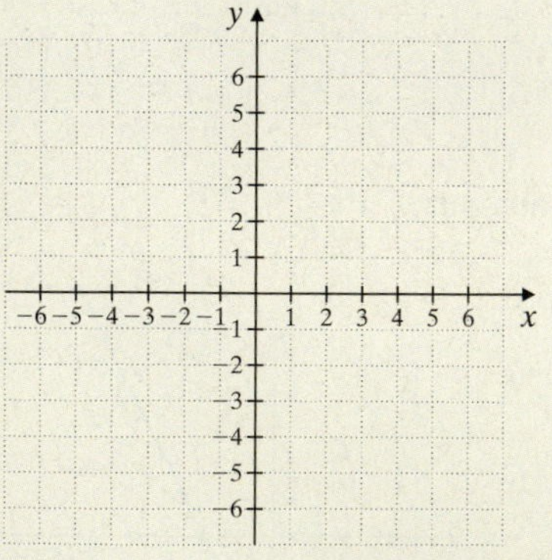

In general, we have the following.

Vertical Lines

The graph of $x = c$, where c is a real number, is a **vertical line** with x-intercept $(c, 0)$.

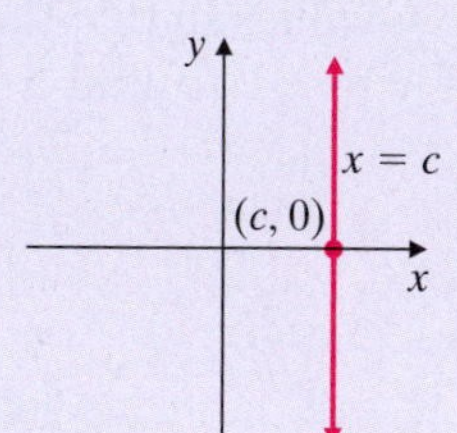

EXAMPLE 7 Graph: $y = -3$

Solution: The equation $y = -3$ can be written as $0x + y = -3$. For any x-value chosen, y is −3. If we choose 4, 1, and −2 as x-values, the ordered pair solutions are (4, −3), (1, −3), and (−2, −3). We use these ordered pairs to graph $y = -3$. The graph is a horizontal line with y-intercept −3 and no x-intercept.

x	y
4	−3
1	−3
−2	−3

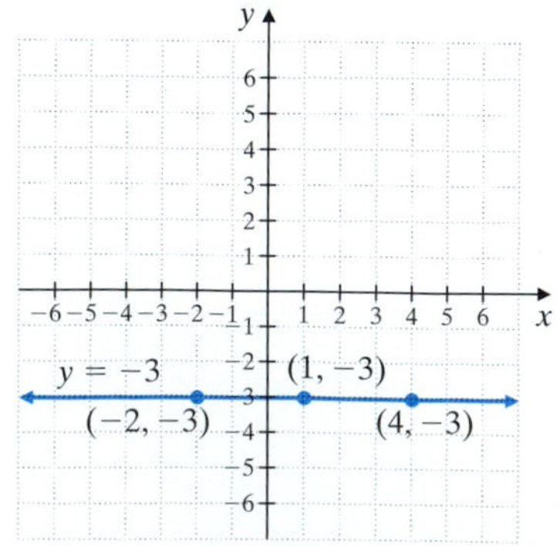

Work Practice Problem 7

Answers

6.

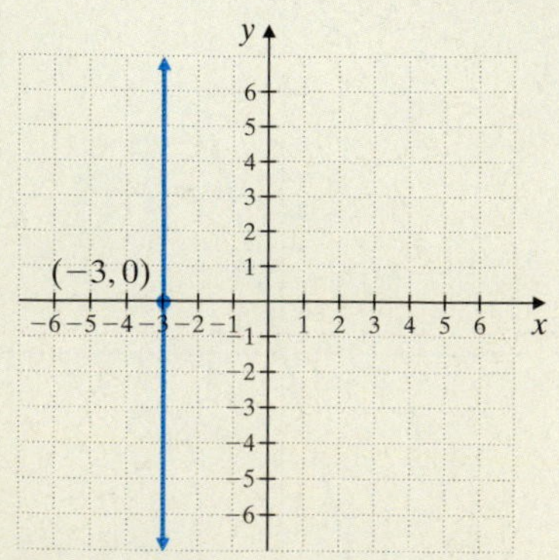

7. See page 975.

In general, we have the following.

Horizontal Lines

The graph of $y = c$, where c is a real number, is a **horizontal line** with y-intercept $(0, c)$.

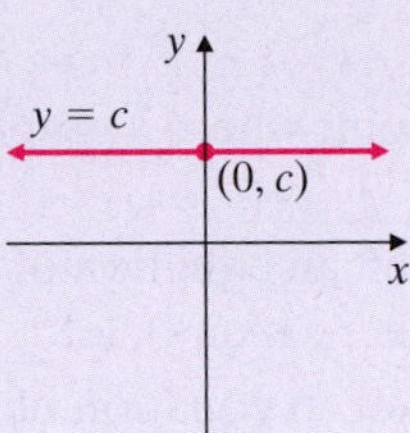

Answer

7.

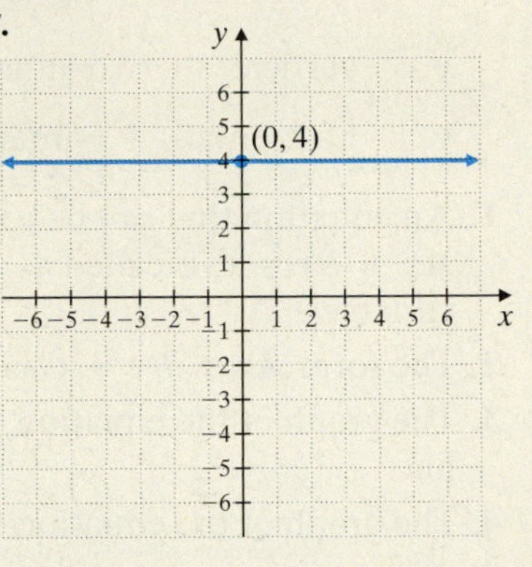

CALCULATOR EXPLORATIONS Graphing

You may have noticed that to use the $\boxed{Y=}$ key on a graphing calculator to graph an equation, the equation must be solved for y. For example, to graph $2x + 3y = 7$, we solve this equation for y.

$$2x + 3y = 7$$

$$3y = -2x + 7 \quad \text{Subtract } 2x \text{ from both sides.}$$

$$\frac{3y}{3} = -\frac{2x}{3} + \frac{7}{3} \quad \text{Divide both sides by 3.}$$

$$y = -\frac{2}{3}x + \frac{7}{3} \quad \text{Simplify.}$$

To graph $2x + 3y = 7$ or $y = -\frac{2}{3}x + \frac{7}{3}$, press the $\boxed{Y=}$ key and enter

$$Y_1 = -\frac{2}{3}x + \frac{7}{3}$$

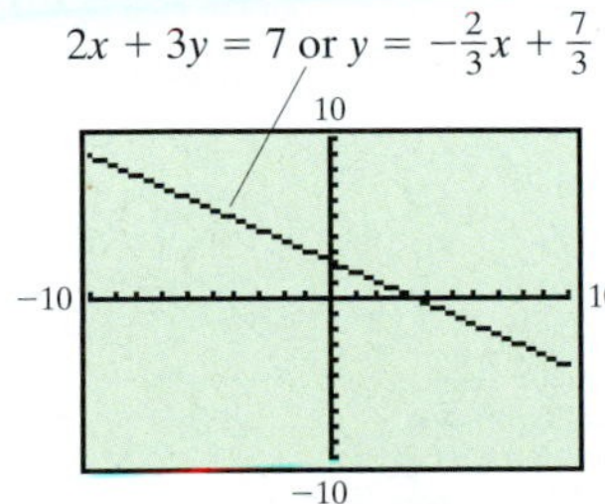

Graph each linear equation.

1. $x = 3.78y$

2. $-2.61y = x$

3. $3x + 7y = 21$

4. $-4x + 6y = 12$

5. $-2.2x + 6.8y = 15.5$

6. $5.9x - 0.8y = -10.4$

Vocabulary and Readiness Check

Use the choices below to fill in each blank. Some choices may be used more than once. Exercises 1 and 2 come from Section 13.2.

x	vertical	x-intercept	linear
y	horizontal	y-intercept	standard

1. An equation that can be written in the form $Ax + By = C$ is called a ________ equation in two variables.
2. The form $Ax + By = C$ is called __________ form.
3. The graph of the equation $y = -1$ is a ___________ line.
4. The graph of the equation $x = 5$ is a _________ line.
5. A point where a graph crosses the y-axis is called a(n) ___________.
6. A point where a graph crosses the x-axis is called a(n) ___________.
7. Given an equation of a line, to find the x-intercept (if there is one), let _____ = 0 and solve for _____.
8. Given an equation of a line, to find the y-intercept (if there is one), let _____ = 0 and solve for _____.

13.3 EXERCISE SET

FOR EXTRA HELP

Student Solutions Manual　PH Math/Tutor Center

CD/Video for Review

MathXL®

MyMathLab

Objective A *Identify the intercepts. See Examples 1 through 3.*

1.

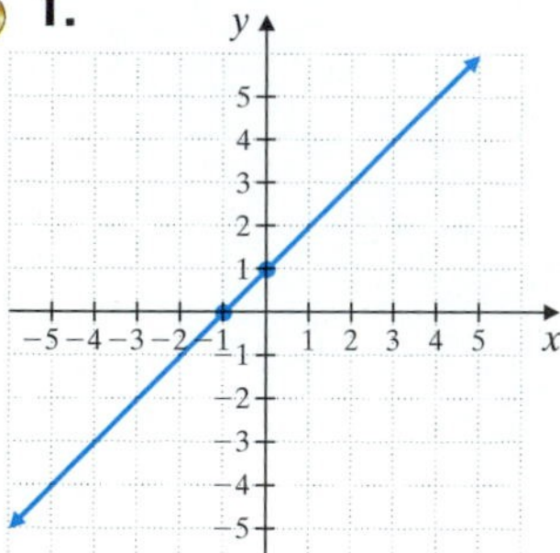

2.

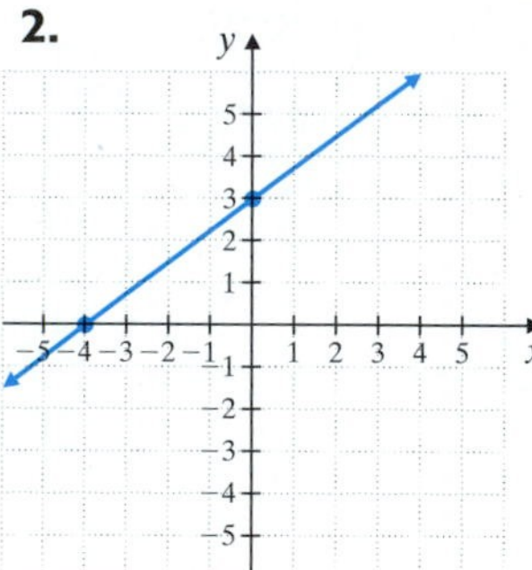

3.

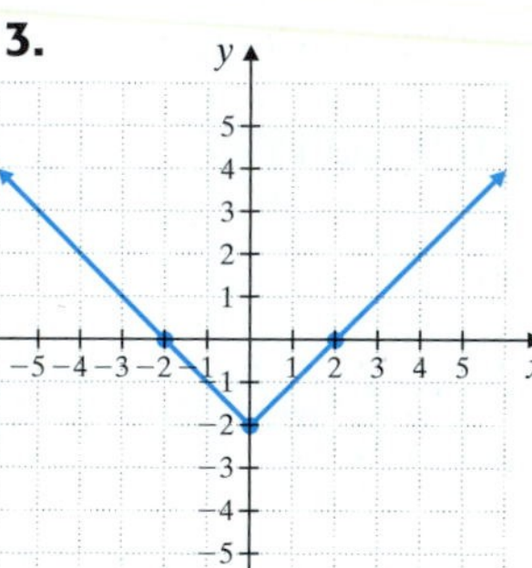

4.

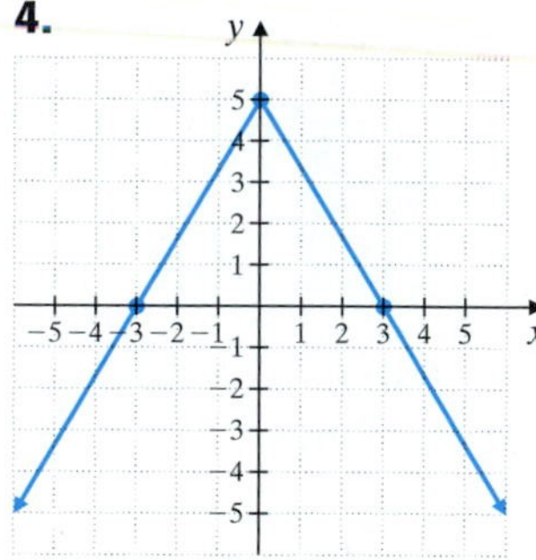

5.

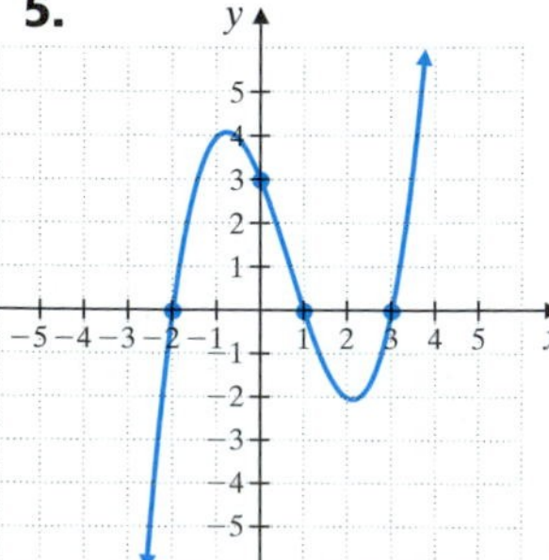

6.

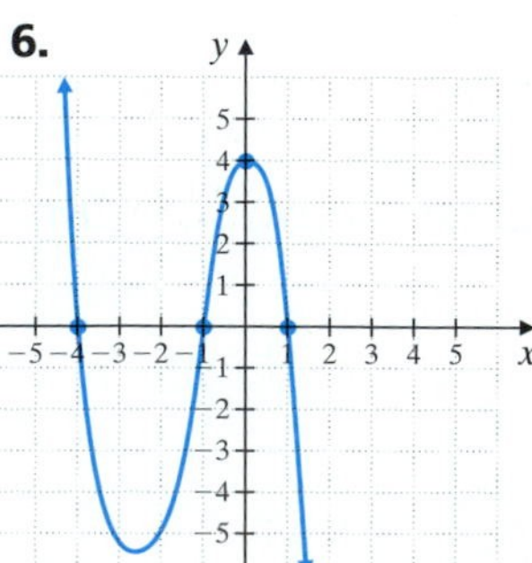

7.

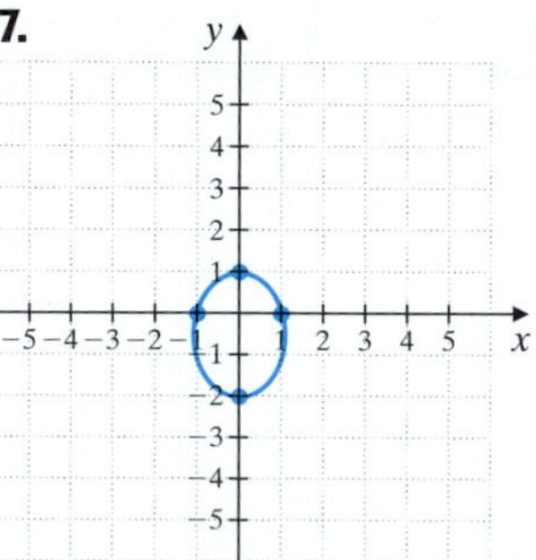

8.

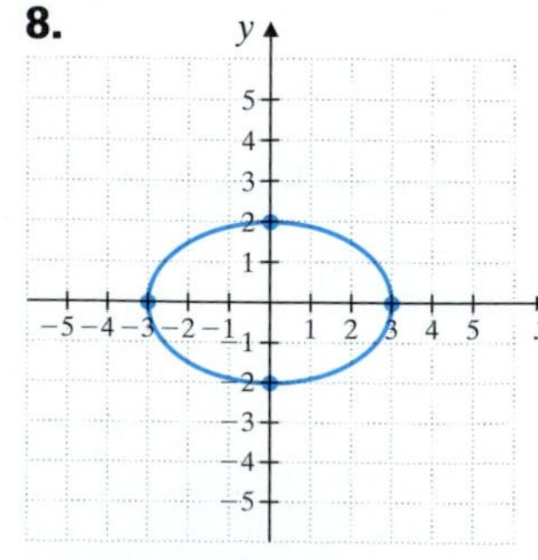

Objective B *Graph each linear equation by finding and plotting its intercepts. See Examples 4 and 5.*

9. $x - y = 3$

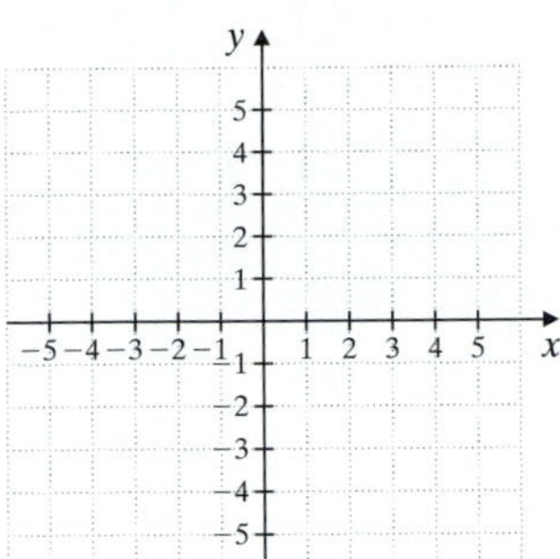

10. $x - y = -4$

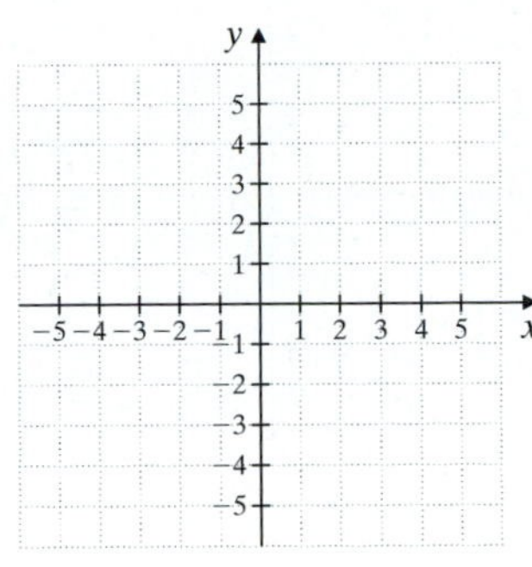

11. $x = 5y$

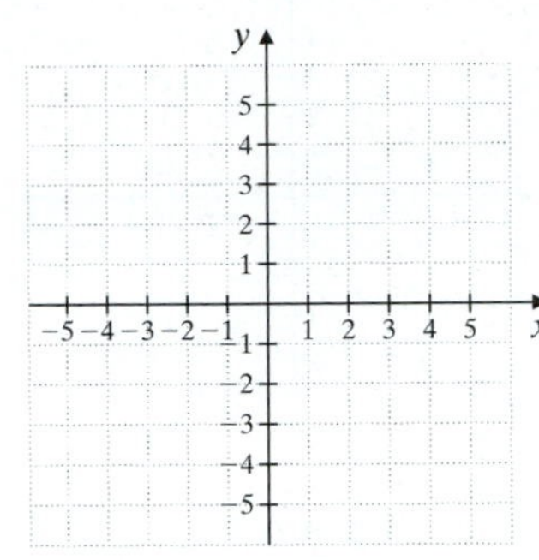

12. $x = 2y$

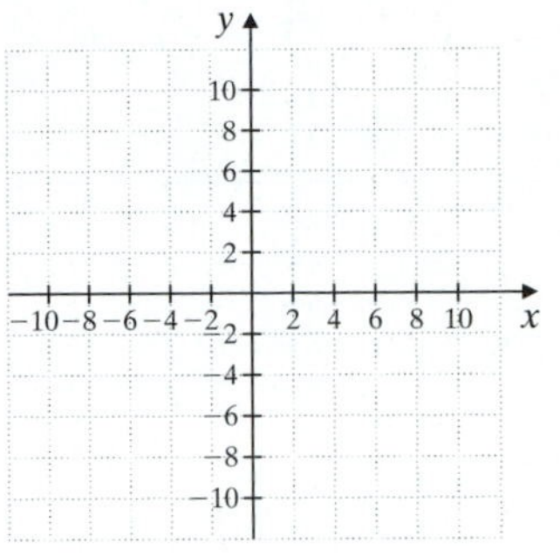

13. $-x + 2y = 6$

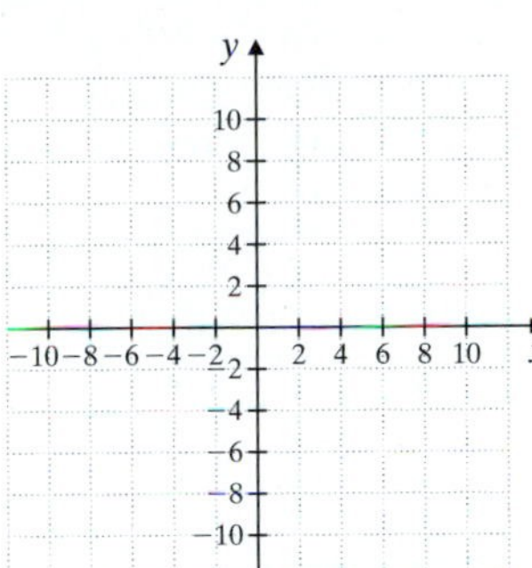

14. $x - 2y = -8$

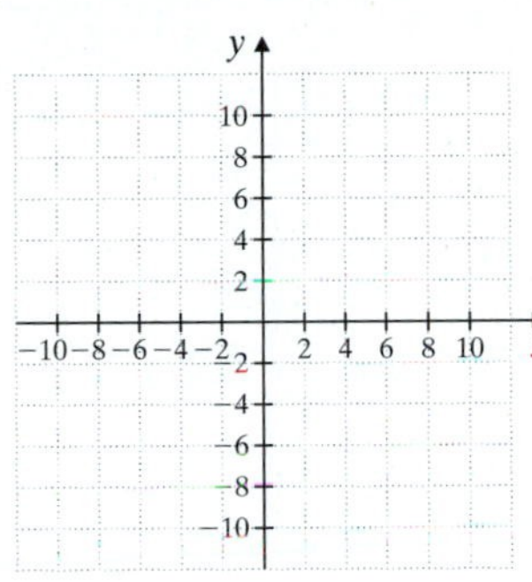

15. $2x - 4y = 8$

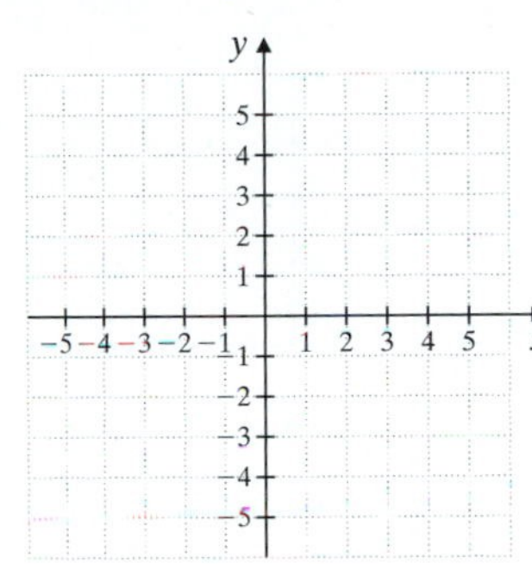

16. $2x + 3y = 6$

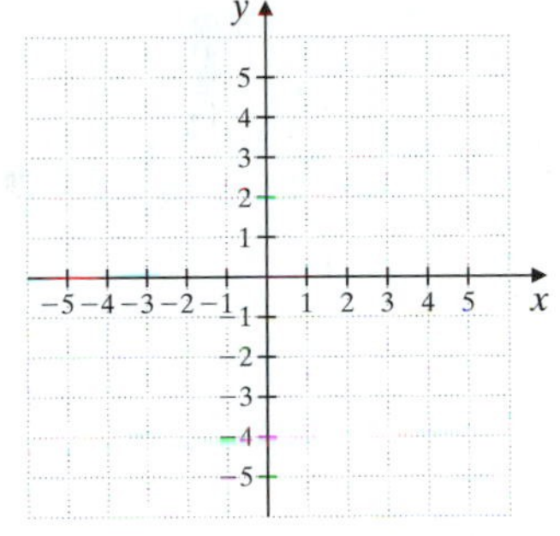

17. $2x - y = 0$

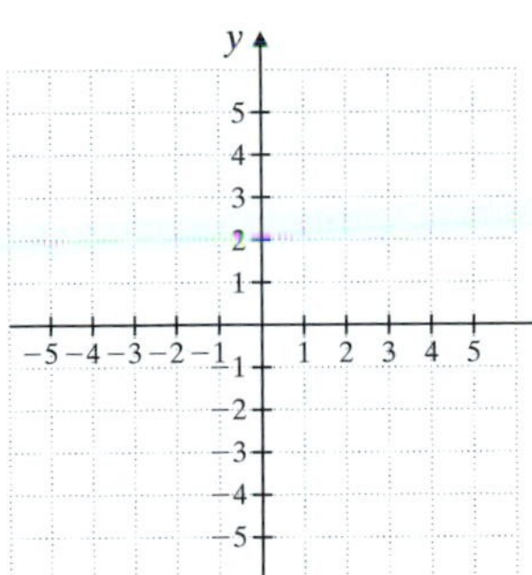

18. $-2x - y = 0$

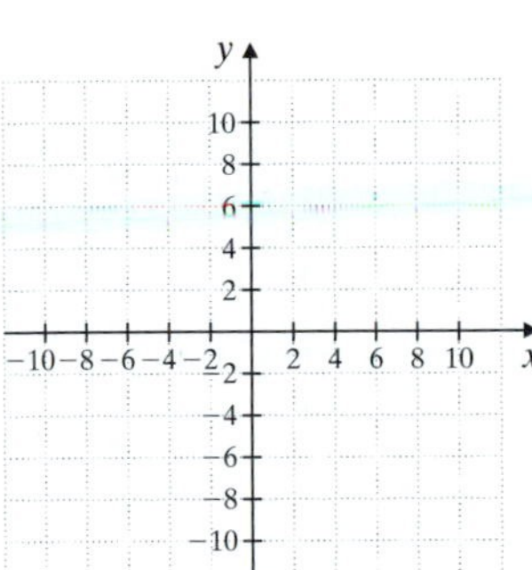

19. $y = 3x + 6$

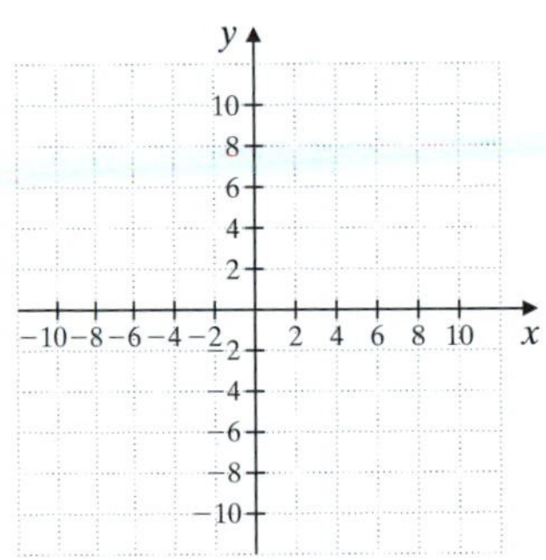

20. $y = 2x + 10$

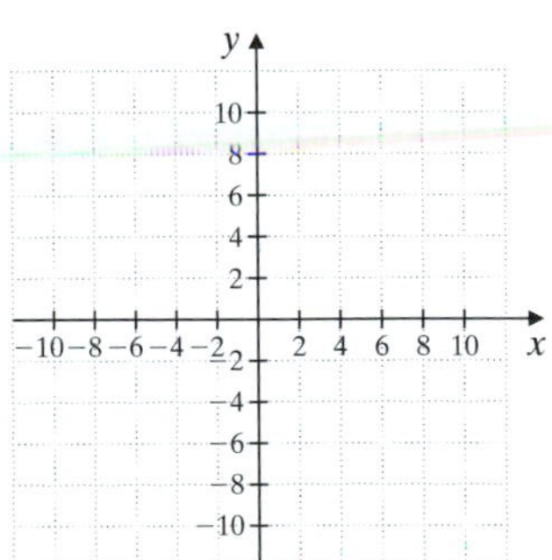

Objective C *Graph each linear equation. See Examples 6 and 7.*

21. $x = -1$

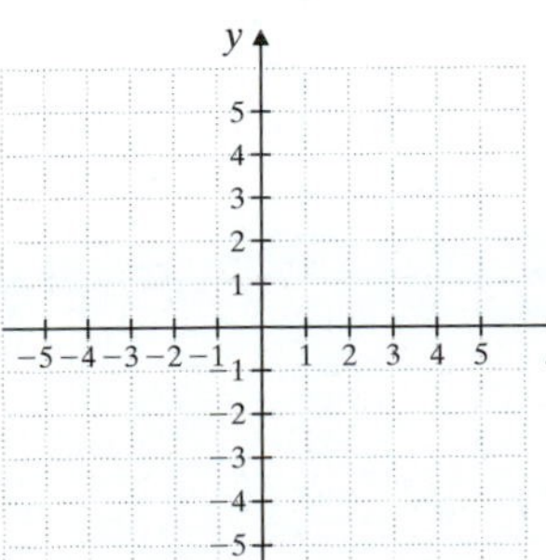

22. $y = 5$

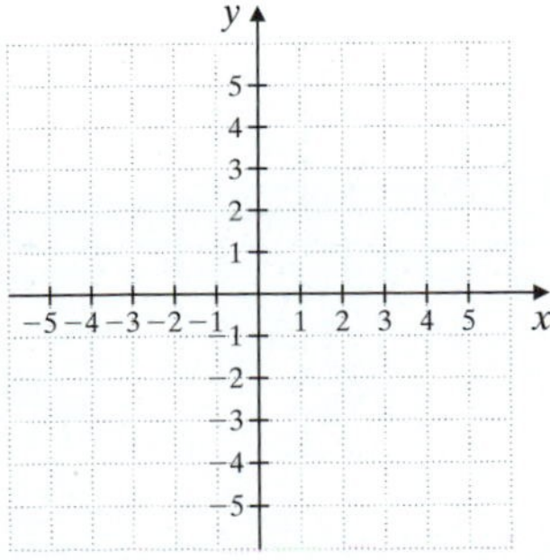

23. $y = 0$

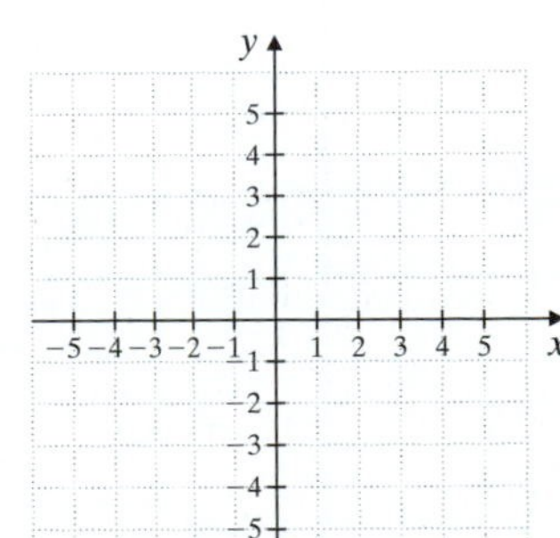

24. $x = 0$

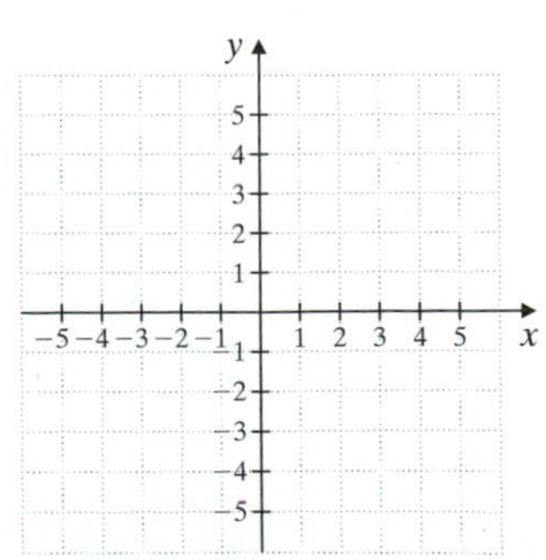

25. $y + 7 = 0$

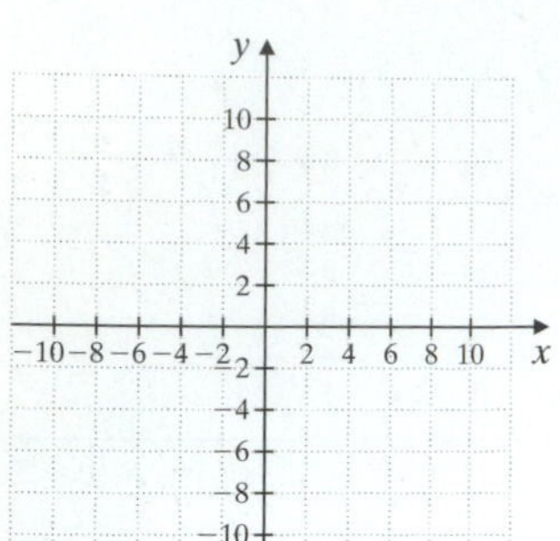

26. $x - 2 = 0$

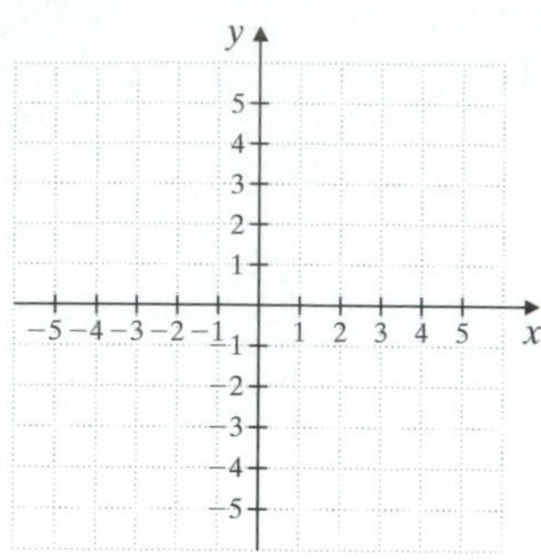

27. $x + 3 = 0$

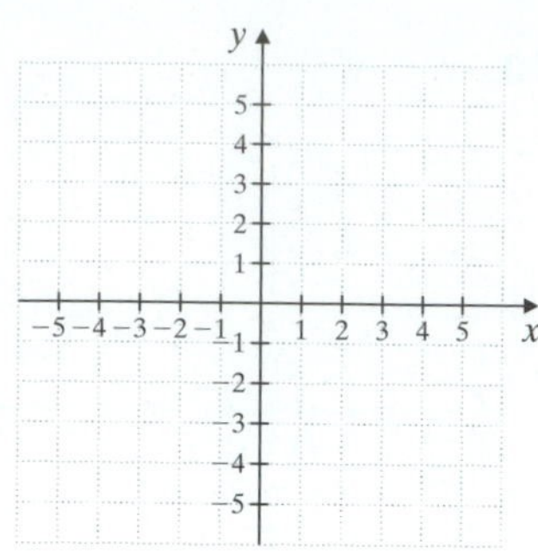

28. $y - 6 = 0$

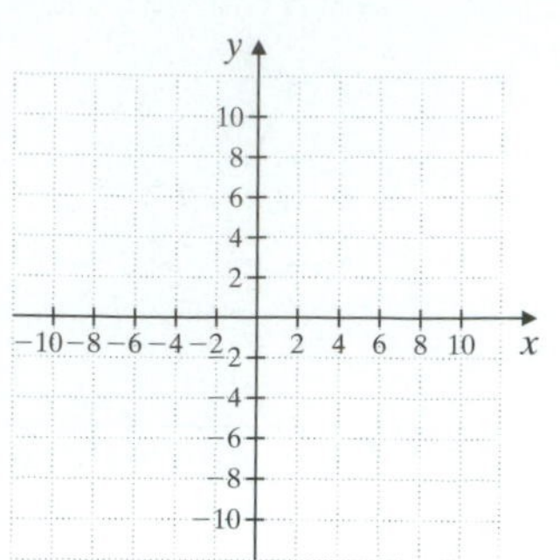

Objectives **B** **C** **Mixed Practice** *Graph each linear equation. See Examples 4 through 7.*

29. $x = y$

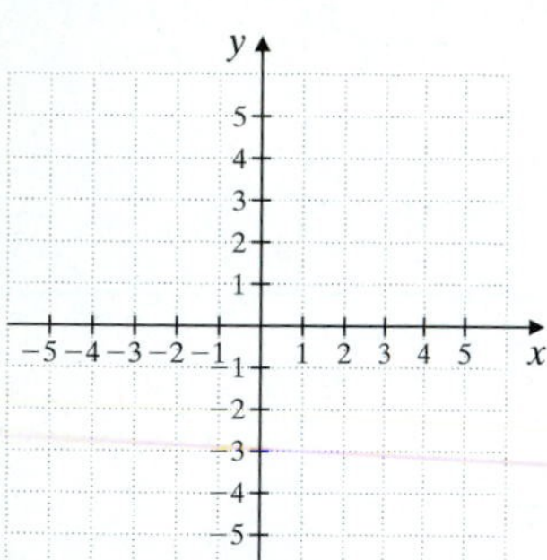

30. $x = -y$

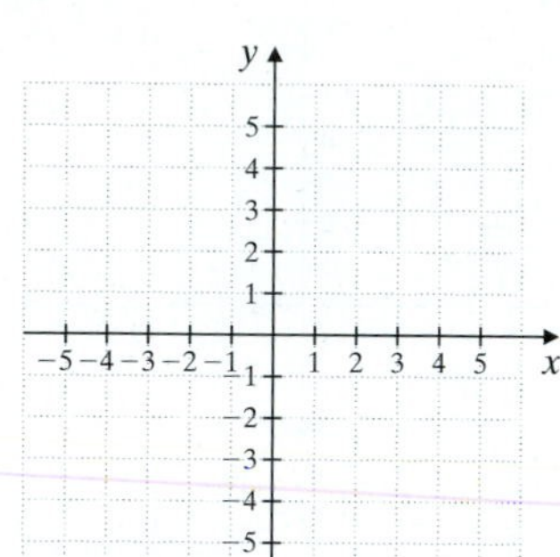

31. $x + 8y = 8$

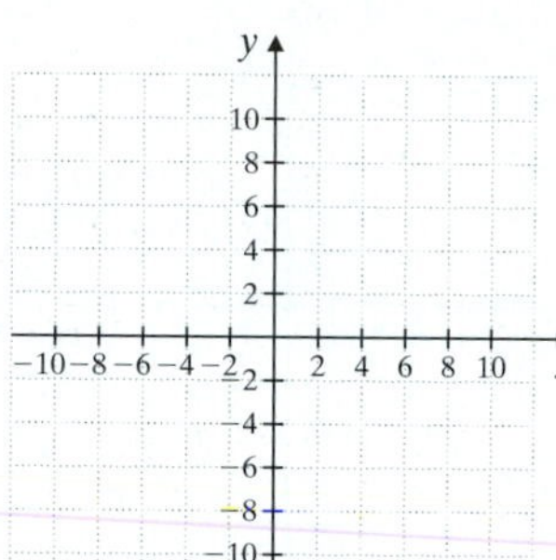

32. $x + 3y = 9$

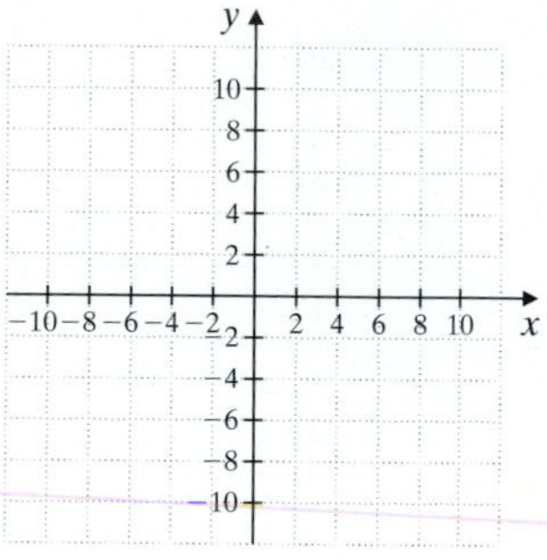

33. $5 = 6x - y$

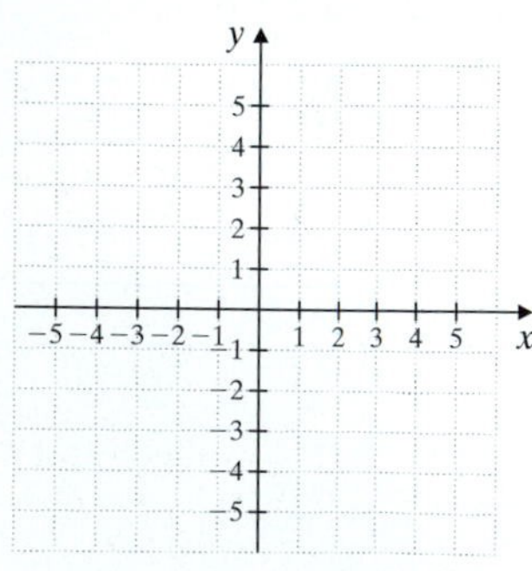

34. $4 = x - 3y$

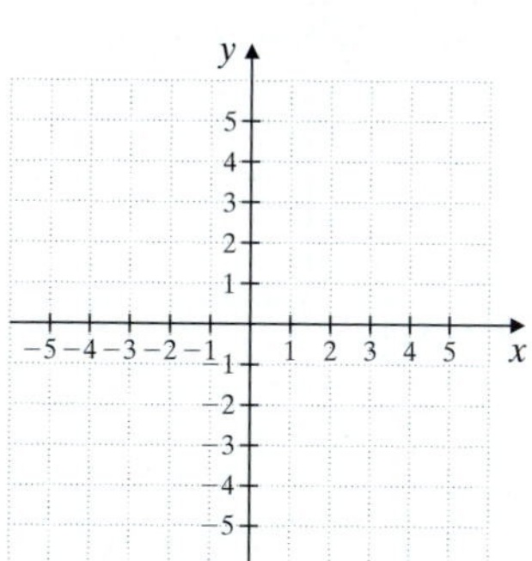

35. $-x + 10y = 11$

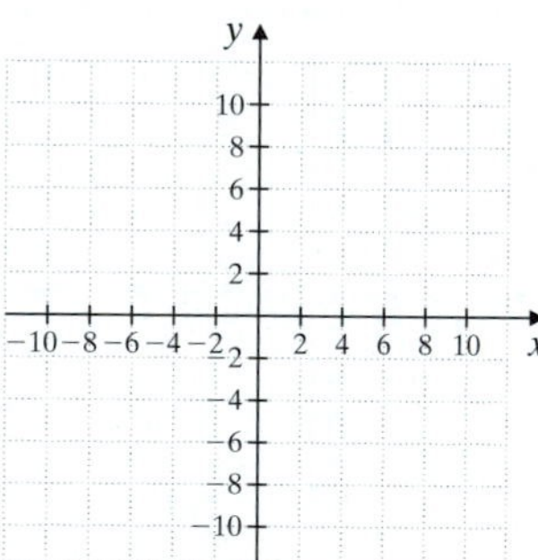

36. $-x + 9y = 10$

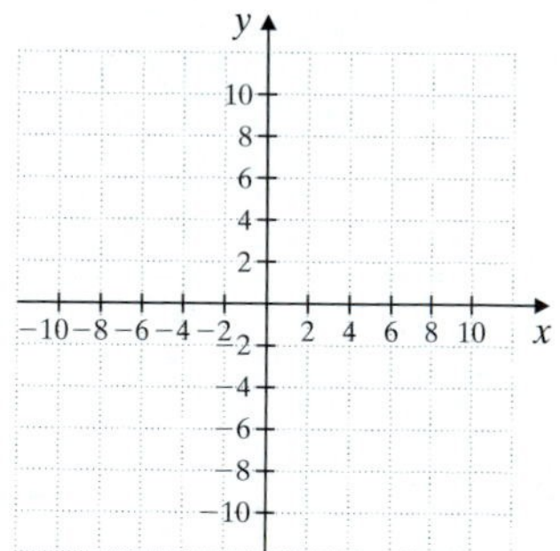

37. $x = -4\frac{1}{2}$

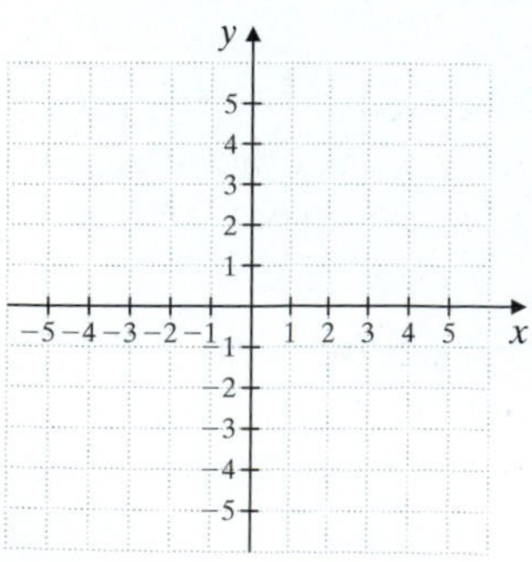

38. $x = -1\frac{3}{4}$

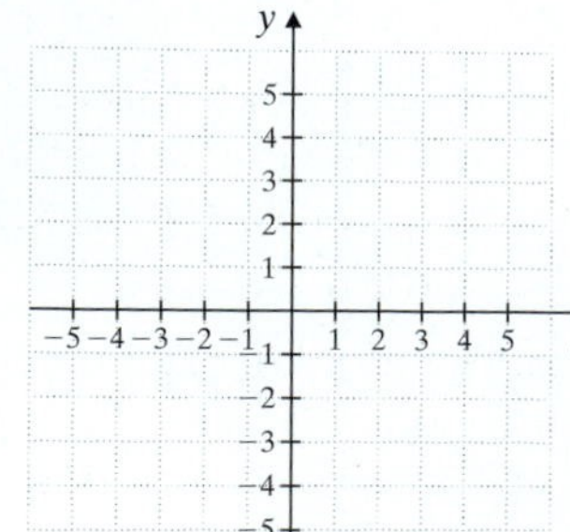

39. $y = 3\frac{1}{4}$

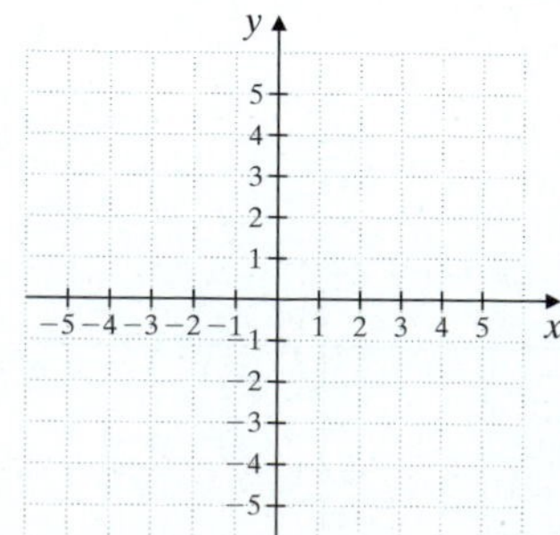

40. $y = 2\frac{1}{2}$

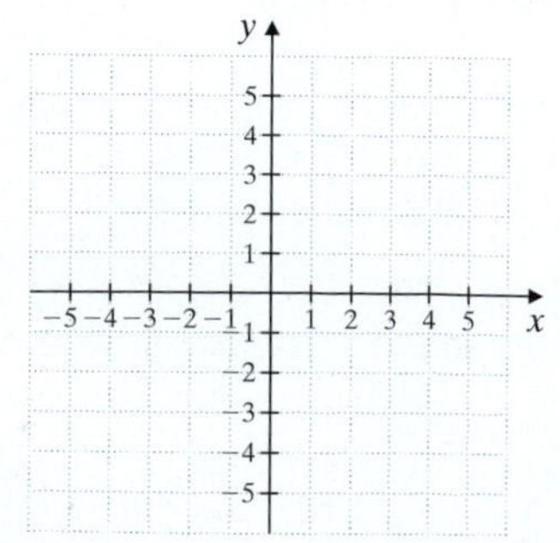

41. $y = -\frac{2}{3}x + 1$

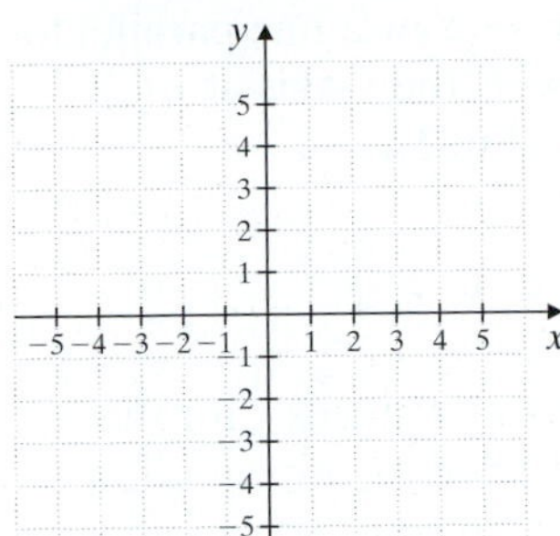

42. $y = -\frac{3}{5}x + 3$

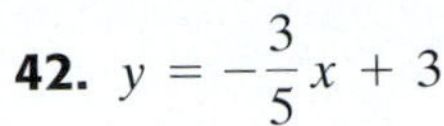

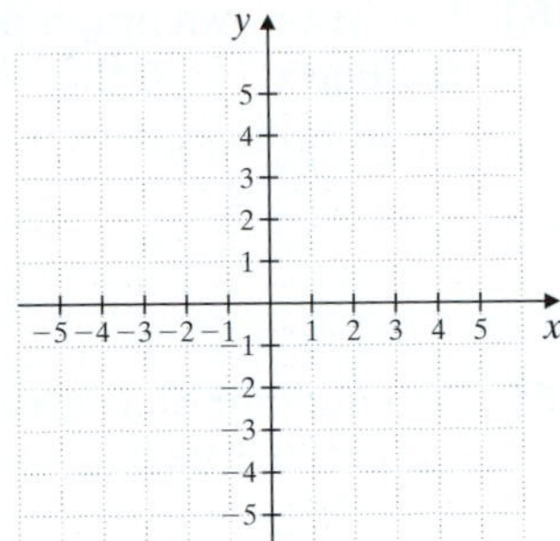

43. $4x - 6y + 2 = 0$

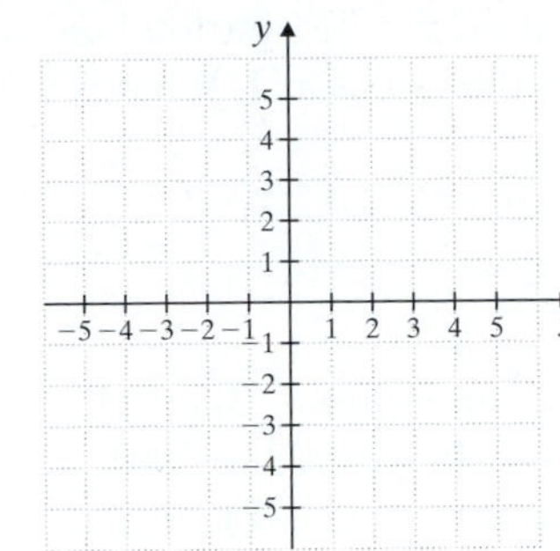

44. $9x - 6y + 3 = 0$

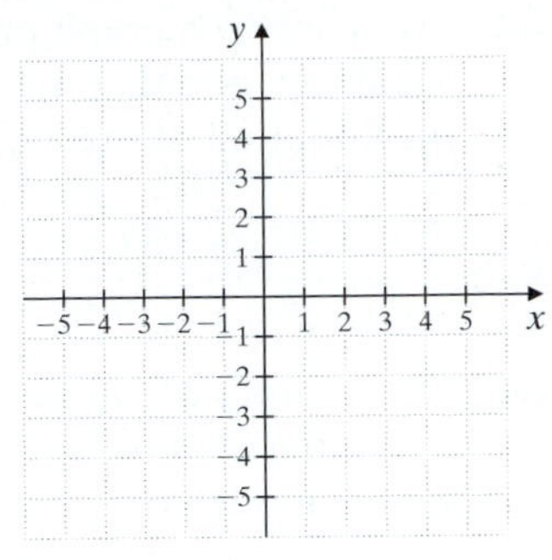

Review

Simplify. See Sections 2.2, 2.3, and 4.2.

45. $\frac{-6 - 3}{2 - 8}$

46. $\frac{4 - 5}{-1 - 0}$

47. $\frac{-8 - (-2)}{-3 - (-2)}$

48. $\frac{12 - 3}{10 - 9}$

49. $\frac{0 - 6}{5 - 0}$

50. $\frac{2 - 2}{3 - 5}$

Concept Extensions

Match each equation with its graph.

51. $y = 3$

52. $y = 2x + 2$

53. $x = 3$

54. $y = 2x + 3$

a.

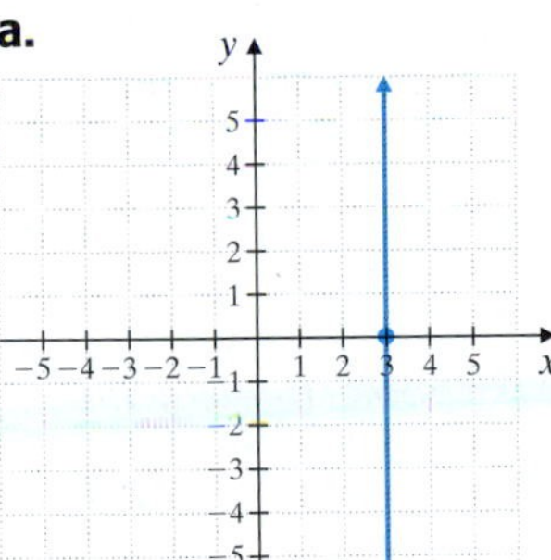

b.

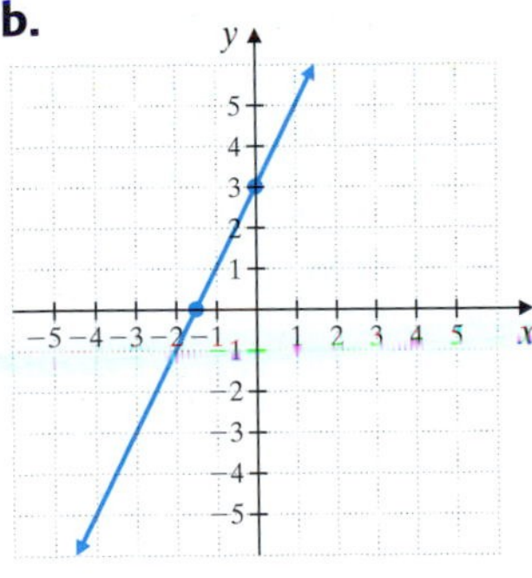

c.

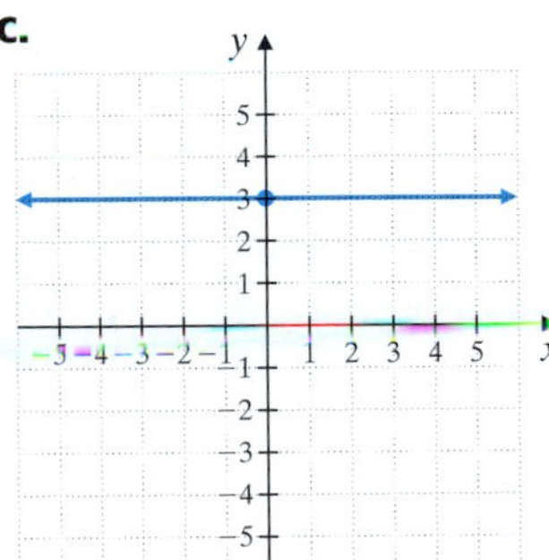

d.

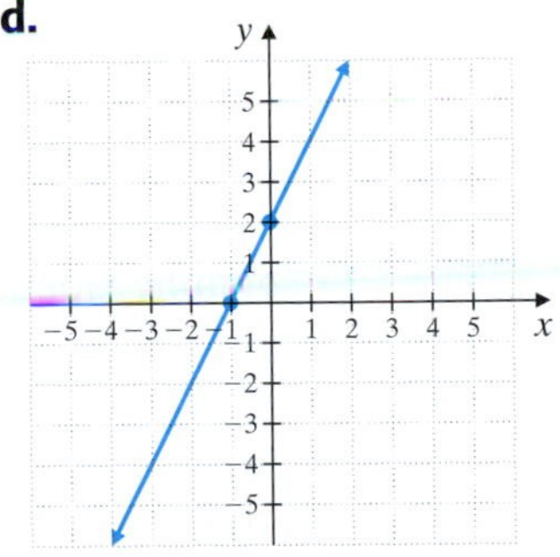

55. What is the greatest number of x- and y-intercepts that a line can have?

56. What is the smallest number of x- and y-intercepts that a line can have?

57. What is the smallest number of x- and y-intercepts that a circle can have?

58. What is the greatest number of x- and y-intercepts that a circle can have?

59. Discuss whether a vertical line ever has a y-intercept.

60. Discuss whether a horizontal line ever has an x-intercept.

61. The production supervisor at Alexandra's Office Products finds that it takes 3 hours to manufacture a particular office chair and 6 hours to manufacture an office desk. A total of 1200 hours is available to produce office chairs and desks of this style. The linear equation that models this situation is $3x + 6y = 1200$, where x represents the number of chairs produced and y the number of desks manufactured.

a. Complete the ordered pair solution (0,) of this equation. Describe the manufacturing situation that corresponds to this solution.

b. Complete the ordered pair solution (,0) of this equation. Describe the manufacturing situation that corresponds to this solution.

c. If 50 desks are manufactured, find the greatest number of chairs that can be made.

Two lines in the same plane that do not intersect are called ***parallel lines.***

62. Use your own graph paper to draw a line parallel to the line $x = 5$ that intersects the x-axis at 1. What is the equation of this line?

63. Use your own graph paper to draw a line parallel to the line $y = -1$ that intersects the y-axis at -4. What is the equation of this line?

Solve.

64. The number of music videos y, in millions, shipped to retailers in the United States from 1998 to 2003 can be modeled by the equation $y = -2.71x + 25$, where x represents the number of years after 1998. (*Source:* Recording Industry Association of America)

a. Find the x-intercept of this equation (round to the nearest tenth).

b. What does this x-intercept mean?

65. The number of a certain chain of stores y for the years 1999–2003 can be modeled by the equation $y = 29.2x + 919$, where x represents the number of years after 1999. (*Source:* Limited Brands)

a. Find the y-intercept of this equation.

b. What does this y-intercept mean?

STUDY SKILLS BUILDER

Are You Familiar with Your Textbook Supplements?

Below is a review of some of the student supplements available for additional study. Check to see if you are using the ones most helpful to you.

- Chapter Test Prep Videos on CD. This material is found with your textbook and is fully explained there. The CD contains video clip solutions to the Chapter Test exercises in this text and are excellent help when studying for chapter tests.
- Lecture Videos on CD-ROM. These video segments are keyed to each section of the text. The material is presented by me, Elayn Martin-Gay, and I have placed a [CD icon] by the exercises in the text that I have worked on the video.
- The *Student Solutions Manual.* This contains worked out solutions to odd-numbered exercises as well as every exercise in the Integrated Reviews, Chapter Reviews, Chapter Tests, and Cumulative Reviews.
- Prentice Hall Tutor Center. Mathematics questions may be phoned, faxed, or emailed to this center.
- MyMathLab, MathXL, and Internet Math. These are computer and Internet tutorials. This supplement may already be available to you somewhere on campus, for example at your local learning resource lab. Take a moment and find the name and location of any such lab on campus.

As usual, your instructor is your best source of information.

Let's see how you are doing with textbook supplements.

1. Name one way the Lecture Videos can be helpful to you.
2. Name one way the Chapter Test Prep Video can help you prepare for a chapter test.
3. List any textbook supplements that you have found useful.
4. Have you located and visited a learning resource lab located on your campus?
5. List the textbook supplements that are currently housed in your campus' learning resource lab.

13.4 SLOPE AND RATE OF CHANGE

Objectives

A Find the Slope of a Line Given Two Points of the Line.

B Find the Slope of a Line Given Its Equation.

C Find the Slopes of Horizontal and Vertical Lines.

D Compare the Slopes of Parallel and Perpendicular Lines.

E Slope as a Rate of Change.

Objective A Finding the Slope of a Line Given Two Points

Thus far, much of this chapter has been devoted to graphing lines. You have probably noticed by now that a key feature of a line is its slant or steepness. In mathematics, the slant or steepness of a line is formally known as its **slope.** We measure the slope of a line by the ratio of vertical change (rise) to the corresponding horizontal change (run) as we move along the line.

On the line below, for example, suppose that we begin at the point (1, 2) and move to the point (4, 6). The vertical change is the change in y-coordinates: $6 - 2$ or 4 units. The corresponding horizontal change is the change in x-coordinates: $4 - 1 = 3$ units. The ratio of these changes is

$$\text{slope} = \frac{\text{change in } y \text{ (vertical change or rise)}}{\text{change in } x \text{ (horizontal change or run)}} = \frac{4}{3}$$

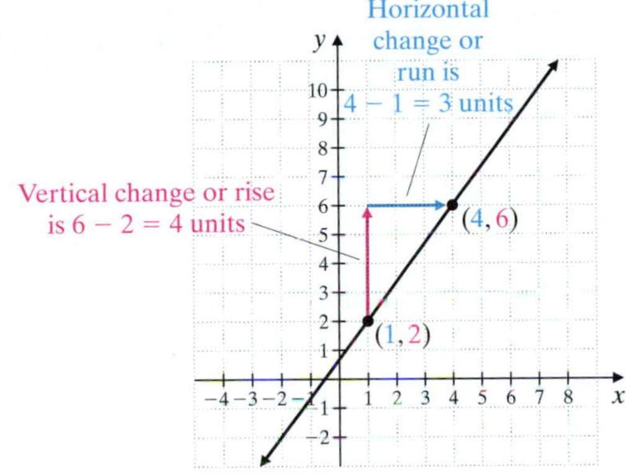

The slope of this line, then, is $\frac{4}{3}$. This means that for every 4 units of change in y-coordinates, there is a corresponding change of 3 units in x-coordinates.

Helpful Hint

It makes no difference what two points of a line are chosen to find its slope. The slope of a line is the same everywhere on the line.

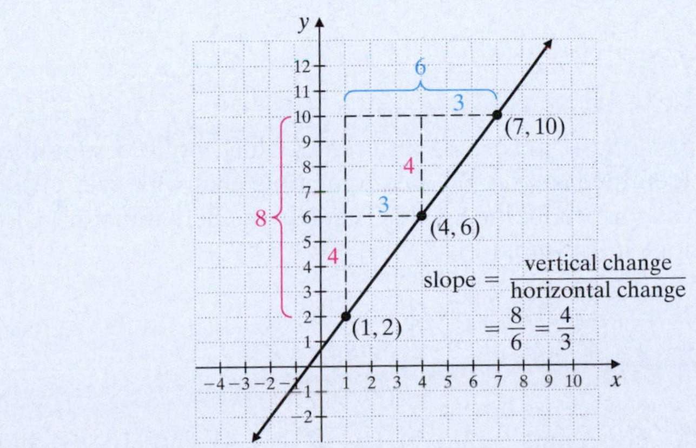

To find the slope of a line, then, choose two points of the line. Label the two x-coordinates of two points x_1 and x_2 (read "x sub one" and "x sub two"), and label the corresponding y-coordinates y_1 and y_2.

The vertical change or **rise** between these points is the difference in the y-coordinates: $y_2 - y_1$. The horizontal change or **run** between the points is the difference of the x-coordinates: $x_2 - x_1$. The slope of the line is the ratio of $y_2 - y_1$ to $x_2 - x_1$, and we traditionally use the letter m to denote slope $m = \frac{y_2 - y_1}{x_2 - x_1}$.

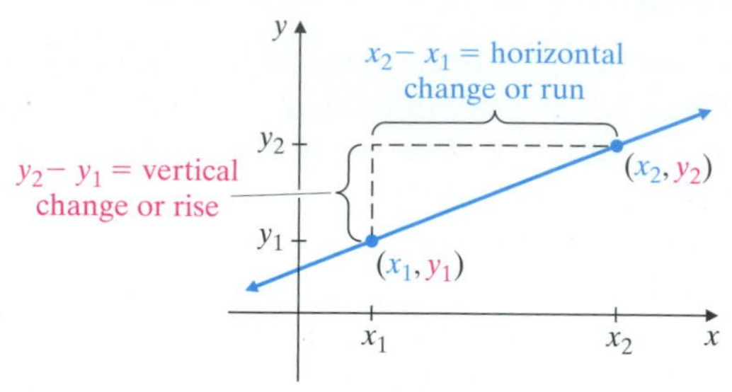

Slope of a Line

The slope m of the line containing the points (x_1, y_1) and (x_2, y_2) is given by

$$m = \frac{\text{rise}}{\text{run}} = \frac{\text{change in } y}{\text{change in } x} = \frac{y_2 - y_1}{x_2 - x_1}, \quad \text{as long as } x_2 \neq x_1$$

PRACTICE PROBLEM 1

Find the slope of the line through $(-2, 3)$ and $(4, -1)$. Graph the line.

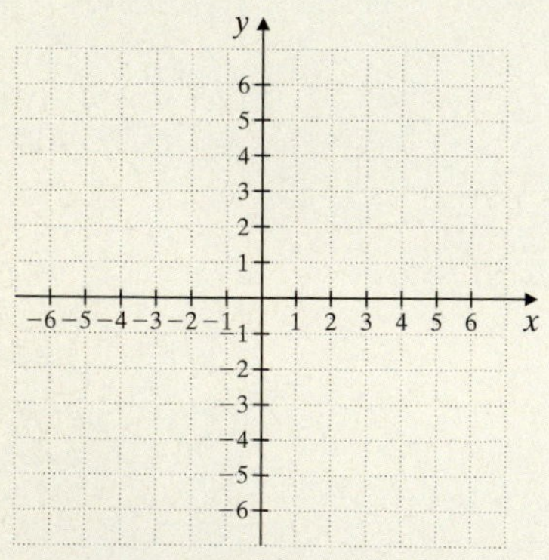

EXAMPLE 1 Find the slope of the line through $(-1, 5)$ and $(2, -3)$. Graph the line.

Solution: Let (x_1, y_1) be $(-1, 5)$ and (x_2, y_2) be $(2, -3)$. Then, by the definition of slope, we have the following.

$$m = \frac{y_2 - y_1}{x_2 - x_1} = \frac{-3 - 5}{2 - (-1)} = \frac{-8}{3} = -\frac{8}{3}$$

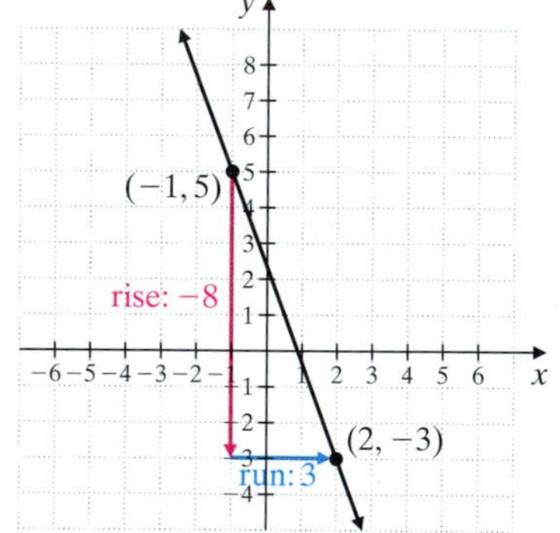

The slope of the line is $-\frac{8}{3}$.

Work Practice Problem 1

Helpful Hint

When finding slope, it makes no difference which point is identified as (x_1, y_1) and which is identified as (x_2, y_2). Just remember that whatever y-value is first in the numerator, its corresponding x-value is first in the denominator. Another way to calculate the slope in Example 1 is

$$m = \frac{y_2 - y_1}{x_2 - x_1} = \frac{5 - (-3)}{-1 - 2} = \frac{8}{-3} \text{ or } -\frac{8}{3} \leftarrow \text{Same slope as found in Example 1.}$$

✔Concept Check The points $(-2, -5)$, $(0, -2)$, $(4, 4)$, and $(10, 13)$ all lie on the same line. Work with a partner and verify that the slope is the same no matter which points are used to find slope.

Answer

1. $-\frac{2}{3}$

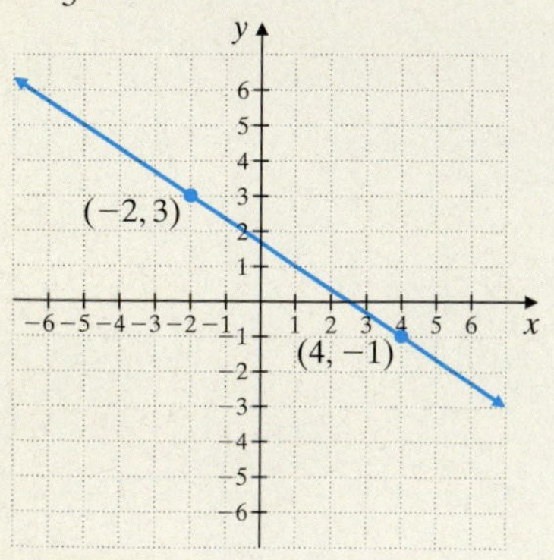

✔ Concept Check Answer

$m = \frac{3}{2}$

EXAMPLE 2 Find the slope of the line through $(-1, -2)$ and $(2, 4)$. Graph the line.

Solution: Let (x_1, y_1) be $(2, 4)$ and (x_2, y_2) be $(-1, -2)$.

$$m = \frac{y_2 - y_1}{x_2 - x_1}$$

$$= \frac{-2 - 4}{-1 - 2} \quad \begin{array}{l}\text{y-value} \\ \text{corresponding x-value}\end{array}$$

$$= \frac{-6}{-3} = 2$$

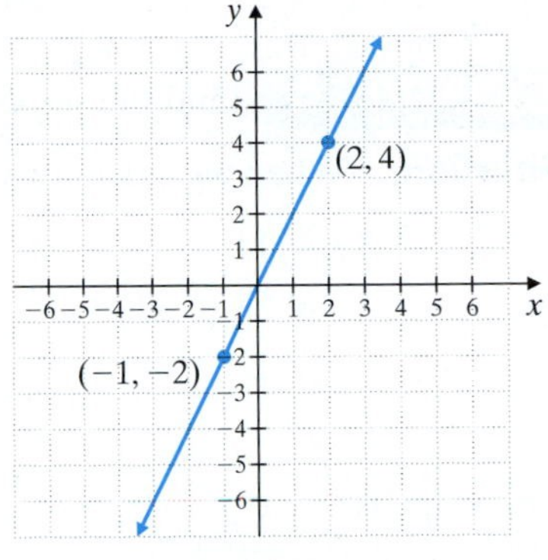

The slope is 2.

Work Practice Problem 2

PRACTICE PROBLEM 2

Find the slope of the line through $(-2, 1)$ and $(3, 5)$. Graph the line.

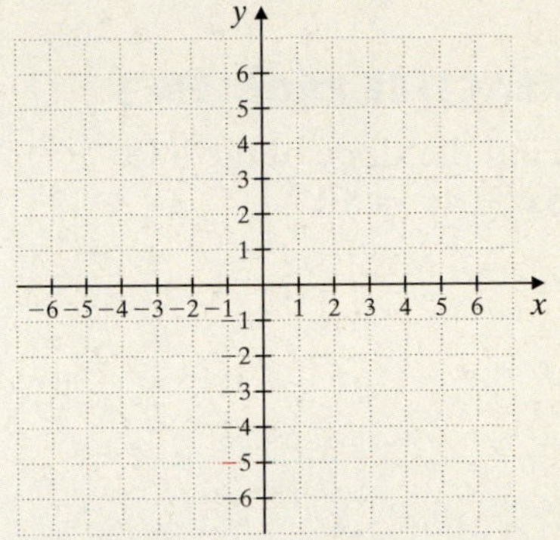

✔Concept Check What is wrong with the following slope calculation for the points $(3, 5)$ and $(-2, 6)$?

$$m = \frac{5 - 6}{-2 - 3} = \frac{-1}{-5} = \frac{1}{5}$$

Notice that the slope of the line in Example 1 is negative, and the slope of the line in Example 2 is positive. Let your eye follow the line with negative slope from left to right and notice that the line "goes down." If you follow the line with positive slope from left to right, you will notice that the line "goes up." This is true in general.

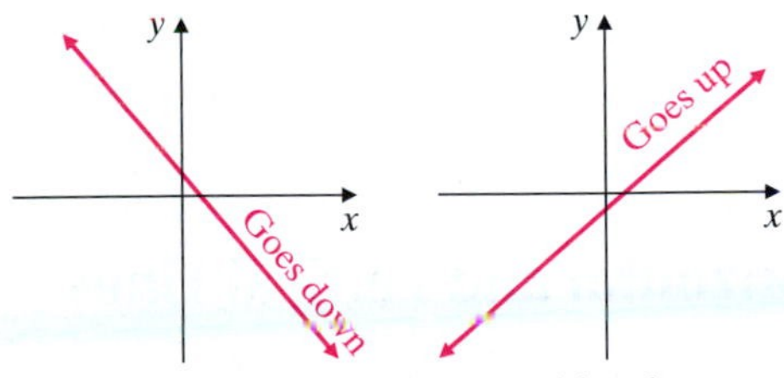

Helpful Hint

To decide whether a line "goes up" or "goes down," always follow the line from left to right.

Objective B Finding the Slope of a Line Given Its Equation

As we have seen, the slope of a line is defined by two points on the line. Thus, if we know the equation of a line, we can find its slope by finding two of its points. For example, let's find the slope of the line

$$y = 3x - 2$$

To find two points, we can choose two values for x and substitute to find corresponding y-values. If $x = 0$, for example, $y = 3 \cdot 0 - 2$ or $y = -2$. If $x = 1$, $y = 3 \cdot 1 - 2$ or $y = 1$. This gives the ordered pairs $(0, -2)$ and $(1, 1)$. Using the definition for slope, we have

$$m = \frac{1 - (-2)}{1 - 0} = \frac{3}{1} = 3 \quad \text{The slope is 3.}$$

Notice that the slope, 3, is the same as the coefficient of x in the equation $y = 3x - 2$. This is true in general.

Answer

2. $\frac{4}{5}$

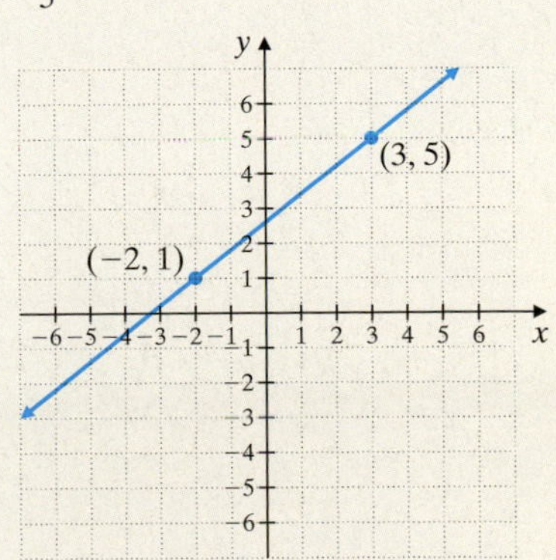

✔ Concept Check Answer

$$m = \frac{5 - 6}{3 - (-2)} = \frac{-1}{5} = -\frac{1}{5}$$

If a linear equation is solved for y, the coefficient of x is the line's slope. In other words, the slope of the line given by $y = mx + b$ is m, the coefficient of x.

$$y = mx + b$$
(m ← slope)

PRACTICE PROBLEM 3

Find the slope of the line $5x + 4y = 10$.

EXAMPLE 3 Find the slope of the line $-2x + 3y = 11$.

Solution: When we solve for y, the coefficient of x is the slope.

$$-2x + 3y = 11$$
$$3y = 2x + 11 \quad \text{Add } 2x \text{ to both sides.}$$
$$y = \frac{2}{3}x + \frac{11}{3} \quad \text{Divide both sides by 3.}$$

The slope is $\frac{2}{3}$.

Work Practice Problem 3

PRACTICE PROBLEM 4

Find the slope of the line $-y = -2x + 7$.

EXAMPLE 4 Find the slope of the line $-y = 5x - 2$.

Solution: Remember, the equation must be solved for y (not $-y$) in order for the coefficient of x to be the slope.

To solve for y, let's divide both sides of the equation by -1.

$$-y = 5x - 2$$
$$\frac{-y}{-1} = \frac{5x}{-1} - \frac{2}{-1} \quad \text{Divide both sides by } -1.$$
$$y = -5x + 2 \quad \text{Simplify.}$$

The slope is -5.

Work Practice Problem 4

Objective C Finding Slopes of Horizontal and Vertical Lines

PRACTICE PROBLEM 5

Find the slope of $y = 3$.

EXAMPLE 5 Find the slope of the line $y = -1$.

Solution: Recall that $y = -1$ is a horizontal line with y-intercept -1. To find the slope, we find two ordered pair solutions of $y = -1$, knowing that solutions of $y = -1$ must have a y-value of -1. We will use $(2, -1)$ and $(-3, -1)$. We let (x_1, y_1) be $(2, -1)$ and (x_2, y_2) be $(-3, -1)$.

$$m = \frac{y_2 - y_1}{x_2 - x_1} = \frac{-1 - (-1)}{-3 - 2} = \frac{0}{-5} = 0$$

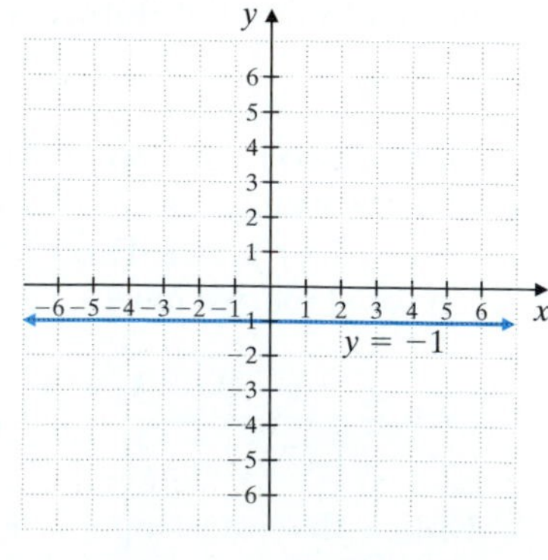

The slope of the line $y = -1$ is 0. Since the y-values will have a difference of 0 for every horizontal line, we can say that all **horizontal lines have a slope of 0.**

Work Practice Problem 5

Answers

3. $-\frac{5}{4}$ **4.** 2 **5.** 0

EXAMPLE 6 Find the slope of the line $x = 5$.

Solution: Recall that the graph of $x = 5$ is a vertical line with x-intercept 5. To find the slope, we find two ordered pair solutions of $x = 5$. Ordered pair solutions of $x = 5$ must have an x-value of 5. We will use $(5, 0)$ and $(5, 4)$. We let $(x_1, y_1) = (5, 0)$ and $(x_2, y_2) = (5, 4)$.

$$m = \frac{y_2 - y_1}{x_2 - x_1} = \frac{4 - 0}{5 - 5} = \frac{4}{0}$$

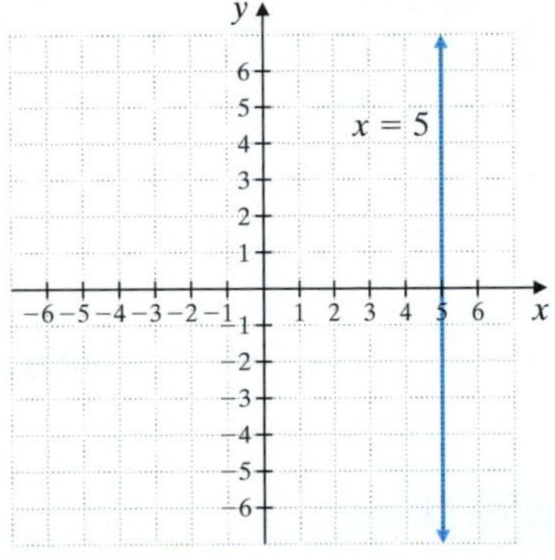

Since $\frac{4}{0}$ is undefined, we say the slope of the vertical line $x = 5$ is undefined.

Since the x-values will have a difference of 0 for every vertical line, we can say that all **vertical lines have undefined slope.**

Work Practice Problem 6

PRACTICE PROBLEM 6

Find the slope of the line $x = -2$.

Helpful Hint

Slope of 0 and undefined slope are not the same. Vertical lines have undefined slope, while horizontal lines have a slope of 0.

Here is a general review of slope.

Summary of Slope

Slope m of the line through (x_1, y_1) and (x_2, y_2) is given by the equation

$$m = \frac{y_2 - y_1}{x_2 - x_1}.$$

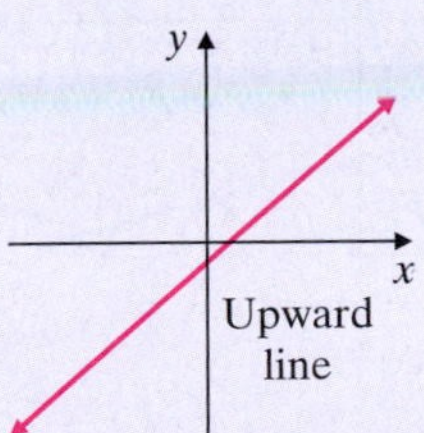

Positive slope: $m > 0$

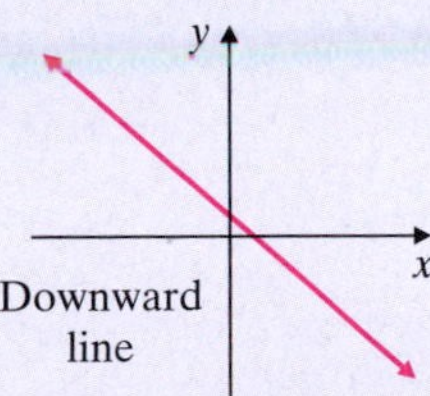

Negative slope: $m < 0$

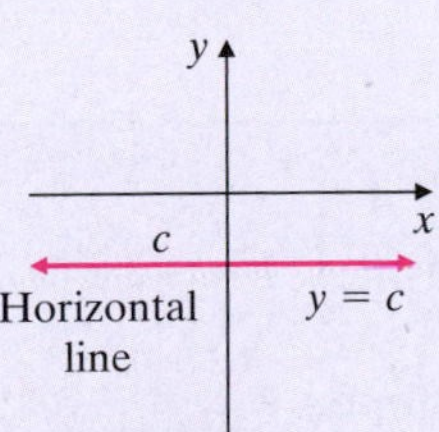

Zero slope: $m = 0$

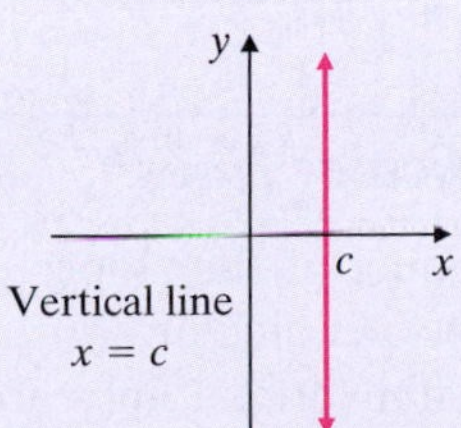

No slope or undefined slope

Objective D Slopes of Parallel and Perpendicular Lines

Two lines in the same plane are **parallel** if they do not intersect. Slopes of lines can help us determine whether lines are parallel. Since parallel lines have the same steepness, it follows that they have the same slope.

Answer

6. undefined slope

For example, the graphs of

$$y = -2x + 4$$

and

$$y = -2x - 3$$

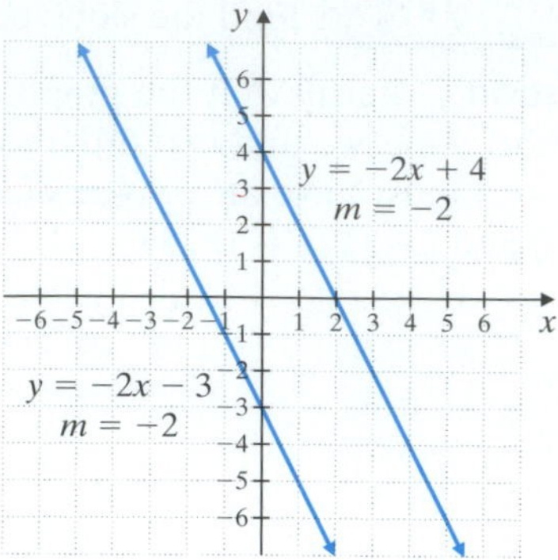

are shown. These lines have the same slope, −2. They also have different y-intercepts, so the lines are parallel. (If the y-intercepts were the same also, the lines would be the same.)

Parallel Lines

Nonvertical parallel lines have the same slope and different y-intercepts.

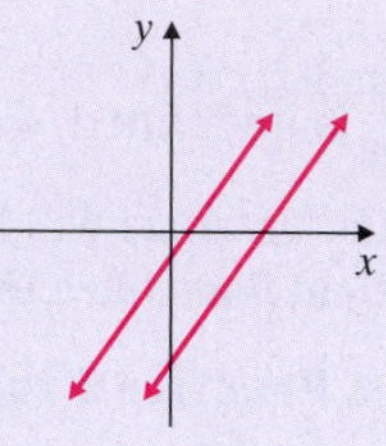

Two lines are **perpendicular** if they lie in the same plane and meet at a 90° (right) angle. How do the slopes of perpendicular lines compare? The product of the slopes of two perpendicular lines is −1.

For example, the graphs of

$$y = 4x + 1$$

and

$$y = -\frac{1}{4}x - 3$$

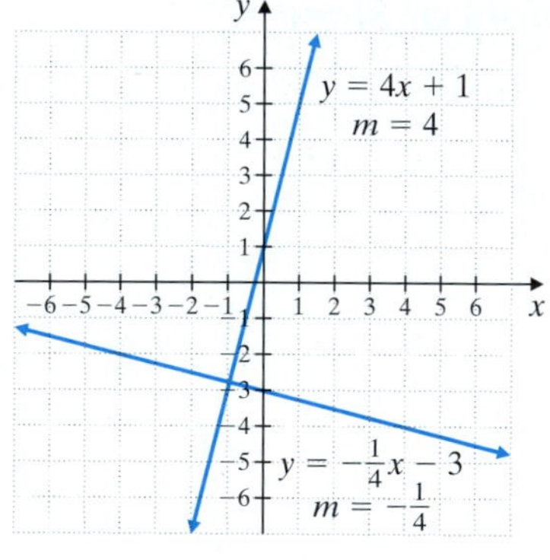

are shown. The slopes of the lines are 4 and $-\frac{1}{4}$. Their product is $4\left(-\frac{1}{4}\right) = -1$, so the lines are perpendicular.

Perpendicular Lines

If the product of the slopes of two lines is −1, then the lines are perpendicular.

(Two nonvertical lines are perpendicular if the slope of one is the negative reciprocal of the slope of the other.)

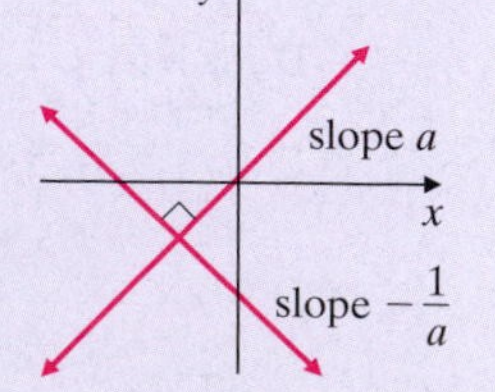

Helpful Hint

Here are examples of numbers that are negative (opposite) reciprocals.

Number	Negative Reciprocal	Their product is −1.
$\frac{2}{3}$	$-\frac{3}{2}$	$\frac{2}{3} \cdot -\frac{3}{2} = -\frac{6}{6} = -1$
-5 or $-\frac{5}{1}$	$\frac{1}{5}$	$-5 \cdot \frac{1}{5} = -\frac{5}{5} = -1$

Here are a few important points about vertical and horizontal lines.

- Two distinct vertical lines are parallel.
- Two distinct horizontal lines are parallel.
- A horizontal line and a vertical line are always perpendicular.

EXAMPLE 7 Determine whether each pair of lines is parallel, perpendicular, or neither.

a. $y = -\frac{1}{5}x + 1$
$2x + 10y = 3$

b. $x + y = 3$
$-x + y = 4$

c. $3x + y = 5$
$2x + 3y = 6$

Solution:

a. The slope of the line $y = -\frac{1}{5}x + 1$ is $-\frac{1}{5}$. We find the slope of the second line by solving its equation for y.

$$2x + 10y = 3$$
$$10y = -2x + 3 \quad \text{Subtract } 2x \text{ from both sides.}$$
$$y = \frac{-2}{10}x + \frac{3}{10} \quad \text{Divide both sides by 10.}$$
$$y = -\frac{1}{5}x + \frac{3}{10} \quad \text{Simplify.}$$

The slope of this line is $-\frac{1}{5}$ also. Since the lines have the same slope and different y-intercepts, they are parallel, as shown in the figure below.

b. To find each slope, we solve each equation for y.

$$x + y = 3 \qquad\qquad -x + y = 4$$
$$y = -x + 3 \qquad\qquad y = x + 4$$

The slope is −1. The slope is 1.

The slopes are not the same, so the lines are not parallel. Next we check the product of the slopes: $(-1)(1) = -1$. Since the product is -1, the lines are perpendicular, as shown in the figure.

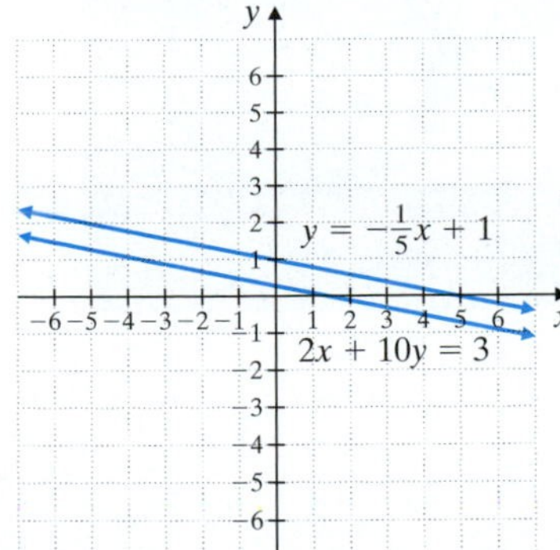

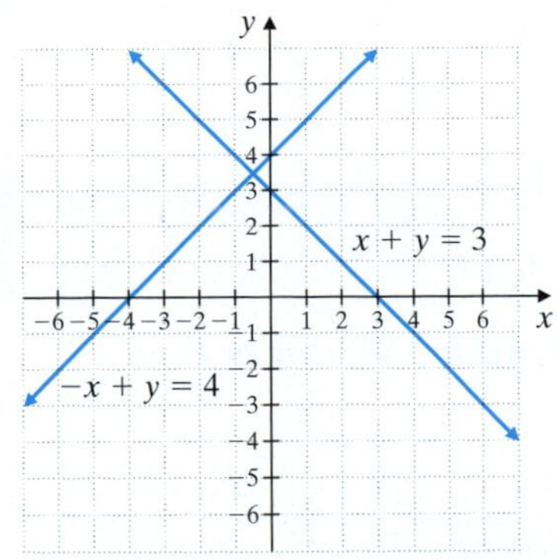

Continued on next page

PRACTICE PROBLEM 7

Determine whether each pair of lines is parallel, perpendicular, or neither.

a. $x + y = 5$
$2x + y = 5$

b. $5y = 2x - 3$
$5x + 2y = 1$

c. $y = 2x + 1$
$4x - 2y = 8$

Answers

7. a. neither **b.** perpendicular **c.** parallel

c. We solve each equation for y to find each slope. The slopes are -3 and $-\frac{2}{3}$. The slopes are not the same and their product is not -1. Thus, the lines are neither parallel nor perpendicular.

Work Practice Problem 7

✔Concept Check Consider the line $-6x + 2y = 1$.

a. Write the equations of two lines parallel to this line.

b. Write the equations of two lines perpendicular to this line.

Objective E Slope as a Rate of Change

$\frac{7}{10}$ pitch

7 feet

10 feet

Slope can also be interpreted as a rate of change. In other words, slope tells us how fast y is changing with respect to x. To see this, let's look at a few of the many real-world applications of slope. For example, the pitch of a roof, used by builders and architects, is its slope. The pitch of the roof on the left is $\frac{7}{10}\left(\frac{\text{rise}}{\text{run}}\right)$. This means that the roof rises vertically 7 feet for every horizontal 10 feet. The rate of change for the roof is 7 vertical feet (y) per 10 horizontal feet (x).

The grade of a road is its slope written as a percent. A 7% grade, as shown below, means that the road rises (or falls) 7 feet for every horizontal 100 feet. $\left(\text{Recall that } 7\% = \frac{7}{100}.\right)$ Here, the slope of $\frac{7}{100}$ gives us the rate of change. The road rises (in our diagram) 7 vertical feet (y) for every 100 horizontal feet (x).

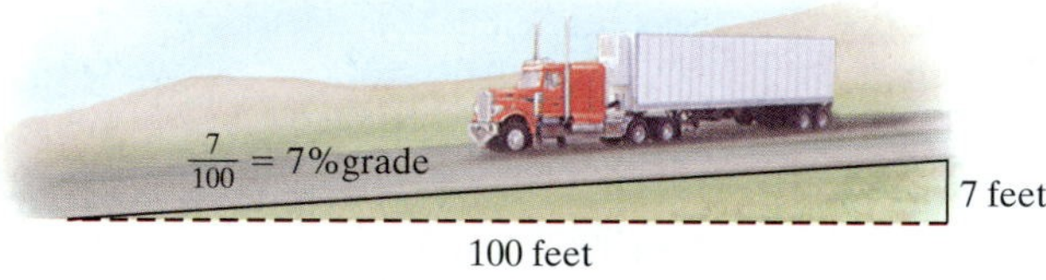

PRACTICE PROBLEM 8

Find the grade of the road shown.

EXAMPLE 8 Finding the Grade of a Road

At one part of the road to the summit of Pike's Peak, the road rises 15 feet for a horizontal distance of 250 feet. Find the grade of the road.

Solution: Recall that the grade of a road is its slope written as a percent.

$$\text{grade} = \frac{\text{rise}}{\text{run}} = \frac{15}{250} = 0.06 = 6\%$$

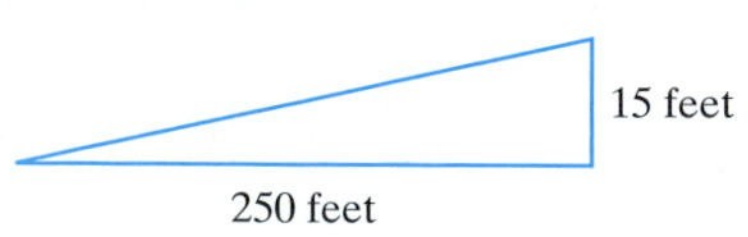

The grade is 6%.

Work Practice Problem 8

Answer

8. 15%

✔ Concept Check Answers

Answers may vary; for example,

a. $y = 3x - 3$, $y = 3x - 1$

b. $y = -\frac{1}{3}x$, $y = -\frac{1}{3}x + 1$

Slope can also be interpreted as a rate of change. In other words, slope tells us how fast y is changing with respect to x.

EXAMPLE 9 Finding the Slope of a Line

The following graph shows the cost y (in cents) of a nationwide long-distance telephone call from Texas with a certain telephone-calling plan, where x is the length of the call in minutes. Find the slope of the line and attach the proper units for the rate of change. Then write a sentence explaining the meaning of slope in this application.

Solution: Use (2, 34) and (6, 62) to calculate slope.

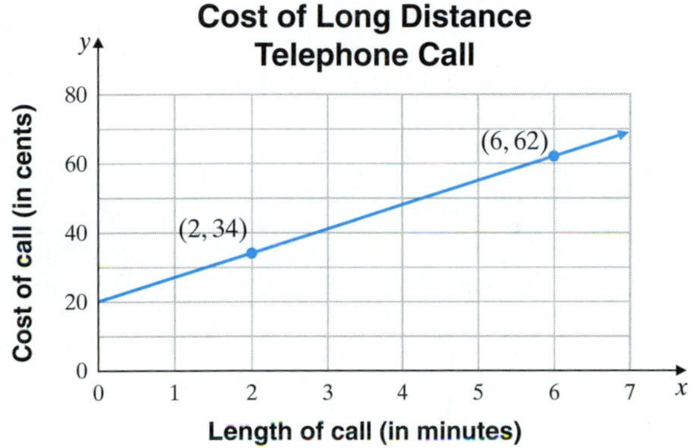

$$m = \frac{62 - 34}{6 - 2} = \frac{28}{4} = \frac{7 \text{ cents}}{1 \text{ minute}}$$

This means that the rate of change of a phone call is 7 cents per 1 minute or the cost of the phone call is 7 cents per minute.

Work Practice Problem 9

PRACTICE PROBLEM 9

Find the slope of the line and write the slope as a rate of change. This graph represents annual food and drink sales y (in billions of dollars) for year x. Write a sentence explaining the meaning of slope in this application.

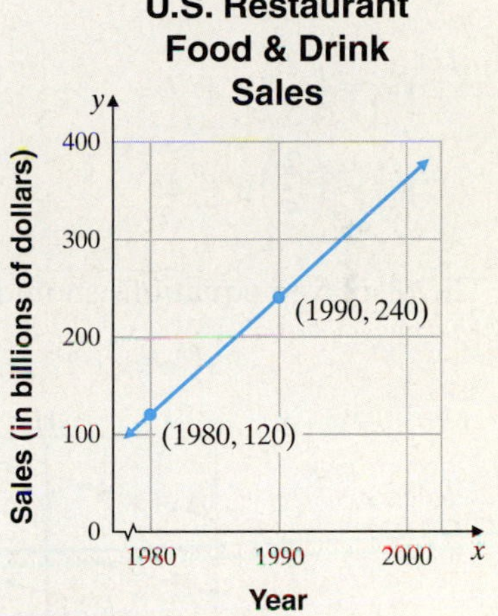

Source: National Restaurant Assn.

Answer

9. $m = 12$; Each year the sales of food and drink from restaurants increases by \$12 billion dollars per year.

CALCULATOR EXPLORATIONS Graphing

It is possible to use a graphing calculator and sketch the graph of more than one equation on the same set of axes. This feature can be used to see parallel lines with the same slope. For example, graph the equations $y = \frac{2}{5}x$, $y = \frac{2}{5}x + 7$, and $y = \frac{2}{5}x - 4$ on the same set of axes. To do so, press the $\boxed{Y=}$ key and enter the equations on the first three lines.

$$Y_1 = \left(\frac{2}{5}\right)x$$

$$Y_2 = \left(\frac{2}{5}\right)x + 7$$

$$Y_3 = \left(\frac{2}{5}\right)x - 4$$

The displayed equations should look like this:

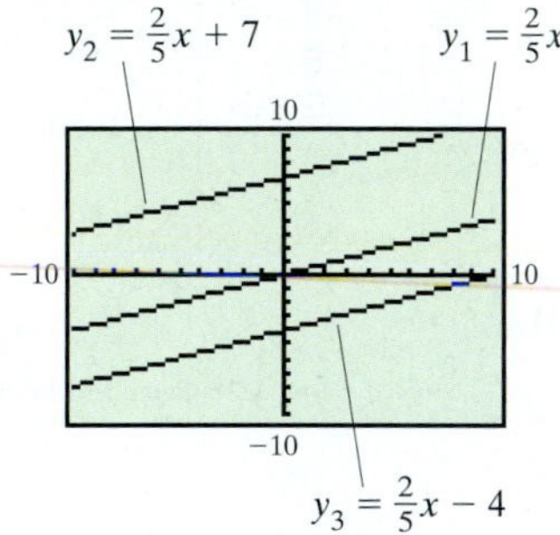

These lines are parallel as expected since they all have a slope of $\frac{2}{5}$. The graph of $y = \frac{2}{5}x + 7$ is the graph of $y = \frac{2}{5}x$ moved 7 units upward with a y-intercept of 7. Also, the graph of $y = \frac{2}{5}x - 4$ is the graph of $y = \frac{2}{5}x$ moved 4 units downward with a y-intercept of −4.

Graph the parallel lines on the same set of axes. Describe the similarities and differences in their graphs.

1. $y = 3.8x$, $y = 3.8x - 3$, $y = 3.8x + 9$

2. $y = -4.9x$, $y = -4.9x + 1$, $y = -4.9x + 8$

3. $y = \frac{1}{4}x$, $y = \frac{1}{4}x + 5$, $y = \frac{1}{4}x - 8$

4. $y = -\frac{3}{4}x$, $y = -\frac{3}{4}x - 5$, $y = -\frac{3}{4}x + 6$

Vocabulary and Readiness Check

Use the choices below to fill in each blank. Not all choices will be used.

m	x	0	positive	undefined
b	y	slope	negative	

1. The measure of the steepness or tilt of a line is called ________.

2. If an equation is written in the form $y = mx + b$, the value of the letter _____ is the value of the slope of the graph.

3. The slope of a horizontal line is _____.

4. The slope of a vertical line is ____________.

5. If the graph of a line moves upward from left to right, the line has __________ slope.

6. If the graph of a line moves downward from left to right, the line has ___________ slope.

7. Given two points of a line, slope $= \dfrac{\text{change in } \_\_\_}{\text{change in } \_\_\_}$.

13.4 EXERCISE SET

FOR EXTRA HELP

Student Solutions Manual

PH Math/Tutor Center

CD/Video for Review

MathXL®

MyMathLab

Objective A *Find the slope of the line that passes through the given points. See Examples 1 and 2.*

1. $(-1, 5)$ and $(6, -2)$

2. $(-1, 16)$ and $(3, 4)$

3. $(1, 4)$ and $(5, 3)$

4. $(3, 1)$ and $(2, 6)$

5. $(5, 1)$ and $(-2, 1)$

6. $(-8, 3)$ and $(-2, 3)$

7. $(5, 4)$ and $(5, 0)$

8. $(-2, -3)$ and $(-2, 5)$

Use the points shown on each graph to find the slope of each line. See Examples 1 and 2.

9.

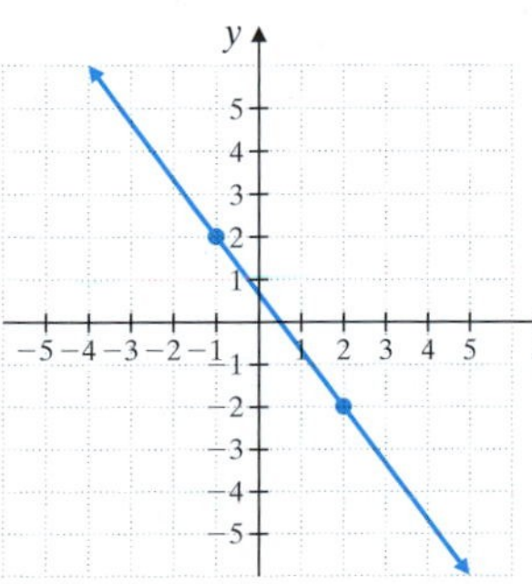

10.

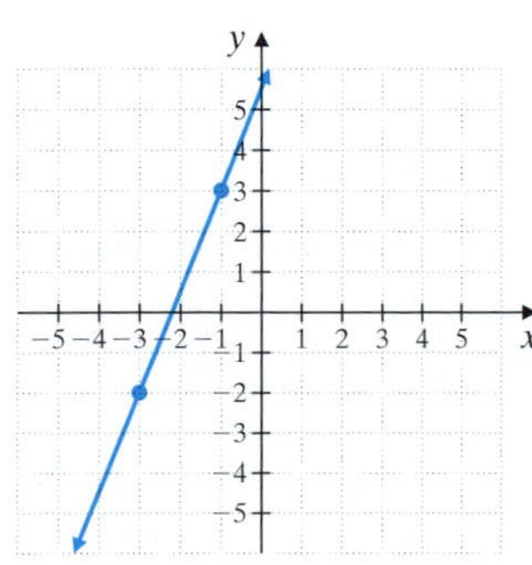

11.

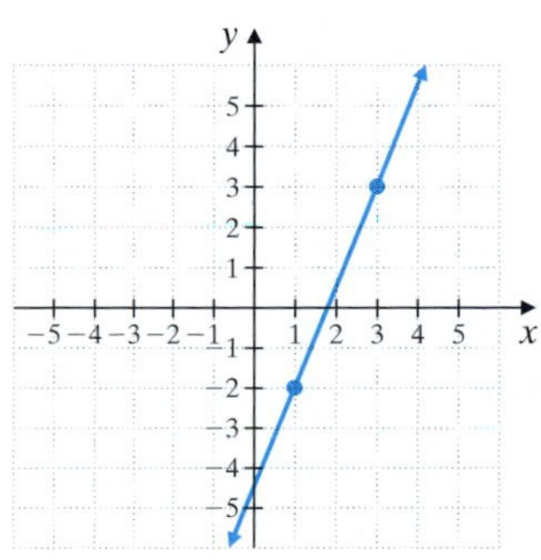

12.

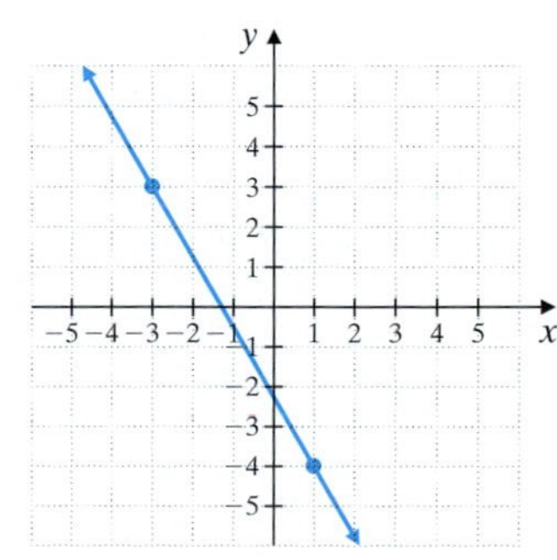

For each graph, determine which line has the greater slope.

13.

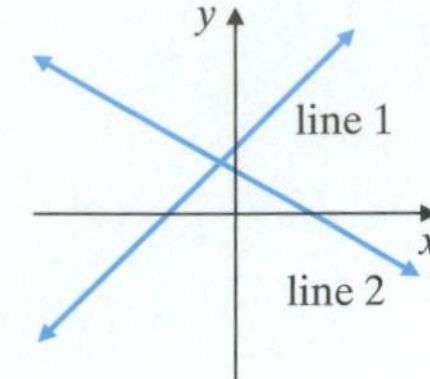

14.

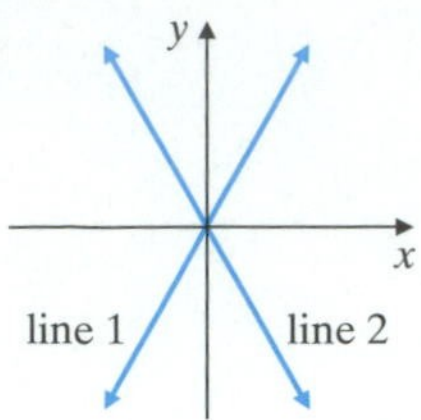

15.

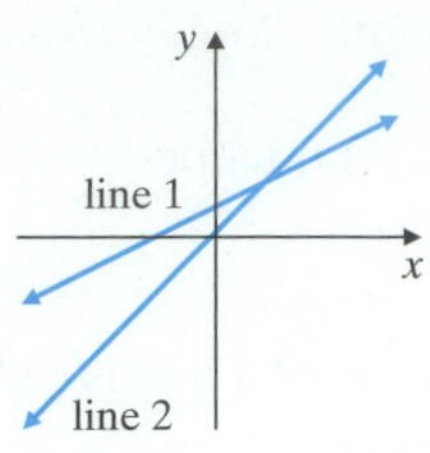

16.

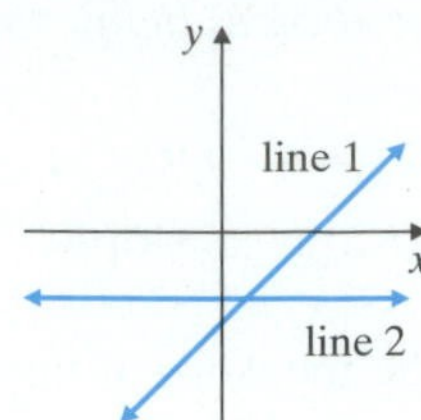

Objectives B C Mixed Practice *Find the slope of each line. See Examples 3 through 6.*

17. $y = 5x - 2$

18. $y = -2x + 6$

19. $y = -0.3x + 2.5$

20. $y = -7.6x - 0.1$

21. $2x + y = 7$

22. $-5x + y = 10$

23.

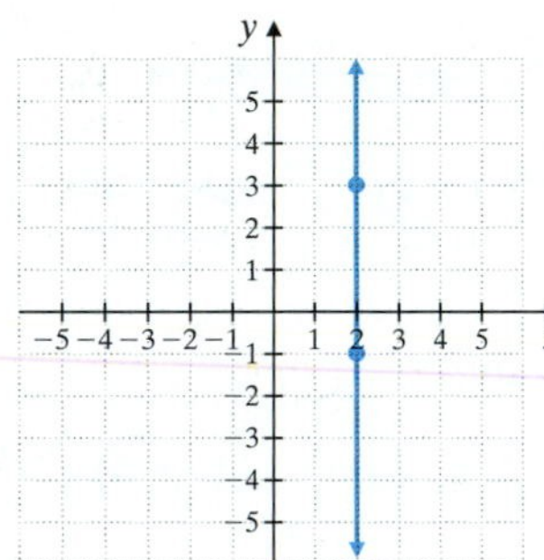

24.

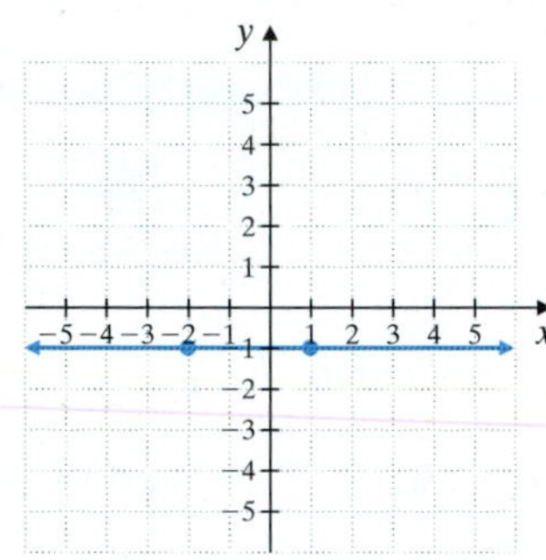

25. $2x - 3y = 10$

26. $3x - 5y = 1$

27. $x = 1$

28. $y = -2$

29. $x = 2y$

30. $x = -4y$

31. $y = -3$

32. $x = 5$

33. $-3x - 4y = 6$

34. $-4x - 7y = 9$

35. $20x - 5y = 1.2$

36. $24x - 3y = 5.7$

△ **Objective D** *Determine whether each pair of lines is parallel, perpendicular, or neither. See Example 7.*

37. $y = \frac{2}{9}x + 3$
$y = -\frac{2}{9}x$

38. $y = \frac{1}{5}x + 20$
$y = -\frac{1}{5}x$

39. $x - 3y = -6$
$y = 3x - 9$

40. $y = 4x - 2$
$4x + y = 5$

41. $6x = 5y + 1$
$-12x + 10y = 1$

42. $-x + 2y = -2$
$2x = 4y + 3$

43. $6 + 4x = 3y$
$3x + 4y = 8$

44. $10 + 3x = 5y$
$5x + 3y = 1$

△ *Find the slope of the line that is* (***a***) *parallel and* (***b***) *perpendicular to the line through each pair of points. See Example 7.*

45. $(-3, -3)$ and $(0, 0)$

46. $(6, -2)$ and $(1, 4)$

47. $(-8, -4)$ and $(3, 5)$

48. $(6, -1)$ and $(-4, -10)$

Objective **E** *The pitch of a roof is its slope. Find the pitch of each roof shown. See Example 8.*

49.

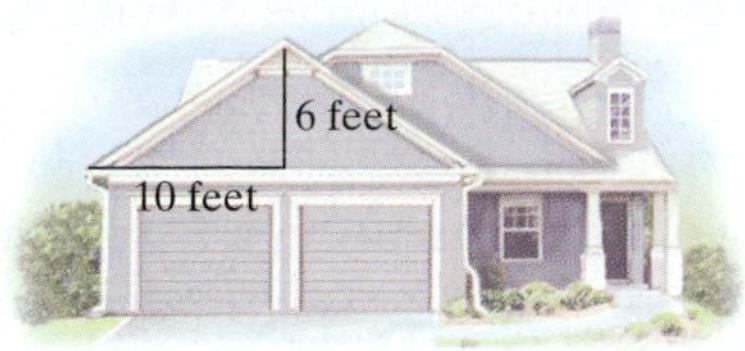

50.

The grade of a road is its slope written as a percent. Find the grade of each road shown. See Example 8.

51.

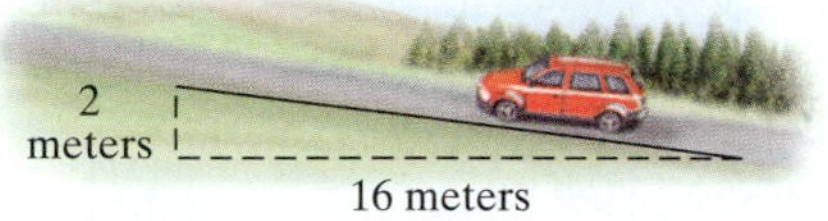

52.

53. One of Japan's superconducting "bullet" trains is researched and tested at the Yamanashi Maglev Test Line near Otsuki City. The steepest section of the track has a rise of 2580 meters for a horizontal distance of 6450 meters. What is the grade of this section of track? (*Source:* Japan Railways Central Co.)

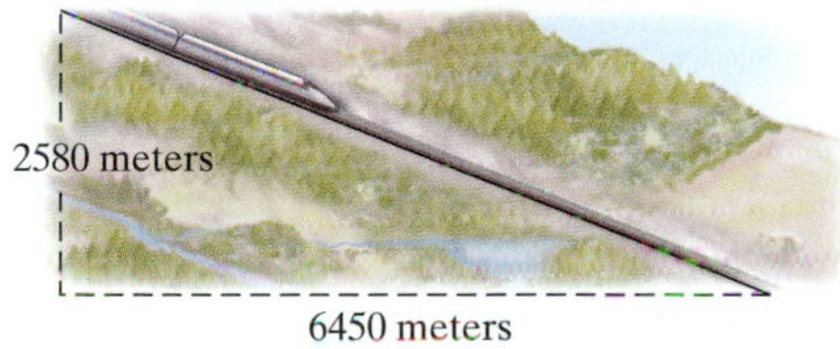

54. Professional plumbers suggest that a sewer pipe should rise 0.25 inch for every horizontal foot. Find the recommended slope for a sewer pipe. Round to the nearest hundredth.

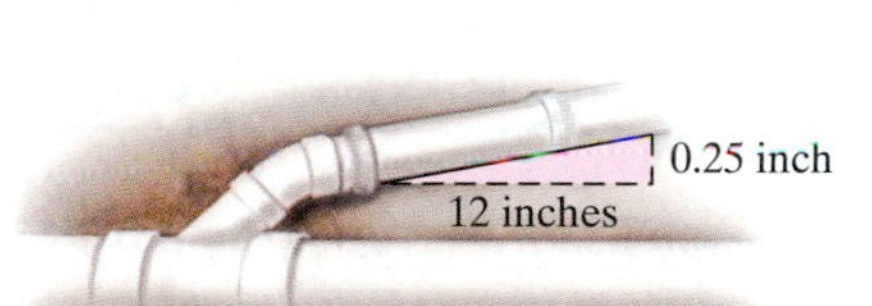

55. The steepest street is Baldwin Street in Dunedin, New Zealand. It has a maximum rise of 10 meters for a horizontal distance of 12.66 meters. Find the grade of this section of road. Round to the nearest whole percent. (*Source: The Guinness Book of Records*)

56. According to federal regulations, a wheelchair ramp should rise no more than 1 foot for a horizontal distance of 12 feet. Write the slope as a grade. Round to the nearest tenth of a percent.

Find the slope of each line and write the slope as a rate of change. Don't forget to attach the proper units. See Example 9.

57. This graph approximates the number of U.S. households that have personal computers y (in millions) for year x.

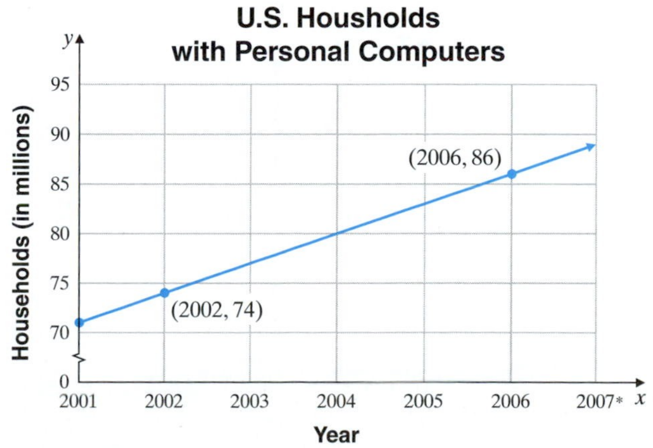

Source: Statistical Abstract of the United States, *projected numbers

58. This graph approximates the total number of cosmetic surgeons for year x.

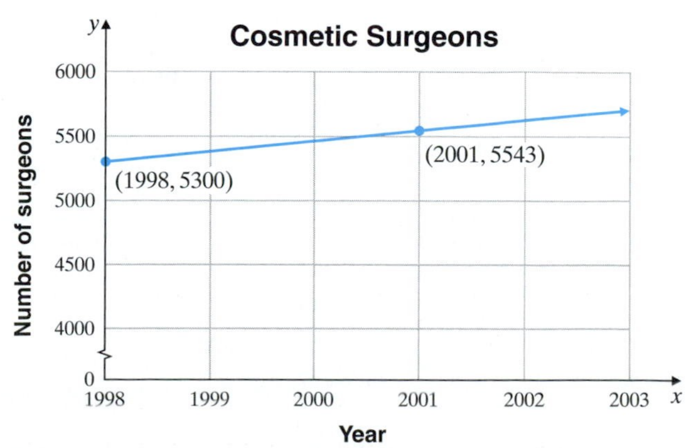

Source: CDC: National Center for Health Statistics

59. The graph below shows the total cost y (in dollars) of owning and operating a compact car where x is the number of miles driven.

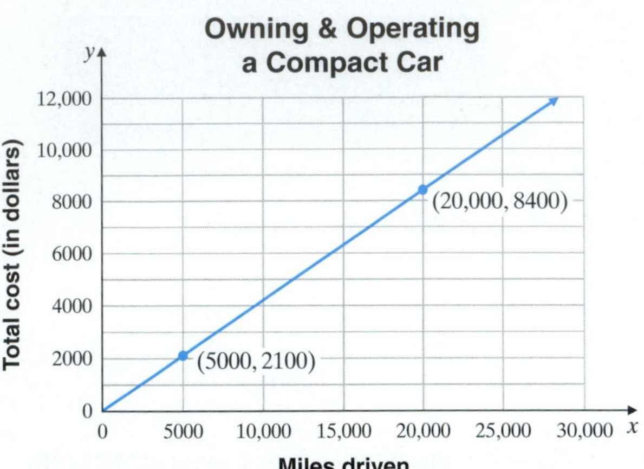

Source: Federal Highway Administration

60. The graph below shows the total cost y (in dollars) of owning and operating a standard pickup truck, where x is the number of miles driven.

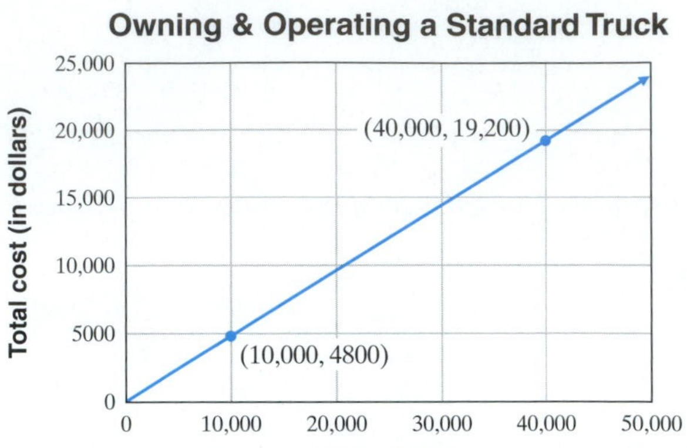

Source: Federal Highway Administration

Review

Solve each equation for y. See Section 9.5.

61. $y - (-6) = 2(x - 4)$

62. $y - 7 = -9(x - 6)$

63. $y - 1 = -6(x - (-2))$

64. $y - (-3) = 4(x - (-5))$

Concept Extensions

Match each line with its slope.

a. $m = 0$

b. undefined slope

c. $m = 3$

d. $m = 1$

e. $m = -\frac{1}{2}$

f. $m = -\frac{3}{4}$

65.

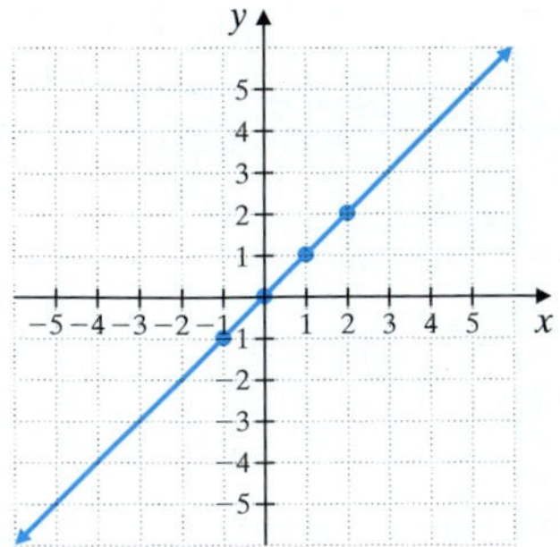

66.

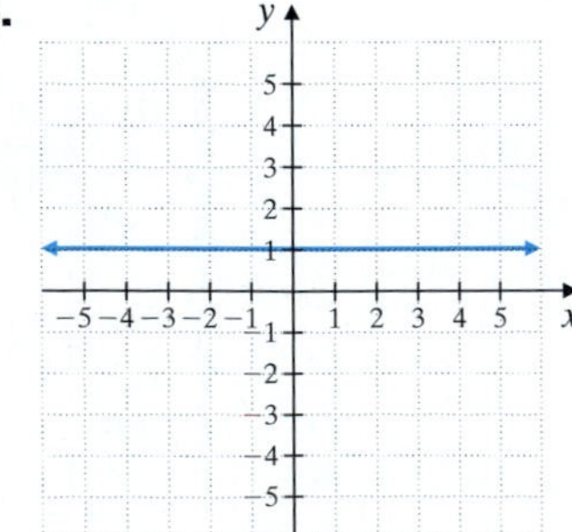

67.

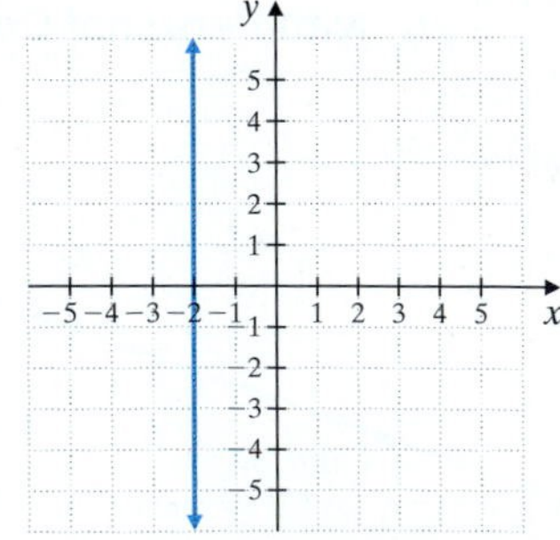

68. 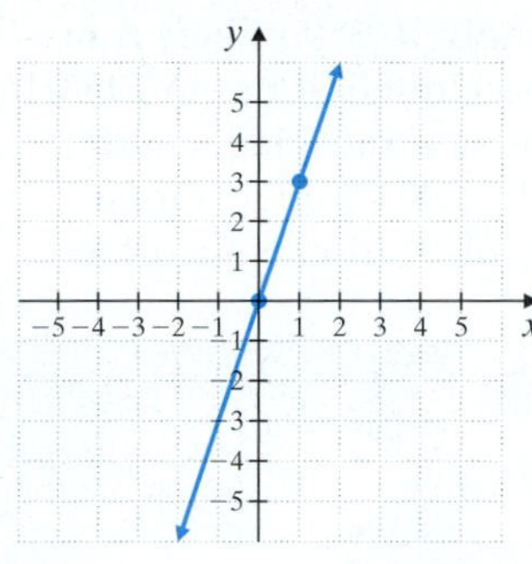

69. 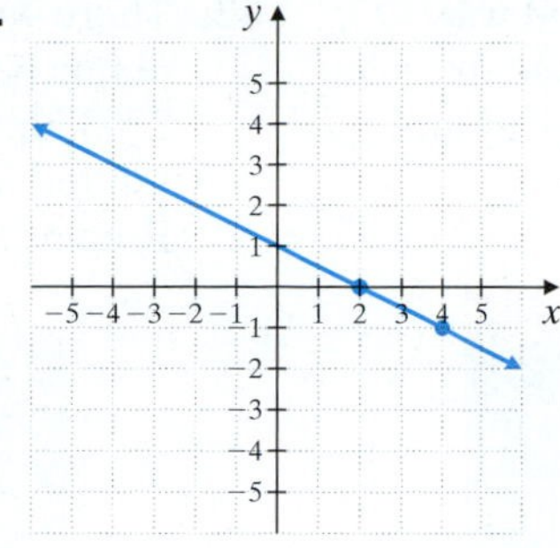

70. 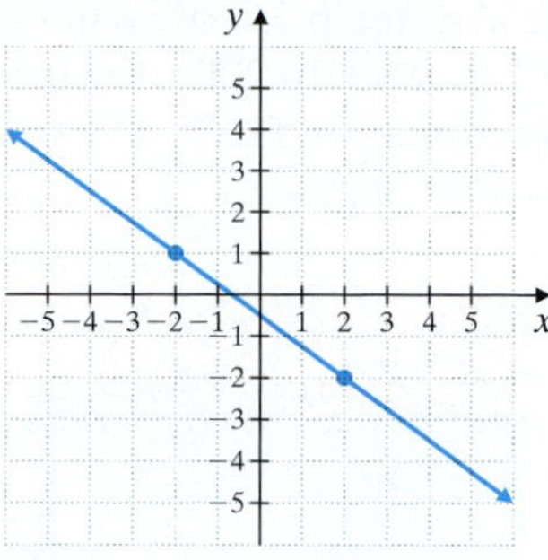

Solve. See a Concept Check in this section.

71. Verify that the points $(2, 1)$, $(0, 0)$, $(-2, -1)$ and $(-4, -2)$ are all on the same line by computing the slope between each pair of points. (See the first Concept Check.)

72. Given the points $(2, 3)$ and $(-5, 1)$, can the slope of the line through these points be calculated by $\frac{1-3}{2-(-5)}$? Why or why not? (See the second Concept Check.)

73. Write the equations of three lines parallel to $10x - 5y = -7$. (See the third Concept Check.)

74. Write the equations of two lines perpendicular to $10x - 5y = -7$. (See the third Concept Check.)

The following line graph shows the average fuel economy (in miles per gallon) by passenger automobiles produced during each of the model years shown. Use this graph to answer Exercises 75 through 79.

75. What was the average fuel economy (in miles per gallon) for automobiles produced during 2001?

76. Find the decrease in average fuel economy for automobiles between the years 1998 to 2000.

77. During which of the model years shown was average fuel economy the lowest?
What was the average fuel economy for that year?

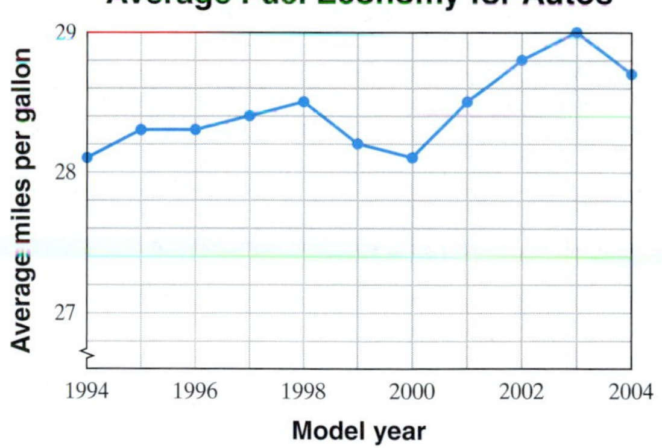

Source: U.S. Environmental Protection Agency, Office of Transportation and Air Quality

78. During which of the model years shown was average fuel economy the highest?
What was the average fuel economy for that year?

79. What line segment has the greatest slope?

80. Find x so that the pitch of the roof is $\frac{1}{3}$.

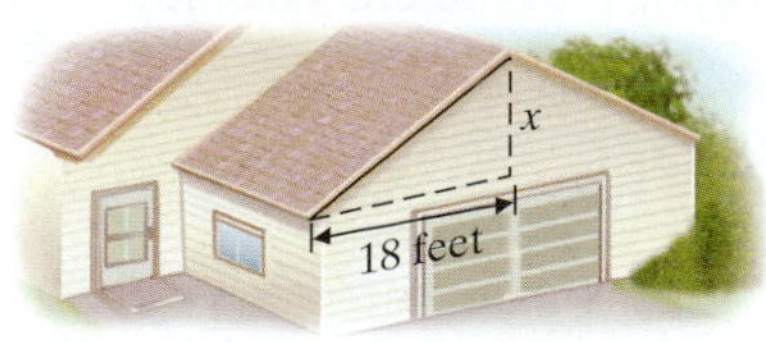

81. Find x so that the pitch of the roof is $\frac{2}{5}$.

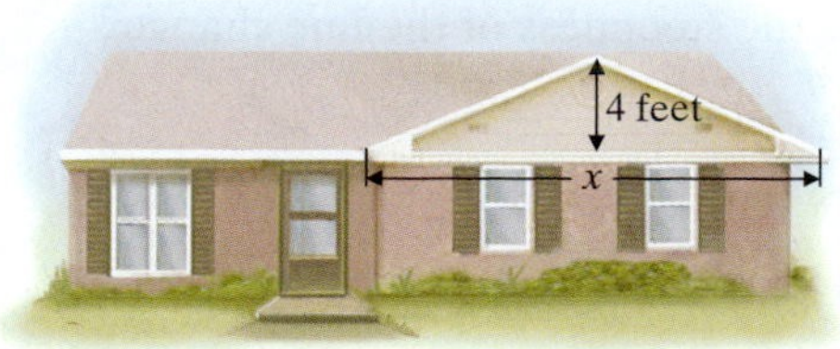

82. The average price of an acre of U.S. farmland was \$974 in 1998. In 2003, the price of an acre rose to approximately \$1275. (*Source:* National Agricultural Statistics Services)

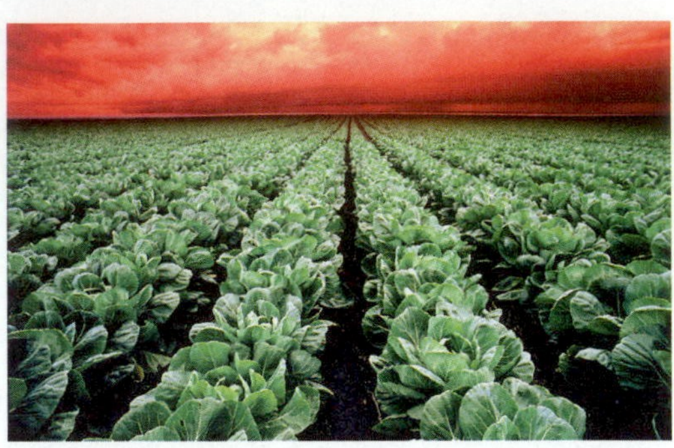

a. Write two ordered pairs of the form (year, price of acre)
b. Find the slope of the line through the two points.
c. Write a sentence explaining the meaning of the slope as a rate of change.

83. There were approximately 10,359 kidney transplants performed in the United States in 1993. In 2003, the number of kidney transplants in the United States rose to 15,138. (*Source:* Organ Procurement and Transplantation Network)

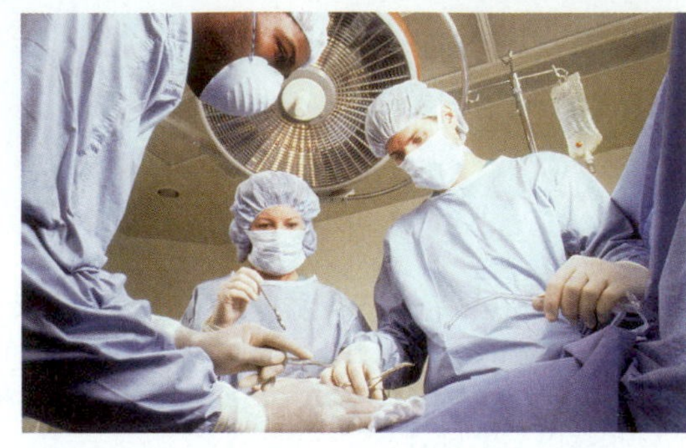

a. Write two ordered pairs of the form (year, number of kidney transplants).
b. Find the slope of the line between the two points.
c. Write a sentence explaining the meaning of the slope as a rate of change.

84. Show that a triangle with vertices at the points $(1, 1)$, $(-4, 4)$, and $(-3, 0)$ is a right triangle.

85. Show that the quadrilateral with vertices $(1, 3)$, $(2, 1)$, $(-4, 0)$, and $(-3, -2)$ is a parallelogram.

Find the slope of the line through the given points.

86. $(2.1, 6.7)$ and $(-8.3, 9.3)$

87. $(-3.8, 1.2)$ and $(-2.2, 4.5)$

88. $(2.3, 0.2)$ and $(7.9, 5.1)$

89. $(14.3, -10.1)$ and $(9.8, -2.9)$

90. The graph of $y = -\frac{1}{3}x + 2$ has a slope of $-\frac{1}{3}$. The graph of $y = -2x + 2$ has a slope of -2. The graph of $y = -4x + 2$ has a slope of -4. Graph all three equations on a single coordinate system. As the absolute value of the slope becomes larger, how does the steepness of the line change?

91. The graph of $y = \frac{1}{2}x$ has a slope of $\frac{1}{2}$. The graph of $y = 3x$ has a slope of 3. The graph of $y = 5x$ has a slope of 5. Graph all three equations on a single coordinate system. As slope becomes larger, how does the steepness of the line change?

13.5 EQUATIONS OF LINES

Objectives

A Use the Slope-Intercept Form to Write an Equation of a Line.

B Use the Slope-Intercept Form to Graph a Linear Equation.

C Use the Point-Slope Form to Find an Equation of a Line Given Its Slope and a Point of the Line.

D Use the Point-Slope Form to Find an Equation of a Line Given Two Points of the Line.

E Use the Point-Slope Form to Solve Problems.

We know that when a linear equation is solved for y, the coefficient of x is the slope of the line. For example, the slope of the line whose equation is $y = 3x + 1$ is 3. In this equation, $y = 3x + 1$, what does 1 represent? To find out, let $x = 0$ and watch what happens.

$y = 3x + 1$

$y = 3 \cdot 0 + 1$ Let $x = 0$.

$y = 1$

We now have the ordered pair $(0, 1)$, which means that 1 is the y-intercept.

This is true in general. To see this, let $x = 0$ and solve for y in $y = mx + b$.

$y = m \cdot 0 + b$ Let $x = 0$.

$y = b$

We obtain the ordered pair $(0, b)$, which means that point is the y-intercept.

The form $y = mx + b$ is appropriately called the *slope-intercept form* of a linear equation. (m: slope; b: y-intercept is $(0, b)$)

Slope-Intercept Form

When a linear equation in two variables is written in **slope-intercept form,**

$y = mx + b$ (m: slope; b: $(0, b)$, y-intercept)

then m is the slope of the line and $(0, b)$ is the y-intercept of the line.

Objective A Using the Slope-Intercept Form to Write an Equation

The slope-intercept form can be used to write the equation of a line when we know its slope and y-intercept.

EXAMPLE 1 Find an equation of the line with y-intercept $(0, -3)$ and slope of $\frac{1}{4}$.

Solution: We are given the slope and the y-intercept. We let $m = \frac{1}{4}$ and $b = -3$ and write the equation in slope-intercept form, $y = mx + b$.

$y = mx + b$

$y = \frac{1}{4}x + (-3)$ Let $m = \frac{1}{4}$ and $b = -3$.

$y = \frac{1}{4}x - 3$ Simplify.

Work Practice Problem 1

PRACTICE PROBLEM 1

Find an equation of the line with y-intercept $(0, -2)$ and slope of $\frac{3}{5}$.

Objective B Using the Slope-Intercept Form to Graph an Equation

We also can use the slope-intercept form of the equation of a line to graph a linear equation.

Answer

1. $y = \frac{3}{5}x - 2$

PRACTICE PROBLEM 2

Graph the equation $y = \frac{2}{3}x - 4$.

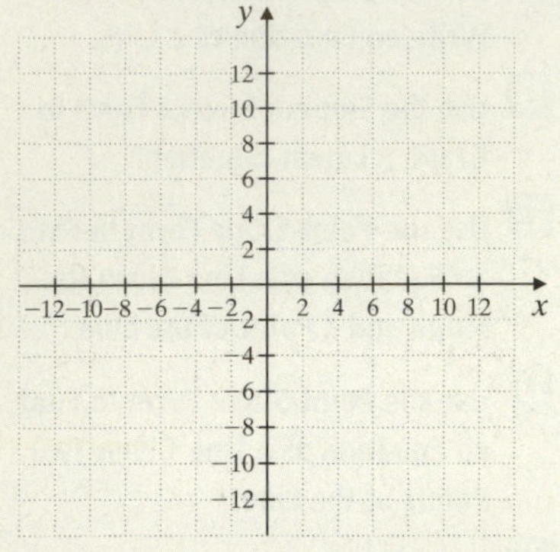

PRACTICE PROBLEM 3

Use the slope-intercept form to graph $3x + y = 2$.

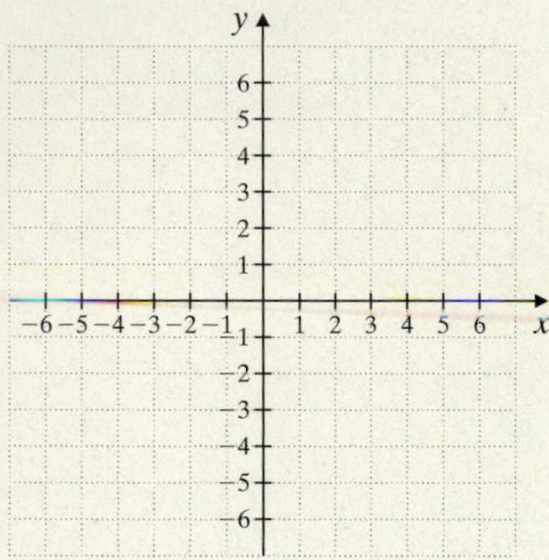

Answers

2.

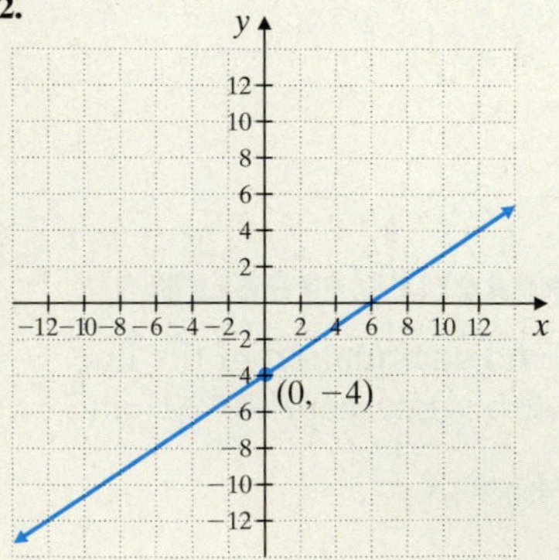

3.

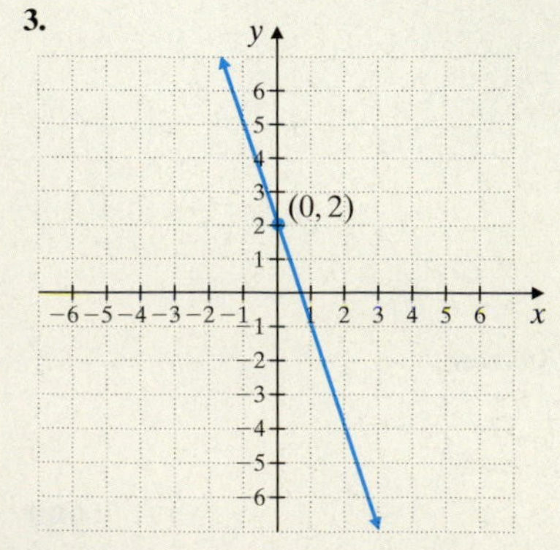

EXAMPLE 2 Use the slope-intercept form to graph the equation

$$y = \frac{3}{5}x - 2$$

Solution: Since the equation $y = \frac{3}{5}x - 2$ is written in slope-intercept form $y = mx + b$, the slope of its graph is $\frac{3}{5}$ and the y-intercept is $(0, -2)$. To graph this equation, we begin by plotting the point $(0, -2)$.

From this point, we can find another point of the graph by using the slope $\frac{3}{5}$ and recalling that slope is $\frac{\text{rise}}{\text{run}}$. We start at the y-intercept and move 3 units up since the numerator of the slope is 3; then move 5 units to the right since the denominator of the slope is 5. We stop at the point $(5, 1)$. The line through $(0, -2)$ and $(5, 1)$ is the graph of $y = \frac{3}{5}x - 2$.

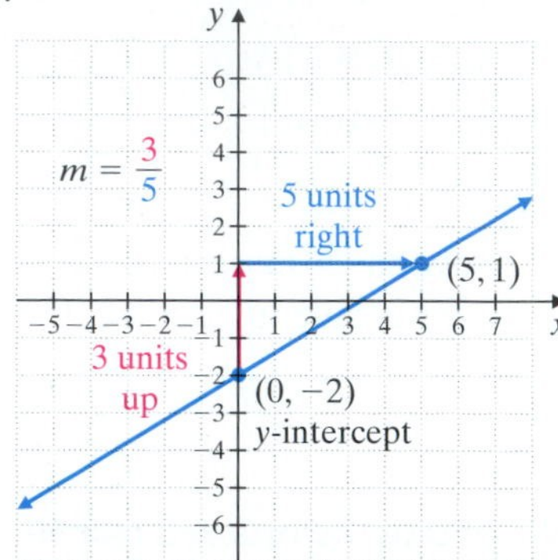

Work Practice Problem 2

EXAMPLE 3 Use the slope-intercept form to graph the equation $4x + y = 1$.

Solution: First we write the given equation in slope-intercept form.

$$4x + y = 1$$
$$y = -4x + 1$$

The graph of this equation will have slope -4 and y-intercept $(0, 1)$. To graph this line, we first plot the point $(0, 1)$. To find another point of the graph, we use the slope -4, which can be written as $\frac{-4}{1}$ $\left(\frac{4}{-1}\text{ could also be used}\right)$. We start at the point $(0, 1)$ and move 4 units down (since the numerator of the slope is -4), and then 1 unit to the right (since the denominator of the slope is 1).

We arrive at the point $(1, -3)$. The line through $(0, 1)$ and $(1, -3)$ is the graph of $4x + y = 1$.

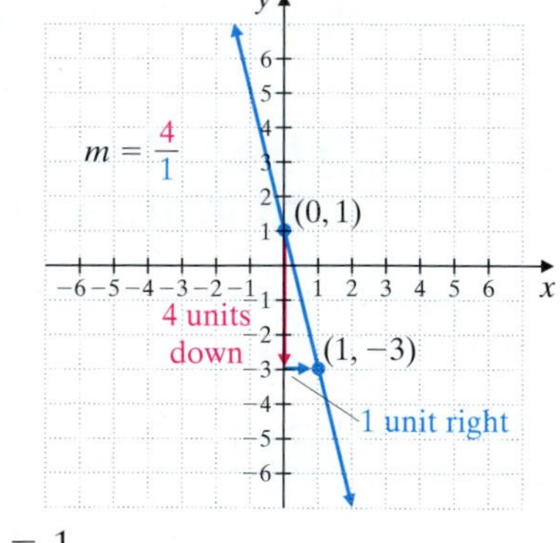

Work Practice Problem 3

Helpful Hint

In Example 3, if we interpret the slope of -4 as $\frac{4}{-1}$, we arrive at $(-1, 5)$ for a second point. Notice that this point is also on the line.

Objective C Writing an Equation Given Its Slope and a Point

Thus far, we have seen that we can write an equation of a line if we know its slope and y-intercept. We can also write an equation of a line if we know its slope and any point on the line. To see how we do this, let m represent slope and (x_1, y_1) represent the point on the line. Then if (x, y) is any other point of the line, we have that

$$\frac{y - y_1}{x - x_1} = m$$

$$y - y_1 = m(x - x_1) \quad \text{Multiply both sides by } (x - x_1).$$

↑ slope

This is the *point-slope form* of the equation of a line.

Point-Slope Form of the Equation of a Line

The **point-slope form** of the equation of a line is $y - y_1 = m(x - x_1)$, where m is the slope of the line and (x_1, y_1) is a point on the line.

EXAMPLE 4 Find an equation of the line with slope -2 that passes through $(-1, 5)$. Write the equation in slope-intercept form, $y = mx + b$, and in standard form, $Ax + By = C$.

Solution: Since the slope and a point on the line are given, we use point-slope form $y - y_1 = m(x - x_1)$ to write the equation. Let $m = -2$ and $(-1, 5) = (x_1, y_1)$.

$$y - y_1 = m(x - x_1)$$
$$y - 5 = -2[x - (-1)] \quad \text{Let } m = -2 \text{ and } (x_1, y_1) = (-1, 5).$$
$$y - 5 = -2(x + 1) \quad \text{Simplify.}$$
$$y - 5 = -2x - 2 \quad \text{Use the distributive property.}$$

To write the equation in slope-intercept form, $y = mx + b$, we simply solve the equation for y. To do this, we add 5 to both sides.

$$y - 5 = -2x - 2$$
$$y = -2x + 3 \quad \text{Slope-intercept form.}$$
$$2x + y = 3 \quad \text{Add } 2x \text{ to both sides and we have standard form.}$$

Work Practice Problem 4

PRACTICE PROBLEM 4

Find an equation of the line with slope -3 that passes through $(2, -4)$. Write the equation in slope-intercept form $y = mx + b$.

Objective D Writing an Equation Given Two Points

We can also find the equation of a line when we are given any two points of the line.

EXAMPLE 5 Find an equation of the line through $(2, 5)$ and $(-3, 4)$. Write the equation in the form $Ax + By = C$.

Solution: First, use the two given points to find the slope of the line.

$$m = \frac{4 - 5}{-3 - 2} = \frac{-1}{-5} = \frac{1}{5}$$

Next we use the slope $\frac{1}{5}$ and either one of the given points to write the equation in point-slope form. We use $(2, 5)$. Let $x_1 = 2$, $y_1 = 5$, and $m = \frac{1}{5}$.

$$y - y_1 = m(x - x_1) \quad \text{Use point-slope form.}$$
$$y - 5 = \frac{1}{5}(x - 2) \quad \text{Let } x_1 = 2,\ y_1 = 5, \text{ and } m = \frac{1}{5}.$$
$$5(y - 5) = 5 \cdot \frac{1}{5}(x - 2) \quad \text{Multiply both sides by 5 to clear fractions.}$$
$$5y - 25 = x - 2 \quad \text{Use the distributive property and simplify.}$$
$$-x + 5y - 25 = -2 \quad \text{Subtract } x \text{ from both sides.}$$
$$-x + 5y = 23 \quad \text{Add 25 to both sides.}$$

Work Practice Problem 5

PRACTICE PROBLEM 5

Find an equation of the line through $(1, 3)$ and $(5, -2)$. Write the equation in the form $Ax + By = C$.

Answers

4. $y = -3x + 2$ **5.** $5x + 4y = 17$

Helpful Hint

When you multiply both sides of the equation from Example 5, $-x + 5y = 23$ by -1, it becomes $x - 5y = -23$.

Both $-x + 5y = 23$ and $x - 5y = -23$ are in the form $Ax + By = C$ and both are equations of the same line.

Objective E Using the Point-Slope Form to Solve Problems

Problems occurring in many fields can be modeled by linear equations in two variables. The next example is from the field of marketing and shows how consumer demand of a product depends on the price of the product.

PRACTICE PROBLEM 6

The Pool Entertainment Company learned that by pricing a new pool toy at \$10, local sales will reach 200 a week. Lowering the price to \$9 will cause sales to rise to 250 a week.

a. Assume that the relationship between sales price and number of toys sold is linear, and write an equation describing this relationship. Write the equation in slope-intercept form. Use ordered pairs of the form (sales price, number sold).

b. Predict the weekly sales of the toy if the price is \$7.50.

EXAMPLE 6 The Whammo Company has learned that by pricing a newly released Frisbee at \$6, sales will reach 2000 Frisbees per day. Raising the price to \$8 will cause the sales to fall to 1500 Frisbees per day.

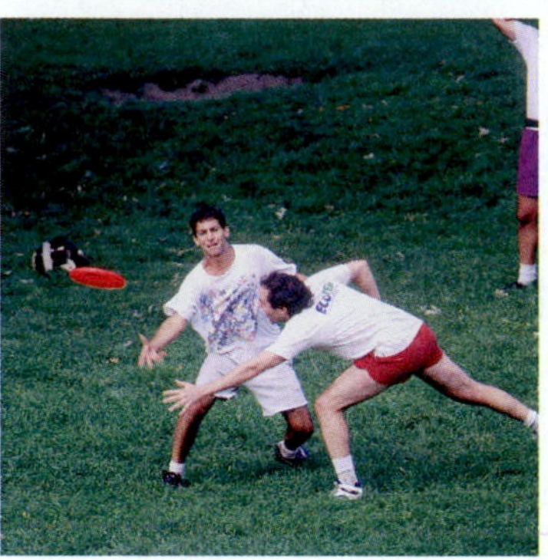

a. Assume that the relationship between sales price and number of Frisbees sold is linear and write an equation describing this relationship. Write the equation in slope-intercept form. Use ordered pairs of the form (sales price, number sold).

b. Predict the daily sales of Frisbees if the price is \$7.50.

Solution:

a. We use the given information and write two ordered pairs. Our ordered pairs are (6, 2000) and (8, 1500). To use the point-slope form to write an equation, we find the slope of the line that contains these points.

$$m = \frac{2000 - 1500}{6 - 8} = \frac{500}{-2} = -250$$

Next we use the slope and either one of the points to write the equation in point-slope form. We use (6, 2000).

$y - y_1 = m(x - x_1)$ Use point-slope form.

$y - 2000 = -250(x - 6)$ Let $x_1 = 6$, $y_1 = 2000$, and $m = -250$.

$y - 2000 = -250x + 1500$ Use the distributive property.

$y = -250x + 3500$ Write in slope-intercept form.

b. To predict the sales if the price is \$7.50, we find y when $x = 7.50$.

$y = -250x + 3500$

$y = -250(7.50) + 3500$ Let $x = 7.50$.

$y = -1875 + 3500$

$y = 1625$

If the price is \$7.50, sales will reach 1625 Frisbees per day.

Work Practice Problem 6

Answers

6. a. $y = -50x + 700$ **b.** 325

We could have solved Example 6 by using ordered pairs of the form (number sold, sales price).

Here is a summary of our discussion on linear equations thus far.

Forms of Linear Equations

$Ax + By = C$	**Standard form** of a linear equation. A and B are not both 0.
$y = mx + b$	**Slope-intercept form** of a linear equation. The slope is m and the y-intercept is $(0, b)$.
$y - y_1 = m(x - x_1)$	**Point-slope form** of a linear equation. The slope is m and (x_1, y_1) is a point on the line.
$y = c$	**Horizontal line** The slope is 0 and the y-intercept is $(0, c)$.
$x = c$	**Vertical line** The slope is undefined and the x-intercept is $(c, 0)$.

Parallel and Perpendicular Lines

Nonvertical parallel lines have the same slope.
The product of the slopes of two nonvertical perpendicular lines is -1.

CALCULATOR EXPLORATIONS Graphing

A graphing calculator is a very useful tool for discovering patterns. To discover the change in the graph of a linear equation caused by a change in slope, try the following. Use a standard window and graph a linear equation in the form $y = mx + b$. Recall that the graph of such an equation will have slope m and y-intercept $(0, b)$.

First graph $y = x + 3$. To do so, press the Y= key and enter $Y_1 = x + 3$. Notice that this graph has slope 1 and that the y-intercept is 3. Next, on the same set of axes, graph $y = 2x + 3$ and $y = 3x + 3$ by pressing Y= and entering $Y_2 = 2x + 3$ and $Y_3 = 3x + 3$.

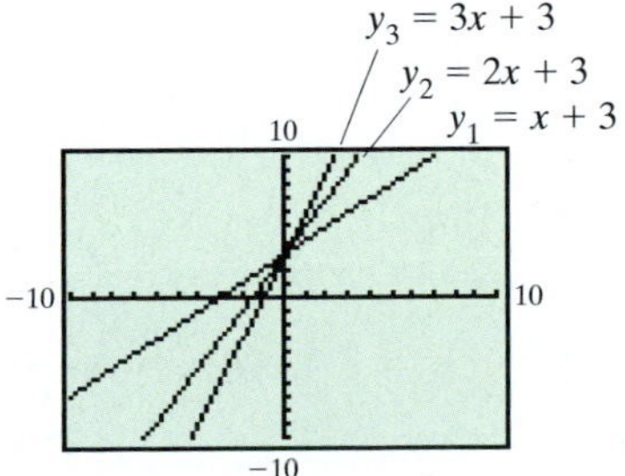

Notice the difference in the graph of each equation as the slope changes from 1 to 2 to 3. How would the graph of $y = 5x + 3$ appear? To see the change in the graph caused by a change in negative slope, try graphing $y = -x + 3$, $y = -2x + 3$, and $y = -3x + 3$ on the same set of axes.

Use a graphing calculator to graph the following equations. For each exercise, graph the first equation and use its graph to predict the appearance of the other equations. Then graph the other equations on the same set of axes and check your prediction.

1. $y = x$; $y = 6x$, $y = -6x$

2. $y = -x$; $y = -5x$, $y = -10x$

3. $y = \frac{1}{2}x + 2$; $y = \frac{3}{4}x + 2$, $y = x + 2$

4. $y = x + 1$; $y = \frac{5}{4}x + 1$, $y = \frac{5}{2}x + 1$

Vocabulary and Readiness Check

Use the choices below to fill in each blank. Some choices may be used more than once and some not at all.

b	(y_1, x_1)	point-slope	vertical	standard
m	(x_1, y_1)	slope-intercept	horizontal	

1. The form $y = mx + b$ is called _______________ form. When a linear equation in two variables is written in this form, _____ is the slope of its graph and $(0, \_\_\_)$ is its y-intercept.

2. The form $y - y_1 = m(x - x_1)$ is called _______________ form. When a linear equation in two variables is written in this form, _____ is the slope of its graph and _________ is a point on the graph.

For Exercises 3, 4, and 7, identify the form that the linear equation in two variables is written in. For Exercises 5 and 6, identify the appearance of the graph of the equation.

3. $y - 7 = 4(x + 3)$; _______________ form

4. $5x - 9y = 11$; _______________ form

5. $y = \frac{1}{2}$; _______________ line

6. $x = -17$; _______________ line

7. $y = \frac{3}{4}x - \frac{1}{3}$; _______________ form

13.5 EXERCISE SET

FOR EXTRA HELP

Student Solutions Manual | PH Math/Tutor Center | CD/Video for Review | MathXL® | MyMathLab

Objective A *Write an equation of the line with each given slope, m, and y-intercept, (0, b). See Example 1.*

1. $m = 5, b = 3$

2. $m = -3, b = -3$

3. $m = -4, b = -\frac{1}{6}$

4. $m = 2, b = \frac{3}{4}$

5. $m = \frac{2}{3}, b = 0$

6. $m = -\frac{4}{5}, b = 0$

7. $m = 0, b = -8$

8. $m = 0, b = -2$

9. $m = -\frac{1}{5}, b = \frac{1}{9}$

10. $m = \frac{1}{2}, b = -\frac{1}{3}$

Objective B *Use the slope-intercept form to graph each equation. See Examples 2 and 3.*

11. $y = 2x + 1$

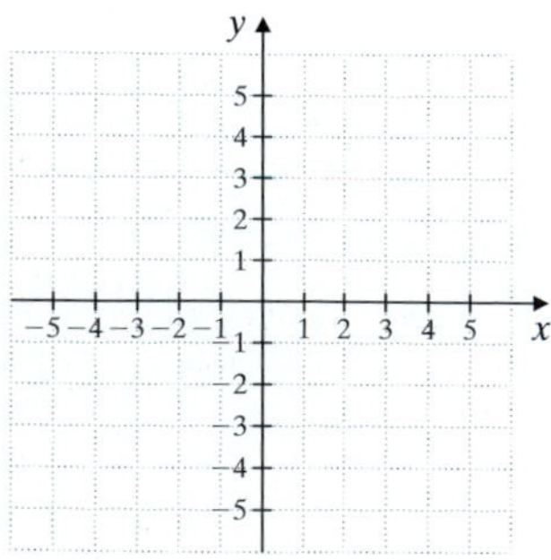

12. $y = -4x - 1$

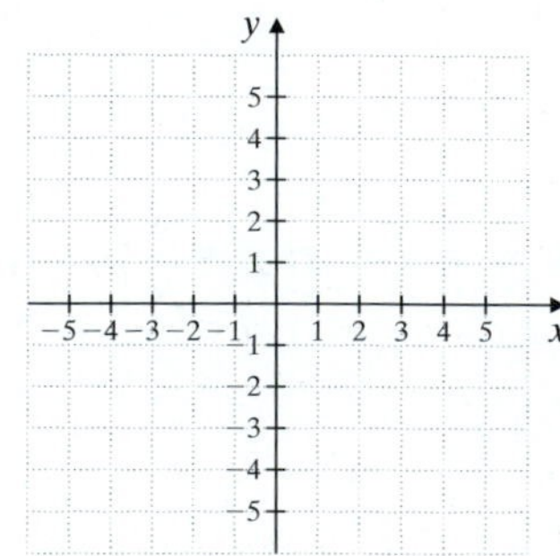

13. $y = \frac{2}{3}x + 5$

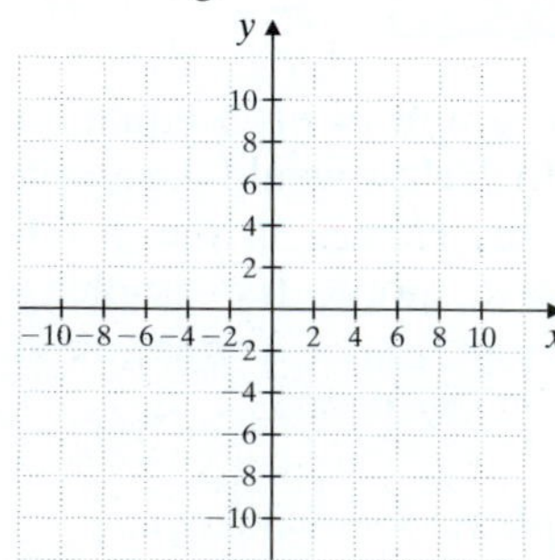

14. $y = \frac{1}{4}x - 3$

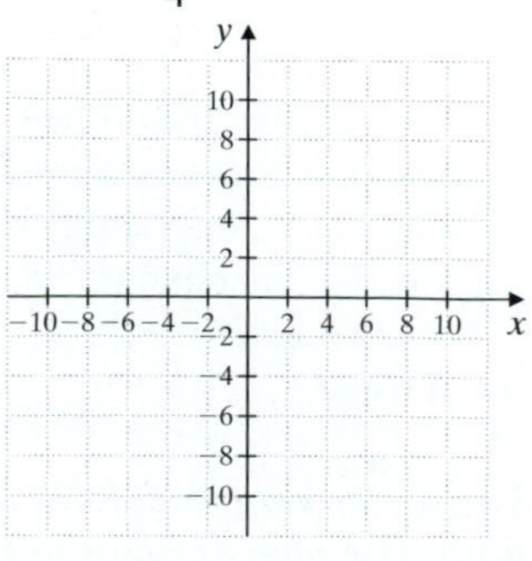

15. $y = -5x$

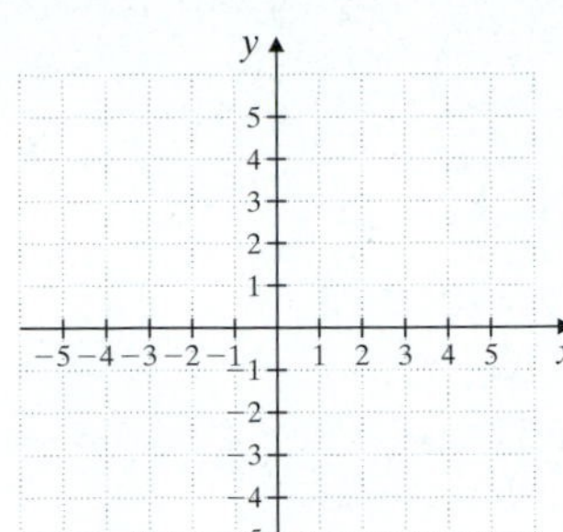

16. $y = -6x$

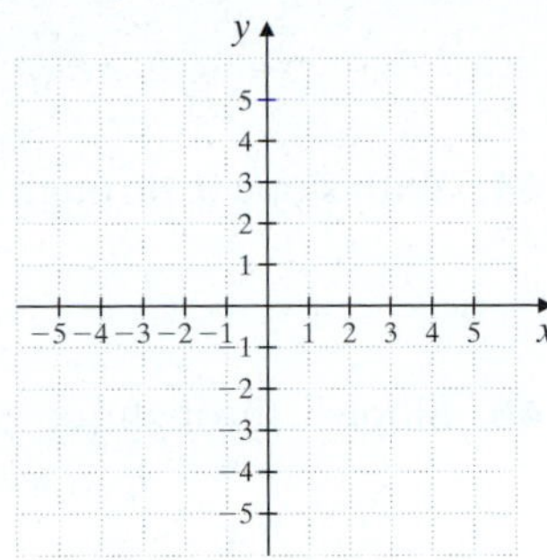

17. $4x + y = 6$

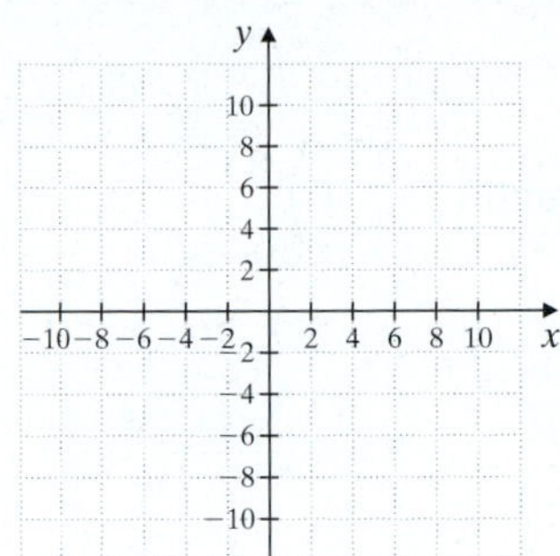

18. $-3x + y = 2$

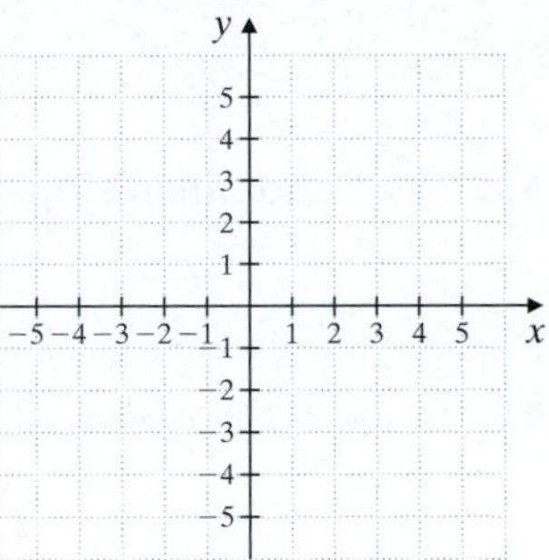

19. $4x - 7y = -14$

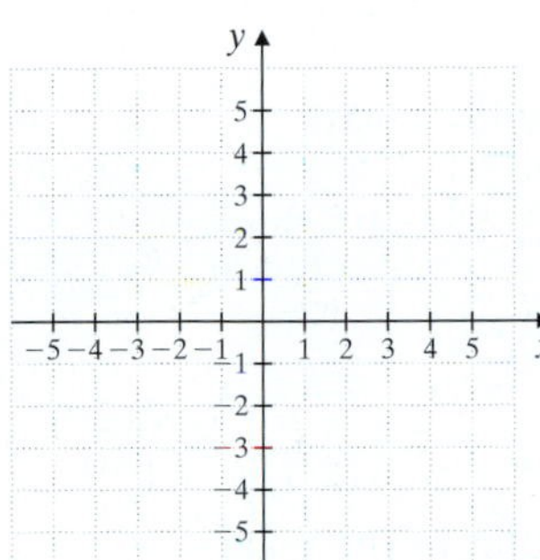

20. $3x - 4y = 4$

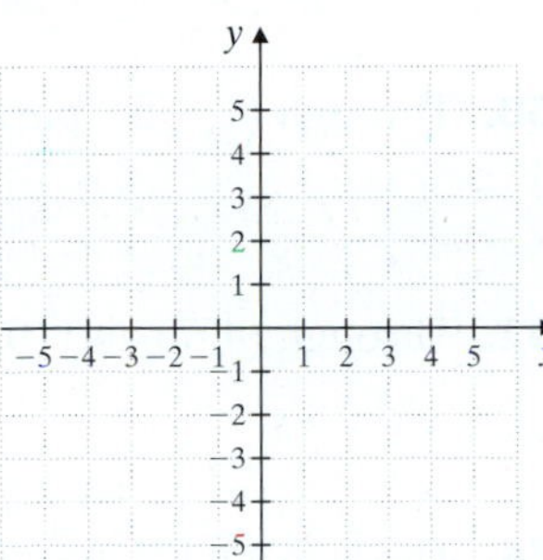

21. $x = \frac{5}{4}y$

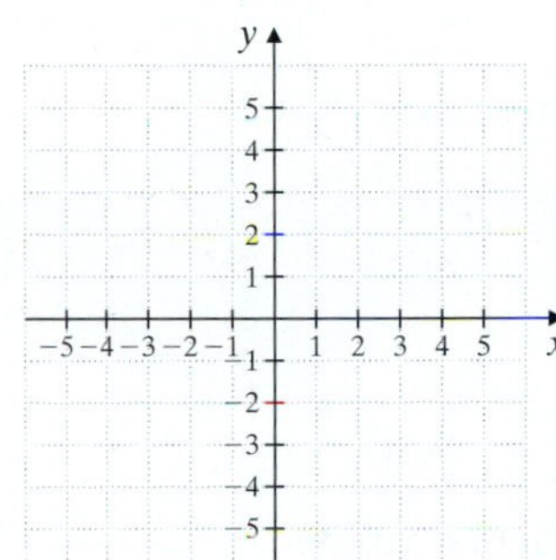

22. $x = \frac{3}{2}y$

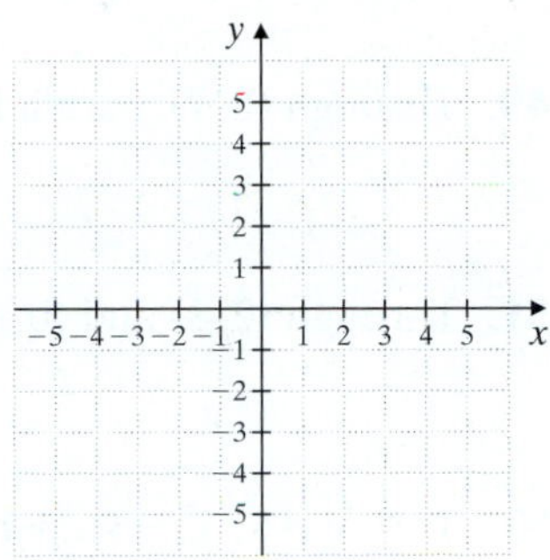

Objective **C** *Find an equation of each line with the given slope that passes through the given point. Write the equation in the form $Ax + By = C$. See Example 4.*

23. $m = 6;\ (2, 2)$

24. $m = 4;\ (1, 3)$

25. $m = -8;\ (-1, -5)$

26. $m = -2;\ (-11, -12)$

27. $m = \frac{3}{2};\ (5, -6)$

28. $m = \frac{2}{3};\ (-8, 9)$

29. $m = -\frac{1}{2};\ (-3, 0)$

30. $m = -\frac{1}{5};\ (4, 0)$

Objective **D** *Find an equation of the line passing through each pair of points. Write the equation in the form $Ax + By = C$. See Example 5.*

31. $(3, 2)$ and $(5, 6)$

32. $(6, 2)$ and $(8, 8)$

33. $(-1, 3)$ and $(-2, -5)$

34. $(-4, 0)$ and $(6, -1)$

35. $(2, 3)$ and $(-1, -1)$

36. $(7, 10)$ and $(-1, -1)$

37. $(0, 0)$ and $\left(-\frac{1}{8}, \frac{1}{13}\right)$

38. $(0, 0)$ and $\left(-\frac{1}{2}, \frac{1}{3}\right)$

Objectives **A** **C** **D** **Mixed Practice** *See Examples 1, 4, and 5. Find an equation of each line described. Write each equation in slope-intercept form when possible.*

39. With slope $-\frac{1}{2}$, through $\left(0, \frac{5}{3}\right)$

40. With slope $\frac{5}{7}$, through $(0, -3)$

41. Through $(10, 7)$ and $(7, 10)$

42. Through $(5, -6)$ and $(-6, 5)$

43. With undefined slope, through $\left(-\frac{3}{4}, 1\right)$

44. With slope 0, through $(6.7, 12.1)$

45. Slope 1, through $(-7, 9)$

46. Slope 5, through $(6, -8)$

47. Slope -5, y-intercept $(0, 7)$

48. Slope -2; y-intercept $(0, -4)$

49. Through $(6, 7)$, parallel to the x-axis

50. Through $(1, -5)$, parallel to the y-axis

51. Through $(2, 3)$ and $(0, 0)$

52. Through $(4, 7)$ and $(0, 0)$

53. Through $(-2, -3)$, perpendicular to the y-axis

54. Through $(0, 12)$, perpendicular to the x-axis

55. Slope $-\frac{4}{7}$, through $(-1, -2)$

56. Slope $-\frac{3}{5}$, through $(4, 4)$

Objective E *Solve. Assume each exercise describes a linear relationship. Write the equations in slope-intercept form. See Example 6.*

57. A rock is dropped from the top of a 400-foot cliff. After 1 second, the rock is traveling 32 feet per second. After 3 seconds, the rock is traveling 96 feet per second.

a. Assume that the relationship between time and speed is linear and write an equation describing this relationship. Use ordered pairs of the form (time, speed).

b. Use this equation to determine the speed of the rock 4 seconds after it was dropped.

58. A Hawaiian fruit company is studying the sales of a pineapple sauce to see if this product is to be continued. At the end of its first year, profits on this product amounted to \$30,000. At the end of the fourth year, profits were \$66,000.

a. Assume that the relationship between years on the market and profit is linear and write an equation describing this relationship. Use ordered pairs of the form (years on the market, profit).

b. Use this equation to predict the profit at the end of 7 years.

59. In 2003 there were approximately 54,000 gas-electric hybrid vehicles sold in the United States. In 2001, there were only 22,000 such vehicles sold. (*Source:* Energy Information Administration, Department of Energy)

a. Write an equation describing the relationship between time and the number of gas-electric hybrid vehicles sold. Use ordered pairs of the form (years past 2001, number of vehicles sold).

b. Use this equation to predict the number of gas-electric hybrid sales in 2006.

60. In 2004, there were approximately 875 thousand restaurants in the United States. In 1972, there were 491 thousand restaurants. (*Source:* National Restaurant Association)

a. Write an equation describing the relationship between time and the number of restaurants. Use ordered pairs of the form (years past 1972, numbers of restaurants in thousands).

b. Use this equation to predict the number of eating establishments in 2012.

61. In 2003 there were approximately 5700 cinema sites in the United States. In 1999 there were 7032 cinema sites. (*Source:* National Association of Theater Owners)

a. Write an equation describing this relationship. Use ordered pairs of the form (years past 1999, number of cinema sites).

b. Use this equation to predict the number of cinema sites in 2007.

62. In 2000, the U.S. population per square mile of land area was 79.6. In 1990, this person per square mile population was 70.3.

a. Write an equation describing the relationship between year and person per square mile. Use ordered pairs of the form (years past 1990, person per square mile).

b. Use this equation to predict the person per square mile population in 2007.

63. In 1997 there were 1509 daily newspapers in the United States. By 2003, there were only 1456 daily newspapers. (*Source:* Statistical Abstract of the United States)

a. Write two ordered pairs of the form (years after 1997, number of daily newspapers) for this situation.

b. The relationship between years after 1997 and numbers of daily newspapers is linear over this period. Use the ordered pairs from part (a) to write an equation for the line relating year after 1997 to numbers of daily newspapers. (Round the slope to one decimal place.)

c. Use the linear equation in part (b) to estimate numbers of daily newspapers in 1999. (Round to the nearest whole.)

64. In 1999, crude oil production by the European Union countries was 3803 thousand barrels per day. In 2002, European Union oil production had decreased to 3482 thousand barrels per day. (*Source:* Energy Information Administration)

a. Write two ordered pairs of the form (years after 1999, crude oil production).

b. Assume that the relationship between years after 1999 and crude oil production is linear over this period. Use the ordered pairs from part (a) to write an equation of the line relating year to crude oil production.

c. Use the linear equation from part (b) to estimate the crude oil production by European Union countries in 2005, if this trend were to continue.

65. The Pool Fun Company has learned that, by pricing a newly released Fun Noodle at \$3, sales will reach 10,000 Fun Noodles per day during the summer. Raising the price to \$5 will cause the sales to fall to 8000 Fun Noodles per day.

a. Assume that the relationship between sales price and number of Fun Noodles sold is linear and write an equation describing this relationship. Use ordered pairs of the form (sales price, number sold).

b. Predict the daily sales of Fun Noodles if the price is \$3.50.

66. The value of a building bought in 1990 may be depreciated (or decreased) as time passes for income tax purposes. Seven years after the building was bought, this value was \$225,000 and 12 years after it was bought, this value was \$195,000.

a. If the relationship between number of years past 1990 and the depreciated value of the building is linear, write an equation describing this relationship. Use ordered pairs of the form (years past 1990, value of building).

b. Use this equation to estimate the depreciated value of the building in 2008.

Review

Find the value of $x^2 - 3x + 1$ *for each given value of x. See Sections 1.7, 1.8, and Chapter 2.*

67. 2 **68.** 5 **69.** -1 **70.** -3

Concept Extensions

Match each linear equation with its graph.

71. $y = 2x + 1$ **72.** $y = -x + 1$ **73.** $y = -3x - 2$ **74.** $y = \frac{5}{3}x - 2$

a.

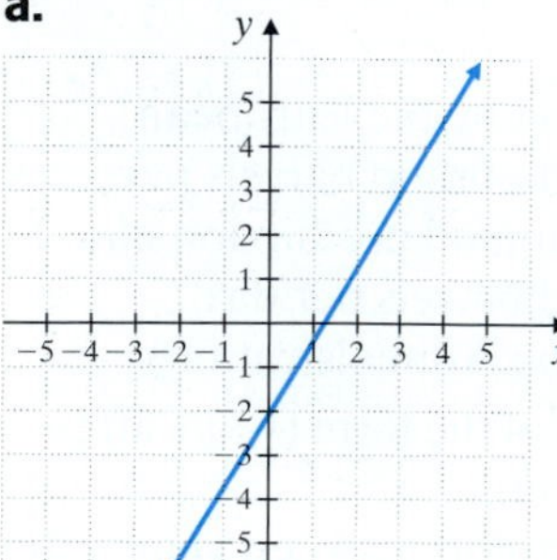

b.

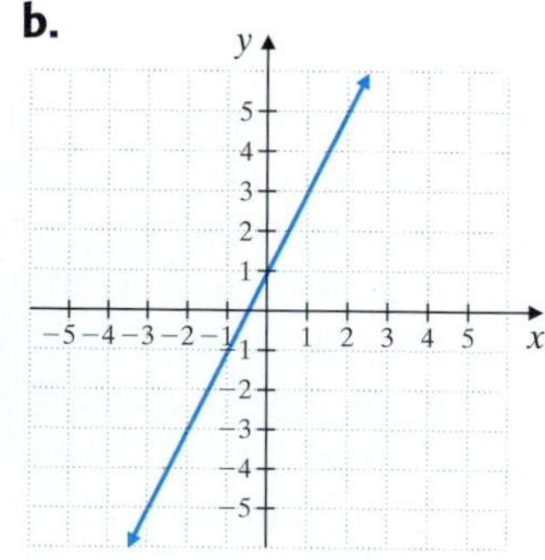

c.

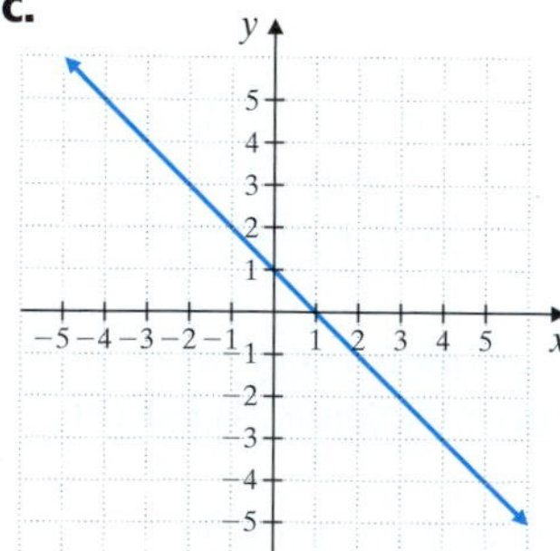

d.

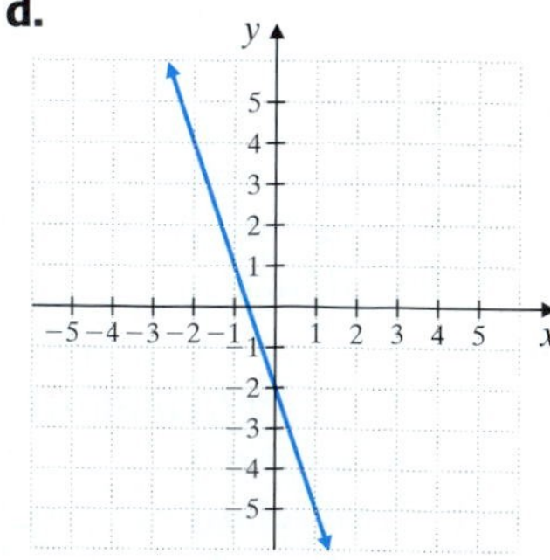

75. Write an equation of the line that contains the point $(-1, 2)$ and has the same slope as the line $y = 3x - 1$.

76. Write an equation of the line that contains the point $(4, 0)$ and has the same slope as the line $y = -2x + 3$.

△ **77.** Write an equation in standard form of the line that contains the point $(-1, 2)$ and is

a. parallel to the line $y = 3x - 1$.

b. perpendicular to the line $y = 3x - 1$.

△ **78.** Write an equation in standard form of the line that contains the point $(4, 0)$ and is

a. parallel to the line $y = -2x + 3$.

b. perpendicular to the line $y = -2x + 3$.

INTEGRATED REVIEW

Sections 13.1–13.5

Summary on Linear Equations

Find the slope of each line.

1.

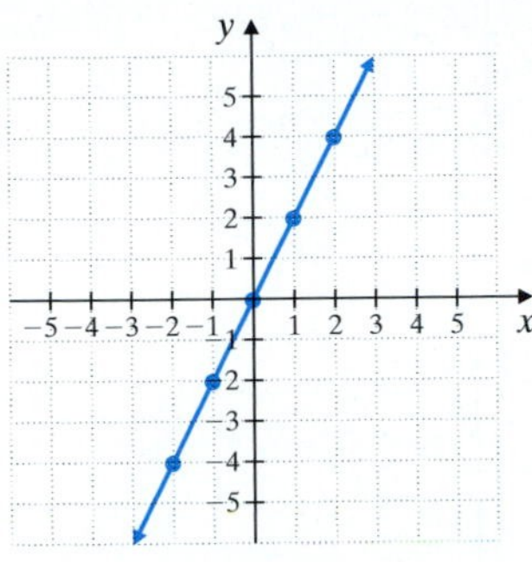

2.

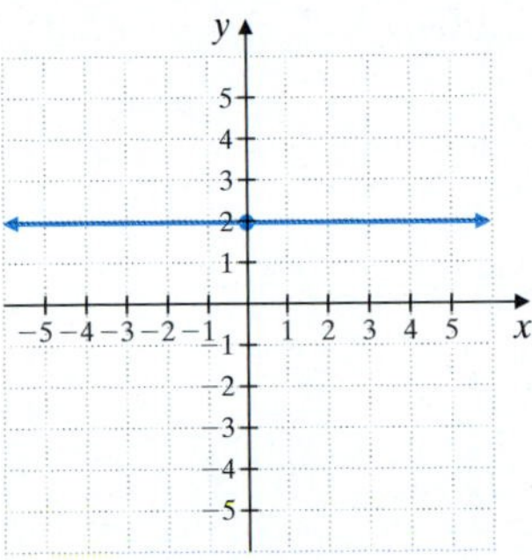

3.

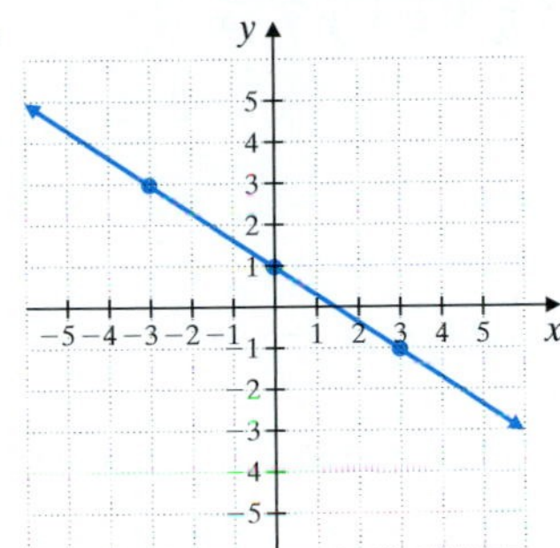

4.

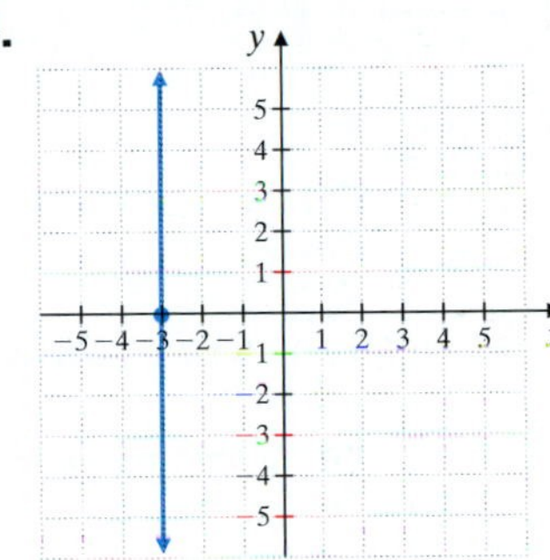

Graph each linear equation. For Exercises 11 and 12, label the intercepts.

5. $y = -2x$

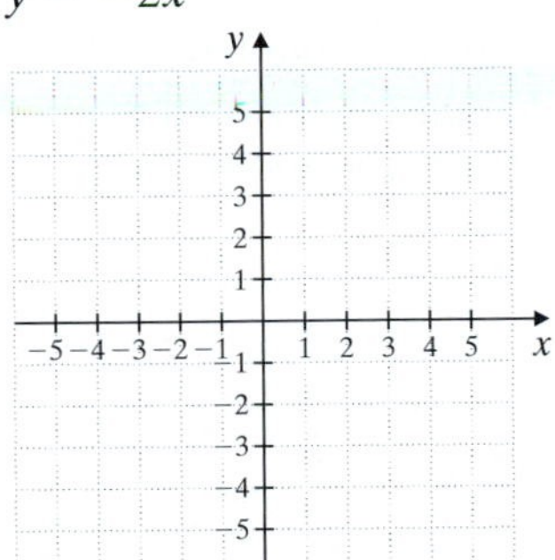

6. $x + y = 3$

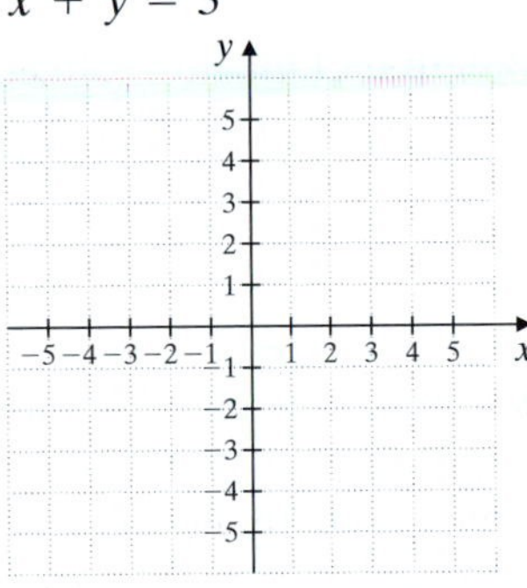

7. $x = -1$

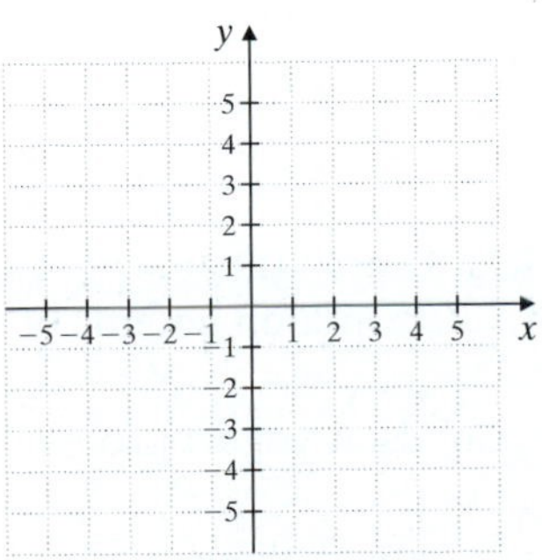

8. $y = 4$

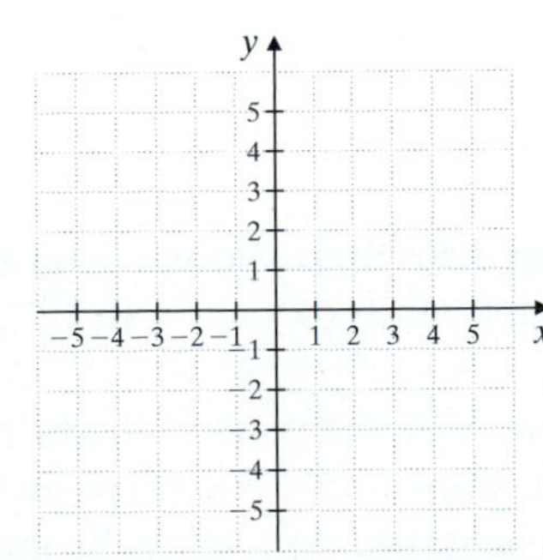

Answers

1. ______

2. ______

3. ______

4. ______

5. see graph

6. see graph

7. see graph

8. see graph

9. see graph ______
10. see graph ______
11. see graph ______
12. see graph ______
13. ______
14. ______
15. ______
16. ______
17. ______
18. ______
19. ______
20. ______
21. ______
22. ______
23. ______
24. a. ______
b. ______
c. ______

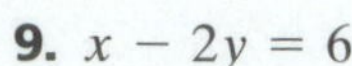

9. $x - 2y = 6$

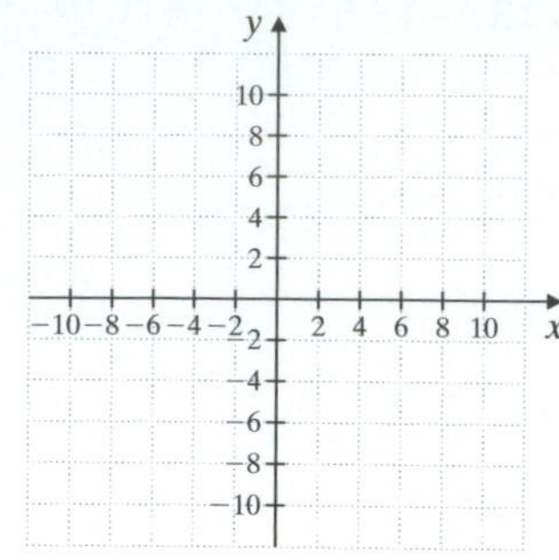

10. $y = 3x + 2$

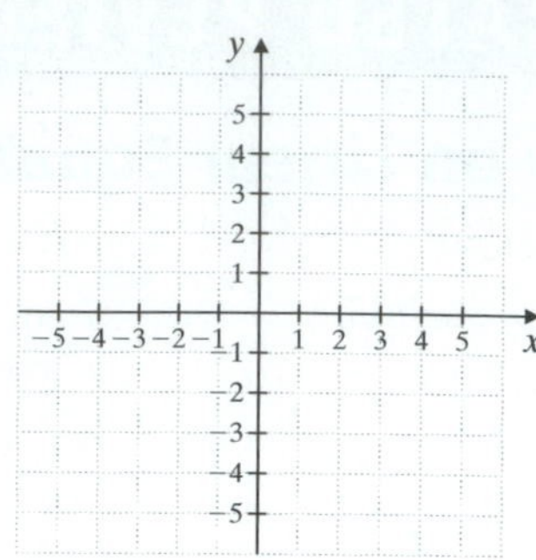

11. $y = -\frac{3}{4}x + 3$

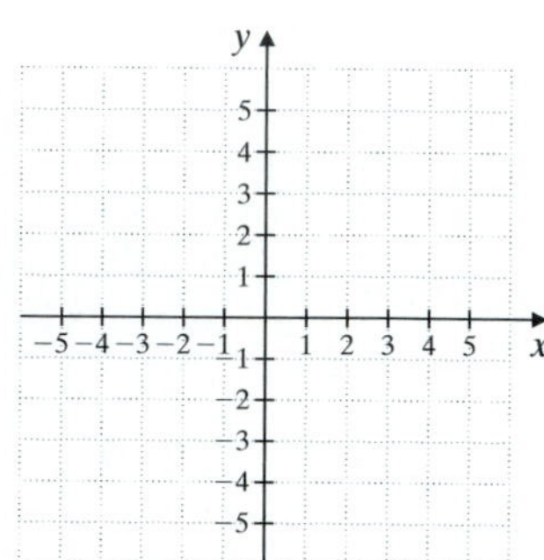

12. $5x - 2y = 8$

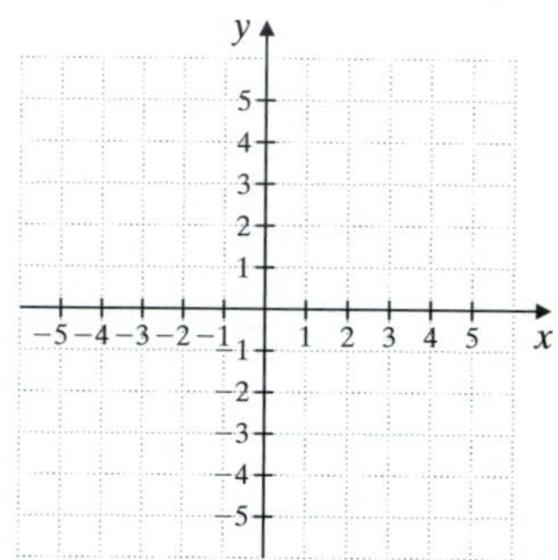

Find the slope of each line by writing the equation in slope-intercept form.

13. $y = 3x - 1$ **14.** $y = -6x + 2$ **15.** $7x + 2y = 11$ **16.** $2x - y = 0$

Find the slope of each line.

17. $x = 2$

18. $y = -4$

19. Write an equation of the line with slope $m = 2$ and y-intercept $\left(0, -\frac{1}{3}\right)$. Write the equation in the form $y = mx + b$.

20. Find an equation of the line with slope $m = -4$ that passes through the point $(-1, 3)$. Write the equation in the form $y = mx + b$.

21. Find an equation of the line that passes through the points $(2, 0)$ and $(-1, -3)$. Write the equation in the form $Ax + By = C$.

Determine whether each pair of lines is parallel, perpendicular, or neither.

22. $6x - y = 7$
$2x + 3y = 4$

23. $3x - 6y = 4$
$y = -2x$

24. Yogurt is an ever more popular food item. In 1998, American Dairy affiliates produced 1639 million pounds of yogurt. In 2002, this number rose to 2135 million pounds of yogurt.

a. Write two ordered pairs of the form (year, millions of pounds of yogurt produced).
b. Find the slope of the line between these two points.
c. Write a sentence explaining the meaning of the slope as a rate of change.

13.6 INTRODUCTION TO FUNCTIONS

Objectives

A Identify Relations, Domains, and Ranges.

B Identify Functions.

C Use the Vertical Line Test.

D Use Function Notation.

Objective A Identifying Relations, Domains, and Ranges

In this chapter, we have studied paired data in the form of ordered pairs. For example, when we list an ordered pair such as $(3, 1)$, we are saying that when x is 3, then y is 1. In other words $x = 3$ and $y = 1$ are related to each other.

For this reason, we call a set of ordered pairs a **relation.** The set of all x-coordinates is called the **domain** of a relation, and the set of all y-coordinates is called the **range** of a relation.

EXAMPLE 1 Find the domain and the range of the relation $\{(0, 2), (3, 3), (-1, 0), (3, -2)\}$.

Solution: The domain is the set of all x-coordinates, or $\{-1, 0, 3\}$, and the range is the set of all y-coordinates, or $\{-2, 0, 2, 3\}$.

Work Practice Problem 1

PRACTICE PROBLEM 1

Find the domain and range of the relation $\{(-3, 5), (-3, 1), (4, 6), (7, 0)\}$.

Objective B Identifying Functions

Paired data occur often in real-life applications. Some special sets of paired data, or ordered pairs, are called *functions*.

Function

A **function** is a set of ordered pairs in which each x-coordinate has exactly one y-coordinate.

In other words, a function cannot have two ordered pairs with the same x-coordinate but different y-coordinates.

EXAMPLE 2 Which of the following relations are also functions?

a. $\{(-1, 1), (2, 3), (7, 3), (8, 6)\}$
b. $\{(0, -2), (1, 5), (0, 3), (7, 7)\}$

Solution:

a. Although the ordered pairs $(2, 3)$ and $(7, 3)$ have the same y-value, each x-value is assigned to only one y-value, so this set of ordered pairs is a function.

b. The x-value 0 is paired with two y-values, -2 and 3, so this set of ordered pairs is not a function.

Work Practice Problem 2

PRACTICE PROBLEM 2

Are the following relations also functions?

a. $\{(2, 5), (-3, 7), (4, 5), (0, -1)\}$
b. $\{(1, 4), (6, 6), (1, -3), (7, 5)\}$

Relations and functions can be described by graphs of their ordered pairs.

Answers

1. domain: $\{-3, 4, 7\}$; range: $\{0, 1, 5, 6\}$
2. a. a function **b.** not a function

PRACTICE PROBLEM 3

Is each graph the graph of a function?

a.

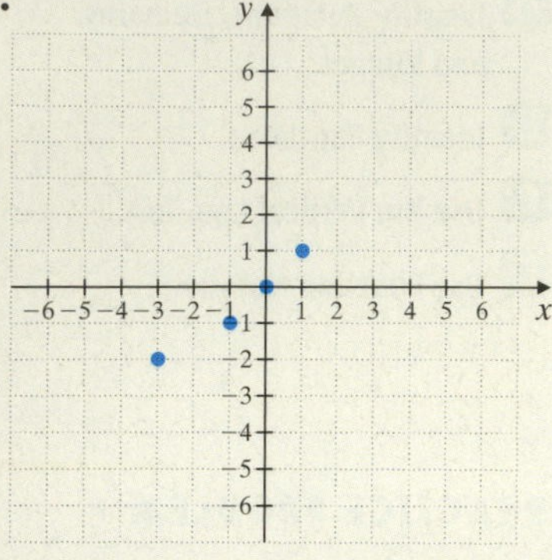

b.

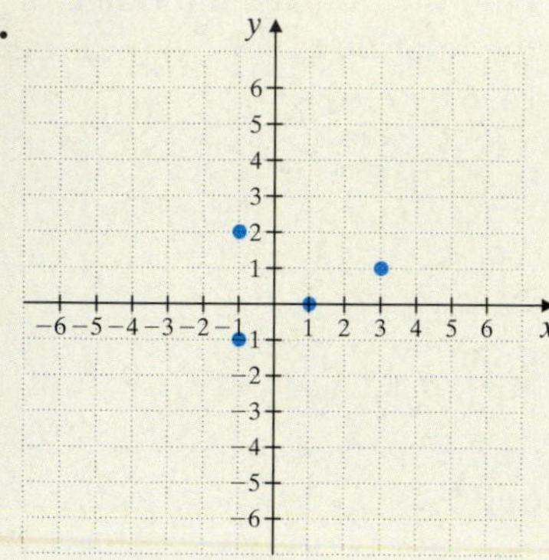

EXAMPLE 3 Which graph is the graph of a function?

a.

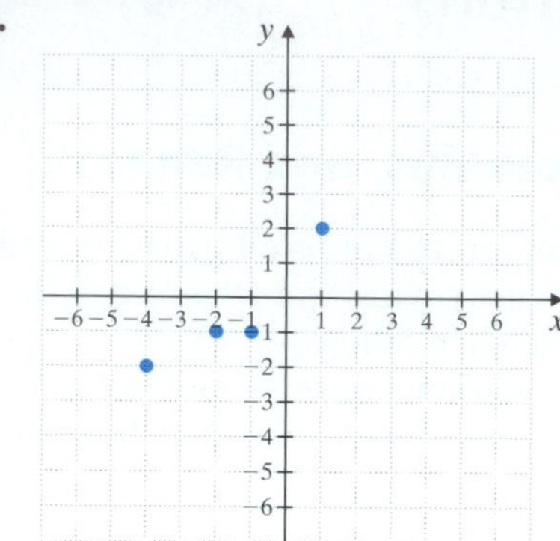

b.

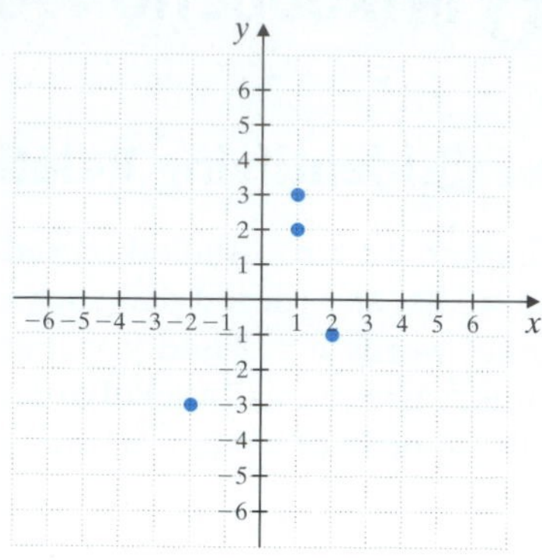

Solution:

a. This is the graph of the relation $\{(-4, -2), (-2, -1), (-1, -1), (1, 2)\}$. Each x-coordinate has exactly one y-coordinate, so this is the graph of a function.

b. This is the graph of the relation $\{(-2, -3), (1, 2), (1, 3), (2, -1)\}$. The x-coordinate 1 is paired with two y-coordinates, 2 and 3, so this is not the graph of a function.

Work Practice Problem 3

Objective C Using the Vertical Line Test

The graph in Example 3(b) was not the graph of a function because the x-coordinate 1 was paired with two y-coordinates, 2 and 3. Notice that when an x-coordinate is paired with more than one y-coordinate, a vertical line can be drawn that will intersect the graph at more than one point. We can use this fact to determine whether a relation is also a function. We call this the vertical line test.

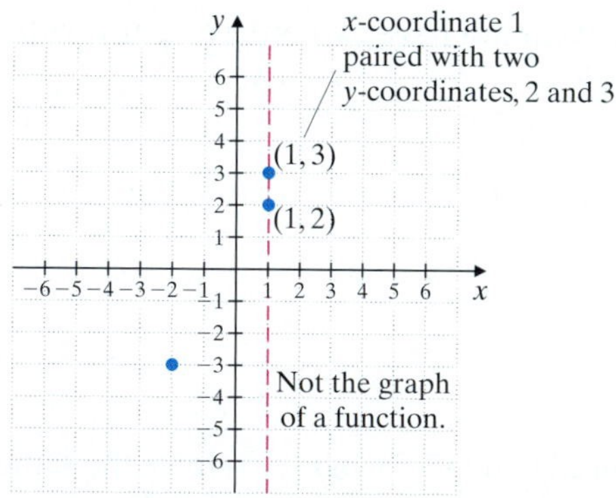

Vertical Line Test

If a vertical line can be drawn so that it intersects a graph more than once, the graph is not the graph of a function. (If no such vertical line can be drawn, the graph is that of a function.)

This vertical line test works for all types of graphs on the rectangular coordinate system.

Answers

3. **a.** a function **b.** not a function

EXAMPLE 4 Use the vertical line test to determine whether each graph is the graph of a function.

a.

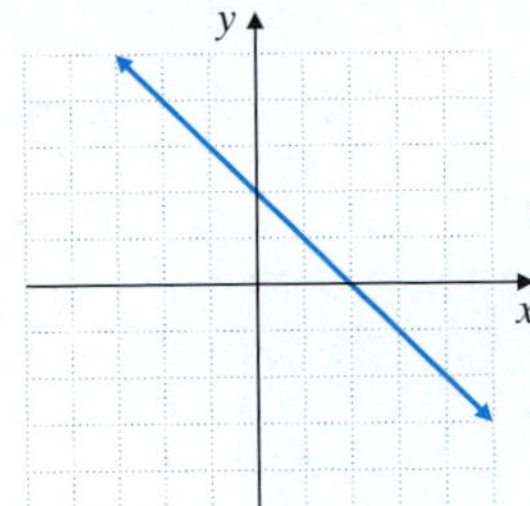

b.

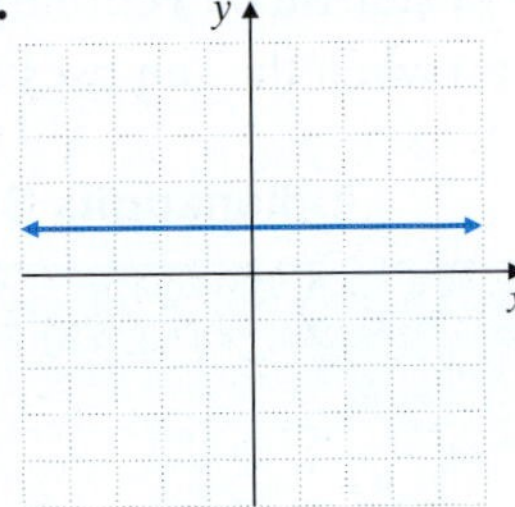

c.

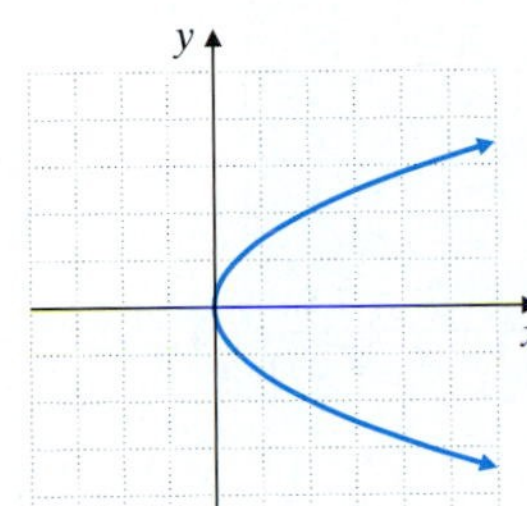

d.

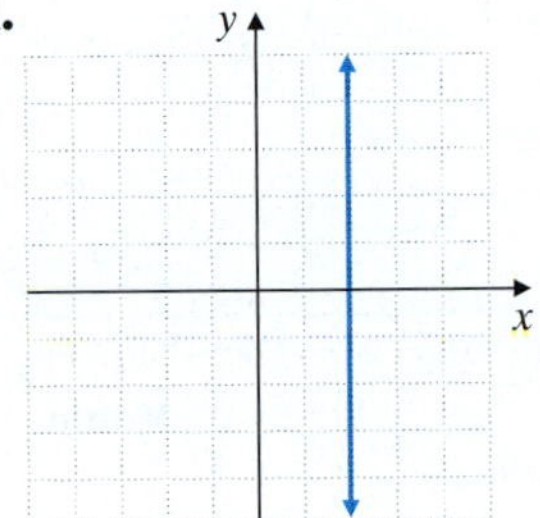

Solution:

a. This graph is the graph of a function since no vertical line will intersect this graph more than once.

b. This graph is also the graph of a function; no vertical line will intersect it more than once.

c. This graph is not the graph of a function. Vertical lines can be drawn that intersect the graph in two points. An example of one is shown.

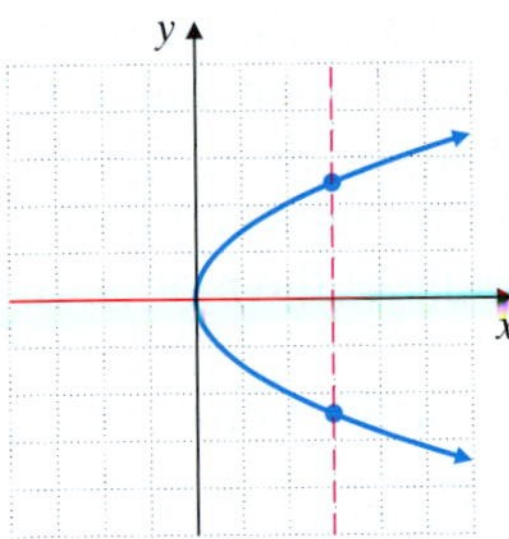

d. This graph is not the graph of a function. A vertical line can be drawn that intersects this line at every point.

Work Practice Problem 4

Examples of functions can often be found in magazines, newspapers, books, and other printed material in the form of tables or graphs such as that in Example 5.

PRACTICE PROBLEM 4

Determine whether each graph is the graph of a function.

a.

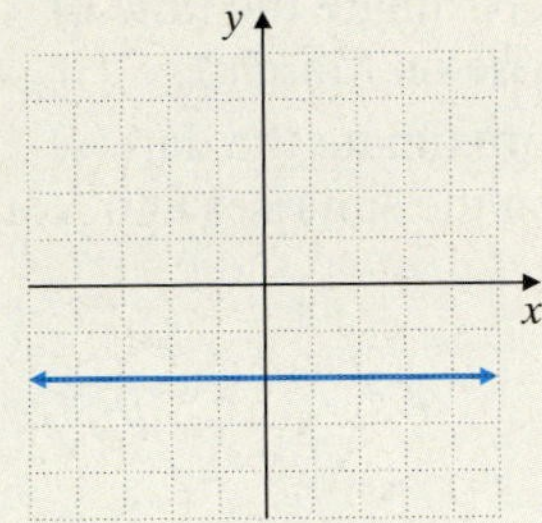

b.

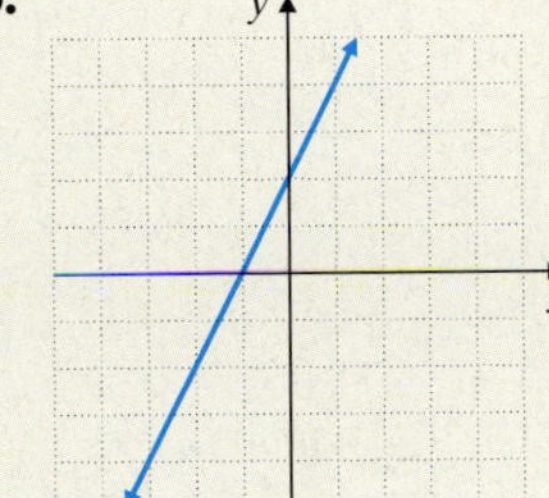

c.

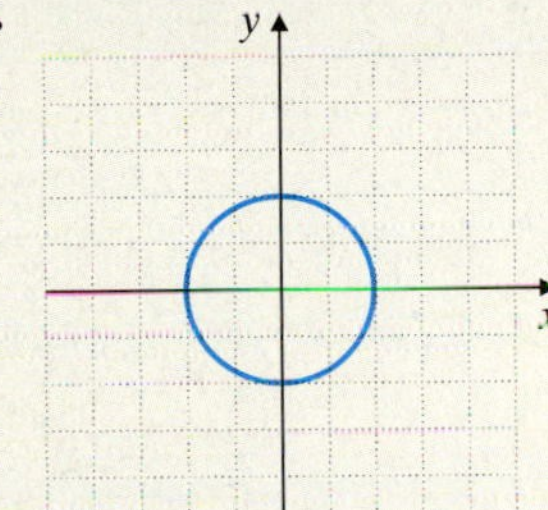

d.

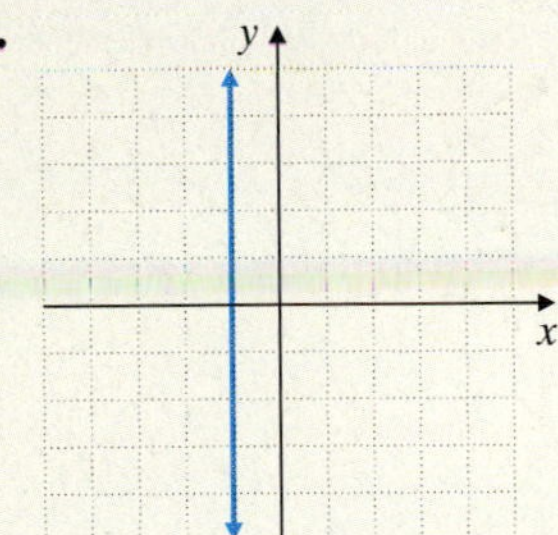

Answers

4. a. a function **b.** a function **c.** not a function **d.** not a function

PRACTICE PROBLEM 5

Use the graph in Example 5 to answer the questions.

a. Approximate the time of sunrise on March 1.

b. Approximate the date(s) when the sun rises at 6 A.M.

EXAMPLE 5 The graph shows the sunrise time for Indianapolis, Indiana, for the year. Use this graph to answer the questions.

a. Approximate the time of sunrise on February 1.

b. Approximate the date(s) when the sun rises at 5 A.M.

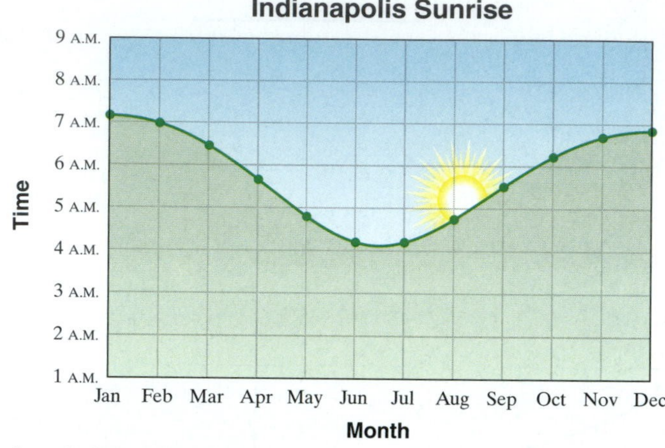

Source: Wolff World Atlas

c. Is this the graph of a function?

Solution:

a. To approximate the time of sunrise on February 1, we find the mark on the horizontal axis that corresponds to February 1. From this mark, we move vertically upward (shown in blue) until the graph is reached. From that point on the graph, we move horizontally to the left until the vertical axis is reached. The vertical axis there reads 7 A.M. as shown below.

b. To approximate the date(s) when the sun rises at 5 A.M., we find 5 A.M. on the time axis and move horizontally to the right (shown in red). Notice that we will hit the graph at two points, corresponding to two dates for which the sun rises at 5 A.M. We follow both points on the graph vertically downward until the horizontal axis is reached. The sun rises at 5 A.M. at approximately the end of the month of April and the middle of the month of August.

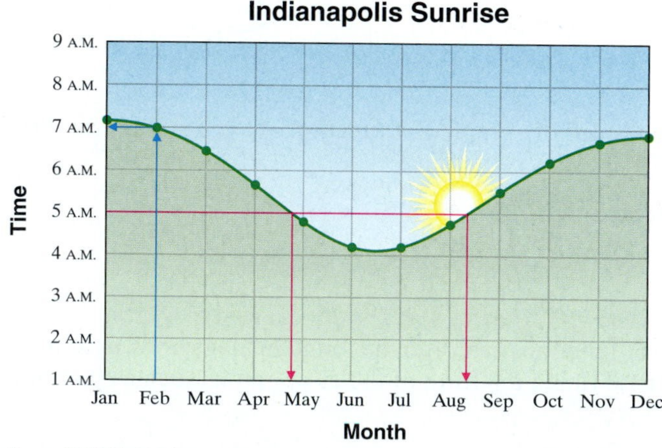

Source: Wolff World Atlas

c. The graph is the graph of a function since it passes the vertical line test. In other words, for every day of the year in Indianapolis, there is exactly one sunrise time.

Work Practice Problem 5

Answers

5. a. 6:30 A.M. **b.** middle of March and middle of September

Objective D Using Function Notation

The graph of the linear equation $y = 2x + 1$ passes the vertical line test, so we say that $y = 2x + 1$ is a function. In other words, $y = 2x + 1$ gives us a rule for writing ordered pairs where every x-coordinate is paired with at most one y-coordinate.

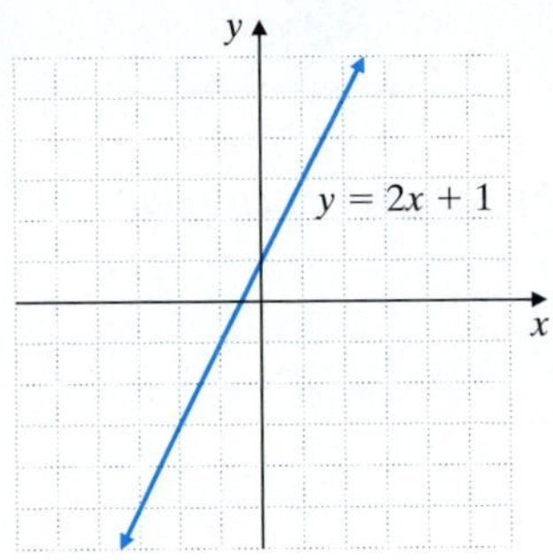

We often use letters such as f, g, and h to name functions. For example, the symbol $f(x)$ means *function of x* and is read "f of x." This notation is called **function notation.** The equation $y = 2x + 1$ can be written as $f(x) = 2x + 1$ using function notation, and these equations mean the same thing. In other words $y = f(x)$.

The notation $f(1)$ means to replace x with 1 and find the resulting y or function value. Since

$$f(x) = 2x + 1$$

then

$$f(1) = 2(1) + 1 = 3$$

This means that, when $x = 1$, y or $f(x) = 3$, and we have the ordered pair (1, 3). Now let's find $f(2)$, $f(0)$, and $f(-1)$.

	$f(x) = 2x + 1$	$f(x) = 2x + 1$	$f(x) = 2x + 1$
	$f(2) = 2(2) + 1$	$f(0) = 2(0) + 1$	$f(-1) = 2(-1) + 1$
	$= 4 + 1$	$= 0 + 1$	$= -2 + 1$
	$= 5$	$= 1$	$= -1$
Ordered Pair:	(2, 5)	(0, 1)	(−1, −1)

Helpful Hint

Note that $f(x)$ is a special symbol in mathematics used to denote a function. The symbol $f(x)$ is read "f of x." It does **not** mean $f \cdot x$ (f times x).

EXAMPLE 6 Given $g(x) = x^2 - 3$, find the following and list the corresponding ordered pair.

a. $g(2)$ **b.** $g(-2)$ **c.** $g(0)$

Solution:

	a.	**b.**	**c.**
	$g(x) = x^2 - 3$	$g(x) = x^2 - 3$	$g(x) = x^2 - 3$
	$g(2) = 2^2 - 3$	$g(-2) = (-2)^2 - 3$	$g(0) = 0^2 - 3$
	$= 4 - 3$	$= 4 - 3$	$= 0 - 3$
	$= 1$	$= 1$	$= -3$
Ordered Pair:	(2, 1)	(−2, 1)	(0, −3)

Work Practice Problem 6

✔Concept Check Suppose that the value of a function f is −7 when the function is evaluated at 2. Write this situation in function notation.

PRACTICE PROBLEM 6

Given $f(x) = x^2 + 1$, find the following and list the corresponding ordered pair.

a. $f(1)$

b. $f(-3)$

c. $f(0)$

Answers

6. a. 2; (1, 2) **b.** 10; (−3, 10) **c.** 1; (0, 1)

✔ Concept Check Answer

$f(2) = -7$

13.6 EXERCISE SET

FOR EXTRA HELP

Student Solutions Manual

PH Math/Tutor Center

CD/Video for Review

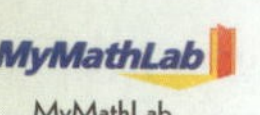
MathXL®

MyMathLab

Objective A *Find the domain and the range of each relation. See Example 1.*

1. $\{(2, 4), (0, 0), (-7, 10), (10, -7)\}$

2. $\{(3, -6), (1, 4), (-2, -2)\}$

3. $\{(0, -2), (1, -2), (5, -2)\}$

4. $\{(5, 0), (5, -3), (5, 4), (5, 3)\}$

Objective B *Determine whether each relation is also a function. See Example 2.*

5. $\{(1, 1), (2, 2), (-3, -3), (0, 0)\}$

6. $\{(11, 6), (-1, -2), (0, 0), (3, -2)\}$

7. $\{(-1, 0), (-1, 6), (-1, 8)\}$

8. $\{(1, 2), (3, 2), (1, 4)\}$

Objectives B C Mixed Practice *Determine whether each graph is the graph of a function. For Exercises 9 through 12, either write down the ordered pairs or use the vertical line test. See Examples 3 and 4.*

9.

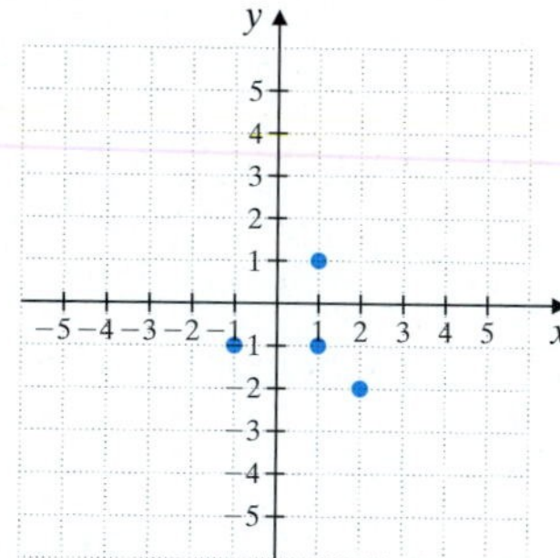

10.

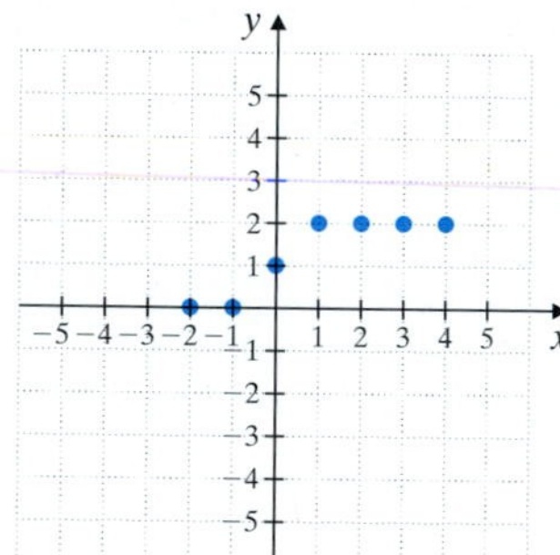

11.

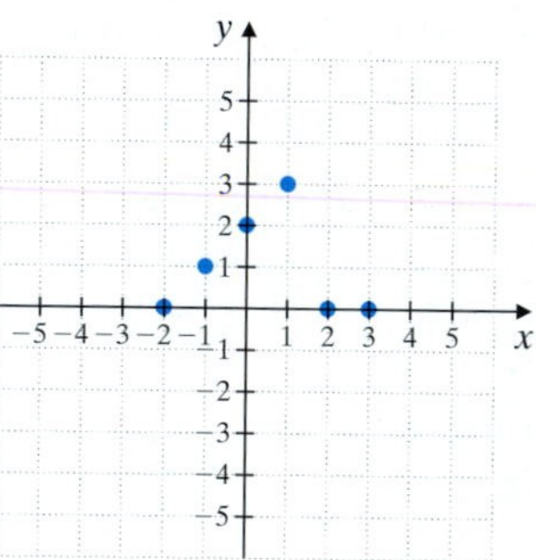

12.

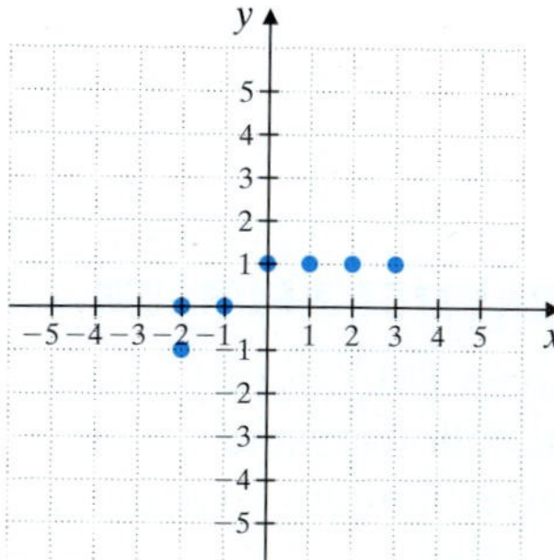

13.

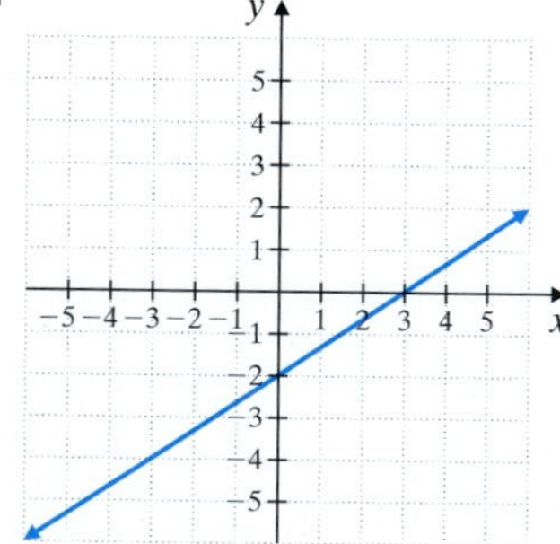

14.

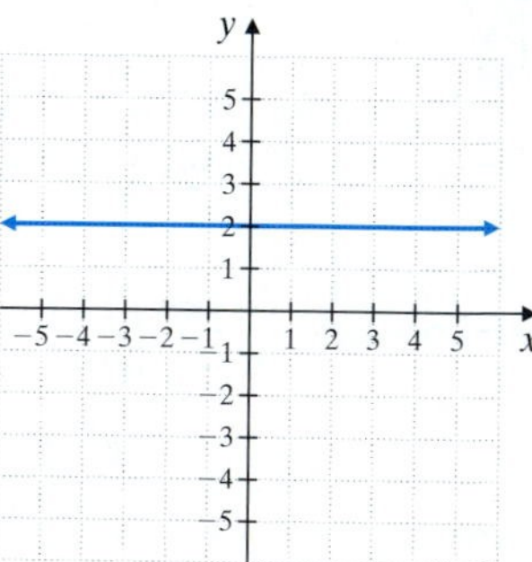

15.

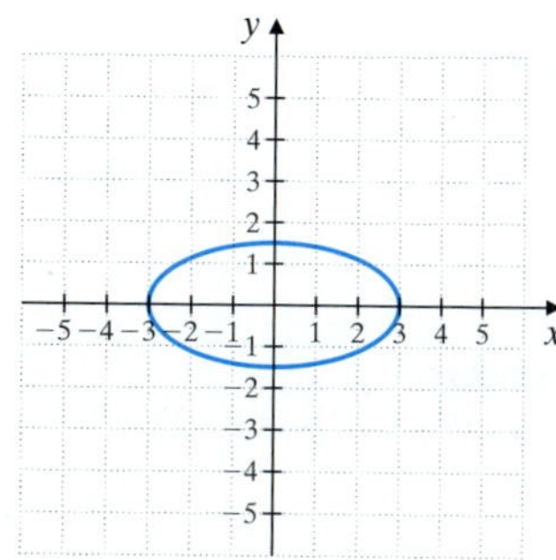

16.

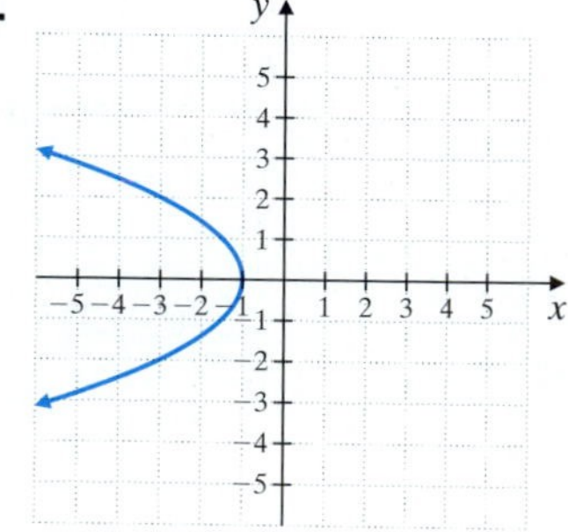

The graph shows the sunset times for Seward, Alaska. Use this graph to answer Exercises 17 through 22.

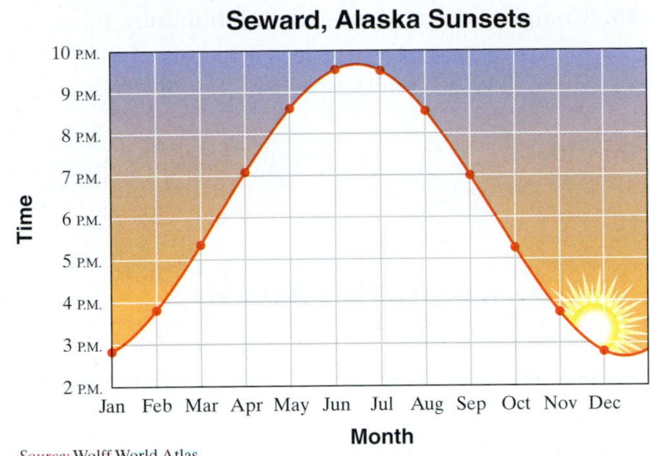

17. Approximate the time of sunset on June 1.

18. Approximate the time of sunset on November 1.

19. Approximate the date(s) when the sunset is at 3 P.M.

20. Approximate the date(s) when the sunset is at 9 P.M.

21. Is this graph the graph of a function? Why or why not?

22. Do you think a graph of sunset times for any location will always be a function? Why or why not?

This graph shows the U.S. hourly minimum wage for each year shown. Use this graph to answer Exercises 23 through 28.

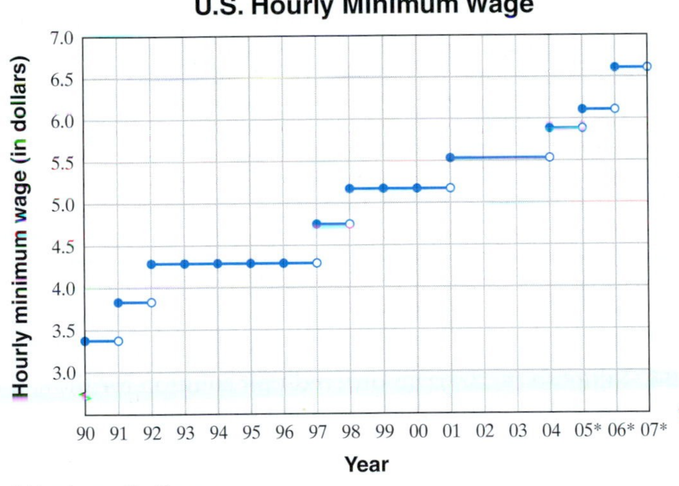

23. Approximate the minimum wage at the beginning of 1997.

24. Approximate the minimum wage at the beginning of 1999.

25. Approximate the year when the minimum wage will increase to over \$5.75 per hour.

26. Approximate the year when the minimum wage increased to over \$5.00 per hour.

27. Is this graph the graph of a function? Why or why not?

28. Do you think that a similar graph of your hourly wage on January 1 of every year (whether you are working or not) will be the graph of a function? Why or why not?

Objective **D** *Find $f(-2)$, $f(0)$, and $f(3)$ for each function. See Example 6.*

29. $f(x) = 2x - 5$

30. $f(x) = 3 - 7x$

31. $f(x) = x^2 + 2$

32. $f(x) = x^2 - 4$

33. $f(x) = 3x$

34. $f(x) = -3x$

35. $f(x) = |x|$

36. $f(x) = |2 - x|$

Find $h(-1)$, $h(0)$, and $h(4)$ for each function. See Example 6.

37. $h(x) = -5x$

38. $h(x) = -3x$

39. $h(x) = 2x^2 + 3$

40. $h(x) = 3x^2$

41. If $f(3) = 6$, write a corresponding ordered-pair solution.

42. If $f(7) = -2$, write a corresponding ordered-pair solution.

Use the graph of f below to answer Exercises 43 through 48.

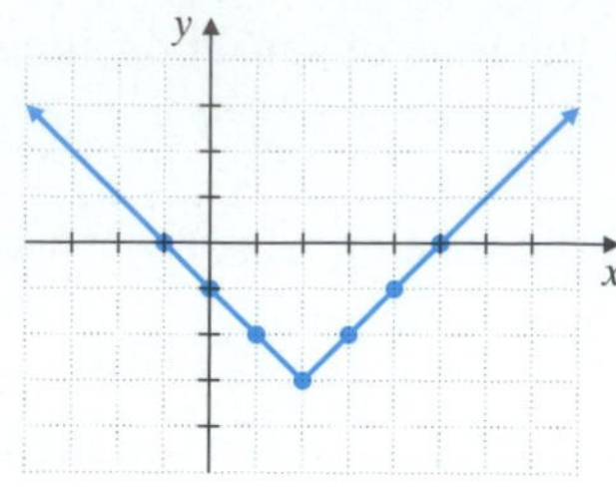

43. Complete the ordered-pair solution for f. (0,)

44. Complete the ordered-pair solution for f. (3,)

45. $f(0) =$ _____?

46. $f(3) =$ _____?

47. If $f(x) = 0$, find the value(s) of x.

48. If $f(x) = -1$, find the value(s) of x.

Review

Solve each inequality. See Section 9.6.

49. $2x + 5 < 7$

50. $3x - 1 \geq 11$

51. $-x + 6 \leq 9$

52. $-2x + 3 > 3$

Find the perimeter of each figure. See Sections 4.4 and 4.5.

△ **53.**

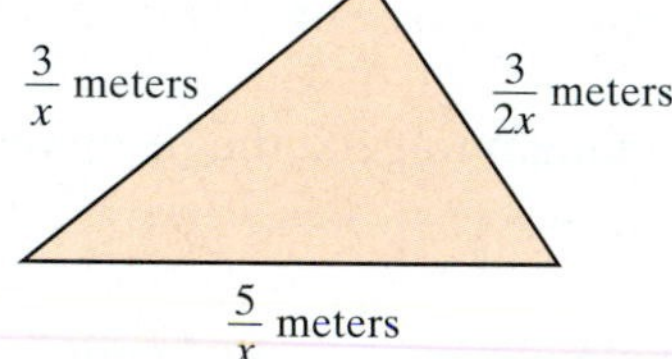

△ **54.**

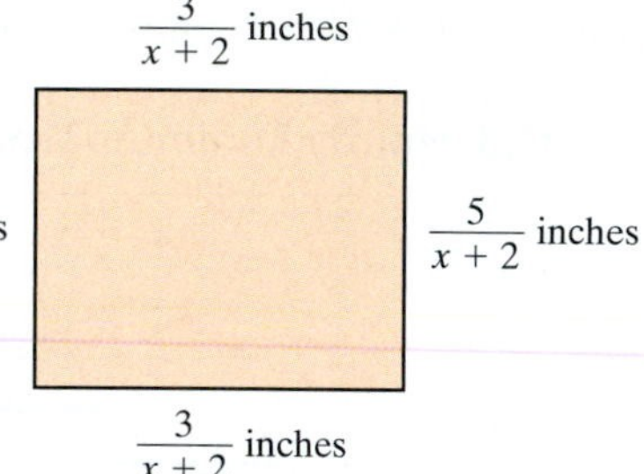

Concept Extensions

Solve. See the Concept Check in this section.

55. If a function f is evaluated at -5, the value of the function is 12. Write this situation using function notation.

56. Suppose (9, 20) is an ordered-pair solution for the function g. Write this situation using function notation.

57. In your own words define (a) function; (b) domain; (c) range.

58. Explain the vertical line test and how it is used.

59. Since $y = x + 7$ is a function, rewrite the equation using function notation.

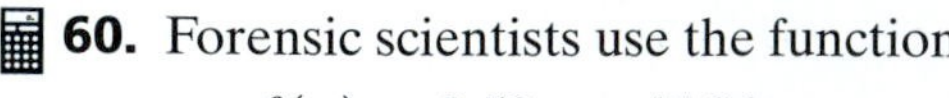

60. Forensic scientists use the function

$$f(x) = 2.59x + 47.24$$

to estimate the height of a woman, in centimeters, given the length x of her femur bone in centimeters.

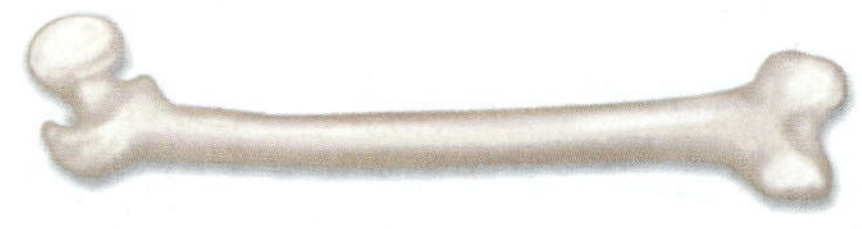

a. Estimate the height of a woman whose femur measures 46 centimeters.

b. Estimate the height of a woman whose femur measures 39 centimeters.

61. The dosage in milligrams of Ivermectin, a heartworm preventive for a dog who weighs x pounds, is given by the function

$$f(x) = \frac{136}{25}x$$

a. Find the proper dosage for a dog that weighs 35 pounds.

b. Find the proper dosage for a dog that weighs 70 pounds.

13.7 GRAPHING LINEAR INEQUALITIES IN TWO VARIABLES

Objectives

A Determine Whether an Ordered Pair is a Solution of a Linear Inequality in Two Variables.

B Graph a Linear Inequality in Two Variables.

Recall that a linear equation in two variables is an equation that can be written in the form $Ax + By = C$, where A, B, and C are real numbers and A and B are not both 0. A **linear inequality in two variables** is an inequality that can be written in one of the forms

$$Ax + By < C \qquad Ax + By \leq C$$
$$Ax + By > C \qquad Ax + By \geq C$$

where A, B, and C are real numbers and A and B are not both 0.

Objective A Determining Solutions of Linear Inequalities in Two Variables

Just as for linear equations in x and y, an ordered pair is a **solution** of an inequality in x and y if replacing the variables with the coordinates of the ordered pair results in a true statement.

EXAMPLE 1 Determine whether each ordered pair is a solution of the inequality $2x - y < 6$.

a. $(5, -1)$ **b.** $(2, 7)$

Solution:

a. We replace x with 5 and y with -1 and see if a true statement results.

$$2x - y < 6$$
$$2(5) - (-1) < 6 \quad \text{Replace } x \text{ with 5 and } y \text{ with } -1.$$
$$10 + 1 < 6$$
$$11 < 6 \quad \text{False}$$

The ordered pair $(5, -1)$ is not a solution since $11 < 6$ is a false statement.

b. We replace x with 2 and y with 7 and see if a true statement results.

$$2x - y < 6$$
$$2(2) - (7) < 6 \quad \text{Replace } x \text{ with 2 and } y \text{ with 7.}$$
$$4 - 7 < 6$$
$$-3 < 6 \quad \text{True}$$

The ordered pair $(2, 7)$ is a solution since $-3 < 6$ is a true statement.

Work Practice Problem 1

PRACTICE PROBLEM 1

Determine whether each ordered pair is a solution of $x - 4y > 8$.

a. $(-3, 2)$

b. $(9, 0)$

Objective B Graphing Linear Inequalities in Two Variables

The linear equation $x - y = 1$ is graphed next. Recall that all points on the line correspond to ordered pairs that satisfy the equation $x - y = 1$.

Notice the line defined by $x - y = 1$ divides the rectangular coordinate system plane into 2 sides. All points on one side of the line satisfy the inequality $x - y < 1$ and all points on the other side satisfy the inequality $x - y > 1$. The graph on the next page shows a few examples of this.

Answers

1. a. no **b.** yes

$x - y < 1$
$1 - 3 < 1$ True
$-2 - 1 < 1$ True
$-4 - (-4) < 1$ True

$x - y > 1$
$4 - 1 > 1$ True
$2 - (-2) > 1$ True
$0 - (-4) > 1$ True

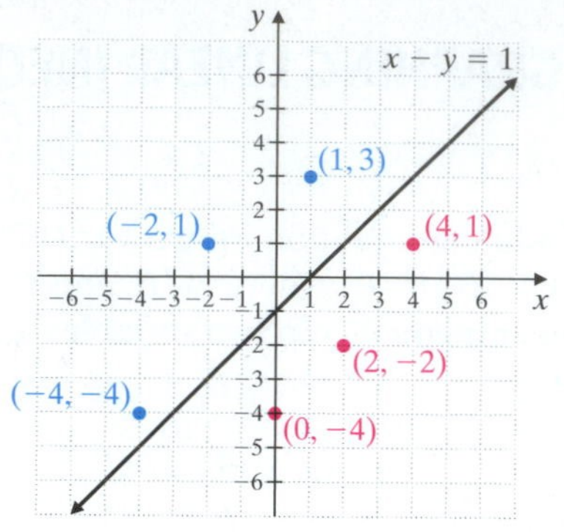

The graph of $x - y < 1$ is the region shaded blue and the graph of $x - y > 1$ is the region shaded red below.

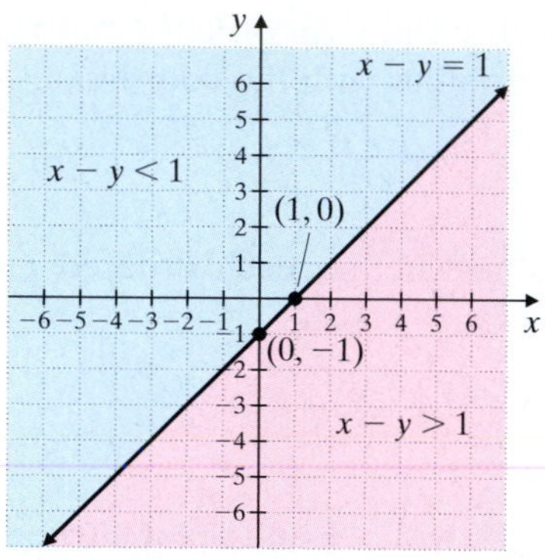

The region to the left of the line and the region to the right of the line are called **half-planes.** Every line divides the plane (similar to a sheet of paper extending indefinitely in all directions) into two half-planes; the line is called the **boundary.**

Recall that the inequality $x - y \le 1$ means

$$x - y = 1 \quad \text{or} \quad x - y < 1$$

Thus, the graph of $x - y \le 1$ is the half-plane $x - y < 1$ along with the boundary line $x - y = 1$.

To Graph a Linear Inequality in Two Variables

Step 1: Graph the boundary line found by replacing the inequality sign with an equal sign. If the inequality sign is $>$ or $<$, graph a dashed boundary line (indicating that the points on the line are not solutions of the inequality). If the inequality sign is $\ge$ or $\le$, graph a solid boundary line (indicating that the points on the line are solutions of the inequality).

Step 2: Choose a point, *not* on the boundary line, as a test point. Substitute the coordinates of this test point into the *original* inequality.

Step 3: If a true statement is obtained in Step 2, shade the half-plane that contains the test point. If a false statement is obtained, shade the half-plane that does not contain the test point.

EXAMPLE 2 Graph: $x + y < 7$

Solution:

Step 1: First we graph the boundary line by graphing the equation $x + y = 7$. We graph this boundary as a *dashed line* because the inequality sign is $<$, and thus the points on the line are not solutions of the inequality $x + y < 7$.

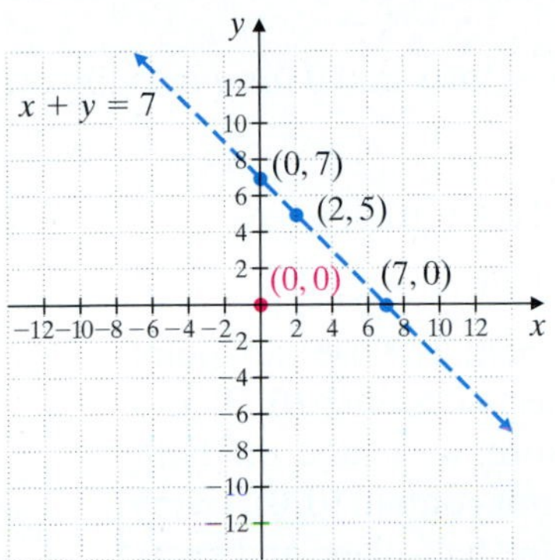

Step 2: Next we choose a test point, being careful *not* to choose a point on the boundary line. We choose (0, 0), and substitute the coordinates of (0, 0) into $x + y < 7$.

$x + y < 7$ Original inequality

$0 + 0 < 7$ Replace x with 0 and y with 0.

$0 < 7$ True

Step 3: Since the result is a true statement, (0, 0) is a solution of $x + y < 7$, and every point in the same half-plane as (0, 0) is also a solution. To indicate this, we shade the entire half-plane containing (0, 0), as shown.

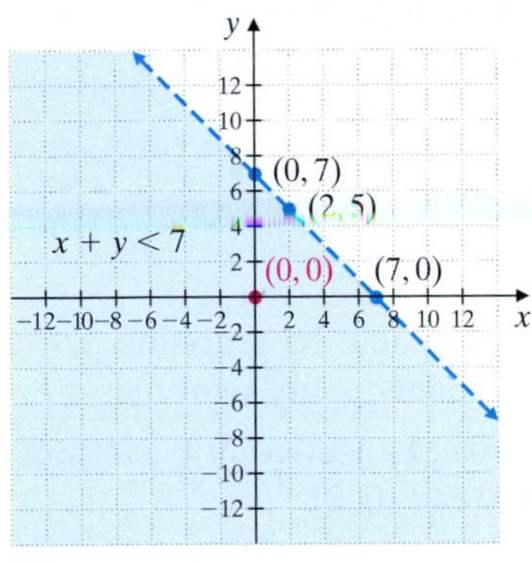

Work Practice Problem 2

✔Concept Check Determine whether (0, 0) is included in the graph of

a. $y \geq 2x + 3$

b. $x < 7$

c. $2x - 3y < 6$

PRACTICE PROBLEM 2

Graph: $x - y > 3$

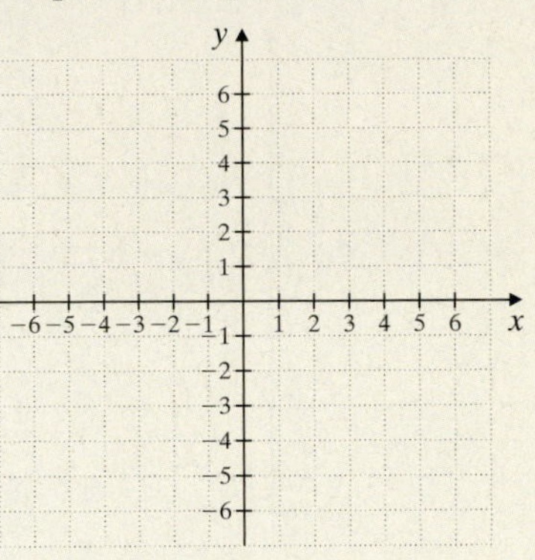

Answer

2.

✔ Concept Check Answers

a. no **b.** yes **c.** yes

PRACTICE PROBLEM 3

Graph: $x - 4y \le 4$

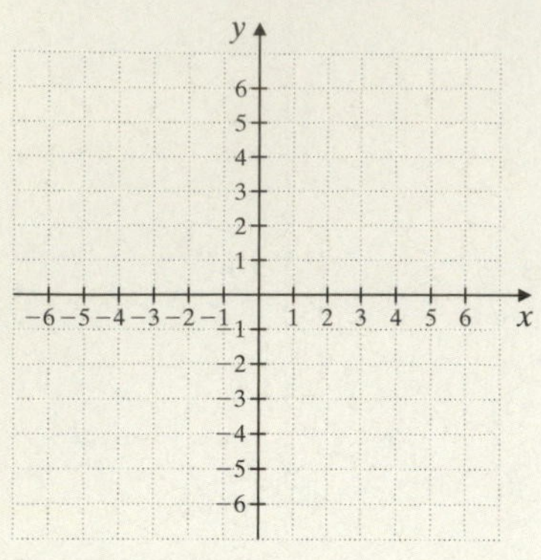

PRACTICE PROBLEM 4

Graph: $y < 3x$

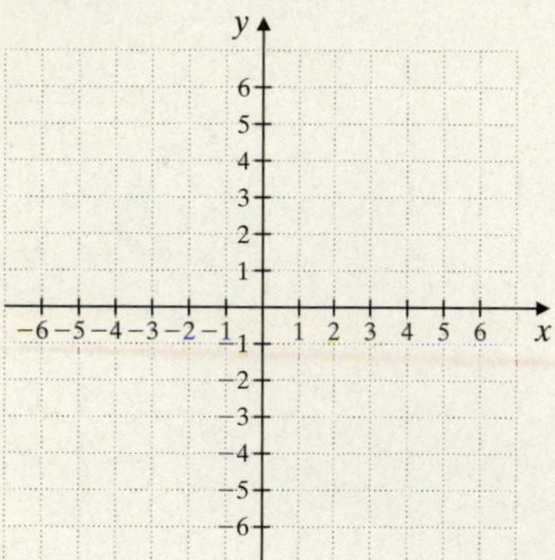

Answers

3.

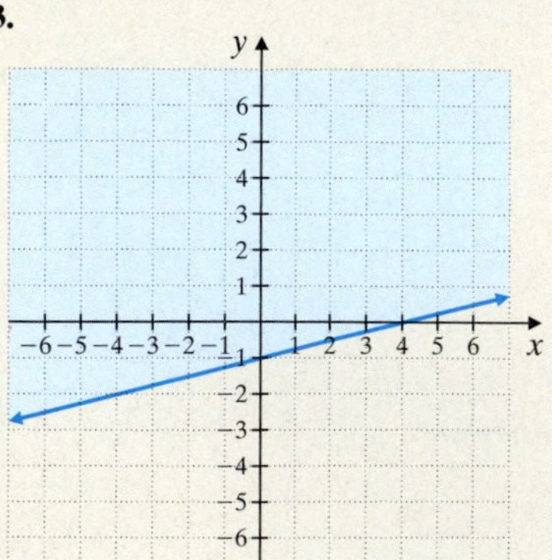

4.

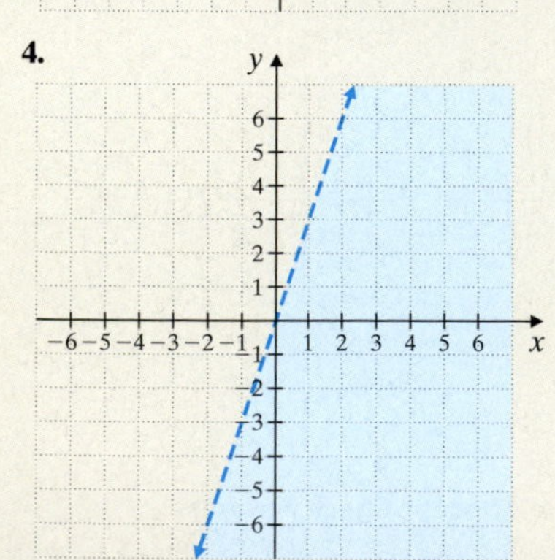

EXAMPLE 3 Graph: $2x - y \ge 3$

Solution:

Step 1: We graph the boundary line by graphing $2x - y = 3$. We draw this line as a solid line because the inequality sign is $\ge$, and thus the points on the line are solutions of $2x - y \ge 3$.

Step 2: Once again, $(0, 0)$ is a convenient test point since it is not on the boundary line.

We substitute 0 for x and 0 for y into the original inequality.

$2x - y \ge 3$

$2(0) - 0 \ge 3$ Let $x = 0$ and $y = 0$.

$0 \ge 3$ False

Step 3: Since the statement is false, no point in the half-plane containing $(0, 0)$ is a solution. Therefore, we shade the half-plane that does not contain $(0, 0)$. Every point in the shaded half-plane and every point on the boundary line is a solution of $2x - y \ge 3$.

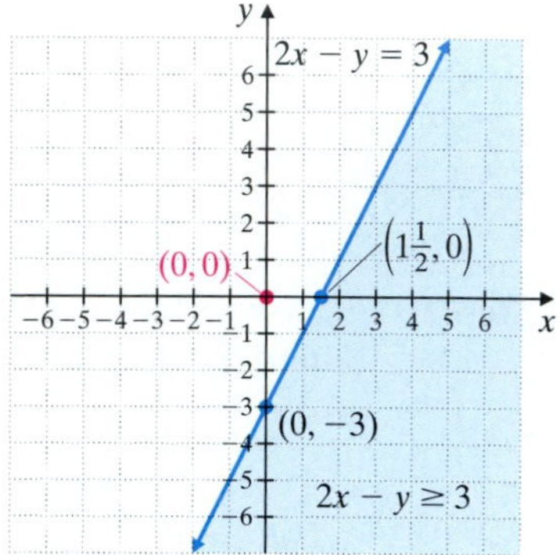

Work Practice Problem 3

Helpful Hint

When graphing an inequality, make sure the test point is substituted into the **original inequality.** For Example 3, we substituted the test point $(0, 0)$ into the **original inequality** $2x - y \ge 3$, *not* $2x - y = 3$.

EXAMPLE 4 Graph: $x > 2y$

Solution:

Step 1: We find the boundary line by graphing $x = 2y$. The boundary line is a dashed line since the inequality symbol is $>$.

Step 2: We cannot use $(0, 0)$ as a test point because it is a point on the boundary line. We choose instead $(0, 2)$.

$x > 2y$

$0 > 2(2)$ Let $x = 0$ and $y = 2$.

$0 > 4$ False

Step 3: Since the statement is false, we shade the half-plane that does not contain the test point $(0, 2)$, as shown.

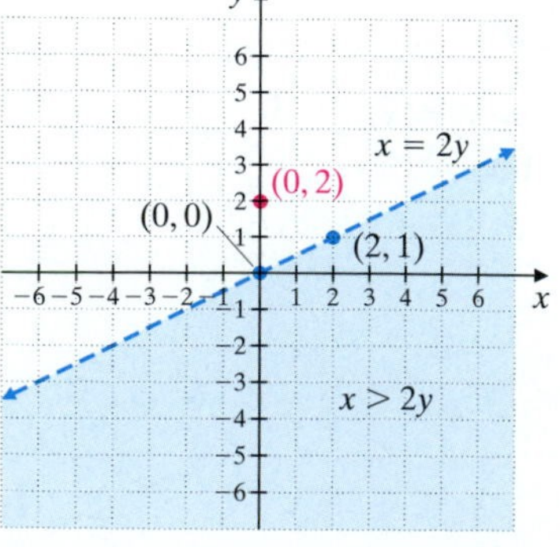

Work Practice Problem 4

EXAMPLE 5 Graph: $5x + 4y \leq 20$

Solution: We graph the solid boundary line $5x + 4y = 20$ and choose (0, 0) as the test point.

$$5x + 4y \leq 20$$

$$5(0) + 4(0) \leq 20 \quad \text{Let } x = 0 \text{ and } y = 0.$$

$$0 \leq 20 \quad \text{True}$$

We shade the half-plane that contains (0, 0), as shown.

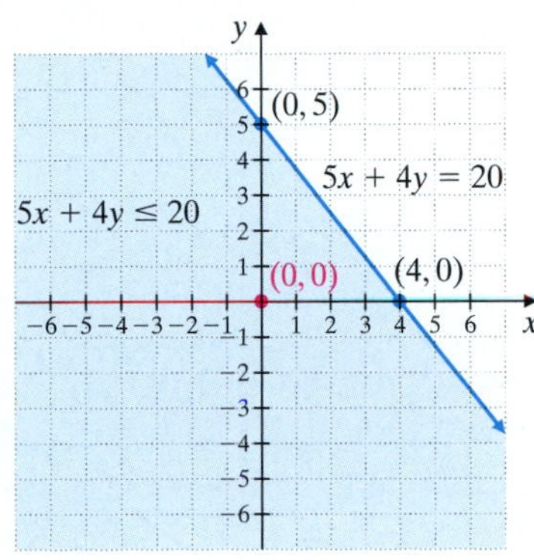

Work Practice Problem 5

EXAMPLE 6 Graph: $y > 3$

Solution: We graph the dashed boundary line $y = 3$ and choose (0, 0) as the test point. (Recall that the graph of $y = 3$ is a horizontal line with y-intercept 3.)

$$y > 3$$

$$0 > 3 \quad \text{Let } y = 0.$$

$$0 > 3 \quad \text{False}$$

We shade the half-plane that does not contain (0, 0), as shown.

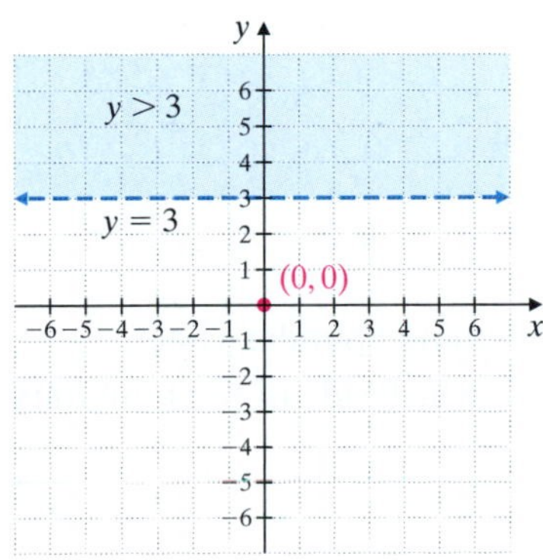

Work Practice Problem 6

PRACTICE PROBLEM 5

Graph: $3x + 2y \geq 12$

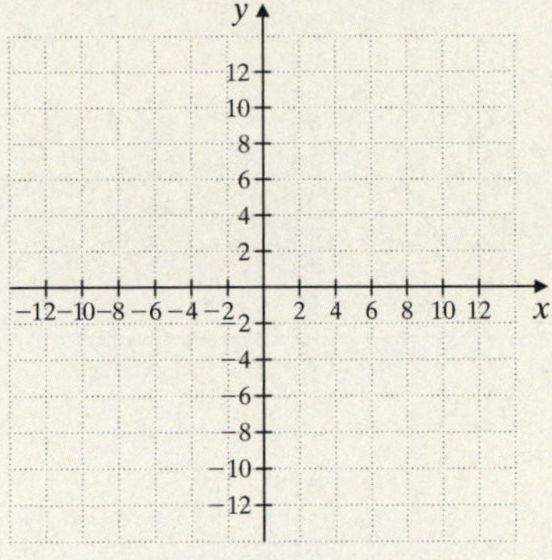

PRACTICE PROBLEM 6

Graph: $x < 2$

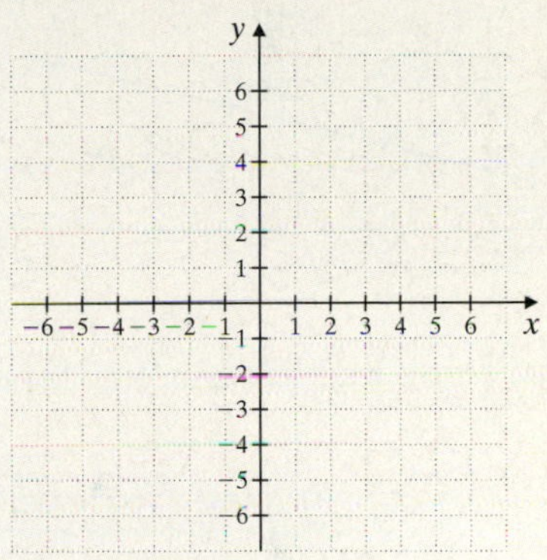

Answers

5.

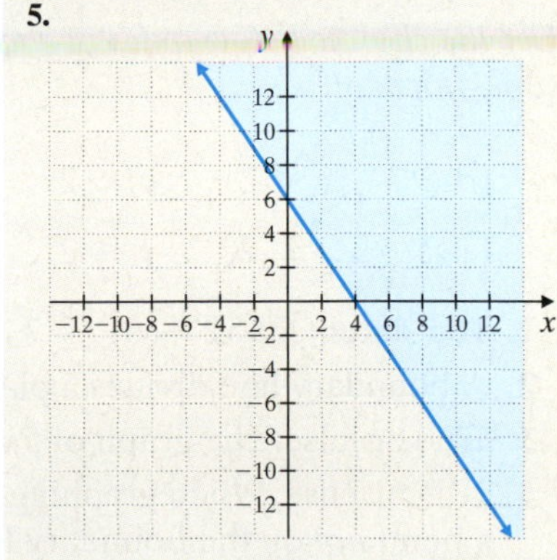

6.

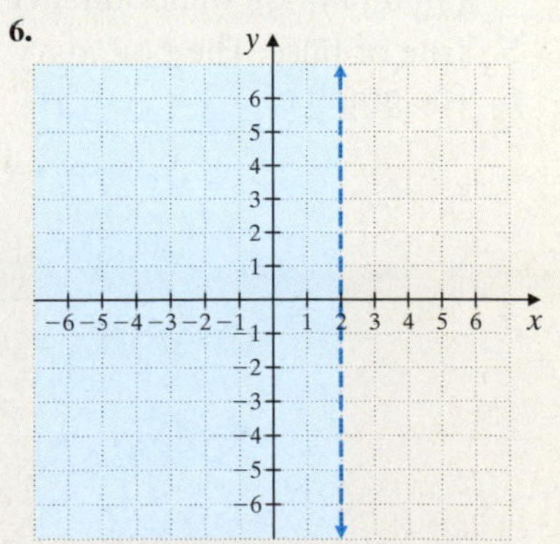

PRACTICE PROBLEM 7

Graph: $y \geq \frac{1}{4}x + 3$

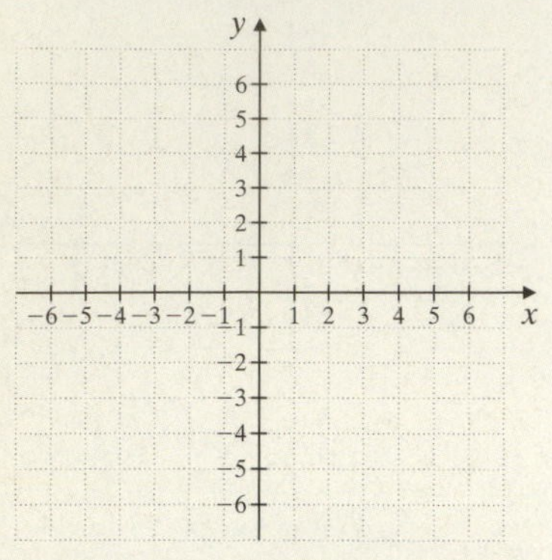

Answer

7.

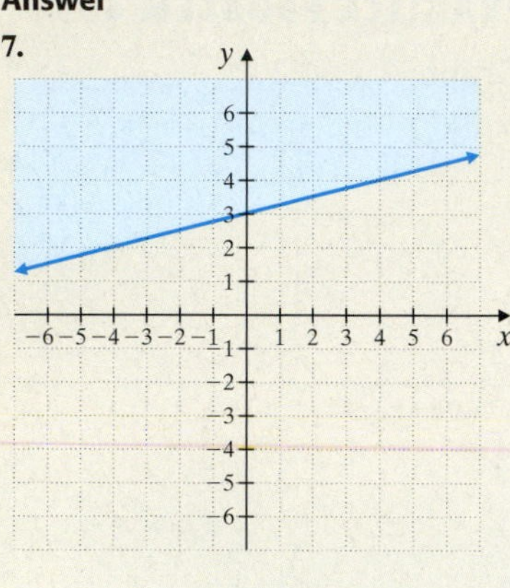

EXAMPLE 7 Graph: $y \leq \frac{2}{3}x - 4$

Solution: Graph the solid boundary line $y = \frac{2}{3}x - 4$. This equation is in slope-intercept form with slope $\frac{2}{3}$ and y-intercept -4.

We use this information to graph the line. Then we choose (0, 0) as our test point.

$$y \leq \frac{2}{3}x - 4$$

$$0 \leq \frac{2}{3} \cdot 0 - 4$$

$$0 \leq -4 \quad \text{False}$$

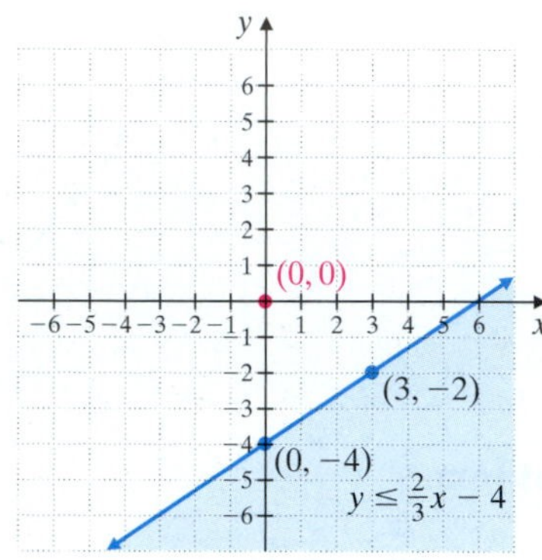

We shade the half-plane that does not contain (0, 0), as shown.

Work Practice Problem 7

Vocabulary and Readiness Check

Use the choices below to fill in each blank. Some choices may be used more than once, and some not at all.

true	$x < 2$	$y < 2$	half-planes
false	$x \leq 2$	$y \leq 2$	linear inequality in two variables

1. The statement $5x - 6y < 7$ is an example of a(n) ______________.

2. A boundary line divides a plane into two regions called ________.

3. True or false: The graph of $5x - 6y < 7$ includes its corresponding boundary line. ______

4. True or false: When graphing a linear inequality, to determine which side of the boundary line to shade, choose a point *not* on the boundary line. ______

5. True or false: The boundary line for the inequality $5x - 6y < 7$ is the graph of $5x - 6y = 7$. ______

6. The graph of ______ is

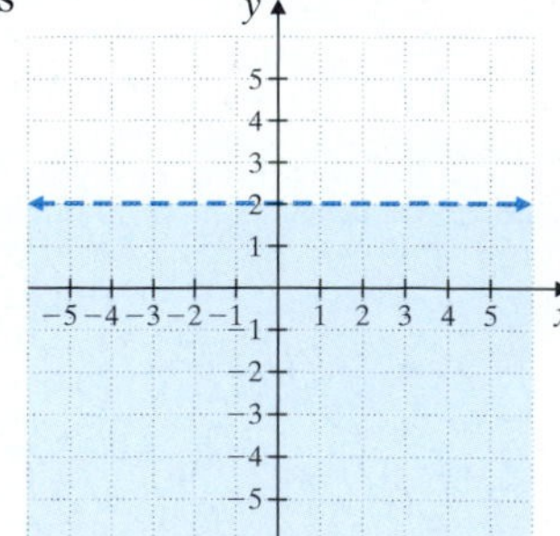

13.7 EXERCISE SET

FOR EXTRA HELP

Student Solutions Manual

PH Math/Tutor Center

CD/Video for Review

MathXL®

MyMathLab

Objective A *Determine whether the ordered pairs given are solutions of the linear inequality in two variables. See Example 1.*

1. $x - y > 3$; $(0, 3)$, $(2, -1)$

2. $y - x < -2$; $(2, 1)$, $(5, -1)$

3. $3x - 5y \leq -4$; $(2, 3)$, $(-1, -1)$

4. $2x + y \geq 10$; $(0, 11)$, $(5, 0)$

5. $x < -y$; $(0, 2)$, $(-5, 1)$

6. $y > 3x$; $(0, 0)$, $(1, 4)$

Objective B *Graph each inequality. See Examples 2 through 7.*

7. $x + y \leq 1$

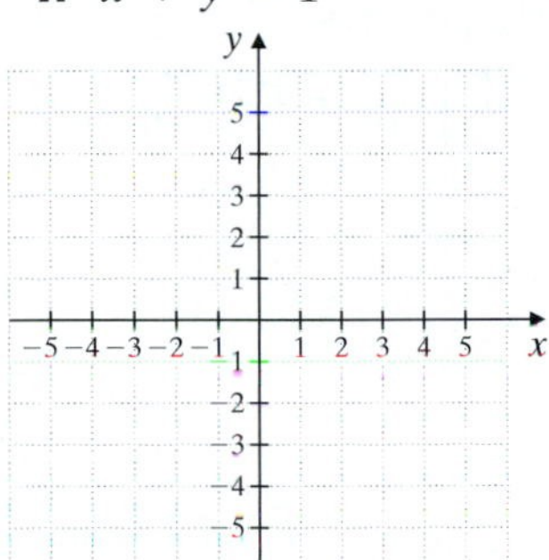

8. $x + y \geq -2$

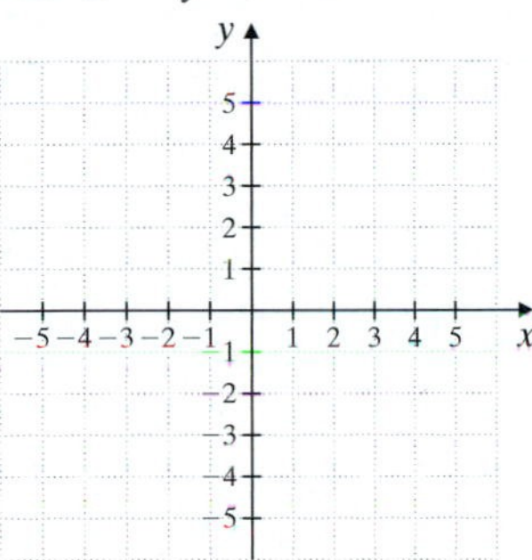

9. $2x - y > -4$

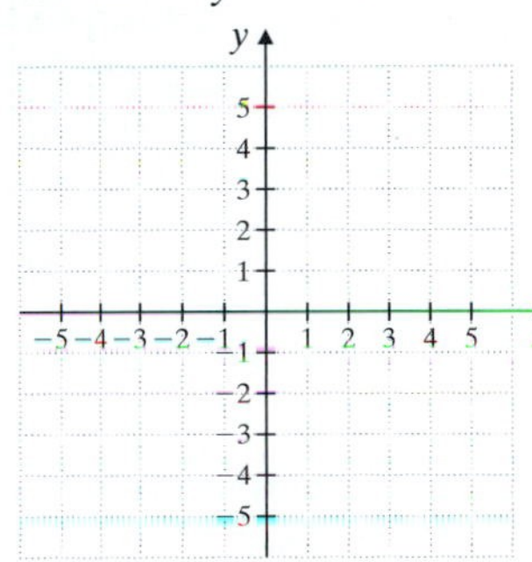

10. $x - 3y < 3$

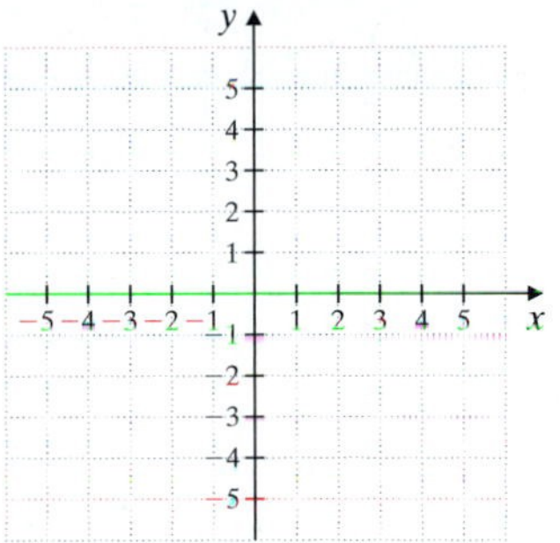

11. $y > 2x$

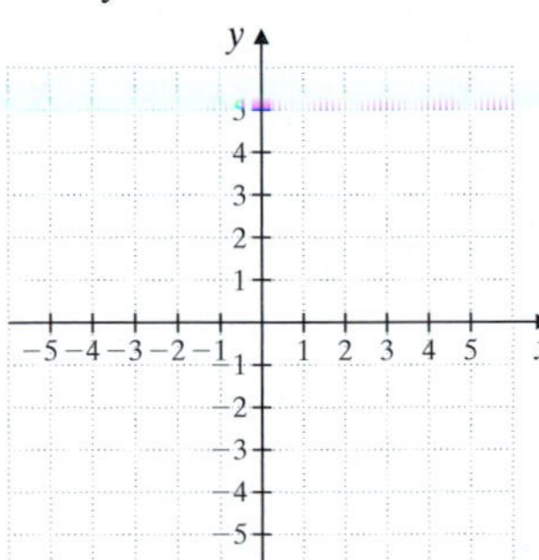

12. $y < 3x$

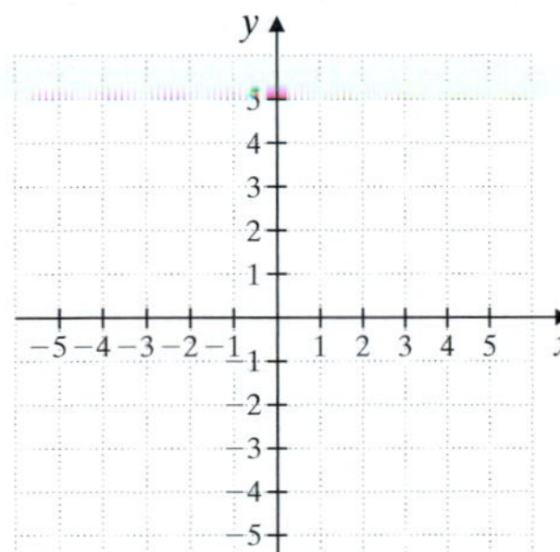

13. $x \leq -3y$

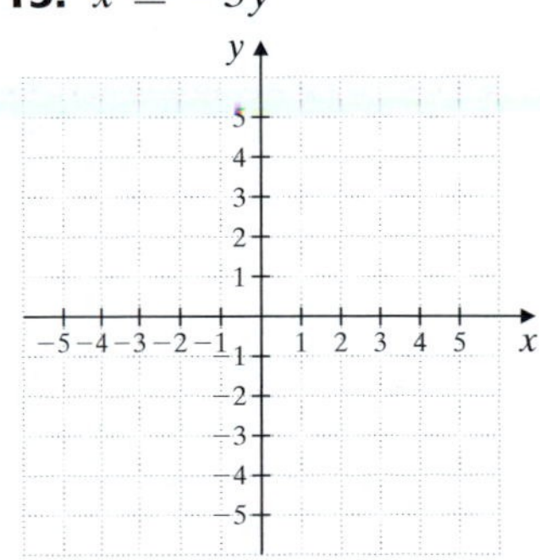

14. $x \geq -2y$

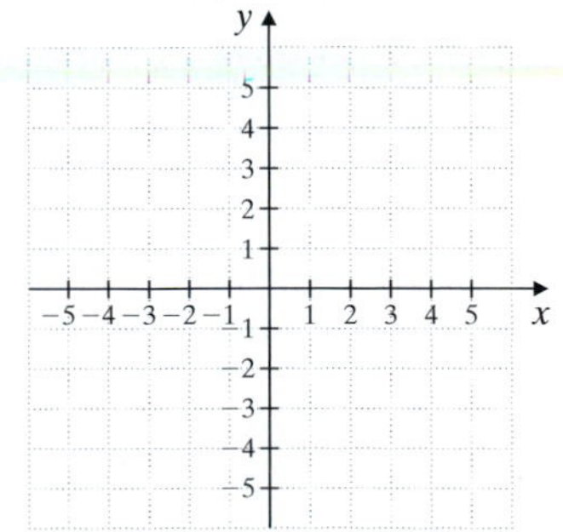

15. $y \geq x + 5$

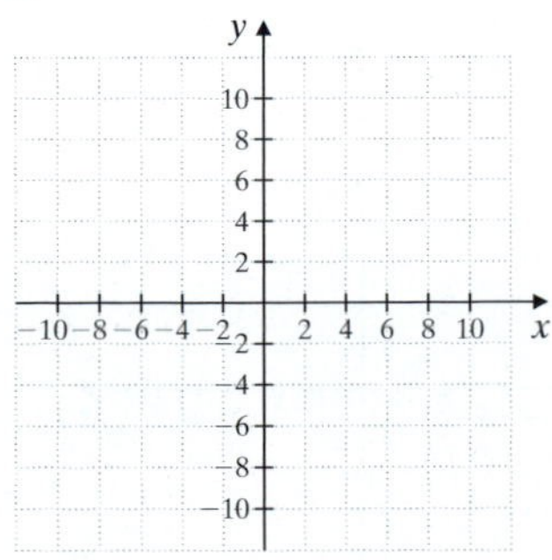

16. $y \leq x + 1$

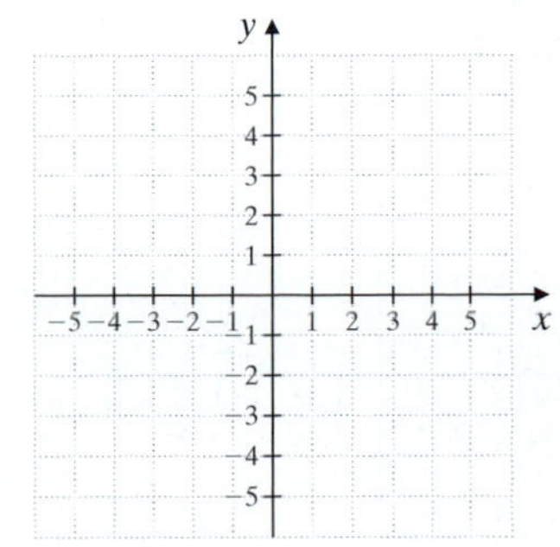

17. $y < 4$

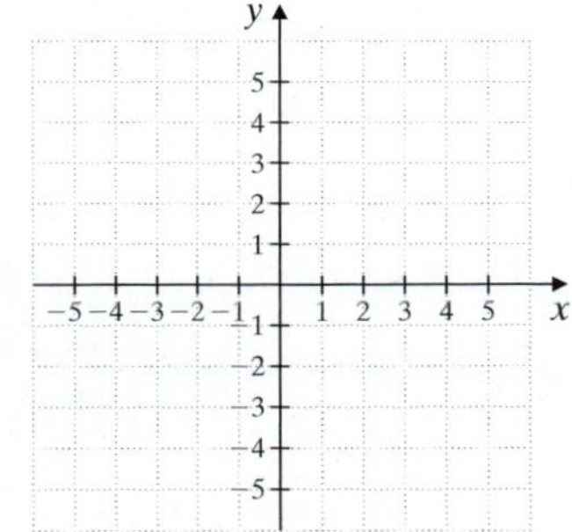

18. $y > 2$

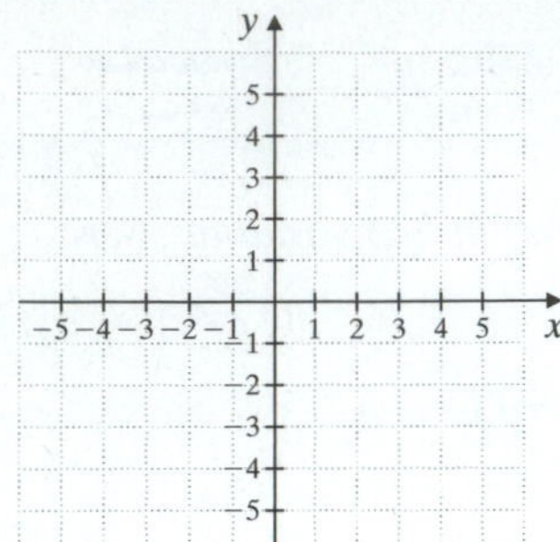

19. $x \geq -3$

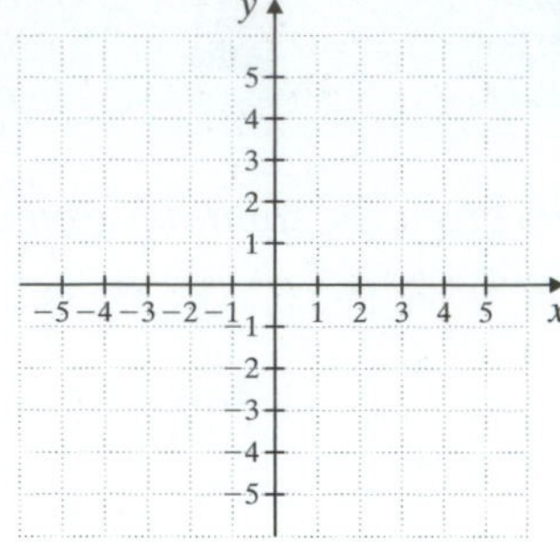

20. $x \leq -1$

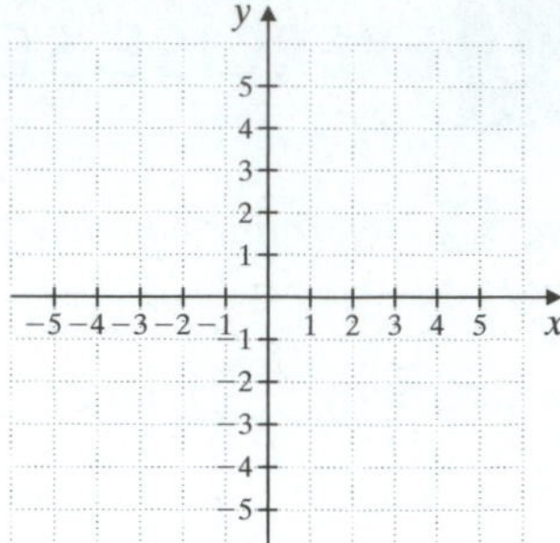

21. $5x + 2y \leq 10$

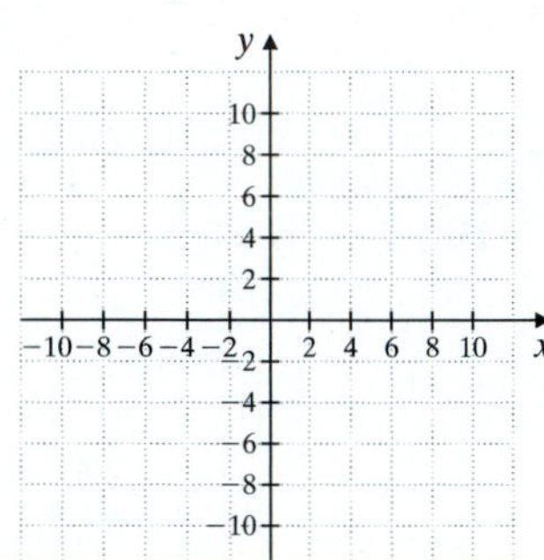

22. $4x + 3y \geq 12$

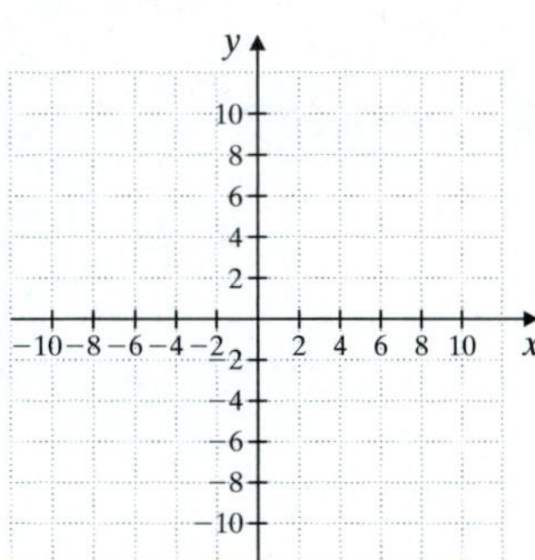

23. $x > y$

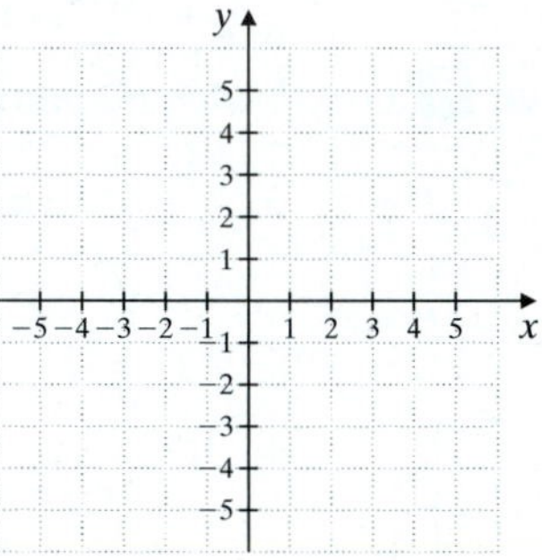

24. $x \leq -y$

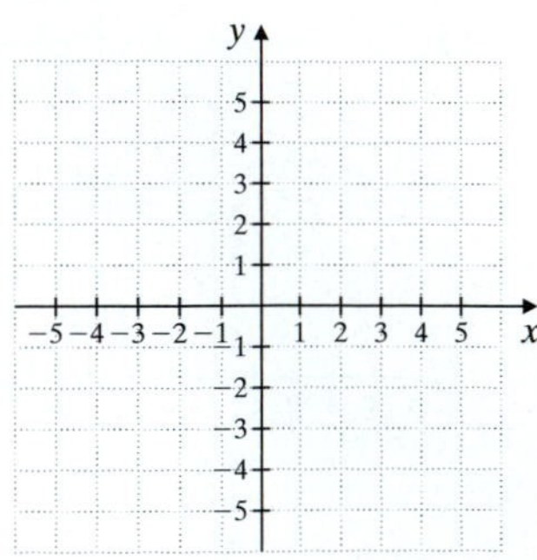

25. $x - y \leq 6$

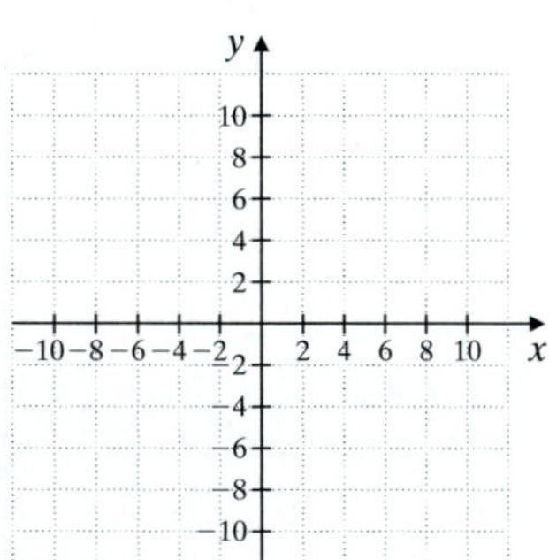

26. $x - y > 10$

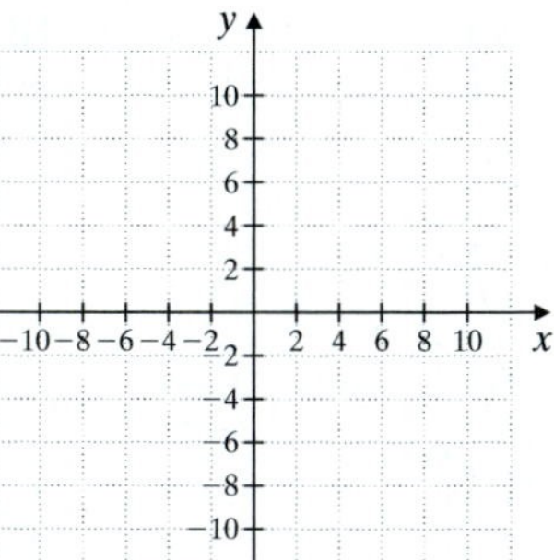

27. $x \geq 0$

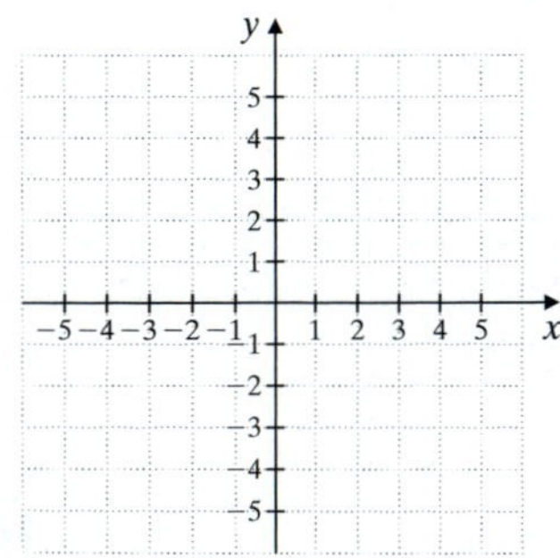

28. $y \leq 0$

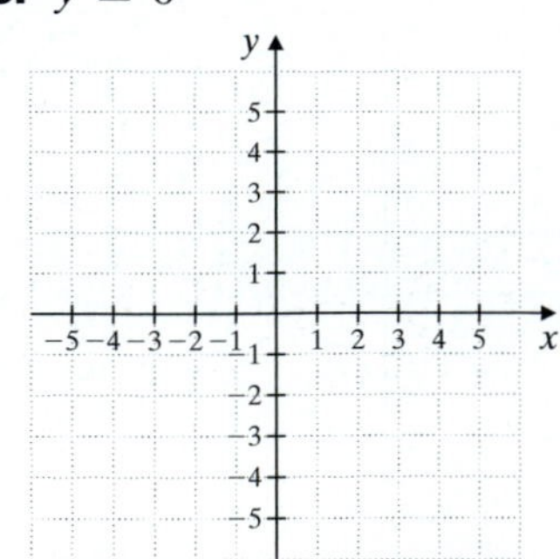

29. $2x + 7y > 5$

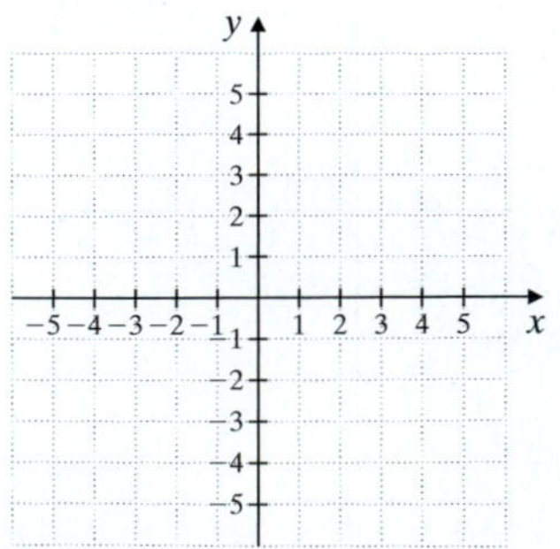

30. $3x + 5y \leq -2$

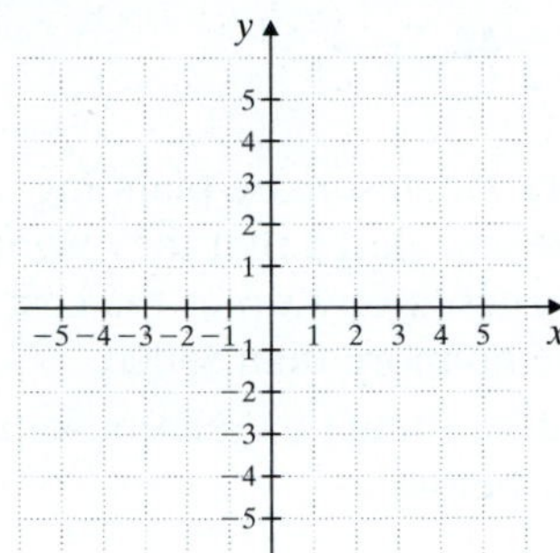

31. $y \geq \frac{1}{2}x - 4$

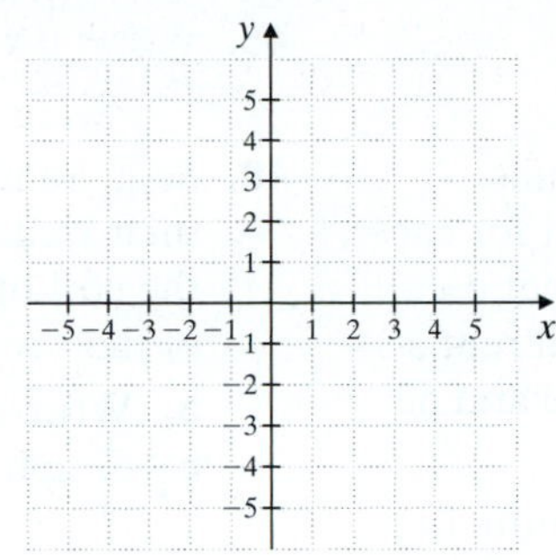

32. $y < \frac{2}{5}x - 3$

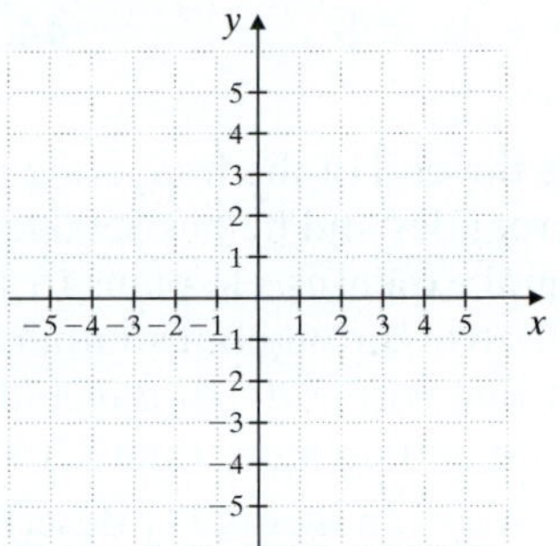

Review

Approximate the coordinates of each point of intersection. See Section 13.1.

33.

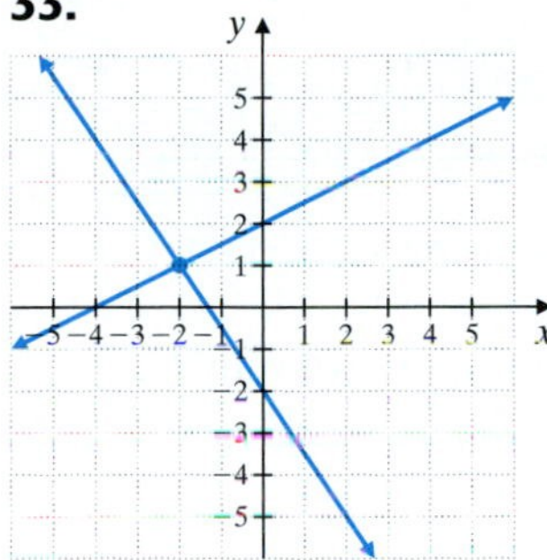

34.

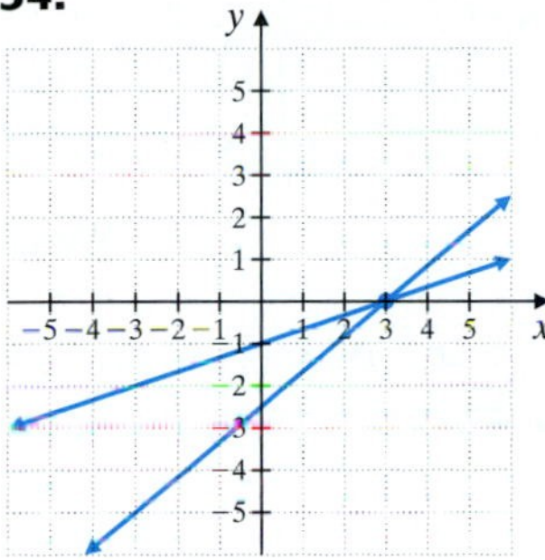

35.

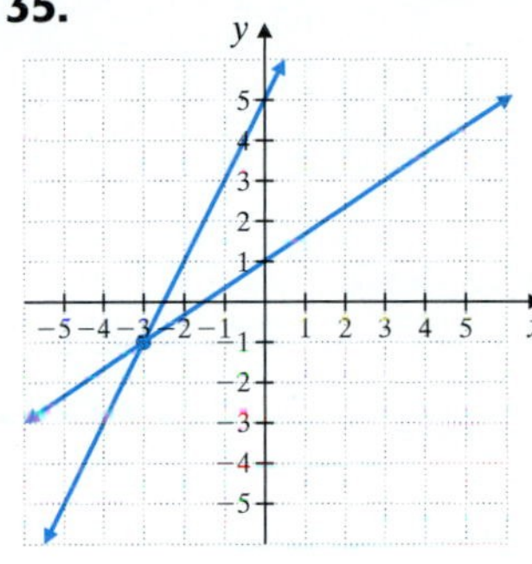

36.

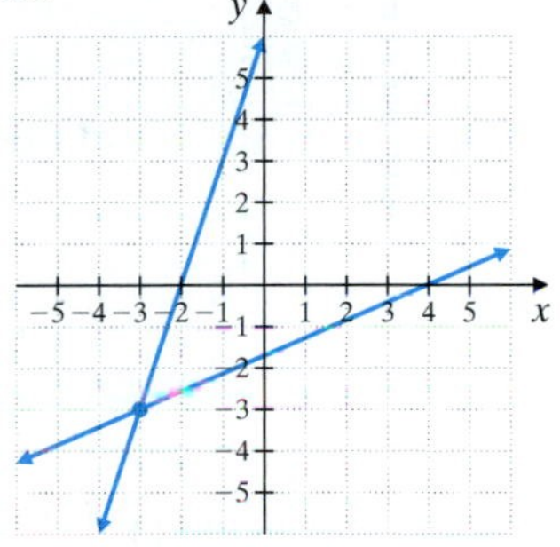

Concept Extensions

Match each inequality with its graph.

a. $x > 2$ **b.** $y < 2$ **c.** $y \leq 2x$ **d.** $y \leq -3x$

37.

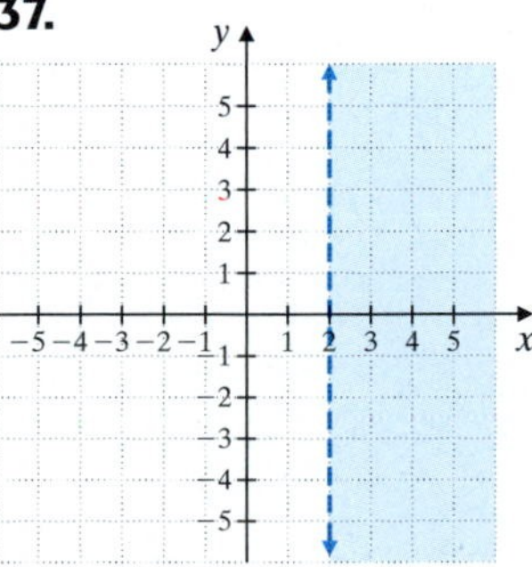

38.

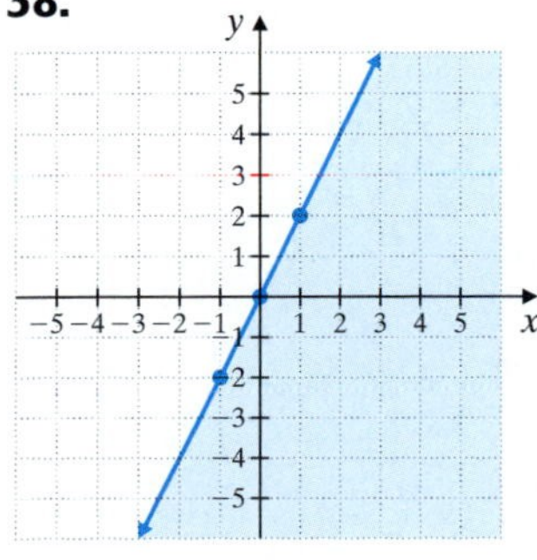

39.

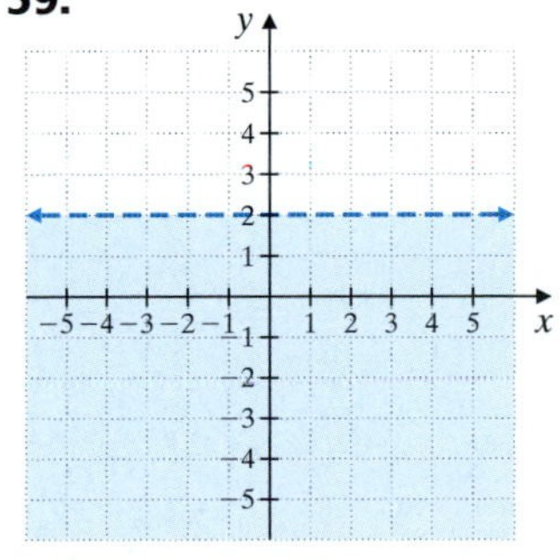

40.

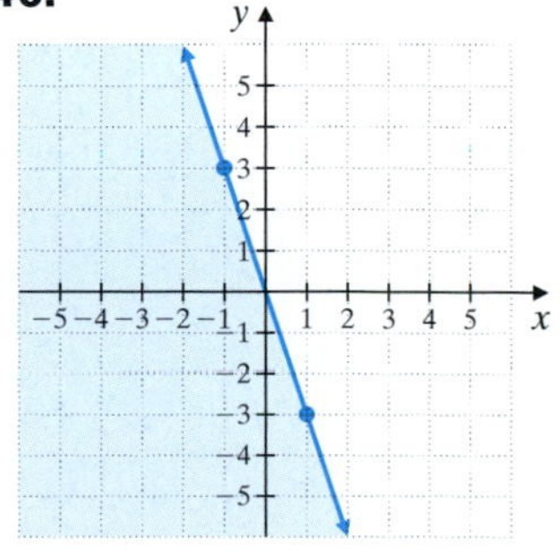

41. Explain why a point on the boundary line should not be chosen as the test point.

42. Write an inequality whose solutions are all points of numbers whose sum is at least 13.

Determine whether $(1, 1)$ *is included in each graph. See the Concept Check in this section.*

43. $3x + 4y < 8$

44. $y > 5x$

45. $y \geq -\frac{1}{2}x$

46. $x > 3$

47. It's the end of the budgeting period for Dennis Fernandes and he has $500 left in his budget for car rental expenses. He plans to spend this budget on a sales trip throughout southern Texas. He will rent a car that costs $30 per day and $0.15 per mile and he can spend no more than $500.

a. Write an inequality describing this situation. Let x = number of days and let y = number of miles.

b. Graph this inequality below.

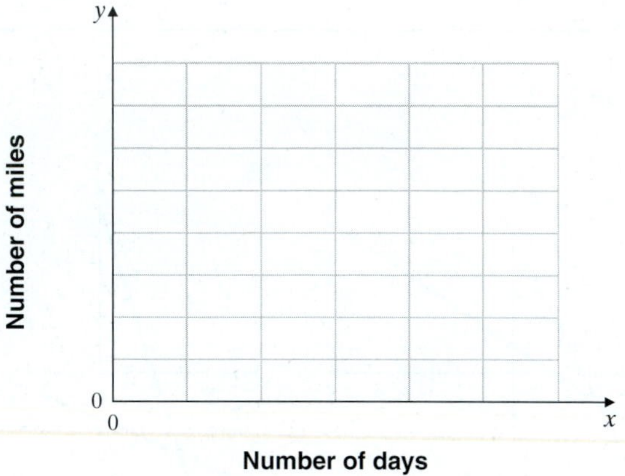

c. Why is the grid showing quadrant I only?

48. Scott Sambracci and Sara Thygeson are planning their wedding. They have calculated that they want the cost of their wedding ceremony x plus the cost of their reception y to be no more than $5000.

a. Write an inequality describing this relationship.

b. Graph this inequality below.

c. Why is the grid showing quadrant I only?

13.8 DIRECT AND INVERSE VARIATION

Objectives

A Solve Problems Involving Direct Variation.

B Solve Problems Involving Inverse Variation.

C Solve Problems Involving Other Types of Direct and Inverse Variation.

D Solve Applications of Variation.

Thus far, we have studied linear equations in two variables. Recall that such an equation can be written in the form $Ax + By = C$, where A and B are not both 0. Also recall that the graph of a linear equation in two variables is a line. In this section, we begin by looking at a particular family of linear equations—those that can be written in the form

$$y = kx$$

where k is a constant. This family of equations is called *direct variation*.

Objective A Solving Direct Variation Problems

Let's suppose that you are earning \$7.25 per hour at a part-time job. The amount of money you earn depends on the number of hours you work. This is illustrated by the following table:

Hours Worked	0	1	2	3	4
Money Earned (before deductions)	0	7.25	14.50	21.75	29.00

and so on

In general, to calculate your earnings (before deductions), multiply the constant \$7.25 by the number of hours you work. If we let y represent the amount of money earned and x represent the number of hours worked, we get the direct variation equation

$$y = 7.25 \cdot x$$

earnings = \$7.25 · hours worked

Notice that in this direct variation equation, as the number of hours increases, the pay increases as well.

Direct Variation

y varies directly as x, or y is directly proportional to x, if there is a nonzero constant k such that

$$y = kx$$

The number k is called the **constant of variation** or the **constant of proportionality.**

In our direct variation example, $y = 7.25x$, the constant of variation is 7.25.

Let's use the previous table to graph $y = 7.25x$. We begin our graph at the ordered-pair solution $(0, 0)$. Why? We assume that the least amount of hours worked is 0. If 0 hours are worked, then the pay is \$0.

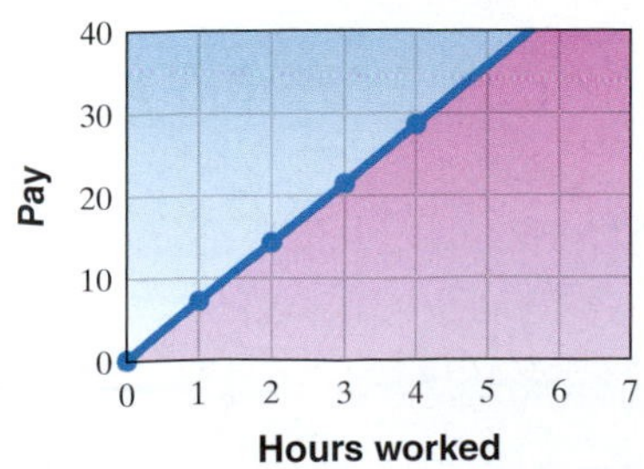

As illustrated in this graph, a direct variation equation $y = kx$ is linear. Also notice that $y = 7.25x$ is a function since its graph passes the vertical line test.

PRACTICE PROBLEM 1

Write a direct variation equation that satisfies:

x	4	$\frac{1}{2}$	1.5	6
y	8	1	3	12

EXAMPLE 1 Write a direct variation equation of the form $y = kx$ that satisfies the ordered pairs in the table below.

x	2	9	1.5	−1
y	6	27	4.5	−3

Solution: We are given that there is a direct variation relationship between x and y. This means that

$$y = kx$$

By studying the given values, you may be able to mentally calculate k. If not, to find k, we simply substitute one given ordered pair into this equation and solve for k. We'll use the given pair $(2, 6)$.

$$y = kx$$
$$6 = k \cdot 2$$
$$\frac{6}{2} = \frac{k \cdot 2}{2}$$
$$3 = k \quad \text{Solve for } k.$$

Since $k = 3$, we have the equation $y = 3x$.
To check, see that each given y is 3 times the given x.

Work Practice Problem 1

Let's try another type of direct variation example.

PRACTICE PROBLEM 2

Suppose that y varies directly as x. If y is 15 when x is 45, find the constant of variation and the direct variation equation. Then find y when x is 3.

EXAMPLE 2 Suppose that y varies directly as x. If y is 17 when x is 34, find the constant of variation and the direct variation equation. Then find y when x is 12.

Solution: Let's use the same method as in Example 1 to find k. Since we are told that y varies directly as x, we know the relationship is of the form

$$y = kx$$

Let $y = 17$ and $x = 34$ and solve for k.

$$17 = k \cdot 34$$
$$\frac{17}{34} = \frac{k \cdot 34}{34}$$
$$\frac{1}{2} = k \quad \text{Solve for } k.$$

Thus, the constant of variation is $\frac{1}{2}$ and the equation is $y = \frac{1}{2}x$.

To find y when $x = 12$, use $y = \frac{1}{2}x$ and replace x with 12.

$$y = \frac{1}{2}x$$
$$y = \frac{1}{2} \cdot 12 \quad \text{Replace } x \text{ with 12.}$$
$$y = 6$$

Thus, when x is 12, y is 6.

Work Practice Problem 2

Let's review a few facts about linear equations of the form $y = kx$.

Answers

1. $y = 2x$ **2.** $k = \frac{1}{3}$; $y = \frac{1}{3}x$; $y = 1$

Direct Variation: y = kx

- There is a direct variation relationship between x and y.
- The graph is a line.
- The line will always go through the origin $(0, 0)$. Why?

  Let $x = 0$. Then $y = k \cdot 0$ or $y = 0$.
- The slope of the graph of $y = kx$ is k, the constant of variation. Why? Remember that the slope of an equation of the form $y = mx + b$ is m, the coefficient of x.
- The equation $y = kx$ describes a function. Each x has a unique y and its graph passes the vertical line test.

EXAMPLE 3 The line is the graph of a direct variation equation. Find the constant of variation and the direct variation equation.

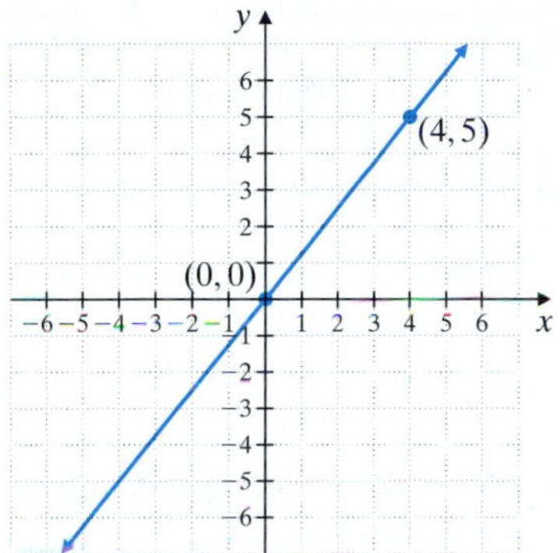

Solution: Recall that k, the constant of variation, is the same as the slope of the line. Thus, to find k, we use the slope formula and find slope.

Using the given points $(0, 0)$, and $(4, 5)$, we have

$$\text{slope} = \frac{5 - 0}{4 - 0} = \frac{5}{4}$$

Thus, $k = \frac{5}{4}$ and the variation equation is $y = \frac{5}{4}x$.

Work Practice Problem 3

PRACTICAL PROBLEM 3

Find the constant of variation and the direct variation equation for the line below.

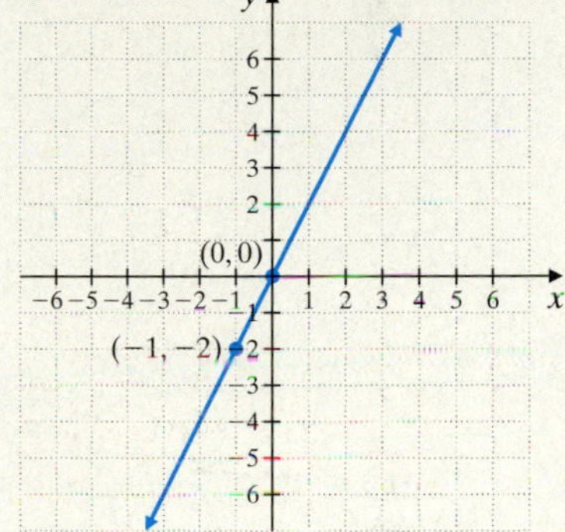

Objective B Solving Inverse Variation Problems

In this section, we introduce another type of variation called inverse variation.

Let's suppose you need to drive a distance of 40 miles. You know that the faster you drive the distance, the sooner you arrive at your destination. Recall that there is a mathematical relationship between distance, rate, and time. It is $d = r \cdot t$. In our example, distance is a constant 40 miles, so we have $40 = r \cdot t$ or $t = \frac{40}{r}$.

For example, if you drive 10 mph, the time to drive the 40 miles is

$$t = \frac{40}{r} = \frac{40}{10} = 4 \text{ hours}$$

If you drive 20 mph, the time is

$$t = \frac{40}{r} = \frac{40}{20} = 2 \text{ hours}$$

Answer

3. $k = 2$; $y = 2x$

Again, notice that as speed increases, time decreases. Below are some ordered-pair solutions of $t = \frac{40}{r}$ and its graph.

Rate (mph)	r	5	10	20	40	60	80
Time (hr)	t	8	4	2	1	$\frac{2}{3}$	$\frac{1}{2}$

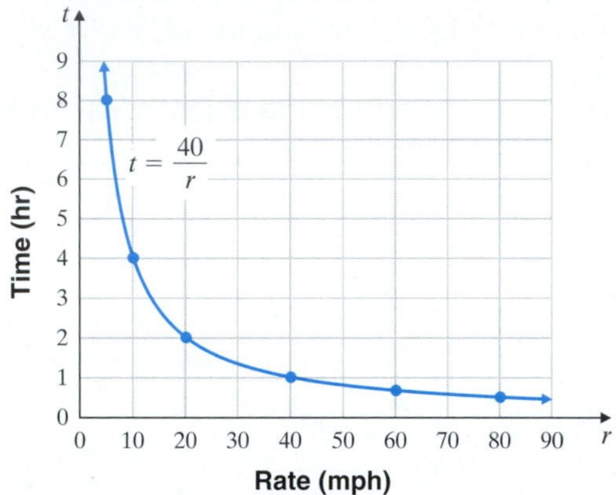

Notice that the graph of this variation is not a line, but it passes the vertical line test so $t = \frac{40}{r}$ does describe a function. This is an example of inverse variation.

Inverse Variation

***y* varies inversely as *x*, or *y* is inversely proportional to *x*,** if there is a nonzero constant k such that

$$y = \frac{k}{x}$$

The number k is called the **constant of variation** or the **constant of proportionality.**

In our inverse variation example, $t = \frac{40}{r}$ or $y = \frac{40}{x}$, the constant of variation is 40.

We can immediately see differences and similarities in direct variation and inverse variation.

Direct Variation	$y = kx$	linear equation	both functions
Inverse Variation	$y = \frac{k}{x}$	rational equation	

Remember from Chapter 12 that $y = \frac{k}{x}$ is a rational equation and not a linear equation. Also notice that because x is in the denominator, x can be any value except 0.

We can still derive an inverse variation equation from a table of values.

PRACTICE PROBLEM 4

Write an inverse variation equation of the form $y = \frac{k}{x}$ that satisfies:

x	4	10	40	−2
y	5	2	$\frac{1}{2}$	−10

EXAMPLE 4 Write an inverse variation equation of the form $y = \frac{k}{x}$ that satisfies the ordered pairs in the table below.

x	2	4	$\frac{1}{2}$
y	6	3	24

Solution: Since there is an inverse variation relationship between x and y, we know that $y = \frac{k}{x}$.

Answer

4. $y = \frac{20}{x}$

To find k, choose one given ordered pair and substitute the values into the equation. We'll use (2, 6).

$$y = \frac{k}{x}$$

$$6 = \frac{k}{2}$$

$$2 \cdot 6 = 2 \cdot \frac{k}{2} \quad \text{Multiply both sides by 2.}$$

$$12 = k \quad \text{Solve.}$$

Since $k = 12$, we have the equation $y = \frac{12}{x}$.

Work Practice Problem 4

Helpful Hint

Multiply both sides of the inverse variation relationship equation $y = \frac{k}{x}$ by x (as long as x is not 0), and we have $xy = k$. This means that if y varies inversely as x, their product is always the constant of variation k. For an example of this, check the table from Example 4:

x	2	4	$\frac{1}{2}$
y	6	3	24

$2 \cdot 6 = 12$ $\quad 4 \cdot 3 = 12$ $\quad \frac{1}{2} \cdot 24 = 12$

EXAMPLE 5 Suppose that y varies inversely as x. If $y = 0.02$ when $x = 75$, find the constant of variation and the inverse variation equation. Then find y when x is 30.

Solution: Since y varies inversely as x, the constant of variation may be found by simply finding the product of the given x and y.

$$k = xy = 75(0.02) = 1.5$$

To check, we will use the inverse variation equation

$$y = \frac{k}{x}$$

Let $y = 0.02$ and $x = 75$ and solve for k.

$$0.02 = \frac{k}{75}$$

$$75(0.02) = 75 \cdot \frac{k}{75} \quad \text{Multiply both sides by 75.}$$

$$1.5 = k \quad \text{Solve for } k.$$

Thus, the constant of variation is 1.5 and the equation is $y = \frac{1.5}{x}$. To find y when $x = 30$, use $y = \frac{1.5}{x}$ and replace x with 30.

$$y = \frac{1.5}{x}$$

$$y = \frac{1.5}{30} \quad \text{Replace } x \text{ with 30.}$$

$$y = 0.05$$

Thus, when x is 30, y is 0.05.

Work Practice Problem 5

PRACTICE PROBLEM 5

Suppose that y varies inversely as x. If y is 4 when x is 0.8, find the constant of variation and the direct variation equation. Then find y when x is 20.

Answer

5. $k = 3.2$; $y = \frac{3.2}{x}$; $y = 0.16$

Objective C Solving Other Types of Direct and Inverse Variation Problems

It is possible for y to vary directly or inversely as powers of x.

Direct and Inverse Variation as nth Powers of x

y varies directly as a power of x if there is a nonzero constant k and a natural number n such that

$$y = kx^n$$

y varies inversely as a power of x if there is a nonzero constant k and a natural number n such that

$$y = \frac{k}{x^n}$$

PRACTICE PROBLEM 6

The area of a circle varies directly as the square of its radius. A circle with radius 7 inches has an area of 49π square inches. Find the area of a circle whose radius is 4 feet.

EXAMPLE 6 The surface area of a cube A varies directly as the square of a length of its sides. If A is 54 when s is 3, find A when $s = 4.2$.

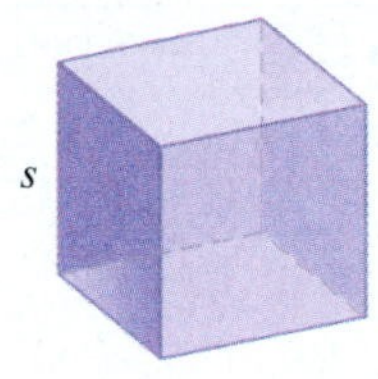

Solution: Since the surface area A varies directly as the square of side s, we have

$A = ks^2$

To find k, let $A = 54$ and $s = 3$.

$A = k \cdot s^2$

$54 = k \cdot 3^2$ Let $A = 54$ and $s = 3$.

$54 = 9k$ $3^2 = 9$.

$6 = k$ Divide by 9.

The formula for surface area of a cube is then

$A = 6s^2$ where s is the length of a side.

To find the surface area when $s = 4.2$, substitute.

$A = 6s^2$

$A = 6 \cdot (4.2)^2$

$A = 105.84$

The surface area of a cube whose side measures 4.2 units is 105.84 sq units.

Work Practice Problem 6

Answer

6. 16π sq ft

Objective D Solving Applications of Variation

There are many real-life applications of direct and inverse variation.

EXAMPLE 7 The weight of a body w varies inversely with the square of its distance from the center of Earth, d. If a person weighs 160 pounds on the surface of Earth, what is the person's weight 200 miles above the surface? (Assume that the radius of Earth is 4000 miles.)

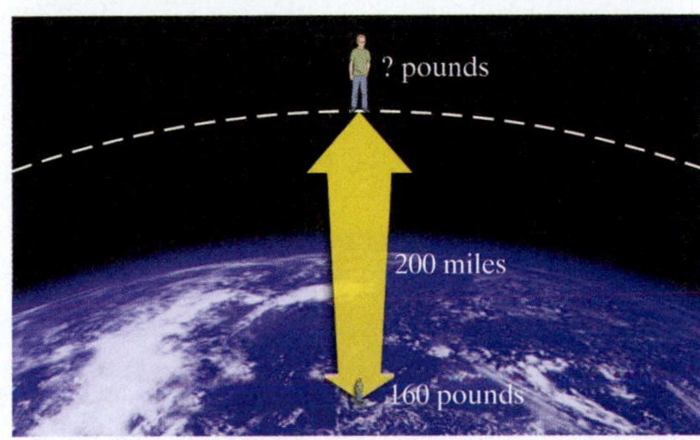

Solution:

1. UNDERSTAND. Make sure you read and reread the problem.
2. TRANSLATE. Since we are told that weight w varies inversely with the square of its distance from the center of Earth, d, we have

$$w = \frac{k}{d^2}$$

3. SOLVE. To solve the problem, we first find k. To do so, use the fact that the person weighs 160 pounds on Earth's surface, which is a distance of 4000 miles from the Earth's center.

$$w = \frac{k}{d^2}$$

$$160 = \frac{k}{(4000)^2}$$

$$2{,}560{,}000{,}000 = k$$

Thus, we have $w = \dfrac{2{,}560{,}000{,}000}{d^2}$.

Since we want to know the person's weight 200 miles above the Earth's surface, we let $d = 4200$ and find w.

$$w = \frac{2{,}560{,}000{,}000}{d^2}$$

$$w = \frac{2{,}560{,}000{,}000}{(4200)^2}$$ A person 200 miles above the Earth's surface is 4200 miles from the Earth's center.

$$w \approx 145$$ Simplify.

4. INTERPRET. *Check:* Your answer is reasonable since the farther a person is from Earth, the less the person weighs. *State:* Thus, 200 miles above the surface of the Earth, a 160-pound person weighs approximately 145 pounds.

Work Practice Problem 7

PRACTICE PROBLEM 7

The distance d that an object falls is directly proportional to the square of the time of the fall, t. If an object falls 144 feet in 3 seconds, find how far the object falls in 5 seconds.

Answer

7. 400 feet

Vocabulary and Readiness Check

Use the choices below to fill in each blank.

direct inverse

State which variation each equation represents.

1. $y = \frac{k}{x}$, where k is a constant. ________

2. $y = kx$, where k is a constant. ________

3. $y = 5x$ ________

4. $y = \frac{5}{x}$ ________

5. $y = \frac{7}{x^2}$ ________

6. $y = 6.5x^4$ ________

13.8 EXERCISE SET

FOR EXTRA HELP

Objective A *Write a direct variation equation, $y = kx$, that satisfies the ordered pairs in each table. See Example 1.*

1.

x	0	6	10
y	0	3	5

2.

x	0	2	−1	3
y	0	14	−7	21

3.

x	−2	2	4	5
y	−12	12	24	30

4.

x	3	9	−2	12
y	1	3	$-\frac{2}{3}$	4

Write a direct variation equation, $y = kx$, that describes each graph. See Example 3.

5.

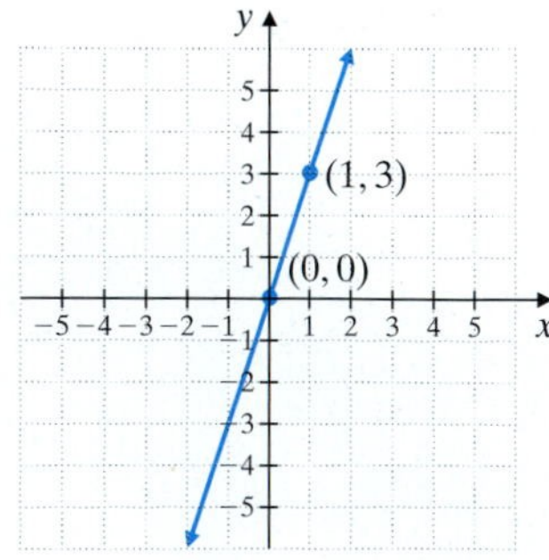

6.

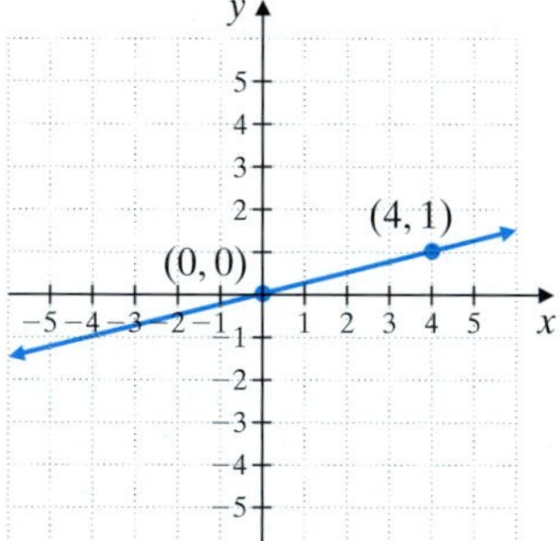

7.

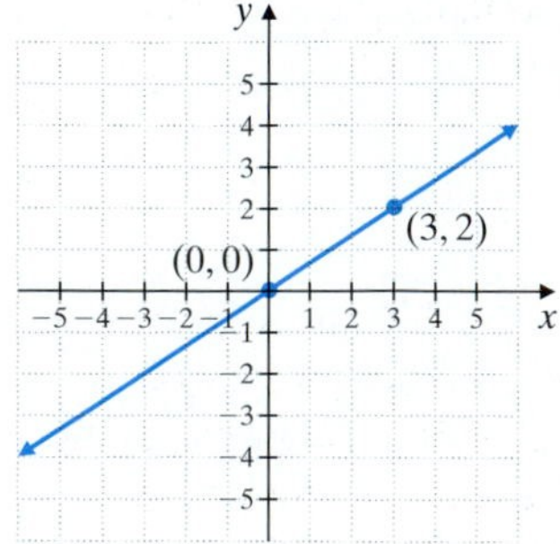

8.

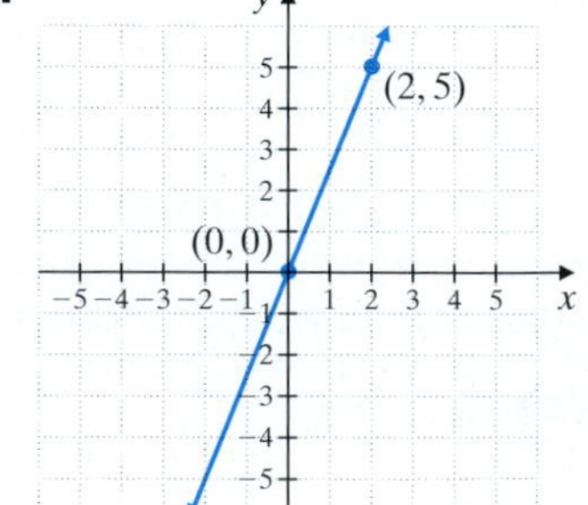

Objective B *Write an inverse variation equation, $y = \frac{k}{x}$, that satisfies the ordered pairs in each table. See Example 4.*

9.

x	1	−7	3.5	−2
y	7	−1	2	−3.5

10.

x	2	−11	4	−4
y	11	−2	5.5	−5.5

11.

x	10	$\frac{1}{2}$	$-\frac{1}{4}$
y	0.05	1	−2

12.

x	4	$\frac{1}{5}$	−8
y	0.1	2	−0.05

Objectives A B Translating *Write an equation to describe each variation. Use k for the constant of proportionality. See Examples 1 through 5.*

13. y varies directly as x

14. a varies directly as b

15. h varies inversely as t

16. s varies inversely as t

17. z varies directly as x^2

18. p varies inversely as x^2

19. y varies inversely as z^3

20. x varies directly as y^4

21. x varies inversely as $\sqrt{y}$

22. y varies directly as d^2

Objectives A B C *Solve. See Examples 2, 5, and 6.*

23. y varies directly as x. If $y = 20$ when $x = 5$, find y when x is 10.

24. y varies directly as x. If $y = 27$ when $x = 3$, find y when x is 2.

25. y varies inversely as x. If $y = 5$ when $x = 60$, find y when x is 100.

26. y varies inversely as x. If $y = 200$ when $x = 5$, find y when x is 4.

27. z varies directly as x^2. If $z = 96$ when $x = 4$, find z when $x = 3$.

28. s varies directly as t^3. If $s = 270$ when $t = 3$, find s when $x = 1$.

29. a varies inversely as b^3. If $a = \frac{3}{2}$ when $b = 2$, find a when b is 3.

30. p varies inversely as q^2. If $p = \frac{5}{16}$ when $q = 8$, find p when $q = \frac{1}{2}$.

Objectives C D *Solve. See Examples 1 through 7.*

31. Your paycheck (before deductions) varies directly as the number of hours you work. If your paycheck is $112.50 for 18 hours, find your pay for 10 hours.

32. If your paycheck (before deductions) is $244.50 for 30 hours, find your pay for 34 hours. (See Exercise 31.)

33. The cost of manufacturing a certain type of headphone varies inversely as the number of headphones increases. If 5000 headphones can be manufactured for $9.00 each, find the cost to manufacture 7500 headphones.

34. The cost of manufacturing a certain composition notebook varies inversely as the number of notebooks increases. If 10,000 notebooks can be manufactured for $0.50 each, find the cost to manufacture 18,000 notebooks. Round your answer to the nearest cent.

35. The distance a spring stretches varies directly with the weight attached to the spring. If a 60-pound weight stretches the spring 4 inches, find the distance that an 80-pound weight stretches the spring.

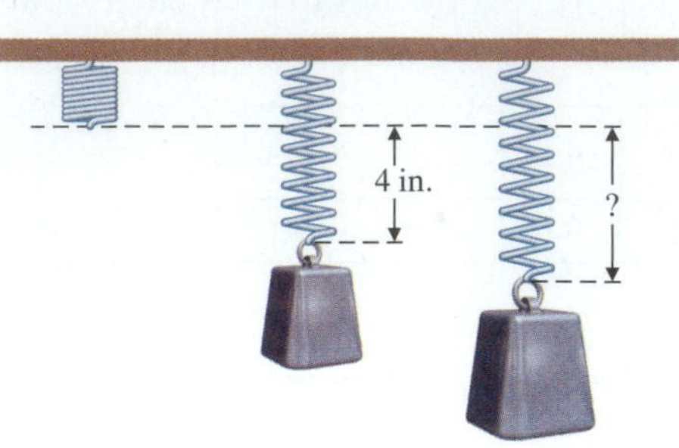

36. If a 30-pound weight stretches a spring 10 inches, find the distance a 20-pound weight stretches the spring. (See Exercise 35.)

37. The weight of an object varies inversely as the square of its distance from the *center* of the Earth. If a person weighs 180 pounds on Earth's surface, what is his weight 10 miles above the surface of the Earth? (Assume that the Earth's radius is 4000 miles and round your answer to one decimal place.)

38. For a constant distance, the rate of travel varies inversely as the time traveled. If a family travels 55 mph and arrives at a destination in 4 hours, how long will the return trip take traveling at 60 mph?

39. The distance d that an object falls is directly proportional to the square of the time of the fall, t. A person who is parachuting for the first time is told to wait ten seconds before opening the parachute. If the person falls 64 feet in 2 seconds, find how far he falls in 10 seconds.

40. The distance needed for a car to stop, d, is directly proportional to the square of its rate of travel, r. Under certain driving conditions, a car traveling 60 mph needs 300 feet to stop. With these same driving conditions, how long does it take a car to stop if the car is traveling 30 mph when the brakes are applied?

Review

Add the equations. See Section 10.4.

41. $\begin{aligned} -3x + 4y &= 7 \\ \underline{3x - 2y} &\underline{= 9} \end{aligned}$

42. $\begin{aligned} x - y &= -9 \\ \underline{-x - y} &\underline{= -14} \end{aligned}$

43. $\begin{aligned} 5x - 0.4y &= 0.7 \\ \underline{-9x + 0.4y} &\underline{= -0.2} \end{aligned}$

44. $\begin{aligned} 1.9x - 2y &= 5.7 \\ \underline{-1.9x - 0.1y} &\underline{= 2.3} \end{aligned}$

Concept Extensions

45. Suppose that y varies directly as x. If x is tripled, what is the effect on y?

46. Suppose that y varies directly as x^2. If x is tripled, what is the effect on y?

47. The period of a pendulum p (the time of one complete back-and-forth swing) varies directly with the square root of its length, ℓ. If the length of the pendulum is quadrupled, what is the effect on the period, p?

48. For a constant distance, the rate of travel r varies inversely with the time traveled, t. If a car traveling 100 mph completes a test track in 6 minutes, find the rate needed to complete the same test track in 4 minutes. (*Hint:* Convert minutes to hours.)

CHAPTER 13 Group Activity

Finding a Linear Model

This activity may be completed by working in groups or individually.

The following table shows the actual number of foreign visitors (in millions) to the United States for the years 2000 through 2003.

Year	Foreign Visitors to the United States (in millions)
2000	50.9
2001	44.8
2002	41.9
2003	40.4

(*Source:* Tourism Industries/International Trade Administration, U.S. Department of Commerce)

1. Make a scatter diagram of the paired data in the table.
2. Use what you have learned in this chapter to write an equation of the line representing the paired data in the table. Explain how you found the equation, and what each variable represents.
3. What is the slope of your line? What does the slope mean in this context?
4. Use your linear equation to predict the number of foreign visitors to the United States in 2010.
5. Compare your linear equation to that found by other students or groups. Is it the same, similar, or different? How?
6. Compare your prediction from question 3 to that of other students or groups. Describe what you find.
7. The number of visitors to the United States for 2004 was estimated to be 45.7 million. If this data point is added to the chart above, how does it affect your results?

Chapter 13 Vocabulary Check

Fill in each blank with one of the words listed below.

y-axis	x-axis	solution	linear	standard	point-slope
x-intercept	y-intercept	y	x	slope	relation
domain	range	direct	inverse	slope-intercept	function

1. An ordered pair is a __________ of an equation in two variables if replacing the variables by the coordinates of the ordered pair results in a true statement.

2. The vertical number line in the rectangular coordinate system is called the ________.

3. A ________ equation can be written in the form $Ax + By = C$.

4. A(n) ____________ is a point of the graph where the graph crosses the x-axis.

5. The form $Ax + By = C$ is called __________ form.

6. A(n) ____________ is a point of the graph where the graph crosses the y-axis.

7. A set of ordered pairs that assigns to each x-value exactly one y-value is called a __________.

8. The equation $y = 7x - 5$ is written in _______________ form.

9. The set of all x-coordinates of a relation is called the __________ of the relation.

10. The set of all y-coordinates of a relation is called the ________ of the relation.

11. A set of ordered pairs is called a __________.

12. The equation $y + 1 = 7(x - 2)$ is written in _____________ form.

13. To find an x-intercept of a graph, let _____ $= 0$.

14. The horizontal number line in the rectangular coordinate system is called the ________.

15. To find a y-intercept of a graph, let _____ $= 0$.

16. The ________ of a line measures the steepness or tilt of a line.

17. The equation $y = kx$ is an example of ________ variation.

18. The equation $y = \frac{k}{x}$ is an example of _________ variation.

Helpful Hint

Are you preparing for your test? Don't forget to take the Chapter 13 Test on page 1049. Then check your answers at the back of the text and use the Chapter Test Prep Video CD to see the fully worked-out solutions to any of the exercises you want to review.

13 Chapter Highlights

DEFINITIONS AND CONCEPTS	EXAMPLES
Section 13.1 The Rectangular Coordinate System	
The **rectangular coordinate system** consists of a plane and a vertical and a horizontal number line intersecting at their 0 coordinate. The vertical number line is called the ***y*-axis** and the horizontal number line is called the ***x*-axis.** The point of intersection of the axes is called the **origin.** To **plot** or **graph** an ordered pair means to find its corresponding point on a rectangular coordinate system. To plot or graph an ordered pair such as $(3, -2)$, start at the origin. Move 3 units to the right and from there, 2 units down. To plot or graph $(-3, 4)$; start at the origin. Move 3 units to the left and from there, 4 units up. An ordered pair is a **solution** of an equation in two variables if replacing the variables with the coordinates of the ordered pair results in a true statement.	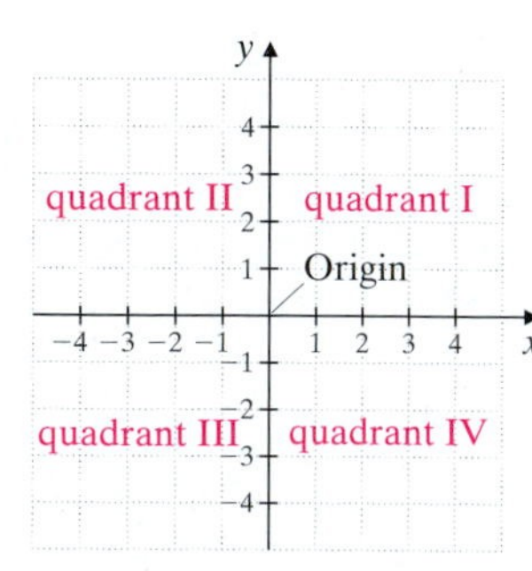 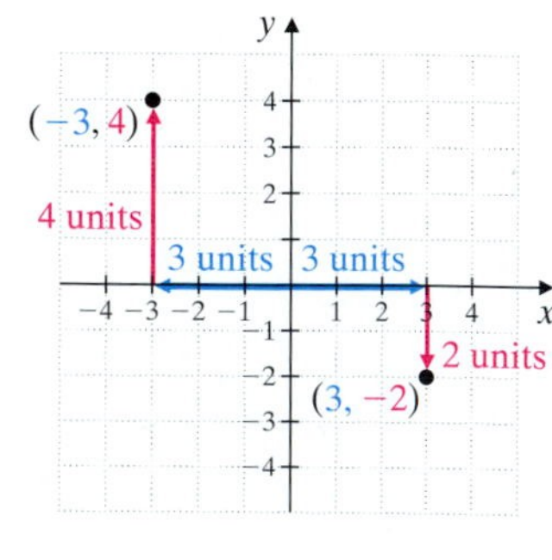

Definitions and Concepts	**Examples**
Section 13.1 The Rectangular Coordinate System (*continued*)	
If one coordinate of an ordered pair solution is known, the other value can be determined by substitution.	Complete the ordered pair (0,) for the equation $x - 6y = 12$. $x - 6y = 12$ $0 - 6y = 12$ Let $x = 0$. $\frac{-6y}{-6} = \frac{12}{-6}$ Divide by -6. $y = -2$ The ordered pair solution is $(0, -2)$.
Section 13.2 Graphing Linear Equations	
A **linear equation in two variables** is an equation that can be written in the form $Ax + By = C$, where A and B are not both 0. The form $Ax + By = C$ is called **standard form.**	$3x + 2y = -6$ $x = -5$ $y = 3$ $y = -x + 10$ $x + y = 10$ is in standard form.
To graph a linear equation in two variables, find three ordered pair solutions. Plot the solution points and draw the line connecting the points.	Graph: $x - 2y = 5$ 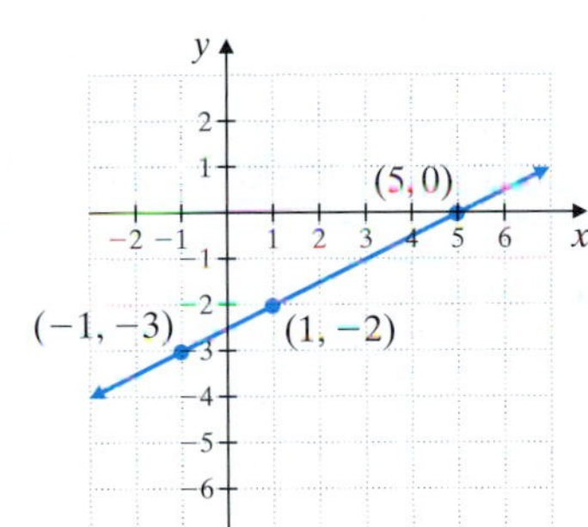
Section 13.3 Intercepts	
An **intercept** of a graph is a point where the graph intersects an axis. If a graph intersects the x-axis at a, then $(a, 0)$ is an ***x*-intercept.** If a graph intersects the y-axis at b, then $(0, b)$ is a ***y*-intercept.**	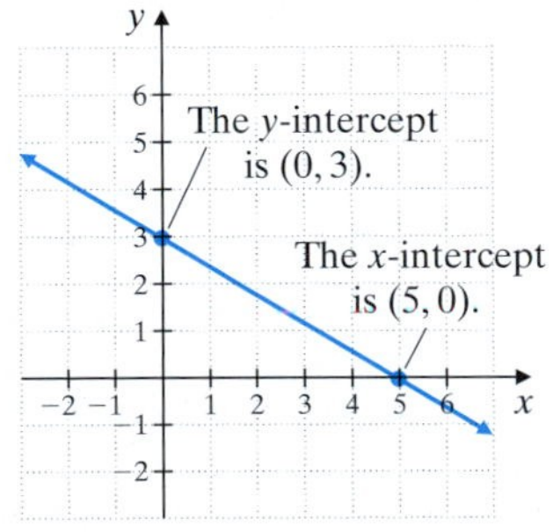
To find the *x*-intercept(s), let $y = 0$ and solve for x. **To find the *y*-intercept(s),** let $x = 0$ and solve for y.	Find the intercepts for $2x - 5y = -10$. If $y = 0$, then: $2x - 5 \cdot 0 = -10$; $2x = -10$; $\frac{2x}{2} = \frac{-10}{2}$; $x = -5$ If $x = 0$, then: $2 \cdot 0 - 5y = -10$; $-5y = -10$; $\frac{-5y}{-5} = \frac{-10}{-5}$; $y = 2$

x	y
5	0
1	−2
−1	−3

continued

DEFINITIONS AND CONCEPTS	EXAMPLES

Section 13.3 Intercepts (*continued*)

The x-intercept is $(-5, 0)$. The y-intercept is $(0, 2)$.

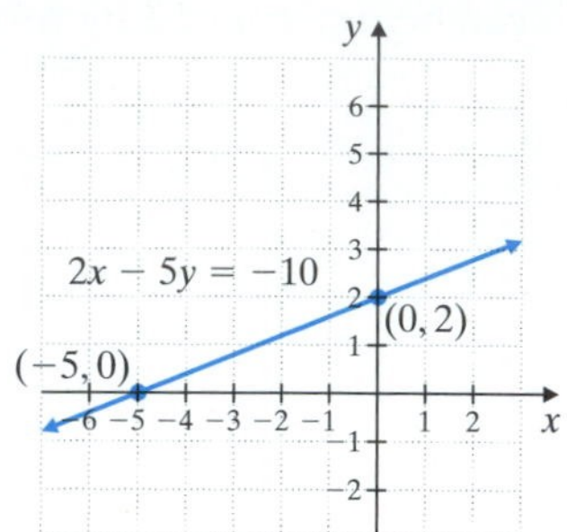

The graph of $x = c$ is a vertical line with x-intercept $(c, 0)$. The graph of $y = c$ is a horizontal line with y-intercept $(0, c)$.

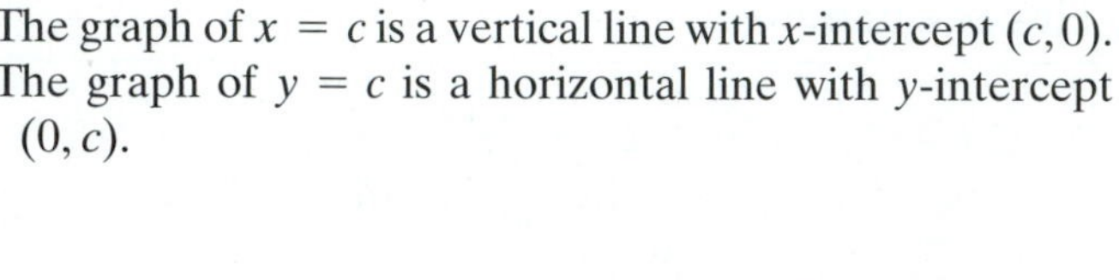

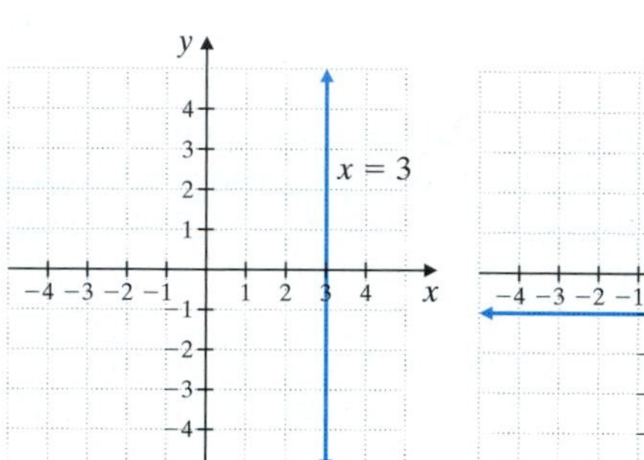

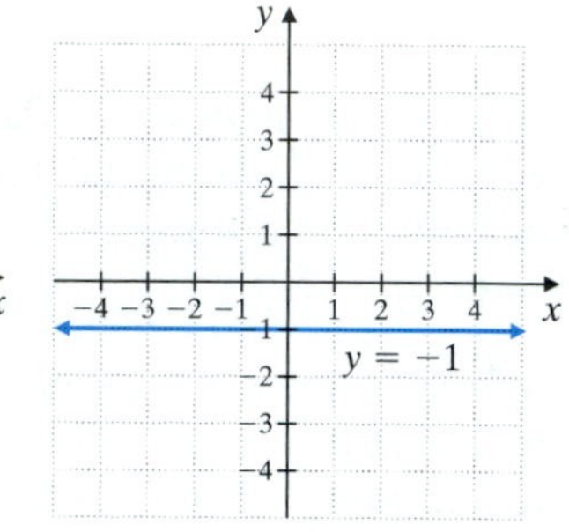

Section 13.4 Slope and Rate of Change

The **slope** m of the line through points (x_1, y_1) and (x_2, y_2) is given by

$$m = \frac{y_2 - y_1}{x_2 - x_1} \quad \text{as long as } x_2 \neq x_1$$

A horizontal line has slope 0.
The slope of a vertical line is undefined.
Nonvertical parallel lines have the same slope.
Two nonvertical lines are perpendicular if the slope of one is the negative reciprocal of the slope of the other.

The slope of the line through points $(-1, 6)$ and $(-5, 8)$ is

$$m = \frac{y_2 - y_1}{x_2 - x_1} = \frac{8 - 6}{-5 - (-1)} = \frac{2}{-4} = -\frac{1}{2}$$

The slope of the line $y = -5$ is 0.
The line $x = 3$ has undefined slope.

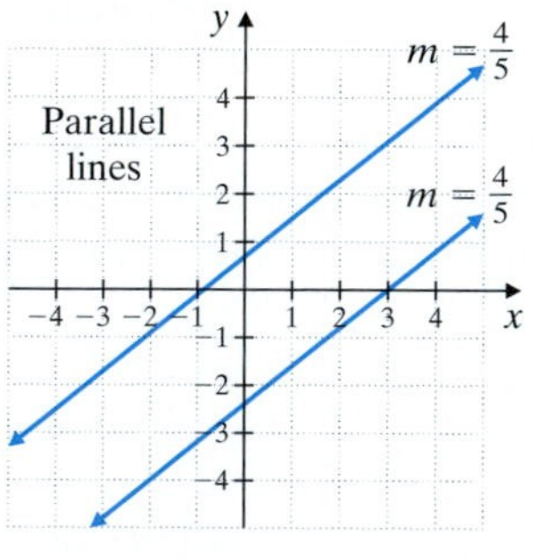

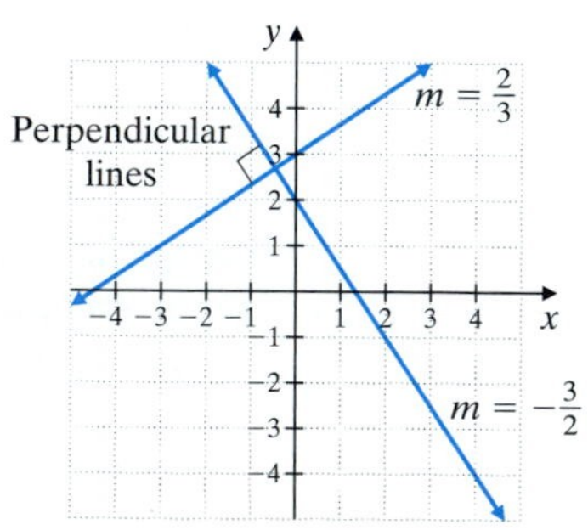

Section 13.5 Equations of Lines

SLOPE-INTERCEPT FORM

$$y = mx + b$$

m is the slope of the line.
$(0, b)$ is the y-intercept.

Find the slope and the y-intercept of the line $2x + 3y = 6$. Solve for y:

$$2x + 3y = 6$$
$$3y = -2x + 6 \quad \text{Subtract } 2x.$$
$$y = -\frac{2}{3}x + 2 \quad \text{Divide by 3.}$$

The slope of the line is $-\frac{2}{3}$ and the y-intercept is $(0, 2)$.

Definitions and Concepts	**Examples**
Section 13.5 Equations of Lines (*continued*)	
Point-Slope Form $y - y_1 = m(x - x_1)$ m is the slope. (x_1, y_1) is a point of the line.	Find an equation of the line with slope $\frac{3}{4}$ that contains the point $(-1, 5)$. $y - 5 = \frac{3}{4}[x - (-1)]$ $4(y - 5) = 3(x + 1)$ Multiply by 4. $4y - 20 = 3x + 3$ Distribute. $-3x + 4y = 23$ Subtract $3x$ and add 20.
Section 13.6 Introduction to Functions	
A set of ordered pairs is a **relation.** The set of all x-coordinates is called the **domain** of the relation and the set of all y-coordinates is called the **range** of the relation.	The domain of the relation $\{(0, 5), (2, 5), (4, 5), (5, -2)\}$ is $\{0, 2, 4, 5\}$. The range is $\{-2, 5\}$.
A **function** is a set of ordered pairs that assigns to each x-value exactly one y-value. **Vertical Line Test** If a vertical line can be drawn so that it intersects a graph more than once, the graph is not the graph of a function. (If no such line can be drawn, the graph is that of a function.)	Which are graphs of functions? 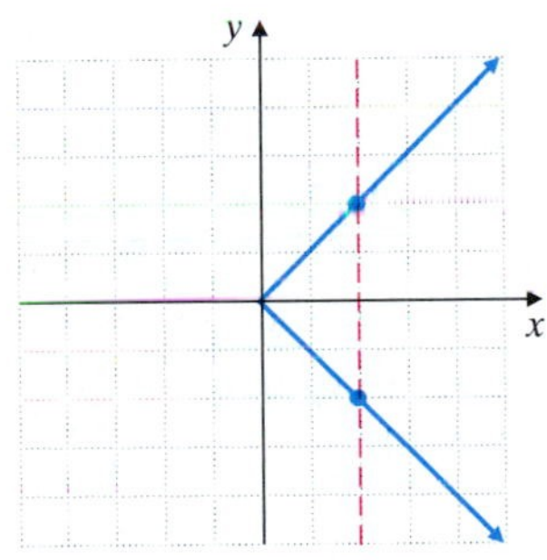 This graph is not the graph of a function. 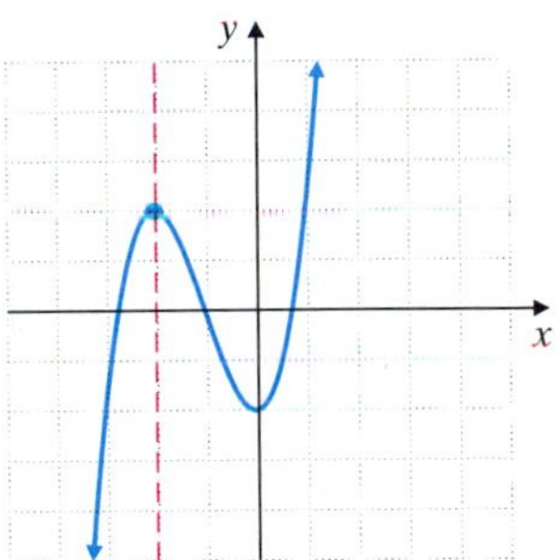 This graph is the graph of a function.
The symbol $f(x)$ means **function of x.** This notation is called **function notation.**	If $f(x) = 3x - 7$, then $f(-1) = 3(-1) - 7$ $= -3 - 7$ $= -10$
Section 13.7 Graphing Linear Inequalities in Two Variables	
A **linear inequality in two variables** is an inequality that can be written in one of these forms: $Ax + By < C$ $Ax + By \le C$ $Ax + By > C$ $Ax + By \ge C$ where A and B are not both 0.	$2x - 5y < 6$ $x \ge -5$ $y > -8x$ $y \le 2$

continued

DEFINITIONS AND CONCEPTS	EXAMPLES

Section 13.7 Graphing Linear Inequalities in Two Variables (*continued*)

TO GRAPH A LINEAR INEQUALITY

1. Graph the boundary line by graphing the related equation. Draw the line solid if the inequality symbol is $\le$ or $\ge$. Draw the line dashed if the inequality symbol is $<$ or $>$.
2. Choose a test point not on the line. Substitute its coordinates into the original inequality.
3. If the resulting inequality is true, shade the half-plane that contains the test point. If the inequality is not true, shade the half-plane that does not contain the test point.

Graph: $2x - y \le 4$

1. Graph $2x - y = 4$. Draw a solid line because the inequality symbol is $\le$.
2. Check the test point (0, 0) in the original inequality, $2x - y \le 4$.

$$2 \cdot 0 - 0 \le 4 \quad \text{Let } x = 0 \text{ and } y = 0.$$
$$0 \le 4 \quad \text{True}$$

3. The inequality is true, so shade the half-plane containing (0, 0) as shown.

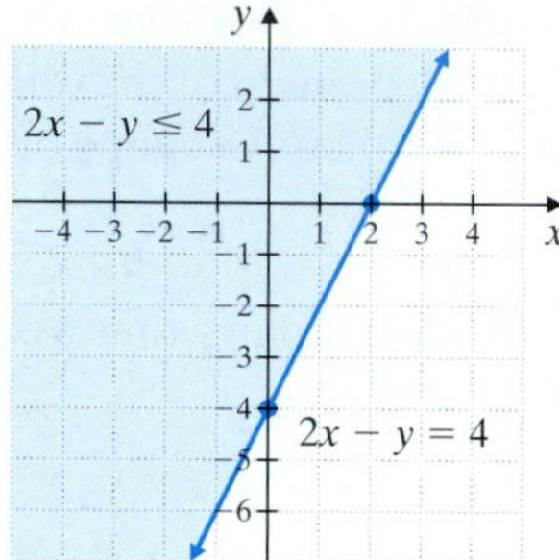

Section 13.8 Direct and Inverse Variation

y **varies directly as** x, or y is **directly proportional to** x, if there is a nonzero constant k such that

$$y = kx$$

y **varies inversely as** x, or y is **inversely proportional to** x, if there is a nonzero constant k such that

$$y = \frac{k}{x}$$

The circumference of a circle C varies directly as its radius r.

$$C = \underbrace{2\pi}_{k} r$$

Pressure P varies inversely with volume V.

$$P = \frac{k}{V}$$

STUDY SKILLS BUILDER

Are You Prepared for a Test on Chapter 13?

Below I have listed some common trouble areas for students in Chapter 13. After studying for your test—but before taking your test—read these.

- If you are having trouble with graphing, you might want to ask your instructor if you can use graph paper on your test. This will save you time and keep your graphs neat.
- Don't forget that the graph of an ordered pair is a *single* point in the rectangular coordinate system.
- Make sure you remember that to find the slope of a linear equation using its equation, *first* solve the equation for y. *Then* the coefficient of x is its slope.

$$2x + 3y = 7$$
$$3y = -2x + 7 \quad \text{Subtract } 2x \text{ from both sides.}$$
$$\frac{3y}{3} = -\frac{2}{3}x + \frac{7}{3} \quad \text{Divide both sides by 3.}$$
$$y = -\frac{2}{3}x + \frac{7}{3} \leftarrow y\text{-intercept}$$

($-\frac{2}{3}$ ← slope)

- Remember that a point that is an x-intercept will have a y-value of 0 and a point that is a y-intercept will have an x-value of 0. Also—the point (0, 0) will be both an x- and y-intercept.
- If you studied functions, remember that $f(x)$ *does not* mean $f \cdot x$. It is a special function notation. If $f(x) = x^2 - 6$, then $f(-3) = (-3)^2 - 6 = 9 - 6 = 3$.

13 CHAPTER REVIEW

(13.1) *Plot each pair on the same rectangular coordinate system.*

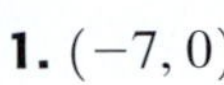

1. $(-7, 0)$

2. $\left(0, 4\frac{4}{5}\right)$

3. $(-2, -5)$

4. $(1, -3)$

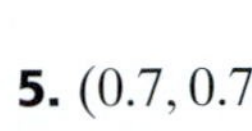

5. $(0.7, 0.7)$

6. $(-6, 4)$

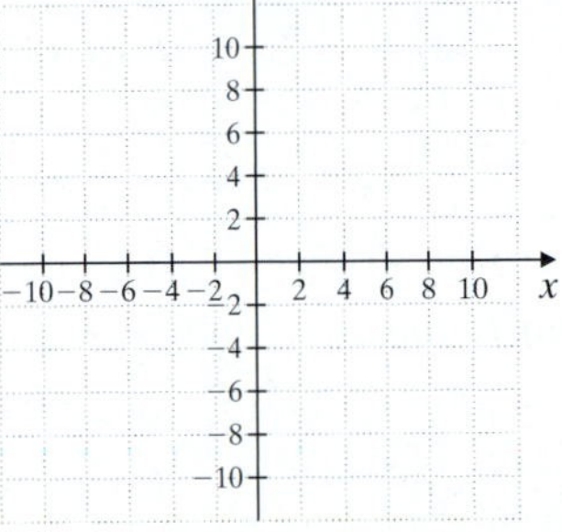

Complete each ordered pair so that it is a solution of the given equation.

7. $-2 + y = 6x$; $(7, \quad)$

8. $y = 3x + 5$; $(\quad, -8)$

Complete the table of values for each given equation.

9. $9 = -3x + 4y$

x	y
	0
	3
9	

10. $y = 5$

x	y
7	
−7	
0	

11. $x = 2y$

x	y
	0
	5
	−5

12. The cost in dollars of producing x compact disc holders is given by $y = 5x + 2000$.

a. Complete the table.

x	1	100	1000
y			

b. Find the number of compact disc holders that can be produced for \$6430.

(13.2) *Graph each linear equation.*

13. $x - y = 1$

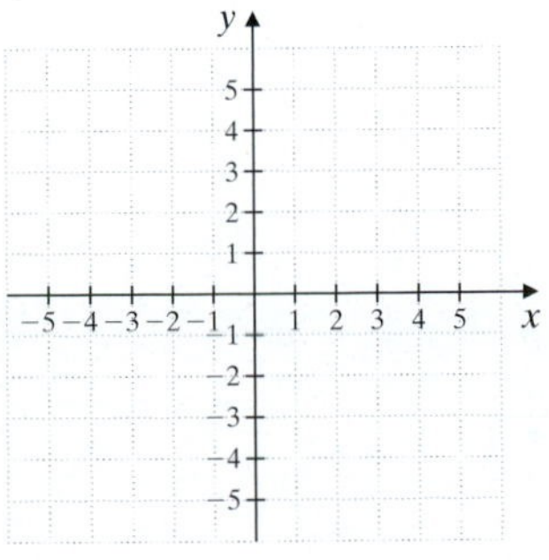

14. $x + y = 6$

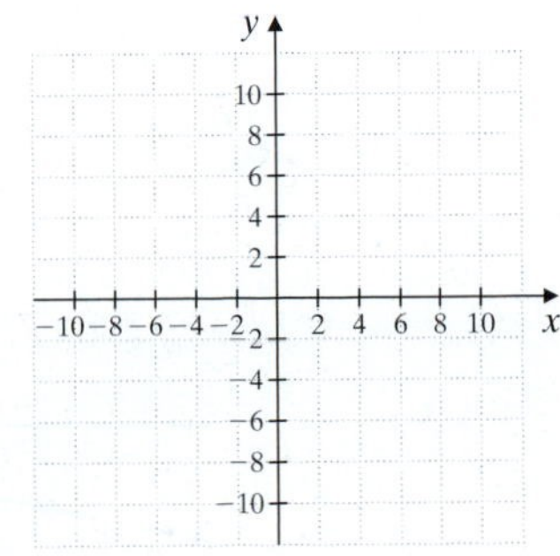

15. $x - 3y = 12$

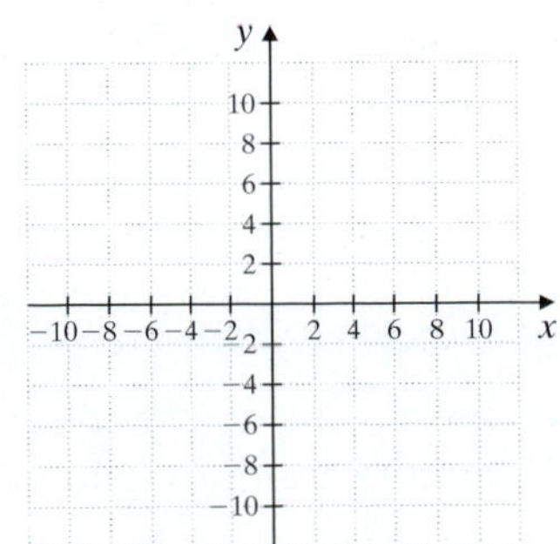

16. $5x - y = -8$

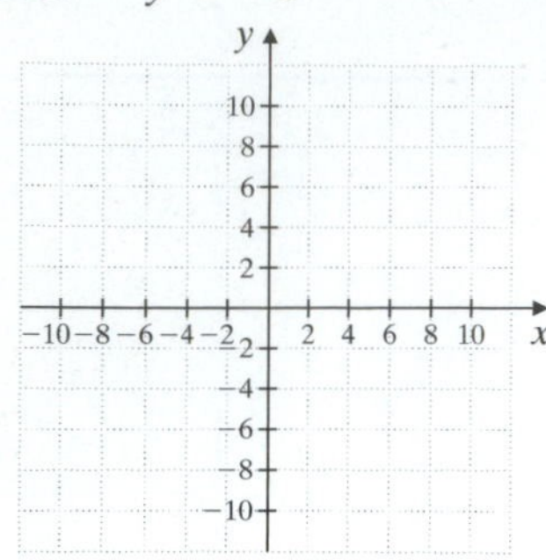

17. $x = 3y$

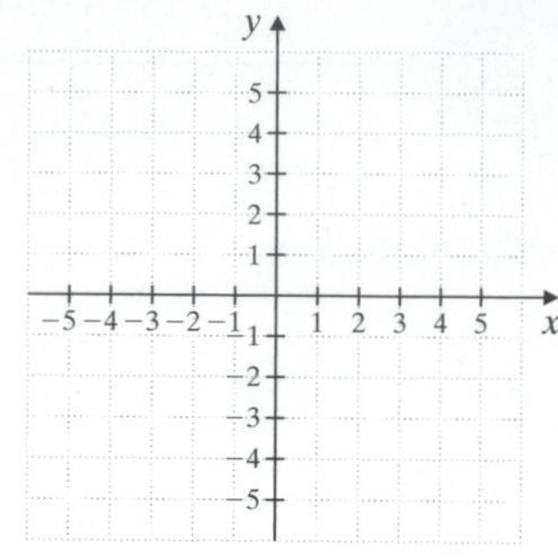

18. $y = -2x$

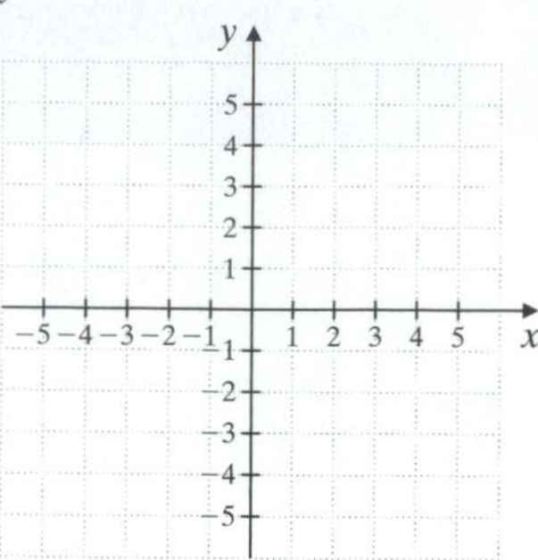

(13.3) *Identify the intercepts in each graph.*

19.

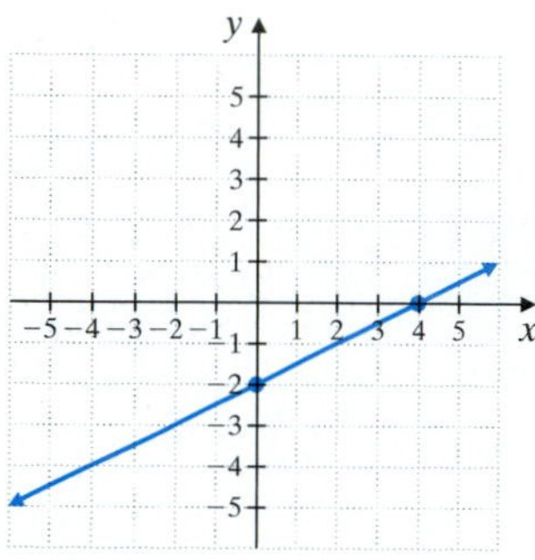

20.

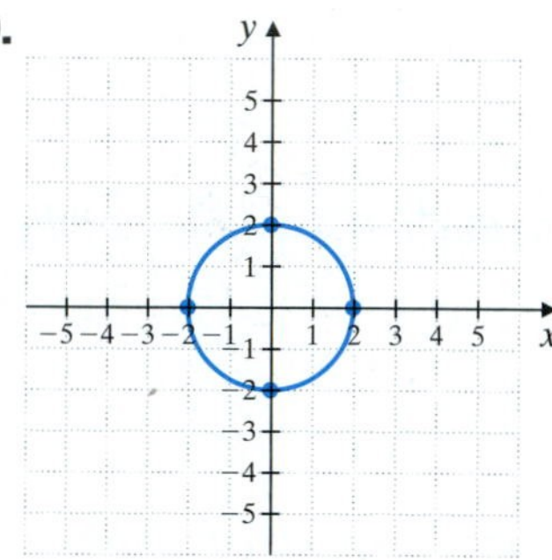

Graph each linear equation.

21. $y = -3$

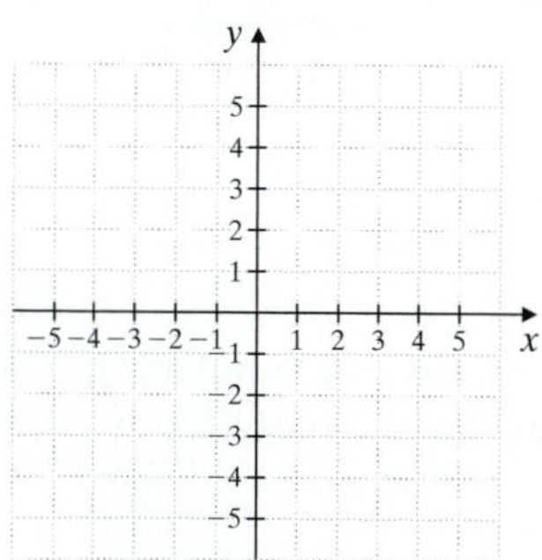

22. $x = 5$

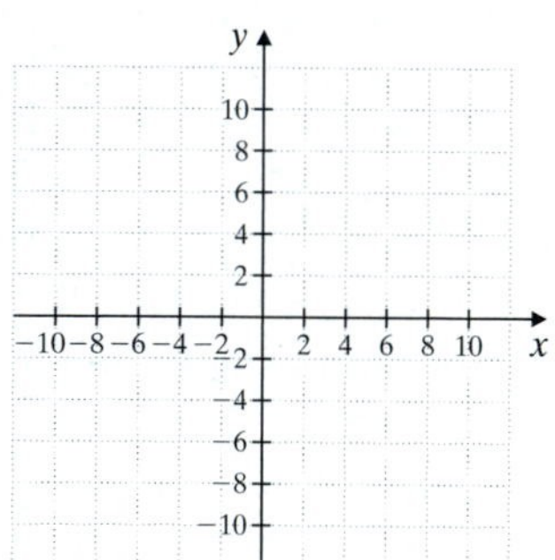

Find the intercepts of each equation.

23. $x - 3y = 12$

24. $-4x + y = 8$

(13.4) *Find the slope of each line.*

25.

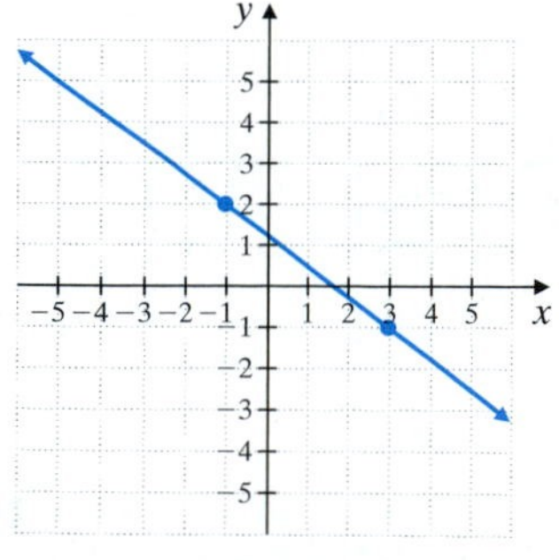

26.

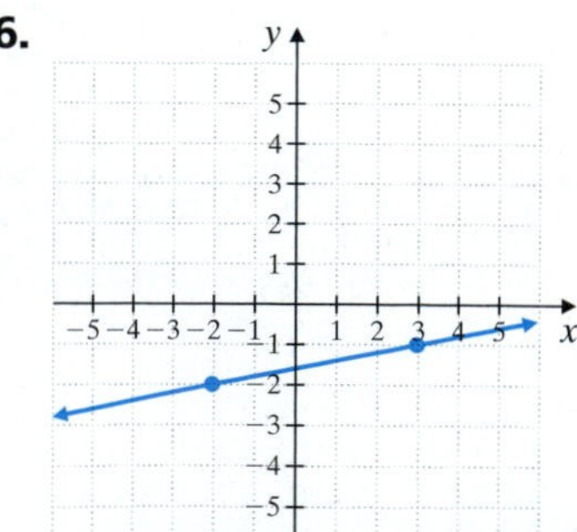

Match each line with its slope.

a.

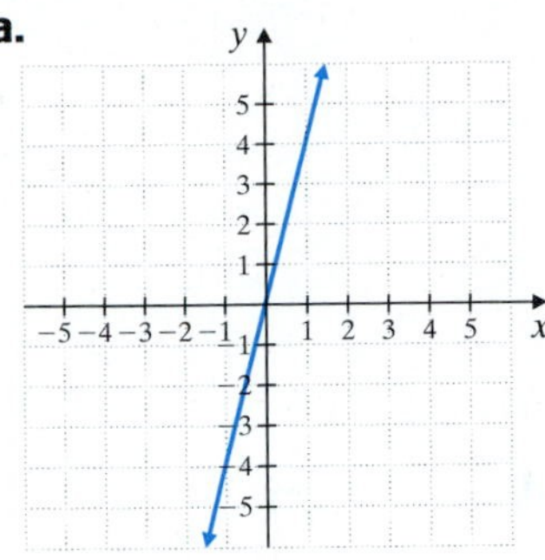

b.

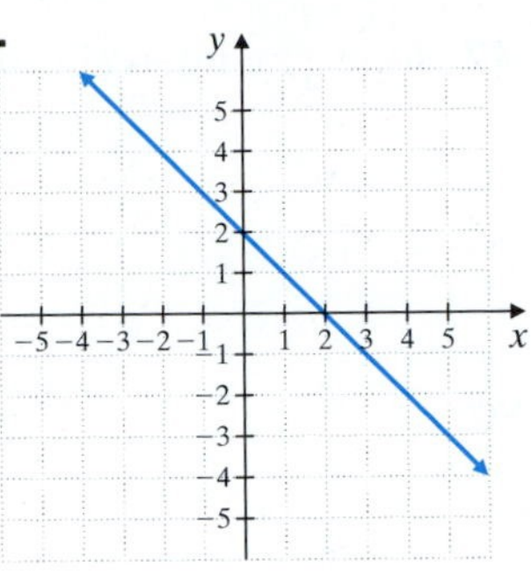

c.

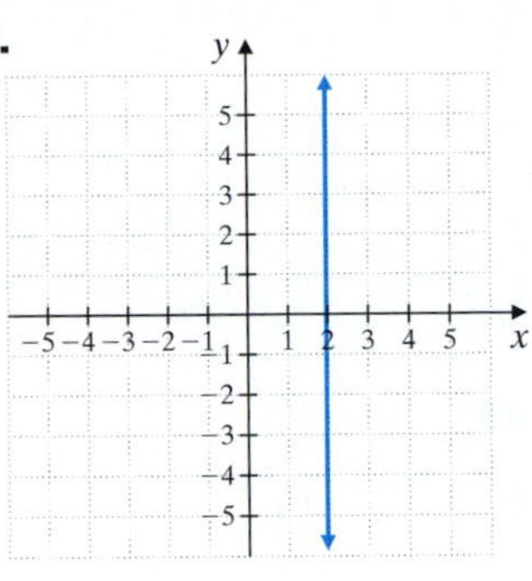

d.

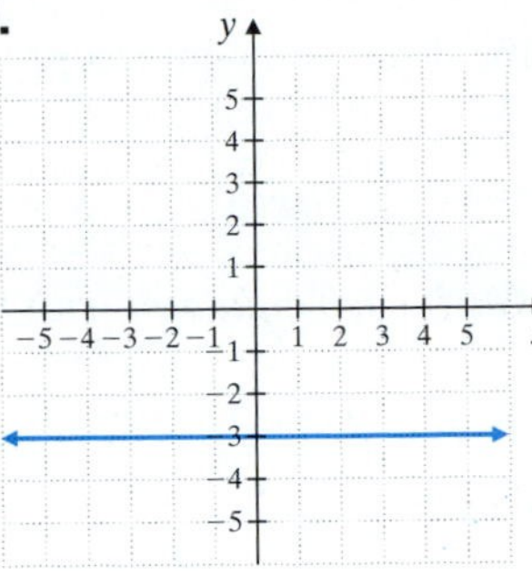

27. $m = 0$

28. $m = -1$

29. undefined slope

30. $m = 4$

Find the slope of the line that passes through each pair of points.

31. $(2, 5)$ and $(6, 8)$

32. $(4, 7)$ and $(1, 2)$

33. $(1, 3)$ and $(-2, -9)$

34. $(-4, 1)$ and $(3, -6)$

Find the slope of each line.

35. $y = 3x + 7$

36. $x - 2y = 4$

37. $y = -2$

38. $x = 0$

△ *Determine whether each pair of lines is parallel, perpendicular, or neither.*

39. $x - y = -6$
$x + y = 3$

40. $3x + y = 7$
$-3x - y = 10$

41. $y = 4x + \frac{1}{2}$
$4x + 2y = 1$

Find the slope of each line and write the slope as a rate of change. Don't forget to attach the proper units.

42. The graph below approximates the number of U.S. persons y (in millions) who have a bachelor's degree or higher per year x.

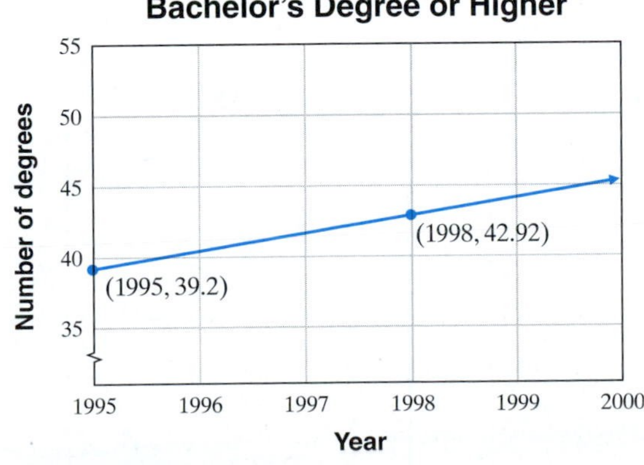

Source: U.S. Census Bureau

43. The graph below approximates the number of U.S. travelers y (in millions) that are vacationing per year x.

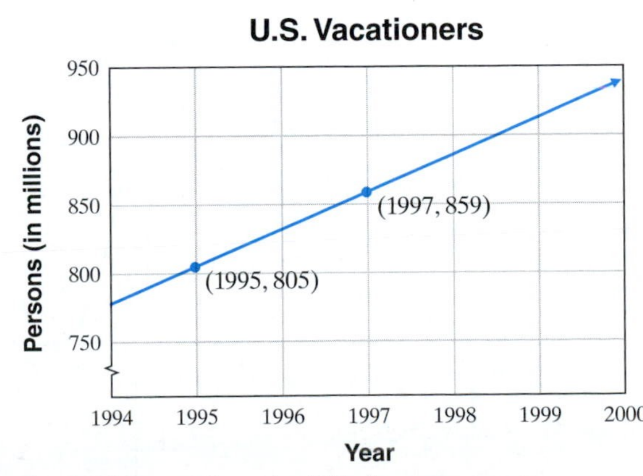

Source: TIA Research Dept., trips of 100 miles or more, one-way

(13.5) *Determine the slope and the y-intercept of the graph of each equation.*

44. $3x + y = 7$

45. $x - 6y = -1$

Write an equation of each line.

46. slope -5; y-intercept $\left(0, \frac{1}{2}\right)$

47. slope $\frac{2}{3}$; y-intercept $(0, 6)$

Match each equation with its graph.

48. $y = 2x + 1$

49. $y = -4x$

50. $y = 2x$

51. $y = 2x - 1$

a.

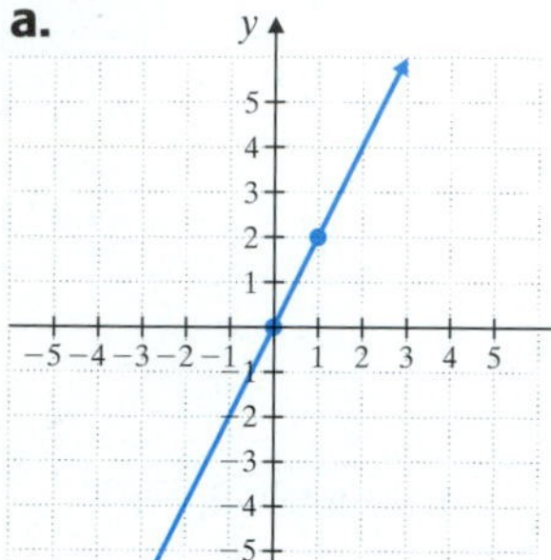

b.

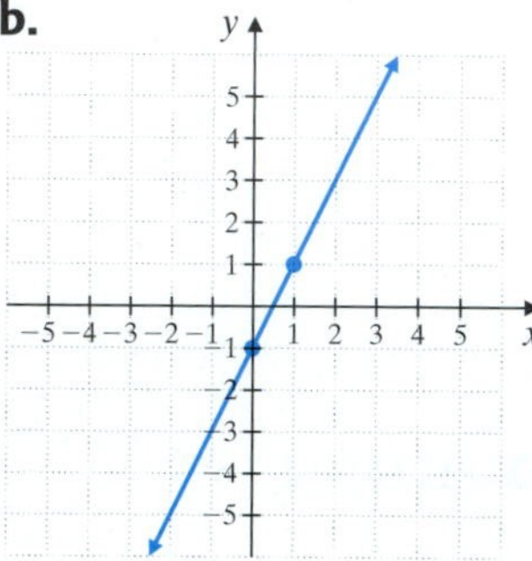

c.

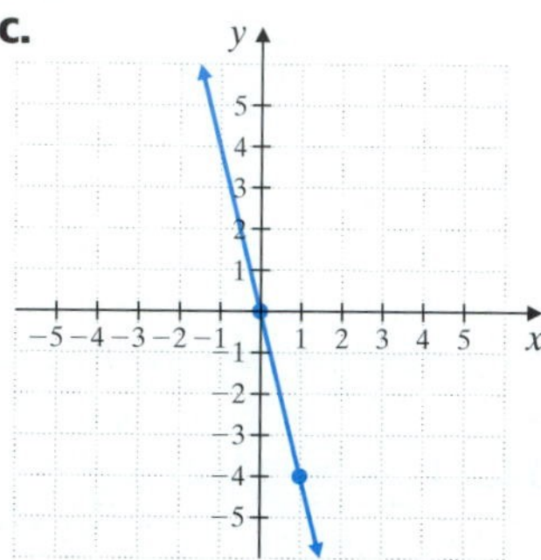

d.

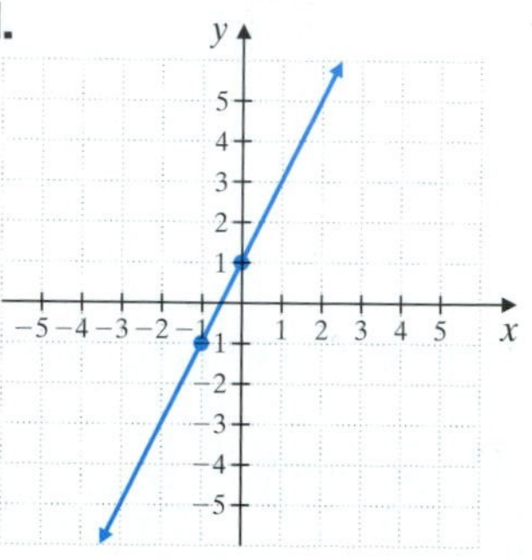

Write an equation of the line with the given slope that passes through the given point. Write the equation in the form $Ax + By = C$.

52. $m = 4$; $(2, 0)$

53. $m = -3$; $(0, -5)$

54. $m = \frac{3}{5}$; $(1, 4)$

55. $m = -\frac{1}{3}$; $(-3, 3)$

Write an equation of the line passing through each pair of points. Write the equation in the form $y = mx + b$.

56. $(1, 7)$ and $(2, -7)$

57. $(-2, 5)$ and $(-4, 6)$

(13.6) *Determine whether each relation or graph is a function.*

58. $\{(7, 1), (7, 5), (2, 6)\}$

59. $\{(0, -1), (5, -1), (2, 2)\}$

60.

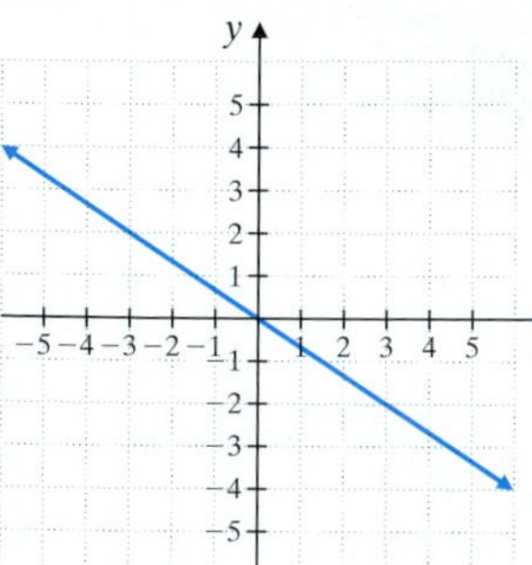

61.

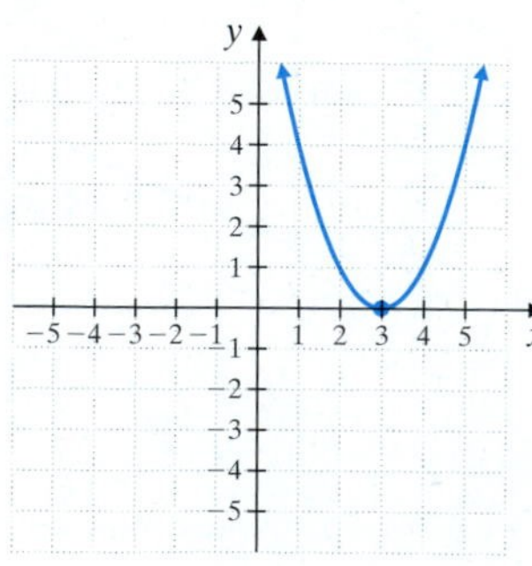

62.

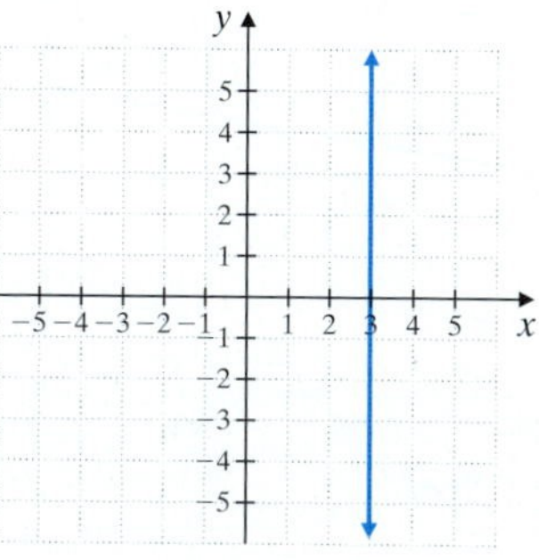

63.

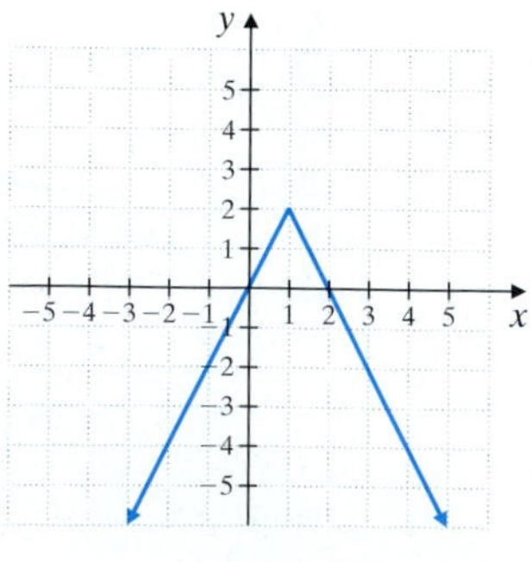

64. Find the indicated function value for the function, $f(x) = -2x + 6$.

a. $f(0)$

b. $f(-2)$

c. $f\left(\frac{1}{2}\right)$

(13.7) *Graph each inequality.*

65. $x + 6y < 6$

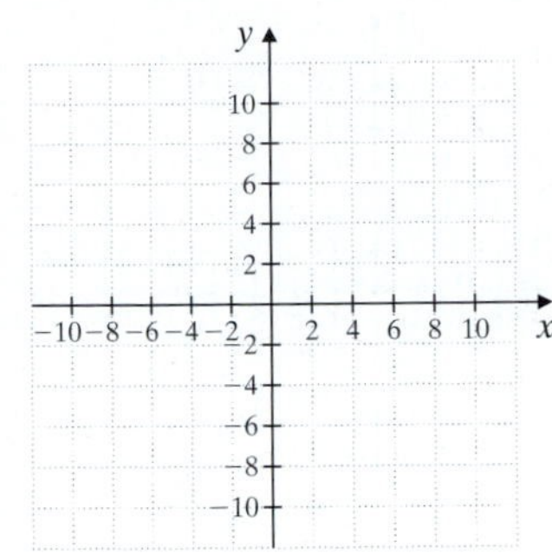

66. $x + y > -2$

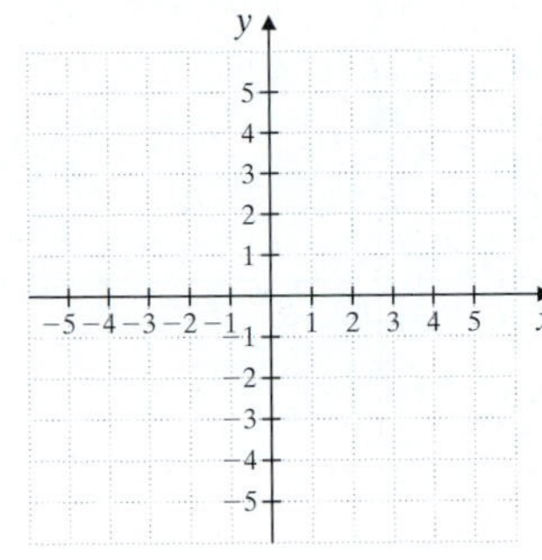

67. $y \geq -7$

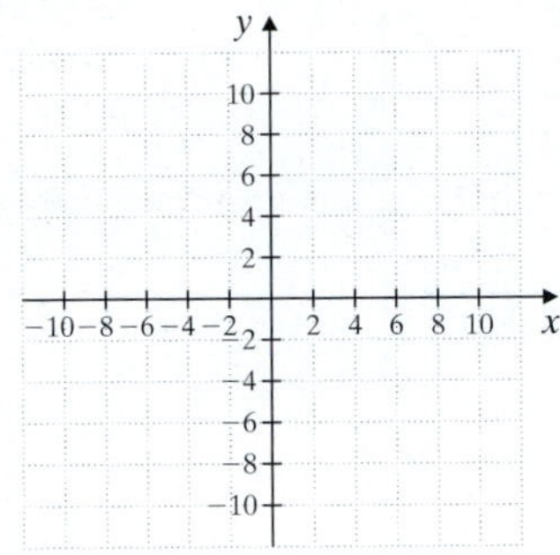

68. $y \leq -4$

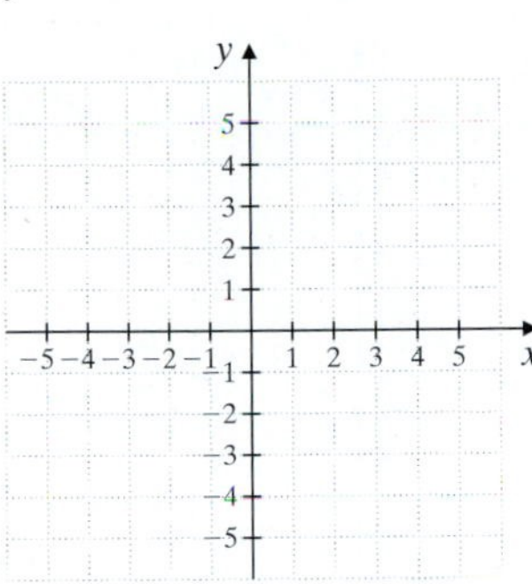

69. $-x \leq y$

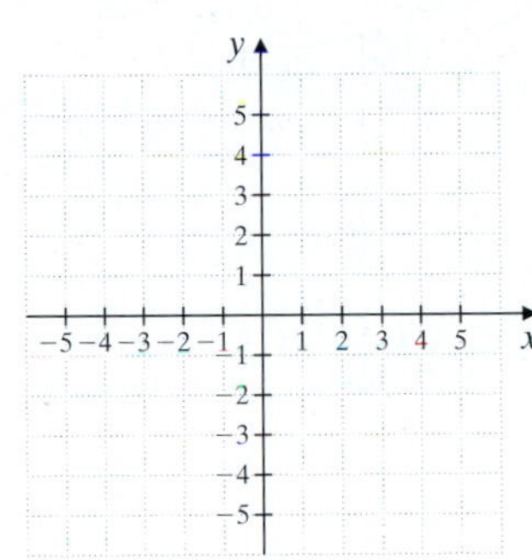

70. $x \geq -y$

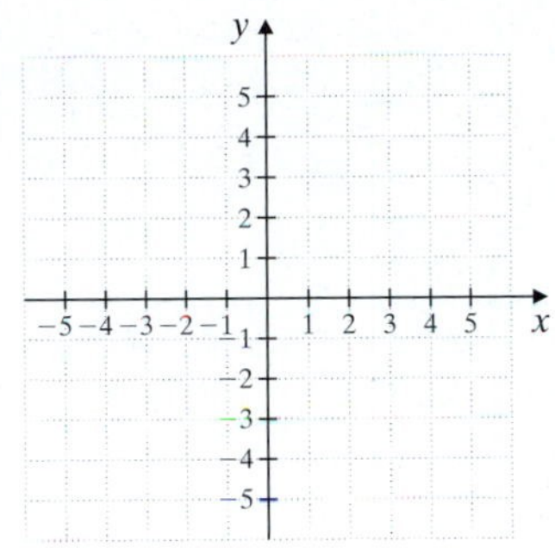

(13.8) *Solve.*

71. y varies directly as x. If $y = 40$ when $x = 4$, find y when x is 11.

72. y varies inversely as x. If $y = 4$ when $x = 6$, find y when x is 48.

73. y varies inversely as x^3. If $y = 12.5$ when $x = 2$, find y when x is 3.

74. y varies directly as x^2. If $y = 175$ when $x = 5$, find y when $x = 10$.

75. The cost of manufacturing a certain medicine varies inversely as the amount of medicine manufactured increases. If 3000 milliliters can be manufactured for \$6600, find the cost to manufacture 5000 milliliters.

76. The distance a spring stretches varies directly with the weight attached to the spring. If a 150-pound weight stretches the spring 8 inches, find the distance that a 90-pound weight stretches the spring.

Mixed Review

Complete the table of values for each given equation.

77. $2x - 5y = 9$

x	y
	1
2	
	−3

78. $x = -3y$

x	y
0	
	1
6	

Find the intercepts for each equation.

79. $2x - 3y = 6$

80. $-5x + y = 10$

Graph each linear equation.

81. $x - 5y = 10$

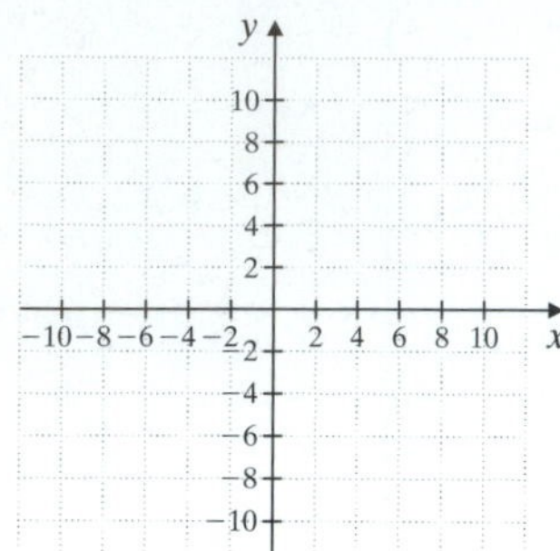

82. $x + y = 4$

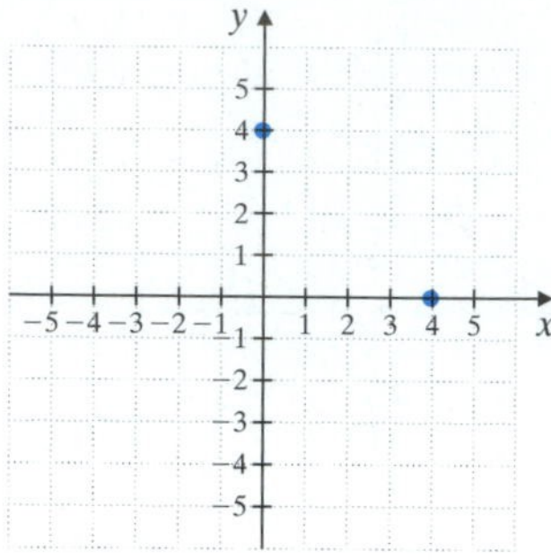

83. $y = -4x$

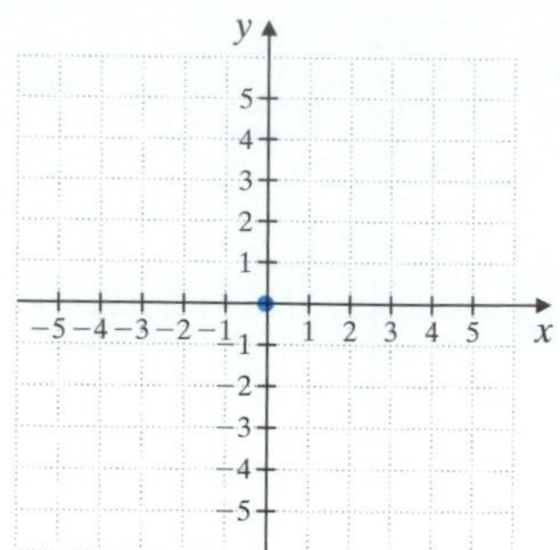

84. $2x + 3y = -6$

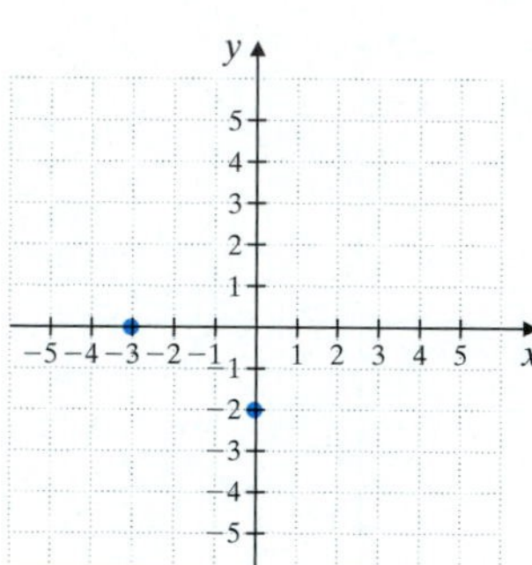

85. $x = 3$

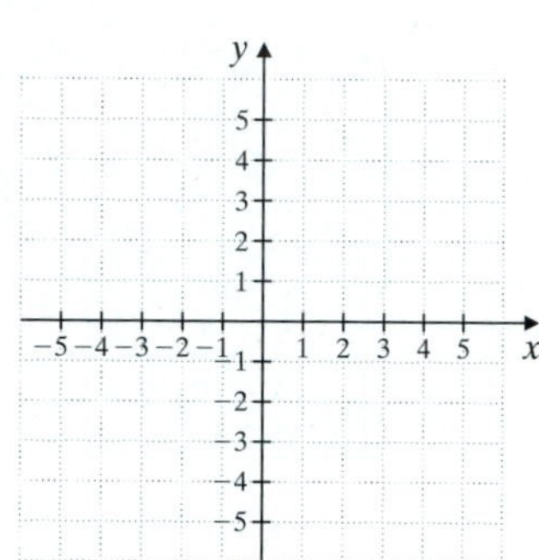

86. $y = -2$

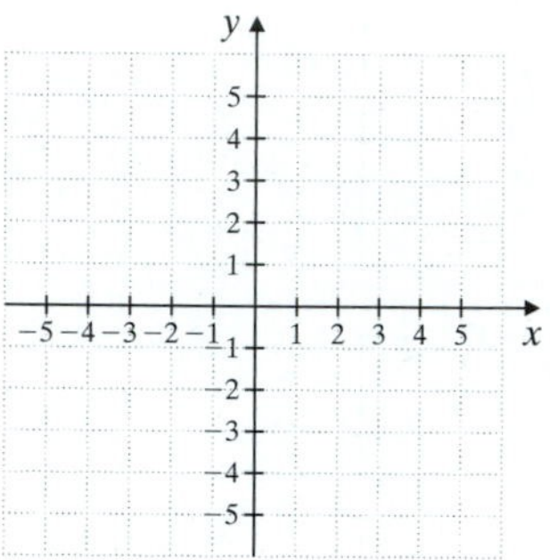

Find the slope of the line that passes through each pair of points.

87. $(3, -5)$ and $(-4, 2)$

88. $(1, 3)$ and $(-6, -8)$

Find the slope of each line.

89.

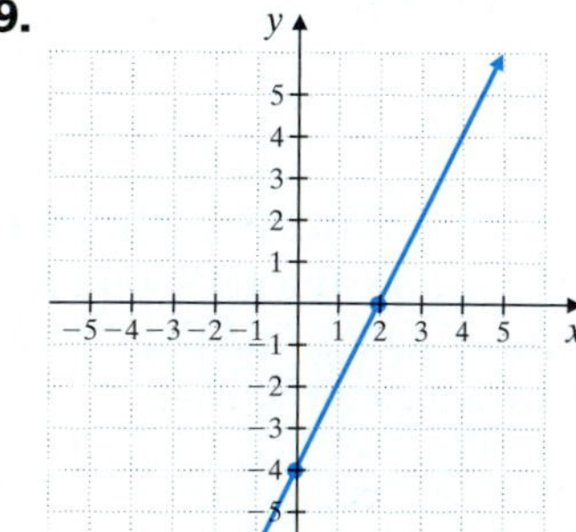

90.

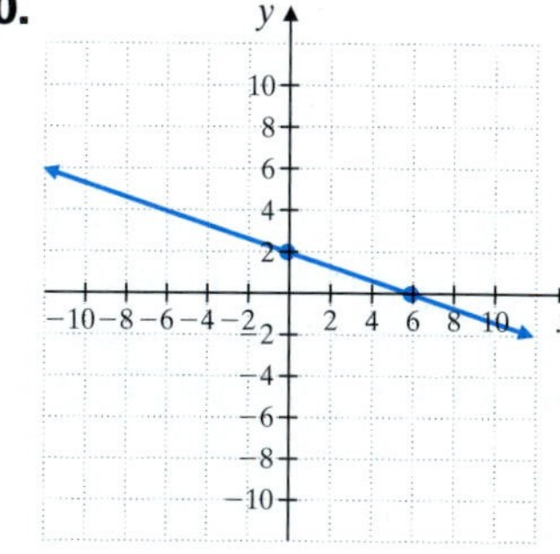

Determine the slope and y-intercept of the graph of each equation.

91. $-2x + 3y = -15$

92. $6x + y - 2 = 0$

Write an equation of the line with the given slope that passes through the given point. Write the equation in the form $Ax + By = C$.

93. $m = -5; (3, -7)$

94. $m = 3; (0, 6)$

Write an equation of the line passing through each pair of points. Write the equation in the form $Ax + By = C$.

95. $(-3, 9)$ and $(-2, 5)$

96. $(3, 1)$ and $(5, -9)$

13 CHAPTER TEST

Remember to use the Chapter Test Prep Video CD to see the fully worked-out solutions to any of the exercises you want to review.

Complete each ordered pair so that it is a solution of the given equation.

1. $12y - 7x = 5; (1, \quad)$

2. $y = 17; (-4, \quad)$

Find the slope of each line.

3.

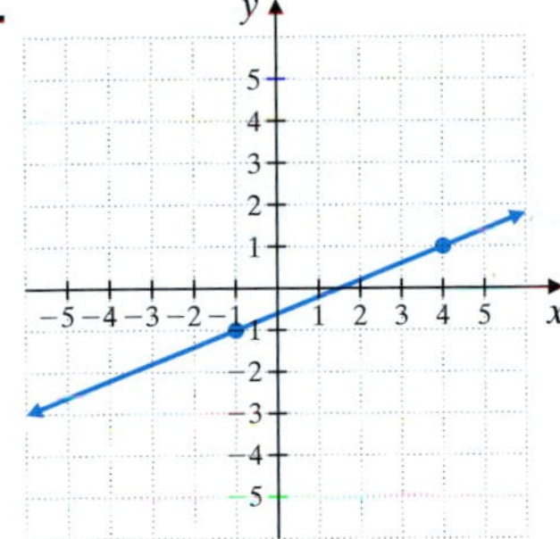

4.

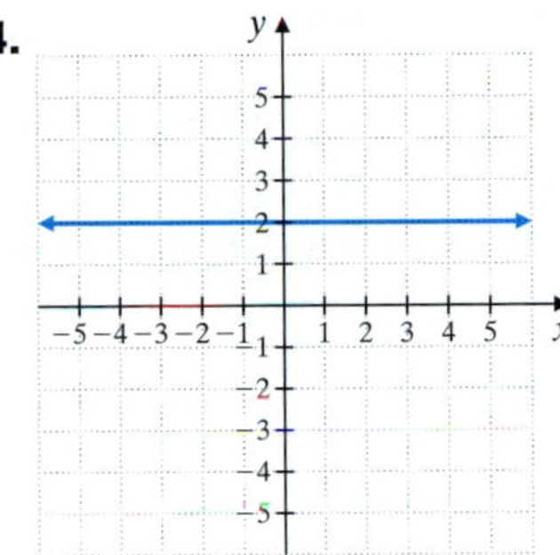

5. Passes through $(6, -5)$ and $(-1, 2)$

6. Passes through $(0, -8)$ and $(-1, -1)$

7. $-3x + y = 5$

8. $x = 6$

Graph.

9. $2x + y = 8$

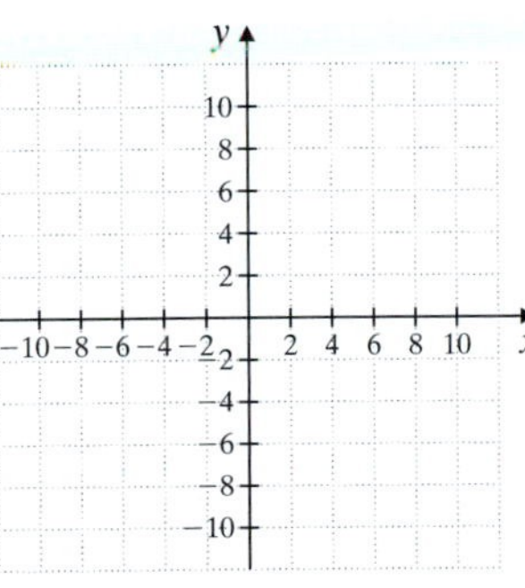

10. $-x + 4y = 5$

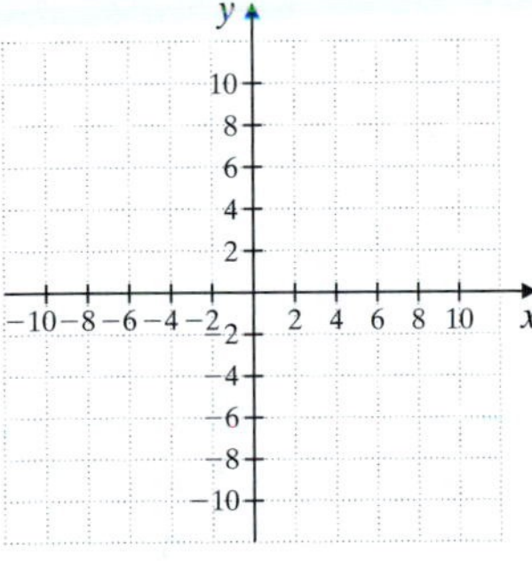

11. $x - y \geq -2$

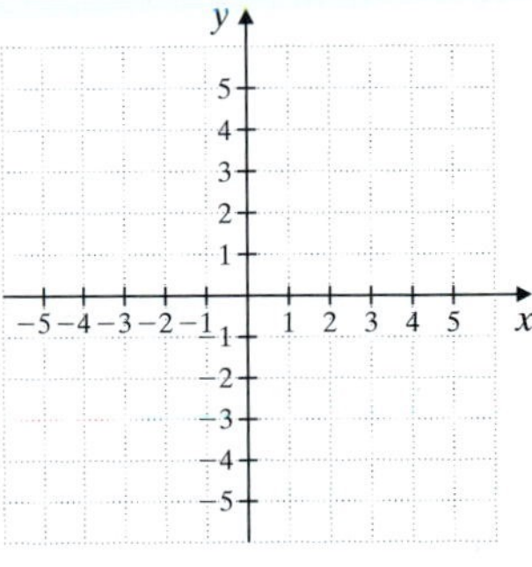

12. $y \geq -4x$

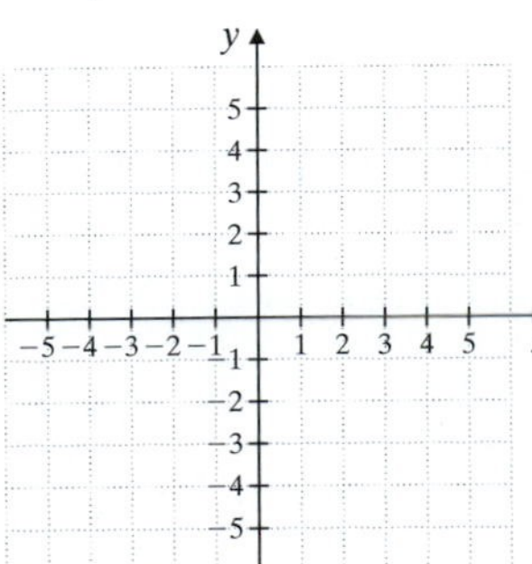

13. $5x - 7y = 10$

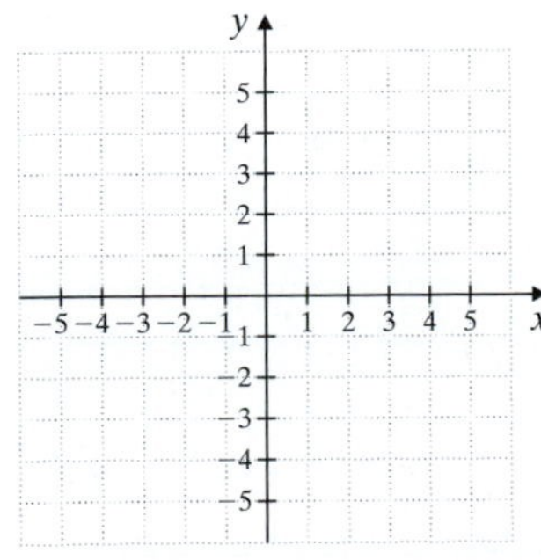

14. $2x - 3y > -6$

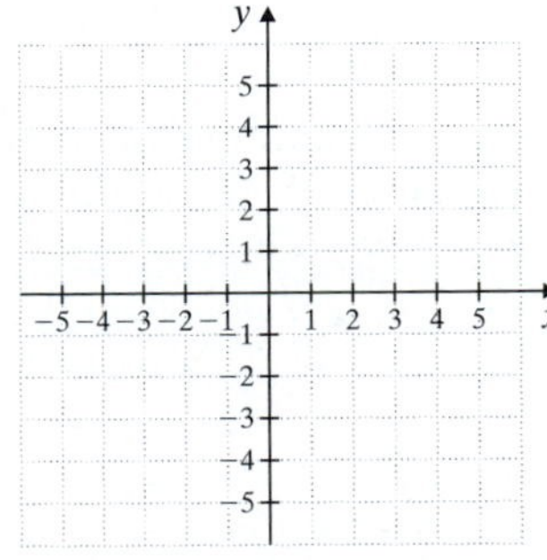

Answers

1. ______

2. ______

3. ______

4. ______

5. ______

6. ______

7. ______

8. ______

9. see graph

10. see graph

11. see graph

12. see graph

13. see graph

14. see graph

15. see graph ______

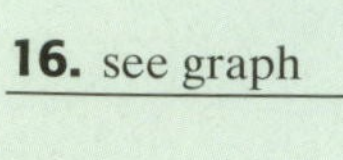

16. see graph ______

17. ______

18. ______

19. ______

20. ______

21. ______

22. ______

23. ______

24. ______

25. ______

26. a. ______

b. ______

c. ______

27. a. ______

b. ______

c. ______

28. ______

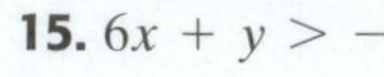

15. $6x + y > -1$

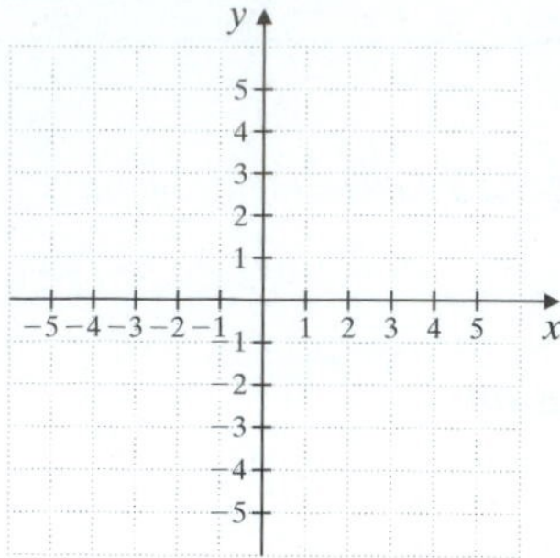

16. $y = -1$

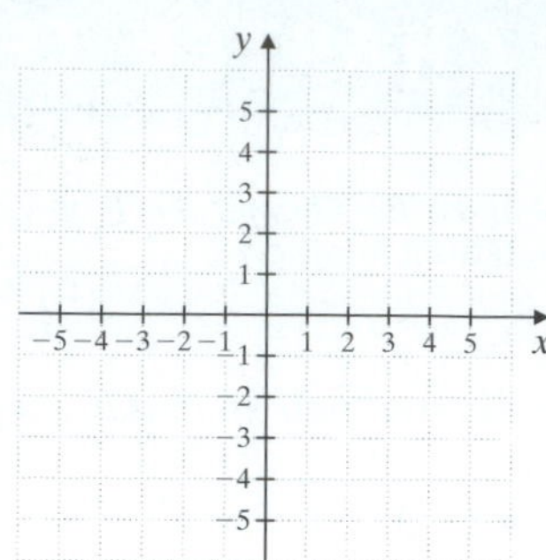

17. Determine whether the graphs of $y = 2x - 6$ and $-4x = 2y$ are parallel lines, perpendicular lines, or neither.

Find the equation of each line. Write the equation in the form $Ax + By = C$.

18. Slope $-\frac{1}{4}$, passes through $(2, 2)$

19. Passes through the origin and $(6, -7)$

20. Passes through $(2, -5)$ and $(1, 3)$

21. Slope $\frac{1}{8}$; y-intercept $(0, 12)$

Determine whether each relation is a function.

22. $\{(-1, 2), (-2, 4), (-3, 6), (-4, 8)\}$

23. $\{(-3, -3), (0, 5), (-3, 2), (0, 0)\}$

24. The graph shown in Exercise 3.

25. The graph shown in Exercise 4.

Find the indicated function values for each function.

26. $f(x) = 2x - 4$

a. $f(-2)$

b. $f(0.2)$

c. $f(0)$

27. $f(x) = x^3 - x$

a. $f(-1)$

b. $f(0)$

c. $f(4)$

△ **28.** The perimeter of the parallelogram below is 42 meters. Write a linear equation in two variables for the perimeter. Use this equation to find x when y is 8.

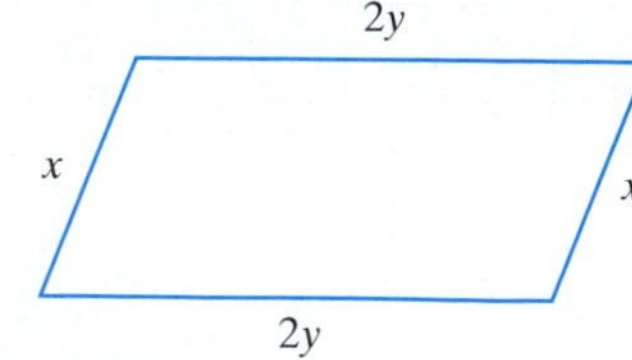

29. The table gives the number of basic cable TV subscribers (in millions) for the years shown. (*Source:* Cisco Systems)

Year	Basic Cable TV Subscribers (in millions)
2000	69.3
2001	70.0
2002	69.9
2003	70.1
2004	70.3
2005	70.5 (estimated)

29. a. ______

a. Write this data as a set of ordered pairs of the form (year, number of basic cable TV subscribers in millions).

b. Create a scatter diagram of the data. Be sure to label the axes properly.

b. see diagram

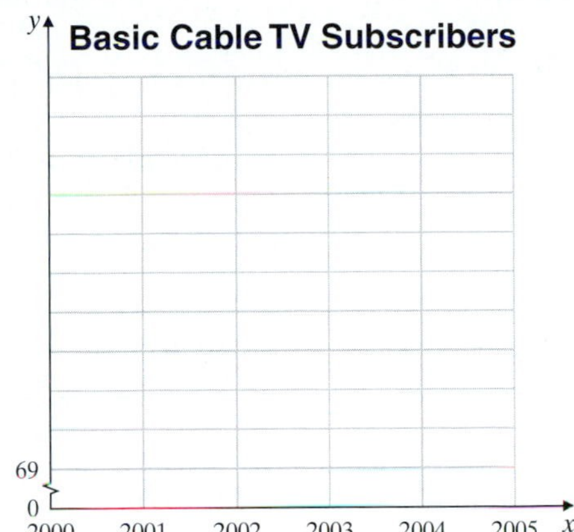

30. ______

30. This graph approximates the movie ticket sales y (in millions) for the year x. Find the slope of the line and write the slope as a rate of change. Don't forget to attach the proper units.

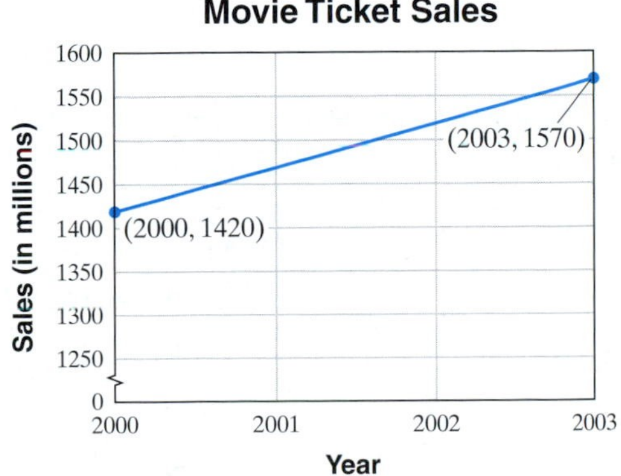

Source: National Association of Theater Owners

31. ______

32. ______

31. y varies directly as x. If $y = 10$ when $x = 15$, find y when x is 42.

32. y varies inversely as x^2. If $y = 8$ when $x = 5$, find y when x is 15.

CUMULATIVE REVIEW

Chapters 1–13

Answers

1. ____
2. ____
3. ____
4. ____
5. ____
6. ____
7. ____
8. ____
9. ____
10. ____
11. ____
12. ____
13. ____
14. ____
15. a. ____
 b. ____
 c. ____
16. a. ____
 b. ____
 c. ____
17. ____
18. ____

△**1.** The state of Colorado is in the shape of a rectangle whose length is about 380 miles and whose width is about 280 miles. Find its area.

2. In a pecan orchard, there are 21 trees in each row and 7 rows of trees. How many pecan trees are there?

3. Add: $1 + (-10) + (-8) + 9$

4. Add: $-2 + (-7) + 3 + (-4)$

5. Write $\frac{8}{3x}$ as an equivalent fraction whose denominator is $24x$.

6. Write $\frac{3}{2c}$ as an equivalent fraction with denominator $8c$.

7. Subtract: $14 - 8\frac{3}{7}$

8. Subtract: $15 - 4\frac{2}{5}$

9. Evaluate $-2x + 5$ for $x = 3.8$.

10. Evaluate $6x - 1$ for $x = -2.1$.

11. Write $\frac{22}{7}$ as a decimal. Round to the nearest hundredth.

12. Write $\frac{37}{19}$ as a decimal. Round to the nearest thousandth.

13. Solve $2x < -4$. Graph the solutions.

−5 −4 −3 −2 −1 0 1 2 3 4 5

14. Solve $3x \leq -9$. Graph the solution set.

−5 −4 −3 −2 −1 0 1 2 3 4 5

Find the degree of each polynomial and tell whether the polynomial is a monomial, binomial, trinomial, or none of these.

15. a. $-2t^2 + 3t + 6$

b. $15x - 10$

c. $7x + 3x^3 + 2x^2 - 1$

16. a. $-7y + 2$

b. $8x - x^2 - 1$

c. $9y^3 - 6y + 2 + y^2$

Perform the indicated operation.

17. Add: $(-2x^2 + 5x - 1)$ and $(-2x^2 + x + 3)$

18. Add: $(9x - 5)$ and $(x^2 - 6x + 5)$

19. Multiply: $(3y+1)^2$

20. Multiply: $(2x-5)^2$

21. Factor: $-9a^5+18a^2-3a$

22. Factor: $2x^5-x^3$

23. Factor: $x^2+4x-12$

24. Factor: $x^2+4x-21$

25. Factor: $8x^2-22x+5$

26. Factor: $15x^2+x-2$

27. Solve: $4x^2-28x=-49$

28. Solve: $x^2-9x=-14$

29. Divide: $\dfrac{2x^2-11x+5}{5x-25}\div\dfrac{4x-2}{10}$

30. Multiply: $\dfrac{3x^2+17x-6}{5x+5}\cdot\dfrac{2x+2}{4x+24}$

Write the rational expression as an equivalent rational expression with the given denominator.

31. $\dfrac{4b}{9a}=\dfrac{\quad}{27a^2b}$

32. $\dfrac{7x}{11y}=\dfrac{\quad}{99x^2y^2}$

33. Add: $1+\dfrac{m}{m+1}$

34. Subtract: $1-\dfrac{m}{m+1}$

35. Solve: $3-\dfrac{6}{x}=x+8$

36. Solve: $2+\dfrac{10}{x}=x+5$

37. Simplify: $\dfrac{\dfrac{x+1}{y}}{\dfrac{x}{y}+2}$

38. Simplify: $\dfrac{\dfrac{x}{2}+2}{\dfrac{x}{2}-2}$

Complete each ordered pair solution so that it is a solution of each equation.

39. $3x+y=12$

a. $(0,\ )$

b. $(\ ,6)$

c. $(-1,\ )$

40. $-x+4y=-20$

a. $(0,\ )$

b. $(\ ,0)$

c. $(\ ,-2)$

19. ______
20. ______
21. ______
22. ______
23. ______
24. ______
25. ______
26. ______
27. ______
28. ______
29. ______
30. ______
31. ______
32. ______
33. ______
34. ______
35. ______
36. ______
37. ______
38. ______
39. a. ______
b. ______
c. ______
40. a. ______
b. ______
c. ______

41. see graph

42. see graph

43.

44.

45.

46.

47. a.

b.

c.

48. a.

b.

c.

41. Graph: $2x + y = 5$

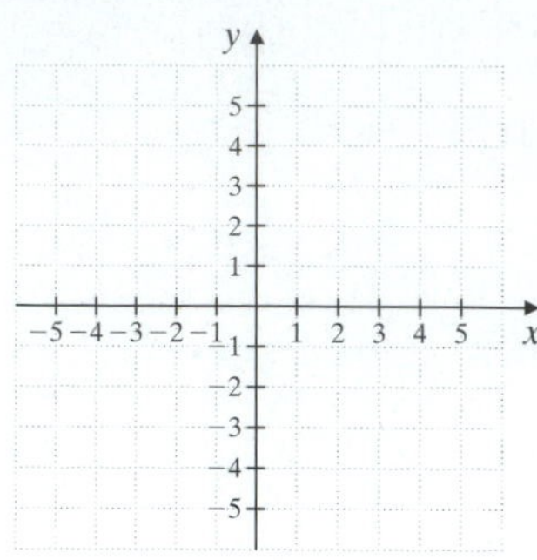

42. Graph: $y = -2x$

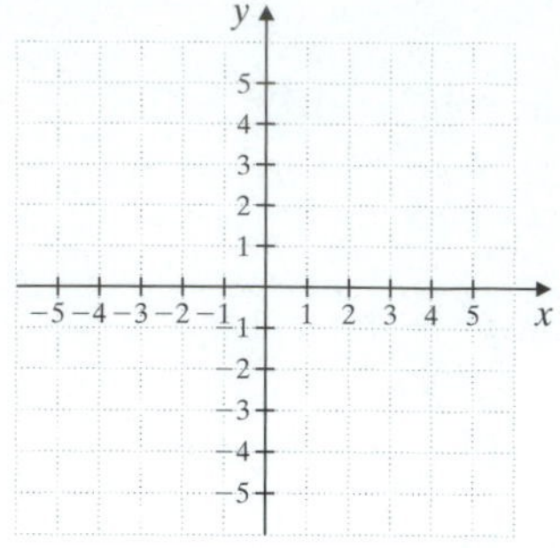

43. Find the slope of the line:
$-2x + 3y = 11$

44. Find the slope of the line:
$7x + 4y = 10$

45. Find an equation of the line passing through $(-1, 5)$ with slope -2. Write the equation in the form $Ax + By = C$.

46. Find an equation of the line passing through $(2, -7)$ with slope -5. Write the equation in the form $Ax + By = C$.

47. Given $g(x) = x^2 - 3$, find each function value and list t[illegible] corresponding ordered pair

a. $g(?$ **b.** $g(-2)$ **c.** $g(0)$

48. Given $f(x) = 3x^2 + 2$, find each function value and list the corresponding ordered pair.

a. $f(0)$ **b.** $f(4)$ **c.** $f(-1)$

14

Systems of Equations

In Chapter 13, we graphed equations containing two variables. As we have seen, equations like these are often needed to represent relationships between two different quantities. There are also many opportunities to compare and contrast two such equations, called a *system of equations*. This chapter presents *linear systems* and ways we solve these systems and apply them to real-life situations.

SECTIONS

CHECK YOUR PROGRESS

Many of the occupations predicted to have the largest increase in number of jobs are in the fields of medicine and computer science. For example, from 2002 to 2012 the job growth predicted for medical assistants is 59% and for computer software engineers is 46%. Although both jobs are growing, they are growing at different rates. In Section 14.3, Exercise 59, we will predict when these occupations might have the same number of jobs.

Job Title	Job Description	Employment (in thousands)	
		2002	2012
Medical Assistant	Perform administrative and clinical tasks to keep the offices of physicians running smoothly.	365	579
Computer Software Engineers	Apply the principles of computer science, engineering, and mathematics to design, test, and evaluate new software and systems for computers.	394	573

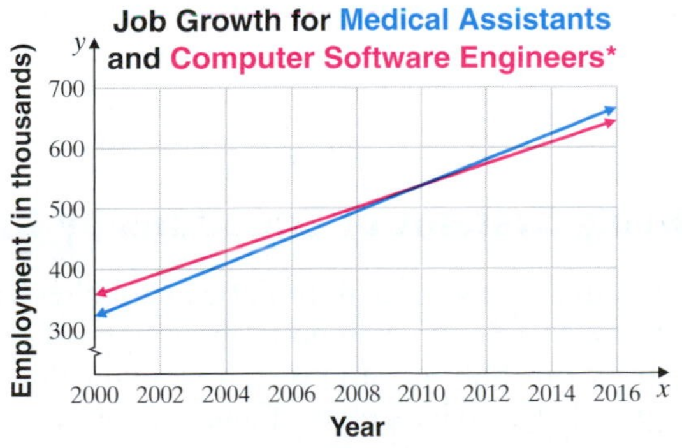

* See the U.S. Bureau of Labor Statistics for more information. Here, we assumed linear growth.

Objectives

A Decide Whether an Ordered Pair Is a Solution of a System of Linear Equations.

B Solve a System of Linear Equations by Graphing.

C Identify Special Systems: Those with no Solution and Those with an Infinite Number of Solutions.

14.1 SOLVING SYSTEMS OF LINEAR EQUATIONS BY GRAPHING

A **system of linear equations** consists of two or more linear equations. In this section, we focus on solving systems of linear equations containing two equations in two variables. Examples of such linear systems are

$$\begin{cases} 3x - 3y = 0 \\ x = 2y \end{cases} \quad \begin{cases} x - y = 0 \\ 2x + y = 10 \end{cases} \quad \begin{cases} y = 7x - 1 \\ y = 4 \end{cases}$$

Objective A Deciding Whether an Ordered Pair Is a Solution

A **solution** of a system of two equations in two variables is an ordered pair of numbers that is a solution of both equations in the system.

PRACTICE PROBLEM 1

Determine whether (3, 9) is a solution of the system

$$\begin{cases} 5x - 2y = -3 \\ y = 3x \end{cases}$$

EXAMPLE 1 Determine whether (12, 6) is a solution of the system

$$\begin{cases} 2x - 3y = 6 \\ x = 2y \end{cases}$$

Solution: To determine whether (12, 6) is a solution of the system, we replace x with 12 and y with 6 in both equations.

$2x - 3y = 6$	First equation	$x = 2y$	Second equation
$2(12) - 3(6) \stackrel{?}{=} 6$	Let $x = 12$ and $y = 6$.	$12 \stackrel{?}{=} 2(6)$	Let $x = 12$ and $y = 6$.
$24 - 18 \stackrel{?}{=} 6$	Simplify.	$12 = 12$	True
$6 = 6$	True		

Since (12, 6) is a solution of both equations, it is a solution of the system.

Work Practice Problem 1

PRACTICE PROBLEM 2

Determine whether (3, −2) is a solution of the system

$$\begin{cases} 2x - y = 8 \\ x + 3y = 4 \end{cases}$$

EXAMPLE 2 Determine whether (−1, 2) is a solution of the system

$$\begin{cases} x + 2y = 3 \\ 4x - y = 6 \end{cases}$$

Solution: We replace x with −1 and y with 2 in both equations.

$x + 2y = 3$	First equation	$4x - y = 6$	Second equation
$-1 + 2(2) \stackrel{?}{=} 3$	Let $x = -1$ and $y = 2$.	$4(-1) - 2 \stackrel{?}{=} 6$	Let $x = -1$ and $y = 2$.
$-1 + 4 \stackrel{?}{=} 3$	Simplify.	$-4 - 2 \stackrel{?}{=} 6$	Simplify.
$3 = 3$	True	$-6 = 6$	False

(−1, 2) is not a solution of the second equation, $4x - y = 6$, so it is not a solution of the system.

Work Practice Problem 2

Objective B Solving Systems of Equations by Graphing

Since a solution of a system of two equations in two variables is a solution common to both equations, it is also a point common to the graphs of both equations. Let's practice finding solutions of both equations in a system—that is, solutions of the system—by graphing and identifying points of intersection.

Answers

1. (3, 9) is a solution of the system, **2.** (3, −2) is not a solution of the system

EXAMPLE 3 Solve the system of equations by graphing.

$$\begin{cases} -x + 3y = 10 \\ x + y = 2 \end{cases}$$

Solution: On a single set of axes, graph each linear equation.

$-x + 3y = 10$

x	y
0	$\frac{10}{3}$
−4	2
2	4

$x + y = 2$

x	y
0	2
2	0
1	1

Helpful Hint The point of intersection gives the solution of the system.

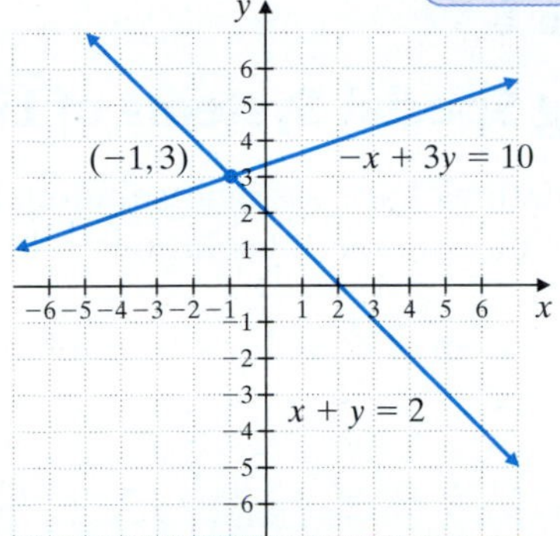

The two lines appear to intersect at the point $(-1, 3)$. To check, we replace x with -1 and y with 3 in both equations.

$-x + 3y = 10$	First equation	$x + y = 2$	Second equation
$-(-1) + 3(3) \stackrel{?}{=} 10$	Let $x = -1$ and $y = 3$.	$-1 + 3 \stackrel{?}{=} 2$	Let $x = -1$ and $y = 3$.
$1 + 9 \stackrel{?}{=} 10$	Simplify.	$2 = 2$	True
$10 = 10$	True		

$(-1, 3)$ checks, so it is the solution of the system.

Work Practice Problem 3

Helpful Hint Neatly drawn graphs can help when "guessing" the solution of a system of linear equations by graphing.

EXAMPLE 4 Solve the system of equations by graphing.

$$\begin{cases} 2x + 3y = -2 \\ x = 2 \end{cases}$$

Solution: We graph each linear equation on a single set of axes.

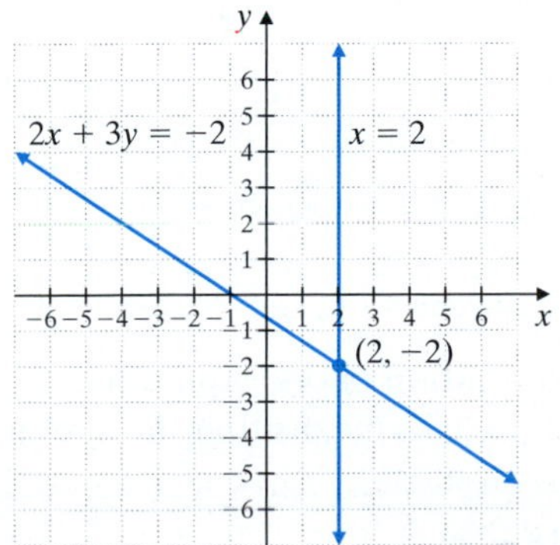

Continued on next page

PRACTICE PROBLEM 3

Solve the system of equations by graphing.

$$\begin{cases} -3x + y = -10 \\ x - y = 6 \end{cases}$$

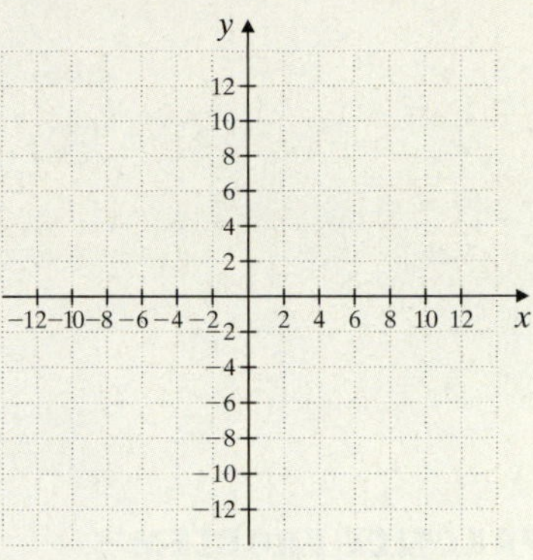

PRACTICE PROBLEM 4

Solve the system of equations by graphing.

$$\begin{cases} x + 3y = -1 \\ y = 1 \end{cases}$$

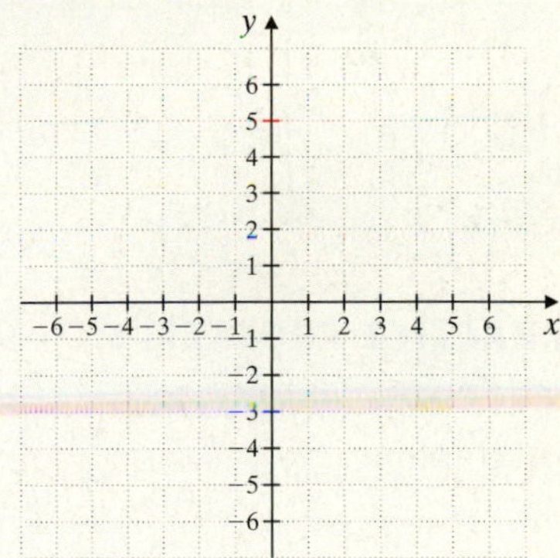

Answers

3. $(2, -4)$,

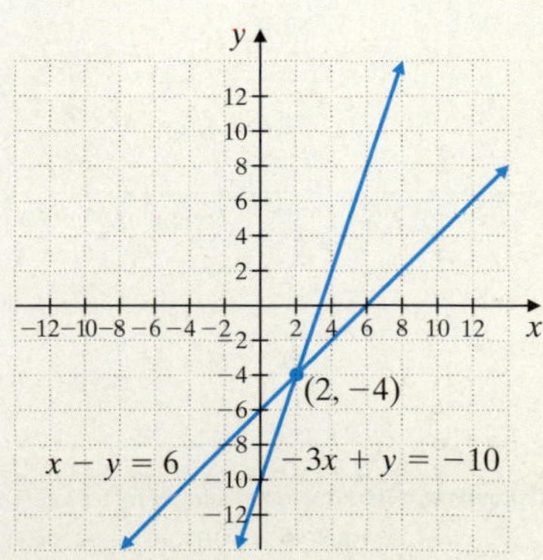

4. See page 1059.

The two lines appear to intersect at the point $(2, -2)$. To determine whether $(2, -2)$ is the solution, we replace x with 2 and y with -2 in both equations.

$2x + 3y = -2$	First equation	$x = 2$	Second equation
$2(2) + 3(-2) \stackrel{?}{=} -2$	Let $x = 2$ and $y = -2$.	$2 \stackrel{?}{=} 2$	Let $x = 2$.
$4 + (-6) \stackrel{?}{=} -2$	Simplify.	$2 = 2$	True
$-2 = -2$	True		

Since a true statement results in both equations, $(2, -2)$ is the solution of the system.

Work Practice Problem 4

Objective C Identifying Special Systems of Linear Equations

Not all systems of linear equations have a single solution. Some systems have no solution and some have an infinite number of solutions.

PRACTICE PROBLEM 5

Solve the system of equations by graphing.

$$\begin{cases} 3x - y = 6 \\ 6x = 2y \end{cases}$$

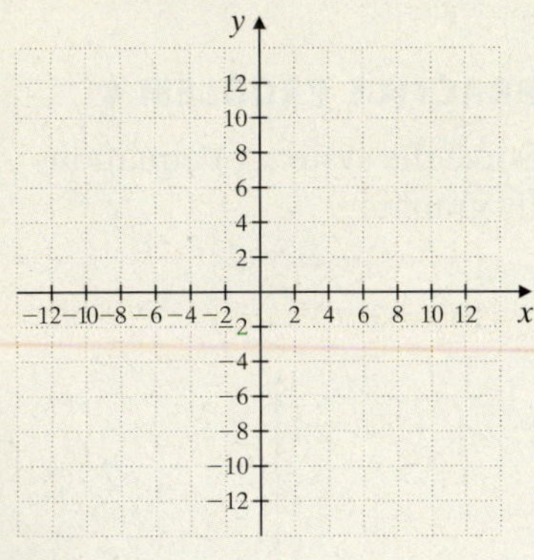

EXAMPLE 5 Solve the system of equations by graphing.

$$\begin{cases} 2x + y = 7 \\ 2y = -4x \end{cases}$$

Solution: We graph the two equations in the system. The equations in slope-intercept form are $y = -2x + 7$ and $y = -2x$. Notice from the equations that the lines have the same slope, -2, and different y-intercepts. This means that the lines are parallel.

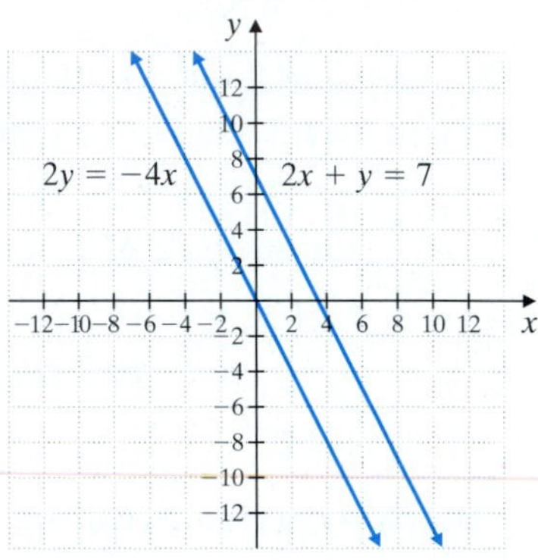

Since the lines are parallel, they do not intersect. This means that the system has *no solution*.

Work Practice Problem 5

PRACTICE PROBLEM 6

Solve the system of equations by graphing.

$$\begin{cases} x + y = -4 \\ -2x - 2y = 8 \end{cases}$$

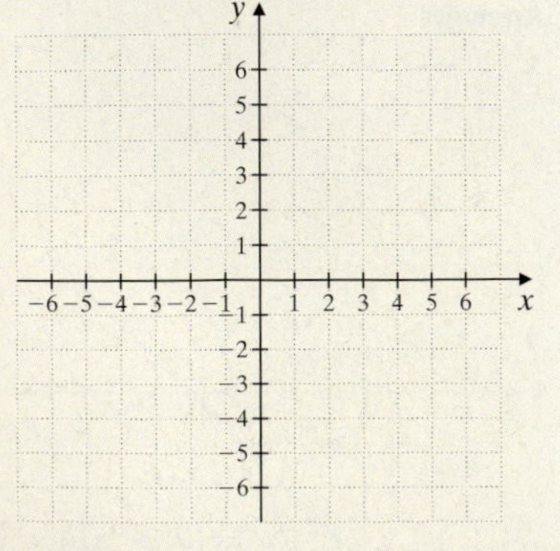

EXAMPLE 6 Solve the system of equations by graphing.

$$\begin{cases} x - y = 3 \\ -x + y = -3 \end{cases}$$

Solution: We graph each equation. The graphs of the equations are the same line. To see this, notice that if both sides of the first equation in the system are multiplied by -1, the result is the second equation.

$x - y = 3$	First equation
$-1(x - y) = -1(3)$	Multiply both sides by -1.
$-x + y = -3$	Simplify. This is the second equation.

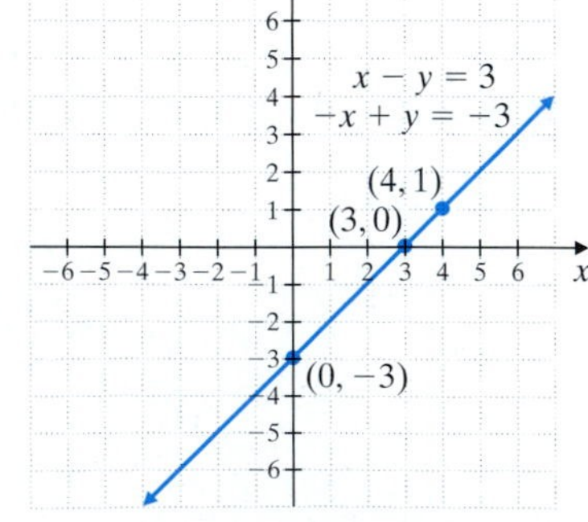

Any ordered pair that is a solution of one equation is a solution of the other and is then a solution of the system. This means that the system has an infinite number of solutions.

Work Practice Problem 6

Answers

5. See page 1059.
6. See page 1059.

Examples 5 and 6 are special cases of systems of linear equations. A system that has no solution is said to be an **inconsistent system.** If the graphs of the two equations of a system are identical, we call the equations **dependent equations.** Thus, the system in Example 5 is an inconsistent system and the equations in the system in Example 6 are dependent equations.

As we have seen, three different situations can occur when graphing the two lines associated with the equations in a linear system. These situations are shown in the figures.

One point of intersection: one solution

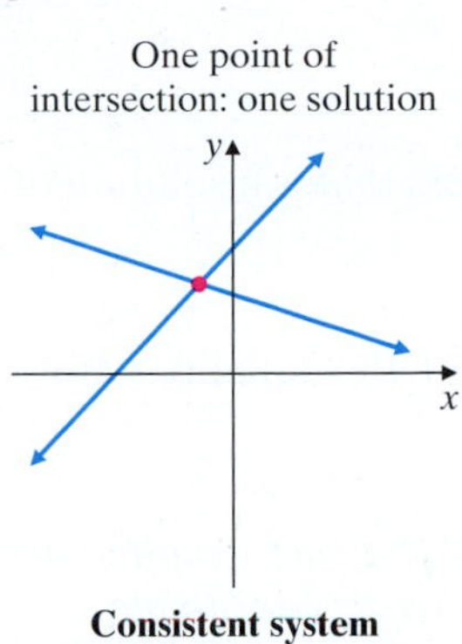

Consistent system
(at least one solution)
Independent equations
(graphs of equations differ)

Parallel lines: no solution

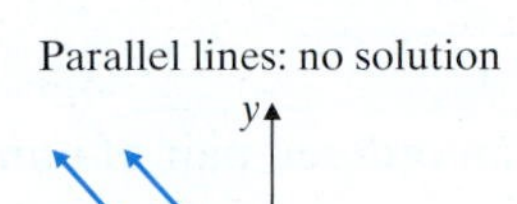
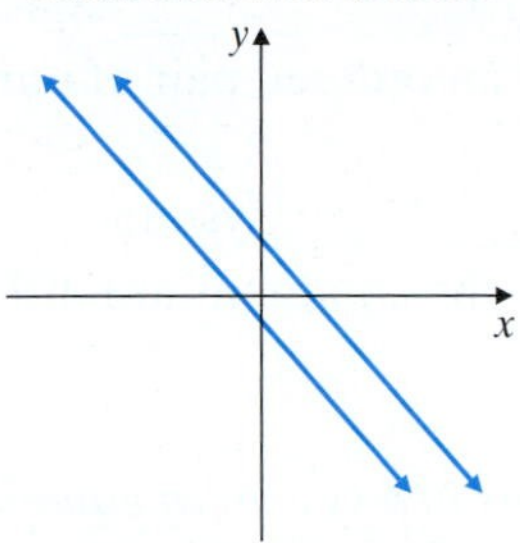

Inconsistent system
(no solution)
Independent equations
(graphs of equations differ)

Same line: infinite number of solutions

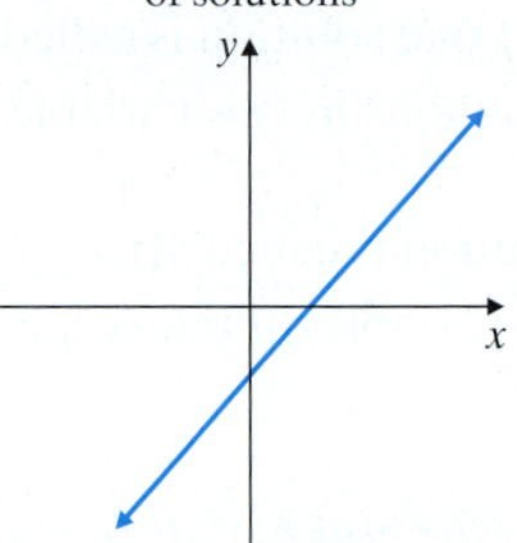

Consistent system
(at least one solution)
Dependent equations
(graphs of equations identical)

Answers

4. $(-4, 1)$,

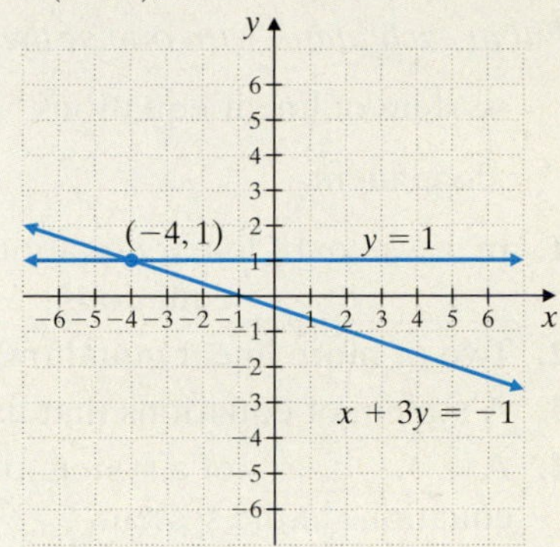

5. no solution,

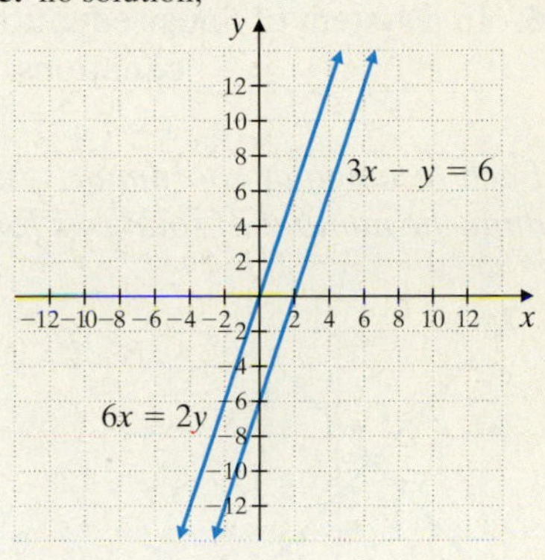

CALCULATOR EXPLORATIONS Graphing

A graphing calculator may be used to approximate solutions of systems of equations. For example, to approximate the solution of the system

$$\begin{cases} y = -3.14x - 1.35 \\ y = 4.88x + 5.25, \end{cases}$$

first graph each equation on the same set of axes. Then use the intersect feature of your calculator to approximate the point of intersection.

The approximate point of intersection is $(-0.82, 1.23)$.

Solve each system of equations. Approximate the solutions to two decimal places.

1. $\begin{cases} y = -2.68x + 1.21 \\ y = 5.22x - 1.68 \end{cases}$

2. $\begin{cases} y = 4.25x + 3.89 \\ y = -1.88x + 3.21 \end{cases}$

3. $\begin{cases} 4.3x - 2.9y = 5.6 \\ 8.1x + 7.6y = -14.1 \end{cases}$

4. $\begin{cases} -3.6x - 8.6y = 10 \\ -4.5x + 9.6y = -7.7 \end{cases}$

6. infinite number of solutions

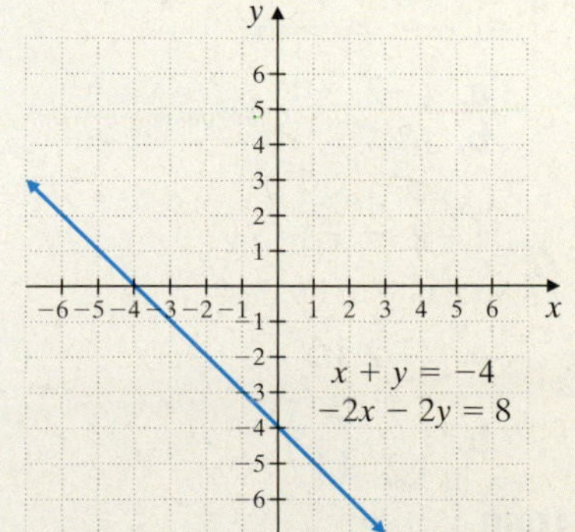

Vocabulary and Readiness Check

Fill in each blank with one of the words or phrases listed below.

system of linear equations	solution	consistent
dependent	inconsistent	independent

1. In a system of linear equations in two variables, if the graphs of the equations are the same, the equations are ____________ equations.
2. Two or more linear equations are called a ________________________.
3. A system of equations that has at least one solution is called a(n) ____________ system.
4. A ____________ of a system of two equations in two variables is an ordered pair of numbers that is a solution of both equations in the system.
5. A system of equations that has no solution is called a(n) ____________ system.
6. In a system of linear equations in two variables, if the graphs of the equations are different, the equations are ____________ equations.

Each rectangular coordinate system shows the graph of the equations in a system of equations. Use each graph to determine the number of solutions for each associated system. If the system has only one solution, give its coordinates.

7.

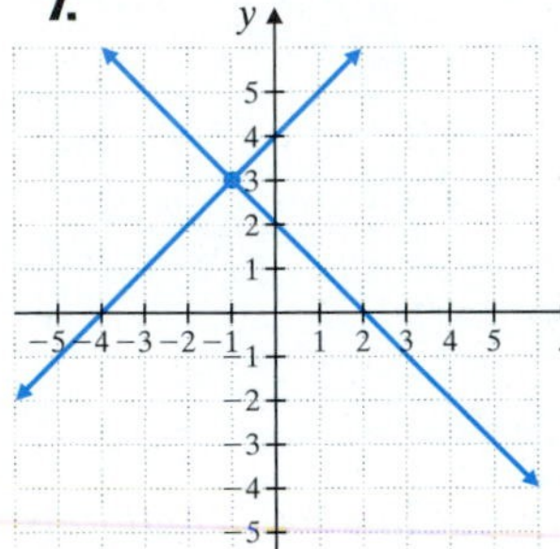

8.

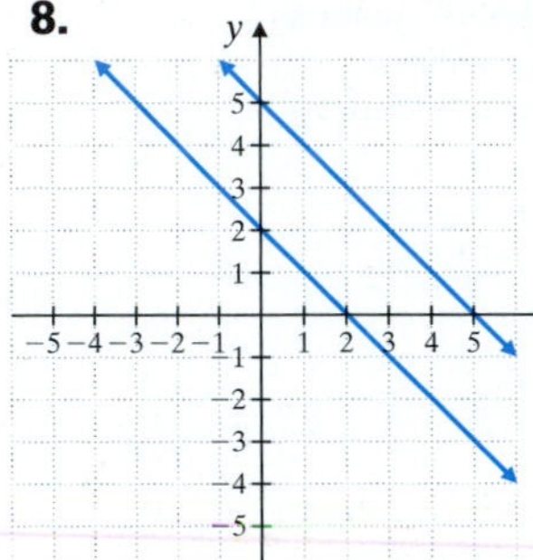

9.

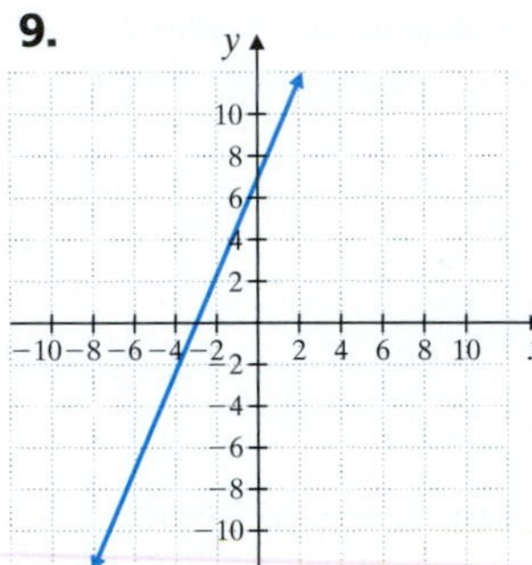

10.

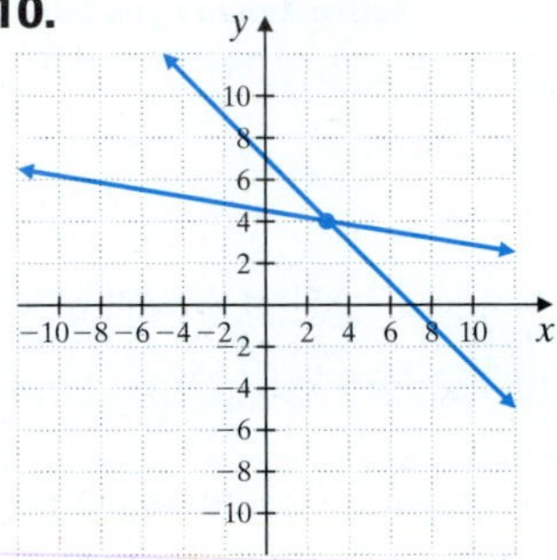

14.1 EXERCISE SET

FOR EXTRA HELP

Student Solutions Manual

PH Math/Tutor Center

CD/Video for Review

MathXL®

MyMathLab

Objective A *Determine whether each ordered pair is a solution of the system of linear equations. See Examples 1 and 2.*

1. $\begin{cases} x + y = 8 \\ 3x + 2y = 21 \end{cases}$

a. $(2, 4)$
b. $(5, 3)$

2. $\begin{cases} 2x + y = 5 \\ x + 3y = 5 \end{cases}$

a. $(5, 0)$
b. $(2, 1)$

3. $\begin{cases} 3x - y = 5 \\ x + 2y = 11 \end{cases}$

a. $(3, 4)$
b. $(0, -5)$

4. $\begin{cases} 2x - 3y = 8 \\ x - 2y = 6 \end{cases}$

a. $(-2, -4)$
b. $(7, 2)$

5. $\begin{cases} 2y = 4x + 6 \\ 2x - y = -3 \end{cases}$

a. $(-3, -3)$
b. $(0, 3)$

6. $\begin{cases} x + 5y = -4 \\ -2x = 10y + 8 \end{cases}$

a. $(-4, 0)$
b. $(6, -2)$

7. $\begin{cases} -2 = x - 7y \\ 6x - y = 13 \end{cases}$

a. $(-2, 0)$
b. $\left(\frac{1}{2}, \frac{5}{14}\right)$

8. $\begin{cases} 4x = 1 - y \\ x - 3y = -8 \end{cases}$

a. $(0, 1)$
b. $\left(\frac{1}{6}, \frac{1}{3}\right)$

Objectives B C Mixed Practice *Solve each system of linear equations by graphing. See Examples 3 through 6.*

9. $\begin{cases} x + y = 4 \\ x - y = 2 \end{cases}$

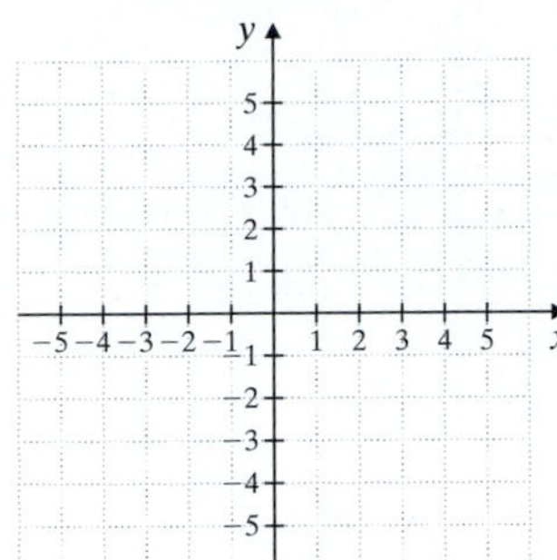

10. $\begin{cases} x + y = 3 \\ x - y = 5 \end{cases}$

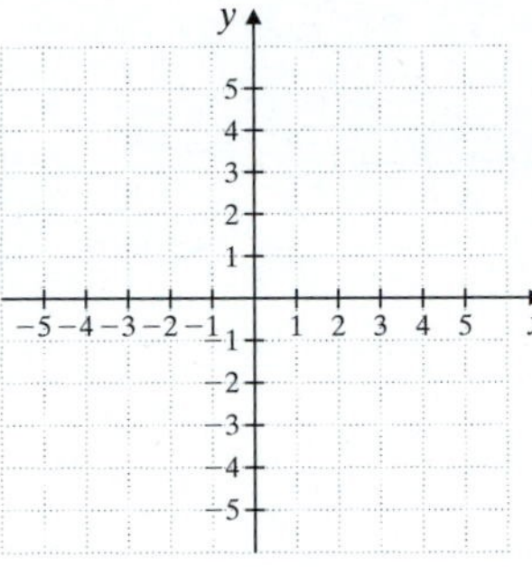

11. $\begin{cases} x + y = 6 \\ -x + y = -6 \end{cases}$

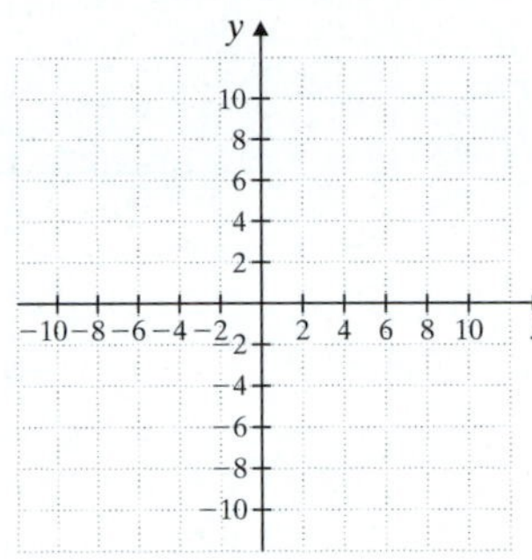

12. $\begin{cases} x + y = 1 \\ -x + y = -3 \end{cases}$

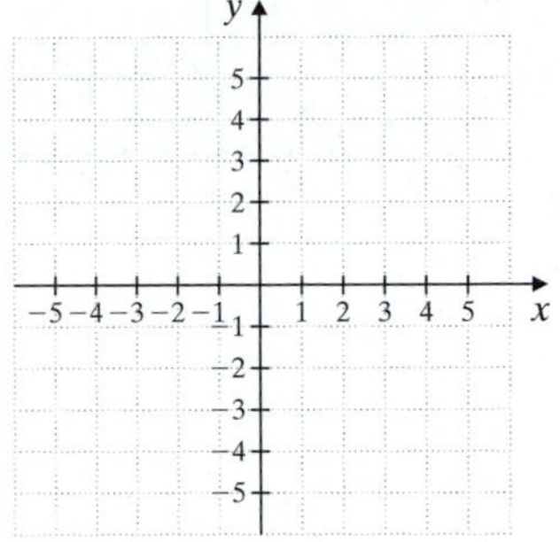

13. $\begin{cases} y = 2x \\ 3x - y = -2 \end{cases}$

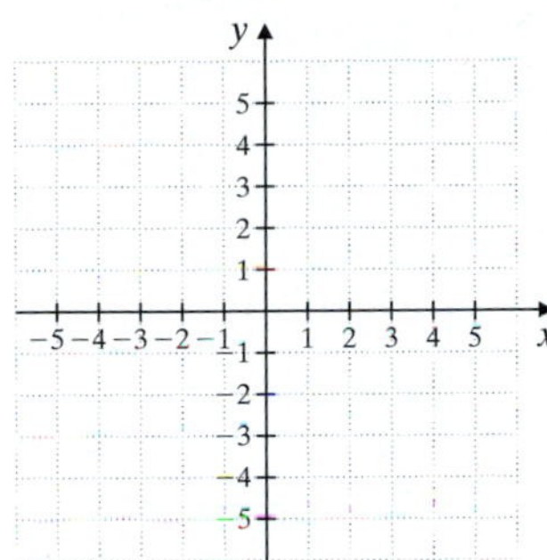

14. $\begin{cases} y = -3x \\ 2x - y = -5 \end{cases}$

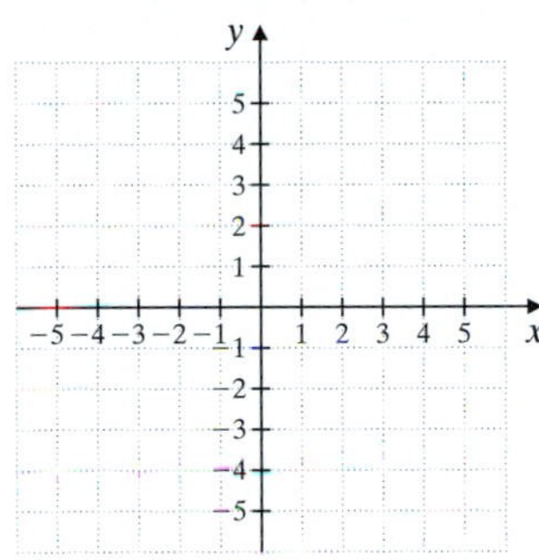

15. $\begin{cases} y = x + 1 \\ y = 2x - 1 \end{cases}$

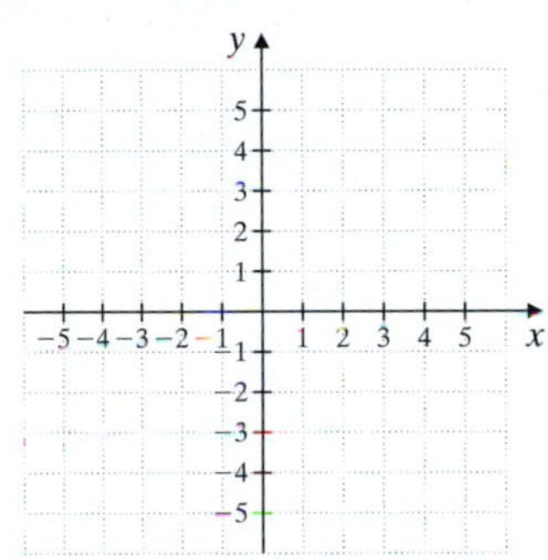

16. $\begin{cases} y = 3x - 4 \\ y = x + 2 \end{cases}$

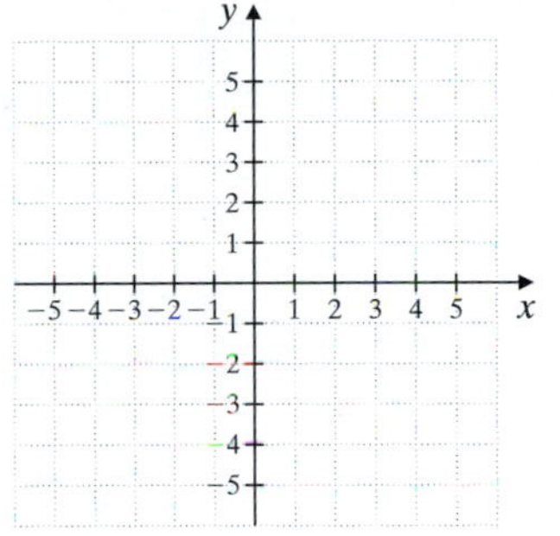

17. $\begin{cases} 2x + y = 0 \\ 3x + y = 1 \end{cases}$

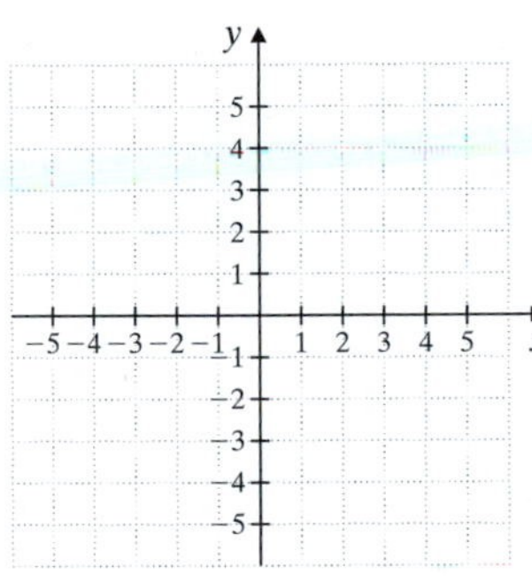

18. $\begin{cases} 2x + y = 1 \\ 3x + y = 0 \end{cases}$

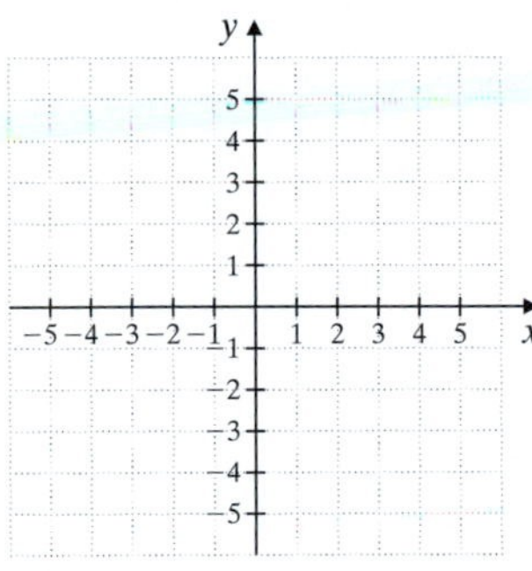

19. $\begin{cases} y = -x - 1 \\ y = 2x + 5 \end{cases}$

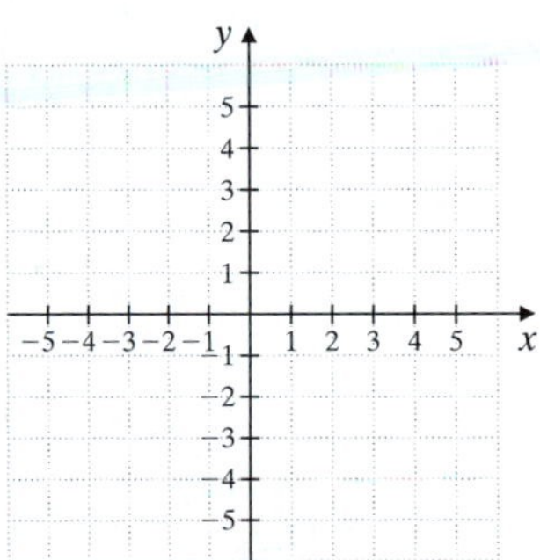

20. $\begin{cases} y = x - 1 \\ y = -3x - 5 \end{cases}$

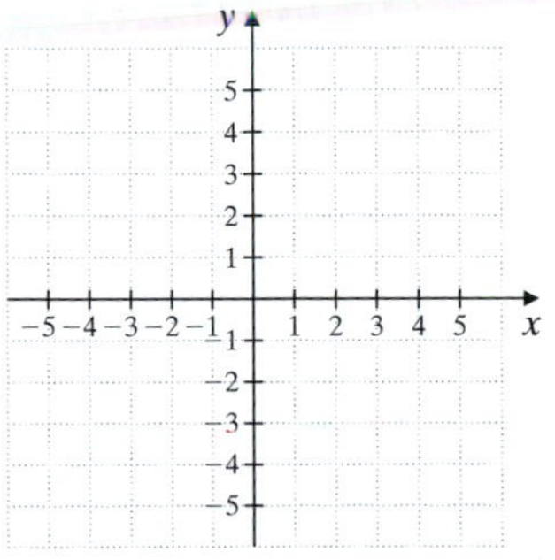

21. $\begin{cases} x + y = 5 \\ x + y = 6 \end{cases}$

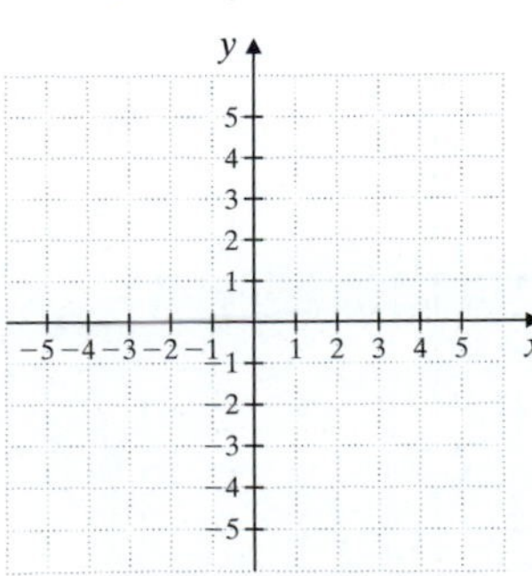

22. $\begin{cases} x - y = 4 \\ x - y = 1 \end{cases}$

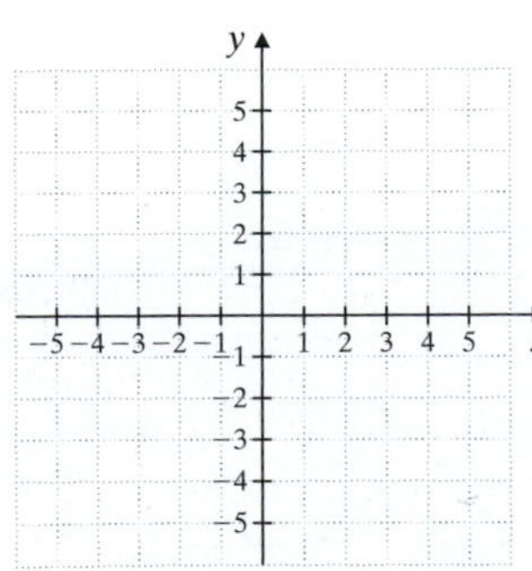

23. $\begin{cases} 2x - y = 6 \\ y = 2 \end{cases}$

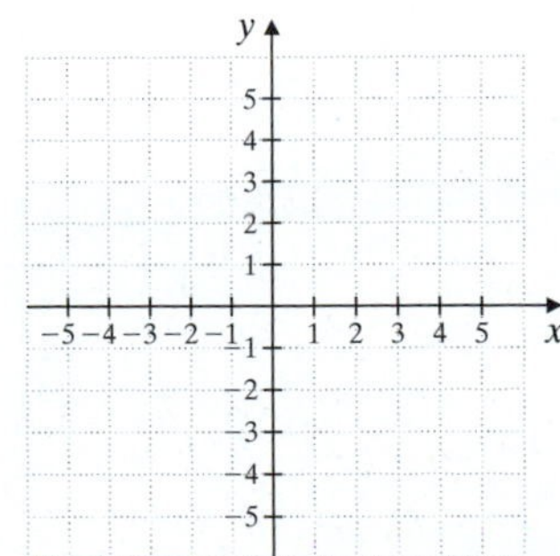

24. $\begin{cases} x + y = 5 \\ x = 4 \end{cases}$

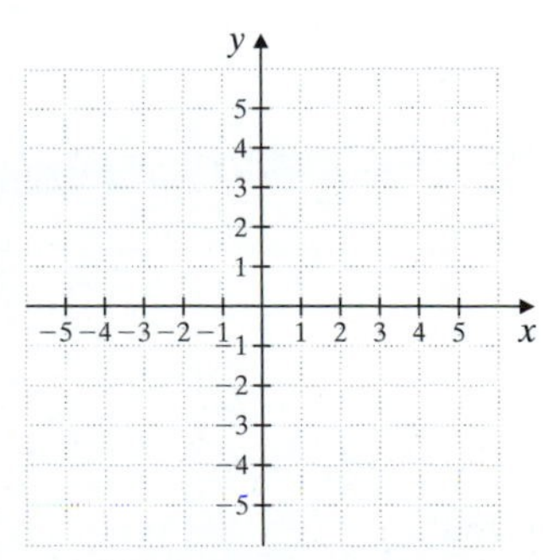

25. $\begin{cases} x - 2y = 2 \\ 3x + 2y = -2 \end{cases}$

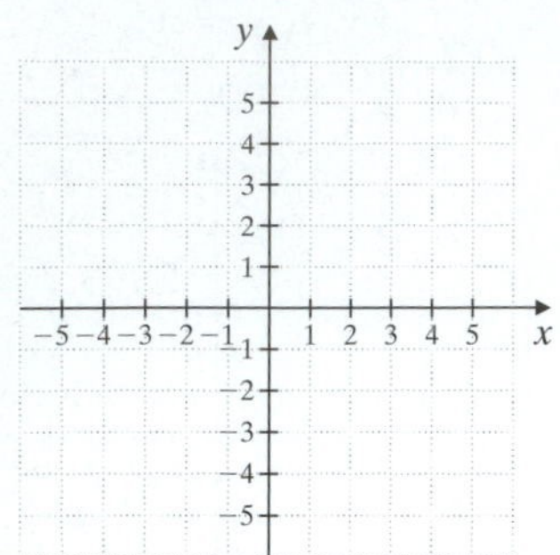

26. $\begin{cases} x + 3y = 7 \\ 2x - 3y = -4 \end{cases}$

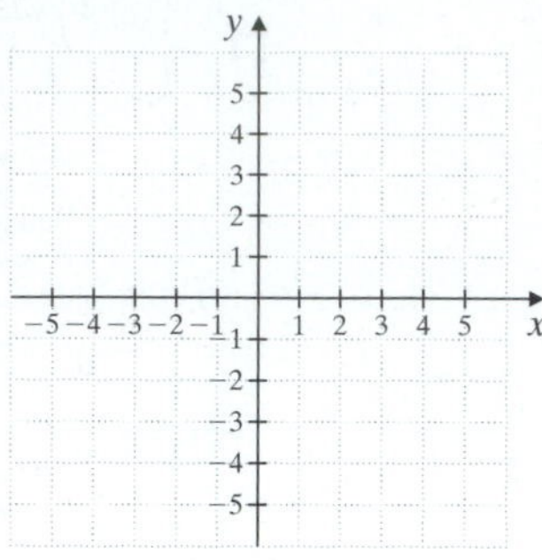

27. $\begin{cases} 2x + y = 4 \\ 6x = -3y + 6 \end{cases}$

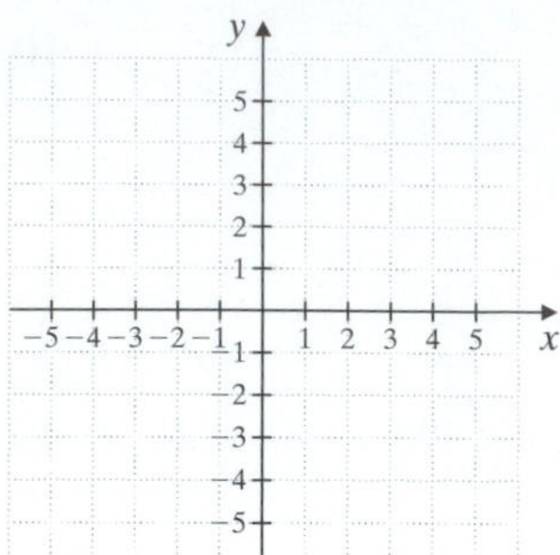

28. $\begin{cases} y + 2x = 3 \\ 4x = 2 - 2y \end{cases}$

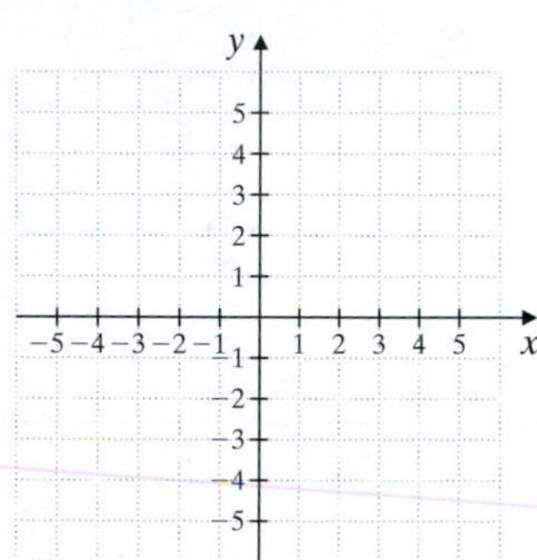

29. $\begin{cases} y - 3x = -2 \\ 6x - 2y = 4 \end{cases}$

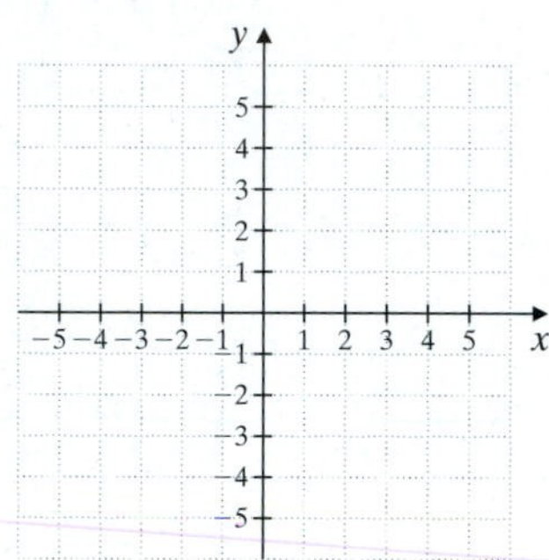

30. $\begin{cases} x - 2y = -6 \\ -2x + 4y = 12 \end{cases}$

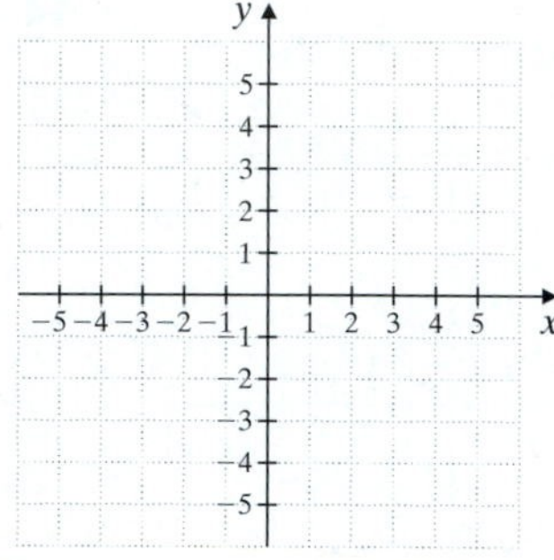

31. $\begin{cases} x = 3 \\ y = -1 \end{cases}$

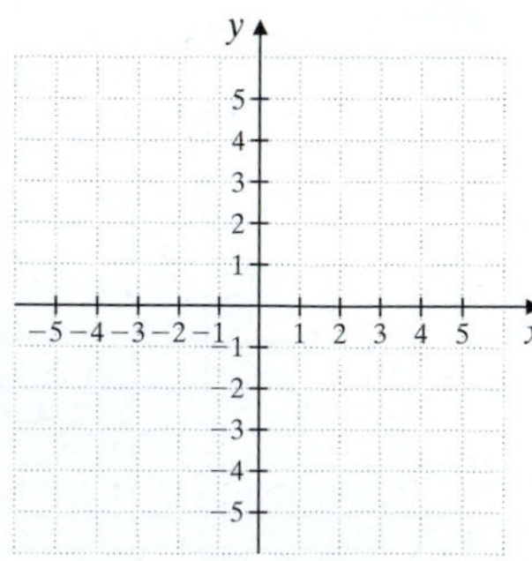

32. $\begin{cases} x = -5 \\ y = 3 \end{cases}$

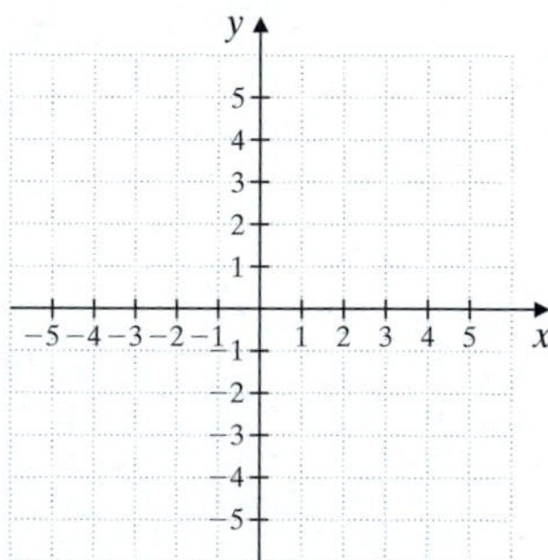

33. $\begin{cases} y = x - 2 \\ y = 2x + 3 \end{cases}$

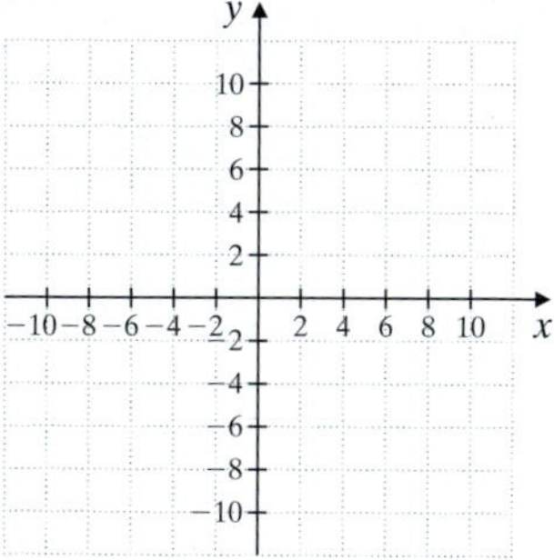

34. $\begin{cases} y = x + 5 \\ y = -2x - 4 \end{cases}$

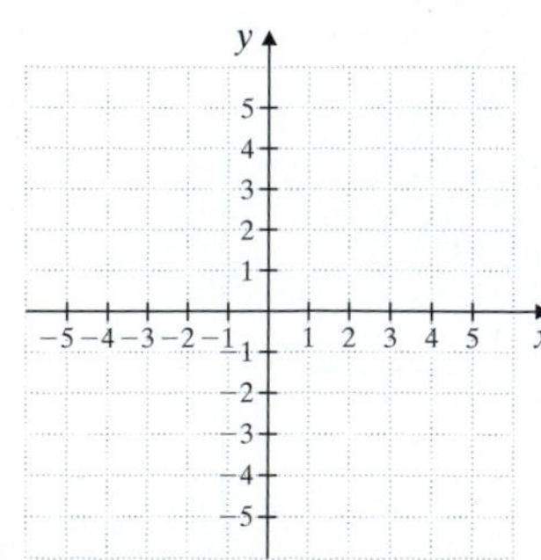

35. $\begin{cases} 2x - 3y = -2 \\ -3x + 5y = 5 \end{cases}$

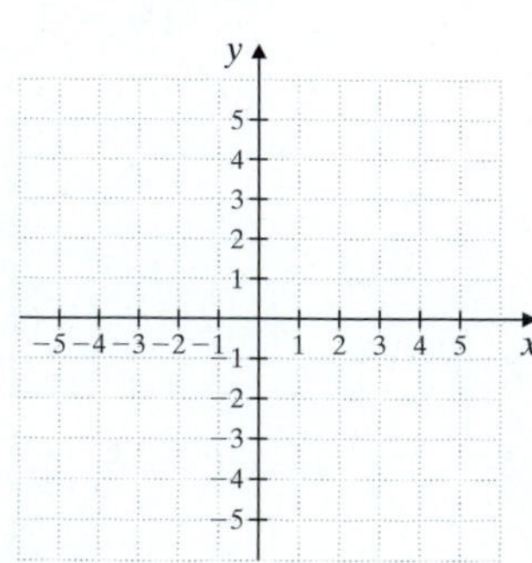

36. $\begin{cases} 4x - y = 7 \\ 2x - 3y = -9 \end{cases}$

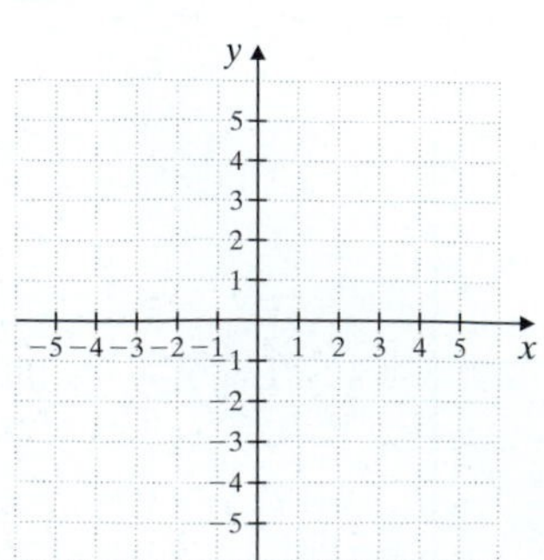

37. $\begin{cases} 6x - y = 4 \\ \frac{1}{2}y = -2 + 3x \end{cases}$

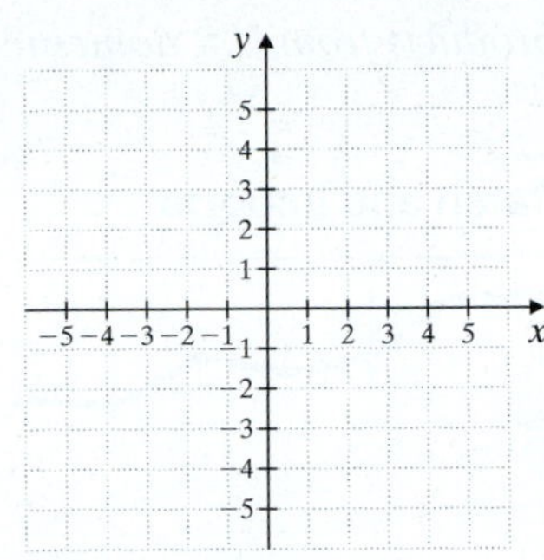

38. $\begin{cases} 3x - y = 6 \\ \frac{1}{3}y = -2 + x \end{cases}$

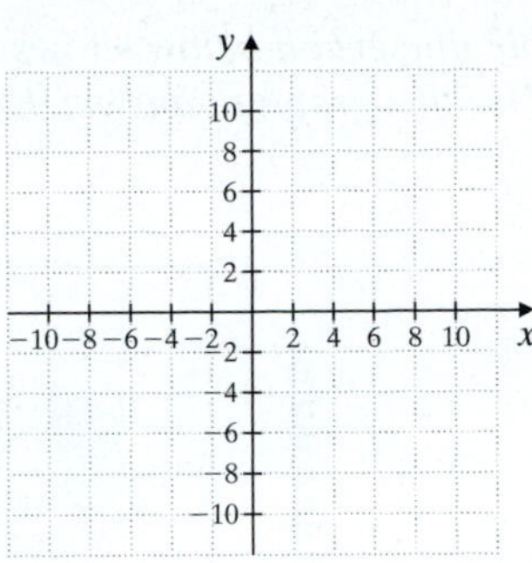

Review

Solve each equation. See Section 9.3.

39. $5(x - 3) + 3x = 1$

40. $-2x + 3(x + 6) = 17$

41. $4\left(\frac{y + 1}{2}\right) + 3y = 0$

42. $-y + 12\left(\frac{y - 1}{4}\right) = 3$

43. $8a - 2(3a - 1) = 6$

44. $3z - (4z - 2) = 9$

Concept Extensions

45. Draw a graph of two linear equations whose associated system has the solution $(-1, 4)$.

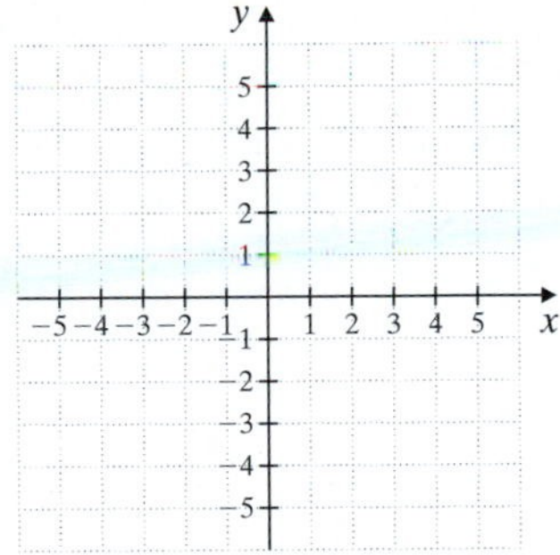

46. Draw a graph of two linear equations whose associated system has the solution $(3, -2)$.

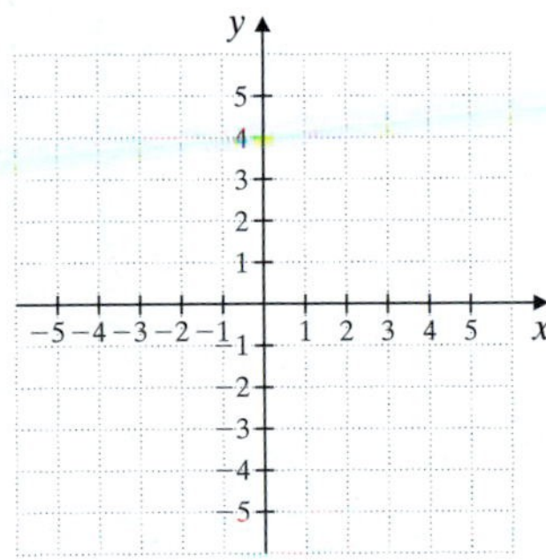

47. Draw a graph of two linear equations whose associated system has no solution.

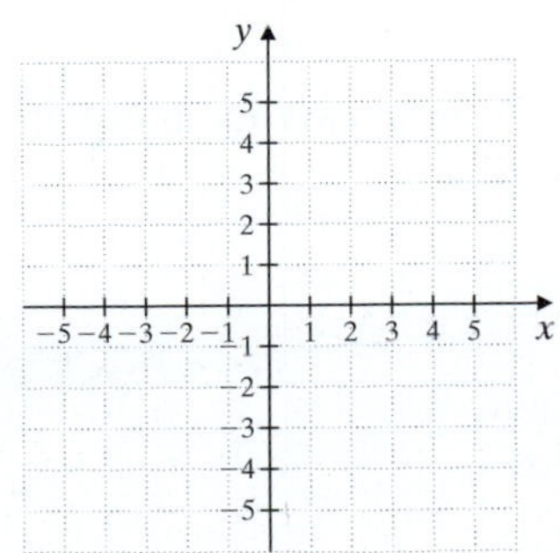

48. Draw a graph of two linear equations whose associated system has an infinite number of solutions.

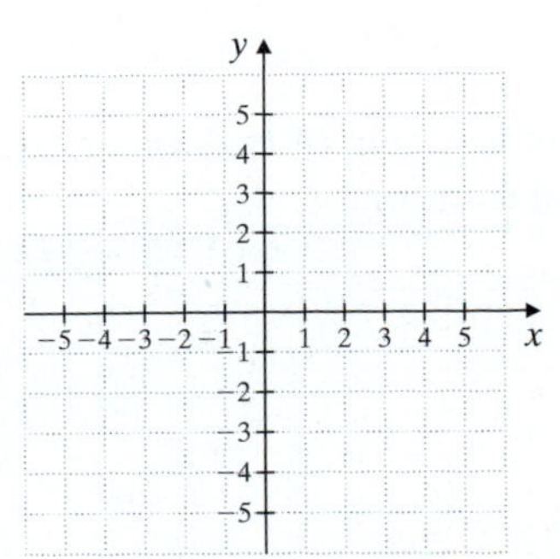

The double line graph below shows the number of pounds of fishery products from U.S. domestic catch and from imports. Use this graph to answer Exercises 49 and 50.

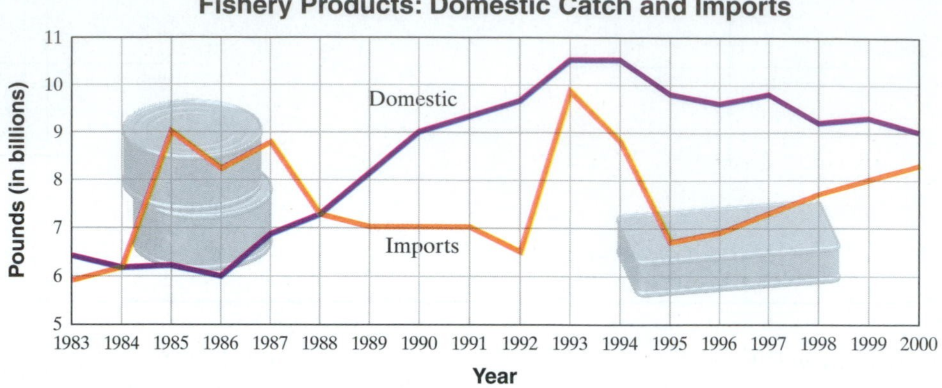

Source: U.S. Bureau of the Census, *Statistical Abstract of the United States*: 2003, 115th ed., Washington, DC, 1995.

49. In what year(s) was the number of pounds of imported fishery products equal to the number of pounds of domestic catch?

50. In what year(s) was the number of pounds of imported fishery products less than or equal to the number of pounds of domestic catch?

*The double line graph below shows the number of Target Stores versus the number of Wal-Mart discount stores. Use this for Exercises 51 and 52. (*Note: *This does not include Wal-Mart Supercenters or Sam's Club) (Sources: Target.com and Walmart.com)*

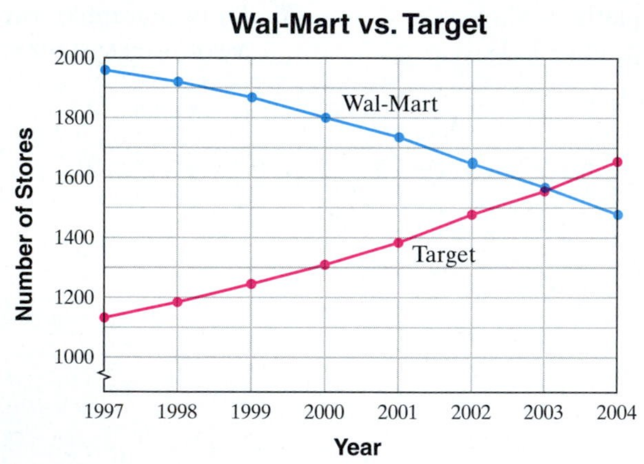

51. In what year(s) was the number of Target stores approximately equal to the number of Wal-Mart discount stores?

52. In what year(s) was the number of Wal-Mart discount stores greater than the number of Target stores?

53. Construct a system of two linear equations that has $(2, 5)$ as a solution.

54. Construct a system of two linear equations that has $(0, 1)$ as a solution.

55. The ordered pair $(-2, 3)$ is a solution of the three linear equations below:

$$x + y = 1$$
$$2x - y = -7$$
$$x + 3y = 7$$

If each equation has a distinct graph, describe the graph of all three equations on the same axes.

56. Below are tables of values for two linear equations.

a. Find a solution of the corresponding system.

b. Graph several ordered pairs from each table and sketch the two lines.

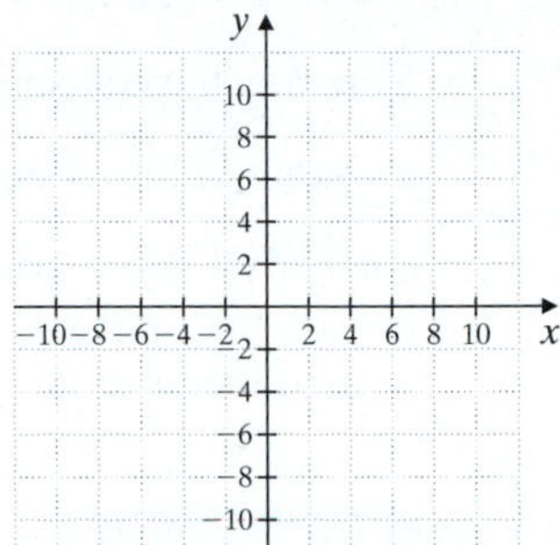

x	y
1	3
2	5
3	7
4	9
5	11

x	y
1	6
2	7
3	8
4	9
5	10

c. Does your graph confirm the solution from part (a)?

57. Explain how to use a graph to determine the number of solutions of a system.

STUDY SKILLS BUILDER

Have You Decided to Successfully Complete This Course?

Hopefully by now, one of your current goals is to successfully complete this course.

If it is not a goal of yours, ask yourself why? One common reason is fear of failure. Amazingly enough, fear of failure alone can be strong enough to keep many of us from doing our best in any endeavor. Another common reason is that you simply haven't taken the time to make successfully completing this course one of your goals.

Anytime you are registered for a course, successfully completing the course should probably be a goal. How do you do this? Start by writing this goal in your mathematics notebook. Then list steps you will take to ensure success. A great first step is to read or reread Section 1.1 and make a commitment to try the suggestions in that section.

Good luck and don't forget that a positive attitude will make a big difference.

Let's see how you are doing.

1. Have you made the decision to make "successfully completing this course" a goal of yours? If no, please list reasons that this has not happened. Study your list and talk to your instructor about this.
2. If your answer to Exercise 1 is yes, take a moment and list, in your notebook, further specific goals that will help you achieve this goal of successfully completing this course. (For example, my goal this semester is not to miss any of my mathematics classes.)
3. Rate your commitment to this course with a number between 1 and 5. Use the diagram below to help.

High Commitment		Average Commitment		Not committed at all
5	4	3	2	1

4. If you have rated your personal commitment level (from the exercise above) as a 1, 2, or 3, list the reasons why this is so. Then determine whether it is possible to increase your commitment level to a 4 or 5.

Objective

A **Use the Substitution Method to Solve a System of Linear Equations.**

14.2 SOLVING SYSTEMS OF LINEAR EQUATIONS BY SUBSTITUTION

Objective A Using the Substitution Method

You may have suspected by now that graphing alone is not an accurate way to solve a system of linear equations. For example, a solution of $\left(\frac{1}{2}, \frac{2}{9}\right)$ is unlikely to be read correctly from a graph. In this section, we discuss a second, more accurate method for solving systems of equations. This method is called the **substitution method** and is introduced in the next example.

PRACTICE PROBLEM 1

Use the substitution method to solve the system:

$$\begin{cases} 2x + 3y = 13 \\ x = y + 4 \end{cases}$$

EXAMPLE 1 Solve the system:

$$\begin{cases} 2x + y = 10 & \text{First equation} \\ x = y + 2 & \text{Second equation} \end{cases}$$

Solution: The second equation in this system is $x = y + 2$. This tells us that x and $y + 2$ have the same value. This means that we may substitute $y + 2$ for x in the first equation.

$2x + y = 10$ First equation

$2(y + 2) + y = 10$ Substitute $y + 2$ for x since $x = y + 2$.

Notice that this equation now has one variable, y. Let's now solve this equation for y.

Helpful Hint Don't forget the distributive property.

$$\begin{aligned} 2(y + 2) + y &= 10 \\ 2y + 4 + y &= 10 && \text{Apply the distributive property.} \\ 3y + 4 &= 10 && \text{Combine like terms.} \\ 3y &= 6 && \text{Subtract 4 from both sides.} \\ y &= 2 && \text{Divide both sides by 3.} \end{aligned}$$

Now we know that the y-value of the ordered pair solution of the system is 2. To find the corresponding x-value, we replace y with 2 in the second equation, $x = y + 2$, and solve for x.

$$\begin{aligned} x &= y + 2 && \text{Second equation.} \\ x &= 2 + 2 && \text{Let } y = 2. \\ x &= 4 \end{aligned}$$

The solution of the system is the ordered pair $(4, 2)$. Since an ordered pair solution must satisfy both linear equations in the system, we could have chosen the equation $2x + y = 10$ to find the corresponding x-value. The resulting x-value is the same.

Check: We check to see that $(4, 2)$ satisfies both equations of the original system.

First Equation		**Second Equation**	
$2x + y = 10$		$x = y + 2$	
$2(4) + 2 \stackrel{?}{=} 10$		$4 \stackrel{?}{=} 2 + 2$	Let $x = 4$ and $y = 2$.
$10 = 10$	True	$4 = 4$	True

Answer

1. $(5, 1)$

The solution of the system is (4, 2).
A graph of the two equations shows the two lines intersecting at the point (4, 2).

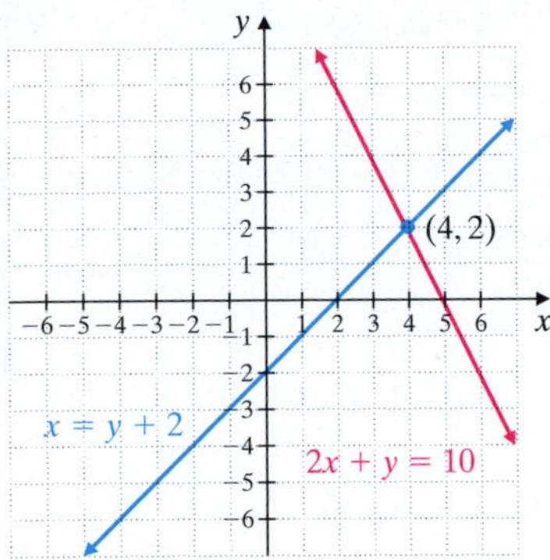

Work Practice Problem 1

EXAMPLE 2 Solve the system:

$$\begin{cases} 5x - y = -2 \\ y = 3x \end{cases}$$

Solution: The second equation is solved for y in terms of x. We substitute $3x$ for y in the first equation.

$5x - y = -2$ First equation

$5x - (3x) = -2$ Substitute $3x$ for y.

Now we solve for x.

$5x - 3x = -2$

$2x = -2$ Combine like terms.

$x = -1$ Divide both sides by 2.

The x-value of the ordered pair solution is -1. To find the corresponding y-value, we replace x with -1 in the second equation $y = 3x$.

$y = 3x$ Second equation

$y = 3(-1)$ Let $x = -1$.

$y = -3$

Check to see that the solution of the system is $(-1, -3)$.

Work Practice Problem 2

PRACTICE PROBLEM 2

Use the substitution method to solve the system:

$$\begin{cases} 4x - y = 2 \\ y = 5x \end{cases}$$

To solve a system of equations by substitution, we first need an equation solved for one of its variables, as in Examples 1 and 2. If neither equation in a system is solved for x or y, this will be our first step.

EXAMPLE 3 Solve the system:

$$\begin{cases} x + 2y = 7 \\ 2x + 2y = 13 \end{cases}$$

Solution: Notice that neither equation is solved for x or y. Thus, we choose one of the equations and solve for x or y. We will solve the first equation for x so that we will not introduce tedious fractions when solving. To solve the first equation for x, we subtract $2y$ from both sides.

$x + 2y = 7$ First equation

$x = 7 - 2y$ Subtract $2y$ from both sides.

Continued on next page

PRACTICE PROBLEM 3

Solve the system:

$$\begin{cases} 3x + y = 5 \\ 3x - 2y = -7 \end{cases}$$

Answers
2. $(-2, -10)$ 3. $\left(\frac{1}{3}, 4\right)$

Since $x = 7 - 2y$, we now substitute $7 - 2y$ for x in the second equation and solve for y.

$$2x + 2y = 13 \quad \text{Second equation}$$
$$2(7 - 2y) + 2y = 13 \quad \text{Let } x = 7 - 2y.$$
$$14 - 4y + 2y = 13 \quad \text{Apply the distributive property.}$$
$$14 - 2y = 13 \quad \text{Simplify.}$$
$$-2y = -1 \quad \text{Subtract 14 from both sides.}$$
$$y = \frac{1}{2} \quad \text{Divide both sides by } -2.$$

Helpful Hint

Don't forget to insert parentheses when substituting $7 - 2y$ for x.

To find x, we let $y = \frac{1}{2}$ in the equation $x = 7 - 2y$.

$$x = 7 - 2y$$
$$x = 7 - 2\left(\frac{1}{2}\right) \quad \text{Let } y = \frac{1}{2}.$$
$$x = 7 - 1$$
$$x = 6$$

To find x, any equation in two variables equivalent to the original equations of the system may be used. We used this equation since it is solved for x.

Check the solution in both equations of the original system. The solution is $\left(6, \frac{1}{2}\right)$.

Work Practice Problem 3

The following steps summarize how to solve a system of equations by the substitution method.

To Solve a System of Two Linear Equations by the Substitution Method

Step 1: Solve one of the equations for one of its variables.

Step 2: Substitute the expression for the variable found in Step 1 into the other equation.

Step 3: Solve the equation from Step 2 to find the value of one variable.

Step 4: Substitute the value found in Step 3 in any equation containing both variables to find the value of the other variable.

Step 5: Check the proposed solution in the original system.

✔**Concept Check** As you solve the system

$$\begin{cases} 2x + y = -5 \\ x - y = 5 \end{cases}$$

you find that $y = -5$. Is this the solution of the system?

EXAMPLE 4 Solve the system:

$$\begin{cases} 7x - 3y = -14 \\ -3x + y = 6 \end{cases}$$

Solution: Since the coefficient of y is 1 in the second equation, we will solve the second equation for y. This way, we avoid introducing tedious fractions.

$$-3x + y = 6 \quad \text{Second equation}$$
$$y = 3x + 6$$

PRACTICE PROBLEM 4

Solve the system:

$$\begin{cases} 5x - 2y = 6 \\ -3x + y = -3 \end{cases}$$

Answer

4. $(0, -3)$

✔ **Concept Check Answer**

no, the solution will be an ordered pair

Next, we substitute $3x + 6$ for y in the first equation.

$$\begin{aligned} 7x - 3y &= -14 && \text{First equation} \\ 7x - 3(3x + 6) &= -14 && \text{Let } y = 3x + 6. \\ 7x - 9x - 18 &= -14 && \text{Use the distributive property.} \\ -2x - 18 &= -14 && \text{Simplify.} \\ -2x &= 4 && \text{Add 18 to both sides.} \\ x &= -2 && \text{Divide both sides by } -2. \end{aligned}$$

To find the corresponding y-value, we substitute -2 for x in the equation $y = 3x + 6$. Then $y = 3(-2) + 6$ or $y = 0$. The solution of the system is $(-2, 0)$. Check this solution in both equations of the system.

Work Practice Problem 4

✔Concept Check To avoid fractions, which of the equations below would you use to solve for x?

a. $3x - 4y = 15$ **b.** $14 - 3y = 8x$ **c.** $7y + x = 12$

Helpful Hint

When solving a system of equations by the substitution method, begin by solving an equation for one of its variables. If possible, solve for a variable that has a coefficient of 1 or -1 to avoid working with time-consuming fractions.

EXAMPLE 5 Solve the system: $\begin{cases} \frac{1}{2}x - y = 3 \\ x = 6 + 2y \end{cases}$

Solution: The second equation is already solved for x in terms of y. Thus we substitute $6 + 2y$ for x in the first equation and solve for y.

$$\begin{aligned} \frac{1}{2}x - y &= 3 && \text{First equation} \\ \frac{1}{2}(6 + 2y) - y &= 3 && \text{Let } x = 6 + 2y. \\ 3 + y - y &= 3 && \text{Apply the distributive property.} \\ 3 &= 3 && \text{Simplify.} \end{aligned}$$

Arriving at a true statement such as $3 = 3$ indicates that the two linear equations in the original system are equivalent. This means that their graphs are identical, as shown in the figure. There is an infinite number of solutions to the system, and any solution of one equation is also a solution of the other.

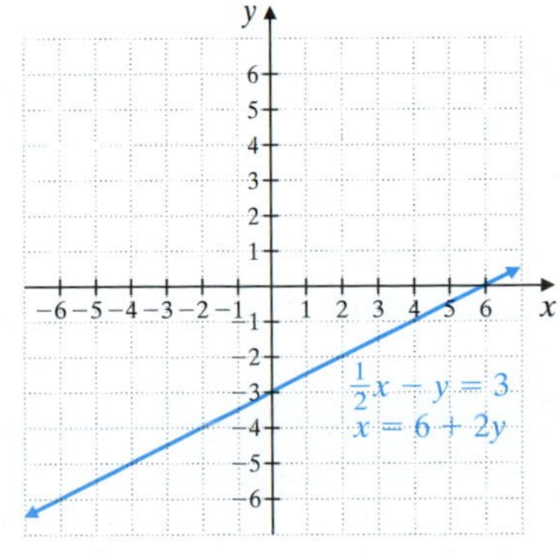

Work Practice Problem 5

PRACTICE PROBLEM 5

Solve the system:

$$\begin{cases} -x + 3y = 6 \\ y = \frac{1}{3}x + 2 \end{cases}$$

Answer

5. infinite number of solutions

✔ Concept Check Answer

c

PRACTICE PROBLEM 6

Solve the system:

$$\begin{cases} 2x - 3y = 6 \\ -4x + 6y = -12 \end{cases}$$

EXAMPLE 6 Solve the system:

$$\begin{cases} 6x + 12y = 5 \\ -4x - 8y = 0 \end{cases}$$

Solution: We choose the second equation and solve for y. (*Note:* Although you might not see this beforehand, if you solve the second equation for x, the result is $x = -2y$ and no fractions are introduced. Either way will lead to the correct solution.)

$$-4x - 8y = 0 \quad \text{Second equation}$$
$$-8y = 4x \quad \text{Add } 4x \text{ to both sides.}$$
$$\frac{-8y}{-8} = \frac{4x}{-8} \quad \text{Divide both sides by } -8.$$
$$y = -\frac{1}{2}x \quad \text{Simplify.}$$

Now we replace y with $-\frac{1}{2}x$ in the first equation.

$$6x + 12y = 5 \quad \text{First equation}$$
$$6x + 12\left(-\frac{1}{2}x\right) = 5 \quad \text{Let } y = -\frac{1}{2}x.$$
$$6x + (-6x) = 5 \quad \text{Simplify.}$$
$$0 = 5 \quad \text{Combine like terms.}$$

The false statement $0 = 5$ indicates that this system has no solution. The graph of the linear equations in the system is a pair of parallel lines, as shown in the figure.

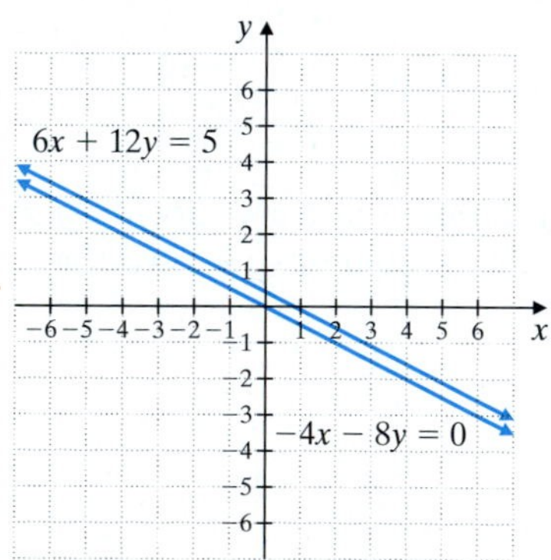

Work Practice Problem 6

✔Concept Check Describe how the graphs of the equations in a system appear if the system has

a. no solution

b. one solution

c. an infinite number of solutions

Answer

6. infinite number of solutions

✔ Concept Check Answers

a. parallel lines **b.** intersect at one point **c.** identical graphs

14.2 EXERCISE SET

FOR EXTRA HELP

Student Solutions Manual

PH Math/Tutor Center

CD/Video for Review

MathXL®

MyMathLab

Objective A *Solve each system of equations by the substitution method. See Examples 1 and 2.*

1. $\begin{cases} x + y = 3 \\ x = 2y \end{cases}$

2. $\begin{cases} x + y = 20 \\ x = 3y \end{cases}$

3. $\begin{cases} x + y = 6 \\ y = -3x \end{cases}$

4. $\begin{cases} x + y = 6 \\ y = -4x \end{cases}$

5. $\begin{cases} y = 3x + 1 \\ 4y - 8x = 12 \end{cases}$

6. $\begin{cases} y = 2x + 3 \\ 5y - 7x = 18 \end{cases}$

7. $\begin{cases} y = 2x + 9 \\ y = 7x + 10 \end{cases}$

8. $\begin{cases} y = 5x - 3 \\ y = 8x + 4 \end{cases}$

Solve each system of equations by the substitution method. See Examples 1 through 6.

9. $\begin{cases} 3x - 4y = 10 \\ y = x - 3 \end{cases}$

10. $\begin{cases} 4x - 3y = 10 \\ y = x - 5 \end{cases}$

11. $\begin{cases} x + 2y = 6 \\ 2x + 3y = 8 \end{cases}$

12. $\begin{cases} x + 3y = -5 \\ 2x + 2y = 6 \end{cases}$

13. $\begin{cases} 3x + 2y = 16 \\ x = 3y - 2 \end{cases}$

14. $\begin{cases} 2x + 3y = 18 \\ x = 2y - 5 \end{cases}$

15. $\begin{cases} 2x - 5y = 1 \\ 3x + y = -7 \end{cases}$

16. $\begin{cases} 3y - x = 6 \\ 4x + 12y = 0 \end{cases}$

17. $\begin{cases} 4x + 2y = 5 \\ -2x = y + 4 \end{cases}$

18. $\begin{cases} 2y = x + 2 \\ 6x - 12y = 0 \end{cases}$

19. $\begin{cases} 4x + y = 11 \\ 2x + 5y = 1 \end{cases}$

20. $\begin{cases} 3x + y = -14 \\ 4x + 3y = -22 \end{cases}$

21. $\begin{cases} x + 2y + 5 = -4 + 5y - x \\ 2x + x = y + 4 \end{cases}$

(*Hint:* First simplify each equation.)

22. $\begin{cases} 5x + 4y - 2 = -6 + 7y - 3x \\ 3x + 4x = y + 3 \end{cases}$

(*Hint:* See Exercise 21.)

23. $\begin{cases} 6x - 3y = 5 \\ x + 2y = 0 \end{cases}$

24. $\begin{cases} 10x - 5y = -21 \\ x + 3y = 0 \end{cases}$

25. $\begin{cases} 3x - y = 1 \\ 2x - 3y = 10 \end{cases}$

26. $\begin{cases} 2x - y = -7 \\ 4x - 3y = -11 \end{cases}$

27. $\begin{cases} -x + 2y = 10 \\ -2x + 3y = 18 \end{cases}$

28. $\begin{cases} -x + 3y = 18 \\ -3x + 2y = 19 \end{cases}$

29. $\begin{cases} 5x + 10y = 20 \\ 2x + 6y = 10 \end{cases}$

30. $\begin{cases} 6x + 3y = 12 \\ 9x + 6y = 15 \end{cases}$

31. $\begin{cases} 3x + 6y = 9 \\ 4x + 8y = 16 \end{cases}$

32. $\begin{cases} 2x + 4y = 6 \\ 5x + 10y = 16 \end{cases}$

33. $\begin{cases} \frac{1}{3}x - y = 2 \\ x - 3y = 6 \end{cases}$

34. $\begin{cases} \frac{1}{4}x - 2y = 1 \\ x - 8y = 4 \end{cases}$

35. $\begin{cases} x = \frac{3}{4}y - 1 \\ 8x - 5y = -6 \end{cases}$

36. $\begin{cases} x = \frac{5}{6}y - 2 \\ 12x - 5y = -9 \end{cases}$

Review

Write equivalent equations by multiplying both sides of each given equation by the given nonzero number. See Section 3.3.

37. $3x + 2y = 6$ by -2

38. $-x + y = 10$ by 5

39. $-4x + y = 3$ by 3

40. $5a - 7b = -4$ by -4

Add the binomials. See Section 10.4.

41. $\begin{array}{r} 3n + 6m \\ \underline{2n - 6m} \end{array}$

42. $\begin{array}{r} -2x + 5y \\ \underline{2x + 11y} \end{array}$

43. $\begin{array}{r} -5a - 7b \\ \underline{5a - 8b} \end{array}$

44. $\begin{array}{r} 9q + p \\ \underline{-9q - p} \end{array}$

Concept Extensions

Solve each system by the substitution method. First simplify each equation by combining like terms.

45. $\begin{cases} -5y + 6y = 3x + 2(x - 5) - 3x + 5 \\ 4(x + y) - x + y = -12 \end{cases}$

46. $\begin{cases} 5x + 2y - 4x - 2y = 2(2y + 6) - 7 \\ 3(2x - y) - 4x = 1 + 9 \end{cases}$

47. Explain how to identify a system with no solution when using the substitution method.

48. Occasionally, when using the substitution method, we obtain the equation $0 = 0$. Explain how this result indicates that the graphs of the equations in the system are identical.

Solve. See a Concept Check in this section.

49. As you solve the system $\begin{cases} 3x - y = -6 \\ -3x + 2y = 7 \end{cases}$, you find that $y = 1$. Is this the solution to the system.

50. As you solve the system $\begin{cases} x = 5y \\ y = 2x \end{cases}$, you find that $x = 0$ and $y = 0$. What is the solution to this system?

51. To avoid fractions, which of the equations below would you use if solving for y? Explain why.

a. $\frac{1}{2}x - 4y = \frac{3}{4}$
b. $8x - 5y = 13$
c. $7x - y = 19$

52. Give the number of solutions for a system if the graphs of the equations in the system are

a. lines intersecting in one point
b. parallel lines
c. same line

Use a graphing calculator to solve each system.

53. $\begin{cases} y = 5.1x + 14.56 \\ y = -2x - 3.9 \end{cases}$

54. $\begin{cases} y = 3.1x - 16.35 \\ y = -9.7x + 28.45 \end{cases}$

55. $\begin{cases} 3x + 2y = 14.04 \\ 5x + y = 18.5 \end{cases}$

56. $\begin{cases} x + y = -15.2 \\ -2x + 5y = -19.3 \end{cases}$

57. For the years 1973 through 2003, the annual percent y of U.S. households that used fuel oil to heat their homes is given by the equation $y = -0.50x + 21.92$, where x is the number of years after 1973. For the same period the annual percent y of U.S. households that used electricity to heat their homes is given by the equation $y = 0.71x + 11.03$, where x is the number of years after 1973. (*Source:* U.S. Census Bureau, American Housing Survey Branch)

a. Use the substitution method to solve this system of equations.

$$\begin{cases} y = -0.50x + 21.92 \\ y = 0.71x + 11.03 \end{cases}$$

(Round your final results to the nearest whole numbers.)

b. Explain the meaning of your answer to part (a).

c. Sketch a graph of the system of equations. Write a sentence describing the use of fuel oil and electricity for heating homes between 1973 and 2003.

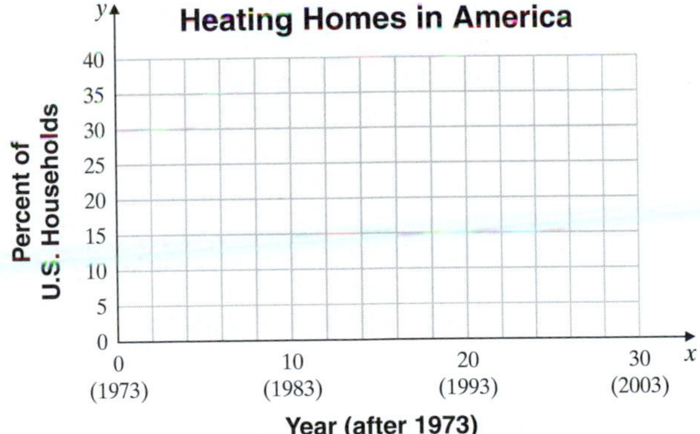

58. The number y of VHS movie format units (in billions) shipped to retailers in the United States from 1998 to 2003 is given by $y = -0.5x + 8.7$, where x is the number of years after 1998. The number y of DVD movie format units (in billions) shipped to retailers in the United States from 1998 to 2003 is given by $y = 1.67x - 0.6$, where x is the number of years after 1998. (*Source:* Business 2.0 e-magazine)

a. Use the substitution method to solve this system of equations:

$$\begin{cases} y = -0.5x + 8.7 \\ y = 1.67x - 0.6 \end{cases}$$

(Round x to the nearest tenth and y to the nearest whole.)

b. Explain the meaning of your answer to part (a).

c. Sketch a graph of the system of equations. Write a sentence describing the trends in the popularity of these two types of movie formats.

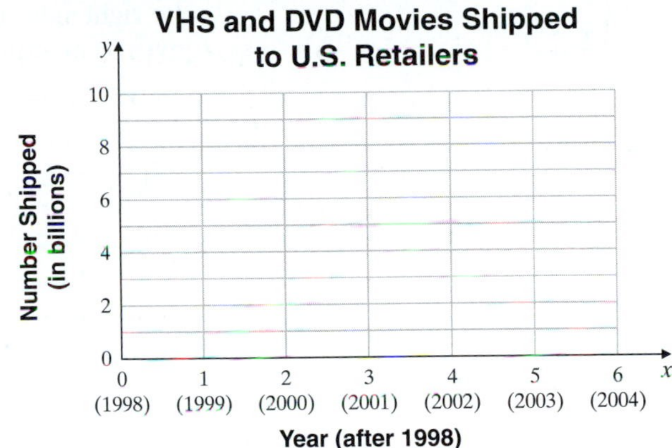

STUDY SKILLS BUILDER

How Are Your Homework Assignments Going?

Remember that it is important to keep up with homework. Why? Many concepts in mathematics build on each other. Often, your understanding of a day's lecture depends on an understanding of the previous day's material.

To complete a homework assignment, remember these 4 things:

- Attempt all of it.
- Check it.
- Correct it.
- If needed, ask questions about it.

Take a moment and review your completed homework assignments. Answer the exercises below based on this review.

1. Approximate the fraction of your homework you have attempted.
2. Approximate the fraction of your homework you have checked (if possible).
3. If you are able to check your homework, have you corrected it when errors have been found?
4. When working homework, if you do not understand a concept, what do you personally do?

Objective

A **Use the Addition Method to Solve a System of Linear Equations.**

14.3 SOLVING SYSTEMS OF LINEAR EQUATIONS BY ADDITION

Objective A Using the Addition Method

We have seen that substitution is an accurate method for solving a system of linear equations. Another accurate method is the **addition** or **elimination method.** The addition method is based on the addition property of equality: Adding equal quantities to both sides of an equation does not change the solution of the equation. In symbols,

$$\text{if } A = B \text{ and } C = D, \text{ then } A + C = B + D$$

To see how we use this to solve a system of equations, study Example 1.

PRACTICE PROBLEM 1

Use the addition method to solve the system:

$$\begin{cases} x + y = 13 \\ x - y = 5 \end{cases}$$

EXAMPLE 1 Solve the system: $\begin{cases} x + y = 7 \\ x - y = 5 \end{cases}$

Solution: Since the left side of each equation is equal to its right side, we are adding equal quantities when we add the left sides of the equations together and the right sides of the equations together. This adding eliminates the variable y and gives us an equation in one variable, x. We can then solve for x.

$x + y = 7$ First equation

$x - y = 5$ Second equation

$2x = 12$ Add the equations to eliminate y.

$x = 6$ Divide both sides by 2.

Helpful Hint

Notice in Example 1 that our goal when solving a system of equations by the addition method is to eliminate a variable when adding the equations.

The x-value of the solution is 6. To find the corresponding y-value, we let $x = 6$ in either equation of the system. We will use the first equation.

$x + y = 7$ First equation

$6 + y = 7$ Let $x = 6$.

$y = 1$ Solve for y.

The solution is (6, 1).

Check: Check the solution in both equations of the original system.

First Equation

$x + y = 7$

$6 + 1 \stackrel{?}{=} 7$ Let $x = 6$ and $y = 1$.

$7 = 7$ True

Second Equation

$x - y = 5$

$6 - 1 \stackrel{?}{=} 5$ Let $x = 6$ and $y = 1$.

$5 = 5$ True

Thus, the solution of the system is (6, 1).

If we graph the two equations in the system, we have two lines that intersect at the point (6, 1) as shown.

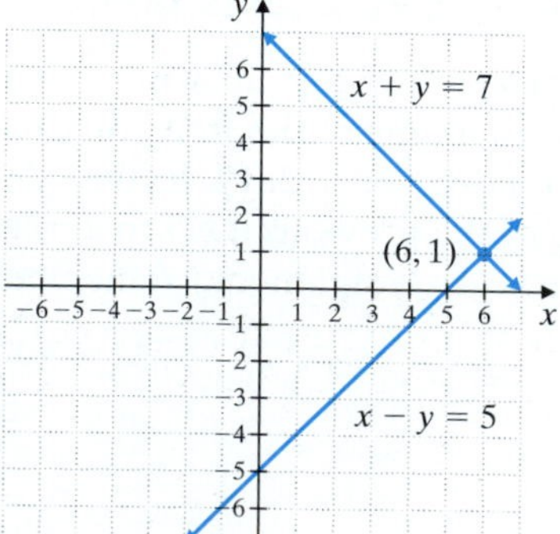

Work Practice Problem 1

Answer

1. (9, 4)

EXAMPLE 2 Solve the system: $\begin{cases} -2x + y = 2 \\ -x + 3y = -4 \end{cases}$

Solution: If we simply add these two equations, the result is still an equation in two variables. However, from Example 1, remember that our goal is to eliminate one of the variables so that we have an equation in the other variable. To do this, notice what happens if we multiply *both sides* of the first equation by -3. We are allowed to do this by the multiplication property of equality. Then the system

$$\begin{cases} -3(-2x + y) = -3(2) \\ -x + 3y = -4 \end{cases} \quad \text{simplifies to} \quad \begin{cases} 6x - 3y = -6 \\ -x + 3y = -4 \end{cases}$$

When we add the resulting equations, the y variable is eliminated.

$$\begin{aligned} 6x - 3y &= -6 \\ -x + 3y &= -4 \\ \hline 5x \quad\quad &= -10 \quad \text{Add.} \\ x &= -2 \quad \text{Divide both sides by 5.} \end{aligned}$$

To find the corresponding y-value, we let $x = -2$ in either of the original equations. We use the first equation of the original system.

$$\begin{aligned} -2x + y &= 2 \quad \text{First equation} \\ -2(-2) + y &= 2 \quad \text{Let } x = -2. \\ 4 + y &= 2 \\ y &= -2 \end{aligned}$$

Check the ordered pair $(-2, -2)$ in both equations of the *original* system. The solution is $(-2, -2)$.

Work Practice Problem 2

Helpful Hint
When finding the second value of an ordered pair solution, any equation equivalent to one of the original equations in the system may be used.

In Example 2, the decision to multiply the first equation by -3 was no accident. **To eliminate a variable** when adding two equations, **the coefficient of the variable in one equation must be the opposite of its coefficient in the other equation.**

Helpful Hint
Be sure to multiply *both sides* of an equation by a chosen number when solving by the addition method. A common mistake is to multiply only the side containing the variables.

EXAMPLE 3 Solve the system: $\begin{cases} 2x - y = 7 \\ 8x - 4y = 1 \end{cases}$

Solution: When we multiply both sides of the first equation by -4, the resulting coefficient of x is -8. This is the opposite of 8, the coefficient of x in the second equation. Then the system

$$\begin{cases} -4(2x - y) = -4(7) \\ 8x - 4y = 1 \end{cases} \quad \text{simplifies to}$$

Helpful Hint
Don't forget to multiply both sides by -4.

$$\begin{aligned} \begin{cases} -8x + 4y = -28 \\ 8x - 4y = 1 \end{cases} \\ \hline 0 = -27 \quad \text{Add the equations.} \end{aligned}$$

Continued on next page

PRACTICE PROBLEM 2

Solve the system:
$\begin{cases} 2x - y = -6 \\ -x + 4y = 17 \end{cases}$

PRACTICE PROBLEM 3

Solve the system:
$\begin{cases} x - 3y = -2 \\ -3x + 9y = 5 \end{cases}$

Answers
2. $(-1, 4)$ **3.** no solution

When we add the equations, both variables are eliminated and we have $0 = -27$, a false statement. This means that the system has no solution. The equations, if graphed, represent parallel lines.

Work Practice Problem 3

PRACTICE PROBLEM 4

Solve the system:

$$\begin{cases} 2x + 5y = 1 \\ -4x - 10y = -2 \end{cases}$$

EXAMPLE 4 Solve the system: $\begin{cases} 3x - 2y = 2 \\ -9x + 6y = -6 \end{cases}$

Solution: First we multiply both sides of the first equation by 3 and then we add the resulting equations.

$$\begin{cases} 3(3x - 2y) = 3(2) \\ -9x + 6y = -6 \end{cases} \quad \text{simplifies to} \quad \begin{cases} 9x - 6y = 6 \\ -9x + 6y = -6 \end{cases} \quad \text{Add the equations.}$$

$$0 = 0$$

Both variables are eliminated and we have $0 = 0$, a true statement. This means that the system has an infinite number of solutions. The equations, if graphed, are the same line.

Work Practice Problem 4

✔ **Concept Check** Suppose you are solving the system

$$\begin{cases} 3x + 8y = -5 \\ 2x - 4y = 3 \end{cases}$$

You decide to use the addition method by multiplying both sides of the second equation by 2. In which of the following was the multiplication performed correctly? Explain.

a. $4x - 8y = 3$ **b.** $4x - 8y = 6$

In the next example, we multiply both equations by numbers so that coefficients of a variable are opposites.

PRACTICE PROBLEM 5

Solve the system:

$$\begin{cases} 4x + 5y = 14 \\ 3x - 2y = -1 \end{cases}$$

EXAMPLE 5 Solve the system: $\begin{cases} 3x + 4y = 13 \\ 5x - 9y = 6 \end{cases}$

Solution: We can eliminate the variable y by multiplying the first equation by 9 and the second equation by 4. Then we add the resulting equations.

$$\begin{cases} 9(3x + 4y) = 9(13) \\ 4(5x - 9y) = 4(6) \end{cases} \quad \text{simplifies to} \quad \begin{cases} 27x + 36y = 117 \\ 20x - 36y = 24 \end{cases}$$

$$47x = 141 \quad \text{Add the equations.}$$

$$x = 3 \quad \text{Solve for } x.$$

To find the corresponding y-value, we let $x = 3$ in one of the original equations of the system. Doing so in any of these equations will give $y = 1$. Check to see that $(3, 1)$ satisfies each equation in the original system. The solution is $(3, 1)$.

Work Practice Problem 5

If we had decided to eliminate x instead of y in Example 5, the first equation could have been multiplied by 5 and the second by -3. Try solving the original system this way to check that the solution is $(3, 1)$.

The following steps summarize how to solve a system of linear equations by the addition method.

Answers

4. infinite number of solutions
5. $(1, 2)$

✔ **Concept Check Answer**

b; answers may vary

To Solve a System of Two Linear Equations by the Addition Method

Step 1: Rewrite each equation in standard form $Ax + By = C$.

Step 2: If necessary, multiply one or both equations by a nonzero number so that the coefficients of a chosen variable in the system are opposites.

Step 3: Add the equations.

Step 4: Find the value of one variable by solving the resulting equation from Step 3.

Step 5: Find the value of the second variable by substituting the value found in Step 4 into either of the original equations.

Step 6: Check the proposed solution in the original system.

✔Concept Check Suppose you are solving the system

$$\begin{cases} -4x + 7y = 6 \\ x + 2y = 5 \end{cases}$$

by the addition method.

a. What step(s) should you take if you wish to eliminate x when adding the equations?

b. What step(s) should you take if you wish to eliminate y when adding the equations?

EXAMPLE 6 Solve the system: $\begin{cases} -x - \dfrac{y}{2} = \dfrac{5}{2} \\ \dfrac{x}{6} - \dfrac{y}{2} = 0 \end{cases}$

Solution: We begin by clearing each equation of fractions. To do so, we multiply both sides of the first equation by the LCD 2 and both sides of the second equation by the LCD 6. Then the system

$$\begin{cases} 2\left(-x - \dfrac{y}{2}\right) = 2\left(\dfrac{5}{2}\right) \\ 6\left(\dfrac{x}{6} - \dfrac{y}{2}\right) = 6(0) \end{cases} \quad \text{simplifies to} \quad \begin{cases} -2x - y = 5 \\ x - 3y = 0 \end{cases}$$

We can now eliminate the variable x by multiplying the second equation by 2.

$$\begin{cases} -2x - y = 5 \\ 2(x - 3y) = 2(0) \end{cases} \quad \text{simplifies to} \quad \begin{cases} -2x - y = 5 \\ 2x - 6y = 0 \end{cases}$$

$$-7y = 5 \quad \text{Add the equations.}$$

$$y = -\frac{5}{7} \quad \text{Solve for } y.$$

To find x, we could replace y with $-\frac{5}{7}$ in one of the equations with two variables. Instead, let's go back to the simplified system and multiply by appropriate factors to eliminate the variable y and solve for x. To do this, we multiply the first equation by -3. Then the system

$$\begin{cases} -3(-2x - y) = -3(5) \\ x - 3y = 0 \end{cases} \quad \text{simplifies to} \quad \begin{cases} 6x + 3y = -15 \\ x - 3y = 0 \end{cases}$$

$$7x = -15 \quad \text{Add the equations.}$$

$$x = -\frac{15}{7} \quad \text{Solve for } x.$$

Check the ordered pair $\left(-\frac{15}{7}, -\frac{5}{7}\right)$ in both equations of the original system. The solution is $\left(-\frac{15}{7}, -\frac{5}{7}\right)$.

Work Practice Problem 6

PRACTICE PROBLEM 6

Solve the system:

$$\begin{cases} -\dfrac{x}{3} + y = \dfrac{4}{3} \\ \dfrac{x}{2} - \dfrac{5}{2}y = -\dfrac{1}{2} \end{cases}$$

Answer

6. $\left(-\frac{17}{2}, -\frac{3}{2}\right)$

✔ Concept Check Answer

a. multiply the second equation by 4

b. possible answer: multiply the first equation by -2 and the second equation by 7

14.3 EXERCISE SET

FOR EXTRA HELP

PH Math/Tutor Center

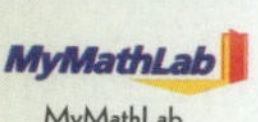

Objective A *Solve each system of equations by the addition method. See Example 1.*

1. $\begin{cases} 3x + y = 5 \\ 6x - y = 4 \end{cases}$

2. $\begin{cases} 4x + y = 13 \\ 2x - y = 5 \end{cases}$

3. $\begin{cases} x - 2y = 8 \\ -x + 5y = -17 \end{cases}$

4. $\begin{cases} x - 2y = -11 \\ -x + 5y = 23 \end{cases}$

Solve each system of equations by the addition method. If a system contains fractions or decimals, you may want to first clear each equation of fractions or decimals. See Examples 2 through 6.

5. $\begin{cases} 3x + y = -11 \\ 6x - 2y = -2 \end{cases}$

6. $\begin{cases} 4x + y = -13 \\ 6x - 3y = -15 \end{cases}$

7. $\begin{cases} 3x + 2y = 11 \\ 5x - 2y = 29 \end{cases}$

8. $\begin{cases} 4x + 2y = 2 \\ 3x - 2y = 12 \end{cases}$

9. $\begin{cases} x + 5y = 18 \\ 3x + 2y = -11 \end{cases}$

10. $\begin{cases} x + 4y = 14 \\ 5x + 3y = 2 \end{cases}$

11. $\begin{cases} x + y = 6 \\ x - y = 6 \end{cases}$

12. $\begin{cases} x - y = 1 \\ -x + 2y = 0 \end{cases}$

13. $\begin{cases} 2x + 3y = 0 \\ 4x + 6y = 3 \end{cases}$

14. $\begin{cases} 3x + y = 4 \\ 9x + 3y = 6 \end{cases}$

15. $\begin{cases} -x + 5y = -1 \\ 3x - 15y = 3 \end{cases}$

16. $\begin{cases} 2x + y = 6 \\ 4x + 2y = 12 \end{cases}$

17. $\begin{cases} 3x - 2y = 7 \\ 5x + 4y = 8 \end{cases}$

18. $\begin{cases} 6x - 5y = 25 \\ 4x + 15y = 13 \end{cases}$

19. $\begin{cases} 8x = -11y - 16 \\ 2x + 3y = -4 \end{cases}$

20. $\begin{cases} 10x + 3y = -12 \\ 5x = -4y - 16 \end{cases}$

21. $\begin{cases} 4x - 3y = 7 \\ 7x + 5y = 2 \end{cases}$

22. $\begin{cases} -2x + 3y = 10 \\ 3x + 4y = 2 \end{cases}$

23. $\begin{cases} 4x - 6y = 8 \\ 6x - 9y = 12 \end{cases}$

24. $\begin{cases} 9x - 3y = 12 \\ 12x - 4y = 18 \end{cases}$

25. $\begin{cases} 2x - 5y = 4 \\ 3x - 2y = 4 \end{cases}$

26. $\begin{cases} 6x - 5y = 7 \\ 4x - 6y = 7 \end{cases}$

27. $\begin{cases} \frac{x}{3} + \frac{y}{6} = 1 \\ \frac{x}{2} - \frac{y}{4} = 0 \end{cases}$

28. $\begin{cases} \frac{x}{2} + \frac{y}{8} = 3 \\ x - \frac{y}{4} = 0 \end{cases}$

29. $\begin{cases} \frac{10}{3}x + 4y = -4 \\ 5x + 6y = -6 \end{cases}$

30. $\begin{cases} \frac{3}{2}x + 4y = 1 \\ 9x + 24y = 5 \end{cases}$

31. $\begin{cases} x - \frac{y}{3} = -1 \\ -\frac{x}{2} + \frac{y}{8} = \frac{1}{4} \end{cases}$

32. $\begin{cases} 2x - \dfrac{3y}{4} = -3 \\ x + \dfrac{y}{9} = \dfrac{13}{3} \end{cases}$

33. $-4(x + 2) = 3y$
$2x - 2y = 3$

34. $-9(x + 3) = 8y$
$3x - 3y = 8$

35. $\begin{cases} \dfrac{x}{3} - y = 2 \\ -\dfrac{x}{2} + \dfrac{3y}{2} = -3 \end{cases}$

36. $\begin{cases} \dfrac{x}{2} + \dfrac{y}{4} = 1 \\ -\dfrac{x}{4} - \dfrac{y}{8} = 1 \end{cases}$

37. $\begin{cases} \dfrac{3}{5}x - y = -\dfrac{4}{5} \\ 3x + \dfrac{y}{2} = -\dfrac{9}{5} \end{cases}$

38. $\begin{cases} 3x + \dfrac{7}{2}y = \dfrac{3}{4} \\ -\dfrac{x}{2} + \dfrac{5}{3}y = -\dfrac{5}{4} \end{cases}$

39. $\begin{cases} 3.5x + 2.5y = 17 \\ -1.5x - 7.5y = -33 \end{cases}$

40. $\begin{cases} -2.5x - 6.5y = 47 \\ 0.5x - 4.5y = 37 \end{cases}$

41. $\begin{cases} 0.02x + 0.04y = 0.09 \\ -0.1x + 0.3y = 0.8 \end{cases}$

42. $\begin{cases} 0.04x - 0.05y = 0.105 \\ 0.2x - 0.6y = 1.05 \end{cases}$

Review

Translating *Rewrite each sentence using mathematical symbols. Do not solve the equations. See Sections 9.3 and 9.4.*

43. Twice a number, added to 6, is 3 less than the number.

44. The sum of three consecutive integers is 66.

45. Three times a number, subtracted from 20, is 2.

46. Twice the sum of 8 and a number is the difference of the number and 20.

47. The product of 4 and the sum of a number and 6 is twice the number.

48. If the quotient of twice a number and 7 is subtracted from the reciprocal of the number, the result is 2.

Concept Extensions

Solve. See a Concept Check in this section.

49. To solve this system by the addition method and eliminate the variable y,

$$\begin{cases} 4x + 2y = -7 \\ 3x - y = -12 \end{cases}$$

by what value would you multiply the second equation? What do you get when you complete the multiplication?

Given the system of linear equations $\begin{cases} 3x - y = -8 \\ 5x + 3y = 2 \end{cases}$

50. Use the addition method and

a. Solve the system by eliminating x.

b. Solve the system by eliminating y.

51. Suppose you are solving the system

$$\begin{cases} 3x + 8y = -5 \\ 2x - 4y = 3. \end{cases}$$

You decide to use the addition method by multiplying both sides of the second equation by 2. In which of the following was the multiplication performed correctly? Explain.

a. $4x - 8y = 3$
b. $4x - 8y = 6$

52. Suppose you are solving the system

$$\begin{cases} -2x - y = 0 \\ -2x + 3y = 6. \end{cases}$$

You decide to use the addition method by multiplying both sides of the first equation by 3, then adding the resulting equation to the second equation. Which of the following is the correct sum? Explain.

a. $-8x = 6$
b. $-8x = 9$

53. When solving a system of equations by the addition method, how do we know when the system has no solution?

54. Explain why the addition method might be preferred over the substitution method for solving the system $\begin{cases} 2x - 3y = 5 \\ 5x + 2y = 6. \end{cases}$

55. Use the system of linear equations below to answer the questions.

$$\begin{cases} x + y = 5 \\ 3x + 3y = b \end{cases}$$

a. Find the value of b so that the system has an infinite number of solutions.
b. Find a value of b so that there are no solutions to the system.

56. Use the system of linear equations below to answer the questions.

$$\begin{cases} x + y = 4 \\ 2x + by = 8 \end{cases}$$

a. Find the value of b so that the system has an infinite number of solutions.
b. Find a value of b so that the system has a single solution.

Solve each system by the addition method.

57. $\begin{cases} 2x + 3y = 14 \\ 3x - 4y = -69.1 \end{cases}$

58. $\begin{cases} 5x - 2y = -19.8 \\ -3x + 5y = -3.7 \end{cases}$

59. Two occupations predicted to greatly increase in number of jobs are medical assistants and computer software engineers. The number of medical assistant jobs predicted for 2002 through 2012 can be approximated by $21.4x - y = -365$. The number of computer software engineer jobs for the same years can be approximated by $17.9x - y = -394$. For both equations, x is the number of years since 2002 and y is the number of jobs in thousands.

a. Use the addition method to solve this system of equations:

$$\begin{cases} 21.4x - y = -365 \\ 17.9x - y = -394 \end{cases}$$

(Round answer to the nearest whole number.)

b. Interpret your solution from part (a).
c. Use the year in your answer to part (b) and estimate the number of medical assistant jobs and computer software engineer jobs in that year.

60. In recent years, the number of daily newspapers printed as morning editions has been increasing and the number of daily newspapers printed as evening editions has been decreasing. The number y of daily morning newspapers in existence from 1993 through 2003 is given by the equation $88x - 5y = -3498$, where x is the number of years since 1993. The number y of daily evening newspapers in existence from 1993 through 2003 is given by the equation $291x + 10y = 9940$, where x is the number of years since 1993. (*Source:* Newspaper Association of America)

a. Use the addition method to solve this system of equations:

$$\begin{cases} 88x - 5y = -3498 \\ 291x + 10y = 9940. \end{cases}$$

(Round to the nearest whole number. Because of rounding, the y-value of your ordered-pair solution may vary.)

b. Interpret your solution from part (a).
c. How many of each type of newspaper were in existence in that year?

INTEGRATED REVIEW Sections 14.1–14.3

Summary on Solving Systems of Equations

Solve each system by either the addition method or the substitution method.

1. $\begin{cases} 2x - 3y = -11 \\ y = 4x - 3 \end{cases}$

2. $\begin{cases} 4x - 5y = 6 \\ y = 3x - 10 \end{cases}$

3. $\begin{cases} x + y = 3 \\ x - y = 7 \end{cases}$

4. $\begin{cases} x - y = 20 \\ x + y = -8 \end{cases}$

5. $\begin{cases} x + 2y = 1 \\ 3x + 4y = -1 \end{cases}$

6. $\begin{cases} x + 3y = 5 \\ 5x + 6y = -2 \end{cases}$

7. $\begin{cases} y = x + 3 \\ 3x = 2y - 6 \end{cases}$

8. $\begin{cases} y = -2x \\ 2x - 3y = -16 \end{cases}$

9. $\begin{cases} y = 2x - 3 \\ y = 5x - 18 \end{cases}$

10. $\begin{cases} y = 6x - 5 \\ y = 4x - 11 \end{cases}$

11. $\begin{cases} x + \frac{1}{6}y = \frac{1}{2} \\ 3x + 2y = 3 \end{cases}$

12. $\begin{cases} x + \frac{1}{3}y = \frac{5}{12} \\ 8x + 3y = 4 \end{cases}$

13. $\begin{cases} x - 5y = 1 \\ -2x + 10y = 3 \end{cases}$

14. $\begin{cases} -x + 2y = 3 \\ 3x - 6y = -9 \end{cases}$

15. $\begin{cases} 0.2x - 0.3y = -0.95 \\ 0.4x + 0.1y = 0.55 \end{cases}$

16. $\begin{cases} 0.08x - 0.04y = -0.11 \\ 0.02x - 0.06y = -0.09 \end{cases}$

17. $\begin{cases} x = 3y - 7 \\ 2x - 6y = -14 \end{cases}$

18. $\begin{cases} y = \frac{x}{2} - 3 \\ 2x - 4y = 0 \end{cases}$

19. $\begin{cases} 2x + 5y = -1 \\ 3x - 4y = 33 \end{cases}$

20. $\begin{cases} 7x - 3y = 2 \\ 6x + 5y = -21 \end{cases}$

21. Which method, substitution or addition, would you prefer to use to solve the system below? Explain your reasoning.

$$\begin{cases} 3x + 2y = -2 \\ y = -2x \end{cases}$$

22. Which method, substitution or addition, would you prefer to use to solve the system below? Explain your reasoning.

$$\begin{cases} 3x - 2y = -3 \\ 6x + 2y = 12 \end{cases}$$

Answers

1. ______
2. ______
3. ______
4. ______
5. ______
6. ______
7. ______
8. ______
9. ______
10. ______
11. ______
12. ______
13. ______
14. ______
15. ______
16. ______
17. ______
18. ______
19. ______
20. ______
21. ______
22. ______

Objective

A Use a System of Equations to Solve Problems.

14.4 SYSTEMS OF LINEAR EQUATIONS AND PROBLEM SOLVING

Objective A Using a System of Equations for Problem Solving

Many of the word problems solved earlier with one-variable equations can also be solved with two equations in two variables. We use the same problem-solving steps that have been used throughout this text. The only difference is that two variables are assigned to represent the two unknown quantities and that the problem is translated into two equations.

Problem-Solving Steps

1. UNDERSTAND the problem. During this step, become comfortable with the problem. Some ways of doing this are to

Read and reread the problem.

Choose two variables to represent the two unknowns.

Construct a drawing.

Propose a solution and check. Pay careful attention to how you check your proposed solution. This will help when writing equations to model the problem.

2. TRANSLATE the problem into two equations.

3. SOLVE the system of equations.

4. INTERPRET the results: *Check* the proposed solution in the stated problem and *state* your conclusion.

PRACTICE PROBLEM 1

Find two numbers whose sum is 50 and whose difference is 22.

EXAMPLE 1 Finding Unknown Numbers

Find two numbers whose sum is 37 and whose difference is 21.

Solution:

1. UNDERSTAND. Read and reread the problem. Suppose that one number is 20. If their sum is 37, the other number is 17 because $20 + 17 = 37$. Is their difference 21? No; $20 - 17 = 3$. Our proposed solution is incorrect, but we now have a better understanding of the problem.

Since we are looking for two numbers, we let

x = first number and

y = second number

2. TRANSLATE. Since we have assigned two variables to this problem, we translate our problem into two equations.

In words:	two numbers whose sum	is	37
	↓	↓	↓
Translate:	$x + y$	$=$	37
In words:	two numbers whose difference	is	21
	↓	↓	↓
Translate:	$x - y$	$=$	21

Answer

1. 36 and 14

3. SOLVE. Now we solve the system.

$$\begin{cases} x + y = 37 \\ x - y = 21 \end{cases}$$

Notice that the coefficients of the variable y are opposites. Let's then solve by the addition method and begin by adding the equations.

$$\begin{aligned} x + y &= 37 \\ x - y &= 21 \\ \hline 2x \quad &= 58 \quad \text{Add the equations.} \\ x &= 29 \quad \text{Divide both sides by 2.} \end{aligned}$$

Now we let $x = 29$ in the first equation to find y.

$$\begin{aligned} x + y &= 37 \quad \text{First equation} \\ 29 + y &= 37 \\ y &= 8 \quad \text{Subtract 29 from both sides.} \end{aligned}$$

4. INTERPRET. The solution of the system is (29, 8).

Check: Notice that the sum of 29 and 8 is $29 + 8 = 37$, the required sum. Their difference is $29 - 8 = 21$, the required difference.

State: The numbers are 29 and 8.

Work Practice Problem 1

EXAMPLE 2 Solving a Problem about Prices

The Cirque du Soleil show Varekai is performing locally. Matinee admission for 4 adults and 2 children is $374, while admission for 2 adults and 3 children is $285.

a. What is the price of an adult's ticket?

b. What is the price of a child's ticket?

c. Suppose that a special rate of $1000 is offered for groups of 20 persons. Should a group of 4 adults and 16 children use the group rate? Why or why not?

Solution:

1. UNDERSTAND. Read and reread the problem and guess a solution. Let's suppose that the price of an adult's ticket is $50 and the price of a child's ticket is $40. To check our proposed solution, let's see if admission for 4 adults and 2 children is $374. Admission for 4 adults is 4($50) or $200 and admission for 2 children is 2($40) or $80. This gives a total admission of $200 + $80 = $280, not the required $374. Again though, we have accomplished the purpose of this process: We have a better understanding of the problem. To continue, we let

A = the price of an adult's ticket and

C = the price of a child's ticket

Continued on next page

PRACTICE PROBLEM 2

Admission prices at a local weekend fair were $5 for children and $7 for adults. The total money collected was $3379, and 587 people attended the fair. How many children and how many adults attended the fair?

Answer

2. 365 children and 222 adults

2. TRANSLATE. We translate the problem into two equations using both variables.

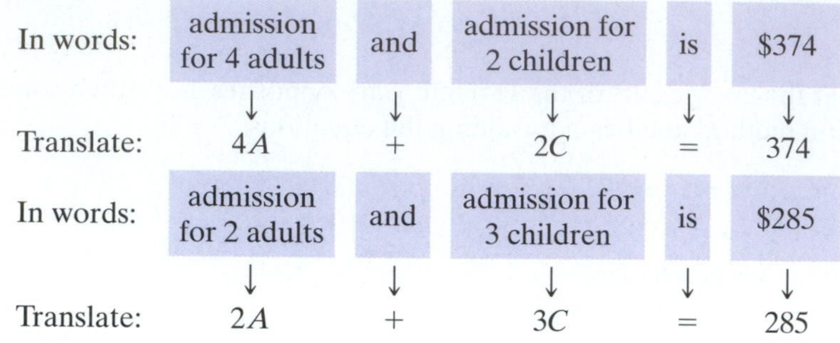

In words:	admission for 4 adults	and	admission for 2 children	is	\$374
	↓	↓	↓	↓	↓
Translate:	$4A$	$+$	$2C$	$=$	374
In words:	admission for 2 adults	and	admission for 3 children	is	\$285
	↓	↓	↓	↓	↓
Translate:	$2A$	$+$	$3C$	$=$	285

3. SOLVE. We solve the system.

$$\begin{cases} 4A + 2C = 374 \\ 2A + 3C = 285 \end{cases}$$

Since both equations are written in standard form, we solve by the addition method. First we multiply the second equation by -2 so that when we add the equations we eliminate the variable A. Then the system

$$\begin{cases} 4A + 2C = 374 \\ -2(2A + 3C) = -2(285) \end{cases} \text{ simplifies to } \begin{cases} 4A + 2C = 374 \\ -4A - 6C = -570 \end{cases}$$

Add the equations.

$$-4C = -196$$

$C = 49$ or \$49, the children's ticket price.

To find A, we replace C with 49 in the first equation.

$4A + 2C = 374$ First equation

$4A + 2(49) = 374$ Let $C = 49$.

$4A + 98 = 374$

$4A = 276$

$A = 69$ or \$69, the adult's ticket price.

4. INTERPRET.

Check: Notice that 4 adults and 2 children will pay

$4(\$69) + 2(\$49) = \$276 + \$98 = \$374$, the required amount. Also, the price for 2 adults and 3 children is $2(\$69) + 3(\$49) = \$138 + \$147 = \$285$, the required amount.

State: Answer the three original questions.

a. Since $A = 69$, the price of an adult's ticket is \$69.

b. Since $C = 49$, the price of a child's ticket is \$49.

c. The regular admission price for 4 adults and 16 children is

$$\begin{aligned} 4(\$69) + 16(\$49) &= \$276 + \$784 \\ &= \$1060 \end{aligned}$$

This is \$60 more than the special group rate of \$1000, so they should request the group rate.

Work Practice Problem 2

EXAMPLE 3 Finding Rates

As part of an exercise program, Louisa and Alfredo start walking each morning. They live 15 miles away from each other. They decide to meet one day by walking toward one another. After 2 hours they meet. If Louisa walks one mile per hour faster than Alfredo, find both walking speeds.

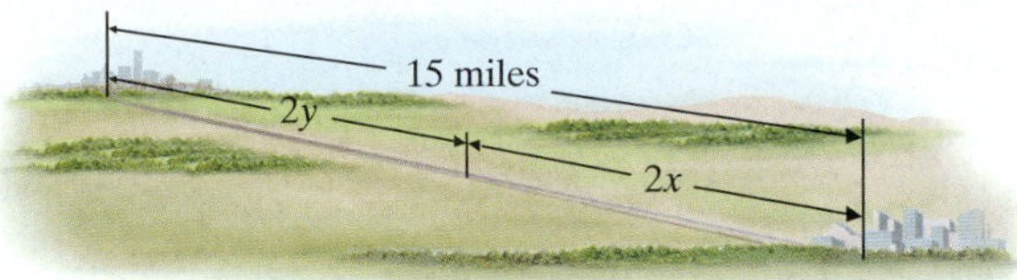

Solution:

1. UNDERSTAND. Read and reread the problem. Let's propose a solution and use the formula $d = r \cdot t$ to check. Suppose that Louisa's rate is 4 miles per hour. Since Louisa's rate is 1 mile per hour faster, Alfredo's rate is 3 miles per hour. To check, see if they can walk a total of 15 miles in 2 hours. Louisa's distance is rate $\cdot$ time $= 4(2) = 8$ miles and Alfredo's distance is rate $\cdot$ time $= 3(2) = 6$ miles. Their total distance is 8 miles + 6 miles = 14 miles, not the required 15 miles. Now that we have a better understanding of the problem, let's model it with a system of equations.

   First, we let

   x = Alfredo's rate in miles per hour and

   y = Louisa's rate in miles per hour

   Now we use the facts stated in the problem and the formula $d = rt$ to fill in the following chart.

	r	· t	= d
Alfredo	x	2	$2x$
Louisa	y	2	$2y$

2. TRANSLATE. We translate the problem into two equations using both variables.

In words:	Alfredo's distance	+	Louisa's distance	=	15 miles
	↓		↓		↓
Translate:	$2x$	+	$2y$	=	15

In words:	Louisa's rate	is	1 mile per hour faster than Alfredo's
	↓	↓	↓
Translate:	y	=	$x + 1$

3. SOLVE. The system of equations we are solving is

$$\begin{cases} 2x + 2y = 15 \\ y = x + 1 \end{cases}$$

Continued on next page

PRACTICE PROBLEM 3

Two cars are 440 miles apart and traveling toward each other. They meet in 3 hours. If one car's speed is 10 miles per hour faster than the other car's speed, find the speed of each car.

	r	· t	= d
Faster Car			
Slower Car			

Answer

3. One car's speed is $68\frac{1}{3}$ mph and the other car's speed is $78\frac{1}{3}$ mph.

Let's use substitution to solve the system since the second equation is solved for y.

$$2x + 2y = 15 \quad \text{First equation}$$

$$2x + 2(x + 1) = 15 \quad \text{Replace } y \text{ with } x + 1.$$

$$2x + 2x + 2 = 15$$

$$4x = 13$$

$$x = \frac{13}{4} = 3\frac{1}{4} \text{ or } 3.25$$

$$y = x + 1 = 3\frac{1}{4} + 1 = 4\frac{1}{4} \text{ or } 4.25$$

4. INTERPRET. Alfredo's proposed rate is $3\frac{1}{4}$ miles per hour and Louisa's proposed rate is $4\frac{1}{4}$ miles per hour.

Check: Use the formula $d = rt$ and find that in 2 hours, Alfredo's distance is (3.25)(2) miles or 6.5 miles. In 2 hours, Louisa's distance is (4.25)(2) miles or 8.5 miles. The total distance walked is 6.5 miles + 8.5 miles or 15 miles, the given distance.

State: Alfredo walks at a rate of 3.25 miles per hour and Louisa walks at a rate of 4.25 miles per hour.

Work Practice Problem 3

PRACTICE PROBLEM 4

Barb Hayes, a pharmacist, needs 50 liters of a 60% alcohol solution. She currently has available a 20% solution and a 70% solution. How many liters of each must she use to make the needed 50 liters of 60% alcohol solution?

EXAMPLE 4 Finding Amounts of Solutions

Eric Daly, a chemistry teaching assistant, needs 10 liters of a 20% saline solution (salt water) for his 2 p.m. laboratory class. Unfortunately, the only mixtures on hand are a 5% saline solution and a 25% saline solution. How much of each solution should he mix to produce the 20% solution?

Solution:

1. UNDERSTAND. Read and reread the problem. Suppose that we need 4 liters of the 5% solution. Then we need 10 − 4 = 6 liters of the 25% solution. To see if this gives us 10 liters of a 20% saline solution, let's find the amount of pure salt in each solution.

	concentration rate	×	amount of solution	=	amount of pure salt
	↓		↓		↓
5% solution:	0.05	×	4 liters	=	0.2 liters
25% solution:	0.25	×	6 liters	=	1.5 liters
20% solution:	0.20	×	10 liters	=	2 liters

Since 0.2 liters + 1.5 liters = 1.7 liters, not 2 liters, our proposed solution is incorrect. But we have gained some insight into how to model and check this problem.

We let

x = number of liters of 5% solution and

y = number of liters of 25% solution

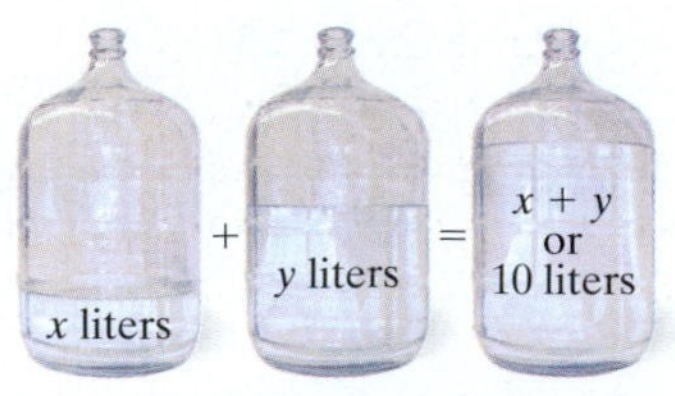

5% saline solution 25% saline solution 20% saline solution

Answer

4. 10 liters of the 20% alcohol solution and 40 liters of the 70% alcohol solution

Now we use a table to organize the given data.

	Concentration Rate	Liters of Solution	Liters of Pure Salt
First Solution	5%	x	$0.05x$
Second Solution	25%	y	$0.25y$
Mixture Needed	20%	10	(0.20)(10)

2. TRANSLATE. We translate into two equations using both variables.

In words:	liters of 5% solution	+	liters of 25% solution	=	10 liters
	↓		↓		↓
Translate:	x	+	y	=	10
In words:	salt in 5% solution	+	salt in 25% solution	=	salt in mixture
	↓		↓		↓
Translate:	$0.05x$	+	$0.25y$	=	(0.20) (10)

3. SOLVE. Here we solve the system

$$\begin{cases} x + y = 10 \\ 0.05x + 0.25y = 2 \end{cases}$$

To solve by the addition method, we first multiply the first equation by −25 and the second equation by 100. Then the system

$$\begin{cases} -25(x + y) = -25(10) \\ 100(0.05x + 0.25y) = 100(2) \end{cases} \quad \text{simplifies to} \quad \begin{cases} -25x - 25y = -250 \\ 5x + 25y = 200 \\ \hline -20x = -50 \quad \text{Add.} \\ x = 2.5 \end{cases}$$

To find y, we let $x = 2.5$ in the first equation of the original system.

$$x + y = 10$$
$$2.5 + y = 10 \quad \text{Let } x = 2.5.$$
$$y = 7.5$$

4. INTERPRET. Thus, we propose that Eric needs to mix 2.5 liters of 5% saline solution with 7.5 liters of 25% saline solution.

Check: Notice that $2.5 + 7.5 = 10$, the required number of liters. Also, the sum of the liters of salt in the two solutions equals the liters of salt in the required mixture:

$$0.05(2.5) + 0.25(7.5) = 0.20(10)$$
$$0.125 + 1.875 = 2$$

State: Eric needs 2.5 liters of the 5% saline solution and 7.5 liters of the 25% saline solution.

Work Practice Problem 4

✔Concept Check Suppose you mix an amount of a 30% acid solution with an amount of a 50% acid solution. Which of the following acid strengths would be possible for the resulting acid mixture?

a. 22% **b.** 44% **c.** 63%

✔ Concept Check Answer

b

14.4 EXERCISE SET

FOR EXTRA HELP

Student Solutions Manual

PH Math/Tutor Center

CD/Video for Review

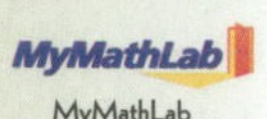
MathXL®

MyMathLab

Without actually solving each problem, choose the correct solution by deciding which choice satisfies the given conditions.

△ **1.** The length of a rectangle is 3 feet longer than the width. The perimeter is 30 feet. Find the dimensions of the rectangle.

a. length = 8 feet; width = 5 feet
b. length = 8 feet; width = 7 feet
c. length = 9 feet; width = 6 feet

△ **2.** An isosceles triangle, a triangle with two sides of equal length, has a perimeter of 20 inches. Each of the equal sides is one inch longer than the third side. Find the lengths of the three sides.

a. 6 inches, 6 inches, and 7 inches
b. 7 inches, 7 inches, and 6 inches
c. 6 inches, 7 inches, and 8 inches

3. Two computer disks and three notebooks cost \$17. However, five computer disks and four notebooks cost \$32. Find the price of each.

a. notebook = \$4; computer disk = \$3
b. notebook = \$3; computer disk = \$4
c. notebook = \$5; computer disk = \$2

4. Two music CDs and four DVDs cost a total of \$40. However, three music CDs and five DVDs cost \$55. Find the price of each.

a. CD = \$12; DVD = \$4
b. CD = \$15; DVD = \$2
c. CD = \$10; DVD = \$5

5. Kesha has a total of 100 coins, all of which are either dimes or quarters. The total value of the coins is \$13.00. Find the number of each type of coin.

a. 80 dimes; 20 quarters
b. 20 dimes; 44 quarters
c. 60 dimes; 40 quarters

6. Samuel has 28 gallons of saline solution available in two large containers at his pharmacy. One container holds three times as much as the other container. Find the capacity of each container.

a. 15 gallons; 5 gallons
b. 20 gallons; 8 gallons
c. 21 gallons; 7 gallons

Objective A *Write a system of equations describing each situation. Do not solve the system. See Example 1.*

7. Two numbers add up to 15 and have a difference of 7.

8. The total of two numbers is 16. The first number plus 2 more than 3 times the second equals 18.

9. Keiko has a total of \$6500, which she has invested in two accounts. The larger account is \$800 greater than the smaller account.

10. Dominique has four times as much money in his savings account as in his checking account. The total amount is \$2300.

Solve. See Examples 1 through 4.

11. Two numbers total 83 and have a difference of 17. Find the two numbers.

12. The sum of two numbers is 76 and their difference is 52. Find the two numbers.

13. A first number plus twice a second number is 8. Twice the first number plus the second totals 25. Find the numbers.

14. One number is 4 more than twice the second number. Their total is 25. Find the numbers.

15. The highest scorer during the WNBA 2004 regular season was Lauren Jackson of the Seattle Storm. Over the season, Jackson scored 36 more points than the second-highest scorer, Lisa Leslie of the Los Angeles Sparks. Together, Jackson and Leslie scored 1232 points during the 2004 regular season. How many points did each player score over the course of the season? (*Source:* Women's National Basketball Association)

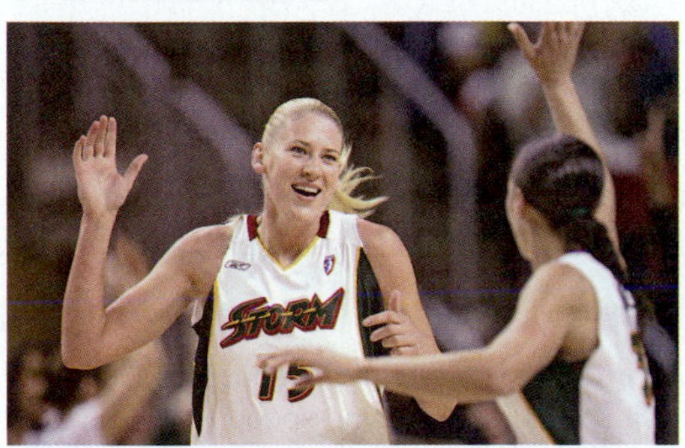

16. Ilya Kovalchuk of the Atlanta Thrashers was tied for the title NHL's leading goal scorer during the 2003–2004 regular season. Bill Guerin of the Dallas Stars, who was ranked ninth for goals, scored 7 fewer goals than Kovalchuk. Together, these two players made a total of 75 goals during the 2003–2004 regular season. How many goals each did Kovalchuk and Guerin make? (*Source:* National Hockey League)

17. Ann Marie Jones has been pricing Amtrak train fares for a group trip to New York. Three adults and four children must pay \$159. Two adults and three children must pay \$112. Find the price of an adult's ticket, and find the price of a child's ticket.

18. Last month, Jerry Papa purchased five DVDs and two CDs at Wall-to-Wall Sound for \$65. This month he bought three DVDs and four CDs for \$81. Find the price of each DVD, and find the price of each CD.

19. Johnston and Betsy Waring have a jar containing 80 coins, all of which are either quarters or nickels. The total value of the coins is \$14.60. How many of each type of coin do they have?

20. Sarah and Keith Robinson purchased 40 stamps, a mixture of 37¢ and 23¢ stamps. Find the number of each type of stamp if they spent \$14.10.

21. Norman and Suzanne Scarpulla own 35 shares of McDonald's stock and 69 shares of The Ohio Art Company stock (makers of Etch A Sketch and other toys). At the close of the markets on a particular day in 2004, their stock portfolio consisting of these two stocks was worth \$1551.00. The closing price of the McDonald's stock was \$25 more per share than the closing price of The Ohio Art Company stock on that day. What was the closing price of each stock on that day? (*Source:* Yahoo finance)

22. Saralee Rose has an investment in Google and Nintendo stock. On a particular day in 2004, Google stock closed at \$169.98 per share, and Nintendo stock closed at \$115.40 per share. Saralee's portfolio made up of these two stocks was worth \$8712.66 at the end of the day. If Saralee owns 16 more shares of Google stock than she owns of Nintendo stock, how many shares of each type of stock does she own?

23. Twice last month, Judy Carter rented a car from Enterprise in Fresno, California, and traveled around the Southwest on business. Enterprise rents its cars for a daily fee, plus an additional charge per mile driven. Judy recalls that her first trip lasted 4 days, she drove 450 miles, and the rental cost her \$240.50. On her second business trip she drove 200 miles in 3 days, and paid \$146.00 for the rental. Find the daily fee and the mileage charge.

24. Joan Gundersen rented a car from Hertz, which rents its cars for a daily fee plus an additional charge per mile driven. Joan recalls that a car rented for 5 days and driven for 300 miles cost her \$178, while a car rented for 4 days and driven for 500 miles cost \$197. Find the daily fee, and find the mileage charge.

25. Pratap Puri rowed 18 miles down the Delaware River in 2 hours, but the return trip took him $4\frac{1}{2}$ hours. Find the rate Pratap can row in still water, and find the rate of the current.

Let x = rate Pratap can row in still water and
y = rate of the current

	d =	r	· t
Downstream		$x + y$	
Upstream		$x - y$	

26. The Jonathan Schultz family took a canoe 10 miles down the Allegheny River in $1\frac{1}{4}$ hours. After lunch it took them 4 hours to return. Find the rate of the current.

Let x = rate the family can row in still water and
y = rate of the current

	d =	r	· t
Downstream		$x + y$	
Upstream		$x - y$	

27. Dave and Sandy Hartranft are frequent flyers with Delta Airlines. They often fly from Philadelphia to Chicago, a distance of 780 miles. On one particular trip they fly into the wind, and the flight takes 2 hours. The return trip, with the wind behind them, only takes $1\frac{1}{2}$ hours. Find the speed of the wind and find the speed of the plane in still air.

28. With a strong wind behind it, a United Airlines jet flies 2400 miles from Los Angeles to Orlando in $4\frac{3}{4}$ hours. The return trip takes 6 hours, as the plane flies into the wind. Find the speed of the plane in still air, and find the wind speed to the nearest tenth of a mile per hour.

29. Jim Williamson began a 186-mile bicycle trip to build up stamina for a triathlete competition. Unfortunately, his bicycle chain broke, so he finished the trip walking. The whole trip took 6 hours. If Jim walks at a rate of 4 miles per hour and rides at 40 miles per hour, find the amount of time he spent on the bicycle.

30. In Canada, eastbound and westbound trains travel along the same track, with sidings to pull onto to avoid accidents. Two trains are now 150 miles apart, with the westbound train traveling twice as fast as the eastbound train. A warning must be issued to pull one train onto a siding or else the trains will crash in $1\frac{1}{4}$ hours. Find the speed of the eastbound train and the speed of the westbound train.

31. Dorren Schmidt is a chemist with Gemco Pharmaceutical. She needs to prepare 12 ounces of a 9% hydrochloric acid solution. Find the amount of a 4% solution and the amount of a 12% solution she should mix to get this solution.

Concentration Rate	Liters of Solution	Liters of Pure Acid
0.04	x	$0.04x$
0.12	y	?
0.09	12	?

32. Elise Everly is preparing 15 liters of a 25% saline solution. Elise has two other saline solutions with strengths of 40% and 10%. Find the amount of 40% solution and the amount of 10% solution she should mix to get 15 liters of a 25% solution.

Concentration Rate	Liters of Solution	Liters of Pure Salt
0.40	x	$0.40x$
0.10	y	?
0.25	15	?

33. Wayne Osby blends coffee for a local coffee café. He needs to prepare 200 pounds of blended coffee beans selling for \$3.95 per pound. He intends to do this by blending together a high-quality bean costing \$4.95 per pound and a cheaper bean costing \$2.65 per pound. To the nearest pound, find how much high-quality coffee bean and how much cheaper coffee bean he should blend.

34. Macadamia nuts cost an astounding \$16.50 per pound, but research by an independent firm says that mixed nuts sell better if macadamias are included. The standard mix costs \$9.25 per pound. Find how many pounds of macadamias and how many pounds of the standard mix should be combined to produce 40 pounds that will cost \$10 per pound. Find the amounts to the nearest tenth of a pound.

△ **35.** Recall that two angles are complementary if the sum of their measures is 90°. Find the measures of two complementary angles if one angle is twice the other.

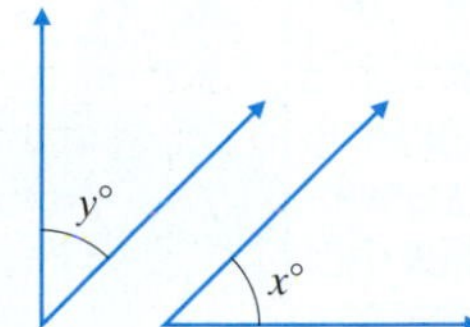

△ **36.** Recall that two angles are supplementary if the sum of their measures is 180°. Find the measures of two supplementary angles if one angle is 20° more than four times the other.

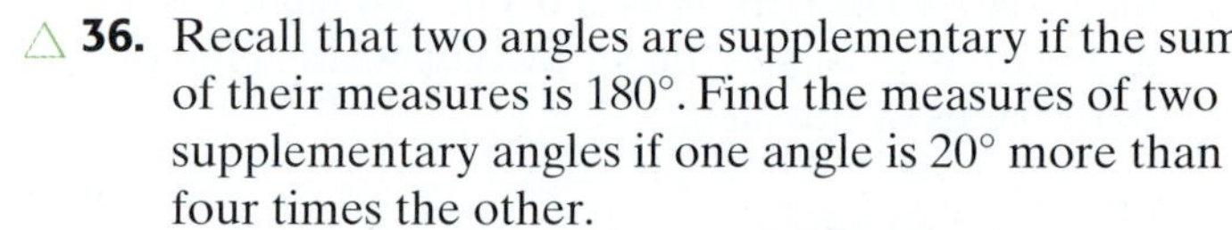
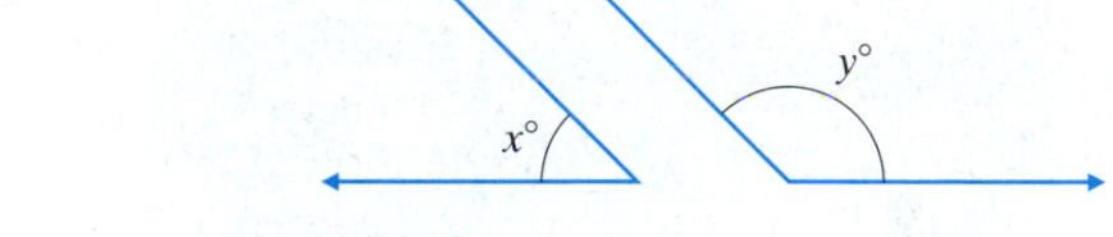

△ **37.** Find the measures of two complementary angles if one angle is 10° more than three times the other.

△ **38.** Find the measures of two supplementary angles if one angle is 18° more than twice the other.

39. Kathi and Robert Hawn had a pottery stand at the annual Skippack Craft Fair. They sold some of their pottery at the original price of \$9.50 each, but later decreased the price of each by \$2. If they sold all 90 pieces and took in \$721, find how many they sold at the original price and how many they sold at the reduced price.

40. A charity fund-raiser consisted of a spaghetti supper where a total of 387 people were fed. They charged \$6.80 for adults and half-price for children. If they took in \$2444.60, find how many adults and how many children attended the supper.

41. The Santa Fe National Historic Trail is approximately 1200 miles between Old Franklin, Missouri, and Santa Fe, New Mexico. Suppose that a group of hikers start from each town and walk the trail toward each other. They meet after a total hiking time of 240 hours. If one group travels $\frac{1}{2}$ mile per hour slower than the other group, find the rate of each group. (*Source:* National Park Service)

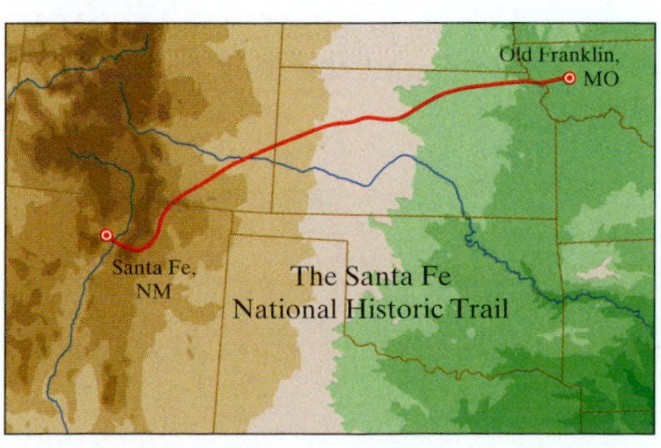

42. California 1 South is a historic highway that stretches 123 miles along the coast from Monterey to Morro Bay. Suppose that two cars start driving this highway, one from each town. They meet after 3 hours. Find the rate of each car if one car travels 1 mile per hour faster than the other car. (*Source:* National Geographic)

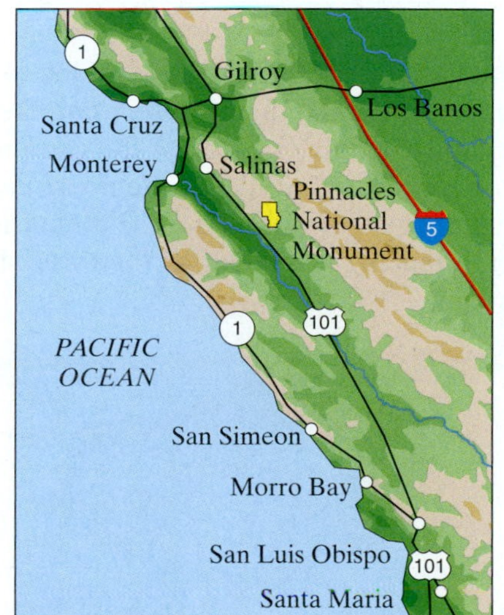

43. A 30% solution of fertilizer is to be mixed with a 60% solution of fertilizer in order to get 150 gallons of a 50% solution. How many gallons of the 30% solution and 60% solution should be mixed?

44. A 10% acid solution is to be mixed with a 50% acid solution in order to get 120 ounces of a 20% acid solution. How many ounces of the 10% solution and 50% solution should be mixed?

45. Traffic signs are regulated by the *Manual on Uniform Traffic Control Devices* (MUTCD). According to this manual, if the sign below is placed on a freeway, its perimeter must be 144 inches. Also, its length is 12 inches longer than its width. Find the dimensions of this sign.

46. According to the MUTCD (see Exercise 45), this sign must have a perimeter of 60 inches. Also, its length must be 6 inches longer than its width. Find the perimeter of this sign.

Review

Find the square of each expression. For example, the square of 7 is 7^2 or 49. The square of 5x is $(5x)^2$ or $25x^2$. See Section 10.1.

47. 4 **48.** 3 **49.** $6x$ **50.** $11y$ **51.** $10y^3$ **52.** $8x^5$

Concept Extensions

Solve. See the Concept Check in the section.

53. Suppose you mix an amount of candy costing \$0.49 a pound with candy costing \$0.65 a pound. Which of the following costs per pound could result?

a. \$0.58 **b.** \$0.72 **c.** \$0.29

54. Suppose you mix a 50% acid solution with pure acid (100%). Which of the following acid strengths are possible for the resulting acid mixture?

a. 25% **b.** 150% **c.** 62% **d.** 90%

△ **55.** Dale and Sharon Mahnke have decided to fence off a garden plot behind their house, using their house as the "fence" along one side of the garden. The length (which runs parallel to the house) is 3 feet less than twice the width. Find the dimensions if 33 feet of fencing is used along the three sides requiring it.

△ **56.** Judy McElroy plans to erect 152 feet of fencing around her rectangular horse pasture. A river bank serves as one side length of the rectangle. If each width is 4 feet longer than half the length, find the dimensions.

CHAPTER 14 Group Activity

Break-Even Point

Sections 14.1, 14.2, 14.3, 14.4

When a business sells a new product, it generally does not start making a profit right away. There are usually many expenses associated with creating a new product. These expenses might include an advertising blitz to introduce the product to the public. These start-up expenses might also include the cost of market research and product development or any brand-new equipment needed to manufacture the product. Start-up costs like these are generally called *fixed costs* because they don't depend on the number of items manufactured. Expenses that depend on the number of items manufactured, such as the cost of materials and shipping, are called *variable costs*. The total cost of manufacturing the new product is given by the cost equation: Total cost = Fixed costs + Variable costs.

For instance, suppose a greeting card company is launching a new line of greeting cards. The company spent \$7000 doing product research and development for the new line and spent \$15,000 on advertising the new line. The company does not need to buy any new equipment to manufacture the cards, but the paper and ink needed to make each card will cost \$0.20 per card. The total cost y in dollars for manufacturing x cards is $y = 22{,}000 + 0.20x$.

Once a business sets a price for the new product, the company can find the product's expected *revenue*. Revenue is the amount of money the company takes in from the sales of its product. The revenue from selling a product is given by the revenue equation: Revenue = Price per item × Number of items sold.

For instance, suppose that the card company plans to sell its new cards for \$1.50 each. The revenue y, in dollars, that the company can expect to receive from the sales of x cards is $y = 1.50x$.

If the total cost and revenue equations are graphed on the same coordinate system, the graphs should intersect. The point of intersection is where total cost equals revenue and is called the *break-even point*. The break-even point gives the number of items x that must be manufactured and sold for the company to recover its expenses. If fewer than this number of items are produced and sold, the company loses money. If more than this number of items are produced and sold, the company makes a profit. In the case of the greeting card company, approximately 16,923 cards must be manufactured and sold for the company to break even on this new card line. The total cost and revenue of producing and selling 16,923 cards is the same. It is approximately \$25,385.

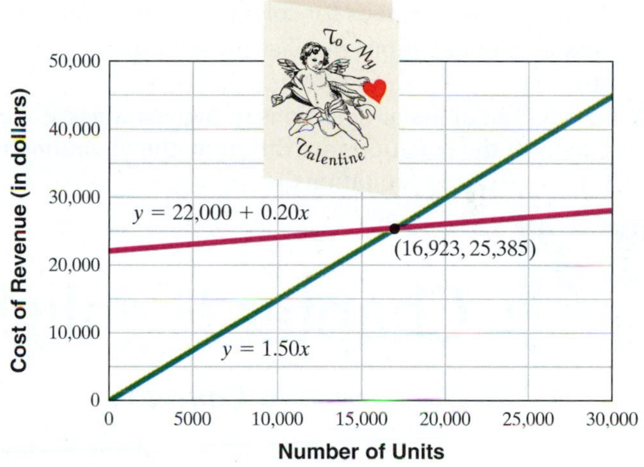

Group Activity

Suppose your group is starting a small business near your campus.

a. Choose a business and decide what campus-related product or service you will provide.

b. Research the fixed costs of starting up such a business.

c. Research the variable costs of producing such a product or providing such a service.

d. Decide how much you would charge per unit of your product or service.

e. Find a system of equations for the total cost and revenue of your product or service.

f. How many units of your product or service must be sold before your business will break even?

Chapter 14 Vocabulary Check

Fill in each blank with one of the words or phrases listed below.

system of linear equations	solution	consistent	independent
dependent	inconsistent	substitution	addition

1. In a system of linear equations in two variables, if the graphs of the equations are the same, the equations are ____________ equations.
2. Two or more linear equations are called a ________________________.
3. A system of equations that has at least one solution is called a(n) ____________ system.
4. A ___________ of a system of two equations in two variables is an ordered pair of numbers that is a solution of both equations in the system.
5. Two algebraic methods for solving systems of equations are __________ and _____________.
6. A system of equations that has no solution is called a(n) _____________ system.
7. In a system of linear equations in two variables, if the graphs of the equations are different, the equations are _____________ equations.

Are you preparing for your test? Don't forget to take the Chapter 14 Test on page 1100. Then check your answers at the back of the text and use the Chapter Test Prep Video CD to see the fully worked-out solutions to any of the exercises you want to review.

14 Chapter Highlights

Definitions and Concepts	**Examples**
Section 14.1 Solving Systems of Linear Equations by Graphing	
A **system of linear equations** consists of two or more linear equations.	$\begin{cases} 2x + y = 6 \\ x = -3y \end{cases}$ $\begin{cases} -3x + 5y = 10 \\ x - 4y = -2 \end{cases}$
A **solution** of a system of two equations in two variables is an ordered pair of numbers that is a solution of both equations in the system.	Determine whether $(-1, 3)$ is a solution of the system. $\begin{cases} 2x - y = -5 \\ x = 3y - 10 \end{cases}$ Replace x with -1 and y with 3 in both equations. $2x - y = -5$; $2(-1) - 3 \stackrel{?}{=} -5$; $-5 = -5$ True. $x = 3y - 10$; $-1 \stackrel{?}{=} 3(3) - 10$; $-1 = -1$ True. $(-1, 3)$ is a solution of the system.
Graphically, a solution of a system is a point common to the graphs of both equations.	Solve by graphing: $\begin{cases} 3x - 2y = -3 \\ x + y = 4 \end{cases}$ 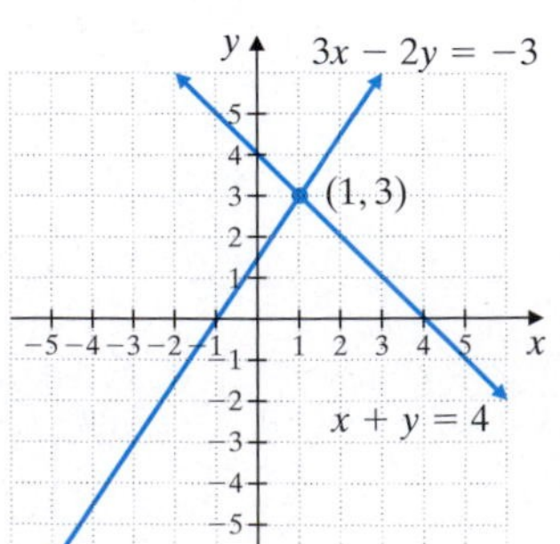

DEFINITIONS AND CONCEPTS	EXAMPLES
Section 14.1 Solving Systems of Linear Equations by Graphing (*continued*)	
Three different situations can occur when graphing the two lines associated with the equations in a linear system.	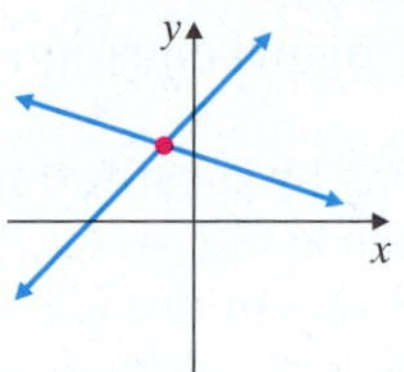 One point of intersection; one solution 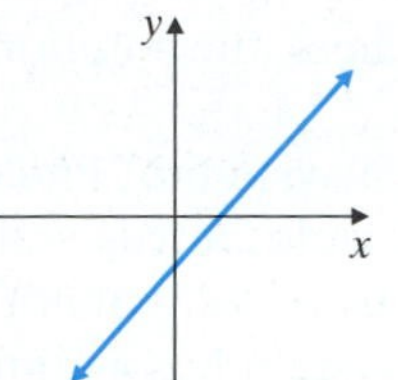 Same line; infinite number of solutions 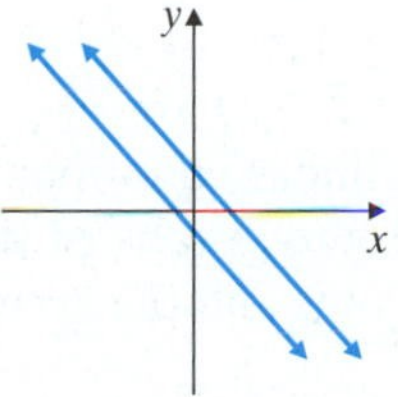 Parallel lines; no solution
Section 14.2 Solving Systems of Linear Equations by Substitution	
TO SOLVE A SYSTEM OF LINEAR EQUATIONS BY THE SUBSTITUTION METHOD **Step 1.** Solve one equation for a variable. **Step 2.** Substitute the expression for the variable into the other equation. **Step 3.** Solve the equation from Step 2 to find the value of one variable. **Step 4.** Substitute the value from Step 3 in either original equation to find the value of the other variable. **Step 5.** Check the solution in both original equations.	Solve by substitution. $\begin{cases} 3x + 2y = 1 \\ x = y - 3 \end{cases}$ Substitute $y - 3$ for x in the first equation. $3x + 2y = 1$ $3(y - 3) + 2y = 1$ $3y - 9 + 2y = 1$ $5y = 10$ $y = 2$ Divide by 5. To find x, substitute 2 for y in $x = y - 3$ so that $x = 2 - 3$ or -1. The solution $(-1, 2)$ checks.
Section 14.3 Solving Systems of Linear Equations by Addition	
TO SOLVE A SYSTEM OF LINEAR EQUATIONS BY THE ADDITION METHOD **Step 1.** Rewrite each equation in standard form $Ax + By = C$. **Step 2.** Multiply one or both equations by a nonzero number so that the coefficients of a variable are opposites. **Step 3.** Add the equations. **Step 4.** Find the value of one variable by solving the resulting equation. **Step 5.** Substitute the value from Step 4 into either original equation to find the value of the other variable.	Solve by addition. $\begin{cases} x - 2y = 8 \\ 3x + y = -4 \end{cases}$ Multiply both sides of the first equation by -3. $\begin{cases} -3x + 6y = -24 \\ 3x + y = -4 \end{cases}$ $7y = -28$ Add. $y = -4$ Divide by 7. To find x, let $y = -4$ in an original equation. $x - 2(-4) = 8$ First equation $x + 8 = 8$ $x = 0$

continued

DEFINITIONS AND CONCEPTS	EXAMPLES

Section 14.3 Solving Systems of Linear Equations by Addition (*continued*)

Step 6. Check the solution in both original equations.

The solution $(0, -4)$ checks.

If solving a system of linear equations by substitution or addition yields a true statement such as $-2 = -2$, then the graphs of the equations in the system are identical and the system has an infinite number of solutions.

Solve: $\begin{cases} 2x - 6y = -2 \\ x = 3y - 1 \end{cases}$

Substitute $3y - 1$ for x in the first equation.

$$\begin{aligned} 2(3y - 1) - 6y &= -2 \\ 6y - 2 - 6y &= -2 \\ -2 &= -2 \quad \text{True} \end{aligned}$$

The system has an infinite number of solutions.

If solving a system of linear equations yields a false statement such as $0 = 3$, the graphs of the equations in the system are parallel lines and the system has no solution.

Solve: $\begin{cases} 5x - 2y = 6 \\ -5x + 2y = -3 \end{cases}$

$$0 = 3 \quad \text{False}$$

The system has no solution.

Section 14.4 Systems of Linear Equations and Problem Solving

PROBLEM-SOLVING STEPS

1. UNDERSTAND. Read and reread the problem.

Two angles are supplementary if the sum of their measures is 180°. The larger of two supplementary angles is three times the smaller, decreased by twelve. Find the measure of each angle. Let

x = measure of smaller angle and

y = measure of larger angle

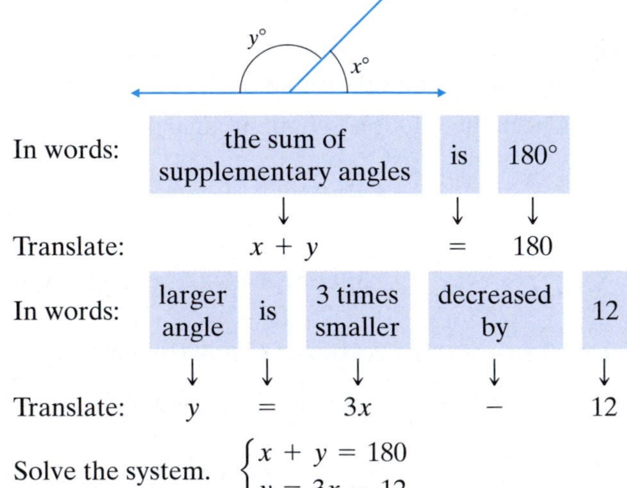

2. TRANSLATE.

In words:	the sum of supplementary angles			is	180°
	↓			↓	↓
Translate:	$x + y$			$=$	180

In words:	larger angle	is	3 times smaller	decreased by	12
	↓	↓	↓	↓	↓
Translate:	y	$=$	$3x$	$-$	12

3. SOLVE.

Solve the system. $\begin{cases} x + y = 180 \\ y = 3x - 12 \end{cases}$

Use the substitution method and replace y with $3x - 12$ in the first equation.

$$\begin{aligned} x + y &= 180 \\ x + (3x - 12) &= 180 \\ 4x &= 192 \\ x &= 48 \end{aligned}$$

4. INTERPRET.

Since $y = 3x - 12$, then $y = 3 \cdot 48 - 12$ or 132.

The solution checks. The smaller angle measures 48° and the larger angle measures 132°.

STUDY SKILLS BUILDER

Are You Prepared for a Test on Chapter 14?

Below I have listed some common trouble areas for students in Chapter 14. After studying for your test—but before taking your test—read these.

- If you are having trouble drawing a neat graph, remember to ask your instructor if you can use graph paper on your test. This will save you time and keep your graphs neat.
- Do you remember how to check solutions of systems of equations? If $(-1, 5)$ is a solution of the system

$$\begin{cases} 3x - y = -8 \\ -x + y = 6 \end{cases}$$

then the ordered pair will make *both* equations a true statement.

$$3x - y = -8$$
$$3(-1) - 5 = -8 \quad \text{Let } x = -1 \text{ and } y = 5.$$
$$-8 = -8 \quad \text{True}$$

$$-x + y = 6$$
$$-(-1) + 5 = 6 \quad \text{Let } x = -1 \text{ and } y = 5.$$
$$6 = 6 \quad \text{True}$$

Remember: This is simply a list of a few common trouble areas. For a review of Chapter 14, see the Highlights and Chapter Review at the end of this chapter.

14 CHAPTER REVIEW

(14.1) *Determine whether each ordered pair is a solution of the system of linear equations.*

1. $\begin{cases} 2x - 3y = 12 \\ 3x + 4y = 1 \end{cases}$

a. $(12, 4)$

b. $(3, -2)$

2. $\begin{cases} 4x + y = 0 \\ -8x - 5y = 9 \end{cases}$

a. $\left(\frac{3}{4}, -3\right)$

b. $(-2, 8)$

3. $\begin{cases} 5x - 6y = 18 \\ 2y - x = -4 \end{cases}$

a. $(-6, -8)$

b. $\left(3, \frac{5}{2}\right)$

4. $\begin{cases} 2x + 3y = 1 \\ 3y - x = 4 \end{cases}$

a. $(2, 2)$

b. $(-1, 1)$

Solve each system of equations by graphing.

5. $\begin{cases} x + y = 5 \\ x - y = 1 \end{cases}$

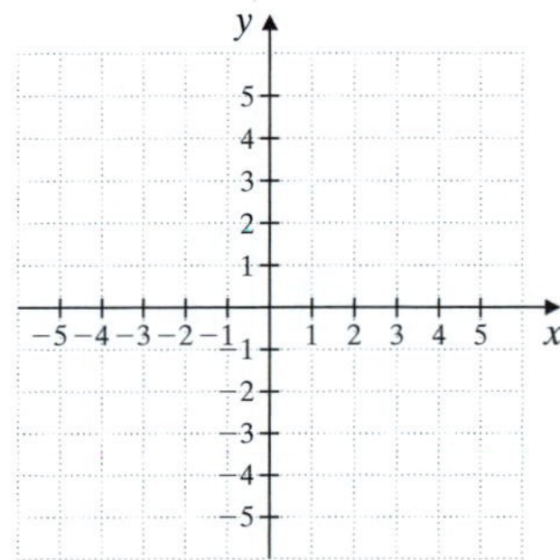

6. $\begin{cases} x + y = 3 \\ x - y = -1 \end{cases}$

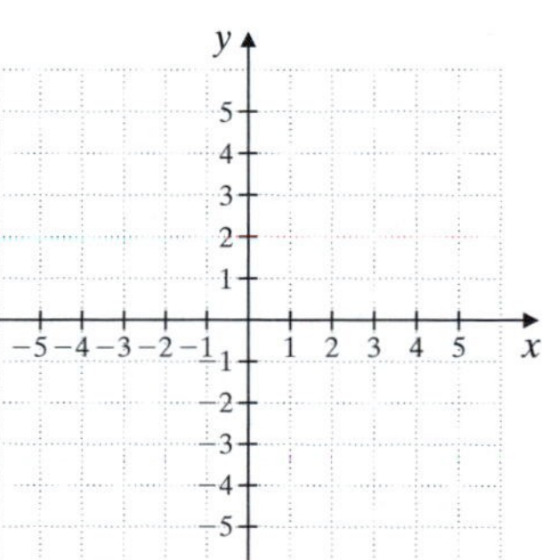

7. $\begin{cases} x = 5 \\ y = -1 \end{cases}$

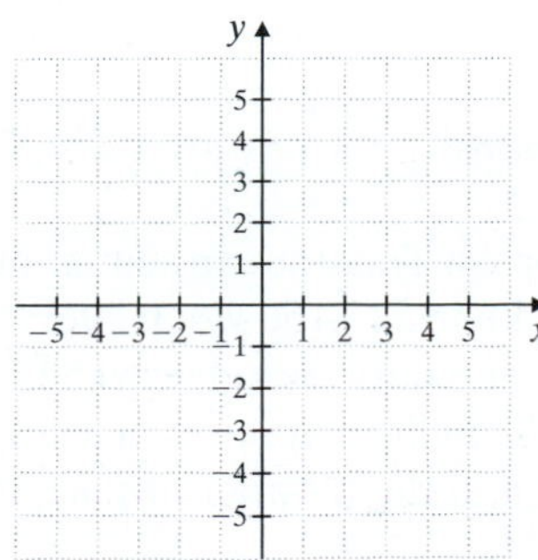

8. $\begin{cases} x = -3 \\ y = 2 \end{cases}$

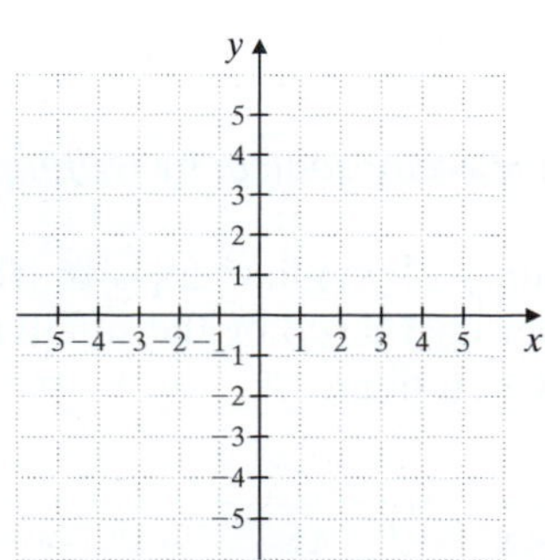

9. $\begin{cases} 2x + y = 5 \\ x = -3y \end{cases}$

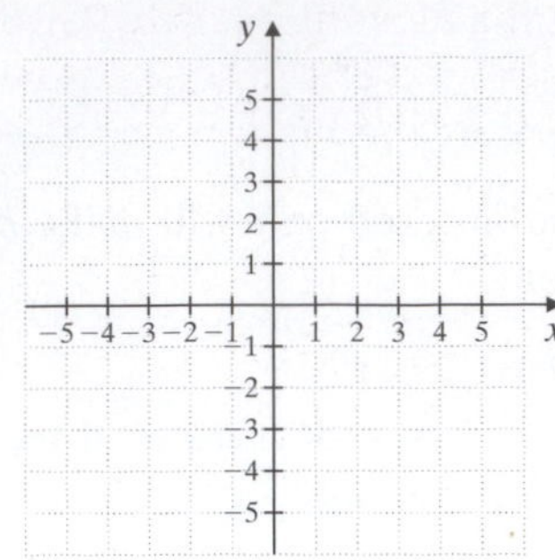

10. $\begin{cases} 3x + y = -2 \\ y = -5x \end{cases}$

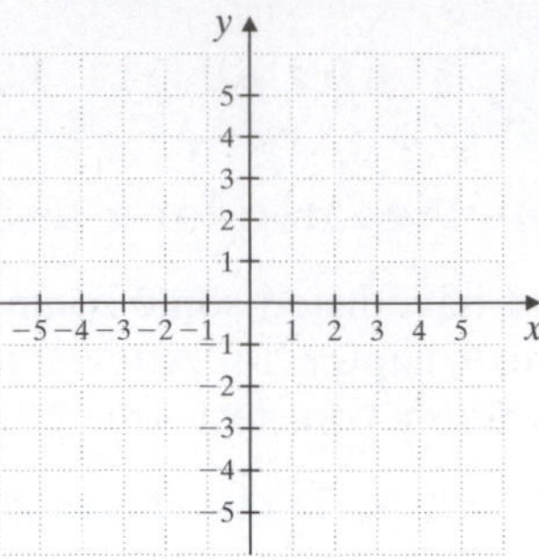

11. $\begin{cases} y = 3x \\ -6x + 2y = 6 \end{cases}$

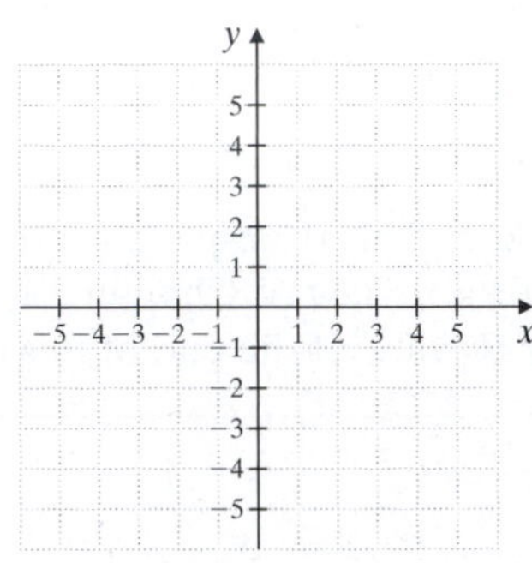

12. $\begin{cases} x - 2y = 2 \\ -2x + 4y = -4 \end{cases}$

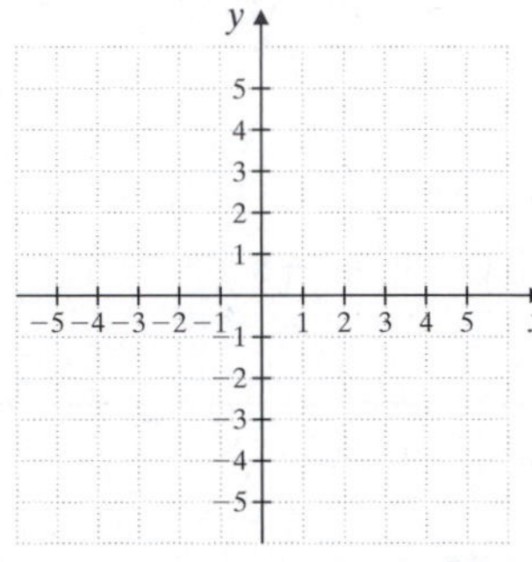

(14.2) *Solve each system of equations by the substitution method.*

13. $\begin{cases} y = 2x + 6 \\ 3x - 2y = -11 \end{cases}$

14. $\begin{cases} y = 3x - 7 \\ 2x - 3y = 7 \end{cases}$

15. $\begin{cases} x + 3y = -3 \\ 2x + y = 4 \end{cases}$

16. $\begin{cases} 3x + y = 11 \\ x + 2y = 12 \end{cases}$

17. $\begin{cases} 4y = 2x + 6 \\ x - 2y = -3 \end{cases}$

18. $\begin{cases} 9x = 6y + 3 \\ 6x - 4y = 2 \end{cases}$

19. $\begin{cases} x + y = 6 \\ y = -x - 4 \end{cases}$

20. $\begin{cases} -3x + y = 6 \\ y = 3x + 2 \end{cases}$

(14.3) *Solve each system of equations by the addition method.*

21. $\begin{cases} 2x + 3y = -6 \\ x - 3y = -12 \end{cases}$

22. $\begin{cases} 4x + y = 15 \\ -4x + 3y = -19 \end{cases}$

23. $\begin{cases} 2x - 3y = -15 \\ x + 4y = 31 \end{cases}$

24. $\begin{cases} x - 5y = -22 \\ 4x + 3y = 4 \end{cases}$

25. $\begin{cases} 2x - 6y = -1 \\ -x + 3y = \dfrac{1}{2} \end{cases}$

26. $\begin{cases} 0.6x - 0.3y = -1.5 \\ 0.04x - 0.02y = -0.1 \end{cases}$

27. $\begin{cases} \dfrac{3}{4}x + \dfrac{2}{3}y = 2 \\ x + \dfrac{y}{3} = 6 \end{cases}$

28. $\begin{cases} 10x + 2y = 0 \\ 3x + 5y = 33 \end{cases}$

(14.4) *Solve each problem by writing and solving a system of linear equations.*

29. The sum of two numbers is 16. Three times the larger number decreased by the smaller number is 72. Find the two numbers.

30. The Forrest Theater can seat a total of 360 people. They take in \$15,150 when every seat is sold. If orchestra section tickets cost \$45 and balcony tickets cost \$35, find the number of seats in the orchestra section and the number of seats in the balcony.

31. A riverboat can head 340 miles upriver in 19 hours, but the return trip takes only 14 hours. Find the current of the river and find the speed of the riverboat in still water to the nearest tenth of a mile.

	d =	r	· t
Upriver		$x - y$	
Downriver		$x + y$	

32. Find the amount of a 6% acid solution and the amount of a 14% acid solution Pat Mayfield should combine to prepare 50 cc (cubic centimeters) of a 12% solution.

33. A deli charges $3.80 for a breakfast of three eggs and four strips of bacon. The charge is $2.75 for two eggs and three strips of bacon. Find the cost of each egg and the cost of each strip of bacon.

34. An exercise enthusiast alternates between jogging and walking. He traveled 15 miles during the past 3 hours. He jogs at a rate of 7.5 miles per hour and walks at a rate of 4 miles per hour. Find how much time, to the nearest hundredth of an hour, he actually spent jogging and how much time he spent walking.

Mixed Review

Solve each system of equations by graphing.

35. $\begin{cases} x - 2y = 1 \\ 2x + 3y = -12 \end{cases}$

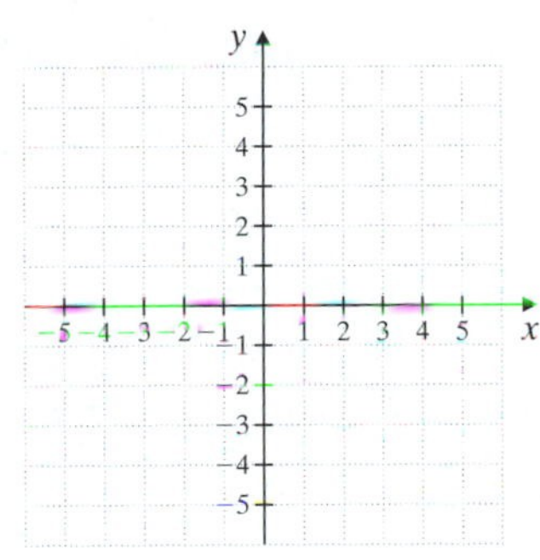

36. $\begin{cases} 3x - y = -4 \\ 6x - 2y = -8 \end{cases}$

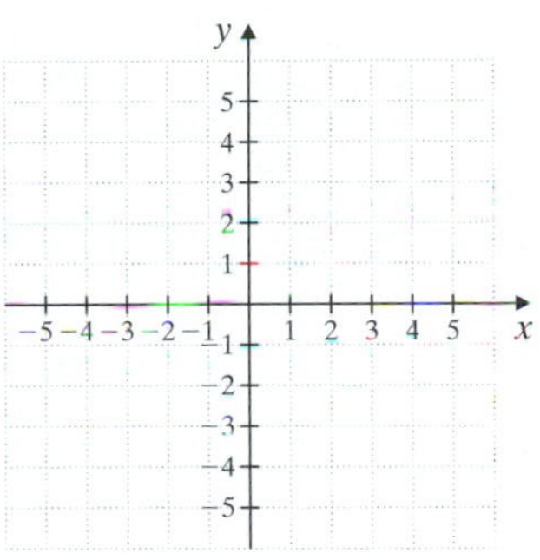

Solve each system of equations.

37. $\begin{cases} x + 4y = 11 \\ 5x - 9y = -3 \end{cases}$

38. $\begin{cases} x + 9y = 16 \\ 3x - 8y = 13 \end{cases}$

39. $\begin{cases} y = -2x \\ 4x + 7y = -15 \end{cases}$

40. $\begin{cases} 3y = 2x + 15 \\ -2x + 3y = 21 \end{cases}$

41. $\begin{cases} 3x - y = 4 \\ 4y = 12x - 16 \end{cases}$

42. $\begin{cases} x + y = 19 \\ x - y = -3 \end{cases}$

43. $\begin{cases} x - 3y = -11 \\ 4x + 5y = -10 \end{cases}$

44. $\begin{cases} -x - 15y = 44 \\ 2x + 3y = 20 \end{cases}$

45. $\begin{cases} 2x + y = 3 \\ 6x + 3y = 9 \end{cases}$

46. $\begin{cases} -3x + y = 5 \\ -3x + y = -2 \end{cases}$

Solve each problem by writing and solving a system of linear equations.

47. The sum of two numbers is 12. Three times the smaller number increased by the larger number is 20. Find the numbers.

48. The difference of two numbers is −18. Twice the smaller decreased by the larger is −23. Find the two numbers.

49. Emma Hodges has a jar containing 65 coins, all of which are either nickels or dimes. The total value of the coins is $5.30. How many of each type does she have?

50. Sarah and Owen Hebert purchased 25 stamps, a mixture of 13¢ and 22¢ stamps. Find the number of each type of stamp if they spent $4.19.

14 CHAPTER TEST

Remember to use the Chapter Test Prep Video CD to see the fully worked-out solutions to any of the exercises you want to review.

Answers

1. ______
2. ______
3. ______
4. ______
5. ______
6. ______
7. see graph
8. see graph
9. ______
10. ______
11. ______
12. ______
13. ______
14. ______

Answer each question true or false.

1. A system of two linear equations in two variables can have exactly two solutions.

2. Although $(1, 4)$ is not a solution of $x + 2y = 6$, it can still be a solution of the system $\begin{cases} x + 2y = 6 \\ x + y = 5 \end{cases}$

3. If the two equations in a system of linear equations are added and the result is $3 = 0$, the system has no solution.

4. If the two equations in a system of linear equations are added and the result is $3x = 0$, the system has no solution.

Is the ordered pair a solution of the given linear system?

5. $\begin{cases} 2x - 3y = 5 \\ 6x + y = 1 \end{cases}$; $(1, -1)$

6. 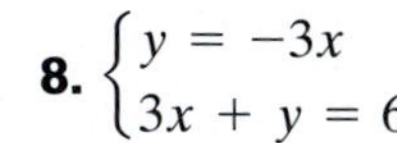 $\begin{cases} 4x - 3y = 24 \\ 4x + 5y = -8 \end{cases}$; $(3, -4)$

Solve each system by graphing.

7. $\begin{cases} x - y = 2 \\ 3x - y = -2 \end{cases}$

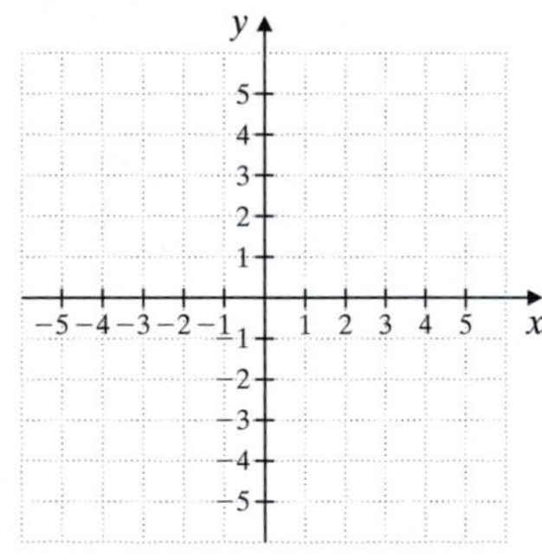

8. $\begin{cases} y = -3x \\ 3x + y = 6 \end{cases}$

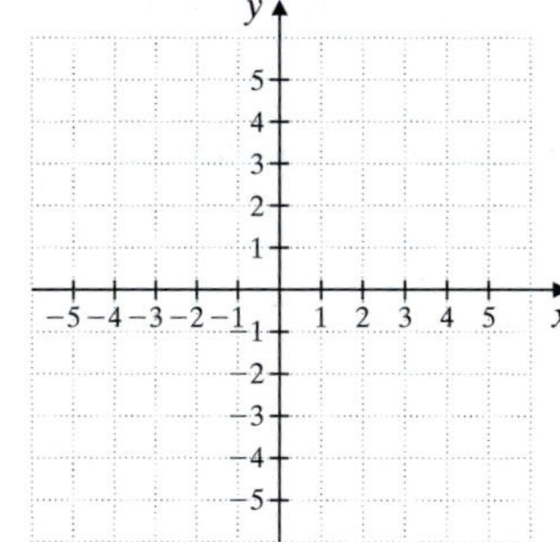

Solve each system by the substitution method.

9. $\begin{cases} 3x - 2y = -14 \\ y = x + 5 \end{cases}$

10. $\begin{cases} \frac{1}{2}x + 2y = -\frac{15}{4} \\ 4x = -y \end{cases}$

Solve each system by the addition method.

11. $\begin{cases} x + y = 28 \\ x - y = 12 \end{cases}$

12. $\begin{cases} 4x - 6y = 7 \\ -2x + 3y = 0 \end{cases}$

Solve each system using the substitution method or the addition method.

13. $\begin{cases} 3x + y = 7 \\ 4x + 3y = 1 \end{cases}$

14. $\begin{cases} 3(2x + y) = 4x + 20 \\ x - 2y = 3 \end{cases}$

15. $\begin{cases} \dfrac{x-3}{2} = \dfrac{2-y}{4} \\ \dfrac{7-2x}{3} = \dfrac{y}{2} \end{cases}$

16. $\begin{cases} 8x - 4y = 12 \\ y = 2x - 3 \end{cases}$

17. $\begin{cases} 0.01x - 0.06y = -0.23 \\ 0.2x + 0.4y = 0.2 \end{cases}$

18. $\begin{cases} x - \dfrac{2}{3}y = 3 \\ -2x + 3y = 10 \end{cases}$

Solve each problem by writing and using a system of linear equations.

19. Two numbers have a sum of 124 and a difference of 32. Find the numbers.

20. Find the amount of a 12% saline solution a lab assistant should add to 80 cc (cubic centimeters) of a 22% saline solution in order to have a 16% solution.

21. Although the number of farms in the United States is still decreasing, small farms are making a comeback. Texas and Missouri are the states with the most number of farms. Texas has 116 thousand more farms than Missouri and the total number of farms for these two states is 336 thousand. Find the number of farms for each state.

22. Two hikers start at opposite ends of the St. Tammany Trails and walk toward each other. The trail is 36 miles long and they meet in 4 hours. If one hiker is twice as fast as the other, find both hiking speeds.

The graph below shows the percent of recorded music purchases that fell within the rap/hip-hop or country music genres for the years shown. Use this graph to answer Exercises 23 and 24.

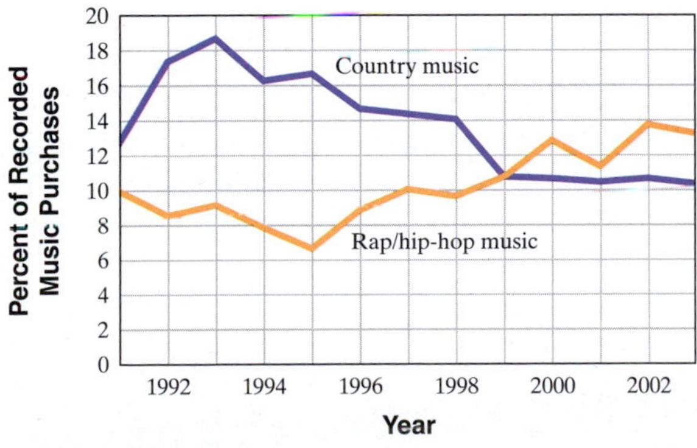

Source: Recording Industry Association of America

23. In what year were purchases of country music equal to purchases of rap/hip-hop music?

24. In what year(s) were there more purchases of country music than rap/hip-hop music?

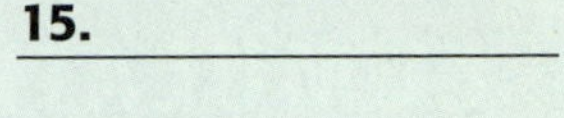
15. ______

16. ______

17. ______

18. ______

19. ______

20. ______

21. ______

22. ______

23. ______

24. ______

CUMULATIVE REVIEW

Chapters 1–14

Answers

1. ______

2. ______

3. ______

4. ______

5. ______

6. ______

7. ______

8. ______

9. ______

10. ______

11. ______

12. ______

13. ______

14. ______

15. ______

16. ______

17. ______

18. ______

19. ______

20. ______

Subtract.

1. $8 - 15$

2. $4 - 7$

3. $-4 - (-5)$

4. $3 - (-2)$

5. Solve: $7x = 6x + 4$

6. Solve: $4x = -2 + 3x$

7. Translate to an equation: 1.2 is 30% of what number?

8. Translate to an equation: 9 is 45% of what number?

9. What percent of 50 is 8?

10. What percent of 16 is 4?

11. Mr. Percy, the principal at Slidell High School, counted 31 freshmen absent during a particular day. If this is 4% of the total number of freshmen, how many freshmen are there at Slidell High School?

12. 2% of the apples in a shipment are rotten. If there are 29 rotten apples, how many apples are in the shipment?

Solve.

13. $-2(x - 5) + 10 = -3(x + 2) + x$

14. $4(4y + 2) = 2(1 + 6y) + 8$

15. Solve $-5x + 7 < 2(x - 3)$. Graph the solution set.

−5 −4 −3 −2 −1 0 1 2 3 4 5

16. Solve $-7x + 4 \leq 3(4 - x)$. Graph the solution set.

−4 −3 −2 −1 0 1 2 3 4

Simplify each expression.

17. $\left(\frac{m}{n}\right)^7$

18. $(-5x^3)(-7x^4)$

19. $\left(\frac{2x^4}{3y^5}\right)^4$

20. $\left(\frac{5x^2}{4y^3}\right)^2$

21. Subtract: $(2x^3 + 8x^2 - 6x) - (2x^3 - x^2 + 1)$

22. Subtract: $(7x + 1) - (-x - 3)$

23. Divide $6x^2 + 10x - 5$ by $3x - 1$.

24. Divide $3x^2 - x - 4$ by $x - 1$.

25. Solve: $x(2x - 7) = 4$

26. Solve: $x(x - 5) = 24$

△ **27.** Find the lengths of the sides of a right triangle if the lengths can be expressed by three consecutive even integers.

28. The sum of a number and its square is 132. Find the number.

29. Subtract: $\dfrac{2y}{2y - 7} - \dfrac{7}{2y - 7}$

30. Add: $\dfrac{x^2 + 3}{x + 9} + \dfrac{9x - 3}{x + 9}$

31. Find the slope of the line $y = -1$.

32. Find the slope of the line $x = 2$.

33. Find an equation of the line through $(2, 5)$ and $(-3, 4)$. Write the equation in the form $Ax + By = C$.

34. Find an equation of the line through $(5, -6)$ and $(-6, 5)$. Write the equation in the form $Ax + By = C$.

35. Find the domain and the range of the relation $\{(0, 2), (3, 3), (-1, 0), (3, -2)\}$.

36. Find the domain and the range of the relation $\{(2, 3), (2, 0), (2, -2), (2, 4)\}$.

21. ________

22. ________

23. ________

24. ________

25. ________

26. ________

27. ________

28. ________

29. ________

30. ________

31. ________

32. ________

33. ________

34. ________

35. ________

36. ________

37. ______

38. ______

39. ______

40. ______

41. ______

42. ______

37. Solve the system: $\begin{cases} x + 2y = 7 \\ 2x + 2y = 13 \end{cases}$

38. Solve the system: $\begin{cases} 3y = x + 6 \\ 4x + 12y = 0 \end{cases}$

39. Solve the system: $\begin{cases} -x - \dfrac{y}{2} = \dfrac{5}{2} \\ \dfrac{x}{6} - \dfrac{y}{2} = 0 \end{cases}$

40. Solve the system: $\begin{cases} x - \dfrac{3y}{8} = -\dfrac{3}{2} \\ x + \dfrac{y}{9} = \dfrac{13}{3} \end{cases}$

41. Find two numbers whose sum is 37 and whose difference is 21.

42. The sum of two numbers is 75 and their difference is 9. Find the two numbers.

15

Roots and Radicals

Having spent the last chapter studying equations, we return now to algebraic expressions. We expand on our skills of operating on expressions—adding, subtracting, multiplying, dividing, and raising to powers—to include finding roots. Just as subtraction is defined by addition and division by multiplication, finding roots is defined by raising to powers. As we master finding roots, we will work with equations that contain roots and solve problems that can be modeled by such equations.

SECTIONS

CHECK YOUR PROGRESS

When we think of pendulums, we often think of grandfather clocks. In fact, pendulums can be used to provide accurate timekeeping. But, did you know that pendulums can also be used to demonstrate that the earth rotates on its axis?

In 1851, French physicist Léon Foucault developed a special pendulum in an experiment to demonstrate that the Earth rotated on its axis. He connected his tall pendulum, capable of running for many hours, to the roof of the Paris Observatory. The pendulum's bob was able to swing back and forth in one plane, but not to twist in other directions. So, when the pendulum bob appeared to move in a circle over time, he demonstrated that it was not the pendulum but the building that moved. And since the building was firmly attached to the earth, it must be the earth rotating which created the apparent circular motion of the bob. In Section 15.1, Exercise 85, roots are used to explore the time it takes Foucault's pendulum to complete one swing of its bob.

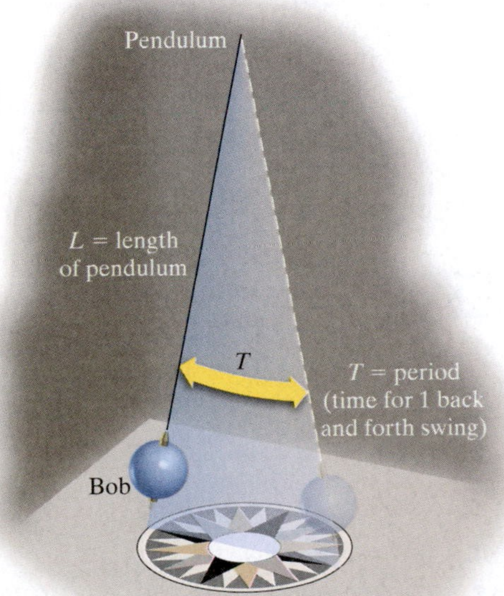

15.1 INTRODUCTION TO RADICALS

Objectives

- **A** Find Square Roots.
- **B** Find Cube Roots.
- **C** Find *n*th Roots.
- **D** Approximate Square Roots.
- **E** Simplify Radicals Containing Variables.

Objective A Finding Square Roots

In this section, we define finding the **root** of a number by its reverse operation, raising a number to a power. We begin with squares and square roots.

The *square* of 5 is $5^2 = 25$.

The *square* of -5 is $(-5)^2 = 25$

The *square* of $\frac{1}{2}$ is $\left(\frac{1}{2}\right)^2 = \frac{1}{4}$.

The reverse operation of squaring a number is finding a **square root** of a number. For example,

A *square root* of 25 is 5, because $5^2 = 25$.

A *square root* of 25 is also -5, because $(-5)^2 = 25$.

A *square root* of $\frac{1}{4}$ is $\frac{1}{2}$, because $\left(\frac{1}{2}\right)^2 = \frac{1}{4}$.

In general, the number b is a square root of a number a if $b^2 = a$.

The symbol $\sqrt{}$ is used to denote the **positive** or **principal square root** of a number. For example,

$\sqrt{25} = 5$ only, since $5^2 = 25$ and 5 is positive.

The symbol $-\sqrt{}$ is used to denote the **negative square root.** For example,

$-\sqrt{25} = -5$

The symbol $\sqrt{}$ is called a **radical** or **radical sign.** The expression within or under a radical sign is called the **radicand.** An expression containing a radical is called a **radical expression.**

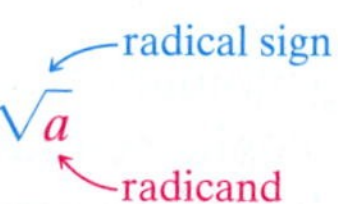

Square Root

If a is a positive number, then

$\sqrt{a}$ is the **positive square root** of a and

$-\sqrt{a}$ is the **negative square root** of a.

Also, $\sqrt{0} = 0$.

EXAMPLES Find each square root.

1. $\sqrt{36} = 6$, because $6^2 = 36$ and 6 is positive.
2. $-\sqrt{64} = -8$. The negative sign in front of the radical indicates the negative square root of 64.
3. $\sqrt{\frac{9}{100}} = \frac{3}{10}$ because $\left(\frac{3}{10}\right)^2 = \frac{9}{100}$ and $\frac{3}{10}$ is positive.
4. $\sqrt{0} = 0$ because $0^2 = 0$.
5. $\sqrt{0.36} = 0.6$ because $(0.6)^2 = 0.36$ and 0.6 is positive.

Work Practice Problems 1–5

PRACTICE PROBLEMS 1–5

Find each square root.

1. $\sqrt{100}$
2. $-\sqrt{36}$
3. $\sqrt{\frac{25}{81}}$
4. $\sqrt{1}$
5. $\sqrt{0.81}$

Answers

1. 10 2. −6 3. $\frac{5}{9}$ 4. 1 5. 0.9

Is the square root of a negative number a real number? For example, is $\sqrt{-4}$ a real number? To answer this question, we ask ourselves, is there a real number whose square is -4? Since there is no real number whose square is -4, we say that $\sqrt{-4}$ is not a real number. In general,

> A square root of a negative number is not a real number.

Study the following table to make sure you understand the differences discussed earlier.

Number	Square Roots of Number	$\sqrt{\text{number}}$	$-\sqrt{\text{number}}$
25	$-5, 5$	$\sqrt{25} = 5$ only	$-\sqrt{25} = -5$
$\frac{1}{4}$	$-\frac{1}{2}, \frac{1}{2}$	$\sqrt{\frac{1}{4}} = \frac{1}{2}$ only	$-\sqrt{\frac{1}{4}} = -\frac{1}{2}$
-9	No real square roots.	$\sqrt{-9}$ is not a real number.	

Objective B Finding Cube Roots

We can find roots other than square roots. For example, since $2^3 = 8$, we call 2 the **cube root** of 8. In symbols, we write

$\sqrt[3]{8} = 2$ The number 3 is called the **index.**

Also,

$\sqrt[3]{-64} = -4$ Since $(-4)^3 = -64$

Notice that unlike the square root of a negative number, the cube root of a negative number is a real number. This is so because while we cannot find a real number whose *square* is negative, we *can* find a real number whose *cube* is negative. In fact, the cube of a negative number is a negative number. Therefore, the cube root of a negative number is a negative number.

EXAMPLES Find each cube root.

6. $\sqrt[3]{1} = 1$ because $1^3 = 1$.
7. $\sqrt[3]{-27} = -3$ because $(-3)^3 = -27$.
8. $\sqrt[3]{\frac{1}{125}} = \frac{1}{5}$ because $\left(\frac{1}{5}\right)^3 = \frac{1}{125}$.

Work Practice Problems 6–8

PRACTICE PROBLEMS 6–8

Find each cube root.

6. $\sqrt[3]{27}$
7. $\sqrt[3]{-8}$
8. $\sqrt[3]{\frac{1}{64}}$

Objective C Finding *n*th Roots

Just as we can raise a real number to powers other than 2 or 3, we can find roots other than square roots and cube roots. In fact, we can take the nth root of a number where n is any natural number. An **nth root** of a number a is a number whose nth power is a.

In symbols, the nth root of a is written as $\sqrt[n]{a}$. Recall that n is called the **index.** The index 2 is usually omitted for square roots.

Helpful Hint

If the index is even, as it is in $\sqrt{}$, $\sqrt[4]{}$, $\sqrt[6]{}$, and so on, the radicand must be nonnegative for the root to be a real number. For example,

$\sqrt[4]{81} = 3$ but $\sqrt[4]{-81}$ is not a real number.

$\sqrt[6]{64} = 2$ but $\sqrt[6]{-64}$ is not a real number.

Answers

6. 3 7. -2 8. $\frac{1}{4}$

✔ **Concept Check** Which of the following is a real number?

a. $\sqrt{-64}$ b. $\sqrt[4]{-64}$ c. $\sqrt[5]{-64}$ d. $\sqrt[6]{-64}$

PRACTICE PROBLEMS 9–12

Find each root.

9. $\sqrt[4]{-16}$
10. $\sqrt[5]{-1}$
11. $\sqrt[4]{81}$
12. $\sqrt[6]{-1}$

EXAMPLES Find each root.

9. $\sqrt[4]{16} = 2$ because $2^4 = 16$ and 2 is positive.
10. $\sqrt[5]{-32} = -2$ because $(-2)^5 = -32$.
11. $-\sqrt[6]{1} = -1$ because $\sqrt[6]{1} = 1$.
12. $\sqrt[4]{-81}$ is not a real number since the index 4 is even and the radicand -81 is negative. In other words, there is no real number that when raised to the 4th power gives -81.

Work Practice Problems 9–12

Objective D Approximating Square Roots

Recall that numbers such as 1, 4, 9, 25, and $\frac{4}{25}$ are called **perfect squares,** since $1^2 = 1, 2^2 = 4, 3^2 = 9, 5^2 = 25$, and $\left(\frac{2}{5}\right)^2 = \frac{4}{25}$. Square roots of perfect square radicands simplify to rational numbers.

What happens when we try to simplify a root such as $\sqrt{3}$? Since 3 is not a perfect square, $\sqrt{3}$ is not a rational number. It cannot be written as a quotient of integers. It is called an **irrational number** and we can find a decimal **approximation** of it. To find decimal approximations, use a calculator or Appendix A.4. (For calculator help, see the next example or the box at the end of this section.)

PRACTICE PROBLEM 13

Use a calculator or Appendix A.4 to approximate $\sqrt{22}$ to three decimal places.

EXAMPLE 13 Use a calculator or Appendix A.4 to approximate $\sqrt{3}$ to three decimal places.

Solution: We may use Appendix A.4 or a calculator to approximate $\sqrt{3}$. To use a calculator, find the square root key $\boxed{\sqrt{\ }}$.

$\sqrt{3} \approx 1.732050808$

To three decimal places, $\sqrt{3} \approx 1.732$.

Work Practice Problem 13

Objective E Simplifying Radicals Containing Variables

Radicals can also contain variables. To simplify radicals containing variables, special care must be taken. To see how we simplify $\sqrt{x^2}$, let's look at a few examples in this form.

If $x = 3$, we have $\sqrt{3^2} = \sqrt{9} = 3$, or x.

If x is 5, we have $\sqrt{5^2} = \sqrt{25} = 5$, or x.

From these two examples, you may think that $\sqrt{x^2}$ simplifies to x. Let's now look at an example where x is a negative number. If $x = -3$, we have $\sqrt{(-3)^2} = \sqrt{9} = 3$, not -3, our original x. To make sure that $\sqrt{x^2}$ simplifies to a nonnegative number, we have the following.

For any real number a,

$$\sqrt{a^2} = |a|.$$

Answers

9. not a real number 10. -1 11. 3 12. not a real number 13. 4.690

✔ **Concept Check Answer**

c

Thus,

$\sqrt{x^2} = |x|,$

$\sqrt{(-8)^2} = |-8| = 8$

$\sqrt{(7y)^2} = |7y|,$ and so on.

To avoid this, for the rest of the chapter we assume that **if a variable appears in the radicand of a radical expression, it represents positive numbers only.** Then

$\sqrt{x^2} = |x| = x$ since x is a positive number.

$\sqrt{y^2} = y$ Because $(y)^2 = y^2$

$\sqrt{x^8} = x^4$ Because $(x^4)^2 = x^8$

$\sqrt{9x^2} = 3x$ Because $(3x)^2 = 9x^2$

EXAMPLES Simplify each expression. Assume that all variables represent positive numbers.

14. $\sqrt{z^2} = z$ because $(z)^2 = z^2$.

15. $\sqrt{x^6} = x^3$ because $(x^3)^2 = x^6$.

16. $\sqrt[3]{27y^6} = 3y^2$ because $(3y^2)^3 = 27y^6$.

17. $\sqrt{16x^{16}} = 4x^8$ because $(4x^8)^2 = 16x^{16}$.

18. $\sqrt[3]{-125a^{12}b^{15}} = -5a^4b^5$ because $(-5a^4b^5)^3 = -125a^{12}b^{15}$.

Work Practice Problems 14–18

PRACTICE PROBLEMS 14–18

Simplify each expression. Assume that all variables represent positive numbers.

14. $\sqrt{z^8}$ **15.** $\sqrt{x^{20}}$

16. $\sqrt{4x^6}$ **17.** $\sqrt[3]{8y^{12}}$

18. $\sqrt[3]{-64x^9y^{24}}$

Answers

14. z^4 **15.** x^{10} **16.** $2x^3$ **17.** $2y^4$ **18.** $-4x^3y^8$

CALCULATOR EXPLORATIONS Simplifying Square Roots

To simplify or approximate square roots using a calculator, locate the key marked $\boxed{\sqrt{}}$. To simplify $\sqrt{25}$ using a scientific calculator, press $\boxed{25}$ $\boxed{\sqrt{}}$. The display should read $\boxed{5}$. To simplify $\sqrt{25}$ using a graphing calculator, press $\boxed{\sqrt{}}$ $\boxed{25}$ $\boxed{\text{ENTER}}$.

To approximate $\sqrt{30}$, press $\boxed{30}$ $\boxed{\sqrt{}}$ (or $\boxed{\sqrt{}}$ $\boxed{30}$). The display should read $\boxed{5.477225575}$. This is an approximation for $\sqrt{30}$. A three-decimal-place approximation is

$\sqrt{30} \approx 5.477$

Is this answer reasonable? Since 30 is between perfect squares 25 and 36, $\sqrt{30}$ is between $\sqrt{25} = 5$ and $\sqrt{36} = 6$. The calculator result is then reasonable since 5.477225575 is between 5 and 6.

Use a calculator to approximate each expression to three decimal places. Decide whether each result is reasonable.

1. $\sqrt{6}$ **2.** $\sqrt{14}$

3. $\sqrt{11}$ **4.** $\sqrt{200}$

5. $\sqrt{82}$ **6.** $\sqrt{46}$

Many scientific calculators have a key, such as $\boxed{\sqrt[x]{y}}$, that can be used to approximate roots other than square roots. To approximate these roots using a graphing calculator, look under the $\boxed{\text{MATH}}$ menu or consult your manual. To use a $\boxed{\sqrt[x]{y}}$ key to find $\sqrt[3]{8}$, press $\boxed{3}$ $\boxed{\sqrt[x]{y}}$ $\boxed{8}$ (press $\boxed{\text{ENTER}}$ if needed.) The display should read $\boxed{2}$.

Use a calculator to approximate each expression to three decimal places. Decide whether each result is reasonable.

7. $\sqrt[3]{40}$ **8.** $\sqrt[3]{71}$

9. $\sqrt[4]{20}$ **10.** $\sqrt[4]{15}$

11. $\sqrt[5]{18}$ **12.** $\sqrt[6]{2}$

15.1 EXERCISE SET

FOR EXTRA HELP

Student Solutions Manual

PH Math/Tutor Center

CD/Video for Review

MathXL®

MyMathLab

Objective A *Find each square root. See Examples 1 through 5.*

1. $\sqrt{16}$ **2.** $\sqrt{64}$ **3.** $\sqrt{\frac{1}{25}}$ **4.** $\sqrt{\frac{1}{64}}$ **5.** $-\sqrt{100}$

6. $-\sqrt{36}$ **7.** $\sqrt{-4}$ **8.** $\sqrt{-25}$ **9.** $-\sqrt{121}$ **10.** $-\sqrt{49}$

11. $\sqrt{\frac{9}{25}}$ **12.** $\sqrt{\frac{4}{81}}$ **13.** $\sqrt{900}$ **14.** $\sqrt{400}$ **15.** $\sqrt{144}$

16. $\sqrt{169}$ **17.** $\sqrt{\frac{1}{100}}$ **18.** $\sqrt{\frac{1}{121}}$ **19.** $\sqrt{0.25}$ **20.** $\sqrt{0.49}$

Objective B *Find each cube root. See Examples 6 through 8.*

21. $\sqrt[3]{125}$ **22.** $\sqrt[3]{64}$ **23.** $\sqrt[3]{-64}$ **24.** $\sqrt[3]{-27}$ **25.** $-\sqrt[3]{8}$

26. $-\sqrt[3]{27}$ **27.** $\sqrt[3]{\frac{1}{8}}$ **28.** $\sqrt[3]{\frac{1}{64}}$ **29.** $\sqrt[3]{-125}$ **30.** $\sqrt[3]{-1}$

Objectives A B C Mixed Practice *Find each root. See Examples 1 through 12.*

31. $\sqrt[5]{32}$ **32.** $\sqrt[4]{81}$ **33.** $\sqrt{81}$ **34.** $\sqrt{49}$

35. $\sqrt[4]{-16}$ **36.** $\sqrt{-9}$ **37.** $\sqrt[3]{-\frac{27}{64}}$ **38.** $\sqrt[3]{-\frac{8}{27}}$

39. $-\sqrt[4]{625}$ **40.** $-\sqrt[5]{32}$ **41.** $\sqrt[6]{1}$ **42.** $\sqrt[5]{1}$

Objective D *Approximate each square root to three decimal places. See Example 13.*

43. $\sqrt{7}$ **44.** $\sqrt{10}$ **45.** $\sqrt{37}$ **46.** $\sqrt{27}$ **47.** $\sqrt{136}$ **48.** $\sqrt{8}$

49. A standard baseball diamond is a square with 90-foot sides connecting the bases. The distance from home plate to second base is $90 \cdot \sqrt{2}$ feet. Approximate $\sqrt{2}$ to two decimal places and use your result to approximate the distance $90 \cdot \sqrt{2}$ feet.

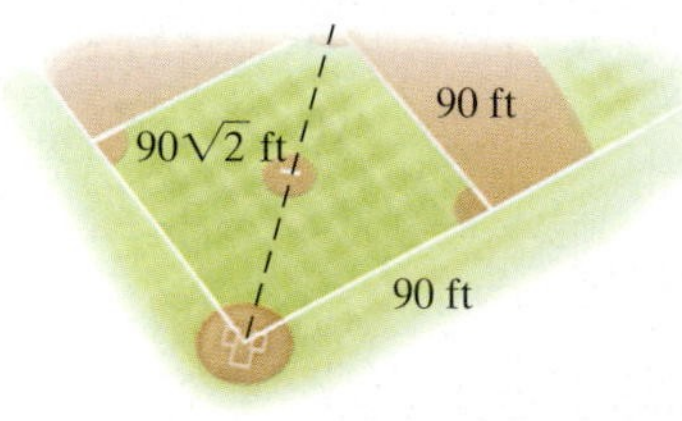

50. The roof of the warehouse shown needs to be shingled. The total area of the roof is exactly $240 \cdot \sqrt{41}$ square feet. Approximate $\sqrt{41}$ to two decimal places and use your result to approximate the area $240 \cdot \sqrt{41}$ square feet. Approximate this area to the nearest whole number.

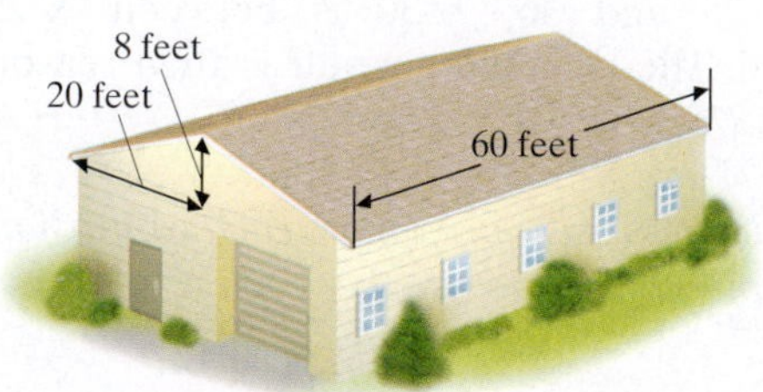

Objective **E** *Find each root. Assume that all variables represent positive numbers. See Examples 14 through 18.*

51. $\sqrt{m^2}$ **52.** $\sqrt{y^{10}}$ **53.** $\sqrt{x^4}$ **54.** $\sqrt{x^6}$

55. $\sqrt{9x^8}$ **56.** $\sqrt{36x^{12}}$ **57.** $\sqrt{81x^2}$ **58.** $\sqrt{100z^4}$

59. $\sqrt{a^2b^4}$ **60.** $\sqrt{x^{12}y^{20}}$ **61.** $\sqrt{16a^6b^4}$ **62.** $\sqrt{4m^{14}n^2}$

63. $\sqrt[3]{a^6b^{18}}$ **64.** $\sqrt[3]{x^{12}y^{18}}$ **65.** $\sqrt[3]{-8x^3y^{27}}$ **66.** $\sqrt[3]{-27a^6b^{30}}$

Review

Write each integer as a product of two integers such that one of the factors is a perfect square. For example, we can write $18 = 9 \cdot 2$, *where 9 is a perfect square. See Section 4.2.*

67. 50 **68.** 8 **69.** 32 **70.** 75

71. 28 **72.** 44 **73.** 27 **74.** 90

Concept Extensions

Solve. See the Concept Check in this section.

75. Which of the following is a real number?
a. $\sqrt[7]{-1}$ **b.** $\sqrt[3]{-125}$
c. $\sqrt[6]{-128}$ **d.** $\sqrt[8]{-1}$

76. a. $\sqrt{-1}$ **b.** $\sqrt[3]{-1}$
c. $\sqrt[4]{-1}$ **d.** $\sqrt[5]{-1}$

The length of a side of a square in given by the expression $\sqrt{A}$, *where A is the square's area. Use this expression for Exercises 77 through 80. Be sure to attach the appropriate units.*

△ **77.** The area of a square is 49 square miles. Find the length of a side of the square.

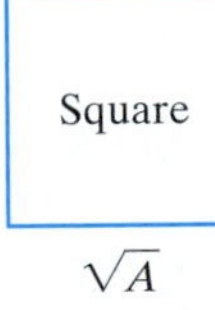

△ **78.** The area of a square is $\frac{1}{81}$ square meters. Find the length of a side of the square.

△ **79.** Sony makes the current smallest mini disc player. It is in the shape of a square with area of 9.0601 square inches. Find the length of a side. (*Source:* SONY)

△ **80.** A parking lot is in the shape of a square with area 2500 square yards. Find the length of a side.

81. Simplify $\sqrt{\sqrt{81}}$.

82. Simplify $\sqrt[3]{\sqrt[3]{1}}$.

83. Simplify $\sqrt{\sqrt{10{,}000}}$.

84. Simplify $\sqrt{\sqrt{1{,}600{,}000{,}000}}$.

85. The formula for calculating the period (one back and forth swing) of a pendulum is $T = 2\pi\sqrt{\frac{L}{g}}$, where T is time of the period of the swing, L is the length of the pendulum, and g is the acceleration of gravity. At the California Academy of Sciences, one can see a Foucault's pendulum with a length $= 30$ ft, and $g = 32$ ft/sec^2. Using $\pi = 3.14$, find the period of this pendulum. (Round to the nearest tenth of a second.)

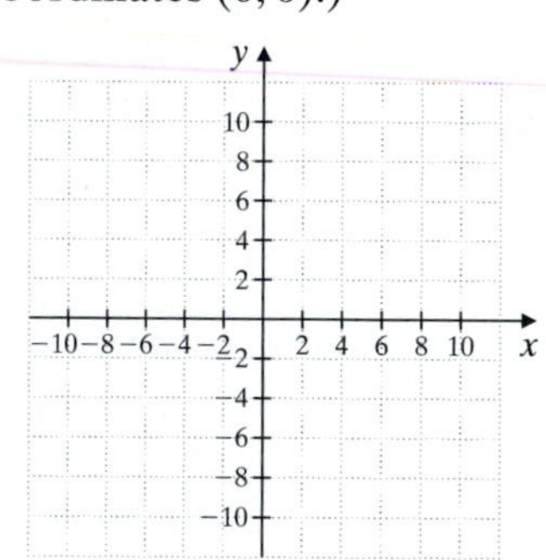

86. If the amount of gold discovered by humankind could be assembled in one place, it would make a cube with a volume of 195,112 cubic feet. Each side of the cube would be $\sqrt[3]{195{,}112}$ feet long. How long would one side of the cube be? (*Source: Reader's Digest*)

87. Explain why the square root of a negative number is not a real number.

88. Explain why the cube root of a negative number is a real number.

89. Graph $y = \sqrt{x}$. (Complete the table below, plot the ordered pair solutions, and draw a smooth curve through the points. Remember that since the radicand cannot be negative, this particular graph begins at the point with coordinates (0, 0).)

x	y
0	0
1	
3	(approximate)
4	
9	

90. Graph $y = \sqrt[3]{x}$. (Complete the table below, plot the ordered pair solutions, and draw a smooth curve through the points.)

x	y
−8	
−2	(approximate)
−1	
0	
1	
2	(approximate)
8	

Use a graphing calculator and graph each function. Observe the graph from left to right and give the ordered pair that corresponds to the "beginning" of the graph. Then tell why the graph starts at that point.

91. $y = \sqrt{x - 2}$

92. $y = \sqrt{x + 3}$

93. $y = \sqrt{x + 4}$

94. $y = \sqrt{x - 5}$

STUDY SKILLS BUILDER

How Well Do You Know Your Textbook?

Let's check to see whether you are familiar with your textbook yet. Remember, for help, see Section 1.1 in this text.

1. What does the icon mean?
2. What does the icon mean?
3. What does the △ icon mean?
4. Where can you find a review for each chapter? What answers to this review can be found in the back of your text?
5. Each chapter contains an overview of the chapter along with examples. What is this feature called?
6. Each chapter contains a review of vocabulary. What is this feature called?
7. There are free CDs in your text. What content is contained on these CDs?
8. What is the location of the section that is entirely devoted to study skills?
9. There are Practice Problems that are contained in the margin of the text. What are they and how can they be used?

15.2 SIMPLIFYING RADICALS

Objective A Simplifying Radicals Using the Product Rule

A square root is simplified when the radicand contains no perfect square factors (other than 1). For example, $\sqrt{20}$ is not simplified because $\sqrt{20} = \sqrt{4 \cdot 5}$ and 4 is a perfect square.

To begin simplifying square roots, we notice the following pattern.

$$\sqrt{9 \cdot 16} = \sqrt{144} = 12$$
$$\sqrt{9} \cdot \sqrt{16} = 3 \cdot 4 = 12$$

Since both expressions simplify to 12, we can write

$$\sqrt{9 \cdot 16} = \sqrt{9} \cdot \sqrt{16}$$

This suggests the following product rule for square roots.

Product Rule for Square Roots

If $\sqrt{a}$ and $\sqrt{b}$ are real numbers, then

$$\sqrt{a \cdot b} = \sqrt{a} \cdot \sqrt{b}$$

In other words, the square root of a product is equal to the product of the square roots.

To simplify $\sqrt{45}$, for example, we factor 45 so that one of its factors is a perfect square factor.

$$\begin{aligned} \sqrt{45} &= \sqrt{9 \cdot 5} && \text{Factor 45.} \\ &= \sqrt{9} \cdot \sqrt{5} && \text{Use the product rule.} \\ &= 3\sqrt{5} && \text{Write } \sqrt{9} \text{ as 3.} \end{aligned}$$

The notation $3\sqrt{5}$ means $3 \cdot \sqrt{5}$. Since the radicand 5 has no perfect square factor other than 1, the expression $3\sqrt{5}$ is in simplest form.

Helpful Hint

A radical expression in simplest form *does not mean* a decimal approximation. The simplest form of a radical expression is an exact form and may still contain a radical.

$$\underbrace{\sqrt{45} = 3\sqrt{5}}_{\text{exact}} \qquad \underbrace{\sqrt{45} \approx 6.71}_{\text{decimal approximation}}$$

EXAMPLES Simplify.

1. $$\begin{aligned} \sqrt{54} &= \sqrt{9 \cdot 6} && \text{Factor 54 so that one factor is a perfect square. 9 is a perfect square.} \\ &= \sqrt{9} \cdot \sqrt{6} && \text{Use the product rule.} \\ &= 3\sqrt{6} && \text{Write } \sqrt{9} \text{ as 3.} \end{aligned}$$

2. $$\begin{aligned} \sqrt{12} &= \sqrt{4 \cdot 3} && \text{Factor 12 so that one factor is a perfect square. 4 is a perfect square.} \\ &= \sqrt{4} \cdot \sqrt{3} && \text{Use the product rule.} \\ &= 2\sqrt{3} && \text{Write } \sqrt{4} \text{ as 2.} \end{aligned}$$

Continued on next page

Objectives

A Use the Product Rule to Simplify Radicals.

B Use the Quotient Rule to Simplify Radicals.

C Use Both Rules to Simplify Radicals Containing Variables.

D Simplify Cube Roots.

PRACTICE PROBLEMS 1–4

Simplify.

1. $\sqrt{40}$
2. $\sqrt{18}$
3. $\sqrt{700}$
4. $\sqrt{15}$

Answers

1. $2\sqrt{10}$ **2.** $3\sqrt{2}$ **3.** $10\sqrt{7}$ **4.** $\sqrt{15}$

3. $\sqrt{200} = \sqrt{100 \cdot 2}$ Factor 200 so that one factor is a perfect square. 100 is a perfect square.

$= \sqrt{100} \cdot \sqrt{2}$ Use the product rule.

$= 10\sqrt{2}$ Write $\sqrt{100}$ as 10.

4. $\sqrt{35}$ The radicand 35 contains no perfect square factors other than 1. Thus $\sqrt{35}$ is in simplest form.

Work Practice Problems 1–4

In Example 3, 100 is the largest perfect square factor of 200. What happens if we don't use the largest perfect square factor? Although using the largest perfect square factor saves time, the result is the same no matter what perfect square factor is used. For example, it is also true that $200 = 4 \cdot 50$. Then

$$\sqrt{200} = \sqrt{4} \cdot \sqrt{50}$$
$$= 2 \cdot \sqrt{50}$$

Since $\sqrt{50}$ is not in simplest form, we continue.

$$\sqrt{200} = 2 \cdot \sqrt{50}$$
$$= 2 \cdot \sqrt{25 \cdot 2}$$
$$= 2 \cdot \sqrt{25} \cdot \sqrt{2}$$
$$= 2 \cdot 5 \cdot \sqrt{2}$$
$$= 10\sqrt{2}$$

PRACTICE PROBLEM 5

Simplify $7\sqrt{75}$.

EXAMPLE 5 Simplify $3\sqrt{8}$.

Solution: Remember that $3\sqrt{8}$ means $3 \cdot \sqrt{8}$.

$3 \cdot \sqrt{8} = 3 \cdot \sqrt{4 \cdot 2}$ Factor 8 so that one factor is a perfect square.

$= 3 \cdot \sqrt{4} \cdot \sqrt{2}$ Use the product rule.

$= 3 \cdot 2 \cdot \sqrt{2}$ Write $\sqrt{4}$ as 2.

$= 6 \cdot \sqrt{2}$ or $6\sqrt{2}$ Write $3 \cdot 2$ as 6.

Work Practice Problem 5

Objective B Simplifying Radicals Using the Quotient Rule

Next, let's examine the square root of a quotient.

$$\sqrt{\frac{16}{4}} = \sqrt{4} = 2$$

Also,

$$\frac{\sqrt{16}}{\sqrt{4}} = \frac{4}{2} = 2$$

Since both expressions equal 2, we can write

$$\sqrt{\frac{16}{4}} = \frac{\sqrt{16}}{\sqrt{4}}$$

This suggests the following quotient rule.

Answer
5. $35\sqrt{3}$

Quotient Rule for Square Roots

If $\sqrt{a}$ and $\sqrt{b}$ are real numbers and $b \neq 0$, then

$$\sqrt{\frac{a}{b}} = \frac{\sqrt{a}}{\sqrt{b}}$$

In other words, the square root of a quotient is equal to the quotient of the square roots.

EXAMPLES Use the quotient rule to simplify.

6. $\sqrt{\frac{25}{36}} = \frac{\sqrt{25}}{\sqrt{36}} = \frac{5}{6}$

7. $\sqrt{\frac{3}{64}} = \frac{\sqrt{3}}{\sqrt{64}} = \frac{\sqrt{3}}{8}$

8. $\sqrt{\frac{40}{81}} = \frac{\sqrt{40}}{\sqrt{81}}$ Use the quotient rule.

$= \frac{\sqrt{4} \cdot \sqrt{10}}{9}$ Use the product rule and write $\sqrt{81}$ as 9.

$= \frac{2\sqrt{10}}{9}$ Write $\sqrt{4}$ as 2.

Work Practice Problems 6–8

PRACTICE PROBLEMS 6–8

Use the quotient rule to simplify.

6. $\sqrt{\frac{16}{81}}$

7. $\sqrt{\frac{2}{25}}$

8. $\sqrt{\frac{45}{49}}$

Objective C Simplifying Radicals Containing Variables

Recall that $\sqrt{x^6} = x^3$ because $(x^3)^2 = x^6$. If a variable radicand has an odd exponent, we write the exponential expression so that one factor is the greatest even power contained in the expression. Then we use the product rule to simplify.

EXAMPLES Simplify each radical. Assume that all variables represent positive numbers.

9. $\sqrt{x^5} = \sqrt{x^4 \cdot x} = \sqrt{x^4} \cdot \sqrt{x} = x^2\sqrt{x}$

10. $\sqrt{8y^2} = \sqrt{4 \cdot 2 \cdot y^2} = \sqrt{4y^2 \cdot 2} = \sqrt{4y^2} \cdot \sqrt{2} = 2y\sqrt{2}$ 4 and y^2 are both perfect square factors so we grouped them under one radical.

11. $\sqrt{\frac{45}{x^6}} = \frac{\sqrt{45}}{\sqrt{x^6}} = \frac{\sqrt{9 \cdot 5}}{x^3} = \frac{\sqrt{9} \cdot \sqrt{5}}{x^3} = \frac{3\sqrt{5}}{x^3}$

12. $\sqrt{\frac{5p^3}{9}} = \frac{\sqrt{5p^3}}{\sqrt{9}} = \frac{\sqrt{p^2 \cdot 5p}}{3} = \frac{\sqrt{p^2} \cdot \sqrt{5p}}{3} = \frac{p\sqrt{5p}}{3}$

Work Practice Problems 9–12

PRACTICE PROBLEMS 9–12

Simplify each radical. Assume that all variables represent positive numbers.

9. $\sqrt{x^{11}}$

10. $\sqrt{18x^4}$

11. $\sqrt{\frac{27}{x^8}}$

12. $\sqrt{\frac{7y^7}{25}}$

Answers

6. $\frac{4}{9}$ 7. $\frac{\sqrt{2}}{5}$ 8. $\frac{3\sqrt{5}}{7}$ 9. $x^5\sqrt{x}$ 10. $3x^2\sqrt{2}$ 11. $\frac{3\sqrt{3}}{x^4}$ 12. $\frac{y^3\sqrt{7y}}{5}$

Objective D Simplifying Cube Roots

The product and quotient rules also apply to roots other than square roots. For example, to simplify cube roots, we look for perfect cube factors of the radicand. Recall that 8 is a perfect cube since $2^3 = 8$. Therefore, to simplify $\sqrt[3]{48}$, we factor 48 as $8 \cdot 6$.

$$\begin{aligned} \sqrt[3]{48} &= \sqrt[3]{8 \cdot 6} && \text{Factor 48.} \\ &= \sqrt[3]{8} \cdot \sqrt[3]{6} && \text{Use the product rule.} \\ &= 2\sqrt[3]{6} && \text{Write } \sqrt[3]{8} \text{ as 2.} \end{aligned}$$

$2\sqrt[3]{6}$ is in simplest form since the radicand 6 contains no perfect cube factors other than 1.

PRACTICE PROBLEMS 13–16

Simplify each radical.

13. $\sqrt[3]{88}$ **14.** $\sqrt[3]{50}$

15. $\sqrt[3]{\frac{10}{27}}$ **16.** $\sqrt[3]{\frac{81}{8}}$

EXAMPLES Simplify each radical.

13. $\sqrt[3]{54} = \sqrt[3]{27 \cdot 2} = \sqrt[3]{27} \cdot \sqrt[3]{2} = 3\sqrt[3]{2}$

14. $\sqrt[3]{18}$ The number 18 contains no perfect cube factors, so $\sqrt[3]{18}$ cannot be simplified further.

15. $\sqrt[3]{\frac{7}{8}} = \frac{\sqrt[3]{7}}{\sqrt[3]{8}} = \frac{\sqrt[3]{7}}{2}$

16. $\sqrt[3]{\frac{40}{27}} = \frac{\sqrt[3]{40}}{\sqrt[3]{27}} = \frac{\sqrt[3]{8 \cdot 5}}{3} = \frac{\sqrt[3]{8} \cdot \sqrt[3]{5}}{3} = \frac{2\sqrt[3]{5}}{3}$

Work Practice Problems 13–16

Answers

13. $2\sqrt[3]{11}$ **14.** $\sqrt[3]{50}$ **15.** $\frac{\sqrt[3]{10}}{3}$ **16.** $\frac{3\sqrt[3]{3}}{2}$

Vocabulary and Readiness Check

Use the choices below to fill in each blank. Not all choices will be used. These exercises come from Sections 15.1 and 15.2.

principal	$a \cdot b$	$\frac{a}{b}$	$\frac{\sqrt{a}}{\sqrt{b}}$	radical sign
index	$\sqrt{a} \cdot \sqrt{b}$	radicand		

1. In the expression $\sqrt[4]{16}$, the number 4 is called the ________, the number 16 is called the ________, and $\sqrt{}$ is called the ________.

2. The symbol $\sqrt{}$ is used to denote the positive, or ________, square root.

3. By the product rule for radicals, $\sqrt{a \cdot b} =$ ________.

4. By the quotient rule for radicals, $\sqrt{\frac{a}{b}} =$ ________.

15.2 EXERCISE SET

FOR EXTRA HELP

Student Solutions Manual

PH Math/Tutor Center

CD/Video for Review

MathXL®

MyMathLab

Objective A *Use the product rule to simplify each radical. See Examples 1 through 4.*

1. $\sqrt{20}$ **2.** $\sqrt{44}$ **3.** $\sqrt{50}$ **4.** $\sqrt{28}$ **5.** $\sqrt{33}$

6. $\sqrt{21}$ **7.** $\sqrt{98}$ **8.** $\sqrt{125}$ **9.** $\sqrt{60}$ **10.** $\sqrt{90}$

11. $\sqrt{180}$ **12.** $\sqrt{150}$ **13.** $\sqrt{52}$ **14.** $\sqrt{75}$

Use the product rule to simplify each radical. See Example 5.

15. $3\sqrt{25}$ **16.** $9\sqrt{36}$ **17.** $7\sqrt{63}$

18. $11\sqrt{99}$ **19.** $-5\sqrt{27}$ **20.** $-6\sqrt{75}$

Objective B *Use the quotient rule and the product rule to simplify each radical. See Examples 6 through 8.*

21. $\sqrt{\frac{8}{25}}$ **22.** $\sqrt{\frac{63}{16}}$ **23.** $\sqrt{\frac{27}{121}}$ **24.** $\sqrt{\frac{24}{169}}$

25. $\sqrt{\frac{9}{4}}$ **26.** $\sqrt{\frac{100}{49}}$ **27.** $\sqrt{\frac{125}{9}}$ **28.** $\sqrt{\frac{27}{100}}$

29. $\sqrt{\frac{11}{36}}$ **30.** $\sqrt{\frac{30}{49}}$ **31.** $-\sqrt{\frac{27}{144}}$ **32.** $-\sqrt{\frac{84}{121}}$

Objective C *Simplify each radical. Assume that all variables represent positive numbers. See Examples 9 through 12.*

33. $\sqrt{x^7}$ **34.** $\sqrt{y^3}$ **35.** $\sqrt{x^{13}}$ **36.** $\sqrt{y^{17}}$

37. $\sqrt{36a^3}$ **38.** $\sqrt{81b^5}$ **39.** $\sqrt{96x^4}$ **40.** $\sqrt{40y^{10}}$

41. $\sqrt{\frac{12}{m^2}}$ **42.** $\sqrt{\frac{63}{p^2}}$ **43.** $\sqrt{\frac{9x}{y^{10}}}$ **44.** $\sqrt{\frac{6y^2}{z^{16}}}$

45. $\sqrt{\frac{88}{x^{12}}}$ **46.** $\sqrt{\frac{500}{y^{22}}}$

Objectives **A** **B** **C** **Mixed Practice** *Simplify each radical. See Examples 1 through 12.*

47. $8\sqrt{4}$

48. $6\sqrt{49}$

49. $\sqrt{\frac{36}{121}}$

50. $\sqrt{\frac{25}{144}}$

51. $\sqrt{175}$

52. $\sqrt{700}$

53. $\sqrt{\frac{20}{9}}$

54. $\sqrt{\frac{45}{64}}$

55. $\sqrt{24m^7}$

56. $\sqrt{50n^{13}}$

57. $\sqrt{\frac{23y^3}{4x^6}}$

58. $\sqrt{\frac{41x^5}{9y^8}}$

Objective **D** *Simplify each radical. See Examples 13 through 16.*

59. $\sqrt[3]{24}$

60. $\sqrt[3]{81}$

61. $\sqrt[3]{250}$

62. $\sqrt[3]{56}$

63. $\sqrt[3]{\frac{5}{64}}$

64. $\sqrt[3]{\frac{32}{125}}$

65. $\sqrt[3]{\frac{23}{8}}$

66. $\sqrt[3]{\frac{37}{27}}$

67. $\sqrt[3]{\frac{15}{64}}$

68. $\sqrt[3]{\frac{4}{27}}$

69. $\sqrt[3]{80}$

70. $\sqrt[3]{108}$

Review

Perform each indicated operation. See Sections 10.4 and 10.5.

71. $6x + 8x$

72. $(6x)(8x)$

73. $(2x + 3)(x - 5)$

74. $(2x + 3) + (x - 5)$

75. $9y^2 - 9y^2$

76. $(9y^2)(-8y^2)$

Concept Extensions

Simplify each radical. Assume that all variables represent positive numbers.

77. $\sqrt{x^6y^3}$

78. $\sqrt{a^{13}b^{14}}$

79. $\sqrt{98x^5y^4}$

80. $\sqrt{27x^8y^{11}}$

81. $\sqrt[3]{-8x^6}$

82. $\sqrt[3]{27x^{12}}$

83. If a cube is to have a volume of 80 cubic inches, then each side must be $\sqrt[3]{80}$ inches long. Simplify the radical representing the side length.

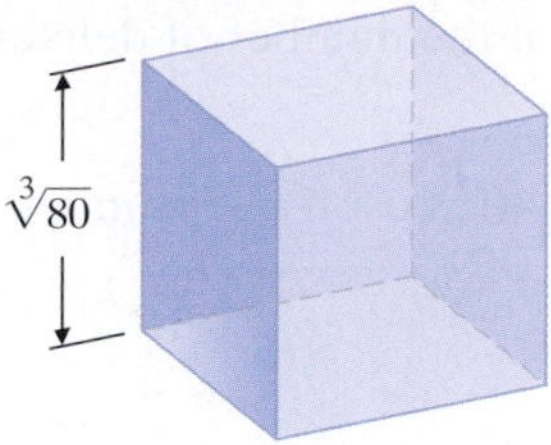

84. Jeannie Boswell is swimming across a 40-foot-wide river, trying to head straight across to the opposite shore. However, the current is strong enough to move her downstream 100 feet by the time she reaches land. (See the figure.) Because of the current, the actual distance she swam is $\sqrt{11{,}600}$ feet. Simplify this radical.

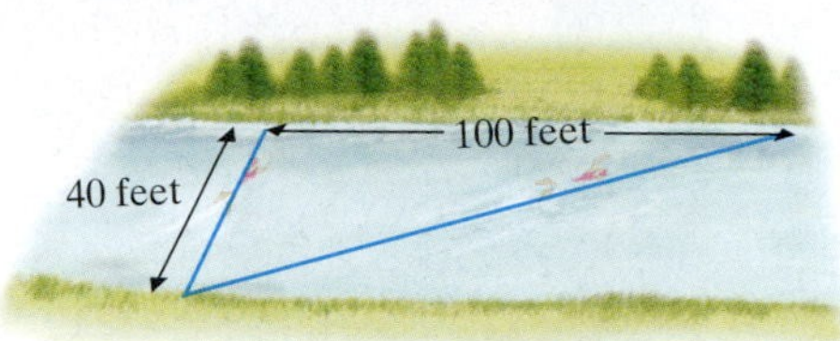

85. By using replacement values for a and b, show that $\sqrt{a^2 + b^2}$ does not equal $a + b$.

86. By using replacement values for a and b, show that $\sqrt{a + b}$ does not equal $\sqrt{a} + \sqrt{b}$.

The length of a side of a cube is given by the expression $\frac{\sqrt{6A}}{6}$, *where A is the cube's surface area. Use this expression for Exercises 87 through 90. Be sure to attach the appropriate units.*

△ **87.** The surface area of a cube is 120 square inches. Find the exact length of a side of the cube.

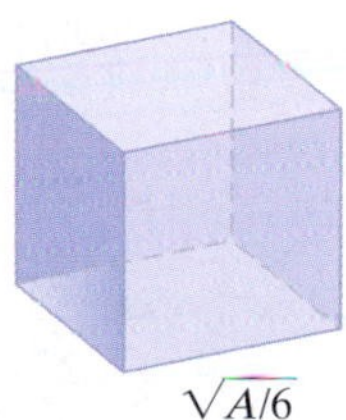

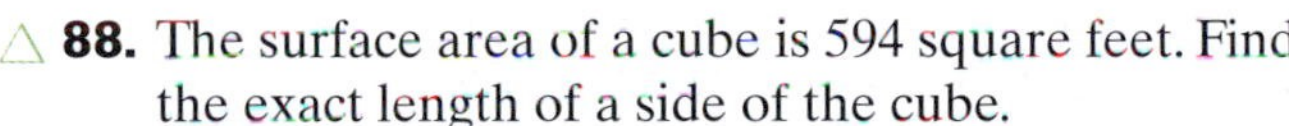

△ **88.** The surface area of a cube is 594 square feet. Find the exact length of a side of the cube.

89. A Guinness World record was set in December 2004, when an electrical engineering student from Johannesburg, South Africa, solved 42 Rubik's cubes in one hour, the most ever in that time. Rubik's cube, named after its inventor, Erno Rubik, was first imagined by him in 1974, and by 1980 was a world-wide phenomenon. These cubes have remained unchanged in size, and a standard Rubik's cube has a surface area of 30.375 square inches. Find the length of one side of a Rubik's cube. (*Source: Guinness World Records*)

△ **90.** The Borg spaceship on *Star Trek: The Next Generation* is in the shape of a cube. Suppose a model of this ship has a surface area of 121 square inches. Find the length of a side of the ship.

The cost C in dollars per day to operate a small delivery service is given by $C = 100\sqrt[3]{n} + 700$, *where n is the number of deliveries per day.*

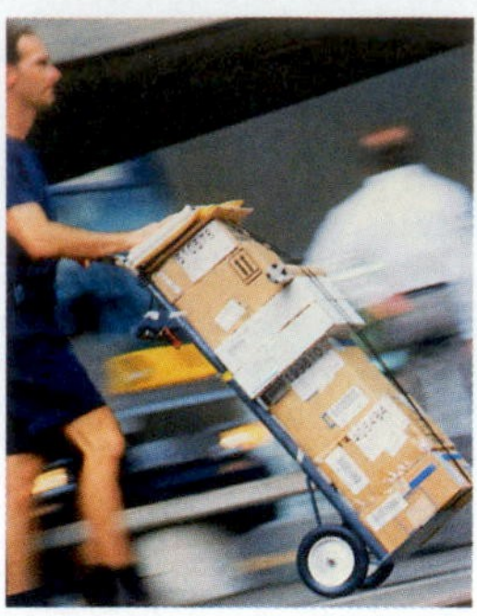

91. Find the cost if the number of deliveries is 1000.

92. Approximate the cost if the number of deliveries is 500.

The Mosteller formula for calculating body surface area is $B = \sqrt{\frac{hw}{3600}}$, *where B is an individual's body surface area in square meters, h is the individual's height in centimeters, and w is the individual's weight in kilograms. Use this formula in Exercises 93 and 94. Round answers to the nearest tenth.*

93. Find the body surface area of a person who is 169 cm tall and weighs 64 kilograms.

94. Approximate the body surface area of a person who is 183 cm tall and weighs 85 kilograms.

15.3 ADDING AND SUBTRACTING RADICALS

Objective A Adding and Subtracting Radicals

Recall that to combine like terms, we use the distributive property.

$$5x + 3x = (5 + 3)x = 8x$$

The distributive property can also be applied to expressions containing the same radicals. For example,

$$5\sqrt{2} + 3\sqrt{2} = (5 + 3)\sqrt{2} = 8\sqrt{2}$$

Also,

$$9\sqrt{5} - 6\sqrt{5} = (9 - 6)\sqrt{5} = 3\sqrt{5}$$

Radical terms such as $5\sqrt{2}$ and $3\sqrt{2}$ are **like radicals,** as are $9\sqrt{5}$ and $6\sqrt{5}$. Like radicals have the same index and the same radicand.

EXAMPLES Add or subtract as indicated.

1. $4\sqrt{5} + 3\sqrt{5} = (4 + 3)\sqrt{5} = 7\sqrt{5}$
2. $\sqrt{10} - 6\sqrt{10} = 1\sqrt{10} - 6\sqrt{10} = (1 - 6)\sqrt{10} = -5\sqrt{10}$
3. $2\sqrt{6} + 2\sqrt{5}$ cannot be simplified further since the radicands are not the same.
4. $\sqrt{15} + \sqrt{15} - \sqrt{2} = 1\sqrt{15} + 1\sqrt{15} - \sqrt{2}$
   $= (1 + 1)\sqrt{15} - \sqrt{2}$
   $= 2\sqrt{15} - \sqrt{2}$

   This expression cannot be simplified further since the radicands are not the same.

Work Practice Problems 1–4

✔Concept Check Which is true?

a. $2 + 3\sqrt{5} = 5\sqrt{5}$
b. $2\sqrt{3} + 2\sqrt{7} = 2\sqrt{10}$
c. $\sqrt{3} + \sqrt{5} = \sqrt{8}$
d. $\sqrt{3} + \sqrt{3} = 3$
e. None of the above is true. In each case, the left-hand side cannot be simplified further.

Objective B Simplifying Square Root Radicals before Adding or Subtracting

At first glance, it appears that the expression $\sqrt{50} + \sqrt{8}$ cannot be simplified further because the radicands are different. However, the product rule can be used to simplify each radical, and then further simplification might be possible.

Objectives

A Add or Subtract Like Radicals.

B Simplify Square Root Radical Expressions, and Then Add or Subtract Any Like Radicals.

C Simplify Cube Root Radical Expressions, and Then Add or Subtract Any Like Radicals.

PRACTICE PROBLEMS 1–4

Add or subtract as indicated.

1. $6\sqrt{11} + 9\sqrt{11}$
2. $\sqrt{7} - 3\sqrt{7}$
3. $\sqrt{2} + \sqrt{2}$
4. $3\sqrt{3} - 3\sqrt{2}$

Answers
1. $15\sqrt{11}$ 2. $-2\sqrt{7}$ 3. $2\sqrt{2}$
4. $3\sqrt{3} - 3\sqrt{2}$

✔ Concept Check Answer
e

PRACTICE PROBLEMS 5–8

Simplify each radical expression.

5. $\sqrt{27} + \sqrt{75}$

6. $3\sqrt{20} - 7\sqrt{45}$

7. $\sqrt{36} - \sqrt{48} - 4\sqrt{3} - \sqrt{9}$

8. $\sqrt{9x^4} - \sqrt{36x^3} + \sqrt{x^3}$

EXAMPLES Simplify each radical expression.

5. $\sqrt{50} + \sqrt{8} = \sqrt{25 \cdot 2} + \sqrt{4 \cdot 2}$ Factor radicands.

$= \sqrt{25} \cdot \sqrt{2} + \sqrt{4} \cdot \sqrt{2}$ Use the product rule.

$= 5\sqrt{2} + 2\sqrt{2}$ Simplify $\sqrt{25}$ and $\sqrt{4}$.

$= 7\sqrt{2}$ Add like radicals.

6. $7\sqrt{12} - 2\sqrt{75} = 7\sqrt{4 \cdot 3} - 2\sqrt{25 \cdot 3}$ Factor radicands.

$= 7\sqrt{4} \cdot \sqrt{3} - 2\sqrt{25} \cdot \sqrt{3}$ Use the product rule.

$= 7 \cdot 2\sqrt{3} - 2 \cdot 5\sqrt{3}$ Simplify $\sqrt{4}$ and $\sqrt{25}$.

$= 14\sqrt{3} - 10\sqrt{3}$ Multiply.

$= 4\sqrt{3}$ Subtract like radicals.

7. $\sqrt{25} - \sqrt{27} - 2\sqrt{18} - \sqrt{16}$

$= 5 - \sqrt{9 \cdot 3} - 2\sqrt{9 \cdot 2} - 4$ Factor radicands and simplify $\sqrt{25}$ and $\sqrt{16}$.

$= 5 - \sqrt{9} \cdot \sqrt{3} - 2\sqrt{9} \cdot \sqrt{2} - 4$ Use the product rule.

$= 5 - 3\sqrt{3} - 2 \cdot 3\sqrt{2} - 4$ Simplify $\sqrt{9}$.

$= 1 - 3\sqrt{3} - 6\sqrt{2}$ Write $5 - 4$ as 1 and $2 \cdot 3$ as 6.

8. $2\sqrt{x^2} - \sqrt{25x^5} + \sqrt{x^5}$

$= 2x - \sqrt{25x^4 \cdot x} + \sqrt{x^4 \cdot x}$ Factor radicands so that one factor is a perfect square. Simplify $\sqrt{x^2}$.

$= 2x - \sqrt{25x^4} \cdot \sqrt{x} + \sqrt{x^4} \cdot \sqrt{x}$ Use the product rule.

$= 2x - 5x^2\sqrt{x} + x^2\sqrt{x}$ Write $\sqrt{25x^4}$ as $5x^2$ and $\sqrt{x^4}$ as x^2.

$= 2x - 4x^2\sqrt{x}$ Add like radicals.

Work Practice Problems 5–8

Objective C Simplifying Cube Root Radicals before Adding or Subtracting

PRACTICE PROBLEM 9

Simplify the radical expression.

$10\sqrt[3]{81p^6} - \sqrt[3]{24p^6}$

EXAMPLE 9 Simplify the radical expression.

$5\sqrt[3]{16x^3} - \sqrt[3]{54x^3}$

$= 5\sqrt[3]{8x^3 \cdot 2} - \sqrt[3]{27x^3 \cdot 2}$ Factor radicands so that one factor is a perfect cube.

$= 5 \cdot \sqrt[3]{8x^3} \cdot \sqrt[3]{2} - \sqrt[3]{27x^3} \cdot \sqrt[3]{2}$ Use the product rule.

$= 5 \cdot 2x \cdot \sqrt[3]{2} - 3x \cdot \sqrt[3]{2}$ Write $\sqrt[3]{8x^3}$ as $2x$ and $\sqrt[3]{27x^3}$ as $3x$.

$= 10x\sqrt[3]{2} - 3x\sqrt[3]{2}$ Write $5 \cdot 2x$ as $10x$.

$= 7x\sqrt[3]{2}$ Subtract like radicands.

Work Practice Problem 9

Answers

5. $8\sqrt{3}$ **6.** $-15\sqrt{5}$ **7.** $3 - 8\sqrt{3}$
8. $3x^2 - 5x\sqrt{x}$ **9.** $28p^2\sqrt[3]{3}$

Vocabulary and Readiness Check

Fill in each blank.

1. Radicals that have the same index and same radicand are called ______________.

2. The expressions $7\sqrt[3]{2x}$ and $-\sqrt[3]{2x}$ are called ______________.

3. $11\sqrt{2} + 6\sqrt{2} =$ ________.
 a. $66\sqrt{2}$ **b.** $17\sqrt{2}$ **c.** $17\sqrt{4}$

4. $\sqrt{5}$ is the same as ________.
 a. $0\sqrt{5}$ **b.** $1\sqrt{5}$ **c.** $5\sqrt{5}$

5. $\sqrt{5} + \sqrt{5} =$ ____
 a. $\sqrt{10}$ **b.** 5 **c.** $2\sqrt{5}$

6. $9\sqrt{7} - \sqrt{7} =$ ____
 a. $8\sqrt{7}$ **b.** 9 **c.** 0

15.3 EXERCISE SET

FOR EXTRA HELP

Student Solutions Manual

PH Math/Tutor Center

CD/Video for Review

MathXL®

MyMathLab

Objective **A** *Add or subtract as indicated. See Examples 1 through 4.*

1. $4\sqrt{3} - 8\sqrt{3}$
2. $\sqrt{5} - 9\sqrt{5}$
3. $3\sqrt{6} + 8\sqrt{6} - 2\sqrt{6} - 5$
4. $12\sqrt{2} - 3\sqrt{2} + 8\sqrt{2} + 10$
5. $6\sqrt{5} - 5\sqrt{5} + \sqrt{2}$
6. $4\sqrt{3} + \sqrt{5} - 3\sqrt{3}$
7. $2\sqrt{3} + 5\sqrt{3} - \sqrt{2}$
8. $8\sqrt{14} + 2\sqrt{14} + \sqrt{5}$
9. $2\sqrt{2} - 7\sqrt{2} - 6$
10. $5\sqrt{7} + 2 - 11\sqrt{7}$

Objective **B** *Add or subtract by first simplifying each radical and then combining any like radicals. Assume that all variables represent positive numbers. See Examples 5 through 8.*

11. $\sqrt{12} + \sqrt{27}$
12. $\sqrt{50} + \sqrt{18}$
13. $\sqrt{45} + 3\sqrt{20}$
14. $5\sqrt{32} - \sqrt{72}$
15. $2\sqrt{54} - \sqrt{20} + \sqrt{45} - \sqrt{24}$
16. $2\sqrt{8} - \sqrt{128} + \sqrt{48} + \sqrt{18}$
17. $4x - 3\sqrt{x^2} + \sqrt{x}$
18. $x - 6\sqrt{x^2} + 2\sqrt{x}$
19. $\sqrt{25x} + \sqrt{36x} - 11\sqrt{x}$
20. $\sqrt{9x} - \sqrt{16x} + 2\sqrt{x}$
21. $\sqrt{\frac{5}{9}} + \sqrt{\frac{5}{81}}$
22. $\sqrt{\frac{3}{64}} + \sqrt{\frac{3}{16}}$
23. $\sqrt{\frac{3}{4}} - \sqrt{\frac{3}{64}}$
24. $\sqrt{\frac{2}{25}} + \sqrt{\frac{2}{9}}$

Objectives **A** **B** **Mixed Practice** *See Examples 1 through 8.*

25. $12\sqrt{5} - \sqrt{5} - 4\sqrt{5}$
26. $\sqrt{6} + 3\sqrt{6} + \sqrt{6}$
27. $\sqrt{75} + \sqrt{48}$
28. $2\sqrt{80} - \sqrt{45}$
29. $\sqrt{5} + \sqrt{15}$
30. $\sqrt{5} + \sqrt{5}$
31. $3\sqrt{x^3} - x\sqrt{4x}$
32. $\sqrt{16x} - \sqrt{x^3}$
33. $\sqrt{8} + \sqrt{9} + \sqrt{18} + \sqrt{81}$
34. $\sqrt{6} + \sqrt{16} + \sqrt{24} + \sqrt{25}$
35. $4 + 8\sqrt{2} - 9$
36. $11 - 5\sqrt{7} - 8$

37. $2\sqrt{45} - 2\sqrt{20}$ **38.** $5\sqrt{18} + 2\sqrt{32}$ **39.** $\sqrt{35} - \sqrt{140}$ **40.** $\sqrt{15} - \sqrt{135}$

41. $6 - 2\sqrt{3} - \sqrt{3}$ **42.** $8 - \sqrt{2} - 5\sqrt{2}$ **43.** $3\sqrt{9x} + 2\sqrt{x}$ **44.** $5\sqrt{x} + 4\sqrt{4x}$

45. $\sqrt{9x^2} + \sqrt{81x^2} - 11\sqrt{x}$ **46.** $\sqrt{100x^2} + 3\sqrt{x} - \sqrt{36x^2}$ **47.** $\sqrt{3x^3} + 3x\sqrt{x}$

48. $x\sqrt{4x} + \sqrt{9x^3}$ **49.** $\sqrt{32x^2} + \sqrt{32x^2} + \sqrt{4x^2}$ **50.** $\sqrt{18x^2} + \sqrt{24x^3} + \sqrt{2x^2}$

51. $\sqrt{40x} + \sqrt{40x^4} - 2\sqrt{10x} - \sqrt{5x^4}$ **52.** $\sqrt{72x^2} + \sqrt{54x} - x\sqrt{50} - 3\sqrt{2x}$

Objective **C** *Simplify each radical expression.*

53. $2\sqrt[3]{9} + 5\sqrt[3]{9} - \sqrt[3]{25}$ **54.** $8\sqrt[3]{4} + 2\sqrt[3]{4} - \sqrt[3]{49}$ **55.** $2\sqrt[3]{2} - 7\sqrt[3]{2} - 6$ **56.** $5\sqrt[3]{11} - 9\sqrt[3]{11} - 5$

57. $\sqrt[3]{81} + \sqrt[3]{24}$ **58.** $\sqrt[3]{32} + \sqrt[3]{4}$ **59.** $2\sqrt[3]{8x^3} + 2\sqrt[3]{16x^3}$ **60.** $3\sqrt[3]{27z^3} + 3\sqrt[3]{81z^3}$

61. $\sqrt{40x} + x\sqrt[3]{40} - 2\sqrt{10x} - x\sqrt[3]{5}$ **62.** $\sqrt{72x^2} + \sqrt[3]{54} - x\sqrt{50} - 3\sqrt[3]{2}$

63. $12\sqrt[3]{y^7} - y^2\sqrt[3]{8y}$ **64.** $19\sqrt[3]{z^{11}} - z^3\sqrt[3]{125z^2}$

Review

Square each binomial. See Section 10.6.

65. $(x + 6)^2$ **66.** $(3x + 2)^2$ **67.** $(2x - 1)^2$ **68.** $(x - 5)^2$

Concept Extensions

69. In your own words, describe like radicals.

70. In the expression $\sqrt{5} + 2 - 3\sqrt{5}$, explain why 2 and -3 cannot be combined.

△ **71.** Find the perimeter of the rectangular picture frame.

△ **72.** Find the perimeter of the plot of land.

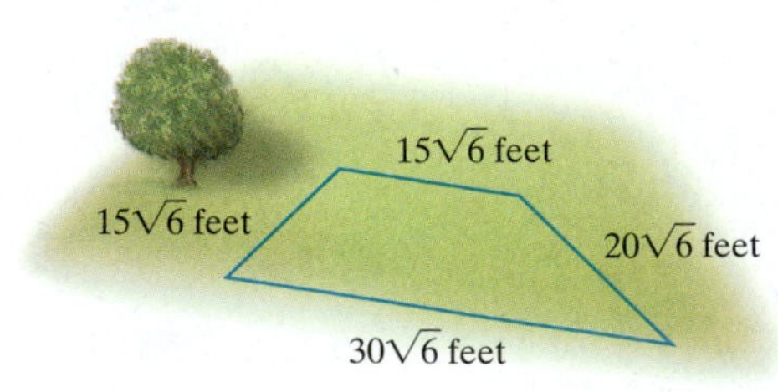

△ **73.** A water trough is to be made of wood. Each of the two triangular end pieces has an area of $\frac{3\sqrt{27}}{4}$ square feet. The two side panels are both rectangular. In simplest radical form, find the total area of the wood needed.

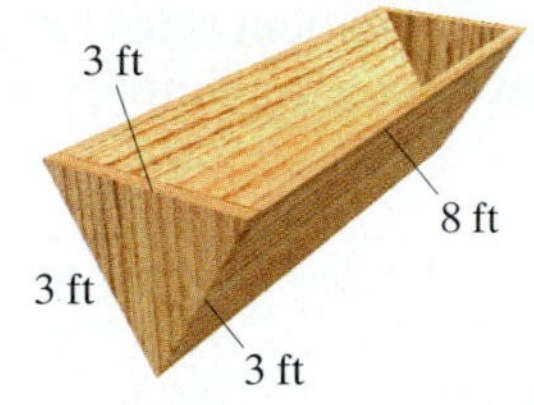

74. Eight wooden braces are to be attached along the diagonals of the vertical sides of a storage bin. Each of four of these diagonals has a length of $\sqrt{52}$ feet, while each of the other four has a length of $\sqrt{80}$ feet. In simplest radical form, find the total length of the wood needed for these braces.

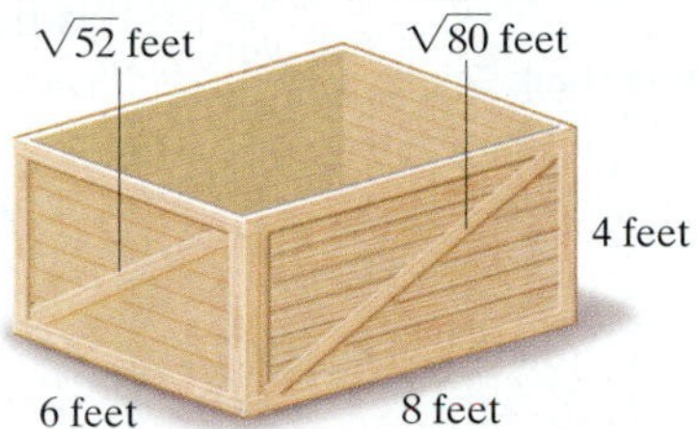

Determine whether each expression can be simplified. If yes, then simplify. See the Concept Check in this section.

75. $4\sqrt{2} + 3\sqrt{2}$

76. $3\sqrt{7} + 3\sqrt{6}$

77. $6 + 7\sqrt{6}$

78. $5x\sqrt{2} + 8x\sqrt{2}$

79. $\sqrt{7} + \sqrt{7} + \sqrt{7}$

80. $6\sqrt{5} - \sqrt{5}$

Simplify.

81. $\sqrt{\frac{x^3}{16}} - x\sqrt{\frac{9x}{25}} + \frac{\sqrt{81x^3}}{2}$

82. $7\sqrt{x^{11}y^7} - x^2y\sqrt{25x^7y^5} + \sqrt{8x^8y^2}$

STUDY SKILLS BUILDER

Learning New Terms?

By now, you have encountered many new terms. It's never too late to make a list of new terms and review them frequently. Remember that placing these new terms (including page references) on 3 × 5 index cards might help you later when you're preparing for a quiz.

Answer the following.

1. How do new terms stand out in this text so that they can be found?
2. Name one way placing a word and its definition on a 3 × 5 card might be helpful.

Objectives

A Multiply Radicals.

B Divide Radicals.

C Rationalize Denominators.

D Rationalize Using Conjugates.

15.4 MULTIPLYING AND DIVIDING RADICALS

Objective A Multiplying Radicals

In Section 15.2, we used the product and quotient rules for radicals to help us simplify radicals. In this section, we use these rules to simplify products and quotients of radicals.

Product Rule for Radicals

If $\sqrt{a}$ and $\sqrt{b}$ are real numbers, then

$$\sqrt{a} \cdot \sqrt{b} = \sqrt{a \cdot b}$$

In other words, the product of the square roots of two numbers is the square root of the product of the two numbers. For example,

$$\sqrt{3} \cdot \sqrt{2} = \sqrt{3 \cdot 2} = \sqrt{6}$$

PRACTICE PROBLEMS 1–4

Multiply. Then simplify each product if possible.

1. $\sqrt{5} \cdot \sqrt{2}$

2. $\sqrt{7} \cdot \sqrt{7}$

3. $\sqrt{6} \cdot \sqrt{3}$

4. $\sqrt{10x} \cdot \sqrt{2x}$

EXAMPLES Multiply. Then simplify each product if possible.

1. $\sqrt{7} \cdot \sqrt{3} = \sqrt{7 \cdot 3}$
$= \sqrt{21}$

2. $\sqrt{3} \cdot \sqrt{3} = \sqrt{3 \cdot 3} = \sqrt{9} = 3$

3. $\sqrt{3} \cdot \sqrt{15} = \sqrt{45}$ Use the product rule.
$= \sqrt{9 \cdot 5}$ Factor the radicand.
$= \sqrt{9} \cdot \sqrt{5}$ Use the product rule.
$= 3\sqrt{5}$ Simplify $\sqrt{9}$.

4. $\sqrt{2x^3} \cdot \sqrt{6x} = \sqrt{2x^3 \cdot 6x}$ Use the product rule.
$= \sqrt{12x^4}$ Multiply.
$= \sqrt{4x^4 \cdot 3}$ Write $12x^4$ so that one factor is a perfect square.
$= \sqrt{4x^4} \cdot \sqrt{3}$ Use the product rule.
$= 2x^2\sqrt{3}$ Simplify.

Work Practice Problems 1–4

From Example 2, we found that

$$\sqrt{3} \cdot \sqrt{3} = 3 \quad \text{or} \quad (\sqrt{3})^2 = 3$$

This is true in general.

If a is a positive number,

$$\sqrt{a} \cdot \sqrt{a} = a \quad \text{or} \quad (\sqrt{a})^2 = a$$

✔Concept Check Identify the true statement(s).

a. $\sqrt{7} \cdot \sqrt{7} = 7$

b. $\sqrt{2} \cdot \sqrt{3} = 6$

c. $(\sqrt{131})^2 = 131$

d. $\sqrt{5x} \cdot \sqrt{5x} = 5x$ (Here x is a positive number.)

When multiplying radical expressions containing more than one term, we use the same techniques we use to multiply other algebraic expressions with more than one term.

Answers

1. $\sqrt{10}$ **2.** 7 **3.** $3\sqrt{2}$ **4.** $2x\sqrt{5}$

✔ Concept Check Answers

a, c, d

EXAMPLE 5 Multiply.

a. $\sqrt{5}(\sqrt{5}-\sqrt{2})$

b. $\sqrt{3x}(\sqrt{x}-5\sqrt{3})$

c. $(\sqrt{x}+\sqrt{2})(\sqrt{x}-\sqrt{7})$

Solution:

a. Using the distributive property, we have

$$\sqrt{5}(\sqrt{5}-\sqrt{2}) = \sqrt{5}\cdot\sqrt{5}-\sqrt{5}\cdot\sqrt{2}$$
$$= 5-\sqrt{10} \quad \text{Since } \sqrt{5}\cdot\sqrt{5}=5 \text{ and } \sqrt{5}\cdot\sqrt{2}=\sqrt{10}$$

b. $\sqrt{3x}(\sqrt{x}-5\sqrt{3}) = \sqrt{3x}\cdot\sqrt{x}-\sqrt{3x}\cdot 5\sqrt{3}$ Use the distributive property.

$= \sqrt{3x\cdot x}-5\sqrt{3x\cdot 3}$ Use the product rule.

$= \sqrt{3\cdot x^2}-5\sqrt{9\cdot x}$ Factor each radicand so that one factor is a perfect square.

$= \sqrt{3}\cdot\sqrt{x^2}-5\cdot\sqrt{9}\cdot\sqrt{x}$ Use the product rule.

$= x\sqrt{3}-5\cdot 3\cdot\sqrt{x}$ Simplify.

$= x\sqrt{3}-15\sqrt{x}$ Simplify.

c. Using the FOIL method of multiplication, we have

$$(\sqrt{x}+\sqrt{2})(\sqrt{x}-\sqrt{7})$$
$$\overset{F}{=\sqrt{x}\cdot\sqrt{x}} - \overset{O}{\sqrt{x}\cdot\sqrt{7}} + \overset{I}{\sqrt{2}\cdot\sqrt{x}} - \overset{L}{\sqrt{2}\cdot\sqrt{7}}$$
$$= x-\sqrt{7x}+\sqrt{2x}-\sqrt{14} \quad \text{Use the product rule.}$$

Work Practice Problem 5

PRACTICE PROBLEM 5

Multiply.

a. $\sqrt{7}(\sqrt{7}-\sqrt{3})$

b. $\sqrt{5x}(\sqrt{x}-3\sqrt{5})$

c. $(\sqrt{x}+\sqrt{5})(\sqrt{x}-\sqrt{3})$

The special product formulas also can be used to multiply expressions containing radicals.

EXAMPLE 6 Multiply.

a. $(\sqrt{5}-7)(\sqrt{5}+7)$

b. $(\sqrt{7x}+2)^2$

Solution:

a. $(\sqrt{5}-7)(\sqrt{5}+7) = (\sqrt{5})^2-7^2$ Recall that $(a-b)(a+b)=a^2-b^2$.

$= 5-49$

$= -44$

b. $(\sqrt{7x}+2)^2$

$= (\sqrt{7x})^2+2(\sqrt{7x})(2)+(2)^2$ Recall that $(a+b)^2=a^2+2ab+b^2$.

$= 7x+4\sqrt{7x}+4$

Work Practice Problem 6

PRACTICE PROBLEM 6

Multiply.

a. $(\sqrt{3}+6)(\sqrt{3}-6)$

b. $(\sqrt{5x}+4)^2$

Objective B Dividing Radicals

To simplify quotients of rational expressions, we use the quotient rule.

Quotient Rule for Radicals

If $\sqrt{a}$ and $\sqrt{b}$ are real numbers and $b \neq 0$, then

$$\frac{\sqrt{a}}{\sqrt{b}} = \sqrt{\frac{a}{b}}$$

Answers

5. a. $7-\sqrt{21}$ b. $x\sqrt{5}-15\sqrt{x}$ c. $x-\sqrt{3x}+\sqrt{5x}-\sqrt{15}$

6. a. -33 b. $5x+8\sqrt{5x}+16$

PRACTICE PROBLEMS 7–9

Divide. Then simplify the quotient if possible.

7. $\dfrac{\sqrt{15}}{\sqrt{3}}$

8. $\dfrac{\sqrt{90}}{\sqrt{2}}$

9. $\dfrac{\sqrt{75x^3}}{\sqrt{5x}}$

EXAMPLES Divide. Then simplify the quotient if possible.

7. $\dfrac{\sqrt{14}}{\sqrt{2}} = \sqrt{\dfrac{14}{2}} = \sqrt{7}$

8. $\dfrac{\sqrt{100}}{\sqrt{5}} = \sqrt{\dfrac{100}{5}} = \sqrt{20} = \sqrt{4 \cdot 5} = \sqrt{4} \cdot \sqrt{5} = 2\sqrt{5}$

9. $\dfrac{\sqrt{12x^3}}{\sqrt{3x}} = \sqrt{\dfrac{12x^3}{3x}} = \sqrt{4x^2} = 2x$

Work Practice Problems 7–9

Objective C Rationalizing Denominators

It is sometimes easier to work with radical expressions if the denominator does not contain a radical. To rewrite the expression so that the denominator does not contain a radical expression, we use the fact that we can multiply the numerator and the denominator of a fraction by the same nonzero number without changing the value of the expression. This is the same as multiplying the fraction by 1. For example, to get rid of the radical in the denominator of $\dfrac{\sqrt{5}}{\sqrt{2}}$, we multiply by 1 in the form of $\dfrac{\sqrt{2}}{\sqrt{2}}$. Then

$$\frac{\sqrt{5}}{\sqrt{2}} = \frac{\sqrt{5}}{\sqrt{2}} \cdot 1 = \frac{\sqrt{5}}{\sqrt{2}} \cdot \frac{\sqrt{2}}{\sqrt{2}} = \frac{\sqrt{5} \cdot \sqrt{2}}{\sqrt{2} \cdot \sqrt{2}} = \frac{\sqrt{10}}{2}$$

This process is called **rationalizing** the denominator.

PRACTICE PROBLEM 10

Rationalize the denominator of $\dfrac{5}{\sqrt{3}}$.

EXAMPLE 10 Rationalize the denominator of $\dfrac{2}{\sqrt{7}}$.

Solution: To rewrite $\dfrac{2}{\sqrt{7}}$ so that there is no radical in the denominator, we multiply by 1 in the form of $\dfrac{\sqrt{7}}{\sqrt{7}}$.

$$\frac{2}{\sqrt{7}} = \frac{2}{\sqrt{7}} \cdot \frac{\sqrt{7}}{\sqrt{7}} = \frac{2 \cdot \sqrt{7}}{\sqrt{7} \cdot \sqrt{7}} = \frac{2\sqrt{7}}{7}$$

Work Practice Problem 10

PRACTICE PROBLEM 11

Rationalize the denominator of $\dfrac{\sqrt{7}}{\sqrt{20}}$.

EXAMPLE 11 Rationalize the denominator of $\dfrac{\sqrt{5}}{\sqrt{12}}$.

Solution: We can multiply by $\dfrac{\sqrt{12}}{\sqrt{12}}$, but see what happens if we simplify first.

$$\frac{\sqrt{5}}{\sqrt{12}} = \frac{\sqrt{5}}{\sqrt{4 \cdot 3}} = \frac{\sqrt{5}}{2\sqrt{3}}$$

To rationalize the denominator now, we multiply by $\dfrac{\sqrt{3}}{\sqrt{3}}$.

$$\frac{\sqrt{5}}{2\sqrt{3}} = \frac{\sqrt{5}}{2\sqrt{3}} \cdot \frac{\sqrt{3}}{\sqrt{3}} = \frac{\sqrt{5} \cdot \sqrt{3}}{2\sqrt{3} \cdot \sqrt{3}} = \frac{\sqrt{15}}{2 \cdot 3} = \frac{\sqrt{15}}{6}$$

Work Practice Problem 11

Answers

7. $\sqrt{5}$ **8.** $3\sqrt{5}$ **9.** $x\sqrt{15}$ **10.** $\dfrac{5\sqrt{3}}{3}$ **11.** $\dfrac{\sqrt{35}}{10}$

EXAMPLE 12 Rationalize the denominator of $\sqrt{\frac{1}{18x}}$.

Solution: First we simplify.

$$\sqrt{\frac{1}{18x}} = \frac{\sqrt{1}}{\sqrt{18x}} = \frac{1}{\sqrt{9} \cdot \sqrt{2x}} = \frac{1}{3\sqrt{2x}}$$

Now to rationalize the denominator, we multiply by $\frac{\sqrt{2x}}{\sqrt{2x}}$.

$$\frac{1}{3\sqrt{2x}} = \frac{1}{3\sqrt{2x}} \cdot \frac{\sqrt{2x}}{\sqrt{2x}} = \frac{1 \cdot \sqrt{2x}}{3\sqrt{2x} \cdot \sqrt{2x}} = \frac{\sqrt{2x}}{3 \cdot 2x} = \frac{\sqrt{2x}}{6x}$$

Work Practice Problem 12

PRACTICE PROBLEM 12

Rationalize the denominator of $\sqrt{\frac{2}{45x}}$.

Objective D Rationalizing Denominators Using Conjugates

To rationalize a denominator that is a sum or a difference, such as the denominator in

$$\frac{2}{4 + \sqrt{3}}$$

we multiply the numerator and the denominator by $4 - \sqrt{3}$. The expressions $4 + \sqrt{3}$ and $4 - \sqrt{3}$ are called conjugates of each other. When a radical expression such as $4 + \sqrt{3}$ is multiplied by its conjugate $4 - \sqrt{3}$, the product simplifies to an expression that contains no radicals.

In general, the expressions $a + b$ and $a - b$ are **conjugates** of each other.

$$(a + b)(a - b) = a^2 - b^2$$
$$(4 + \sqrt{3})(4 - \sqrt{3}) = 4^2 - (\sqrt{3})^2 = 16 - 3 = 13$$

Then

$$\frac{2}{4 + \sqrt{3}} = \frac{2(4 - \sqrt{3})}{(4 + \sqrt{3})(4 - \sqrt{3})} = \frac{2(4 - \sqrt{3})}{13}$$

EXAMPLE 13 Rationalize the denominator of $\frac{2}{1 + \sqrt{3}}$.

Solution: We multiply the numerator and the denominator of this fraction by the conjugate of $1 + \sqrt{3}$, that is, by $1 - \sqrt{3}$.

$$\frac{2}{1 + \sqrt{3}} = \frac{2(1 - \sqrt{3})}{(1 + \sqrt{3})(1 - \sqrt{3})}$$
$$= \frac{2(1 - \sqrt{3})}{1^2 - (\sqrt{3})^2}$$
$$= \frac{2(1 - \sqrt{3})}{1 - 3}$$
$$= \frac{2(1 - \sqrt{3})}{-2}$$
$$= -\frac{2(1 - \sqrt{3})}{2} \qquad \frac{a}{-b} = -\frac{a}{b}$$
$$= -1(1 - \sqrt{3}) \qquad \text{Simplify.}$$
$$= -1 + \sqrt{3} \qquad \text{Multiply.}$$

Helpful Hint Don't forget that $(\sqrt{3})^2 = 3$.

Work Practice Problem 13

PRACTICE PROBLEM 13

Rationalize the denominator of $\frac{3}{2 + \sqrt{7}}$.

Answers

12. $\frac{\sqrt{10x}}{15x}$ **13.** $-2 + \sqrt{7}$

PRACTICE PROBLEM 14

Rationalize the denominator of $\frac{\sqrt{2}+5}{\sqrt{2}-1}$.

EXAMPLE 14 Rationalize the denominator of $\frac{\sqrt{5}+4}{\sqrt{5}-1}$.

Solution:

$$\frac{\sqrt{5}+4}{\sqrt{5}-1} = \frac{(\sqrt{5}+4)(\sqrt{5}+1)}{(\sqrt{5}-1)(\sqrt{5}+1)}$$ Multiply the numerator and denominator by $\sqrt{5}+1$, the conjugate of $\sqrt{5}-1$.

$$= \frac{5+\sqrt{5}+4\sqrt{5}+4}{5-1}$$ Multiply.

$$= \frac{9+5\sqrt{5}}{4}$$ Simplify.

Work Practice Problem 14

PRACTICE PROBLEM 15

Rationalize the denominator of $\frac{7}{2-\sqrt{x}}$.

EXAMPLE 15 Rationalize the denominator of $\frac{3}{1+\sqrt{x}}$.

Solution:

$$\frac{3}{1+\sqrt{x}} = \frac{3(1-\sqrt{x})}{(1+\sqrt{x})(1-\sqrt{x})}$$ Multiply the numerator and denominator by $1-\sqrt{x}$, the conjugate of $1+\sqrt{x}$.

$$= \frac{3(1-\sqrt{x})}{1-x}$$

Work Practice Problem 15

Answers

14. $7+6\sqrt{2}$ **15.** $\frac{7(2+\sqrt{x})}{4-x}$

Vocabulary and Readiness Check

Fill in each blank.

1. $\sqrt{7}\cdot\sqrt{3} =$ ______
2. $\sqrt{10}\cdot\sqrt{10} =$ ______
3. $\dfrac{\sqrt{15}}{\sqrt{3}} =$ ______
4. The process of eliminating the radical in the denominator of a radical expression is called ______.
5. The conjugate of $2 + \sqrt{3}$ is ______.

15.4 EXERCISE SET

FOR EXTRA HELP

Student Solutions Manual

PH Math/Tutor Center

CD/Video for Review

MathXL®

MyMathLab

Objective A *Multiply and simplify. Assume that all variables represent positive real numbers. See Examples 1 through 6.*

1. $\sqrt{8}\cdot\sqrt{2}$
2. $\sqrt{3}\cdot\sqrt{12}$
3. $\sqrt{10}\cdot\sqrt{5}$
4. $\sqrt{2}\cdot\sqrt{14}$
5. $(\sqrt{6})^2$
6. $(\sqrt{10})^2$
7. $\sqrt{2x}\cdot\sqrt{2x}$
8. $\sqrt{5y}\cdot\sqrt{5y}$
9. $(2\sqrt{5})^2$
10. $(3\sqrt{10})^2$
11. $(6\sqrt{x})^2$
12. $(8\sqrt{y})^2$
13. $\sqrt{3x^5}\cdot\sqrt{6x}$
14. $\sqrt{21y^7}\cdot\sqrt{3y}$
15. $\sqrt{2xy^2}\cdot\sqrt{8xy}$
16. $\sqrt{18x^2y^2}\cdot\sqrt{2x^2y}$
17. $\sqrt{6}(\sqrt{5} + \sqrt{7})$
18. $\sqrt{10}(\sqrt{3} - \sqrt{7})$
19. $\sqrt{10}(\sqrt{2} + \sqrt{5})$
20. $\sqrt{6}(\sqrt{3} + \sqrt{2})$
21. $\sqrt{7y}(\sqrt{y} - 2\sqrt{7})$
22. $\sqrt{5b}(2\sqrt{b} + \sqrt{5})$
23. $(\sqrt{3} + 6)(\sqrt{3} - 6)$
24. $(\sqrt{5} + 2)(\sqrt{5} - 2)$
25. $(\sqrt{3} + \sqrt{5})(\sqrt{2} - \sqrt{5})$
26. $(\sqrt{7} + \sqrt{5})(\sqrt{2} - \sqrt{5})$
27. $(2\sqrt{11} + 1)(\sqrt{11} - 6)$
28. $(5\sqrt{3} + 2)(\sqrt{3} - 1)$
29. $(\sqrt{x} + 6)(\sqrt{x} - 6)$
30. $(\sqrt{y} + 5)(\sqrt{y} - 5)$
31. $(\sqrt{x} - 7)^2$
32. $(\sqrt{x} + 4)^2$
33. $(\sqrt{6y} + 1)^2$
34. $(\sqrt{3y} - 2)^2$

Objective B *Divide and simplify. Assume that all variables represent positive real numbers. See Examples 7 through 9.*

35. $\dfrac{\sqrt{32}}{\sqrt{2}}$
36. $\dfrac{\sqrt{40}}{\sqrt{10}}$
37. $\dfrac{\sqrt{21}}{\sqrt{3}}$
38. $\dfrac{\sqrt{55}}{\sqrt{5}}$
39. $\dfrac{\sqrt{90}}{\sqrt{5}}$
40. $\dfrac{\sqrt{96}}{\sqrt{8}}$
41. $\dfrac{\sqrt{75y^5}}{\sqrt{3y}}$
42. $\dfrac{\sqrt{24x^7}}{\sqrt{6x}}$
43. $\dfrac{\sqrt{150}}{\sqrt{2}}$
44. $\dfrac{\sqrt{120}}{\sqrt{3}}$
45. $\dfrac{\sqrt{72y^5}}{\sqrt{3y^3}}$
46. $\dfrac{\sqrt{54x^3}}{\sqrt{2x}}$
47. $\dfrac{\sqrt{24x^3y^4}}{\sqrt{2xy}}$
48. $\dfrac{\sqrt{96x^5y^3}}{\sqrt{3x^2y}}$

Objective C *Rationalize each denominator and simplify. Assume that all variables represent positive real numbers. See Examples 10 through 12.*

49. $\frac{\sqrt{3}}{\sqrt{5}}$

50. $\frac{\sqrt{2}}{\sqrt{3}}$

51. $\frac{7}{\sqrt{2}}$

52. $\frac{8}{\sqrt{11}}$

53. $\frac{1}{\sqrt{6y}}$

54. $\frac{1}{\sqrt{10z}}$

55. $\sqrt{\frac{5}{18}}$

56. $\sqrt{\frac{7}{12}}$

57. $\sqrt{\frac{3}{x}}$

58. $\sqrt{\frac{5}{x}}$

59. $\sqrt{\frac{1}{8}}$

60. $\sqrt{\frac{1}{27}}$

61. $\sqrt{\frac{2}{15}}$

62. $\sqrt{\frac{11}{14}}$

63. $\sqrt{\frac{3}{20}}$

64. $\sqrt{\frac{3}{50}}$

65. $\frac{3x}{\sqrt{2x}}$

66. $\frac{5y}{\sqrt{3y}}$

67. $\frac{8y}{\sqrt{5}}$

68. $\frac{7x}{\sqrt{2}}$

69. $\sqrt{\frac{y}{12x}}$

70. $\sqrt{\frac{x}{20y}}$

Objective D *Rationalize each denominator and simplify. Assume that all variables represent positive real numbers. See Examples 13 through 15.*

71. $\frac{3}{\sqrt{2}+1}$

72. $\frac{6}{\sqrt{5}+2}$

73. $\frac{4}{2-\sqrt{5}}$

74. $\frac{2}{\sqrt{10}-3}$

75. $\frac{\sqrt{5}+1}{\sqrt{6}-\sqrt{5}}$

76. $\frac{\sqrt{3}+1}{\sqrt{3}-\sqrt{2}}$

77. $\frac{\sqrt{3}+1}{\sqrt{2}-1}$

78. $\frac{\sqrt{2}-2}{2-\sqrt{3}}$

79. $\frac{5}{2+\sqrt{x}}$

80. $\frac{9}{3+\sqrt{x}}$

81. $\frac{3}{\sqrt{x}-4}$

82. $\frac{4}{\sqrt{x}-1}$

Review

Solve each equation. See Sections 9.3 and 11.6.

83. $x + 5 = 7^2$

84. $2y - 1 = 3^2$

85. $4z^2 + 6z - 12 = (2z)^2$

86. $16x^2 + x + 9 = (4x)^2$

87. $9x^2 + 5x + 4 = (3x + 1)^2$

88. $x^2 + 3x + 4 = (x + 2)^2$

Concept Extensions

△ **89.** Find the area of a rectangular room whose length is $13\sqrt{2}$ meters and width is $5\sqrt{6}$ meters.

△ **90.** Find the volume of a microwave oven whose length is $\sqrt{3}$ feet, width is $\sqrt{2}$ feet, and height is $\sqrt{2}$ feet.

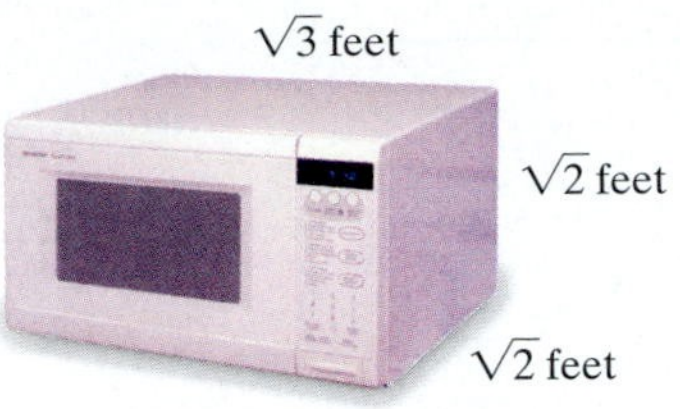

△ **91.** If a circle has area A, then the formula for the radius r of the circle is

$$r = \sqrt{\frac{A}{\pi}}$$

Rationalize the denominator of this expression.

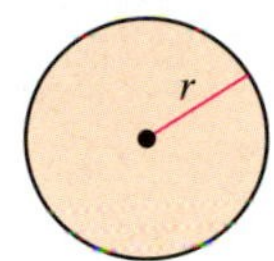

△ **92.** If a round ball has volume V, then the formula for the radius r of the ball is

$$r = \sqrt[3]{\frac{3V}{4\pi}}$$

Simplify this expression by rationalizing the denominator.

Identify each statement as true or false. See the Concept Check in this section.

93. $\sqrt{5} \cdot \sqrt{5} = 5$

94. $\sqrt{5} \cdot \sqrt{3} = 15$

95. $\sqrt{3x} \cdot \sqrt{3x} = 2\sqrt{3x}$

96. $\sqrt{3x} + \sqrt{3x} = 2\sqrt{3x}$

97. $\sqrt{11} + \sqrt{2} = \sqrt{13}$

98. $\sqrt{11} \cdot \sqrt{2} = \sqrt{22}$

99. When rationalizing the denominator of $\frac{\sqrt{2}}{\sqrt{3}}$, explain why both the numerator and the denominator must be multiplied by $\sqrt{3}$.

100. In your own words, explain why $\sqrt{6} + \sqrt{2}$ cannot be simplified further, but $\sqrt{6} \cdot \sqrt{2}$ can be.

101. When rationalizing the denominator of $\frac{\sqrt[3]{2}}{\sqrt[3]{3}}$, explain why both the numerator and the denominator must be multiplied by $\sqrt[3]{9}$.

102. When rationalizing the denominator of $\frac{5}{1 + \sqrt{2}}$, explain why multiplying by $\frac{\sqrt{2}}{\sqrt{2}}$ will not accomplish this, but multiplying by $\frac{1 - \sqrt{2}}{1 - \sqrt{2}}$ will.

It is often more convenient to work with a radical expression whose numerator is rationalized. Rationalize the numerator of each expression by multiplying the numerator and denominator by the conjugate of the numerator.

103. $\frac{\sqrt{3} + 1}{\sqrt{2} - 1}$

104. $\frac{\sqrt{2} - 2}{2 - \sqrt{3}}$

Answers

1. ______
2. ______
3. ______
4. ______
5. ______
6. ______
7. ______
8. ______
9. ______
10. ______
11. ______
12. ______
13. ______
14. ______
15. ______
16. ______
17. ______
18. ______
19. ______
20. ______
21. ______
22. ______

INTEGRATED REVIEW

Sections 15.1–15.4

Simplifying Radicals

Simplify. Assume that all variables represent positive numbers.

1. $\sqrt{36}$

2. $\sqrt{48}$

3. $\sqrt{x^4}$

4. $\sqrt{y^7}$

5. $\sqrt{16x^2}$

6. $\sqrt{18x^{11}}$

7. $\sqrt[3]{8}$

8. $\sqrt[4]{81}$

9. $\sqrt[3]{-27}$

10. $\sqrt{-4}$

11. $\sqrt{\frac{11}{9}}$

12. $\sqrt[3]{\frac{7}{64}}$

13. $-\sqrt{16}$

14. $-\sqrt{25}$

15. $\sqrt{\frac{9}{49}}$

16. $\sqrt{\frac{1}{64}}$

17. $\sqrt{a^8b^2}$

18. $\sqrt{x^{10}y^{20}}$

19. $\sqrt{25m^6}$

20. $\sqrt{9n^{16}}$

Add or subtract as indicated.

21. $5\sqrt{7} + \sqrt{7}$

22. $\sqrt{50} - \sqrt{8}$

23. $5\sqrt{2} - 5\sqrt{3}$

24. $2\sqrt{x} + \sqrt{25x} - \sqrt{36x} + 3x$

Multiply and simplify if possible.

25. $\sqrt{2} \cdot \sqrt{15}$

26. $\sqrt{3} \cdot \sqrt{3}$

27. $(2\sqrt{7})^2$

28. $(3\sqrt{5})^2$

29. $\sqrt{3}(\sqrt{11} + 1)$

30. $\sqrt{6}(\sqrt{3} - 2)$

31. $\sqrt{8y} \cdot \sqrt{2y}$

32. $\sqrt{15x^2} \cdot \sqrt{3x^2}$

33. $(\sqrt{x} - 5)(\sqrt{x} + 2)$

34. $(3 + \sqrt{2})^2$

Divide and simplify if possible.

35. $\dfrac{\sqrt{8}}{\sqrt{2}}$

36. $\dfrac{\sqrt{45}}{\sqrt{15}}$

37. $\dfrac{\sqrt{24x^5}}{\sqrt{2x}}$

38. $\dfrac{\sqrt{75a^4b^5}}{\sqrt{5ab}}$

Rationalize each denominator.

39. $\sqrt{\dfrac{1}{6}}$

40. $\dfrac{x}{\sqrt{20}}$

41. $\dfrac{4}{\sqrt{6} + 1}$

42. $\dfrac{\sqrt{2} + 1}{\sqrt{x} - 5}$

23. ____________

24. ____________

25. ____________

26. ____________

27. ____________

28. ____________

29. ____________

30. ____________

31. ____________

32. ____________

33. ____________

34. ____________

35. ____________

36. ____________

37. ____________

38. ____________

39. ____________

40. ____________

41. ____________

42. ____________

Objectives

A Solve Radical Equations by Using the Squaring Property of Equality Once.

B Solve Radical Equations by Using the Squaring Property of Equality Twice.

15.5 SOLVING EQUATIONS CONTAINING RADICALS

Objective A Using the Squaring Property of Equality Once

In this section, we solve **radical equations** such as

$$\sqrt{x+3} = 5 \quad \text{and} \quad \sqrt{2x+1} = \sqrt{3x}$$

Radical equations contain variables in the radicand. To solve these equations, we rely on the following squaring property.

The Squaring Property of Equality

If $a = b$, then $a^2 = b^2$.

Unfortunately, this squaring property does not guarantee that all solutions of the new equation are solutions of the original equation. For example, if we square both sides of the equation

$$x = 2$$

we have

$$x^2 = 4$$

This new equation has two solutions, 2 and -2, while the original equation $x = 2$ has only one solution. For this reason, we must **always check proposed solutions of radical equations in the original equation.**

PRACTICE PROBLEM 1

Solve: $\sqrt{x-2} = 7$

EXAMPLE 1 Solve: $\sqrt{x+3} = 5$

Solution: To solve this radical equation, we use the squaring property of equality and square both sides of the equation.

$$\begin{aligned} \sqrt{x+3} &= 5 \\ \left(\sqrt{x+3}\right)^2 &= 5^2 && \text{Square both sides.} \\ x+3 &= 25 && \text{Simplify.} \\ x &= 22 && \text{Subtract 3 from both sides.} \end{aligned}$$

Check: We replace x with 22 in the original equation.

$$\begin{aligned} \sqrt{x+3} &= 5 && \text{Original equation} \\ \sqrt{22+3} &\stackrel{?}{=} 5 && \text{Let } x = 22. \\ \sqrt{25} &\stackrel{?}{=} 5 \\ 5 &= 5 && \text{True} \end{aligned}$$

Since a true statement results, 22 is the solution.

Helpful Hint

Don't forget to check the proposed solutions of radical equations in the original equation.

Work Practice Problem 1

PRACTICE PROBLEM 2

Solve: $\sqrt{6x-1} = \sqrt{x}$

EXAMPLE 2 Solve: $\sqrt{x} = \sqrt{5x-2}$

Solution: Each radical is by itself on one side of the equation. Let's begin solving by squaring both sides.

$$\begin{aligned} \sqrt{x} &= \sqrt{5x-2} && \text{Original equation} \\ (\sqrt{x})^2 &= \left(\sqrt{5x-2}\right)^2 && \text{Square both sides.} \\ x &= 5x-2 && \text{Simplify.} \\ -4x &= -2 && \text{Subtract } 5x \text{ from both sides.} \\ x &= \frac{-2}{-4} = \frac{1}{2} && \text{Divide both sides by } -4 \text{ and simplify.} \end{aligned}$$

Answers

1. $x = 51$ **2.** $x = \frac{1}{5}$

Check: We replace x with $\frac{1}{2}$ in the original equation.

$$\sqrt{x} = \sqrt{5x - 2} \quad \text{Original equation}$$
$$\sqrt{\frac{1}{2}} \stackrel{?}{=} \sqrt{5 \cdot \frac{1}{2} - 2} \quad \text{Let } x = \frac{1}{2}.$$
$$\sqrt{\frac{1}{2}} \stackrel{?}{=} \sqrt{\frac{5}{2} - 2} \quad \text{Multiply.}$$
$$\sqrt{\frac{1}{2}} \stackrel{?}{=} \sqrt{\frac{5}{2} - \frac{4}{2}} \quad \text{Write 2 as } \frac{4}{2}.$$
$$\sqrt{\frac{1}{2}} = \sqrt{\frac{1}{2}} \quad \text{True}$$

This statement is true, so the solution is $\frac{1}{2}$.

Work Practice Problem 2

EXAMPLE 3 Solve: $\sqrt{x} + 6 = 4$

Solution: First we write the equation so that the radical is by itself on one side of the equation.

$$\sqrt{x} + 6 = 4$$
$$\sqrt{x} = -2 \quad \text{Subtract 6 from both sides to get the radical by itself.}$$

Normally we would now square both sides. Recall, however, that $\sqrt{x}$ is the principal or nonnegative square root of x so that $\sqrt{x}$ cannot equal -2 and thus this equation has no solution. We arrive at the same conclusion if we continue by applying the squaring property.

$$\sqrt{x} = -2$$
$$(\sqrt{x})^2 = (-2)^2 \quad \text{Square both sides.}$$
$$x = 4 \quad \text{Simplify.}$$

Check: We replace x with 4 in the original equation.

$$\sqrt{x} + 6 = 4 \quad \text{Original equation}$$
$$\sqrt{4} + 6 \stackrel{?}{=} 4 \quad \text{Let } x = 4.$$
$$2 + 6 = 4 \quad \text{False}$$

Since 4 *does not* satisfy the original equation, this equation has no solution.

Work Practice Problem 3

PRACTICE PROBLEM 3

Solve: $\sqrt{x} + 9 = 2$

Example 3 makes it very clear that we *must* check proposed solutions in the original equation to determine if they are truly solutions. If a proposed solution does not work, we say that the value is an **extraneous solution.**

The following steps can be used to solve radical equations containing square roots.

To Solve a Radical Equation Containing Square Roots

Step 1: Arrange terms so that one radical is by itself on one side of the equation. That is, isolate a radical.

Step 2: Square both sides of the equation.

Step 3: Simplify both sides of the equation.

Step 4: If the equation still contains a radical term, repeat Steps 1 through 3.

Step 5: Solve the equation.

Step 6: Check all solutions in the original equation for extraneous solutions.

Answer

3. no solution

PRACTICE PROBLEM 4

Solve: $\sqrt{9y^2 + 2y - 10} = 3y$

EXAMPLE 4 Solve: $\sqrt{4y^2 + 5y - 15} = 2y$

Solution: The radical is already isolated, so we start by squaring both sides.

$$\sqrt{4y^2 + 5y - 15} = 2y$$
$$\left(\sqrt{4y^2 + 5y - 15}\right)^2 = (2y)^2 \quad \text{Square both sides.}$$
$$4y^2 + 5y - 15 = 4y^2 \quad \text{Simplify.}$$
$$5y - 15 = 0 \quad \text{Subtract } 4y^2 \text{ from both sides.}$$
$$5y = 15 \quad \text{Add 15 to both sides.}$$
$$y = 3 \quad \text{Divide both sides by 5.}$$

Check: We replace y with 3 in the original equation.

$$\sqrt{4y^2 + 5y - 15} = 2y \quad \text{Original equation}$$
$$\sqrt{4 \cdot 3^2 + 5 \cdot 3 - 15} \stackrel{?}{=} 2 \cdot 3 \quad \text{Let } y = 3.$$
$$\sqrt{4 \cdot 9 + 15 - 15} \stackrel{?}{=} 6 \quad \text{Simplify.}$$
$$\sqrt{36} \stackrel{?}{=} 6$$
$$6 = 6 \quad \text{True}$$

This statement is true, so the solution is 3.

Work Practice Problem 4

PRACTICE PROBLEM 5

Solve: $\sqrt{x + 1} - x = -5$

EXAMPLE 5 Solve: $\sqrt{x + 3} - x = -3$

Solution: First we isolate the radical by adding x to both sides. Then we square both sides.

$$\sqrt{x + 3} - x = -3$$
$$\sqrt{x + 3} = x - 3 \quad \text{Add } x \text{ to both sides.}$$
$$\left(\sqrt{x + 3}\right)^2 = (x - 3)^2 \quad \text{Square both sides.}$$
$$x + 3 = x^2 - 6x + 9 \quad \text{Simplify.}$$

Helpful Hint

Don't forget that $(x - 3)^2 = (x - 3)(x - 3) = x^2 - 6x + 9$.

To solve the resulting quadratic equation, we write the equation in standard form by subtracting x and 3 from both sides.

$$x + 3 = x^2 - 6x + 9$$
$$3 = x^2 - 7x + 9 \quad \text{Subtract } x \text{ from both sides.}$$
$$0 = x^2 - 7x + 6 \quad \text{Subtract 3 from both sides.}$$
$$0 = (x - 6)(x - 1) \quad \text{Factor.}$$
$$0 = x - 6 \quad \text{or} \quad 0 = x - 1 \quad \text{Set each factor equal to zero.}$$
$$6 = x \qquad 1 = x \quad \text{Solve for } x.$$

Check: We replace x with 6 and then x with 1 in the original equation.

Let $x = 6$.

$$\sqrt{x + 3} - x = -3$$
$$\sqrt{6 + 3} - 6 \stackrel{?}{=} -3$$
$$\sqrt{9} - 6 \stackrel{?}{=} -3$$
$$3 - 6 \stackrel{?}{=} -3$$
$$-3 = -3 \quad \text{True}$$

Let $x = 1$.

$$\sqrt{x + 3} - x = -3$$
$$\sqrt{1 + 3} - 1 \stackrel{?}{=} -3$$
$$\sqrt{4} - 1 \stackrel{?}{=} -3$$
$$2 - 1 \stackrel{?}{=} -3$$
$$1 = -3 \quad \text{False}$$

Since replacing x with 1 resulted in a false statement, 1 is an extraneous solution. The only solution is 6.

Work Practice Problem 5

Answers

4. $y = 5$ **5.** $x = 8$

Objective B Using the Squaring Property of Equality Twice

If a radical equation contains two radicals, we may need to use the squaring property twice.

EXAMPLE 6 Solve: $\sqrt{x-4} = \sqrt{x} - 2$

Solution:

$$\sqrt{x-4} = \sqrt{x} - 2$$

$$\left(\sqrt{x-4}\right)^2 = \left(\sqrt{x} - 2\right)^2 \quad \text{Square both sides.}$$

$$x - 4 = x - 4\sqrt{x} + 4$$

$$-8 = -4\sqrt{x} \quad \text{To get the radical term alone, subtract } x \text{ and 4 from both sides.}$$

$$2 = \sqrt{x} \quad \text{Divide both sides by } -4.$$

$$4 = x \quad \text{Square both sides again.}$$

Check the proposed solution in the original equation. The solution is 4.

Work Practice Problem 6

Helpful Hint

Don't forget:

$$\begin{aligned}(\sqrt{x} - 2)^2 &= (\sqrt{x} - 2)(\sqrt{x} - 2)\\ &= \sqrt{x} \cdot \sqrt{x} - 2\sqrt{x} - 2\sqrt{x} + 4\\ &= x - 4\sqrt{x} + 4\end{aligned}$$

PRACTICE PROBLEM 6

Solve: $\sqrt{x} + 3 = \sqrt{x + 15}$

Answer

6. $x = 1$

15.5 EXERCISE SET

FOR EXTRA HELP

Student Solutions Manual

PH Math/Tutor Center

CD/Video for Review

MathXL®

MyMathLab

Objective A *Solve each equation. See Examples 1 through 3.*

1. $\sqrt{x} = 9$

2. $\sqrt{x} = 4$

3. $\sqrt{x + 5} = 2$

4. $\sqrt{x + 12} = 3$

5. $\sqrt{x} - 2 = 5$

6. $4\sqrt{x} - 7 = 5$

7. $3\sqrt{x} + 5 = 2$

8. $3\sqrt{x} + 8 = 5$

9. $\sqrt{x} = \sqrt{3x - 8}$

10. $\sqrt{x} = \sqrt{4x - 3}$

11. $\sqrt{4x - 3} = \sqrt{x + 3}$

12. $\sqrt{5x - 4} = \sqrt{x + 8}$

Solve each equation. See Examples 4 and 5.

13. $\sqrt{9x^2 + 2x - 4} = 3x$

14. $\sqrt{4x^2 + 3x - 9} = 2x$

15. $\sqrt{x} = x - 6$

16. $\sqrt{x} = x - 2$

17. $\sqrt{x + 7} = x + 5$

18. $\sqrt{x + 5} = x - 1$

19. $\sqrt{3x + 7} - x = 3$

20. $x = \sqrt{4x - 7} + 1$

21. $\sqrt{16x^2 + 2x + 2} = 4x$

22. $\sqrt{4x^2 + 3x + 2} = 2x$

23. $\sqrt{2x^2 + 6x + 9} = 3$

24. $\sqrt{3x^2 + 6x + 4} = 2$

Objective B *Solve each equation. See Example 6.*

25. $\sqrt{x - 7} = \sqrt{x} - 1$

26. $\sqrt{x - 8} = \sqrt{x} - 2$

27. $\sqrt{x} + 2 = \sqrt{x + 24}$

28. $\sqrt{x} + 5 = \sqrt{x + 55}$

29. $\sqrt{x + 8} = \sqrt{x} + 2$

30. $\sqrt{x} + 1 = \sqrt{x + 15}$

Objectives A B Mixed Practice *Solve each equation. See Examples 1 through 6.*

31. $\sqrt{2x + 6} = 4$

32. $\sqrt{3x + 7} = 5$

33. $\sqrt{x + 6} + 1 = 3$

34. $\sqrt{x + 5} + 2 = 5$

35. $\sqrt{x + 6} + 5 = 3$

36. $\sqrt{2x - 1} + 7 = 1$

37. $\sqrt{16x^2 - 3x + 6} = 4x$

38. $\sqrt{9x^2 - 2x + 8} = 3x$

39. $-\sqrt{x} = -6$

40. $-\sqrt{y} = -8$

41. $\sqrt{x + 9} = \sqrt{x} - 3$

42. $\sqrt{x} - 6 = \sqrt{x + 36}$

43. $\sqrt{2x + 1} + 3 = 5$

44. $\sqrt{3x - 1} + 1 = 4$

45. $\sqrt{x} + 3 = 7$

46. $\sqrt{x} + 5 = 10$

47. $\sqrt{4x} = \sqrt{2x + 6}$

48. $\sqrt{5x + 6} = \sqrt{8x}$

49. $\sqrt{2x + 1} = x - 7$

50. $\sqrt{2x + 5} = x - 5$

51. $x = \sqrt{2x - 2} + 1$

52. $\sqrt{2x - 4} + 2 = x$

53. $\sqrt{1 - 8x} - x = 4$

54. $\sqrt{2x + 5} - 1 = x$

Review

Translating *Translate each sentence into an equation and then solve. See Section 9.4.*

55. If 8 is subtracted from the product of 3 and x, the result is 19. Find x.

56. If 3 more than x is subtracted from twice x, the result is 11. Find x.

57. The length of a rectangle is twice the width. The perimeter is 24 inches. Find the length.

58. The length of a rectangle is 2 inches longer than the width. The perimeter is 24 inches. Find the length.

Concept Extensions

Solve each equation.

59. $\sqrt{x-3}+3=\sqrt{3x+4}$

60. $\sqrt{2x+3}=\sqrt{x-2}+2$

61. Explain why proposed solutions of radical equations must be checked in the original equation.

62. Is 8 a solution of the equation $\sqrt{x-4}-5=\sqrt{x+1}$? Explain why or why not.

63. The formula $b=\sqrt{\frac{V}{2}}$ can be used to determine the length b of a side of the base of a square-based pyramid with height 6 units and volume V cubic units.

a. Find the length of the side of the base that produces a pyramid with each volume. (Round to the nearest tenth of a unit.)

V	20	200	2000
b			

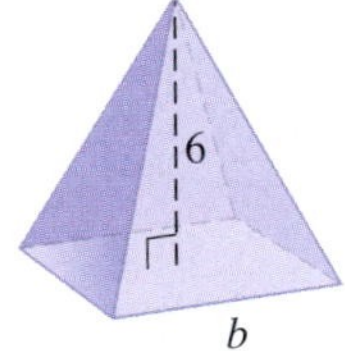

b. Notice in the table that volume V has been increased by a factor of 10 each time. Does the corresponding length b of a side increase by a factor of 10 each time also?

64. The formula $r=\sqrt{\frac{V}{2\pi}}$ can be used to determine the radius r of a cylinder with height 2 units and volume V cubic units.

a. Find the radius needed to manufacture a cylinder with each volume. (Round to the nearest tenth of a unit.)

V	10	100	1000
r			

b. Notice in the table that volume V has been increased by a factor of 10 each time. Does the corresponding radius increase by a factor of 10 each time also?

Graphing calculators can be used to solve equations. To solve $\sqrt{x-2}=x-5$, for example, graph $y_1=\sqrt{x-2}$ and $y_2=x-5$ on the same set of axes. Use the Trace and Zoom features or an Intersect feature to find the point of intersection of the graphs. The x-value of the point is the solution of the equation. Use a graphing calculator to solve the equations below. Approximate solutions to the nearest hundredth.

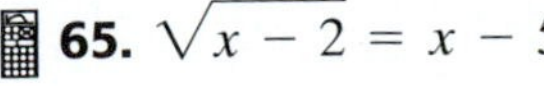

65. $\sqrt{x-2}=x-5$

66. $\sqrt{x+1}=2x-3$

67. $-\sqrt{x+4}=5x-6$

68. $-\sqrt{x+5}=-7x+1$

THE BIGGER PICTURE Simplifying Expressions and Solving Equations

Now we continue our outline from Sections 9.6, 10.7, 11.6, 12.4, and 12.5. Although suggestions are given, this outline should be in your own words. Once you complete this new portion, try the exercises below.

I. Simplifying Expressions

- **A. Exponents (Sections 10.1, 10.2)**
- **B. Polynomials**
  1. Add (Section 10.4)
  2. Subtract (Section 10.4)
  3. Multiply (Section 10.5)
  4. Divide (Section 10.7)
- **C. Factoring Polynomials** (Chapter 11 Integrated Review)
- **D. Rational Expressions**
  1. Simplify (Section 12.1)
  2. Multiply (Section 12.2)
  3. Divide (Section 12.2)
  4. Add or Subtract (Section 12.4)
- **E. Radicals**
  1. Simplify square roots: If possible, factor the radicand so that one factor is a perfect square. Then use the product rule, and simplify.

     $\sqrt{75} = \sqrt{25 \cdot 3} = \sqrt{25} \cdot \sqrt{3} = 5\sqrt{3}$

  2. Add or subtract: Only like radicals (same index and radicand) can be added or subtracted.

     $$8\sqrt{10} - \sqrt{40} + \sqrt{5}$$
     $$= 8\sqrt{10} - 2\sqrt{10} + \sqrt{5}$$
     $$= 6\sqrt{10} + \sqrt{5}$$

  3. Multiply or divide:

     $$\sqrt{a} \cdot \sqrt{b} = \sqrt{ab}; \frac{\sqrt{a}}{\sqrt{b}} = \sqrt{\frac{a}{b}}.$$
     $$\sqrt{11} \cdot \sqrt{3} = \sqrt{33};$$
     $$\frac{\sqrt{140}}{\sqrt{7}} = \sqrt{\frac{140}{7}} = \sqrt{20} = \sqrt{4 \cdot 5} = 2\sqrt{5}$$

  4. Rationalizing the denominator:

     a. If the denominator is one term,

     $$\frac{5}{\sqrt{11}} = \frac{5 \cdot \sqrt{11}}{\sqrt{11} \cdot \sqrt{11}} = \frac{5\sqrt{11}}{11}$$

     b. If the denominator has two terms, multiply by 1 in the form of $\frac{\text{conjugate of denominator}}{\text{conjugate of denominator}}$.

     $$\frac{13}{3 + \sqrt{2}} = \frac{13}{3 + \sqrt{2}} \cdot \frac{3 - \sqrt{2}}{3 - \sqrt{2}}$$
     $$= \frac{13(3 - \sqrt{2})}{9 - 2} = \frac{13(3 - \sqrt{2})}{7}$$

II. Solving Equations and Inequalities

- **A. Linear Equations** (Section 9.3)
- **B. Linear Inequalities** (Section 9.6)
- **C. Quadratic and Higher Degree Equations** (Section 11.6)
- **D. Equations with Rational Expressions** (Section 12.5)
- **E. Equations with Radicals** To solve, isolate a radical, then square both sides. You may have to repeat this. Check possible solution in the original equation.

$\sqrt{x + 49} + 7 = x$	
$\sqrt{x + 49} = x - 7$	Subtract 7 from both sides.
$x + 49 = x^2 - 14x + 49$	Square both sides.
$0 = x^2 - 15x$	Set terms equal to 0.
$0 = x(x - 15)$	Factor.
~~$x = 0$~~ or $x = 15$	Set each factor equal to 0 and solve.

Perform indicated operations and simplify. If necessary, rationalize the denominator.

1. $\sqrt{56}$
2. $\sqrt{\dfrac{20x^5}{49}}$
3. $(-5x^{12}y^{-3})(3x^{-7}y^{14})$
4. $\sqrt{\dfrac{10}{11}}$
5. $\dfrac{8}{\sqrt{5} - 1}$
6. $\dfrac{1}{2}(6x^2 - 4) + \dfrac{1}{3}(6x^2 - 9) - 14$

Solve each equation or inequality.

7. $9x - 7 = 7x - 9$
8. $\dfrac{x}{5} = \dfrac{x - 3}{11}$
9. $-5(2y + 1) \le 3y - 2 - 2y + 1$
10. $x(x + 1) = 42$
11. $\dfrac{-6}{x - 7} + \dfrac{8}{x} = \dfrac{-4}{x - 7}$
12. $1 + 4(x - 2) = x(x - 6) - x^2 + 13$

15.6 RADICAL EQUATIONS AND PROBLEM SOLVING

Objectives

A **Use the Pythagorean Theorem to Solve Problems.**

B **Solve Problems Using Formulas Containing Radicals.**

Objective A Using the Pythagorean Theorem

Applications of radicals can be found in geometry, finance, science, and other areas of technology. Our first application involves the Pythagorean theorem, which gives a formula that relates the lengths of the three sides of a right triangle. We first studied the Pythagorean theorem in Chapter 11 and we review it here.

The Pythagorean Theorem

If a and b are lengths of the legs of a right triangle and c is the length of the hypotenuse, then $a^2 + b^2 = c^2$.

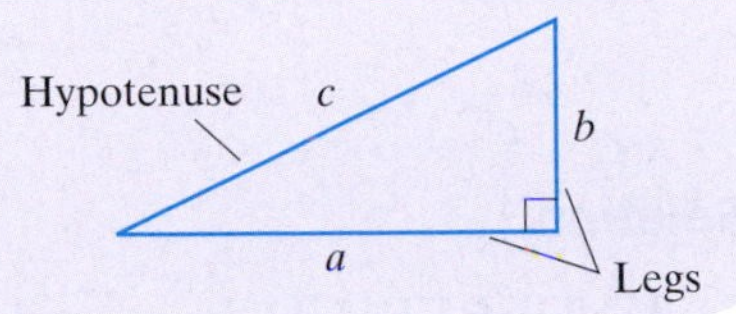

EXAMPLE 1 Find the length of the hypotenuse of a right triangle whose legs are 6 inches and 8 inches long.

Solution: Because this is a right triangle, we use the Pythagorean theorem. We let $a = 6$ inches and $b = 8$ inches. Length c must be the length of the hypotenuse.

$a^2 + b^2 = c^2$ Use the Pythagorean theorem.

$6^2 + 8^2 = c^2$ Substitute the lengths of the legs.

$36 + 64 = c^2$ Simplify.

$100 = c^2$

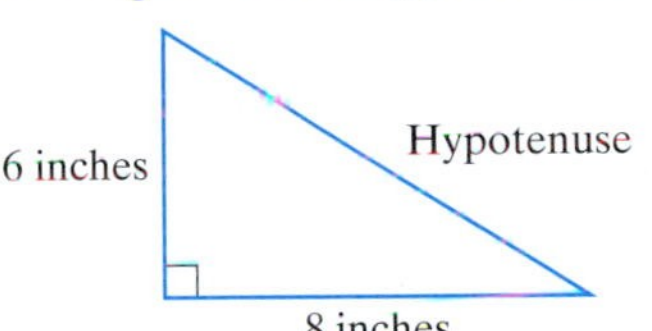

Since c represents a length, we know that c is positive and is the principal square root of 100.

$100 = c^2$

$\sqrt{100} = c$ Use the definition of principal square root.

$10 = c$ Simplify.

The hypotenuse has a length of 10 inches.

Work Practice Problem 1

PRACTICE PROBLEM 1

Find the length of the hypotenuse of the right triangle shown.

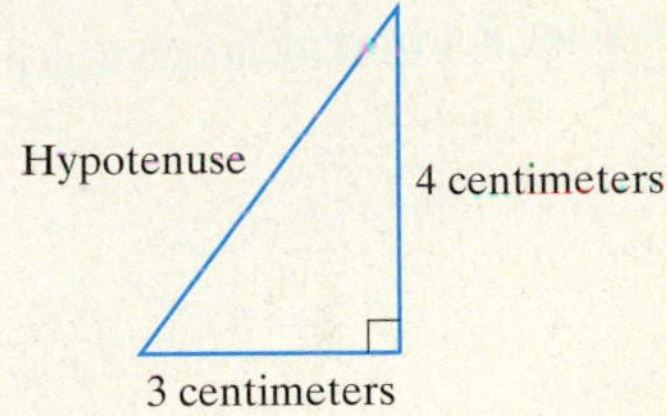

EXAMPLE 2 Find the length of the leg of the right triangle shown. Give the exact length and a two-decimal-place approximation.

Solution: We let $a = 2$ meters and b be the unknown length of the other leg. The hypotenuse is $c = 5$ meters.

2 meters

5 meters

Leg

$a^2 + b^2 = c^2$ Use the Pythagorean theorem.

$2^2 + b^2 = 5^2$ Let $a = 2$ and $c = 5$.

$4 + b^2 = 25$

$b^2 = 21$

$b = \sqrt{21} \approx 4.58$ meters

The length of the leg is exactly $\sqrt{21}$ meters or approximately 4.58 meters.

Work Practice Problem 2

PRACTICE PROBLEM 2

Find the length of the leg of the right triangle shown. Give the exact length and a two-decimal-place approximation.

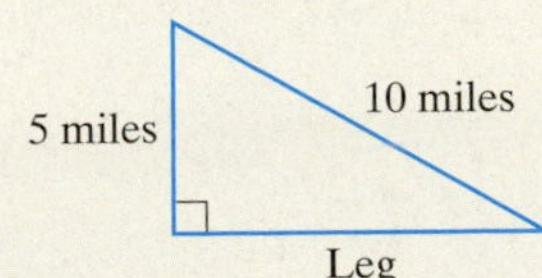

Answers

1. 5 cm **2.** $5\sqrt{3}$ mi; 8.66 mi

PRACTICE PROBLEM 3

Evan Saacks wants to determine the distance at certain points across a pond on his property. He is able to measure the distances shown on the following diagram. Find how wide the pond is to the nearest tenth of a foot.

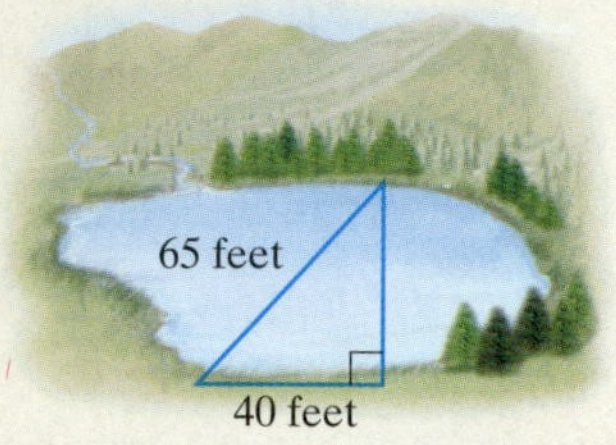

Answer

3. 51.2 feet

EXAMPLE 3 Finding a Distance

A surveyor must determine the distance across a lake at points P and Q as shown in the figure. To do this, she finds a third point R perpendicular to line PQ. If the length of $\overline{PR}$ is 320 feet and the length of $\overline{QR}$ is 240 feet, what is the distance across the lake? Approximate this distance to the nearest whole foot.

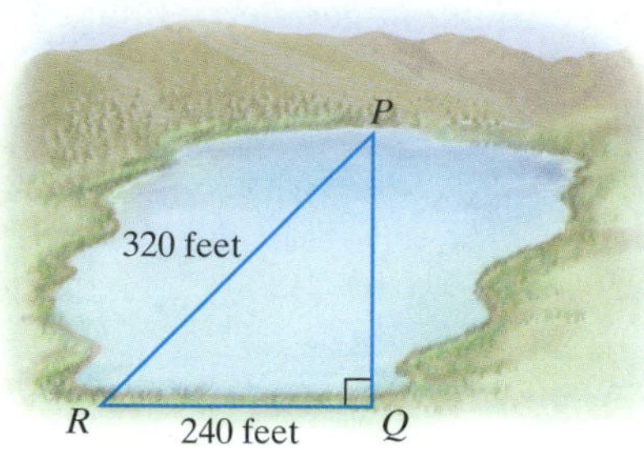

Solution:

1. UNDERSTAND. Read and reread the problem. We will set up the problem using the Pythagorean theorem. By creating a line perpendicular to line PQ, the surveyor deliberately constructed a right triangle. The hypotenuse, $\overline{PR}$, has a length of 320 feet, so we let $c = 320$ in the Pythagorean theorem. The side $\overline{QR}$ is one of the legs, so we let $a = 240$ and $b =$ the unknown length.

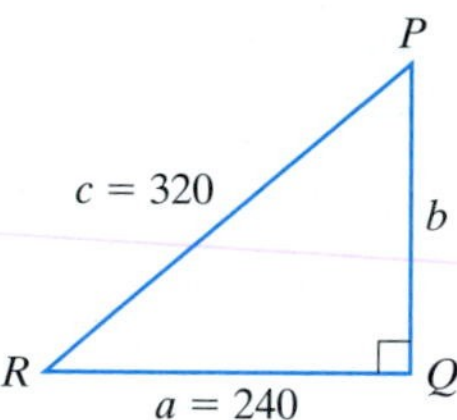

2. TRANSLATE.

$a^2 + b^2 = c^2$ Use the Pythagorean theorem.

$240^2 + b^2 = 320^2$ Let $a = 240$ and $c = 320$.

3. SOLVE.

$57{,}600 + b^2 = 102{,}400$

$b^2 = 44{,}800$ Subtract 57,600 from both sides.

$b = \sqrt{44{,}800}$ Use the definition of principal square root.

$= 80\sqrt{7}$ Simplify.

4. INTERPRET.

Check: See that $240^2 + (\sqrt{44{,}800})^2 = 320^2$.

State: The distance across the lake is *exactly* $\sqrt{44{,}800}$ or $80\sqrt{7}$ feet. The surveyor can now use a calculator to find that $80\sqrt{7}$ feet is *approximately* 211.6601 feet, so the distance across the lake is roughly 212 feet.

Work Practice Problem 3

Objective B Using Formulas Containing Radicals

The Pythagorean theorem is an extremely important result in mathematics and should be memorized. But there are other applications involving formulas containing radicals that are not quite as well known, such as the velocity formula used in the next example.

EXAMPLE 4 Finding the Velocity of an Object

A formula used to determine the velocity v, in feet per second, of an object after it has fallen a certain height (neglecting air resistance) is $v = \sqrt{2gh}$, where g is the acceleration due to gravity and h is the height the object has fallen. On Earth, the acceleration g due to gravity is approximately 32 feet per second per second. Find the velocity of a person after falling 5 feet.

Solution: We are told that $g = 32$ feet per second per second. To find the velocity v when $h = 5$ feet, we use the velocity formula.

$$v = \sqrt{2gh} \quad \text{Use the velocity formula.}$$
$$= \sqrt{2 \cdot 32 \cdot 5} \quad \text{Substitute known values.}$$
$$= \sqrt{320}$$
$$= 8\sqrt{5} \quad \text{Simplify the radicand.}$$

The velocity of the person after falling 5 feet is *exactly* $8\sqrt{5}$ feet per second, or *approximately* 17.9 feet per second.

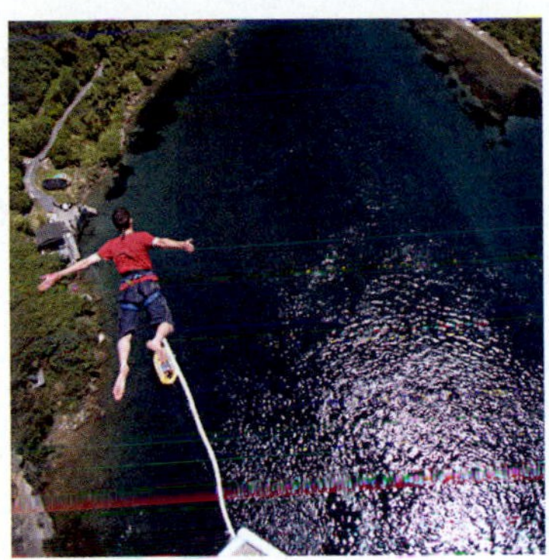

Work Practice Problem 4

PRACTICE PROBLEM 4

Use the formula from Example 4 and find the velocity of an object after it has fallen 20 feet.

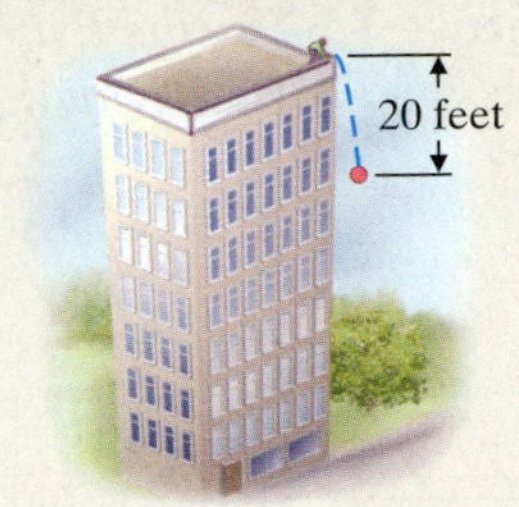

Answer

4. $16\sqrt{5}$ ft per sec ≈ 35.8 ft per sec

STUDY SKILLS BUILDER

Are You Prepared for a Test on Chapter 15?

Below I have listed some *common trouble areas* for students in Chapter 15. After studying for your test—but before taking your test—read these.

- Do you understand the difference between $\sqrt{3} \cdot \sqrt{2}$ and $\sqrt{3} + \sqrt{2}$?

  $$\sqrt{3} \cdot \sqrt{2} = \sqrt{3 \cdot 2} = \sqrt{6}$$

  $\sqrt{3} + \sqrt{2}$ cannot be simplified further. The terms are unlike terms.

- Do you understand the difference between rationalizing the denominator of $\frac{\sqrt{3}}{\sqrt{7}}$ and rationalizing the denominator of $\frac{\sqrt{3}}{\sqrt{7}+1}$?

  $$\frac{\sqrt{3}}{\sqrt{7}} = \frac{\sqrt{3} \cdot \sqrt{7}}{\sqrt{7} \cdot \sqrt{7}} = \frac{\sqrt{21}}{7}$$

  $$\frac{\sqrt{3}}{\sqrt{7}+1} = \frac{\sqrt{3}(\sqrt{7}-1)}{(\sqrt{7}+1)(\sqrt{7}-1)}$$
  $$= \frac{\sqrt{3}(\sqrt{7}-1)}{7-1} = \frac{\sqrt{3}(\sqrt{7}-1)}{6}$$

- To solve an equation containing a radical, don't forget to first isolate the radical.

  $$\sqrt{x} - 10 = -4$$
  $$\sqrt{x} = 6 \quad \text{Isolate the radical.}$$
  $$(\sqrt{x})^2 = 6^2 \quad \text{Square both sides.}$$
  $$x = 36 \quad \text{Simplify.}$$

Make sure you check the proposed solution in the original equation.

Remember: This is simply a listing of a few common trouble areas. For a review of Chapter 15, see the Highlights and Chapter Review at the end of the chapter.

15.6 EXERCISE SET

FOR EXTRA HELP

Objective A *Use the Pythagorean theorem to find the length of the unknown side of each right triangle. Give an exact answer and a two-decimal-place approximation. See Examples 1 and 2.*

1.

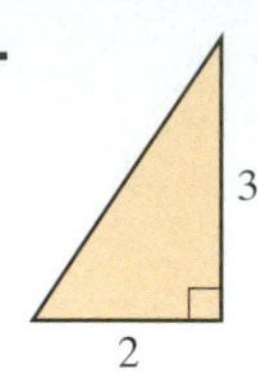

2.

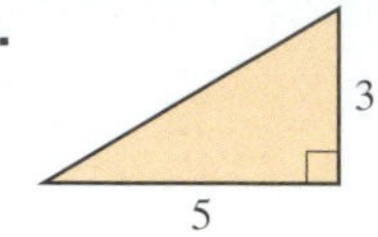

3.

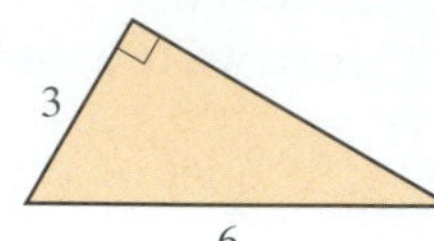

4.

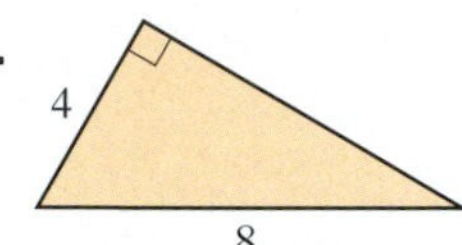

5.

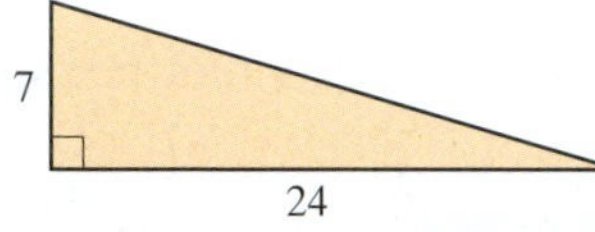

6.

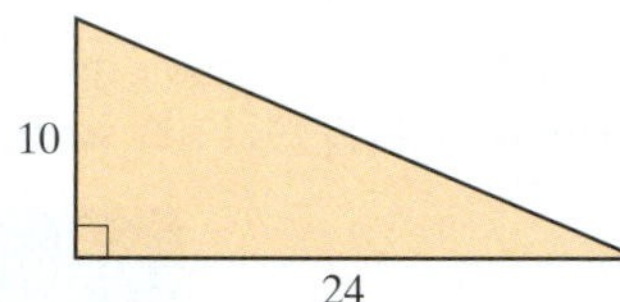

7.

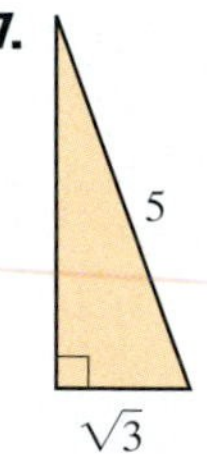

8.

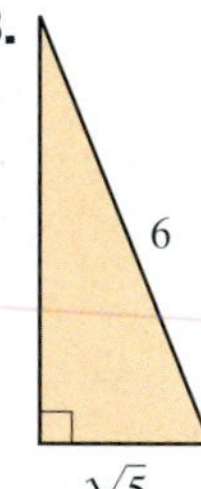

9.

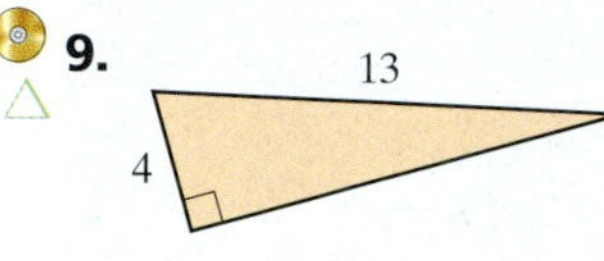

10.

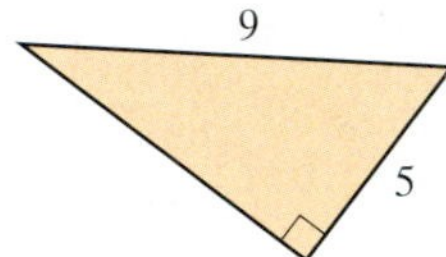

Find the length of the unknown side of each right triangle with sides a, b, and c, where c is the hypotenuse. See Examples 1 and 2. Give an exact answer and a two-decimal-place approximation.

11. $a = 4, b = 5$

12. $a = 2, b = 7$

13. $b = 2, c = 6$

14. $b = 1, c = 5$

15. $a = \sqrt{10}, c = 10$

16. $a = \sqrt{7}, c = \sqrt{35}$

Solve each problem. See Example 3.

17. A wire is used to anchor a 20-foot-tall pole. One end of the wire is attached to the top of the pole. The other end is fastened to a stake five feet away from the bottom of the pole. Find the length of the wire, to the nearest tenth of a foot.

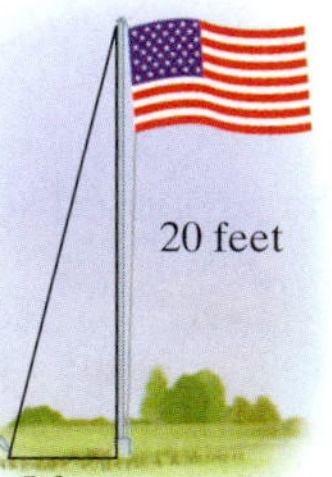

18. Jim Spivey needs to connect two underground pipelines, which are offset by 3 feet, as pictured in the diagram. Neglecting the joints needed to join the pipes, find the length of the shortest possible connecting pipe rounded to the nearest hundredth of a foot.

△ **19.** Robert Weisman needs to attach a diagonal brace to a rectangular frame in order to make it structurally sound. If the framework is 6 feet by 10 feet, find how long the brace needs to be to the nearest tenth of a foot.

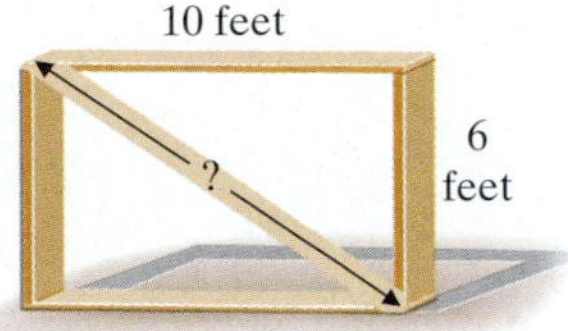

△ **20.** Elizabeth Kaster is flying a kite. She let out 80 feet of string and attached the string to a stake in the ground. The kite is now directly above her brother Mike, who is 32 feet away from the stake. Find the height of the kite to the nearest foot.

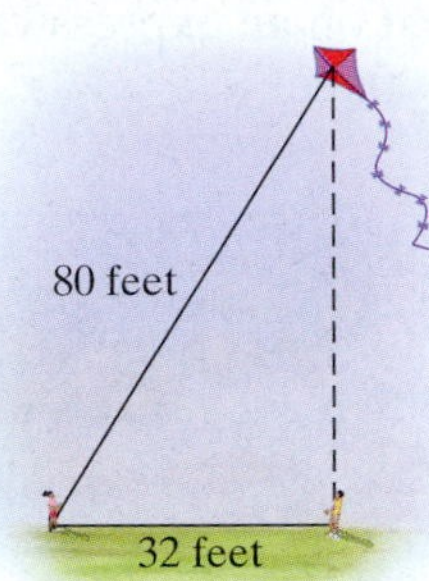

Objective **B** *Solve each problem. See Example 4.*

△ **21.** For a square-based pyramid, the formula $b = \sqrt{\frac{3V}{h}}$ describes the relationship between the length b of one side of the base, the volume V, and the height h. Find the volume if each side of the base is 6 feet long, and the pyramid is 2 feet high.

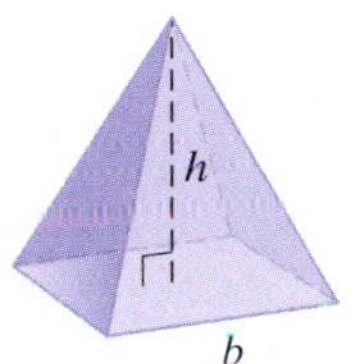

22. The formula $t = \frac{\sqrt{d}}{4}$ relates the distance d, in feet, that an object falls in t seconds, assuming that air resistance does not slow down the object. Find how long, to the nearest hundredth of a second, it takes an object to reach the ground from the top of the Sears Tower in Chicago, a distance of 1730 feet. (*Source:* Council on Tall Buildings and Urban Habitat)

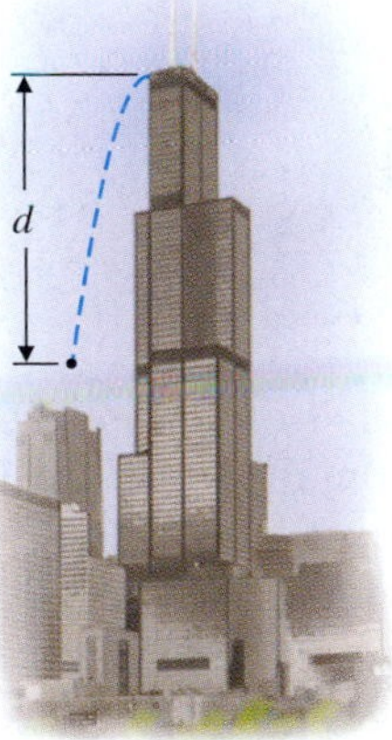

23. Police use the formula $s = \sqrt{30fd}$ to estimate the speed s of a car just before it skidded. In this formula, the speed s is measured in miles per hour, d represents the distance the car skidded in feet and f represents the coefficient of friction. The value of f depends on the type of road surface, and for wet concrete f is 0.35. Find how fast a car was moving if it skidded 280 feet on wet concrete. Round your result to the nearest mile per hour.

24. The coefficient of friction of a certain dry road is 0.95. Use the formula in Exercise 23 to find how far a car will skid on this dry road if it is traveling at a rate of 60 mph. Round the length to the nearest foot.

25. The formula $v = \sqrt{2.5r}$ can be used to estimate the maximum safe velocity, v, in miles per hour, at which a car can travel if it is driven along a curved road with a **radius of curvature** r in feet. Find the maximum safe speed to the nearest whole number if a cloverleaf exit on an expressway has a radius of curvature of 300 feet.

26. Use the formula from Exercise 25 to find the radius of curvature if the safe velocity is 30 mph.

△ **27.** The maximum distance d in kilometers that you can see from a height of h meters is given by $d = 3.5\sqrt{h}$. Find how far you can see from the top of the Bank One Tower in Indianapolis, a height of 285.4 meters. Round to the nearest tenth of a kilometer. (*Source: World Almanac and Book of Facts*, 2001)

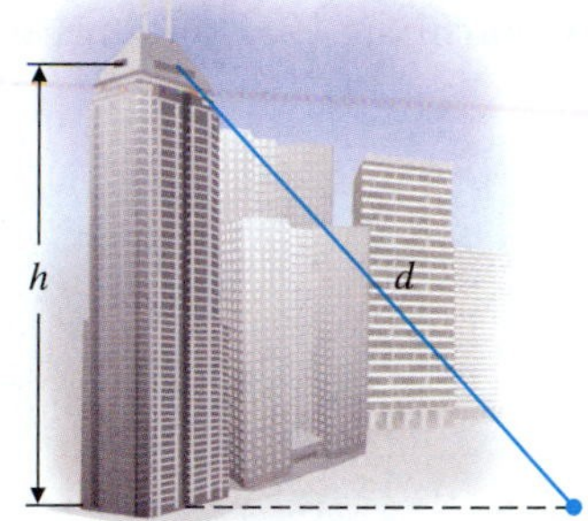

△ **28.** Use the formula from Exercise 27 to find how far you can see from the top of the Chase Tower Building in Houston, Texas, a height of 305 meters. Round to the nearest tenth of a kilometer. (*Source:* Council on Tall Buildings and Urban Habitat)

29. Use the formula from Exercise 27 to find how far you can see from the top of the First Interstate Tower in Houston, Texas, a height of 295.7 meters. Round to the nearest tenth of a kilometer. (*Source:* Council on Tall Buildings and Urban Habitat)

30. Use the formula from Exercise 27 to find how far you can see from the top of the Gas Company Tower in Los Angeles, California, a height of 228.3 m. Round to the nearest tenth of a kilometer. (*Source:* Council on Tall Buildings and Urban Habitat)

Review

Find two numbers whose square is the given number. See Section 15.1.

31. 9 **32.** 25 **33.** 100

34. 49 **35.** 64 **36.** 121

Concept Extensions

For each triangle, find the length of y, then x.

△ **37.**

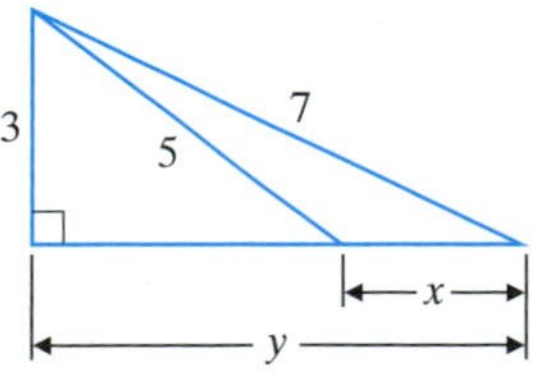

△ **38.**

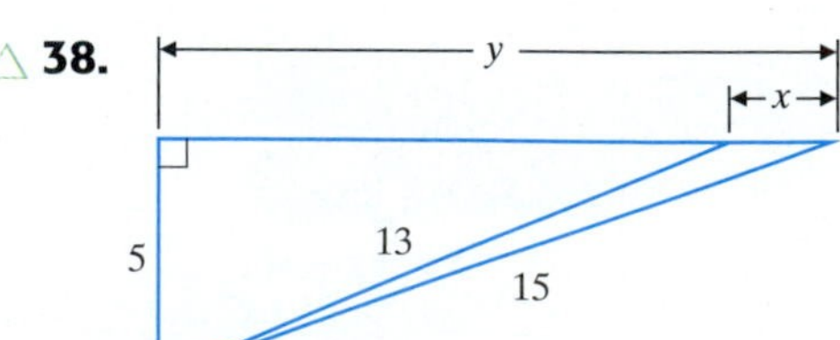

△ **39.** Mike and Sandra Hallahan leave the seashore at the same time. Mike drives northward at a rate of 30 miles per hour, while Sandra drives west at 60 mph. Find how far apart they are after 3 hours to the nearest mile.

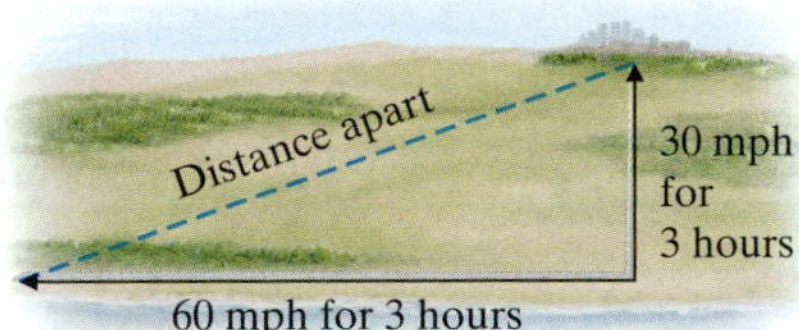

△ **40.** Railroad tracks are invariably made up of relatively short sections of rail connected by expansion joints. To see why this construction is necessary, consider a single rail 100 feet long (or 1200 inches). On an extremely hot day, suppose it expands 1 inch in the hot sun to a new length of 1201 inches. Theoretically, the track would bow upward as pictured.

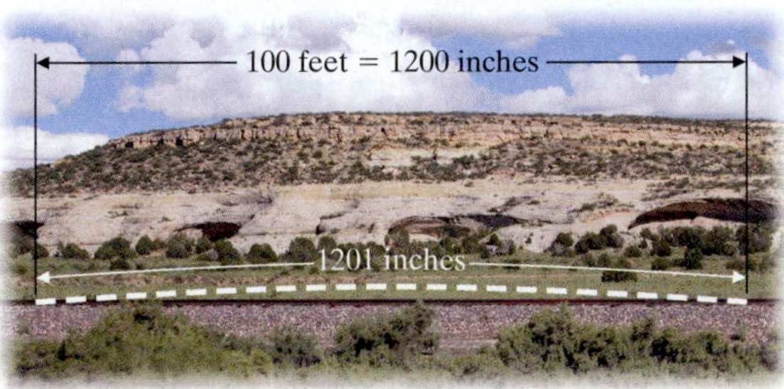

Let us approximate the bulge in the railroad this way.

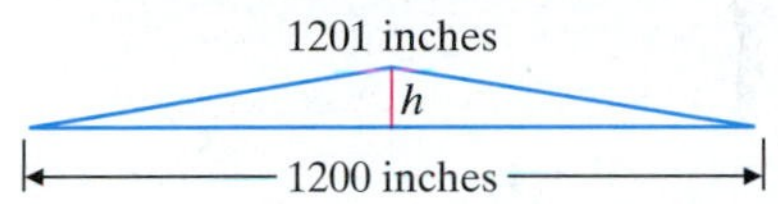

Calculate the height h of the bulge to the nearest tenth of an inch.

41. Based on the results of Exercise 40, explain why railroads use short sections of rail connected by expansion joints.

CHAPTER 15 Group Activity

Graphing and the Distance Formula

One application of radicals is finding the distance between two points in the coordinate plane. This can be very useful in graphing.

The distance d between two points with coordinates (x_1, y_1) and (x_2, y_2) is given by the **distance formula** $d = \sqrt{(x_2 - x_1)^2 + (y_2 - y_1)^2}$.

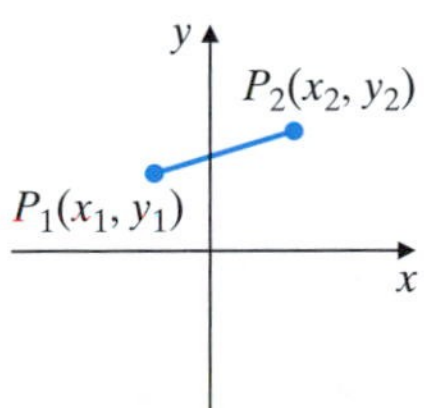

Suppose we want to find the distance between the two points $(-1, 9)$ and $(3, 5)$. We can use the distance formula with $(x_1, y_1) = (-1, 9)$ and $(x_2, y_2) = (3, 5)$. Then we have

$$\begin{aligned} d &= \sqrt{(x_2 - x_1)^2 + (y_2 - y_1)^2} \\ &= \sqrt{[3 - (-1)]^2 + (5 - 9)^2} \\ &= \sqrt{(4)^2 + (-4)^2} \\ &= \sqrt{16 + 16} \\ &= \sqrt{32} = 4\sqrt{2} \end{aligned}$$

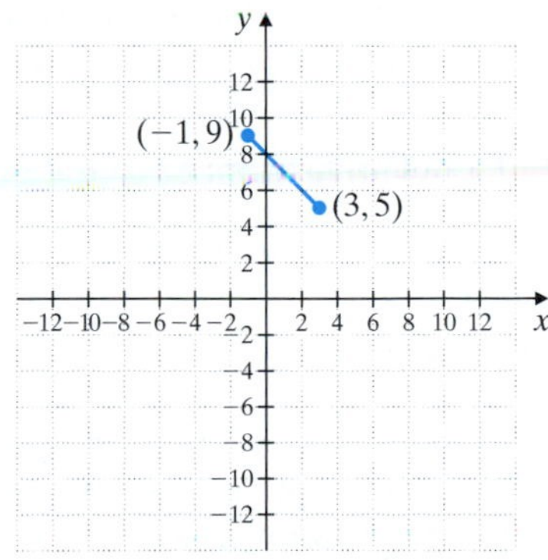

The distance between the two points is exactly $4\sqrt{2}$ units or approximately 5.66 units.

Group Activity

Brainstorm to come up with several disciplines or activities in which the distance formula might be useful. Make up an example that shows how the distance formula would be used in one of the activities on your list. Then present your example to the rest of the class.

Chapter 15 Vocabulary Check

Fill in each blank with one of the words or phrases listed below.

index	radicand	like radicals
rationalizing the denominator	conjugate	
principal square root	radical	

1. The expressions $5\sqrt{x}$ and $7\sqrt{x}$ are examples of ______________.
2. In the expression $\sqrt[3]{45}$ the number 3 is the _________, the number 45 is the ___________, and $\sqrt{}$ is called the __________ sign.
3. The _____________ of $a + b$ is $a - b$.
4. The ______________________ of 25 is 5.
5. The process of eliminating the radical in the denominator of a radical expression is called ______________________________.

Helpful Hint

Are you preparing for your test? Don't forget to take the Chapter 15 Test on page 1156. Then check your answers at the back of the text and use the Chapter Test Prep Video CD to see the fully worked-out solutions to any of the exercises you want to review.

15 Chapter Highlights

Definitions and Concepts	Examples
Section 15.1 Introduction to Radicals	
The **positive or principal square root** of a positive number a is written as $\sqrt{a}$. The **negative square root** of a is written as $-\sqrt{a}$. $\sqrt{a} = b$ only if $b^2 = a$ and $b > 0$.	$\sqrt{25} = 5$ $\sqrt{100} = 10$ $-\sqrt{9} = -3$ $\sqrt{\frac{4}{49}} = \frac{2}{7}$
A square root of a negative number is not a real number.	$\sqrt{-4}$ is not a real number.
The **cube root** of a real number a is written as $\sqrt[3]{a}$ and $\sqrt[3]{a} = b$ only if $b^3 = a$. The ***n*th root** of a number a is written as $\sqrt[n]{a}$ and $\sqrt[n]{a} = b$ only if $b^n = a$. In $\sqrt[n]{a}$, the natural number n is called the **index,** the symbol $\sqrt{}$ is called a **radical,** and the expression within the radical is called the **radicand.** (*Note:* If the index is even, the radicand must be nonnegative for the root to be a real number.)	$\sqrt[3]{64} = 4$ $\sqrt[3]{-8} = -2$ $\sqrt[4]{81} = 3$ $\sqrt[5]{-32} = -2$ index ↓ $\sqrt[n]{a}$ ↑ radicand
Section 15.2 Simplifying Radicals	
Product Rule for Radicals If $\sqrt{a}$ and $\sqrt{b}$ are real numbers, then $\sqrt{a \cdot b} = \sqrt{a} \cdot \sqrt{b}$ A square root is in **simplified form** if the radicand contains no perfect square factors other than 1. To simplify a square root, factor the radicand so that one of its factors is a perfect square factor.	$\sqrt{45} = \sqrt{9 \cdot 5}$ $= \sqrt{9} \cdot \sqrt{5}$ $= 3\sqrt{5}$

Definitions and Concepts	**Examples**
Section 15.2 Simplifying Radicals (*continued*)	
Quotient Rule for Radicals If $\sqrt{a}$ and $\sqrt{b}$ are real numbers and $b \neq 0$, then $\sqrt{\frac{a}{b}} = \frac{\sqrt{a}}{\sqrt{b}}$	$\sqrt{\frac{18}{x^6}} = \frac{\sqrt{9 \cdot 2}}{\sqrt{x^6}} = \frac{\sqrt{9} \cdot \sqrt{2}}{x^3} = \frac{3\sqrt{2}}{x^3}$
Section 15.3 Adding and Subtracting Radicals	
Like radicals are radical expressions that have the same index and the same radicand.	$5\sqrt{2}, -7\sqrt{2}, \sqrt{2}$
To **combine like radicals** use the distributive property.	$2\sqrt{7} - 13\sqrt{7} = (2 - 13)\sqrt{7} = -11\sqrt{7}$ $\sqrt{8} + \sqrt{50} = 2\sqrt{2} + 5\sqrt{2} = 7\sqrt{2}$
Section 15.4 Multiplying and Dividing Radicals	
The product and quotient rules for radicals may be used to simplify products and quotients of radicals.	Perform each indicated operation and simplify. Multiply. $\sqrt{2} \cdot \sqrt{8} = \sqrt{16} = 4$ $(\sqrt{3x} + 1)(\sqrt{5} - \sqrt{3})$ $= \sqrt{15x} - \sqrt{9x} + \sqrt{5} - \sqrt{3}$ $= \sqrt{15x} - 3\sqrt{x} + \sqrt{5} - \sqrt{3}$ Divide. $\frac{\sqrt{20}}{\sqrt{2}} = \sqrt{\frac{20}{2}} = \sqrt{10}$
The process of eliminating the radical in the denominator of a radical expression is called **rationalizing the denominator.**	Rationalize the denominator. $\frac{5}{\sqrt{11}} = \frac{5 \cdot \sqrt{11}}{\sqrt{11} \cdot \sqrt{11}} = \frac{5\sqrt{11}}{11}$
The **conjugate** of $a + b$ is $a - b$.	The conjugate of $2 + \sqrt{3}$ is $2 - \sqrt{3}$.
To rationalize a denominator that is a sum or difference of radicals, multiply the numerator and the denominator by the conjugate of the denominator.	Rationalize the denominator. $\frac{5}{6 - \sqrt{5}} = \frac{5(6 + \sqrt{5})}{(6 - \sqrt{5})(6 + \sqrt{5})}$ $= \frac{5(6 + \sqrt{5})}{36 - 5}$ $= \frac{5(6 + \sqrt{5})}{31}$

Definitions and Concepts	**Examples**
Section 15.5 Solving Equations Containing Radicals	
To Solve a Radical Equation Containing Square Roots **Step 1.** Get one radical by itself on one side of the equation. **Step 2.** Square both sides of the equation. **Step 3.** Simplify both sides of the equation. **Step 4.** If the equation still contains a radical term, repeat Steps 1 through 3. **Step 5.** Solve the equation. **Step 6.** Check solutions in the original equation.	Solve: $\sqrt{2x-1} - x = -2$ $\sqrt{2x-1} = x - 2$ $(\sqrt{2x-1})^2 = (x-2)^2$ Square both sides. $2x - 1 = x^2 - 4x + 4$ $0 = x^2 - 6x + 5$ $0 = (x-1)(x-5)$ Factor. $x - 1 = 0$ or $x - 5 = 0$ $x = 1$ $x = 5$ Solve. Check both proposed solutions in the original equation. Here, 5 checks but 1 does not. The only solution is 5.
Section 15.6 Radical Equations and Problem Solving	
Problem-Solving Steps **1.** UNDERSTAND. Read and reread the problem.	A rain gutter is to be mounted on the eaves of a house 15 feet above the ground. A garden is adjacent to the house so that the closest a ladder can be placed to the house is 6 feet. How long a ladder is needed for installing the gutter? Let $x =$ the length of the ladder. 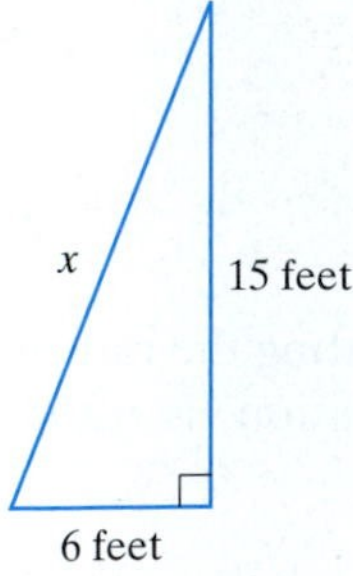
2. TRANSLATE.	Here, we use the Pythagorean theorem. The unknown length x is the hypotenuse. In words: $(\text{leg})^2 + (\text{leg})^2 = (\text{hypotenuse})^2$ Translate:
3. SOLVE.	$6^2 + 15^2 = x^2$ $36 + 225 = x^2$ $261 = x^2$ $\sqrt{261} = x$ or $x = 3\sqrt{29}$
4. INTERPRET.	Check and state. The ladder needs to be $3\sqrt{29}$ feet or approximately 16.2 feet long.

15 CHAPTER REVIEW

(15.1) *Find each root.*

1. $\sqrt{81}$ **2.** $-\sqrt{49}$ **3.** $\sqrt[3]{27}$ **4.** $\sqrt[4]{81}$

5. $-\sqrt{\frac{9}{64}}$ **6.** $\sqrt{\frac{36}{81}}$ **7.** $\sqrt[4]{16}$ **8.** $\sqrt[3]{-8}$

9. Which radical(s) is not a real number?

a. $\sqrt{4}$ **b.** $-\sqrt{4}$ **c.** $\sqrt{-4}$ **d.** $\sqrt[3]{-4}$

10. Which radical(s) is not a real number?

a. $\sqrt{-5}$ **b.** $\sqrt[3]{-5}$ **c.** $\sqrt[4]{-5}$ **d.** $\sqrt[5]{-5}$

Find each root. Assume that all variables represent positive numbers.

11. $\sqrt{x^{12}}$ **12.** $\sqrt{x^8}$ **13.** $\sqrt{9y^2}$ **14.** $\sqrt{25x^4}$

(15.2) *Simplify each expression using the product rule. Assume that all variables represent positive numbers.*

15. $\sqrt{40}$ **16.** $\sqrt{24}$ **17.** $\sqrt{54}$ **18.** $\sqrt{88}$ **19.** $\sqrt{x^5}$

20. $\sqrt{y^7}$ **21.** $\sqrt{20x^2}$ **22.** $\sqrt{50y^4}$ **23.** $\sqrt[3]{54}$ **24.** $\sqrt[3]{88}$

Simplify each expression using the quotient rule. Assume that all variables represent positive numbers.

25. $\sqrt{\frac{18}{25}}$ **26.** $\sqrt{\frac{75}{64}}$ **27.** $-\sqrt{\frac{50}{9}}$ **28.** $-\sqrt{\frac{12}{49}}$

29. $\sqrt{\frac{11}{x^2}}$ **30.** $\sqrt{\frac{7}{y^4}}$ **31.** $\sqrt{\frac{y^5}{100}}$ **32.** $\sqrt{\frac{x^3}{81}}$

(15.3) *Add or subtract by combining like radicals.*

33. $5\sqrt{2} - 8\sqrt{2}$

34. $\sqrt{3} - 6\sqrt{3}$

35. $6\sqrt{5} + 3\sqrt{6} - 2\sqrt{5} + \sqrt{6}$

36. $-\sqrt{7} + 8\sqrt{2} - \sqrt{7} - 6\sqrt{2}$

Add or subtract by simplifying each radical and then combining like terms. Assume that all variables represent positive numbers.

37. $\sqrt{28} + \sqrt{63} + \sqrt{56}$

38. $\sqrt{75} + \sqrt{48} - \sqrt{16}$

39. $\sqrt{\frac{5}{9}} - \sqrt{\frac{5}{36}}$

40. $\sqrt{\frac{11}{25}} + \sqrt{\frac{11}{16}}$

41. $\sqrt{45x^2} + 3\sqrt{5x^2} - 7x\sqrt{5} + 10$

42. $\sqrt{50x} - 9\sqrt{2x} + \sqrt{72x} - \sqrt{3x}$

(15.4) *Multiply and simplify if possible. Assume that all variables represent positive numbers.*

43. $\sqrt{3} \cdot \sqrt{6}$

44. $\sqrt{5} \cdot \sqrt{15}$

45. $\sqrt{2}(\sqrt{5}-\sqrt{7})$

46. $\sqrt{5}(\sqrt{11}+\sqrt{3})$

47. $(\sqrt{3}+2)(\sqrt{6}-5)$

48. $(\sqrt{5}+1)(\sqrt{5}-3)$

49. $(\sqrt{x}-2)^2$

50. $(\sqrt{y}+4)^2$

Divide and simplify if possible. Assume that all variables represent positive numbers.

51. $\dfrac{\sqrt{27}}{\sqrt{3}}$

52. $\dfrac{\sqrt{20}}{\sqrt{5}}$

53. $\dfrac{\sqrt{160}}{\sqrt{8}}$

54. $\dfrac{\sqrt{96}}{\sqrt{3}}$

55. $\dfrac{\sqrt{30x^6}}{\sqrt{2x^3}}$

56. $\dfrac{\sqrt{54x^5y^2}}{\sqrt{3xy^2}}$

Rationalize each denominator and simplify.

57. $\dfrac{\sqrt{2}}{\sqrt{11}}$

58. $\dfrac{\sqrt{3}}{\sqrt{13}}$

59. $\sqrt{\dfrac{5}{6}}$

60. $\sqrt{\dfrac{7}{10}}$

61. $\dfrac{1}{\sqrt{5x}}$

62. $\dfrac{5}{\sqrt{3y}}$

63. $\sqrt{\dfrac{3}{x}}$

64. $\sqrt{\dfrac{6}{y}}$

65. $\dfrac{3}{\sqrt{5}-2}$

66. $\dfrac{8}{\sqrt{10}-3}$

67. $\dfrac{\sqrt{2}+1}{\sqrt{3}-1}$

68. $\dfrac{\sqrt{3}-2}{\sqrt{5}+2}$

69. $\dfrac{10}{\sqrt{x}+5}$

70. $\dfrac{8}{\sqrt{x}-1}$

(15.5) *Solve each radical equation.*

71. $\sqrt{2x}=6$

72. $\sqrt{x+3}=4$

73. $\sqrt{x}+3=8$

74. $\sqrt{x}+8=3$

75. $\sqrt{2x+1}=x-7$

76. $\sqrt{3x+1}=x-1$

77. $\sqrt{x}+3=\sqrt{x+15}$

78. $\sqrt{x-5}=\sqrt{x}-1$

(15.6) *Use the Pythagorean theorem to find the length of each unknown side. Give an exact answer and a two-decimal-place approximation.*

 **79.**

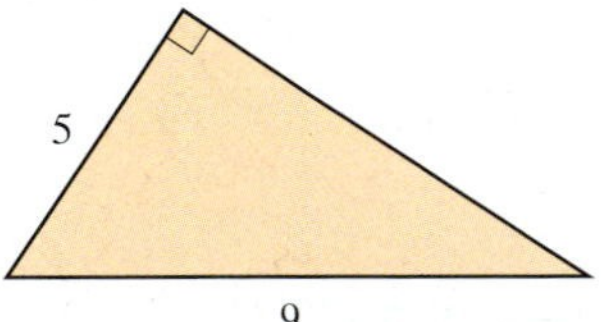

△ **80.**

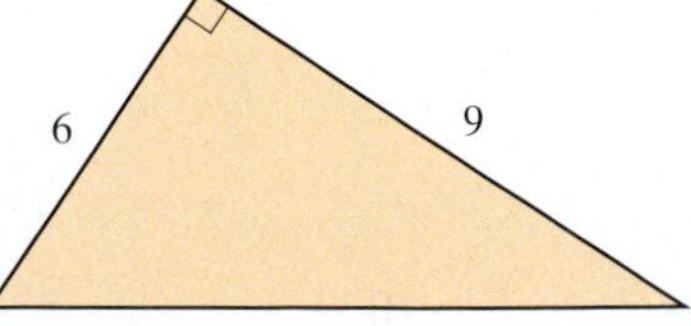

△ **81.** Romeo is standing 20 feet away from the wall below Juliet's balcony during a school play. Juliet is on the balcony, 12 feet above the ground. Find how far apart Romeo and Juliet are.

△ **82.** The diagonal of a rectangle is 10 inches long. If the width of the rectangle is 5 inches, find the length of the rectangle.

Use the formula $r = \sqrt{\frac{S}{4\pi}}$, *where* r = *the radius of a sphere and* S = *the surface area of the sphere, for Exercises 83 and 84.*

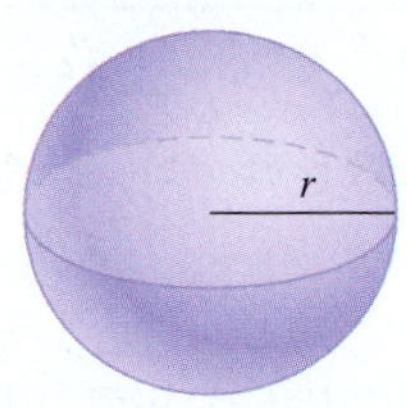

△ **83.** Find the radius of a sphere to the nearest tenth of an inch if the surface area is 72 square inches.

△ **84.** Find the exact surface area of a sphere if its radius is 6 inches. (Do not approximate π.)

Mixed Review

Find each root. Assume all variables represent positive numbers.

85. $\sqrt{144}$

86. $-\sqrt[3]{64}$

87. $\sqrt{16x^{16}}$

88. $\sqrt{4x^{24}}$

Simplify each expression. Assume all variables represent positive numbers.

89. $\sqrt{18x^7}$

90. $\sqrt{48y^6}$

91. $\sqrt{\frac{y^4}{81}}$

92. $\sqrt{\frac{x^9}{9}}$

Add or subtract by simplifying and then combining like terms. Assume all variables represent positive numbers.

93. $\sqrt{12} + \sqrt{75}$

94. $\sqrt{63} + \sqrt{28} - \sqrt{9}$

95. $\sqrt{\frac{3}{16}} - \sqrt{\frac{3}{4}}$

96. $\sqrt{45x^3} + x\sqrt{20x} - \sqrt{5x^3}$

Multiply and simplify if possible. Assume all variables represent positive numbers.

97. $\sqrt{7} \cdot \sqrt{14}$

98. $\sqrt{3}(\sqrt{9} - \sqrt{2})$

99. $(\sqrt{2} + 4)(\sqrt{5} - 1)$

100. $(\sqrt{x} + 3)^2$

Divide and simplify if possible. Assume all variables represent positive numbers.

101. $\frac{\sqrt{120}}{\sqrt{5}}$

102. $\frac{\sqrt{60x^9}}{\sqrt{15x^7}}$

Rationalize each denominator and simplify.

103. $\sqrt{\frac{2}{7}}$

104. $\frac{3}{\sqrt{2x}}$

105. $\frac{3}{\sqrt{x} - 6}$

106. $\frac{\sqrt{7} - 5}{\sqrt{5} + 3}$

Solve each radical equation.

107. $\sqrt{4x} = 2$

108. $\sqrt{x - 4} = 3$

109. $\sqrt{4x + 8} + 6 = x$

110. $\sqrt{x - 8} = \sqrt{x} - 2$

111. Use the Pythagorean theorem to find the length of the unknown side. Give an exact answer and a two-decimal-place approximation.

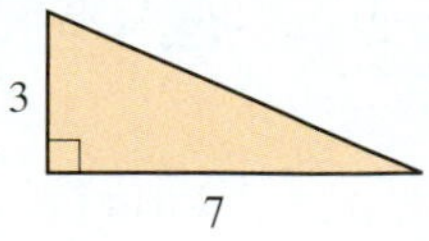

112. The diagonal of a rectangle is 6 inches long. If the width of the rectangle is 2 inches, find the length of the rectangle.

Answers

1. ____

2. ____

3. ____

4. ____

5. ____

6. ____

7. ____

8. ____

9. ____

10. ____

11. ____

12. ____

13. ____

14. ____

15. ____

16. ____

17. ____

18. ____

19. ____

15 CHAPTER TEST

Remember to use the Chapter Test Prep Video CD to see the fully worked-out solutions to any of the exercises you want to review.

Simplify each radical. Indicate if the radical is not a real number. Assume that x represents a positive number.

1. $\sqrt{16}$

2. $\sqrt[3]{125}$

3. $\sqrt[4]{81}$

4. $\sqrt{\frac{9}{16}}$

5. $\sqrt[4]{-81}$

6. $\sqrt{x^{10}}$

Simplify each radical. Assume that all variables represent positive numbers.

7. $\sqrt{54}$

8. $\sqrt{92}$

9. $\sqrt{y^7}$

10. $\sqrt{24x^8}$

11. $\sqrt[3]{27}$

12. $\sqrt[3]{16}$

13. $\sqrt{\frac{5}{16}}$

14. $\sqrt{\frac{y^3}{25}}$

Perform each indicated operation. Assume that all variables represent positive numbers.

15. $\sqrt{13} + \sqrt{13} - 4\sqrt{13}$

16. $\sqrt{18} - \sqrt{75} + 7\sqrt{3} - \sqrt{8}$

17. $\sqrt{\frac{3}{4}} + \sqrt{\frac{3}{25}}$

18. $\sqrt{7} \cdot \sqrt{14}$

19. $\sqrt{2}(\sqrt{6} - \sqrt{5})$

20. $(\sqrt{x}+2)(\sqrt{x}-3)$

21. $\dfrac{\sqrt{50}}{\sqrt{10}}$

22. $\dfrac{\sqrt{40x^4}}{\sqrt{2x}}$

Rationalize each denominator. Assume that all variables represent positive numbers.

23. $\sqrt{\dfrac{2}{3}}$

24. $\dfrac{8}{\sqrt{5y}}$

25. $\dfrac{8}{\sqrt{6}+2}$

26. $\dfrac{1}{3-\sqrt{x}}$

Solve each radical equation.

27. $\sqrt{x}+8=11$

28. $\sqrt{3x-6}=\sqrt{x+4}$

29. $\sqrt{2x-2}=x-5$

△ **30.** Find the length of the unknown leg of the right triangle shown. Give an exact answer.

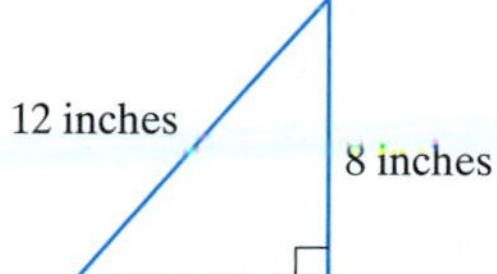

△ **31.** The formula $r=\sqrt{\dfrac{A}{\pi}}$ can be used to find the radius r of a circle given its area A. Use this formula to approximate the radius of the given circle. Round to two decimal places.

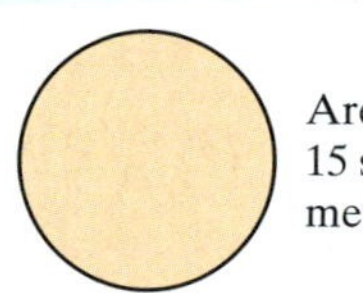

20. ______________

21. ______________

22. ______________

23. ______________

24. ______________

25. ______________

26. ______________

27. ______________

28. ______________

29. ______________

30. ______________

31. ______________

CUMULATIVE REVIEW

Chapters 1–15

Answers

1. ______
2. ______
3. ______
4. ______
5. ______
6. ______
7. ______
8. ______
9. ______
10. ______
11. a. ______ b. ______ c. ______ d. ______
12. a. ______ b. ______ c. ______ d. ______
13. ______
14. ______
15. ______
16. ______
17. ______
18. ______
19. a. ______ b. ______ c. ______
20. a. ______ b. ______ c. ______

1. Round 736.2359 to the nearest tenth.

2. Round 328.174 to the nearest tenth.

3. Add: $23.85 + 1.604$

4. Add: $12.762 + 4.29$

5. Is -9 a solution of the equation $3.7y = -3.33$?

6. Is 6 a solution of the equation $2.8x = 16.8$?

7. Find: $\sqrt{\frac{1}{36}}$

8. Find: $\sqrt{\frac{4}{25}}$

9. Solve: $4(2x - 3) + 7 = 3x + 5$

10. Solve: $3(2 - 5x) + 24x = 12$

11. Write the following numbers in standard notation, without exponents.

a. 1.02×10^5
b. 7.358×10^{-3}
c. 8.4×10^7
d. 3.007×10^{-5}

12. Write the following numbers in standard notation, without exponents.

a. 8.26×10^4
b. 9.9×10^{-2}
c. 1.002×10^5
d. 8.039×10^{-3}

13. Multiply: $(3x + 2)(2x - 5)$

14. Multiply: $(5x - 1)(4x + 1)$

15. Factor $xy + 2x + 3y + 6$ by grouping.

16. Factor: $16x^3 - 28x^2 + 12x - 21$

17. Factor: $3x^2 + 11x + 6$

18. Factor: $9x^2 - 5x - 4$

Are there any values for x for which each expression is undefined?

19. a. $\frac{x}{x - 3}$

b. $\frac{x^2 + 2}{x^2 - 3x + 2}$

c. $\frac{x^3 - 6x^2 - 10x}{3}$

20. a. $\frac{x - 3}{x}$

b. $\frac{x + 1}{5}$

c. $\frac{x^2 - 3}{x^2 - 4}$

21. Simplify: $\frac{x^2 + 4x + 4}{x^2 + 2x}$

22. Simplify: $\frac{16x^2 - 4y^2}{4x - 2y}$

Perform each indicated operation.

23. a. $\frac{a}{4} - \frac{2a}{8}$

b. $\frac{3}{10x^2} + \frac{7}{25x}$

24. a. $\frac{x}{5} - \frac{3x}{10}$

b. $\frac{9}{12a^2} + \frac{5}{16a}$

25. Solve: $\frac{4x}{x^2 + x - 30} + \frac{2}{x - 5} = \frac{1}{x + 6}$

26. Solve: $\frac{3}{x + 3} = \frac{12x + 19}{x^2 + 7x + 12} - \frac{5}{x + 4}$

27. Graph $y = -3$.

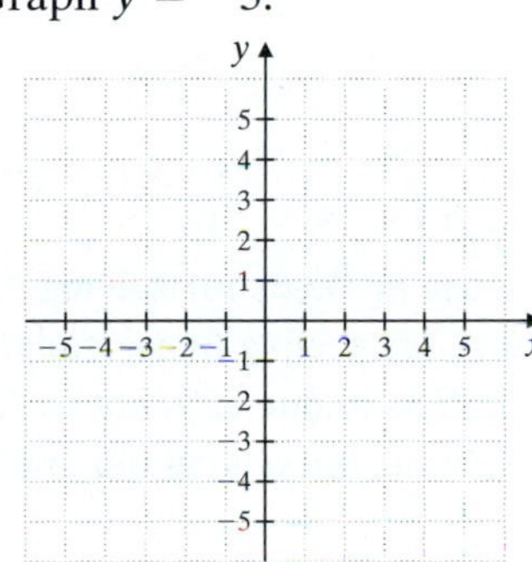

28. Graph $x = 2$.

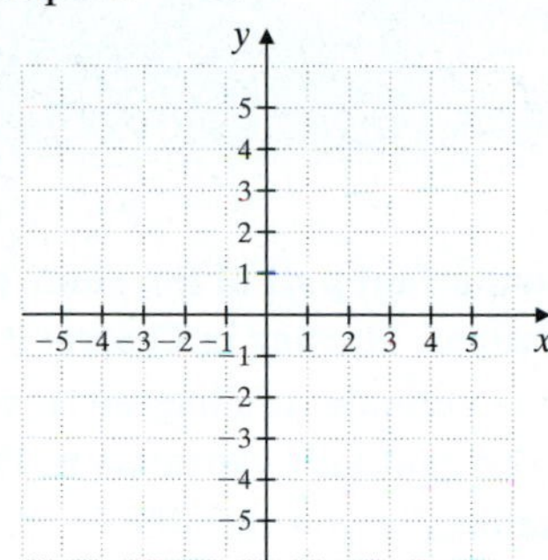

Find each cube root.

29. $\sqrt[3]{1}$

30. $\sqrt[3]{8}$

31. $\sqrt[3]{-27}$

32. $\sqrt[3]{-8}$

33. $\sqrt[3]{\frac{1}{125}}$

34. $\sqrt[3]{\frac{27}{64}}$

Simplify.

35. $\sqrt{54}$

36. $\sqrt{40}$

37. $\sqrt{200}$

38. $\sqrt{125}$

Add or subtract by first simplifying each radical.

39. $7\sqrt{12} - 2\sqrt{75}$

40. $\sqrt{75} + \sqrt{48}$

41. $2\sqrt{x^2} - \sqrt{25x^5} + \sqrt{x^5}$

42. $5\sqrt{x^2} + \sqrt{36x} + \sqrt{49x^2}$

Rationalize each denominator.

43. $\frac{2}{\sqrt{7}}$

44. $\frac{4}{\sqrt{5}}$

45. Solve: $\sqrt{x} = \sqrt{5x - 2}$

46. Solve: $\sqrt{x + 5} = x - 1$

21. ______
22. ______
23. a. ______
b. ______
24. a. ______
b. ______
25. ______
26. ______
27. see graph
28. see graph
29. ______
30. ______
31. ______
32. ______
33. ______
34. ______
35. ______
36. ______
37. ______
38. ______
39. ______
40. ______
41. ______
42. ______
43. ______
44. ______
45. ______
46. ______

16 Quadratic Equations

An important part of the study of algebra is learning to use methods for solving equations. Starting in Chapter 2, we presented techniques for solving linear equations in one variable. In Chapter 11, we solved quadratic equations in one variable by factoring the quadratic expressions. We now present other methods for solving quadratic equations in one variable.

SECTIONS

CHECK YOUR PROGRESS

Your heart is a muscular pump that continuously circulates blood throughout your body. Coronary heart disease and congestive heart failure are two examples of cardiovascular diseases that affect more than 64 million Americans and accounted for about 900,000 deaths in a recent year. Unfortunately, many deaths occur while severe heart failure patients wait for donor hearts.

In 2001, history was made when the AbioCor® artificial heart, developed by ABIOMED, was successfully implanted in a 58-year-old telephone company employee and teacher. This artificial heart is completely self-contained and is the first artificial heart to be used in nearly 20 years. In Section 16.3, Exercise 70, we will use a method called the quadratic formula to predict when the net income of the ABIOMED Corporation will reach a certain level.

Percent of Active Heart Registrants at End of Calendar Year by Time Waiting, 1996–2003

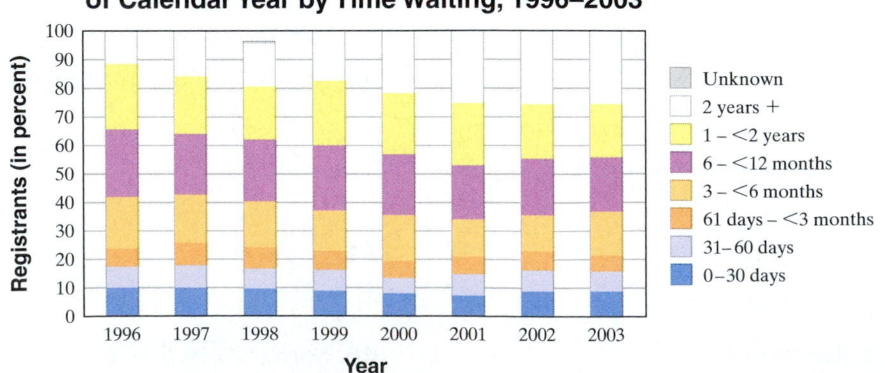

Source: 2004 OPTN/SRTR Annual Report, Additional Analyses.

16.1 SOLVING QUADRATIC EQUATIONS BY THE SQUARE ROOT PROPERTY

Recall that a quadratic equation is an equation that can be written in the form

$$ax^2 + bx + c = 0$$

where a, b, and c are real numbers and $a \neq 0$.

Objective A Solving Quadratic Equations by Factoring

Recall from Section 11.6 that to solve quadratic equations by factoring, we use the **zero factor property:** If the product of two numbers is zero, then at least one of the two numbers is zero. Examples 1 and 2 review the process of solving quadratic equations by factoring.

EXAMPLE 1 Solve: $x^2 - 4 = 0$

Solution:

$$x^2 - 4 = 0$$

$$(x + 2)(x - 2) = 0 \quad \text{Factor.}$$

$$x + 2 = 0 \quad \text{or} \quad x - 2 = 0 \quad \text{Use the zero factor property.}$$

$$x = -2 \qquad\qquad x = 2 \quad \text{Solve each equation.}$$

The solutions are -2 and 2.

Work Practice Problem 1

EXAMPLE 2 Solve: $3y^2 + 13y = 10$

Solution: Recall that to use the zero factor property, one side of the equation must be 0 and the other side must be factored.

$$3y^2 + 13y = 10$$

$$3y^2 + 13y - 10 = 0 \quad \text{Subtract 10 from both sides.}$$

$$(3y - 2)(y + 5) = 0 \quad \text{Factor.}$$

$$3y - 2 = 0 \quad \text{or} \quad y + 5 = 0 \quad \text{Use the zero factor property.}$$

$$3y = 2 \qquad\qquad y = -5 \quad \text{Solve each equation.}$$

$$y = \frac{2}{3}$$

The solutions are $\frac{2}{3}$ and -5.

Work Practice Problem 2

Objective B Using the Square Root Property

Consider solving Example 1, $x^2 - 4 = 0$, another way. First, add 4 to both sides of the equation.

$$x^2 - 4 = 0$$

$$x^2 = 4 \quad \text{Add 4 to both sides.}$$

Now we see that the value for x must be a number whose square is 4. Therefore $x = \sqrt{4} = 2$ or $x = -\sqrt{4} = -2$. This reasoning is an example of the square root property.

Objectives

A Review Factoring to Solve Quadratic Equations.

B Use the Square Root Property to Solve Quadratic Equations.

C Use the Square Root Property to Solve Applications.

PRACTICE PROBLEM 1

Solve: $x^2 - 25 = 0$

PRACTICE PROBLEM 2

Solve: $2x^2 - 3x = 9$

Answers

1. 5 and -5 **2.** $-\frac{3}{2}$ and 3

Square Root Property

If $x^2 = a$ for $a \geq 0$, then

$$x = \sqrt{a} \quad \text{or} \quad x = -\sqrt{a}$$

PRACTICE PROBLEM 3

Use the square root property to solve $x^2 - 16 = 0$.

EXAMPLE 3 Use the square root property to solve $x^2 - 9 = 0$.

Solution: First we solve for x^2 by adding 9 to both sides.

$$x^2 - 9 = 0$$
$$x^2 = 9 \quad \text{Add 9 to both sides.}$$

Next we use the square root property.

$$x = \sqrt{9} \quad \text{or} \quad x = -\sqrt{9}$$
$$x = 3 \qquad\qquad x = -3$$

Check:

$$x^2 - 9 = 0 \quad \text{Original equation}$$
$$3^2 - 9 \stackrel{?}{=} 0 \quad \text{Let } x = 3.$$
$$0 = 0 \quad \text{True}$$

$$x^2 - 9 = 0 \quad \text{Original equation}$$
$$(-3)^2 - 9 \stackrel{?}{=} 0 \quad \text{Let } x = -3.$$
$$0 = 0 \quad \text{True}$$

The solutions are 3 and -3.

Work Practice Problem 3

PRACTICE PROBLEM 4

Use the square root property to solve $3x^2 = 11$.

EXAMPLE 4 Use the square root property to solve $2x^2 = 7$.

Solution: First we solve for x^2 by dividing both sides by 2. Then we use the square root property.

$$2x^2 = 7$$
$$x^2 = \frac{7}{2} \quad \text{Divide both sides by 2.}$$
$$x = \sqrt{\frac{7}{2}} \quad \text{or} \quad x = -\sqrt{\frac{7}{2}} \quad \text{Use the square root property.}$$
$$x = \frac{\sqrt{7}\cdot\sqrt{2}}{\sqrt{2}\cdot\sqrt{2}} \qquad x = -\frac{\sqrt{7}\cdot\sqrt{2}}{\sqrt{2}\cdot\sqrt{2}} \quad \text{Rationalize the denominator.}$$
$$x = \frac{\sqrt{14}}{2} \qquad x = -\frac{\sqrt{14}}{2} \quad \text{Simplify.}$$

Remember to check both solutions in the original equation. The solutions are $\frac{\sqrt{14}}{2}$ and $-\frac{\sqrt{14}}{2}$.

Work Practice Problem 4

PRACTICE PROBLEM 5

Use the square root property to solve $(x - 4)^2 = 49$.

EXAMPLE 5 Use the square root property to solve $(x - 3)^2 = 16$.

Solution: Instead of x^2, here we have $(x - 3)^2$. But the square root property can still be used.

$$(x - 3)^2 = 16$$
$$x - 3 = \sqrt{16} \quad \text{or} \quad x - 3 = -\sqrt{16} \quad \text{Use the square root property.}$$
$$x - 3 = 4 \qquad x - 3 = -4 \quad \text{Write } \sqrt{16} \text{ as 4 and } -\sqrt{16} \text{ as } -4.$$
$$x = 7 \qquad x = -1 \quad \text{Solve.}$$

Answers

3. 4 and -4 4. $\frac{\sqrt{33}}{3}$ and $-\frac{\sqrt{33}}{3}$
5. 11 and -3

Check:

$(x-3)^2 = 16$	Original equation	$(x-3)^2 = 16$	Original equation
$(7-3)^2 \stackrel{?}{=} 16$	Let $x = 7$.	$(-1-3)^2 \stackrel{?}{=} 16$	Let $x = -1$.
$4^2 \stackrel{?}{=} 16$	Simplify.	$(-4)^2 \stackrel{?}{=} 16$	Simplify.
$16 = 16$	True	$16 = 16$	True

Both 7 and −1 are solutions.

Work Practice Problem 5

EXAMPLE 6 Use the square root property to solve $(x+1)^2 = 8$.

Solution: $(x+1)^2 = 8$

$$x + 1 = \sqrt{8} \quad \text{or} \quad x + 1 = -\sqrt{8} \qquad \text{Use the square root property.}$$
$$x + 1 = 2\sqrt{2} \qquad x + 1 = -2\sqrt{2} \qquad \text{Simplify the radical.}$$
$$x = -1 + 2\sqrt{2} \qquad x = -1 - 2\sqrt{2} \qquad \text{Solve for } x.$$

Check both solutions in the original equation. The solutions are $-1 + 2\sqrt{2}$ and $-1 - 2\sqrt{2}$. This can be written compactly as $-1 \pm 2\sqrt{2}$. The notation ± is read as "plus or minus."

Work Practice Problem 6

EXAMPLE 7 Use the square root property to solve $(x-1)^2 = -2$.

Solution: This equation has no real solution because the square root of −2 is not a real number.

Work Practice Problem 7

EXAMPLE 8 Use the square root property to solve $(5x-2)^2 = 10$.

Solution: $(5x-2)^2 = 10$

$$5x - 2 = \sqrt{10} \quad \text{or} \quad 5x - 2 = -\sqrt{10} \qquad \text{Use the square root property.}$$
$$5x = 2 + \sqrt{10} \qquad 5x = 2 - \sqrt{10} \qquad \text{Add 2 to both sides.}$$
$$x = \frac{2 + \sqrt{10}}{5} \qquad x = \frac{2 - \sqrt{10}}{5} \qquad \text{Divide both sides by 5.}$$

Check both solutions in the original equation. The solutions are $\frac{2 + \sqrt{10}}{5}$ and $\frac{2 - \sqrt{10}}{5}$, which can be written as $\frac{2 \pm \sqrt{10}}{5}$.

Work Practice Problem 8

Helpful Hint

For some applications and graphing purposes, decimal approximations of exact solutions to quadratic equations may be desired.

Exact solutions from Example 8		Decimal approximations
$\frac{2 + \sqrt{10}}{5}$	≈	1.032
$\frac{2 - \sqrt{10}}{5}$	≈	−0.232

PRACTICE PROBLEM 6

Use the square root property to solve $(x-5)^2 = 18$.

Helpful Hint

read "plus or minus" ↓

The notation $-1 \pm \sqrt{5}$, for example, is just a shorthand notation for both $-1 + \sqrt{5}$ and $-1 - \sqrt{5}$.

PRACTICE PROBLEM 7

Use the square root property to solve $(x+3)^2 = -5$.

PRACTICE PROBLEM 8

Use the square root property to solve $(4x+1)^2 = 15$.

Answers

6. $5 \pm 3\sqrt{2}$ 7. no real solution
8. $\frac{-1 \pm \sqrt{15}}{4}$

Objective C Using the Square Root Property to Solve Applications

Many real-world applications are modeled by quadratic equations. In the next example, we use the quadratic formula $h = 16t^2$. This formula gives the distance h traveled by a free-falling object in time t. One important note is that this formula does not take into account any air resistance.

PRACTICE PROBLEM 9

Use the formula $h = 16t^2$ (see Example 9) to find how long, to the nearest tenth of a second, it takes a free-falling body to fall 650 feet.

EXAMPLE 9 Finding the Length of Time of a Dive

The record for the highest dive into a lake was made by Harry Froboess of Switzerland. In 1936 he dove 394 feet from the airship Hindenburg into Lake Constance. To the nearest tenth of a second, how long did his dive take? (*Source: Guinness World Records*)

Solution:

1. UNDERSTAND. To approximate the time of the dive, we use the formula* $h = 16t^2$ where t is time in seconds and h is the distance in feet traveled by a free-falling body or object. For example, to find the distance traveled in 1 second, or 3 seconds, we let $t = 1$ and then $t = 3$.

   If $t = 1$, $h = 16(1)^2 = 16 \cdot 1 = 16$ feet
   If $t = 3$, $h = 16(3)^2 = 16 \cdot 9 = 144$ feet

   Since a body travels 144 feet in 3 seconds, we now know the dive of 394 feet lasted longer than 3 seconds.
2. TRANSLATE. Use the formula $h = 16t^2$, let the distance $h = 394$, and we have the equation $394 = 16t^2$.
3. SOLVE. To solve $394 = 16t^2$ for t, we will use the square root property.

$$394 = 16t^2$$

$$\frac{394}{16} = t^2 \quad \text{Divide both sides by 16.}$$

$$24.625 = t^2 \quad \text{Simplify.}$$

$$\sqrt{24.625} = t \quad \text{or} \quad -\sqrt{24.625} = t \quad \text{Use the square root property.}$$

$$5.0 \approx t \quad \text{or} \quad -5.0 \approx t \quad \text{Approximate.}$$

4. INTERPRET.

Check: We reject the solution -5.0 since the length of the dive is not a negative number.

State: The dive lasted approximately 5 seconds.

Work Practice Problem 9

*The formula $h = 16t^2$ does not take into account air resistance.

Answer

9. 6.4 sec

16.1 EXERCISE SET

FOR EXTRA HELP

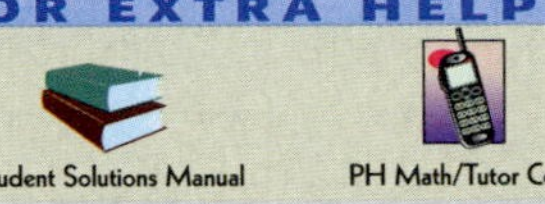

Objective A *Solve each equation by factoring. See Examples 1 and 2.*

1. $k^2 - 49 = 0$

2. $k^2 - 9 = 0$

3. $m^2 + 2m = 15$

4. $m^2 + 6m = 7$

5. $2x^2 - 32 = 0$

6. $2x^2 - 98 = 0$

7. $4a^2 - 36 = 0$

8. $7a^2 - 175 = 0$

9. $x^2 + 7x = -10$

10. $x^2 + 10x = -24$

Objective B *Use the square root property to solve each quadratic equation. See Examples 3 and 4.*

11. $x^2 = 64$

12. $x^2 = 121$

13. $x^2 = 21$

14. $x^2 = 22$

15. $x^2 = \frac{1}{25}$

16. $x^2 = \frac{1}{16}$

17. $x^2 = -4$

18. $x^2 = -25$

19. $3x^2 = 13$

20. $5x^2 = 2$

21. $7x^2 = 4$

22. $2x^2 = 9$

23. $x^2 - 2 = 0$

24. $x^2 - 15 = 0$

Use the square root property to solve each quadratic equation. See Examples 5 through 8.

25. $(x - 5)^2 = 49$

26. $(x + 2)^2 = 25$

27. $(x + 2)^2 = 7$

28. $(x - 7)^2 = 2$

29. $\left(m - \frac{1}{2}\right)^2 = \frac{1}{4}$

30. $\left(m + \frac{1}{3}\right)^2 = \frac{1}{9}$

31. $(p + 2)^2 = 10$

32. $(p - 7)^2 = 13$

33. $(3y + 2)^2 = 100$

34. $(4y - 3)^2 = 81$

35. $(z - 4)^2 = -9$

36. $(z + 7)^2 = -20$

37. $(2x - 11)^2 = 50$

38. $(3x - 17)^2 = 28$

39. $(3x - 7)^2 = 32$

40. $(5x - 11)^2 = 54$

Use the square root property to solve. See Examples 3 through 8.

41. $x^2 - 2 = 0$

42. $x^2 - 15 = 0$

43. $(x + 6)^2 = 24$

44. $(x + 5)^2 = 20$

45. $\frac{1}{2}n^2 = 5$

46. $\frac{1}{5}y^2 = 2$

47. $(4x - 1)^2 = 5$

48. $(7x - 2)^2 = 11$

49. $3z^2 = 36$

50. $3z^2 = 24$

51. $(8 - 3x)^2 - 45 = 0$

52. $(10 - 9x)^2 - 75 = 0$

Objective C *Solve. For Exercises 53 through 56, use the formula* $h = 16t^2$. *See Example 9. Round each answer to the nearest tenth of a second.*

53. The highest regularly performed dives are made by professional divers from La Quebrada. If this cliff in Acapulco has a height of 87.6 feet, determine the time of a dive. (*Source: Guinness World Records*)

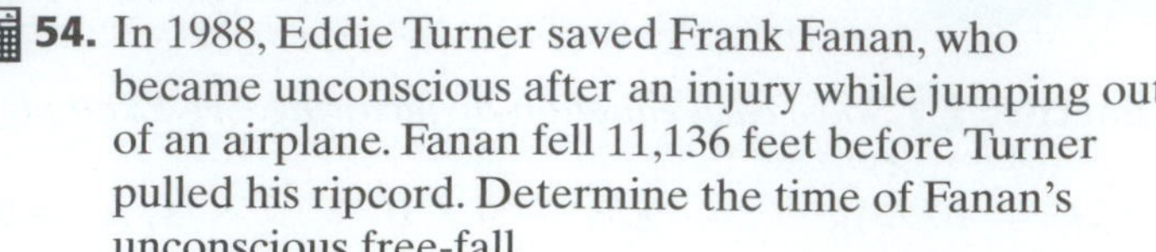

54. In 1988, Eddie Turner saved Frank Fanan, who became unconscious after an injury while jumping out of an airplane. Fanan fell 11,136 feet before Turner pulled his ripcord. Determine the time of Fanan's unconscious free-fall.

55. In 1997, stuntman Stig Gunther of Denmark jumped from a height of 343 feet off a crane onto an airbag. Determine the time of Gunther's stunt fall. (*Source: Guinness World Records,* 2005)

56. Eugene Andreev holds the official Federation Aeronautique Internationale (FAI) world's record for the longest free-fall jump. On November 1, 1962, he fell 80,380 feet before opening his parachute. How long did Andreev free-fall? (*Source: Guinness World Records,* 2005)

The formula for area of a square is $A = s^2$ *where s is the length of a side. Use this formula for Exercises 57 through 60. For each exercise, give an exact answer and a two-decimal-place approximation.*

57. If the area of a square is 20 square inches, find the length of a side.

58. If the area of a square is 32 square meters, find the length of a side.

59. The "Water Cube" National Swimming Center is being constructed in Beijing for the 2008 Summer Olympics. Its square base has an area of 31,329 sq meters. Find the length of a side of this building. (*Source:* ARUP East Asia)

60. The Washington Monument has a square base whose area is approximately 3039 square feet. Find the length of a side. (*Source: The World Almanac*)

Review

Factor each perfect square trinomial. See Section 11.5.

61. $x^2 + 6x + 9$

62. $y^2 + 10y + 25$

63. $x^2 - 4x + 4$

64. $x^2 - 20x + 100$

Concept Extensions

65. Explain why the equation $x^2 = -9$ has no real solution.

66. Explain why the equation $x^2 = 9$ has two solutions.

Solve each quadratic equation by first factoring the perfect square trinomial on the left side. Then apply the square root property.

67. $x^2 + 4x + 4 = 16$

68. $y^2 - 10y + 25 = 11$

△ **69.** The area of a circle is found by the equation $A = \pi r^2$. If the area A of a certain circle is 36π square inches, find its radius r.

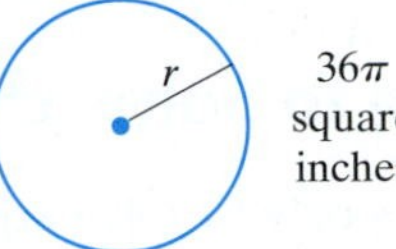

70. Neglecting air resistance, the distance d in feet that an object falls in t seconds is given by the equation $d = 16t^2$. If a sandblaster drops his goggles from a bridge 400 feet from the water below, find how long it takes for the goggles to hit the water.

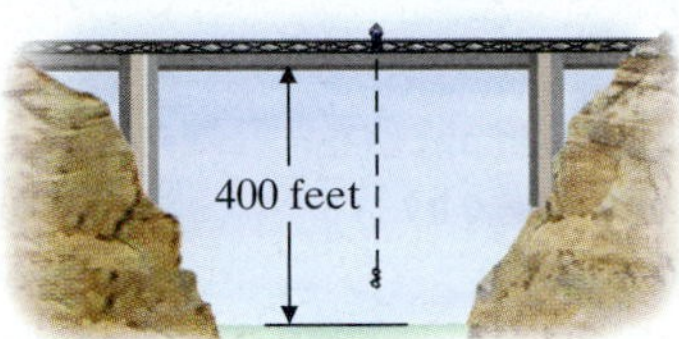

Solve each quadratic equation by using the square root property. Use a calculator and round each solution to the nearest hundredth.

71. $x^2 = 1.78$

72. $(x - 1.37)^2 = 5.71$

73. The number y of Barnes & Noble Booksellers open for business from 1999 through 2003 is given by the equation $y = -0.07(x - 192.5)^2 + 3135$, where $x = 0$ represents the year 1999. Assume that this trend continues, and find the first year in which there will be 727 stores open for business. (*Hint:* Replace y with 727 in the equation and solve for x. Round to the nearest year.) (*Source:* Based on data from Barnes & Noble Corporation)

74. World cotton production y (in millions of bales) from 2002 to 2004 can be represented by the equation $y = 6.4(x + 0.0065)^2 + 88.31$, where $x = 0$ represents the year 2002. Assume that this trend continues and find the year in which there will be 249 millions of bales produced. (*Hint:* Replace y with 249 in the equation and solve for x. Round to the nearest year.) (*Source:* Based on data from U.S. Department of Agriculture, Foreign Agricultural Service)

STUDY SKILLS BUILDER

Are You Preparing for Your Final Exam?

Let's review the tips for preparing for your final exam. To prepare for your final exam, try the following study techniques:

- Review the material that you will be responsible for on your exam. This includes material from your textbook, your notebook, and any handouts from your instructor.
- Review any formulas that you may need to memorize.
- Check to see if your instructor or mathematics department will be conducting a final exam review.
- Check with your instructor to see whether final exams from previous semesters/quarters are available to students for review.
- Use your previously taken exams as a practice final exam. To do so, rewrite the test questions in mixed order on blank sheets of paper. This will help you prepare for exam conditions.
- If you are unsure of a few concepts, see your instructor or visit a learning lab for assistance. Also, view the video segment of any troublesome sections.
- If you need further exercises to work, try the Cumulative Reviews at the end of the chapters.

Once again, good luck! I hope you are enjoying this textbook and your mathematics course.

Objectives

A **Solve Quadratic Equations of the Form $x^2 + bx + c = 0$ by Completing the Square.**

B **Solve Quadratic Equations of the Form $ax^2 + bx + c = 0$ by Completing the Square.**

16.2 SOLVING QUADRATIC EQUATIONS BY COMPLETING THE SQUARE

Objective A Completing the Square to Solve $x^2 + bx + c = 0$

In the last section, we used the square root property to solve equations such as

$$(x + 1)^2 = 8 \quad \text{and} \quad (5x - 2)^2 = 3$$

Notice that one side of each equation is a quantity squared and that the other side is a constant. To solve

$$x^2 + 2x = 4$$

notice that if we add 1 to both sides of the equation, the left side is a perfect square trinomial that can be factored.

$$x^2 + 2x + 1 = 4 + 1 \quad \text{Add 1 to both sides.}$$
$$(x + 1)^2 = 5 \quad \text{Factor.}$$

Now we can solve this equation as we did in the previous section by using the square root property.

$$x + 1 = \sqrt{5} \quad \text{or} \quad x + 1 = -\sqrt{5} \quad \text{Use the square root property.}$$
$$x = -1 + \sqrt{5} \quad\quad x = -1 - \sqrt{5} \quad \text{Solve.}$$

The solutions are $-1 \pm \sqrt{5}$.

Adding a number to $x^2 + 2x$ to form a perfect square trinomial is called **completing the square** on $x^2 + 2x$.

In general, we have the following:

Completing the Square

To complete the square on $x^2 + bx$, add $\left(\frac{b}{2}\right)^2$. To find $\left(\frac{b}{2}\right)^2$, **find half the coefficient of x, and then square the result.**

PRACTICE PROBLEM 1

Solve $x^2 + 8x + 1 = 0$ by completing the square.

EXAMPLE 1 Solve $x^2 + 6x + 3 = 0$ by completing the square.

Solution: First we get the variable terms alone by subtracting 3 from both sides of the equation.

$$x^2 + 6x + 3 = 0$$
$$x^2 + 6x = -3 \quad \text{Subtract 3 from both sides.}$$

Next we find half the coefficient of the x-term, and then square it. We add this result to *both sides* of the equation. This will make the left side a perfect square trinomial. The coefficient of x is 6, and half of 6 is 3. So we add 3^2 or 9 to both sides.

$$x^2 + 6x + 9 = -3 + 9 \quad \text{Complete the square.}$$
$$(x + 3)^2 = 6 \quad \text{Factor the trinomial } x^2 + 6x + 9.$$
$$x + 3 = \sqrt{6} \quad \text{or} \quad x + 3 = -\sqrt{6} \quad \text{Use the square root property.}$$
$$x = -3 + \sqrt{6} \quad\quad x = -3 - \sqrt{6} \quad \text{Subtract 3 from both sides.}$$

Check by substituting $-3 + \sqrt{6}$ and $-3 - \sqrt{6}$ in the original equation. The solutions are $-3 \pm \sqrt{6}$.

Work Practice Problem 1

Answer

1. $-4 \pm \sqrt{15}$

Helpful Hint Remember, when completing the square, add the number that completes the square to **both sides of the equation.**

EXAMPLE 2 Solve $x^2 - 10x = -14$ by completing the square.

Solution: The variable terms are already alone on one side of the equation. The coefficient of x is -10. Half of -10 is -5, and $(-5)^2 = 25$. So we add 25 to both sides.

$$x^2 - 10x = -14$$
$$x^2 - 10x + 25 = -14 + 25$$

Helpful Hint Add 25 to *both* sides of the equation.

$(x - 5)^2 = 11$ Factor the trinomial and simplify $-14 + 25$.

$x - 5 = \sqrt{11}$ or $x - 5 = -\sqrt{11}$ Use the square root property.

$x = 5 + \sqrt{11}$ $\quad x = 5 - \sqrt{11}$ Add 5 to both sides.

The solutions are $5 \pm \sqrt{11}$.

Work Practice Problem 2

PRACTICE PROBLEM 2

Solve $x^2 - 14x = -32$ by completing the square.

Objective B Completing the Square to Solve $ax^2 + bx + c = 0$

The method of completing the square can be used to solve *any* quadratic equation whether the coefficient of the squared variable is 1 or not. When the coefficient of the squared variable is not 1, we first divide both sides of the equation by the coefficient of the squared variable so that the new coefficient is 1. Then we complete the square.

EXAMPLE 3 Solve $4x^2 - 8x - 5 = 0$ by completing the square.

Solution: Since the coefficient of x^2 is 4, not 1, we first divide both sides of the equation by 4 so that the coefficient of x^2 is 1.

$$4x^2 - 8x - 5 = 0$$

$x^2 - 2x - \frac{5}{4} = 0$ Divide both sides by 4.

$x^2 - 2x = \frac{5}{4}$ Get the variable terms alone on one side of the equation.

The coefficient of x is -2. Half of -2 is -1, and $(-1)^2 = 1$. So we add 1 to both sides.

$$x^2 - 2x + 1 = \frac{5}{4} + 1$$

$(x - 1)^2 = \frac{9}{4}$ Factor $x^2 - 2x + 1$ and simplify $\frac{5}{4} + 1$.

$x - 1 = \sqrt{\frac{9}{4}}$ or $x - 1 = -\sqrt{\frac{9}{4}}$ Use the square root property.

$x = 1 + \frac{3}{2}$ $\quad x = 1 - \frac{3}{2}$ Add 1 to both sides and simplify the radical.

$x = \frac{5}{2}$ $\quad x = -\frac{1}{2}$ Simplify.

Both $\frac{5}{2}$ and $-\frac{1}{2}$ are solutions.

Work Practice Problem 3

PRACTICE PROBLEM 3

Solve $4x^2 - 16x - 9 = 0$ by completing the square.

Answers

2. $7 \pm \sqrt{17}$ **3.** $\frac{9}{2}$ and $-\frac{1}{2}$

The following steps may be used to solve a quadratic equation in x by completing the square.

To Solve a Quadratic Equation in x by Completing the Square

Step 1: If the coefficient of x^2 is 1, go to Step 2. If not, divide both sides of the equation by the coefficient of x^2.

Step 2: Get all terms with variables on one side of the equation and constants on the other side.

Step 3: Find half the coefficient of x and then square the result. Add this number to both sides of the equation.

Step 4: Factor the resulting perfect square trinomial.

Step 5: Use the square root property to solve the equation.

PRACTICE PROBLEM 4

Solve $2x^2 + 10x = -13$ by completing the square.

EXAMPLE 4 Solve $2x^2 + 6x = -7$ by completing the square.

Solution: The coefficient of x^2 is not 1. We divide both sides by 2, the coefficient of x^2.

$$2x^2 + 6x = -7$$

$$x^2 + 3x = -\frac{7}{2} \quad \text{Divide both sides by 2.}$$

$$x^2 + 3x + \frac{9}{4} = -\frac{7}{2} + \frac{9}{4} \quad \text{Add } \left(\frac{3}{2}\right)^2 \text{ or } \frac{9}{4} \text{ to both sides.}$$

$$\left(x + \frac{3}{2}\right)^2 = -\frac{5}{4} \quad \text{Factor the left side and simplify the right.}$$

There is no real solution to this equation since the square root of a negative number is not a real number.

Work Practice Problem 4

PRACTICE PROBLEM 5

Solve $2x^2 = -6x + 5$ by completing the square.

EXAMPLE 5 Solve $2x^2 = 10x + 1$ by completing the square.

Solution: First we divide both sides of the equation by 2, the coefficient of x^2.

$$2x^2 = 10x + 1$$

$$x^2 = 5x + \frac{1}{2} \quad \text{Divide both sides by 2.}$$

Next we get the variable terms alone by subtracting $5x$ from both sides.

$$x^2 - 5x = \frac{1}{2}$$

$$x^2 - 5x + \frac{25}{4} = \frac{1}{2} + \frac{25}{4} \quad \text{Add } \left(-\frac{5}{2}\right)^2 \text{ or } \frac{25}{4} \text{ to both sides.}$$

$$\left(x - \frac{5}{2}\right)^2 = \frac{27}{4} \quad \text{Factor the left side and simplify the right side.}$$

$$x - \frac{5}{2} = \sqrt{\frac{27}{4}} \quad \text{or} \quad x - \frac{5}{2} = -\sqrt{\frac{27}{4}} \quad \text{Use the square root property.}$$

$$x - \frac{5}{2} = \frac{3\sqrt{3}}{2} \qquad x - \frac{5}{2} = -\frac{3\sqrt{3}}{2} \quad \text{Simplify.}$$

$$x = \frac{5}{2} + \frac{3\sqrt{3}}{2} \qquad x = \frac{5}{2} - \frac{3\sqrt{3}}{2}$$

The solutions are $\dfrac{5 \pm 3\sqrt{3}}{2}$.

Work Practice Problem 5

Answers

4. no real solution and -2 **5.** $\dfrac{-3 \pm \sqrt{19}}{2}$

Vocabulary and Readiness Check

Use the choices below to fill in each blank. Not all choices will be used, and these exercises come from Sections 16.1 and 16.2.

$\sqrt{a}$	linear equation	zero	$\left(\frac{b}{2}\right)^2$	$\frac{b}{2}$	6	3
$\pm\sqrt{a}$	quadratic equation	one	completing the square		9	

1. By the zero factor property, if the product of two numbers is zero, then at least one of these two numbers must be _______.

2. If a is a positive number, and if $x^2 = a$, then $x =$ _________.

3. An equation that can be written in the form $ax^2 + bx + c = 0$ where a, b, and c are real numbers and a is not zero is called a(n) _________________.

4. The process of solving a quadratic equation by writing it in the form $(x + a)^2 = c$ is called ___________________.

5. To complete the square on $x^2 + 6x$, add ____.

6. To complete the square on $x^2 + bx$, add _______.

16.2 EXERCISE SET

FOR EXTRA HELP

Student Solutions Manual

PH Math/Tutor Center

CD/Video for Review

MathXL®

MyMathLab

Objective A *Solve each quadratic equation by completing the square. See Examples 1 and 2.*

1. $x^2 + 8x = -12$

2. $x^2 - 10x = -24$

3. $x^2 + 2x - 7 = 0$

4. $z^2 + 6z - 9 = 0$

5. $x^2 - 6x = 0$

6. $y^2 + 4y = 0$

7. $z^2 + 5z = 7$

8. $x^2 - 7x = 5$

9. $x^2 - 2x - 1 = 0$

10. $x^2 - 4x + 2 = 0$

11. $y^2 + 5y + 4 = 0$

12. $y^2 - 5y + 6 = 0$

Objective B *Solve each quadratic equation by completing the square. See Examples 3 through 5.*

13. $3x^2 - 6x = 24$

14. $2x^2 + 18x = -40$

15. $5x^2 + 10x + 6 = 0$

16. $3x^2 - 12x + 14 = 0$

17. $2x^2 = 6x + 5$

18. $4x^2 = -20x + 3$

19. $2y^2 + 8y + 5 = 0$

20. $4z^2 - 8z + 1 = 0$

Objectives A B Mixed Practice *Solve each quadratic equation by completing the square. See Examples 1 through 5.*

21. $x^2 + 6x - 25 = 0$

22. $x^2 - 6x + 7 = 0$

23. $x^2 - 3x - 3 = 0$

24. $x^2 - 9x + 3 = 0$

25. $2y^2 - 3y + 1 = 0$

26. $2y^2 - y - 1 = 0$

27. $x(x + 3) = 18$

28. $x(x - 3) = 18$

29. $3z^2 + 6z + 4 = 0$

30. $2y^2 + 8y + 9 = 0$

31. $4x^2 + 16x = 48$

32. $6x^2 - 30x = -36$

Review

Simplify each expression. See Section 15.2.

33. $\frac{3}{4} - \sqrt{\frac{25}{16}}$

34. $\frac{3}{5} + \sqrt{\frac{16}{25}}$

35. $\frac{1}{2} - \sqrt{\frac{9}{4}}$

36. $\frac{9}{10} - \sqrt{\frac{49}{100}}$

Simplify each expression. See Section 15.4.

37. $\frac{6 + 4\sqrt{5}}{2}$

38. $\frac{10 - 20\sqrt{3}}{2}$

39. $\frac{3 - 9\sqrt{2}}{6}$

40. $\frac{12 - 8\sqrt{7}}{16}$

Concept Extensions

41. In your own words, describe a perfect square trinomial.

42. Describe how to find the number to add to $x^2 - 7x$ to make a perfect square trinomial.

43. Write your own quadratic equation to be solved by completing the square. Write it in the form

perfect square trinomial = a number that is not a perfect square

$$x^2 + 6x + 9 = 11$$

For example,

a. Solve $x^2 + 6x + 9 = 11$.

b. Solve your quadratic equation by completing the square.

44. Follow the directions of Exercise 43, except write your equation in the form

perfect square trinomial = negative number

Solve your quadratic equation by completing the square.

45. Find a value of k that will make $x^2 + kx + 16$ a perfect square trinomial.

46. Find a value of k that will make $x^2 + kx + 25$ a perfect square trinomial.

47. Retail sales y (in millions of dollars) for bookstores in the United States from 2001 through 2003 can be represented by the equation $y = 10x^2 + 513x + 15{,}743$. In this equation x is the number of years after 2001. Assume that this trend continues and predict the years after 2001 in which the retail sales for U.S. bookstores will be \$20,487 million. (*Source:* Based on data from the U.S. Bureau of the Census, Monthly Retail Surveys Branch)

48. The average price of gold y (in dollars per ounce) from 2000 through 2003 is given by the equation $y = 12x^2 - 10x + 278$. Assume that this trend continues and find the year after 2000 in which the price of gold will be \$1620 per ounce. (*Source:* Based on data from U.S. Geological survey, Minerals Information)

Recall that a graphing calculator may be used to solve an equation. For example, to solve $x^2 + 8x = -12$ (Exercise 1), graph

$y_1 = x^2 + 8x$ *(left side of equation) and*

$y_2 = -12$ *(right side of equation)*

The x-coordinate of the point of intersection of the graphs is the solution. Use a graphing calculator and solve each equation. Round solutions to the nearest hundredth.

49. Exercise 1

50. Exercise 2

51. Exercise 17

52. Exercise 8

16.3 SOLVING QUADRATIC EQUATIONS BY THE QUADRATIC FORMULA

Objectives

A Use the Quadratic Formula to Solve Quadratic Equations.

B Approximate Solutions to Quadratic Equations.

Objective A Using the Quadratic Formula

We can use the technique of completing the square to develop a formula to find solutions of any quadratic equation. We develop and use the **quadratic formula** in this section.

Recall that a quadratic equation in **standard form** is

$$ax^2 + bx + c = 0, \quad a \neq 0$$

To develop the quadratic formula, let's complete the square for this quadratic equation in standard form.

First we divide both sides of the equation by the coefficient of x^2 and then get the variable terms alone on one side of the equation.

$$x^2 + \frac{b}{a}x + \frac{c}{a} = 0$$ Divide by a; recall that a cannot be 0.

$$x^2 + \frac{b}{a}x = -\frac{c}{a}$$ Get the variable terms alone on one side of the equation.

The coefficient of x is $\frac{b}{a}$. Half of $\frac{b}{a}$ is $\frac{b}{2a}$ and $\left(\frac{b}{2a}\right)^2 = \frac{b^2}{4a^2}$. So we add $\frac{b^2}{4a^2}$ to both sides of the equation.

$$x^2 + \frac{b}{a}x + \frac{b^2}{4a^2} = -\frac{c}{a} + \frac{b^2}{4a^2}$$ Add $\frac{b^2}{4a^2}$ to both sides.

$$\left(x + \frac{b}{2a}\right)^2 = -\frac{c}{a} + \frac{b^2}{4a^2}$$ Factor the left side.

$$\left(x + \frac{b}{2a}\right)^2 = -\frac{4ac}{4a^2} + \frac{b^2}{4a^2}$$ Multiply $-\frac{c}{a}$ by $\frac{4a}{4a}$ so that the terms on the right side have a common denominator.

$$\left(x + \frac{b}{2a}\right)^2 = \frac{b^2 - 4ac}{4a^2}$$ Simplify the right side.

Now we use the square root property.

$$x + \frac{b}{2a} = \sqrt{\frac{b^2 - 4ac}{4a^2}} \quad \text{or} \quad x + \frac{b}{2a} = -\sqrt{\frac{b^2 - 4ac}{4a^2}}$$ Use the square root property.

$$x + \frac{b}{2a} = \frac{\sqrt{b^2 - 4ac}}{2a} \qquad x + \frac{b}{2a} = -\frac{\sqrt{b^2 - 4ac}}{2a}$$ Simplify the radical.

$$x = -\frac{b}{2a} + \frac{\sqrt{b^2 - 4ac}}{2a} \qquad x = -\frac{b}{2a} - \frac{\sqrt{b^2 - 4ac}}{2a}$$ Subtract $\frac{b}{2a}$ from both sides.

$$x = \frac{-b + \sqrt{b^2 - 4ac}}{2a} \qquad x = \frac{-b - \sqrt{b^2 - 4ac}}{2a}$$ Simplify.

The solutions are $\frac{-b \pm \sqrt{b^2 - 4ac}}{2a}$. This final equation is called the **quadratic formula** and gives the solutions of any quadratic equation.

Quadratic Formula

If a, b, and c are real numbers and $a \neq 0$, a quadratic equation written in the standard form $ax^2 + bx + c = 0$ has solutions

$$x = \frac{-b \pm \sqrt{b^2 - 4ac}}{2a}$$

Helpful Hint

Don't forget that to correctly identify a, b, and c in the quadratic formula, you should write the equation in standard form.

Quadratic Equations in Standard Form

$5x^2 - 6x + 2 = 0$	$a = 5, b = -6, c = 2$
$4y^2 - 9 = 0$	$a = 4, b = 0, c = -9$
$x^2 + x = 0$	$a = 1, b = 1, c = 0$
$\sqrt{2}x^2 + \sqrt{5}x + \sqrt{3} = 0$	$a = \sqrt{2}, b = \sqrt{5}, c = \sqrt{3}$

PRACTICE PROBLEM 1

Solve $2x^2 - x - 5 = 0$ using the quadratic formula.

EXAMPLE 1 Solve $3x^2 + x - 3 = 0$ using the quadratic formula.

Solution: This equation is in standard form with $a = 3$, $b = 1$, and $c = -3$. By the quadratic formula, we have

$$x = \frac{-b \pm \sqrt{b^2 - 4ac}}{2a}$$

$$x = \frac{-1 \pm \sqrt{1^2 - 4 \cdot 3 \cdot (-3)}}{2 \cdot 3} \quad \text{Let } a = 3, b = 1, \text{ and } c = -3.$$

$$= \frac{-1 \pm \sqrt{1 + 36}}{6} \quad \text{Simplify.}$$

$$= \frac{-1 \pm \sqrt{37}}{6}$$

Check both solutions in the original equation. The solutions are $\frac{-1 + \sqrt{37}}{6}$ and $\frac{-1 - \sqrt{37}}{6}$.

Work Practice Problem 1

PRACTICE PROBLEM 2

Solve $3x^2 + 8x = 3$ using the quadratic formula.

EXAMPLE 2 Solve $2x^2 - 9x = 5$ using the quadratic formula.

Solution: First we write the equation in standard form by subtracting 5 from both sides.

$$2x^2 - 9x = 5$$

$$2x^2 - 9x - 5 = 0$$

Next we note that $a = 2$, $b = -9$, and $c = -5$. We substitute these values into the quadratic formula.

$$x = \frac{-b \pm \sqrt{b^2 - 4ac}}{2a}$$

$$x = \frac{-(-9) \pm \sqrt{(-9)^2 - 4 \cdot 2 \cdot (-5)}}{2 \cdot 2} \quad \text{Substitute in the formula.}$$

$$= \frac{9 \pm \sqrt{81 + 40}}{4} \quad \text{Simplify.}$$

$$= \frac{9 \pm \sqrt{121}}{4} = \frac{9 \pm 11}{4}$$

Helpful Hint

Notice that the fraction bar is under the entire numerator $-b \pm \sqrt{b^2 - 4ac}$.

Answers

1. $\frac{1 + \sqrt{41}}{4}$ and $\frac{1 - \sqrt{41}}{4}$
2. $\frac{1}{3}$ and -3

Then,

$$x = \frac{9 - 11}{4} = -\frac{1}{2} \quad \text{or} \quad x = \frac{9 + 11}{4} = 5$$

Check $-\frac{1}{2}$ and 5 in the original equation. Both $-\frac{1}{2}$ and 5 are solutions.

Work Practice Problem 2

The following steps may be useful when solving a quadratic equation by the quadratic formula.

To Solve a Quadratic Equation by the Quadratic Formula

Step 1: Write the quadratic equation in standard form: $ax^2 + bx + c = 0$.

Step 2: If necessary, clear the equation of fractions to simplify calculations.

Step 3: Identify a, b, and c.

Step 4: Replace a, b, and c in the quadratic formula with the identified values, and simplify.

✓Concept Check For the quadratic equation $2x^2 - 5 = 7x$, if $a = 2$ and $c = -5$ in the quadratic formula, the value of b is which of the following?

a. $\frac{7}{2}$ **b.** 7 **c.** -5 **d.** -7

EXAMPLE 3 Solve $7x^2 = 1$ using the quadratic formula.

Solution: First we write the equation in standard form by subtracting 1 from both sides.

$$7x^2 = 1$$
$$7x^2 - 1 = 0$$

Helpful Hint $7x^2 - 1 = 0$ can be written as $7x^2 + 0x - 1 = 0$. This form helps you see that $b = 0$.

Next we replace a, b, and c with the identified values: $a = 7, b = 0, c = -1$.

$$x = \frac{0 \pm \sqrt{0^2 - 4 \cdot 7 \cdot (-1)}}{2 \cdot 7} \quad \text{Substitute in the formula.}$$

$$= \frac{\pm\sqrt{28}}{14} \quad \text{Simplify.}$$

$$= \frac{\pm 2\sqrt{7}}{14}$$

$$= \pm\frac{2\sqrt{7}}{2 \cdot 7}$$

$$= \pm\frac{\sqrt{7}}{7}$$

The solutions are $\frac{\sqrt{7}}{7}$ and $-\frac{\sqrt{7}}{7}$.

Work Practice Problem 3

PRACTICE PROBLEM 3

Solve $5x^2 = 2$ using the quadratic formula.

Answer

3. $\frac{\sqrt{10}}{5}$ and $-\frac{\sqrt{10}}{5}$

✓ Concept Check Answer

d

Notice that we could have solved the equation $7x^2 = 1$ in Example 3 by dividing both sides by 7 and then using the square root property. We solved the equation by the quadratic formula to show that this formula can be used to solve any quadratic equation.

PRACTICE PROBLEM 4

Solve $x^2 = -2x - 3$ using the quadratic formula.

EXAMPLE 4 Solve $x^2 = -x - 1$ using the quadratic formula.

Solution: First we write the equation in standard form.

$$x^2 + x + 1 = 0$$

Next we replace a, b, and c in the quadratic formula with $a = 1, b = 1$, and $c = 1$.

$$x = \frac{-1 \pm \sqrt{1^2 - 4 \cdot 1 \cdot 1}}{2 \cdot 1} \quad \text{Substitute in the formula.}$$

$$= \frac{-1 \pm \sqrt{-3}}{2} \quad \text{Simplify.}$$

There is no real number solution because $\sqrt{-3}$ is not a real number.

Work Practice Problem 4

PRACTICE PROBLEM 5

Solve $\frac{1}{3}x^2 - x = 1$ using the quadratic formula.

EXAMPLE 5 Solve $\frac{1}{2}x^2 - x = 2$ using the quadratic formula.

Solution: We write the equation in standard form and then clear the equation of fractions by multiplying both sides by the LCD, 2.

$$\frac{1}{2}x^2 - x = 2$$

$$\frac{1}{2}x^2 - x - 2 = 0 \quad \text{Write in standard form.}$$

$$x^2 - 2x - 4 = 0 \quad \text{Multiply both sides by 2.}$$

Here, $a = 1, b = -2$, and $c = -4$, so we substitute these values into the quadratic formula.

$$x = \frac{-(-2) \pm \sqrt{(-2)^2 - 4 \cdot 1 \cdot (-4)}}{2 \cdot 1}$$

$$= \frac{2 \pm \sqrt{20}}{2} = \frac{2 \pm 2\sqrt{5}}{2} \quad \text{Simplify.}$$

$$= \frac{2(1 \pm \sqrt{5})}{2} = 1 \pm \sqrt{5} \quad \text{Factor and simplify.}$$

The solutions are $1 - \sqrt{5}$ and $1 + \sqrt{5}$.

Work Practice Problem 5

Notice that in Example 5, although we cleared the equation of fractions, the coefficients $a = \frac{1}{2}, b = -1$, and $c = -2$ will give the same results.

Answers

4. no real solution
5. $\frac{3 + \sqrt{21}}{2}$ and $\frac{3 - \sqrt{21}}{2}$

Helpful Hint

When simplifying an expression such as

$$\frac{3 \pm 6\sqrt{2}}{6}$$

first factor out a common factor from the terms of the numerator and then simplify.

$$\frac{3 \pm 6\sqrt{2}}{6} = \frac{3(1 \pm 2\sqrt{2})}{2 \cdot 3} = \frac{1 \pm 2\sqrt{2}}{2}$$

Objective B Approximate Solutions to Quadratic Equations

Sometimes approximate solutions for quadratic equations are appropriate.

EXAMPLE 6 Approximate the exact solutions of the quadratic equation in Example 1. Round the approximations to the nearest tenth.

Solution: From Example 1, we have exact solutions $\frac{-1 \pm \sqrt{37}}{6}$. Thus,

$$\frac{-1 + \sqrt{37}}{6} \approx 0.847127088 \approx 0.8 \text{ to the nearest tenth.}$$

$$\frac{-1 - \sqrt{37}}{6} \approx -1.180460422 \approx -1.2 \text{ to the nearest tenth.}$$

Thus approximate solutions to the quadratic equation in Example 1 are 0.8 and −1.2.

Work Practice Problem 6

PRACTICE PROBLEM 6

Approximate the exact solutions of the quadratic equation in Practice Problem 1. Round the approximations to the nearest tenth.

Answer

6. $\frac{1 + \sqrt{41}}{4} \approx 1.9, \frac{1 - \sqrt{41}}{4} \approx -1.4$

Vocabulary and Readiness Check

Fill in each blank.

1. The quadratic formula is ________________.

Identify the values of a, b, and c in each quadratic equation.

2. $5x^2 - 7x + 1 = 0$; $a =$ ___, $b =$ ____, $c =$ ___

3. $x^2 + 3x - 7 = 0$ $a =$ ___, $b =$ ___, $c =$ ____

4. $x^2 - 6 = 0$ $a =$ ___, $b =$ ___, $c =$ ____

16.3 EXERCISE SET

FOR EXTRA HELP

Student Solutions Manual

PH Math/Tutor Center

CD/Video for Review

MathXL®

MyMathLab

Objective A *Use the quadratic formula to solve each quadratic equation. See Examples 1 through 4.*

1. $x^2 - 3x + 2 = 0$

2. $x^2 - 5x - 6 = 0$

3. $3k^2 + 7k + 1 = 0$

4. $7k^2 + 3k - 1 = 0$

5. $4x^2 - 3 = 0$

6. $25x^2 - 15 = 0$

7. $5z^2 - 4z + 3 = 0$

8. $3x^2 + 2x + 1 = 0$

9. $y^2 = 7y + 30$

10. $y^2 = 5y + 36$

11. $2x^2 = 10$

12. $5x^2 = 15$

13. $m^2 - 12 = m$

14. $m^2 - 14 = 5m$

15. $3 - x^2 = 4x$

16. $10 - x^2 = 2x$

17. $6x^2 + 9x = 2$

18. $3x^2 - 9x = 8$

19. $7p^2 + 2 = 8p$

20. $11p^2 + 2 = 10p$

21. $x^2 - 6x + 2 = 0$

22. $x^2 - 10x + 19 = 0$

23. $2x^2 - 6x + 3 = 0$

24. $5x^2 - 8x + 2 = 0$

25. $3x^2 = 1 - 2x$

26. $5y^2 = 4 - y$

27. $4y^2 = 6y + 1$

28. $6z^2 = 2 - 3z$

29. $20y^2 = 3 - 11y$

30. $2z^2 = z + 3$

31. $x^2 + x + 2 = 0$

32. $k^2 + 2k + 5 = 0$

Use the quadratic formula to solve each quadratic equation. See Example 5.

33. $\frac{m^2}{2} = m + \frac{1}{2}$

34. $\frac{m^2}{2} = 3m - 1$

35. $3p^2 - \frac{2}{3}p + 1 = 0$

36. $\frac{5}{2}p^2 - p + \frac{1}{2} = 0$

37. $4p^2 + \frac{3}{2} = -5p$

38. $4p^2 + \frac{3}{2} = 5p$

39. $5x^2 = \frac{7}{2}x + 1$

40. $2x^2 = \frac{5}{2}x + \frac{7}{2}$

41. $x^2 - \frac{11}{2}x - \frac{1}{2} = 0$

42. $\frac{2}{3}x^2 - 2x - \frac{2}{3} = 0$

43. $5z^2 - 2z = \frac{1}{5}$

44. $9z^2 + 12z = -1$

Objectives **A** **B** **Mixed Practice** *Use the quadratic formula to solve each quadratic equation. Find the exact solutions; then approximate these solutions to the nearest tenth. See Example 6.*

45. $3x^2 = 21$

46. $2x^2 = 26$

47. $x^2 + 6x + 1 = 0$

48. $x^2 + 4x + 2 = 0$

49. $x^2 = 9x + 4$

50. $x^2 = 7x + 5$

51. $3x^2 - 2x - 2 = 0$

52. $5x^2 - 3x - 1 = 0$

Review

Graph the following linear equations in two variables. See Sections 13.2 and 13.3.

53. $y = -3$

54. $x = 4$

55. $y = 3x - 2$

56. $y = 2x + 3$

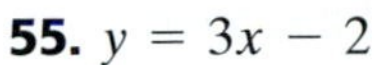
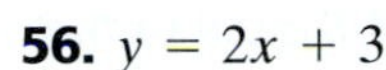

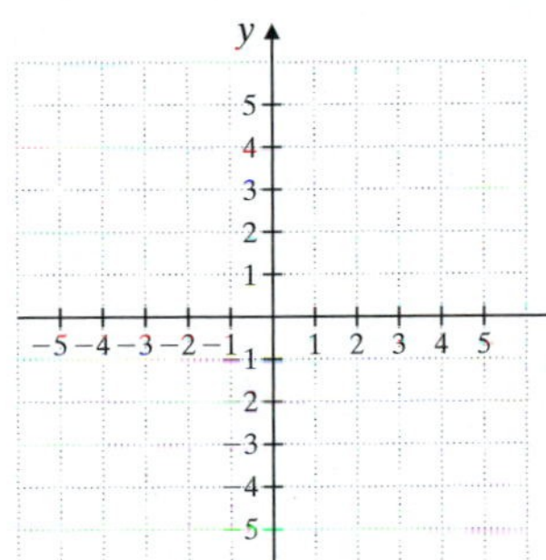

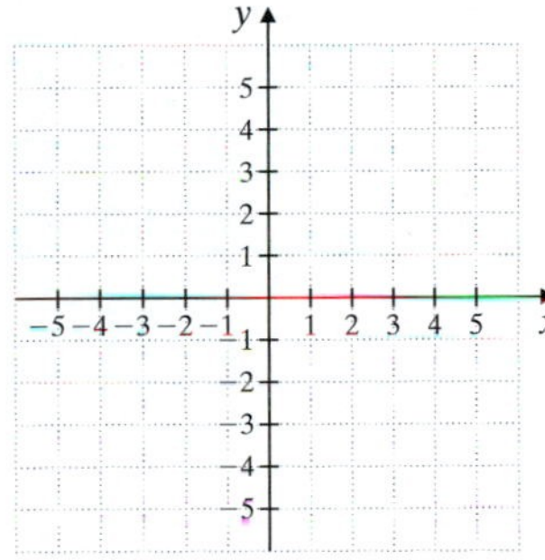

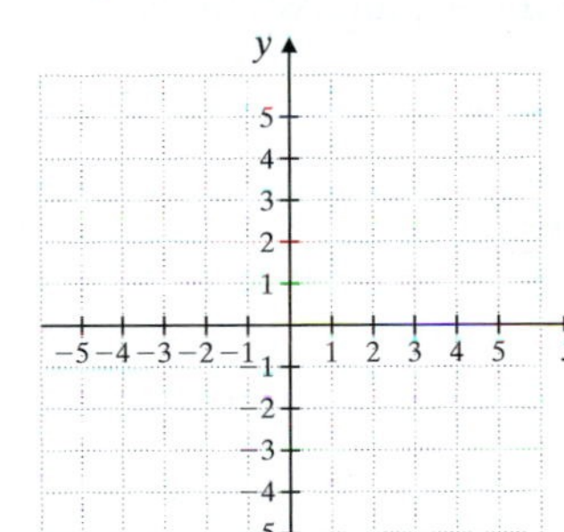

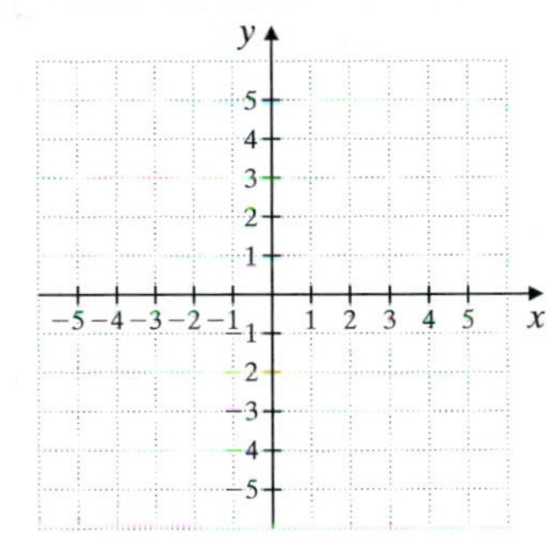

Concept Extensions

Solve. See the Concept Check in this section. For the quadratic equation $5x^2 + 2 = x$, *if* $a = 5$,

57. What is the value of b?

a. $\frac{1}{5}$ **b.** 0 **c.** -1 **d.** 1

58. What is the value of c?

a. 5 **b.** x **c.** -2 **d.** 2

For the quadratic equation $7y^2 = 3y$, *if* $b = 3$,

59. What is the value of a?

a. 7 **b.** -7 **c.** 0 **d.** 1

60. What is the value of c?

a. 7 **b.** 3 **c.** 0 **d.** 1

△ **61.** The largest chocolate bar was a 5026-lb scaled-up model of a Novi chocolate bar, made by the Elah-Dufour United Food Company in 2000. The bar had a base area of 50.8 square feet and its length was 0.5 feet longer than twice its width. Find the length and the width of the bar, rounded to one decimal place. (*Source: Guinness World Records*, 2005)

△ **62.** The area of a rectangular conference room table is 95 square feet. If its length is six feet longer than its width, find the dimensions of the table. Round each dimension to the nearest tenth.

Solve each equation using the quadratic formula.

63. $x^2 + 3\sqrt{2}x - 5 = 0$

64. $y^2 - 2\sqrt{5}y - 1 = 0$

65. Explain how the quadratic formula is developed and why it is useful.

Use the quadratic formula and a calculator to solve each equation. Round solutions to the nearest tenth.

66. $1.2x^2 - 5.2x - 3.9 = 0$

67. $7.3z^2 + 5.4z - 1.1 = 0$

A rocket is launched from the top of an 80-foot cliff with an initial velocity of 120 feet per second. The height, h, of the rocket after t seconds is given by the equation $h = -16t^2 + 120t + 80$.

68. How long after the rocket is launched will it be 30 feet from the ground? Round to the nearest tenth of a second.

69. How long after the rocket is launched will it strike the ground? Round to the nearest tenth of a second. (*Hint:* The rocket will strike the ground when its height $h = 0$.)

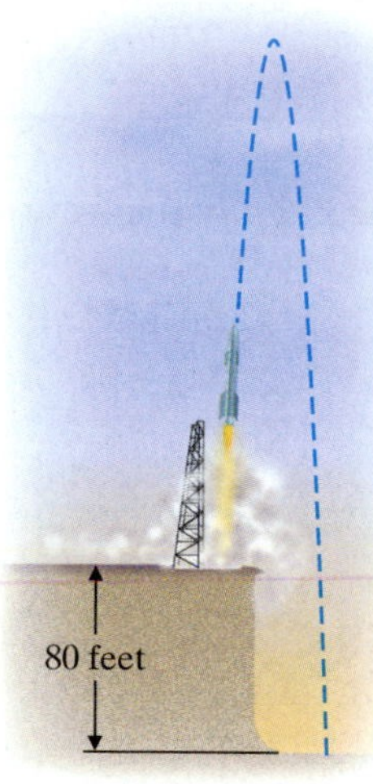

70. The net revenues y (in thousands of dollars) of ABIOMED, Inc., maker of the AbioCor® artificial heart, from 2002 through 2004, can be modeled by the equation $y = 1351.5x^2 + 4670.5x - 24{,}193$, where $x = 0$ represents 2002. Assume that this trend continues and predict the year in which ABIOMED's net revenues will be \$16,113 thousand. (*Source:* Based on data from ABIOMED Corporation)

71. The average annual salary y (in dollars) for NFL players for the years 2000 through 2003 is approximated by $y = 21{,}400x^2 - 16{,}100x + 1{,}111{,}000$, where $x = 0$ represents the year 2000. Assume that this trend continues and predict the year in which the average NFL salary will be approximately \$2,351,800. (*Source:* Based on data from NFL Players Association and *USA Today*)

THE BIGGER PICTURE Simplifying Expressions and Solving Equations

Now we continue our outline from Sections 9.6, 10.7, 11.6, 12.4, 12.5, and 15.5. Although suggestions are given, this outline should be in your own words. Once you complete this new portion, try the exercises below.

I. Simplifying Expressions

- **A. Exponents** (Sections 10.1, 10.2)
- **B. Polynomials**
  1. Add (Section 10.4)
  2. Subtract (Section 10.4)
  3. Multiply (Section 10.5)
  4. Divide (Section 10.7)
- **C. Factoring Polynomials** (Chapter 11 Integrated Review)
- **D. Rational Expressions**
  1. Simplify (Section 12.1)
  2. Multiply (Section 12.2)
  3. Divide (Section 12.2)
  4. Add or Subtract (Section 12.4)
- **E. Radicals**
  1. Simplify (Section 15.2)
  2. Add or Subtract (Section 15.3)
  3. Multiply or Divide (Section 15.4)
  4. Rationalize the denominator (Section 15.4)

II. Solving Equations and Inequalities

- **A. Linear Equations** (Section 9.3)
- **B. Linear Inequalities** (Section 9.6)
- **C. Quadratic and Higher Degree Equations** (Section 11.6)
  1. If in the form $x^2 = a$, solve by the Square Root Property. If not, write the equation in standard form (one side is 0).
  2. If the polynomial on one side factors, solve by factoring.
  3. If the polynomial does not factor, solve by the quadratic formula.
- **D. Equations with Rational Expressions** (Section 12.5)
- **E. Equations with Radicals** (Section 15.5)

By factoring:

$$x^2 + x = 6$$
$$x^2 + x - 6 = 0$$
$$(x - 2)(x + 3) = 0$$
$$x - 2 = 0 \quad \text{or} \quad x + 3 = 0$$
$$x = 2 \quad \text{or} \quad x = -3$$

By Square Root Property:

$$9x^2 = 2$$
$$x^2 = \frac{2}{9}$$
$$x = \pm\sqrt{\frac{2}{9}} = \frac{\pm\sqrt{2}}{3}$$

By quadratic formula:

$$x^2 + x = 5$$
$$x^2 + x - 5 = 0$$
$$a = 1, b = 1, c = -5$$
$$x = \frac{-1 \pm \sqrt{1^2 - 4(1)(-5)}}{2 \cdot 1}$$
$$x = \frac{-1 \pm \sqrt{21}}{2}$$

Perform the indicated operations and simplify.

1. $7.9 - 9.7$
2. $5 + (-3) + (-7)$
3. $(-4)^2 - 5^2$
4. $7x - 2 + \frac{1}{3}(9x - 3) + 5$
5. $\left(\frac{1}{2}x + 5\right)\left(\frac{1}{2}x - 5\right)$
6. $\dfrac{9x^2y + 3xy - 12y}{3xy}$
7. $\dfrac{x^2}{(x-5)(x-4)} - \dfrac{3x + 10}{(x-5)(x-4)}$
8. $\dfrac{x}{x - 10} + \dfrac{5}{x + 3}$
9. $\sqrt{50}$
10. $\dfrac{\sqrt{30a^2b^3}}{\sqrt{3ab}}$
11. $\sqrt{\dfrac{2}{3}}$
12. $\dfrac{7x - 14}{x^2 - 4} \cdot \dfrac{x^2 + 5x + 6}{49}$

Solve.

13. $x^2 + 3x - 5 = 0$
14. $x^2 + x = x^2 + 6$
15. $-2x \le 5.6$
16. $2x^2 + 15x = 8$
17. $\sqrt{x + 2} + 4 = x$
18. $\dfrac{5}{x} - \dfrac{3}{x - 4} = \dfrac{7 + x}{x(x - 4)}$

Answers

1.
2.
3.
4.
5.
6.
7.
8.
9.
10.
11.
12.
13.
14.
15.
16.
17.
18.
19.
20.

INTEGRATED REVIEW

Sections 16.1–16.3

Summary on Quadratic Equations

An important skill in mathematics is learning when to use one technique in favor of another. We now practice this by deciding which method to use when solving quadratic equations. Although both the quadratic formula and completing the square can be used to solve any quadratic equation, the quadratic formula is usually less tedious and thus preferred. The following steps may be used to solve a quadratic equation.

To Solve a Quadratic Equation

Step 1: If the equation is in the form $ax^2 = c$ or $(ax + b)^2 = c$, use the square root property and solve. If not, go to Step 2.

Step 2: Write the equation in standard form: $ax^2 + bx + c = 0$.

Step 3: Try to solve the equation by the factoring method. If not possible, go to Step 4.

Step 4: Solve the equation by the quadratic formula.

Choose and use a method to solve each equation.

1. $5x^2 - 11x + 2 = 0$

2. $5x^2 + 13x - 6 = 0$

3. $x^2 - 1 = 2x$

4. $x^2 + 7 = 6x$

5. $a^2 = 20$

6. $a^2 = 72$

7. $x^2 - x + 4 = 0$

8. $x^2 - 2x + 7 = 0$

9. $3x^2 - 12x + 12 = 0$

10. $5x^2 - 30x + 45 = 0$

11. $9 - 6p + p^2 = 0$

12. $49 - 28p + 4p^2 = 0$

13. $4y^2 - 16 = 0$

14. $3y^2 - 27 = 0$

15. $x^2 - 3x + 2 = 0$

16. $x^2 + 7x + 12 = 0$

17. $(2z + 5)^2 = 25$

18. $(3z - 4)^2 = 16$

19. $30x = 25x^2 + 2$

20. $12x = 4x^2 + 4$

21. $\frac{2}{3}m^2 - \frac{1}{3}m - 1 = 0$

22. $\frac{5}{8}m^2 + m - \frac{1}{2} = 0$

23. $x^2 - \frac{1}{2}x - \frac{1}{5} = 0$

24. $x^2 + \frac{1}{2}x - \frac{1}{8} = 0$

25. $4x^2 - 27x + 35 = 0$

26. $9x^2 - 16x + 7 = 0$

27. $(7 - 5x)^2 = 18$

28. $(5 - 4x)^2 = 75$

29. $3z^2 - 7z = 12$

30. $6z^2 + 7z = 6$

31. $x = x^2 - 110$

32. $x = 56 - x^2$

33. $\frac{3}{4}x^2 - \frac{5}{2}x - 2 = 0$

34. $x^2 - \frac{6}{5}x - \frac{8}{5} = 0$

35. $x^2 - 0.6x + 0.05 = 0$

36. $x^2 - 0.1x - 0.06 = 0$

37. $10x^2 - 11x + 2 = 0$

38. $20x^2 - 11x + 1 = 0$

39. $\frac{1}{2}z^2 - 2z + \frac{3}{4} = 0$

40. $\frac{1}{5}z^2 - \frac{1}{2}z - 2 = 0$

41. Explain how you will decide what method to use when solving quadratic equations.

21. ______

22. ______

23. ______

24. ______

25. ______

26. ______

27. ______

28. ______

29. ______

30. ______

31. ______

32. ______

33. ______

34. ______

35. ______

36. ______

37. ______

38. ______

39. ______

40. ______

41. ______

Objectives

A Graph Quadratic Equations of the Form $y = ax^2$.

B Graph Quadratic Equations of the Form $y = ax^2 + bx + c$.

PRACTICE PROBLEM 1

Graph: $y = -3x^2$

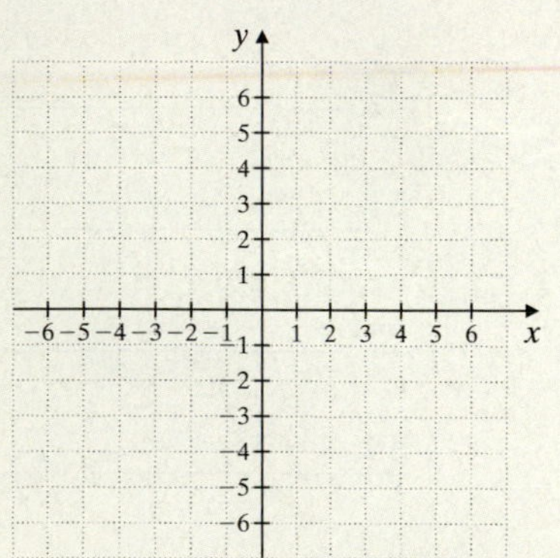

Answer

1.

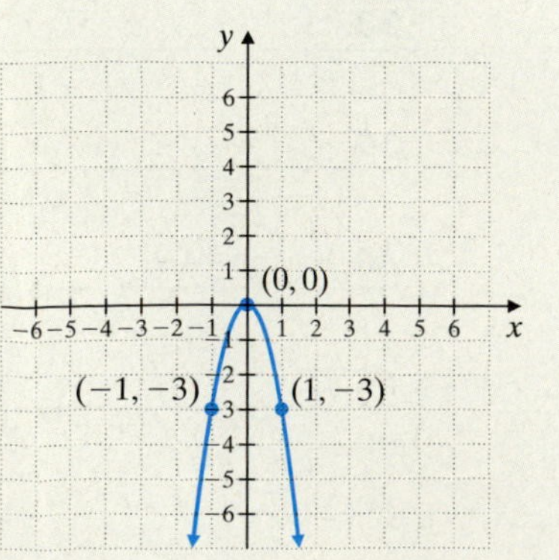

16.4 GRAPHING QUADRATIC EQUATIONS IN TWO VARIABLES

Recall from Section 13.2 that the graph of a linear equation in two variables $Ax + By = C$ is a straight line. In this section, we will find that the graph of a quadratic equation in the form $y = ax^2 + bx + c$ is a parabola.

Objective A Graphing $y = ax^2$

We begin our work by graphing $y = x^2$. To do so, we will find and plot ordered pair solutions of this equation. Let's select a few values for x, find the corresponding y-values, and record them in a table of values to keep track. Then we can plot the points corresponding to these solutions on a coordinate plane.

If $x = -3$, then $y = (-3)^2$, or 9.
If $x = -2$, then $y = (-2)^2$, or 4.
If $x = -1$, then $y = (-1)^2$, or 1.
If $x = 0$, then $y = 0^2$, or 0.
If $x = 1$, then $y = 1^2$, or 1.
If $x = 2$, then $y = 2^2$, or 4.
If $x = 3$, then $y = 3^2$, or 9.

x	y
−3	9
−2	4
−1	1
0	0
1	1
2	4
3	9

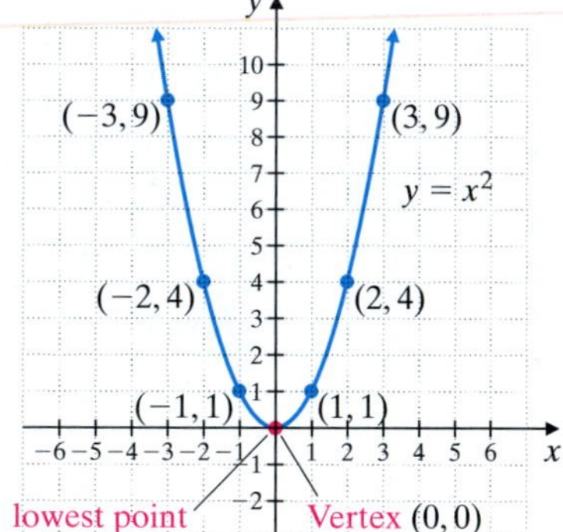

The graph of $y = x^2$ is a smooth curve through the plotted points. This curve is called a **parabola.** The lowest point on a parabola opening upward is called the **vertex.** The vertex is (0, 0) for the parabola $y = x^2$. If we fold the graph paper along the y-axis, the two pieces of the parabola match perfectly. For this reason, we say the graph is **symmetric about the y-axis,** and we call the y-axis the **line of symmetry.**

Notice that the parabola that corresponds to the equation $y = x^2$ opens upward. This happens when the coefficient of x^2 is positive. In the equation $y = x^2$, the coefficient of x^2 is 1. Example 1 shows the graph of a quadratic equation where the coefficient of x^2 is negative.

EXAMPLE 1 Graph: $y = -2x^2$

Solution: We begin by selecting x-values and calculating the corresponding y-values. Then we plot the ordered pairs found and draw a smooth curve through those points. Notice that when the coefficient of x^2 is negative, the corresponding parabola opens downward. When a parabola opens downward, the vertex is the highest point of the parabola. The vertex of this parabola is (0, 0).

$y = -2x^2$

x	y
0	0
1	−2
2	−8
3	−18
−1	−2
−2	−8
−3	−18

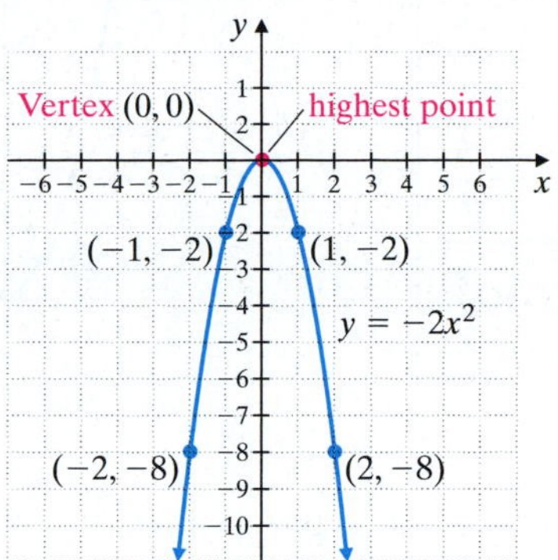

■ **Work Practice Problem 1**

Objective B Graphing $y = ax^2 + bx + c$

Just as for linear equations, we can use x- and y-intercepts to help graph quadratic equations. Recall from Chapter 13 that an x-intercept is the point where the graph crosses the x-axis. A y-intercept is the point where the graph crosses the y-axis. We find intercepts just as we did in Chapter 13.

Recall that:

To find x-intercepts, let $y = 0$ and solve for x.
To find y-intercepts, let $x = 0$ and solve for y.

EXAMPLE 2 Graph: $y = x^2 - 4$

Solution: If we write this equation as $y = x^2 + 0x + (-4)$, we can see that it is in the form $y = ax^2 + bx + c$. To graph it, we first find the intercepts. To find the y-intercept, we let $x = 0$. Then

$$y = 0^2 - 4 = -4$$

To find x-intercepts, we let $y = 0$.

$$0 = x^2 - 4$$
$$0 = (x - 2)(x + 2)$$
$$x - 2 = 0 \quad \text{or} \quad x + 2 = 0$$
$$x = 2 \qquad\qquad x = -2$$

Thus far, we have the y-intercept $(0, -4)$ and the x-intercepts $(2, 0)$ and $(-2, 0)$. Now we can select additional x-values, find the corresponding y-values, plot the points, and draw a smooth curve through the points.

PRACTICE PROBLEM 2

Graph: $y = x^2 - 9$

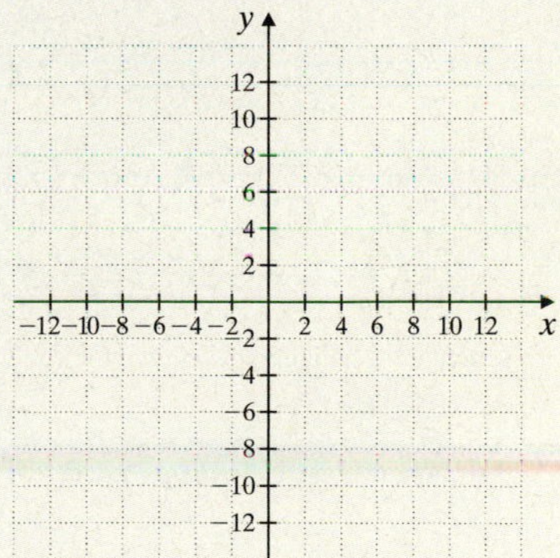

Answer

2.

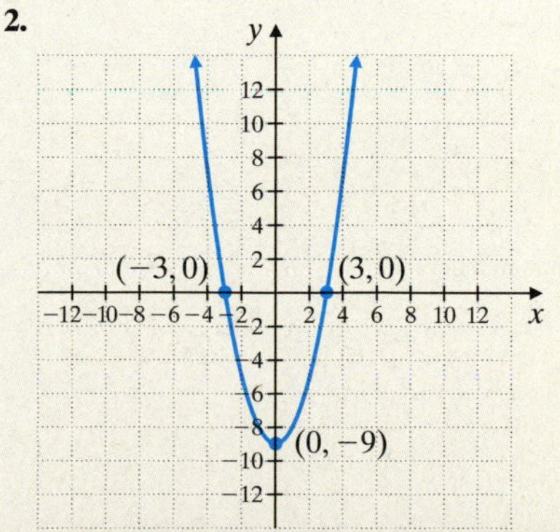

$y = x^2 - 4$

x	y
0	−4
1	−3
2	0
3	5
−1	−3
−2	0
−3	5

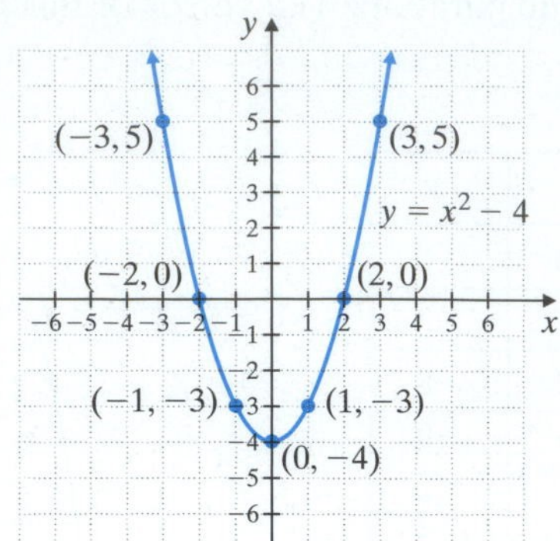

Notice that the vertex of this parabola is $(0, -4)$.

Work Practice Problem 2

✔Concept Check Tell whether the graph of each equation opens upward or downward.

a. $y = 2x^2$ **b.** $y = 3x^2 + 4x - 5$ **c.** $y = -5x^2 + 2$

For the graph of $y = ax^2 + bx + c$,

If a is positive, the parabola opens upward.
If a is negative, the parabola opens downward.

✔Concept Check For which of the following graphs of $y = ax^2 + bx + c$ would the value of a be negative?

a.

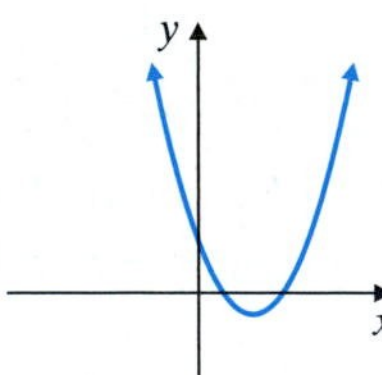

b.

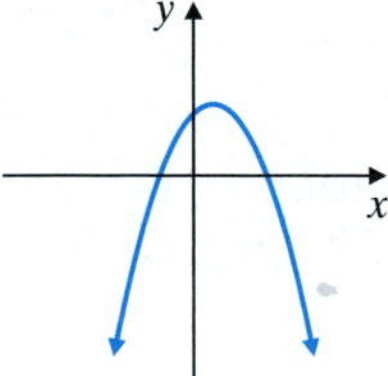

Thus far, we have accidentally stumbled upon the vertex of each parabola that we have graphed. However, our choice of values for x may not yield an ordered pair for the vertex of the parabola. It would be helpful if we could first find the vertex of a parabola. Next we would determine whether the parabola opens upward or downward. Finally we would calculate additional points such as x- and y-intercepts as needed. In fact, there is a formula that may be used to find the vertex of a parabola.

Vertex Formula

The vertex of the parabola $y = ax^2 + bx + c$ has x-coordinate

$$\frac{-b}{2a}$$

The corresponding y-coordinate of the vertex is obtained by substituting the x-coordinate into the equation and finding y.

✔ Concept Check Answer
a. upward, **b.** upward, **c.** downward; b

One way to develop this formula is to notice that the x-value of the vertex of the parabolas that we are considering lies halfway between its x-intercepts. Another way to develop this formula is to complete the square on the general form of a quadratic equation: $y = ax^2 + bx + c$. We will not show the development of this formula here.

EXAMPLE 3 Graph: $y = x^2 - 6x + 8$

Solution: In the equation $y = x^2 - 6x + 8$, $a = 1$ and $b = -6$.

Vertex: The x-coordinate of the vertex is

$$\frac{-b}{2a} = \frac{-(-6)}{2 \cdot 1} = 3 \quad \text{Use the vertex formula, } \frac{-b}{2a}.$$

To find the corresponding y-coordinate, we let $x = 3$ in the original equation.

$$y = x^2 - 6x + 8 = 3^2 - 6 \cdot 3 + 8 = -1$$

The vertex is $(3, -1)$ and the parabola opens upward since a is positive. We now find and plot the intercepts.

Intercepts: To find the x-intercepts, we let $y = 0$.

$$0 = x^2 - 6x + 8$$

We factor the expression $x^2 - 6x + 8$ to find $(x - 4)(x - 2) = 0$. The x-intercepts are $(4, 0)$ and $(2, 0)$.

If we let $x = 0$ in the original equation, then $y = 8$ gives us the y-intercept $(0, 8)$. Now we plot the vertex $(3, -1)$ and the intercepts $(4, 0)$, $(2, 0)$, and $(0, 8)$. Then we can sketch the parabola.

These and two additional points are shown in the table.

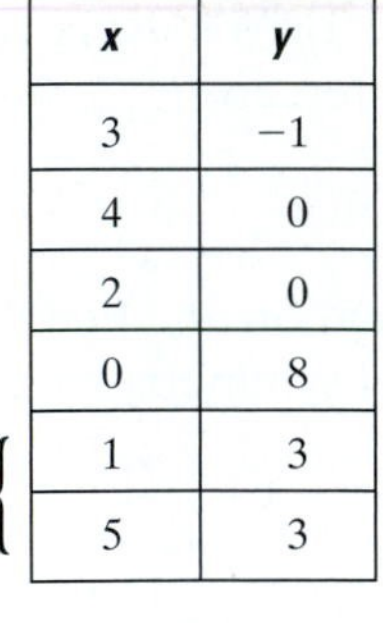

	x	y
	3	−1
	4	0
	2	0
	0	8
Additional points:	1	3
	5	3

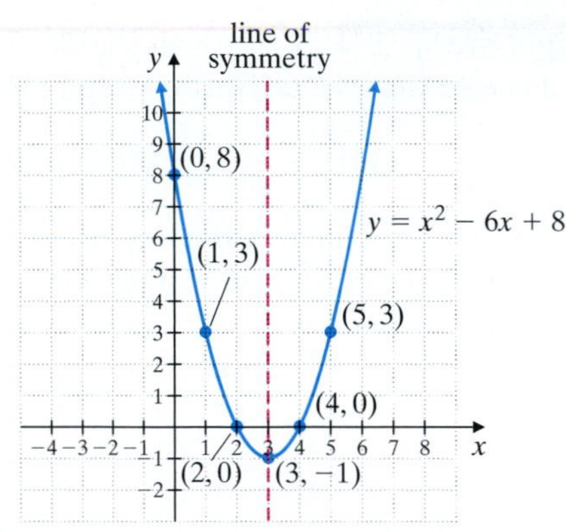

Work Practice Problem 3

PRACTICE PROBLEM 3

Graph: $y = x^2 - 2x - 3$

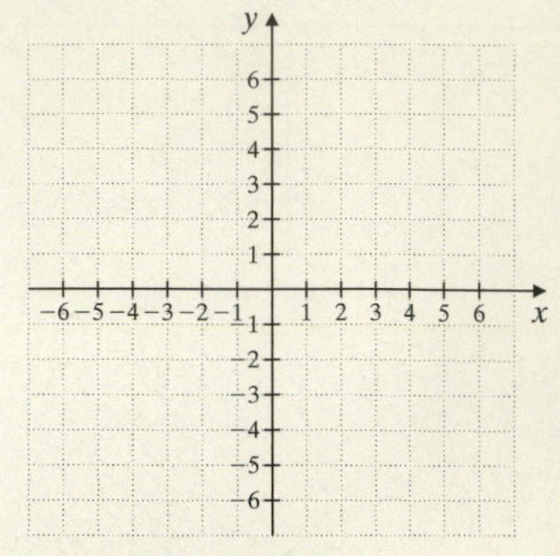

Study Example 3 and let's use it to write down a general procedure for graphing quadratic equations.

Graphing Parabolas Defined by $y = ax^2 + bx + c$

1. **Find the vertex by using the formula $x = -\frac{b}{2a}$.** Don't forget to find the y-value of the vertex.
2. **Find the intercepts.**
   - Let $x = 0$ and solve for y to find the y-intercept. There will be only one.
   - Let $y = 0$ and solve for x to find any x-intercepts. There may be 0, 1, or 2.
3. **Plot the vertex and the intercepts.**
4. **Find and plot additional points on the graph.** Then draw a smooth curve through the plotted points. Keep in mind if $a > 0$, the parabola opens up and if $a < 0$, the parabola opens down.

Answer

3.

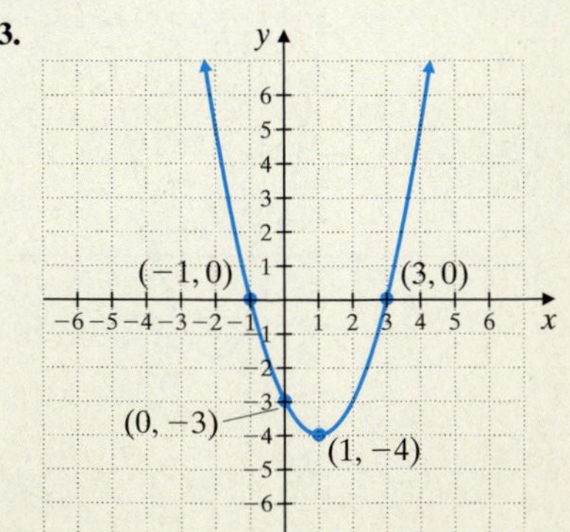

PRACTICE PROBLEM 4

Graph: $y = x^2 - 4x + 1$

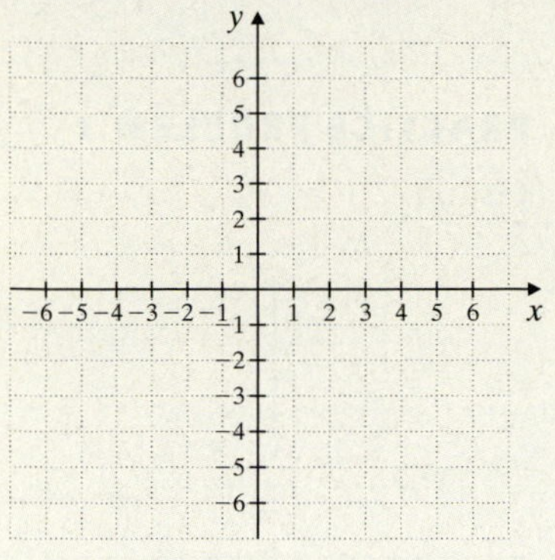

EXAMPLE 4 Graph: $y = x^2 + 2x - 5$

Solution: In the equation $y = x^2 + 2x - 5$, $a = 1$ and $b = 2$. Using the vertex formula, we find that the x-coordinate of the vertex is

$$x = \frac{-b}{2a} = \frac{-2}{2 \cdot 1} = -1$$

The y-coordinate is

$$y = (-1)^2 + 2(-1) - 5 = -6$$

Thus the vertex is $(-1, -6)$.
To find the x-intercepts, we let $y = 0$.

$$0 = x^2 + 2x - 5$$

This cannot be solved by factoring, so we use the quadratic formula.

$$x = \frac{-2 \pm \sqrt{2^2 - 4(1)(-5)}}{2 \cdot 1} \quad \text{Let } a = 1, b = 2, \text{ and } c = -5.$$

$$x = \frac{-2 \pm \sqrt{24}}{2}$$

$$x = \frac{-2 \pm 2\sqrt{6}}{2} \quad \text{Simplify the radical.}$$

$$x = \frac{2\left(-1 \pm \sqrt{6}\right)}{2} = -1 \pm \sqrt{6}$$

The x-intercepts are $\left(-1 + \sqrt{6}, 0\right)$ and $\left(-1 - \sqrt{6}, 0\right)$. We use a calculator to approximate these so that we can easily graph these intercepts.

$$-1 + \sqrt{6} \approx 1.4 \quad \text{and} \quad -1 - \sqrt{6} \approx -3.4$$

To find the y-intercept, we let $x = 0$ in the original equation and find that $y = -5$. Thus the y-intercept is $(0, -5)$. You will find because of symmetry, that $(-2, -5)$ is also an ordered-pair solution.

x	y
-1	-6
$-1 + \sqrt{6} \approx 1.4$	0
$-1 - \sqrt{6} \approx -3.4$	0
0	-5
-2	-5

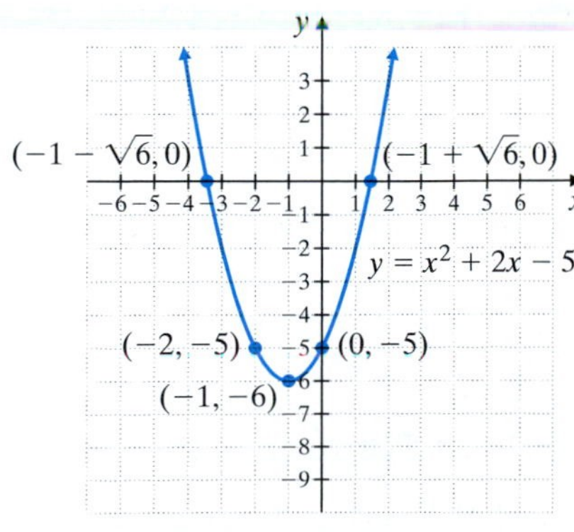

Work Practice Problem 4

Answer

4.

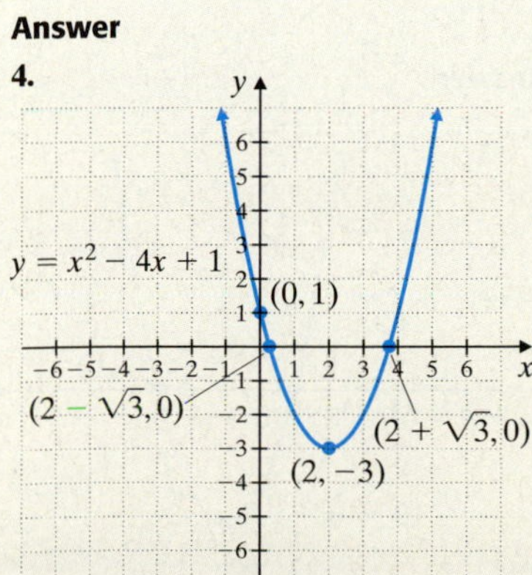

Helpful Hint

Notice that the number of x-intercepts of the graph of the parabola $y = ax^2 + bx + c$ is the same as the number of real solutions of $0 = ax^2 + bx + c$.

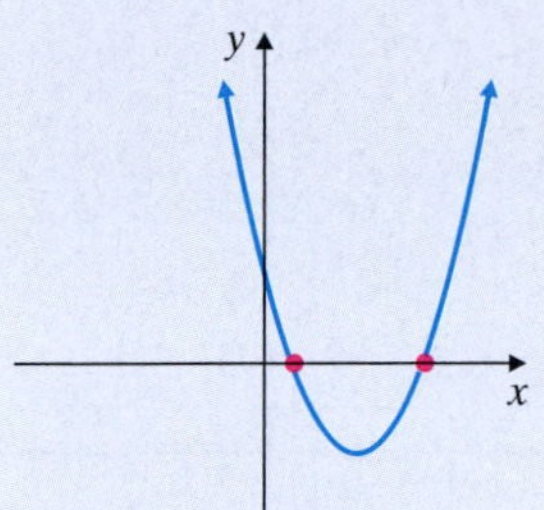

$y = ax^2 + bx + c$
$a > 0$

Two x-intercepts
Two real solutions of
$0 = ax^2 + bx + c$

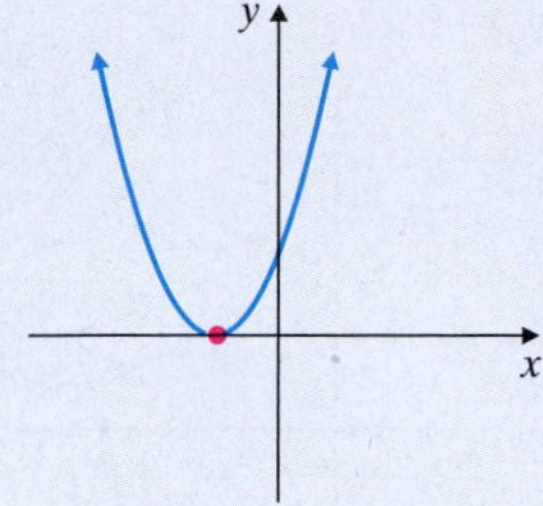

$y = ax^2 + bx + c$
$a > 0$

One x-intercept
One real solution of
$0 = ax^2 + bx + c$

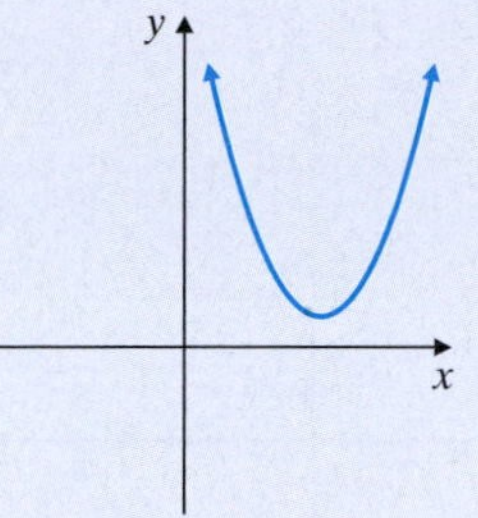

$y = ax^2 + bx + c$
$a > 0$

No x-intercepts
No real solutions of
$0 = ax^2 + bx + c$

CALCULATOR EXPLORATIONS Graphing

Recall that a graphing calculator may be used to solve quadratic equations. The x-intercepts of the graph of $y = ax^2 + bx + c$ are solutions of $0 = ax^2 + bx + c$. To solve $x^2 - 7x - 3 = 0$, for example, graph $y_1 = x^2 - 7x - 3$. The x-intercepts of the graph are the solutions of the equation.

Use a graphing calculator to solve each quadratic equation. Round solutions to two decimal places.

1. $x^2 - 7x - 3 = 0$

2. $2x^2 - 11x - 1 = 0$

3. $-1.7x^2 + 5.6x - 3.7 = 0$

4. $-5.8x^2 + 2.3x - 3.9 = 0$

5. $5.8x^2 - 2.6x - 1.9 = 0$

6. $7.5x^2 - 3.7x - 1.1 = 0$

16.4 EXERCISE SET

FOR EXTRA HELP

Objective A *Graph each quadratic equation by finding and plotting ordered pair solutions. See Example 1.*

1. $y = 2x^2$

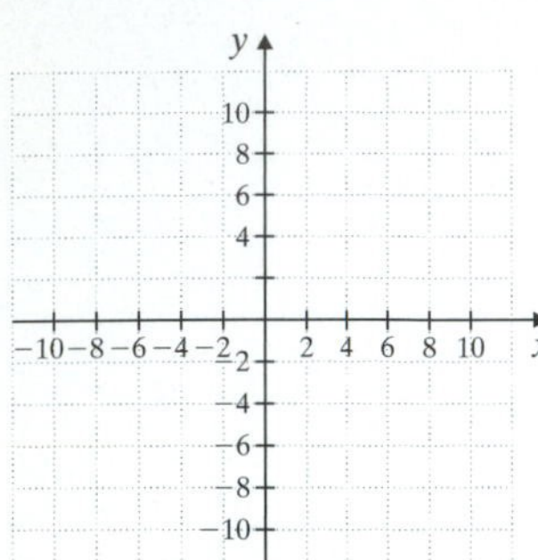

2. $y = 3x^2$

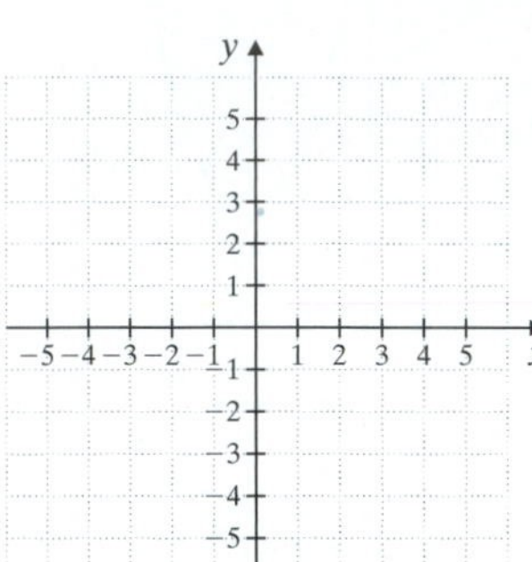

3. $y = -x^2$

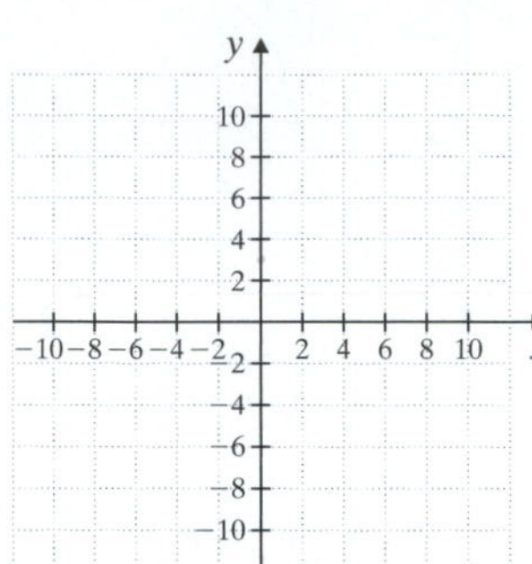

4. $y = -4x^2$

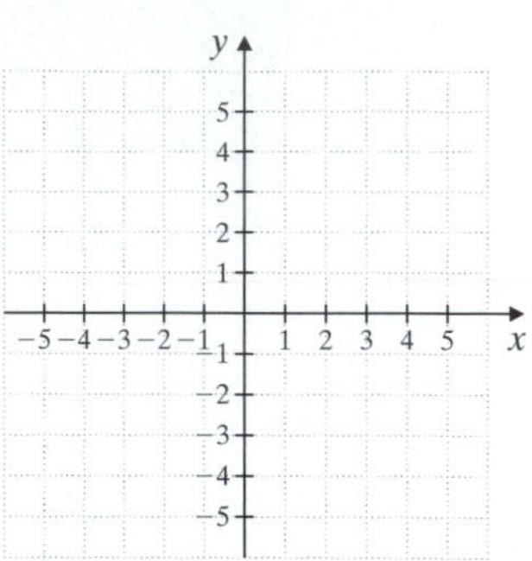

Objective B *Sketch the graph of each equation. Label the vertex and the intercepts. See Examples 2 through 4.*

5. $y = x^2 - 1$

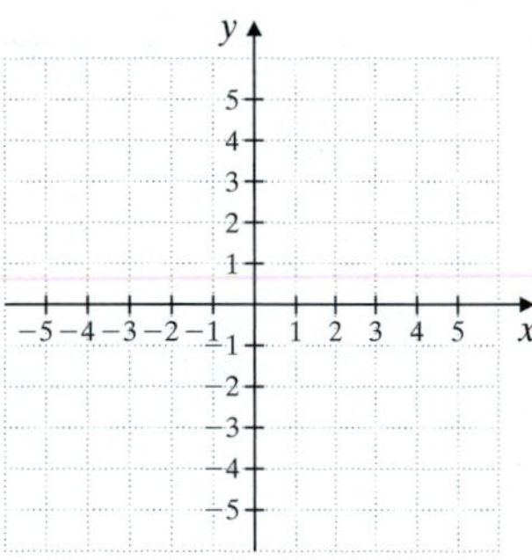

6. $y = x^2 - 16$

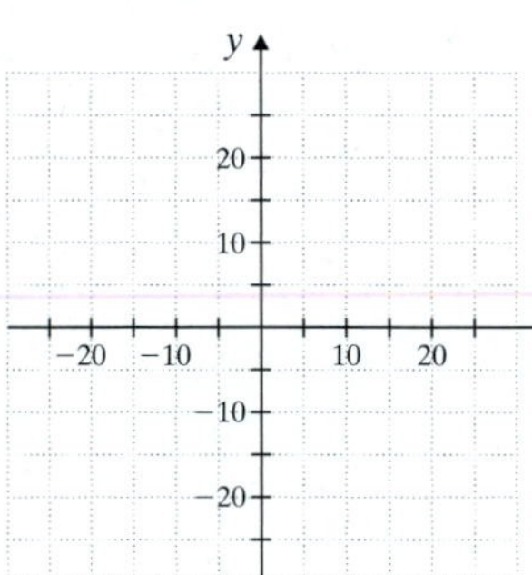

7. $y = x^2 + 4$

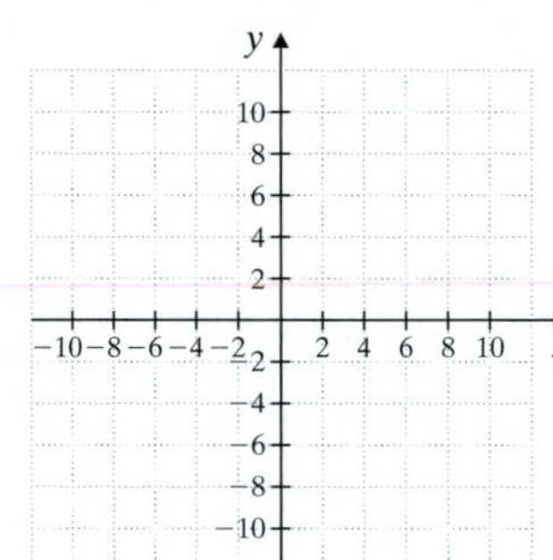

8. $y = x^2 + 9$

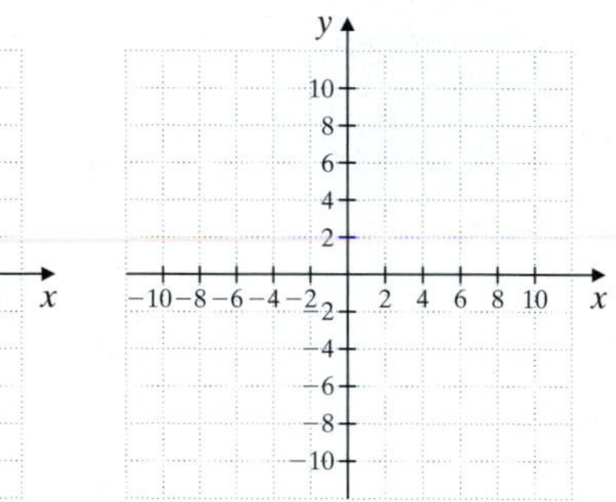

9. $y = -x^2 + 4x - 4$

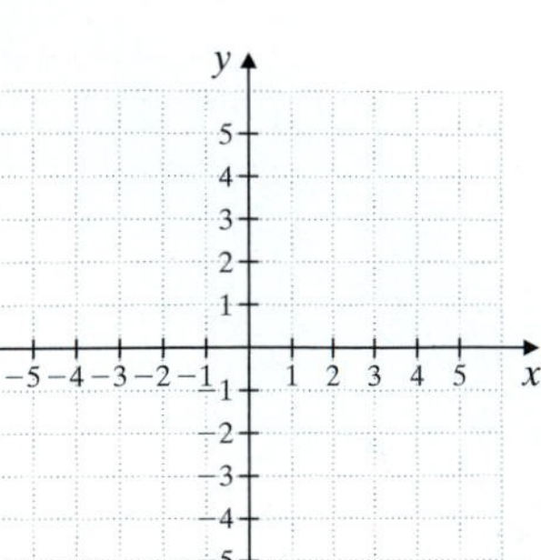

10. $y = -x^2 - 2x - 1$

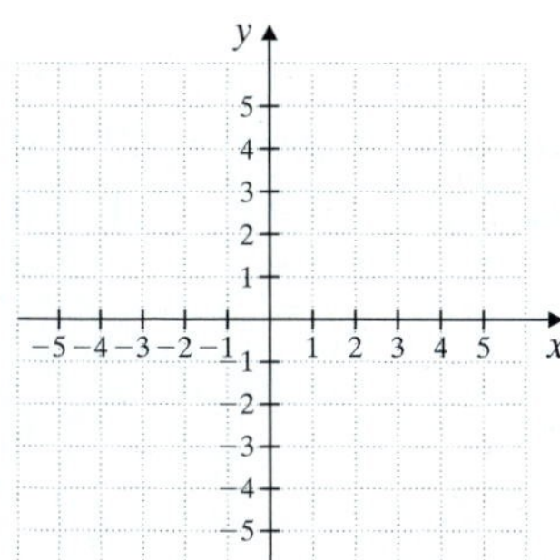

11. $y = x^2 + 5x + 4$

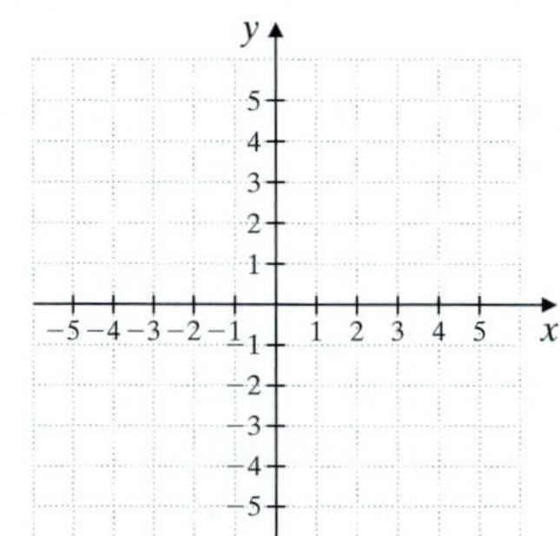

12. $y = x^2 + 7x + 10$

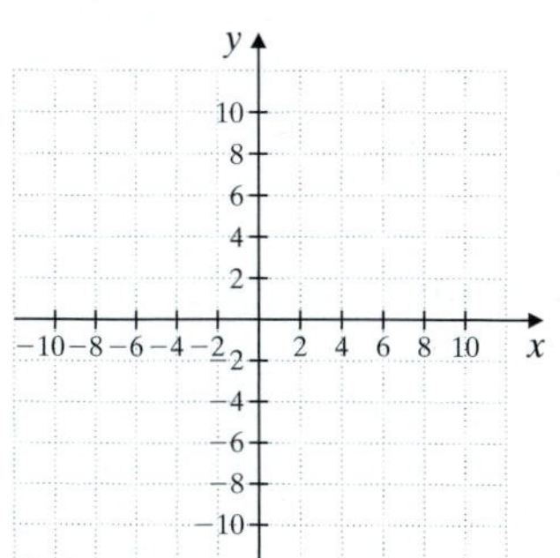

13. $y = x^2 - 4x + 5$

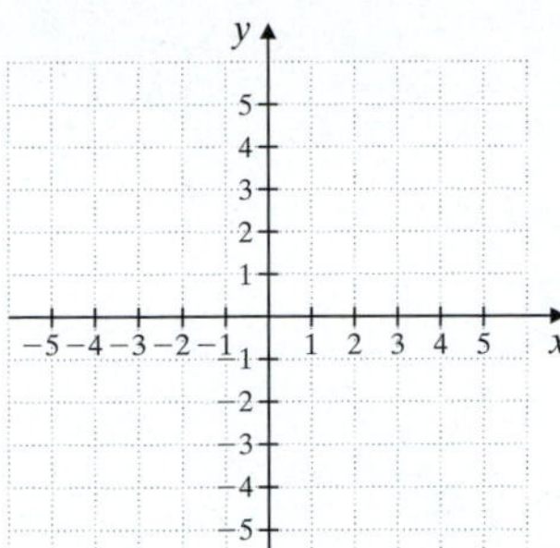

14. $y = x^2 - 6x + 10$

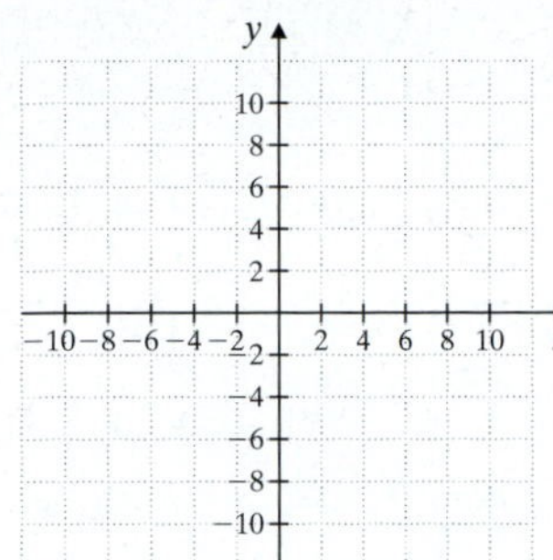

15. $y = 2 - x^2$

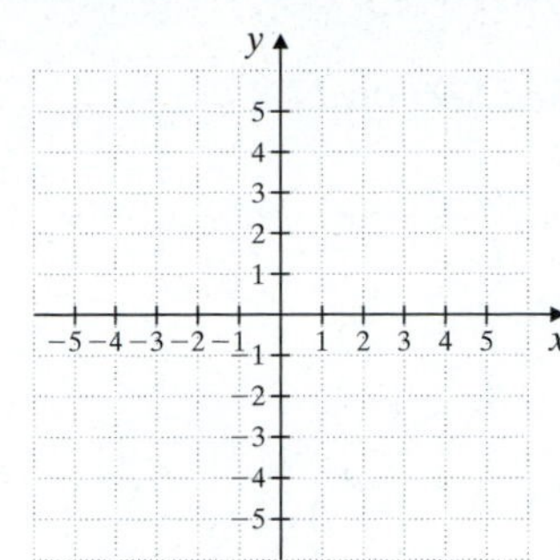

16. $y = 3 - x^2$

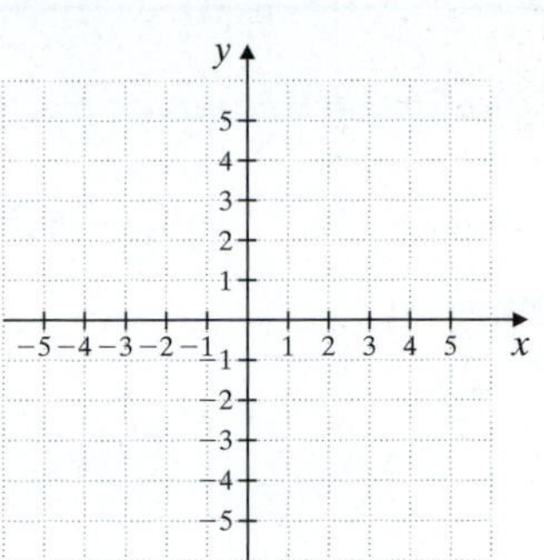

Objectives **A B Mixed Practice** *Sketch the graph of each equation. Label the vertex and the intercepts. See Examples 1 through 4.*

17. $y = \frac{1}{3}x^2$

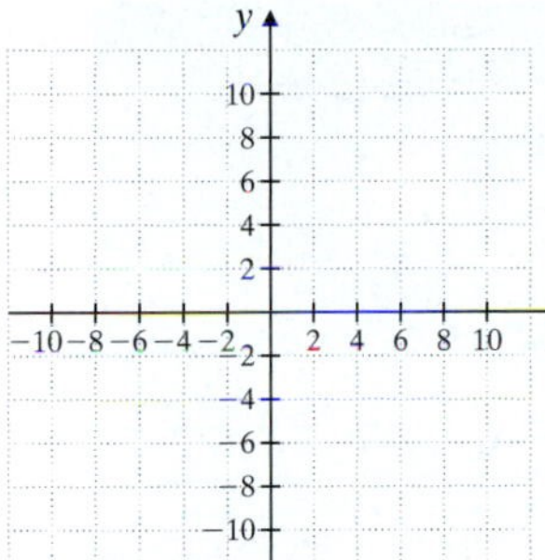

18. $y = \frac{1}{2}x^2$

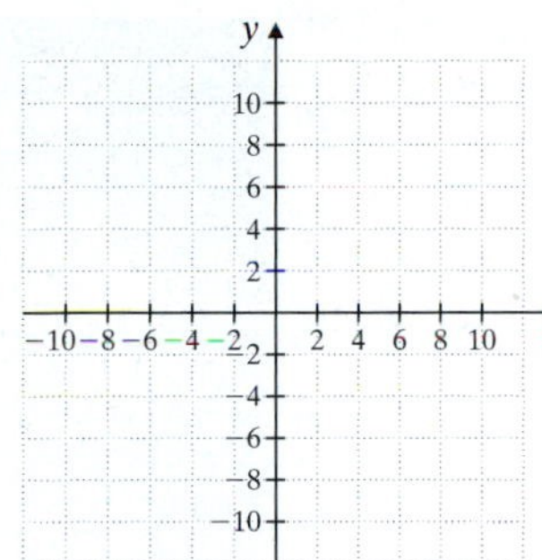

19. $y = x^2 + 6x$

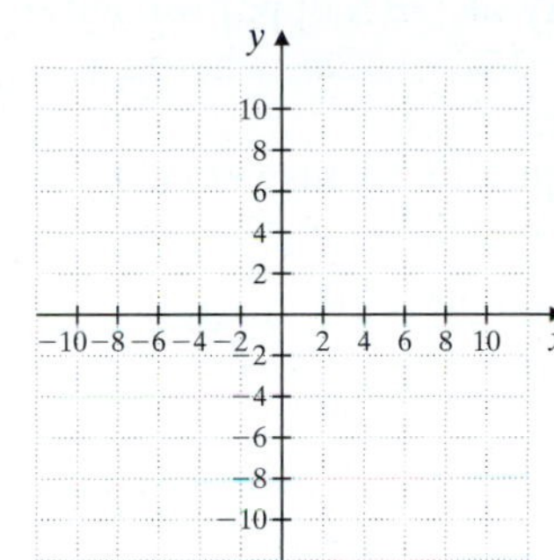

20. $y = x^2 - 4x$

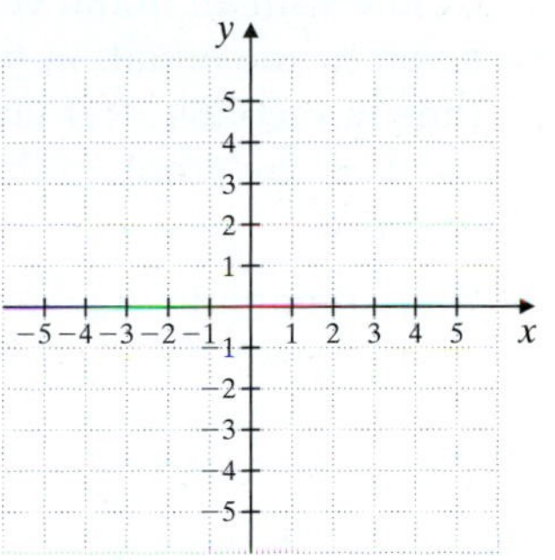

21. $y = x^2 + 2x - 8$

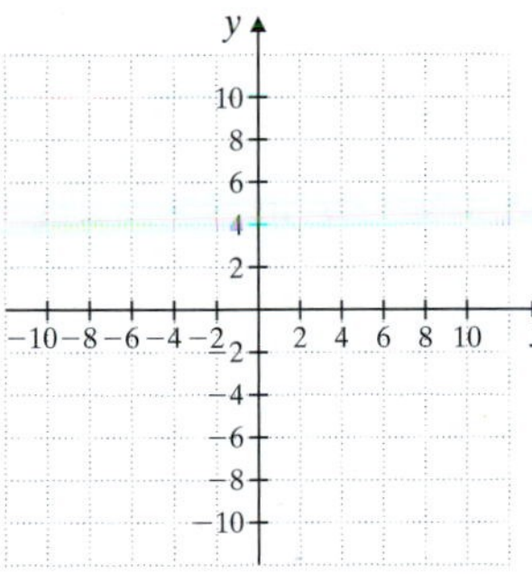

22. $y = x^2 - 2x - 3$

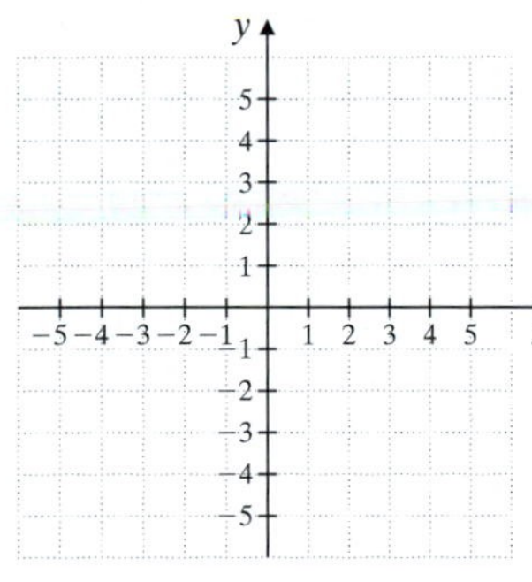

23. $y = -\frac{1}{2}x^2$

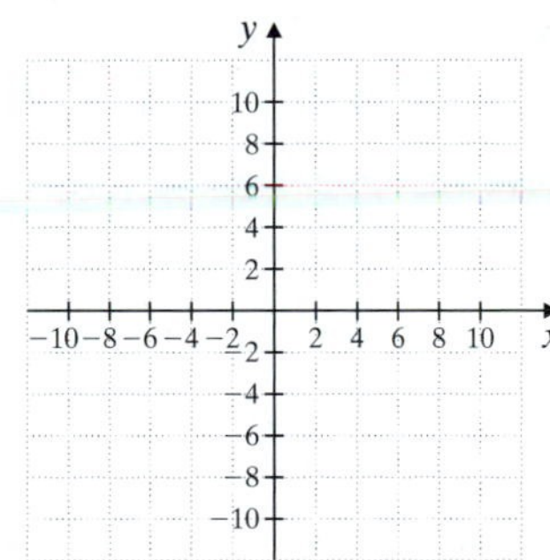

24. $y = -\frac{1}{3}x^2$

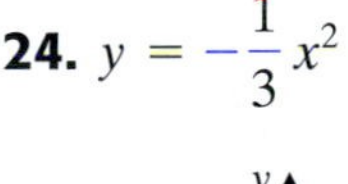

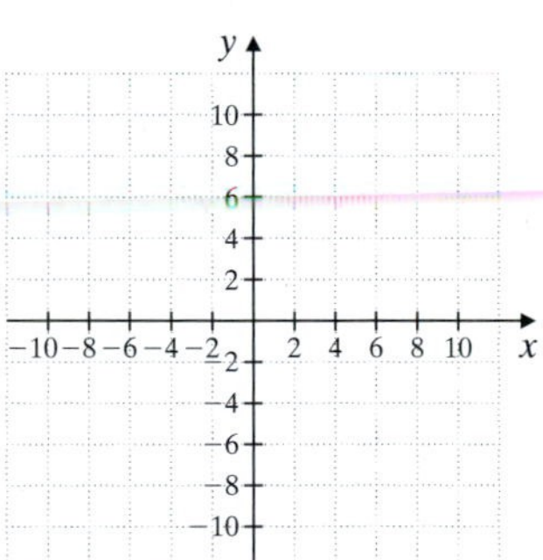

25. $y = 2x^2 - 11x + 5$

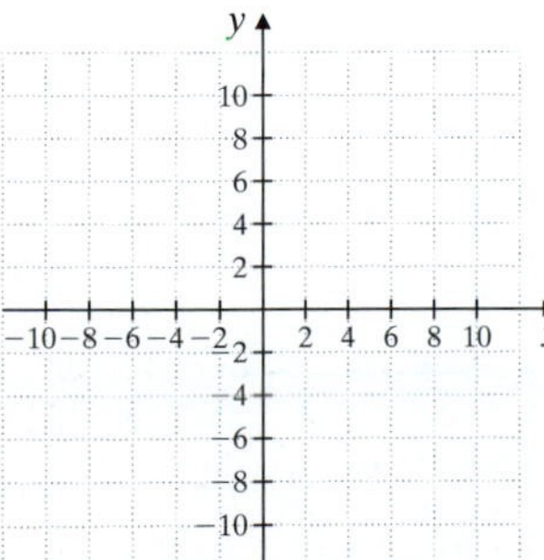

26. $y = 2x^2 + x - 3$

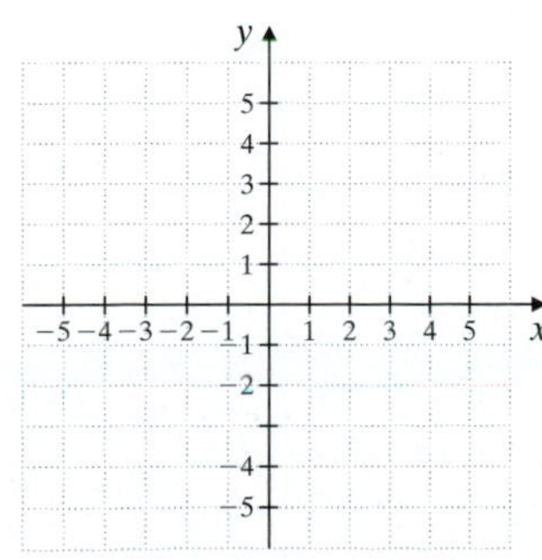

27. $y = -x^2 + 4x - 3$

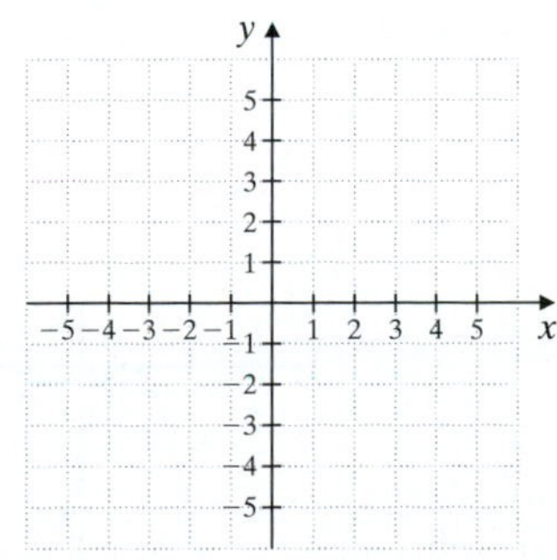

28. $y = -x^2 + 6x - 8$

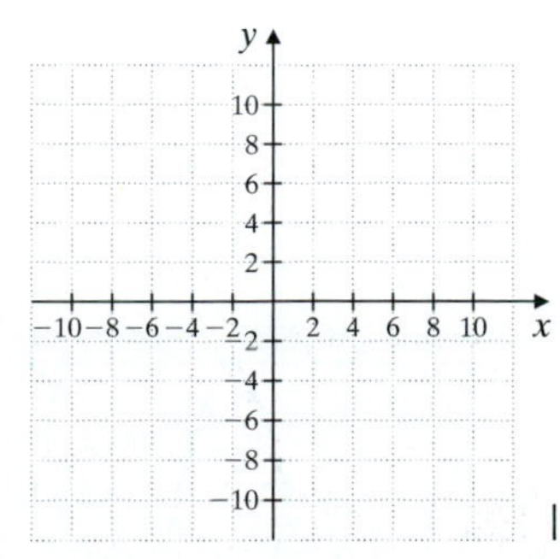

Review

Simplify each complex fraction. See Section 12.7.

29. $\dfrac{\frac{1}{7}}{\frac{2}{5}}$ **30.** $\dfrac{\frac{3}{8}}{\frac{1}{7}}$ **31.** $\dfrac{\frac{1}{x}}{\frac{2}{x^2}}$ **32.** $\dfrac{\frac{x}{5}}{\frac{2}{x}}$

33. $\dfrac{2x}{1 - \frac{1}{x}}$ **34.** $\dfrac{x}{x - \frac{1}{x}}$ **35.** $\dfrac{\frac{a-b}{2b}}{\frac{b-a}{8b^2}}$ **36.** $\dfrac{\frac{2a^2}{a-3}}{\frac{a}{3-a}}$

Concept Extensions

37. The height h of a fireball launched from a Roman candle with an initial velocity of 128 feet per second is given by the equation $h = -16t^2 + 128t$, where t is time in seconds after launch.

Use the graph of this equation to answer each question.

a. Estimate the maximum height of the fireball.
b. Estimate the time when the fireball is at its maximum height.
c. Estimate the time when the fireball returns to the ground.

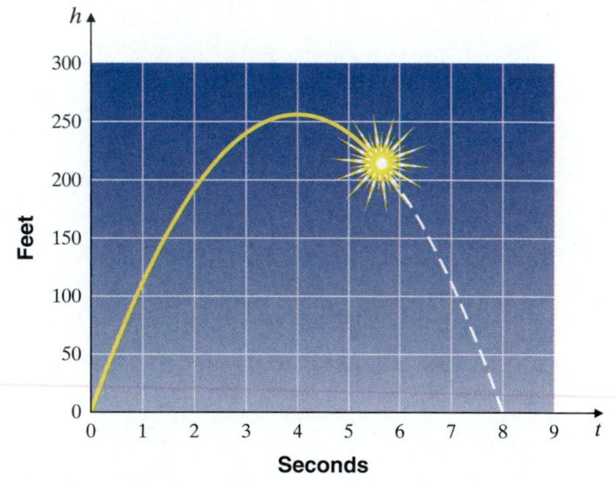

38. Determine the maximum number and the minimum number of x-intercepts for a parabola. Explain your answers.

Match the graph of each equation of the form $y = ax^2 + bx + c$ with the given description.

39. $a > 0$, two x-intercepts **40.** $a < 0$, one x-intercept **41.** $a < 0$, no x-intercept

42. $a > 0$, no x-intercept **43.** $a > 0$, one x-intercept **44.** $a < 0$, two x-intercepts

A
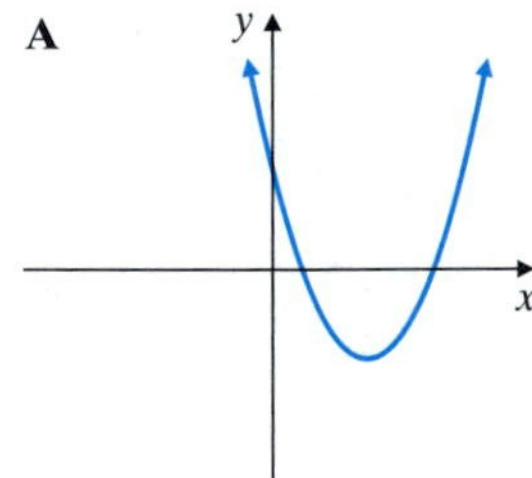

B
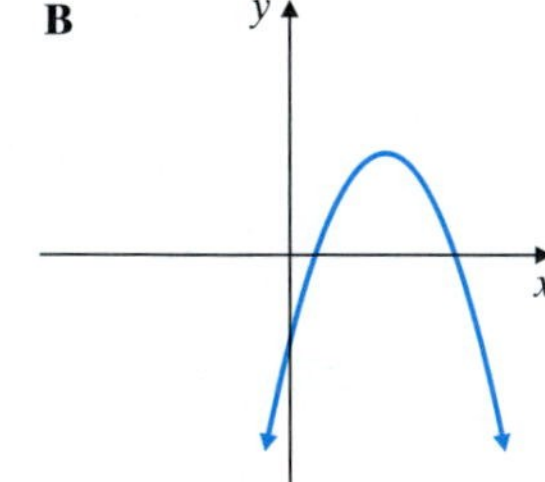

C
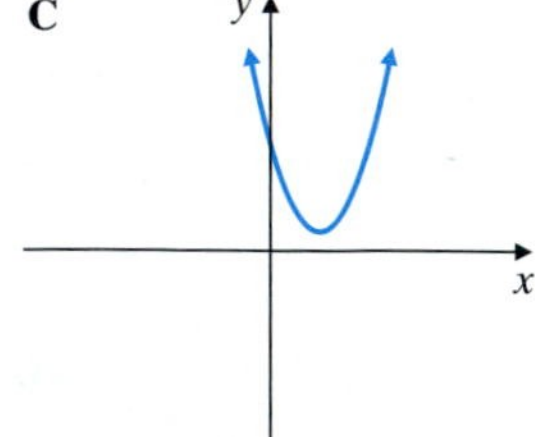

D
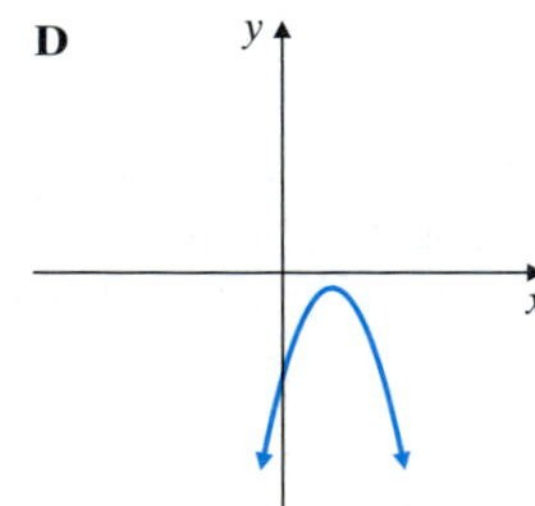

E
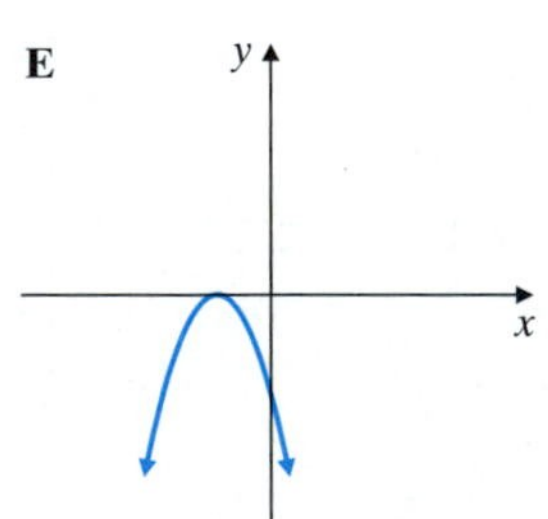

F
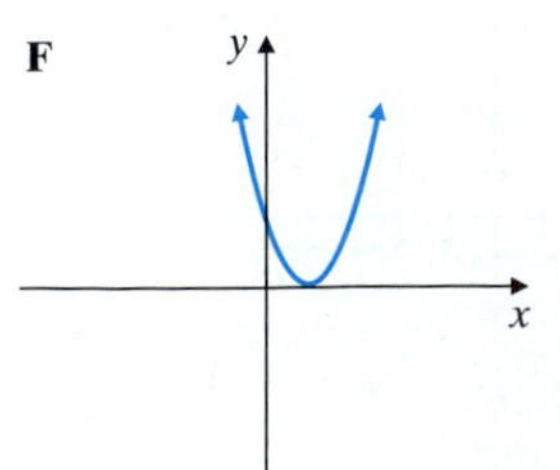

CHAPTER 16 Group Activity

Uses of Parabolas

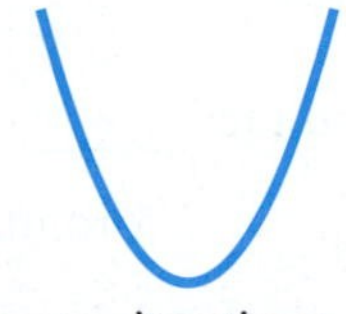

In this chapter, we learned that the graph of a quadratic equation in two variables of the form $y = ax^2 + bx + c$ is a shape called a **parabola.** The figure to the right shows the general shape of a parabola.

The shape of a parabola shows up in many situations, both natural and human-made, in the world around us.

Natural Situations

- **Hurricanes** The paths of many hurricanes are roughly shaped like a parabola. In the northern hemisphere, hurricanes generally begin moving to the northwest. Then, as they move farther from the equator, they swing around to head in a northeastern direction.

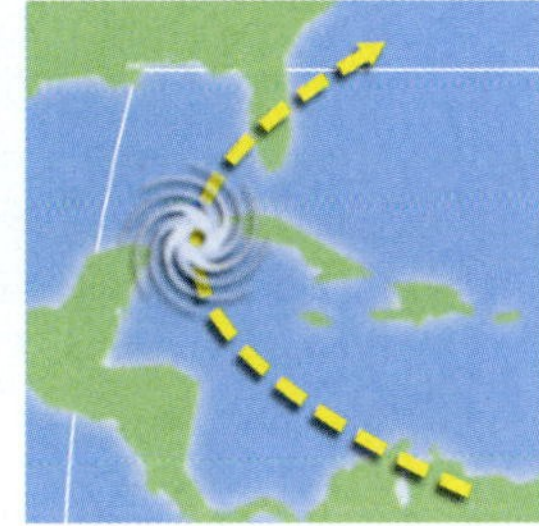

- **Projectiles** The force of the earth's gravity acts on a projectile launched into the air. The resulting path of the projectile, anything from a bullet to a football, is generally shaped like a parabola.

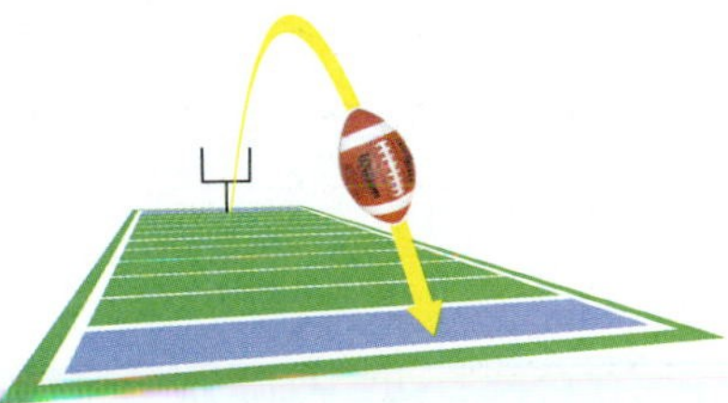

- **Orbits** There are several different possible shapes of orbits of satellites, planets, moons, and comets in outer space. One of the possible types of orbits is in the shape of a parabola. A parabolic orbit is most often seen with comets.

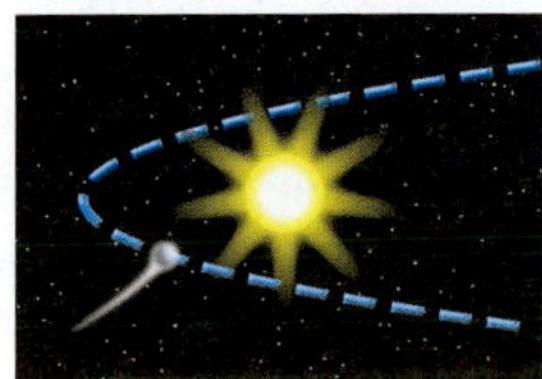

Human-Made Situations

- **Telescopes** Because a parabola has nice reflecting properties, its shape is used in many kinds of telescopes. The largest non-steerable radio telescope is Arecibo Observatory in Puerto Rico. This telescope consists of a huge parabolic dish built into a valley. The dish is about 1000 feet across.

- **Training Astronauts** Astronauts must be able to work in zero-gravity conditions on missions in space. However, it's nearly impossible to escape the force of gravity on earth. To help astronauts train to work in weightlessness, a specially modified jet can be flown in a parabolic path. At the top of the parabola, weightlessness can be simulated for up to 30 seconds at a time.

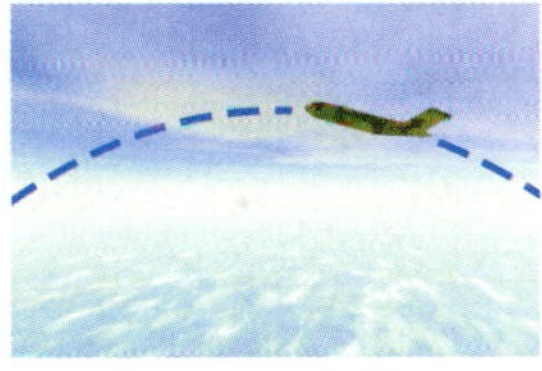

- **Architecture** The reinforced concrete arches used in many modern buildings are based on the shape of a parabola.

- **Music** The design of the modern flute incorporates a parabolic head joint.

Group Activity

There are many other physical applications of parabolas. For example, satellite dishes often have parabolic shapes. Choose a physical example of a parabola given here or use one of your own and write a report (with diagrams).

Chapter 16 Vocabulary Check

Fill in each blank with one of the words or phrases listed below.

square root	vertex
completing the square	quadratic

1. If $x^2 = a$, then $x = \sqrt{a}$ or $x = -\sqrt{a}$. This property is called the ______________ property.
2. The formula $\frac{-b}{2a}$ where $y = ax^2 + bx + c$ is called the __________ formula.
3. The process of solving a quadratic equation by writing it in the form $(x + a)^2 = c$ is called ______________________.
4. The formula $x = \frac{-b \pm \sqrt{b^2 - 4ac}}{2a}$ is called the ____________ formula.

Helpful Hint

Are you preparing for your test? Don't forget to take the Chapter 16 Test on page 1200. Then check your answers at the back of the text and use the Chapter Test Prep Video CD to see the fully worked-out solutions to any of the exercises you want to review.

16 Chapter Highlights

DEFINITIONS AND CONCEPTS	EXAMPLES
Section 16.1 Solving Quadratic Equations by the Square Root Property	
SQUARE ROOT PROPERTY If $x^2 = a$ for $a \geq 0$, then $x = \sqrt{a}$ or $x = -\sqrt{a}$.	Solve the equation. $(x - 1)^2 = 15$ $x - 1 = \sqrt{15}$ or $x - 1 = -\sqrt{15}$ $x = 1 + \sqrt{15}$ $\quad x = 1 - \sqrt{15}$
Section 16.2 Solving Quadratic Equations by Completing the Square	
TO SOLVE A QUADRATIC EQUATION BY COMPLETING THE SQUARE **Step 1.** If the coefficient of x^2 is not 1, divide both sides of the equation by the coefficient. **Step 2.** Get all terms with variables alone on one side. **Step 3.** Complete the square by adding the square of half of the coefficient of x to both sides. **Step 4.** Factor the perfect square trinomial. **Step 5.** Use the square root property to solve.	Solve $2x^2 + 12x - 10 = 0$ by completing the square. $\frac{2x^2}{2} + \frac{12x}{2} - \frac{10}{2} = \frac{0}{2}$ Divide by 2. $x^2 + 6x - 5 = 0$ Simplify. $x^2 + 6x = 5$ Add 5. The coefficient of x is 6. Half of 6 is 3 and $3^2 = 9$. Add 9 to both sides. $x^2 + 6x + 9 = 5 + 9$ $(x + 3)^2 = 14$ Factor. $x + 3 = \sqrt{14}$ or $x + 3 = -\sqrt{14}$ $x = -3 + \sqrt{14}$ $\quad x = -3 - \sqrt{14}$

Definitions and Concepts	**Examples**

Section 16.3 Solving Quadratic Equations by the Quadratic Formula

Quadratic Formula

If a, b, and c are real numbers and $a \neq 0$, the quadratic equation $ax^2 + bx + c = 0$ has solutions

$$x = \frac{-b \pm \sqrt{b^2 - 4ac}}{2a}$$

To Solve a Quadratic Equation by the Quadratic Formula

Step 1. Write the equation in standard form: $ax^2 + bx + c = 0$.

Step 2. If necessary, clear the equation of fractions.

Step 3. Identify a, b, and c.

Step 4. Replace a, b, and c in the quadratic formula with the identified values, and simplify.

Identify a, b, and c in the quadratic equation

$$4x^2 - 6x = 5$$

First, subtract 5 from both sides.

$$4x^2 - 6x - 5 = 0$$

$a = 4$, $b = -6$, and $c = -5$.

Solve $3x^2 - 2x - 2 = 0$.

In this equation, $a = 3$, $b = -2$, and $c = -2$.

$$x = \frac{-(-2) \pm \sqrt{(-2)^2 - 4(3)(-2)}}{2 \cdot 3}$$

$$= \frac{2 \pm \sqrt{4 - (-24)}}{6}$$

$$= \frac{2 \pm \sqrt{28}}{6} = \frac{2 \pm \sqrt{4 \cdot 7}}{6} = \frac{2 \pm 2\sqrt{7}}{6}$$

$$= \frac{2(1 \pm \sqrt{7})}{2 \cdot 3} = \frac{1 \pm \sqrt{7}}{3}$$

Section 16.4 Graphing Quadratic Equations in Two Variables

The graph of a quadratic equation $y = ax^2 + bx + c$, $a \neq 0$, is called a **parabola.** The lowest point on a parabola opening upward or the highest point on a parabola opening downward is called the **vertex.** The vertical line through the vertex is the **line of symmetry.**

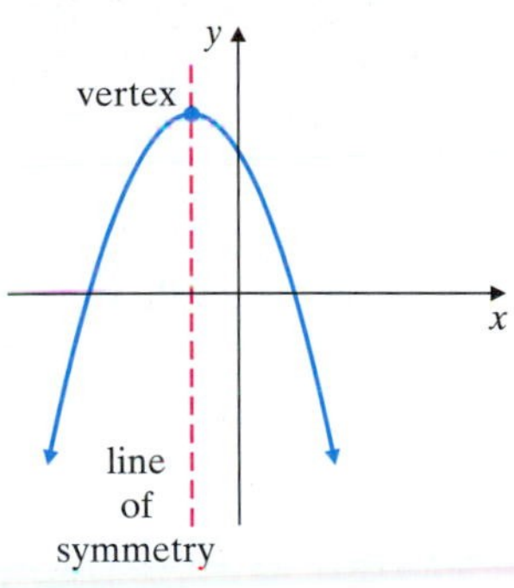

Vertex Formula

The vertex of the parabola $y = ax^2 + bx + c$ has x-coordinate $\frac{-b}{2a}$. To find the corresponding y-coordinate, substitute the x-coordinate into the original equation and solve for y.

Graph: $y = 2x^2 - 6x + 4$

The x-coordinate of the vertex is

$$x = \frac{-b}{2a} = \frac{-(-6)}{2(2)} = \frac{6}{4} = \frac{3}{2}$$

The y-coordinate is

$$y = 2\left(\frac{3}{2}\right)^2 - 6\left(\frac{3}{2}\right) + 4 = 2\left(\frac{9}{4}\right) - 9 + 4 = -\frac{1}{2}$$

The vertex is $\left(\frac{3}{2}, -\frac{1}{2}\right)$.

The y-intercept is

$$y = 2 \cdot 0^2 - 6 \cdot 0 + 4 = 4$$

The x-intercepts are

$$0 = 2x^2 - 6x + 4$$
$$0 = 2(x - 2)(x - 1)$$
$$x = 2 \quad \text{or} \quad x = 1$$

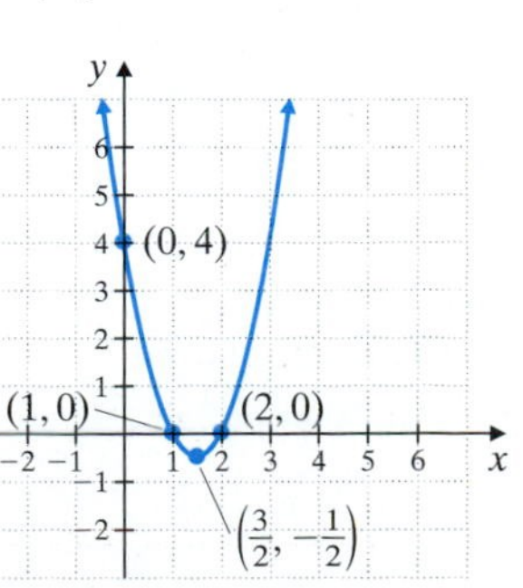

16 CHAPTER REVIEW

(16.1) *Solve each quadradic equation by factoring.*

1. $x^2 - 121 = 0$

2. $y^2 - 100 = 0$

3. $3m^2 - 5m = 2$

4. $7m^2 + 2m = 5$

Use the square root property to solve each quadratic equation.

5. $x^2 = 36$

6. $x^2 = 81$

7. $k^2 = 50$

8. $k^2 = 45$

9. $(x - 11)^2 = 49$

10. $(x + 3)^2 = 100$

11. $(4p + 5)^2 = 41$

12. $(3p + 7)^2 = 37$

Solve. For Exercises 13 and 14, use the formula $h = 16t^2$*, where h is the height in feet at time t seconds.*

13. If Kara Washington dives from a height of 100 feet, how long before she hits the water?

14. How long does a 5-mile free-fall take? Round your result to the nearest tenth of a second. (*Hint:* 1 mi = 5280 ft)

(16.2) *Solve each quadratic equation by completing the square.*

15. $x^2 - 9x = -8$

16. $x^2 + 8x = 20$

17. $x^2 + 4x = 1$

18. $x^2 - 8x = 3$

19. $x^2 - 6x + 7 = 0$

20. $x^2 + 6x + 7 = 0$

21. $2y^2 + y - 1 = 0$

22. $y^2 + 3y - 1 = 0$

(16.3) *Use the quadratic formula to solve each quadratic equation.*

23. $9x^2 + 30x + 25 = 0$

24. $16x^2 - 72x + 81 = 0$

25. $7x^2 = 35$

26. $11x^2 = 33$

27. $x^2 - 10x + 7 = 0$

28. $x^2 + 4x - 7 = 0$

29. $3x^2 + x - 1 = 0$

30. $x^2 + 3x - 1 = 0$

31. $2x^2 + x + 5 = 0$

32. $7x^2 - 3x + 1 = 0$

For the Exercise numbers given, approximate the exact solutions to the nearest tenth.

33. Exercise 29

34. Exercise 30

35. The average price of silver (in cents per ounce) from 2001 to 2003 is modeled by the equation $y = 38x^2 - 43x + 446$. In this equation, x is the number of years since 2001. Assume that this trend continues and find the year after 2001 in which the price of silver will be 1556 cents per ounce. (*Source:* U.S. Geological Survey, Minerals Information)

36. The average price of platinum (in dollars per ounce) from 2001 to 2004 is modeled by the equation $y = 64x^2 - 87x + 545$. In this equation, x is the number of years since 2001. Assume that this trend continues and find the year after 2001 in which the price of platinum will be 2327 dollars per ounce. (*Source:* U.S. Geological Survey, Minerals Information)

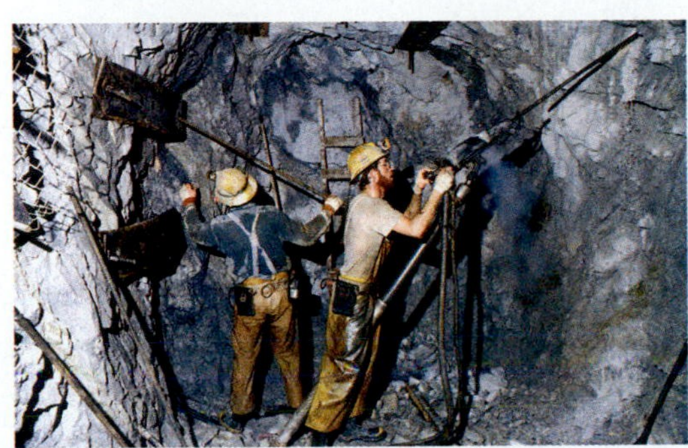

(16.4) *Graph each quadratic equation and find and plot any intercept points.*

37. $y = 5x^2$

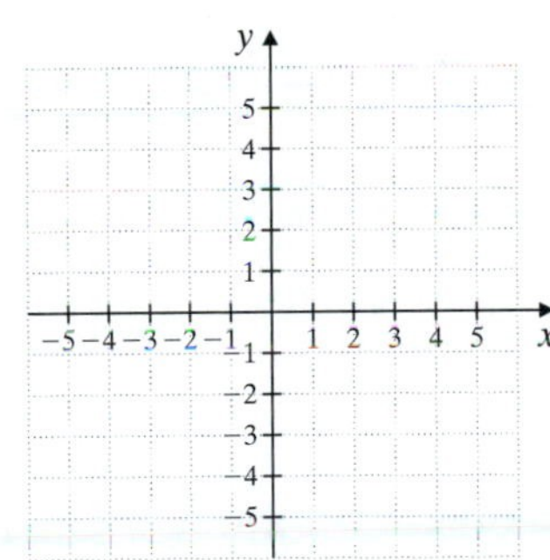

38. $y = -\frac{1}{2}x^2$

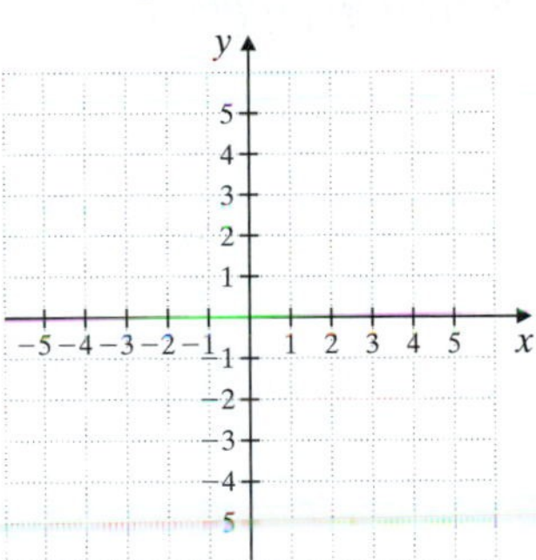

Graph each quadratic equation. Label the vertex and the intercept points with their coordinates.

39. $y = x^2 - 25$

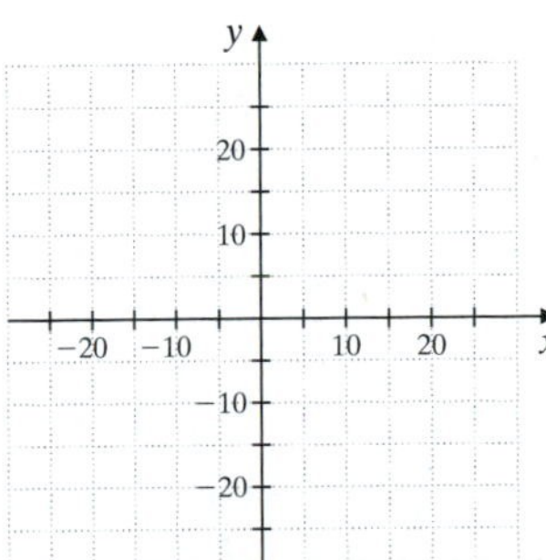

40. $y = x^2 - 36$

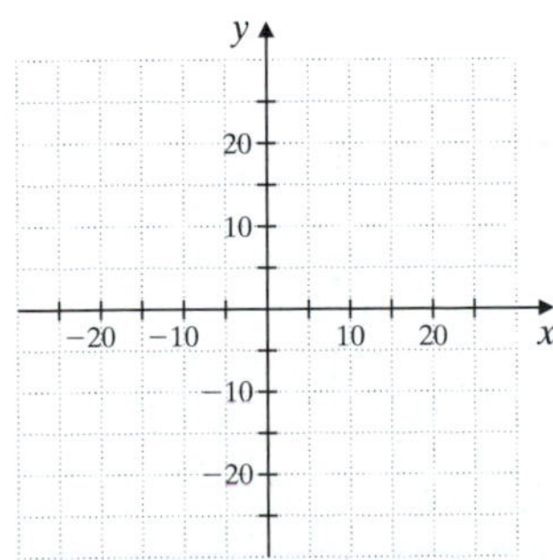

41. $y = x^2 + 3$

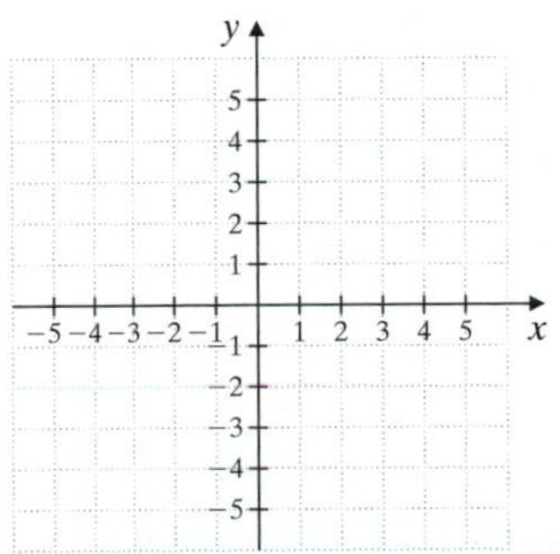

42. $y = x^2 + 8$

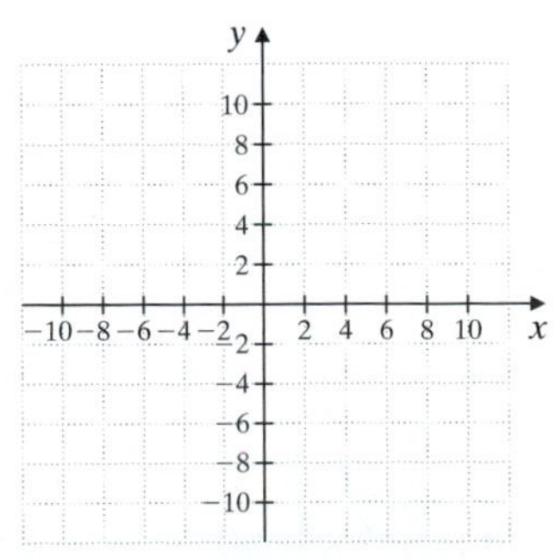

43. $y = -4x^2 + 8$

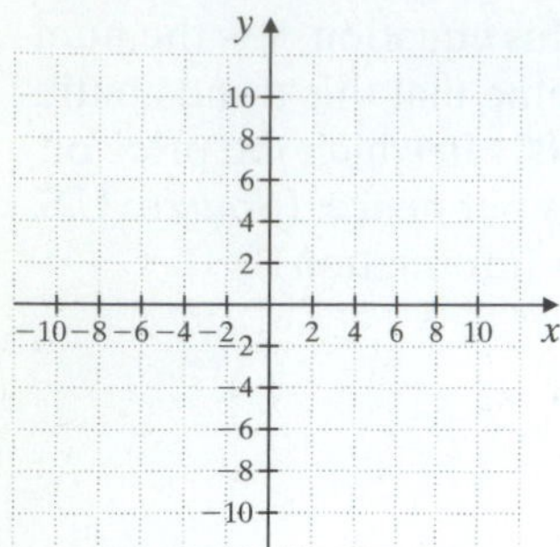

44. $y = -3x^2 + 9$

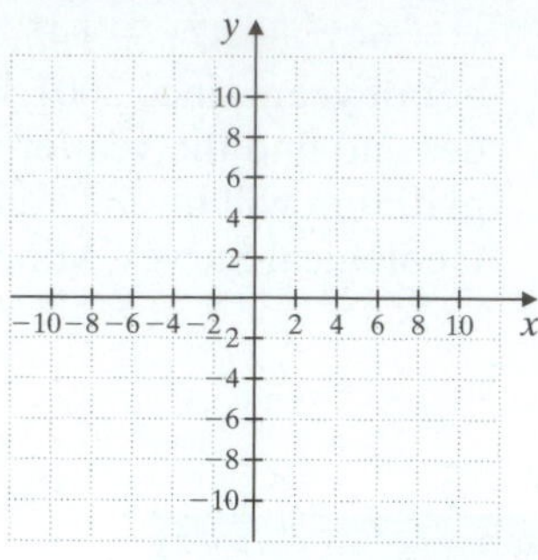

45. $y = x^2 + 3x - 10$

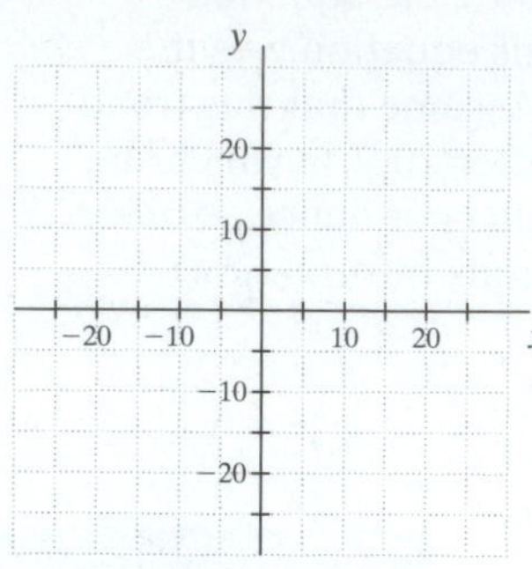

46. $y = x^2 + 3x - 4$

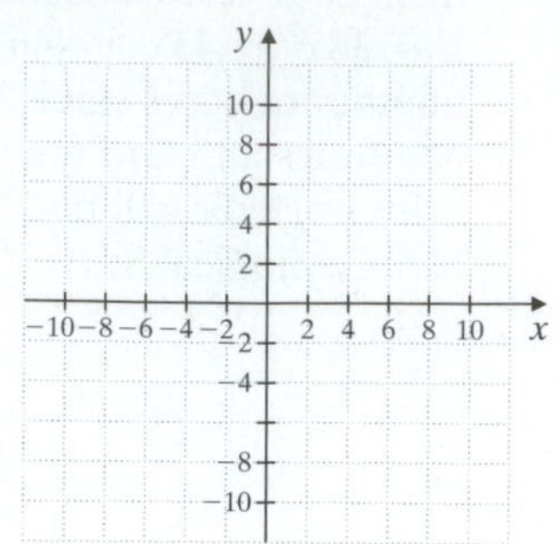

47. $y = -x^2 - 5x - 6$

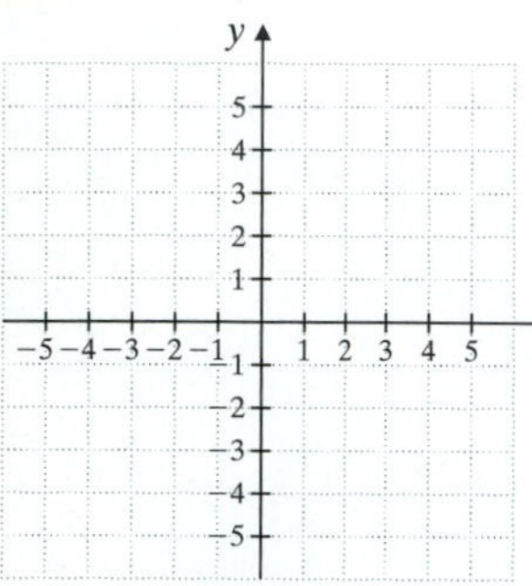

48. $y = 3x^2 - x - 2$

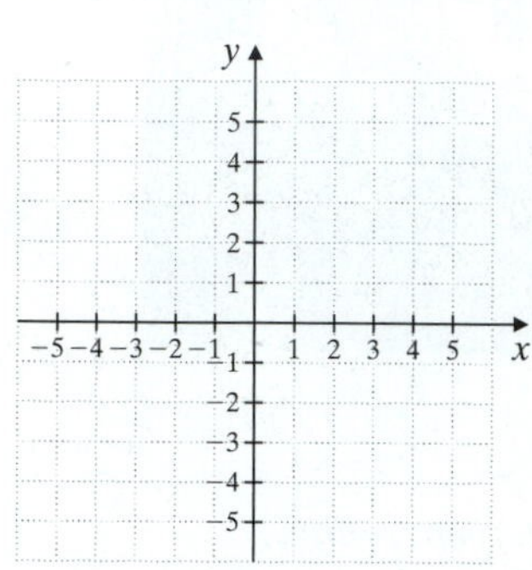

49. $y = 2x^2 - 11x - 6$

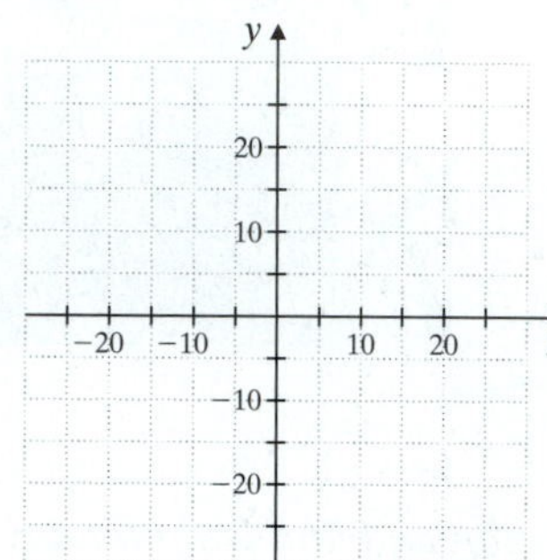

50. $y = -x^2 + 4x + 8$

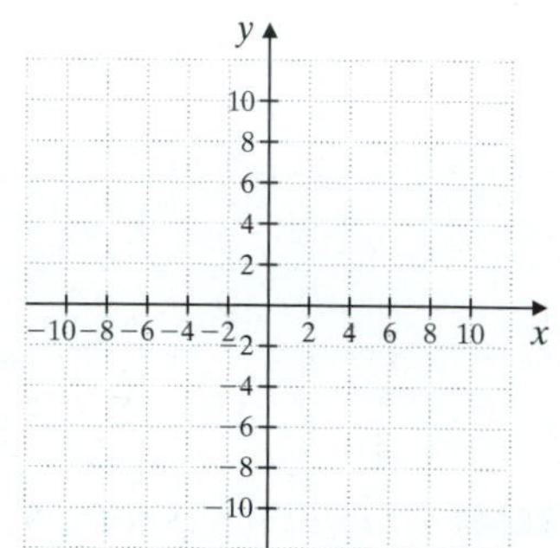

Match each quadratic equation with its graph.

51. $y = 2x^2$

52. $y = -x^2$

53. $y = x^2 + 4x + 4$

54. $y = x^2 + 5x + 4$

A

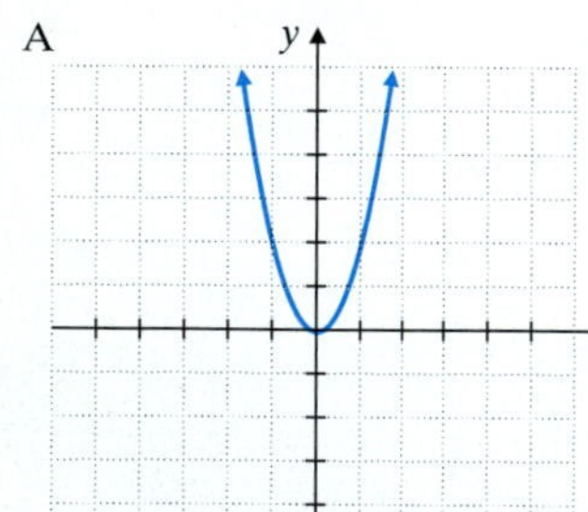

B

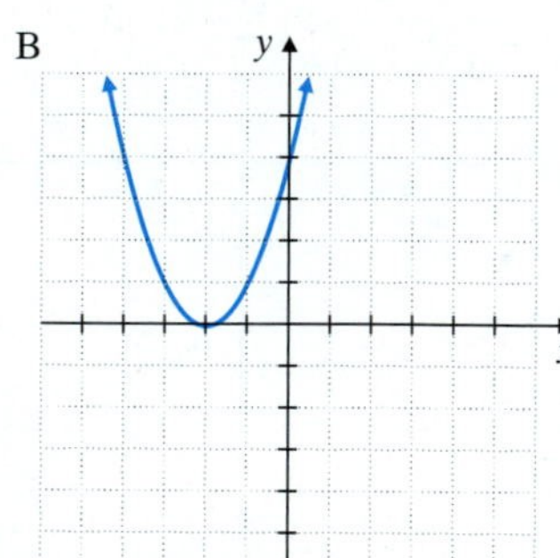

C

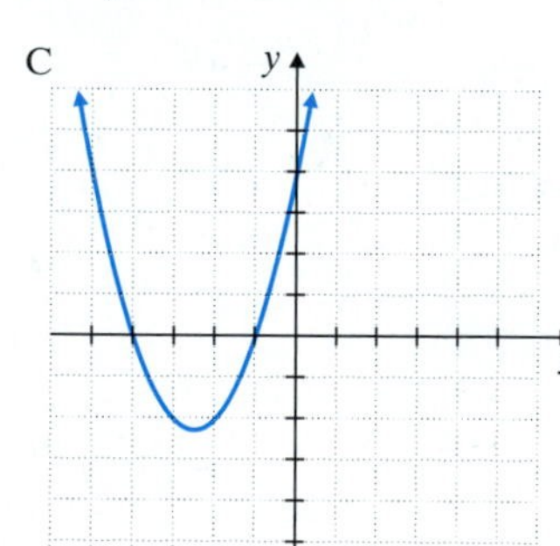

D

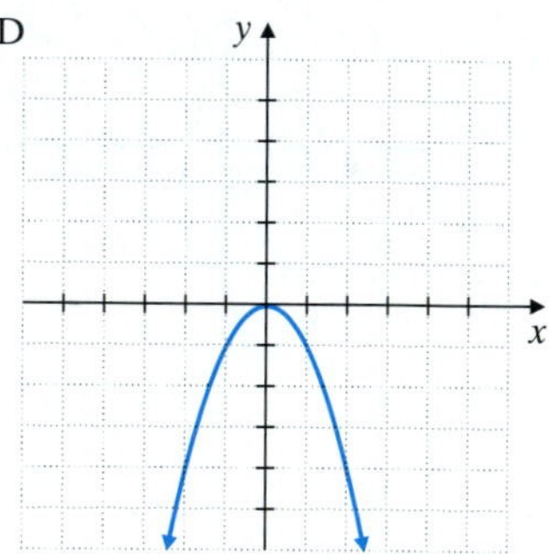

Quadratic equations in the form $y = ax^2 + bx + c$ are graphed below. Determine the number of real solutions for the related equation $0 = ax^2 + bx + c$ from each graph.

55.

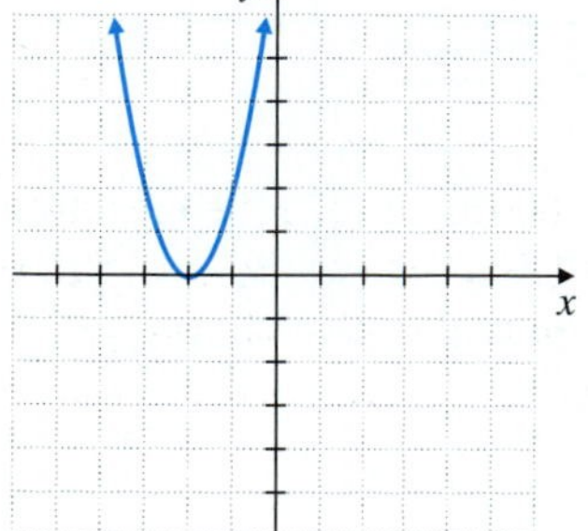

56.

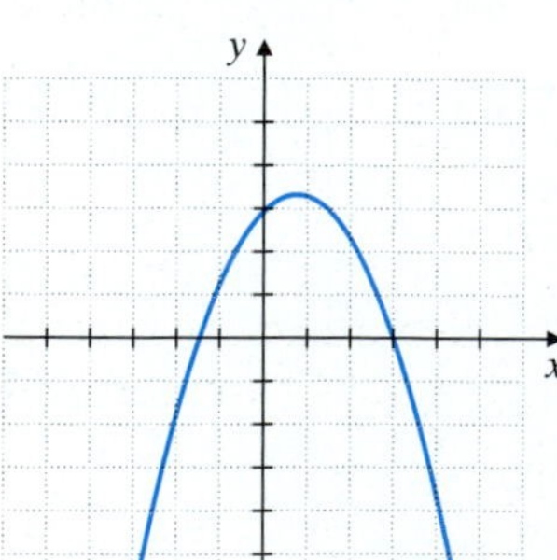

57.

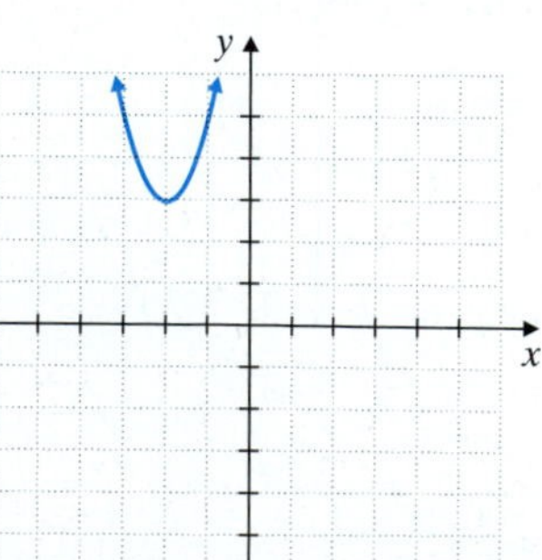

58.

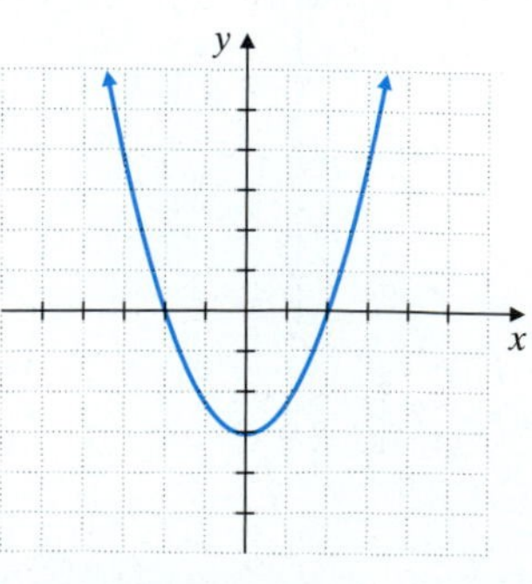

Mixed Review

Use the square root property to solve each quadratic equation.

59. $x^2 = 49$

60. $y^2 = 75$

61. $(x - 7)^2 = 64$

Solve each quadratic equation by completing the square.

62. $x^2 + 4x = 6$

63. $3x^2 + x = 2$

64. $4x^2 - x - 2 = 0$

Use the quadratic formula to solve each quadratic equation.

65. $4x^2 - 3x - 2 = 0$

66. $5x^2 + x - 2 = 0$

67. $4x^2 + 12x + 9 = 0$

68. $2x^2 + x + 4 = 0$

Graph each quadratic equation. Label the vertex and the intercept points with their coordinates.

69. $y = 4 - x^2$

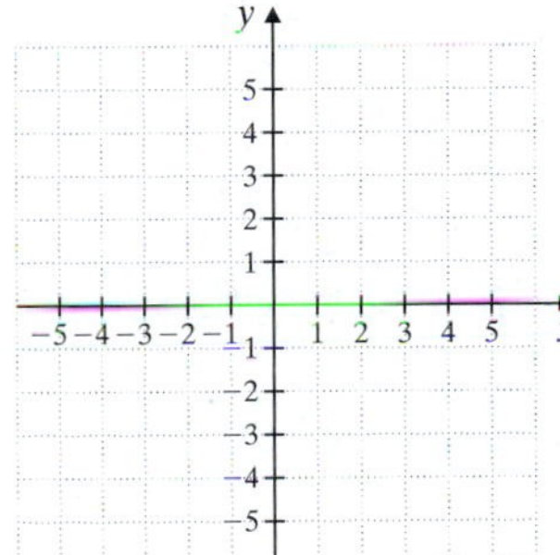

70. $y = x^2 + 4$

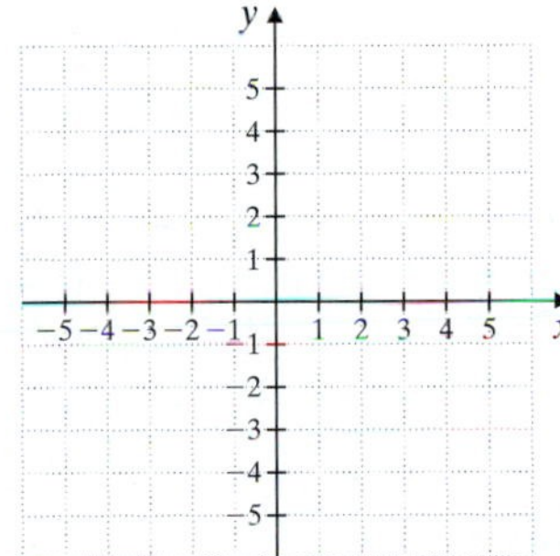

71. $y = x^2 + 6x + 8$

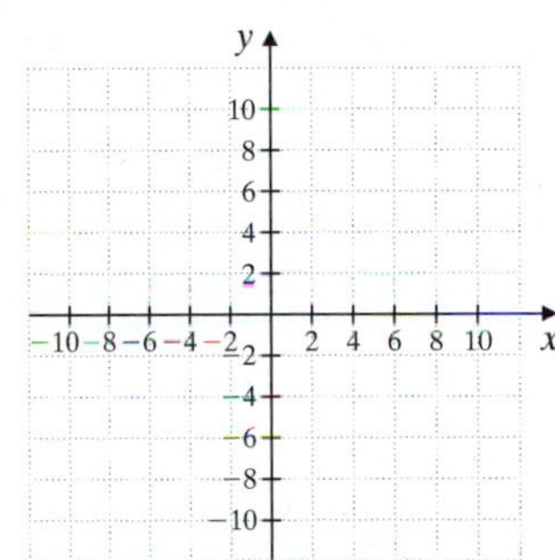

72. $y = x^2 - 2x - 4$

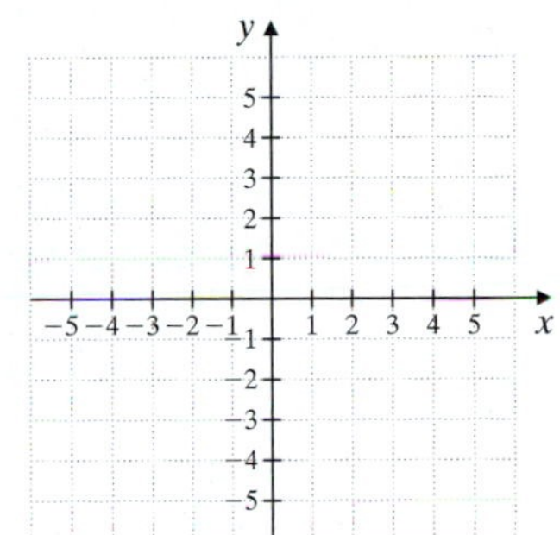

STUDY SKILLS BUILDER

Are You Prepared for a Test on Chapter 16?

Below I have listed some common trouble areas for students in Chapter 16. After studying for your test—but before taking your test—read these.

- Don't forget that to use the square root property, one side of your equation should be a squared variable or variable expression.

Solve: $3x^2 = 15$

$x^2 = 5$ — Divide both sides by 3 to isolate x^2.

$x = \sqrt{5}$ or $x = -\sqrt{5}$ — Use the square root property.

- Remember that to identify a, b, and c for the quadratic formula, write the quadratic equation in standard form: $ax^2 + bx + c = 0$

Solve: $x^2 = -x + 1$

$x^2 + x - 1 = 0$ — Write in standard form.

Here, $a = 1, b = 1$, and $c = -1$.

$$x = \frac{-1 \pm \sqrt{1^2 - 4(1)(-1)}}{2(1)} = \frac{-1 \pm \sqrt{5}}{2}$$

Remember: This is simply a listing of a few common trouble areas. For a review of Chapter 16, see the Highlights and Chapter Review at the end of this chapter.

16 CHAPTER TEST

Answers

Remember to use the Chapter Test Prep Video CD to see the fully worked-out solutions to any of the exercises you want to review.

Solve by factoring.

1. $x^2 - 400 = 0$

2. $2x^2 - 11x = 21$

Solve using the square root property.

3. $5k^2 = 80$

4. $(3m - 5)^2 = 8$

Solve by completing the square.

5. $x^2 - 26x + 160 = 0$

6. $3x^2 + 12x - 4 = 0$

Solve using the quadratic formula.

7. $x^2 - 3x - 10 = 0$

8. $p^2 - \frac{5}{3}p - \frac{1}{3} = 0$

Solve by the most appropriate method.

9. $(3x - 5)(x + 2) = -6$

10. $(3x - 1)^2 = 16$

11. $3x^2 - 7x - 2 = 0$

12. $x^2 - 4x - 5 = 0$

13. $3x^2 - 7x + 2 = 0$

14. $2x^2 - 6x + 1 = 0$

△ **15.** The height of a triangle is 4 times the length of the base. The area of the triangle is 18 square feet. Find the height and base of the triangle.

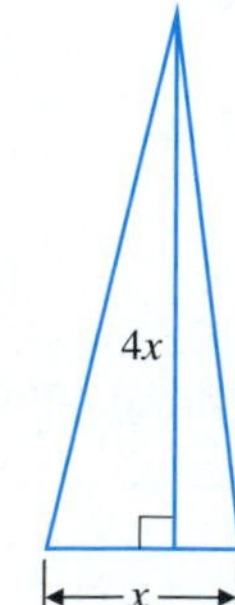

1. ______

2. ______

3. ______

4. ______

5. ______

6. ______

7. ______

8. ______

9. ______

10. ______

11. ______

12. ______

13. ______

14. ______

15. ______

Graph each quadratic equation. Label the vertex and the intercept points with their coordinates.

16. $y = -5x^2$

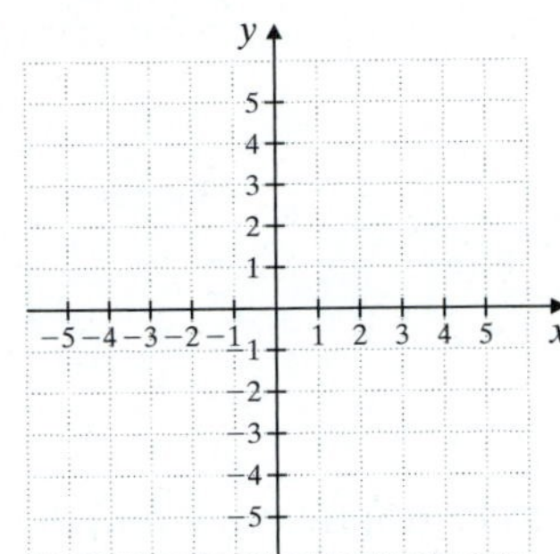

17. $y = x^2 - 4$

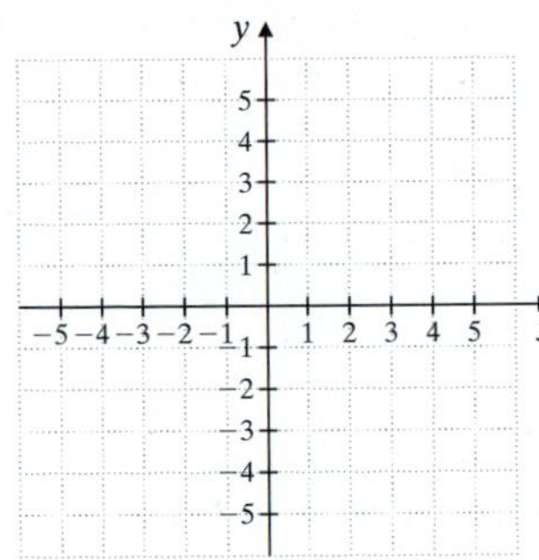

18. $y = x^2 - 7x + 10$

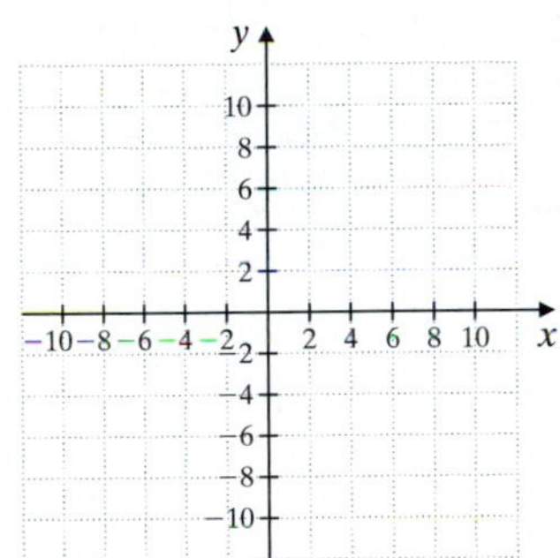

19. $y = 2x^2 + 4x - 1$

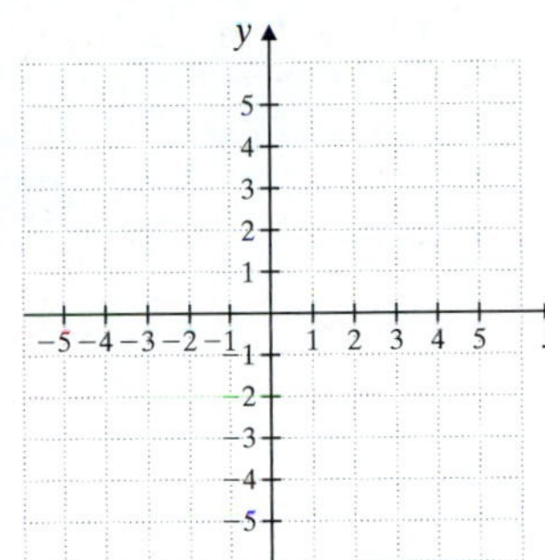

△ **20.** The number of diagonals d that a polygon with n sides has is given by the formula

$$d = \frac{n^2 - 3n}{2}$$

Find the number of sides of a polygon if it has 9 diagonals.

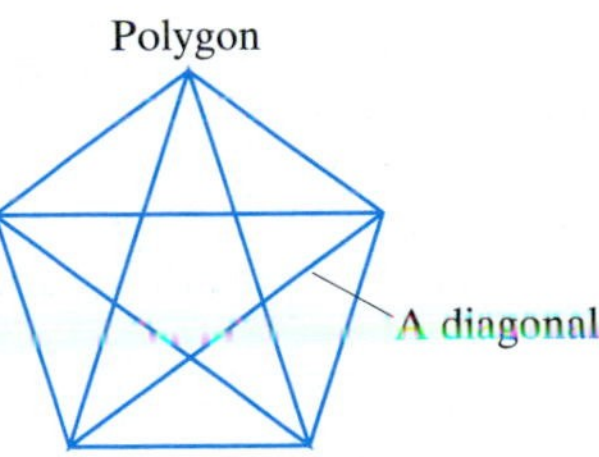

Solve.

 **21.** The highest dive from a diving board by a woman was made by Lucy Wardle of the United States. She dove from a height of 120.75 feet at Ocean Park, Hong Kong, in 1985. To the nearest tenth of a second, how long did the dive take? Use the formula $h = 16t^2$.

22. The value of Washington State's mineral production y (in millions of dollars) from 2000 through 2003 is modeled by the equation $y = 26x^2 - 136x + 607$. In this equation, $x = 0$ represents the year 2000. Assume that this trend continues and find the year after 2000 in which the value of mineral production is $727 million. (*Source:* U.S. Geological Survey, Minerals Information)

16. see graph

17. see graph

18. see graph

19. see graph

20. ______

21. ______

22. ______

CUMULATIVE REVIEW

Chapters 1–16

Divide.

Answers

1. ______
2. ______
3. ______
4. ______
5. ______
6. ______
7. ______
8. ______
9. ______
10. ______
11. ______
12. ______
13. ______
14. ______
15. ______
16. ______
17. ______
18. ______
19. ______
20. ______

1. $\frac{786.1}{1000}$

2. $\frac{818}{1000}$

3. $-\frac{0.12}{10}$

4. $-\frac{5.03}{100}$

5. Using the circle graph below, determine the percent of Americans that have one or more working computers at home.

6. Using the circle graph below, find the percent of Americans that have fewer than 3 working computers at home.

Number of Working Computers at Home

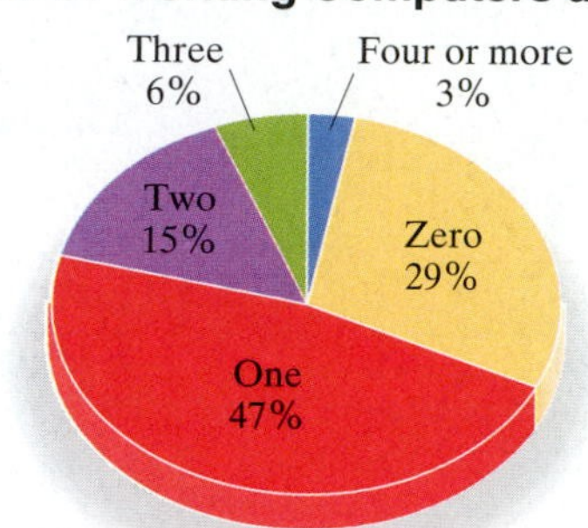

Source: UCLA Center for Communication Policy, 2003

△ **7.** Find the perimeter of a rectangle with a length of 11 inches and a width of 3 inches.

△ **8.** Find the perimeter of a triangular yard whose sides are 6 feet, 8 feet, and 11 feet.

△ **9.** Find the area of the parallelogram.

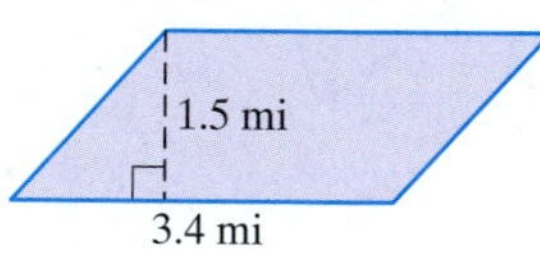

△ **10.** Find the area of the triangle.

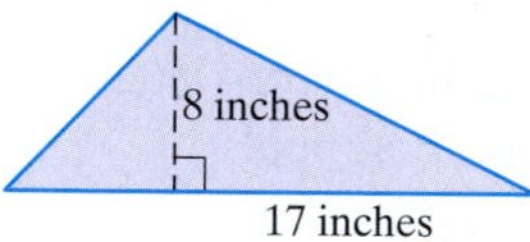

11. Convert 3210 ml to liters.

12. Convert 4321 cl to liters.

13. Solve: $8(2 - t) = -5t$

14. Solve: $\frac{5}{2}x - 1 = x + \frac{1}{4}$

Simplify the following expressions.

15. 3^0

16. -2^0

17. Factor: $r^2 - r - 42$

18. Factor: $y^2 + 3y - 70$

19. Factor: $10x^2 - 13xy - 3y^2$

20. Factor: $72x^2 - 35xy + 3y^2$

21. Factor: $8x^2 - 14x + 5$ by grouping.

22. Factor: $15x^2 - 4x - 4$ by grouping.

23. Factor completely.

a. $4x^3 - 49x$

b. $162x^4 - 2$

24. Factor completely.

a. $9x^3 - x$

b. $5x^4 - 5$

25. Solve: $(5x - 1)(2x^2 + 15x + 18) = 0$

26. Solve: $(x + 4)(40x^2 - 34x + 3) = 0$

27. Simplify: $\dfrac{x^2 + 8x + 7}{x^2 - 4x - 5}$

28. Simplify: $\dfrac{x^2 - 6x + 5}{x^2 + 6x - 7}$

29. The quotient of a number and 6, minus $\frac{5}{3}$ is the quotient of the number and 2. Find the number.

30. Four times a number, added to 5 is divided by 6. The result is $\frac{7}{2}$. Find the number.

31. Complete the table for the equation $y = 3x$.

	x	y
a.	−1	
b.		0
c.		−9

32. Complete the table for the equation $y = -2x + 7$.

	x	y
a.	0	
b.		0
c.	5	

Determine whether each pair of lines is parallel, perpendicular, or neither.

33. a. $y = -\frac{1}{5}x + 1$
$2x + 10y = 3$

b. $x + y = 3$
$-x + y = 4$

c. $3x + y = 5$
$2x + 3y = 6$

34. a. $y = -2x + 3$
$y = -2x + 5$

b. $-2x + y = 3$
$x + 2y = 9$

c. $3x - 2y = -8$
$3x + 2y = 1$

35. Which of the following relations are also functions?

a. $\{(-1, 1), (2, 3), (7, 3), (8, 6)\}$

b. $\{(0, -2), (1, 5), (0, 3), (7, 7)\}$

36. Which of the following relations are also functions?

a. $\{(-2, 3), (-5, 7), (9, 3), (0, 0)\}$

b. $\left\{(9, -1), \left(0, \frac{1}{2}\right), (2, -1), (9, 0)\right\}$

37. Solve the system:

$$\begin{cases} 2x + y = 10 \\ x = y + 2 \end{cases}$$

38. Solve the system:

$$\begin{cases} 8x - 3y = -4 \\ y = 7x - 3 \end{cases}$$

21. ______
22. ______
23. a. ______
b. ______
24. a. ______
b. ______
25. ______
26. ______
27. ______
28. ______
29. ______
30. ______
31. see table
32. see table
33. a. ______
b. ______
c. ______
34. a. ______
b. ______
c. ______
35. a. ______
b. ______
36. a. ______
b. ______
37. ______
38. ______

39. ______

40. ______

41. ______

42. ______

43. ______

44. ______

45. ______

46. ______

47. ______

48. ______

Find each square root.

39. $\sqrt{36}$ **40.** $\sqrt{81}$ **41.** $\sqrt{\dfrac{9}{100}}$ **42.** $\sqrt{\dfrac{16}{25}}$

Rationalize the denominator.

43. $\dfrac{2}{1 + \sqrt{3}}$ **44.** $\dfrac{7}{\sqrt{5} - 2}$

Use the square root property to solve.

45. $(x - 3)^2 = 16$ **46.** $(x + 4)^2 = 9$

47. Solve $\frac{1}{2}x^2 - x = 2$ by using the quadratic formula.

48. Solve $2x^2 = \frac{5}{2}x + \frac{7}{2}$ by using the quadratic formula.

Appendix

A Tables

A.1 TABLE OF GEOMETRIC FIGURES

Plane Figures Have Length and Width but No Thickness or Depth		
Name	**Description**	**Figure**
Polygon	Union of three or more coplanar line segments that intersect with each other only at each endpoint, with each endpoint shared by two segments.	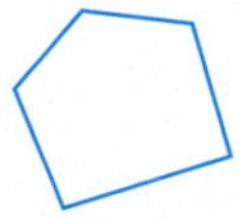
Triangle	Polygon with three sides (sum of measures of three angles is 180°).	
Scalene Triangle	Triangle with no sides of equal length.	
Isosceles Triangle	Triangle with two sides of equal length.	
Equilateral Triangle	Triangle with all sides of equal length.	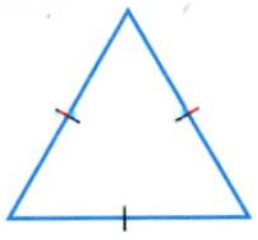
Right Triangle	Triangle that contains a right angle.	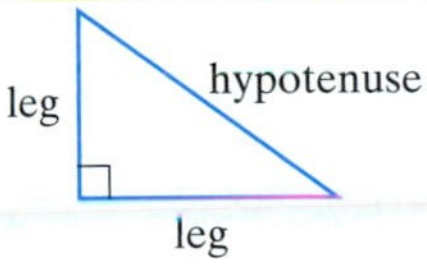 leg, hypotenuse, leg
Quadrilateral	Polygon with four sides (sum of measures of four angles is 360°).	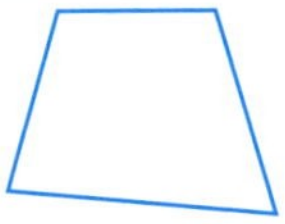
Trapezoid	Quadrilateral with exactly one pair of opposite sides parallel.	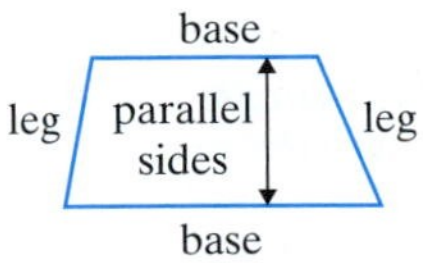 base, leg, parallel sides, leg, base
Isosceles Trapezoid	Trapezoid with legs of equal length.	
Parallelogram	Quadrilateral with both pairs of opposite sides parallel.	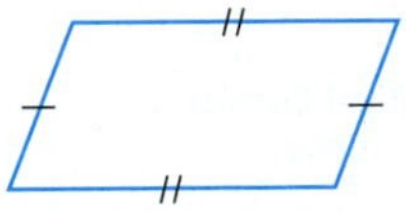
Rhombus	Parallelogram with all sides of equal length.	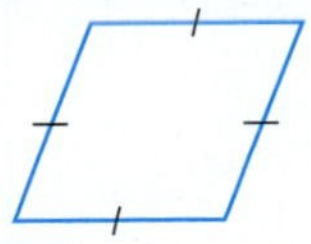

(*Continued*)

Plane Figures Have Length and Width but No Thickness or Depth (continued)		
Name	**Description**	**Figure**
Rectangle	Parallelogram with four right angles.	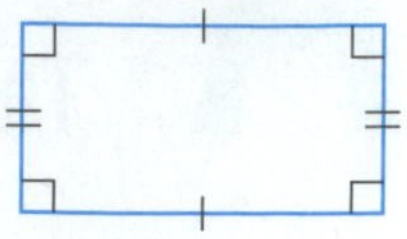
Square	Rectangle with all sides of equal length.	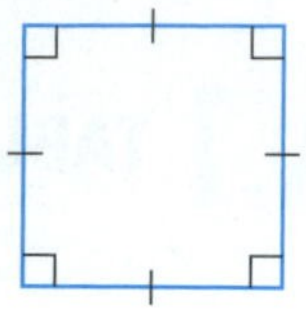
Circle	All points in a plane the same distance from a fixed point called the **center.**	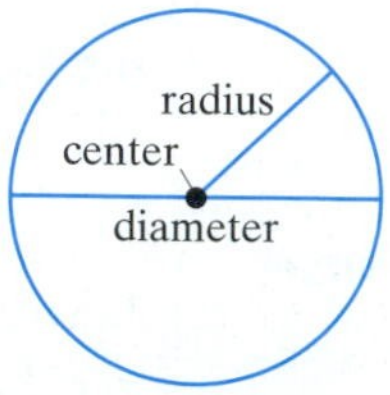 radius, center, diameter

Solid Figures Have Length, Width, and Height or Depth		
Name	**Description**	**Figure**
Rectangular Solid	A solid with six sides, all of which are rectangles.	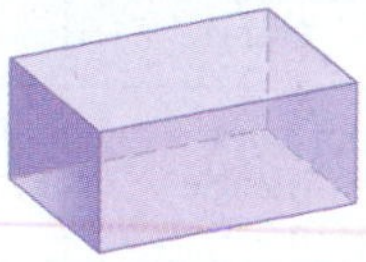
Cube	A rectangular solid whose six sides are squares.	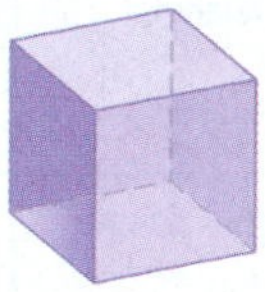
Sphere	All points the same distance from a fixed point, called the **center.**	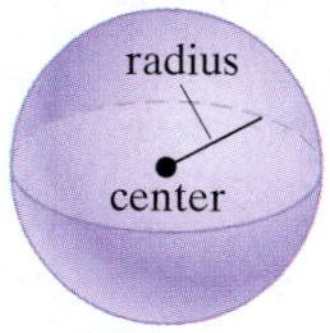 radius, center
Right Circular Cylinder	A cylinder having two circular bases that are perpendicular to its altitude.	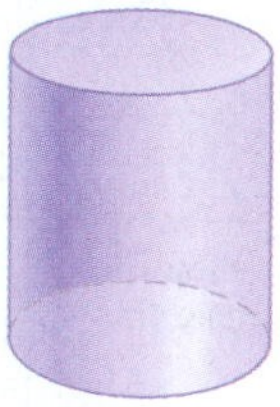
Right Circular Cone	A cone with a circular base that is perpendicular to its altitude.	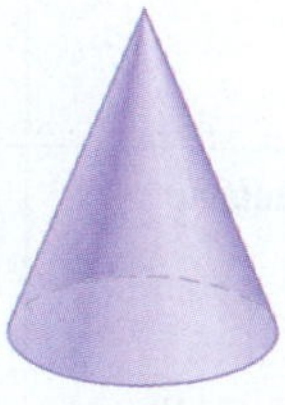

A.2 TABLE OF PERCENTS, DECIMALS, AND FRACTION EQUIVALENTS

Percent	Decimal	Fraction
1%	0.01	$\frac{1}{100}$
5%	0.05	$\frac{1}{20}$
10%	0.1	$\frac{1}{10}$
12.5% or $12\frac{1}{2}$%	0.125	$\frac{1}{8}$
$16.\overline{6}$% or $16\frac{2}{3}$%	$0.1\overline{6}$	$\frac{1}{6}$
20%	0.2	$\frac{1}{5}$
25%	0.25	$\frac{1}{4}$
30%	0.3	$\frac{3}{10}$
$33.\overline{3}$% or $33\frac{1}{3}$%	$0.\overline{3}$	$\frac{1}{3}$
37.5% or $37\frac{1}{2}$%	0.375	$\frac{3}{8}$
40%	0.4	$\frac{2}{5}$
50%	0.5	$\frac{1}{2}$
60%	0.6	$\frac{3}{5}$
62.5% or $62\frac{1}{2}$%	0.625	$\frac{5}{8}$
$66.\overline{6}$% or $66\frac{2}{3}$%	$0.\overline{6}$	$\frac{2}{3}$
70%	0.7	$\frac{7}{10}$
75%	0.75	$\frac{3}{4}$
80%	0.8	$\frac{4}{5}$
$83.\overline{3}$% or $83\frac{1}{3}$%	$0.8\overline{3}$	$\frac{5}{6}$
87.5% or $87\frac{1}{2}$%	0.875	$\frac{7}{8}$
90%	0.9	$\frac{9}{10}$
100%	1.0	1
110%	1.1	$1\frac{1}{10}$
125%	1.25	$1\frac{1}{4}$
$133.\overline{3}$% or $133\frac{1}{3}$%	$1.\overline{3}$	$1\frac{1}{3}$
150%	1.5	$1\frac{1}{2}$
$166.\overline{6}$% or $166\frac{2}{3}$%	$1.\overline{6}$	$1\frac{2}{3}$
175%	1.75	$1\frac{3}{4}$
200%	2.0	2

A.3 TABLE ON FINDING COMMON PERCENTS OF A NUMBER

Common Percent Equivalences*	Shortcut Method for Finding Percent	Example
$1\% = 0.01 \left(\text{or } \frac{1}{100}\right)$	To find 1% of a number, multiply by 0.01. To do so, move the decimal point 2 places to the left.	1% of 210 is 2.10 or 2.1. 1% of 1500 is 15. 1% of 8.6 is 0.086.
$10\% = 0.1 \left(\text{or } \frac{1}{10}\right)$	To find 10% of a number, multiply by 0.1, or move the decimal point of the number one place to the left.	10% of 140 is 14. 10% of 30 is 3. 10% of 17.6 is 1.76.
$25\% = \frac{1}{4}$	To find 25% of a number, find $\frac{1}{4}$ of the number, or divide the number by 4.	25% of 20 is $\frac{20}{4}$ or 5. 25% of 8 is 2. 25% of 10 is $\frac{10}{4}$ or $2\frac{1}{2}$.
$50\% = \frac{1}{2}$	To find 50% of a number, find $\frac{1}{2}$ of the number, or divide the number by 2.	50% of 64 is $\frac{64}{2}$ or 32. 50% of 1000 is 500. 50% of 9 is $\frac{9}{2}$ or $4\frac{1}{2}$.
$100\% = 1$	To find 100% of a number, multiply the number by 1. In other words, 100% of a number is the number.	100% of 98 is 98. 100% of 1407 is 1407. 100% of 18.4 is 18.4.
$200\% = 2$	To find 200% of a number, multiply the number by 2.	200% of 31 is $31 \cdot 2$ or 62. 200% of 750 is 1500. 200% of 6.5 is 13.

*See Appendix A.2.

A.4 TABLE OF SQUARES AND SQUARE ROOTS

n	n^2	$\sqrt{n}$	n	n^2	$\sqrt{n}$
1	1	1.000	51	2601	7.141
2	4	1.414	52	2704	7.211
3	9	1.732	53	2809	7.280
4	16	2.000	54	2916	7.348
5	25	2.236	55	3025	7.416
6	36	2.449	56	3136	7.483
7	49	2.646	57	3249	7.550
8	64	2.828	58	3364	7.616
9	81	3.000	59	3481	7.681
10	100	3.162	60	3600	7.746
11	121	3.317	61	3721	7.810
12	144	3.464	62	3844	7.874
13	169	3.606	63	3969	7.937
14	196	3.742	64	4096	8.000
15	225	3.873	65	4225	8.062
16	256	4.000	66	4356	8.124
17	289	4.123	67	4489	8.185
18	324	4.243	68	4624	8.246
19	361	4.359	69	4761	8.307
20	400	4.472	70	4900	8.367
21	441	4.583	71	5041	8.426
22	484	4.690	72	5184	8.485
23	529	4.796	73	5329	8.544
24	576	4.899	74	5476	8.602
25	625	5.000	75	5625	8.660
26	676	5.099	76	5776	8.718
27	729	5.196	77	5929	8.775
28	784	5.292	78	6084	8.832
29	841	5.385	79	6241	8.888
30	900	5.477	80	6400	8.944
31	961	5.568	81	6561	9.000
32	1024	5.657	82	6724	9.055
33	1089	5.745	83	6889	9.110
34	1156	5.831	84	7056	9.165
35	1225	5.916	85	7225	9.220
36	1296	6.000	86	7396	9.274
37	1369	6.083	87	7569	9.327
38	1444	6.164	88	7744	9.381
39	1521	6.245	89	7921	9.434
40	1600	6.325	90	8100	9.487
41	1681	6.403	91	8281	9.539
42	1764	6.481	92	8464	9.592
43	1849	6.557	93	8649	9.644
44	1936	6.633	94	8836	9.695
45	2025	6.708	95	9025	9.747
46	2116	6.782	96	9216	9.798
47	2209	6.856	97	9409	9.849
48	2304	6.928	98	9604	9.899
49	2401	7.000	99	9801	9.950
50	2500	7.071	100	10,000	10.000

Appendix

B The Bigger Picture

OUTLINE: PART 1

Operations on Sets of Numbers and Solving Equations

To see the sections that formed this outline, see Sections 1.6, 1.7, 2.4, 3.3, 4.3, 4.7, 5.4, 6.1, 6.3, 6.4, and 7.3.

I. Operations on Sets of Numbers

A. Whole Numbers

1. Add or Subtract: (Sec. 1.3)

$$\begin{array}{r} 14 \\ +\,39 \\ \hline 53 \end{array} \qquad \begin{array}{r} 300 \\ -\,27 \\ \hline 273 \end{array}$$

2. Multiply or Divide: (Sec. 1.5, 1.6)

$$\begin{array}{r} 238 \\ \times\,47 \\ \hline 1666 \\ 9520 \\ \hline 11{,}186 \end{array} \qquad \begin{array}{r} 127\text{ R2} \\ 7\overline{)891} \\ -7 \\ \hline 19 \\ -14 \\ \hline 51 \\ -49 \\ \hline 2 \end{array}$$

3. Exponent: (Sec. 1.7)

$$3^4 = \overbrace{3 \cdot 3 \cdot 3 \cdot 3}^{\text{4 factors of 3}} = 81$$

4. Order of Operations: (Sec. 1.7)

$24 \div 3 \cdot 2 - (2 + 8) = 24 \div 3 \cdot 2 - (10)$ Simplify within parentheses.

$= 8 \cdot 2 - 10$ Multiply or divide from left to right.

$= 16 - 10$ Multiply or divide from left to right.

$= 6$ Add or subtract from left to right.

5. Square Root: (Sec. 7.3)

$\sqrt{25} = 5$ *because* $5 \cdot 5 = 25$ and 5 is a positive number.

B. Integers

1. Add: (Sec. 2.2) $-5 + (-2) = -7$ Adding like signs.
Add absolute value. Attach the common sign.

$-5 + 2 = -3$ Adding unlike signs.
Subtract absolute values. Attach the sign of the number with the larger absolute value.

2. Subtract: Add the first number to the opposite of the second number. (Sec. 2.3)

$7 - 10 = 7 + (-10) = -3$

3. **Multiply or Divide:** Multiply or divide as usual. If the signs of the two numbers are the same, the answer is positive. If the signs of the two numbers are different, the answer is negative. (Sec. 2.4)

$$-5 \cdot 5 = -25, \quad \frac{-32}{-8} = 4$$

C. **Fractions**

1. **Simplify:** Factor the numerator and denominator. Then divide out factors of 1 by dividing out common factors in the numerator and denominator. (Sec. 4.2)

Simplify: $\frac{20}{28} = \frac{4 \cdot 5}{4 \cdot 7} = \frac{5}{7}$

2. **Multiply:** Numerator times numerator over denominator times denominator. (Sec. 4.3)

$$\frac{5}{9} \cdot \frac{2}{7} = \frac{10}{63}$$

3. **Divide:** First fraction times the reciprocal of the second fraction. (Sec. 4.3)

$$\frac{2}{11} \div \frac{3}{4} = \frac{2}{11} \cdot \frac{4}{3} = \frac{8}{33}$$

4. **Add or Subtract:** Must have same denominators. If not, find the LCD, and write each fraction as an equivalent fraction with the LCD as denominator. (Sec. 4.4, 4.5)

$$\frac{2}{5} + \frac{1}{15} = \frac{2}{5} \cdot \frac{3}{3} + \frac{1}{15} = \frac{6}{15} + \frac{1}{15} = \frac{7}{15}$$

D. **Decimals**

1. **Add or Subtract:** Line up decimal points. (Sec. 5.2)

$$\begin{array}{r} 1.27 \\ +\ 0.6 \\ \hline 1.87 \end{array}$$

2. **Multiply:** (Sec. 5.3)

$$\begin{array}{rl} 2.56 & \text{2 decimal places} \\ \times\ 3.2 & \text{1 decimal place} \\ \hline 512 & 2 + 1 = 3 \\ 7680 & \\ \hline 8.192 & \text{3 decimal places} \end{array}$$

3. **Divide:** (Sec. 5.4)

$$8\overline{)5.6} = 0.7 \qquad 0.6\overline{)0.786} = 1.31$$

II. **Solving Equations**

A. **Equations in General:** Simplify both sides of the equation by removing parentheses and adding any like terms. Then use the Addition Property to write variable terms on one side, constants (or numbers) on the other side. Then use the Multiplication Property to solve for the variable by dividing both sides of the equation by the coefficient of the variable. (Sec. 3.3)

Solve: $2(x - 5) = 80$

$$\begin{array}{rll} 2x - 10 &= 80 & \text{Use the distributive property.} \\ 2x - 10 + 10 &= 80 + 10 & \text{Add 10 to both sides.} \\ 2x &= 90 & \text{Simplify.} \\ \frac{2x}{2} &= \frac{90}{2} & \text{Divide both sides by 2.} \\ x &= 45 & \text{Simplify.} \end{array}$$

B. **Proportions:** Set cross products equal to each other. Then solve. (Sec. 6.1)

$\frac{14}{3} = \frac{2}{n}$, *or* $14 \cdot n = 3 \cdot 2$, *or* $14 \cdot n = 6$, *or* $n = \frac{6}{14} = \frac{3}{7}$

C. Percent Problems

1. Solved by Equations: Remember that "of" means multiplication and "is" means equals. (Sec. 6.3)

"12% of some number is 6" translates to

$$12\% \cdot n = 6 \textit{ or } 0.12 \cdot n = 6 \textit{ or } n = \frac{6}{0.12} \textit{ or } n = 50$$

2. Solved by Proportions: Remember that percent, p, is identified by % or percent, base, b, usually appears after "of" and amount, a, is the part compared to the whole. (Sec. 6.4)

"12% of some number is 6" translates to

$$\frac{6}{b} = \frac{12}{100} \textit{ or } 6 \cdot 100 = b \cdot 12 \textit{ or } \frac{600}{12} = b \textit{ or } 50 = b$$

OUTLINE: PART 2

To see the sections that formed this outline, see Sections 9.6, 10.7, 11.6, 12.4, 12.5, 15.5, and 16.3.

Simplifying Expressions and Solving Equations

I. Simplifying Expressions

A. Exponents (Sec. 10.1, 10.2)

$$x^7 \cdot x^5 = x^{12};\ (x^7)^5 = x^{35};\ \frac{x^7}{x^5} = x^2;\ x^0 = 1;\ 8^{-2} = \frac{1}{8^2} = \frac{1}{64}$$

B. Polynomials

1. Add: Combine like terms. (Sec. 10.4)

$$\begin{aligned}(3y^2 + 6y + 7) + (9y^2 - 11y - 15) &= 3y^2 + 6y + 7 + 9y^2 - 11y - 15\\ &= 12y^2 - 5y - 8\end{aligned}$$

2. Subtract: Change the sign of the terms of the polynomial being subtracted, then add. (Sec. 10.4)

$$\begin{aligned}(3y^2 + 6y + 7) - (9y^2 - 11y - 15) &= 3y^2 + 6y + 7 - 9y^2 + 11y + 15\\ &= -6y^2 + 17y + 22\end{aligned}$$

3. Multiply: Multiply each term of one polynomial by each term of the other polynomial. (Sec. 10.5)

$$\begin{aligned}(x + 5)(2x^2 - 3x + 4) &= x(2x^2 - 3x + 4) + 5(2x^2 - 3x + 4)\\ &= 2x^3 - 3x^2 + 4x + 10x^2 - 15x + 20\\ &= 2x^3 + 7x^2 - 11x + 20\end{aligned}$$

4. Divide: (Sec. 10.7)

a. To divide by a monomial, divide each term of the polynomial by the monomial.

$$\frac{8x^2 + 2x - 6}{2x} = \frac{8x^2}{2x} + \frac{2x}{2x} - \frac{6}{2x} = 4x + 1 - \frac{3}{x}$$

b. To divide by a polynomial other than a monomial, use long division.

$$
\begin{array}{r}
x - 6 + \dfrac{40}{2x+5} \\
2x + 5\overline{)2x^2 - 7x + 10} \\
\underline{2x^2 + 5x} \\
-12x + 10 \\
\underline{-12x - 30} \\
40
\end{array}
$$

C. Factoring Polynomials

See the Chapter 11 Integrated Review for steps. (Sec. 11.5)

$$
\begin{aligned}
3x^4 - 78x^2 + 75 &= 3(x^4 - 26x^2 + 25) && \text{Factor out GCF—always first step.} \\
&= 3(x^2 - 25)(x^2 - 1) && \text{Factor trinomial.} \\
&= 3(x + 5)(x - 5)(x + 1)(x - 1) && \text{Factor further—each difference of squares.}
\end{aligned}
$$

D. Rational Expressions

1. **Simplify:** Factor the numerator and denominator. Then divide out factors of 1 by dividing out common factors in the numerator and denominator. (Sec. 12.1)

$$\frac{x^2 - 9}{7x^2 - 21x} = \frac{(x + 3)(x - 3)}{7x(x - 3)} = \frac{x + 3}{7x}$$

2. **Multiply:** Multiply numerators, then multiply denominators. (Sec. 12.2)

$$\frac{5z}{2z^2 - 9z - 18} \cdot \frac{22z + 33}{10z} = \frac{5 \cdot z}{(2z + 3)(z - 6)} \cdot \frac{11(2z + 3)}{2 \cdot 5 \cdot z} = \frac{11}{2(z - 6)}$$

3. **Divide:** First fraction times the reciprocal of the second fraction. (Sec. 12.2)

$$\frac{14}{x + 5} \div \frac{x + 1}{2} = \frac{14}{x + 5} \cdot \frac{2}{x + 1} = \frac{28}{(x + 5)(x + 1)}$$

4. **Add or subtract:** Must have same denominator. If not, find the LCD and write each fraction as an equivalent fraction with the LCD as denominator. (Sec. 12.4)

$$
\begin{aligned}
\frac{9}{10} - \frac{x + 1}{x + 5} &= \frac{9(x + 5)}{10(x + 5)} - \frac{10(x + 1)}{10(x + 5)} \\
&= \frac{9x + 45 - 10x - 10}{10(x + 5)} = \frac{-x + 35}{10(x + 5)}
\end{aligned}
$$

E. Radicals

1. **Simplify square roots:** If possible, factor the radicand so that one factor is a perfect square. Then use the product rule and simplify. (Sec. 15.2)

$$\sqrt{75} = \sqrt{25 \cdot 3} = \sqrt{25} \cdot \sqrt{3} = 5\sqrt{3}$$

2. **Add or subtract:** Only like radicals (same index and radicand) can be added or subtracted. (Sec. 15.3)

$$8\sqrt{10} - \sqrt{40} + \sqrt{5} = 8\sqrt{10} - 2\sqrt{10} + \sqrt{5} = 6\sqrt{10} + \sqrt{5}$$

3. **Multiply or divide:** $\sqrt{a} \cdot \sqrt{b} = \sqrt{ab}$; $\dfrac{\sqrt{a}}{\sqrt{b}} = \sqrt{\dfrac{a}{b}}$. (Sec. 15.4)

$$\sqrt{11} \cdot \sqrt{3} = \sqrt{33}; \frac{\sqrt{140}}{\sqrt{7}} = \sqrt{\frac{140}{7}} = \sqrt{20} = \sqrt{4 \cdot 5} = 2\sqrt{5}$$

4. Rationalizing the denominator: (Sec. 15.4)

a. If denominator is one term,

$$\frac{5}{\sqrt{11}} = \frac{5 \cdot \sqrt{11}}{\sqrt{11} \cdot \sqrt{11}} = \frac{5\sqrt{11}}{11}$$

b. If denominator is two terms, multiply by 1 in the form of $\frac{\text{conjugate of denominator}}{\text{conjugate of denominator}}$.

$$\frac{13}{3+\sqrt{2}} = \frac{13}{3+\sqrt{2}} \cdot \frac{3-\sqrt{2}}{3-\sqrt{2}} = \frac{13(3-\sqrt{2})}{9-2} = \frac{13(3-\sqrt{2})}{7}$$

II. Solving Equations and Inequalities

A. Linear Equations: Power on variable is 1 and there are no variables in denominator. (Sec. 9.3)

$7(x-3) = 4x+6$ Linear equation. (If fractions, multiply by LCD.)
$7x - 21 = 4x + 6$ Use the distributive property.
$7x = 4x + 27$ Add 21 to both sides.
$3x = 27$ Subtract $4x$ from both sides.
$x = 9$ Divide both sides by 3.

B. Linear Inequalities: Same as linear equation except if you multiply or divide by a negative number, then reverse direction of inequality. (Sec. 9.6)

$-4x + 11 \le -1$ Linear inequality.
$-4x \le -12$ Subtract 11 from both sides.
$\frac{-4x}{-4} \ge \frac{-12}{-4}$ Divide both sides by -4 and reverse the direction of the inequality symbol.
$x \ge 3$ Simplify.

C. Quadratic and Higher Degree Equations: Solve: first write the equation in standard form (one side is 0.)

1. If the polynomial on one side factors, solve by factoring. (Sec. 11.6)
2. If the polynomial does not factor, solve by the quadratic formula. (Sec. 16.3)

By factoring:

$$x^2 + x = 6$$
$$x^2 + x - 6 = 0$$
$$(x-2)(x+3) = 0$$
$$x - 2 = 0 \text{ or } x + 3 = 0$$
$$x = 2 \text{ or } x = -3$$

By quadratic formula:

$$x^2 + x = 5$$
$$x^2 + x - 5 = 0$$
$$a = 1, b = 1, c = -5$$
$$x = \frac{-1 \pm \sqrt{1^2 - 4(1)(-5)}}{2 \cdot 1}$$
$$x = \frac{-1 \pm \sqrt{21}}{2}$$

D. Equations with Rational Expressions: Make sure the proposed solution does not make the denominator 0. (Sec. 12.5)

$\frac{3}{x} - \frac{1}{x-1} = \frac{4}{x-1}$ Equation with rational expressions.

$x(x-1) \cdot \frac{3}{x} - x(x-1) \cdot \frac{1}{x-1} = x(x-1) \cdot \frac{4}{x-1}$ Multiply through by $x(x-1)$.

$3(x - 1) - x \cdot 1 = x \cdot 4$	Simplify.
$3x - 3 - x = 4x$	Use the distributive property.
$-3 = 2x$	Simplify and move variable terms to right side.
$-\dfrac{3}{2} = x$	Divide both sides by 2.

E. Equations with Radicals: To solve, isolate a radical, then square both sides. You may have to repeat this. Check possible solution(s) in the original equation. (Sec. 15.5)

$\sqrt{x + 49} + 7 = x$	
$\sqrt{x + 49} = x - 7$	Subtract 7 from both sides.
$x + 49 = x^2 - 14x + 49$	Square both sides.
$0 = x^2 - 15x$	Set terms equal to 0.
$0 = x(x - 15)$	Factor.
$x = 0$ or $x = 15$	Set each factor equal to 0 and solve.

Appendix

C Factoring Sums and Differences of Cubes

Although the sum of two squares usually does not factor, the sum or difference of two cubes can be factored and reveal factoring patterns. The pattern for the sum of cubes can be checked by multiplying the binomial $x + y$ and the trinomial $x^2 - xy + y^2$. The pattern for the difference of two cubes can be checked by multiplying the binomial $x - y$ by the trinomial $x^2 + xy + y^2$.

Sum or Difference of Two Cubes

$$a^3 + b^3 = (a + b)(a^2 - ab + b^2)$$

$$a^3 - b^3 = (a - b)(a^2 + ab + b^2)$$

EXAMPLE 1 Factor $x^3 + 8$.

Solution:

First, write the binomial in the form $a^3 + b^3$.

$$x^3 + 8 = x^3 + 2^3 \quad \text{Write in the form } a^3 + b^3.$$

If we replace a with x and b with 2 in the formula above, we have

$$x^3 + 2^3 = (x + 2)[x^2 - (x)(2) + 2^2]$$

$$= (x + 2)(x^2 - 2x + 4)$$

Helpful Hint

When factoring sums or differences of cubes, notice the sign patterns.

same sign

$$x^3 + y^3 = (x + y)(x^2 - xy + y^2)$$

opposite sign always positive

same sign

$$x^3 - y^3 = (x - y)(x^2 + xy + y^2)$$

opposite sign always positive

EXAMPLE 2 Factor $y^3 - 27$.

Solution:

$$\begin{aligned} y^3 - 27 &= y^3 - 3^3 && \text{Write in the form } a^3 - b^3. \\ &= (y - 3)[y^2 + (y)(3) + 3^2] \\ &= (y - 3)(y^2 + 3y + 9) \end{aligned}$$

EXAMPLE 3 Factor $64x^3 + 1$.

Solution:

$$\begin{aligned} 64x^3 + 1 &= (4x)^3 + 1^3 \\ &= (4x + 1)[(4x)^2 - (4x)(1) + 1^2] \\ &= (4x + 1)(16x^2 - 4x + 1) \end{aligned}$$

EXAMPLE 4 Factor $54a^3 - 16b^3$.

Solution: Remember to factor out common factors first before using other factoring methods.

$$\begin{aligned} 54a^3 - 16b^3 &= 2(27a^3 - 8b^3) && \text{Factor out the GCF 2.} \\ &= 2[(3a)^3 - (2b)^3] && \text{Difference of two cubes.} \\ &= 2(3a - 2b)[(3a)^2 + (3a)(2b) + (2b)^2] \\ &= 2(3a - 2b)(9a^2 + 6ab + 4b^2) \end{aligned}$$

C EXERCISE SET

FOR EXTRA HELP

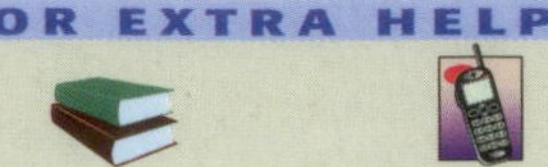

MyMathLab

Factor the sum or difference of two cubes. See Examples 1 through 4.

1. $a^3 + 27$

2. $b^3 - 8$

3. $8a^3 + 1$

4. $64x^3 - 1$

5. $5k^3 + 40$

6. $6r^3 - 162$

7. $x^3y^3 - 64$

8. $8x^3 - y^3$

9. $x^3 + 125$

10. $a^3 - 216$

11. $24x^4 - 81xy^3$

12. $375y^6 - 24y^3$

Factor the binomials completely.

13. $27 - t^3$

14. $125 + r^3$

15. $8r^3 - 64$

16. $54r^3 + 2$

17. $t^3 - 343$

18. $s^3 + 216$

19. $s^3 - 64t^3$

20. $8t^3 + s^3$

Appendix D Systems of Linear Inequalities

A **solution of a system of linear inequalities** is an ordered pair that satisfies each inequality in the system. The set of all such ordered pairs is the solution set of the system. Graphing this set gives us a picture of the solution set. We can graph a system of inequalities by graphing each inequality in the system and identifying the region of overlap.

EXAMPLE 1 Graph the solutions of the system: $\begin{cases} 3x \geq y \\ x + 2y \leq 8 \end{cases}$

Solution: We begin by graphing each inequality on the *same* set of axes. The graph of the solutions of the system is the region contained in the graphs of both inequalities. In other words, it is their intersection.

First let's graph $3x \geq y$. The boundary line is the graph of $3x = y$. We sketch a solid boundary line since the inequality $3x \geq y$ means $3x > y$ or $3x = y$. The test point $(1, 0)$ satisfies the inequality, so we shade the half-plane that includes $(1, 0)$.

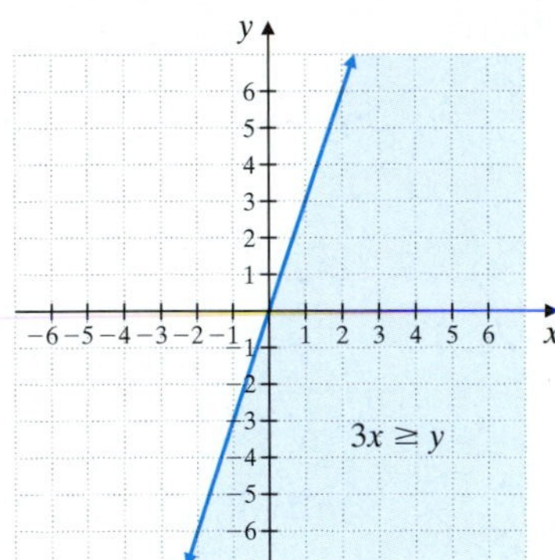

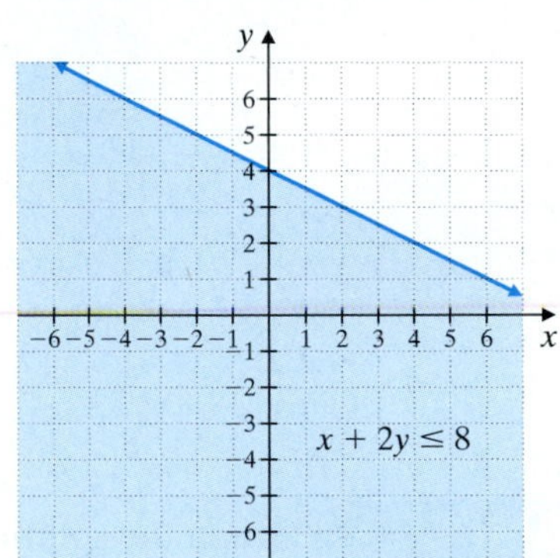

Next we sketch a solid boundary line $x + 2y = 8$ on the same set of axes. The test point $(0, 0)$ satisfies the inequality $x + 2y \leq 8$, so we shade the half-plane that includes $(0, 0)$. (For clarity, the graph of $x + 2y \leq 8$ is shown here on a separate set of axes.) An ordered pair solution of the system must satisfy both inequalities. These solutions are points that lie in both shaded regions. The solution of the system is the darkest shaded region. This solution includes parts of both boundary lines.

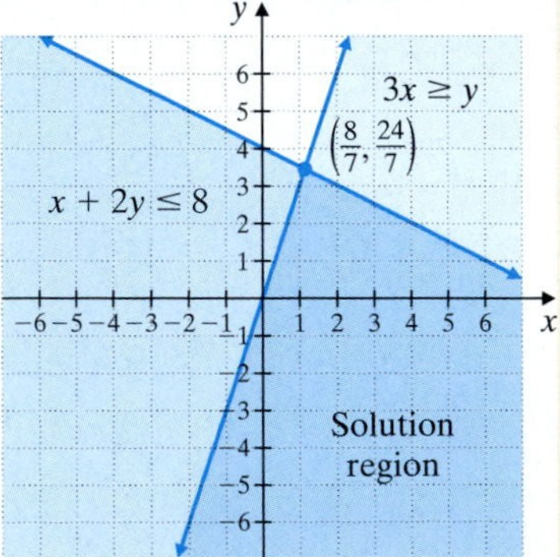

In linear programming, it is sometimes necessary to find the coordinates of the **corner point:** the point at which the two boundary lines intersect. To find the corner point for the system of Example 1, we solve the related linear system

$$\begin{cases} 3x = y \\ x + 2y = 8 \end{cases}$$

using either the substitution or the elimination method. The lines intersect at $\left(\frac{8}{7}, \frac{24}{7}\right)$, the corner point of the graph.

Graphing the Solutions of a System of Linear Inequalities

Step 1: Graph each inequality in the system on the same set of axes.

Step 2: The solutions of the system are the points common to the graphs of all the inequalities in the system.

EXAMPLE 2 Graph the solutions of the system: $\begin{cases} x - y < 2 \\ x + 2y > -1 \end{cases}$

Solution: Graph both inequalities on the same set of axes. Both boundary lines are dashed lines since the inequality symbols are $<$ and $>$. The solutions of the system are the regions shown by the darkest shading. In this example, the boundary lines are not a part of the solution.

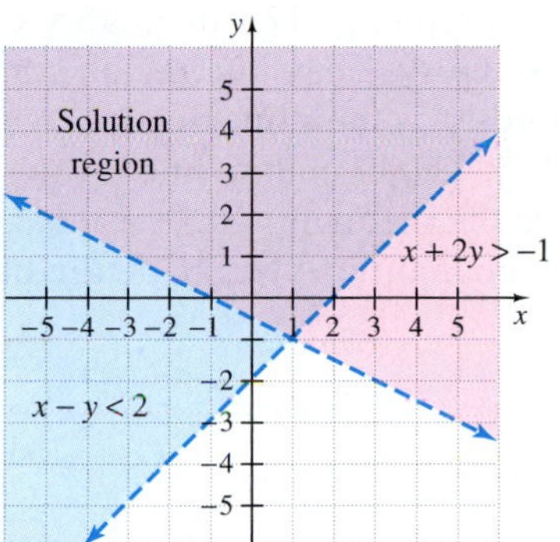

EXAMPLE 3 Graph the solutions of the system $\begin{cases} -3x + 4y < 12 \\ x \geq 2 \end{cases}$

Solution: Graph both inequalities on the same set of axes.

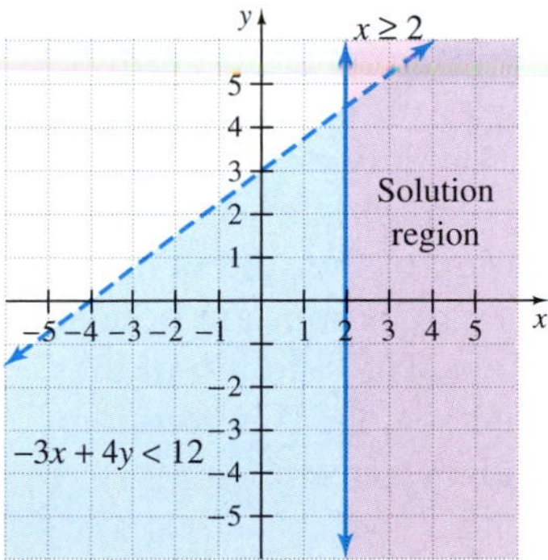

The solutions of the system are the darkest shaded regions, including a portion of the line $x = 2$.

D EXERCISE SET

MathXL®

Graph the solutions of each system of linear inequalities. See Examples 1 through 3.

1. $\begin{cases} y \geq x + 1 \\ y \geq 3 - x \end{cases}$

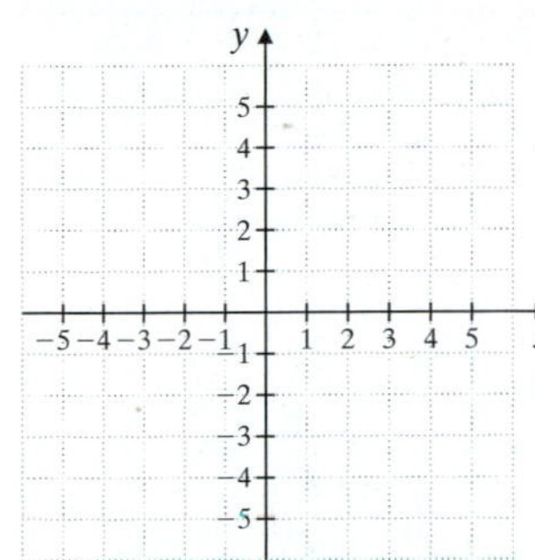

2. $\begin{cases} y \geq x - 3 \\ y \geq -1 - x \end{cases}$

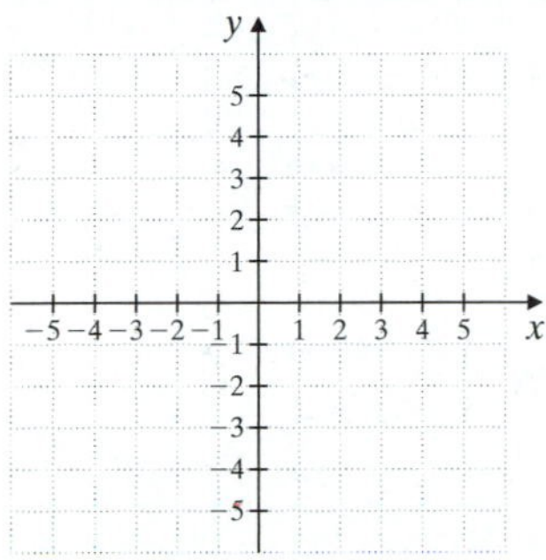

3. $\begin{cases} y < 3x - 4 \\ y \leq x + 2 \end{cases}$

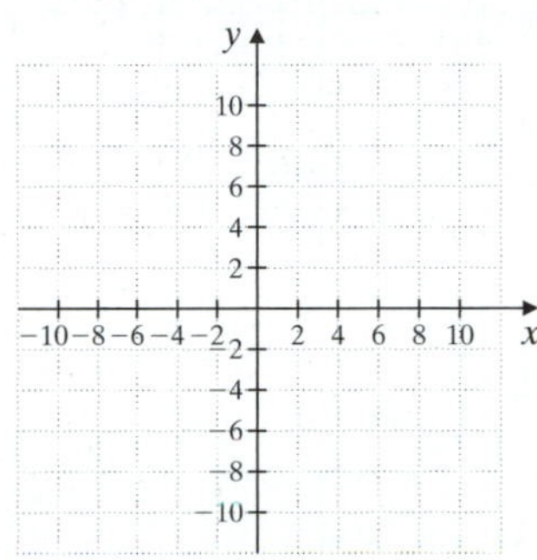

4. $\begin{cases} y \leq 2x + 1 \\ y > x + 2 \end{cases}$

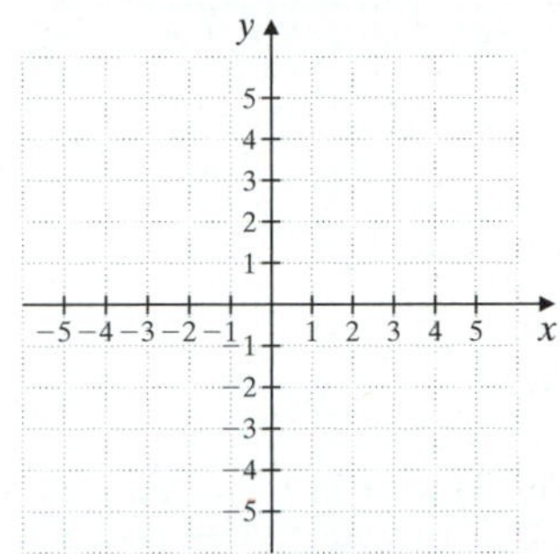

5. $\begin{cases} y < -2x - 2 \\ y > x + 4 \end{cases}$

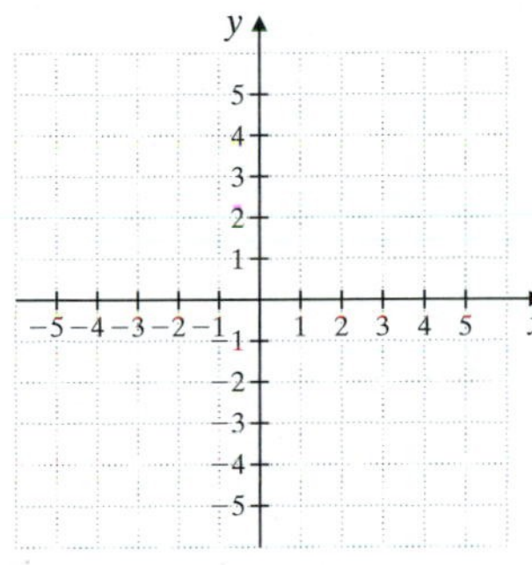

6. $\begin{cases} y \leq 2x + 4 \\ y \geq -x - 5 \end{cases}$

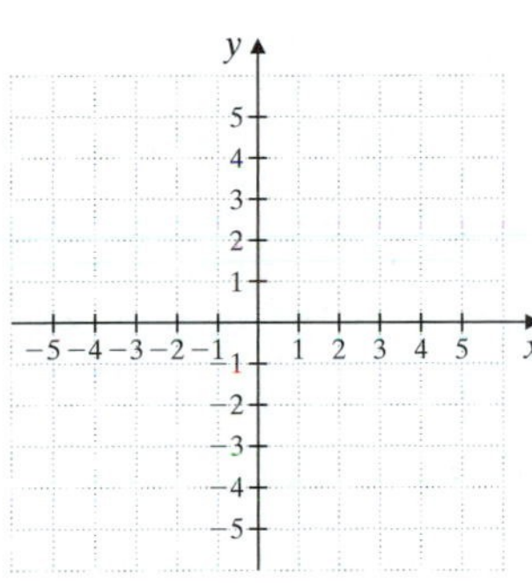

7. $\begin{cases} y \geq -x + 2 \\ y \leq 2x + 5 \end{cases}$

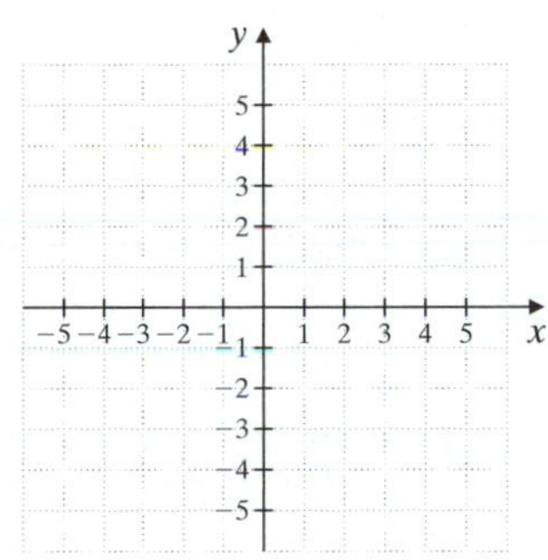

8. $\begin{cases} y \geq x - 5 \\ y \leq -3x + 3 \end{cases}$

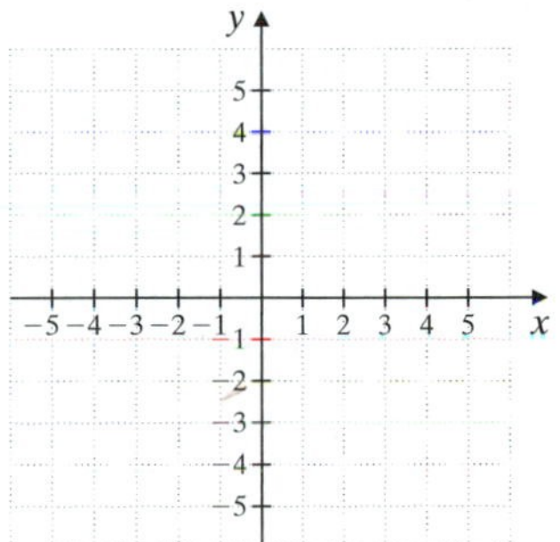

9. $\begin{cases} x \geq 3y \\ x + 3y \leq 6 \end{cases}$

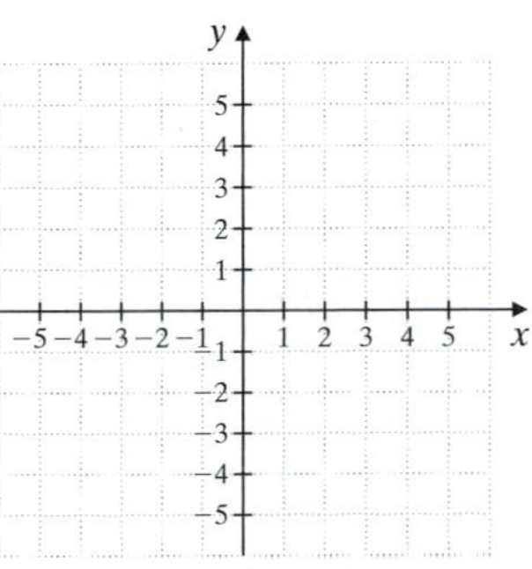

10. $\begin{cases} -2x < y \\ x + 2y < 3 \end{cases}$

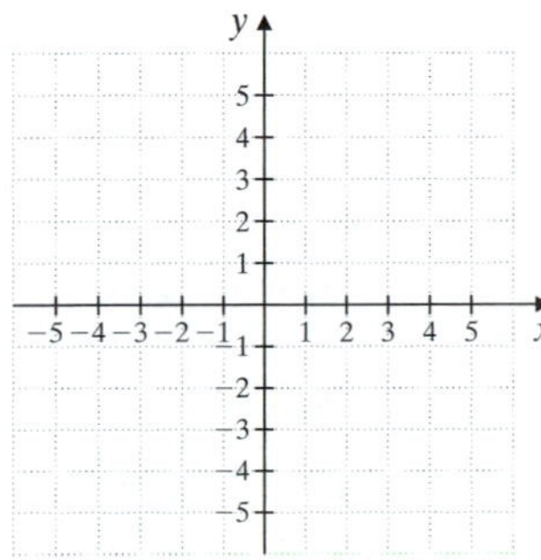

11. $\begin{cases} y + 2x \geq 0 \\ 5x - 3y \leq 12 \end{cases}$

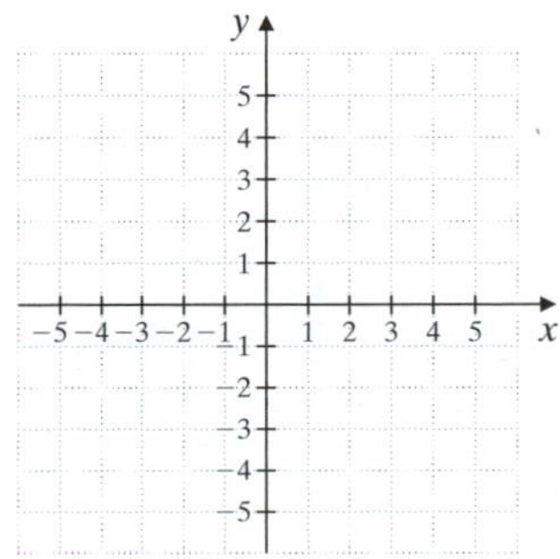

12. $\begin{cases} y + 2x \leq 0 \\ 5x + 3y \geq -2 \end{cases}$

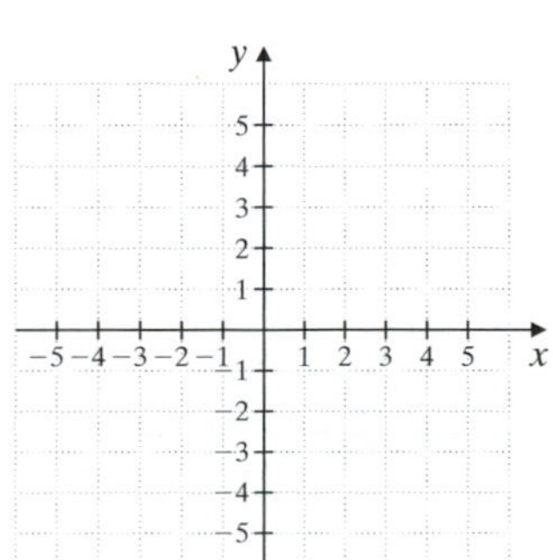

Graph the solutions of each system of linear inequalities. See Examples 1 through 3.

13. $\begin{cases} 3x - 4y \geq -6 \\ 2x + y \leq 7 \end{cases}$

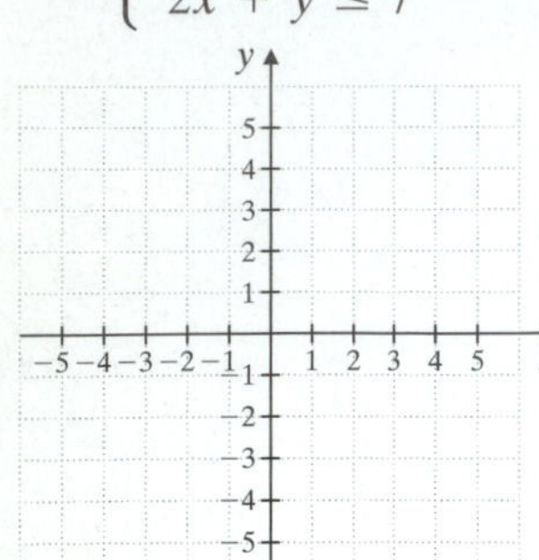

14. 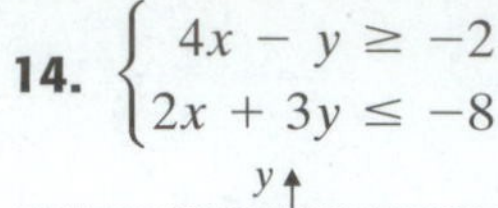
$\begin{cases} 4x - y \geq -2 \\ 2x + 3y \leq -8 \end{cases}$

15. $\begin{cases} x \leq 2 \\ y \geq -3 \end{cases}$

16. $\begin{cases} x \geq -3 \\ y \geq -2 \end{cases}$

17. $\begin{cases} y \geq 1 \\ x < -3 \end{cases}$

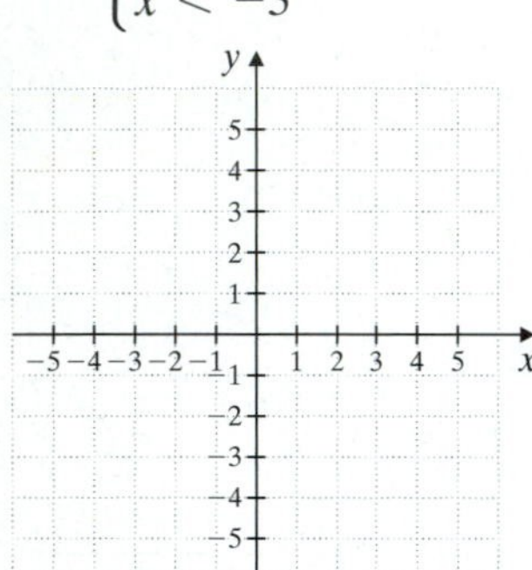

18. $\begin{cases} y > 2 \\ x \geq -1 \end{cases}$

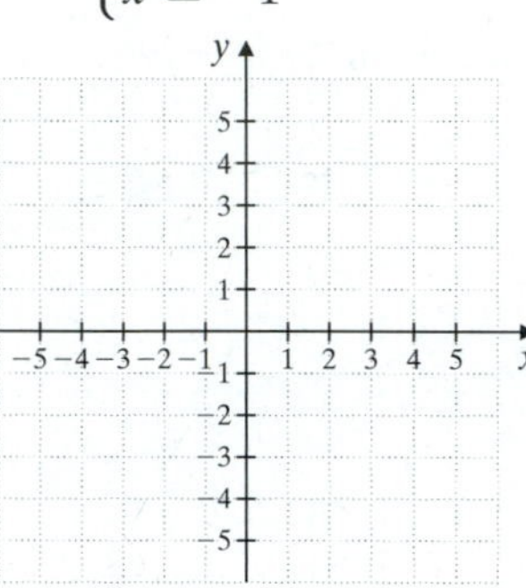

19. $\begin{cases} 2x + 3y < -8 \\ x \geq -4 \end{cases}$

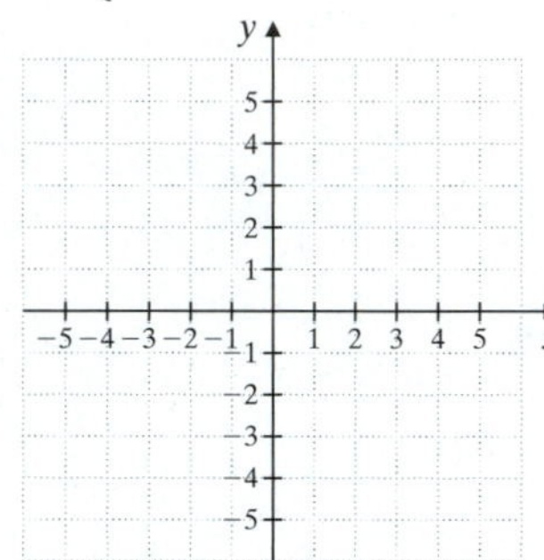

20. $\begin{cases} 3x + 2y \leq 6 \\ x < 2 \end{cases}$

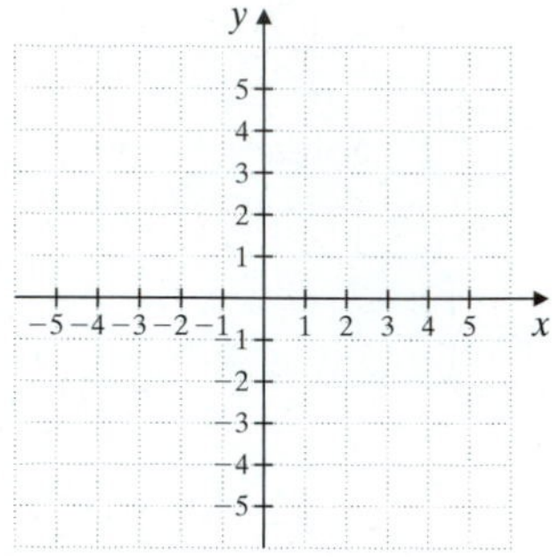

21. $\begin{cases} 2x - 5y \leq 9 \\ y \leq -3 \end{cases}$

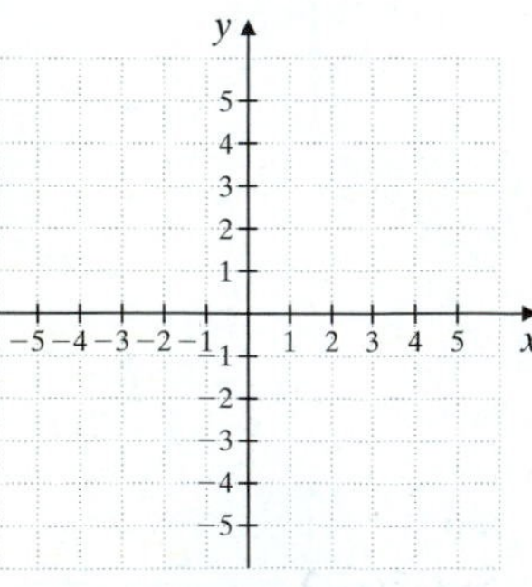

22. $\begin{cases} 2x + 5y \leq -10 \\ y \geq 1 \end{cases}$

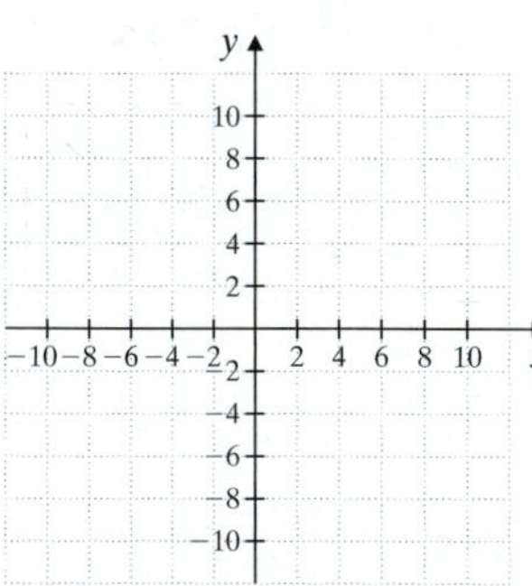

23. $\begin{cases} y \geq \frac{1}{2}x + 2 \\ y \leq \frac{1}{2}x - 3 \end{cases}$

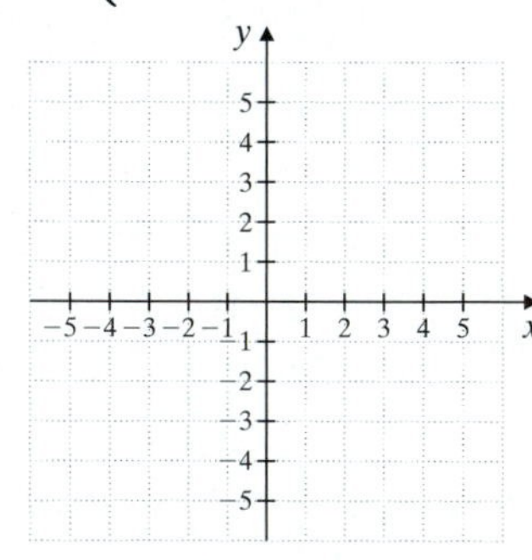

24. $\begin{cases} y \geq \frac{-3}{2}x + 3 \\ y < \frac{-3}{2}x + 6 \end{cases}$

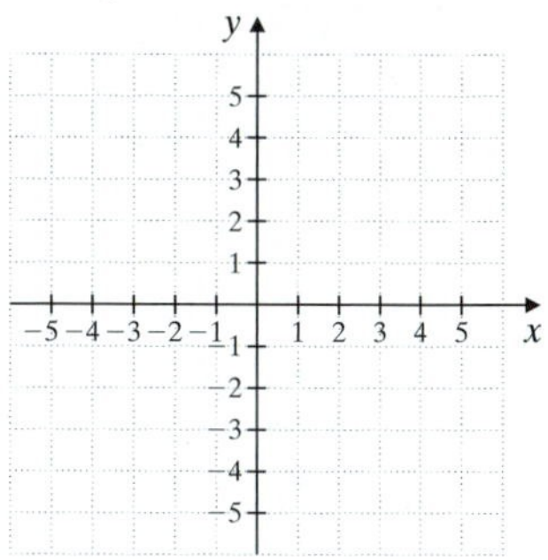

Match each system of inequalities to the corresponding graph.

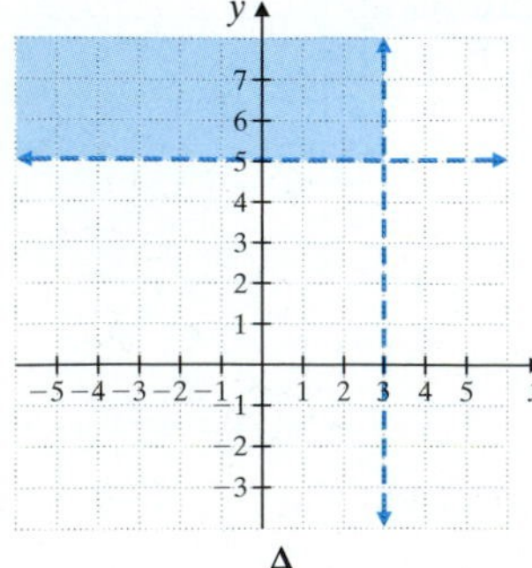

A

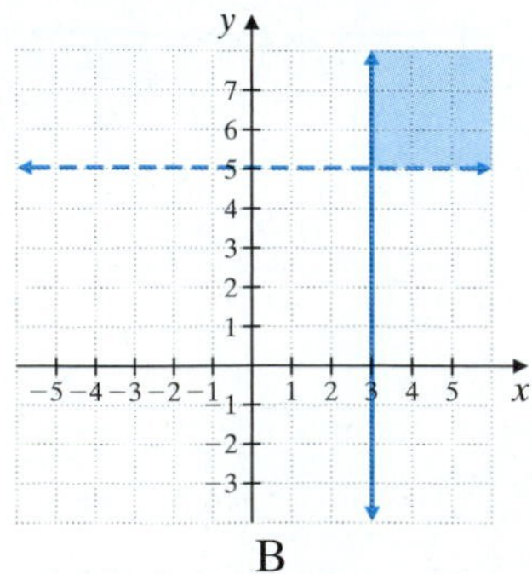

B

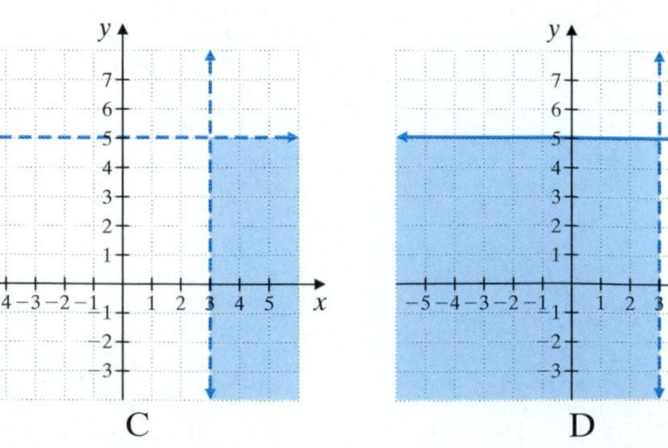

C

D

25. $\begin{cases} y < 5 \\ x > 3 \end{cases}$

26. $\begin{cases} y > 5 \\ x < 3 \end{cases}$

27. $\begin{cases} y \leq 5 \\ x < 3 \end{cases}$

28. $\begin{cases} y > 5 \\ x \geq 3 \end{cases}$

Appendix

E Geometric Formulas

Rectangle

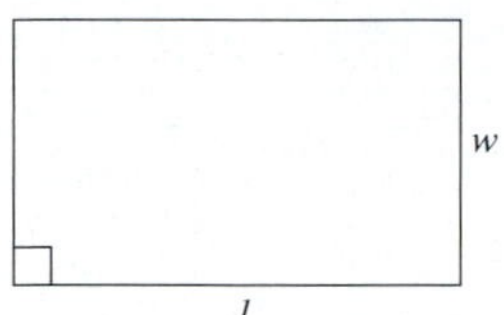

Perimeter: $P = 2l + 2w$
Area: $A = lw$

Square

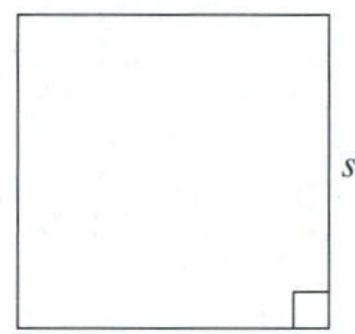

Perimeter: $P = 4s$
Area: $A = s^2$

Triangle

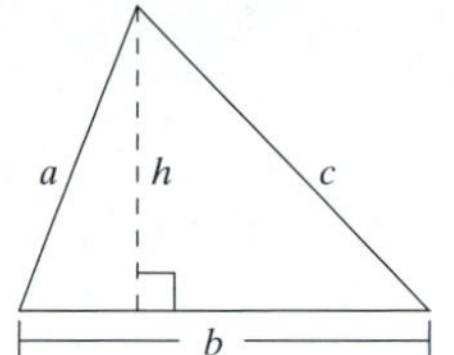

Perimeter: $P = a + b + c$
Area: $A = \frac{1}{2}bh$

Sum of Angles of Triangle

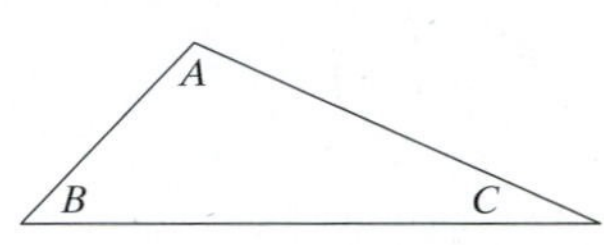

$A + B + C = 180°$
The sum of the measures of the three angles is $180°$.

Pythagorean Theorem (for right triangles)

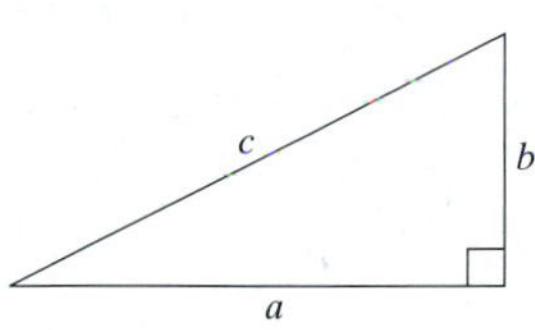

$a^2 + b^2 = c^2$
One $90°$ (right) angle

Isosceles Triangle

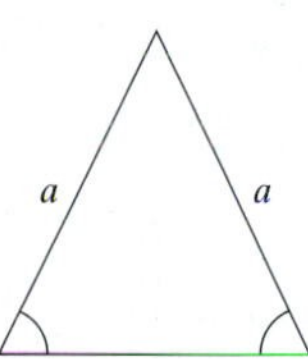

Triangle has:
two equal sides and
two equal angles.

Equilateral Triangle

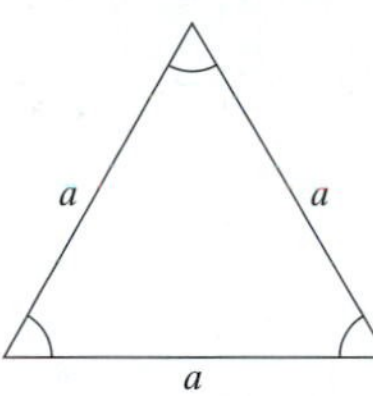

Triangle has:
three equal sides and
three equal angles.
Measure of each angle is $60°$.

Trapezoid

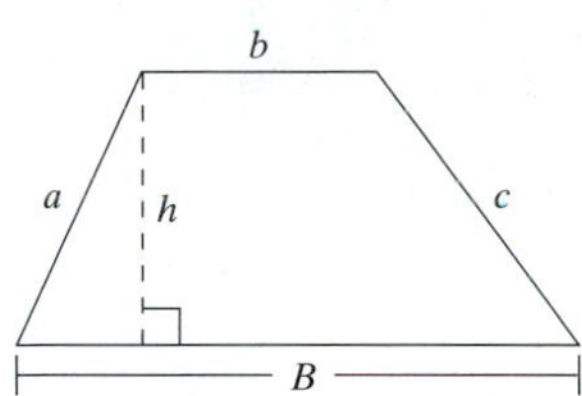

Perimeter:
$P = a + b + c + B$
Area: $A = \frac{1}{2}h(B + b)$

Parallelogram

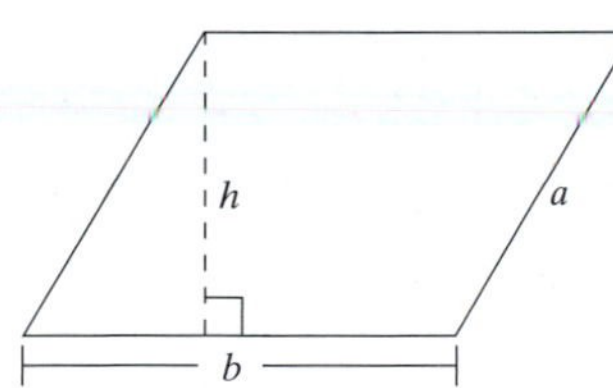

Perimeter: $P = 2a + 2b$
Area: $A = bh$

Circle

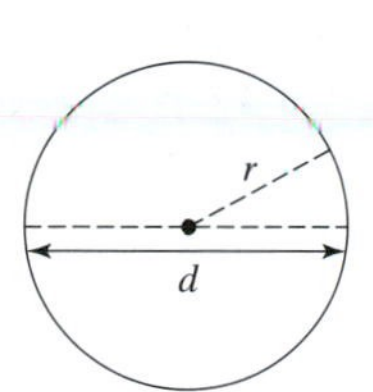

Circumference: $C = \pi d$
$C = 2\pi r$
Area: $A = \pi r^2$

Rectangular Solid

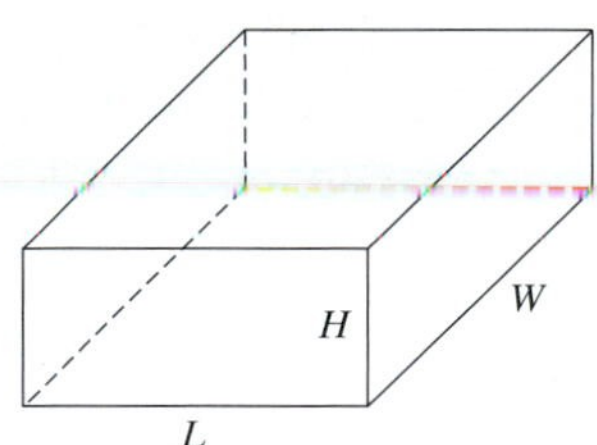

Volume: $V = LWH$
Surface Area:
$S = 2LW + 2HL + 2HW$

Cube

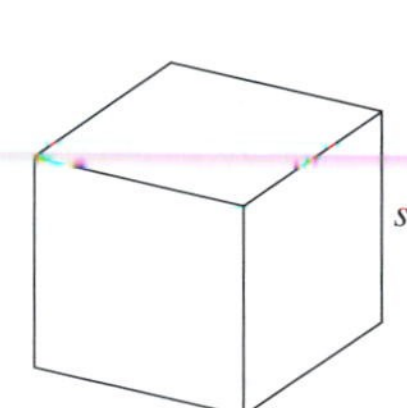

Volume: $V = s^3$
Surface Area: $S = 6s^2$

Cone

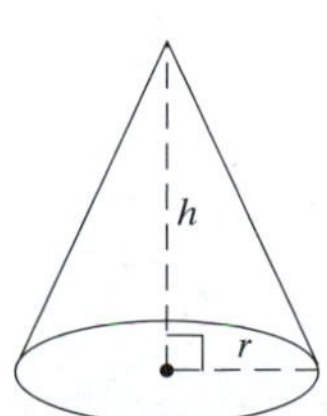

Volume: $V = \frac{1}{3}\pi r^2 h$

Right Circular Cylinder

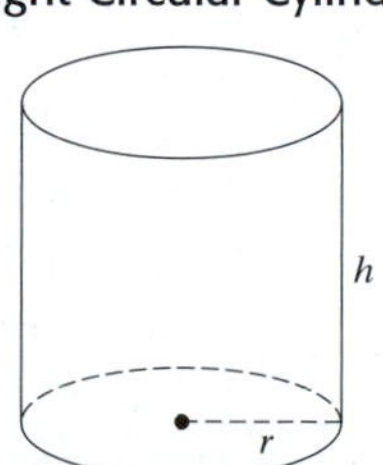

Volume: $V = \pi r^2 h$
Surface Area:
$S = 2\pi r^2 + 2\pi rh$

Sphere

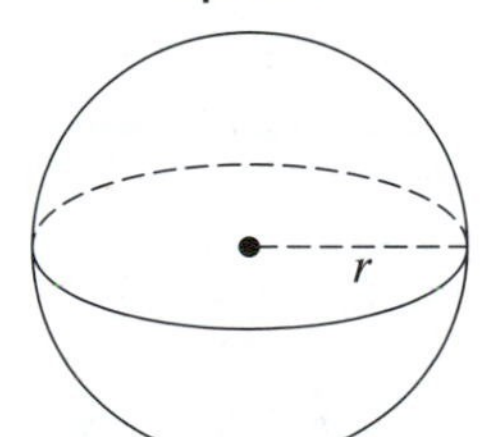

Volume: $V = \frac{4}{3}\pi r^3$
Surface Area: $S = 4\pi r^2$

Square-Based Pyramid

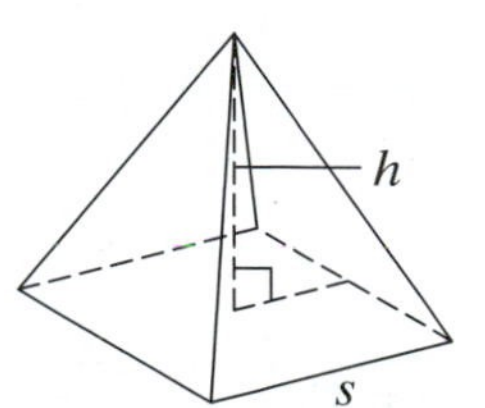

Volume: $V = \frac{1}{3} \cdot s^2 \cdot h$

ANSWERS TO SELECTED EXERCISES

CHAPTER 1 The Whole Numbers

Vocabulary and Readiness Check **1.** whole **3.** words **5.** period

Exercise Set 1.2 **1.** tens **3.** thousands **5.** hundred-thousands **7.** millions **9.** three hundred fifty-four **11.** eight thousand, two hundred seventy-nine **13.** twenty-six thousand, nine hundred ninety **15.** two million, three hundred eighty-eight thousand **17.** twenty-four million, three hundred fifty thousand, one hundred eighty-five **19.** sixty-five thousand, seven hundred seventy-three **21.** one thousand, six hundred seventy-nine **23.** two million, eight hundred thousand **25.** eleven thousand, two hundred thirty-nine **27.** two hundred two thousand, seven hundred **29.** 6587 **31.** 59,800 **33.** 13,601,011 **35.** 7,000,017 **37.** 260,997 **39.** 353 **41.** 484,235 **43.** 54,500,000 **45.** 714 **47.** 200 + 9 **49.** 3000 + 400 + 70 **51.** 80,000 + 700 + 70 + 4 **53.** 60,000 + 6000 + 40 + 9 **55.** 30,000,000 + 9,000,000 + 600,000 + 80,000 **57.** 5532: five thousand, five hundred thirty-two **59.** 5000 + 400 + 90 + 2 **61.** Mt. Washington **63.** Chihuahua **65.** Labrador retriever; one hundred forty-six thousand, seven hundred fourteen **67.** 25 pounds **69.** 9861 **71.** no; one hundred five **73.** answers may vary **75.** 135,500,000,000,000

Calculator Explorations **1.** 134 **3.** 340 **5.** 2834 **7.** 770 **9.** 109 **11.** 8978

Vocabulary and Readiness Check **1.** number **3.** 0 **5.** minuend, subtrahend, difference **7.** order, commutative

Exercise Set 1.3 **1.** 36 **3.** 292 **5.** 49 **7.** 5399 **9.** 209,078 **11.** 25 **13.** 212 **15.** 11,926 **17.** 16,717 **19.** 35,901 **21.** 632,389 **23.** 25 **25.** 600 **27.** 288 **29.** 168 **31.** 5723 **33.** 504 **35.** 79 **37.** 32,711 **39.** 5041 **41.** 31,213 **43.** 1034 **45.** 9 **47.** 8518 **49.** 22,876 **51.** 25 ft **53.** 24 in. **55.** 29 in. **57.** 44 m **59.** 2093 **61.** 266 **63.** 20 **65.** 544 **67.** 72 **69.** 88 **71.** 2028 thousand **73.** $619 **75.** 264,000 sq ft **77.** 283,000 sq ft **79.** 340 ft **81.** 264 pages **83.** 813,109 **85.** 100 dB **87.** 58 dB **89.** 9042 **91.** 124 ft **93.** California **95.** 421 stores **97.** Florida and Georgia **99.** 5894 mi **101.** minuend: 48; subtrahend: 1 **103.** minuend: 70; subtrahend: 7 **105.** answers may vary **107.** correct **109.** incorrect; 530 **111.** incorrect: 685 **113.** correct **115.** $\begin{array}{r} 5269 \\ -2385 \\ \hline 2884 \end{array}$ **117.** answers may vary **119.** no; 1089 more pages

Vocabulary and Readiness Check **1.** graph **3.** 70; 60

Exercise Set 1.4 **1.** 420 **3.** 640 **5.** 2800 **7.** 500 **9.** 21,000 **11.** 34,000 **13.** 328,500 **15.** 36,000 **17.** 39,990 **19.** 30,000,000 **21.** 5280; 5300; 5000 **23.** 9440; 9400; 9000 **25.** 14,880; 14,900; 15,000 **27.** 19,000 **29.** 38,000 points **31.** $68,000,000,000 **33.** $3,000,000 **35.** US 208,000,000; India 69,000,000 **37.** 130 **39.** 80 **41.** 5700 **43.** 300 **45.** 11,400 **47.** incorrect **49.** correct **51.** correct **53.** $3400 **55.** 900 mi **57.** 6000 ft **59.** 400,000 people **61.** 3,000 children **63.** $339,000,000; $340,000,000; $300,000,000 **65.** $179,000,000; $180,000,000; $200,000,000 **67.** 5723, for example **69. a.** 8550 **b.** 8649 **71.** answers may vary **73.** 140 m

Calculator Explorations **1.** 3456 **3.** 15,322 **5.** 272,291

Vocabulary and Readiness Check **1.** 0 **3.** product; factor **5.** grouping; associative **7.** length

Exercise Set 1.5 **1.** 24 **3.** 0 **5.** 0 **7.** 87 **9.** $6 \cdot 3 + 6 \cdot 8$ **11.** $4 \cdot 3 + 4 \cdot 9$ **13.** $20 \cdot 14 + 20 \cdot 6$ **15.** 512 **17.** 3678 **19.** 1662 **21.** 6444 **23.** 1157 **25.** 24,418 **27.** 24,786 **29.** 15,600 **31.** 0 **33.** 6400 **35.** 48,126 **37.** 142,506 **39.** 2,369,826 **41.** 64,790 **43.** 3,949,935 **45.** 63 sq m **47.** 680 sq ft **49.** 240,000 **51.** 300,000 **53.** c **55.** c **57.** 880 **59.** 4200 **61.** 4480 **63.** 375 cal **65.** $3290 **67. a.** 20 **b.** 100 **c.** 2000 lb **69.** 8800 sq ft **71.** 56,000 sq ft **73.** 5828 pixels **75.** 2100 characters **77.** 1280 cal **79.** $10, $60; $10, $200; $12, $36; $12, $36; $372 **81.** 21,700,000 qt **83.** 135 **85.** 2144 **87.** 23 **89.** 15 **91.** $5 \cdot 6$ or $6 \cdot 5$ **93. a.** 5 + 5 + 5 or 3 + 3 + 3 + 3 + 3 **b.** answers may vary **95.** $\begin{array}{r} 203 \\ \times\ 14 \\ \hline 812 \\ 2030 \\ \hline 2842 \end{array}$ **97.** $\begin{array}{r} 42 \\ \times\ 93 \\ \hline \end{array}$ **99.** answers may vary **101.** 506 windows

Calculator Explorations **1.** 53 **3.** 62 **5.** 261 **7.** 0

Vocabulary and Readiness Check **1.** quotient; dividend; divisor **3.** 1 **5.** undefined

Exercise Set 1.6 **1.** 6 **3.** 12 **5.** 0 **7.** 31 **9.** 1 **11.** 8 **13.** undefined **15.** 1 **17.** 0 **19.** 9 **21.** 29 **23.** 74 **25.** 338 **27.** undefined **29.** 9 **31.** 25 **33.** 68 R 3 **35.** 236 R 5 **37.** 38 R 1 **39.** 326 R 4 **41.** 13 **43.** 49 **45.** 97 R 8 **47.** 209 R 11 **49.** 506 **51.** 202 R 7 **53.** 54 **55.** 99 R 100 **57.** 202 R 15 **59.** 579 R 72 **61.** 17 **63.** 511 R 3 **65.** 2132 R 32 **67.** 6080 **69.** 23 R 2 **71.** 5 R 25 **73.** 20 R 2 **75.** 33 students **77.** 165 lb **79.** 310 yd **81.** 89 bridges **83.** 11 light poles **85.** 5 mi **87.** 1760 yd **89.** 20 **91.** 387 **93.** 79 **95.** 74° **97.** 9278 **99.** 15,288 **101.** 679 **103.** undefined **105.** 9 R 12 **107.** c **109.** b **111.** $732,000,000 **113.** increase; answers may vary **115.** no; answers may vary **117.** answers may vary

The Bigger Picture **1.** 121 **2.** 43 **3.** 3198 **4.** 89 R 11 **5.** 0 **6.** 0 **7.** 26 **8.** undefined **9.** 82 **10.** 1844

Integrated Review **1.** 194 **2.** 6555 **3.** 4524 **4.** 562 **5.** 67 **6.** undefined **7.** 1 **8.** 5 **9.** 0 **10.** 0 **11.** 0 **12.** 3 **13.** 63 **14.** 9 **15.** 138 **16.** 276 **17.** 1169 **18.** 9826 **19.** 182 R 4 **20.** 79,317 **21.** 1099 R 2 **22.** 111 R 1 **23.** 663 R 24 **24.** 1076 R 60 **25.** 1037 **26.** 9899 **27.** 30,603 **28.** 47,500 **29.** 71 **30.** 558 **31.** 6 R 8 **32.** 53 **33.** 183 **34.** 231 **35.** 9740; 9700; 10,000 **36.** 1400; 1000 **37.** 20,800; 20,800; 21,000 **38.** 432,200; 432,200; 432,000 **39.** perimeter: 24 ft; area: 36 sq ft **40.** perimeter: 42 in.; area: 98 sq in. **41.** 28 mi **42.** 26 m **43.** 24 **44.** 124 **45.** Lake Pontchartrain Bridge: 2175 ft **46.** $6246

Calculator Explorations **1.** 729 **3.** 1024 **5.** 2048 **7.** 2526 **9.** 4295 **11.** 8

Vocabulary and Readiness Check **1.** base; exponent **3.** addition **5.** division

Exercise Set 1.7 **1.** 4^3 **3.** 7^6 **5.** 12^3 **7.** $6^2 \cdot 5^3$ **9.** $9 \cdot 8^2$ **11.** $3 \cdot 2^4$ **13.** $3 \cdot 2^4 \cdot 5^5$ **15.** 64 **17.** 125 **19.** 32 **21.** 1 **23.** 7 **25.** 128 **27.** 256 **29.** 256 **31.** 729 **33.** 144 **35.** 100 **37.** 20 **39.** 729 **41.** 192 **43.** 162 **45.** 21 **47.** 7 **49.** 5 **51.** 16 **53.** 46 **55.** 8 **57.** 64 **59.** 83 **61.** 2 **63.** 48 **65.** 4 **67.** undefined **69.** 59 **71.** 52 **73.** 44 **75.** 12 **77.** 21 **79.** 3 **81.** 43 **83.** 8 **85.** 16 **87.** 49 sq m **89.** 529 sq mi **91.** true **93.** false **95.** $(2 + 3) \cdot 6 - 2$ **97.** $24 \div (3 \cdot 2) + 2 \cdot 5$ **99.** 1260 ft **101.** 6,384,814 **103.** answers may vary; $(20 - 10) \cdot 5 \div 25 + 3$

The Bigger Picture **1.** 216 **2.** 48 **3.** 200 **4.** 15 **5.** 37 **6.** 200 **7.** 799 **8.** 2160 **9.** 10 R 34 **10.** 27

Vocabulary and Readiness Check **1.** expression **3.** expression; variables **5.** equation

Exercise Set 1.8 **1.** 28; 14; 147; 3 **3.** 152; 152; 0; undefined **5.** 57; 55; 56; 56 **7.** 9 **9.** 8 **11.** 6 **13.** 5 **15.** 117 **17.** 94 **19.** 5 **21.** 34 **23.** 20 **25.** 4 **27.** 4 **29.** 0 **31.** 33 **33.** 125 **35.** 121 **37.** 100 **39.** 60 **41.** 4 **43.** 16; 64; 144; 256 **45.** yes **47.** no **49.** no **51.** yes **53.** no **55.** yes **57.** 12 **59.** 6 **61.** 4 **63.** none **65.** 11 **67.** $x + 8$ **69.** $x + 8$ **71.** $20 - x$ **73.** $512x$ **75.** $\frac{6}{x}$ **77.** $5x + (17 + x)$ **79.** $5x$ **81.** $11 - x$ **83.** $x - 5$ **85.** $6 \div x$ or $\frac{6}{x}$ **87.** $50 - 8x$ **89.** 274,657 **91.** 777 **93.** $5x$ **95.** As t gets larger $16t^2$ gets larger.

Chapter 1 Vocabulary Check **1.** whole numbers **2.** perimeter **3.** place value **4.** exponent **5.** area **6.** digits **7.** variable **8.** equation **9.** expression **10.** solution **11.** set **12.** sum **13.** divisor **14.** dividend **15.** quotient **16.** factor **17.** product **18.** minuend **19.** subtrahend **20.** difference **21.** addend

Chapter Review **1.** tens **2.** ten-millions **3.** seven thousand, six hundred forty **4.** forty-six million, two hundred thousand, one hundred twenty **5.** 3000 + 100 + 50 + 8 **6.** 400,000,000 + 3,000,000 + 200,000 + 20,000 + 5000 **7.** 81,900 **8.** 6,304,000,000 **9.** 1,953,631 **10.** 3,844,829 **11.** San Jose, CA **12.** New York, NY **13.** 67 **14.** 67 **15.** 65 **16.** 304 **17.** 449 **18.** 840 **19.** 3914 **20.** 7908 **21.** 4211 **22.** 1967 **23.** 1334 **24.** 886 **25.** 17,897 **26.** 34,658 **27.** 7523 mi **28.** $197,699 **29.** 216 ft **30.** 66 km **31.** 111,863 **32.** 54,269 **33.** May **34.** August **35.** $110 **36.** $240 **37.** 40 **38.** 50 **39.** 880 **40.** 500 **41.** 3800 **42.** 58,000 **43.** 40,000,000 **44.** 800,000 **45.** 7300 **46.** 4100 **47.** 2700 mi **48.** 300,000 **49.** 2208 **50.** 1396 **51.** 2280 **52.** 2898 **53.** 560 **54.** 900 **55.** 0 **56.** 0 **57.** 16,994 **58.** 8954 **59.** 113,634 **60.** 44,763 **61.** 411,426 **62.** 636,314 **63.** 1500 **64.** 4920 **65.** $898 **66.** $96,400 **67.** 91 sq mi **68.** 500 sq cm **69.** 7 **70.** 4 **71.** 5 R 2 **72.** 4 R 2 **73.** undefined **74.** 0 **75.** 33 R 2 **76.** 19 R 7 **77.** 24 R 2 **78.** 35 R 15 **79.** 506 R 10 **80.** 907 R 40 **81.** 2793 R 140 **82.** 2012 R 60 **83.** 18 R 2 **84.** 21 R 2 **85.** 27 boxes **86.** 13 miles **87.** 51 **88.** 59 **89.** 64 **90.** 125 **91.** 405 **92.** 400 **93.** 16 **94.** 10 **95.** 15 **96.** 7 **97.** 12 **98.** 9 **99.** 42 **100.** 33 **101.** 9 **102.** 2 **103.** 6 **104.** 29 **105.** 40 **106.** 72 **107.** 5 **108.** 7 **109.** 49 sq m **110.** 9 sq in. **111.** 5 **112.** 17 **113.** undefined **114.** 0 **115.** 121 **116.** 2 **117.** 4 **118.** 20 **119.** $x - 5$ **120.** $x + 7$ **121.** $10 \div x$ **122.** $5x$ **123.** yes **124.** no **125.** no **126.** yes **127.** 11 **128.** 175 **129.** 14 **130.** none **131.** 417 **132.** 682 **133.** 2196 **134.** 2516 **135.** 1101 **136.** 1411 **137.** 458 R 8 **138.** 237 R 1 **139.** 70,848 **140.** 95,832 **141.** 1644 **142.** 8481 **143.** 840 **144.** 300,000 **145.** 12 **146.** 6 **147.** no **148.** yes **149.** 53 full boxes with 18 left over **150.** $86

Chapter Test **1.** eighty-two thousand, four hundred twenty-six **2.** 402,550 **3.** 141 **4.** 113 **5.** 14,880 **6.** 766 R 42 **7.** 200 **8.** 98 **9.** 0 **10.** undefined **11.** 33 **12.** 21 **13.** 48 **14.** 36 **15.** 5,698,000 **16.** 82 **17.** 52,000 **18.** 13,700 **19.** 1600 **20.** 92 **21.** 122 **22.** 1605 **23.** 7 R 2 **24.** $17 **25.** $126 **26.** 360 cal **27.** $7905 **28.** 20 cm; 25 sq cm **29.** 60 yd; 200 sq yd **30.** 30 **31.** 1 **32.** **a.** $\frac{x}{17}$ **b.** $2x - 20$ **33.** yes **34.** 10

CHAPTER 2 Integers and Introduction to Solving Equations

Vocabulary and Readiness Check **1.** integers **3.** inequality symbols **5.** is less than; is greater than **7.** absolute value

Exercise Set 2.1 **1.** −1235 **3.** +14,433 **5.** +118 **7.** −13,000 **9.** −10,458 million **11.** −160, −147; Guillermo **13.** −81

15. 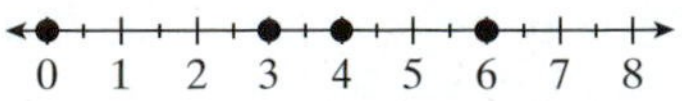

17. 

19. 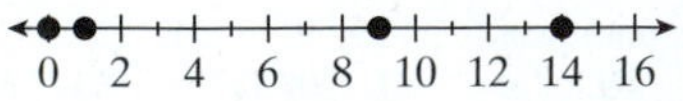

21. 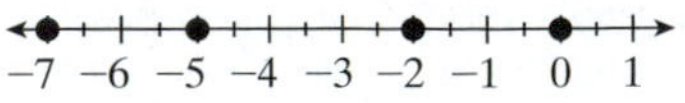

23. > **25.** < **27.** > **29.** < **31.** 5 **33.** 8 **35.** 0 **37.** 55 **39.** −5 **41.** 4 **43.** −23 **45.** 85 **47.** 7 **49.** −20 **51.** −3 **53.** 43 **55.** 15 **57.** 33 **59.** 6 **61.** −2 **63.** 32 **65.** −7 **67.** < **69.** < **71.** = **73.** < **75.** > **77.** < **79.** > **81.** < **83.** 31; −31 **85.** 28; 28 **87.** Caspian Sea

89. Lake Superior **91.** Mars **93.** Uranus **95.** 13 **97.** 35 **99.** 360 **101.** $-|-8|, -|3|, 2^2, -(-5)$ **103.** $-|-6|, -|1|, |-1|, -(-6)$ **105.** $-10, -|-9|, -(-2), |-12|, 5^2$ **107.** a, d **109.** 8 **111.** false **113.** true **115.** false **117.** answers may vary **119.** no; answers may vary

Calculator Explorations **1.** -159 **3.** 44 **5.** $-894{,}855$

Vocabulary and Readiness Check **1.** 0 **3.** *a*

Exercise Set 2.2

1. 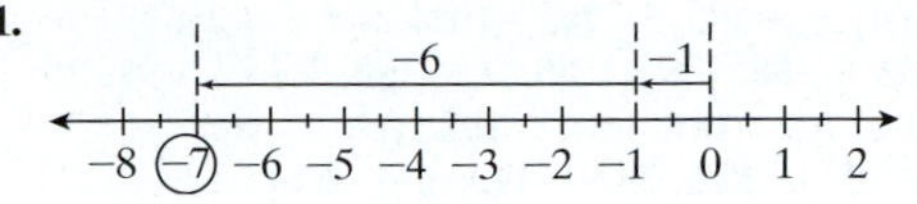

3. 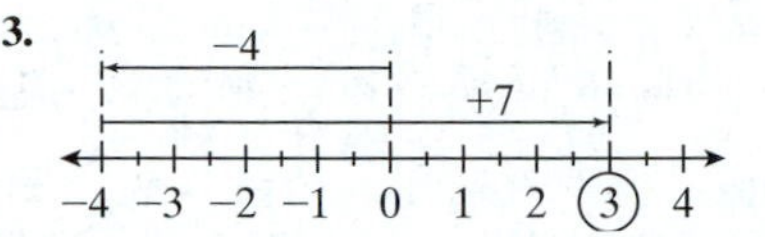

5. 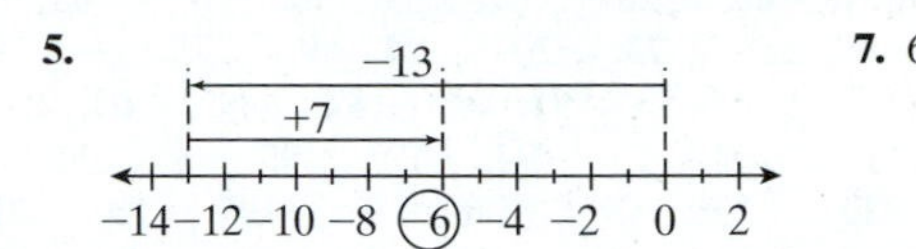

7. 67

9. -10 **11.** 0 **13.** 4 **15.** -6 **17.** -2 **19.** -9 **21.** -24 **23.** -840 **25.** 7 **27.** -3 **29.** -30 **31.** 40 **33.** -20 **35.** -125 **37.** -7 **39.** -246 **41.** 16 **43.** 13 **45.** -28 **47.** -11 **49.** 20 **51.** -34 **53.** -1 **55.** 0 **57.** -42 **59.** -70 **61.** 3 **63.** -21 **65.** 19 **67.** -10 **69.** $0 + (-215) + (-16) = -231$; 231 ft below the surface **71.** Sorenstam: -1; Hurst: $+3$ **73.** \$65,000,000 **75.** \$40,000,000 **77.** 2°C **79.** \$13,609 **81.** -47°F **83.** -7679 m **85.** 44 **87.** 141 **89.** answers may vary **91.** -3 **93.** 10 **95.** true **97.** false **99.** answers may vary

Vocabulary and Readiness Check **1.** b **3.** d

Exercise Set 2.3 **1.** 0 **3.** 3 **5.** -5 **7.** 22 **9.** 3 **11.** -20 **13.** -12 **15.** -13 **17.** 508 **19.** -14 **21.** -4 **23.** -12 **25.** -42 **27.** -19 **29.** 14 **31.** -56 **33.** -5 **35.** -145 **37.** -37 **39.** 3 **41.** 1 **43.** -1 **45.** -31 **47.** 44 **49.** -32 **51.** -9 **53.** 14 **55.** -11 **57.** 31 **59.** 12 **61.** 20 **63.** 15°F **65.** 4°F **67.** 265°F **69.** \$49 **71.** -10°C **73.** 154 ft **75.** 69 ft **77.** 652 ft **79.** 144 ft **81.** 1197°F **83.** $-$\$782 billion **85.** $-5 + x$ **87.** $-20 - x$ **89.** 5 **91.** 1058 **93.** answers may vary **95.** 16 **97.** -20 **99.** -4 **101.** 0 **103.** -14 **105.** false **107.** answers may vary

Vocabulary and Readiness Check **1.** negative **3.** positive **5.** 0 **7.** undefined

Exercise Set 2.4 **1.** 12 **3.** -36 **5.** -81 **7.** 0 **9.** 48 **11.** -12 **13.** 80 **15.** 0 **17.** -15 **19.** -9 **21.** -27 **23.** -36 **25.** -64 **27.** -8 **29.** -5 **31.** 7 **33.** 0 **35.** undefined **37.** -14 **39.** 0 **41.** -15 **43.** -63 **45.** 42 **47.** -24 **49.** 49 **51.** -5 **53.** -9 **55.** -6 **57.** 120 **59.** -1080 **61.** 0 **63.** -6 **65.** -7 **67.** 3 **69.** -1 **71.** -32 **73.** 180 **75.** 1 **77.** -30 **79.** -1104 **81.** -2870 **83.** -56 **85.** -18 **87.** 35 **89.** -1 **91.** undefined **93.** 6 **95.** 16; 4 **97.** 0; 0 **99.** -6 **101.** 252 **103.** $-71 \cdot x$ or $-71x$ **105.** $-16 - x$ **107.** $-29 + x$ **109.** $\frac{x}{-33}$ or $x \div (-33)$ **111.** $3 \cdot (-4) = -12$ yd; a loss of 12 yd **113.** $5 \cdot (-20) = -100$; a depth of 100 feet **115.** -210°C **117.** -189°C **119.** $-$\$1828 million **121. a.** -576 million or 576 million less **b.** -144 million or 144 million less **123.** 109 **125.** 8 **127.** -19 **129.** -28 **131.** -8 **133.** negative **135.** $(-5)^{17}, (-2)^{17}, (-2)^{12}, (-5)^{12}$ **137.** answers may vary

The Bigger Picture **1.** 5 **2.** 55 **3.** -6 **4.** -12 **5.** -12 **6.** 81 **7.** -81 **8.** -34 **9.** 10 **10.** -30 **11.** 30 **12.** -11 **13.** -30 **14.** -80 **15.** 15 **16.** 45

Integrated Review **1.** -50; $+122$ or 122 **2.** −4 −3 −2 −1 0 1 2 3 4 **3.** $>$ **4.** $<$ **5.** $<$ **6.** $>$ **7.** 3 **8.** -4 **9.** 9 **10.** 5 **11.** -11 **12.** 3 **13.** -64 **14.** 0 **15.** 12 **16.** -20 **17.** -48 **18.** -9 **19.** 10 **20.** -2 **21.** 106 **22.** -3 **23.** 0 **24.** 4 **25.** 42 **26.** 6 **27.** 19 **28.** -900 **29.** -12 **30.** -19 **31.** undefined **32.** 0 **33.** -4 **34.** -44 **35.** 125 **36.** 20 **37.** $\frac{x}{-17}$ or $x \div (-17)$ **38.** $-3 + x$ **39.** $x - (-18)$ **40.** $-7 \cdot x$ or $-7x$ **41.** 9 **42.** -15 **43.** 27 **44.** 33 **45.** -15 **46.** -4

Calculator Explorations **1.** 48 **3.** -258

Vocabulary and Readiness Check **1.** division **3.** average **5.** subtraction

Exercise Set 2.5 **1.** -125 **3.** -64 **5.** 32 **7.** -8 **9.** -11 **11.** -43 **13.** -8 **15.** 17 **17.** -1 **19.** 4 **21.** -3 **23.** 16 **25.** 13 **27.** -77 **29.** 80 **31.** 256 **33.** 53 **35.** 4 **37.** -64 **39.** 4 **41.** 16 **43.** -27 **45.** 34 **47.** 65 **49.** -7 **51.** 36 **53.** -117 **55.** 30 **57.** -3 **59.** -30 **61.** 1 **63.** -12 **65.** 0 **67.** -20 **69.** 9 **71.** -16 **73.** -128 **75.** 1 **77.** -50 **79.** -2 **81.** -19 **83.** 28 **85.** 2 **87.** no; answers may vary **89.** 4050 **91.** 45 **93.** 32 in. **95.** 30 ft **97.** $2 \cdot (7 - 5) \cdot 3$ **99.** $-6 \cdot (10 - 4)$ **101.** answers may vary **103.** answers may vary **105.** 20,736 **107.** 8900 **109.** 9

Vocabulary and Readiness Check **1.** expression **3.** equation; expression **5.** solution **7.** addition

Exercise Set 2.6 **1.** yes **3.** no **5.** yes **7.** yes **9.** 18 **11.** -12 **13.** 9 **15.** -17 **17.** 4 **19.** -4 **21.** -14 **23.** -17 **25.** 0 **27.** 1 **29.** -7 **31.** -50 **33.** -25 **35.** 36 **37.** 21 **39.** 12 **41.** -80 **43.** -2 **45.** $x - (-2)$ **47.** $-6 \cdot x$ or $-6x$ **49.** $-15 + x$ **51.** $-8 \div x$ or $\frac{-8}{x}$ **53.** 41,574 **55.** -409 **57.** answers may vary **59.** answers may vary

Chapter 2 Vocabulary Check **1.** opposites **2.** absolute value **3.** integers **4.** negative **5.** positive **6.** inequality symbols **7.** solution **8.** average **9.** expression **10.** equation **11.** is less than; is greater than **12.** addition **13.** multiplication

Chapter 2 Review **1.** −1572 **2.** +11,239 **3.** −7 −6 −5 −4 −3 −2 −1 0 1 2 3 4 5 6 7 **4.** −7 −6 −5 −4 −3 −2 −1 0 1 2 3 4 5 6 7 **5.** 11 **6.** 0 **7.** −8 **8.** 9 **9.** −16 **10.** 2 **11.** > **12.** < **13.** > **14.** > **15.** 18 **16.** −42 **17.** false **18.** true **19.** true **20.** true **21.** 2 **22.** 3 **23.** −5 **24.** −10 **25.** Elevator D **26.** Elevator B **27.** 2 **28.** 14 **29.** 4 **30.** 17 **31.** −23 **32.** −22 **33.** −21 **34.** −70 **35.** 0 **36.** 0 **37.** −151 **38.** −606 **39.** −20°C **40.** −150 ft **41.** −7 **42.** +1 **43.** 8 **44.** −16 **45.** −24 **46.** −10 **47.** 20 **48.** 8 **49.** 0 **50.** −32 **51.** 0 **52.** 7 **53.** −10 **54.** −17 **55.** 692 ft **56.** −25 **57.** −14 or 14 feet below ground **58.** 82 feet **59.** true **60.** false **61.** 21 **62.** −18 **63.** −64 **64.** 60 **65.** 25 **66.** −1 **67.** 0 **68.** 24 **69.** −5 **70.** 3 **71.** 0 **72.** undefined **73.** −20 **74.** −9 **75.** 38 **76.** −5 **77.** $(-5)(2) = -10$ **78.** $(-50)(4) = -200$ **79.** $-1024 \div 4 = -256$ **80.** $-45 \div 9 = -5$ **81.** 49 **82.** −49 **83.** 0 **84.** −8 **85.** −16 **86.** 35 **87.** −32 **88.** −8 **89.** 7 **90.** −14 **91.** 39 **92.** −117 **93.** −2 **94.** −12 **95.** −3 **96.** −35 **97.** −5 **98.** 5 **99.** −1 **100.** −7 **101.** no **102.** yes **103.** −13 **104.** −20 **105.** −3 **106.** −9 **107.** −13 **108.** −31 **109.** 44 **110.** −26 **111.** −19 **112.** 38 **113.** 6 **114.** −5 **115.** −15 **116.** −19 **117.** 48 **118.** −21 **119.** 21 **120.** −5 **121.** −$9 **122.** 6°C **123.** 13,118 ft **124.** −27°C **125.** 2 **126.** 3 **127.** −5 **128.** −25 **129.** −20 **130.** 17 **131.** −21 **132.** −17 **133.** 12 **134.** −9 **135.** −200 **136.** 3

Chapter Test **1.** 3 **2.** −6 **3.** −100 **4.** 4 **5.** −30 **6.** 12 **7.** 65 **8.** 5 **9.** 12 **10.** −6 **11.** 50 **12.** −2 **13.** −11 **14.** −46 **15.** −117 **16.** 3456 **17.** 28 **18.** −213 **19.** −2 **20.** 2 **21.** −5 **22.** −32 **23.** −17 **24.** 1 **25.** −1 **26.** 88 feet below sea level **27.** 45 **28.** 31,642 **29.** 3820 ft below sea level **30.** −4 **31. a.** $17x$ **b.** $20 - 2x$ **32.** 5 **33.** −28 **34.** −20 **35.** −4

Cumulative Review **1.** hundred-thousands (Sec. 1.2, Ex. 1) **2.** hundreds (Sec. 1.2) **3.** thousands (Sec. 1.2, Ex. 2) **4.** thousands (Sec. 1.2) **5.** ten-millions (Sec. 1.2, Ex. 3) **6.** hundred-thousands (Sec. 1.2) **7. a.** < **b.** > **c.** > (Sec. 1.2, Ex. 3) **8. a.** > **b.** < **c.** > (Sec. 1.2) **9.** 39 (Sec. 1.3, Ex. 3) **10.** 39 (Sec. 1.3) **11.** 7321 (Sec. 1.3, Ex. 6) **12.** 3013 (Sec. 1.4) **13.** 36,184 mi (Sec. 1.3, Ex. 11) **14.** $525 (Sec. 1.3) **15.** 570 (Sec. 1.4, Ex. 1) **16.** 600 (Sec. 1.4) **17.** 1800 (Sec. 1.4, Ex. 5) **18.** 5000 (Sec. 1.4) **19. a.** $5 \cdot 6 + 5 \cdot 5$ **b.** $20 \cdot 4 + 20 \cdot 7$ **c.** $2 \cdot 7 + 2 \cdot 9$ (Sec. 1.5, Ex. 2) **20. a.** $5 \cdot 2 + 5 \cdot 12$ **b.** $9 \cdot 3 + 9 \cdot 6$ **c.** $4 \cdot 8 + 4 \cdot 1$ (Sec. 1.5) **21.** 78,875 (Sec. 1.5, Ex. 5) **22.** 31,096 (Sec. 1.5) **23. a.** 6 **b.** 8 **c.** 7 (Sec. 1.6, Ex. 1) **24. a.** 7 **b.** 8 **c.** 12 (Sec. 1.6) **25.** 741 (Sec. 1.6, Ex. 4) **26.** 456 (Sec. 1.6) **27.** 2 cards each; 10 cards left over (Sec. 1.6, Ex. 11) **28.** $9 (Sec. 1.6) **29.** 81 (Sec. 1.7, Ex. 5) **30.** 125 (Sec. 1.7) **31.** 6 (Sec. 1.7, Ex. 6) **32.** 4 (Sec. 1.7) **33.** 180 (Sec. 1.7, Ex. 8) **34.** 56 (Sec. 1.7) **35.** 2 (Sec. 1.7, Ex. 16) **36.** 5 (Sec. 1.7) **37.** 14 (Sec. 1.8, Ex. 1) **38.** 14 (Sec. 1.8) **39. a.** 9 **b.** 3 **c.** 0 (Sec. 2.1, Ex. 4) **40. a.** 4 **b.** 7 (Sec. 2.1) **41.** 3 (Sec. 2.2, Ex. 7) **42.** 5 (Sec. 2.2) **43.** 22 (Sec. 2.3, Ex. 12) **44.** 5 (Sec. 2.3) **45.** −21 (Sec. 2.4, Ex. 1) **46.** −10 (Sec. 2.4) **47.** 0 (Sec. 2.4, Ex. 3) **48.** −54 (Sec. 2.4) **49.** −16 (Sec. 2.5, Ex. 8) **50.** −27 (Sec. 2.5)

CHAPTER 3 Solving Equations and Problem Solving

Vocabulary and Readiness Check **1.** expression; term **3.** combine like terms **5.** variable; constant **7.** associative **9.** numerical coefficient

Exercise Set 3.1 **1.** $8x$ **3.** $-n$ **5.** $-2c$ **7.** $-6x$ **9.** $12a - 5$ **11.** $42x$ **13.** $-33y$ **15.** $72a$ **17.** $2y + 6$ **19.** $3a - 18$ **21.** $-12x - 28$ **23.** $2x + 1$ **25.** $15c + 3$ **27.** $-21n + 20$ **29.** $7w + 15$ **31.** $11x - 8$ **33.** $-2y + 16$ **35.** $-2y$ **37.** $-7z$ **39.** $8d - 3c$ **41.** $6y - 14$ **43.** $-q$ **45.** $2x + 22$ **47.** $-3x - 35$ **49.** $-3z - 15$ **51.** $-6x + 6$ **53.** $3x - 30$ **55.** $-r + 8$ **57.** $-7n + 3$ **59.** $9z - 14$ **61.** −6 **63.** $-4x + 10$ **65.** $2xy + 20$ **67.** $7a + 12$ **69.** $3y + 5$ **71.** $(14y + 22)$ m **73.** $(11a + 12)$ ft **75.** $(-25x + 55)$ in. **77.** $36y$ sq in. **79.** $(32x - 64)$ sq km **81.** $(60y + 20)$ sq mi **83.** 2000 sq ft **85.** 64 ft **87.** $(3x + 6)$ ft **89.** −3 **91.** 8 **93.** 0 **95.** incorrect; $15x - 10$ **97.** correct **99.** distributive **101.** associative **103.** $(20x + 16)$ sq mi **105.** $4824q + 12{,}274$ **107.** answers may vary **109.** answers may vary

Vocabulary and Readiness Check **1.** equivalent **3.** simplifying **5.** addition

Exercise Set 3.2 **1.** 6 **3.** 8 **5.** −4 **7.** 6 **9.** −1 **11.** 18 **13.** −8 **15.** −50 **17.** 3 **19.** −22 **21.** −6 **23.** 24 **25.** −30 **27.** −4 **29.** 4 **31.** 3 **33.** −3 **35.** 1 **37.** 5 **39.** −9 **41.** −4 **43.** −3 **45.** −1 **47.** 0 **49.** 3 **51.** −6 **53.** −35 **55.** 10 **57.** −2 **59.** 0 **61.** 28 **63.** −5 **65.** −28 **67.** −28 **69.** 5 **71.** −4 **73.** $-7 + x$ **75.** $x - 11$ **77.** $-13x$ **79.** $\frac{x}{-12}$ or $-\frac{x}{12}$ **81.** $-11x + 5$ **83.** $-10 - 7x$ **85.** $4x + 7$ **87.** $2x - 17$ **89.** $-6(x + 15)$ **91.** $\frac{45}{-5x}$ **93.** $\frac{17}{x} + (-15)$ **95.** 2005 **97.** 35,000 **99.** answers may vary **101.** no; answers may vary **103.** answers may vary **105.** 67,896 **107.** −48 **109.** 42

Integrated Review **1.** expression **2.** equation **3.** equation **4.** expression **5.** simplify **6.** solve **7.** $8x$ **8.** $-4y$ **9.** $-2a - 2$ **10.** $5a - 26$ **11.** $-8x - 14$ **12.** $-6x + 30$ **13.** $5y - 10$ **14.** $15x - 31$ **15.** $(12x - 6)$ sq m **16.** $(2x + 9)$ ft **17.** −4 **18.** −3 **19.** −10 **20.** 6 **21.** −15 **22.** −120 **23.** −5 **24.** −13 **25.** −24 **26.** −54 **27.** 12 **28.** −42 **29.** 2 **30.** 2 **31.** −3 **32.** 5 **33.** −5 **34.** 6 **35.** $x - 10$ **36.** $-20 + x$ **37.** $10x$ **38.** $\frac{10}{x}$ **39.** $-2x + 5$ **40.** $-4(x - 1)$

Calculator Explorations **1.** yes **3.** no **5.** yes

Vocabulary and Readiness Check **1.** $3x - 9 + x - 16$; $5(2x + 6) - 1 = 39$ **3.** addition **5.** distributive

Exercise Set 3.3 **1.** -12 **3.** -3 **5.** 1 **7.** -45 **9.** -9 **11.** 6 **13.** 8 **15.** 5 **17.** 0 **19.** -5 **21.** -22 **23.** 6 **25.** -11 **27.** -7 **29.** -5 **31.** 270 **33.** 5 **35.** 3 **37.** 9 **39.** -6 **41.** 11 **43.** 3 **45.** 4 **47.** -4 **49.** 3 **51.** -1 **53.** -4 **55.** -5 **57.** 0 **59.** 4 **61.** 1 **63.** -30 **65.** $-42 + 16 = -26$ **67.** $-5(-29) = 145$ **69.** $3(-14 - 2) = -48$ **71.** $\frac{100}{2(50)} = 1$ **73.** 97 million returns **75.** 20 million returns **77.** 33 **79.** -37 **81.** b **83.** a **85.** $6x - 10 = 5x - 7$; $6x = 5x + 3$; $x = 3$ **87.** -81 **89.** 891 **91.** no; answers may vary

The Bigger Picture **1.** -5 **2.** 11 **3.** -7 **4.** 103 **5.** 75 **6.** -11 **7.** -6 **8.** 9

Exercise Set 3.4 **1.** $-5 + x = -7$ **3.** $3x = 27$ **5.** $-20 - x = 104$ **7.** $2x = 108$ **9.** $5(-3 + x) = -20$ **11.** $9 + 3x = 33; 8$ **13.** $3 + 4 + x = 16; 9$ **15.** $x - 3 = \frac{10}{5}; 5$ **17.** $30 - x = 3(x + 6); 3$ **19.** $5x - 40 = x + 8; 12$ **21.** $3(x - 5) = \frac{108}{12}; 8$ **23.** $4x = 30 - 2x; 5$ **25.** California: 55 votes; Florida: 27 **27.** falcon: 185 mph; pheasant: 37 mph **29.** India: 8407; US: 5758 **31.** Xbox: \$420; games: \$140 **33.** 5470 mi **35.** Michigan Stadium: 107,501; Neyland Stadium: 102,854 **37.** China: 140 million; Spain: 70 million **39.** Spain: 6609 cars per day; Germany: 13,218 cars per day **41.** \$225 **43.** 78 **45.** 590 **47.** 1000 **49.** 3000 **51.** answers may vary **53.** \$8250 **55.** \$5

Chapter 3 Vocabulary Check **1.** simplified; combined **2.** like **3.** variable **4.** algebraic expression **5.** terms **6.** numerical coefficient **7.** evaluating the expression **8.** constant **9.** equation **10.** solution **11.** distributive **12.** multiplication **13.** addition

Chapter Review **1.** $10y - 15$ **2.** $-6y - 10$ **3.** $-6a - 7$ **4.** $-8y + 2$ **5.** $2x + 10$ **6.** $-3y - 24$ **7.** $11x - 12$ **8.** $-4m - 12$ **9.** $-5a + 4$ **10.** $12y - 9$ **11.** $16y - 5$ **12.** $x - 2$ **13.** $(4x + 6)$yd **14.** $20y$ m **15.** $(6x - 3)$ sq yd **16.** $(45x + 8)$ sq cm **17.** -2 **18.** 10 **19.** -7 **20.** 5 **21.** -12 **22.** 45 **23.** -6 **24.** -1 **25.** -25 **26.** -8 **27.** -2 **28.** -2 **29.** -8 **30.** -45 **31.** 5 **32.** -5 **33.** -63 **34.** -15 **35.** 5 **36.** 12 **37.** -6 **38.** 4 **39.** $-5x$ **40.** $x - 3$ **41.** $-5 + x$ **42.** $\frac{-2}{x}$ **43.** $2x + 11$ **44.** $-5x - 50$ **45.** $\frac{70}{x + 6}$ **46.** $2(x - 13)$ **47.** 21 **48.** -10 **49.** 2 **50.** 2 **51.** 11 **52.** -5 **53.** -15 **54.** 10 **55.** -2 **56.** -6 **57.** -1 **58.** 1 **59.** 0 **60.** 20 **61.** $20 - (-8) = 28$ **62.** $5[2 + (-6)] = -20$ **63.** $\frac{-75}{5 + 20} = -3$ **64.** $-2 - 19 = -21$ **65.** $2x - 8 = 40$ **66.** $6x = x + 2$ **67.** $\frac{x}{2} - 12 = 10$ **68.** $x - 3 = \frac{x}{4}$ **69.** 5 **70.** -16 **71.** 2386 votes **72.** 84 DVDs **73.** $-11x$ **74.** $-35x$ **75.** $22x - 19$ **76.** $-9x - 32$ **77.** -1 **78.** -25 **79.** 13 **80.** -6 **81.** -22 **82.** -6 **83.** -15 **84.** 18 **85.** -5 **86.** 2 **87.** 11 **88.** -1 **89.** 0 **90.** -6 **91.** 5 **92.** 1 **93.** Hawaii: 4309 mi; Delaware: 5894 mi **94.** North Dakota: 86,782 mi; South Dakota: 83,688 mi

Chapter Test **1.** $-5x + 5$ **2.** $-6y - 14$ **3.** $-8z - 20$ **4.** $(15x + 15)$ in. **5.** $(12x - 4)$ sq m **6.** -6 **7.** -6 **8.** 24 **9.** -2 **10.** 6 **11.** 3 **12.** -2 **13.** 0 **14.** 4 **15.** $-23 + x$ **16.** $-2 - 3x$ **17.** $2 \cdot 5 + (-15) = -5$ **18.** $3x + 6 = -30$ **19.** -2 **20.** 8 free throws **21.** 244 women

Cumulative Review **1.** three hundred eight million, sixty-three thousand, five hundred fifty-seven (Sec. 1.2, Ex. 7) **2.** two hundred seventy-six thousand, four (Sec. 1.2) **3.** 13 in. (Sec. 1.3, Ex. 9) **4.** 18 in. (Sec. 1.3) **5.** 726 (Sec. 1.3, Ex. 8) **6.** 9585 (Sec. 1.3) **7.** 249,000 (Sec. 1.4, Ex. 3) **8.** 844,000 (Sec. 1.4) **9.** 200 (Sec. 1.5, Ex. 3a) **10.** 29,230 (Sec. 1.5) **11.** 208 (Sec. 1.6, Ex. 5) **12.** 86 (Sec. 1.6) **13.** 7 (Sec. 1.7, Ex. 9) **14.** 35 (Sec. 1.7) **15.** 26 (Sec. 1.8, Ex. 4) **16.** 10 (Sec. 1.8) **17.** 20 is a solution. (Sec. 1.8, Ex. 7) **18. a.** $<$; **b.** $>$ (Sec. 2.1) **19.** 3 (Sec. 2.2, Ex. 1) **20.** -7 (Sec. 2.2) **21.** -25 (Sec. 2.2, Ex. 5) **22.** -4 (Sec. 2.2) **23.** 3 (Sec. 2.2, Ex. 7) **24.** 17 (Sec. 2.2) **25.** -14 (Sec. 2.3, Ex. 2) **26.** -5 (Sec. 2.3) **27.** 11 (Sec. 2.3, Ex. 3) **28.** 29 (Sec. 2.3) **29.** -4 (Sec. 2.3, Ex. 4) **30.** -3 (Sec. 2.3) **31.** -2 (Sec. 2.4, Ex. 10) **32.** 6 (Sec. 2.4) **33.** 5 (Sec. 2.4, Ex. 11) **34.** -13 (Sec. 2.4) **35.** -16 (Sec. 2.4, Ex. 12) **36.** -10 (Sec. 2.4) **37.** 9 (Sec. 2.5, Ex. 1) **38.** -32 (Sec. 2.5) **39.** -9 (Sec. 2.5, Ex. 2) **40.** 25 (Sec. 2.5) **41.** $6y + 2$ (Sec. 3.1, Ex. 2) **42.** $3x + 9$ (Sec. 3.1) **43.** not a solution (Sec. 2.6, Ex. 1) **44.** solution (Sec. 2.6) **45.** 3 (Sec. 2.6, Ex. 7) **46.** -5 (Sec. 2.6) **47.** 12 (Sec. 3.2, Ex. 7) **48.** -2 (Sec. 3.2) **49.** software: \$420; computer system: \$1680 (Sec. 3.4, Ex. 4) **50.** 11 (Sec. 3.4)

CHAPTER 4 Fractions and Mixed Numbers

Vocabulary and Readiness Check **1.** fraction; denominator; numerator **3.** improper; proper; mixed number

Exercise Set 4.1 **1.** numerator: 1; denominator: 2; proper **3.** numerator: 10; denominator: 3; improper **5.** numerator: 15; denominator: 15; improper **7.** $\frac{1}{3}$ **9. a.** $\frac{11}{4}$ **b.** $2\frac{3}{4}$ **11. a.** $\frac{23}{6}$ **b.** $3\frac{5}{6}$ **13.** $\frac{7}{12}$ **15.** $\frac{3}{7}$ **17.** $\frac{4}{9}$ **19. a.** $\frac{4}{3}$ **b.** $1\frac{1}{3}$ **21. a.** $\frac{11}{2}$ **b.** $-5\frac{1}{2}$ **23.** $\frac{1}{6}$ **25.** $\frac{5}{8}$ **27.** 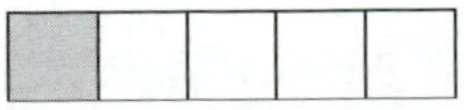 **29.**  **31.** 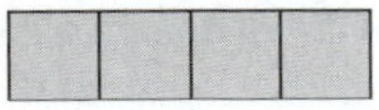 **33.** $\frac{42}{131}$ **35. a.** 89 **b.** $\frac{89}{131}$ **37.** $\frac{8}{43}$ **39.** $\frac{15}{28}$ of the tropical storms **41.** $\frac{11}{31}$ of the month **43.** $\frac{10}{31}$ of the class **45. a.** $\frac{33}{50}$ of the states **b.** 17 states **c.** $\frac{17}{50}$ of the states **47. a.** $\frac{21}{50}$ **b.** 29 **c.** $\frac{29}{50}$ **49.** 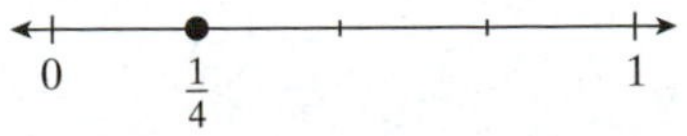

51. 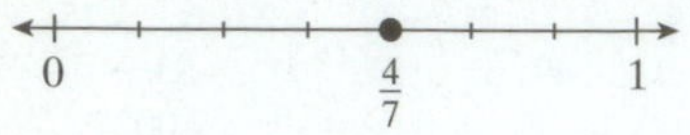 number line from 0 to 1 with point at $\frac{4}{7}$ **53.** 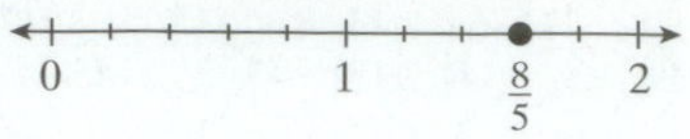 number line from 0 to 2 with point at $\frac{8}{5}$ **55.** 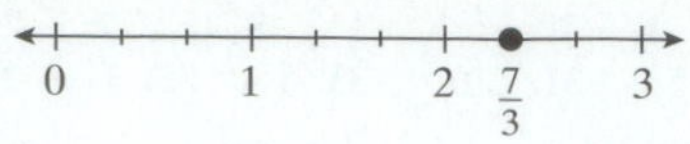 number line from 0 to 3 with point at $\frac{7}{3}$ **57.** 1

59. −5 **61.** 0 **63.** 1 **65.** undefined **67.** 3 **69.** $\frac{7}{3}$ **71.** $\frac{18}{5}$ **73.** $\frac{53}{8}$ **75.** $\frac{83}{7}$ **77.** $\frac{187}{20}$ **79.** $\frac{500}{3}$ **81.** $3\frac{2}{5}$ **83.** $4\frac{5}{8}$

85. $3\frac{2}{15}$ **87.** 15 **89.** $1\frac{7}{175}$ **91.** $6\frac{65}{112}$ **93.** 9 **95.** 125 **97.** $\frac{-11}{2}; \frac{11}{-2}$ **99.** $\frac{13}{-15}; -\frac{13}{15}$ **101.** answers may vary

103.  ▲ ▲ △ △ △ △ **105.** 6 **107.** $\frac{57}{171}$ of the licensees **109.** $\frac{5}{13}$ of the training centers

Calculator Explorations **1.** $\frac{4}{7}$ **3.** $\frac{20}{27}$ **5.** $\frac{8}{15}$ **7.** $\frac{2}{9}$

Vocabulary and Readiness Check **1.** prime factorization **3.** prime **5.** equivalent

Exercise Set 4.2 **1.** $2^2 \cdot 5$ **3.** $2^4 \cdot 3$ **5.** 3^4 **7.** $2 \cdot 3^4$ **9.** $2 \cdot 5 \cdot 11$ **11.** $5 \cdot 17$ **13.** $2^4 \cdot 3 \cdot 5$ **15.** $2^2 \cdot 3^2 \cdot 23$ **17.** $\frac{1}{4}$ **19.** $\frac{2x}{21}$ **21.** $\frac{7}{8}$ **23.** $\frac{2}{3}$ **25.** $\frac{7}{10}$ **27.** $-\frac{7}{9}$ **29.** $\frac{5x}{6}$ **31.** $\frac{27}{64}$ **33.** $\frac{5x}{8}$ **35.** $-\frac{5}{8}$ **37.** $\frac{3x^2y}{2}$ **39.** $\frac{3}{4}$ **41.** $\frac{5}{8z}$ **43.** $\frac{3}{14}$ **45.** $-\frac{11}{17y}$ **47.** $\frac{7}{8}$ **49.** $14a^2$ **51.** equivalent **53.** not equivalent **55.** equivalent **57.** equivalent **59.** not equivalent **61.** not equivalent **63.** $\frac{1}{4}$ of a shift **65.** $\frac{1}{2}$ mi **67. a.** $\frac{3}{10}$ **b.** 35 states **c.** $\frac{7}{10}$ **69.** $\frac{5}{12}$ of the wall **71. a.** 28 **b.** $\frac{14}{25}$ **73.** $\frac{291}{449}$ of individuals **75.** −3 **77.** −14 **79.** answers may vary **81.** $\frac{3}{5}$ **83.** $\frac{9}{25}$ **85.** $\frac{1}{25}$ **87.** $2^2 \cdot 3^5 \cdot 5 \cdot 7$ **89.** answers may vary **91.** no; answers may vary **93.** $\frac{2}{25}$ **95.** answers may vary **97.** 2235, 105, 900, 1470 **99.** 15; answers may vary

Vocabulary and Readiness Check **1.** $\frac{a \cdot c}{b \cdot d}$ **3.** $\frac{2 \cdot 2 \cdot 2}{7}; \frac{2}{7} \cdot \frac{2}{7} \cdot \frac{2}{7}$ **5.** $\frac{a \cdot d}{b \cdot c}$

Exercise Set 4.3 **1.** $\frac{18}{77}$ **3.** $-\frac{5}{28}$ **5.** $\frac{1}{15}$ **7.** $\frac{18x}{55}$ **9.** $\frac{3a^2}{4}$ **11.** $\frac{x^2}{y}$ **13.** 0 **15.** $-\frac{17}{25}$ **17.** $\frac{1}{56}$ **19.** $\frac{1}{125}$ **21.** $\frac{4}{9}$ **23.** $-\frac{4}{27}$ **25.** $\frac{4}{5}$ **27.** $-\frac{1}{6}$ **29.** $-\frac{16}{9x}$ **31.** $\frac{121y}{60}$ **33.** $-\frac{1}{6}$ **35.** $\frac{x}{25}$ **37.** $\frac{10}{27}$ **39.** $\frac{18x^2}{35}$ **41.** $\frac{10}{9}$ **43.** $-\frac{1}{4}$ **45.** $\frac{9}{16}$ **47.** xy^2 **49.** $\frac{77}{2}$ **51.** $-\frac{36}{x}$ **53.** $\frac{3}{49}$ **55.** $-\frac{19y}{7}$ **57.** $\frac{4}{11}$ **59.** $\frac{8}{3}$ **61.** $\frac{15x}{4}$ **63.** $\frac{8}{9}$ **65.** $\frac{1}{60}$ **67.** b **69.** $\frac{2}{5}$ **71.** $-\frac{5}{3}$ **73. a.** $\frac{1}{3}$ **b.** $\frac{12}{25}$ **75. a.** $-\frac{36}{55}$ **b.** $-\frac{44}{45}$ **77.** yes **79.** no **81.** 50 **83.** 20 **85.** 128 **87.** 868 mi **89.** $\frac{3}{16}$ in. **91.** $1838 **93.** 20 contestants **95.** $\frac{1}{14}$ sq ft **97.** 3840 mi **99.** 2400 mi **101.** 201 **103.** 196 **105.** answers may vary **107.** 5 **109.** 34,897,200 people **111.** 37

The Bigger Picture **1.** $\frac{16}{27}$ **2.** $-\frac{3}{4}$ **3.** $\frac{4}{3}$ **4.** $\frac{12}{5}$ **5.** 8 **6.** 72 **7.** $\frac{1}{24}$ **8.** 40 **9.** 35 **10.** 24 **11.** 0 **12.** −6

Vocabulary and Readiness Check **1.** like; unlike **3.** $-\frac{a}{b}$ **5.** least common denominator (LCD)

Exercise Set 4.4 **1.** $\frac{7}{11}$ **3.** $\frac{2}{3}$ **5.** $-\frac{1}{4}$ **7.** $-\frac{1}{2}$ **9.** $\frac{2}{3x}$ **11.** $-\frac{x}{9}$ **13.** $\frac{6}{11}$ **15.** $\frac{3}{4}$ **17.** $-\frac{3}{y}$ **19.** $-\frac{19}{33}$ **21.** $-\frac{1}{3}$ **23.** $\frac{7a-3}{4}$ **25.** $\frac{9}{10}$ **27.** $-\frac{13x}{14}$ **29.** $\frac{9x+1}{15}$ **31.** $-\frac{x}{2}$ **33.** $-\frac{2}{3}$ **35.** $\frac{3x}{4}$ **37.** $\frac{5}{4}$ **39.** $\frac{2}{5}$ **41.** 1 in. **43.** 2 m **45.** $\frac{7}{25}$ **47.** $\frac{1}{50}$ **49.** $\frac{7}{10}$ mi **51.** $\frac{2}{5}$ **53.** 45 **55.** 72 **57.** 150 **59.** $24x$ **61.** 126 **63.** 168 **65.** 14 **67.** 20 **69.** 25 **71.** $56x$ **73.** $8y$ **75.** $20a$ **77.** $\frac{89}{100}; \frac{90}{100}; \frac{96}{100}; \frac{92}{100}; \frac{68}{100}; \frac{91}{100}; \frac{80}{100}; \frac{69}{100}; \frac{8}{100}; \frac{70}{100}$ **79.** India **81.** 9 **83.** 125 **85.** 49 **87.** 24 **89.** $\frac{2}{7} + \frac{9}{7} = \frac{11}{7}$ **91.** answers may vary **93.** 1; answers may vary **95.** $814x$ **97.** answers may vary **99.** a, b, and d

Calculator Explorations **1.** $\frac{37}{80}$ **3.** $\frac{95}{72}$ **5.** $\frac{394}{323}$

Vocabulary and Readiness Check **1.** equivalent; least common denominator **3.** $\frac{4}{24}; \frac{15}{24}; \frac{19}{24}$ **5.** expression; equation

Exercise Set 4.5 **1.** $\frac{5}{6}$ **3.** $\frac{1}{6}$ **5.** $-\frac{4}{33}$ **7.** $-\frac{3}{14}$ **9.** $\frac{3x}{5}$ **11.** $\frac{24-y}{12}$ **13.** $\frac{11}{36}$ **15.** $-\frac{44}{7}$ **17.** $\frac{89a}{99}$ **19.** $\frac{4y-1}{6}$ **21.** $\frac{x+6}{2x}$ **23.** $-\frac{8}{33}$ **25.** $\frac{3}{14}$ **27.** $\frac{11y-10}{35}$ **29.** $-\frac{11}{36}$ **31.** $\frac{1}{20}$ **33.** $\frac{33}{56}$ **35.** $\frac{17}{16}$ **37.** $-\frac{2}{9}$ **39.** $-\frac{11}{30}$ **41.** $\frac{15+11y}{33}$ **43.** $-\frac{53}{42}$

45. $\frac{7x}{8}$ **47.** $\frac{11}{18}$ **49.** $\frac{44a}{39}$ **51.** $-\frac{11}{60}$ **53.** $\frac{5y+9}{9y}$ **55.** $\frac{19}{20}$ **57.** $\frac{56}{45}$ **59.** $\frac{40+9x}{72x}$ **61.** $-\frac{5}{24}$ **63.** $\frac{37x-20}{56}$ **65.** $<$ **67.** $>$ **69.** $>$ **71.** $\frac{13}{12}$ **73.** $\frac{1}{4}$ **75.** $\frac{11}{6}$ **77.** $\frac{34}{15}$ or $2\frac{4}{15}$ cm **79.** $\frac{17}{10}$ or $1\frac{7}{10}$ m **81.** $x+\frac{1}{2}$ **83.** $-\frac{3}{8}-x$ **85.** $\frac{5}{8}$ in. **87.** $\frac{7}{100}$ mph **89.** $\frac{47}{32}$ in. **91.** $\frac{49}{100}$ of students **93.** $\frac{19}{25}$ **95.** $\frac{11}{50}$ **97.** $\frac{21}{25}$ **99.** -20 **101.** -6 **103.** $\frac{3}{5}+\frac{4}{5}=\frac{7}{5}$. **105.** $\frac{223}{540}$ **107.** $\frac{49}{44}$ **109.** answers may vary **111.** standard mail

Integrated Review **1.** $\frac{3}{7}$ **2.** $\frac{5}{4}$ or $1\frac{1}{4}$ **3.** $\frac{73}{85}$ **4.**  **5.** -1 **6.** 17 **7.** 0 **8.** undefined **9.** $5\cdot 13$ **10.** $2\cdot 5\cdot 7$ **11.** $3^2\cdot 5\cdot 7$ **12.** $3^2\cdot 7^2$ **13.** $\frac{1}{7}$ **14.** $\frac{6}{5}$ **15.** $-\frac{14}{15}$ **16.** $-\frac{9}{10}$ **17.** $\frac{2x}{5}$ **18.** $\frac{3}{8y}$ **19.** $\frac{11z^2}{14}$ **20.** $\frac{7}{11ab^2}$ **21.** not equivalent **22.** equivalent **23. a.** $\frac{1}{25}$ **b.** 48 **c.** $\frac{24}{25}$ **24. a.** $\frac{1}{5}$ **b.** 368 **c.** $\frac{4}{5}$ **25.** 30 **26.** 14 **27.** 90 **28.** $\frac{28}{36}$ **29.** $\frac{55}{75}$ **30.** $\frac{40}{48}$ **31.** $\frac{4}{5}$ **32.** $-\frac{2}{5}$ **33.** $\frac{3}{25}$ **34.** $\frac{1}{3}$ **35.** $\frac{4}{5}$ **36.** $\frac{5}{9}$ **37.** $-\frac{1}{6y}$ **38.** $\frac{3x}{2}$ **39.** $\frac{1}{18}$ **40.** $\frac{4}{21}$ **41.** $-\frac{7}{48z}$ **42.** $-\frac{9}{50}$ **43.** $\frac{37}{40}$ **44.** $\frac{11}{36}$ **45.** $\frac{11}{18}$ **46.** $\frac{25y+12}{50}$ **47.** $\frac{2}{3}\cdot x$ or $\frac{2}{3}x$ **48.** $x\div\left(-\frac{1}{5}\right)$ **49.** $-\frac{8}{9}-x$ **50.** $\frac{6}{11}+x$ **51.** 1020 **52.** 24 lots **53.** $\frac{3}{4}$ ft

Vocabulary and Readiness Check **1.** complex **3.** division **5.** addition

Exercise Set 4.6 **1.** $\frac{1}{6}$ **3.** $\frac{7}{3}$ **5.** $\frac{x}{6}$ **7.** $\frac{23}{22}$ **9.** $\frac{2x}{13}$ **11.** $\frac{17}{60}$ **13.** $\frac{5}{8}$ **15.** $\frac{35}{9}$ **17.** $-\frac{17}{45}$ **19.** $\frac{11}{8}$ **21.** $\frac{29}{10}$ **23.** $\frac{27}{32}$ **25.** $\frac{1}{100}$ **27.** $\frac{9}{64}$ **29.** $\frac{7}{6}$ **31.** $-\frac{2}{5}$ **33.** $-\frac{2}{9}$ **35.** $\frac{5a}{2}$ **37.** $\frac{7}{2}$ **39.** $\frac{9}{20}$ **41.** $-\frac{13}{2}$ **43.** $\frac{9}{25}$ **45.** $-\frac{5}{32}$ **47.** 1 **49.** $-\frac{2}{5}$ **51.** $-\frac{11}{40}$ **53.** $\frac{x+6}{16}$ **55.** $\frac{7}{2}$ or $3\frac{1}{2}$ **57.** $\frac{49}{6}$ or $8\frac{1}{6}$ **59.** no; answers may vary **61.** $\frac{5}{8}$ **63.** $\frac{11}{56}$ **65.** halfway between *a* and *b* **67.** false **69.** false **71.** true **73.** no; answers may vary **75.** addition, multiplication, subtraction, division **77.** subtraction, multiplication, division, addition **79.** $\frac{17}{7}$ **81.** $\frac{18}{5}$

Calculator Explorations **1.** $\frac{280}{11}$ **3.** $\frac{3776}{35}$ **5.** $26\frac{1}{14}$ **7.** $92\frac{3}{10}$

Vocabulary and Readiness Check **1.** mixed number **3.** round

Exercise Set 4.7 **1.**

3.
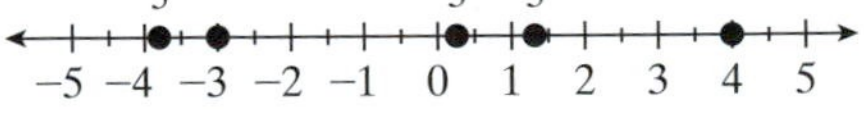

5. b **7.** a **9.** $\frac{8}{21}$ **11.** $4\frac{3}{8}$ **13.** $\frac{77}{10}$ or $7\frac{7}{10}$; 8 **15.** $\frac{836}{35}$ or $23\frac{31}{35}$; 24 **17.** $\frac{25}{2}$ or $12\frac{1}{2}$ **19.** $5\frac{1}{2}$ **21.** $18\frac{2}{3}$ **23.** a **25.** b **27.** $6\frac{2}{3}$; 7 **29.** $13\frac{11}{14}$; 14 **31.** $17\frac{7}{25}$ **33.** $47\frac{53}{84}$ **35.** $25\frac{5}{14}$ **37.** $13\frac{13}{24}$ **39.** $2\frac{3}{5}$; 3 **41.** $7\frac{5}{14}$; 7 **43.** $\frac{24}{25}$ **45.** $3\frac{5}{9}$ **47.** $15\frac{3}{4}$ **49.** 4 **51.** $5\frac{11}{14}$ **53.** $6\frac{2}{9}$ **55.** $\frac{25}{33}$ **57.** $35\frac{13}{18}$ **59.** $2\frac{1}{2}$ **61.** $72\frac{19}{30}$ **63.** $\frac{11}{14}$ **65.** $5\frac{4}{7}$ **67.** $13\frac{16}{33}$ **69.** $-5\frac{2}{7}-x$ **71.** $1\frac{9}{10}\cdot x$ **73.** $9\frac{2}{5}$ in. **75.** $7\frac{13}{20}$ in. **77.** $\frac{7}{2}$ or $3\frac{1}{2}$ sq yd **79.** $\frac{15}{16}$ sq in. **81.** $21\frac{5}{24}$ m **83.** no; she will be $\frac{1}{12}$ ft short **85.** $3\frac{3}{16}$ mi **87.** $4\frac{2}{3}$ m **89.** $9\frac{7}{12}$ min **91.** $1\frac{4}{5}$ min **93.** $-10\frac{3}{25}$ **95.** $-24\frac{7}{8}$ **97.** $-13\frac{59}{60}$ **99.** $-\frac{1}{2}$ **101.** $-1\frac{23}{24}$ **103.** $\frac{73}{1000}$ **105.** $1x$ or x **107.** $1a$ or a **109.** a, b, c **111.** Incorrect; to divide mixed numbers, first write each mixed number as an improper fraction. **113.** answers may vary **115.** Incorrect; $3\frac{2}{3}\cdot 1\frac{1}{7}=\frac{11}{3}\cdot\frac{8}{7}=\frac{88}{21}$ or $4\frac{4}{21}$ **117.** answers may vary

The Bigger Picture **1.** $\frac{5}{17}$ **2.** $\frac{4}{5}$ **3.** $\frac{29}{30}$ **4.** $\frac{1}{24}$ **5.** $1\frac{33}{40}$ **6.** $\frac{27}{64}$ **7.** $12\frac{3}{7}$ **8.** $9\frac{13}{24}$ **9.** $\frac{16}{33}$ **10.** $\frac{34}{27}$ or $1\frac{7}{27}$

Exercise Set 4.8 **1.** $-\frac{2}{3}$ **3.** $\frac{1}{13}$ **5.** $\frac{4}{5}$ **7.** $\frac{11}{12}$ **9.** $-\frac{7}{10}$ **11.** $\frac{11}{16}$ **13.** $\frac{11}{18}$ **15.** $\frac{2}{7}$ **17.** 12 **19.** -27 **21.** $\frac{27}{8}$ **23.** $\frac{1}{21}$ **25.** $\frac{2}{11}$ **27.** $-\frac{3}{10}$ **29.** 1 **31.** $\frac{15}{2}$ **33.** -1 **35.** -15 **37.** $\frac{3x-28}{21}$ **39.** $\frac{y+10}{2}$ **41.** $\frac{7x}{15}$ **43.** $\frac{4}{3}$ **45.** 2 **47.** $\frac{21}{10}$

49. $-\frac{1}{14}$ **51.** -3 **53.** $-\frac{1}{9}$ **55.** 50 **57.** $-\frac{3}{7}$ **59.** 4 **61.** $\frac{3}{5}$ **63.** $-\frac{1}{24}$ **65.** $-\frac{5}{14}$ **67.** 4 **69.** -36 **71.** 57,200 **73.** 330 **75.** answers may vary **77.** -2

Chapter 4 Vocabulary Check **1.** reciprocals **2.** composite number **3.** equivalent **4.** improper fraction **5.** prime number **6.** simplest form **7.** proper fraction **8.** mixed number **9.** numerator; denominator **10.** prime factorization **11.** undefined **12.** 0 **13.** like **14.** least common denominator **15.** complex fraction **16.** cross products

Chapter Review **1.** $\frac{2}{6}$ **2.** $\frac{4}{7}$ **3.** $\frac{7}{3}$ or $2\frac{1}{3}$ **4.** $\frac{13}{4}$ or $3\frac{1}{4}$ **5.** $\frac{11}{12}$ **6.** $\frac{108}{131}$ **7.** -1 **8.** 1 **9.** 0 **10.** undefined **11.** number line from 0 to 1 with point at $\frac{7}{9}$ **12.** number line from 0 to 1 with point at $\frac{4}{7}$ **13.** number line from 0 to 2 with point at $\frac{5}{4}$ **14.** number line from 0 to 2 with point at $\frac{7}{5}$ **15.** $3\frac{3}{4}$ **16.** 3 **17.** $\frac{11}{5}$ **18.** $\frac{35}{9}$ **19.** $\frac{3}{7}$ **20.** $\frac{5}{9}$ **21.** $-\frac{1}{3x}$ **22.** $-\frac{y^2}{2}$ **23.** $\frac{29}{32c}$ **24.** $\frac{18z}{23}$ **25.** $\frac{5x}{3y^2}$ **26.** $\frac{7b}{5c^2}$ **27.** $\frac{2}{3}$ of a foot **28.** $\frac{3}{5}$ of the cars **29.** no **30.** yes **31.** $\frac{3}{10}$ **32.** $-\frac{5}{14}$ **33.** $\frac{9}{x^2}$ **34.** $\frac{y}{2}$ **35.** $-\frac{1}{27}$ **36.** $\frac{25}{144}$ **37.** -2 **38.** $\frac{15}{4}$ **39.** $\frac{27}{2}$ **40.** $-\frac{5}{6y}$ **41.** $\frac{12}{7}$ **42.** $-\frac{63}{10}$ **43.** $\frac{77}{48}$ sq ft **44.** $\frac{4}{9}$ sq m **45.** $\frac{10}{11}$ **46.** $\frac{2}{3}$ **47.** $-\frac{1}{3}$ **48.** $\frac{4x}{5}$ **49.** $\frac{4y-3}{21}$ **50.** $-\frac{1}{15}$ **51.** $3x$ **52.** 24 **53.** 20 **54.** 35 **55.** $49a$ **56.** $45b$ **57.** 40 **58.** 10 **59.** $\frac{3}{4}$ of his homework **60.** $\frac{3}{2}$ mi **61.** $\frac{11}{18}$ **62.** $\frac{7}{26}$ **63.** $-\frac{1}{12}$ **64.** $-\frac{5}{12}$ **65.** $\frac{25x+2}{55}$ **66.** $\frac{4+3b}{15}$ **67.** $\frac{7y}{36}$ **68.** $\frac{11x}{18}$ **69.** $\frac{4y+45}{9y}$ **70.** $-\frac{15}{14}$ **71.** $\frac{91}{150}$ **72.** $\frac{5}{18}$ **73.** $\frac{19}{9}$ m **74.** $\frac{3}{2}$ ft **75.** $\frac{21}{50}$ of the donors **76.** $\frac{1}{4}$ yd **77.** $\frac{4x}{7}$ **78.** $\frac{3y}{11}$ **79.** -2 **80.** $-7y$ **81.** $\frac{15}{4}$ **82.** $-\frac{5}{24}$ **83.** $\frac{8}{13}$ **84.** $-\frac{1}{27}$ **85.** $\frac{29}{110}$ **86.** $-\frac{1}{7}$ **87.** $20\frac{7}{24}$ **88.** $2\frac{51}{55}$; 3 **89.** $\frac{26}{5}$ or $5\frac{1}{5}$; 6 **90.** $\frac{21}{4}$ or $5\frac{1}{4}$ **91.** 22 mi **92.** $\frac{110}{3}$ or $36\frac{2}{3}$ g **93.** each measurement is $4\frac{1}{4}$ in. **94.** $\frac{7}{10}$ yd **95.** $-27\frac{5}{14}$ **96.** $1\frac{5}{27}$ **97.** $-3\frac{15}{16}$ **98.** $-\frac{33}{40}$ **99.** $\frac{5}{6}$ **100.** $-\frac{9}{10}$ **101.** -10 **102.** -6 **103.** 15 **104.** 1 **105.** 5 **106.** -4 **107.** $\frac{1}{4}$ **108.** $\frac{y^2}{2x}$ **109.** $\frac{1}{5}$ **110.** $-\frac{7}{12x}$ **111.** $\frac{11x}{12}$ **112.** $-\frac{23}{55}$ **113.** $-\frac{32}{5}$ or $-6\frac{2}{5}$ **114.** $12\frac{3}{8}$; 13 **115.** $2\frac{19}{35}$; 2 **116.** $\frac{22}{7}$ or $3\frac{1}{7}$ **117.** $-\frac{1}{12}$ **118.** $-\frac{9}{14}$ **119.** $-\frac{4}{9}$ **120.** -19 **121.** $44\frac{1}{2}$ yd **122.** $\frac{81}{2}$ or $40\frac{1}{2}$ sq ft

Chapter Test **1.** $\frac{7}{16}$ **2.** $\frac{23}{3}$ **3.** $18\frac{4}{3}$ **4.** $\frac{4}{35}$ **5.** $-\frac{3x}{5}$ **6.** not equivalent **7.** equivalent **8.** $2^2 \cdot 3 \cdot 7$ **9.** $3^2 \cdot 5 \cdot 11$ **10.** $\frac{4}{3}$ **11.** $-\frac{4}{3}$ **12.** $\frac{8x}{9}$ **13.** $\frac{x-21}{7x}$ **14.** y^2 **15.** $\frac{16}{45}$ **16.** $\frac{9a+4}{10}$ **17.** $-\frac{2}{3y}$ **18.** $\frac{1}{a^2}$ **19.** $\frac{3}{4}$ **20.** $14\frac{1}{40}$ **21.** $16\frac{8}{11}$ **22.** $\frac{64}{3}$ or $21\frac{1}{3}$ **23.** $\frac{45}{2}$ or $22\frac{1}{2}$ **24.** $-\frac{5}{3}$ or $-1\frac{2}{3}$ **25.** $\frac{9}{16}$ **26.** $\frac{3}{8}$ **27.** $\frac{11}{12}$ **28.** $\frac{3}{4x}$ **29.** $\frac{76}{21}$ **30.** -2 **31.** -4 **32.** 1 **33.** $\frac{5}{2}$ **34.** $\frac{4}{31}$ **35.** $3\frac{3}{4}$ ft **36.** $\frac{23}{50}$ **37.** $\frac{13}{50}$ **38.** $2820 **39.** perimeter: $3\frac{1}{3}$ ft; area: $\frac{2}{3}$ sq ft **40.** 24 mi

Cumulative Review **1.** five hundred forty-six (Sec. 1.2, Ex. 5) **2.** one hundred fifteen (Sec. 1.2) **3.** twenty-seven thousand, thirty-four (Sec 1.2, Ex. 6) **4.** six thousand five hundred seventy-three (Sec. 1.2) **5.** 759 (Sec. 1.3, Ex. 1) **6.** 631 (Sec. 1.3) **7.** 514 (Sec. 1.3, Ex. 7) **8.** 933 (Sec. 1.3) **9.** 278,000 (Sec. 1.4, Ex. 2) **10.** 1440 (Sec. 1.4) **11.** 57,600 megabytes (Sec. 1.5, Ex. 7) **12.** 1305 mi (Sec. 1.5) **13.** 7089 R 5 (Sec. 1.6, Ex. 7) **14.** 379 R 10 (Sec. 1.6) **15.** 7^3 (Sec. 1.7, Ex. 1) **16.** 7^2 (Sec. 1.7) **17.** $3^4 \cdot 9^3$ (Sec. 1.7, Ex. 4) **18.** $9^4 \cdot 5^2$ (Sec 1.7) **19.** 6 (Sec. 1.8, Ex. 2) **20.** 52 (Sec. 1.8) **21.** -6824 (Sec. 2.1, Ex. 1) **22.** -21 (Sec. 2.1) **23.** -4 (Sec. 2.2, Ex. 3) **24.** 5 (Sec. 2.2) **25.** 3 (Sec. 2.3, Ex. 9) **26.** 10 (Sec. 2.3) **27.** 25 (Sec. 2.4, Ex. 8) **28.** -16 (Sec. 2.4) **29.** -16 (Sec. 2.5, Ex. 8) **30.** 25 (Sec. 2.5) **31.** $6y + 2$ (Sec. 3.1, Ex. 2) **32.** $6x + 9$ (Sec. 3.1) **33.** -14 (Sec. 3.2, Ex. 4) **34.** -18 (Sec. 3.2) **35.** -1 (Sec. 3.3, Ex. 2) **36.** -11 (Sec. 3.3) **37.** $\frac{2}{5}$ (Sec. 4.1, Ex. 3) **38.** $2^2 \cdot 3 \cdot 13$ (Sec. 4.2) **39. a.** $\frac{38}{9}$ **b.** $\frac{19}{11}$ (Sec. 4.1, Ex. 20) **40.** $\frac{39}{5}$ (Sec. 4.1) **41.** $\frac{7x}{11}$ (Sec. 4.2, Ex. 5) **42.** $\frac{2}{3y}$ (Sec. 4.2) **43.** $\frac{35}{12}$ or $2\frac{11}{12}$ (Sec. 4.7, Ex. 2) **44.** $\frac{8}{3}$ or $2\frac{2}{3}$ (Sec. 4.7) **45.** $\frac{5}{12}$ (Sec. 4.3, Ex. 11) **46.** $\frac{11}{56}$ (Sec. 4.7)

CHAPTER 5 Decimals

Vocabulary and Readiness Check **1.** words; standard form **3.** decimals **5.** tenths; tens

Exercise Set 5.1 **1.** five and sixty-two hundredths **3.** sixteen and twenty-three hundredths **5.** negative two hundred five thousandths **7.** one hundred sixty-seven and nine thousandths **9.** three thousand and four hundredths **11.** one hundred five and six tenths **13.** two and forty-three hundredths

15.

Preprinted Name
Preprinted Address
Current date
DATE

PAY TO THE
ORDER OF *R.W. Financial* $ *321.42*
Three hundred twenty-one and 42/100 DOLLARS

FOR ______ *Signature*

17.

Preprinted Name
Preprinted Address
Current date
DATE

PAY TO THE
ORDER OF *Bell South* $ *59.68*
Fifty-nine and 68/100 DOLLARS

FOR ______ *Signature*

19. 2.8 **21.** 9.08 **23.** −705.625 **25.** 0.0046 **27.** $\frac{7}{10}$ **29.** $\frac{27}{100}$ **31.** $\frac{2}{5}$ **33.** $5\frac{2}{5}$ **35.** $-\frac{29}{500}$ **37.** $7\frac{1}{125}$ **39.** $15\frac{401}{500}$ **41.** $\frac{601}{2000}$ **43.** 0.8; $\frac{8}{10}$ or $\frac{4}{5}$ **45.** seventy-seven thousandths; $\frac{77}{1000}$ **47.** $<$ **49.** $<$ **51.** $<$ **53.** $=$ **55.** $<$ **57.** $>$ **59.** $<$ **61.** $>$ **63.** 0.6 **65.** 98,210 **67.** −0.23 **69.** 0.594 **71.** 3.1 **73.** 3.142 **75.** $27 **77.** $0.20 **79.** 10.4 cm **81.** 1.73 hr **83.** $48 **85.** 24.623 hr **87.** 5766 **89.** 35 **91.** b **93.** a **95.** answers may vary **97.** 7.12 **99.** $\frac{26,849,576}{100,000,000,000}$ **101.** answers may vary **103.** answers may vary **105.** 0.0612, 0.0586 **107.** 0.01, 0.0839, 0.09, 0.1 **109.** $20

Calculator Explorations **1.** 328.742 **3.** 5.2414 **5.** 865.392

Vocabulary and Readiness Check **1.** last **3.** like **5.** vertically

Exercise Set 5.2 **1.** 7.8 **3.** 10.35 **5.** 27.0578 **7.** −8.57 **9.** 10.33 **11.** 465.56; $\begin{array}{r} 230 \\ +\,230 \\ \hline 460 \end{array}$ **13.** 115.123; $\begin{array}{r} 100 \\ 6 \\ +\;\;9 \\ \hline 115 \end{array}$ **15.** 50.409 **17.** 4.4 **19.** 15.3 **21.** 598.23 **23.** 1.83; 6 − 4 = 2 **25.** 876.6; $\begin{array}{r} 1000 \\ -\,100 \\ \hline 900 \end{array}$ **27.** 194.4 **29.** −6.32 **31.** −6.15 **33.** 3.1 **35.** 2.9988 **37.** 16.3 **39.** 3.1 **41.** −5.62 **43.** 363.36 **45.** −549.8 **47.** 861.6 **49.** 115.123 **51.** 0.088 **53.** −180.44 **55.** −1.1 **57.** 3.81 **59.** 3.39 **61.** 1.61 **63.** no **65.** yes **67.** no **69.** $6.9x + 6.9$ **71.** $-10.97 + 3.47y$ **73.** $0.17 **75.** 28.56 m **77.** 11.2 in. **79.** $7.52 **81.** 285.8 mph **83.** $3.4 billion **85.** $1498.49 million **87.** 240.8 in. **89.** 67.44 ft **91.** 18.193 mph **93.** Switzerland **95.** 8.1 lb

97.

Country	Pounds of Chocolate per Person
Switzerland	22.0
Norway	16.0
Germany	15.8
United Kingdom	14.5
Belgium	13.9

99. 138 **101.** $\frac{1}{125}$ **103.** 6.08 in. **105.** $1.20 **107.** 1 nickel, 1 dime, and 2 pennies; 3 nickels and 2 pennies; 1 dime and 7 pennies **109.** answers may vary **111.** 0.777 mi **113.** $-109.544x + 15.604$

Vocabulary and Readiness Check **1.** sum **3.** right; zeros **5.** circumference

Exercise Set 5.3 **1.** 1.36 **3.** 0.6 **5.** −17.595 **7.** 55.008 **9.** 28.56; $7 \times 4 = 28$ **11.** 0.1041 **13.** 8.23854; $\begin{array}{r} 1 \\ \times\,8 \\ \hline 8 \end{array}$ **15.** 11.2746 **17.** 65 **19.** 0.83 **21.** −7093 **23.** 709.3 **25.** 0.0983 **27.** 0.02523 **29.** 0.0492 **31.** 14,790 **33.** 1.29 **35.** −9.3762 **37.** 0.5623 **39.** 36.024 **41.** 1,500,000,000 **43.** 49,800,000 **45.** −0.6 **47.** 17.3 **49.** no **51.** yes **53.** 10π cm $\approx$ 31.4 cm **55.** 18.2π yd $\approx$ 57.148 yd **57.** 24.8 g **59.** 7.2 sq in. **61.** 250π ft $\approx$ 785 ft **63.** 135π m $\approx$ 423.9 m **65.** 64.9605 in. **67.** $715.20 **69. a.** 62.8 m and 125.6 m **b.** yes **71.** $340 **73.** 77,800.5 yen **75.** 715.59 euros **77.** 8 **79.** −9 **81.** 3.64 **83.** 3.56 **85.** −0.1105 **87.** 3,831,600 mi **89.** answers may vary **91.** answers may vary

Calculator Explorations **1.** not reasonable **3.** reasonable

Vocabulary and Readiness Check **1.** quotient; divisor; dividend **3.** left; zeros

Exercise Set 5.4 **1.** 4.6 **3.** 0.094 **5.** 300 **7.** 5.8 **9.** 6.6; $6\overline{)36}$ with quotient 6 **11.** 0.413 **13.** −600 **15.** 7 **17.** 4.8 **19.** 2100 **21.** 5.8 **23.** 5.5 **25.** 9.8; $7\overline{)70}$ with quotient 10 **27.** 9.6 **29.** 45 **31.** 54.592 **33.** 0.0055 **35.** 23.87 **37.** 114.0 **39.** 0.83397 **41.** 2.687 **43.** −0.0129 **45.** 12.6 **47.** 1.31 **49.** 0.045625 **51.** 0.413 **53.** −7 **55.** −4.8 **57.** 2100 **59.** 30 **61.** −58,000 **63.** −0.69 **65.** 0.024 **67.** 65 **69.** −5.65 **71.** −7.0625 **73.** yes **75.** no **77.** 11 qt **79.** 5.1 m **81.** 11.4 boxes **83.** 24 tsp **85.** 8 days **87.** 248.1 mi **89.** $5109.62 per hr **91.** $\frac{21}{50}$ **93.** $-\frac{1}{10}$ **95.** 4.26 **97.** 1.578 **99.** −26.66 **101.** 904.29 **103.** c **105.** b **107.** 85.5 **109.** 8.6 ft **111.** answers may vary **113.** 65.2–82.6 knots **115.** 319.64 m

The Bigger Picture **1.** 22.172 **2.** 3.951 **3.** 9133.2 **4.** −6.8 **5.** 1.404 **6.** 8.66 **7.** 0.051 **8.** 2.14 **9.** $\frac{2}{15}$ **10.** $-\frac{16}{75}$

Integrated Review **1.** 2.57 **2.** 4.05 **3.** 8.9 **4.** 3.5 **5.** 0.16 **6.** 0.24 **7.** 0.27 **8.** 0.52 **9.** −4.8 **10.** 6.09 **11.** 75.56 **12.** 289.12 **13.** −24.974 **14.** −43.875 **15.** −8.6 **16.** 5.4 **17.** −280 **18.** 1600 **19.** 224.938 **20.** 145.079 **21.** 0.56 **22.** −0.63 **23.** 27.6092 **24.** 145.6312 **25.** 5.4 **26.** −17.74 **27.** −414.44 **28.** −1295.03 **29.** −34 **30.** −28 **31.** 116.81 **32.** 18.79 **33.** 156.2 **34.** 1.562 **35.** 25.62 **36.** 5.62 **37.** 200 mi **38.** $0.90 **39.** $24.88 billion or $24,880,000,000

Vocabulary and Readiness Check **1.** False **3.** False

Exercise Set 5.5 **1.** 0.2 **3.** 0.68 **5.** 0.75 **7.** −0.08 **9.** 2.25 **11.** $0.91\overline{6}$ **13.** 0.425 **15.** 0.45 **17.** $-0.\overline{3}$ **19.** 0.4375 **21.** $0.\overline{63}$ **23.** 5.85 **25.** 0.624 **27.** −0.33 **29.** 0.44 **31.** 0.6 **33.** 0.62 **35.** 0.71 **37.** 0.02 **39.** < **41.** = **43.** < **45.** < **47.** < **49.** > **51.** < **53.** < **55.** 0.32, 0.34, 0.35 **57.** 0.49, 0.491, 0.498 **59.** 5.23, $\frac{42}{8}$, 5.34 **61.** 0.612, $\frac{5}{8}$, 0.649 **63.** 0.59 **65.** −3 **67.** 5.29 **69.** 9.24 **71.** 0.2025 **73.** −1.29 **75.** −15.4 **77.** −3.7 **79.** 25.65 sq in. **81.** 0.248 sq yd **83.** 5.76 **85.** 5.7 **87.** 3.6 **89.** $\frac{77}{50}$ **91.** $\frac{5}{2}$ **93.** = 1 **95.** > 1 **97.** < 1 **99.** 0.189 **101.** 6000 stations **103.** answers may vary

Exercise Set 5.6 **1.** 5.9 **3.** −0.43 **5.** 10.2 **7.** −4.5 **9.** 4 **11.** 0.45 **13.** 4.2 **15.** −4 **17.** 1.8 **19.** 10 **21.** 7.6 **23.** 60 **25.** −0.07 **27.** 20 **29.** 0.0148 **31.** −8.13 **33.** 1.5 **35.** −1 **37.** −7 **39.** 7 **41.** 53.2 **43.** $3x - 16$ **45.** $\frac{3}{5x}$ **47.** $\frac{32}{55}$ **49.** 3.7 **51.** $6x - 0.61$ **53.** $-2y + 6.8$ **55.** 9.1 **57.** −3 **59.** $-4z + 16.67$ **61.** 15.7 **63.** 5.85 **65.** −2.1 − 10.1 **67.** answers may vary **69.** 7.683 **71.** 4.683

Vocabulary and Readiness Check **1.** average **3.** mean (or average) **5.** grade point average

Exercise Set 5.7 **1.** mean: 21; median: 23; no mode **3.** mean: 8.1; median: 8.2; mode: 8.2 **5.** mean: 0.5; median: 0.5; mode: 0.2 and 0.5 **7.** mean: 370.9; median: 313.5; no mode **9.** 1495 ft **11.** 1415 ft **13.** answers may vary **15.** 2.79 **17.** 3.64 **19.** 6.8 **21.** 6.9 **23.** 88.5 **25.** 73 **27.** 70 and 71 **29.** 9 rates **31.** $\frac{1}{3}$ **33.** $\frac{3}{5y}$ **35.** $\frac{11}{15}$ **37.** 35, 35, 37, 43 **39.** yes; answers may vary

Chapter 5 Vocabulary Check **1.** decimal **2.** numerator; denominator **3.** vertically **4.** and **5.** sum **6.** mode **7.** circumference **8.** median; mean **9.** mean **10.** standard form

Chapter Review **1.** tenths **2.** hundred-thousandths **3.** negative twenty-three and forty-five hundredths **4.** three hundred forty-five hundred-thousandths **5.** one hundred nine and twenty-three hundredths **6.** two hundred and thirty-two millionths **7.** 8.06 **8.** −503.102 **9.** 16,025.0014 **10.** 14.011 **11.** $\frac{4}{25}$ **12.** $-12\frac{23}{1000}$ **13.** 0.00231 **14.** 25.25 **15.** > **16.** = **17.** < **18.** > **19.** 0.6 **20.** 0.94 **21.** −42.90 **22.** 16.349 **23.** 887,000,000 **24.** 600,000 **25.** 18.1 **26.** 5.1 **27.** −7.28 **28.** −12.04 **29.** 320.312 **30.** 148.74236 **31.** 1.7 **32.** 2.49 **33.** −1324.5 **34.** −10.136 **35.** 65.02 **36.** 199.99802 **37.** 52.6 mi **38.** −5.7 **39.** 22.2 in. **40.** 38.9 ft **41.** 72 **42.** 9345 **43.** −78.246 **44.** 73,246.446 **45.** 14π m ≈ 43.96 m **46.** 20π in. ≈ 62.8 in. **47.** 0.0877 **48.** 15.825 **49.** 70 **50.** −0.21 **51.** 8.059 **52.** 30.4 **53.** 0.02365 **54.** −9.3 **55.** 7.3 m **56.** 45 mo **57.** 0.8 **58.** −0.923 **59.** $2.\overline{3}$ or 2.333 **60.** $0.21\overline{6}$ or 0.217 **61.** = **62.** < **63.** < **64.** < **65.** 0.832, 0.837, 0.839 **66.** $\frac{5}{8}$, 0.626, 0.685 **67.** 0.42, $\frac{3}{7}$, 0.43 **68.** $\frac{19}{12}$, 1.63, $\frac{18}{11}$ **69.** −11.94 **70.** 3.89 **71.** −129 **72.** 0.81 **73.** 55 **74.** 7.26 **75.** 6.9 sq ft **76.** 5.46 sq in. **77.** 0.3 **78.** 92.81 **79.** 8.6 **80.** −80 **81.** 1.98 **82.** −1.5 **83.** −20 **84.** 1 **85.** mean: 17.8; median: 14; no mode **86.** mean: 58.1; median: 60; mode: 45 and 86 **87.** mean: 24,500; median: 20,000; mode: 20,000 **88.** mean: 447.3; median: 420; mode: 400 **89.** 3.25 **90.** 2.57 **91.** two hundred and thirty-two ten-thousandths **92.** −16.09 **93.** $\frac{847}{10,000}$ **94.** 0.75, $\frac{6}{7}$, $\frac{8}{9}$ **95.** −0.07 **96.** 0.1125 **97.** 51.057 **98.** > **99.** < **100.** 86.91 **101.** 3.115 **102.** $123.00 **103.** $3646.00 **104.** −1.7 **105.** 5.26 **106.** −12.76 **107.** −14.907 **108.** 8.128 **109.** −7.245 **110.** 4900 **111.** 23.904 **112.** 9600 sq ft **113.** yes **114.** 0.1024 **115.** 3.6 **116.** mean: 74.4; median: 73; mode: none **117.** mean: 619.17; median: 647.5; mode: 327

Chapter Test **1.** forty-five and ninety-two thousandths **2.** 3000.059 **3.** 17.595 **4.** −51.20 **5.** −20.42 **6.** 40.902 **7.** 0.037 **8.** 34.9 **9.** 0.862 **10.** < **11.** < **12.** $\frac{69}{200}$ **13.** $-24\frac{73}{100}$ **14.** −0.5 **15.** 0.941 **16.** 1.93 **17.** −6.2 **18.** $0.5x - 13.4$ **19.** −3 **20.** 3.7 **21.** mean: 38.4; median: 42; no mode **22.** mean: 12.625; median: 12.5; mode: 12 and 16 **23.** 3.07 **24.** 4,583,000,000 **25.** 2.31 sq mi **26.** 18π mi, 56.52 mi **27. a.** 9904 sq ft **b.** 198.08 oz **28.** 54 mi

Cumulative Review Chapters 1–5 **1.** seventy-two (Sec. 1.2, Ex. 4) **2.** one hundred seven (Sec. 1.2) **3.** five hundred forty-six (Sec. 1.2, Ex. 5) **4.** five thousand, twenty-six (Sec. 1.2) **5.** 759 (Sec. 1.3, Ex. 1) **6.** 19 in. (Sec. 1.3) **7.** 514 (Sec. 1.3, Ex. 7) **8.** 121 R 1 (Sec. 1.6) **9.** 278,000 (Sec. 1.4, Ex. 2) **10.** 1, 2, 3, 5, 6, 10, 15, 30 (Sec. 4.2) **11.** 20,296 (Sec. 1.5, Ex. 4) **12.** 0 (Sec. 1.5) **13. a.** 7 **b.** 12 **c.** 1 **d.** 1 **e.** 20 **f.** 1 (Sec. 1.6, Ex. 2) **14.** 25 (Sec. 1.6) **15.** 7 (Sec. 1.7, Ex. 9) **16.** 49 (Sec. 1.7) **17.** 81 (Sec. 1.7, Ex. 5) **18.** 125 (Sec. 1.7) **19.** 81 (Sec. 1.7, Ex. 7) **20.** 1000 (Sec. 1.7) **21.** 2 (Sec. 1.8, Ex. 3) **22.** 6 (Sec. 1.8) **23. a.** −13 **b.** 2 **c.** 0 (Sec. 2.1, Ex. 5) **24. a.** 7 **b.** −4 **c.** 1 (Sec. 2.1) **25.** −23 (Sec. 2.2, Ex. 4) **26.** −22 (Sec. 2.2) **27.** 180 (Sec. 1.7, Ex. 8) **28.** 32 (Sec. 1.7) **29.** −49 (Sec. 2.4, Ex. 9) **30.** −32 (Sec. 2.4) **31.** 25 (Sec. 2.4, Ex. 8) **32.** −9 (Sec. 2.4) **33.** $\frac{4}{3}$ or $1\frac{1}{3}$ (Sec. 4.1, Ex. 10) **34.** $\frac{11}{4}$ or $2\frac{3}{4}$ (Sec. 4.1) **35.** $\frac{11}{4}$ or $2\frac{3}{4}$ (Sec. 4.1, Ex. 11) **36.** $\frac{14}{3}$ or $4\frac{2}{3}$ (Sec. 4.1) **37.** $2^2 \cdot 3^2 \cdot 7$ (Sec. 4.2, Ex. 3) **38.** 62 (Sec. 1.3) **39.** $-\frac{36}{13}$ (Sec. 4.2, Ex. 8) **40.** $\frac{79}{8}$ (Sec. 4.1) **41.** equivalent (Sec. 4.2, Ex. 10) **42.** > (Sec. 4.5) **43.** $\frac{10}{33}$ (Sec. 4.3, Ex. 1) **44.** $\frac{3}{2}$ or $1\frac{1}{2}$ (Sec. 4.7) **45.** $\frac{1}{8}$ (Sec. 4.3, Ex. 2) **46.** 37 (Sec. 4.7) **47.** −24 (Sec. 3.2, Ex. 3) **48.** −8 (Sec. 3.2) **49.** 829.6561 (Sec. 5.2, Ex. 2) **50.** 230.8628 (Sec. 5.2) **51.** 18.408 (Sec. 5.3, Ex. 1) **52.** 28.251 (Sec. 5.3)

CHAPTER 6 Percent

Vocabulary and Readiness Check **1.** proportion, ratio

Exercise Set 6.1 **1.** $\frac{2}{15}$ **3.** $\frac{5}{6}$ **5.** $\frac{5}{12}$ **7.** $\frac{1}{10}$ **9.** $\frac{7}{20}$ **11.** $\frac{19}{18}$ **13.** $\frac{47}{25}$ **15.** $\frac{17}{40}$ **17.** $\frac{4}{9}$ **19.** $\frac{15}{1}$ **21.** answers may vary **23.** 4 **25.** $\frac{50}{9}$ **27.** $\frac{21}{4}$ **29.** 7 **31.** −3 **33.** $\frac{14}{9}$ **35.** 5 **37.** no solution **39.** 123 lb **41.** 165 cal **43.** 120 cal **45. a.** 18 tsp **b.** 6 tbsp **47.** 6 people **49.** 112 ft; 11-in. difference **51.** 102.9 mg **53.** 1248 ft **55.** 434 emergency room visits **57.** 28 workers **59.** 2.4 c **61. a.** 0.1 gal **b.** 13 fl oz **63. a.** 2062.5 mg **b.** no **65.** $2^3 \cdot 5^2$ **67.** 2^5 **69.** 0.8 ml **71.** 1.25 ml

The Bigger Picture **1.** $\frac{1}{4}$ **2.** $\frac{7}{200}$ **3.** $\frac{7}{2}$ or $3\frac{1}{2}$ **4.** $\frac{9}{20}$ **5.** 7.62 **6.** 0.152 **7.** 8 **8.** $\frac{5}{2}$ or $2\frac{1}{2}$ **9.** 0.004 **10.** $\frac{35}{12}$ or $2\frac{11}{12}$

Vocabulary and Readiness Check **1.** percent **3.** percent **5.** 0.01

Exercise Set 6.2 **1.** 96% **3.** football; 37% **5.** 50% **7.** 0.41 **9.** 0.06 **11.** 1.00 or 1 **13.** 0.736 **15.** 0.028 **17.** 0.006 **19.** 3.00 or 3 **21.** 0.3258 **23.** $\frac{2}{25}$ **25.** $\frac{1}{25}$ **27.** $\frac{9}{200}$ **29.** $\frac{7}{4}$ or $1\frac{3}{4}$ **31.** $\frac{1}{16}$ **33.** $\frac{31}{300}$ **35.** $\frac{179}{800}$ **37.** 0.6% **39.** 22% **41.** 530% **43.** 5.6% **45.** 22.28% **47.** 300% **49.** 70% **51.** 70% **53.** 80% **55.** 68% **57.** $37\frac{1}{2}\%$ **59.** $33\frac{1}{3}\%$ **61.** 450% **63.** 190% **65.** 81.82% **67.** 26.67% **69.** 0.6; $\frac{3}{5}$; $23\frac{1}{2}\%$; $\frac{47}{200}$; 80%; 0.8; $0.333\overline{3}$; $\frac{1}{3}$; 87.5%; 0.875; 0.075; $\frac{3}{40}$ **71.** 2; 2; 280%; $2\frac{4}{5}$; 7.05; $7\frac{1}{20}$; 454%; 4.54 **73.** 0.38; $\frac{19}{50}$ **75.** 0.252; $\frac{63}{250}$ **77.** 0.322; $\frac{161}{500}$ **79.** 0.004; $\frac{1}{250}$ **81.** 0.12; $\frac{3}{25}$ **83.** 44.2% **85.** 0.7% **87.** $\frac{39}{1000}$ **89.** $\frac{11}{36}$ **91.** $1\frac{5}{6}$ **93. a.** 52.9% **b.** 52.86% **95.** b, d **97.** 4% **99.** 75% **101.** greater **103.** 0.266; 26.6% **105.** network systems and data communication analysts **107.** 0.49 **109.** answers may vary

Vocabulary and Readiness Check **1.** is **3.** amount; base; percent **5.** greater

Exercise Set 6.3 **1.** $18\% \cdot 81 = x$ **3.** $20\% \cdot x = 105$ **5.** $0.6 = 40\% \cdot x$ **7.** $x \cdot 80 = 3.8$ **9.** $x = 9\% \cdot 43$ **11.** $x \cdot 250 = 150$ **13.** 3.5 **15.** 28.7 **17.** 10 **19.** 600 **21.** 110% **23.** 34% **25.** 1 **27.** 645 **29.** 500 **31.** 5.16% **33.** 25.2 **35.** 35% **37.** 35 **39.** 0.624 **41.** 0.5% **43.** 145 **45.** 63% **47.** 4% **49.** 30 **51.** $3\frac{7}{11}$ **53.** $\frac{17}{12} = \frac{x}{20}$ **55.** $\frac{8}{9} = \frac{14}{x}$ **57.** c **59.** b **61.** What number equals thirty-three and one-third percent of twenty-four? **63.** a **65.** c **67.** a **69.** a **71.** answers may vary **73.** 686.625 **75.** 12,285

The Bigger Picture **1.** $\frac{19}{45}$ **2.** 63 **3.** −0.021 **4.** 8 **5.** 48 **6.** 1.6 **7.** 250% **8.** 45.6 or $45\frac{3}{5}$ **9.** 28% **10.** 180

Vocabulary and Readiness Check **1.** amount; base; percent **3.** amount

Exercise Set 6.4 **1.** $\frac{a}{45} = \frac{98}{100}$ **3.** $\frac{a}{150} = \frac{4}{100}$ **5.** $\frac{14.3}{b} = \frac{26}{100}$ **7.** $\frac{84}{b} = \frac{35}{100}$ **9.** $\frac{70}{400} = \frac{p}{100}$ **11.** $\frac{8.2}{82} = \frac{p}{100}$ **13.** 26 **15.** 18.9 **17.** 600 **19.** 10 **21.** 120% **23.** 28% **25.** 37 **27.** 1.68 **29.** 1000 **31.** 210% **33.** 55.18 **35.** 45% **37.** 75 **39.** 0.864 **41.** 0.5% **43.** 140 **45.** 9.6 **47.** 113% **49.** $-\frac{7}{8}$ **51.** $3\frac{2}{15}$ **53.** 0.7 **55.** 2.19 **57.** answers may vary **59.** no; $a = 16$ **61.** no; 300 or 300% **63.** answers may vary **65.** 80.0%

The Bigger Picture **1.** $\frac{19}{45}$ **2.** 63 **3.** −0.021 **4.** 8 **5.** 48 **6.** 1.6 **7.** 250% **8.** $45\frac{3}{5}$ or 45.6 **9.** 28% **10.** 180

Integrated Review **1.** $\frac{9}{10}$ **2.** $\frac{9}{25}$ **3.** $\frac{43}{50}$ **4.** $\frac{8}{23}$ **5.** $\frac{2}{3}$ **6.** $\frac{13}{4}$ **7.** 25 **8.** 35 **9.** 86 weeks **10.** 7 boxes **11.** 94% **12.** 17% **13.** 37.5% **14.** 350% **15.** 470% **16.** 800% **17.** 45% **18.** 106% **19.** 675% **20.** 325% **21.** 2% **22.** 6% **23.** 0.71 **24.** 0.31 **25.** 0.03 **26.** 0.04 **27.** 2.24 **28.** 7 **29.** 0.029 **30.** 0.066 **31.** 0.07; $\frac{7}{100}$ **32.** 0.05; $\frac{1}{20}$ **33.** 0.068; $\frac{17}{250}$ **34.** 0.1125; $\frac{9}{80}$ **35.** 0.74; $\frac{37}{50}$ **36.** 0.45; $\frac{9}{20}$ **37.** 0.163; $\frac{49}{300}$ **38.** 0.127; $\frac{19}{150}$ **39.** 13.5 **40.** 100 **41.** 350 **42.** 120% **43.** 28% **44.** 76 **45.** 34 **46.** 130% **47.** 46% **48.** 37.8 **49.** 150 **50.** 62

Exercise Set 6.5 **1.** 8.8 lb **3.** 1600 bolts **5.** 14% **7.** 545 **9.** 6.5% **11.** 496 chairs; 5704 chairs **13.** 93,870 physician assistants **15.** 50% **17.** 12.5% **19.** 29.2% **21.** $175,000 **23.** 31.2 hr **25.** $867.87; $20,153.87 **27.** 2224 **29.** 28 million; 63 million **31.** 30; 60% **33.** 52; 80% **35.** 2; 25% **37.** 120; 75% **39.** 44% **41.** 1.3% **43.** 155.2% **45.** 2% **47.** 5.6% **49.** 20.1% **51.** 51.9% **53.** 92.7% **55.** 4.56 **57.** 11.18 **59.** $-\frac{19}{24}$ **61.** $\frac{28}{39}$ **63.** The increased number is double the original number. **65.** percent increase $= \frac{30}{150} = 20\%$ **67.** False; the percents are different.

Vocabulary and Readiness Check **1.** sales tax **3.** commission **5.** sale price

Exercise Set 6.6 **1.** $7.50 **3.** $858.93 **5.** 7% **7. a.** $120 **b.** $130.20 **9.** $177; $1917 **11.** $485 **13.** 6% **15.** $16.10; $246.10 **17.** $53,176.04 **19.** 14% **21.** $4888.50 **23.** $185,500 **25.** $8.90; $80.10 **27.** $98.25; $98.25 **29.** $143.50; $266.50 **31.** $3255; $18,445 **33.** $45; $255 **35.** $27.45; $332.45 **37.** $3.08; $59.08 **39.** $7074 **41.** 8% **43.** 1200 **45.** 132 **47.** 16 **49.** d **51.** $4.00; $6.00; $8.00 **53.** $7.20; $10.80; $14.40 **55.** a discount of 60% is better; answers may vary **57.** $26,838.45

Calculator Explorations **1.** $936.31 **3.** $9674.77 **5.** $634.49

Vocabulary and Readiness Check **1.** simple **3.** compound **5.** total amount

Exercise Set 6.7 **1.** $32 **3.** $73.60 **5.** $750 **7.** $33.75 **9.** $700 **11.** $101,562.50 **13.** $5562.50 **15.** $14,280 **17.** $46,815.37 **19.** $2327.14 **21.** $58,163.65 **23.** $2915.75 **25.** $2938.66 **27.** $2971.89 **29.** 32 yd **31.** 35 m **33.** $\frac{9x}{20}$ **35.** $-\frac{131}{225}$ **37.** answers may vary **39.** answers may vary

Chapter 6 Vocabulary Check **1.** of **2.** is **3.** percent **4.** compound interest **5.** $\frac{\text{amount}}{\text{base}}$ **6.** 100% **7.** 0.01 **8.** $\frac{1}{100}$ **9.** base; amount **10.** percent of decrease **11.** percent of increase **12.** sales tax **13.** total price **14.** commission **15.** amount of discount **16.** sale price **17.** proportion **18.** ratio

Chapter Review **1.** $\frac{1}{5}$ **2.** $\frac{2}{3}$ **3.** 6 **4.** 500 **5.** 312.5 **6.** 50 **7.** 9 **8.** −17 **9.** 3 **10.** −19 **11.** 675 parts **12.** $33.75 **13.** 37% **14.** 77% **15.** 0.26 **16.** 0.75 **17.** 0.035 **18.** 0.015 **19.** 2.75 **20.** 4.00 or 4 **21.** 0.4785 **22.** 0.8534 **23.** 160% **24.** 5.5% **25.** 7.6% **26.** 8.5% **27.** 71% **28.** 65% **29.** 600% **30.** 900% **31.** $\frac{7}{100}$ **32.** $\frac{3}{20}$ **33.** $\frac{1}{4}$ **34.** $\frac{17}{200}$ **35.** $\frac{51}{500}$ **36.** $\frac{1}{6}$ **37.** $\frac{1}{3}$ **38.** $1\frac{1}{10}$ **39.** 40% **40.** 70% **41.** $58\frac{1}{3}\%$ **42.** $166\frac{2}{3}\%$ **43.** 125% **44.** 60% **45.** 6.25% **46.** 62.5% **47.** 100,000 **48.** 8000 **49.** 23% **50.** 114.5 **51.** 108.8 **52.** 150% **53.** 418 **54.** 300 **55.** 159.6 **56.** 180% **57.** 110% **58.** 165 **59.** 66% **60.** 16% **61.** 20.9% **62.** 106.25% **63.** $206,400 **64.** $13.23 **65.** $263.75 **66.** $1.15 **67.** $5000 **68.** $300.38 **69.** discount: $900; sale price: $2100 **70.** discount: $9; sale price: $81 **71.** $160 **72.** $325 **73.** $30,104.61 **74.** $17,506.54 **75.** $180.61 **76.** $33,830.10 **77.** 0.038 **78.** 1.245 **79.** 54% **80.** 9520% **81.** $\frac{47}{100}$ **82.** $\frac{7}{125}$ **83.** $37\frac{1}{2}\%$ **84.** 120% **85.** 268.75 **86.** 110% **87.** 708.48 **88.** 134% **89.** 300% **90.** 38.4 **91.** 560 **92.** 325% **93.** 26% **94.** $6786.50 **95.** $617.70 **96.** $3.45 **97.** 12.5% **98.** $1491 **99.** $11,687.50

Chapter Test **1.** 0.85 **2.** 5 **3.** 0.006 **4.** 5.6% **5.** 610% **6.** 35% **7.** $1\frac{1}{5}$ **8.** $\frac{77}{200}$ **9.** $\frac{1}{500}$ **10.** 55% **11.** 37.5% **12.** 175% **13.** 20% **14.** $\frac{16}{25}$ **15.** 33.6 **16.** 1250 **17.** 75% **18.** 38.4 lb **19.** $56,750 **20.** $358.43 **21.** 5% **22.** discount: $18; sale price: $102 **23.** $395 **24.** 1% **25.** $647.50 **26.** $2005.63 **27.** $427 **28.** 5.8% **29.** $\frac{15}{2}$ **30.** −6 **31.** 18 bulbs

Cumulative Review **1.** 20,296 (Sec. 1.5, Ex. 4) **2.** 31,084 (Sec. 1.5) **3.** −10 (Sec. 2.3, Ex. 8) **4.** 10 (Sec. 2.3) **5.** 1(Sec. 2.6, Ex. 2) **6.** −1 (Sec. 2.6) **7.** 2 (Sec. 3.3, Ex. 4) **8.** 5 (Sec. 3.3) **9.** $\frac{21}{7}$(Sec. 4.4, Ex. 20) **10.** $\frac{40}{5}$ (Sec. 4.4)

11. $\frac{-10}{27}$ (Sec. 4.2, Ex. 6) **12.** $\frac{5y}{16}$ (Sec. 4.2) **13.** $\frac{5}{12}$ (Sec. 4.3, Ex. 13) **14.** $\frac{-4}{7}$ (Sec. 4.3) **15.** $-\frac{1}{2}$ (Sec. 4.4, Ex. 9) **16.** $\frac{1}{5}$ (Sec. 4.4) **17.** $\frac{1}{28}$ (Sec. 4.5, Ex. 5) **18.** $\frac{16}{45}$ (Sec. 4.5) **19.** $\frac{3}{2}$ (Sec. 4.6, Ex. 2) **20.** $\frac{50}{9}$ (Sec. 4.6) **21.** 3 (Sec. 4.8, Ex. 9) **22.** 4 (Sec. 4.8) **23. a.** $\frac{38}{9}$ **b.** $\frac{19}{11}$ (Sec. 4.1, Ex. 20) **24. a.** $\frac{17}{5}$ **b.** $\frac{44}{7}$ (Sec. 4.1) **25.** $\frac{1}{8}$ (Sec. 5.1, Ex. 9) **26.** $\frac{17}{20}$ (Sec. 5.1) **27.** $-105\frac{83}{1000}$ (Sec. 5.1, Ex. 11) **28.** $17\frac{3}{200}$ (Sec. 5.1) **29.** 67.69 (Sec. 5.2, Ex. 6) **30.** 27.94 (Sec. 5.2) **31.** 76.8 (Sec. 5.3, Ex. 5) **32.** 1248.3 (Sec. 5.3) **33.** −76,300 (Sec. 5.3, Ex. 7) **34.** −8537.5 (Sec. 5.3) **35.** 50 (Sec. 5.4, Ex. 10) **36.** no (Sec. 5.4) **37.** 80.5 (Sec. 5.7, Ex. 4) **38.** 48 (Sec. 5.7) **39.** $\frac{50}{63}$ (Sec. 6.1, Ex. 2) **40.** $\frac{29}{38}$ (Sec. 6.1) **41.** 63 (Sec. 6.1, Ex. 4) **42.** −11 (Sec. 6.1) **43.** 17.5 mi (Sec. 6.1, Ex. 6) **44.** 35 problems (Sec. 6.1) **45.** $\frac{19}{1000}$ (Sec. 6.2, Ex. 9) **46.** $\frac{23}{1000}$ (Sec. 6.2) **47.** $\frac{1}{3}$ (Sec. 6.2, Ex. 11) **48.** $1\frac{2}{25}$ (Sec. 6.2) **49.** 21 (Sec. 6.3, Ex. 7) **50.** 35.7 (Sec. 6.3)

CHAPTER 7 Graphs and Triangle Applications

Vocabulary and Readiness Check **1.** bar **3.** line

Exercise Set 7.1 **1.** Washington **3.** 25 million bushels **5.** California and Pennsylvania **7.** about 5% increase **9.** 90,000 **11.** 120,000 **13.** 12,000 **15.** 76,000 wildfires/year **17.** September **19.** 27 **21.** $\frac{1}{3}$ **23.** Tokyo, Japan; about 35.2 million or 35,200,000 **25.** New York; 21.9 million, or 21,900,000 **27.** approximately 2 million

29.

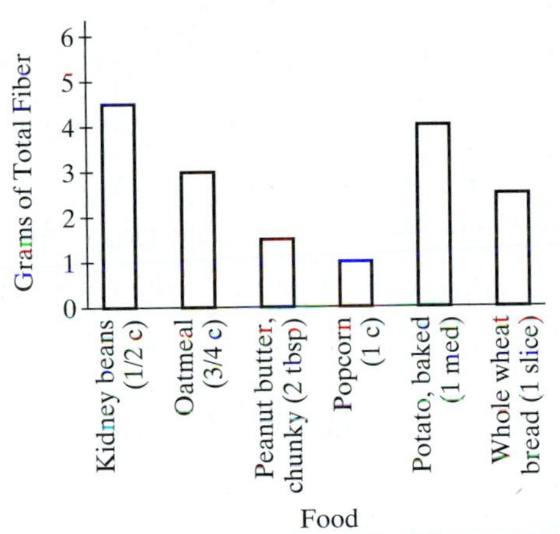

31.

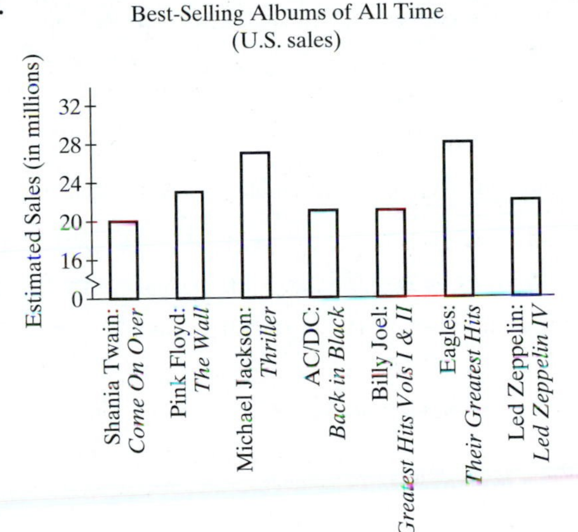

33. 15 adults **35.** 61 adults **37.** 24 adults **39.** 12 adults **41.** $\frac{9}{100}$ **43.** 45–54 **45.** 21 million householders **47.** 44 million householders **49.** 4 million householders **51.** |; 1 **53.** 𝍸 |||; 8 **55.** 𝍸 |; 6 **57.** 𝍸 |; 6 **59.** ||; 2 **63.** 2.7 goals **65.** 1982 **67.** decrease **69.** 1990, 2006 **71.** 3.6 **73.** 6.2 **75.** 25% **77.** 34% **79.** 83°F **81.** Sunday; 68°F **83.** Tuesday; 13°F **85.** answers may vary

61.

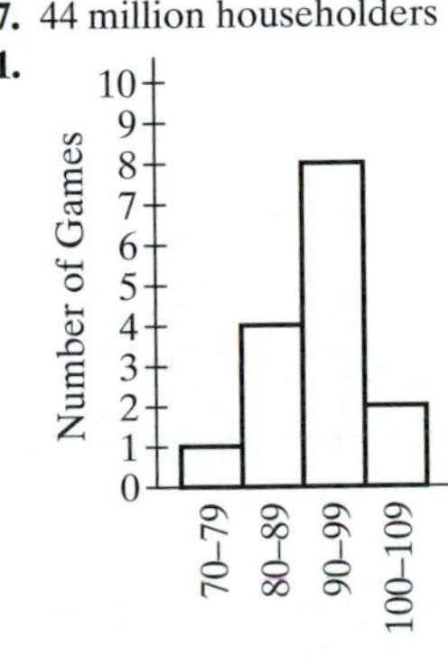

Vocabulary and Readiness Check **1.** circle **3.** 360

Exercise Set 7.2 **1.** parent or guardian's home **3.** $\frac{9}{35}$ **5.** $\frac{9}{16}$ **7.** Asia **9.** 37% **11.** 17,100,000 sq mi **13.** 2,850,000 sq mi **15.** 55% **17.** nonfiction **19.** 31,400 books **21.** 27,632 books **23.** 25,120 books

25.

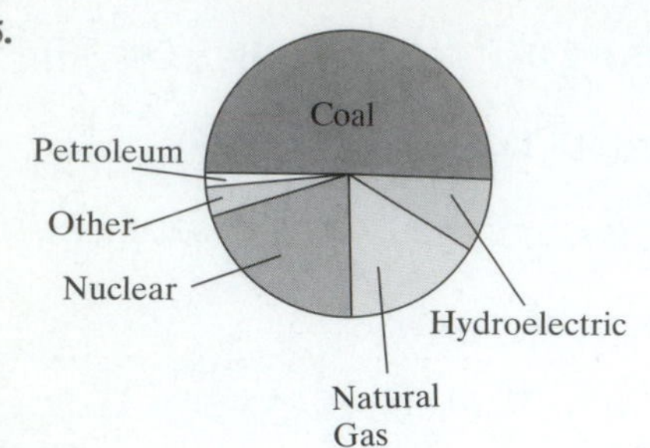

27.

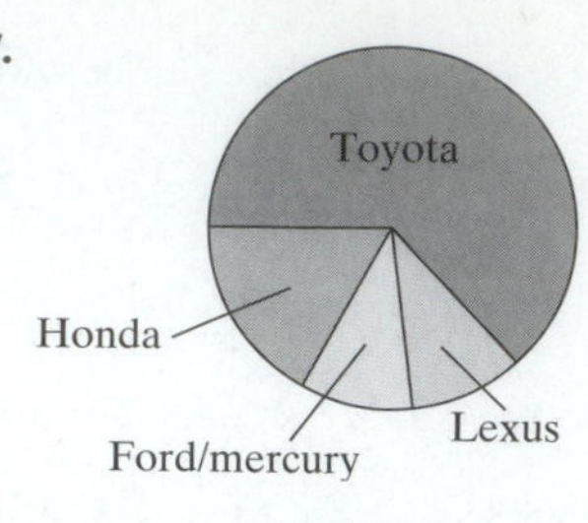

29. $2^2 \times 5$ **31.** $2^3 \times 5$ **33.** 5×17 **35.** answers may vary **37.** 129,600,002 sq km **39.** 55,542,858 sq km **41.** 602 respondents **43.** 2198 respondents **45.** $\frac{301}{837}$ **47.** no; answers may vary

Integrated Review **1.** 500,000 **2.** 550,000 **3.** Retail salespeople **4.** Waitstaff **5.** Oroville Dam; 755 ft **6.** New Bullards Bar Dam; 635 ft **7.** 15 ft **8.** 4 dams **9.** Thursday and Saturday; 100°F **10.** Monday; 82°F **11.** Sunday, Monday, and Tuesday **12.** Wednesday, Thursday, Friday, and Saturday **13.** 70 qt containers **14.** 52 qt containers **15.** 2 qt containers **16.** 6 qt containers **17.** ||; 2 **18.** |; 1 **19.** |||; 3 **20.** 卌 |; 6 **21.** 卌; 5 **22.**

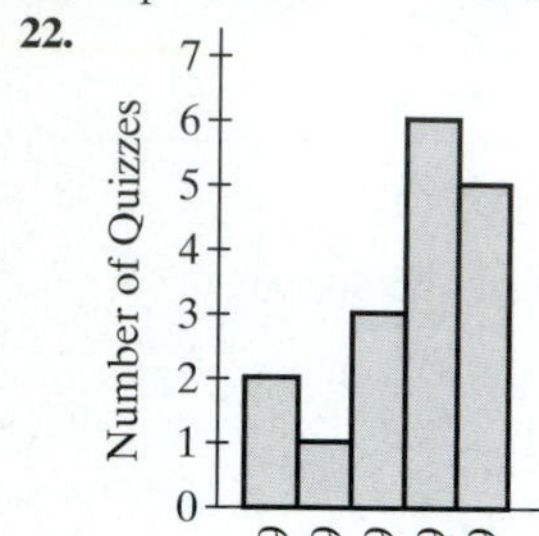

Calculator Explorations **1.** 32 **3.** 3.873 **5.** 9.849

Vocabulary and Readiness Check **1.** 10; −10 **3.** radical **5.** perfect squares **7.** c^2; b^2

Exercise Set 7.3 **1.** 2 **3.** 11 **5.** $\frac{1}{9}$ **7.** $\frac{4}{8} = \frac{1}{2}$ **9.** 1.732 **11.** 3.873 **13.** 5.568 **15.** 5.099 **17.** 16 **19.** 9.592 **21.** $\frac{7}{12}$ **23.** 8.426 **25.** 13 in. **27.** 6.633 cm **29.** 52.802 m **31.** 117 mm **33.** 5 **35.** 12 **37.** 17.205 **39.** 44.822 **41.** 42.426 **43.** 1.732 **45.** 8.5 **47.** 141.42 yd **49.** 25.0 ft **51.** 340 ft **53.** $\frac{5}{6}$ **55.** $\frac{x}{30}$ **57.** $\frac{3x-5}{9}$ **59.** $\frac{9x}{64}$ **61.** 6, 7 **63.** 10, 11 **65.** answers may vary **67.** no

The Bigger Picture **1.** 9 **2.** $\frac{2}{7}$ **3.** 15 **4.** 8 **5.** 24 **6.** 33 **7.** 28 **8.** 42 **9.** $\frac{3}{2}$ or $1\frac{1}{2}$ **10.** $\frac{17}{16}$ or $1\frac{1}{16}$

Vocabulary and Readiness Check **1.** false **3.** true **5.** false

Exercise Set 7.4 **1.** congruent **3.** not congruent **5.** congruent **7.** congruent **9.** $\frac{2}{1}$ **11.** $\frac{3}{2}$ **13.** 4.5 **15.** 6 **17.** 5 **19.** 13.5 **21.** 17.5 **23.** 10 **25.** 28.125 **27.** 10 **29.** 500 ft **31.** approx 560 ft high **33.** 14.4 ft **35.** 52 neon tetras **37.** 381 ft **39.** 4.01 **41.** −1.23 **43.** $3\frac{8}{9}$ in.; no **45.** 32.7

Vocabulary and Readiness Check **1.** outcome **3.** probability **5.** 0

Exercise Set 7.5 **1.**

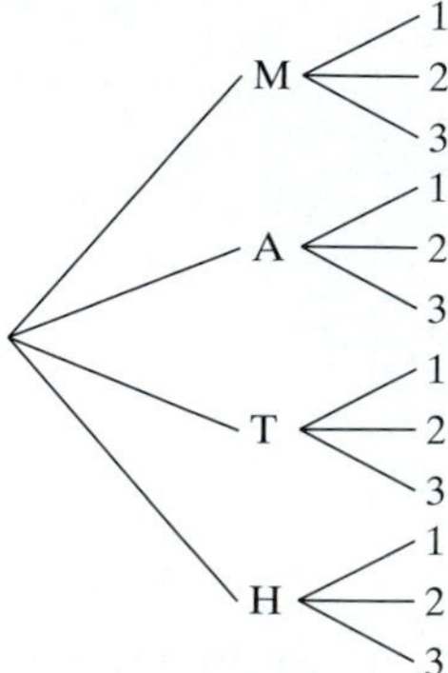

3.

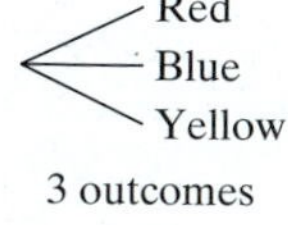

5.

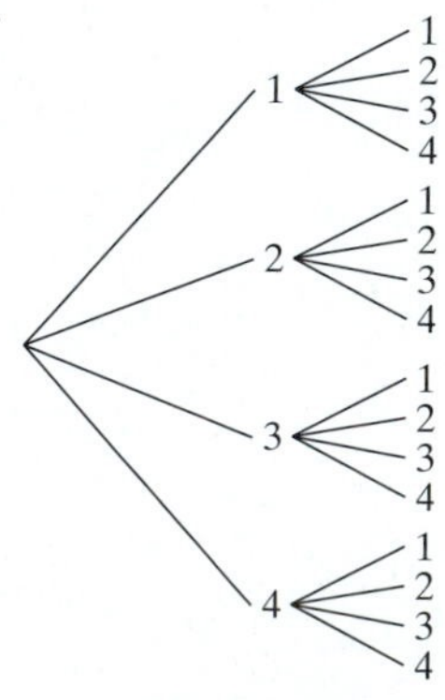

7.

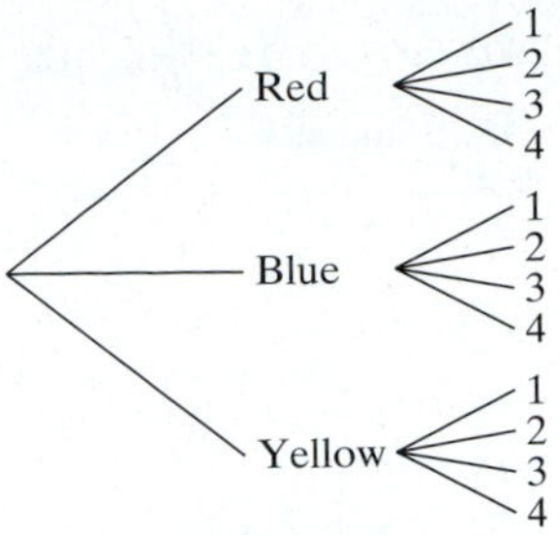

12 outcomes

9.

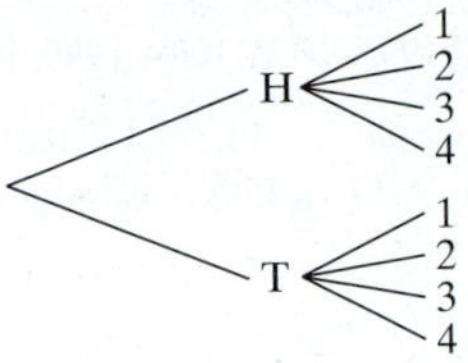

8 outcomes

11. $\frac{1}{6}$ **13.** $\frac{1}{3}$ **15.** $\frac{1}{2}$ **17.** $\frac{2}{3}$ **19.** $\frac{1}{3}$ **21.** 1 **23.** $\frac{2}{3}$ **25.** $\frac{1}{7}$ **27.** $\frac{2}{7}$ **29.** $\frac{4}{7}$ **31.** $\frac{19}{100}$ **33.** $\frac{1}{20}$ **35.** $\frac{5}{6}$ **37.** $\frac{1}{6}$
39. $\frac{20}{3}$ or $6\frac{2}{3}$ **41.** $\frac{1}{52}$ **43.** $\frac{1}{13}$ **45.** $\frac{1}{4}$ **47.** $\frac{1}{2}$ **49.** $\frac{5}{36}$ **51.** 0 **53.** answers may vary

Chapter 7 Vocabulary Check **1.** congruent **2.** similar **3.** leg **4.** leg **5.** hypotenuse **6.** right **7.** Pythagorean **8.** bar **9.** outcomes **10.** pictograph **11.** tree diagram **12.** experiment **13.** circle **14.** probability **15.** histogram; class interval; class frequency

Chapter Review **1.** 4,000,000 **2.** 2,400,000 **3.** South **4.** Northeast **5.** South, West **6.** Northeast, Midwest **7.** 7.5% **8.** 2003 **9.** 1980, 1990, 2000, 2003 **10.** answers may vary **11.** 68 **12.** 78 **13.** 84 **14.** 6 **15.** 27 **16.** 4 employees **17.** 1 employee **18.** 9 employees **19.** 18 employees **20.** 𝍸; 5 **21.** |||; 3 **22.** ||||; 4

23.

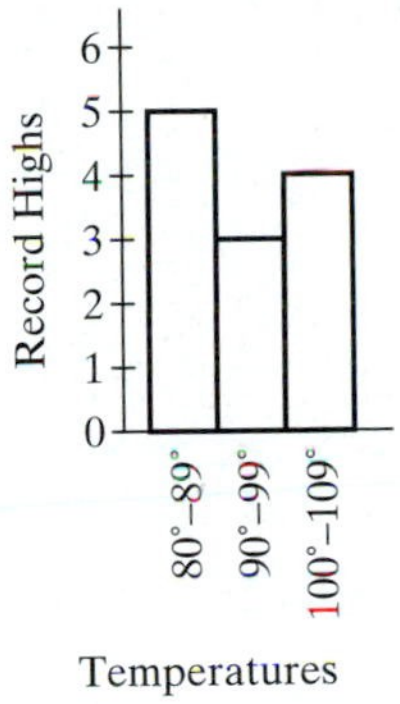

24. mortgage payment **25.** utilities **26.** $1225 **27.** $700 **28.** $\frac{39}{160}$ **29.** $\frac{7}{40}$ **30.** $\frac{5}{7}$ **31.** 20 states **32.** 11 states **33.** 17 states **34.** 2 states **35.** 8 **36.** 12 **37.** 3.464 **38.** 3.873 **39.** 0 **40.** 1 **41.** 7.071 **42.** 8.062 **43.** $\frac{2}{5}$ **44.** $\frac{1}{10}$ **45.** 13 **46.** 29 **47.** 10.7 **48.** 86.6 **49.** 28.28 cm **50.** 88.2 ft **51.** congruent: ASA **52.** not congruent **53.** $13\frac{1}{3}$ **54.** 17.4 **55.** approximately 33 ft
56. $x = \frac{5}{6}$ in.; $y = 2\frac{1}{6}$ in.

57.

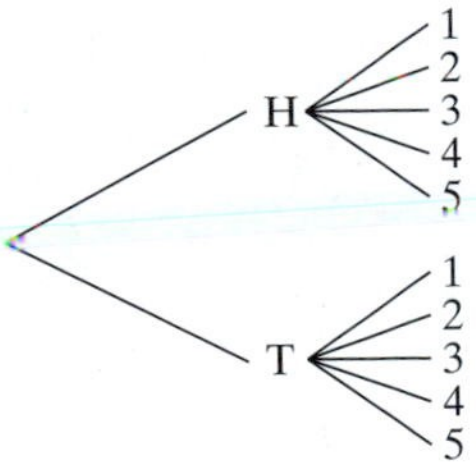

10 outcomes

58.

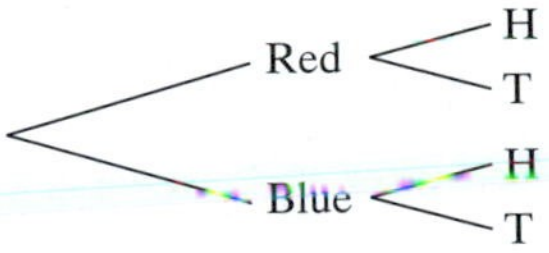

4 outcomes

59.

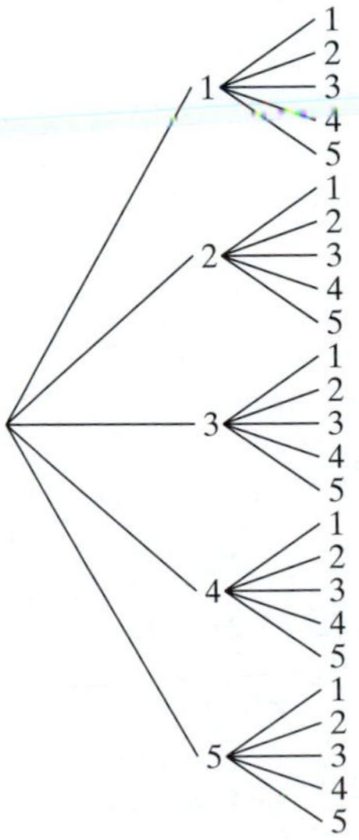

25 outcomes

60.

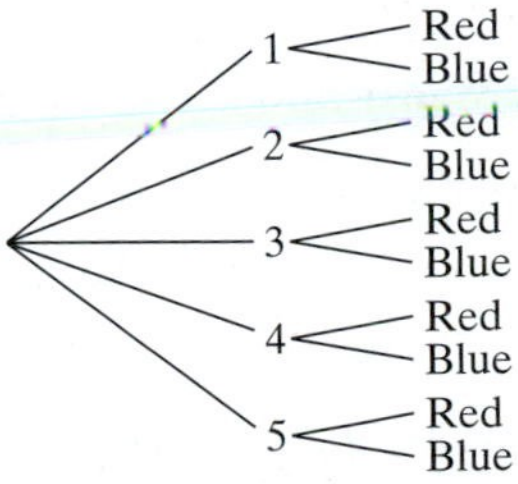

10 outcomes

61. $\frac{1}{6}$ **62.** $\frac{1}{6}$ **63.** $\frac{1}{5}$ **64.** $\frac{1}{5}$ **65.** $\frac{3}{5}$ **66.** $\frac{2}{5}$ **67.** $\frac{1}{4}$ **68.** $\frac{3}{8}$ **69.** $\frac{1}{4}$ **70.** $\frac{1}{8}$ **71.** Insight 2WD Manual; 66 mpg **72.** RX 400h 4WD eCVT; 25 mpg **73.** Insight 2WD CVT; 55 mpg **74.** Camry 2WD ECT-1; 37 mpg **75.** 6 **76.** $\frac{4}{9}$ **77.** 10.247 **78.** 5.657 **79.** 86.6 **80.** 20.8 **81.** 12 **82.** $6\frac{1}{2}$

Chapter Test **1.** $225 **2.** 3rd week; $350 **3.** $1100 **4.** June, August, September **5.** February; 3 cm **6.** March and November

7.

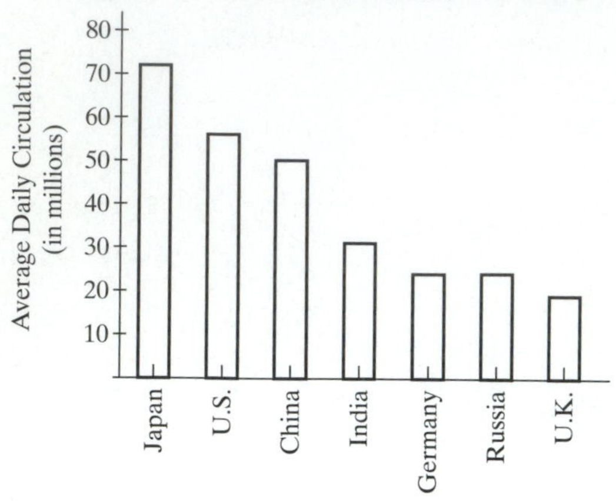
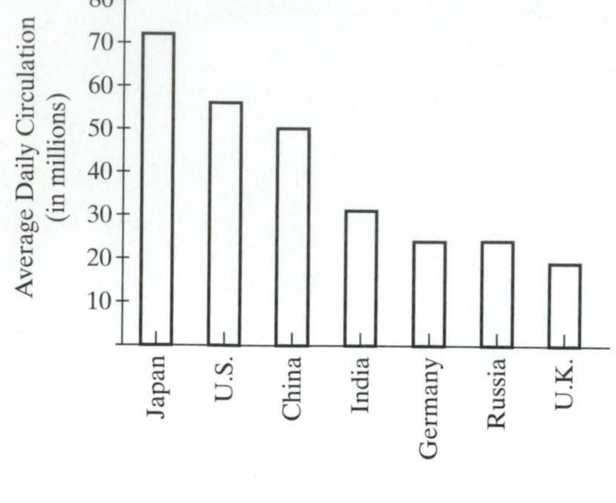

8. 1.5% **9.** 1990, 1991, 2000 **10.** 1994–1995, 1995–1996, 1998–1999, 1999–2000, 2002–2003 **11.** $\frac{17}{40}$ **12.** $\frac{31}{22}$ **13.** 40,920,000 people **14.** 21,120,000 people **15.** 9 students **16.** 11 students **17.** |; 1; |||; 3; ||||; 4; 𝍸; 5; 𝍸 |||; 8; ||||; 4

18.

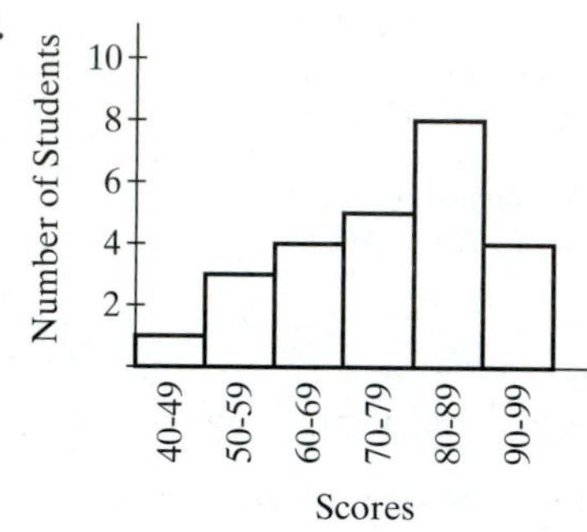

19. 7 **20.** 12.530 **21.** $\frac{4}{5}$ **22.** 5.66 cm **23.** 7.5 **24.** 69 ft

25.

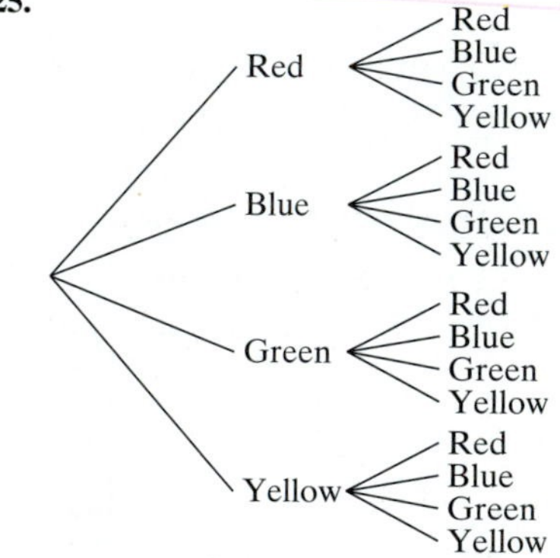

26.

H — H, T
T — H, T

27. $\frac{1}{10}$ **28.** $\frac{1}{5}$

Cumulative Review **1.** 47; Sec. 1.7, Ex. 12 **2.** −70; Sec. 1.7 **3.** −12; Sec. 2.3, Ex. 11 **4.** 9; Sec. 2.3 **5.** −3; Sec. 3.2, Ex. 2 **6.** −6; Sec. 3.2 **7.** 2; Sec. 4.8, Ex. 7 **8.** −10; Sec. 4.8 **9.** $7\frac{17}{24}$; Sec. 4.7, Ex. 9 **10.** $8\frac{3}{20}$; Sec. 4.7 **11.** $5\frac{9}{10}$; Sec. 5.1, Ex. 8 **12.** $2\frac{4}{5}$; Sec. 5.1 **13.** 3.432; Sec. 5.2, Ex. 5 **14.** 7.327; Sec. 5.2 **15.** 0.8496; Sec. 5.3, Ex. 2 **16.** 0.0294; Sec. 5.3 **17.** −0.052; Sec. 5.4, Ex. 3 **18.** 0.136; Sec. 5.4 **19.** 4.09; Sec. 5.5, Ex. 13 **20.** 7.29; Sec. 5.5 **21.** 0.25; Sec. 5.5, Ex. 1 **22.** 0.375; Sec. 5.5 **23.** 0.7; Sec. 5.6, Ex. 5 **24.** 1.68; Sec. 5.6 **25.** 8.944; Sec. 7.3, Ex. 7b **26.** 7.746; Sec. 7.3 **27.** $\frac{1}{6}$; Sec. 7.3, Ex. 5 **28.** $\frac{4}{7}$; Sec. 7.3 **29.** 4.90 in.; Sec. 7.3, Ex. 10 **30.** 16.12; Sec. 7.3 **31.** $\frac{31}{2}$; Sec. 6.1, Ex. 5 **32.** 16; Sec. 6.1 **33.** $\frac{12}{19}$; Sec. 7.4, Ex. 2 **34.** $\frac{4}{9}$; Sec. 7.4 **35.** 0.046; Sec. 6.2, Ex. 4 **36.** 0.32; Sec. 6.2 **37.** 0.0074; Sec. 6.2, Ex. 6 **38.** 0.027; Sec. 6.2 **39.** 21; Sec. 6.3, Ex. 7 **40.** 14.4; Sec. 6.3 **41.** 52; Sec. 6.4, Ex. 9 **42.** 38; Sec. 6.4 **43.** 8.5%; Sec. 6.6, Ex. 2 **44.** 6.5%; Sec. 6.6 **45.** \$6772.12; Sec. 6.7, Ex. 5 **46.** \$144.05; Sec. 6.7 **47.** 80.5; Sec. 5.7, Ex. 4 **48.** 48.5; Sec. 5.7 **49.** $\frac{1}{3}$; Sec. 7.5, Ex. 4 **50.** $\frac{1}{2}$; Sec. 7.5

CHAPTER 8 Geometry and Measurement

Vocabulary and Readiness Check **1.** plane **3.** space **5.** ray **7.** straight **9.** acute **11.** parallel; intersecting **13.** degrees **15.** vertical

Exercise Set 8.1 **1.** line; line CD or line l or $\overleftrightarrow{CD}$ **3.** line segment; line segment MN or $\overline{MN}$ **5.** angle; $\angle GHI$ or $\angle IHG$ or $\angle H$ **7.** ray; ray UW or $\overrightarrow{UW}$ **9.** $\angle CPR$, $\angle RPC$ **11.** $\angle TPM$, $\angle MPT$ **13.** straight **15.** right **17.** obtuse **19.** acute **21.** 67° **23.** 163° **25.** 32° **27.** 30° **29.** $\angle MNP$ and $\angle RNO$; $\angle PNQ$ and $\angle QNR$ **31.** $\angle SPT$ and $\angle TPQ$; $\angle SPR$ and $\angle RPQ$; $\angle SPT$ and $\angle SPR$; $\angle TPQ$ and $\angle QPR$ **33.** 27° **35.** 132° **37.** $x = 30°, y = 150°, z = 30°$ **39.** $m\angle x = 77°; m\angle y = 103°; m\angle z = 77°$ **41.** $m\angle x = 100°; m\angle y = 80°; m\angle z = 100°$ **43.** $m\angle x = 134°; m\angle y = 46°; m\angle z = 134°$ **45.** $\angle ABC$ or $\angle CBA$ **47.** $\angle DBE$ or $\angle EBD$ **49.** 15° **51.** 50° **53.** 65° **55.** 95° **57.** $\frac{9}{8}$ or $1\frac{1}{8}$ **59.** $\frac{7}{32}$ **61.** $\frac{5}{6}$ **63.** $\frac{4}{3}$ or $1\frac{1}{3}$ **65.** 54.8° **67.** false; answers may vary **69.** true **71.** $m\angle a = 60°; m\angle b = 50°; m\angle c = 110°; m\angle d = 70°; m\angle e = 120°$ **73.** no; answers may vary **75.** no; answers may vary

Vocabulary and Readiness Check **1.** perimeter **3.** π **5.** $\frac{22}{7}$(or 3.14); 3.14$\left(\text{or } \frac{22}{7}\right)$

Exercise Set 8.2 **1.** 64 ft **3.** 120 cm **5.** 21 in. **7.** 48 ft **9.** 42 in. **11.** 155 cm **13.** 21 ft **15.** 360 ft **17.** 346 yd **19.** 22 ft **21.** $55 **23.** 72 in. **25.** 28 in. **27.** $36.12 **29.** 96 m **31.** 66 ft **33.** 74 cm **35.** 17π cm; 53.38 cm **37.** 16π mi; 50.24 mi **39.** 26π m; 81.64 m **41.** $C = 15\pi$ ft; 47.1 ft **43.** 12,560 ft **45.** 30.7 mi **47.** $14\pi \approx 43.96$ cm **49.** 40 mm **51.** 84 ft **53.** 23 **55.** 1 **57.** 10 **59.** 216 **61. a.** width: 30 yd; length: 40 yd **b.** 140 yd **63.** b **65. a.** 62.8 m; 125.6 m **b.** yes **67.** $44 + 10\pi \approx 75.4$ m **69.** 26.9 ft

Vocabulary and Readiness Check **1.** surface area **3.** area **5.** square

Exercise Set 8.3 **1.** 7 sq m **3.** $9\frac{3}{4}$ sq yd **5.** 15 sq yd **7.** 2.25π sq in. ≈ 7.065 sq in. **9.** 36.75 sq ft **11.** 28 sq m **13.** 22 sq yd **15.** $36\frac{3}{4}$ sq ft **17.** $22\frac{1}{2}$ sq in. **19.** 25 sq cm **21.** 86 sq mi **23.** 24 sq cm **25.** 36π sq in. $\approx 113\frac{1}{7}$ sq in. **27.** $V = 72$ cu in.; $SA = 108$ sq in. **29.** $V = 512$ cu cm; $SA = 384$ sq cm **31.** $V = 4\pi$ cu yd $\approx 12\frac{4}{7}$ cu yd; $SA = (2\sqrt{13}\pi + 4\pi)$ sq yd ≈ 35.20 sq yd **33.** $V = \frac{500}{3}\pi$ cu in. $\approx 523\frac{17}{21}$ cu in.; $SA = 100\pi$ sq in. $\approx 314\frac{2}{7}$ sq in. **35.** $V = 9\pi$ cu in. $\approx 28\frac{2}{7}$ cu in. **37.** $V = 75$ cu cm **39.** $2\frac{10}{27}$ cu in. **41.** $V = 8.4$ cu ft; $SA = 26$ sq ft **43.** 113,625 sq ft **45.** 168 sq ft **47.** 960 cu cm **49.** 9200 sq ft **51.** $V = \frac{1372}{3}\pi$ cu in. or $457\frac{1}{3}\pi$ cu in.; $SA = 196\pi$ sq in. **53.** 381 sq ft **55.** $V = 36\pi$ cu in. ≈ 113.04 cu in. **57.** 4π sq ft ≈ 12.56 sq ft **59.** $V = 4.5\pi$ cu m; 14.13 cu m **61.** 168 sq ft **63.** $10\frac{5}{6}$ cu in. **65.** 25 **67.** 9 **69.** 5 **71.** 20 **73.** perimeter **75.** area **77.** area **79.** perimeter **81.** 12-in. pizza **83.** 26,696.5 cu ft **85.** no; answers may vary **87.** 7.74 sq in. **89.** 298.5 sq m **91.** no; answers may vary

Integrated Review **1.** 153°; 63° **2.** $m\angle x = 75°; m\angle y = 105°; m\angle z = 75°$ **3.** $m\angle x = 128°; m\angle y = 52°; m\angle z = 128°$ **4.** $m\angle x = 52°$ **5.** 4.6 in. **6.** $4\frac{1}{4}$ in. **7.** 20 m; 25 sq m **8.** 12 ft; 6 sq ft **9.** 10π cm ≈ 31.4 cm; 25π sq cm ≈ 78.5 sq cm **10.** 32 mi; 44 sq mi **11.** 54 cm; 143 sq cm **12.** 62 ft; 238 sq ft **13.** $V = 64$ cu in.; $SA = 96$ sq in. **14.** $V = 30.6$ cu ft; $SA = 63$ sq ft **15.** $V = 400$ cu cm; **16.** $V = 4\frac{1}{2}\pi$ cu mi $\approx 14\frac{1}{7}$ cu mi

Vocabulary and Readiness Check **1.** meter **3.** yard **5.** feet **7.** feet

Exercise Set 8.4 **1.** 5 ft **3.** 36 ft **5.** 8 mi **7.** 102 in. **9.** $3\frac{1}{3}$ yd **11.** 33,792 ft **13.** 4.5 yd **15.** 0.25 ft **17.** 13 yd 1 ft **19.** 7 ft 1 in. **21.** 1 mi 4720 ft **23.** 62 in. **25.** 26 ft **27.** 84 in. **29.** 11 ft 2 in. **31.** 22 yd 1 ft **33.** 6 ft 5 in. **35.** 7 ft 6 in. **37.** 14 ft 4 in. **39.** 83 yd 1 ft **41.** 6000 cm **43.** 4.0 cm **45.** 0.5 km **47.** 1.7 m **49.** 15 m **51.** 42,000 cm **53.** 7000 m **55.** 83 mm **57.** 0.201 dm **59.** 40 mm **61.** 8.94 m **63.** 2.94 m or 2940 mm **65.** 1.29 cm or 12.9 mm **67.** 12.640 km or 12,640 m **69.** 54.9 m **71.** 1.55 km **73.** $348\frac{2}{3}$; 12,552 **75.** $11\frac{2}{3}$; 420 **77.** 5000; 0.005; 500 **79.** 0.065; 65; 0.000065 **81.** 342,000; 342,000,000; 34,200,000 **83.** 10 ft 6 in. **85.** 5100 ft **87.** 4.8 times **89.** 26.7 mm **91.** 15 ft 9 in. **93.** 3.35 m **95.** $105\frac{1}{3}$ yd **97.** $\frac{21}{100}$ **99.** 0.13 **101.** 0.25 **103.** no **105.** yes **107.** no **109.** Estimate: 13 yd **111.** answers may vary; for example, $1\frac{1}{3}$ yd or 48 in. **113.** answers may vary **115.** 334.89 sq m

Vocabulary and Readiness Check **1.** mass **3.** gram **5.** 2000

Exercise Set 8.5 **1.** 32 oz **3.** 10,000 lb **5.** 9 tons **7.** $3\frac{3}{4}$ lb **9.** $1\frac{3}{4}$ tons **11.** 204 oz **13.** 9800 lb **15.** 76 oz **17.** 1.5 tons **19.** $\frac{1}{20}$ lb **21.** 92 oz **23.** 161 oz **25.** 5 lb 9 oz **27.** 53 lb 10 oz **29.** 8 tons 750 lb **31.** 3 tons 175 lb **33.** 8 lb 11 oz

35. 31 lb 2 oz **37.** 1 ton 700 lb **39.** 0.5 kg **41.** 4000 mg **43.** 25,000 g **45.** 0.048 g **47.** 0.0063 kg **49.** 15,140 mg **51.** 6250 g **53.** 350,000 cg **55.** 13.5 mg **57.** 5.815 g or 5815 mg **59.** 1850 mg or 1.850 g **61.** 1360 g or 1.360 kg **63.** 13.52 kg **65.** 2.125 kg **67.** 200,000; 3,200,000 **69.** $\frac{269}{400}$ or 0.6725; 21,520 **71.** 0.5; 0.0005; 50 **73.** 21,000; 21,000,000; 2,100,000 **75.** 8.064 kg **77.** 30 mg **79.** 5 lb 8 oz **81.** 35 lb 14 oz **83.** 6 lb 15.4 oz **85.** 144 mg **87.** 6.12 kg **89.** 130 lb **91.** 211 lb **93.** 0.16 **95.** 0.875 **97.** no **99.** yes **101.** no **103.** answers may vary **105.** true **107.** answers may vary

Vocabulary and Readiness Check **1.** capacity **3.** fluid ounces **5.** cups **7.** quarts

Exercise Set 8.6 **1.** 4 c **3.** 16 pt **5.** $3\frac{1}{2}$ gal **7.** 5 pt **9.** 8 c **11.** $3\frac{3}{4}$ qt **13.** $10\frac{1}{2}$ qt **15.** 9 c **17.** 23 qt **19.** $\frac{1}{4}$ pt **21.** 14 gal 2 qt **23.** 4 gal 3 qt 1 pt **25.** 22 pt **27.** 13 gal 2 qt **29.** 4 c 4 fl oz **31.** 1 gal 1 qt **33.** 2 gal 3 qt 1 pt **35.** 17 gal **37.** 4 gal 3 qt **39.** 5000 ml **41.** 0.00016 kl **43.** 5.6 L **45.** 320 cl **47.** 0.41 kl **49.** 0.064 L **51.** 160 L **53.** 3600 ml **55.** 19.3 L **57.** 4.5 L or 4500 ml **59.** 8410 ml or 8.41 L **61.** 16,600 ml or 16.6 L **63.** 3840 ml **65.** 162.4 L **67.** 336; 84; 168 **69.** $\frac{1}{4}$; 1; 2 **71.** 1.59 L **73.** 18.954 L **75.** 4.3 oz **77.** yes **79.** \$0.316 **81.** $\frac{4}{5}$ **83.** $\frac{3}{5}$ **85.** $\frac{9}{10}$ **87.** no **89.** no **91.** answers may vary **93.** answers may vary **95.** answers may vary **97.** 1.5 cc **99.** 2.7 cc **101.** 54 u or 0.54 cc **103.** 86 u or 0.86 cc

Exercise Set 8.7 **1.** 25.57 fl oz **3.** 218.44 cm **5.** 40 oz **7.** 57.66 mi **9.** 3.77 gal **11.** 13.5 kg **13.** 1.5; $1\frac{2}{3}$; 150; 60 **15.** 54.9; 5486.4; 180; 2160 **17.** 3.94 in. **19.** 80.5 kph **21.** 0.007 oz **23.** yes **25.** 2795 mi **27.** 90 mm **29.** 112.5 g **31.** 104 mph **33.** 26.24 ft **35.** 3 mi **37.** 8 fl oz **39.** b **41.** b **43.** c **45.** d **47.** d **49.** 25°C **51.** 40°C **53.** 122°F **55.** 239°F **57.** −6.7°C **59.** 61.2°C **61.** 197.6°F **63.** 54.3°F **65.** 56.7°C **67.** 80.6°F **69.** 21.1°C **71.** 244.4°F **73.** 7232°F **75.** 29 **77.** 36 **79.** yes **81.** no **83.** no **85.** yes **87.** 2.13 sq m **89.** 1.19 sq m **91.** 1.69 sq m **93.** 510,000,000°C **95.** answers may vary

Chapter 8 Vocabulary Check **1.** weight **2.** mass **3.** meter **4.** unit fractions **5.** gram **6.** liter **7.** line segment **8.** complementary **9.** line **10.** perimeter **11.** angle; vertex **12.** area **13.** ray **14.** transversal **15.** straight **16.** volume **17.** vertical **18.** adjacent **19.** obtuse **20.** right **21.** acute **22.** supplementary **23.** surface area

Chapter Review **1.** right **2.** straight **3.** acute **4.** obtuse **5.** 65° **6.** 75° **7.** 58° **8.** 98° **9.** 90° **10.** 25° **11.** $\angle a$ and $\angle b$; $\angle b$ and $\angle c$; $\angle c$ and $\angle d$; $\angle d$ and $\angle a$ **12.** $\angle x$ and $\angle w$; $\angle y$ and $\angle z$ **13.** $m\angle x = 100°$; $m\angle y = 80°$; $m\angle z = 80°$ **14.** $m\angle x = 155°$; $m\angle y = 155°$; $m\angle z = 25°$ **15.** $m\angle x = 53°$; $m\angle y = 53°$; $m\angle z = 127°$ **16.** $m\angle x = 42°$; $m\angle y = 42°$; $m\angle z = 138°$ **17.** 69 m **18.** 30.6 cm **19.** 36 m **20.** 90 ft **21.** 32 ft **22.** 440 ft **23.** 5.338 in. **24.** 31.4 yd **25.** 240 sq ft **26.** 140 sq m **27.** 600 sq cm **28.** 189 sq yd **29.** 49π sq ft $\approx$ 153.86 sq ft **30.** 82.81 sq m **31.** 119 sq in. **32.** 1248 sq cm **33.** 144 sq m **34.** 432 sq ft **35.** 130 sq ft **36.** $V = 15\frac{5}{8}$ cu in.; $SA = 37\frac{1}{2}$ sq in. **37.** $V = 84$ cu ft; $SA = 136$ sq ft **38.** $V = 20{,}000\pi$ cu cm $\approx$ 62,800 cu cm **39.** $V = \frac{1}{6}\pi$ cu km $\approx \frac{11}{21}$ cu km **40.** $2\frac{2}{3}$ cu ft **41.** 307.72 cu in. **42.** $7\frac{1}{2}$ cu ft **43.** 0.5π cu ft or $\frac{1}{2}\pi$ cu ft **44.** 9 ft **45.** 24 yd **46.** 7920 ft **47.** 18 in. **48.** 17 yd 1 ft **49.** 3 ft 10 in. **50.** 4200 cm **51.** 820 mm **52.** 0.01218 m **53.** 0.00231 km **54.** 21 yd 1 ft **55.** 3 ft 8 in. **56.** 9.5 cm or 95 mm **57.** 2.74 m or 274 cm **58.** 169 yd 2 ft **59.** 258 ft 4 in. **60.** 108.5 km **61.** 0.24 sq m **62.** 4.125 lb **63.** 4600 lb **64.** 3 lb 4 oz **65.** 5 tons 300 lb **66.** 0.027 g **67.** 40,000 g **68.** 21 dag **69.** 0.0003 dg **70.** 3 lb 9 oz **71.** 33 lb 8 oz **72.** 21.5 mg **73.** 0.6 kg or 600 g **74.** 4 lb 4 oz **75.** 9 tons 1075 lb **76.** 14 qt **77.** 5 c **78.** 7 pt **79.** 72 c **80.** 4 qt 1 pt **81.** 3 gal 3 qt **82.** 3800 ml **83.** 1.4 kl **84.** 3060 cl **85.** 1 gal 1 qt **86.** 7 gal **87.** 736 ml or 0.736 L **88.** 15.5 L or 15,500 ml **89.** 2 gal 3 qt **90.** 6 fl oz **91.** 10.88 L **92.** yes **93.** 22.96 ft **94.** 10.55 m **95.** 4.55 gal **96.** 8.27 qt **97.** 425.25 g **98.** 10.35 kg **99.** 2.36 in. **100.** 180.4 lb **101.** 107.6°F **102.** 320°F **103.** 5.2°C **104.** 26.7°C **105.** 1.7°C **106.** 329°F **107.** 108° **108.** 89° **109.** 95° **110.** 57° **111.** 27.3 in. **112.** 194 ft **113.** 1624 sq m **114.** 9π sq m $\approx$ 28.26 sq m **115.** $346\frac{1}{2}$ cu in. **116.** $V = 140$ cu in.; $SA = 166$ sq in. **117.** 75 in. **118.** 4 tons 200 lb **119.** 500 cm **120.** 0.000286 km **121.** 1.4 g **122.** 27 qt **123.** 186.8°F **124.** 11°C **125.** 9117 m or 9.117 km **126.** 35.7 L or 35,700 ml **127.** 8 gal 1 qt **128.** 12.8 kg

Chapter Test **1.** 12° **2.** 56° **3.** 50° **4.** $\angle x = 118°$; $\angle y = 62°$; $\angle z = 118°$ **5.** $\angle x = 73°$; $\angle y = 73°$; $\angle z = 73°$ **6.** 6.2 m **7.** 10 in. **8.** circumference $= 18\pi \approx$ 56.52 in.; area $= 81\pi \approx$ 254.34 sq in. **9.** perimeter = 24.6 yd; area = 37.1 sq yd **10.** perimeter = 68 in.; area = 185 sq in. **11.** $62\frac{6}{7}$ cu in. **12.** 30 cu ft **13.** 16 in. **14.** 18 cu ft **15.** 62 ft; \$115.94 **16.** 23 ft 4 in. **17.** 10 qt **18.** 1.875 lb **19.** 5600 lb **20.** $4\frac{3}{4}$ gal **21.** 0.04 g **22.** 2400 g **23.** 36 mm **24.** 0.43 g **25.** 830 ml **26.** 1 gal 2 qt **27.** 3 lb 13 oz **28.** 8 ft 3 in. **29.** 2 gal 3 qt **30.** 66 mm or 6.6 cm **31.** 2.256 km or 2256 m **32.** 28.9°C **33.** 54.7°F **34.** 5.6 m **35.** 4 gal 3 qt **36.** 91.4 m **37.** 16 ft 6 in.

Cumulative Review **1.** 5; Sec. 3.3, Ex. 1 **2.** 6; Sec. 3.3 **3. a.** $\frac{16}{625}$ **b.** $\frac{1}{16}$; Sec. 4.3, Ex. 10 **4. a.** $-\frac{1}{27}$ **b.** $\frac{9}{49}$; Sec. 4.3 **5.** $9\frac{3}{10}$; Sec. 4.7, Ex. 11 **6.** $9\frac{11}{15}$; Sec. 4.7 **7.** $20x - 10.9$; Sec. 5.2, Ex. 13 **8.** $1.2y + 1.8$; Sec. 5.2 **9.** 28.4405; Sec. 5.5, Ex. 14 **10.** 2.16; Sec. 5.5 **11.** =; Sec. 5.5, Ex. 9 **12.** >; Sec. 5.5 **13.** −1.3; Sec. 5.6, Ex. 6 **14.** 30; Sec. 5.6 **15.** 424 ft; Sec. 7.3, Ex. 11 **16.** 236 ft; Sec. 7.3 **17. a.** $\frac{5}{7}$; **b.** $\frac{7}{24}$; Sec. 6.1, Ex. 3 **18. a.** $\frac{1}{4}$; **b.** $\frac{4}{9}$; Sec. 6.1 **19.** no; Sec. 2.6, Ex. 1 **20.** yes; Sec. 2.6

21. -6; Sec. 2.6; Ex. 8 **22.** 1.02; Sec. 2.6 **23.** 22.4 cc; Sec. 6.1, Ex. 7 **24.** 7.5 c; Sec. 6.1 **25.** 17%; Sec. 6.2, Ex. 1 **26.** 38%; Sec. 6.2 **27.** 200; Sec. 6.3, Ex. 10 **28.** 1200; Sec. 6.3 **29.** 2.7; Sec. 6.4, Ex. 7 **30.** 12.6; Sec. 6.4 **31.** 32%; Sec. 6.5, Ex. 5 **32.** 27%; Sec. 6.5 **33.** sales tax: \$6.41, total price: \$91.91; Sec. 6.6, Ex. 1 **34.** sales tax: \$30; total price: \$405; Sec. 6.6 **35.** 57; Sec. 5.7; Ex. 3 **36.** 83; Sec. 5.7 **37.** $\frac{1}{4}$; Sec. 7.5, Ex. 5 **38.** $\frac{2}{7}$; Sec. 7.5 **39.** 42°; Sec. 8.1, Ex. 4 **40.** 43°; Sec. 8.1 **41.** 96 in.; Sec. 8.4, Ex. 1 **42.** 21 feet; Sec. 8.4 **43.** 4 tons 1650 lb; Sec. 8.5, Ex. 4 **44.** 18 lb 2 oz; Sec. 8.5 **45.** 15°C; Sec. 8.7, Ex. 6 **46.** 30°C; Sec. 8.7

CHAPTER 9 Equations, Inequalities, and Problem Solving

Vocabulary and Readiness Check **1.** whole **3.** inequality **5.** real

Exercise Set 9.1 **1.** $<$ **3.** $>$ **5.** $=$ **7.** $<$ **9.** $32 < 212$ **11.** $30 \leq 45$ **13.** true **15.** false **17.** true **19.** false **21.** $20 \leq 25$ **23.** $6 > 0$ **25.** $-12 < -10$ **27.** $7 < 11$ **29.** $5 \geq 4$ **31.** $15 \neq -2$ **33.** 14,494; -282 **35.** $-43{,}413$ **37.** 475; -195 **39.** 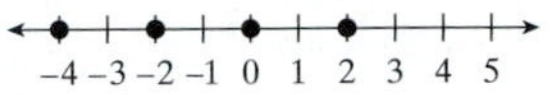

41. 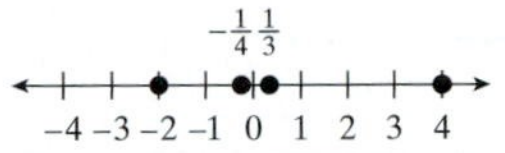

43. 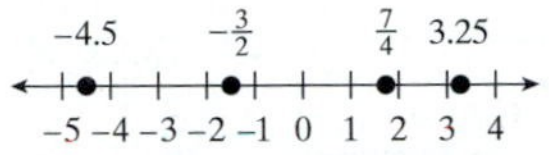

45. whole, integers, rational, real **47.** integers, rational, real **49.** natural, whole, integers, rational, real **51.** rational, real **53.** false **55.** true **57.** false **59.** false **61.** $>$ **63.** $=$ **65.** $<$ **67.** $<$ **69.** $762 < 1548$ **71.** went down by 261 or -261 **73.** $-0.04 > -26.7$ **75.** Sun **77.** Sun **79.** answers may vary

Vocabulary and Readiness Check **1.** commutative property of addition **3.** distributive property **5.** associative property of addition **7.** opposites or additive inverses

Exercise Set 9.2 **1.** $16 + x$ **3.** $y \cdot (-4)$ **5.** yx **7.** $13 + 2x$ **9.** $x \cdot (yz)$ **11.** $(2 + a) + b$ **13.** $(4a) \cdot b$ **15.** $a + (b + c)$ **17.** $17 + b$ **19.** $24y$ **21.** y **23.** $26 + a$ **25.** $-72x$ **27.** s **29.** $-\frac{5}{2}x$ **31.** $4x + 4y$ **33.** $9x - 54$ **35.** $6x + 10$ **37.** $28x - 21$ **39.** $18 + 3x$ **41.** $-2y + 2z$ **43.** $-y - \frac{5}{3}$ **45.** $5x + 20m + 10$ **47.** $8m - 4n$ **49.** $-5x - 2$ **51.** $-r + 6 + 7p$ **53.** $3x + 4$ **55.** $-x + 3y$ **57.** $6r + 8$ **59.** $-36x - 70$ **61.** $-1.6x - 2.5$ **63.** $4(1 + y)$ **65.** $11(x + y)$ **67.** $-1(5 + x)$ **69.** $30(a + b)$ **71.** commutative property of multiplication **73.** associative property of addition **75.** commutative property of addition **77.** associative property of multiplication **79.** identity element for addition **81.** distributive property **83.** multiplicative inverse property **85.** identity element for multiplication **87.** 2 **89.** -23 **91.** -25 **93.** -8; $\frac{1}{8}$ **95.** $-x$; $\frac{1}{x}$ **97.** $2x$; $-2x$ **99.** false **101.** no **103.** yes **105.** yes **107.** yes **109.** **a.** commutative property of addition **b.** commutative property of addition **c.** associative property of addition **111.** answers may vary **113.** answers may vary

Calculator Explorations **1.** solution **3.** not a solution **5.** solution

Exercise Set 9.3 **1.** -6 **3.** 3 **5.** 1 **7.** $\frac{3}{2}$ **9.** 0 **11.** -1 **13.** 4 **15.** -4 **17.** -3 **19.** 2 **21.** 50 **23.** 1 **25.** $\frac{7}{3}$ **27.** 0.2 **29.** all real numbers **31.** no solution **33.** no solution **35.** all real numbers **37.** 18 **39.** $\frac{19}{9}$ **41.** $\frac{14}{3}$ **43.** 13 **45.** 4 **47.** all real numbers **49.** $-\frac{3}{5}$ **51.** -5 **53.** 10 **55.** no solution **57.** 3 **59.** -17 **61.** $(6x - 8)$ m **63.** $-8 - x$ **65.** $-3 + 2x$ **67.** $9(x + 20)$ **69.** **a.** all real numbers **b.** answers may vary **c.** answers may vary **71.** a **73.** b **75.** c **77.** answers may vary **79.** **a.** $x + x + x + 2x + 2x = 28$ **b.** $x = 4$ **c.** $x = 4$ cm; $2x = 8$ cm **81.** answers may vary **83.** 15.3 **85.** -0.2

Integrated Review **1.** whole, integers, rational, real **2.** natural, whole, integers, rational, real **3.** rational, real **4.** natural, whole, integers, rational, real **5.** integers, rational, real **6.** rational, real **7.** rational, real **8.** irrational, real **9.** $7d - 11$ **10.** $9z + 48$ **11.** 16 **12.** $-12x - 15$ **13.** 5 **14.** -1 **15.** -2 **16.** -2 **17.** $-\frac{5}{6}$ **18.** $\frac{1}{6}$ **19.** 1 **20.** 6 **21.** 4 **22.** 1 **23.** $\frac{9}{5}$ **24.** $-\frac{6}{5}$ **25.** all real numbers **26.** all real numbers **27.** 0 **28.** -1.6 **29.** $\frac{4}{19}$ **30.** $-\frac{5}{19}$ **31.** $\frac{7}{2}$ **32.** $-\frac{1}{4}$

Exercise Set 9.4 **1.** -25 **3.** $-\frac{3}{4}$ **5.** 234, 235 **7.** Belgium: 32; France: 33; Spain: 34 **9.** 3 in.; 6 in.; 16 in. **11.** 1st piece: 5 in.; 2nd piece: 10 in.; 3rd piece: 25 in. **13.** Governor of California: \$175,000; Governor of Florida: \$124,575 **15.** 172 mi **17.** 25 mi **19.** 1st angle: 37.5°; 2nd angle: 37.5°; 3rd angle: 105° **21.** A: 60°; B: 120°; C: 120°; D: 60° **23.** 5 ft, 12 ft **25.** 1997: 15.1 million prescriptions; 2001: 20.6 million prescriptions **27.** 45°, 135° **29.** 58°, 60°, 62° **31.** 1 **33.** 280 mi **35.** Johnson: 4932; Kenseth: 5022 **37.** Montana: 56 counties; California: 58 counties **39.** Neptune: 8 moons; Uranus: 21 moons; Saturn: 18 moons **41.** -16 **43.** Sahara: 3,500,000 sq mi; Gobi: 500,000 sq mi **45.** Korea: 9, Italy: 10, France: 11 **47.** Brown: 66,362; Randall: 53,074 **49.** Illinois **51.** Texas: \$29.4 million; Florida: \$27.2 million **53.** answers may vary **55.** 34 **57.** 225π **59.** 15 ft by 24 ft **61.** 720 blinks per hour; 11,520 blinks per day; 4,204,800 blinks per year **63.** answers may vary **65.** answers may vary

Exercise Set 9.5 **1.** $h = 3$ **3.** $h = 3$ **5.** $h = 20$ **7.** $c = 12$ **9.** $r = 2.5$ **11.** $h = \frac{f}{5g}$ **13.** $w = \frac{V}{lh}$ **15.** $y = 7 - 3x$ **17.** $R = \frac{A - P}{PT}$ **19.** $A = \frac{3V}{h}$ **21.** $a = P - b - c$ **23.** $h = \frac{S - 2\pi r^2}{2\pi r}$ **25. a.** area: 103.5 sq ft; perimeter: 41 ft **b.** baseboard: perimeter; carpet: area **27. a.** area: 480 sq in.; perimeter: 120 in. **b.** frame: perimeter; glass: area **29.** 70 ft **31.** −10°C **33.** 6.25 hr **35.** length: 78 ft; width: 52 ft **37.** 18 ft, 36 ft, 48 ft **39.** 137.5 mi **41.** 96 piranhas **43.** 2 bags **45.** one 16-in. pizza **47.** 4.65 min **49.** 13 in. **51.** 2.25 hr **53.** 12,090 ft **55.** 50°C **57.** 515,509.5 cu in. **59.** 449 cu in. **61.** 333°F **63.** 0.32 **65.** 2.00 or 2 **67.** 17% **69.** 720% **71.** $V = G(N - R)$ **73.** multiplies the volume by 8; answers may vary **75.** −40° **77.** $\frac{\triangle - \square}{\blacksquare}$ **79.** 44.3 sec **81.** $P = 3{,}200{,}000$ **83.** $V = 113.1$

Vocabulary and Readiness Check **1.** expression **3.** inequality **5.** equation **7.** −5 **9.** 4.1

Exercise Set 9.6 **1.** −1 **3.** $\frac{1}{2}$ **5.** 4 **7.** −2 **9.** −1 3 **11.** 0 2 **13.** $\{x \mid x \geq -5\}$ −5 **15.** $\{y \mid y < 9\}$ 9 **17.** $\{x \mid x > -3\}$ −3 **19.** $\{x \mid x \leq 1\}$ 1 **21.** $\{x \mid x < -3\}$ −3 **23.** $\{x \mid x \geq -2\}$ −2 **25.** $\{x \mid x < 0\}$ 0 **27.** $\left\{y \mid y \geq -\frac{8}{3}\right\}$ $-\frac{8}{3}$ **29.** $\{y \mid y > 3\}$ 3 **31.** $\{x \mid x > -15\}$ **33.** $\{x \mid x \geq -11\}$ **35.** $\left\{x \middle| x > \frac{1}{4}\right\}$ **37.** $\{y \mid y \geq -12\}$ **39.** $\{z \mid z < 0\}$ **41.** $\{x \mid x > -3\}$ **43.** $\left\{x \middle| x \geq -\frac{2}{3}\right\}$ **45.** $\{x \mid x \leq -2\}$ **47.** $\{x \mid x > -13\}$ **49.** $\{x \mid x \leq -8\}$ **51.** $\{x \mid x > 4\}$ **53.** $\left\{x \middle| x \leq \frac{5}{4}\right\}$ **55.** $\left\{x \middle| x > \frac{8}{3}\right\}$ **57.** $\{x \mid x \geq 0\}$ **59.** all numbers greater than −10 **61.** 35 cm **63.** 193 **65.** 86 people **67.** 35 min **69.** 81 **71.** 1 **73.** $\frac{49}{64}$ **75.** about 120 **77.** 2003 and 2004 **79.** 2001 **81.** > **83.** ≥ **85.** when multiplying or dividing by a negative number **87.** final exam score ≥ 78.5

The Bigger Picture **1.** −3 **2.** $\{x \mid x < -3\}$ **3.** $\frac{2}{9}$ **4.** $-\frac{1}{4}$ **5.** $\{x \mid x \geq -15\}$ **6.** no solution **7.** 7 **8.** $\{x \mid x < 37\}$ **9.** all real numbers **10.** $\frac{41}{29}$

Chapter 9 Vocabulary Check **1.** linear equation in one variable **2.** equivalent equations **3.** formula **4.** linear inequality in one variable **5.** all real numbers **6.** no solution **7.** the same **8.** reversed **9.** opposites **10.** reciprocals

Chapter Review **1.** < **2.** > **3.** > **4.** > **5.** < **6.** > **7.** = **8.** = **9.** > **10.** < **11.** $4 \geq -3$ **12.** $6 \neq 5$ **13.** $0.03 < 0.3$ **14.** $400 > 155$ **15. a.** 1, 3 **b.** 0, 1, 3 **c.** −6, 0, 1, 3 **d.** $-6, 0, 1, 1\frac{1}{2}, 3, 9.62$ **e.** π **f.** all numbers in set **16. a.** 2, 5 **b.** 2, 5 **c.** −3, 2, 5 **d.** $-3, -1.6, 2, 5, \frac{11}{2}, 15.1$ **e.** $\sqrt{5}, 2\pi$ **f.** all numbers in set **17.** Friday **18.** Wednesday **19.** commutative property of addition **20.** multiplicative identity property **21.** distributive property **22.** additive inverse property **23.** associative property of addition **24.** commutative property of multiplication **25.** distributive property **26.** associative property of multiplication **27.** multiplicative inverse property **28.** additive identity property **29.** −4 **30.** −4 **31.** 2 **32.** −3 **33.** no solution **34.** no solution **35.** $\frac{3}{4}$ **36.** $-\frac{8}{9}$ **37.** 20 **38.** 0.25 **39.** $\frac{23}{7}$ **40.** $-\frac{6}{23}$ **41.** 6665.5 in. **42.** short piece: 4 ft; long piece: 8 ft **43.** Kellogg: 35 plants; Keebler: 18 plants **44.** −39, −38, −37 **45.** 3 **46.** −4 **47.** $w = 9$ **48.** $h = 4$ **49.** $m = \frac{y - b}{x}$ **50.** $s = \frac{r + 5}{vt}$ **51.** $x = \frac{2y - 7}{5}$ **52.** $y = \frac{2 + 3x}{6}$ **53.** $\pi = \frac{C}{d}$ **54.** $\pi = \frac{C}{2r}$ **55.** 15 m **56.** 18 ft by 12 ft **57.** 1 hr and 20 min **58.** 40°C **59.** −2 **60.** 0 5 **61.** $\{x \mid x \leq 1\}$ **62.** $\{x \mid x > -5\}$ **63.** $\{x \mid x \leq 10\}$ **64.** $\{x \mid x < -4\}$ **65.** $\{x \mid x < -4\}$ **66.** $\{x \mid x \leq 4\}$ **67.** $\{y \mid y > 9\}$ **68.** $\{y \mid y \geq -15\}$ **69.** $\left\{x \middle| x < \frac{7}{4}\right\}$ **70.** $\left\{x \middle| x \leq \frac{19}{3}\right\}$ **71.** at least \$2500 **72.** score must be less than 83 **73.** $x = 4$ **74.** $y = -14$ **75.** $a = -\frac{3}{2}$ **76.** $x = 21$ **77.** all real numbers **78.** no solution **79.** −13 **80.** shorter piece: 4 in.; longer piece: 19 in. **81.** $h = \frac{3V}{A}$ **82.** $\{x \mid x > 9\}$ 9 **83.** $\{x \mid x > -4\}$ −4 **84.** $\{x \mid x \leq 0\}$ 0

Chapter Test **1.** $|-7| > 5$ **2.** $(9 + 5) \geq 4$ **3. a.** 1, 7 **b.** 0, 1, 7 **c.** −5, −1, 0, 1, 7 **d.** $-5, -1, \frac{1}{4}, 0, 1, 7, 11.6$ **e.** $\sqrt{7}, 3\pi$ **f.** all numbers in set **4.** associative property of addition **5.** commutative property of multiplication **6.** distributive property **7.** multiplicative inverse property **8.** 9 **9.** −3 **10.** $10y + 1$ **11.** $-2x + 10$ **12.** 8 **13.** no solution **14.** 0 **15.** 27 **16.** 3 **17.** 0.25 **18.** $\frac{25}{7}$ **19.** 21 **20.** 7 gal **21.** $x = 6$ **22.** $h = \frac{V}{\pi r^2}$ **23.** $y = \frac{3x - 10}{4}$ **24.** $\{x \mid x < -2\}$ −2 **25.** $\{x \mid x \leq -8\}$ **26.** $\{x \mid x \geq 11\}$ **27.** $\left\{x \middle| x > \frac{2}{5}\right\}$ **28.** New York: 1077; Indiana: 427

Cumulative Review **1. a.** 4 **b.** −5 **c.** −6; Sec. 2.1, Ex. 6 **2. a.** < **b.** = **c.** > **d.** < **e.** >; Sec. 2.1 **3.** −19; Sec. 2.5, Ex. 10 **4.** $\frac{8}{3}$; Sec. 2.5 **5.** −8; Sec. 2.2, Ex. 9 **6.** −19; Sec. 2.2 **7.** 4; Sec. 2.2, Ex. 10 **8.** 8; Sec. 2.2 **9.** −10; Sec. 2.2, Ex. 14 **10.** −0.3; Sec. 5.2 **11.** $-\frac{8}{33}$; Sec. 4.5, Ex. 4 **12.** $-\frac{23}{40}$; Sec. 4.5 **13.** $\frac{6-x}{3}$; Sec. 4.5, Ex. 6 **14.** $\frac{10+x}{2}$; Sec. 4.5 **15.** $-\frac{8}{9}$ **16.** $\frac{11}{12}$; Sec. 4.6 **17.** $-5\frac{3}{5}$; Sec. 4.7, Ex. 21 **18.** $6\frac{1}{4}$; Sec. 4.7 **19.** $\frac{14}{15}$; Sec. 4.7, Ex. 22 **20.** $-1\frac{2}{25}$; Sec. 4.7 **21.** 4.09; Sec. 5.5, Ex. 13 **22.** −2.4; Sec. 5.5 **23.** 7; Sec. 7.3, Ex. 1 **24.** 8; Sec. 7.3 **25.** $\frac{2}{5}$; Sec. 7.3, Ex. 6 **26.** $\frac{3}{10}$; Sec. 7.3 **27.** 2; Sec. 5.6, Ex. 3 **28.** −5; Sec. 5.6 **29.** true; Sec. 9.1, Ex. 3 **30.** true; Sec. 9.1 **31.** true; Sec. 9.1, Ex. 4 **32.** true; Sec. 9.1 **33.** false; Sec. 9.1, Ex. 5 **34.** true; Sec. 9.1 **35.** true; Sec. 9.1, Ex. 6 **36.** false; Sec. 9.1 **37.** $15 - 10z$; Sec. 9.2, Ex. 8 **38.** $-8x + 4$; Sec. 9.2 **39.** $3x + 17$; Sec. 9.2, Ex. 12 **40.** $10x + 21$; Sec. 9.2 **41.** 0; Sec. 9.3, Ex. 4 **42.** $\frac{30}{7}$; Sec. 9.3 **43.** Republicans: 223; Democrats: 208; Sec. 9.6, Ex. 4 **44.** 1; Sec. 9.4 **45.** 79.2 yr; Sec. 9.5, Ex. 1 **46.** $w = \frac{V}{lh}$; Sec. 9.5 **47.** 

; Sec. 9.6, Ex. 2 **48.** $\{x \mid x > 10\}$; Sec. 9.6 **49.** $\{x \mid x \geq 1\}$; Sec. 9.6, Ex. 9 **50.** $\{x \mid x > 4\}$; Sec. 9.6

CHAPTER 10 Exponents and Polynomials

Vocabulary and Readiness Check **1.** exponent **3.** add **5.** 1

Exercise Set 10.1 **1.** 49 **3.** −5 **5.** −16 **7.** 16 **9.** $\frac{1}{27}$ **11.** 112 **13.** 4 **15.** 135 **17.** 150 **19.** $\frac{80}{7}$ **21.** x^7 **23.** $(-3)^{12}$ **25.** $15y^5$ **27.** $x^{19}y^6$ **29.** $-72m^3n^8$ **31.** $-24z^{20}$ **33.** $20x^5$ sq ft **35.** x^{36} **37.** p^8q^8 **39.** $8a^{15}$ **41.** $x^{10}y^{15}$ **43.** $49a^4b^{10}c^2$ **45.** $\frac{r^9}{s^9}$ **47.** $\frac{m^9p^9}{n^9}$ **49.** $\frac{4x^2z^2}{y^{10}}$ **51.** $64z^{10}$ sq dm **53.** $27y^{12}$ cu ft **55.** x^2 **57.** −64 **59.** p^6q^5 **61.** $\frac{y^3}{2}$ **63.** 1 **65.** 1 **67.** −7 **69.** 2 **71.** −81 **73.** $\frac{1}{64}$ **75.** b^6 **77.** a^9 **79.** $-16x^7$ **81.** $a^{11}b^{20}$ **83.** $26m^9n^7$ **85.** z^{40} **87.** $64a^3b^3$ **89.** $36x^2y^2z^6$ **91.** z^8 **93.** $3x$ **95.** 1 **97.** $81x^2y^2$ **99.** 40 **101.** $\frac{y^{15}}{8x^{12}}$ **103.** $2x^2y$ **105.** −2 **107.** 5 **109.** −7 **111.** c **113.** e **115.** answers may vary **117.** answers may vary **119.** 343 cu m **121.** volume **123.** answers may vary **125.** answers may vary **127.** x^{9a} **129.** a^{5b} **131.** x^{5a}

Calculator Explorations **1.** 5.31 EE 3 **3.** 6.6 EE −9 **5.** 1.5×10^{13} **7.** 8.15×10^{19}

Vocabulary and Readiness Check **1.** $\frac{1}{x^3}$ **3.** scientific notation

Exercise Set 10.2 **1.** $\frac{1}{64}$ **3.** $\frac{7}{x^3}$ **5.** −64 **7.** $\frac{5}{6}$ **9.** p^3 **11.** $\frac{q^4}{p^5}$ **13.** $\frac{1}{x^3}$ **15.** z^3 **17.** $\frac{4}{9}$ **19.** $\frac{1}{9}$ **21.** $-p^4$ **23.** −2 **25.** x^4 **27.** p^4 **29.** m^{11} **31.** r^6 **33.** $\frac{1}{x^{15}y^9}$ **35.** $\frac{1}{x^4}$ **37.** $\frac{1}{a^2}$ **39.** $4k^3$ **41.** $3m$ **43.** $-\frac{4a^5}{b}$ **45.** $-\frac{6}{7y^2}$ **47.** $\frac{27a^6}{b^{12}}$ **49.** $\frac{a^{30}}{b^{12}}$ **51.** $\frac{1}{x^{10}y^6}$ **53.** $\frac{z^2}{4}$ **55.** $\frac{x^{11}}{81}$ **57.** $\frac{49a^4}{b^6}$ **59.** $-\frac{3m^7}{n^4}$ **61.** $a^{24}b^8$ **63.** 200 **65.** x^9y^{19} **67.** $-\frac{y^8}{8x^2}$ **69.** $\frac{25b^{33}}{a^{16}}$ **71.** $\frac{27}{z^3x^6}$ cu in. **73.** 7.8×10^4 **75.** 1.67×10^{-6} **77.** 6.35×10^{-3} **79.** 1.16×10^6 **81.** 1.36×10^4 **83.** 0.0000000008673 **85.** 0.033 **87.** 20,320 **89.** 700,000,000 **91.** 9.4×10^8 **93.** 1,230,000,000,000 **95.** Yahoo!: 115,000,000: 1.15×10^8; eBay: 58,000,000; 5.8×10^7 **97.** 0.000036 **99.** 0.0000000000000000028 **101.** 0.0000005 **103.** 200,000 **105.** 2.7×10^9 gal **107.** $-2x + 7$ **109.** $2y - 10$ **111.** $-x - 4$ **113.** $9a^{13}$ **115.** −5 **117.** answers may vary **119. a.** 1.3×10^1 **b.** 4.4×10^7 **c.** 6.1×10^{-2} **121. a.** false **b.** true **c.** false **123.** $\frac{1}{x^{9s}}$ **125.** a^{4m+5}

Vocabulary and Readiness Check **1.** binomial **3.** trinomial **5.** constant

Exercise Set 10.3 **1.** 1; $-3x$; 5 **3.** −5; 3.2; 1; −5 **5.** 1; binomial **7.** 3; none of these **9.** 6; trinomial **11.** 4; binomial **13. a.** −6 **b.** −11 **15. a.** −2 **b.** 4 **17. a.** −15 **b.** −10 **19.** 184 ft **21.** 595.84 ft **23.** 1044 thousand, or 1,044,000 visitors **25.** $-11x$ **27.** $23x^3$ **29.** $16x^2 - 7$ **31.** $12x^2 - 13$ **33.** $7s$ **35.** $-1.1y^2 + 4.8$ **37.** $\frac{5}{6}x^4 - 7x^3 - 19$ **39.** $\frac{3}{20}x^3 + 6x^2 - \frac{13}{20}x - \frac{1}{10}$ **41.** 2, 1, 1, 0; 2 **43.** 4, 0, 4, 3; 4 **45.** $9ab - 11a$ **47.** $4x^2 - 7xy + 3y^2$ **49.** $-3xy^2 + 4$ **51.** $14y^3 - 19 - 16a^2b^2$ **53.** $7x^2 + 0x + 3$ **55.** $x^3 + 0x^2 + 0x - 64$ **57.** $5y^3 + 0y^2 + 2y - 10$ **59.** $2y^4 + 0y^3 + 0y^2 + 8y + 0y^0$ or $2y^4 + 0y^3 + 0y^2 + 8y + 0$ **61.** $6x^5 + 0x^4 + x^3 + 0x^2 - 3x + 15$ **63.** $4x^2 + 7x + x^2 + 5x$; $5x^2 + 12x$ **65.** $5x + 3 + 4x + 3 + 2x + 6 + 3x + 7x$; $21x + 12$ **67.** $10x + 19$ **69.** $-x + 5$ **71.** answers may vary **73.** answers may vary **75.** x^{13} **77.** a^3b^{10} **79.** $2y^{20}$ **81.** answers may vary **83.** answers may vary **85.** $11.1x^2 - 7.97x + 10.76$

Exercise Set 10.4 **1.** $12x + 12$ **3.** $-3x^2 + 10$ **5.** $-3x^2 + 4$ **7.** $-y^2 - 3y - 1$ **9.** $7.9x^3 + 4.4x^2 - 3.4x - 3$ **11.** $\frac{1}{2}m^2 - \frac{7}{10}m + \frac{13}{16}$ **13.** $8t^2 - 4$ **15.** $15a^3 + a^2 - 3a + 16$ **17.** $-x + 14$ **19.** $5x^2 + 2y^2$ **21.** $-2x + 9$ **23.** $2x^2 + 7x - 16$ **25.** $2x^2 + 11x$ **27.** $-0.2x^2 + 0.2x - 2.2$ **29.** $\frac{2}{5}z^2 - \frac{3}{10}z + \frac{7}{20}$ **31.** $-2z^2 - 16z + 6$ **33.** $2u^5 - 10u^2 + 11u - 9$ **35.** $5x - 9$ **37.** $4x - 3$ **39.** $11y + 7$ **41.** $-2x^2 + 8x - 1$ **43.** $14x + 18$ **45.** $3a^2 - 6a + 11$ **47.** $3x - 3$ **49.** $7x^2 - 4x + 2$ **51.** $7x^2 - 2x + 2$ **53.** $4y^2 + 12y + 19$ **55.** $-15x + 7$ **57.** $-2a - b + 1$ **59.** $3x^2 + 5$ **61.** $6x^2 - 2xy + 19y^2$ **63.** $8r^2s + 16rs - 8 + 7r^2s^2$ **65.** $(x^2 + 7x + 4)$ ft **67.** $\left(\frac{19}{2}x + 3\right)$ units **69.** $(3y^2 + 4y + 11)$ m **71.** $-6.6x^2 - 1.8x - 1.8$ **73.** $6x^2$ **75.** $-12x^8$ **77.** $200x^3y^2$ **79.** 2; 2 **81.** 4; 3; 3; 4 **83.** b **85.** e **87. a.** $4z$ **b.** $3z^2$ **c.** $-4z$ **d.** $3z^2$; answers may vary **89. a.** m^3 **b.** $3m$ **c.** $-m^3$ **d.** $-3m$; answers may vary **91.** $-0.325x^2 + 10.14x + 83.58$

Vocabulary and Readiness Check **1.** distributive **3.** $(5y - 1)(5y - 1)$

Exercise Set 10.5 **1.** $24x^3$ **3.** x^4 **5.** $-28n^{10}$ **7.** $-12.4x^{12}$ **9.** $-\frac{2}{15}y^3$ **11.** $-24x^8$ **13.** $6x^2 + 15x$ **15.** $7x^3 + 14x^2 - 7x$ **17.** $-2a^2 - 8a$ **19.** $6x^3 - 9x^2 + 12x$ **21.** $12a^5 + 45a^2$ **23.** $-6a^4 + 4a^3 - 6a^2$ **25.** $6x^5y - 3x^4y^3 + 24x^2y^4$ **27.** $-4x^3y + 7x^2y^2 - xy^3 - 3y^4$ **29.** $4x^4 - 3x^3 + \frac{1}{2}x^2$ **31.** $x^2 + 7x + 12$ **33.** $a^2 + 5a - 14$ **35.** $x^2 + \frac{1}{3}x - \frac{2}{9}$ **37.** $12x^4 + 25x^2 + 7$ **39.** $12x^2 - 29x + 15$ **41.** $1 - 7a + 12a^2$ **43.** $4y^2 - 16y + 16$ **45.** $x^3 - 5x^2 + 13x - 14$ **47.** $x^4 + 5x^3 - 3x^2 - 11x + 20$ **49.** $10a^3 - 27a^2 + 26a - 12$ **51.** $49x^2y^2 - 14xy^2 + y^2$ **53.** $12x^2 - 64x - 11$ **55.** $2x^3 + 10x^2 + 11x - 3$ **57.** $2x^4 + 3x^3 - 58x^2 + 4x + 63$ **59.** $8.4y^7$ **61.** $-3x^3 - 6x^2 + 24x$ **63.** $2x^2 + 39x + 19$ **65.** $x^2 - \frac{2}{7}x - \frac{3}{49}$ **67.** $9y^2 + 30y + 25$ **69.** $a^3 - 2a^2 - 18a + 24$ **71.** $(4x^2 - 25)$ sq yd **73.** $(6x^2 - 4x)$ sq in. **75.** $25x^2$ **77.** $9y^6$ **79. a.** $6x + 12$ **b.** $9x^2 + 36x + 35$; answers may vary **81.** $13x - 7$ **83.** $30x^2 - 28x + 6$ **85.** $-7x + 5$ **87.** $x^2 + 3x$ **89.** $x^2 + 5x + 6$ **91.** $11a$ **93.** $25x^2 + 4y^2$ **95. a.** $a^2 - b^2$ **b.** $4x^2 - 9y^2$ **c.** $16x^2 - 49$ **d.** answers may vary

Exercise Set 10.6 **1.** $x^2 + 7x + 12$ **3.** $x^2 + 5x - 50$ **5.** $5x^2 + 4x - 12$ **7.** $4y^2 - 25y + 6$ **9.** $6x^2 + 13x - 5$ **11.** $6y^3 + 4y^2 + 42y + 28$ **13.** $x^2 + \frac{1}{3}x - \frac{2}{9}$ **15.** $0.08 - 2.6a + 15a^2$ **17.** $2x^2 + 9xy - 5y^2$ **19.** $x^2 + 4x + 4$ **21.** $4x^2 - 4x + 1$ **23.** $9a^2 - 30a + 25$ **25.** $x^4 + x^2 + 0.25$ **27.** $y^2 - \frac{4}{7}y + \frac{4}{49}$ **29.** $4a^2 - 12a + 9$ **31.** $25x^2 + 90x + 81$ **33.** $9x^2 - 42xy + 49y^2$ **35.** $16m^2 + 40mn + 25n^2$ **37.** $25x^8 - 15x^4 + 9$ **39.** $a^2 - 49$ **41.** $x^2 - 36$ **43.** $9x^2 - 1$ **45.** $x^4 - 25$ **47.** $4y^4 - 1$ **49.** $16 - 49x^2$ **51.** $9x^2 - \frac{1}{4}$ **53.** $81x^2 - y^2$ **55.** $4m^2 - 25n^2$ **57.** $a^2 + 9a + 20$ **59.** $a^2 - 14a + 49$ **61.** $12a^2 - a - 1$ **63.** $x^2 - 4$ **65.** $9a^2 + 6a + 1$ **67.** $4x^2 + 3xy - y^2$ **69.** $a^2 - \frac{1}{4}y^2$ **71.** $6b^2 - b - 35$ **73.** $x^4 - 100$ **75.** $16x^2 - 25$ **77.** $25x^2 - 60xy + 36y^2$ **79.** $4r^2 - 9s^2$ **81.** $(4x^2 + 4x + 1)$ sq ft **83.** $\frac{5b^5}{7}$ **85.** $-\frac{2a^{10}}{b^5}$ **87.** $\frac{2y^8}{3}$ **89.** c **91.** d **93.** 2 **95.** $(24x^2 - 32x + 8)$ sq m **97.** answers may vary

Integrated Review **1.** $35x^5$ **2.** $-32y^9$ **3.** -16 **4.** 16 **5.** $2x^2 - 9x - 5$ **6.** $3x^2 + 13x - 10$ **7.** $3x - 4$ **8.** $4x + 3$ **9.** $7x^6y^2$ **10.** $\frac{10b^6}{7}$ **11.** $144m^{14}n^{12}$ **12.** $64y^{27}z^{30}$ **13.** $16y^2 - 9$ **14.** $49x^2 - 1$ **15.** $\frac{y^{45}}{x^{63}}$ **16.** $\frac{1}{64}$ **17.** $\frac{x^{27}}{27}$ **18.** $\frac{r^{58}}{16s^{14}}$ **19.** $2x^2 - 2x - 6$ **20.** $6x^2 + 13x - 11$ **21.** $2.5y^2 - 6y - 0.2$ **22.** $8.4x^2 - 6.8x - 4.2$ **23.** $2y^2 - 6y - 1$ **24.** $6z^2 + 2z + \frac{11}{2}$ **25.** $x^2 + 8x + 16$ **26.** $y^2 - 18y + 81$ **27.** $2x + 8$ **28.** $2y - 18$ **29.** $7x^2 - 10xy + 4y^2$ **30.** $-a^2 - 3ab + 6b^2$ **31.** $x^3 + 2x^2 - 16x + 3$ **32.** $x^3 - 2x^2 - 5x - 2$ **33.** $6x^2 - x - 70$ **34.** $20x^2 + 21x - 5$ **35.** $2x^3 - 19x^2 + 44x - 7$ **36.** $5x^3 + 9x^2 - 17x + 3$ **37.** $4x^2 - \frac{25}{81}$ **38.** $144y^2 - \frac{9}{49}$

Exercise Set 10.7 **1.** $12x^3 + 3x$ **3.** $4x^3 - 6x^2 + x + 1$ **5.** $5p^2 + 6p$ **7.** $-\frac{3}{2x} + 3$ **9.** $-3x^2 + x - \frac{4}{x^3}$ **11.** $-1 + \frac{3}{2x} - \frac{7}{4x^4}$ **13.** $x + 1$ **15.** $2x + 3$ **17.** $2x + 1 + \frac{7}{x - 4}$ **19.** $3a^2 - 3a + 1 + \frac{2}{3a + 2}$ **21.** $4x + 3 - \frac{2}{2x + 1}$ **23.** $2x^2 + 6x - 5 - \frac{2}{x - 2}$ **25.** $x + 6$ **27.** $x^2 + 3x + 9$ **29.** $-3x + 6 - \frac{11}{x + 2}$ **31.** $2b - 1 - \frac{6}{2b - 1}$ **33.** $ab - b^2$ **35.** $4x + 9$ **37.** $x + 4xy - \frac{y}{2}$ **39.** $2b^2 + b + 2 - \frac{12}{b + 4}$ **41.** $y^2 + 5y + 10 + \frac{24}{y - 2}$ **43.** $-6x - 12 - \frac{19}{x - 2}$ **45.** $x^3 - x^2 + x$ **47.** 3 **49.** -4 **51.** $3x$ **53.** $9x$ **55.** $(3x^3 + x - 4)$ ft **57.** $(2x + 5)$ m **59.** answers may vary **61.** c

The Bigger Picture **1.** -5.93 **2.** $-\frac{2}{5}$ **3.** $5x^9y^4$ **4.** $\frac{1}{8a^4}$ **5.** $6y^3 - 2y^2 - 6y$ **6.** $8y^2 - 3y - 7$ **7.** $4x^3 - 13x^2 + 10x - 21$ **8.** $36m^2 - 60m + 25$ **9.** $4n - 1 + \frac{2}{n}$ **10.** $2x - 6 + \frac{14}{3x - 1}$ **11.** -0.6 **12.** $\{x \mid x > 0.6\}$ **13.** $\{x \mid x \le 2\}$ **14.** $\frac{2}{3}$

Vocabulary Check **1.** term **2.** FOIL **3.** trinomial **4.** degree of polynomial **5.** binomial **6.** coefficient **7.** degree of a term **8.** monomial **9.** polynomials

Chapter Review **1.** base: 3; exponent: 2 **2.** base: −5; exponent: 4 **3.** base: 5; exponent: 4 **4.** base: x; exponent: 6 **5.** 512 **6.** 36 **7.** −36 **8.** −65 **9.** 1 **10.** 1 **11.** y^9 **12.** x^{14} **13.** $-6x^{11}$ **14.** $-20y^7$ **15.** x^8 **16.** y^{15} **17.** $81y^{24}$ **18.** $8x^9$ **19.** x^5 **20.** z^7 **21.** a^4b^3 **22.** x^3y^5 **23.** $\frac{4}{x^3y^4}$ **24.** $\frac{x^6y^6}{4}$ **25.** $40a^{19}$ **26.** $36x^3$ **27.** 3 **28.** 9 **29.** b **30.** c **31.** $\frac{1}{49}$ **32.** $-\frac{1}{49}$ **33.** $\frac{2}{x^4}$ **34.** $\frac{1}{16x^4}$ **35.** 125 **36.** $\frac{9}{4}$ **37.** $\frac{17}{16}$ **38.** $\frac{1}{42}$ **39.** x^8 **40.** z^8 **41.** r **42.** y^3 **43.** c^4 **44.** $\frac{x^3}{y^3}$ **45.** $\frac{1}{x^6y^{13}}$ **46.** $\frac{a^{10}}{b^{10}}$ **47.** 2.7×10^{-4} **48.** 8.868×10^{-1} **49.** 8.08×10^7 **50.** -8.68×10^5 **51.** 1.124×10^8 **52.** 1.5×10^5 **53.** 867,000 **54.** 0.00386 **55.** 0.00086 **56.** 893,600 **57.** 1,431,280,000,000,000 cu km **58.** 0.0000000001 m **59.** 0.016 **60.** 400,000,000,000 **61.** 5 **62.** 2 **63.** 5 **64.** 6 **65.** 22; 78; 154.02; 400 **66.** $2a^2$ **67.** $-4y$ **68.** $15a^2 + 4a$ **69.** $22x^2 + 3x + 6$ **70.** $-6a^2b - 3b^2 - q^2$ **71.** cannot be combined **72.** $8x^2 + 3x + 6$ **73.** $2x^5 + 3x^4 + 4x^3 + 9x^2 + 7x + 6$ **74.** $-7y^2 - 1$ **75.** $-6m^7 - 3x^4 + 7m^6 - 4m^2$ **76.** $-x^2 - 6xy - 2y^2$ **77.** $-5x^2 + 5x + 1$ **78.** $-2x^2 - x + 20$ **79.** $6x + 30$ **80.** $9x - 63$ **81.** $8a + 28$ **82.** $54a - 27$ **83.** $-7x^3 - 35x$ **84.** $-32y^3 + 48y$ **85.** $-2x^3 + 18x^2 - 2x$ **86.** $-3a^3b - 3a^2b - 3ab^2$ **87.** $-6a^4 + 8a^2 - 2a$ **88.** $42b^4 - 28b^2 + 14b$ **89.** $2x^2 - 12x - 14$ **90.** $6x^2 - 11x - 10$ **91.** $4a^2 + 27a - 7$ **92.** $42a^2 + 11a - 3$ **93.** $x^4 + 7x^3 + 4x^2 + 23x - 35$ **94.** $x^6 + 2x^5 + x^2 + 3x + 2$ **95.** $x^4 + 4x^3 + 4x^2 - 16$ **96.** $x^6 + 8x^4 + 16x^2 - 16$ **97.** $x^3 + 21x^2 + 147x + 343$ **98.** $8x^3 - 60x^2 + 150x - 125$ **99.** $x^2 + 14x + 49$ **100.** $x^2 - 10x + 25$ **101.** $9x^2 - 42x + 49$ **102.** $16x^2 + 16x + 4$ **103.** $25x^2 - 90x + 81$ **104.** $25x^2 - 1$ **105.** $49x^2 - 16$ **106.** $a^2 - 4b^2$ **107.** $4x^2 - 36$ **108.** $16a^4 - 4b^2$ **109.** $(9x^2 - 6x + 1)$ sq m **110.** $(5x^2 - 3x - 2)$ sq mi **111.** $\frac{1}{7} + \frac{3}{x} + \frac{7}{x^2}$ **112.** $-a^2 + 3b - 4$ **113.** $a + 1 + \frac{6}{a - 2}$ **114.** $4x + \frac{7}{x + 5}$ **115.** $a^2 + 3a + 8 + \frac{22}{a - 2}$ **116.** $3b^2 - 4b - \frac{1}{3b - 2}$ **117.** $2x^3 - x^2 + 2 - \frac{1}{2x - 1}$ **118.** $-x^2 - 16x - 117 - \frac{684}{x - 6}$ **119.** $\left(5x - 1 + \frac{20}{x^2}\right)$ ft **120.** $(7a^3b^6 + a - 1)$ units **121.** $-\frac{1}{8}$ **122.** $4x^4y^7$ **123.** $\frac{2x^6}{3}$ **124.** $\frac{27a^{12}}{b^6}$ **125.** $\frac{x^{16}}{16y^{12}}$ **126.** $9a^2b^8$ **127.** $11x - 5$ **128.** $2y^2 - 10$ **129.** $5y^2 - 3y - 1$ **130.** $5x^2 + 3x - 2$ **131.** $28x^3 + 12x$ **132.** $6x^2 + 11x - 10$ **133.** $x^3 + x^2 - 18x + 18$ **134.** $28x^2 - 71x + 18$ **135.** $25x^2 + 40x + 16$ **136.** $36x^2 - 9$ **137.** $4a - 1 + \frac{2}{a^2} - \frac{5}{2a^3}$ **138.** $x - 3 + \frac{25}{x + 5}$ **139.** $2x^2 + 7x + 5 + \frac{19}{2x - 3}$

Chapter Test **1.** 32 **2.** 81 **3.** −81 **4.** $\frac{1}{64}$ **5.** $-15x^{11}$ **6.** y^5 **7.** $\frac{1}{r^5}$ **8.** $\frac{y^{14}}{x^2}$ **9.** $\frac{1}{6xy^8}$ **10.** 5.63×10^5 **11.** 8.63×10^{-5} **12.** 0.0015 **13.** 62,300 **14.** 0.036 **15. a.** 4, 3; 7, 3; 1, 4; −2, 0 **b.** 4 **16.** $-2x^2 + 12x + 11$ **17.** $16x^3 + 7x^2 - 3x - 13$ **18.** $-3x^3 + 5x^2 + 4x + 5$ **19.** $x^3 + 8x^2 + 3x - 5$ **20.** $3x^3 + 22x^2 + 41x + 14$ **21.** $6x^4 - 9x^3 + 21x^2$ **22.** $3x^2 + 16x - 35$ **23.** $9x^2 - \frac{1}{25}$ **24.** $16x^2 - 16x + 4$ **25.** $64x^2 + 48x + 9$ **26.** $x^4 - 81b^2$ **27.** 1001 ft; 985 ft; 857 ft; 601 ft **28.** $(4x^2 - 9)$ sq in. **29.** $\frac{x}{2y} + \frac{1}{4} - \frac{7}{8y}$ **30.** $x + 2$ **31.** $9x^2 - 6x + 4 - \frac{16}{3x + 2}$

Cumulative Review **1.** 81; Sec. 1.7, Ex. 5 **2.** 125; Sec. 1.7 **3.** 81; Sec. 1.7, Ex. 7 **4.** 27; Sec. 1.7 **5. a.** $7 + x$ **b.** $15 - x$ **c.** $2x$ **d.** $\frac{x}{5}$ **e.** $x - 2$; Sec. 1.8, Ex. 8 **6. a.** $x + 3$ **b.** $3x$ **c.** $2x$ **d.** $10 - x$ **e.** $5x + 7$; Sec. 1.8 **7.** $14x - 9$; Sec. 3.1, Ex. 11 **8.** $5x + 10$; Sec. 3.1 **9.** $6x + 18$; Sec. 3.1, Ex. 12 **10.** $-2y - 0.6z + 2$; Sec. 3.1 **11.** −3; Sec. 3.2, Ex. 1 **12.** 140; Sec. 3.2 **13.** −4; Sec. 3.3, Ex. 3 **14.** 19; Sec. 3.3 **15.** 15; Sec. 4.8, Ex. 3 **16.** 24; Sec. 4.8 **17.** $-\frac{2}{33}$; Sec. 4.8, Ex. 5 **18.** $-\frac{1}{25}$; Sec. 4.8 **19.** $\frac{1}{8}$; Sec. 5.1, Ex. 9 **20.** $\frac{1}{4}$; Sec. 5.1 **21.** $43\frac{1}{2}$; Sec. 5.1, Ex. 10 **22.** $10\frac{3}{4}$; Sec. 5.1 **23.** $-105\frac{83}{1000}$; Sec. 5.1, Ex. 11 **24.** $-31\frac{7}{100}$; Sec. 5.1 **25.** 1.88; Sec. 5.2, Ex. 11 **26.** −8.8; Sec. 5.2 **27.** 1.012; Sec. 5.3, Ex. 12 **28.** −3.05; Sec. 5.3 **29. a.** 11, 112 **b.** 0, 11, 112 **c.** −3, −2, 0, 11, 112 **d.** $-3, -2, 0, \frac{1}{4}, 11, 112$ **e.** $\sqrt{2}$ **f.** $-2, 0, \frac{1}{4}, 112, -3, 11, \sqrt{2}$; Sec. 9.1, Ex. 11 **30.** rational numbers, real numbers; Sec. 9.1 **31.** 4, Sec. 9.3, Ex. 5 **32.** 5; Sec. 9.3 **33.** 10; Sec. 9.4, Ex. 1 **34.** 2; Sec. 9.4 **35.** 40 ft; Sec. 9.5, Ex. 2 **36.** 32 ft; Sec. 9.5 **37.** $\{x \mid x \le 4\}$; Sec. 9.6, Ex. 7 **38.** $\{x \mid x \ge 10\}$; Sec. 9.6 **39. a.** x^{11} **b.** $\frac{t^4}{16}$ **c.** $81y^{10}$; Sec. 10.1, Ex. 33 **40. a.** y^6 **b.** $\frac{8}{27}$ **c.** $64x^6$; Sec. 10.1

41. $\frac{b^3}{27a^6}$; Sec. 10.2, Ex. 10 **42.** $\frac{y^2}{4x^6}$ (Sec. 10.2) **43.** $\frac{1}{25y^6}$; Sec. 10.2. Ex. 14 **44.** $\frac{1}{27x^{21}}$ (Sec. 10.2) **45.** $10x^3$; Sec. 10.3, Ex. 8

46. $8x^3$ (Sec. 10.3) **47.** $5x^2 - 3x - 3$; Sec. 10.3, Ex. 9 **48.** $-5x^2 + 7x$ (Sec. 10.3) **49.** $7x^3 + 14x^2 + 35x$; Sec. 10.5, Ex. 4

50. $-2x^3 + 2x^2 - 2x$ (Sec. 10.5) **51.** $3x^3 - 4 + \frac{1}{x}$; Sec. 10.7; Ex. 2 **52.** $2x^6 - 6x + 1$ (Sec. 10.7)

CHAPTER 11 Factoring Polynomials

Vocabulary and Readiness Check **1.** factors **3.** least **5.** false

Exercise Set 11.1 **1.** 4 **3.** 6 **5.** 1 **7.** y^2 **9.** z^7 **11.** xy^2 **13.** 7 **15.** $4y^3$ **17.** $5x^2$ **19.** $3x^3$ **21.** $9x^2y$ **23.** $10a^6b$ **25.** $3(a + 2)$ **27.** $15(2x - 1)$ **29.** $x^2(x + 5)$ **31.** $2y^3(3y + 1)$ **33.** $2x(16y - 9x)$ **35.** $4(x - 2y + 1)$ **37.** $3x(2x^2 - 3x + 4)$ **39.** $a^2b^2(a^5b^4 - a + b^3 - 1)$ **41.** $5xy(x^2 - 3x + 2)$ **43.** $4(2x^5 + 4x^4 - 5x^3 + 3)$ **45.** $\frac{1}{3}x(x^3 + 2x^2 - 4x^4 + 1)$ **47.** $(x^2 + 2)(y + 3)$ **49.** $(y + 4)(z + 3)$ **51.** $(z^2 - 6)(r + 1)$ **53.** $-1(x + 7)$ **55.** $-1(2 - z)$ **57.** $-1(-3a + b - 2)$ **59.** $(x + 2)(x^2 + 5)$ **61.** $(x + 3)(5 + y)$ **63.** $(3x - 2)(2x^2 + 5)$ **65.** $(5m^2 + 6n)(m + 1)$ **67.** $(y - 4)(2 + x)$ **69.** $(2x + 1)(x^2 + 4)$ **71.** $(x - 2y)(4x - 3)$ **73.** $(5q - 4p)(q - 1)$ **75.** $2(2y - 7)(3x^2 - 1)$ **77.** $x^2 + 7x + 10$ **79.** $b^2 - 3b - 4$ **81.** 2, 6 **83.** −1, −8 **85.** −2, 5 **87.** −8, 3 **89.** b **91.** factored **93.** not factored **95.** **a.** 8684 thousand barrels per day **b.** 9022 thousand barrels per day **c.** $-13(x^2 - 17x - 652)$ or $13(-x^2 + 17x + 652)$ **97.** $4x^2 - \pi x^2$; $x^2(4 - \pi)$ **99.** $(x^3 - 1)$ units **101.** answers may vary **103.** answers may vary

Vocabulary and Readiness Check **1.** true **3.** false

Exercise Set 11.2 **1.** $(x + 6)(x + 1)$ **3.** $(y - 9)(y - 1)$ **5.** $(x - 3)(x - 3)$ or $(x - 3)^2$ **7.** $(x - 6)(x + 3)$ **9.** $(x + 10)(x - 7)$ **11.** prime **13.** $(x + 5y)(x + 3y)$ **15.** $(a^2 - 5)(a^2 + 3)$ **17.** $(m + 13)(m + 1)$ **19.** $(t - 2)(t + 12)$ **21.** $(a - 2b)(a - 8b)$ **23.** $2(z + 8)(z + 2)$ **25.** $2x(x - 5)(x - 4)$ **27.** $(x - 4y)(x + y)$ **29.** $(x + 12)(x + 3)$ **31.** $(x - 2)(x + 1)$ **33.** $(r - 12)(r - 4)$ **35.** $(x + 2y)(x - y)$ **37.** $3(x + 5)(x - 2)$ **39.** $3(x - 18)(x - 2)$ **41.** $(x - 24)(x + 6)$ **43.** prime **45.** $(x - 5)(x - 3)$ **47.** $6x(x + 4)(x + 5)$ **49.** $4y(x^2 + x - 3)$ **51.** $(x - 7)(x + 3)$ **53.** $(x + 5y)(x + 2y)$ **55.** $2(t + 8)(t + 4)$ **57.** $x(x - 6)(x + 4)$ **59.** $2t^3(t - 4)(t - 3)$ **61.** $5xy(x - 8y)(x + 3y)$ **63.** $3(m - 9)(m - 6)$ **65.** $-1(x - 11)(x - 1)$ **67.** $\frac{1}{2}(y - 11)(y + 2)$ **69.** $x(xy - 4)(xy + 5)$ **71.** $2x^2 + 11x + 5$ **73.** $15y^2 - 17y + 4$ **75.** $9a^2 + 23ab - 12b^2$ **77.** $x^2 + 5x - 24$ **79.** answers may vary **81.** $2x^2 + 28x + 66$; $2(x + 3)(x + 11)$ **83.** $-16(t - 5)(t + 1)$ **85.** $\left(x + \frac{1}{4}\right)\left(x + \frac{1}{4}\right)$ or $\left(x + \frac{1}{4}\right)^2$ **87.** $(x + 1)(z - 10)(z + 7)$ **89.** 15; 28; 39; 48; 55; 60; 63; 64 **91.** 9; 12; 21 **93.** $(x^n + 10)(x^n - 2)$

Exercise Set 11.3 **1.** $x + 4$ **3.** $10x - 1$ **5.** $4x - 3$ **7.** $(2x + 3)(x + 5)$ **9.** $(y - 1)(8y - 9)$ **11.** $(2x + 1)(x - 5)$ **13.** $(4r - 1)(5r + 8)$ **15.** $(5x + 1)(2x + 3)$ **17.** $(3x - 2)(x + 1)$ **19.** $(3x - 5y)(2x - y)$ **21.** $(3m - 5)(5m + 3)$ **23.** $(x - 4)(x - 5)$ **25.** $(2x + 11)(x - 9)$ **27.** $(7t + 1)(t - 4)$ **29.** $(3a + b)(a + 3b)$ **31.** $(7p + 1)(7p - 2)$ **33.** $(6x - 7)(3x + 2)$ **35.** prime **37.** $(8x + 3)(3x + 4)$ **39.** $x(3x + 2)(4x + 1)$ **41.** $3(7b + 5)(b - 3)$ **43.** $(3z + 4)(4z - 3)$ **45.** $2y^2(3x - 10)(x + 3)$ **47.** $(2x - 7)(2x + 3)$ **49.** $3(x^2 - 14x + 21)$ **51.** $(4x + 9y)(2x - 3y)$ **53.** $-1(x - 6)(x + 4)$ **55.** $x(4x + 3)(x - 3)$ **57.** $(4x - 9)(6x - 1)$ **59.** $b(8a - 3)(5a + 3)$ **61.** $2x(3x + 2)(5x + 3)$ **63.** $2y(3y + 5)(y - 3)$ **65.** $5x^2(2x - y)(x + 3y)$ **67.** $-1(2x - 5)(7x - 2)$ **69.** $p^2(4p - 5)(4p - 5)$ or $p^2(4p - 5)^2$ **71.** $-1(2x + 1)(x - 5)$ **73.** $-4(12x - 1)(x - 1)$ **75.** $(2t^2 + 9)(t^2 - 3)$ **77.** prime **79.** $a(6a^2 + b^2)(a^2 + 6b^2)$ **81.** $x^2 - 16$ **83.** $x^2 + 4x + 4$ **85.** $4x^2 + 4x + 1$ **87.** no **89.** $4x^2 + 21x + 5$; $(4x + 1)(x + 5)$ **91.** $\left(2x + \frac{1}{2}\right)\left(2x + \frac{1}{2}\right)$ or $\left(2x + \frac{1}{2}\right)^2$ **93.** $(y - 1)^2(4x^2 + 10x + 25)$ **95.** 2; 14 **97.** 2 **99.** answers may vary

Exercise Set 11.4 **1.** $(x + 3)(x + 2)$ **3.** $(y + 8)(y - 2)$ **5.** $(8x - 5)(x - 3)$ **7.** $(5x^2 - 3)(x^2 + 5)$ **9.** **a.** 9.2 **b.** $9x + 2x$ **c.** $(2x + 3)(3x + 1)$ **11.** **a.** −20, −3 **b.** $-20x - 3x$ **c.** $(3x - 4)(5x - 1)$ **13.** $(3y + 2)(7y + 1)$ **15.** $(7x - 11)(x + 1)$ **17.** $(5x - 2)(2x - 1)$ **19.** $(2x - 5)(x - 1)$ **21.** $(2x + 3)(2x + 3)$ or $(2x + 3)^2$ **23.** $(2x + 3)(2x - 7)$ **25.** $(5x - 4)(2x - 3)$ **27.** $x(2x + 3)(x + 5)$ **29.** $2(8y - 9)(y - 1)$ **31.** $(2x - 3)(3x - 2)$ **33.** $3(3a + 2)(6a - 5)$ **35.** $a(4a + 1)(5a + 8)$ **37.** $3x(4x + 3)(x - 3)$ **39.** $y(3x + y)(x + y)$ **41.** prime **43.** $6(a + b)(4a - 5b)$ **45.** $p^2(15p + q)(p + 2q)$ **47.** $(7 + x)(5 + x)$ or $(x + 7)(x + 5)$ **49.** $(6 - 5x)(1 - x)$ or $(5x - 6)(x - 1)$ **51.** $x^2 - 4$ **53.** $y^2 + 8y + 16$ **55.** $81z^2 - 25$ **57.** $16x^2 - 24x + 9$ **59.** $10x^2 + 45x + 45$; $5(2x + 3)(x + 3)$ **61.** $(x^n + 2)(x^n + 3)$ **63.** $(3x^n - 5)(x^n + 7)$ **65.** answers may vary

Calculator Explorations

	$x^2 - 2x + 1$	$x^2 - 2x - 1$	$(x - 1)^2$
$x = 5$	16	14	16
$x = -3$	16	14	16
$x = 2.7$	2.89	0.89	2.89
$x = -12.1$	171.61	169.61	171.61
$x = 0$	1	−1	1

Vocabulary and Readiness Check **1.** perfect square trinomial **3.** perfect square trinomial **5.** $(x + 5y)^2$ **7.** false

Exercise Set 11.5 **1.** yes **3.** no **5.** yes **7.** no **9.** no **11.** yes **13.** $(x + 11)^2$ **15.** $(x - 8)^2$ **17.** $(4a - 3)^2$ **19.** $(x^2 + 2)^2$ **21.** $2(n - 7)^2$ **23.** $(4y + 5)^2$ **25.** $(xy - 5)^2$ **27.** $m(m + 9)^2$ **29.** prime **31.** $(3x - 4y)^2$ **33.** $(x + 2)(x - 2)$ **35.** $(9 + p)(9 - p)$ or $-1(p + 9)(p - 9)$ **37.** $-1(2r + 1)(2r - 1)$ **39.** $(3x + 4)(3x - 4)$ **41.** prime **43.** $(-6 + x)(6 + x)$ or $-1(6 + x)(6 - x)$ **45.** $(m^2 + 1)(m + 1)(m - 1)$ **47.** $(x + 13y)(x - 13y)$ **49.** $2(3r + 2)(3r - 2)$ **51.** $x(3y + 2)(3y - 2)$ **53.** $16x^2(x + 2)(x - 2)$ **55.** $xy(y - 3z)(y + 3z)$ **57.** $4(3x - 4y)(3x + 4y)$ **59.** $9(4 - 3x)(4 + 3x)$ **61.** $(5y - 3)(5y + 3)$ **63.** $(11m + 10n)(11m - 10n)$ **65.** $(xy - 1)(xy + 1)$ **67.** $\left(x - \frac{1}{2}\right)\left(x + \frac{1}{2}\right)$ **69.** $\left(7 - \frac{3}{5}m\right)\left(7 + \frac{3}{5}m\right)$ **71.** $(9a + 5b)(9a - 5b)$ **73.** $(x + 7y)^2$ **75.** $2(4n^2 - 7)^2$ **77.** $x^2(x^2 + 9)(x + 3)(x - 3)$ **79.** $pq(8p + 9q)(8p - 9q)$ **81.** $x = 6$ **83.** $m = -2$ **85.** $z = \frac{1}{5}$ **87.** $\left(x - \frac{1}{3}\right)^2$ **89.** $(x + 2 + y)(x + 2 - y)$ **91.** $(b - 4)(a + 4)(a - 4)$ **93.** $(x + 3 + 2y)(x + 3 - 2y)$ **95.** $(x^n + 10)(x^n - 10)$ **97.** 8 **99.** answers may vary **101.** $(x + 6)$ **103.** $a^2 + 2ab + b^2$ **105. a.** 777 ft **b.** 441 ft **c.** 7 sec **d.** $(29 + 4t)(29 - 4t)$ **107. a.** 1456 feet **b.** 816 feet **c.** 10 seconds **d.** $16(10 + t)(10 - t)$

Integrated Review **1.** $(x - 3)(x + 4)$ **2.** $(x - 8)(x - 2)$ **3.** $(x + 2)(x - 3)$ **4.** $(x + 1)^2$ **5.** $(x - 3)^2$ **6.** $(x + 2)(x - 1)$ **7.** $(x + 3)(x - 2)$ **8.** $(x + 3)(x + 4)$ **9.** $(x - 5)(x - 2)$ **10.** $(x - 6)(x + 5)$ **11.** $2(x - 7)(x + 7)$ **12.** $3(x - 5)(x + 5)$ **13.** $(x + 3)(x + 5)$ **14.** $(y - 7)(3 + x)$ **15.** $(x + 8)(x - 2)$ **16.** $(x - 7)(x + 4)$ **17.** $4x(x + 7)(x - 2)$ **18.** $6x(x - 5)(x + 4)$ **19.** $2(3x + 4)(2x + 3)$ **20.** $(2a - b)(4a + 5b)$ **21.** $(2a + b)(2a - b)$ **22.** $(x + 5y)(x - 5y)$ **23.** $(4 - 3x)(7 + 2x)$ **24.** $(5 - 2x)(4 + x)$ **25.** prime **26.** prime **27.** $(3y + 5)(2y - 3)$ **28.** $(4x - 5)(x + 1)$ **29.** $9x(2x^2 - 7x + 1)$ **30.** $4a(3a^2 - 6a + 1)$ **31.** $(4a - 7)^2$ **32.** $(5p - 7)^2$ **33.** $(7 - x)(2 + x)$ **34.** $(3 + x)(1 - x)$ **35.** $3x^2y(x + 6)(x - 4)$ **36.** $2xy(x + 5y)(x - y)$ **37.** $3xy(4x^2 + 81)$ **38.** $2xy^2(3x^2 + 4)$ **39.** $2xy(1 + 6x)(1 - 6x)$ **40.** $2x(x - 3)(x + 3)$ **41.** $(x + 6)(x + 2)(x - 2)$ **42.** $(x - 2)(x + 6)(x - 6)$ **43.** $2a^2(3a + 5)$ **44.** $2n(2n - 3)$ **45.** $(3x - 1)(x^2 + 4)$ **46.** $(x - 2)(x^2 + 3)$ **47.** $6(x + 2y)(x + y)$ **48.** $2(x + 4y)(6x - y)$ **49.** $(x + y)(5 + x)$ **50.** $(x - y)(7 + y)$ **51.** $(7t - 1)(2t - 1)$ **52.** prime **53.** $-1(3x + 5)(x - 1)$ **54.** $-1(7x - 2)(x + 3)$ **55.** $(1 - 10a)(1 + 2a)$ **56.** $(1 + 5a)(1 - 12a)$ **57.** $(x + 3)(x - 3)(x - 1)(x + 1)$ **58.** $(x + 3)(x - 3)(x + 2)(x - 2)$ **59.** $(x - 15)(x - 8)$ **60.** $(y + 16)(y + 6)$ **61.** prime **62.** $(4a - 7b)^2$ **63.** $(5p - 7q)^2$ **64.** $(7x + 3y)(x + 3y)$ **65.** $-1(x - 5)(x + 6)$ **66.** $-1(x - 2)(x - 4)$ **67.** $(3r - 1)(s + 4)$ **68.** $(x - 2)(x^2 + 1)$ **69.** $(x - 2y)(4x - 3)$ **70.** $(2x - y)(2x + 7z)$ **71.** $(x + 12y)(x - 3y)$ **72.** $(3x - 2y)(x + 4y)$ **73.** $(x^2 + 2)(x + 4)(x - 4)$ **74.** $(x^2 + 3)(x + 5)(x - 5)$ **75.** answers may vary **76.** yes; $9(x^2 + 9y^2)$

Vocabulary and Readiness Check **1.** quadratic **3.** 3, −5

Exercise Set 11.6 **1.** 2, −1 **3.** 6, 7 **5.** −9, −17 **7.** 0, −6 **9.** 0, 8 **11.** $-\frac{3}{2}, \frac{5}{4}$ **13.** $\frac{7}{2}, -\frac{2}{7}$ **15.** $\frac{1}{2}, -\frac{1}{3}$ **17.** −0.2, −1.5 **19.** 9, 4 **21.** −4, 2 **23.** 0, 7 **25.** 0, −20 **27.** 4, −4 **29.** 8, −4 **31.** −3, 12 **33.** $\frac{7}{3}, -2$ **35.** $\frac{8}{3}, -9$ **37.** $0, -\frac{1}{2}, \frac{1}{2}$ **39.** $\frac{17}{2}$ **41.** $\frac{3}{4}$ **43.** $-\frac{1}{2}, \frac{1}{2}$ **45.** $-\frac{3}{2}, -\frac{1}{2}, 3$ **47.** −5, 3 **49.** $-\frac{5}{6}, \frac{6}{5}$ **51.** $2, -\frac{4}{5}$ **53.** $-\frac{4}{3}, 5$ **55.** −4, 3 **57.** 0, 8, 4 **59.** −7 **61.** $0, \frac{3}{2}$ **63.** 0, 1, −1 **65.** $-6, \frac{4}{3}$ **67.** $\frac{6}{7}, 1$ **69.** $\frac{47}{45}$ **71.** $\frac{17}{60}$ **73.** $\frac{7}{10}$ **75.** didn't write equation in standard form; should be $x = 4$ or $x = -2$ **77.** answers may vary, for example, $(x - 6)(x + 1) = 0$ **79.** answers may vary, for example, $x^2 - 12x + 35 = 0$ **81. a.** 300; 304; 276; 216; 124; 0; −156 **b.** 5 sec **c.** 304 ft **83.** $0, \frac{1}{2}$ **85.** 0, −15

The Bigger Picture **1.** −34 **2.** x^{22} **3.** $-4x^3 - 6x^2 + 8$ **4.** $y - 1 + \frac{3}{y^2}$ **5.** $10x(x + 5)(x - 5)$ **6.** $(x - 1)(x - 35)$ **7.** $3(2y + 5)(x - 1)$ **8.** $x(5y - 7)(y + 1)$ **9.** $5, -\frac{1}{2}$ **10.** 1 **11.** −2, 14 **12.** $\frac{33}{17}$

Exercise Set 11.7 **1.** width: x; length: $x + 4$ **3.** x and $x + 2$ if x is an odd integer **5.** base: x; height: $4x + 1$ **7.** 11 units **9.** 15 cm, 13 cm, 22 cm, 70 cm **11.** base: 16 mi; height: 6 mi **13.** 5 sec **15.** width: 5 cm; length: 6 cm **17.** 54 diagonals **19.** 10 sides **21.** −12 or 11 **23.** 14, 15 **25.** 13 feet **27.** 5 in. **29.** 12 mm, 16 mm, 20 mm **31.** 10 km **33.** 36 ft **35.** 9.5 sec **37.** 20% **39.** length: 15 mi; width: 8 mi **41.** 105 units **43.** 11,250 thousand acres **45.** 10,750 thousand acres **47.** 1995 **49.** answers may vary **51.** 8 m **53.** 10 and 15 **55.** width of pool: 29 m; length of pool: 35 m

Chapter 11 Vocabulary Check **1.** quadratic equation **2.** factoring **3.** greatest common factor **4.** perfect square trinomial

Chapter Review **1.** $2x - 5$ **2.** $2x^4 + 1 - 5x^3$ **3.** $5(m + 6)$ **4.** $4x(5x^2 + 3x + 6)$ **5.** $(2x + 3)(3x - 5)$ **6.** $(x + 1)(5x - 1)$ **7.** $(x - 1)(3x + 2)$ **8.** $(3x + 5)(2x - 1)$ **9.** $(a + 3b)(3a + b)$ **10.** $(x + 4)(x + 2)$ **11.** $(x - 8)(x - 3)$ **12.** prime **13.** $(x - 6)(x + 1)$ **14.** $(x + 4)(x - 2)$ **15.** $(x + 6y)(x - 2y)$ **16.** $(x + 5y)(x + 3y)$ **17.** $2(3 - x)(12 + x)$ **18.** $4(8 + 3x - x^2)$ **19.** $5y(y - 6)(y - 4)$ **20.** −48, 2 **21.** factor out the GCF, 3 **22.** $(2x + 1)(x + 6)$ **23.** $(2x + 3)(2x - 1)$ **24.** $(3x + 4y)(2x - y)$ **25.** prime **26.** $(2x + 3)(x - 13)$ **27.** $(6x + 5y)(3x - 4y)$ **28.** $5y(2y - 3)(y + 4)$ **29.** $5x^2 - 9x - 2$; $(5x + 1)(x - 2)$ **30.** $16x^2 - 28x + 6$; $2(4x - 1)(2x - 3)$ **31.** yes **32.** no **33.** no **34.** yes **35.** yes **36.** no **37.** yes **38.** no **39.** $(x + 9)(x - 9)$ **40.** $(x + 6)^2$ **41.** $(2x + 3)(2x - 3)$ **42.** $(3t + 5s)(3t - 5s)$ **43.** prime **44.** $(n - 9)^2$ **45.** $3(r + 6)^2$ **46.** $(3y - 7)^2$ **47.** $5m^6(m + 1)(m - 1)$ **48.** $(2x - 7y)^2$ **49.** $3y(x + y)^2$ **50.** $(4x^2 + 1)(2x + 1)(2x - 1)$ **51.** −6, 2 **52.** $0, -1, \frac{2}{7}$ **53.** $-\frac{1}{5}, -3$ **54.** −7, −1 **55.** −4, 6 **56.** −5 **57.** 2, 8 **58.** $\frac{1}{3}$ **59.** $-\frac{2}{7}, \frac{3}{8}$ **60.** 0, 6 **61.** 5, −5 **62.** $x^2 - 9x + 20 = 0$ **63.** c **64.** d **65.** 9 units **66.** 8 units, 13 units, 16 units, 10 units **67.** width: 20 in.; length: 25 in. **68.** 36 yd **69.** 19 and 20 **70. a.** 17.5 sec and 10 sec; answers may vary **b.** 27.5 sec **71.** 32 cm **72.** $6(x + 4)$ **73.** $7(x - 9)$ **74.** $(4x - 3)(11x - 6)$ **75.** $(x - 5)(2x - 1)$ **76.** $(3x - 4)(x^2 + 2)$ **77.** $(y + 2)(x - 1)$

78. $2(x+4)(x-3)$ **79.** $3x(x-9)(x-1)$ **80.** $(2x+9)(2x-9)$ **81.** $2(x+3)(x-3)$ **82.** $(4x-3)^2$ **83.** $5(x+2)^2$ **84.** $-\frac{7}{2}, 4$ **85.** $-3, 5$ **86.** $0, -7, -4$ **87.** $3, 2$ **88.** $0, 16$ **89.** 19 in.; 8 in.; 21 in. **90.** length: 6 in.; width: 2 in.

Chapter Test **1.** $3x(3x-1)$ **2.** $(x+7)(x+4)$ **3.** $(7+m)(7-m)$ **4.** $(y+11)^2$ **5.** $(x^2+4)(x+2)(x-2)$ **6.** $(a+3)(4-y)$ **7.** prime **8.** $(y-12)(y+4)$ **9.** $(a+b)(3a-7)$ **10.** $(3x-2)(x-1)$ **11.** $5(6+x)(6-x)$ **12.** $3x(x-5)(x-2)$ **13.** $(6t+5)(t-1)$ **14.** $(x-7)(y-2)(y+2)$ **15.** $x(1+x^2)(1+x)(1-x)$ **16.** $(x+12y)(x+2y)$ **17.** $3, -9$ **18.** $-7, 2$ **19.** $-7, 1$ **20.** $0, \frac{3}{2}, -\frac{4}{3}$ **21.** $0, 3, -3$ **22.** $-3, 5$ **23.** $0, \frac{5}{2}$ **24.** 17 ft **25.** width: 6 units; length: 9 units **26.** 7 sec **27.** hypotenuse: 25 cm; legs: 15 cm, 20 cm **28.** 8.25 sec

Cumulative Review **1.** -3; Sec. 2.5, Ex. 13 **2.** -14; Sec. 2.5 **3.** -9; Sec. 2.5, Ex. 14 **4.** -72; Sec. 2.5 **5.** -9; Sec. 2.6, Ex. 3 **6.** -9; Sec. 2.6 **7.** $6y+2$; Sec. 3.1, Ex. 2 **8.** $-4a-1$; Sec. 3.1 **9.** $-x+5$; Sec. 3.1, Ex. 4 **10.** $7.3x-6$; Sec. 3.1 **11.** -15; Sec. 3.2, Ex. 6 **12.** -2; Sec. 3.2 **13.** 2; Sec. 3.3, Ex. 4 **14.** 0; Sec. 3.3 **15.** $\frac{31}{2}$; Sec. 6.1, Ex. 5 **16.** -17; Sec. 6.1 **17. a.** $9 \le 11$ **b.** $8 > 1$ **c.** $3 \ne 4$; Sec. 9.1, Ex. 7 **18. a.** $5 \ge 1$; **b.** $2 \ne -4$; Sec. 9.1 **19.** every real number; Sec. 9.3, Ex. 7 **20.** no solution; Sec. 9.3 **21.** $l = \frac{V}{wh}$; Sec. 9.5, Ex. 5 **22.** $x = \frac{2y+5}{3}$; Sec. 9.5 **23.** 5^{18}; Sec. 10.1, Ex. 16 **24.** 7^{18}; Sec. 10.1 **25.** y^{16}; Sec. 10.1; Ex. 17 **26.** x^{33}; Sec. 10.1 **27.** x^6; Sec. 10.2; Ex. 9 **28.** y^{22}; Sec. 10.2 **29.** $\frac{y^{18}}{z^{36}}$; Sec. 10.2, Ex. 11 **30.** $\frac{y^8}{x^2}$; Sec. 10.2 **31.** $\frac{1}{x^{19}}$; Sec. 10.2, Ex. 13 **32.** y^{13}; Sec. 10.2 **33.** $4x$; Sec. 10.3, Ex. 6 **34.** $2y$; Sec. 10.3 **35.** $13x^2-2$; Sec. 10.3, Ex. 7 **36.** $12y-2y^2$; Sec. 10.3 **37.** $4x^2-4xy+y^2$; Sec. 10.5, Ex. 8 **38.** $9x^2+6x+1$; Sec. 10.5 **39.** t^2+4t+4; Sec. 10.6, Ex. 5 **40.** $x^2-8x+16$; Sec. 10.6 **41.** $x^4-14x^2y+49y^2$; Sec. 10.6, Ex.8 **42.** $x^4+14x^2y+49y^2$; Sec. 10.6 **43.** $2xy-4+\frac{1}{2y}$; Sec. 10.7, Ex. 3 **44.** $4ab^2-1+\frac{2}{a}$; Sec. 10.7 **45.** $(x+3)(5+y)$; Sec. 11.1, Ex. 9 **46.** $(y-2)(9+x)$; Sec. 11.1 **47.** $(x^2+2)(x^2+3)$; Sec. 11.2, Ex. 7 **48.** $(x^2+1)(x^2-5)$; Sec. 11.2 **49.** $2(x-2)(3x+5)$; Sec. 11.4, Ex. 2 **50.** $5(x+2)(2x+1)$; Sec. 11.4 **51.** 3 sec; Sec. 11.7, Ex. 1 **52.** $-12, 10$; Sec. 11.7

CHAPTER 12 Rational Expressions

Vocabulary and Readiness Check **1.** rational expression **3.** -1 **5.** 2 **7.** $\frac{-a}{b}, \frac{a}{-b}$

Exercise Set 12.1 **1.** $\frac{7}{4}$ **3.** $-\frac{8}{3}$ **5.** $-\frac{11}{2}$ **7. a.** \$403 **b.** \$7 **c.** decrease; answers may vary **9.** $x=0$ **11.** $x=-2$ **13.** $x=\frac{5}{2}$ **15.** $x=0, x=-2$ **17.** none **19.** $x=6, x=-1$ **21.** $x=-2, x=-\frac{7}{3}$ **23.** 1 **25.** -1 **27.** $\frac{1}{4(x+2)}$ **29.** $\frac{1}{x+2}$ **31.** can't simplify **33.** -5 **35.** $\frac{7}{x}$ **37.** $\frac{1}{x-9}$ **39.** $5x+1$ **41.** $\frac{x^2}{x-2}$ **43.** $7x$ **45.** $\frac{x+5}{x-5}$ **47.** $\frac{x+2}{x+4}$ **49.** $\frac{x+2}{2}$ **51.** $-(x+2)$ **53.** $\frac{x+1}{x-1}$ **55.** $x+y$ **57.** $\frac{5-y}{2}$ **59.** $\frac{2y+5}{3y+4}$ **61.** $\frac{-(x-10)}{x+8}; \frac{-x+10}{x+8}; \frac{x-10}{-(x+8)}; \frac{x-10}{-x-8}$ **63.** $\frac{-(5y-3)}{y-12}; \frac{-5y+3}{y-12}; \frac{5y-3}{-(y-12)}; \frac{5y-3}{-y+12}$ **65.** correct **67.** correct **69.** $\frac{3}{11}$ **71.** $\frac{4}{3}$ **73.** $\frac{117}{40}$ **75.** correct **77.** incorrect; $\frac{1+2}{1+3}=\frac{3}{4}$ **79.** answers may vary **81.** answers may vary **83.** 400 mg **85.** 45.5% **87.** 85.9

Vocabulary and Readiness Check **1.** reciprocals **3.** $\frac{a \cdot b}{b \cdot c}$ **5.** $\frac{6}{7}$

Exercise Set 12.2 **1.** $\frac{21}{4y}$ **3.** x^4 **5.** $-\frac{b^2}{6}$ **7.** $\frac{x^2}{10}$ **9.** $\frac{1}{3}$ **11.** $\frac{m+n}{m-n}$ **13.** $\frac{x+5}{x}$ **15.** $\frac{(x+2)(x-3)}{(x-4)(x+4)}$ **17.** $\frac{2x^4}{3}$ **19.** $\frac{12}{y^6}$ **21.** $x(x+4)$ **23.** $\frac{3(x+1)}{x^3(x-1)}$ **25.** m^2-n^2 **27.** $-\frac{x+2}{x-3}$ **29.** $\frac{x+2}{x-3}$ **31.** $\frac{5}{6}$ **33.** $\frac{3x}{8}$ **35.** $\frac{3}{2}$ **37.** $\frac{3x+4y}{2(x+2y)}$ **39.** $\frac{2(x+2)}{x-2}$ **41.** $-\frac{y(x+2)}{4}$ **43.** $\frac{(a+5)(a+3)}{(a+2)(a+1)}$ **45.** $\frac{5}{x}$ **47.** $\frac{2(n-8)}{3n-1}$ **49.** 1440 **51.** 5 **53.** 81 **55.** 73 **57.** 56.7 **59.** 411,755 sq yd **61.** 1364 feet per second **63.** 1 **65.** $-\frac{10}{9}$ **67.** $-\frac{1}{5}$ **69.** true **71.** false; $\frac{x^2+3x}{20}$ **73.** $\frac{2}{9(x-5)}$ sq ft **75.** $\frac{x}{2}$ **77.** $\frac{5a(2a+b)(3a-2b)}{b^2(a-b)(a+2b)}$ **79.** answers may vary **81.** 1543.81 euros

Vocabulary and Readiness Check **1.** $\frac{9}{11}$ **3.** $\frac{a+c}{b}$ **5.** $\frac{5-(6+x)}{x}$

Exercise Set 12.3 **1.** $\frac{a+9}{13}$ **3.** $\frac{3m}{n}$ **5.** 4 **7.** $\frac{y+10}{3+y}$ **9.** $5x+3$ **11.** $\frac{4}{a+5}$ **13.** $\frac{1}{x-6}$ **15.** $4x^3$ **17.** $8x(x+2)$ **19.** $(x+3)(x-2)$ **21.** $3(x+6)$ **23.** $5(x-6)^2$ **25.** $6(x+1)^2$ **27.** $x-8$ or $8-x$ **29.** $(x-1)(x+4)(x+3)$ **31.** $(3x+1)(x+1)(x-1)(2x+1)$ **33.** $2x^2(x+4)(x-4)$ **35.** $\frac{6x}{4x^2}$ **37.** $\frac{24b^2}{12ab^2}$ **39.** $\frac{9y}{2y(x+3)}$ **41.** $\frac{9ab+2b}{5b(a+2)}$ **43.** $\frac{x^2+x}{x(x+4)(x+2)(x+1)}$ **45.** $\frac{18y-2}{30x^2-60}$ **47.** $2x$ **49.** $\frac{x+3}{2x-1}$ **51.** $x+1$ **53.** $\frac{1}{x^2-8}$ **55.** $\frac{6(4x+1)}{x(2x+1)}$ **57.** $\frac{29}{21}$ **59.** $-\frac{5}{12}$ **61.** $\frac{7}{30}$ **63.** d **65.** $\frac{20}{x-2}$ m **67.** answers may vary **69.** 3 packages hot dogs and 2 packages buns **71.** answers may vary **73.** answers may vary

Exercise Set 12.4 **1.** $\frac{5}{x}$ **3.** $\frac{75a+6b^2}{5b}$ **5.** $\frac{6x+5}{2x^2}$ **7.** $\frac{11}{x+1}$ **9.** $\frac{x-6}{(x-2)(x+2)}$ **11.** $\frac{35x-6}{4x(x-2)}$ **13.** $-\frac{2}{x-3}$ **15.** 0 **17.** $-\frac{1}{x^2-1}$ **19.** $\frac{5+2x}{x}$ **21.** $\frac{6x-7}{x-2}$ **23.** $-\frac{y+4}{y+3}$ **25.** $\frac{-5x+14}{4x}$ or $-\frac{5x-14}{4x}$ **27.** 2 **29.** $\frac{9x^4-4x^2}{21}$ **31.** $\frac{x+2}{(x+3)^2}$ **33.** $\frac{9b-4}{5b(b-1)}$ **35.** $\frac{2+m}{m}$ **37.** $\frac{x^2+3x}{(x-7)(x-2)}$ or $\frac{x(x+3)}{(x-7)(x-2)}$ **39.** $\frac{10}{1-2x}$ **41.** $\frac{15x-1}{(x+1)^2(x-1)}$ **43.** $\frac{x^2-3x-2}{(x-1)^2(x+1)}$ **45.** $\frac{a+2}{2(a+3)}$ **47.** $\frac{y(2y+1)}{(2y+3)^2}$ **49.** $\frac{x-10}{2(x-2)}$ **51.** $\frac{2x+21}{(x+3)^2}$ **53.** $\frac{-5x+23}{(x-2)(x-3)}$ **55.** $\frac{7}{2(m-10)}$ **57.** $\frac{2x^2-2x-46}{(x+1)(x-6)(x-5)}$ **59.** $\frac{n+4}{4n(n-1)(n-2)}$ **61.** 10 **63.** 2 **65.** $\frac{25a}{9(a-2)}$ **67.** $\frac{x+4}{(x-2)(x-1)}$ **69.** $x=\frac{2}{3}$ **71.** $x=-\frac{1}{2}, x=1$ **73.** $x=-\frac{15}{2}$ **75.** $\frac{6x^2-5x-3}{x(x+1)(x-1)}$ **77.** $\frac{4x^2-15x+6}{(x-2)^2(x+2)(x-3)}$ **79.** $\frac{-2x^2+14x+55}{(x+2)(x+7)(x+3)}$ **81.** $\frac{2x-16}{(x+4)(x-4)}$ in. **83.** $\frac{P-G}{P}$ **85.** answers may vary **87.** $\left(\frac{90x-40}{x}\right)^\circ$ **89.** answers may vary

The Bigger Picture **1.** −17.7 **2.** 78.26 **3.** 28 **4.** $4x^4-x^2-17$ **5.** $\frac{x-1}{5}$ **6.** $\frac{14}{x+1}$ **7.** $-\frac{11}{18}$ **8.** $\frac{-4x-27}{45}$ or $-\frac{4x+27}{45}$ **9.** $x(9x-11)(x+1)$ **10.** $(4y-7)(3x+1)$ **11.** 12 **12.** $\left\{x \middle| x>\frac{1}{2}\right\}$ **13.** $-\frac{8}{3}$ **14.** −4, 6

Exercise Set 12.5 **1.** 30 **3.** 0 **5.** −2 **7.** −5, 2 **9.** 5 **11.** 3 **13.** 1 **15.** 5 **17.** no solution **19.** 4 **21.** −8 **23.** 6, −4 **25.** 1 **27.** 3, −4 **29.** −3 **31.** 0 **33.** −2 **35.** 8, −2 **37.** no solution **39.** 3 **41.** −11, 1 **43.** $I=\frac{E}{R}$ **45.** $B=\frac{2U-TE}{T}$ **47.** $W=\frac{Bh^2}{705}$ **49.** $G=\frac{V}{N-R}$ **51.** $r=\frac{C}{2\pi}$ **53.** $x=\frac{3y}{3+y}$ **55.** $\frac{1}{x}$ **57.** $\frac{1}{x}+\frac{1}{2}$ **59.** $\frac{1}{3}$ **61.** 5 **63.** 100°, 80° **65.** 22.5°, 67.5° **67.** no; multiplying both terms in the expression by 4 changes the value of the original expression.

The Bigger Picture **1.** $12x^3-11x^2-13x+10$ **2.** $4x^2-4xy+y^2$ **3.** $4y^3(2-5y^2)$ **4.** $(9m-2n)(m-n)$ **5.** −35 **6.** $\frac{16x-70}{x(x-10)}$ or $\frac{2(8x-35)}{x(x-10)}$ **7.** $-12x^{13}$ **8.** 2 **9.** $-\frac{1}{2}$ **10.** $\{y \mid y \geq 4\}$ **11.** $-\frac{27}{23}$ **12.** $\frac{a^{14}}{b^{14}}$

Integrated Review **1.** expression; $\frac{3+2x}{3x}$ **2.** expression; $\frac{18+5a}{6a}$ **3.** equation; 3 **4.** equation; 18 **5.** expression; $\frac{x-1}{x(x+1)}$ **6.** expression; $\frac{3(x+1)}{x(x-3)}$ **7.** equation; no solution **8.** equation; 1 **9.** expression; 10 **10.** expression; $\frac{z}{3(9z-5)}$ **11.** expression; $\frac{5x+7}{x-3}$ **12.** expression; $\frac{7p+5}{2p+7}$ **13.** equation; 23 **14.** equation; 3 **15.** expression; $\frac{25a}{9(a-2)}$ **16.** expression; $\frac{9}{4(x-1)}$ **17.** expression; $\frac{3x^2+5x+3}{(3x-1)^2}$ **18.** expression; $\frac{2x^2-3x-1}{(2x-5)^2}$ **19.** expression; $\frac{4x-37}{5x}$ **20.** equation; $-\frac{7}{3}$ **21.** equation; $\frac{8}{5}$ **22.** expression; $\frac{29x-23}{3x}$ **23.** answers may vary **24.** answers may vary

Vocabulary and Readiness Check **1.** c

Exercise Set 12.6 **1.** 2 **3.** −3 **5.** $2\frac{2}{9}$ hr **7.** $1\frac{1}{2}$ min **9.** trip to park rate: r; to park time: $\frac{12}{r}$; return trip rate: r; return time: $\frac{18}{r}=\frac{12}{r}+1$; $r=6$ mph **11.** 1st portion: 10 mph; cooldown: 8 mph **13.** 2 **15.** $108.00 **17.** 20 mph **19.** 5 **21.** 217 mph **23.** 8 mph **25.** 2.2 mph; 3.3 mph **27.** 3 hr **29.** $666\frac{2}{3}$ mi **31.** 20 hr **33.** car: 70 mph; motorcycle: 60 mph **35.** $5\frac{1}{4}$ hr **37.** $\frac{1}{2}$ **39.** $\frac{3}{7}$ **41.** first pump: 28 min; second pump: 84 min **43.** answers will vary **45.** $R=\frac{D}{T}$ **47.** 3.75 min

Exercise Set 12.7 **1.** $\frac{2}{3}$ **3.** $\frac{2}{3}$ **5.** $\frac{1}{2}$ **7.** $-\frac{21}{5}$ **9.** $\frac{27}{16}$ **11.** $\frac{4}{3}$ **13.** $\frac{1}{21}$ **15.** $-\frac{4x}{15}$ **17.** $\frac{m-n}{m+n}$ **19.** $\frac{2x(x-5)}{7x^2+10}$ **21.** $\frac{1}{y-1}$ **23.** $\frac{1}{6}$ **25.** $\frac{x+y}{x-y}$ **27.** $\frac{3}{7}$ **29.** $\frac{a}{x+b}$ **31.** $\frac{7(y-3)}{8+y}$ **33.** $\frac{3x}{x-4}$ **35.** $-\frac{x+8}{x-2}$ **37.** $\frac{s^2+r^2}{s^2-r^2}$ **39.** $\frac{(x-6)(x+4)}{x-2}$ **41.** Steffi Graf **43.** Martina Navratilova and Steffi Graf **45.** answers may vary **47.** $\frac{13}{24}$ **49.** $\frac{R_1R_2}{R_2+R_1}$ **51.** $\frac{2x}{2-x}$ **53.** $\frac{1}{y^2-1}$ **55.** 12 hr

Chapter 12 Vocabulary Check **1.** ratio **2.** proportion **3.** cross products **4.** rational expression **5.** complex fraction **6.** rate

Chapter Review **1.** $x = 2, x = -2$ **2.** $x = \frac{5}{2}, x = -\frac{3}{2}$ **3.** $\frac{4}{3}$ **4.** $\frac{11}{12}$ **5.** $\frac{2}{x}$ **6.** $\frac{3}{x}$ **7.** $\frac{1}{x-5}$ **8.** $\frac{1}{x+1}$ **9.** $\frac{x(x-2)}{x+1}$ **10.** $\frac{5(x-5)}{x-3}$ **11.** $\frac{x-3}{x-5}$ **12.** $\frac{x}{x+4}$ **13.** $\frac{x+a}{x-c}$ **14.** $\frac{x+5}{x-3}$ **15.** $\frac{3x^2}{y}$ **16.** $-\frac{9x^2}{8}$ **17.** $\frac{x-3}{x+2}$ **18.** $\frac{-2x(2x+5)}{(x-6)^2}$ **19.** $\frac{x+3}{x-4}$ **20.** $\frac{4x}{3y}$ **21.** $(x-6)(x-3)$ **22.** $\frac{2}{3}$ **23.** $\frac{1}{2}$ **24.** $\frac{3(x+2)}{3x+y}$ **25.** $\frac{1}{x+2}$ **26.** $\frac{1}{x-3}$ **27.** $\frac{2x-10}{3x^2}$ **28.** $\frac{2x+1}{2x^2}$ **29.** $14x$ **30.** $(x-8)(x+8)(x+3)$ **31.** $\frac{10x^2y}{14x^3y}$ **32.** $\frac{36y^2x}{16y^3x}$ **33.** $\frac{x^2-3x-10}{(x+2)(x-5)(x+9)}$ **34.** $\frac{3x^2+4x-15}{(x+2)^2(x+3)}$ **35.** $\frac{4y-30x^2}{5x^2y}$ **36.** $\frac{-2x+10}{(x-3)(x-1)}$ **37.** $\frac{-2x-2}{x+3}$ **38.** $\frac{5x+5}{(x+4)(x-2)(x-1)}$ **39.** $\frac{x-4}{3x}$ **40.** $-\frac{x}{x-1}$ **41.** 30 **42.** 3, −4 **43.** no solution **44.** 5 **45.** $\frac{9}{7}$ **46.** −6, 1 **47.** 3 **48.** 2 **49.** fast car speed: 30 mph; slow car speed: 20 mph **50.** 20 mph **51.** $17\frac{1}{2}$ hr **52.** $8\frac{4}{7}$ days **53.** $-\frac{7}{18y}$ **54.** $\frac{6}{7}$ **55.** $\frac{3y-1}{2y-1}$ **56.** $-\frac{7+2x}{2x}$ **57.** $\frac{1}{2x}$ **58.** $\frac{x(x-3)}{x+7}$ **59.** $\frac{x-4}{x+4}$ **60.** $\frac{(x-9)(x+8)}{(x+5)(x+9)}$ **61.** $\frac{1}{x-6}$ **62.** $\frac{2x+1}{4x}$ **63.** $\frac{2}{(x+3)(x-2)}$ **64.** $-\frac{3x}{(x+2)(x-3)}$ **65.** $\frac{1}{2}$ **66.** no solution **67.** 1 **68.** $1\frac{5}{7}$ days **69.** $\frac{3}{10}$ **70.** $\frac{2}{3}$

Chapter Test **1.** $x = -1, x = -3$ **2. a.** \$115 **b.** \$103 **3.** $\frac{3}{5}$ **4.** $\frac{1}{x+6}$ **5.** −1 **6.** $-\frac{1}{x+y}$ **7.** $\frac{2m(m+2)}{m-2}$ **8.** $\frac{a+2}{a+5}$ **9.** $\frac{(x-6)(x-7)}{(x+7)(x+2)}$ **10.** 15 **11.** $\frac{y-2}{4}$ **12.** $-\frac{1}{2x+5}$ **13.** $\frac{3a-4}{(a-3)(a+2)}$ **14.** $\frac{3}{x-1}$ **15.** $\frac{2(x+3)(x+5)}{x(x^2+4x+1)}$ **16.** $\frac{x^2+2x+35}{(x+9)(x+2)(x-5)}$ **17.** $\frac{4y^2+13y-15}{(y+5)(y+1)(y+4)}$ **18.** $\frac{30}{11}$ **19.** −6 **20.** no solution **21.** no solution **22.** −2, 5 **23.** $\frac{xz}{2y}$ **24.** $b-a$ **25.** $\frac{5y^2-1}{y+2}$ **26.** $x = 1$ and $x = 5$ **27.** 30 mph **28.** $6\frac{2}{3}$ hr

Cumulative Review **1.** 35%; Sec. 6.2, Ex. 17 **2.** 80%; Sec. 6.2 **3.** $66\frac{2}{3}$%; Sec. 6.2, Ex. 18 **4.** $11\frac{1}{9}$%; Sec. 6.2 **5.** 225%; Sec. 6.2, Ex. 19 **6.** 375%; Sec. 6.2 **7.** commutative property of multiplication; Sec. 9.2, Ex. 18 **8.** commutative property of addition; Sec. 9.2 **9.** associative property of addition; Sec. 9.2, Ex. 16 **10.** associative property of multiplication; Sec. 9.2 **11.** shorter piece, 2 ft; longer piece, 8 ft; Sec. 9.4, Ex. 3 **12.** 16 ft, 16 ft, 13 ft; Sec. 9.4 **13.** $\frac{y-b}{m} = x$; Sec. 9.5, Ex. 6 **14.** $b = y - mx$; Sec. 9.5 **15.** $x \le -10$;  −10 ; Sec. 9.6, Ex. 4 **16.** $\{x|x > -9\}$; Sec. 9.6 **17.** x^3; Sec. 10.1, Ex. 24 **18.** x^8; Sec. 10.1 **19.** $4^4 = 256$; Sec. 10.1, Ex. 25 **20.** 7^8; Sec. 10.1 **21.** −27; Sec. 10.1, Ex. 26 **22.** 16; Sec. 10.1 **23.** $2x^4y$; Sec. 10.1, Ex. 27 **24.** $13ab$; Sec. 10.1 **25.** $\frac{2}{x^3}$; Sec. 10.2, Ex. 2 **26.** $\frac{9}{x^2}$; Sec. 10.2 **27.** $\frac{1}{16}$; Sec. 10.2, Ex. 4 **28.** $-\frac{1}{27}$; Sec. 10.2 **29.** $10x^4 + 30x$; Sec. 10.5, Ex. 5 **30.** $12y^3 - 6y$; Sec. 10.5 **31.** $-15x^4 - 18x^3 + 3x^2$; Sec. 10.5, Ex. 6 **32.** $-35y^3 + 15y^2 - 5y$; Sec. 10.5 **33.** $4x^2 - 4x + 6 + \frac{-11}{2x+3}$; Sec. 10.7, Ex. 7 **34.** $3x - 5 + \frac{9}{2x+1}$; Sec. 10.7 **35.** $(x+3)(x+4)$; Sec. 11.2, Ex. 1 **36.** $(x+7)(x+10)$; Sec. 11.2 **37.** $(5x+2y)^2$; Sec. 11.5, Ex. 5 **38.** $4(3a-2b)^2$; Sec. 11.5 **39.** $x = 11, x = -2$; Sec. 11.6, Ex. 4 **40.** $x = 3, x = -5$; Sec. 11.6 **41.** $\frac{2}{5}$; Sec. 12.2, Ex. 2 **42.** $\frac{6x}{5}$; Sec. 12.2 **43.** $3x - 5$; Sec. 12.3, Ex. 3 **44.** 4; Sec. 12.3 **45.** $\frac{3}{x-2}$; Sec. 12.4, Ex. 2 **46.** $\frac{7x+6}{x^2-9}$; Sec. 12.4 **47.** $t = 5$; Sec. 12.5, Ex. 2 **48.** $y = \frac{2}{3}$; Sec. 12.5 **49.** $2\frac{1}{10}$ hr; Sec. 12.6, Ex. 2 **50.** $7\frac{1}{5}$ hr; Sec. 12.6 **51.** $\frac{3}{z}$; Sec. 12.7, Ex. 3 **52.** $\frac{x-3}{9}$; Sec. 12.7

CHAPTER 13 Graphing Equations and Inequalities

Vocabulary and Readiness Check **1.** x-axis **3.** origin **5.** x-coordinate; y-coordinate **7.** solution

Exercise Set 13.1 **1.**

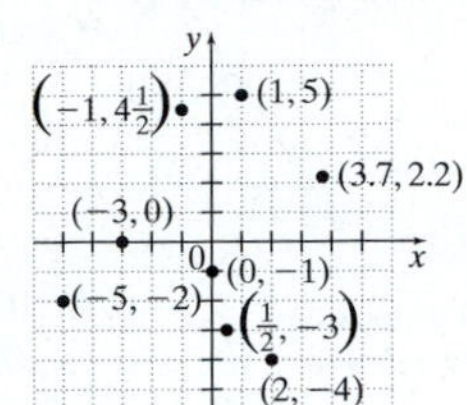

(1, 5) and (3.7, 2.2) are in quadrant I, $\left(-1, 4\frac{1}{2}\right)$ is in quadrant II, (−5, −2) is in quadrant III, (2, −4) and $\left(\frac{1}{2}, -3\right)$ are in quadrant IV, (−3, 0) lies on the x-axis, (0, −1) lies on the y-axis **3.** (0, 0) **5.** (3, 2) **7.** (−2, −2) **9.** (2, −1) **11.** (0, −3) **13.** (1, 3) **15.** (−3, −1) **17. a.** (2001, 28.5), (2002, 29.5), (2003, 32.4), (2004, 34.3) **b.** In the year 2004, $34.3 billion was spent on pet-related expenditures. **c.**

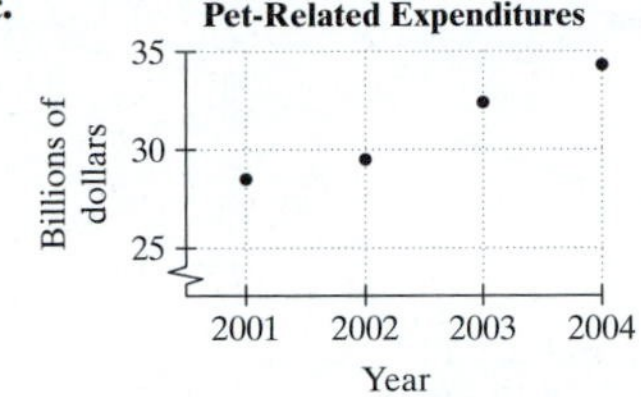

d. Pet-related expenditures increased every year.

19. a. (0.50, 10), (0.75, 12), (1.00, 15), (1.25, 16), (1.50, 18), (1.50, 19), (1.75, 19), (2.00, 20) **b.** When Minh studied 1.25 hours, her quiz score was 16. **c.**

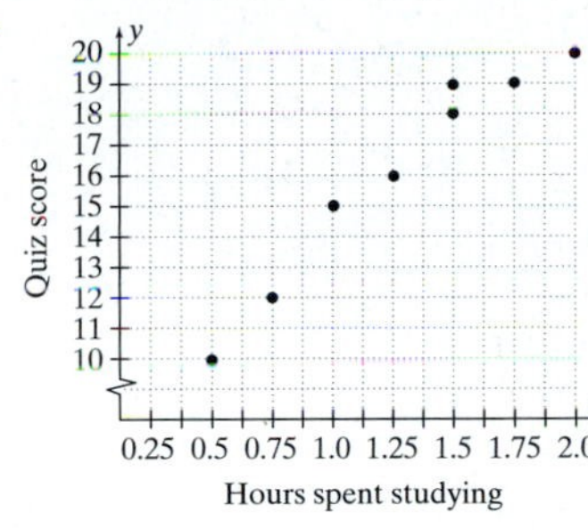

d. answers may vary **21.** (−4, −2), (4, 0) **23.** (−8, −5), (16, 1) **25.** $0; 7; -\frac{2}{7}$ **27.** 2; 2; 5 **29.** 0; −3; 2 **31.** 2; 6; 3 **33.** −12; 5; −6 **35.** $\frac{5}{7}; \frac{5}{2}; -1$ **37.** 0; −5; −2

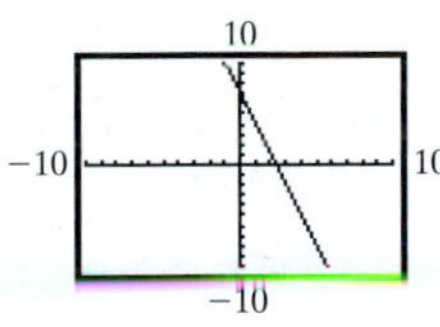

39. 2; 1; −6

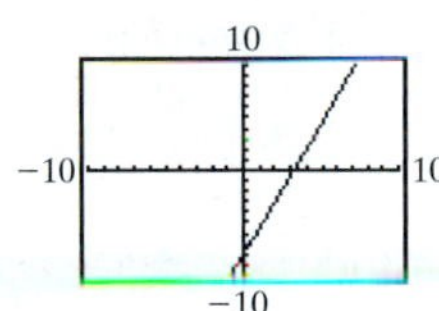

41. a. 13,000; 21,000; 29,000 **b.** 45 desks **43. a.** 10.6; 5.8; 1 **b.** 2002 **45.** $y = 5 - x$ **47.** $y = \frac{5 - 2x}{4}$ **49.** $y = -2x$ **51.** false **53.** true **55.** negative; negative **57.** positive; negative **59.** 0; 0 **61.** y **63.** no; answers may vary **65.** answers may vary **67.** (4, −7) **69.** 26 units **71.** $25 million, $26 million, $25 million, $27 million **73.** answers may vary

Calculator Explorations **1.**

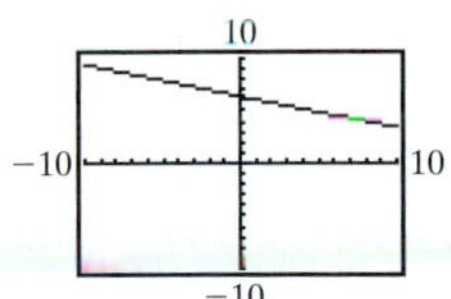

3. **5.**

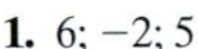

Exercise Set 13.2 **1.** 6; −2; 5

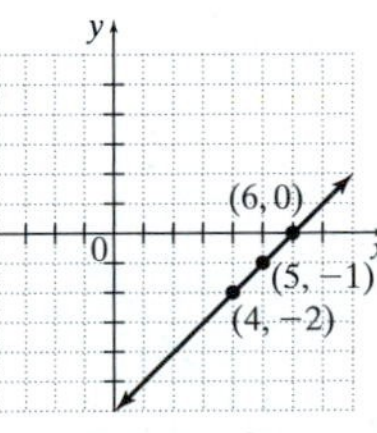

3. −4; 0; 4

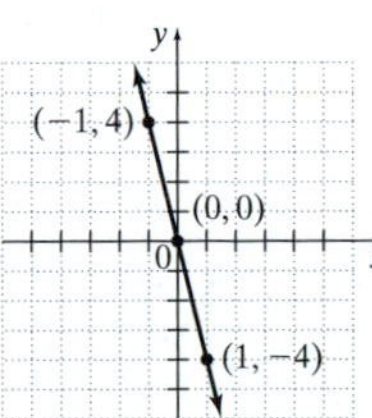

5. 0; 2; −1

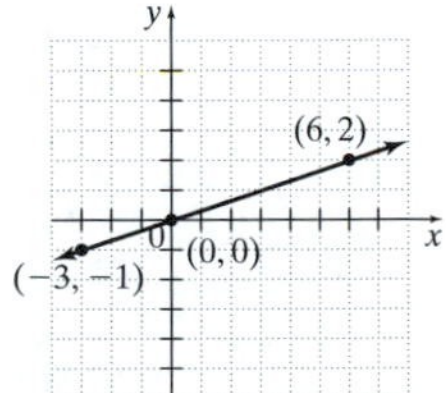

7. 3; −1; −5

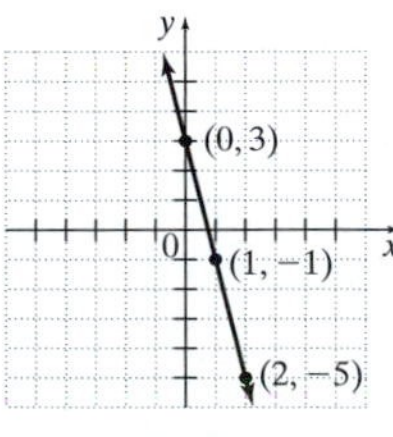

9.

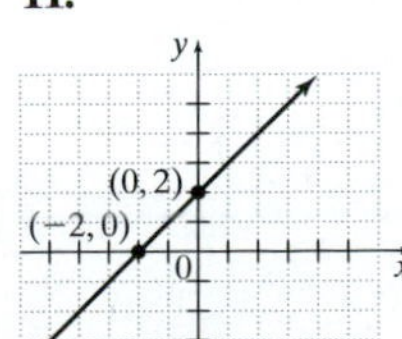

11.

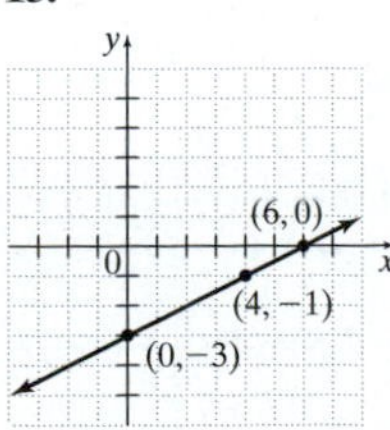

13.

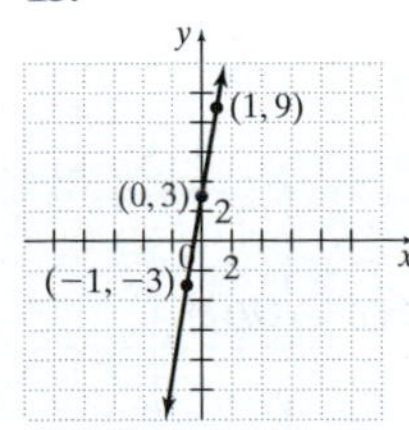

15.

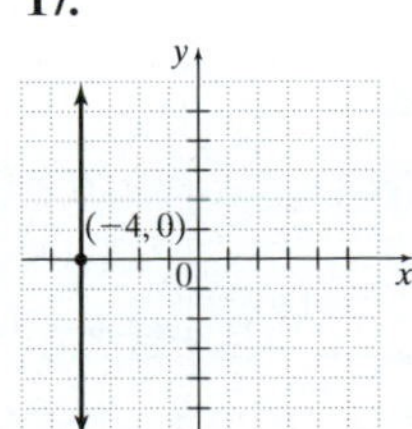

17. **19.**

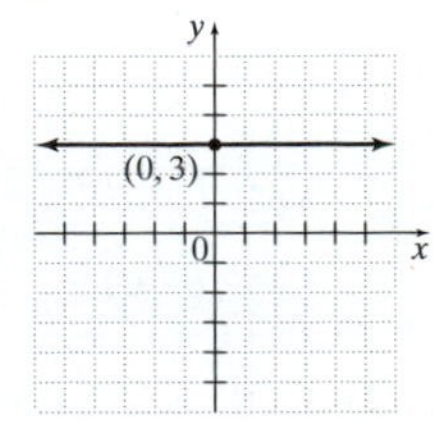

21.

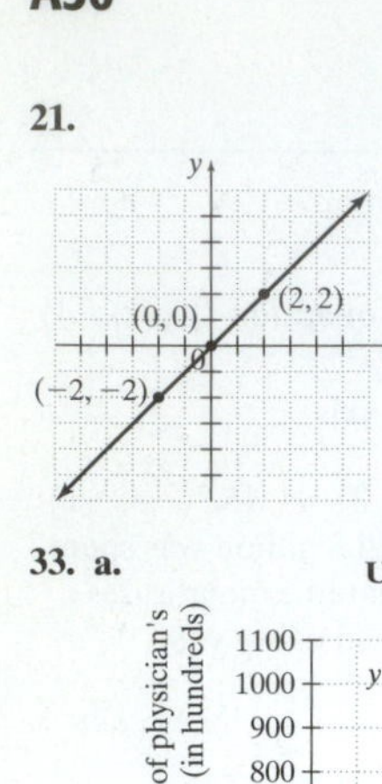

23.

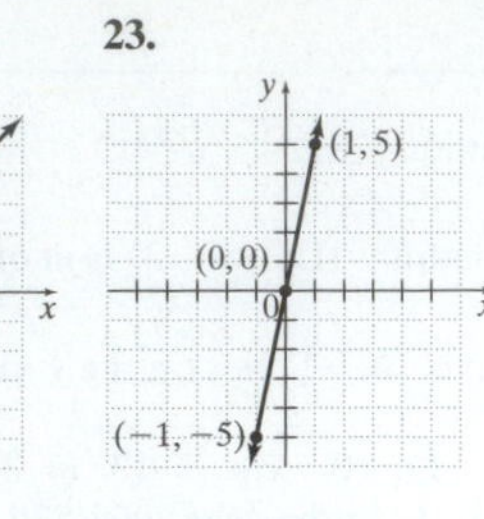

25.

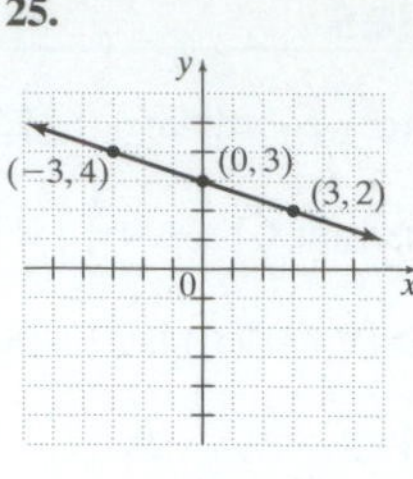

27.

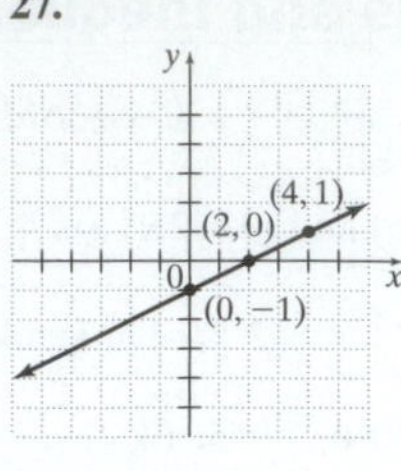

29.

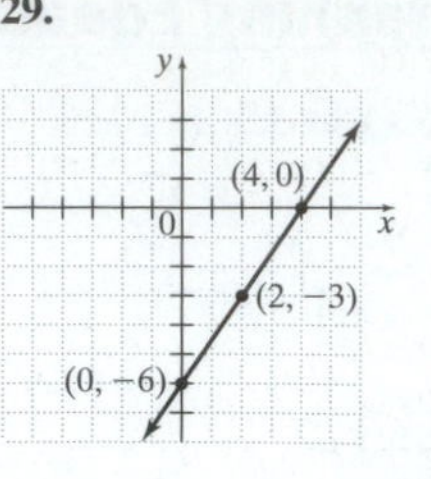

31.

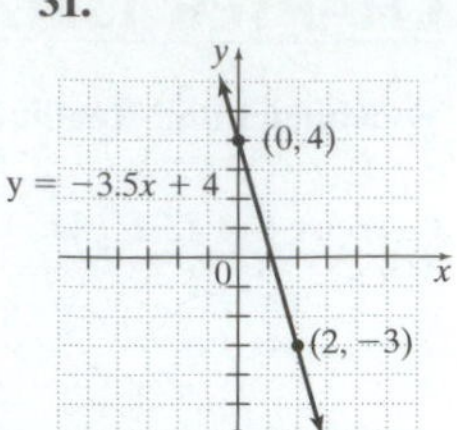

33. a.

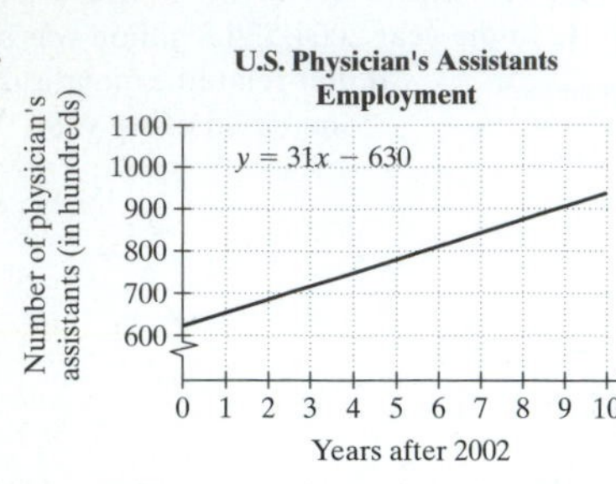

b. yes; answers may vary **35. a.**

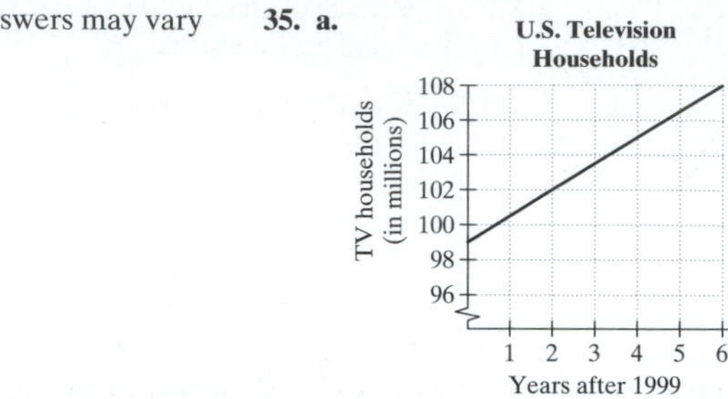

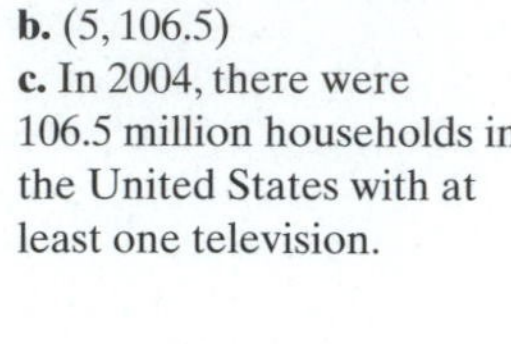

b. (5, 106.5) **c.** In 2004, there were 106.5 million households in the United States with at least one television.

37. (4, −1) **39.** 3; −3 **41.** 0; 0 **43.**

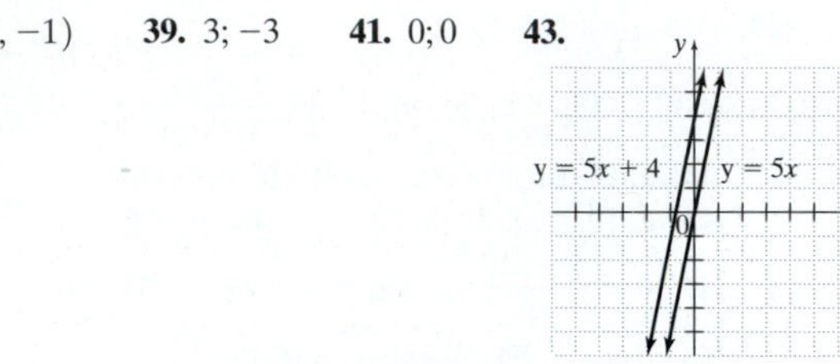

45.

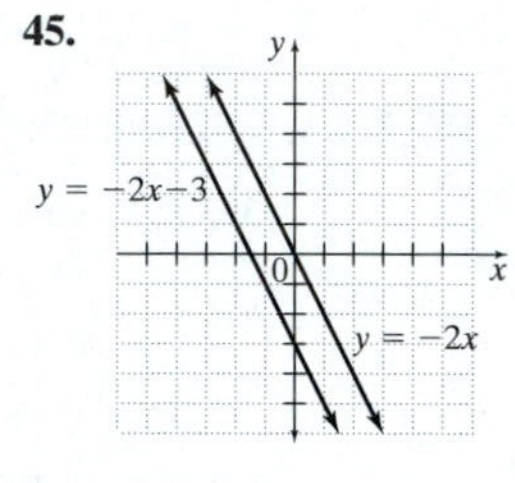

47. 0; 1; 1; 4; 4

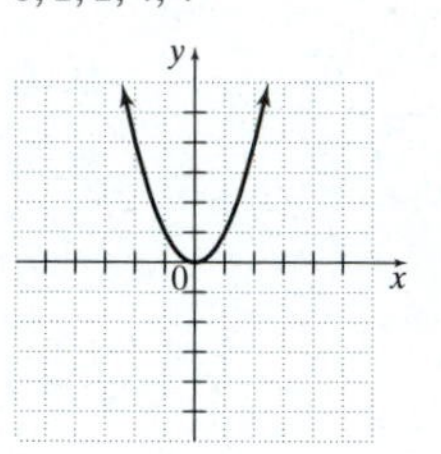

49. $x + y = 12$; 9 cm **51.** yes; answers may vary

Calculator Explorations 1.

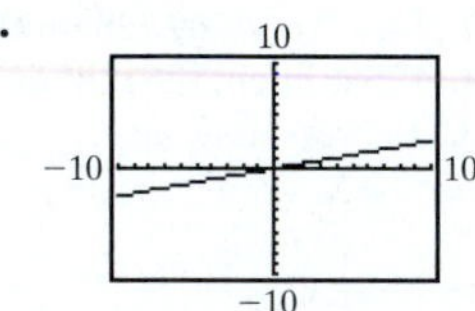

5.

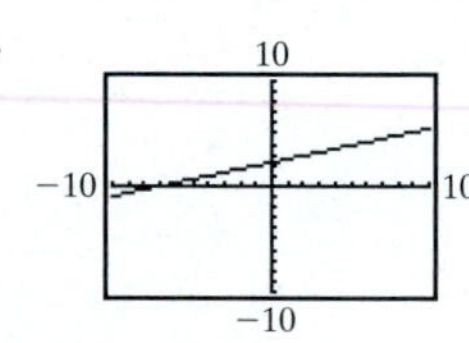

Vocabulary and Readiness Check 1. linear **3.** horizontal **5.** y-intercept **7.** y; x

Exercise Set 13.3 1. (−1, 0); (0, 1) **3.** (−2, 0); (2, 0); (0, −2) **5.** (−2, 0); (1, 0); (3, 0); (0, 3) **7.** (−1, 0); (1, 0); (0, 1); (0, −2)

9.

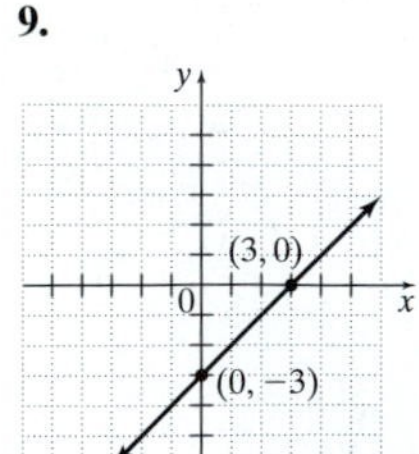

11.

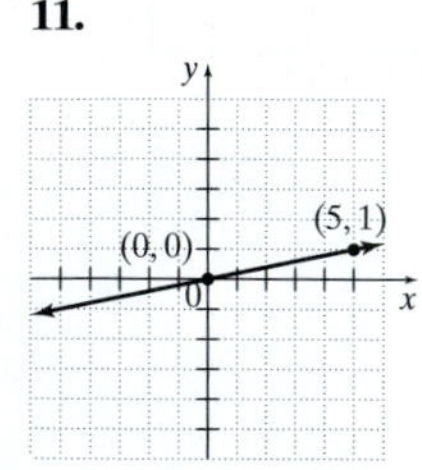

13.

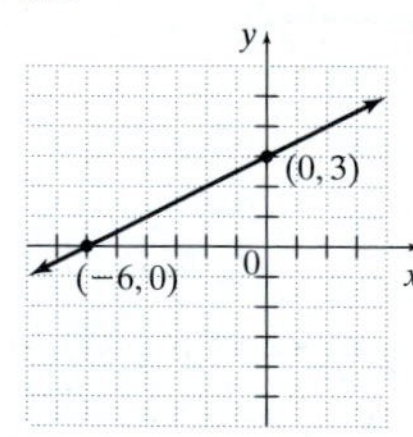

15.

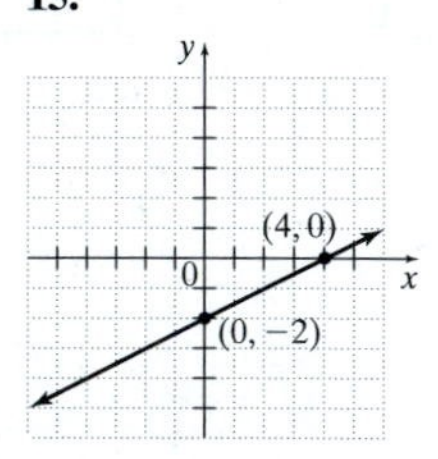

17.

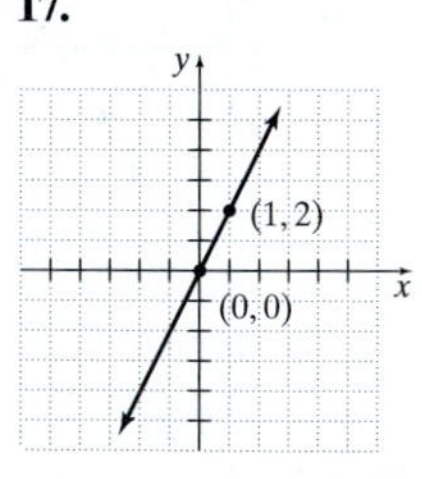

19.

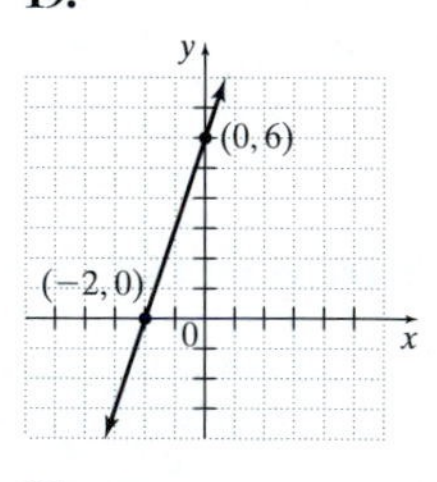

21.

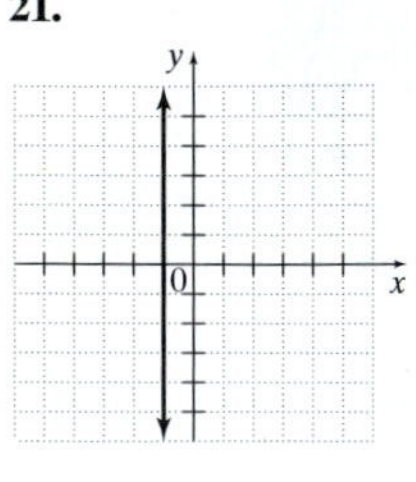

23.

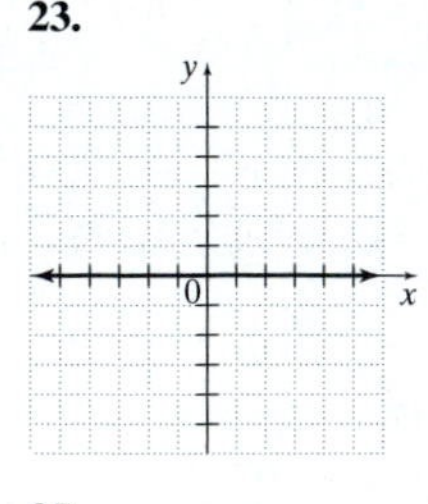

25.

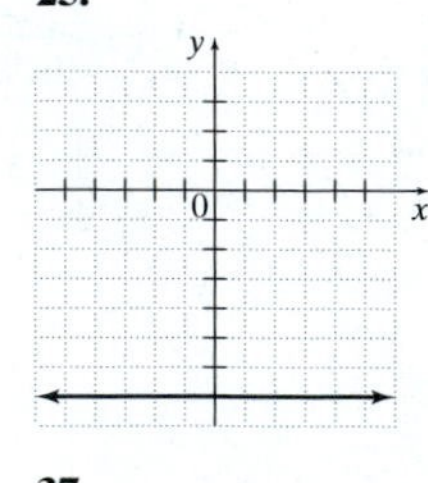

27.

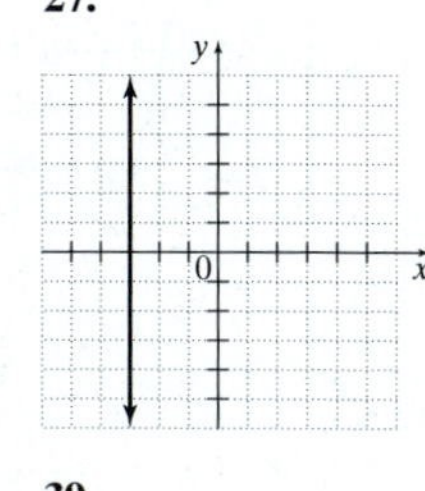

29.

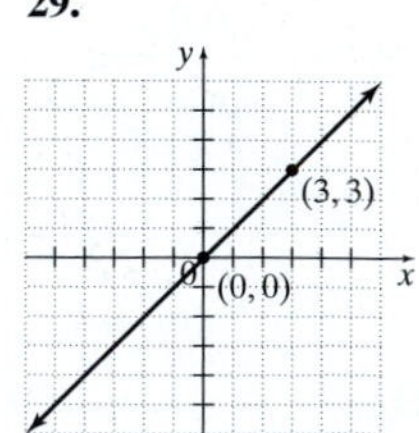

31.

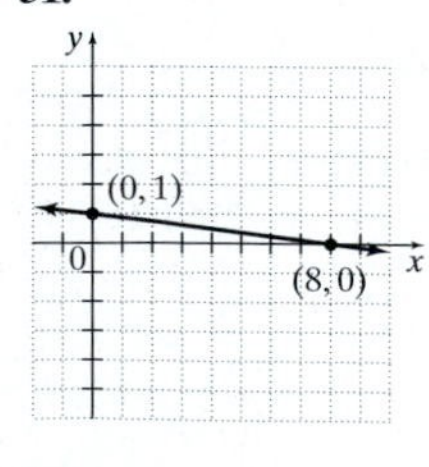

33.

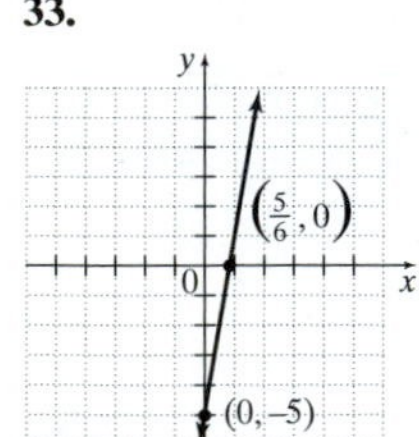

35.

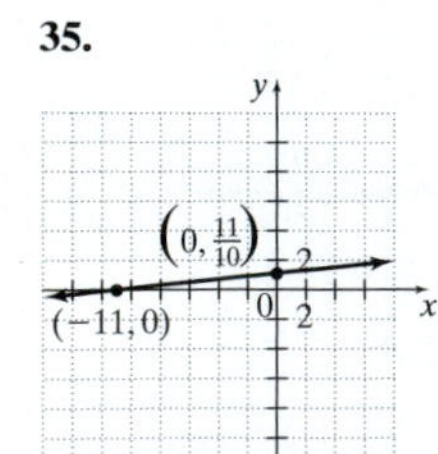

37.

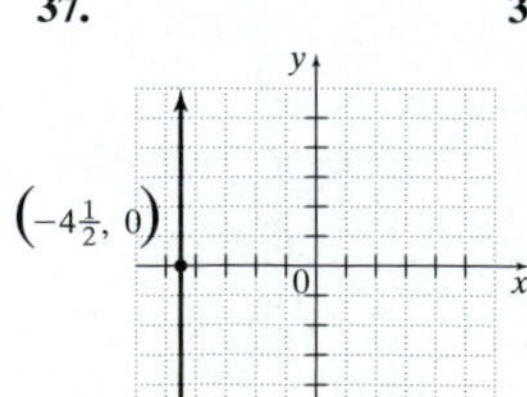

39.

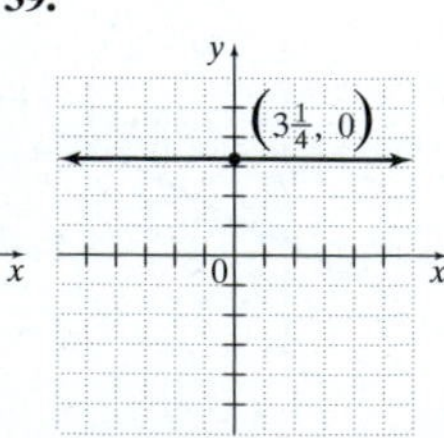

41.

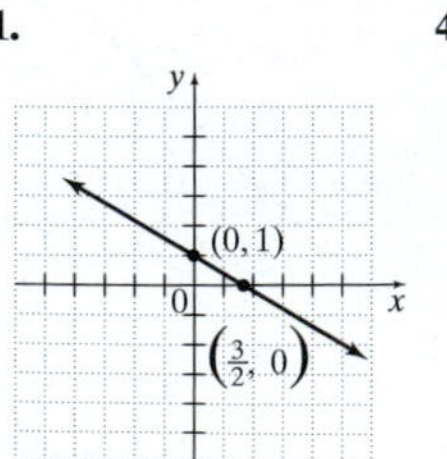

43.

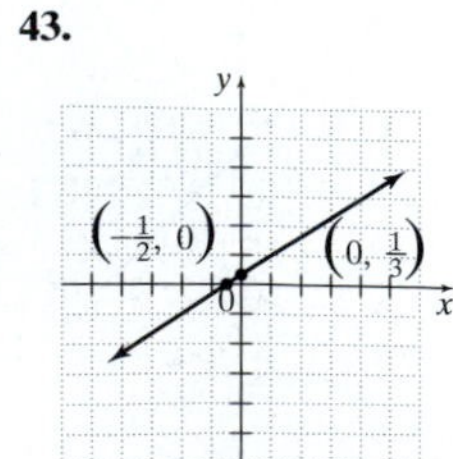

45. $\frac{3}{2}$ **47.** 6 **49.** $-\frac{6}{5}$ **51.** c **53.** a **55.** infinite **57.** 0 **59.** answers may vary
61. a. (0, 200); no chairs and 200 desks are manufactured. **b.** (400, 0); 400 chairs and no desks are manufactured. **c.** 300 chairs **63.** $y = -4$
65. a. (0, 919) **b.** In 1999, the number of stores was 919.

Calculator Explorations **1.**

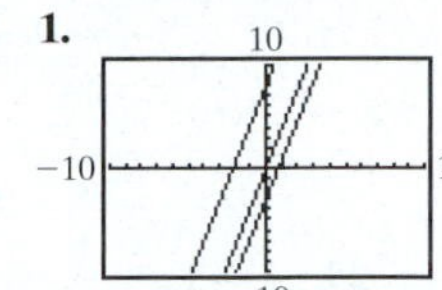

3.

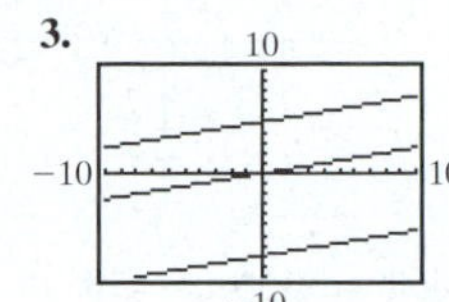

Vocabulary and Readiness Check **1.** slope **3.** 0 **5.** positive **7.** y; x

Exercise Set 13.4 **1.** $m = -1$ **3.** $m = -\frac{1}{4}$ **5.** $m = 0$ **7.** undefined slope **9.** $m = -\frac{4}{3}$ **11.** $m = \frac{5}{2}$ **13.** line 1 **15.** line 2
17. $m = 5$ **19.** $m = -0.3$ **21.** $m = -2$ **23.** undefined slope **25.** $m = \frac{2}{3}$ **27.** undefined slope **29.** $m = \frac{1}{2}$ **31.** $m = 0$
33. $m = -\frac{3}{4}$ **35.** $m = 4$ **37.** neither **39.** neither **41.** parallel **43.** perpendicular **45. a.** 1 **b.** −1 **47. a.** $\frac{9}{11}$ **b.** $-\frac{11}{9}$
49. $\frac{3}{5}$ **51.** 12.5% **53.** 40% **55.** 79% **57.** $m = 3$; Every 1 year, there are/should be 3 million more U.S. households with personal computers.
59. $m = 0.42$; It costs \$0.42 per 1 mile to own and operate a compact car. **61.** $y = 2x - 14$ **63.** $y = -6x - 11$ **65.** d **67.** b **69.** e
71. $m = \frac{1}{2}$ **73.** answers may vary **75.** 28.5 **77.** 1994 and 2000; 28.1 miles per gallon **79.** from 2000 to 2001 **81.** $x = 20$
83. a. (1993, 10,359); (2003, 15,139) **b.** 478 **c.** For the years 1993 through 2003, the number of kidney transplants increased at a rate of 478 per year.
85. Opposite sides are parallel since their slopes are equal, so the figure is a parallelogram. **87.** 2.0625 **89.** −1.6 **91.** the line becomes steeper

Calculator Explorations **1.**

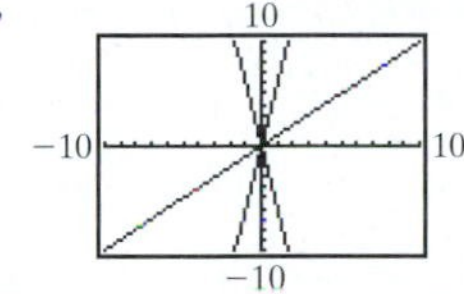

3.

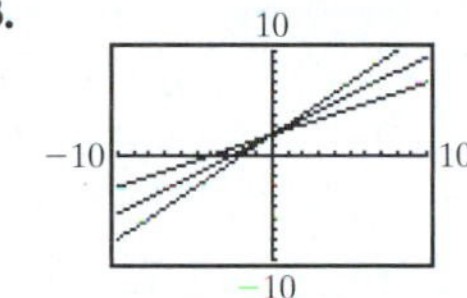

Vocabulary and Readiness Check **1.** slope-intercept; m; b **3.** point-slope **5.** horizontal **7.** slope- intercept

Exercise Set 13.5 **1.** $y = 5x + 3$ **3.** $y = -4x - \frac{1}{6}$ **5.** $y = \frac{2}{3}x$ **7.** $y = -8$ **9.** $y = -\frac{1}{5}x + \frac{1}{9}$

11.

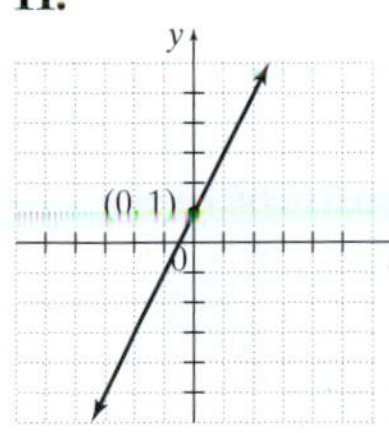

13.

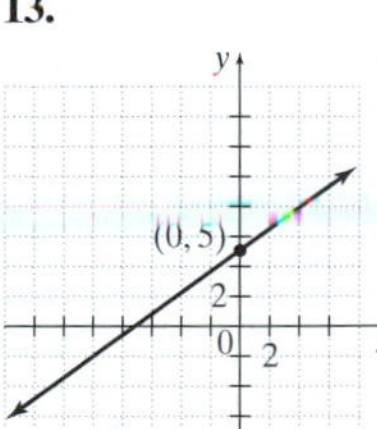

15.

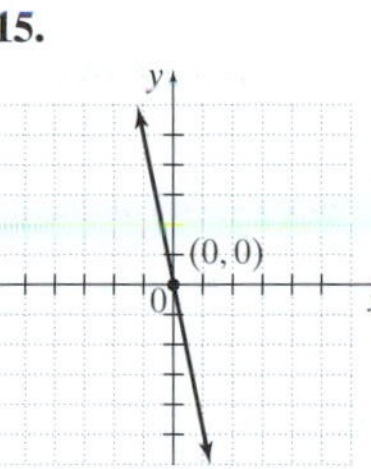

17.

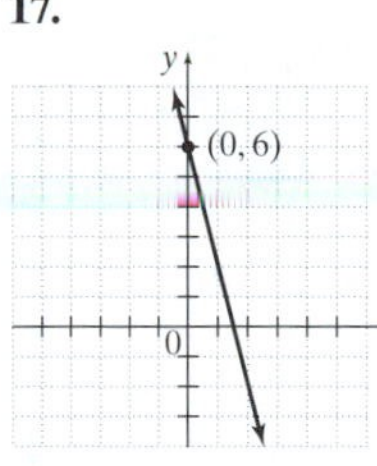

19.

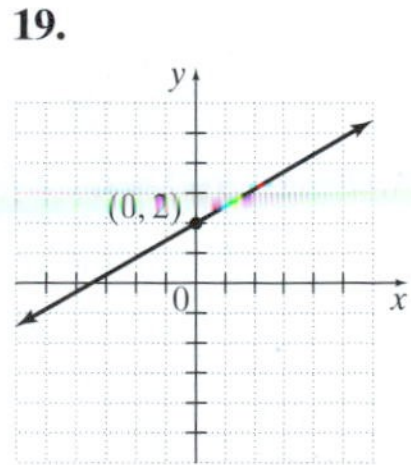

21.

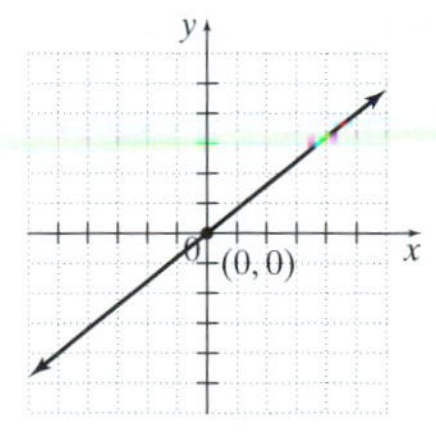

23. $-6x + y = -10$ **25.** $8x + y = -13$ **27.** $3x - 2y = 27$ **29.** $x + 2y = -3$ **31.** $2x - y = 4$ **33.** $8x - y = -11$
35. $4x - 3y = -1$ **37.** $8x + 13y = 0$ **39.** $y = -\frac{1}{2} + \frac{5}{3}$ **41.** $y = -x + 17$ **43.** $x = -\frac{3}{4}$ **45.** $y = x + 16$ **47.** $y = -5x + 7$
49. $y = 7$ **51.** $y = \frac{3}{2}x$ **53.** $y = -3$ **55.** $y = -\frac{4}{7}x - \frac{18}{7}$ **57. a.** $s = 32t$ **b.** 128 ft/sec
59. a. $y = 16{,}000x + 22{,}000$ **b.** 102,000 vehicles **61. a.** $y = -333x + 7032$ **b.** 4368 cinema sites
63. a. (0, 1509); (6, 1456) **b.** $y = -8.8x + 1509$ **c.** 1491 daily newspapers **65. a.** $S = -1000p + 13{,}000$ **b.** 9500 Fun Noodles **67.** −1
69. 5 **71.** b **73.** d **75.** $3x - y = -5$ **77. a.** $3x - y = -5$ **b.** $x + 3y = 5$

Integrated Review **1.** $m = 2$ **2.** $m = 0$ **3.** $m = -\frac{2}{3}$ **4.** slope is undefined

5.

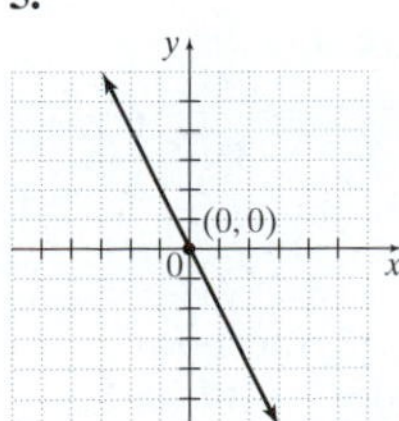

6.

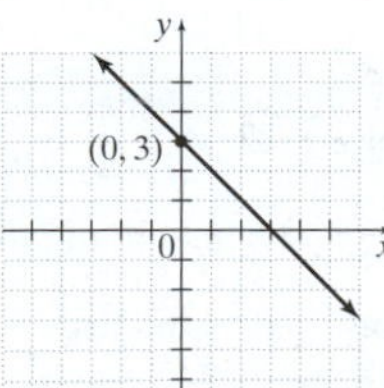

7.

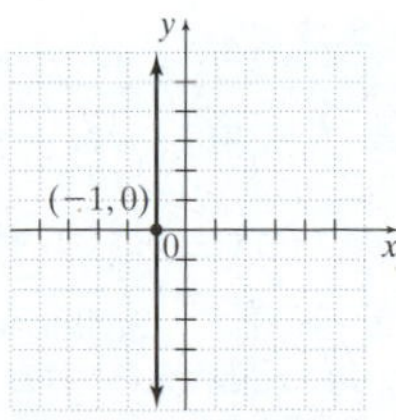

8.

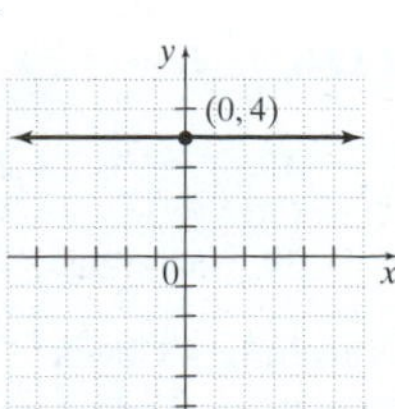

9.

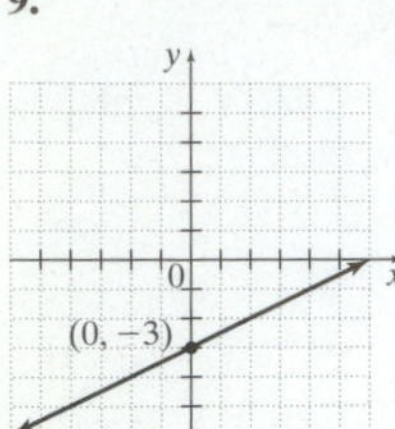

10.

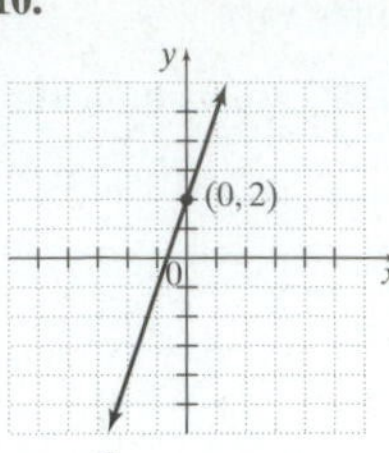

11.

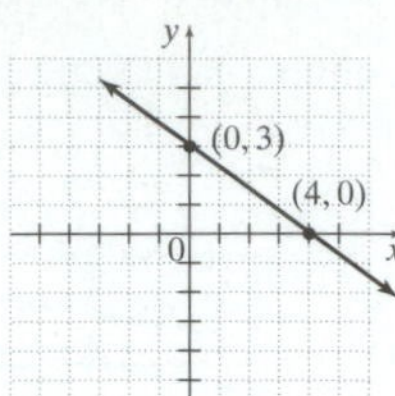

12.

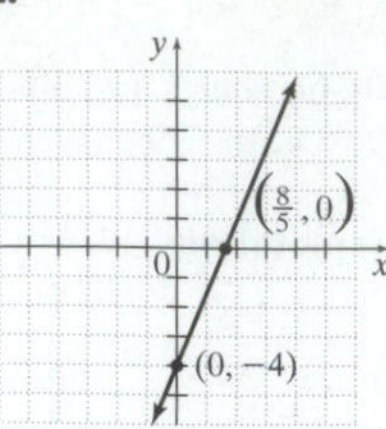

13. $m = 3$ **14.** $m = -6$ **15.** $m = -\frac{7}{2}$ **16.** $m = 2$ **17.** undefined slope **18.** $m = 0$ **19.** $y = 2x - \frac{1}{3}$ **20.** $y = -4x - 1$
21. $-x + y = -2$ **22.** neither **23.** perpendicular **24. a.** (1998, 1639); (2002, 2135) **b.** 124 **c.** For the years 1998 through 2002, the amount of yogurt produced increased at a rate of 124 million pounds per year.

Exercise Set 13.6 **1.** $\{-7, 0, 2, 10\}$; $\{-7, 0, 4, 10\}$ **3.** $\{0, 1, 5\}$; $\{-2\}$ **5.** yes **7.** no **9.** no **11.** yes **13.** yes **15.** no **17.** 9:30 P.M. **19.** January 1 and December 1 **21.** yes; it passes the vertical line test **23.** $4.75 per hour **25.** 2005 **27.** yes; answers may vary **29.** −9, −5, 1 **31.** 6, 2, 11 **33.** −6, 0, 9 **35.** 2, 0, 3 **37.** 5, 0, −20 **39.** 5, 3, 35 **41.** (3, 6) **43.** −1 **45.** −1 **47.** −1, 5 **49.** $x < 1$ **51.** $x \geq -3$ **53.** $\frac{19}{2x}$ m **55.** $f(-5) = 12$ **57.** answers may vary **59.** $f(x) = x + 7$
61. a. 190.4 mg **b.** 380.8 mg

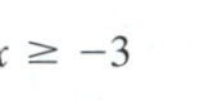

Vocabulary and Readiness Check **1.** linear inequality in two variables **3.** false **5.** true

Exercise Set 13.7 **1.** no; no **3.** yes; no **5.** no; yes

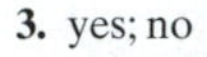

7.

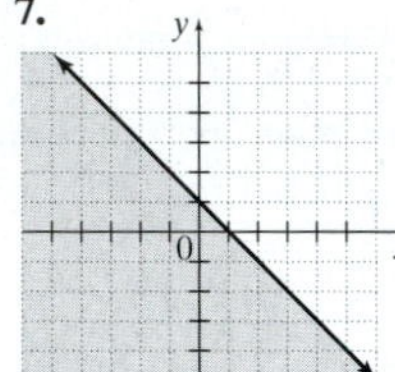

9.

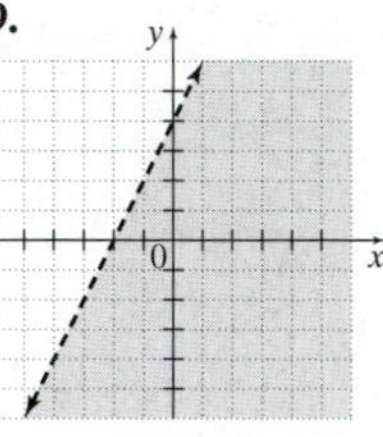

11.

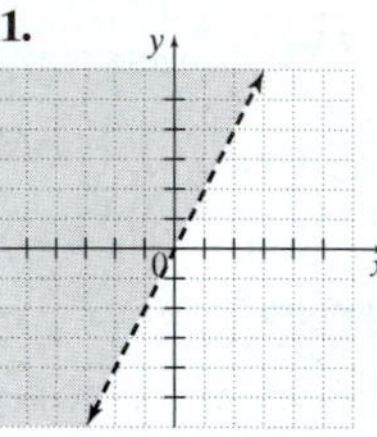

13.

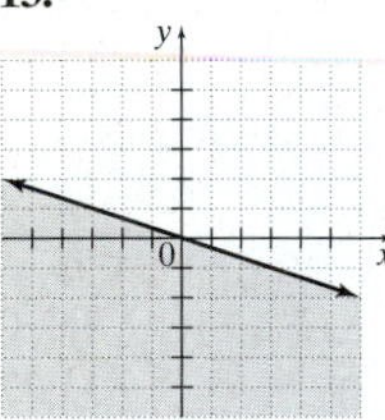

15.

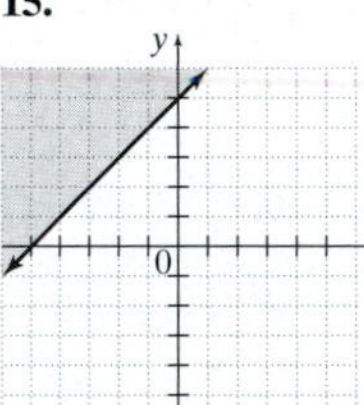

17.

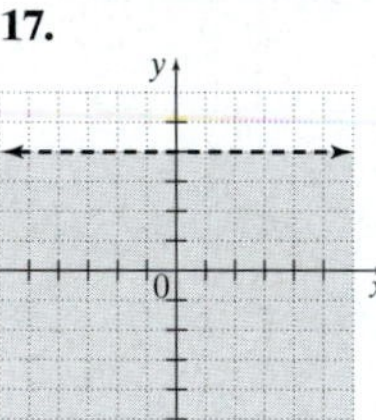

19.

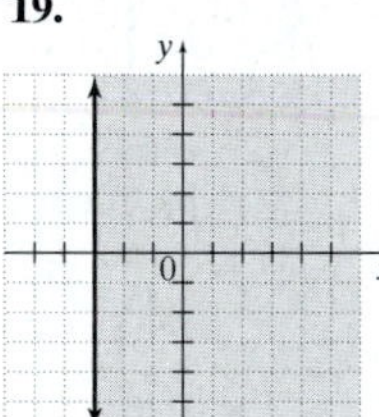

21.

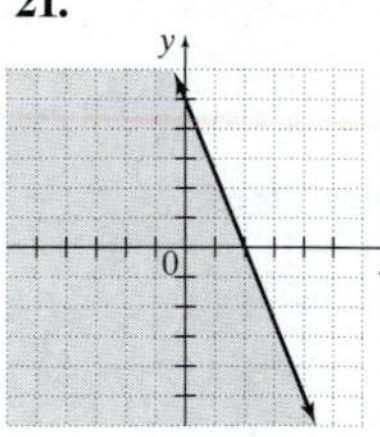

23.

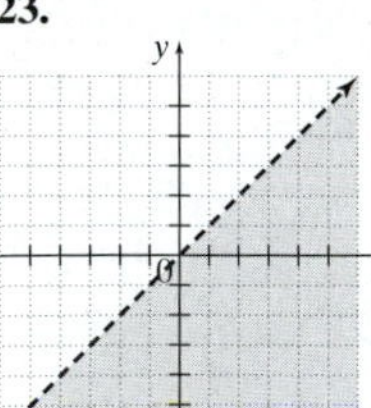

25.

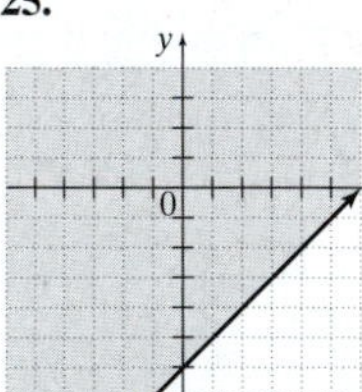

27.

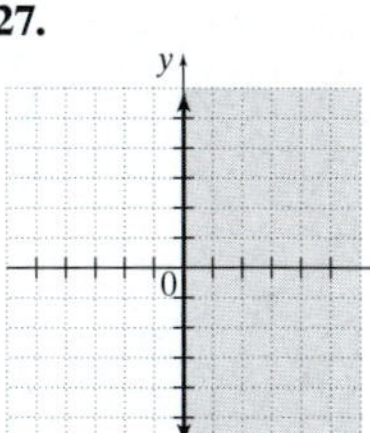

29.

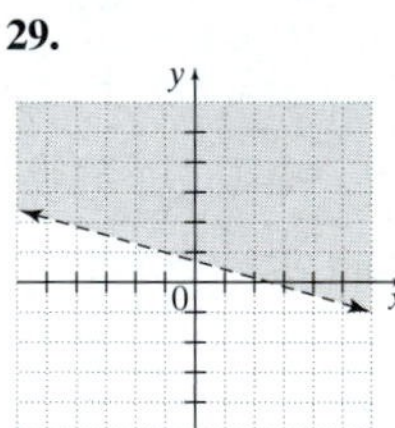

31.

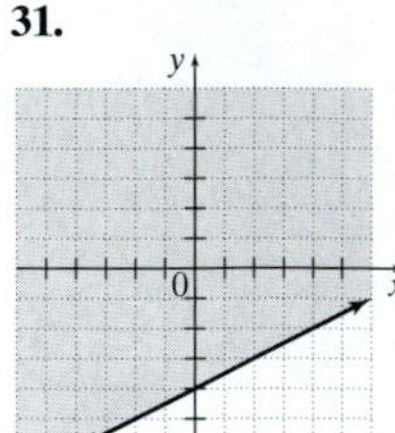

33. (−2, 1) **35.** (−3, −1) **37.** a **39.** b **41.** answers may vary **43.** yes **45.** yes
47. a. $30x + 0.15y \leq 500$ **b.**

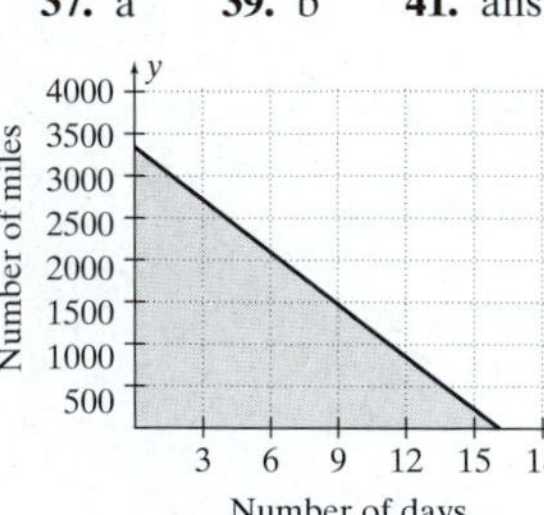

c. answers may vary

Vocabulary and Readiness Check **1.** inverse **3.** direct **5.** inverse

Exercise Set 13.8 **1.** $y = \frac{1}{2}x$ **3.** $y = 6x$ **5.** $y = 3x$ **7.** $y = \frac{2}{3}x$ **9.** $y = \frac{7}{x}$ **11.** $y = \frac{0.5}{x}$ **13.** $y = kx$ **15.** $h = \frac{k}{t}$
17. $z = kx^2$ **19.** $y = \frac{k}{z^3}$ **21.** $x = \frac{k}{\sqrt{y}}$ **23.** $y = 40$ **25.** $y = 3$ **27.** $z = 54$ **29.** $a = \frac{4}{9}$ **31.** $62.50 **33.** $6 **35.** $5\frac{1}{3}$ in.
37. 179.1 lb **39.** 1600 feet **41.** $2y = 16$ **43.** $-4x = 0.5$ **45.** multiplied by 3 **47.** it is doubled

Chapter 13 Vocabulary Check **1.** solution **2.** y-axis **3.** linear **4.** x-intercept **5.** standard **6.** y-intercept **7.** function **8.** slope-intercept **9.** domain **10.** range **11.** relation **12.** point-slope **13.** y **14.** x-axis **15.** x **16.** slope **17.** direct **18.** inverse

Chapter Review **1–6.**

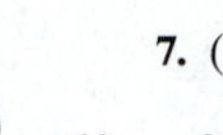

7. (7, 44) **8.** $\left(-\frac{13}{3}, -8\right)$ **9.** −3; 1; 9 **10.** 5; 5; 5 **11.** 0; 10; −10 **12. a.** 2005; 2500; 7000 **b.** 886 compact disc holders

13. **14.** 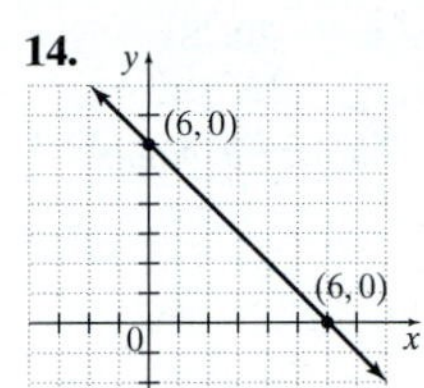 **15.** 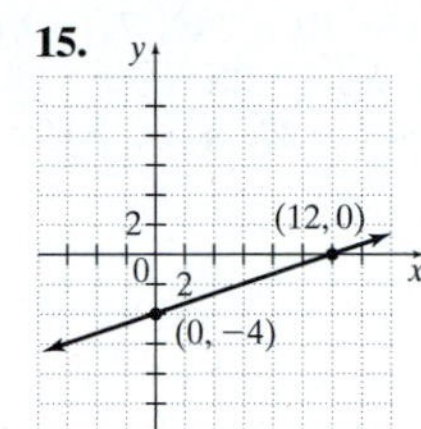 **16.** 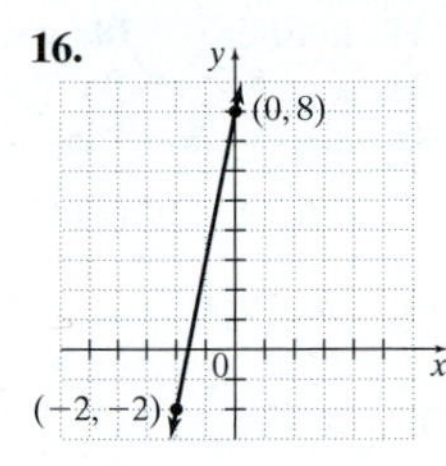 **17.** 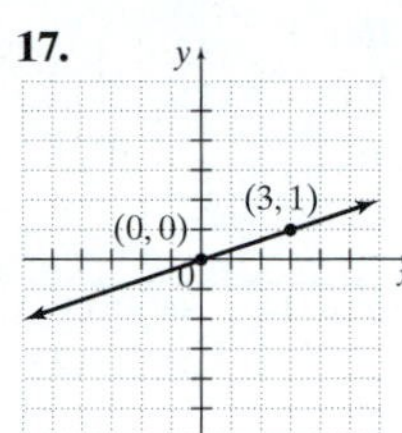 **18.** 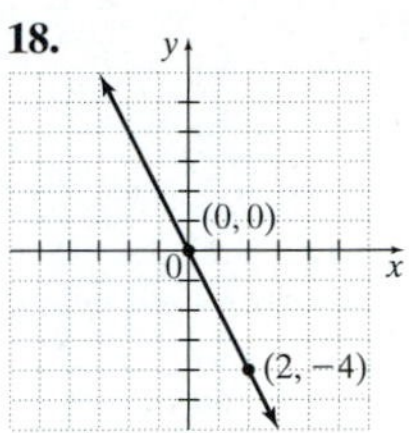

19. (4, 0); (0, −2) **20.** (−2, 0); (2, 0); (0, 2); (0, −2) **21.**  **22.** 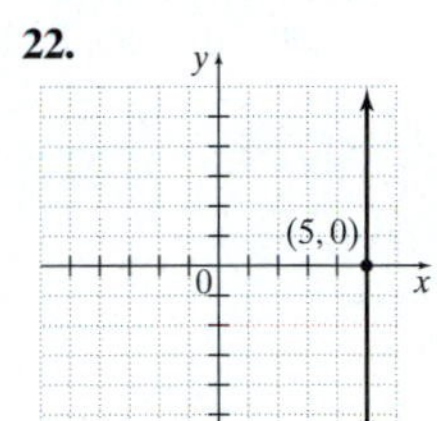 **23.** (12, 0), (0,−4)

24. (−2, 0), (0, 8) **25.** $m = -\frac{3}{4}$ **26.** $m = \frac{1}{5}$ **27.** d **28.** b **29.** c **30.** a **31.** $\frac{3}{4}$ **32.** $\frac{5}{3}$ **33.** 4 **34.** −1 **35.** 3 **36.** $\frac{1}{2}$ **37.** 0 **38.** undefined slope **39.** perpendicular **40.** parallel **41.** neither **42.** $m = 1.24$; Every 1 year, 1.24 million more persons have a bachelor's degree or higher. **43.** $m = 27$; Every 1 year, 27 million more people go on vacations. **44.** $m = -3$; (0, 7) **45.** $m = \frac{1}{6}; \left(0, \frac{1}{6}\right)$ **46.** $y = -5x + \frac{1}{2}$ **47.** $y = \frac{2}{3}x + 6$ **48.** d **49.** c **50.** a **51.** b **52.** $-4x + y = -8$ **53.** $3x + y = -5$ **54.** $-3x + 5y = 17$ **55.** $x + 3y = 6$ **56.** $y = -14x + 21$ **57.** $y = -\frac{1}{2}x + 4$ **58.** no **59.** yes **60.** yes **61.** yes **62.** no **63.** yes **64. a.** 6 **b.** 10 **c.** 5

65. 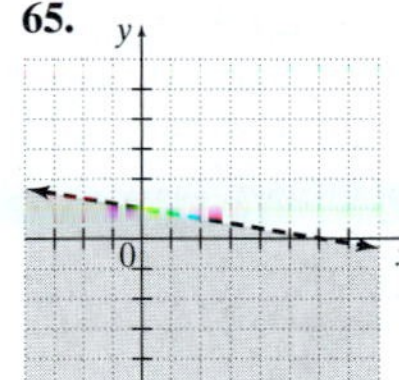 **66.** 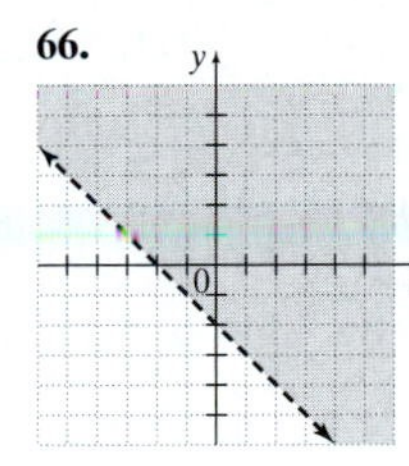 **67.** 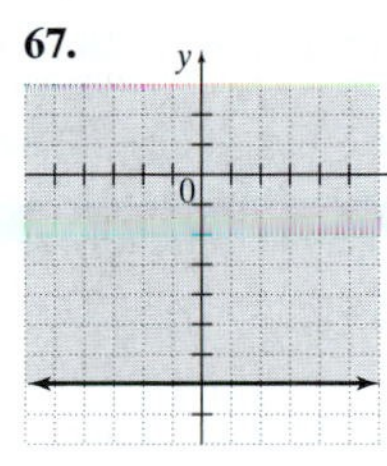 **68.** 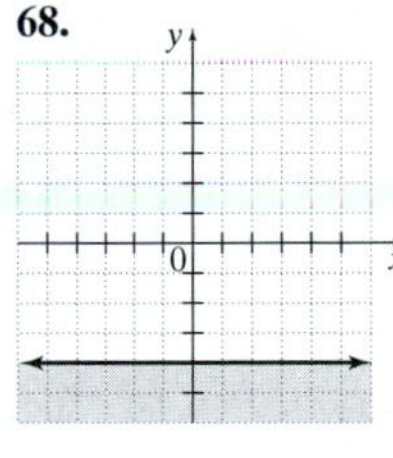 **69.** 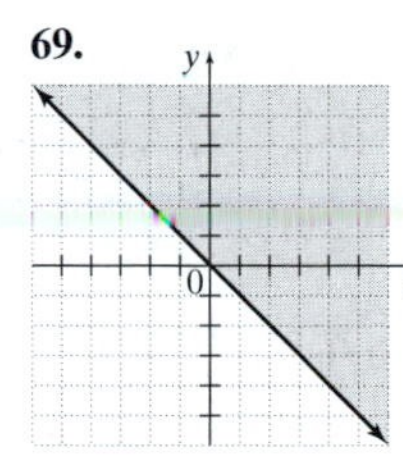 **70.** 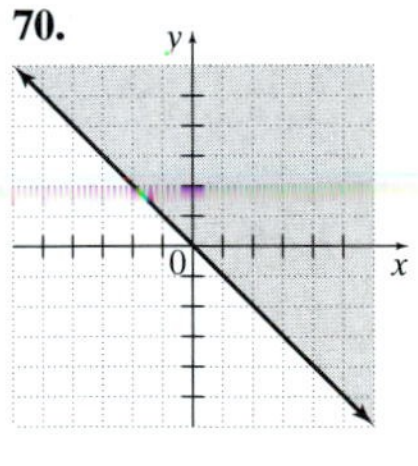

71. $y = 110$ **72.** $y = \frac{1}{2}$ **73.** $y = \frac{100}{27}$ **74.** $y = 700$ **75.** \$3960 **76.** $4\frac{4}{5}$ in. **77.** 7; −1; −3 **78.** 0; −3; −2 **79.** (3, 0); (0, −2) **80.** (−2, 0); (0, 10) **81.** 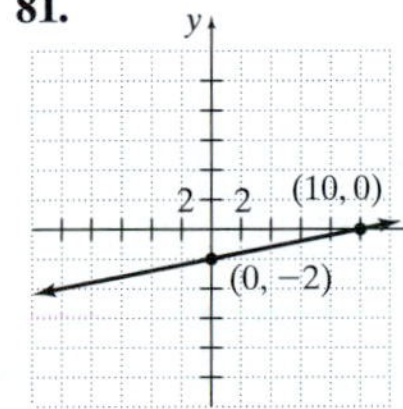 **82.** 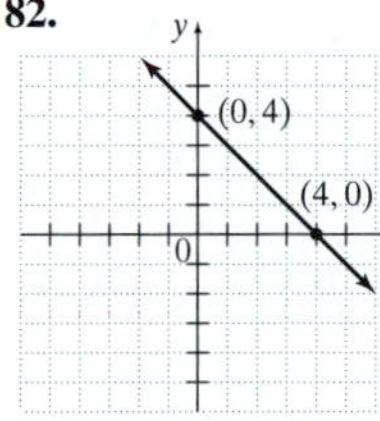 **83.** 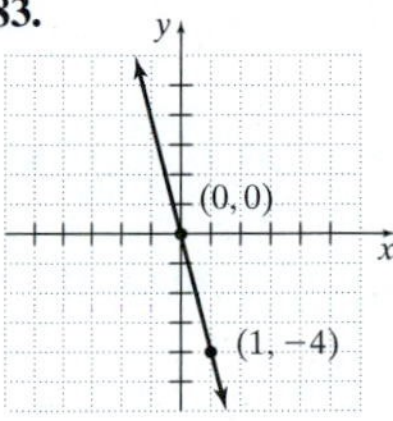 **84.** 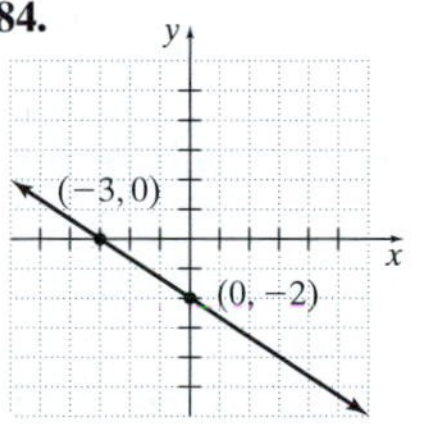

85. 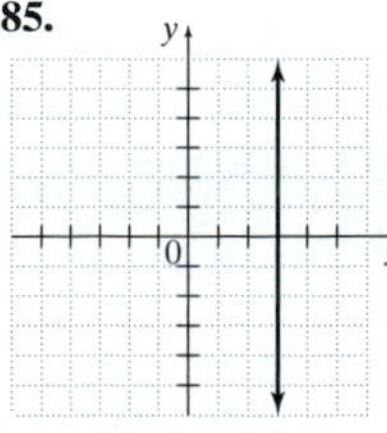 **86.** 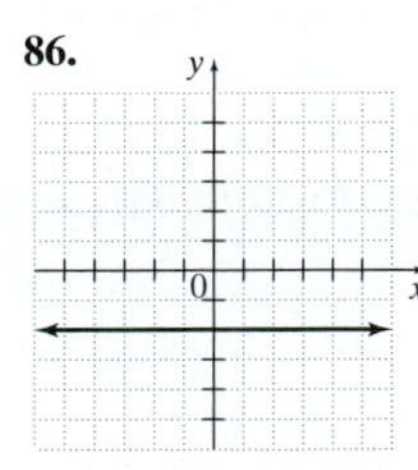 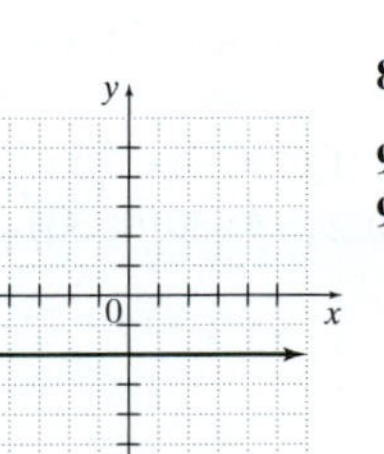

87. −1 **88.** $\frac{11}{7}$ **89.** 2 **90.** $-\frac{1}{3}$ **91.** $m = \frac{2}{3}$; (0, −5) 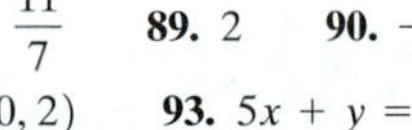 **92.** $m = -6$; (0, 2) **93.** $5x + y = 8$ **94.** $3x - y = -6$ **95.** $4x + y = -3$ **96.** $5x + y = 16$

Chapter Test **1.** (1, 1) **2.** (−4, 17) **3.** $m = \frac{2}{5}$ **4.** $m = 0$ **5.** $m = -1$ **6.** $m = -7$ **7.** $m = 3$ **8.** undefined slope

9.

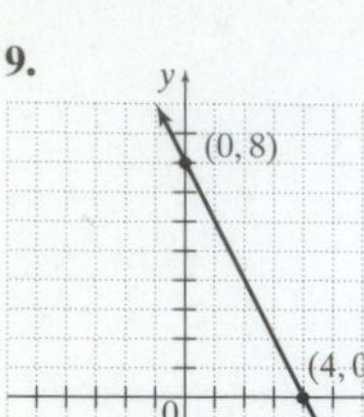

10.

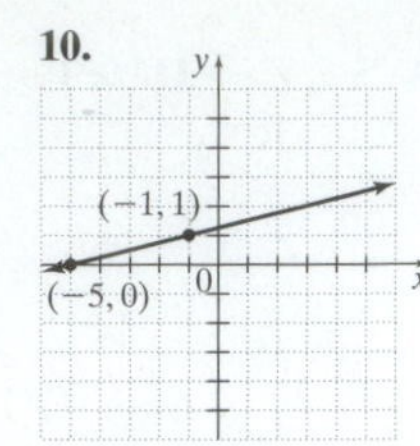

11.

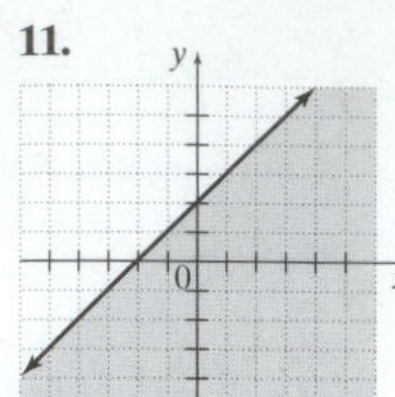

12.

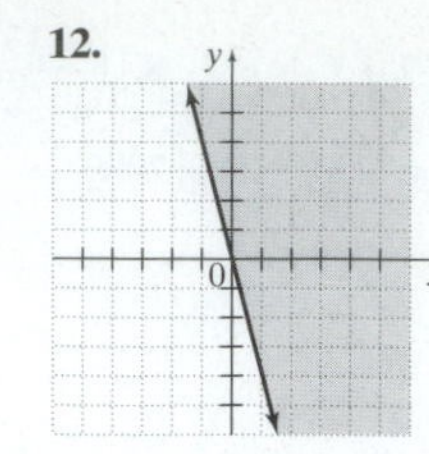

13.

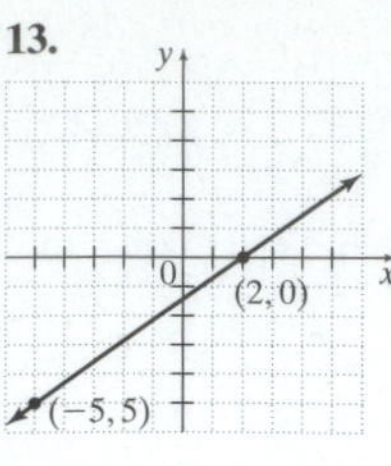

14.

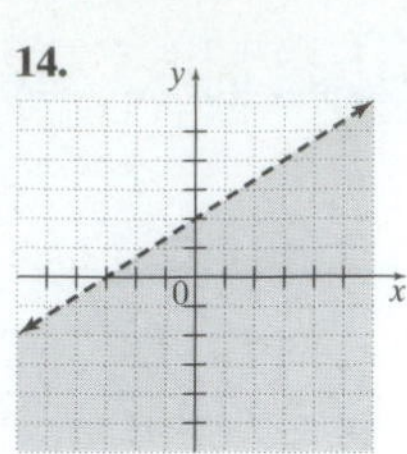

15.

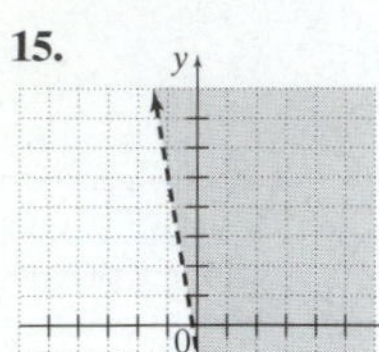

16.

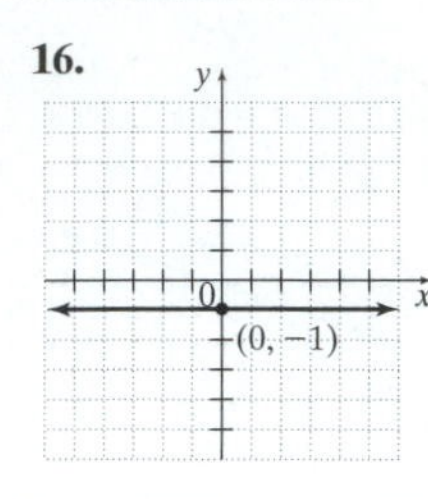

17. neither **18.** $x + 4y = 10$ **19.** $7x + 6y = 0$ **20.** $8x + y = 11$ **21.** $x - 8y = -96$ **22.** yes **23.** no **24.** yes **25.** yes **26. a.** -8 **b.** -3.6 **c.** -4 **27. a.** 0 **b.** 0 **c.** 60 **28.** $x + 2y = 21$; $x = 5$ m

29. a. (2000, 69.3); (2001, 70.0); (2002, 69.9); (2003, 70.1); (2004, 70.3); (2005, 70.5)
b.

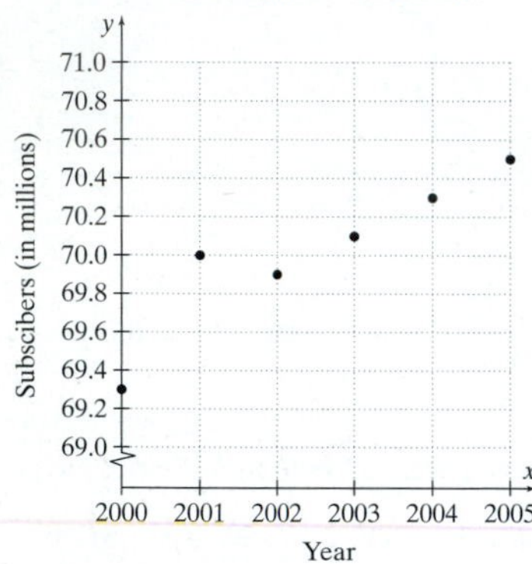

30. $m = 50$; Every 1 year, 50 million more movie tickets are sold. **31.** 28 **32.** $\frac{8}{9}$

Cumulative Review **1.** 106,400 sq mi; Sec. 1.5, Ex. 6 **2.** 147 trees; Sec. 1.5 **3.** -8; Sec. 2.2, Ex. 15 **4.** -10; Sec. 2.2 **5.** $\frac{64}{24x}$; Sec. 4.4, Ex. 21 **6.** $\frac{12}{8c}$; Sec. 4.4 **7.** $5\frac{4}{7}$; Sec. 4.7, Ex. 14 **8.** $10\frac{3}{5}$; Sec. 4.7 **9.** -2.6; Sec. 5.5, Ex. 16 **10.** -13.6; Sec. 5.5 **11.** 3.14; Sec. 5.5, Ex. 4 **12.** 1.947; Sec. 5.5 **13.** $\{x|x < -2\}$; Sec. 9.6, Ex. 6 **14.** $\{x|x \le -3\}$; Sec. 9.6 **15. a.** 2; trinomial **b.** 1; binomial **c.** 3; none of these; Sec. 10.3, Ex. 3 **16. a.** 1; binomial **b.** 2; trinomial **c.** 3; none of these; Sec. 10.3 **17.** $-4x^2 + 6x + 2$; Sec. 10.4, Ex. 2 **18.** $x^2 + 3x$; Sec. 10.4 **19.** $9y^2 + 6y + 1$; Sec. 10.6, Ex. 4 **20.** $4x^2 - 20x + 25$; Sec. 10.6 **21.** $3a(-3a^4 + 6a - 1)$; Sec. 11.1, Ex. 5 **22.** $x^3(2x^2 - 1)$; Sec. 11.1 **23.** $(x - 2)(x + 6)$; Sec. 11.2, Ex. 3 **24.** $(x - 3)(x + 7)$; Sec. 11.2 **25.** $(4x - 1)(2x - 5)$; Sec. 11.3, Ex. 2 **26.** $(5x + 2)(3x - 1)$; Sec. 11.3 **27.** $\frac{7}{2}$; Sec. 11.6, Ex. 5 **28.** 7, 2; Sec. 11.6 **29.** 1; Sec. 12.2, Ex. 7 **30.** $\frac{3x - 1}{10}$; Sec. 12.2 **31.** $\frac{12ab^2}{27a^2b}$; Sec. 12.3, Ex. 9a **32.** $\frac{63x^3y}{99x^2y^2}$; Sec. 12.3 **33.** $\frac{2m + 1}{m + 1}$; Sec. 12.4, Ex. 5 **34.** $\frac{1}{m + 1}$; Sec. 12.4 **35.** $-3, -2$; Sec. 12.5, Ex. 3 **36.** $-5, 2$; Sec. 12.5 **37.** $\frac{x + 1}{x + 2y}$; Sec. 12.7, Ex. 5 **38.** $\frac{x + 4}{x - 4}$; Sec. 12.7 **39. a.** (0, 12) **b.** (2, 6) **c.** (−1, 15); Sec. 13.1, Ex. 3 **40. a.** (0, −5) **b.** (20, 0) **c.** (12, −2); Sec. 13.1

41.

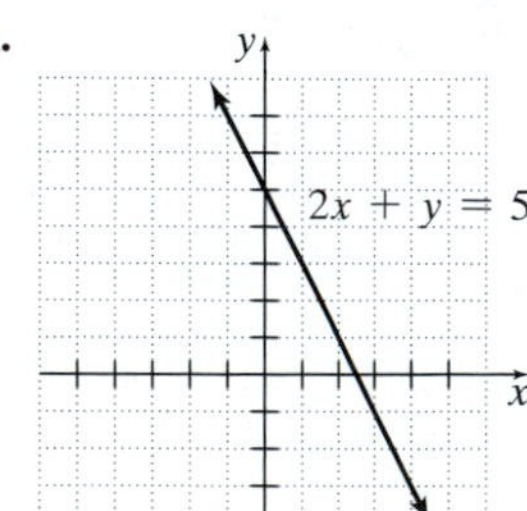

; Sec. 13.2, Ex. 1 **42.** ; Sec. 13.2 **43.** $\frac{2}{3}$; Sec. 13.4, Ex. 3 **44.** $-\frac{7}{4}$; Sec. 13.4 **45.** $2x + y = 3$; Sec. 13.5, Ex.4 **46.** $5x + y = 3$; Sec. 13.5 **47. a.** 1; (2, 1) **b.** 1; (−2, 1) **c.** −3; (0, −3); Sec. 13.6, Ex. 6 **48. a.** 2; (0, 2) **b.** 50; (4, 50) **c.** 5; (−1, 5); Sec. 13.6

CHAPTER 14 Systems of Equations

Calculator Explorations **1.** (0.37, 0.23) **3.** (0.03, −1.89)

Vocabulary and Readiness Check **1.** dependent **3.** consistent **5.** inconsistent **7.** 1 solution, (−1, 3) **9.** infinite number of solutions.

Exercise Set 14.1 **1. a.** no **b.** yes **3. a.** yes **b.** no **5. a.** yes **b.** yes **7. a.** no **b.** no

9.
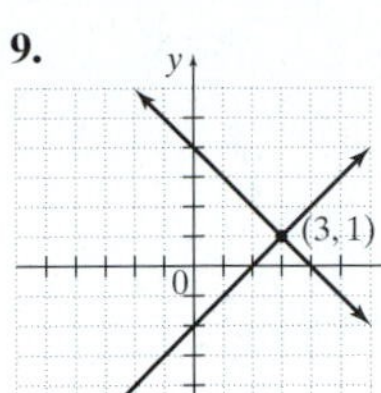

11.
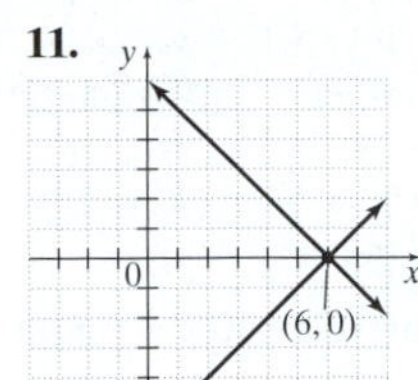

13.
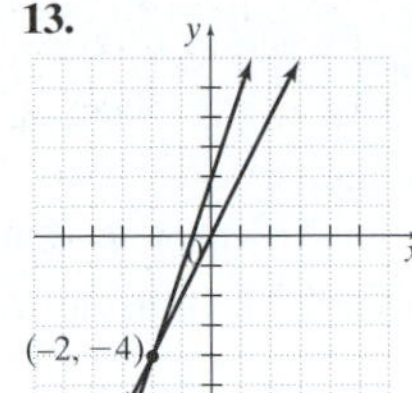

15.
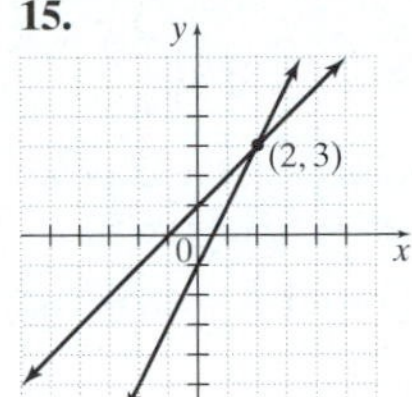

17.
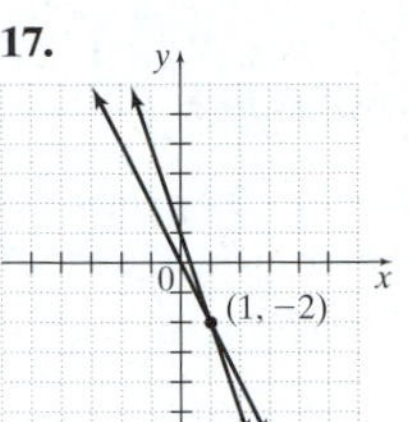

19.
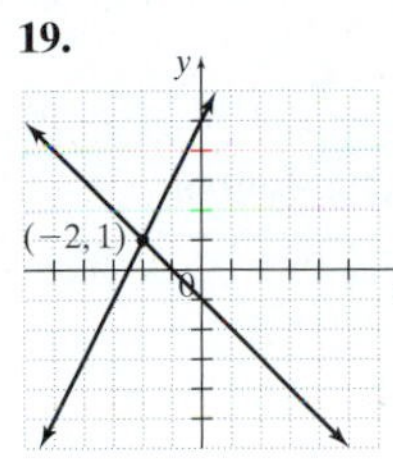

21. no solution
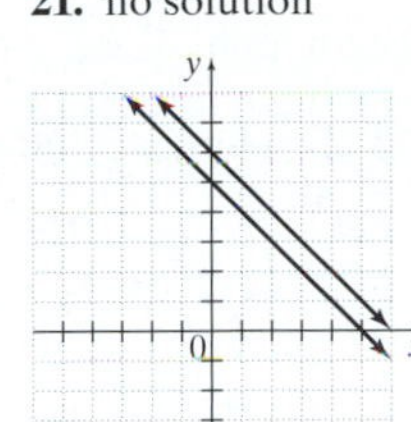

23.
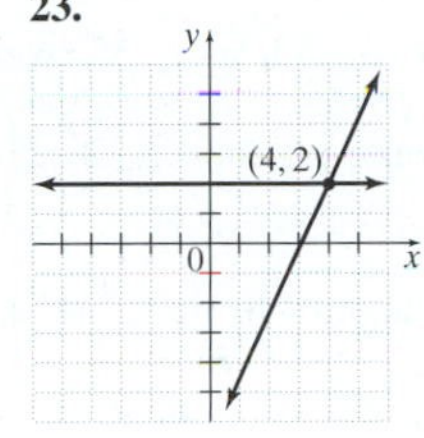

25.
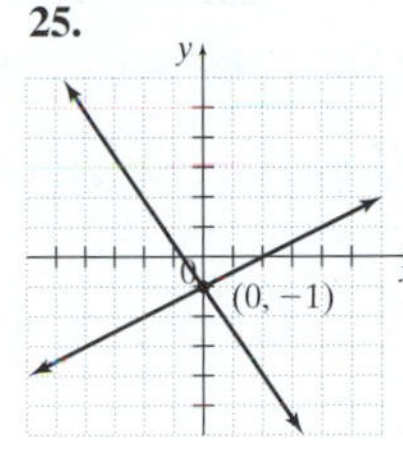

27. no solution
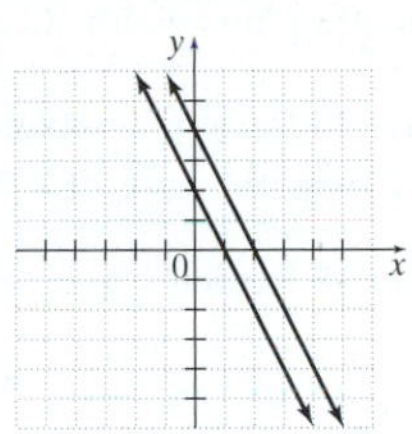

29. infinite number of solutions
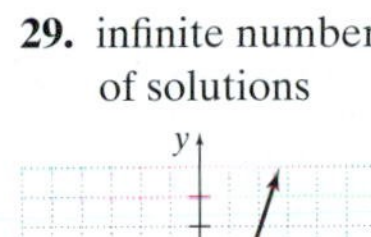
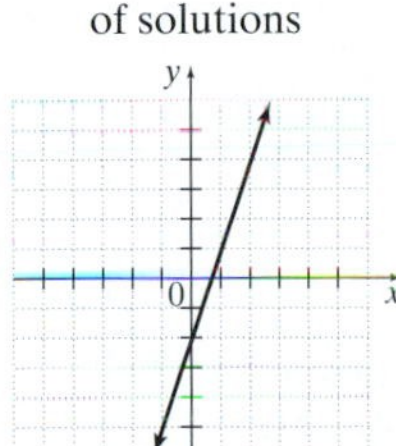

31.
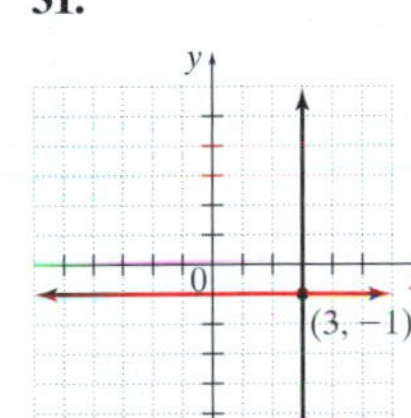

33.
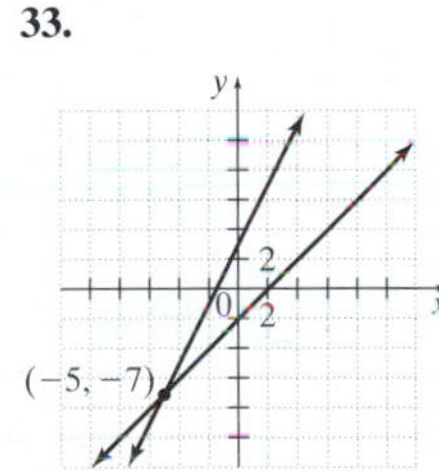

35.
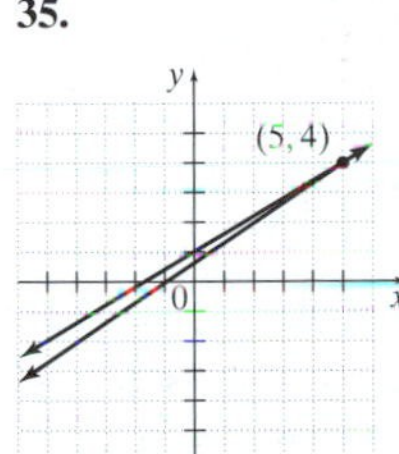

37. infinite number of solutions
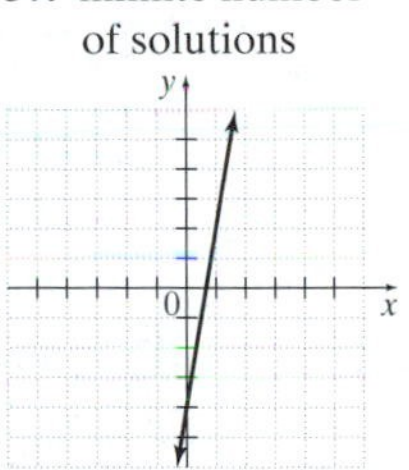

39. $x = 2$ **41.** $y = -\frac{2}{5}$ **43.** $a = 2$ **45.** answers may vary **47.** answers may vary **49.** 1984, 1988 **51.** 2003 **53.** answers may vary **55.** answers may vary **57.** answers may vary

Exercise Set 14.2 **1.** (2, 1) **3.** (−3, 9) **5.** (2, 7) **7.** $\left(-\frac{1}{5}, \frac{43}{5}\right)$ **9.** (2, −1) **11.** (−2, 4) **13.** (4, 2) **15.** (−2, −1) **17.** no solution **19.** (3, −1) **21.** (3, 5) **23.** $\left(\frac{2}{3}, -\frac{1}{3}\right)$ **25.** (−1, −4) **27.** (−6, 2) **29.** (2, 1) **31.** no solution **33.** infinite number of solutions **35.** $\left(\frac{1}{2}, 2\right)$ **37.** $-6x - 4y = -12$ **39.** $-12x + 3y = 9$ **41.** $5n$ **43.** $-15b$ **45.** (1, −3) **47.** answers may vary **49.** no; answers may vary **51.** c; answers may vary **53.** (−2.6, 1.3) **55.** (3.28, 2.1) **57. a.** (9, 17) **b.** In 1973 + 9 = 1982, the percent of households that used fuel oil and electricity for heat was the same, 17%. **c.**

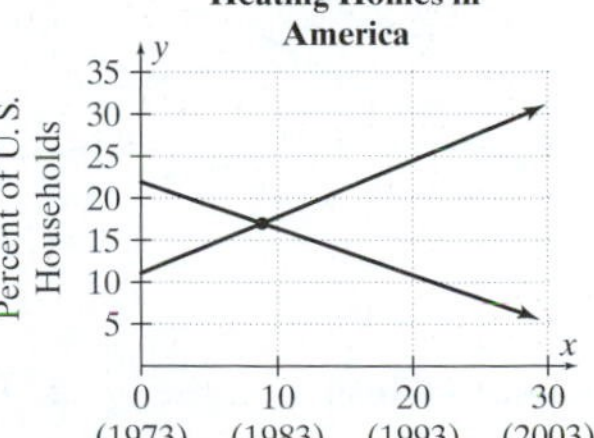

Exercise Set 14.3 **1.** (1, 2) **3.** (2, −3) **5.** (−2, −5) **7.** (5, −2) **9.** (−7, 5) **11.** (6, 0) **13.** no solution **15.** infinite number of solutions **17.** $\left(2, -\frac{1}{2}\right)$ **19.** (−2, 0) **21.** (1, −1) **23.** infinite number of solutions **25.** $\left(\frac{12}{11}, -\frac{4}{11}\right)$ **27.** $\left(\frac{3}{2}, 3\right)$ **29.** infinite number of solutions **31.** (1, 6) **33.** $\left(-\frac{1}{2}, -2\right)$ **35.** infinite number of solutions **37.** $\left(-\frac{2}{3}, \frac{2}{5}\right)$ **39.** (2, 4) **41.** (−0.5, 2.5)

43. $2x + 6 = x - 3$ **45.** $20 - 3x = 2$ **47.** $4(x + 6) = 2x$ **49.** 2; $6x - 2y = -24$ **51.** b; answers may vary **53.** answers may vary **55. a.** $b = 15$ **b.** any real number except 15 **57.** $(-8.9, 10.6)$ **59. a.** (8, 536) or (8, 537) **b.** In 2010 (2002 + 8), the number of medical assistant jobs equals the number of computer software engineer jobs. **c.** 536 thousand or 537 thousand

Integrated Review **1.** (2, 5) **2.** (4, 2) **3.** (5, −2) **4.** (6, −14) **5.** (−3, 2) **6.** (−4, 3) **7.** (0, 3) **8.** (−2, 4) **9.** (5, 7) **10.** (−3, −23) **11.** $\left(\frac{1}{3}, 1\right)$ **12.** $\left(-\frac{1}{4}, 2\right)$ **13.** no solution **14.** infinite number of solutions **15.** (0.5, 3.5) **16.** (−0.75, 1.25) **17.** infinite number of solutions **18.** no solution **19.** (7, −3) **20.** (−1, −3) **21.** answers may vary **22.** answers may vary

Exercise Set 14.4 **1.** c **3.** b **5.** a **7.** $\begin{cases} x + y = 15 \\ x - y = 7 \end{cases}$ **9.** $\begin{cases} x + y = 6500 \\ x = y + 800 \end{cases}$ **11.** 33 and 50 **13.** 14 and −3 **15.** Jackson: 634 points; Leslie: 598 points **17.** child's ticket: \$18; adult's ticket: \$29 **19.** quarters: 53; nickels: 27 **21.** McDonald's: \$31.50; The Ohio Art Company: \$6.50 **23.** daily fee: \$32; mileage charge: \$0.25 per mi **25.** distance downstream = distance upstream = 18 mi; time downstream: 2 hr; time upstream: $4\frac{1}{2}$ hr; still water: 6.5 mph; current: 2.5 mph **27.** still air: 455 mph; wind: 65 mph **29.** $4\frac{1}{2}$ hr **31.** 12% solution: $7\frac{1}{2}$ oz; 4% solution: $4\frac{1}{2}$ oz **33.** \$4.95 beans: 113 lb; \$2.65 beans: 87 lb **35.** 60°, 30° **37.** 20°, 70° **39.** number sold at \$9.50: 23; number sold at \$7.50: 67 **41.** $2\frac{1}{4}$ mph and $2\frac{3}{4}$ mph **43.** 30%: 50 gal; 60%: 100 gal **45.** length: 42 in.; width: 30 in. **47.** 16 **49.** $36x^2$ **51.** $100y^6$ **53.** a **55.** width: 9 ft; length: 15 ft

Chapter 14 Vocabulary Check **1.** dependent **2.** system of linear equations **3.** consistent **4.** solution **5.** addition; substitution **6.** inconsistent **7.** independent

Chapter 14 Review **1. a.** no **b.** yes **2. a.** yes **b.** no **3. a.** no **b.** no **4. a.** no **b.** yes

5.
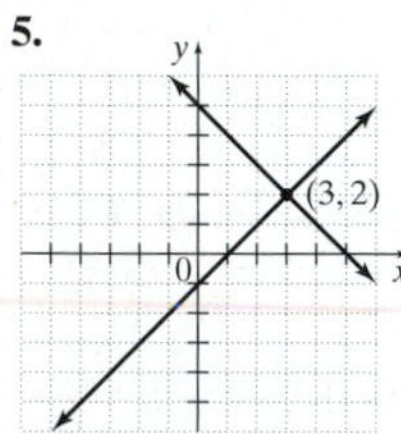

6.
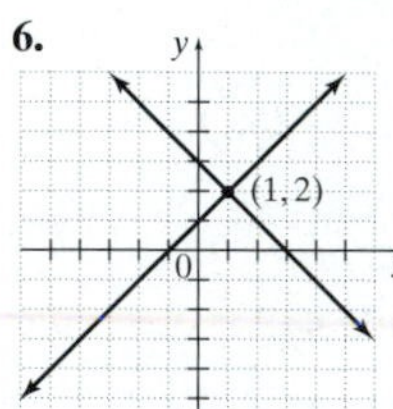

7.
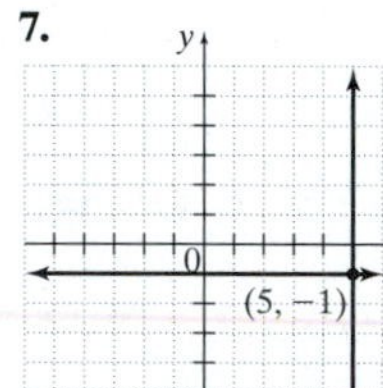

8.
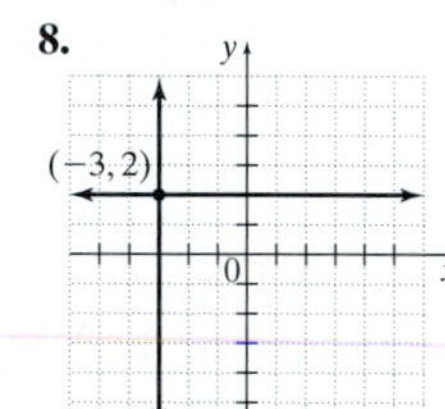

9.
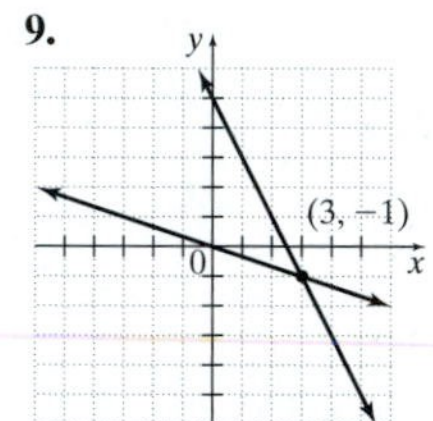

10.
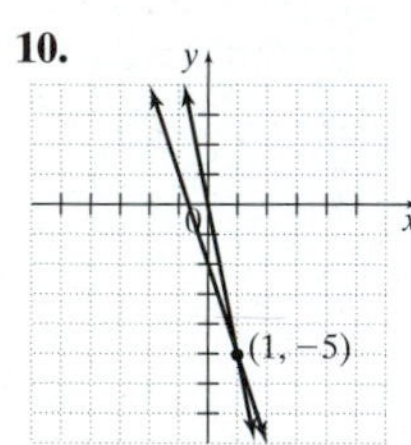

11. no solution
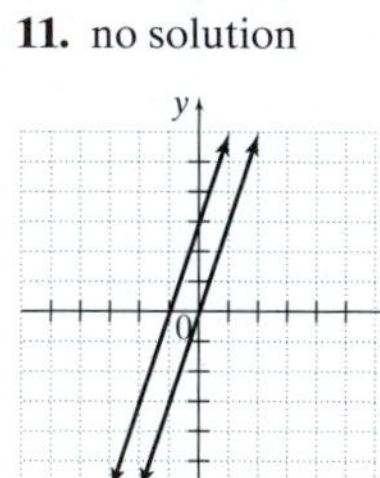

12. infinite number of solutions

13. (−1, 4) **14.** (2, −1) **15.** (3, −2) **16.** (2, 5) **17.** infinite number of solutions **18.** infinite number of solutions **19.** no solution **20.** no solution **21.** (−6, 2) **22.** (4, −1) **23.** (3, 7) **24.** (−2, 4) **25.** infinite number of solutions **26.** infinite number of solutions **27.** (8, −6) **28.** $\left(-\frac{3}{2}, \frac{15}{2}\right)$ **29.** −6 and 22 **30.** orchestra: 255 seats; balcony: 105 seats **31.** current of river: 3.2 mph; speed in still water: 21.1 mph **32.** 6% solution: $12\frac{1}{2}$ cc; 14% solution: $37\frac{1}{2}$ cc **33.** egg: \$0.40; strip of bacon: \$0.65 **34.** jogging: 0.86 hr; walking: 2.14 hr

35.
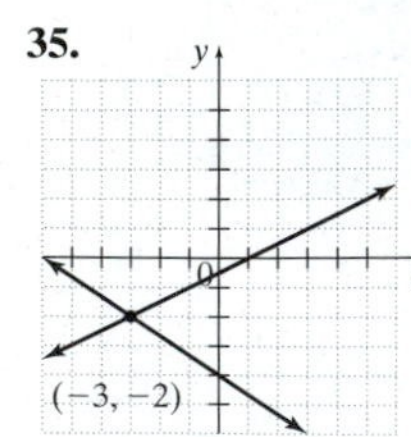

36. infinite number of solutions
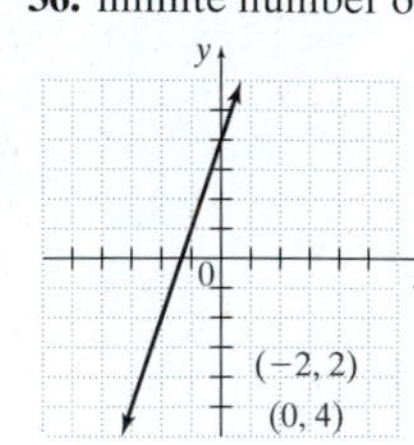
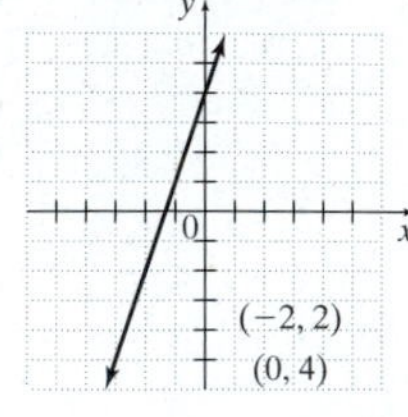

37. (3, 2) **38.** (7, 1) **39.** $\left(1\frac{1}{2}, -3\right)$ **40.** no solution **41.** infinite number of solutions **42.** (8, 11) **43.** (−5, 2) **44.** (16, −4) **45.** infinite number of solutions **46.** no solution **47.** 4 and 8 **48.** −5 and 13 **49.** 24 nickels and 41 dimes **50.** 13¢ stamps: 17; 22¢ stamps: 9

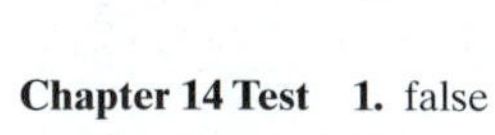

Chapter 14 Test **1.** false **2.** false **3.** true **4.** false **5.** no **6.** yes

7.
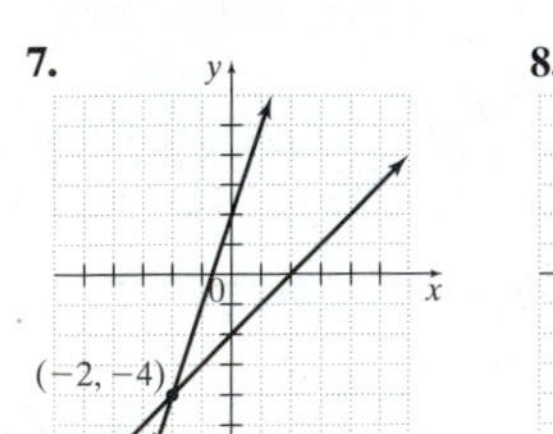

8.
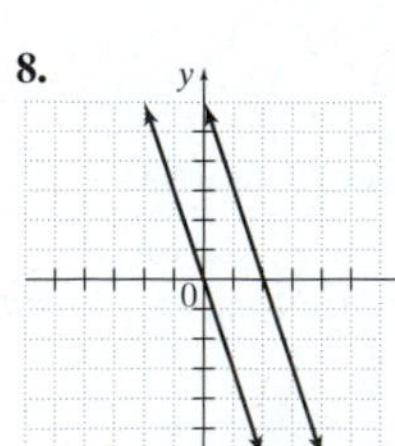

9. (−4, 1) **10.** $\left(\frac{1}{2}, -2\right)$ **11.** (20, 8) **12.** no solution **13.** (4, −5) **14.** (7, 2) **15.** (5, −2) **16.** infinite number of solutions **17.** (−5, 3) **18.** $\left(\frac{47}{5}, \frac{48}{5}\right)$ **19.** 78, 46 **20.** 120 cc **21.** Texas: 226 thousand; Missouri: 110 thousand **22.** 3 mph; 6 mph **23.** 1999 **24.** 1991–1999

Cumulative Review **1.** −7; Sec. 2.3, Ex. 6 **2.** −3; Sec. 2.3 **3.** 1; Sec. 2.3, Ex. 7 **4.** 5; Sec. 2.3 **5.** 4; Sec. 2.6, Ex. 4 **6.** −2; Sec. 2.6 **7.** $1.2 = 30\% \cdot x$; Sec. 6.3, Ex. 2 **8.** $9 = 45\% \cdot x$; Sec. 6.3 **9.** 16%; Sec. 6.4, Ex.10 **10.** 25%; Sec. 6.4 **11.** 775 freshmen; Sec. 6.5, Ex. 3 **12.** 1450 apples; Sec. 6.5 **13.** no solution; Sec. 9.3, Ex. 6 **14.** $\frac{1}{2}$; Sec. 9.3 **15.** $\left\{x \middle| x > \frac{13}{7}\right\}$; 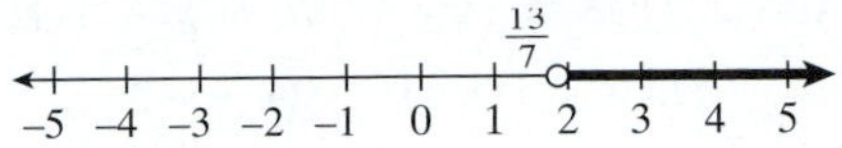

; Sec. 9.6, Ex. 8 **16.** $\{x | x \geq -2\}$; 

; Sec. 9.6

17. $\frac{m^7}{n^7}$, $n \neq 0$; Sec. 10.1, Ex. 22 **18.** $35x^7$; Sec. 10.1 **19.** $\frac{16x^{16}}{81y^{20}}$, $y \neq 0$; Sec. 10.1, Ex. 23 **20.** $\frac{25x^4}{16y^6}$; Sec. 10.1

21. $9x^2 - 6x - 1$; Sec. 10.4, Ex. 5 **22.** $8x + 4$; Sec. 10.4 **23.** $2x + 4 - \frac{1}{3x - 1}$; Sec. 10.7, Ex. 5 **24.** $3x + 2 - \frac{2}{x - 1}$; Sec. 10.7 **25.** $-\frac{1}{2}$, 4; Sec. 11.6, Ex. 6 **26.** 8, −3; Sec. 11.6 **27.** 6, 8, 10; Sec. 11.7, Ex. 5 **28.** −12 and 11; Sec. 11.7 **29.** 1; Sec. 12.3, Ex. 2 **30.** x; Sec. 12.3 **31.** $m = 0$; Sec. 13.4, Ex. 5 **32.** undefined; Sec. 13.4 **33.** $-x + 5y = 23$; Sec. 13.5, Ex. 5 **34.** $x + y = -1$; Sec. 13.5 **35.** domain: {−1, 0, 3}; range: {−2, 0, 2, 3}; Sec. 13.6, Ex. 1 **36.** domain: {2}; range: {3, 0, −2, 4}; Sec. 13.6 **37.** $\left(6, \frac{1}{2}\right)$; Sec. 14.2, Ex. 3 **38.** (−3, 1); Sec. 14.2 **39.** $\left(-\frac{15}{7}, -\frac{5}{7}\right)$; Sec. 14.3, Ex. 6 **40.** (3, 12); Sec. 14.3 **41.** 29 and 8; Sec. 14.4, Ex. 1 **42.** 42 and 33; Sec. 14.4

CHAPTER 15 Roots and Radicals

Calculator Explorations **1.** 2.449 **3.** 3.317 **5.** 9.055 **7.** 3.420 **9.** 2.115 **11.** 1.783

Exercise Set 15.1 **1.** 4 **3.** $\frac{1}{5}$ **5.** −10 **7.** not a real number **9.** −11 **11.** $\frac{3}{5}$ **13.** 30 **15.** 12 **17.** $\frac{1}{10}$ **19.** 0.5 **21.** 5 **23.** −4 **25.** −2 **27.** $\frac{1}{2}$ **29.** −5 **31.** 2 **33.** 9 **35.** not a real number **37.** $-\frac{3}{4}$ **39.** −5 **41.** 1 **43.** 2.646 **45.** 6.083 **47.** 11.662 **49.** $\sqrt{2} \approx 1.41$; 126.90 ft **51.** m **53.** x^2 **55.** $3x^4$ **57.** $9x$ **59.** ab^2 **61.** $4a^3b^2$ **63.** a^2b^6 **65.** $-2xy^9$ **67.** $25 \cdot 2$ **69.** $16 \cdot 2$ or $4 \cdot 8$ **71.** $4 \cdot 7$ **73.** $9 \cdot 3$ **75.** a, b **77.** 7 mi **79.** 3.01 in. **81.** 3 **83.** 10 **85.** $T = 6.1$ seconds **87.** answers may vary **89.** 1; 1.7; 2; 3

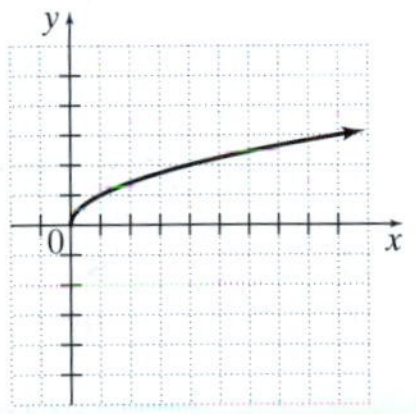

91. (2, 0) **93.** (−4, 0)

Vocabulary and Readiness Check **1.** index, radicand, radical sign **3.** $\sqrt{a} \cdot \sqrt{b}$

Exercise Set 15.2 **1.** $2\sqrt{5}$ **3.** $5\sqrt{2}$ **5.** $\sqrt{33}$ **7.** $7\sqrt{2}$ **9.** $2\sqrt{15}$ **11.** $6\sqrt{5}$ **13.** $2\sqrt{13}$ **15.** 15 **17.** $21\sqrt{7}$ **19.** $-15\sqrt{3}$ **21.** $\frac{2\sqrt{2}}{5}$ **23.** $\frac{3\sqrt{3}}{11}$ **25.** $\frac{3}{2}$ **27.** $\frac{5\sqrt{5}}{3}$ **29.** $\frac{\sqrt{11}}{6}$ **31.** $-\frac{\sqrt{3}}{4}$ **33.** $x^3\sqrt{x}$ **35.** $x^6\sqrt{x}$ **37.** $6a\sqrt{a}$ **39.** $4x^2\sqrt{6}$ **41.** $\frac{2\sqrt{3}}{m}$ **43.** $\frac{3\sqrt{x}}{y^5}$ **45.** $\frac{2\sqrt{22}}{x^6}$ **47.** 16 **49.** $\frac{6}{11}$ **51.** $5\sqrt{7}$ **53.** $\frac{2\sqrt{5}}{3}$ **55.** $2m^3\sqrt{6m}$ **57.** $\frac{y\sqrt{23y}}{2x^3}$ **59.** $2\sqrt[3]{3}$ **61.** $5\sqrt[3]{2}$ **63.** $\frac{\sqrt[3]{5}}{4}$ **65.** $\frac{\sqrt[3]{23}}{2}$ **67.** $\frac{\sqrt[3]{15}}{4}$ **69.** $2\sqrt[3]{10}$ **71.** $14x$ **73.** $2x^2 - 7x - 15$ **75.** 0 **77.** $x^3y\sqrt{y}$ **79.** $7x^2y^2\sqrt{2x}$ **81.** $-2x^2$ **83.** $2\sqrt[3]{10}$ in. **85.** answers may vary **87.** $2\sqrt{5}$ in. **89.** 2.25 in. **91.** \$1700 **93.** 1.7 sq m

Vocabulary and Readiness Check **1.** like radicals **3.** $17\sqrt{2}$ **5.** $2\sqrt{5}$

Exercise Set 15.3 **1.** $-4\sqrt{3}$ **3.** $9\sqrt{6} - 5$ **5.** $\sqrt{5} + \sqrt{2}$ **7.** $7\sqrt{3} - \sqrt{2}$ **9.** $-5\sqrt{2} - 6$ **11.** $5\sqrt{3}$ **13.** $9\sqrt{5}$ **15.** $4\sqrt{6} + \sqrt{5}$ **17.** $x + \sqrt{x}$ **19.** 0 **21.** $\frac{4\sqrt{5}}{9}$ **23.** $\frac{3\sqrt{3}}{8}$ **25.** $7\sqrt{5}$ **27.** $9\sqrt{3}$ **29.** $\sqrt{5} + \sqrt{15}$ **31.** $x\sqrt{x}$ **33.** $5\sqrt{2} + 12$ **35.** $8\sqrt{2} - 5$ **37.** $2\sqrt{5}$ **39.** $-\sqrt{35}$ **41.** $6 - 3\sqrt{3}$ **43.** $11\sqrt{x}$ **45.** $12x - 11\sqrt{x}$ **47.** $x\sqrt{3x} + 3x\sqrt{x}$ **49.** $8x\sqrt{2} + 2x$ **51.** $2x^2\sqrt{10} - x^2\sqrt{5}$ **53.** $7\sqrt[3]{9} - \sqrt[3]{25}$ **55.** $-5\sqrt[3]{2} - 6$ **57.** $5\sqrt[3]{3}$ **59.** $4x + 4x\sqrt[3]{2}$ **61.** $x\sqrt[3]{5}$ **63.** $10y^2\sqrt[3]{y}$ **65.** $x^2 + 12x + 36$ **67.** $4x^2 - 4x + 1$ **69.** answers may vary **71.** $8\sqrt{5}$ in. **73.** $\left(48 + \frac{9\sqrt{3}}{2}\right)$ sq ft **75.** yes; $7\sqrt{2}$ **77.** no **79.** yes; $3\sqrt{7}$ **81.** $\frac{83x\sqrt{x}}{20}$

Vocabulary and Readiness Check **1.** $\sqrt{21}$ **3.** $\sqrt{\frac{15}{3}}$ or $\sqrt{5}$ **5.** $2 - \sqrt{3}$

Exercise Set 15.4 **1.** 4 **3.** $5\sqrt{2}$ **5.** 6 **7.** $2x$ **9.** 20 **11.** $36x$ **13.** $3x^3\sqrt{2}$ **15.** $4xy\sqrt{y}$ **17.** $\sqrt{30} + \sqrt{42}$ **19.** $2\sqrt{5} + 5\sqrt{2}$ **21.** $y\sqrt{7} - 14\sqrt{y}$ **23.** -33 **25.** $\sqrt{6} - \sqrt{15} + \sqrt{10} - 5$ **27.** $16 - 11\sqrt{11}$ **29.** $x - 36$ **31.** $x - 14\sqrt{x} + 49$ **33.** $6y + 2\sqrt{6y} + 1$ **35.** 4 **37.** $\sqrt{7}$ **39.** $3\sqrt{2}$ **41.** $5y^2$ **43.** $5\sqrt{3}$ **45.** $2y\sqrt{6}$ **47.** $2xy\sqrt{3y}$ **49.** $\frac{\sqrt{15}}{5}$ **51.** $\frac{7\sqrt{2}}{2}$ **53.** $\frac{\sqrt{6y}}{6y}$ **55.** $\frac{\sqrt{10}}{6}$ **57.** $\frac{\sqrt{3x}}{x}$ **59.** $\frac{\sqrt{2}}{4}$ **61.** $\frac{\sqrt{30}}{15}$ **63.** $\frac{\sqrt{15}}{10}$ **65.** $\frac{3\sqrt{2x}}{2}$ **67.** $\frac{8y\sqrt{5}}{5}$ **69.** $\frac{\sqrt{3xy}}{6x}$ **71.** $3\sqrt{2} - 3$ **73.** $-8 - 4\sqrt{5}$ **75.** $5 + \sqrt{30} + \sqrt{6} + \sqrt{5}$ **77.** $\sqrt{6} + \sqrt{3} + \sqrt{2} + 1$ **79.** $\frac{10 - 5\sqrt{x}}{4 - x}$ **81.** $\frac{3\sqrt{x} + 12}{x - 16}$ **83.** $x = 44$ **85.** $z = 2$ **87.** $x = 3$ **89.** $130\sqrt{3}$ sq m **91.** $\frac{\sqrt{A\pi}}{\pi}$ **93.** true **95.** false **97.** false **99.** answers may vary **101.** answers may vary **103.** $\frac{2}{\sqrt{6} - \sqrt{2} - \sqrt{3} + 1}$

Integrated Review **1.** 6 **2.** $4\sqrt{3}$ **3.** x^2 **4.** $y^3\sqrt{y}$ **5.** $4x$ **6.** $3x^5\sqrt{2x}$ **7.** 2 **8.** 3 **9.** -3 **10.** not a real number **11.** $\frac{\sqrt{11}}{3}$ **12.** $\frac{\sqrt[3]{7}}{4}$ **13.** -4 **14.** -5 **15.** $\frac{3}{7}$ **16.** $\frac{1}{8}$ **17.** a^4b **18.** x^5y^{10} **19.** $5m^3$ **20.** $3n^8$ **21.** $6\sqrt{7}$ **22.** $3\sqrt{2}$ **23.** cannot be simplified **24.** $\sqrt{x} + 3x$ **25.** $\sqrt{30}$ **26.** 3 **27.** 28 **28.** 45 **29.** $\sqrt{33} + \sqrt{3}$ **30.** $3\sqrt{2} - 2\sqrt{6}$ **31.** $4y$ **32.** $3x^2\sqrt{5}$ **33.** $x - 3\sqrt{x} - 10$ **34.** $11 + 6\sqrt{2}$ **35.** 2 **36.** $\sqrt{3}$ **37.** $2x^2\sqrt{3}$ **38.** $ab^2\sqrt{15a}$ **39.** $\frac{\sqrt{6}}{6}$ **40.** $\frac{x\sqrt{5}}{10}$ **41.** $\frac{4\sqrt{6} - 4}{5}$ **42.** $\frac{\sqrt{2x} + 5\sqrt{2} + \sqrt{x} + 5}{x - 25}$

Exercise Set 15.5 **1.** 81 **3.** -1 **5.** 49 **7.** no solution **9.** 4 **11.** 2 **13.** 2 **15.** 9 **17.** -3 **19.** $-1, -2$ **21.** no solution **23.** $0, -3$ **25.** 16 **27.** 25 **29.** 1 **31.** 5 **33.** -2 **35.** no solution **37.** 2 **39.** 36 **41.** no solution **43.** $\frac{3}{2}$ **45.** 16 **47.** 3 **49.** 12 **51.** 3, 1 **53.** -1 **55.** $3x - 8 = 19$; $x = 9$ **57.** $2(2x + x) = 24$; length = 8 in. **59.** 4, 7 **61.** answers may vary **63.** **a.** 3.2, 10, 31.6 **b.** no **65.** 7.30 **67.** 0.76

The Bigger Picture

1. $2\sqrt{14}$ **2.** $\frac{2x^2\sqrt{5x}}{7}$ **3.** $-15x^5y^{11}$ **4.** $\frac{\sqrt{110}}{11}$ **5.** $2(\sqrt{5} + 1)$ or $2\sqrt{5} + 2$ **6.** $5x^2 - 19$ **7.** -1 **8.** $-\frac{5}{2}$ **9.** $\left\{y \middle| y \geq -\frac{4}{11}\right\}$ **10.** $6, -7$ **11.** $\frac{28}{3}$ **12.** 2

Exercise Set 15.6 **1.** $\sqrt{13}$; 3.61 **3.** $3\sqrt{3}$; 5.20 **5.** 25 **7.** $\sqrt{22}$; 4.69 **9.** $3\sqrt{17}$; 12.37 **11.** $\sqrt{41}$; 6.40 **13.** $4\sqrt{2}$; 5.66 **15.** $3\sqrt{10}$; 9.49 **17.** 20.6 ft **19.** 11.7 ft **21.** 24 cu ft **23.** 54 mph **25.** 27 mph **27.** 59.1 km **29.** 60.2 km **31.** $3, -3$ **33.** $10, -10$ **35.** $8, -8$ **37.** $y = 2\sqrt{10}$; $x = 2\sqrt{10} - 4$ **39.** 201 miles **41.** answers may vary

Chapter 15 Vocabulary Check **1.** like radicals **2.** index; radicand; radical **3.** conjugate **4.** principal square root **5.** rationalizing the denominator

Chapter Review **1.** 9 **2.** -7 **3.** 3 **4.** 3 **5.** $-\frac{3}{8}$ **6.** $\frac{2}{3}$ **7.** 2 **8.** -2 **9.** c **10.** a, c **11.** x^6 **12.** x^4 **13.** $3y$ **14.** $5x^2$ **15.** $2\sqrt{10}$ **16.** $2\sqrt{6}$ **17.** $3\sqrt{6}$ **18.** $2\sqrt{22}$ **19.** $x^2\sqrt{x}$ **20.** $y^3\sqrt{y}$ **21.** $2x\sqrt{5}$ **22.** $5y^2\sqrt{2}$ **23.** $3\sqrt[3]{2}$ **24.** $2\sqrt[3]{11}$ **25.** $\frac{3\sqrt{2}}{5}$ **26.** $\frac{5\sqrt{3}}{8}$ **27.** $-\frac{5\sqrt{2}}{3}$ **28.** $-\frac{2\sqrt{3}}{7}$ **29.** $\frac{\sqrt{11}}{x}$ **30.** $\frac{\sqrt{7}}{y^2}$ **31.** $\frac{y^2\sqrt{y}}{10}$ **32.** $\frac{x\sqrt{x}}{9}$ **33.** $-3\sqrt{2}$ **34.** $-5\sqrt{3}$ **35.** $4\sqrt{5} + 4\sqrt{6}$ **36.** $-2\sqrt{7} + 2\sqrt{2}$ **37.** $5\sqrt{7} + 2\sqrt{14}$ **38.** $9\sqrt{3} - 4$ **39.** $\frac{\sqrt{5}}{6}$ **40.** $\frac{9\sqrt{11}}{20}$ **41.** $10 - x\sqrt{5}$ **42.** $2\sqrt{2x} - \sqrt{3x}$ **43.** $3\sqrt{2}$ **44.** $5\sqrt{3}$ **45.** $\sqrt{10} - \sqrt{14}$ **46.** $\sqrt{55} + \sqrt{15}$ **47.** $3\sqrt{2} - 5\sqrt{3} + 2\sqrt{6} - 10$ **48.** $2 - 2\sqrt{5}$ **49.** $x - 4\sqrt{x} + 4$ **50.** $y + 8\sqrt{y} + 16$ **51.** 3 **52.** 2 **53.** $2\sqrt{5}$ **54.** $4\sqrt{2}$ **55.** $x\sqrt{15x}$ **56.** $3x^2\sqrt{2}$ **57.** $\frac{\sqrt{22}}{11}$ **58.** $\frac{\sqrt{39}}{13}$ **59.** $\frac{\sqrt{30}}{6}$ **60.** $\frac{\sqrt{70}}{10}$ **61.** $\frac{\sqrt{5x}}{5x}$ **62.** $\frac{5\sqrt{3y}}{3y}$ **63.** $\frac{\sqrt{3x}}{x}$ **64.** $\frac{\sqrt{6y}}{y}$ **65.** $3\sqrt{5} + 6$ **66.** $8\sqrt{10} + 24$ **67.** $\frac{\sqrt{6} + \sqrt{2} + \sqrt{3} + 1}{2}$ **68.** $\sqrt{15} - 2\sqrt{3} - 2\sqrt{5} + 4$ **69.** $\frac{10\sqrt{x} - 50}{x - 25}$ **70.** $\frac{8\sqrt{x} + 8}{x - 1}$ **71.** 18 **72.** 13 **73.** 25 **74.** no solution **75.** 12 **76.** 5 **77.** 1 **78.** 9 **79.** $2\sqrt{14}$; 7.48 **80.** $\sqrt{117}$; 10.82 **81.** $4\sqrt{34}$ ft; 23.33 ft

82. $5\sqrt{3}$ in.; 8.66 in. **83.** 2.4 in. **84.** 144π sq in. **85.** 12 **86.** -4 **87.** $4x^8$ **88.** $2x^{12}$ **89.** $3x^3\sqrt{2x}$ **90.** $4y^3\sqrt{3}$ **91.** $\frac{y^2}{9}$ **92.** $\frac{x^4\sqrt{x}}{3}$ **93.** $7\sqrt{3}$ **94.** $5\sqrt{7}-3$ **95.** $-\frac{\sqrt{3}}{4}$ **96.** $4x\sqrt{5x}$ **97.** $7\sqrt{2}$ **98.** $3\sqrt{3}-\sqrt{6}$ **99.** $\sqrt{10}-\sqrt{2}+4\sqrt{5}-4$ **100.** $x+6\sqrt{x}+9$ **101.** $2\sqrt{6}$ **102.** $2x$ **103.** $\frac{\sqrt{14}}{7}$ **104.** $\frac{3\sqrt{2x}}{2x}$ **105.** $\frac{3\sqrt{x}+18}{x-36}$ **106.** $\frac{\sqrt{35}-3\sqrt{7}-5\sqrt{5}+15}{-4}$ **107.** 1 **108.** 13 **109.** 14 **110.** 9 **111.** $\sqrt{58}$; 7.62 **112.** $4\sqrt{2}$ in.; 5.66 in.

Chapter Test **1.** 4 **2.** 5 **3.** 3 **4.** $\frac{3}{4}$ **5.** not a real number **6.** x^5 **7.** $3\sqrt{6}$ **8.** $2\sqrt{23}$ **9.** $y^3\sqrt{y}$ **10.** $2x^4\sqrt{6}$ **11.** 3 **12.** $2\sqrt[3]{2}$ **13.** $\frac{\sqrt{5}}{4}$ **14.** $\frac{y\sqrt{y}}{5}$ **15.** $-2\sqrt{13}$ **16.** $\sqrt{2}+2\sqrt{3}$ **17.** $\frac{7\sqrt{3}}{10}$ **18.** $7\sqrt{2}$ **19.** $2\sqrt{3}-\sqrt{10}$ **20.** $x-\sqrt{x}-6$ **21.** $\sqrt{5}$ **22.** $2x\sqrt{5x}$ **23.** $\frac{\sqrt{6}}{3}$ **24.** $\frac{8\sqrt{5y}}{5y}$ **25.** $4\sqrt{6}-8$ **26.** $\frac{3+\sqrt{x}}{9-x}$ **27.** 9 **28.** 5 **29.** 9 **30.** $4\sqrt{5}$ in. **31.** 2.19 m

Cumulative Review **1.** 736.2; Sec. 5.1, Ex. 15 **2.** 328.2; Sec. 5.1 **3.** 25.454; Sec. 5.2, Ex. 1 **4.** 17.052; Sec. 5.1 **5.** no; Sec. 5.3, Ex. 13 **6.** yes; Sec. 5.3 **7.** $\frac{1}{6}$; Sec. 7.3, Ex. 5 **8.** $\frac{2}{5}$; Sec. 7.3 **9.** 2; Sec. 9.3, Ex. 1 **10.** $\frac{2}{3}$; Sec. 9.3 **11. a.** 102,000 **b.** 0.007358 **c.** 84,000,000 **d.** 0.00003007; Sec. 10.2, Ex. 18 **12. a.** 82,600 **b.** 0.099 **c.** 100,200 **d.** 0.008039; Sec. 10.2 **13.** $6x^2-11x-10$; Sec. 10.5, Ex. 7b **14.** $20x^2+x-1$; Sec. 10.5 **15.** $(y+2)(x+3)$; Sec. 11.1, Ex. 10 **16.** $(4x^2+3)(4x-7)$; Sec. 11.1 **17.** $(3x+2)(x+3)$; Sec. 11.3, Ex. 1 **18.** $(9x+4)(x-1)$; Sec. 11.3 **19. a.** $x=3$ **b.** $x=2, x=1$ **c.** none; Sec. 12.1, Ex. 2 **20. a.** $x=0$ **b.** none **c.** $x=2, x=-2$; Sec. 12.1 **21.** $\frac{x+2}{x}$; Sec. 12.1, Ex. 5 **22.** $2(2x+y)$; Sec. 12.1 **23. a.** 0 **b.** $\frac{15+14x}{50x^2}$; Sec. 12.4, Ex. 1 **24. a.** $-\frac{x}{10}$ **b.** $\frac{12+5a}{16a^2}$; Sec. 12.4 **25.** $-\frac{17}{5}$; Sec. 12.5, Ex. 4 **26.** 2; Sec. 12.5 **27.** Sec. 13.3, Ex. 7 **28.** ; Sec. 13.3

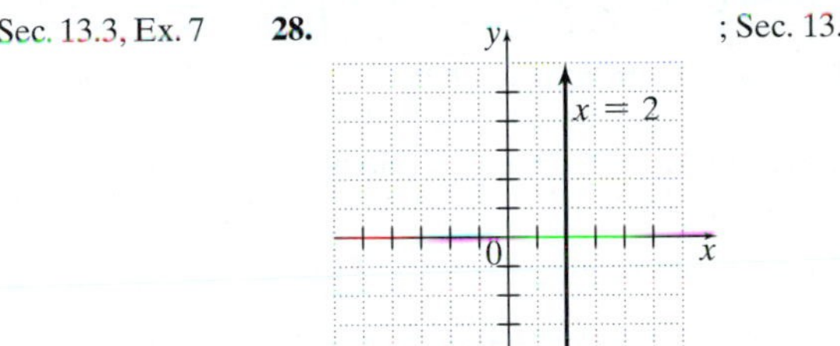

29. 1; Sec. 15.1, Ex. 6 **30.** 2; Sec. 15.1 **31.** -3; Sec. 15.1, Ex. 7 **32.** -2; Sec. 15.1 **33.** $\frac{1}{5}$; Sec. 15.1, Ex. 8 **34.** $\frac{3}{4}$; Sec. 15.1 **35.** $3\sqrt{6}$; Sec. 15.2, Ex. 1 **36.** $2\sqrt{10}$; Sec. 15.2 **37.** $10\sqrt{2}$; Sec. 15.2, Ex. 3 **38.** $5\sqrt{5}$; Sec. 15.2 **39.** $4\sqrt{3}$; Sec. 15.3, Ex. 6 **40.** $9\sqrt{3}$; Sec. 15.3 **41.** $2x-4x^2\sqrt{x}$; Sec. 15.3, Ex. 8 **42.** $12x+6\sqrt{x}$; Sec. 15.3 **43.** $\frac{2\sqrt{7}}{7}$; Sec. 15.4, Ex. 10 **44.** $\frac{4\sqrt{5}}{5}$; Sec. 15.4 **45.** $\frac{1}{2}$; Sec. 15.5, Ex. 2 **46.** 4; Sec. 15.5

CHAPTER 16 Quadratic Equations

Exercise Set 16.1 **1.** ± 7 **3.** $-5, 3$ **5.** ± 4 **7.** ± 3 **9.** $-5, -2$ **11.** ± 8 **13.** $\pm\sqrt{21}$ **15.** $\pm\frac{1}{5}$ **17.** no real solution **19.** $\pm\frac{\sqrt{39}}{3}$ **21.** $\pm\frac{2\sqrt{7}}{7}$ **23.** $\pm\sqrt{2}$ **25.** 12, -2 **27.** $-2\pm\sqrt{7}$ **29.** 1, 0 **31.** $-2\pm\sqrt{10}$ **33.** $\frac{8}{3}, -4$ **35.** no real solution **37.** $\frac{11\pm 5\sqrt{2}}{2}$ **39.** $\frac{7\pm 4\sqrt{2}}{3}$ **41.** $\pm\sqrt{2}$ **43.** $-6\pm 2\sqrt{6}$ **45.** $\pm\sqrt{10}$ **47.** $\frac{1\pm\sqrt{5}}{4}$ **49.** $\pm 2\sqrt{3}$ **51.** $\frac{-8\pm 3\sqrt{5}}{-3}$ or $\frac{8\pm 3\sqrt{5}}{3}$ **53.** 2.3 sec **55.** 4.6 seconds **57.** $2\sqrt{5}$ in. $\approx$ 4.47 in. **59.** 177 meters **61.** $(x+3)^2$ **63.** $(x-2)^2$ **65.** answers may vary **67.** 2, -6 **69.** $r=6$ in. **71.** ± 1.33 **73.** $x=7$, which is 2006

Vocabulary and Readiness Check **1.** zero **3.** quadratic equation **5.** 9

Exercise Set 16.2 **1.** $-6, -2$ **3.** $-1\pm 2\sqrt{2}$ **5.** 0, 6 **7.** $\frac{-5\pm\sqrt{53}}{2}$ **9.** $1\pm\sqrt{2}$ **11.** $-1, -4$ **13.** $-2, 4$ **15.** no real solution **17.** $\frac{3\pm\sqrt{19}}{2}$ **19.** $-2\pm\frac{\sqrt{6}}{2}$ **21.** $-3\pm\sqrt{34}$ **23.** $\frac{3\pm\sqrt{21}}{2}$ **25.** $\frac{1}{2}, 1$ **27.** $-6, 3$ **29.** no real solution **31.** 2, -6 **33.** $-\frac{1}{2}$ **35.** -1 **37.** $3+2\sqrt{5}$ **39.** $\frac{1-3\sqrt{2}}{2}$ **41.** answers may vary **43. a.** $-3\pm\sqrt{11}$ **b.** answers may vary **45.** $k=8$ or $k=-8$ **47.** 8 years, or 2009 **49.** $-6, -2$ **51.** $\approx -0.68, 3.68$

Vocabulary and Readiness Check **1.** $x=\frac{-b\pm\sqrt{b^2-4ac}}{2a}$ **3.** 1; 3; -7

Exercise Set 16.3 **1.** $x = 2, 1$ **3.** $k = \frac{-7 \pm \sqrt{37}}{6}$ **5.** $x = \pm\frac{\sqrt{3}}{2}$ **7.** no real solution **9.** $y = 10, -3$ **11.** $x = \pm\sqrt{5}$ **13.** $m = -3, 4$ **15.** $x = -2 \pm \sqrt{7}$ **17.** $x = \frac{-9 \pm \sqrt{129}}{12}$ **19.** $p = \frac{4 \pm \sqrt{2}}{7}$ **21.** $x = 3 \pm \sqrt{7}$ **23.** $x = \frac{3 \pm \sqrt{3}}{2}$ **25.** $x = \frac{1}{3}, -1$ **27.** $y = \frac{3 \pm \sqrt{13}}{4}$ **29.** $y = \frac{1}{5}, -\frac{3}{4}$ **31.** no real solution **33.** $m = 1 \pm \sqrt{2}$ **35.** no real solution **37.** $p = -\frac{1}{2}, -\frac{3}{4}$ **39.** $x = \frac{7 \pm \sqrt{129}}{20}$ **41.** $x = \frac{11 \pm \sqrt{129}}{4}$ **43.** $z = \frac{1 \pm \sqrt{2}}{5}$ **45.** $\pm\sqrt{7}$; $-2.6, 2.6$ **47.** $-3 \pm 2\sqrt{2}$; $-5.8, -0.2$ **49.** $\frac{9 \pm \sqrt{97}}{2}$; $9.4, -0.4$ **51.** $\frac{1 \pm \sqrt{7}}{3}$; $1.2, -0.5$ **53.** **55.**

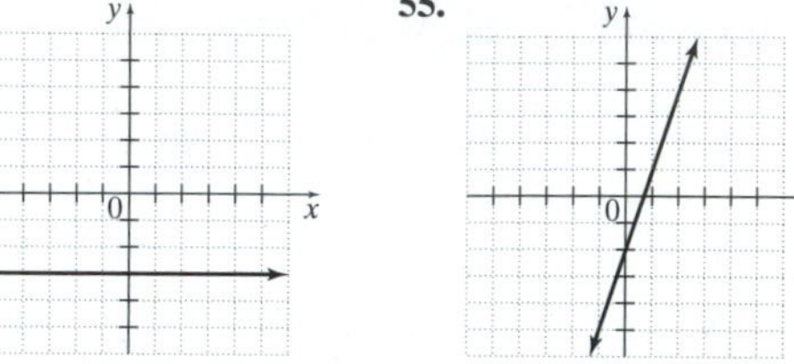

57. c **59.** b **61.** 10.3 ft by 4.9 ft **63.** $x = \frac{-3\sqrt{2} \pm \sqrt{38}}{2}$ **65.** answers may vary **67.** $-0.9, 0.2$ **69.** 8.1 sec **71.** 2008

The Bigger Picture **1.** -1.8 **2.** -5 **3.** -9 **4.** $10x + 2$ **5.** $\frac{1}{4}x^2 - 25$ **6.** $3x + 1 - \frac{4}{x}$ **7.** $\frac{x+2}{x-4}$ **8.** $\frac{x^2 + 8x - 50}{(x-10)(x+3)}$ **9.** $5\sqrt{2}$ **10.** $b\sqrt{10a}$ **11.** $\frac{\sqrt{6}}{3}$ **12.** $\frac{x+3}{7}$ **13.** $\frac{-3 \pm \sqrt{29}}{2}$ **14.** 6 **15.** $\{x \mid x \geq -2.8\}$ **16.** $\frac{1}{2}, -8$ **17.** 7 **18.** 27

Integrated Review **1.** $x = 2, \frac{1}{5}$ **2.** $x = \frac{2}{5}, -3$ **3.** $x = 1 \pm \sqrt{2}$ **4.** $x = 3 \pm \sqrt{2}$ **5.** $a = \pm 2\sqrt{5}$ **6.** $a = \pm 6\sqrt{2}$ **7.** no real solution **8.** no real solution **9.** $x = 2$ **10.** $x = 3$ **11.** $p = 3$ **12.** $p = \frac{7}{2}$ **13.** $y = \pm 2$ **14.** $y = \pm 3$ **15.** $x = 1, 2$ **16.** $x = -3, -4$ **17.** $z = 0, -5$ **18.** $z = \frac{8}{3}, 0$ **19.** $x = \frac{3 \pm \sqrt{7}}{5}$ **20.** $x = \frac{3 \pm \sqrt{5}}{2}$ **21.** $m = \frac{3}{2}, -1$ **22.** $m = \frac{2}{5}, -2$ **23.** $x = \frac{5 \pm \sqrt{105}}{20}$ **24.** $x = \frac{-1 \pm \sqrt{3}}{4}$ **25.** $x = 5, \frac{7}{4}$ **26.** $x = 1, \frac{7}{9}$ **27.** $x = \frac{7 \pm 3\sqrt{2}}{5}$ **28.** $x = \frac{5 \pm 5\sqrt{3}}{4}$ **29.** $z = \frac{7 \pm \sqrt{193}}{6}$ **30.** $z = \frac{-7 \pm \sqrt{193}}{12}$ **31.** $x = 11, -10$ **32.** $x = 7, -8$ **33.** $x = 4, -\frac{2}{3}$ **34.** $x = 2, -\frac{4}{5}$ **35.** $x = 0.5, 0.1$ **36.** $x = 0.3, -0.2$ **37.** $x = \frac{11 \pm \sqrt{41}}{20}$ **38.** $x = \frac{11 \pm \sqrt{41}}{40}$ **39.** $z = \frac{4 \pm \sqrt{10}}{2}$ **40.** $z = \frac{5 \pm \sqrt{185}}{4}$ **41.** answers may vary

Calculator Explorations **1.** $x = -0.41, 7.41$ **3.** $x = 0.91, 2.38$ **5.** $x = -0.39, 0.84$

Exercise Set 16.4 **1.** **3.** **5.** **7.**

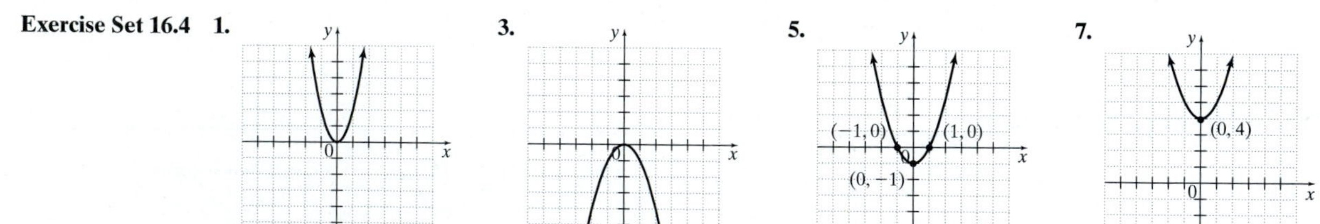

9. **11.** **13.** **15.** **17.**

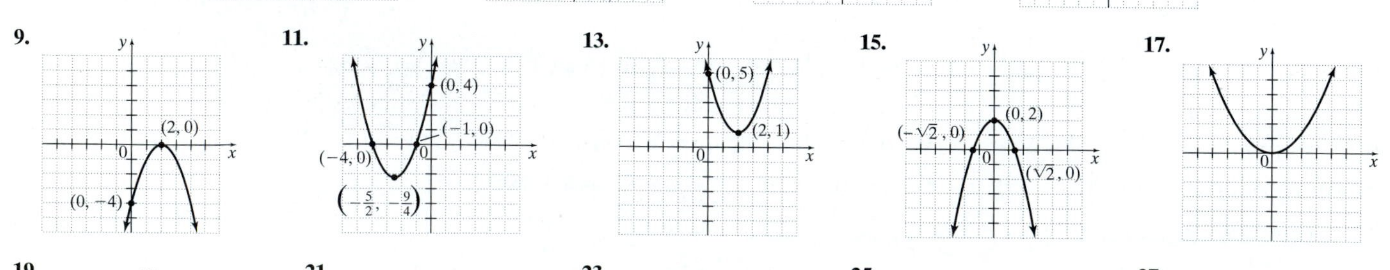

19. **21.** **23.** **25.** **27.**

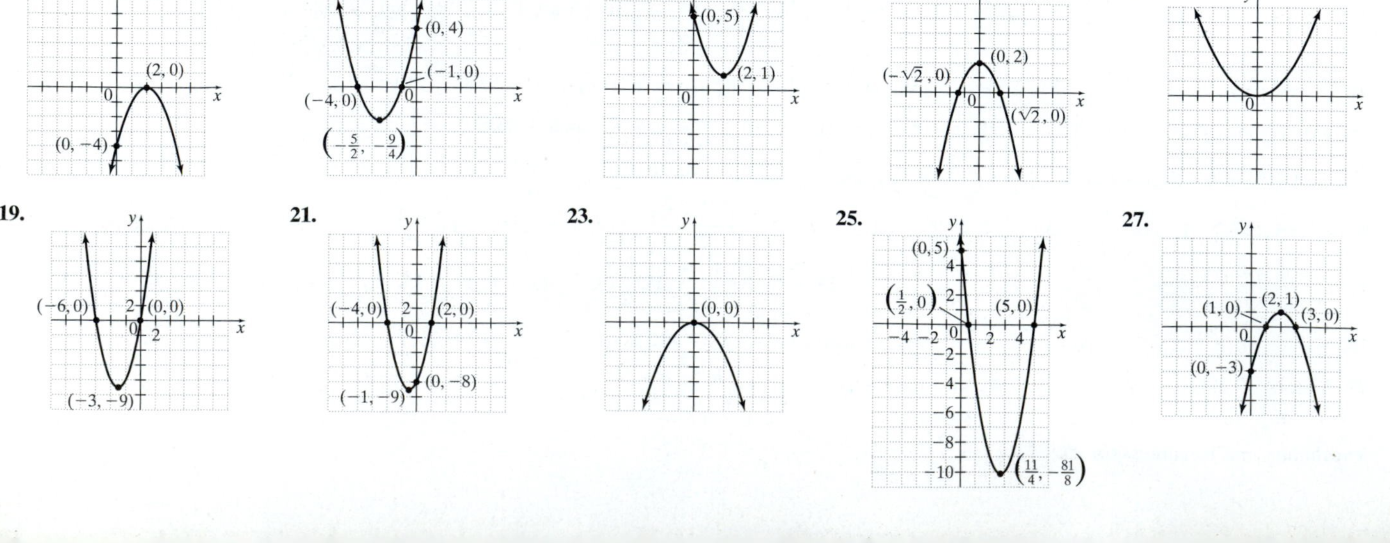

29. $\frac{5}{14}$ **31.** $\frac{x}{2}$ **33.** $\frac{2x^2}{x-1}$ **35.** $-4b$ **37. a.** 256 ft **b.** $t = 4$ sec **c.** $t = 8$ sec **39.** A **41.** D **43.** F

Chapter 16 Vocabulary Check **1.** square root **2.** vertex **3.** completing the square **4.** quadratic

Chapter Review **1.** $x = \pm 11$ **2.** $y = \pm 10$ **3.** $m = -\frac{1}{3}, 2$ **4.** $m = \frac{5}{7}, -1$ **5.** $x = \pm 6$ **6.** $x = \pm 9$ **7.** $k = \pm 5\sqrt{2}$ **8.** $k = \pm 3\sqrt{5}$ **9.** $x = 4, 18$ **10.** $x = 7, -13$ **11.** $p = \frac{-5 \pm \sqrt{41}}{4}$ **12.** $p = \frac{-7 \pm \sqrt{37}}{3}$ **13.** 2.5 sec **14.** 40.6 sec **15.** $x = 1, 8$ **16.** $x = -10, 2$ **17.** $x = -2 \pm \sqrt{5}$ **18.** $x = 4 \pm \sqrt{19}$ **19.** $x = 3 \pm \sqrt{2}$ **20.** $x = -3 \pm \sqrt{2}$ **21.** $y = \frac{1}{2}, -1$ **22.** $y = \frac{-3 \pm \sqrt{13}}{2}$ **23.** $x = -\frac{5}{3}$ **24.** $x = \frac{9}{4}$ **25.** $x = \pm\sqrt{5}$ **26.** $\pm\sqrt{3}$ **27.** $x = 5 \pm 3\sqrt{2}$ **28.** $x = -2 \pm \sqrt{11}$ **29.** $x = \frac{-1 \pm \sqrt{13}}{6}$ **30.** $x = \frac{-3 \pm \sqrt{13}}{2}$ **31.** no real solution **32.** no real solution **33.** 0.4, −0.8 **34.** 0.3, −3.3 **35.** 2007 **36.** 2007

37.
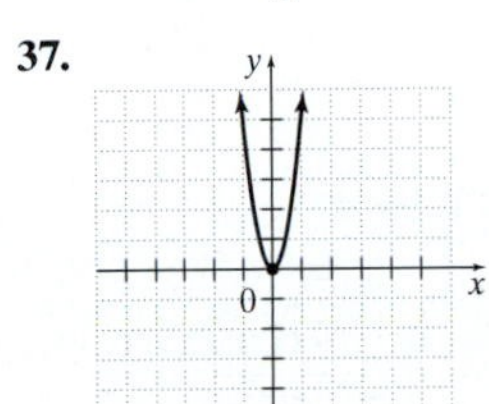

38.
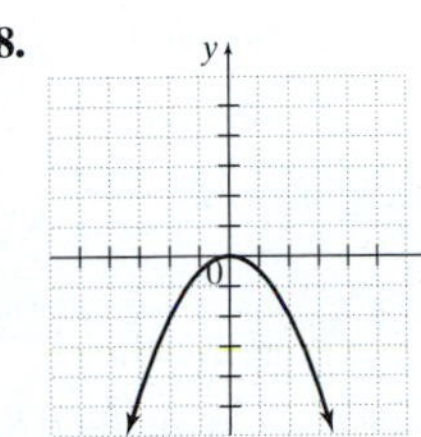

39.
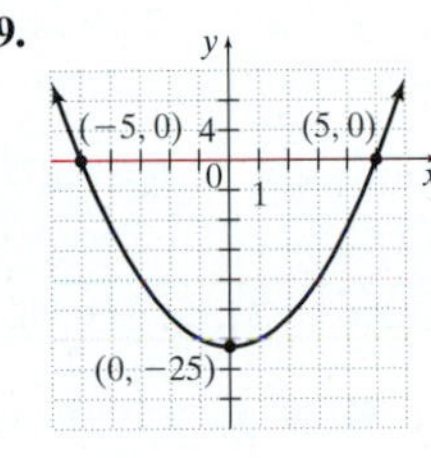

40.
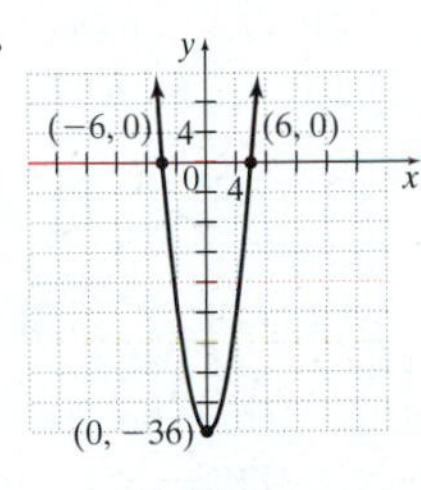

41.
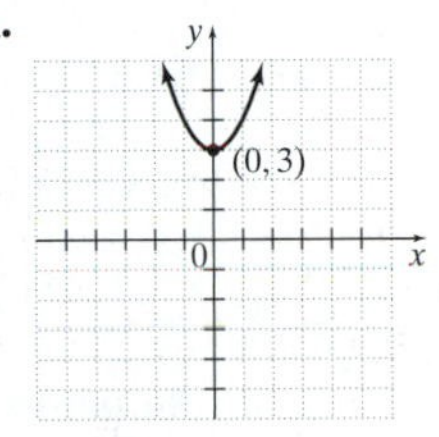

42.
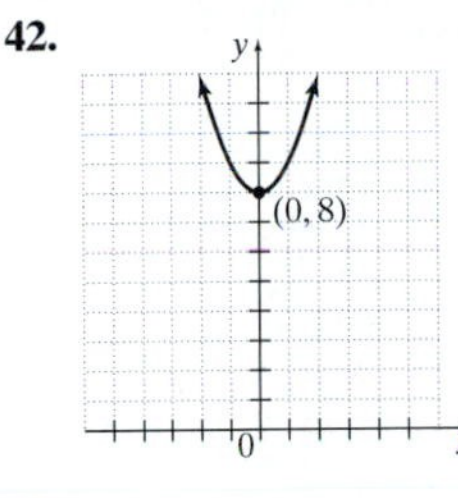

43.
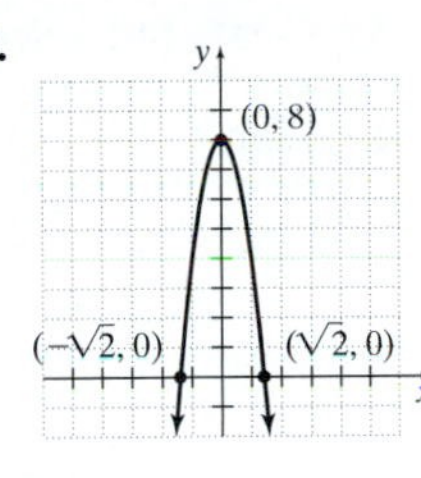

44.
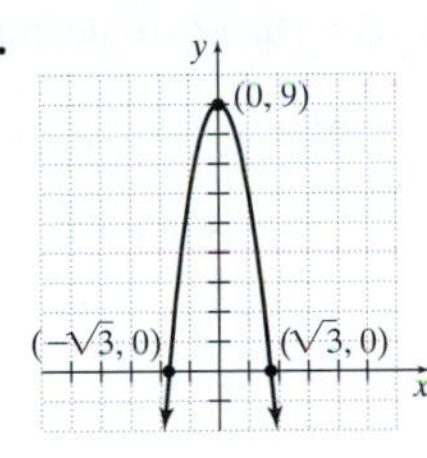

45.
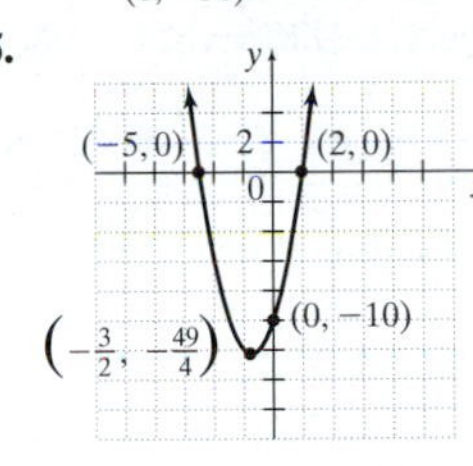

46.
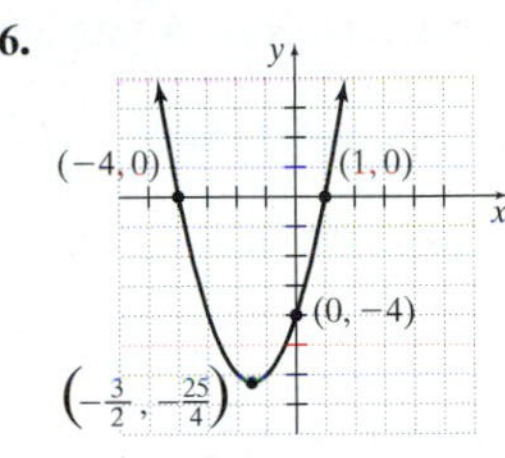

47.
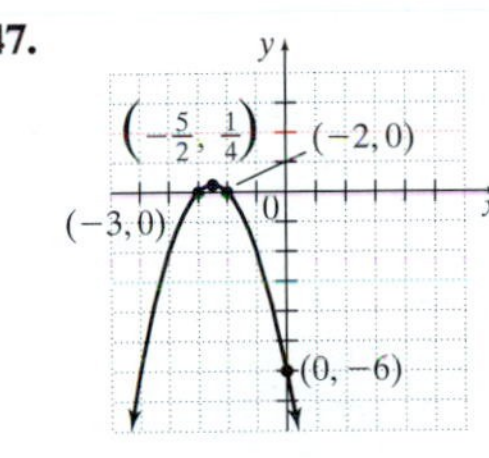

48.
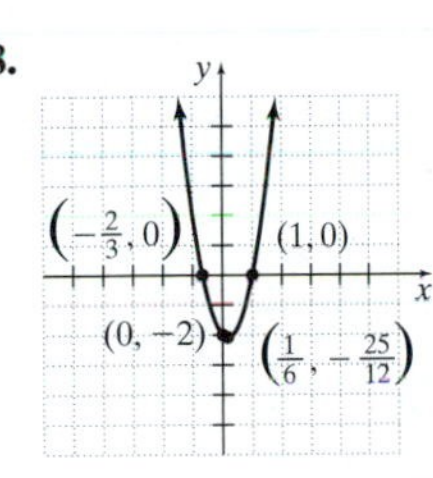

49.
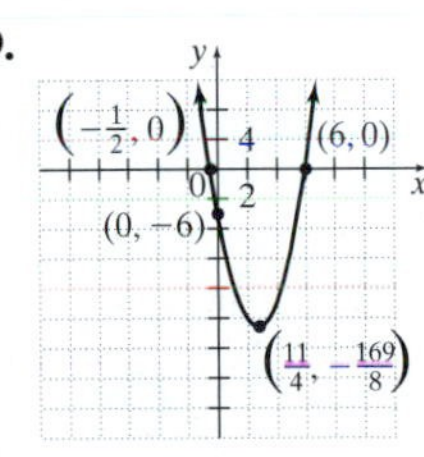

50.
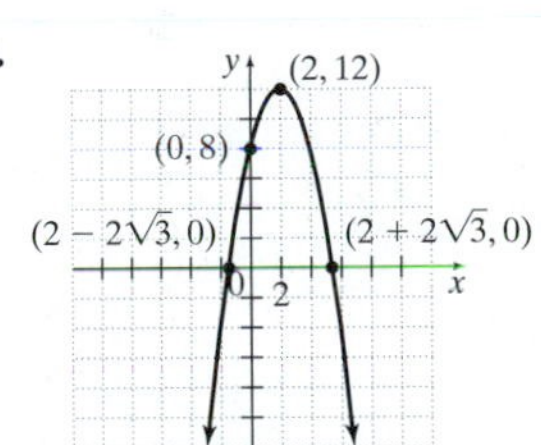

51. A **52.** D **53.** B **54.** C **55.** one real solution **56.** two real solutions **57.** no real solution **58.** two real solutions **59.** $x = \pm 7$ **60.** $x = \pm 5\sqrt{3}$ **61.** $x = 15, -1$ **62.** $x = -2 \pm \sqrt{10}$ **63.** $x = \frac{2}{3}, -1$ **64.** $x = \frac{1 \pm \sqrt{33}}{8}$ **65.** $x = \frac{3 \pm \sqrt{41}}{8}$ **66.** $x = \frac{-1 \pm \sqrt{41}}{10}$ **67.** $x = -\frac{3}{2}$ **68.** no real solution

69.
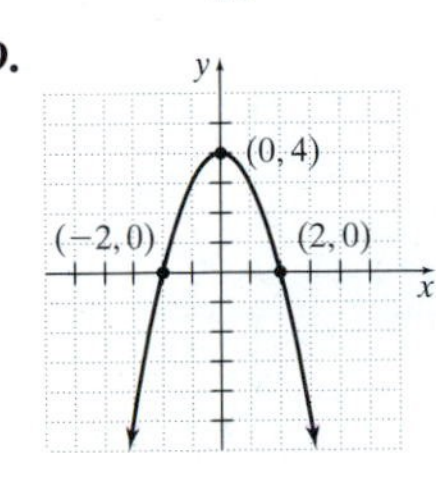

70.
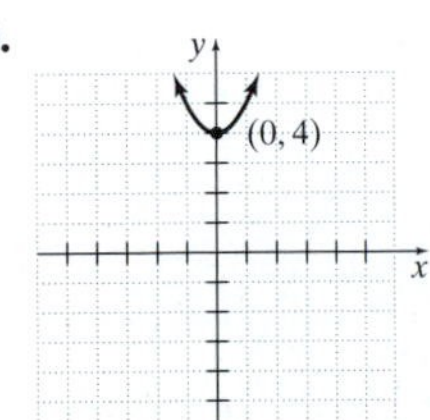

71.
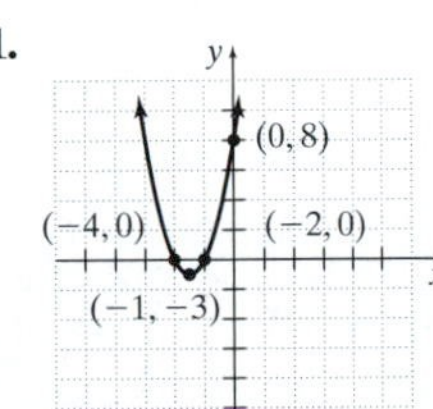

72.
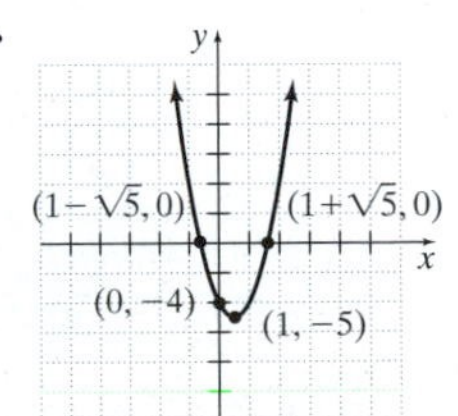

Chapter Test **1.** $x = \pm 20$ **2.** $x = -\frac{3}{2}, 7$ **3.** $k = \pm 4$ **4.** $m = \frac{5 \pm 2\sqrt{2}}{3}$ **5.** $x = 10, 16$ **6.** $x = \frac{-6 \pm 4\sqrt{3}}{3}$ **7.** $x = -2, 5$ **8.** $p = \frac{5 \pm \sqrt{37}}{6}$ **9.** $x = 1, -\frac{4}{3}$ **10.** $x = -1, \frac{5}{3}$ **11.** $x = \frac{7 \pm \sqrt{73}}{6}$ **12.** $x = -1, 5$ **13.** $x = 2, \frac{1}{3}$ **14.** $x = \frac{3 \pm \sqrt{7}}{2}$ **15.** base: 3 ft; height: 12 ft

16. (0, 0)

17.
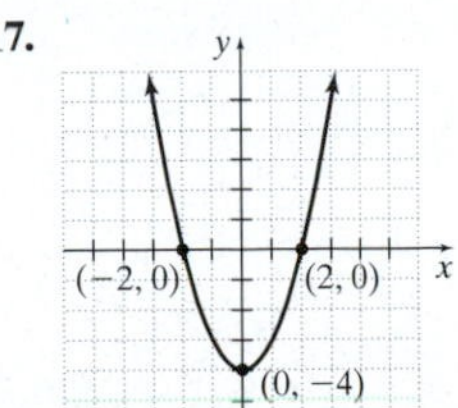

18.
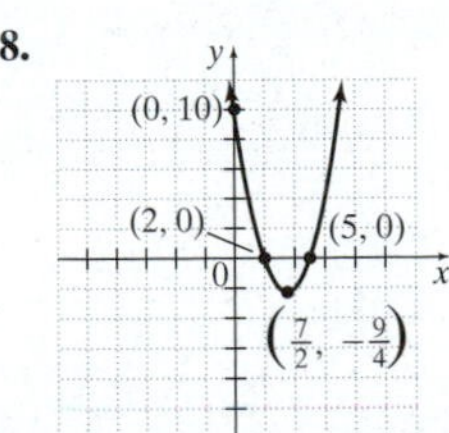

19.

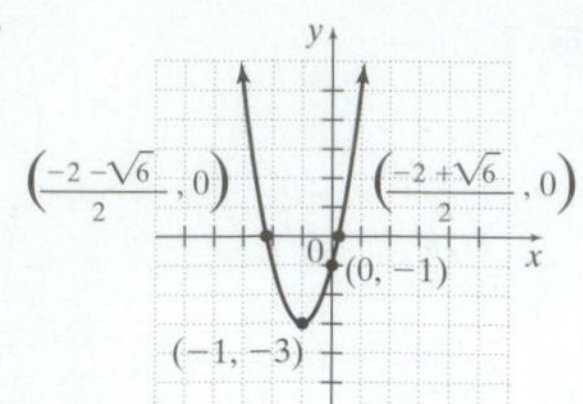

20. 6 sides **21.** 2.7 sec **22.** 2006

Cumulative Review **1.** 0.7861; Sec. 5.4, Ex. 8 **2.** 0.818; Sec. 5.4 **3.** −0.012; Sec. 5.4, Ex. 9 **4.** −0.0503; Sec. 5.4 **5.** 71%; Sec. 7.2, Ex. 2 **6.** 91%; Sec. 7.2 **7.** 28 in.; Sec. 8.2, Ex. 2 **8.** 25 ft; Sec. 8.2 **9.** 5.1 sq mi; Sec. 8.3, Ex. 2 **10.** 68 sq in.; Sec. 8.3 **11.** 3.21 L; Sec. 8.6, Ex. 5 **12.** 43.21 L; Sec. 8.6 **13.** $\frac{16}{3}$; Sec. 9.3, Ex. 2 **14.** $\frac{5}{6}$; Sec. 9.3 **15.** 1; Sec. 10.1, Ex. 28 **16.** −1; Sec. 10.1 **17.** $(r + 6)(r - 7)$; Sec. 11.2, Ex. 4 **18.** $(y - 7)(y + 10)$; Sec. 11.2 **19.** $(2x - 3y)(5x + y)$; Sec. 11.3, Ex. 4 **20.** $(9x - y)(8x - 3y)$; Sec. 11.3 **21.** $(2x - 1)(4x - 5)$; Sec. 11.4, Ex. 1 **22.** $(3x - 2)(5x + 2)$; Sec. 11.4 **23. a.** $x(2x + 7)(2x - 7)$; Sec. 11.5, Ex. 16 **b.** $2(9x^2 + 1)(3x + 1)(3x - 1)$; Sec. 11.5, Ex. 17 **24. a.** $x(3x - 1)(3x + 1)$ **b.** $5(x^2 + 1)(x - 1)(x + 1)$; Sec. 11.5 **25.** $\frac{1}{5}, -\frac{3}{2}, -6$; Sec. 11.6, Ex. 8 **26.** $-4, \frac{3}{4}, \frac{1}{10}$; Sec. 11.6 **27.** $\frac{x + 7}{x - 5}$; Sec. 12.1, Ex. 4 **28.** $\frac{x - 5}{x + 7}$; Sec. 12.1 **29.** −5; Sec. 12.6, Ex. 1 **30.** 4; Sec. 12.6 **31. a.** −3 **b.** 0 **c.** −3; Sec. 13.1, Ex. 4 **32. a.** 7 **b.** $\frac{7}{2}$ **c.** −3; Sec. 13.1 **33. a.** parallel **b.** perpendicular **c.** neither; Sec. 13.4, Ex. 7 **34. a.** parallel **b.** perpendicular **c.** neither; Sec. 13.4 **35. a.** function **b.** not a function; Sec. 13.6, Ex. 2 **36. a.** function **b.** not a function; Sec. 13.6 **37.** (4, 2); Sec. 14.2, Ex. 1 **38.** (1, 4); Sec. 14.2 **39.** 6; Sec. 15.1, Ex. 1 **40.** 9; Sec. 15.1 **41.** $\frac{3}{10}$; Sec. 15.1, Ex. 3 **42.** $\frac{4}{5}$; Sec. 15.1 **43.** $-1 + \sqrt{3}$; Sec. 15.4, Ex. 13 **44.** $7(\sqrt{5} + 2)$; Sec. 15.4 **45.** 7, −1; Sec. 6.1, Ex. 5 **46.** −1, −7; Sec. 16.1 **47.** $1 \pm \sqrt{5}$; Sec. 16.3, Ex.5 **48.** $\frac{5 \pm \sqrt{137}}{8}$; Sec. 16.3

APPENDIX

Exercise Set Appendix C **1.** $(a + 3)(a^2 - 3a + 9)$ **3.** $(2a + 1)(4a^2 - 2a + 1)$ **5.** $5(k + 2)(k^2 - 2k + 4)$ **7.** $(xy - 4)(x^2y^2 + 4xy + 16)$ **9.** $(x + 5)(x^2 - 5x + 25)$ **11.** $3x(2x - 3y)(4x^2 + 6xy + 9y^2)$ **13.** $(3 - t)(9 + 3t + t^2)$ **15.** $8(r - 2)(r^2 + 2r + 4)$ **17.** $(t - 7)(t^2 + 7t + 49)$ **19.** $(s - 4t)(s^2 + 4st + 16t^2)$

Exercise Set Appendix D **1.**

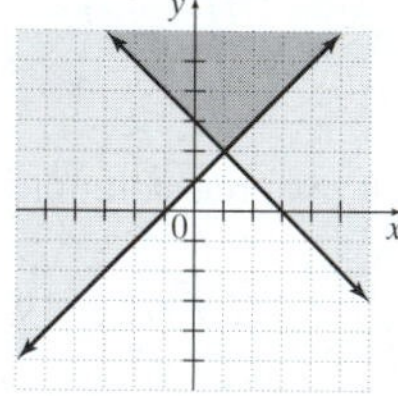

3.

5.

7.

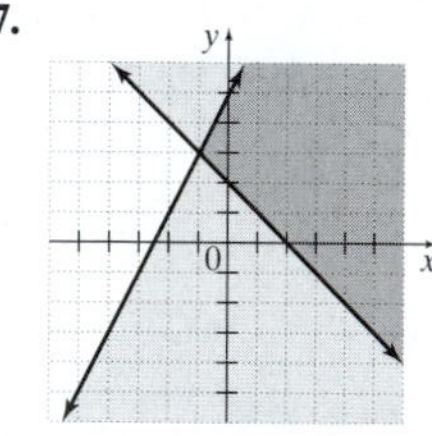

9.

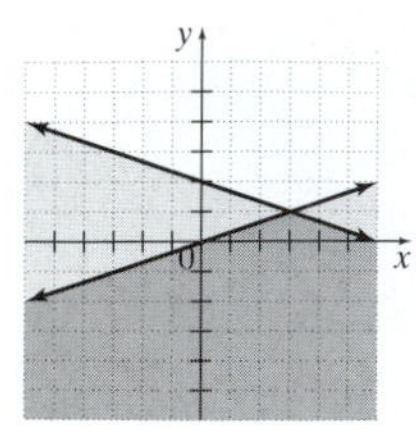

11.

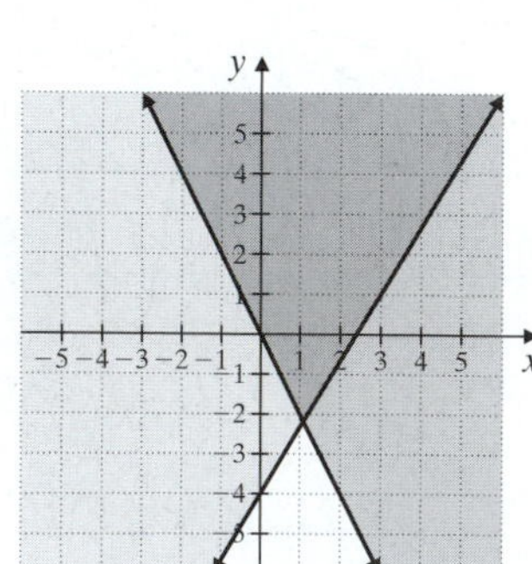

13.

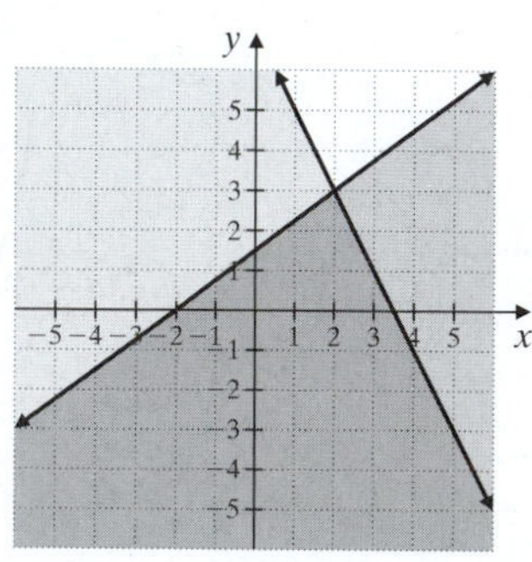

15.

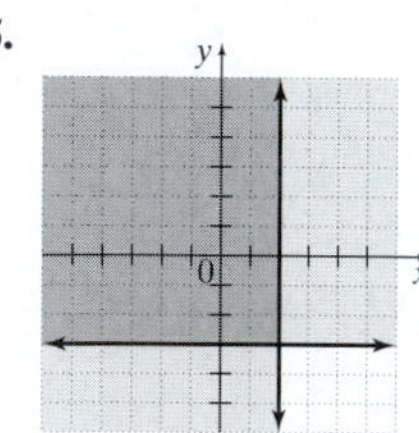

17.

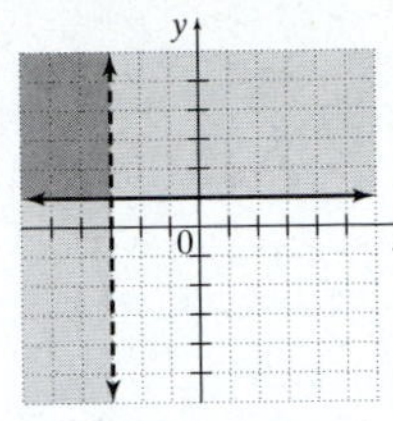

19.

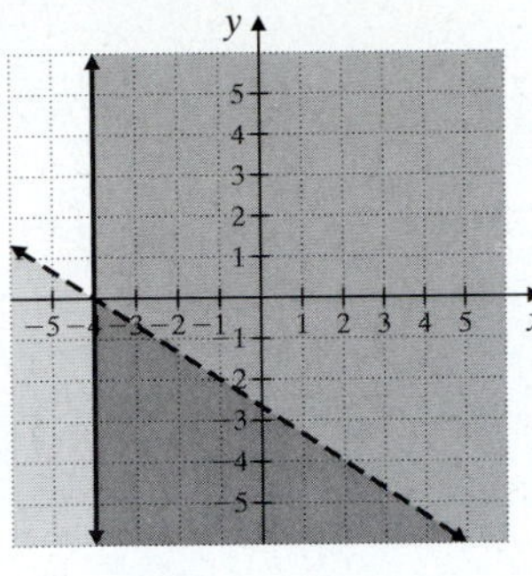

21.

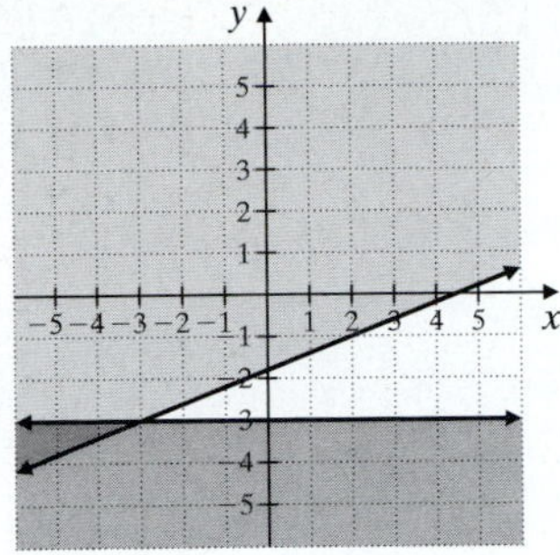

23.

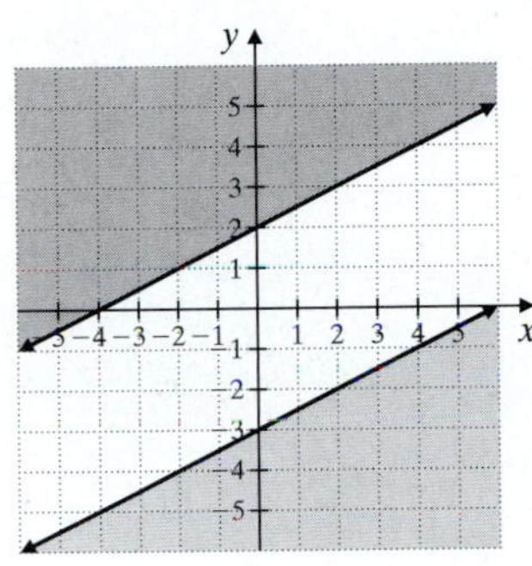

25. C **27.** D

INDEX

Photo Credits

Table of Contents Norbert Rosing/National Geographic Image Collection; © SuperStock; Rachel Epstein/PhotoEdit Inc.; NASA/SIPA Press; Jeff Greenberg/PhotoEdit Inc.; © 2006 American Honda Motor Co., Inc.; Tom Bean/Tom & Susan Bean, Inc.; Kevin Dietsch/UPI/Landov LLC; David Young Wolff/Photo Edit; AP/Wide World Photos; AP/Wide World Photos; ©Corbis. All Rights Reserved; ©Jose Luis Pelaez/CORBIS All Rights Reserved; Javier Larrea/AGE Fotostock America, Inc.; Wyman Ira/Corbis/Sygma

Chapter 1 CO Norbert Rosing/National Geographic Image Collection **Page 2** Sepp Seitz/Woodfin Camp & Associates **Page 12** NASA/AP Wide World Photos **Page 21** Sisse Brimberg/National Geographic Image Collection **Page 35** Copyright © David Nunuk/Photo Researchers, Inc. **Page 44** Max Alexander/© Dorling Kindersley **Page 47** Hyatt Corporation

Chapter 2 **Page 101** John Elk/Bruce Coleman Inc.; © Derek Croucher/Corbis/Bettmann **Page 109** Image by © Royalty-Free/Corbis **Page 125** Andrew Redington/Getty Images, Inc. **Page 149** "Melancholia (engraving)" by Albrecht Durer (141–1528). Guildhall Library, Corporation of London, UK/Bridgeman Art Library International **Page 153** Gillianne Tedder/Photolibrary.com; Greg Vaughan/PacificStock.com

Chapter 3 CO © SuperStock **Page 195** Mark Richards/PhotoEdit Inc. **Page 197** D. Robert & Lorri Franz/© Corbis. All rights reserved; Jim Zipp/PhotoResearchers, Inc. **Page 198** Steve Morrell/AP Wide World Photos; F. Rangel/The Image Works **Page 209** AP Photos/Robert F. Bukaty

Chapter 4 CO Rachel Epstein/PhotoEdit Inc. **Page 238** Keith Brofsky/Getty Images, Inc./PhotoDisc **Page 246** Kevin Horan/Getty Images, Inc./Stone Allstock **Page 302** Rev. Ronald Royer/Photo Researchers, Inc.

Chapter 5 CO NASA/SIPA Press **Page 331** AP Wide World Photos **Page 338** Kaku Kurita/Time Life Pictures/Getty Images; Bill Ross/Corbis-NY **Page 341** Stan Honda/AFP/Getty Images **Page 351** Noah Berger/Landov LLC **Page 362** Angelo Hornak/Corbis/Bettmann **Page 372** Tony Latham/Getty Images Inc./Stone Allstock

Chapter 6 CO Jeff Greenberg/PhotoEdit Inc. **Page 425** C. Borland/Getty Images, Inc./PhotoDisc **Page 454** Getty Images/PhotoDisc **Page 455** © Jeff Greenberg/Omni-Photo Communications, Inc. **Page 458** StockByte/Getty Images, Inc./StockByte **Page 461** David Young-Wolff; Novastock **Page 468** Susan Van Etten; Michael St. Maur Sheil/Corbis

Chapter 7 CO © 2006 American Honda Motor Co., Inc.

Chapter 8 CO Tom Bean/Tom & Susan Bean, Inc. **Page 563** AP Wide World Photos; Will & Deni McIntyre/Photo Researchers, Inc. **Page 571** © Tom and Susan Bean, Inc./DRK PHOTO **Page 587** William West/AFP; Rich Iwasaki **Page 592** Theo Allofs/Image Bank/Getty Images **Page 627** Bill Bachmann/© Index Stock Imagery, Inc.; © Jon Feingersh/Corbis/Bettmann

Chapter 9 CO Kevin Dietsch/UPI/Landov LLC; **Page 651** Tim Davis/Corbis RF; Rick Wilking/Reuters/Corbis/Bettmann; **Page 679** ©Rolf Bruderer/Masterfile; **Page 681** U.S. Congress, Office of Technology Assessment; **Page 682** ©Rick Gomez/Masterfile; AP Wide World Photos; **Page 687** NASA/Science Source/Photo Researchers, Inc.; **Page 689** Bill Bachmann/Stock Boston; **Page 700** Colin BraleyCorbis/Reuters America LLC; **Page 701** ©2005 Norbert Wu/www.norbertwu.com; **Page 718** Catherine Karnow/Woodfin Camp & Associates

Chapter 10 CO David Young Wolff/Photo Edit; **Page 745** Brad Lewis/SeaPics.com; Photo Researchers, Inc.; **Page 763** Terrence Moore/Woodfin Camp & Associates; Masterfile Corporation; **Page 792** David Olsen/Getty Images Inc./Stone Allstock

Chapter 11 CO AP/Wide World Photos; **Page 810** Robert Harding/Digital Vision/Getty Images; AP/Wide World Photos; **Page 837** AP/Wide World Photos; John Freeman/Stone/Getty Images; **Page 849** AP/Wide World Photos; **Page 869** AP/Wide World Photos

Chapter 12 CO AP/Wide World Photos; **Page 879** Chuck Keeler/Frozen Images/The Image Works; **Page 886** Matthew McVay/Stock Boston; Tebo Photography; **Page 918** ©Charles O'Rear/CORBIS; **Page 924** Gavin Lawrence/Getty Images, Inc.

Chapter 13 CO © Corbis. All Rights Reserved; **Page 964** ©Michael Newman/Photo Edit; **Page 969** AP/Wide World Photos; SuperStock, Inc.; **Page 988** Betty Crowell/Faraway Places; **Page 996** Miles Ertman/Masterfile Corporation; Jim Pickerell/The Stock Connection; **Page 1000** Mike Yamashita/Woodfin Camp & Associates; **Page 1005** Mary Kate Denny/Photo Edit; **Page 1006** Ann States Photography

Chapter 14 CO ©Jose Luis Pelaez/CORBIS All Rights Reserved; **Page 1073** Lloyd Sutton/Masterfile Corporation; **Page 1083** AP/Wide World Photos; **Page 1089** AP/Wide World Photos; Bruce Bennett Studios/Getty Images, Inc.; **Page 1090** AP/Wide World Photos; **Page 1101** AP/Wide World Photos

Chapter 15 CO Javier Larrea/AGE Fotostock America, Inc.; **Page 1112** California Academy of Sciences, Geology; **Page 1119** Jessica Wecker/Photo Researchers, Inc.; Photofest; **Page 1120** Getty Images, Inc./Stone; **Page 1145** Dallas and John Heaton/The Stock Connection

Chapter 16 CO Wyman Ira/Corbis/Sygma; **Page 1164** Photofest; **Page 1166** AP/Wide World Photos; **Page 1193** National Astronomy and Ionosphere Center; **Page 1197** Ted Clutter/Photo Researchers, Inc.

ELAYN MARTIN-GAY: PREALGEBRA & INTRODUCTORY ALGEBRA, 2E, CHAPTER TEST PREP VIDEO CD

0-13-231930-6

Pearson Prentice Hall
Pearson Education, Inc.
Upper Saddle River, NJ 07458

YOU SHOULD CAREFULLY READ THE TERMS AND CONDITIONS BEFORE USING THE CD-ROM PACKAGE. USING THIS CD-ROM PACKAGE INDICATES YOUR ACCEPTANCE OF THESE TERMS AND CONDITIONS.

Pearson Education, Inc. provides this program and licenses its use. You assume responsibility for the selection of the program to achieve your intended results, and for the installation, use, and results obtained from the program. This license extends only to use of the program in the United States or countries in which the program is marketed by authorized distributors.

LICENSE GRANT
You hereby accept a nonexclusive, nontransferable, permanent license to install and use the program ON A SINGLE COMPUTER at any given time. You may copy the program solely for backup or archival purposes in support of your use of the program on the single computer. You may not modify, translate, disassemble, decompile, or reverse engineer the program, in whole or in part.

TERM
The License is effective until terminated. Pearson Education, Inc. reserves the right to terminate this License automatically if any provision of the License is violated. You may terminate the License at any time. To terminate this License, you must return the program, including documentation, along with a written warranty stating that all copies in your possession have been returned or destroyed.

LIMITED WARRANTY
THE PROGRAM IS PROVIDED "AS IS" WITHOUT WARRANTY OF ANY KIND, EITHER EXPRESSED OR IMPLIED, INCLUDING, BUT NOT LIMITED TO, THE IMPLIED WARRANTIES OF MERCHANTABILITY AND FITNESS FOR A PARTICULAR PURPOSE. THE ENTIRE RISK AS TO THE QUALITY AND PERFORMANCE OF THE PROGRAM IS WITH YOU. SHOULD THE PROGRAM PROVE DEFECTIVE, YOU (AND NOT PEARSON EDUCATION, INC. OR ANY AUTHORIZED DEALER) ASSUME THE ENTIRE COST OF ALL NECESSARY SERVICING, REPAIR, OR CORRECTION. NO ORAL OR WRITTEN INFORMATION OR ADVICE GIVEN BY PEARSON EDUCATION, INC., ITS DEALERS, DISTRIBUTORS, OR AGENTS SHALL CREATE A WARRANTY OR INCREASE THE SCOPE OF THIS WARRANTY.

SOME STATES DO NOT ALLOW THE EXCLUSION OF IMPLIED WARRANTIES, SO THE ABOVE EXCLUSION MAY NOT APPLY TO YOU. THIS WARRANTY GIVES YOU SPECIFIC LEGAL RIGHTS AND YOU MAY ALSO HAVE OTHER LEGAL RIGHTS THAT VARY FROM STATE TO STATE.

Pearson Education, Inc. does not warrant that the functions contained in the program will meet your requirements or that the operation of the program will be uninterrupted or error-free. However, Pearson Education, Inc. warrants the CD-ROM(s) on which the program is furnished to be free from defects in material and workmanship under normal use for a period of ninety (90) days from the date of delivery to you as evidenced by a copy of your receipt. The program should not be relied on as the sole basis to solve a problem whose incorrect solution could result in injury to person or property. If the program is employed in such a manner, it is at the user's own risk and Pearson Education, Inc. explicitly disclaims all liability for such misuse.

LIMITATION OF REMEDIES
Pearson Education, Inc.'s entire liability and your exclusive remedy shall be:
1. the replacement of any CD-ROM not meeting Pearson Education, Inc.'s "LIMITED WARRANTY" and that is returned to Pearson Education, or
2. if Pearson Education is unable to deliver a replacement CD-ROM that is free of defects in materials or workmanship, you may terminate this agreement by returning the program.

IN NO EVENT WILL PEARSON EDUCATION, INC. BE LIABLE TO YOU FOR ANY DAMAGES, INCLUDING ANY LOST PROFITS, LOST SAVINGS, OR OTHER INCIDENTAL OR CONSEQUENTIAL DAMAGES ARISING OUT OF THE USE OR INABILITY TO USE SUCH PROGRAM EVEN IF PEARSON EDUCATION, INC. OR AN AUTHORIZED DISTRIBUTOR HAS BEEN ADVISED OF THE POSSIBILITY OF SUCH DAMAGES, OR FOR ANY CLAIM BY ANY OTHER PARTY.

SOME STATES DO NOT ALLOW FOR THE LIMITATION OR EXCLUSION OF LIABILITY FOR INCIDENTAL OR CONSEQUENTIAL DAMAGES, SO THE ABOVE LIMITATION OR EXCLUSION MAY NOT APPLY TO YOU.

GENERAL
You may not sublicense, assign, or transfer the license of the program. Any attempt to sublicense, assign or transfer any of the rights, duties, or obligations hereunder is void.

This Agreement will be governed by the laws of the State of New York.

Should you have any questions concerning this Agreement, you may contact Pearson Education, Inc. by writing to:
ESM Media Development
Higher Education Division
Pearson Education, Inc.
1 Lake Street
Upper Saddle River, NJ 07458

Should you have any questions concerning technical support, you may write to:
New Media Production
Higher Education Division
Pearson Education, Inc.
1 Lake Street
Upper Saddle River, NJ 07458

YOU ACKNOWLEDGE THAT YOU HAVE READ THIS AGREEMENT, UNDERSTAND IT, AND AGREE TO BE BOUND BY ITS TERMS AND CONDITIONS. YOU FURTHER AGREE THAT IT IS THE COMPLETE AND EXCLUSIVE STATEMENT OF THE AGREEMENT BETWEEN US THAT SUPERSEDES ANY PROPOSAL OR PRIOR AGREEMENT, ORAL OR WRITTEN, AND ANY OTHER COMMUNICATIONS BETWEEN US RELATING TO THE SUBJECT MATTER OF THIS AGREEMENT.

System Requirements
Windows
Pentium II 300-MHz processor-based computer
Windows 2000 (Service Pack 4) or XP
64 MB RAM
7.2 MB available hard drive space (optional—for minimum QuickTime installation)
800 x 600 resolution
8x CD drive (required)
QuickTime 7.x
Sound card
Internet Explorer 6.0 or Netscape 6.23, 7.2

Macintosh
Power PC G3 233 MHz
Mac OS 10.x
64 MB RAM
19 MB on OS X (optional—if QuickTime installation is needed)
800 x 600 resolution monitor
8x CD drive (required)
QuickTime 6 or 7
Internet Explorer 5.2, Safari 1.2, or Netscape 7.2

Support Information
If you are having problems with this software, please call our Media Support Line (800) 677-6337 Monday through Friday 8:00 a.m. to 8:00 p.m. and Sunday 5:00 p.m. to 12:00 a.m. EST. You can also get support by filling out the web form located at: http://247.prenhall.com/mediaform

Our technical staff will need to know certain things about your system in order to help us solve your problems more quickly and efficiently. If possible, please be at your computer when you call for support. You should have the following information ready:

- Textbook ISBN
- CD-ROM ISBN
- corresponding product and title
- computer make and model
- Operating System (Windows or Macintosh) and Version
- RAM available
- hard disk space available
- Sound card? Yes or No
- printer make and model
- network connection
- detailed description of the problem, including the exact wording of any error messages.

NOTE: Pearson does not support and/or assist with the following:

- third-party software (i.e. Microsoft including Microsoft Office suite, Apple, Borland, etc.)
- homework assistance
- Textbooks and CD-ROMs purchased used are not supported and are non-replaceable. To purchase a new CD-ROM, contact Pearson Individual Order Copies at 1-800-282-0693.